A TEXTBOOK OF

INORGANIC CHEMISTRY

for JEE Main & Advanced

A TEXTBOOK OF

INORGANIC CHEMISTRY

for JEE Main & Advanced

Dr. RK Gupta

ARIHANT PRAKASHAN (Series), MEERUT

ARIHANT PRAKASHAN (Series), MEERUT

卐 **ADMINISTRATIVE & PRODUCTION OFFICES**

Regd. Office
'Ramchhaya' 4577/15, Agarwal Road, Darya Ganj, New Delhi -110002
Tele: 011- 47630600, 43518550; Fax: 011- 23280316

Head Office
Kalindi, TP Nagar, Meerut (UP) - 250002
Tel: 0121-2401479, 2512970, 4004199; Fax: 0121-2401648

卐 **SALES & SUPPORT OFFICES**

Agra, Ahmedabad, Bengaluru, Bareilly, Chennai, Delhi, Guwahati, Hyderabad, Jaipur, Jhansi, Kolkata, Lucknow, Meerut, Nagpur & Pune

卐 **ISBN : 978-93-13192-08-4**

卐 **Price : ₹ 800.00**

Published by Arihant Publications (India) Ltd.

PREFACE

In the recent years, the format and pattern of JEE Main & Advanced have been constantly changing. In pursuant to these latest changes, I present a new text book the Inorganic Chemistry to help the students master the Inorganic Chemistry. The book covers all the aspects of Inorganic at the level of Classes XI-XII with a view to master the syllabus of all Engineering Entrances.

Features of the Book

- Text has been devised in such a way as to be assimilable, without a prior knowledge of the Inorganic Chemistry.
- The basic thrust of the NCERT books has been followed and most of the problems from the NCERT books have been included in each chapter.
- The whole text has been divided into twenty chapters, out of which seven devoted to General section, the backbone of Chemistry from which almost 40% questions are being asked in JEE Main & Advanced every year.
- *p*-block elements have been sub-divided into six separate blocks to make it more relevant to NCERT pattern and examination oriented.
- Every chapter starts with Quick-Points that is the idea in nut-shell of the concepts/basics for application in the current unit and thus helpful in creating confidence among aspirants.
- Target-Exercises at the end of each Section are meant to check your ability at the spot before you proceed to the next Section. This is followed by Practice-Exercises on NCERT-based questions.
- There are lots of other exercises (with complete solutions) for testing analytical-skills. They include Comprehension-Based Questions, True-False, Fill-in-blanks, Assertion-Reason, Matrix-Matching, Integer Answer Types, MCQs with One and More than One Correct Options, Very Short and Analytical Questions compiled Sections-wise .
- Brain –Twisters are provide to sharp your knowledge specially of Olympiad Aspirants.
- Deep – Focus are scattered throughout the text. These are the warning concepts to catch your eyes for spot revision before examination.
- Essential Inorganic CHEMISTRY is thought to be non-conceptual, nevertheless, I hope this text will be able to debunk this misconception and will be immensely helpful.

Acknowledgement

Finally, there are many faculty and students, who have helped in the development of contents by their valuable suggestions and unsolicited comments .They know who they are, and I hope that they will accept my sincere thanks for all they did.

It is also a pleasure to acknowledge the many people for their friendly assistance in completing this book, especially Shri Y C Jain, Chairman, Arihant for encouraging me at each step, and for being an exemplary Publisher, and good friend too. My special thanks are also to Daya Sindh Computers, Meerut for completing the book well in time. I am also indebted to MS Rawat, Keshav Agarwal, Harsh Kumar and Narendra Singh who left no stone unturned for completing this project successfully.

I must thank Almighty God for His inspiration and guidance, and my family members for their unquestioning timely encouragement without which this work would not be possible.

I would be glad to from the faculty and other of any errors which may have escaped. Any suggestion whereby the book can be improved will be heartly welcomed.

Dr. RK Gupta

raghvesh@gmail.com

CONTENTS

Chemistry touches almost every aspect of our lives, our culture, and our environment. Its scope encompasses the air we breathe, the food we eat, the fluids we drink, our clothings, dwellings, transportation and fuel supplies, and our fellow creatures.

1

Fundamental Concepts *of* Inorganic Chemistry

Chemistry is the science that describes matter—its chemical and physical properties, the chemical and physical changes it undergoes, and the energy changes that accompany those processes.

Matter includes everything that is tangible, from our bodies and the stuff of our everyday lives to the grandest objects in the universe. Some call chemistry as the central science. It rests on the foundations of mathematics and physics and in turn underlies the life sciences biology and medicine.

All living matter contains carbon combined with hydrogen. The chemistry of compounds of carbon and hydrogen is called **Organic Chemistry**. The study of substances that do not contain carbon combined with hydrogen is called the **Inorganic Chemistry**. The branch of chemistry that is concerned with the detection or identification of substances present in a sample (*qualitative analysis*) or with the amount of each that is present (*quantitative analysis*) is called **Analytical Chemistry. Physical Chemistry** applies the **mathematical theories** and methods of physics to the properties of matter and to the study of chemical processes and the accompanying energy changes. As its name suggests, **Biochemistry** is the study of the chemistry of processes in living organisms.

Learning Framework of the Chapter ...

- Matter and Energy
- Physical and Chemical Changes
- Laws of Chemical Combination
- Dalton and Constitution of Matter
- Atomic Number, Mass Number and Isotopes
- Equivalent Weight
- The Mole
- Per cent Composition and Empirical Formula

Matter and Energy

Matter is anything that has mass and occupies space. Mass is a measure of the quantity of matter in a sample of any material. The more massive an object is, the more force is required to put it in the motion. All bodies consist of matter. Our senses of sight and touch usually tell us that an object occupies space. In the case of colourless, odourless, and tasteless gases (such as air), our senses may fail us.

Energy is defined as the capacity to do work or to transfer heat. Energy can be in the form of

(i) mechanical energy
(ii) light energy
(iii) electrical energy
(iv) heat energy

Energy is classified into two principal types:

Kinetic Energy

A body in motion, such as rolling boulder, possesses energy because of its motion. Such energy is called **kinetic energy.** It represents the capacity for doing work directly. It is easily transferred between objects.

Potential Energy

It is the energy an object possesses because of its position, condition or composition. Coal, for example, possesses chemical energy, a form of potential energy, because of its composition. Thermal plants burn coal, producing heat which is converted to electrical energy. A boulder located at top of a mountain possesses potential energy because of its height. It can roll down the mountain side, and potential energy is converted into kinetic energy.

Different States of Matter

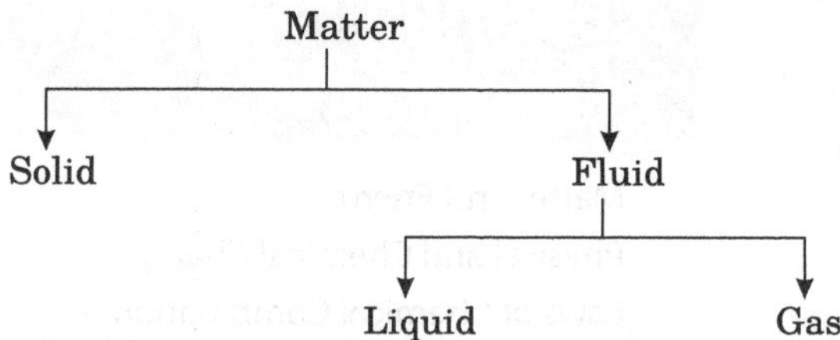

A molecule is the smallest particle of any kind of matter, element or compound, which can exist in the free state and can still preserve the character of that kind of matter. **The properties of a substance,** are the properties of its molecules.

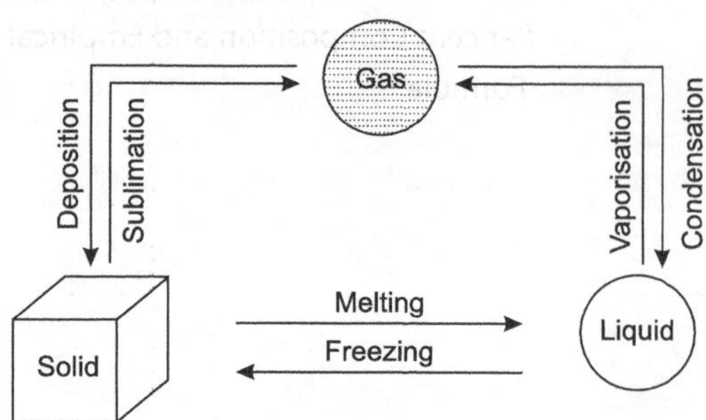

Fig. 1.1

Solids

In the solid state, the intermolecular force is very strong. The molecules in a solid species, therefore, cling together, with a great force. Thus, solids have definite shape and volume.

Liquid

In the liquid state, the intermolecular forces are very feeble. They are still perceptible. Thus, liquid has no definite shape but takes the shape of that part of the vessel in which it is stored.

Gases

In the gaseous state, the intermolecular force is negligible and molecules lie far apart from one another. They move in different directions with different velocities colliding among themselves and against sides of the container.

Properties	Ice (Solid H_2O)	Liquid H_2O	Steam (Gaseous H_2O)
Rigidity	Rigid (definite shape and volume)	Flows and assumes shape of container (no definite shape but definite volume)	Fills any container completely (no definite volume and shape)
Expansion on heating	Slight	Slight	Expands infinitely
Compressibility	Slight	Slight	Easily
Melting point	solid >	liquid >	Compressed gas

We know that when water is boiled, it disappears; when a candle burns, both the wick and the wax entirely disappear, practically all its weight is lost. Again a piece of magnesium wire, on being strongly heated in air, burns with a brilliant light and leaves a white residue, which when weighed, will be found to be heavier than the piece of magnesium taken.

When we try to follow these, we are confronted with some questions which naturally come to our mind? Is water actually destroyed? Is there is destruction of the substance of which the candle is decomposed? Whence does the additional weight of magnesium come? Can matter be created? We can certainly answer these questions one by one.

- When water is boiled, it only changes its state and is not destroyed.

$$H_2O(l) \longrightarrow H_2O(g)$$

The candle loses weight due to combustion and the products are colourless gases that are disappearing into the atmosphere.

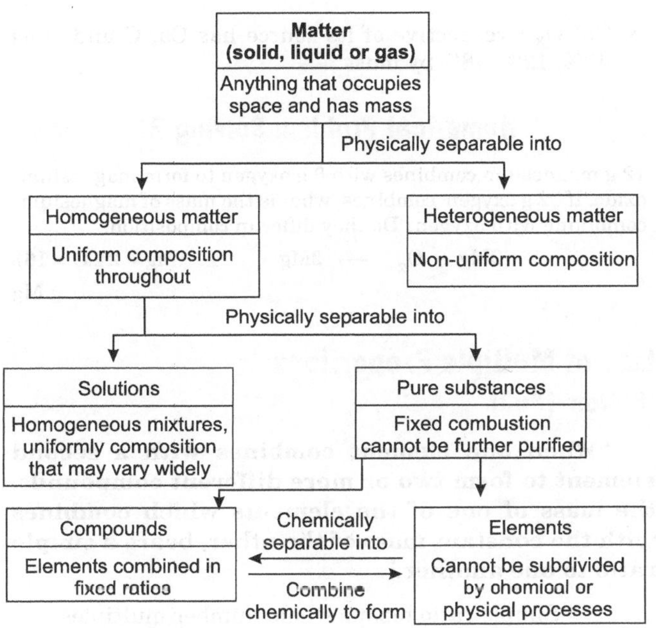

Fig.1.2 Classification of matter

Magnesium gains weight due to its reaction with the oxygen of the atmosphere and gives white residue (of magnesium oxide). But in no chemical action any kind of matter is destroyed or created.

Physical and Chemical Changes

Matter is capable of undergoing variety of changes as

Physical Change

If there is change in some of the specific properties like the state (solid to liquid or *vice-versa*), texture, electrical or magnetic without bringing about any change in the composition or constitution of its molecules, then this type of change is termed as physical change.

Original state can be restored by change in experimental condition.

Illustration :

H_2O (liquid) is converted into H_2O vapour (steam) on heating, and steam into H_2O (liquid) on condensation. H_2O (liquid) is converted into ice at freezing point, and ice melts to $H_2O(l)$ at melting point.

$$H_2O(s) \rightleftharpoons H_2O(l) \rightleftharpoons H_2O(g).$$

Thus, it is completely a physical change.

- When an electric current is passed into copper wire wrapped over iron, it is magnetised. When flow of electric current stops, magnetic behaviour is also lost. Thus, it is a physical change.

Chemical Change

If there is change in the specific properties of the matter due to change in the composition or constitution of its molecules, this is termed as chemical change. **There is formation of new molecules as a result of a chemical change.**

Illustration :

- Acidified water is converted into hydrogen and oxygen due to electric decomposition

$$2H_2O \longrightarrow 2H_2(g) + O_2(g)$$

This is a chemical change, since H_2O is converted into new compounds H_2 and O_2.

- When iron piece is exposed to air for a long time, resulting takes place. Rust appears in the form of brown-coloured layer and is formed when iron reacts with moist air.

$$Fe \xrightarrow{O_2} FeO \xrightarrow{O_2} \underset{\text{rust}}{Fe_2O_3 \cdot xH_2O}$$

Think Yourself !

When a candle (of wax) burns, wax melts. After sometime wax is completely converted into CO_2 and H_2O. What is the type of this change?

We know that wax is made of carbon, hydrogen and oxygen when wax melts, it is a **physical change.** But when it is converted into CO_2 and H_2O due to burning, it is a **chemical change.**

Table 1.1 **Distinction between Physical and Chemical Change**

	Physical change	Chemical change
Example	$H_2O(s) \underset{\text{f.p.}}{\overset{\text{m.p.}}{\rightleftharpoons}} H_2O(l)$	$CaCO_3(s) \xrightarrow{\text{Heat}} CaO(s) + CO_2(g)$
Composition	no change, water in solid and liquid state have identical composition H and O being in 2 : 1 ratio	Changes due to formation of new compounds. $CaCO_3(s)$, $CaO(s)$ and $CO_2(g)$ have completely different composition.
Properties	Physical properties differ but chemical properties are identical.	Physical properties and chemical properties of the reactants and products are quite different.
Stability	A temporary change, above equilibrium is reversible. Original state is restored by change in exp- erimental condition.	A permanent change, reaction is spontaneous going to completion. Original state is not restored.
Thermal effect	Physical change can take place in the absence or presence of heat. Above change absorbs heat in forward side and loses heat in backward side.	Chemical change initially endothermic (takes place with absorption of heat). Above change requires heat.

Laws of Chemical Combination

Law of Conservation of Mass

(Lavoisier–1789)

"As a result of chemical change, the total mass of the reactants is equal to the total mass of the products."

Thus, of masses m and n of reactants A and B react to form masses x and y of the products C and D

$$\underset{m}{A} + \underset{n}{B} \longrightarrow \underset{x}{C} + \underset{y}{D}$$

Then, $m + n = x + y$

$$\underset{4g}{2H_2(g)} + \underset{32g}{O_2(g)} \longrightarrow \underset{36g}{2H_2O(g)}$$

$$\underset{100g}{CaCO_3(s)} \longrightarrow \underset{56g}{CaO(s)} + \underset{44g}{CO_2(g)}$$

$$\underset{30g}{CH_3COOH} + \underset{23g}{C_2H_5OH} \longrightarrow \underset{44g}{CH_3COOC_2H_5} + \underset{9g}{H_2O}$$

Numerical Problem Solving 1

$$\underset{17g}{AgNO_3} + \underset{7.45\,g}{KCl} \longrightarrow \underset{14.35\,g}{AgCl} + \underset{?}{KNO_3}$$

What is mass of KNO_3?

Law of Definite or Constant Composition

(Proust–1799)

"A chemical substance always consists of the same elements combined together in the same proportions by mass."

If the elements A and B combine together to form the compound AB, then in whatever manner AB is formed, then it is always composed of these two elements and mass percentage of A and B remains constant.

AB can be formed by **two methods.**

$$A + B \longrightarrow AB$$

First method a g b g $\frac{a}{b}$ thus, ratio of A and B in AB by mass

Second method x g y g $\frac{x}{y}$

thus, $\frac{a}{b} = \frac{x}{y}$

- H_2O irrespective of its source has H and O in 1 : 8 mass ratio in pure state.
- $CaCO_3$ irrespective of its source has Ca, C and O as 40%, 12%, 48% by mass.

Numerical Problem Solving 2

12 g magnesium combines with 8 g oxygen to form magnesium oxide. If 32 g oxygen combines, what is the mass of magnesium combining with oxygen? Do they differ in composition.

$$2Mg + O_2 \longrightarrow 2Mg \qquad (Mg = 24, O = 16)$$

...... g Mg

Law of Multiple Proportions

(Dalton-1803)

"When one element combines with a second element to form two or more different compounds, the mass of one of the elements which combines with the constant mass of the other, bears a simple ratio to one another."

Simple ratio means whole number multiples

A combines with B to form AB and AB_2. Then ratio of B (by mass in AB and AB_2) that combines with a definite mass of A is in a simple multiple ratio.

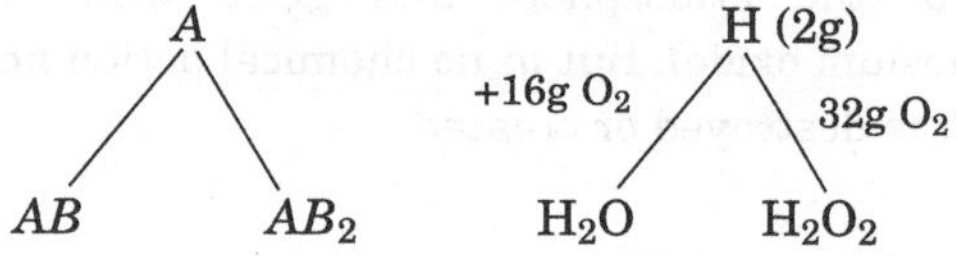

- 2 g H_2 combines with 16 g O_2 to form 18 g H_2O
  2g H_2 combines with 32 g O_2 to form 34 g H_2O_2
  Thus, ratio of oxygen (by mass) that combines with 1 g H_2 is $8 : 16 = 1 : 2$

Numerical Problem Solving 3

(i) Carbon and oxygen form two oxides carbon monoxide and carbon dioxide. What is the ratio of oxygen that combines with 1 g carbon in each?

$$2C + O_2 \longrightarrow 2CO$$

$$C + O_2 \longrightarrow CO_2$$

Ratio = ...

(ii) Consider the following reactions.

(a) $2C + H_2 \longrightarrow C_2H_2$ (b) $2C + 2H_2 \longrightarrow C_2H_4$

(c) $2C + 3H_2 \longrightarrow C_2H_6$ (d) $C + 2H_2 \longrightarrow CH_4$

Is law of multiple proportions followed.

Yes/No

Law of Reciprocal Proportions

(Richter-1792)

"The masses of two or more different elements which separately combine with a definite mass of another element are either the same as or are simple multiples of the masses of these different elements when they combine amongst themselves."

A combines with *C* to form *AC*

B combines with *C* to form *BC*

A combines with *B* to form *AB*

Then ratio of *A* and *B* by mass in *AB* is same as ratio of *A* and *B*, and *AC* and *BC*.

(a) $C + 2H_2 \longrightarrow CH_4$

(b) $O_2 + 2H_2 \longrightarrow 2H_2O(O_2H_4)$

(c) $C + O_2 \longrightarrow CO_2$

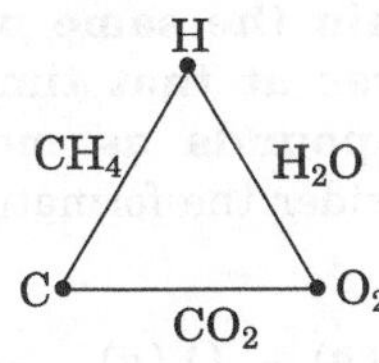

4 g H_2 combines with 12 g carbon in (a)

4 g H_2 combines with 32 g oxygen in (b)

Thus, ratio of carbon and oxygen in (a) and (b) by mass that combines with a definite mass of H_2 (in this case 4 g) = 12 : 32 = **3 : 8** in (c) 12 g carbon combines with 32 g oxygen thus ratio of carbon and oxygen (by mass) in CO_2 is = **3 : 8**

Compound	Elements	Mass ratio
CH_4	C and H	3 : 1
H_2O	O and H	8 : 1
CO_2	C and O	3 : 8

Numerical Problem Solving 4

Phosphorus forms PH_3 and PCl_3 with H_2 and Cl_2 respectively.

$$\underset{31\,g}{P} + \underset{3\,g}{\frac{3}{2}H_2} \longrightarrow PH_3$$

$$\underset{31\,g}{P} + \underset{3\times 35.5\,g}{\frac{3}{2}Cl_2} \longrightarrow PCl_3$$

In which ratio by mass H_2 and Cl_2 combine to form HCl?

Ans. ....

Law of Equivalent Proportions

(Richter–1792)

"Elements combine in the ratio of their equivalent mass or in simple multiple of them. (Equivalent mass – Refer in this unit)"

Equivalent mass of hydrogen = 1

Equivalent mass of oxygen = 8

Equivalent mass of magnesium = 12

Equivalent mass of calcium = 20

Equivalent mass of chlorine = 35.5

$$H_2 + Cl_2 \longrightarrow 3HCl$$

Ratio by mass 2g 2 × 35.5 g 2 × 36.5 g

H_2 and Cl_2 combine in the mass ratio = 2 : 2 × 35.5 = 1 : 35.5

Thus, ratio is also the ratio of their equivalent masses.

$$\underset{4\,g}{2H_2} + \underset{32\,g}{O_2} \longrightarrow \underset{36\,g}{2H_2O}$$

H_2 and O_2 combine in the mass ratio = 4 : 32 = 1 : 8

This is also the ratio of their equivalent masses.

Law of **Reciprocal Proportion is a special case of Law of Equivalent Proportion**

Numerical Problem Solving 5

Consider the following reactions (Mg = 24, Cl = 35.5, Ca = 40)

(a) $Mg + H_2 \longrightarrow MgH_2$

(b) $Mg + Cl_2 \longrightarrow MgCl_2$

(c) $Ca + Cl_2 \longrightarrow CaCl_2$

Equivalent mass of hydrogen is 1, derive equivalent mass of magnesium, chlorine and calcium.

Ans. Mg = ..., Cl = ... Ca = ...

Law of Gaseous Volume or Gay-Lussac's Law (Gay-Lussac–1808)

When gases combine together they do so in simple ratio by volume and the volume of the product formed (if gaseous), also bear a simple relation to the volume of the reacting gases, measured under the same conditions of temperature and pressure.

(1 mole of every gas at NTP = 22.4 L and equal moles occupy equal volumes)

	Gaseous Reactions					R of V Reactants	P
(a)	H_2 (1 mol, 1 volume)	+	Cl_2 (1 mol, 1 volume)	$\longrightarrow$	2HCl (2 mol, 2 volumes)	1:1	2
(b)	N_2 (1 mol, 1 volume)	+	$3H_2$ (3 mol, 3 volumes)	$\longrightarrow$	$2NH_3$ (2 mol, 2 volumes)	1:3	2
(c)	$2H_2$ (2 mol, 2 volumes)	+	O_2 (1 mol, 1 volume)	$\longrightarrow$	$2H_2O$ (2 mol, 2 volumes)	2:1	2

***R of V** = Ratio of volume, ****P** = Product

Numerical Problem Solving 6

Nitrogen combines with oxygen in the following reaction

$$N_2(g) + O_2(g) \longrightarrow 2NO(g)$$

What is ratio of the reactants and products?

(a) By mass (b) By equivalent
(c) By moles (d) By volume

Ans.	N_2	O_2	NO
(a)	....	....	....
(b)	....	....	....
(c)	....	....	....
(d)	....	....	....

Dalton and Constitution of Matter

Dalton Atomic Theory (1808)

- All elements are composed of minute particles called atoms, which do not undergo subdivision during chemical reactions and cannot be created or destroyed by any chemical process. (Taylor)
- Atoms of the same element are similar to one another in all respects and are equal in weights.
- Atoms of different elements have different masses (called atomic masses) and different properties.
- Chemical combination between two or more elements takes place by interaction of their atoms in simple numerical proportions (as 1 : 1; 1 : 2; 1 : 3; etc.)

Reaction	Ratio
$H_2 + Cl_2 \longrightarrow 2HCl$	1 : 1
$N_2 + 3H_2 \longrightarrow 2NH_3$	1 : 3
$C + O_2 \longrightarrow CO_2$	1 : 1

- Combining masses of the elements represent the combining masses of the atoms.
Dalton atomic, theory can explain various laws of chemical combination.

Limitations and Defects of Dalton's Atomic Theory

- It can explain the laws of chemical combination by weight, but it fails to explain Gay-Lussac's law of gaseous volume.
- Distinction between the ultimate particles of an element and the smallest particle of a compound, both of which are made of atoms, could not be made.
- It assumes that atoms of same element are like and have same mass. But this is not true for isotopes (two or more atoms with same atomic number but different atomic mass are called isotopes).
- $^{12}_{6}C$, $^{13}_{6}C$, $^{14}_{6}C$, have atoms of different masses. $^{1}_{1}H$, $^{2}_{1}H$, $^{3}_{1}H$ have atoms of different masses.

Avogadro's Hypothesis

Berzelius assumed **"equal volumes of different gases under identical conditions of temperature and pressure contain the same number of atoms (The word atom was at that time used for both elements and compounds as molecule was not thought of).** We consider the formation of $H_2O(g)$ from $H_2(g)$ and $O_2(g)$

	$2H_2(g)$ +	$O_2(g)$	$\longrightarrow$ $2H_2O(g)$
	2 volume	1 volume	2 volume
Berzelius	2 atoms	1 atom	2 atoms
	1 atom	$\frac{1}{2}$ atom	1 atom

1 atom of H_2 combines with 1/2 atom of O_2. This allows atom to be divisible. This is against Dalton's atomic theory. Hence, Berzelius hypothesis was discarded. Failure led to assume that gases exist in polyatomic state (diatomic, triatomic etc.) and ultimate particles which can exsit in the free state were defined as **molecule.** Avogadro's hypothesis is based on molecules :

"Equal volume of all gases under the same conditions of temperature and pressure contain the same number of molecule".

Avogadro's hypothesis led to

- the word molecule and made the distinction between atoms and molecules.
- conclude that molecules of hydrogen, chlorine, oxygen, nitrogen and fluorine are diatomic.
- conclude that molecular weight = 2 × vapour density.
- Conclude that one gram-molecular mass (molecular weight) has volume 22.4 L (22.4×10^{-3} m^3) at NTP (pressure = 1 atm and T = 273.15 K)

Number of molecules present in amount equal to molecular weight (which we shall call one mole) are 6.02×10^{23} (called Avogadro's number)

H_2	+	Cl_2	$\longrightarrow$	$2HCl$
1 volume		1 volume		2 volumes
1 mole		1 mole		2 moles
6.02×10^{23} molecules		6.02×10^{23} molecules		$2 \times 6.02 \times 10^{23}$ molecules

Dalton's Theory as Modified by Avogadro's Hypothesis

1. Substances (elements and compounds) consist of molecules which are composed of indivisible particles called atoms. Molecules are capable of independent existence and are divisible into atoms.
2. Molecules of a particular substance are similar to one another; with identical properties and mass. Molecules of one substance are different from those of other substances.
3. When there is chemical reaction between two or more substances, the molecules of each substance are first disintegrated into atoms which combine together to produce new molecules of the new substances.

$$N_2(g) + 3H_2(g) \longrightarrow 2NH_3$$
$$\downarrow \qquad \downarrow$$
$$2N \;+\; 6H \longrightarrow 2NH_3$$

Atomic Number, Mass Number and Isotopes

- All atoms can be identified by the number of protons and neutrons they contain. *The atomic number (Z) is the number of protons in the nucleus of each atom of an element.*
- In a neutral atom the number of protons is equal to the number of electrons present in the atom.
- The chemical identity of an atom can be determined solely from its atomic number. For example atomic number of oxygen is 8. This means each neutral oxygen atom has 8 protons and 8 electrons.
- The *mass number (A) is the total number of neutrons and protons present in the nucleus of an atom of an element.*
- Except for hydrogen (which has only one proton), all atomic nuclei contain both protons and neutrons.
- The mass number is given by

mass number $(A) = P + N = Z + N$

= atomic number + no. of neutrons

Thus, N (number of neutrons) $= A - Z$

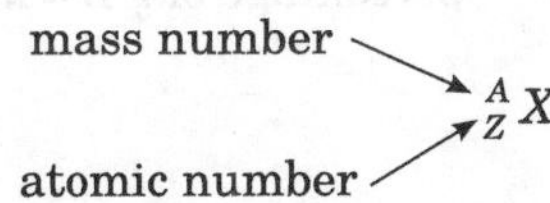

- Atomic number, number of neutrons, and mass number all must be positive integers (whole numbers).
- Atoms of a given element do not all have the same mass. Most elements have two or more **isotopes** (atoms that have the same atomic number but different mass numbers). There are three isotopes of hydrogen

^1_1H	hydrogen	$1P$	$0N$
^2_1H	deuterium	$1P$	$1N$
^3_1H	tritium	$1P$	$2N$

Thus, isotopes have different number of neutrons.

- The chemical properties of an element are determined primarily by the protons and electrons in its atoms; neutrons do not take part in chemical changes under normal conditions. Therefore, isotopes of the same element have similar chemical reactions.

Atomic Weight

The atomic weight of an element is the smallest weight of it present in the molecular weights of its compounds.

It may also be defined thus —

The atomic weight of an element is a number which expresses how many times the weight of one atom of the element is greater than the weight of one atom of hydrogen.

At. wt. of an element

$$= \frac{\text{wt. of 1 atom of the element}}{\text{wt. of 1 atom of hydrogen}}$$

Thus, the atomic weights are not the actual weights of atoms but the relative weights of the atoms of various elements expressed in terms of hydrogen as unit. Hence, when we say that the atomic weight of nitrogen is 14, we mean that an atom of nitrogen is 14 times heavier than an atom of hydrogen; that is to say, if the weight of one atom of hydrogen is taken as unit, the weight of 1 atom of nitrogen will be 14 times the weight of 1 atom of hydrogen.

Now a days the standard for atomic weight is C^{12}. On this basis the atomic weight may be defined as :

The atomic weight of an element is a number which indicates as to how many times an element of that element is heavier as compared with 1/12 of the mass of an atom of C-12 ($^{12}_6C$).

$$\text{At. wt. of an element} = \frac{\text{wt. of an atom}}{\frac{1}{12} \times \text{wt. of an atom of } ^{12}C}$$

The gram atomic weight (or the gram-atom) of an element is its atomic weight expressed in grams. Thus, 1 g-atom of magnesium stands for 24 g of Mg.

The French chemists Dulong and Petit in 1819 found out that the product of the specific heat of an element in the solid state and its atomic weight is constant, and is equal to 6.4.

Specific heat × atomic weight = 6.4 (approx.)

Atomic Mass Unit

An early observation regarding atomic weight was that carbon and hydrogen have relative atomic masses, also traditionally called **atomic weight,** AW, of approximately 12 and 1, respectively. Thousands of experiments on the composition of compounds have resulted in the establishment of scale of relative weights based on the atomic mass unit (amu), which is defined as.

"Exactly 1/12 of the mass of an atom of a particular kind of carbon atom, called carbon-12. Its value is 1.66054×10^{-24} g.

On this scale, the atomic weight of hydrogen is 1.00794 amu

Sodium is 22.989768 amu – 23 times of H
Magnesium is 24.3050 amu – 24 times of H

- amu is also called dalton (Da).
- Mass of one ${}^{12}_{6}$C atom = 12 amu (exactly)

$$1 \text{ amu} = \frac{\text{mass of one } {}^{12}_{6}\text{C atom}}{12}$$

$$= 1.66054 \times 10^{-24} \text{ g}$$

- Thus, mass of one H-atom $= \frac{1}{N_0} = 1$ amu

mass of one O-atom $= \frac{16}{N_0} = 16$ amu

Average Atomic Weight

Average mass number of natural occurring element containing two or more isotopes is given by

$$\bar{A} = \frac{\Sigma AX}{\Sigma X} = \frac{A_1X_1 + A_2X_2 + \ldots}{X_1 + X_2 + \ldots}$$

where, $A_1, A_2, \ldots$ are mass numbers of the different isotopes with percentage or ratios as $X_1, X_2, \ldots$

If ${}^{35}_{17}$Cl and ${}^{37}_{17}$Cl are two isotopes of chlorine in the ratio of 3 : 1, then average mass number of Cl is

$$\bar{A}\,(\text{Cl}) = \frac{35 \times 3 + 37 \times 1}{3 + 1} = 35.5$$

Illustration 1 *A divalent cation (Atomic number Z) is isoelectronic of CO_2 and has (Z + 2) neutrons. What is ionic mass of divalent cation?*

Solution A divalent cation (M^{2+}) has as many as electrons as $CO_2 = 6 + 16 = 22$

Hence, (Z) atomic number of

$$M^{2+} = 22 + 2 = 24$$

or number of protons (P) in $M^{2+} = 24$

number of neutrons (N) in $M^{2+} = Z + 2 = 24 + 2 = 26$

$\therefore$ ionic mass of $M^{2+} = P + N = 24 + 26 =$ **50**

Illustration 2 *Ionic mass of X^{2-} is 16. It has 8 neutrons. Determine the number of its (a) protons (b) number of electrons and (c) atomic number Z.*

Solution Ionic mass of $X^{2-} = 16$, neutrons = 8

(a) protons = 16 – 8 = **8**

$\therefore$ electrons in neutral $X = 8$

(b) X gains two electrons to form X^{2-},

hence, number of electrons is $X^{2-} = 8 + 2 =$ **10**

(c) Z = atomic number = **8**

Illustration 3 *Write the appropriate symbol for each of the following isotopes.*
(*a*) $Z = 74, A = 186$ (*b*) $Z = 80, A = 201$

Solution For any isotope, system is

mass number → ${}^{A}_{Z}X$ ← atomic number

(a) $Z = 74, A = 186$ ${}^{186}_{74}X$
(b) $Z = 80, A = 201$ ${}^{201}_{80}X$

Illustration 4 ${}^{10}_{5}$B *and* ${}^{11}_{5}$B *are two isotopes of boron. If average mass number is 10.2, what is the percentage of each?*

Solution Let the percentage of ${}^{10}_{5}$B $= x$

then percentage of ${}^{11}_{5}$B $= (100 - x)$

$$\text{Average mass number } \bar{A} = \frac{A_1X_1 + A_2X_2}{X_1 + X_2}$$

$$\therefore \quad 10.2 = \frac{10x + 11(100 - x)}{x + 100 - x}$$

$$10.2 = \frac{10x + 1100 - 11x}{100}$$

$$x = 80$$

Thus, percentage of ${}^{10}_{5}$B = **80%**

and percentage of ${}^{11}_{5}$B = **20%**

Target Practice 1

Avogadro's number = 6.02×10^{23} atoms/mol

1 mole of a gas at STP occupies 22.4 L

1. How many atoms of oxygen are there in 18 g of water? ....
2. How many atoms of hydrogen are there in 18 g of water?
3. How many molecules of H_2O are there in 18 g of water? ......
4. What is the mass of 1 mole of O_2? ....
5. What is the mass of 1 molecule of O_2? ....
6. What is the mass of 2 moles of H_2SO_4? ....
7. What is the density (mass/voume) of O_2 at STP? ....
8. 3 L of a gas weighs 2 g. What is the molecular mass? ....
9. What volume does 22 g of CO_2 at STP occupy? ....
10. How many atoms of hydrogen are in 67.2 L of H_2 at STP? ....
11. You are given 1.0 g samples of He, Fe and C. Which contains the largest number of atoms? Which contains the smallest? (He = 4, Fe = 56, C = 12)

Note : Avogadro's number = 6.02×10^{23} atoms/mol

1 mole of a gas at STP occupies 22.4 L

Answers

1. 6.02×10^{23} **2.** 1.204×10^{24} **3.** 6.02×10^{23} **4.** 32 g **5.** 5.32×10^{-23} g **6.** 196 g **7.** 1.43 g/L **8.** 14.9 g/L **9.** 11.2 L **10.** 3.612×10^{24}

11. $He=\frac{N_0}{4}$, $Fe=\frac{N_0}{56}$, $C=\frac{N_0}{12}$

He (Largest) Fe (smallest)

Equivalent Weight

It is also called the chemical equivalent (or simply equivalent) or the combining weight.

The weight of an element which can combine with or displace from a compound **1 part by weight of hydrogen,** or **8 parts by weight of oxygen** or **35.5 parts by weight of chlorine** is called the equivalent weight of the element.

1 part by weight of hydrogen is actually the reference weight.

- $\underset{2\,g}{H_2} + \underset{(35.5\times 2)\,g}{Cl_2} \longrightarrow \underset{2\times 36.5\,g}{2HCl}$

$\Rightarrow$ 2 g hydrogen is equivalent (≡) to 71 g chlorine (Cl_2)

thus, 1 g hydrogen is equivalent to (≡) 35.5 g Cl_2

Thus, equivalent weight of Cl_2 = 35.5 g equivalent^{-1}

and also equivalent weight of HCl = 36.5 g equivalent^{-1}

$\Rightarrow \quad \underset{4\,g}{2H_2} + \underset{32\,g}{O_2} \longrightarrow \underset{36\,g}{2H_2O}$

4 g hydrogen ≡ 32 g oxygen

Thus, 1 g hydrogen ≡ 8 g oxygen

Thus, equivalent weight of oxygen

= 8 g equivalent^{-1}

Thus equivalent weight of water = 9 g equivalent^{-1}

$\Rightarrow \quad \underset{28\,g}{N_2} + \underset{6\,g}{3H_2} \longrightarrow \underset{34\,g}{2NH_3}$

6 g hydrogen ≡ 28 g nitrogen and 34 g ammonia

Thus, 1 g hydrogen

$\equiv \frac{28}{6}$ (= 4.66 g) nitrogen and $\frac{34}{6}$ (5.66 g) ammonia.

Thus, equivalent weight of nitrogen = 4.66 g equivalent^{-1}

and equivalent weight of ammonia = 5.66 g equivalent^{-1}

$\Rightarrow \quad \underset{24\,g}{Mg} + \underset{98\,g}{H_2SO_4} \longrightarrow \underset{120\,g}{MgSO_4} + \underset{2\,g}{H_2}$

2 g H_2 is displaced by 24 g Mg

Thus, 1 g H_2 is displaced by 12 g Mg

Thus, equivalent weight of magnesium

= 12 g equivalent^{-1}

Also, 2 g H_2 is displaced by 98 g H_2SO_4

Thus, 1 g H_2 is displaced by 49 g H_2SO_4

Thus, equivalent weight of H_2SO_4 = 49 g equivalent.

DEEP Focus

Equivalent weight is not a fixed quantity. It varies reaction to reaction.

$$N_2 + 3H_2 \longrightarrow 2NH_3$$

Equivalent weight of nitrogen (N_2) as equivalent to 1 g H_2 = 4.66 g equivalent^{-1}

$$\underset{28\,g}{N_2} + \underset{32\,g}{O_2} \longrightarrow 2NO$$

32 g oxygen ≡ 28 g nitrogen

Thus, 8 g oxygen ≡ 7 g nitrogen

Thus, equivalent weight of nitrogen

= **7 g equivalent^{-1}**

In acid-base reactions, one **equivalent weight,** or **equivalent (eq),** of an acid is defined as the mass of acid (in grams) that will furnish 6.022×10^{23} hydrogen ions (1 mole) or that will react with 6.022×10^{23} hydroxide ions (1 mole). One mole of an acid contains 6.022×10^{23} formula units of the acid.

$$\underset{\substack{1\text{ mol}\\36.50\text{ g}\\6.022\times10^{23}\text{ FU}}}{HCl} \xrightarrow{H_2O} \underset{\substack{1\text{ mol}\\1.00\text{ g}\\6.022\times10^{23}\text{ FU}}}{H^+(aq)} + \underset{\substack{1\text{ mol}\\35.50\text{ g}\\6.022\times10^{23}\text{ FU}}}{Cl^-(aq)}$$

We observe that one mole of HCl can produce 6.022×10^{23} H^+, hence one mole of HCl is one equivalent. The same is true for all monoprotic acids (with one ionisable H^+)

Sulphuric acid (H_2SO_4) is a diprotic acid. One molecule of H_2SO_4 can furnish $2H^+$ ions

$$\underset{\substack{1\text{ mol}\\6.022\times10^{23}\text{ FU}}}{H_2SO_4} \xrightarrow{H_2O} \underset{\substack{2\text{ mol}\\2\times(6.022\times10^{23})\text{ FU}}}{2H^+(aq)} + \underset{\substack{1\text{ mol}\\6.022\times10^{23}\text{ FU}}}{SO_4^{2-}(aq)}$$

FU = Formula Unit

Thus, equation shows that one mole H_2SO_4 can produce $2(6.022 \times 10^{23})$ H^+ ions therefore, one mole of H_2SO_4 is two equivalent weights in all reactions in which both acidic hydrogen ions react.

One **equivalent weight** of a base is defined as the mass of the base (expressed in grams) that will furnish 6.022×10^{23} hydroxide ions or the mass of the base that will react with 6.022×10^{23} hydrogen ions.

Detailed study of Equivalent weights based on **redox reaction** are taken **in unit-3**

Numerical Problem Solving 7

(i) Taking equivalent weight of hydrogen as (1) and that of oxygen as (8) derive equivalent weights of the underlined.

(a) $Mg + H_2 \longrightarrow \underline{MgH_2}$ ....

(b) $Mg + \frac{1}{2}O_2 \longrightarrow \underline{MgO}$ ....

(c) $C + O_2 \longrightarrow \underline{CO_2}$ ....

(d) $\frac{1}{2}N_2 + O_2 \longrightarrow \underline{NO_2}$ ....

(ii) How many equivalents are in (based on ionization)?

(a) 3.65 g HCl ($\rightleftharpoons H^+ + Cl^-$) ....

(b) 9.8 g H_2SO_4 ($\rightleftharpoons 2H^+ + SO_4^{2-}$).........

(c) 9.8 g of H_3PO_4 ($\rightleftharpoons 3H^+ + PO_4^{3-}$) ....

(iii) Cobalt has three radioactive isotopes used in medical studies. Atoms of these isotopes have 30, 31 and 33 neutrons, respectively. Give the symbol for each of these isotopes ....

The Mole

You are already familiar with the dozen (12 items) and the gross (144 items).

The SI unit for amount is the mole, abbreviated mol. It is defined as the amount of substance that contains as many entities (atoms, molecules, or other particles) as there are atoms in exactly 0.012 kg of pure carbon-12 atoms. Many experiments have refined the number, and the currently accepted value is

$$1 \text{ mole} = 6.0221367 \times 10^{23} \text{ particles}$$

One mole is always associated with a chemical formula and refers to Avogadro's number of particles. It is designated by N_0

$$N_0 = 6.022 \times 10^{23} \text{ mol}^{-1}$$

$$16 \text{ g oxygen atom} = 1 \text{ mole of oxygen atom}$$
$$= 6.022 \times 10^{23} \text{ atoms}$$

$$32 \text{ g oxygen gas} = 1 \text{ mole of oxygen } (O_2) \text{ gas}$$
$$= 6.022 \times 10^{23} \text{ oxygen molecules}$$
$$= 2 \times 6.022 \times 10^{23} \text{ oxygen atoms}$$

DEEP Focus

When to use 1.008 g/mol for the molar mass of hydrogen and when to use 2.016 g/mol?

The statement, "a mole of hydrogen" is ambiguous; you should always say either a mole of hydrogen *atoms* or a mole of hydrogen *molecules.* Better still write **1 mol H** or **1 mol H_2.** If you do this is, then you will see that the molar masses should be expressed as **1.008 g H/mol H** and **2.016 g H_2/mol H_2.** This distinction is very much like the distinction between one dozen socks and one dozen pairs of socks (that is, the H atom is analogous to a single sock and the H_2 molecule to a pair of socks).

The mass of one mole of atoms of pure element in grams is numerically equal to atomic weight of that element in amu. This is also called **molar mass** of the element; its **units are grams/mole**

Molar mass of a substance is the mass in gram of one mole of that substance.

$$\text{Moles of substance} = \frac{\text{mass in grams}}{\text{molar mass in grams/mole}}$$

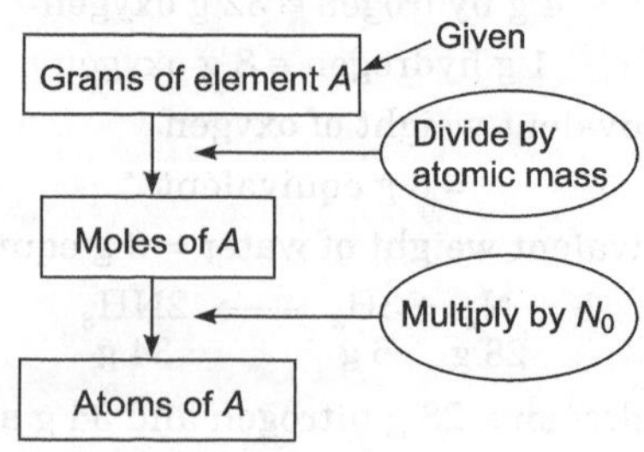

Fig. 1.3

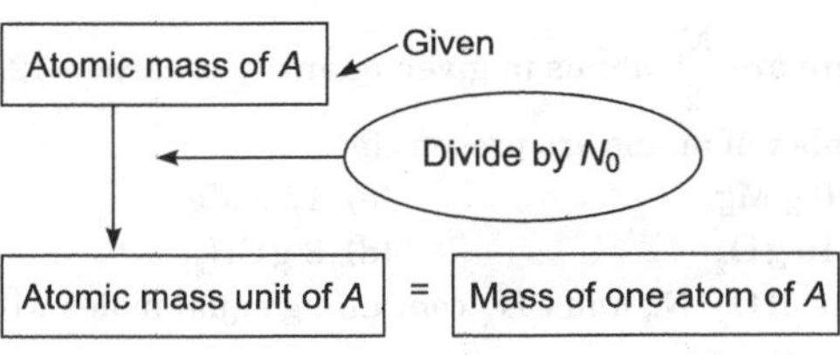

Fig. 1.4

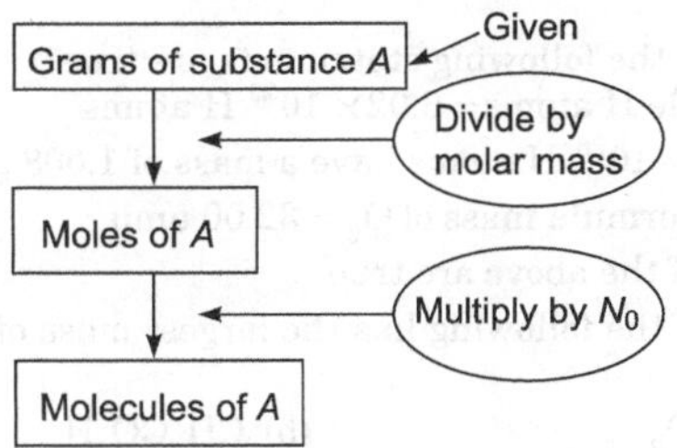

Fig. 1.5

Table 1.2 **One Mole of Some Ionic Compounds**

Compound	Formula weight	A sample with a mass of 1 mole	Contains
Sodium chloride	58.44	58.44 g NaCl	6.022×10^{23} Na^+ ions or 1 mole of Na^+ ions 6.022×10^{23} Cl^- ions or 1 mole of Cl^- ions
Calcium chloride	111.0	111.0 g $CaCl_2$	6.022×10^{23} Ca^{2+} ions or 1 mole of Ca^{2+} ions $2(6.022\times 10^{23})$ Cl^- ions or 2 moles of Cl^- ions
Aluminium sulphate	342.1	342.1 g $Al_2(SO_4)_3$	$2(6.022\times 10^{23})$ Al^{3+} ions or 2 moles of Al^{3+} ions $3(6.022\times 10^{23})$ SO_4^{2-} ions or 3 moles of SO_4^{2-} ions

Illustration 4 *Sulphur is a non-metallic element. Its presence in coal gives rise to the acid-rain phenomenon. How many sulphur atoms are in 10 kg of coal sample containing 1% sulphur.* (S = 32 *g mol*$^{-1}$)

Solution 10 kg coal = 10×10^3 g coal

$$= \frac{10\times 10^3 \times 1}{100} \text{ g sulphur}$$

$$= 10^2/32 \text{ mol sulphur}$$

$$= (10^2/32)\times 6.022\times 10^{23} \text{ atoms of sulphur}$$

$$= \mathbf{1.882\times 10^{24}} \textbf{ sulphur atoms}$$

Illustration 5 *How many H atoms are present in 25.6 g of urea* $[(NH_2)_2CO)]$*, having molar mass of 60 g mol*$^{-1}$?

Solution 25.6 g urea $= \frac{25.6}{60}$ mol urea

$$= \frac{25.6}{60}\times 6.022\times 10^{23} \text{ molecules}$$

$$= \frac{25.6}{60}\times 6.022\times 10^{23}\times 4\text{H atoms}$$

$$= \mathbf{1.0274\times 10^{24}} \textbf{ H atoms}$$

Illustration 6 *Mass of one atom of an element X is* 2.32481×10^{-23} *g. Express this in amu.*

Solution Let atomic mass of the element be = A

Number of atoms in one mol of $X = N_0$

Hence, mass of one atom of $X = \frac{A}{N_0}$

$$\therefore \quad \frac{A}{N_0} = 2.32481\times 10^{-23} \text{ g}$$

$$\therefore \quad A = 2.32481\times 10^{-23}\times N_0$$

$$= 2.32481\times 10^{-23}\times 6.022\times 10^{23}$$

$$= 14$$

Hence, mass of one atom of $X = \frac{14}{N_0} = \mathbf{14\ amu}$

Target Practice 2

1. Which of the following has the greatest mass?
   (a) 200 molecules of water
   (b) 100 atoms of Fe
   (c) 200 molecules of O_2
   (d) 200 molecules of CH_4
2. If the fertilizers listed below are priced according to their nitrogen content, which will be the least expensive per 50 kg bag?
   (a) Urea $(NH_2)_2CO$
   (b) Ammonia NH_3
   (c) Ammonium nitrate NH_4NO_3
   (d) Guanidine $HNC(NH_2)_2$
3. One atom of an element has a mass of 9.123×10^{-23} g. There is only one isotope of the element. The element is
   (a) chlorine (b) chromium
   (c) magnesium (d) manganese
4. One mole of $(NH_4)_2Cr_2O_7$ contains
   (a) one mole of nitrogen (b) four moles of hydrogen
   (c) one mole of chromium (d) seven moles of oxygen

5. Which of the following statements are true?
(a) 1 mole H atoms $= 6.02\times 10^{23}$ H atoms
(b) 6.02×10^{23} H atoms have a mass of 1.008 g
(c) The formula mass of $O_2 = 32.00$ amu
(d) All of the above are true

6. Which of the following has the largest mass of carbon per gram?
(a) H_2CO_3 (b) CH_3CO_2H
(c) CH_3OH (d) CH_3CH_2OH

7. 1 g urea (NH_2CONH_2), 1 g acetic acid (CH_3COOH) and 1 g formaldehyde (HCHO) will have H-atoms in the ratio
(a) 2 : 2 : 1 (b) 1 : 1 : 2
(c) 1 : 1 : 1 (d) 1 : 2 : 1

8. There are $\frac{N_0}{2}$ atoms in given quantity of carbon-12. Same number of atoms are present in
(a) 6 g Mg (b) 12 g Mg
(c) 16 g O_2 (d) 8 g CH_4

9. Mass of O_2, N_2 and CH_4 containing equal moles will be in the ratio of
(a) 16 : 14 : 16 (b) 8 : 7 : 4
(c) 1 : 1 : 1 (d) 4 : 7 : 4

10. Which has maximum volume at STP? (1 mole of a gas = 22.4 L at STP)
(a) 1.6 g CH_4 (b) 6.02×10^{22} N_2 molecules
(c) 4.4 g CO_2 (d) Equal

Answers

1. (a) 200 molecules $H_2O = \frac{200}{N_0}$ mol $H_2O = \frac{3600}{N_0}$ g H_2O

(b) 100 atoms Fe $= \frac{100}{N_0}$ mol Fe $= \frac{100}{N_0} \times 56 = \frac{5600}{N_0}$ g Fe

(c) 200 molecules $O_2 = \frac{200}{N_0}$ mol $O_2 = \frac{6400}{N_0}$ g O_2

(d) 200 molecules CH_4

$= \frac{200}{N_0} \times 10$ g $CH_4 = \frac{3200}{N_0}$ g CH_4

Thus, (c)

2. (c)

3. $9.123\times 10^{-23} \times 6.02\times 10^{23}$ g mol$^{-1} \approx 55$
Mn. Thus, (d)

4. (d) **5.** (d)

6. (a) H_2CO_3, $\frac{12}{62}$ g carbon (b) CH_3CO_2H, $\frac{24}{60}$ g carbon

(c) CH_3OH, $\frac{12}{32}$ g carbon (d) CH_3CH_2OH, $\frac{24}{40}$ g carbon (max.)

Thus, (d)

7. 1 g urea $= \frac{1}{60}$ mol $= \frac{4N_0}{60}$ H-atom

1 g acetic acid $= \frac{1}{60}$ mol $= \frac{4N_0}{60}$ H-atom

1 g formaldehyde $= \frac{1}{30}$ mol $= \frac{2N_0}{30}$ H-atom $= \frac{4N_0}{60}$ H-atom

Thus, 1 : 1 : 1

Thus, (c)

8. (b) **9.** (b) **10.** (d)

Practice Exercise 1

1. Substances *X* and *Y* are colourless gases obtained by combination of sulphur and oxygen. Substances *X* results from combination of 6.00 g of sulphur with 6.00 g of oxygen and substance *Y* results from combination of 8.60 g of sulphur with 12.9 g of oxygen. Show that the mass ratios in the two substances are simple multiples of each other.

2. In addition to carbon monoxide (CO) and carbon dioxide (CO_2), there is a third compound of carbon called carbon suboxide. If a 2.500 g sample of carbon suboxide contains 1.32 g of C and 1.18 g of O, show that the law of multiple proportions is followed. What is possible formula of carbon suboxide?

3. In which one of the following pairs do the two species resemble each other most closely in chemical properties? Explain.
(a) ${}^1_1H$ and ${}^1_1H^+$ (b) ${}^{14}_7N$ and ${}^{14}_7N^{3-}$
(c) ${}^{12}_6C$ and ${}^{13}_6C$

4. One isotope of a metallic element has mass number 65 and 35 neutrons in the nucleus. The cation derived from the isotope has 28 electrons. Write the symbol for this cation.

5. Which of the following symbols provides more information about the atom : ${}^{23}Na$ or ${}_{11}Na$? Explain.

6. Fill in the blanks in the following table :

Symbol		${}^{54}_{26}Fe$			
Protons	5	—	—	79	86
Neutrons	6	—	16	117	136
Electrons	5	—	18	79	
Net charge	—	—	−3		0

7. How many H atoms are in 6.0 g of isopropanol C_3H_8O?

8. How many moles of cobalt atoms are there in 6.022×10^9 Co atoms?

9. The density of water is 1.00 g/mL at 4°C. How many water molecules are present in 2.56 mL water at this temperature?

10. Aspirin has the formula $C_9H_8O_4$. How many moles of aspirin are there in a tablet weighing 500 mg? How many molecules?

11. What is molecular weight of chloroform if 0.0275 mol weighs 3.28 g?

12. An average cup of coffee contains about 125 mg caffeine, $C_8H_{10}N_4O_2$. How many moles of caffeine are in a cup? How many molecules?

13. Select correct alternate. Only **one** alternate is correct.
(A) In which case purity of the substance is 100%?
(a) 1 mole of $CaCO_3$ gave 11.2 L CO_2 (at STP)

(b) 1 mole of $MgCO_3$ gave 40.0 g MgO
(c) 1 mole of $NaHCO_3$ gave 4 g H_2O
(d) 1 mole of $Ca(HCO_3)_2$ gave 1 mole CO_2

(B) Consider following laws of chemical combination with examples:
I : Law of multiple proportion : N_2O, NO, NO_2
II : Law of reciprocal proportion : H_2O, SO_2, H_2S
Which is correct with examples?
(a) I and II (b) I only
(c) II only (d) None of these

(C) H_2S contains 94.11% sulphur; SO_2 contains 50% oxygen and H_2O contains 11.11% hydrogen. Thus,
(a) law of multiple proportion is followed
(b) law of reciprocal proportion is followed
(c) law of conservation of mass is followed
(d) all of the above are followed

(D) Sodium combines with $^{35}_{17}Cl$ and $^{37}_{17}Cl$ to give two samples of sodium chloride. Their formation follows the law of
(a) gaseous diffusion (b) conservation of mass
(c) reciprocal proportion (d) none of these

(E) According to Dalton's atomic theory, the smallest particle in which matter can exist is called
(a) an atom (b) an ion
(c) an electron (d) a molecule

(F) M^{2+} ion is isoelectronic of SO_2 and has $(Z + 2)$ neutrons (Z is atomic number of M). Thus, ionic mass of M^{2+} is
(a) 70 (b) 66
(c) 68 (d) 64

(G) X^+, Y^{2+} and Z^- are isoelectronic of CO_2. Increasing order of protons in X^+, Y^{2+} and Z^- is
(a) $X^+ = Y^{2+} = Z^-$ (b) $X^+ < Y^{2+} < Z^-$
(c) $Z^- < X^+ < Y^{2+}$ (d) $Y^{2+} < X^+ < Z^-$

(H) X^-, Y^{2-} and Z^{3-} are isotonic and isoelectronic. Thus, increasing order of atomic number of X, Y and Z is
(a) $X < Y < Z$ (b) $Z < Y < X$
(c) $X = Y = Z$ (d) $Z < X < Y$

(I) If each O-atom has two equivalents, volume of one equivalent of O_2 gas at STP is
(a) 22.4 L (b) 11.2 L
(c) 5.6 L (d) 44.8 L

(J) Each drop of H_2O has 0.018 mL at room temperature. Number of H_2O molecules in one drop is
(a) 1×10^{-3} (b) 6.02×10^{20}
(c) 22.4×10^{-3} (d) $6.02 \times 3 \times 10^{2}$

(K) 1 g CH_4 and 4 g of compound X have equal number of moles. Thus, molar mass of X is
(a) 16 g mol^{-1} (b) 32 g mol^{-1}
(c) 4 g mol^{-1} (d) 64 g mol^{-1}

Answers

1. Ratio S : O
X : SO_2 — 1 : 2
Y : SO_3 — 1 : 3
Thus, O atom in SO_2 and SO_3 are simple multiple of each other.

2. Carbon suboxide C_3O_2

$$O{=}C{=}C{=}C{=}O$$

3. (c) **4.** $^{65}_{30}M^{2+}$ **5.** ^{23}Na

6.

Symbol	$^{11}_{5}B$	$^{54}_{26}Fe$	$^{31}_{15}P^{3-}$	$^{196}_{79}Au$	$^{222}_{86}Rn$
Protons	5	26	15	79	86
Neutrons	6	28	16	117	136
Electrons	5	26	18	79	86
Net charge	zero	zero	–3	zero	zero

7. 4.8176×10^{23} H atoms
8. 10^{-14} mol cobalt atoms
9. 8.56×10^{22} water molecules
10. 2.78×10^{-3} mol, 1.67×10^{21} molecules
11. 119.27 g mol^{-1}
12. 6.44×10^{-4} mol, 3.88×10^{20} N_0 molecules
13. A. (b) B. (a)
C. (b) D. (d)
E. (a) F. (a)
G. (c) H. (b)
I. (c) J. (b)
K. (d)

Per cent Composition and Empirical Formula

- The per cent composition of a compound is expressed by identifying the elements present and giving the mass per cent of each.
- By knowing per cent composition of compound makes it possible to calculate the chemical formula of the compounds. The method involved is to find the relative number of mole of each element in the compound and then use & numbers to establish the mole ratios of the elements.
- A formula derived up to above stage from data about per cent composition, is called an **empirical formula** because it tells the ratios of atoms in a compound. The **simplest** or **empirical formula** for a compound is the smallest whole number ratio of atoms present.

For **molecular compounds** the **molecular formula** indicates the actual number of atoms present in a molecule of the compound. It may be the same as the simplest formula or elase whole-number multiple of it. Simplest and molecular formulae of water are both H_2O. However for hydrogen peroxide they are **HO** and $\mathbf{H_2O_2}$ respectively.

$$\text{Multiple } n = \frac{\text{molecular weight}}{\text{empirical formula weight}}$$

- Molecular formula = (empirical formula)$_n$

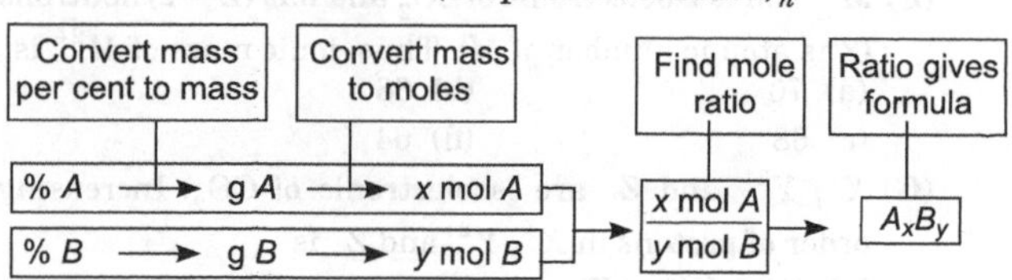

Illustration 1 *What is simplest formula of a salt that contains 40% calcium, 12% carbon and 48% oxygen?*

Solution

Element	Relative mass	Atomic weight (At. wt.)	Relative number of atoms	Divide by smallest number	Smallest whole number ratio of atoms
Ca	40.0	40.0	$\frac{40}{40} = 1.0$	$\frac{1}{1} = 1$ (Ca)	1
C	12.0	12.0	$\frac{12}{12} = 1.0$	$\frac{1}{1} = 1$ (C)	1
O	48.0	16.0	$\frac{48}{16} = 3.0$	$\frac{3}{1} = 3$ (O)	3

Thus, simplest formula = $\mathbf{CaCO_3}$

Illustration 2 *Glucose or blood sugar has the molecular formula $C_6H_{12}O_6$. What is the empirical formula and what is the per cent composition of glucose?*

Solution Molecular formula : $C_6H_{12}O_6$

Mol ratio of C H O

6 12 6

simplest mol ratio 1 2 3

hence, empirical formula = $\mathbf{CH_2O}$

$$\text{mass of 1 mole glucose} = 12 \times 6 + 1 \times 12 + 16 \times 6 = 180 \text{ g mol}^{-1}$$

$$\% \text{ of carbon} = \frac{12 \times 6}{180} \times 100 = \mathbf{40.0\%}$$

$$\% \text{ of hydrogen} = \frac{1 \times 12}{180} \times 100 = \mathbf{6.67\%}$$

$$\% \text{ of oxygen} = \frac{16 \times 6}{180} \times 100 = \mathbf{53.33\%}$$

Illustration 3 *Analysis of a pure compound of an oxide of sulphur gave 50.1% sulphur and 49.9% oxygen by mass. What is the simplest formula of the compound?*

Solution

Element	Relative mass	Atomic weight (At. wt.)	Relative Number of atoms (Divide by At. wt.)	Divide by smallest number	Smallest whole number ratio of atom
S	50.1	32.1	$\frac{50.1}{32.1} = 1.56$	$\frac{1.56}{1.56} = 1$ (S)	1
O	49.9	16.0	$\frac{49.9}{16.0} = 3.12$	$\frac{3.12}{1.56} = 2$ (O)	2

Thus, simplest formula = $\mathbf{SO_2}$

Problem Solving TiPs

(Known common fractions in decimal form)

As Illustration (1) illustrates, sometimes we must convert a fraction to a whole number by multiplication by the correct integer. But we must first recognise which fraction is represented by a non-zero part of a number. The decimal equivalents of the following fractions may be useful.

Fraction	Decimal equivalent (to 2 places)	To convert to integer multiply by
1/2	0.50	2
1/3	0.33	3
2/3	0.67	3
1/4	0.25	4
3/4	0.75	4
1/5	0.20	5
2/5	0.40	5
3/5	0.60	5
4/5	0.80	5

Table 1.3 **A Summary of Terms Used**

Term	Definition or usage
Atomic weight standard	An atomic weight of 12.00000 is arbitrarily assigned to $^{12}_{6}C$.
Isotopic mass (nuclidic mass)	The mass, in atomic mass units, u, of a single atom, on a scale in which the mass of an atom of $^{12}_{6}C$ is arbitrarily defined as 12.00000 u (*e.g.*, $^{35}_{17}Cl$ has an isotopic mass of 34.968852 u).
Atomic weight (relative atomic weight)	A dimensionless (pure) number that expresses the mass of the naturally occurring mixture of isotopes of an element, relative to an arbitrarily assigned atomic weight of 12.00000 for carbon-12. These are the values listed in a table of atomic weights (*e.g.*, atomic weight of Cl = 35.4527).
Formula weight (relative formula weight)	A dimensionless (pure) number that expresses the mass of a formula unit of a compound, relative to the atomic weight standard, carbon-12 (*e.g.*, formula weight of NaCl = 58.44). Tabulated atomic weights are used in computing formula weights.
Molecular weight (relative molecular weight)	A dimensionless (pure) number that compares the mass of a molecule to the atomic weight standard, carbon-12 (*e.g.*, molecular weight of CCl_4 = 153.82). Tabulated atomic weights are used in computing molecular weights.
Mole	An amount of substance containing the same number of elementary units (6.02214×10^{23}) as there are $^{12}_{6}C$ atoms in 12.00000 g $^{12}_{6}C$.
Molar mass (molar weight, mole weight)	The mass of one mole of a substance, whether the substance is composed of individual atoms (*e.g.*, 35.45 g Cl/mol Cl), formula units (*e.g.*, 58.44 g NaCl/mol NaCl), or molecules (*e.g.*, 153.82 g CCl_4/mol CCl_4).

**Where ever the term "weight" occurs, the term "mass" is equally (or even more) appropriate, e.g., atomic weight = atomic mass.*

Practice Exercise 2

1. Write the atomic masses of the following in amu.

(a) $^{23}_{11}Na$ (b) $^{14}_{6}C$

(c) $^{14}_{7}N$ (d) $^{27}_{13}Al$

(e) $_{12}Mg^{2+}$ containing 10 electrons

(f) $_{8}O^{2-}$ containing 10 electrons

2. Which is heavier in each pair?

(a) 1 amu of ^{12}C or 1 amu of ^{13}C

(b) 1 mol of $^{23}_{11}Na$ or 1 mol of $^{24}_{11}Na$

(c) 0.5 mol of $^{4}_{2}He$ or 1 mol of $^{2}_{1}H$

3. The "lead" in lead pencils is actually almost pure carbon and the mass of a period mark made by a lead pencil is about 0.0001 g. How many carbon atoms are in the period?

4. A binary compound of zinc and sulphur contains 67.1% zinc by mass. What is the ratio of zinc and sulphur atoms in the compound? (Zn = 65; S = 32)

5. What is the empirical formula and what is the per cent composition of dimethylhydrazine, $C_2H_8N_2$, a colourless liquid used as a rocket fuel?

6. What is the empirical formula of an ingredient in bufferin tablets that has the per cent composition?
C : 14.25% O : 56.93%, Mg : 28.83% by mass

7. Convert the following per cent composition into molecular formulae:

(a) **Diborane** H : 21.86%, B : 78.14%,
Mol. wt. : 28 g mol^{-1}

(b) **Trioxan** C : 40.00%, H : 6.71%,
O : 53.28%, Mol wt. : 90 g mol^{-1}

8. $Al_2(SO_4)_3 \cdot xH_2O$ has 8.20 per cent Al by mass. Calculate value of x.

9. Select correct alternate. Only **one** alternate is correct.

(A) If the equivalent weight of an element is 32, then the percentage of oxygen in its oxide is

(a) 16 (b) 40

(c) 32 (d) 20

(B) A hydrocarbon has 3 g carbon per gram of hydrogen, hence simplest formula is

(a) CH_4 (b) C_6H_6

(c) C_3H_8 (d) CH_2

(C) Molar ratio of Na_2SO_3 and H_2O is 1 : 7 in $Na_2SO_3 \cdot xH_2O$. Hence, mass percentage is

(a) 12.5 : 87.5 (b) 87.5 : 12.5

(c) 50 : 50 (d) 75 : 25

Answers

1. (a) 23 amu (b) 14 amu (c) 14 amu (d) 27 amu
(e) 24 amu (f) 16 amu

2. (a) 1 amu of ^{13}C (b) 1 mol of $^{24}_{11}Na$ (c) equal mass

3. 5.018×10^{18} atoms

4. Zn : S :: 1 : 1

5. Empirical formula : CH_4N, C : 40%, H : 13.3%, N : 46.7%

6. CO_3Mg **7.** (a) B_2H_6 (b) $C_3H_6O_3$ **8.** $x = 18$

9. **A.** (d) **B.** (a) **C.** (c)

Total Practice Set

(Read, Plan & Solve!)

Problem 1. Analysis of chlorophyll shows that it contains 2.68 per cent magnesium. How many atoms of magnesium does 100 g of chlorophyll contain?

Plan *First we determine moles of magnesium in* 1 g *of chlorophyll and*

$$\text{number of Mg atoms} = mol \times N_0$$

Solution Mg in chlorophyll is 2.68%

Thus, $100 \text{ g chlorophyll} = 100 \times \dfrac{2.68}{100} \text{ g Mg}$

$$= \frac{2.68}{24} \text{ mol Mg}$$

$$= \frac{2.68 N_0}{24} \text{ atoms of Mg}$$

$$= \frac{2.68 \times 6.022 \times 10^{23}}{24}$$

$$= \mathbf{6.7246 \times 10^{22} \text{ atoms}}$$

Problem 2. Carbohydrates are compounds containing carbon, hydrogen and oxygen in which the hydrogen to oxygen ratio is 2 : 1. A certain carbohydrate contain 40% carbon by mass. Derive molecular formula of the compound if molar mass is 180 g mol^{-1}.

Plan *Molar ratio of hydrogen and oxygen and mass per cent of carbon have been given. Molecular weight and molecular formula are thus derived.*

Solution Let carbohydrate be $C_x(H_2O)_y$ where x and y are whole numbers.

$\therefore \quad 12x + 18y = 180$

Also $\quad$ 180 g of the carbohydrate has = $12x$ carbon

$\therefore \quad$ 100 g of the carbohydrate has

$$= \frac{12x \times 100}{180} \% \text{ carbon}$$

$$\therefore \quad \frac{1200x}{180} = 40$$

$$\therefore \quad x = 6$$

$$\therefore \quad 12 \times 6 + 18y = 180$$

$$\therefore \quad y = 6$$

$\therefore \quad$ Carbohydrate is $C_x(H_2O)_y = C_6(H_2O)_6$

$$= \mathbf{C_6H_{12}O_6}$$

Problem 3. Myoglobin stores oxygen for metabolic processes in muscle. Chemical analysis shows that it contains 0.34% Fe by mass. What is the molar mass of myoglobin? (There is one Fe atom per molecule).

Plan *Every myoglobin molecule has one* Fe *atom. Every* 100 g *of it has* 0.34 g Fe *hence, molar mass of myoglobin containing* 56 g Fe *per mol can be determined.*

Solution If 0.34 g Fe is present then myoglobin = 100 g

If 56 g of Fe is present then myoglobin $= \dfrac{100}{0.34} \times 56$

$$= \mathbf{16470.6 \text{ g mol}^{-1}}$$

Problem 4. A compound X contains 63.3 per cent manganese (Mn) and 36.7 per cent oxygen by mass. When X is heated, oxygen gas is evolved and a new compound Y containing 72.0 per cent Mn and 28.0 per cent O is formed. (a) Determine the empirical formula of X and Y. (b) Write a balanced equation for the conversion of X to Y. (Mn = 55, O = 16)

Plan *From the percentage of* Mn *and* O, *empirical formula of X and Y are derived.*

Solution **For the compound X**

	%	mol = $\dfrac{\%}{\text{atomic mass}}$	Simplest ratio
Mn	63.3	1.15	1
O	36.7	2.29	2

Empirical formula : $\mathbf{MnO_2}$

For the compound Y

	%	mol = $\dfrac{\%}{\text{atomic mass}}$	Simplest ratio	Whole number ratio
Mn	72.0	1.31	1	3
O	28.0	1.75	1.33	4

Empirical formula : $\mathbf{Mn_3O_4}$

$$3MnO_2 \longrightarrow Mn_3O_4 + O_2$$

Problem 5. Ferrocene, a substance proposed for use as a gasoline additive, has the per cent composition 5.42% H, 64.56% C, and 30.02% Fe. What is the empirical formula of ferrocene?

Plan *Per cent composition is to be converted to moles, then to molar ratio and then into integral mole ratio (if it is fraction).*

Solution

Element	%	Mol = $\frac{\%}{\text{atomic mass}}$	mole ratio
^{12}C	64.56	5.38	10
^{1}H	5.42	5.42	10
^{56}Fe	30.02	0.54	1

Empirical formula is $\mathbf{C_{10}H_{10}Fe}$.

Problem 5. Phosphoric acid (H_3PO_4) is a colourless, syrupy liquid used in detergents, fertilizer, tooth pastes, and in carbonated beverages for a "tangy" flavour. Calculate the per cent composition by mass of H, P and O in this compound. (H= 1, P= 31, O= 16)

Plan *In one mole (= 98 g) amounts of* H, P *and* O *are known, hence their mass per cent can be calculated.*

Solution One mole of H_3PO_4 = 1 mole × 98 g mol^{-1}

$$= 98 \text{ g}$$

H present in one mole = 3 g

$$\therefore \quad \% \text{ of H in } H_3PO_4 = \frac{3}{98} \times 100 = \mathbf{3.06\%}$$

P present in one mole = 31 g

$$\therefore \quad \% \text{ of P in } H_3PO_4 = \frac{31}{98} \times 100 = \mathbf{31.63\%}$$

O present in one mole = 64 g

$$\therefore \quad \% \text{ of O in one mole} = \frac{64}{98} \times 100 = \mathbf{65.31\%}$$

Problem 7. What is the ratio of the masses of oxygen that are combined with 1.08 g of nitrogen in the compounds N_2O_3 and NO?

Plan *First we calculate the mass of* O *that combines with one gram of* N *in each compound. Then we determine the ratio of the values of* $\frac{g(O)}{g(N)}$ *for two compounds.*

Solution In N_2O_3: $\frac{?g(O)}{g(N)} = \frac{48.0\ g(O)}{28.0\ g(N)} = \frac{1.71\ g(O)}{g(N)}$

In NO : $\frac{?\ g(O)}{gN} = \frac{16.\ g(O)}{14.0\ g(N)} = \frac{1.14\ g(O)}{g(N)}$

The ratio is $\frac{g(O)}{g(N)}$ in (N_2O_3)

$$\frac{1.71\ g(O)/g(N)}{1.14\ g(O)/g(N)} = \frac{1.5}{1.0} = \frac{3}{2}$$

$\frac{g(O)}{g(N)}$ in (NO)

Thus, we observe that the ratio is 3 mass units g(O) in (N_2O_3) to 2 mass units of O (in NO or N_2O_2). This represents law of multiple proportions.

Master Exercises

Exercise 1 (Stage 1 : Learning)

Short Answer Type Questions

1. What is the name of the following process?

$$\text{gas} \longrightarrow \text{solid}$$

2. What is the valency of carbonate if its equivalent weight is 30 g $equiv^{-1}$?
3. 24 g of magnesium combines with oxygen forming 40 g of oxide. What is equivalent weight of magnesium?
4. 27 g of aluminium combines with hydrogen forming 30 g of hydride. What is equivalent weight of aluminium?
5. H_3PO_2 only forms NaH_2PO_2. What is its equivalent weight of acid?
6. H_3PO_3 forms NaH_2HPO_3 and Na_2HPO_3. What is equivalent weight of acid at each stage of salt formation?
7. H_3PO_4 forms NaH_2PO_4, Na_2HPO_4 and Na_3PO_4. What is equivalent weight of acid at each stage of salt formation?
8. When a piece of sulphur is burnt in air, it disappears. How could you prove that it is not destroyed?
9. Taking VD of hydrogen as 1, determine
   (a) VD of CH_4 (b) VD of O_2
   (c) VD of CO_2
10. Calculate number of equivalents in
   (a) 1.6 g O_2 gas (b) 0.49 g H_2SO_4
   (c) 1.80 g H_2O (d) 1.6 g CH_4
11. Calculate equivalent weights of the underlined based on following reaction : (Take molecular weight = M)
   (a) $\underline{N_2} + 3H_2 \longrightarrow 2\underline{NH_3}$
   (b) $Ca(OH)_2 + \underline{H_3PO_4} \longrightarrow CaHPO_4$
   (c) $\underline{Zn} + H_2SO_4 \longrightarrow ZnSO_4 + H_2$
   (d) $\underline{2FeSO_4} + H_2SO_4 + O \longrightarrow Fe_2(SO_4)_3 + H_2O$
12. What is the relation between specific heat and atomic weight?
13. The specific heat of a metal M is 0.25. Its equivalent weight is 12.
   (a) What is its atomic weight?
   (b) What is the formulas of its (i) chloride, (ii) sulphate and (iii) phosphate?
14. Calculate number of moles and atoms of each in 1.8 g $C_6H_{12}O_6$.

Analytical Questions

Law of Chemical Combination

1. Benzene, ethane and ethylene are just three of a large number of hydrocarbons—compounds that contain only carbon and hydrogen. Show how the following data are consistent with the law of multiple proportions.

Compound	Mass of carbon in 5.00 g sample	Mass of hydrogen in 5.00 g sample
Benzene	4.61 g	0.39 g
Ethane	4.00 g	1.00 g
Ethylene	4.29 g	0.71 g

2. Ammonia, NH_3 and hydrazine, N_2H_4, are both binary compounds of nitrogen and hydrogen. Based on the law of multiple proportions, how many grams of hydrogen would you expect 2.34 g of nitrogen to combine with to yield ammonia? To yield hydrazine.
3. If 3.670 g of nitrogen combine with 0.5275 g of hydrogen to yield compound X, how many grams of nitrogen would combine with 1.575 g of hydrogen to make the same compound? Is X ammonia or hydrazine?

Mole

4. The table below gives number of electrons, protons and neutrons in atoms or ions of a number of elements. Answer the following :

Atom or ion of element	A	B	C	D	E	F	G
Number of electrons	5	10	18	28	36	5	9
Number of protons	5	7	19	30	35	5	9
Number of neutrons	5	7	20	36	46	6	10

(a) Which of the species are neutral?
(b) Which are negatively charged?
(c) Which are positively charged?
(d) What are the conventional symbols for all the species?

5. The carat is the unit of mass used by jewellers. One carat is exactly 200 mg. How many carbon atoms are present in a 24 carat diamond?

6. The natural abundance of the two stable isotopes of hydrogen (hydrogen and deuterium) are ${}^1_1H : 99.985\%$ and ${}^2_1H : 0.015\%$. Assume that water exists as either H_2O or D_2O. Calculate the number of D_2O molecules in exactly 400 mL water. (Density = 1.00 g mL^{-1})

7. A solution containing 10.0 millimol of $CaCl_2$ is diluted to 1 L. Calculate the number of grams of $CaCl_2 \cdot 2H_2O$ per mL of final solution.

Percentage Composition and Empirical Formula

8. All of the substance listed below are fertilizers that contribute nitrogen to the soil. Which of these is the richest source of nitrogen on a mass percentage basis?
(a) Urea, $(NH_2)_2CO$
(b) Ammonium nitrate, NH_4NO_3
(c) Guanidine, $HNC(NH_2)_2$
(d) Ammonia, NH_3

9. The aluminium sulphate hydrate $[Al_2(SO_4)_3 \cdot xH_2O]$ contains 8.20 per cent Al by mass. Calculate x, that is, the number of water molecules associated with each $Al_2(SO_4)_3$ units.

10. Calculate the per cent composition by mass of all the elements in calcium phosphate $Ca_3(PO_4)_2$.

11. Platinum forms two different compounds with chlorine. One contains 26.7 per cent Cl by mass, and the other contains 42.1 per cent Cl by mass. Determine the empirical formulae of the two compounds. (Pt = 195)

12. Disilane, Si_2H_x, is analysed and found to contain 90.28% by weight silicon. What is the value of x ? (Si = 28)

13. Cytochrome *C* is an iron-containing enzyme found in the cells of all aerobic organisms. If cytochrome *C* is 0.43% Fe by weight, what is its minimum molecular weight?

14. What is empirical formula of stannous fluoride, a compound added to tooth paste to protect teeth against decay? Its mass per cent composition is 24.25% F, 75.75% Sn.

15. Analysis of a metal chloride XCl_3 shows that it contains 67.2% Cl by mass. Calculate the molar mass of *X* and identify the element.

16. A sample of a compound of nitrogen (N) and oxygen (O) contains 1.52 g of N and 3.47 g of O. The molar mass of this compound is known to be between 90 g and 95 g. Determine the molecular formula and molar mass of this compound. (N = 14, O = 16)

NCERT Problems

1. Calculate the molar mass of
(i) H_2O (ii) CO_2
(iii) CH_4

2. Determine the empirical formula of an oxide of iron which has 69.9% iron and 30.1% oxygen (Fe = 55.85 amu; O = 16.00 amu).

3. How much copper can be obtained from 100 g of anhydrous copper sulphate ($CuSO_4$)? Cu = 63.5; S = 32, O = 16 (all in amu).

4. Calculate the following in three moles of ethane gas (C_2H_6)
(i) moles of carbon atoms
(ii) moles of hydrogen atoms
(iii) molecules of ethane

5. Which one of the following will have largest number of atoms? [Au = 197, Na = 23, Li = 7, Cl = 35.5 in amu]
(i) 1g Au(*s*) (ii) 1 g Na(*s*)
(iii) 1g Li(*s*) (iv) 1 g $Cl_2(g)$

6. What will be the mass of one ${}^{12}C$ atom in gram?

7. Calculate the number of atoms in each of the following :
(i) 52 moles He (ii) 52 u of He
(iii) 52 g He

Exercise 2

(Stage 2 : High Skill Problem Solving)

Only One Option Correct

1. In which of the following pairs do the two species resemble each other most closely in chemical properties?
(a) ${}^1_1H$ and ${}^2_1H$ (b) ${}^{16}_8O$ and ${}^{16}_8O^{2-}$
(c) ${}^{24}_{12}Mg$ and ${}^{24}_{12}Mg^{2+}$ (d) ${}^{14}_7N$ and ${}^{14}_7N^{3-}$

2. One isotope of a non-metallic element has mass number 127 and 74 neutrons in the nucleus. The anion derived from the isotope has 54 electrons. Hence, symbol for the anion is
(a) ${}^{127}_{54}X^-$ (b) ${}^{127}_{53}X^-$
(c) ${}^{74}_{53}X^-$ (d) ${}^{74}_{54}X^-$

3. Which of the following is the richest source of ammonia on a mass percentage basis?
(a) NH_4NO_3 (b) NH_2CONH_2
(c) NH_3 (d) $HNC(NH_2)_2$

4. Which of the following substances contains greatest mass of chlorine?
(a) 5.0 g Cl_2 (b) 0.5 mol Cl_2
(c) 0.10 mol KCl (d) 30.0 g $MgCl_2$

5. When 0.273 g of Mg is heated strongly in a nitrogen (N_2) atmosphere, 0.378 g of the compound is formed. Hence, compound formed is
(a) Mg_3N_2 (b) Mg_3N
(c) Mg_2N_3 (d) MgN

6. A certain metal sulphide, MS_2, is used extensively as a high temperature lubricant. If MS_2 is 40.06% by mass sulphur, metal M has atomic mass
(a) 160 amu (b) 64 amu
(c) 40 amu (d) 96 amu

7. The molar mass of a compound if 0.372 mol of it has a mass of 186 g is
(a) 200 g (b) 372 g
(c) 500 g (d) 186 g

8. Which of the following has maximum number of C atoms?
(a) 4.4 g CO_2 (b) 3.0 g C_2H_6
(c) 4.4 g C_3H_8 (d) 1.3 g C_6H_6

9. Number of atoms present in 1.8 g H_2O, 1.7 g NH_3 and 1.6 g CH_4, in the increasing order is
(a) $H_2O < NH_3 < CH_4$ (b) $CH_4 < NH_3 < H_2O$
(d) $CH_4 = NH_3 = H_2O$ (d) $NH_3 < CH_3 < H_2O$

10. Number of atoms in 20 g Ca is equal to number of atoms is
(a) 20 g Mg (b) 1.6 g CH_4
(c) 1.8 g H_2O (d) 1.7 g NH_3

11. $Na_2SO_3 \cdot xH_2O$ has 50% H_2O by mass. Hence, x is
(a) 4 (b) 5
(c) 6 (d) 7

12. Mass of one atom of X is 6.66×10^{-23} g. Hence, number of mole of atom X in 40 kg is
(a) 10^3 mol (b) 10^{-3} mol
(c) $\frac{40 \times 10^3}{6.66 \times 10^{-23}}$ mol (d) $\frac{40 \times 10^3}{6.022 \times 10^{23}}$ mol

13. To make 0.01 mol, which of the following has maximum mass?
(a) $NaHCO_3$ (b) Na_2CO_3
(c) Na_2SO_4 (d) $Na_2C_2O_4$

14. In a glass-tube, there is 18 g of glucose. 0.08 mol of glucose is taken. Glucose left in the glass-tube is
(a) 0.10 g (b) 17.92 g
(c) 3.60 mol (d) 3.60 g

15. Rest mass of an electron is 9.11×10^{-31} kg. Molar mass of the electron is
(a) 1.50×10^{-31} kg mol^{-1} (b) 9.11×10^{-31} kg mol^{-1}
(c) 5.5×10^{-7} kg mol^{-1} (d) 6.02×10^{23} kg mol^{-1}

16. A sample of ammonium phosphate $(NH_4)_3PO_4$ contains 3.18 moles of H atoms. The number of moles of O atoms in the sample is
(a) 0.265 (b) 0.795
(c) 1.06 (d) 3.18

17. A sample of $CuSO_4 \cdot 5H_2O$ contains 3.782 g of Cu. How many grams of oxygen are in this sample? (Cu = 63.5)
(a) 0.952 g (b) 3.80 g
(c) 4.761 g (d) 8.576 g

18. Cortisone is a molecular substance containing 21 atoms of carbon per molecule. The mass percentage of carbon in cortisone is 69.98%. Its molar mass is
(a) 176.5 (b) 252.2
(c) 287.6 (d) 360.1

19. In a hydrocarbon C is 3 g per g of hydrogen. Hence, hydrocarbon is
(a) CH_4 (b) C_2H_6
(c) C_3H_8 (d) C_4H_{10}

20. Which has maximum percentage of Cl?
(a) $C_6H_6Cl_6$ (b) $CHCl_3$
(c) C_6H_5Cl (d) CH_3Cl

21. A spherical ball of radius 7 cm contains 56% iron. If density is 1.4 g/cm^3, number of moles of Fe present approximately is
(a) 10 (b) 15
(c) 20 (d) 25

22. In the following final result is ... 0.1 mol CH_4 + 3.01×10^{23} molecules CH_4 – 9.6 g CH_4 = x mol H atoms
(a) 0 mol H atom (b) 0.2 mol H atom
(c) 0.3 mol H atom (d) 0.4 mol H atom

23. If we assume that $N_0 = 1.2\times 10^{23}$ mol^{-1}, then molar mass of O_2 will be taken as
(a) 32 g mol^{-1} (b) $\frac{32}{6}$ g mol^{-1}
(c) 32×10^{23} g mol^{-1} (d) $\frac{1\times 10^{23}}{32}$ g mol^{-1}

24. CO, CO_2, C_2O_3 follow
(a) law of definite proportion
(b) law of multiple proportion
(c) law of conservation of mass
(d) all of the above

25. In a gas S and O are 50% by mass, hence, their mol ratio is
(a) 1 : 1 (b) 1 : 2
(c) 2 : 1 (d) 3 : 1

One or More Than One Options Correct

1. Equal moles of the substance are present in
(a) 1.6 g CH_4 (b) 1.7 g NH_3
(c) 1.8 g H_2O (d) 1.2 g Mg

2. In which cases law of conservation of mass are followed?
(a) 10 g $CaCO_3$ on heating gives 0.1 mole CaO and 2.24 L CO_2 at NTP
(b) 1 mole N_2 and 1 mole H_2 combine to form 2 moles of NH_3
(c) 0.2 mole SO_2 combines with 3.2 g O_2 to form 0.2 mole SO_3
(d) 1 g H_2 combines with 0.5 g O_2 to form 1.5 g H_2O

3. One formula unit of a substance (underlined) is present in each of
(a) 200 g $\underline{CaCO_3}$ with 50% impurity
(b) 38 g $\underline{H_2O}$ and D_2O with 1 mole D_2O
(c) 360 g $\underline{C_{12}H_{22}O_{11}}$ and $C_6H_{12}O_6$ with 0.1 mole $C_6H_{12}O_6$
(d) 46 g $\underline{N_2H_4CO}$ with 60% purity

4. Which represents one mole?
(a) 22.4 L He at STP
(b) 16 g O_2
(c) 158 g $KMnO_4$
(d) 392 g $FeSO_4\cdot(NH_4)_2SO_4\cdot 6H_2O$

5. Select correct alternate(s).
(a) 23.5 g SO_2 would contain the same mass of oxygen as is contained in 33.7 g of As_2O_5 (As = 75)
(b) 46 g $CaCl_2$ would contain the same number of chloride ions as are contained in 48.6 g NaCl
(c) 17 g NH_3 would contain the same volume of H_2 at NTP as are contained in 18 g H_2O at NTP
(d) 1 mole CO would combine with same mole of O_2 to form CO_2 as required by 1 mole of SO_2 to form SO_3

6. Equal mass of water is lost on heating of 1 mole of each of
(a) $NaHCO_3$ (b) $Ca(HCO_3)_2$
(c) $KHCO_3$ (d) $CuSO_4\cdot 5H_2O$

7. Empirical formula will be same in with following compounds
(a) CH_3CO_2H (b) $C_6(H_2O)_6$
(c) HCHO (d) $C_{12}(H_2O)_{11}$

8. Consider following species
I : CO; II : NO;
III : N_2; IV : HCHO
Select species with equal number of neutrons, protons and electrons in each.
(a) CO, N_2 (b) CO, HCHO
(c) NO, HCHO (d) N_2, NO, CO

9. Which of the following may contain two neutrons and two protons?
(a) He (b) D_2
(c) He_2^{2+} (d) H_2

10. ${}^{12}_{6}C$ and ${}^{14}_{6}C$ resemble in
(a) number of protons (b) number of electrons
(c) number of neutrons (d) chemical properties

Brain Twisters

1. What is equivalent weight of H_3O^+?

2. Equivalent weight of
(a) $H_3PO_2 = \frac{\text{molecular weight}}{1}$
(b) $H_3PO_2 = \frac{\text{molecular weight}}{2}$
(c) $H_3PO_3 = \frac{\text{molecular weight}}{3}$
Explain the difference.

3. Determine equivalent weight of carbon in the following :
(a) $C + O_2 \longrightarrow CO_2$
(b) $C + 2H_2 \longrightarrow CH_4$
(c) $2C + H_2 \longrightarrow C_2H_2$
(d) $2C + 2H_2 \longrightarrow C_2H_4$
(e) $2C + 3H_2 \longrightarrow C_2H_6$
(f) $3C + 4H_2 \longrightarrow C_3H_8$
What do you concludes about equivalent weight?

4. Equivalent weight of two oxides A and B of sulphur are 16 and 13.33 g equivalent^{-1} respectively. What are these oxides?

5. Equivalent weights of two hydrides of nitrogen are 5.66 and 8. Derive formulae of hydride.

6. A solution contains 200 g ppm $CaCO_3$. Express this concentration in
(a) millimoles per gram
(b) number of $CaCO_3$ molecules per gram

7. How many carbon atoms are present in a diamond weighing 200 mg and of 20 carat purity?

8. 100 g of ethyl alcohol-water mixture contains 1 mole of ethyl alcohol. What is molar ratio of ethyl alcohol and water in the mixture?

9. Students A and B derived equivalent weights of KHC_2O_4 quite different. Explain.

10. 0.24 g of magnesium was burnt in the atmosphere of oxygen when residue formed weighed 0.40 g. Derive equivalent weight of magnesium.

11. Cost of a fertilizer is based on the percentage of nitrogen in it. Which of the fertilizers should be priced maximum?

(a) NH_4NO_3 (b) NH_2CONH_2
(c) KNO_3 (d) NH_4Cl

12. One equivalent of calcium phosphate $Ca_3(PO_4)_2$ is dissolved in HCl

$$Ca_3(PO_4)_2 + 6HCl \longrightarrow 3CaCl_2 + 2H_3PO_4$$

(a) How many moles of $CaCl_2$ are formed?
(b) How many moles of H_3PO_4 are formed?

Passage Comprehension Questions

Passage 1

Study the following observations and answer the questions at the end of it.

The following is a crude but effective method for estimating the order of magnitude of Avogadro's number using stearic acid ($C_{18}H_{36}O_2$). When stearic acid is added to water, its molecules collect at the surface and form a monolayer; that is, the layer is only one molecule thick. The cross-sectional area of each stearic acid molecule has been measured to be 0.21 nm^2. In one experiment it is found that 1.4×10^{-4} g of stearic acid is needed to form a monolayer over water in a dish of diameter 20 cm. (the area of a circle of radius r is πr^2.)

1. Based on these measurements value of Avogadro's number is

(a) 3×10^{23} (b) 6×10^{23}
(c) 4×10^{23} (d) 1×10^{23}

2. What is equivalent of 1g H in amu for this value of Avogadro's number?

(a) 1.66×10^{-24} g (b) 3.33×10^{-24} g
(c) 2.5×10^{-24} g (d) 1×10^{-23} g

Passage 2

Study the following passage regarding fertilizer K_2O and answer the questions at the end of it.

Potash is any potassium mineral that is used for its potassium content. Most of the potash produced in the United States goes into fertilizer. The major sources of potash are potassium chloride (KCl) and potassium sulphate (K_2SO_4). Potash production is often reported as the potassium oxide (K_2O) equivalent or the amount of K_2O that could be made from a given mineral. KCl costs ₹ 50 per kg.

1. What is the cost of K per mol of the KCl sample?

(a) ₹ 13.42 mol^{-1} (b) ₹ 3.73 mol^{-1}
(c) ₹ 1.00 mol^{-1} (d) ₹ 2.00 mol^{-1}

2. For what price must K_2SO_4 be sold in order to supply the same amount of potassium as in KCl?

(a) ₹ 58.40 kg^{-1} (b) ₹ 50.00 kg^{-1}
(c) ₹ 42.82 kg^{-1} (d) ₹ 25.00 kg^{-1}

3. What mass (in kg) of K_2O contains the same number of moles of K atoms as 1.00 kg KCl?

(a) 0.158 kg (b) 0.315 kg
(c) 1.262 kg (d) 0.631 kg

Passage 3

Study the following observations and answer the questions at the end of it.

The mass spectrum (given below) of magnesium has three peaks, which indicates that magnesium has three isotopes. The questions given below are based on this mass spectrum.

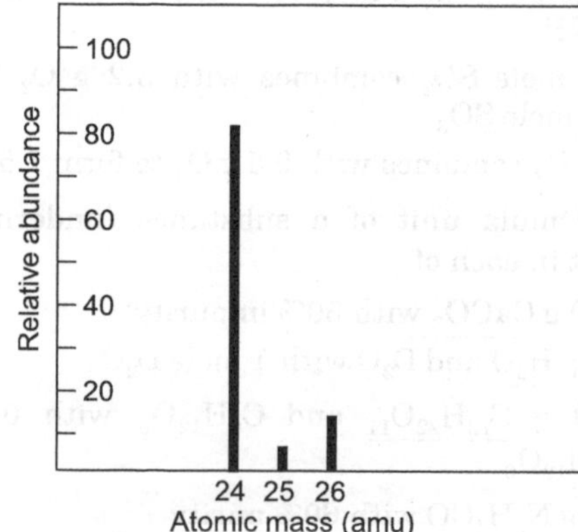

1. Which isotope has maximum number of atoms per gram of each?

(a) Mg-24 (b) Mg-25
(c) Mg-26 (d) Equal

2. Number of atoms in one mole of each isotope is placed in increasing order

(a) $^{24}Mg < ^{25}Mg < ^{26}Mg$
(b) $^{26}Mg < ^{25}Mg < ^{24}Mg$

(c) $^{26}Mg = ^{25}Mg = ^{24}Mg$
(d) given data is insufficient

3. Average atomic mass of Mg is approximately
(a) 25.0 (b) 24.3
(c) 25.2 (d) 25.8

Assertion & Reason

Codes :
(a) Both **A** and **R** are true and **R** is the correct explanation of **A.**
(b) Both **A** and **R** are true but **R** is not the correct explanation of **A.**
(c) **A** is true but **R** is false.
(d) **A** is false but **R** is true.

1. **Assertion (A) :** When 10 g of $CaCO_3$ is decomposed, 5.6 g of residue is left and 4.4 g of CO_2 escapes.
Reason (R) : Law of mass conservation is followed.

2. **Assertion (A) :** SO_2 and SO_3 obey **"law of multiple proportion".**
Reason (R) : Every sample of SO_2 contains 1 part sulphur and 1 part oxygen by mass.

3. **Assertion (A) :** SO_2 from oxidation of sulphur or H_2S contains sulphur and oxygen in the mass ratio of 1 : 2.
Reason (R) : Different samples of a pure chemical substance always contain the same proportions of elements by mass.

4. **Assertion (A) :** In Fe^{2+}, there are 24 electrons and 30 neutrons and thus, ionic mass is 56.
Reason (R) : Ionic mass = neutron + proton and proton = electrons in neutral species

True & False

1. If there are 10 electrons and 8 neutrons in X^{2-}, then its ionic mass is 18.
2. SO_2 contains sulphur and oxygen in 1:1 mass ratio.
3. 20 g $CaCO_3$ (50% pure) on strongly heating gives 5.6 g residue.
4. Mass of 1 mole of electrons is 5.48×10^{-7} kg.
5. Number of atoms in 1.6 g CH_4, 1.7 g NH_3 and 1.8 g H_2O are equal.
6. If 1 mole of H_2 is taken out from 2 g H_2, 1 g H_2 is left.
7. 1 amu is equal to 1.66×10^{-24} g.
8. Out of CH_3Cl, $C_6H_6Cl_6$, and CCl_3CHO, maximum percentage of chlorine is in $C_6H_6Cl_6$.
9. 1 ppm is equal to 1 mg per kg.
10. Equivalent mass of NH_3 is 17.

Fill in the Blanks

1. NO and NO_2 follow the law of ....
2. SO_3 contains sulphur and oxygen in the mass ratio of ....
3. Hydrocarbon containing 4 g carbon per gram hydrogen is ....
4. If X^{2+} is isotonic (same number of neutrons) of H_2O and contains 10 electrons then ionic mass of X^{2+} is ....
5. 1 mole of barium nitrate and 1 mole of ammonium phosphate form .... mole of barium phosphate.
6. Percentage of sodium sulphite (Na_2SO_3) in hydrated salt is 50. Hence, molecular formula of hydrated salt is ....
7. Concentration 1 ppm = ... ppb.
8. Concentration 1 microgram per gram = ... ppm.
9. H_2O, H_2S and SO_2 follow law of ....
10. Atomic mass unit is also called ....

Matrix-Match Type Questions

I. Only One Correct

1. Match % of carbon (in List I) with the compound (in List II).

List I	List II
A. 20%	1. CH_4
B. 52.2%	2. CO_2
C. 75%	3. C_2H_6O
D. 27.3%	4. CN_2OH_4
E. 12%	5. $CaCO_3$

2. Match Laws of chemical combinations (in List I) with the examples followed (in List II).

List I	List II
A. Law of conservation of mass	1. CH_4 has carbon and hydrogen in 3 : 1 mass ratio.
B. Law of multiple proportion	2. 10 mL N_2 combines with 30 mL of H_2 to form 20 mL of NH_3.
C. Law of definite proportion	3. S and O_2 combine to form SO_2 and SO_3.
D. Law of reciprocal proportion	4. In H_2S and SO_2 mass ratio of H and O w.r.t. sulphur is 1: 16, hence in H_2O, mass ratio of H and O is 1: 8.
E. Gay Lussac's law	5. 4.2 g $MgCO_3$ gives 2.0 g residue on heating.

II. One or More Correct

1. **Match the item in Column I with the compounds in Column II.**

	Column I		Column II
A.	Same % of C (40%)	1.	CH_3COOH
B.	Same % of H (6.67%)	2.	NH_2CONH_2
C.	Same % of O (53.3%)	3.	HCHO
D.	Same empirical formula	4.	$CH_3CH_2CH_2OH$
		5.	$HCOOCH_3$

2. **Match the item in Column I with the given amount in Column II.**

	Column I		Column II
A.	1 equivalent	1.	8 g oxygen (O)
B.	0.5 mole	2.	1 g hydrogen (H)
C.	$\frac{N_0}{2}$ atoms	3.	35.5 g chlorine (Cl_2)
D.	N_0 molecules	4.	66.0 g H_3PO_2
		5.	49 g H_2SO_4

Integer Answer Types

This section contains 8 questions. The answer to each of the questions is a single digit integer, ranging from 0 to 9. The appropriate bubbles below the respectively question numbers in the ORS have to be darkened. For example, if the correct answers to question number *X, Y, Z* and *W* (say) are 6, 0, 9 and 2, respectively, then the correct darkening of bubbles will like the following.

X	Y	Z	W
0	0	0	0
1	1	1	1
2	2	2	2
3	3	3	3
4	4	4	4
5	5	5	5
6	6	6	6
7	7	7	7
8	8	8	8
9	9	9	9

Set I

1. **1 g of hydrogen has *x* g of carbon in methane. This *x* is ......**

2. **$Na_2SO_3 \cdot yH_2O$ has 50% H_2O by mass. Thus, *y* is ......**

3. **If Avogadro's number would have been 1×10^{10}, instead of 6.02×10^{23}, then mass of one atom of H would be ...... amu.**

4. **Sum of neutrons and electrons in H_2^{2+} is ......**

5. **When neutron changes to proton, atomic number changes by ...... unit.**

6. **When 1 mole of Na_2CO_3 is heated, CO_2 lost is ...... mole.**

7. **Empirical formula of substance with molar mass 180 is CH_2O. Thus, ratio of molecular formula to empirical formula is ......**

8. **When 2.495 g of $CuSO_4 \cdot xH_2O$ (molar mass 249.5) is heated 0.05 mole of H_2O is lost. Thus, *x* is ......**

Set II

1. **In Japan, radioactive water is leaking into sea from nuclear reactors (April 04, 2011). 0.060 kg of radioactive water has number of moles = ...**

2. **$^{35}_{17}Cl$ and $^{37}_{17}Cl$ are two isotopes of chlorine. If atomic mass of chlorine is 35.5, then mass ratio of these isotopes is $x:1$ where $x =$ ....**

3. **Mass of one atom of element is 6.66×10^{-23} g. Number of gram atoms in 0.320 kg of element would be ....**

4. **X^{2-} is isoelectronic of $_{10}Ne$. 1.6 g of X^{2-} has 0.1 mole. Thus, there are .... neutrons is X^{2-}.**

5. **In the following reaction,**

$$SO_2(g) + 2H_2S(g) \longrightarrow 3S(s) + 2H_2O(l)$$

2 moles each of SO_2 and H_2S forms sulphur = ... moles

6. **In the lightest isotope of hydrogen, number of neutrons present is ....**

7. **Total number of ions in aluminium phosphate is ....**

8. **11.2 g of Fe completely react with 3.36 L of O_2 gas under standard condition to form Fe_xO_y. Thus, $(x + y)$ is ....**

Answers & Solutions

Numerical Problem Solving

1. Mass of reactants = mass of products

$$17 + 7.45 = 14.35 + KNO_3(x)$$

$$\therefore \quad x = 10.10 \text{ g}$$

2. $Mg + \frac{1}{2}O_2 \longrightarrow MgO$

24 g	$\left(\frac{32}{2}\right)$ g
12 g	16 g
?	32 g

Mg = 24 g. MgO has constant composition.

3. (i) $C + \frac{1}{2}O_2 \longrightarrow CO_2$

12 g	16 g
1 g	$\frac{16}{12}$ g

$C + O_2 \longrightarrow CO_2$

12 g	32 g
1 g	$\frac{32}{12}$ g

Ratio of oxygen in CO and $CO_2 = \frac{16}{12} : \frac{32}{12} = 1:2$

(ii)

Carbon	Hydrogen	Amount of hydrogen combines with 12 g carbon	Ratio
24 g	2 g	1 g	1
24 g	4 g	2 g	2
24 g	6 g	3 g	3
12 g	4 g	4 g	4

Ratio of hydrogen in C_2H_2, C_2H_4, C_2H_6 and CH_4 that combines with 12 g carbon is = 1 : 2 : 3 : 4.

Thus, law of multiple proportion is followed.

4. 1 : 35.5

5. (a) $Mg = \frac{24}{2} = 12$ g equivalent^{-1}

(b) $Cl_2 = 35.5$ g equivalent^{-1}

(c) $Ca = \frac{40}{2} = 20$ g equivalent^{-1}

6.

	$N_2(g)$ +	$O_2(g)$ $\longrightarrow$	$2NO(g)$
mass	28 g	32g	2 × 30 g
mole	1 mol	1 mol	2 mol
equivalent	4	4	4
	22.4 L	22.4 L	2 × 22.4 L at NTP

$$32\text{g } CO_2 \equiv 28 \text{ g } N_2$$

$$\therefore \quad 8 \text{ g } O_2 \equiv 7\text{g } N_2$$

Thus, equivalent mass of N_2 = 7 g equivalent^{-1}

Hence, equivalent of N_2 used $= \frac{28}{7} = 4$

Equivalent mass of O_2 = 8 g equivalent^{-1}

Hence, equivalent used of $O_2 = \frac{32}{8} = 4$

$$32 \text{ g } O_2 = 2 \times 30 \text{ g NO}$$

$$8 \text{ g } O_2 = 15 \text{ g NO}$$

Thus, equivalent mass of NO = 15 g equivalent^{-1}

Hence, equivalent of NO formed = 4

	N_2	O_2	NO
(a)	7	8	15
(b)	1	1	1
(c)	1	1	2
(d)	1	1	2

7. (i) (a) $2 \text{ g } H_2 \equiv 26 \text{ g } MgH_2$

$1 \text{ g } H_2 \equiv$ **13 g MgH_2**

(b) $16 \text{ g } O_2 \equiv 40 \text{ g MgO}$

$8 \text{ g } O_2 \equiv$ **20 g MgO**

(c) $32 \text{ g } O_2 \equiv 44 \text{ g } CO_2$

$8 \text{ g } O_2 \equiv$ **11 g CO_2**

(d) $32 \text{ g } O_2 \equiv 46 \text{ g } NO_2$

$8 \text{ g } O_2 \equiv$ **11.5 g NO_2**

(ii) (a) Equivalent of weight HCl = 36.5

hence, equivalents of HCl $= \frac{3.65}{36.5} = 0.1$

(b) Equivalent weight of $H_2SO_4 = 49$

hence, equivalent of $H_2SO_4 = \frac{9.8}{49} = 0.2$

(c) Equivalent weight of $H_3PO_4 = \frac{98}{3}$

hence, equivalents of $H_3PO_4 = \frac{9.8}{98/3} = 0.3$

(iii) ${}^{57}_{27}Co \,|\, {}^{58}_{27}Co \,|\, {}^{60}_{27}Co$ $(Z(Co) = 27)$

Master Exercises

Exercise 1

Short Answer Type Questions

1. Deposition
2. Since, carbonate is = 60 g mol^{-1}

hence, valency $= \frac{\text{molar mass}}{\text{equivalent mass}} = \frac{60}{30} = 2$

3. Mass of magnesium = 24 g

Mass of oxide = 40 g

Oxygen combined = 40 − 24 = 16 g

16 g oxygen ≡ 24 g magnesium

$\therefore$ 8 g oxygen $\equiv \frac{24}{16} \times 8 = 12$ g

$\therefore$ Equivalent weight of Mg = 12

4. Mass of aluminium = 27 g

Mass of hydride = 30 g

Hydrogen combined = 3 g

3 g hydrogen ≡ 27 g aluminium

$\therefore$ 1 g hydrogen $\equiv \frac{27}{3} \times 1 = 9$ g

$\therefore$ Equivalent weight of Al = 9

5. $H_3PO_2 + NaOH \longrightarrow NaH_2PO_2$

$H_3PO_2 \equiv 1\ NaOH$

Thus, equivalent weight = molecular weight

= 66 g $equivalent^{-1}$

6. $H_3PO_3 \longrightarrow NaH_2PO_3 \longrightarrow Na_2HPO_3$

$H_3PO_3 \equiv 1\ H^+\ (NaH_2PO_3)$

Equivalent weight = molecular weight

= 82 g $equivalent^{-1}$

$H_3PO_3 \equiv 2\ H^+\ (Na_2HPO_3)$

Thus, equivalent weight $= \frac{\text{molecular weight}}{2}$

= 41 g $equivalent^{-1}$

7. | | Equivalent weight |
|---|---|
| $H_3PO_4 \equiv NaH_2PO_4 \equiv 1H^+$ | $= M = 98$ |
| $H_3PO_4 \equiv Na_2HPO_4 \equiv 2H^+$ | $= \frac{M}{2} = 49$ |
| $H_3PO_4 \equiv Na_3PO_4 \equiv 3H^+$ | $= \frac{M}{3} = 32.66$ |

8. Sulphur changes to SO_2.
9. (a) 8 (b) 16 (c) 22
10. (a) 1.6 g $O_2 = \frac{1.6}{8} = 0.2$ equivalent

(b) $\frac{0.49}{49} = 0.01$ (equivalent weight of $H_2SO_4 = 49$)

(c) $\frac{1.80}{9} = 0.2$ (equivalent weight of $H_2O = 9$)

(d) $\frac{1.6}{4} = 0.4$ (equivalent weight of $CH_4 = 4$)

11. (a) $N_2 \equiv 6H$

$\therefore$ $1\ H = \frac{N_2}{6} = \frac{28}{6} = \frac{M}{6}$

$2NH_3 \equiv 6H$

$\therefore$ $1H = \frac{2NH_3}{6} = \frac{NH_3}{3} = \frac{17}{3} = \frac{M}{3}$

(b) $H_3PO_4 \equiv 2OH^-$

$\therefore$ $1\ OH^- \equiv \frac{H_3PO_4}{2} = \frac{M}{2}$

(c) $Zn \equiv 2H^+$

$\therefore$ Equivalent weight of $Zn = \frac{M}{2}$

(d) $2FeSO_4 \equiv 1\ (O) \equiv 16\ g \equiv 2 \times 8$ g

$\therefore$ 8g (O) ≡ 1 $FeSO_4$

$\therefore$ Equivalent weight of $FeSO_4 = M$

12. Specific heat × atomic weight = 6.4

(Dulong and Petit's law)

13. (a) Atomic weight × specific heat = 6.4

Atomic weight $= \frac{6.4}{0.25} = 25.6$

$\therefore$ Valency $= \frac{\text{atomic weight}}{\text{equivalent weight}} = \frac{25.6}{12.0} = 2.13 \approx 2$

(b) Thus, (i) MCl_2 (ii) MSO_4 (iii) $M_2(PO_4)_2$

14. 1.8 g $C_6H_{12}O_6 = \frac{1.8}{180}$ mole $C_6H_{12}O_6$

= 0.01 mole

= 0.01 × 6 moles of carbon

= 0.01 × 6 × N_0 atoms of carbon

= 0.01 × 12 moles of hydrogen

= 0.01 × 12 × N_0 atoms of hydrogen

= 0.01 × 6 moles of oxygen

= 0.01 × 6 × N_0 atoms of oxygen

Analytical Questions

	C : H
1. Benzene	4.61 : 0.39
	12 : 1
Ethane	4 : 1
Ethylene	6 : 1

Thus, mass ratio of C (w.r.t. H = 1 g) in benzene, ethane and ethylene is 6 : 2 : 3

2. For NH_3, 0.5014 g H; for N_2H_4, 0.3342 g H.

3. $N = 3.6700$ g

$H = 0.5275$ g

Total mass of compound $X = 4.1975$ g

N(%) = 87.43%

H(%) = 12.57%

	%		Ratio
N	87.43	6.24	1
H	12.57	12.57	2

Thus, X is N_2H_4

N in Y = 10.9578 g, X is N_2H_4

4. (a) Neutral species : A, F, G

(b) Negatively charged : B, E

(c) Positively charged : C, D

(d) Conventional symbols

$^{10}_{5}A$, $^{14}_{7}B^{3-}$, $^{39}_{19}C^{+}$, $^{66}_{30}D^{2+}$, $^{81}_{35}E^{-}$, $^{11}_{5}F$, $^{19}_{9}G$

5. One carat = 200 mg diamond

24 carat = $200 \times 24 \times 10^{-3}$ g

$$= \frac{200 \times 24 \times 10^{-3}}{12} \text{ g atom (mol)}$$

$= 0.4$ g atom

$= 0.4 \times 6.02 \times 10^{23}$ atom

$= \mathbf{2.408 \times 10^{23}}$ **atoms**

6. Water = 400 mL = 400 g

$$\therefore \quad D_2O = 400 \times \frac{0.015}{100} \text{ g}$$

$$= \frac{400 \times 0.015}{20 \times 100} \text{ moles}$$

$= 1.8066 \times 10^{21}$ molecules

7. 10 millimol $CaCl_2 \cdot 2H_2O$ in 1000 mL

$$\therefore \text{ Millimol per mL} = \frac{10}{1000} = 0.01$$

$= 0.01 \times 10^{-3}$ mol

$= 1 \times 10^{-5}$ mol

Mass of $CaCl_2 \cdot 2H_2O$ per mL = 147×10^{-5} g mL^{-1}

$= 1.47 \times 10^{-3}$ g mL^{-1}

8. (d) NH_3

9. Molar of $Al_2(SO_4)_3 \cdot xH_2O = 342 + 18x$

$$\% \text{ of Al} = \frac{54 \times 100}{342 + 18x} = 8.2$$

x = **18**

10. Ca : 38.71 O : 41.29 P : 20.00

11. I II

%	Mol	Ratio	%	Mol	Ratio
Pt = 73.3	0.38	1	Pt = 57.9	0.30	1
Cl = 26.7	0.75	2	Cl = 42.1	1.18	4
Compound :	$PtCl_2$			$PtCl_4$	

12. $x = 6$

13. Let molar mass be = M

If 0.43 g Fe then molar mass = 100 g

$$56 \text{ g Fe than molar mass} = \frac{100 \times 56}{0.43}$$

$= \mathbf{13023.26\ g\ mol^{-1}}$

14. SnF_2

15. Cr : 52 g mol^{-1}

16. N_2O_4, 92 g mol^{-1}

NCERT Problems

1. (i) 18 g mol^{-1} (ii) 44 g mol^{-1} (iii) 16 g mol^{-1}

2.

	%	Ratio	Integer ratio
Fe	69.9	$\frac{69.9}{55.85} = 1.25$	1 = 2
O	30.1	$\frac{30.1}{16.00} = 1.88$	1.5 = 3

Thus, Fe_2O_3

3. $CuSO_4$ = 159.5 g mol ≡ 63.5 g Cu

$$\therefore \text{ Cu in 100 g } CuSO_4 = \frac{63.5}{159.5} \times 100$$

$= 39.81$ g

4. (i) 6 moles carbon

(ii) 18 moles hydrogen

(iii) $3N_0$ molecules ethane

5.

	moles	Atoms
moles in (i)	$\frac{1}{197}$	$\frac{N_0}{197}$
(ii)	$\frac{1}{23}$	$\frac{N_0}{23}$
(iii)	$\frac{1}{7}$	$\frac{N_0}{7}$ (max)
(iv)	$\frac{1}{71}$	$\frac{N_0}{71}$

6. 6.02×10^{23} atoms = 12 g carbon

$$\therefore \quad 1 \text{ atom} = \frac{12}{6.02 \times 10^{23}} \text{ g}$$

7. (i) $52N_0$ (ii) $\frac{52}{N_0} \times N_0 = 52$

(iii) $\frac{52}{4} \times N_0$

Exercise 2

Only One Option Correct

1. (a) **2.** (b) **3.** (c) **4.** (b)

5. Mg $= 0.273$ g

Compound of Mg with N $= 0.378$ g

N-reacted $= 0.105$ g

	Gram	Mole	Ratio	
Mg	0.273	0.011375	1.51	3
N	0.105	0.0075	1	2

Thus, Mg_3N_2 **(a)**

6. $$\underset{\substack{M+60\\100}}{MS_2} \equiv \underset{\substack{64\\40.06}}{2S}$$

$$\frac{M+64}{100} = \frac{64}{40.06}$$

$$\therefore \quad M = 96$$

Thus, **(d)**

7. (c) **8.** (c) **9.** (a) **10.** (b)

11. $Na_2SO_3 \cdot xH_2O = 126 + 18x$

$$\frac{18x}{126+18x} = 0.5$$

$$x = 7$$

Thus, **(d)**

12. Mass of one atom $= 6.66 \times 10^{-23}$ g

$\therefore$ Atomic mass $= 6.66 \times 10^{-23} \times 6.02 \times 10^{23}$

$= 40 \text{ g mol}^{-1}$

$\therefore$ Number of moles in 40 kg $= \dfrac{40000 \text{ g}}{40 \text{ g mol}^{-1}} = \mathbf{1000\ mol}$

Thus, **(a)**

13. (c) **14.** (d) **15.** (c)

16. Number of moles of O atoms $= \dfrac{1}{3}$ moles of H

$$= \frac{3.18 \text{ mol}}{3} = 1.06 \text{ mol}$$

Thus, (c)

17. (d)

18. Let molar mass be $= M$

Mass of 21 carbon atoms $= 252$

$$\% \text{ of carbon} = \frac{252 \times 100}{M} = 69.98$$

$$\therefore \quad M = 360.1$$

Thus, **(d)**

19. (a) **20.** (b)

21. Volume of spherical ball $= \dfrac{4}{3}\pi r^3 = \dfrac{4}{3} \times \dfrac{22}{7} \times (7)^3$

$$\text{Mass} = V \times d = \frac{4}{3} \times \frac{22}{7} \times (7)^3 \times 1.4 \text{ g}$$

$$\text{Fe } (56\%) = \frac{4}{3} \times \frac{22}{7} \times (7)^3 \times 1.4 \times \frac{56}{100} \text{ g} = 1126.4$$

Moles $= 20$

Thus, **(c)**

22. (a) **23.** (a) **24.** (b) **25.** (b)

One or More Than One Options Correct

1. (a, b, c) **2.** (a, c) **3.** (a, b, c)
4. (a, c, d) **5.** (a, b, d) **6.** (a, c)
7. (a, b, c) **8.** (d) **9.** (a, b) **10.** (a, b, d)

Brain Twisters

1. $H_3O^+ \rightleftharpoons H_2O + H^+$

$$1\,H^+ \equiv 1\,H_3O^+$$

$\therefore$ Equivalent weight of H_3O^+ = molecular weight

= 19 g equivalent^{-1}

2. Equivalent weight in acid is based on number of ionisable protons which is called Basicity. **Ionisable hydrogen is attached to electronegative atom as O, and X (X = F, Cl, Br, I)**

(a) H_3PO_2 has one ionisable hydrogen (monobasic acid)

$$H_3PO_2 \rightleftharpoons H_2PO_2^- + H^+$$

hence, equivalent weight ≡ molecular weight

(b) $\underset{\text{(dibasic)}}{H_3PO_3} \rightleftharpoons HPO_3^{2-} + 2H^+$

$$2H^+ \equiv H_3PO_3$$

$$\therefore \quad 1\,H^+ \equiv \frac{H_3PO_3}{2}$$

Hence, equivalent weight $= \dfrac{\text{molecular weight}}{2}$

(c) $\underset{\text{(tribasic)}}{H_3PO_4} \rightleftharpoons PO_4^{3-} + 3H^+$

$$3H^+ \equiv H_3PO_4$$

$$1\,H^+ \equiv \frac{H_3PO_4}{3}$$

hence, equivalent weight $= \dfrac{\text{molecular weight}}{3}$

(In the Unit-Chemical Bonding, structures have been given; they can be used to predict ionisable hydrogens).

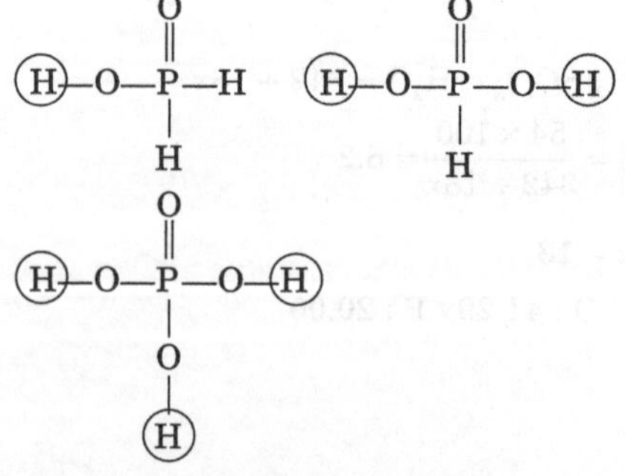

Ionisable hydrogens (circled) are attached to oxygen atoms.

3. (a) $32\text{ g } O_2 \equiv 12\text{ g C}$

$8\text{ g } O_2 \equiv \mathbf{3\text{ g C}}$

(b) $4\text{ g } H_2 \equiv 12\text{ g C}$

$1\text{ g H} \equiv \mathbf{3\text{ g C}}$

(c) $2\text{ g } H_2 \equiv 24\text{ g C}$

$1\text{ g } H_2 \equiv \mathbf{12\text{ g C}}$

(d) $4\text{ g } H_2 \equiv 24\text{ g C}$

$1\text{ g } H_2 \equiv \mathbf{6\text{ g C}}$

(e) $6\text{ g } H_2 \equiv 24\text{ g C}$

$1\text{ g } H_2 \equiv \mathbf{4\text{ g C}}$

(f) $8\text{ g } H_2 = 36\text{ g C}$

$1\text{ g } H_2 \equiv \mathbf{4.5\text{ g C}}$

Equivalent weight is not a definite property. It depends on the reaction and product formed.

4. Molecular weight

Let A be $= S_xO_y$ $\quad 32x + 16y$

and B be $= S_pO_q$ $\quad 32p + 16q$

In *A* $\quad 16\,y\text{ g oxygen} \equiv (32x + 16y)\text{ g } S_xO_y$

hence, $8\text{ g oxygen} \equiv \dfrac{32x + 16y}{2y} = 16$ (given)

$$\frac{32x}{y} + 8 = 16$$

$$\frac{x}{y} = \frac{1}{2} = \frac{\text{number of S atom}}{\text{number of O atom}}$$

Thus, A is SO_2

In *B* $\quad 16\,q\text{ g oxygen} \equiv (32p + 16q)\text{ g } S_pO_q$

$8\text{ g oxygen} \equiv \dfrac{32p + 16q}{2q} = 13.33$ (given)

$$\frac{16p}{q} = 5.33$$

$$\frac{p}{q} = \frac{5.33}{16} = \frac{1}{3} = \frac{\text{number of S-atom}}{\text{number of O-atom}}$$

Thus, B is SO_3

5. Hydrides

(I) N_xH_y

molar mass $= (14x + y)$

$y\text{ gH} = (14x + y)\text{ g } N_xH_y$

$1\text{ gH} \equiv \dfrac{14x + y}{y} = 5.66$

$$\frac{14x}{y} = 4.66$$

$$\frac{x}{y} = \frac{1}{3} = \frac{\text{number of nitrogen atom}}{\text{number of hydrogen atom}}$$

Hence, hydrogen is NH_3

(II) $\dfrac{14x + y}{y} = 8$ (given)

$$\frac{14x}{y} = 7$$

$$\frac{x}{y} = \frac{1}{2} = \frac{\text{number of nitrogen atom}}{\text{number of hydrogen atom}}$$

Thus, hydride appears NH_2 (unstable).

Thus, actual hydride is N_2H_4 (hydrazine).

6. (a) 200 g parts per million (out of 10^6 g)

$= \dfrac{200}{100}$ moles $CaCO_3$ per million gram

$= 2000$ millimoles $CaCO_3$ per million gram

$\equiv \dfrac{2000}{10^6} = 2 \times 10^{-3}$ millimoles per gram

(b) 2 moles $CaCO_3$

$= 2 \times N_0$ molecules $CaCO_3$ per million gram

$= \dfrac{2 \times N_0}{10^6}$ molecules per gram

7. Purity $= \dfrac{20}{24} \times 100 = \dfrac{250}{3}$ g

(24 carat is assumed to be 100%)

Pure diamond in 200 mg sample

$$= \frac{200}{1000} \times \frac{250}{3} \times \frac{1}{100} = \frac{1}{6}\text{ g}$$

Number of moles $= \dfrac{1}{6} \times \dfrac{1}{12}$ g carbon

Number of carbon atoms

$$= \frac{6.02 \times 10^{23}}{72} = 8.33 \times 10^{21}\text{ atom}$$

8. 1 mole ethyl alcohol = 46 g ethyl alcohol (CH_3CH_2OH)

Water present $= 100 - 46 = 54$ g

Moles of $CH_3CH_2OH = \dfrac{46}{46} = 1$

Moles of $H_2O = \dfrac{54}{18} = 3$, $\quad$ Ratio $= 1 : 3$

9. $KHC_2O_4 \rightleftharpoons K^+ + HC_2O_4^-$

$HC_2O_4^- \rightleftharpoons H^+ + C_2O_4^{2-}$

One of the students (say A) would have derived equivalent weight based on H^+ which is = molecular weight

Another student (say B) would have derived equivalent weight based on $C_2O_4^{2-}$ (oxalate)

10. Magnesium = 0.24 g

Magnesium oxide (residue) = 0.40 g

Oxygen combined = 0.16 g

0.16 g oxygen ≡ 0.24 g Mg

Thus, $\quad$ 8 g oxygen = **12 g Mg**

11. (a) $NH_4NO_3 = \dfrac{28 \times 100}{80}$

(b) $NH_2CONH_2 = \dfrac{28 \times 100}{60}$ Max

(c) $KNO_3 = \dfrac{14}{101} \times 100$ least

(d) $NH_4Cl = \dfrac{14 \times 100}{53.5}$

12. $Ca_3(PO_4) \rightleftharpoons 3Ca^{2+} + 2PO_4^{3-}$

Based on valency, equivalent weight of

$$Ca_3(PO_4)_2 = \frac{\text{molecular weight}}{6}$$

Hence, one equivalent $Ca_3(PO_4)_2 = \frac{\text{mole}}{6}$

1 mole $Ca_3(PO_4)_2 \equiv 3$ moles $CaCl_2 \equiv 2$ moles H_3PO_4

$\therefore \frac{1}{6}$ mole $Ca_3(PO_4)_3 = \frac{1}{2}$ mole $CaCl_2 \equiv \frac{1}{3}$ mole H_3PO_4

Passage Comprehension Questions

Passage 1

1. (a) Dish of 20 cm diameter (10 cm = 0.1 m radius) has area $= \pi r^2 = \pi \times 0.01\ m^2$

$0.21 \times 10^{-18}\ m^2$ is the area of = 1 molecule

$\therefore\ \pi \times 0.01\ m^2$ is the area of $= \frac{\pi \times 0.01}{0.21 \times 10^{-18}}$ molecules

1.4×10^{-4} g stearic acid $= \frac{1.4 \times 10^{-4}}{284}$ mol

$$= \frac{1.4 \times 10^{-4}}{284} N_0 \text{ molecules}$$

$$\frac{1.4 \times 10^{-4} N_0}{284} = \frac{\pi \times 0.01}{0.21 \times 10^{-18}}$$

$$N_0 = 3.03 \times 10^{23}$$

2. (b) $\frac{1}{N_0} = \frac{1}{3.03 \times 10^{23}} = 3.33 \times 10^{-24}$ g

Passage 2

1. (b) Rs. 50 per kg = Rs. 50 per $\left(\frac{1000}{74.5}\right)$ mol

$$\text{cost} = \frac{\text{Rs. } 50}{\frac{1000}{74.5}} \text{ per mol} = \text{Rs. } 3.73 \text{ per mol}$$

2. (c) 1 kg KCl $= \frac{1000}{74.5}$ mol K^+ costing Rs. 50/ –

Re 1 gives $= \frac{1000}{74.5 \times 50}$ mol of K^+ (in KCl)

1 kg $K_2SO_4 = \frac{1000}{174}$ mol

$= \frac{2000}{174}$ mol K^+ costing Rs. x

Re 1 gives $= \frac{1000 \times 2}{174x}$ mol of K^+ (in K_2SO_4)

$$\therefore \frac{1000}{74.5 \times 50} = \frac{2000}{174x}$$

$$x = \textbf{Rs. 42.82 kg}^{-1}$$

3. (d) Let mass of $K_2O = x$ g

$= \frac{x}{94}$ mol $K_2O = \frac{2x}{94}$ mol K^+

1000 g KCl $= \frac{1000}{74.5}$ mol KCl $= \frac{1000}{74.5}$ mol K^+

$$\therefore \frac{2x}{94} = \frac{1000}{74.5}$$

$$x = 631 \text{ g} = 0.631 \text{ kg}$$

Passage 3

1. (a) **2.** (c)

3. (b) For spectrum, approximate percentage of

$^{24}Mg = 80$

$^{25}Mg = 8$

$^{26}Mg = 12$

$$\text{Average atomic mass} = \frac{M_1X_1 + M_2X_2 + M_3X_3}{100} = 24.32$$

Assertion & Reason

1. (a) **2.** (b) **3.** (d) **4.** (a)

True & False

1. F(16) **2.** T **3.** T **4.** T **5.** F (moles one equal)
6. F (no H_2 is left) **7.** T **8.** T **9.** T
10. F (it is $\frac{17}{3}$ g)

Fill in the Blanks

1. multiple proportion **2.** 2 : 3 **3.** C_2H_6 **4.** 20
5. 0.33 **6.** $Na_2S_2O_3 \cdot 7H_2O$ **7.** 10^3 **8.** 1
9. reciprocal proportions **10.** Dalton (Da)

Matrix-Match Type Questions

I. Only One Correct

1. (A) — (4); (B) — (3); (C) — (1); (D) — (2); (E) — (5)

2. (A) — (5); (B) — (3); (C) — (1); (D) — (4); (E) — (2)

II. One or More Correct

1. (A) — (1,3,4,5); (B) — (1,2,3,5); (C) — (1,3,5); (D) — (1,3,5)

2. (A) — (1,2,3,4,5); (B) — (1,3,5); (C) — (1); (D) — (4)

Integer Answer Types

Set I

Questions →	1	2	3	4	5	6	7	8
Answers →	3	7	1	0	1	0	6	5

1. $CH_4 = 16 \text{ g mol}^{-1}$

4g H has = 12 g carbon

Thus, 1 g H has $= \dfrac{12}{4} = 3$ g

Thus, $x = \mathbf{3}$

2. $\underbrace{Na_2SO_3}_{\downarrow} \cdot \underbrace{yH_2O}_{\downarrow}$

126 g 18 y

$$\frac{18y}{126 + 18y} = \frac{50}{100}$$

$$y = 7$$

3. 1 amu

4. H_2^{2+} H_2 has = 2 electron, 0 neutron

H_2^{2+} has = 0 electron, 0 neutron

5. ${}^1_0n \longrightarrow {}^1_1H + {}^{0}_{-1}e$

due to emission of electron, atomic number increases by 1 unit

6. Na_2CO_3 is thermally stable, no effect of heating.

7. Empirical formula weight = 30

Thus, $\dfrac{180}{30} = 6$

8. $\dfrac{2.495}{249.5} = 0.01$ mol

$= 0.01x\ H_2O$

$= 0.05$ mol

$\therefore$ $x = \mathbf{5}$

Set II

Questions →	1	2	3	4	5	6	7	8
Answers →	3	3	8	8	3	0	2	5
	⓪	⓪	⓪	⓪	⓪	⓪	⓪	⓪
	①	①	①	①	①	①	①	①
	②	②	②	②	②	②	②	②
	③	③	③	③	③	③	③	③
	④	④	④	④	④	④	④	④
	⑤	⑤	⑤	⑤	⑤	⑤	⑤	⑤
	⑥	⑥	⑥	⑥	⑥	⑥	⑥	⑥
	⑦	⑦	⑦	⑦	⑦	⑦	⑦	⑦
	⑧	⑧	⑧	⑧	⑧	⑧	⑧	⑧
	⑨	⑨	⑨	⑨	⑨	⑨	⑨	⑨

1. Radioactive water is heavy water with heavy isotope (${}^2_1H$) of hydrogen. Thus, molar mass of heavy water
$= 2 \times 2 + 16$
$= 20 \text{ g mol}^{-1}$
$0.060 \text{ kg} = 0.060 \times 1000 \text{ g} = 60 \text{ g} = \dfrac{60}{20} = 3$ moles

Thus, **3**

2. Average atomic mass $= \dfrac{A_1X_1 + A_2X_2}{x_1 + x_2}$

$= \dfrac{35x + 37}{x + 1} = 35.5$

Thus, $x = \mathbf{3}$

3. One gram atom has N_0 atoms $= 6.02 \times 10^{23}$ atoms

If atomic mass $= A$

then mass of one atom $= \dfrac{A}{N_0} = 6.66 \times 10^{-23}$ g

$$\frac{A}{N_0} = 6.66 \times 10^{-23}$$

$\therefore$ $A = 6.66 \times 10^{-23} \times 6.02 \times 10^{23}$

$= 40 \text{ g mol}^{-1}$

Number of gram atoms (moles) in 0.320 kg

$= \dfrac{0.320 \times 1000}{40} = \mathbf{8}$

4. X^{2-} is isoelectronic with Ne.

Thus, electrons in $X^{2-} = 10$

Proton (= atomic number) in $X^{2-} = 8$

Also mole $= \dfrac{\text{Mass}}{\text{atomic mass}}$

$0.1 = \dfrac{1.6}{A}$

$\therefore$ $A = 16$

Also, neutrons + protons = Atomic mass = 16

$\therefore$ neutrons $= 16 - 8 = \mathbf{8}$

5. $SO_2(g) + 2H_2S(g) \longrightarrow 3S(s) + 2H_2O(l)$

	$SO_2(g)$	$2H_2S(g)$	$3S(s)$	$2H_2O(l)$
As per equation	1 mol	2 mol	3 mol	2 mol
taken :	2 mol	2 mol	?	

If one takes 2 moles $SO_2(g)$ then 4 moles of $H_2S(g)$ are required. But 2 moles of $H_2S(g)$ have been taken. Hence, $H_2S(g)$ is **limiting reagent** that forms 3 moles of sulphur.

Thus, **3**

6. ${}^1_1H$ neutrons = 0

7. $AlPO_4 \rightleftharpoons Al^{3+} + PO_4^{3-}$ (two ions)

8. $\underset{x \text{ mol}}{xFe} + \underset{y \text{ mol}}{yO_2} \longrightarrow Fe_xO_y$

$= (22.4\ y\ \text{L})$ $\quad \dfrac{x}{0.2} = \dfrac{22.4y}{3.36}$

$11.2 \text{ g} = \dfrac{11.2}{56} = 0.2$ mol

$$\frac{x}{0.2} = \frac{22.4y}{3.36}$$

$$\frac{x}{y} = \frac{2}{3}$$

Thus, Fe_2O_3 $x + y = \mathbf{5}$

2

Chemical Bonding

The theory of the chemical bond .... is still far from perfect. Most of the principles that have been developed are crude, and only rarely can they be used in making an accurate quantitative prediction. However, they are the best that we have, as yet, and I agree with Poincare's that "it is far better to forsee even without certainty than not to forsee at all".

Linus Pauling

Quick Points

- Formal charge is given by

$$F = v - \frac{s}{2} - u$$

where v = valence electron

s = shared electrons forming bonds

u = unshared electrons

- 1 pm $= 1 \times 10^{-12}$ m $= 1 \times 10^{-10}$ cm
- Charge on electron $= 4.8 \times 10^{-10}$ esu $= 1.6 \times 10^{-19}$ C
- In a complex, $[ML_n]$, M is the metal and L the ligand and n the number of ligands.
- If ligand is weak, (as H_2O) arrangement of electrons in valence shell is not affected but with strong ligand, (as CN^-) unpaired electrons get paired.

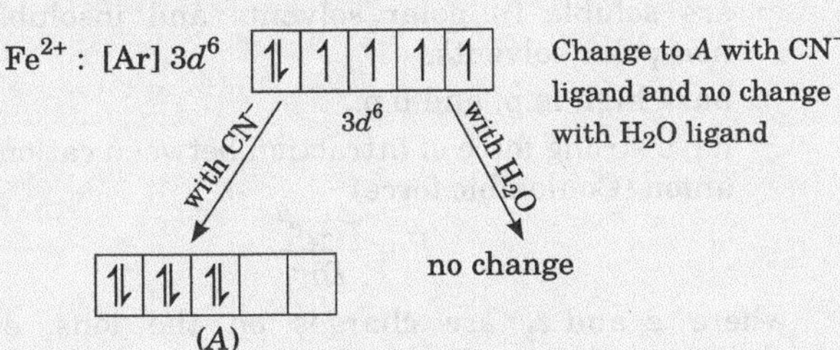

- If there are N unpaired electrons, magnetic moment is given by

$$\mu = \sqrt{N(N+2)} \text{ BM}$$

Learning Framework of the Chapter ...

- Ionic Bonding
- Covalent Bonding
- Coordinate Bonding
- Formal charge
- VSEPR Theory
- Assessing Molecular Shape
- Sigma and Pi Bonds
- Valence Bond Theory
- Hybridisation
- Molecular Orbital Theory
- Bonding in Complexes
- Dipole Moment
- Fajan's Rule
- Hydrogen Bonding
- Resonance
- Metallic Bonding

A molecule will only be formed, if it is more stable, and has a lower energy than the individual atoms. Normally only electrons in the outermost shell of an atom are involved in forming bonds and in this process each atom attains a stable electronic configuration of inert gas. We divide elements into three classes :

(A) Electropositive elements, whose atoms give up one or more electrons easily. They have low ionisation potentials.

(B) Electronegative elements, which can gain electrons. They have higher value of electronegativity.

(C) Elements which have little tendency to lose or gain electrons.

Types of Bonds

	Atoms involved	Bonding
(i)	$A + B$	Ionic
(ii)	$B + B$	Covalent
(iii)	$A + A$	Metallic
(iv)	Electron deficient molecule or ion (Lewis acid) and electron-rich molecule or ion (Lewis base)	Coordinate
(v)	H and electronegative element (N, O, F)	Hydrogen

Ionic Bonding

- An ionic bond is formed when a **metal atom** transfers one or more electrons to a **non-metal atom.**

$$Na \longrightarrow Na^+ + e^-$$
$$Cl + e^- \longrightarrow Cl^-$$

As a result of this transfer the metal atom becomes **cation** and the non-metal the **anion.** IA(1), IIA(2) and VIA(16) , VIIA(17) elements are involved (Table 2.1).

- We can understand this exchange of electrons by showing electron configuration.

	3s	3p			3s	3p
$_{11}$Na [Ne]	[↑]		→	Na^+ [Ne]	___	___
$_{17}$Cl [Ne]	[↑↓]	[↑↓][↑↓][↑]		Cl^- [Ne]	[↑↓]	[↑↓][↑↓][↑↓]

Table 2.1 **Simple Binary Ionic Compounds**

Group: Metal		Group: Non-metal		General formula	Ions present	Example	Melting point (°C)
IA(1)	+	VIIA(17)	$\longrightarrow$	MX	(M^+, X^-)	LiBr	547
IIA(2)	+	VIIA(17)	$\longrightarrow$	MX_2	$(M^{2+}, 2X^-)$	$MgCl_2$	708
IIIA(13)	+	VIIA(17)	$\longrightarrow$	MX_3	$(M^{3+}, 3X^-)$	GaF_3	800 (subl)
IA(1)	+	VIA(16)	$\longrightarrow$	M_2X	$(2M^+, X^{2-})$	Li_2O	> 1700
IIA(2)	+	VIA(16)	$\longrightarrow$	MX	(M^{2+}, X^{2-})	CaO	2580
IIIA(13)	+	VIA(16)	$\longrightarrow$	M_2X_3	$(2M^{3+}, 3X^{2-})$	Al_2O_3	2045
IA(1)	+	VA(15)	$\longrightarrow$	M_3X	$(3M^+, X^{3-})$	Li_3N	840
IIA(2)	+	VA(15)	$\longrightarrow$	M_3X_2	$(3M^{2+}, 2X^{3-})$	Ca_3P_2	~1600
IIIA(13)	+	VA(15)	$\longrightarrow$	MX	(M^{3+}, X^{3-})	AlP	

- Cations and anions attain stable noble gas configuration with outer shell octet of electrons. In some cases cations may not have octet but different outer-shell (see Tables 2.2 and 2.3)
- In the solid state each cation surrounds itself with anions, and each anion with cations. These very large number of ions are arranged in an orderly network called **ionic crystal.**
- Ionic solids
  - are good conductor of electricity in fused state and in aqueous solution,
  - are soluble in polar solvents and insoluble in non-polar solvents,
  - have high m.p. and b.p.,
  - have strong force of attraction between cation and anion (Coulombic force)

$$F = \frac{z_1 z_2 e^2}{Dr^2}$$

where z_1 and z_2 are charges on the ions, e the electronic charge; D the dielectric constant of the medium and r the inter-ionic distance (*i.e.*, the bond length).

Table 2.2 **Electronic Configuration of Some Metal Ions**

"Octet"	"18"	"18 + 2"	Variable	
Na^+, Mg^{2+}	Cu^+ Zn^{2+}	In^+	Cr^{2+}	[Ar] $3d^4$
K^+, Ca^{2+}	Ag^+ Cd^{2+}	Tl^+	Cr^{3+}	[Ar] $3d^3$
Rb^+, Sr^{2+}	Au^+ Hg^{2+}	Sn^{2+}	Mn^{2+}	[Ar] $3d^5$
Cs^+, Ba^{2+}	Ga^{3+}	Pb^{2+}	Mn^{3+}	[Ar] $3d^4$
Fr^+, Ra^{2+}	Sb^{3+}	Sb^{3+}	Fe^{2+}	[Ar] $3d^6$
Al^{3+}	Tl^{3+}	Bi^{3+}	Fe^{3+}	[Ar] $3d^5$
Sc^{3+}			Co^{2+}	[Ar] $3d^7$
Y^{3+}			Co^{3+}	[Ar] $3d^6$
La^{3+}			Ni^{2+}	[Ar] $3d^8$
			Ni^{3+}	[Ar] $3d^7$

Table 2.3 **Lewis Dot Formulae for Representative Elements**

Group	IA (1)	IIA (2)	IIIA (13)	IVA (14)	VA (15)	VIA (16)	VIIA (17)	0
Number of electrons in valence shell	1	2	3	4	5	6	7	8 (except He)
Period 1	H•							He:
Period 2	Li•	Be:	B̈•	·C̈·	·N̈·	·Ö:	·F̤̈:	:N̤̈e:
Period 3	Na•	Mg:	Äl•	·S̈i·	·P̈·	·S̈:	·C̤̈l:	:Ä̤r:
Period 4	K•	Ca:	G̈a•	·G̈e·	·Äs·	·S̈e:	·B̤̈r:	:K̤̈r:
Period 5	Rb•	Sr:	Ïn•	·S̈n·	·S̈b·	·T̈e:	·Ï:	:Ẍ̤e:
Period 6	Cs•	Ba:	T̈l•	·P̈b·	·B̈i·	·P̈o:	·Ä̤t:	:R̤̈n:
Period 7	Fr•	Ra:						

Illustration 1 *Explain the formation of* Li_2O *based on electrons configuration.*

Solution The Lewis dot formula for the atoms and ions are

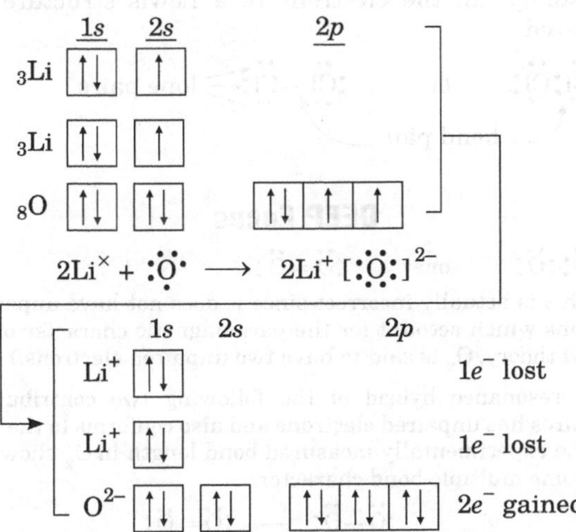

Solubility of Ionic Solids

By the Coulombic force of attraction (F) between cation and anion.

$$F = \frac{z_1 z_2 e^2}{Dr^2}$$

we can study the solubility of ionic solids in different solvents.

DEEP Focus

- Greater the force of attraction between ions, smaller the tendency of the ions to go into solution and thus, smaller the solubility.
  - For a given solute, greater the value of D (Dielectric constant of the solvent) then smaller the force of attraction between positive and negative ions, hence greater the solubility.

    $$D(H_2O) > D(CH_3CH_2OH) > D(CH_3OCH_3)$$

  hence solubility of ionic solid (say NaCl) in

  $$H_2O > CH_3CH_2OH > CH_3OCH_3$$

  - Greater the interionic distance (r) between positive and negative ions, smaller the force of attraction between positive and negative ions hence greater the solubility in the given solvent

    LiF LiCl LiBr LiI $\longrightarrow$ (max)
    interionic distance increases

    LiF NaF KF RbF CsF $\longrightarrow$ (max)
    interionic distance increases

    (maximum towards the direction of arrow)

    (max) $\longleftarrow$ force of attraction increases

    $\longrightarrow$ (max) solubility in water increases

  However, if one of the ions is very big, the solubility increases with decreasing size of the other ion.

  CsF > CsCl > CsBr > CsI (very big cation)
  LiI > NaI > KI > CsI (very big anion)
  $LiNO_3 > NaNO_3 > KNO_3 > CsNO_3$ (very big anion)

  - Also, greater the charge (z_1 and z_2) then greater the force of attraction between positive and negative ions hence smaller the tendency of the ions to go into the solution hence smaller the solubility.

    Na^+ Cl^- (NaCl) Ba^{2+} Cl^- ($BaCl_2$) Al^{3+} Cl^- $AlCl_3$

    charge increases max $\longrightarrow$

    max $\longleftarrow$ solubility increases

    Na^+ Cl^- NaCl Na^+ SO_4^{2-} Na_2SO_4 Na^+ PO_4^{3-} Na_3PO_4

    charge increases max $\longrightarrow$

    max $\longleftarrow$ solubility increases

Energetics of Ionic Bond Formation

The energy given off when gaseous positive and negative ions come together to form 1 mole of the solid ionic compound is called **lattice energy** (U).

$$Na^+(g) + Cl^-(g) \longrightarrow NaCl(s), \quad \Delta H = -U$$

For the reverse process

$$NaCl(s) \longrightarrow Na^+(g) + Cl^-(g), \quad \Delta H = +U$$

Lattice energy can be calculated using **Born-Haber cycle**

DEEP Focus

(I) $M(s) + \frac{1}{2}X_2(g) \longrightarrow M^+X^-(s), \qquad \Delta H_f^\circ = Q = \Delta H_1$

$$M(s) \longrightarrow M(g); \qquad \Delta H = S$$

$$M(g) \longrightarrow M^+(g) + e^-; \qquad \Delta H = I$$

$$\frac{1}{2}X_2(g) \longrightarrow X(g); \qquad \Delta H = \frac{D}{2}$$

$$X(g) + e^- \longrightarrow X^-(g); \qquad \Delta H = -E$$

$$M^+(g) + X^-(g) \longrightarrow M^+X^-(s); \qquad \Delta H = -U$$

(II) $M(s) + \frac{1}{2}X_2(g) \longrightarrow M^+X^-(s);$

$$\Delta H_2 = S + I + \frac{D}{2} - E - U$$

where ΔH_f° is the standard heat of formation of $MX(s)$,

S = the heat of sublimation of $M(s)$,

I = the ionisation energy of $M(g)$

D = the dissociation energy of $X_2(g)$,

E = the electron-affinity of $X(g)$

I and II represent the same chemical change hence,

$$\Delta H_1 = \Delta H_2$$

$$\therefore \quad Q = S + I + \frac{D}{2} - E - U$$

$$\therefore \quad U = S + I + \frac{D}{2} - E - Q$$

(a) $M^+X^-(s) + aq \longrightarrow M^+(aq) + X^-(aq)$

enthalpy change = ΔH_{soln}

(b) $M^+(g) + X^-(g) + aq \longrightarrow M^+(aq) + X^-(aq)$

enthalpy change = $\Delta H_{Hydration}$

(c) $M^+X^-(s) \longrightarrow M^+(g) + X^-(g)$

enthalpy change = U

We find $\quad (a) = (b) + (c)$

$$\therefore \quad \Delta H_{soln} = \Delta H_{Hydration} + U$$

Both lattice energy and hydration energies decrease with increase in ionic size.

DEEP Focus

- If the anion and the cation are of comparable size, the cationic radius will influence the lattice energy. Since lattice energy decreases much more than the hydration energy with increasing ionic size, solubility will increase as we go down the group.

$$\underset{\text{solubility increases} \qquad \text{max}}{\xrightarrow{(BeOH)_2 < Mg(OH)_2 < Ca(OH)_2 < Sr(OH)_2 < Ba(OH)_2}}$$

- If the anion is large compared to the cation, the lattice energy will remain almost constant within a particular group. Since the hydration energies decrease down a group, solubility will decrease as found for alkaline earth metal carbonates and sulphates.

Covalent Bonding

If duplet (2) or octet (8) is completed by sharing of electrons between two electronegative elements, the bond formed is called **covalent bond.**

The Lewis Theory

The Octet rule

For many light atoms a stable arrangement is attained when the atom is surrounded by eight electrons—the octet rule. (In case of H_2, duplet is completed)

- Number of bonds between two atoms

$$= \frac{\text{electrons shared}}{2}$$

Molecule	Electrons shared	Bonds	
H : H	2	1 (single)	H—H
:Ö: :Ö:	4	2 (double)	O=O
:N: :N:	6	3 (triple)	N≡N
:Ö:ₓH with H below (H₂O)	2 (between H and O)	1 (two single bonds)	O—H, with H below O
Hₓ N̈ ₓH with H below (NH₃)	2 (between H and N)	1 (three single bonds)	H—N—H, with H below N
H above, Hₓ C ₓH, H below (CH₄)	2 (between H and C)	1 (four single bonds)	H—C—H, with H above and below C

- All the valence (outer-shell) electrons of the atoms in a Lewis structure must be shown in the structure.
- Usually, all the electrons in a Lewis structure are paired.
- :C̈l:C̈l: or :C̈l—C̈l: ← lone pairs; bond pair (the shared pair in :Cl:Cl: corresponds to the bond —)

DEEP Focus

- :Ö::Ö: or :Ö=Ö:

This is actually incorrect since it does not have unpaired electrons which account for the paramagnetic character of O_2 (by MO theory, O_2 is said to have two unpaired electrons.)

A resonance hybrid of the following two contributing structures has unpaired electrons and also conforms to the fact that the experimentally measured bond length in O_2 shows to have some multiple bond character.

:Ö—Ö: ⟷ :Ö=Ö:

- If there are odd number of electrons (15 as in NO), then there is at least one unpaired electron somewhere in the structure.
- **At least one atom lacks a completed octet of electrons and species is paramagnetic.**

complete octet of N and O

unpaired electron ⟶ $\cdot\ddot{N}=\ddot{O}:$ ⟷ $\cdot\ddot{N}::\ddot{O}:$

electrons shared between N and O = 5

number of bonds = 2.5

Octet of electrons around each oxygen atom (four in double bonds and four in lone pairs)

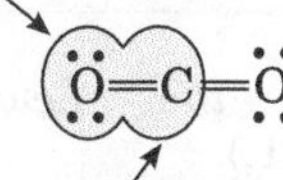

Octet of electrons around carbon atom (four in each of two double bonds)

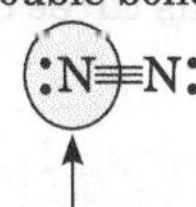

Octet of electrons around each N atom (Six in triple bond and two in lone pair)

- A mathematical relationship can be used to determine total number of electrons shared :

$$S = N - A$$

where S is the total number of electrons shared in the molecule or polyatomic ion; N is the number of valence shell electrons needed by all the atoms in the molecule or ion to achieve noble gas configurations ($N = 8 \times$ number of atoms (H excluded) $+ 2 \times$ number of H atoms); A is the number of electrons available in the valence shell of all of the representative atoms. This is equal to the sum of their periodic group numbers.

In CO_2, A for each O atom is 6 and for C atom is 4.

Thus, $A = 4 + 2 \times 6 = 16$

$N = 8 \times 3 = 24$

Thus, $S = 24 - 16 = 8.$

Thus, there are, eight electrons shared to form bonds

$:\ddot{O}::C::\ddot{O}:$

Illustration 1 *Calculate total number of shared electrons in* H_2SO_4 *and thus, assign structure.*

Solution Let S be the total number of shared electrons then

$$S = N - A$$

where N is the number of valence shell electrons needed by all atoms in a molecule or ion and A is the number of electrons available in the valence shell of all of the respresentative atoms

$N = 8 \times$ number of atoms (excluding H) $+ 2 \times$ number of H-atoms

A = sum of the periodic group number of atoms.

In H_2SO_4.

$N = 8 \times 5$ (one sulphur and four oxygen) $+ 2 \times 2$(H)

$= 44$

$A = 2$(for two H - atoms) $+ 6$ (for sulphur atom) $+ 24$(for four oxygen atoms)

$= 32$

$\therefore S = 44 - 32 = 12$

Thus, H_2SO_4 has six σ-bonds (each bond is formed by sharing of one pair of electrons)

```
        O
        |
  H—O—S—O—H
        |
        O
```

Exceptions to the Octet Rule

In so many cases, the octet rule is violated :

- In BeF_2, octet of Be is not complete

octet incomplete

$:\ddot{F}\times Be\times\ddot{F}:$

octet complete

- In BF_3, octet of B is not complete

$:\ddot{F}:$

octet incomplete [$B\times\ddot{F}:$] octet complete

$:\ddot{F}:$

- The octet rule is also violated where atoms have an extra energy level which is close in energy to the p level, and may accept electrons and be used for bonding. PF_3 obeys the octet rule, but PF_5 does not. PF_5 has ten outer electrons, and uses one $3s$, three $3p$ and one $3d$ orbitals

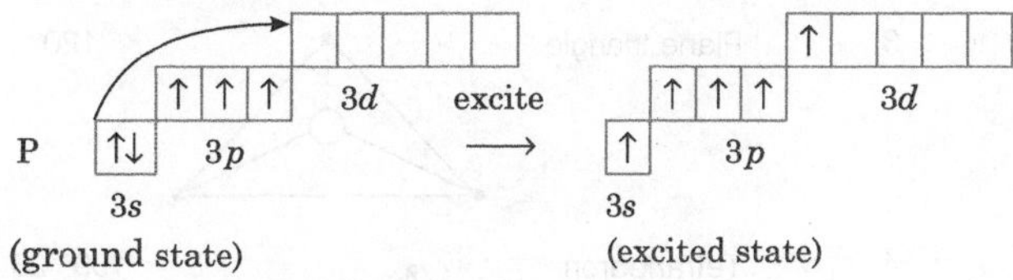

Any compound with more than four covalent bonds must violate the octet rule, and these become very common in the elements after the first two periods of eight elements in the Periodic Table.

In SF_6, there are six S—F bonds having in all 12 electrons in sulphur.

```
     F
  F  |
   \ |
  F—S—F
     | \
     |  F
     F
```

In PCl_5, there are ten electrons in phosphorus. In such cases, **Sugden** introduced **singlet linkage** in which one electron is donated (instead of one pair of electrons) to the

electron deficient atom so that octet rule is not violated. This is indicated by (—). Thus, PCl_5 and SF_6 are represented as

PCl_5: Lewis dot structure (P surrounded by five :Cl: atoms) or Cl—P—Cl with two Cl above and one Cl below

SF_6: Lewis dot structure (S surrounded by six :F: atoms) or S bonded to six F atoms

Sidgwick-Powell Theory

Depending on the number of electron pairs in the outer-shell of the central atom, shape of the molecule can be decided based on this theory. In this concept bond pairs and lone pairs are taken as equivalent since all electron pairs take up some space, and also their tendency of repulsion. This repulsion is minimised if the electron pairs are oriented in space as far apart as possible. Molecular shapes predicted by Sidgwick–Powell theory have been summarised in Table 2.4.

Table 2.4 **Molecular Shapes Predicted By Sidgwick-Powell Theory**

Number of electron-pairs in outer-shell	Shapes of molecules	Bond angles
2	Linear	180°
3	Plane triangle	120°
4	Tetrahedron	109° 28′
5	Trigonal bipyramid	120° and 90°
6	Octahedron	90°
7	Pentagonal bipyramidal	72° and 90°

Illustration 2 *Draw the Lewis structure for aluminium tri-iodide* (AlI_3).

Solution **The outer-shell electron configuration of Al is $3s^2 3p^1$. The Al atom forms three covalent bonds with the I atoms as follows :**

:I:—Al bonded to three I atoms (one to the left, one above, one below)

We find that the octet rule is satisfied for the I atom, there are only six valence electrons around Al atom. The molecule is an example of the incomplete octet. Thus, AlI_3 is a Lewis acid—an electron pair acceptor.

Illustration 3 *Draw the Lewis structure for* PF_5 *in which all five* F *atoms are bonded directly to* P *atom.*

Solution **The outer-shell electron configuration for**

$$P: 3s^2 3p^3, \quad F: 2s^2\, 2p^5$$

Total number of valence electrons = $7 \times 5 + 5 = 40$

P is a third-period element and it can have an expanded octet. The Lewis structure of PF_5 is

:F:—P bonded to five :F: atoms (one above, one below, one to the left, two to the right)

Illustration 4 *In the vapour phase, beryllium chloride consists of discrete* $BeCl_2$ *molecules. Is the octet rule satisfied for* Be *in the compound. If not, can you form an octet around* Be *by drawing another resonance structure? How plausible is this structure?*

Solution **In single unit, octet of Be is short of four electrons.**

:Cl:Be:Cl:

Octet around Be is formed in the following resonance structure but it is not plausible.

$$\overset{+}{\ddot{\underset{\circ\circ}{Cl}}}=\overset{2-}{Be}=\ddot{\underset{\circ\circ}{Cl}}^{+}$$

Coordinate Bonding

A covalent bond results from the sharing of pair of electrons between two atoms where each atom contributes one electron to the bond. It is also possible to have an electron pair bond where both electrons originate from one atom and none from the other. Such bonds are called **coordinate bonds** or **dative bonds.** Since in coordinate bonds, two electrons are shared by two atoms, they differ from normal covalent bonds only in the way they are formed, and once formed they are identical to normal covalent bonds.

- It is represented as : $\longrightarrow$
- Atom/ion/molecule donating electron pair is called **DONOR** or **LEWIS BASE**
- Atom/ion/molecule accepting electron pair is called **ACCEPTOR** or **LEWIS ACID;** (→) points donor to acceptor.
- NH_4^+ : NH_3 has three (N—H) bonds and one lone pair on N-atom. In NH_4^+ formation this lone pair is donated to H^+ (having no electron)

$$\underset{\text{Lewis base}}{NH_3} + \underset{\text{Lewis acid}}{H^+} \longrightarrow NH_4^+$$

$$\left[\begin{array}{c} H \\ | \\ H—\ddot{N}\rightarrow H^+ \\ | \\ H \end{array}\right] \text{ or } \left[\begin{array}{c} H \\ | \\ H—N\rightarrow H \\ | \\ H \end{array}\right]^{\oplus}$$

- BF_4^- : In BF_3, B has incomplete octet. F^- can donate its one of the lone pairs to B forming BF_4^-

$$\left[:\ddot{\underset{\cdot\cdot}{F}}:\rightarrow \begin{array}{c} F \\ | \\ B—F \\ | \\ F \end{array}\right] \text{ or } \left[F\rightarrow \begin{array}{c} F \\ | \\ B—F \\ | \\ F \end{array}\right]^-$$

- SO_3 $\quad O=\overset{\overset{O}{\uparrow}}{S}\rightarrow O$
- PCl_6^- $\quad \left[\begin{array}{c} Cl \\ | \\ \begin{array}{l}Cl\\Cl\\Cl\end{array}\!\!>P\leftarrow Cl \\ | \\ Cl \end{array}\right]^-$
- SbF_6^- $\quad \left[\begin{array}{c} Cl \\ | \\ \begin{array}{l}F\\F\\F\end{array}\!\!>Sb\leftarrow F \\ | \\ Cl \end{array}\right]^-$
- NH_3 and BF_3 $\quad \left[\begin{array}{ccc} H & & F \\ | & & | \\ H—\ddot{N} & \rightarrow & B—F \\ | & & | \\ H & & F \end{array}\right]$
- $[Fe(H_2O)_6]^{2+}$ $\quad \left[\begin{array}{ccc} H_2O & OH_2 & OH_2 \\ \searrow & \downarrow & \swarrow \\ & Fe & \\ \nearrow & \uparrow & \nwarrow \\ H_2O & OH_2 & OH_2 \end{array}\right]^{2+}$

Properties of the coordinate compounds are intermediates of ionic and covalent compounds. A comparison is summarised in Table 2.5.

Table 2.5 **Comparison of Ionic, Covalent and Coordinate Compounds**

	Ionic	Covalent	Coordinate
(1) Binding force	between ions strong (coulombic)	between molecules smaller (van der Waals')	in between
(2) m.p./b.p.	high	less than ionic	in between
(3) Conduction	conductor of electricity in fused state and in aqueous solution.	bad conductor	greater than covalent
(4) Solubility in polar solvent (H_2O)	high	less	in between
(5) Solubility in non polar solvents (ether)	low	high	in between
(6) Physical state	generally solid	liquid and gas	solid, liquid, gas

Illustration 1 *Classify the following bonds as ionic, polar covalent or covalent, and give your reasons :*

(*a*) SiSi *bond in* $Cl_3SiSiCl_3$
(*b*) SiCl *bond in* $Cl_3SiSiCl_3$
(*c*) CaF *bond in* CaF_2
(*d*) NH *bond in* NH_3

Solution (a) Covalent due to identical electronegativity.

(b) One electrons pair is shared between Si and Cl and thus, covalent bond is expected but electronegativity of Cl is greater than that of Si and some polarity develops giving polar-covalent nature.

(c) Ionic since Ca completes its octet by transfer of two outer-shell electrons and two F atoms gain these two electrons thus, completing their octets

$$Ca : [Ar]\,4s^2, \quad F : [He]\,2s^2\,2p^5$$

(d) Polar covalent, explanation as in (b).

Illustration 2 *In the gas phase, aluminium chloride exists as a dimer (a unit of two) with the formula* Al_2Cl_6*. Its skeletal structure is given by*

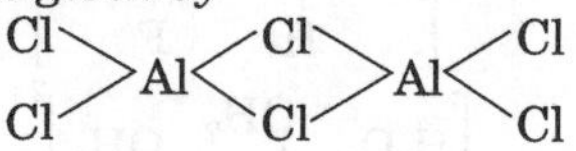

Complete the Lewis structure and indicate the coordinate covalent bonds in the molecule.

Solution

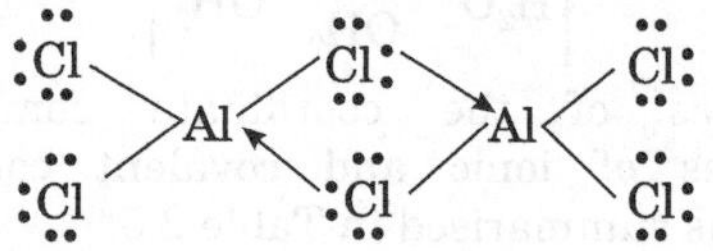

Formal Charge

The formal charge of an atom in a molecule is the hypothetical charge the atom would have if we could redistribute the electrons in the bonds evenly between the atoms in the bond. Formal charge (F) is determined using

$$F = v - \frac{s}{2} - u$$

where, v = valence electrons
s = shared electrons (forming bonds)
u = unshared electrons

- In $BrCl_3$ Cl—Br—Cl (with Cl bonded below Br)

	v	s	u	Formal charge
Br	7	6	4	0
Cl	7	2	6	0

- In ICl_4^-

	v	s	u	Formal charge
I	7	8	4 (one extra)	–1
Cl	7	2	6	0

- In NH_4^+

	v	s	u	Formal charge
N	5	8	0	+1
H	1	2	0	0

- In many cases, we can draw more than one structure for a molecule, with different arrangement of atoms. We can use the following guidelines involving formal charge to decide between two or more possible structures for a molecule or ion :

DEEP Focus

(a) A molecular structure that gives a Lewis structure in which all formal charges are zero is preferable to one in which some formal charges are not zero.

(b) If the Lewis structure must have non-zero formal charges, the structure with the lowest number of non-zero formal charge is preferable and a Lewis structure with no more than one large formal charge (–2, +2 and so on) is preferable to a structure with several formal charges.

(c) Lewis structure should have adjacent formal charges of zero or of opposite sign.

(d) When we must choose among several Lewis structures with similar distribution of formal charges on the more electronegative atom is preferable.

Thiocyanate ion can be assigned three structures :

	Formal charges			Total charge
	N	C	S	
I [:N≡C—S̤̈:]⁻	–1	0	0	–1
II [:C̈=N=S̈:]⁻	+1	–2	0	–1
III [:C̈=S=N̈:]⁻	–1	–2	+2	–1

In all these cases total formal charge is –1. But we prefer structure I because it has the lowest number of atoms with non-zero formal charge [guideline (b)].

- Generally the most stable atomic skeleton is also that for which the sum of the electronegativity differences of adjacent atoms is maximum. Thus, for the arrangements HOCl and HClO, the electronegativities of the atoms yield :

$$\overset{2.2}{H}—\overset{3.44}{O}—\overset{3.16}{Cl}; \quad \Sigma = (3.44 - 2.20) + (3.44 - 3.16) = 1.52$$

$$\overset{2.2}{H}—\overset{3.16}{Cl}—\overset{3.44}{O}; \quad \Sigma = (3.16 - 2.20) + (3.44 - 3.16) = 1.24$$

Thus, H—O—Cl is the most stable structure. This is also consistent with formal charge distribution and Pauling's electroneutrality principle: in a molecule, electrons are so distributed that the residual charge on each atom becomes zero or nearly zero; hydrogen and the most electropositive metals may acquire partial positive charge and the most electronegative atom may acquire the negative charge.

Illustration 1 *Calculate the formal charge on each atom in the following electron-dot structure for* SO_2 :

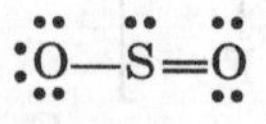

Solution Let v = valence electrons, F = formal charge, s = shared electrons, u = unshared electrons.

Atom	v	s	u	$F = v - \frac{s}{2} - u$
Sulphur	6	6	2	+1
Single-bonded oxygen	6	2	6	–1
Double-bonded oxygen	6	4	4	0

SO_2 may be written as

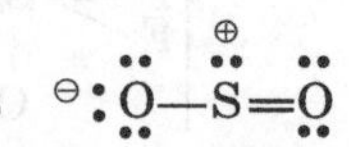

Illustration 2 *We can write two electron-dot structures for N_2O, one of which has a negative formal charge on the oxygen atom and the other of which has a negative formal charge on the terminal nitrogen atom.*

$$:N{=}N{-}\ddot{\underset{\cdot\cdot}{O}}:^{\ominus} \longleftrightarrow {}^{\ominus}:\ddot{N}{=}\overset{\oplus}{N}{=}\ddot{O}$$

(I) (II)

Which of the two electron-dot structures gives a more accurate representation of the molecule?

Solution Since oxygen is a more electronegative element than nitrogen, the structure that places a negative formal charge on oxygen is probably lower in energy than the structure that has a negative formal charge on nitrogen. Thus, (I) is more favourable than (II).

Target Practice 1

1. Which one would better represent bonding in the molecule HNO?
   I : $H{-}\ddot{N}{=}\ddot{O}:$ .... and II : $H{-}\ddot{O}{=}\ddot{N}$
2. In each of the following Lewis structure, *Z* represents a main-group element. Name the group to which *Z* belongs in each case and give an example of a such a compound or ion that actually exists.

	Group of Element Z	Example
(a) $:\ddot{O}{=}\underset{\cdot\cdot}{\ddot{Z}}{=}\ddot{O}:$	....	....
(b) $[\ddot{\underset{\cdot\cdot}{O}}{=}\ddot{Z}{-}\ddot{\underset{\cdot\cdot}{O}}:]^-$	....	....
(c) $H{-}\ddot{\underset{\cdot\cdot}{O}}{-}Z(:\ddot{O}:)_2{-}\ddot{O}:$	....	....

3. Draw Lewis structure for (i) ClO_3^- (chlorate ion) and (ii) NO_2^+ (nitronium ion).
   (i) .... (ii) ....

Answers

1. I is better as octets of N and O are complete and duplet of H is also complete.
2. **Group of Element Z** **Example**
   (a) Zero (18) $:\ddot{O}{=}\underset{\cdot\cdot}{\ddot{Xe}}{=}\ddot{O}:$
   (b) V (15) $[\ddot{\underset{\cdot\cdot}{O}}{=}\ddot{N}{-}\ddot{\underset{\cdot\cdot}{O}}:]^-$
   (c) VII (17) $H{-}\ddot{\underset{\cdot\cdot}{O}}{-}Cl(:\ddot{O}:)_2{-}\ddot{\underset{\cdot\cdot}{O}}:$
3. (i) $[:\ddot{\underset{\cdot\cdot}{O}}{-}\overset{:\ddot{O}:}{\underset{\cdot\cdot}{Cl}}{-}\ddot{\underset{\cdot\cdot}{O}}:]^-$ (ii) $[\ddot{O}{=}\overset{+}{N}{=}\ddot{O}]$

Practice Exercise 1

1. Atomic number of the four elements *A*, *B*, *C* and *D* are respectively $Z-1, Z, Z+1$ and $Z+2$. If $Z = 9$, what is type of bonding between :
   (a) *A* and *D* (b) *B* and *D*
   (c) *A* and *A* (d) *B* and *B*
2. Valence electrons in the elements *X* and *Y* are 3 and 6. What is the formula of the compound formed from *X* and *Y*?
3. Classify Lewis acid and Lewis base out of :
   (a) H_2O (b) NH_3
   (c) H^+ (d) BF_3
   (e) AlH_3 (f) *R*OH
   (g) CH_3NH_2 (h) CN^-
   (i) $^+CH_3$
4. Use your knowledge of electronegativities, to arrange the following bonds in terms of increasing ionic character
   C—H, F—H, Na—Cl, Br—H, K—F
5. Represent each of the following ionic compounds by an appropriate Lewis structures
   (a) KIO_3
   (b) $Ca(OCl)_2$
   (c) NH_4ClO_4
6. Propose Lewis structures for the following ionic species containing sulphur-to-sulphur bonds
   (a) S_2^{2-} (b) S_3^{2-}
   (c) S_4^{2-} (d) S_5^{2-}
7. Explain the formation of following compounds
   (a) CaO (b) Mg_3N_2
8. Which of the following atoms would be expected to form negative ions and which would be expected to form positive ions in binary ionic compounds?
   (a) P, I, Mg, Cl, In, Cs, O, Pb, Co
   (b) Ca, Br, Na, N, F, Al, Sn, S, Cd
9. Calculate the lattice energy of a salt *MX* (*s*) from the data given below :

Heat of formation of *MX*	$(\Delta H) = -550$ kJ mol^{-1}
Heat of sublimation of *M*	$(S) = 80$ kJ mol^{-1}
Heat of dissociation of X_2	$(D) = 155$ kJ mol^{-1}
Ionisation energy of *M*	$(I) = 374$ kJ mol^{-1}
Electron affinity of *X*	$(E) = -343$ kJ mol^{-1}

10. Arrange the following in increasing lattice energy :
    (a) $BeCO_3$, $MgCO_3$, $CaCO_3$, $SrCO_3$, $BaCO_3$
    (b) Li_2CO_3, Na_2CO_3, K_2CO_3

(c) LiCl, NaCl, KCl
(d) LiF, LiBr, LiI

11. The lattice energy of CaO (s) is –3460 kJ mol^{-1}; the lattice energy of K_2O is – 2240 kJ mol^{-1}. Account for this difference.
12. Specify which compound in the following pairs of ionic compounds has the higher lattice energy?
(a) KCl or MgO (b) LiF or LiBr
(c) Mg_3N_2 or NaCl
13. Bond length can be calculated by merely adding covalent bond radii which are H = 0.28 Å, C = 0.77 Å, N = 0.70Å, O = 0.66Å, Cl = 0.99Å, (C=) = 0.67Å, (C≡) = 0.61Å, (N≡) = 0.55 Å. Calculate bond lengths in:
(a) NH_3 (b) CH_2Cl_2
(c) HOCl (d) HCN
(e) $CH_2{=}CH_2$ (f) CH≡CH
14. In which molecule is the van der Waals' force likely to be the most important in determining the melting point and boiling point?
ICl, Br_2, HCl, H_2S, CO
15. Write Lewis structure of :
(a) C_3^{4-} (b) PF_6^-
(c) ICl_4^-
16. The chlorine-to-oxygen bond distance in ClO_4^- is 1.44 Å. What do you conclude about the structure of this ion?
17. Show that by using an expanded octet for the sulphur atom in thionyl chloride ($SOCl_2$), you can write a Lewis structure with no formal charge.
18. Write Lewis structures for the following molecules :
(a) H_2NOH (b) N_2F_2
(c) HONO (d) H_2NNO_2
19. The skeletal structure of acetic acid shown below is correct, but some of the bonds are wrong.
(a) Identify the incorrect bonds and explain what is wrong with them
(b) Write the correct Lewis structure.

```
      H  :O:
      |   |
   H=C—C—O—H
      |    ··
      H
```

20. Give an example of an ion or molecule containing Al that
(a) obeys the octet rule
(b) has an expanded octet
(c) has an incomplete octet
21. Why does NH_3 forms coordination complexes, whereas the isoelectronic species CH_4 does not ?
22. PF_3 can act as a donor molecule while NF_3 shows little tendency to act as a donor. Account for this observation.
23. Indicate the coordinate covalent bond (or bonds) in the product :
(a) $Al^{3+} + 6H_2O \longrightarrow Al(OH)_6^{3-} + 6H^+$
(b) $CO + BF_3 \longrightarrow OCBF_3$
(c) $Sn^{2+} + 3Cl^- \longrightarrow SnCl_3^-$
(d) $H_2O + H^+ \longrightarrow H_3O^+$
(e) $SO_3 + O^{2-} \longrightarrow SO_4^{2-}$
24. Write structures of the following indicating coordinate bonds wherever exist :
(a) N_2O (b) $Cr_2O_7^{2-}$
(c) CrO_5 (d) Al_2Cl_6
(e) H_3PO_4 (f) H_3PO_3
(g) H_3PO_2 (h) NO_2
25. For each of the following pairs of elements, state whether the binary compound they form is likely to be ionic or covalent.
(a) B and F (b) K and Br

Answers

1. (a) Ionic (b) Ionic (c) Covalent (d) Covalent
2. X_2Y_3
3. Lewis acid : (c), (d), (e), (i); Lewis base : (a), (b), (f), (g), (h)
4. C—H < Br—H < F—H < Na—Cl < K—F
5. (a) K^+ $[:\ddot{O}—\ddot{I}—\ddot{O}:]^-$ (with :Ö: below I)

(b) $[:\ddot{O}—\ddot{Cl}:]^-$ $Ca^{2+}[:\ddot{O}—\ddot{Cl}:]^-$

(c) [H—N(H)(H)—H]$^+$ [:Ö—Cl(:Ö:)(:Ö:)—Ö:]$^-$

6. (a) $[:\ddot{S}—\ddot{S}:]^{2-}$ similarly others.
7. (a) Ca : [Ar] $4s^2$ $\longrightarrow$ Ca^{2+} [Ar]
O : $1s^22s^22p^4$ $\longrightarrow$ O^{2-} $1s^22s^22p^6$
Ca^{2+} $[:\ddot{O}:]^{2-}$
(b) $3Mg\times_\times + 2\cdot\dot{N}\cdot \longrightarrow 3Mg^{2+}, 2[:\ddot{N}\times]^{3-}$ or by electron configuration.
8. Refer Table 2.1
9. –738.5 kJ mol^{-1}
10. (a) $BaCO_3 < SrCO_3 < CaCO_3 < MgCO_3 < BeCO_3$
(b) $K_2CO_3 < Na_2CO_3 < Li_2CO_3$ (c) KCl < NaCl < LiCl
(d) LiI < LiBr < LiF
11. $K^+ > Ca^{2+}$ (size), $K^+ < Ca^{2+}$ (charge)
O^{2-} is identical, polarising power of $K^+ < Ca^{2+}$ thus, K_2O is more ionic than CaO thus, lattice energy of CaO > K_2O.
12. (a) MgO (b) LiF (c) Mg_3N_2
13. (a) N—H 0.98 Å (b) C—Cl 1.76 Å
C—H 1.05 Å
(c) O—H 0.94 Å (d) H—C 1.05 Å
O—Cl 1.65 Å C≡N 1.16 Å
(e) C—H 1.05 Å (f) C—H 1.05 Å
C=C 1.34 Å C≡C 1.22 Å

14. Br_2, other molecules have a dipole in addition to van der Waals' forces.

15.

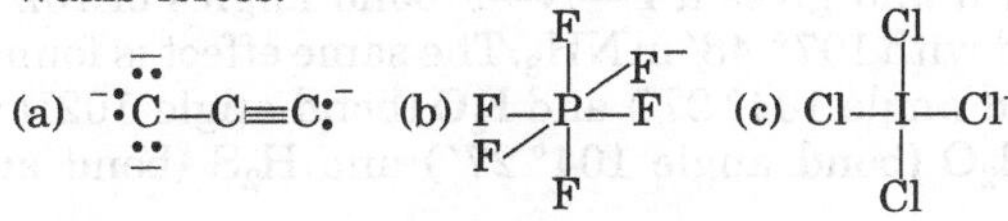

16. There must be considerable double bond character in the bonds.

17.

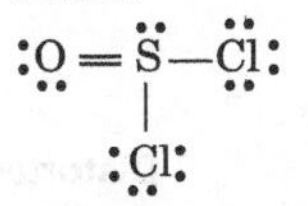

Formal charge = valence electrons $-\dfrac{\text{bonding electrons}}{2}$ – lone pair electrons

$$\text{Oxygen} = 6 - \frac{4}{2} - 4 = 0$$

$$\text{Sulphur} = 6 - \frac{8}{2} - 2 = 0$$

$$\text{Chlorine} = 7 - \frac{2}{2} - 6 = 0$$

18.

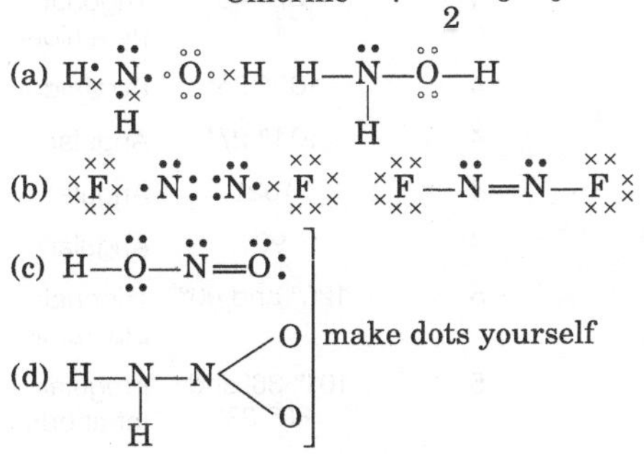
make dots yourself

19.

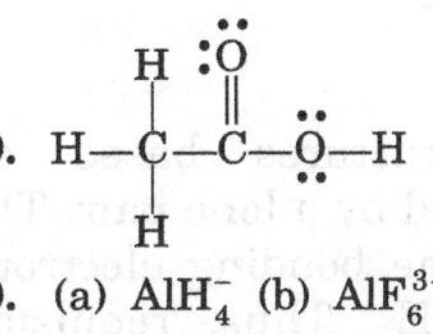

20. (a) AlH_4^- (b) AlF_6^{3-} (c) $AlCl_3$

21. Due to lone-pair on N

22. In PF_3 lone-pair on P is easily donated compared to that of N.

23. Lewis bases (donor) are H_2O (OH^-), CO, Cl^-, H_2O, O^{2-}

24.

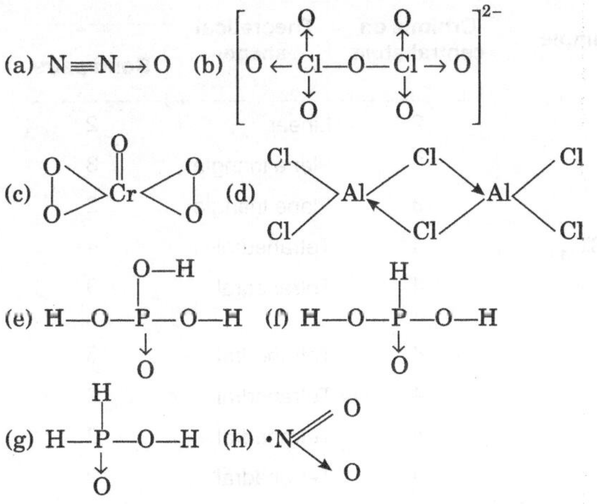

25. (a) Covalent (b) Ionic

Valence Shell Electron Pair Repulsion (VSEPR) Theory (Gillespie Theory)

- The shape of the molecule is determined by repulsions between all of the electron pairs present in the valence shell.
- There can be
  — (lone pair–lone pair)
  — (lone pair–bond pair)
  — (bond pair–bond pair) repulsions.
- Repulsion is in order

$$(lp - lp) > (lp - bp) > (bp - bp)$$

- The magnitude of repulsion between pairs of electrons depends on the electronegativity difference between the central atom and the other atom.
- Double bonds cause more repulsion than single bonds, and triple bonds cause more repulsion than a double bond.
- From the Table 2.6, it is quite evident that presence of lone pair will change geometry of the molecule (which is not shown in Table 2.4).

DEEP Focus

To find which geometry applies, we determine the **Steric number,** S_N, of the central atom, which is defined as

$$S_N = \begin{pmatrix}\text{Number of atom bonded}\\ \text{to central atom}\end{pmatrix} + \begin{pmatrix}\text{Number lone pair}\\ \text{on central atom}\end{pmatrix}$$

The steric number is used to predict molecular geometries. If there are no lone pairs on the central atom then

S_N = number of atoms bonded to central atom = n

There is no repulsion and geometry is retained when lone-pairs are present, the situation changes slightly due to various types of repulsion as shown. Double-bonded or triple-bonded atoms count the same as single-bonded atoms in determining the steric number.

In CO_2, two double-bonded oxygen atoms are attached to central carbon and there are no lone pairs on that atom, thus $S_N = 2$

O=C=O
↑ ↑
Consider this double-bonded atom as single atom.

Effect of Electronegativity

NF_3 and NH_3 both have structures based on tetrahedron with one corner occupied by a lone pair. The high electronegativity of F pulls the bonding electrons further away from N than in NH_3. Thus, repulsion between bond pairs is less in NF_3 than NH_3. Hence, the lone pair in NF_3 causes a greater distortion from tetrahedral and gives a F—N—F bond angles of 102° 30′ compared with 107° 48′ in NH_3. The same effect is found in H_2O (bond angle 104° 27′) and F_2O (bond angle 102°) and also in H_2O (bond angle 104° 27′) and H_2S (bond angle 90°).

Table 2.6 **The Effect of Bonding and Lone Pairs on Geometry and Bond Angles**

Example	Orbitals on central atom	Theoretical shape	Number of		VSEPR notation	S_N	Bond angle	Distorted geometry due to repulsion
			Bond pairs	Lone pairs				
$BeCl_2$	2	Linear	2	0	AX_2	2	180°	Linear
BF_3	3	Plane triangle	3	0	AX_3	3	120°	Plane triangle
SO_2	3	Plane triangle	2	1	AX_2E	3	119°	Angular
CH_4, SiH_4	4	Tetrahedral	4	0	AX_4	4	109° 28′	Tetrahedral
NH_3	4	Tetrahedral	3	1	AX_3E	4	107° 48′	Trigonal pyramidal
NF_3	4	Tetrahedral	3	1	AX_3E	4	102° 35′	Pyramidal
H_2O	4	Tetrahedral	2	2	AX_2E_2	4	104° 27′	Angular
F_2O	4	Tetrahedral	2	2	AX_2E_2	4	102°	Angular
H_2S	4	Tetrahedral	2	2	AX_2E_2	4	90°	Angular
PCl_5	5	Trigonal bipyramidal	5	0	AX_5	5	120° and 90°	Trigonal bipyramidal
XeO_2F_2, SF_4, IF_4^+	5	Trigonal bipyramidal	4	1	AX_4E	5	101° 36′ and 86° 33′	Irregular tetrahedral
ClF_3	5	Trigonal bipyramidal	3	2	AX_3E_2	5	87° 40′	T-shaped
I_3^-, XeF_2	5	Trigonal bipyramidal	2	3	AX_2E_3	5	180°	Linear (also see text)
SF_6	6	Octahedral	6	0	AX_6	6	90°	Octahedral
$XeOF_4$, BrF_5	6	Octahedral	5	1	AX_5E	6	84° 30′	Square pyramidal
ICl_4^-, XeF_4	6	Octahedral	4	2	AX_4E_2	6	90°	Square planar
IF_7	7	Pentagonal bipyramidal	7	0	AX_7	7	72°, 90°	Pentagonal bipyramidal

E stands for lone pair.

Isoelectronic Principle

Isoelectronic species usually have the same structure. This may be extended to species with the same number of valence electrons.

Species	Steric number	Structures
CH_4, NH_4^+, BF_4^-	4	Tetrahedral
CO_3^{2-}, NO_3^-, SO_3	3	Planar triangle
CO_2, N_3^-, NO_2^+	2	Linear
PCl_5	5	Trigonal bipyramidal
SF_6	6	Octahedral

Applications

(a) **Ammonia** (NH_3) There are (*lp* – *bp*) and (*bp* – *bp*) repulsions. The steric number is 4, and the electron pairs arrange themselves into an approximately tetrahedral structure. The lone-pair is not identical to three bonding pairs, however, so there is no reason for the electron pair structure to be exactly tetrahedral. It is found that lone pair tends to occupy more space than bonding **pairs hence bond angles are reduced** to 107° 48′ and structure is pyramidal.

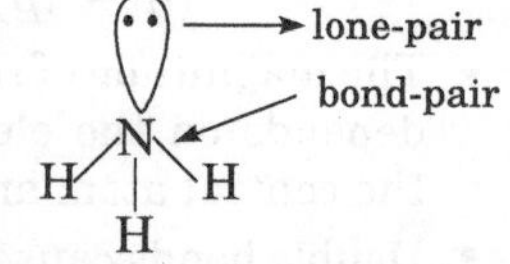

> **DEEP Focus**
>
> The presence of the lone pair causes slight distortion from 109°28′ to

(b) Water (H_2O) : There are two bond pairs and two lone pairs. Presence of two lone pairs distorts the bond angle from 109° 28′ to 104° 27′.

(c) Phosphorus pentachloride (PCl_5) : There is no lone pair so the structure is not distorted. However, a trigonal bipyramidal is not a completely regular structure, since some bond angles are 90° and others 120°.

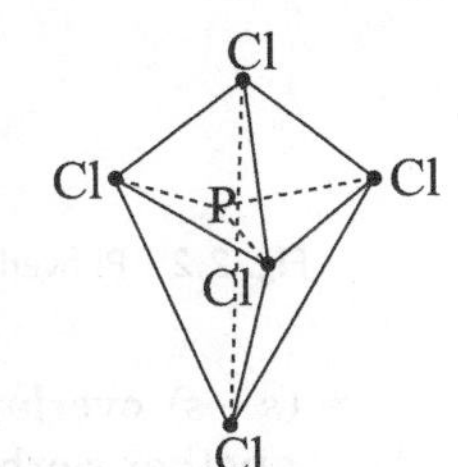

Symmetrical structures are usually more stable than asymmetrical ones. Thus, PCl_5 is highly unstable, and in the solid state it splits into $[PCl_4]^+$ and $[PCl_6]^-$ ions.

$$\underbrace{2PCl_5}_{\text{unsymmetrical}} \longrightarrow \underbrace{\underset{\text{tetrahedral}}{[PCl_4]^+} + \underset{\text{octahedral}}{[PCl_6]^-}}_{\text{symmetrical}}$$

(d) Chlorine trifluoride (ClF_3) There are three bond pairs and two lone pairs. As a general rule, if lone pairs occur in a trigonal bipyramidal they will be located in the equatorial positions (round the middle) rather than the apical positions (top and bottom), since this arrangement minimises the repulsive forces.

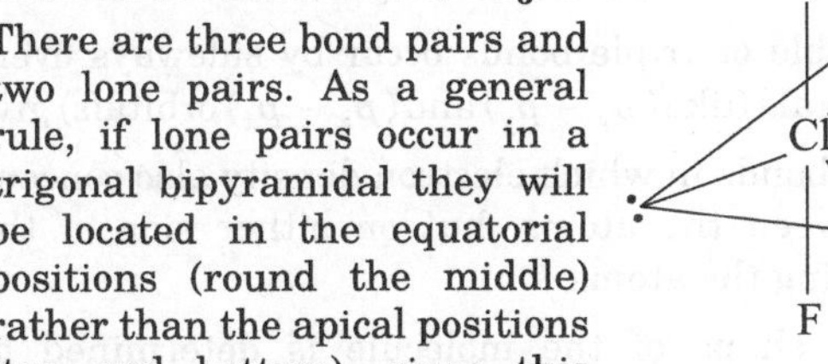

(e) Sulphur tetrafluoride (SF_4) There are four bond pairs and one lone pair. To minimise the repulsive forces the lone pair occupies an equatorial position, and F atoms are located at the other four corners.

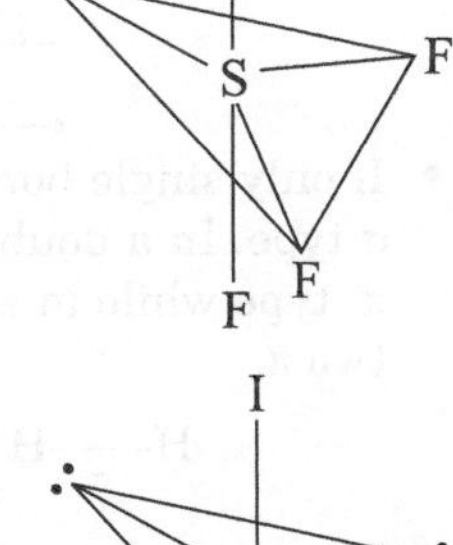

(f) Triiodide ion (I_3^-) I_3^- ion is made up of an I_2 molecule with an I^- bonded to it by means of a coordinate bond in which I_2 is **lone pair acceptor (Lewis acid) and I^- the lone pair donor (Lewis base).** There are two bond pairs and three lone pairs in the outer-shell of central atom. To minimise the repulsive forces the three lone pairs occupy the equatorial position. The ion is therefore linear in shape with a bond angle of exactly 180°.

$$\underset{\text{Lewis acid}}{I_2} + \underset{\text{Lewis base}}{I^-} \longrightarrow [I - I \leftarrow I]^-$$

(g) Iodine heptafluoride (IF_7) This is the only common example of a non-transition element using seven orbitals for bonding giving a pentagonal bipyramid.

(h) ClO_3^+ The central chlorine atom has all of its valence electrons (Cl loses one electron in the formation of Cl^+) involved in bonds to three oxygen atoms and no lone pair exists. Steric number is three and structure is thus **trigonal planar.**

(i) ClO_2^+ Chlorine (being positively charged) has six valence electrons. Two pairs (in all four electrons) are involved in bonding with two oxygen atoms and thus one lone pair exists. Thus, steric number $(2 + 1 = 3)$ is three. The predicted geometry is bent with an angle some what less than 120°.

Illustration 1 *Use the VSEPR model to predict the geometry of the following*

(*a*) AsH_3 (*b*) OF_2 (*c*) $AlCl_4^-$ (*d*) C_2H_4 (*e*) ClF_2^+

Solution

Species	Structure	Repulsion/ angle	Geometry
(a) AsH_3	H—$\ddot{As}$—H, with H below As	*lp* – *bp* *bp* – *bp* bond angle less than 109.5°	trigonal pyramidal
(b) OF_2	:F̤—Ö—F̤:	*lp* – *lp* *lp* – *bp* *bp* – *bp* much less than that of AsH_3	bent shape (like H_2O), since electronegativity of F > O, hence bond angle is less than that of H_2O
(c) $AlCl_4^-$	$[\,$:Cl: above; :Cl—Al—Cl:; :Cl: below$\,]^-$	*bp* – *bp* (no lone pair) 109.5°	tetrahedral
(d) C_2H_4	H₂C=CH₂ (H>C=C<H)	*bp* – *bp* (no lone pair) 120°	trigonal planar
(e) ClF_2^+	:Cl—F: with :F: below Cl	as in (b)	angular

Illustration 2 *The triiodide ion (I_3^-) in which the I atoms are arranged in a straight line is stable, but the corresponding F_3^- ion does not exist. Explain.*

Solution F_3^- and I_3^- are of same group. I can expand its octet but F cannot and thus, in I_3^- octet rule is not violated, but in F_3^- octet rule is violated.

Assessing Molecular Shape

DEEP Focus

By use of VSEPR theory, we can decide the shape of the molecule based on four-step procedure.

- Draw a plausible Lewis structure of the species (molecule or polyatomic ion)
- Determine the number of electron pairs around the central atom and identify each pair as being either a bond-pair or a lone pair.
- Establish the geometrical orientation of the electron pairs around the central atom (linear, trigonal planar, tetrahedral, trigonal bipyramidal, or octahedral).
- Use the positions around the central atom that are occupied by other atoms (not lone-pair electrons) to specify the geometrical shape using Table 2.6.

ClO_3^-

- Number of valence electrons is

from Cl		from O		due to charge		
7	+	3 × 6	+	1	=	26

- Less electronegative atom is central atom in the Lewis structure. Thus, plausible structure is

$$:\ddot{O}-\ddot{Cl}-\ddot{O}:^{-}$$
$$\quad\quad\quad |$$
$$\quad\quad :\ddot{O}:$$

- There are three bond pairs (—) and one lone pair (. .) on chlorine atom total being four.
- By Table 2.6, the orientation of four electron pairs is tetrahedral.
- In VSEPR notation the ClO_3^- ion is AX_3E (E stands for lone pair of electrons — Table 2.6). The ion has a pyramidal shape.

ICl_4^-

- Number of valence electrons

from I		from Cl		due to charge		
7	+	7 × 4	+	1	=	36

- I is the central atom being less electronegative and four Cl atoms around it.
- To complete octet, total electrons required = 32 that means four electrons are extra and lie on I atom as two lone pairs. Thus, plausible structure is

:Cl: and :Cl: (top), :Cl: and :Cl: (bottom) bonded to $\ddot{I}$ (with lone pairs)
[four bond pairs (—)
two lone pairs (:)]

- By Table 2.6, the orientation of six electron-pairs around the central atom is octahedral.
- The ICl_4^- anion is of the type AX_4E_2 and thus, it is **square planar.**

Sigma and Pi Bonds

There can be following types of overlapping along the axes (end to end) :

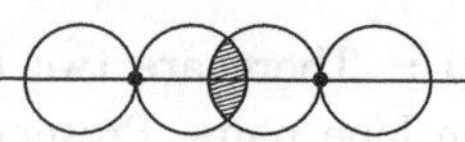

Fig. 2.1 Sigma overlap (lobes point along nuclei)

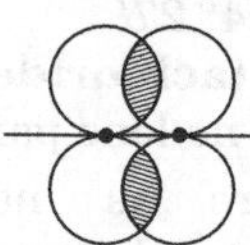

Fig. 2.2 Pi overlap (lobes are at right angles to the line joining the nuclei)

- $(s - s)$ overlapping when s-orbital overlaps with another s-orbital
- $(s - p)$ overlapping, (p can be p_x or p_y or p_z)
- $(p_x - p_x)$ overlapping
- any of the hybrid orbitals overlaps with another hybrid orbital or s or p orbital. (to be taken later)

Bond formed in this manner is called **sigma** (σ) bond in which electron density is concentrated in between the two atoms, and on a line joining the two atoms.

- Double or triple bonds occur by sideways overlap of orbitals (like $(p_y - p_y)$ and $(p_z - p_z)$ orbitals) giving **pi** (π) bonds in which electron density also concentrates between the atoms, but on either side of the line joining the atoms.
- The shape of the molecule is determined by the σ-bonds (and lone pairs) but not by the π bonds. Pi bonds merely shorten bond length. Thus,

$$(C{\equiv}C) < (C{=}C) < (C{-}C)$$
$$\xrightarrow{\text{bond length}}$$
$$\xleftarrow{\text{bond energy}}$$

- If only single bonds exist in a molecule, all will be of σ type. In a double bond, one will be σ and the other π type while in a triple bond, one will be σ and other two π.

$H \overset{}{\underset{\sigma}{—}} H$ $\quad$ $O \overset{\sigma}{\underset{\pi}{=}} O$ $\quad$ $N \overset{\sigma}{\underset{2\pi}{\equiv}} N$

$C(C{\equiv}N)_4$: N≡C—C(—C≡N)(C≡N)—C≡N
(8π, 8σ)

Benzene C_6H_6 (H—C ring with alternating double bonds)
(3π, 12σ)

- In a reaction, it is the π bond that actually takes part

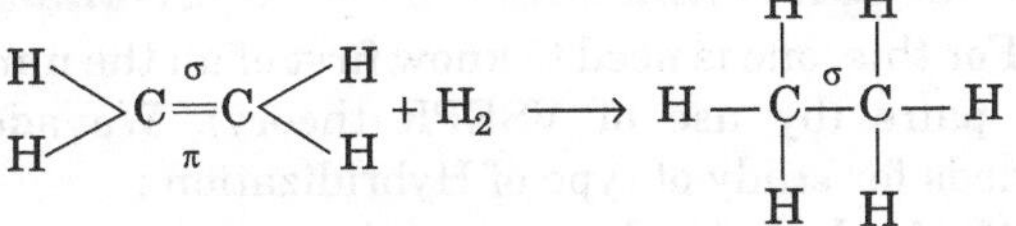

Thus, π-bond is more reactive than σ bond.

Valence Bond Theory

The covalent bond is a region of high electron charge-density (high electron-probability) that results from the overlap of atomic orbitals between two atoms. In general, the greater the amount of overlap between two orbitals, the stronger the bond. For each bond there is a condition of maximum atomic orbital overlap leading to maximum bond strength at a particular internuclear distance (bond length). This is called **Valence Bond Approach.**

- This is a localised electron model of bonding.
- Most of the electrons retain the same orbital locations as in a separated atoms, and the bonding electrons are localised (fixed) in the region of atomic orbital overlap.
- In H_2 **molecules,** H—H σ bond is by *s-s* overlapping

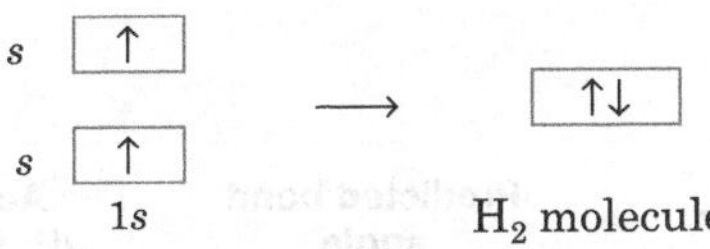

- **In F_2 molecule** F—F σ bond is by axial overlapping of two *p*-orbitals.

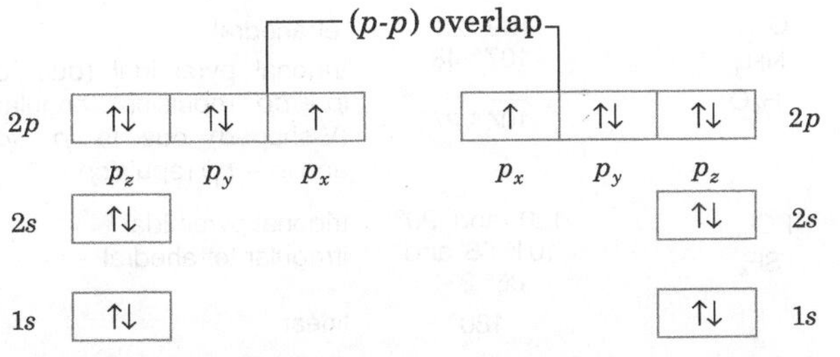

- **In O_2 molecule** One σ bond is by axial overlap of *p*-orbitals and π bond is by lateral overlap of *p*-orbitals.

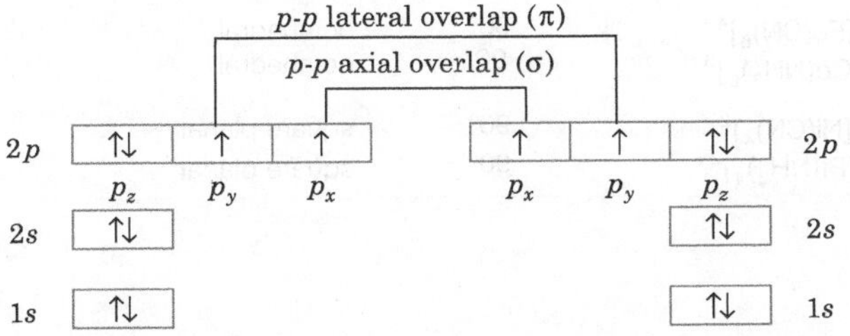

- **In N_2 molecule** One σ bond is by axial overlap of *p*-orbitals and two π bonds are by lateral overlap of *p*-orbitals.

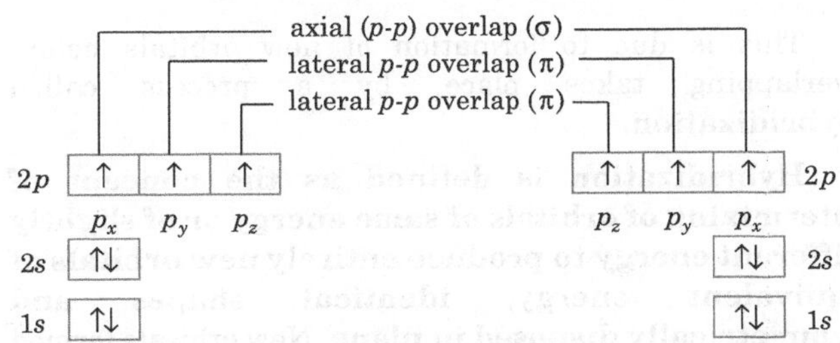

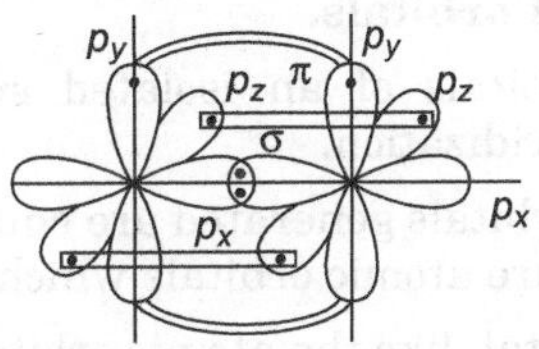

Fig. 2.3 Structure of N_2 by valence bond method

- **In NH_3 :** Three σ bonds are by $(s - p)$ overlap of H and N-atoms

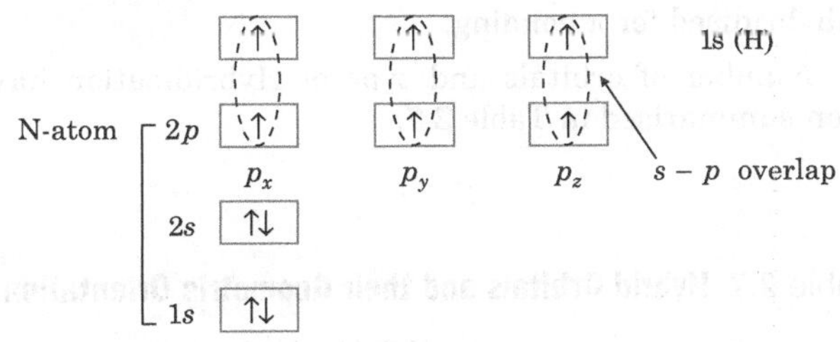

Hybridisation

We consider formation of CH_4 molecule based on **valence bond approach.** To account for tetravalent nature of carbon, we take it in excited state in which there are four unpaired electrons and four C—H σ bonds are supposed to be by overlap of $s - s, s - p_x, s - p_y$ and $s - p_z$ types.

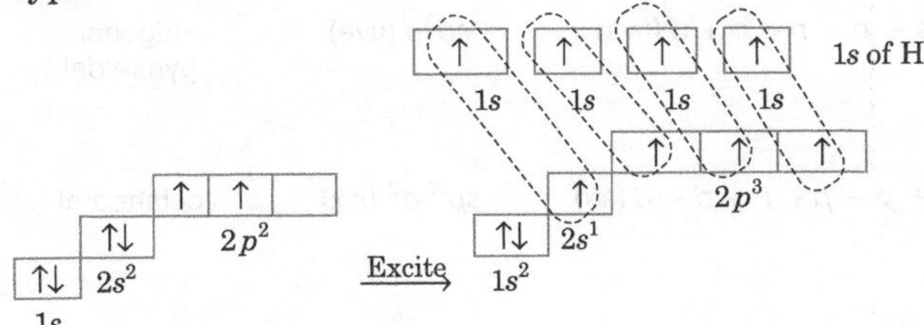

What do you think about these (C—H) *bonds?*

- are they equivalent?
- are they having different bond lengths and bond energies (due to different types of overlap)?

our answer is :

- they are equivalent;
- they have equal bond energy and bond length.

This is due to formation of new orbitals before overlapping takes place, by a process called **Hybridization.**

Hybridization is defined as the concept of intermixing of orbitals of same energy or of slightly different energy to produce entirely new orbitals of equivalent energy, identical shapes and symmetrically disposed in plane. New orbitals formed are called **hybrid orbitals.**

- Only the orbitals of an isolated single atom can undergo Hybridization.
- The hybrid orbitals generated are equal in number to that of the pure atomic orbitals which mix up.
- A hybrid orbital, like the atomic orbitals, cannot have more than two electrons of opposite spins.

Hybrid orbitals do not make π bonds. If there are π-bonds, equal number of atomic orbitals must be left unhybridised for π bonding.

Number of orbitals and type of Hybridization have been summarised in Table 2.7.

How to Determine the Type of Hybridization

For this, one is need to know first of all the numbers of lone pairs (by use of VSEPR theory). We adopt two methods for study of type of Hybridization :

Method I : Count number of atoms directly attached to central atom + lone pairs + single electrons

Method II : Count σ bonds + ⟶ (coordinate) bonds + lone pairs + single electrons

This number comes out to be

2 ⇒ *sp* hybridization

3 ⇒ sp^2 hybridization

4 ⇒ sp^3 hybridization or dsp^2 Hybridization

(decided by way of orbital used)

5 ⇒ sp^3d hybridization

6 ⇒ sp^3d^2 hybridization or d^2sp^3 hybridization

(based on outer *d* or inner *d*-orbitals are used).

We illustrate methods in Table 2.8.

Table 2.7 **Hybrid Orbitals and their Geometric Orientation**

Atomic orbitals (number)	Hybrid orbitals (number)	Orientation	Example	Predicted bond angle	Actual shape (Refer table 2.6)
$s + p$ (two)	*sp* (two)	linear	$BeCl_2$, CO_2, C_2H_2	180°	linear
$s + p + p$ (three)	sp^2 (three)	trigonal planar	BF_3, C_2H_4, SO_2, SO_3	120°	trigonal planar
$s + p + p + p$ (four)	sp^3 (four)	tetrahedral	CH_4 NH_3 H_2O	109° 28′ 107° 48′ 104° 27′	tetrahedral trigonal pyramidal (due to *lp* – *bp* repulsion) angular (V-shaped) due to *lp* – *lp* and *lp* – *bp* repulsion
$s + p + p + p + d$ (five)	sp^3d (five)	trigonal pyramidal	PCl_5 SF_4 I_3^-	120° and 90° 101° 36′ and 86° 33′ 180°	trigonal pyramidal irregular tetrahedral linear
$s + p + p + p + d + d$ (six)	sp^3d^2 (six)	octahedral	SF_6 $[Fe(H_2O)_6]^{3+}$ XeF_4	90° 90° 90°	octahedral octahedral square planar (with two lone pairs)
$d + d + s + p + p + p$ (six)	d^2sp^3 (six)	octahedral	$[Fe(CN)_6]^{4-}$ $[Co(NH_3)_6]^{3+}$	90° 90°	octahedral octahedral
$d + s + p + p$ (four)	dsp^2 (four)	square planar	$[Ni(CN)_4]^{2-}$ $[Pt(NH_3)_4]^{2+}$	90° 90°	square planar square planar

Table 2.8 **Method of Determination of Type of Hybridization**

Molecule or Ion	Central atom	Number of electrons attached to central atom					Total hybrid orbitals		Type
		atoms (*a*)	σ bonds (*b*)	*lp* (*c*)	single electron (*d*)	→ bonds (*e*)	Method I (*a* + *b* + *c*)	Method II (*b* + *c* + *d* + *e*)	
H—Be—H	Be	2	2	—	—	—	2	2	*sp*
BH_3 (H₂B—H)	B	3	3	—	—	—	3	3	sp^2
CH_4	C	4	4	—	—	—	4	4	sp^3
O=C=O	C	2	2	—	—	—	2	2	*sp*
H—C(H)=C(H)—H	C (any)	3	3	—	—	—	3	3	sp^2
H—C≡C—H	C (any)	2	2	—	—	—	2	2	*sp*
$\ddot{N}H_3$	N	3	3	1	—	—	4	4	sp^3
$H_2\ddot{O}$:	O	2	2	2	—	—	4	4	sp^3
PCl_5	P	5	5	—	—	—	5	5	sp^3d
SF_6	S	6	6	—	—	—	6	6	sp^3d^2
•NO_2	N	2	1	—	1	1	3	3	sp^2
:N≡N:	N (any)	1	1	1	—	—	2	2	*sp*
or $\ddot{S}O_2$	S	2	1	1	—	1	3	3	sp^2
H_3C—C*(=O)—O—H	C*	3	3	—	—	—	3	3	sp^2
CH_2=CH—CH=CH_2	C (any)	3	3	—	—	—	3	3	sp^2
H—O—S(→O)(→O)—O—H	S	4	2	—	—	2	4	4	sp^3
H—O—S(=O)(=O)—O—H	S	4	4	—	—	—	4	4	sp^3
:N_α≡N*=$\ddot{O}$:	N*	2	2	—	—	—	2	2	*sp*
	O	1	1	2	—	—	3	3	sp^2
	N_α	1	1	1	—	—	2	2	*sp*

Applications

(a) BeH_2

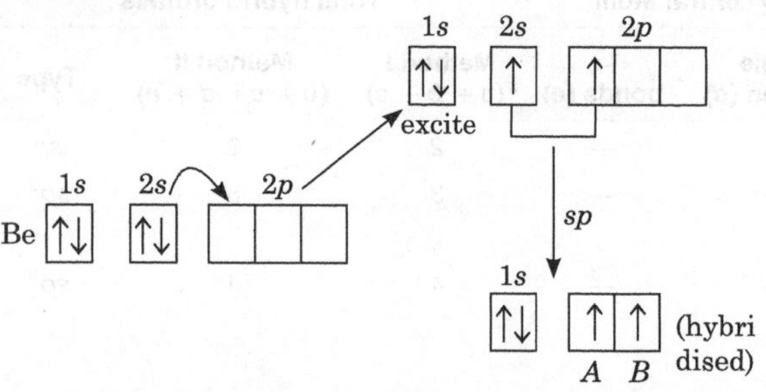

A and *B* *sp*-hybrid orbitals from σ bonds with H by (*sp-s*) overlap.

(b) BH_3

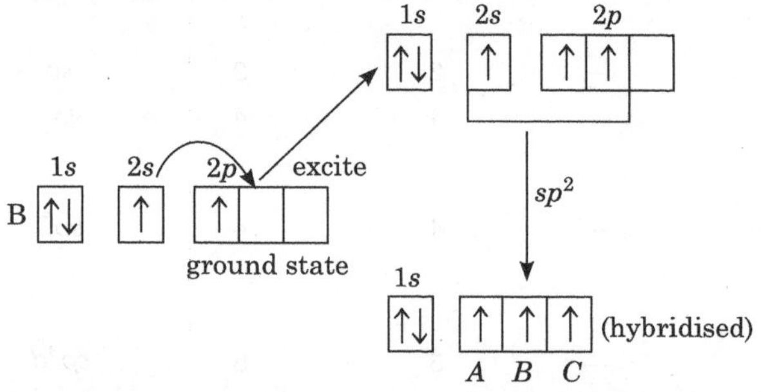

A, *B* and *C* sp^2-hybrid orbitals form σ bonds with H by ($sp^2 - s$) overlap.

(c) CH_4

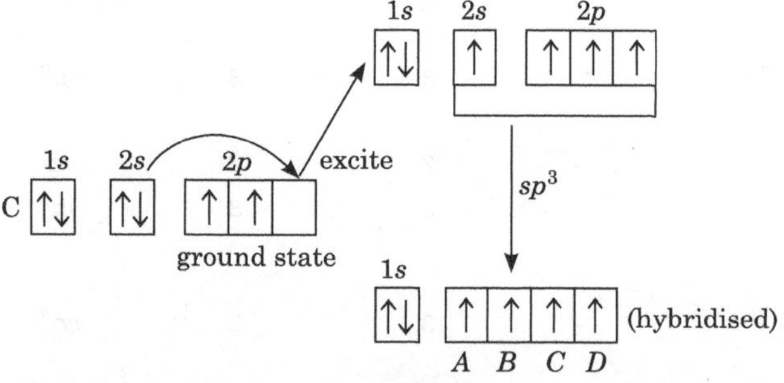

A, *B*, *C* and *D* sp^3-hybrid orbitals form σ bonds with H by ($sp^3 - s$) overlap.

(d) PCl_5

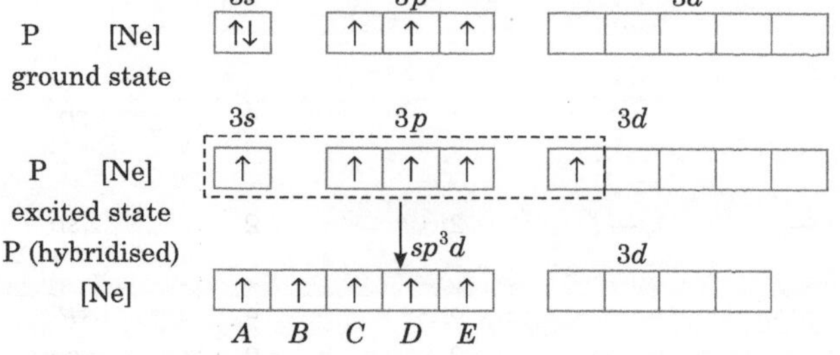

sp^3d hybrid orbitals form σ-bonds with Cl atoms

(e) SF_6

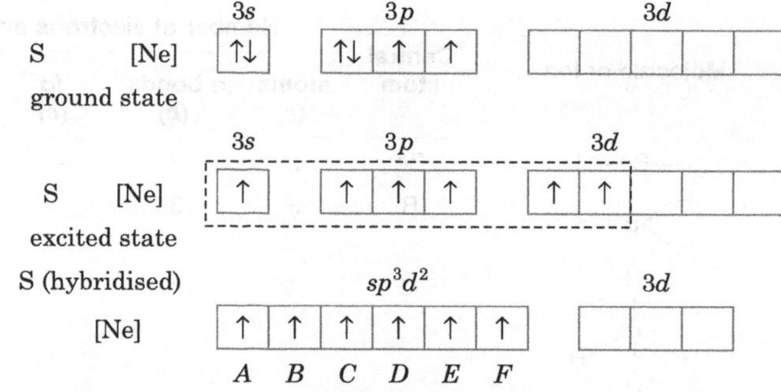

sp^3d^2 hybrid orbitals form σ-bonds with F-atoms

DEEP Focus

There, are also two types of new hybrid orbitals
(i) an ***sd*** hybrid orbital
and (ii) an a ***pd*** hybrid orbital.
(i) is formed by interaction of one *s*- and one d_{z^2}- orbital

$$s + d_{z^2} \longrightarrow \text{two } \textbf{sd} \text{ hybrid orbital}$$

Combination of d_{z^2} with *s* increases the (+) lobes of d_{z^2} and diminishes the negative one. Angular hybrid results from combination of *s* with other *d*-orbitals.

$s + d_{z^2}$ (+) + (+ +) → (+ +) (–)

$s + d_{z^2}$ (+) + (+ +) → (– –) +

$s + d_{x^2-y^2}$ (+) + (+ + – –) → (+ +)

$s + d_{x^2-y^2}$ (+) + (+ + – –) → (– – +)

(ii) $p_z \pm d_{z^2} \rightarrow$ Each combination increases one lobe of d_{z^2} and diminishes the other.

(+ +) + (+ + –) → (– +)

$p_z d_{z^2}$

- While determining type of hybridization on the atom, π bonds are never taken into account, but lone pairs are always considered.

H—C—H with H above and below (σ on each bond): four σ bonds (sp^3)

H—C=C—H with H on each C (σ, σ/π): three σ bonds with each C (sp^2)

H—C≡C—H (σ, 2π): two σ with each C (sp)

N—N̈—H with H below: three σ bonds and one lone pair on N (sp^3)

H—Ö—H: two σ bonds and two lone pairs on O (sp^3)

:N≡N: one σ bond and one lone pair on each N (sp)

Bond energies and bond lengths of different types of bonds have been summarised in Table 2.9.

Table 2.9 **Some Representative Bond Energies and Bond Lengths**

Bond	Bond energy kJ mol^{-1}	Bond length pm	Bond	Bond energy kJ mol^{-1}	Bond length pm	Bond	Bond energy kJ mol^{-1}	Bond length pm
H—H	435	74	C=C	611	134	N—N	163	145
C—H	414		C≡C	837	120	N=N	418	123
sp^3 - H		110	C—N	305	147	N≡N	946	109
sp^2 - H		109	C=N	615	128	N—O	230	136
sp - H		108	C≡N	891	116	N=O	590	115
H—N	389	100	C—O	360	143	O—O	142	145
H—O	464	97	C=O	728	123	O=O	498	121
H—S	368	132	C—Cl	326	177	F—F	159	128
H—F	569	101				Cl—Cl	243	199
H—Cl	431	136				Br—Br	192	228
H—Br	368	151				I—I	151	266
H—I	297	170						
C—C	347							
sp^3 -sp^3		154						
sp^3 -sp^2		150						
sp^2 -sp^2		147						
sp^3 -sp		146						
sp^2 -sp		143						
sp-sp		137						

Target Practice 2

1. Steric number of the central atom in IF_4^- is ....
2. Steric number and predicted geometry of AX_6 is ....
3. Which of the following in each pair will have the larger bond angle?

	I	II	
(a)	CH_4,	NH_3	....
(b)	OF_2,	OCl_2	....
(c)	NH_3,	NF_3	....
(d)	PH_3,	NH_3	....

4. Using the general principle of isoelectric groups, identify the species of line (II) that are "isoelectronic" with those of line (I). Each entry might have more than one "isoelectronic" partner in the other line.

(I) $(H_2N)_2CO$, $HONO_2$, OCO $CO(CN)_2$ ClO_2^+ $Si_3O_9^{6-}$

(II) H_2CCO, $(HO)_2CO$, ONN, H_2CNN, H_3CNO_2, F_2CO, BF, B_2O_2, $(CH_3)_2CO$, CH_2CN^-, SO_2, Cyclic $(SO_3)_3$

Answers

1. Two lone pairs + four bonding atoms = 6

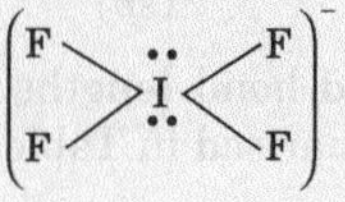

$$\left(\begin{matrix} F & & F \\ & \ddot{\underset{..}{I}} & \\ F & & F \end{matrix}\right)^-$$

2. Six, octahedral
3. (a) CH_4 (b) OCl_2 (c) NH_3 (d) NH_3
4. For each horizontal line shown the molecules, ions, or the groups underlined are "isoelectronic".

(I)	(II)
$HONO_2$	CH_3NO_2
$(H_2N)_2CO$	$(HO)_2CO, F_2CO, (CH_3)_2CO$
$(CN)_2$	B_2O_2 (2 × 9 electron)
ClO_2^+	SO_2 (18 electrons)
$Si_3O_9^{6-}$	$(SO_3)_3$ (3 × 40 electrons)

Practice Exercise 2

1. The H—P—H bond angles in PH_3 are smaller than the H—N—H angles in NH_3. Explain.
2. PCl_5 has the shape of a trigonal bipyramid whereas IF_5 has the shape of square pyramid. Explain.
3. Which one of each of the following pairs is expected to have the larger bond angle?
(a) H_2O and NH_3 (b) SF_2 and BeF_2
(c) BF_3 and BF_4^- (d) NH_3 and NF_3
4. Which of the following molecules are linear?
ICl_2^-, IF_2^+, OF_2, SnI_2, $CdBr_2$
5. Which molecule out of AX_3, AX_4, AX_5 and AX_6, is most likely to have a trigonal bipyramidal structure? If the central atom, *A*, has no lone pairs, what type of Hybridization will it have?
6. Two of the following species have the same shape. Which are these two ? What is their shape ? What are the shapes of the other two? NI_3, I_3^-, SO_3^{2-}, NO_3^-?
7. Use the VSEPR model to predict the shape of
(a) OSF_2 (b) O_2SF_2
(c) XeF_4 (d) ClO_4^-
(e) I_3^-
8. Predict the geometrical shapes of the following molecules and then predict which molecules you would expect to have resultant dipole moments : (also refer sec-dipole moment)
(a) SO_2 (b) NH_3
(c) H_2S (d) C_2H_4
(e) SF_6 (f) CH_2Cl_2
9. Match the terms in *A* with those in *B*

(A)	*(B)*
(a) NH_3 molecule	(i) Heitler London
(b) SF_6	(ii) sp^3d^2
(c) Metallic bond	(iii) Pyramidal shape
(d) VB approach	(iv) Mobile electrons
(e) MO theory	(v) Bond order
(f) VSEPR theory	(vi) Bond angles

10. Two different bond lengths are observed in PF_5 molecule, but only one bond length is observed in SF_6. Explain.
11. Draw an electron dot structure for Br_3^-. Deduce an approximate value for the bond angle and explain your deduction.
12. Predict the shapes of
(a) $TeCl_4$ (b) ClO_3
(c) ClO_2^+ (d) ClO_3^+
13. Match each of the following species with one of these Hybridization schemes

(I)	(II)
(A) *sp*	(a) SF_6
(B) sp^2	(b) CS_2
(C) sp^3	(c) $SnCl_4$
(D) sp^3d	(d) NO_3^-
(E) sp^3d^2	(e) AsF_5

14. The ion ClF_2^- is linear, but the ion ClF_2^+ is bent. Describe Hybridization schemes for the central Cl atom to account for this difference in structure.
15. SbF_3 reacts with XeF_4 and XeF_6 to form ionic compounds $XeF_3^+SbF_6^-$ and $XeF_5^+SbF_6^-$.
Describe the geometries of these two compounds.
16. What are the Hybridization states of the C and N atoms in the following molecule?

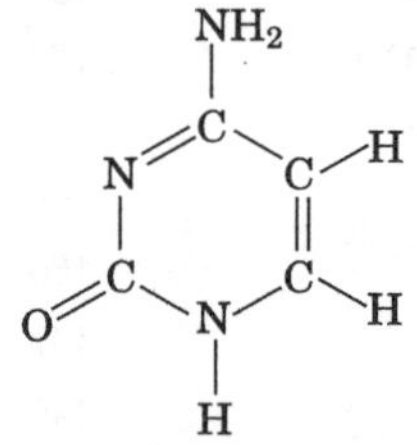

17. What are the total number of σ and π bonds in
(a) methyl isocyanate CH_3NCO ?
(b) tetracyanomethane $C(CN)_4$?
(c) naphthalene $C_{10}H_8$
18. When ethene reacts with H_2, ethane is formed. Which bond is affected?
19. Although both carbon and silicon are in group IVA (14), very few Si = Si bonds are known. Explain.
20. Aspirin has the following connections among atoms. Complete the electron dot structure for aspirin. Also give the σ and π bonds and Hybridization of each atom.

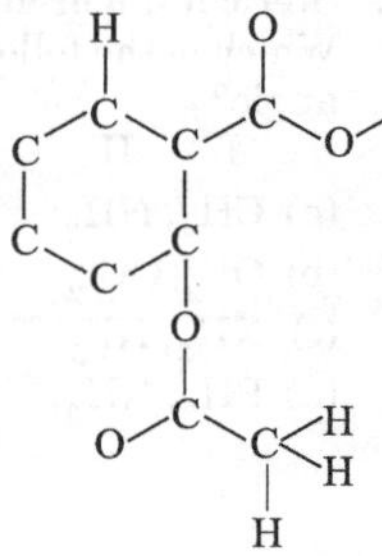

Answers

1. See text **2.** See text

3. (a) NH_3 (b) BeF_2 (c) BF_3 (d) NH_3 **4.** ICl_2^-, $CdBr_2$

5. AX_5, sp^3d

6. NI_3 and SO_3^{2-} are trigonal pyramidal; I_3^- is linear; NO_3^- is trigonal planar.

7. (a) trigonal pyramidal (b) tetrahedral (c) sqare planar (d) tetrahedral (e) linear.

8. Resultant dipole moment in (a), (b), (c) and (f).

9. (a)—(iii) (b)—(ii) (c)—(iv) (d)—(i) (e)—(v) (f)—(vi)

10. PCl_5 has trigonal bipyramidal structure in which bond angles of 90° and 120° exist, hence, two types of bond length, while in case of SF_6 there is ony one type of angle (90°) in octahedral structure.

11. The molecule should be linear, using two axial dsp^3 orbital, with three other dsp^3 orbitals occupied by lone pairs.

12. (a) seesaw shaped, (b) trigonal pyramid with bond angles less than 109.5°, (c) angular, with bond angloo less than 109.5° (d) trigonal.

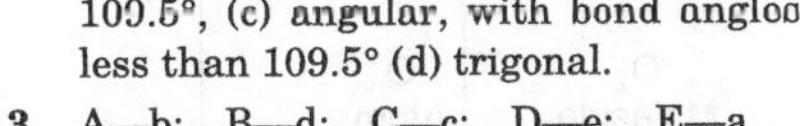

13. A—b; B—d; C—c; D—e; E—a

14. ClF_2^- — sp^3d (three lone pairs on Cl at three corners and two F at vertices)

ClF_2^+ — sp^3 (two lone pairs on Cl)

15. XeF_3^+ T-shaped
XeF_5^+ Square pyramidal
SbF_6^- Octahedral

16. Each C and N, both are sp^2 hybridised.

17. (a) $H-CH_2-N=C=O$ 6σ, 2π (b) $N\equiv C-CH(C\equiv N)-C\equiv N$ 8σ, 8π

(c)

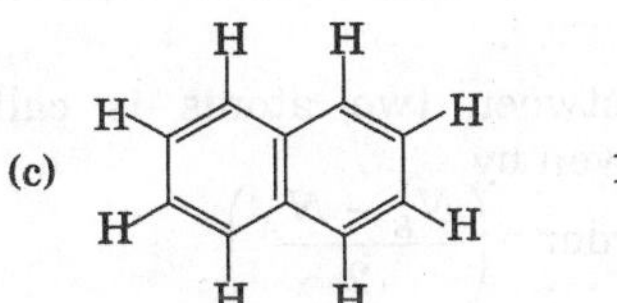

19σ, 5π

18. π-bond of C=C is affected, thus π bond is more reactive than σ-bond.

19. Larger size of silicon (Si) results in poor sideway overlap of *p*-orbitals to form π-bond.

20. 5π, 21σ, C and O with multiple bonds $-sp^2$, C and O with single bonds $-sp^3$

Molecular Orbital Theory

- There is **Linear Combination of Atomic Orbitals (LCAO)** to form **Molecular Orbitals (MOs)**
- The number of MO produced is equal to the number of atomic orbitals that are combined.
- Two types of MOs are formed :

* Bonding MO at a lower energy than the original atomic orbital

* Anti-bonding MO at a higher energy

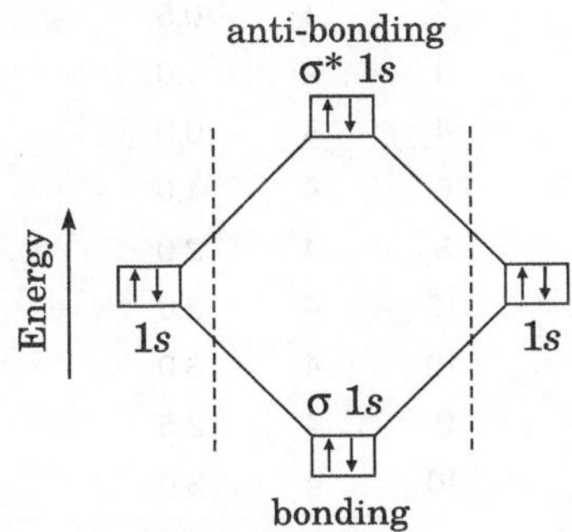

Following MOs are formed when atomic orbitals overlap :

Atomic orbitals	Molecular orbitals	
	Bonding	Anti-bonding
1s + 1s	σ 1s	σ * 1s
2s + 2s	σ 2s	σ * 2s
$2p_x + 2p_x$	σ $2p_x$	σ * $2p_x$
$2p_y + 2p_y$	π $2p_y$	π * $2p_y$
$2p_z + 2p_z$	π $2p_z$	π * $2p_z$

(σ before MO indicates that bond formed is σ type and π before it indicates bond formed is π. *indicates anti-bonding MO)

- Electrons are filled in the increasing energy of the MO which is in order (Aufbau rule)

(a) $\sigma\, 1s, \sigma^*\, 1s, \sigma\, 2s, \sigma^*\, 2s, \sigma\, 2p_x, \begin{vmatrix} \pi\, 2p_y \\ \pi\, 2p_z \end{vmatrix} \begin{vmatrix} \pi^*\, 2p_y \\ \pi^*\, 2p_z \end{vmatrix}, \sigma^*\, 2p_x$

increasing energy (for electrons > 14)

(b) $\sigma\, 1s, \sigma^*\, 1s, \sigma\, 2s, \sigma^*\, 2s, \begin{vmatrix} \pi\, 2p_y \\ \pi\, 2p_z \end{vmatrix}, \sigma\, 2p_x,$

increasing energy (for electrons ≤ 14)

DEEP Focus

- $\pi 2p_y$, $\pi 2p_z$ and also $\pi^* 2p_y$, $\pi^* 2p_z$ are degenerate levels (with same energy). Electrons enter MO of identical energies singly before they pair up (Hund's rule).
- The maximum number of electrons that can be assigned to given MO is two (Pauli's exclusion principle).
- Number of bonds between two atoms is called **bond-order** and is given by

$$\text{Bond-order} = \left(\frac{N_B - N_A}{2}\right)$$

where N_B = number of electrons in bonding MO

N_A = number of electrons in anti-bonding MO

- For a stable molecule/ion, $N_B > N_A$
- Given molecule or ion is paramagnetic if it has unpaired electrons in MO, otherwise diamagnetic.

- MOs formation for H_2 and O_2 are represented in Figs. 2.4 and 2.5 respectively.

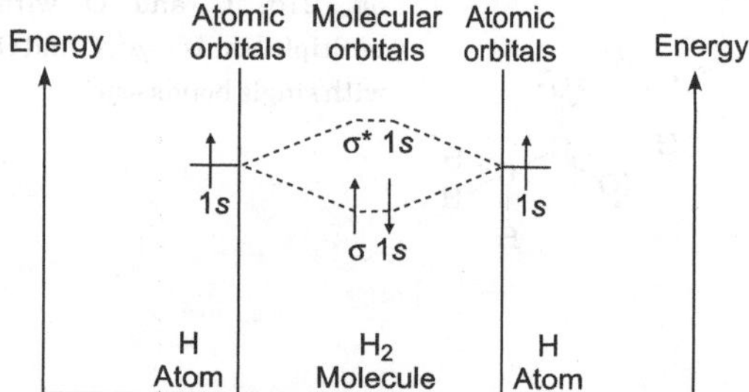

Fig. 2.4 Electronic configuration, atomic and molecular orbitals for hydrogen

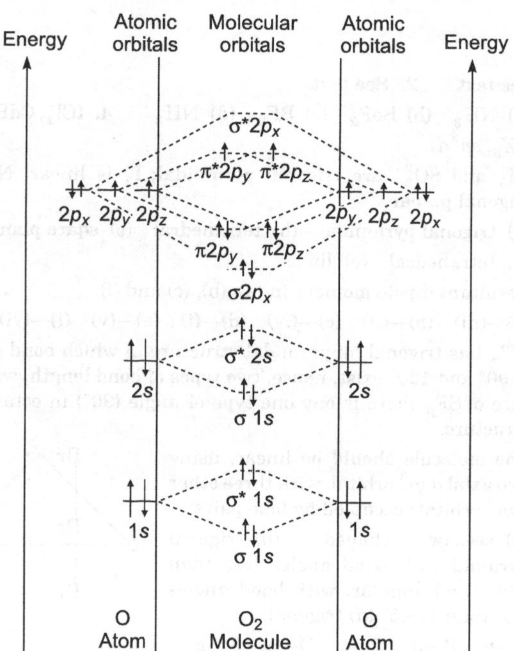

Fig. 2.5 Electronic configuration, atomic and molecular orbitals for oxygen

- Various other species have been summarised in Tables 2.10 and 2.11.

Table 2.10 **MO Electronic Configuration of Diatomic Species up to 14 Electrons**

S. no.	Molecule or ion	Total no. of electrons	Electronic configuration							N_B	N_A	Bond order	Paramagnetic (P) diamagnetic (D)
			$\sigma 1s$	$\sigma^* 1s$	$\sigma 2s$	$\sigma^* 2s$	$\pi 2p_y$	$\pi 2p_z$	$\sigma 2p_x$				
1.	H_2	2	2							2	0	1.0	D
2.	H_2^+	1	1							1	0	0.5	P
3.	He_2	4	2	2						2	2	0.0	D
4.	He_2^+	3	2	1						2	1	0.5	P
5.	Li_2	6	2	2	2					4	2	1.0	D
6.	Be_2	8	2	2	2	2				4	4	0.0	D
7.	B_2	10	2	2	2	2	1	1		6	4	1.0	P
8.	C_2	12	2	2	2	2	2	2		8	4	2.0	D
9.	N_2	14	2	2	2	2	2	2	2	10	4	3.0	D
10.	O_2^{2+}	14	(same as N_2)							10	4	3.0	D
11.	N_2^+	13	2	2	2	2	2	2	1	9	4	2.5	P
12.	CN^-	14	(same as N_2)							10	4	3.0	D

Table 2.11 **MO Electronic Configuration of Diatomic Species of More Than 14 Electrons**

S. No.	Molecule or ion	Total no. of electrons	$\sigma 1s$	$\sigma^* 1s$	$\sigma 2s$	$\sigma^* 2s$	$\sigma 2p_x$	$\pi 2p_y$	$\pi 2p_z$	$\pi^* 2p_y$	$\pi^* 2p_z$	$\sigma^* 2p_x$	N_B	N_A	Bond order	Paramagnetic (P) Diamagnetic (D)
			Electronic configuration													
1.	O_2	16	2	2	2	2	2	2	2	1	1		10	6	2.0	P
2.	O_2^+	15	2	2	2	2	2	2	2	1			10	5	2.5	P
3.	NO	15	2	2	2	2	2	2	2	1			10	5	2.5	P
4.	O_2^- (superoxide ion)	17	2	2	2	2	2	2	2	2	1		10	7	1.5	P
5.	O_2^{2-} (peroxide ion)	18	2	2	2	2	2	2	2	2	2		10	8	1.0	D
6.	F_2	18	2	2	2	2	2	2	2	2	2		10	8	1.0	D
7.	Ne_2	20	2	2	2	2	2	2	2	2	2	2	10	10	0.0	D

- MO electronic configuration is written similar to that of atomic orbitals. For example for He_2 (4 electrons) we write

  $\sigma 1s^2 \sigma^* 1s^2$ or KK^* (**K indicates first bonding MO and K^* anti-bonding MO and are core electrons**)

  and for O_2 we write

  $$KK^* \sigma 2s^2 \sigma^* 2s^2 \sigma 2p_x^2 \; \pi 2p_y^2 \, \pi 2p_z^2 \, \pi^* 2p_y^1 \, \pi^* 2p_z^1$$

  In O_2, $\pi^* 2p_y$ and $\pi^* 2p_z$ each has one electron being degenerate levels (of equal energy) hence it is paramagnetic.

- Greater the bond order, larger the bond length and greater the bond energy. Thus for O_2, O_2^+, O_2^-, O_2^{2-}

  O_2^+ O_2 O_2^- O_2^{2-}

  bond length →

  ← bond order

  ← bond energy

Bonding in Complexes

Valence Bond Theory

- The metal M loses requisite number of electrons to form cation. Number of electrons lost corresponds to the **oxidation number** of the metal ion.
- The metal ion makes available a number of empty orbitals equal to its **coordination number** for the formation of coordinate bonds with the ligand orbitals.
- A weak ligand (like H_2O) will not affect the electronic configuration of the metal/metal ion.
- A strong ligand (like CN^-, OH^- etc.) will make unpaired electrons paired. (Also refer chapter-18)

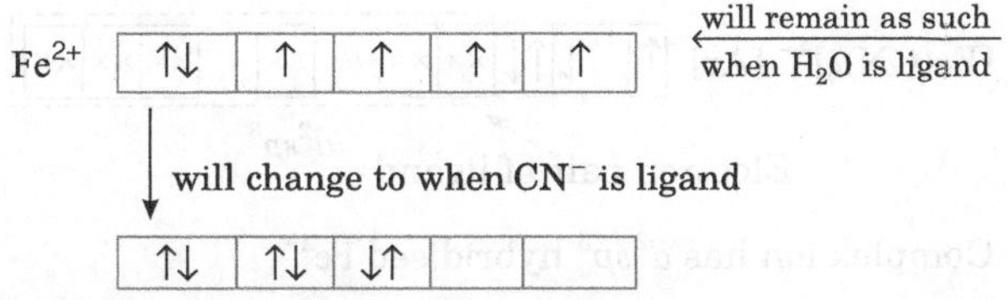

- The metal ion orbitals hybridise to form a new set of equivalent hybridised orbitals with definite stereochemical directions and summarised in Table 2.12.

Table 2.12 **Common Type of Hybridization and Geometry of Complexes**

Coordination number (CN)	Hybridization	Geometry	Examples
2	sp	Linear	$[Ag(NH_3)_2]^+$
3	sp^2	Trigonal planar	$[HgI_3]^-$, $[Cu(PMe_3)_3]^-$
4	sp^3	Tetrahedral	$[VO_4]^{3-}$, $[MnO_4]^-$, $[NiCl_4]^{2-}$, $Ni(CO)_4$
4	dsp^2	Square planar	$[Cu(NH_3)_4]^{2+}$, $[Ni(CN)_4]^{2-}$, $[Pt(NH_3)_4]^{2+}$
5	sp^3d, dsp^3	Trigonal bipyramidal	$[CuCl_5]^{3-}$, $[Ni(CN)_5]^{3-}$, $[Fe(CO)_5]$
6	d^2sp^3, sp^3d^2	Octahedral	$[Fe(CN)_6]^{4-}$, $[Fe(H_2O)_6]^{3+}$

DEEP Focus

- If there are unpaired electrons the complex ion is said to be **paramagnetic** and magnetic moment is = $\sqrt{N(N+2)}$ BM where N is the number of unpaired electrons.
- The complex with so many unpaired electrons is **high spin complex** and that with paired electrons and one or two unpaired electrons is called **low-spin complex.**
- If inner *d*-orbital is involved in hybridisation, it is called **inner-*d* complex** as in

$$[Fe(CN)_6]^{4-};\ d^2sp^3$$

and if outer *d*-orbital is involved then it is called **outer-*d* complex** as in

$$[Fe(H_2O)_6]^{3+};\ sp^3d^2$$

(i) $[Fe(CN)_6]^{4-}$

oxidation state of Fe = +2

C.N. of Fe = 6

[*in all six empty hybrid orbitals are required by six CN^- ligands (a strong ligand)*]

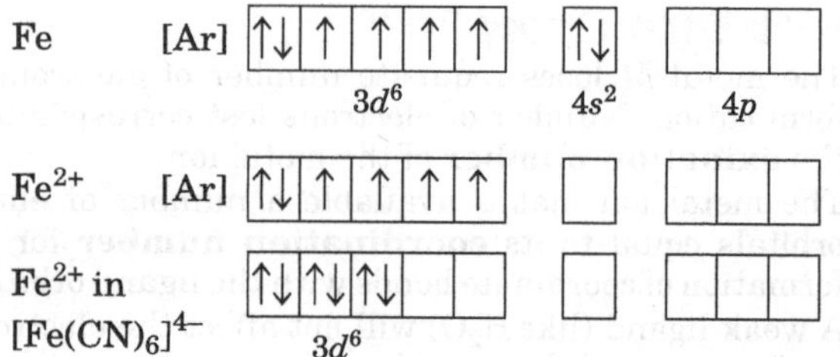

(CN^- is a strong ligand making unpaired electrons paired)

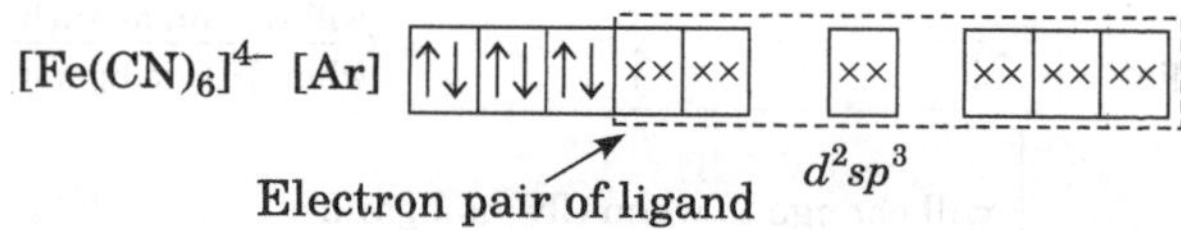

- Complex ion has d^2sp^3 hybridised Fe^{2+}.
- Inner-*d* complex (inner *d*-orbital takes part).
- Diamagnetic (since no electron unpaired).
- Magnetic moment = 0 (zero spin complex or spin-paired complex).
- Octahedral geometry, bond angle 90° (Table 7.8).

(ii) $[Fe(CN)_6]^{3-}$

oxidation number of Fe = +3

If we proceed as in $[Fe(CN)_6]^{4-}$ we have

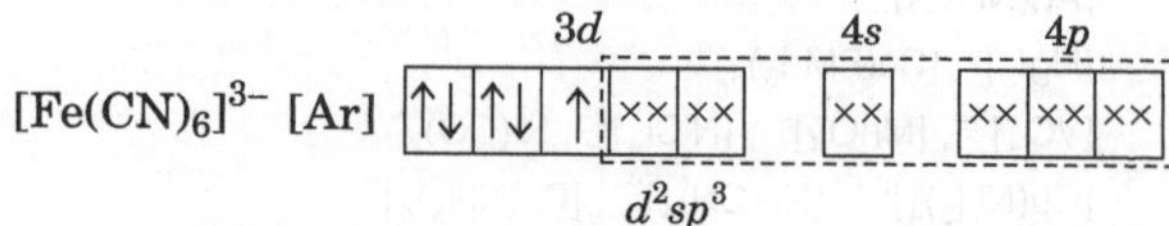

- It has one unpaired electron hence it is paramagnetic with magnetic moment $\sqrt{3}$ BM.
- All other characteristics remain as in $[Fe(CN)_6]^{4-}$.

(iii) $[Fe(H_2O)_6]^{3+}$

oxidation number of Fe = +3

C.N. of Fe = 6

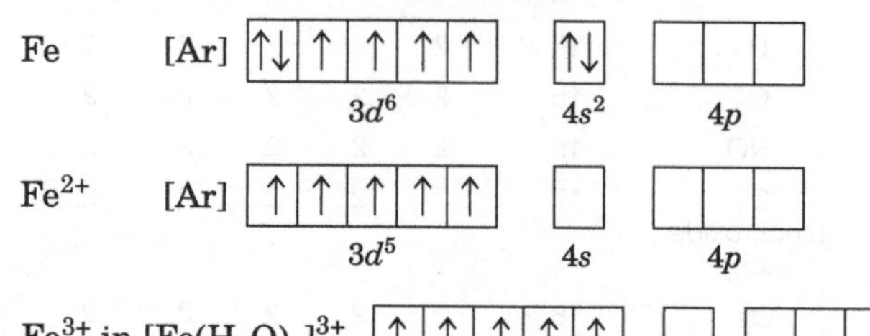

Fe^{3+} in $[Fe(H_2O)_6]^{3+}$ [Ar] ↑ ↑ ↑ ↑ ↑

(H_2O is a weak ligand hence no change in electron-spin. Outer *d*-orbitals will be used for taking electron pairs from H_2O)

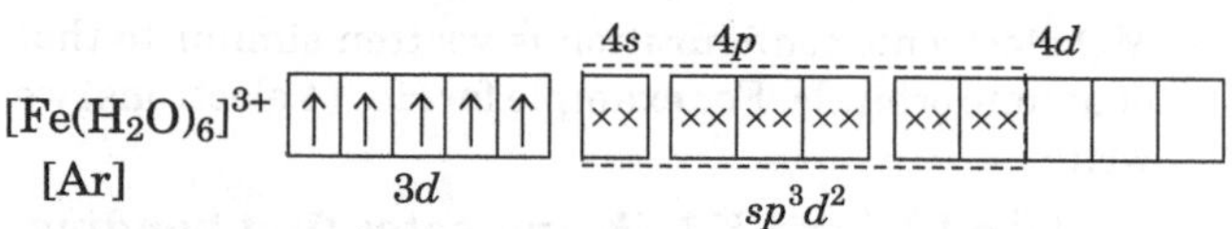

- sp^3d^2 hybridised Fe
- Outer–*d* complex, H_2O being a weak ligand does not affect electron spin in Fe^{3+}.
- A high spin complex with five unpaired electrons, paramagnetic; magnetic moment is $\sqrt{35}$ BM.
- Octahedral geometry, bond angle 90°.

(iv) $[FeF_6]^{3-}$

F^- is a weak ligand like H_2O hence structure is very similar to $[Fe(H_2O)_6]^{3+}$

(v) $[Cr(NH_3)_6]^{3+}$

oxidation number of Cr = +3

C.N. of Cr = 6

As discussed above

3d 4s 4p

$[Cr(NH_3)_6]^{3+}$ [Ar] ↑ ↑ ↑ ×× ×× ×× ×× ×× ××

d^2sp^3

- d^2sp^3 hybridised Cr
- Inner –*d* complex
- Highly paramagnetic, magnetic moment $\sqrt{15}$ BM.
- Octahedral geometry, bond angle 90°.

(vi) $[Ni(H_2O)_4]^{2+}$

Oxidation number of Ni = +2

CN of Ni = 4

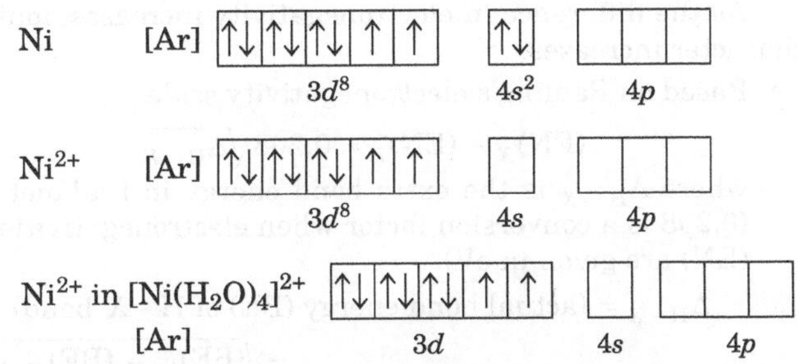

(no change, H_2O being a weak ligand)

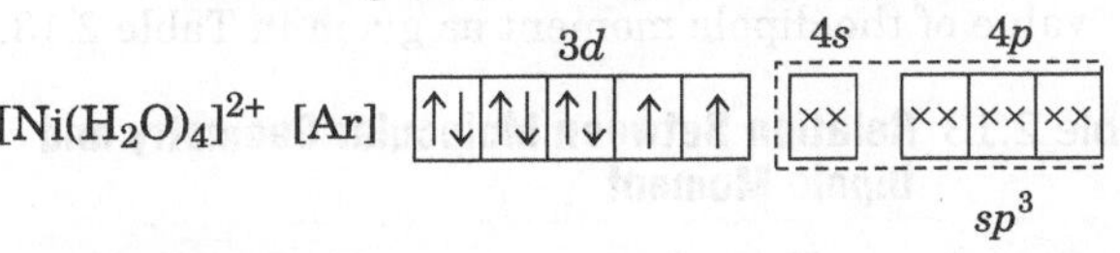

- sp^3 hybridised Ni
- Paramagnetic with two unpaired electrons, magnetic moment is $2\sqrt{2}$ BM
- Tetrahedral geometry, bond angle 109° 28′

(vii) $[Ni(CN)_4]^{2-}$

CN^- is a strong ligand, hence unpaired electrons in Ni^{2+} are paired making one inner *d*-orbital vacant for the ligand.

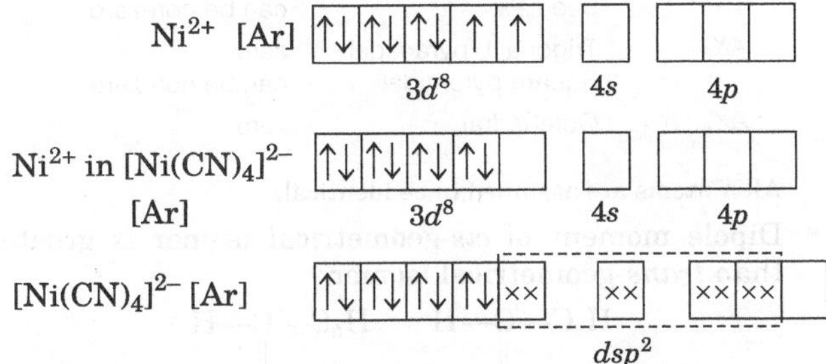

- dsp^2 hybridised Ni, inner *d*-orbital complex
- Diamagnetic
- Square planar geometry, bond angle 90°

(viii) $[Ni(CO)_4]$ oxidation number of Ni = 0

C.N. of Ni = 4

(CO (carbonyl) is a strong ligand hence electronic configuration is affected)

Ni [Ar] | ↑↓ | ↑↓ | ↑↓ | ↑ | ↑ | ↑↓ | | | |

$3d^8$ $4s^2$ $4p$

Ni in $[Ni(CO)_4]$ [Ar] | ↑↓ | ↑↓ | ↑↓ | ↑↓ | ↑↓ | | | | |

$3d^{10}$ $4s$ $4p$

Unpaired electrons are paired and outer 4 *s* electrons are accommodated in 3*d*.

$[Ni(CO)_4]$ [Ar] | ↑↓ | ↑↓ | ↑↓ | ↑↓ | ↑↓ | ×× | ×× | ×× | ×× |

$3d^{10}$ 4s 4p sp^3

- sp^3 hybridised Ni,
- Diamagnetic,
- Tetrahedral geometry, bond angle 109° 28′.

Dipole Moment

In H_2, there is no displacement of the electric charge due to same electronegativity of both H-atoms and the **bond is non-polar.** In HCl, the Cl atom has a more electronegativity than does the H atom. Electronic charge distribution is shifted towards the Cl atom. The **H—Cl bond is said to be polar.**

$$H{:}\,H \qquad H^{\delta+}{:}\ddot{\underset{\cdot\cdot}{Cl}}{:}^{\delta-}$$

The magnitude of the charge displacement in a polar covalent bond is measured through a quantity called **dipole moment** μ. It is the product of the magnitude of charges (δ) and the distance separating them (*d*). (Here the symbol (δ) suggests a small magnitude of charge, less than the charge on an electron)

$$\mu = \delta \times d$$

- If $\delta = 4.8 \times 10^{-10}$ esu and $d = 1$ Å $= 1 \times 10^{-8}$ cm then $\mu = 4.8 \times 10^{-10} \times 1 \times 10^{-8}$ $= 4.8 \times 10^{-18}$ esu cm

  A molecule will have a dipole moment of 1 Debye (1D) if charges of 1.0×10^{-10} esu are separated by a distance of 1 Å

  Thus, 4.8×10^{-10} esu cm = 4.8 D
- In S.I. unit, $1\text{ D} = 3.33 \times 10^{-30}$ coulomb metre (when charge $\delta = 3.33 \times 10^{-20}$ C and $d = 1 \times 10^{-10}$ m).
- In diatomic molecule,

  $$\mu = \delta \times d$$

  but in polyatomic molecule with angle θ, resultant dipole moment is the vector summation of the vector moments.

  $$\mu_{\text{resultant}} = 2\mu_{AB} \cos\left(\frac{\theta}{2}\right)$$

  (B with two A atoms, each at angle θ/2 from the resultant arrow)

  where μ_{AB} is the vector moment of the side *AB*.
- Also $\mu_{\text{resultant}} \propto \cos\left(\frac{\theta}{2}\right) \propto \frac{1}{\theta}$

  Thus, greater the value of θ, smaller the value of μ.
- Symmetrical molecules without lone pair of electrons will have $\mu = 0$

  $$\mu_2 \; H\!-\!B\!-\!H \; \mu_1 \qquad \mu = \vec{\mu}_1 + \overleftarrow{\mu}_2 = \vec{\mu}_1 - \vec{\mu}_2 = 0$$

  In this case resultant μ_2 of H—B—H on the left is cancelled by μ_1 of the right.

- O=C=O $\theta = 180°$, $\mu = 0$
- (benzene, C_6H_6) $\theta = 120°$, planar, $\mu = 0$
- $\mu_{H_2O}(\theta = 104.5°) > \mu_{NH_3}(\theta = 107°) > \mu_{SO_2}(\theta = 120°)$

	ortho	meta	para
	$\theta = 60°$	$\theta = 120°$	$\theta = 180°$
Theoretical	$\mu = 6.30$ D	$\mu = 3.80$ D	$\mu = 0.00$ D
Experimental	$\mu = 6.00$ D	$\mu = 3.79$ D	$\mu = 0.00$ D

In this case μ (*o*-isomer) > μ (*m*-isomer) > μ (*p*-isomer) but experimental value of *o*-isomer < theoretical value.

This is due to dipole-dipole repulsion in *o*-isomer that increases bond angle greater than 60° and μ decreases.

o-OH, H	o-Cl, Cl	o-CH₃, CH₃
attraction due to H-bonding θ decreases, μ increases	repulsion increases θ, μ decreases	no repulsion or attraction no change in actual value.

- % ionic character in a molecule

$$= \frac{\text{observed value of } \mu}{\text{theoretical value of } \mu} \times 100$$

Molecule	Dipole moment (Debye)	Bond length (pm)	Per cent ionic character
HF	1·91	101	39.4%
HCl	1·03	136	16.0%
HBr	0.78	151	10.8%
HI	0.38	170	4.7%

Thus, HF is having only approximately 40% ionic character, the order being

$$HF > HCl > HBr > HI$$

- Ionic character can also be decided based on electronegativity difference. Molecule H—*X* is 50% ionic if electronegativity difference of *X* and H is 1.7 eV.

Chemists have found that a correlation exists between the difference in **electronegativity** of bonded atoms and **degree of ionicity** (expressed as "% ionic character")

As the difference in electronegativity increases, ionic character increases.

- Based on Pauling's electronegativity scale

$$(EN)_X - (EN)_H = 0.208\sqrt{\Delta_{H-X}}$$

where Δ_{H-X} is the extra bond energy in kcal mol^{-1} (0.208 is a conversion factor when electronegativities (EN) are given in eV).

$$\Delta_{H-X} = \text{(actual bond energy (BE) of H}-X\text{ bond)} - \sqrt{(BE)_{H-H}\,(BE)_{X-X}}$$

- Geometry of the molecule can be decided based on the value of the dipole moment as given in Table 2.13.

Table 2.13 **Relation Between Molecular Geometry and Dipole Moment**

Formula	Molecular geometry	Dipole moment
AX	Linear	can be non-zero
AX$_2$	Linear	zero
	Bent	can be non-zero
AX$_3$	Trigonal planar	zero
	Trigonal pyramidal	can be non-zero
	T-shaped	can be non-zero
AX$_4$	Tetrahedral	zero
	Square planar	zero
	See saw	can be non-zero
AX$_5$	Trigonal bipyramidal	zero
	Square pyramidal	can be non-zero
AX$_6$	Octahedral	zero

All *X* atoms are assumed to be identical.

- Dipole moment of *cis*-geometrical isomer is greater than *trans*-geometrical isomer

$H_3C—C—H$ / $H_3C—C—H$ (cis), $\mu = +ve$ $H_3C—C—H$ / $H—C—CH_3$ (trans), $\mu = 0$

Illustration 1 *Arrange the following compounds in order of increasing dipole moment:*

(a) 1,2,3-trichlorobenzene   (b) 1,4-dichlorobenzene   (c) 1,2,4-trichlorobenzene   (d) 1,3,5-trichlorobenzene

Solution (b) and (d) are symmetrical and thus resultant dipole moment is zero.

In (a) bond moments are towards the same direction but in (c) there is net dipole at the C_2 position. Thus, dipole moment of (a) is maximum.

(b) = (d) < (c) < (a)

(a) [1,2,3-trichlorobenzene structure] (c) [1,2,4-trichlorobenzene structure]

Illustration 2 *Predict whether each of the following molecules has a dipole moment :*

(*a*) IBr (*b*) BF_3 (*c*) CH_2Cl_2

Solution (a) Since IBr molecule is diatomic, it has a linear geometry. Bromine is more electronegative than iodine, hence IBr is polar with bromine at the negative end. Thus, the molecule does have a dipole moment.

(b) Since fluorine is more electronegative than boron, each B—F bond in BF_3 is polar and three bond moments are equal. However, the symmetry of a trigonal planar shape means that the three bond moments exactly cancel one another.

(c) CH_2Cl_2 is similar to CH_4 (tetrahedral), but it has three different bond angles : HCH, HCCl, and ClCCl, which are close to but not equal to 109.5°. Since Cl is more electronegative than C and which is more electronegative than H, the bond moments do not cancel and the molecule possesses a dipole moment. CH_2Cl_2 is thus a polar molecule.

Resultant dipole moment is indicated by $\leftrightarrow$

Illustration 3 *Predict whether or not the following molecules will have non-zero dipole moments.*

O_3, CS_2, SCl_2, CH_2O, FHC=CHF (trans)

Solution **In O_3,** all of the atoms are the same, the central oxygen is **hypervalent,** sharing three instead of the usual two electron pairs, and hence is electron-poor and must bear a partial positive charge. VSEPR rules predict a bent molecule, making μ **non-zero.**

:S=C=S: **CS_2,** like CO_2 has zero dipole moment.

:Cl—S—Cl: **SCl_2,** has two lone pair on S atom, and shows a non-zero μ due to noncancelling S—Cl dipoles.

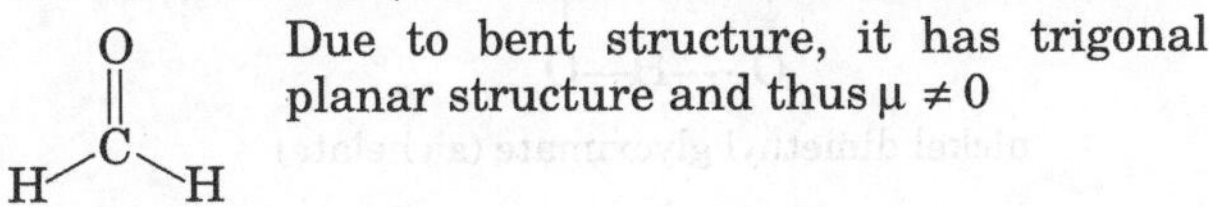

Due to bent structure, it has trigonal planar structure and thus $\mu \neq 0$

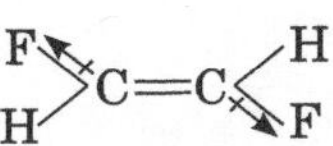

It is the *trans*-isomer. Dipole moments of C—F bonds cancel one each other and thus, μ = 0.

Fajan's Rule

This rule is used to decide relative ionic and covalent character in a molecule. A molecule is predominant covalent if

— smaller the size of cation
— larger the size of anion
— greater the charge on cation and anion
— ion does not have inert gas configuration.

We can also say that these factors favour polarisation.

- Greater the covalent nature
  — lower the m.p./b.p.
  — larger the lattice energy
  — smaller conducting nature
  — lower solubility in polar solvent
  — higher solubility in non-polar solvent.

(a) LiCl, NaCl, KCl, RbCl, CsCl

Size of cation is the deciding factor (charge of each ion and size of Cl is same)

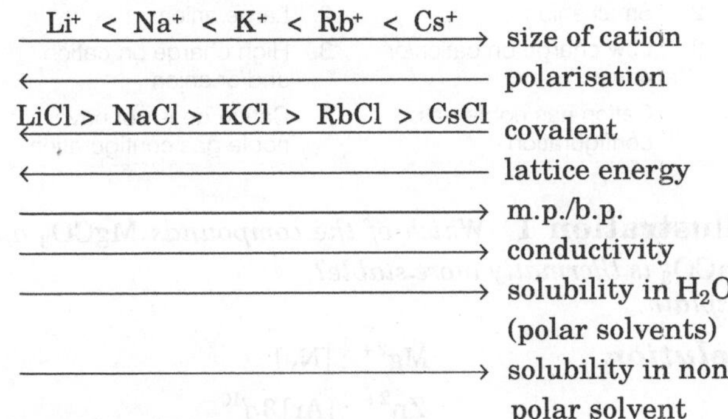

(b) NaF, NaCl, NaBr, NaI

Deciding factor is the size of the anion. Larger size of the anion, greater the polarisation, hence greater the covalent nature.

$F^- < Cl^- < Br^- < I^- \longrightarrow$ size of anion

$F^- < Cl^- < Br^- < I^- \longrightarrow$ polarisation

$NaF < NaCl < NaBr < NaI \longrightarrow$ covalent nature

$\longleftarrow$ lattice energy

(c) NaCl $MgCl_2$ $AlCl_3$ $SiCl_4$ PCl_5

$Na^+ > Mg^{2+} > Al^{3+} > Si^{4+} > P^{5+}$ $\xleftarrow{\text{max}}$ size of cation

$\xrightarrow{\text{max}}$ charge on cation

Greater the charge, smaller the size then greater the polarisation

$Na^+ < Mg^{2+} < Cl^{3+} < Si^{4+} < P^{5+} \longrightarrow$ polarisation

$NaCl < MgCl_2 < AlCl_3 < SiCl_4 < PCl_5 \longrightarrow$ covalent nature

(d) NaF, Na_2O Na_3N

$$\xrightarrow{F^- < O^{2-} < N^{3-}} \text{size of anion}$$

$$\xrightarrow{F^- < O^{2-} < N^{3-}} \text{charge on anion}$$

$$\xrightarrow{F^- < O^{2-} < N^{3-}} \text{polarisation}$$

$$\xrightarrow{NaF < Na_2O < Na_3N} \text{covalent nature}$$

(e) CuCl and NaCl

$[Cu^+]$ $[Ar]3d^{10}$

$[Na^+]$ $[Ne]$

Cations with 18-electron shell have greater polarising power than the 8-electron-shell ions with the same charge and size. This is due to the increased electronegativity of the 18-electron shell ions as the inner electrons have poor shielding effect on the nucleus. Thus, CuCl is covalent and NaCl is ionic.

DEEP Focus

The conditions predicting the nature of bonds are :

Ionic	Covalent
1. Large cation	1. Small cation
2. Small anion	2. Large anion
3. Low charge on cation/or anion	3. High charge on cation and/or anion
4. Cation has noble gas configuration	4. Cation does not have noble gas configuration

Illustration 1 *Which of the compounds* $MgCO_3$ *and* $ZnCO_3$ *is thermally more stable? Explain.*

Solution Mg^{2+} : [Ne]

Zn^{2+} : $[Ar]3d^{10}$

Mg^{2+} has noble-gas electron configuration while Zn^{2+} has an 18-electron configuration of the valence-shell. Thus, polarising power of Zn^{2+} on an anion is larger than a cation of the same size and charge. Thus, lattice energy of $ZnCO_3$ is greater than that of $MgCO_3$ and thus, $ZnCO_3$ is thermally more stable.

Hydrogen Bonding

- The concept of the hydrogen bonding was introduced by **Latimer** and **Rodebush** (1920).
- Hydrogen bonding is said to be formed when slightly acidic hydrogen attached to a strongly electronegative atom such as F, N and O, is held with weak electrostatic forces by the non-bonded pair of electrons of another atom. The coordination number of hydrogen in such cases is two. It acts as a bridge between two atoms, to one of which it is covalently bonded, and to the other attached through electrostatic forces, also called **Hydrogen Bond.**

DEEP Focus

- Of all the electronegative donor atoms, only F, N and O atoms enter into stable hydrogen bond formation.
- The weak electrostatic interaction leading to the hydrogen bond formation is shown by dotted (....) lines. Thus *X*—H....*Y* represents hydrogen bonding between hydrogen and *Y* atom.

Nature of Hydrogen-Bonding

Since the hydrogen bond is formed by the most electronegative elements only, it is considered that its nature is dipole-ion, dipole-induced dipole or dipole-dipole interaction of the type :

$$\overset{\delta-}{A}—\overset{\delta+}{H}...B^- \quad \text{or} \quad \overset{\delta-}{A}—\overset{\delta+}{H}..\overset{\delta-}{B}—\overset{\delta+}{R}$$

(R is the part of the molecule *B*—*R*)

Bond energy of H—bonding in $(\overset{\delta+}{H}...\overset{\delta-}{F}—\overset{\delta+}{H})$ is 113.4 kJ mol^{-1} due to dipole-dipole interactions.

- Energies of H—bonds formed by neutral molecules have been summarised in Table 2.14.

Table 2.14 **Energies of Hydrogen Bonds Formed by Neutral Molecules**

Bond	Energy of the bond is kJ mol^{-1}
F—H....O	45
F—H....F	23 – 25
O—H....O	20 – 30
O—H....S	15 – 20
O—H....F	12 – 15
N—H....N	12 – 25
N—H....O	25

Intramolecular H-Bonding

This type of H-bonding occurs when polar H and electronegative atom are present in the same molecule.

(a) *o*-hydroxy benzaldehyde

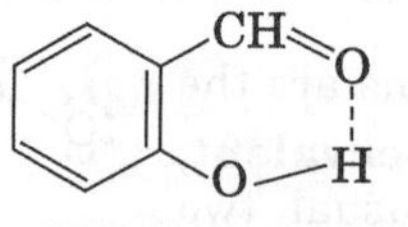

(b) *o*-nitro phenol

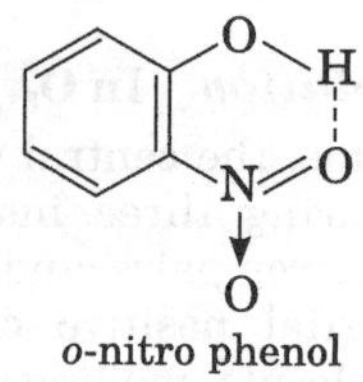

(c) nickel dimethyl glyoximate (a chelate)

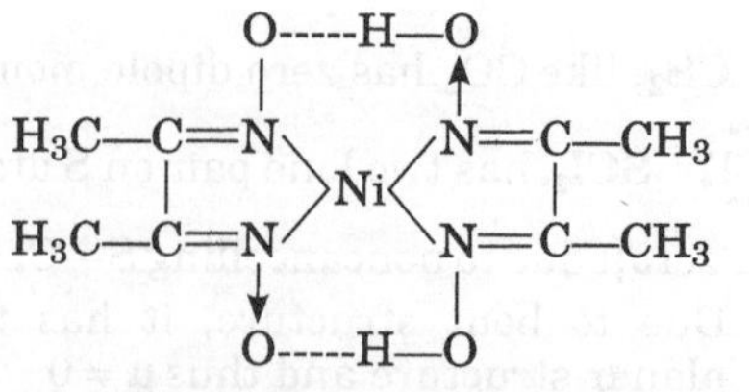

The necessary conditions for the formation of intramolecular hydrogen-bonding are :

(a) the ring formed as a result of hydrogen bonding should be planar.

(b) a 5- or 6- membered ring should be formed.

(c) interacting atoms should be placed in such a way that there is minimum strain during the ring closure.

In case of *anti*-form of pyridine-2-carboxaldoxime there is intramolecular hydrogen bonding but it is not shown in *syn*-form

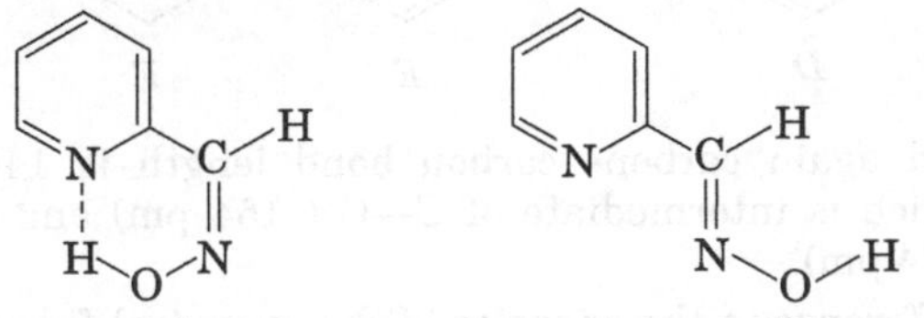

anti *syn*

Intermolecular H-Bonding

This type of H-bonding takes place between H and electronegative element present in the different molecules of the same substance (as in between H_2O and H_2O) or different substances (as in between H_2O and NH_3)

(a) **In water molecules**

Due to polar nature of H_2O, there is association of water molecules giving a liquid state of abnormally high boiling point.

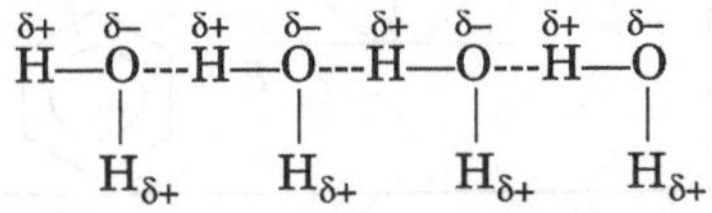

$$\overset{\delta+}{H}-\overset{\delta-}{O}\cdots\overset{\delta+}{H}-\overset{\delta-}{O}\cdots\overset{\delta+}{H}-\overset{\delta-}{O}\cdots\overset{\delta+}{H}-\overset{\delta-}{O}$$

If we actually observe, one water molecule is joined to four water molecules — two with H-atoms and other two with O-atoms. Thus coordination number of water molecule in water is **four.**

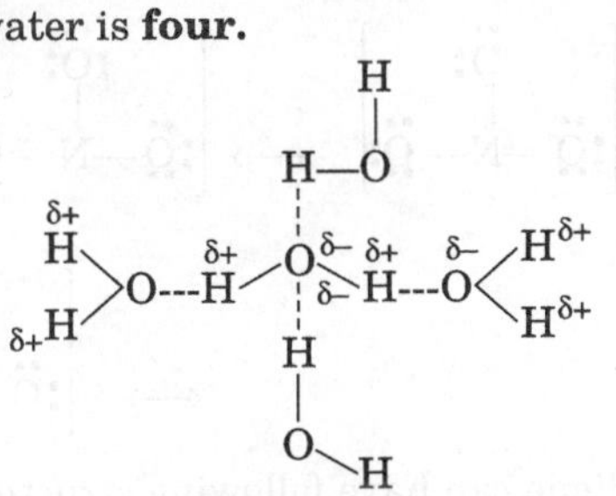

DEEP Focus

- When ice is formed from liquid water, some air gap is formed (in tetrahedral packing of water molecules). **Due to this volume of ice is greater than liquid water and thus ice is lighter than water.** We can say that density decreases when ice is formed. Reversely when ice melts, density increases but only up to 4°C, after this intermolecular H-bonding between water molecules breaks hence volume increases and hence density decreases. **Thus, water has maximum density at 4°C.**

(b) In hydrofluoric acid (HF), there is again association by H-bonding.

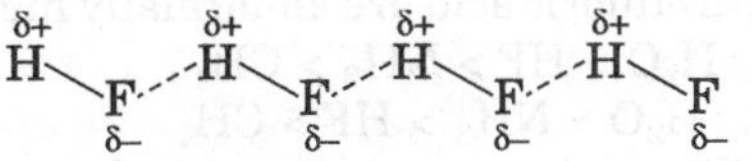

$$\overset{\delta+}{H}-\overset{\delta-}{F}\cdots\overset{\delta+}{H}-\overset{\delta-}{F}\cdots\overset{\delta+}{H}-\overset{\delta-}{F}\cdots\overset{\delta+}{H}-\overset{\delta-}{F}$$

However in the gaseous state, several polymeric forms of the HF molecules exist in which the monomers are held together through H-bonding. A pentagonal arrangement of H—F molecules is shown below :

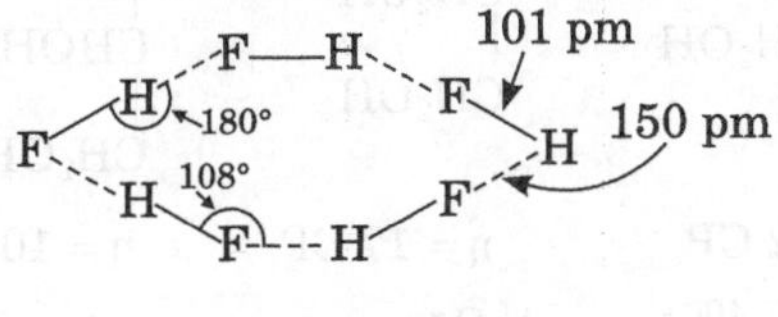

101 pm, 150 pm, 180°, 108°

(c) There is also similar H-bonding in alcohol (R—OH) ammonia (NH_3) and phenol (C_6H_5OH) molecules.

(d) Carboxylic acid dimerises in gaseous state due to H-bonding

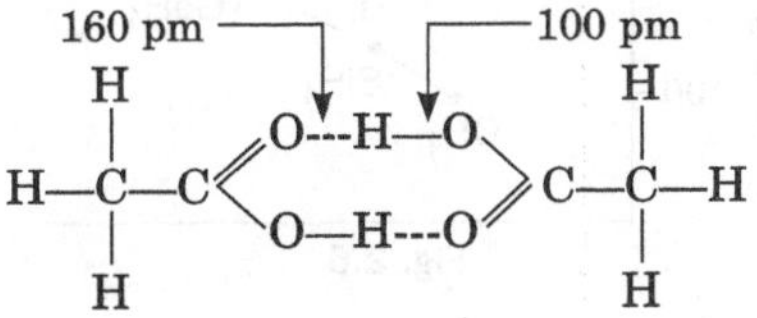

160 pm, 100 pm

(e) Alcohol is said to be highly soluble in water due to *crossed* intermolecular H-bonding (between H_2O and R—OH molecules).

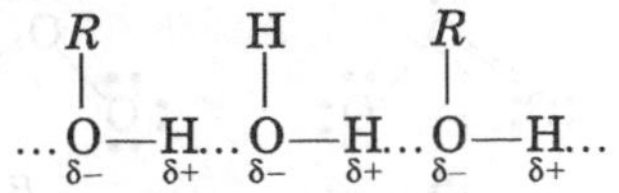

$$\cdots\overset{\delta-}{O}(R)-\overset{\delta+}{H}\cdots\overset{\delta-}{O}(H)-\overset{\delta+}{H}\cdots\overset{\delta-}{O}(R)-\overset{\delta+}{H}\cdots$$

However isomeric ether is less soluble in water due to its (ether) non-polar nature.

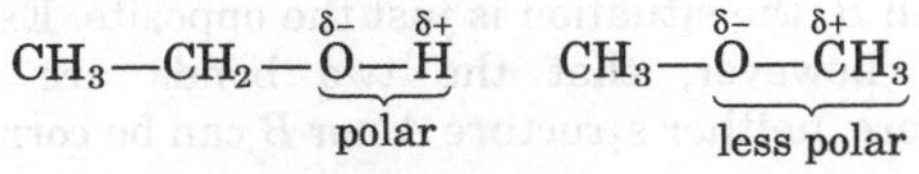

$$\underbrace{CH_3-CH_2-\overset{\delta-}{O}-\overset{\delta+}{H}}_{\text{polar}} \qquad \underbrace{CH_3-\overset{\delta-}{O}-\overset{\delta+}{CH_3}}_{\text{less polar}}$$

DEEP Focus

(f) Though the hydrogen atoms in a methyl groups are not polarised, if an electronegative group like chloro, carbonyl, nitro or cyano is attached to it, the C—H bond gets polarised due to the inductive effect and the hydrogen atom becomes slightly acidic resulting in the formation of weak hydrogen bonds.

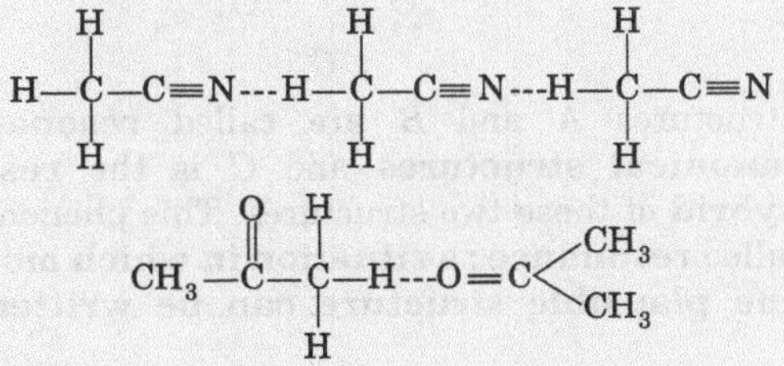

$$H-CH_2-C{\equiv}N\cdots H-CH_2-C{\equiv}N\cdots H-CH_2-C{\equiv}N$$

$$CH_3-\overset{O}{\overset{\|}{C}}-CH_2-H\cdots O{=}C(CH_3)_2$$

Consequences of the Hydrogen-Bonding

(a) Due to H-bonding boiling points of water, ammonia, hydrofluoric acid are abnormally high (Fig. 2.6.)

(b) B.P. : $H_2O > HF > NH_3 > CH_4$

M.P. : $H_2O > NH_3 > HF > CH_4$

(c) Solubility of the organic compounds in water is due to H-bond formation.

(d) Due to hydrogen-bonding, viscosity (η) of the liquid increases as given below (in centipoise (CP).

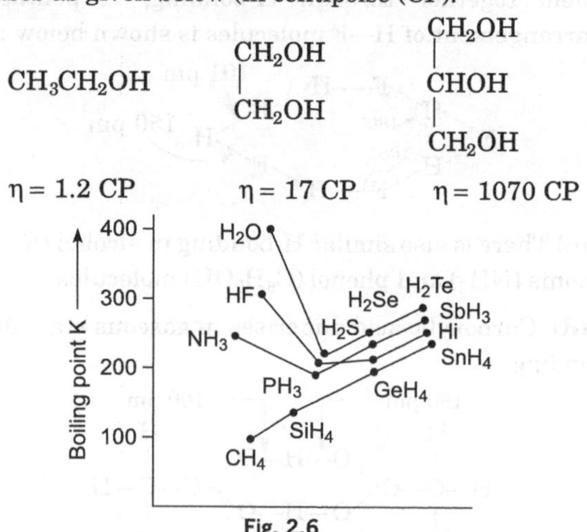

Fig. 2.6

Resonance

We can write two electronic structures of ozone, O_3.

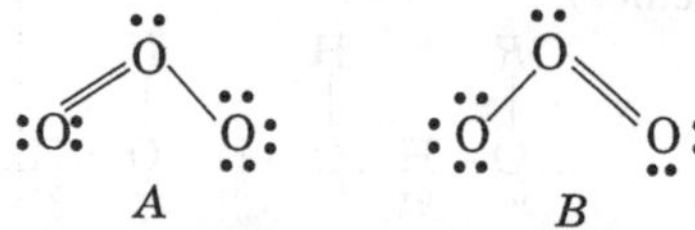

In *A*, the oxygen—oxygen bond on the left is a double bond and the oxygen—oxygen bond on the right is a single bond. In *B*, the situation is just the opposite. Experiment shows, however, that the two bonds are identical. Therefore, neither structure *A* nor *B* can be correct.

- One of the bonding pairs in ozone is spread over the region of all three atoms rather than associated with a particular oxygen—oxygen bond. This is called **delocalised-bonding**— a type of bonding in which a bonding pair of electrons is spread over a number of atoms rather than localised between two.

C

- Structures *A* and *B* are called **resonating** or **canonical structures** and *C* is the **resonance hybrid** of these two structures. This phenomenon is called **resonance : a situation in which more than one plausible structure can be written for a species, and in which the true structure cannot be written at all.**
- O=O bond length is 121 pm and O—O bond length is 145 pm but in ozone experimental measured O—O bond length is 128 pm suggesting that resonance hybrid is a true representation.
- In benzene, *D* and *E* are extreme resonating structure, while *F* is a resonance hybrid of these two structures,

D ⟷ *E* ≡ *F*

and again carbon—carbon bond length is 140 pm which is intermediate of C—C (154 pm) and C=C (134 pm).

- Difference in the energies of the canonical forms and resonance hybrid form is called **resonance stabilisation energy** (Δ*E*), and provides stability to the molecule.

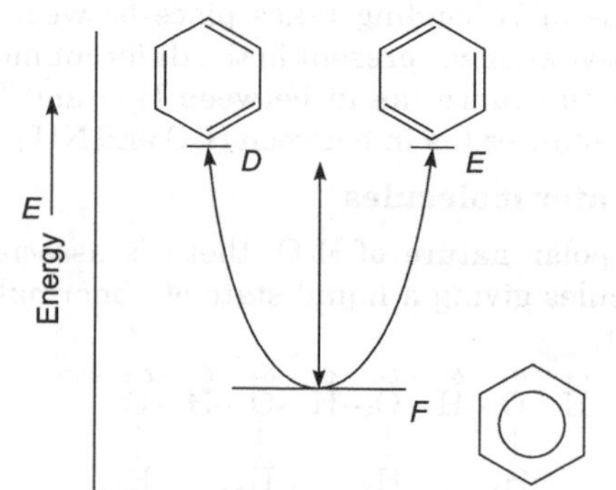

- In nitrate ion, nitrogen–oxygen bond lengths are equal due to resonance,

$$[O{=}N(O^-)_2]^- \longleftrightarrow [O{=}N(O^-)_2]^- \longleftrightarrow [O{=}N(O^-)_2]^-$$

- 1, 3-butadiene can have following structures:

$$CH_2{=}CH{-}CH{=}CH_2 \longleftrightarrow CH_2{=}CH{-}\overset{\oplus}{C}H{-}\overset{\ominus}{C}H_2$$

$$\longleftrightarrow \overset{\oplus}{C}H_2{-}CH{=}CH{-}\overset{\ominus}{C}H_2 \longleftrightarrow \ldots$$

- Vinyl chloride also has following resonating structures :

$$CH_2{=}CH{-}\ddot{\underset{..}{Cl}}\!: \longleftrightarrow \overset{\ominus}{C}H_2{-}CH{=}\overset{\oplus}{Cl}$$

and C—Cl bond is stable.

- C—Cl bond in chlorobenzene is stable due to resonance.

Vinyl chloride and chlorobenzene do not give S_N reactions due to resonance stabilisation of C—Cl bond.

- Phenoxide ($C_6H_5O^-$) ion is stable due to resonance.

- Thus phenol (which is also an oxo-acid) is more acidic than an alcohol because of resonance charge delocalisation of the phenoxide ion.
- Carbon-oxygen bond lengths in carboxylate ion are equal due to resonance, and thus stability of the carboxylate ion is increased.

Conditions for Resonance

- The canonical forms should not differ in the atomic arrangements. Only positions of the electrons can change. Different canonical forms are separated by $\longleftrightarrow$ (double headed arrow)
- The number of unpaired electrons should be the same in all the canonical forms. Thus, following structure of benzene is incorrect:

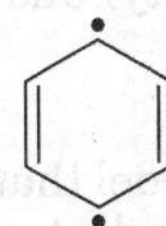

- The positive charge should reside as far as possible, on the less electronegative element and *vice-versa*. Thus, H—F can be represented as

$$\text{H—F} \longleftrightarrow \underset{\text{correct}}{\overset{+\ \ -}{\text{HF}}} \longleftrightarrow \underset{\text{incorrect}}{\overset{-\ \ +}{\text{HF}}}$$

- Like charges should not reside on adjacent atoms. For HN_3, the structure (II) becomes unstable because of the positive charges on the adjacent nitrogen atoms.

$$\underset{\text{(I)}}{\text{H—N}=\overset{+}{\text{N}}=\overset{-}{\text{N}}} \longleftrightarrow \underset{\text{(II)}}{\text{H—}\overset{+}{\text{N}}\text{—}\overset{+}{\text{N}}=\text{N}^{2-}} \longleftrightarrow \underset{\text{(III)}}{\text{H—}\overset{-}{\text{N}}\text{—}\overset{+}{\text{N}}\equiv\text{N}}$$

- The larger the number of the contributing structures, the greater the stability of the molecule.
- Greater numbers of the covalent bonds add to the stability of the canonical forms.

In BF_3, the actual bond order of B—F bond is more than 1 and is found to have 54% π character. Thus, its resonating structures can be written as:

(I) $\longleftrightarrow$ (II) $\longleftrightarrow$ etc.

Thus, (II) is significant due to the additional bond formed.

Illustration 1 *Draw two resonance structures for diazomethane* (CH_2N_2). *Show formal charges.*

Solution The skeletal structure of CH_2N_2 is,

$$\text{H}_2\text{C—N—N}$$

Two resonance structures are:

$$\text{H}_2\text{C}=\overset{+}{\text{N}}=\text{N}^- \longleftrightarrow \text{H}_2\overset{-}{\text{C}}\text{—}\overset{+}{\text{N}}\equiv\text{N:}$$

Illustration 2 *Draw three resonance structures for the molecule* N_2O *in which the atoms are arranged in the order* NNO. *Also indicate formal charges.*

Solution Resonance structures are shown below:

$$^-\text{N}=\overset{+}{\text{N}}=\text{O} \longleftrightarrow \text{:N}\equiv\overset{+}{\text{N}}\text{—O:}^- \longleftrightarrow {}^{2-}\text{:N—}\overset{+}{\text{N}}\equiv\overset{+}{\text{O}}$$

Metallic Bonding

- Most metals crystallise in close-packed structures. The ability of metals to conduct electricity and heat must result from strong electronic interactions among 8 to 12 nearest neighbours (which is also called coordination number). Bonding in metals is called **metallic bonding.** *It results from the electrical attractions among positively charged metal ions and mobile, delocalised electrons belonging to the crystal as a whole*
- Two models are considered to explain metallic bonding:

(a) Band model

(b) Electron-sea model

Band Model

- The interaction of two atomic orbitals, say the 3*s*-orbitals of two sodium atoms, produces two molecular orbitals, one bonding orbital and one anti-bonding orbital. If *N* atomic orbitals interact, *N* molecular orbitals are formed. Atoms interact more strongly with nearby atoms than with those farther away. The energy that separates bonding and anti-bonding molecular orbitals resulting from two given atomic orbitals decreases as the interaction (overlap) between the atomic orbitals decreases. When we consider all the possible interactions among one mole of Na atoms, there is formation of series of very closely spaced molecular orbitals (formally) ($3\sigma s$ and $3\sigma * s$). This consists of a nearly continuous **band** of orbitals belonging to the crystal as a whole. One mole of Na atoms contributes one mole (6.02×10^{23}) of valence electrons (Fig. 2.7) thus, 6.02×10^{23} orbitals in the band are half-filled.

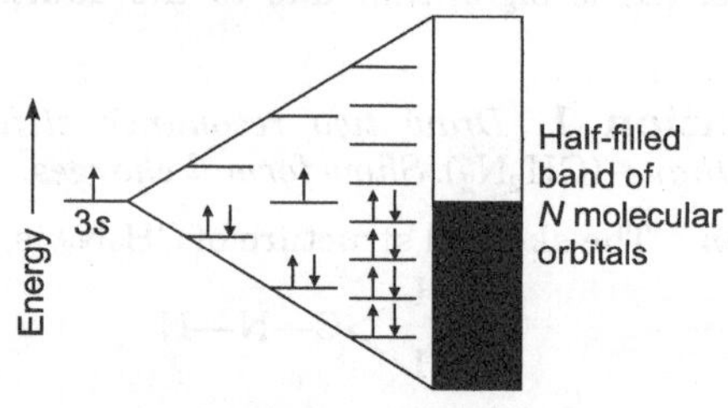

Fig. 2.7 The band of orbitals resulting from interaction of the 3*s*-orbitals in a crystal of sodium

- The empty 3*p* atomic orbitals of Na atoms also interact to form a wide band of $3\times 6.02\times 10^{23}$ orbitals. The 3*s* and 3*p* atomic orbitals are quite close in energy, so that these bands of molecular orbitals overlap (Fig. 2.8). The two overlapping bands contain $4\times 6.02\times 10^{23}$ orbitals and only 6.023×10^{23} electrons. Because each orbital can hold two electrons, the resulting combination of bands is only one-eighth full.

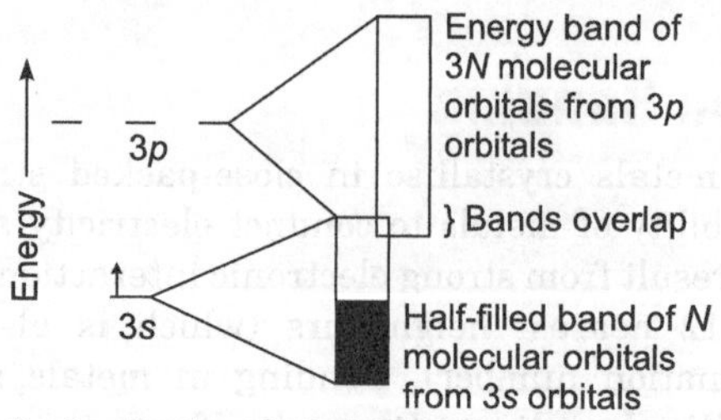

Fig. 2.8 Overlapping of a half-filled "3s" band with an empty "3p" band of Na_N crystal

- According to band theory, the highest-energy electrons of metallic crystals occupy either a partially filled band or a filled band that overlaps an empty band. A band within which (or into which) electrons must move to allow electrical conduction is called a **conduction band.** The electrical conductivity of a metal decreases as temperature increases. The increase in temperature causes thermal agitation of the metal ions. This impedes the flow of electrons when an electric field is applied.
- Crystalline non-metals, such as diamond and phosphorus, are **insulators**—they do not conduct electricity. It is due to the fact that their highest-energy electrons occupy filled bands of molecular orbitals that are separated from the lowest empty band (conduction band) by an energy difference called the **band gap.** In an **insulator**, this band gap is an energy difference that is too large for electrons to jump to get to the conduction band. (Fig. 2.9).

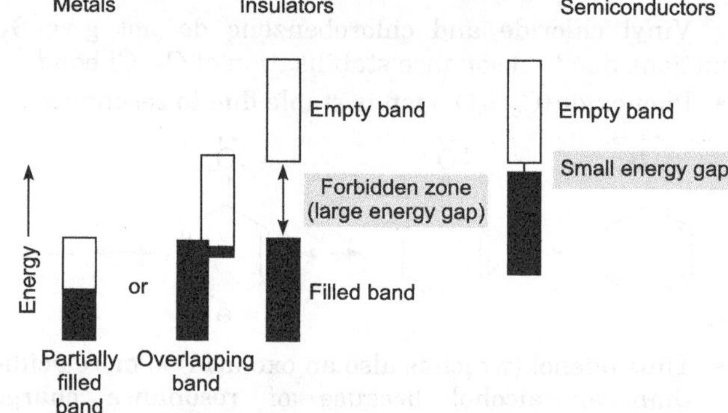

Fig. 2.9 Distinction among metals, insulators and semiconductors. In each case an unshaded area represents a conduction band.

- Elements that are **semiconductors** have filled bands that are only slightly below, but do not overlap with empty bands. They do not conduct electricity at low temperatures, but a small increase in temperature is sufficient to excite some of the highest-energy electrons into the empty conduction band.

Electron-Sea Model

Metals have

- ability to conduct electricity
- ability to conduct heat
- ease of deformation [that is, the ability to be flattened into sheets (malleability) and to be drawn into wires (ductility)]
- lustrous appearance

One over simplified model that can account for some of these properties is the electron-sea model. The metal is pictured (Fig. 2.10) as a network of positive ions immersed in a "sea of electrons". In lithium, the ions would be Li^+ and one electron per atom would be contributed to the sea. These free electrons account for the characteristic metallic properties. If the ends of a bar of metal are connected to a source of electric current, electrons from the external source enter the bar at one end. Free electrons pass through the metal and leave the other end at the same rate.

In thermal conductivity no electrons leave or enter the metal, but those in the region being heated gain kinetic energy and transfer this to other electrons.

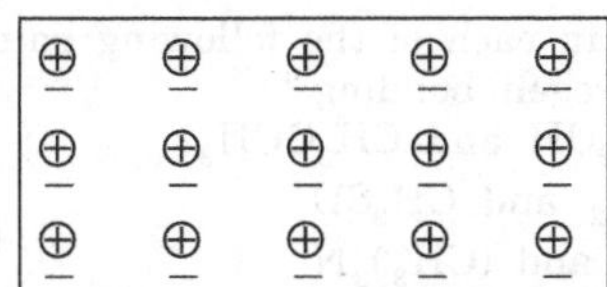

Fig. 2.10 Metallic Bonding in electron-sea model

According to the electron-sea model, the case of deformation of metals can be thought of in this way: If one layer of metal ions is forced across another, perhaps by hammering, the internal structure remains unchanged as the sea of electrons rapidly adjusts to the new situation (Fig. 2.11)

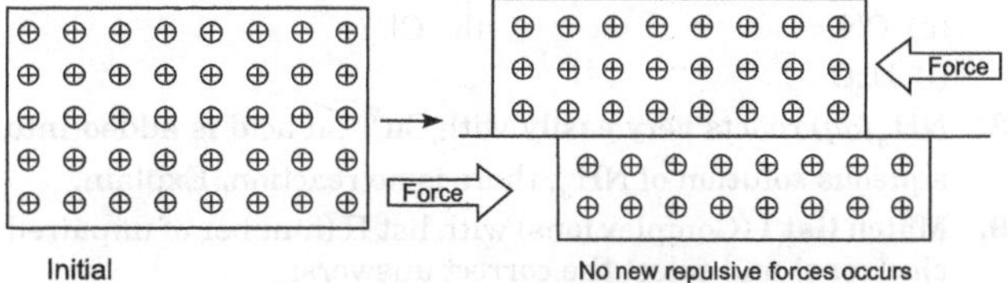

Fig. 2.11 Effect of distortion (by hammering) on metal sheet (assuming electron-sea model)

Target Practice 3

1. The molecular ion $S_3N_3^-$ has the cyclic structure

All (S—N) bonds are equivalent
Give at least two equivalent resonance structures

....
I II

2. For each of the following bond pairs, decide which is the more polar and indicate the negative and positive poles
   (a) B—F and B—Cl ....
   (b) Si—O and P—P ....
3. What is per cent ionic character in the molecule ($M—X$) if electronegativity difference is 1.7 eV? ....
4. Arrange the following in increasing values of dipole moment (i) HF, HCl, HBr, HI and H_2 ....
   (ii) H_2O, H_2S, SO_2, CO_2
5. Select the species with zero dipole moment and maximum dipole moment.
   $CH_4, CH_3Cl, CH_2Cl_2, CHCl_3, CCl_4$

Answers

1. $\longleftrightarrow$

2. (a) B—F > B—Cl $\overset{\delta+}{B}—\overset{\delta-}{F}$
   (b) Si—O > P—P (non-polar) $\overset{\delta+}{Si}—\overset{\delta-}{O}$

3. 50%
4. (i) $H_2 < HI < HBr < HCl < HF$
   (ii) $CO_2 < H_2S < SO_2 < H_2O$
5. Zero : CCl_4, CH_4
   Maximum : CH_3Cl
   Order being : $CH_4 = CCl_4 < CHCl_3 < CH_2Cl_2 < CH_3Cl$

Practice Exercise 3

1. Which of the following is paramagnetic?
   (a) O_2^{2-} (b) O_2^-
   (c) BN
2. N_2 has a greater dissociation energy than N_2^+, whereas O_2 has a lower dissociation energy than O_2^+. Explain.
3. What are the bond-order in each of:
   (a) CN^+ (b) CN
   (c) CN^-?
   Which of these species should have the shortest bond length?
4. For each of the following species : H_2^-; C_2^+ and F_2^+
   (a) write the molecular orbital diagram
   (b) determine the bond order
   (c) indicate whether the species is paramagnetic or diamagnetic.
5. Arrange the following species in order of increasing stability: Li_2, Li_2^+, Li_2^-
   Justify your answer using M.O. theory.
6. Explain :
   (a) $[Fe(H_2O)_6]^{2+}$ has magnetic moment (μ) of 4.90 BM, but $[Fe(CN)_6]^{4-}$ has μ = 0.
   (b) $[Co(CN)_6]^{3-}$ is stable but $[Co(CN)_6]^{4-}$ is unstable.
   (c) $[NiCl_4]^{2-}$ is paramagnetic but $[Ni(CN)_4]^{2-}$ is diamagnetic.
   (d) $[Co(NH_3)_6]^{3+}$ is diamagnetic but $[CoF_6]^{3-}$ is paramagnetic.

7. Classify weak and strong ligands out of :

(a) OH^- (b) F^-
(c) I^- (d) NH_3
(c) CH_3COO^- (f) en
(g) CN^- (h) Cl^-
(i) H_2O

8. NH_3 (*aq*) reacts very easily with Cu^{2+}; if acid is added into aqueous solution of NH_3, there is no reaction. Explain.

9. Match **list I** (Complex ions) with list II (number of unpaired electrons) and select the correct answers:

	List I		List II
A	$[CrF_6]^{4-}$	1	one
B	$[MnF_6]^{4-}$	2	two
C	$[Fe(CN)_6]^{4-}$	3	zero
D	$[Mn(CN)_6]^{4-}$	4	four
		5	five

	A	B	C	D
(a)	4	1	2	5
(b)	2	5	3	1
(c)	4	5	3	1
(d)	2	1	3	5

10. Select the species which is best described by the statement to the right :

(a) CO_2, NH_3, CO has a zero dipole moment.
(b) CH_4, NH_3, HF has highest boiling point.
(c) Cl_2, Br_2, I_2 has the lowest boiling point.
(d) HOI, HOBr, HOCl is the weakest acid.

11. Which of these molecules would you expect to be polar and which are to be non-polar?
Cl_2, ICl, BF_3, NO, SO_2, XeF_4, H_2CCl_2, OCS.

12. Calculate the dipole moment of HCl molecule if the bond length is 1.2476 Å.

13. If the observed dipole moment of LiF is 6.32 D, calculate percentage ionic character; given Li—F bond length is 0.156 nm.

14. Does the molecule OCS have a higher or lower dipole moment than CS_2.

15. Which has larger lattice energy in :
(I) Al_2O_3 or Al_2Se_3 (II) ZnO or NaCl

16. Aqueous solution of a mixture contains LiCl, NaCl and KCl. How is LiCl separated ?

17. Arrange the following in increasing solubility in H_2O :
(a) NaF, MgO, AlN
(b) MgF_2, CaF_2, AgF
(c) AgCl, AgBr, AgI, AgF

18. Which of the ion Ca^{2+} or Cd^{2+} has a stronger polarising action on anions?

19. In going from CsF to CsI, the melting point of the crystal decreases. Explain.

20. Which one in each of the following pairs is expected to exhibit hydrogen- bonding ?
(a) CH_3CH_2OH and CH_3OCH_3
(b) CH_3NH_2 and CH_3SH
(c) CH_3OH and $(CH_3)_3N$

21. Which is expected to have the highest melting point?
PH_3, NH_3, $(CH_3)_3N$

22. Explain:
(a) Maleate acid ion is more stable than fumarate acid ion.
(b) Dipole moment of *o*-chlorophenol is less than *o*-fluorophenol.

23. In each of the following pairs, select the species having the greater resonance stabilisation :
(a) HNO_3 and NO_3^-
(b) H—C(=O)—H and H—C(=O)—O^-
(c) H_2CO_3 and CO_3^{2-}

24. Draw all possible octet structural formulae for N_3^-.

25. The structure of 1,3-butadiene is often written as CH_2=CH—CH=CH_2. The distance between the central carbon atoms is 1.46 Å. Comment on the adequacy of the assigned structure.

26. Carbon suboxide has the formula C_3O_2. C—C distances are found to be 130 pm and C—O distances are 120 pm. Write various Lewis structures.

27. How many energy levels are present in the 3*s* conduction band of a single crystal of sodium weighing 26.8 mg? How many electrons are present in this band?

28. Which of the following factors are especially important in determining whether an element has metallic properties?
(a) Atomic number
(b) Atomic weight
(c) Number of valence electrons
(d) Number of vacant atomic orbitals
(e) Total number of electronic shells in the atom.

29. What is the energy gap in band theory? Compare its size in conductors, semiconductors and insulators.

30. Compare the temperature dependence of electrical conductivity of a metal with that of a typical metalloid.

31. What single factor accounts for the ability of metals to conduct both heat and electricity in the solid state? Why are ions solids poor conductors of heat and electricity even though they are composed of charged particles?

Answers

1. (b) O_2^- with one unpaired electron as decided by M.O. theory.

2. N_2 loses bonding electron hence, destablished; O_2 loses anti-bonding electron hence, stabilised.

3. (a) 2 (b) 2.5 (c) 3
CN^- has the shortest bond-length.

4. H_2^- : B.O. = 0.5, paramagentic
C_2^+ : B.O. = 1.5, paramagnetic
F_2^+ : B.O. = 1.5, paramagnetic
(Refer Table 2.10, 2.11)

5. $Li_2^+ = Li_2^- < Li_2$

6. See text

7. F^-, I^-, Cl^- and H_2O are weak ligands, other are strong ligands.
8. NH_3 forms deep blue coloured complex with Cu^{2+} and colour change appears rapidly. Acid neutralises NH_4OH (aq NH_3) but reaction appears bit slowly.
9. (c)
10. (a) CO_2 (b) HF (c) Cl_2 (d) HOI
11. Polar : ICl, NO, SO_2, CH_2Cl_2, OCS (due to electronegativity difference between the bonded atoms)
    Non-polar : Cl_2 (due to identical atom)
    BF_3, XeF_4 (due to symmetry)
12. 5.99 Debye **13.** 84.4% **14.** Higher dipole moment of OCS
15. (I) Al_2O_3 (II) ZnO
16. Ether extracts LiCl (covalent) from mixture.
17. (a) AlN < MgO < NaF (b) CaF_2 < MgF_2 < AgF
    (c) AgI < AgBr < AgCl < AgF
18. Cd^{2+} **19.** Due to decrease in ionic character.
20. (a) CH_3CH_2OH (b) CH_3NH_2 (c) CH_3OH
21. NH_3,
22. (a) H—C(—C(=O)—O$^\ominus$)=C(—C(=O)—O—H)—H, with the H of —OH bonded to the $O^\ominus$: stable due to intramolecular H-bonding

    (maleate acid ion)

    $^-$O—C(=O)—C(H)=C(H)—C(=O)—OH: no H-bonding possible

    (fumarate acid ion)

    (b) o-chlorophenol (Cl and OH on adjacent ring carbons) (no H-bonding), o-fluorophenol (F - - - - - - H—O)

    H-bonding in *o*-fluorophenol decrease angle (< 60°) hence, μ is increased.
23. (a) NO_3^- (with three equivalent resonating structure)
    (b) H—C(=O)—O$^-$ (with two equivalent resonating structures)
    (c) CO_3^{2-}
24. $\overset{-}{\ddot{N}}:\overset{+}{N}::\overset{-}{\ddot{N}} \longleftrightarrow :N\vdots\vdots\overset{+}{N}:\ddot{N}:^{2-} \longleftrightarrow {}^{2-}:\ddot{N}:\overset{+}{N}\vdots\vdots N:$
25. It is due to resonanting structures that all C—C bond lengths are equal.
26. O=C=C=C=O, $\bar{O}$—C≡C—C≡O$^+$
27. 7.015×10^{20} energy levels in the 3s conduction band.
    7.015×10^{20} electrons in the band.
28. (c) and (a) **29.** See text **30.** See text **31.** See text

Total Practice Set

(Read, Plan & Solve!)

Problem 1. The platinum-chlorine distance has been found to be 2.32 Å in several crystalline compounds. If this value applies to both of the compounds shown in Fig., what is the chlorine-chlorine distance in (a) structure I and (b) structure II?

(I): NH_3 (top), Cl—Pt—Cl, NH_3 (bottom)

(II): NH_3 (top), Cl—Pt—NH_3, Cl (bottom)

Plan *I and II are geometrical isomers and based on position of* Cl*-atoms, distance between them can be determined.*

Solution (a) Given (Pt—Cl) distance = 2.32 Å

Since both Cl and Pt atoms are in straight line hence in (I) Cl—Cl distance

$$= 2 \times 2.32 \text{ Å}$$

$$= \mathbf{4.64 \text{ Å}}$$

(b) In (II), Cl—Cl distance

$$= \sqrt{(2.32)^2 + (2.32)^2}$$

$$= 2.32\sqrt{2}$$

$$= 2.32 \times 1.414$$

$$= \mathbf{3.28 \text{ Å}}$$

Cl —2.32 Å— Pt, 2.32 Å, Cl

Problem 2. The C—C single-bond distance is 1.54Å. What is the distance between the terminal carbons in propane? Assume that the four bonds of any carbon atom are pointed towards the corners of a regular tetrahedron.

Plan *Structure of propane is assumed to be tetrahedral with bond angle of* $\theta = 109°28'$. *Thus, from the value of* $\left(\frac{\theta}{2}\right)$ *and sin values, distance between terminal* C*-atoms can be determined.*

Solution Two terminal carbons can be assumed to be at A and B, while the central carbon at O. Then,

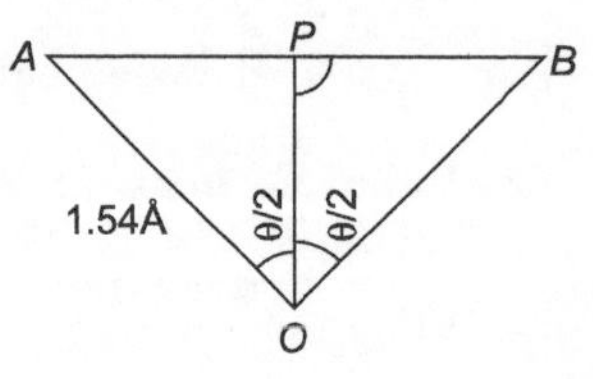

$$AB = 2AP$$

But $\frac{AP}{AO} = \sin\left(\frac{\theta}{2}\right)$

$$= \sin\left(\frac{109°\,28'}{2}\right) \because \text{(in tetrahedral structure } \theta = 109°\,28')$$

$$= \sin(54°\,44')$$

$\therefore$ $AP = AO \sin(54°\,44')$

$$= 1.54 \times 0.82 = 1.26\text{Å}$$

$\therefore$ $AB = 2AP = 2.52\text{Å}$

Problem 3. The dipole moment of HBr is 2.60×10^{-30} C.m. and the interatomic spacing is 1.41 Å. What is the per cent ionic character of HBr ?

Plan *Per cent ionic character* $= \frac{\textit{observed value}}{\textit{theoretical value}} \times 100$

Solution Theoretical value of dipole moment of a 100% ionic character $= e \times d$

$$= (1.60 \times 10^{-19} \text{ C})\,(1.41 \times 10^{-10} \text{ m})$$

$$= 2.26 \times 10^{-29} \text{ Cm}$$

observed value of dipole moment $= 2.60 \times 10^{-30}$ Cm

$$\text{per cent ionic character} = \frac{\text{observed value}}{\text{theoretical value}} \times 100$$

$$= 2.60 \times \frac{2.60 \times 10^{-30}}{2.26 \times 10^{-29}} \times 100 = \mathbf{11.5\%}$$

Problem 4. A diatomic molecule has a dipole moment of 1.2 D. If its bond distance is 1.0 Å, what fraction of an electronic charge exists on each atom?

Plan *Based on dipole moment of 1.2 D, electronic charge is calculated. This value is compared to theoretical value of charge* (= 4.8×10^{-10} esu).

Solution For a diatomic molecule dipole moment $\mu = e \times d$

$$\therefore \quad \text{electronic charge} = \frac{\mu}{d} = \frac{1.2 \times 10^{-18} \text{ esu cm}}{1.0 \times 10^{-8} \text{ cm}}$$

$$= 1.2 \times 10^{-10} \text{ esu}$$

actual value of electronic charge $= 4.8 \times 10^{-10}$ esu

$$\therefore \quad \text{fraction of the electronic charge} = \frac{1.2 \times 10^{-10}}{4.8 \times 10^{-10}} = \mathbf{0.25}$$

Problem 5. At 300 K and 1.00 atm, the density of gaseous HF is 3.17 g/L. Explain the observation, and support you explanation by calculations.

Plan *HF molecules are associated by H-bonding. Overall molecular weight of the associated molecule is determined, hence number of* HF *molecules associated can be predicted.*

Solution $PV = nRT = \frac{w}{m}RT$

$$\therefore \quad P = \frac{w}{Vm}RT = \frac{dRT}{m}$$

$$\therefore \quad m \text{ (molecular weight)} = \frac{dRT}{P} = \frac{3.17 \times 0.0821 \times 300}{1}$$

$$= 78.08 \text{ g mol}^{-1}$$

But normal molecular weight of HF = 19 g mol^{-1}.

Thus, four molecules of HF are associated in gaseous phase by hydrogen bonding, forming tetramer of molecular weight 78.08 g mol^{-1}.

Problem 6. Dipole moment of H_2O is 1.85 D. If bond angle is 105° and O—H bond length is 0.94 Å, determine magnitude of the charge on the oxygen atom in the water molecule.

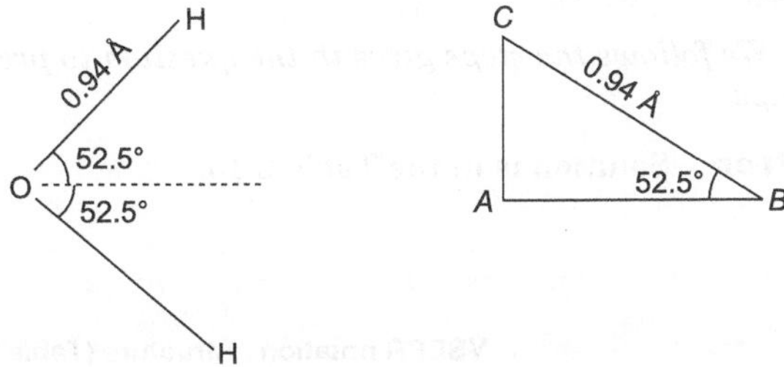

Plan *From the bond angle and vector moment* $\mu = e \times l$
Thus, e (charge) can be determined.

Solution

$$AB = BC \cos 52.5°$$

$$= 0.94 \times 0.609 = 0.572 \text{ Å}$$

$$= 0.572 \times 10^{-8} \text{ cm}$$

Also $\mu = e \times AB$

$$\therefore \quad e = \frac{\mu}{AB} = \frac{1.85 \times 10^{-18} \text{ esu cm}}{0.572 \times 10^{-8} \text{ cm}}$$

$$= \mathbf{3.23 \times 10^{-10} \text{ esu}}$$

Problem 7. Draw all the geometric isomers of PBr_2Cl_3 molecule. State whether each isomer has a dipole moment.

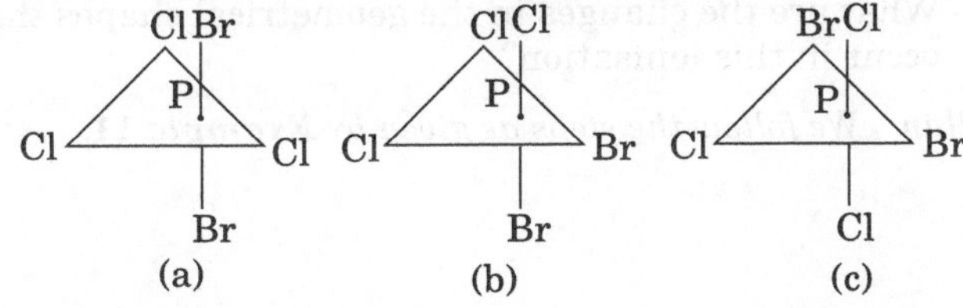

Plan *Symmetrical molecules will have net dipole zero while unsymmetrical molecules will have dipoles.*

Solution (a) is symmetrical hence no dipole.
(b) and (c) have dipoles.

Problem 8. Azobenzene, $(C_6H_5)_2N_2$, is an important starting material in the production of dyes.

(a) Describe the hybridisation scheme for the atoms marked with (*).

(b) Indicate the value of the bond angles labelled α and β.

Plan *From the type of hybridisation, bond angles can be predicted.*

Solution (a) Each sp^2 (b) $\alpha = \beta =$ **120°**

Problem 9. Write Lewis structure for the following compounds :

(a) BaO (b) $MgCl_2$ (c) K_2S

Plan *Based on valence-shell electron-configuration, Lewis structure can be written.*

Solution (a) Ba = $[Xe]6s^2$: two electrons excess to inert gas configuration; can lose two electrons.
O = $1s^2 2s^2 2p^4$: two electrons less than octet; can gain two electrons

$$Ba + O \longrightarrow [Ba]^{2+} [O]^{2-}$$

Lewis structure

(b) Mg = $[Ne]3s^2$: two electrons excess to inert gas configuration; can lose two electrons.
Cl = $[Ne]3s^2 3p^5$: one electron less than octet; two chlorine atoms gain two electrons lost by magnesium

$$Mg + 2Cl \longrightarrow [Cl]^- [Mg^{2+}] [Cl]^-$$

Lewis structure

(c) $$2K + S \longrightarrow [K]^+ [S]^{2-} [K]^+$$

Lewis structure

Problem 10. Applying the VSEPR theory, predict the shape of methyl isocyanate CH_3NCO.

Plan *Using valence-shell electrons and Lewis structure, shape of* CH_3NCO *is predicted.*

Solution Number of valence electrons

$$\begin{matrix} C & H_3 & N & C & O \\ \downarrow & \downarrow & \downarrow & \downarrow & \downarrow \\ 4 & +3 & +5 & +4 & +6 = 22 \end{matrix}$$

We assign structure I with six bond pairs (–) and five lone pairs (..)

(I) $H-\overset{H}{\underset{H}{C}}-\ddot{N}-\ddot{C}-\ddot{O}:$

But this structure lacks octet of N and C, hence by shift of electrons of lone pairs to bonding pairs, we write structure II in which each of atoms has completed octet (H with duplet).

(II) $H-\overset{H}{\underset{H}{C}}-\ddot{N}=C=\ddot{O}:$

Two bonding pairs one lone pair (trigonal planar)

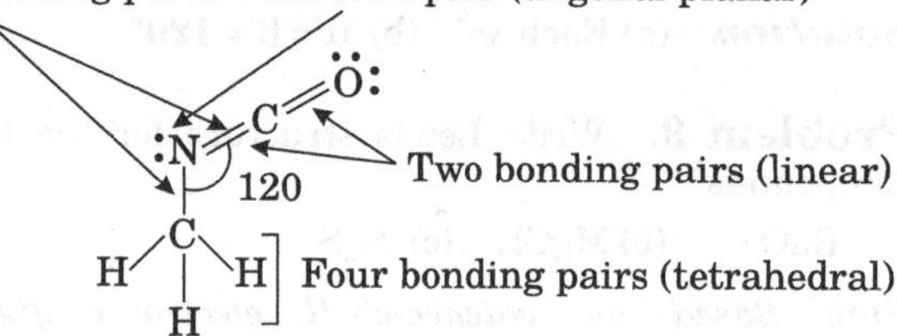

Again note here, we have not taken π bonds into account to decide shape of the molecule.

Problem 11. By the following method you can predict the geometrical shape of species with only one central atom, without first drawing the Lewis structure :

1. Total number of electron pairs
   = (number of valence electrons
   ± electrons (for ionic) charge) / 2
2. Number of bond electron pairs = number of atoms – 1
3. Number of electron pairs around central atom
   = total number of electron pairs
   –3 [number terminal atoms (except H)]
4. Number lone pair
   = (number of central electron pairs – number bond pair)

What is the shape of the following based on this method?

(a) PCl_5 (b) NH_3 (c) H_2O
(d) ClO_3^- (e) ICl_4^-

Plan *We follows the steps gives in the question to predict the shape.*

Solution Solution is in the Table 2.15.

Table 2.15 **Solution of Example 11**

Species	Steps 1	2(X)	3	4(E)	VSEPR notation	Structure (Table 2.6)
(a) PCl_5	$\frac{5+(5\times 7)}{2}=20$	5	20 – 15 = 5	5 – 5 = 0	AX_5	trigonal bipyramidal
(b) NH_3	$\frac{5+3}{2}=4$	3	4 – 0 = 4	4 – 3 = 1	AX_3E	trigonal pyramidal
(c) H_2O	$\frac{2+6}{2}=4$	2	4 – 0 = 4	4 – 2 = 2	AX_2E_2	angular
(d) ClO_3^-	$\frac{7+18+1}{2}=13$	3	13 – 9 = 4	4 – 3 = 1	AX_3E	pyramidal
(e) ICl_4^-	$\frac{7+28+1}{2}=18$	4	18 – 12 = 6	6 – 4 = 2	AX_4E_2	square planar

All the structures determined by this new method is similar to that used by VSEPR theory and is widely acceptable.

Problem 12. In certain polar solvents PCl_5 undergoes an ionisation reaction in which Cl^- ion leaves one PCl_5 molecule and attaches itself to another.

$$2PCl_5 \rightleftharpoons PCl_4^+ + PCl_6^-$$

What are the changes in the geometrical shapes that occur in this ionisation?

Plan *We follow the steps as given by* ***Example*** **11.**

Solution

	Steps	PCl_5	PCl_4^+	PCl_6^-
(1)	Total number of electron pairs	$\frac{5+35}{2}=20$	$\frac{5+28-1}{2}=16$	$\frac{5+42+1}{2}=24$
(2)	Number of bonded electron pairs	5	4	6
(3)	Number of electron pairs around central atom	20 – 15 = 5	16 – 12 = 4	24 – 18 = 6
(4)	Number of lone pairs	0	0	0
(5)	VSEPR notation	AX_5	AX_4	AX_6
(6)	Structure	trigonal bipyramidal	tetrahedral	octahedral

***If you* manage to determine type of hybridisation of P atom in each, then structure can be decided, but this will also require whether lone pair is present or not.**

PCl_5	sp^3d	**trigonal bipyramidal**
PCl_4^+	sp^3	**tetrahedral**
PCl_6^-	sp^3d^2	**octahedral**

Problem 13. The HF_2^- ion exists in the solid state and also in liquid HF solution, but not in dilute aqueous solution. Explain.

Solution In the solid state and in liquid HF, the HF_2^- ion is held together by hydrogen bonding. In aqueous solutions, there is hydrogen bonding, but each HF molecule being stronger acid than H_2O, forms H_3O^+ and F^-.

$$H_2F_2 \rightleftharpoons HF_2^- + H^+ \quad \text{(in liquid HF)}$$

$$HF + H_2O \rightleftharpoons H_3O^+ + F^- \quad \text{(in solution)}$$

Problem 14. Copper (II) iodide is unstable even at ordinary temperature. Explain.

Plan *Explanation can be made in terms of polarisation of cation by anion and vice-versa as predicted by Fajan's rule.*

Solution The Cu^{2+} ion with a 17-electron outer-shell and a comparatively small radius (0.08 nm) has a strong polarising power, while the iodide ion with a larger radius (r = 0.22 nm) has a high polarisability. Hence, the polarisation of the I^- by the Cu^{2+} leads to the complete transition of an electron from the anion to the cation : the Cu^{2+} ion is thus reduced to Cu^+ ion and I^- is oxidised to free iodine. The compound CuI_2 therefore, does not exist.

$$2CuI_2 \longrightarrow 2CuI + I_2$$

$$2I^- \longrightarrow I_2 + 2e^-$$

$$Cu^{2+} + e^- \longrightarrow Cu$$

Problem 15. Assuming the additivity of covalent radii in the C—I bond, what would be the iodine-iodine distance in each of the diiodibenzene? Assume that the ring is regular hexagon and that each C—I bond lies on a line through the centre of the hexagon. (Take C—C bond-length = 1.40 Å radius of iodine atom = 1.33 Å, radius of carbon atom = 0.77 Å)

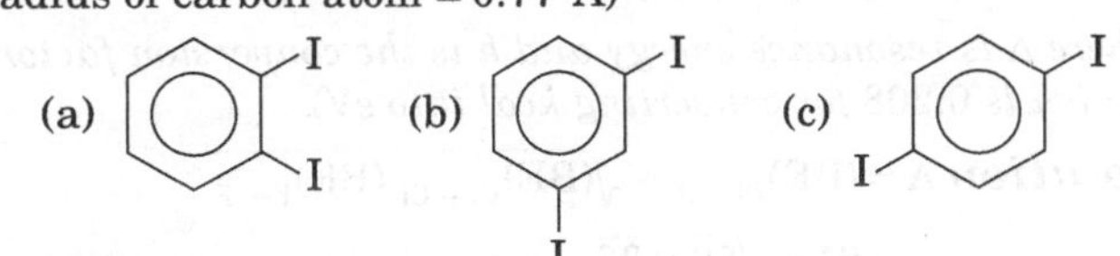

Plan *In o- m- and p-isomers, bond angles are respectively 60°, 120° and 180° hence, distance between two-atoms can be calculated.*

Solution (a) I-atoms are at C and D positions. CD = ?

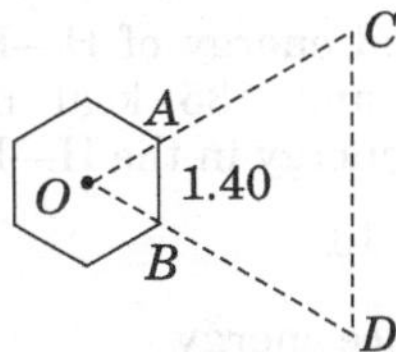

Also, ΔOCD is equilateal triangle.

$$CD = OC = OA + AC$$
$$= 1.40 + 1.33 + 0.77$$
$$\mathbf{= 3.50\ Å}$$

(AC is the sum of radii of carbon and iodine and $OA = AB$)

(b) $CD = 2DE$

But $\frac{DE}{OD} = \sin 60°$

$$DE = OD \sin 60°$$
$$= (OB + BD) \sin 60°$$
$$= (1.40 + 0.77 + 1.33) \sin 60°$$

$$= 3.50 \times \frac{\sqrt{3}}{2} = 3.03 \text{ Å}$$

$\therefore$ $CD = \mathbf{6.06\ Å}$

($BD = AC$ as in (a))

(c) $OC = OD = 3.50$ Å

$\therefore$ $CD = 2 \times 3.50 = \mathbf{7.0\ Å}$

C

O

D

Problem 16. Calculate the electronegativity of chlorine from the bond energy of Cl—F bond (61 kcal mol^{-1}), F—F bond (38 kcal mol^{-1}) and Cl—Cl bond (58 kcal mol^{-1}), and electronegativity of fluorine 4.0 eV.

Plan *Based on Pauling's scale :*

$$(EN)_F - (EN)_{Cl} = k\sqrt{\Delta} = 0.208\sqrt{\Delta}$$

where Δ *is resonance energy and k is the conversion factor (which is* 0.208 *for converting kcal into eV).*

Solution $\Delta = (BE)_{Cl-F} - \sqrt{(BE)_{Cl-Cl}\,(BE)_{F-F}}$

$$= 61 - \sqrt{58 \times 38}$$

$$= 61 - 46.95 = 14.05 \text{ kcal}$$

$$(EN)_{Cl} = (EN)_F - 0.208\sqrt{\Delta}$$

$$= 4.0 - 0.208\sqrt{14.05}$$

$$= 4.0 - 0.78 = \mathbf{3.22\ eV}$$

Problem 17. Bond energy of H—H, F—F and H—F bonds are 104, 38 and 135 kcal mol^{-1} respectively. Calculate resonance energy in the H—F molecule.

Plan *As in Problem* 16.

Solution Resonance energy

$$\Delta_{H-F} = (BE)_{H-F} - \sqrt{(BH)_{H_2}(BE)_{F_2}}$$

$$= 135 - \sqrt{104 \times 38}$$

$$= 135 - 62.86$$

$$= \mathbf{72.14\ kcal\ mol^{-1}}$$

Problem 18. Calculate the steric numbers for iodine in (IF_4^-) and for bromine in (BrO_4^-).

Plan S_N = *(Bonding atom + Lone pairs)*

Solution The central I^- has eight valence electrons

:I:ˣ⁻ × (gained electrons)

Each *F* atom has seven valence electrons of its own and needs to one of the electrons from I^- to attain noble-gas configuration ($s^2 p^6$). Thus, four, of the I^- valence electrons take part in covalent bonds, leaving the remaining four to form two lone pairs. Thus,

:F:
:F—I—F:
:F:

S_N = 4 (bonded atoms) + 2 (lone pairs) = 6

The central Br^- has also eight electrons. In BrO_4^-, each oxygen atom needs to share two of electrons from Br^- to attain noble-gas configurations thus all the electrons are used in bonding and no lone pair exists.

Thus, S_N = 4 (bonded pair) + 0 (lone pair) = 4.

Master Exercises

Exercise 1
(Stage 1 : Learning)

Short Answer Type Questions

1. Predict whether the molecules NH_3 and SF_6 will have dipole moments.
2. Acrolein is the starting material for certain plastics

$$H-\overset{H}{\overset{|}{C}}=\overset{H}{\overset{|}{C}}-\overset{H}{\overset{|}{C}}=\ddot{O}:$$

   (a) Which bonds in the molecule are polar and which are non-polar?
   (b) Which is the most polar bond in the molecule? Which is the most negative atom of this bond?
3. Consider the nitrogen—oxygen ond lengths is NO_2^+, NO_2^- and NO_3^-. In which ion is the bond predicted to be longest? In which is it predicted to the shortest?
4. Explain why the species I in each set has larger bond angle than II?

	I	II
(a)	CH_4	NH_3
(b)	OCl_2	OF_2
(c)	NH_3	NF_3
(d)	NH_3	PH_3

5. Explain why the valence bond structure of carbon monoxide is best described as

$$:C\equiv O$$

   with a dative bond from oxygen atom to the less electronegative carbon atom.
6. The bond dissociation energy of C_2 (599 kJ mol^{-1}) decreases slightly on forming C_2^+ (513 kJ mol^{-1}) and increases greatly on forming C_2^- (818 kJ mol^{-1}). Why is the change much greater for the addition of an electron?

Analytical Questions

Ionic, Covalent and Coordinate Bonding

1. Indicate what is wrong with each of the following Lewis structures? Replace each with a more acceptable structure.

   (a) $[:\ddot{S}-C=\ddot{N}:]^-$ (b) $[:\ddot{C}l]^+ [:\ddot{O}:]^{2-} [\ddot{C}l:]^+$

   (c) $:\ddot{O}=N=\ddot{O}:$ (d) $:\ddot{C}l-\underset{:\ddot{C}l:}{\underset{|}{\ddot{N}}}=\ddot{C}l:$

2. Assign formal charges to the following species. If there are no formal charges present for certains of these species, so indicate

   (a) $\left[:\ddot{O}-\overset{:\ddot{O}:}{\overset{|}{Cl}}-\ddot{C}l:\right]^-$ (b) $:\ddot{O}-\ddot{S}=\ddot{O}:$ (S bonded to both O atoms)

   (c) $\left[:\ddot{F}-\underset{:\ddot{F}:}{\underset{|}{\overset{:\ddot{F}:}{\overset{|}{B}}}}-\ddot{F}:\right]^-$ (d) $[:\ddot{N}=N=\ddot{N}:]^-$

   (e) $[:\ddot{N}-N\equiv N:]^-$ (f) $[:\ddot{O}=N=\ddot{O}:]^+$

3. In which of the following molecules would you expect the nitrogen-to-nitrogen bond to be the shortest?
   (a) N_2H_4 (b) N_2
   (c) N_2O_4 (d) N_2O
4. The common forms of N and P are N_2 and P_4 respectively.

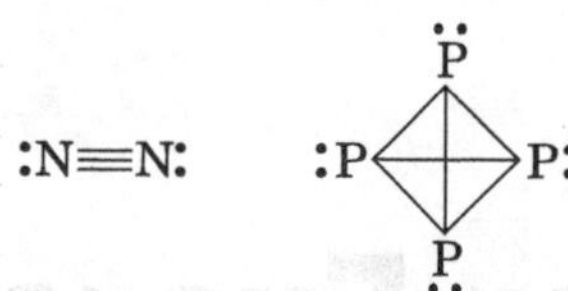

Using the following data, explain why the stable forms are N_2, not N_4; and P_4 not P_2?
Mean bond enthalpy, ($kJ\ mol^{-1}$).

N≡N	946	P≡P	481
N—N	160	P—P	215

5. A sample of lead (II) bromide, $PbBr_2$, was placed in a crucible. The ends of two carbon electrodes were put in the solid and connected to a lamp and power pack. At first the lamp did not light, but when the bromide was heated and began to melt the lamp began to glow. Explain these observations.
6. What are two types of bonds present in B_2H_6?
7. The experimentally determined N—F bond length in NF_3 is greater than the sum of the single bond covalent bond radii of N and O. Explain.
8. The correct order of increasing C—O bond length of CO, CO_3^{2-}, CO_2 is
   (a) $CO_3^{2-} < CO_2 < CO$ (b) $CO_2 < CO_3^{2-} < CO$
   (c) $CO < CO_3^{2-} < CO_2$ (d) $CO < CO_2 < CO_3^{2-}$
9. In the dichromate dianion
   (a) 4 Cr—O bonds are equivalent
   (b) 6 Cr—O bonds are equivalent
   (c) all Cr—O bonds are equivalent
   (d) all Cr—O bonds are non-equivalent

Hybridization/Shape of Molecules

10. Two of the following species have the same shape. Which are these two ? What is their shape ? What are the shapes of the other two ?
    NI_3, I_3^-, SO_3^{2-}, NO_3^-
11. Write plausible Lewis structures for the following molecules
    (a) H_2NOH (b) N_2F_2
    (c) HONO (d) H_2NNO_2
12. PF_5 and BrF_5 appear similar formulae, but structures are different, explain.
13. ClF_2^- ion is linear but ClF_2^+ is bent. Describe hybridisation schemes for the central Cl atom to account for this structure difference.
14. The hyponitrite ion, $^-$O—N=N—O$^-$, exists in solid compounds as the *trans* isomer. Using VB theory, explain why *cis – trans* isomers might be expected for this ion?
15. A molecular compound is composed of 60.4% Xe, 21.1% O and 17.5% F, by mass. If the molecular weight is 217.3 u, what is the molecular formula? Predict the molecular geometry using the VSEPR model. Describe the bonding, using valence bond theory.
16. I. Which one of the following molecules is planar?
    (a) NF_3 (b) NCl_3
    (c) PH_3 (d) BF_3
    II. What is the type and number of bonds between the two carbon atoms in CaC_2?
17. Each of the following compounds has a nitrogen-nitrogen bond
    N_2, N_2H_4, N_2F_2. Match each compound with one of the following bond lengths
    1.10 Å, 1.22 Å, 1.45 Å. Describe the geometry about one of the N atoms in each compound. Why hybrid orbitals are needed to describe the bonding in valence bond theory?
18. "EI" is the general symbol for a representative element. In each case, name the probable element EI.

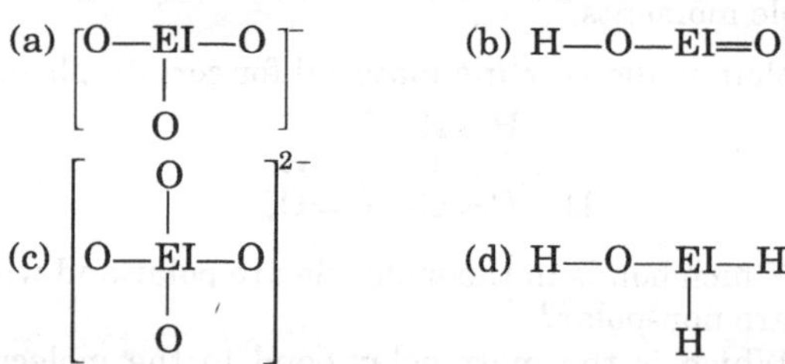

19. A monomeric anhydrous complex of Co gives the following results :

Co	: 21.24%	SO_4^{2-}	: 34.65%
NH_3	: 24.77%	H_2O	: 6.53%
Cl^-	: 12.81%		

(Co : 58 , Cl= 35.5 , S= 32, N= 14); Z (Co = 27)

The compound is diamagnetic and gives white ppt with $BaCl_2$. What is its empirical formula ? What is its structure ?

[**Hint** : Determine molar ratio. Since complex is diamagnetic, hence, cobalt has no unpaired electron. SO_4^{2-} is outside coordinate sphere since it gives white ppt with $BaCl_2$.]

20. $N(CH_3)_3$ is pyramidal but $N(SiH_3)_3$ is planar. Explain.
21. Fill in the blanks
    (a) The shape of CH_3^+ is ....
    (b) The valence atomic orbitals on C in silver acetylide is .... hybridised.
22. Using the VSEPR theory, identify the type of hybridisation and draw the structure of OF_2. What are the oxidation states of O and F?
23. Among the following species identify the isostructural pairs
    NF_3, NO_3^-, BF_3, H_3O^+, NH_3

24. I. Select correct alternate
Among KO_2 , AlO_2^-, BaO_2 and NO_2^+, unpaired electron is present in :
(a) NO_2^+ and BaO_2 (b) KO_2 and AlO_2^-
(c) KO_2 only (d) BaO_2 only
II. Among N_2O, SO_2, I_3^+ and I_3^-, the linear species are .... and ....
III. The angle P—P—P in P_4 molecule is ....

25. What is the geometry and the type of hybrid orbital present about the central atom in BF_3?

26. Interpret the non-linear shape of H_2S molecule and non-linear shape of PCl_3 using VSEPR theory.

27. Discuss the hybridisation of carbon atoms in allene (C_3H_4) and show that π-orbital overlap.

Select the correct alternate (Q. 28 to 29)

28. The geometry of $Ni(CO)_4$ and $Ni(PH_3)_2Cl_2$ are
(a) both square planar
(b) tetrahedral and square planar
(c) both tetrahedral
(d) square planar and tetrahedral

29. In the compound $CH_2{=}CH{-}CH_2{-}CH_2{-}C{\equiv}CH$ the C_2—C_3 bond is of the type
(a) sp-sp^2 (b) sp^3-sp^3
(c) sp-sp^3 (d) sp^2-sp^3

30. Deduce the structure of $[NiCl_4]^{2-}$ and $[Ni(CN)_4]^{2-}$ considering the hybridisation of the metal ion. Calculate the magnetic moment (spin only) of the species.

MO Theory

31. Suggest reasons why the following do not exist as stable molecules?
(a) H_3 (b) HHe
(c) He_2 (d) H_3O

32. Write MO diagram of (a) CN^+ and (b) BN.

33. What is the molecular orbital configuration of HeH^+? Do you expect the ion to be stable.

34. (a) When N_2 goes to N_2^+, the N—N bond distance ... and when O_2 goes to O_2^+, O—O bond distance ...
(b) Write MO electron distribution of O_2. Specific its bond order and magnetic property.

Dipole Moment

35. Estimate the per cent ionic character of the HBr molecule, given that the dipole moment (μ) is 0.79 D and HBr bond length 151 pm.

36. (a) Draw Lewis structure for nitrosyl fluoride FNO.
(b) What is the geometrical shape ?
(c) Explain why the measured resultant dipole moment for FNO, 1.81 D, is so much larger than the value for nitryl fluoride, NO_2F (0.47 D)?

37. Explain how the dipole moment could be used to distinguish between the *cis* and *trans* isomers of 1,2-dibromoethene?

(Br)(H)C=C(Br)(H) — *cis*   (H)(Br)C=C(Br)(H) — *trans*

38. The length of the dipole of HF molecule is 4×10^{-11} m. Calculate the dipole moment in coulomb-metres and in debyes.

39. Two substances *A* and *B* having the same molecular formula, C_4H_8O , were examined in the gaseous state by electron diffraction. The carbon-oxygen distance was found to be 1.43 Å in the compound *A* and 1.24 Å in compound *B*. What can you conclude about the structure of these two compounds ?

40. The N_2F_2 molecule can exist in either of the following two forms:

(a) What is the hybridisation of N in the molecule?
(b) Which structure has a dipole moment?

41. The dipole moment of HCN molecule is 2.9 D. Calculate the length of the dipole.

42. The dipole moment of the NH_3 molecule is 1.48 D. Calculate the length of the dipole. Can the molecule be assumed to have the shape of a regular triangle.

43. I. Arrange (A) toluene, (B) *m*-dichlorobenzene, (C) *o*-dichlorobenzene and (D) *p*-dichlorobenzene in order of increasing dipole moment.
II. Select the correct alternate :
The geometry of H_2S and its dipole moment are
(a) angular and non-zero (b) angular and zero
(c) linear and non-zero (d) linear and zero

Fajan's Rule/Lattice Energy

44. Compare the stability (in the solid state) of the following pairs of compounds:
(a) LiF and LiF_2 (b) Cs_2O and CsO
(c) $CaBr_2$ and $CaBr_3$

45. Assume the following bonds in order of increasing ionic character:
C to H, F to H, Br to H, Na to I, K to F, Li to Cl

46. Explain.
(a) $AuCl_3$ is less stable than AuCl and $PbCl_4$ is less stable than $PbCl_2$.
(b) K_2CO_3 melts at 890°C without decomposing while Ag_2CO_3 decomposes at 220°C.
(c) $BaCl_2$ ionises in aqueous solution while $HgCl_2$ does not (or very smaller).

(d) The ionic radii of Na^+ and Cu^+ are the same (0.098 nm) but melting point of NaCl (801°C) is much higher than that of CuCl (430°C).

47. Anhydrous $AlCl_3$ is covalent. From the data given below, predict whether it would remain covalent or become ionic in aqueous solution.

Ionisation energy for Al = 5137 kJ mol^{-1}
$\Delta H_{hydration}$ (Al^{3+}) = − 4665 kJ mol^{-1}
$\Delta H_{hydration}$ (Cl^-) = − 381 kJ mol^{-1}

Hydrogen Bonding/Metallic Bonding/General

48. I. There is no S—S bond in :
(a) $S_2O_4^{2-}$ (b) $S_2O_5^{2-}$
(c) $S_2O_3^{2-}$ (d) $S_2O_7^{2-}$

II. Out of $Ni(CO)_4$, $Ni(CN)_4^{2-}$ and $NiCl_4^{2-}$ select paramagnetic and diamagnetic.

III. What is the type of hybridisation of C_2—C_3 single bond in HC≡C—CH=CH_2 ?

IV. Out of $SnCl_2$, NCO^-, CS_2, NO_2^+ and SO_2 select the one with linear structure.

V. What is increasing strength of hydrogen bonding (X.... H—X) (X = O, S, F, Cl, N)?

49. I. What is type of hybrid orbitals used by the chloride in ClO_2^-?

II. The cyanide ion, CN^- and N_2 are isoelectronic. But in contrast to CN^-, N_2 is chemically inert because of ....

III. The maximum possible number of hydrogen bonds a water molecule can form is ....

IV. Which of the following have identical bond order

CN^-, O_2^-, NO^+, CN^+ ?

V. Which has highest dipole moment out of
(a) 2, 2-dimethyl propane
(b) *trans*-2-pentene
(c) *cis*-3-hexene
(d) 2,2,3,3-tetramethyl butane?

50. (I) The structure of the enolic form of CH_3—CO—CH_2—CO—CH_3 with intra-molecular H-bonding is .... :

(II) Pick out the isoelectronic structure from the following:

CH_3^+, H_3O^+, NH_3, CH_3^-

51. Explain why *o*-hydroxybenzaldehyde is a liquid at room temperature while *p*-hydroxybenzaldehyde is high melting solid?

Resonance

52. NO_2 gas is paramagnetic at room temperature. On cooling below 0°C, its molecular weight increases and it loses its paramagnetism. When it is reheated, the behaviour is reversed. Explain. How does this phenomenon differ from resonance?

53. The central C—C bond in 1,3-butadiene is shorter than that of *n*-butane. Explain.

54. Select the correct alternate :
The most unlikely representation of resonance structure of *p*-nitrophenoxide is:

(a) (b)

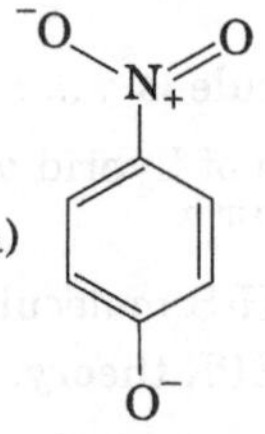

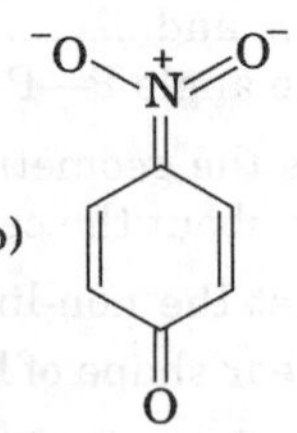

(c) 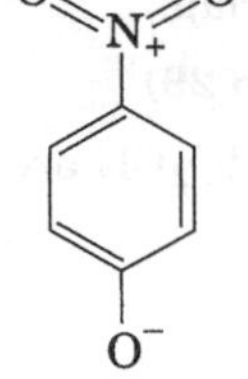 (d)

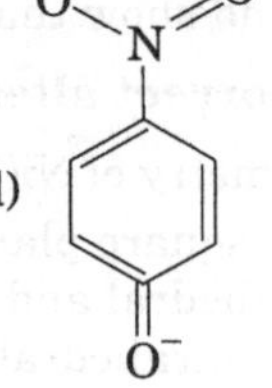

NCERT Problems

1. Two *p*-orbitals from one atom and two *p* orbitals from another atom are combined to form molecular orbitals. How many MOs will result from this combination? Explain.

2. Show the shapes of bonding and anti-bonding MOs formed by combination of (a) two *s* orbitals (b) two *p* orbitals (side to side).

3. How do the bonding and anti-bonding MOs formed from a given pair of AOs compare to each other with respect to (a) energy (b) presence of nodes (c) internuclear electron density?

4. Arrange the following species in order of increasing stability : Li_2, Li_2^+, Li_2^-. Justify your choice with a molecular orbital energy level diagram.

5. Use molecular orbital theory to explain why the Be_2 molecule does not exist?

6. Explain why the bond order of N_2 is greater than N_2^+, but the bond order of O_2 is less than that of O_2^+ ?

7. Compare the relative stability of the following species and indicate their magnetic properties (diamagnetic or paramagnetic); O_2, O_2^+, O_2^- (superoxide), O_2^{2-} (peroxide ion).

8. Explain the significance of bond order. Can bond order be used for quantitative comparisons of the strength of chemical bonds.

9. What is the energy gap in band theory? Compare its size in conductors, semiconductors and insulators.

10. Which of the following substances will exhibit H-bonding? Draw the H bonds between two molecules of the substance where appropriate.

(a) CH_3CH_2OH (b) $CH_3—\overset{\overset{O}{||}}{C}—OH$

(c) $CH_3—\overset{\overset{O}{||}}{C}—CH_3$ (d) $CH_3—\overset{\overset{O}{||}}{C}—NH_2$

11. How can one non-polar induce a dipole in a nearby non-polar molecule?

12. What type(s) of intermolecular forces exist between the following pairs? (a) HBr and H_2S, (b) Cl_2 and CBr_4, (c) I_2 and NO_3^-, and (d) NH_3 and C_6H_6 ?

[Hint : To identify intermolecular forces, it is useful to classify the species being considered as (1) non-polar molecules, (2) polar molecules, and (3) ions. Keep in mind that dispersion forces exist between all species.]

Exercise 2
(Stage 2 : High Skill Problem Solving)

Only One Option Correct

1. In which of the following molecules would you expect the nitrogen–to–nitrogen bond to be the shortest?
(a) N_2H_4 (b) N_2
(c) N_2O_4 (d) N_2O

2. One of the following ions has a trigonal planar shape
(a) SO_3^{2-} (b) PO_4^{3-}
(c) CN^- (d) CO_3^{2-}

3. Which of the following molecules or ions is not linear?
(a) $BeCl_2$ (b) ICl_2^-
(c) CS_2 (d) ICl_2^+

4. The formal charge of the O atoms in the ion $[:\ddot{O}=N=\ddot{O}:]^+$ is
(a) –2 (b) –1
(c) 0 (d) +1

5. All the following molecules are linear except one
(a) SO_2 (b) CO_2
(c) HCN (d) NO

6. All the following molecules are polar except one
(a) BCl_3 (b) CH_2Cl_2
(c) NH_3 (d) FNO

7. Correct Lewis structure is
(a) $[:\ddot{\underset{..}{O}}—C\equiv\ddot{N}:]^-$ (b) $[C\equiv C:]^{2-}$
(c) $[:\ddot{\underset{..}{Cl}}—\ddot{\underset{..}{O}}:]^-$ (d) $:\ddot{N}=\ddot{O}:$

8. In carbon-hydrogen-oxygen compounds
(a) all O—to—H bonds are π bonds
(b) and C—to—H bonds are σ bonds
(c) all C—to—C bonds are σ and π bonds
(d) all C—to—C bonds are π bonds

9. The hybridisation scheme for the central atom includes a *d*-orbital contribution in
(a) I_3^- (b) PCl_3
(c) NO_3^- (d) H_2Se

10. Delocalised molecular orbitals are found in
(a) H_2 (b) HS^-
(c) CH_4 (d) CO_3^{2-}

11. A substance in which large energy gap separates the valence and conduction band is a
(a) metal (b) metalloid
(c) semi conductor (d) insulator

12. Which of the following has the highest percentage of ionic character in its bonding?
(a) LiI (b) $MgCl_2$
(c) CsF (d) CsI

13. In which of the following sets do all the three compounds have bonds that are mainly ionic?
(a) NaCl, NCl_3, CCl_4 (b) CsBr, $BaBr_2$, SrO
(c) CsF, BF_3, NH_3 (d) Al_2O_3, CaO, SO_2

14. Which of the following salts is not soluble in water?
(a) K_2CO_3 (b) Rb_2SO_4
(c) $BaCO_3$ (d) $NaIO_3$

15. Which of the following is an electron-deficient compound?
(a) $NaBH_4$ (b) CO_2
(c) B_2H_6 (d) Al_2Cl_6

16. Which of the following is non-polar but contains polar bonds?
(a) HCl (b) H_2O
(c) NH_3 (d) NO_2

17. Following consists of discrete polar molecules at room temperatures
(a) CS_2 (b) $RaCl_2$
(c) HCN (d) F_2

18. One of the following is least volatile
(a) HCN (b) H—F
(c) H_2O (d) CS_2

19. Which of the following does not obey 18-electron rule?
(a) $Cr(CO)_6$ (b) $Fe(CO)_5$
(c) $V(CO)_6$ (d) $Mn_2(CO)_{10}$

20. Which of the following will have the molecular shape of a trigonal bipyramid?
(a) PF_3Cl_2 (b) IF_5
(c) BrF_5 (d) SbF_5^{2-}

21. Which of the following diatomic molecules would be stabilised by the removal of an electron?
(a) C_2 (b) CN
(c) N_2 (d) O_2

22. The expected spin-only magnetic moments (BM) for $[Fe(CN)_6]^{4-}$ and $[FeF_6]^{3-}$ respectively are
(a) 1.73 and 1.73 (b) 1.73 and 5.92
(c) 0.0 and 1.73 (d) 0.00 and 5.92

23. Which of the following species/molecules has a planar geometry?
(a) $Ni(CO)_4$ (b) SF_4
(c) $CoCl_4^{2-}$ (d) XeF_4

24. O_2 molecule is paramagnetic due to presence of
(a) two unpaired electrons in anti-bonding MO
(b) one unpaired electron in anti-bonding MO
(c) two unpaired electrons in bonding MO
(d) one unpaired electron in bonding MO

25. Resonance is not observed in
(a) CO_2 (b) NH_3
(c) SO_3 (d) NO_2

26. Which of the following can provide an electron pair for the formation of a coordinate covalent bond?
(a) H^+ (b) H^-
(c) H_2 (d) He^{2+}

27. Allyl cyanide has
(a) 9 sigma bonds and 4 pi bonds
(b) 9 sigma bonds, 3 pi bonds and 1 lone pair
(c) 8 sigma bonds and 5 pi bonds
(d) 8 sigma bonds, 3 pi bonds

28. (C—Cl) bond in $CH_2{=}CH{-}Cl$ (vinyl chloride) is stabilised in the same way as in
(a) benzyl chloride (b) benzoyl chloride
(c) chlorobenzene (d) allyl chloride

29. Arrange the following compounds in increasing dipole moment : toluene (I), *m*-chlorobenzene (II), *o*-dichlorobenzene (III), *p*-dichlorobenzene (IV)
(a) I< IV< II< III (b) IV< I< II< III
(c) IV< I< III< II (d) IV< II< I< III

30. When N_2^+ is formed from N_2 bond-order .... and when O_2^+ is formed from O_2 bond-order .... :
(a) increases
(b) decreases
(c) increases, decreases
(d) decreases, increases

31. Which of the following is most polar bond ?
(a) Cl—Cl (b) N—F
(c) C—F (d) O—F

32. The molecules that will have dipole moments are
(a) 1,2-dichloro benzene (b) *cis*-3-hexene
(c) *trans*-2-pentene (d) all of these

33. Which of the following have identical bond order?
(I) CN^- (II) O_2^-
(III) NO^+ (IV) CN^+
(a) I, III (b) II, IV
(c) I, II, III (d) I, IV

34. $C_2{-}C_3$ sigma single bond in vinyl acetylene $CH_2{=}CH{-}C{\equiv}CH$ is due to overlapping of
(a) *sp*—*sp* (b) sp^2—sp^2
(c) *sp*—sp^2 (d) sp^2—*sp*

35. C—H bond distance is the longest in
(a) C_2H_2 (b) C_2H_4
(c) C_2H_6 (d) $C_2H_2Br_2$

36. Among the following species, identify the isostructural pairs

NF_3, NO_3^-, BF_3, H_3O^+, HN_3

(a) [NF_3, NO_3^-] and [BF_3, H_3O^+]
(b) [NF_3, HN_3] and [NO_3^-, BF_3]
(c) [NF_3, H_3O^+] and [NO_3^-, BF_3]
(d) [NF_3, H_3O^+] and [HN_3, BF_3]

37. Mg_2C_3 reacts with water forming propyne. C_3^{4-} has
(a) two sigma and two pi bonds
(b) three sigma and one pi bonds
(c) two sigma and one pi bonds
(d) two sigma and three pi bonds

38. Which of the following statements is correct for $CsBr_3$?
(a) It is covalent compound
(b) It contains Cs^{3+} and Br^-
(c) It contains Cs^+ and Br_3^-
(d) It contains Cs^+, Br^- and lattice Br_2 molecule

39. Among KO_2, AlO_2^-, BaO_2 and NO_2^+, unpaired electron is present in
(a) KO_2 only (b) NO_2^+ and BaO_2
(c) KO_2 and AlO_2^- (d) BaO_2 only

40. In which case there is change in hybridisation when
(a) NH_3 is protonated
(b) AlH_3 combines with H^-
(c) in both cases
(d) in none of the above case

41. Molecular shapes of SF_4, CF_4 and XeF_4 are
(a) the same, with 2, 0 and 1 lone pair of electrons respectively
(b) the same, with 1, 1 and 1 lone pair of electrons respectively
(c) different with 0, 1 and 2 lone pair of electrons respectively
(d) different with 1, 0 and 2 lone pair of electrons respectively

42. The hybridisation of atomic orbitals of N in NO_2^+, NO_3^- and NH_4^+ are respectively
(a) sp, sp^2, sp^3 (b) sp, sp^3, sp^2
(c) sp^2, sp, sp^3 (d) sp^2, sp^3, sp

43. Amongst H_2O, H_2S, H_2Se and H_2Te, the one with the highest boiling point is
(a) H_2O because of hydrogen bonding
(b) H_2Te because of higher molecular weight
(c) H_2S because of hydrogen bonding
(d) H_2Se because of lower molecular weight

44. Which is paramagnetic and coloured?
(a) $K_2Cr_2O_7$ (b) $(NH_4)_2[TiCl_6]$
(c) $VOSO_4$ (d) $K_3[Cu(CN)_4]$

45. $MgSO_4$ is soluble while $BaSO_4$ is insoluble in H_2O. This is because
(a) lattice energy of $BaSO_4$ is greater than $MgSO_4$
(b) $BaSO_4$ is more covalent than $MgSO_4$
(c) hydration energy of Mg^{2+} is greater than Ba^{2+}
(d) lattice energy of $MgSO_4$ is greater than $BaSO_4$

46. The strength of bonds by overlapping of atomic orbitals is in order
(a) $s-s > s-p > p-p$ (b) $s-s < p-p < s-p$
(c) $s-p < s-s < p-p$ (d) $p-p < s-s < s-p$

47. Correct order of dipole moment is

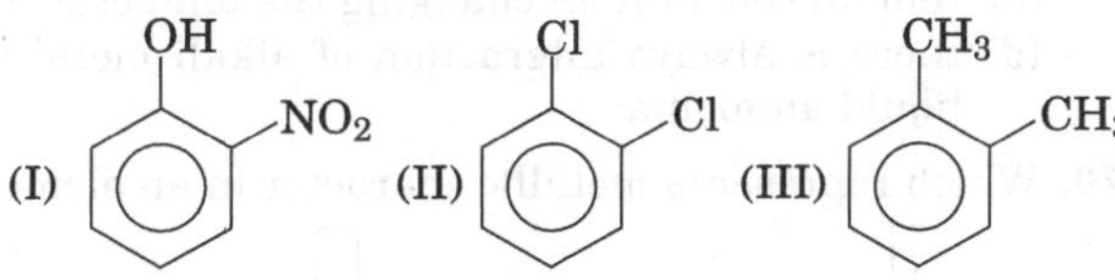

(a) I = II ≡ III (b) I< II< III
(c) I> II> III (d) II< III< I

48. A : tetracyanomethane B : carbon dioxide
C : benzene D : 1,3-buta-di-ene
Ratio of σ and π bonds is in order
(a) $A = B < C < D$
(b) $A = B < D < C$
(c) $A = B = C = D$
(d) $C < D < A < B$

49. Which of the following molecules has the shortest nitrogen-to-carbon bond?
(a) C_2H_4 (b) C_2H_6
(c) C_2H_2 (d) C_2Cl_6

50. Which of the following species are isoelectronic?
(a) NH_4^+, CH_4 (b) N_2, CO
(c) $C_6H_6, B_3N_3H_6$ (d) All of these

51. Which of the following molecules does not have a dipole moment?
(a) IBr (b) $CHCl_3$
(c) CH_2Cl_2 (d) BF_3

52. Which has maximum dipole moment?
(a) (b)
(c) (d)

53. Bond order of O_2, O_2^+, O_2^- and O_2^{2-} is in order
(a) $O_2^- < O_2^{2-} < O_2 < O_2^+$
(b) $O_2^{2-} < O_2^- < O_2 < O_2^+$
(c) $O_2^+ < O_2 < O_2^- < O_2^{2+}$
(d) $O_2 < O_2^+ < O_2^- < O_2^{2-}$

54. Central O^- atom in O_3 is ...... hybridised.
(a) sp (b) sp^2
(c) sp^3 (d) dsp^2

55. Which of the following species contain at least one atom that violates the octet rule?
(a) O—Cl—O (b) F—Xe—F
(c) Both of these (d) None of these

56. $ZnCO_3$ is thermally more stable than $MgCO_3$, because
(a) $Mg(OH)_2$ exhibits only basic properties while $Zn(OH)_2$ is amphoteric
(b) polarising action of Zn^{2+} with 18-electron configuration on the anion is larger than that of Mg^{2+} with a noble-gas electron configuration and of same size and charge.
(c) both (a) and (b) are correct
(d) none of the above is correct

57. Polarising action of Cd^{2+} on anions is stronger than that of Ca^{2+} because
(a) the charges of the ions are same
(b) their radii are same (Ca^{2+} = 0.104 nm; Cd^{2+} = 0.099 nm)
(c) the Ca^{2+} ion has a noble-gas electron configuration, and the Cd^{2+} ion, an 18-electron configuration of its outer shell
(d) all of the above are correct

58. Which of the following is not stable?
(a) H_2^+ (b) H_2
(c) He_2 (d) HHe

59. Which of the following has the smallest bond length?
(a) NO^+ (b) NO
(c) NO^- (d) Equal

60. Some of the following properties are important in determining whether an element has metallic properties

I : atomic number
II : atomic weight
III : number of valence electrons
IV : number of vacant atomic orbitals
V : total number of electronic shells in the atom

Select correct properties from the codes given below

(a) I, II, III, IV (b) I, III, IV, V
(c) III, IV (d) III, IV, V

61. Azide ion (N_3^-) exhibits on N—N bond order of 2 and may be represented by resonance structures I, II and III given below

$\overset{-}{:\ddot{N}}=\overset{+}{N}=\overset{-}{\ddot{N}:} \longleftrightarrow {}^{2-}:\ddot{N}-\overset{+}{N}\equiv N: \longleftrightarrow :N\equiv\overset{+}{N}-\ddot{\underset{..}{N}}:^{2-}$

I II III

Select correct statements.

(a) Structures I and II make greater contributions than III
(b) Structures II and III make greater contributions than I
(c) Structures I and III make greater contributions than II
(d) All the above three structures make equal contributions

62. Which of the following molecules are expected to exhibit intermolecular H-bonding?

I : Acetic acid II : *o*-nitrophenol
III : *m*-nitrophenol IV : *o*-boric acid

Select correct alternate.

(a) I, II, III (b) I, II, IV
(c) I, III, IV (d) II, III, IV

63. In terms of polar character, which one of the following order is correct?

(a) $NH_3 < H_2O < HF < H_2S$
(b) $H_2S < NH_3 < H_2O < HF$
(c) $H_2O < NH_3 < H_2S < HF$
(d) $HF < H_2O < NH_3 < H_2S$

64. How many σ and π-bonds are there in the molecule of tetracyano-ethylene?

(a) 4σ, 14π (b) 5σ, 13π
(c) 8σ, 10π (d) 9σ, 9π

65. Match **List I** with **List II** and select the correct answer

List I (species)	**List II (O—N—O angle)**
A. NO_2^+	1. 180°
B. NO_2	2. 132°
C. NO_2^-	3. 120°
D. NO_3^-	4. 115°
	5. 109°

	A	B	C	D		A	B	C	D
(a)	5	4	3	2	(b)	5	2	4	3
(c)	1	2	4	3	(d)	1	4	3	2

66. Metallic bonding is explained by

(a) band model (b) electron-sea model
(c) both (a) and (b) (d) None of these

67. Energy gaps between the highest filled band and the lowest empty bands in elements *A, B, C* and *D* are 0, 70, 530 and 110 kJ mol^{-1}. Thus

(a) element *A* is metal
(b) element *C* is insulator
(c) element *B* and *D* can be semiconductors
(d) all are correct conclusion about elements *A, B, C* and *D*

68. Select correct statements.

(a) The delocalised electrons in the π-orbitals of graphite cause this allotropic form to conduct electricity
(b) Pauling assigns "metallic valences" equal to their group number to the elements of IA to VIA
(c) Diatomic lithium molecules Li_2 are stable towards dissociation
(d) All the above are correct statements

69. Solutions of alkali metals in liquid ammonia at −33°C conduct electricity without chemical reaction at the electrodes. Hence, conductivity should

(a) increase as temperature is lowered
(b) decrease as temperature is lowered
(c) remain constant on changing the temperature
(d) there is always interaction of alkali metal with liquid ammonia

70. Which represents metallic character in an element?

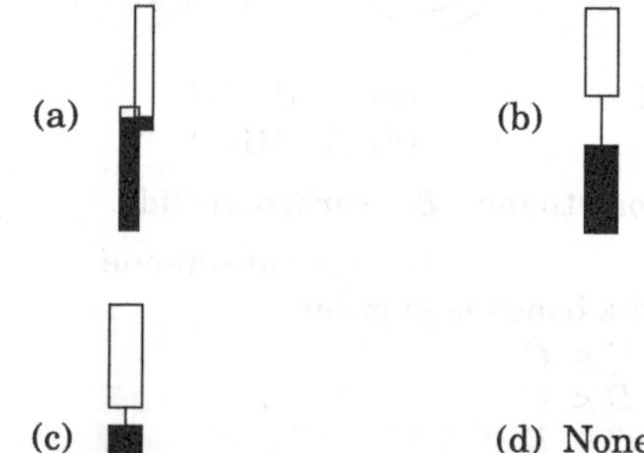

71. Isoelectronic species are

(a) CO, CN^-, NO^+, N_2, BF
(b) $-CH_3, -\ddot{N}H_2, -\ddot{\underset{..}{O}}H, -\ddot{\underset{..}{F}}:$
(c) $>CH_2, >\ddot{N}H, >\ddot{O}:$
(d) all of the above

72. Select, isoelectronic species out of

CO, $\gt\!\!-CH$, $\gt\!\!-\ddot{N}$:, $=CH_2$, $=\ddot{\underset{..}{O}}$, $\gt\!\!-\ddot{O}^{\oplus}$, $=\ddot{N}H$, $>\ddot{N}H$

(a) $\gt\!\!-CH$ $\gt\!\!-\ddot{N}$: $\gt\!\!-\ddot{O}^{\oplus}$ (b) $=CH_2, =\ddot{N}H, =\ddot{\underset{..}{O}}$
(c) both (a) and (b) (d) None of these

73. Some, group in species in Column I are isoelectronic with that in II

Column I	Column II
$HONO_2$, OCO, CO	$(OH)_2CO$, F_2CO, ONN, BF, $(CH_3)_2CO$

Select isoelectronic groups in Column I to Column II

	Column I	Column II
(a)	$(H_2N)_2CO$	$(HO)_2CO, F_2CO, (CH_3)_2CO$
(b)	$HONO_2$	CH_3NO_2
(c)	CO	BF
(d)	all of the above	

74. Decreasing bond-order are expected to be
(a) $ClO_4^- > IO_4^-$
(b) $ClO_3^- > BrO_3^- > IO_3^-$
(c) both (a) & (b)
(d) None of these

75. Out of NO, NO^+, O_2, NO^-, O_2^+ select species with same bond-order and paramagnetic.

		Electron	Bond-order	
(a)	NO	15	2.5	Paramagnetic
	O_2^+	15	2.5	
(b)	O_2	16	2	Paramagnetic
	NO^-	16	2	

(c) Both (a) and (b)
(d) None of the above

76. Formation of anion from a neutral atom *X* is favoured by
(a) high electron affinity
(b) large size of *X*
(c) low ionisation potential
(d) high charge on anion *X*

77. When a chemical bond is formed, there is decrease in
(a) kinetic energy
(b) potential energy
(c) repulsive force
(d) attractive force

78. Which of the following bonds is the weakest?
(a) Coordinate bond
(b) Hydrogen bond
(c) van der Waals' forces
(d) Covalent bond

79. An ionic compound A^+B^- is most likely to be formed from *A* and *B* when
(a) ionisation energy of *A* is low
(b) electron affinity of *B* is low
(c) electronegativity of *B* is low
(d) ionisation energy of *B* is low

80. In I_3^-, Lewis base is
(a) I_2
(b) I^-
(c) I_2^+
(d) I_2^-

81. Which set contains no ionic species?
(a) NH_4Cl, OF_2, H_2S
(b) CO_2, CCl_4, Cl_2
(c) BF_3, AlF_3, TlF_3
(d) I_2, CaO, CH_3Cl

82. Which combination will give the strongest ionic bond?
(a) K^+ and Cl^-
(b) K^+ and O^{2-}
(c) Ca^{2+} and Cl^-
(d) Ca^{2+} and O^{2-}

83. Ionisation energies (IE) and electron affinities (EA) of F and Cl are given below:

Element	IE	EA
F	17.4 eV	3.45 eV
Cl	13.0 eV	3.61 eV

Compound formed between F and Cl is predominantly
(a) F^+Cl^-
(b) Cl^+F^-
(c) F—Cl
(d) Cl<(F)(F)

84. Elements *A*, *B* and *C* have respectively 1,3 and 5 electrons in valence shell. Ionic compound is not formed from
(a) *A* and *C*
(b) *A* and *B*
(c) *B* and *B*
(d) *B* and *C*

85. $Mg^{2+}O^{2-}$ is formed since
(a) IE of Mg is low
(b) EA of O is low
(c) IE of Mg is high
(d) IE of O is low

86. Solubility of KCl is maximum in
(a) CH_3CH_2OH
(b) CH_3OCH_3
(c) H_2O
(d) CH_3COOH

87. Which pair is not correct order of lattice energy?
(a) KCl > MgO
(b) AlN > MgO
(c) $BeCO_3 > MgCO_3$
(d) $BeCO_3 = MgCO_3$

88. Solubility of NaCl, Na_2SO_4 and Na_3PO_4 in water in increasing order is
(a) $NaCl < Na_2SO_4 < Na_3PO_4$
(b) $Na_3PO_4 < Na_2SO_4 < NaCl$
(c) $NaCl < Na_3PO_4 < Na_2SO_4$
(d) $Na_2SO_4 < NaCl < Na_3PO_4$

89. Select correct statement.
(a) LiF and MgO are isostructural and also isodimensional, but a crystal of MgO is much harder than one of LiF
(b) The thermal stability of the isomorphous sulphates of Ca^{2+}, Sr^{2+} and Ba^{2+} w.r.t. decomposition into metal oxide and sulphur trioxide increases in the order $CaSO_4$, $SrSO_4$, $BaSO_4$
(c) Both (a) & (b)
(d) None of the above

90. Select correct statement.
(a) Both lattice energy and hydration energies decrease with ionic size
(b) Lattice energy can be calculated using Born-Haber cycle
(c) If the anion is large compared to the cation, the lattice energy will remain almost constant within a particular group
(d) All the above are correct statements

91. Which of the following compounds has both ionic and covalent bonding?
(a) NaBr (b) $Ba(CN)_2$
(c) PCl_5 (d) CH_3CH_2OH

92. Compound *X* is highly volatile and insoluble in water. Bonding in *X* is
(a) ionic (b) covalent
(c) polar covalent (d) coordinate

93. The maximum covalency is equal to the number of
(a) paired *p*-electrons
(b) unpaired *s*-electrons
(c) unpaired *s*- and *p*-electrons
(d) *s*- and *p*-electrons in the valence shell

94. Select incorrect statement.
(a) Double bond is shorter than a single bond
(b) σ-bond is weaker than a π-bond
(c) Double bond is stronger than a single bond
(d) Covalent bond is stronger than a hydrogen bond

95. The As—Cl bond distance in $AsCl_3$ is 2.20 Å. If radius of chlorine atom is 0.99 Å then radius of As atom is
(a) 0.24 Å (b) 3.19 Å
(c) 1.21 Å (d) 2.09 Å

96. Covalency of carbon in the CO molecule is three because
(a) an unexcited carbon atom has two unpaired electrons
(b) the carbon atom can be an acceptor of an electron pair
(c) the carbon atom has four valence electrons
(d) maximum covalency of carbon is three

97. The number of electron dots in Lewis structure indicates
(a) number of valence electrons in an atom
(b) number of protons in the nucleus of the atom
(c) total number of electrons in an atom
(d) total number of electrons and protons in an atom

98. If there are five electron pairs in outer shell, then structure and bond angle as predicted by Sidgwick-Powell theory is
(a) octahedron, 90°
(b) trigonal bipyramidal, 120° and 90°
(c) pentagonal bipyramidal, 72° and 90°
(d) tetrahedron, 109° 28′

99. Octet rule is not followed in
(a) CCl_4, N_2O_4 and N_2O_5 (b) $BF_3, BeCl_2$ and NO_2
(c) $NaCl, MgCl_2, MgO$ (d) PCl_3, NH_3, H_2O

100. Which has maximum number of lone-pairs of electrons on the central atom?
(a) XeF_2 (b) H_3O^+
(c) XeF_4 (d) XeF_6

101. Geometrical configuration of BF_3 and NF_3 molecules is
(a) the same because of same covalency of the central atom
(b) different because BF_3 is polar and NF_3 is non-polar
(c) different because BF_3 is non-polar and NF_3 is polar
(d) none of the above is correct

102. Coordinate covalent bond is absent in
(a) $[Fe(CN)_6]^{4-}$
(b) adduct of NH_3 and BH_3
(c) H_3O^+
(d) PCl_3

103. The compound NH_3—BF_3 can be easily separated into its compounds because
(a) BF_3 is highly reactive
(b) NH_3 is highly reactive
(c) BF_3 and NH_3 are unstable
(d) BF_3 and NH_3 have their independent existence

104. Consider following compounds
I : $K_4[Fe(CN)_6]$, II : NH_4Cl, III : H_2SO_4
Ionic, covalent and coordinate bonds are present in
(a) I, II and III (b) I and III
(c) II and III (d) I and II

105. Which statement is not correct about NO_2?
(a) It is paramagnetic
(b) It forms dimer and paramagnetism is lost
(c) It has one coordinate bonds
(d) It has nitrogen oxygen triple bond

106. In ammonia, H—N—H bond angle is
(a) 106.7° (b) 104.5°
(c) 109.5° (d) 120°

107. The angle between the bonding orbitals of a molecule AX_3 with zero dipole moment is
(a) 120° (b) 109°
(c) 104° (d) 180°

108. Number of bonding pairs (*X*) and lone pairs (*Y*) around the central atom in the I_3^- ion are

	X	*Y*		*X*	*Y*
(a)	2	2	(b)	2	3
(c)	3	2	(d)	4	3

109. The nitrogen atoms in NH_3, NH_2^- and NH_4^+ are all surrounded by eight electrons. When these species are arranged in increasing order of H—N—H bond angle, correct order is
(a) NH_3, NH_2^-, NH_4^+ (b) NH_4^+, NH_2^-, NH_3
(c) NH_3, NH_4^+, NH_2^- (d) NH_2^-, NH_3, NH_4^+

110. Which of these molecules have non-bonding electron pairs on the central atom?
I : SF_4; II : ICl_3; III : SO_2
(a) II only (b) I and II only
(c) I and III only (d) I, II and III

111. Which species has the same shape as the NO_3^- ion?
(a) SO_3 (b) SO_3^{2-}
(c) ClF_3 (d) ClO_3^-

112. Which one of the following compounds has the electron-pair geometry as the trigonal bipyramidal with three equatorial positions occupied by lone pairs of electrons?
(a) $[AlCl_3]$ (b) XeF_2
(c) $[Pt(NH_3)_2Cl_2]$ (d) $CH_3—Mg—Br$

113. Electron pairs occupy localised orbitals. Their orbitals are oriented in such a way that the repulsion between electron clouds becomes
(a) zero (b) minimum
(c) maximum (d) infinite

114. When the number of electron pairs on the central atom is six, then geometry of the molecule is
(a) octahedral (b) trigonal bipyramidal
(c) equilateral triangle (d) linear

115. PCl_5 has a shape of trigonal bipyramid whereas IF_5 has the shape of a square pyramid. It is due to
(a) presence of unshared electron pair on I which is oriented so as to minimise repulsion while P in PCl_5 has no unshared pair
(b) octet of P is complete while that of I is incomplete
(c) P and I are of different groups
(d) F and Cl have different extent of repulsion

116. Select correct statement about valence-bond approach.
(a) Each bond is formed by maximum overlap for its maximum stability
(b) It represents localised electron model of bonding
(c) Most of the electrons retain the same orbital locations as in a separated atoms
(d) All the above are correct statements

117. Which has a maximum repulsive interaction?
(a) *bp-bp* (b) *lp-lp*
(c) *lp-bp* (d) Equal

118. A molecule of the type AX_5 has square pyramidal geometry. Hence, number of lone pairs on *A* is
(a) 4 (b) 3
(c) 2 (d) 1

119. The bond angle in H_2O is 105° and in H_2S it is 90°. It is due to
(a) the larger size of S atom as compared to O atom which minimises repulsion and allows the bonds in H_2S to be purely *p*-type
(b) liquid state of H_2O as compared to gaseous state of SO_2
(c) both (a) & and (b)
(d) none of the above is correct

120. True structure is predicted by
(a) valence-bond approach
(b) Sidgwick approach
(c) hybrid formation
(d) none of the above

121. The hybrid orbital of the central atom in AlF_4^- is
(a) *sp* (b) sp^2
(c) sp^3 (d) dsp^2

122. What hybrid orbitals are employed by carbon atoms 1, 2 and 3, respectively as labelled in the compound shown?

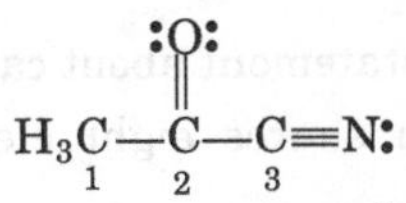

(a) sp^3, *sp*, *sp* (b) sp^3, sp^2, *sp*
(c) sp^3, *sp*, sp^2 (d) sp^3, sp^2, sp^2

123. The geometry of the atoms in the species PCl_4^+ is best described as
(a) tetrahedral (b) see-saw
(c) square (d) trigonal bipyramidal

124. In vinyl acetylene $CH≡C—CH=CH_2$, type of overlapping in ($C_2σ C_3$) bond is
(a) sp^2-*sp* (b) *sp* - sp^2
(c) sp^3 - sp^3 (d) sp^3 - sp^2

125. In which case, hybridisation of the central atom is affected when
(a) NH_3 changes to NH_4^+ (b) AlH_3 changes to AlH_4^-
(c) in both cases (d) in none case

126. Hybridisation of the nitrogen atom and electron geometry around nitrogen atom in pyridine is

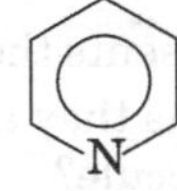

(a) sp^3, pyramidal (b) sp^2, planar trigonal
(c) sp^2, linear (d) sp^3, tetrahedral

127. Which pair of substances will have the most similar geometry?
(a) SO_3, SO_3^{2-} (b) SO_3, SO_4^{2-}
(c) SO_3, CO_3^{2-} (d) SO_4^{2-}, CO_3^{2-}

128. Hybridisation of the underlined atom is affected when
(a) $\underline{C}H_3COOH$ is decarboxylated
(b) $CH_3\underline{C}H_2OH$ is dehydrated
(c) $CH_3\underline{C}H_3$ is chlorinated
(d) $\underline{C}_6H_6$ is nitrated

129. Hybridisation on carbon in short lived species CH_2^{2+} is
(a) *sp*-linear (b) sp^2-trigonal
(c) sp^3-tetrahedral (d) none is correct

130. Hybridisation on carbon in carbanion $\overset{\ominus}{C}H_3$ is
(a) sp^3 (b) sp^3d^2
(c) sp^2 (d) sp^3d

131. The cyanide ion CN^- and the nitrogen molecule N_2 are isoelectronic. However, in contrast to CN^-, N_2 is chemically inert due to
(a) unsymmetrical electron distribution
(b) low bond energy
(c) absence of bond polarity
(d) presence of greater number of electrons in bonding

132. Select correct statement about carbonium ion $\overset{\oplus}{CH_5}$
(a) This cation shares eight electrons among five bonds
(b) There is no empty orbital
(c) It is not electron deficient
(d) All the above are correct statements

133. Select correct statement about carbenium ion $\overset{\oplus}{CH_3}$
(a) It is planar with empty *p*-orbital
(b) There are three sigma (C—H) bonds
(c) It is electron deficient with six electron in outer-shell
(d) All the above are correct statements

134. Which of the following has maximum bond energy?
(a) O_2^- (b) O_2^+
(c) O_2 (d) O_2^{2-}

135. In which pair or pairs is the stronger bond found in the first species?
I : O_2^{2-}, O_2; II : N_2, N_2^+; III : NO^+, NO^-
(a) I only (b) II only
(c) I and III only (d) II and III only

136. Which plot best represents the potential energy (E) of two hydrogen atoms as they approach one another to form a hydrogen molecule?

(a) 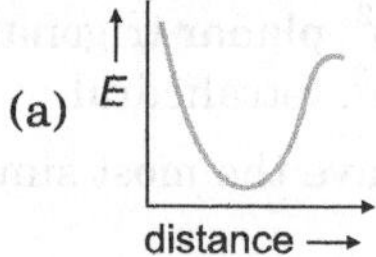

(b) 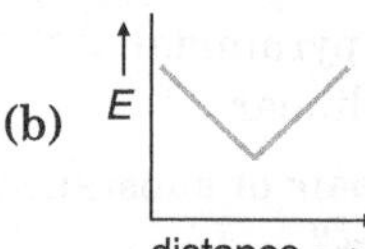

(c) 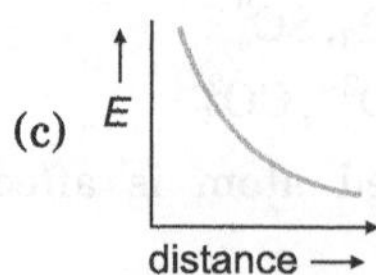

(d) 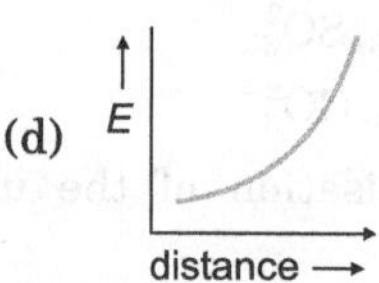

137. Number of unpaired electrons in $[V(H_2O)_6]^{3+}$ is
(a) 2 (b) 3
(c) 4 (d) 5

138. dsp^2 hybridisation is in
(a) $[Ni(CN)_4]^{2-}$ (b) $[Pt(en)_2]^{2+}$
(c) both (a) and (b) (d) none of these

139. Select outer *d*-complexes
I : $[Fe(CN)_6]^{3-}$; II : $[Fe(H_2O)_6]^{3+}$
III : $[CoF_6]^{3-}$
(a) II, III (b) I, II
(c) I, III (d) I, II, III

140. A set of inner *d*-complex is
(a) $[Fe(CN)_6]^{4-}, [Ni(CN)_4]^{2-}, [Pt(en)_2]^{2+}$
(b) $[Fe(H_2O)_6]^{3+}, [CoF_6]^{3-}, [Cr(CN)_6]^{3-}$
(c) both (a) and (b)
(d) none of the above

One or More Than One Options Correct

1. Which of the following have identical bond-order?
(a) CN^- (b) O_2^-
(c) NO^+ (d) CN^+

2. Select correct statement(s) about C_3^{4-} ion
(a) It reacts with H_2O forming C_3H_6
(b) It reacts with H_2O forming C_3H_4
(c) It has two sigma and two pi bonds
(d) It has three sigma and one pi bonds

3. The planar shape of $N(SiH_3)_3$ is explained by the
(a) type of hybrid orbitals of nitrogen
(b) additional $d\pi$-$p\pi$ overlap along the N—Si bond
(c) higher electronegativity of nitrogen
(d) higher electronegativity of silicon

4. Which are the species in which sulphur undergoes sp^3 hybridisation?
(a) SF_4 (b) SCl_2
(c) SO_4^{2-} (d) H_2S

5. Which of the following are non-polar?
(a) SiF_4 (b) XeF_4
(c) SF_4 (d) BF_3

6. In which of the following pairs, molecules/ions have similar shape?
(a) CO_2 and H_2O (b) BF_3 and $(CH_3)_3C^{\oplus}$
(c) CCl_4 and $PtCl_4$ (d) NH_3 and BF_3

7. In BrF_3 molecule, the lone pairs occupy equatorial position to minimize
(a) lone pair-bond pair repulsion only
(b) bond pair-bond pair repulsion only
(c) lone pair-lone pair repulsion and lone pair-bond pair repulsion
(d) lone pair-lone pair repulsion only

8. H_2O is dipolar, whereas BeF_2 is not. It is because
(a) the electronegativity of F is greater than that of O
(b) H_2O involves hydrogen bonding whereas BeF_2 is a discrete molecule
(c) H_2O is linear and BeF_2 is angular
(d) H_2O is angular and BeF_2 is linear

9. Among the following the pair in which the two species are not isostructural is
(a) SiF_4 and SF_4
(b) IO_3^- and XeO_3
(c) BH_4^- and NH_4^+
(d) PF_6^- and SF_6

10. $PbCl_4$ exists but $PbBr_4$ and PbI_4 do not because of
(a) Br^- and I^- ions are bigger in size
(b) chlorine is a gas
(c) Br_2 and I_2 are more electronegative
(d) inability of bromine and iodine to oxidise Pb^{2+} to Pb^{4+}

11. Which is/are correct statements?
(a) A solute will dissolve in water if hydration energy is greater than lattice energy
(b) If the anion is large compared to the cation, the lattice energy wil remain almost constant
(c) Solubility of II A hydroxide is in order.
$$Be(OH)_2 < Mg(OH)_2 < Ca(OH)_2 < Sr(OH)_2$$
(d) None of the above is correct

12. In which molecule is the van der Waals' force likely to be most important in determining melting point and boiling point?
(a) ICl (b) Br_2
(c) H_2S (d) CO

13. Which have linear structure?
(a) BeF_2 (b) $Ag[CN]_2^-$
(c) CO_2 (d) XeF_2

14. In I_3^-
(a) I_2 is Lewis acid and I^- as Lewis base
(b) I_2 is Lewis base and I^- as Lewis acid
(c) I_3^- itself behaves as amphoteric anion
(d) I_3^- does not exist

15. In the following case (s), hybridisation of the underlined atom is affected
(a) $\underline{P}Cl_5$ (solid) dissociates into PCl_4^+ and PCl_6^-
(b) LiH reacts with $\underline{Al}H_3$ forming $LiAlH_4$
(c) $\underline{N}H_3$ is protonated
(d) $H_3\underline{P}O_2$ is heated forming PH_3 and H_3PO_3

16. The correct order of the O—O bond length in O_2, H_2O_2 and O_3 is
(a) $H_2O_2 > O_3 > O_2$
(b) $O_2 > O_3 > H_2O_2$
(c) $O_2 > H_2O_2 > O_3$
(d) $O_3 > H_2O_2 > O_2$

17. Which combinations of the compounds and their geometry are correct?
(a) $HgCl_2$—linear (b) ClF_3—V-shaped
(c) ClF_3—T-shaped (d) ICl_4^-—square planar

18. Which combination of the compounds with their magnetic moments are correct?
(a) $K_3Mn(C_2O_4)_3 \cdot 3H_2O$ $\mu = 4.90$ BM
(b) $K_3Fe(C_2O_4)_3 \cdot 3H_2O$ $\mu = 5.92$ BM
(c) $Ni(CO)_4$ $\mu = 0.00$ BM
(d) $[Ni(CN)_4]^{2-}$ $\mu = 0.00$ BM

19. The correct structural representation of diborane is
(a) $[BH_2]^+[BH_4]^-$
(b) HB—BH with two H atoms above and two H atoms below (H–B bonds drawn vertically)
(c) H₂B bridged with two H atoms to BH₂ (bridged structure)
(d) $H_2\left[\begin{matrix}H \\ H\end{matrix} > B{=}B < \begin{matrix}H \\ H\end{matrix}\right]$

20. Which of the following is/are electron deficient compounds?
(a) $NaBH_4$
(b) B_2H_6
(c) $AlCl_3$
(d) C_3^{4-}

Passage Comprehension Questions

Passage 1

A paper published in the research journal *Science* in 2007 (S. Vallina and R. Simo, *Science*, 305, 506 Jan 26, 2007) reported studies of dimethyl sulphide (DMS, an important green house gas that is released by marine phytoplankton. This gas "represents the largest natural source of atmosphere sulphur and a major precusor of hygroscopic (*i.e.*, cloud forming) particles in clean air over the remote oceans, thereby acting to reduce the amount of solar radiation that crosses the atmosphere and is absorbed by ocean."

Answer the following questions

1. Steric number of sulphur in DMS is
(a) 2 (b) 3
(c) 4 (d) 5

2. Select correct statement(s).
(a) Bond pair-bond pair repulsions of the two (S—CH_3) bonds in DMS would be greater than the bond-pair bond-pair repulsions of the S—CH_3 and S—H bonds in thiol.
(b) Bond angle $\angle$C—S—C in DMS is greater than bond angle $\angle$C—S—H in thiol
(c) Both (a) & (b) are correct
(d) None of the above is correct

3. The mean sea water concentration of DMS in the ocean is 2.7 nm (nanomolar). How many molecules of DMS are present in 1.0 m^3 of sea water?
(a) 16.254×10^{17} molecules
(b) 2.7×10^{17} molecules
(c) 2.7×10^{-6} molecules
(d) 16.257×10^{9} molecules

Passage 2

Assume following structure of N_2O and answer the questions.

$N{=}N{=}O$ (I) $\quad$ $:N{-}N{\equiv}O:$ (II) $\quad$ $:N{\equiv}N{-}O:$ (III)

$N{-}N{=}O$ (IV) $\quad$ $:N{=}N{=}O$ (V) $\quad$ $:N{=}N{=}O:$ (VI)

1. Which is most favourable structure?
(a) I (b) III
(c) V (d) VI

2. Select correct statement(s).
(a) Structure I is not acceptable as there are 10 electron on N atom and 18 electrons total
(b) Structure IV is not acceptable as there are only three bonds
(c) Structure VI is not acceptable as there are 10 electrons on oxygen atom
(d) All the above are correct statements

3. Structure III is represented as
(a) $:N{\equiv}\overset{+}{N}{-}\ddot{O}:^{-}$ (b) $:N{\equiv}\overset{-}{N}{-}\ddot{O}:^{+}$
(c) $:\overset{+}{N}{=}N{-}\ddot{O}:^{-}$ (d) $:\overset{-}{N}{\equiv}N-\ddot{O}:^{+}$

Passage 3

We consider dissociation energy of different species.

Species	Dissociation energy (kJ mol^{-1})
O_2^+	642.9
O_2	493.6
O_2^-	395.0
N_2^+	840.7
N_2	941.7
N_2^-	765.0
NO^+	1046.9
NO	026.9
NO^-	487.8

Answer the following questions :

1. In which case bond-order increases?
(a) O_2 changes to O_2^+ (b) N_2 changes to N_2^+
(c) NO changes to NO^- (d) In all cases

2. In which case electron from anti-bonding MO (HOMO) is removed? (HOMO : Highest Occupied Molecular Orbital)
(a) O_2 to O_2^+ (b) N_2 to N_2^+
(c) N_2 to N_2^- (d) In all cases

3. In which case electron from bonding MO is removed?
(a) O_2 to O_2^+ (b) N_2 to N_2^+
(c) NO to NO^+ (d) O_2 to O_2^-

Passage 4

Consider the following molecules and answer the questions given :

I : H_2O; II : CO_2; III : NH_3
IV : CCl_4; V : ClF

1. In which compound are the bonds most polar?
(a) H_2O (b) CO_2
(c) CCl_4 (d) ClF

2. Which compounds are not polar?
(a) H_2O, CO_2 (b) CO_2, NH_3
(c) H_2O, CCl_4 (d) CO_2, CCl_4

3. Which atoms in H_2O and ClF are negatively charged?
(a) O in H_2O and Cl in ClF
(b) H in H_2O and F in ClF
(c) O in H_2O and F in ClF
(d) None of the above is correct

Passage 5

The platinum-chlorine distance has been found to be 2.32 Å in several crystalline compounds. This value applies to both of the compounds shown in figure,

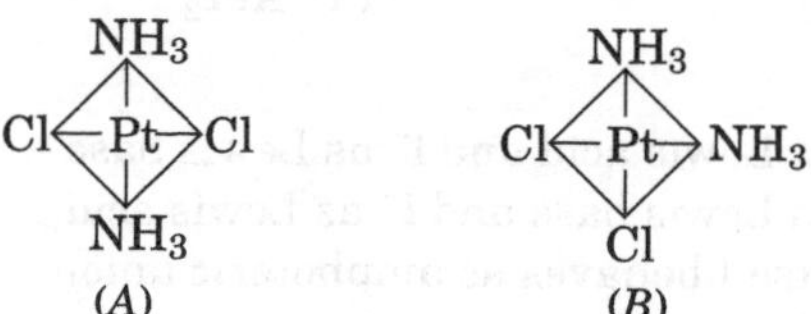

Based on the above structures, answer the following questions

1. Cl—Cl distance in structure (*A*) is
(a) 2.32 Å (b) 4.64 Å
(c) 1.16 Å (d) 9.28 Å

2. Cl—Cl distance in structure (*B*) is
(a) 2.32 Å (b) 1.52 Å
(c) 2.15 Å (d) 3.28 Å

3. *A* is
(a) *cis*-isomer (b) *trans*-isomer
(c) chiral isomer (d) none of these

4. *B* is
(a) *cis*-isomer (b) nuclear isomer
(c) chiral isomer (d) coordinate isomer

Passage 6

By the following method you can predict the geometrical shape of species with only one central atom, without first drawing the Lewis structure:

1. Total number of electron pairs
= (number of valence electrons ± electrons (for ionic charge)/2

2. Number of bond electron pairs = number of atoms −1

3. Number of electron pairs around central atom = total number of electron pairs − 3 [number terminal atoms (except H)]

4. Number of lone pair = (number of central electron pairs − number bond pairs)

Read the above method and answer the following questions:

1. If A is the central element of the molecule containing A and X element and E the number of electron pairs round it then VSEPR notation AX_3E will be for the molecules

(a) PCl_5, ICl_4^- (b) NH_3, H_2O
(c) NH_3, ClO_3^- (d) ICl_4^-, ClO_3^-

2. VSEPR notation of PCl_5, H_2O and ICl_4^- are

	PCl_5	H_2O	ICl_4^-
(a)	AX_5	AX_2E_2	AX_4E_2
(b)	AX_5	AX_2E	AX_4E
(c)	AX_5	AX_2	AX_4E
(d)	AX_5	AX_2E_2	AX_4E

3. Square planar structure is predicted for

(a) ICl_4^-, ClO_3^-
(b) PCl_4^+, PCl_6^-
(c) ICl_4^-, PCl_4^+
(d) $ICl_4^-, [Ni(CN)_4^{2-}]$

4. Based on above method, structure of the some of the molecules have been matched. Which is the incorrect matching?

(a) PCl_5 — trigonal bipyramidal
(b) ClO_3^- — square planar
(c) ICl_4^- — square planar
(d) PCl_4^+ — tetrahedral

Passage 7

Read the following short write-up and answer the questions at the end of it.

In certain polar solvents PCl_5 undergoes an ionisation reaction in which Cl^- ion leaves one PCl_5 molecule and attaches itself to another.

$$2PCl_5 \rightleftharpoons PCl_4^+ + PCl_6^-$$

1. Select **incorrect** statement(s).

(a) Dissociation is a redox reaction
(b) Hybridisation changes from sp^3d to sp^3d^2 (PCl_6^-) and sp^3(PCl_4^+)
(c) Structure changes from trigonal bipyramidal to tetrahedral (PCl_4^+) and octahedral (PCl_6^-)
(d) None of the above is incorrect

2. Number of lone pairs around phosphorus in PCl_5, PCl_4^+ and PCl_6^- are respectively

(a) 0, 1, 2 (b) 0, 0, 0
(c) 1, 2, 3 (d) 0, 2, 1

Passage 8

Read the following short write-up and answer the questions at the end of it.

"Each of the following compounds has a nitrogen-nitrogen bond"

(A) N_2, (B) N_2H_4
(C) N_2F_2

Molecules may be polar or non-polar, that is dependent on the electronegativity difference between two atoms. Their bond-lengths (between N-atoms) have been 1.45 Å, 1.10 Å and 1.22 Å but not in correct order.

1. Compounds have been matched with their bond-length (between N-atoms). Which is not the correct matching?

(a) A – 1.10 Å (b) B – 1.45 Å
(c) C – 1.22 Å (d) None of these

2. Select incorrect statement(s).

(a) N_2 is an inert gas
(b) N_2F_2 has sp^2-hybridised N-atom
(c) N_2H_4 is an inert solvent
(d) N_2H_4 is used as rocket fuel

Passage 9

Read the following short write up and answer the questions at the end of it.

Two substances having the same molecular formula, C_4H_8O, were examined in the gaseous state by electron diffraction. The carbon-oxygen distance was found to be 143 pm in the compound A and 124 pm in compound B.

C—O bond length = 143 pm
C=O bond length = 123 pm

1. Which of the above compounds is inert?

(a) A (b) B
(c) Both (a) and (b) (d) None of these

2. Which of the above can show functional isomerism?

(a) A (b) B
(c) Both (a) and (b) (d) None of these

3. Keto-enol tautomerism occurs in

(a) A (b) B
(c) both (a) and (b) (d) none of these

4. Distinction between two functional isomers of the same main functional group can be made by

(a) Iodoform test (b) Tollen's test
(c) Br_2 water (d) Phthalein test

5. Probable structures of *A* and *B* can be

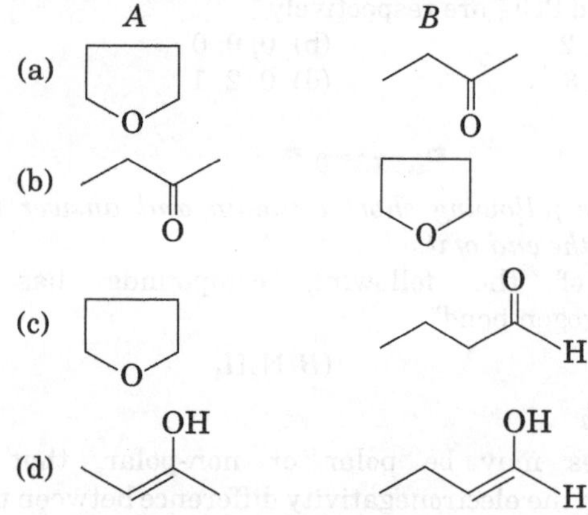

Passage 10

Read the following experimental fact and answer the question at the end of it.

A sample of lead (II) bromide, $PbBr_2$, was placed in a crucible. The ends of two carbon electrodes were put in the solid and connected to a lamp and power pack. At first the lamp did not light, but when the bromide was heated and began to melt the lamp began to glow.

1. Select correct statement(s).
(a) $PbBr_2$ is non-conducting in solid state
(b) $PbBr_2$ is conducting in aqueous solution
(c) $PbBr_2$ is conducting in fused state
(d) $PbBr_2$ is non-conducting in molten state

Passage 11

Read the following short-write up and answer the questions at the end of it.

"Methyl isocyanate is used to make certain pesticides. In December 1984, water leaked into a tank containing this substance at a chemical plant, producing a toxic cloud that killed thousands of people in Bhopal.

1. Which is not the resonance structure of methyl isocyanate?
(a) $CH_3—\ddot{N}=C=\ddot{O}:$ (b) $CH_3—\overset{\ominus}{\ddot{N}}—C\equiv\overset{\oplus}{O}:$
(c) $CH_3—N\equiv\underset{\oplus}{C}—\ddot{O}:^{\ominus}$ (d) $CH_3—\overset{\oplus}{N}\equiv C—\ddot{O}:^{\ominus}$

2. Which of the atoms in CH_3NCO has maximum *s*-character?
(a) N (b) O
(c) C(in CH_3) (d) C (bonded to O and N)

3. Total number of lone pairs are
(a) 2 (b) 3
(c) 4 (d) 5

Assertion & Reason

Codes :
(a) Both **A** and **R** are true and **R** is the correct explanation of **A.**
(b) Both **A** and **R** are true but **R** is not the correct explanation of **A.**
(c) **A** is true but **R** is false.
(d) **A** is false but **R** is true.

1. Assertion (A) : On cooling, the brown colour of nitrogen dioxide disappears.
Reason (R) : On cooling NO_2 undergoes dimerisation resulting in the pairing of odd electrons of NO_2.

2. Assertion (A) : Tin (Sn) exhibits +2 and +4 oxidation states only and not +3 oxidation state.
Reason (R) : Inert pair effect of electron pair in *s*-orbital is responsible.

3. Assertion (A) : $PbCl_2$ is more stable than $PbCl_4$.
Reason (R) : $PbCl_4$ is powerful oxidising agent.

4. Assertion (A) : BF_3 is a weaker Lewis acid than BCl_3.
Reason (R) : The planar BF_3 molecule is stabilised to a greater extent than BCl_3 by B—*X* π-bonding.

5. Assertion (A) : O_2 is paramagnetic in nature.
Reason (R) : Bond-order of oxygen is two.

6. Assertion (A) : KO_2 is paramagnetic.
Reason (R) : K^+ ion has unpaired electron.

7. Assertion (A) : Dipole moment of *o*-dichloro benzene is larger than the theoretical value.
Reason (R) : Due to dipole-dipole repulsion of two Cl-atoms in *ortho* position bond angle is greater than 60°.

8. Assertion (A) : N-atom in NH_3 is sp^3 hybridised and bond angle is 107°.
Reason (R) : *lp-bp* repulsion (VSEPR) decreases bond angle to 107°.

9. Assertion (A) : P_4 exists but N_4 does not exist.
Reason (R) : In P_4 form P-atoms are joined by single bonds while in N_4, two N-atoms are joined by double and other two by single bonds.

10. Assertion (A) : PCl_5 exists but NCl_5 does not exist.
Reason (R) : *d*-orbital in P-atom takes part in chemical bonding; this lacks in N-atom.

11. Asserton (A) : The C—C bond distance in methyl acetylene is unusally short (145 pm) for a single bond.

Reason (R) : Hyperconjugation involving "no bond" structure,

$$H-\ddot{C}=C=C\langle^{H}_{H}$$

increases the bond order of the C — CH_3 bond.

12. Assertion (A) : Bond angle in NH_3 is greater than that in PH_3.

Reason (R) : P—H bonds are longer and the lower electronegativity of P permits electron density to be displaced towards H to a greater extent than in the case of NH_3.

13. Assertion (A) : Following structure is acceptable

$$:N\equiv N-\ddot{\underset{..}{O}}:$$

Reason (R) : Octet rule is followed for each atom central N has positive charge and oxygen has negative charge.

True & False

1. Solid sodium chloride does not conduct electricity.
2. Dipole moment of HF is less than that of HCl because of shorter bond length of HF.
3. Bond energy of all the P—Cl bonds in PCl_3 and PCl_5 is same.
4. The covalent molecule $BeCl_2$ has permanent dipole moment.
5. The H—N—H bond angle in NH_3 is greater than the H—As—H bond angle in AsH_3.
6. All the molecules with polar bonds have dipole moment.
7. Sugden introduced singlet linkage in which one electron is donated instead of one pair of electrons.
8. PCl_5 in solid state dissociates into PCl_4^+ and PCl_6^-.
9. O_2, O_2^- and O_2^{2-} are paramagnetic species.
10. Dipole moment of BH_3 is zero.
11. Dipole moment of $CHCl_3$ is larger than that of HCl.
12. All the B—H bonds in B_2H_6 are equivalent.
13. Hybridisation of sulphur in SO_2 is *sp*.
14. Bond length of (C=O) bond is larger than (C—O) bond in CO_3^{2-}.
15. The highest occupied energy band is called the valence band and the lowest unoccupied energy band is called the conduction band.
16. NaCl conducts electricity only in aqueous solution.
17. C_2 and C_3 atoms in CH≡C—CH=CH$_2$ are respectively *sp* and *sp*^{2} hybridised.
18. Structures of $[Ni(H_2O)_4]^{2+}$ and $[Ni(CN)_4]^{2-}$ are different.
19. Dipole moment of $[Ni(CN)_4]^{2-}$ is zero.
20. Dipole moment of *o*-dichloro benzene is smaller than the theoretical value, while in case of *o*-nitrophenol, value is larger than the theoretical value.

Fill in the Blanks

1. The complex ion $[Pt(NH_3)_4]^{2+}$ has a .... structure.
2. $AlCl_3$ is covalent while AlF_3 is ionic. The above fact can be justified on the basis of ....
3. Water has maximum density at .... °C.
4. O_2 is .... in nature because it has two unpaired electrons.
5. The abnormally large specific heat of water molecule can be explained on the basis of ....
6. The bond formed between two atoms one of low ionisation potential and the other of high electron affinity is ....
7. By the axial overlapping of two *p*-orbitals of same energy, a strong .... bond is formed.
8. When NH_3 and BF_3 are bonded, NH_3 is a .... and BF_3 is a ....
9. Types of bonding in NH_4Cl are .... and ....
10. NO_2 is .... but N_2O_4 is ....
11. Bond orders of oxygen, superoxide ion and peroxide ion are .... and .... and of these .... and .... are paramagnetic.
12. Hybridisation of N atom in NO_2, NO_3^- and NH_3 are respectively .... and ....
13. Ni atom in $[Ni(CN)_4]^{2-}$ is .... hybridised and in $Ni(CO_4)$ is .... hybridised.
14. Magnetic moment of iron in $[Fe(H_2O)_6]^{2+}$ is ....
15. A molecule like H—*X* is 50% ionic if electronegativity difference of H and *X* is .... eV.
16. Out of NH_3, PH_3 and AsH_3 boiling point of .... is maximum and that of .... is minimum.
17. Coordination number of H_2O in liquid state is ....
18. When N_2 changes to N_2^+, the N—N bond distance .... and bond order...........
19. When O_2 changes to O_2^- the O—O bond distance .... and bond order ....
20. CN^- and N_2 are isoelectronic. But in contrast to CN^-, N_2 is chemically inert because of ....

21. Unsymmetrical $PCl_5(s)$ splits into .... and ....

22. In I_3^-, I_2 is a .... and I^- is a ....

23. Metallic bonding is explained by .... and ....

24. In the case of insulators, the energy gap is .... while in case of semiconductors energy gap is ....

25. The highest occupied energy band is called the .... and lowest unoccupied energy band is called the ....

Matrix-Match Type Questions

I. Only One Correct

1. Match the compounds (in List I) with the structure (in List II)

List I	List II
A. XeF_4	1. Distorted octahedral
B. XeF_6	2. Tetrahedral
C. XeO_3	3. Square planar
D. XeO_4	4. Pyramidal

2. Match the type of species (in List I) with the compound (in List II).

List I	List II
A. Ionic species	1. HCl
B. Non-polar covalent	2. $LiCH_3$
C. Polar covalent species having sigma bond only	3. KCH_3
	4. NO

3. Match the compounds (in List I) with the type of hybridisation on the central atom/ion (in List II).

List I	List II
A. BCl_3	1. sp^3
B. NH_3	2. sp^3d^2
C. $[Cu(NH_3)_4]^{2+}$	3. sp^2
D. $[Fe(H_2O)_6]^{2+}$	4. dsp^2

4. Match the compounds (in List I) with the structures (in List II).

List I	List II
A. $(CH_3)_3B$	1. Square planar
B. NH_4Cl	2. Trigonal bipyramid
C. $[ICl_4]^-$	3. Tetrahedral
D. PCl_5	4. Trigonal

5. Match type of bond (given in List I) with method of formation (given in List II).

List I	List II
A. σ-bond	1. Lateral overlapping
B. Covalent bond	2. Sharing of electrons
C. Ionic bond	3. Transfer of electrons
D. π-bond	4. Donating an electron
	5. Accepting an electron
	6. Axial overlapping

6. Match the hybridization (in List I) with the respective geometrical shape of the hybrid orbitals (in List II).

List I	List II
A. *sp*	1. Tetrahedral
B. sp^2	2. Linear
C. sp^3	3. Trigonal
D. dsp^2	4. Square planar
E. d^2sp^3	5. Pyramidal
	6. Octahedral

7. Match the list of type of bonds (List I) with list of compounds (List II).

List I	List II
A. Ionic	1. H_2O
B. Single covalent	2. NH_3
C. Coordinate covalent	3. $CaCl_2$
D. Hydrogen bond	4. C_2H_2
E. Multiple covalent	5. CH_4
	6. NH_4^+

8. Match the species (in List I) with the bond order (in List I).

List I	List II
A. N_2	1. 1.0
B. O_2	2. 2.0
C. F_2	3. 2.5
D. O_2^+	4. 3.0

9. Match the items under List (I) with items under List (II).

List (I) Molecule	List (II) Shape
A. PCl_5	1. V-shaped
B. F_2O	2. Triangular planar
C. BCl_3	3. Trigonal bipyramidal
D. NH_3	4. Trigonal pyramidal
	5. Tetrahedral

10. Match List I (Hybrid bond orbitals) with List II in (species) and select the correct answer.

List I	List II
A. sp^3	1. ICl_4^-
B. dsp^2	2. $TeCl_4$
C. sp^3d	3. MnO_4^-
D. sp^3d^2	4. $Ni(CN)_4^{2-}$

11. Match the species (in List I) with bond-order (in List II) and number of unpaired electrons (in List III).

List I (Species)	List II (Bond-order)	List III (Unpaired electron)
Be_2^+	2	2
O_2	2.5	0
O_2^+	0.5	1
O_2^{2-}	1.0	1

12. Match the species (in List I) with Bond-order (in List II) and isoelectronic species (in List III).

List I (Species)	List II (Bond-order)	List III (Isoelectronic species)
NO^+	2.5	O_2^+
NO	3.0	O_2
NO^-	2.0	N_2

II. One or More Correct

1. Match the species in Column I with corresponding character(s) in Column II.

Column I	Column II
A. $\overset{\oplus}{C}H_3$	1. Bond order 3
B. $\overset{\oplus}{C}H_5$	2. Paramagnetic
C. $NO^{\oplus}$	3. sp^2
D. O_2	4. Carbocation
E. N_2	5. Carbonium ion
	6. Octet rule violated

2. Match the species in Column I with the corresponding character(s) in Column II.

Column I	Column II
A. CO_2	1. Zero dipole moment
B. CH_4	2. Expansion of octet
C. $ClF_2^{\oplus}$	3. sp^3 hybridisation
D. PCl_5	4. sp^3d hybridisation
E. $ClF_2^{\ominus}$	5. Linear

Integer Answer Types

This section contains 8 questions. The answer to each of the questions is a single digit integer, ranging from 0 to 9. The appropriate bubbles below the respectively question numbers in the *ORS* have to be darkened. For example, if the correct answers to question number *X, Y, Z* and *W* (say) are 6, 0, 9 and 2, respectively

X	Y	Z	W
0	0	0	0
1	1	1	1
2	2	2	2
3	3	3	3
4	4	4	4
5	5	5	5
6	6	6	6
7	7	7	7
8	8	8	8
9	9	9	9

Set I

1. Steric number of CH_4 is ....

2. Steric number of H_2O is ....

3. Lone pairs in I_3^- are ....

4. Bond-order in NO^+ is ....

5. Number of hybrid orbitals in XeF_4 is ....

6. Number of unpaired electron(s) in O_2^- is/are ....

7. Formal charge on sulphur in SO_2 is ....

8. In a compound of Xe, F and O

$$Xe = 60.4\% \ (131.3)$$
$$F = 17.5\% \ (19.0)$$

Thus, number of oxygen atom ....

Set II

This section contains 9 questions.

1. A slightly polar molecule *AB* has dipole moment of 0.24 D. If bond-length is 1 Å, ionic character is ......%

2. In a very ideal imaginary condition, a sample of ice was taken, and its osmotic pressure was found to be 0.274 atm at 300 K. (density = 1 g/mL). Thus, number of water units attached to one water unit is ....

3. In O_3, central oxygen atom is hypervalent. It shares .... electron pairs.

4. Total number of orbitals involved in the formation of hybrid orbitals in ClF_2^- is ....

5. Total number of orbitals involved in the formation of hybrid orbitals in ClF_2^+ is ....

6. AX_4 forms square planar type structure. Distance between *A* and *X* atoms is 4 units then distance between two *X* atoms in *trans* position is ....

7. Out of CH_2Cl_2, CH_4, CCl_4, H_2O, $CHCl_3$, *d*-dichlorobenzene, *o*-cresol, *p*-xylene, SCl_2, BF_3, IBr and CH_2O, non-zero value of dipole moment is of ....

8. Superoxide ion has .... electrons in anti-bonding molecular orbitals.

9. Difference of σ and π bonds in $C(CN)_4$ is ....

Answers & Solutions
Master Exercises

Exercise 1

Short Answer Type Questions

1. The NH_3 molecule has a dipole moment because the three N—H bond dipoles add to give a net dipole pointing upward to the N atom from the base of the pyramid that defines the NH_3 structure.

The SF_6 molecule has no dipole moment because each S—F bond dipole is balanced by one of equal magnitude pointing in opposite direction of the other side of the molecule.

2. (a) CH and CO bonds are polar.

(b) The CO bond is most polar, and O-atom is the most negative atom.

$$CH_2{=}CH{-}CH{=}O \longleftrightarrow \overset{\oplus}{C}H_2{-}CH{=}CH{-}O^{\ominus}$$

3. Bond-order NO is

2 in NO^{2+}
1.5 in NO_2^+
1.33 in NO_3^-

Thus, longest NO bond in NO_3^- and shortest in NO_2^+

Fact : Larger the bond-order, shorter the bond-length

NO_2^+ NO_2^- NO_3^-

Bond order $\xleftarrow{\text{max}}$

Bond length $\xrightarrow{\text{max}}$

4. (a) The bond angle is smaller in NH_3 than in CH_4 because of the repulsion between the lone pair on N and the bonding pairs.

(b) The bond angle is greater in OCl_2 than in OF_2 because there is some π interactions in OCl_2 involving donation from the filled *p*-orbitals on oxygen into the empty *d*-orbitals on chlorine No π-bonding is possible for OF_2 because all orbitals are filled on both atoms.

(c) The bond angle in NH_3 is larger than that in NF_3 because the N—F bonds are larger than N—H bonds and the electron density is displaced towards the more electronegative F, both effects diminish the bonding pair-bonding pair repulsion.

(d) The bond angle in NH_3 is larger than in PH_3 because the P—H bonds are longer and the lower electronegativity of P permits electron-density to be displaced towards hydrogen to a greater extent than in the case of NH_3. Both of these effects diminish the repulsion due to bond-pairs.

5. C : $1s^2 2s^2 2p^2$.

We consider formation of two *sp*-hybrid orbitals, one directed toward oxygen and directed 180° opposite. This contains carbon lone pair, leaving empty *sp* hybrid orbitals with a filled oxygen σ-orbital.

The two unhybridised singly occupied carbon 2*p*-orbitals form π bonds by overlap with singly occupied oxygen 2*p* orbitals.

6. C_2 (12 *e*) has molecular orbital electronic configuration as:

$(\sigma 1s)^2(\sigma^*1s)^2(\sigma 2s)^2(\sigma^*2s)^2(\pi 2p_y^2)(\pi 2p_z^2)\sigma 2p_x^0$

Bond-order

$$= \frac{\text{Electrons in (bonding MO – anti-bonding MO)}}{2}$$

$$= \frac{8-4}{2} = 2$$

C_2^-**(13*e*) :** $(\sigma 1s)^2(\sigma^*1s)^2(\sigma 2s)^2(\sigma^*2s)^2(\pi 2p_y^2 \pi 2p_z^2)\,\sigma^2 p_x^1$

$$\text{Bond-order} = \frac{9-4}{2} = 2.5$$

Electron is added to higher energy orbital ($\sigma 2p_x$) and bond-order increases.

C_2^+**(11*e*) :** $(\sigma 1s)^2\,(\sigma^*1s)^2(\sigma 2s)^2(\sigma^*2s)^2(\pi 2p_y^2 \pi 2p_z^1)$

$$\text{Bond-order} = \frac{7-4}{2} = 1.5$$

Electron is removed from lower-energy orbital and bond-order decreases. This bond-dissociation energy of C_2^+ is much lower than of C_2^- in which bond-order is higher and also electron is added to higher energy orbital.

Analytical Questions

1. (a) Two valence electrons missing, S and C have incomplete octet, $[\ddot{\text{:S}}{=}C{=}\ddot{\text{N}}\text{:}]^-$,

(b) Correct,

(c) One valence electron missing; odd electron species with structure $\ddot{\text{:O}}{=}\dot{\text{N}}{=}\ddot{\text{O}}\text{:}$

(d) N cannot have an expanded octet, correct structure is :

$$\text{:}\ddot{\underset{\cdot\cdot}{Cl}}{-}\ddot{N}{-}\ddot{\underset{\cdot\cdot}{Cl}}\text{:}$$
$$|$$
$$\text{:}\underset{\cdot\cdot}{Cl}\text{:}$$

2.

		v	s	u	$F = v - \frac{s}{2} - u$
(a)	O	6	2	6	−1
	Cl	7	6	2	+2
(b)	S	6	8	2	0
	O	6	4	4	0
(c)	B	3	8	0	−1
	F	7	2	6	0
(d)	N (centre)	5	8	0	1
	N (terminal)	5	4	4	−1
(e)	N (centre)	5	8	0	+1
	N (left)	5	2	6	−2
	N (right)	5	6	2	0
(f)	N	5	8	0	+1
	O	6	4	4	0

3. (b) Structure with the highest bond order and the shortest bond length (N≡N)

4. BE of fission of 6 (P—P) bond in P_4 is greater than 6 (N—N) bond in N_4; and BE of(N≡N) is greater than that of (P≡P).

5. The glowing of lamp tells us that the solid conducts when in molten state in which there are ions which are free to move. This also suggest that $PbBr_2$ is ionic.

6. (a) Which are formed by two electrons.

(b) Which are formed by one electron.

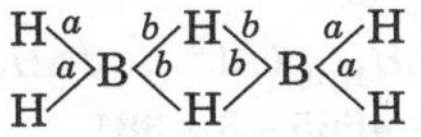

$$\underset{H}{\overset{H}{}}\!\!{}^{a}_{}\!\!>\!B\!<\!\!{}^{b}_{b}\!\!\underset{H}{\overset{H}{}}\!\!>\!B\!<\!\!{}^{a}_{a}\underset{H}{\overset{H}{}}$$

7. N and F both have high electron density due to smaller atomic radii. Thus, $lp-bp$ repulsion increases bond-length.

8. Bond-length is inversely proportional to bond-order

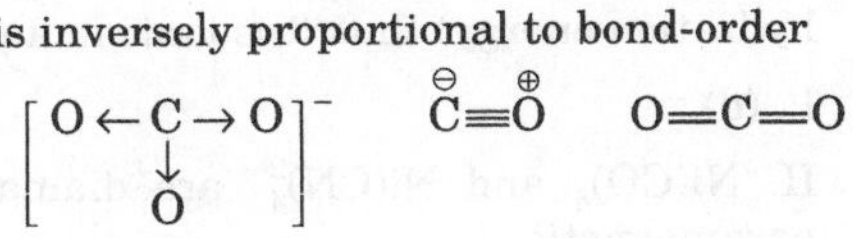

$$\left[O \leftarrow C \rightarrow O\right]^- \text{ (with } C \downarrow O\text{)} \qquad \overset{\ominus}{C}\equiv\overset{\oplus}{O} \qquad O=C=O$$

Bond-order : between 2 and 3   3   2

Thus, **(d)**

9. 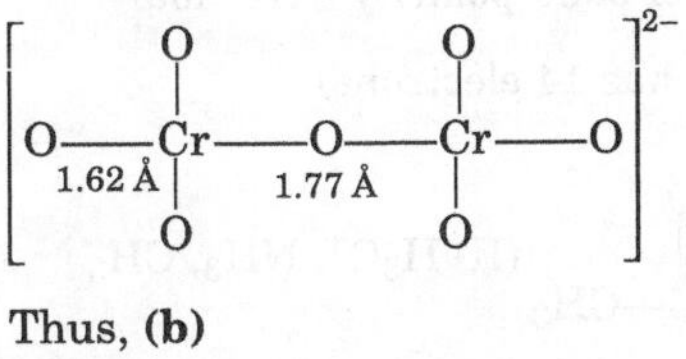 $\left[O—Cr(=O)_2—O—Cr(=O)_2—O \right]^{2-}$ (Cr—O terminal 1.62 Å; Cr—O bridging 1.77 Å)

Thus, **(b)**

10. NI_3 and SO_3^{2-} are trigonal pyramidal; I_3^- is linear and NO_3^- is trigonal planar.

11. 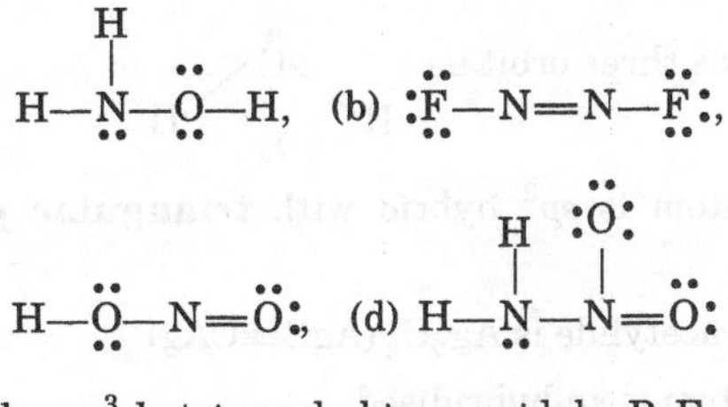 (a) H—N(H)—O—H, (b) :F—N=N—F:,

(c) H—O—N=O:, (d) H—N(H)—N=O:

12. PCl_5 : sp^3d trigonal bipyramidal, BrF_5 sp^3d^2 square based pyramid.

13. $[[:F—Cl—F:]^+ \; sp^3d]$ $\quad [[:F—Cl—F:]^+ \; sp^3]$

14. 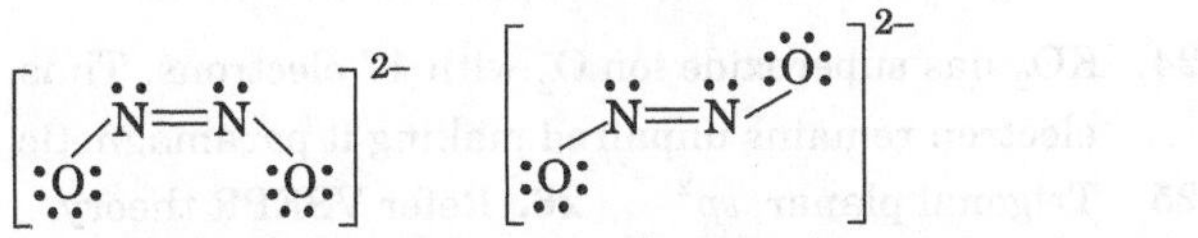 $\left[{}^{:}O{}^{:}—N=N—{}^{:}O{}^{:} \right]^{2-}$ (*cis-*) $\quad \left[{}^{:}O{}^{:}—N=N—{}^{:}O{}^{:} \right]^{2-}$ (*trans-*)

Since the π-bond between the nitrogen atoms must be broken to interconvert these two forms, hence, this ion will show *cis-trans* isomerisation.

15.

Atom	%	Atomic mass	Mol	Molar ratio
Xe	60.4	131.3	0.46	1
O	22.1	16.0	1.38	3
F	17.5	19.0	0.92	2

Thus, empirical formula = XeO_3F_2

Empirical formula weight = 217.3

Molar mass = 217.3

Molecular formula = XeO_3F_2

16. I. (a) II. one σ, two π

17. N_2 : 1.10 Å linear, sp hybrid

N_2F_2 : 1.22 Å trigonal planar, sp^2 hybrid

N_2H_2 : 1.45 Å tetrahedral, sp^3 hybrid

18. (a) Cl (b) N (c) S (d) N

19.

Atom	%	Atomic mass/ Molar masses	Mol	Molar ratio
Co	21.24	58	0.367	1
NH_3	24.77	17	1.46	4
SO_4^{2-}	34.65	96	0.36	1
H_2O	6.53	18	0.36	1
Cl	12.81	35.5	0.36	1

Empirical formula $Co(NH_3)_4SO_4 \cdot H_2OCl$

The compound gives white ppt with $BaCl_2$

Hence, SO_4 is outside coordination sphere $[Co(NH_3)_4(Cl)(H_2O)]SO_4$

20. In $N(CH_3)_3$, lone pair on N remains on it causing lp-bp repulsion while in $N(SiH_3)_3$, lone pair is transferred to vacant d-orbital of Si.

21. (a) $\overset{+}{C}H_3$ has three orbitals 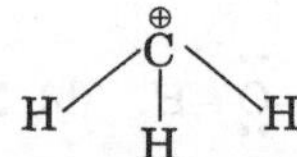

Thus, C-atom is sp^2 hybrid with **triangular planar** structure

(b) Silver acetylide is Ag_2C_2 ($AgC{\equiv}CAg$)

Thus, C-atom is *sp*-hybridised.

22. sp^3, +2, F—Ö—F (bent, with two lone pairs on O)

23. (NF_3, H_3^+O) and (NO_3^-, BF_3) pairs isostructural.

24. KO_2 has superoxide ion O_2^- with 17 electrons. Thus, one electron remains unpaired making it paramagnetic.

25. Trigonal planar, sp^2 **26.** Refer VSEPR theory.

27. Allene is $\underset{sp^2}{CH_2}{=}\underset{sp}{C}{=}\underset{sp^2}{CH_2}$ **28.** (b) **29.** (d)

30. Refer text **31.** Refer text

32. (a) and (b) both have 12 electrons.

KK^* $\sigma 2s$ [↑↓] $\sigma^* 2s$ [↑↓] $\pi 2p_y$ [↑↓] $\pi 2p_z$ [↑↓]

33. Total electrons = 2 [↑↓] $\sigma 1s^2$

bond order = 1, stable.

34. (a) Refer Table 2.10 :

Species	Bond order	Comparison of bond order	Comparison of bond distance
N_2	3	$N_2 > N_2^+$	$N_2 < N_2^+$
N_2^+	2.5		increases
O_2	2	$O_2^+ > O_2$	$O_2^+ < O_2$
O_2^+	2.5		decreases

35. Calculated value of $\mu = e \times l$

$$= 4.8 \times 10^{-10} \text{ esu} \times 1.51 \times 10^{-10} \text{ cm}$$
$$= 7.248 \times 10^{-18} \text{ esu cm}$$

$(1 \times 10^{-18}$ esu-cm = 1 Debye) = 7.248 Debye

Observed value of μ = 0.79 Debye

$$\% \text{ ionic character} = \frac{\text{observed value} \times 100}{\text{calculated value}}$$
$$= \frac{0.79}{7.248} \times 100$$
$$= 10.9\%$$

36. (a) $:\ddot{F}—\ddot{N}{=}\ddot{O}:$ (b) angular (bent)

(c) larger μ is a result of the molecule being asymmetric (bent) and having rather larger (EN) difference. NO_2F molecule has a symmetric shape.

37. $\mu_{cis} > \mu_{trans} (= 0)$

38. l = length $= 4 \times 10^{-11}$ m $= 4 \times 10^{-9}$ cm

e = charge $= 4.8 \times 10^{-10}$ esu.

$\therefore \quad \mu = e \times l = 4.8 \times 10^{-10} \times 4 \times 10^{-9}$ esu-cm

$= 1.92 \times 10^{-18}$ esu-cm = 1.92 D

Since, 1 D $= 3.3 \times 10^{-30}$ C-m

$\mu = 1.92 \times 3.33 \times 10^{-30}$ C-m

$\therefore \quad = 6.4 \times 10^{-30}$ C-m

39. C—O single bond length = 0.77 + 0.66 = 1.43 Å

C=O double bond length = 1.22 Å (given). It is very near to predicted for a double bond. Hence, structure are :

(tetrahydrofuran ring with O) and (butan-2-one, C=O)

40. (a) sp^2 (b) II

41. $\mu = e \times l$

$\mu = 2.9$ D $= 2.9 \times 10^{-10}$ esu-cm

$e = 4.8 \times 10^{-10}$ esu.

$$\therefore \quad l = \frac{\mu}{e} = \frac{2.9 \times 10^{-18} \text{ esu-cm}}{4.8 \times 10^{-10} \text{ esu}}$$
$$= 6.04 \times 10^{-9} \text{ cm} = \mathbf{0.0604 \text{ nm}}$$

42. 0.0308 nm, No since $\mu \neq 0$

43. I. $D < A < B < C$ II. (a)

44. $LiF > LiF_2$, $CsO > Cs_2O$, $CaBr_2 > CaBr_3$

45. C—H < Br—H < Na—I < F—H < Li—Cl < K—F

46. Refer text

47. Hydration energy of

$$AlCl_3 = \Delta H_{Hyd.}(Al^{3+}) + 3\Delta H_{Hyd}(Cl^-)$$
$$= -4665 - 3 \times 381$$
$$= -5808 \text{ kJ mol}^{-1}$$

IE of $AlCl_3 = 5137$ kJ mol^{-1}

Since, hydration energy > IE, hence, $AlCl_3$ ionises (by hydration energy) thus it is ionic in aqueous solution.

48. I. (d)

II. $Ni(CO)_4$ and $Ni(CN)_4^{2-}$ are diamagnetic, $NiCl_4^{2-}$ is paramagnetic.

III. sp^3-sp IV. NCO^-, CS_2, NO_2^+ V. S < Cl < N < O < F

49. I. sp^3 II. absence of bond polarity III. four

IV. CN^-, NO^+ (each has 14 electrons)

50. (c) (I) CH_3—C(—O—H···O=)C—CH_3 with CH bridge (intramolecular H-bonded enol form) (II) H_3O^+, NH_3, CH_3^-

51. There is intramolecular H-bonding in *o*-hydroxybenzaldehyde (weak) while intermolecular H-bonding in *p*-isomer (strong).

52. NO_2 has one unpaired electron, hence paramagnetic. On cooling, thee is formation of new N—N bond (in N_2O_4) using this unpaired electron; hence, paramagnetism is lost and molecular weight increases.

$$2NO_2 \rightleftharpoons N_2O_4$$

Resonance forms are only multiple representation of the same structure.

53. Refer sec. 7.15. Due to resonance in conjugate diene

54. (c)

NCERT Problems

Refer Text

Exercise 2

Only One Option Correct

1. (b)	**2.** (d)	**3.** (d)	**4.** (c)	**5.** (a)	**6.** (a)	**7.** (c)	**8.** (b)	**9.** (a)	**10.** (d)
11. (d)	**12.** (c)	**13.** (b)	**14.** (c)	**15.** (c)	**16.** (a)	**17.** (b)	**18.** (c)	**19.** (c)	**20.** (a)
21. (b)	**22.** (d)	**23.** (a)	**24.** (a)	**25.** (b)	**26.** (b)	**27.** (b)	**28.** (c)	**29.** (b)	**30.** (d)
31. (c)	**32.** (d)	**33.** (a)	**34.** (d)	**35.** (c)	**36.** (c)	**37.** (a)	**38.** (c)	**39.** (a)	**40.** (b)
41. (d)	**42.** (a)	**43.** (a)	**44.** (c)	**45.** (c)	**46.** (a)	**47.** (d)	**48.** (a)	**49.** (c)	**50.** (d)
51. (d)	**52.** (a)	**53.** (b)	**54.** (b)	**55.** (c)	**56.** (b)	**57.** (c)	**58.** (c)	**59.** (a)	**60.** (c)
61. (b)	**62.** (c)	**63.** (b)	**64.** (d)	**65.** (c)	**66.** (c)	**67.** (d)	**68.** (d)	**69.** (a)	**70.** (a)
71. (d)	**72.** (c)	**73.** (d)	**74.** (c)	**75.** (c)	**76.** (a)	**77.** (b)	**78.** (c)	**79.** (a)	**80.** (b)
81. (b)	**82.** (d)	**83.** (b)	**84.** (c)	**85.** (a)	**86.** (c)	**87.** (a)	**88.** (b)	**89.** (c)	**90.** (d)
91. (b)	**92.** (b)	**93.** (d)	**94.** (b)	**95.** (c)	**96.** (b)	**97.** (a)	**98.** (b)	**99.** (b)	**100.** (a)
101. (c)	**102.** (d)	**103.** (d)	**104.** (a)	**105.** (d)	**106.** (a)	**107.** (a)	**108.** (b)	**109.** (d)	**110.** (d)
111. (a)	**112.** (b)	**113.** (b)	**114.** (a)	**115.** (a)	**116.** (d)	**117.** (b)	**118.** (d)	**119.** (a)	**120.** (c)
121. (c)	**122.** (b)	**123.** (a)	**124.** (a)	**125.** (b)	**126.** (b)	**127.** (c)	**128.** (b)	**129.** (a)	**130.** (a)
131. (c)	**132.** (d)	**133.** (d)	**134.** (b)	**135.** (d)	**136.** (a)	**137.** (a)	**138.** (c)	**139.** (a)	**140.** (a)

One or More Than One Options Correct

1. (a, c) **2.** (b, c) **3.** (a, b) **4.** (b, c, d)
5. (a, b, d) **6.** (b) **7.** (c) **8.** (d) **9.** (a)
10. (a) **11.** (a, b, c) **12.** (b) **13.** (a, b, c)
14. (a) **15.** (a), (b) **16.** (a) **17.** (a, c, d)
18. (a, b, c, d) **19.** (c) **20.** (b, c)

Passage Comprehension Questions

Passage 1

1. (c) **2.** (c) **3.** (a)

Passage 2

1. (b) **2.** (d) **3.** (a)

Passage 3

1. (a) **2.** (a) **3.** (b)

Passage 4

1. (a) **2.** (d) **3.** (c)

Passage 5

1. (b) **2.** (d) **3.** (b) **4.** (a)

Passage 6

1. (c) **2.** (a) **3.** (d) **4.** (b)

Passage 7

1. (a) **2.** (b)

Passage 8

1. (d) **2.** (c)

Passage 9

1. (a) **2.** (c) **3.** (b) **4.** (a), (b) **5.** (a), (c)

Passage 10

1. (c)

Passage 11

1. (c) **2.** (d) **3.** (b)

Assertion & Reason

1. (a) **2.** (a) **3.** (b) **4.** (a) **5.** (b) **6.** (c)
7. (d) **8.** (a) **9.** (c) **10.** (a) **11.** (a) **12.** (b)
13. (a)

True & False

1. T **2.** F **3.** F **4.** F **5.** T **6.** F **7.** T
8. T **9.** F **10.** T **11.** T **12.** F **13.** F **14.** F
15. T **16.** F **17.** F **18.** T **19.** T **20.** T

Fill in the Blanks

1. square planar **2.** Fajan's rule **3.** 4°
4. paramagnetic **5.** H-bonding **6.** ionic **7.** sigma
8. Lewis base, Lewis acid
9. ionic (between NH_4^+ and Cl^-), covalent (between N and H), coordinate (between H^+ and NH_3).
10. paramagnetic, diamagnetic
11. 2, 1.5, 1 oxygen, superoxide ion **12.** sp^2, sp^2, sp^3
13. dsp^2, sp^3 **14.** $\sqrt{24}$ BM **15.** 1.7 **16.** NH_3, PH_3
17. 4 **18.** increases, decreases **19.** increases, decreases
20. large N≡N bond energy **21.** PCl_4^+, PCl_6^-
22. Lewis acid, Lewis base
23. electron sea model, band model
24. very large, very short
25. valence band, conduction band

Matrix-Match Type Questions

I. Only One Correct

1. (A)—(3); (B)—(1); (C)—(4); (D)—(2)
2. (A)—(3); (B)—(2); (C)—(1)
3. (A)—(3); (B)—(1); (C)—(4); (D)—(2)
4. (A)—(4); (B)—(3); (C)—(1); (D)—(2)
5. (A)—(6); (B)—(2); (C)—(3); (D)—(1)
6. (A)—(2); (B)—(3); (C)—(1); (D)—(4); (E)—(6)
7. (A)—(3); (B)—(5); (C)—(6); (D)—(1); (E)—(4);
8. (A)—(4); (B)—(2); (C)—(1); (D)—(3)
9. (A)—(3); (B)—(1); (C)—(2); (D)—(4)
10. (A)—(3); (B)—(4); (C)—(2); (D)—(1)

11.

List I	List II	List III
Be_2^+	0.5	1
O_2	2.0	2
O_2^+	2.5	1
O_2^{2-}	1.0	0

12.

List I	List II	List III
NO^+	3.0	N_2
NO	2.5	O_2^+
NO^-	2.0	O_2

II. One or More Correct

1. (A)—(3,4); (B)—(5,6); (C)—(1); (D)—(2,3); (E)—(1)
2. (A)—(1,5); (B)—(1,3); (C)—(3); (D)—(2,4); (E)—(3,5)

Integer Answer Types

Set I

Questions →	1	2	3	4	5	6	7	8
Answers →	4	4	3	3	6	1	0	3
	0	0	0	0	0	0	0	0
	1	1	1	1	1	1	1	1
	2	2	2	2	2	2	2	2
	3	3	3	3	3	3	3	3
	4	4	4	4	4	4	4	4
	5	5	5	5	5	5	5	5
	6	6	6	6	6	6	6	6
	7	7	7	7	7	7	7	7
	8	8	8	8	8	8	8	8
	9	9	9	9	9	9	9	9

Set II

Questions →	1	2	3	4	5	6	7	8	9
Answers →	5	4	3	5	4	8	7	7	0
	0	0	0	0	0	0	0	0	0
	1	1	1	1	1	1	1	1	1
	2	2	2	2	2	2	2	2	2
	3	3	3	3	3	3	3	3	3
	4	4	4	4	4	4	4	4	4
	5	5	5	5	5	5	5	5	5
	6	6	6	6	6	6	6	6	6
	7	7	7	7	7	7	7	7	7
	8	8	8	8	8	8	8	8	8
	9	9	9	9	9	9	9	9	9

> **The members of the same group of elements (arranged as every eighth in a group) stand to each other in the same relation as the extremities of one or more octaves in music.** John Newlands

3 Periodicity of Elements

Quick Points

Electrons in various subshells of a shell (orbit) are filled based on
—Aufbau rule
—Hund's rule
—Pauli's exclusion rule

- α-particle is positively charged helium nucleus
  Its emission causes decrease in atomic number by two units and atomic mass by four units

$$^{226}_{88}\text{Ra} \longrightarrow {}^{222}_{86}\text{Rn} + {}^{4}_{2}\text{He}$$

Emission of β-particle increases atomic number by one unit resulting in the formation of isobar.

$$^{200}_{79}\text{Au} \longrightarrow {}^{200}_{80}\text{Hg} + {}^{0}_{-1}e$$

This is due to conversion of neutron to proton

$$^{1}_{0}n \longrightarrow {}^{1}_{1}\text{H} + {}^{0}_{-1}e$$

- ns^2 pair in p-block elements is inert if it can't be unpaired (inert-pair effect). If a p-block element M^{x+} does not have noble gas configuration (n^2np^6 or $1s^2$ (He)) then it is expected to show inert-pair effect where x is the number of valence electrons.

Atom	Z	E.C.	x	M^{x+}	E.C.	Inert pair effect
B	5	$1s^22s^22p^1$	3	B^{3+}	$1s^2$	No
Al	13	$1s^22s^22p^63s^23p^1$	3	Al^{3+}	$1s^22s^22p^6$	No
Ga	31	[Ar] $3d^{10}4s^24p^1$	3	Ga^{3+}	[Ar] $3d^{10}$	Yes
P	15	$1s^22s^22p^63s^23p^3$	5	P^{5+}	$1s^22s^22p^6$	No
As	33	[Ar] $3d^{10}4s^24p^3$	5	As^{5+}	[Ar] $3d^{10}$	Yes

Learning Framework of the Chapter ...

- The Origin of The Periodic Table
- Classification of the Elements in Blocks
- Periodicity
- Valency and Oxidation Number
- Shielding (Screening) Effect and Effective Nuclear Charge
- Size of an Atom and Ion
- Ionisation Energy
- Electron Affinity
- Electronegativity
- Metallic Nature
- Hydration and Hydration Energy
- Acid-Base Character of Oxides
- Hydrides
- Melting Points and Boiling Points
- Density
- Atomic Volume

Ga [Ar] $3d^{10}$ (ground state) — ns^2: ↑↓; np^1: ↑ — valency +1 if ns^2 pair is inert.

Ga (first excited state) [Ar] $3d^{10}$ — ns^1: ↑; np: ↑ ↑ — valency +3 if ns^2 pair is excited, and thus three unpaired electrons take part in chemical bonding.

(For detail refer Inert-Pair Effect in Chemical-Bonding Chapter-2)

The concept of chemical periodicity is central to the study of Inorganic Chemistry. Chemical periodicity and the Periodic Table now find their natural interpretation in the detailed electronic structure of atom; indeed they played a major role at the turn of the century in elucidating the mysterious phenomenon of radioactivity and the quantum effects which led ultimately to Bohr's theory of the H-atom.

The Origin of The Periodic Table

- A brief historical survey of various development in the study of Periodic Table is given in Table 3.1. There is no single best form of the Periodic Table since the choice depends on the purpose for which the table is used. The most convenient form for various studies is "long form" with separate panels for the actinide and lanthanide elements.
- There has always been confusion over the numbering system to be used for the individual groups in the table. We adopt the 1-18 numbering scheme recommended by IUPAC in 1988. The 'long form' of the Periodic Table is given in Table 3.2. For comparison, numbering in Roman digits I, II, ... is also given.
- One advantage of the IUPAC system is that the groups headed by Fe, Co, and Ni have separate designations (8, 9 and 10, respectively) instead of all being placed together in VIII B. Also, for the transition elements

  group number = number of electrons in $(n-1)d$ and ns orbitals

  and for the p-block elements

  group number = 10 + outer-shell electrons

Table 3.1 Mendeleef's Periodic Table Precursors and Successors

1772 L.B.G. de Morveau made the first table of "chemically simple" substances. A. L. Lavoisier used this in his *traité Elémentaire de Chimie* published in 1789.

de Morveau's Table of Chemically Simple Substances

de Morveau's table of chemically simple substances (updated with modern representations by Mazurs) :

Non Decomposable Substances				
Simple substances	Simple combustible substances	Metallic substances	Simple earths	Alkalies
	N	As Fe	SiO_2	NaOH
	C	Mo Sn	Al_2O_3	KOH
O	S	W Pb	BaO	NH_4OH
H	P	Mn Cu	CaO	
	Cl	Ni Hg	MgO	
	B	Co Ag		
	F	Bi Pt		
	etc.	Sb Au		
		Zn		

1817-29 J.W. Döbereiner discovered many triads of elements and compounds, the combining weight of the central component being the average of its partners (*e.g.* CaO, SrO, BaO, and NiO, CuO, ZnO).

1850 Johann Dobereiner's Triads (1780-1849)

Triads are found with sequence of three similar elements, where the middle element has a mass equal to the average of the least and most massive. The diagram below, updated from **here,** uses mid-nineteenth century atomic mass information rather than modern data. If atomic number (Z) are used (a property unknown in 1850), the triads are exact :

Atomic mass (1850)		Atomic number	
Li 7, Na 23, K 39	$\longrightarrow \frac{7+39}{2} = 23$	Li 3, Na 11, K 19	$\longrightarrow \frac{3+19}{2} = 11$
Ca 40, Sr 87, Ba 137	$\longrightarrow \frac{40+137}{2} = 88.5$	Ca 20, Sr 38, Ba 56	$\longrightarrow \frac{20+56}{2} = 38$
P 31, As 75, Sb 122	$\longrightarrow \frac{31+122}{2} = 76.5$	P 15, As 33, Sb 51	$\longrightarrow \frac{15+51}{2} = 33$
S 32, Se 78, Te 128	$\longrightarrow \frac{32+128}{2} = 80$	S 16, Se 34, Te 52	$\longrightarrow \frac{16+52}{2} = 34$
Cl 35.5, Br 80, I 127	$\longrightarrow \frac{35.5+127}{2} = 81.25$	Cl 17, Br 35, I 53	$\longrightarrow \frac{17+53}{2} = 35$

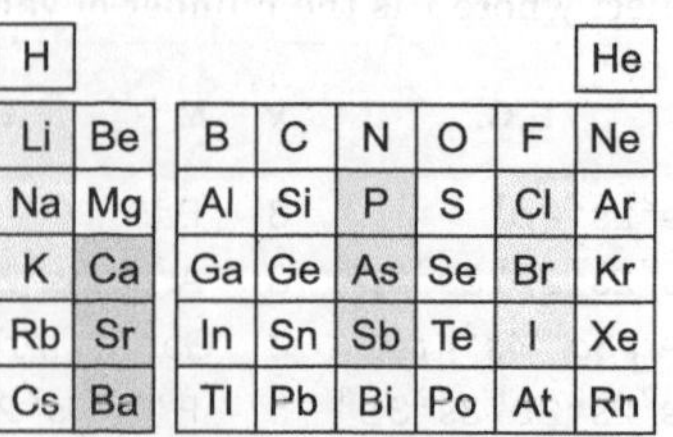

1843 L. Gmelin included a V-shaped arrangement of 16 triads in the 4th edition of his *Handbuch der Chemie.*

O N H
F Cl Br I Li Na K
S Se Te Mg Ca Sr Ba
P As Sb Be Ce La
C B Bi Zr Th Al
Ti Ta W Sn Cd Zn
Mo V Cr U Mn Ni Fe
Bi Pb Ag Hg Cu
Os Ir Rh Pt Pd Au

1857 J. B. Dumas published a rudimentary table of 32 elements in 8 columns indicating their relationships.

1862 A.E.B. de Chancourtois first arranged the elements in order of increasing atomic weight; he located similar elements in this way and published a helical form in 1863.

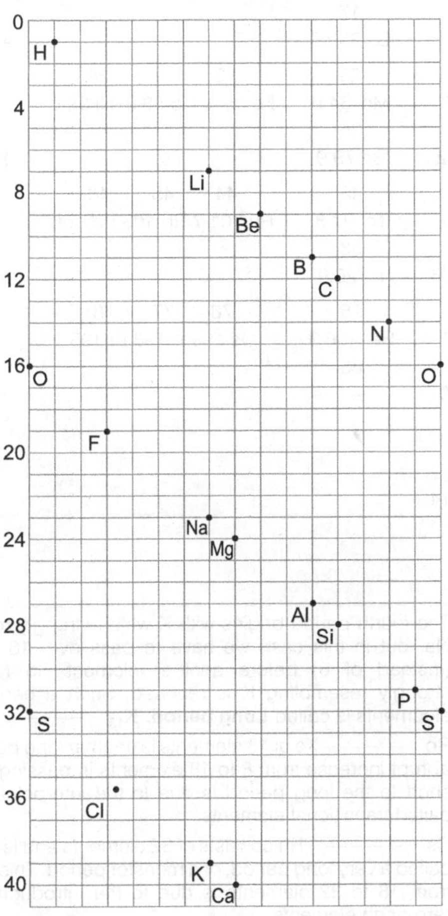

1864 L. Meyer published a table of valencies for 49 elements.

1864 W. Odling drew up an almost correct table with 17 vertical columns and including 57 elements.

1865 J.A.R. Newlands propounded his law of octaves after several partial classifications during the preceding 2y; he also correctly predicted the atomic weight of the undiscovered element germanium.

1868-9 L. Meyer drew up an atomic volume curve and a Periodic Table, but this latter was not published until 1895.

1869 D.I. Mendeleef enunciated his periodic law that "the properties of the elements are a periodic function of their atomic weights". He published several forms of Periodic Table, one containing 63 elements.

1871 D. I. Mendeleef modified and improved his tables and predicted the discovery of 10 elements (now known as Sc, Ga, Ge, Tc, Re, Po, Fr, Ra, Ac and Pa). He fully described with amazing prescience the properties of 4 of these (Sc, Ga, Ge, Po). Note, however, that it was not possible to predict the existence of the noble gases or the number of lanthanide elements.

1894-8 Lord Rayleigh, W. Ramsay and M. W. Travers detected and then isolated the noble gases (He, Ne, Ar, Kr, Xe).

1913 N. Bohr explained the form of the Periodic Table on the basis of his theory of atomic structure and showed that there could be only 14 lanthanide elements.

1913 H.G.J. Moseley observed regularities in the characteristic X-ray spectra of the elements; he thereby discovered atomic numbers Z and provided justification for the ordinal sequence of the elements.

1940 E. McMillan and P. Abelson synthesized the first transuranium element $_{93}Np$. Others were synthesized by G.T. Seaborg and his colleagues during the next 15 years.

1944 G. T. Seaborg proposed the actinide hypothesis and predicted 14 elements (up to $Z = 103$) in this group.

The Mayan Periodic Table

The Mayan Periodic Table of Elements, named for its similarity to the ancient Mesoamerican calendar, is based on electron shells.

Main Features of Mendeleef's Periodic Classification

Study and explanation of the table

In the table, the elements are arranged in order of their increasing atomic weights in horizontal series, in such a way that elements having similar properties are placed in the same vertical column. The horizontal series are generally written as periods, and the nine vertical columns-headed by the Roman numerals from I to VIII and zero (0) are called groups. The periods are short or long.

Periodic Table

Table 3.1 **With Atomic numbers written above and Atomic weights written by the side of the Symbols**

Periods	Group I	Group II	Group III	Group IV	Group V	Group VI	Group VII	Group VIII	Group 0
	R_2O RH	R_2O_2 RH_2	R_2O_3 RH_3	R_2O_4 RH_4	R_2O_5 RH_3	R_2O_6 RH_2	R_2O_7 RH	R_2O_8 ← formula of oxides ← formula of hydrides.	or Gr. VIII B
	A B	A B	A B	A B	A B	A B	A B	A	
1							1 H 1.008		2 He 4
2	3 Li 6.94	4 Be 9.02	5 B 10.82	6 C 12	7 N 14.01	8 O 16	9 F 19		10 Ne 20.2
3	11 Na 23	12 Mg 24.3	13 Al 26.97	14 Si 28.06	15 P 31.04	16 S 32	17 Cl 35.46		18 Ar 39.94
4	19 K 39.1	20 Ca 40	21 Sc 45.1	22 Ti 48.1	23 V 51	24 Cr 52	25 Mn 54.9	26 27 28 Fe 55.8 Co 58.9 Ni 58.6	
	29 Cu 63.5	30 Zn 65.3	31 Ga 70	32 Ge 72.5	33 As 74.9	34 Sc 79.2	35 Br 79.9		36 Kr 83.7
5	37 Rb 85.4	38 Sr 87.63	39 Y 88.9	40 Zr 91.2	41 Nb 92.9	42 Mo 96	43 Tc 97.8	44 45 46 Ru 101.7 Rh 103 Pd 106	
	47 Ag 107.8	48 Cd 112	49 In 114.8	50 Sn 119	51 Sb 121.76	52 Te 127.6	53 I 126.9		54 Xe 131
6	55 Cs 132.9	56 Ba 137.37	57 La 139 58-71 Rare Earths	72 Hf 178.6	73 Ta 180.8	74 W 184	75 Re 186.3	76 77 78 Os 191 Ir 198 Pt 195	
	79 Au 197.2	80 Hg 200.6	81 Tl 204	82 Pb 207.2	83 Bi 209	84 Po 210	85 —		86 Rn 222
7	87 —	88 Ra 226	89 Ac 227	90 Th 232.12	91 Pa 231	92 U 238.14			

Short and Long Periods

Period	Elements
1	H He (Short period)
2	Li (Electropositive) Be B C N O F (Electronegative) Ne (Inert) There is a regular change from the strongly electropositive element Li to the strongly electronegative element F, which is followed by the inert element Ne.
3	Na Mg Al Si P S Cl A There is same periodicity as in second period. Na resembles Li; Mg resembles Be; Al resembles B and so on. Such periods of eight elements are called short periods.
4	The fourth period begins with K which strongly resembles Na; but in this case we have to pass over 18 elements (instead of 8) before another element Rb (rubidium) strongly resembling K is reached. Such a period of 18 elecments is called **Long period.** K_{19} ——— Kr (36)
5	Rb ——— Xe of 18 elements is another long period. The abrupt increase from 8 to 18 elements in passing from the short to the long period is due to the interpolation of so called transitional elements.
6	Cs ——— Rn consists of 32 elements and is therefore called a very long period, or a monster period. This increase from 18 to 32 elements is due to the introduction of 14 rare-earth elements.
7	It is fragmentary. It starts like a long period, but soon comes to an end. Uranium (at fifth place) is the element of highest atomic weight known at that time.

Intervals of Perodicity

Thus, we observe that intervals of periodicity are $2n^2$ (n = period), *i.e.*, 2, 8, 18, 18, 32

Even series The first seven elements of a long period form the **even series.**

Odd series The last seven elements of a long period form the **odd series.**

Group Zero The inert gases like helium, argon, etc. present in the air were discovered afterwards. They were placed in zero group. They being inert and havig no valency, stands between the highly electropositive alkali metals and the strongly electornegative halogens.

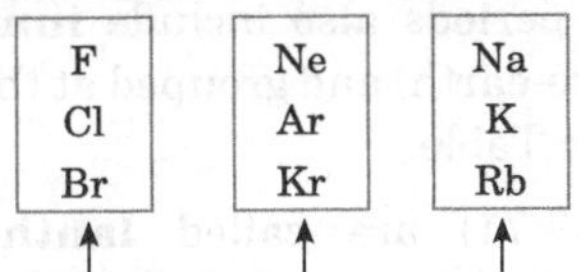

Partington considered group zero as sub group B of group VIII.

Subgroups Each group is divided into subgroup A and B. The elements belonging to the even series of their respective long periods are placed to the left (sub group A), while those belonging to the odd series are arranged on the right-hand side (subgroup B) of each vertical column. The members belonging to a subgroup are more closely related than the other members of the same group.

The Rare Earths Lanthanum in Group III and in the 6th period forms a cluster with 14 other elements (atomic number 58-71) which have got atomic weights differing by one or two or four; they are analogous in chemical properties, and it is very difficult to separate them in analysis. They are known as the **elements** of the **rare earths.**

The Radioactive Elements There is also a class of elements, placed in last period of the Periodic Table, which differ from the rest in as much as they are *radioactive, i.e.,* they constantly liberate energy, whch appears to be derived from the disintegration of atoms.

Periodicity in the properties of the elements, when arranged in the Periodic Table is shown in terms of.

1. Valency The valencies towards hydrogen regularly increase up to the middle of a short period and then gradually fall and come back to unity. The valencies towards oxygen, on the other hand increase regularly from 1 to 7, and then begin with 1. Elements of group zero have zero valency.

LiH	BeH_2	BH_3	CH_4	NH_3	OH_2	FH
1	2	3	4	3	2	1
Na_2O	MgO	Al_2O_3	SiO_2	N_2O_5	SO_3	Cl_2O_7
1	2	3	4	5	6	7

2. Electrochemical Character In group I, elements are highly electropositive. In between there are non-metals (less electropositive) followed by electronegative elements as halogens

Li	Be	B	C	N	O	F
Na	Mg	Al	Si	P	S	Cl

3. Other Properties The same periodicity in the properties can be observed with regard to the

(a) melting point
(b) boiling point
(c) specific gravity
(d) atomic volume

In atomic heat periodicity is not observed.

Applications of the Periodic Law

1. In classification of the elements

(a) We can form an idea of the properties of an element and its compounds by knowing the group it belongs to

(b) We can compare the properties of different elements and their compounds belonging to different groups.

2. In the prediction of new elements When Mendeleef first prepared the Periodic Table, many blank spaces were automatically left in it. Space of the elements (not known at that time) Sc, Ga, Ge were left blank. However he predicted their properties by naming

eka-boron (below boron)
eka-aluminium (below aluminium)
eka-silicon (below silicon)

	B	Al	Si
Named by Mendeleef]	Eka-B,	Eka-Al	Eka-Si
Named afterwards]		Ga	Ge

The following table summarises the predicted (15 years before discovery by Mendeleef in 1871) and observed (discovered by Winkler in 1886)

Property	Ekosilicon (1871)	Germanium (discovered in 1886 by Winkler)
At. wt.	72	72.3
Sp. wt.	5.5	5.469
At. vol.	13	13.2
Oxide	EsO_2	GeO_2
Sp. wt. oxide	4.7	4.703
Chloride	$EsCl_4$	$GeCl_4$
Boiling point chloride	< 100°	86°
Density chloride	1.9	1.887
Fluoride	EsF_4	$GeF_4 \cdot 3H_2O$
Not gaseous		white solid mass
Ethyl compound	$EsEt_4$	$Ge(C_2H_5O)_4$
Boiling point ethyl compound	160°	160°
Sp. wt. ethyl compound	0.96	a little < 1

*Eka a prefix being the sanskrit numeral one

3. In correcting doubtful atomic weight An element has got only one proper place in the Periodic Table. Position is fixing by comparing its properties with those of neighbouring elements.

Elements 113 and 115 Discovered

Two super heavy elements, element 113 and element 115, have been discovered by scientists. The team observed atomic decay patterns or chains, that confirm the existence of these two elements.

In these decay chains, element 113 is produced *via* the alpha decay of element 115. The observations were made in experiments conducted at the JINR (Joint Institute for Nuclear Research in Russia) U400 cyclotron with the Dubna gas-filled separator between July 14 and August 10, 2003. Scientists from the Glenn T. Seaborg Institute and the Chemical Biology and Nuclear Science Division at Lawrence Livermore National Lab, worked in collaboration with researchers from the JINR. The results have been published in *Physical Review C.*

Modern Periodic Law

The properties of elements are periodic functions of their atomic numbers. If element are arranged in increasing order of their atomic numbers, there is repetition of properties after 2, 8, 8, 18, 18 and 32 elements. The numbers 2, 8, 18 and 32 are also referred as **magic numbers.**

Repetition of properties after these numbers is due to repetition of electronic configurations of the outermost orbit.

H_1	$1s^1$	H to Fr have been placed together [in group 1 (IA)] due to the fact that outermost orbit has ns^1 configuration.
Li_3	$1s^2, 2s^1$	
Na_{11}	$1s^2, 2s^2 2p^6, 3s^1$	
K_{19}	$1s^2, 2s^2 2p^6, 3s^2 3p^6, 4s^1$	
Rb_{37}	[Kr] $5s^1$	
Cs_{55}	[Xe] $6s^1$	
Fr_{87}	[Rn] $7s^1$	

Long form (Modern form) of Periodic Table

It is annexed in Table 3.2. The elements have been arranged in the increasing order of **atomic number.**

- It can explain why certain elements resemble one another and also why they differ from other elements in their properties. The arrangement of elements in this table is also in keeping with their electronic structures.
- This table also explains the cause of periodicity of properties (discussed in following sections) and the reason why similar properties recur at certain regular intervals.
- The horizontal rows constitute **periods** while the vertical rows constitute **groups** or **families.**
- 4th, 5th, 6th and 7th periods also include **transition** elements.
- 6th and 7th periods also include **inner transition** elements (rare-earth) and grouped at the bottom of the main Periodic Table.
- Elements (58-71) are called **lanthanides** while elements (90-103) are called **actinides.** Each is a group of 14 elements and grouped in IIIB. (3rd period)

Explanation for Aufbau Principle

The arrangement of the long Periodic Table may be explained from the distribution of electors in various quantum levels. Each period starts with an element having one valence electron in the *s*-orbital as given below :

Element	Valence electron
H	$1s^1$
Li	$2s^1$
Na	$3s^1$
K	$4s^1$
Rb	$5s^1$
Cs	$6s^1$
Fr	$7s^1$

Electrons in various subshells have been filled based on Aufbau principle *i.e.,*

$$1s < 2s < 2p < 3s < 3p < 4s < 3d < 4p < 5s < 4d < 5p < 6s < 4f < 6d < 6p < 7s < 5f \ldots$$

Thus, 3*d* orbital which is filled after 4*s* is a part of fourth period. Thus one can determine number of elements in a given period assuming two elements in *s*, six in -*p*, ten in -*d* and fourteen in *f*-suborbits.

Period	Suborbit	Elements	Total elements	Comment
1	1*s*	2	2	
2	2*s*	2	8	
	2*p*	6		
3	3*s*	2	8	
	3*p*	6		
4	4*s*	2	18	3*d* is in 4th period
	3*d*	10		
	4*p*	6		
5	5*s*	2	18	4*d* is in 5th period
	4*d*	10		
	5*p*	6		
6	6*s*	2	32	4*f* and 5*d* are in 6th period
	4*f*	14		
	5*d*	10		
	6*p*	6		

The sixth period is further complicated by the filling of 4*f*-orbitals after 4*d*-orbitals. As 4*d*-suborbit may accommodate maximum of 14 electrons, we have the first series of 14 inner-transition elements, each related to the entry of one electron in the 4*f*-orbital. If these are placed in the Periodic Table, it would require 14 extra group; this would make the table unnecessarily long. Hence, these elements have been placed separately at the bottom of the Periodic Table. Similarly elements with 5*f*-orbitals form a separate block.

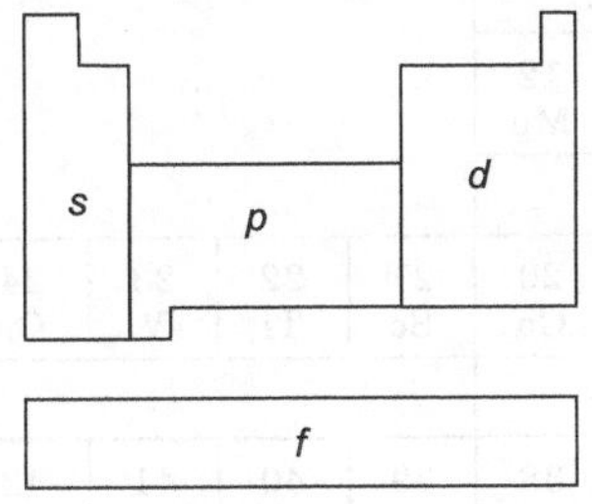

- The eight families belonging to the *s*- and *p*-blocks are often called the representative elements; those to the *d*-block are called the transition elements and members of the *f*-block are termed the inner-transition elements.

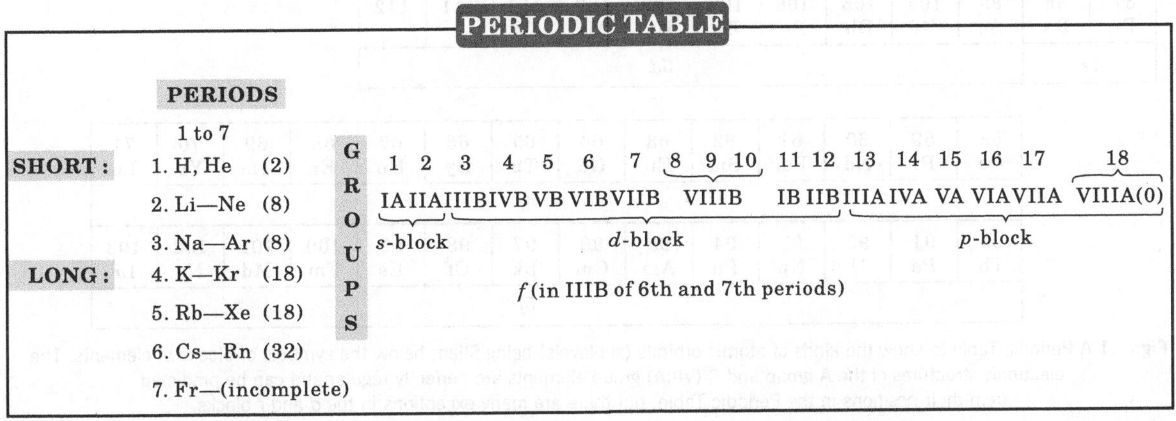

Group	IA	IIA	IIIB	IVB	VB	VIB	VIIB	VIII			IB	IIB	IIIA	IVA	VA	VIA	VIIA	0
	(1)	(2)	(3)	(4)	(5)	(6)	(7)	(8)	(9)	(10)	(11)	(12)	(13)	(14)	(15)	(16)	(17)	(18)
Period	1 H																1 H	2 He
1	*1s*																*1s*	
	3 Li	4 Be											5 B	6 C	7 N	8 O	9 F	10 Ne
2	*2s*												*2p*					
	11 Na	12 Mg											13 Al	14 Si	15 P	16 S	17 Cl	18 Ar
3	*3s*												*3p*					
	19 K	20 Ca	21 Sc	22 Ti	23 V	24 Cr	25 Mn	26 Fe	27 Co	28 Ni	29 Cu	30 Zn	31 Ga	32 Ge	33 As	34 Se	35 Br	36 Kr
4	*4s*		*3d*										*4p*					
	37 Rb	38 Sr	39 Y	40 Zr	41 Nb	42 Mo	43 Tc	44 Ru	45 Rh	46 Pd	47 Ag	48 Cd	49 In	50 Sn	51 Sb	52 Te	53 I	54 Xe
5	*5s*		*4d*										*5p*					
	55 Cs	56 Ba	57 La	72 Hf	73 Ta	74 W	75 Re	76 Os	77 Os	78 Pt	79 Au	80 Hg	81 Tl	82 Pb	83 Bi	84 Po	85 At	86 Rn
6	*6s*		*5d*										*6p*					
	87 Fr	88 Ra	89 Ac	104 Rf	105 Db	106 Sg	107 Bh	108 Hs	109 Mt	110 Ds	111	112						
7	*7s*		*6d*															

Period														
	58 Ce	59 Pr	60 Nd	61 Pm	62 Sm	63 Eu	64 Gd	65 Tb	66 Dy	67 Ho	68 Er	69 Tm	70 Yb	71 Lu
6	*4f*													
	90 Th	91 Pa	92 U	93 Np	94 Pu	95 Am	96 Cm	97 Bk	98 Cf	99 Es	100 Fm	101 Md	102 No	103 Lr
7	*5f*													

Fig. 3.1 A Periodic Table to show the kinds of atomic orbitals (sublevels) being filled, below the symbols of blocks of elements. The electronic structures of the A group and O (VIIIA) group elements are perfectly regular and can be predicted from their positions in the Periodic Table, but there are many exceptions in the *d* and *f* blocks.

Table 3.2 **Long Form of Periodic Table**

Main-Group Elements (groups 1–2) · *Transition Metals* (groups 3–12) · *Main-Group Elements* (groups 13–18)

Key: 1 ← Atomic number; H ← Symbol; 1.00794 ← Atomic weight

PERIODS	1 IA	2 IIA	3 IIIB	4 IVB	5 VB	6 VIB	7 VIIB	8	9 VIIIB	10	11 IB	12 IIB	13 IIIA	14 IVA	15 VA	16 VIA	18 VIIA	18 VIII A
1	1 H 1.00794																	2 He 4.002602
2	3 Li 6.941	4 Be 9.012182											5 B 10.811	6 C 12.011	7 N 14.00674	8 O 15.9994	9 F 18.9984032	10 Ne 20.1797
3	11 Na 22.989768	12 Mg 24.3050											13 Al 26.981539	14 Si 28.0855	15 P 30.973762	16 S 32.066	17 Cl 35.4527	18 Ar 39.948
4	19 K 39.0983	20 Ca 40.078	21 Sc 44.955910	22 Ti 47.88	23 V 50.9415	24 Cr 51.9961	25 Mn 54.93805	26 Fe 55.847	27 Co 58.93320	28 Ni 58.6934	29 Cu 63.546	30 Zn 65.39	31 Ga 69.723	32 Ge 72.61	33 As 74.92159	34 Se 78.96	35 Br 79.904	36 Kr 83.80
5	37 Rb 85.4678	38 Sr 87.62	39 Y 88.90585	40 Zr 91.224	41 Nb 92.90638	42 Mo 95.94	43 Tc (98)	44 Ru 101.07	45 Rh 102.9055	46 Pd 106.42	47 Ag 107.8682	48 Cd 112.411	49 In 114.82	50 Sn 118.710	51 Sb 121.757	52 Te 127.60	53 I 126.90447	54 Xe 132.29
6	55 Cs 132.90543	56 Ba 137.327	57 La* 138.9055	72 Hf 178.49	73 Ta 180.9479	74 W 183.85	75 Re 186.207	76 Os 190.2	77 Ir 192.22	78 Pt 195.08	79 Au 196.96654	80 Hg 200.59	81 Tl 204.3833	82 Pb 207.2	83 Bi 207.98037	84 Po (209)	85 At (210)	86 Rn (222)
7	87 Fr (223)	88 Ra (226)	89 Ac** (227)	104 Rf (261.11)	105 Db (262.114)	106 Sg (263.118)	107 Bh (262.12)	108 Hs (265)	109 Mt (266)	110 Ds (269)				114 Uuq				

Inner-Transition Metals

*Lanthanides	58 Ce 140.115	59 Pr 140.90765	60 Nd 144.24	61 Pm (145)	62 Sm 150.36	63 Eu 151.965	64 Gd 157.25	65 Tb 158.92534	66 Dy 162.50	67 Ho 164.93032	68 Er 167.26	69 Tm 168.93421	70 Yb 173.04	71 Lu 174.967
**Actinides	90 Th 232.0381	91 Pa (231)	92 U 238.0289	93 Np (237)	94 Pu (244)	95 Am (243)	96 Cm (247)	97 Bk (247)	98 Cf (251)	99 Es (252)	100 Fm (257)	101 Md (258)	102 No (259)	103 Lr (262)

Classification of The Elements in Blocks

In Periodic Table, the elements have been arranged in different **periods** and **groups** based on electronic configuration. Classification of the elements into following **four blocks** is decided by the differentiating electron (last filling electron). Block is that in which orbital differentiating electron goes (Aufbau rule)

(a) *s*-block element (b) *p*-block element

(c) *d*-block element (d) *f*-block element

Element	Atomic no.	E.C.	Last electron goes into	Block
Na	11	[Ne] $3s^1$	*s*	*s*-block
Al	13	[Ne] $3s^2 3p^1$	*p*	*p*-block
Sc	21	[Ar] $4s^2 3d^1$	*d*	*d*-block
Ce	58	[Xe] $6s^2 4f^2$	*f*	*f*-block

Blocks have been represented in Fig. 3.1 and Table 3.3.

s-Block Elements

If shells up to $(n-1)$ are completely filled and the last electron enters the *s*-orbital of the outermost (*n*th) shell, the elements of this class are called **s-block** elements.

- Thus differentiating electron enters the *ns*-orbital.
- General electronic configuration is

$$[\text{inert gas}]\ ns^x \qquad (x = 1, 2)$$

- If $x = 1$, elements constitute a group of **alkali metals** placed in **first group** (IA).
- If $x = 2$, elements constitute a group of **alkaline earth metals** placed in **second group** (IIA).
- Alkali metals form **monovalent** cations while alkaline earth metals form **divalent** cations, each with nearest stable configuration of inert gas.
- Period of the element is the value of x (number of shells).

Group IA (First) alkali metals (*A*)	Electronic configuration of			Group IIA (Second) alkaline earth metals (*B*)
	Metal *A*	A^+ or B^{2+}	Metal *B*	
Li_3	[He] $2s^1$	[He]	[He] $2s^2$	Be_4
Na_{11}	[Ne] $3s^1$	[Ne]	[Ne] $3s^2$	Mg_{12}
K_{19}	[Ar] $4s^1$	[Ar]	[Ar] $4s^2$	Ca_{20}
Rb_{37}	[Kr] $5s^1$	[Kr]	[Kr] $5s^2$	Sr_{38}
Cs_{55}	[Xe] $6s^1$	[Xe]	[Xe] $6s^2$	Ba_{56}
Fr_{87}	[Rn] $7s^1$	[Rn]	[Rn] $7s^2$	Ra_{88}

Table 3.3 **The *s*, *p*, *d* and *f*-Block of the Periodic Table**

	IA (1)	IIA (2)	IIIB (3)		IVB (4)	VB (5)	VIB (6)	VIIB (7)	VIII (8)	VIII (9)	VIII (10)	IB (11)	IIB (12)	IIIA (13)	IVA (14)	VA (15)	VIA (16)	VIIA (17)	0 (18)
	s-orbital block			*f*-orbital block	*d*-orbital block									*p*-orbital block					
	s^1	s^2	d^1s^2		d^2s^2	d^3s^2	d^5s^1	d^5s^2	d^6s^2	d^7s^2	d^8s^2	$d^{10}s^1$	$d^{10}s^2$	s^2p^1	s^2p^2	s^2p^3	s^2p^4	s^2p^5	s^2p^6
$\overset{*}{n}=1$	1 H																		s^2 2 He
$n=2$	3 Li	4 Be												5 B	6 C	7 N	8 O	9 F	10 Ne
$n=3$	11 Na	12 Mg												13 Al	14 Si	15 P	16 S	17 Cl	18 Ar
$n=4$	19 K	20 Ca	21 Sc		22 Ti	23 V	24 Cr	25 Mn	26 Fe	27 Co	28 Ni	29 Cu	30 Zn	31 Ga	32 Ge	33 As	34 Se	35 Br	36 Kr
$n=5$	37 Rb	38 Sr	39 Y		40 Zr	41 Nb d^4s^1	42 Mo	43 Tc	44 Ru d^7s^1	45 Rh d^8s^1	46 Pd	47 Ag	48 Cd	49 In	50 Sn	51 Sb	52 Te	53 I	54 Xe
$n=6$	55 Cs	56 Ba	57 La	58 Ce → 71 Lu	72 Hf	73 Ta	74 W d^4s^2	75 Re	76 Os	77 Ir	78 Pt d^9s^1	79 Au	80 Hg	81 Tl	82 Pb	83 Bi	84 Po	85 At	86 Rn
$n=7$	87 Fr	88 Ra	89 Ac	90 Th → 103 Lr	104 Rf	105 Db	106 Sg	107 Bh	108 Hs	109 Mt	110 Ds	111	112						

$n=6$	Lanthanide Series	58 Ce	59 Pr	60 Nd	61 Pm	62 Sm	63 Eu	64 Gd	65 Tb	66 Dy	67 Ho	68 Er	69 Tm	70 Yb	71 Lu	4*f* subshell being filled
$n=7$	Actinide Series	90 Th	91 Pa	92 U	93 Np	94 Pu	95 Am	96 Cm	97 Bk	98 Cf	99 Es	100 Fm	101 Md	102 No	103 Lr	5*f* subshell being filled

**n* is the principal quantum number. The $d^1s^2, d^2s^2, \ldots$ designations represent known configurations. They refer to $(n-1)d$ and ns orbitals.

p-Block Elements

If shells up to $(n-1)$ are completely filled and differentiating electron enters the *p*-orbital of the *n*th orbit, elements of this class are called ***p*-block** elements.

- General E.C. is
  $ns^2\, np^x$ $\quad (x = 1 \text{ to } 6)$
- Group is $(2 + x)$ A based on old system IIIA, IVA, ... and $(12 + x)$ based on new IUPAC system.
- The *s*-block and *p*-block elements together are called **representative elements or main group elements.**

Period	Elements					
2	B	C	N	O	F	Ne
3	Al	Si	P	S	Cl	Ar
4	Ga	Ge	As	Se	Br	Kr
5	In	Sn	Sb	Te	I	Xe
6	Tl	Pb	Bi	Po	At	Rn
E.C.	ns^2np^1	ns^2np^2	ns^2np^3	ns^2np^4	ns^2np^5	ns^2np^6
x	1	2	3	4	5	6
old	IIIA	IVA	VA	VIA	VIIA	VIIIA (0)
new	13	14	15	16	17	18

Illustration 1 *Group the following electron configurations in pairs that would represent similar chemical properties of their atoms :*

(*a*) $1s^2\ 2s^2\ 2p^6\ 3s^2$
(*b*) $1s^2\ 2s^2\ 2p^3$
(*c*) $1s^2 2s^2 2p^6 3s^2 3p^6\ 4s^2 3d^{10}\ 4p^6$
(*d*) $1s^2\ 2s^2$
(*e*) $1s^2\ 2s^2\ 2p^6$
(*f*) $1s^2\ 2s^2\ 2p^6\ 3s^2\ 3p^3$

Solution If there is repetition of electron configuration (E.C.) of the valence shell, then chemical properties are also repeated. Thus pairs with similar chemical properties are :

		Valence shell EC
I :	(a) and (d)	[inert gas] ns^2
II :	(b) and (f)	$ns^2\ np^3$-nitrogen family of *p*-block
III :	(c) and (e)	$ns^2\ np^6$-inert gas

d-Block Elements

If outermost (*n*th) and penultimate $[(n-1)\text{th}]$ are incompletely filled and differentiating electron enters the

$(n-1)d$ orbitals (*i.e.*, *d*-orbital of penultimate shell) then elements of this class are called ***d*-block** elements.

- *d*-block elements are placed between *s*-block and *p*-block elements and thus also called transition elements.
- EC is $(n-1)d^x ns^y$

where x = number of electrons in $(n-1)d$ orbital
x = 1 to 10 (except 4 and 9)
y = number of electrons in *ns*-orbital.
y = 2 in all cases except in groups 6 (VIB) and 11 (IB) y = 1

- IUPAC (new pattern) group = $(x+y)$ but old group is decided by
  - group is $(x+y)$B if $x < 6$
  - group is VIIIB if x = 6, 7, 8
  - group is yB if x = 10

Element	Z (atomic number)	EC	x	y	Group	
					New	Old
Sc	21	[Ar] $3d^1 4s^2$	1	2	3	III B
Ti	22	[Ar] $3d^2 4s^2$	2	2	4	IV B
V	23	[Ar] $3d^3 4s^2$	3	2	5	V B
Cr	24	[Ar] $3d^5 4s^1$	5	1	6	VI B
Mn	25	[Ar] $3d^5 4s^2$	5	2	7	VII B
Fe	26	[Ar] $3d^6 4s^2$	6	2	8	VIII B
Co	27	[Ar] $3d^7 4s^2$	7	2	9	VIII B
Ni	28	[Ar] $3d^8 4s^2$	8	2	10	VIII B
Cu	29	[Ar] $3d^{10} 4s^1$	10	1	11	I B
Zn	30	[Ar] $3d^{10} 4s^2$	10	2	12	II B

- *d*-block elements are further classified into following three transition series on the basis of which $(n-1)d$ is being filled:

Series	Elements	$(n-1)$ *d* being filled
3*d*	Sc_{21}—Zn_{30}	3*d*
4*d*	Y_{39}—Cd_{48}	4*d*
5*d*	La_{57}, Hf_{72}—Hg_{80}	5*d*

Illustration 2 *Specify the group of the Periodic Table in which each of the following elements is found?*
(*a*) [Ne]$3s^1$ (*b*) [Ne]$3s^2\,3p^3$ (*c*) [Ne]$3s^2\,3p^6$
(*d*) [Al]$4s^2\,3d^8$

Solution

EC	Group	Element
(a) [Ne]$3s^1$	IA (1)	Na
(b) [Ne]$3s^2\,3p^3$	VA (15)	P
(c) [Ne]$3s^2\,3p^6$	VIIIA (18)	Ar
(d) [Ne] $4s^2\,3d^8$	VIIIB (10)	Ni

Illustration 3 *A M^{2+} ion derived from a metal in the first transition metal series has four electrons in 3d subshell. What element might M be?*

Solution EC of M^{2+} : $[Ar]_{18}\,4s^0\,3d^4$
$\therefore$ EC of M : $[Ar]4s^1\,3d^5$ (and not $4s^2\,3d^4$)
Total electrons = 24
Hence, *M* is **Cr.**

f-Block Elements

If n, $(n-1)$ and $(n-2)$ are incompletely filled and last electron enters into *f*-orbital of ante penultimate *i.e.*, $(n-2)$ th shell, elements of this class are called ***f*-block** elements.

- Thus differentiating electron enters $(n-2)f$ orbital.
- General E.C. is $(n-2)f^{2-14}(n-1)d^{0,1}ns^2$
- If n =6, elements belong to **lanthanides** with E.C. $4f^{2-14}5d^{0,1}6s^2$ of the elements with atomic number Ce (58) to Lu (71)

Ce_{58} : $[Xe]_{54}\,4f^2 5d^0 6s^2$ Gd_{64} : $[Xe]_{54}\,4f^7 5d^1 6s^2$
Lu_{71} : $[Xe]_{54}\,4f^{14} 5d^1 6s^2$

- If n =7, elements belong to **actinides** with E.C. $5f^{2-14}6d^{0,1}7s^2$ of the elements with atomic number Th (90) to Lr (103).

Th_{90} : $[Rn]_{86}\,5f^2 6d^0 7s^2$
Cm_{96} : $[Rn]_{86}\,5f^7 6d^1 7s^2$
Lr_{103} : $[Rn]_{86}\,5f^{14} 6d^1 7s^2$

- All the elements of *f*-block (4*f* series—Lanthanides and 5*f* series—Actinides) are placed in 3rd group (IIIB) and form a separate block at the bottom of the Periodic Table.
- They are also called **inner-transition** elements.

DEEP Focus

- They were also called **rare-earth elements** since their oxides (earth) were rare in earlier days.

Nomenclature of the Elements

Elements with atomic numbers greater than 92 (atomic number of uranium) are called the transuranium elements. All these elements are man-made through artificial nuclear reactions.

$$^{238}_{92}U + ^{1}_{0}n \longrightarrow ^{239}_{92}U + \gamma$$
$$^{239}_{92}U \longrightarrow ^{239}_{93}Np + ^{0}_{-1}e$$
$$^{239}_{93}Np \longrightarrow ^{239}_{94}Pu + ^{0}_{-1}e$$

- At present, transuranium elements with atomic numbers up to 110 have been reported.
- Very recently, on August 16,2003, IUPAC approved the name for the element of atomic number 110, as **darmstadtium**, with symbol **Ds**.

- **CNIC (Commission on Nomenclature of Inorganic Chemistry)** appointed by IUPAC in 1994, approved official names for elements with atomic number 104 to 109. With $Z > 109$, IUPAC recommended a nomenclature to be followed for naming these elements until their names are officially recognised. This nomenclature is based on the Latin words for the atomic numbers of the elements.
- The names are derived by using roots for the three digits in the atomic number of the element and adding the ending –ium. The roots for the numbers are :

Digit	Name	Abbreviation
0	nil	n
1	un	u
2	bi	b
3	tri	t
4	quad	q
5	pent	p
6	hex	h
7	sept	s
8	oct	o
9	enn	e

Thus element with atomic number 109 will be named as **une** (**u** for 1, **n** for 0 and **e** for 9).

Table 3.4 summarises the names of the elements with atomic number above 100.

Table 3.4 **IUPAC Nomenclature of Elements with Atomic Number above 100**

Atomic number	Name	Symbol	IUPAC official name	IUPAC symbol
101	Un-nil-unnium	Unu	Mendelevium	Md
102	Un-nil-bium	Unb	Nobelium	No
103	Un-nil-trium	Unt	Lawrencium	Lr
104	Un-nil-quadium	Unq	Rutherfordium	Rf
105	Un-nil-pentium	Unp	Dubnium	Db
106	Un-nil-hexium	Unh	Seaborgium	Sg
107	Un-nil-septium	Uns	Bohrium	Bh
108	Un-nil-octium	Uno	Hassnium	Hs
109	Un-nil-ennium	Une	Meitnerium	Mt
110	Un-un-nillium	Uun	Darmstadtium***	
111	Un-un-unnium	Uuu	**	
112	Un-un-bium	Uub	**	
113	Un-un-trium	Uut	α	
114	Un-un-quadium	Uuq	**	
115	Un-un-pentium	Uup	α	
116	Un-un-hexium	Uuh	*	
117	Un-un-septium	Uus	*	
118	Un-un-octium	Uuo	*	

* Elements yet to be discovered,
** IUPAC names yet to be announced.
*** IUPAC approved the name on August 16, 2003.
α Discovered in 2003.

Target Practice 1

1. What are the blocks of the following?
   (a) Sc^{3+} (18*e*) ....
   (b) Na^{+} (10*e*) ....
   (c) Al^{3+} (10*e*) ....
2. Lanthanides (58-71) have .... elements.
3. In the absence of Aufbau rule, what are the blocks of
   (a) K(19) .... (b) Sc (21) ....
   (c) Ca(20) ....
4. What are atomic numbers of the following elements?
   (a) Uut .... (b) Uuo ....
   (c) Unt ....
5. What are symbols for the following having atomic numbers?
   (a) 117 .... (b) 115 ....
   (c) 104 ....

Answers

1. Blocks are based on atomic number and not on the number of electrons. Thus,
   (a) Sc^{3+} (18*e*, $z = 21$)[Ar] $4s^2 3d^1$ — *d*-block
   (b) Na^{+}(10*e*, $z = 11$)[Ne] $3s^1$ — *s*-block
   (c) Al^{3+}(10*e*, $z = 13$)[Ne] $3s^2 3p^1$ — *p*-block
2. 14
3. If Aufbau rule is not followed, electrons are filled in the available orbitals.
   (a) K(19) $1s^2 2s^2 2p^6 3s^2 3p^6 3d^1$ — *d*-block
   (b) Sc (21) $1s^2 2s^2 2p^6 3s^2 3p^6 3d^3$ — *d*-block
   (c) Ca(20) $1s^2 2s^2 2p^6 3s^2 3p^6 3d^2$ — *d*-block
4. (a) Uut 113
   ↑↑↑
   1 1 3
   (b) Uuo 118
   (c) Unt 103
5. (a) 117 uus (b) 115 uup (c) 104 uuq

Practice Exercise 1

1. If each orbital can take maximum of three electrons (elements), what is the number of elements in
(a) 2nd period, (b) 4th period?

2. By use of electron-configuration, decide the group and period of the following elements with atomic number
13, 35, 56, 78, 65

3. (a) If **Aufbau rule** is not taken into account, determine the number of elements in each period.
(b) Assume total number of 112 elements, then how many periods would be required in the Periodic Table without separate position of lanthanides and actinides?

4. What are the groups and periods of the elements with atomic numbers 37 and 56?

5. Without writing EC can you decide whether the element is alkali metal or alkaline earth metals of the given two types of elements.

6. Without using Aufbau rule which of the following elements would be placed in *s*-block?

$Z = 19, 38, 29$

7. If it can happen that each orbital can accommodate 3 electrons (for example EC of Na is $1s^3 2s^3 2p^5$), what is block of the elements with $Z = 12, 19, 37$ when Aufbau rule
(i) is also valid, (ii) is not valid?

8. What is the most important relationship among elements in the same group in the Periodic Table?

9. What would be the atomic number of the as-yet-undiscovered alkaline earth element of period 8?

10. What are the blocks if EC in the excited states are :
(A) : $1s^2, 2s^2 2p^6, 3s^1 3p^3 3d^2$ (B) : $1s^2, 2s^1 2p^1$
(C) : $1s^2, 2s^1 2p^3$

11. The *p*-block elements show two types of oxidation state due to inert pair effect
n (valence electrons)
$(n - 2)$
Also stability of *n*th state decreases and that of $(n-2)$ th increases as one goes down the group. What are different oxidation states of groups 13, 14, 15 starting from 4th period? Also write order of their stabilities.

13	14	15
Ga	Ge	As
In	Sn	Sb
Tl	Pb	Bi

12. Following are the *p*-block elements:

Ge	As	Se
32	33	34

If each orbital can take maximum of three electrons and in the absence of Aufbau rule, specify the
(i) block
(ii) group
(iii) period of the above elements.

13. A neutral atom of a certain element has 15 electrons. Without consulting a Periodic Table, answer the following questions:
(a) What is ground state electron-configuration?
(b) How should the element be classified?
(c) Are the atoms of the element diamagnetic or paramagnetic.

14. With reference to the Periodic Table, name
(a) a halogen element in the fourth period
(b) an element similar to phosphorus in chemical properties
(c) the most reactive metal in the fifth period
(d) an element that has an atomic number smaller than 20 and is similar to strontium.

15. Elements *A, B, C, D* and *E* have the following electron configurations:
$A : 1s^2, 2s^2 2p^1$ $B : 1s^2, 2s^2 2p^6, 3s^2 3p^1$
$C : 1s^2, 2s^2 2p^6, 3s^2 3p^3$ $D : 1s^2, 2s^2 2p^6, 3s^2 3p^5$
$E : 1s^2, 2s^2 2p^6, 3s^2 3p^6, 4s^2$
Which among these will belong to the same group in the Periodic Table?

16. By definition, *d*-block elements are those in which *n* and $(n-1)$ shells are incomplete and last electron is being filled into $(n-1)d$-orbitals. For
Zn(30) $[Ar]_{18}\, 3d^{10}\, 4s^2$
Cd(48) $[Kr]_{36}\, 4d^{10}\, 5s^2$
Hg(80) $[Xe]_{54}\, 4f^{14}\, 5d^{10}\, 6s^2$
You find, $(n-1)d$ is completely filled. Are their positions in *d*-block justified.

17. What are the blocks, groups and period of the following iso-electronic ions/species ?
(a) Na^+, Mg^{2+}, Al^{3+} (b) O^{2-}, Na^+, Ne
(c) Fe^{2+}, Mn^+, Cr

18. If Aufbau rule is not followed, what are the blocks of the following elements ?
(a) K (19) (b) Fe (26)
(c) Ga (31) (d) Sn (50)

19. A metal ion with +3 charge has five electrons in the 3*d*-subshell. Identify the metal.

20. EC of Gd (64) is written as:

$$[Xe]_{54}\, 4f^7 5d^1 6s^2$$

and not as $[Xe]_{54}\, 4f^8 6s^2$. Explain.

21. Last electron in Lu(71) goes into 5*d*, but it is studied in *f*-block. Explain.

22. Match the block of the elements with their atomic number:

Z	Block
55	*p*
45	*f*
81	*s*
64	*d*

23. Write EC of Cm (96). What is its
(a) group (b) period
(c) block and (d) magnetic moment?

24. Identify the group, family, and/or other Periodic Table location of each element with the outer electron configuration :
(a) $ns^2\, np^3$
(b) ns^1
(c) $ns^2\, (n-1)\, d^{0-2}\, (n-2)\, f^{1-14}$
(d) $ns^2\, (n-1)\, d^{1-10}$

Answers

1. **(a)** Total orbitals available in 2nd period :

one in 2*s* ⎤
three in 2*p* ⎦ = 4 hence, number of elements = 12

(b) Total orbitals available in 4th period :

one in 4*s* ⎤
five in 3*d* ⎥ = 9 hence, number of elements = 27
three in 4*p* ⎦

2.

Z	Group	Period
13	13 (IIIA)	3
35	17 (VIIA)	4
56	2 (IIA)	6
78	10 (VIIIB)	6
65	3 (IIIB)	6

3. **(a)** Each group would be having number of elements = $2n^2$, where *n* stands for number of period.

n	Elements
1	2
2	8
3	18
4	32
5	50
6	72
7	98

(b) Total number of six periods, sixth period having only 2 elements [based on part (a)].

4.

Z	Period	Group
37	4th	IA (1)
56	5th	IIA (2)

5. Yes : Atomic number of alkali metals is not divisible by 2 while that of alkaline earth metals, it is divisible by 2.

e.g., Na — 11
Mg — 12

6.

Z	(EC)	Block
19	$1s^2 2s^2 2p^6 3s^2 3p^6 3d^1$	*d*-block
38	$1s^2 2s^2 2p^6 3s^2 3p^6 3d^{10} 4s^2 4p^6 4d^2$	*d*-block
29	$1s^2 2s^2 2p^6 3s^2 3p^6 3d^{10} 4s^1$	*s*-block

7.

Atomic no.	(EC)			
	Aufbau rule is valid	Block	Aufbau rule is not valid	Block
12	$1s^3 2s^3 2p^6$	*p*	$1s^3 2s^3 2p^6$	*p*
19	$1s^3 2s^3 2p^9 3s^3 3p^1$	*p*	$1s^3 2s^3 2p^9 3s^3 3p^1$	*p*
37	$1s^3 2s^3 2p^9 3s^3 3p^9 4s^3 3d^7$	*d*	$1s^3 2s^3 2p^9 3s^3 3p^9 3d^{10}$	*d*

8. They have similar valence shell E.C.

9. Atomic number = 120

10. (A) *p*-block (B) *s*-block (C) *p*-block

11. Ga, In, Tl +3, +1 $Ga^+ < In^+ < Tl^+$
Ge, Sn, Pb +4, +2 $Ge^{2+} < Sn^{2+} < Pb^{2+}$
As, Sb, Bi +5, +3 $As^{3+} < Sb^{3+} < Bi^{3+}, Pb^{2+} > Pb^{4+}$, $Sn^{2+} < Sn^{4+}$

12.

Element	(EC)	Block	Group	Period
Ge_{32}	$1s^3 2s^3 2p^9 3s^3 3p^9 3d^5$	*d*	17	3
As_{33}	$1s^3 2s^3 2p^9 3s^3 3p^9 3d^6$	*d*	18	3
Se_{34}	$1s^3 2s^3 2p^9 3s^3 3p^9 3d^7$	*d*	19	3

13. (a) $1s^2\ 2s^2\ 2p^6\ 3s^2\ 3p^3$ (b) *p*-block, group VA (15), period 3
(c) paramagnetic since there are (three) unpaired electrons

14. (a) Br (b) N (C) Rb (d) Mg **15.** *A* and *B*

16. Zn, Cd, Hg, with $(n-1)\ d^{10}\ ns^2$ electronic configuration can't be considered in *d*-block due to completely filled $(n-1)d$ orbital. Although these elements form +2 cations, they do not have much in common with Be, Mg and Ca to Ra group except for some resemblance between Zn, Mg and Be.
2(IIA) group elements form +2 cation with inert gas configuration while 12(IIB) group elements form +2 cation with $(n-1)d^{10}$ configuration which makes study of these elements in *d*-block.

17. It is atomic number and not the number of electrons that decide position in the Periodic Table.

	Ions/species	Group	Period	Block
(a)	Na^+	1(IA)	3	*s*
	Mg^{2+}	2(IIA)	3	*s*
	Al^{3+}	13(IIIA)	3	*p*
(b)	O^{2-}	16 (VIA)	2	*p*
	Na^+	1 (IIA)	3	*s*
	Ne	18 (VIIIA)	2	*p*
(c)	Fe^{2+}	8 (VIIIB)	4	*d*
	Mn^+	7 (VIIB)	4	*d*
	Cr	6 (VIB)	4	*d*

18. In the absence of Aufbau rule, (EC) and last filling electron decides the block

(i) K_{19} $1s^2 2s^2 2p^6 3s^2 3p^6 3d^1$
(ii) Fe_{26} $1s^2 2s^2 2p^6 3s^2 3p^6 3d^8$
(iii) Ga_{31} $1s^2 2s^2 2p^6 3s^2 3p^6 3d^{10} 4s^2 4p^1$
(iv) Sn_{50} $1s^2 2s^2 2p^6 3s^2 3p^6 3d^{10} 4s^2 4p^6 4d^{10} 4f^4$

	Changed block	Original block
(i)	*d*	*s*
(ii)	*d*	*d*
(iii)	*p*	*p*
(iv)	*f*	*p*

Thus, block would change in case of K and Sn if EC is written without Aufbau rule.

19. Fe (Z = 26) : $[Ar]_{18}\ 4s^2, 3d^6$
Fe^{3+} : $[Ar]_{18}\ 4s^0, 3d^5$

20. In this EC all the electrons in $4f^7$ are unpaired and thus is stable.

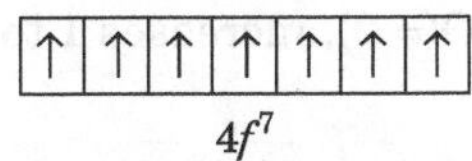

$4f^7$

21. Lu_{71} is studied in *f*-block based on its properties.

22.

Z	Block
55	*s*
45	*d*
81	*p*
64	*f*

23. Cm(96)

EC $[Rn]_{86}\ 5f^7\ 6d^1\ 7s^2$

$[Rn]_{86}$ ↑ ↑ ↑ ↑ ↑ ↑ ↑ | ↑ | ↑↓

$5f^7$ $6d^1$ $7s^2$

(a) Group = 3 (IIIB)

(b) Period = 7th

(c) Block = *f*

(d) Magnetic moment $= \sqrt{N(N+2)}$ BM $= \sqrt{80}$ BM

N = number of unpaired electrons

24.

Group	Family
(a) VA (15)	nitrogen
(b) IA (1)	alkali
(c) IIIB (3)	inner-transition metals
(d) —	transition metals

Periodicity

The recurrence of similar properties of elements after certain regular intervals when they are arranged in the order of increasing atomic number is called **periodicity.**

- Periodicity is due to recurrence of similar outer electronic configurations of their atoms at certain regular intervals. (Table 3.5)
- Recurrence of properties is after 2, 8, 8, 18, 18 and 32 elements (these number are called magic numbers).

Table 3.5 **E.C. of Elements of Group 1, 17 and 18**

1 (Alkali metals)		17 (Halogen family)	18 (Noble gases)
Li	[He] $2s^1$	F : $1s^2 2s^2 2p^5$	He : $1s^2$
Na	[Ne] $3s^1$	Cl : ... $3s^2 3p^5$	Ne : $2s^2 2p^6$
K	[Ar] $4s^1$	Br : ... $4s^2 4p^5$	Ar : ... $3s^2 3p^6$
Rb	[Kr] $5s^1$	I : ... $5s^2 5p^5$	Kr : ... $4s^2 4p^6$
Cs	[Xe] $6s^1$	At : ... $6s^2 6p^5$	Xe : ... $5s^2 5p^6$
Fr	[Rn] $7s^1$		Rn : ... $6s^2 6p^6$

From left to right across a period there is a transition from metals to metalloids to non-metals. Consider the third-period elements from Na to Ar.

DEEP Focus

- Sodium, the first element in the third period, is a very reactive metal, whereas chlorine, the second-to-last element of that period, is a very reactive non-metal. In between, the elements show a gradual transition from metallic properties to non-metallic properties.
- Na, Mg and Al all have extensive three-dimensional atomic networks, which are held together by forces characteristic of the metallic state.
- Si is a metalloid; it has a giant three-dimensional structure in which the Si atoms are held together very strongly. Starting with phosphorus, the elements exist in simple, discrete molecular units (P_4, S_8, Cl_2 and Ar) which have low melting points (m.p.) and boiling points (b.p.).

- Within a periodic group the physical properties vary more predictably, especially if the elements are in same physical state. For example, the m.p. of Ar and Xe are −189.2°C and −111.9°C, respectively. We can estimate the m.p. of the intermediate element (which is Kr) by taking the average of these two values as follows :

$$\text{m.p. of Kr} = \frac{(-189.2) + (-111.9)}{2}$$

$$= -150.55°\,C$$

This value is quite close to the actual m.p. of −156.6°C.

Valency and Oxidation Number

Valency and oxidation number of the elements are determined along a group and period.

Variation in a Group

- For a group, valency is constant.
- $\text{Valency} = N \quad \text{if} \quad N \le 4$
  $= (N - 8) \quad \text{if} \quad N \ge 4$

where N is the number of valence electrons.

	1 (IA)	2 (IIA)	13 (IIIA)	14 (IVA)	15 (VA)	16 (VIA)	17 (VIIA)	18 (VIIIA)
	Li	Be	B	C	N	O	F	Ne
Z	3	4	5	6	7	8	9	10
N	1	2	3	4	5	6	7	8
valency	1	2	3	4 −4	−3	−2	−1	0

In CO_2, C is +4 and in CH_4, C is −4.

Due to **Inert-pair effect** (refer *p*-block elements), *p*-block elements in 4th and higher periods have two types of oxidation states/valencies, N and $(N-2)$. Stability of higher Nth state decreases and that of lower $(N-2)$th state increases going down the group.

Group	Element	N	Oxidation state	Stability order
13 (IIIA)	Ga, In, Tl	3	+1, +3	$Ga^{3+} > In^{3+} > Tl^{3+}$ $Ga^{+} < In^{+} < Tl^{+}$ $Ga^{3+} > Ga^{+}$ $Tl^{3+} < Tl^{+}$
14 (IVA)	Ge, Sn, Pb	4	+2, +4	$Ge^{4+} > Sn^{4+} > Pb^{4+}$ $Ge^{2+} < Sn^{2+} < Pb^{2+}$ $Sn^{4+} > Sn^{2+}$ $Pb^{4+} < Pb^{2+}$
15 (VA)	As, Sb, Bi	5	+3, +5	$As^{5+} > Sb^{5+} > Bi^{5+}$ $As^{3+} < Sb^{3+} < Bi^{3+}$ $As^{5+} > As^{3+}$ $Bi^{5+} < Bi^{3+}$

Variation in Period

- **Valency, w.r.t. O(= 2), increases 1 to 7 in a period left to right**

	Na	Mg	Al	Si	P	S	Cl
Compounds with oxygen	Na_2O	MgO	Al_2O_3	SiO_2	P_2O_5	SO_3	Cl_2O_7
Valency	1	2	3	4	5	6	7

- **Valency, w.r.t. H (= 1), increases 1 to 4 and then falls to 1.**

	Li	Be	B	C	N	O	F
Compounds with hydrogen	LiH	BeH_2	BH_3	CH_4	NH_3	H_2O	HF
Valency (numerical value)	1	2	3	4	3	2	1

- **Inert gases [group 18 (VIIIA)], have zero valency.**

Oxidation states/valencies of the elements have been given in Fig. 3.2.

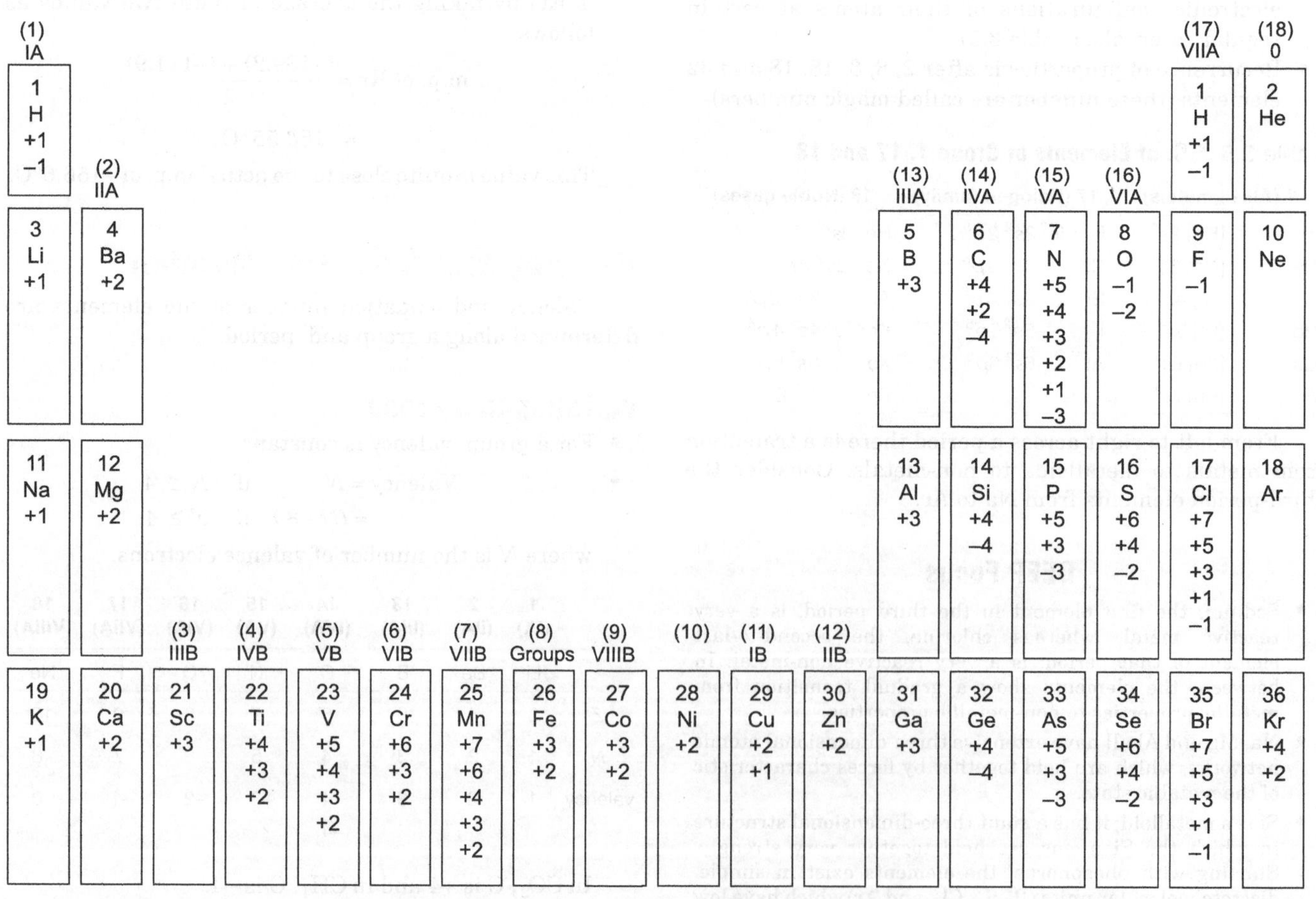

Fig. 3.2 Some non-zero oxidation states (number)

DEEP Focus

Based on stability

$$Sn^{2+} \longrightarrow Sn^{4+} + 2e^- \quad \Delta G^\circ < 0$$
$$Pb^{2+} \longrightarrow Pb^{4+} + 2e^- \quad \Delta G^\circ > 0$$
$$Tl^{3+} + 2e^- \longrightarrow Tl^{+} \quad \Delta G^\circ < 0$$
$$Bi^{5+} + 2e^- \longrightarrow Bi^{3+} \quad \Delta G^\circ < 0$$

(Note : If ΔG° is negative, reaction is spontaneous)

Shielding (Screening) Effect and Effective Nuclear Charge

A valence-electron in a multi-electron atom is attracted by the nucleus and repelled by the electrons of inner-shells. The combined effect of this attractive and repulsive force acting on the valence-electron is that the valence-electron experiences less attraction from the nucleus. This is called **shielding** or **screening effect.**

DEEP Focus

- According to **J.C. Slater,** *effective nuclear charge Z* due to screening is not the actual charge Z of the nucleus of the atom, but depends on the type of orbital in which the electron is housed, and on the ability of the other electrons in more penetrating orbitals to screen (or shield) the electron in question from the nucleus.*
- The relative extent to which the various orbitals penetrate the electron clouds of other orbitals is $s > p > d > f$. Thus, for any given principal quantum number n, an electron will experience the greatest effective nuclear charge when in an s-orbital, than a p-orbital and so on.
- The electron is more stable in the more penetrating orbital.

Effective nuclear charge Z^* due to screening is given by

$$Z^* = Z - \sigma$$

where Z is the atomic number and σ Slater's screening constant. Values of σ are given below for various types of orbitals. In all cases $0 < \sigma < Z$.

Electrons in orbitals ↓ (of shell) →	σ per electron of orbit		
	n	$(n-1)$	$(n-2), (n-3)\ldots$
s-or *p*-orbital	0.35	0.85	1.00
d-or *f*-orbital	0.35	1.00*	1.00

*for all groups on left, including s or p-electrons of the same n.

- For **1s** electron, $\sigma = 0.30$ (in case of **two-electron system)**

– For H-atom ($Z = 1$), there is no screening, being one-electron system. Thus

$$\sigma = 0 \quad \text{and} \quad Z^* = Z = 1$$

– For He-atom ($Z = 2$), $1s^2$, there is screening by one electron on the other electron.

$$\sigma = 0.30 \text{ (of 1s electron)}$$
$$\therefore \quad Z^* = 2 - 0.30 = 1.70$$

– For Li-atom, there is screening of outermost electron by 2 electrons in 1st orbit, $(n-1)$.

Thus σ for two electrons in $(n-1) = 2 \times 0.85 = 1.70$

$$\therefore \quad Z^* = 3 - 1.70 = 1.3$$

DEEP Focus

- In all cases, screening by $(Z-1)$ electrons is to be considered.
- One electron system will have $\sigma = 0$ and thus $Z^* = Z$.

  Thus in case of H, He^+, Li^{2+}, Be^{3+}... with one electron Z^* = atomic number.

- **In a Period**

Element	Z	Value of σ for (Z – 1) electrons in			$Z^* = Z - \sigma$
		1st orbit	2nd orbit	Total	
Li	3	2 × 0.85		1.70	1.30
Be	4	2 × 0.85	1 × 0.35	2.05	1.95
B	5	2 × 0.85	2 × 0.35	2.40	2.60
C	6	2 × 0.85	3 × 0.35	2.75	3.25
N	7	2 × 0.85	4 × 0.35	3.10	3.90
O	8	2 × 0.85	5 × 0.35	3.45	4.55
F	9	2 × 0.85	6 × 0.35	3.80	5.20
Ne	10	2 × 0.85	7 × 0.35	4.15	5.85

Thus, in a period (left to right), value of Z^* increases, increment being of 0.65 per element.

DEEP Focus

For second period with n valence electron

$$Z^* = 0.65[1 + n]$$

For third period with n valences electron

$$Z^* = 1.55 + 0.65n$$

- For higher periods with d- and f-orbitals, variation may differ.

- **In a Group**

Element	Z	n	Value of σ for (Z – 1) electrons in orbit							Total	$Z^* = Z - \sigma$
			n	(n – 1)	(n – 2)	(n – 3)	(n – 4)	(n – 5)	(n – 6)		
Li	3	2	0	1.70	—	—	—	—	—	1.70	1.30
Na	11	3	0	6.80	2.00	—	—	—	—	8.8	2.20
K	19	4	0	6.80	8.00	2.00	—	—	—	16.8	2.20
Rb	37	5	0	6.80	8.00	18.00	2.00	—	—	34.8	2.20
Cs	55	6	0	6.80	18.00	18.00	8.00	2.00	—	52.8	2.20
Fr	87	7	0	6.80	32.00	18.00	18.00	8.00	2.00	84.8	2.20

- Electron in nth orbit is screened by electrons in inner-shells, hence $\sigma = 0$ for electrons in nth orbit in the above case.

DEEP Focus

- *We find amazing effective nuclear charge in a group. Except for* Li *(2nd period) all other elements have constant value of* Z^* *(2.2 in case of alkali metals). Thus while studying variation of any property in a group, not Z but principal quantum number, n will be the* deciding factor. As we shall see in the coming sections, variation of radius in a group

$$r_n = \frac{n^2 a_0}{Z} \qquad (a_0 = \text{Bohr's radius})$$

will be decided by n and not by Z (that will be actually Z^* and which we have seen is constant).

Thus while studying periodicity in a group, do not consider effect of Z.

- One way to illustrate electron shielding is to consider the amount of energy required to remove the two electrons from the helium atom.

$$He(g) \longrightarrow He^+(g) + e^- \quad E = 2373 \text{ kJ mol}^{-1}$$

$$He^+(g) \longrightarrow He^{2+}(g) + e^- \quad E = 5231 \text{ kJ mol}^{-1}$$

The reason it takes so much more energy to remove the second electron is that with only one electron present, there is no shielding, and the electron feels the full effect of the +2 nuclear charge.

Illustration 1 *One method for calculating Z_{eff} is to use the equation :*

$$Z_{eff} = \sqrt{\frac{(E)(n^2)}{1312 \text{ kJ/mol}}}$$

where E is the energy necessary to remove an electron from an atom and n is the principal quantum number of the electron. Use this equation to calculate Z_{eff} values for the highest-energy electron in potassium (E = 418.8 kJ mol^{-1}) and krypton (E = 1350.7 kJ mol^{-1}).

Solution For K, Z = 19 $1s^2, 2s^2\, 2p^6, 3s^2\, 3p^6, 4s^1$

$4s^1$ is the electron of highest energy, $n = 4$

$$Z_{eff} = \sqrt{\frac{(418.8)(4^2)}{1312}} = \mathbf{2.26}$$

For Kr, Z = 36, [Ar] $4s^2\, 3d^{10}\, 4p^6$, electron of highest energy is in $4p$, $n = 4$

$$Z_{eff} = \sqrt{\frac{(1350.7)(16)}{1312}} = \mathbf{4.06}$$

Size of an Atom and Ion

Atomic radius is taken as the distance from the centre of the nucleus to the outermost shell of the electrons.

- Based on probability concept, an atom does not have well defined boundary. Probability of finding the electron is never zero even at large distance from the nucleus. Hence, exact value of the atomic radius can't be evaluated. Atomic radius is taken as the **effective size** which is the distance of the closest approach of one atom to another atom in a given bonding situation.
- Atomic radius can be

– **Covalent radius** : It is one-half of the distance between the centres of two nuclei (of like atoms) bonded by a single covalent bond.

For atoms known to form single bonds in their elemental state, the experimental bond length may be halved to obtain the respective covalent radius. For example

Element	Bond length (pm)	Atomic radius (pm)
H_2	74	37*
F_2	142	71*
Cl_2	199	99
C (diamond)	154	77

*Accepted values are H : 29 pm; F : 64 pm.

Atomic radii of elements which do not form single bonds (in their elemental form) may be calculated from observed single bond lengths in compounds with other elements of known atomic radii.

The set of covalent radii obtained in the manner fit experimental bond lengths quite satisfactorily, as may be observed from following table for the halides of carbon, silicon and germanium. The experimental bond lengths (in pm) given in parenthesis show fair agreement (without ten per cent) except some fluorides. The atomic radius of fluorine has been taken as 64 pm because this value fits more closely with experimental data on a large number of fluorine compounds. The correction is empirical in nature.

Similarly, the radius of the hydrogen atom is taken to be 29 pm. These adjusted values for some elements are shown below :

Molecule	M=C	M=Si	M=Ge
MF_4	141 (148)	181 (154)	186 (167)
MCl_4	176 (176)	216 (201)	221 (208)
MBr_4	191 (191)	231 (215)	236 (231)
MI_4	210 (211)	250 (243)	255 (250)

Covalent atomic radii (single bond), pm

H	29	Be	89	B	80	C	77	N	70	O	66	F	64
				Al	126	Si	117	P	110	S	104	Cl	99
		Zn	131	Ga	126	Ge	122	As	121	Se	117	Br	114
		Cd	148	In	144	Sn	140	Sb	141	Te	137	I	113
		Hg	148	Tl	147	Pb	146	Bi	151				

In a heteronuclear, diatomic molecule of *AB* type, if the bonding is purely covalent, then the bond length r_{A-B} is equal to the sum of the covalent radii of *A* and *B*, *i.e.*,

$$r_{A-B} = r_A + r_B$$

Thus, covalent radii are additive. Thus known the value of r_{A-B} and r_A (or r_B), value of r_B (or r_A) can be calculated. If there is difference in the electronegativities of the bonded atoms, it causes shortening of the bonds. Schomaker and Stevenson gave following relation to calculate internuclear distance of the molecule *AB* with different electronegativity X_A and X_B of atom *A* and *B* respectively.

$$r_{A-B} = r_A + r_B - 0.07(X_A - X_B)^2$$

DEEP Focus

Multiplicity of the bond also causes a shortening of the bond. Usually a double bond is about 0.86 times and a triple bond about 0.78 times the single bond length for the second period elements.

One important contribution is likely to come from the attraction between the opposite partial charge built up on the atoms due to their electronegativity difference. The gradual decrease in C—F bond length with increasing substitution by fluorine in the following compounds supports this view.

Compound	C—F distance (pm)
CH_3F	139.1
CH_2F_2	135.8
CHF_3	133.2
CF_4	132

When two atoms are joined by multiple bonds, the bond length is shorter than the sum of their single bond covalent radii. From the observed bond-lengths, it is possible to assign "double bond radii" or "triple bond radii" to the atoms. Thus the carbon-carbon bond length of 135 pm in ethylene leads to a double bond radius of carbon equal to 67 pm. Some typical values are given below.

	B	C	N	O
Double bond radius	71	67	62	62
Triple bond radius	64	60	55	

The periodic variation in covalent radii of atoms follows the same trend as the atomic volume. The radii decrease from left to right along a period with increase in nuclear charge. The size should also increase downward along any group with increasing principal quantum number of the outermost shell.

– **van der Waals' radius** : It is one-half of the internucleus distance between two adjacent atoms in two nearest neighbouring molecules of the substance in solid state.

This is obtained from the shortest distance to which two non-bonded atoms can approach before repulsion between their electron clouds come into play.

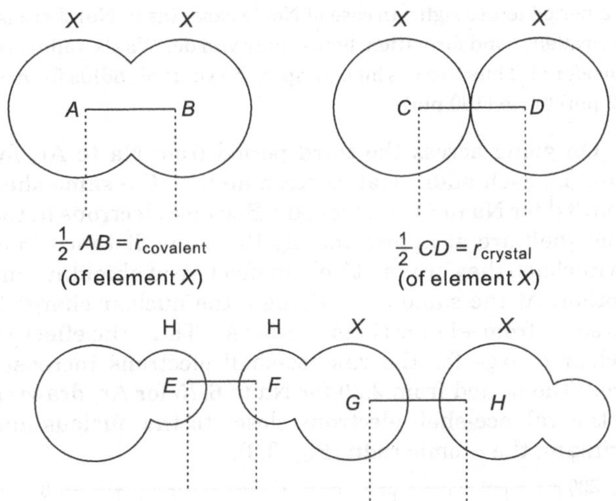

$$\frac{1}{2}EF = r_{\text{van der Waals'}} \text{ of hydrogen in HX molecule}$$

$$\frac{1}{2}GH = r_{\text{van der Waals'}} \text{ of } X \text{ in HX molecule.}$$

$$r_{\text{covalent}} < r_{\text{crystal}} < r_{\text{van der Waals'}}$$

– **Crystal radius** : It is one-half of the distance between the nuclei of two adjacent metal atoms in the metallic closed packed crystal lattice.

- Atomic radius/ionic radius in the *n*th orbit is given by

$$r_n = \frac{n^2 a_0}{Z^*}$$

where *n* is principal quantum number (*i.e.*, number of shell), a_0 the Bohr's radius of H-atom (= 0.529Å) and Z^*, the effective nuclear charge.

Radius is also dependent on the extent of force of attraction which pulls outer shell inward.

Variation in a Period

	Li	Be	B	C	N	O	F	Ne
Z	3	4	5	6	7	8	9	10
Z^*	1.30	1.95	2.60	3.25	3.90	4.55	5.20	5.85
n	2	2	2	2	2	2	2	2
r_n (pm) (covalent)	123	90	80	77	75	74	72	160 ← (van der Waals')

In a period left to right :

- Z (atomic number) increases (by one unit)
- Z^* (effective nuclear charge due to screening) also increases (but by 0.65 unit)
- n (number of shells) remains constant
- Electrons are pulled close to the nucleus by the increased Z^*.

Thus, $r_n \propto \dfrac{1}{Z^*}$

DEEP Focus

Thus atomic radii decrease with increase in atomic number in a period left to right. In case of Noble gases (as in Ne) there is no covalent bond formation, hence, only van der Waals' radius is considered. Thus there is high jump in the value of radius from F (72 pm) to Ne (160 pm).

On going across the third period from Na to Ar, for example, each additional electron adds to the same shell (from $3s^1$ for Na to $3s^2\ 3p^6$ for Ar). Because electrons in the same shell are at approximately the same distance from the nucleus, they are relatively ineffective at shielding one another. At the same time, though, the nuclear charge Z increases from +11 for Na to +18 for Ar. Thus, the effective nuclear charge for the valence-shell electrons increases across the period from 2.20 for Na to 6.75 for Ar, drawing all the valence-shell electrons closer to the nucleus and shrinking the atomic radii (Fig. 3.3).

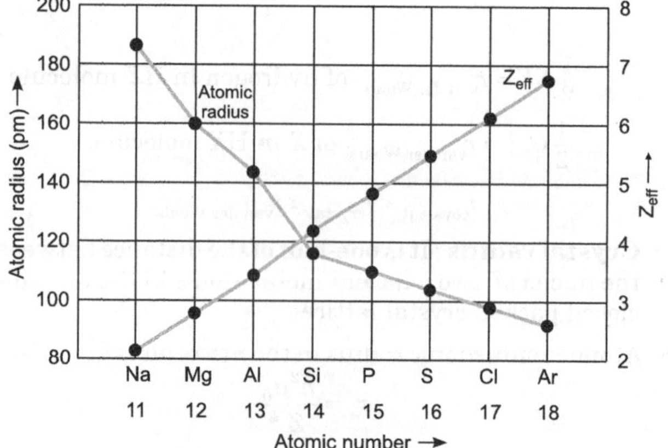

Fig. 3.3 A plot of both atomic radius and Z_{eff} for the valence-shell electrons *versus* atomic number. As Z_{eff} increases, the valence-shell electrons are attracted more strongly to the nucleus, and the atomic radius therefore decreases.

Variation in a Group

We consider 1st group (alkali metals) elements :

Element	Z	Z*	n	Radius (pm)
Li	3	1.3	2	123
Na	11	2.2	3	157
K	19	2.2	4	203
Rb	37	2.2	5	216
Cs	55	2.2	6	235
Fr	87	2.2	7	—
	increases	constant	increases	increases

DEEP Focus

In a group, top to bottom,

- **Z increases**
- **but Z^* (effective nuclear charge due to screening) almost remains constant**
- **n increases**

Since $r_n \propto \dfrac{n^2}{Z^*}$, hence $r_n \propto n^2$ (Z^* remaining constant) thus r_n increases moving along a group top to bottom.

Fig. 3.4 gives the variation of covalent radii of the elements (H to K). In the case of the noble gases, van der Waals' radii have been plotted.

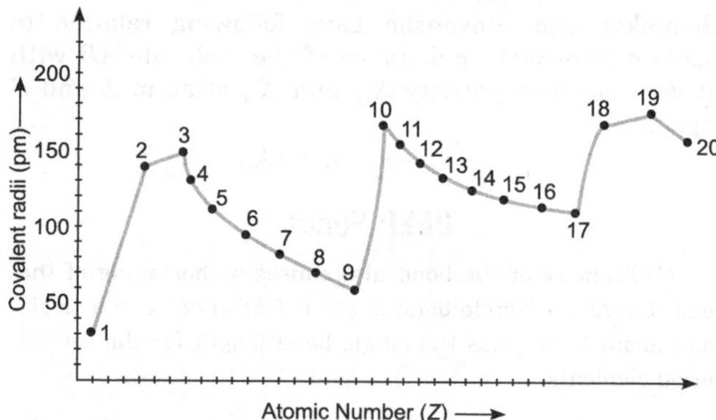

Fig. 3.4 Covalent radii of first 20 elements.

Ionic Radius

The effective distance from the nucleus of the ion up to which it has an influence in the ionic bond is called ionic radius.

- **Metal ions (positive ions) are smaller than the atoms from which they were formed** due to two factors :

  (i) The whole of the outer shell of electrons is usually removed.

  (ii) In an atom, the number of positive charges on the nucleus is exactly the same as the number of orbital electrons.

$$Na \,)))\, \longrightarrow Na^+ \,))\, + \,)$$
2 8 1 — 2 8 — 1

When a positive ion is formed, the number of positive charges on the nucleus exceeds the number of orbital electrons, and the effective nuclear charge is increased. This results in the remaining electrons being more strongly attracted by the nucleus. These electrons are pulled in further reducing the size. **A positive ion is thus always smaller than the corresponding atom and the more electrons which are removed (that is greater the charge on the ion), the smaller the ion becomes.**

Thus, $Mg > Mg > Mg^{2+}$

$Fe > Fe^{2+} > Fe^{3+}$

- **The negative ion is always larger than that of the corresponding atom.** This is due to :
  (i) negative ion is formed by gain of one or more electrons in the neutral atom and thus number of electrons increases but magnitude of nuclear charge remains the same.
  (ii) nuclear charge per electron is thus reduced and the electron cloud is held less tightly by the nucleus resulting in the expansion of the outer-shell. Thus size of negative ion is increased.

$$O^{2-} > O^- > O, \quad F^- > F, \quad I^- > I > I^+$$

- The expansion, that occurs when a group 17 (VIIA, halogen) atom gains an electron to form an anion, can be explained by the change in the quantum number of the valence shell since the added electron simply completes an already occupied *p*-subshell [Ne] : $3s^2\,3p^5$ for a Cl atom to [Ne] $3s^2\,3p^6$ for Cl^- anion. Thus, the expansion is due entirely to a decrease in effective nuclear charge and to an increase in electron-electron repulsion that occurs when an extra electron is added.
- *Iso*-electronic species (containing equal number of electrons) will have different size due to different values of *Z*. Thus, N^{3-}, O^{2-}, F^-, Na^+, Mg^{2+}, Al^{3+}, Si^{4+} (each having 10*e*) are *iso*-electronic but their radii differ and $\propto \frac{1}{Z}$

	Z	Electron	e/Z
N^{3-}	7	10	1.43
O^{2-}	8	10	1.25
F^-	9	10	1.11
Na^+	11	10	0.91
Mg^{2+}	12	10	0.83
Al^{3+}	13	10	0.77
Si^{4+}	14	10	0.71

Thus, greater the value of (*e*/*Z*), greater the size of iso-electronic species.

- In the main groups, radii increase on descending the group because extra shell is added.

Li^+	Na^+	K^+
76 pm	102 pm	138 pm

- The ionic radii decrease moving from left to right across any period in the Periodic Table.

Na^+	Mg^{2+}	Al^{3+}
102 pm	72 pm	53.5 pm

DEEP Focus

The *d* and *f*-orbital do not shield the nuclear charge very effectively. Thus there is significant reduction in the size of the ions just after *d* or *f*-orbitals have been filled completely. This is called **lanthanide contraction** (refer *d*-block elements). Zr (160 pm) and Hf (159 pm) thus have equal size due to lanthanide contraction.

Ionisation Energy

- Ionisation energy (*IE*), **sometimes also called ionisation potential (*IP*),** of an element is defined as the amount of energy required to remove an electron from an isolated gaseous atom of that element resulting in the formation of positive ion.

$$M(g) \xrightarrow{(IE)} M^+(g) + e^-$$

- (*IE*) is thus a direct measure of the ease with which an atom can change into cation. The smaller the ionisation energy, the easier it is for the neutral atom to change into a positive ion.
- $(IE)_1, (IE)_2, (IE)_3 \ldots$ are respectively first, second, third, ... ionisation energies required to remove first, second, third, ... electron from the isolated gaseous atoms.

$$M(g) \xrightarrow{(IE)_1} M^+(g) + e^-$$

$$M^+(g) \xrightarrow{(IE)_2} M^{2+}(g) + e^-$$

$$M^{2+}(g) \xrightarrow{(IE)_3} M^{3+}(g) + e^-$$

Thus, second ionisation energy, $(IE)_2$, causes ionisation of $M^+(g)$ to form $M^{2+}(g)$ and third ionisation energy, $(IE)_3$ causes ionisation of $M^{2+}(g)$ to form $M^{3+}(g)$ and so on ...

- $(IE)_3$ is also called first ionisation energy of $M^{2+}(g)$ and second ionisation energy of $M^+(g)$. $(IE)_2$ is also called first ionisation energy of $M^+(g)$.
- **If an electron has been removed from an atom, it becomes increasingly difficult to remove the second and subsequent electrons from the resulting positively charged ions on account of electrostatic attraction. This is due to the fact that after the removal of an electron, the number of electrons decreases while the nuclear charge remains the same. Consequently, the remaining electrons are held**

more tightly by the nucleus and it becomes difficult to remove second electron. Similarly it is further difficult to remove electron from $M^{2+}(g)$. Thus

$$(IE)_1 < (IE)_2 < (IE)_3 < \ldots$$

Note : 1 eV = 1.602×10^{-19} J, 1 J = 6.242×10^{18} eV

Successive ionisation energies (kJ mol^{-1}) of some elements are given below :

Element	Electronic configuration	$(IE)_1$	$(IE)_2$	$(IE)_3$
Li	$1s^2 2s^1$	520.1	7297	11810
Be	$1s^2 2s^2$	899.3	1758	14810
B	$1s^2 2s^2 2p^1$	800.6	2428	25026
C	$1s^2 2s^2 2p^2$	1086.2	2353	4618
N	$1s^2 2s^2 2p^3$	1402.1	2855	4577
O	$1s^2 2s^2 2p^4$	1313.7	3388	5297
F	$1s^2 2s^2 2p^5$	1680.8	3375	6045
Ne	$1s^2 2s^2 2p^6$	2080.4	3963	6130

Factors Influencing Ionisation Energy

(IE) variation in a period and group may or may not be regular and can be influenced by :

(1) Size of the Atom

In a small atom, the electrons are tightly held, while in a larger atom, the electrons are less strongly held, the Coulombic force of attraction *F* being

$$F = \frac{k\,Ze^2}{r^2}$$

where Ze is charge on nucleus, *e* the electronic charge, *r* the atomic size and *k* the constant.

Thus, the ionisation energy decreases as the size of the atom increases.

(2) Nuclear Charge

As given above, the force of attraction between the nucleus and the outermost electron increases with increase in nuclear charge. Greater the nuclear charge, greater the energy required to pull the electron from the atom. Here *(IE)* increases with increase in nuclear charge.

(3) Shielding Effect

The electrons in the inner-shells act as a screen or shield between the nucleus and the electron in the outermost-shell. This is called **shielding** or **screening** effect. The larger the number of electrons in the inner-shells, greater is the screening effect and smaller the force of attraction and thus *(IE)* decreases.

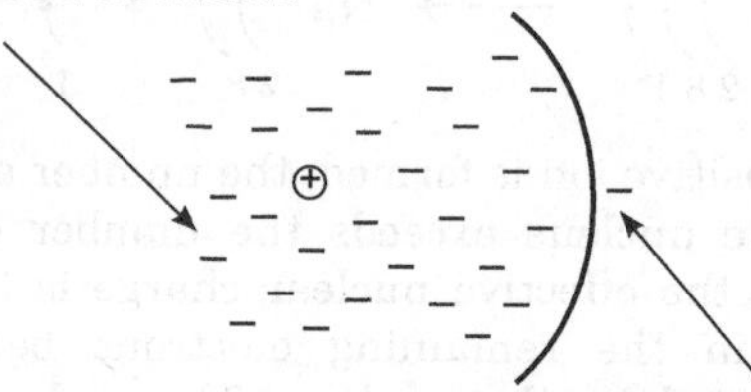

Shielding is most effective whenever there is a full shell (or shells) of electrons between the outermost electron and the nucleus, as in case of noble gases (group VIII A). Hence, there is a sharp decrease in *(IE)* going from noble gas to alkali metals as given below:

	He ⟶ Li		Ne ⟶ Na		Ar ⟶ K	
IE (kJ mol^{-1})	2372	520	2081	513	1521	419

The shielding is not so important across a period as it is down a group. One reason is that an electron in, say, a p_x orbital has little shielding effect on an electron in a p_y or p_z orbital (p_x, p_y and p_z are mutually at right angles to one another hence less shielding). *s*-orbitals are spherically symmetrical, hence filling of 1*s* or 2*s* will experience shielding effect. This comparison is shown in Fig. 3.5.

(4) Penetration Effect

The ionisation energy also depends on the type of electron which is removed. *s, p, d,* and *f*-electrons have orbitals with different shape. An *s*-electron penetrates nearer to the nucleus, and is therefore more tightly held than a *d*-electron, and a *d* electron is more tightly held than an *f*-electron. Other factors being equal, ionisation energies are in the order *s* > *p* > *d* > *f*. Thus, the increase in *(IE)* is not quite smooth on moving from left to right in the Periodic Table. $(IE)_1$ of group 13 (III A) elements (where a *p*-electron is being removed) is actually less than adjacent group 2 (IIA) (where an *s*-electron is being removed). Also refer Fig. 3.5.

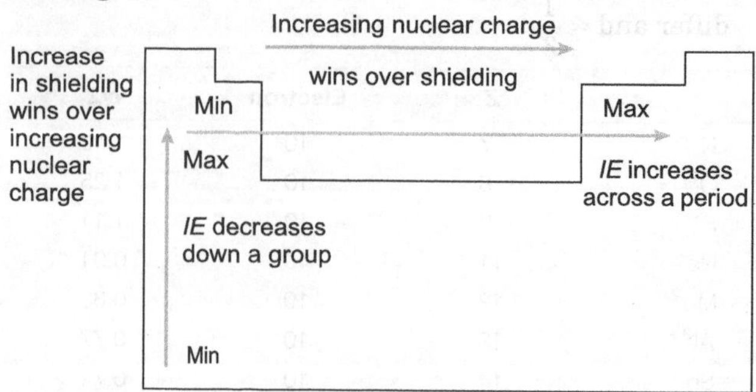

Fig. 3.5 Comparison of shielding effect across a period and a group

(5) Electronic Configuration

If an atom has fully-filled or half-filled orbitals, its (*IE*) is higher than expected normally from its position in the Periodic Table.

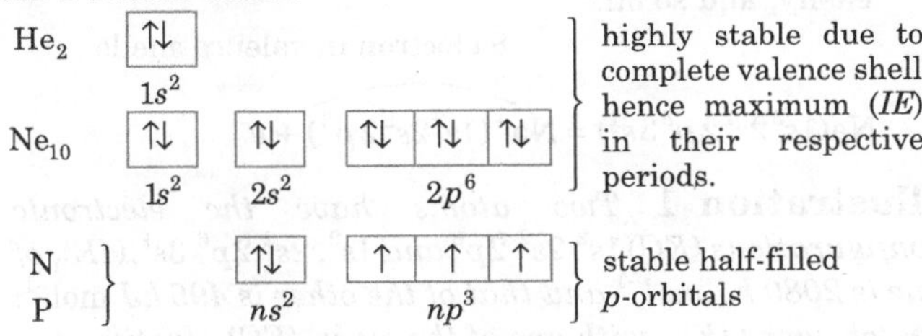

Variation of (*IE*) in a Group

On moving down a group :

(i) nuclear charge increases

(ii) Z^* (effective nuclear charge) due to screening is almost constant.

(iii) number of shells increases, hence atomic size increases.

(iv) there is increase in the number of inner electrons which shield the valence electrons from the nucleus.

Thus, force of attraction between electrons and nucleus decreases and tendency to remove the valence electron increases. Hence (*IE*) decreases on moving down the group. Variations of $(IE)_1$ of alkali metals (group I) and boron family (group 13) are given in Fig. 3.6.

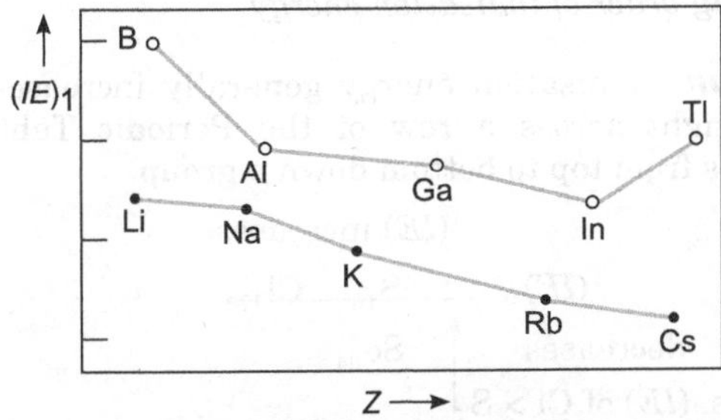

Fig. 3.6 Variation of $(IE)_1$ of groups 1 (•) and 13 (o), (not according to scale)

DEEP Focus

In general the first ionisation energy decreases in a regular way on descending the main groups. A departure from this trend occurs in group 13 (boron family), where the expected decrease occurs between B and Al, but the values for the remaining elements Ga, In and Tl do not continue the trend and are irregular. Ga with completely filled 3*d*-orbitals (no 3*d* in Al) decreases screening thus Ga is smaller in size than it would otherwise be.

	$(IE)_1$ Group 1 (in kJ mol^{-1})		$(IE)_1$ Group 13 (in kJ mol^{-1})	
Li	520	B	801	
Na	496	Al	577	decrease
K	419	Ga	579	constant
Rb	403	In	558	decrease
Cs	376	Tl	589	increase
	decrease			

As expected the first ionisation energies decrease down the groups in case of the main group elements in the Periodic Table. But in case of transition elements opposite trends are observed. Thus, $(IE)_1$ of the corresponding elements of 3*d* and 4*d*-series are almost similar but these an smaller than the $(IE)_1$ of the 5*d*- series elements. Certainly, the higher values of ionisation energies of the 5*d*-transition elements are consistent with the relatively small size of the atoms.

Variation of (IE) in a Period

On moving across a period, the atomic size decreases and nuclear charge increases and therefore the force of attraction exerted by the nucleus on the electron in outer most shell increases. Hence **(*IE*) increases along a period** from left to right. Fig. 3.7 explains general variation of (*IE*) in a group and period. Fig. 3.8 shows the pattern of the ionisation energies of the elements H to Ca.

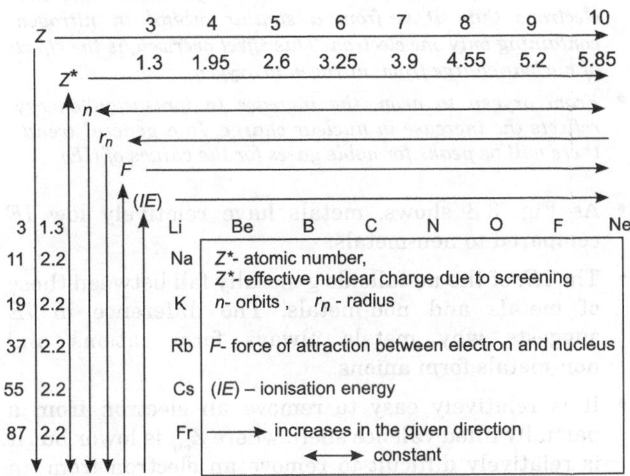

Fig. 3.7 Variation of (*IE*) in a group and period

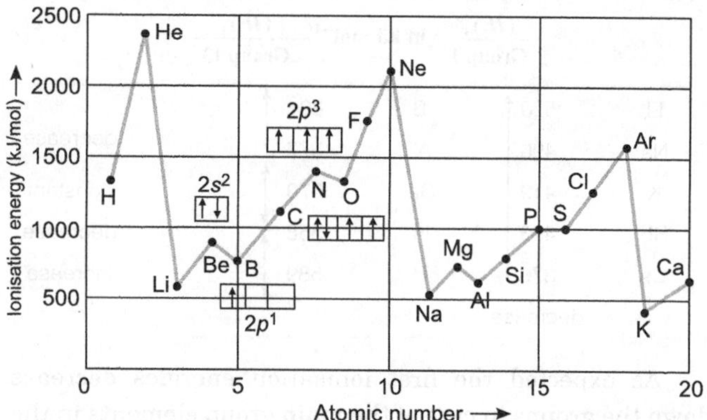

Fig. 3.8 Variation of *IE* with atomic number showing apparent anomalies of B and O, and Al and S

DEEP Focus

- *There are two bumps, or dips, around beryllium / boron and nitrogen / oxygen. The dip from beryllium to boron occurs because the extra electron of boron enters a 2p-orbital, and this electron will feel the effects of the shielding by the pair of electrons in 2s-orbital. The shielding has the effect of lowering of (IE). The increase from boron to carbon reflects the increased nuclear charge of carbon. Also, because the two 2p-electrons of carbon are in different orbitals, they have little shielding effect on one another. A similar state of affairs accounts for the rise from carbon to nitrogen.*
- *The dip from nitrogen to oxygen is a different matter. Compared to nitrogen, the oxygen atom's extra electron must go into a 2p-orbital that already has one electron in it. Putting two electrons into the same orbital brings about repulsion between them. For this reason it is easier to remove an electron from the 2p-orbital in oxygen containing two electrons than it is from a similar orbital in nitrogen containing only one electron. This effect overweighs the effect of nuclear charge from nitrogen to oxygen.*
- *From oxygen to neon, the increase in ionisation energy reflects the increase in nuclear charge. In a general trend, there will be peaks for noble gases for the values of (IE).*

- As Fig. 3.8 shows, metals have relatively low *IE* compared to non-metals.
- The *IE*, of the metalloids generally fall between those of metals and non-metals. The difference in *IE* suggests why metals always form cations and non-metals form anions.
- It is relatively easy to remove an electron from a partially filled valence shell, where Z_{eff} is lower but it is relatively difficult to remove an electron from an atom or ion that has a filled valence shell, when Z_{eff} is higher. In other words, there is a large amount of stability associated with filled *s* and *p*-subshells (a noble gas electron configuration), which corresponds to having eight electrons (an octet) in the valence shell of an atom or ion. Thus,
  - sodium ([Ne] $3s^1$) loses one electron easily
  - magnesium ([Ne] $3s^2$) loses only two electrons easily,
  - aluminium ([Ne] $3s^2\ 3p^1$ loses only three electrons easily, and so on.

8 electron in valence shells

$$Na(1s^2 2s^2 2p^6 3s^1) = Na^+(1s^2 2s^2 2p^6) + e^-$$

Illustration 1 *Two atoms have the electronic configurations (EC)* $1s^2\ 2s^2\ 2p^6$ *and* $1s^2, 2s^2\ 2p^6\ 3s^1$*.* $(IE)_1$ *of one is 2080 kJ* mol^{-1} *and that of the other is 496 kJ* mol^{-1}*. Match each* $(IE)_1$*, with one of the given (EC). Justify your choice.*

Solution

Element	(EC)				
A	$1s^2\ 2s^2\ 2p^6$	↑↓ ($1s^2$)	↑↓ ($2s^2$)	↑↓ ↑↓ ↑↓ ($2p^6$)	
		stable inert gas *EC* and removal of electron requires higher energy			
B	$1s^2\ 2s^2\ 2p^6\ 3s^1$	↑↓ ($1s^2$)	↑↓ ($2s^2$)	↑↓ ↑↓ ↑↓ ($2p^6$)	↑ ($3s^1$)
		has unpaired electron, removal of electron requires lower energy			

Thus $(IE)_1$ of *A* is 2080 kJ mol^{-1} and that of *B* is 496 kJ mol^{-1}.

Illustration 2 *Arrange the elements* Se, Cl *and* S *in increasing order of ionisation energy.*

Solution Ionisation energy generally increases from left to right across a row of the Periodic Table and decreases from top to bottom down a group.

(*IE*) increases

(*IE*) decreases: S_{16} → Cl_{17}; Se_{34} ↑

Thus, (*IE*) of Cl > S
and (*IE*) of Se < S
Thus, the order is : Se < S < Cl.

Illustration 3 *Which has the larger fifth ionisation energy,* Ge *or* As?

Solution

		Group
Ge (32) :	[Ar] $4s^2\ 3d^{10}\ 4p^2$	IVA (14)
As (33) :	[Ar] $4s^2\ 3d^{10}\ 4p^3$	VA (15)

As their positions in the Periodic Table indicate, the group IVA (14) element Ge has four valence-shell electrons and thus four relatively low ionisation energies, whereas the group VA (15) element. As has five valence-shell electrons and five low ionisation energies. Ge thus has larger fifth ionisation energy than As.

Target Practice 2

1. Arrange the following atoms in order of increasing ionisation energy :

   Si, K, P, Ca ....

2. Answer each of the following questions.

   (a) Of the elements S, Se and Cl which has the largest atomic radius? ....

   (b) Which has the layer radius, Br or Br^- ? ....

   (c) Which should have the largest difference between the first and second ionisation energy : Si, Na, P or Mg? ....

   (d) Which has the largest ionisation energy? ....

   N, P, As

   (e) Which of the following has the largest radius? ....

   O^{2-}, N^{3-} or F^-

3. Select the atom or ion in each pair that has the larger radius

   (a) Cl or Cl^- ....

   (b) Al or O ....

   (c) I or I^+ ....

4. If screening effect is not considered, and thus effective nuclear charge is the atomic number, then

   (a) larger radius out of Li, Be is ....

   (b) larger ionisation potential out of Li, Na is ....

   (c) larger radius out of Li, Na is ....

Answers

1. K < Ca < Si < P
2. (a) Se (b) Br^-

   (c) Na (since Na^+ attains stable inert gas configuration)

   (d) N (e) N^{3-}
3. (a) Cl^- (b) Al (c) I
4. (a)

$$r_n = \frac{n^2 a_0}{2}$$

$$r_n\,(Li) = \frac{490}{3}$$

$$r_n\,(Be) = \frac{490}{4}$$

Thus, $r_n\,(Be) < r_n\,(Li)$

Thus, Li

(b) $(IE)_Z = (IE)_H \dfrac{Z^2}{n^2}$

$$(IE)_{Li} = (IE)_H \left(\frac{9}{4}\right) = (2.25)\,(IE)_H$$

$$(IE)_{Na} = (IE)_H \left(\frac{121}{9}\right) > (IE)_{Li}$$

Thus, Na

(c) $r_n\,(Na) = \dfrac{990}{11} < r_n\,(Li)$

(as in Cl (a))

Thus, Li

Practice Exercise 2

1. There is recurrence of properties after 2, 8, 8, 18, 18 and 32 elements in the Periodic Table. Explain.
2. What would be the number of elements of recurrence of properties if each orbital can take three electrons instead of two?
3. Group the following electronic configurations in pairs that would represent similar chemical properties of their atoms :

   (a) $1s^2, 2s^2 2p^5$

   (b) $1s^2, 2s^1$

   (c) $1s^2, 2s^2 2p^6$

   (d) $1s^2, 2s^2 2p^6, 3s^2 3p^5$

   (e) $1s^2, 2s^2 2p^6, 3s^2 3p^6, 4s^1$

   (f) $1s^2, 2s^2 2p^6, 3s^2 3p^6, 4s^2 3d^{10} 4p^6$
4. The boiling points of neon and krypton are –245.9°C and –152.9°C respectively. Using these data estimate the boiling point of argon.
5. Which of the following properties show a clear periodic variation along a group?

   (a) First ionisation energy

   (b) Molar mass of the elements

   (c) Number of isotopes of an element

   (d) Atomic radius.
6. Pb^{4+} salts are better oxidising agents while Ge^{2+} salts are better reducing agents. Explain.
7. $BiCl_5$ does not exist. Explain.
8. Which of the following reactions are not possible/possible ?

   (i) $Sn^{2+} + 2Hg^{2+} \longrightarrow Hg_2^{2+} + Sn^{4+}$

   (ii) $Ge^{2+} + Pb^{4+} \longrightarrow Pb^{2+} + Ge^{4+}$

   (iii) $Tl^+ + Ga^{3+} \longrightarrow Ga^+ + Tl^{3+}$
9. Determine value of effective nuclear charge of

   (i) Mg ($Z = 12$) (ii) Sc ($Z = 21$)

   (iii) Cu ($Z = 29$)
10. He and He^+ have different size though each has same value of Z and same number of orbit ($n = 1$). Explain. You may use equation $r_n = \dfrac{n^2 a_0}{Z}$
11. Ionisation potential (IP) of the element/ion with atomic number Z is given by

$$(IP)_Z = 13.6 \left(\frac{Z^2}{n^2}\right) \text{eV.}$$

    Compare values of IP of He and He^+.
12. Ionisation potential of Be^{x+} is found to be 217.6 eV when in the first orbit. What is the value of x?
13. Compare atomic radii of third period elements while moving left to right.

14. As we shall be taking in further chapters, covalent radii of Cr(24) to Cu(29) do not vary appreciably. Explain.

	Cr_{24}	Mn_{25}	Fe_{26}	Co_{27}	Ni_{28}	Cu_{29}
radius (pm)	117	117	117	116	115	117

15. Referring to the Periodic Table, arrange the following atoms in order of increasing radius
(a) P, Si, N (b) Na^+, Cl^-, O^{2-}, Ne

16. On the basis of their positions in the Periodic Table, select the atom with the larger atomic radius in each of the following pairs :
(a) Na, Cs (b) Be, Ba
(c) N, Sb (d) F, Br
(e) Ne, Xe

17. Calculate atomic radius of Li and Na atoms
(i) without consideration of screening
(ii) with consideration of screening.
Which gives better result ?

18. What is effective nuclear charge on Li if its radius is 1.23 Å ?

19. Compare the ionic size of :
(a) Fe^{2+} and Fe^{3+} (b) Cu^+ and Cu^{2+}
(c) H^+, H^- and H

20. Arrange the following in increasing atomic radius :
(i) Cr_{24}, Mn_{25}, Fe_{26} (ii) Ni_{28}, Cu_{29}, Zn_{30}
(iii) Ge_{32}, As_{33}, Se_{34} (iv) Tl_{22}, Zr_{40}, Hf_{72}

21. Arrange the following in increasing radius :
(i) Li^+, Na^+, K^+ (ii) K^+, Ca^{2+}, Ar
(iii) Mg, Mg^+, Mg^{2+} (iv) O^{2-}, N^{3-}, F^-
(v) O, O^-, O^{2-}

22. Calculate ionic radius of Li^{2+}, Be^{3+} and He^+ each containing one electron in the ground state.

23. For each of the following pairs, indicate which one of the two species is larger?
(a) N^{3-} or F^- (b) Mg^{2+} or Ca^{2+}
(c) Fe^{2+} or Fe^{3+} (d) Na^+ or Ne
(e) Au^+ or Au^{3+} (f) P^{3-} or N^{3-}

24. The atomic radius of F is 72 pm and that of F^- is 136 pm. Calculate the per cent increase in volume that occurs when F(g) is converted to $F^-(g)$.

25. *IE* (in kJ mol^{-1}) of Li and Be are given below :

	Z	$(IE)_1$	$(IE)_2$	$(IE)_3$
Li	3	520.1	7297	11810
Be	4	899.3	1758	14810

$(IE)_2$ of Li is much higher than that of its $(IE)_1$ but in case of Be $(IE)_3$ is much higher than that of its $(IE)_1$. Explain.

26. Can you identify the element of third period whose successive (*IEs*) are given below :

Element	$(IE)_1$	$(IE)_2$	$(IE)_3$
X	513	4562	6920
Y	738	1451	7733
Z	1521	2666	3931

27. (*IE*) of Na is 513 kJ mol^{-1}. What is value in eV $molecule^{-1}$?

28. Which element in each of the following pairs has the larger ionisation energy (*IE*)?
(a) K or Br (b) S or Te
(c) Ga or Se (d) Ne or Sr

29. For one-electron species, (*IE*) is given by

$$(IE) = 13.6\left(\frac{Z^2}{n^2}\right)\text{eV}$$

Calculate (*IE*) of He^+, Li^{2+} and Be^{3+}.

30. $(IE)_1 < (IE)_2$ for any case. Can you compare (*IE*) of He and He^+ using equation in Q. 29 above ?

31. $(IE)_1$ of Cs is 376 kJ mol^{-1}. What is the maximum number of Cs^+ ions that can be produced per joule of energy absorbed by a sample of gaseous Cs atoms ?

32. Which atom should have a smaller value of $(IE)_1$?
(a) Oxygen or sulphur (b) Nitrogen or phosphorus

33. Which of the following atoms should have a smaller value of second (*IE*) : Na or Mg?

34. (a) Which has the larger third ionisation energy, Be or N?
(b) Which has the larger fourth ionisation energy, Ga or Ge?

35. Three atoms have the following electron configurations:
(i) $1s^2\,2s^2\,2p^6\,3s^2\,3p^1$ (ii) $1s^2\,2s^2\,2p^6\,3s^2\,3p^5$
(iii) $1s^2\,2s^2\,2p^6\,3s^2\,3p^6\,4s^1$
(a) Which of the three has the largest $(IE)_1$?
(b) Which has the largest $(IE)_4$?

36. Which has larger (*IE*) in each pair?
(a) N or O^+ (b) Be^+ or B^+
(c) Be or B (d) Na^+ or Ne

37. $(IE)_1$ of Be is greater than that of (B) but $(IE)_2$ of (B) is greater than that of Be. Explain.

38. Match the electronic configuration with the values of (*IE*) of the elements in the *n*th period ($n \neq 1$) :

	EC		(*IE*) kJ mol^{-1}
1.	ns^2	I	4000
2.	ns^2np^1	II	1500
3.	ns^2np^3	III	1200
4.	ns^2np^6	IV	1400

39. In general, ionisation energy increases from left to right across a given period. Aluminium, however, has a lower ionisation energy than magnesium. Explain.

40. $(IE)_1$ and $(IE)_2$ in kJ mol^{-1} of K and Ca have been given as

	$(IE)_1$	$(IE)_2$
K	419	3052
Ca	590	1145

Compare their values and comment on the differences.

41. Arrange the following isoelectronic species in order of
(a) increasing ionic radius and
(b) increasing (*IE*)
O^{2-}, F^-, Na^+, Mg^{2+}

42. The first four ionisation energies of an element (in kJ mol^{-1}) are approximately 738, 1450, 7.7×10^3 and 1.1×10^4. To which periodic group does this element belong? Why?

43. Which element in each of the following sets has the smallest first ionisation energy, and which has the largest?
(a) Li, Ba, K (b) B, Be, Cl
(c) Ca, C, Cl

44. The $(IE)_1$ and the $(IE)_2$ in kJ mol^{-1} of a few elements designated by Roman numerals are shown below :

Element	$(IE)_1$	$(IE)_2$
I	2372	5251
II	520	7300
III	900	1760
IV	1680	3380

Which of the above elements is likely to be a
(a) reactive metal
(b) reactive non-metal
(c) noble gas
(d) metal that forms a stable binary halide of the formula AX_2?

Answers

1. Electrons in various suborbits are filled in the increasing order of their energies as decided by Aufbau rule. 3*d*-sub-orbit is filled after 4*s*; we can say that 3*d* is part of 4th period. Similarly 4*d* is a part of 5th period and so on.

Period		Orbitals	Elements
1	$1s^2$	1	2
2	$2s^2 2p^6$	4	8
3	$3s^2 3p^6$	4	8
4	$4s^2 3d^{10} 4p^6$	9	18
5	$5s^2 4d^{10} 5p^6$	9	18
6	$6s^2 4f^{14} 6d^{10} 6p^6$	16	32

Thus, there are respectively 2, 8, 8, 18, 18 and 32 elements in first, second, third, fourth, fifth and sixth period. Hence, recurrence of properties takes place after these number of elements.

2. If each orbital can take three electrons, number of elements of recurrence of properties would be 3, 12, 12, 27, 27 and 48 elements [see practice exercise 2 (above)]

3. (a) and (d), (b) and (e), (c) and (f)

4. – 199.4°C **5.** (a) and (d)

6. By inert pair effect, stabilities of

$$Pb^{4+} < Pb^{2+} \quad \text{and} \quad Ge^{4+} > Ge^{2+}$$

hence $\quad Pb^{4+} + 2e^- \longrightarrow Pb^{2+}$

and $\quad Ge^{2+} \longrightarrow Ge^{4+} + 2e^-$

are spontaneous towards more stable ions ($\Delta G° < 0$). Pb^{4+} can be easily reduced to Pb^{2+} and thus is an oxidising agent; Ge^{2+} can be easily oxidised and thus, is a reducing agent.

7. Due to inert pair effect, stability of

$$Bi^{3+} > Sb^{3+} > As^{3+} \quad \text{and} \quad Bi^{5+} < Sb^{5+} < As^{5+}$$

hence Bi^{5+} salts are least stable; hence $BiCl_5$ does not exist.

8. (i) possible, stability of $Sn^{4+} > Sn^{2+}$
(ii) possible, stability of $Ge^{4+} > Ge^{2+}$
(iii) not possible, stability of $Ga^{3+} > Ga^+$, $Tl^{3+} < Tl^+$

9. (i) Mg (*Z* = 12)

EC	$1s^2$	$2s^2 2p^6$	$3s^2$
	$(n-2)$	$(n-1)$	n
σ of	2*e* in $(n-2)$th	8*e* in $(n-1)$th	1*e* in *n*th orbit
	2.0 +	6.8 +	0.35 = 9.15

$$\therefore \quad Z^* = 12 - 9.15 = 2.85$$

Remember Z* increases by 0.65 unit per element in a period. $Z^*(\text{Na}) = 2.20$

(ii) Sc(*Z* = 21)

$$\text{EC} \quad \underset{(n-3)}{1s^2} \quad \underset{(n-2)}{2s^2 2p^6} \quad \underbrace{3s^2 3p^6 \; 3d^1}_{(n-1)} \quad \underset{n}{4s^2}$$

$$\sigma = 2 + 8 + \underbrace{8 \times 0.85}_{\text{for } (s,\, p)} + \underbrace{1}_{\text{for } (d)} + 0.35 = 18.15$$

$$Z^* = 21 - 18.15 = 2.85$$

(iii) Cu (*Z* = 29)

$$\text{EC} \quad \underset{(n-3)}{1s^2} \quad \underset{(n-2)}{2s^2 2p^6} \quad \underset{(n-1)}{3s^2 3p^6 3d^{10}} \quad \underset{n}{4s^1}$$

$$\sigma \quad 2 + 8 + \underset{(\text{for } s,\, p)}{6.8} + \underset{(\text{for } d)}{10} = 24.8$$

$$Z^* = 29 - 24.8 = 4.2$$

10. For the $(1s^2)$ there is screening by one electron, hence

$$Z^* = Z - \sigma = 1.70$$

$$\therefore \quad r_n = \frac{(1)^2 a_0}{1.70} = 0.5882\, a_0$$

In $He^+ (1s^1)$ there is no screening, hence $Z^* = Z - 0 = 2$

$$\therefore \quad r_1' = \frac{(1)^2 a_0}{4.0} = 0.2500\, a_0$$

Thus, radius of He> He^+

11. For He, $Z^* = 1.70$

$\therefore (IP)_{He} = 13.6 \times (1.70)^2 = 39.30$ eV] (for comparison only)

For He^+, $Z^* = Z = 2$

$$(IP)_{He^+} = 13.6 \times (2)^2 = 54.40 \text{ eV}$$

Thus, $\quad (IP)_{He}^{+} > (IP)_{He}$

12. $x = 3$ (since $Z = Z^* = 4$, *i.e.*, no screening when it is one-electron system).

13. Atomic radius decreases Na > Mg > Al > Si > P > S > Cl << Ar but there is sharp jump in the value of Ar (in which van der Waals' radius is taken).

14. It is due to the fact that attractive force just balances the repulsive force due to screening.

15. (a) N < P < Si; (b) $Na^+ < Ne < O^{2-} < Cl^-$

16. (a) Cs (b) Ba (c) Sb (d) Br (e) Xe

17.

	Z	*Z**	*n*	$r_n = n^2 a_0 / Z$	$r_n = n^2 a_0 / Z^*$
Li	3	1.3	2	0.71 Å	1.63 Å
Na	11	2.2	3	0.43 Å	2.16 Å

Since $r_{Li} < r_{Na}$ (experimentally) hence results are better when screening is considered.

18.

$$r_n = \frac{n^2 a_0}{Z^*}$$

$$1.23 = \frac{(2)^2 \times 0.529}{Z^*}$$

$$Z^* = 1.72$$

19. (a) $Fe^{3+} < Fe^{2+}$ (b) $Cu^+ > Cu^{2+}$ (c) $H^+ < H < H^-$

20. (i) Cr = Mn = Fe (repulsive force just balances attractive force)
(ii) Ni < Cu < Zn (repulsive force overcomes attractive force)
(iii) Se < As < Ge (normal gradation)
(iv) Ti < Zr = Hf (due to lanthanide contraction)

21. (i) $Li^+ < Na^+ < K^+$ (ii) $Ca^{2+} < K^+ < Ar$
(iii) $Mg^{2+} < Mg^+ < Mg$ (iv) $F^- < O^{2-} < N^{3-}$
(v) $O < O^- < O^{2-}$

22. Radius of $Li^{2+} = 0.1763$ Å

$Be^{3+} = 0.1323$ Å, $He^+ = 0.2645$ Å — use equation $r_n = \frac{n^2 a_0}{Z}$

In this case, $Z = Z^*$ since each contains one electron and no screening takes place, *i.e.*, $\sigma = 0$.

23. (a) N^{3-} (b) Ca^{2+} (c) Fe^{2+} (d) Ne (e) Au^+ (f) P^{3-}

24. $5.74 \times 10^2\%$

25. $(IE)_2$ of Li involves removal of electron from $Li^+ (1s^2)$ which has inert gas (stable) helium configuration, hence $(IE)_2$ of Li >>> $(IE)_1$. $(IE)_3$ of Be involves removal of electron from $Be^{2+}(1s^2)$ which is again stable hence $(IE)_3$ of Be >>> $(IE)_1$

26. X : Na its $(IE)_2$ is very high
Y : Mg its $(IE)_3$ is very high
Z : Ar its $(IE)_1$ is very high

27. $513 \text{ kJ mol}^{-1} = \frac{513000 \text{ J}}{6.02 \times 10^{23}} \text{ molecule}^{-1}$

$= 8.522 \times 10^{-19} \text{ J molecule}^{-1}$

$= \frac{8.522 \times 10^{-19}}{1.6 \times 10^{-19}} \text{ eV} = 5.326 \text{ eV}$

28. (a) Br (b) S (c) Se (d) Ne

29.

Species	(IE)	
He^+	54.4 eV	Being single electron species, $Z^* = Z$ (there is no screening effect)
Li^{2+}	122.4 eV	
Be^{3+}	217.6 eV	

30. In case of He, there is screening by one electron hence $Z^* = 1.7$

$(IE)_{He} = 13.6 \times (1.7)^2 = 39.4$ eV,

$(IE)_{He^+} = 54.4$ eV (in Q. 29)

31. 376 kJ mol^{-1}

$376 \times 10^3 \text{ J mol}^{-1}$

376×10^3 J ionise = 1 mol

$= 6.02 \times 10^{23}$ atoms

1 J ionises $= \frac{6.02 \times 10^{23}}{376 \times 10^3}$ atoms

$= 1.6 \times 10^{18}$ atoms

32. (a) S (b) P **33.** Mg

34. (a) Be (b) Ga

35. (a) largest $(IE)_1$ = (ii)
(b) largest $(IE)_4$ = (i)

36. (a) O^+ (b) B^+ (c) Be (d) Na^+

37. Be [↑↓] $2s^2$; B [↑↓] $2s^2$ [↑| |] $2p^1$

Unpaired electron in B is easier to remove than that paired electron in Be. Hence $(IE)_1$ of Be > $(IE)_1$ of B, and by $(IE)_1$ Be^+ and B^+ are formed.

Be^+ [↑] $2s^1$; B^+ [↑↓] $2s^2$

Now reverse is the case; removal of unpaired electron from Be^+ is easier than that of paired electron from B^+. Hence $(IE)_2$ of B > $(IE)_2$ of Be.

38.

EC	*(IE)*
1	III
2	IV
3	II
4	I

39. Refer Text

40. $(IE)_2$ of K >>> $(IE)_1$ of K since K^+ (after one electron is removed) has stable inert gas configuration as compared to corresponding values for Ca in which Ca^+ can lose second electron easily.

41. (a) $Mg^{2+} < Na^+ < F^- < O^{2-}$ (b) $O^{2-} < F^- < Na^+ < Mg^{2+}$

42. Alkaline earth elements.

43.

	Min	Max
(a)	K	Li
(b)	B	Cl
(c)	Ca	Cl

44. (a) II (b) IV (c) I (d) III

Electron Affinity (EA)

Electron affinity (*EA*) is the amount of energy associated with the gain of electrons by the gaseous atom.

$$F(g) + e^- \longrightarrow F^-(g), \quad EA = -328 \text{ kJ mol}^{-1}$$

Electron affinity is also called **electron gain enthalpy.** The sign of the enthalpy change indicates that this is an exothermic process; however the electron affinity of fluorine is assigned a value of + 328 kJ mol^{-1}. Thus, we can think of electron affinity as the energy that must be supplied to remove an electron from a negative ion.

$$F^-(g) \longrightarrow F(g) + e^- \quad \Delta H = +328 \text{ kJ mol}^{-1}.$$

DEEP Focus

Two features of electron affinity based on above line to remember are :

(1) The electron affinity of an element is equal to the enthalpy change that accompanies the ionisation process of its ion.

(2) A large positive electron affinity means that the negative ion is very stable (that is, the atom has greater tendency to accept an electron).

- The greater the energy released in the process of taking up the extra electron, the greater is the electron affinity.
- The electron affinity of an atom measures the tightness with which it binds an additional electron to itself.

- In terms of orbital diagram:

F [↑↓] [↑↓] [↑↓|↑↓|↑] $+ e^- \rightarrow$ F$^-$: [↑↓] [↑↓] [↑↓|↑↓|↑↓] $EA = -328$ kJ mol^{-1}
1s 2s 2p — 1s 2s 2p

In becoming F$^-$, fluorine acquires the electron configuration of noble gas (Ne).

- By the line of reasoning outlined above, we might expect that even metal atoms can form negative ions in the gaseous state, and this happens too.

Li [↑↓] [↑] $+ e^- \rightarrow$ Li$^-$: [↑↓] [↑↓] $EA = -59.6$ kJ mol^{-1}
1s 2s — 1s 2s

- There are also cases when addition of electron experience repulsion; in such cases (*EA*) is a positive quantity. This is the case for the noble gases [group 18 (VIII A)], where the added electron goes into $(n + 1)s$ orbital of higher energy.

Ne [↑↓] [↑↓] [↑↓|↑↓|↑↓] $+ e^- \longrightarrow$
1s 2s 2p

Ne$^-$: [↑↓] [↑↓] [↑↓|↑↓|↑↓] [↑] $EA = +29$ kJ mol^{-1}
1s 2s 2p 3s

- The group 2 (IIA) elements with filled *s*-subshell (ns^2), also have positive (*EA*).
- Even better examples of positive electron affinities are those associated with gaining a second electron. Here the electron to be added is approaching not a neutral atom but a negative ion. A strong repulsion is felt and the energy of the system increases.

$$O^- + e^- \longrightarrow O^{2-} \qquad EA = 702 \text{ kJ mol}^{-1}$$
$$S^- + e^- \longrightarrow S^{2-} \qquad EA = 332 \text{ kJ mol}^{-1}$$

This process [$O^-(g) + e^- \longrightarrow O^{2-}(g)$] is unfavourable in the gas phase because the resulting increase in electron-electron repulsion overweighs the stability gained by achieving the noble gas configuration. However, note that O^{2-} is common in ionic compounds; in solids, O^{2-} is stabilised by neighbouring cations.

- (EA) values (kJ mol^{-1}) of some elements are given below :

1	H -72								He (21)
2	Li -60	Be (241)		B -23	C -122	N 0	O -142	F -328	Ne (29)
3	Na -53	Mg (231)	Cu -123	Al -44	Si -119	P -74	S -200	Cl -348	Ar (35)
4	K -48	Ca (156)	Ag -125	Ga (-36)	Ge -116	As -77	Se -194	Br -323	Kr (39)
5	Rb -47	Sr (119)	Au -222	In (-34)	Sn -120	Sb -101	Te -190	I -295	Xe (40)
6	Cs -45	Ba (52)		Tl (-48)	Pb -101	Bi -101	Po (-173)	At (-270)	Rn (40)
7	Fr (-44)								

*Estimated value are in parentheses

Variation of (*EA*) in a Period

- On moving across a period, the atomic size decreases and hence, the force of attraction exerted by the nucleus on the electrons increases. Consequently, the atom has a greater tendency to attract additional electron, *i.e.*, its electron affinity increases.
- Also (*EA*) values of metals are low while those of non-metals are high.
- Halogens have high electron affinities. This is due to their strong tendency to gain an additional electron to change into the stable ns^2np^6 configuration.

Variation of (*EA*) in a Group

On moving down a group, the atomic size increases and, therefore, the effective nuclear attraction decreases and thus, electron affinity decreases.

DEEP Focus

- The fact that fluorine has a less negative electron affinity than chlorine seems to be due to the relatively greater effectiveness of 2*p*-electrons in the small F atom to repel the additional electron entering the atom than do 3*p*-electrons in the larger Cl atom.

- The results have been represented in Fig. 3.9.

The positive value of (*EA*) of Be, Mg and N are also interesting. The stability of the filled *s*-shell (in Be and Mg) and half-filled *p*-shell (in N) and the resulting reluctance of these atoms to add any extra electron is indicated.

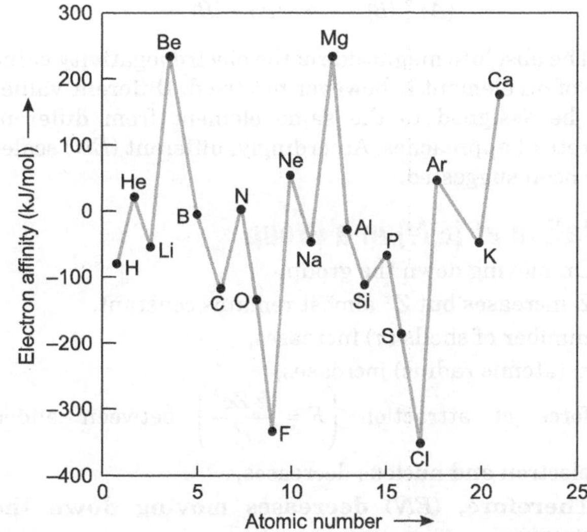

Fig. 3.9 A plot of electron affinity *versus* atomic number for the first 20 elements. The general horizontal trend is that electron affinities become more negative (more energy is released as an extra electron is added) from group IA (1) through group VIIA (17) for a given period. Exceptions occur at the IIA (2) and VA (15) elements.

Illustration 1 *Why does nitrogen have a less favourable (more positive) electron-affinity than its neighbours on either side, C and O?*

Solution Electron configurations of these elements are :

$$C_6 : 1s^2\, 2s^2\, 2p_x^1\, 2p_y^1\, 2p_z^0$$
$$N_7 : 1s^2\, 2s^2\, 2p_x^1\, 2p_y^1\, 2p_z^1$$
$$O_8 : 1s^2\, 2s^2\, 2p_x^2\, 2p_y^1\, 2p_z^1$$

Carbon has only two electrons in its $2p$-subshell and can readily accept another in its vacant $2p_z$-orbital. Nitrogen, however, has a half-filled $2p$-subshell, and the additional electron must pair up in a $2p$-orbital, when it feels a repulsion from the electron already present.

Thus the electron affinity of nitrogen is less favourable than that of carbon. Oxygen also must add an electron to an orbital that already has one electron, but the additional stabilising effect of increased effective nuclear charge (Z_{eff}), across the Periodic Table counteracts the effect of electron repulsion, resulting in a more favourable electron-affinity for O than for N.

Electronegativity (EN)

Electronegativity is a measure of the tendency of an element to attract electrons to itself.

$$X(g) + e^- \longrightarrow X^-(g)$$

In a molecule, tendency of the atom to attract bonding pair towards itself is its electronegativity. *B* is said to be more electronegative than *A* if it pulls bonding pair towards itself.

$$(A) \overset{\bullet}{\times} (B) \longrightarrow \overset{\delta+}{(A)}\text{—}\overset{\delta-}{(B)}$$

The absolute magnitude of the electronegativity value (*EN*) of an element is however not fixed. Different values may be assigned to the same element from different theoretical approaches. Accordingly, different (*EN*) scales have been suggested.

Variation of (*EN*) in a Group

- On moving down the group
  - Z increases but Z^* almost remains constant,
  - number of shells (n) increases,
  - r_n (atomic radius) increases,
  - force of attraction $\left(F = \dfrac{k\,Ze^2}{r_n^2}\right)$ between added electron and nucleus decreases,

Therefore, (*EN*) decreases moving down the group.

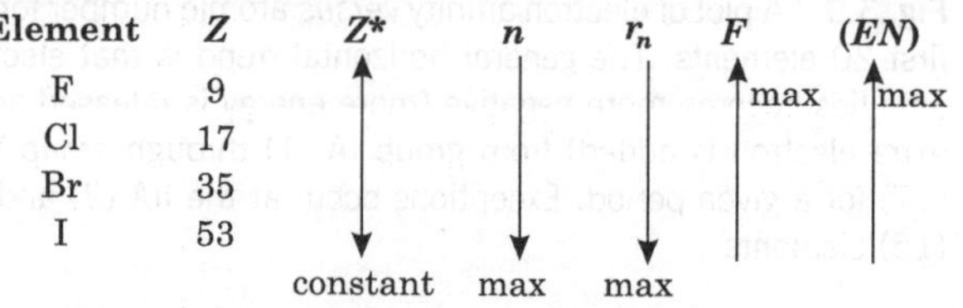

Element	Z	Z*	n	r_n	F	(EN)
F	9				max	max
Cl	17					
Br	35					
I	53	constant	max	max		

Variation of (*EN*) in a Period

- While moving across a period left to right
  - Z, Z^* increases,
  - n (number of shells) remains constant,
  - r_n decreases,
  - $F\left(= k\dfrac{Ze^2}{r_n^2}\right)$ increases

hence, (EN) increases along a period.

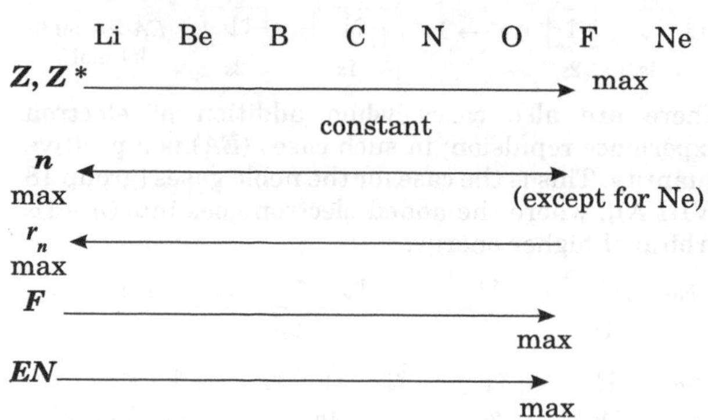

- **Fluorine is the most electronegative element in the Periodic Table.**

Fig. 3.10 represents variation of electronegativity of elements with atomic number. In this :

- the **strongest non-metals, the halogens, appear at the peaks;**

  the **alkali metals mark the** troughs. This is largely a result of new shell of electrons starting with noble gases **which have zero values of electronegativities.**

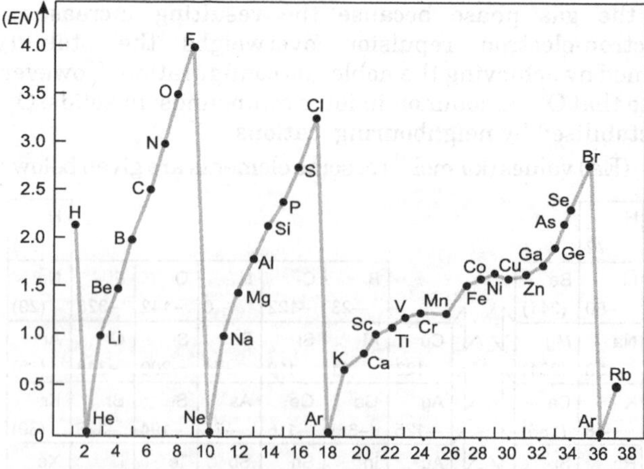

Fig. 3.10 Electronegativities of the elements H(1) to Rb (37). The (*EN*) of the noble gases have been set at zero.

- the group 1 (IA) metals have their nuclei quite strongly shielded. Hence, they show little tendency to gain new electron.

  the electronegativities of transition metals like their (*IE*), do not change greatly from one another.

Diagonal Relationship and Electronegativity

- Moving one place to the right across period, the increased nuclear charge holds the electrons more tightly to the atom. Moving down one place in a group, the extra shell of electrons decreases the attraction of the nucleus for the outer electrons. A **diagonal** move means that these effects tend to compensate for one another.
- The electronegativity increases as we go from Li to Be but it decreases as we move from Be to Mg. As we move diagonally, these two effects partly cancel each other and there is no marked change in electronegativity and thus Li and Mg have close values of electronegativities. Because of this, Li and Mg show similarity in properties— a diagonal relationship (between elements lying diagonally in Periodic Table).
- Elements' pairs showing this relationship are :

Li—Mg, Be—Al, B—Si

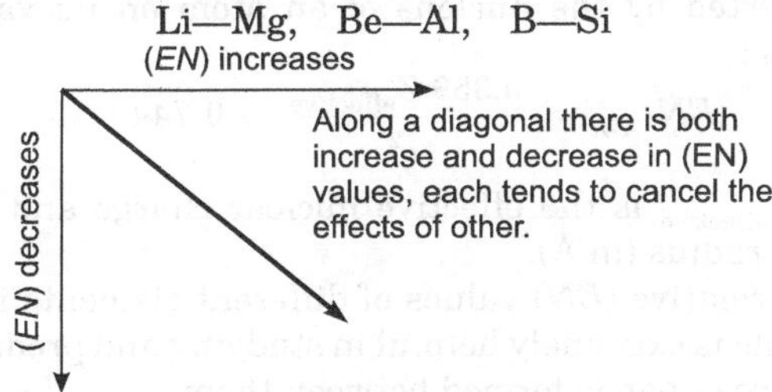

DEEP Focus

Electronegativity of an element depends on its valence state and hence it will change from one compound or one bonding situation to another. An atom will have a greater attraction for electrons when its oxidation state is high, rather than when it is in a low oxidation state — increase of positive oxidation state or formal charge of an atom increases its *(EN)*. Thus, *(EN)* value of sulphur in

$$SF_6 > SCl_2$$

Fe (II)	1.83
Fe (III)	1.96
Tl (I)	1.62
Tl (III)	2.04
Sn (II)	1.80
Sn (IV)	1.96
Pb (II)	1.87
Pb (IV)	2.33

DEEP Focus

Hybrid orbitals with greater *s*-character should be more electronegative than hybrid orbitals with less *s*-character. As *s*-character increases, *(EN)* value increases hence acidic character increases.

	CH_4	C_2H_4	C_2H_2
hybridization of C-atom	sp^3	sp^2	sp
s-character	25%	33.3%	50%
(EN)	⟶		
acidic nature	⟶		

The estimated values (in Pauling equivalents) for C and N in different hybridization states are :

Atom	sp^3	sp^2	sp
C	2.5	2.75	3.3
N	3.7	3.9	4.7

Thus, carbon in C_2H_2 is more electronegative than chlorine; *sp*-hybridised nitrogen is more electronegative than fluorine.

Pauling's Electronegativity

Let us consider two covalent molecules H—H and *X*—*X*

$$H—H + X—X \longrightarrow 2H—X$$

The above reaction is generally exothermic, the bond H—*X* is thus more stable and hence stronger than either of the bonds H—H or *X*—*X*. If the bond H—*X* were purely covalent (like the bonds H—H of *X*—*X*, *i.e.,* the electron pair forming the bond would have been equally shared by H and *X*, the bond H—*X* would not be so. Pauling suggested that extra stability was derived from unequal sharing of the bonding electrons between *A* and *B,* giving rise to some ionic character. The extent of this ionic character, as measured by the extra stability of the bond, is dependent on unequal sharing of the bonding electrons, which depends on the *(EN)* difference of H and *X*. Extra stability of the bond H—*X* has been correlated to the difference in (EN).

This scale determines electronegativity of the element *X* w.r.t. hydrogen H

$$\chi_X - \chi_H = 0.208\sqrt{\Delta_{H-X}}$$

where χ (chi) represents electronegativity, Δ_{H-X} is the extra bond energy in kcal mol^{-1}

and $$\Delta_{H-X} = E_{H-X} - \sqrt{E_{H-H}\,E_{X-X}}$$

E_{H-X}, E_{H-H} and E_{X-X} represent bond energy of $H-X, H-H$ and $X-X$ bonds respectively.

Converting the equation to SI units

$$\chi_X - \chi_H = 0.1017\sqrt{\Delta_{H-X}}$$

where Δ_{H-X} is measured in kJ mol^{-1}.

- $\chi_H = 2.05$ (and not zero) to avoid any negative value of the element.

Some other elements have following values of electronegativity based on Pauling's scale :

Metals / Non-metals / Metalloids

Period	(1) IA	(2) IIA	(3) IIIB	(4) IVB	(5) VB	(6) VIB	(7) VIIB	(8) VIIIB	(9) VIIIB	(10) VIIIB	(11) IB	(12) IIB	(13) IIIA	(14) IVA	(15) VA	(16) VIA	(17) VIIA	(18) VIIIA
1	1 H 2.1																	2 He
2	3 Li 1.0	4 Be 1.5											5 B 2.0	6 C 2.5	7 N 3.0	8 O 3.5	9 F 4.0	10 Ne
3	11 Na 1.0	12 Mg 1.2											13 Al 1.5	14 Si 1.8	15 P 2.1	16 S 2.5	17 Cl 3.0	18 Ar
4	19 K 0.9	20 Ca 1.0	21 Sc 1.3	22 Ti 1.4	23 V 1.5	24 Cr 1.6	25 Mn 1.6	26 Fe 1.7	27 Co 1.7	28 Ni 1.8	29 Cu 1.8	30 Zn 1.6	31 Ga 1.7	32 Ge 1.9	33 As 2.1	34 Se 2.4	35 Br 2.8	36 Kr
5	37 Rb 0.9	38 Sr 1.0	39 Y 1.2	40 Zr 1.3	41 Nb 1.5	42 Mo 1.6	43 Tc 1.7	44 Ru 1.8	45 Rh 1.8	46 Pd 1.8	47 Ag 1.6	48 Cd 1.6	49 In 1.6	50 Sn 1.8	51 Sb 1.9	52 Te 2.1	53 I 2.5	54 Xe
6	55 Cs 0.8	56 Ba 1.0	57 La 1.1 *	72 Hf 1.3	73 Ta 1.4	74 W 1.5	75 Re 1.7	76 Os 1.9	77 Ir 1.9	78 Pt 1.8	79 Au 1.9	80 Hg 1.7	81 Tl 1.6	82 Pb 1.7	83 Bi 1.8	84 Po 1.9	85 At 2.1	86 Rn
7	87 Fr 0.8	88 Ra 1.0	89 Ac 1.1 †															

*	58 Ce 1.1	59 Pr 1.1	60 Nd 1.1	61 Pm 1.1	62 Sm 1.1	63 Eu 1.1	64 Gd 1.1	65 Tb 1.1	66 Dy 1.1	67 Ho 1.1	68 Er 1.1	69 Tm 1.1	70 Yb 1.0	71 Lu 1.2
†	90 Th 1.2	91 Pa 1.3	92 U 1.5	93 Np 1.3	94 Pu 1.3	95 Am 1.3	96 Cm 1.3	97 Bk 1.3	98 Cf 1.3	99 Es 1.3	100 Fm 1.3	101 Md 1.3	102 No 1.3	103 Lr 1.5

Electronegativity values are given at the bottoms of the boxes.
The noble gases are not included in this discussion.

- **If two atoms have similar electronegativity, the bond between H and *X* will be predominantly covalent.**
- **A large difference in electronegativity leads to a bond with a high degree of polar character, and bond will be predominantly ionic.**
- Ionic character of a bond varies with the difference in electronegativity as given in Fig. 8.11. This is based on the ionic character; HI : 4%, HBr : 11%, HCl : 19% and HF : 45% which are known from dipole moment measurements.

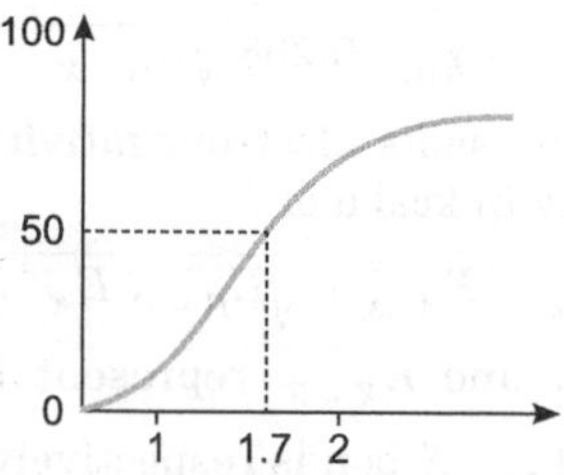

Fig. 3.11 Electronegativity difference

- 50% ionic character is shown in the cases when electronegativity difference is about 1.7. Thus,

$(\chi_X - \chi_H) = 1.7$ 50% ionic

$(\chi_X - \chi_H) > 1.7$ predominantly ionic

$(\chi_X - \chi_H) < 1.7$ predominantly covalent

Mulliken-Jaffe Electronegativity

Electronegativity (*EN*) can be regarded as the average of the ionisation energy (*IE*) and the electron affinity (*EA*) of an atom.

$$(EN) = \frac{IE + EA}{2}$$

If both (*EA*) and (*IE*) are determined in eV units then Pauling's electronegativity $(EN)_P$ is related to Mulliken's electronegativity $(EN)_M$ by the equation :

$$(EN)_P = 0.336[(EN)_M - 0.615]$$

- If (*EA*) and (*IE*) are in eV, then (*EN*) in eV is given by
  $0.187\ (IE + EA) + 0.17$
- **Mulliken's values were about 2.8 times larger than the Pauling's values.**

Allred-Rochow's Electronegativity

Allred and Rochow defined electronegativity as the force exerted by the nucleus of an atom on its valence electrons :

$$(EN)_{A\text{–}R} = \frac{0.359\ Z_{\text{effective}}}{r^2} + 0.744$$

where $Z_{\text{effective}}$ is the effective nuclear charge and *r* the covalent radius (in Å).

The relative (*EN*) values of different elements in the same scale is extremely helpful in studying and predicting the nature of bonds formed between them.

Metallic Nature

Metals have the tendency to form cations by loss of electrons and this property makes the elements as **electropositive** elements or metals.

$$M(g) \longrightarrow M^+(g) + e$$

- The tendency of an element to lose electron is closely connected to the (*IE*) of the element. **The smaller the (*IE*) of an element, the greater will be its tendency to lose electrons** and **thus greater will be its metallic character.**
- Tendency to oxidise itself provides reducing property to the elements thus, smaller the (*IE*), greater the metallic character, hence greater the reducing nature : **(*IE*) increases moving** along a **period left to right** and decreases down the group, hence metallic and reducing nature decrease along the period and increase down the group.

Nuclear charge	Shielding		Tendency to make +ve ions	Reducing power	Metallic nature
increases	increases	But the shielding effect wins so ⇒	increases	increases	increases

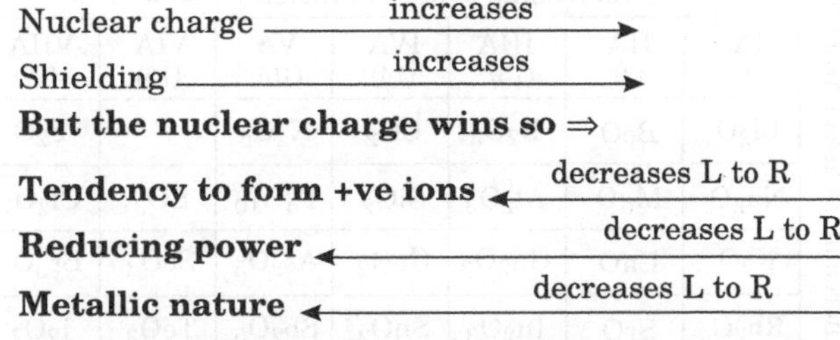

Notice that the most reactive metals (the top of the activity series) are on the left of the Periodic Table, whereas the least reactive metals (the bottom of the activity series) are in the transition metal groups closer to the right side of the table as shown in Fig. 3.12.

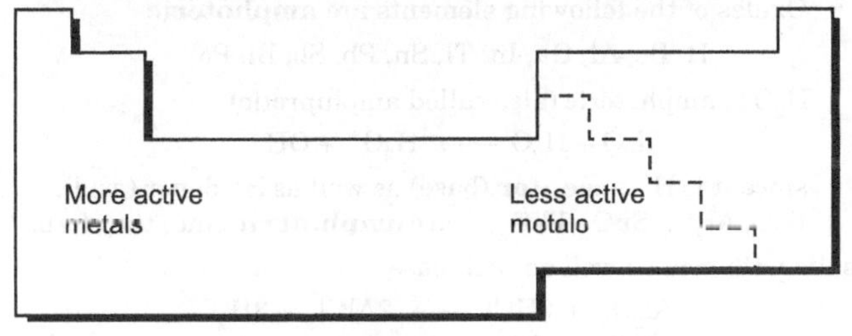

Fig. 3.12

Redox reactions are common for almost every element in the Periodic Table except for the noble gas elements of group VIIIA (18). In general, metals act as reducing agents, and reactive non-metals such as O_2 and the halogens act as oxidizing agents as given in Fig. 3.13.

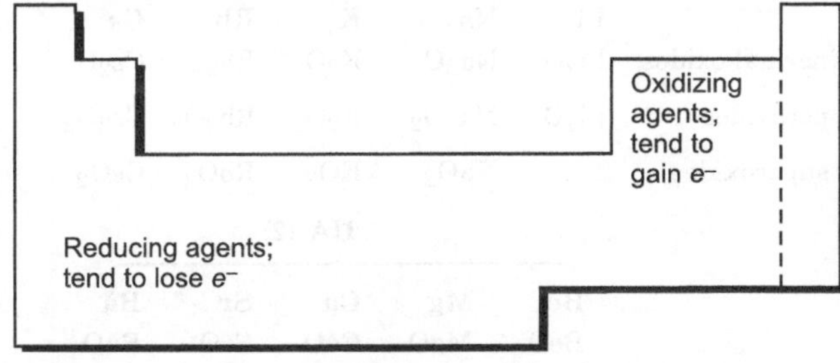

Fig. 3.13

Hydration and Hydration Energy

Hydration energy is the enthalpy change that accompanies the dissolving of one mole of gaseous ions in water.

$$Li^+(g) + H_2O \longrightarrow [Li(H_2O)]^+, \quad \Delta H = -806 \text{ kJ mol}^{-1}$$

- Size of the ion and its charge determines extent of hydration.
- Greater the charge, smaller the size of the ion, greater the attraction for the lone pair of O of H_2O, hence greater the extent of hydration and hence greater the hydration energy
  - size of the hydrated ions increases,
  - ionic mobility decreases [heavier (hydrated) ions moves slower].

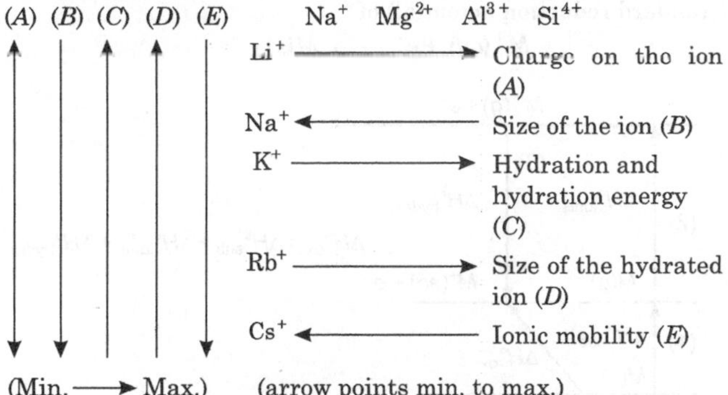

Table 3.6 compares extent of hydration and other related properties of 1(IA) group cations.

Table 3.6 **Hydration and Ionic Mobilities**

Ion	Ionic radius (Å)	Ionic mobilities*	Radius of hydrated ion (Å)	Hydration energy** $\Delta H^\circ_{hyd.}$	Oxidation energy** $\Delta H^\circ_{ox.}$	$E^\circ_{redn.}$ (V)
Li^+	0.76	33.5	3.40	−544	169	− 3.045
Na^+	1.02	43.5	2.76	−435	208	− 2.714
K^+	1.38	64.5	2.32	−352	191	− 2.925
Rb^+	1.52	67.5	2.28	−326	200	− 2.925
Cs^+	1.67	68.0	2.28	−293	196	− 2.923

* Values in 10^{-8} m^2 V^{-1} s^{-1}

** Values in kJ mol^{-1}

- Oxidation of an alkali metal M to ions $M^+(aq)$ involves the **hypothetical** steps :
  1. Sublimation of the solid metal to the gaseous state, $M(g)$,
  2. Ionisation of the gaseous metal to the gaseous metal ions, $M^+(g)$,
  3. Hydration of the gaseous metal ions to produce $M^+(aq)$.

DEEP Focus

From Table 3.6 we see that they are positive quantities, but the value is smallest for Li. This is primarily because of the less negative ΔH for the hydration of Li^+. Thus, it appears that Li(s) should be easiest to oxidise of the alkali metal and is thus best reducing agent of the alkali metals but in aqueous solution which is also confirmed by its most negative value of $E^\circ_{\text{redn.}}$ (standard reduction potential) of

$$M^+(aq) + e^- \longrightarrow M(s)$$

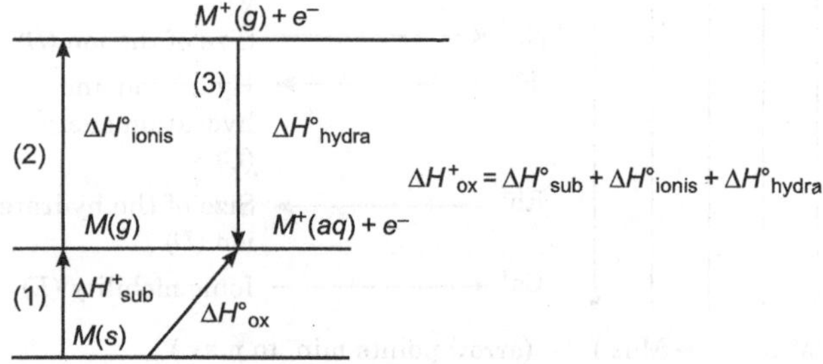

Acid-Base Character of Oxides

On moving across a period, the basic character of the oxides gradually changes first into amphoteric and finally into acidic character.

- On moving down the group, reverse behaviour is observed, *i.e.*, from more acidic to more basic.

  Oxides of the element M in H_2O produce MOH

  – If electronegativities difference of M and O is greater than that of H and O in H_2O then MOH is acidic due to formation of H_3O^+

  $$M\text{—O—H} + H_2O \longrightarrow H_3O^+ + MO^-$$

  – If electronegativities difference of M and O is less than that of H and O in H_2O then MOH is basic due to formation of OH^-

  $$M\text{—O—H} + \text{H—O—H} \longrightarrow [MOH_2]^+ + OH^-$$

- Stability of oxides decreases across a period.
- Behaviour of the elements is shown in Fig. 3.14.

Increasing acidic character ⟶

Increasing base character ↓

IA (1)	IIA (2)	IIIA (13)	IVA (14)	VA (15)	VIA (16)	VIIA (17)
Li_2O	BeO	B_2O_3	CO_2	N_2O_5		F_2O
Na_2O	MgO	Al_2O_3	SiO_2	P_4O_{10}	SO_3	Cl_2O_7
K_2O	CaO	Ga_2O_3	GeO_2	As_2O_5	SeO_3	Br_2O_7
Rb_2O	SrO	In_2O_3	SnO_2	Sb_2O_5	TeO_3	I_2O_7
Cs_2O	BaO	Tl_2O_3	PbO_2	Bi_2O_5	PoO_3	At_2O_7

DEEP Focus

- Oxides of the following elements are **amphoteric**

  H, Be, Al, Ga, In, Tl, Sn, Pb, Sb, Bi, Po

  H_2O is amphoteric (also called amphiprotic)

  $$H_2O + H_2O \longrightarrow H_3O^+ + OH^-$$

  since it is H^+ **acceptor** (base) as well as H^+ donor (acid).

  BeO, Al_2O_3, SnO_2, PbO_2, ... are **amphoteric** since they form salts with acid as well as with base

  $$\underset{\text{base}}{Al_2O_3} + \underset{\text{acid}}{6HCl} \longrightarrow 2AlCl_3 + 3H_2O$$

  $$\underset{\text{acid}}{Al_2O_3} + \underset{\text{base}}{2NaOH} + 3H_2O \longrightarrow 2Na[Al(OH)_4]$$

  – oxide is acidic if it reacts with a base.

  – oxide is basic if it reacts with an acid.

IA (1)

	Li	Na	K	Rb	Cs
normal oxides	Li_2O	Na_2O	K_2O	Rb_2O	Cs_2O
peroxides	Li_2O	Na_2O_2	K_2O_2	Rb_2O_2	Cs_2O_2
superoxides		NaO_2	KO_2	RbO_2	CsO_2

IIA (2)

Be	Mg	Ca	Sr	Ba
BeO	MgO	CaO	SrO	BaO
		CaO_2	SrO_2	BaO_2

1	2	3	4	5	6	7	8	9	10	11	12	13	14	15	16	17	18
H																	
Li	Be											B	C	N	O	F	
Na	Mg											Al	Si	P	B	Cl	
K	Ca	Sc	Ti	V	Cr	Mn	Fe	Co	Ni	Cu	Zn	Ga	Ge	As	Se	Br	
Rb	Sr	Y	Zr	Nb	Mo	Tc	Ru	Rh	Pd	Ag	Cd	In	Sn	Sb	Te	I	
Cs	Ba	La	Hf	Ta	W	Re	Os	Ir	Pt	Au	Hg	Tl	Pb	Bi	Po	At	
Fr	Ra	Ac															

Transition metals

Predominantly base oxides

Acidic, covalent oxides

Oxides with amphoteric nature generally ionic

Fig. 3.14 The nature of oxides of the elements

Illustration 1 *Arrange the following oxides in order of increasing molecular (acidic) character:*

$$SO_3 , Cl_2O_7 , CaO \text{ and } PbO_2$$

Solution Acidic character of oxides increases as non-metallic character of the element that is combined with oxygen increases

increasing non-metallic character

	Ca <	Pb <	S <	Cl
Periodic group :	IIA	IVA	VIA	VIIA
	(2)	(12)	(16)	(17)

increasing (molecular) acidic character

$$CaO < PbO_2 < SO_3 < Cl_2O_7$$

Hydrides

- Hydrogen combines with a number of other elements including metals and non-metals to form compounds callod hydrides (with H^-).
- Covalent nature of hydrides increases across a period and decreases down the group.
- Ionic hydride are better reducing agents than covalent hydride and reducing nature of hydride decreases across a period and increases down the group.
- Covalent and ionic hydrides are classified in Fig. 3.15.

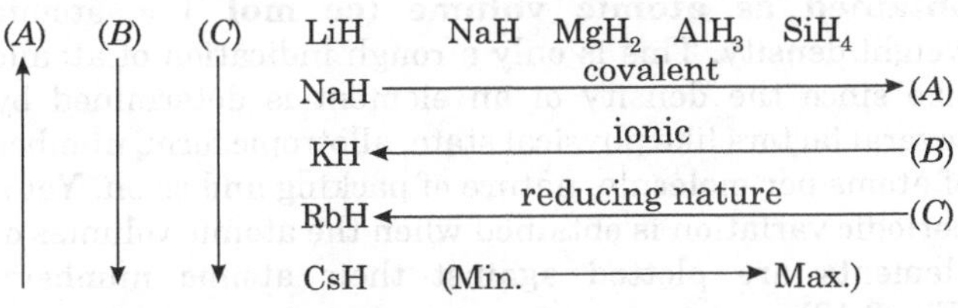

IA 1	IIA	IIIB	IVB	VB	VIB	VIIB	VIII			IB	IIB	IIIA	IVA	VA	VIA	VIIA	VIIIA 18
H	2											13	14	15	16	17	
Li	Be				Transition metals							B	C	N	O	F	
Na	Mg	3	4	5	6	7	8	9	10	11	12	Al	Si	P	S	Cl	
K	Ca	Sc	Ti	V	Cr	Mn	Fe	Co	Ni	Cu	Zn	Ga	Ge	As	Se	Br	
Rb	Sr	Y	Zr	Nb	Mo	Tc	Ru	Rh	Pd	Ag	Cd	In	Sn	Sb	Te	I	
Cs	Ba	La	Hf	Ta	W	Re	Os	Ir	Pt	Au	Hg	Tl	Pb	Bi	Po	At	
Fr	Ra	Ac															

Mainly ionic with often of uncertain formula :

Covalent, molecular structures fixed formula

absorbed into metal lattice

Fig. 3.15 The nature of hydrides of the elements

Melting Points and Boiling Points

A solid melts when the forces holding its constituent units in position is overcome by thermal energy as a consequence of increase in temperature. The melting temperature is determined by a number of different factors :

(1) Nature of packing of the atoms, ions, or molecules,

(2) The lattice energy,

(3) Association through hydrogen bond, etc.

Boiling involves complete detachment of the particles of a liquid. Both the melting and boiling points of an element thus indicate the nature and extent of binding between the atoms of that element. In spite of the contribution of a large number of variable factors in controlling the melting and boiling points, a striking periodicity is observed in their values. Fig. 3.16

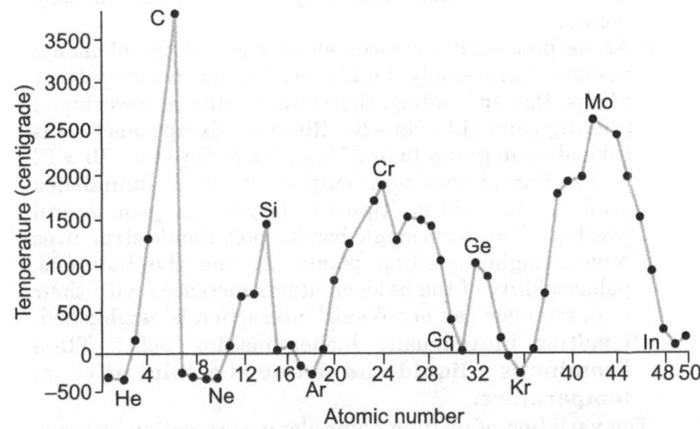

Fig. 3.16 Periodic variation of melting point of elements

represents a plot of melting points of some elements against their respective atomic numbers. The shape of the curve indicates periodic variation. It may be further observed that :

(i) The noble gases have very low melting points.

(ii) The melting point of the elements in any period increases gradually to group 14 where it becomes maximum; iot then decreases to the next noble gas.

(iii) The melting points of elements occurring in a group usually decrease downwards, *i.e.*, as the elements bear higher atomic numbers.

DEEP Focus

(i) The noble gas atoms contain a completed electron shell. So there exists little tendency for these atoms to combine between themselves. This is why they occupy the lowest range of melting points.

(ii) The alkali metals have only one electron in their outermost shells. Their contribution to metallic bonding is, therefore, weak, resulting in low melting points.

(iii) As the number of electrons participating in metallic bonding increases, the melting point also increases : Be > Li; Mg > Na.

(iv) Boron, carbon and silicon form giant three-dimensional structures; breaking of these arrangements require large amount of energy. Hence, they have very high melting points.

(v) Nitrogen, oxygen, fluorine—all form stable diatomic molecules containing triple, double and single bonds (between their atoms) respectively. They do not form giant three-dimensional lattices—may be due to greater repulsion between the non-bonding electrons and the greater stabilities of these multiple bonds. At the same time, very poor binding exists among the molecules themselves. Consequently they have very low melting points.

(vi) As we proceed downwards along a group, the elements involve increasingly highly orbital for binding their atoms. Poorer bonding, therefore, results in lowering of melting point (Li—Na—K—Rb—Cs). Exceptions to this take place in group 15 to 17 (*e.g.*, P > N; S > O; I > Br > Cl > F). Phosphorus and sulphur have a diminished tendency to bond multiplicity (larger size poor lateral overlap). They form single bonded polyatomic structures having higher melting points. Among the halogens, polarisability of the halogen atoms increases with their size; stronger van der Waals' interaction is established, resulting in gradually higher melting point. **Thus bromine is a liquid and iodine is a solid at room temperature.**

The variation of melting point along a transition series is irregular. The *d*-orbitals and the electrons there in contribute to metallic bonding to a varying degree. A sharp drop in melting point is observed just after the completion of a transition series, *e.g.*, Cu : 1356 K *vs.* Zn : 692.5 K; Ag : 1233.8 K *vs.* Cd : 593.9 K. **It appears that the completed *d*-orbitals do no longer effectively participate in metallic bonding.**

Density

The densities of the elements in the solid state also exhibit periodic variation when plotted against their atomic numbers. It may be observed from Fig. 3.17 that the densities of the alkali metals occupy the minima in the curve; the density first rises and then decreases along a period, the elements in the middle having the highest density values.

Here again the underlying reason is largely complicated by the nature of packing of the atoms. A simple interpretation may be provided by remembering that **stronger and compact binding leads to high density. The binding force is again dependent on the number of valence electrons provided by each atom.** This is why the density of the central elements in any period is highest.

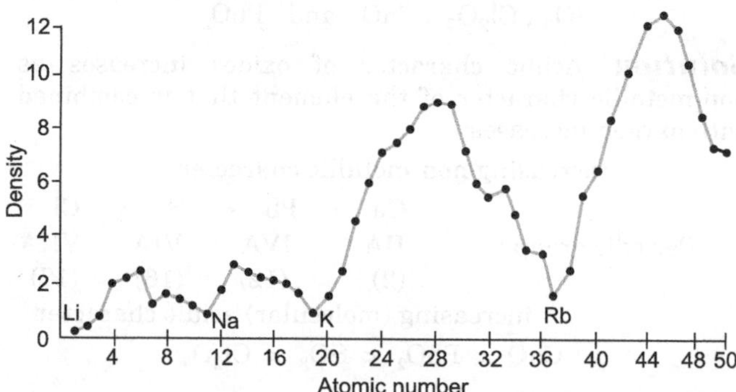

Fig. 3.17 Periodic variation of density with atomic number

The increase of density down any group in the Periodic Table is similarly explained from the increasing number of neutrons and protons in the nuclei.

Atomic Volume

Lothar Meyer, in 1870, independently introduced the idea of periodicity in properties of the elements from study of their atomic volumes. **The atomic volume is the volume of one mole of atoms of an element; it was obtained as atomic volume (cc mol^{-1})** = atomic weight/density. This is only a rough indication of atomic size since the density of an element is determined by several factors like physical state, allotropic form, number of atoms per molecule, nature of packing and so on. Yet a periodic variation is obtained when the atomic volumes of elements are plotted against their atomic numbers. (Fig. 3.18)

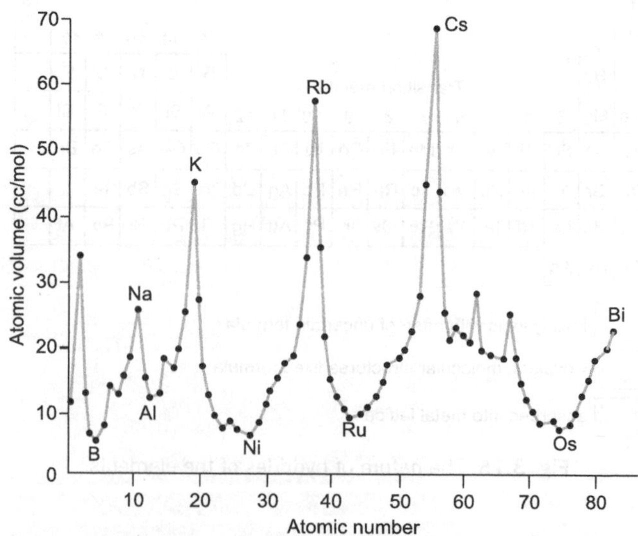

Fig. 3.18 Periodic variation of atomic volume

Target Practice 3

1. Any salt MCl_2 is soluble in water. Relate lattice energy and hydration energy for MCl_2 to be water soluble ....
2. Arrange hydroxide of alkaline earth metals in increasing solubility in water. ....
3. Arrange solubility of sulphates of
   (a) alkali metals
   (b) alkaline earth metals
   in increasing order ....
4. Which is more stable (thermally) in each pair
   (a) $BaCO_3, SrCO_3$ (b) $BeCO_3, CaCO_3$
   (c) $MgCO_3, BeCO_3$ (d) Na_2CO_3, Li_2CO_3
   (e) $NaHCO_3, Na_2CO_3$
5. What is enthalpy change in the formation of MgO?
   (i) $Mg(s) \longrightarrow Mg(g)$; $\Delta H = S$
   (ii) $Mg(g) \longrightarrow Mg^+(g) + e^-$; $\Delta H = I_1$
   (iii) $Mg^+(g) \longrightarrow Mg^{2+}(g) + e^-$; $\Delta H = I_2$
   (iv) $O_2(g) \longrightarrow 2O(g)$; $\Delta H = D$
   (v) $O(g) + e^- \longrightarrow O^-(g)$; $\Delta H = -E_1$
   (vi) $O^-(g) + e^- \longrightarrow O^{2-}(g)$; $\Delta H = E_2$
   (vii) $Mg^{2+}(g) + O^{2-}(g) \longrightarrow MgO(s)$; $\Delta H = -U$

Answers

1. For a salt to be water soluble, hydration energy should exceed the lattice energy
   hydration energy > lattice energy
2. $Be(OH)_2 < Mg(OH)_2 < Ca(OH)_2 < Sr(OH)_2 < Ba(OH)_2$
   For salts of small anions (as fluorides, hydroxides), the lattice energy decreases from Be to Ba due to increase in size. Hydration energy also decrease but this decrease is less than decrease in lattice energy. As a result hydration energy increases. Hence, solubility of hydroxides increases.
3. (a) $Li_2SO_4 < Na_2SO_4 < K_2SO_4 < Rb_2SO_4$
   (b) $BaSO_4 < CrSO_4 < CaSO_4 < MgSO_4 < BeSO_4$
4. (a) $BaCO_3 > SrCO_3$
   (b) $CaCO_3 > BeCO_3$
   (c) $MgCO_3 > BeCO_3$
   (d) $Na_2CO_3 > Li_2CO_3$
   (e) $Na_2CO_3 > NaHCO_3$
5. Add all steps (i) to (vii) (with (iv)/2)
   $$Mg(s) + \frac{1}{2}O_2(g) \longrightarrow MgO(s)$$
   Net enthalpy change is
   $$\Delta H = S + I_1 + I_2 + \frac{D}{2} - E_1 + E_2 - U$$

Practice Exercise 3

1. Which has greater (EA) value in each pair?
   (a) F or Cl^- (b) O or O^-
   (c) Ne^+ or F (d) F or F^-
2. EA of Cu is 123 kJ mol^{-1} but that of Zn is –87 kJ mol^{-1}. Explain.
3. We consider formation of NaCl(s) using Born-Haber cycle which involves following steps :
   (I) $Na(s) + \frac{1}{2}Cl_2 \longrightarrow NaCl(s)$ $\Delta H = Q$
   (II) $Na(s) \longrightarrow Na(g)$ $\Delta H = S$
   (III) $Na(g) \longrightarrow Na^+(g) + e^-$ $\Delta H = I$
   (IV) $\frac{1}{2}Cl_2(g) \longrightarrow Cl\ (g)$ $\Delta H = \frac{D}{2}$
   (V) $Cl(g) + e^- \longrightarrow Cl^-\ (g)$ $\Delta H = ?$
   (VI) $Na^+(g) + Cl^-(g) \longrightarrow NaCl\ (s)$ $\Delta H = -U$
   Can you derive value of (EA) of Cl(g) atom.
4. Why are the electron affinities of the alkaline earth metals positive?
5. Is it likely that Ar will form the anion Ar^-.
6. Considering their electrons affinities, do you think it is possible for alkali metals to form an anion like M^- where M represents an alkali metal.
7. What is relationship between the electron affinity of a univalent cation such as Na^+ and the ionisation energy of the neutral atom?
8. Calculate effective nuclear charge that added electron would experience in F-atom in the formation of negative ion.
9. (a) Explain Na_2CO_3 is thermally stable, but Li_2CO_3 and $MgCO_3$ both decompose on heating.
   (b) $BeCl_2$ and $AlCl_3$ are Lewis acids.
   (c) B and Si resemble each other in their properties.
10. Which is more electronegative in each pair?
    (a) F^- or Na^+ (b) Ne or F^-
    (c) F or Cl (d) H^- or He
11. Electronegativity of F on Pauling's scale is 4.0. What is the value on Mulliken's scale ?
12. A^+B^- and A^-B^+ can be formed from elements (A) and (B). Explain their formation based on relative values of (EN), (EA) and (IE).
13. Covalent radius of F is 0.72 Å. Calculate its Allred-Rochow's electronegativity.
14. Covalent radius of oxygen is 0.74 Å and its A.R. electronegativity is 3.5. What is effective nuclear charge experienced by oxygen ?
15. Make a **Born-Haber** cycle for the preparation of $MgBr_2(s)$.
16. Calculate the electronegativity of carbon from the following data :
    $E_{H-H} = 104.2$ kcal mol^{-1}; $E_{C-C} = 83.1$ kcal mol^{-1}
    $E_{C-H} = 98.8$ kcal mol^{-1}

17. (*IE*) and (*EA*) of fluorine are respectively 17.418 eV and –9.278 eV. Calculate (*EN*) of fluorine atom.

18. Four atoms are arbitrarily labelled *D, E, F* and *G*. Their electronegativities are as follows $D = 3.8$, $E = 3.3$, $F = 2.8$ and $G = 1.3$. If the atoms of these elements form the molecules *DE, DG, EG*, and *DF*, how would you arrange these molecules in order of increasing covalent bond character.

19. Tendency to lose electrons shows reducing property of the element. Arrange the following in order of reducing property:

(a) Na, K, Rb (b) Na, Mg, Al

(c) F^-, Cl^-, Br^-, I^- (d) Mg, Ca, Sr

20. Li is better reducing agent than Na. Explain.

21. Which of the following reactions are possible ?

(a) $3Mg + 2Al^{3+} \longrightarrow 2Al + 3Mg^{2+}$

(b) $Cl_2 + 2I^- \longrightarrow I_2 + 2Cl^-$

(c) $K^+ + Na \longrightarrow K + Na^+$

22. Arrange the following ions in increasing

$$Na^+, Mg^{2+}, Al^{3+}$$

(i) extent of hydration

(ii) hydration energy

(iii) size of the hydrated ions

(iv) ionic mobility

(v) standard reduction potential ($E^\circ_{redn.}$),

$$M^{n+}(aq) + ne^- \longrightarrow M(s)$$

(vi) enthalpy of oxidation (ΔH°_{ox}) of *M(s)*.

23. In the preparation of hydrocarbon by Kolbe's electrolysis, generally *R*COONa or *R*COOK are taken but not *R*COOLi. Explain.

24. LiCl is hydrated but NaCl is always anhydrous. Explain.

25. Arrange NH_3, PH_3 and AsH_3 in increasing reducing nature.

26. Fill in the blanks :

(i) NaH is .... hydride.

(ii) B_2H_6 is called ....

(iii) SiH_4 is called ....

(iv) Most widely used mixed hydride in organic synthesis is ....

27. Write the formulae and names of the hydride of the following second period elements :

Li, C, N, O, F.

28. SnO_2 forms two types of colloidal sols, positively charged in an acidic medium and negatively charged in a basic medium. Explain.

29. Arrange following oxides in increasing acidic nature

$$Li_2O, BeO, B_2O_3$$

30. Explain:

(i) Aqueous CO_2 can cause rusting.

(ii) Al_2O_3 is amphoteric while B_2O_3 is acidic.

(iii) BeO and Al_2O_3 both are amphoteric.

31. Which oxide is more basic, MgO or BaO? Why?

32. Write the formulae and names of the oxides of second-period elements (Li to N). Identify the oxides as acidic, basic or amphoteric.

Answers

1. (i) F (ii) O^- (more positive) (iii) Ne^+ (iv) F

2. Cu (Ar) $3d^{10}\ 4s^1 + e^- \longrightarrow Cu^-$ (Ar) $3d^{10}\ 4s^2$

In case of Cu, added electron goes into empty 4*s*-orbital where it can get repulsion to a certain extent giving a positive value of (*EA*). In case of Zn, added electron can be accommodated into 4*p*-orbital giving a negative value of (*EA*).

3. (v) involves addition of electron to Cl(*g*)

$$Cl(g) + e^- \longrightarrow Cl^-(g) \qquad EA = x$$

Energy released is (*EA*)

Adding (ii) to (vi) gives

$$Na(s) + \frac{1}{2}Cl_2(g) \longrightarrow NaCl(s)$$

which is also the step (i), hence

$$S + I + \frac{D}{2} + x(EA) - U = Q$$

hence, *x* is calculated.

4. $$M(g) + e^- \longrightarrow M^-(g)$$

The extra electron must enter *np*-subshell, which is effectively shielding the two *ns*-electrons and the inner electrons. Consequently, the alkaline earth metals have little tendency to pick up an extra electron.

5. Added electron enters 4*s* orbital of higher energy where it gets repulsion. Thus, (*EA*) value is positive. Thus, Ar^- is not formed.

6. $$M(g) + e^- \longrightarrow M^-(g) \qquad \Delta H = -\text{ ve}$$

$$ns^1 \uparrow \qquad\qquad ns^2 \uparrow\downarrow$$

Yes, since $M^-(g)$ has paired *ns*-electrons

7. $$Na^+(g) + e^- \longrightarrow Na(g) \qquad \Delta H = -(EA)$$

$$Na(g) \longrightarrow Na^+(g) + e^- \qquad \Delta H = +(IE)$$

$$\therefore \quad (IE) = -(EA)$$

8. $Z^* = Z - S = 5.85$

9. (i) Due to diagonal relationship, Li and Mg resemble.

(ii) Due to diagonal relationship.

(iii) Due to diagonal relationship.

10. (a) both zero (b) both zero (c) F (d) both zero

11. 11.2 (4.8 times larger)

12. $$A \longrightarrow A^+ + e^- \text{ if } (IE)_A < (IE)_B$$

$$B + e^- \longrightarrow B^- \text{ if } (EA)_A < (EA)_B$$

Thus, A^+B^- is formed if $(IE)_A + (EA)_A < (IE)_B + (EA)_B$

or $$\frac{(IE)_A + (EA)_A}{2} < \frac{(IE)_B + (EA)_B}{2}$$

or $$(EN)_A < (EN)_B$$

Thus, A^+B^- exists if electronegativity of $A < B$. Similarly, A^-B^+ is formed if electronegativity of $A > B$.

13. 4.7952 **14.** 4.2039

15.

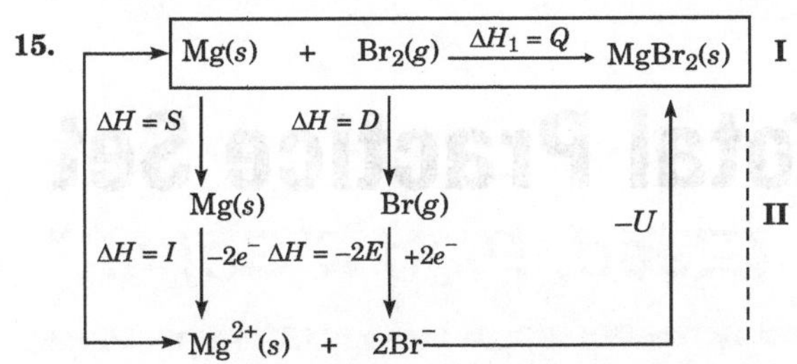

$$\Delta H_{II} = S + I + D - 2E - U = \Delta H_1 = Q$$

S = heat of sublimation of Mg(s)

I = ionisation energy of Mg(g)

D = dissociation energy of $Br_2(g)$

$-E$ = electron affinity of Br(g)

$-U$ = lattice energy of $MgBr_2(s)$

Q = heat of formation of $MgBr_2(s)$

16. Using Pauling's scale :

$$\chi_C = \chi_H + 0.182\sqrt{\Delta_{C-H}}$$

$$\chi_H = 2.05$$

$$\Delta_{C-H} = E_{C-H} - \sqrt{E_{H-H} \cdot E_{C-C}}$$

$$= 98.8 - \sqrt{104.2 \times 83.1} = 5.75$$

$$\chi_C = 2.49$$

17. 4.07 eV

18. Smaller the (*EN*) difference, larger the covalent nature.

$$DG < DF < EG < DE$$

increasing covalent bond character ⟶

19. (a) Na < K < Rb (b) Na > Mg > Al (c) $F^- < Cl^- < Br^- < I^-$

(d) Mg < Ca < Sr

20. The large hydration energy is responsible for the case with which lithium ion will be made in solution. The electrode potential of Li reflects this tendency for Li to convert into hydrated ion, rather than its inherent ability as a reducing agent. *If lithium takes part in reactions that do not involve water, then it does show less reducing power than the other members of the group.*

21. (a) and (b)

22. (i) $Na^+ < Mg^{2+} < Al^{3+}$, (ii) $Na^+ < Mg^{2+} < Al^{3+}$

(iii) $Na^+ < Mg^{2+} < Al^{3+}$, (iv) $Al^{3+} < Mg^{2+} < Na^+$

(v) $Al^{3+} < Mg^{2+} < Na^+$ (E°_{red} = –ve), (vi) Al < Mg < Na

23. Ionic mobility of $Li^+ << Na^+$ or K^+ (Refer Table 3.6)

Thus, Na^+ or K^+ reaches cathode faster than Li^+ and thus is discharged at a faster rate. (This is due to greater hydration of Li^+ than other ions)

24. Li^+ (smaller size) is more hydrated than Na^+ (larger size).

25. $NH_3 < PH_3 < AsH_3$

26. (i) an ionic (ii) diborane (iii) silane (iv) $LiAlH_4$

27. LiH, CH_4, NH_3, H_2O, HF

28. SnO_2 forms Sn^{4+} in acidic medium and SnO_3^{2-} in basic medium. These ions are preferentially adsorbed on SnO_2 forming colloidal sol.

$$SnO_2 + 4H^+ \longrightarrow Sn^{4+} + 2H_2O$$

$$SnO_2 + Sn^{4+} \longrightarrow \underset{\text{positively charged}}{[SnO_2]Sn^{4+}}$$

$$SnO_2 + 2OH^- \longrightarrow SnO_3^{2-} + H_2O$$

$$SnO_2 + SnO_3^{2-} \longrightarrow \underset{\text{negatively charged}}{[SnO_2]\,SnO_3^{2-}}$$

29. $\underset{\text{basic}}{Li_2O} < \underset{\text{amphoteric}}{BeO} < \underset{\text{acidic}}{B_2O_3}$

30. (i) CO_2 is an acidic oxide

$$CO_2 + H_2O \longrightarrow H_2CO_3 \rightleftharpoons H^+ + HCO_3^-$$

H^+ can react with Fe forming Fe^{2+} and thus cause rusting.

$$Fe + 2H^+ \longrightarrow Fe^{2+} + H_2$$

(ii) While going down the group, basic nature increases and acidic nature decreases, hence

B_2O_3 is an acidic oxide.

Al_2O_3 is an amphoteric oxide.

(iii) Due to diagonal relationship.

31. BaO

32.

Li_2O	basic
BeO	amphoteric
B_2O_3	acidic
CO_2	acidic
N_2O_5	acidic

Total Practice Set
(Read, Plan & Solve!)

Problem 1. Draw a graph that shows how the second (IE) (given in kJ mol^{-1}) of the elements helium to sodium change. Explain the shape of the graph.

	He	Li	Be	B	C	N	O	F	Ne	Na
$(IE)_2$	525	729	175	242	235	285	333	347	395	456
	0	8	7	7	3	6	8	1	2	2

Solution Once Li has lost an electron, it has noble gas electron structure, *i.e.*, a filled set of electron shells. Like the noble gas it has high $(IE)_2$. When Be has lost one electron, it is isoelectronic of Li. Its single outer electron is shielded from the nucleus by a filled shell of electron. This gives a low $(IE)_2$. In similar fashion, there is a trend upwards across the period, with dips at places when an electron is outside a filled *s*-orbital, or where two electrons begin to fill a single *p*-orbital. Thus, behaviour of monovalent cation (formed after one electron is lost) with atomic number *Z*, is just parallel to the atom with atomic number (*Z* –1).

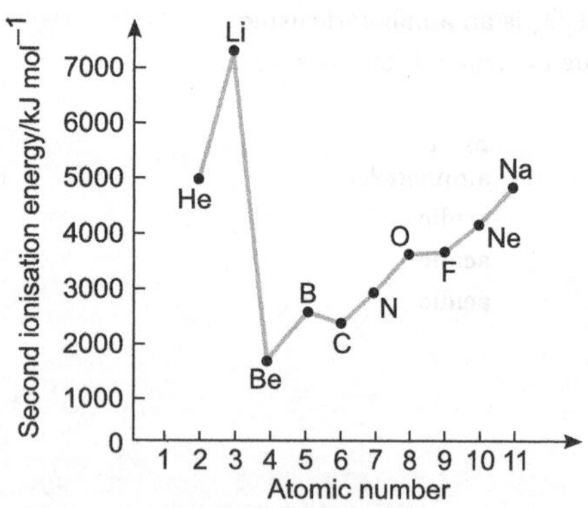

A graph of second ionisation energy against atomic number for the elements helium to sodium

For $(IE)_2$ behaviour is similar to ⟶	He^+	Li^+	Be^+	B^+	C^+	N^+	O^+	F^+	Ne^+	Na^+
	H	He	Li	Be	B	C	N	O	F	Ne

Problem 2. Of the following pairs, which would you expect to have

(a) the lower first (*IE*), Sr or Be?
(b) the more non-metallic behaviour, Ga or P?
(c) the larger atomic radius, Mg or Sr?
(d) the more negative (*EA*), Br or I?

Solution In the Periodic Table (*IE*), (*EA*) and non-metallic behaviour decrease down the group and increase across a period while radius increases down the group and decreases along a period.

(a) Sr has a lower (*IE*) than Be.
(b) P has more non-metallic behaviour than Ga.
(c) Sr has larger atomic radius than Mg.
(d) Br has more (*EA*) than I.

Problem 3. In the modern Periodic Table, the elements in groups IB(11) and IIB(12) are shown in *d*-block. This block is the place where the transition elements are found. Why do you think some chemists argue that the Group IB(11) and IIB(12) elements should not be considered as *d*-block elements?

Solution In the *d*-block, the 3*d*, 4*d*, or 5*d* orbitals are being filled. The properties of the transition elements are a result of their having unfilled *d*-orbitals. However IB and IIB elements have full *d*-orbitals.

IB (11) — $(n-1)d^{10}\ ns^1$ *e.g.* Cu_{29} : [Ar] $3d^{10}\ 4s^1$
IIB (12) — $(n-1)d^{10}\ ns^2$ Zn_{30} : [Ar] $3d^{10}\ 4s^2$

Indeed, these elements have *s*-orbitals filling. This is one reason why some chemists think them misplaced tacked on the end of the transition elements. However, the chemical properties of the IB metals make to include them among the transition metals, *viz.*, Cu^{2+} is coloured, forms complexes etc. The IIB metals are best excluded.

Problem 4. There are certain exceptions to the properties of the elements.

(i) Which metal is liquid at room temperature ?
(ii) Which non-metal is a solid with an extremely high melting point?
(iii) Which non-metal is a liquid at room temperature?
(iv) Which two elements would you expect to combine in the most violent fashion?

Solution (i) Hg
(ii) Carbon (diamond or graphite)
(iii) Br
(iv) Cs and F

Problem 5. The heats of formation (ΔH_f°) of the oxides of the third period, sodium to chlorine, are in kJ mol^{-1}

Na_2O	MgO	Al_2O_3	SiO_2	P_4O_{10}	SO_3	Cl_2O_7
−416	−602	−1676	−911	−2984	−395	+250

Divide these values by the number of oxygen atoms in the formula of the oxide. The resulting figure is a measure of the strength of the oxide. What is the link between the figures and the structures of the oxides?

Solution

Oxide	ΔH_f° (kJ mol^{-1})	Number of O-atoms = x	$\Delta H_f^\circ / x$
Na_2O	−416	1	−416
MgO	−602	1	−602
Al_2O_3	−1676	3	−558.7
SiO_2	−911	2	−455.5
P_4O_{10}	−2984	10	−298.4
SO_3	−395	3	−131.7
Cl_2O_7	+250	7	35.7

More negative the value of $(\Delta H_f^\circ / x)$, greater the stability. Thus, MgO is most stable and Cl_2O_7 is least stable, stability being :

$$Cl_2O_7 < SO_3 < P_4O_{10} < Na_2O < SiO_2 < Al_2O_3 < MgO$$

In general, stability decreases across the period from left to right (except Na_2O). This is shown in given fig.

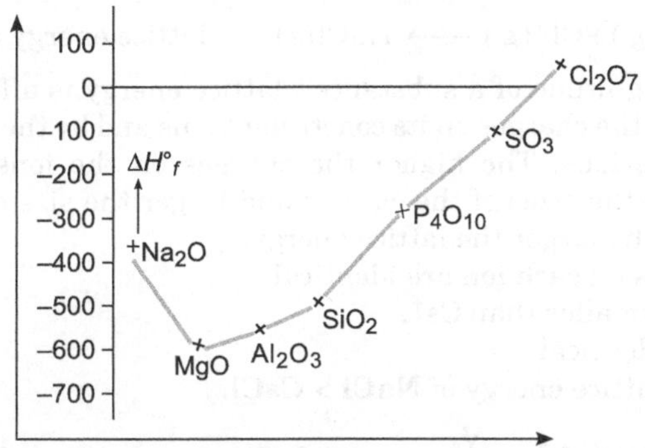

Problem 6. Match the columns in (*A*) with those given in (*B*).

(*A*)	(*B*)
(a) Inner transition elements	(i) Diagonal relationship
(b) *d*-block elements	(ii) *p*-block elements
(c) Elements with unpaired electrons	(iii) Typical elements
(d) Be—Al	(iv) Block deciding rule
(e) *s* and *p*-block elements	(v) Transition elements
(f) Elements of 3rd period	(vi) Paramagnetic
(g) ns^2np^1 to ns^2np^5	(vii) Normal elements
(h) Aufbau principle	(viii) *f*-block

Solution

A	B
(a)	(viii)
(b)	(v)
(c)	(vi)
(d)	(i)
(e)	(vii)
(f)	(iii)
(g)	(ii)
(h)	(iv)

Problem 7. What is effective nuclear charge at the periphery of nitrogen atom when an extra electron is added in the formation of an anion. Compare this value when the atom is ionised.

Solution $N(7) = 1s^2\, 2s^2 2p^3$

Added electron will be screened by five ($2s^2 2p^3$) electrons in **2nd orbit** and two ($1s^2$) electrons in **1st orbit.**

Added electron is screened

$(n-1)$ n

The screening constant, $\sigma = 5 \times 0.35$ in nth orbit
$= 2 \times 0.85$ in $(n-1)$th orbit

Total $\sigma = 3.45$

$\therefore$ Effective nuclear charge $= 7 - 3.45 =$ **3.55**

In the formation of positive ion, electron in the outermost orbit is screened by four electrons in ***n*th orbit** and two electrons in **(*n* − 1)th orbit**

This electron is now screened

$(n-1)$ n

Screening constant, $\sigma = 4 \times 0.35$ in nth orbit
$= 2 \times 0.85$ in $(n-1)$th orbit

Total value of $\sigma = 3.1$

$\therefore$ Effective nuclear charge $= 7 - 3.1 =$ **3.9**

Problem 8. The sums of first and second ionisation energies and those of third and fourth ionisation energies (in MJ mol^{-1}) of nickel and platinum are :

	$(IE)_1 + (IE)_2$	$(IE)_3 + (IE)_4$
Ni	2.49	8.80
Pt	2.66	6.70

Based on this information, write

(i) the most common oxidation states of Ni and Pt

(ii) name of the metal (Ni or Pt) which can more easily form compounds in its +4 oxidation state.

Solution (i) Ni = +2, Pt = +4 since $(IE)_1 + (IE)_2$ of Ni is less than its $(IE)_3 + (IE)_4$ and reverse is the case in Pt.

(ii) Platinum forms more stable complexes in +4 **state** due to its higher stability than +2 state.

Problem 9. Which atom should have a higher second *IE* : Li or Be?

Solution Ionisation of Li and Be can be represented as

$$\underset{1s^2\,2s^1}{\text{Li}(g)} \longrightarrow \underset{1s^2}{\text{Li}^+(g)} \longrightarrow \underset{1s^1}{\text{Li}^{2+}(g)}$$

$$\underset{1s^2\,2s^2}{\text{Be}(g)} \longrightarrow \underset{1s^2\,2s^1}{\text{Be}^+(g)} \longrightarrow \underset{1s^2}{\text{Be}^{2+}(g)}$$

Since, 1*s*-electrons shield 2*s*-electrons much more effectively than they shield each other, we predict that it should be much easier to remove a 2*s* electron from Be^+ than to remove a 1*s*-electron from Li^+ which has inert gas [He] configuration.

Problem 10. A technique called photoelectron spectroscopy is used to measure the *IE* of atoms. A sample is irradiated with UV light, and electrons are ejected from the valence shell. The kinetic energies of the ejected electrons are measured. Since the energy of the UV photon and the kinetic energy of the ejected electrons are known, we can write

$$h\nu = (IE) + \frac{1}{2}mu^2$$

where ν is the frequency of the UV light, and *m* and *u* are the mass and velocity of the electron, respectively. In one experiment the kinetic energy of the ejected electron from potassium is found to be 5.34×10^{-19} J using UV source of wavelength 162 nm. Calculate *IE* of potassium ($h = 6.62 \times 10^{-34}$ J s, $c = 3 \times 10^8$ m s^{-1}).

Solution

$$h\nu = (IE) + \frac{1}{2}mu^2$$

$$IE = h\nu - \frac{1}{2}mu^2$$

$$= \frac{hc}{\lambda} - \frac{1}{2}mu^2$$

$$= \left(\frac{6.62 \times 10^{-34} \times 3 \times 10^8}{162 \times 10^{-9}}\right) - (5.34 \times 10^{-19})$$

$$= \mathbf{6.91 \times 10^{-19}\ J\ atom^{-1}}$$

$$= \mathbf{416.54 \times 10^3\ J\ mol^{-1}}$$

$$= \mathbf{416.54\ kJ\ mol^{-1}}$$

Problem 11. The energy needed for the following process is 1.96×10^4 kJ mol^{-1}

$$\text{Li}(g) \longrightarrow \text{Li}^{3+}(g) + 3e^-$$

If the first ionisation energy of lithium is 520 kJ mol^{-1}, calculate the second ionisation energy of lithium, that is the energy required for the process

$$\text{Li}^+(g) \longrightarrow \text{Li}^{2+}(g) + e^-$$

Solution Stepwise ionisation of Li(*g*) is written as :

$$\text{Li}(g) \longrightarrow \text{Li}^+(g) + e^- \quad (IE)_1 = 520\ \text{kJ mol}^{-1}$$

$$\text{Li}^+(g) \longrightarrow \text{Li}^{2+}(g) + e^- \quad (IE)_2 = ? = x$$

$$\text{Li}^{2+}(g) \longrightarrow \text{Li}^{3+}(g) + e^- \quad (IE)_3 = ? = y$$

$$\text{Li}(g) \longrightarrow \text{Li}^{3+}(g) + 3e^-$$

$$(IE)_{\text{Total}} = 1.96 \times 10^4\ \text{kJ mol}^{-1}$$

$Li^{2+}(g)$ is isoelectronic of H(*g*), hence

$$(IE)_3 = (2.18 \times 10^{-18})(Z^2)\left(\frac{1}{n^2}\right)\ \text{J atom}^{-1}$$

$$= 2.18 \times 10^{-18} \times 9\ \text{J atom}^{-1}$$

$$= 2.18 \times 10^{-18} \times 9 \times 6.02 \times 10^{23}\ \text{J mol}^{-1}$$

$$\therefore \quad y = 11.81 \times 10^3\ \text{kJ mol}^{-1}$$

$$(IE)_{\text{Total}} = (IE)_1 + (IE)_2 + (IE)_3$$

$$1.96 \times 10^4 = 520 + x + 11.81 \times 10^3$$

$$\therefore \quad x = (IE)_2 = \mathbf{7.269 \times 10^3\ kJ\ mol^{-1}}$$

Problem 12. Which has the larger lattice energy, NaCl or CsCl?

Solution Lattice energy is defined as the energy when gaseous ions combine to form ionic solid.

$$\text{Na}^+(g) + \text{Cl}^-(g) \longrightarrow \text{NaCl}(s), \quad \text{lattice energy} = -U$$

The magnitude of a substance's lattice energy is affected both by the charges on its constituent ions and by the sizes of those ions. The higher the charges on the ions and smaller the sizes of the cations and larger the size of the anion, the larger the lattice energy.

Charges on each ion are identical.

Na^+ is smaller than Cs^+.

Cl^- is identical

Thus, lattice energy of **NaCl > CsCl.**

Problem 13. $\frac{N_0}{2}$ atoms of $X(g)$ are converted into $X^+(g)$ by energy ΔH_1. $\frac{N_0}{2}$ atoms of $X(g)$ are converted into $X^+(g)$ and $\frac{N_0}{2}$ atoms if $X(g)$ are converted into $X^-(g)$ by energy ΔH_2. Calculate (*IE*) and (*EA*) of $X(g)$.

Solution Let (*IE*) of $X(g) = I$ per atom

and (*EA*) of $X(g) = -E$ per atom

(i) $X(g) \longrightarrow X^+(g) + e^-$

Energy required to ionise $\frac{N_0}{2}$ atom of $(X)g = \frac{N_0 I}{2}$

$$\frac{N_0 I}{2} = \Delta H_1$$

$$\therefore \quad I = \frac{2\Delta H_1}{N_0}$$

(ii) $X(g) \longrightarrow X^+(g) + e^-$,

$$\text{Energy} = \frac{N_0 I}{2} \text{ for } \frac{N_0}{2} \text{ atoms}$$

$X(g) + e^- \longrightarrow X^-(g)$,

$$\text{Energy} = \frac{-N_0 E}{2} \text{ for } \frac{N_0}{2} \text{ atoms}$$

$$\therefore \quad \frac{N_0 I}{2} - \frac{N_0 E}{2} = \Delta H_2$$

$$N_0 \frac{2\Delta H_1}{2N_0} - \frac{N_0 E}{2} = \Delta H_2$$

$$\frac{-N_0 E}{2} = \Delta H_2 - \Delta H_1$$

$$E = \frac{2(\Delta H_2 - \Delta H_1)}{N_0}$$

$\therefore$ **Ionisation Energy (*IE*)** $= \frac{2\Delta H_1}{N_0}$ **atom**$^{-1}$,

Electron Affinity (*EA*)

$$= \frac{2(\Delta H_2 - \Delta H_1)}{N_0} \text{ atom}^{-1}$$

Problem 14. Calculate the electronegativity of fluorine from the following data :

$$E_{H-H} = 104.2 \text{ kcal mol}^{-1}$$
$$E_{F-F} = 36.6 \text{ kcal mol}^{-1}$$
$$E_{H-F} = 134.6 \text{ kcal mol}^{-1}$$

Electronegativity of H is 2.05.

Solution By Pauling's electonegativity scale

$$\chi_F - \chi_H = 0.182\sqrt{\Delta_{H-F}}$$

(Values of BE are in kcal mol^{-1})

$$\Delta_{H-F} = E_{HF} - \sqrt{E_{H-H} \cdot E_{F-F}}$$
$$= 134.6 - \sqrt{104.2 \times 36.6}$$
$$= 72.84 \text{ kcal}$$
$$\chi_F - \chi_H = 0.182\sqrt{72.84} = 1.5534$$
$$\therefore \quad \chi_F = \chi_H + 1.5534$$
$$= \mathbf{3.6034}$$

Problem 15. $(IE)_1$ and $(IE)_2$ of Mg(g) are 740 and 1450 kJ mol^{-1}. Calculate percentage of $Mg^+(g)$ and $Mg^{2+}(g)$ if 1 g of Mg(g) absorbs 50.0 kJ of energy.

Solution Number of moles of Mg(g) $= \frac{1}{24} = 0.0417$,

Energy absorbed in the ionisation of Mg(g) to $Mg^+(g)$

$$= 0.0417 \times 740$$
$$= 30.83 \text{ kJ}$$

Energy unused = 50 – 30.83 = 19.17 kJ

19.17 kJ will be used in the ionisation of $Mg^+(g)$ to $Mg^{2+}(g)$. Hence, number of moles of $Mg^+(g)$ converted to $Mg^{2+}(g)$

$$= \frac{19.17}{1450} = 0.0132$$

$Mg^+(g)$ left as $Mg^+(g)$ = 0.0417 – 0.0132 = 0.0285 mol

$$\therefore \quad \text{Percentage of } Mg^+(g) = \frac{0.0285}{0.0417} \times 100 = \mathbf{68.35\%}$$

and percentage of $Mg^{2+}(g)$ = **31.65%**

Master Exercises

Exercise 1
(Stage 1 : Learning)

Short Answer Type Questions

1. Comment on the following melting point (given in °C) sequences.

Cu	1083	Zn	419.5
Ag	961	Cd	320.8
Au	1064	Hg	–38.9

2. For the effective nuclear charge for the last valence electron (n) of the second period

$$Z^* = 0.65 + 0.65n = 0.65[1 + n]$$

Calculate Z^* for Li, Be, B and C.

3. Compute the oxidation numbers of each element in the ions

(a) $H_2IO_6^{3-}$ (b) $S_3O_6^{2-}$

(c) Na_2O_2

4. Rank the 5s, 5p, 5d, 5f and 5g orbitals in order of increasing

(a) penetrating ability

(b) shielding ability and

(c) degree to which they are shielded by inner electrons

5. The single-bonded metallic radius of Na is 157 pm. Give a reasonable estimate of the following radii, assuming that the increment between radii of different magnitudes is 60 pm

(a) the covalent radius of Na

(b) the cationic radius of Na

(c) the anionic radius of Na

(d) the van der Waals' radius of Na

6. Without the use of tables, arrange each of the following sets in order of increasing size

(a) Li, C, F, Ne (b) Be, Ca, Ba, Ra

(c) B, Al, Ga, In, Tl (d) V, Nb, Ta, Db

7. Fill in each blank with one of the following; increase; decrease; stay the same; more rapidly; less rapidly.

(a) Electronegativities of atoms .... from left to right across a period.

(b) Electronegativities of atoms .... from top to bottom down a group.

(c) Electronegativities of atoms change .... when crossing the d-block of elements than when crossing the p-block.

(d) Theoretical radii of atoms $< r_{max} >$ .... from left to right across the f-block in the sixth period, but real radii .... across the f-block .... than they do across the p-block.

(e) Radii of atoms .... from top to bottom going down group 2; radii .... from top to bottom going down Group 13(III).

Analytical Questions

1. Arrange the following species in order of increasing size :

$$Ar, K^+, Cl^-, S^{2-}, Ca^{2+}$$

2. Refer to the Periodic Table and identify :

(a) an element that is both in group 15 (IIIA) and in the fifth period

(b) an element similar to sulphur

(c) a highly reactive metal in the sixth period

(d) the halogen element in the fifth period

(e) an element with atomic number greater than 50 that is similar to the element with atomic number 18.

3. Match each of the lettered items in the column on the left with an appropriate numbered items in the column on the right. Some of the numbered items may be used more than once, and some not at all.

(a) Tl	1. an alkaline-earth metal
(b) $Z = 70$	2. element in the fifth period and group 15 (VA)
(c) Ni	3. largest atomic radius of all the elements
(d) [Ar]4s^2	4. an element in the fourth period and group 16(VIA)
(e) a metalloid	5. 3d^8
(f) a non-metal	6. one p-electron in the shell of highest n
	7. d-block element
	8. f-block element

4. For each of the following pairs, indicate the atom that has the larger size :

(a) Br or As (b) Sr or Mg
(c) Ca or Cs (d) Ne or Xe
(e) C or O (f) Hg or Cl

5. Arrange the following in the increasing value of $(IE)_1$:

Sr, Cs, S, F, As

6. Listed below are the electronic configuration of five elements. Arrange the elements in the increasing order of metallic character :

(a) $[Ar]\,3d^{10}\,4s^2$ (b) $[Ar]\,4s^2$
(c) $[Ar]\,3d^{10}\,4s^2\,4p^6\,5s^2$ (d) $[Ar]\,3d^{10}\,4s^2\,4p^5$
(e) $[Ar]\,3d^{10}\,4s^2\,4p^6\,5s^1$

7. With reference only to the Periodic Table, indicate which of the atoms Bi, S, Ba, As and Ca :

(a) is most metallic;
(b) is most non-metallic;
(c) has the intermediate value when the five are arranged in order of increasing first ionisation energy.

8. Arrange the following elements in order of decreasing metallic character : Sc, Fe, Rb, Br, O, Ca, F, Te.

9. Which of the following species would you expect to be diamagnetic and which paramagnetic : K^+, Cr^{3+}, Zn^{2+}, Cd, Co^{3+}, Sn^{2+}, Br.

10. Assuming that the seventh period is 32 members long, what should be the atomic number of the noble gas following radon (Rn)? Of the alkali metal following francium (Fr)? What would you expect their approximate atomic weights to be?

11. Use the following system of naming elements in which first alphabets of the digits are written collectively,

0	1	2	3	4	5	6	7	8	9
nil	uni	bi	tri	quad	pent	hex	sept	oct	enn

to write three-letter symbol for the elements with atomic number **101** to **109**.

[Example : 101 is Unu ...]

12. Match the following lettered items in the column on the left with an appropriate numbered items in the column on the right. All the numbered items should be used at least once, and some must be used more than once :

(a) $Z = 37$ 1. two unpaired p-electrons
(b) $Z = 9$ 2. diamagnetic
(c) $Z = 16$ 3. more negative electron affinity than elements on either side of it in the same period.
(d) $Z = 30$
(e) $Z = 82$ 4. lower $(IE)_1$ than Ca but greater than Cs.
(f) $Z = 12$

13. Listed below are two atomic properties of the element germanium. Refer to the Periodic Table and indicate probable values for each of the following elements, expressed as greater than, about equal to, or less than the value for Ge.

	Covalent radius	$(IE)_1$
(a) Ge	122 pm	762 kJ mol^{-1}
(b) Al		
(c) In	$> \approx <$	$> \approx <$
(d) Se		

14. For the following groups of elements, select the one that has the property noted

(a) The largest atom : H, Ar, Ag, Ba, Te, Au
(b) The lowest $(IE)_1$: B, Sc, Al, Br, Mg, Pb
(c) The smallest (most negative) (EA)
: Na, I, Ba, Se, Cl, P
(d) The largest number of unpaired electrons
: F, N, S^{2-}, Mg^{2+}, Sc^{3+}, Ti^{3+}

15. An element is oxidised by fluorine and not by chlorine. Could the element be sodium? Aluminum? Sulphur? Oxygen? Explain your answer.

16. Explain why the most negative oxidation state of an atom of group 15 (VA) element is equal to the number of electrons required to fill the valence shell of the neutral atom?

17. Predict which member of each of the following pairs will

(a) conduct electricity	: Ca or S?
(b) form a negative ion	: Co or Cl?
(c) form a solution of an acid in water	N_2O_3 or CaO?
(d) form an ionic compound with fluorine	: Al or P?
(e) form a covalent compound with O	: K or N ?
(f) have a higher electronegativity	: Ba or Br
(g) form a solution of base in water	: CO_2 or BaO
(h) a base	: CsOH or BrOH
(i) an acid	: $In(OH)_3$ or $B(OH)_3$
(j) an oxidising agent	: S or Na
(k) an ionic compound	: RbCl or BrCl
(l) more easily reduced	: Ca or O
(m) react as a base	: $P(OH)_3$ or $Al(OH)_3$
(n) contain ions	: KNO_2 or $ClNO_2$
(o) easily hydrated	: Li^+ or Na^+
(p) greater $(IE)_1$	: Be or B
(q) greater (EA)	: F or Cl^-
(r) positive (EA)	: F or O^-
(s) greater $(IE)_2$	: N or O

(t) max. $(IE)_1$: Na^+, Ne^+, F^+, O^+
(u) zero (*EN*) : Ne or Na
(v) liquid element : Hg or Zn
(w) amphoteric oxide : B_2O_3, Al_2O_3
(x) acidic oxide : B_2O_3, Al_2O_3
(y) basic oxide : N_2O_5, In_2O_3
(z) harder element : carbon (diamond) or lead

18. Match each of the following elements on the right with its description on the left :

(a) A dark-red liquid I : Ca
(b) A colourless gas that burns in oxygen gas II : Au
(c) A reactive metal that attacks water III : H_2
(d) A totally inert gas IV : Ar
(e) A shiny metal that is used in jewellery V : Br_2

19. You are given four substances: a fuming red liquid, a dark metallic-looking solid, a pale yellow gas, and a yellow-green gas that attacks glass. You are told that these substances are the first four members of group VIIA. Name each.

20. A student is given samples of three elements, *X*, *Y*, and *Z*, which could be an alkali metal, a member of group IVA, and a member of group VA. She makes the following observations :

Element *X* has a metallic lusture and conducts electricity. It reacts slowly with *aq* HCl to produce H_2 gas. Element *Y* is a light-yellow solid and does not conduct electricity. Element *Z* has a metallic lusture and conducts electricity. When exposed to air, it slowly forms a white powder aqueous solution of which is basic. What can you conclude about the elements from these observations?

21. Listed below are the locations of certain elements in groups and periods of the Periodic Table. Arrange these elements in the expected order of increasing first ionisation energy.

(a) element in the fourth period and group IVA
(b) element in the third period and group VIA
(c) element in the sixth period and group IIIA
(d) element in the second period and group VIIIA
(e) element in the fourth period and group VIA

22. The $(IE)_1$ and the $(IE)_2$ (in kJ mol^{-1}), of three elements I, II and III are given below :

	I	II	III
$(IE)_1$	403	549	1142
$(IE)_2$	2640	1060	2080

Identify the element which is likely to be :

(a) a non-metal
(b) an alkali metal
(c) an alkaline-earth metal

23. The element 119 has not been discovered. What would be the IUPAC name and the symbol for this element? Write the formula of the most stable chloride and oxide.

Exercise 2
(Stage 2 : High Skill Problem Solving)

Only One Option Correct

1. If Aufbau rule is not followed, K-19 will be placed in .... block.

(a) *s* (b) *p*
(c) *d* (d) *f*

2. First, second and third IP values are 100 eV, 150 eV and 1500 eV. Element can be

(a) Be (b) B
(c) F (d) Na

3. Which has maximum ionisation potential?

(a) N (b) O
(c) O^+ (d) Na

4. $N_0/2$ atoms of *X*(*g*) are converted into $X^+(g)$ by energy E_1. $N_0/2$ atoms of *X*(*g*) are converted into $X^-(g)$ by energy E_2. Hence, ionisation potential and electron affinity of *X* (*g*) are

(a) $\frac{2E_1}{N_0}, \frac{2(E_1 - E_2)}{N_0}$ (b) $\frac{2E_1}{N_0}, \frac{2E_2}{N_0}$
(c) $\frac{(E_1 - E_2)}{N_0}, \frac{2E_2}{N_0}$ (d) none is correct

5. Which is maximum hydrated?

(a) NaCl (b) $MgCl_2$
(c) $AlCl_3$ (d) $SiCl_4$

6. Which has maximum polarising power in cation ?

(a) Li^+ (b) Mg^{2+}
(c) Al^{3+} (d) O^{2-}

7. Electron affinity is positive when

(a) O^- is formed from O
(b) O^{2-} is formed from O^-
(c) O^+ is formed from O
(d) electron affinity is always a negative value

8. Following triads have approximately equal size
(a) Na^+, Mg^{2+}, Al^{3+} (*iso*-electronic)
(b) F^-, Ne, O^{2-} (*iso*-electronic)
(c) Fe, Co, Ni
(d) Mn^+, Fe^{2+}, Cr (*iso*-electronic)

9. Ionic radii of
(a) $Ti^{4+} < Mn^{7+}$ (b) $^{35}Cl^- < ^{37}Cl^-$
(c) $K^+ > Cl^-$ (d) $P^{3+} > P^{5+}$

10. Which of the following ions has the smallest radius?
(a) Ti^{2+} (b) Pt^{2+}
(c) Ni^{2+} (d) Zr^{2+}

11. When the following five anions are arranged in order of decreasing ionic radius, the correct sequence is
(a) $Se^{2-}, I^-, Br^-, O^{2-}, F^-$ (b) $I^-, Se^{2-}, O^{2-}, Br^-, F^-$
(c) $Se^{2-}, I^-, Br^-, F^-, O^{2-}$ (d) $I^-, Se^{2-}, Br^-, O^{2-}, F^-$

12. For which of the following crystals would you expect the assumption of anion-anion contact to be valid ?
(a) CsBr (b) NaF
(c) KCl (d) NaI

13. Which of the following ions has the largest heat of hydration?
(a) Na^+ (b) Al^{3+}
(c) F^- (d) Sr^{2+}

14. Which of the following anions is most easily polarized?
(a) Cl^- (b) Se^{2-}
(c) Br^- (d) Te^{2-}

15. Of the four ΔH values needed to calculate a lattice energy using the Born-Haber cycle, the one that is most difficult to measure is
(a) the heat of sublimation of the metal
(b) the heat of formation of gaseous atoms of the non-metal
(c) the ionization energy of the metal
(d) the electron affinity of the non-metal

16. A molecule H—*X* will be 50% ionic if electronegativity difference of H and *X* is
(a) 1.2 eV (b) 1.4 eV
(c) 1.5 eV (d) 1.7 eV

17. Which pair is different from the others?
(a) Li—Mg (b) Na—K
(c) Ca—Mg (d) B—Al

18. Which pair is different from the others?
(a) Li—Mg (b) B—Si
(c) Be—Al (d) Li—Na

19. Representative elements belong to
(a) *s*-and *p*-block (b) *d*-block
(c) *d*- and *f*-block (d) *f*-block

20. Covalency is favoured in the following cases
(a) a smaller cation
(b) a larger anion
(c) large charges on cation or anion
(d) in all the above cases

21. Inert pair effect is shown by
(a) *s*-block (b) *p*-block
(c) *d*-block (d) *f*-block

22. Stability of ions of Ge, Sn and Pb will be in the order
(a) $Ge^{2+} < Sn^{2+} < Pb^{2+}$ (b) $Ge^{4+} > Sn^{4+} > Pb^{4+}$
(c) $Sn^{4+} > Sn^{2+}$ (d) all are correct

23. Which is incorrect statement?
(a) Tl^{3+} salts are better oxidising agents
(b) Ga^+ salts are better reducing agents
(c) Pb^{4+} salts are better oxidising agents
(d) As^{5+} salts are better oxidising agents

24. The first element of a group in many ways differs from the other heavier members of the group. This is due to
(a) the small size
(b) the high electronegativity and high ionisation potential
(c) the unavailability of *d*-orbitals
(d) all of the above

25. Catenation properties of C, Si, Ge, Sn, Pb are in order
(a) C >> Si > Ge ≈ Sn >> Pb
(b) C < Si < Ge < Sn < Pb
(c) C > Si > Sn > Ge > Pb
(d) None of the above is correct

26. Melting points of NaCl, NaBr, NaI and NaF will be in order
(a) NaI < NaBr < NaCl < NaF
(b) NaF < NaCl < NaBr < NaI
(c) NaBr < NaF < NaCl < NaI
(d) NaCl < NaI < NaF < NaBr

27. Melting point is maximum for
(a) LiCl (b) NaCl
(c) KCl (d) RbCl

28. Which is/are amphoteric oxides?
(a) BeO (b) SnO
(c) ZnO (d) All of these

29. The correct order of increasing ionic character is
(a) $BeCl_2 < MgCl_2 < CaCl_2 < BaCl_2$
(b) $BeCl_2 < CaCl_2 < MgCl_2 < BaCl_2$
(c) $BaCl_2 < CaCl_2 < MgCl_2 < BeCl_2$
(d) $MgCl_2 < CaCl_2 < BeCl_2 < BaCl_2$

30. Atomic number 64 will have electronic configuration
(a) $[Xe]_{54}\, 6s^2\, 4f^8$ (b) $[Xe]_{54}\, 6s^2\, 4f^7\, 5d^1$
(c) $[Xe]_{54}\, 4f^{10}$ (d) $[Xe]_{54}\, 6s^2\, 4f^7\, 5p^1$

31. $^{238}_{92}$U(IIIB) changes to $^{234}_{90}$Th by emission of α-particle. Daughter element will be in
(a) III B (b) I B
(c) V B (d) II A

32. Which is correct order of size of O^-, O^{2-}, F^- and F?
(a) $O^{2-} > O^- > F^- > F$ (b) $O^- > O^{2-} > F^- > F$
(c) $O^{2-} > F^- > F > O^-$ (d) $O^{2-} > F^- > O^- > F$

33. The electronegativities of N, C, Si and P are such that
(a) P < Si < C < N (b) Si < P < N < C
(c) Si < P < C < N (d) P < Si < N < C

34. The electron affinities of N, O, S and Cl are such that
(a) N < O < S < Cl (b) O < N < Cl < S
(c) O ≈ Cl < N ≈ S (d) O < S < Cl < N

35. Fluorine has the highest electronegativity among the $ns^2\ np^5$ group on the Pauling scale, but the electron affinity of fluorine is less than that of chlorine because
(a) the atomic number of fluorine is less than that of chlorine
(b) fluorine being the first member of the family behaves in an unusual manner
(c) chlorine can accommodate an electron better than fluorine by utilising its vacant 3*d*-orbital
(d) small size, high electron density and an increased electron repulsion makes addition of an electron to fluorine less favourable than that in the case of chlorine.

36. (*A*), (*B*), (*C*) are elements in the third short period. Oxide of (*A*) is ionic, that of (*B*) is amphoteric and of (*C*) a giant molecule. (*A*), (*B*) and (*C*) will have atomic number in the order
(a) $(A) < (B) < (C)$ (b) $(C) < (B) < (A)$
(c) $(A) < (C) < (B)$ (d) $(B) < (A) < (C)$

37. Going down in a group from F to I, which of the following properties decreases?
(a) Ionic radius (b) Ionisation energy
(c) Oxidising power (d) Electronegativity

38. Which of the following will have maximum electron affinity?
(a) $1s^2 2s^2 2p^5$ (b) $1s^2 2s^2 2p^6$
(c) $1s^2 2s^2 2p^6 3s^2 3p^5$ (d) $1s^2 2s^2 2p^6 3s^2 3p^6$

39. Size of cation is smaller than that of the atom because of
(a) the whole of the outer shell of electrons is removed
(b) effective nuclear charge increases
(c) due to gain of electrons
(d) statement, that cation is smaller than atom, is wrong

40. The sizes of the second and third row transition elements being almost the same. This is due to
(a) *d*-and *f*-orbitals do not shield the nuclear charge very effectively
(b) lanthanide contraction
(c) both (a) and (b) are true
(d) None of the above is true

41. The factors that influence the ionisation energies are
(a) the size of the atom
(b) the charge on the nucleus
(c) how effectively the inner electron shell screen the nuclear charge
(d) all of the above

42. Following are the values of the electron affinities of (in kJ mol^{-1}) the formation of O^- and O^{2-} from O
(a) –142, – 702 (b) –142, 702
(c) 142, 702 (d) –142, –142

43. The ionic radii of O^{2-}, F^-, Na^+ and Mg^{2+} are 1.35, 1.34, 0.95 and 0.66 Å respectively. The radius of the Ne atom is
(a) 1.39 Å (b) 1.12 Å
(c) 0.85 Å (d) 0.50 Å

44. The element which does not exist in liquid state at room temperature are
(a) Na (b) Br
(c) Hg (d) Ga

45. The statement that is not correct for the periodic classification of elements is
(a) the properties of elements are the periodic functions of their atomic numbers
(b) non-metallic elements are lesser in number than metallic elements
(c) the first ionisation energies of elements along a period do not vary in a regular manner with increase in atomic number
(d) for transition elements the *d*-subshells are filled with electrons monotonically with increase in atomic number

46. Which one of the following has the maximum number of unpaired electrons?
(a) Mg^{2+} (b) Ti^{3+}
(c) V^{3+} (d) Fe^{2+}

47. The following acids have been arranged in the order of decreasing acid strength. Identify the correct order
ClOH (I) BrOH (II) IOH (III)
(a) I > II > III (b) II > I > III
(c) III > II > I (d) I > III > II

48. The statement is not true for the long form of the Periodic Table
(a) It reflects the sequence of filling the electrons in the order of the sub-energy shells *s, p, d* and *f*
(b) It helps to predict the stable valency states of the elements
(c) It reflects trends in physical and chemical properties of the elements
(d) It helps to predict the relative ionicity of the bond between any two elements

49. The electronic configuration $1s^2, 2s^2 2p^6, 3s^2 3p^6 3d^{10}, 4s^2 4p^6 4d^{10}, 5s^2$ is for
(a) *f*-block element (b) *d*-block element
(c) *p*-block element (d) *s*-block element

50. Which of the following statement is false?
(a) Elements of IB and IIB groups are transition elements
(b) Elements of VB group do not contain metalloids

(c) Elements of IA and IIA groups are normal elements

(d) Elements of IVB group are neither strongly electronegative nor strongly electropositive.

51. Transition metals are characterised by the properties

(a) variable valency and tendency to form complex

(b) coloured compounds

(c) high melting and boiling points

(d) all of the above

52. Following graph shows variation of ionisation potential (IP) with atomic number in second period (Li—Ne). Value of ionisation potential (IP) of Na (11) will be

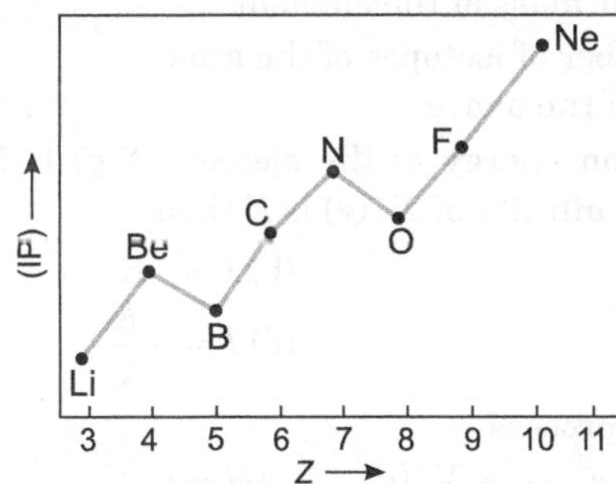

(a) above Ne (b) below Ne but above O

(c) below Li (d) between N and O

53. Based on the fig. in Q. 52, which is incorrect statement

(a) I_1 of Be $> I_1$ of B but I_2 of Be $< I_2$ of B

(b) I_1 of Be $< I_1$ of B but I_2 of Be $< I_2$ of B

(c) I_2 of $Be^+ < I_1$ of B^+

(d) I_1 of Be^{2+} is abnormally high

54. Which is/are correct order of ionic mobility?

(a) $Li^+ < Na^+ < K^+$ (b) $Al^{3+} < Mg^{2+} < Na^+$

(c) Both (a) and (b) (d) None of these

55. $M(g) \longrightarrow M^+(g) + e^-, \Delta H = 100$ eV

$M(g) \longrightarrow M^{2+}(g) + 2e^-, \Delta H = 250$ eV

which is incorrect statement?

(a) I_1 of $M(g)$ is 100 eV (b) I_1 of $M^+(g)$ is 150 eV

(c) I_2 of $M(g)$ is 250 eV (d) I_2 of $M(g)$ is 150 eV

56. AB is predominantly ionic as A^+B^- if

(a) $(IP)_A < (IP)_B$ (b) $(EA)_A < (EA)_B$

(c) $(EN)_A < (EN)_B$ (d) $(IP)_B < (IP)_A$

(IP stands for ionisation potential, EA for electron affinity and EN for electronegativity).

57. EN of the element (A) is E_1 and IP is E_2. Hence, EA will be

(a) $2E_1 - E_2$ (b) $E_1 - E_2$

(c) $E_1 - 2E_2$ (d) $(E_1 + E_2)/2$

58. Stability order of group IIIA (boron family) element is

(a) $Ga^+ < In^+ < Tl^+$ (b) $Ga^{3+} > Ga^+$

(c) both (a) & (b) (d) None of these

59. The metal which gives no amphoteric oxide is

(a) Zn (b) Cu

(c) Sn (d) Al

60. The first ionisation potential (in eV) of N, O atoms are

(a) 14.6, 13.6 (b) 13.6, 14.6

(c) 13.6, 13.6 (d) 14.6, 14.6

61. Atomic radii of F and Ne in Å are given by

(a) 0.72, 1.6 (b) 1.6, 1.6

(c) 0.72, 0.72 (d) none of these

62. Match list-I (atomic number of elements) with list-II (position of elements in Periodic Table) and select the correct answer using the codes given below the lists

List I	List II
A. 19	1. *p*-block
B. 22	2. *f*-block
C. 32	3. *d*-block
D. 64	4. *s*-block

Codes

	A	B	C	D		A	B	C	D
(a)	1	2	3	4	(b)	4	3	1	2
(c)	4	1	3	2	(d)	2	1	3	4

63. Fluorine does not form any polyhalide as other halogens because

(a) it has maximum ionic character

(b) it has low F—F bond energy (38.5 kcal mol^{-1})

(c) of the absence of *d*-orbitals in the valence shell of fluorine

(d) it brings about maximum coordination number in other elements

64. Consider the following statements

1. Cs^+ is more highly hydrated than the other alkali metal ions.
2. Among the alkali metals Li, Na, K and Rb, lithium has the highest melting point.
3. Among the alkali metals only lithium forms a stable nitride by direct combination.

Of these statements

(a) 1, 2 and 3 are correct (b) 1 and 2 are correct

(c) 1 and 3 are correct (d) 2 and 3 are correct

65. Match list I (atomic number of the elements) with list II (position in the Periodic Table) and select the correct answer using the codes given below the lists:

List I	List II
A. 52	1. *s*-block
B. 56	2. *p*-block
C. 57	3. *d*-block
D. 60	4. *f*-block

Codes

	A	B	C	D		A	B	C	D
(a)	2	1	3	4	(b)	2	1	4	3
(c)	1	2	3	4	(d)	1	2	4	3

66. If Aufbau and Hund's rule are not used, then incorrect statement is
(a) K^+ would be coloured ion
(b) Na will be in same *s*-block (if these rules are true)
(c) Cu would be *s*-block element
(d) magnetic moment of Cr(24) would be zero

67. Which set has all the coloured ions?
(a) Cu^+, Cu^{2+}, Ni^{2+} (b) Cu^{2+}, Fe^{2+}, Co^{2+}
(c) Cu^{2+}, Co^{2+}, Sc^{3+} (d) Na^+, Mg^{2+}, Al^{3+}

68. For the element (*X*), student (*A*) measured its radius as 102 nm, student (*B*) as 103 nm and (*C*) as 100 nm using same apparatus. Their teacher explained that measurements were correct by saying that recorded values by (*A*), (*B*) and (*C*) are
(a) crystal, van der Waals' and covalent radii
(b) covalent, crystal and van der Waals' radii
(c) van der Waals', ionic and covalent
(d) none of the above is correct

69. Which set shows inert pair effect?
(a) Fe, Co, Ni (b) B, Al, Ga
(c) Ge, Sn, Pb (d) He, Ne, Ar

70. Gd (64) has .... unpaired electrons with sum of spin ....
(a) 7, 3.5 (b) 8, 3
(c) 6, 3 (d) 8, 4

71. Element Unq has atomic number of
(a) 102 (b) 103
(c) 104 (d) 112

72. The relative thermal stabilities of alkali metal halides are such that
(a) CsCl > RbCl > KCl > NaCl > LiCl
(b) LiCl > NaCl > KCl > RbCl > CsCl
(c) CsCl > RbCl < KCl > NaCl < LiCl
(d) CsCl < RbCl > KCl < NaCl > LiCl

73. Consider the isoelectronic ions

K^+, S^{2-}, Cl^- and Ca^{2+}

The radii of these ionic species follow the order
(a) $Ca^{2+} > K^+ > Cl^- > S^{2-}$ (b) $Cl^- > S^{2-} > K^+ > Ca^{2+}$
(c) $S^{2-} > Cl^- > K^+ > Ca^{2+}$ (d) $K^+ > Ca^{2+} > S^{2-} > Cl^-$

74. Astatine is the last element in halogen group (VII). It is expected to be
(a) more electronegative than iodine
(b) composed of diatomic molecules
(c) limited to an oxidation number of −1 in its compounds
(d) a solid at room temperature

75. The first ionisation potential of Al is smaller than that of Mg because
(a) the atomic size of Al > Mg
(b) the atomic size of Al < Mg
(c) Al has one unpaired electron in *p*-orbital
(d) the atomic number of Al > Mg

76. The non-metallic cation is in
(a) CrO_2Cl_2 (b) VOCl
(c) NH_4Cl (d) PCl_3

77. Which of the following properties shows a clear periodic variation?
(a) First ionisation energy
(b) Molar mass of the element
(c) Number of isotopes of the atom
(d) All of the above

78. Ionisation energy of the element *X*(*g*) is *I* and the electron affinity of $X^+(g)$ is *E* then
(a) $I = E$ (b) $I = -E$
(c) $I = \frac{E}{2}$ (d) $I = -\frac{E}{2}$

79. For the process

$$X(g) + e^- \longrightarrow X^-(g), \quad \Delta H = x$$

and $X^-(g) \longrightarrow X(g) + e^-$, $\Delta H = y$

Select correct alternate
(a) Ionisation energy of $X^-(g)$ is *y*
(b) Electron affinity of *X*(*g*) is *x*
(c) Electron affinity of *X*(*g*) is −*y*
(d) All the above are correct statements

80. The most metallic of the following elements is
(a) Mg (b) Li
(c) K (d) Ca

81. The most negative electron affinity of the following elements is that of
(a) Br (b) Sn
(c) Ba (d) Li

82. The highest first ionisation energy of the following elements is that of
(a) Cs (b) Cl
(c) I (d) Li

83. The heat of hydration of Ca^{2+}, Sr^{2+} and Ba^{2+} in the decreasing order is
(a) $Ca^{2+} > Sr^{2+} > Ba^{2+}$ (b) $Ca^{2+} > Ba^{2+} > Sr^2$
(c) $Sr^{2+} > Ba^{2+} > Ca^{2+}$ (d) $Ba^{2+} > Sr^{2+} > Ca^{2+}$

84. The compound of vanadium has magnetic moment of 1.73 BM. The vanadium chloride has the formula
(a) VCl_2 (b) VCl_3
(c) VCl_4 (d) VCl_5

85. Which of the following does not represent the correct order of the properties indicated?
(a) $Ni^{2+} > Cr^{2+} > Fe^{2+} > Mn^{2+}$ (size)
(b) Sc > Ti > Cr > Mn (size)
(c) $Ni^{2+} < Co^{2+} < Fe^{2+} < Mn^{2+}$ (unpaired electron)
(d) $Fe^{2+} > Co^{2+} > Ni^{2+} > Cu^{2+}$ (unpaired electron)

86. Which of the following order is wrong?
(a) $NH_3 < PH_3 < AsH_3$ — acidic
(b) Li < Be < B < C — $(IE)_1$
(c) $Al_2O_3 < MgO < Na_2O < K_2O$ — basic
(d) $Li^{2+} < Na^+ < K^+ < Cs^+$ — ionic radius

87. The ions O^{2-}, F^-, Na^+, Mg^{2+} and Al^{3+} are isoelectronic. Their ionic radii show
(a) an increase from O^{2-} to F^- and then decrease from Na^+ to Al^{3+}
(b) a decrease from O^{2-} to F^- and then increase from Na^+ to Al^{3+}
(c) a significant increase from O^{2-} to Al^{3+}
(d) a significant decrease from O^{2-} to Al^{3+}

88. Ionic radii are
(a) inversely proportional to effective nuclear charge
(b) inversely proportional to square of effective nuclear charge
(c) directly proportinal to effective nuclear charge
(d) directly proportional to square of effective nuclear charge

89. Sodium forms Na^+ ion but it does not form Na^{2+} because
(a) very low value of $(IE)_1$ and $(IE)_2$
(b) very high value of $(IE)_1$ and $(IE)_2$
(c) low value of $(IE)_1$ and low value of $(IE)_2$
(d) low value of $(IE)_1$ and high value of $(IE)_2$

90. The correct decreasing order of ionic radius is
(a) $C^{4-} > N^{3-} > O^{2-} > F^-$ (b) $F^- > O^{2-} > N^{3-} > C^{4-}$
(c) $O^{2-} > F^- > C^{4-} > N^{3-}$ (d) $N^{3-} > O^{2-} > F^- > C^{4-}$

91. First long period contains .... elements.
(a) 8 (b) 18
(c) 32 (d) 2

92. Recently (in 2003) element with atomic number 110 has been named by IUPAC as
(a) Hs (b) Mt
(c) Ds (d) Sg

93. Recently (in Aug 2003) two new elements have been discovered with atomic numbers
(a) 113, 114 (b) 114, 115
(c) 115, 116 (d) 113, 115

94. Element 113 is produced via
(a) α-decay of element 115
(b) β-decay of element 114
(c) α-decay of element 111
(d) β-decay of element 112

95. Numbering of groups as 1, 2, ... 18 was adopted by IUPAC in
(a) 1986 (b) 1906
(c) 1908 (d) 1988

96. Element with valence shell-electronic configuration as $(n-1)d^5 ns^1$ is placed in
(a) IA, *s*-block (b) VIA, *s*-block
(c) VIB, *s*-block (d) VIB, *d*-block

97. Which set does not show correct matching?
(a) Sc^{3+} [Ne] $3s^2 3p^6$ zero group
(b) Fe^{2+} [Ar] $3d^6$ VIII group
(c) Cr [Ar] $3d^5\ 4s^1$ VIB group
(d) All of the above

98. Two new elements (discovered in Aug 2003) with atomic number 113 and 115 are to be placed in
(a) *s*-block (b) *p*-block
(c) *d*-block (d) *f*-block

99. Most stable cation of element 113 will be
(a) M^{3+} (b) M^{2+}
(c) M^+ (d) M_2^{2+}

100. Element with atomic number 115 has configuration as ...... and with most stable cation as ......
(a) [Rn] $7s^2 5d^{10} 4f^{14} 7p^3$ M^{3+}
(b) [Rn] $7s^2 5d^{10} 4f^{14} 7p^3$ M^{5+}
(c) [Rn] $7s^2 5d^{10} 4f^{14} 7p^3$ M^+
(d) [Rn] $5d^{10} 4f^{14} 7p^5$ M^{5+}

101. Main group elements constitute:
(a) *s*-and *p*-block (b) *p*-and *d*-block
(c) *s*-and *d*-block (d) *d*-and *f*-block

102. Elements *X* and *Y* have valence shell electron configuration as
$X : ns^2 np^5$; $Y : ns^1$
Which compound is likely formed from *X* and *Y* ?
(a) X_3Y_5 (b) Y_3X_5
(c) XY (d) YX

103. Recently discovered elements (August 2003) with atomic number 113 and 115 have valence electrons in
(a) *s*-orbital (b) *p*-orbital
(c) *d*-orbital (d) *f*-orbital

104. Chalcogens are elements of
(a) group 16 (b) *p*-block
(c) $ns^2 np^4$ configuration (d) all are correct

105. M^{3+} has electronic configuration as [Ar] $3d^{10}$, hence it lies in
(a) *s*-block (b) *p*-block
(c) *d*-block (d) *f*-block

106. Transition elements have vacant
(a) *s*-orbitals (b) *p*-orbitals
(c) *d*-orbitals (d) *f*-orbitals

107. Of the following pairs, the one containing examples of metalloid elements in the Periodic Table is
(a) Na and K (b) F and Cl
(c) Cu and Ag (d) B and Si

108. Pick out the property which is not shown by transition elements.

(a) Show variable oxidation state
(b) Impart colour to flame
(c) Are paramagnetic in nature
(d) Act as catalytic agents

109. Which of the following is a transition element?
(a) Al (b) As
(c) Ni (d) Rb

110. Which of the following has the electronic configuration [Ar] $3d^5$?
(a) Cr (b) Fe^{3+}
(c) Mn (d) V

111. Which of the following forms a stable +4 oxidation state?
(a) Lanthanum (b) Cerium
(c) Europium (d) Gadolinium

112. The lanthanides have electron configuration with $6s^2$ in common but with variable occupation of the
(a) 6*p*-level (b) 5*p*-level
(c) 5*d*-level (d) 4*f*-level

113. Which group of elements is analogous to the lanthanides?
(a) Halides (b) Actinides
(c) Chalcogenides (b) Borides

114. EC of Gd (64) is written as
(a) $[Xe]_{54}\ 4f^7 5d^1 6s^2$ (b) $[Xe]_{54}\ 4f^8 6s^2$
(c) $[Xe]_{54}\ 4f^9 6s^1$ (d) $[Xe]_{54}\ 4f^{10}$

115. Recently discovered element with atomic number 115 is
(a) Uun (b) Uub
(c) Uup (d) Uus

116. In Periodic Table, metallic elements appear
(a) in the left-hand columns
(b) in the top rows
(c) in the right-hand columns
(d) in the bottom rows

117. The atoms of the elements belonging to the same group of the Periodic Table will have
(a) the same number of protons
(b) the same number of electrons in the valence-shell
(c) the same number of neutrons
(d) the same number of electrons

118. Which will have graded property similar to EC $1s^2 2s^2 2p^6 3s^2 3p^6 4s^1$?
(a) $[Ar]\,3d^{10} 4s^1$ (b) $[Kr]\,4d^{10} 5s^1$
(c) $[Kr]\,5s^1$ (d) All of these

119. Recently discovered element with atomic number 115 has configuration
(a) ns^2np^3 (d) ns^2np^1
(c) ns^2np^5 (d) ns^2np^2

120. Element with atomic number 113 has been reported in August 2003. Its electronic configuration is similar to that of
(a) Si (b) Ga
(c) Bi (d) At

121. Which has maximum stability?
(a) $AsCl_3$ (b) $SbCl_3$
(c) $BiCl_3$ (d) Equal

122. Which reaction is most spontaneous?
(a) $Ga^{3+} + 2e^- \longrightarrow Ga^+$ (b) $In^{3+} + 2e^- \longrightarrow In^+$
(c) $Tl^{3+} + 2e^- \longrightarrow Tl^+$ (d) Cannot be predicted

123. With respect to oxygen maximum valency is shown by
(a) halogen family (b) oxygen family
(c) nitrogen family (d) boron family

124. Which is best oxidising agent?
(a) Ge^{4+} (b) Pb^{4+}
(c) Sn^{4+} (d) Sn^{2+}

125. Valence electrons in the element *A* are 3 and that in element *B* are 6. Most probable compound formed from *A* and *B* is
(a) A_2B (b) AB_2
(c) A_6B_3 (d) A_2B_3

126. The relative extent to which the various orbitals penetrate the electron clouds of other orbitals is
(a) $s > p > d > f$ (b) $s < p < d < f$
(c) $s < d < p < f$ (d) $d < s < p < f$

127. Screening effect is not observed in
(a) He^+ (b) Li^{2+}
(c) Be^{3+} (d) in all cases

128. In which case effective nuclear charge is minimum?
(a) Be (b) Be^{2+}
(c) Be^{3+} (d) Equal

129. For Be, $Z_{eff} = 1.95$ and for Be^{x+}, $Z_{eff} = 2.30$. Hence, ion is
(a) Be^+ (b) Be^{2+}
(c) Be^{3+} (d) Be^0

130. For Cu, effective nuclear charge felt by a 4*s*-electron is
(a) 2.2 (b) 2.95
(c) 1.0 (d) 1.65

131. State, which one of the following has the largest atomic radius?
(a) Cs (b) Mg
(c) Ba (d) Cr

132. Which one of the following has the smallest atomic radius?
(a) F (b) Cl
(c) Cs (d) Mg

133. Covalent radius of nitrogen is 70 pm. Hence, covalent radius of boron is about
(a) 60 pm (b) 110 pm
(c) 50 pm (d) 40 pm

134. In which case bond length is shortened?
(a) When multiplicity occurs between atoms
(b) When electronegativities are different
(c) In both cases
(d) In none of the cases

135. Select correct statement about radius of an atom
(a) Values of van der Waals' radii are larger than those of covalent radii because the van der Waals' forces are much weaker than the forces operating between atoms in a covalently bonded molecule
(b) The metallic radii are smaller than the van der Waals' radii, since the bonding forces in the metallic crystal lattice are much stronger than the van der Waals' forces
(c) Both (a) & (b) are correct
(d) None of the above is correct

136. The separation of lanthanides in ion-exchange method is based on
(a) basicity of the hydroxides
(b) size of the hydrated ions
(c) size of the unhydrated ion
(d) the solubility of their nitrates

137. Select correct statement(s).
(a) Across a transition series, there is only a small decrease in atomic radius from one element to another due to very small increase in effective nuclear charge
(b) The rate of decrease in the size across the lanthanide series is less than the across the first transition series
(c) Both (a) & (b) are correct statements
(d) None of the above statement is correct

138. Select incorrect order of size of ions/atoms
(a) $I^+ < I < I^-$ (b) Fe = Co = Ni
(c) Ni < Cu < Zn (d) None is incorrect

139. Select correct alternate.
(a) Due to lanthanide contraction Zr and Hf have almost equal size
(b) Due to completion of 3*d*-subshell, the electronic charge denisty in this subshell becomes very high which increases the inter-electronic repulsion hence, size increases
(c) Both (a) & (b) are correct statements
(d) Both (a) & (b) are incorrect statements

140. The lanthanides contraction refers to
(a) ionic radius of the series
(b) valence electrons of the series
(c) the density of the series
(d) electronegativity

141. The second ionisation energy is always higher than the first ionisation energy because the
(a) ion becomes more stable attaining an octet or duplet configurations
(b) electron is more tightly bound to the nucleus in an ion
(c) electron is attracted more by the core electrons
(d) none of the above is the correct explanation

142. SI unit of *IE* is
(a) $kJ\ mol^{-1}$ (b) $eV\ atom^{-1}$
(c) $J\ mol^{-1}$ (d) $kcal\ mol^{-1}$

143. Which has maximum *IE*?
(a) Mg (b) Mg^+
(c) Mg^{2+} (d) Equal

144. Which pair represents incorrect first (*IE*)?
(a) Be > B (b) N > O
(c) Li > Na (d) He > He^+

145. Which represents alkali metals based on $(IE)_1$ and $(IE)_2$ values?

		$(IE)_1$	$(IE)_2$
(a)	*X*	100	110
(b)	*Y*	95	120
(c)	*Z*	195	500
(d)	*M*	200	250

146. Which element has the highest first (*IE*)?
(a) N (b) Ne
(c) He (d) H

147. *IE* of an element does not depend on
(a) its nuclear charge (b) the shielding effect
(c) electron neutrality (d) penetration effect

148. The dominant factor in determining the *IE* of the elements on moving down the groups is its
(a) atomic radius
(b) effective nuclear charge
(c) both (a) and (b)
(d) none of the above

149. Select correct statement.
(a) $(IE)_1$ of the corresponding elements of 3*d* and 4*d* series are almost similar
(b) $(IE)_1$ of 5*d*-series elements are smaller than that of 3*d* and 4*d*-series elements
(c) both (a) & (b) are correct
(d) both (a) & (b) are incorrect

150. Higher values of ionisation energies of the 5*d*-transition elements are consistent with the
(a) relatively smaller effective nuclear charge
(b) relatively smaller size of their atoms
(c) relatively smaller penetration
(d) all the above are correct

One or More Than One Options Correct

1. Stability of ions of Ge, Sn and Pb will be in order
(a) $Ge^{2+} < Sn^{2+} < Pb^{2+}$ (b) $Pb^{4+} < Sn^{4+} < Ge^{4+}$
(c) $Sn^{2+} < Sn^{4+}$ (d) $Pb^{4+} < Pb^{2+}$

2. The first element of a group in many ways differs from the other heavier members of the group. This is due to
(a) the smaller size
(b) the high electronegativity and high ionisation potential
(c) the unavailability of *d*-orbitals
(d) the higher shielding effect

3. The factors that influence the ionisation energies are
(a) the size of the atom
(b) the charge on the nucleus
(c) how effectively the inner electrons screen the nuclear charge
(d) the atomic mass

4. Which is/are correct order of ionic mobility?
(a) $Li^+ < Na^+ < K^+$ (b) $Na^+ < Mg^{2+} < Al^{3+}$
(c) $Al^{3+} < Mg^{2+} < Na^+$ (d) $K^+ < Na^+ < Li^+$

5. Consider following ionisation steps:

$$M(g) \longrightarrow M^+(g) + e^-; \quad \Delta H = 100\,\text{eV}$$
$$M(g) \longrightarrow M^{2+}(g) + 2e^-; \quad \Delta H = 250\,\text{eV}$$

Select correct statement(s).
(a) $(IE)_1$ of $M(g)$ is 100 eV
(b) $(IE)_1$ of $M^+(g)$ is 150 eV
(c) $(IE)_2$ of $M(g)$ is 250 eV
(d) $(IE)_2$ of $M(g)$ is 150 eV

6. An increase in both atomic and ionic radii with atomic number occurs in any group of the Periodic Table and in accordance with this the ionic radii of Ti(IV) and Zr(IV) ions are 0.68 Å and 0.74 Å respectively; but for Hf(IV) ion the ionic radius is 0.75 Å which is almost the same as that for Zr(IV) ion. This is due to
(a) greater degree of covalency in compound of Hf^{4+}
(b) lanthanide contraction
(c) difference in the coordination number of Zr^{4+} and Hf^{4+} in their compounds
(d) actinide contraction

7. When an isotope undergoes K-capture, its mass number
(a) remains the same while the atomic number increases by one
(b) remains the same while the atomic number increases by two
(c) remains the same while the atomic number decreases by one
(d) as well as the atomic number decreases by one

8. The melting point of RbBr is 682°C while that of NaF is 988°C. The principal reason that the melting point of NaF is much higher than that of RbBr is that
(a) the molar mass of NaF is smaller than that of RbBr
(b) the bond in RbBr has more covalent character than the bond in NaF
(c) the difference in electronegativity between Rb and Br is smaller than the difference between Na and F.
(d) the internuclear distance, $r_c + r_a$ is greater for RbBr than for NaF

9. Select correct statement(s).
(a) Cs^+ is more highly hydrated than the other alkali metal ions
(b) Among the alkali metals Li, Na, K and Rb, lithium has the highest melting point
(c) Ionic mobility of Li^+ is maximum among Li^+, Na^+ and K^+ ions
(d) Ionisation potential of Li is smaller than that of Na

10. Select correct statements.
(a) Ce, Pr and Nd are *f*-block elements
(b) Cu, Pd and Ni are *d*-block elements
(c) Cu([Ar]$3d^{10}\,4s^1$) and K([Ar]$4s^1$) have been placed in *s*-block
(d) Si, Ge and As are metalloids

Brain Twisters

1. Let us suppose that you have discovered a new element of atomic number 162, which you choose to name khalidium (Kh). Assume that the normal order of filling of orbitals still prevails in the eighth period.
(a) Write the full electron configuration of this element.
(b) Write the valence electron configuration of this element and of its +2 ion.
(c) Note that a new type of orbital is used in Kh that is not used in any known atom. How many nodal planes will there be in a sketch of this type of orbital? How many nodal spheres will there be? Will the electrons in this type of orbital be good or poor at shielding the valence electrons of Kh? Will electrons in this type of orbital be good or poor at penetrating the inner orbitals of Kh?
(d) Calculate Z^* fo the outermost valence electron of Kh.
(e) Using an estimated covalent radius of 150 pm, calculate the Allred-Rochow electronegativity of Kh. Do you think that the Pauling electronegativity of Kh will be higher, lower, or about the same as this value?

2. How many joules of energy must be absorbed be convert to Li^+ all the atoms present in 1.00 mg of gaseous Li? $(IE)_1$ of Li is 320.3 kJ mol^{-1}. (Li = 7)

3. How much energy in kJ must be absorbed to ionise completely all the third shell electrons in a mole of gaseous phosphorus atoms?
$(IE)_1 = 1012$ $(IE)_2 = 1903$ $(IE)_3 = 2912$
$(IE)_4 = 4957$ $(IE)_t = 6274$ (values in kJ mol^{-1})
Also express result in eV per atom.

4. Use (*IE*) and (*EA*) listed below to determine whether the following process is endothermic or exothermic :

$$Mg(g) + 2F(g) \longrightarrow Mg^{2+}(g) + 2F^-(g)$$
$$(IE)_1 \text{ of } Mg(g) = 737.7 \text{ kJ mol}^{-1}$$
$$(IE)_2 \text{ of } Mg(g) = 1451 \text{ kJ mol}^{-1}$$
$$(EA) \text{ of } F(g) = -328 \text{ kJ mol}^{-1}$$

5. What is the value in kJ mol^{-1} which is equal to 1 eV $atom^{-1}$?

6. A mixture contains F and Cl atoms. The removal of an electron from each atom of the sample requires 284 kJ while the addition of an electron to each atom of the mixture releases 68.8 kJ. Determine the mole % composition of the mixture.

	(*IE*) per atom	(*EA*) per atom
F	27.91×10^{-22} kJ	-5.53×10^{-22} kJ
Cl	20.77×10^{-22} kJ	-5.78×10^{-22} kJ

7. A hydrogen like ion is an ion containing only one electron. The energies of the electron in a hydrogen like ion are given by

$$E_n = -(2.18\times 10^{-18})\, Z^2 \left(\frac{1}{n^2}\right) \text{joule}$$

where, *n* is the principal quantum number and *Z* is the atomic number of the element. Calculate the ionisation energy (in kJ mol^{-1}) of the He^+ ion.

8. Plasma is a state of matter consisting of positive gaseous ions and electrons. In the plasma state, a mercury atom could be stripped of its electrons and therefore would exist as Hg^{80+}.

Calculate energy required for the step

$$Hg^{79+}(g) \longrightarrow Hg^{80+}(g) + e^-$$

9. The atomic radius of K is 216 pm and that of K^+ is 133 pm. Calculate the per cent decrease in volume that occurs when $K(g)$ is converted to $K^+(g)$.

Passage Comprehension Questions

Passage 1

Following questions are based on Sc($Z = 21$)

1. Out of Sc^{3+}, Sc^{2+} and Sc^+ paramagnetic as well coloured ions are

(a) Sc^+, Sc^{2+}
(b) Sc^+, Sc^{3+}
(c) Sc^{2+}, Sc^{3+}
(d) all being *d*-block element

2. For any 1*s* electron, Z^* (effective nuclear charge is)

(a) 21.0 (b) 20.65
(c) 16.85 (d) 9.75

3. For 3*d*-electron, Z^* is

(a) 20.65 (b) 16.85
(c) 9.75 (d) 3.00

Passage 2

Following rule in general, classifies *p*-block elements into metals, non-metals and metalloids (where *P* is the period and *N* the valence electrons)

$(P+1) > N$, the element would be metal
$(P+1) < N$, the element would be non-metal
$(P+1) = N$, the element would be metalloid

Answer the following questions

1. Metalloid will be out of elements with atomic number 13, 14, 15, 16

(a) 13 (b) 14
(c) 15 (d) 16

2. Metals will be out of 13, 31, 35, 54

(a) 13, 31 (b) 31, 35
(c) 35, 54 (d) 13, 54

3. Non-metals will be out of 13, 31, 35, 54

(a) 13, 31 (b) 31, 35
(c) 35, 54 (d) 13, 54

4. Which will find use in semi-conductors?

(a) 5, 13 (b) 5, 14
(c) 13, 14 (d) 14, 16

Passage 3

In the following table I_1, I_2 and I_3 of the main group elements in 2nd period have been given.

	Li	Be	B	C	N	O	F	Ne
I_1	513	899	801	1086	1402	1314	1681	2080
I_2	7298	1757	2427	2352	2856	3388	3374	3952
I_3	11815	14848	3660	4620	4578	5300	6050	6122

Answer the following question based on the above table.

1. There is increase in ionisation energy along a period going Li to Ne, but I_1 of B is less than that of Be. This is due to

(a) valence electron in B is in 2*p*-orbital of higher energy than that of Be in 2*s*-orbital of lower energy
(b) removal of paired electron in Be requires higher energy than that of removal of unpaired electron in B.
(c) both (a) & (b) above
(d) None of the above

2. We observe that I_1 of Be > I_1 of B

but I_2 of Be < I_2 of B.

Select correct explanation(s).

(a) Electron in Be in paired while in B, electron is unpaired, hence I_1 of Be > I_1 of B
(b) Electron in Be^+ is unpaired while in B^+, electron is paired, hence I_2 of Be < I_2 of B.
(c) Both (a) & (b) above
(d) None of the above

3. These occurs a fall while going from N to O. It is due to
(a) stable configuration of N ↑ | ↑ | ↑
(b) higher atomic number of O
(c) stable configuration of O ↑↓ | ↑↓ | ↑ | ↑
(d) lone pairs on O.

Passage 4

The sums of first and second ionisation energies and those of third and fourth ionisation energies (in MJ mol^{-1}) of nickel and platinum are

	$(IE)_1 + (IE)_2$	$(IE)_3 + (IE)_4$
Ni	2.49	8.80
Pt	2.66	6.70

Based on this information, answer the following questions. Name of the metal (Ni or Pt) which can more easily form compounds in its +4 oxidation state.

1. Most common oxidation states of Ni and Pt are respectively
(a) +2, +2 (b) +4, +2
(c) +2, +4 (d) +4, +4

2. Which is most stable complex of Ni?
(a) $[Ni(CN)_4]^{2-}$ (b) $[Ni(CN)_6]^{2-}$
(c) $[Ni(CN)_3]^-$ (d) $[Ni(CO)_6]$

3. Which is least stable complex of Pt?
(a) $K_2[PtCl_6]$ (b) $[Pt(en)_3]Cl_4$
(c) $K_2[PtCl_4]$ (d) $K_2[Pt(C_2O_4)_3]$

Passage 5

The $(IE)_1$ and the $(IE)_2$ in kJ mol^{-1} of a few elements designated by Roman numerals are shown below:

Element	$(IE)_1$	$(IE)_2$
A	2372	5251
B	520	7300
C	900	1760
D	1680	3380

Based on above information, answer the following questions:

1. Which of the above elements is likely to be a reactive metal?
(a) *A* (b) *B*
(c) *C* (d) *D*

2. Which of the above elements is likely to a reactive non-metal?
(a) *A* (b) *B*
(c) *C* (d) *D*

3. Which represents a noble gas?
(a) *A* (b) *B*
(c) *C* (d) *D*

4. Which of the above elements forms a stable binary halide of the formula MX_2?
(a) *A* (b) *B*
(c) *C* (d) *D*

Passage 6

The heats of formation (ΔH_f°) of the oxides of the third period, sodium to chlorine, are in kJ mol^{-1}

Na_2O	MgO	Al_2O_3	SiO_2	P_4O_{10}	SO_3	Cl_2O_7
− 416	− 602	− 1676	− 911	− 2984	− 395	+250

Based on these data, answer the following questions

1. Which oxide has maximum negative heat of formation per oxygen atom?
(a) P_4O_{10} (b) Al_2O_3
(c) Na_2O (d) MgO

2. Most stable and least stable oxides are
(a) P_4O_{10}, Cl_2O_7 (b) Na_2O, Cl_2O_7
(c) MgO, Cl_2O_7 (d) Cl_2O_7, MgO

3. Stability order of the above oxides is
(a) $Cl_2O_7 < SO_3 < P_4O_{10} < SiO_2 < Al_2O_3 < MgO < Na_2O$
(b) $Cl_2O_7 < SO_3 < P_4O_{10} < Na_2O < SiO_2 < Al_2O_3 < MgO$
(c) $Cl_2O_7 < SO_3 < Na_2O < MgO < SiO_2 < Al_2O_3 < P_4O_{10}$
(d) $P_4O_{10} < Al_2O_3 < SiO_2 < MgO < Na_2O < SO_3 < Cl_2O_7$

Passage 7

Consider following table comparing ionic radius

Ion →	N^{3-}	O^{2-}	F^-	Na^+	Mg^{2+}
Number of electron	10	10	10	10	10
Number of nuclear protons	7	8	9	11	12
Ionic radius (pm)	146	140	133	98	79

Answer the following questions.

1. Select correct alternate(s) in terms of size.
(a) $Na > Na^+$ (b) $Mg > Mg^+ > Mg^{2+}$
(c) $F^- > F$ (d) $O^{2-} > O^- > O$

2. All the above species are isoelectronic but they differ in size. It is due to
(a) increase in number of protons
(b) removal of valence shell
(c) decrease in repulsive force
(d) addition of additional shell

Passage 8

The singly-bonded metallic radius of Na is 157 pm. Assume that the increment between radii of different magnitudes is 60 pm. Answer the following questions.

1. The covalent radius of Na is
(a) 157 pm (b) 97 pm
(c) 217 pm (d) 267 pm

2. Ionic radius of Na^+ is
(a) 157 pm (b) 97 pm
(c) 217 pm (d) 267 pm

3. Ionic radius of Na^- is
(a) 157 pm (b) 97 pm
(c) 217 pm (d) 267 pm

4. van der Waals' radius of Na is
(a) 157 pm (b) 97 pm
(c) 217 pm (d) 267 pm

Assertion & Reason

Codes :
(a) Both **A** and **R** are true and **R** is the correct explanation of **A.**
(b) Both **A** and **R** are true but **R** is not the correct explanation of **A.**
(c) **A** is true but **R** is false.
(d) **A** is false but **R** is true.

1. Assertion (A) : Tin exhibits +2 and +4 oxidation states only and not +3 oxidation state.

Reason (R) : Inert pair effect of electrons in the *s*-orbital is responsible.

2. Assertion (A) : Na^+ and Al^{3+} are isoelectronic but the magnitude of ionic radius of Al^{3+} is less than that of Na^+.

Reason (R) : The magnitude of effective nuclear charge of the outershell electrons in Al^{3+} is greater than that in Na^+.

3. Assertion (A) : The atomic radii of the elements of the oxygen family are smaller than the atomic radii of the corresponding elements of the nitrogen family.

Reason (R) : The members of the oxygen family are more electronegative and thus have lower values of nuclear charge than those of the nitrogen family.

4. Assertion (A) : $PbCl_2$ is more stable than $PbCl_4$.

Reason (R) : $PbCl_4$ is powerful oxidising agent.

5. Assertion (A) : The mobility of sodium ion is lower than that of potassium ion.

Reason (R) : The ionic mobilities depend upon the effective radius of the ion.

6. Assertion (A) : Sc$(Z = 21)$ is placed in *d*-block elements.

Reason (R) : Last filling electron goes into 3*d*-suborbit.

7. Assertion (A) : Ionisation energy decreases on moving down the group.

Reason (R) : Force of attraction between electrons and nucleus decreases on moving down the group.

8. Assertion (A) : Li and Mg shows diagonal relationship.

Reason (R) : Li and Mg are diagonal to each other in the Periodic Table.

9. Assertion (A) : Fluorine has a less negative electron affinity than chlorine.

Reason (R) : There is relatively greater effectiveness of 2*p*-electrons in the small F atom to repel the additional electron entering the atom than to 3*p*-electrons in the larger Cl atom.

10. Assertion (A) : $SnCl_2$ is a reducing agent.

Reason (R) : $SnCl_2$ reduces $FeCl_3$ to $FeCl_2$ and $HgCl_2$ to Hg.

True & False

1. Electron affinity of Cl is greater than that of F.

2. Al_2O_3 in aqueous NaOH forms $Na[Al(OH)_4]$.

3. CO is an acidic oxide.

4. PbO_2 is a reducing agent.

5. Ionic mobility of Li^+ is greater than that of Na^+.

6. B_2O_3 is an amphoteric oxide.

7. ZnO is amphiprotic.

8. LiCl can be used to control humidity.

9. $(IE)_2$ of H is less than that of He.

10. First long period has 18 elements.

11. Na_2CO_3 is thermally stable but Li_2CO_3 is not.

12. Atomic size of Fe > Co > Ni.

13. Atomic size of Ne > F.

14. $(IE)_2$ of Li is maximum in the Periodic Table.

15. The noble gas core together with $(n-1)d^{10}$ electrons is called pseudo-noble gas core.

16. Helium is the most unreactive element of the Periodic Table.

Fill in the Blanks

1. .... is the lightest of all metals.
2. .... is the only non-metallic element which is liquid at room temperature.
3. Two atoms of similar electronegativity would be expected to form a .... compound.
4. Of the halide ions, ... is the most powerful reducing agent.
5. As the atomic radius increases, ionisation energy ....
6. The element having electronic configuration $1s^2 2s^2 2p^6 3s^2 3p^2$ has valency ....
7. The substance which loses electrons is said to be .... and the substance which gains electrons is said to be ....
8. Elements of group .... have greater tendency to form positive ions than elements of group IIA.
9. An element which can exist as a positive ion in acidic solution and also as a negative ion in basic solution is said to be ....
10. On Mulliken scale, the average of ionisation energy and electron affinity is called ....
11. .... is the most electronegative element in the Periodic Table.
12. Ionic mobility of Al^{3+} is .... than that of Na^+.
13. $(IE)_1$ of Be is .... than that of B but $(IE)_2$ of Be is .... than that of B.
14. Electron affinity of F .... than that of Cl.
15. In aqueous solution .... is the best reducing agent among alkali metals.

Matrix-Match Type Questions

I. Only One Correct

1. Match atomic number of elements (in List I) with position of elements in the Periodic Table (in List II).

List I		List II
A.	19	1. *p*-block
B.	22	2. *f*-block
C.	32	3. *d*-block
D.	64	4. *s*-block

2. Match the type of elements (in List I) with the corresponding electronic configuration (in List II).

List I	List II
A. Inert gas elements	1. ns^1 to ns^2np^5
B. Main group elements	2. $1s^2$ to ns^2np^6
C. Transition elements	3. $(n-2)f^{1-14}$ $(n-1)s^2p^6d^{10}ns^2$
D. Inner transition elements	4. $(n-1)d^{1-9}ns^2$
	5. $(n-1)d^{10}ns^2$

3. Match the classification (in List I) with the element (in List II).

List I	-List I
A. Alkaline earth metal	1. Rb
B. Radioactive element	2. As
C. Alkali metal	3. Mg
D. Metalloid	4. Fr

4. Match the element (in List I) with its unique properties (in List II).

List I	List II
A. F	1. Maximum ionisation energy
B. Cl	2. Maximum electronegativity
C. Fe	3. Maximum electron-affinity
D. He	4. Recently named by IUPAC
E. Ds	5. Variable valency

5. Match the atomic number (in List I) with its IUPAC nomenclature (in List II).

List I	List II
A. 105	1. Uuh
B. 107	2. Uun
C. 109	3. Uns
D. 110	4. Unp
E. 116	5. Une

II. One or More Correct

1. Match the element in Column I with its corresponding property (ies) in Column II.

Column I	Column II
A. F	1. Maximum electron-affinity
B. Cl	2. Maximum electronegativity
C. Br	3. Exists as X_2 in liquid state
D. I	4. Exists as X_2 in solid state
E. At	5. X_2 sublimates on heating
	6. X_2 is the best oxidising agent
	7. Radioactive element

2. Match the oxide in Column II with its corresponding property (ies) in Column II.

	Column I		Column II
A.	Na_2O	1.	Soluble in HCl
B.	MgO	2.	Soluble in NaOH
C.	Al_2O_3	3.	Turns blue litmus red
D.	SiO_2	4.	Turns red litmus blue
E.	P_2O_5	5.	Giant molecule
F.	SO_3	6.	Acid-rain

Integer Answer Types

This section contains 8 questions. The answer to each of the questions is a single digit integer, ranging from 0 to 9. The appropriate bubbles below the respectively question numbers in the *ORS* have to be darkened. For example, if the correct answers to question number *X*, *Y*, *Z* and *W* (say) are 6, 0, 9 and 2, respectively

X	Y	Z	W
0	0	0	0
1	1	1	1
2	2	2	2
3	3	3	3
4	4	4	4
5	5	5	5
6	6	6	6
7	7	7	7
8	8	8	8
9	9	9	9

Set I

1. Magnetic moment of an element is 1.73. Thus, unpaired electron can be ....

2. Maximum oxidation state shown by Mn is ....

3. Electronegativity $= \frac{(IP) + (EA)}{N}$.

This $N =$ ....

4. Stability of Tl is maximum when its valency is ....

5. Catention is maximum for the element with atomic number ....

6. Lanthanides have been placed in period ....

7. Actinides have been placed in period ....

8. Number of unpaired electron in Gd(64) is ....

Set II

1. Effective nuclear charge of an element with three valence electrons in 2.60. Thus, atomic number of the element is ....

2. Number of amphoteric oxides out of Na_2O, MgO, Al_2O_3, SnO_2, BeO_2 and CaO is ....

3. Maximum electron affinity is shown by the species at serial number.

(1) O (2) F (3) Cl^- (4) He (5) N

4. Diatomic molecule is formed by the element at serial number (1) and at serial number ....

(1) $[Ne]3s^2 3p^1$ (2) $[Ne]3s^1$
(3) $[Ne]3s^2$ (4) $[He]2s^2 2p^3$
(5) $[Ne]3s^2 3p^6$ (6) [Ne]

5. An ionic compound is formed of the type *XY* from *X* with electronic configuration ns^1 and *Y* with valence shell electrons ....

6. Maximum oxidation state shown by chalcogens is ....

7. Ionisation energy of M^{3+} is 217.6 eV compared to that of He^+ (54.4 eV). Thus, atomic number of M^{3+} is ....

8. Semi-conductor with minimum value of atomic number is ....

Test Yourself on Periodic Table

1. Test your knowledge of the Periodic Table by indicating where on the blank outline elements that meet the following descriptions appear:

(a) Elements with the valence-shell ground-state electron configuration $ns^2\ np^5$

(b) An elements whose fourth shell contains two *p*-electrons

(c) An element with the ground-state electron configuration $[Ar]\ 4s^2\ 3d^{10}$

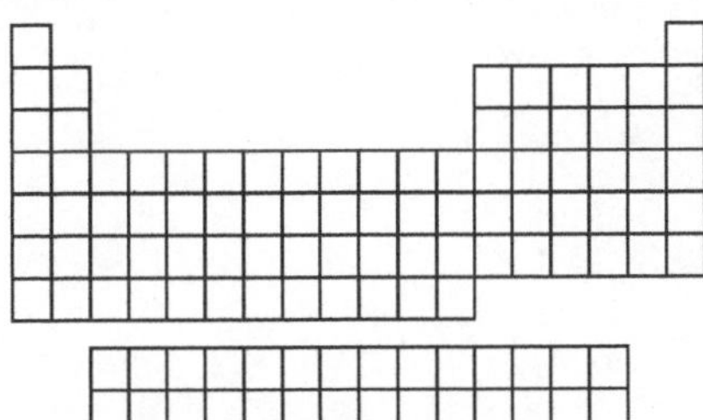

2. Circle the approximate part or parts of the Periodic Table where the following elements appear :

(a) Elements with the smallest values of $(IE)_1$

(b) Elements with the largest atomic radii

(c) Elements with the most negative values of EA

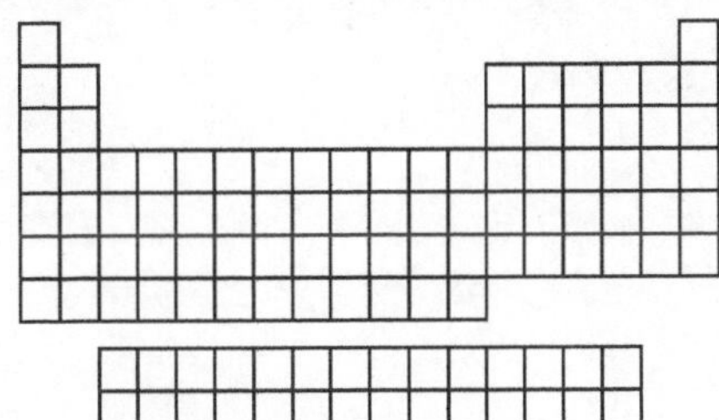

3. Test your knowledge of the Periodic Table by indicating approximately where on the blank outline the following elements appear :

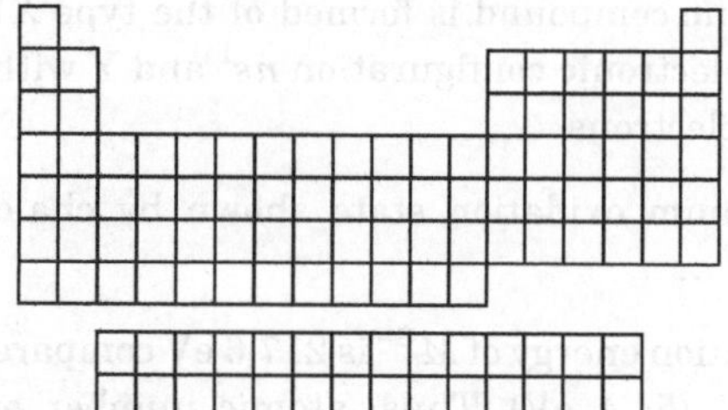

(a) Main groups (b) Halogens
(c) Alkali metals (d) Noble gases
(e) Alkaline earths metals (f) Group 3A elements
(g) Lanthanides

4. Look at the location of elements *A*, *B*, *C*, and *D* in the following Periodic Table :

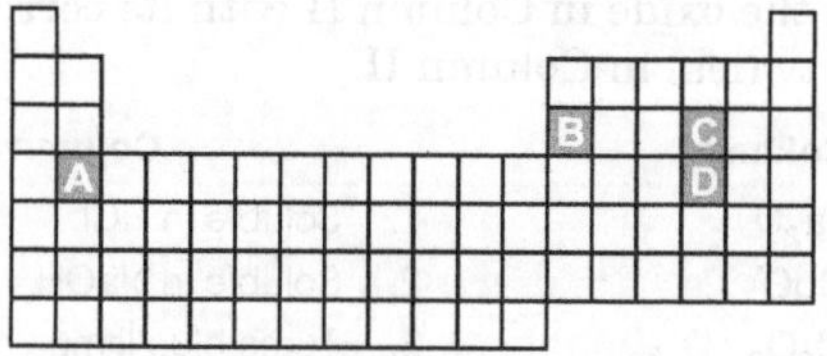

(a) Write the formula of the oxide that has each of these elements in its highest oxidation state.

(b) Classify each oxide as basic, acidic, or amphoteric.

(c) Which oxide is the most ionic? Which is the most covalent?

(d) Which of these oxides are molecular? Which are solids with an infinitely extended three-dimensional crystal structure?

(e) Which of these oxides has the highest melting point?
Which has the lowest melting point?

5. Look at the location of elements *A*, *B*, *C*, and *D* in the following Periodic Table:

(a) Write the formula of the simplest binary hydride of each element.

(b) Classify of these hydride as ionic, covalent or interstitial.

(c) What are the oxidation states of hydrogen and other elements in the hydrides of *A*, *C* and *D*?

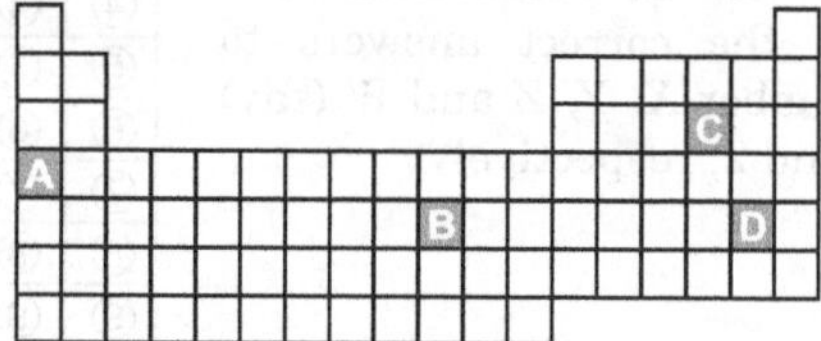

Answers & Solutions
Master Exercises

Exercise 1

Short Answer Type Questions

1. Melting point is a reflection of the attractive force which holds the constituent particles of the solid element. In metals, this force depends primarily on the size and nature of packing of the atoms in the crystal and on the contribution of valence shell electrons in bonding. The lower melting point of Ag suggests that *d*-electrons in metallic bonding are involved to a smaller extent compared to Cu and Au. This is confirmed by the fact that common oxidation of Ag is +1 while that of Cu and Au are +2 and +3 respectively.

   The lower melting points of Zn—Cd—Hg in comparison to Cu—Ag—Au also indicate that there is contraction of *d*-orbital from increasing nuclear charge; in Hg this is reinforced by still poor shielding by the *f*-electrons.

2. This is general formula to calculate Z^* directly by use of valence electrons *n*.

Element	Z	EC	n (valence electrons)	$Z^* = 0.65[1 + n]$
Li	3	$1s^2 2s^1$	1	1.30
Be	4	$1s^2 2s^2$	2	1.95
B	5	$1s^2 2s^2 2p^1$	3	2.60
C	6	$1s^2 2s^2 2p^2$	4	3.25

3. (a) H being least electronegative, its oxidation number is taken as $= +1$.

   and O being most electronegative, its oxidation number is taken as $= -2$. Take oxidation number of $I = x$

   Thus, $H_2IO_6^{3-}$

   $$+2 + x - 12 = -3$$

   $\therefore$ $x = +7$

   (b) Oxygen being more electronegative than sulphur is assigned oxidation number $= -2$

   and oxidation number of sulphur $= x$

   Thus, $S_3O_6^{2-}$

   $$3x - 12 = -2$$
   $$x = 3.33$$

   (c) Na_2O_2

   $$+2 + 2x = 0$$
   $$x = -1$$

4. (a), (b) $5g < 5f < 5d < 5p < 5s$

   (c) $5s < 5p < 5d < 5f < 5g$

5. (a) 157 pm (b) 97 pm (c) 217 pm (d) 217 pm

6. (a) F < C < Li < Ne (Ne has van der Waal's radius)

   (b) Be < Ca < Ba < Ra

   (c) B < Al ≈ Ga < In ≈ Tl

   (d) V < Nb = Ta ≈ Db

7. (a) increase (b) decrease (c) less rapidly

   (d) stay the same; decreases; less rapidly

   (e) increase; increase, less rapidly

Analytical Questions

1. $Ca^{2+} < K^+ < Ar < Cl^- < S^{2-}$
2. (a) In (b) C (c) Cs (d) I (e) Xe
3. (a) 6 (b) 8 (c) 7 (d) 1 (e) 2 (f) 4
4. (a) As (b) Sr (c) Cs (d) Xe (e) C (f) Hg
5. Cs < Sr < As < S < F
6. (d) < (a) < (b) < (c) < (e)
7. (a) most metallic : Ba (b) most non-metallic : S

   (c) intermediate $(IE)_1$: Bi
8. Rb > Ca > Sc > Fe > Te > Br > O > F
9. Diamagnetic (which does not have unpaired electrons) : K^+, Zn^{2+}, Cd, Sn^{2+}

   Paramagnetic (which has unpaired electrons) : Cr^{3+}, Co^{3+}
10. Noble gas has $Z = 118$ and alkali metal $Z = 119$; atomic weight should be about 310.
11.

101	102	103	104	105	106	107	108	109
Unu	Unb	Unt	Unq	Unp	Unh	Uns	Uno	Une

12. (a) 4 (b) 3 (c) 1 (d) 2 (e) 1 (f) 2
13.

		Covalent radius	$(IE)_1$
(b)	Al	125 pm > Ge	577 < Ge
(c)	In	150 pm > Ge	558 < Ge
(d)	Se	114 pm < Ge	941 > Ge

14. (a) Ba (b) Al (c) Cl (d) N
15. Oxygen is more electronegative than chlorine and cannot be oxidised by it.

16. Outer shell electronic configuration : ns^2np^3

(Three electrons less than octet)

Number of electrons required to fill the valence shell = 3

Hence, X^{3-} has configuration = ns^2np^6

X^{3-} has stable nearest inert gas configuration hence most negative oxidation state = – 3

17. (a) Ca (b) Cl (c) N_2O_3 (d) Al (e) N (f) Br (g) BaO (h) CsOH (i) $B(OH)_3$ (j) S (k) RbCl (l) O (m) $Al(OH)_3$ (n) KNO_2 (o) Li^+ (p) Be (q) F (r) O^- (s) O (t) Na^+ (u) Ne (v) Hg (w) Al_2O_3 (x) B_2O_3 (y) In_2O_3 (z) carbon (diamond)

18. (a) – V (b) — III (c) — I (d) — IV (e) — II

19. a fuming red liquid — Br_2

a dark metallic looking solid — I_2

a pale yellow gas — Cl_2

yellow green gas — F_2

20. X : Sn or Pb, Y : P, Z : alkali metal

21. (a) Ge, (b) S, (c) Tl, (d) Ne, (e) Se $c < a < e < b < d$

22. (a) a non-metal : III

(b) an alkali metal : I

(c) an alkaline earth metal : II

23. IUPAC name : Uue, Chloride : MCl, Oxide : M_2O

Exercise 2

Only One Option Correct

1. K-19 (in absence of Aufbau rule) :

$$1s^2 2s^2 2p^6 3s^2 3p^6 \underset{\uparrow}{\underline{3d^1}}$$

last-filling electron goes into d-orbital. Thus, d-block element.

Thus, **(c)**

2. I_3 is very high indicating +2 oxidation state is most stable. Thus, element is of group 2.

Thus, Be.

Thus, **(a)**

3. $\left.\begin{array}{l}N(7)\\ \text{and } O^+(8)\end{array}\right] 1s^2 2s^2 \quad \underbrace{2p_x^1 2p_y^1 2p_z^1}_{\text{Maximum multiplicity most stable}}$

O] $1s^2 2s^2 2p_x^2 2p_y^1 2p_z^1$

(IP) is dependent on the value of Z.

Z of $O^+(8) > Z$ of N(7)

Thus, (IP) of O^+ > N

Na with one valence electron has least IP

Thus, (c)

4. Let $IP = I$

$EA = E$

$$X(g) \longrightarrow X^+(g) + e^-$$

$$\frac{N_0}{2} I = E_1$$

$$\therefore \quad I = \frac{2E_1}{N_0}$$

$$X(g) + e^- \longrightarrow X^-(g)$$

$$\frac{N_0 E}{2} = E_2$$

$$\therefore \quad E = \frac{2E_2}{N_0}$$

Thus, **(b)**

5. Smaller cation, layer charge, larger hydration

Size of cation $Na^+ > Mg^{2+} > Al^{3+} > Si^{4+}$ Minimum size

Charge ⟶ Maximum charge

Thus, $SiCl_4$ is maximum hydrated.

Thus, **(d)**

6. (c) **7.** (b)

8. Repulsive force are balanced by attractive force, thus, **(c)**

9. (d) **10.** (c)

11.

O	F	$F < O$	$I > Br > F$
S	Cl	Thus, $F^- < O^{2-}$	$I^- > Br^- > F^-$
Se	Br	$Se > O$	
	I	$Se^{2-} > O^{2-}$	

$I > Se$

$\therefore \quad I^- > Se^{2-}$

Thus, $I^- > Se^{2-} > Br^- > O^{2-} > F^-$

Thus, **(d)**

12. (d)

13. Maximum hydrated ion is Al^{3+}.

Thus, **(b)**

14. Smaller cation, larger anion, maximum polarising power

Thus, Te^{2-}

Thus, **(d)**

15. (d) **16.** (d)

17. (a) Li—Mg diagonal relationship

(b) Na—K, (c) Ca—Mg, (d) B—Al — same group

Thus, **(a)**

18. Except (d) all show diagonal relationship.

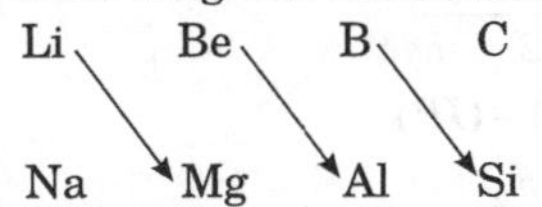

Thus, (d)

19. (a) **20.** (d) **21.** (b) **22.** (d)

23. (a) $Tl^{3+} + 2e^- \longrightarrow \underset{\text{stable}}{Tl^+}$

Tl^{3+} is an oxidising agent thus correct.

(b) $Ga^+ \longrightarrow \underset{\text{stable}}{Ga^{3+}} + 2e^-$

Ga^+ is a reducing agent thus correct.

(c) $Pb^{4+} + 2e^- \longrightarrow \underset{\text{stable}}{Pb^{2+}}$

Pb^{4+} is an oxidising agent thus correct.

(d) $\underset{\text{stable}}{As^{5+}} + 2e^- \longrightarrow As^{3+}$

As^{5+} cannot be reduced

Thus, As^{5+} salts are not oxidising agent.

Thus, **(d)**

24. (d) **25.** (a)

26. Smaller cation, larger anion, larger charge (on cation or anion)

greater the polarising power,

larger the covalent nature

smaller the melting point

Na^+ (same), charge on cation, anion–same

Thus, size of anion will decide polarising power.

Size : $F^- < Cl^- < Br^- < F^-$

Polarising power : $F^- < Cl^- < Br^- < I^-$

Covalent NaF < NaCl < NaBr < NaI

Melting point NaF > NaCl > NaBr > NaF

Thus, **(a)**

27. (b)

28. BeO, SnO, ZnO, all are soluble in alkalis as well as in acids, thus are amphoteric oxides.

$$MO + 2HCl \longrightarrow MCl_2 + H_2O$$

$$MO + 2NaOH \longrightarrow Na_2MO_2 + H_2O$$

Thus, **(d)**

29. Polarising power of $Be^{2+} > Mg^{2+} > Ca^{2+} > Ba^{2+}$

Thus, ionic nature of $BeCl_2 < MgCl_2 < CaCl_2 < BaCl_2$

Thus, **(a)**

30. Stable configuration with all unpaired electron in *f*-orbital is $[Xe]\,6s^2 4f^7 5d^1$

Thus, **(b)**

31. Element (58-71) and (90-103) are placed in IIIB.

$$\underset{\text{IIIB}}{^{238}_{92}U} \longrightarrow \underset{\text{IIIB}}{^{234}_{90}Th} + \alpha$$

Thus, **(a)**

32. O > F

$O^- > F^-$ Thus, $O^{2-} > O^- > F^- > F$

$O^{2-} > F^-$

Thus, **(a)**

33. (c) **34.** (a)

35. Due to empty *d*-orbital, screening effect decreases, hence tendency to attract electron increases.

Thus, **(c)**

36. Oxides of *C* is a giant molecule indicating covalent nature

Oxide of *B* is amphoteric

Oxide of *A* is ionic

Thus, *A* is largest in size.

Thus, atomic number of *A* is least

$$A < B < C$$

Thus, **(a)**

37.

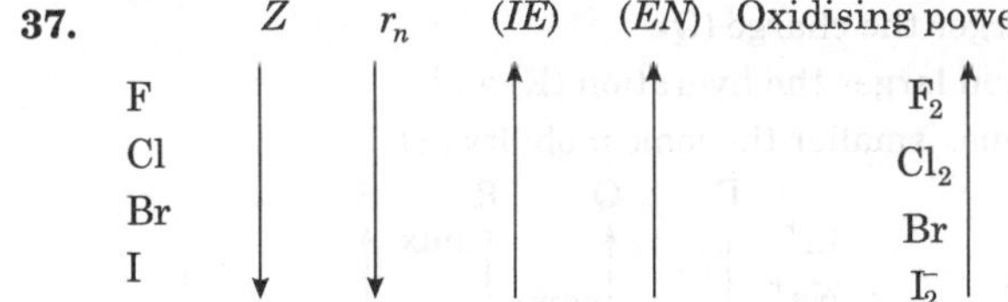

Thus, **(a)**

38. (a) —F (b) —Ne zero (*EA*) (c) —Cl

(d) —Ar zero (*EA*)

(c) Cl > (a) F due to empty *d*-orbital in Cl

Thus, **(c)**

39. (b) **40.** (c) **41.** (d)

42. $O(g) + e^- \longrightarrow O^-(g)$ *EA* = – ve

$O^-(g) + e^- \longrightarrow O^{2-}(g)$ *EA* = + ve

Thus, **(b)**

43. Ionic radii of $O^{2-} > F^- > Na^+ > Mg^{2+}$

We take van der Waals' radius of Ne which should have maximum size.

Thus, **(a)**

44. (a) **45.** (d)

46. Mg^{2+} : $1s^2 2s^2 2p^6$ unpaired electron zero

Ion	Configuration	3d	Unpaired
Ti^{3+}	$1s^2 2s^2 2p^6 3s^2 3p^6 3d^1$	↑	1
V^{3+}	$1s^2 2s^2 2p^6 3s^2 3p^6 3d^2$	↑ ↑	2
Fe^{2+}	$1s^2 2s^2 2p^6 3s^2 3p^6 3d^6$	⇅ ↑ ↑ ↑ ↑	4

47. I > II > III

Thus, (a)

48. (d) **49.** (b)

50. (IV) B contains C, Si, Ge, Sn, Pb

Ge, Sn, Pb are strongly electropositive.

Thus, **(d)**

51. (d)

52. I.P. of Li > Na
Thus, Na falls below Li
Thus, **(c)**

53.

		Be	B
I_1 of Be > I_1 of B		$1s^2 2s^2$	$1s^2 2s^2 2p^1$
I_2 of Be < I_2 of B		$\downarrow I_1$	$\downarrow I_1$
I_3 of Be > I_3 of B		$1s^2 2s^1 (Be^+)$	$1s^2 2s^2 (B^+)$
		$\downarrow I_2$	$\downarrow I_2$
		$1s^2 (Be^{2+})$	$1s^2 2s^1 (B^{2+})$
		$\downarrow I_3$	$\downarrow I_3$
		$1s^1 (Be^{3+})$	$1s^2 (B^{3+})$

Thus, (b) is incorrect.
Thus, **(b)**

54. smaller the size (P) of the ion
larger the charge (Q)
then larger the hydration (R) and
thus, smaller the ionic mobility (S)

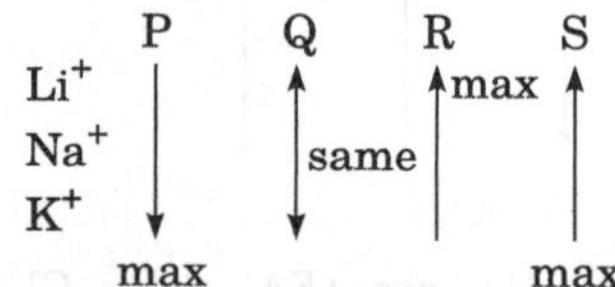

Thus, (a) is true

Na^+ Mg^{2+} Al^{3+}

P ← max

Q → max

R → max

S → max

Thus, (b) is also true
Thus, **(c)**

55.

		ΔH
	$M(s) \xrightarrow{I_1} M^+(g) + e^-$	100 eV
	$M(g) \xrightarrow{I_1 + I_2} M^{2+}(g) + 2e^-$	250 eV
Thus,	$M^+(g) \xrightarrow{I_2} M^{2+}(g) + e^-$	150 eV

Thus, (a) is true, (b) is also true (I_1 of $M^+(g)$ is I_2 of $M(g)$)
Thus, **(c)** is incorrect

56. $A \longrightarrow A^+ + e^-$ if $(IP)_A < (IP)_B$
$B + e^- \longrightarrow B^-$ if $(EA)_B > (EA)_A$

$$(EN) = \frac{(IP) + (EA)}{2}$$

A^+B^- is formed
if $(EN)_A < (EN)_B$
Thus, **(c)**

57. $(EN) = \dfrac{(IP) + (EA)}{2}$
$(EA) = 2(EN) - (IP)$
$= 2E_1 - E_2$
Thus, **(a)**

58. (c) due to inert-pair effect.

59. Copper forms only basic oxide
Thus, **(b)**

60. *(IP)* of N > O
Thus, (a)

61. In case of Ne we take van der Waals' radius
Thus, **(a)**

62. A-19 *s*-block
B-22 *d*-block
C-32 *p*-block
D-64 *f*-block
Thus, **(b)**

63. (c) **64.** (d) **65.** (a)

66. (a) K(19) $1s^2 2s^2 2p^6 3s^2 3p^6 3d^1$
$K^+(19)$ $1s^2 2s^2 2p^6 3s^2 3p^6$
colourless being no unpaired electron in *d*-orbital
Thus, **(a)** is incorrect.
(b) Na(11) $1s^2 2s^2 2p^6 3s^1$ *s*-block — true
(c) Cu(29) $1s^2 2s^2 2p^6 3s^2 3p^6 3d^{10} 4s^1$ *s*-block —true
(d) Cr(24) $1s^2 2s^2 2p^6 3s^2 3p^6 3d^6$ [↑↓|↑↓|↑↓| |]
Hund's rule absent
Thus, (d) is true
Only (a) is incorrect.

67. Cu^+ is colourless
Sc^{3+} is colourless
Na^+, Mg^{3+}, Al^{3+} — colourless
as they have no electron unpaired in *d*-orbitals.
Thus, **(b)**

68. Crystal > van der Waals' > covalent
Thus, **(a)**

69. *p*-block element of **higher periods**
Ge, Sn, Pb
Thus, **(c)**

70. Gd(64) $4f^7 5d^1$
eight unpaired electron
Sum of spin $= 8 \times \frac{1}{2} = 4$
Thus, **(d)**

71. U n q
↓ ↓ ↓
1 0 4
Thus, **(c)**

72. Size of cation decides lattice energy and thermal stability.

Smaller size, larger polarising power thus larger the lattice energy.

Thus, $LiCl > NaCl > KCl > RbCl > CsCl$

Thus, **(b)**

73.

	Z	electron
K^+	19	18
S^{2-}	16	18
Cl^-	17	18
Ca^{2+}	20	18

Above are isoelectronic species, hence size $\propto \frac{1}{Z}$

Thus, $S^{2-} > Cl^- > K^+ > Ca^{2+}$

Thus, **(c)**

74. (d)

75. Mg $1s^2\,2s^2\,2p^6\,3s^2$ $3s^2$ [↑↓]

Al $1s^2\,2s^2\,2p^6\,3s^2\,3p^1$ [↑↓] [↑| |]

Thus, **(c)**

76. (c) NH_4^+ is from non-metal N

77. (a)

78. $X(g) \longrightarrow X^+(g) + e^-; \quad \Delta H = I$

$X^+(g) + e^- \longrightarrow X(g); \quad \Delta H = E = -I$

Thus, **(b)**

79. $X(g) + e^- \longrightarrow X^-(g); \quad \Delta H = x = EA \text{ of } X(g)$

$X^-(g) \longrightarrow (X)g + e^-; \quad \Delta H = y = (IE) \text{ of } X^-(g) = -(EA) \text{ of } X(g)$

Thus, (a), (b), (c) are true

Thus, **(d)**

80. (c) **81.** (a) **82.** (b)

83. Smaller size, larger charge ⇒ larger hydration and hydration energy

Size $Ca^{2+} < Si^{2+} < Ba^{2+}$

hydration $Ca^{2+} > Sr^{2+} > Ba^{2+}$

Thus, **(a)**

84. Unpaired electrons = N

$$\text{Magnetic moment} = \sqrt{N(N+2)}\text{ BM}$$

$$\sqrt{N(N+2)} = 1.73$$

$$N(N+2) = (1.73)^2 = 3$$

$$N^2 + 2N - 3 = 0$$

$$(N+3)(N-1) = 0$$

Thus, $N = 1$ (one unpaired electron)

$V(23) : 1s^2 2s^2 2p^6 3s^2 3p^6 3d^3 4s^2$

$V^{4+} : 1s^2 2s^2 2p^6 3s^2 3p^6 3d^1$

Thus, $\mathbf{VCl_4}$

Thus, **(c)**

85. Ni^{2+} is smallest thus, (a)

86. $(IE)_1$ of Be > B

Thus, (b) is incorrect

87. Size $\propto \frac{1}{Z}$ for isoelectronic ion.

Thus, $O^{2-} > F^- > Na^+ > Al^{3+}$

(Z) 8 9 11 13

Thus, **(d)**

81. $r_n = \frac{n^2 a_0}{Z^*}$

$r_n \propto \frac{1}{Z^*}$

Thus, **(a)**

89. (d)

90. Size $\propto \frac{1}{Z}$ for isoelectronic ion

Size : $C^{4-} > N^{3-} > O^{2-} > F^-$

Z: 6 7 8 9

Thus, **(a)**

91. (b) **92.** (c) **93.** (d)

94. ${}_{115}X \longrightarrow {}_{113}Y + {}^4_2He$

Thus, **(a)**

95. (d)

96. $(n-1)d^5ns^1$ represents group VIB (transition metals) of d-block

Thus, **(d)**

97. $Sc^{2+}(18e)[Ne]:3s^2 3p^6$

$Sc[Ne]:3s^2 3p^6 4s^2 3d^1$ Thus, d-block IIIB group

Thus, given matching is incorrect.

Thus, **(a)**

98. $113 : [Rn]\, 7s^2 5d^{10} 4f^{14} 7p^1$

$115\ 7s^2 7p^3 : [Rn]\, 7s^2 5d^{10} 4f^{14} 7p^3$

Thus, p-block element

Thus, **(b)**

99. Valence electrons in $M(113) = 3$

Thus, M^{3+} is the stable cation but due to inert pair effect M^+ is most stable

Thus, **(c)**

100. $M(115) : [Rn]\, 7s^2 5d^{10} 4f^{14} 7p^3$

Valence shell is $7s^2 7p^3$ with five electrons.

Thus, +5, +3 are variable oxidation state. But due to inert-pass effects $M^{5+} < M^{3+}$ (stable).

Thus, **(a)**

101. (a)

102. $X : ns^2np^5$ electronegative element

valency = −1

Y : electropositive alkali metal

valency = +1

$Y \longrightarrow Y^+ + e^-$ $\quad Y^+$

$X + e^- \longrightarrow X^-$ $\quad X^-$ have inert gas configuration

Compound is YX

Thus, **(d)**

103. (b)

104. Oxygen family is chalcogens.

Thus, **(d)**

105. M^{3+} : [Ar] $3d^{10}$

M : [Ar] $3d^{10}4s^14p^1$

Thus, p-block

Thus, **(b)**

106. (c) **107.** (d)

108. Alkali (Li, Na, K) and alkaline earth metals (Ca, Sr, Ba) impart colour in flame

Thus, **(b)**

109. (c)

110. Fe : [Ar] $3d^64s^2$

Fe^{2+} : [Ar] $3d^5$

Thus, **(b)**

111. (b)

112. Lanthanides are f-block elements.

Thus, **(d)**

113. Actinides and lanthanides constitute f-blocks

Thus, **(b)**

114. Gd(64) [Xe] $6s^2$

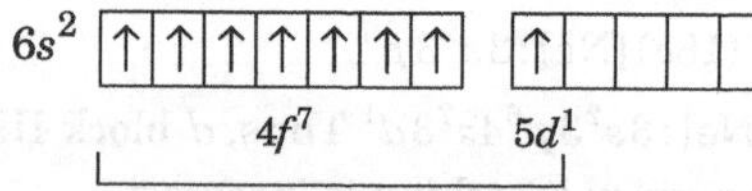

all electrons, unpaired, hence stable

Thus, **(a)**

115. 1 1 5
↑ ↑ ↑
U u p

Thus, **(c)**

116. (a) **117.** (b)

118. $1s^22s^22p^63s^23p^64s^1$: it is a case of alkali metals (K).

Thus, graded properties will be shown by element with identical configuration of the valence shell.

Thus, **(c)**

119. (a)

120. Element (113) lies below Tl (group 13, *i.e.*, IIIA)

Thus, it resembles **Ga**

Thus, **(b)**

121. Due to inert-pair effect

Stability of $Al^{3+} < Sb^{3+} < Bi^{3+}$

Thus, **(c)**

122. Stability of $Ga^+ > In^+ > Tl^+$

and $Ga^+ < In^{3+} < Tl^{3+}$

and $Tl^+ > Tl^{3+}$

Thus, Tl^{3+} salts are easily reduced to Tl^+. Most spontaneous reaction is

$$Tl^{3+} + + 2e^- \longrightarrow Tl^+$$

Thus, **(c)**

123.

Element		Compound with oxygen	Valency
Halogen	:	Cl_2O_7	+7
Oxygen	:	O_3	0
Nitrogen	:	N_2O_5	+5
Boron	:	B_2O_3	+3

Thus, **(a)**

124. Stability of $Pb^{2+} > Sn^{2+} > Ge^{2+}$

$Pb^{4+} < Sn^{4+} < Ge^{4+}$

$Pb^{2+} > Pb^{4+}$

$Sn^{4+} > Sn^{2+}$

$Ge^{4+} > Ge^{2+}$

Thus, $Pb^{2+} + 2e^- \longrightarrow Pb^{2+}$

Thus, Pb^{4+} salts are better oxidising agent being easily reduced to Pb^{2+}.

Thus, **(b)**

125. Valence electrons in $A = 3$

Valence electrons in $B = 6$

Thus A is electropositive and B is electronegative; A can lose three electrons and B can gain two electrons to attain stable configuration.

Thus, A exists as A^{3+}

and B as B^{2-}

Thus, compound is A_2B_3

Thus, **(d)**

126. (a)

127. He^+, Li^{2+} and Be^{3+} each has one electron, hence no screening.

Thus, **(d)**

128. Effective nuclear charge $Z_{eff} = Z - S$

where Z = atomic number

and S = screening constant

= 0.35 per electron for electron in nth orbit.

= 0.85 per electron for electrons in $(n-1)$th orbit

= 1.00 per electron for electrons in $(n-2)$th, $(n-3)$th, $(n-4)$th orbit

= 0.30 per electron in 1s-orbital (when alone)

Be $1s^2 2s^2$ one valence-electron in $2s$ is screened by one electron in $2s$-orbital (nth orbit) and two electrons in $1s$ orbital (($n-1$) th orbit)

$$\therefore \quad S = 0.35 + 2 \times 0.85 = 2.05$$

$$\therefore \quad Z_{eff} = 4 - 2.05 = 1.95 \text{ (given)}$$

Be^+ $1s^2 2s^1$ $2s$-electron is screened by two electrons in $1s$-orbital (($n-1$) th orbital)

$$S = 2 \times 0.85 = 1.70$$

$$\therefore \quad Z_{eff} = 4 - 1.70 = 2.30$$

Be^{2+} $1s^2$ one electron in $1s$-orbital (alone exists) is screened by another electron in same orbit. Thus,

$$S = 0.30$$

$$\therefore \quad Z_{eff} = 4 - 0.30 = 3.70$$

Be^{3+} $1s^1$ no-screening (single electron)

Thus, $Z_{eff} = 4.$

Thus, $Z_{effective}$: $Be < Be^+ < Be^{2+} < Be^{3+}$

Thus, **(a)**

129. Refer question 128.

Thus, **(a)**

130. Cu(29)

$$\underbrace{1s^2}_{(n-3)} \quad \underbrace{2s^2 2p^6}_{(n-2)} \quad \underbrace{3s^2 3p^6 3d^{10}}_{(n-1)} \quad \underbrace{4s^1}_{n}$$

↑ This electron is screened by 18 electrons in $(n-1)$, 8-electrons in $(n-2)$ and 2-electrons in $(n-3)$

$$\text{Thus, } \sigma = \underbrace{8 \times 0.85}_{(n-1)s^2p^6} + \underbrace{1 \times 10}_{(n-1)d^{10}} + \underbrace{8 \times 1}_{(n-2)d^{10}} + \underbrace{2 \times 1}_{(n-3)s^2}$$

$$= 26.8$$

Thus, $Z^* = 29 - 26.8 = 2.2$

Thus, **(a)**

131.

	Atomic number (Z)	Period (n)
(a) Cs	55	6
(b) Mg	12	3
(c) Ba	56	6
(d) Cr	24	4

$r_n \propto n^2$, but in case of Cs and Ba, $r_n \propto \dfrac{1}{Z}$

Thus, **(a)**

132. (a)

133. $r_n \propto \dfrac{1}{Z}$ along a period.

Thus, B(5) > C(6) > N(7)

Thus, **(b)**

134. (c) **135.** (c) **136.** (b) **137.** (c) **138.** (d)
139. (c) **140.** (a) **141.** (b) **142.** (a)

143. As electron is lost, effective nuclear charge increases, force of attraction increases. Hence,

$$(IE)_1 < (IE)_2 < (IE)_3$$

Thus, Mg^{2+} has maximum (IE)

Thus, **(c)**

144. $He \xrightarrow{I_1} He^+ \xrightarrow{I_2} He^{2+}$

$I_2 > I_1$, hence, (d) is incorrect.

Thus, **(d)**

145. For alkali metals

$$(IE)_1 << (IE)_2$$

Thus, **(c)**

146. (c) **147.** (c)

148. Along a group Z^* (effective nuclear charge) is almost constant. Thus, (IE) depends mainly on r_n.

Thus, **(a)**

149. (a) **150.** (b)

One or More Than One Options Correct

1. (a), (b), (c), (d) **2.** (a), (b), (c) **3.** (a), (b), (c)
4. (a), (c) **5.** (a), (b), (d) **6.** (b) **7.** (c) **8.** (d)
9. (b) **10.** (a), (b), (d)

Brain Twisters

1. (a) $1s^2 2s^2 2p^6 3s^2 3p^6 4s^2 3d^{10} 4p^6 5s^2 4d^{10} 5p^6 6s^2 4f^{14} 5d^{10} 6p^6 7s^2 5f^{14} 6d^{10} 7p^6 8s^2 5g^{18} 6f^{14} 7d^{10}$

(b) $8s^2 7d^{10}$, $7d^{10}$

(c) Four nodal planes, zero nodal sphere poor shielders, poor penetrators.

(d) $Z^* = 4.35$

(e) $\chi_{A-R} = 0.744 + [3590 \times (4.35 - 0.35)]/(150)^2$
$= 0.744 + 0.64 = 1.384$

The value should be much higher.

2. $1.00 \text{ mg} = 1 \times 10^{-3} \text{ g} = \dfrac{1 \times 10^{-3}}{7} \text{ g mol Li}$

1 mol of Li is converted by = 520.3 kJ

$\therefore \quad \dfrac{1 \times 10^{-3}}{7}$ mol is converted to Li^+ by

$$= \frac{520 \times 1 \times 10^{-3}}{7} \text{ kJ}$$

$$= 0.0743 \text{ kJ}$$

3.
$$(IE)_{Total} = (IE)_1 + \dots + (IE)_5$$
$$= 1.7058 \times 10^4 \text{ kJ mol}^{-1}$$
$$= 1.7058 \times 10^7 \text{ J mol}^{-1}$$

Since, $1.6 \times 10^{-19} \text{ J} = 1 \text{ eV}$

$$\therefore \quad 1.7058 \times 10^7 \text{ J} = \frac{1.7058 \times 10^7}{1.6 \times 10^{-19}} \text{ eV mol}^{-1}$$

$$= 1.0661 \times 10^{26} \text{ eV mol}^{-1}$$

$$= \frac{1.0661 \times 10^{26}}{6.02 \times 10^{23}}$$

$$= \mathbf{177.1\ eV\ atom^{-1}}$$

4.

		ΔH
$Mg(g) \longrightarrow Mg^+(g) + e^-$		737.7 kJ
$Mg^+(g) \longrightarrow Mg^{2+}(g) + e^-$		1451 kJ
$2F(g) + 2e^- \longrightarrow 2F^-(g)$		-328×2 kJ
$Mg(g) + 2F(g) \longrightarrow Mg^{2+}(g) + 2F^-(g)$	$\Delta H =$	1532.7 kJ

Thus, **endothermic**

5. $1\ eV\ atom^{-1} = 1.6 \times 10^{-19}\ J\ atom^{-1}$

$$= 1.6 \times 10^{-19} \times 6.02 \times 10^{23}\ J\ mol^{-1}$$

$$= 96.32 \times 10^3\ J\ mol^{-1}$$

$$= \mathbf{96.32\ kJ\ mol^{-1}}$$

6. IE of F $= 27.91 \times 10^{-22}$ kJ atom^{-1}

$$= 27.91 \times 10^{-22} \times 6.02 \times 10^{23}\ kJ\ mol^{-1}$$

$$= 1680.2\ kJ\ mol^{-1}$$

IE of Cl $= 20.77 \times 10^{-22} \times 6.02 \times 10^{23}$ kJ mol^{-1}

$$= 1250.35\ kJ\ mol^{-1}$$

EA of F $= -5.53 \times 10^{-22} \times 6.02 \times 10^{23}$ kJ mol^{-1}

$$= -332.91\ kJ\ mol^{-1}$$

EA of Cl $= -5.78 \times 10^{-22} \times 6.02 \times 10^{23}$ kJ mol^{-1}

$$= -347.96\ kJ\ mol^{-1}$$

Let F $= n_1$ moles and Cl $= n_2$ moles

For the formation of cation

$$1680.2n_1 + 1250.35n_2 = 284$$

or $\quad n_1 + 0.744n_2 = 0.1690 \quad$ ...(i)

and for the formation of anion

$$332.91n_1 + 347.96n_2 = 68.8$$

or, $\quad n_1 + 1.045n_2 = 0.207 \quad$ ...(ii)

From (i) and (ii)

$$n_1 = 0.0785\ mol$$

$$n_2 = 0.123\ mol$$

Total moles $= 0.123 + 0.0785 = 0.2015$

$\because$ Mol % of F $= \dfrac{0.123}{0.2015} \times 100 = 61.04\%$

Mol % F = 38.96%

7. $(IE)_{He^+} = (-E_n)$

$$= \frac{2.18 \times 10^{-18}\ Z^2}{n^2}\ J\ atom^{-1}$$

$$= \frac{2.18 \times 10^{-18} \times (2)^2}{1}\ J\ atom^{-1} \text{ for } He^+\ (n = 1, Z = 2)$$

$$= 2.18 \times 10^{-18} \times 4 \times 6.02 \times 10^{23}\ J\ mol^{-1}$$

$$= 5.2494 \times 10^6\ J\ mol^{-1}$$

$$= \mathbf{5.2494 \times 10^3\ kJ\ mol^{-1}}$$

8. $Hg^{79+}(g) \longrightarrow Hg^{80+}(g) + e^-$

$Hg^{79+}(g)$ is isoelectronic of hydrogen (electron = 1) ($Z = 79, n = 1$)

From Q. 7.

$$\therefore (IE)_{Hg^{79+}} = \frac{2.18 \times 10^{-18} \times (79)^2}{1}\ J\ atom^{-1}$$

$$= 1.36 \times 10^{-17}\ J\ atom^{-1}$$

$$= \frac{2.18 \times 10^{-18} \times (79)^2 \times 6.02 \times 10^{23}}{1000}\ kJ\ mol^{-1}$$

$$= \mathbf{8.19 \times 10^6\ kJ\ mol^{-1}}$$

9. Volume of K atom $= \dfrac{4}{3}\pi r^3 = \dfrac{4}{3}\pi(216)^3$

Volume of K$^+$ ion $= \dfrac{4}{3}\pi(133)^3$

Decrease in volume $= \dfrac{4}{3}\pi[(216)^3 - (133)^3]$

$$\% \text{ decrease} = \frac{\frac{4}{3}\pi[(216)^3 - (133)^3]}{\frac{4}{3}\pi(216)^3} \times 100$$

$$= \mathbf{76.66\%}$$

Passage Comprehension Questions

Passage 1

1. (a) **2.** (b) **3.** (d)

Passage 2

1. (b) **2.** (a) **3.** (c) **4.** (b)

Passage 3

1. (c) **2.** (c) **3.** (a)

Passage 4

1. (c) **2.** (a) **3.** (c)

Passage 5

1. (b) **2.** (d) **3.** (a) **4.** (c)

Passage 6

1. (d) **2.** (c) **3.** (b)

Passage 7

1. (a), (b), (c), (d) **2.** (a)

Passage 8

1. (a) **2.** (b) **3.** (c) **4.** (c)

Assertion & Reason

1. (b) **2.** (a) **3.** (c) **4.** (b) **5.** (b) **6.** (a)
7. (a) **8.** (b) **9.** (a) **10.** (b)

True & False

1. T : Due to shielding effect

2. T : $2NaOH + Al_2O_3 + 3H_2O \longrightarrow 2Na[Al(OH)_4]$

3. F : Neutral oxide

4. F : PbO_2 (oxidation number of Pb = +4) is an oxidising agent since Pb^{2+} is more stable than Pb^{4+} due to inert pair effect.

$$Pb^{4+} + 2e^- \longrightarrow Pb^{2+}$$

5. F : Smaller the size of the cation, larger the hydration and hence smaller the ionic mobility. Li^+ is smaller than Na^+ hence, it is more hydrated than Na^+ hence, its ionic mobility is less than that of Na^+.

6. F : B_2O_3 is an acidic oxide.

$$\underset{\text{acid}}{B_2O_3} + \underset{\text{base}}{2NaOH} \longrightarrow 2NaBO_2 + H_2O$$

B_2O_3 is not soluble in acid.

7. F : ZnO is amphoteric but not amphiprotic.

Amphoteric oxide is that which can be an acid as well as a base. Amphiprotic oxide is that which is H^+ ion donor as well as its acceptor.

$$H_2O + H_2O \longrightarrow H_3O^+ + OH^-$$

H_2O is an amphiprotic.

8. T : LiCl can absorb moisture due to greater tendency of hydration of Li^+ hence, LiCl is humidity controller.

9. F : Hydrogen has only one electron. Its $(IE)_2$ is simply imagination.

10. T

11. T : $Li_2CO_3 \xrightarrow{\Delta} Li_2O + CO_2$

12. F : They have approximately equal size.

13. T : Atomic size $\propto \frac{1}{Z}$ but in case of Ne, **van der Waals'** radius is considered hence atomic radius of Ne > F.

14. T : $Li \xrightarrow{(IE)_1} Li^+ + e^-$

$$Li^+ \xrightarrow{(IE)_2} Li^{2+} + e^-$$

Li^+ is isoelectronic of He. Hence, after losing one electron, Li^+ attains He electronic configuration. Hence, its (IE), (which is $(IE)_2$ of Li), is maximum.

15. T

16. T : Its $(IE)_1$ is maximum in the Periodic Table.

Fill in the Blanks

1. Li **2.** Br **3.** covalent **4.** I^- **5.** decreases **6.** four **7.** reducing agent, oxidsing agent **8.** IA **9.** amphoteric **10.** electronegativity **11.** F **12.** less **13.** greater, less **14.** less **15.** Li.

Matrix-Match Type Questions

I. Only One Correct

1. (A)—(4); (B)—(3); (C)—(1); (D)—(2)
2. (A)—(2); (B)—(1); (C)—(4); (D)—(3)
3. (A)—(3); (B)—(4); (C)—(1); (D)—(2)
4. (A)—(2); (B)—(3); (C)—(5); (D)—(1); (E)—(4)
5. (A)—(4); (B)—(3); (C)—(5); (D)—(2); (E)—(1)

II. One or More Correct

1. (A)—(2,6); (B)—(1); (C)—(3); (D)—(4,5); (E)—(7)
2. (A)—(1,4); (B)—(1,4); (C)—(1,2); (D)—(1,2,5); (E)—(2,3); (F)—(2,3,6)

Integer Answer Types

Set I

Questions →	1	2	3	4	5	6	7	8
Answers →	1	7	2	1	6	6	7	8
	⓪	⓪	⓪	⓪	⓪	⓪	⓪	⓪
	①	①	①	①	①	①	①	①
	②	②	②	②	②	②	②	②
	③	③	③	③	③	③	③	③
	④	④	④	④	④	④	④	④
	⑤	⑤	⑤	⑤	⑤	⑤	⑤	⑤
	⑥	⑥	⑥	⑥	⑥	⑥	⑥	⑥
	⑦	⑦	⑦	⑦	⑦	⑦	⑦	⑦
	⑧	⑧	⑧	⑧	⑧	⑧	⑧	⑧
	⑨	⑨	⑨	⑨	⑨	⑨	⑨	⑨

Set II

Questions →	1	2	3	4	5	6	7	8
Answers →	5	3	2	4	7	6	4	5
	⓪	⓪	⓪	⓪	⓪	⓪	⓪	⓪
	①	①	①	①	①	①	①	①
	②	②	②	②	②	②	②	②
	③	③	③	③	③	③	③	③
	④	④	④	④	④	④	④	④
	⑤	⑤	⑤	⑤	⑤	⑤	⑤	⑤
	⑥	⑥	⑥	⑥	⑥	⑥	⑥	⑥
	⑦	⑦	⑦	⑦	⑦	⑦	⑦	⑦
	⑧	⑧	⑧	⑧	⑧	⑧	⑧	⑧
	⑨	⑨	⑨	⑨	⑨	⑨	⑨	⑨

Test Yourself On Periodic Table

1.

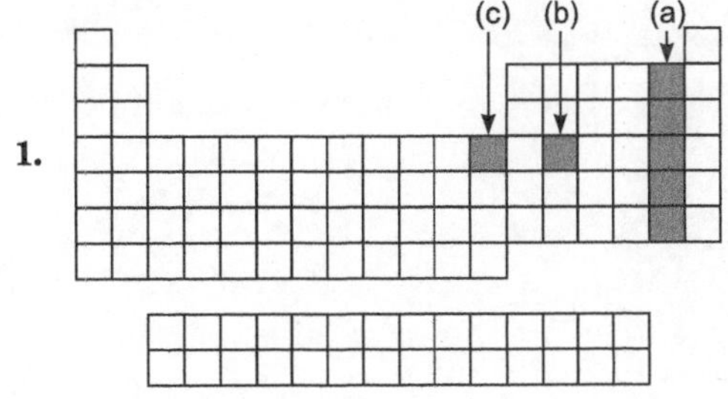

2.

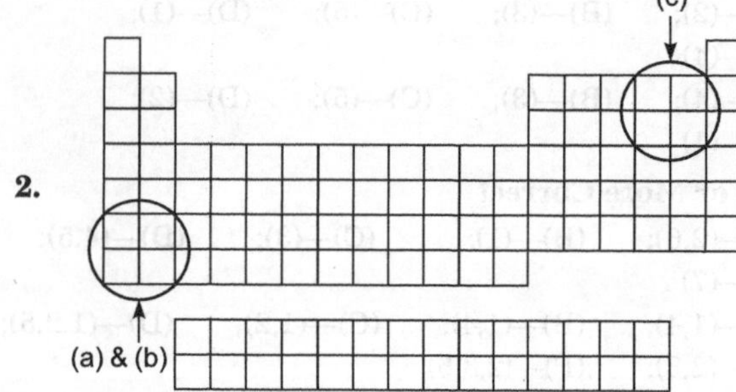

3.

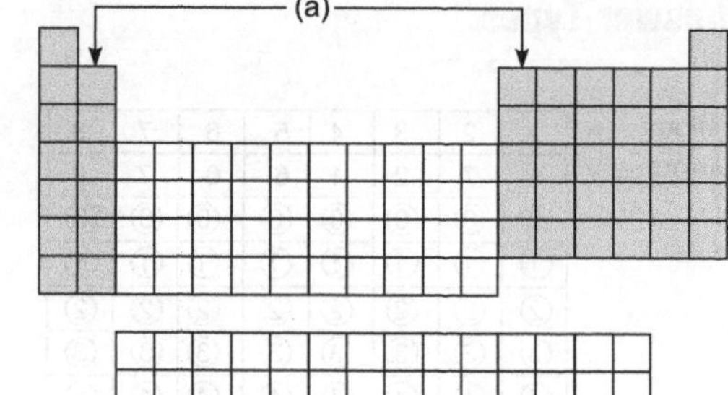

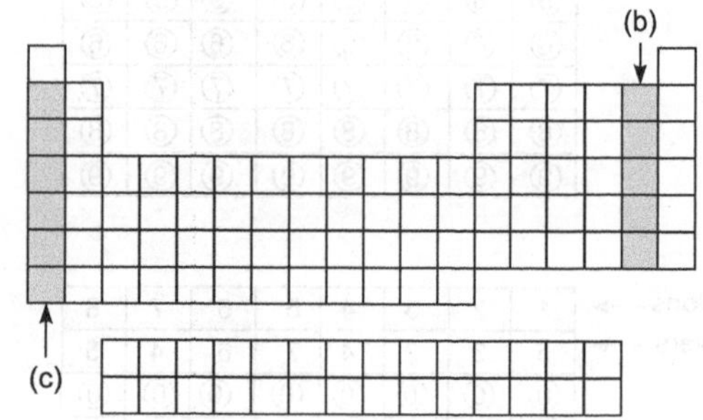

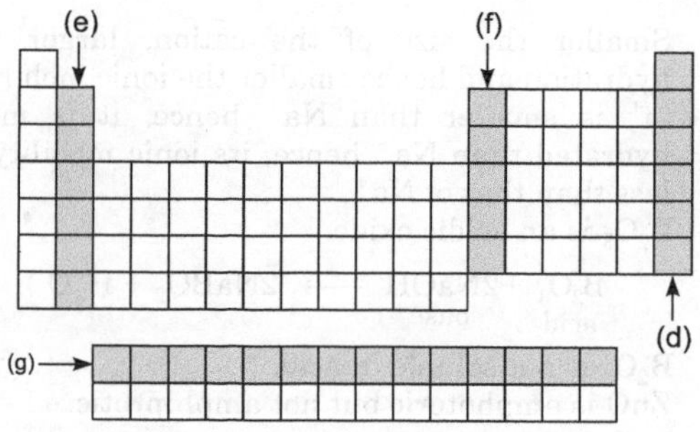

4. (a) A : CaO, B : Al_2O_3, C : SO_3, D : SeO_3

(b) basic : CaO (c) most ionic : CaO

acidic : SO_3, SeO_3 most covalent : SO_3

amophoteric; Al_2O_3

(d) CaO, Al_2O_3 : three-dimensional structure

SO_3, SeO_3 : molecular

(e) CaO : highest melting point

SO_3 : lowest melting point

5. (a) A : NaH (b) ionic (c) $Na = +1, H = -1$

B : PdH_x	interstitial	
C : H_2S	covalent	$S = -2, H = +1$
D : HI	covalent	$I = -1, H = +1$

4

Oxidation-Reduction

A fair business of gain of electrons (reduction) and loss of electrons (oxidation) is a redox reaction. How much is lost in a half-reaction, is also gained in equal amount (electrons) — actually the best fair deal.

Learning Framework of the Chapter ...

- Oxidation Number and Oxidation State
- Oxidation and Reduction
- How to Balance a Redox Reaction
- Balanced Redox Reactions
- Equivalent Weights

Quick Points

- Number of ionisable H atoms (attached to electronegative atom O) determines basicity of an acid.
  HNO_3, H_3PO_2 (monobasic acid) — one H attached to O-atom
  H_2SO_4, H_3PO_3 (dibasic acid) — two H attached to O-atoms
  H_3PO_4 (tribasic acid) — three H attached to O-atoms
- Superoxide ion is O_2^-
- Peroxide ion is O_2^{2-}
- $\text{Mole} = \dfrac{\text{mass}}{\text{molar mass}}$
- N_0 (Avogadro's number) $= 6.02 \times 10^{23}\ \text{mol}^{-1}$
- $\Delta G^\circ = -nFE^\circ$
  where, n = number of electrons exchanged
  F = number of faradays
  E° = standard electrode potential
  ΔG° = standard Gibb's free energy change
- If different number of electrons are exchanged in a series of reactions then

$$\Delta G_3^\circ = \Delta G_1^\circ + \Delta G_2^\circ$$

or

$$E_3^\circ = \frac{n_1E_1^\circ + n_2E_2^\circ}{n_3}$$

$$E = E^\circ - \frac{0.0591}{n} \log Q$$

where, Q = reaction quotient.

$$E^\circ = \frac{2.303RT}{nF} \log K_{eq}$$

at equilibrium when $E = 0$

and Q = equilibrium constant = K_{eq}

Oxidation Number and the Oxidation State

The oxidation number (O.N.) of an atom in a molecule or in a polyatom ion is a hypothetical charge the atom would have if the electrons in each bond were located on the more electronegative atom.

The oxidation number (or oxidation state) of an element in a compound is determined as the number of electrons that have passed from one atom of a given element to other atoms (positive oxidation) or to one atom of a given element from other atom (negative oxidation) Oxidation states simplify the whole process of working out what is being oxidised and what is being reduced in redox reactions.

- Oxidation state is the oxidation number per atom.
- An unreacted element has oxidation number of zero.
- A monoatomic ion has oxidation number equal to its charge.

Oxidation Number of Elements in Covalent Compounds and in Ions

Element/Ion	O.N.	Element/Ion	O.N.
H_2	0	Mg^{2+}	+2
O_2	0	Fe^{3+}	+3
N_2	0	Mn^{7+}	+7
Na	0	O_2^-	– 0.5 (per atom)
Mg	0	O^{2-}	–2
H^+	+1	O_2^{2-}	–1 (per atom)
H^-	–1		

Vanadium forms a number of different ions — for example, V^{2+} and V^{3+}. If you think about how these might be produced from vanadium metal, the 2+ ion will be formed by oxidising the metal by removing two electrons :

$$V \longrightarrow V^{2+} + 2e^-$$

The vanadium is now said to be in an oxidation state of +2.

Removal of another electron gives the V^{3+} ion

$$V^{2+} \longrightarrow V^{3+} + e^-$$

The vanadium now has an oxidation state of +3.

Removal of another electron gives a more unusual looking ion, VO^{2+}.

$$V^{3+} + H_2O \longrightarrow VO^{2+} + e^- + 2H^+$$

The vanadium is now in an oxidation state of +4. Notice that the oxidation state isn't simply counting the charge on the ion (that was true for the first two cases but not for this one).

The positive oxidation state is counting the total number of electrons which have had to be removed-starting from the element.

It is also possible to remove a fifth electron to give another ion (easily confused with the one before). The oxidation state of the vanadium is now +5.

Every time you oxidise the vanadium by removing another electron from it, its oxidation state increases by 1.

Fairly obviously, if you start adding electrons again the oxidation state will fall. You could eventually get back to the element vanadium which would have an oxidation state of zero.

What if you kept on adding electrons to the element? You can't actually do that with vanadium, but you can with an element like sulphur.

Oxidation states are represented by integers, which can be positive, zero, or negative. In some cases average oxdidation state of element is a fraction such as 8/3 for iron in magnetite.

Working Out the Oxidation States

You don't work out oxidation states by counting the numbers of electrons transferred. It would take far too long. Instead you learn some simple rules, and do some very simple sums.

- The oxidation state of an uncombined element is zero. That's obviously so, because it hasn't been either oxidised or reduced yet. This applies whatever the structure of the element-whether it is, for example Xe or Cl_2 or S_8, or whether it has a giant structure like carbon or silicon.
- The sum of the oxidation states of all the atoms or ions in a neutral compound is zero.
- The sum of the oxidation states of all the atoms in an ion is equal to the charge on the ion.
- The more electronegative element in a substance is given a negative oxidation state. The less electronegative one is given a positive oxidation state. **Remember that fluorine is the most electronegative element with oxygen second.**
- Some elements almost always have the same oxidation states in their compounds

Element	Usual oxidation state	Exceptions
Group 1 metals	always +1	
Group 2 metals	always +2	
Oxygen	usually –2	except in peroxides and F_2O
Hydrogen	usually +1	except in metal hydrides where it is –1
Fluorine	always –1	
Chlorine	usually –1	except in compounds with O or F

The reasons for the exceptions

(i) *Hydrogen in the metal hydrides* Metal hydrides include compounds like sodium hydride, NaH. In this, the hydrogen is present as a hydride ion, H^-. The oxidation state of a simple ion like hydride is equal to the charge on the ion-in this case, –1.

Alternatively, you can think of it that the sum of the oxidation states in a neutral compound is zero. Since Group 1 metals always have an oxidation state of +1 in their compounds, it follows that the hydrogen must have an oxidation state of –1 (+1 – 1 = 0).

$$\underset{\substack{\uparrow \\ 1+x=0 \quad \therefore \quad x=-1}}{NaH}$$

(ii) *Oxygen in peroxides* Peroxides include hydrogen peroxide, H_2O_2. This is an electrically neutral compound and so the sum of the oxidation states of the hydrogen and oxygen must be zero.
Since, each hydrogen has an oxidation state of +1, each oxygen must have an oxidation state of –1 to balance it.

$$\underset{\substack{\uparrow \\ 2+2x=0}}{H_2O_2}$$
$$x=-1$$

(iii) *Oxygen in* F_2O The problem here is that oxygen isn't the most electronegative element. The fluorine is more electronegative and has an oxidation state of –1. In this case, the oxygen has an oxidation state of +2.

(iv) *Chlorine in compounds with fluorine or oxygen* There are so many different oxidation states that chlorine can have in these, that it is safer to simply remember that the chlorine doesn't have an oxidation state of –1 in them, and work out its actual oxidation state when you need it. You will find an example of this below.

- While naming the ion or compound, it is desirable to **put oxidation number per atom** (*i.e.*, oxidation state) in Roman figures within brackets (anion end with ate) (Table 4.1).

Table 4.1 **Naming the Acid and its Ions**

Oxoanions	Formula	Acids	Formula
Bromate (I)	BrO^-	Bromic (I)	HBrO
Bromate (III)	BrO_2^-	Bromic (III)	$HBrO_2$
Bromate (V)	BrO_3^-	Bromic (V)	$HBrO_3$
Bromate (VII)	BrO_4^-	Bromic (VII)	$HBrO_4$
Chromate (VI)	CrO_4^{2-}	Chromic (VI)	H_2CrO_4
Nitrate (V)	NO_3^-	Nitric	HNO_3
Nitrate (III)	NO_2^-	Nitrous	HNO_2
Phosphate (V)	PO_4^{3-}	Phosphoric	H_3PO_4
Sulphate (IV)	SO_3^{2-}	Sulphurous	H_2SO_3
Sulphate (VI)	SO_4^{2-}	Sulphuric	H_2SO_4
Peroxodisulphate (VI)	$S_2O_8^{2-}$	Peroxy-disulphuric	$H_2S_2O_8$

Elements with Multiple Oxidation States

Most elements have more than one possible oxidation state. Carbon has nine integer oxidation states, and there are also molecules in which the average degree of oxidation of several carbons is fractional.

Integer oxidation states of carbon with examples

1. **– 4** : CH_4
2. **– 3** : C_2H_6
3. **– 2** : CH_3Cl
4. **–1** : C_2H_2
5. **0** : CH_2Cl_2
6. **+1** : $CHCl_2—CHCl_2$
7. **+2** : $CHCl_3$
8. **+3** : C_2Cl_6
9. **+4** : CCl_4

Examples of fractional oxidation states for carbon

1. **- (6/5)** : $C_5H_5^-$
2. **- (6/7)** : $C_7H_7^+$
3. **- (10/8)** : $C_8H_8^{2-}$

Oxygen has 8 different oxidation states

1. **-2 in oxides, *e.g.*,** ZnO, CO_2, H_2O
2. **-1 in perioxides, *e.g.*,** H_2O_2
3. **- (1/2) in superoxides, *e.g.*,** KO_2
4. **- (1/3) in inorganic ozonides, *e.g.*,** RbO_3
5. **0 in** O_2
6. **+ (1/2) in dioxygenyl, *e.g.*, dioxygenyl hexfluoroarsenate** $O_2^+[AsF_6]^-$
7. **+1 in** O_2F_2
8. **+2 in** OF_2

Note that since fluorine is more electronegative than oxygen, O_2F_2 and OF_2 are considered as fluorides, rather than as a peroxide and an oxide.

Fractional Oxidation States

Fractional oxidation states are often used to represent the average oxidation states of several atoms in a structure. For example, in KO_2, the supoeroxide ion has a charge of –1 divided among two equivalent atoms, so each oxygen is assigned an oxidation state of –1/2. This ion is decribed as a resonance hybrid of two Lewis structures, and each oxygen has oxidation state 0 in one structure and –1 in the other.

For the cyclopentadienyl ion $C_5H_5^-$, the oxidation state of C is (–1) + (–1/5) = – 6/5. The –1 occurs because each C is bonded to one less electronegative H, and the –1/5 because the total ionic charge is divided among five equivalent C.

If the average refers to atoms which are not equivalent, the average oxidation state may not be representative of each atom. This is true in magnetic Fe_3O_4 whose formula leads to an average oxidation state of +8/3. In fact **two thirds of the iron ions are Fe^{3+} and one third Fe^{2+}.**

Similarly the ozonide ion O_3^- has an average oxidation state of –1/3. This ion is V-shaped with a central oxygen which is not equivalent to the two others and cannot be assumed to have the same oxidation state.

You will have come across names like iron (II) sulphate and iron (III) chloride. The (II) and (III) are the oxidation states of the iron in the two compounds: +2 and +3 respectively. That tells you that they contain Fe^{2+} and Fe^{3+} ions.

This can also be extended to the negative ion. Iron (II) sulphate is $FeSO_4$. There is also a compound $FeSO_3$ with the old name of iron (II) sulphite. The modern names reflect the oxidation states of the sulphur in the two compounds.

The sulphate ion is SO_4^{2-}. The oxidation state of the sulphur is +6. The ion is more properly called the sulphate (VI) ion.

The sulphite ion is SO_3^{2-}. The oxidation state of the sulphur is +4. This ion is more properly called the sulphate(IV) ion. The **ate ending simply shows that the sulphur is in a negative ion.**

So, $FeSO_4$ is properly called iron (II) sulphate(VI), and $FeSO_3$ is iron(II) sulphate (IV). In fact, because of the easy confusion between these names, the old names sulphate and sulphite are normally still used in introductory chemistry courses.

Oxidation numbers of underlined atoms have been determined by known values of the O.N. of other atoms in it.

(1) (a) $\underline{Cr}O_4^{2-}$ $\quad x - 8 = -2 \quad x = 6$

(b) $\underline{Cr}_2O_7^{2-}$ $\quad 2x - 14 = -2, \quad 2x = 12$ or $x = 6$ (per atom)

(2) (a) $H_3\underline{P}O_4$ $\quad 3 + x - 8 = 0, \quad x = +5$

(b) $H_3\underline{P}O_3$ $\quad 3 + x - 6 = 0, \quad x = +3$

(c) $H_3\underline{P}O_2$ $\quad 3 + x - 4 = 0, \quad x = +1$

(3) (a) $H_2\underline{S}O_3$ $\quad 2 + x - 6 = 0 \quad x = +4$

(b) $H_2\underline{S}O_4$ $\quad 2 + x - 8 = 0 \quad x = +6$

(c) $H_2\underline{S}_2O_3$ $\quad 2 + 2x - 6 = 0 \quad x = +2$

(4) (a) $H\underline{Cl}O_4$ $\quad 1 + x - 8 = 0 \quad x = +7$

(b) $H\underline{Cl}O_3$ $\quad 1 + x - 6 = 0 \quad x = +5$

(c) $H\underline{Cl}O_2$ $\quad 1 + x - 4 = 0 \quad x = +3$

(d) $H\underline{Cl}O$ $\quad 1 + x - 2 = 0 \quad x = +1$

(e) $H\underline{Cl}$ $\quad 1 + x = 0 \quad x = -1$

— **In (2), (3) and (4) oxyacids have atoms with different O.N.**

— **Greater the O.N. of the element forming oxyacids, greater the acid strength except for oxyacids of phosphorus :** Thus, acid strength increases.

In (3) $H_2S_2O_3 < H_2SO_3 < H_2SO_4$

(4) $HCl < HClO < HClO_2 < HClO_3 < HClO_4$

} order of acid strength

If acid is strong its conjugate base (formed when H^+ is lost from an acid) is weak and *vice-versa*. Thus, basic strength of conjugate bases of oxyacids of Cl will be in order:

$$ClO_4^- < ClO_3^- < ClO_2^- < ClO^-$$

Hence, greater the O.N. of the element in conjugate bases, smaller the basic strength.

Table 4.2 gives oxidation number of underlined atom.

Table 4.2 **Oxidation Number of the Atom in Different Species (Ion/Molecule)**

Species	O.N.	Remarks	Species	O.N.
$[\underline{Ni}(CN)_4]^{2-}$	+2		$H\underline{N}O_2$	+3
$\underline{Ni}(CO)_4$	0		$H\underline{N}O_3$	+5
$[\underline{Fe}(H_2O)]^{2+}$	+2		$H_2\underline{N}_2O_2$	+1
$\underline{Mn}O_4^-$	+7		$\underline{C}_6H_{12}O_6$	0
$\underline{Mn}O_4^{2-}$	+6		$\underline{C}O_2$	+4
$\underline{Mn}O_2$	+4		$\underline{N}H_3$	–3
$\underline{N}O_2$	+4		$\underline{N}_2H_4$	–2
$\underline{N}H_2OH$	–1		$\underline{N}_2O$	+1
$\underline{N}O$	+2			
$\underline{N}_2O_5$	+5 (per atom)			
$\underline{N}_2O_3$	+3 (per atom)			
$K\underline{O}_2$	–0.5 (per atom)			
$\underline{Cr}O_5$	+6	It has two peroxy linkages $CrO(O_2)_2$		
$H_2\underline{S}O_5$	+6	It has one peroxy linkage H—O—S(=O)(=O)—O—O—H		
$H_2\underline{S}_2O_8$	+6 (per atom)	It has one peroxy linkage		
$[\underline{Fe}(CN)_6]^{4-}$	+2			
$[\underline{Fe}(CN)_6]^{3-}$	+3			
$[\underline{Fe}(CN)_5(NO)^+]^{2-}$	+2			
$\underline{P}OCl_3$	+5	O as well as Cl are attached to P directly hence ON of O = – 2, and that of Cl = – 1		
$P\underline{O}Cl_3$	–2			
$PO\underline{Cl}_3$	–1 (per atom)			
$[\underline{Cr}(en)_3]^{3+}$	+3	en = CH_2NH_2–CH_2NH_2 is neutral		

Table 4.3 gives common oxidation numbers for group A elements.

Table 4.3 Common Oxidation Number (States) for Group A Elements in Compounds and Ions

Element (s)	Common O.N.	Examples	Other O.N.
H	+1	H_2O, CH_4, NH_4Cl	−1 in metal hydrides, *e.g.*, NaH, CaH_2
Group IA	+1	KCl, NaH, $RbNO_3$, K_2SO_4	None
Group IIA	+2	$CaCl_2$, MgH_2, $Ba(NO_3)_2$, $SrSO_4$	None
Group IIIA	+3	$AlCl_3, BF_3, Al(NO_3)_3$, GaI_3	None in common compounds
Group IVA	+2 +4	CO, PbO, $SnCl_2$, $Pb(NO_3)_2$, CCl_4, SiO_2, SiO_3^{2-}, $SnCl_4$	Many others are also seen for C and Si
Group VA	−3 in binary compounds with metals −3 in NH_4^+, binary compounds with H	Mg_3N_2, Na_3P, Ca_3As NH_3, PH_3, AsH_3, NH_4^+	+3, *e.g.*, NO_2^-, PCl_3, +5, *e.g.*, NO_3^-, PO_4^{3-}, AsF_5, P_4O_{10}
Oxygen	−2	H_2O, P_4O_{10}, Fe_2O_3, CaO, ClO_3^-	+2 in OF_2, −1 in peroxides, *e.g.*, H_2O_2, Na_2O_2, $-\frac{1}{2}$ in superoxides, *e.g.*, KO_2, RbO_2
Group VIA (other than oxygen)	−2 in binary compound s with metals and H −2 in binary compound s with NH_4^+	H_2S, CaS, Fe_2S_3, Na_2Se, $(NH_4)_2S$, $(NH_4)_2Se$	+4 with O and the lighter halogens, *e.g.*, SO_2, SeO_2, Na_2SO_3, SO_3^{2-}, SF_4 +6 with O and the lighter halogens, *e.g.*, SO_3, TeO_3, H_2SO_4, SO_4^{2-}, SF_6
Group VIIA	−1 in binary compound s with metals and H, −1 in binary compound s with NH_4^+	MgF_2, KI, $ZnCl_2$, $FeBr_3$, NH_4Cl, NH_4Br	Cl, Br, or I with O or with a lighter halogen +1, *e.g.*, BrF, ClO^-, BrO^-; +3, *e.g.*, ICl_3; ClO_2^-, BrO_2^-; +5, *e.g.*, BrF_5; ClO_3^-, BrO_3^-; +7, *e.g.*, IF_7, ClO_4^-, BrO_4^-

Table 4.4 gives some non-zero oxidation numbers/states.

Do you know ......

By convention, oxidation numbers are represented as $+n$ and $-n$, while ionic charges as n^+ and n^-; as +3, Fe^{3+}, −3, X^{3-} usually, the element with thze positive oxidation number is written first. $(\overset{+1}{Cl}\overset{-2}{O})^{-1}$, $(\overset{+4}{C}\overset{-2}{O_2})$

However, for historic reasons, in compounds containing nitrogen and hydrogen, such as NH_3, and many other compounds containing carbon and hydrogen, such as CH_4, hydrogen is written last, although it has a positive oxidation state.

Most transition metals and the metals of Group IIIA (boron family), IVA (carbon family) and VA (nitrogen family), exhibit more than one oxidation number. These metals may form two or more binary compounds with the same non-metal. To distinguish among all possibilities, the oxidation number of the metal is indicated by a Roman numeral in parentheses following its name. This method can be applied to any binary compound of a metal and a non-metal, whether the compound is ionic or molecule. An older method still in use but not recommended by the IUPAC uses "ous" and "ic" suffixes to indicate lower and higher oxidation number respectively.

Formula	Oxidation number of metal	IUPAC name	Common name
Cu_2O	+1	Copper (I) oxide	Cuprous oxide
CuF_2	+2	Copper (II) fluoride	Cupric fluoride
FeS	+2	Iron (II) sulphide	Ferrous sulphide
Fe_2O_3	+3	Iron (III) oxide	Ferric oxide
$SnCl_2$	+2	Tin (II) chloride	Stannous chloride
$SnCl_4$	+4	Tin (IV) chloride	Stannic chloride
PbO	+2	Lead (II) oxide	Plumbous oxide
PbO_2	+4	Lead (IV) oxide	Plumbic oxide
Hg_2Cl_2	+1	Mercury (I) chloride	Mercurous chloride
$HgCl_2$	+2	Mercury (II) chloride	Mercuric chloride

Formula	Oxidation number		Name
H_2SO_3	+4	2-units	Sulphurous acid
H_2SO_4	+6		Sulphuric acid
HNO_2	+3	2-units	Nitrous acid
HNO_3	+5		Nitric acid
$HBrO_2$	+3	2-units	Bromous acid
$HBrO_3$	+5		Bromic acid

DEEP Focus

Ternary acids that have one fewer O atoms than the "-ous" acids (two fewer O atoms than the "-ic" acids) are named using the prefix "hypo-" and the suffix "-ous".

Formula	Oxidation number	Name
HClO	+1	hypochlorous acid
H_3PO_2	+1	hypophosphorus acid
HIO	+1	hypoiodous acid
$H_2N_2O_2$	+1	hyponitrous acid

DEEP Focus

Acids containing one more oxygen atom per central non-metal atom than the normal "-ic acid" are named "-per" acids.

Formula	Oxidation number	Name
$HClO_4$	+7	perchloric acid
$HBrO_4$	+7	perbromic acid
HIO_4	+7	periodic acid

Ternary acids (oxoacids) are compounds of hydrogen, oxygen, and usually a non-metal. Non-metals that exhibit more than one oxidation state form more than one ternary acid. These ternary acids differ in the number of oxygen atoms they contain. The suffixes "-ous" and "-ic" indicate lower and higher oxidation states, respectively. The common ternary "-ic" acids are shown below.

Acids containing one fewer oxygen atom per central atom are named in the same way except that the "-ic" suffix is changed to "ous". **The oxidation number of the central element is lower by 2 in the "-ous"** acid than in the "-ic" acid.

Thus, oxoacids of chlorine can be named as

Formula	Oxidation number	Name
HClO	+1	hypochlorous acid
$HClO_2$	+3	chlorous acid
$HClO_3$	+5	chloric acid
$HClO_4$	+7	perchloric acid

Problem-Solving Tip Naming ternary acids and their anions.

Decreasing oxidation number of central atom ↓

Ternary acid	Anion
PerXXXic acid	PerXXXate
XXXic acid	XXXate
XXXous acid	XXXite
hypoXXXous acid	hypoXXXite

Decreasing number of oxygen atom of central atom ↓

The stem (XXX) represents the stem of the name *e.g.*, "nitr", "sulphur" or "chlor"

Do you know ......

Oxidation state of the central atom is equal to its periodic group number, except for halogen

IIIA	IVA	VA	VIA	VIIA
$H_3\overset{+3}{B}O_3$ boric acid	$H_2\overset{+4}{C}O_3$ carbonic acid	$H\overset{+5}{N}O_3$ nitric acid		
	$H_4\overset{+4}{Si}O_4$ silicic acid	$H_3\overset{+5}{P}O_4$ phosphoric acid	$H_2\overset{+6}{S}O_4$ sulphuric acid	$H\overset{+5}{Cl}O_3$ chloric acid
		$H_3\overset{+5}{As}O_4$ arsenic acid	$H_2\overset{+6}{Se}O_4$ selenic acid	$H\overset{+5}{I}O_3$ iodic acid

Illustration 1 *What is the name of* ClO^- *ion?*

Solution We have

Oxidation number of Cl = x

Oxidation number of O = − 2

Total charge on ion = − 1

Thus, O.N.(Cl) + O.N.(O) = total charge

$$x - 2 = -1 \quad \therefore \quad x = +1$$

Since, it is an oxoanion, hence it ends with ate. Thus, it is chlorate (I) ion.

Note : Its common name is hypochlorite.

Illustration 2 *What is the formula of*

(a) *iron (II) chloride* (b) *iron (III) chloride*

(c) *copper (II) nitrate*

Solution (a) $FeCl_2$ (b) $FeCl_3$ (c) $Cu(NO_3)_2$

Illustration 3 *Thiosulphuric acid (a dibasic acid) contains one sulphur and three oxygen atoms directly joined to central sulphur atom. What is oxidation number of sulphur atom?*

Solution Thiosulphuric acid is a dibasic acid represented as

$$H-O-\underset{\underset{S}{\|}}{\overset{\overset{O}{\|}}{S}}-O-H$$

The difference in the environment of two S-atoms is shown by radioactive tracer studies using ^{35}S (a radioactive isotope of sulphur)

$$SO_3^{2-} + S^* \rightleftharpoons S^*SO_3^{2-} \rightleftharpoons S^* + SO_3^{2-}$$

Thiosulphuric acid $H_2S_2O_3$ decomposes into sulphurous acid (H_2SO_3) and sulphur and the entire radioactivity is found associated with sulphur atom.

$$H_2S_2O_3 \rightleftharpoons H_2SO_3 + S$$

Thus, oxidation number of terminal S atom in $H_2S_2O_3 = -2$

and oxidation number of central S-atom in $H_2S_2O_3 = 6$

net oxidation number of two S-atom =4

oxidation number per sulphur atom = **+2**

Table 4.4 **Some Non-zero Oxidation States* (Numbers)**

Groups

IA	IIA	IIIB	IVB	VB	VIB	VIIB	(VIIIB)			IB	IIB	IIIA	IVA	VA	VIA	VIIA	0
1 H **+1** **−1**																1 H **+1** **−1**	2 He
3 Li **+1**	4 Be **+2**											5 B **+3**	6 C **+4** +2 −4	7 N **+5** +4 +3 +2 +1 **−3**	8 O −1 **−2**	9 F **−1**	10 Ne
11 Na **+1**	12 Mg **+2**											13 Al **+3**	14 Si +4 −4	15 P **+5** +3 −3	16 S **+6** +4 **−2**	17 Cl +7 +5 +3 +1 **−1**	18 Ar
19 K **+1**	20 Ca **+2**	21 Sc **+3**	22 Ti **+4** +3 +2	23 V **+5** **+4** +3 +2	24 Cr **+6** **+3** +2	25 Mn **+7** +6 **+4** +3 **+2**	26 Fe **+3** **+2**	27 Co **+3** **+2**	28 Ni **+2**	29 Cu **+2** +1	30 Zn **+2**	31 Ga **+3**	32 Ge **+4** −4	33 As +5 **+3** −3	34 Se +6 +4 −2	35 Br +7 +5 +3 +1 **−1**	36 Kr +4 +2

O.N. shown in bold are common

Illustration 4 *Determine change in oxidation number in of the underlined of the following conversions:*
(a) <u>Cl</u>$^-$ *changes to* ClO_4^- (b) <u>Mn</u>O_4^- *changes to* MnO_2
(c) <u>C</u>$_2O_4^{2-}$ *changes to* CO_2 (d) <u>Cr</u>$_2O_7^{2-}$ *to* Cr^{3+}

Solution

	Oxidation number of the underlined		
	in reactant	in product	change
(a)	Cl^- = −1	ClO_4^- = +7	8
(b)	MnO_4^- = +7	MnO_2 = +4	3
(c)	$C_2O_4^{2-}$ = +3	CO_2 = +4	1 (per atom)
(d)	$Cr_2O_7^{2-}$ = +6	Cr^{3+} = +3	3 (per atom)

Illustration 5 H_2O_2 *changes to* O_2 *or* H_2O *depending on the medium. What is change in oxidation number of oxygen per atom?*

Solution (i) $H_2O_2 \longrightarrow O_2$

−2 0] for every two oxygen atoms

Total change in oxidation number = 2 units

Change in oxidation number per atom = 1 unit

(ii) $H_2O_2 \longrightarrow 2H_2O$

coefficient 2 is used to balance O

−2 −4] oxidation number of two O-atoms

Total change in oxidation number = 2 units

Change in oxidation number per atom = 1 unit

Illustration 6 SO_2 *under atmospheric condition changes to* SO_x^{2-}. *If oxidation number of S in* SO_x^{2-} *is +6, what is* SO_x^{2-} *?*

Solution Oxidation number of S in SO_x^{2-} = + 6

Oxidation number of O = − 2

Thus, $\underset{\uparrow}{S}\ \underset{\uparrow}{O_x^{2-}}$

$$6 - 2x = -2$$
$$2x = 8$$
$\therefore$ $$x = 4$$

Thus, SO_x^{2-} is $\mathbf{SO_4^{2-}}$.

Illustration 7 *Match the species with correct oxidation number of the underlined:*

Species	Oxidation number
A : $\underline{Cr}O_2Cl_2$	I : +7
B : $\underline{Mn}O_3Cl$	II : +1
C : $\underline{N}H_2OH$	III : +6
D : $\underline{O}F_2$	IV : +2
E : $\underline{N}_2O$	V : –1

Solution

Species	O.N.
A	+6
B	+7
C	–1
D	+2
E	+1

Illustration 8 *What is the oxidation state of hydrogen in* $LiAlH_4$*? Write an equation for the reaction of hydrogen in that oxidation state with water.*

Solution $LiAlH_4$

$$+1 + 3 + 4x = 0$$
$$x = -1$$

Thus, hydrogen is in the form of hydride (H^-) in $LiAlH_4$.

$$H^- + H_2O \longrightarrow H_2 + OH^-$$

Illustration 9 *What is the oxidation number of* Cl *in*
(*a*) $Ca(ClO_2)_2$ (*b*) $Ba(ClO_3)_2$ (*c*) $CaOCl_2$?

Solution (a) $Ca(ClO_2)_2$ has chlorite ion ClO_2^-

$$ClO_2^-$$
$$x - 4 = -1$$
$$x = \mathbf{+3}$$

(b) $Ba(ClO_3)_2$ has chlorate ion ClO_3^-

$$ClO_3^-$$
$$x - 6 = -1$$
$$x = \mathbf{+5}$$

(c) $CaOCl_2$ is bleaching powder having two types of chlorine, Cl^- and ClO^- (hypochlorite ion)

oxidation number of Cl in Cl^- = –1

and oxidation number of Cl in ClO^- = +1

Thus, in combined state oxidation number = **0**

Illustration 10 *In qualitative analysis, a confirmatory test for chromate ion involves the formation of blue transient peroxo species,* CrO_5*, which is better represented as* $CrO(O_2)_2$*. It is formed when* H_2O_2 *is added to an acidic solution containing* $Cr_2O_7^{2-}$*. What is oxidation state of* Cr *in* $CrO(O_2)_2$*? Write the equation of formation of* CrO_5.

Solution

$$Cr_2O_7^{2-} + 2H^+ + 4H_2O_2 \longrightarrow 2CrO(O_2)_2 + 5H_2O$$

Oxidation state of oxygen in peroxide = – 1

Oxidation state of oxygen in oxide = – 2

Hence in, $CrO(O_2)_2$

$$x - 2 - 4 = 0$$
$$x = \mathbf{+6}$$

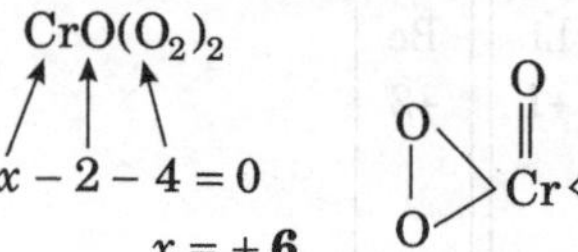

Illustration 11 *The complex* $[Fe(H_2O)_5NO]^{2+}$ *is formed in the **'ring-test'** for nitrate ion when freshly prepared* $FeSO_4$ *solution is added to aqueous solution of* NO_3^- *followed by addition of conc.* H_2SO_4*. If* NO *exists as nitrosyl* NO^+*, what is oxidation number of iron? Explain. Also calculate magnetic moment of* Fe.

Solution Let oxidation state of Fe = x

$$[Fe(H_2O)_5NO]^{2+}$$
$$x + 0 + 1 = 2$$
$\therefore$ $$x = +1$$

Thus, oxidation state of iron = +1

Iron is added as $FeSO_4$ with oxidation state of iron = +2. It is due to charge transfer from NO to Fe^{2+}. NO changes to NO^+ and Fe^{2+} to Fe^+.

$$NO \longrightarrow NO^+ + e^-$$
$$Fe^{2+} + e^- \longrightarrow Fe^+$$

[Ar] | ↑↓ | ↑ | ↑ | ↑ | ↑ |

Electronic configuration of Fe^{2+} = [Ar] $3d^6$

Fe^{2+} gains one electron from NO hence, it has now three unpaired electrons ($N = 3$)

[Ar] | ↑↓ | ↑↓ | ↑ | ↑ | ↑ |
(from NO)

$$\text{Magnetic moment} = \sqrt{N(N+2)}\ \text{BM} = \sqrt{3 \times 5}\ \text{BM}$$
$$= \sqrt{15} = \mathbf{3.873\ BM}$$

Illustration 12 *The tetrathionate anion,* $S_4O_6^{2-}$*, is a sulphur oxoanion derived from the compound tetrathionic acid,* $H_4S_4O_6$*. Anion can be viewed as the adduct formed from the reaction between Lewis base* S_2^{2-} *and Lewis acid* SO_3.

$$S_2^{2-} + 2SO_3 \longrightarrow S_4O_6^{2-}$$

Two of the sulphur atoms are present in the oxidation state 0 and two are in oxidation state +5.

Alternately it can be explained as

$$\underset{-2}{\overset{\uparrow}{S_2^{2-}}} + \underset{+12}{\overset{\uparrow}{2SO_3}} \longrightarrow S_4O_6^{2-}$$

Oxidation number of two sulphur atom in: -2, $+12$

Thus, oxidation number per sulphur atom $= \frac{10}{4} = 2.5$

Note $\quad 2S_2O_3^{2-} \longrightarrow S_4O_6^{2-} + 2e^-$

oxidation number per atom: 2, 2.5

Target Practice 1

1. Oxidation number of Cr in CrO_2Cl_2 is +6. What is oxidation number (per atom) of oxygen? ....
2. $H_2SO_5 + H_2O \longrightarrow H_2SO_4 + H_2O_2$.
   Derive oxidation number of sulphur in H_2SO_5. ....
3. Write the common and IUPAC names of the following compounds.

	Common	IUPAC
(a) HIO_3	....	....
(b) HClO	....	....
(c) $FeCl_2$	....	....
(d) Hg_2Cl_2	....	....
(e) $Fe(NO_3)_3$	....	....

4. Chloride of a copper metal is colourless. What is its IUPAC name?
5. Compare basic strength of the following
   (a) $H_2PO_4^-$ ☐ $H_2PO_3^-$
   (b) HPO_4^{2-} ☐ HPO_3^{2-}
   (c) ClO^- ☐ ClO_2^-
   (d) NH_3 ☐ NH_2^-

Answers

1. CrO_2Cl_2

   $6 + 2x - 2 = 0$

   Thus, x (oxidation number of oxygen) $= -2$
2. Formation of H_2O_2 indicates that there is one peroxy linkage in H_2SO_5.

   $$HO-\overset{\overset{O}{\|}}{\underset{\underset{O}{\|}}{S}}-O-O-H$$

   $$H_2SO_3(O_2)$$

   $$+2 + x - 6 - 2 = 0$$

   $\therefore$ $x = 6$. Thus, oxidation number of sulphur = + 6
3.

	Common	IUPAC
(a)	Iodic acid	hydroiodic (V) acid
(b)	hypochlorous acid	hydrochloric (I) acid
(c)	ferrous chloride	iron (II) chloride
(d)	mercurous chloride	mercury (I) chloride
(e)	ferric nitrate	iron (III) nitrate

4. Since copper salt is colourless, it means there is no electron unpaired in *d*-orbital.

   Cu : [Ar] $3d^{10}4s^1$

   Cu^+ : [Ar] $3d^{10}$

   Thus, copper is +1

   Thus, it is Cu_2Cl_2 or CuCl : Copper (I) chloride
5. (a) $H_2PO_4^- < H_2PO_3^-$ (b) $HPO_4^{2-} < HPO_3^{2-}$
   (c) $ClO^- > ClO_2^-$ (d) $NH_3 < NH_2^-$

Oxidation and Reduction

Oxidation and Reduction in Terms of Oxygen Transfer

- Oxidation is gain of oxygen.
- Reduction is loss of oxygen.

  For example, in the extraction of iron from its ore :

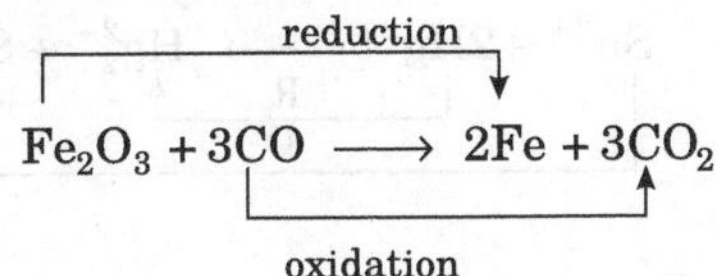

Because both reduction and oxidation are going on side-by-side, this is known as a **redox** reaction.

Oxidising and Reducing Agents

An oxidising agent is substance which oxidises something else. In the above example, the iron (III) oxide is the oxidising agent.

A reducing agent reduces something else. In the equation, the carbon monoxide is the reducing agent.

- Oxidising agents give oxygen to another substance.
- Reducing agents remove oxygen from another substance.

Oxidation and Reduction in Terms of Hydrogen Transfer

- Oxidation is loss of hydrogen.
- Reduction is gain of hydrogen.

Notice that these are exactly the opposite of the oxygen definitions.

For example, ethanol can be oxidised to ethanal :

$$CH_3CH_2OH \longrightarrow CH_3CHO$$

oxidation by loss of hydrogen

You would need to use an oxidising agent to remove the hydrogen from the ethanol. A commonly used oxidising agent is potassium dichromate (VI) solution acidified with dilute sulphuric acid.

Note The equation for this is rather complicated for this introductory page. If you are interested, you will find a similar example (ethanol to ethanoic acid) on the page dealing with writing equations for redox reactions.

Ethanal can also be reduced back to ethanol again by adding hydrogen to it. A possible reducing agent is sodium tetrahydridoborate, $NaBH_4$. Again the equation is too complicated to be worth bothering about at this point.

reduction by gain of hydrogen

$$CH_3CHO \longrightarrow CH_3CH_2OH$$

DEEP Focus

- Oxidising agents give oxygen to another substance or remove hydrogen from it.
- Reducing agents remove oxygen from another substance or give hydrogen to it.

Oxidation and Reduction in Terms of Electron Transfer

- Oxidation is loss of electrons.
- Reduction is gain of electrons.

Oil	Ring
oxidation is loss	reduction is gain

A simple example The equation shows a simple redox reaction which can obviously be described in terms of oxygen transfer.

$$CuO + Mg \longrightarrow Cu + MgO$$

Copper (II) oxide and magnesium oxide are both ionic. The metals obviously aren't. If you rewrite this as an ionic equation, it turns out that the oxide ions are spectator ions and you are left with:

reduction by gain of electrons

$$Cu^{2+} + Mg \longrightarrow Cu + Mg^{2+}$$

oxidation by loss of electrons

If you look at the equation above, the magnesium is reducing the copper(II) ions by giving them electrons to neutralise the charge. Magnesium is a reducing agent.

Looking at it the other way round, the copper (II) ions are removing electrons from the magnesium to create the magnesium ions. The copper (II) ions are acting as an oxidising agent.

Writing Ionic Equations for Redox Reactions

Electron-Half-Equations

What is an electron-half-equation?

When magnesium reduces hot copper (II) oxide to copper, the ionic equation for the reaction is :

$$Cu^{2+} + Mg \longrightarrow Cu + Mg^{2+}$$

You can split the ionic equation into two parts, and look at it from the point of view of the magnesium and of the copper (II) ions separately. This shows clearly that the magnesium has lost two electrons, and the copper (II) ions have gained them.

$$Mg \longrightarrow Mg^{2+} + 2e^-$$

$$Cu^{2+} + 2e^- \longrightarrow Cu$$

These two equations are described as **"electron-half-equations" or "half-equations" or "ionic-half-equations" or "half-reactions"**

Any redox reaction is made up of two half-reactions : in one of them electrons are being lost (an oxidation process) and in the other one those electrons are being gained (a reduction process).

DEEP Focus

- An oxidising agent oxidises something else.
- Oxidation is loss of electrons (OIL RIG), that means that an oxidising agent takes electrons from that other substance.
- So an oxidising agent must gain electrons.

Or you could think it out like this :

- An oxidising agent oxidises something else, that means that the oxidising agent must be being reduced.
- Reduction is gain of electrons (OIL RIG). so an oxidising agent must gain electrons.

A complete reaction showing oxidation and reduction together is called a **redox- reaction.**

$$Fe + Cu^{2+} \longrightarrow Cu + Fe^{2+}$$

R (Cu²⁺ → Cu), O (Fe → Fe²⁺)

Some examples of redox reactions are:

(a) $$Sn^{2+} + 2Hg^{2+} \longrightarrow Hg_2^{2+} + Sn^{4+}$$

R, O

(b) $MnO_4^- + 5Fe^{2+} + 8H^+ \longrightarrow 5Fe^{3+} + Mn^{2+} + 4H_2O$ (O: $Fe^{2+} \to Fe^{3+}$; R: $MnO_4^- \to Mn^{2+}$)

(c) $Cr_2O_7^{2-} + 6Fe^{2+} + 14H^+ \rightarrow 6Fe^{3+} + 2Cr^{3+} + 7H_2O$ (O: $Fe^{2+} \to Fe^{3+}$; R: $Cr_2O_7^{2-} \to Cr^{3+}$)

(d) $3Cu + 2NO_3^- + 8H^+ \longrightarrow 2NO + 3Cu^{2+} + 4H_2O$ (R: $NO_3^- \to NO$; O: $Cu \to Cu^{2+}$)

DEEP Focus

If one of the half-reactions does not take place, other half will also not take place. We can say oxidation and reduction go side by side.

In the above examples, two substances react — one of them is an oxidising agent and the other is a reducing agent. Such reactions relate to intermolecular oxidation-reduction ones.

Consider the following example:

$$\underset{\text{O.N.}=0}{3Cl_2} + 6OH^- \longrightarrow 5\underset{-1}{Cl^-} + \underset{+5}{ClO_3^-} + 3H_2O$$

(R: $Cl_2 \to Cl^-$; O: $Cl_2 \to ClO_3^-$)

In this case we find that Cl_2 has been oxidised as well as reduced. Such type of redox reaction is called **disproportionation** (autoxidation-autoreduction) reaction. Examples are:

- $2Cu^+ \xrightarrow{\text{Disproportionates}} Cu + Cu^{2+}$ (R, O)
- $2HCHO + OH^- \longrightarrow CH_3OH + HCOO^-$ (R, O)
- $4ClO_3^- \longrightarrow 3ClO_4^- + Cl^-$ (O, R)
- $3MnO_4^{2-} + 2H_2O \longrightarrow MnO_2 + 2MnO_4^- + 4OH^-$ (R, O)
- $\begin{matrix}CHO\\|\\CHO\end{matrix} + OH^- \longrightarrow \begin{matrix}CH_2OH\\|\\COO^-\end{matrix}$
- $2H_2O_2 \longrightarrow 2H_2O + O_2$

In disproportionation reactions, compounds are simultaneously formed that contain a given element in a more oxidised and more reduced state than the initial one. The initial substance displays functions of both an oxidising and a reducing agent. Such reactions are possible if the relevant element is an intermediate oxidation in the initial compound. In the following disproportionation reaction

$$4H_3PO_3 \longrightarrow 3H_3PO_4 + PH_3,$$

Oxidation number of P (in H_3PO_3) is +3 which is the intermediate of +5 (in H_3PO_4) and –3 (in PH_3). Such substances have **oxidation- reduction duality.**

The reaction between chlorine and cold dilute sodium hydroxide solution is :

$$Cl_2 + 2OH^- \longrightarrow Cl^- + ClO^- + H_2O$$

Obviously the chlorine has changed oxidation state because it has ended up in compounds starting from the original element. Checking all the oxidation states shows :

$$\underset{0}{Cl_2} + 2\underset{-2\ +1}{OH^-} \longrightarrow \underset{-1}{Cl^-} + \underset{+1\ -2}{ClO^-} + \underset{+1\ -2}{H_2O}$$

The chlorine is the only thing to have changed oxidation state. One atom has been reduced because its oxidation state has fallen. The other has been oxidised.

This is a good example of disproportionation reaction. A disproportionation reaction is one in which a single substance is both oxidised and reduced.

Illustration 1 *Write unbalanced oxidation and reduction half-reactions for the following processes:*

(a) $Te(s) + NO_3^-(aq) \longrightarrow TeO_2(s) + NO(g)$

(b) $H_2O_2(aq) + Fe^{2+}(aq) \longrightarrow Fe^{3+}(aq) + H_2O(l)$

(c) $Mn(s) + NO_3^-(aq) \longrightarrow Mn^{2+}(aq) + NO_2(g)$

(d) $Mn^{3+}(aq) \longrightarrow MnO_2(s) + Mn^{2+}(aq)$

Solution (a) $\underset{0}{Te}(s) + \underset{+5}{N}O_3^-(aq) \longrightarrow \underset{+4}{Te}O_2(s) + \underset{+2}{N}O(g)$

Oxidation : $Te(s) \longrightarrow TeO_2(s)$

Reduction : $NO_3^-(aq) \longrightarrow NO(g)$

(b) $H_2\underset{-1}{O_2}(aq) + \underset{+2}{Fe^{2+}}(aq) \longrightarrow \underset{+3}{Fe^{3+}}(aq) + H_2\underset{-2}{O}(l)$

Oxidation : $Fe^{2+}(aq) \longrightarrow Fe^{3+}(aq)$

Reduction : $H_2O_2(aq) \longrightarrow H_2O(l)$

(c) **Oxidation :** $Mn(s) \longrightarrow Mn^{2+}(aq)$

Reduction : $NO_3^-(aq) \longrightarrow NO_2(g)$

(d) **Oxidation :** $Mn^{3+}(aq) \longrightarrow MnO_2(s)$

Reduction : $Mn^{3+}(aq) \longrightarrow Mn^{2+}(aq)$

In this reaction, Mn^{3+} has been oxidised to $MnO_2(s)$ as well as reduced to $Mn^{2+}(aq)$. Thus, it is a **disproportionation** reaction.

Illustration 2 *Identify the species oxidised and reduced in the following*:

(a) 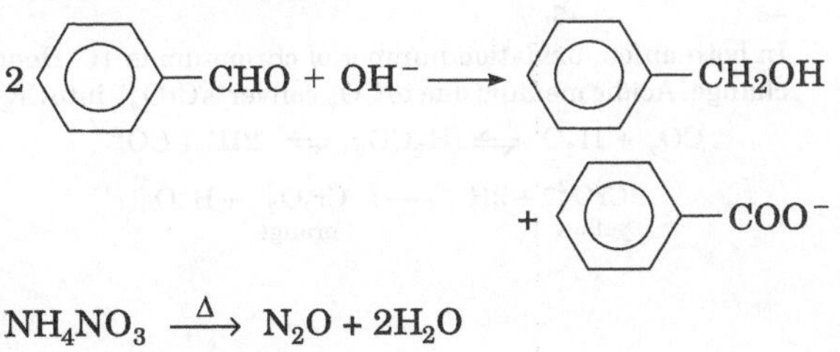 $2\,C_6H_5{-}CHO + OH^- \longrightarrow C_6H_5{-}CH_2OH + C_6H_5{-}COO^-$

(b) $NH_4NO_3 \xrightarrow{\Delta} N_2O + 2H_2O$

(c) $PCl_5 \xrightarrow{\Delta} PCl_3 + Cl_2$

(d) CHO, NO_2 (benzene ring) $\longrightarrow$ COOH, NH_2 (benzene ring)

Solution (a) It is a case of Cannizzaro reaction. RCHO has been oxidised to RCOO$^-$ as well as reduced to RCH$_2$OH in a proton-hydride transfer reaction. Thus, it is a disproportionation reaction.

CHO (benzene ring) $\xrightarrow{\text{oxidation}}$ COO^- (benzene ring)

CHO (benzene ring) $\xrightarrow{\text{reduction}}$ CH_2OH (benzene ring)

(b) $NH_4NO_3 \xrightarrow{\Delta} N_2O + 2H_2O$

There are two types of nitrogen, one in NH_4^+ ion (oxidation number = −3) and the other in NO_3^- (oxidation number = + 5). On decomposition, N_2O (oxidation number = + 1) is formed. Thus,

NH_4^+ is oxidised to N_2O and NO_3^- is reduced to N_2O.

But it is not the disproportionation reaction.

Both the elements are in the same initial substance. Reactions of this kind are known as **intramolecular oxidation reduction reactions.**

(c)
$$\underset{(\text{O.N.})\ +5\ -1}{PCl_5} \xrightarrow{\Delta} \underset{+3\ -1}{PCl_3} + \underset{0}{Cl_2}$$

Phosphorus (V) of PCl_5 is thus reduced to P(III) of PCl_3. Cl (−1) is oxidised to Cl_2(0).

(d) —CHO group is oxidised to —COOH and —NO_2 group is reduced to —NH_2.

Thus, it is **intramolecular redox-reaction.**

Target Practice 2

1. I^- changes to I_3^- in a reaction. Is it oxidation or reduction?

$$I^- \longrightarrow I_3^-$$

2. On passing CO_2 into aqueous yellow coloured CrO_4^{2-} solution, an orange coloured $Cr_2O_7^{2-}$ solution is obtained. This is oxidation or reduction?

$$\underset{\text{yellow}}{CrO_4^{2-}(aq)} \xrightarrow{CO_2} Cr_2O_7^{2-}$$

3. Anhydride of an acid is formed by dehydration and oxidation number of central atom remains unchanged.

Acid : $H_2N_2O_2, HNO_2, HNO_3$

Anhydride : N_2O_5, N_2O, NO, N_2O_3

Match acid with anhydride.

Acid	Anhydride
$H_2N_2O_2$	....
HNO_2	....
HNO_3	....

4. NH_4NO_2 changes to N_2 on heating.

$$NH_4NO_2 \xrightarrow{\Delta} N_2 + 2H_2O$$

Indicate the species (i) oxidised (ii) reduced

5. Select oxidant and reductant in the following :

(a) $Sn^{2+} + 2Hg^{2+} \longrightarrow Hg_2^{2+} + Sn^{4+}$

(b) $Zn + 2OH^- \longrightarrow ZnO_2^{2-} + H_2$

(c) $MnO_2 \longrightarrow Mn^{2+} + MnO_4^-$

Oxidant	Reductant
....	....
....	....
....	....

Answers

1. $2I^- \longrightarrow I_2 \xrightarrow{I^-} I_3^-$

I^- is oxidised to I_2 which is soluble in KI forming KI_3 (I_3^-). Thus, I_2 is formed by oxidation.

2. $\underset{+6}{CrO_4^{2-}} \longrightarrow \underset{+6}{Cr_2O_7^{2-}}$

In both anion, oxidation number of chromium is +6. Hence, no change. Acidic medium due to CO_2 converts CrO_4^{2-} into $Cr_2O_7^{2-}$.

$$CO_2 + H_2O \rightleftharpoons H_2CO_3 \rightleftharpoons 2H^+ + CO_3^{2-}$$

$$\underset{\text{yellow}}{2CrO_4^{2-}} + 2H^+ \longrightarrow \underset{\text{orange}}{Cr_2O_7^{2-}} + H_2O$$

3. $H_2N_2O_2 \longrightarrow N_2O$
$HNO_2 \longrightarrow N_2O_3$
$HNO_3 \longrightarrow N_2O_5$

4. $\underset{-3}{NH_4^+} \longrightarrow \underset{+1}{N_2O}$ **Oxidation**

$\underset{+5}{NO_3^-} \longrightarrow \underset{+1}{N_2O}$ **Reduction**

5.

	Oxidant	Reductant
(a)	Hg^{2+}	Sn^{2+}
(b)	OH^-	Zn
(c)	MnO_2	MnO_2

Practice Exercise 1

1. What is oxidation number of carbon in CN^-?
2. What is the change in oxidation number of the underlined atom in following conversions?
   (a) $NaH\underline{C}O_3 + HCl \longrightarrow NaCl + H_2O + CO_2$
   (b) $2NaOH + \underline{Br_2} \longrightarrow NaBr + NaBrO + H_2O$
   (c) $K\underline{I} \longrightarrow KIO_3$
   (d) $H_2\underline{S}O_5 + 2K\underline{I} \longrightarrow K_2SO_4 + H_2O + I_2$
   (e) $2H\underline{N}O_3 \longrightarrow H_2O + N_2O_5$
   (f) $CaO\underline{Cl_2} + H_2O \longrightarrow Ca(OH)_2 + Cl_2$
   (g) $\underline{Mn}O_4^- + 3OH^- \longrightarrow MnO_2 + H_2O$
   (dil.)
   (h) $\underline{As_2}S_3 + OH^- \longrightarrow AsS_4^{3-}$
3. In $H_2S_2O_8$ oxidation state of sulphur is six? Explain by structure.
4. Write a formula for
   (a) sulphate of iron with Fe in the oxidation state +3.
   (b) an oxoacid of nitrogen with N in the oxidation state +3.
   (c) an oxide of chlorine with Cl in the oxidation state +7.
5. Which is the anhydride of HNO_3 out of N_2O, NO, N_2O_5?
6. Which is stronger base in each pair?
   (a) HSO_4^-, HSO_3^- (b) NO_2^-, NO_3^-
   (c) Cl^-, ClO^-
7. VO_2^{2+} and VO^{2+} are known as vanadyl ion. Determine oxidation number of vanadium in each.
8. Assign oxidation numbers to each atom in the following substances :
   (a) ethane (C_2H_6)
   (b) borax ($Na_2B_4O_7$)
   (c) silicate mineral of magnesium ($Mg_2Si_2O_6$)
9. (a) What is the type of linkage in $(NH_3)_3CrO_4$ if oxidation number of Cr is +4?
   (b) $K_3Cr(O_2)_x$ has peroxy linkage with oxidation number of Cr as +5. What is value of x ?
10. Determine the oxidation state of the underlined element in :
    (a) $K_4\underline{P_2}O_7$ (b) $K\underline{Au}Cl_4$
    (c) $Rb_4Na[H\underline{V_{10}}O_{28}]$ (d) $\underline{I}Cl$
    (e) $Ba(\underline{Cl}O_2)_2$ (f) $\underline{O}F_2$,
    (g) $Ba_2\underline{Xe}O_6$
11. Identify oxidant and reductant in the following redox reactions :
    (a) $2MnO_2 + 4KOH + O_2 \longrightarrow 2K_2MnO_4 + 2H_2O$
    (b) $2K_2MnO_4 + Cl_2 \longrightarrow 2KCl + 2KMnO_4$
    (c) $2MnO_4^{2-} + 2H_2O \xrightarrow{\text{current}} 2MnO_4^- + 2OH^- + H_2$
    (d) $P_4 + 5O_2 \longrightarrow P_4O_{10}$
    (e) $Hg^{2+} + NO_2^- + H_2O \longrightarrow Hg + NO_3^- + 2H^+$
12. Identify disproportionation reactions in the following:
    (a) $3MnO_4^{2-} + 4H^+ \longrightarrow MnO_2 + 2MnO_4^- + 2H_2O$
    (b) $2KClO_3 \longrightarrow 2KCl + 3O_2$
    (c) $2C_6H_5CHO + OH^- \longrightarrow C_6H_5COO^- + C_6H_5CH_2OH$
    (d) $2CH_3CHO \xrightarrow{Al(OEt)_3} CH_3COOCH_2CH_3$
    (e) $4H_3PO_3 \xrightarrow{\Delta} 3H_3PO_4 + PH_3$
    (f) $P_4 + 3NaOH + 3H_2O \xrightarrow{\Delta} 3NaH_2PO_2 + PH_3$
    (g) $NH_4NO_3 \xrightarrow{\Delta} N_2O + 2H_2O$
13. Underline the species which are oxidised and reduced in the following (unbalanced) reactions:
    (a) $CN^- + OCl^- \longrightarrow N_2 + HCO_3^- + Cl^-$
    (b) $C_6H_5OH + O_3 \longrightarrow CO_2 + H_2O + O_2$
    (c) $I^- + O_3 \longrightarrow O_2 + I_2$
    (d) $S_2O_3^{2-} + Cl_2 \longrightarrow HSO_4^- + Cl^-$
    (e) $C + ZnO \longrightarrow Zn + CO$
14. Classify each of the following unbalanced half-reactions as either an oxidation or a reduction:
    (a) $O_2(g) \longrightarrow OH^-(aq)$
    (b) $H_2O_2(aq) \longrightarrow O_2(g)$
    (c) $MnO_4^-(aq) \longrightarrow MnO_4^{2-}(aq)$
    (d) $CH_3OH(aq) \longrightarrow CH_2O(aq)$

Answers

1. +2 (**Hint:** In the cyanide ion, CN^-, N is more electronegative than C and taking the common negative oxidation state of N as −3, oxidation state of C = + 2)
2. (a) No change
   (b) Decreases to −1, w.r.t. NaBr and increases to +1 w.r.t. NaBrO
   (c) 6
   (d) No change in S of H_2SO_5 (since its peroxy linkage is affected) and 1 in KI
   (e) No change
   (f) There are two types of Cl in $CaOCl_2.Cl^-$ (oxidation number = − 1) and Cl in ClO^- (oxidation number = +1) Cl^- is oxidised to Cl_2 and ClO^- is reduced to Cl_2 we find overall change is zero (since in $CaOCl_2$ total oxidation number is 0)
   (g) 3 (h) 2
3. $H_2S_2O_8$ (called Marshall's acid) is peroxy disulphuric acid as shown:

```
        O           O
        ↑           ↑
  H—O—S—O—O—S—OH
        ↓    ↑      ↓
        O    |      O
       peroxy linkage
```

with oxidation state of
O = − 1 (2 for two oxygen atoms)

Other six oxygen atoms have oxidation state of −2 each, hence, **oxidation state of sulphur is six :**

$$2x + 2(\text{for two H}) - 12(\text{for six O}) - 2\ (\text{peroxy}) = 0$$

$\therefore\ x = 6$ per sulphur atom

4. (a) $Fe_2(SO_4)_3$ (b) HNO_2 (c) Cl_2O_7
5. N_2O_5 (oxidation number of the element (N) in oxyacid and its anhydride remains same).
6. (a) HSO_3^- (b) NO_2^- (c) Cl^-
   Smaller the oxidation number, stronger the conjugate base.
7. +6 in VO_2^{2+} and +4 in VO^{2+}

8. (a) C = − 3, H = + 1 (b) Na = + 1, B = + 3, O = − 2
(c) Mg = + 2, Si = +4, O = −2

9. (a) $(NH_3)_3CrO_4$, take oxidation number of O = x

$$\underset{0}{(NH_3)_3}\ \underset{+4}{Cr}\ \underset{+4x\,=\,0}{O_4}$$

$x = -1$

Thus, oxygen atoms are in the form of peroxy linkage. The compound can be written as $(NH_3)_3Cr(O_2)_2$.

(b) $K_3Cr(O_2)_x$ has $Cr(O_2)_x^{3-}$ ion

$$Cr(O_2)_x^{3-}$$

$$+5 - 2x = -3$$

$$-2x = -8$$

$$x = 4$$

It is $K_3Cr(O_2)_4$

10. (a) P = + 5 (b) Au = + 3 (c) V = + 5 (d) I = + 1
(e) Cl = + 3 (f) O = + 2 (g) Xe = +8

11.

	Oxidant	Reductant
(a)	O_2	MnO_2
(b)	Cl_2	K_2MnO_4
(c)	H_2O	MnO_4^{2-}
(d)	O_2	P_4
(e)	Hg^{2+}	NO_2^-

12. (a) MnO_4^{2-} is reduced to MnO_2 as well as oxidised to MnO_4^- hence, disproportionation.

(b) Cl of $KClO_3$ (O.N. = + 5) is reduced to Cl^- and O of $KClO_3$ (O.N. = −2) is oxidised to O_2 hence, redox but not disproportionation.

(c) C_6H_5CHO is oxidised to $C_6H_5COO^-$ as well as reduced to $C_6H_5CH_2OH$ **(Cannizzaro reaction)** hence disproportionation.

(d) $CH_3COOC_2H_5$ (ester) is formed from CH_3COOH (oxidation product of CH_3CHO) and CH_3CH_2OH (reduction product of CH_3CHO), hence CH_3CHO disproportionates in presence of aluminium ethoxide **(Tischenko reaction)**.

(e) H_3PO_3 (O.N. of P = +3) disproportionates to H_3PO_4 (O.N. of P = +5) and PH_3 (O.N. of P = −3).

(f) P_4 (O.N. = 0) disproportionates to NaH_2PO_2 (O.N. of P = +1) and PH_3 (O.N. of P = −3).

(g) NH_4^+ (O.N. of N = −3) oxidises to N_2O (O.N. of N = +1). NO_3^- (O.N. of N = +5) redues to N_2O; hence redox but not disproportionation.

13. (a) Oxidation of C(O.N. = + 2) and N (O.N. = − 3) of CN^- into HCO_3^- (O.N. of C = +4) and N_2 (O.N. of N = 0) respectively and reduction of OCl^- (O.N. of Cl = +1) into Cl^- (O.N. of Cl = −1).

(b) Oxidation of C_6H_5OH into CO_2 by O_3

(c) Oxidation of I^- into I_2 by O_3

(d) Oxidation of $S_2O_3^{2-}$ into HSO_4^- by Cl_2

(e) ZnO is reduced to Zn

14. (a) Reduction (b) Oxidation (c) Reduction
(d) Oxidation

Do you Know ......

A redox predominance diagram shows the predominant (thermodynamically most stable) oxidation state and chemical form of an element at any given potential. In this type of diagram, elements with no redox chemistry (such as Ar) will predominate at all potentials and will cover the entire diagram. **Among those with more than one thermodynamically stable oxidation state, more strongly oxidising stable chemical forms are confined to higher** regions of redox predominance diagrams. Correspondingly strongly reducing stable chemical forms will be low in the predominance diagrams Fig 4.1 represents redox predominance diagram of iron. Of the chemical forms of iron shown, the most strongly oxidising is the ferrate ion, FeO_4^{2-}; the most strongly reducing is iron metal, Fe.

Oxidation state	Half-reaction	$E°/V$	
+6	$FeO_4^{2-} + 8H^+ + 3e^- \rightleftharpoons Fe^{3+} + 4H_2O$	+2.20 V	FeO_4^{2-}
			2.20
+3	$Fe^{3+} + e^- \rightleftharpoons Fe^{2+}$	+0.77 V	Fe^{3+}
			0.77
+2	$Fe^{2+} + 2e^- \rightleftharpoons Fe(s)$	−0.44 V	Fe^{2+}
			−0.44
			Fe

Fig. 4.1 Redox predominance diagram for the element iron, with the corresponding reduction half-reactions

How to Balance a Redox Reaction?

Working out electron-half-equations and using them to build ionic equations.

Illustration 1 *The reaction between chlorine and iron (II) ions*

Chlorine gas oxidises iron (II) ions to iron (III) ions. In the process, the chlorine is reduced to chloride ions.

You would have to know this, or be told it by an examiner. In building equations, there is quite a lot that you can work out as you go along, but you have to have somewhere to start from!

You start by writing down what you know for each of the half-reactions. In the chlorine case, you know that chlorine (as molecules) turns into chloride ions :

$$Cl_2 \longrightarrow Cl^-$$

The first thing to do is to balance the atoms that you have got as far as you possibly can :

$$Cl_2 \longrightarrow 2Cl^-$$

Always check that you have the existing atoms balanced before you do anything else. If you forget to do this, everything else that you do after wards is a complete waste of time!

Now you have to add things to the half-equation in order to make it balance completely.

All you are allowed to add are :

- electrons
- water

- hydrogen ions (unless the reaction is being done under alkaline conditions in which case, you can add hydroxide ions instead)

In the chlorine case, all that is wrong with the existing equation that we've produced so far is that the charges don't balance. The left-hand side of the equation has no charge, but the right-hand side carries 2 negative charges.

That's easily put right by adding two electrons to the left-hand side. The final version of the half-reaction is :

$$Cl_2 + 2e^- \longrightarrow 2Cl^-$$

Now you repeat this for the iron (II) ions. You know (or are told) that they are oxidised to iron (III) ions. Write this down :

$$Fe^{2+} \longrightarrow Fe^{3+}$$

The atoms balance, but the charges don't. There are 3 positive charges on the right-hand side, but only 2 on the left.

You need to reduce the number of positive charges on the right-hand side. That's easily done by adding an electron to that side :

$$Fe^{2+} \longrightarrow Fe^{3+} + e^-$$

It is obvious that the iron reaction will have to happen twice for every chlorine molecule that reacts. Allow for that, and then add the two half-equations together.

$$\begin{array}{rcl} 2\times[Fe^{2+} & \longrightarrow & Fe^{3+} + e^- \\ Cl_2 + 2e^- & \longrightarrow & 2Cl^- \\ \hline 2Fe^{2+} + Cl_2 & \longrightarrow & 2Fe^{3+} + 2Cl^- \end{array}$$

But don't stop there!! Check that everything balances-atoms and charges. It is very easy to make small mistakes, especially if you are trying to multiply and add up more complicated equations.

Illustration 2 *The reaction between hydrogen peroxide and manganate (VII) ions*

The first example was a simple bit of chemistry which you may well have come across. The technique works just as well for more complicated (and perhaps unfamiliar) chemistry.

Manganate (VII) ions, MnO_4^-, oxidise hydrogen peroxide, H_2O_2, to oxygen gas. The reaction is done with potassium manganate (VII) solution and hydrogen peroxide solution acidified with dilute sulphuric acid.

During the reaction, the manganate (VII) ions are reduced to manganese (II) ions.

Let's start with the hydrogen peroxide half-equation. What we know is :

$$H_2O_2 \longrightarrow O_2$$

The oxygen is already balanced. What about the hydrogen?

All you are allowed to add to this equation are water, hydrogen ions and electrons. If you add water to supply the extra hydrogen atoms needed on the right-hand side, you will mess up the oxygens again-that's obviously wrong!

Add two hydrogen ions to the right-hand side.

$$H_2O_2 \longrightarrow O_2 + 2H^+$$

Now all you need to do is balance the charges. You would have to add 2 electrons to the right-hand side to make the overall charge on both sides zero.

$$H_2O_2 \longrightarrow O_2 + 2H^+ + 2e^-$$

Now for the manganate(VII) half-equation:

You know (or are told) that the manganate(VII) ions turn into manganese(II) ions. Write that down.

$$MnO_4^- \longrightarrow Mn^{2+}$$

The manganese balances, but you need four oxygens on the right-hand side. These can only come from water-that's the only oxygen-containing thing you are allowed to write into one of these equations in acid conditions.

$$MnO_4^- \longrightarrow Mn^{2+} + 4H_2O$$

By doing this, we've introduced some hydrogens. To balance these, you will need 8 hydrogen ions on the left-hand side.

$$MnO_4^- + 8H^+ \longrightarrow Mn^{2+} + 4H_2O$$

Now that all the atoms are balanced, all you need to do is balance the charges. At the moment there are a net 7+ charges on the left-hand side (1– and 8+), but only 2+ on the right. Add 5 electrons to the left-hand side to reduce the 7+ to 2+.

$$MnO_4^- + 8H^+ + 5e^- \longrightarrow Mn^{2+} + 4H_2O$$

This is the typical sort of half-equation which you will have to be able to work out. The sequence is usually :

- Balance the atoms apart from oxygen and hydrogen.
- Balance the oxygens by adding water molecules.
- Balance the hydrogens by adding hydrogen ions.
- Balance the charges by adding electrons.

Combining the half-reactions to make the ionic equation for the reaction

The two half-equations we've produced are :

This reaction needs 5 electrons

$$MnO_4^- + 8H^+ + 5e^- \longrightarrow Mn^{2+} + 4H_2O$$

$$H_2O_2 \longrightarrow O_2 + 2H^+ + 2e^-$$

but this is only producing 2 electrons

You have to multiply the equations so that the same number of electrons are involved in both. In this case, everything would work out well if you transferred 10 electrons.

$$\begin{array}{rcl} 2\times(MnO_4^- + 8H^+ + 5e^- & \longrightarrow & Mn^{2+} + 4H_2O) \\ 5\times(H_2O_2 & \longrightarrow & O_2 + 2H^+ + 2e^-) \\ \hline 2MnO_4^- + 16H^+ + 5H_2O_2 & \longrightarrow & 2Mn^{2+} + 8H_2O + 5O_2 \\ & & + 10H^+ \end{array}$$

But this time, you haven't quite finished. During the checking of the balancing, you should notice that there are hydrogen ions on both sides of the equation :

hydrogen ions on both sides of the equation

$$2MnO_4^- + 16H^+ + 5H_2O_2 \longrightarrow 2Mn^{2+} + 8H_2O + 5O_2 + 10H^+$$

You can simplify this down by subtracting 10 hydrogen ions from both sides to leave the final version of the ionic equation-but don't forget to check the balancing of the atoms and charges!

$$2MnO_4^- + 6H^+ + 5H_2O_2 \longrightarrow 2Mn^{2+} + 8H_2O + 5O_2$$

You will often find that hydrogen ions or water molecules appear on both sides of the ionic equation in complicated cases built up in this way. Always check, and then simplify where possible.

Working out oxidation number and using them to balance equation

In the entrance examination it is never mentioned what method is to be used. **We adopt here "quick" method that will certainly be a time-saving method.**

Acidic Medium

Illustration 1 We want to balance following equation:

$$Cu + NO_3^- \longrightarrow NO + Cu^{2+}$$

$$\underset{0}{Cu} + \underset{+5}{NO_3^-} \longrightarrow \underset{+2}{NO} + \underset{+2}{Cu^{2+}}$$

+3e⁻ ; −(2e⁻)

ALERT

- Affected atoms are balanced.
- Take total O.N. of the affected atom on each side.

(a) Cross multiply by change in O.N.

$$3Cu + 2NO_3^- \longrightarrow 2NO + 3Cu^{2+}$$

(b) Balance oxygen by H_2O

$$3Cu + 2NO_3^- \longrightarrow 2NO + 3Cu^{2+} + 4H_2O$$

(c) Balance hydrogen by H^+

$$3Cu + 2NO_3^- + 8H^+ \longrightarrow 2NO + 3Cu^{2+} + 4H_2O$$

Illustration 2 $MnO_4^- + Fe^{2+} \longrightarrow Fe^{3+} + Mn^{2+}$

$$MnO_4^- + Fe^{2+} \longrightarrow Fe^{3+} + Mn^{2+}$$

$-e^-$; $+5e^-$

step (a) $MnO_4^- + 5Fe^{2+} \longrightarrow 5Fe^{3+} + Mn^{2+}$

step (b) $MnO_4^- + 5Fe^{2+} \longrightarrow 5Fe^{3+} + Mn^{2+} + 4H_2O$

step (c)

$$MnO_4^- + 5Fe^{2+} + 8H^+ \longrightarrow 5Fe^{3+} + Mn^{2+} + 4H_2O$$

Illustration 3 $Cr_2O_7^{2-} + Fe^{2+} \longrightarrow Fe^{3+} + Cr^{3+}$

In this Cr atoms on RHS are not balanced. Hence, first step is to balance Cr^{3+} and take O.N. of two Cr atoms

$$\underset{+12}{Cr_2O_7^{2-}} + \underset{+2}{Fe^{2+}} \longrightarrow \underset{+3}{Fe^{3+}} + \underset{+6}{2Cr^{3+}}$$

$-e$; $+6e^-$

By taking steps of illustration (1), balanced equation is

$$Cr_2O_7^{2-} + 6Fe^{2+} + 14H^+ \longrightarrow 6Fe^{3+} + 2Cr^{3+} + 7H_2O$$

Illustration 4 $MnO_4^- + C_2O_4^{2-} \longrightarrow 2CO_2 + Mn^{2+}$

$$\underset{+7}{MnO_4^-} + \underset{+6}{C_2O_4^{2-}} \longrightarrow \underset{+8}{2CO_2} + \underset{+2}{Mn^{2+}}$$

$-2e^-$; $+5e^-$

Carbon atoms are taken balanced and taking usual steps:

$$2MnO_4^- + 5C_2O_4^{2-} + 16H^+ \longrightarrow 10CO_2 + 2Mn^{2+} + 8H_2O$$

Illustration 5 $H_2O_2 + MnO_4^- \longrightarrow Mn^{2+} + O_2$

$$\underset{-2}{H_2O_2} + \underset{+7}{MnO_4^-} \longrightarrow \underset{+2}{Mn^{2+}} + \underset{0}{O_2}$$

R ; O

H_2O_2 has been oxidised to O_2. Thus, oxygen atoms undergoing oxidation are balanced. Taking O.N. of both oxygen and as by usual steps, balanced equation is

$$5H_2O_2 + 2MnO_4^- + 6H^+ \longrightarrow 2Mn^{2+} + 5O_2 + 8H_2O$$

Total H on LHS = (double of H_2O) − (H of H_2O_2)

Illustration 6 ***We may face some difficulties when two or more atoms are oxidised or reduced. In the following example***

$$As_2S_3 + NO_3^- + H^+ \longrightarrow AsO_4^{3-} + S + NO_2 + H_2O$$

As^{3+} as well as S^{2-} have been oxidised to AsO_4^{3-} and S respectively. In such case take total change (increase or decrease) in O.N. for cross-multiplication.

$$\underset{+6}{2As^{3+}} + \underset{-6}{3S^{2-}} + \underset{+5}{NO_3^-} \longrightarrow \underset{+4}{NO_2} + \underset{0}{3S} + \underset{+10}{2AsO_4^{3-}}$$

R ; O ; O

Total change in O.N. of oxidised part = 10

Hence, by usual steps, balance equation is

$$As_2S_3 + 10NO_3^- + 4H^+ \longrightarrow 10NO_2 + 2AsO_4^{3-} + 3S + 2H_2O$$

Basic Medium

We adopt here same method as in acidic medium, difference is only in balancing of oxygen.

— Cross-multiply by change in O.N.

— Add H_2O on excess oxygen side.

— Add OH^- twice of H_2O on the opposite side.

Illustration 1 *Br_2 disproportionates into Br^- and BrO_3^- in alkaline medium.*

$$\underset{0}{\frac{1}{2}Br_2} + \underset{0}{\frac{1}{2}Br_2} \longrightarrow \underset{-1}{Br^-} + \underset{+5}{BrO_3^-}$$

(R: 0 → −1; O: 0 → +5)

Cross multiplication by change in O.N. gives

$$3Br_2 \longrightarrow 5Br^- + BrO_3^-$$

(We have taken $\frac{1}{2}$ Br_2 for oxidation and reduction to balance affected atoms.)

Three O-atoms are extra on RHS, hence, add three H_2O molecules (excess O-side) on RHS, and add $6OH^-$ on LHS

$$3Br_2 + 6OH^- \longrightarrow 5Br^- + BrO_3^- + 3H_2O$$

Illustration 2 $\underset{+7}{Mn}O_4^- + H_2\underset{-2}{O_2} \xrightarrow{\text{Basic}} \underset{+4}{Mn}O_2 + \underset{0}{O_2}$

Affected atoms are balanced; by cross-multiplication

$$\underbrace{2MnO_4^- + 3H_2O_2}_{14\ (O)} \longrightarrow \underbrace{2MnO_2 + 3O_2}_{10\ (O)}$$

add H_2O on LHS (excess O-side)

$$2MnO_4^- + 3H_2O_2 + 4H_2O \longrightarrow 2MnO_2 + 3O_2$$

add OH^- twice of H_2O on RHS

$$2MnO_4^- + 3H_2O_2 + 4H_2O \longrightarrow 2MnO_2 + 3O_2 + 8OH^-$$

We have proceeded in a right way, but we have not considered H due to H_2O_2. Hence, H-atoms as well as charge are not balanced. Hence, add $6H^+$ on RHS which will neutralise OH^- forming $6H_2O$ and thus, final result will be

$$2MnO_4^- + 3H_2O_2 \longrightarrow 2MnO_2 + 3O_2 + 2OH^- + 2H_2O$$

Illustration 3 *A purple solution of aqueous potassium permanganate ($KMnO_4$) reacts with aqueous sodium sulphite (Na_2SO_3) in basic solution to yield the green manganate ion (MnO_4^{2-}) and sulphate ion (SO_4^{2-}). The unbalanced net ionic equation is:*

$$MnO_4^-(aq) + SO_3^{2-}(aq) \longrightarrow MnO_4^{2-}(aq) + SO_4^{2-}(aq)$$

Balance the equation.

change in O.N.

$$\underset{+7}{MnO_4^-}(aq) \xrightarrow{\text{reduction}} \underset{+6}{MnO_4^{2-}}(aq) \quad 1 \text{ unit}$$

$$\underset{+4}{SO_3^{2-}}(aq) \longrightarrow \underset{+6}{SO_4^{2-}}(aq) \quad 2 \text{ units}$$

Cross-multiplication followed by addition gives

$$\underbrace{2MnO_4^-(aq) + SO_3^{2-}(aq)}_{11 \text{ oxygen}} \longrightarrow \underbrace{2MnO_4^{2-}(aq) + SO_4^{2-}(aq)}_{12 \text{ oxygen}}$$

Add H_2O on extra oxygen side and add twice OH^- on opposite side

$$2MnO_4^-(aq) + SO_3^{2-}(aq) + 2OH^-(aq) \longrightarrow 2MnO_4^{2-}(aq) + SO_4^{2-}(aq) + H_2O(aq)$$

Illustration 4 *Write unbalanced half-reactions for the following net ionic equations:*

(a) $Mn^{2+}(aq) + ClO_3^-(aq) \longrightarrow MnO_2(s) + ClO_2(aq)$

(b) $Cr_2O_7^{2-}(aq) + Fe^{2+}(aq) \longrightarrow Cr^{3+}(aq) + Fe^{3+}(aq)$

(c) $Cl_2(g) + OH^-(aq) \longrightarrow Cl^-(aq) + ClO_3^-(aq)$

Oxidation	Reduction
(a) $Mn^{2+}(aq) \longrightarrow MnO_2(s)$	$ClO_3^-(aq) \longrightarrow ClO_2(aq)$
(b) $Fe^{2+}(aq) \longrightarrow Fe^{3+}(aq)$	$Cr_2O_7^{2-}(aq) \longrightarrow Cr^{3+}(aq)$
(c) $Cl_2(g) \longrightarrow ClO_3^-(aq)$	$Cl_2(g) \longrightarrow Cl^-(aq)$

Illustration 5 *Balance the following half-reactions in acidic medium. You may balance charges by putting electrons.*

(*a*) $VO^{2+}(aq) \longrightarrow V^{3+}(aq)$

(*b*) $NO_3^-(aq) \longrightarrow NO_2(g)$

(*c*) $IO_3^-(aq) \longrightarrow I_3^-(aq)$

(a) $$\underset{+4}{VO^{2+}}(aq) \longrightarrow \underset{+3}{V^{3+}}(aq)$$

VO^{2+} is reduced to V^{3+} by gain of one electron.

$$VO^{2+}(aq) \longrightarrow V^{3+}(aq)$$

Balance O : $VO^{2+}(aq) \longrightarrow V^{3+}(aq) + H_2O(l)$

Balance H : $VO^{2+}(aq) + 2H^+(aq) \longrightarrow V^{3+}(aq) + H_2O(l)$

Balance charge :

$$VO^{2+}(aq) + 2H^+(aq) + e^- \longrightarrow V^{3+}(aq) + H_2O(l)$$

(b) $$\underset{+5}{NO_3^-}(aq) \longrightarrow \underset{+4}{NO_2}(g)$$

$$NO_3^-(aq) + 2H^+(aq) + e^- \longrightarrow NO_2(g) + H_2O(l)$$

(c) $$IO_3^-(aq) \longrightarrow I_3^-(aq)$$

$IO_3^-(aq)$ is first oxidised to I_2 which dissolves in I^- forming I_3^-

$$\underset{+5}{IO_3^-}(aq) \longrightarrow \frac{1}{2}\underset{0}{I_2}(s)$$

$$IO_3^-(aq) + 6H^+ + 5e^- \longrightarrow \frac{1}{2}I_2(s) + 3H_2O(l)$$

Note decrease in oxidation number = electrons added on LHS
increase in oxidation number = electrons added on RHS

Illustration 6 *Balance the following half-reactions in a basic medium :*

(a) $CrO_4^{2-}(aq) \longrightarrow Cr(OH)_4^-(aq)$

(b) $Bi^{3+}(aq) \longrightarrow BiO_3^-(aq)$

(c) $ClO^-(aq) \longrightarrow Cl^-(aq)$

(a) $$\underset{+6}{CrO_4^{2-}}(aq) \longrightarrow \underset{+3}{Cr(OH)_4^-}(aq)$$

$$CrO_4^{2-}(aq) + 4H_2O(l) + 3e^- \longrightarrow Cr(OH)_4^-(aq) + 4OH^-(aq)$$

(b) $$\underset{+3}{Bi^{3+}}(aq) \longrightarrow \underset{+5}{BiO_3^-}(aq)$$

$$Bi^{3+}(aq) + 6OH^- \longrightarrow BiO_3^-(aq) + 3H_2O(l) + 2e^-$$

(c) $\underset{+1}{ClO^-}(aq) \longrightarrow \underset{-1}{Cl^-}(aq) + 2H_2O$

$ClO^-(aq) + H_2O(l) + 2e^- \longrightarrow Cl^-(aq) + 2OH^-(aq)$

Illustration 7 *Write balanced net ionic equations for the following reactions in basic solution:*

(*a*) $MnO_4^-(aq) + IO_3^-(aq) \longrightarrow MnO_2(s) + IO_4^-(aq)$
(*b*) $Cu(OH)_2(s) + N_2H_4(aq) \longrightarrow Cu(s) + N_2(g)$
(*c*) $Fe(OH)_2(s) + CrO_4^{2-}(aq) \longrightarrow Fe(OH)_3(s) + Cr(OH)_4^-(aq)$
(*d*) $H_2O_2(aq) + ClO_4^-(aq) \longrightarrow ClO_2^-(aq) + O_2(g)$

Solution

(a) $2MnO_4^-(aq) + 3IO_3^-(aq) + H_2O(l) \longrightarrow 2MnO_2(s) + 3IO_4^-(aq) + 2OH^-(aq)$
(b) $2Cu(OH)_2(s) + N_2H_4(aq) \longrightarrow 2Cu(s) + N_2(g) + 4H_2O(l)$
(c) $3Fe(OH)_2(s) + CrO_4^{2-}(aq) + 4H_2O(l) \longrightarrow 3Fe(OH)_3(s) + Cr(OH)_4^-(aq) + OH^-(aq)$
(d) $ClO_4^-(aq) + 2H_2O_2(aq) \longrightarrow ClO_2^-(aq) + 2H_2O(l) + 2O_2(g)$

Illustration 8 *Write balanced net ionic equations for the following reactions in acidic solution:*

(*a*) $Zn(s) + VO^{2+}(aq) \longrightarrow Zn^{2+}(aq) + V^{3+}(aq)$
(*b*) $Ag(s) + NO_3^-(aq) \longrightarrow Ag^+(aq) + NO_2(g)$
(*c*) $Mg(s) + VO_4^{3-}(aq) \longrightarrow Mg^{2+}(aq) + V^{2+}(aq)$
(*d*) $I^-(aq) + IO_3^-(aq) \longrightarrow I_3^-(aq)$

Solution

(a) $Zn(s) + 2VO^{2+}(aq) + 4H^+(aq) \longrightarrow Zn^{2+}(aq) + 2V^{3+}(aq) + 2H_2O(l)$
(b) $Ag(s) + 2H^+(aq) + NO_3^-(aq) \longrightarrow Ag^+(aq) + NO_2(g) + H_2O(l)$
(c) $3Mg(s) + 2VO_4^{3-}(aq) + 16H^+(aq) \longrightarrow 3Mg^{2+}(aq) + 2V^{2+}(aq) + 8H_2O(l)$
(d) $IO_3^-(aq) + 8I^-(aq) + 6H^+(aq) \longrightarrow 3I_3^-(aq) + 3H_2O(l)$

Balanced Redox Reactions

Some balanced redox reactions are in the medium specified:

1. $3S_2O_4^{2-}$ (dithionite ion) $+ 2CrO_4^{2-} + 4H_2O \longrightarrow 6SO_3^{2-} + 2Cr(OH)_3 + 2H^+$
2. $3CN^- + 2MnO_4^- + H_2O \longrightarrow 2MnO_2 + 3CNO^- + 2OH^-$
3. $4[Fe(CN)_6]^{3-} + N_2H_4 + 4OH^- \longrightarrow 4[Fe(CN)_6]^{4-} + N_2 + 4H_2O$
4. $4Fe(OH)_2 + O_2 + 2H_2O \longrightarrow 4Fe(OH)_3$
5. $3C_2H_5OH + 4MnO_4^- \longrightarrow 3CH_3COO^- + 4MnO_2 + 4H_2O + OH^-$
6. $S_2O_3^{2-} + 5H_2O + 4Cl_2 \longrightarrow 2SO_4^{2-} + 8Cl^- + 10H^+$
7. $2NH_4^+ + Cr_2O_7^{2-} \longrightarrow Cr_2O_3 + N_2 + 4H_2O$
8. $P_4 + 8H^+ + 20NO_3^- + 8H_2O \longrightarrow 12H_2PO_4^- + 20NO$
9. $2MnO_4^- + 6H^+ + 5NO_2^- \longrightarrow 2Mn^{2+} + 5NO_3^- + 3H_2O$
10. $S_8 + 12OH^- \longrightarrow 4S^{2-} + 2S_2O_3^{2-} + 6H_2O$
11. $5S_2O_3^{2-} + 8MnO_4^- + 14H^+ \longrightarrow 10SO_4^{2-} + 8Mn^{2+} + 7H_2O$
12. $2HS^- + 4HSO_3^- \longrightarrow 3S_2O_3^{2-} + 3H_2O$
13. $4Fe^{3+} + 2NH_2OH \longrightarrow 4Fe^{2+} + N_2O + 4H^+$
14. $4O_2^-$ (superoxide ion) $+ 2H_2O \longrightarrow 4OH^- + 3O_2$
15. $2[Fe(CN)_6]^{4-}$ (ferrocyanide ion) $+ H_2O_2 + 2H^+ \longrightarrow 2[Fe(CN)_6]^{3-}$ (ferricyanide ion) $+ 2H_2O$
16. $H_2C_2O_4 + MnO_2 + 2H^+ \longrightarrow Mn^{2+} + 2H_2O + 2CO_2$
17. $IO_3^- + 3HSO_3^- \longrightarrow I^- + 3H^+ + 3SO_4^{2-}$
18. $5I^- + IO_3^- + 6H^+ \longrightarrow 3H_2O + 3I_2$
19. $3MnO_4^{2-} + 2H_2O \longrightarrow MnO_2 + 2MnO_4^- + 4OH^-$
20. $2Fe_2S_3 + 6H_2O + 3O_2 \longrightarrow 4Fe(OH)_3 + 6S$
21. $3IBr + 2BrO_3^- + 3H_2O \longrightarrow 3IO_3^- + 5Br^- + 6H^+$
22. $As_2S_3 + 12OH^- + 14H_2O_2 \longrightarrow 2AsO_4^{3-} + 20H_2O + 3SO_4^{2-}$
23. $2CrI_3 + 27H_2O_2 + 10OH^- \longrightarrow 2CrO_4^{2-} + 6IO_4^- + 32H_2O$
24. $4Ag + 8CN^- + O_2 + 2H_2O \longrightarrow 4[Ag(CN)_2]^- + 4OH^-$
25. $B_2Cl_4 + 6OH^- \longrightarrow 2BO_2^- + 4Cl^- + 2H_2O + H_2$
26. $2Ag_2S + 8CN^- + O_2 + H_2O \longrightarrow 4[Ag(CN)_2]^- + 2S + 4OH^-$
27. $H_2O_2 + 2H^+ + 2Fe^{2+} \longrightarrow 2H_2O + 2Fe^{3+}$
28. $CrO + 2ClO^- + 2OH^- \longrightarrow CrO_4^{2-} + 2Cl^- + H_2O$
29. $5As_2S_3 + 4MnO_4^- + 9H_2O + 12H^+ \longrightarrow 10H_3AsO_4 + 15S + 4Mn^{2+}$
30. $O_3 + 2I^- + H_2O \longrightarrow O_2 + I_2 + 2OH^-$
31. $S_2O_3^{2-} + 4Cl_2 + 5H_2O \longrightarrow 2HSO_4^- + 8H^+ + 8Cl^-$
32. $2CN^- + 5ClO^- + H_2O \longrightarrow 2HCO_3^- + 5Cl^- + N_2$
33. $6ClO_2 + 6OH^- \longrightarrow 5ClO_3^- + Cl^- + 3H_2O$
34. $4Zn + 10H^+ + NO_3^- \longrightarrow 4Zn^{2+} + NH_4^+ + 3H_2O$
35. $4Zn + 10H^+ + 2NO_3^- \longrightarrow 4Zn^{2+} + N_2O + 5H_2O$
36. $2MnO + 5PbO_2 + 8H^+ \longrightarrow 2MnO_4^- + 5Pb^{2+} + 4H_2O$
37. $2OCl^- + 2OH^- + Bi_2O_3 \longrightarrow 2BiO_3^- + H_2O + 2Cl^-$
38. $8Al + 3OH^- + 3HSO_4^- \longrightarrow 4Al_2O_3 + 3H_2O + 3S^{2-}$
39. $4Ag + 8CN^- + O_2 + 2H_2O \longrightarrow 4Ag(CN)_2^- + 4OH^-$
40. $P_4 + 4OH^- + 2H_2O \longrightarrow 2PH_3 + 2HPO_3^{2-}$

Target Practice 3

Balance the following redox reactions

1. $SO_2(g) + HNO_2(aq) \longrightarrow H_2SO_4(aq) + NO(g)$
2. $Al(s) + H_2SO_4(aq) \longrightarrow Al_2(SO_4)_3(aq) + H_2(g)$
3. $Au^{3+}(aq) + I^-(aq) \longrightarrow Au(s) + I_2(s)$
4. $S^{2-}(aq) + I_2(s) \longrightarrow SO_4^{2-}(aq) + I^-(aq)$
5. $H_2O_2(aq) + ClO_4^-(aq) \longrightarrow O_2(g) + ClO_2^-(aq)$
6. $Br_2(aq) + OH^-(aq) \longrightarrow Br^-(aq) + BrO_3^-(aq)$
7. $Mn(s) + HNO_3(aq) \longrightarrow Mn^{2+}(aq) + NO_2(g)$
8. $I_2(s) + OCl^-(aq) \longrightarrow IO_3^-(aq) + Cl^-(aq)$
9. $Cr_2O_7^{2-}(aq) + HNO_2(aq) \longrightarrow Cr^{3+}(aq) + NO_3^-(aq)$
10. $CrO_4^{2-}(aq) + S^{2-}(aq) \longrightarrow Cr(OH)_3(s) + S(s)$

Answers

1. $SO_2(g) + 2HNO_2(aq) \longrightarrow H_2SO_4(aq) + 2NO(g)$
2. $2Al(s) + 3H_2SO_4(aq) \longrightarrow Al_2(SO_4)_3(aq) + 3H_2(g)$
3. $2Au^{3+}(aq) + 6I^-(aq) \longrightarrow 2Au(s) + 3I_2(s)$
4. $S^{2-}(aq) + 4I_2(s) + 4H_2O(l) \longrightarrow 8I^-(aq) + SO_4^{2-}(aq) + 8H^+(aq)$
5. $2H_2O_2(aq) + ClO_4^-(aq) \longrightarrow 2O_2(g) + ClO_2^-(aq) + 2H_2O(l)$
6. $3Br_2(aq) + 6OH^-(aq) \longrightarrow 5Br^-(aq) + BrO_3^-(aq) + 3H_2O(l)$
7. $Mn(s) + 2NO_3^-(aq) + 4H^+(aq) \longrightarrow Mn^{2+}(aq) + 2NO_2(g) + 2H_2O$
8. $I_2(s) + 5OCl^-(aq) + H_2O(l) \longrightarrow IO_3^-(aq) + 5Cl^-(aq) + 2H^+(aq)$
9. $Cr_2O_7^{2-}(aq) + 6NO_2^-(aq) + 8H^+(aq) \longrightarrow 2Cr^{3+}(aq) + 6NO_3^-(aq) + 4H_2O(l)$
10. $2CrO_4^{2-}(aq) + 3S^{2-}(aq) + 8H_2O(l) \longrightarrow 2Cr(OH)_3(s) + 3S(s) + 10OH^-(aq)$

Equivalent Weights

In **unit 1,** detailed study of equivalent weight has been given. To determine number of equivalents or equivalent weight we can compare the molecular weight to

$\equiv 1\text{ g H} = 1\text{ mol of } H^+$

$\equiv 17\text{ g } OH^- = 1\text{ mol of } OH^-$

$\equiv 12\text{ g C}$

$\equiv 35.5\text{ g Cl}$

$\equiv 8\text{ g O}$

Equivalent Weight of Non-metals

Equivalent weight of non-metals can be determined by

(i) Direct combination with hydrogen

	Non-metal	Compound with hydrogen	Mass of hydrogen	Equivalent Mass = of non-metal combining with 1 g H
(a)	Carbon–12	CH_4	4 g	3 g
(b)	Boron–10.8	BH_3	3 g	3.6 g
(c)	Nitrogen–14	NH_3	3 g	4.66 g

(ii) Conversion into oxide

Non-metal	Compound with oxygen	Oxygen combined	Mass of non-metal combining with 8 g oxygen = equivalent weight
Carbon–12	CO_2	32 g	$\frac{12}{32} \times 8 = 3$ g
Nitrogen–14	NO	16 g	$\frac{14}{16} \times 8 = 7$ g
Nitrogen–14	NO_2	32 g	$\frac{14}{32} \times 8 = 3.5$ g
Nitrogen–14	N_2O	16 g	$\frac{28}{16} \times 8 = 14$ g

In the above example, it is clear that depending upon the compound formed **nitrogen** has different equivalent weight.

Equivalent Weight of Metals

1. Replacement of hydrogen from acids

We compare the mass of metal that displaces 1 g (H_2)

$$\underset{(M)\text{ g}}{Zn} + 2HCl \longrightarrow ZnCl_2 + \underset{2\text{ g}}{H_2}$$

$$H_2 \equiv Zn$$

$$2\text{ g } H_2 \equiv M\text{ g metal}$$

$$\therefore \quad 1\text{ g } H_2 \equiv \frac{M}{2}$$

$$\therefore \quad \text{Equivalent weight of Zn} = \frac{\text{Atomic weight}}{2}$$

(b) $\underset{(2M)\text{ g}}{2Al} + 6HCl \longrightarrow AlCl_3 + \underset{6\text{ g}}{3H_2}$

$$3H_2 \equiv 2\,Al$$
$$6\text{ g } H_2 \equiv 2M\text{g}$$
$$1\text{ g } H_2 \equiv \frac{2M}{6} = \frac{M}{3}$$

$\therefore$ Equivalent weight of Al $= \dfrac{\text{atomic weight}}{3}$

2. Conversion of Metals into Oxide

We compare the mass of metal that combines with 8 g oxygen.

$$2M + xO_2 \longrightarrow M_2O_x$$
$$2M \equiv 32x\text{ g} \quad (x = \text{valency of metal } M)$$
$$32x\text{ g } O_2 \equiv 2\text{ atomic weight}$$
$$8\text{ g } O_2 \equiv \frac{2 \times \text{atomic weight}}{32x} \times 8$$
$$= \frac{\text{atomic weight}}{2x}$$

We can also calculate equivalent weight by taking weight of metal and metal oxide.

Let weight of metal $= a$ g

Weight of metal oxide $= b$ g

Oxygen combined $= (b - a)$ g

$(b - a)$ g oxygen combined with $= a$ g metal

$\therefore$ 8 g oxygen combined with $= \dfrac{a}{b-a} \times 8$ g

$\therefore$ Equivalent weight of metal $= \dfrac{8a}{(b-a)}$

(a) $$2Na + \frac{1}{2}O_2 \longrightarrow Na_2O$$
$$x = \frac{1}{2}$$

Equivalent weight of sodium = atomic weight
$= 23$ g equivalent^{-1}

(b) $$2Mg + O_2 \longrightarrow 2MgO$$
$$x = 1$$

Equivalent weight of magnesium $= \dfrac{\text{atomic weight}}{2}$

$$= \frac{24}{2}$$
$$= 12 \text{ g equivalent}^{-1}$$

(c) $$2Al + \frac{3}{2}O_2 \longrightarrow Al_2O_3$$
$$x = \frac{3}{2}$$

Equivalent weight of aluminium $= \dfrac{\text{atomic weight}}{3}$

$$= \frac{27}{3}$$
$$= 9 \text{ g equivalent}^{-1}$$

3. Conversion of Metals into Chlorides

We compare the mass of metal that combines with 35.5 g of chlorine.

Let mass of metal $= a$ g

Mass of metal chloride $= b$ g

Chlorine combined $= (b - a)$ g

$(b - a)$ g chlorine $\equiv a$ g metal

$\therefore$ 35.5 g chlorine $\equiv \dfrac{a}{(b-a)} \times 35.5$ g

$\therefore$ Equivalent weight of metal $= \dfrac{35.5a}{(b-a)}$

Alternatively

$M + \dfrac{x}{2}Cl_2 \longrightarrow MCl_x$ equivalent weight of M

$= \dfrac{\text{atomic weight}}{x}$

$Li + \dfrac{1}{2}Cl_2 \longrightarrow LiCl$ $x = 1$; equivalent weight of Li = atomic weight

$Mg + Cl_2 \longrightarrow MgCl_2$ $x = 2$; equivalent weight of Mg $= \dfrac{\text{atomic weight}}{2}$

$Al + \dfrac{3}{2}Cl_2 \longrightarrow AlCl_3$ $x = 3$; equivalent weight of Al $= \dfrac{\text{atomic weight}}{3}$

Illustration 1 (*a*) *0.24 g of magnesium combined with oxygen forming 0.40 g magnesium oxide. Derive equivalent weight of magnesium.*

(*b*) *0.27 g aluminium combined with hydrogen forming 0.30 g aluminium hydride. Derive equivalent weight of aluminium.*

(*c*) *1.95 g of platinum forms 3.37 g of chlorides on reaction with chlorine. Derive equivalent weight of platinum.*

Solution (a) Mass of magnesium = 0.24 g

Mass of magnesium oxide = 0.40 g

Oxygen combined = 0.40 − 0.24
= 0.16 g

0.16 g oxygen combined with = 0.24 g magnesium

$\therefore$ 8 g oxygen combined with $= \dfrac{0.24}{0.16} \times 8 = 12$ g

$\therefore$ Equivalent weight of magnesium

= 12 g equivalent^{-1}

(b) Mass of aluminium = 0.27 g

Mass of aluminium hydride = 0.30 g

Hydrogen combined = 0.03 g

0.03 g hydrogen combines with = 0.27 g aluminium

$\therefore$ 1 g hydrogen combines with $= \frac{0.27}{0.03} \times 1\text{ g} = 9\text{ g}$

$\therefore$ Equivalent weight of aluminium

$= \mathbf{9\text{ g equivalent}^{-1}}$

(c) Mass of platinum = 1.95 g

Mass of platinum chloride = 3.37 g

Mass of chlorine combined = 3.37 − 1.95

= 1.42 g

1.42 g chlorine combines with = 1.95 g platinum

$\therefore$ 35.5 g chlorine combines with $= \frac{1.95 \times 35.5}{1.42}$

$= \frac{195}{4} = 48.75\text{ g}$

$\therefore$ Equivalent weight of platinum

$= \mathbf{48.75\text{ g equivalent}^{-1}}$

Equivalent Weight of Acids and Bases

In a Neutralisation Reaction

We compare molecular weight w.r.t. ionisable H^+ (in acids, called basicity) or OH^- (in bases, called acidity).

		M
$HCl \rightleftharpoons H^+ + Cl^-$	$E = M/1$	36.5
$H_3PO_2 \rightleftharpoons H^+ + H_2PO_2^-$	$E = M/1$	66.0
$HCO_3^- \rightleftharpoons H^+ + CO_3^{2-}$	$E = M/1$	61.0
$H_2SO_4 \rightleftharpoons 2H^+ + SO_4^{2-}$	$E = M/2$	98.0
$H_3PO_3 \rightleftharpoons 2H^+ + HPO_3^{2-}$	$E = M/2$	82.0
$H_2C_2O_4 \rightleftharpoons 2H^+ + C_2O_4^{2-}$	$E = M/2$	90.0
$H_3PO_4 \rightleftharpoons 3H^+ + PO_4^{3-}$	$E = M/3$	98.0
$NaOH \rightleftharpoons Na^+ + OH^-$	$E = M$	40.0
$Ca(OH)_2 \rightleftharpoons Ca^{2+} + 2OH^-$	$E = M/2$	74.0
$Al(OH)_3 \rightleftharpoons Al^{3+} + 3OH^-$	$E = M/3$	78.0

Facts : It must be noted that equivalent weight is not a definite property but varies reaction to reaction.

H_3PO_4 is a tribasic acid. But in the following reaction only two H^+ ions have been neutralised:

$$H_3PO_4 + Ca(OH)_2 \longrightarrow CaHPO_4 + 2H_2O$$

Hence, equivalent weight of $H_3PO_4 = M/2 = 49$

and if reaction proceeds completely

$$H_3PO_4 + 3NaOH \longrightarrow Na_3PO_4 + 3H_2O$$

$$\text{Equivalent weight} = \frac{M}{3} = \frac{98}{3} = 32.66$$

Illustration 1 *Calculate moles of* $Ca(OH)_2$ *required to carry out following conversion taking one mole in each case:*

(*a*) $\begin{matrix} COOH \\ | \\ COOH \end{matrix}$ *into* $\begin{matrix} COO \\ | \\ COO \end{matrix}\!\!>\!Ca$ (*b*) H_3PO_4 *into* $CaHPO_4$

(*c*) NH_4Cl *into* NH_3 (*d*) $NaHCO_3$ *into* $CaCO_3$

Solution (a) $\begin{matrix} COOH \\ | \\ COOH \end{matrix}$ is a dibasic acid

$$\underset{1\text{ mol}}{\begin{matrix} COOH \\ | \\ COOH \end{matrix}} + \underset{1\text{ mol}}{Ca(OH)_2} \longrightarrow \begin{matrix} COOH \\ | \\ COOH \end{matrix}\!\!>\!Ca$$

$Ca(OH)_2$ required = **1 mol**

(b) $H_3PO_4 + Ca(OH)_2 \longrightarrow CaHPO_4 + 2H_2O$

1 mol of $H_3PO_4 \equiv 2H^+$ neutralised by 1 mol of $Ca(OH)_2$

$Ca(OH)_2$ required = **1 mol**

(c) $2NH_4Cl + Ca(OH)_2 \longrightarrow CaCl_2 + 2NH_3 + 2H_2O$

2 mol $NH_4Cl \equiv 1$ mol $Ca(OH)_2$

1 mol $NH_4Cl \equiv$ **0.5 mol $Ca(OH)_2$**

(d) $2NaHCO_3 + Ca(OH)_2 \longrightarrow Na_2CO_3 + CaCO_3 + 2H_2O$

2 mol $NaHCO_3 \equiv 1$ mol $Ca(OH)_2$

1 mol $NaHCO_3 \equiv$ **0.5 mol $Ca(OH)_2$**

Illustration 2 *Calculate equivalent weight of the underlined:*

(*a*) $Ca(OH)_2 + 2\underline{H_2SO_4} \longrightarrow Ca(HSO_4)_2 + 2H_2O$

(*b*) $\underline{[Al(OH)_3(H_2O)_3]} + OH^- \longrightarrow [Al(OH)_4(H_2O)_2^-] + H_2O$

(*c*) $\underline{[Al(OH)_3(H_2O)_3]} + H_3O^+ \longrightarrow [Al(OH)_2(H_2O)_4^+] + H_2O$

Solution (a) H_2SO_4 changes to HSO_4^- *i.e.*, only one H^+ reacts

$$H_2SO_4 \equiv H^+$$

$\therefore$ Equivalent weight of H_2SO_4 = molecular weight = **98**

(b) $[Al(OH)_3(H_2O)_3] \equiv 1\ OH^-$

$\therefore$ equivalent weight = molecular weight

(c) $[Al(OH)_3(H_2O)_3] \equiv 1\ H_3O^+$

$\therefore$ equivalent weight = molecular weight

In a Redox Reaction

Depending on the nature of products of a redox reaction, equivalent weight is determined.

- Compare molecular weight w.r.t. 8 g (O) if oxygen is gained or lost. For this convert every (O) atom gained or lost into **two parts** :

(a) $Cl_2 + H_2O \longrightarrow 2HCl + O$

$\therefore$ $Cl_2 \equiv 1\ (O)$ taking weight as 16

$\equiv 2\ (O)$ taking weight as 8

$\frac{Cl_2}{2} \equiv 1\ (O)$

$\therefore$ Equivalent weight of $Cl_2 = \frac{M}{2}$

(b) $KMnO_4$ in acidic medium loses 2.5 oxygen atoms

$$2KMnO_4 + 3H_2SO_4 \longrightarrow K_2SO_4 + 2MnSO_4 + 3H_2O + 5(O)$$

Thus, $KMnO_4 \equiv 2.5\ (O)$ atoms taking weight as 16

$\equiv 5\ (O)$ atoms taking weight as 8

$\frac{KMnO_4}{5} \equiv 1\ (O)$

$\therefore$ Eq. wt. of $KMnO_4$ in acidic medium $= \frac{M}{5}$

(c) $KMnO_4$ in basic and neutral medium loses 1.5 oxygen atoms

$$2KMnO_4 + H_2O \longrightarrow 2MnO_2 + 2KOH + 3(O)$$

Thus, $KMnO_4 \equiv 1.5\ (O)$ atoms taking weight as 16

$\equiv 3\ (O)$ atoms taking weight as 8

$\frac{KMnO_4}{3} \equiv 1\ (O)$

$\therefore$ Eq. wt. of $KMnO_4$ in basic medium $= \frac{M}{3}$

(d) $KMnO_4$ in strongly alkaline medium loses 0.5 oxygen atom

$KMnO_4 \equiv 0.5\ (O)$ atom taking weight as 16

$\equiv 1\ (O)$ taking weight at 8.

$\therefore$ Eq. wt. of $KMnO_4$ in neutral medium $= \frac{M}{1}$

Thus, eq. wt. is not a definite property. It also depends on medium. Eq. wt. of $KMnO_4$ as we shall above is quite different in three media, acidic, basic and neutral. (mol. wt. = 158)

$$\text{Acidic} = \frac{M}{5} = 31.6$$

$$\text{Basic and neutral} = \frac{M}{3} = 52.66$$

$$\text{Strongly alkaline} = \frac{M}{1} = 158.0$$

- In certain cases I_2 is formed or used. In such cases compare molecular weight with one I atom.

(a) $2CuSO_4 + 4KI \longrightarrow I_2 + Cu_2I_2 + 2K_2SO_4$

$2CuSO_4 \equiv I_2 \equiv 2I$

$\therefore$ $I \equiv CuSO_4$

$\therefore$ Eq. wt. of $CuSO_4 = M$

(b) $2Na_2S_2O_3 + I_2 \longrightarrow 2NaI + Na_2S_4O_6$

$2Na_2S_2O_3 \equiv I_2 \equiv 2I$

$\therefore$ Eq. wt. of $Na_2S_2O_3 = M$

(c) $MnO_2 + 4HCl \longrightarrow MnCl_2 + Cl_2 + 2H_2O$

$Cl_2 + 2KI \longrightarrow 2KCl + I_2$

$MnO_2 \equiv I_2 \equiv 2I$

$\therefore$ Eq. wt. of $MnO_2 = \frac{M}{2}$

(d) $Cr_2O_7^{2-} + 6I^- + 14H^+ \longrightarrow 3I_2 + 2Cr^{3+} + 7H_2O$

$Cr_2O_7^{2-} \equiv 3I_2 \equiv 6I$

$\therefore$ Eq. wt. of $Cr_2O_7^{2-} = \frac{M}{6}$

DEEP Focus

- Easiest method of determination of eq. wt. in a redox reaction is by

$$\text{Eq. wt.} = \frac{\text{formula weight (stoichiometric)}}{\text{change in O.N. or electron exchanged}}$$

We have used the term **"formula weight"** to simplify the definition. While using this term, check

– affected atoms are balanced

– take total O.N. of the affected atom of each side.

Illustration 1

(a) $\underset{+7}{MnO_4^-} \longrightarrow \underset{+2}{Mn^{2+}}$

$$\text{Eq. wt.} = \frac{\text{mol. wt. of one unit of } MnO_4^-}{5}$$

(b) $\underset{-2}{H_2O_2} \longrightarrow \underset{0}{O_2}$ O.N. of two (O) on each side

$$\text{Eq. wt.} = \frac{M}{2}$$

(c) $Cr_2O_7^{2-} \longrightarrow Cr^{3+}$ check here Cr atoms are not balanced

$\underset{12}{Cr_2O_7^{2-}} \longrightarrow \underset{6}{2Cr^{3+}}$ now balanced O.N. of two Cr atoms

$$\text{Eq. wt.} = \frac{M}{6}$$

We can also take it as follows :

$Cr_2O_7^{2-} \longrightarrow Cr^{3+}$

$\underset{6}{\frac{1}{2}Cr_2O_7^{2-}} \longrightarrow \underset{3}{Cr^{3+}}$ Cr atoms are balanced O.N. of one Cr atom on each side

$$\text{Eq. wt.} = \frac{\left(\frac{1}{2}\right)\text{formula weight}}{3} = \frac{M/2}{3} = \frac{M}{6}$$

(half molecule is on LHS)

(d) $\underset{0}{N_2 + 3H_2} \longrightarrow \underset{-6}{2NH_3}$ O.N. of two N-atoms

$$\text{Eq. wt. of } NH_3 = \frac{2\text{ formula weight of } NH_3}{6} = \frac{M}{3} = \frac{17}{3}$$

$$\text{Eq. wt. of } N_2 = \frac{M}{6} = \frac{28}{6} = \frac{14}{3}$$

(e) $2MnO_4^{2-} \longrightarrow MnO_2 + MnO_4^-$

MnO_4^{2-} disproportionates to MnO_2 and MnO_4^-. If E_R is the equivalent weight in the reduction half-reaction and E_o is the equivalent weight in the oxidation half-reaction then net equivalent weight E

$$E = E_R + E_o$$

$$\underset{+6}{MnO_4^{2-}} \longrightarrow \underset{+4}{MnO_2} \quad E_R = \frac{M}{2}$$

$$\underset{+6}{MnO_4^{2-}} \longrightarrow \underset{+7}{MnO_4^-} \quad E_o = \frac{M}{1}$$

$$E(\text{net}) = \frac{M}{2} + \frac{M}{1} = \mathbf{1.5\ M}$$

(f) $$Cl_2 + OH^- \longrightarrow Cl^- + ClO_2^-$$

$$\frac{1}{2}\underset{0}{Cl_2} \longrightarrow \underset{-1}{Cl^-} \quad E_R = \frac{M}{2}$$

$$\frac{1}{2}\underset{0}{Cl_2} \longrightarrow \underset{+3}{ClO_2^-} \quad E_o = \frac{M}{6}$$

(0.5 mol of Cl_2 is used in each half-reaction)

$$E_{(\text{net})} = \frac{M}{2} + \frac{M}{6} = \frac{2}{3}M$$

Illustration 2 *Equivalent weight of* Br_2 *is 96 in the following disproportionation reaction*:

$Br_2 + OH^- \longrightarrow Br^- + H_2O + ?$ *(oxidised product)*

Identify the oxidised product. (Br = 80)

Solution Molar mass of Br_2 = 160 g mol^{-1}

$$\frac{1}{2}\underset{0}{Br_2} \xrightarrow{\text{reduction}} \underset{-1}{Br^-}$$

$$\text{Equivalent weight of } Br_2 = \frac{\text{formula weight}}{1} = \frac{80}{1} = 80$$

Net equivalent weight (given) = 96

∴ Equivalent weight of Br_2 in oxidation process
$$= 96 - 80 = 16$$

Let oxidised product may contain one bromine

$$\frac{1}{2}\underset{0}{Br_2} \longrightarrow \underset{n}{\text{oxidised part}}$$

$$\text{Equivalent weight} = 16 = \frac{80}{\text{change in O.N.}} = \frac{80}{n}$$

$$n = \frac{80}{16} = 5$$

Thus, oxidation number of Br in oxidised part is +5. It is $\mathbf{BrO_3^-}$

Illustration 3 HCHO *disproportionates to* $HCOO^-$ *and* CH_3OH *in presence of* OH^- *(Cannizzaro reaction).*

$$2HCHO + OH^- \longrightarrow HCOO^- + CH_3OH$$

Calculate equivalent weight of HCHO.

Solution $$\underset{\text{O.N.}=0}{HCHO} \xrightarrow{\text{oxidation}} \underset{2}{HCOO^-}$$

$$\text{Equivalent weight of HCHO} = \frac{\text{molecular weight}}{\text{change in O.N.}} = \frac{M}{2}$$

$$\underset{0}{HCHO} \xrightarrow{\text{reduction}} \underset{2}{CH_3OH}$$

$$\text{Equivalent weight of HCHO} = \frac{M}{2}$$

In a disproportionation reaction, net equivalent weight = sum of equivalent weights of oxidation and reduction reactions

$$= \frac{M}{2} + \frac{M}{2} = M = \mathbf{30}$$

Target Practice 4

1. In the following redox reactions, write values of **a** to **f**

I. $\mathbf{a}MnO_4^- + \mathbf{b}Fe^{2+} + \mathbf{c}H^+ \longrightarrow \mathbf{d}Mn^{2+} + \mathbf{e}Fe^{3+} + \mathbf{f}H_2O$

II. $\mathbf{a}Cr_2O_7^{2-} + \mathbf{b}Fe^{2+} + \mathbf{c}H^+ \longrightarrow \mathbf{d}Cr^{3+} + \mathbf{e}Fe^{3+} + \mathbf{f}H_2O$

III. $\mathbf{a}MnO_4^- + \mathbf{b}H_2O_2 + \mathbf{c}H^+ \longrightarrow \mathbf{d}Mn^{2+} + \mathbf{e}O_2 + \mathbf{f}H_2O$

	a	**b**	**c**	**d**	**e**	**f**
I						
II						
III						

2. Molecular weight of $KMnO_4$ is 158. $KMnO_4$ can be reduced to $MnSO_4$, K_2MnO_4 and MnO_2 and equivalent weight of $KMnO_4$ comes out to be 158, 52.66, 31.6. Match the equivalent weights with compounds formed by reduction.

Equivalent weight of $KMnO_4$	Compound formed
158	....
52.66	....
31.6	....

3. MnO_4^- is an oxidant for Fe^{2+}, $C_2O_4^{2-}$ etc. Mention moles of MnO_4^- required by 1 mol each of

	MnO_4^- required
(a) Ferrous oxalate	....
(b) Ferric sulphate	....
(c) Ferrous chloride	....

4. Write half reactions using electrons
(a)
$$2MnO_4^- + 5C_2O_4^{2-} + 16H^+ \longrightarrow 10CO_2 + 2Mn^{2+} + 8H_2O$$
Oxidation ....
Reduction ....
(b) $Cr_2O_7^{2-} + 6Fe^{2+} + 14H^+ \longrightarrow 2Cr^{3+} + 6Fe^{3+} + 7H_2O$
Oxidation ....
Reduction ....
(c) $3Cl_2 + 6NaOH \longrightarrow 5NaCl + NaClO_3 + 3H_2O$
Oxidation ....
Reduction ....

5. $KHCO_3$ and KHC_2O_4 can be estimated by base or/and MnO_4^-. Derive equivalent weight

	Using base	Using MnO_4^-
$KHCO_3$	....	....
KHC_2O_4	....	....

Answers

1.

	a	b	c	d	e	f
I	1	5	8	1	5	4
II	1	6	14	2	6	7
III	2	5	6	2	5	8

2.

Equivalent weight	Compound formed
158	K_2MnO_4
52.66	MnO_2
31.6	$MnSO_4$

3. (a) $FeC_2O_4 \rightleftharpoons Fe^{2+} + C_2O_4^{2-}$
$$5Fe^{2+} + MnO_4^- \longrightarrow Mn^{2+} + 5Fe^{3+}$$
$$5C_2O_4^{2-} + 2MnO_4^- \longrightarrow 2Mn^{2+} + 10CO_2$$
$$5FeC_2O_4 \equiv 3MnO_4^-$$
$\therefore$ $1\ FeC_2O_4 \equiv \mathbf{0.6\ MnO_4^-}$
$\therefore$ $MnO_4^- = 0.6$ mol

(b) Ferric sulphate $Fe_2(SO_4)_3$ is already has iron in +3 state. Hence, no further oxidation by MnO_4^-
$\therefore$ MnO_4^- = Zero

(c) $FeCl_2 \rightleftharpoons Fe^{2+}$
$$5Fe^{2+} \equiv MnO_4^-$$
$$1\ Fe^{2+} \equiv \mathbf{0.2\ mol\ MnO_4^-}$$

4. (a) **Oxidation :** $Fe^{2+} \longrightarrow Fe^{3+} + e^-$
Reduction :
$$MnO_4^- + 8H^+ + 5e^- \longrightarrow Mn^{2+} + 4H_2O$$
(b) **Oxidation :** $Fe^{2+} \longrightarrow Fe^{3+} + e^-$
Reduction :
$$Cr_2O_7^{2-} + 14H^+ + 6e^- \longrightarrow 2Cr^{3+} + 7H_2O$$
(c) **Oxidation :** $\frac{1}{2}Cl_2 + 6OH^- \longrightarrow ClO_3^- + 3H_2O + 5e^-$
Reduction : $\frac{1}{2}Cl_2 + e^- \longrightarrow Cl^-$

5.

Compound	Using base	Using MnO_4^-
$KHCO_3$	$E = M$	cannot be used
KHC_2O_4	$E = M$	$E = \frac{M}{2}$

Practice Exercise 2

1. Balance the following redox reactions in acidic solution:
(a) $VO_4^{3-} + Fe^{2+} \longrightarrow Fe^{3+} + VO^{2+}$
(b) $BrO_3^- + N_2H_4 \longrightarrow Br^- + N_2$
(c) $UO_2 + Cr_2O_7^{2-} + H^+ \longrightarrow UO_2^{2+} + Cr^{3+} + H_2O$

2. Balance the following redox reaction in basic medium:
(a) $Mn^{2+} + H_2O_2 \longrightarrow MnO_2 + H_2O$
(b) $MnO_4^- + NO_2^- \longrightarrow MnO_2 + NO_3^-$
(c) $Mn^{2+} + ClO_3^- \longrightarrow MnO_2 + ClO_2$
(d) $MnO_4^- + NO_2 \longrightarrow MnO_2 + NO_3^-$

3. Balance the following redox reactions in acidic or basic aqueous solution as indicated:
(a) $H_2S + NO_3^- \longrightarrow NO_2 + S_8$ (acidic)
(b) $MnO_4^- + SO_2 \longrightarrow SO_4^{2-} + Mn^{2+}$ (acidic)
(c) $Bi(OH)_3 + Sn(OH)_3^- \longrightarrow Sn(OH)_6^{2-} + Bi$ (basic)
(d) $MnO_4^- + I^- \longrightarrow MnO_2 + IO_3^-$ (basic)
(e) $H_2O_2 + MnO_4^- \longrightarrow O_2 + MnO_2$ (basic)
(f) $Al + NO_3^- \longrightarrow Al(OH)_4^- + NH_3$ (basic)
(g) $Zn + NO_3^- \longrightarrow Zn^{2+} + N_2$ (acidic)

4. Balance the following equation taking place in basic solution:
$$Fe(OH)_2(s) + O_2(g) \longrightarrow Fe(OH)_3(s)$$

5. Write balanced redox reactions of the following:
(a) $H_2O_2(aq)$ reduces $Cr_2O_7^{2-}(aq)$ to green coloured $Cr^{3+}(aq)$ in acidic medium.
(b) $H_2O_2(aq)$ oxidises $Mn^{2+}(aq)$ to MnO_2 in basic medium.
(c) 4 moles of $Sn^{2+}(aq)$ reduces $IO_4^-(aq)$ to $I^-(aq)$.

6. A given amount of Fe^{2+} is oxidised by x moles of MnO_4^- in acidic medium. Calculate number of moles of $Cr_2O_7^{2-}$ required to oxidise same amount in Fe^{2+} in acidic medium.

7. $C_2O_4^{2-}$ can be oxidised to CO_2 by MnO_4^- in acidic as well as in basic medium. What is ratio of moles of MnO_4^- used per mol of $C_2O_4^{2-}$?

8. How many moles of KI are oxidised by number of moles of KIO_3 formed when 1 mole of I_2 is boiled with excess of KOH?
$$6KOH + 3I_2 \longrightarrow 5KI + KIO_3 + 3H_2O$$
$$KI + KIO_3 \xrightarrow{H^+} I_2 + H_2O$$

9. Calculate number of moles of MnO_4^- required to oxidise 1 mole of $Fe(HC_2O_4)_2$ in acidic medium. How many moles of $Ca(OH)_2$ are required to react with 1 mole of $Fe(HC_2O_4)_2$?

10. $H_2O_2(aq)$ reduces $ClO_4^-(aq)$ in acidic medium. If 1 mole of H_2O_2 reduces 0.5 mole of $ClO_4^-(aq)$, what is the reduced species?

11. $MnO_4^-(aq)$ can oxidise $IO_3^-(aq)$ to $IO_4^-(aq)$ and itself gets reduced to $MnO_2(s)$. Calculate number of moles of $IO_3^-(aq)$ that can reduce 0.1 mole of $MnO_4^-(aq)$.

12. Calculate equivalent weight of the underlined :

(a) $\underline{Na_2CO_3} + H_2SO_4 \longrightarrow Na_2SO_4 + H_2O + CO_2$

(b) $\underline{NaHCO_3} + HCl \longrightarrow NaCl + H_2O + CO_2$

(c) $\underline{2C_6H_5OH} + 2Na \longrightarrow 2C_6H_5ONa + H_2$

(d) $\underline{\begin{matrix} CH_2NH_2 \\ | \\ CH_2NH_2 \end{matrix}} + 2HCl \longrightarrow \begin{matrix} CH_2NH_3^+Cl^- \\ | \\ CH_2NH_3^+Cl^- \end{matrix}$

(e) $\underline{CH\equiv CH} + 2Na \longrightarrow NaC\equiv CNa + H_2$

13. Mercuric oxide, HgO, can be analysed by reaction with iodide and then titration with an acid:

$$HgO + 4I^- \longrightarrow HgI_4^{2-} + 2OH^-$$

What is its equivalent weight?

14. Calculate the equivalent weight of the following substances as acids or bases:

(a) $KH(IO_3)_2$ (b) $KHC_2O_4 \cdot H_2C_2O_4$
(c) $Al(OH)_4^-$ (d) $SO_2(OH)_2$
(e) $PO(OH)_3$ (f) $Zn(OH)_4^{2-}$

15. Calculate equivalent weight of KHC_2O_4 (potassium acid oxalate)

(a) when it is neutralised by KOH, and

(b) when it is oxidised by MnO_4^- in basic solution.

16. 0.24 g of a metal on heating in air forms 0.40 g of metal oxide. What is equivalent weight of metal?

17. Equivalent weight of $KMnO_4$ in acidic medium is 31.6. What is corresponding value in basic medium?

18. Determine equivalent weight of underlined species taking its molecular weight as *M*

(a) $\underline{4H_3PO_3} \xrightarrow{\Delta} \underline{3H_3PO_4} + \underline{PH_3}$

(b) $\underline{2H_2O_2} \xrightarrow{\Delta} 2H_2O + \underline{O_2}$

(c) $\underline{MnO_2} \xrightarrow{OH^-} \underline{MnO_4^-} + Mn^{2+}$

(d) $\underline{2H_2O} \xrightarrow{\Delta} \underline{2H_2} + \underline{O_2}$

19. Calculate the grams of one equivalent each of the following for the indicated reaction:

(a) $FeSO_4(\longrightarrow Fe^{3+})$ (b) $H_2S(\longrightarrow S^0)$
(c) $H_2O_2(\longrightarrow O_2)$ (d) $H_2O_2(\longrightarrow H_2O)$

Answers

1. (a) $VO_4^{3-} + Fe^{2+} + 6H^+ \longrightarrow Fe^{3+} + VO^{2+} + 3H_2O$

(b) $2BrO_3^- + 3N_2H_4 \longrightarrow 2Br^- + 3N_2 + 6H_2O$

(c) $3UO_2 + Cr_2O_7^{2-} + 14H^+ \longrightarrow 3UO_2^{2+} + 2Cr^{3+} + 7H_2O$

2. (a) $Mn^{2+} + H_2O_2 + 2OH^- \longrightarrow MnO_2 + 2H_2O$

(b) $2MnO_4^- + 3NO_2^- + H_2O \longrightarrow MnO_2 + 3NO_3^- + 2OH^-$

(c) $Mn^{2+} + 2ClO_3^- \longrightarrow MnO_2 + 2ClO_2$

(d) $MnO_4^- + 3NO_2 + 2OH^- \longrightarrow MnO_2 + 3NO_3^- + H_2O$

3. (a) $8H_2S + 16NO_3^- + 16H^+ \longrightarrow 16NO_2 + S_8 + 16H_2O$

(b) $2MnO_4^- + 5SO_2 + 2H_2O \longrightarrow 2Mn^{2+} + 5SO_4^{2-} + 4H^+$

(c) $2Bi(OH)_3 + 3Sn(OH)_3^- + 3OH^- \longrightarrow 2Bi + 3Sn(OH)_6^{2-}$

(d) $2MnO_4^- + I^- + H_2O \longrightarrow 2MnO_2 + IO_3^- + 2OH^-$

(e) $3H_2O_2 + 2MnO_4^- \longrightarrow 3O_2 + 2MnO_2 + 2OH^- + 2H_2O$

(f) $8Al + 3NO_3^- + 5OH^- + 18H_2O \longrightarrow 8Al(OH)_4^- + 3NH_3$

(g) $5Zn + 2NO_3^- + 12H^+ \longrightarrow 5Zn^{2+} + N_2 + 6H_2O$

4. $4Fe(OH)_2(s) + 2H_2O(l) + O_2(g) \longrightarrow 4Fe(OH)_3(s)$

5. (a) $Cr_2O_7^{2-} + 3H_2O_2 + 8H^+ \longrightarrow 2Cr^{3+} + 3O_2 + 7H_2O$

(b) $Mn^{2+} + H_2O_2 + 2OH^- \longrightarrow MnO_2 + 2H_2O$

(c) $4Sn^{2+} + IO_4^- + 8H^+ \longrightarrow 4Sn^{4+} + I^- + 4H_2O$

6. $6Fe^{2+} + Cr_2O_7^{2-} \xrightarrow{H^+} 2Cr^{3+} + H_2O + 6Fe^{3+}$

$$5Fe^{2+} + MnO_4^- \xrightarrow{H^+} Mn^{2+} + H_2O + 5Fe^{3+}$$

$\frac{\text{mole of } MnO_4^-}{\text{mole of } Cr_2O_7^{2-}} = \frac{5}{6}$, $\frac{x}{\text{mole of } Cr_2O_7^{2-}} = \frac{5}{6}$ (ratio of Fe^{2+} oxidised)

$$\text{mole of } Cr_2O_7^{2-} = 1.2x$$

7. $5C_2O_4^{2-} + 2MnO_4^- \xrightarrow{H^+} 10CO_2 + 2Mn^{2+}$

$$3C_2O_4^{2-} + 2MnO_4^- \xrightarrow{OH^-} 6CO_2 + 2MnO_2$$

$$1\ C_2O_4^{2-} \equiv \frac{2}{5} MnO_4^- \text{ (acidic)}$$

$$1\ C_2O_4^{2-} \equiv \frac{2}{3} MnO_4^- \text{ (basic)}$$

ratio of mole of MnO_4^- used = 3 : 5

8. By given reaction

$$3I_2 \equiv KIO_3$$

$$1\ I_2 = \frac{1}{3} KIO_3 \equiv \frac{1}{3} KI$$

Hence, $\frac{1}{3}$ mol.

9. $Fe(HC_2O_4)_2 \rightleftharpoons Fe^{2+} + 2HC_2O_4^-$

MnO_4^- oxidises Fe^{2+} to Fe^{3+} and $HC_2O_4^-$ to CO_2.

$$MnO_4^- + 5Fe^{2+} \longrightarrow 5Fe^{3+} + Mn^{2+}$$

$$2MnO_4^- + 5HC_2O_4^- \longrightarrow 10\ CO_2 + 2Mn^{2+}$$

$$5Fe^{2+} \equiv 1\ MnO_4^-$$

$$1Fe^{2+} \equiv \frac{1}{5} MnO_4^-$$

$$5HC_2O_4^- \equiv 2MnO_4^-$$

$$2HC_2O_4^- \equiv \frac{4}{5} MnO_4^-$$

Total MnO_4^- required by 1 mol of $Fe(HC_2O_4)_2$

$$= \frac{1}{5} + \frac{4}{5} = 1 \text{ mol}$$

$Fe(HC_2O_4)_2$ has two acidic H^+

$$Fe(HC_2O_4)_2 \rightleftharpoons Fe^{2+} + 2HC_2O_4^-$$
$$\downarrow\uparrow$$
$$2H^+ + 2C_2O_4^{2-}$$
$$2H^+ + Ca(OH)_2 \longrightarrow Ca^{2+} + 2H_2O$$

Thus, $Ca(OH)_2$ required by 1 mol of $Fe(HC_2O_4)_2$ = 1 mol

10. $\underset{-2}{H_2O_2} + \underset{+7}{ClO_4^-} \longrightarrow \underset{0}{O_2}$ + reduced species (Z) O.N. = x

Balancing O.N.

$$(7-x)\,H_2O_2 + 2ClO_4^- \longrightarrow (7-x)\,O_2 + 2Z$$

$(7-x)$ mol H_2O_2 reduces = 2 mol ClO_4^-

$$1 \text{ mol } H_2O_2 \text{ reduces} = \frac{2}{7-x} \text{ mol } ClO_4^-$$

$$\frac{2}{7-x} = 0.5 \text{ mol } ClO_4^- \text{ (given)}$$

$$4 = 7 - x$$
$$x = 3$$

Oxidation number of Cl in Z is = + 3, Z is ClO_2^-.

11. $\underset{+7}{2MnO_4^-} + \underset{+5}{3IO_3^-} \longrightarrow \underset{+7}{3IO_4^-} + \underset{+4}{2MnO_2}$

2 mol $MnO_4^- \equiv$ 3 mol IO_3^-

0.1 mol $MnO_4^- \equiv$ 0.15 mol IO_3^-

12. Equivalent weight of

(a) $Na_2CO_3 = \frac{\text{mol. wt.}}{2} = 53$

(b) $NaHCO_3 = \frac{\text{mol. wt.}}{1} = 84$

(c) $C_6H_5OH = \frac{\text{mol. wt.}}{1} = 94$

(d) $\begin{matrix} CH_2NH_2 \\ | \\ CH_2NH_2 \end{matrix} = \frac{\text{mol. wt.}}{2} = 30$

(e) $CH \equiv CH = \frac{\text{mol. wt.}}{2} = 13$

13. $HgO \equiv 2OH^- \equiv 2H^+$

$\therefore\ 1H^+ \equiv \frac{HgO}{2}$

Equivalent weight of HgO $= \frac{\text{molecular weight}}{2}$

14. If M is the molecular weight and E the equivalent weight, then

(a) $E = M$ (b) $E = \frac{M}{3}$ (c) $E = \frac{M}{4}$

(d) $E = \frac{M}{2}$ (e) $E = \frac{M}{3}$ (f) $E = \frac{M}{4}$

15. (a) $KHC_2O_4 + KOH \longrightarrow K_2C_2O_4 + H_2O$

$KHC_2O_4 \equiv 1\,H^+ \equiv 1\,OH^-$ (H^+ ion is neutralised)

$$\text{Equivalent weight} = \frac{\text{mol. wt.}}{1} = 128$$

(b) $KHC_2O_4 + MnO_4^- \longrightarrow MnO_2 + 2CO_2$

$\uparrow$ 6 (for two C) under KHC_2O_4; $\uparrow$ +7 under MnO_4^-; $\uparrow$ +4 under MnO_2; $\uparrow$ 8 (for two C) under $2CO_2$

($C_2O_4^{2-}$ ion is oxidised)

Equivalent weight of $KHC_2O_4 = \frac{\text{mol wt.}}{2} = 64$

16. 12 g/equivalent.

17. $31.6 \times \frac{5}{3} = 52.66$ g / equivalent.

18. (a) $\underset{+3}{H_3PO_3} \longrightarrow \underset{+5}{H_3PO_4}$, Eq. wt. $H_3PO_3 = H_3PO_4 = \frac{M}{2}$;

$\underset{+3}{H_3PO_3} \longrightarrow \underset{-3}{PH_3}$, $H_3PO_3 = PH_3 = \frac{M}{6}$

In total reaction, equivalent weight of the species undergoing disproportionation

= sum of equivalent weight of half reactions

$$= \frac{M}{2} + \frac{M}{6} = \frac{2}{3}M$$

(b) $2H_2O_2 \longrightarrow 2H_2O + O_2$, $H_2O_2 = \frac{M}{2}$, $O_2 = \frac{M}{4}$

(c) $MnO_2 \longrightarrow MnO_4^- + Mn^{2+}$ $MnO_2 = \frac{M}{2}$, $MnO_4^- = \frac{M}{3}$,

(d) $2H_2O \longrightarrow 2H_2 + O_2$, $O_2 = \frac{M}{4}$, $H_2 = \frac{M}{2}$

19. (a) $\underset{+2}{Fe^{2+}} \longrightarrow \underset{+3}{Fe^{3+}}$, 56 g

(b) $\underset{-2}{H_2S} \longrightarrow \underset{0}{S}$, 17 g

(c) $\underset{-2}{H_2O_2} \longrightarrow \underset{0}{O_2}$, 17 g } O.N. of two O –atoms have been considered

(d) $\underset{-2}{H_2O_2} \longrightarrow \underset{-4}{2H_2O}$, 17 g } O.N. of two O-atoms have been considered

Total Practice Set

(Read, Plan & Solve!)

Problem 1. If 10 g V_2O_5 is dissolved in acid and reduced to V^{2+} by treatment with zinc metal, how many moles of I_2 could be reduced by the resulting V^{2+} solution, as it is oxidised to V^{4+}? (V= 51)

$$V_2O_5 + 10H^+ + 6e^- \longrightarrow 2V^{2+} + 5H_2O$$
$$V^{2+} + I_2 + H_2O \longrightarrow 2I^- + VO^{2+} + 2H^+$$

Plan *On comparison of stoichiometric coefficients, we find*

$$1 \text{ mol } V_2O_5 \equiv 2 \text{ mol } I_2$$

Solution From the above equations,

1 mol of V_2O_5 is reduced to 2 mol of V^{2+}

1 mol of V^{2+} reduces 1 mol of I_2

Thus, 1 mol of V_2O_5 reduces 2 mol of I_2

hence, 10 g of $V_2O_5 = \left(= \frac{10}{182} \text{ mol}\right)$ reduces

$$= 2 \times \frac{10}{182} \text{ mol } I_2$$
$$= \mathbf{0.11 \text{ mol } I_2}$$

Problem 2. Which of the following are examples of disproportionation reactions?

(a) $Ag(NH_3)_2^+ + 2H^+ \longrightarrow Ag^+ + 2NH_4^+$

(b) $Cl_2 + 2OH^- \longrightarrow Cl^- + ClO^- + H_2O$

(c) $CaCO_3 \longrightarrow CaO + CO_2$

(d) $2HgO \longrightarrow 2Hg + O_2$

(e) $Cu_2O + 2H^+ \longrightarrow Cu + Cu^{2+} + H_2O$

(f) $CuS + O_2 \longrightarrow Cu + SO_2$

(g) $2HCuCl_2 \longrightarrow Cu + Cu^{2+} + 4Cl^- + 2H^+$

Plan *A disproportionation reaction is that in which the same species is oxidised as well as reduced.*

$$Cu^+ \xrightarrow{\text{Oxidation}} Cu^{2+}$$
$$Cu^+ \xrightarrow{\text{Reduction}} Cu$$

Solution (a) $\underset{+1\ \ -3}{Ag(NH_3)_2^+} + 2H^+ \longrightarrow \underset{+1}{Ag^+} + \underset{-3}{NH_4^+}$

no change, neither oxidation nor reduction

(b) $\underset{0}{Cl_2} + 2OH^- \longrightarrow \underset{-1}{Cl^-} + \underset{+1}{ClO^-} + H_2O$

Cl_2 is reduced to Cl^- as well as oxidised to ClO^- hence, $\mathbf{Cl_2}$ **undergoes** disproportionation.

(c) $\underset{+2\ \ +4}{CaCO_3} \longrightarrow \underset{+2}{CaO} + \underset{+4}{CO_2}$

no change

(d) $\underset{+2\ \ -2}{2HgO} \longrightarrow \underset{0}{2Hg} + \underset{0}{O_2}$

There is oxidation of O and reduction of Hg. It is a redox reaction but not disproportionation.

(e) $\underset{+1}{Cu_2O} + 2H^+ \longrightarrow \underset{0}{Cu} + \underset{+2}{Cu^{2+}} + H_2O$

There is oxidation of Cu^+ to Cu^{2+} as well as reduction to Cu, hence, Cu^+ **disproportionates.**

(f) $\underset{+2\ \ -2}{CuS} + \underset{0}{O_2} \longrightarrow \underset{0}{Cu} + \underset{+4\ \ -2}{SO_2}$

This is also a redox but not disproportionation reaction.

(g) $\underset{+1\ \ -1}{2HCuCl_2} \longrightarrow \underset{0}{Cu} + \underset{+2}{Cu^{2+}} + \underset{-1}{4Cl^-} + 2H^+$

Cu^+ disproportionates to Cu^{2+} and Cu.

Problem 3. MnO_4^- can oxidise NO_2^- to NO_3^- in basic medium. How many moles of NO_2^- are oxidised by 1 mole of MnO_4^-?

Plan *On comparison of change of oxidation number,*

$$2 \text{ mol of } MnO_4^- \equiv 3 \text{ mol of } NO_2^-$$

Solution NO_2^- is oxidised to NO_3^- by MnO_4^- (in basic medium) which is reduced to MnO_2

$$\underset{+7}{MnO_4^-} + \underset{+3}{NO_2^-} \longrightarrow \underset{+5}{NO_3^-} + \underset{+4}{MnO_2}$$

Oxidation ($NO_2^- \to NO_3^-$); Reduction ($MnO_4^- \to MnO_2$)

Thus, $MnO_4^- \longrightarrow MnO_2$ oxidation number decreases by 3-units

$NO_2^- \longrightarrow NO_3^-$ oxidation number increases by 2-units

Thus,
$$2MnO_4^- \equiv 3NO_2^-$$
$$1\,MnO_4^- \equiv \frac{3}{2}\,NO_2^-$$
$$= \mathbf{1.5\ mol\ NO_2^-}$$

Problem 4. In certain laboratory procedures it is necessary to remove NH_4^+ ions from solution by heating with HNO_3. Write balanced ionic reaction if NH_4^+ ion is oxidised to N_2 by HNO_3, which in turn is reduced to NO_2. Calculate moles of HNO_3 required by 1 mole of $(NH_4)_2SO_4$.

$$6NO_3^- + 2NH_4^+ + 4H^+ \longrightarrow N_2 + 6NO_2 + 6H_2O$$

Plan *From stoichiometric coefficients :*

$$6\text{ mol } HNO_3(NO_3^-) \equiv 2\text{ mol } NH_4^+$$
$$\equiv 1\text{ mol } (NH_4)_2SO_4$$

Solution In the product, we find there are two types of nitrogen (as N_2 and as NO_2). To know which nitrogen is from which reactant, isotopes of nitrogen are used. If N of NH_4^+ is labelled by its certain isotope, we find that nitrogen certainly appears in N_2

$$(NH_4)_2SO_4 \equiv 2NH_4^+ \equiv 6NO_3^-$$

hence, **one mole of $(NH_4)_2SO_4$ is oxidised by six moles of HNO_3.**

Problem 5. Calculate number of moles of (a) MnO_4^- and (b) $Cr_2O_7^{2-}$ to oxidise 1 mole of FeC_2O_4 (ferrous oxalate) in acidic medium.

Plan Fe^{2+} *and* $C_2O_4^{2-}$ *(in* FeC_2O_4*) both are oxidised by* MnO_4^-/H^+ *as well as by* $Cr_2O_7^{2-}/H^+$*. Thus, on comparison of stoichiometric coefficients, number of moles of* $Cr_2O_7^{2-}$ *and that of* MnO_4^- *required by 1 mole of* FeC_2O_4 *can be calculated.*

Solution Fe^{2+} is oxidised to Fe^{3+} and $C_2O_4^{2-}$ is oxidised to CO_2 both by MnO_4^- and $Cr_2O_7^{2-}$ in acidic medium.

(a) $5Fe^{2+} + MnO_4^- + 8H^+ \longrightarrow 5Fe^{3+} + Mn^{2+} + 4H_2O$

$5C_2O_4^{2-} + 2MnO_4^- + 16H^+ \longrightarrow 10CO_2 + 8H_2O + Mn^{2+}$

Thus, $\quad 5FeC_2O_4 \equiv 3MnO_4^-$

$\therefore \quad 1\text{ mol } FeC_2O_4 \equiv \mathbf{0.6\ mol\ MnO_4^-}$

(b) $6Fe^{2+} + Cr_2O_7^{2-} + 14H^+ \longrightarrow 6Fe^{3+} + 2Cr^{3+} + 7H_2O$

$6C_2O_4^{2-} + 2Cr_2O_7^{2-} + 28H^+ \longrightarrow 12CO_2 + 4Cr^{3+} + 14H_2O$

Thus, $\quad 6FeC_2O_4 \equiv 3Cr_2O_7^{2-}$

$\therefore \quad 1\text{ mol } FeC_2O_4 \equiv \mathbf{0.5\ mol\ of\ Cr_2O_7^{2-}}$

Problem 6. The following is a completely balanced equation:

$$3Sn + 12HCl + 4HNO_3 \longrightarrow 3SnCl_4 + 4NO + 8H_2O$$

In this reaction, what fraction of the formula weights are the equivalent weights of Sn and HNO_3, respectively?

Plan *Equivalent weight* $= \dfrac{\textit{formula weight}}{\textit{change in oxidation number}}$

or $\quad = \dfrac{\textit{formula weight}}{\textit{number of electrons exchanged}}$

Solution The half-reactions are

oxidation : $\quad Sn + 4Cl^- \longrightarrow SnCl_4 + 4e^-$

reduction : $\quad NO_3^- + 4H^+ + 3e^- \longrightarrow NO + 2H_2O$

Equivalent weight of $Sn = \dfrac{\text{formula weight}}{4}$

Thus, in the oxidation, there are 4 equivalents Sn per mol Sn.

Equivalent weight of $HNO_3 = \dfrac{\text{formula weight}}{3}$

Thus, in the reduction, there are 3 equivalents HNO_3 per mol HNO_3.

Problem 7. Aqueous solution of sodium hypochlorite (NaOCl) is a household bleach and is a strong oxidising agent that reacts with chromite ion $[Cr(OH)_4^-]$ in basic solution to yield chromate (CrO_4^{2-}) and chloride ion. Determine

(a) number of equivalents of each reactant

(b) number of moles of NaOCl required by 1 mole of $[Cr(OH)_4^-]$.

Plan *(a) Number of equivalents = number of moles × change in oxidation number*

(b) If change in oxidation number of each reactant is known, stoichiometric coefficients are also known. Hence, number of moles of NaOCl *can be determined.*

Solution

$$ClO^-(aq) + Cr(OH)_4^-(aq) \longrightarrow CrO_4^{2-}(aq) + Cl^-(aq)$$

Oxidation numbers: ClO^-: +1; $Cr(OH)_4^-$: +3; CrO_4^{2-}: +6; Cl^-: −1. Cr: 3 units; Cl: 2 units.

Thus,

$$3ClO^-(aq) + 2Cr(OH)_4^-(aq) \longrightarrow 2CrO_4^{2-}(aq) + 3Cl^-(aq)$$

(a) Equivalents of $ClO^-(aq)$ used $= 3 \times 2 = \mathbf{6}$

Equivalents of $Cr(OH)_4^-(aq)$ used $= 2 \times 3 = \mathbf{6}$

(b) $2 \text{ mol } [Cr(OH)_4^-] = 3 \text{ mol } ClO^-$

$1 \text{ mol } [Cr(OH)_4^-] = \frac{3}{2} = \mathbf{1.5 \text{ mol } ClO^-}$

Note : Remember, in a redox reaction, number of equivalents of each reactant is always same.

Problem 8. Alcohol levels in the blood can be determined by a redox titration with dichromate solution in acidic medium according to the balanced equation.

$$C_2H_5OH(aq) + 2Cr_2O_7^{2-}(aq) + 16H^+(aq) \longrightarrow 2CO_2(g) + 4Cr^{3+}(aq) + 11H_2O(l)$$

What is the blood alcohol level in mass per cent if 8.76 mL of 0.05 M $Cr_2O_7^{2-}$ solution is required for titration of a 10.00 g sample of blood?

Plan *Based on stoichiometric coefficients, number of millimol (= molarity × mL) of $Cr_2O_7^{2-}$ are to be related to millimol of alcohol hence, mass per cent of alcohol level in blood can be determined.*

Solution 2 millimol of $Cr_2O_7^{2-}$ solution

$= 1$ millimol of C_2H_5OH

8.76×0.05 millimol of $Cr_2O_7^{2-}$ solution

$= \frac{1}{2} \times 8.76 \times 0.05$ millimol of C_2H_5OH

$= 0.219$ millimol

$= 0.219 \times 10^{-3}$ mol

$= 0.219 \times 10^{-3} \times 46$ g C_2H_5OH

$= 0.01$ g C_2H_5OH

Thus, 10 g blood sample has $= 0.01$ g C_2H_5OH

hence, alcohol level (mass per cent) $= \frac{0.01}{10} \times 100$

$= \mathbf{0.10\%}$

Master Exercises

Exercise 1
(Stage 1 : Learning)

Short Answer Type Questions

1. When Cl^- is converted into Cl_2, what is change in oxidation number?
2. What is oxidation number of oxygen in superoxide ion O_2^- ?
3. Equivalent weight of MnO_4^- is equal to molecular weight, when it changes into new species X. What is X? It is by oxidation or reduction.
4. What is equivalent weight of $NaHC_2O_4$ when
   (a) it behaves as an acid?
   (b) it behaves as a reducing agent?
5. Oxidation number of sulphur in H_2SO_5 is +6. Explain by structure.
6. Oxidation number of sulphur in $H_2S_2O_8$ is +6. Explain by structure.
7. Derive equivalent weight of
   (i) H_3PO_2 (ii) H_3PO_3
   (iii) H_3PO_4
   at every stage of neutralisation by NaOH.
8. When CO_2 is passed into aqueous K_2CrO_4 solution (yellow), it changes to orange coloured solution. Explain by reaction.
9. When SO_2 gas is passed into aqueous $K_2Cr_2O_7$ solution (orange), it changes to green coloured solution. Explain by reaction.
10. Fe^{2+} is oxidised to Fe^{3+} by MnO_4^- or $Cr_2O_7^{2-}$ in acidic medium. Explain by reaction with change in colour.

Analytical Questions

Oxidation Number

1. The ion A^{n+} is oxidised to AO_3^- by MnO_4^-, changing to Mn^{2+} in acidic solution. Given that 2.68×10^{-3} mol of A^{n+} requires 1.61×10^{-3} mol of MnO_4^-. What is the value of n?
2. Arrange the following sulphur containing anions in order of increasing oxidation state of sulphur:
   $SO_3^{2-}; S_2O_3^{2-}; S_2O_8^{2-}; HSO_4^-; HS^-; S_4O_6^{2-}$
3. Nitrogen forms five different compounds with oxygen. Write appropriate formulae for these compounds if the oxidation state of N in them is +1, +2, +3, +4 and +5.
4. A compound is made of Ba (oxidation number = + 2) Xe(oxidation number = 8) and oxygen. Xe and O are present in the atomic ratio of 1:6, what is the compound?
5. $KMnO_4$ oxidises X^{n+} ion into XO_3^-, itself changing to Mn^{2+} in acid solution. 2.68×10^{-3} mol of X^{n+} requires 1.61×10^{-3} mol of MnO_4^-. What is atomic weight of X, if the weight of 1 g equivalent of XCl_n is 56?

Redox Reactions

6. Balance the following redox reactions. Also underline disproportionation reactions:
   (a) $MnO_4^- + I^- + H^+ \longrightarrow Mn^{2+} + I_2 + H_2O$
   (b) $Cl_2 + I^- + H_2O \longrightarrow IO_3^- + Cl^- + H^+$
   (c) $BrO_3^- + N_2H_4 \longrightarrow Br^- + N_2 + H_2O$
   (d) $VO_4^{3-} + Fe^{2+} + H^+ \longrightarrow VO^{2+} + Fe^{3+} + H_2O$
   (e) $UO^{2+} + Cr_2O_7^{2-} + H^+ \longrightarrow UO_2^{2+} + Cr^{3+} + H_2O$
   (f) $Pb(NO_3)_2 \longrightarrow PbO + NO_2 + O_2$
   (g) $IO_3^- + HSO_3^- + H^+ \longrightarrow I_2 + SO_4^{2-} + H_2O$
   (h) $S_2O_4^{2-} + H_2O \longrightarrow S_2O_3^{2-} + HSO_3^-$
   (i) $Cl_2 + OH^- \longrightarrow ClO^- + Cl^- + H_2O$
   (j) $C_6H_5CHO + OH^- \longrightarrow C_6H_5COO^- + C_6H_5CH_2OH$
   (k) $S_2O_3^{2-} + H^+ \longrightarrow H_2O + SO_2 + S$
   (l) $Cl^- + NO_3^- + H^+ \longrightarrow NO_2 + Cl_2 + H_2O$
   (m) $ClO^- + H_2O \longrightarrow ClO_2^- + Cl_2 + OH^-$
   (n) $Cl_2 + S^{2-} \longrightarrow S + Cl^-$

(o) $O_3 + Pb^{2+} + H_2O \longrightarrow PbO_2 + H^+ + O_2$

(p) HO_2^- (hydrogen peroxide ion) $+ Br^- + H_2O \longrightarrow OH^- + Br_2$

(q) $Au + CN^- + O_2 + H_2O \longrightarrow [Au(CN)_2]^- + OH^-$

7. Identify the oxidising agent and the reducing agent in the following redox reactions:

(a) $5ClO_3^- + 3As_2S_3 + 9H_2O \longrightarrow 5Cl^- + 6H^+ + 6H_2AsO_4^- + 9S\downarrow$

(b) $7CN^- + 2OH^- + 2Cu(NH_3)_4^{2+} \longrightarrow 2Cu(CN)_3^{2-} + 8NH_3 + CNO^- + H_2O$

8. H_2O_2 can be used as oxidising agent or reducing agent depending on the medium. Explain, why this behaviour is possible? Give one example of each.

9. Balance the following redox reactions:

(a) $Zn + NO_3 + OH \longrightarrow NH_3 + ZnO_2^{2-}$

(b) $HgS + Cl^- + NO_3^- + H^+ \longrightarrow HgCl_4^{2-} + NO + S$

(c) $MnO_4^- + H^+ + H_2O_2 \longrightarrow Mn^{2+} + O_2$

(d) $P_2H_4 \longrightarrow PH_3 + P_4H_2$

(e) $ClO_2 + O_2^{2-} \longrightarrow ClO_2^- + O_2$

(f) $Cl_2 + IO_3^- + OH^- \longrightarrow IO_4^- + Cl^-$

10. Calculate the number of moles of MnO_4^- that will be needed to react with one mole of SO_3^{2-} (sulphite ion) in acidic solution.

11. H_2O_2 acts both as an oxidising agent and as a reducing agent in alkaline solution towards first row transition metal ions. Illustrate both these properties of H_2O_2 using chemical equations.

12. Give an example of oxidation of one halide by another halogen. Explain the feasibility of the reaction.

Equivalent Weight

13. Determine the equivalent weight of the underlined by taking its molecular weight as M:

(a) $2KMnO_4 + 5H_2SO_4 + 10FeSO_4 \longrightarrow 5Fe_2(SO_4)_3 + K_2SO_4 + 2MnSO_4 + 5H_2O$

(b) $K_2Cr_2O_7 + 7H_2SO_4 + 6FeSO_4 \longrightarrow K_2SO_4 + Cr_2(SO_4)_3 + 3Fe_2(SO_4)_3 + 7H_2O$

(c) $\underline{CH_3CHO} + 2[Ag(NH_3)_2]^+ + 3OH^- \longrightarrow CH_3COO^- + 2Ag + 4NH_3 + 2H_2O$

(d) $\underline{2KClO_3} \longrightarrow 2KCl + 3O_2$

(e) $3Cl_2 + 6NaOH \longrightarrow 5NaCl + NaClO_3 + 3H_2O$

(f) $\underline{NH_4NO_3} \longrightarrow N_2O + 3H_2O$

(g) $\underline{PCl_5} \longrightarrow PCl_3 + Cl_2$

(h) $\underline{2CuSO_4} + 4KI \longrightarrow Cu_2I_2 + 2K_2SO_4 + I_2$

(i) $\underline{S_2O_3^{2-}} + 5H_2O + 4Cl_2 \longrightarrow 2SO_4^{2-} + 8Cl^- + 10H^+$

(j) $\underline{2HS^-} + \underline{4HSO_3^-} \longrightarrow 3S_2O_3^{2-} + 3H_2O$

(k) $4Fe^{2+} + \underline{2NH_2OH} \longrightarrow 4Fe^{3+} + N_2O + 4H^+$

(l) $\underline{IO_3^-} + \underline{3HSO_3^-} \longrightarrow I^- + 3H^+ + 3SO_4^{2-}$

(m) $\underline{3MnO_2} + 6KOH + KClO_3 \longrightarrow 3K_2MnO_4 + KCl + 3H_2O$

(n) $\underline{H_2C_2O_4} + MnO_2 + 2H^+ \longrightarrow Mn^{2+} + 2H_2O + 2CO_2$

14. Photosynthesis of a carbohydrate in plants is given by chemical equation

$$6CO_2 + 12H_2O \longrightarrow C_6H_{12}O_6 + 6H_2O + 6O_2$$

Calculate:

(a) equivalents of O_2 formed per mol of CO_2 consumed.

(b) equivalents of $C_6H_{12}O_6$ per mol of H_2O used.

(c) equivalents of CO_2 used per mol of H_2O used.

(d) If you have net result as:

$$6CO_2 + 6H_2O \longrightarrow C_6H_{12}O_6 + 6O_2$$

It is correct or incorrect? Explain.

Exercise 2
(Stage 2 : High Skill Problem Solving)

Only One Option Correct

1. A sulphur-containing species that can't be a reducing agent is

(a) SO_2 (b) SO_3^{2-}
(c) H_2SO_4 (d) S^{2-}

2. Which of the following changes requires a reducing agent?

(a) $CrO_4^{2-} \longrightarrow Cr_2O_7^{2-}$
(b) $BrO_3^- \longrightarrow BrO^-$
(c) $H_3AsO_3 \longrightarrow HAsO_4^{2-}$
(d) $Al(OH)_3 \longrightarrow Al(OH)_4^-$

3. In balancing the half-reaction

$$CN^- \longrightarrow CNO^- \text{ (skeletal)}$$

the number of electrons that must be added is

(a) 0
(b) 1 on the right
(c) 1 on the left
(d) 2 on the right

4. For the reaction between MnO_4^- and $C_2O_4^{2-}$ in basic solution, the unbalanced equation is

$$MnO_4^- + C_2O_4^{2-} \longrightarrow MnO_2(s) + CO_3^{2-}$$

In a balanced equation, the number of OH^- ions is
(a) 0 (b) 4 on the right
(c) 4 on the left (d) 2 on the left

5. Which of the following is disproportionation reaction?
(a) $CaCO_3 + 2H^+ \longrightarrow Ca^{2+} + H_2O + CO_2$
(b) $2CrO_4^{2-} + 2H^+ \longrightarrow Cr_2O_7^{2-} + H_2O$
(c) $Cr_2O_7^{2-} + 2OH^- \longrightarrow 2CrO_4^{2-} + H_2O$
(d) $Cu_2O + 2H^+ \longrightarrow Cu + Cu^{2+} + H_2O$

6. When MnO_4^- and I^- react in a strongly basic solution, the products will most likely be
(a) Mn and I_2 (b) MnO_4^{2-}, IO_3^-
(c) MnO_2, O_2, IO^- (d) Mn^{2+}, I_2

7. Of the following acids, the one that is strongest is
(a) $HBrO_4$ (b) HOCl
(c) HNO_2 (d) H_3PO_3

8. When PbS(*s*) is treated with warm dilute HNO_3, the products most likely will be
(a) Pb^{2+}, S, NO_2 (b) Pb^{2+}, S, NO
(c) PbO, S, NO (d) PbO_2, SO_4^{2-}, N_2

9. In an experiment 20 g of vanadium (V) oxide (molar mass = 182) was reduced by excess of zinc dust in acidic solution to vanadium (II) ions. The required number of moles of iodine to reoxidise vanadium (II) to VO^{2+} is
(a) 0.22 (b) 0.11
(c) 0.30 (d) 0.23

10. For a titration of 100 cm^3 of 0.1 M Sn^{2+} to Sn^{4+}, 50 cm^3 of 0.40 M Ce^{4+} solution was required. The oxidation state of cerium in the reduction product is
(a) +1 (b) +2
(c) +3 (d) 0

11. In the following reaction (unbalanced), equivalent weight of As_2S_3 is related to molecular weight *M* by

$$As_2S_3 + H^+ + NO_3^- \longrightarrow NO + H_2O + AsO_4^{3-} + SO_4^{2-}$$

(a) $\frac{M}{2}$ (b) $\frac{M}{4}$
(c) $\frac{M}{28}$ (d) $\frac{M}{24}$

12. 1 mole of FeC_2O_4 is oxidised by *x* mole of $Cr_2O_7^{2-}$ in acidic medium, *x* is
(a) 3 (b) 1.5
(c) 0.5 (d) 1.0

13. 1 mole of MnO_4^- will oxidise *x* mole of ferric oxalate in acidic medium, *x* is
(a) $\frac{5}{6}$ (b) $\frac{6}{5}$
(c) 5 (d) 6

14. In the reaction, 4 moles of electrons are transferred to 1 mole of HNO_3. The possible product obtained due to reduction
(a) 0.4 mol of N_2
(b) 0.5 mol of N_2O
(c) 1 mol of NO_2
(d) 1 mol of NH_3

15. 1 mole of N_2H_4 loses 10 mol of electrons to form a new compound *Y*. Assuming that all the nitrogen appear in the new compound, what is the oxidation state of nitrogen in *Y*? (no change in the oxidation state of H)
(a) −1 (b) −3
(c) +3 (d) +5

16. Oxidation number of P in $Ba(H_2PO_2)_2$ is
(a) +3 (b) +2
(c) +1 (d) −1

17. Which combination appears odd w.r.t. oxidation number per atom of the underlined?
(a) $H_2\underline{S}O_5, H_2\underline{S}_2O_8, K_2\underline{Cr}_2O_7$
(b) $\underline{Cr}O_5, \underline{Cr}O_4^{2-}, \underline{S}O_4^{2-}$
(c) Both (a) and (b)
(d) None of the above

18. In $[Cr(O_2)(NH_3)_4(H_2O)]Cl_2$ oxidation number of Cr is +3, then O_2 will be in the form
(a) dioxo (b) peroxo
(c) superoxo (d) oxo

19. Strongest conjugate base is
(a) ClO^- (b) ClO_2^-
(c) ClO_3^- (d) ClO_4^-

20. Which is not a disproportionation reaction?
(a) $2\,C_6H_5CHO \xrightarrow{Al(OEt)_3} C_6H_5COOCH_2C_6H_5$
(b) $\begin{matrix} CHO \\ | \\ COOH \end{matrix} + OH^- \longrightarrow \begin{matrix} CH_2OH \\ | \\ COO^- \end{matrix} + \begin{matrix} COO^- \\ | \\ COO^- \end{matrix}$
(c) $NaH + H_2O \longrightarrow NaOH + H_2$
(d) All of the above

21. Oxidation number of Cr in the following compounds are *x*, *y*, *z* and *w*.

$\underset{x}{K[CrO(O_2)(OH)]}$, $\underset{y}{K_3[Cr(O_2)_4]}$ $\underset{z}{(NH_3)_3[Cr(O_2)_2]}$, $\underset{w}{CrO_2Cl_2}$

These values are
(a) 6, 13, 5, 6 (b) 4, 4, 4, 6
(c) 4, 5, 4, 6 (d) 4, 5, 6, 6

22. If a small amount of $KMnO_4$ is added to conc. H_2SO_4, a green coloured solution with oxidation number of Mn as +7 is formed. Probable compound is

(a) MnO_3HSO_4 (b) K_2MnO_4
(c) MnO_2HSO_4 (d) MnO_2

23. Equivalent weight of Mn^{3+} in the following reaction is (Mn = 55)

$$Mn^{3+} \longrightarrow Mn^{2+} + MnO_2$$

(a) 27.5 (b) 55
(c) 110 (d) 165

24. Which of the following is the example of a disproportionation reaction?

(a) $CaCO_3 \longrightarrow CaO + CO_2$
(b) $2HCuCl_2 \longrightarrow Cu + Cu^{2+} + 4Cl^- + 2H^+$
(c) $PCl_5 \longrightarrow PCl_3 + Cl_2$
(d) $2H_2S + SO_2 \longrightarrow 3S + 2H_2O$

25. Out of the following redox reactions

I : $NH_4NO_3 \xrightarrow{\Delta} N_2O + 2H_2O$

II : $NH_4NO_2 \xrightarrow{\Delta} N_2 + 2H_2O$

III : $PCl_5 \xrightarrow{\Delta} PCl_3 + Cl_2$

disproportionation is not shown in

(a) I, II (b) II, III
(c) I, III (d) I, II and III

26. Bleaching powder ($CaOCl_2$) has two types of chlorine atoms with oxidation number of +1 and –1. It can be represented as:

(a) $Ca^{2+}(ClO_2^-)Cl^-$
(b) $Ca^{2+}(ClO_3^-)Cl^-$
(c) $Ca^{2+}(ClO^+)Cl^-$
(d) $Ca^{2+}(ClO^-)Cl^-$

27. In the following redox reaction

$$Zn(s) + NO_3^-(aq) + H^+(aq) \longrightarrow Zn^{2+}(aq) + NH_4^+(aq)$$

$Zn(s)$ and $NO_3^-(aq)$ respectively are:

(a) oxidant, reductant
(b) reductant, oxidant
(c) both oxidant
(d) both reductant

28. Out of the following select species with different oxidation states of the underlined atoms

I : $CaO\underline{Cl_2}$; II : $H_2\underline{S_2}O_8$; III. $Na_2\underline{S_4}O_6$

IV : $\underline{Cr}O_5$; V : $\underline{N_2}H_4O_3$

(a) All except I
(b) All except II
(c) All except III
(d) All except IV

29. Photographic paper is developed with alkaline hydroquinone

$$2AgBr(s) + 2OH^-(aq) + \text{(hydroquinone: OH, OH)} \longrightarrow 2Ag(s) + 2H_2O(l) + 2Br^-(aq) + \text{(quinone: O, O)}$$

Select correct statement.

(a) Hydroquinone is the oxidant
(b) Ag^+ is the oxidant
(c) Br^- is the oxidant
(d) Ag^+ is the reductant

30. In the following redox reaction

$$Cu(OH)_2(s) + N_2H_4(aq) \longrightarrow Cu(s) + N_2(g)$$

number of moles of $Cu(OH)_2$ reduced by one mole of N_2H_4 is

(a) 1 (b) 2
(c) 3 (d) 4

31. In the following half-reaction in basic medium

$$Bi^{3+}(aq) \longrightarrow BiO_3^-(aq)$$

one mole of Bi^{3+} is equivalent to

(a) 1 (b) 2
(c) 3 (d) 4

32. The complex $[Fe(H_2O)_5NO]^{2+}$ is formed in the ring-test for nitrate when freshly prepared $FeSO_4$ solution is added to aqueous solution of NO_3^- followed by addition of conc. H_2SO_4. This complex is formed by charge transfer in which

(a) Fe^{2+} changes to Fe^{3+} and NO^+ changes to NO
(b) Fe^{2+} changes to Fe^{3+} and NO changes to NO^+
(c) Fe^{2+} changes to Fe^+ and NO changes to NO^+
(d) no charge transfer takes place

33. When KCN is added to $CuSO_4$ solution there is formation of the stable water soluble complex. This complex is

(a) $K_4[Cu(CN)_6]$ (b) $K_3[Cu(CN)_4]$
(c) $K_2[Cu(CN)_4]$ (d) $K[Cu(CN)_3]$

34. In cyanide method, silver metal is obtained as

$$2K[Ag(CN)_2] + Zn \longrightarrow K_2[Zn(CN)_4] + 2Ag$$

In this

(a) Ag has been oxidised and Zn has been reduced
(b) Ag^+ has been reduced and Zn has been oxidised
(c) both the metals have been oxidised
(d) both the metals have been reduced.

35. In the following unbalanced redox reaction

$$H_2S(g) + SO_2(g) \longrightarrow S(s) + H_2O$$

total number of equivalent of SO_2 is
(a) 4 (b) 3
(c) 2 (d) 1

36. In the above reaction of question (35), one equivalent of $H_2S(g)$ will reduce
(a) 1 mol of SO_2 (b) 0.5 mol of SO_2
(c) 0.25 mol of SO_2 (d) 2 mol of SO_2

37. Which has maximum number of equivalent per mol of the oxidant?
(a) $Zn(s) + VO^{2+}(aq) \longrightarrow Zn^{2+}(aq) + V^{3+}(aq)$
(b) $Ag(s) + NO_3^-(aq) \longrightarrow Ag^+(aq) + NO_2(g)$
(c) $Mg(s) + VO_4^{3-}(aq) \longrightarrow Mg^{2+}(aq) + V^{2+}(aq)$
(d) $I^-(aq) + IO_3^-(aq) \longrightarrow I_3^-(aq)$

38. Which has maximum oxidation number of the underlined atom in the following?
(a) $\underline{Mn}O_4^{2-}$ (b) $\underline{Cr}O_5$
(c) $\underline{Cr}O_2Cl_2$ (d) Equal

39. Which has least number of equivalent per mol of the reactant?
(a) MnO_4^- changes to MnO_2
(b) MnO_4^- changes to Mn^{2+}
(c) MnO_4^- changes to MnO_4^{2-}
(d) MnO_4^- changes to Mn_2O_3

40. Oxidation number of Cr is +5 in
(a) K_3CrO_8 (b) $(NH_3)_3CrO_4$
(c) K_2CrO_4 (d) $[Cr(NH_3)_5(H_2O)]Cl_3$

41. Sulphur atoms are of two environments in $S_4O_6^{2-}$, difference in oxidation states of two types of sulphur is
(a) 5 (b) 6
(c) 7 (d) 10

42. When Br_2 is passed into hot aqueous solution of Na_2CO_3,
(a) Br^- is formed by reduction
(b) BrO_3^- is formed by oxidation
(c) CO_2 is evolved
(d) all of the above are correct

43. Consider following unbalanced reactions
I : $Zn + \text{dil. } H_2SO_4 \longrightarrow ZnSO_4 + H_2$
II : $Zn + \text{conc. } H_2SO_4 \longrightarrow ZnSO_4 + SO_2 + H_2O$
III : $Zn + HNO_3 \longrightarrow Zn(NO_3)_2 + NH_4NO_3 + H_2O$
Oxidation number of hydrogen changes in
(a) I, II, III (b) I, II
(c) II, III (d) I

44. NH_3 is oxidised to NO by O_2 (air) in basic medium. Number of equivalent of NH_3 oxidised by 1 mole of O_2 is
(a) 4 (b) 5
(c) 6 (d) 7

45. In the following redox reactions, NH_3 appears either in reactant or product. In which case equivalent weight of NH_3 is maximum?
(a) $N_2 + 3H_2 \longrightarrow 2NH_3$
(b) $4NH_3 + 5O_2 \longrightarrow 4NO + 6H_2O$
(c) $2NH_3 + 2Na \longrightarrow 4NaNH_2 + H_2$
(d) Equal in the above all cases

46. In the following unbalanced redox reaction,
$$Cu_3P + Cr_2O_7^{2-} \longrightarrow Cu^{2+} + H_3PO_4 + Cr^{3+}$$
Equivalent weight of H_3PO_4 is
(a) $\frac{M}{3}$ (b) $\frac{M}{6}$
(c) $\frac{M}{7}$ (d) $\frac{M}{8}$

47. In the following reaction
$$Zn + 2OH^- + 2H_2O \longrightarrow Zn(OH)_4^{2-} + H_2$$
Species which has been reduced is
(a) Zn (b) OH^-
(c) H_2O (d) none of these

48. $H_2S_2O_8$ and H_2SO_5 both have +6 oxidation state of sulphur. It is due to
(a) presence of peroxy group
(b) presence of superoxo group
(c) presence of neutral O_2
(d) presence of ozone

49. H_2O_2 oxidises MnO_2 to MnO_4^- in basic medium. H_2O_2 and MnO_2 react in the molar ratio of
(a) 1:1 (b) 2:1
(c) 2:3 (d) 3:2

50. Equivalent weight of H_3PO_2 in a reaction is found to be half of its molecular weight. It can be due to its
(a) reaction of its two H^+ ions
(b) oxidation to H_3PO_3
(c) both (a) and (b)
(d) None of the above

51. Oxidation number of Cl atoms in $CaOCl_2$ are

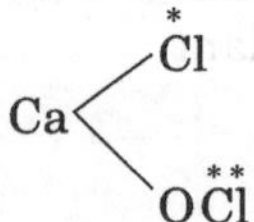

(a) zero on each
(b) −1 on Cl* and +1 on Cl**
(c) +1 on Cl* and −1 on Cl**
(d) +1 on each

52. Prussian blue has two types of iron with oxidation number as shown
$$\overset{III}{Fe}[\overset{II}{Fe}(CN)_6]$$
What is the net charge on Prussian blue?
(a) −1 (b) +1
(c) 0 (d) −2

53. Consider following redox reactions

I : $(NH_4)_2Cr_2O_7 \longrightarrow N_2 + Cr_2O_3 + H_2O$

II : $NH_4NO_3 \longrightarrow N_2O + 2H_2O$

III : $2KClO_3 \longrightarrow 2KCl + 3O_2$

Select intramolecular redox reactions

(a) I, II (b) II, III
(c) I, III (d) I, II, III

54. 0.05 mole of $Ca(OH)_2$ can neutralise H_2SO_4. This H_2SO_4 can be neutralised by

(a) 0.05 mol of NaOH
(b) 0.10 mol of NaOH
(c) 0.05 mol of $POH(OH)_2$
(d) None of the above is correct

55. $NaHC_2O_4$ is neutralised by NaOH and can also be oxidised by $KMnO_4$ (in acidic medium). Equivalent weight is related to molecular weight (M) of $NaH_2C_2O_4$ in these reactions as

(a) M, M (b) $\frac{M}{2}, \frac{M}{2}$
(c) $\frac{M}{2}, M$ (d) $M, \frac{M}{2}$

56. If $OX_1 + ae^- \rightleftharpoons Red_1, \quad E_1°$

$OX_2 + be^- \rightleftharpoons Red_2, \quad E_2°$

Then potential at the equivalence point is

(a) $E = \frac{aE_1° + bE_2°}{(a+b)}$ (b) $E = \frac{aE_1° - bE_2°}{(a+b)}$
(c) $E = \frac{bE_2° - aE_1°}{(a+b)}$ (d) $E = \frac{aE_1° - bE_2°}{(a-b)}$

57. In the following redox reaction,

$5Fe^{2+} + MnO_4^- + 8H^+ \rightleftharpoons Mn^{2+} + 5Fe^{3+} + 4H_2O$

Given : $Fe^{3+} + e^- \longrightarrow Fe^{2+}, \quad E_1°$

$MnO_4^- + 8H^+ + 5e^- \longrightarrow Mn^{2+} + 4H_2O, \quad E_2°$

Potential at the equivalence point is

(a) $E = (E_1° + E_2°) - 0.08\,pH$
(b) $E = \frac{E_1° + 5E_2°}{6} - 0.08\,pH$
(c) $E = (E_1° - E_2°) - 0.08\,pH$
(d) $E = \frac{E_1° - 5E_2°}{6} + 0.08\,pH$

58. Which reaction is suitably matched?

(a) $\begin{matrix} CHO \\ | \\ CHO \end{matrix} + OH^- \longrightarrow \begin{matrix} CH_2OH \\ | \\ COO^- \end{matrix}$
(crossed Cannizzaro reaction)

(b) $3H_3PO_2 \longrightarrow 2H_3PO_3 + PH_3$
(disproportionation reaction)

(c) $\begin{matrix} CHO \\ | \\ COOH \end{matrix} + OH^- \longrightarrow \begin{matrix} COO^- \\ | \\ COO^- \end{matrix} + \begin{matrix} CH_2OH \\ | \\ COO^- \end{matrix}$
(Intramolecular Cannizzaro reaction)

(d) $MnO_2 \xrightarrow{OH^-} MnO_4^- + Mn^{2+}$ (Redox reaction)

59. If CO_2 is passed into waste water containing CrO_4^{2-} (yellow) then

(a) solution turn green due to formation of Cr^{3+}
(b) solution turns blue due to formation of CrO_5
(c) solution turns orange due to formation of $Cr_2O_7^{2-}$
(d) no effect takes place

60. In the following reaction

$VO_3^{2-} + MnO_4^- \longrightarrow Mn^{2+} + VO_4^{3-}$

1 mole of VO_3^{2-} is oxidised by MnO_4^-

(a) 0.2 mol (b) 0.4 mol
(c) 0.8 mol (d) 1.0 mol

One or More Than One Options Correct

1. One mole of ferric oxalate is titrated with acidified potassium permanganate solution. In this

(a) 3 mol of potassium permanganate are required
(b) 1.2 mol of potassium permanganate are required.
(c) oxalate ion is oxidised to carbon dioxide.
(d) ferric ion is oxidised to ferrate (VI) ion

2. Consider following reaction :

$\begin{matrix} CHO \\ | \\ CHO \end{matrix} + OH^- \longrightarrow \begin{matrix} COO^- \\ | \\ CH_2OH \end{matrix}$

(a) It is a disproportionation reaction
(b) It is intramolecular redox reaction
(c) OH^- is a reducing as well as oxidising agent
(d) $\begin{matrix} CHO \\ | \\ CHO \end{matrix}$ is a reducing as well as oxidising agent

3. Out of FeO_4^{2-}, $FeCl_3$, $FeCl_2$ and Fe,

(a) best reducing agent is Fe
(b) best oxidising agent is FeO_4^{2-}
(c) best oxidising agent is $FeCl_3$
(d) best reducing agent is FeO_4^{2-}

4. 0.1 mol of $NaHC_2O_4$ is

(a) neutralised by 0.1 mol of NaOH
(b) neutralised by 0.05 mol of $Ca(OH)_2$
(c) oxidised by 0.04 mol of $KMnO_4$ is acidic medium
(d) oxidised by 0.02 mol of K_2MnO_4 is basic medium

5. Select correct statements.
Equivalent weight of $KMnO_4$ (molar mass = M) is
(a) $\frac{M}{5}$ in acidic medium
(b) $\frac{M}{3}$ in neutral basic medium
(c) M in strongly alkaline medium
(d) $\frac{M}{2}$ in acidic, basic and neutral media

6. Intramolecular redox reactions are
(a) $PCl_5 \longrightarrow PCl_3 + Cl_2$
(b) $2KClO_3 \longrightarrow 2KCl + 3O_2$
(c) $\begin{matrix} CHO \\ | \\ CHO \end{matrix} + OH^- \longrightarrow \begin{matrix} COO^- \\ | \\ CH_2OH \end{matrix}$
(d) $NH_4NO_2 \longrightarrow N_2 + 2H_2O$

7. Out of $\underset{(I)}{H_2SO_4}, \underset{(II)}{H_2S_2O_8}, \underset{(III)}{H_2SO_5}, \underset{(IV)}{H_2S_2O_3}$ and $\underset{(V)}{H_2S_2O_7}$
(a) oxidation number of sulphur is +6 in all except IV
(b) peroxy linkage is in I, II and III
(c) peroxy linkage is in II and III
(d) oxidation number of sulphur is –2 in I and V

8. Select correct statement(s).
(a) Equivalent weight of $Ca(HC_2O_4)_2$ is $\frac{M}{2}$ when it is a reducing agent
(b) Equivalent weight of $Ca(HC_2O_4)_2$ is M when it behaves as an acid
(c) $Ca(HC_2O_4)_2$ can be estimated by MnO_4^-/H^+
(d) $Ca(HC_2O_4)_2$ can be estimated by an acid

9. Consider following reaction
$Br_2 + OH^- \longrightarrow BrO_3^- + Br^- + H_2O$ (unbalanced)
Select correct statement.
(a) Equivalent weight of Br_2 when it is reduced to Br^- is 80
(b) Equivalent weight of Br_2 when it is oxidised to BrO_3^- is 96
(c) Net equivalent weight of Br_2 is 96
(d) It is a disproportionation reaction

10. Following is/are disproportionation reaction(s)
(a) $2CCl_3CHO + NaOH \longrightarrow CCl_3COONa + CCl_3CH_2OH$
(b) $NH_4NO_3 \longrightarrow N_2O + 2H_2O$
(c) $2KClO_3 \longrightarrow 2KCl + 3O_2$
(d) $5Cl_2 + 6OH^- \longrightarrow ClO_3^- + 5Cl^- + 3H_2O$

Passage Comprehension Questions

Passage 1

Consider following **Redox Predominance Diagram** for iron in different oxidation state

Oxidation state	Half-reaction	$E°/V$	
+6	$FeO_4^{2-} + 8H^+ + 3e^- \rightleftharpoons Fe^{3+} + 4H_2O$	+2.20 V	FeO_4^{2-} 2.20
+3	$Fe^{3+} + e^- \longrightarrow Fe^{2+}$	+0.77 V	Fe^{3+} 0.77
+2	$Fe^{2+} + 2e^- \rightleftharpoons Fe$	–0.44 V	Fe^{2+} – 0.44 Fe

Answers the following questions.

1. Select best oxidising and reducing agent
(a) Fe, Fe^{2+}
(b) FeO_4^{2-}, Fe^{3+}
(c) FeO_4^{2-}, Fe^{2+}
(d) FeO_4^{2-}, Fe

2. Most thermodynamically stable species is
(a) FeO_4^{2-}
(b) Fe^{3+}
(c) Fe^{2+}
(d) Fe

3. Least thermodynamically species is
(a) FeO_4^{2-}
(b) Fe^{3+}
(c) Fe^{2+}
(d) Fe

Passage 2

Every disproportionation reaction is a redox reaction but every redox reaction may not be disproportionation. Disproportionation reaction may be intramolecular if same species is oxidised and reduced.

Based on the above statement answer the following questions.

1. A: $2C_6H_5CHO + KOH \longrightarrow C_6H_5COOK + C_6H_5CH_2OH$
B: $3Cl_2 + 6KOH \longrightarrow 5KCl + KClO_3 + 3H_2O$
Select correct statement.
(a) Both are redox reactions
(b) Both are disproportionation reaction
(c) A is a disproportionation and B is a redox reaction
(d) A is a redox reaction, B is a disproportionation reaction

2. Following reaction is used to produce Br_2
$$5Br^- + BrO_3^- + 6H^+ \longrightarrow 3Br_2 + 3H_2O$$
In this reaction.
(a) Br^- disproportionates to Br_2
(b) BrO_3^- disproportionates to Br_2
(c) Br^- has been oxidised and BrO_3^- has been reduced
(d) Br^- has been reduced and BrO_3^- has been oxidised.

Passage 3

Following passage is for the drunken drivers on roads which are becoming fatal by their wrong driving.

Every year in our country about 25,000 people are killed and 500,000 more are injured as a result of drunk driving. In spite of efforts to educate the public about the dangers of driving while intoxicated and stiffer penalties for drunk driving offenses, law enforcement agencies still have to devote a great deal of work to remove drunk drivers from roads.

The police often use a device called a breath analyser to test drivers suspected to being drunk. The chemical basis of this device is a redox reaction. A sample of the driver's breath is drawn into the breath analyser, where it is treated with an acidic solution of potassium dichromate.

$$3CH_3CH_2OH + 2K_2Cr_2O_7 + 8H_2SO_4 \longrightarrow 3CH_3COOH + 2Cr_2(SO_4)_3 + 2K_2SO_4 + 11H_2O$$

The driver's blood alcohol level can be determined readily by measuring the degree of this colour change (read from a calibrated meter on the instrument). The current legal limit of blood alcohol content is 0.1 per cent by mass. Anything higher constitutes intoxication.

Read the passage and answer the questions given below:

1. Above test is based on
(a) smell of ethanol and ethanoic acid
(b) change of colour yellow to blue
(c) change of colour orange to green
(d) change of colour orange to colourless

2. Equivalent mass of $K_2Cr_2O_7$ is
(a) $\frac{\text{molar mass}}{6}$ (b) $\frac{\text{molar mass}}{3}$
(c) $\frac{\text{molar mass}}{2}$ (d) $\frac{\text{molar mass}}{1}$

3. Equivalent mass of CH_3CH_2OH is
(a) $\frac{\text{molar mass}}{1}$
(b) $\frac{\text{molar mass}}{2}$
(c) $\frac{\text{molar mass}}{3}$
(d) $\frac{\text{molar mass}}{4}$

4. The alcohol content in a 10.0 g sample of blood from a driver required 4.23 mL of 0.07654 M $K_2Cr_2O_7$ for titration. Should the police prosecute the individual for drunken driving.
(a) yes
(b) no
(c) nothing can be predicted
(d) None of the above

Assertion & Reason

Codes :
(a) Both **A** and **R** are true and **R** is the correct explanation of **A**
(b) Both **A** and **R** are true but **R** is not the correct explanation of **A**
(c) **A** is true but **R** is false
(d) **A** is false but **R** is true

1. Assertion (A) : NH_4NO_3 undergoes a redox reaction to N_2O and H_2O.

Reason (R) : It is a disproportionation reaction.

2. Assertion (A) : Oxidation number of sulphur in H_2SO_5 is +6.

Reason (R) : It has one peroxy linkage.

3. Assertion (A) : Oxidation number of chromium in CrO_5 is +6.

Reason (R) : Oxidation number of each oxygen atom is –1.5.

4. Assertion (A) : Br_2 is oxidised to BrO_3^- in basic medium.

Reason (R) : When Br_2 is passed into hot NaOH, it disproportionates to Br^- and BrO_3^-.

5. Assertion (A) : 1 mole of FeC_2O_4 is oxidised by 0.6 mole of MnO_4^- in acidic medium.

Reason (R) : MnO_4^- oxidises both Fe^{2+} as well as $C_2O_4^{2-}$.

True & False

1. Lewis structures should have adjacent formal charges of zero or of opposite sign.

2. Oxidation number of sulphur in H_2SO_5 is +6.

3. Oxidation number of sulphur in $H_2S_2O_8$ is +7.

4. Following reaction is disproportionation reaction,

$$NH_4^+ + NO_3^- \longrightarrow N_2O + 2H_2O$$

5. SO_3^{2-} is named is sulphate (IV).

6. HCHO disproportionates in presence of OH^-.

7. Oxidation number of iron in brown complex (obtained in **ring-test** of NO_3^-) is +1.

$$[Fe(H_2O)_5NO]^{2+}$$

8. Equivalent weight of MnO_4^- when it changes to MnO_2 in concentrated basic medium is one-fifth of molar mass.

9. Oxidation number of chromium in CrO_2^{2+} is +6.

10. Oxidation number of each chlorine in $CaOCl_2$ is –1.

Fill in the Blanks

1. Oxidation number of Mn in MnO_2, MnO_4^{2-} and MnO_4^- are......, ... and .... respectively.

2. Oxidation number of Mn in MnO_3Cl is ....

3. Br_2 disproportionates to .... and .... in cold Na_2CO_3 solution.

4. Equivalent weight of Cl_2 in the following reaction is ....

$$Cl_2 \longrightarrow ClO_3^- + Cl^-$$

5. Oxidation number of Fe in $[Fe(H_2O)_5NO]$ is +1 due to charge transfer ....

6. CrO_5 has oxidation number of Cr as +6 hence, it has ....

7. (o-nitrobenzaldehyde: benzene ring with –CHO and NO_2) $\longrightarrow$ (o-aminobenzoic acid: benzene ring with –COOH and NH_2)

This reaction is .... redox-reaction.

8. 1 mole of $Cr_2O_7^{2-}$ oxidises .... moles of FeC_2O_4 in acidic medium.

9. In copper extraction, final stage reaction is ....

10. $$\begin{matrix} CHO \\ | \\ CHO \end{matrix} \xrightarrow{OH^-} \begin{matrix} COO^- \\ | \\ CH_2OH \end{matrix}$$

is a type of .... reaction.

Matrix-Match Type Questions

I. Only One Correct

1. Match the different compounds of Mn (in List I) with oxidation number (in List II).

	List I		List II
A.	MnO_3HSO_4	1.	+4
B.	K_2MnO_4	2.	+7
C.	MnO_2	3.	+6
D.	Mn_2O_3	4.	+2
E.	$Mn(HCO_3)_2$	5.	+3

2. Match the half-reactions (in List I) with change in oxidation number (in List II).

	List I		List II
A.	$Cl^- \longrightarrow ClO_4^-$	1.	2
B.	$Cr^{3+} \longrightarrow CrO_5$	2.	8
C.	$H_2O_2 \longrightarrow O_2$	3.	0
D.	$CrO_2^{2+} \longrightarrow CrO_4^{2-}$	4.	3

II. One or More Correct

1. Match the compounds in Column I with the matter in Column II.

	Column I		Column II
A.	$H_2S_2O_8$	1.	Peroxy linkage
B.	$\underline{Cr}O_5$	2.	Oxidation number of the underlined atom 6
C.	$H_2\underline{S}O_5$	3.	H_2O_2 is the hydrolysis product
D.	$K_2\underline{Cr}O_4$	4.	Caro's acid
E.	$(NH_3)_3\overset{IV}{\underline{Cr}}O_4$	5.	Marshall's acid

2. Match the compounds/Ions having underlined atoms of different oxidation states in Column I with values in Column II.

	Column I		Column II
A.	$CaO\underline{Cl_2}$	1.	6
B.	$Na_2\underline{S_2}O_3$	2.	3
C.	$\underline{N_2}H_4O_2$	3.	2
D.	$[\underline{Fe_2}(CN)_6]^-$ (Prussian blue)	4.	–1
E.	$Cr\underline{O_5}$	5.	1
		6.	–2
		7.	–3

Integer Answer Types

This section contains 8 questions. The answer to each of the questions is a single digit integer, ranging from 0 to 9. The appropriate bubbles below the respectively question numbers in the ORS have to be darkened. For example, if the correct answers to question number *X, Y, Z* and *W* (say) are 6, 0, 9 and 2, respectively, then

X	Y	Z	W
⓪	⓪	⓪	⓪
①	①	①	①
②	②	②	②
③	③	③	③
④	④	④	④
⑤	⑤	⑤	⑤
⑥	⑥	⑥	⑥
⑦	⑦	⑦	⑦
⑧	⑧	⑧	⑧
⑨	⑨	⑨	⑨

1. $Cr_2O_7^{2-}$ changes to Cr^{3+} in acidic medium. Thus, molecular weight of $Cr_2O_7^{2-} = x \times$ equivalent weight. What is the value of x?

2. In the following reaction

$$2NH_3 \longrightarrow N_2 + 2NH_3$$

equivalent weight of NH_3 is $= \dfrac{\text{molecular weight}}{y}$

What is value of y?

3. In a Cannizzaro reaction of HCHO with NaOH

$$2HCHO + NaOH \longrightarrow HCOONa + CH_3OH$$

Molecular weight of HCHO = $z \times$ equivalent weight. What is value of z?

4. Oxidation number of Cr in $(NH_3)_3Cr(O_2)_2$ is ......

5. Number of peroxy linkage in $K_3Cr(O_2)_x$ having oxidation number of Cr as +5 is ....

6. Difference in oxidation states of two types of sulphur in $S_4O_6^{2-}$ is ....

7. Based on redox predominance diagram, out of FeO_4^{2-}, Fe^{3+}, Fe^{2+} and Fe, best oxidising agent is the species with oxidation number of iron as ....

8. Maximum oxidation state of chlorine is ....

Answers & Solutions

Master Exercises

Exercise 1

Short Answer Type Questions

1. $\underset{-1}{Cl^-} \longrightarrow \frac{1}{2}\underset{0}{Cl_2}$

Change in oxidation number = 1 unit

2. O_2^-, If oxidation number of oxygen = x

then $2x = -1 \quad \therefore \quad x = -\frac{1}{2}$

3. MnO_4^- (oxidation number of Mn = + 7)

It is maximum value of oxidation number. In the product formed, oxidation number should be less than +7.

Since, equivalent weight = molecular weight, hence oxidation number of Mn = + 6 in new species formed by reduction.

This is MnO_4^{2-}

4. $NaHC_2O_4 \rightleftharpoons Na^+ + HC_2O_4^- \rightleftharpoons H^+ + C_2O_4^{2-}$

(a) It has one ionisable H^+ thus equivalent weight = molecular weight

(b) $C_2O_4^{2-}$ is a reducing agent, itself oxidised to CO_2.

$$\underset{+6}{\underset{\uparrow}{C_2O_4^{2-}}} \longrightarrow \underset{+8}{\underset{\uparrow}{2CO_2}} + 2e^- \quad \text{(for two C-atoms)}$$

Change in oxidation number = 2 units

Thus, equivalent weight = $\dfrac{\text{molecular weight}}{2}$

5. It has one peroxy linkage in which oxidation number of each oxygen is –1

$$H_2SO_5 + H_2O \longrightarrow H_2SO_4 + H_2O_2$$

```
         O ← -2
         ‖
   HO—S—O—O—H
         ‖  └┬┘
         O   -2
         ↑
        -2
```

6. $H_2S_2O_8 + 2H_2O \longrightarrow 2H_2SO_4 + H_2O_2$

It has also one peroxy linkage

```
        O           O
        ‖           ‖
  HO—S—O—O—S—OH
        ‖           ‖
        O           O
```

7. (i) $\underset{\text{monobasic acid}}{H_3PO_2} \rightleftharpoons H^+ + H_2PO_2^- \rightleftharpoons$ no further ionisation

Thus, equivalent weight = M

(ii) $\underset{\text{dibasic}}{H_3PO_3} \rightleftharpoons H_2PO_3^- + H^+ \qquad E = M$

$H_3PO_3 \rightleftharpoons HPO_3^{-} + 2H^+ \qquad E = \dfrac{M}{2}$

(iii) $\underset{\text{tribasic acid}}{H_3PO_4} \rightleftharpoons H_2PO_4^- + H^+ \qquad E = M$

$H_3PO_4 \rightleftharpoons HPO_4^{2-} + 2H^+ \qquad E = \dfrac{M}{2}$

$H_3PO_4 \rightleftharpoons PO_4^{3-} + 3H^+ \qquad E = \dfrac{M}{3}$

8. $CO_2 + H_2O \rightleftharpoons H_2CO_3 \rightleftharpoons H^+ + HCO_3^-$

CO_2 makes the solution acidic, which converts CrO_4^{2-} into $Cr_2O_7^{2-}$

$$\underset{\text{yellow}}{2CrO_4^{2-}} + 2H^+ \rightleftharpoons \underset{\text{orange}}{Cr_2O_7^{2-}} + H_2O$$

9. $\underset{\text{orange}}{Cr_2O_7^{2-}} + SO_2 \longrightarrow SO_4^{2-} + \underset{\text{green}}{Cr^{3+}}$

(oxidation: $SO_2 \to SO_4^{2-}$; reduction: $Cr_2O_7^{2-} \to Cr^{3+}$)

10. $5Fe^{2} + \underset{\text{pinkish violet}}{MnO_4^-} + 8H^+ \longrightarrow \underset{\text{colourless}}{Mn^{2+}} + 5Fe^{3+} + 4H_2O$

$6Fe^{2+} + \underset{\text{orange}}{Cr_2O_7^{2-}} + 14H^+ \longrightarrow 6Fe^{3+} + \underset{\text{green}}{2Cr^{3+}} + 7H_2O$

Analytical Questions

1.

				change
	A^{n+}	$\longrightarrow$	AO_3^-	
O.N. →	$+n$		$+5$	$(5-n)$
	MnO_4^-	$\longrightarrow$	Mn^{2+}	
O.N. →	$+7$		$+2$	5

Balancing (O.N.)

$$(5-n)MnO_4^- + 5A^{n+} \longrightarrow 5AO_3^- + (5-n)\,Mn^{2+}$$

Thus, $\dfrac{\text{moles of } MnO_4^-}{\text{moles of } A^{n+}} = \dfrac{1.61\times10^{-3}}{2.68\times10^{-3}}$

$\therefore \quad \dfrac{5-n}{5} = \dfrac{1.61\times10^{-3}}{2.68\times10^{-3}}$

$(5-n) = 3$

$\therefore \quad n = 2$

2. $HS^- < S_2O_3^{2-} < SO_3^{2-} < S_4O_6^{2-} < S_2O_8^{2-} = HSO_4^-$

Oxidation state is the oxidation number per atom $S_2O_8^{2-}$ has one peroxy linkage.

3. $N_2O(+1), NO(+2), HNO_2(+3), NO_2(+4), HNO_3(+5)$

4. Ba Xe O

n : 1 : 6

$2n + 8 - 12 = 0$

$\therefore \quad n = 2$

$\therefore \quad Ba_2XeO_6$

5. As in 11.

$$n = 2$$

$$\therefore \quad \underset{+2}{X^{2+}} \longrightarrow \underset{+5}{XO_3^-} \quad \underset{3}{\text{change}}$$

$\therefore$ Equivalent mass of $XCl_2 = \frac{M}{3} = 56$

$$\therefore \quad M = 56 \times 3 = 168$$

$$XCl_2 = M + 35.5 \times 2 = 168$$

$$\therefore \quad M = 97$$

6. Refer Text (h), (l), (j), (k), (m) are disproportionation reaction.

7. **Oxidising agent** (a) ClO_3^- (b) $Cu(NH_3)_4^{2+}$

Reducing agent (a) Ag_2S_3 (b) CN^-

8. $2H_2O_2 \longrightarrow 2H_2O + O_2$

Since, H_2O_2 disproportionates to H_2O and O_2.

9. (a) $4\,Zn + NO_3^- + 7OH^- \longrightarrow NH_3 + 4ZnO_2^{2-} + 2H_2O$

(b) $3HgS + 12Cl^- + 2NO_3^- + 8H^+ \longrightarrow 3HgCl_4^{2-} + 2NO + 3S + 4H_2O$

(c) $2MnO_4^- + 6H^+ + 5H_2O_2 \longrightarrow 2Mn^{2+} + 5O_2 + 8H_2O$

(d) $5P_2H_4 \longrightarrow 6PH_3 + P_4H_2$

(e) $2ClO_2 + O_2^{2-} \longrightarrow 2ClO_2^- + O_2$

(f) $Cl_2 + IO_3^- + 2OH^- \longrightarrow IO_4^- + 2Cl^- + H_2O$

10. 0.4 mol $(2MnO_4^- + 5SO_3^- \longrightarrow 5SO_4^{2-} + 2Mn^{2+})$

11. $2H_2O_2 \longrightarrow 2H_2O + O_2$

Since, H_2O_2 disproportionates to H_2O (by reduction) and O_2 (by oxidation) hence acts both as an oxidising agent and as a reducing agent. Examples are

H_2O_2 as oxidising agent

$$2MnO_2 + 3H_2O_2 + 2OH^- \longrightarrow 2MnO_4^- + 4H_2O$$

H_2O_2 as reducing agent

$$MnO_2 + H_2O_2 \longrightarrow Mn^{2+} + O_2 + 2OH^-$$

12. $2I^- + Cl_2 \longrightarrow 2Cl^- + I_2$

oxidising power is in order:

$I_2 < Br_2 < Cl_2 < F_2$ based on electrochemical series.

13. (a) $E = \frac{M}{5}$ (b) $E = \frac{M}{6}$ (c) $E = M$ (d) $E = \frac{M}{6}$

(e) $E = \frac{3M}{5}$ (f) $E = \frac{M}{4}$ (g) $E = \frac{M}{2}$ (h) $E = M$

(i) $E = \frac{M}{8}$ (j) $E(HS^-) = \frac{M}{4}$ (k) $E = \frac{M}{2}$

(l) $E(HSO_3^-) = \frac{M}{2}$, $E(IO_3^-) = \frac{M}{6}$, $E(HSO_3^-) = \frac{M}{2}$

(m) $E = M$ (n) $E = \frac{M}{2}$

14. (a) $6CO_2 \equiv 6O_2$

1 mol $CO_2 \equiv$ 1 mol $O_2 \equiv$ 32 g $O_2 \equiv$ 4 equivalents

(b) 12 mol H_2O used = 1 mol $C_6H_{12}O_6$

$\equiv$ 24 equivalents $C_6H_{12}O_6$

1 mol H_2O used $\equiv$ 2 equivalents $C_6H_{12}O_6$

$$\underset{O.N = 24}{6CO_2} \longrightarrow \underset{= 0}{C_6H_{12}O_6}$$

change = 24 units

Equivalent weight of $CO_2 = \frac{6\text{ mol}}{24} = \frac{\text{mol. wt.}}{4}$

$\therefore$ 1 mol CO_2 = 4 equivalents CO_2

Equivalent weight of $C_6H_{12}O_6 = \frac{\text{mol. wt.}}{24}$

$\therefore$ 1 mol $C_6H_{12}O_6$ = 24 equivalents of $C_6H_{12}O_6$

(c) 2 equivalents of CO_2

(d) No, O_2 (product) is from H_2O (reactant), hence 12 H_2O is taken as reactant. It is confirmed by taking isotopic O^{18} when all O^{18} appear in O_2.

$$6CO_2 + 12H_2O^{18} \longrightarrow C_6H_{12}O_6 + 6H_2O + 6O_2^{18}$$

Exercise 2

Only One Option Correct

1. If the species is a reducing agent, it means it can be oxidised easily thus it should have an O.N. less than maximum values of O.N.

	O.N.	
SO_2	4	SO_2, SO_3^{2-} and S^{2-} can be reducing agent
SO_3^{2-}	4	
SO_4^{2-}	6	
S^{2-}	−2	
but		SO_4^{2-} can't be.

Thus, **(c)**

2. (a,d) no change

(b) reduction, (c) oxidation

Thus, (b)

3. $CN^- \longrightarrow CNO^-$

O.N. of nitrogen $4 + x = -1$ $\quad 4 + x - 2 = -1$

$x = -5 \quad x = -3$

CN^- is thus oxidised. Thus, two electrons are lost. Thus, **(d)**

4. (c) **5.** (d) **6.** (b)

7. Strongest acid is that which has maximum value of O.N. of the central atom

$HBrO_4$	O.N. of Br = + 7
HOCl	O.N. of Cl = + 1
HNO_2	O.N. of N = + 3
H_3PO_3	O.N. of P = + 3

Thus, **(a)**

8. (b)

9. $$V_2O_5 \longrightarrow V^{2+} \longrightarrow VO^{2+}$$

$$\underset{\substack{20\text{ g}\\0.11\text{ mol}}}{V_2O_5} \longrightarrow \underset{0.22\text{ mol}}{2V^{2+}}$$

$$\underset{+2}{V^{2+}} + \underset{0}{I_2} \longrightarrow \underset{+4}{VO^{2+}} + \underset{-2}{2I^-}$$

Thus, $\underset{0.11\text{ mol}}{V_2O_5} \equiv 2V^{2+} = 2VO^{2+} \equiv \underset{0.22\text{ mol}}{2I_2}$

Thus, **(a)**

10. $Sn^{2+} \longrightarrow Sn^{4+} + 2e^-$ increase in O.N. = 2

$Ce^{4+} + ne^- \longrightarrow Ce^{(4-n)}$ decrease in O.N. = n

To balance

$$nSn^{2+} + 2Ce^{4+} \longrightarrow nSn^{4+} + 2Ce^{(4-n)}$$

Millimoles of $Sn^{2+} = 100 \times 0.1 = 10$

Millimoles of $Ce^{4+} = 50 \times 0.4 = 20$

Thus, $\frac{n}{2} = \frac{10}{20}$

$\therefore$ $n = 1$

Thus, oxidation state of cerium in the reduced product $= (Ce^{4-1} = Ce^{3+}) = +3$

Thus, **(c)**

11. $$As_2S_3 \longrightarrow AsO_4^{3-} + 3SO_4^{3-}$$

O.N. $2As = +6$, $3S = -6$ (Net = 0); $2 \times 5 = 10$, $3 \times 6 = 18$ (Total = 28); Net change = 28

Thus, equivalent mass of $A_2S_3 = \frac{M}{28}$

Thus, **(c)**

12. $$FeC_2O_4 \rightleftharpoons Fe^{2+} + C_2O_4^{2-}$$

Fe^{2+} is oxidised to Fe^{3+} and $C_2O_4^{2-}$ to CO_2 by $Cr_2O_7^{2-}$

$$Fe^{2+} \longrightarrow Fe^{3+} + e^-] \times 6$$

$$Cr_2O_7^{2-} + 14H^+ + 6e^- \longrightarrow 2Cr^{3+} + 7H_2O$$

Thus, $6Fe^{2+} \equiv 1\,Cr_2O_7^{2-}$

$$1\,Fe^{2+} \equiv \frac{1}{6}Cr_2O_7^{2-}$$

$$C_2O_4^{2-} \longrightarrow 2CO_2 \times 2e^-] \times 3$$

$$Cr_2O_7^{2-} + 14H^+ + 6e^- \longrightarrow 2Cr^{3+} + 7H_2O$$

$$3C_2O_4^{2-} \longrightarrow Cr_2O_7^{2-}$$

$$1C_2O_4^{2-} \longrightarrow \frac{1}{3}Cr_2O_7^{2-}$$

Thus, total $Cr_2O_7^{2-}$ required $= \frac{1}{6} + \frac{1}{3} = \frac{1}{2} = 0.5$ mol

Thus, **(c)**

13. MnO_4^- will only oxidise oxalate ($C_2O_4^{2-}$), ferric is already in oxidised state

$$2MnO_4^- + 5C_2O_4^- \longrightarrow 10CO_2 + 2MnO_4^-$$

$$Fe_2(C_2O_4)_3 \rightleftharpoons 2Fe^{3+} + 3C_2O_4^{2-}$$

$\therefore$ 2 mol of MnO_4^- oxidises = 5 mol of $C_2O_4^{2-}$

1 mol of MnO_4^- oxidises $= \frac{5}{2}$ mol of $C_2O_4^{2-}$

$= \frac{5}{6}$ mol of $Fe_2(C_2O_4)_3$

Thus, **(a)**

14. $$\underset{+5}{HNO_3} + 4e^- \longrightarrow \text{Product has O.N.} = 1$$

In N_2O oxidation number of N = + 1. Thus, 0.5 mol of N_2O are formed by reduction of 1 mol of HNO_3 by 4 mol of electrons thus, **(b)**

15. $$N_2H_4 \longrightarrow N_2(---) + 10e^-$$

O.N. two N-atoms of = −4 O.N. of two N-atoms in oxidised species = − 4 + 10 = + 6

Thus, oxidation state of N in new compound = +3.

Thus, **(c)**

16. (c)

17. H_2SO_5 (peroxy sulphuric acid) H—O—O—S(=O)(=O)—O—H (peroxy linkage)

$H_2S_2O_8$ (peroxy disulphuric acid)

H—O—S(=O)(=O)—O—O—S(=O)(=O)—O—H

All of the above have **O.N. = 6** in sulphur, peroxy linkage (—O—O—) has −1 O.N. for oxygen, CrO_5 has two peroxy linkage, thus O.N. of Cr = +6

[Structure of CrO_5: Cr with one =O and two O—O peroxy rings]

$K_2Cr_2O_7$ and CrO_4^{2-} both has +6 oxidation number for Cr.

Thus, **(d)**

18. $[Cr(O_2)(NH_3)_4(H_2O)]Cl_2$

Cr: +3, O_2: $+x$, NH_3: +0, H_2O: +0, Cl_2: −2 = 0

$\therefore$ $x = -1$

Thus, (O_2) exists as O_2^- (superoxide ion)

Thus, **(c)**

19. If acid is weak, its conjugate base is strong. Greater the O.N. of Cl, stronger the acid and thus weaker the conjugate base

	HClO	$HClO_2$	$HClO_3$	$HClO_4$
O.N. of Cl	+1	+3	+5	+7
acid strength ⟶				
conjugate base	ClO^-	ClO_2^-	ClO_3^-	ClO_4^-
base strength ⟵				

Thus, **(a)**

20. $$NaOH + H_2O \longrightarrow NaOH + H_2$$

(H in NaOH: −1; H in H_2O: +1; H in product NaOH: +1; H_2: +0)

H^- changes to H_2 — oxidation

H^+ changes to H_2 — reduction
Thus, different H are involved.
Thus, it is a redox but not disproportionation.
Thus, **(c)**

21. $K[CrO](O_2)(OH)]$

$+1 + x\ 2 - 2 - 1 = 0$

$\therefore\quad x = 4$

$K_3[Cr(O_2)_4]$

$3 + y - 8 = 0,\quad y = 5$

$(NH_3)_3[Cr(O_2)_2]$

$0\quad +z\quad -4 = 0$ thus, $z = 4$

CrO_2Cl_2

$w - 4 - 2 = 0,\quad w = 6$

Thus, **(c)**

22. (a) MnO_3HSO_4

$x - 6 - 1 = 0$

$x = +7$

Thus, **(a)**

23. $\underset{+3}{Mn^{3+}} + \underset{+3}{Mn^{3+}} \longrightarrow \underset{+2}{Mn^{2+}} + \underset{+4}{MnO_2}$

red, oxid

Mn^{3+} disproportionates to Mn^{2+} and MnO_2

Thus, equivalent mass of $Mn^{3+} = \dfrac{2 \text{ units of } Mn^{3+}}{2}$

$= \dfrac{M}{1} = 55$

Thus, **(b)**

24. (a) $\underset{+2}{Ca}\overset{-2}{C}\underset{+4}{O_3} \longrightarrow \underset{+2}{Ca}\overset{-2}{O} + \underset{+4}{C}\overset{-2}{O_2}$

No change,

(b) $2HCuCl_2 \longrightarrow Cu + Cu^{2+} + 4Cl^- + 2H^+$
$+1\ +1\ -1 \qquad 0 \qquad +2 \qquad -1 \qquad +1$

Thus, Cu^+ disproportionates to Cu and Cu^{2+}.

Thus, **(b)**

25. (I) $\underset{-3}{NH_4^+} + \underset{+5}{NO_3^-} \longrightarrow \underset{+1}{N_2O} + 2H_2O$

reduction, oxidation

It is a redox but not disproportionation. Since, different N-species are involved.

(II) $\underset{-3}{NH_4^+} + \underset{+3}{NO_2^-} \longrightarrow \underset{0}{N_2}$

Same as (I)

(III) $PCl_5 \longrightarrow PCl_3 + Cl_2$
$+5\ -1 \qquad +3\ -1 \qquad 0$

P^{+5} is reduced to P^{3+}

Cl^- is oxidised to Cl_2

Thus, different species are involved. Thus, it is also not a disproportionation reaction.

Thus, **(d)**

26. $Ca^{2+}(ClO^-)(Cl^-)$

Thus, **(d)**

27. $Zn \longrightarrow Zn^{2+}$

Zn is oxidised hence, it is a reducing agent (reductant)

$\underset{+5}{NO_3^-} \longrightarrow \underset{-3}{NH_4^+}$

NO_3^- is reduced hence, it is oxidising agent (oxidant)

Thus, **(b)**

28. I : Ca^{2+} $Cl^- \leftarrow -1$, $OCl^- \leftarrow +1$

II : +6 both and S

III : 0 and 5 IV –2, 1

V : $\underset{-3}{NH_4^+}$ $\underset{+5}{NO_3^-}$

Thus, (b)

29. Ag^+ is reduced hence, it is oxidant. Hydroquinone is oxidised hence, it is reductant.

Thus, **(b)**

30. $\underset{+2}{2Cu^{2+}} + \underset{-4}{N_2H_4} \longrightarrow \underset{0}{Cu} + \underset{0}{N_2} + 4H^+$

$\therefore$ Thus 1 mol $N_2H_4 \equiv 2$ mol Cu^{2+}

Thus, **(b)**

31. $\underset{+3}{Bi^{3+}} \longrightarrow \underset{+5}{BiO_3^-}$

Change in oxidation number = 2 units

Thus, number of equivalents in one mole $Bi^{3+} = 2$

Thus, **(b)**

32. $Fe^{2+} + e^- \longrightarrow Fe^+$

$NO \longrightarrow NO^+ + e^-$

Based on magnetic properties of the complex, it is found that iron has three unpaired electrons thus it is as Fe^+ formed by reduction of Fe^{2+} by NO which is oxidised to NO^+.

Thus, **(c)**

33. (b) **34.** (b)

35. $2H_2S(g) + SO_2(g) \longrightarrow 3S(s) + 2H_2O(l)$

H_2S: -2; SO_2: $+4$; S: 0

Change in oxidation number = 4 units

Thus, number of equivalent of $SO_2 = 4$

Thus, **(a)**

36. H_2S changes to S

Thus, change in oxidation number = 2 units

Thus, one equivalents $H_2S = 2$

$$\underset{\substack{2\text{ mol}\\4\text{ equivalent}}}{2H_2S} + \underset{\substack{1\text{ mol}\\1\text{ mol}}}{SO_2} \longrightarrow \underset{3\text{ mol}}{3S} + 2H_2O$$

1 equivalent of H_2S will reduce 0.25 mol of SO_2

Thus, **(c)**

37.

Reaction	Change in O.N.	Number of equivalent
(a) $VO^{2+}(aq) \longrightarrow V^{3+}(aq)$; $x-2=+2$, $x=+4$ → $+3$	1 unit	1
(b) $NO_3^-(aq) \longrightarrow NO_2(g)$; $+5$ → $+4$	1 unit	1
(c) $VO_4^{3-}(aq) \longrightarrow V^{2+}(aq)$; $+5$ → $+2$	3 unit	3
(d) $IO_3^-(aq) \longrightarrow I_3^-(aq)$; $+5$ → $-\frac{1}{3}$	$\frac{16}{3}$	$\frac{16}{3}$

Thus, **(d)**

38. (a) $\underline{Mn}O_4^{2-}$ $x = 6$

(b) $\underline{Cr}O_5$ $x = 6$

(c) $\underline{Cr}O_2Cl_2$ $x = 6$

Thus, **(d)**

39.

		Change in O.N.	Equivalent per mol of MnO_4^-
(a)	$\underset{+7}{MnO_4^-} \longrightarrow \underset{+4}{MnO_2}$	3	3
(b)	$\longrightarrow \underset{+2}{Mn^{2+}}$	5	5
(c)	$\longrightarrow \underset{+6}{MnO_4^{2-}}$	1	1
(d)	$\longrightarrow \frac{1}{2}\underset{+3}{Mn_2O_3}$	4	4

Thus, **(c)**

40. K_3CrO_8 has $[Cr(O_2)_4]^{3-}$ with four peroxy linkage

$\therefore$ $[Cr(O_2)_4]^{3-}$

$x - 8 = -3$

$x = +5$

Thus, **(a)**

41. (a)

42. (d) $Na_2CO_3 + H_2O \longrightarrow 2NaOH + H_2CO_3$

$H_2CO_3 \longrightarrow H_2O + CO_2$

$$6NaOH + Br_2 \longrightarrow \underset{\text{by reduction}}{5NaBr} + \underset{\text{by oxidation}}{NaBrO_3} + 3H_2O$$

Thus, **(d)**

43. I : $Zn + \underset{+1}{2H^+} \longrightarrow \underset{0}{H_2}$ changes

II : $Zn + \underset{+1}{H^+} + SO_4^{2-} \longrightarrow Zn^{2+} + SO_2 + \underset{+1}{H_2O}$ no change

III : $Zn + \underset{+1}{H^+} + NO_3^- \longrightarrow Zn^{2+} + \underset{+1}{NH_4^+} + \underset{+1}{H_2O}$ no change

Thus, in (I) only

Thus, **(d)**

44. $4\underset{-3}{NH_3} + 5O_2 \longrightarrow 4\underset{+2}{NO} + 6H_2O$

1 mol O_2 oxidises $= \frac{4}{5}$ mol of NH_3

$= \frac{4}{5} \times 5$ equivalent of $NH_3 = 4$

Thus, **(a)**

45. (a) $3H_2 \equiv 2NH_3$

$1\,H \equiv \frac{NH_3}{3}$

$\therefore$ Equivalent weight of $NH_3 = \frac{M}{3}$

(b) $4NH_3 \equiv 5O_2$

$\equiv 10(O)$ atoms taking weight as 16

$\equiv 20(O)$ atoms taking weight as 8

Thus, $1(O) \equiv \frac{1}{5}\,NH_3$

$\therefore$ Equivalent weight of $NH_3 = \frac{M}{5}$

(c) $2NH_3 \equiv H_2$

$\therefore$ $1\,H \equiv 1\,NH_3$

Equivalent weight of $NH_3 = M$

Thus, **(c)**

46. $Cu_3\underset{-3}{P} \longrightarrow H_3\underset{+5}{P}O_4$

Change in oxidation number = 8 units of phosphorus

$\therefore$ Equivalent weight of $H_3PO_4 = \frac{M}{8}$

Thus, **(d)**

47. $\underset{-2\,+1}{OH^-} \longrightarrow \underset{0}{H_2}$

Thus, OH^- is reduced to H_2

Thus, **(b)**

48. $H_2S_2O_8$ HO—S(→O)(→O)—O—O—S(→O)(→O)—OH (peroxy linkage)

H_2SO_5 HO—S(→O)(→O)—O—O—H

Thus, **(a)**

49.
$$\underset{-2}{H_2O_2} + \underset{+4}{MnO_2} \longrightarrow \underset{+7}{MnO_4^-} + \underset{-4}{2H_2O}$$

$3e^-$; $+2e^-$

(ox. no. of two O atoms taken)

Balancing of O.N.

$$3H_2O_2 + 2MnO_2 \longrightarrow MnO_4^- + 6H_2O$$

Thus, $H_2O_2 : MnO_2 = 3:2$

Thus, **(d)**

50. (a) $H_3PO_2 \equiv HPO_2^- + 2H^+$

But H_3PO_2 is a monobasic acid, hence, false

(b) $H_3PO_2 \longrightarrow H_3PO_3$

O.N. +1 +3

$\therefore$ Equivalent weight $= \dfrac{M}{2}$

51. Ca(Cl)(OCl) Cl is –1

OCl is –1

Thus, OCl

$-2 + x = -1$

$\therefore x = +1$

Thus, **(b)**

52. $\overset{III}{Fe}[\overset{II}{Fe}(CN)_6]$

$+3 + 2 - 6 = -1$

Thus, **(a)**

53. Intramolecular redox reactions are in which different atoms are oxidised and reduced in same substance.

Substance	Oxidised	Reduced
I. $(NH_4)_2Cr_2O_7$	NH_4^+ to N_2	$Cr_2O_7^{2-}$ to Cr^{3+}
II. NH_4NO_3	NH_4^+ to N_2O	NO_3^- to N_2O
III. $KClO_3$	O^- to O_2	$ClO_3^-(+5)$ to Cl^-

Thus, **(d)**

54. $\underset{0.05\ mol}{Ca(OH)_2} + H_2SO_4 \longrightarrow CaSO_4 + 2H_2O$

Thus, $1\ mol\ Ca(OH)_2 \equiv 1\ mol\ H_2SO_4$

hence, $0.05\ mol\ Ca(OH)_2 \equiv 0.05\ mol\ H_2SO_4$

$$\underset{0.05}{H_2SO_4} + \underset{0.10\ mol}{2NaOH} \longrightarrow Na_2SO_4 + 2H_2O$$

Thus, **(b)**

55. $NaHC_2O_4 + NaOH \longrightarrow Na_2C_2O_4 + H_2O$

$$NaHC_2O_4 \equiv HC_2O_4^- \equiv H^+ \equiv OH^-$$

Thus, $E(NaHC_2O_4) = M$

$$\underset{+6}{HC_2O_4^-} \equiv C_2O_4^{3-} \equiv \underset{+8}{2CO_2} \text{] for two C-atoms}$$

Thus, $E(NaHC_2O_4) = \dfrac{M}{2}$

Thus, **(d)**

56. At the equivalent point, all species are in unit concentration and

$$\Delta G^\circ(\text{net}) = \Delta G_1^\circ + \Delta G_2^\circ$$

Total electrons exchanged $= (a + b)$

$\therefore$ $\Delta G^\circ = -nFE^\circ$

$$-(a+b)FE_3^\circ = -aFE_1^\circ - bFE_2^\circ$$

$\therefore$ $E_3^\circ = \dfrac{aE_1^\circ + bE_2^\circ}{a+b}$

Thus, **(a)**

57. $\Delta G^\circ(\text{net}) = \Delta G_1^\circ + \Delta G_2^\circ$

$$-6FE_3^\circ = -FE_1^\circ - 5E_2^\circ$$

$\therefore$ $E_3^\circ = \dfrac{E_1^\circ + 5E_2^\circ}{6}$

Except H^+, all other species = 1 M

$\therefore$ $E = E_3^\circ - \dfrac{0.0591}{6} \log\left(\dfrac{1}{H^+}\right)^8$

$$= \frac{E_1^\circ + 5E_2^\circ}{6} - \frac{0.0591}{6} \times 8 \log\left(\frac{1}{H^+}\right)$$

$$= \frac{E_1^\circ + 5E_2^\circ}{6} - 0.08\ \text{pH}$$

Thus, **(b)**

58. (a) Intramolecular Cannizzaro

(b) Disproportionation thus, true

Thus, **(b)**

59. $CO_2 + H_2O \rightleftharpoons H_2CO_3 \rightleftharpoons 2H^+$

$$\underset{\text{yellow}}{2CrO_4^{2-}} + 2H^+ \longrightarrow H_2O + \underset{\text{orange}}{Cr_2O_7^{2-}}$$

Thus, **(c)**

60. $$\underset{+4}{VO_3^{2-}} + \underset{+7}{MnO_4^-} \longrightarrow \underset{+2}{Mn^{2+}} + \underset{+5}{VO_4^{3-}}$$

5 unit ; 1 unit

Thus, $5VO_3^{2-} + MnO_4^- \longrightarrow Mn^{2+} + 5VO_4^{3-}$

$$5\ mol\ VO_3^{2-} \equiv 1\ mol\ MnO_4^-$$
$$1\ mol\ VO_3^{2-} \equiv 0.2\ mol\ MnO_4^-$$

Thus, **(a)**

One or More Than One Options Correct

1. $Fe_2(C_2O_4)_3 \rightleftharpoons 2Fe^{3+} + 3C_2O_4^{2-}$
$\quad$ 1 mol $\qquad\qquad\qquad$ 3 mol

Fe^{3+} is already in the oxidised state
$C_2O_4^{2-}$ is oxidised to CO_2 by MnO_4^- in acidic medium.

$$2MnO_4^- + 5C_2O_4^{2-} \longrightarrow 10CO_2 + 2Mn^{2+}$$

$$5 \text{ mol } C_2O_4^{2-} \equiv 2 \text{ mol } MnO_4^-$$

$$\therefore \quad 3 \text{ mol } C_2O_4^{2-} \equiv \frac{2\times 3}{5} = 1.2 \text{ mol}$$

Thus, (b) is correct
$C_2O_4^{2-}$ is oxidised to CO_2
Thus, (c) is correct
Thus, (b) and (c)

2. One CHO is oxidised to COO^-
and one CHO is reduced to CH_2OH
Thus, it is not a disproportionation reaction. It is intramolecular redox reaction. Thus, (b) is true and $\begin{matrix} CHO \\ | \\ CHO \end{matrix}$ is reducing as well as oxidising agent.
Thus, (d) is also true.
Thus, (b) and (d)

3. Refer redox predominance diagram (3.1). Species which has least oxidation number is oxidised and thus is the best reducing agent. Thus, (a) is true
Species which has maximum oxidation number is reduced and thus is the best oxidising agent. Thus, (b) is also true.
Thus, (a) and (b)

4. $NaHC_2O_4 \rightleftharpoons Na^+ + H^+ + C_2O_4^{2-}$
$\quad$ 0.1 $\qquad\qquad$ 0.1 (↑ neutralised by OH^-) $\qquad$ 0.1 (↑ oxidised by MnO_4^-)

$$\underset{0.1\text{ mol}}{1H^+} \equiv \underset{0.1\text{ mol}}{1OH^-} \equiv \underset{0.1\text{ mol}}{NaOH} \equiv \frac{1}{2}\underset{0.05\text{ mol}}{Ca(OH)_2}$$

Thus, (a) is true
Thus, (b) is also true.

$$5C_2O_4^{2-} + 2MnO_4^- \longrightarrow 2Mn^{2+} + 10CO_2$$

$$5C_2O_4^{2-} \equiv 2MnO_4^-$$

$$\therefore \quad 0.1 \text{ mol } C_2O_4^{2-} \equiv 0.04 \text{ mol } MnO_4^-$$

Thus, (c) is also true
Thus, (a), (b) and (c) are true

5. $\qquad\qquad\qquad\qquad\qquad\qquad\qquad\qquad$ Equivalent weight

(a) $\underset{+7}{MnO_4^-} \longrightarrow \underset{+2}{Mn^{2+}}$ in acidic medium $= \dfrac{M}{5}$

(b) $\underset{+7}{MnO_4^-} \longrightarrow \underset{+4}{MnO_2}$ in basic neutral medium $= \dfrac{M}{3}$

(c) $\underset{+7}{MnO_4^-} \longrightarrow \underset{+6}{MnO_4^{2-}}$ in strongly alkaline medium $= \dfrac{M}{1}$

Thus, (a), (b) and (c) are true.

6. (a) $\underset{+5\ -1}{PCl_5} \longrightarrow \underset{+3}{PCl_3} + \underset{0}{Cl_2}$

(b) $\underset{+5\ -2}{2KClO_3} \longrightarrow \underset{-1}{KCl} + \underset{0}{O_2}$

(c) $\begin{matrix} CHO \\ | \\ CHO \end{matrix}$ —oxidation→ COO^- ; —reduction→ CH_2OH

(d) $\underset{-3\ +4}{NH_4NO_2} \longrightarrow \underset{0}{N_2}$

Thus, (a),(b),(c) and (d)

7.

Acid	Structure	Oxidation number of sulphur	Peroxy linkage
I. H_2SO_4 sulphuric acid	O ‖ HO—S—OH ‖ S	+6	No
II. $H_2S_2O_8$ peroxydi-sulphuric acid	O $\qquad$ O ‖ $\qquad$ ‖ HO—S—O—O—S—OH ‖ $\qquad$ ‖ S $\qquad$ S	+6	Yes
III. H_2SO_5 peroxy-sulphuric acid	O ‖ HO—S—O—O—H ‖ S	+6	Yes
IV. $H_2S_2O_3$ thiosulph-uric acid	O ‖ HO—S—OH ‖ S	+2 0	No
V. $H_2S_2O_7$ (oleum)	SO_3 dissolved in H_2SO_4	+6	No

(a) is true
(b) is not true
(c) is true
(d) is not true
Thus, (a) and (c)

8. $\underset{+6}{C_2O_4^{2-}} \longrightarrow \underset{+4}{2CO_2}$

Equivalent weight $= \dfrac{M}{2}$

Thus, (a) is true

$$Ca(HC_2O_4)_2 \rightleftharpoons 2HC_2O_4^-$$

$$HC_2O_4^- \rightleftharpoons H^+ + C_2O_4^{2-}$$

Equivalent weight $= M$
Thus, (b) is true
$C_2O_4^{2-}$ can be estimated by MnO_4^-/H^+
Thus, (c) is also true

$Ca(HC_2O_4)_2$ cannot be estimated by acid

Thus, (d) is not true

Thus, (a), (b) and (c) are true.

9. $Br_2 + OH^- \longrightarrow BrO_3^- + Br^- + H_2O$

(a) $\underset{0}{\frac{1}{2}Br_2} \longrightarrow \underset{-1}{Br^-}$

Equivalent weight $= \frac{M}{2} = 80$

Thus, (a) is true.

(b) $\underset{0}{\frac{1}{2}Br_2} \longrightarrow \underset{+5}{BrO_3^-}$

Equivalent weight $= \frac{M}{10} = \frac{160}{10} = 16$

Thus, (b) is not true

(c) Net equivalent weight = Equivalent (in oxidation part + in reduction part)

$= 16 + 80 = 96$

Thus, (c) is true

(d) Br_2 disproportionates to Br^- (reduction) and BrO_3^- (oxidation)

Thus, (d) is also true.

Thus, (a), (c) and (d) true.

10. (a) $CCl_3CHO \longrightarrow \underset{\text{by oxidation}}{CCl_3COO^-} + \underset{\text{by reduction}}{CCl_3CH_2OH}$

Thus, (a) is true

(b) $\underset{-3}{NH_4^+} + \underset{+5}{NO_3^-} \longrightarrow \underset{+1}{N_2O} + 2H_2O$

(reduction: NO_3^- → N_2O; oxidation: NH_4^+ → N_2O)

Different ions are involved, hence (b) is not true

(c) $\underset{+5\ -2}{ClO_3^-} \longrightarrow \underset{-1}{Cl^-} + \underset{0}{O_2}$

(d) Different atoms are involved, hence (c) is not true (d) is true (refer question 9(d)).

Thus, (a) and (d) are true

Passage Comprehension Questions

Passage 1

1. (d) **2.** (a) **3.** (d)

Passage 2

1. (b) **2.** (c)

Passage 3

1. (c) **2.** (a)

3. (b) $\underset{-2}{CH_3CH_2OH} \longrightarrow \underset{0}{CH_3COOH}$

Change in O.N. = 2 units

Thus, $E = \frac{M}{2}$

4. (a) 4.23×0.07654 millimol $K_2Cr_2O_7$

$\equiv 4.23 \times 0.07654 \times 6$ milliequivalent of $K_2Cr_2O_7$

$\equiv 4.23 \times 0.07654 \times 6$ milliequivalent of CH_3CH_2OH

$\equiv \frac{4.23 \times 0.07654 \times 6 \times 23}{1000}$ g CH_3CH_2OH

$\equiv 0.045$ g in 10 g sample

Thus, % of alcohol content $= \frac{0.045}{10} \times 100$

$= 0.45 > 0.10$

Thus, **(a)**

Assertion & Reason

1. (c) **2.** (a) **3.** (c) **4.** (d) **5.** (a)

True & False

1. T **2.** T **3.** F **4.** F **5.** T **6.** T **7.** T
8. F **9.** T **10.** F

Fill in the Blanks

1. +4, +6, +7 **2.** +7 **3.** Br^-, BrO^- **4.** 42.6
5. from NO to Fe^{2+} **6.** two peroxy linkages
7. intramolecular **8.** 2 **9.** disproportionation
10. intramolecular disproportionation.

Matrix-Match Type Questions

I. Only One Correct

1. (A) — (2); (B) — (3); (C) — (1); (D) — (5); (E) — (4)

2. (A) — (2); (B) — (4); (C) — (1); (D) — (3);

II. One or More Correct

1. (A) — (1,2,3,5); (B) — (1,2); (C) — (1,2,3); (D) — (2); (E)—(1)

2. (A) — (4,5); (B) — (1,6); (C) — (2,7); (D) — (2,3); (E)—(4,6)

Integer Answer Types

Questions →	1	2	3	4	5	6	7	8
Answers →	6	3	1	4	4	5	6	7
	⓪	⓪	⓪	⓪	⓪	⓪	⓪	⓪
	①	①	①	①	①	①	①	①
	②	②	②	②	②	②	②	②
	③	③	③	③	③	③	③	③
	④	④	④	④	④	④	④	④
	⑤	⑤	⑤	⑤	⑤	⑤	⑤	⑤
	⑥	⑥	⑥	⑥	⑥	⑥	⑥	⑥
	⑦	⑦	⑦	⑦	⑦	⑦	⑦	⑦
	⑧	⑧	⑧	⑧	⑧	⑧	⑧	⑧
	⑨	⑨	⑨	⑨	⑨	⑨	⑨	⑨

"If the radiance of a thousand suns were to be burst into the sky, that would be like the splendour of the mighty one — yet when the sinister and gigantic clouds rose up ....

I am become death, the shatters of world ......"

A passge from Bhagvad Gita was quoted by Robert Jungk in his book while describing the first testing of Atom Bomb in Japan.

5

Radioactivity

Learning Framework of the Chapter ...

- Radioactivity
- Units of Radioactivity
- Theory of Radioactive Disintegration
- Nuclear Stability
- Nuclear Reactions
- Nuclear Fission
- Nuclear Fusion
- Artificial Radioactivity
- Binding Energy and Packing Fraction
- Applications of Radioactive Isotopes
- Radiations and Matter : Detection and Biological Effects

Quick Points

- Elements in the Periodic Table, have been placed in the increasing order of atomic number.
- In the element ${}_{p}^{A}X$,

  A = atomic mass = proton + neutron = $p + n$

  p = proton = atomic number = electrons in neutral atom
- $E = mc^2$ by Einstein equation, where m is the mass and c the velocity of light.
- By first-order kinetics,

$$\text{rate}, -\frac{dN}{dt} = \lambda N$$

  where λ = disintegration constant (also called rate constant) in time^{-1} unit

$$\lambda = \frac{2.303}{t} \log_{10} \frac{N_0}{N}$$

  where N_0 is the number of atoms disintegrating at start and N is the number of atoms disintegrating at time t.
- When 50% disintegration has taken place

$$N = \frac{N_0}{2}, t = t_{50} \text{ called half-life period}$$

$$t_{50} = \frac{2.303 \log 2}{\lambda} = \frac{0.693}{\lambda}$$

  Thus, t_{50} is independent of initial number of atoms.

- Also, $N = N_0 \left(\frac{1}{2}\right)^y$

  where $y = \frac{\text{total time}}{t_{50}}$

Radioactivity

Radioactivity was discovered by **Henri Becquerel** in **1896,** while investigating any possible relation between fluorescence and X-rays. A crystal of potassium uranyl sulphate was used as a source of fluorescence material. On exposure to sunlight, it emitted radiations which could penetrate paper, glass and aluminium sheets and ionise gases. These types of radiations were also observed in darkness. Nature and origin of these radiations were investigated by Becquerel, Rutherford, Pierre Curie and Marie Curie. Uranium changed to thorium during the emission of these radiations.

Radioactivity is the phenomenon of spontaneous disintegration of certain nuclei to form new elements with the emission of **"active radiation"**, they are called **active** as they affect photographic plates and interact with electric and magnetic fields. Active radiations are summarised in Table 5.1.

α-particles : $^{238}_{92}\text{U} \longrightarrow {}^{234}_{90}\text{Th} + {}^{4}_{2}\text{He}(\alpha)$

β-particles : $^{234}_{90}\text{Th} \longrightarrow {}^{234}_{91}\text{Pa} + {}_{-1}e^{0}\ (\beta)$

γ-rays : $^{230}_{90}\text{Th}^* \longrightarrow {}^{230}_{90}\text{Th} + \gamma$

Table 5.1 **Types of Radioactive Decay**

Types of decay	Radiation	Equivalent process	Nuclear change: Atomic number	Nuclear change: Mass number	Usual nuclear condition
Alpha emission (α)	^4_2He	–	– 2	– 4	$Z > 83$
Beta emission (β)	$^{0}_{-1}e$	$^1_0n \rightarrow {}^1_1p + {}^{0}_{-1}e$	+ 1	0	$\frac{N}{Z}$ too large
Positron emission (β^+)	^0_1e	$^1_1p \rightarrow {}^1_0n + {}^0_1e$	– 1	0	$\frac{N}{Z}$ too small
Electron capture* (E.C.)	X rays	$^1_1p + {}^{0}_{-1}e \rightarrow {}^1_0n$	– 1	0	$\frac{N}{Z}$ too small
Gamma emission (γ)	$^0_0\gamma$	–	0	0	Excited nucleus

(N = neutrons, Z = atomic number)

*Also called K capture-an electron from the K-shell ($n = 1$) is captured by the nucleus

- β-rays are the same as cathode-rays but β-electrons travel with much higher speeds than the cathode-ray electrons. Also the velocity of cathode-ray electrons is constant while the β-electrons have a wide velocity range.
- Emission of α- and β-radiations do not cause health hazard to a greater extent due to their low penetrating power (see Table 5.2).
- γ-rays are extremely penetrating and represent a greater health hazard to those who work with radioactive chemicals :
  Characteristics of α-, β- and γ-radiations are summarised in Table 5.2

Table 5.2 **Properties of Alpha, Beta and Gamma Rays**

Type	Alternate name	Charge	Mass	Speed (ms^{-1})	Penetrating power	Stopped by
Alpha (α)	Helium nucleus (2 protons + 2 neutrons)	+2 (deflection towards –ve plate in electric field)	$4m_H$	2×10^7	Low $\left(\frac{1}{100}\text{ mm of Al}\right)$	Air or paper
Beta (β)	Electron	–1 (deflection towards +ve plate in electric field)	m_e	2×10^8	Medium $\left(\frac{1}{10}\text{ mm of Al}\right)$	Thin aluminium
Gamma (γ)	High energy electro-magnetic radiation	0 (no effect in electric or magnetic field)	0	3×10^8	High 25 cm of steel/ 5 cm of lead	Thick lead
Positron	Positive electron	+1	m_e	–	Medium	Thin aluminium

Illustration 1 *Write balanced nuclear equations for each of the following*:

(*a*) *α-emission from curium*-242
(*b*) *β-emission from magnesium*-28
(*c*) *positron emission from xenon*-118

Solution The key to writing nuclear equations is to make sure that the number of nucleons is the same on both sides of the equation and that the number of elementary and nuclear charges is the same.

(a) $^{242}_{96}\text{Cm} \longrightarrow {}^4_2\text{He} + {}^{238}_{94}\text{Pu}$ (b) $^{28}_{12}\text{Mg} \longrightarrow {}^{0}_{-1}e + {}^{28}_{13}\text{Al}$

(c) $^{118}_{54}\text{Xe} \longrightarrow {}^{0}_{+1}e + {}^{118}_{53}\text{I}$

Table 5.3 **Distinction between a Radioactive Change and a Chemical Change**

Radioactive change	Chemical change
(1) New elements are produced $^{226}_{88}Ra \longrightarrow ^{222}_{86}Rn + ^{4}_{2}He$	(1) One compound can be converted to other $CaCO_3 \xrightarrow{\Delta} CaO + CO_2$
(2) It is irreversible $^{222}_{86}Rn + ^{4}_{2}He \not\longrightarrow$	(2) It can be reversible or irreversible $N_2 + 3H_2 \rightleftharpoons 2NH_3$
(3) Rate of a reaction is independent of imposed external conditions like temperature, pressure, etc.	(3) Rate is affected by change in temperature, pressure etc.
(4) It is spontaneous and needs not be initiated by external means.	(4) It is made spontaneous by external source as heat, pressure $2H_2O \xrightarrow{Heat} 2H_2 + O_2$ $C + O_2 \xrightarrow{Ignite} CO_2$
(5) Large amount of heat is liberated about one million times greater than that liberated in a chemical change.	(5) Change may be exothermic or endothermic. Heat change is generally very low.

Radio Elements and the Periodic Table

Elements formed due to emission of α- and β^--particles are placed in groups other than parent element. **This is by Soddy and Fajan's group displacement law (1913).**

- *When an α-particle is emitted by a radioactive atom, the product has an atomic number two units less than that the parent element and it occupies a position two places to the left of the parent element in the Periodic Table.*

$$\underset{\text{II A}}{^{226}_{88}Ra} \longrightarrow \underset{\text{Zero}}{^{222}_{86}Rn} + ^{4}_{2}He$$

$$\underset{\text{Zero}}{^{222}_{86}Rn} \longrightarrow \underset{\text{VIA}}{^{218}_{84}Po} + ^{4}_{2}He$$

$$\underset{\text{VIA}}{^{218}_{84}Po} \longrightarrow \underset{\text{IVA}}{^{214}_{82}Pb} + ^{4}_{2}He$$

- *When a β-particle is emitted by a radioactive atom, the atomic number of the product becomes one unit greater than that of its parent; the product occupies a position one place to the right of the parent element in the Periodic Table.*

β-emission involves the conversion of a neutron into proton and electron

$$^{1}_{0}n \longrightarrow ^{1}_{1}H + ^{0}_{-1}e + \text{nutrino}$$

The electron escapes the atom as β-particle, leaving one proton extra in the nucleus, due to which atomic number increases by one unit. The nutrinos are energy particles.

$$\underset{\text{IVA}}{^{214}_{82}Pb} \longrightarrow \underset{\text{VA}}{^{214}_{83}Bi} + ^{0}_{-1}e$$

Illustration 1 ***Elements with atomic number 89 to 103 all have been placed in group IIIB (actinides). What are the groups of the elements formed in the following disintegration?***

$$^{238}_{92}U \xrightarrow{-\alpha} A \xrightarrow{-\alpha} B \xrightarrow{-\alpha} C \xrightarrow{-\beta} D \xrightarrow{-\beta} E$$

Solution **$^{238}_{92}U$ belongs to actinide series hence, is of IIIB. Other elements are placed in the groups as shown:**

Elements	Atomic Number	Group
A	90	IIIB
B	88	IIA (alkaline-earth)
C	86	Zero (noble gas)
D	87	IA (alkali)
E	88	IIA (alkaline-earth)

Units of Radioactivity

Various types of units are used for the radioactivity, each one named after the scientist using it. These are summarised in Table 5.4.

Table 5.4 **Units of Radioactivity**

Name	Symbol	Defined to be	Quantity measured
Curie	Ci	3.7×10^{10} disintegration per second (dps)	Decay events
Rutherford	Ru	1×10^6 dps	Decay events
Becquerel	Bq (SI unit)	1 dps	Decay events
Gray	Gy	1 kg tissue receiving 1 J of energy	Energy absorbed
Rad	Rad	1 rad = 0.01 Gy	Energy absorbed
Sievert	Sv	gray × quality number of radiation (*e.g.*, 20 for alpha and 1 for beta)	Tissue damage
Rem	Rem	1 rem = 0.01 Sv	Tissue damage

$$1 \text{ millicurie (mCi)} = 3.7 \times 10^7 \text{ dps}$$

$$1 \text{ microcurie } (\mu Ci) = 3.7 \times 10^4 \text{ dps}$$

$$1 \text{ Ci} = 3.7 \times 10^{10} \text{ Bq} = 37 \text{ GBq } (G = \text{giga}, 1 G = 10^9)$$

Specific activity of a radionuclide is its activity per kilogram (or dm³) of the radioactive material.

(In some cases specific activity can also be taken as the activity per gram)

Theory of Radioactive Disintegration

According to this theory (Rutherford, 1903) the atoms of a radioactive element are inherently unstable. They disintegrate spontaneously, emitting α- or β-particles, and resulting in the formation of a new element which is physically and chemically different from its parent element. The new element formed may, in turn, be again radioactive and will disintegrate in a similar fashion. The process will continue until the atoms of stable element are reached. Following radioactive disintegration series have been identified among the naturally occurring radioactive elements and are summarised in Table 5.5. **Series is decided by dividing the atomic mass of the parent element by 4. Series will be $(4n + x)$, where x is the remainder.**

Neptunium series is artificial series while the other are natural series. The thorium, neptunium and actinium series are so named because an isotope of these elements has the largest half-life in the respective series. The actinium series was once thought to originate with actinium-227—the isotope having the longest half-life amongst the series members. However, it has been investigated that the series actually initiates with a very long-lived uranium-235 isotope.

There is no cross-linking between the four series. Thus, an isotope of one series will not decay to an isotope belonging to another series.

Table 5.5 **Radioactive Disintegration Series**

Series		Starts with	Ends with
$4n$	– Thorium series	$^{232}_{90}Th$	$^{208}_{82}Pb$
$(4n + 1)$	– Neptunium series	$^{241}_{94}Pu$	$^{209}_{83}Bi$
$(4n + 2)$	– Uranium series	$^{238}_{92}U$	$^{206}_{82}Pb$
$(4n + 3)$	– Actinium series	$^{235}_{92}U$	$^{207}_{82}Pb$

Illustration 1 *In Table 5.5, four disintegration series have been given. Calculate number of α- and β-particles emitted in each series.*

Solution

Series		Net disintegration reaction	Number of particles	
			α	β
$4n$	– Thorium series	$^{232}_{90}Th \longrightarrow ^{208}_{82}Pb$	6	4
$(4n + 1)$	– Neptunium series	$^{241}_{94}Pu \longrightarrow ^{209}_{83}Bi$	8	5
$(4n + 2)$	– Uranium series	$^{238}_{92}U \longrightarrow ^{206}_{82}Pb$	8	6
$(4n + 3)$	– Actinium series	$^{235}_{92}U \longrightarrow ^{207}_{82}Pb$	7	4

Rate of Radioactive Disintegration

Radioactive decay law states :"*The rate of disintegration of a radioactive material (the decay rate) is directly proportional to the number of atoms present.*"

$$\therefore \qquad -\frac{dN}{dt} \propto N$$

$$-\frac{dN}{dt} = \lambda N$$

where $-\frac{dN}{dt}$ represents rate of decay in disintegration of atoms per unit time (say second), N the number of atoms disintegrating at that time and λ the disintegration constant (also called decay constant).

$$\lambda = \frac{\text{rate of decay}}{N}$$

Radioactive decay is a **first-order** process*.

If N_0 is the number of atoms at the start of decay and N the number of atoms at a given time then by first order kinetics,

$$-\int \frac{dN}{dt} = \lambda \int N$$

$$\lambda = \frac{2.303}{t} \log_{10}\left(\frac{N_0}{N}\right)$$

$$N = N_0 e^{-\lambda t}$$

If T_{50} (or $t_{1/2}$, both notations have been used in further parts) is the half-life which is the time for half of the substance to disappear.

thus, at $t = t_{1/2}$, $N = \frac{N_0}{2}$ then $\lambda = \frac{2.303}{t_{1/2}} \log 2$

$$\therefore \qquad t_{1/2} = \frac{2.303 \log 2}{\lambda} = \frac{0.693}{\lambda}$$

Thus, half-life period is independent of number of atoms present at a given time.

Half-life provides an easy way to estimate the time required before a radioactive element is no longer a health hazard. Strontium-90, for example, is a β-emitter with a half-life of 29.1 years. Significant quantities of strontium-90 were dispersed into the environment in atmospheric nuclear bomb tests in the 1960s and 1970s, and from the half-life, we know that a little less than half is still around. Sr-90 is taken into the body and deposited in bone, taking the place of calcium. Radiation damage by Sr-90 in bone has been directly linked to bone-related cancers.

If $y = \frac{t}{t_{1/2}}$ = number of half-lives in a time t

then $$N = N_0\left(\frac{1}{2}\right)^y$$

and in concentration terms $C = C_0\left(\frac{1}{2}\right)^y$

The average-life period (t_{av}) is related to $t_{1/2}$ and disintegration constant by equation:

$$t_{av} = \frac{1}{\lambda} = \frac{t_{1/2}}{0.693} = 1.443\ t_{1/2}$$

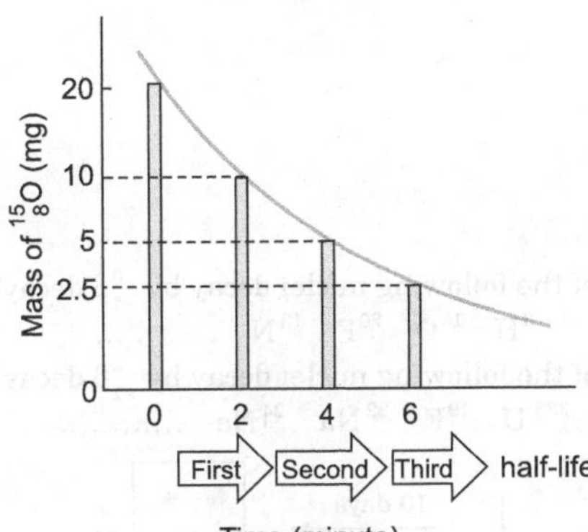

Fig. 5.1 Decay of 20.0 mg of oxygen-15. After each half-life period of 2.0 min, the mass of oxygen-15 decreases by one half

Illustration 1 *Phosphorus-32, a radioactive isotope used in leukemia- therapy has a half-life of 14.28 days. What per cent of a sample disintegrated after 35.0 days?*

Solution Let sample initially present = N_0

and after 35.0 days sample = N

$$\text{decay constant } \lambda = \frac{0.693}{T_{50}} = \frac{0.693}{14.28} = 0.0485 \text{ day}^{-1}$$

$$\lambda = \frac{2.303}{t} \log \frac{N_0}{N}$$

$$0.0485 = \frac{2.303}{35} \log \frac{N_0}{N}$$

$$\therefore \quad \log \frac{N_0}{N} = \frac{35 \times 0.0485}{2.303} = 0.7375$$

$$\frac{N_0}{N} = 5.464$$

$$\frac{N}{N_0} = 0.183$$

% of sample left = 0.183×100 = **18.3**

% of sample disintegrated = 100 − 18.3

= **81.7%**

Alternate Method

Let initial amount of sample = $N_0 = 100$

and sample left after time $t = N$

then $N = N_0 \left(\frac{1}{2}\right)^y$

where $y = \frac{\text{total time } (t)}{T_{50}} = \frac{35.0}{14.28} = 2.451$

$$N = 100\left(\frac{1}{2}\right)^{2.451} = \mathbf{18.3\%}$$

$\therefore$ Sample disintegrated = $N_0 - N$ = 100 − 18.3

= **81.7%**

Illustration 2 *The decay constant for sodium-24, a radioisotope used medically in blood-study, is* $4.63 \times 10^{-2}\ h^{-1}$. *What is the half-life of* ^{24}Na?

Solution Half-life (T_{50}) is related to the decay constant λ of the radioactive disintegration (which follows first-order kinetics) by equation

$$T_{50} = \frac{0.693}{\lambda}$$

$$\therefore \quad T_{50} = \frac{0.693}{4.63 \times 10^{-2}\ h^{-1}} = 14.97 \text{ h}$$

The half-life of sodium-24 is **14.97 h**

Illustration 3 *The half-life of radon-222, a radioactive gas of concern as a health hazard is 3.823 days. What is the decay constant when it undergoes first-order radioactive disintegration?*

Solution As given in above example

$$T_{50} = \frac{0.693}{\lambda}$$

or

$$\lambda = \frac{0.693}{T_{50}} = \frac{0.693}{3.823} \text{ days}$$

$$= 0.181 \text{ day}^{-1}$$

Radioactive Equilibrium

According to the theory of radioactive disintegration, the product of a radioactive disintegration may also be radioactive, and the rate of disintegration of the daughter element depends upon the amount of it present at different time. When steady state is reached

$$\left(\frac{dN}{dt}\right)_{\text{parent}} = \left(\frac{dN}{dt}\right)_{\text{daughter}}$$

or

$$\lambda_1 N_1 = \lambda_2 N_2$$

where λ_1 and λ_2 are the disintegration constants for the decay of parent and daughter radioactive species and N_1 and N_2 are the number of atoms disintegrating at a given time.

Thus, $\frac{N_1}{N_2} = \frac{\lambda_2}{\lambda_1} = \frac{t_1 \text{ (half-life)}}{t_2 \text{ (half-life)}}$

If the parent (say A) has smaller λ than the daughter (B), but none are very small, a steady state is reached at a stage when

$$\frac{N_B}{N_A} = \frac{\lambda_A}{\lambda_B - \lambda_A}$$

This is called a **transient equilibrium,** which is in fact a steady state.

Illustration 1 *Radioactive equilibrium is set up between uranium and radium atoms in a uranium mineral in the ratio of 28 : 1.* T_{50} *for radium is 1620 years. Calculate the disintegration constant of uranium.*

Solution When radioactive equilibrium is set up

$$\lambda_1 N_1 = \lambda_2 N_2$$

$$\frac{\lambda_U}{\lambda_{Ra}} = \frac{N_{Ra}}{N_U}$$

$$\frac{\lambda_U}{\frac{0.693}{(T_{50})_{Ra}}} = \frac{1}{28}$$

$$\therefore \quad \lambda_U = \frac{1}{28} \times \left(\frac{0.693}{1620 \text{ years}}\right)$$

$$= \mathbf{1.528 \times 10^{-5}\ yr^{-1}}$$

Target Practice 1

1. Complete the following nuclear equations. Write the mass number and atomic number for the remaining particle, as well as its symbol.
   (a) $^{54}_{26}Fe + ^{4}_{2}He \longrightarrow 2^{1}_{0}n + ?$ ....
   (b) $^{27}_{13}Al + ^{4}_{2}He \longrightarrow ^{30}_{15}P +$ ....
   (c) $^{32}_{16}S + ^{1}_{0}n \longrightarrow ^{1}_{1}H +$ ....
   (d) $^{18}_{9}F \longrightarrow ^{18}_{8}O +$ ....
   (e) $? \longrightarrow ^{24}_{12}Mg + ^{0}_{-1}e$
2. What particle is emitted in the following nuclear reactions?
   (a) Gold-198 decays to mercury-198
   (b) Radon-222 decays to polonium-218
   (c) Cesium-137 decays to barium-137
   (d) Indium-110 decays to cadmium-110
3. Which of the following nuclei decay by $^{0}_{-1}\beta$ decay?
   ^{3}H ^{16}O ^{20}F ^{13}N ....
4. Which of the following nuclei decay by $^{0}_{+1}\beta$ decay?
   ^{238}U ^{19}F ^{22}Na ^{24}Na ....
5. 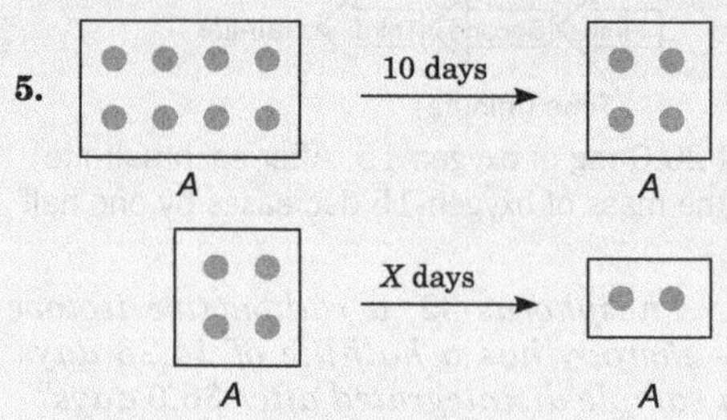

   Specify X...........

Answers

1. (a) $^{56}_{28}Ni$ (b) $^{1}_{0}n$ (c) $^{32}_{15}P$ (d) $^{0}_{1}e$ (e) $^{24}_{11}Na$
2. (a) $^{198}_{79}Au \longrightarrow ^{198}_{80}Hg + ^{0}_{-1}e$
   (b) $^{222}_{86}Rn \longrightarrow ^{218}_{84}Po + ^{4}_{2}He$
   (c) $^{137}_{55}Cs \longrightarrow ^{137}_{56}Ba + ^{0}_{-1}e$
   (d) $^{110}_{49}In \longrightarrow ^{110}_{48}Cd + ^{0}_{1}e$
3. Refer text
4. Refer text
5. $X = 10$ (Half-life of first-order disintegration is independent of initial radioactivity)

Practice Exercise 1

1. Although a neutron within a nucleus is stable, a neutron in free space decays into a proton and an electron (β-particle). Write its equation.
2. What are approximate atomic weight of electron and positron per mol?
3. $^{234}_{90}Th$ disintegrates to give $^{206}_{82}Pb$ as the final product. How many α- and β-particles are emitted during the process?
4. Fill in the blanks :
   (a) Alpha-rays are fast moving .... nuclei.
   (b) Emission of one α-and two β-particles will result in the formation of ....
   (c) .... rays cause health hazards to maximum extent.
   (d) Nuclear fusion is a .... reaction.
   (e) A radioactive change is always ....
   (f) The rate of radioactive change is independent of pressure and ....
   (g) New elements are produced in ....
5. Identify chemical change and radioactive change in the following :
   (a) $^{14}_{6}C \longrightarrow ^{14}_{7}N + ^{0}_{-1}e$
   (b) $^{27}_{13}Al + ^{4}_{2}He \longrightarrow ^{30}_{15}P + ^{1}_{0}n$
   (c) $^{27}_{13}Al + 3^{1}_{1}H \xrightarrow{\Delta} AlH_3$
   (d) $^{23}_{11}Na + ^{2}_{1}H \longrightarrow ^{24}_{11}Na + ^{1}_{1}H$
6. $^{238}_{92}U$ (group III B) emits one α particle. What is the group of new element formed?
7. Assign groups to the elements A, B, C, D and E starting from element X (group II A)

   $$\underset{\text{(II A)}}{X} \xrightarrow{-\alpha} A \xrightarrow{-\beta} B \xrightarrow{-\beta} C \xrightarrow{-\alpha} D$$

   What is relation between A and D?
   What is relation between A, B and C?
8. If a radioactive element $_{79}Au$ emits β-particle, what is the new group? Is there any change in the number of moles of daughter element? $_{79}Au$ belongs to group I B (copper family).

9. What is relation between following units:
(a) Bq and Ci (b) Bq and Ru
(c) Ru and Ci

10. $\left(-\dfrac{dN}{dt} = \lambda N\right)$ represents the rate of decay in disintegration of atoms per unit time, where N is the number of atoms disintegrating at that time and λ the disintegration constant. Calculate number of curies when number of atoms equal to one Avogadro's number disintegrates in the case when $\lambda = 0.6146\times 10^{-10}\ s^{-1}$.

11. The decay constant for mercury-197, a radioisotope used medically in kidney-scans, is $1.08\times 10^{-2}\ h^{-1}$. What is the half-life of mercury-197?

12. The half-life of carbon-14 is 5715 yr. What is decay constant?

13. Write balanced nuclear equations for each of the following:
(a) β-emission from ruthenium–106
(b) α-emission from bismuth–189
(c) electron capture by polonium–204.

Answers

1. $^1_0n \longrightarrow {}^1_1p + {}^0_{-1}\beta$
2. atomic weight of electron or positron $= 9.1\times 10^{-28} \times 6.00\times 10^{23}\ g\ mol^{-1} = 5.48\times 10^{-4}\ g\ mol^{-1}$
3. 7α and 6β
4. (a) helium; (b) an isotope; (c) γ (d) thermonuclear (e) spontaneous (f) temperature (g) radioactive change.
5. (a), (b) and (d) radioactive change; (c) chemical change.
6. III B (Z = 90 to 104 are in III B)
7. *A* (Zero), *B* (IA), *C*(IIA), *D* (Zero)
A and *D* are isotopes;
A, *B* and *C* are isobars.
8. $_{79}Au \longrightarrow {}_{80}Hg + \beta^-$. $_{80}Hg$ belongs to II B. (zinc family)
There is no change in number of moles since atomic mass remains unchanged.
9. (a) 1 Ci = 3.7×10^{10} Bq (b) 1 Bq = 1×10^{6} Ru
(c) 1 Ci = 3.7×10^{4} Ru
10. 1000 Ci 11. 64.17 h 12. $1.212\times 10^{-4}\ yr^{-1}$
13. (a) $^{106}_{44}Ru \longrightarrow {}^{0}_{-1}e + {}^{106}_{45}Rh$
(b) $^{189}_{83}Bi \longrightarrow {}^{4}_{2}He + {}^{185}_{81}Tl$
(c) $^{204}_{84}Po + {}^{0}_{-1}e \longrightarrow {}^{204}_{83}Bi$

Nuclear Stability

A number of basic questions have probably occurred to you as we have been describing nuclear decay processes.

Why do some radioactive nuclei decay by α-emission and some by β^--emission ?

Why do the lighter elements have so few naturally occurring radioactive nuclides, whereas those of the heavier elements all seem to be radioactive?

From the Fig. 5.2, it is clear that nuclides with even number of protons and neutrons have higher binding energies than their neighbours and are stable. Experimentally, it is found that nuclei with certain number of protons or neutrons appear to be very stable. These numbers called **magic numbers** and associated with specially stable nuclei, were later explained by the **shell model.** According to this theory, **a magic number is the number of nuclear particles in a completed shell of protons or neutrons.**

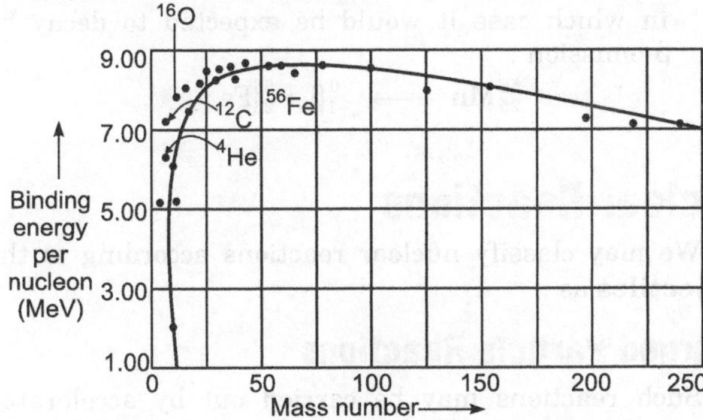

Fig. 5.2 Binding energy per nucleon

- Magic number for protons : 2, 8, 20, 28, 50, 82, 114
- Magic number for neutrons : 2, 8, 20, 28, 50, 82, 126
- ^4_2He (α-particle) is stable. It contains two protons and two neutrons hence, it contains magic number of protons (2), neutrons (2).

$$^{238}_{92}U \longrightarrow {}^{206}_{82}Pb \text{ (stable)}$$

It ($^{206}_{82}Pb$) has 82 protons (magic number)
Table 5.6 gives number of stable isotopes.

Table 5.6 **Number of Stable Isotopes with Even and Odd Numbers of Protons and Neutrons**

	Number of stable isotopes			
	157	52	50	5
Number of protons	Even	Even	Odd	Odd
Number of neutrons	Even	Odd	Even	Odd

- Finally when we plot each stable nuclide on a graph between the number of protons (*Z*) and the number of neutrons (*N*), these stable nuclides fall in a certain region, or band, of graph. (Fig. 5.3). **The band of stability is the region in which stable nuclides lie in a plot of number of protons against number of neutrons.** From this graphs it is clear that
- Stable nuclei are those for which number of neutrons and protons are equal and are generally upto 20.

$$({}^4_2He\ ,\ {}^{12}_{6}C\ ,\ {}^{16}_{8}O\ ,{}^{28}_{14}Si\ ,{}^{40}_{20}Ca)$$

- For higher atomic numbers, because of increasing repulsive forces between protons, large number of neutrons must be present to stabilise a nucleus and the neutron/proton (n/p) ratio increases.

For bismuth the ratio is about 1.5 : 1

- Isotopes of all the known elements with $Z > 83$ are radioactive.

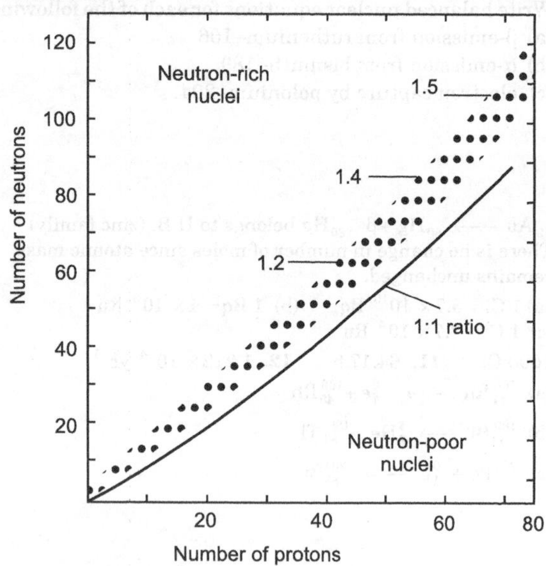

Fig. 5.3 Plot of the number of neutrons against the number of protons in stable nuclei (shown by dots)

Mechanism of Radioactive Decay

The emission of an α-particle by a nucleus is not difficult to visualise. A bundle of four nucleons (two protons, two neutrons) is ejected from a nucleus, and the nucleus becomes energetically more stable. In the emission of β-particle, neutron is converted into proton.

$${}^{238}_{92}U \longrightarrow {}^{234}_{90}Th + {}^{4}_{2}He \quad \alpha\text{-emission}$$

$${}^{1}_{0}n \longrightarrow {}^{1}_{1}p + {}^{0}_{-1}e \quad \beta^-\text{-emission}$$

$${}^{1}_{1}p \longrightarrow {}^{1}_{0}n + {}^{0}_{+1}e \quad \beta^+\text{-emission}$$

$${}^{1}_{1}p + {}_{-1}e^{0} \longrightarrow {}^{1}_{0}n \quad \text{electron capture}$$

- In general, if a nuclide lies above the belt of stability (Fig. 5.3), the $\left(\frac{n}{p}\right)$ ratio is too high.

A neutron is converted to a proton and a β^- particle is emitted.

$${}^{14}_{6}C \longrightarrow {}^{14}_{7}N + {}_{-1}e^{0}$$

$$\frac{n}{p} = 1.33 \quad \frac{n}{p} = 1 \text{ (stable)}$$

- If a nuclide lies below the belt of stability, the $\left(\frac{n}{p}\right)$ ratio is too low.

Either a proton is converted to a neutron, followed by β^+ emission or electron capture occurs.

$${}^{38}_{19}K \longrightarrow {}^{38}_{18}Ar + {}^{0}_{+1}e \text{ (or } {}^{0}_{+1}\beta)$$

$${}^{15}_{8}O \longrightarrow {}^{15}_{7}N + {}^{0}_{+1}\beta$$

$${}^{106}_{47}Ag + {}^{0}_{-1}e \longrightarrow {}^{106}_{46}Pd$$

$${}^{37}_{18}Ar + {}^{0}_{-1}e \longrightarrow {}^{37}_{17}Cl$$

- All nuclides with atomic number greater than 83 are beyond the band of stability and are radioactive. Many of these decay by α-emission.

$${}^{226}_{88}Ra \longrightarrow {}^{222}_{86}Rn + {}^{4}_{2}He$$

$${}^{210}_{84}Po \longrightarrow {}^{206}_{82}Pb + {}^{4}_{2}He$$

Illustration 1 *Consider the following radioactive change*

$${}^{22}_{11}Na \longrightarrow {}^{22}_{10}Ne$$

What is the type of particle captured or emitted?

Solution ${}^{22}_{11}Na \longrightarrow {}^{22}_{10}Ne$

There is change in atomic number of one unit, hence, it can

(i) capture an electron

$${}^{22}_{11}Na + {}^{0}_{-1}e \longrightarrow {}^{22}_{10}Ne$$

(ii) emit a positron ${}^{22}_{11}Na \longrightarrow {}^{22}_{10}Ne + {}^{0}_{+1}\beta$

Illustration 2 *Identify probable mode(s) of decay for each isotope and write an equation for the decay process.* (*a*) *oxygen-15*, ${}^{15}_{8}O$ (*b*) *uranium-234* ${}^{234}_{92}U$ (*c*) *fluorine-20*, ${}^{20}_{9}F$ (*d*) *manganese-56*, ${}^{56}_{25}Mn$

Solution (a) Oxygen-15 has 7 neutrons and 8 protons, so the n/p ratio is less than 1-too low for ${}^{15}O$ to be stable. Nuclei with too few neutrons are expected to decay by either positron emission or electron capture. In this instance, the process is ${}^{0}_{+1}\beta$ emission, and the equation is ${}^{15}_{8}O \longrightarrow {}^{0}_{+1}\beta + {}^{15}_{7}N$.

(b) Alpha emission is a common mode of decay for isotopes of elements with atomic numbers higher than 83. The decay of uranium-234 is one example.

$${}^{234}_{92}U \longrightarrow {}^{230}_{90}Th + {}^{4}_{2}\alpha$$

(c) Fluorine-20 has 11 neutrons and 9 protons, a high n/p ratio. The ratio is lowered by β emission :

$${}^{20}_{9}F \longrightarrow {}^{0}_{-1}\beta + {}^{20}_{10}Ne$$

(d) The atomic weight of manganese is 54.85. The higher mass number, 56, suggests that this radioactive isotope has an excess of neutrons, in which case it would be expected to decay by β-emission :

$${}^{56}_{25}Mn \longrightarrow {}^{0}_{-1}\beta + {}^{56}_{26}Fe$$

Nuclear Reactions

We may classify nuclear reactions according to the **projectiles** as :

Charged Particle Reactions

Such reactions may be carried out by accelerated protons, deutrons or alpha particles:

Projectile	Reaction type
Proton	$(p, n), (p, \alpha), (p, \gamma), (p, d)$
Deutron	$(d, p), (d, n), (d, \alpha), (d, \gamma)$
Alpha particle	$(\alpha, n), (\alpha, p)$

- **By writing (p, n), it means, proton is absorbed and neutron is emitted.**

Examples :

(i) p, n $\quad {}^{24}_{11}Na + {}^{1}_{1}H \longrightarrow {}^{24}_{12}Mg + {}^{1}_{0}n$

(ii) p, α $\quad {}^{7}_{3}Li + {}^{1}_{1}H \longrightarrow {}^{4}_{2}He + {}^{4}_{2}He$

(iii) p, γ $\quad {}^{27}_{13}Al + {}^{1}_{1}H \longrightarrow {}^{28}_{14}Si + \gamma$

(iv) d, p $\quad {}^{27}_{13}Al + {}^{2}_{1}H \longrightarrow {}^{28}_{13}Al + {}^{1}_{1}H$

(v) d, n $\quad {}^{27}_{13}Al + {}^{2}_{1}H \longrightarrow {}^{28}_{14}Si + {}^{1}_{0}n$

(vi) d, α $\quad {}^{6}_{3}Li + {}^{2}_{1}H \longrightarrow {}^{4}_{2}He + {}^{4}_{2}He$

(vii) α, n $\quad {}^{9}_{4}Be + {}^{4}_{2}He \longrightarrow {}^{12}_{6}C + {}^{1}_{0}n$

(viii) α, p $\quad {}^{14}_{7}N + {}^{4}_{2}He \longrightarrow {}^{17}_{8}O + {}^{1}_{1}H$

- Reaction (ii) is the first artificial nuclear reaction carried out with artificially accelerated particles.
- Reaction (vi) gives one of the highest non-fission energies (22.2 MeV) recorded for light atoms.
- Neutron was discovered through the reaction (vii).
- Reaction (viii) is the first artificial nuclear reaction studied.

Neutral Particle Reactions

Neutrons are generally used as projectiles in such reactions. Neutrons of high energy are obtained from different artificial nuclear reactions. These are slowed down through **moderators** like graphite, heavy water, paraffin, etc. Much of the energies of the neutron is transferred to these substances through elastic scattering.

Example :

(i) n, γ $\quad {}^{238}_{92}U + {}^{1}_{0}n \longrightarrow {}^{239}_{92}U + \gamma$

(ii) n, p $\quad {}^{14}_{7}N + {}^{1}_{0}n \longrightarrow {}^{14}_{6}C + {}^{1}_{1}H$

(iii) n, α $\quad {}^{6}_{3}Li + {}^{1}_{0}n \longrightarrow {}^{4}_{2}He + {}^{3}_{1}H$

- Reaction (i) shows the formation of transuranium elements.

Reaction (iii) shows the production of tritium.

Reactions initiated by photons may also be considered as neutral particle reactions.

$${}^{2}_{1}H + \gamma \longrightarrow {}^{1}_{1}H + {}^{1}_{0}n$$

$${}^{9}_{4}Be + \gamma \longrightarrow {}^{8}_{4}Be + {}^{1}_{0}n$$

Nuclear Fission

It is a nuclear reaction in which a heavy nucleus splits into lighter nuclei and energy is released.

- Californium -252 decays both by alpha emission (97%) and by spontaneous fission (3%). During spontaneous fission, the nucleus splits into two stabler, lighter nuclei plus several neutrons.

$${}^{252}_{98}Cf \longrightarrow {}^{142}_{56}Ba + {}^{106}_{42}Mo + 4{}^{1}_{0}n$$

- When a neutron strikes the ${}^{235}_{92}U$ nucleus, the nucleus splits into roughly equal parts, giving off several neutrons.

$${}^{1}_{0}n + {}^{235}_{92}U \longrightarrow \begin{cases} {}^{142}_{54}Xe + {}^{90}_{38}Sr + 4({}^{1}_{0}n) \\ {}^{139}_{56}Ba + {}^{94}_{36}Kr + 3({}^{1}_{0}n) \\ {}^{144}_{55}Cs + {}^{90}_{37}Rb + 2({}^{1}_{0}n) \end{cases}$$

- If the neutrons from each nuclear fission are absorbed by other uranium -235 nuclei, these nuclei split and release even more neutrons. Thus, **a chain reaction can occur.** *A nuclear chain reaction is a self-sustaining series of nuclear fissions caused by the absorption of neutrons released from previous nuclear fissions.* This is shown in Fig. 5.4.

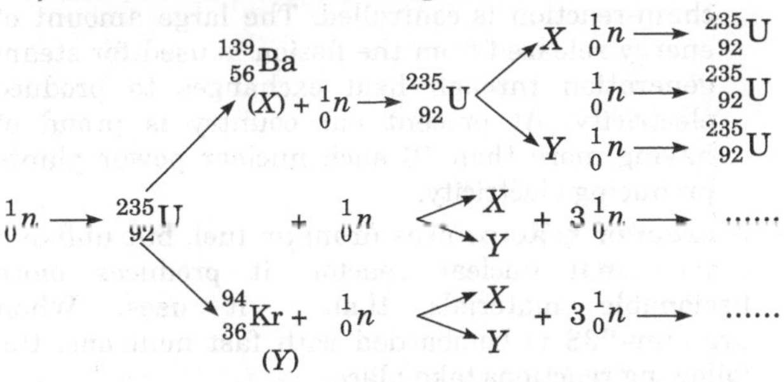

Fig. 5.4 Representation of nuclear chain reaction of nuclear fissions

- There should be a **critical size** of the fissionable material to maintain a fission chain. This in turn, requires a minimum **critical mass** of the fissionable material. It is the *smallest mass of the fissionable material in which a chian reaction can be sustained.* If the mass is much larger than this (a supercritical) mass, the number of nuclei that split multiply rapidly. *An atomic bomb is detonated with a small amount of chemical explosive that pushes together two or more masses of fissionable material to get a supercritical mass.*

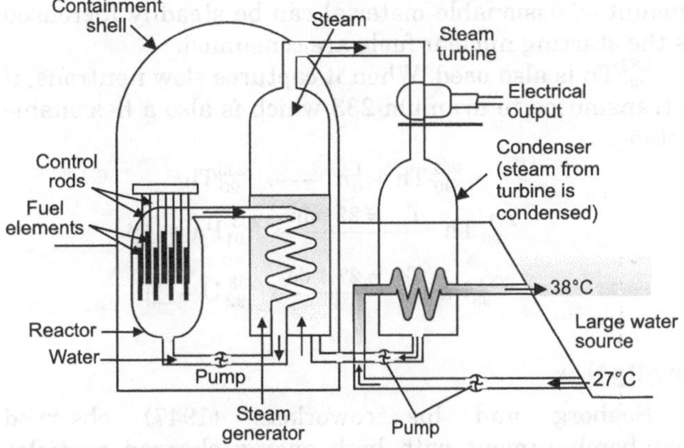

Fig. 5.5 A diagram of a nuclear power plant. Heat produced in the reactor core is transferred by coolant circulating in a closed loop to a steam generator, and the steam then drives a turbine to generate electricity.

A nuclear fission reactor (Fig. 5.5) is a device that permits a controlled chain nuclear-fissions. **Control rods** made of elements such as boron and cadmium, absorb additional neutrons and can therefore, slow the chain reactions.

A nuclear reactor consists of three components:

(a) A fissile material (uranium enriched in $^{235}_{92}U$, say 2.3%)

(b) A moderator (graphite or heavy water D_2O) to slow the neutrons so that they are captured and become effective to bring about fission reaction.

(c) Control rods made of boron steel or cadmium, which are capable of absorbing neutrons and are used to ensure that neutron flux is under control and thus chain-reaction is controlled. The large amount of energy released from the fission is used for steam generation through heat exchanges to produce electricity. At present our country is proud of having more than 12 such nuclear power plants producing electricity.

- A **breeder reactor** uses uranium fuel, but unlike a conventional nuclear reactor, it produces more fissionable materials than it uses. When uranium-238 is bombarded with fast neutrons, the following reactions take place:

$$^{238}_{92}U + ^{1}_{0}n \longrightarrow ^{239}_{92}U$$

$$^{239}_{92}U \xrightarrow{T_{50} = 23.4 \text{ min}} ^{239}_{93}Np + ^{0}_{-1}\beta$$

$$^{239}_{93}Np \xrightarrow{T_{50} = 2.35 \text{ day}} ^{239}_{94}Pu + ^{0}_{-1}\beta$$

In this manner the non-fissionable uranium-238 is transmuted into the fissionable isotope plutonium-239.

In a typical breeder reactor, nuclear fuel containing uranium-235 or plutonium-239 is mixed with uranium-238 so that breeding takes place within the core. For every uranium-235 (or plutonium-239) nucleus undergoing fission, more than one neutron is captured by uranium-238 to generate plutonium-239. Thus, the amount of fissionable material can be steadily increased as the starting nuclear fuels are consumed.

$^{232}_{90}Th$ is also used. When it captures slow neutrons, it is transmuted to uranium-233 which is also a fissionable isotope.

$$^{232}_{90}Th + ^{1}_{0}n \longrightarrow ^{233}_{90}Th$$

$$^{233}_{90}Th \xrightarrow{T_{50} = 22 \text{ min}} ^{233}_{91}Pa + ^{0}_{-1}\beta$$

$$^{233}_{93}Pa \xrightarrow{T_{50} = 27.4 \text{ day}} ^{233}_{92}U + ^{0}_{-1}\beta$$

Spallation

Seaborg and his coworkers (1947) observed that bombardment with high energy charged particles (*e.g.,* 400 MeV protons or 200 MeV deutrons) can break some target nuclei into several smaller nuclei with the emission of a large number of nucleons (10 to 20 or more). These are called spallation reactions.

$$^{75}_{33}As + ^{2}_{1}H \longrightarrow ^{56}_{25}Mn + 9^{1}_{1}H + 12^{1}_{0}n$$

$$^{238}_{92}U + ^{4}_{2}He \longrightarrow ^{187}_{74}W + 20^{1}_{1}H + 35^{1}_{0}n$$

Nuclear Fusion

In this type of nuclear reaction, certain light nuclei may fuse together with the liberation of tremendous amount of energy. To achieve this, the colliding nuclei must possess enough kinetic energy to overcome the initial repulsion between the positively charged cores. This energy may be made available by raising the temperature of the reacting system to several million degrees. Such reactions are therefore, also **known as thermonuclear reactions.**

$$^{2}_{1}H + ^{3}_{1}H \longrightarrow ^{4}_{2}He + ^{1}_{0}n + 17.6 \text{ MeV}$$

The energy of a fusion process is due to mass defect (converted to binding energy). The high temperature required to initiate such reactions may be attained initially through a fission process.

Spectroscopic evidence indicates that the sun is a tremendous fusion reactor consisting of 73% H, 26% He and 1% other elements. Following fusion reactions are believed to take place:

$$^{1}_{1}H + ^{2}_{1}H \longrightarrow ^{3}_{2}He$$

$$^{3}_{2}He + ^{3}_{2}He \longrightarrow ^{4}_{2}He + 2^{1}_{1}H$$

$$^{1}_{1}H + ^{1}_{1}H \longrightarrow ^{2}_{1}H + ^{0}_{+1}e$$

A major concern in choosing the proper nuclear fusion process for energy production is the temperature necessary to carry out the process. Some promising reactions are:

Reactions	Energy released
$^{2}_{1}H + ^{2}_{1}H \longrightarrow ^{3}_{1}H + ^{1}_{1}H$	6.3×10^{-13} J
$^{2}_{1}H + ^{3}_{1}H \longrightarrow ^{4}_{2}He + ^{1}_{0}n$	2.8×10^{-12} J
$^{6}_{3}Li + ^{2}_{1}H \longrightarrow 2^{4}_{2}He$	3.6×10^{-12} J

Artificial Radioactivity

Bombardment of stable elements with high energy α-particles, protons, neutrons, deutrons or γ-rays produce radioactive nuclides. These radio-nuclides do not occur naturally and may be called man-made or artificial. The radioactivity exhibited by these artificial radio-nuclides is referred to as **artificial radioactivity** or as **induced radioactivity.** (I. Curie and F. Joliot–1934)

The decay resulting from artificial radioactivity is usually a one-step process, *i.e.,* the radioisotope disintegrates into a stable isotope which does not show any further tendency to decay.

(a) The first artificial breakdown of the atomic nucleus was achieved by Rutherford in 1919. He bombarded nitrogen with α-particles and obtained oxygen and protons

$$^{14}_{7}N + ^{4}_{2}He \longrightarrow ^{17}_{8}O + ^{1}_{1}H$$

(b) The newly formed isotope undergoes positron emission.

$$^{27}_{13}Al + ^{4}_{2}He \longrightarrow ^{30}_{15}P + ^{1}_{0}n$$

$$^{30}_{15}P \xrightarrow{T_{50} = 3.25 \text{ min}} ^{30}_{14}Si + ^{0}_{1}e$$

artificial radioactive stable

$$^{24}_{12}Mg + ^{2}_{1}H \longrightarrow ^{22}_{11}Na + ^{4}_{2}He$$

$$^{22}_{11}Na \xrightarrow{T_{50} = 2.6 \text{ yr}} ^{22}_{10}Ne + ^{0}_{1}e$$

$$^{24}_{12}Mg + ^{4}_{2}He \longrightarrow ^{27}_{14}Si + ^{1}_{0}n$$

$$^{27}_{14}Si \longrightarrow ^{27}_{13}Al + ^{0}_{1}e$$

$$^{10}_{5}B + ^{4}_{2}He \longrightarrow ^{13}_{7}N + ^{1}_{0}n$$

$$^{13}_{7}N \longrightarrow ^{13}_{6}C + ^{0}_{1}e$$

(c) The newly formed isotope undergoes electron emission (β-decay)

$$^{23}_{11}Na + ^{1}_{0}n \longrightarrow ^{24}_{11}Na \xrightarrow{T_{50} - 15 \text{ h}} ^{24}_{12}Mg + ^{0}_{-1}e$$

$$^{9}_{4}Be + ^{2}_{1}H \longrightarrow ^{10}_{4}Be + ^{1}_{1}H$$

$$^{10}_{4}Be \longrightarrow ^{10}_{5}B + ^{0}_{-1}e$$

$$^{31}_{15}P + ^{1}_{0}n \longrightarrow ^{28}_{13}Al + ^{4}_{2}He$$

$$^{28}_{13}Al \longrightarrow ^{28}_{14}Si + ^{0}_{-1}e$$

$$^{16}_{8}O + ^{1}_{0}n \longrightarrow ^{13}_{6}C + ^{4}_{2}He$$

(d) The newly formed isotope undergoes K-electron capture (electron from K-shell or 1*s* energy level). In this process energy is emitted in the form of X-rays.

$$^{48}_{22}Ti + ^{2}_{1}H \longrightarrow ^{49}_{23}V + ^{1}_{0}n$$

$$^{49}_{23}V + ^{0}_{-1}e\text{(K-shell)} \longrightarrow ^{49}_{22}Ti + \text{X-rays}$$

(e) There can be formation of isomer in artificial nuclear reactions.

$$^{79}_{35}Br + ^{1}_{0}n \longrightarrow ^{80}_{35}Br$$

$$^{80}_{35}Br \longrightarrow ^{80}_{36}Kr + ^{0}_{-1}e$$

Thus ^{80}Br exists as two nuclear-isomers.

Binding Energy and Packing Fraction

The nearest integer to the mass of a nuclide (in the atomic mass scale) was called the mass number of that isotope. The difference between the actual isotopic mass and mass number was termed as **mass defect** of the nuclide.

Thus, Δm (mass defect) = isotopic mass (m) − mass number (A)

Aston introduced a term "packing fraction" for each nuclide to compare their mass excess, it was defined as:

$$\text{Packing fraction } (f) = \frac{\Delta m}{A} \times 10^4$$

- The packing fraction does not have any precise theoretical significance but it gives an indication about the stability of the nucleus.
- A negative packing fraction indicates that the isotopic mass is less than the nearest whole number. This suggests that a small fraction of the mass has been converted into energy of formation of that nucleus. **Hence, the more negative the packing fraction of an element, the more stable should be the nucleus.**
- On the other hand, a positive packing fraction indicates that the concerned nucleus should be less stable.
- Generalisations are shown in the Fig. 5.6.

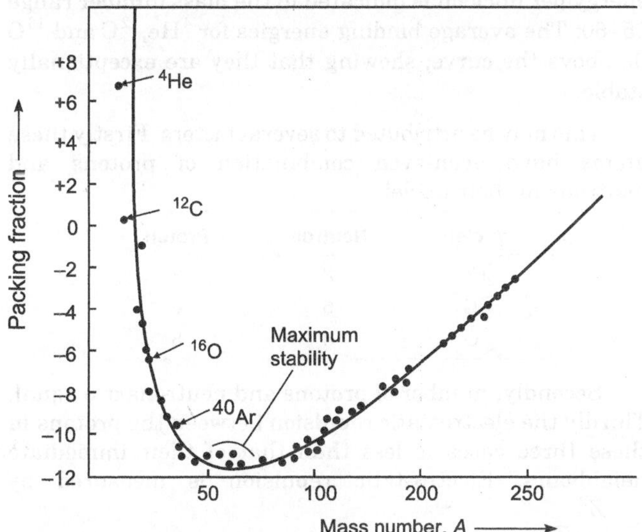

Fig. 5.6 Plot of packing fraction against mass number

This shows:

– The packing fractions of the stable nuclides lie more or less on a smooth curve.

– The curve passes through a minima in the mass number range 50–60, implying the maximum stability.

– After the minima, packing fraction steadily increases and ultimately becomes positive for elements of high mass number.

This is consistent with the observed radioactivity, *i.e.*, inherent instability, of these elements.

$$\text{Binding energy } (B) = \Delta m \times c^2$$

Binding energy of the nucleus is equal to the energy required to split the nucleus into its component nucleons.

Nuclear binding energy is generally expressed in the non-SI unit of MeV (mega electron volt).

Since, $1 \text{ amu} = 1.66 \times 10^{-27}$ kg

$\therefore$ (B) for one amu $= 1.66 \times 10^{-27} \times (3 \times 10^8)^2$ J

$$= \frac{1.66 \times 10^{-27} \times (3 \times 10^8)^2 \text{ J}}{1.602 \times 10^{-19} \text{ J/eV}}$$

$$= 932.6 \times 10^6 \text{ eV} = 932.6 \text{ MeV}$$

$$\therefore \quad (B) = \Delta m \text{ (in amu)} \times 931 \text{ MeV}$$

$$\text{Binding energy per nucleon } (\overline{B}) = \frac{B}{A}$$

$$= \frac{\text{total binding energy}}{\text{total number of nucleons}}$$

A plot of average binding energy per nucleon against mass number gives a nearly smooth curve (Fig 5.2). The curve is practically an inversion of the curve obtained by plotting packing fractions (Fig 5.6). Maximum binding energy per nucleon is indicated in the mass number range 55–60. The average binding energies for ^4He, ^{12}C and ^{16}O lie above the curve, showing that they are exceptionally stable.

This may be attributed to several factors. Firstly these atoms have even-even combination of protons and neutrons in their nuclei.

Nuclei	Neutron	Proton
^4_2He	2	2
$^{12}_6C$	6	6
$^{16}_8O$	8	8

Secondly, number of protons and neutrons are equal. Thirdly the electrostatic repulsion between the protons in these three cases is less than that of their immediate neighbours. Electrostatic repulsion is measured by

$$= \frac{Z^2}{(A)^{1/3}}$$

where Z is the number of protons and A the mass number. We compare this type of force with the immediate neighbours.

Table 5.7 **Electrostatic Force of Different Nuclei**

Nuclei	Z	A	Electrostatic force
^4_2He	2	4	$\frac{(2)^2}{(4)^{1/3}} = 2.53$
^6_3Li	3	6	4.97
$^{12}_6C$	6	12	15.78
$^{14}_7N$	7	14	20.33
$^{16}_8O$	8	16	25.39
$^{19}_9F$	9	19	30.45

In case of helium-4 and oxygen-16, there is an additional reason for exceptional stability. In each case number of neutrons (2 and 8) corresponds to magic number.

Note: *To convert* amu *into* MeV *conversion factor is generally taken* 931.

Illustration 1 *Helium-6 is a radioactive isotope with* $T_{50} = 0.81$ s. *Calculate the binding energy if mass defect is* 0.03141 *g mol*$^{-1}$. *Also calculate binding energy of helium-4 whose mass defect is* 0.03038 *g mol*$^{-1}$. *Is a* ^6He *nucleus more stable or less stable than a* ^4He *nucleus?*

Solution Binding energy of $^6He = \Delta m \times c^2$

$$= \left(\frac{0.03141}{1000} \text{ kg mol}^{-1}\right)(3 \times 10^8 \text{ ms}^{-1})^2$$

$$= 2.8269 \times 10^{12} \text{ J mol}^{-1}$$

Binding energy of

$$^4He = \left(\frac{0.03038}{1000} \text{ kg mol}^{-1}\right)(3 \times 10^8 \text{ m s}^{-1})^2$$

$$= 2.7342 \times 10^{12} \text{ J mol}^{-1}$$

$$\text{Binding energy of } ^6He \text{ per nucleon} = \frac{2.8269 \times 10^{12}}{6}$$

$$= 4.7115 \times 10^{11} \text{ J per nucleon}$$

$$\text{Binding energy of } ^4He \text{ per nucleon} = \frac{2.7342 \times 10^{12}}{4}$$

$$= 6.8355 \times 10^{11} \text{ per nucleon}$$

Thus, **^6He nucleus is less stable than a ^4He nucleus.**

Applications of Radioactive Isotopes

Carbon-Dating

Radiocarbon-dating of archaeological artifacts depends on the slow and constant production of radioactive carbon-14 in the upper atomosphere by neutron bombardment of nitrogen atoms (The neutrons come from the bombardment of other atoms by cosmic rays).

$$^{14}_7N + ^1_0n \longrightarrow ^{14}_6C + ^1_1H$$

$$^{14}_6C + O_2 \longrightarrow ^{14}_6CO_2$$

Carbon-14 atoms produced in the upper atomosphere combine with oxygen to yield $^{14}CO_2$, which slowly diffuses into the lower atmosphere, where it mixes with ordinary $^{12}CO_2$ and is taken up by plants during photosynthesis. When these plants are eaten, carbon-14 enters the food chain and is ultimately distributed evenly throughout all living organisms.

As long as a plant or animal is alive, a dynamic equilibrium exists in which an organism excretes or exhales the same amount of ^{14}C that it takes in and the ratio of ^{14}C to ^{12}C in the living organism is the same as that in the upper atmosphere. When the plant or animal dies, however, it no longer takes in more ^{14}C, and the ratio of ^{14}C to ^{12}C slowly decreases due to decay of ^{14}C

$$^{14}_6C \longrightarrow ^{14}_7N + ^0_{-1}e$$

Thus, from the activity of ^{14}C in the dead organism and that of the fresh sample (in the atmosphere), age can be decided.

$$\lambda(\text{decay constant}) = \frac{2.303}{t(\text{age})}\log\frac{N_0\ (\text{original activity})}{N\ (\text{final activity})}$$

Also, $$\lambda = \frac{0.693}{T_{50}}\ (\text{half-life})$$

Thus, t (age) can be determined.

or $$N = N_0\left(\frac{1}{2}\right)^y$$

where $$y = \frac{\text{total time}\,(t)}{\text{half-life}\,(T_{50})}$$

Rock (Geological) Dating

Uranium-238 has a half-life of 4.46×10^9 years and decays through a series of events to yield lead-206.

$$^{238}_{92}U \longrightarrow\ ^{206}_{82}Pb$$

The age of uranium containing rock can, therefore, be determined by measuring the $^{238}U/^{206}Pb$ ratio.

It is based on the kinetics of radioactive decay. It is assumed that no lead was initially present in the sample and the whole of it came from the uranium.

Initial radioactivity = [U] + [Pb] = N_0 in terms of gram atoms

Final radioactivity = [U] = N

We have assumed that due to high value of T_{50} of uranium, its amount remains unchanged.

$$\frac{N_0}{N} = \frac{[U]+[Pb]}{[U]} = 1 + \frac{[Pb]}{[U]}$$

Thus, $$\lambda = \frac{2.303}{T_{50}}\log\left(\frac{N_0}{N}\right) = \frac{2.303}{T_{50}}\log\left[1 + \frac{[Pb]}{[U]}\right]$$

or $$\left[1 + \frac{[Pb]}{[U]}\right] = (2)^y$$

where $y = \dfrac{t}{T_{50}}$ hence, t is calculated.

Thus, knowing the ratio of uranium to lead or thorium to lead in a particular minerals and knowing the half-lives of the radioisotopes, the age of the natural substance may be estimated.

Similarly, potassium-40 has a half-life of 1.26×10^9 years and decays through electron capture and through positron emission to yield argon-40.

$$^{40}_{19}K + ^{0}_{-1}e \longrightarrow\ ^{40}_{18}Ar$$

$$^{40}_{19}K \longrightarrow\ ^{40}_{18}Ar + ^{0}_{1}e$$

The age of a rock can be determined by crushing a sample, measuring the amount of ^{40}Ar gas that escapes, and comparing the amount of argon-40 with the amount of ^{40}K remaining in the sample

Medical Uses of Radioactivity

- French physician Henri Danlos first used radium in the treatment of tuberculous skin lesion. Now, uses of radioactivity have become a crucial part of modern medical care, both diagnostic and therapeutic
- Various techniques used in medical treatment are grouped into three classes:
  (i) *in vivo* procedures,
  (ii) therapeutic procedures and
  (iii) imaging procedures.
- In *vivo* studies—those take place inside the body—are carried out to assess the functioning of a particular organ or body system. A radio-pharmaceutical agent is administered, and its path in the body—whether it is absorbed, excreted, diluted or concentrated is determined by analysis of blood or urine samples. ^{51}Cr radioisotope is used in this type of study.
- Therapeutic procedures —those in which radiation is used as a weapon to kill diseased tissue—can involve either external or internal source of radiation.

  Cobalt-60 used in external radiation therapy for the treatment of cancer. It is embedded in the cancerous tissue and emitts both β- and γ-rays

  $$^{60}_{27}Co \longrightarrow\ ^{60}_{28}Ni + ^{0}_{-1}e + \gamma\ (\gamma \text{ rays destroys cancer cells})$$

  Sodium-24 injected alongwith salt solution to trace the flow of blood.

  Iodine-131 used in internal radiation therapy for the treatment of thyroid disease.

  Phosphorus-32 used for leukemia therapy which controls *polycythemiavera*-a condition in which the bone marrow produces excess red blood cells. The radioactive P-32 (in the form of phosphate) gets preferentially absorbed in the bone marrow and emits γ-rays, which destroy excessive red blood corpuscles.
- Imaging procedures gives diagnostic information about the health of body organs by analysing the distribution pattern of radioisotopes introduced into the body.

  Iodine-131 employed to study the structure and activity of thyroid glands.

  Iodine-123 used in imaging the brain.

  Technetium-99 *M* used in bone scans.
- **MRI** (magnetic resonance imaging) uses radio-waves to stimulate certain nuclei in the presence of an extremely powerful magnetic field.

Power Generation

The fission of uranium-235 nucleus by slow neutrons, releases energy to the tune of 200 MeV

$$^{235}_{92}U + ^{1}_{0}n \longrightarrow\ ^{141}_{56}Ba + ^{92}_{36}Kr + 2\text{-}3\ ^{1}_{0}n + 200\ \text{MeV}$$

which may be converted into electrical energy. Power produced by nuclear energy is unmatched as compared to any other energy resource. Theoretically, one tonne of uranium can give as much electrical energy as 3 million tonne of coal or 12 million barrels of oil. Further, not only the reserves of coal and oil are exhausting fastly, but burning of these fossil fuels pollutes the environment. Nuclear energy is a clean and abundant resource for power generation.

Agricultural Applications

Isotopes can be used to measure the uptake of nutrients. Plants take up phosphorus containing compounds from the soil through their roots. By adding a small amount of radioactive ^{32}P, a β-emitter (with a half-life of 14.3 days), to a fertiliser and then measuring the rate at which the radioactivity appears in the leaves, plant biologists can determine the rate at which phosphorus is taken up. The outcome can assist agriculture-scientists in identification of hybrid strain of plants that can absorb phosphorus quickly, resulting in faster-maturing crops, better yields per acre, and more food or fibre at lower cost. To measure pesticide level, a pesticide can be tagged with a radioisotope and then applied to a test field. By counting the disintegrations of radioactive tracer, information can be obtained about how much pesticide is in soil; is taken up by the plant, and is carried off in runoff surface water. **(Radioactive isotope used to carry out these investigation has short half-life and decays to harmless levels in a few days)**

Isotopes Dilution Technique is used to determine, the volume of blood in an animal subject. A small amount of radioactive isotope is injected into the bloodstream. After a period of time to allow the isotope to become distributed throughout the body, a blood sample is taken and its radioactivity measured.

Chemical Investigation

Structural Determination

When more than one atom of a particular element are present in a molecule, it needs to be decided whether all of these are equivalent or not. Radioisotope labelling settles the issue.

- To decide nature of bonding in PCl_5.

In order to determine whether all five chlorides in phosphorus (V) chloride are in a chemically similar environment, the compound may be synthesised using labelled chlorine molecules.

$$PCl_3 + \dot{C}l_2 \longrightarrow P\dot{C}l_5$$

The two labelled chlorine would rapidly exchange with three non-labelled ones, if all five chlorines are in a similar environment, and as a result all five chlorines would have radioactive character. If above synthesised PCl_5 is hydrolysed.

$$P\dot{C}l_5 + H_2O \longrightarrow P\dot{O}Cl_3 + 2H\dot{C}l$$

HCl is found to be radioactive and $POCl_3$ is found to be non-radioactive. Thus, two chlorines are different from the other three.

- To decide structure of $S_2O_3^{2-}$.

The formula of thiosulphate ion is $S_2O_3^{2-}$. Following reaction is found to be reversible depending on the experimental conditions :

$$SO_3^{2-}(aq) + S(s) \underset{}{\overset{H^+}{\rightleftharpoons}} S_2O_3^{2-}(aq)$$

If one starts with elemental sulphur enriched with the radioactive sulphur-35, the isotope acts as a "label" for S atoms. All the labels are found in the sulphur precipitated when $S_2O_3^{2-}(aq)$ is treated with dilute acids. Thus, two sulphur atoms are not structurally equivalent as would be case if the structure were:

$$[:\ddot{O}-\ddot{S}-\ddot{O}-\ddot{S}-\ddot{O}:]^{2-}$$

Otherwise, the radioisotope would be present in both the sulphur atoms. Based on this thiosulphate has the structure:

$$\left[\begin{array}{c} :\ddot{O}: \\ | \\ :\ddot{O}-S-\ddot{S}: \\ | \\ :\ddot{O}: \end{array}\right]^{2-}$$

Oxidation number of central S atom is **+6** and that of terminal is **–2**, thus net oxidation number is **4** (for two S-atoms) and difference is **8**.

Study of photosynthesis also uses radioisotope ^{18}O to determine the source of O_2

$$6CO_2 + 12H_2{}^{18}O \longrightarrow C_6H_{12}O_6 + 6O_2^{18} + 6H_2O$$

On starting with ^{18}O enriched water ($H_2{}^{18}O$) we find that only O_2 has radioisotope indicating that O_2 comes only from H_2O and not from CO_2.

- **To study the mechanism of the following reaction :**

$$CH_3-\overset{\overset{O}{\|}}{C}-O-H + C_2H_5\overset{18}{O}H \longrightarrow CH_3-\overset{\overset{O}{\|}}{C}-\overset{18}{O}C_2H_5 + H_2O$$

O of C_2H_5OH is enriched with radioisotope ^{18}O. After esterification, ^{18}O is found in the ester and not in H_2O indicating —OH of CH_3COOH is replaced by $—OC_2H_5$ of C_2H_5OH which acts as an acid in this reaction.

Illustration 1 *What is the age of a bone fragment that shows an average of 2.9 disintegrations per minute (dpm) per gram of carbon? The carbon in living organisms undergoes an average of 15.3 dpm/min per gram, and the half-life of* ^{14}C *is 5715 years.*

Solution T_{50} of ^{14}C = 5715 yr

$$\therefore \text{ decay constant } \lambda = \frac{0.693}{T_{50}} = \frac{0.693}{5715 \text{ yr}}$$

$$= 1.2126 \times 10^{-4} \text{ yr}^{-1}$$

Also, $$\lambda = \frac{2.303}{t\,(\text{age})} \log \frac{N_0}{N}$$

where N_0 the initial activity = 15.3 dpm and N the activity after time t = 2.9 dpm

$$\therefore \quad 1.2126 \times 10^{-4} = \frac{2.303}{t} \log \frac{15.3}{2.9}$$

$$t = \mathbf{13718 \text{ yr}}$$

Illustration 2 *A* 1.00 *mL solution containing* 0.240 μCi *of tritium is injected into a dog's bloodstream. After a period of time to allow the isotope to be dispersed, a* 1.00 *mL sample of blood is drawn. The radioactivity of this sample is found to be* 4.3×10^{-4} μCi/*mL. What is the total volume of blood in the dog?*

Solution Let total volume = X mL

$$\text{Activity} = 0.240\,\mu\text{Ci}$$

$$\text{Activity per mL} = 0.240\,\mu\text{Ci}/X$$

After dilution in the blood stream

$$\text{activity per mL} = 4.3 \times 10^{-4}\,\mu\text{Ci/mL}$$

Thus, $$0.240\,\mu\text{Ci}/X = 4.3 \times 10^{-4}\,\mu\text{Ci/mL}$$

$$\therefore \quad X = \mathbf{558.14 \text{ mL}}$$

Radiations and Matter Detection and Biological Effects

Radiations from nuclear processes affect matter in part by dissipating energy in it. In some cases, these radiations may also excite electrons in matter. When these electrons undergo transitions back to their ground states, light is emitted. The ions, free electrons and light produced in matter can be used to detect nuclear radiations. Because nuclear radiations can ionise molecules and break chemical bonds, they adversely affect biological organisms.

Radiation Counters

A Geiger counter, a kind of ionisation counter used to count particles emitted by radioactive nuclei consists of a metal tube fitted with gas such as argon.

Alpha and beta-particles can be detected by a Geiger counter. To detect neutrons, boron trifluoride is added to the gas in the tube. Neutrons react with boron–10 nuclei to produce alpha particles, which can then be detected.

$$^1_0n + ^{10}_5\text{B} \longrightarrow ^7_3\text{Li} + ^4_2\text{He}$$

A scintillation counter is a device that detects nuclear radiations from flashes of light generated in a material by the radiation. A phosphor is a substance that emits flashes of light when struck by radiation. The gamma rays are detected by this counter– a sodium iodide crystal containing thallium (I) iodide is used as a phosphor for gamma radiations. Rutherford used zinc sulphide as a phosphor to detect alpha particles.

Biological Effects and Radiation Dosage

Quantity of radiation intermixing with biological organisms might be small but the effects are quite adverse because important chemical bonds may be broken; DNA is specially affected.

Effect of nuclear radiations on biological tissues is expressed in terms of RAD (radiation adsorbed dose) *which is the dosage of radiation that deposits* 1×10^{-2} J *of energy per kilogram of tissue.*

- Neutrons are more destructive than gamma rays of the same radiation dosage measured in rads.

A REM (roentgen equivalent for man) is a unit of radiation dosage used to relate various kinds of radiation in terms of biological destruction.

$$\text{Rem} = \text{rad} \times \text{RBE}$$

where RBE stands for the **relative biological effectiveness.**

- For beta and gamma-radiations, RBE ≈ 1 radiation.
- For neutron, RBE ≈ 5
- For alpha-radiation, RBE ≈ 10.

The biological effects of different radiation-doses are given in Table 5.8

Table 5.8 **Biological Effects of Short-Term Radiation on Humans**

Dose (rem)	Biological effect
0–25	No detectable effects
25–100	Temporary decrease in white blood cell count
100–200	Nausea, vomitting, longer-term decrease in white blood cells
200–300	Vomitting, diarrhoea, loses of apetite, restlessness
300–600	Vomitting, diarrhoea, hemorrhaging, eventually death in some cases
above 600	Eventually death in nearly all cases.

Target Practice 2

1. What is Einstein mass-energy relationship? ....
2. When an electron is emitted, mass number remains unchanged but atomic number increases by one unit. Write nuclear reaction. ....
3. Write full name of MRI.
4. What is application of radioactive sodium chloride containing sodium-24?
5. Which radioactive isotope is used to control **polycythemia vera?**
6. Cobalt-60 is used to irradiate tumors with γ-rays.

$$^{60}_{27}\text{Co} \longrightarrow {}^{60}_{28}\text{Ni} + {}^{0}_{-1}e + \gamma$$

Which is responsible for destroying cancer-cells?

7. A museum wishes to analyse a piece of ruby for chromium content. What should be preferred method of analysis?

Answers

1. $E = mc^2$
2. ${}^{1}_{0}n \longrightarrow {}^{1}_{1}\text{H} + {}^{0}_{-1}e$
3. Magnetic Resonance Imaging
4. It is used to study obstructive vascular diseases in the extremities of the body.
5. Phosphorus-32
6. γ-rays,
7. Neutron activation analysis.

Practice Exercise 2

1. The half-life periods of two elements *A* and *B* are 5 days and 5 months, respectively. Which of the two is more stable ?
2. Which of the following are not stable ?
(i) ${}^{14}_{6}\text{C}$ (ii) ${}^{13}_{6}\text{C}$
(iii) ${}^{14}_{7}\text{N}$ (iv) ${}^{24}_{11}\text{Na}$
3. A proposed series of reactions (known as the carbon-nitrogen cycle) that could be important in every hottest region of the interior of the sun is:

$$^{12}\text{C} + {}^{1}\text{H} \longrightarrow A + \gamma$$
$$A \longrightarrow B + {}^{0}_{+1}e$$
$$B + {}^{1}\text{H} \longrightarrow C + \gamma$$
$$C + {}^{1}\text{H} \longrightarrow D + \gamma$$
$$D \longrightarrow E + {}^{0}_{+1}e$$
$$E + {}^{1}\text{H} \longrightarrow {}^{12}\text{C} + F$$

Identify the series labelled *A* to *F*.

4. Give product of the following nuclear reactions:
(a) ${}^{109}_{47}\text{Ag} + {}^{4}_{2}\text{He} \longrightarrow ?$
(b) ${}^{10}_{5}\text{B} + {}^{4}_{2}\text{He} \longrightarrow ? + {}^{1}_{0}n$
(c) ${}^{246}_{96}\text{Cm} + {}^{12}_{6}\text{C} \longrightarrow ? + 4\,{}^{1}_{0}n$
(d) ${}^{250}_{98}\text{Cf} + {}^{11}_{5}\text{B} \longrightarrow ? + 4\,{}^{1}_{0}n$
5. Write balanced nuclear equations for the following processes :
(a) Alpha emission of ${}^{162}\text{Re}$
(b) Electron capture of ${}^{138}\text{Sm}$
(c) Beta emission of ${}^{188}\text{W}$
(d) Positron emission of ${}^{165}\text{Ta}$
6. Write the abbreviated notation for the following bombardment reaction, in which neutron was first discovered:

$$^{9}_{4}\text{Be} + {}^{4}_{2}\text{He} \longrightarrow {}^{12}_{6}\text{C} + {}^{1}_{0}n$$

7. Write the nuclear equation for the bombardment reaction denoted as ${}^{27}_{13}\text{Al}\,(p,d)^{26}_{13}\,\text{Al}$.
8. Complete the following nuclear reactions :
(i) ${}^{40}_{20}\text{Ca}\,(n, \alpha)$ (ii) ${}^{24}_{12}\text{Mg}\,(d, \alpha)$
(iii) ${}^{3}_{1}\text{H}\,(\alpha, n)$ (iv) ${}^{197}_{79}\text{Au}\,(n, \gamma)$
(v) ${}^{27}_{13}\text{Al}\,(\gamma, n)$ (vi) ${}^{9}_{4}\text{Be}\,(\alpha, n)$
9. What are the types of reactions?
(a) ${}^{14}_{7}\text{Ca} + {}^{4}_{2}\text{He} \longrightarrow {}^{17}_{8}\text{O} + {}^{1}_{1}\text{H}$
(b) ${}^{6}_{3}\text{Li} + {}^{1}_{0}n \longrightarrow {}^{4}_{2}\text{He} + {}^{3}_{1}\text{H}$
(c) ${}^{2}_{1}\text{H} + \gamma \longrightarrow {}^{1}_{1}\text{H} + {}^{1}_{0}n$
10. How a radioactive nucleus can emit a β^- particle (electron) if there are no electrons present in the nucleus?
11. Why both α and β-particles are emitted in a radioactive decay series?
12. Why is ${}^{238}_{92}\text{U}$ not suitable for a chain reaction ?
13. A typical neutron-initiated fission of ${}^{235}_{92}\text{U}$ yields ${}^{97}_{42}\text{Mo}$, two neutrons and an isotope of what element?
14. If an atom of ${}^{235}_{92}\text{U}$, after the absorption of a slow neutron undergoes fission to form two fragments ${}^{139}_{54}\text{Xe}$ and ${}^{94}_{38}\text{Sr}$, what other particles are produced and how many?
15. The binding energy per nucleon for ${}^{235}_{92}\text{U}$ is 7.5 MeV and 8.5 MeV for nuclei of half that mass. If ${}^{235}_{92}\text{U}$ undergoes nuclear fission in two such nuclei of equal size, how much energy is released in the process?
16. The overall reaction in the sun responsible for the energy it radiates is

$$4\,({}^{1}_{1}\text{H}) \longrightarrow {}^{4}_{2}\text{He} + 2\,{}^{0}_{1}e\,(\beta^+)$$

It has been estimated that the sun produces 3×10^{22} kJ s^{-1}. How many moles of H-atoms are converted to He in 1 second, to produce this energy ? (Take binding energy 27 MeV)

17. Fill in the blanks :
(a) .... is a thermonuclear reaction.

(b) The energy production in the sun and stars could be explained on the basis of .... by means of which hydrogen is converted into helium.

(c) .... was awarded the 1967 Physics Nobel Prize for his theory of energy production in the sun and the stars.

(d) Hydrogen bomb is based on ....

(e) Instrument used to accelerate nuclear projectiles is called ....

(f) .... was awarded the 1939 Physics Nobel Prize for invention and development of cyclotron.

(g) ${}^{9}_{4}Be + {}^{4}_{2}He \longrightarrow {}^{12}_{6}C$ led to the discovery of .... by Chadwick.

18. Make distinction between nuclear fusion and nuclear fission.

19. An artificial element shows activity of 12.5 cps (counts per second) at its formation as recorded in a G.M. counter. After 10 min activity is 7.5 cps. In the absence of element, G.M. counter records activity of 2.5 cps. What is its half-life period and activity after 20 days?

20. Give one example each of (i) α-particle-induced reaction (ii) proton-induced reaction and (iii) neutron-induced reaction.

21. Packing fraction of element *A*, *B*, *C* are –20, 27, +18, respectively. Arrange them in the increasing order of stability.

Answers

1. Element B is more stable.

2. (i) ${}^{14}_{6}C$ (ii) ${}^{13}_{6}C$ and (iv) ${}^{24}_{11}Na$ are not stable.

3. A : ${}^{13}_{7}N$ B : ${}^{13}_{6}C$ C : ${}^{14}_{7}N$ D : ${}^{15}_{8}O$ E : ${}^{15}_{7}N$
F : ${}^{4}_{2}He$

4. (a) ${}^{110}_{49}In$ (b) ${}^{10}_{7}N$ (c) ${}^{254}_{102}No$ (d) ${}^{257}_{103}Lr$

5. (a) ${}^{162}_{75}Re \longrightarrow {}^{158}_{73}Ta + {}^{4}_{2}He$

(b) ${}^{138}_{62}Sm + {}^{0}_{-1}e \longrightarrow {}^{138}_{61}Pm$

(c) ${}^{188}_{74}W \longrightarrow {}^{188}_{75}Re + {}^{0}_{-1}e$

(d) ${}^{165}_{73}Ta \longrightarrow {}^{165}_{72}Hf + {}^{0}_{1}e$

6. ${}^{9}_{4}Be\ (\alpha, n)\ {}^{12}_{6}C$ **7.** ${}^{27}_{13}Al + {}^{1}_{1}H \longrightarrow {}^{26}_{13}Al + {}^{2}_{1}H$

8. (i) ${}^{40}_{20}Ca + {}^{1}_{0}n \longrightarrow {}^{37}_{18}Ar + {}^{4}_{2}He$

(ii) ${}^{24}_{12}Mg + {}^{2}_{1}H \longrightarrow {}^{22}_{11}Na + {}^{4}_{2}He$

(iii) ${}^{3}_{1}H + {}^{4}_{2}He \longrightarrow {}^{6}_{3}Li + {}^{1}_{0}n$

(iv) ${}^{197}_{79}Au + {}^{1}_{0}n \longrightarrow {}^{198}_{79}Au + \gamma$

(v) ${}^{27}_{13}Al + \gamma \longrightarrow {}^{26}_{13}Al + {}^{1}_{0}n$

(vi) ${}^{9}_{4}Be + {}^{4}_{2}He \longrightarrow {}^{12}_{6}C + {}^{1}_{0}n$

9. (a) α, *p* ; (b) *n*, α ; (c) γ, *n*. **10.** See text

11. See text **12.** Due to less stability **13.** ${}^{137}_{50}Sn$

14. Three neutrons **15.** 235 MeV

16. 4.61×10^{13} mol H atoms s^{-1}.

17. (a) nuclear fusion (b) carbon-cycle (c) H.A. Bethe (d) nuclear fusion (e) cyclotron (f) E.O. Lawrence (g) neutron

18. See Text

19. half-life = 10 days, 5 cps (2.5 cps is zero error of the G.M. counter.)

20. ${}^{14}_{7}N\ (\alpha, n)\ {}^{17}_{9}F \longrightarrow {}^{17}_{8}O + {}^{0}_{+1}e$

${}^{11}_{5}B\ (p, n)\ {}^{11}_{6}C \longrightarrow {}^{11}_{5}B + {}^{0}_{+1}e$

${}^{23}_{11}Na\ (n, \gamma)\ {}^{24}_{11}Na \longrightarrow {}^{24}_{12}Mg + {}^{0}_{-1}e$

21. $C < A < B$.

Total Practice Set
(Read, Plan & Solve!)

Problem 1. Calculate the nuclear binding energy per nucleon of $^{12}_{6}C$ from the following data:
Given mass of proton $(m_p) = 1.00758$ amu
mass of neutron $(m_n) = 1.00893$ amu
mass of electron $(m_e) = 0.00055$ amu
Isotopic mass of $^{12}C = 12.00000$ amu. Take 1 amu = 931 MeV

Plan *From the mass defect Δm (amu),*

$$\text{binding energy} = \Delta m \times 931 \text{ MeV}$$

$$\text{binding energy per nucleon} = \frac{\Delta m \times 931}{\text{number of nucleons}}$$

Solution $^{12}_{6}C$ contains 6 protons, 6 neutrons and 6 electrons hence, calculated atomic mass of $^{12}_{6}C$

$$= 6\, m_p + 6\, m_n + 6\, m_e$$
$$= 6 \times 1.00758 + 6 \times 1.00893 + 6 \times 0.00055$$
$$= 12.10236$$

$\therefore$ mass defect = 12.10236 − 12.00000
= 0.10236 amu

binding energy = 0.10236 × 931 = 95.30 MeV

$$\text{binding energy per nucleon} = \frac{95.30}{12} = \mathbf{7.94\ MeV}$$

Problem 2. The sun radiates energy at the rate of 4×10^{26} J s^{-1}. If the energy of fusion process

$$4\,^1_1H \longrightarrow \,^4_2He + 2\,^0_1e$$

is 27 MeV, calculate amount of hydrogen that would be consumed per day for the given process.

Plan *Energy radiated by the sun per second is to be converted per day. Energy of the fusion process is known, hence, amount of ^1_1H can be calculated.*

Solution $27 \text{ MeV} = 27 \times 10^6 \times 1.6 \times 10^{-19}$ J
$= 43.2 \times 10^{-13}$ J

and 4×10^{26} J s^{-1} $= 4 \times 10^{26} \times 24 \times 3600$ J day^{-1}
$= 34.56 \times 10^{30}$ J day^{-1}

43.2×10^{-13} J of energy is obtained from
= 4 amu of H

$$= \frac{4}{6.02 \times 10^{23}} \text{ g of H}$$

34.56×10^{30} J day^{-1} of energy is obtained from

$$= \frac{4}{6.02 \times 10^{23}} \times \frac{34.56 \times 10^{30}}{43.2 \times 10^{-13}} \text{ g H}$$

$$= \mathbf{53.16 \times 10^{18}\ g}$$

Problem 3. A chemist prepares 1.00 g of pure $^{11}_{6}C$. This isotope has half-life of 21 min, decaying by equation :

$$^{11}_{6}C \longrightarrow \,^{11}_{5}B + \,^0_1e$$

(a) What is rate of disintegration per second (dps) at start?

(b) How much of this isotope ($^{11}_{6}C$) is left after 24 h of its preparation?

Plan $\frac{-dN}{dt} = \lambda N_0$, *where N_0 is the number of atoms of $^{11}_{6}C$ at start in 1 g.*

Use $$\lambda = \frac{2.303}{t} \log \frac{N_0}{N}$$

or $$C = C_0 \left(\frac{1}{2}\right)^y \text{ to calculate amount left after 24 h}$$

Solution (a) $$-\frac{dN}{dt} = \lambda N_0$$

where λ (disintegration constant) $= \frac{0.693}{T_{50}}$ and N_0 the number of atoms present initially

$$N_0 = \frac{\text{weight}}{\text{atomic weight}} \times \text{Avogadro's number}$$

$$-\frac{dN}{dt} = \frac{0.693}{21 \times 60} \times \frac{1}{11} \times 6.02 \times 10^{23} \text{ dps}$$

$$= \mathbf{3 \times 10^{19}\ dps}$$

(b) Using equation : $C = C_0 \left(\frac{1}{2}\right)^y$

where, $$y = \frac{\text{total time}}{T_{50}} = \frac{24 \times 60 \text{ min}}{21 \text{ min}} = 68.57$$

C_0 = initial amount = 1 g
C = amount of radioactive substance left after y **half-lives**

$$\therefore \quad C = 1\left(\frac{1}{2}\right)^{68.57}$$

Taking log and then antilog, $C = \mathbf{2.28 \times 10^{-21}}$ **g ($^{11}_{6}$C).**

Problem 4. Calculate the weight in grams of 1 Ci of ^{14}C (T_{50} = 5000 yr), (1 yr = 31.536×10^6 s).

Plan $$-\frac{dN}{dt} = \lambda N = \lambda \frac{w}{A} N_0$$

where w is the weight of the element of mass number A; N_0 is the Avogadro's number.

Solution Let weight required = w g

$$= \frac{w}{14} \times 6.02 \times 10^{23} \text{ atoms}$$

$$\text{Disintegration constant } (\lambda) = \frac{0.693}{5000 \times 31.536 \times 10^6} \text{ s}^{-1}$$

and 1 Ci = 3.7×10^{10} dps

$$\therefore \quad -\frac{dN}{dt} = \lambda N$$

$$3.7 \times 10^{10} = \frac{0.693 \times w \times 6.02 \times 10^{23}}{5000 \times 31.536 \times 10^6 \times 14}$$

$$\therefore \quad w = \mathbf{0.196\ g}$$

Problem 5. How many mL of radon (Rn) under standard condition of temperature and pressure are in equilibrium with 1 g of radium (Ra). Given half-life period of ^{226}Ra = 1590 yr and for ^{222}Rn = 3.82 days.

Plan At radioactive equilibrium :

$$N_1\lambda_1 = N_2\lambda_2 \quad \therefore \quad \frac{N_1}{N_2} = \frac{\lambda_2}{\lambda_1}$$

Solution $$\frac{N_1(\text{Rn})}{N_2(\text{Ra})} = \frac{\lambda_2(\text{Ra})}{\lambda_1(\text{Rn})} = \frac{T_{50}\ (\text{Rn})}{T_{50}\ (\text{Ra})}$$

$$\frac{N_1(\text{Rn})}{(1/226)\ \text{mol}} = \frac{3.82 \text{ days}}{365 \times 1590 \text{ days}}$$

$$N_1(\text{Rn}) = 2.91 \times 10^{-8} \text{ mol}$$

$$\therefore \quad \text{mL of Rn at STP} = \mathbf{6.52 \times 10^{-4}\ mL}$$

Problem 6. Estimate the age of a rock which is found to contain 5×10^{-5} mL of helium at STP and 3×10^{-7} g of uranium. (T_{50} of $^{238}_{92}U$ is 4.51×10^9 yr and each $^{238}_{92}U$ is converted into $^{206}_{82}Pb$).

Plan $$^{238}_{92}U \longrightarrow {}^{206}_{82}Pb + 6\beta + 8\alpha \text{ (helium)}$$

We take the equation $$\left[\frac{[\text{Pb}]}{[\text{U}]} + 1\right] = (2)^y$$

where $\frac{[\text{Pb}]}{[\text{U}]}$ *is the molar ratio,* $y = \frac{t}{T_{50}}$

Solution Number of moles of He at STP

$$= \frac{5 \times 10^{-5}}{22400} \text{ mol}$$

$$[\text{Pb}] = \frac{1}{8}[\text{He}] = \frac{5 \times 10^{-5}}{8 \times 22400} \text{ mol}$$

$$= 2.79 \times 10^{-10} \text{ mol}$$

$$[\text{U}] = \frac{3 \times 10^{-7}}{238} = 12.6 \times 10^{-10} \text{ mol}$$

$$\therefore \quad \left[\frac{2.79 \times 10^{-10}}{12.6 \times 10^{-10}} + 1\right] = (2)^y$$

$$1.2214 = (2)^y$$

taking log, we have y = 0.2886

but y = total number of half-life = $\frac{t}{T_{50}}$

$$\therefore \quad t = y \times T_{50}$$

$$= 0.2886 \times 4.5 \times 10^9$$

$$= \mathbf{1.30 \times 10^9\ yr}$$

Problem 7. 1 g of radium produces 0.043 cm^3 of helium gas at STP in one year. The number of α-particles produced in the same period is 11.6×10^{17}. Calculate the value of Avogadro's number.

Plan One mole = 22400 cm^3 at STP = N_0
Thus, from the volume of the gas and number of particles, N_0 can be determined.

Solution $$_{88}Ra \longrightarrow {}_{86}Rn + {}^4_2He$$

thus, 0.043 cm^3 of helium gas at STP

$$= 11.6 \times 10^{17} \ \alpha\text{-particles}$$

$$\therefore \ 22400 \text{ cm}^3 \text{ of helium at STP} = \frac{11.6 \times 10^{17}}{0.043} \times 22400$$

$$N_0 = \mathbf{6.043 \times 10^{23}\ mol^{-1}}$$

Problem 8. The curie was originally intended to be the activity of one gram of radium-226, [T_{50} = 1620 a, here 'a' stands for annum (year) in SI unit]. Calculate actual weight of radium which is equivalent to one curie. given 1 a = 3.1536×10^7 s

Plan $$-\frac{dN}{dt} = -\lambda N = -\lambda \frac{w}{A} N_0$$

Thus, w (mass of the radioactive substance) can be calculated.

Solution $$\lambda = \frac{0.693}{1620} = \frac{0.693}{1620 \times 3.1536 \times 10^7} \text{ s}^{-1}$$

but by definition, 1 Ci (curie) = 3.7×10^{10} dps

$$\therefore \quad -\frac{dN}{dt} = \lambda N$$

$$3.7 \times 10^{10} = \frac{0.693}{1620 \times 3.1536 \times 10^7} N$$

$$\therefore \quad N = 2.72766 \times 10^{21} \text{ atoms}$$

$$\text{moles of Ra-226} = \frac{2.72766 \times 10^{21}}{6.02 \times 10^{23}}$$

$$= 0.004531$$

$$\text{amount in g} = \text{mol} \times 226$$

$$= \mathbf{1.024\ g}$$

Problem 9. The half-life of ^{32}P is 14.3 days. Calculate the specific activity of a specimen of phosphorus containing 1.00 part per million ^{32}P.(atomic wt. of P = 31)

Plan *Determine rate of disintegration per second (dps) for the given isotope. Activity of* 1 curie = 3.7×10^{10} *dps. Thus, activity in the given isotope can be calculated.*

Solution The specific activity of a radionuclide is its activity per kg of the radioactive material.

$$1 \text{ kg} = 1000 \text{ g}\ ^{31}P = \frac{1000}{31} N_0 \text{ atoms of } ^{31}P$$

But ^{32}P content is 1 ppm (1 atm ^{32}P out of 10^6 atoms)

$$\therefore \quad ^{32}P = \frac{1000\, N_0}{31} \times \frac{1}{10^6} = \frac{1000\times 6.02\times 10^{23}}{31\times 10^6}$$

$$= 1.942\times 10^{19} \text{ atoms of } ^{32}P$$

$$\text{also} \quad -\frac{dN}{dt} = \lambda N, \quad \left(\lambda = \frac{0.693}{T_{50}}\right)$$

$$-\frac{dN}{dt} = \frac{0.693}{14.3\times 24\times 3600}\times 1.942\times 10^{19} \text{ dps}$$

$$= 1.089\times 10^{13} \text{ s}^{-1}.$$

hence, specific activity (number of curies in 1 kg of ^{32}P)

$$= \frac{1.089\times 10^{13}}{3.7\times 10^{10}} = \mathbf{294.32\ Ci\ kg^{-1}}$$

Problem 10. An archaeological specimen containing ^{14}C gives 40 counts in 5 min per gram of carbon. A specimen of freshly cut wood gives 20.3 counts per gram of carbon per minute. The counter used recorded a background count of 5 counts per minute in absence of any ^{14}C-containing sample. What is the age of the specimen? [T_{50} of $^{14}C = 5668\ a$]

Plan *Initial and final activities (in terms of counts per min) have been given. Net value are determined (since instruments has zero-error). Thus, age can be determined.*

Solution Counter has zero-error of 5 counts per min (cpm)

$$\therefore \quad \text{Initial activity} = 20.3 - 5 = 15.3 \text{ cpm}$$

$$\text{Activity after time } t = \left(\frac{40}{5} - 5\right) = 3.0 \text{ cpm}$$

$$\therefore \quad \lambda = \frac{2.303}{t}\log\frac{15.3}{3.0}$$

$$\frac{0.693}{5668} = \frac{2.303}{t}\log\frac{15.3}{3.0}$$

$$\therefore \quad \mathbf{t = 13328\ yr}$$

Problem 11. Calculate the energy released in the reaction :

$$^{35}_{17}Cl + {}^{1}_{0}n \longrightarrow {}^{35}_{16}S + {}_{1}H^{1}$$

[(Cl-35) = 34.9688 amu, (S-35) = 34.9690 amu, (H-1) = 1.0078 amu, $^{1}_{0}n$ = 1.0087 amu]. Assume 1 amu = 931 MeV

Plan *From the mass defect* (Δm)
energy (in MeV) = (Δm *in amu* × 931)

Solution Mass defect

$$\Delta m = [m(Cl) + m(n)] - [m(S) + m(H)]$$

$$= [34.9688 + 1.0087] - [34.9690 + 1.0078]$$

$$= 0.0007 \text{ amu}$$

$$\therefore \quad \text{energy released} = 0.0007\times 931$$

$$= \mathbf{0.6517\ MeV}$$

Problem 12. Calculate the maximum energy of a β-particle emitted from a ^{14}C nucleus which does not emit any γ-radiation. The masses of C–14 and N–14 isotopes are 14.003242 amu and 14.003074 amu respectively.

$$^{14}_{6}C \longrightarrow {}^{14}_{7}N + {}_{-1}e^{0}$$

Plan *In solving the above problem, the mass of emitted β-particle is not taken into account. This is because the atom N-14 produced contains one more extra nuclear electron than the C-14 atom. But when a positron is emitted two electron masses have to be taken into account*
(i) one for the positron emitted and
(ii) one for the extra electron that has to leave the extra nuclear shells to maintain electrical neutrality. Thus in the reaction :

$$^{23}_{12}Mg \longrightarrow {}^{23}_{11}Na + {}_{1}e^{0}$$

$$\Delta m = [\text{mass of (Mg - 23)}] - [\text{mass of (Na - 23)} + 2 \text{ mass of electron}]$$

Solution $\Delta m = m(\text{C-14}) - m(\text{N-14}) = 0.000168$ amu

$$\therefore \quad \text{energy released in the process} = 0.000168\times 931$$

$$= \mathbf{0.1564\ MeV}$$

This is maximum energy of β-particle.

Problem 13. A sample of $^{90}_{38}Sr$ originally had an activity of 0.500 mCi (millicurie) (a) What is the specific activity of the sample? (b) What is the activity of the sample after 30.0 years? T_{30} of $^{90}_{38}Sr$ is 19.9 years.

Plan *Initial activity* (N_0) *is known, hence activity* (N) *after time t can be calculated using*

$$t = \frac{2.303}{t}\log\frac{N_0}{N}$$

Solution The specific activity of a radio-nuclide is its activity per kg of the radioactive material.
Hence, first we calculate the weight of $^{90}_{38}Sr$ which has an activity of 0.500 mCi

$$\therefore \quad -\frac{dN}{dt} = \lambda N = \lambda\frac{wN_0}{A}$$

$$1 \text{ mCi} = 3.7\times 10^7 \text{ dps}$$

$$\therefore \quad 0.500 \text{ mCi} = 0.5\times 3.7\times 10^7 \text{ dps}$$

$$\therefore \quad 0.5 \times 3.7 \times 10^7 = \frac{0.693 \times w \times 6.02 \times 10^{23}}{19.9 \times 365 \times 24 \times 60 \times 60 \times 90}$$

$$\therefore \quad w = 2.5 \times 10^{-6} \text{ g} = 2.5 \times 10^{-9} \text{ kg}$$

$$\therefore \quad \text{Specific activity} = \frac{0.5 \times 3.7 \times 10^7}{2.5 \times 10^{-9}}$$

$$= \mathbf{7.4 \times 10^{15}\ dis/kg\text{-}s}$$

(b) using $N = N_0 \left(\frac{1}{2}\right)^y$

where $y = \frac{t}{T_{50}} = \frac{30.0}{19.9} = 1.5075$

$$N = 0.5 \left(\frac{1}{2}\right)^{1.5075} \text{ mCi}$$

taking log $\log N = \log 0.5 - 1.5075 \log 2$

$$= -0.3010 - 1.5075 \times 0.3010$$

$$= -0.7548$$

$$\therefore \quad N = 0.176 \text{ mCi}$$

$$= 0.176 \times 3.7 \times 10^7 \text{ dps}$$

$$= \mathbf{6.51 \times 10^6\ dps}$$

Problem 14. In order to determine the volume of blood in an animal without killing it, a 1.00 mL sample of an aqueous solution containing tritium is injected into the animal's blood-stream. The sample injected has an activity of 1.8×10^6 cps (counts per second). After sufficient time for the sample to be completely mixed with the animals blood due to normal blood circulation, 2.00 mL of blood are withdrawn from animal, and the activity of the blood sample withdrawn is found to be 1.2×10^4 cps. Calculate volume of animal's blood.

Plan Compare the activity with volume of the blood sample to calculate volume of animal blood.

Solution Let volume of the blood in the animal = V mL

$$\text{Total activity before mixing} = 1.8 \times 10^6 \text{ cps}$$

$$\text{Total activity after mixing in } V \text{ mL} = \frac{1.2 \times 10^4}{2} \times V$$

$$= 0.6 \times 10^4\ V \text{ cps}$$

Thus, $0.6 \times 10^4\ V = 1.8 \times 10^6$

(assuming activity remains constant)

$$V = \frac{1.8 \times 10^6}{0.6 \times 10^4} = \mathbf{300\ mL}$$

Problem 15. A sample of uraninite, a uranium containing mineral, was found on analysis to contain 0.214 g of lead-206 for every g uranium. If all the lead came from the radioactive disintegration of the uranium and assume that all isotopes of uranium other than ^{238}U can be neglected, estimate the date when the mineral was formed in the earth's crust? (The half-life of ^{238}U is 4.5×10^9 years).

Plan Use $1 + \frac{[\text{Pb}]}{[\text{U}]} = 2^y$ *where* $y = \frac{t}{T_{50}}$.

Solution If 0.214 g of ^{206}Pb is present, then 1 g of ^{238}U is also present.

$$\therefore \quad [\text{Pb}] = \frac{0.214}{206} = 0.00104 \text{ mol}, \quad [\text{U}] = \frac{1}{238} = 0.0042 \text{ mol}$$

$$\therefore \quad \frac{[\text{Pb}]}{[\text{U}]} = 0.247$$

∴ For 'rock dating' analysis, we use

$$\left[1 + \frac{[\text{Pb}]}{[\text{U}]}\right] = 2^y \quad \text{where } y = \frac{t}{T_{50}}$$

$$[1 + 0.247] = (2)^y$$

Taking log, we get $y = 0.3188$

$$\therefore \quad \frac{t}{T_{50}} = 0.3188$$

$$\therefore \quad t = 0.3188 \times T_{50}$$

$$= 0.3188 \times 4.5 \times 10^9$$

$$= \mathbf{1.4344 \times 10^9\ years}$$

Problem 16. Given 2.00 kg of $^{238}_{92}$U ($T_{50} = 4.50 \times 10^9$ years). The ultimate decay product in this series is $^{206}_{82}$Pb. (a) What weight of $^{238}_{92}$U is left after 3.00×10^9 years? (b) What weight of $^{206}_{82}$Pb is produced in this time?

Plan Use $C = C_0 \left(\frac{1}{2}\right)^y$ *where* C_0 *is the initial amount, C after time t and* $y = \frac{t}{T_{50}}$.

Solution (a) $^{238}_{92}\text{U} \longrightarrow {}^{206}_{82}\text{Pb} + \underbrace{8(\alpha) + 6(\beta)}_{\text{particles}}$

taking $C = C_0 \left(\frac{1}{2}\right)^y$ where $y = \frac{3.00 \times 10^9}{4.5 \times 10^9} = 0.667$

$$C = 2 \left(\frac{1}{2}\right)^{0.667}$$

taking log on both sides

or $\log C = \log 2 - 0.667 \log 2$

$$= 0.3010 - 0.666 \times 0.3010 = 0.1003$$

$$\therefore \quad \mathbf{C = 1.26\ kg}$$

∴ (b) ^{92}U disintegrated = 2.00 − 1.26

$$= 0.74 \text{ kg} = 740 \text{ g}$$

$$= \frac{740}{238} = 3.109 \text{ mol}$$

$\therefore$ Pb formed = 3.109 mol

$= 3.109 \times 206 = 640.5$ g

$= \mathbf{0.6405\ kg}$

Problem 17. A sample of $^{14}CO_2$ was to be mixed with ordinary CO_2 for a biological tracer experiment. In order that 10 cm^3 of the diluted gas at STP should have 10^4 dis/min, how many μCi of radioactive carbon–14 are needed to prepare 0.06 m^3 of the diluted gas ?

Plan *Given volume is to be related to the dps (activity) and thus, number of microcuries (μCi) can be calculated.*

Solution 10 cm^3 of the diluted gas has activity

$$= 10^4 \text{ dis/min} = \frac{10^4}{60} \text{ dps}$$

$\therefore$ 0.06 m^3 (= 60,000 cm^3) of the diluted gas has

$$\text{activity} = \frac{10^4}{60} \times \frac{60000}{10} \text{ dps}$$

hence, number of μCi needed $= 10^4 \times \frac{60000}{600 \times 3.7 \times 10^4}$

$(1\,\mu\text{Ci} = 3.7 \times 10^4 \text{ dps})$

$= \mathbf{27.03\ \mu Ci}$

Problem 18. The half-life of ^{212}Pb is 10.6 h, that of its daughter ^{212}Bi is 60.5 min. How long will it take for a maximum daughter activity to grow in freshly separated ^{212}Pb ?

Plan *We define time of maximum activity by equation*

$$t_{max} = \frac{2.303}{(\lambda_d - \lambda_p)} \log \frac{\lambda_d}{\lambda_p}$$

where λ_d is the disintegration constant of daughter element and λ_p that of parent element.

Solution $\lambda_d = \frac{0.693}{T_{50}\ (^{212}\text{Bi})} = \frac{0.693}{60.5 \text{ min}}$

$= 0.01145 \text{ min}^{-1}$

$\lambda_p = \frac{0.693}{T_{50}\ (^{212}\text{Pb})} = \frac{0.693}{10.6 \times 60}$

$= 0.00109 \text{ min}^{-1}$

$\therefore$ $t_{max} = \frac{2.303}{(0.01145 - 0.00109)} \log\left(\frac{0.01145}{0.00109}\right)$

$= \mathbf{227\ min}$

Problem 19. An isotopic species of lithium hydride $^6Li^2H$ is a potential nuclear fuel, on the basis of the reaction

$$^6_3\text{Li} + ^2_1\text{H} \longrightarrow 2^4_2\text{He}$$

Calculate the expected power production in MW, associated with 1.00 g of $^6Li^2H$ per day (process is 100% efficient).

$[^6_3\text{Li} = 6.01512 \text{ u}, ^2_1\text{H} = 2.01410 \text{ u}, ^4_2\text{He} = 4.00260 \text{ u.}]$, u is amu and 1 u = 931 MeV

Plan *Energy* $= \Delta m \times c^2$, 1 MW $= 10^6$ W $= 10^6$ J s^{-1} *Thus, energy in MW can be calulated.*

Solution Δm = mass decay $= m(^6_3\text{Li}) + m(^2_1\text{H}) - 2m(^4_2\text{He})$

$= (6.01512 + 2.01410) - 2 \times 4.0026$

$= 0.02402$ u

thus, energy of one atomic event (one mass decay)

$= 0.0240 \times 931 \times 10^6$ eV

$= 0.0240 \times 931 \times 10^6 \times 1.6 \times 10^{-19}$ J

$= 3.59 \times 10^{-12}$ J

energy per mol of LiH $= 3.59 \times 10^{-12} \times 6.02 \times 10^{23}$

$= 2.16 \times 10^{12}$ J mol^{-1}

energy per g of LiH $= \frac{2.16 \times 10^{12}}{8}$ J g^{-1}

energy per g of LiH per second $= \frac{2.16 \times 10^{12}}{8 \times 24 \times 3600}$ J g^{-1} s^{-1}

$= 3.125 \times 10^6$ J s^{-1}g^{-1}

$= 3.125 \times 10^6$ W g^{-1} (Js^{-1} = 1 W)

$= 3.125$ MW g^{-1} (1MW = 10^6 W)

Problem 20. Tritium, or hydrogen-3, is formed in the upper atmosphere by cosmic rays, similar to the formation of C-14. Tritium has been used to determine the age of wines. A certain wine that has been aged in a bottle has a tritium content only 78% of that in a similar wine of the same mass that has just been bottled. How long has the aged wine been in the bottle ? T_{50} of tritium is 12.2 years.

Plan *Use* $\lambda = \frac{2.303}{t} \log \frac{N_0}{N}$ *and* $\lambda = \frac{0.693}{T_{50}}$

Thus, time t can be calculated.

Solution $\lambda = \frac{2.303}{t} \log \frac{N_0}{N}$

$$\frac{0.693}{T_{50}} = \frac{2.303}{t} \log \frac{N_0}{N}$$

$$\frac{0.693}{12.2} = \frac{2.303}{t} \log \frac{100}{78}$$

$\therefore$ $t = \mathbf{4.375\ years}$

Problem 21. When a positron and an electron collide, they are annihilated and two gamma photons of equal energy are emitted. Calculate the wavelength corresponding to this gamma emission:

Plan *Energy per photon* $= m_e c^2$

$\therefore$ $E = h\nu = \frac{hc}{\lambda}$

Thus, λ can be calculated.

Solution ${}_1e^0 + {}_{-1}e^0 \longrightarrow$ 2 gamma photons of equal energy

The energy of the above process is

$$E = 2\, m_e c^2$$
$$= 2\,(9.11\times 10^{-31}\ \text{kg})\,(3\times 10^8\ \text{m s}^{-1})^2$$
$$= 1.6398\times 10^{-13}\ \text{J}$$

$\therefore$ energy of one photon $= 8.199\times 10^{-14}$ J

taking $E = \dfrac{hc}{\lambda}$

$$\lambda = \frac{hc}{E} = \frac{6.62\times 10^{-34}\times 3\times 10^8}{8.199\times 10^{-14}}\ \text{m}$$
$$= 2.42\times 10^{-12}\ \text{m} = \mathbf{2.42\ pm}$$

Problem 22. Natural nitrogen is 99.63% ^{14}N with mass 14.00307 amu and 0.37% ^{15}N, with mass number 15.00011 amu. Which isotope is more stable, ^{14}N or ^{15}N?

Plan *Binding energy per nucleon* $= \dfrac{\Delta m\,(amu)\times 931}{number\ of\ nucleons}$

Isotope with larger binding energy per nucleon is more stable.

Solution There are $7n$, $7p$ and $7e^-$ in ^{14}N

$$\text{mass of nucleons} = 7\times 1.00782505 + 7\times 1.008664904$$
$$= 14.11543\ \text{amu}$$
$$\text{mass defect} = 14.11543 - 14.00307$$
$$= 0.11236\ \text{amu}$$
$$\text{binding energy per nucleon} = \frac{0.11236\times 931}{14}\ \text{MeV}$$
$$= 7.472\ \text{MeV}$$

There are $8n$, $7p$ and $7e^-$ in ^{15}N

$$\text{mass of nucleons} = 8\times 1.008664904 + 7\times 1.00782505$$
$$= 15.12409\ \text{amu}$$
$$\text{mass defect} = 15.12409 - 15.00011$$
$$= 0.12398\ \text{amu}$$
$$\text{binding energy per nucleon} = \frac{0.12398\times 931}{15}\ \text{MeV}$$
$$= 7.695\ \text{MeV}$$

As binding energy per nucleon is larger for ^{15}N, hence, ^{15}N is more stable.

Problem 23. What is the change in mass (in grams) when 2 mol of hydrogen atoms combine to form 1 mol of hydrogen molecule?

$$2H \longrightarrow H_2 \qquad \Delta E = -436\ \text{kJ}$$

Plan *The problem asks us to calculate the mass defect Δm when the energy change ΔE is known. To do this, we have to rearrange the Einstein equation to solve for mass, remembering that $1\ J = 1\ kg\ m^2/s^2$.*

Solution $\Delta E = \Delta mc^2$

$$\Delta m = \frac{\Delta E}{c^2} = \frac{-436\ \text{kJ}}{(3\times 10^8)^2\ \text{m}^2\ \text{s}^{-2}}$$
$$= \frac{-436\times 10^3\ \text{J}}{9\times 10^{16}\ \text{m}^2\ \text{s}^{-2}}$$
$$= \frac{-436\times 10^3\ \text{kg m}^2\ \text{s}^{-2}}{9\times 10^{16}\ \text{m}^2\ \text{s}^{-2}}$$
$$= \mathbf{-4.84\times 10^{-12}\ kg}$$

Problem 24. Calculate the volume of H_2 gas at STP which would be produced per minute from radiolysis of water in a 1000 kW reactor, assuming that 100 eV of energy produces 1 molecule of hydrogen

$$2\,H_2O \longrightarrow H_2 + H_2O_2 \quad (1\ \text{eV} = 1.6\times 10^{-19}\ \text{J})$$

Plan *Energy given in kW is to be converted into J s^{-1} and thus, volume of H_2 gas produced in this reactor can be calculated.*

Solution Given $1\ \text{eV} = 1.6\times 10^{-19}$ J

$\therefore$ $100\ \text{eV} = 100\times 1.6\times 10^{-19}$ J
$= 1.6\times 10^{-17}$ J

Also $1000\ \text{kW} = 1000\times 1000$ W
$= 1000\times 1000\ \text{J s}^{-1}$
$= 10^6\ \text{J s}^{-1}$
$= 10^6\times 60$ J in one minute (as required)

1.6×10^{-17} J produces = 1 molecule of H_2

$\therefore$ 60×10^6 J produces $= \dfrac{60\times 10^6}{1.6\times 10^{-17}}$ molecules
$= 37.5\times 10^{23}$ molecules
$= \dfrac{37.5\times 10^{23}}{6.02\times 10^{23}}$ mol H_2
$= 6.23$ mol H_2

1 mol at STP has volume = 22.4 L

$\therefore$ 6.23 mol at STP has volume = **139.55 L**

Master Exercises

Exercise 1
(Stage 1 : Learning)

Analytical Questions

Theory of Radioactive Disintegration

1. Half-life of ^{221}Fr is 4.8 min. If 1 mg of the isotope is isolated, how much will remain after (a) 19.2 min (b) 48 min?
2. 0.1 mg of ^{239}Pu emits 1.4×10^7 particles per minute. What is its half-life?
3. A radioactive isotope has a half-life of 14.8 hr. How much time it will take for the activity to drop to (a) 10 per cent, (b) 5 per cent of its original value?
4. Radioactive equilibrium is established in a uranium mineral between radium and uranium atoms in the ratio 1 : 28. If half-life period of radium is 1620 years, calculate the disintegration constant of uranium.
5. 1 g of ^{226}Ra emits 11.6×10^{17} α-particles in a year. If $\lambda = 4.353\times 10^{-4}$ year^{-1}, calculate the value of Avogadro's number.
6. Calculate the mass of ^{14}C (T_{50} = 5720 years) atom which has radioactivity of 1 mCi.
7. The half-life of ^{40}K is 1.28×10^9 years. What mass of this nuclide has an activity of 1.00μCi ?
8. $^{210}_{84}Po$ decays with α-particle to $^{206}_{82}Pb$ with a half-life of 138.4 days. If 1.0 g of $^{210}_{84}Po$ is placed in a sealed tube how much of helium will accumulate in 69.2 days? Express the answer in cm^3 at STP.
9. 10 g atoms of an α-active radioactive isotope are disintegrating in a sealed container. In one hour, the He gas collected at STP is 11.2 cm^3. Calculate half-life of radioactive isotope.
10. 0.1 g atom of radioactive isotope ^A_ZX (half-life 5 days) is taken. How many number of atoms will decay during eleventh day.
11. A certain radioisotope $^A_ZX(T_{50} = 10 \text{ days})$ decays to give $^{A-4}_{Z-2}Y$. If one g atom of ^A_ZX is kept in a sealed vessel, how much He will accumulate in 20 days at STP?
12. The disintegration rate of a certain radioactive sample at any instant is 4750 dpm (disintegration per minute). Five minutes later, the rate becomes 2700 dpm. Calculate half-life of sample.
13. Chlorine-34 has a half-life of only 1.53 s. How long does it take for 99.99% of a ^{34}Cl sample to decay?

Nuclear stability and Nuclear Reactions

14. Calculate the number of α and β-particles in the change :

$$^{235}_{92}U \longrightarrow \,^{207}_{82}Pb$$

15. 1 g of $^{198}_{79}Au$ (T_{50} = 65 h.) decays by β-emission to produce stable Hg.
    (a) Write nuclear reaction for process.
    (b) How much Hg will be present after 260 h?
16. $^{238}_{92}U$ is radioactive and it emits α and β-particles to form $^{206}_{82}Pb$. Calculate the number of α and β-particles emitted in this conversion. An ore of $^{238}_{92}U$ is found to contain $^{238}_{92}U$ and $^{206}_{82}Pb$ in the weight ratio of 1 : 0.1. The half-life period of $^{238}_{92}U$ is 4.5×10^9 years. Calculate the age of the ore.
17. (a) The super heavy element ^{287}Uuq was made by firing a beam of ^{40}Ca ions at ^{240}Pu. Three neutrons were ejected in the reaction. Write a balanced nuclear equation for the synthesis of ^{287}Uuq.
    (b) Boron is an effective absorber of neutrons. When boron-10 adds a neutron, an α-particle is emitted. Write an equation for this nuclear reaction.

Binding Energy

18. Calculate the mass defect and binding energy in
    (a) $^{10}_5B$ (isotopic mass = 10.01019 amu) and
    (b) ^7_3Li (isotopic mass = 7.01436 amu).

19. The alpha-particles emitted by radium have energies of 4.795 MeV and 4.611 MeV. What is the wavelength of the gamma ray associated with the decay?

20. What is the wavelength (in nm) of gamma-rays whose energy is 1.50 eV?

21. Calculate the mass defect (in g/mol) and the binding energy (in MeV/nucleon) for the following nuclides. Which of the two is more stable?
(a) ^{50}Cr (atomic mass = 49.94605 amu)
(b) ^{64}Zn (atomic mass = 63.92915 amu)

Applications

22. A sample of uranium ore was found to contain 0.28 g ^{206}Pb and 1.67 g of ^{238}U. Assuming that all Pb came from the decay of ^{238}U, what is the age of the ore ? (Given T_{50} of ^{238}U is 4.51×10^9 years).

23. A sample of ^{238}U (half-life = 4.5×10^9 years) ore is found to contain 23.8 g of ^{238}U and 20.6 g of ^{206}Pb. Calculate the age of the ore.

NCERT Problems

1. What is meant by nuclear binding energy? Calculate the binding energy per nucleon of Li isotope, which has the isotopic mass of 7.016 m_u. The individual masses of neutron and proton are 1.008665 m_u and 1.007277 m_u, respectively and the mass of electron = 0.000548 m_u.

2. The atomic mass of ^{16}O is 15.995 m_u while the individual masses of proton and neutron are 1.0073 and 1.0087 m_u. The mass of electron = 0.000548 m_u. Calculate the binding energy of the oxygen nucleus in joules.

3. The isotopic masses of $^2_1$H and $^4_2$He are 2.0141 and 4.0026 m_u respectively and the velocity of light in vacuum is 2.998×10^8 ms^{-1}. Calculate the quantity of energy (in J) liberated when two moles of $^2_1$H undergoes fusion to form one mole of $^4_2$He.

4. The radioactive isotope $^{60}_{27}$Co which has now replaced radius in the treatment of cancer can be made by a (n, p) or (n, γ) reaction. For each reaction, indicate the appropriate target nucleus. If the half-life of $^{60}_{27}$Co is 7 yr, evaluate the decay constant in s^{-1}.

5. A piece of wood from an archaeological source shows a ^{14}C activity which is 60% of the activity found in fresh wood today. Calculate the age of the archaeological sample. ($t_{1/2}$ = C = 5770 yr)

6. Calculate the mass of ^{140}La in a sample whose activity is 3.7×10^{10} Bq (1 becquerel, Bq = 1 disintegration per second) given that its $t_{1/2}$ is 40 hr.

7. Calculate the binding energy per nucleon for ^{12}C, ^{14}N, ^{16}O, and comment on their relative magnitudes. Masses of proton and neutron are 1.0078 and 1.0087 m_u respectively. (1 m_u = 931 MeV)

Exercise 2
(Stage 2 : High Skill Problem Solving)

Only One Option Correct

1. In the chain reaction
$$^{238}_{92}\text{U} \longrightarrow \text{Ba} + \text{Kr} + 3\,^1_0n + \text{energy } E$$
neutrons and energy produced at nth step will be:
(a) $3n, nE$ (b) $3^n, nE$
(c) $3^n, 3^{n-1}E$ (d) $3^{n-1}, nE$

2. A radioactive element decays by the sequence, and with half-lives, given below :
$$X \xrightarrow[30\text{ min}]{\alpha} Y \xrightarrow[2\text{ days}]{2\beta} Z$$
Which of the following statements about this system are correct ?
(a) After two hours, less than 10% of the initial X is left
(b) Maximum amount of Y present at any time before 30 min is less than 50% of the initial amount of X
(c) Atomic number of X and Z are same
(d) All of the above

3. $^{27}_{13}$Al is a stable isotope. $^{29}_{13}$Al is expected to disintegrate by
(a) α-emission (b) β-emission
(c) positron emission (d) neutron emission

4. The number of α and β particles lost when $^{238}_{92}$U changes to $^{206}_{82}$Pb are
(a) 8α, 6β (b) 6α, 8β
(c) 6α, 6β (d) 8α, 8β

5. Assuming that only particles emitted from atoms during natural radioactive decay are α and β-particles, which of the following atoms could not possibly result from the natural decay of $^{235}_{92}$U
(a) $^{231}_{89}$Ac (b) $^{227}_{89}$Ac
(c) $^{225}_{89}$Ac (d) $^{207}_{82}$Pb

6. Of the following types of radiations, the only one to be deflected in a magnetic field is
(a) X-ray (b) γ-ray
(c) β-ray (d) neutrons

7. A process that produces a one-unit increase in atomic number is
(a) electron capture (b) β^--emission
(c) α-emission (d) γ-ray emission

8. Of the following nuclides, the one most likely to be radioactive is
(a) ${}^{31}_{15}P$ (b) ${}^{66}_{30}Zn$
(c) ${}^{14}_{7}N$ (d) ${}^{14}_{6}C$

9. One of the following elements has eight naturally occurring stable isotopes
(a) Ra (b) Au
(c) Cd (d) Br

10. Of the following nuclides, the one most likely to decay by positron (β^+) emission is
(a) ${}^{59}Cu$ (b) ${}^{63}Cu$
(c) ${}^{67}Cu$ (d) ${}^{68}Cu$

11. Among the following nuclides, the highest binding energy per nucleon is found for
(a) ${}^{3}_{1}H$ (b) ${}^{16}_{8}O$
(c) ${}^{56}_{26}Fe$ (d) ${}^{235}_{92}U$

12. The most radioactive of the isotopes of an element is the one with the largest value of its
(a) half-life (b) neutron number
(c) atomic number (d) decay constant

13. The radiation that produces the greatest number of ions as it passes through matter is
(a) α (b) γ
(c) β^- (d) β^+

14. The type of radiation that has the greatest penetrating power through matter is
(a) α (b) γ
(c) β^- (d) visible light

15. In uranium mineral, the atomic ratio N_{U-238}/N_{Pb-206} is nearly equal to one. The age (in years) of the mineral is nearly (half-life period of U-238 is 4.5×10^9 years)
(a) 3.0×10^8 (b) 4.5×10^9
(c) 3.0×10^9 (d) 4.5×10^8

16. For an α-emitting isotope, the value of disintegration constant is 0.49×10^{-10} per year. The amount of the isotope of a given sample will reduce to half its value after a period (in years) of nearly
(a) 0.45×10^{10} (b) 0.9×10^{10}
(c) 1.41×10^{10} (d) 2.82×10^{10}

17. In nuclear reaction

$${}^{7}_{3}Li + {}^{1}_{1}H \longrightarrow 2\,{}^{4}_{2}He$$

the mass loss is nearly 0.02 amu. Hence, the energy released (in units of million kcal/mol) in the process is approximately
(a) 430 (b) 220
(c) 120 (d) 50

18. Consider the following nuclear reactions
1. ${}^{14}_{7}N + {}^{4}_{2}He \longrightarrow {}^{17}_{8}O + {}^{1}_{1}H$
2. ${}^{9}_{4}Be + {}^{1}_{1}H \longrightarrow {}^{6}_{3}Li + {}^{4}_{2}He$
3. ${}^{24}_{12}Mg + {}^{4}_{2}He \longrightarrow {}^{27}_{14}Si + {}^{1}_{0}n$
4. ${}^{10}_{5}B + {}^{4}_{2}He \longrightarrow {}^{13}_{7}N + {}^{1}_{0}n$

Examples of induced radioactivity would include reactions:
(a) 3 and 4 (b) 1 and 2
(c) 1, 3 and 4 (d) 1, 2, 3 and 4

19. Match list I with list II and select the correct answer, using the codes given :

List I	List II
A. ${}^{32}P$	1. Location of tumour in brain
B. ${}^{24}Na$	2. Location of blood clots and circulatory disorders
C. ${}^{60}Co$	3. Radio-therapy
D. ${}^{131}I$	4. Agriculture research

Codes

	A	B	C	D
(a)	4	1	2	3
(b)	4	3	2	1
(c)	4	2	3	1
(d)	3	1	2	4

20. Aluminium-25 decays by emitting a positron. The species immediately produced has
(a) $12p$, $13n$, $13e^-$ (b) $13p$, $12n$, $13e^-$
(c) $12p$, $13n$, $12e^-$ (d) $14p$, $11n$, $14e^-$

21. The half-life of ${}^{99}Tc$ is 6.0 h. Hence, average life is
(a) 4.17 h (b) 3.0 h
(c) 8.66 h (d) 8.00 h

22. Tritium, ${}^{3}_{1}H$ has a half-life of 12.26 years. A 5.00 mL sample of tritiated water has an activity of 2.40×10^9 cpm. How many years will it take for the activity to fall to 3.00×10^8 cpm?
(a) 6.13 (b) 12.26
(c) 24.52 (d) 36.78

23. In a series of three steps in radioactive disintegration sequence starting with ${}^{228}_{88}Ra$, the particles emitted are, successively, β^-, β^- and α-particles. The resulting product is an isotope of
(a) ${}_{92}U$ (b) ${}_{90}Th$
(c) ${}_{88}Ra$ (d) ${}_{86}Rn$

24. 1 g of ${}^{200}_{79}Au$ (T_{50} = 13 days) emits a β-particle forming a stable Hg atom. Hg formed at the end of 52 days
(a) 0.0625 g (b) 0.9375 g
(c) 0.7500 g (d) 0.2500 g

25. At radioactive equilibrium, the ratio between the atoms of two radioactive elements A and B was found to be $3.1\times 10^9 : 1$ respectively. If T_{50} of the element A is 2×10^{10} yrs, then T_{50} of the element B is

(a) 6.2×10^9 yr (b) 6.45 yr

(c) 2×10^{10} yr (d) 3.1×10^9 yr

26. Weight of ^{14}C to have radioactivity 1 curie [λ (disintegration constant) = 4.4×10^{-12} s^{-1}] is

(a) 2×10^{-4} kg (b) 0.9×10^{-4} kg

(c) 1.7×10^{-4} kg (d) 3.7×10^{-10} kg

27. If half-life period is 100 yr, average life is nearly

(a) 100 yr (b) 70 yr

(c) 144 yr (d) 90 yr

28. There are 0.618 μg of ^{206}Pb and 0.238 μg of ^{238}U in a rock. If T_{50} of ^{238}U is 1.5×10^9 yr, age of the rock is

(a) 1.5×10^9 yr (b) 3.0×10^9 yr

(c) 4.5×10^9 yr (d) 0.75×10^9 yr

29. $A \xrightarrow[T_A]{k_A}$ product, $B \xrightarrow[T_B]{k_B}$ product

A and B are two radioactive elements with half-life periods T_A and T_B (in years) and k_A(yr^{-1}) and k_B(atom^{-1} year^{-1}). If half-life periods are equal, disintegration rate at the start of disintegration would be

(a) $k_A T_A$ (b) 0.693

(c) both (a) and (b) (d) None of these

30. Which of the following nuclear reactions is an example of nuclear fusion?

(a) $^{12}_{6}C + ^{1}_{1}H \longrightarrow ^{13}_{7}N + \gamma$

(b) $^{14}_{7}N + ^{1}_{0}n \longrightarrow ^{12}_{6}C + ^{1}_{1}H$

(c) $^{2}_{1}H + ^{3}_{1}H \longrightarrow ^{4}_{2}He + ^{1}_{0}n$

(d) $^{235}_{92}U + ^{1}_{0}n \longrightarrow ^{142}_{56}Ba + ^{91}_{36}Kr + 3^{1}_{0}n$

31. Match list I with list II and select the correct answer:

List I (Isotope)	List II (Characteristic)
A. $^{40}_{20}$Ca	1. unstable, α-emitter
B. $^{133}_{53}$I	2. unstable, β-emitter
C. $^{121}_{53}$I	3. unstable, positron emitter
D. $^{232}_{90}$Th	4. stable

	A	B	C	D
(a)	1	2	3	4
(b)	1	3	2	4
(c)	4	3	2	1
(d)	4	2	3	1

32. In the esterification of an organic acid R—COOH by alcohol R'OH, O of acid is isotopic.

$$R\text{—}\overset{\overset{\displaystyle O}{\|}}{C}\text{—}\overset{18}{O}H + R'\text{—}O\text{—}H \longrightarrow R\text{—}\overset{\overset{\displaystyle O}{\|}}{C}\text{—}OR' + H_2O$$

Which of the following statement is correct?

(a) ^{18}O is in ester

(b) ^{18}O is in water

(c) ^{18}O is in both

(d) ^{18}O is in none

33. Thiosulphate ion, $S_2O_3^{2-}$ on acidification changes to SO_2 along with precipitation of sulphur

$$^{35}S^{32}SO_3^{2-} + 2H^+ \longrightarrow H_2O + SO_2 + S$$

Which is correct statement?

(a) ^{35}S is in sulphur (b) ^{35}S is in SO_2

(c) ^{35}S is in both (d) ^{35}S is in none

34. Water used as moderator in nuclear reactor is called

(a) heavy water (b) hard water

(c) nuclear water (d) critical water

35. $^{14}_{7}N + ^{1}_{0}n \longrightarrow ^{14}_{6}C + ^{1}_{1}H$ is written as

(a) $^{14}_{7}N\ (n, e)\ ^{1}_{1}H$ (b) $^{14}_{7}N\ (p, n)\ ^{14}_{6}C$

(c) $^{14}_{7}N\ (n, p)\ ^{14}_{6}C$ (d) $^{14}_{6}C\ (p, n)\ ^{14}_{7}N$

36. $^{238}_{92}$U (III B) emits two α-particles; new element will be placed in

(a) VII B (b) VII A

(c) II A (d) II B

37. After the emission of a β-particle followed by α-particle from $^{214}_{83}$Bi, the number of neutrons in the atom is

(a) 130 (b) 129

(c) 128 (d) 127

38. Which of the following are fissile isotopes?

I. $^{238}_{92}$U II. $^{233}_{92}$U

III. $^{239}_{94}$Pu IV. $^{232}_{90}$U

Select the correct answer from the following

(a) I and II (b) II and III

(c) I and IV (d) All of these

39. ^{23}Na is the more stable isotope of Na. $^{24}_{11}$Na can undergo radioactive decay by

(a) β^--emission

(b) α-emission

(c) β^+-emission

(d) K-electron capture

40. The missing fission product in the reaction

$$^{235}_{92}U + ^{1}_{0}n \longrightarrow ^{146}_{57}La + \ldots + 3^{1}_{0}n \text{ is}$$

(a) $^{86}_{35}$Br (b) $^{87}_{35}$Br

(c) $^{89}_{35}$Br (d) $^{89}_{32}$Ge

41. Which is the correct combination?

Emission	Change in Atomic number	Mass number	Neutron number
(a) α	−2	−2	−2
(b) β^-	+1	0	−1
(c) γ	0	0	−1
(d) β^+	−1	0	0

42. Match the radioisotopes with their applications from the alternates given.

Radioisotopes	Applications
I. Cobalt-60	1. Leukemia therapy
II. Potassium-40	2. Thyroid therapy
III. Iodine-123	3. Geological dating
IV. Phosphorus-32	4. Cancer therapy
V. Carbon-14	5. Archaeological dating

	I	II	III	IV	V
(a)	1	2	3	4	5
(b)	5	4	3	2	1
(c)	4	3	2	1	5
(d)	4	3	2	5	1

43. Constituents of wine are carbon, hydrogen and oxygen. Half-lives of ^{13}C, ^{15}O and ^{3}H are respectively 5730 years, 124 s and 12.5 years. A bottle of wine was sealed about 6 years ago. To confirm its age, which of the isotopes would you choose to determine its age?
(a) $^{13}_{6}C$ (b) $^{15}_{8}O$
(c) $^{3}_{1}H$ (d) Any of these

44. 'Fat man' relates to
(a) Pu bomb
(b) U bomb
(c) Th bomb
(d) literary word from a book

45. 'Little boy' relates to
(a) Pu bomb (b) U bomb
(c) H bomb (d) He bomb

46. Thorium is found in
(a) pitchblende (b) uranite
(c) monazite (d) malachite green

47. Consider following statements
I. 'Carbon-dating' is based on the measurement of activity of ^{14}C
II. ^{40}K is used to determine age of the objects upto 1 million years old
III. The uranium-lead method is based on the natural uranium-238 decay series which ends up with the production of stable lead-206.
Select correct statements.
(a) I, II (b) II, III
(c) I, III (d) I, II, III

48. Beta-emission takes place
(a) from the elements above the band of stability
(b) when neutron is converted to proton
(c) with shifting of the new element one group towards right
(d) following all the facts given above

49. Mass defect of 1 g gives energy equal to
(a) 9×10^{13} J
(b) 5.625×10^{32} eV
(c) both (a) and (b)
(d) None of the above

50. Select correct alternate.
(a) $^{1}_{1}p \longrightarrow {^{1}_{0}n}$ is positron emission
(b) $^{1}_{1}p \longrightarrow {^{1}_{0}n}$ is K-electron capture
(c) both (a) and (b)
(d) None of the above

One or More Than One Options Correct

1. Carbon-14 dating method is based on the fact that
(a) carbon-14 fraction is same in all objects
(b) carbon-14 is highly insoluble
(c) ratio of carbon-14 and carbon-12 remains constant during disintegration
(d) carbon-14 is highly soluble

2. Select correct statement(s).
(a) ^{131}I is used for the treatment of thyroid cancer
(b) ^{59}Co cannot be used for treatment of cancer
(c) ^{33}P is used for treatment of leukemia
(d) excessive use of radioactive elements cause cancer

3. Radioactive disintegration rate is affected by
(a) temperature (b) pressure
(c) electric field (d) None of these

4. $^{238}_{92}U$ (III B) undergoes follows emissions

$$^{238}_{92}U \xrightarrow{-\alpha} A \xrightarrow{-\alpha} B \xrightarrow{-\beta} C$$

Which is/are correct statements?
(a) *A* will be of IB group
(b) *A* will be of IIIB group
(c) *B* will be of IIA (alkaline earth metal) group
(d) *C* will be of IIIA (boron family) group

5. Beta-emission takes place
(a) from the elements above the band of stability
(b) when neutron is converted to proton
(c) such that new element remains in the same group
(d) such that atomic number of the element formed increases by one unit

6. Mass defect of 1 g gives energy equal to
(a) 9×10^{13} J (b) 5.625×10^{32} eV
(c) 2.15×10^{10} kcal (d) 9×10^{6} ergs

7. Target nucleus A is converted to product nucleus B by (p, n) as: $A(p, n)\ B$
In this case
(a) A and B are isotopes
(b) A and B are isobars
(c) A and B are isotones
(d) B has higher atomic number than that of A

8. In a nuclear reactor, oxides of which of the following metals are used as fuel material?
(a) Uranium (b) Thorium
(c) Actinium (d) Plutonium

9. Two radioactive substances X and Y have disintegration constant in the ratio 2 : 1. If half-life of Y is 4 days then after 12 days starting with equal mole fraction of each in mixture of X and Y
(a) mole fraction of X and Y remains unchanged
(b) mole fraction of X is larger than that of Y
(c) mole fraction of X is smaller than that of Y
(d) half-life of X is smaller than that of Y

10. Select correct statement(s).
(a) Neutron-proton ratio after a nuclide, ${}^{238}_{92}U$ loses an α-particle is 1.6
(b) ${}^{27}_{13}Al$ can be converted to ${}^{30}_{15}P$ by (α, n) reaction
(c) Nuclear fusion reactions are known as thermonuclear reactions
(d) Larger the value of disintegration constant, greater the stability of radioactive element

Brain Twisters

1. An atomic battery operated with watch uses β-particles from ${}^{147}_{61}Pm$ ($T_{50} = 2.62$ yr) as the energy source. How long would it take for the rate of β-emission in the battery to be reduced to 5% of its original values?

2. If 1 g of ${}^{226}Ra$ emits 11.6×10^{17} α-particles in a year. If $\lambda = 4.353 \times 10^{-4}$ yr^{-1}, calculate value of Avogadro's number.

3. One of the hazards of nuclear explosion is the generation of ${}^{90}Sr$ and its subsequent incorporation in bones. This nuclide has a half-life of 28.1 years. Suppose one microgram (10^{-6} g) was absorbed by a new-born child, how much ${}^{90}Sr$ will remain in his bones after 20 years?

4. The activity of 30 µg of ${}^{247}Cm$ is 2.8 nCi (10^{-9} curie). Calculate the disintegration rate constant and the half-life of ${}^{247}Cm$.

5. Upon irradiating californium with neutrons, a scientist discovered a new nuclide having mass number of 250 and a half-life of 0.50 h. Three hours after the irradiation, the observed radioactivity due to nuclide was 10 dis/min. How many atoms of the nuclide were prepared initially?

6. What mass of ${}^{14}C$ with $T_{50} = 5730$ years has activity equal to one curie?

7. What is the wavelength (in nm) of gamma-rays whose energy is 1.50 eV?

8. A positron has the same mass as an electron (9.109×10^{-31} kg) but an opposite charge. When the two particles encounter each other, annihilation occurs and only gamma-rays are produced. How much energy (in kJ mol^{-1}) is produced?

9. Radiolysis of water yields hydrogen according to the reaction:

$$2H_2O \longrightarrow H_2 + H_2O_2$$

Calculate the volume of hydrogen produced per hour in a 200 kW reactor. The yield of hydrogen is 1.7 molecules per 100 eV of energy absorbed.

10. The nuclide ${}^{50}_{21}Sc$, mass 49.9516 amu, is neutron rich. It decays to form ${}^{50}_{22}Ti$, mass 49.94479 amu. If the emitted β^- particle has a kinetic energy of 0.80 MeV, what is the kinetic energy of the neutrino emitted simultaneously ?

11. How much of heat would be developed per hour from 1 Ci ${}^{14}C$ source if all the energy of the β^- decay were imprisoned ?

12. In a thorium mineral the ratio by weight of ${}^{208}Pb : {}^{232}Th$ is 0.2. The decay constant of ${}^{232}Th$ is 4.98×10^{-11} year^{-1}. Calculate the age of mineral.

13. The activity of hair of an Egyptian mummy is 10 dismin^{-1} g^{-1} of ${}^{14}C$. What is the age of the mummy? T_{50} of ${}^{14}C$ is 5730 years and disintegration rate of fresh sample of ${}^{14}C$ is 15 dis min^{-1} g^{-1}.

14. 1 g of ${}^{14}C$ from an archaeological sample gives 9.4 counts min^{-1}. 1 g of ${}^{14}C$ from a recently cut tree gives 15.3 counts min^{-1} under the same conditions. How old is the archaeological sample? (T_{50} of ${}^{14}C$ is 5730 years).

15. Potassium-40 decays with a half-life of 1.3×10^9 yr. What is the age of lunar rock sample that contains these isotopes in the ratio 34.1 atoms ${}^{40}K$: 65.9 atoms ${}^{40}Ar$? Assume that no argon was originally in the sample and that none has been lost by weathering.

16. ${}^{238}_{92}U$ by successive radioactive decays changes to ${}^{206}_{82}Pb$. A sample of uranium ore was analysed and found to contain 1.0 g ${}^{238}U$ and 0.1 g ${}^{206}Pb$. Assuming that ${}^{206}Pb$ has accumulated due to decay of ${}^{238}U$, find out the age of the ore. T_{50} of ${}^{238}U$ is 4.5×10^9 years.

17. As a result of being exposed to the radiation released during the Japan nuclear accident in very recent Sunami, the dose of iodine-131 in a person's body is 7.4 mCi (1 mCi = 1×10^{-3} Ci). Calculate the number of atoms of iodine-131 (T_{50} = 8.1 days).

Passage Comprehension Questions

Passage 1

Back in 1953 the U.S. Army started an experimental programme of food irradiation so that deployed troops could have fresh food without refrigeration. The procedure is a simple one. Food is exposed to high levels of radiation to kill insects and harmful bacteria. It is then packaged in airtight containers, in which it can be stored for months without deterioration. The radiation sources for most food preservation are cobalt-60 and cesium-137, both of which are γ-emitters, although X-rays and electron beams can also be used to irradiate food.

Read the above passage and answer the following questions.

1. The benefits of food irradiations are
   (a) it reduces energy demand by eliminating the need for refrigerations
   (b) it prolongs shelf life of various foods
   (c) both (a) and (b) are correct
   (d) None of the above is correct
2. There have been certain objections against food preservation by irradiations. These are:
   (a) irradiated food may itself become radioactive
   (b) irradiation can destroy the nutrients such as vitamins and amino acids
   (c) ionising radiation produces reactive species such as the hydroxyl radical which then react with organic molecules to produce potentially harmful substances.
   (d) irradiated food may be much costlier
3. Select correct statement(s).
   (a) A **rad** is a measure of the amount of high energy radiation absorbed by matter
   (b) 1 **rad** of α-rays has the same effect on tissue as 1 **rad** of γ-rays
   (c) Iodine-131 can be used for scanning the thyroid for disorders
   (d) Strontium-90 is useful isotope for food radiation

Passage 2

In a series of extraordinary events, a natural nuclear fission reactor operated at the time when the first life forms appeared on Earth. As is often the case in scientific endeavors, humans are not necessarily the innovators but merely the imitators of nature.

Read the above passage and answer the following questions.

1. Nuclear-fission reaction is
   (a) chain reaction
   (b) induced by neutrons
   (c) related to Pu-bomb
   (d) related to U-bomb
2. Select correct statements.
   (a) Sr-90 and Cs-137 are the other by-products of nuclear fission released through bomb testing
   (b) Sr-90 and Cs-137 can produce cancer of the bone
   (c) Sr-90 like Ca is incorporated into bone and thus produces bone-cancer
   (d) Cs-137 through radioactive, is rapidly eliminated from the tissue being quite soluble like K

Passage 3

Read the following passage and answer the questions at the end.

A person received an anonymous gift of a decorative cube which he placed on his desk. A few months later he became ill and died shortly afterward. After investigation, the cause of his death was linked to the box. The box was air-tight and had no toxic chemicals on it. What might have killed the man?

1. (a) α-rays emitting material
   (b) β-rays emitting material
   (c) γ-rays emitting material
   (d) KCN
2. Am-241 is used in smoke detectors because it has a long half-life (458 y). Given the schematic diagram of a smoke detector below, explain how it works?

   (a) it emits α-particles which are energetic enough to ionise air molecules
   (b) it emits β-particles which can complete the circuit
   (c) it emits γ-particles which have high penetration
   (d) it emits α- and β-particles

Passage 4

Radiolysis of water yields H_2 *as given below*

$$2H_2O \longrightarrow H_2O_2 + H_2$$

In the analysis it is found that 1.7 molecules of H_2 are yielded per 100 eV of energy absorbed. A nuclear power reactor of 200 kW capacity has been installed based on this reaction.

Read the above passage and answer the following questions.

1. Above reaction is an example of
   (a) disproportionation
   (b) redox
   (c) decomposition
   (d) displacement
2. What mass of heavy water is required to yield 448 kL of heavy hydrogen at STP?
   (a) 996 kg (b) 448 kg
   (c) 800 kg (d) 720 kg
3. Volume of H_2 produced per minute in the above reactor is
   (a) 140 L (b) 70 L
   (c) 35 L (d) 28 L

Passage 5

To detect bombs that may be smuggled into airplanes, the Federal Aviation Administration (FAA) will soon require all major airports in the United States to install thermal neutron analysers. The thermal neutron analyser will bombard baggage with low-energy neutrons, converting some of the nitrogen-14 nuclei to nitrogen-15, with simultaneous emission of γ-rays. Because nitrogen content is usually high in explosives, detection of a high dosage of γ-rays will suggest that a bomb may be present.

Read the above passage and answer the following questions.

1. Nuclear process involves reaction
   (a) ${}^{14}_{7}N + {}^{1}_{0}n \longrightarrow {}^{15}_{7}N$
   (b) ${}^{14}_{7}N + {}^{1}_{0}n \longrightarrow {}^{14}_{6}C + {}^{1}_{1}H$
   (c) ${}^{14}_{7}N \longrightarrow {}^{14}_{6}C + {}^{0}_{1}e$
   (d) ${}^{14}_{7}N \longrightarrow {}^{15}_{7}N + {}^{-1}_{0}e$
2. Consider the following nuclear reactions
   1. ${}^{14}_{7}N + {}^{4}_{2}He \longrightarrow {}^{17}_{8}O + {}^{1}_{1}H$
   2. ${}^{9}_{4}Be + {}^{1}_{1}H \longrightarrow {}^{6}_{3}Li + {}^{4}_{2}He$
   3. ${}^{24}_{12}Mg + {}^{4}_{2}He \longrightarrow {}^{27}_{14}Si + {}^{1}_{0}n$
   4. ${}^{10}_{5}B + {}^{4}_{2}He \longrightarrow {}^{13}_{7}N + {}^{1}_{0}n$

   Examples of induced radioactivity would include reactions
   (a) 3 and 4 (b) 1 and 2
   (c) 1, 3 and 4 (d) 1, 2, 3 and 4
3. The sensitivity of the neutral activation analysis depends on the
   (a) neutron flux available for irradiation
   (b) availability of nucleus to absorb a neutron
   (c) energy of decay process
   (d) All of the above are correct
4. Select correct statement(s).
   (a) Photon or particle is ionised and collected by the positive electrode in a Geiger counter
   (b) A scintillation counter contains a phosphor, a substance that emits a flash of light when struck by a photon or particle
   (c) Both of the above
   (d) None of the above

Passage 6

One of the products of radiation fall out from atmospheric bomb testing or nuclear accidents as at Chernobyl is iodine-131. People living in a region where iodine-131 was known to have been deposited were encouraged to use table salt enriched with non-radioactive isotope of iodine.

Read the above short write-up and answer the following questions.

1. Non-radioactive isotope prescribed in the table-salt is
   (a) I-130 (b) I-127
   (c) I-132 (d) I-129
2. The rate of iodine uptake by the thyroid is
   (a) dependent upon the iodine concentration
   (b) dependent upon pressure of iodine
   (c) follows second-order kinetics
   (d) independent of iodine concentration
3. By use of non-radioactive iodine isotope in common salt
   (a) rate of iodine-131 uptake is sharply reduced compared to uptake of iodine-127
   (b) a new non-radioactive isotope is formed
   (c) taste of table salt becomes better
   (d) iodine-131 being water soluble is eliminated from tissue

Passage 7

Read the following figure in paragraph and answer the questions at the end of it.

A number of basic questions have probably occurred to you as we have been describing nuclear decay processes.

Why do some radioactive nuclei decay by α-emission and some by β^--emission?

Why do the lighter elements have so few naturally occurring radioactive nuclides, whereas those of the heavier elements all seem to be radioactive?

Experimentally, it is found that nuclei with certain number of protons or neutrons appear to be very stable. These numbers called **magic numbers** and associated with specially stable nuclei, were later explained by the **shell model.**

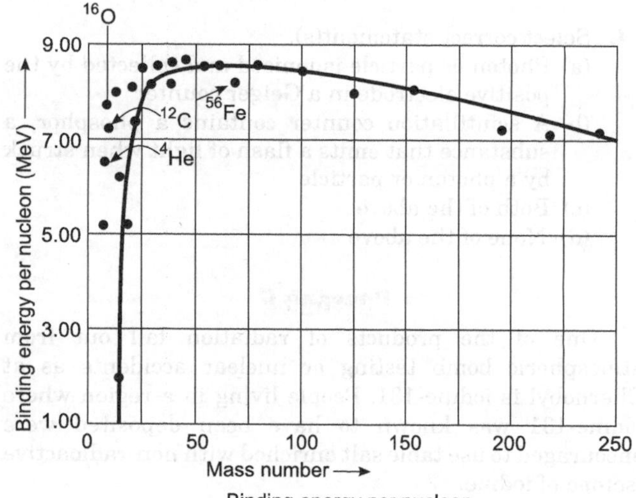

Binding energy per nucleon

1. Which have comparatively higher energy?
   (a) Even number of protons and even number of neutrons
   (b) Even number of protons but odd number of neutrons
   (c) Even number of neutrons but odd number of protons
   (d) Odd number of protons as well as odd number of neutrons
2. Magic number for protons are
   (a) 2, 8, 20, 28, 50, 126
   (b) 2, 8, 20, 28, 50, 82, 114
   (c) 2, 8, 20, 50, 82, 126
   (d) 1, 3, 5, 7, 9, 11
3. Magic number for neutrons are
   (a) 1, 3, 9, 11, 13
   (b) 11, 19, 37, 55, 87
   (c) 2, 8, 20, 28, 50, 82, 126
   (d) 2, 8, 18, 36, 54, 86

Passage 8

Don't Press the Panic Button.

Despite the crisis in Japan's Plants, India must not take nuclear power off the table.

Answer the following questions

1. In view of recent nuclear blasts due to Sunami in Japan countries are
   (a) going to stop nuclear-energy plan
   (b) devising methods to avoid this type of disaster
   (c) stressing thermal energy plan
   (d) stressing hydro-electric power
2. Nuclear plan can't be taken off the table because of
   (a) unusual toxicity of the radioactive substance
   (b) technology outweighs such risks
   (c) nuclear power remains the only economically viable renewable source of energy
   (d) alternates (b) and (c)

Assertion & Reason

Codes :

(a) Both **A** and **R** are true and **R** is the correct explanation of **A**.
(b) Both **A** and **R** are true but **R** is not the correct explanation of **A**.
(c) **A** is true but **R** is false.
(d) **A** is false but **R** is true.

1. **Assertion (A) :** When there is emission of β-particle atomic number increases by one unit.
   $$^{14}_{6}C \longrightarrow ^{14}_{7}N + ^{0}_{-1}e$$
   Reason (R) : Neutron is converted to proton
   $$^{1}_{0}n \longrightarrow ^{1}_{1}p + ^{0}_{-1}e$$
2. **Assertion (A) :** When there is an emission of α-particle, new element is displaced two groups of the parent element.
   Reason (R) : Atomic mass decreases by 2 units due to α-particle emission.
3. **Assertion (A) :** Half-life period of the radioactive disintegration is independent of concentration.
   Reason (R) : 75% disintegration takes place in a time twice the 50% disintegration.
4. **Assertion (A) :** γ-rays produce a greater health hazard to those who work with radioactive element.
   Reason (R) : γ-rays are extremely penetrating high energy electromagnetic radiations.
5. **Assertion (A) :** Element formed by the successive emission of 1α-particle and 2β-particles is placed in the same group as the parent element.
   Reason (R) : α-particle is $[^{4}_{2}He]^{2+}$ while β-particle is $^{0}_{-1}e$.
6. **Assertion (A) :** Nuclear-fission is a chain reaction.
   Reason (R) : Extra electrons formed in the intermediate steps further attack the nuclei.

True & False

1. A radioactive change is always irreversible and independent of temperature and pressure.
2. Thorium (90) formed from emission of α-particle from uranium (92, group IIIB) has been placed in group IIIB.
3. Increase in atomic number by one unit due to emission of β-particle is as a result of conversion of neutron into proton.
4. Successive emission of 1α and 2β-particles results in the formation of isotopes.
5. γ-rays are not affected in magnetic or electric field.
6. γ-rays have least penetrating power.

7. Nuclear-fusion chain initiation is endothermic and further reaction is also endothermic.
8. D_2O or graphite is a moderator in nuclear reactor.
9. Elements with atomic number greater than 92 are called actinides.
10. IUPAC representation of atomic number 108 is Uno.
11. Ds is the latest IUPAC approved name of the element of atomic number 110.
12. The more positive the packing fraction of an element, the more stable should be nucleus.
13. Energy equivalent of 1 amu is 931 MeV.
14. Age of earth is determined by Geological dating.
15. MRI uses radio-waves.
16. Iodine-131 is used in imaging the brain.
17. ${}^{23}_{11}Na$ changes to ${}^{24}_{12}Mg$ by bombardment of ${}^{1}_{0}n$ and successive emission of β-particle.
18. 'Fat man' relates to literary word from a book.
19. ${}^{40}K$ is used to determine age of objects upto 1 million years old.
20. 'Little boy' relates to U-bomb.

Fill in the Blanks

1. SI unit of radioactivity is .... which is equal to .... dps.
2. Disintegration of 3.7×10^{10} atoms per second (3.7×10^{10} dps) is called ....
3. Successive emission of 1 α- and 2 β-particles results in the formation of ....
4. There is an increase in the atomic number by 1 unit due to ....
5. ${}^{234}_{90}Th$ changes to ${}^{206}_{82}Pb$ by emission of .... α-particles and .... β-particles.
6. Series | Starts with | Ends with

Series	Starts with	Ends with
(a) $4n$	...	...
(b) $4n+1$	...	...
(c) $4n+2$	...	...
(d) $4n+3$	...	...

7. Radioactive disintegration follows .... order kinetic.
8. Radioactive change is .... and independent of .... and ....
9. k^{-1}(k is radioactive disintegration constant) is called ....
10. Concentration of the radioactive element after one average life is .... of the original concentration.
11. ${}^{24}_{11}Na \xrightarrow{(p,\,n)} {}^{24}_{12}Mg$ reaction is written as ....
12. Nuclear reaction ${}^{14}_{7}N + {}^{4}_{2}He \longrightarrow {}^{17}_{8}O + {}^{1}_{1}H$ is written as ....
13. There should be a .... of the fissionable material to maintain a fission chain.
14. Graphite is used in the nuclear reactor as a ....
15. IUPAC representation of the element of atomic number 115 is ....
16. The overall reaction in the sun responsible for the energy it radiates is ....
17. Elements with atomic numbers greater than 92 are called the .... elements.
18. The more .... the packing fraction of an element, the more stable should be the nucleus.
19. Energy equivalent of 1 amu is .... MeV.
20. Age of dead bodies is determined by ....
21. Cobalt-60 used in the treatment of ....
22. .... is a device that detects nuclear radiations from flashes of light generated in a matter by the radiation.
23. A REM is a unit of .... used to relate various kinds of radiation in terms of biological destruction.
24. Unit of radioactive disintegration constant is ....
25. Element formed from ${}^{238}_{92}U$ (group IIIB) by emission of 1 α-particle is placed in group ....

Matrix-Match Type Questions

I. Only One Correct

1. Match the atomic number of the transuranium element (in List I) with their IUPAC representation in (List II).

List I	List II
A. 100	1. Unp
B. 101	2. Unh
C. 102	3. Unq
D. 103	4. Unt
E. 106	5. Unu
	6. Unb
	7. Unn

2. Match the name of scientist (in List I) with the discovery or law (in List II).

List I	List II
A. Henry-Becquerel	1. Artificial radioactivity
B. Soddy-Fajan	2. Cyclotron
C. Rutherford	3. Group displacement law
D. I. Curie, F. Jolit	4. Radioactive disintegration
E. E.O. Lawrence	5. Radioactivity

3. **Match the series (in List I) with the end product (in List II).**

List I	List II
A. $4n$	1. $^{207}_{82}Pb$
B. $(4n + 1)$	2. $^{212}_{83}Bi$
C. $(4n + 2)$	3. $^{209}_{83}Bi$
D. $(4n + 3)$	4. $^{206}_{82}Pb$
	5. $^{208}_{82}Pb$
	6. $^{207}_{83}Bi$

4. **Match the series (in List I) with starting element (in List II) and end element (in List III).**

List I	List II	List III
A. $4n$	1. $^{241}_{94}Pu$	I. $^{207}_{87}Pb$
B. $(4n + 1)$	2. $^{232}_{92}Th$	II. $^{209}_{83}Bi$
C. $(4n + 2)$	3. $^{235}_{92}U$	III. $^{209}_{83}Pb$
D. $(4n + 3)$	4. $^{238}_{92}U$	IV. $^{206}_{82}Pb$

II. One or More Correct

1. **Match the Column I with the properties (one or more) in Column.**

Column I	Column II
A. Determination of age of rock	1. $N = N_0 \left(\frac{1}{2}\right)^y$
B. Carbon-dating	2. $\frac{0.693}{\lambda}$
C. Half-life	3. $^{238}_{92}U \longrightarrow {}^{206}_{82}Pb$
D. α-particle	4. $^{14}_{6}C \longrightarrow {}^{14}_{7}N$
E. β-particle	5.

2. **Match the Column I giving particles with the reaction involving these particles (on or more) in Column II.**

Column I	Column II
A. $^{4}_{2}He$	1. $^{14}_{7}N \longrightarrow {}^{17}_{8}O$
B. $^{1}_{1}H$	2. $^{27}_{13}Al \longrightarrow {}^{30}_{15}P$
C. $^{1}_{0}n$	3. $^{30}_{15}P \longrightarrow {}^{30}_{14}Si$
D. $^{0}_{-1}e$	4. $^{23}_{11}Na \longrightarrow {}^{24}_{11}Na$
E. $^{0}_{1}e$	5. $^{24}_{11}Na \longrightarrow {}^{24}_{12}Mg$

Integer Answer Types

This section contains 8 questions. The answer to each of the questions is a single digit integer, ranging from 0 to 9. The appropriate bubbles below the respectively question numbers in the *ORS* have to be darkened. For example, if the correct answers to question number *X*, *Y*, *Z* and *W* (say) are 6, 0, 9 and 2, respectively, then the correct darkening of bubbles will like the following :

X	Y	Z	W
⓪	⓪	⓪	⓪
①	①	①	①
②	②	②	②
③	③	③	③
④	④	④	④
⑤	⑤	⑤	⑤
⑥	⑥	⑥	⑥
⑦	⑦	⑦	⑦
⑧	⑧	⑧	⑧
⑨	⑨	⑨	⑨

Set I

1. **In the following disintegration**

$$^{238}_{92}U \longrightarrow {}^{206}_{82}Pb + \ldots$$

number of β-particles emitted is ....

2. **In question (1) above, number of α-particles emitted is ....**

3. **16 g X changes to 1 g (half life 1 day) after .... days**

4. **Atomic number of positron is taken as ....**

5. **When $^{7}_{4}Be$ captures one electron, mass number of the element formed is ....**

6. **Charge on α-particles is X^+; X is ....**

7. **Charge or γ-particles is ....**

8. **10 g of a radioactive element loses 8.75 g in 9 days. Thus, half-life is .... days.**

Set II

1. **1.00 mL solution containing 0.24 μCi of tritium is injected into a dog's blood stream. After a period of time to allow the isotope to be dispersed, a 1.00 mL sample of blood is drawn. The radioactivity of this sample is found to be 0.04 μCi/mL. What is the total volume of blood in the dog?**

2. **How many neutrons are contained in A formed by the reaction?**

$$^{9}_{4}Be\,(\alpha, n) \longrightarrow A$$

3. **How many neutrons are emitted in the reaction?**

$$^{236}U \longrightarrow {}^{141}Ba + {}^{92}Kr$$

4. **In the following decay process, half-life is .... days**

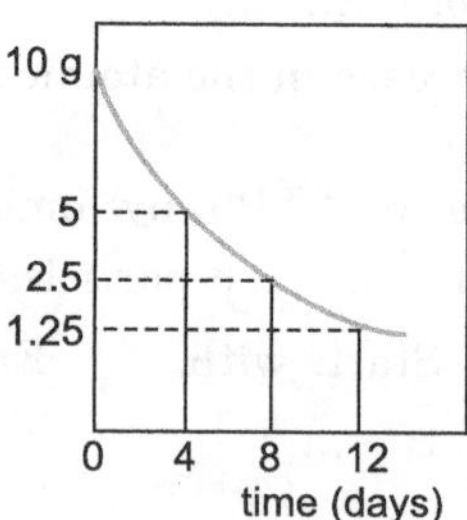

5. **What is the number of particles left in the following decay process?**

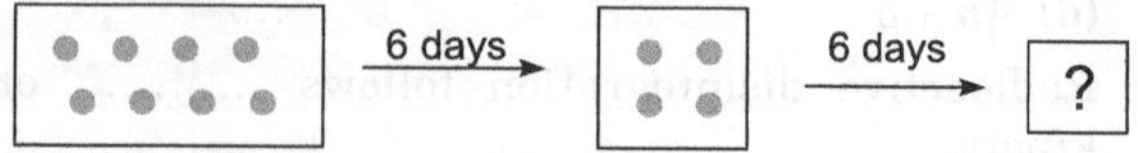

6. **In the following reaction,**

$$CH_3\overset{\overset{\displaystyle O}{\|}}{C}OH + CH_3CH_2\overset{18}{O}H \longrightarrow CH_3\overset{\overset{\displaystyle O}{\|}}{C}OC_2H_5 + H_2O$$

molar mass difference between water formed in the reaction and ordinary water is .... unit.

7. **How many curies are contained in the radioactive change with activity 18.5×10^{10} dps?**

8. **Carbon-dating uses carbon isotope with number of neutrons ....**

Answers & Solutions

Master Exercises

Exercise 1

Analytical Questions

1. Half-life = 4.8 min

(a) time = 19.2 min

$$y = \text{number of half-life} = \frac{19.2}{4.8} = 4$$

$$\therefore \quad C = C_0\left(\frac{1}{2}\right)^y$$

$$= 1\text{ mg}\left(\frac{1}{2}\right)^4$$

$$= \frac{1}{16}\text{ mg}$$

$$= \mathbf{0.0625\ mg}$$

(b) time = 48 min

$$y = \frac{48}{4.8} = 10$$

time of 99.9% reaction = 10 times of half-life
hence, 99.9% reacted

$$\text{unreacted} = 0.1\% = 1 \times \frac{0.1}{100}$$

$$= \mathbf{1 \times 10^{-3}\ mg}$$

2. $-\dfrac{dN}{dt} = \lambda N$

$$0.1\text{ mg } ^{239}\text{Pu} = 0.1 \times 10^{-3}\text{ g} = \frac{0.1 \times 10^{-3}}{239}\text{ mol}$$

$$\therefore \quad N = \frac{0.1 \times 10^{-3} \times 6.02 \times 10^{23}}{239}\text{ atoms}$$

$$\therefore \quad 1.4 \times 10^7 = \frac{\lambda \times 0.1 \times 10^{-3} \times 6.02 \times 10^{23}}{239}$$

$$\therefore \quad \lambda = 5.56 \times 10^{-11}\text{ min}^{-1}$$

$$\therefore \quad t_{50} = \frac{0.693}{\lambda} = \mathbf{7.481 \times 10^{11}\ s}$$

$$= \mathbf{2.372\ years}$$

3.

$$\lambda = \frac{2.303}{t}\log\frac{C_0}{C}$$

$$t = \frac{2.303}{\lambda}\log\frac{C_0}{C}$$

$$= \frac{2.303 \times t_{50}}{0.693}\log\left(\frac{C_0}{C}\right)$$

(a)

$$t = \frac{2.303 \times 14.8}{0.693}\log\frac{100}{10}$$

$$= \mathbf{49.184\ h}$$

(b)

$$t = \frac{2.303 \times 14.8}{0.693}\log\frac{100}{5} = \mathbf{64\ h}$$

4.

$$N_1\lambda_1 = N_2\lambda_2$$

radium uranium

$$\frac{\lambda_1}{\lambda_2} = \frac{N_2}{N_1} = \frac{28}{1}$$

$$\frac{\left(\dfrac{0.693}{t_{50}}\right)_1}{\left(\dfrac{0.693}{t_{50}}\right)_2} = \frac{(t_{50})_2}{(t_{50})_1} = \frac{28}{1}$$

$$(t_{50})_2 = 1620 \times 28\text{ years} = 1.43047 \times 10^{12}\text{ s}$$

$$\therefore \quad \lambda_2 = \frac{0.693}{t_{50}} = \mathbf{4.844 \times 10^{-13}\ s^{-1}}$$

5.

$$1\text{ g radium} = \frac{N_0}{226}\text{ atoms}$$

$$-\frac{dN}{dt} = \frac{\lambda N_0}{226}$$

$$11.6 \times 10^{17} = \frac{4.353 \times 10^{-4}\ N_0}{226}$$

$$\therefore \quad N_0 = \mathbf{6.02 \times 10^{23}\ mol^{-1}}$$

6. $1\text{ mCi} = 0.001\text{ Ci} = 3.7 \times 10^7\text{ dps}$

$$t_{50} = 5720\text{ years}$$

$$= 5720 \times 3.1536 \times 10^7\text{ s}$$

$$(1\text{ yr} = 3.1536 \times 10^7\text{ s})$$

$$\lambda = \frac{0.693}{t_{50}} = \frac{0.693}{5720 \times 3.1336 \times 10^7}\text{ s}^{-1}$$

$$\frac{dN}{dt} = \lambda\left(\frac{w}{14}\right) \times 6.02 \times 10^{23}\text{ atom s}^{-1}$$

$$3.7 \times 10^7 = \frac{0.693}{3.1536 \times 10^7 \times 5720} \times \frac{w \times 6.02 \times 10^{23}}{14}$$

$$w = \mathbf{2.24 \times 10^{-4}\ g}$$

7.

$$1.00\ \mu\text{Ci} = 1.0 \times 10^{-6}\text{ Ci}$$

$$= 3.7 \times 10^{10} \times 1.0 \times 10^{-6}\text{ dps}$$

$$\therefore \quad \frac{-dN}{dt} = 3.7 \times 10^4\text{ dps}$$

$$1\text{ year} = 3.1536 \times 10^7\text{ s}$$

$$\lambda = \frac{0.693}{t_{50}} = \frac{0.693}{1.28 \times 10^9 \times 3.1536 \times 10^7}$$

$$= 1.72 \times 10^{-17}\text{ s}^{-1}$$

$$-\frac{dN}{dt} = \lambda N = \frac{\lambda w N_0}{40}$$

$$3.7\times10^4=\frac{1.72\times10^{-17}\times w\times6.02\times10^{23}}{40}$$

$w=\mathbf{0.143\ g}$

8. $^{210}_{84}Po \longrightarrow {}^{4}_{2}He + {}^{206}_{82}Pb$

$1\ g\ Po=\frac{1}{210}\ mol\ Po$

Po left in 69.2 days $=C=C_0\left(\frac{1}{2}\right)^y$

where, $y=\frac{\text{total time}}{t_{50}}=\frac{69.2}{138.4}=\left(\frac{1}{2}\right)$

$C=1\left(\frac{1}{2}\right)^{1/2}$

$=0.7071\ g$

Po changed to helium $=1-0.7071\ g$

$=0.2929\ g$

$=1.3947\times10^{-3}\ mol$

Helium formed $=31.24$ cc at STP

9. Initial amount of an α-active isotope $=N_0=10$ g atom

Helium formed $=11.2$ cc at STP

$=\frac{11.2}{22400}$ g atom at STP

$=5\times10^{-4}$ g atom

∴ Element left $=(10-5\times10^{-4})$

$=9.9995$ g atom

$\lambda=\frac{2.303}{t}\log\frac{N_0}{N}$

$=2.303\log\frac{10}{9.9995}=5\times10^{-5}\ h^{-1}$

$t_{50}=\frac{0.693}{\lambda}=\frac{0.693}{5\times10^{-3}}\ h$

$=13857\ h=\mathbf{1.582\ years}$

10. Amount left after 10 days = amount disintegrating on 11th day.

$C=C_0\left(\frac{1}{2}\right)^y \qquad y=\frac{\text{total time}}{t_{50}}=2$

$=0.1\left(\frac{1}{2}\right)^2$

$=0.1\times\frac{1}{4}$

$=0.025$ g atm after 10 days

Amount left on 11th day $\left(y=\frac{1}{5}\text{ half-life}\right)$

when $C_0=0.025$ g atom

$C=0.025\left(\frac{1}{2}\right)^{1/5}$ g atom

$=0.0218$ g atom

Amount disintegrating on 11th day $=C_0-C$

$=0.025-0.0218$

$=3.2\times10^{-3}$ g atom

$=1.926\times10^{21}$ atom

11. ${}^{A}_{Z}X \longrightarrow {}^{A-4}_{Z-2}Y + {}^{4}_{2}He$

$t_{50}=10$ days

∴ $y=\frac{t}{t_{50}}=\frac{20}{2}=2$

X left after 20 days

$C=C_0\left(\frac{1}{2}\right)^y=\frac{1}{4}$ g atom

∴ Helium formed = amount of X decomposed

$=1-\frac{1}{4}=\frac{3}{4}$ mol

∴ Volume at STP $=\frac{3}{4}\times22.4\ L$

$=\mathbf{16.8\ L}$

12. $-\frac{dN}{dt}=\lambda N_0$ at start

$-\frac{dN}{dt}=\lambda N$ after time t

$4750=\lambda N_0$

$2700=\lambda N$

∴ $\frac{N_0}{N}=\frac{4750}{2700}$

Also $\lambda=\frac{2.303}{t}\log\frac{N_0}{N}$

$=\frac{2.303}{5}\log\frac{4750}{200}=0.113\ min^{-1}$

∴ $t_{50}=\frac{0.693}{\lambda}=\frac{0.693}{0.113}=\mathbf{6.1328\ min}$

13. Let initial amount $=10=N_0$

Amount decomposed $=99.99\%$

$N=$ amount left $=100-99.99$

$=0.01$

∴ $\lambda=\frac{2.303}{t}\log\frac{N_0}{N}$

$\frac{0.693}{1.53}=\frac{2.303}{t}\log\frac{100}{0.01}$

$t=\mathbf{20.34\ s}$

14. ${}^{235}_{92}U \longrightarrow {}^{207}_{82}Pb;\ 7-\alpha,\ 4-\beta$

15. ${}^{198}_{79}Au \longrightarrow {}^{198}_{80}Hg + {}^{0}_{-1}e(\beta)$

Au and Hg are **isobars.**

Half-life = 65 days, $t=260$ days

$y=\frac{\text{total time}}{\text{hal-life}}=4$

Au left $C=C_0\left(\frac{1}{2}\right)^y=1\left(\frac{1}{2}\right)^4=\frac{1}{16}$ g

Hg formed $=C_0-C=1-\frac{1}{16}=\frac{15}{16}=\mathbf{0.9375\ g}$

16. $\left[1+\frac{[Pb]}{[U]}\right]=1+\left(\frac{\frac{0.1}{206}}{\frac{1}{238}}\right)=1.1155=(2)^y$

$y=\frac{t}{t_{50}}$

$$y \log 2 = \log 1.1155$$
$$y = 0.1578$$
$$\therefore \quad t = 4.5 \times 10^9 \times 0.1578$$
$$= \mathbf{7.1 \times 10^8 \text{ years}}$$

17. (a) $^{242}_{94}Pu + ^{40}_{20}Ca \longrightarrow ^{287}_{114}Uuq + ^{1}_{0}n$

Uuq has atomic number **114**

↑↑↑
1 1 4

(b) $^{10}_{5}B + ^{1}_{0}n \longrightarrow ^{7}_{3}Li + \underset{(\alpha)}{^{4}_{2}He}$

18. (a) 69.9 MeV (b) 42.6 MeV

19. $$\Delta E = 4.795 - 4.611 = 0.184 \text{ MeV}$$
$$= 0.184 \times 10^6 \text{ eV}$$
$$= 0.184 \times 10^6 \times 1.6 \times 10^{-19} \text{ J}$$
$$= 2.944 \times 10^{-14} \text{ J}$$
$$\Delta E = \frac{hc}{\lambda}$$
$$\therefore \quad \lambda = \frac{hc}{\Delta E} = \frac{6.62 \times 10^{-34} \times 3 \times 10^8}{2.944 \times 10^{-14}}$$
$$= 6.746 \times 10^{-12} \text{ m}$$
$$= \mathbf{6.746 \text{ pm}}$$

20. 827.5 nm

21.

Mass defect g mol^{-1}	BE MeV/neutron
(a) 0.467	8.72
(b) 0.60015	8.76

22. Refer Q. 22.

23. $$\left(1 + \frac{[Pb]}{[U]}\right) = 2^y$$
$$\left(1 + \frac{20.6/206}{23.8/238}\right) = 2^y$$
$$2 = 2^y$$
$$y = 1$$
$$\frac{t}{t_{50}} = 1$$
$$t = 4.5 \times 10^9 \text{ years}$$

NCERT Problems

1. Mass defect
$$= 7.016 - (4 \times 1.008665 + 3 \times 1.007277 + 3 \times 0.000548)$$
$$= 7.016 - 7.058135$$
$$= -0.042135 \text{ units}$$
$$\therefore \quad BE = 0.042135 \times 931 \text{ MeV} = 39.228 \text{ MeV}$$
$$\therefore \quad \text{BE per nucleon} = \left(\frac{BE}{7}\right) = 5.604 \text{ MeV/nucleon}$$

2. Mass defect $= 15.995 - [8(1.0073 + 1.0087 + 0.000548)]$
$$= 15.995 - 16.132384 = 0.137384 \text{ unit}$$
$$\therefore \quad BE = 0.137384 \times 931 \text{ MeV}$$
$$= 127.9045 \text{ MeV}$$
$$= 127.9045 \times 10^6 \text{ eV}$$
$$= 127.9045 \times 10^8 \times 1.6 \times 10^{-19} \text{ J}$$
$$= \mathbf{2.046 \times 10^{-11} \text{ J per molecule}}$$
$$= 2.046 \times 10^{-11} \times 6.02 \times 10^{23}$$
$$= \mathbf{1.232 \times 10^{13} \text{ J mol}^{-1}}$$

3. $2\,^{2}_{1}H \longrightarrow ^{4}_{2}He$

$$\text{Mass defect} = m(^{4}_{2}He) - 2m(^{2}_{1}H)$$
$$= 4.0026 - 2 \times 2.0141$$
$$= -0.0256 \text{ units}$$
$$= -0.0256 \times 10^{-3} \text{ kg unit}$$
$$\therefore \quad BE = \Delta mc^2$$
$$= 0.0256 \times 10^{-3} \times (2.998 \times 10^8)^2$$
$$= \mathbf{2.3 \times 10^{12} \text{ J}}$$

4. $$^{60}_{28}Ni + ^{1}_{0}n \xrightarrow{(n,\,p)} ^{60}_{27}Co + ^{1}_{1}H$$

Target nucleus $^{60}_{28}Ni$

$$^{59}_{27}Co + ^{1}_{0}n \xrightarrow{n,\,\gamma} ^{60}_{27}Co + \gamma$$

Target nucleus $= ^{59}_{27}Co$

Half-life of $^{60}_{27}Co$ = 7 years

$$t_{50} = 7 \times 3.1536 \times 10^7 \text{ s}$$
$$\therefore \quad \lambda = \frac{0.693}{t_{50}} = \frac{0.693}{7 \times 3.1536 \times 10^7} \text{ s}^{-1}$$
$$= \mathbf{3.14 \times 10^{-9} \text{ s}^{-1}}$$

5. $$\lambda = \frac{2.303}{t} \log \frac{C_0}{C}$$
$$\frac{0.693}{t_{50}} = \frac{2.303}{t} \log \frac{C_0}{C}$$
$$\frac{0.693}{5770} = \frac{2.303}{t} \log \frac{100}{60}$$
$$t = \mathbf{4254 \text{ years}}$$

6. $$-\frac{dN}{dt} = \lambda N$$
$$3.7 \times 10^{10} = \frac{0.693 \times w \times 6.02 \times 10^{23}}{40 \times 60 \times 60 \times 140}$$
$$\therefore \quad w = 1.8 \times 10^{-6} \text{ g}$$

7. Refer text.

Exercise 2

Only One Option Correct

1. In the chain reaction,

	Energy	Neutrons
1st step	E	3
2nd step	$3E$	9
3rd step	$9E$	27
nth step	$3^{n-1}E$	3^n

Thus, **(d)**

2. (d) 3. (d) 4. (a) 5. (c) 6. (c) 7. (b)
8. (d) 9. (c) 10. (a) 11. (c) 12. (d) 13. (a)
14. (b)
15. To determine age by rock-dating

$$1+\frac{[Pb]}{[U]}=2^y$$
$$(1+1)=(2)^y$$
$$\therefore \quad y=1$$
$$\therefore \quad \frac{t}{T_{50}}=y=1$$
$$\therefore \quad t=T_{50}=4.5\times10^9 \text{ year}$$

Thus, **(b)**

16. Disintegration constant $k=0.49\times10^{-10}$ year^{-1}

$$\therefore \quad T_{50}=\frac{0.693}{k}=\frac{0.693}{0.49\times10^{-10}}\text{ year}$$
$$=1.41\times10^{10}\text{ year}$$

Thus, **(c)**

17. $$\text{Mass defect}=0.02\text{ amu}$$
$$=\frac{0.02}{N_0}\text{ g atom}^{-1}$$
$$=\frac{0.02}{N_0}\times N_0\text{ g mol}^{-1}$$
$$=0.02\text{ g mol}^{-1}$$
$$E=mc^2=0.02\times9\times10^{20}\text{ erg mol}^{-1}$$
$$=\frac{0.02\times9\times10^{20}\times0.002}{8.314\times10^7\times10^6}\text{ million kcal mol}^{-1}$$
$$\approx430 \quad \begin{pmatrix}1\text{ million}=10^6\\ 8.314\times10^7\text{ erg}=0.002\text{ kcal}\end{pmatrix}$$

Thus, **(a)**

18. (d) 19. (c)
20. Positron is emitted with proton changes to neutron

$$^1_1\text{H}\longrightarrow{}^1_0n+{}^0_{+1}e$$
Thus, $$^{25}_{13}\text{Al}\longrightarrow{}^{25}_{12}X+{}^1_0e$$

Number of electrons remained = 13 (as in Al)

Thus, **(a)** $\quad ^{25}_{12}X$ has $12p$, $13n$, $13e$.

21. (c)
22. $$C=C_0\left(\frac{1}{2}\right)^y$$
$$(3.00\times10^8)=(2.40\times10^9)\left(\frac{1}{2}\right)^y$$
$$\left(\frac{1}{2}\right)^y=\left(\frac{1}{8}\right)=\left(\frac{1}{2}\right)^3$$
$$\therefore \quad y=3$$
$$\text{or} \quad \frac{t}{T_{50}}=3$$
$$\therefore \quad t=3\times T_{50}=36.78\text{ years}$$

Thus, **(d)**

23. (c)
24. β-particle emission gives isobar.

$$^{200}_{79}\text{Au}\longrightarrow{}^{200}_{80}\text{Hg}+{}^0_{-1}e$$
$$C=C_0\left(\frac{1}{2}\right)^y$$
$$y=\frac{52}{13}=4$$
$$C=1\left(\frac{1}{2}\right)^4=\frac{1}{16}$$
$$\therefore \quad \text{Au left}=\frac{1}{16}\text{ g}$$
$$\text{Au changed to Hg}=1-\frac{1}{16}=\frac{15}{16}=0.9375\text{ g}$$

Thus, **(b)**

25. $$\frac{N_A}{N_B}=\frac{(T_{50})_A}{(T_{50})_B}$$
$$3.1\times10^9=\frac{2\times10^{10}}{(T_{50})_B}$$
$$(T_{50})_B=6.45\text{ year}$$

Thus, **(b)**

26. $$-\frac{dN}{dt}=kN=k\frac{w}{A}N_0$$
$$3.7\times10^{10}=4.4\times10^{-12}\times\frac{w}{14}\times6.02\times10^{23}$$
$$w=2\times10^{-4}\text{ kg} \quad (1\text{ Ci}=3.7\times10^{10}\text{ dps})$$

Thus, **(a)**

27. (c)
28. $$[\text{Pb}]=\frac{0.618\times10^{-6}}{206}\text{ mol}=3\times10^{-9}\text{ mol}$$
$$[\text{U}]=\frac{0.238\times10^{-6}}{238}\text{ mol}=1\times10^{-9}\text{ mol}$$
$$\text{For rock-dating,} \quad \left(1+\frac{[\text{Pb}]}{[\text{U}]}\right)=(2)^y$$
$$\therefore \quad y=2=\frac{t}{T_{50}}$$
$$\therefore \quad t=3.0\times10^9\text{ years}$$

Thus, **(b)**

29. $T_A = \frac{0.693}{k_A}$ by first order rate law (based on unit of k_A)

$T_B = \frac{1}{k_B a}$ by second order rate law (based on unit of k_B)

$$T_A = T_B \quad \text{(given)}$$

$$\therefore \quad \frac{0.693}{k_A} = \frac{1}{k_B a} \quad (a = \text{initial activity})$$

$$\therefore \quad \frac{k_A}{k_B} = 0.693a$$

$$(\text{Rate})_A = k_A(a)$$

$$(\text{Rate})_B = k_B(a)^2$$

$$\therefore \quad \frac{(\text{Rate})_A}{(\text{Rate})_B} = \frac{k_A}{k_B a} = 0.693 = k_A T_A$$

Thus, **(c)**

30. (c) **31.** (d) **32.** (b) **33.** (a) **34.** (a)
35. (c)

36. $$^{238}_{92}\text{U} \xrightarrow{-2\alpha} {}^{230}_{88}X$$

Element of atomic number 88 belongs to IIA
Thus, **(c)**

37. (c) **38.** (c) **39.** (a) **40.** (b) **41.** (b)
42. (c) **43.** (c) **44.** (a) **45.** (b) **46.** (c)
47. (d) **48.** (d) **49.** (c) **50.** (c)

One or More Than One Options Correct

1. (c) **2.** (a),(b),(d) **3.** (a) **4.** (b),(c),(d)
5. (a),(b),(d) **6.** (a),(b),(c) **7.** (b),(d) **8.** (a),(b),(d)
9. (c),(d) **10.** (a),(b),(c)

Brain Twisters

1. $$\lambda = \frac{2.303}{t} \log \frac{100}{5}$$

$$\frac{0.693}{t_{50}} = \frac{2.303}{t} \log 20$$

$$\therefore \quad \frac{0.693}{2.62} = \frac{2.303}{t} \log 20$$

$$t = \mathbf{11.33\ years}$$

2. $$1 \text{ g radium} = \frac{N_0}{226} \text{ atoms}$$

$$-\frac{dN}{dt} = \frac{\lambda N_0}{226}$$

$$11.6 \times 10^{17} = \frac{4.353 \times 10^{-4} N_0}{226}$$

$$\therefore \quad N_0 = \mathbf{6.02 \times 10^{23}\ mol^{-1}}$$

3. $$t_{50} = 28.1 \text{ years}$$

$$\lambda = \frac{2.303}{t} \log \frac{C_0}{C}$$

$$\frac{0.693}{t_{50}} = \frac{2.303}{t} \log \frac{C_0}{C}$$

$$\frac{0.693}{28.1} = \frac{2.303}{20} \log \frac{1 \times 10^{-6}}{C}$$

$$C = 0.6107 \times 10^{-6} \text{ g}$$

$$= 0.6107\ \mu\text{g}$$

4. $$2.8 \text{ nCi} = 2.8 \times 10^{-9} \text{ Ci}$$

$$= 2.8 \times 10^{-9} \times 3.7 \times 10^{10} \text{ dps}$$

$$= 103.6 \text{ dps}$$

$$103.6 = -\frac{dN}{dt} = \lambda \times \frac{30 \times 10^{-6} \times 6.02 \times 10^{23}}{247}$$

$$\lambda = 1.417 \times 10^{-15} \text{ s}^{-1}$$

$$\therefore \quad t_{50} = \frac{0.693}{\lambda} = \mathbf{4.891 \times 10^{14}\ s}$$

$$= \mathbf{1.551 \times 10^7\ years}$$

5. $$\lambda = \frac{0.693}{t_{50}} = \frac{0.693}{0.5} \text{ h}^{-1}$$

$$-\frac{dN}{dt} = \lambda N$$

$$-\frac{dN}{dt} = 10 \text{ dis min}^{-1} = 10 \times 60 \text{ dis h}^{-1}$$

$$10 \times 60 = \frac{0.693}{0.5} \text{N}$$

$$N = 207.9 \text{ atoms}$$

6. $1 \text{ Ci} = 3.7 \times 10^{10}$ dps

$$t_{50} = 5730 \text{ years} = 5730 \times 3.1536 \times 10^7 \text{ s}$$

$$\therefore \quad \lambda = \frac{0.693}{t_{50}} = \frac{0.693}{5730 \times 3.1536 \times 10^7} \text{ s}^{-1}$$

$$-\frac{dN}{dt} = \lambda N = \frac{\lambda w \times 6.02 \times 10^{23}}{14}$$

$$3.7 \times 10^{10} = \frac{0.693 \times w \times 6.02 \times 10^{23}}{5730 \times 3.1536 \times 10^7 \times 14}$$

$$w = 0.2244 \text{ g}$$

7. 8.27×10^{-4} nm

8. Refer solved example 5.21

9. $$\text{Energy} = 200 \text{ kWh} = 200 \times 10^3 \text{ watt}$$

$$= 200 \times 10^3 \text{ J s}^{-1}$$

$$= 200 \times 10^3 \times 3600 \text{ J h}^{-1}$$

$$= \frac{200 \times 10^3 \times 3600}{1.6 \times 10^{-19}} \text{ eV}$$

$$= 4.5 \times 10^{27} \text{ eV}$$

$$\text{Molecules of } H_2 \text{ produced} = \frac{4.5 \times 10^{27} \times 1.7}{100}$$

$$= 7.65 \times 10^{25} \text{ molecules}$$

$$= \frac{7.65 \times 10^{25}}{6.02 \times 10^{23}} \text{ mol } H_2$$

$$= 127.076 \text{ mol } H_2$$

$$= \mathbf{2846.5\ L} \text{ at STP}$$

10. $$\text{Mass defect} = 49.9516 - 49.94479 = 6.81\times10^{-3}\text{ amu}$$

$$\text{Binding energy} = 6.81\times10^{-3}\times931\text{ MeV} = 6.34011\text{ MeV}$$

Kinetic energy of the emitted β^- particle = 0.80 V

∴ Kinetic energy of neutrino emitted = **5.54 MeV**

11. $${}^{14}_{6}\text{C} \longrightarrow {}^{14}_{7}\text{N} + {}^{0}_{-1}e$$

$$\text{Mass defect} = {}^{14}_{6}\text{C} - {}^{14}_{7}\text{N} = 14.00324 - 14.00307 = 1.7\times10^{-4}\text{ amu}$$

$$\text{Binding energy} = 1.7\times10^{-4}\times931\text{ MeV} = 0.1583\text{ MeV per nucleon}$$

$$1\text{ Ci} = 3.7\times10^{10}\text{ dps} = 3.7\times10^{10}\times3600\text{ dph}$$

$$\therefore\quad \text{Total heat} = 3.7\times10^{10}\times3600\times0.1583\text{ MeV} = 2.1086\times10^{13}\text{ MeV} = 2.1086\times10^{19}\text{ eV} = 2.1086\times10^{19}\times1.6\times10^{-19}\text{ J}$$

12. $$\frac{[\text{Pb}]}{[\text{Th}]} = \left(\frac{0.2}{208}\times232\right) = 0.2231$$

$$\therefore\quad (2)^y = 1 + \frac{[\text{Pb}]}{[\text{Th}]} = 1.2231$$

$$\therefore\quad y\log2 = 0.08745$$

$$y = 0.2905$$

$$\frac{t}{t_{50}} = 0.2905$$

$$\therefore\quad t = 0.2905\times T_{50} = 0.2905\times\frac{0.693}{4.98\times10^{-11}}$$

$$= \mathbf{4.04\times10^9\ years}$$

13. $$\lambda = \frac{2.303}{t}\log\frac{C_0}{C}$$

$$\frac{0.693}{5730} = \frac{2.303}{t}\log\frac{15}{10}$$

$$\therefore\quad t = 3353\text{ years}$$

14. $$\lambda = \frac{2.303}{t}\log\frac{C_0}{C}$$

$$\frac{0.693}{5730} = \frac{2.303}{t}\log\frac{15.3}{9.4}$$

$$t = 4028.62\text{ year}$$

15. $$(\text{Ar}) = 65.9\text{ atoms} = N_0$$

$$(\text{K}) = 39.1\text{ atoms} = N$$

$$\therefore\quad \lambda = \frac{2.303}{t}\log\frac{N_0}{N}$$

$$\frac{0.693}{t_{50}} = \frac{2.303}{t}\log\frac{N_0}{N}$$

$$\frac{0.693}{1.3\times10^9} = \frac{2.303}{t}\log\frac{65.9}{34.1}$$

$$\therefore\quad t = \mathbf{1.236\times10^9\ years}$$

17. $$7.4\text{ mCi} = 7.4\times10^{-3}\text{ Ci} = 7.4\times10^{-3}\times3.7\times10^{10}\text{ dps} = 2.738\times10^8\text{ dps}$$

$$\therefore\quad -\frac{dN}{dt} = \lambda N$$

$$t_{50} = 8.1\text{ days} = 8.1\times24\times3600\text{ s}$$

$$\therefore\quad \lambda = \frac{0.693}{t_{50}}$$

$$2.738\times10^8 = \frac{0.693N}{8.1\times24\times3600}$$

$$N = 2.765\times10^{14}\text{ atoms}$$

Passage Comprehension Questions

Passage 1

1. (c) **2.** (a),(b),(c) **3.** (a)

Passage 2

1. (a),(b),(c),(d) **2.** (a),(c),(d)

Passage 3

1. (c) **2.** (a)

Passage 4

1. (b) **2.** (c) **3.** (d)

Passage 5

1. (a) **2.** (d) **3.** (d) **4.** (c)

Passage 6

1. (b) **2.** (a) **3.** (a)

Passage 7

1. (a) **2.** (b) **3.** (c)

Passage 8

1. (b) **2.** (d)

Assertion & Reason

1. (a) **2.** (c) **3.** (b) **4.** (a) **5.** (b) **6.** (a)

True & False

1. T **2.** T **3.** T **4.** T **5.** T **6.** F **7.** F
8. T **9.** F **10.** T **11.** T **12.** F **13.** T **14.** T
15. T **16.** F **17.** T **18.** F **19.** T **20.** T

Fill in the Blanks

1. Bq (Becquerel), 1 **2.** one curie **3.** isotope
4. conversion of neutron into proton **5.** 7, 6
6. (a) ${}^{232}_{90}\text{Th}$, ${}^{208}_{82}\text{Pb}$ (b) ${}^{241}_{94}\text{Pu}$, ${}^{209}_{83}\text{Bi}$ (c) ${}^{238}_{92}\text{U}$, ${}^{206}_{82}\text{Pb}$
(d) ${}^{235}_{92}\text{U}$, ${}^{207}_{82}\text{Pb}$
7. first **8.** irreversible, temperature, pressure

9. average life **10.** $\frac{1}{e}$

11. $^{24}_{11}\text{Na} + ^{1}_{1}\text{H} \longrightarrow ^{24}_{12}\text{Mg} + ^{1}_{0}n$

12. $^{14}_{7}\text{N}(\alpha, p)\,^{17}_{8}\text{O}$ **13.** critical size **14.** moderator

15. Uup **16.** $4\,^{1}_{1}\text{H} \longrightarrow ^{4}_{2}\text{He} + 2\,^{0}_{1}e(\beta^{+})$ **17.** transuranium

18. negative **19.** 931 **20.** carbon-dating **21.** cancer

22. scintillation counter **23.** radiation dosage

24. time^{-1} **25.** IIIB

Matrix-Match Type Questions

I. Only One Correct

1. (A)—(7); (B)—(5); (C)—(6); (D)—(4); (E)—(2);
2. (A)—(5); (B)—(3); (C)—(4); (D)—(1); (E)—(2);
3. (A)—(5); (B)—(3); (C)—(4); (D)—(1);
4. (A)—(2)—(III); (B)—(1)—(II); (C)—(4)—(IV); (D)—(3)—(I);

II. One or More Correct

1. (A)—(1,2,3); (B)—(1,2,4); (C)—(1,2); (D)—(3); (E)—(3,4)
2. (A)—(1,2); (B)—(1); (C)—(2,4); (D)—(5); (E)—(3)

Integer Answer Types

Set I

Questions →	1	2	3	4	5	6	7	8
Answers →	6	8	4	1	7	2	0	3
	0	0	0	0	0	0	0	0
	1	1	1	1	1	1	1	1
	2	2	2	2	2	2	2	2
	3	3	3	3	3	3	3	3
	4	4	4	4	4	4	4	4
	5	5	5	5	5	5	5	5
	6	6	6	6	6	6	6	6
	7	7	7	7	7	7	7	7
	8	8	8	8	8	8	8	8
	9	9	9	9	9	9	9	9

Set II

Questions →	1	2	3	4	5	6	7	8
Answers →	6	6	3	4	2	0	5	8
	0	0	0	0	0	0	0	0
	1	1	1	1	1	1	1	1
	2	2	2	2	2	2	2	2
	3	3	3	3	3	3	3	3
	4	4	4	4	4	4	4	4
	5	5	5	5	5	5	5	5
	6	6	6	6	6	6	6	6
	7	7	7	7	7	7	7	7
	8	8	8	8	8	8	8	8
	9	9	9	9	9	9	9	9

Chemical equations represent a very precise, yet a very versatile, language that describes chemical changes.

6

Chemical Equations and Reactions

Learning Framework of the Chapter ...

- Chemical Equations
- Stoichiometry
- Limiting Reactants
- Yields of Chemical Reactions
- Sequential Reactions

Chemical Equations

Chemical reactions always involve changing one or more substances into one or more different substances. That is, they involve regrouping atoms or ions to form other substances.

A chemical equation represents the final change in composition when two or more reactants react together to form new substances.

Thus, chemical equations are used to

— describe chemical reactions,

— and they show

(1) the substances that react, called **reactants**

(2) the substances formed, called **products** and

(3) the relative amounts of the substances involved.

Take following chemical equation which represents a reaction of stream of chlorine gas, Cl_2, with solid phosphorus, P_4 giving a liquid phosphorus trichloride (which appears as a flame).

$$\underbrace{P_4(s) + 6Cl_2(g)}_{\text{Reactants}} \longrightarrow \underbrace{4PCl_3(l)}_{\text{Products}}$$

In a chemical equation, the formulas for

— the reactants are written to the left of the arrow.

— and that of the products are written to the right of the arrow.

— the physical states of the reactants are also indicated by

(*s*) for solid

(*l*) for liquid

(*g*) for gas

A substance dissolved in water, that is, an aqueous solution of a substance, is indicated by (*aq*)

Based on the **law of conservation of matter,** which states that **matter can neither be created nor destroyed,** if the total mass of the reactants is 10 g, and if the reaction completely converts the reactants to products, you must end up with 10 g of the products. This also means that if 100 atoms (say) of a particular element are contained in the reactants, then those of 100 atoms must also appear in the products in same fashion.

When we apply law of conservation of matter to the following reaction

$$P_4(s) + 6Cl_2(g) \longrightarrow 4PCl_3(l)$$

it tells that 1 molecule of phosphorus, P_4 (with 4 phosphorus atoms) and 6 molecules of diatomic chlorine (Cl_2, with 12 chlorine atoms) are required to produce four molecules of PCl_3. Because each PCl_3 molecules contains 1 P atom and 3 Cl atom, hence in all four PCl_3 molecules are needed.

(Cl-atoms) $6 \times 2 = 12$ on the reactant side; $4 \times 3 = 12$ on the product side.

$$P_4(s) + 6Cl_2(g) \longrightarrow 4PCl_3(l)$$

(P-atom) 4 on the reactant side; 4 on the product side

The numbers in front of formulae in balanced chemical equations are required by the law of conservation of matter. The relationship between the quantities of chemical reactants and products is called **stoichiometry** and the coefficients in a balanced equation are **stoichiometric coefficients**

$$\mathbf{2}Fe(s) + \mathbf{3}Cl_2(g) \longrightarrow \mathbf{2}FeCl_3(l)$$

(2, 3, 2: stoichiometric coefficients)

This equation is read as :

2 atoms of iron (Fe) combine with 3 molecules of chlorine (Cl_2) to form 2 molecules of ferric chloride ($FeCl_3$)

This equation also represents that 2 moles of iron combine with 3 moles of chlorine (Cl_2) to form 2 moles of $FeCl_3$.

Balancing the Chemical Equations

Balancing a chemical equation ensures that the same number of atoms of each element appears on both sides of the equation. Many chemical eqautions can be balanced by trial and error, although some will involve more trial than others when balancing chemical equations,

— Formulae for reactants and products must be correct

— Subscripts in the formulae of reactants and products cannot be changed to balance equations. Thus, CO_2 cannot be changed to CO

— Look for elements that appear in only one place on each side of the equation and balance these elements first.

— If free, uncombined elements appear on either side, balance them last.

Illustration 1 *Al reacts with* HCl *to form* $AlCl_3$ *and* H_2. *Unbalanced equation is*

$$Al + HCl \longrightarrow AlCl_3 + H_2$$

Solution In this form, the **equation does not satisfy** the **Law of conservation of matter** as except Al, other atoms are not balanced (Cl one on the left and three on the right, H one on the left and two on the right)

	Steps	Reaction stage	Atoms of Al Cl H (Reactants)	Al Cl H (Products)	Conclusion
I	Balance Cl by placing 3 before HCl	$Al + HCl \longrightarrow AlCl_3 + H_2$ $Al + 3HCl \longrightarrow AlCl_3 + H_2$	1 1 1 1 3 3	1 3 2 1 3 2	Unbalanced chlorine balanced but reaction unbalanced
II	3 H appears in LHS and 2 H on RHS. Least common multiple is 6. Multiply HCl on LHS by 2 and H_2 on RHS by 3	$Al + 6HCl \longrightarrow AlCl_3 + 3H_2$	1 6 6	1 3 6	H balanced but reaction is unbalanced
III	Balance Cl by multiplying $AlCl_3$ on RHS by 2	$Al + 6HCl \longrightarrow 2AlCl_3 + 3H_2$	1 6 6	2 6 6	All balanced but Al is not balanced
IV	Balance Al on LHS by multiplying Al on LHS by 2	$2Al + 6HCl \longrightarrow 2AlCl_3 + 3H_2$	2 6 6	2 6 6	all balanced
V	Specify physical state	$2Al(s) + 6HCl(aq) \longrightarrow 2AlCl_3(aq) + 3H_2(g)$			

Illustration 2 *We now balance the following reaction with same strategy* :

$$\underbrace{C_3H_8(g)+O_2(g)}_{\text{Reactants}} \longrightarrow \underbrace{CO(g)+H_2O(l)}_{\text{Products}}$$

Solution ***Step I Balance the* C-atoms** Multiply CO_2 on RHS by **3.**

$$C_3H_8(g)+O_2(g) \longrightarrow 3CO_2(g)+H_2O(l)$$

unbalanced in terms of H and O

***Step II Balance the* H-atoms** Multiply H_2O on RHS by **4**

$$C_3H_8(g)+O_2(g) \longrightarrow 3CO_2(g)+4H_2O(l)$$

unbalanced in terms of O

Step III **There are in all 10 oxygen atoms in RHS Balance O on LHS by multiplying by 5**

$$C_3H_8(g)+5O_2(g) \longrightarrow 3CO_2(g)+4H_2O(l)$$

balanced

Numerical Problem Solving 1

Balance following reactions

(a) $KClO_3(s) \longrightarrow KCl(s)+O_2(s)$

(b) $P_4(s)+O_2(g) \longrightarrow P_4O_{10}(s)$

(c) $Zn(s)+HNO_3(aq) \longrightarrow Zn(NO_3)_2(aq)+NO_2(g) + H_2O(l)$

(d) $Al(s)+NaOH(aq) \longrightarrow NaAlO_2(aq)+H_2(g)+H_2O(l)$

Different Types of Chemical Reactions

Reactions in Aqueous Solutions

A solution is a homogeneous mixture of two or more substances. One substance is generally considered the **solvent,** the medium in which another substance-the **solute-** is dissolved.

There are many important chemical reactions occuring in aqueous solutions where water is the solvent.

- **In formula unit equations,** we show complete formulae for all compounds.

$$2AgNO_3(aq)+Cu(s) \longrightarrow 2Ag(s)+Cu(NO_3)_2(aq)$$

- **In total ionic equations,** formulae are written to show the (predominant) form in which each substance exists when it is in contact with aqueous solution. Ions that have common source or that remain in solution after the reaction in complete are shown in brackets in total ionic equations

$$2[Ag^+(aq)]+NO_3^-(aq)+Cu(s) \longrightarrow 2Ag(s) + [Cu^{2+}(aq)+2NO_3^-(aq)]$$

We observe that NO_3^- ions do not participate in the reaction. Because they do not change, they are thus simply the **"spectator"** ions.

- The net ionic equation shows only the species that react. It does not have the spectator ions.

$$2Ag^+(aq)+Cu(s) \longrightarrow 2Ag(s)+Cu^{2+}(aq)$$

Table 6.1 Can be used to represent a chemical equation in aqueous solution many of the reactions are **exchange reactions** in which the ions of the reactants change partners

$$A^+B^- + C^+D^- \longrightarrow A^+D^- + C^+B^-$$

Take the reaction of aqueous silver nitrate with aqueous potassium chloride.

$$AgNO_3(aq)+KCl(aq) \longrightarrow AgCl(s)+KNO_3(aq)$$

From the table 6.1, we observe $AgNO_3$, KNO_3 and KCl both are water soluble and thus can give ions in aqueous solution. AgCl is insoluble and is thus precipitated.

$$Ag^+(aq)+NO_3^-(aq)+K^+(aq)+Cl^-(aq) \longrightarrow AgCl(s)+K^+(aq)+NO_3^-(aq)$$

Net reaction taking place is thus between Ag^+ and Cl^- ion in aqueous solution

$$Ag^+(aq)+Cl^-(aq) \longrightarrow AgCl(s)$$

Acid-Base Reactions

- Acid is a substance that gives H^+ ion in aqueous solution

$$HCl \rightleftharpoons H^+ + Cl^-$$

- Base is a substance that gives OH^- ion in aqueous solution

$$NaOH \rightleftharpoons Na^+ + OH^-$$

Acid-base reactions are among the most important kinds of chemical reactions.

Many acid-base reactions occur in nature in both plants and animals.

The reaction of an acid with a metal hydroxide (base) produces a salt and water. Such reactions are called **neutralisation reactions** due to the fact that acidic and basic properties are lost.

DEEP Focus

Acid is H^+ ion donor base is OH^- ion donor.

In nearly all neutralisation reactions, the driving force is the combination of $H^+(aq)$ from an acid and $OH^-(aq)$ from a base to form H_2O.

$$\underset{\text{strong acid}}{HCl(aq)}+\underset{\text{strong base}}{NaOH(aq)} \longrightarrow \underset{\text{weak}}{H_2O(l)}+\underset{\text{strong electrolyte}}{NaCl(aq)}$$

By ionisation

$$[H^+(aq)+Cl^-(aq)]+[Na^+(aq)+OH^-(aq)] \longrightarrow H_2O(l)+[Na^+(aq)+Cl^-(aq)]$$

Cancelling common ion,

$$H^+(aq)+OH^-(aq) \longrightarrow H_2O(l)$$

This is the net ionic equation for all reactions of strong acids with strong soluble bases that form soluble salts and water.

DEEP Focus

A weak acid is that one that is not ionised completely in aqueous solution. Neutralisation of weak monobasic acid by a strong base results in the formation of soluble salt and H_2O

CH_3COOH (acetic acid) is such type one of the weak acids :

$$\underset{\text{weak acid}}{CH_3COOH(aq)} + \underset{\text{strong base}}{KOH(aq)} \longrightarrow \underset{\text{soluble salt}}{CH_3COOK(aq)} + \underset{\text{weak}}{H_2O(l)}$$

Thus, we write

$$CH_3COOH(aq) + [K^+(aq) + OH^-(aq)] \longrightarrow [CH_3COO^-(aq) + K^+(aq)] + H_2O(l)$$

Net reaction is thus :

$$CH_3COOH(aq) + OH^-(aq) \longrightarrow CH_3COO^-(aq) + H_2O(l)$$

Numerical Problem Solving 2

Write net reaction when a weak acid (as CH_3COOH) is neutralised by a weak base (as NH_4OH)

Table 6.1 **Solubility of Common Ionic Compounds in Water**

Generally soluble	Exceptions
Common inorganic acids (HCl, H_2SO_4, HNO_3, H_3PO_4)	No common exceptions
Low molecular weight organic acid [CH_3COOH, HCOOH, $(COOH)_2$]	No common exceptions
Na^+, K^+, NH_4^+ compounds	No common exceptions
Chlorides (Cl^-)	Insoluble, AgCl, Hg_2Cl_2 soluble in hot water : $PbCl_2$
Bromide (Br^-)	Insoluble : AgBr, Hg_2Br_2, $PbBr_2$ Moderately soluble : $HgBr_2$
Iodides (I^-)	Insoluble : Many heavy metal iodides (AgI, PbI_2, CdI_2)
Sulphates (SO_4^{2-})	Insoluble : $BaSO_4$, $PbSO_4$, $HgSO_4$ Moderately soluble : $CaSO_4$, $SrSO_4$, Ag_2SO_4
Nitrates (NO_3^-), Nitrites (NO_2^-)	Moderately soluble : $AgNO_2$
Chlorates (ClO_3^-), Perchlorates (ClO_4^-), Permanganate (MnO_4^-), Acetates (CH_3COO^-)	Moderately soluble : CH_3COOAg
Sulphides (S^{2-})	Soluble those of NH_4^+, Na^+, K^+, Mg^{2+}, Ca^{2+}
Oxides (O^{2-}), Hydroxides (OH^-)	Soluble : Li_2O, Na_2O, K_2O, BaO (dissolve in H_2O with release of heat *i.e.,* exothermic reactions forming soluble hydroxide LiOH, NaOH, KOH, $Ba(OH)_2$) Moderately soluble : CaO, SrO (dissolve in H_2O with release of heat, *i.e.,* exothermic reactions forming $Ca(OH)_2$, $Sr(OH)_2$)

Generally soluble	Exceptions
Carbonates (CO_3^{2-}) Phosphate (PO_4^{3-}) Oxalate ($C_2O_4^{2-}$) Chromate (CrO_4^{2-}) Arsenate (AsO_4^{3-})	Soluble : those of NH_4^+, Na^+, K^+

Precipitation Reactions

In precipitation reactions an insoluble solid, a **precipitate** forms and then settles out of solution.

DEEP Focus

The driving force for these reactions is the strong attraction between cations and anions. This results in the removal of ions from solution by the formation of precipitate.

Both silver nitrate and potassium chloride are water-soluble ionic compounds. When combined in water, they undergo an exchange reaction to produce insoluble silver chloride and soluble potassium nitrate

$$AgNO_3(aq) + KCl(aq) \longrightarrow AgCl(s) + KNO_3(aq)$$

Reactants		**Products**
$Ag^+(aq) + NO_3^-(aq)$	$\longrightarrow$	Insoluble $AgCl(s)$
$K^+(aq) + Cl^-(aq)$		$K^+(aq) + NO_3^-(aq)$

Predicting the Outcome of a Precipitation Reaction

Whether the species in the product is precipitated or not, can be predicted by solubility guidelines as given in Table 6.1

⇒ Lead nitrate (soluble) is mixed with potassium chromate (soluble) to form potassium nitrate and lead chromate. In Table 6.1, potassium nitrate is soluble and lead chromate is insoluble and thus is precipitated.

$$Pb(NO_3)_2(aq) + K_2CrO_4(aq) \longrightarrow PbCrO_4(s) + 2KNO_3(aq)$$

Reactants		**Products**
$Pb^{2+}(aq) + 2NO_3^-(aq)$	$\longrightarrow$	$PbCrO_4(s)$ – insoluble
$2K^+(aq) + CrO_4^{2-}(aq)$		$2K^+(aq) + 2NO_3^-(aq)$

$K^+(aq)$ and $NO_3^-(aq)$ appear on both sides and thus final outcome is the reaction between $Pb^{2+}(aq)$ and $CrO_4^{2-}(aq)$ ions.

$$Pb^{2+}(aq) + CrO_4^{2-}(aq) \longrightarrow PbCrO_4(s)$$

Numerical Problem Solving 3

Write the net ionic reactions of

(a) $K_2CrO_4(aq) + AgNO_3(aq) \longrightarrow$

(b) $BaCl_2(aq) + K_2SO_4(aq) \longrightarrow$

(c) $CaCl_2(aq) + Na_2CO_3(aq) \longrightarrow$

Do You Know ---

Our teeth and bones were formed by very slow precipitation reactions in which mostly calcium phosphate, $Ca_3(PO_4)_2$ was deposited in the correct geometric arrangements.

Sea shells, which are formed in a slow precipitation reactions, are mostly calcium carbonate, ($CaCO_3$), a white compound. Traces of transition metal ions give them colour.

Stoichiometry

A basic question raised in the chemical laboratory is, "How much product will be formed from specific amounts of reactants?" Or in some cases, question may be "How much starting material (reactant) must be used to obtain a specific amount of product. **Stoichiometry** (Greek words : **stoicheion** – meaning "first principle or element", and **metron** – meaning "measure") describes the quantitative relationships among element in compounds (**composition stoichiometry**) and among substances as they undergo chemical changes (reaction stoichiometry).

- Two moles of potassium chlorate ($KClO_3$) on heating (called thermal decomposition) give two moles of potassium chloride (KCl) and three moles of oxygen (O_2) gas

$$2KClO_3(s) \longrightarrow 2KCl(s) + 3O_2(g)$$

- Two moles silver nitrate ($AgNO_3$) and one mole potassium chromate (K_2CrO_4) form one mole silver chromate (Ag_2CrO_4) and two moles potassium nitrate (KNO_3) is aqueous solution

$$2AgNO_3(aq) + K_2CrO_4(aq) \longrightarrow Ag_2CrO_4(s) + 2KNO_3(aq)$$

Problem Solving Tip

For stoichiometry

— Write correct formulae for all reactants and products, and balance the resulting equation. A balanced chemical equation must always include the same number of each kind of atom on both sides of the equation.

— Convert the quantities of same or all given or known substances (usually reactants) into moles.

— Use the coefficients in the balanced equation to calculate the number of moles of the unknown quantities (usually products)

— Using the calculated number of moles and molar masses, convert the unknown quantities to the desired units.

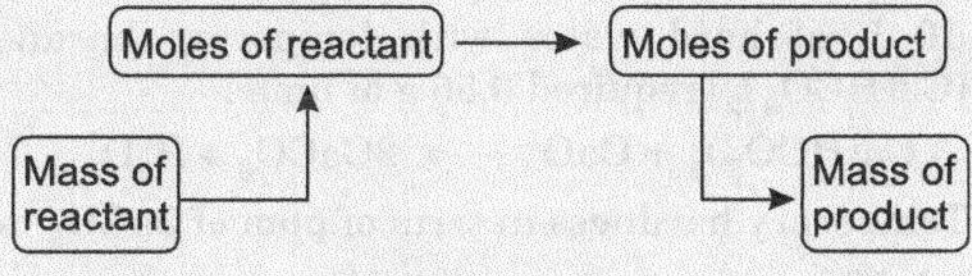

DEEP Focus

Take quantity of each reactant in terms of moles, for stoichiometric calculations as coefficients in a balanced chemical reaction represent moles.

	A_2	+ $3B_2$	$\longrightarrow$ $2AB_2$
Correct	1 mol	3 mol	2 mol
Incorrect	1 g	3 g	2 g

Mass relations in chemical reactions can be illustrated by :

Illustration 1 *Glucose, $C_6H_{12}O_6(s)$ on reaction with $O_2(g)$ gives $CO_2(g)$ and $H_2O(l)$*

$$C_6H_{12}O_6(s) + 6O_2(g) \longrightarrow 6CO_2(g) + 6H_2O(l)$$

Suppose we have 18 g of glucose. Then often quantity of O_2, CO_2, and H_2O can be determined based on stoichiometric factor. **Strategy** is that using a balanced equation, we can perform the stoichiometric calculations.

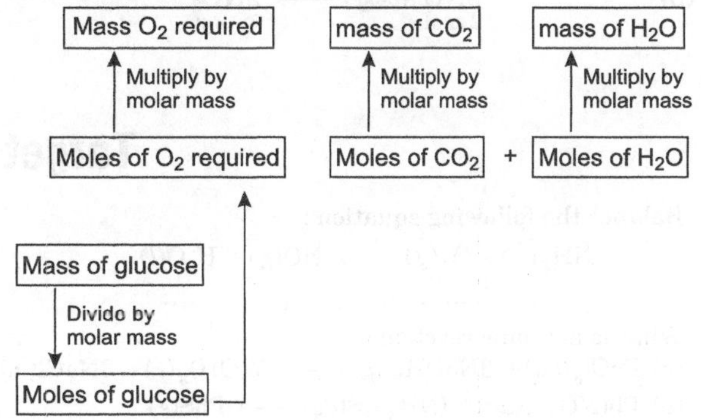

Step I **:** Balanced equation is

$$C_6H_{12}O_6(s) + 6O_2(g) \longrightarrow 6CO_2(g) + 6H_2O(l)$$

Step II **:** Convert amount of glucose into moles :

$$18 \text{ g glucose} = \frac{18 \text{ g}}{180 \text{ g mol}^{-1}} = 0.1 \text{ mole}$$

Step III **: Stoichiometrically**

$$\frac{\text{moles of } C_6H_{12}O_6}{\text{moles of } O_2} = \frac{1}{6}$$

$$\therefore \text{ moles of } O_2 \text{ required} = 6 \times \text{ moles of } C_6H_{12}O_6 = 0.6 \text{ mole} = 0.6 \times 32 = \mathbf{19.2\ g}$$

Step IV **:** Also $\dfrac{\text{moles of } CO_2 \text{ formed}}{\text{moles of } C_6H_{12}O_6 \text{ reacts}} = \dfrac{6}{1}$

$$\therefore \text{ Moles of } CO_2 \text{ formed} = 0.6 \text{ mole} = 0.6 \times 44 = \mathbf{26.4\ g}$$

Step V **:** Also, $\dfrac{\text{moles of } H_2O \text{ formed}}{\text{moles of } C_6H_{12}O_6 \text{ reacted}} = \dfrac{6}{1}$

$$\therefore \text{ Moles of } H_2O \text{ formed} = 0.6 \text{ mole} = \mathbf{10.8\ g}$$

Results can be summarised in the tabular form :

Equation :	$C_6H_{12}O_6(s)$ +	$6O_2(g)$ $\longrightarrow$	$6CO_2(g)$ +	$6H_2O(l)$
Initial amount (in moles)	0.1	0.6	0	0
change	–0.1	–0.6	+0.6	0.6
After reaction	0	0	0.6	0.6
moles			0.6×44 **26.4 g**	0.6×18 **10.8 g**

Illustration 2 *The reaction between nitric oxide (NO) and oxygen to form nitrogen dioxide (NO_2) is a key step in photochemical smog formation:*

$$2NO(g) + O_2(g) \longrightarrow 2NO_2(g)$$

(a) How many moles of NO_2 are formed by the complete reaction of 0.25 mole of O_2?

(b) How many gram of NO_2 are formed by the complete reaction of 1.20 g of NO?

Solution (a)

$$\underset{1\ \text{mol}}{2NO + O_2} \longrightarrow \underset{2\ \text{mol}}{2NO_2}$$

1 mol O_2 gives = 2 mol NO_2

hence, 0.25 mol O_2 gives = **0.50 mol NO_2**

(b)

$$\underset{2\ \text{mol}}{2NO} + O_2 \longrightarrow \underset{2\ \text{mol}}{2NO_2}$$

1 mol NO gives = 1 mol NO_2

hence, $1.20 \text{ g NO} = \dfrac{1.20}{30}$ mol gives

$$= 1 \times \frac{1.20}{30} \text{ mol } NO_2$$

$$= \frac{1.20}{30} \times 46 \text{ g } NO_2$$

$$= \mathbf{1.84\ g\ NO_2}$$

Target Practice 1

1. Balance the following equation :

$$NH_3(g) + O_2(g) \longrightarrow NO(g) + H_2O(l)$$

....

2. What is net ionic reaction of
   (a) $FeCl_3(aq) + 3NaOH(aq) \longrightarrow Fe(OH)_3(s) + 3NaCl(aq)$
   (b) $Pb(NO_3)_2(aq) + (NH_4)_2S(aq) \longrightarrow PbS(s) + 2NH_3NO_3(aq)$

3. What is the difference in the net reactions of the following :

$$HCOOH(aq) + KOH(aq) \longrightarrow HCOOK(aq) + H_2O(l)$$

and $HNO_3(aq) + KOH(aq) \longrightarrow KNO_3(aq) + H_2O(l)$

....

4. Net reaction of

$KOH(aq) + HCl(aq) \longrightarrow KCl(aq) + H_2O(l)$ is

$OH^-(aq) + H^+(aq) \longrightarrow H_2O(l)$

but net reaction of

$NH_4OH(aq) + HCl(aq) \longrightarrow NH_4Cl(aq) + H_2O(l)$

is $NH_4OH(aq) + H^+(aq) \longrightarrow NH_4^+(aq) + H_2O(l)$

Explain ......

5. Which compound or compounds in each of the following groups is (are) expected to be soluble in water?
   (a) $CuO, CuCl_2, FeCO_3$ ....
   (b) $AgI, Ag_3PO_4, AgNO_3$ ....
   (c) $K_2CO_3, KI, KMnO_4$ ....

Answers

1. $4NH_3(g) + 5O_2(g) \longrightarrow 4NO(g) + 6H_2O(l)$
2. (a) $Fe^{3+}(aq) + 3OH^-(aq) \longrightarrow Fe(OH)_3(s)$
   (b) $Pb^{2+}(aq) + S^{2-}(aq) \longrightarrow PbS(s)$
3. $\underset{\text{weak acid}}{HCOOH(aq)} + OH^-(aq) \longrightarrow HCOO^-(aq) + H_2O(l)$

   $H^+(aq) + OH^-(aq) \longrightarrow H_2O(l)$
4. Since NH_4OH is a weak base
5. (a) $CuCl_2$ (b) $AgNO_3$ (c) K_2CO_3, KI, $KMnO_4$

Practice Exercise 1

1. Consider the following reaction taking place completely

$$2Na + 2H_2O \longrightarrow 2NaOH + H_2$$

   (a) How many moles of H_2 will be formed from 0.46 g of Na?
   (b) How many gram of NaOH will be formed by 0.01 mole of Na? (Na = 23, H = 1, O = 16)

2. Consider the following reaction:

$$3Mg + N_2 \longrightarrow Mg_3N_2$$

   (a) Calculate amount of Mg required to prepare 10 g of Mg_3N_2 in 100% reaction.
   (b) How many moles of Mg_3N_2 are formed from 5.6 g of N_2 with excess of Mg in 100% reaction? (Mg = 14, N = 14)

3. One of the reactions that occurs in a blast furnace, where iron ore is converted to cast iron, is

$$Fe_2O_3 + 3CO \longrightarrow 2Fe + 3CO_2$$

Suppose that 1.64×10^3 kg of Fe are obtained from 2.62×10^3 kg sample Fe_2O_3. Assuming that the reaction goes to completion, what is the per cent purity of Fe_2O_3 in the original sample?

4. Select correct alternate. Only **one** alternate is correct.

   (A) 10 L of hard water with temporary hardness of $[Ca(HCO_3)_2]$ required 0.56 g of lime :

$$Ca(HCO_3)_2 + CaO \longrightarrow 2CaCO_3 + H_2O$$

   Temporary hardness in term of ppm of $CaCO_3$ is

   (a) 56 (b) 1/2
   (c) 100 (d) 200

   (B) 5.0 g of $KClO_3$ gave 0.03 mole of O_2. Hence, percentage purity of $KClO_3$ is :

$$2KClO_3 \longrightarrow 2KCl + 3O_2$$

   (a) 49% (b) 50%
   (c) 95% (d) 98%

(C) $CaCO_3$ is decomposed by HCl (density 1.825 g/c.c.)

$$CaCO_3 + 2HCl \longrightarrow CaCl_2 + H_2O + CO_2$$

Volume of HCl required to decompose 10 g of 50% pure $CaCO_3$ is

(a) 1.825 mL (b) 3.65 mL
(c) 0.9125 mL (d) 2 mL

Answers

1. (a) 0.01 mol (b) 0.40 g **2.** (a) 7.2 g (b) 0.2 mol
3. 89.42%
4. (A) (d) (B) (a) (C) (d)

Limiting Reactants

Reactants are often carried out with an excess of one reactant over that required by stoichiometry. This is usually done to ensure that one of the reactants in the reaction is consumed completely, even though some of another reactant remains unused. The extent to which a reaction takes place depends on the amount of the reactants.

— If the reactants are present exactly in the same stoichiometric ratio as required, then no reactant is left unused.

— If one of the reactants is present in smaller quantity than as required stoichiometricallly, then other reactant is left unreacted and amount of product formed is determined by the reactant which is present in smaller quantity and is called **Limiting Reactant.** The other reactant is said to be **Excess Reactant.**

Illustration 1 *In the following general example*

$$A + 3B \longrightarrow 2C$$

1 mole A reacts with 3 moles B to give 2 moles C and 1 mole D. Thus, stoichiometric requirement for the reaction to go completely, is that molar ratio of A and B should be 3 : 1

$$\frac{n_A}{n_B} = 3$$

Solution

Case I : If $n_A = 1$

$n_B = 2$

i.e., $\frac{n_A}{n_B} = \frac{1}{2}$

then B is in smaller quantity, and amount of the products is determined by B. Thus, B is limiting reactant. We present the calculation as

	A +	$3B$ ⟶	$2C$
Reaction says	1 mol	3 mol	2 mol
We take	1 mol	2 mol	?

1 mole A forms = 2 moles C

3 moles B form = 2 moles C

Thus 2 moles B form $= \frac{4}{3} =$ **1.33 moles** $C <$ 2 moles C

Also, 3 moles B react with = 1 mole A

2 moles B react with $= \frac{2}{3}$ mole A = 0.66 mole A

Thus A is in excess and B reacts completely.
B is thus **limiting reactant.**

Problem Solving Tips

- If you cannot decide how to solve a limiting reactant problem, do the entire calculation based on each reactant. **The smallest answer is always right**
- Moles of product formed are always determined by the starting moles of limiting reactant.

Illustration 2 *We perform the following reaction with four moles of* CO *and* 3 *moles* O_2.

$$2CO(g) + O_2(g) \longrightarrow 2CO_2(g)$$

	2CO(g)	$O_2(g)$	$2CO_2(g)$
Stoichiometry	2 mol	1 mol	2 mol
We take	4	3	?

We determine the amount of CO_2 *(product) based on each reactant. The reactant giving the smaller amount of* CO_2 *will be the* ***limiting reactant.***

2 moles CO give = 2 moles CO_2
4 moles CO give = 4 moles CO_2
1 mole O_2 gives = 2 moles CO_2
3 moles O_2 gives = 6 moles CO_2

Thus, smaller quantity of the product (CO_2) is from CO, which is thus **limiting reactant** and O_2 is **excess reactant.**

How much excess reactant (O_2) is left, is decided by stoichiometric ratio of the reactants or from the limiting reactant

2 moles CO consume = 1 mole O_2

Thus, 4 moles CO consume = 2 moles O_2

Thus, $\frac{n_{CO}}{n_{O_2}} = \frac{2}{1} = \frac{4}{2}$

Thus, O_2 left unreacted = 3 − 2 = 1 mole

After reaction is completed, reaction mixture contains 4 moles CO_2 (product) and 1 mole O_2 (unreacted reactant)

We can represent the above change as

$$2CO(g) + O_2(g) \longrightarrow 2CO_2(g)$$

	$2CO(g)$	$O_2(g)$	$2CO_2(g)$
Stoichiometry	2	1	2
Initial moles	4	3	0
Change	–4	–2	+4
After complete reaction	0	1	4

Illustration 3 *We want to calculate the amount of* CO_2 *formed by the reaction of* 16 g CH_4 *with* 48 g O_2.

$$CH_4 + 2O_2 \longrightarrow CO_2 + 2H_2O$$

	CH_4	$2O_2$	CO_2	$2H_2O$
Stoichiometry (moles)	1	2	1	2
We take	16 g	48 g	?	?
	= 1 mole	1.5 mole	?	?

We do the calculation based on mole ratio :

Required ratio $\dfrac{1 \text{ mole } CH_4}{2 \text{ moles } O_2} = \dfrac{0.5 \text{ mole } CH_4}{1 \text{ mole } O_2}$

Each mole of O_2 would require exactly 0.50 mole of CH_4 to be completely used up

Available ratio

$$\frac{1 \text{ mole } CH_4}{1.5 \text{ moles } O_2} = \frac{0.67 \text{ mole } CH_4}{1 \text{ mole } O_2}$$

We have 0.67 mole of CH_4 for each mole of O_2. Thus, there is O_2 in small quantity and CO_2 formed is decided by moles of O_2

2 moles of give = **1 mole CO_2**

1.5 moles O_2 gives = **0.75 mole CO_2**

We do the calculation based on each reactant, and limiting reactant would be that give smaller amount of CO_2.

1 mole CH_4 gives = 1 mole CO_2

2 moles O_2 gives = 1 mole CO_2

1.5 moles O_2 gives = 0.75 mole CO_2 (smaller)

Thus, O_2 is the **limiting reactant.**

How much methane (CH_4) is used :

2 moles O_2 consumes = 1 mole CH_4

1.5 moles O_2 consumes = 0.75 mole CH_4

Thus, CH_4 left unreacted = 1 – 0.75

= **0.25 mole CH_4**

= **4 g CH_4**

Numerical Problem Solving 4

In the following reaction, calculate the amount formed in each case

	$N_2(g)$ +	$3H_2(g)$ ⟶	$2NH_3$
Case I	1	3	?
Case II	2	3	?
Case III	2	4	?
Case IV	1	1	?

Illustration 4 *Urea* $(NH_2)_2CO$, *is prepared by reacting ammonia with carbon dioxide.*

$$2NH_3(g) + CO_2(g) \longrightarrow (NH_2)_2CO(aq) + H_2O(l)$$

In one process, 637.2 g *of* NH_3 *are allowed to react with* 1142 g *of* CO_2 *(a) Which of the two reactants is the limiting reagent? (b) Calculate the mass of* $(NH_2)_2CO$ *formed (c) How much of the excess reagent (in gram) is left at the end of the reaction?*

Solution Number of moles of

NH_3 taken $= \dfrac{637.2 \text{ g}}{17 \text{ g mol}^{-1}} = 37.482$ mol

Number of moles of CO_2 taken $= \dfrac{1142 \text{ g}}{44 \text{ g mol}^{-1}}$

= 25.955 mol

$$2NH_3 + CO_2 \longrightarrow (NH_2)_2CO + H_2O$$

(a) Limiting reagent is that one which is in smaller quantity stoichiometrically.

2 mol NH_3 react with = 1 mol CO_2

37.482 mol NH_3 react with $= \dfrac{37.482}{2}$

= 18.741 mol CO_2

Reversely,

1 mol CO_2 reacts with = 2 mol NH_3

25.955 mol CO_2 reacts with = 51.910 mol NH_3

Thus, NH_3 is present in **limiting quantity** and CO_2 is present in excess.

(b) Quantity of the product is always determined from the limiting reactant.

2 mol NH_3 gives = 1 mol $(NH_2)_2CO$

37.482 mol NH_3 gives $= \dfrac{37.482}{2}$

= **18.741 mol $(NH_2)_2CO$**

(c) CO_2 required by 37.482 mol of NH_3 = 18.741 mol

Thus, CO_2 left = 25.955 – 18.741

= 7.214 mol = 7.21 × 44 g

= **317.416 g**

Illustration 5 *Consider following reaction :*

$$BaCl_2(aq) + H_2SO_4(aq) \longrightarrow BaSO_4(s) + 2HCl(aq)$$

50 mL of 20.8% (w/V) $BaCl_2(aq)$ *and 100 mL of 9.8% (w/V)* $H_2SO_4(aq)$ *solutions are mixed. What is the amount of* $BaSO_4(s)$ *formed? Which reactant is left over after the reaction?* (Ba = 137, Cl = 35.5, S = 32, H = 1, O = 16)

Solution Number of grams of $BaCl_2$ present in 50 mL aqueous solution which is 20.8% $(w/V) = \dfrac{20.8}{100} \times 50$

= 10.4 g

moles of $BaCl_2 = \dfrac{10.4}{208} = 0.05$ mol

Number of grams of H_2SO_4 present in 100 mL aqueous solution which is 9.8% w/V = 9.8 g

$$\text{moles of } H_2SO_4 = \frac{9.8}{98} = 0.1 \text{ mol}$$

$$\underset{\substack{1\text{ mol}\\0.05\text{ mol}}}{BaCl_2(aq)} + \underset{\substack{1\text{ mol}\\0.10\text{ mol}}}{H_2SO_4(aq)} \longrightarrow \underset{\substack{1\text{ mol}\\?}}{BaSO_4(s)} + 2HCl(aq)$$

1 mol $BaCl_2$ gives = 1 mol $BaSO_4$

$\therefore$ 0.05 mol $BaCl_2$ gives = 0.05 mol $BaSO_4$

= 0.05 × 233 g

= **11.65 g $BaSO_4$**

1 mol $BaCl_2$ reacts with = 1 mol H_2SO_4

0.05 mol $BaCl_2$ reacts with = 0.05 mol H_2SO_4

But, H_4SO_4 taken = 0.10 mol H_2SO_4

Thus, H_2SO_4 left unreacted = 0.10 – 0.05 = 0.05 mol

= 0.05 × 98 g = 4.9 g

Volume of $H_2SO_4(aq)$ solution left = **50 mL** which is 9.8% (w/V).

Yields of Chemical Reactions

The maximum mass of product that can be obtained from a chemical reaction is the **theoretical yield.** Generally, loss of product occurs during the isolation and purification steps. In addition, some reactions do not go completely to products, and reactions are sometimes complicated by giving more than one set of products. For all these reasons, the **actual yield** of the product – the mass of material that is actually obtained in the laboratory or a chemical plant – is always less than the theoretical yield. In all such cases, it is desirable to report a **per cent yield,** which specifies how much of the theoretical yield was obtained, is defined as

$$\text{Per cent yield} = \frac{\text{actual yield}}{\text{theoretical yield}} \times 100\%$$

Illustration 1 $KClO_3(s)$ *decomposes into* $KCl(s)$ *and* $O_2(g)$

There is complete conversion of $KClO_3$ *into the products*

$$KClO_3(s) \longrightarrow 2KCl(s) + 3O_2(g)$$

	$KClO_3(s)$	$2KCl(s)$	$3O_2(g)$
Initial	2 mol	0	0
Change	–2	+2	+3
After	0	2	3

Thus, yield is 100%

There can be side reactions

$$2KClO_3(s) \xrightarrow[\text{reaction}]{\text{main}} 2KCl(s) + 3O_2(g)$$

$$4KClO_3(s) \xrightarrow[\text{reaction}]{\text{side}} 3KClO_4(s) + KCl$$

Take 2 moles of $KClO_3(s)$ and assume product is 2 moles $O_2(g)$ due to side reaction, then per cent yield of $O_2(g)$ is

$$= \frac{\text{actual yield}}{\text{theoretical yield}} \times 100$$

$$= \frac{2}{3} \times 100$$

$$= 66.67\%$$

Illustration 2 *Acetaldehyde* (CH_3CHO) *can be obtained by the oxidation or ethyl alcohol* (CH_3CH_2OH).

There is no side reaction and all CH_3CH_2OH is converted to CH_3CHO

$$CH_3CH_2OH \longrightarrow CH_3CHO + H_2$$

Thus, yield is 100%

There can be side reaction giving acetic acid (CH_3COOH)

$$\underset{1\text{ mol}}{CH_3CH_2OH} \longrightarrow \underset{1\text{ mol}}{CH_3CHO} \longrightarrow CH_3COOH$$

due to no side reaction

If there is also formation of 0.2 mol CH_3COOH, then yield of CH_3CHO would be 0.8 mol

$$\text{Per cent yield of } CH_3CHO = \frac{\text{actual yield}}{\text{theoretical yield}} \times 100$$

$$= \frac{0.8}{1} \times 100$$

$$= 80\%$$

Numerical Problem Solving 5

Acetyle chloride (CH_3COCl) can be reduced to acetaldehyde (CH_3CHO) in presence of $BaSO_4$/Pd — reaction is called Rosenmund reaction. Some ethyl alcohol (CH_3CH_2OH) is also obtained due to further reduction of CH_3CHO (into CH_3CH_2OH) if experimental conditions are not efficient. In an experiment one mole of CH_3COCl gave 2.3 g CH_3CH_2OH along with CH_3CHO (main product). What is per cent yield of CH_3CHO?

Illustration 3 *Diethyl ether* ($C_4H_{10}O$), *the "ether" used medically as an anaesthetic, is prepared from ethyl alcohol* (C_2H_5OH) *in presence of an acid*

$$2C_2H_5OH(l) \xrightarrow[\text{acid}]{140\,°C} C_2H_5OC_2H_5(l) + H_2O(l)$$

Calculate the per cent yield if 46.0 g of ethyl alcohol gives 30.0 g of ether.

Solution

$$\underset{\substack{2\text{ mol}\\92\text{ g}}}{2C_2H_5OH(l)} \longrightarrow \underset{\substack{1\text{ mol}\\74\text{ g}}}{C_2H_5OC_2H_5(l)} + \underset{\substack{1\text{ mol}\\18\text{ g}}}{H_2O(l)}$$

Theoretical yield

92 g of C_2H_5OH gives = 74 g ether with 100% yield

$\therefore$ 46g of C_2H_5OH gives $= \frac{74}{92} \times 46 = 37$ g ether with 100% yield

But, actual yield is = 30.0 g

$\therefore$ per cent yield $= \frac{\text{actual yield}}{\text{theoretical yield}} \times 100$

$= \frac{30}{37} \times 100 = \mathbf{81.08\%}$

Illustration 4 *A 15.6 g sample of C_6H_6 (benzene) is mixed with excess HNO_3 in presence of H_2SO_4. We isolate 18.0 g of $C_6H_5NO_2$ (nitrobenzene). What is the per cent yield of $C_6H_5NO_2$ in this reaction?*

Solution

$$C_6H_6 + HNO_3 \xrightarrow{H_2SO_4} C_6H_5NO_2 + H_2O$$

$$\underset{\substack{1\text{ mol}\\78\text{ g}\\15.6\text{ g}}}{C_6H_6} + HNO_3 \xrightarrow{H_2SO_4} \underset{\substack{1\text{ mol}\\123\text{ g}\\?\ (\text{theoretical yield})}}{C_6H_5NO_2} + H_2O$$

Theoretical yield

78 g of C_6H_6 gives = 123 g $C_6H_5NO_2$

$\therefore$ 15.6 g of C_6H_6 gives $= \frac{123}{78} \times 15.6$ $C_6H_5NO_2$

$= 24.6$ g

Actual yield = 18.0 g

$\therefore$ per cent yield $= \frac{\text{actual yield}}{\text{theoretical yield}} \times 100$

$= \frac{18.0}{24.6} \times 100 = \mathbf{73.17\%}$

Illustration 5 *The per cent yield for the following reaction carried out in carbon tetrachloride (CCl_4) solution is 67%.*

$$Br_2 + Cl_2 \longrightarrow 2BrCl$$

(a) *What amount of BrCl would be formed from the reaction of 0.025 mole Br_2 and 0.025 mole Cl_2?*

(b) *What amount of Br_2 is left unchanged?*

Solution

$$\underset{\substack{1\text{ mol}\\0.025}}{Br_2} + \underset{\substack{1\text{ mol}\\0.025}}{Cl_2} \longrightarrow \underset{\substack{2\text{ mol}\\0.050\text{ mol}}}{2BrCl}$$

(a) Theoretical yield of BrCl = 0.050 mol

actual yield of BrCl $= 0.050 \times \frac{67}{100}$

$= \mathbf{0.0335\ mol}$

(b) Br_2 reacted $= 0.025 \times \frac{67}{100} = 0.01675$ mol

Br_2 left unreacted = 0.025 − 0.01675 = **0.00825 mol**

Sequential Reactions

Often more than one reaction is required to change starting materials into the desired product. This is true for many reactions that we carry out in the laboratory and for many industrial process. These are called sequential reactions. The amount of desired product from each reaction is taken as the starting material for the next reaction. We consider following illustrations.

Illustration 1 *I* : $2KClO_3 \longrightarrow 2KCl + 3O_2$

II : $4Al + 3O_2 \longrightarrow 2Al_2O_3$

$KClO_3$ decomposes in step I to give O_2, which in turn, is used by Al to form Al_2O_3 in step II. First we determine O_2 formed in step I and then Al used by this O_2 in step II. Both reactions can be added to determine amount of $KClO_3$ that can give required amount of O_2 needed for Al.

Net: $2KClO_3 + 4Al \longrightarrow 2KCl + 2Al_2O_3$

Thus, $KClO_3 \equiv 2Al$.

Illustration 2 I : $CaO + 3C \longrightarrow CaC_2 + CO$

II : $CaC_2 + 2H_2O \longrightarrow Ca(OH)_2 + C_2H_2$

CaC_2 (calcium carbide) is prepared in step I. It is used to prepare acetylene (C_2H_2) in step II. Suppose we want to determine amount of CaO that can give enough CaC_2 to be converted into required amount of C_2H_2. Amount of CaO is determined in step I and then amount of C_2H_2 in step II. We can relate CaO and C_2H_2 stoichiometrically by writing net reaction which is

$$CaO + 3C + 2H_2O \longrightarrow Ca(OH)_2 + C_2H_2 + CO$$

Thus, $CaO \equiv C_2H_2$

Illustration 3 *Purification of* Ni *in Mond's process is based on the conversion of* Ni *(impure) into pure* $Ni(CO)_4$ *using* CO *(step II) which in turn is obtained in step* I

Step I : $C + H_2O \xrightarrow{\text{red hot}} CO + H_2$

Step II : $Ni + 4CO \longrightarrow Ni(CO)_4$

If we combine both steps then

$$4C + 4H_2O + Ni \longrightarrow Ni(CO)_4 + 4H_2$$

Numerical Problem Solving 6

Take illustration (3) given above. What mass of $Ni(CO)_4$ could be obtained from the CO produced by the reaction of 72 g carbon? Assume 100% reaction and 100% recovering in both cases. (Ni = 58.7)

Illustration 4 *Consider the formation of cyanogen, C_2N_2, and its subsequent decomposition in water given by the equations :*

$$2Cu^{2+} + 6CN^- \longrightarrow 2[Cu(CN)_2]^- + C_2N_2$$

$$C_2N_2 + H_2O \longrightarrow HCN + HOCN$$

How much hydrocyanic acid, HCN, can be produced from 20.0 g of KCN, assuming 100% yield?
(H = 1, C = 12, N = 14, K = 39)

Solution

Step I: $2Cu^{2+} + 6CN^- \longrightarrow 2[Cu(CN)_2]^- + C_2N_2$

Step II: $C_2N_2 + H_2O \longrightarrow HCN + HOCN$

Net: $2Cu^{2+} + 6CN^- + H_2O \longrightarrow 2[Cu(CN)_2]^- + HCN + HOCN$

Thus, 6 mol KCN gives = 1 mol HCN

$\therefore \frac{20}{65}$ mol KCN gives $= \frac{20}{65 \times 6}$ mol HCN

= 0.05128 mol HCN

= 0.05128 × 27 g HCN

= **1.38462 g HCN**

Illustration 5 *What mass of potassium chlorate* ($KClO_3$) *would be required to supply the proper amount of oxygen needed to burn 33.2 g of methane* (CH_4)?

$$2KClO_3 \longrightarrow 2KCl + 3O_2$$
$$CH_4 + 2O_2 \longrightarrow CO_2 + 2H_2O$$

(K = 39, O = 16, C = 12, H = 1, Cl = 35.5)

Solution **Step I:** $2KClO_3 \longrightarrow 2KCl + 3O_2$

Step II: $CH_4 + 2O_2 \longrightarrow CO_2 + 2H_2O$.

$KClO_3$ gives O_2, and O_2 in turn is used to combust CH_4. Thus, O_2 is the intermediate reactant. To cancel out O_2 multiply step II reaction by 1.5, thus,

Net reaction is

$$2KClO_3 + 1.5\, CH_4 \longrightarrow 2KCl + 1.5\, CO_2 + 3H_2O$$

1.5 mol of CH_4 is combusted by O_2 supplied by = 2 mol $KClO_3$

hence, $\frac{33.2}{16}$ mol of CH_4 is combusted by O_2 supplied by

$= \frac{2}{1.5} \times \frac{33.2}{16}$ mol $KClO_3$

= 2.767 mol $KClO_3$

= **338.92 g**

Target Practice 2

1. Which is the limiting reactant in the following?

$$\underset{14\text{ g}}{N_2(g)} + \underset{3\text{g}}{3H_2(g)} \longrightarrow 2NH_3(g)$$

2. What is the amount of the product, NH_3 in the question (1) react?
3. Carbon monoxide (CO) and hydrogen (H_2) react to form methanol (CH_3OH)

$$CO(g) + 2H_2(g) \longrightarrow CH_3OH(l)$$

How much $CH_3OH(l)$ obtained from 0.01 mole $CO(g)$ and 0.08 g $H_2(g)$?

4. 50 g of calcium carbonate sample, $CaCO_3$, (lime stone) gave 1.12 L carbon dioxide gas measured at STP. (1 mole of every ideal gas at STP has 22.4 L volume)
What is per cent purity of $CaCO_3$?
5. Take following sequential reactions :

$$2H_2O(l) \longrightarrow 2H_2(g) + O_2(g)$$
$$N_2(g) + 3H_2(g) \longrightarrow 2NH_3(g)$$
$$CaCO_3(s) \longrightarrow CaO(s) + CO_2(g)$$
$$NH_3(g) + CO_2(g) + H_2O(l) \longrightarrow NH_4HCO_3(aq)$$

We want to produce 0.10 mol NH_4HCO_3.
$H_2O(l)$ required is ....
$N_2(g)$ required is ....

Answers

1. $N_2(g) + 3H_2(g) \longrightarrow 2NH_3$

N_2	$3H_2$	$2NH_3$
1 mol	3 mol	2 mol
28 g	6 g	34 g
14 g	3 g	17 g

Both $N_2(g)$ and $H_2(g)$ are in exact quantity.

2. 17 g (as in question (1))
3. $CO(g) + 2H_2(g) \longrightarrow CH_3OH(l)$

$CO(g)$	$2H_2(g)$	$CH_3OH(l)$
1 mol	2 mol	1 mol
0.01 mol	0.02 mol	0.01 mol
0.02 mol	0.04 mol	0.02 mol

$CO(g) = 0.01$ mol

$H_2(g) = 0.08$ g = 0.04 mol

Thus, $CO(g)$ is limiting reagent

$CH_3OH(l)$ formed = 0.01 mol = **0.32 g**

4. $CaCO_3 \longrightarrow CaO + CO_2$

$CaCO_3$	CO_2
1 mol (100 g)	1 mol = 22.4 L at STP
50 g	11.2 L CO_2

But actual yield = 1.12 L CO_2 from 5.0 g pure $CaCO_3$

$\therefore$ Per cent purity $= \frac{5}{50} \times 100 =$ **10% pure**

5. (a) $2H_2O(l) \longrightarrow 2H_2(g) + O_2(g)] \times 0.75$

(b) $N_2(g) + 3H_2(g) \longrightarrow 2NH_3(g)] \times 0.50$

(c) $CaCO_3(s) \longrightarrow CaO(s) + CO_2(g)]$

$NH_3(g) + CO_2(g) + H_2O(l) \longrightarrow NH_4HCO_3(aq)$

0.75(a) + 0.50(b) gives

$$0.5\, N_2(g) + 1.5\, H_2O(l) \longrightarrow NH_3(g) + 0.75\, O_2(g)$$

Thus, $N_2(g)$ required = **0.05 mol** for 0.1 mol NH_4HCO_3

$H_2O(l)$ required = 0.15 mol in (a) and (b) + 0.10 mol in final reaction = **0.25 mol**

Practice Exercise 2

1. The per cent yield for the reaction

$$PCl_3 + Cl_2 \longrightarrow PCl_5$$

is 83.2%. What mass of PCl_5 would be expected from the reaction of 56.7 g of PCl_3 with excess Cl_2? (P = 31, Cl = 35.5)

2. Calcium carbide is made in an electric-furnace by the reaction :

$$CaO + 3C \longrightarrow CaC_2 + CO$$

The crude product is usually 85% CaC_2 and 15% unreacted CaO. (a) How much CaO should we start with to produce 450 kg of crude product? (b) How much CaC_2 would this crude product contain? (Ca = 40, C = 12, O = 16)

3. If 1.87 g of acetic acid (CH_3COOH) reacts with 2.31 g of isopentyl alcohol ($C_5H_{12}O$) to give 2.96 g isopentyl acetate ($C_7H_{14}O_2$), what is the per cent yield of the reaction?

$$CH_3COOH + C_5H_{12}O \longrightarrow C_7H_{14}O_2 + H_2O$$

4. Hydrogen, obtained by the electrical decomposition of water, was combined with chlorine to produce 62.5 g of hydrogen chloride. Calculate the mass of water decomposed.

$$2H_2O \longrightarrow 2H_2 + O_2$$
$$H_2 + Cl_2 \longrightarrow 2HCl$$

5. Urea, NH_2CONH_2, is formed when CO_2 reacts with NH_3

$$CO_2 + 2NH_3 \longrightarrow NH_2CONH_2 + H_2O$$

NH_3 and CO_2 are obtained in the following steps

I : $N_2 + 3H_2 \xrightarrow{90\%} 2NH_3$

II : $CaCO_3 \xrightarrow{50\%} CaO + CO_2$

Calculate amount of N_2 and $CaCO_3$ to produce 60 kg of urea.

6. The reaction between aluminium and iron (III) oxide can generate temperature approaching 3000°C and is used in welding metals:

$$2Al + Fe_2O_3 \longrightarrow Al_2O_3 + 2Fe$$

In one process 124 g Al are reacted with 601 g of Fe_2O_3.
(a) Calculate the number of moles of Al_2O_3 formed.
(b) How much of the excess reagent is left at the end of the reaction?(Al = 27, Fe = 56, O = 16)

7. H_2O can be obtained from by the reaction

$$2H_2 + O_2 \longrightarrow 2H_2O$$

If we take 1 mol O_2 and 1 g H_2,
(a) Which is the limiting reactant?
(b) How much of the excess reactant is left unreacted?
(c) How much of H_2O is formed?

8. Nitric oxide (NO) reacts instantly with oxygen gas to form nitrogen dioxide (NO_2), a dark-brown gas

$$2NO(g) + O_2(g) \longrightarrow 2NO_2(g)$$

In one experiment 1.0 mol of NO is mixed with 1.0 mol of O_2
(a) Calculate which of the two reactants is the limiting reagent.
(b) Calculate also the number of mol of NO_2 produced.

9. *Select correct alternate. Only one alternate is correct.*

(A) In the following reaction :

$$MnO_2 + 4HCl \longrightarrow MnCl_2 + 2H_2O + Cl_2$$

2 moles MnO_2 reacts with 4 moles of HCl to form 11.2 L Cl_2 at STP. Thus, per cent yield of Cl_2 is
(a) 25% (b) 50%
(c) 100% (d) 75%

(B) 2 moles of 50% pure $Ca(HCO_3)_2$ on heating forms 1 mole of CO_2. Thus, per cent yield of CO_2 is

$$Ca(HCO_3)_2 \xrightarrow{\Delta} CaO + H_2O + 2CO_2$$

(a) 50% (b) 75%
(c) 80% (d) 100%

(C) Mixture of 1 mole of Na_2CO_3 and 2 moles of $NaHCO_3$ forms 1 mole of CO_2 on heating. Thus, per cent yield of CO_2 is
(a) 25% (b) 50%
(c) 75% (d) 100%

(D) Per cent yield of NH_3 in the following reaction is 80%

$$NH_2CONH_2 + 2NaOH \xrightarrow{\Delta} Na_2CO_3 + 2NH_3$$

6 g NH_2CONH_2 reacts with 8 g NaOH to form NH_3
(a) 3.4 g (b) 2.72 g
(c) 4.25 g (d) 11.2 g

(E) 100 mL solution of NaOH (containing 4 g NaOH per litre) and 50 mL of HCl (containing 7.3 g HCl per litre) react as:

$$NaOH(aq) + HCl(aq) \longrightarrow NaCl(aq) + H_2O(l)$$

0.5 g of NaCl is formed. Thus, unreacted NaOH is
(a) 0.058 g (b) 3.66 g
(c) 10.8 g (d) 0.63 g

(F) Al and $KClO_3$ react together to form Al_2O_3 according to

$$2KClO_3 \longrightarrow 2KCl + 3O_2$$
$$4Al + 3O_2 \longrightarrow 2Al_2O_3$$

4 moles of $KClO_3$ (50% pure) on reaction with excess of Al forms Al_2O_3
(a) 2 mol (b) 4 mol
(c) 6 mol (d) 8 mol

(G) NaOH is formed according to reaction

$$2Na + \frac{1}{2}O_2 \longrightarrow Na_2O$$

$$Na_2O + H_2O \longrightarrow 2NaOH$$

To make 4 g NaOH, Na required is
(a) 4.6 g (b) 4.0 g
(c) 2.3 g (d) 0.23 g

(H) NH_3 is formed in the following steps

I : $Ca + 2C \longrightarrow CaC_2$ 50% yield

II : $CaC_2 + N_2 \longrightarrow CaCN_2 + C$ 100% yield

III : $CaCN_2 + 3H_2O \longrightarrow 2NH_3 + CaCO_3$ 50% yield

To obtain 2 moles NH_3, calcium required is

(a) 1 mol (b) 2 mol
(c) 3 mol (d) 4 mol

(I) $MnO_2 + 2KClO_3 \longrightarrow K_2MnO_4 + Cl_2 + 2O_2$

$Cl_2 + K_2MnO_4 \longrightarrow 2KCl + MnO_2 + O_2$

Each reaction takes place to the extent of 50%. 11.2 L of O_2 at STP is obtained from $KClO_3$ using

(a) 0.67 mol (b) 1.12 mol
(c) 1.33 mol (d) 2.66 mol

(J) NaOH can be prepared by two methods each with two steps having 100% extent.

I : $2Na + 2H_2O \longrightarrow 2NaOH + H_2$

$2H_2 + O_2 \longrightarrow 2H_2O$

II : $2Na + \frac{1}{2}O_2 \longrightarrow Na_2O$

$Na_2O + H_2O \longrightarrow 2NaOH$

Which gives better yield?

(a) I (b) II
(c) Both equal (d) None is suitable

Answers

1. 71.53 g **2.** (a) 393.75 kg (b) 382.50 kg
3. 86.74% **4.** 15.411 g
5. 31.111 kg N_2, 200 kg $CaCO_3$
6. (a) 2.296 mol (b) 1.46 mol
7. (a) H_2 (b) O_2 left = 24 g = 0.75 mol (c) 9 g H_2O
8. (a) NO is the limiting reactant (b) 1 mol NO_2
9. (A) (b) (B) (a) (C) (d) (D) (b) (E) (a)
(F) (a) (G) (c) (H) (d) (I) (a) (J) (c)

Total Practice Set

(Read, Plan & Solve!)

Problem 1. Assume that you have 1.39 moles of N_2 and 3.44 moles of H_2. (a) How many grams of ammonia (NH_3) can you make? (b) How many grams of which reactant will be left over? (c) How many grams of the limiting reactant will be required to react with the reactant present in larger amount?

$$N_2 + 3H_2 \longrightarrow 2NH_3$$

$$\underset{1\text{ mol}}{N_2} + \underset{3\text{ mol}}{3H_2} \longrightarrow \underset{2\text{ mol}}{2NH_3}$$

Plan *N_2 and H_2 react in 1:3 mol ratio. Reactant present in smaller quantity (in terms of mol) stoichiometrically is the limiting reactant and the other present in larger quantity is left over. Thus, amount of NH_3, reactant left over and the reactant required for complete reaction can be calculated.*

Solution Number of moles of N_2 = 1.39 mol

Number of moles of H_2 taken = 3.44 mol

Number of moles of H_2 required by 1.39 moles of $N_2 = 1.39 \times 3 = 4.17$ mol

Thus, H_2 is present in **smaller quantity** and is thus the **limiting reactant.**

(a) 3 mol of H_2 gives = 2 mol NH_3

$$3.44 \text{ mol gives} = \frac{2}{3} \times 3.44$$

$$= \mathbf{2.293 \text{ mol } NH_3}$$

(b) 3 mol of H_2 react with = 1 mol N_2

$$3.44 \text{ mol } H_2 \text{ react with} = \frac{1}{3} \times 3.44 = 1.147 \text{ mol } N_2$$

Thus, number of N_2 left over = 1.390 – 1.147

$$= \mathbf{0.243 \text{ mol}}$$

(c) Number of moles of

H_2 required by 1.39 mol N_2 = 4.17 mol

Number of moles of H_2 actually present = 3.44 mol

Additional H_2 required = 4.17 – 3.44

$$= \mathbf{0.73 \text{ mol}}$$

Problem 2. Cisplatin, an anticancer agent used for the treatment of solid tumours, is prepared by the reaction of ammonia with potassium tetra chloroplatinate:

$$\underset{\substack{\text{potassium}\\\text{tetrachloroplatinate}}}{K_2PtCl_6} + 2NH_3 \longrightarrow \underset{\text{cisplatin}}{Pt(NH_3)_2Cl_2} + 2KCl$$

Assume that 10.0 g of K_2PtCl_4 and 10.0 g of NH_3 are allowed to react.

(a) Which reactant is limiting and which is in excess?

(b) How many grams of the excess reactant are consumed, and how many gram remain?

(c) How many grams of cisplatin are formed?
(Pt = 195, K = 39, Cl = 35.5, N = 14, H = 1)

Plan *Given amounts are converted into moles. Limiting reactant is that which is present stoichiometrically in smaller amount. Quantity of the product is always determined from the limiting reactant.*

Solution Number of moles of

$$K_2PtCl_4 = \frac{10 \text{ g}}{415 \text{ g mol}^{-1}} = 0.0241 \text{ mol}$$

$$\text{Number of moles of } NH_3 = \frac{10}{17} = 0.5882 \text{ mol}$$

Moles of NH_3 required by 0.0241 moles of K_2PtCl_4

$$= 0.0482 \text{ mol}$$

but, NH_3 taken = 0.5882 mol >> 0.0482 mol

(a) Thus, **K_2PtCl_4 is the limiting reactant** and **NH_3 is in excess.**

(b) NH_3 consumed = 0.0482 mol = 0.0482×17 g

$$= \mathbf{0.8194 \text{ g}}$$

NH_3 left = 10 – 0.8194 = **9.1806 g**

(c) 1 mole of K_2PtCl_4 gives = 1 mol of $\underset{\text{cisplatin}}{Pt(NH_3)_2Cl_2}$

∴ 0.0241 mole K_2PtCl_4 gives = 0.0241mol $Pt(NH_3)_2Cl_2$

$$= 0.0241 \times 300 \text{ g}$$

$$= \mathbf{7.23 \text{ g}}$$

Problem 3. At high temperatures carbon reacts with water to produce a mixture of carbon monoxide, CO and hydrogen, H_2.

$$C + H_2O \xrightarrow{\text{red heat}} CO + H_2$$

CO is separated from H_2, and then used to separate nickel from cobalt by forming a volatile compound, nickel tetracarbonyl, $Ni(CO)_4$.

$$Ni + 4CO \longrightarrow Ni(CO)_4$$

What mass of $Ni(CO)_4$ could be obtained from the CO produced by the reaction of 75.0 g of carbon? Assume 100% reaction and 100% recovery in both steps. (Ni= 59)

Plan *We determine the number of moles of* CO *formed in the first reaction.*

From the number of moles of CO *produced in the first reaction, we calculate the number of grams of* $Ni(CO)_4$ *that would be formed in the second reaction. Or, if we combine both the reactions*

$$4C + 4H_2O + Ni \longrightarrow Ni(CO)_4 + 4H_2$$

we find that 4 moles carbon gives = 1 mole of $Ni(CO)_4$.

$$g\ C \longrightarrow mol\ C \longrightarrow mol\ of\ CO \longrightarrow mol\ of\ Ni(CO)_4$$

Solution Number of mol of carbon

$$= \frac{75\ g}{12\ g\ mol^{-1}} = 6.25\ mol$$

4 mol of carbon give = 1 mol of $Ni(CO)_4$

hence, 6.25 mol of carbon give $= \frac{6.25}{4}$ mol of $Ni(CO)_4$

$$= \frac{6.25}{4} \times 171\ g$$

$$= \mathbf{267.2\ g}$$

Problem 4. Phosphoric acid, (H_3PO_4) is widely used to make fertilizers, and can be prepared in a two-step process.

Step I : $P_4 + 5O_2 \longrightarrow P_4O_{10}$

Step II : $P_4O_{10} + 6H_2O \longrightarrow 4H_3PO_4$

We allow 272 g of phosphorus to react with excess oxygen, which forms tetraphosphorus decoxide (P_4O_{10}) in 89.5% yield. In the step II reaction, 96.8% yield of H_3PO_4 is obtained. What mass of H_3PO_4 is obtained? (P= 31, H= 1, O= 16)

Plan *We convert amount of* P_4 *given into number of moles. Also* 1 *mole* $P_4 \equiv 1$ *mole* P_4O_{10} *thus, number of moles (with* 89.5% *yield) of* P_4O_{10} *are known.*

Also 1 *mole of* P_4O_{10} *gives* 4 *moles of* H_3PO_4 *in step II. Thus, amount of* H_3PO_4 *is also known (with 96.8% yield).*

Solution ***In Step I*** **:** Number of mol of P_4

$$= \frac{272\ g}{4 \times 31\ g\ mol^{-1}} = 2.1935$$

Theoretical P_4O_{10} formed = 2.1935 mol

$$\text{Actual } P_4O_{10} \text{ formed} = 2.1935 \times \frac{89.5}{100}$$

$$= 1.9632\ mol$$

In Step II **:** Theoretical H_3PO_4 formed

$$= 4 \times \text{moles of } P_4$$

$$= 4 \times 1.9632\ mol\ H_3PO_4$$

$$\text{Actual } H_3PO_4 \text{ formed} = 4 \times 1.9632 \times \frac{96.8}{100}$$

$$= 7.6016\ mol$$

$$= 7.6016\ mol \times 98\ g\ mol^{-1}$$

$$= \mathbf{744.96\ g}$$

Problem 5. Lime stone $(CaCO_3)$ reacts with hydrochloric acid according to the equation:

$$CaCO_3 + 2HCl \longrightarrow CaCl_2 + H_2O + CO_2$$

If 1.00 mole of CO_2 has a volume of 22.4 L under the reaction conditions, how many litres of gas can be formed by the reaction of 2.35 g of $CaCO_3$ with 2.35 g of HCl? Which reactant is limiting?

Plan *Quantity of the product is always determined from the reactant present stoichiometrically in smaller amount (expressed in terms of moles)*

Reactant present in smaller amount is the limiting reactant. (Ca = 40, C = 12, O = 16, Cl = 35.5, H = 1)

Solution $\underset{1\ mol}{CaCO_3} + \underset{2\ mol}{2HCl} \longrightarrow CaCl_2 + H_2O + \underset{\substack{1\ mol \\ 22.4\ L}}{CO_2}$

$$CaCO_3 = \frac{2.35}{100} = 0.0235\ mol$$

$$HCl = \frac{2.35}{36.5} = 0.0644\ mol$$

1 mol $CaCO_3$ requires = 2 mol HCl

$\therefore$ 0.0235 mol $CaCO_3$ requires = 0.0470 mol HCl

HCl taken = 0.0644 mol

Thus, HCl is present in excess and amount of the product is determined from $CaCO_3$, which is the **limiting reactant.**

1 mol $CaCO_3$ gives = 1 mol CO_2 = 22.4 L CO_2

0.0235 mol $CaCO_3$ gives = **0.5264 L** $\mathbf{CO_2}$

Problem 6. What is the per cent free SO_3 in an oleum $(H_2S_2O_7)$ that is labelled "109% H_2SO_4" ?

Plan *Such a label refers to the total mass of pure* H_2SO_4, 109 g, *that would be present after dilution of* 100 g *of the oleum, when all free* SO_3 *would combine with water to form* H_2SO_4.

Solution 9 g of H_2O is mixed with 100 g oleum

$$H_2O + SO_3 \longrightarrow H_2SO_4$$

Since, 18 g H_2O combines with 80 g SO_3

9 g H_2O combines with 40 g SO_3

thus, 100 g of oleum has 40 g SO_3

40% free SO_3 in oleum

Problem 7. A 0.7160 g of a hydrated metallic sulphate of formula $M_x(SO_4)_y \cdot 18H_2O$ precipitated 0.7002 g of $BaSO_4$. When mixed with K_2SO_4 and crystallised, the metallic sulphate yielded an alum isomorphous with potash alum. Find the atomic weight of the metal.

Plan *$M_x(SO_4)_y \cdot 18H_2O$ is isomorphous with alum with formula $K_2SO_4\ Al_2(SO_4)_3 \cdot 24H_2O$ thus, $x = 2$, and $y = 3$. From the stoichiometric amount of $BaSO_4$, and given amount of $BaSO_4$, atomic weight of M is determined.*

Solution Crystallised substance formed with K_2SO_4 is isomorphous with alum,

$$K_2SO_4 \cdot Al_2(SO_4)_3 \cdot 24\,H_2O$$

Thus, M is trivalent, M^{3+} and hydrated metallic sulphate is $M_2(SO_4)_3.18H_2O$

$$M_2(SO_4)_3 \cdot 18H_2O + 3BaCl_2 \longrightarrow 3BaSO_4 + ...$$

1 mol of hydrated salt gives = 3 mol of $BaSO_4$

or $(2M + 612)$ g of hydrated salt gives = 3×233 g $BaSO_4$

0.7160 g of hydrated salt gives

$$= \frac{3 \times 233}{(2M + 612)} \times 0.7160 \text{ g } BaSO_4$$

$$\therefore \quad \frac{3 \times 233 \times 0.7160}{(2M + 612)} = 0.7002 \text{ g (given)}$$

$$(2M + 612) = 714.77$$

$$2M = 102.77$$

$$M = \mathbf{51.38}$$

Problem 8. A 0.4852-g sample of iron-ore is dissolved in acid, and iron is oxidised to the +3 state and then precipitated as the hydrous oxide, $Fe_2O_3 \cdot xH_2O$. The precipitate is filtered, washed and ignited to Fe_2O_3, which is found to weigh 0.2481 g. Calculate the percentage of iron (Fe) in the sample. (Fe = 56, O = 16)

Plan *Only Fe(II) in the pure state after oxidation to Fe(III) is precipitated as $Fe_2O_3 \cdot xH_2O$ which gives Fe_2O_3. From the weight of Fe_2O_3, percentage of iron in iron-ore can be determined.*

Solution Let iron in iron-ore be = w g

$$\underset{\substack{2 \times 56 \text{ g} \\ w \text{ g}}}{2Fe(II) \longrightarrow 2Fe(III)} \longrightarrow Fe_2O_3 \cdot xH_2O \longrightarrow \underset{\substack{160 \text{ g} \\ 0.2481 \text{ g}}}{Fe_2O_3}$$

$$\text{Thus,} \quad \frac{w}{2 \times 56} = \frac{0.2481}{160}$$

$$\therefore \quad w = \frac{0.2481}{160} \times 2 \times 56 \text{ g pure Fe(II)}$$

$$= 0.1737 \text{ g in } 0.4852 \text{ g sample}$$

$$\text{Thus, percentage of Fe(II)} = \frac{0.1737 \times 100}{0.4852}$$

$$= \mathbf{35.80}$$

Master Exercises

Exercise 1
(Stage 1 : Learning)

Short Answer Type Questions

1. Classify following as soluble or insoluble salts in water. FeS, PbS, $CaSO_4$, NaOH, CaO, $CaCO_3$, CaC_2O_4
2. What is driving force in a precipitation reaction?
3. What is driving force in a neutralisation reaction?
4. One mole of HCl(*aq*) is neutralised by one mole of NaOH(*aq*). If we take $Ca(OH)_2(aq)$, then how many moles of $Ca(OH)_2$ are required?
5. Calculate moles of H_2SO_4 required to complete reaction of
   (i) 1 mole $Ca(OH)_2$
   (ii) 1 mole KOH
6. Consider the reaction
   $$2NO + Br_2 \longrightarrow 2NOBr$$
   For every 5.00 moles of bromine that reacts, how many moles of
   (a) NO react and
   (b) NOBr are produced?
7. What mass of ClO_2 is required to produce 8.36 kg of $HClO_3$ by the following reaction?
   $$2ClO_2 + H_2O \longrightarrow HClO_3 + HClO_2$$
8. We allow 32.0 g of methane, CH_4, to react as completely as possible with excess of O_2, to form CO_2 and H_2O.
   (a) Write the balanced equation
   (b) What mass of oxygen will react?
9. What mass of pentane, C_5H_{12}, produce 3.01×10^{22} CO_2 molecules when burnt is excess oxygen O_2.
10. Lime, CaO, dissolved in muriatic acid, HCl, to form calcium chloride, $CaCl_2$, and water, H_2O
    (a) How many moles of HCl would be required to dissolve 8.8 moles of CaO
    (b) How many moles of H_2O would be formed?
11. What mass of potassium can be produced by the reaction of 150.0 g of Na with 150.0 g of KCl?
    $$Na + KCl \xrightarrow{\Delta} NaCl + K$$
12. How much 71.1% pure Na_2SO_4 could be produced from 245 g of 93.9% pure NaCl?
    $$2NaCl + H_2SO_4 \longrightarrow Na_2SO_4 + 2HCl$$

Analytical Questions

Stoichiometry

1. Dichloromethane (CH_2Cl_2), used as a solvent, is prepared by reaction of methane (CH_4) with chlorine (Cl_2). How many grams of CH_2Cl_2 result from reaction of 1.85 kg of methane if the yield is 43.1%?
2. 4 g of a mixture of dry sodium carbonate and sodium bicarbonate containing 75% of the latter was heated until there was no loss in the weight. Calculate volume of CO_2 evolved at STP.
3. Find the volume of hydrogen which would be evolved at STP by reacting zinc with 50 mL of 40% pure dil. H_2SO_4 (density = 1.3 g / mL).

Limiting Reactants

4. Lithium oxide, (Li_2O) is used aboard the space shuttle to remove water from the air supply according to the equation
   $$Li_2O(s) + H_2O(g) \longrightarrow 2LiOH(s)$$
   If there are 80.0 kg of water to be removed and 65 kg of Li_2O available, which reactant is limiting? How many kilograms of the excess reactant remain? (Li = 7, O = 16, H = 1)
5. Consider the reaction
   $$MnO_2 + 4HCl \longrightarrow MnCl_2 + Cl_2 + 2H_2O$$

If 0.86 mole of MnO_2 and 48.2 g of HCl react, which reagent will be used up first? How many gram of Cl_2 will be produced? (Mn = 55, O = 16, Cl = 35.5)

Yields of Chemical Reactions

6. Ethyl alcohol (C_2H_5OH) is prepared industrially by the reaction of ethylene (C_2H_4) with water. What is the per cent yield of the reaction if 9.6 g of ethylene gives 4.7 g of ethyl alcohol?

$$C_2H_4(g) + H_2O(l) \longrightarrow C_2H_5OH(l)$$

7. Industrially, vanadium metal can be obtained by reacting vanadium (V) oxide with calcium at high temperatures

$$5Ca + V_2O_5 \longrightarrow 5CaO + 2V$$

In one process, 1.54×10^3 g of V_2O_5 react with 1.96×10^3 g of Ca, (a) Calculate the theoretical yield of V. (b) Calculate the per cent yield if 803 g of V are obtained. (Ca = 40, V = 51, O = 16)

Sequential Reactions

8. The chief ore of zinc is the sulphide, ZnS. The ore is concentrated by floatation process and then heated in air, which converts the ZnS to ZnO.

I. $2ZnS + 3O_2 \xrightarrow{90.6\%} 2ZnO + 2SO_2$

The ZnO is then treated with dilute H_2SO_4

II. $2ZnO + 2H_2SO_4 \xrightarrow{100\%} 2ZnSO_4 + 2H_2O$

to produce $ZnSO_4(aq)$ which produces Zn metal on electrolysis.

III. $2ZnSO_4 + 2H_2O \xrightarrow{98.2\%} 2Zn + 2H_2SO_4 + O_2$

What mass of Zn will be obtained from an ore containing 225 kg of ZnS? Efficiencies of the processes have been indicated above the arrow mark. (Zn = 65, S = 32, O = 16, H = 1)

9. How many grams of calcium carbonate are required to produce enough carbon dioxide which can convert 0.5 mole of sodium carbonate into sodium bicarbonate?

Exercise 2
(Stage 2 : High Skill Problem Solving)

Only One Option Correct

1. What mass of CO_2 could be formed by the reaction of 16 g CH_4 with 48 g of O_2?

$$CH_4 + 2O_2 \longrightarrow CO_2 + 2H_2O$$

(a) 44 g (b) 33 g
(c) 16 g (d) 24 g

2. For the formation of 5.00 mol of water, which reaction uses the most nitric acid?

(a) $3Cu + 8HNO_3 \longrightarrow 3Cu(NO_3)_2 + 2NO + 4H_2O$
(b) $Al_2O_3 + 6HNO_3 \longrightarrow 2Al(NO_3)_3 + 3H_2O$
(c) $4Zn + 10HNO_3 \longrightarrow 4Zn(NO_3)_2 + NH_4NO_3 + 3H_2O$
(d) $Cu + 4HNO_3 \longrightarrow Cu(NO_3)_2 + 2NO_2 + 2H_2O$

3. One mole of SO_2Cl_2 is dissolved in water and $Ca(OH)_2$ is added to neutralise acidic solution. Moles of $Ca(OH)_2$ required are

(a) 2 (b) 3
(c) 4 (d) 5

4. 0.5 mole of $BaCl_2$ is mixed with 0.2 mole of $(NH_4)_3PO_4$. Maximum number of moles of barium phosphate formed in this reaction is

(a) 0.1 (b) 0.2
(c) 0.3 (d) 0.5

5. Reactant present in excess (with extra number of moles of that reactant) in question (4) above is

(a) $BaCl_2$, 0.3 mol (b) $(NH_4)_3PO_4$, 0.2 mol
(c) $(NH_4)_3PO_4$, 0.1 mol (d) $BaCl_2$, 0.2 mol

6. In the following reaction,

$$2H_2S + SO_2 \longrightarrow 2H_2O + 3S$$

1 mol each of SO_2 and H_2S will give sulphur

(a) 1 mol (b) 3 mol
(c) 1.5 mol (d) 2 mol

7. One mole BaF_2 is treated with two moles of H_2SO_4. To make the resulting mixture neutral, NaOH is added. NaOH required in this process is

(a) 4 mol NaOH (b) 2 mol NaOH
(c) 3 mol NaOH (d) 1 mol NaOH

8. Temporary hardness is due to HCO_3^- of Mg^{2+} and Ca^{2+}. It is removed by addition of CaO.

$$Ca(HCO_3)_2 + CaO \longrightarrow 2CaCO_3 + H_2O$$

Mass of CaO required to precipitate 2 g $CaCO_3$ is

(a) 2.00 g (b) 0.56 g
(c) 0.28 g (d) 1.12 g

9. Hardness of water is measured in terms of ppm (parts per million) of $CaCO_3$. It is the amount (in g) of $CaCO_3$ present in 10^6 g H_2O. In a sample of water, 10 L

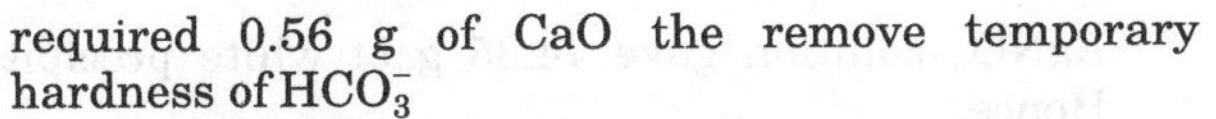

required 0.56 g of CaO the remove temporary hardness of HCO_3^-

$$Ca(HCO_3)_2 + CaO \longrightarrow 2CaCO_3 + H_2O$$

Temporary hardness is

(a) 200 ppm $CaCO_3$ (b) 100 ppm $CaCO_3$
(c) 50 ppm $CaCO_3$ (d) 25 ppm $CaCO_3$

10. Permanent hardness is due to SO_4^{2-} and Cl^- of Ca^{2+} and Mg^{2+} and is removed by the addition of Na_2CO_3

$$CaSO_4 + Na_2CO_3 \longrightarrow CaCO_3 + Na_2SO_4$$
$$CaCl_2 + Na_2CO_3 \longrightarrow CaCO_3 + 2NaCl$$

If hardness is 100 ppm $CaCO_3$, amount of Na_2CO_3 required to soften 10 L of hard water is

(a) 2.12 g (b) 0.106 g
(c) 10.6 g (d) 1.06 g

11. All the oxygen in a 0.5434-g sample of a pure oxide of iron is removed by reduction in a stream of H_2. The loss in weight is 0.1210 g. Hence, formula of the iron oxide is : (Fe – 56)

(a) FeO (b) Fe_2O_3
(c) Fe_3O_4 (d) FeO_2

12. A certain sample of phosphate rock contains 26.26% P_2O_5. A 0.5428 g sample is analysed by precipitating $MgNH_4PO_4 \cdot 6H_2O$ and igniting the precipitate to $Mg_2P_2O_7$. Thus, $Mg_2P_2O_7$ obtained is

(a) 0.8486 g (b) 0.1424 g
(c) 0.3648 g (d) 0.2228 g

13. The calcium in a 0.8432-g sample is precipitated as CaC_2O_4. The precipitate is washed, ignited to $CaCO_3$ and found to weigh 0.3462 g. Percentage of CaO in the sample is

(a) 11.5% (b) 23%
(c) 18% (d) 46%

14. Number of millilitres of a 1.6% $BaCl_2$ (w/V) solution which required to precipitate the sulphur as $BaSO_4$ in a 0.60-g sample that contains 12% S is

(a) 58.50 mL (b) 14.63 mL
(c) 29.25 mL (d) 21.00 mL

15. An iron sample contains 18% Fe_3O_4. What is the amount so that it is precipitated as Fe_2O_3 which weighs 0.40 g?

(a) 2.15 g (b) 1.075 g
(c) 4.30 g (d) 2.01 g

16. An aqueous solution of glucose is 10%. The volume in which 1 g mol of it is dissolved will be

(a) 0.9 L (b) 9 L
(c) 1.8 L (d) 18 L

17. In diammonium phosphate $(NH_4)_2HPO_4$, the percentage of P_2O_5 is

(a) 35.87 (b) 46.44
(c) 51.99 (d) 53.78

18. The oxide of an element contains 67.67% of oxygen. Equivalent weight of the element is

(a) 2.46 (b) 3.82
(c) 4.36 (d) 4.96

19. Which pair has same percentage of carbon?

(a) CH_3COOH and $C_6H_{12}O_6$
(b) CH_3COOH and $C_{12}H_{22}O_{11}$
(c) CH_3COOH and C_2H_5OH
(d) $C_6H_{12}O_6$ and $C_{12}H_{22}O_{11}$

20. The mass of 70% H_2SO_4 required for neutralization of 1 mole of NaOH is

(a) 70 g (b) 35 g
(c) 30 g (d) 95 g

21. 1000 g $CaCO_3$ solution contain 10 g carbonate. The concentration of solution as carbonate is

(a) 10 ppm (b) 100 pm
(c) 1000 ppm (d) 10000 ppm

22. 800 g of a 40% solution by weight was cooled. 100 g of solute precipitated. The percentage composition of remaining solution is

(a) 31.4% (b) 20.0%
(c) 23.0% (d) 24%

23. The hydrogen phosphate of certain metal has $MHPO_4$ formula. The formula of metal chloride is

(a) MCl (b) M_2Cl
(c) MCl_3 (d) MCl_2

24. A bivalent metal has 37.2 equivalent weight. The molecular weight of its chloride is

(a) 216.6 (b) 148.8
(c) 145.4 (d) 172.8

25. A body drinks 500 mL of 9% glucose solution. The number of glucose molecules he has consumed are [mol. wt. of glucose = 180]

(a) 0.5×10^{23} (b) 1.0×10^{23}
(c) 1.5×10^{23} (d) 2.0×10^{23}

One or More Than One Options Correct

1. 1 mole aqueous solution of SO_2Cl_2 is neutralised by

(a) 2 mol of $Ca(OH)_2$ (b) 1 mol of $Ca(OH)_2$
(c) 2 mol KOH (d) 4 mol KOH

2. The molar mass of haemoglobin is about 65000 g mol^{-1}. Every haemoglobin contains 4 iron atoms. Thus

(a) iron content in haemoglobin is 0.35% by mass
(b) 1 mole of haemoglobin contains 56 g iron
(c) 1 mole of haemoglobin contains 224 g iron
(d) if iron content is increased to 0.56%, molar mass of haemoglobin would be higher than 65000 g mol^{-1}.

3. In one process giving below

$$2NH_3(g) + CO_2(g) \longrightarrow NH_2CONH_2(aq) + H_2O(l)$$

637.2 g of NH_3 are allowed to react with 1142 g of CO_2. In this

(a) NH_3 is the limiting reactant and CO_2 is in excess
(b) NH_3 is in excess and CO_2 is the limiting reactant
(c) 1124 g of urea is formed
(d) 319 g of CO_2 is left unreacted

4. If 0.80 mole of MnO_2 and 146 g of HCl react

$$MnO_2 + 4HCl \longrightarrow MnCl_2 + Cl_2 + 2H_2O$$

then,

(a) 0.80 mole of Cl_2 is formed
(b) 0.80 mole of HCl remains unreacted
(c) MnO_2 is completely reacts
(d) MnO_2 is the limiting reactant

5. When 0.273 g of Mg is heated strongly in a nitrogen (N_2) atmosphere, a chemical reaction occurs. The product of the reaction weighs 0.378 g. Empirical formula of the compound

(a) MgN (b) Mg_3N_2
(c) MgN_2 (d) Mg_2N

6. We have 1.6 g CH_4, 1.7 g NH_3 and 1.8 g H_2O select correct alternate(s)

(a) There are equal number of moles of each reactant
(b) Total number of atoms in $CH_4 > NH_3 > H_2O$
(c) Total number of H-atoms are in the ratio of 4 : 3 : 2
(d) Total number of C-atoms in CH_4 < that of N-atoms in NH_3 < that of O-atoms in H_2O

7. Density of H_2O is 1 g mL^{-1}. If we have 1 mL H_2O then

(a) it is 55.55 mol L^{-1}
(b) it has 3.33×10^{22} H_2O molecules
(c) it has 3.33×10^{22} H-atoms
(d) it has 6.66×10^{22} O-atoms

8. In diammonium hydrogen phosphate, $(NH_4)_2HPO_4$, percentage as

(a) P_2O_5 is 53.78% (b) NH_3 is 25.76%
(c) P is maximum (d) N is maximum

9. Mole fraction of ethanol (C_2H_5OH) is ethanol-water system is 0.25. Thus it has

(a) 25% ethanol by weight of solution
(b) 75% water by weight of solution
(c) 46% ethanol by weight of solution
(d) 54% water by weight of solution

10. 12.8 g of mixture of *M*Cl (volatile) and KCl (non-volatile) on reaction with excess of aqueous $AgNO_3$ solution gave 28.7 g white precipitate. 12.8 g of same mixture on a heating gave a gas that on passing into $AgNO_3$ solution gave 14.35 g of white precipitate. Hence,

(a) ionic mass of M^+ is 18 g mol^{-1}
(b) mixture has equal mole fraction of *M*Cl and KCl
(c) KCl and *M*Cl when separately treated with $AgNO_3$ solution would give same number of moles of while precipitates
(d) AgCl is water soluble

Brain Twisters

1. $CaCO_3(s) + 2HCl(aq) \longrightarrow CaCl_2(aq) + H_2CO_3(aq)$

$CaCO_3(s) + 2H^+(aq) + 2Cl^-(aq) \longrightarrow$
$[Ca^{2+}(aq) + 2Cl^-(aq)] + H_2CO_3(aq)$

Thus, net reaction is

$$CaCO_3(s) + 2H^+(aq) \longrightarrow Ca^{2+}(aq) + H_2CO_3(aq)$$

This is an exothermic reaction (in which heat is evolved) and thus there is thermal decomposition of carbonic acid into gaseous carbon dioxide and water. Explain

$$H_2CO_3(aq) \longrightarrow H_2O(l) + CO_2(g)$$

2. For each of the following write balanced

(i) formulae unit (ii) total ionic, and
(iii) net ionic equations

(Refer Table 6.1)

(a) Black-and-white photographic film contains some silver bromide, which can be formed by the reaction of sodium bromide with silver nitrate.
(b) Barium sulphate is used when X-rays of gastrointestinal tract are made. It can be prepared by reacting barium chloride with dilute suphuric acid.
(c) In water purification, small solid particles are often "trapped" as aluminium hydroxide precipitates and falls to the bottom of the sedimentation pool. Aluminium sulphate reacts with calcium hydroxide (from lime) to form aluminium hydroxide and calcium sulphate. (Do not use NaOH, why?)
(d) Our bones are mostly calcium phosphate. Calcium chloride reacts with potassium phosphate to form calcium phosphate and calcium sulphate.

3. Aluminium building materials have a hard transparent protective coating of aluminium oxide, formed by reaction with oxygen in air. The sulphuric acid, H_2SO_4, in **acid rain** dissolves this protective coating and forms aluminium sulphate and water.

(a) How many moles of H_2SO_4 are required to react with 4.0 moles of Al_2O_3?
(b) How many moles of $Al_2(SO_4)_3$ are formed in (a)

4. Calcium carbide is made in an electric furnace by the reaction.

$$CaO + 3C \longrightarrow CaC_2 + CO$$

The crude product is usually 85% CaC_2 and 15% unreacted CaO.

(a) How much CaO should we start with to produce 450 kg of crude product

(b) How much CaC_2 would this crude product contain?

5. A mixture of $\mathbf{CuSO_4}$ and $\mathbf{CuSO_4 \cdot 5H_2O}$ has a mass of 1.245 g. After heating ot drive off all the water, the mass is only 0.832 g. What is the mass per cent of $CuSO_4 \cdot 5H_2O$ in the mixture?

6. Nickel forms a compound with carbon monoxide, $Ni_x(CO)_y$. To determine its formula, you carefully heat a 0.0973 g sample in air to convert the nickel to 0.0426 g of NiO and the CO to 0.100 g CO_2. What is the empirical formulae of $Ni_x(CO)_y$ [Ni = 58.7]

7. Sodium azide, the explosive chemical used in automobile airbags, is made by the following reaction.

$$NaNO_3 + 3NaNH_2 \longrightarrow NaN_3 + 3NaOH + NH_3$$

If you combine 15.0 g $NaNO_3$ with 15.0 g $NaNH_2$, what mass of NaN_3 is produced?

8. Silicon and hydrogen form a series of compounds with general formula Si_xH_y. A 6.22 g sample of the compound is burnt in oxygen. All the Si is converted to 11.64 g of SiO_2, and all of the H is converted to 6.980 g of H_2O. What is the empirical formula of the silicon compound? (Si = 28)

9. Copper metal can be prepared from copper ore is following steps :

I : $Cu_2S(s) + O_2(g) \longrightarrow 2Cu(s) + SO_2(g)$

II : $CuS(s) + O_2(g) \longrightarrow Cu(s) + SO_2(g)$

Suppose an ore sample contains 11.0% impurity in addition to a mixture of CuS and Cu_2S. Heating 100.0 g of the mixture produces 75.4 g of copper metal with a purity of 89.5% what is the weight per cent of CuS and Cu_2S in the ore?

10. In some countries, a person will receive a "driving while intoxicated" (DWI) ticket of the blood alcohol level (BAL) is 100 mg per decilitre (dL) of blood or higher. Suppose a person is found to have a BAL of 0.033 mole of ethanol (C_2H_5OH) per litre of blood. Will a person receive a DWI ticket?

Passage Comprehension Questions

Passage 1

When sulphuric acid dissolves in water, the following reactions take place

$$H_2SO_4 \longrightarrow H^+ + HSO_4^- \quad (100\% \text{ ionisation})$$

$$H_2SO_4^- \longrightarrow H^+ + SO_4^{2-} \quad (10\% \text{ ionisation})$$

We start with 0.2 M aqueous solution of H_2SO_4. Answer the following questions.

1. $[SO_4^{2-}]$ in the above case is

(a) 0.1 M (b) 0.01 M
(c) 0.2 M (d) 0.02 M

2. $[HSO_4^-]$ in the above case is

(a) 0.1 M (b) 0.09 M
(c) 0.18 M (d) 0.08 M

3. $[H_3O^+]$ in aqueous solution is

(a) 0.28 M (b) 0.22 M
(c) 0.02 M (d) 0.10 M

4. Maximum concentration is species

(a) H_2SO_4 (b) H_3O^+
(c) SO_4^{2-} (d) HSO_4^-

Passage 2

About half of the world's production of pigments for paints involves the formation of white TiO_2. In the united states, it is made on a large scale by the chloride process, starting with ores containing only small amounts of rutile, TiO_2. The ore is treated with chlorine and carbon (coke). This produces $TiCl_4$ and gaseous products.

$$2TiO_2 + 3C + 4Cl_2 \xrightarrow{70.0\%} 2TiCl_4 + CO_2 + 2CO$$

The $TiCl_4$ is then converted into TiO_2 of high purity.

$$TiCl_4 + O_2 \xrightarrow{93.0\%} TiO_2 + 2Cl_2$$

Efficiencies have been indicated above the arrow mark. (Ti = 48, Cl = 35.5, O = 16, C = 12)
Answer the following questions

1. If we start with 100 kg of ore (containing 7.5% rutile), pure TiO_2 obtained is

(a) 4.8825 kg (b) 7.3238 kg
(c) 9.7650 kg (d) 2.4413 kg

2. Cl_2 required in this reaction is

(a) 0.1221 kmol (b) 8.666 kg
(c) 4.333 kg (d) 0.1221 kg

Passage 3

A fluorine disposal plant was constructed to carryout the reactions

$$F_2 + 2NaOH \longrightarrow \frac{1}{2}O_2 + 2NaF + H_2O$$

$$2NaF + CaO + H_2O \longrightarrow CaF_2 + 2NaOH$$

As the plant operated, excess lime was added to bring about complete precipitation of the fluoride as CaF_2. Over a period of operation, 2000 kg of fluorine were fed into a plant and 10,000 kg of lime were required. (F = 19, Ca = 40, O = 16, H = 1, Na = 23)

Answer the following questions

1. What was the percentage utilisation of lime (CaO)?

(a) 29.5% (b) 14.75%
(c) 59.0% (d) 60.0%

2. CaO used by given quantity of F_2 is

(a) 1475 kg (b) 2947.4 kg
(c) 5900 kg (d) 6000 kg

3. CaF_2 precipitated is

(a) 37.79 kmol (b) 263.16 kmol
(c) 35.71 kmol (d) 26.32 kmol

Passage 4

11.2 g of mixture of *M*Cl (volatile) and NaCl gave 28.7 g of white precipitate with excess of $AgNO_3$ solution. 11.2 g of same mixture on heating gave a gas that on passing into $AgNO_3$ solution gave 14.35 g of white precipitate

Answer the following questions

1. Ionic mass of M^+ is

(a) 18 (b) 9
(c) 36 (d) 23

2. *M*Cl and NaCl are the molar ratio

(a) 1:1 (b) 1:2
(c) 2:1 (d) 1:3

3. In terms of moles, mixture has

(a) 0.2 mol (b) 0.1 mol
(c) 0.15 mol (d) 0.01 mol

Passage 5

Cisplatin, an anticancer agent used for the treatment of solid tumors, is prepared by the reaction of ammonia with potassium tetra chloroplatinate

$$\underset{\text{potassium tetrachloroplatinate}}{K_2PtCl_4} + 2NH_3 \longrightarrow \underset{\textit{cis}\text{-platin}}{[Pt(NH_3)_2]Cl_2} + 2KCl$$

Assume that 10.0 g of K_2PtCl_4 and 10.0 of NH_3 are allowed to react. (K= 39, Pt= 195, Cl = 35.5)

Answer the following questions

1. Which reactant is limiting and which is in excess?

	Limiting	Excess
(a)	K_2PtCl_4	NH_3
(b)	NH_3	K_2PtCl_4

(c) Exact molar ratio
(d) Would be decided by the quantity of the product formed

2. Number of moles of K_2PtCl_6 consumed is

(a) 0.048 (b) 0.024
(c) 0.012 (d) 0.096

3. Number of moles of NH_3 consumed is

(a) 0.048 (b) 0.024
(c) 0.096 (d) 0.192

4. Number of moles of *cis*-platin formed is

(a) 0.012 (b) 0.024
(c) 0.048 (d) 0.192

5. Number of moles of excess reactant which remains unreacted is

(a) 0.024 (b) 0.34
(c) 0.54 (d) 0.56

Passage 6

Phosphoric acid, H_3PO_4, is widely used to make fertilisers, and can be prepared in a two-step process.

Step I : $P_4 + 5O_2 \longrightarrow P_4O_{10}$

Step II : $P_4O_{10} + 6H_2O \longrightarrow 4H_3PO_4$

We allow 310 g of phosphorus to react with excess oxygen, which forms tetraphosphorus decoxide, P_4O_{10}, in 50% yield. In the step II reaction, 25% yield of H_3PO_4 is obtained. (P = 31, H= 1, O= 16)

Answer the following questions

1. Assume 100% yield then number of moles of H_3PO_4 would be obtained

(a) 1.25 (b) 10.0
(c) 5.0 (d) 2.5

2. Actual number of moles of H_3PO_4 obtained based on experimental conditions is

(a) 5.0 (b) 2.5
(c) 1.25 (d) 0.3125

3. H_3PO_4 thus formed is converted into calcium phosphate by treating with 1 mole of calcium hydroxide. Calcium phosphate obtained is

(a) 2.5 (b) 0.33
(c) 0.1563 (d) 1.25

Passage 7

Chile salt peter is a natural source of $NaNO_3$ which also contains $NaIO_3$. The $NaIO_3$ can be used as a source of iodine produced in the following reactions

$$IO_3^- + 3HSO_3^- \longrightarrow I^- + 3H^+ + 3SO_4^{2-} \quad ...(i)$$

$$5I^- + IO_3^- + 6H^+ \longrightarrow 3I_2(s) + 3H_2O \quad ...(ii)$$

1.00 L of the starting solution which contains 5.80 g $NaIO_3$/L, is treated with stoichiometric quantity of $NaHSO_3$. Then a further quantity of the starting solution is added to the reaction mixture to bring about the second reaction. [I = 127, Na = 23, S = 32]

Answer the following questions

1. How many grams of $NaHSO_3$ are required in step (i)?
   (a) 5.80 g (b) 17.4 g
   (c) 9.14 g (d) 3.0 g
2. What additional volume of the starting solution must be added in step (ii)?
   (a) 0.2 L (b) 0.4 L
   (c) 0.8 L (d) 1.0 L
3. How many equivalents of IO_3^- are present in one mole of $NaIO_3$ in step (i)?
   (a) 2 (b) 4
   (c) 6 (d) 8
4. How many equivalents of IO_3^- are used in step (ii)?
   (a) 6 (b) 5
   (c) 0.06 (d) 0.035

Passage 8

The depletion of ozone (O_3) in the stratosphere has been a matter of great concern among scientists in recent years. It believed that ozone can react with nitric oxide (NO) that is discharged from the high altitude jet plane. The reaction is

$$O_3 + NO \longrightarrow O_2 + NO_2$$

Answer the following questions

1. If 0.740 g of O_3 reacts with 0.670 g of NO, how many gram of NO_2 will be produced?
   (a) 0.71 g (b) 0.74 g
   (c) 0.68 g (d) 0.81 g
2. Which compound is the limiting reactant?
   (a) NO
   (b) O_3
   (c) Both are in equimolar ratio
   (d) Both are in stoichiometric ratio
3. Number of moles of the excess reactant remaining at the end of the reaction is
   (a) 0.007 mol O_3 (b) 0.014 mol O_3
   (c) 0.007 mol NO (d) 0.014 mol NO

Passage 9

Let us explore a reaction with a limiting reactant. Here zinc metal is added to a flask containing aqueous HCl and H_2 gas is a product

$$Zn(s) + 2HCl(aq) \longrightarrow ZnCl_2(aq) + H_2(g)$$

Each of three flasks contain 0.10 mol HCl. Zinc is added to each flask in the following quantities :

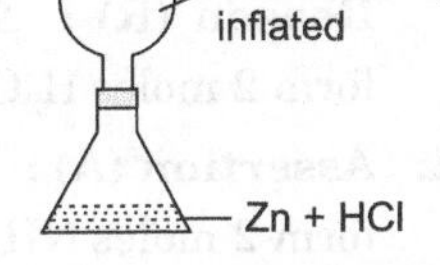

Flask 1 : 7.00 g Zn
Flask 2 : 3.27 g Zn
Flask 3 : 1.31 g Zn

When the reactants are combined, the H_2 inflates the balloon attached to the flask.

Based on the above experiment, answer the following questions

1. Balloon is inflated completely in
   (a) flask 1 (b) flask 2
   (c) Both (a) and (b) (d) None of these
2. Zinc reacts completely in
   (a) flask 1 (b) flask 2
   (c) flask 3 (d) flask 2 and 3
3. Consider following statements
   I : In flask 1, balloon inflates completely, but some Zn remains when inflation ceases
   II : In flask 2, balloon inflates completely, but no Zn remains
   III : In flask 3, balloon does not inflate completely and no zinc remains

   Select correct statements.
   (a) I, II, III (b) I, II
   (c) I, III (d) II, III

Assertion & Reason

Codes :

(a) Both **A** and **R** are true and **R** is the correct explanation of **A**.
(b) Both **A** and **R** are true but **R** is not the correct explanation of **A**.
(c) **A** is true but **R** is false.
(d) **A** is false but **R** is true.

1. **Assertion (A) :** In the following reaction
   $$SO_2 + 2H_2S \longrightarrow 3S + 2H_2O$$
   1 mole SO_2 and 1 mole H_2S form 3 moles sulphur.
   Reason (R) : 1 mole SO_2 and 2 moles H_2S form 3 moles sulphur and 2 moles H_2O.
2. **Assertion (A) :** 20 g of 50% pure $CaCO_3$ on heating gave 2.24 L CO_2 at STP and 15.6 g non-volatile residue.
   $$CaCO_3 \xrightarrow{\Delta} CaO + CO_2$$

Reason (R) : Pure $CaCO_3$ is 10 g which gave 4.4 g CO_2 (= 0.1 mol = 2.24 L at STP) and 15.6 g residue is left.

3. Assertion (A) : $2H_2 + O_2 \longrightarrow 2H_2O$

2 g H_2 react with 1 g O_2 to form 2 g H_2O.

Reason (R) : 2 moles H_2 react with 1 mole O_2 to form 2 moles H_2O.

4. Assertion (A) : 2 moles N_2 react with 3 moles H_2 to form 2 moles NH_3.

$$N_2 + 3H_2 \longrightarrow 2NH_3$$

Reason (R) : N_2 is excess and H_2 is the limiting reactant.

5. Assertion (A) : 4 moles of $KClO_3$ (50% pure) gave 3 moles of O_2 on heating strongly and thus yield is 100%

$$2KClO_3 \longrightarrow 2KCl + 3O_2$$

Reason (R) : 3 moles of O_2 are obtained from 2 moles of $KClO_3$.

True & False

1. 1 mole of Ag_2CO_3 on strongly heating gave 1 mole of residue.
2. 1 mole of Na_2CO_3 and 2 moles of $NaHCO_3$ on heating gave 2 moles residue, 1 mole CO_2 and 1 mole H_2O.
3. 1 mole SO_2 and 1 mole H_2S form 3 moles S.
   $$SO_2 + 2H_2S \longrightarrow 3S + 2H_2O$$
4. In the reaction between 2 g H_2 and 3 g O_2, H_2 is the limiting reactant and O_2 is in excess.
   $$2H_2 + O_2 \longrightarrow 2H_2O$$
5. Percentage yield in the following reaction is 50%.
   $$NH_4NO_3 \xrightarrow{\Delta} N_2O + 2H_2O$$
   1 mole of 50% pure NH_4NO_3 gave 11.2 L N_2O at STP.
6. 100 mL of 20.8% $BaCl_2$ (molar mass = 208) and 50 mL of 17.4% K_2SO_4 (molar mass = 174) form 0.5 moles of $BaSO_4$ (molar mass = 233)
7. 1 g $AgNO_3$ reacts with 1 g HCl to form 1 g AgCl.
8. 2 moles of PCl_5 are 50% dissociated at 400 K. After reaction, there are total of 3 moles.
9. In the following reactions,
   $$2KClO_3 \longrightarrow 2KCl + 3O_2$$
   $$2Al + \frac{3}{2}O_2 \longrightarrow Al_2O_3$$
   1 mole of $KClO_3$ is required to form 1 mole Al_2O_3.
10. Gravimetry is studied under stoichiometry.

Fill in the Blanks

1. 16.8 g of mixture of $NaHCO_3$ and Na_2CO_3 (50% each) on strongly heating gives .... L CO_2 at STP.
2. In the following reaction,
   $$MnO_2 + 4HCl \longrightarrow MnCl_2 + Cl_2$$
   1 mole each of MnO_2 and HCl forms .... g Cl_2.
3. $$C(s) + H_2O \longrightarrow CO + H_2$$
   $$Ni + 4CO \longrightarrow Ni(CO)_4$$
   To prepare 1 mole of $Ni(CO)_4$, C(s) required is .... mole.
4. In the reaction between 1 mole each of H_2S and $AsCl_3$.
   $$2AsCl_3 + 3H_2S \longrightarrow As_2S_3 + 6HCl$$
   limiting reagent is ....
5. In the following reaction,
   $$\underset{1\,g}{AgNO_3} + \underset{1\,g}{HCl} \longrightarrow AgCl + HNO_3$$
   unreacted reagent is ....

Matrix-Match Type Questions

I. Only One Correct

1. Match the number of moles (in List I) with the compound (in List II).

List I	List II
A. 0.5 mol	1. 92 g CH_3CH_2OH
B. 1.0 mol	2. 66 g CO_2
C. 1.5 mol	3. 54 g H_2O
D. 2.0 mol	4. 15 g HCHO
E. 3.0 mol	5. 44 g CH_3CHO

2. Match the reaction (in List I) with the maximum yield of the product (in List II).

List I	List II
A. $\underset{1\,g}{2H_2} + \underset{1\,g}{O_2} \longrightarrow 2H_2O$	1. 1.028 g
B. $\underset{1\,g}{3H_2} + \underset{1\,g}{N_2} \longrightarrow 2NH_3$	2. 1.333 g
C. $\underset{1\,g}{H_2} + \underset{1\,g}{Cl_2} \longrightarrow 2HCl$	3. 1.125 g
D. $\underset{1\,g}{2H_2} + \underset{1\,g}{C} \longrightarrow CH_4$	4. 1.214 g

II. One or More Correct

1. Match Column I (reactant) with the amount in Column II.

Column I	Column II
A. $\underset{1\ \text{mol}}{CH_3COOH} + NaOH \longrightarrow CH_3COONa + H_2O$ NaOH used to react completely with CH_3COOH	1. 1 mole
B. $\underset{84\ g}{MgCO_3} \xrightarrow{\Delta} MgO + CO_2$ Residue formed	2. 1 equivalent
C. $\underset{2\ \text{mol}}{2SO_2} + \underset{64\ g}{O_2} \longrightarrow 2SO_3$ Reactant in excess under standard conditions	3. 40 g
D. $2H_2 + O_2 \longrightarrow 2H_2O$ O_2 used to form 36 g H_2O	4. 22.4 L at STP
E. $N_2H_4 + O_2 \longrightarrow N_2 + 2H_2O$ N_2 formed from 32 g each reactant	5. 4 equivalents

Integer Answer Types

This section contains 8 questions. The answer to each of the questions is a single digit integer, ranging from 0 to 9. The appropriate bubbles below the respectively question numbers in the ORS have to be darkened. For example, if the correct answers to question number *X, Y, Z* and *W* (say) are 6, 0, 9 and 2, respectively, then the correct darkening of bubbles will like the following

X	*Y*	*Z*	*W*
0	0	0	0
1	1	1	1
2	2	2	2
3	3	3	3
4	4	4	4
5	5	5	5
6	6	6	6
7	7	7	7
8	8	8	8
9	9	9	9

Set I

1. What mass (in grams) of CH_4 produces 3.01×10^{23} H_2O molecules when burnt to excess O_2 ?
2. How many moles of H_2O are formed by 10 moles of HNO_3 and 10 moles Cu in the following reaction?
$$3Cu + 8HNO_3 \longrightarrow 3Cu(NO_3)_2 + 2NO + 4H_2O$$
3. 20.8 g of MCl_x ($M = 137$) or reaction with excess of K_2SO_4 yields 23.3 g of sulphate of metal *M*. Calculate value of *x*.
4. Free SO_3 in one mole of pure oleum is ...... mole.
5. The molar mass of haemoglobin is about 65000 g mol^{-1}. Haemoglobin contains 0.35% Fe by mass. Thus, iron atoms present in haemoglobin molecules are ......
6. How many moles of oxygen are required to react completely with 24 g of CH_4 ?
7. How many moles of $KClO_3$ would be required to supply the proper amount of oxygen needed to burn 24 g of CH_4 completely?
$$2KClO_3 \longrightarrow 2KCl + 3O_2$$
$$CH_4 + 2O_2 \longrightarrow CO_2 + 2H_2O$$
8. In the following sequence of reactions
$$CaOCl_2 + H_2O \longrightarrow Ca(OH)_2 + Cl_2$$
$$2KI + Cl_2 \longrightarrow 2KCl + I_2$$
$$I_2 + 2Na_2S_2O_3 \longrightarrow Na_2S_4O_6 + 2NaI$$
Calculate moles of $Na_2S_2O_3$ required for complete reduction of I_2 obtained from 127 g $CaOCl_2$ sample (50% purity).

Set II

1. How many species form 0.05 mole from 3 g of $CH_3COOH, CH_3CH_2CH_2OH, NH_2CONH_2, HNO_3$?
2. How many substances underlined have equivalent weight equal to molecular weight based on the reactions?
$$\underline{H_3PO_4} + Ca(OH)_2 \longrightarrow CaHPO_4 + 2H_2O$$
$$\underline{H_3PO_2} + NaOH \longrightarrow NaH_2PO_2 + H_2O$$
$$\underline{NaHC_2O_4} + NaOH \longrightarrow Na_2C_2O_4 + H_2O$$
3. 1 mole of C_2H_x is oxidised to CO_2 and H_2O by 3 moles O_2. Thus, 140 g C_2H_x has .... moles.
4. Out of the following, how many substances have underlined atoms of different nature?
$Na_2\underline{S_4}O_6, Na_2\underline{S_2}O_3, CaO\underline{Cl_2}, \underline{N_2}H_4O_3, \underline{Fe_3}O_4, K\underline{I_3}, Cr\underline{O_5}, K_2\underline{Cr_2}O_7$
5. 6 g urea sample on heating with NaOH give 0.224 L NH_3 at STP.
$$NH_2CONH_2 + 2NaOH \longrightarrow 2NH_3 + Na_2CO_3$$
Thus, percentage yield of NH_3 is ....
6. Sample of $CaCO_3$ of 50% purity forms 0.56 g residue. Thus, sample weighs .... g.
7. Pure nitrogen is obtained by decomposition of nitride $M_x(N_3)_y$ where $x + y = ?$
8. In the following reaction
$$2K_2MnO_4 + Cl_2 \longrightarrow 2KCl + 2KMnO_4$$
2 moles Cl_2 and 2 moles K_2MnO_4 form .... moles $KMnO_4$.

Answers & Solutions

Numerical Problems Solving

1. (a) $2KClO_3(s) \longrightarrow 2KCl(s) + 3O_2(g)$

(b) $P_4(s) + 5O_2(g) \longrightarrow P_4O_{10}(s)$

(c) $Zn(s) + 4HNO_3(aq) \longrightarrow Zn(NO_3)_2(aq) + 2NO_2(g) + 2H_2O(l)$

(d) $2Al(s) + 2NaOH(aq) + 2H_2O(l) \longrightarrow 2NaAlO_2(aq) + 3H_2(g)$

2. $CH_3COOH(aq) + NH_4OH(aq) \longrightarrow CH_3COO^-(aq) + NH_3^+(aq) + H_2O(l)$

3. (a) $CrO_4^{2-}(aq) + 2Ag^+(aq) \longrightarrow Ag_2CrO_4(s)$

(b) $Ba^{2+}(aq) + SO_4^{2-}(aq) \longrightarrow BaSO_4(s)$

(c) $Ca^{2+}(aq) + CO_3^{2-}(aq) \longrightarrow CaCO_3(s)$

4. I : 2 mol NH_3

II : 2 mol, N_2 is in excess

III : 2.66 mol, N_2 is in excess

IV : 0.66 mol, H_2 is limiting reagent.

5. $\underset{1\text{ mol}}{CH_3COCl} + H_2 \longrightarrow \underset{\substack{1\text{ mol}\\44\text{ g}}}{CH_3CHO} \xrightarrow{+H_2} \underset{\substack{1\text{ mol}\\46\text{ g}}}{CH_3CH_2OH}$

If no side reaction, $CH_3CHO = 44$ g

but same CH_3CH_2OH is also formed.

$CH_3CH_2OH = 2.3$ g for 4.4 g CH_3CHO

Thus, actual yield of $CH_3CHO = 44.0 - 4.4 = 39.6$ g

$$\% \text{ yield} = \frac{39.6}{44.0} \times 100 = \mathbf{90.0\%}$$

6. $C(s) + H_2O \longrightarrow CO(g) + H_2(g)$

$Ni(s) + 4CO(g) \longrightarrow Ni(CO)_4$

After balancing

$$\underset{4 \times 12\text{ g}}{4C(s)} \longrightarrow \underset{170.7\text{ g}}{Ni(CO)_4}$$

$$72\text{ g} \longrightarrow \frac{170.7}{48} \times 72\text{ g}$$

$$= \mathbf{256.05\text{ g}}$$

Master Exercises

Exercise 1

Short Answer Type Questions

1. Soluble : NaOH, CaO

Partial soluble : $CaSO_4$

Insoluble : FeS, PbS, $CaCO_3$, CaC_2O_4

2. Interaction between cation and anion that leads to formation of insoluble salts

3. $H^+(aq) + OH^-(aq) \longrightarrow H_2O(aq)$

4. $\underset{1\text{ mol}}{HCl(aq)} + \underset{1\text{ mol}}{NaOH(aq)} \longrightarrow NaCl(aq) + H_2O$

$\underset{\substack{2\text{ mol}\\1\text{ mol}}}{2HCl(aq)} + \underset{\substack{1\text{ mol}\\0.5\text{ mol}}}{Ca(OH)_2(aq)} \longrightarrow CaCl_2(aq) + 2H_2O$

5. $\underset{1\text{ mol}}{Ca(OH)_2} + \underset{1\text{ mol}}{H_2SO_4} \longrightarrow CaSO_4 + 2H_2O$

$\underset{\substack{2\text{ mol}\\1\text{ mol}}}{2KOH} + \underset{\substack{1\text{ mol}\\0.5\text{ mol}}}{H_2SO_4} \longrightarrow K_2SO_4 + 2H_2O$

6. NO = 10 moles

NOBr = 10 moles

7. $\underset{\substack{2\text{ mol}\\135\text{ g}}}{2ClO_2} \longrightarrow \underset{\substack{1\text{ mol}\\84.5\text{ g}}}{HClO_3} + HClO_2$

$$8.36\text{ kg } HClO_3 \text{ is from} = \frac{135}{84.5} \times 8.36\ CO_2$$

$$= \mathbf{13.356\text{ kg } ClO_2}$$

8. $\underset{\substack{1\text{ mol}\\16\text{ g}\\32\text{ g}}}{CH_4} + \underset{\substack{2\text{ mol}\\64\text{ g}\\128\text{ g}}}{2O_2} \longrightarrow CO_2 + 2H_2O$

9. $\underset{\substack{1\text{ mol}}}{C_5H_{12}(g)} + 8O_2(g) \longrightarrow \underset{5\text{ mol}}{5CO_2(g)} + 6H_2O(l)$

1 mole = 72 g $\qquad$ 5 N_0 molecules

$5 \times 6.02 \times 10^{23}$ molecules of CO_2 are from = 72 g C_5H_{12}

Thus, 3.01×10^{22} molecules of CO_2 are from

$$= \frac{72 \times 3.01 \times 10^{22}}{5 \times 6.02 \times 10^{23}} = 0.72\text{ g } C_5H_{12}$$

10. $CaO + 2HCl \longrightarrow CaCl_2 + H_2O$
1 mol, 2 mol, 1 mol, 1 mol
8.8 mol, 8.8 × 2 mol, 8.8 mol

11. $Na + KCl \longrightarrow NaCl + K$
1 mol, 1 mol, 1 mol
23.0 g, 74.5 g, 39 g
150.0 g, 150.0 g, 78.5 g, 254.35 g

KCl is a limiting quantity and thus, K formed = 78.5 g

12. $2NaCl + H_2SO_4 \longrightarrow Na_2SO_4 + 2HCl$
2 mol, 1 mol
2 × 58.5 g, 142 g

117 g NaCl yield = 142 g pure Na_2SO_4

$$\left(245 \times \frac{93.9}{100}\right) \text{ g NaCl yield} = \frac{142 \times 245 \times 93.9}{117 \times 100} \text{ g pure } Na_2SO_4$$

$= 279.21$ g pure Na_2SO_4

$= 279.21 \times \frac{100}{74.1}$ of 74.1% purity

$= \mathbf{376.80\ g}$

Analytical Questions

1. 4236 g

2. $2NaHCO_3 \xrightarrow{\Delta} Na_2CO_3 + H_2O + CO_2$
2 mol, 1 mol = 22.4 L at STP

$Na_2CO_3 \xrightarrow{\Delta} X$

$NaHCO_3$ in mixture $= 4 \times \frac{75}{100} = 3\text{ g} = 0.0357$ mol

CO_2 formed $= \frac{0.0357}{2}$ mol

$= \frac{0.0357}{2} \times 22.4$ L CO_2 at STP

$= 0.4$ L = **400 mL**

3. 5.943 L **4.** 41 kg H_2O, Li_2O is limiting reactant

5. $MnO_2 + 4HCl \longrightarrow MnCl_2 + 2H_2O + Cl_2$
1 mol, 4 mol, 1 mol
0.86 mol, 48.2 g = 1.32 mol

HCl required by 0.86 mol MnO_2 = 3.44 mol
but HCl taken = 1.32 mol

Thus, HCl is the limiting reactant.

Thus, Cl_2 formed $= \frac{1}{4} \times 1.32$

$= \mathbf{0.33\ mol\ Cl_2}$

$= 23.43$ g

6. 29.80% **7.** (a) **863.08 g** (b) **93.04%**

8. $ZnS \longrightarrow ZnO \longrightarrow ZnSO_4 \longrightarrow Zn$
97 g, 65 g

$\therefore \frac{225 \times 90.6 \times 100 \times 98.2}{100 \times 100 \times 100}$ kg gives Zn = **134.14 kg**

9. $CaCO_3 \longrightarrow CaO + CO_2$

$Na_2CO_3 + H_2O + CO_2 \longrightarrow 2NaHCO_3$
50 g

Exercise 2

Only One Option Correct

1. $CH_4 + 2O_2 \longrightarrow CO_2 + 2H_2O$
1 mol, 2 mol, 44 g
16 g, 64 g

given 16 g 48 g = 1.5 mol

Thus, O_2 is the limiting reactant

$\therefore$ 2 mol O_2 give = 44 g CO_2

1.5 mol O_2 give $= \frac{44}{2} \times 1.5 = 33$ g CO_2

Thus, **(b)**

2. Ratio of HNO_3 per mol of H_2O

(a) $\frac{8}{4} = 2$ (b) $\frac{6}{3} = 2$

(c) $\frac{10}{3} = 3.33$ (d) $\frac{4}{2} = 2$

Thus, **(c)**

3. $SO_2Cl_2 + 2H_2O \longrightarrow H_2SO_4 + 2HCl$
1 mol, 1 mol, 2 mol

$H_2SO_4 + Ca(OH)_2 \longrightarrow CaSO_4 + 2H_2O$

$2HCl + Ca(OH)_2 \longrightarrow CaCl_2 + 2H_2O$

Thus, total $Ca(OH)_2$ required = 2 mol

Thus, **(a)**

4. $3BaCl_2 + 2(NH_4)_3PO_4 \longrightarrow Ba_3(PO_4)_2 + 6NH_4Cl$
3 mol, 2 mol, 1 mol
given 0.5 mol, 0.2 mol, 0.1 mol

Thus, $(NH_4)_3PO_4$ is the limiting reactant giving 0.1 mol $Ba_3(PO_4)_2$

Thus, **(a)**

5. $BaCl_2$ required by 0.2 mol of $(NH_4)_3PO_4$ = 0.3 mol

Thus, $BaCl_2$ in excess = 0.5 − 0.3
= 0.2 mol

Thus, **(d)**

6. $2H_2S + SO_2 \longrightarrow 2H_2O + 3S$
2 mol, 1 mol, 3 mol
given 1 mol, 1 mol, ?

Thus, H_2S is the limiting reactant forming 1.5 mol S

Thus, **(c)**

7. $BaF_2 + H_2SO_4 \longrightarrow BaSO_4 + H_2F_2$

1 mol 1 mol 1 mol 1 mol

given 1 mol 2 mol

Thus, unreacted $H_2SO_4 = 1$ mol

H_2F_2 formed = 1 mol

$H_2SO_4 + 2NaOH \longrightarrow Na_2SO_4 + 2H_2O$
1 mol 2 mol

$H_2F_2 + 2NaOH \longrightarrow Na_2F_2 + 2H_2O$
1 mol 2 mol

Thus, total NaOH required by 1 mol H_2F_2 (formed) and 1 mol H_2SO_4 (unreacted) = 4 mol

Thus, **(a)**

8. (b)

9. $Ca(HCO_3)_2 + CaO \longrightarrow 2CaCO_3 + H_2O$

56 g 200 g

0.56 g 2.00 g in 10 L water

10 L H_2O (= 10^4 mL) has $CaCO_3 = 2$ g

10^6 mL (ppm) has $CaCO_3 = 200$ g

Thus, 200 ppm $CaCO_3$

Thus, **(a)**

10. $Na_2CO_3 \equiv CaCO_3$

106 g 100 g

Hardness in 10^6 g H_2O = 100 g $CaCO_3$

$\therefore$ Hardness in 10 L (10^4 mL = 10^4 g) H_2O = 1 g $CaCO_3$

100 g $CaCO_3$ required = 106 g Na_2CO_3

$\therefore$ 1 g $CaCO_3$ required = 1.06 g Na_2CO_3

Thus, **(d)**

11. Iron oxide = 0.5434 g

Oxygen lost as H_2O = 0.1210 g

Iron = 0.5434 − 0.1210 = 0.4224 g

Element	Amount	%	mole	Ratio
Iron	0.4224 g	77.73	1.39	1
Oxygen	0.1210 g	22.26	1.39	1

Thus, FeO **(a)**

12. $P_2O_5 \longrightarrow Mg_2P_2O_7$

1 mol 1 mol

142 g 222 g

P_2O_5 in the mineral $= 0.5428 \times \frac{26.26}{100}$

$= 0.1425 \text{ g} = 1 \times 10^{-3}$ mol

Thus, $Mg_2P_2O_7$ obtained $= 1 \times 10^{-3}$ mol = 0.2228 g

Thus, **(d)**

13. $CaO \longrightarrow CaC_2O_4 \longrightarrow CaCO_3$

56 g 100 g

? 0.3462 g

Pure CaO in the mineral $= \frac{56}{100} \times 0.3462$

= 0.1938 g

$\therefore$ % of pure CaO $= \frac{0.1938}{0.8432} \times 100$

= 23%

Thus, **(b)**

14. $S \longrightarrow BaSO_4 \longrightarrow BaCl_2$

32 208 g

$0.60 \times 0.12 = 0.072$ g ?

$BaCl_2$ required by 0.072 g sulphur to be precipitated as

$BaSO_4 = \frac{208}{32} \times 0.072$

= 0.468 g

Thus, volume of $BaCl_2$ solution $= \frac{0.468 \times 100}{1.6}$

= 29.25 mL

Thus, **(c)**

15. $Fe_3O_4 \longrightarrow 1.5\,Fe_2O_3$

1 mol 1.5 mol

232 g 240 g

? 0.40 g

Fe_3O_4 required by 0.40 g $Fe_2O_3 = \frac{232}{240} \times 0.4$

= 0.3867 g (pure)

$= 0.3867 \times \frac{100}{18}$ g

= 2.15 g impure Fe_3O_4

Thus, **(a)**

16. 10 g glucose is in = 100 mL

$\therefore$ 180 g glucose is in $= \frac{100 \times 180}{10}$ mL

= 1800 mL = 1.8 L

Thus, **(c)**

17. $2(NH_4)_2HPO_4 = P_2O_5$

2×132 g = 142 g

Thus, % of $P_2O_5 = \frac{142}{264} \times 100 = 53.78$

Thus, **(d)**

18. Oxygen = 67.6%

Metal = 32.33%

Thus, 67.67 g oxygen combines with = 32.33 g metal

hence, 8 g oxygen combines with $= \frac{32.33 \times 8}{67.67}$

= 3.82

Thus, **(b)**

19. (a)

20. $2NaOH + H_2SO_4 \longrightarrow Na_2SO_4 + 2H_2O$

2 mol 1 mol

1 mol 0.5 mol = 49 g H_2SO_4 (pure)

Thus, 70% H_2SO_4 required $= 49 \times \frac{100}{70} = 70$ g

Thus, **(a)**

21. (d)

22. Solute present in 800 g solution $= 800 \times \dfrac{40}{100}$

$= 320$ g

Solute precipitated $= 100$ g

Solute left $= 220$ g

$\therefore$ Total weight of solution $= 800 - 100 = 700$ g

700 g solution has solute $= 220$

100 g solution has solute $= \dfrac{220}{700} \times 100$

$= 31.43\%$

Thus, **(a)**

23. $MHPO_4 \rightleftharpoons M^{2+} + HPO_4^{2-}$

Thus, metal is divalent forming MCl_2

Thus, **(d)**

24. Equivalent weight of M^{2+} – 37.2

Thus, molar mass of $M^{2+} = 37.2 \times 2 = 74.4$

Thus, molar mass of $MCl_2 = 74.4 + 71 = 145.5$

Thus, **(c)**

25. Glucose in 500 mL solution $= \dfrac{9 \times 500}{100}$

$= 45$ g

$= \dfrac{45}{80} = 0.25$ mol

$= 0.25 \times 6.02 \times 10^{23}$ molecules

$= 1.5 \times 10^{23}$

Thus, **(c)**

One or More Than One Options Correct

1. $\underset{1\text{ mol}}{SO_2Cl_2} + 2H_2O \longrightarrow \underset{1\text{ mol}}{H_2SO_4} + \underset{2\text{ mol}}{2HCl}$

$1 \text{ mol } H_2SO_4 \equiv 1 \text{ mol } Ca(OH)_2 \equiv 2 \text{ mol KOH}$

$1 \text{ mol HCl} \equiv \dfrac{1}{2} \text{ mol } Ca(OH)_2 \equiv 1 \text{ mol KOH}$

$2 \text{ mol HCl} \equiv 1 \text{ mol } Ca(OH)_2 \equiv 2 \text{ mol KOH}$

Thus, mixture of 1 mol H_2SO_4 and 2 mol HCl requires 2 mol $Ca(OH)_2$ thus, (a) is correct

4 mol KOH thus, (d) is correct

2. Iron in haemoglobin $= \dfrac{4 \times 56}{65000} \times 100 = 0.35$

Thus, (a) is true

Iron in mole haemoglobin $= 224$ g

Thus, (b) is false and (c) is true

Molar mass of haemoglobin if iron content is 0.56%

$= \dfrac{100 \times 4 \times 56}{0.56} = 40000 < 65000$

Thus, molar mass is decreased

Thus, (d) is also false

Thus, (a), (c)

3. (a), (c), (d)

4. 146 g HCl $= \dfrac{146}{36.5} = 4$ mol

$$\underset{1\text{ mol}}{MnO_2} + \underset{4\text{ mol}}{4HCl} \longrightarrow MnCl_2 + \underset{1\text{ mol}}{Cl_2} + 2H_2O$$

	MnO_2	HCl	Cl_2
Initial	0.8 mol	3.2 mol	0 mol
after	0 mol	0.8 mol	0.8 mol

Thus, MnO_2 (less) is limiting reagent thus, (d) is true HCl (more) is in excess.

Amount of product is decided by MnO_2

Cl_2 formed $= 0.8$ mol thus (a) is true

HCl unreacted $= 0.8$ mol thus, (b) is true

Thus, (a), (b), (c) and (d) are correct alternates

5. Mg $= 0.273$ g

Product formed with nitrogen $= 0.378$ g

Nitrogen reacted $= 0.105$ g

Element	Amount	Mole	Rate	Whole number
Mg	0.273 g	0.011	1.5	3
N	0.105	0.0075	1.0	2

Thus, Mg_3N_2

Thus, (b)

6.

Species	Amount	Mole	Total atoms	H-atoms
CH_4	1.6 g	0.1	$0.5\,N_0$	$0.4\,N_0$
NH_3	1.7 g	0.1	$0.4\,N_0$	$0.3\,N_0$
H_2O	1.8 g	0.1	$0.3\,N_0$	$0.2\,N_0$

They have equal number of moles thus, (a) is true Number of atoms is $CH_4 > NH_3 > H_2O$ thus, (b) is also true

Ratio of H-atom 4 : 3 : 2 thus, (c) is also true.

Thus, (a), (b) and (c)

7. $H_2O = 1 \text{ g mL}^{-1} = 1000 \text{ gL}^{-1}$

$= \dfrac{1000}{18} = 55.55 \text{ mol L}^{-1}$

Thus, (a) is true

$1 \text{ g} = \dfrac{1}{18} \text{ mol} = \dfrac{6.02 \times 10^{23}}{18}$ H_2O molecules

$= 3.33 \times 10^{22}$ H_2O molecules

Thus, (b) is true

It has 6.06×10^{22} H-atoms thus (c) is false

It has 3.33×10^{22} O-atoms thus (d) is also false

Thus, (a) and (b)

8. 1 mole of $(NH_3)_2HPO_4$ has 0.5 mole P_2O_5 and 2 mole of NH_3.

Thus, 1 mole (= 132 g) $(NH_4)_2HPO_4$ has = 0.5 mole (= 71 g) P_2O_5

$$\% \text{ of } P_2O_5 = \frac{71}{132} \times 100 = 53.78\%$$

Thus, (a) is correct.

Also, 132 g $(NH_4)_2HPO_4$ has = 34 g NH_3

Thus, $\% \text{ as } NH_3 = \frac{34}{132} \times 100 = 25.76\%$

Thus, (b) is correct.

Also, 132 g has = 28 g N, 9g H and 64 g O

Thus, % of O is maximum

Thus, (c) and (d) are incorrect.

Thus, (a) and (b)

9. Mole fraction of ethanol $(C_2H_5OH) = 0.25 = \frac{n_1}{n_1 + n_2}$

Mole fraction of water $(H_2O) = 0.75 = \frac{n_2}{n_1 + n_2}$

$$\therefore \quad \frac{n_1}{n_2} = \frac{0.25}{0.75} = \frac{1}{3}$$

$$C_2H_5OH = n_1 = 1 \text{ mole} = 46 \text{ g}$$

$$H_2O = n_2 = 3 \text{ moles} = 54 \text{ g}$$

$$\text{Total mass} = 46 + 54 = 100 \text{ g}$$

Thus, it is 46% by weight of solution and 54% by weight of solution.

Thus, (c) and (d) are correct.

10. $\underset{(M + 35.5)}{MCl} + AgNO_3 \longrightarrow \underset{143.5}{AgCl} \downarrow$

$\underset{74.5}{KCl} + AgNO_3 \longrightarrow \underset{143.5}{AgCl} \downarrow$

Let $MCl = x$ g

than $KCl = (12.8 - x)$ g

$$AgCl \text{ due to } MCl = \frac{143.5x}{(M + 35.5)}$$

$$AgCl \text{ due to } KCl = \frac{143.5}{74.5}(12.8 - x)$$

$$\frac{143.5x}{(M + 35.5)} + \frac{143.5(12.8 - x)}{74.5} = 28.7$$

Since, MCl is volatile, in second case AgCl due to MCl only.

Thus, $\left(\frac{143.5x}{M + 35.5}\right) = 14.35$

$$\therefore \quad 14.35 + \frac{143.5(12.8 - x)}{74.5} = 28.7$$

$$1 + \frac{10(12.8 - x)}{74.5} = 2$$

$$(12.8 - x) = 7.45$$

$$x = 5.35 \text{ g}$$

Thus, $\frac{143.5 \times 5.35}{(M + 35.5)} = 14.35$

$$\therefore \quad M = 18$$

Thus, ionic mass of $M^+ = 18$, thus (a) is true and molar mass of $MCl = 53.5$

Thus, moles of $MCl = \frac{5.35}{53.5} = 0.1$

Moles of $KCl = \left(\frac{12.80 - 5.35}{74.5}\right) = 0.1$

Thus, (b) is also true

Thus, (c) is also true

AgCl is water insoluble

Thus, (d) is false

Thus, (a), (b) and (c)

Brain Twisters

1. Heat evolved in the reaction (exothermic) causes decomposition of $H_2CO_3(aq)$ into $CO_2(g)$ and $H_2O(l)$

$$H_2CO_3 \longrightarrow H_2O(l) + CO_2(g)$$

Thus, H_2CO_3 is unstable.

2. (a)

(i) $AgNO_3(aq) + NaBr(aq) \longrightarrow AgBr(s) + NaNO_3(aq)$

(ii) $Ag^+(aq) + NO_3^-(aq) + Na^+(aq) + Br^-(aq) \longrightarrow AgBr(s) + Na^+(aq) + NO_3^-(aq)$

(iii) $Ag^+(aq) + Br^-(aq) \longrightarrow AgBr(s)$

(b) (i)

$BaCl_2(aq) + H_2SO_4(aq) \longrightarrow BaSO_4(s) + 2HCl(aq)$

(ii) $Ba^{2+}(aq) + 2Cl^-(aq) + 2H^+(aq) + SO_4^{2-}(aq) \longrightarrow BaSO_4(s) + 2H^+(aq) + 2Cl^-(aq)$

(iii) $Ba^{2+}(aq) + SO_4^{2-}(aq) \longrightarrow BaSO_4(s)$

(c) (i) $Al_2(SO_4)_3(aq) + 3Ca(OH)_2(aq) \longrightarrow 2Al(OH)_3(s) + 3CaSO_4(aq)$

(ii) $2Al^{3+}(aq) + 3SO_4^{2-}(aq) + 3Ca^{2+}(aq) + 6OH^-(aq) \longrightarrow 2Al(OH)_3(s) + 3Ca^{2+}(aq) + 3SO_4^{2-}(aq)$

(iii) $2Al^{3+} + 6OH^-(aq) \longrightarrow 2Al(OH)_3(s)$

NaOH dissolves $Al(OH)_3$ hence can't be used

(d) (i) $CaCl_2(aq) + 2K_3PO_4(aq) \longrightarrow Ca_3(PO_4)_2(s) + 6KCl(aq)$

(ii) $3Ca^{2+}(aq) + 6Cl^-(aq) + 6K^+(aq) + 2PO_4^{3-}(aq) \longrightarrow Ca_3(PO_4)_2(s) + 6K^+(aq) + 6Cl^-(aq)$

(iii) $3Ca^{3+} + 2PO_4^{3-} \longrightarrow Ca_3(PO_4)_2(s)$

3. $\underset{\substack{1 \text{ mol} \\ 4 \text{ mol}}}{Al_2O_3} + \underset{\substack{3 \text{ mol} \\ 12 \text{ mol}}}{3H_2SO_4} \longrightarrow \underset{\substack{1 \text{ mol} \\ 4 \text{ mol}}}{Al_2(SO_4)_3} + 3H_2O$

$$H_2SO_4 \text{ required} = 12 \text{ mol}$$

$$Al_2(SO_4)_3 \text{ formed} = 4 \text{ mol}$$

4. $\underset{56 \text{ g}}{CaO} + 3C \longrightarrow \underset{64 \text{ g}}{CaC_2} + CO$

Crude produce is = 85% CaC_2 + 15% CaO (unreacted)

450 kg of crude product has 15% CaO $= 450 \times \frac{15}{100}$

$= 67.50$ kg CaO

and $85\%\ CaC_2 = 450 \times \frac{85}{100}$

$= 382.5$ kg CaC_2

64 kg CaC_2 is from = 56 g CaO

Thus, 382.5 kg CaC_2 is from $= \frac{56}{64} \times 382.5$

$= 334.69$ kg CaO

Thus, total CaO required $= 67.50 + 334.69$

$= 402.19$ kg

(b) 382.5 kg CaC_2

5. $CuSO_4 = x$ g

$CuSO_4 \cdot 5H_2O = (1.245 - x)$

$$\underbrace{CuSO_4 \cdot 5H_2O}_{249.5} \longrightarrow \underbrace{CuSO_4}_{159.5\ g} + 5H_2O(g)\uparrow$$

$CuSO_4$ in $(1.245 - x)$ g

$CuSO_4 \cdot 5H_2O = \frac{159.5}{249.5}(1.245 - x)$ g

Thus, $\frac{159.5}{249.5}(1.245 - x) + x = 0.832$

$x = 0.1$ g

$\therefore$ $CuSO_4 \cdot 5H_2O = 1.145$ g

6.
$$\underset{\substack{(58.7x+30y)\\0.0973\ g}}{Ni_x(CO)_y} \longrightarrow \underset{\substack{74.7x\\0.0426}}{xNiO} + \underset{\substack{44y\\0.10}}{yCO_2}$$

$$\frac{74.7x}{44y} = \frac{0.0426}{0.10}$$

$\therefore$ $\frac{x}{y} = 0.25 = \frac{1}{4}$

Thus, empirical formula is $= Ni(CO)_4$

7. 3 mol $NaNH_2$ form = 1 mol NaN_3

0.0769 mol $NaNH_2$ forms = 0.0256 NaN_3

Thus, $NaNH_2$ is limiting reactant and NaN_3 obtained = 0.0256 mol

7.
$$\underset{\substack{1\ mol\\85\ g\\15\ g}}{NaNO_3} + \underset{\substack{3\ mol\\(3\times 39)\ g\\15g}}{3NaNH_2} \longrightarrow \underset{\substack{1\ mol\\65\ g\\?}}{NaN_3}$$

8.33 g (based as $NaNH_2$)

11.47 g (based on $NaNO_3$)

Limiting reagent is **$NaNH_2$**

and thus **NaN_3 obtained is 8.33 g**

8.
$$\underset{(28x+y)}{Si_xH_y} + O_2 \longrightarrow \underset{\substack{60x\\11.64\ g}}{xSiO_2} + \underset{\substack{9y\ H_2O\\6.980\ g}}{\frac{y}{2}H_2O}$$

$$\frac{60x}{9y} = \frac{11.64}{6.980}$$

$$\frac{x}{y} = 0.25 = \frac{1}{4}$$

Thus, Si_xH_y is **SiH_4**

9.
$$\underset{159\ g}{Cu_2S} \longrightarrow \underset{127\ g}{2Cu}$$

$$\underset{95.5\ g}{CuS} \longrightarrow \underset{63.5\ g}{Cu}$$

There is 11.0% impurity. Thus, mixture has 89.0 g of Cu_2S and CuS.

Let $Cu_2S = x$ g

then CuS $= (89.0 - x)$ g

Thus, Cu obtained from x g $Cu_2S = \left(\frac{127x}{159}\right)$ g

and from $(89.0 - x)$ g CuS $= \frac{63.5(89.0 - x)}{95.5}$ g

Total Cu obtained

$$\frac{127x}{159} + \frac{63.5(89.0 - x)}{95.5} = \frac{75.4 \times 89.5}{100} \text{ (given)}$$

$= 67.483$ g

This gives $x = Cu_2S = 62.45$

% of Cu_2S = **62.45%**

CuS = **27.55%**

10. 0.033 mol C_2H_5OH per litre $= 0.033 \times 46 \times 1000$ mg/L

$= 1518$ mg/L

$= 151.8$ mg/dL

This excess prescribed limit, hence **yes DWI ticket**

Passage Comprehension Questions

Passage 1

Step I $H_2SO_4 \longrightarrow H^+ + HSO_4^-$ 100%

Step II $HSO_4^- \longrightarrow H^+ + SO_4^{2-}$ 10%

$H^+ + H_2O \longrightarrow H_3O^+$

SO_4^{2-} is from step II

Step I is 100% thus $[H^+] = 0.2$ M

and $[HSO_4^-] = 0.2$ M

Step II is 10% thus $[H^+] = 0.2 \times 0.1 = 0.02$ M

and $[SO_4^{2-}] = 0.02$ M

total $[H_3O^+] = 0.2 + 0.02 = 0.22$ M

$[SO_4^{2-}] = 0.02$ M

and $[HSO_4^-] = 0.20 - 0.02 = 0.18$ M

1. (d) **2.** (c) **3.** (b) **4.** (b)

Passage 2

1. (a) **2.** (a,b)

Passage 3

1. (a) **2.** (b) **3.** (a)

Passage 4

1. Let MCl $= x$ g, NaCl $= (11.2 - x)$ g

$$MCl + AgNO_3 \longrightarrow MNO_3 + \underset{\text{white ppt}}{AgCl\downarrow}$$

$(M + 35.5)$ g MCl gives $= 143.5$ g AgCl

$$x \text{ g } MCl \text{ gives} = \frac{143.5x}{(M+35.5)} \text{ g AgCl} \quad ...(i)$$

$$\underset{(58.5\text{ g})}{NaCl} + AgNO_3 \longrightarrow NaNO_3 + \underset{(143.5\text{ g})}{AgCl\downarrow}$$

$$(11.2 - x) \text{ g NaCl gives} = \frac{143.5(11.2-x)}{58.5} \text{ g AgCl} \quad ...(ii)$$

$$\frac{143.5x}{M+35.5} + \frac{143.5(11.2-x)}{58.5} = 28.7 \text{ (given)} \quad ...(iii)$$

On heating MCl vaporises (being volatile)

Thus, AgCl is due to MCl only

$$\text{Thus,} \quad \frac{143.5x}{M+35.5} = 14.35 \quad ...(iv)$$

Thus, from (III)

$$\frac{143.5(11.2-x)}{58.5} = 14.35$$

Thus, gives $\quad x = 5.35$

From (IV) $\quad M = 18$

MCl in mixture $= 5.35$ g

NaCl in mixture $= 5.85$ g

Thus, ionic mass of M^+ is 18.

Also, molar mass of MCl $= 53.5$ g mol^{-1}

$$\text{moles of } MCl = \frac{5.35}{53.5} = 0.1$$

$$\text{moles of NaCl} = \frac{5.85}{58.5} = 0.1$$

Total moles $= 0.2$

Mole fraction of MCl $= 0.5$

Mole fraction of MCl $= 0.5$

Thus, **1.** (a) **2.** (b) **3.** (a)

Passage 5

1. (a) $10 \text{ g } K_2PtCl_6 = \frac{10}{417} \text{ mol} = 0.024 \text{ mol}$ (limiting)

$10 \text{ g } NH_3 = \frac{10}{17} = 0.588 \text{ mol}$ (excess)

K_2PtCl_6 consumed $= 0.024$ mol $=$ *cis*-platin formed

NH_3 consumed $= 0.048$ mol

NH_3 unreacted $= 0.588 - 0.048$

$= 0.54$ mol

2. (b) **3.** (a) **4.** (b) **5.** (c)

Passage 6

1. (b) $310 \text{ g } P_4 = \frac{310}{124} \text{ mol } P_4 = 10 \text{ mol } H_3PO_4$

2. (c) 2.5 mol P_4O_{10} (100% yield)

$= 2.5 \times \frac{50}{100}$ (50% yield)

$= 2.5 \times 0.5 \times 0.25 \times 4$ (25% yield)

$= 1.25$ mol H_3PO_4

3. (b)

$$\underset{3\text{ mol}}{3Ca(OH)_2} + \underset{2\text{ mol}}{2H_3PO_4} \longrightarrow \underset{1\text{ mol}}{Ca_3(PO_4)_2}$$

$$1 \text{ mol } Ca(OH)_2 = \frac{1}{3} \text{ mol } Ca_3(PO_4)_2$$

$$2 \text{ mol } H_3PO_4 = 1 \text{ mol } Ca_3(PO_4)_2$$

$$1.25 \text{ mol } H_3PO_4 = \frac{1.25}{2} \text{ mol.}$$

$Ca(OH)_2$ is limiting reagent.

Passage 7

1. (c) $5.8 \text{ g } NaIO_3 = \frac{5.8}{198} \text{ mol } NaIO_3$

$\therefore \quad NaHSO_3 = 3 \times \frac{5.8}{198}$ mol

$= \frac{3 \times 5.8}{198} \times 104$ g $NaHSO_3$

$= 9.14$ g

2. (a) I^- formed in step I $= \frac{5.8}{198}$ mol

IO_3^- required in step II $= \frac{5.9}{198 \times 5}$ mol

$= \frac{5.8 \times 198}{198 \times 5}$ g $NaIO_3$

$= 1.16$ g $NaIO_3$

$\therefore \quad \text{Volume} = \frac{1.16}{5.8} = \mathbf{0.2\ L}$

3. (c) $\underset{+5}{IO_3^-} \longrightarrow \underset{-1}{I^-}$

Change in oxidation number $= 6$ units

Thus, one mole $NaIO_3 = 6$ equivalents

4. (d) $1.16 \text{ g } NaIO_3 = \frac{1.16}{198}$ mol

$= \frac{6.96}{198}$ equivalent (since $IO_3^- = 6I$)

$= 0.035$ equivalent

Passage 8

1. (a) $0.74 \text{ g } O_3 = \frac{0.74}{48} = 0.0154 \text{ mol } O_3$

$0.67 \text{ g NO} = \frac{0.67}{30} = 0.0223 \text{ mol NO}$

O_3 is the limiting reagent and NO is in excess

$= 0.0223 - 0.0154 = 0.007$ mol

Thus, O_3 taken $= NO_2$ formed

$= 0.0154$ mol NO_2

$= 0.0154 \times 46$ g NO_2

$= 0.71$ g

2. (b) **3.** (c)

Passage 9

1. (c) 2. (d) 3. (a)

Assertion & Reason

1. (d) 2. (a) 3. (d) 4. (a) 5. (b)

True & False

1. F 2. T 3. F 4. F 5. F 6. T 7. F
8. T 9. T 10. T

Fill in the Blanks

1. 1.12 2. 17.75 3. 4 4. H_2S 5. HCl

Matrix-Match Type Questions

I. Only One Correct

1. (A) — (4); (B) — (5); (C) — (2);
(D) — (1); (E) — (3)
2. (A) — (3); (B) — (4); (C) — (1); (D) — (2);

II. One or More Correct

1. (A) — (1,2,3); (B) — (1,3); (C) — (1,4,5);
(D) — (1,4,5); (E) — (1,4,5)

Integer Answer Types

Set I

Questions →	1	2	3	4	5	6	7	8
Answers →	4	5	2	1	4	6	2	1
	⓪	⓪	⓪	⓪	⓪	⓪	⓪	⓪
	①	①	①	①	①	①	①	①
	②	②	②	②	②	②	②	②
	③	③	③	③	③	③	③	③
	④	④	④	④	④	④	④	④
	⑤	⑤	⑤	⑤	⑤	⑤	⑤	⑤
	⑥	⑥	⑥	⑥	⑥	⑥	⑥	⑥
	⑦	⑦	⑦	⑦	⑦	⑦	⑦	⑦
	⑧	⑧	⑧	⑧	⑧	⑧	⑧	⑧
	⑨	⑨	⑨	⑨	⑨	⑨	⑨	⑨

Set II

Questions →	1	2	3	4	5	6	7	8
Answers →	3	2	4	7	5	2	1	2
	⓪	⓪	⓪	⓪	⓪	⓪	⓪	⓪
	①	①	①	①	①	①	①	①
	②	②	②	②	②	②	②	②
	③	③	③	③	③	③	③	③
	④	④	④	④	④	④	④	④
	⑤	⑤	⑤	⑤	⑤	⑤	⑤	⑤
	⑥	⑥	⑥	⑥	⑥	⑥	⑥	⑥
	⑦	⑦	⑦	⑦	⑦	⑦	⑦	⑦
	⑧	⑧	⑧	⑧	⑧	⑧	⑧	⑧
	⑨	⑨	⑨	⑨	⑨	⑨	⑨	⑨

Phenolphthalein is colourless in acidic solution and pink in alkaline solution. Methyl-red is red in acidic solution and yellow in alkaline solution. What is an idea? A drop of phenolphthalein or methyl red can indicate completion of acid-alkali reaction by colour change in volumetric titration.

7

Volumetric

Quick Points

- Mole = $\dfrac{\text{mass}}{\text{molar mass}}$
- N_0 (Avogadro's number) = 6.02×10^{23} mol^{-1}

  1 equivalent of an acid ≡ 1 equivalent of a base in an acid-base reaction

  1 equivalent of a reducing agent ≡ 1 equivalent of an oxidising agent in a redox reaction

  Equivalent weight in a redox reaction

  $$= \frac{\text{molar mass (one unit)}}{\text{change in O.N. (equal on both sides)}}$$

  $$\underset{+7}{MnO_4^-} \longrightarrow \underset{+2}{Mn^{2+}} \qquad E(MnO_4^-) = \frac{M}{5}$$

  $$\underset{\underset{12}{\uparrow}}{Cr_2O_7^{2-}} \longrightarrow 2\underset{\underset{6}{\uparrow}}{Cr}^{3+} \qquad E(Cr_2O_7^{2-}) = \frac{M}{6}$$

- $$\Delta G^\circ = -nFE^\circ$$

  For redox reactions such that $\Delta G_1^\circ + \Delta G_2^\circ = \Delta G_3^\circ$

  $$E_3^\circ = \frac{n_1E_1^\circ + n_2E_2^\circ}{n_3}$$

  where n_i is the number of electrons exchanged in a reaction with standard electrode potential E_i°, F is number of coulombs (1 F = 96500 coulombs)

  Electrode potential is given by Nernst equation

  $$E = E^\circ - \frac{2.303RT}{nF} \log Q$$

Learning Framework of the Chapter ...

- Titrimetry
- Methods of Expressing Concentration of a Solution
- Mole Fraction
- Per cent Concentration
- Molarity (M)
- Molality (m)
- Equivalent Mass and Normality
- Volumetric Titrations
- Acid-Base Titrations
- Redox Titrations
- Iodometric/Iodimetric Titrations

where, Q is reaction quotient.

At equilibrium, $E = 0$ and $Q = K_{eq}$

$$\therefore \quad E^\circ = \frac{2.303RT}{nF} \log K_{eq}$$

$$E^\circ = \frac{0.0591}{n} \log K_{eq}$$

$$\left(\text{at 298 K when } \frac{2.303RT}{F} = 0.0591 \text{ V}\right)$$

Titrimetry

Titrimetric methods include a large and powerful group of quantitative methods that are based on measuring the amount of a reagent of known concentration that is consumed in an analysis. Titrimetric can be

Volumetric	Gravimetric	Coulometric
involves measuring the volume of a solution of known concentration that is needed to react with the species under investigation	involves measurement of mass involved instead of its volume	involves measurement of electric current required to bring about a particular chemical change.

Present unit describes stoichiometric based on **Volumetric Titrimetry.**

Terms to be used in Volumetric Titrimetry

A **standard solution** (or a **standard titrant**) is a reagent of known concentration that is used to carry out a titrimetric analysis.

A **titration** is performed by slowly adding a standard solution from a burette to a solution under examination (analyte) until the reaction between the two is judged complete. The volume or mass of reagent needed to complete the titration is determined from the difference between the initial and final readings.

It is sometimes necessary to add an excess of the standard titrant and then determine the excess amount by **back-titration** with a second standard titrant.

Primary and Secondary Standards

A **primary standard** is a highly purified compound that serves as a reference material in volumetric and mass titrimetric methods. The accuracy of a method is mainly dependent on the properties of this compound. A primary standard must be

1. of high purity
2. of atmospheric stability, (no oxidation, reduction or decomposition
3. anhydrous so that composition of the solid does not change with variation in humidity
4. of reasonable solubility in the titration medium
5. of reasonably large molar mass so that the relative error associated with weighing the standard is minimised.

In most of the cases, above requirements are not met and their standard solutions cannot be prepared by dissolving a weighed amount in a given volume of solution. They are termed as **secondary standard.**

Standard Solutions

The ideal standard solution (of known concentration) must

- be sufficiently stable so that it is necessary to determine its concentration only once.
- react rapidly with the analyte.
- give sharp end-point with the analyte.

- If concentration can be known simply by dissolving the known amount in a given volume of solutions, they are termed as **primary standard solutions.** In most cases, concentration can be determined by standarisation against primary standard solution. Such solutions are termed as **secondary standard solution.**

Equivalence Points and End Points

The equivalent point in a titration is a theoretical point reached when the amount of added titrant is chemically equivalent to the amount of analyte in the sample. Equivalent point cannot be determined experimentally. We can only estimate its position by observing same physical change associated with the condition of equivalence. This is called end-point of titration.

Equivalent point is reached when 1 mole of HCl is neutralised by 1 mole of NaOH. But we cannot detect this equivalent point. If one drop of suitable indicator (in this case phenolphthalein) is added in the HCl (analyte) then additional drop of NaOH (from burette) will give light pink colour. This is the end-point. Thus volume of titrant at the end-point is larger than that at the equivalent point.

Sometimes even rough estimate of the concentration is not known. In such cases excess of reagent is added so that complete reaction of the analyte takes place. Suppose we have been given NH_3 solution. We allow its reaction with excess of HCl or H_2SO_4 of known concentration. Acid reacts equivalent to NH_3 and unreacted acid is determined by titration with NaOH. Estimation of unreacted acid by NaOH is **called back titration.**

Explanation of Terms

We can determine the concentration of oxalic acid ($H_2C_2O_4 \cdot 2H_2O$), potassium dichromate ($K_2Cr_2O_7$), copper sulphate ($CuSO_4 \cdot 5H_2O$) by dissolving the known amount in a given volume of solution. These are stable to atmospheric condition and their concentration does not change. They are thus, **primary standard.** Ferrous salts (Fe^{2+} as $FeSO_4$), potassium permanganate ($KMnO_4$), mineral acids (HCl, HNO_3, H_2SO_4), sodium hydroxide (NaOH) etc, are not stable to atmospheric condition. Their solutions are prepared by dissolving a known amount or volume but they have to be standarised using primary standard. Thus, all these are secondary standard.

Secondary standard	Standarised against primary standard
NaOH	$H_2C_2O_4$ (oxalic acid)
$KMnO_4$	$H_2C_2O_4$
$FeSO_4 \cdot 7H_2O$	$K_2Cr_2O_7$
$Na_2S_2O_3 \cdot 5H_2O$ (sodium thiosulphate)	$CuSO_4 \cdot 5H_2O$
HCl, HNO_3, H_2SO_4	NaOH (standarised with $H_2C_2O_4$)
I_2	$Na_2S_2O_3 \cdot 5H_2O$ (standarised with $CuSO_4 \cdot 5H_2O$)

Mineral acids are standarised using sodium hydroxide which is turn is standarised first using oxalic acid.

Iodine is standarised using sodium thiosulphate which in turn is standarised first using copper sulphate solution.

Methods of Expressing Concentration of a Solution

The concentration of a chemical **solution** refers to the amount of **solute** that is dissolved in a **solvent.** We normally think of a solute as a solid that is added to a solvent (for example NaCl in H_2O) but the solute could exist in another phase. For example, if we add a small amount of ethanol (C_2H_5OH) to water, then the ethanol is solute and the water is the solvent. If we add a smaller amount of water to a larger amount of ethanol, then water could be solute.

The amount of solute in a solution can be expressed in terms of two system of units:

(i) **'Group A'** units specify the amount of solute in a given volume of solution, such as density, molarity, etc.

(ii) **'Group B'** units specify the amount of solute for a given mass of solvent or solution, such as mass%, mole fraction, molality, etc.

The advantage of 'group A' units is the ease of solution preparation and that of 'group B' units is temperature independence.

In a given solution, let solute be represented by (1) and the solvent by (2)

w_1 = mass of solute

w_2 = mass of solvent

m_1 = molar mass of solute

m_2 = molar mass of solvent

n_1 = moles of solute $= \dfrac{w_1}{m_1}$

n_2 = moles of solvent $= \dfrac{w_2}{m_2}$

$w_1 + w_2$ = mass of solution

$(n_1 + n_2)$ = moles of solution

d = density in g cm^{-3}

V = volume of solution in cm^3

E_1 = equivalent weight of solute

Expressing Concentration in Groups *A* and *B*

Following are the terms given in Tables 7.1 and 7.2 widely used to express concentration of solution.

Table 7.1 **Expressing Concentration in "Group *A*"**

Name	Symbol	Definition	Formula	Usual SI unit
Volume per cent	C	amount of solute present in 100 mL of solution	$C = \dfrac{100w_1}{V}$	kg m^{-3}
Volume per litre	C	amount of solute present in 1000 mL of solution	$C = \dfrac{1000w_1}{V}$ gL^{-1}	
Molarity	M	number of moles of solute present in 1 L (1 dm^3, 1000 cm^3) of solution	$M = \dfrac{1000w_1}{m_1V}$	mol dm^{-3}
Normality	N	number of g-equivalent of solute present in 1 L (1 dm^3, 1000 cm^3) of solution	$N = \dfrac{1000w_1}{E_1V}$	g equiv dm^{-3}

Table 7.2 **Expressing Concentrations in "Group *B*"**

Name	Symbol	Definition	Formula	Usual SI unit
% concentration by mass of solution	%	amount of solute present in 100 g of solution.	$\% = \dfrac{100w_1}{w_1 + w_2}$	dimensionless
% concentration by mass of solvent	%	amount of solute present in 100 g of solvent	$\% = \dfrac{100w_1}{w_2}$	dimensionless
Molality	m	number of moles of solute present in 1 kg solvent	$m = \dfrac{1000w_1}{m_1w_2}$	mol kg^{-1}
Mole fraction	X	fraction of a component out of total moles present in solution	$X_1 = \dfrac{n_1}{n_1 + n_2}$, $X_2 = \dfrac{n_2}{n_1 + n_2}$, $X_1 + X_2 = 1$	dimensionless

Standard notation for molality is m and should not be confused with m_1 (molar mass of solute) or m_2 (molar mass of solvent)

Closely related to weight per cent and particularly useful for dilute solutions are the concentration units parts per million (ppm) and parts per billion (ppb)

$$\text{Parts per million (ppm)} = \frac{\text{mass of solute component}}{\text{total mass of solution}} \times 10^6$$

$$\text{Parts per billion (ppb)} = \frac{\text{mass of solute component}}{\text{total mass of solution}} \times 10^9$$

Mole Fraction

Fraction of the substance in the mixture expressed in terms of mole is called its mole fraction.

n_A = moles of A in mixture of A and B

n_B = moles of B in same mixture

$$\text{mole fraction of } A \text{ in the mixture} = X_A = \frac{n_A}{n_A + n_B}$$

$$\text{mole fraction of } B \text{ in the mixture} = X_B = \frac{n_B}{n_A + n_B}$$

This is thus, the number of moles of a compound divided by the total number of moles of all chemical species in the solution. Keep in mind, the sum of all mole fractions in a solution always equals 1.

Thus,
$$X_A + X_B = 1$$
$$X_A = 1 - X_B$$

Illustration 1 *What are the mole fractions of the components of the solution formed when 92 g glycerol (92 g mol^{-1}) is mixed with 90 g water (18 g mol^{-1})?*

Solution

Components	Step 1	Step 2	Step 3
Glycerol	convert amount into moles $= \frac{\text{amount}}{\text{molar mass}}$ $= \frac{92}{92} = 1\,\text{mol}$	Take total moles = 1 + 5 = 6	Mole fraction = $\frac{\text{mole of the species}}{\text{total moles}}$ $X_{\text{glycerol}} = \frac{1}{6}$
Water	mole = $\frac{90}{18}$ = 5 mol		$X_{\text{water}} = \frac{5}{6}$

Check: Sum of mole fraction is always equal to one

$$X_{\text{glycerol}} + X_{\text{Water}} = \frac{1}{6} + \frac{5}{6} = 1$$

Per cent Concentration

In a solution, let there be w_1 g solute (molar mass = m_1)

w_2 g solvent (molar mass = m_2)

V mL of solution of density d g/mL

There are following types of expressions used to denote % concentration:

(a) mass per cent

$$\left(\frac{w}{W}\right) = \frac{\text{mass of solute}}{\text{mass of solution}} \times 100\%$$
$$= \left(\frac{w_1}{w_1 + w_2} \times 100\right)$$

(b) volume per cent

$$\left(\frac{v}{V}\right) = \frac{\text{volume of solute}}{\text{volume of solution}} \times 100\%$$
$$= \frac{v}{V} \times 100$$

(c) mass/volume per cent

$$\left(\frac{w}{V}\right) = \frac{\text{mass of solute, g}}{\text{volume of solution, mL}} \times 100\%$$
$$= \frac{w}{V} \times 100$$

- Note that the denominator in each of these expressions refers to the solution rather than to the solvent.
- (a) and (b) do not depend on the unit employed while (c) has unit of mass/volume (say g/mL).
- Of the three expressions, only mass per cent is **temperature independent.**
- To avoid confusion, we generally use terms like % by **mass of solution** which means amount of solute (w_1) g present in 100 g solution
$$= \frac{w_1}{w_1 + w_2} \times 100$$
and % by **mass of solvent** which means amount of solute (w_1) g present in 100 g solvent
$$= \frac{w_1}{w_2} \times 100$$
- For every dilute solutions, parts per million, 10^6 (ppm) is a convenient way to express concentration.
$$\text{Concentration (ppm)} = \frac{\text{mass of solute}}{\text{mass of solution}} \times 10^6$$
- For even more dilute solutions, parts per billion, 10^9 (ppb) rather than 10^6 (ppm) is used.
$$\text{Concentration (ppb)} = \frac{\text{mass of solute}}{\text{mass of solution}} \times 10^9$$
- Concentration in parts per thousand (ppt) is also commonly used.
$$\text{Concentration (ppt)} = \frac{\text{mass of solute}}{\text{mass of solution}} \times 10^3$$

Illustration 1 *Aqueous urea solution is 20% by mass of solution. Calculate percentage by mass of solvent.*

Solution If solute $w_1 = 20$ g

then solution $w_1 + w_2 = 100$ g

$\therefore$ solvent = 100 − 20 = 80 g

Hence, percentage by mass of solvent $= \frac{20}{80} \times 100 = \mathbf{25\%}$

Alternate method Given, $\frac{w_1}{w_1 + w_2} \times 100 = 20$

$$\frac{w_1}{w_1 + w_2} = \frac{1}{5}$$

Thus, $w_1 : (w_1 + w_2) = 1:5$

or $w_1 : w_2 = 1:4$

$\therefore$ $\frac{w_1}{w_2} \times 100 = \mathbf{25\%}$

Illustration 2 *Equal moles of water and urea are taken in a flask. What is mass percentage of urea in the solution?*

Solution If $H_2O = x \text{ mol} = 18x \text{ g}$

then urea $= x \text{ mol} = 60x \text{ g}$

$$\text{total mass of the solution} = 18x + 60x = 78x \text{ g}$$

$$\text{mass percentage of urea} = \frac{18x}{78x} \times 100 = \mathbf{23.077\%}$$

Molarity (*M*)

Number of moles of solute present per litre of the solution is its molarity.

$$\text{Molarity } (M) = \frac{\text{moles of solute}}{\text{litres of solution}}$$

If 0.1 mol of a solute is present in 100 mL (0.1 L) solution then molarity

$$M = \frac{0.1}{0.1} = 1 \text{ M}$$

DEEP Focus

moles of solute = molarity × litres of solution

millimoles of solute = molarity × mL of solution

Analytical Molarity

The analytical molarity of a solution gives the total number of moles of a solute in one litre of the solution (or the total number of millimoles in one millilitre). Thus, analytical molarity specifies the amount by which the solution can be prepared.

- Sulphuric acid solution that has analytical concentration of 1 mol per litre can be prepared by dissolving 1.0 mol or 98 g of H_2SO_4 in one litre of solution.
- Sodium hydroxide solution that has analytical molarity of 0.1 M has 0.1 mol or 4 g of it in one litre of solution.

Illustration 1 *What is the molarity of solution made when 6 g urea (molar mass 60 g mol^{-1}) is dissolved in water to make 100 mL of solution?*

Solution Solute $= 6 \text{ g} = \frac{6}{60} = 0.1 \text{ mol}$

$$\text{Volume} = 100 \text{ mL} = \frac{100}{1000} \text{ L} = 0.1 \text{ L}$$

$$\text{Molarity} = \frac{0.1 \text{ mol}}{0.1 \text{ L}} = 1 \text{ mol L}^{-1}$$

Equilibrium Molarity

The equilibrium molarity expresses the molar concentration of a particular species in a solution at equilibrium. In order to specify the equilibrium molarity of a particular species, it is necessary to know how the solute behaves when it is dissolved in a solvent.

- If **analytical molarity** of NaOH is 0.1 M, then equilibrium molarity of NaOH, [NaOH] = 0.0 M (assuming 100% ionisation).

  that of OH^-, $[OH^-] = 0.1 \text{ M}$

  and that of Na^+, $[Na^+] = 0.1 \text{ M}$

  $$NaOH \rightleftharpoons Na^+ + OH^-$$

- If H_2SO_4 ionises as $H_2SO_4 \rightleftharpoons H^+ + HSO_4^-$ then 1.0 M H_2SO_4 (analytical) has equilibrium concentration as (assume 100% ionisation)

  $$[H_2SO_4] = 0.0 \text{ M}$$
  $$[HSO_4^-] = 1.0 \text{ M}$$
  $$H^+ \text{ as } [H_3O^+] = 1.0 \text{ M}$$

- However, there is no difference between analytical and equilibrium molarity when solute dissolved is non-electrolyte or no ionisation takes place in solution. 1 M glucose (non-electrolyte), 1 M urea (non-electrolyte), etc., are also 1 M as equilibrium molarity.

Illustration 2 *Concentration of glucose* ($C_6H_{12}O_6$) *in normal blood is approximately* 90 *mg per* 100 *mL. What is the molarity of the glucose solution in blood?*

Solution Amount of glucose $= 90 \text{ mg} = \frac{90}{1000} = 0.09 \text{ g}$

$$= \frac{0.09}{180} = 0.0005 \text{ mol}$$

$$\text{Volume of solution} = 100 \text{ mL} = 0.10 \text{ L}$$

$$\text{Molarity} = \frac{\text{moles of solute}}{\text{litres of solution}} = \frac{0.0005}{0.10}$$

$$= \mathbf{0.005 \text{ M}}$$

Illustration 3 *Hydrochloric acid is sold commercially as a* 12.0 *M solution. How many moles and how many grams of* HCl *are in* 300.0 *mL of* 12.0 *M solution?*

Solution $$\text{Molarity} = \frac{\text{moles of solute}}{\text{litres of solution}}$$

$$12.0 = \frac{\text{moles of solute (HCl)}}{0.30 \text{ L}}$$

$$\text{moles of HCl} = 12 \times 0.30$$
$$= \mathbf{3.6\ mol\ HCl}$$
$$= 3.6 \times 36.5\ \text{g HCl}$$
$$= \mathbf{131.4\ g\ HCl}$$

Illustration 4 *How many moles of solute are present in each of the following solutions?*

(a) 35.0 *mL of* 0.2 *M* HNO_3

(b) 175 *mL of* 0.5 M *glucose* ($C_6H_{12}O_6$)

Solution Molarity × volume in mL = millimoles of solute

(a) 35.0 mL × 0.2 M = 7.0 millimol = 0.007 mol

(b) 175 mL × 0.5 M = 87.5 millimol = 0.0875 mol

Illustration 5 *How many millilitres of 0.45 M* $BaCl_2$ *solution contain* 15.0 g *of* $BaCl_2$ (208 *g* mol^{-1}) ?

Solution Let volume of the solution containing 15.0 g $BaCl_2$ be *V* mL

$$\therefore\quad V \times 0.45 = 0.45V\ \text{millimol} = 0.45V \times 10^{-3}\ \text{mol}$$

$$\therefore\quad 0.45V \times 10^{-3} = \frac{15}{208}$$

$$\therefore\quad V = \frac{15}{208} \times \frac{1000}{0.45} = \mathbf{160.3\ mL}$$

Illustration 6 *Calculate the molar concentration of a solution that is 50% in* NaOH (*w/W*) *and has a specific gravity of* 1.52.

Solution $\text{Specific gravity} = \dfrac{\text{mass}}{\text{volume}} = 1.52$

(w.r.t. H_2O = 1)

50% NaOH (*w/W*) means

100 g NaOH solution has = 50 g NaOH

$$\frac{100}{1.52}\ \text{mL NaOH solution has} = 50\ \text{g NaOH}$$

$$\therefore\quad \frac{1}{15.2}\ \text{L NaOH solution has} = \frac{50}{40} = 1.25\ \text{mol NaOH}$$

$$\therefore\quad \text{Molarity of NaOH} = 1.25 \times 15.2\ \text{M}$$
$$= \mathbf{19\ M}$$

Illustration 7 HCl *is 80% ionised in* 0.01 *M aqueous solution. What is its equilibrium molarity? What are other species present? What are their molar concentrations?*

Solution Initial [HCl] = 0.01 M

$$HCl \rightleftharpoons H^+ + Cl^-$$
$$H^+ + H_2O \rightleftharpoons H_3O^+$$

or
$$HCl + H_2O \rightleftharpoons H_3O^+ + Cl^-$$

HCl ionised = 80% of 0.01 M

$$= 0.01 \times \frac{80}{100} = 0.008\ \text{M}$$

Equilibrium molarity

$$[HCl] = 0.010 - 0.008 = \mathbf{0.002\ M}$$
$$[H_3O^+] = \mathbf{0.008\ M}$$
$$[Cl^-] = \mathbf{0.008\ M}$$

Illustration 8 *What is molarity of* K^+ *in aqueous solution that contains* 17.4 *ppm* K_2SO_4 (174 *g* mol^{-1})

Solution K_2SO_4 is 17.4 ppm *i.e.*,

$$10^6\ \text{g } (\approx \text{mL}) \text{ solution has} = 17.4\ \text{g } K_2SO_4$$

$$10^3\ \text{mL solution has} = \frac{17.4 \times 10^3}{10^6}\ \text{g } K_2SO_4$$
$$= 0.0174\ \text{g L}^{-1}\ K_2SO_4$$

$$\therefore\quad [K_2SO_4] = \frac{0.0174}{174} = 1 \times 10^{-4}\ \text{M}$$

$$K_2SO_4 \rightleftharpoons 2K^+ + SO_4^{2-}$$

$$\therefore\quad \text{Molarity of } K^+ = \mathbf{2 \times 10^{-4}\ M}$$

Illustration 9 *How many millimoles of solute are contained in*

(*a*) 2.00 *L of* 2.76×10^{-3} M $KMnO_4$?

(*b*) 250 *mL of a solution that contains* 4.20 *ppm* $CuSO_4$?

Solution (a) Since,

volume in mL × molarity = millimol

$$\therefore\quad 2.00\ \text{L} (= 2000\ \text{mL}) \text{ of } 2.76 \times 10^{-3}\ \text{M } KMnO_4$$
$$= 2.76 \times 10^{-3} \times 2000$$
$$= \mathbf{5.52\ millimol}$$

(b) 4.20 ppm means 10^6 mL solution has 4.20 g $CuSO_4$

$$= \frac{4.20}{159.5}\ \text{mol}$$

$$\text{hence,}\quad 250\ \text{mL solution has} = \frac{4.20}{159.5} \times \frac{250}{10^6}\ \text{mol } CuSO_4$$
$$= 6.58 \times 10^{-6}\ \text{mol}$$
$$= \mathbf{6.58 \times 10^{-3}\ millimol}$$

Molality (*m*)

- The **molality of a solution** is defined as the **number of moles** of solute per **kilogram of solvent** ($\mathbf{mol\ kg^{-1}}$)

$$\text{Molality } (m) = \frac{\text{moles of solute}}{\text{mass of solvent (kg)}}$$

- The main advantage of using molality is that it is **temperature independent** because masses do not change when substances are heated or cooled.
- The disadvantages of using molality are that amount of solutions are measured by mass rather than by volume and that the density of the solution must be known to convert molality into molarity.

DEEP Focus

- Molality (*m*) is related to molarity (*M*) by relation

$$m = \frac{1000M}{1000d - Mm_1}$$

where *d* is the density (g mL^{-1}) and m_1 is the molar mass of the solute

Because the density of water at 25°C is about 1 kilogram per litre, molality is approximately equal to molarity for dilute aqueous solutions at this temperature.

Remember ! This is only for water (solvent) and if solution is not dilute and at other temperature, this approximation is not true.

Thus, 18 M H_2SO_4 ($d = 1.8\ \text{mL}^{-1}$, $m_1 = 98\ \text{g mol}^{-1}$) is **500 molal.**

$$m = \frac{1000 \times 18}{(1000 \times 1.8 - 18 \times 98)} = 500$$

DEEP Focus

Molality (m) expressed in terms of mole fraction of solute (X_1) is given by relation:

$$m = \frac{1000 X_1}{(1 - X_1) m_2}$$

where m_2 is the molar mass of the solvent.

Illustration 1. *100 g solution of urea in water has 40 g urea (molar mass = 60 g mol^{-1}). What is molality of urea solution? What is mole fraction of urea in solution?*

Solution Mass of urea and water = 100 g

$$\text{mass of urea} = 40\ \text{g}$$

$$\text{moles of urea} = \frac{40}{60} = \frac{2}{3}$$

$$= 0.667\ \text{mol}$$

$$\text{mass of } H_2O = 60\ \text{g}$$

$$\text{moles of } H_2O = \frac{60}{18} = 3.333\ \text{mol}$$

$$\text{kg of solvent } (H_2O) = \frac{60}{1000} = 0.06\ \text{kg}$$

$$\text{molality } (m) = \frac{\text{moles of urea}}{\text{kg of } H_2O} = \frac{0.667}{0.06}$$

$$= \mathbf{11.12\ molal}$$

Total moles of urea and water = 3.333 + 0.667 = 4.00

$$\therefore \quad \text{mole fraction of urea} = \frac{0.667}{4.00} = \mathbf{0.167}$$

Alternatively

If molality of urea solution is **11.12 mol kg^{-1}** then using relation:

$$\text{molality } (m) = \frac{1000 X_1}{(1 - X_1)\, m_2}$$

$$11.12 = \frac{1000 X_1}{(1 - X_1) 18}$$

$$\frac{1 - X_1}{X_1} = \frac{1000}{11.12 \times 18} = 5$$

$$\frac{1}{X_1} = 1 + 5 = 6$$

$$X_1 = \frac{1}{6} = \mathbf{0.167}$$

Equivalent Mass and Normality

Equivalent mass

- Equivalent mass of a substance is the mass which can combine or displace. (For more details refer unit 2 and unit 3)

1 g hydrogen ≡ 8 g oxygen ≡ 35.5 g chlorine ≡ 80 g bromine ≡ 127 g iodine ≡ 12 g carbon ≡ 17 g OH^-

$$\underset{\substack{1\ \text{mol} \\ 28\ \text{g}}}{N_2} + \underset{\substack{3\ \text{mol} \\ 6\ \text{g}}}{3H_2} \longrightarrow \underset{\substack{2\ \text{mol} \\ 34\ \text{g}}}{2NH_3}$$

$$6\ \text{g}\ H_2 \equiv 28\ \text{g}\ N_2$$

$$1\ \text{g}\ H_2 \equiv \frac{28}{6}\ \text{g}\ N_2$$

$$\therefore \quad \text{equivalent mass of } N_2 = \frac{14}{3}\ \text{g equiv}^{-1}$$

Similarly, equivalent mass of

$$NH_3 = \frac{34}{6} = \frac{17}{3}\ \text{g equiv}^{-1}$$

- $$\underset{\substack{2\ \text{mol} \\ 4\ \text{g}}}{2H_2} + \underset{\substack{1\ \text{mol} \\ 32\ \text{g}}}{O_2} \longrightarrow \underset{\substack{2\ \text{mol} \\ 36\ \text{g}}}{2H_2O}$$

$$32\ \text{g}\ O_2 \equiv 36\ \text{g}\ H_2O$$

$$\therefore \quad 8\text{g}\ O_2 = 9\text{g}\ H_2O$$

Thus, equivalent mass of H_2O = 9 g

(determined w.r.t. oxygen)

Also, $4\text{g}\ H_2 \equiv 36\ \text{g}\ H_2O$

$1\text{g}\ H_2 \equiv 9\ \text{g}\ H_2O$

Thus, w.r.t. hydrogen equivalent mass of H_2O is also 9 g equiv^{-1}

- $$\text{Number of equivalent} = \frac{\text{mass}}{\text{equivalent mass}}$$

Normality (*N*) Concentration of the solution expressed in number of equivalent dissolved per litre of solution is called normality.

An equivalent is a measure of the reactive capacity of a given molecule. **Normality is the only concentration unit that is reaction dependent.** 1 M sulphuric acid is 2 N for acid-base reactions because each mole of sulphuric acid produce 2 moles of H^+ ions. On the other hand, 1 M sulphuric acid is 1 N for sulphate ion, since 1 mole sulphuric acid provides one mole of sulphate ion.

$$H_2SO_4 \rightleftharpoons 2H^+ + SO_4^{2-}$$

H_3PO_4 (*ortho* phosphoric acid) can be neutralised stepwise in three steps and depending upon the number of H^+ ions neutralised, normality would be different (say we take 1 M H_3PO_4)

$$H_3PO_4 + OH^- \longrightarrow H_2PO_4^- + H_2O$$

$$H^+ \text{ neutralised} = 1$$

$$\text{Equivalent mass} = \frac{\text{molar mass}}{1}$$

$$\text{Normality} = 1\,N$$

$$H_3PO_4 + 2OH^- \longrightarrow HPO_4^{2-} + 2H_2O$$

$$H^+ \text{ neutralised} = 2$$

$$\text{Equivalent mass} = \frac{\text{molar mass}}{2}$$

$$\text{Normality} = 2\,N$$

$$H_3PO_4 + 3OH^- \longrightarrow PO_4^{3-} + 3H_2O$$

$$H^+ \text{ neutralised} = 3$$

$$\text{Equivalent mass} = \frac{\text{molar mass}}{3}$$

$$\text{Normality} = 3\,N$$

$$\text{Normality}\,(N) = \frac{\text{number of equivalent}}{\text{litre of solution}}$$

If 0.40 g NaOH is dissolved in 100 mL (= 0.1 L) solution,

then, $\quad \text{number of equivalent of NaOH} = \frac{0.40}{40} = 0.01$

$\therefore \quad \text{normality of NaOH solution} = \frac{0.01}{0.10} = 0.1\,N$

Also, $NV = \text{milliequivalent}\,(x) = \frac{x}{1000}\,\text{equivalent.}$

$NE = \text{concentration in g L}^{-1}$.

Illustration 1 *Calculate equivalents of the underlined based on redox reactions / acid-base reactions.*

(*a*) $\underline{N_2H_4} + 2I_2 \longrightarrow N_2(g) + 4I^- + 4H^+$

(*b*) $\underline{5H_2O_2} + 2MnO_4^- + 6H^+ \longrightarrow 2Mn^{2+} + 5O_2 + 8H_2O$

(*c*) $\underline{B_4O_7^{2-}} + 2H^+ + 5H_2O \longrightarrow 4H_3BO_3$

(*d*) $\underline{S}(s) + CN^- \longrightarrow SCN^-$

$2SCN^- + 3IO_3^- + 2H^+ + 6Cl^- \longrightarrow 2SO_4^{2-} + 2CN^- + 3ICl_2^- + H_2O$

Solution (a) $N_2H_4 \longrightarrow N_2$

O.N. -4 $\qquad$ 0 (for two N-atoms)

change in O.N. = 4

$\therefore \quad 1 \text{ mol } N_2H_4 = 4 \text{ equivalents } N_2H_4$

(b) $H_2O_2 \longrightarrow O_2$

$-2 \qquad 0$ (for two O-atoms)

change in O.N. = 2

$\therefore \quad 1 \text{ mol } H_2O_2 \equiv 2 \text{ equivalents } H_2O_2$

(c) $\quad B_4O_7^{2-} \equiv 2H^+$

$\therefore \quad 1 \text{ mol } B_4O_7^{2-} \equiv 2 \text{ equivalents}$

(d) $\quad \underset{0}{S} \longrightarrow SCN^- \longrightarrow \underset{+6}{SO_4^{2-}}$

$1 \text{ mol S} = 6 \text{ equivalents of S}$

Illustration 2 *Determine equivalent mass of the underlined:*

(a) $\underline{H_3PO_4} + Ca(OH)_2 \longrightarrow CaHPO_4 + 2H_2O$

(b) $\underline{H_3PO_4} + 3NaOH \longrightarrow Na_3PO_4 + 3H_2O$

(c) $\underline{H_3PO_4} + NaOH \longrightarrow NaH_2PO_4 + H_2O$

(d) $2H_2 + \underline{O_2} \longrightarrow 2H_2O$

(e) $\underline{2CuSO_4} + 4KI \longrightarrow Cu_2I_2 + K_2SO_4 + I_2$

Solution (a) H_3PO_4 is a tribasic acid but in this case only two equivalent H^+ are neutralised by $Ca(OH)_2$.

$$H_3PO_4 \equiv Ca(OH)_2 \equiv 2OH^-$$

$$\therefore \quad 1OH^- \equiv \frac{H_3PO_4}{2}$$

Thus, equivalent mass of $H_3PO_4 = \frac{\text{molar mass}}{2} = \frac{98}{2}$

$= \mathbf{49\ g\ equiv^{-1}}$

(b) Similarly, $H_3PO_4 \equiv 3NaOH \equiv 3OH^-$

$$\frac{H_3PO_4}{3} = 1\,OH^-$$

equivalent mass of $H_3PO_4 = \frac{98}{3}$ **g equiv^{-1}**

(c) $\quad H_3PO_4 \equiv NaOH \equiv 1\,OH^-$

equivalent mass of H_3PO_4 = **98 g equiv^{-1}**

(d) $\quad 2H_2 \equiv O_2$

$4H \equiv O_2$

$1H \equiv \frac{O_2}{4}$

$\therefore$ equivalent mass of $O_2 = \frac{\text{molar mass}}{4}$ = **8 g equiv^{-1}**

(e) $\quad 2CuSO_4 \equiv I_2 \equiv 2I$

$CuSO_4 \equiv I$

equivalent mass of $CuSO_4$ = molar mass

= **159.5 g equiv^{-1}**

Illustration 3 *At a given condition,* 1 *mole of* O_2 *occupies volume* 30 L. *What is volume occupied by* 1 *equivalent of* O_2?

Solution Molar mass of O_2 (= 1 mol) = 32 g

then equivalent mass of O_2 = 8 g

hence, $\quad 1 \text{ mol } O_2 = 4 \text{ equivalent } O_2$

If $\quad 1 \text{ mol } O_2$ occupies = 30 L

then $\quad 1 \text{ equivalent } O_2$ occupies $= \frac{30}{4}$ = **7.5 L**

Illustration 4 0.024 *g of* Mg *on burning in oxygen gave* 0.40 *g of* MgO. *What is equivalent mass of* Mg?

Solution Mass of Mg = 0.24 g, mass of MgO = 0.40 g, oxygen combined = 0.16 g

$\therefore \quad$ 0.16 g oxygen ≡ 0.24 g Mg

$\therefore \quad$ 8 g oxygen ≡ 12 g Mg

Thus, equivalent mass of Mg = **12 g equiv^{-1}**

Illustration 5 *When 6 g of a monobasic acid is dissolved in 100 mL solution, it is 1 N. What is equivalent mass of the acid?*

Solution $$\text{Normality} = \frac{\text{equivalent of substance}}{\text{litre of solution}}$$

$$1\text{ N} = \frac{6}{E}\Big/0.1$$

$$\frac{6}{E} = 0.1$$

$$E = \mathbf{60\ g\ equiv^{-1}}$$

Illustration 6 *Oxalic acid, $H_2C_2O_4$, forms two series of salt $HC_2O_4^-$ and $C_2O_4^{2-}$. 1.5 g of $H_2C_2O_4$ is dissolved in 250 mL solution. Determine its normality at each stage of salt formation.*

Solution $$H_2C_2O_4 \rightleftharpoons H^+ + HC_2O_4^-$$

$$H_2C_2O_4 \equiv 1\text{ H}$$

Hence, equivalent mass of

$$H_2C_2O_4 = \text{molar mass} = 90\text{ g equiv}^{-1}$$

$$N = \frac{\text{equivalent}}{\text{volume (L)}} = \frac{1.5/90}{0.250} = \mathbf{0.067\ N}$$

$$H_2C_2O_4 \rightleftharpoons 2H^+ + C_2O_4^{2-}$$

$$H_2C_2O_4 \equiv 2\text{H}$$

∴ Equivalent mass of

$$H_2C_2O_4 = \frac{\text{molar mass}}{2} = 45\text{ g equiv}^{-1}$$

$$N = \frac{1.5/45}{0.250} = \mathbf{0.133\ N}$$

Illustration 7 *10 g of MnO_2 on reaction with conc. HCl liberated 0.1 equivalent of Cl_2. What is per cent purity of MnO_2? (Mn = 55, O = 16, Cl = 35.5)*

$$MnO_2 + 4HCl \longrightarrow MnCl_2 + Cl_2 + 2H_2O$$

Solution $$\underset{87\text{ g}}{MnO_2} \equiv \underset{71\text{ g}}{Cl_2}$$

$$MnO_2 \equiv Cl_2 \equiv 2Cl$$

$$\therefore \quad 1\text{ Cl} = \frac{MnO_2}{2} = \frac{87}{2} = 43.5\text{ g}$$

$$1\text{ equivalent of } Cl_2 = 43.5\text{ g } MnO_2$$

$$0.1\text{ equivalent of } Cl_2 = 4.35\text{g } MnO_2 \text{ (pure)}$$

But, MnO_2 taken = 10 g

$$\therefore \text{ percentage of pure } MnO_2 = \frac{4.35}{10} \times 100 = \mathbf{43.5\%}$$

Target Practice 1

1. How many grams of Na^+ are contained is 25 mL of 1 M Na_2SO_4 ....
2. How many moles and millimoles of benzoic acid are contained in 10 mL of 0.5 M benzoic acid ....
3. Calculate molar concentration of ethanol (C_2H_5OH) in an aqueous solution that contains 2.3 g of C_2H_5OH in 3.5 L solution ....
4. Describe preparation of 2.00 L of 0.108 M $BaCl_2$ from $BaCl_2 \cdot 2H_2O$ (Ba = 137) ....
5. Describe the preparation of 500 mL of 0.01 M Cl^- solution from solid $BaCl_2 \cdot 2H_2O$ (Ba = 137) ....
6. Calculate the resulting molarity when 145 mL of 6.00 M H_2SO_4 is mixed with 245 mL of 3.00 M H_2SO_4 ....
7. A solution contains 0.10 mole per litre of each of the following acids : HCl, H_2SO_4, H_3PO_4.
   (a) Is the molarity the same for each acid? ....
   (b) Is the number of molecules per litre the same for each acid? ....
   (c) Is the normality the same for each acid? ....
8. Calculate the volume of 0.0600 M $Ba(OH)_2$ solution that contains the same number of moles of $Ba(OH)_2$ as 180 mL of 0.09 M $Ba(OH)_2$ solution? ....
9. Calculate the volume of 4.00 M NaOH solution required to prepare 200 mL of a 0.750 M solution of NaOH ....
10. Complete the table for the following aqueous solution

	Solute	Mass solute (g)	Volume (litres)	Molarity
(a)	HCl	0.365	0.01	(I) ...
(b)	H_2SO_4	0.98	(II) ...	1.0
(c)	H_3PO_4	(III) ...	1.00	0.1
(d)	NaOH	(IV) ...	2.00	0.25
(e)	$Ca(OH)_2$	7.40	1.00	(V) ...

11. What is normality if molarity of the underlined solute is 1 M?
   (a) $\underline{H_3PO_4} + Ca(OH)_2 \longrightarrow CaHPO_4 + 2H_2O$ ....
   (b) $\underline{H_3PO_3} + NaOH \longrightarrow NaH_2PO_3 + H_2O$ ....
   (c) $\underline{H_3PO_2} + Ca(OH)_2 \longrightarrow Ca(OH)H_2PO_2 + H_2O$ ....
   (d) $\underline{(NH_4)_2SO_4} + 2NaOH \longrightarrow Na_2SO_4 + 2NH_3 + 2H_2O$ ....
   (e) $\underline{Ba(OH)_2} + 2HCl \longrightarrow BaCl_2 + 2H_2O$ ....

Answers

1. $Na_2SO_4 \rightleftharpoons 2Na^+ + SO_4^{2-}$

One mole furnishes two moles of Na^+ ions

millimoles = volume (mL) × molarity

= 25 × 1

= 25 millimol of Na_2SO_4

= 50 millimol of Na^+

$$\therefore \quad \text{Moles of } Na^+ = \frac{50}{1000} = 0.050$$

Mole × molar mass = 0.050 × 23 = 1.15 g Na^+

2. Benzoic acid = 10×0.5 millimol

$= 5$ millimol $= 0.005$ mol

3. $\text{Molarity} = \dfrac{\text{moles}}{\text{volume in litre}}$

$2.3 \text{ g } C_2H_5OH = \dfrac{2.3}{46} = 0.05 \text{ mol}$

$\therefore \quad \text{Molarity} = \dfrac{0.05}{3.5} = 0.0143 \text{ mol L}^{-1}$

4. $BaCl_2 \cdot 2H_2O = 2 \times 0.108$

$= 0.216$ mol

$= 0.216 \times 244$ g

$= 52.704$ g

We dissolve 52.704 g $BaCl_2 \cdot 2H_2O$ in 2.00 L solution

5. $0.01 \text{ M } Cl^- = \dfrac{0.01}{2} \text{ M } BaCl_2 \cdot 2H_2O$

$= 0.005 \text{ M } BaCl_2 \cdot 2H_2O$

$= 0.005 \text{ mol L}^{-1} \; BaCl_2 \cdot 2H_2O$

$= 0.0025$ mol in 500 mL

Dissolve 0.0025 mol (= 0.61 g)$BaCl_2 \cdot 2H_2O$ in 500 mL solution.

6. $\text{Resulting molarity} = \dfrac{\Sigma MV}{\Sigma V}$

$$= \frac{M_1V_1 + M_2V_2}{V_1 + V_2}$$

$$= \frac{6 \times 145 + 3 \times 245}{145 + 245}$$

$= \mathbf{4.0125 \text{ M}}$

7. $\text{Molarity} = \dfrac{\text{Mol}}{\text{L}}$

	HCl	H_2SO_4	H_3PO_4	
(a)	0.1 M	0.1 M	0.1 M	Molarity same-Yes
(b)	$0.1 \times N_0$	$0.1 \times N_0$	$0.1 \times N_0$	Molecules same-Yes
(c)	0.1 N	0.2 N	0.3 N	Normality same-No

(N_0 = Avogadro's number)

8. It is a case of dilution

$M_1V_1 = M_2V_2$

$0.06 \times V_1 = 0.09 \times 180$

$V_1 = \mathbf{270 \text{ mL}}$

9. It is a case of dilution

$M_1V_1 = M_2V_2$

$4 \times V_1 = 0.75 \times 200$

$V_1 = \mathbf{37.5 \text{ mL}}$

10. I — 1 M

II — 0.01 L

III — 9.8 g

IV — 20.0 g

V — 0.1 M

11.

		Normality
(a) Two H^+ ions are neutralised	—	2 N
(b) One H^+ ion is neutralised	—	1N
(c) One H^+ ion is neutralised	—	1 N
(d) One mole $(NH_4)_2SO_4$ is neutralised by two (OH^-) ions	—	2 N
(e) One mole $Ba(OH)_2$ is neutralised by two HCl	—	2 N

Practice Exercise 1

1. How many equivalents of hydrogen are in 1 equivalent of H_3PO_4?

2. 0.400 g of iron (III) sulphate are present in 100 mL of solution. Express concentration of iron (III) and sulphate ions in terms of

(i) ppm (ii) molarity

3. H_2SO_4 ($d = 1.80 \text{ g mL}^{-1}$) is 98% by weight of solution. Express concentration in terms of molarity of

(i) hydrogen ion, (ii) sulphate ion

4. MCl_x is 100% ionised in 0.01 M aqueous solution. If $[Cl^-] = 0.03$ M, what is the formula? Also calculate $[M^{x+}]$ and $[MCl_x]$.

5. A solution of Na_2SO_4 is 50% ionised. What is analytical molarity of Na_2SO_4 if $[Na^+] = 1.0$ M?

6. 6 g of an urea sample (molar mass = 60 g mol^{-1}) was dissolved in 100 mL aqueous solution and was 0.8 M. Comment.

7. The same amount of a metal combines with 0.4 g of oxygen and 4 g of a halogen. Determine equivalent mass of halogen.

8. (a) The equivalent mass of a metal is double that of oxygen. How many times is the mass of its oxide greater than the mass of the metal?

(b) 1.575 g of a diabasic acid $H_2C_2O_4 \cdot xH_2O$ in 250 mL solution is 0.1 N. What is the value of x?

9. Urea solution is 20% by weight of solution. What is percentage concentration by weight of solvent?

10. Calculate normality of:

(a) 0.01 M $Ba(OH)_2$ (b) 0.01 M H_3PO_3

(c) 0.01 M H_3PO_2 (d) 0.01 M H_3PO_4 solutions.

11. What is weight of sucrose in 1 kg solution of 1 molal concentration?

12. One litre of 1 M solution of following are to be prepared. Which has maximum weight of solute?

(Mn = 55, K = 39, Cr = 52, Fe = 56)

(a) $KMnO_4$ (b) $K_2Cr_2O_7$

(c) $FeSO_4 \cdot (NH_4)_2SO_4 \cdot 6H_2O$

13. 2.78 g $FeSO_4 \cdot xH_2O$ in 100 mL solution gives 0.1 M solution. What is the value of x?

14. CH_3CH_2OH solution in water is 46% by weight of solution. What is mole fraction of each component? What is molality as water solvent? What is molality as ethanol solvent?

15. Calculate volume of water that must be added to 250 mL of 1.25 N solution to make it 0.500 N.

16. What volumes of 12.0 N and 3.00 N HCl must be mixed to give 1.00 L of 6.00 N HCl?

Answers

1. 1 equivalent of H_3PO_4 has 1 equivalent of H^+ (as well as 1 equivalents of PO_4^{3-})

2. $Fe_2(SO_4)_3 \longrightarrow \underset{\text{iron (III)}}{2Fe^{3+}} + \underset{\text{sulphate}}{3SO_4^{2-}}$

Molar mass of $Fe_2(SO_4)_3 = 400 \text{ g mol}^{-1}$

Moles of $Fe_2(SO_4)_3 = \dfrac{0.400}{400} = 0.001$ mol

Volume of solution = 100 mL = 0.1 L

$\therefore$ Molarity of $Fe_2(SO_4)_3 = \dfrac{0.001}{0.1} = \mathbf{0.01\ M}$

(i) $Fe^{3+} = 0.001 \times 2 = 0.002 \text{ mol} = 0.002 \times 56$ g

$= 0.112$ g in 100 mL

$= \dfrac{0.112 \times 10^6}{100}$ g in 10^6 mL

$= 1120$ ppm

$SO_4^{2-} = 0.001 \times 3$ mol

$= 0.001 \times 3 \times 96$ g in 100 mL

$= \dfrac{0.001 \times 3 \times 96 \times 10^6}{100}$ g in 10^6 mL

$= 2880$ ppm

(ii) Molarity : $[Fe^{3+}] = 0.01 \times 2 = \mathbf{0.02\ M}$

$[SO_4^{2-}] = 0.01 \times 3 = \mathbf{0.03\ M}$

3. H_2SO_4 = 98% by weight of solution

= 98 g in 100 g solution

$= 98$ g in $\dfrac{100}{1.8}$ mL solution $\quad \left(\text{Volume} = \dfrac{\text{mass}}{\text{density}}\right)$

$\therefore$ Moles of $H_2SO_4 = \dfrac{98}{98} = 1$ mol

$\text{Volume} = \dfrac{100}{1.8 \times 1000} \text{ L} = \dfrac{1}{18} \text{ L}$

$\therefore$ Molarity of $H_2SO_4 = \dfrac{\text{mole}}{\text{litre}} = \dfrac{1}{1/18} = 18$ M

$\therefore \quad H_2SO_4 \rightleftharpoons 2H^+ + SO_4^{2-}$

$[H^+] = 36$ M

$[SO_4^{2-}] = 18$ M

4. $\underset{0.01\text{ M}}{MCl_x} \rightleftharpoons \underset{0.01\text{ M}}{M^{x+}} + \underset{0.01x\text{ M}}{xCl^-}$

There is 100% ionisation

$[Cl^-]$ in $\quad MCl_x = 0.03$ M

$\therefore \quad 0.01x = 0.03$

$x = 3$

Thus, solute is MCl_3

$\therefore \quad [M^{3+}] = 0.01$ M

$[MCl_3] = 0.00$ M $\quad$ (none is left)

5. $Na_2SO_4 \rightleftharpoons 2Na^+ + SO_4^{2-}$

$[Na_2SO_4] = \dfrac{[Na^+]}{2}$ if 100% ionised

$= [Na^+]$ if 50% ionised

$= 1$ M

6. Moles of urea $= \dfrac{6}{60} = 0.1$ mol

Volume = 100 mL = 0.1 L

$\text{Molarity} = \dfrac{\text{mole}}{\text{litre}} = \dfrac{0.1}{0.1} = 1 \text{ M}$

But it is 0.8 M (< 1 M)

Thus, urea is 80% pure.

7. Equivalent of oxygen = Equivalent of halogen

$\dfrac{0.4}{8} = \dfrac{4}{E}$

$E = \dfrac{4 \times 8}{0.4} = 80 \text{ g equivalent}^{-1}$

8. (a) Equivalent mass of oxygen = 8

Equivalent mass of metal = 16

Equivalent mass of metal oxide = 24

Thus, $\dfrac{\text{equivalent mass of metal oxide}}{\text{equivalent mass of metal}} = \dfrac{24}{16} = \mathbf{1.5}$

(b) Molar mass of $H_2C_2O_4 \cdot xH_2O = (90 + 18x)$

Moles of $H_2C_2O_4 \cdot xH_2O = \dfrac{1.575}{(90 + 18x)}$

Volume of solution = 250 mL = 0.250 L

$\text{Molarity} = \dfrac{\text{mole}}{\text{litre}} = \dfrac{\dfrac{1.575}{(90+18x)}}{0.250} = \dfrac{6.3}{(90 + 18x)}$

Being a dibasic acid

$\text{normality} = 2 \times \text{molarity} = \dfrac{2 \times 6.3}{(90 + 18x)}$

$= 0.1$ N (given)

$\dfrac{2 \times 6.3}{(90 + 18x)} = 0.1$

$\therefore \quad x = 2$

9. $w_1 = 20$ g solute

w_2 = mass of solvent

$w_1 + w_2 = 100$ g

$\therefore \quad w_2 = 80$ g

$\therefore \quad$ % by weight of solvent $= \dfrac{20}{80} \times 100 = 25\%$

10. (a) 0.02 N $\quad [Ba(OH)_2 \rightleftharpoons Ba^{2+} + 2OH^-]$

(b) 0.02 N $\quad$ (H_3PO_3 has only two ionisable H^+ ions)

(c) 0.01 N $\quad$ (H_3PO_2 has only one ionisable H^+ ion)

(d) 0.03 N $\quad$ (H_3PO_4 has three ionisable H^+ ions)

Focus Hydrogen ion attached to oxygen atom in oxyacids is ionisable. This decides basicity of an acid.

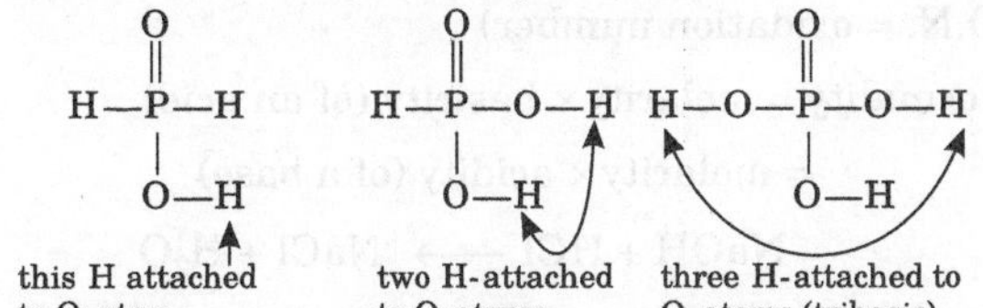

11. 1 molal sucrose means 1 mol sucrose (= 342 g) in 1 kg water.

Total weight of solution = 1000 + 342

= 1342 g

1342 g solution has sucrose = 342 g

Thus, 1000 g solution has sucrose $= \frac{342}{1342} \times 1000 = \mathbf{254.84\ g}$

12.

Solute	Molecular weight	Moles in one litre of 1 M solution
$KMnO_4$	158 g mol^{-1}	1 mol
$K_2Cr_2O_7$	294 g mol^{-1}	1 mol
$FeSO_4 \cdot (NH_4)_2SO_4 \cdot 6H_2O$	392 g mol^{-1}	1 mol

Thus, maximum weight is of $\mathbf{FeSO_4 \cdot (NH_4)_2SO_4 \cdot 6H_2O}$

13. Molar mass of $FeSO_4 \cdot xH_2O = 152 + 18x$

Moles of $FeSO_4 \cdot xH_2O = \frac{2.78}{(152 + 184)}$

Volume = 100 mL = 0.1 L

Molarity $= \frac{2.78/(152 + 18x)}{0.1}$

Thus, $\frac{2.78}{152 + 18x} \Big/ 0.1 = 0.1$ (given)

$\therefore \quad x = 7$

14. CH_3CH_2OH = 46% by weight of solution

Thus, CH_3CH_2OH = 46 g = 1 mol = 0.046 kg

H_2O = 54 g = 3 mol = 0.054 kg

Total moles = 4

Mole fraction of $CH_3CH_2OH = X_{CH_3CH_2OH} = \frac{1}{4}$

Mole fraction of $H_2O = X_{H_2O} = \frac{3}{4}$

Molality as water solvent $= \frac{\text{mole of solute } (CH_3CH_2OH)}{\text{kg of solvent } (H_2O)}$

$= \frac{1}{0.054}$

$= \mathbf{18.52\ molal}$

Molality as ethanol (CH_3CH_2OH) solvent

$= \frac{\text{mole of solute } (H_2O)}{\text{kg of solvent } (CH_3CH_2OH)}$

$= \frac{3}{0.046}$

$= \mathbf{65.22\ molal}$

15. It is a case of dilution :

$N_1V_1 = N_2V_2$

$0.5 \times V_1 = 1.25 \times 250$

$V_1 = 625$ mL

Thus, additional water to be added = 625 − 250

= **375 mL**

16. If volume of 12.00 N HCl = x L

then volume of 3.00 N = $(1 - x)$ L

Final normality = 6.00 N

$N \text{ (final)} = \frac{N_1V_1 + N_2V_2}{V_1 + V_2}$

$6 = \frac{12x + 3(1 - x)}{1}$

$x = 0.333$ L

Thus, volume of 12 N HCl = 0.333 L

and volume of 3 N HCl = 0.667 L

Volumetric Titrations

While diluting a solution from one concentration to another, we can use

$$M_1V_1 = M_2V_2$$

and

$$N_1V_1 = N_2V_2$$

(concentrated solution) (diluted solution)

- For a redox reaction or neutralisation reaction

  $N_1V_1 = N_2V_2$ is always true

  but $M_1V_1 = M_2V_2$ is not always true
- Normality = change in O.N. × molarity in a redox reaction

  (O.N. = oxidation number)
- Normality = molarity × basicity (of an acid)

  = molarity × acidity (of a base)

$$NaOH + HCl \longrightarrow NaCl + H_2O$$

Here, $N_1V_1 = N_2V_2$

and also $M_1V_1 = M_2V_2$

$$2NaOH + H_2SO_4 \longrightarrow Na_2SO_4 + 2H_2O$$

Here, $N_1V_1(NaOH) = N_2V_2(H_2SO_4)$

but $M_1V_1(NaOH) = 2M_2V_2(H_2SO_4)$

$$I_2 + 2Na_2S_2O_3 \longrightarrow Na_2S_4O_6 + 2NaI$$

$N_1V_1(I_2) = N_2V_2(Na_2S_2O_3)$

but $2M_1V_1(I_2) = M_2V_2(Na_2S_2O_3)$

$$MnO_4^- + 5Fe^{2+} + 8H^+ \longrightarrow Mn^{2+} + 4H_2O + 5Fe^{3+}$$

Here, $N_1V_1(Fe^{2+}) = N_2V_2(MnO_4^-)$

but $M_1V_1(Fe^{2+}) = 5M_2V_2(MnO_4^-)$

DEEP Focus

Thus, it is always better to convert molarity factor into normality factor by multiplying it with change in O.N. or acidity or basicity.

$(MnO_4^- \longrightarrow Mn^{2+})$, 1 M MnO_4^- = 5 N MnO_4^-

$(Fe^{2+} \longrightarrow Fe^{3+})$, 1 M Fe^{2+} = 1 N Fe^{2+}

$(Cr_2O_7^{2-} \longrightarrow 2Cr^{3+})$, 1 M $Cr_2O_7^{2-}$ = 6 N $Cr_2O_7^{2-}$

$(H_2SO_4 \longrightarrow 2H^+ + SO_4^{2-})$, 1 M H_2SO_4 = 2 N H_2SO_4

- Normality of a mixture of two or more acids or bases (strong)

$$N_{mix} = \frac{N_1V_1 + N_2V_2 + \dots}{V_1 + V_2 + \dots} = \frac{\Sigma NV}{\Sigma V}$$

where $N_1, N_2, \dots$ are normalities and $V_1, V_2 \dots$ are volumes of acids or bases.

- **Normality of mixture of acid and bases (strong)**
  let N_1V_1 = milliequivalent of acid
  N_2V_2 = milliequivalent of base
  - If $N_1V_1 = N_2V_2$ mixture is **neutral**
  - If $N_1V_1 > N_2V_2$, then mixture is **acidic** and resultant normality of acidic solution is:

$$N(\text{resultant}) = \frac{N_1V_1 - N_2V_2}{V_1 + V_2}$$

  - If $N_1V_1 < N_2V_2$, then mixture is **basic** and resultant normality of basic solution is:

$$N(\text{resultant}) = \frac{N_2V_2 - N_1V_1}{V_1 + V_2}$$

Acid-Base Titrations

Acid or acid mixture can be titrated against a suitable base or *vice-versa* using a suitable indicator. These are summarised in Table 7.3

Table 7.3 **Acid-Base Titrations**

Combination	Suitable indicator	Colour change	
		in acid	in alkali
1. Strong acid/ strong base	Methyl orange Bromothymol blue	Red Yellow	Yellow Blue
2. Strong acid/ weak base	Methyl red	Red	Yellow
3. Weak acid/ strong base	Phenolphthalein	Colourless	Purple
4. Weak acid/ weak base	Not suitable for a titration		

- In all cases : equivalent of an acid ≡ equivalent of a base
- When neutralisation point is reached $N_1V_1 = N_2V_2$

End point is the point of completion of the reaction indicated by suitable indicator. Hence, *it has additional drop of titrating reagent* but, we use

$$N_1V_1 = N_2V_2$$

$$NaOH + HCl \longrightarrow NaCl + H_2O$$

$$\begin{aligned}\text{1 equivalent of NaOH} &\equiv \text{1 equivalent of HCl}\\ &\equiv \text{1 mol of HCl}\\ &\equiv \text{1 equivalent of } H_2SO_4\\ &\equiv \text{0.5 mol of } H_2SO_4\end{aligned}$$

$$Na_2CO_3 + 2HCl \longrightarrow 2NaCl + H_2O + CO_2$$

$$Na_2CO_3 + H_2SO_4 \longrightarrow Na_2SO_4 + H_2O + CO_2$$

$$\begin{aligned}\text{1 equivalent of } Na_2CO_3 &\equiv \text{0.5 mol of } Na_2CO_3\\ &= \text{1 equivalent of HCl}\\ &\equiv \text{1 mol of HCl}\\ &\equiv \text{1 equivalent of } H_2SO_4\\ &\equiv \text{0.5 mol of } H_2SO_4\end{aligned}$$

Titration of Mixture of Bases with Two Indicators

Every indicator has a working range

Indicator	pH range	Behaving as
Phenolphthalein	8–10	weak organic acid
Methyl orange	3–4.4	weak organic base

Thus, methyl orange with lower pH range can indicate complete neutralisation of all types of bases.

Extent of reaction of different bases with acid (HCl) using these two indicators is summarised below :

Mixture	Phenolphthalein	Methyl orange
NaOH	100% reaction is indicated $NaOH + HCl \longrightarrow NaCl + H_2O$	100% reaction is indicated, $NaOH + HCl \longrightarrow NaCl + H_2O$
Na_2CO_3	50% reaction up to $NaHCO_3$ stage is indicated $Na_2CO_3 + HCl \longrightarrow NaHCO_3 + NaCl$	100% reaction is indicated, $Na_2CO_3 + 2HCl \longrightarrow 2NaCl + H_2O + CO_2$
$NaHCO_3$	No reaction is indicated	100% reaction is indicated $NaHCO_3 + HCl \longrightarrow NaCl + H_2O + CO_2$

Suppose volume of given standard acid solution (say HCl) required

for complete reaction of $Na_2CO_3 = x$ mL

for complete reaction of $NaHCO_3 = y$ mL

for complete reaction of $NaOH = z$ mL

There may be different combination of mixture of bases. We may adopt two methods:

Method I: We carry two titrations separately with two different indicators.

Method II: We carry single titration but adding second indicator after first end-point is reached.

These have been summarised in Table 7.4

Table 7.4 **Results with Two Indicators**

Mixture	Method I: Volume of HCl used with indicator		Method II: Volume of HCl used with indicator	
	Phenolphthalein	Methyl orange	Phenolphthalein	Methyl orange added after first end-point is reached
1. NaOH + Na_2CO_3	$\left(z+\frac{x}{2}\right)$ ↑ ↑ 100% 50%	$(x + z)$ 100% each	$\left(z+\frac{x}{2}\right)$ ↑ ↑ 100% 50%	$\frac{x}{2}$ (remaining 50% Na_2CO_3 is indicated)
2. NaOH + $NaHCO_3$	$z + 0$ 100% no change	$(y + z)$ 100% each	$(z + 0)$	y (remaining 100% $NaHCO_3$ is indicated)
3. Na_2CO_3 + $NaHCO_3$	$\frac{x}{2} + 0$ ↑ ↑ 50% no change	$(x + y)$ 100% each	$\left(\frac{x}{2}+0\right)$	$\left(\frac{x}{2}+y\right)$ (remaining 50% of Na_2CO_3 and 100% of $NaHCO_3$ are indicated.)

- Bromocresol-green also indicates complete reaction of Na_2CO_3, $NaHCO_3$ and NaOH, thus its behaviour is similar to methyl orange.
- Bromocresol can also be used as an indicator for the determination of phosphates when reaction of acid (say HCl) up to $H_2PO_4^-$ is indicated.

$$PO_4^{3-} + 2H^+ \longrightarrow H_2PO_4^-$$

- Thymolphthalein indicates reaction upto the formation of HPO_4^{2-}.

$$PO_4^{3-} + H^+ \longrightarrow HPO_4^{2-}$$

Illustration 1 *Calculate the molar concentration of a dilute HCl solution if*

(a) *titration of a 50 mL of* 0.04 *M* $Ba(OH)_2$ *required* 20 *mL of the acid.*

(b) 50 *mL of* HCl *yielded* 0.6010 *g* AgCl

Solution (a) $Ba(OH)_2 + 2HCl \longrightarrow BaCl_2 + 2H_2O$

$$0.04\ M\ Ba(OH)_2 = 0.08N\ Ba(OH)_2$$

$$\underset{HCl}{N_1V_1} = \underset{Ba(OH)_2}{N_2V_2}$$

$$N_1 \times 20 = 0.08 \times 50$$

$$N_1(HCl) = 0.20\ N$$

∴ molar concentration = **0.20 M**

(b) $$\text{Ag}^+ + \underset{\text{(from HCl)}}{\text{Cl}^-} \longrightarrow \text{AgCl}$$

1 mol   1 mol = 143.5 g

?   0.6010 g

143.5 g AgCl is obtained from = 1 mol HCl

∴ 0.6010 g AgCl is obtained from $= \frac{0.6010}{143.5}$ mol HCl

$= 4.188 \times 10^{-3}$ mol HCl

This amount of HCl is contained in 50 mL solution (= 0.050 L)

∴ molar concentration of HCl $= \frac{4.188 \times 10^{-3}}{0.050}$

$=$ **0.08376 M**

Illustration 2 *A* 1.00 *g sample of a metal X (that is known to form* X^{2+} *ions) was added to* 0.10 *L of* 0.50 *M* H_2SO_4*. After all the metal had reacted, the remaining acid required* 0.0334 *L of* 0.50 *M* NaOH *solution for neutralisation. Calculate molar mass of the metal and identify the element.*

Solution $X + H_2SO_4 \longrightarrow XSO_4 + H_2$

or $X + 2H^+ \longrightarrow X^{2+} + H_2$

$$2H^+ + 2OH^- \longrightarrow 2H_2O$$

1 mol H_2SO_4 ≡ 1 mol of X = molar mass of X

Let unused acid = V L

then V L of 0.5 M H_2SO_4 ≡ 0.0334 L of 0.5 M NaOH

or V L of 1 N H_2SO_4 ≡ 0.0334 L of 0.5 N NaOH

∴ V = 0.0167 L of H_2SO_4

H_2SO_4 used = 0.10 – 0.0167 = 0.0833 L

Number of moles of H_2SO_4 = 0.0833 × 0.5 mol

≡ 1 g metal X

$\therefore$ $$1 \text{ mol of } H_2SO_4 = \frac{1}{0.0833 \times 0.5} \text{ g metal } X$$
$$= \mathbf{24.01 \text{ g mol}^{-1}}$$

Metal is **Mg**.

Note : This is a case of **back-titration.**

Kjeldahl Method

Kjeldahl method is used to determine amount of N-containing compounds as NH_3, NH_2CONH_2, NH_4Cl, etc. Compound is decomposed by alkali (NaOH, $Ca(OH)_2$, etc.) and NH_3 formed is absorbed by excess of standard HCl or H_2SO_4 solution. Unreacted acid is determined by **back-titration.**

If V is the volume of acid or normality (N) required to absorb NH_3 from w g of the compound then,

$$NH_3 \equiv H^+$$

$$\text{Equivalent of acid} \equiv \text{equivalent of } NH_3 = \frac{NV}{1000}$$

$$NH_3 \text{ present} = \frac{17NV}{1000} \text{ g}$$

$$w \text{ g of the compound has} = \frac{17NV}{1000} \text{ g } NH_3$$

$$100 \text{ g of the compound has} = \frac{17NV \times 100}{1000w} NH_3$$

$$\% \text{ of } NH_3 = \frac{1.7\ NV}{w}$$

$$\% \text{ in terms of N} = \frac{1.4\ NV}{w} \quad \text{since } N \equiv NH_3 \equiv H^+$$

$$\% \text{ in terms of } NH_4Cl = \frac{5.35\ NV}{w} \quad \text{since } NH_4Cl \equiv NH_3 \equiv H^+$$

$$\% \text{ in terms of } (NH_4)_2SO_4 = \frac{6.6\ NV}{w}$$

$$\text{since } (NH_4)_2SO_4 \equiv 2NH_3 \equiv 2H^+$$

$$\% \text{ in terms of } NH_2CONH_2 = \frac{3.0\ NV}{w}$$

$$\text{since } NH_2CONH_2 \equiv 2NH_3 \equiv 2H^+$$

Illustration 3 *A 1.047 g sample of canned tuna was analysed by the Kjeldahl method; 24.61 mL of 0.1181 M HCl were required to titrate the liberated ammonia. Calculate the percentage of nitrogen in the sample.*

Solution $\text{N(in the sample)} \longrightarrow NH_3 \xrightarrow{HCl} NH_4Cl$

$$\text{moles of HCl} \equiv \text{moles of } NH_3 \equiv \text{moles of N}$$
$$= \frac{24.61 \times 0.1180}{1000}$$
$$= 2.9064 \times 10^{-3}$$

$$\therefore \text{ N in the given sample} = 2.9064 \times 10^{-3} \text{ mol}$$
$$= 2.9064 \times 10^{-3} \times 14 \text{ g}$$
$$= 0.0407 \text{ g}$$

$$\% \text{ of nitrogen} = \frac{0.0407}{1.047} \times 100 = \mathbf{3.89\%}$$

Note: $\% \text{ of nitrogen} = \frac{1.4\ NV}{w} = \frac{1.4 \times 0.1181 \times 24.61}{1.047}$

where, N = normality of acid

V = volume of acid required to neutralise NH_3

w = weight of sample

Illustration 4 *A 0.9092 g sample of a wheat flour was analysed by the Kjeldahl procedure. The ammonia formed was distilled into 50 mL of 0.05 M HCl. A 7.50 mL of 0.05 M NaOH was required in back titration. Calculate percentage of nitrogen in the flour.*

Solution Nitrogen content (protein) of the flour was converted into NH_3 which was absorbed into HCl.

$$N \longrightarrow NH_3 \xrightarrow{HCl} NH_4Cl$$

Let unreacted HCl be = V mL

$$V \text{ mL of 0.05 HCl} \equiv 7.50 \text{ mL of 0.05 M NaOH}$$
$$\therefore \quad V = 7.50 \text{ mL}$$
$$\text{used HCl} = 50.00 - 7.50 = 42.50 \text{ mL}$$

$$\text{moles of HCl} = \text{moles of } NH_3 = \text{moles of N} = \frac{42.50 \times 0.05}{1000}$$
$$= 2.125 \times 10^{-3} \text{ mol}$$

$$\therefore \text{ nitrogen in the flour sample} = 2.125 \times 10^{-3} \times 14 \text{ g}$$
$$= 0.02975 \text{ g}$$

$$\% \text{ of nitrogen} = \frac{0.02975}{0.9092} \times 100 = \mathbf{3.272\%}$$

Redox Titrations

Reducing agents (R.A.) like Fe, Fe^{2+}, H_2O_2, $CaOCl_2$, MnO_2, etc. can be determined by suitable oxidising agents (O.A.) like MnO_4^-, $Cr_2O_7^{2-}$, etc. in acidic or basic or neutral medium. These are called **redox titrations.** For example, Fe^{2+} is oxidised to Fe^{3+} by MnO_4^- which in turn is reduced to Mn^{2+} in acidic medium.

$$5Fe^{2+} + MnO_4^- + 8H^+ \longrightarrow Mn^{2+} + 5Fe^{3+} + 4H_2O$$

Oxidising agents (O.A.) and reducing agents (R.A.) are summarised in Table 7.5.

This will certainly help to choose a particular O.A. or R.A. in a particular redox titration.

In all cases, again, we remind

$$N_1V_1 = N_2V_2$$

but $$M_1V_1 \neq M_2V_2$$

and $$\text{Eq. wt.} = \frac{\text{mol. wt.}}{\text{total change in O.N.}}$$
$$= \frac{\text{mol. wt.}}{\text{number of electrons gained or lost}}$$

Thus, in acidic medium MnO_4^- is changed to Mn^{2+} by gain of 5 electrons or there is change of 5 units in O.N. hence,

Eq. wt. of MnO_4^{2-} (acidic medium) = $M/5$

Following example will certainly help, why to take total change in O.N. while determining equivalent weight

$Cr_2O_7^{2-}$ is changed to Cr^{3+}

$$Cr_2O_7^{2-} \longrightarrow 2Cr^{3+}$$

O.N. of two Cr atoms = 12 on LHS (6 per atom)

O.N. of two Cr atoms = 6 on RHS (3 per atom)

Hence, eq. wt. = $\left(\frac{M}{6}\right)$

We can take it like this also

$$\frac{1}{2}(Cr_2O_7^{2-}) \longrightarrow Cr^{3+}$$

O.N. 6 3

change in O.N. = 3

$$\text{eq. wt.} = \frac{\text{(formula wt. as given in equation)}}{\text{change in O.N.}}$$

$$= \frac{M/2}{3} = \frac{M}{6}$$

Again note this point that it will cause error in result if not considered.

Again take the following examples :

(a) $C_2O_4^{2-} \longrightarrow 2CO_2 + 2e^-$

O.N. = 6 4 for two carbon atoms

change in O.N. = 2

$\therefore$ eq. wt. of $C_2O_4^{2-} = M/2$

(b) H_2O_2 (reducing agent) $\longrightarrow O_2 + 2H^+ + 2e^-$

O.N. = −2 0 for two oxygen atoms

change in O.N. = 2

eq. wt. of $H_2O_2 = \frac{M}{2}$

(c) H_2O_2 (oxidising agent) $\longrightarrow 2e^- + 2H^+ + 2H_2O$

O.N. = −2 −4 for two oxygen atoms

change in O.N. = 2

$\therefore$ eq. wt of $H_2O_2 = \frac{M}{2}$

Table 7.5 **Oxidising Agents (O.A.)/Reducing Agents (R.A.) with Equivalent Weight**

	Species	Changed to	Reaction	Electrons exchanged or change in O.N.	Eq. wt.
1.	MnO_4^- (O.A.)	Mn^{2+} in acidic medium	$MnO_4^- + 8H^+ + 5e^- \longrightarrow Mn^{2+} + 4H_2O$	5	$E = \frac{M}{5}$
2.	MnO_4^- (O.A.)	MnO_2 in dilute basic* medium	$MnO_4^- + 3e^- + 2H_2O \longrightarrow MnO_2 + 4OH^-$	3	$E = \frac{M}{3}$
3.	MnO_4^- (O.A.)	MnO_4^{2-} in neutral medium	$MnO_4^- + e^- \longrightarrow MnO_4^{2-}$	1	$E = \frac{M}{1}$
4.	$Cr_2O_7^{2-}$ (O.A.)	Cr^{3+} in acidic medium	$Cr_2O_7^{2-} + 14H^+ + 6e^- \longrightarrow 2Cr^{3+} + 7H_2O$	6	$E = \frac{M}{6}$
5.	MnO_2 (O.A.)	Mn^{2+} in acidic medium	$MnO_2 + 4H^+ + 2e^- \longrightarrow Mn^{2+} + 2H_2O$	2	$E = \frac{M}{2}$
6.	Cl_2 (O.A.) (in bleaching powder)	Cl^-	$Cl_2 + 2e^- \longrightarrow 2Cl^-$	2	$E = \frac{M}{2}$
7.	$CuSO_4$ (O.A.) (in iodometric titration)**	Cu^+	$Cu^{2+} + e^- \longrightarrow Cu^+$	1	$E = \frac{M}{1}$
8.	$S_2O_3^{2-}$ (R.A.)	$S_4O_6^{2-}$	$2S_2O_3^{2-} \longrightarrow S_4O_6^{2-} + 2e^-$	2 (for two ions)	$E = \frac{2M}{2} = M$
9.	H_2O_2 (O.A.)	H_2O	$H_2O_2 + 2H^+ + 2e^- \longrightarrow 2H_2O$	2	$E = \frac{M}{2}$
10.	H_2O_2 (R.A.)	O_2	$H_2O_2 \longrightarrow O_2 + 2H^+ + 2e^-$ (O.N. of oxygen is H_2O_2 is (−1) per atom)	2	$E = \frac{M}{2}$
11.	Fe^{2+} (R.A.)	Fe^{3+}	$Fe^{2+} \longrightarrow Fe^{3+} + e^-$	1	$E = \frac{M}{1}$

*In strong basic medium (which is not practically taken) $4MnO_4^- + 4OH^- \longrightarrow 4MnO_4^{2-} + O_2 + 2H_2O$, $E = \frac{M}{1}$

In $S_2O_3^{2-}$, there are two types of sulphur, terminal one with oxidation number −2 and central one with oxidation number +6, thus net value of oxidation number of +4. In $S_4O_6^{2-}$, oxidation number of two types of sulphur are **0 (zero) and **five** and thus net value of +5. When $S_2O_3^{2-}$ is oxidised to $S_4O_6^{2-}$, oxidation number increase by two units for two S-atoms.

DEEP Focus

"**Volume**" concentration of H_2O_2 can be expressed in terms of normality or molarity.

$$\underset{2\text{ mol}}{2H_2O_2} \longrightarrow 2H_2O + \underset{22400\text{ mL at STP}}{O_2}$$

If H_2O_2 is **"x volume"** it means 1 mL of H_2O_2 at STP gives x mL O_2.

From the above equation,

22400 mL of O_2 at STP are from = 2 mol H_2O_2

$\therefore$ x mL O_2 at STP are from $= \dfrac{2x}{22400}$ mol H_2O_2

$\dfrac{2x}{22400}$ mol H_2O_2 are in 1 mL H_2O_2 solution.

Hence, molarity of "x volume" $H_2O_2 = \dfrac{\text{mol}}{\text{litre}}$

$$= \frac{2x}{22400} \times 1000 = \frac{x}{11.2}\text{ M}$$

Normality of "x volume" $H_2O_2 = \dfrac{2x}{11.2} = \dfrac{x\text{ N}}{5.6}$

Redox titrations (excluding iodometric and iodimetric) are summarised in Table 7.6

Table 7.6 **Redox Titrations (Excluding Iodometric/Iodimetric)**

Estimation of	By titrating with	Reactions	Relation* between O.A. and R.A.
1. Fe^{2+}	MnO_4^-	$Fe^{2+} \longrightarrow Fe^{3+} + e^-$ $MnO_4^- + 8H^+ + 5e^- \longrightarrow Mn^{2+} + 4H_2O$	$5Fe^{2+} \equiv MnO_4^-$ Eq. wt. $Fe^{2+} = M/1$ Eq. wt. $MnO_4^- = M/5$
2. Fe^{2+}	$Cr_2O_7^{2-}$	$Fe^{2+} \longrightarrow Fe^{3+} + e^-$ $Cr_2O_7^{2-} + 14H^+ + 6e^- \longrightarrow 2Cr^{3+} + 7H_2O$	$6Fe^{2+} \equiv Cr_2O_7^{2-}$ Eq. wt. $Cr_2O_7^{2-} = M/6$
3. $C_2O_4^{2-}$	MnO_4^-	$C_2O_4^{2-} \longrightarrow 2CO_2 + 2e^-$ $MnO_4^- + 8H^+ + 5e^- \longrightarrow Mn^{2+} + 4H_2O$	$5C_2O_4^{2-} \equiv 2MnO_4^-$ Eq. wt. $C_2O_4^{2-} = M/2$ Eq. wt. $MnO_4^- = M/5$
4. H_2O_2	MnO_4^-	$H_2O_2 \longrightarrow 2H^+ + O_2 + 2e^-$	$5H_2O_2 \equiv 2MnO_4^-$ Eq. wt. $H_2O_2 = M/2$ Eq. wt. $MnO_4^- = M/5$
5. As_2O_3	MnO_4^-	$As_2O_3 + 5H_2O \longrightarrow 2AsO_4^{3-} + 10H^+ + 4e^-$	Eq. wt. $As_2O_3 = M/4$
6. AsO_3^{3-}	BrO_3^-	$AsO_3^{3-} + H_2O \longrightarrow AsO_4^{3-} + 2H^+ + 2e^-$ $BrO_3^- + 6H^+ + 6e^- \longrightarrow Br^- + 3H_2O$	Eq. wt. of $AsO_3^{3-} = M/2$ Eq. wt. of $BrO_3^- = M/6$

*M is molar mass of O.A. or R.A.

Illustration 1 *A solution of* $KHC_2O_4 \cdot H_2C_2O_4 \cdot 2H_2O$ *is* 0.2 *N as an acid. What is its normality as reducing agent?*

Solution $KHC_2O_4 \cdot H_2C_2O_4 \rightleftharpoons K^+ + 3H^+ + 2C_2O_4^{2-}$ H^+ ion is neutralised by OH^-

$$H^+ + OH^- \longrightarrow H_2O$$

$C_2O_4^{2-}$ ion is oxidised hence, it is a reducing agent

$$C_2O_4^{2-} \longrightarrow 2CO_2 + 2e^-$$

Since, there are three ionisable H^+ ions hence,

$$N_1\text{ (acid), normality} = M \times 3$$

and since there are two $C_2O_4^{2-}$ ions exchanging four electrons, hence,

N_2 (reducing agent), normality = molarity $(M) \times 4$

As an acid $\quad 3M = 0.2$ N (given)

$$\therefore \quad M = \frac{0.2}{3} = \frac{\text{M}}{15}$$

$$\therefore \quad N_2\text{ (reducing agent)} = 4 \times \frac{1}{15} = \frac{4}{15} = \mathbf{0.267\ M}$$

Illustration 2 *The concentration of a hydrogen peroxide solution can be conveniently determined by titration against a standardised potassium permanganate solution in an acidic medium according to the following equation:*

$$2MnO_4^- + 5H_2O_2 + 6H^+ \longrightarrow 5O_2 + 2Mn^{2+} + 8H_2O$$

If 36.44 mL of 0.01652 M $KMnO_4$ solution are required to completely oxidise 25 mL of a H_2O_2 solution, calculate the molarity of the H_2O_2 solution.

Solution **Method I**

$\underset{+7}{MnO_4^-} \xrightarrow{\text{reduction}} \underset{+2}{Mn^{2+}}$ **Change in O.N. Normality**

5 units $\quad = 5 \times M$

(for two O atoms) $\underset{-2}{H_2O_2} \longrightarrow \underset{0}{O_2}$ 2 units $\quad = 2 \times M$

Thus, 25.00 mL of

$$M_1\ H_2O_2 \equiv 36.44\text{ mL of }0.01652\text{ M }MnO_4^-$$

Using, $\quad N_1V_1 = N_2V_2$

$$25.00 \times 2M_1 \ H_2O_2 = 36.44 \times 5 \times 0.01652 \ M \ MnO_4^-$$

$$\therefore \quad M_1 = \frac{36.44 \times 5 \times 0.01652}{25 \times 2}$$

$$= \mathbf{0.0602\ M}$$

Method II Number of moles of

$$MnO_4^- \text{ used} = \frac{36.44 \times 0.01652}{1000}$$

$$= 0.602 \times 10^{-3} \text{ mol}$$

$$\text{Number of moles of } H_2O_2 = 2.5 \times \text{mol of } MnO_4^-$$

$$= 1.505 \times 10^{-3} \text{ mol}$$

Let molarity of $H_2O_2 = M_1$

then number of moles of $H_2O_2 = \dfrac{25 \times M_1}{1000}$

$$\therefore \quad \frac{25 \times M_1}{1000} = 1.505 \times 10^{-3}$$

$$\therefore \quad M_1 = \frac{1.505 \times 10^{-3} \times 1000}{25} = \mathbf{0.0602\ M}$$

Illustration 3 $KBrO_3$ *can be used to determine the concentration of* As(III)

$$3H_3AsO_3 + BrO_3^- \longrightarrow Br^- + 3H_3AsO_4$$

What is the molar concentration of As(III) *in a solution if* 22.35 *mL of* 0.10 M $KBrO_3$ *is needed to titrate* 50.00 *mL of* As(III) *solution?*

Solution Let molar concentration of a solution containing As(III) = M_1

	Change in O.N.	Normality
$\underset{+3}{H_3AsO_3} \xrightarrow{\text{oxidation}} \underset{+5}{H_3AsO_4}$	2 units	2 × molarity
$\underset{+5}{BrO_3^-} \xrightarrow{\text{reduction}} \underset{-1}{Br^-}$	6 units	6 × molarity

$$\underset{\text{As(III)}}{N_1V_1} = \underset{BrO_3^-}{N_2V_2}$$

$$2 \times M_1 \times 50 = 6 \times 0.10 \times 22.35$$

$$M_1 = \frac{6 \times 0.10 \times 22.35}{100}$$

$$= \mathbf{0.1341\ M}$$

Illustration 4 *Titration of* 0.1467 *g of primary standard* $Na_2C_2O_4$ *required* 28.85 *mL of* $KMnO_4$ *solution. Calculate the molar concentration of* $KMnO_4$ *solution.*

Solution

$$\underset{+6}{5C_2O_4^{2-}} + \underset{+7}{2MnO_4^-} + 16H^+ \longrightarrow \underset{+2}{2Mn^{2+}} + \underset{8}{10CO_2} + 8H_2O$$

Method I (Based on number of equivalent)

In a reaction (redox/neutralisation) number of equivalents of different reactants are same.

$$\text{Equivalent of } C_2O_4^{2-} = \text{equivalents of } MnO_4^-$$

$$\text{Equivalent weight of } C_2O_4^{2-} = \frac{\text{molecular weight}}{2}$$

$$\therefore \quad \text{Molarity} = 2 \times \text{normality}$$

$$\text{Equivalent weight of } MnO_4^- = \frac{\text{molecular weight}}{5}$$

$$\therefore \quad \text{Molarity} = 5 \times \text{normality}$$

Let molarity of $MnO_4^- = M_1$

$$\text{normality} = 5M_1$$

$$\text{Number of equivalents of } MnO_4^- = \frac{28.85 \times 5M_1}{1000}$$

$$\text{Number of equivalents of } Na_2C_2O_4 = \frac{0.1467}{67}$$

$$\therefore \quad \frac{28.85 \times 5M_1}{1000} = \frac{0.1467}{67}$$

$$\therefore \quad M_1 = \mathbf{0.01518\ M}$$

Method II (Based on stoichiometric molar coefficients)

$$\text{Moles of } MnO_4^- = \frac{28.85M_1}{1000}$$

$$\text{Moles of } Na_2C_2O_4 = \frac{0.1467}{134}$$

$$5 \text{ mol } C_2O_4^{2-} \equiv 2 \text{ mol } MnO_4^-$$

$$\therefore \quad \frac{0.1467}{134} \text{ mol } C_2O_4^{2-} = \frac{2}{5} \times \frac{0.1467}{134} \text{ mol } MnO_4^-$$

$$\therefore \quad \frac{2}{5} \times \frac{0.1467}{134} = \frac{28.85M_1}{1000}$$

$$\therefore \quad M_1 = \mathbf{0.01518\ M}$$

Illustration 5 *A* 100 *mL of sample of water was treated to convert any iron present to* Fe^{2+}. *Addition of* 25 *mL of* 0.002 *M* $K_2Cr_2O_7$ *resulted in the reaction :*

$$6Fe^{2+} + Cr_2O_7^{2-} + 14H^+ \longrightarrow 6Fe^{3+} + 2Cr^{3+} + 7H_2O$$

The excess $K_2Cr_2O_7$ *was back-titrated with* 7.50 *mL of* 0.01 *M* Fe^{2+} *solution. Calculate the parts per million of iron in the sample*

Solution In the given reaction equivalents of $Cr_2O_7^{2-}$ ≡ equivalents of Fe^{2+}

Unreacted $Cr_2O_7^{2-}$ has been determined by titration with standard 0.01 M Fe^{2+} solution.

Method 1 Let unreacted $Cr_2O_7^{2-} = V$ mL

$$0.01\,M\,Fe^{2+} = 0.01\,N\,Fe^{2+}$$

$$0.002\,M\,Cr_2O_7^{2-} = 0.012N\,Cr_2O_7^{2-}$$

$$\therefore \quad \underset{Cr_2O_7^{2-}}{N_1V_1} = \underset{Fe^{2+}}{N_2V_2}$$

$$0.012 \times V = 0.01 \times 7.50$$

$$\therefore \quad V \text{ (unreacted } Cr_2O_7^{2-}) = \frac{0.01 \times 7.50}{0.012} = 6.25 \text{ mL}$$

$$\therefore \quad \text{Volume of } Cr_2O_7^{2-} \text{ used by } Fe^{2+} \text{ (in water sample)}$$

$$= 25.00 - 6.25$$

$$= 18.75 \text{ mL}$$

$\therefore$ Equivalents of $Cr_2O_7^{2-}$ used $= \dfrac{18.75 \times 0.012}{1000}$

$= 2.25 \times 10^{-4}$

$\therefore$ Equivalents of Fe^{2+} in water $= 2.25 \times 10^{-4}$

Thus, Fe present in 100 mL water $= 2.25 \times 10^{-4} \times 56$ g

$= 0.0126$ g

Fe parts per million $= \dfrac{0.0126}{100} \times 10^6$

$=$ **126 ppm**

Method II Volume of $Cr_2O_7^{2-}$ used = 18.75 mL

Moles of $Cr_2O_7^{2-}$ used $= \dfrac{18.75 \times 0.002}{1000}$

$= 3.75 \times 10^{-5}$ mol

Moles of Fe^{2+} (based on reaction) $= 6 \times Cr_2O_7^{2-}$

$= 6 \times 3.75 \times 10^{-5}$ mol

$\therefore$ Fe present in 100 mL $H_2O = 2.25 \times 10^{-4}$ mol

$= 2.25 \times 10^{-4} \times 56$ g

$= 0.0126$ g

Fe parts per million = **126 ppm**

Illustration 6 *Treatment of hydroxylamine* (NH_2OH) *with an excess of* Fe(III) *results in the formation of* N_2O *and an equivalent amount of* Fe(II)

$$2NH_2OH + 4Fe^{3+} \longrightarrow N_2O(g) + 4Fe^{2+} + 4H^+ + H_2O$$

Calculate the molar concentration of an NH_2OH *solution if* Fe(II) *produced from* 50 *mL solution required* 23.61 *mL of* 0.0217 *M* $K_2Cr_2O_7$.

Solution Fe^{2+} obtained by reduction of Fe^{3+} by NH_2OH is estimated by $Cr_2O_7^{2-}$.

$$2NH_2OH + 4Fe^{3+} \longrightarrow N_2O(g) + 4Fe^{2+} + 4H^+ + H_2O$$

$$6Fe^{2+} + Cr_2O_7^{2-} + 14H^+ \longrightarrow 6Fe^{3+} + 2Cr^{3+} + 7H_2O$$

Based on stoichiometric coefficients :

$$1\,Cr_2O_7^{2-} \equiv 6Fe^{2+} \equiv 3\,NH_2OH$$

or $\quad 1 \text{ mol } Cr_2O_7^{2-} \equiv 3 \text{ mol } NH_2OH$

$\therefore \quad \dfrac{23.61 \times 0.0217}{1000}$ mol $Cr_2O_7^{2-}$

$= \dfrac{3 \times 23.61 \times 0.0217}{1000}$ mol NH_2OH

$= 1.537 \times 10^{-3}$ mol NH_2OH in 50 mL solution

$\therefore$ Molar concentration of $NH_2OH = \dfrac{1.537 \times 10^{-3}}{0.050}$

$=$ **0.0307 M**

Illustration 7 *The* SO_2 *present in air is mainly responsible for the acid-rain phenomenon. Its concentration can be determined by titration against a standard permanganate solution as follows :*

$$5SO_2 + 2MnO_4^- + 2H_2O \longrightarrow 5SO_4^{2-} + 2Mn^{2+} + 4H^+$$

Calculate the number of grams of SO_2 *in a sample of air if* 7.37 *mL of* 0.008 *M* $KMnO_4$ *solution are required for the titration.*

Solution Number of moles of MnO_4^- required

$= \dfrac{7.37 \times 0.008}{1000} = 5.896 \times 10^{-5}$ mol

Also, number of moles of

$SO_2 = 2.5 \times$ number of moles of MnO_4^-

$= 1.474 \times 10^{-4}$ mol

SO_2 in air sample $= 1.474 \times 10^{-4} \times 64$ g

$=$ **9.434×10^{-3} g**

Iodometric/Iodimetric Titrations

Compounds containing iodine are widely used in titrations.

(i) Iodide ions can be oxidised to I_2 by suitable oxidising agents

$$2I^-(aq) \longrightarrow I_2(s) + 2e^-$$

(ii) Iodate (V) ions, IO_3^-, will oxidise I^- to I_2

$$IO_3^-(aq) + 5I^-(aq) + 6H^+(aq) \longrightarrow 3I_2(s) + 3H_2O(l)$$

(iii) Thiosulphate ions, ($S_2O_3^{2-}$) can reduce iodine to iodide ions.

$$\underset{\text{colourless}}{2S_2O_3^{2-}(aq)} + \underset{\text{black}}{I_2(s)} \longrightarrow \underset{\text{colourless}}{S_4O_6^{2-}} + 2I^-$$

Although solid iodine is black (deep violet))and insoluble in water, it is converted into soluble triiodide ions (I_3^-)

$$\underset{\text{black}}{I_2(s)} + I^- \rightleftharpoons \underset{\text{dark brown}}{I_3^-(aq)}$$

Small amount of iodine molecules (I_2), gives blue colour with starch, hence, completion of the reaction can be detected when blue colour disappears at the end-point. It should be noted that *starch, being colloidal, can adsorb* I_2 *hence, it should be added when there is very small amount of* I_2 *left (indicated by appearance of pale yellow colour) otherwise exact point is not detectable.*

In all the titrations, KI is added to oxidising agent and I_2 formed is estimated by sodium thiosulphate solution. These are summarised in Table 7.7.

Table 7.7 **Iodometric Titration (Titration Solution is $Na_2S_2O_3 \cdot 5H_2O$)**

S. No.	Estimation of	Reaction	Relation between O.A. and R.A.
1.	I_2	$I_2 + 2Na_2S_2O_3 \longrightarrow 2NaI + Na_2S_4O_6^{2-}$ or $I_2 + 2S_2O_3^{2-} \longrightarrow 2I^- + S_4O_6^{2-}$	$I_2 \equiv 2I \equiv 2Na_2S_2O_3$ Eq. wt. of $Na_2S_2O_3 = M/1$
2.	$CuSO_4$	$2CuSO_4 + 4KI \longrightarrow Cu_2I_2 + 2K_2SO_4 + I_2$ or $Cu^{2+} + 4I^- \longrightarrow \underset{\text{white ppt.}}{Cu_2I_2} + I_2$	$2CuSO_4 \equiv I_2 \equiv 2I \equiv 2Na_2S_2O_3$ Eq. wt. of $CuSO_4 = M/1$
3.	$CaOCl_2$	$CaOCl_2 + H_2O \longrightarrow Ca(OH)_2 + Cl_2$ $Cl_2 + 2KI \longrightarrow 2KCl + I_2$ $Cl_2 + 2I^- \longrightarrow 2Cl^- + I_2$	$CaOCl_2 \equiv Cl_2 \equiv I_2 \equiv 2I \equiv 2Na_2S_2O_3$ Eq. wt. of $CaOCl_2 = M/2$
4.	MnO_2	$MnO_2 + 4HCl(conc.) \xrightarrow{\Delta} MnCl_2 + Cl_2 + 2H_2O$ $Cl_2 + 2KI \longrightarrow 2KCl + I_2$ or $MnO_2 + 4H^+ + 2Cl^- \longrightarrow Mn^{2+} + 2H_2O + Cl_2$ $Cl_2 + 2I^- \longrightarrow I_2 + 2Cl^-$	$MnO_2 \equiv Cl_2 \equiv I_2 \equiv 2I \equiv 2Na_2S_2O_3$ Eq. wt. of $MnO_2 = M/2$
5.	IO_3^-	$IO_3^- + 5I^- + 6H^+ \longrightarrow 3I_2 + 3H_2O$	$IO_3^- \equiv 3I_2 \equiv 6I \equiv 6Na_2S_2O_3$ Eq. wt. of $IO_3^- = M/6$
6.	H_2O_2	$H_2O_2 + 2I^- + 2H^+ \longrightarrow I_2 + 2H_2O$	$H_2O_2 \equiv I_2 \equiv 2I \equiv 2Na_2S_2O_3$ Eq. wt. of $H_2O_2 = M/2$
7.	Cl_2	$Cl_2 + 2I^- \longrightarrow 2Cl^- + I_2$	$Cl_2 \equiv I_2 \equiv 2I \equiv 2Na_2S_2O_3$ Eq. wt. of $Cl_2 = M/2$
8.	O_3	$O_3 + 6I^- + 6H^+ \longrightarrow 3I_2 + 3H_2O$	$O_3 \equiv 3I_2 \equiv 6I \equiv 6Na_2S_2O_3$ Eq. wt. of $O_3 = M/6$
9.	ClO^-	$ClO^- + 2I^- + 2H^+ \longrightarrow H_2O + Cl^- + I_2$	$ClO^- \equiv I_2 \equiv 2I \equiv 2Na_2S_2O_3$ Eq. wt. of $ClO^- = M/2$
10.	$Cr_2O_7^{2-}$	$Cr_2O_7^{2-} + 14H^+ + 6I^- \longrightarrow 3I_2 + 2Cr^{3+} + 7H_2O$	$Cr_2O_7^{2-} \equiv 3I_2 \equiv 6I$ Eq. wt. of $Cr_2O_7^{2-} = M/6$
11.	MnO_4^-	$2MnO_4^- + 10I^- + 16H^+ \longrightarrow 2MnO_4^- + 5I_2 + 8H_2O$	$2MnO_4^- \equiv 5I_2 \equiv 10I$ Eq. wt. of $MnO_4^- = M/5$
12.	BrO_3^-	$BrO_3^- + 6I^- + 6H^+ \longrightarrow Br^- + 3I_2 + 3H_2O$	$BrO_3^- \equiv 3I_2 \equiv 6I$ Eq. wt. of $BrO_3^- = M/6$
13.	As(V)	$H_3AsO_4 + 2I^- + 2H^+ \longrightarrow H_3AsO_3 + H_2O + I_2$	$H_3AsO_4 \equiv I_2 \equiv 2I$ Eq. wt. of $H_3AsO_4 = M/2$
14.	HNO_2	$2HNO_2 + 2I^- \longrightarrow I_2 + 2NO + H_2O$	$2HNO_2 \equiv I_2 \equiv 2I$ Eq. wt. of $HNO_2 = M/1$
15.	HClO	$HClO + 2I^- + H^+ \longrightarrow Cl^- + I_2 + H_2O$	Eq. wt. of $HClO = M/2$

In **iodimetric** (different from iodometric) titration, I_2 solution is taken in burette and reducing agents like sodium thiosulphate (As_2O_3) etc. are taken in flask.

$$I_2 + 2Na_2S_2O_3 \longrightarrow 2NaI + Na_2S_4O_6$$

$$2I_2 + As_2O_3 + 5H_2O \longrightarrow 2AsO_4^{3-} + 4I^- + 10H^+$$

These titrations are usually performed in neutral or mildly alkaline (pH 8) to weakly acid solutions. If the pH is too alkaline, I_2 will disproportionate to hypoiodate and iodide ions.

$$I_2 + 2OH^- \longrightarrow IO^- + I^- + H_2O$$

Weakly acidic medium is preferred since

- Starch used for the end-point detection tends to hydrolyse or decompose in strong acid.
- Reducing power of several reducing agents is increased in neutral solution.
- I^- produced in the reaction tends to be oxidised by dissolved O_2 in acid solution:

$$4I^- + O_2 + 4H^+ \longrightarrow 2I_2 + 2H_2O$$

The pH for the titration of arsenic (III) with I_2 can be maintained neutral by adding $NaHCO_3$. Iodimetric titrations have been summarised in Table **7.8.**

Table 7.8 **Iodimetric Titration (Titrating Solution is I_2 in KI)**

S. No.	Estimation	Reaction	Relation between O.A. and R.A.
1.	H_2S (in acidic medium)	$H_2S + I_2 \longrightarrow S + 2I^- + 2H^+$	$H_2S \equiv I_2 \equiv 2I$ Eq. wt. of $H_2S = M/2$
2.	SO_3^{2-} (in acidic medium)	$SO_3^{2-} + I_2 + H_2O \longrightarrow SO_4^{2-} + 2I^- + 2H^+$	$SO_3^{2-} \equiv I_2 = 2I$ Eq. wt. of $SO_3^{2-} = M/2$
3.	Sn^{2+} (in acidic medium)	$Sn^{2+} + I_2 \longrightarrow Sn^{4+} + 2I^-$	$Sn^{2+} \equiv I_2 \equiv 2I$ Eq. wt. of $Sn^{2+} = M/2$
4.	As (III) (at pH 8)	$H_2AsO_3^- + I_2 + H_2O \longrightarrow HAsO_4^{2-} + 2I^- + 3H^+$	$H_2AsO_3^- \equiv I_2 \equiv 2I$ Eq. wt. of $H_2AsO_3^- = M/2$
5.	N_2H_4	$N_2H_4 + 2I_2 \longrightarrow N_2 + 4H^+ + 4I^-$	$N_2H_4 \equiv 2I_2 \equiv 4I$ Eq. wt. of $N_2H_4 = M/4$

Illustration 1 *A 0.200 g sample containing copper is analysed iodometrically. Copper (II) is reduced to copper (I) by iodide.*

$$2Cu^{2+} + 4I^- \longrightarrow 2CuI + I_2$$

What is the per cent copper in the sample if 20.0 mL of 0.10 M $Na_2S_2O_3$ *is required for titration of the liberated* I_2*? (Cu = 63.5 g mol*$^{-1}$*)*

Solution $2Cu^{2+} + 4I^- \longrightarrow \underset{\text{white ppt.}}{2CuI\downarrow} + I_2$

$$I_2 + 2S_2O_3^{2-} \longrightarrow S_4O_6^{2-} + 2I^-$$

By stoichiometric coefficients:

$$2Cu^{2+} \equiv I_2 \equiv 2S_2O_3^{2-}$$

$\therefore$ $Cu^{2+} \equiv S_2O_3^{2-}$

20 mL of 0.10 M $S_2O_3^{2-}$ $= 20 \times 0.1$ millimol $= 2 \times 10^{-3}$ mol $S_2O_3^{2-}$

$\therefore$ Pure $Cu^{2+} = 2 \times 10^{-3}$ mol $= 2 \times 10^{-3} \times 63.5$ g

% of pure Cu in 0.20 g sample $= \dfrac{2 \times 10^{-3} \times 63.5 \times 100}{0.2}$

$= \mathbf{63.5\%}$

Illustration 2 *A solution of* $Na_2S_2O_3$ *is standardised iodometrically against a 0.1262 g of pure* $KBrO_3$*, requiring 45.0 mL* $Na_2S_2O_3$ *solution. What is the molarity of* $Na_2S_2O_3$ *solution? (*$KBrO_3$ *= 167 g mol*$^{-1}$*)*

Solution The reactions are

$$BrO_3^- + 6I^- + 6H^+ \longrightarrow Br^- + 3I_2 + 3H_2O$$

$$3I_2 + 6S_2O_3^{2-} \longrightarrow 6I^- + 3S_4O_6^{2-}$$

Let, molarity of $Na_2S_2O_3$ solution $= M_1$

moles of $Na_2S_2O_3$ used $= \dfrac{M_1 \times 45}{1000} = 0.045\,M_1$

moles of $KBrO_3$ taken $= \dfrac{0.1262}{167} = 7.56 \times 10^{-4}$

Based on stoichiometric coefficients:

6 mol of $S_2O_3^{2-}$ = 1 mol of BrO_3^-

$\therefore$ $0.045\,M_1$ mol of $S_2O_3^{2-} = \dfrac{0.045M_1}{6}$ mol of BrO_3^-

$\therefore$ $\dfrac{0.045M_1}{6} = 7.56 \times 10^{-4}$

$\therefore$ $M_1 = \mathbf{0.10\ M}$

$\therefore$ Molarity of $Na_2S_2O_3$ solution $=$ **0.10 M**

Target Practice 2

1. 100 mL of $NaHC_2O_4$ (*A*) is neutralised by 100 mL of 1 M NaOH. Thus 100 mL of *A* will be oxidised by .... mL of 1 M $KMnO_4$ in acidic medium.
2. $KMnO_4$ is reduced to $MnCl_2$ in HCl medium, to MnO_2 in KOH medium and to K_2MnO_4 in H_2O medium.
   Thus, (a) in 1 M $KMnO_4$ = ... N in HCl medium
   (b) 1 M $KMnO_4$ = ... N in KOH medium
   (c) 1 M $KMnO_4$ = ... N in H_2O medium
3. We have 100 mL of 1 M $KMnO_4$. In acidic medium it is to be used. Calculate volume of water to be added so that it is 0.5 N. ....
4. Calculate normality of the underlined if each reactant is 1 M.
   (a) $\underline{2CuSO_4} + 4KI \longrightarrow Cu_2I_2 + 2K_2SO_4 + I_2$ ....
   (b) $\underline{2Na_2S_2O_3} + I_2 \longrightarrow 2NaI + Na_2S_4O_6$ ....
   (c) $\underline{H_2O_2} + 2KI \longrightarrow 2KOH + I_2$ ....
   (d) $Fe^{2+} \longrightarrow Fe^{3+} + e^-$ ....

5. 10 mL of 1 M Fe^{2+} is oxidised by each of 1 M MnO_4^- and 1 M $Cr_2O_7^{2-}$ in separate titration in acidic medium.

Thus, volume of MnO_4^- required = ... mL

Volume of $Cr_2O_7^{2-}$ required = ... mL

6. 0.1 M HCl solution is being titrated by two students using 0.1 M NaOH.

Student A : 20 mL of 0.1 M HCl + 40 mL H_2O

Student B : 20 mL of 0.1 M HCl + 80 mL of H_2O.

Calculate volume of NaOH in both cases.

(a) ....

(b) ....

7. A 100 mL sample of brackish water was made ammoniacal and sulphide it contained was titrated with 16.0 mL of 0.025 M $AgNO_3$ solution.

$$2Ag^+ + S^{2-} \longrightarrow Ag_2S(s)$$

Calculate concentration of H_2S in water in ppm. ....

Answers

1. 100 mL of $NaHC_2O_4$ is neutralised by 100 mL of 1 M NaOH. This is an acid-base titration. $NaHC_2O_4$ has one ionisable H^+ ion, hence,

$$100 \times M_1 (= N_1) = 100 \times 1 \text{ M NaOH}$$

$$\therefore \quad M_1 = N_1 \ (NaHC_2O_4) = 1$$

$NaHC_2O_4$ is oxidised to CO_2 by $KMnO_4$ in acidic medium.

$$C_2O_4^{2-} \longrightarrow 2CO_2 + 2e^-$$

Thus, 1 M $NaHC_2O_4$ = 1 N as acid

= 2 N as reducing agent

1 M $KMnO_4$ = 5 N as oxidising agent

$$\therefore \quad 100 \times 2N\ (A) = V \times 5N\ KMnO_4$$

$$\therefore \quad V = \mathbf{40\ mL}$$

2. (a) 5 N (b) 3 N (c) 1 N

3. 1 M $KMnO_4$ = 5 N in acidic medium

$$100 \times 5\ N = V \times 0.5\ N$$

$$V = 1000\ mL$$

Thus, water to be added = **900 mL**

4. (a) $\underline{2\,CuSO_4} \equiv I_2 \equiv 2I$,

$CuSO_4 \equiv I \ \therefore\ 1\ M = 1\ N$

(b) $2Na_2S_2O_3 \equiv I_2 \equiv 2I$

$\therefore\ Na_2S_2O_3 \equiv I \ \therefore\ 1\ M = 1\ N$

(c) $H_2O_2 \equiv I_2 \equiv 2I$

$\therefore\ I = \dfrac{H_2O_2}{2} \ \therefore\ 1\ M = 2\ N$

(d) $Fe^{2+} \equiv 1\ e^-$

$\therefore\ 1\ M = 1\ N$

5. $Fe^{2+} \longrightarrow Fe^{3+} + e^-$

$$MnO_4^- + 8H^+ + 5e^- \longrightarrow Mn^{2+} + 4H_2O$$

10 mL of 1 M Fe^{2+} ≡ 10 mL of 1 N Fe^{2+}

V mL of 1 M MnO_4^- ≡ V mL of 5 N MnO_4^- in acidic medium

$$\therefore \quad V \times 5\ N = 10 \times 1\ N$$

$$V = 2\ mL$$

$$Cr_2O_7^{2-} + 14H^+ + 6e^- \longrightarrow 2Cr^{3+} + 7H_2O$$

V mL of 1 M $Cr_2O_7^{2-}$ ≡ V mL of 6 N $Cr_2O_7^{2-}$

$$\therefore \quad V \times 6\ N = 10 \times 1\ N$$

$$\therefore \quad V = \frac{10}{6} = \mathbf{1.67\ mL}$$

6. In both case same volume of NaOH is used.

(A) N_{HCl} (resultant) $= \dfrac{20 \times 0.1}{60} = \dfrac{N}{30}$

(B) N_{HCl} (resultant) $= \dfrac{20 \times 0.1}{100} = \dfrac{N}{50}$

Thus, resultant volume in A = 60 mL

$$60\ mL\ of\ \frac{N}{30} \equiv V\ mL\ of\ 0.1\ N\ NaOH$$

$$\therefore \quad V = \mathbf{20\ mL}$$

Resultant volume of B = 100 mL

$$100\ mL\ of\ \frac{N}{50} \equiv V\ mL\ of\ 0.1\ N\ NaOH$$

$$\therefore \quad V = \mathbf{20\ mL}$$

7. 16.0 mL of 0.025 M $AgNO_3$ ≡ 100 mL of H_2O containing H_2S

$$\text{Moles of } AgNO_3 = \frac{16 \times 0.025}{1000}$$

$$= 4 \times 10^{-4}\ mol$$

$$\therefore \quad \text{Moles of } H_2S = \frac{1}{2} \text{ mol of } AgNO_3$$

$$= 2 \times 10^{-4}\ mol\ H_2S$$

$$= 2 \times 10^{-4} \times 34\ g\ in\ 100\ mL\ H_2O$$

$$= \frac{68 \times 10^{-4} \times 10^6}{100}\ g\ in\ 10^6\ mL\ (ppm)\ H_2O$$

$$= \mathbf{68\ ppm}$$

Alternatively :

$$1\ H_2S \equiv 2Ag^+$$

Thus, Normality of H_2S = 2 × Molarity

$$100 \times 2 \times M\ (H_2S) = 16.0 \times 0.025$$

$$M(H_2S) = 2 \times 10^{-3}\ mol\ L^{-1}$$

$$= 68 \times 10^{-3}\ gL^{-1}$$

$$= 68 \times 10^{-6}\ g\ mL^{-1}$$

$$= 68 \times 10^{-6} \times 10^6\ g\ ppm$$

$$= \mathbf{68\ ppm}$$

Practice Exercise 2

1. Household ammonia, used as a window cleaner and for other cleaning purposes, is $NH_3(aq)$. 31.08 mL of 0.9928 M HCl (*aq*) is required to neutralise the NH_3 present in a 5.00 mL sample,
$$NH_3 + H^+ \longrightarrow NH_4^+$$
What is the molarity of NH_3 in the sample?
2. What volume of a 0.232 N HCl solution contains
(a) 3.17 meq. of solute
(b) 6.5 eq. of solute?
3. Calculate the normality of each of the following solution:
(a) 7.88 g of HNO_3 per litre solution
(b) 26.5 g of Na_2CO_3 per litre solution (if acidified to form CO_2).
4. How many (a) meq. (b) equivalent, (c) mole and (d) gram are present in 60 mL of 4.0 N H_2SO_4?
5. What is the purity of conc. H_2SO_4 ($d = 1.8$ g / mL) if 5.00 mL is neutralised by 84.6 mL of 2.00 N NaOH?
6. Urea sample is 60% pure. Calculate volume of 1 N HCl required to absorb NH_3 from 1.0 g of urea sample on decomposition with NaOH.
7. $Mn^{2+}(aq)$ can be determined by titration with $KMnO_4(aq)$
$$3Mn^{2+} + 2MnO_4^- + 2H_2O \longrightarrow 5MnO_2 + 4H^+$$
A 25.00 mL sample of $Mn^{2+}(aq)$ requires 35 mL of 0.06 M $KMnO_4(aq)$ for its titration. What is the molarity of the $Mn^{2+}(aq)$?
8. A particular acid-rain water has SO_3^{2-}. If a 25.00 mL sample of this water requires 34.08 mL of 0.01964 M $KMnO_4$ for its titration, what is the molarity of SO_3^{2-} in acid-rain?
$$2MnO_4^- + 5SO_3^{2-} + 6H^+ \longrightarrow 5SO_4^{2-} + 2Mn^{2+} + 3H_2O$$
9. What is molarity of a MnO_4^- solution if 32.00 mL of the solution is required to titrate 40.00 mL of 0.400 N Fe^{2+}?
$$MnO_4^- + 5Fe^{2+} + 8H^+ \longrightarrow Mn^{2+} + 5Fe^{3+} + 4H_2O$$
10. In a reaction, $Cr_2O_7^{2-}$ is reduced to Cr^{3+}. What is concentration of 0.1 M $K_2Cr_2O_7$ in equivalent per litre?
$$Cr_2O_7^{2-} + 6H^+ + 6e^- \longrightarrow 2Cr^{3+} + 7H_2O$$
(Here we have given half reaction; if not given be careful in balancing affected atoms and in taking oxidation number of two Cr atoms)
11. What mass of phosphoric acid (H_3PO_4) is required to make 550 mL of 0.400 N solution?
(a) assuming complete neutralisation of the acid
$$H_3PO_4 + 3OH^- \longrightarrow 3H_2O + PO_4^{3-}$$
(b) assuming reduction to HPO_3^{2-}.
$$H_3PO_4 + 2e^- \longrightarrow HPO_3^{2-} + H_2O$$
12. The concentration of Fe^{2+} ion in aqueous solution can be determined by redox titration with bromate ion, BrO_3^-, according to net ionic equation:
$$6Fe^{2+} + BrO_3^- + 6H^+ \longrightarrow 6Fe^{3+} + Br^- + 3H_2O$$
What is the molar concentration of Fe^{2+} if 31.50 mL of 0.10 M $KBrO_3$ is required for complete reaction with 10.0 mL of the Fe^{2+} solution?
13. A 1.100 g sample of copper ore is dissolved and the $Cu^{2+}(aq)$ is treated with excess KI. The liberated I_2 requires 12.12 mL of 0.10 M $Na_2S_2O_3$ solution for titration. What is % copper by mass in the ore?
14. What mass of $Na_2S_2O_3 \cdot 5H_2O$ is needed to make 500 mL of 0.200 N solution for the reaction?
$$2S_2O_3^{2-} + I_2 \longrightarrow S_4O_6^{2-} + 2I^-$$

Answers

1. $M(NH_3) \times 5 = 0.9928 \times 31.08$ HCl
$M(NH_3) = 6.17$ M

2. (a) $N \times V$ (mL) = milliequiv.
$0.232 \times V = 3.17$
$\therefore \quad V = \dfrac{3.17}{0.232} = 13.66$ mL
(b) $\dfrac{N \times V\text{ (mL)}}{1000} = \text{equiv.}$
$\dfrac{0.232 \times V}{1000} = 6.5$
$V = 2.8017 \times 10^4$ mL = **28.017 L**

3. (a) $7.88 \text{ g L}^{-1} \text{ HNO}_3 = \dfrac{7.88}{63} \text{ mol L}^{-1}$
$= 0.125$ M
$= 0.125$ N (being monobasic acid)
(b) $Na_2CO_3 + 2HCl \longrightarrow 2NaCl + CO_2 + H_2O$
$Na_2CO_3 \equiv 2H^+$
$\therefore$ Normality of $Na_2CO_3 = 2 \times$ molarity
26.5 Na_2CO_3 per litre $= \dfrac{26.5}{106} \text{ mol L}^{-1} = 0.25$ M
$\therefore$ Normality = **0.50 N**

4. (a) meq = normality × volume (mL)
= 4.0 × 60
= 240
(b) eq = 0.240
(c) mol $= \dfrac{0.240}{2} = 0.120$ mol
(d) gram = 0.120 × 98 = **11.76 g**

5. 5.00 mL of H_2SO_4 = 84.6 mL of 2.00 N NaOH
$\therefore \quad N(H_2SO_4) = 33.84$ N
$\therefore \quad M(H_2SO_4) = 16.92$ M
1000 mL H_2SO_4 has = 16.92 mol H_2SO_4
since, mass/volume = density
$\therefore$ 1000 × 1.8 g H_2SO_4 has = 16.92 × 98 g H_2SO_4
$\therefore \quad 100 \text{ g } H_2SO_4 \text{ has} = \dfrac{16.92 \times 98 \times 100}{1000 \times 1.8}$
= **92.12%**

6. $NH_2CONH_2 + 2NaOH \longrightarrow 2NH_3 + CO_2 + H_2O$
$2NH_3 + 2HCl \longrightarrow 2NH_4Cl$
1 mol $NH_2CONH_2 \equiv 2$ mol HCl

1.0 g impure NH_2CONH_2 = 0.6 g pure NH_2CONH_2

= 0.01 mol

∴ HCl required by 0.01 mol urea = 0.02 mol

= 20 millimol

$1 \times V = 20$

∴ $V = 20$ **mL**

7. $Mn^{2+} \longrightarrow MnO_2$ **Change in O.N.** **Normality**

		Change in O.N.	Normality
$Mn^{2+} \longrightarrow MnO_2$			
+2	+4	2-unit	2 × molarity
$MnO_4^- \longrightarrow MnO_2$			
+7	+4	3-unit	3 × molarity

25 mL of $2M_1$ $Mn^{2+} \equiv 35 \times 0.06 \times 3$ N MnO_4^-

∴ M_1 = **0.126 M** Mn^{2+}

8.

		Change in O.N.	Normality
$SO_3^{2-} \longrightarrow SO_4^{2-}$			
+4	+6	2-unit	2 × molarity
$MnO_4^- \longrightarrow Mn^{2+}$			
+7	+2	5-unit	5 × molarity

$25 \times 2M_1$ (SO_3^{2-}) in $H_2O \equiv 34.08 \times 0.01964 \times 5\, MnO_4^-$

∴ M_1 = **0.0669 M** SO_3^{2-}

9. MnO_4^- Fe^{2+}

$32 \times N_1 \equiv 40 \times 0.4$ N

N_1 (MnO_4^-) = 0.5 N

∴ M_1 (MnO_4^-) = $\frac{0.5}{5}$ = **0.1 M**

10. 0.1 M $K_2Cr_2O_7$ ≡ 0.6 N $K_2Cr_2O_7$

11. (a) All the H^+ ions are neutralised.

Thus, 0.400 N = $\frac{0.400}{3}$ M H_3PO_4

Thus, moles of H_3PO_4 in 550 mL = $\frac{0.400 \times 550}{3 \times 1000}$

∴ Mass of H_3PO_4 = 0.073 mol

= **7.1867 g**

(b) It is reduced to HPO_3^{2-}

∴ 1 $H_3PO_4 \equiv 2e^-$

∴ $1\ e^- \equiv \frac{H_3PO_4}{2}$

∴ 0.400 N = $\frac{0.400}{2}$ M H_3PO_4

Thus, moles of H_3PO_4 in 550 mL

= $\frac{0.400 \times 550}{2 \times 1000}$ mol

= 0.11 mol

∴ = 10.78 g

12. $Fe^{2+} \longrightarrow Fe^{3+} + e^-$ Normality = molarity
+2 +3

$BrO_3^- \longrightarrow Br^-$ Normality = 6 × molarity
+5 −1

Fe^{2+} BrO_3^-

$10.00 \times N_1 \equiv 31.50 \times 0.10 \times 6$ N BrO_3^-

∴ $N_1(Fe^{2+})$ = 1.89 N

= **1.89 M** Fe^{2+}

13. $2Cu^{2+} + 4I^- \longrightarrow Cu_2I_2 + I_2$

$2S_2O_3^{2-} + I_2 \longrightarrow S_4O_6^{2-} + 2I^-$

$2Cu^{2+} \equiv I_2 \equiv 2I$

∴ $1\ Cu^{2+} \equiv I \equiv S_2O_3^{2-}$

∴ 1 mol Cu^{2+} = 1 equivalent Cu^{2+}

Also, $2S_2O_3^{2-} \equiv I_2 \equiv 2I$

∴ $1\ S_2O_3^{2-} \equiv I$

∴ 1 mol $S_2O_3^{2-}$ ≡ 1 equivalent $S_2O_3^{2-} \equiv Cu^{2+}$

Moles of Cu^{2+} ≡ moles of $S_2O_3^{2-}$

= $\frac{12.12 \times 0.1}{1000}$

= 1.212×10^{-3} mol

= $1.212 \times 10^{-3} \times 63.5$ g

= 0.076962 g

It is present in 1.1 g impure Cu ore

% of Cu^{2+} in sample = $\frac{0.076962}{1.100} \times 100$

= 6.996%

14. As in Q. 13

500 mL of 0.200 N $Na_2S_2O_3 \cdot 5H_2O$

≡ 500 × 0.200 millimol

= 0.1 mol

= **24.8 g** $Na_2S_2O_3 \cdot 5H_2O$

Total Practice Set
(Read, Plan & Solve!)

Problem 1. What are the molar concentrations of the solutes listed below **when** dissolved in water?
(a) 3.28 mol C_2H_5OH in 7.16 L of solution
(b) 3.67×10^{-3} mol CH_3OH in 50.00 mL of solution
(c) 22.3 g $(CH_3)_2CO$ in 125 mL solution
(d) 14.5 mL of pure glycerol; $C_3H_8O_3$ ($d = 1.26$ g / cm^3) in 425 mL solution

Plan $Molarity = mol\ L^{-1} = \dfrac{g}{g\ mol^{-1}}\ L^{-1} = \dfrac{g \times 1000}{g\ mol^{-1} \times mL}$

$mass = volume \times density$

Solution (a) 3.28 mol C_2H_5OH is 7.16 L of solution

$\therefore \quad M(C_2H_5OH) = \dfrac{3.28}{7.16}$ mol L^{-1} = **0.458 M**

(b) 3.67×10^{-3} mol CH_3OH in 50.00 mL (= 0.05 L) solution

$\therefore \quad M(CH_3OH) = \dfrac{3.67\times 10^{-3}}{0.05}$ mol L^{-1} = **0.0734 M**

(c) 22.3 g $(CH_3)_2CO$ in 125 mL solution

$= \dfrac{22.3}{58}$ mol $(CH_3)_2CO$ in 0.125 L solution

$\therefore \quad M[(CH_3)_2CO] = \dfrac{\frac{22.3}{58}}{0.125}$ mol L^{-1} = **3.076 M**

(d) 14.5 mL of pure glycerol = 14.5×1.26 g

$= \dfrac{14.5\times 1.26}{92}$ mol in 0.425 L solution

$\therefore \quad M(\text{glycerol})$ = **0.467 M**

Problem 2. Water is evaporated from 135 mL of 0.224 M $MgSO_4$ solution until the solution volume becomes 105 mL. What is the molarity of $MgSO_4$ in the solution that results?

Plan *When solution is evaporated, volatile solvent (like* H_2O*) evaporates, but non-volatile solute remains unaffected. Due to change in volume of solution, molarity also changes.*

Solution 135 mL of 0.224 M $MgSO_4$ solution
$= 135\times 0.224$ millimol of $MgSO_4$
$= \dfrac{135\times 0.224}{1000}$ mol $MgSO_4$
$= 0.0302$ mol $MgSO_4$

Same amount is present in 105 mL solution after evaporation of water.

Thus, 105 mL solution has = 0.0302 mol $MgSO_4$
1000 mL solution has = $\dfrac{0.0302}{105}\times 1000$ = **0.288 M**

Alternate method $\underset{\text{initial}}{M_1V_1} = \underset{\text{final}}{M_2V_2}$

$0.224\times 135 = M_2 \times 105$
M_2 = **0.228 M**

Problem 3. Find the volume of a 0.2 M solution of MnO_4^- which will react with 50.0 mL of 0.1 M solution of $C_2O_4^{2-}$ in acidic medium.

Plan *Normality of* MnO_4^- $= 5\times$ *molarity*
and *normality of* $C_2O_4^{2-}$ $= 2\times$ *molarity*
use $N_1V_1 = N_2V_2$
Milliequivalents of reacting species are equal.

Solution

$$\underset{+7}{2MnO_4^-} + \underset{+6}{5C_2O_4^{2-}} + 16H^+ \longrightarrow \underset{+8}{10CO_2} + \underset{+2}{2Mn^{2+}} + 8H_2O$$

(+6 → +8: Oxidation; +7 → +2: Reduction)

Note: We have taken oxidation number of two C atoms in $C_2O_4^{2-}$ (true definition of O.N.) hence, we also take O.N. of two C atoms in CO_2.

Thus, $N(MnO_4^-)$ = change in O.N. $\times M$
$N = 5\times 0.2 = 1$ N
$N(C_2O_4^{2-}) = 2\times M = 2\times 0.1 = 0.2$ N

using $N_1V_1(MnO_4^-) = N_2V_2(C_2O_4^{2-})$
$1.0\times V_1 = 0.2\times 50$
V_1 = **10 mL 0.2 M MnO_4^-**

Note: We can't use $M_1V_1 = M_2V_2$ when change in O.N. or acidity and basicity are different.

Alternate method Let volume of $MnO_4^- = V$ mL

V mL of 0.2 M MnO_4^- = $0.2V$ millimol of MnO_4^-
50 mL of 0.1 M $C_2O_4^{2-}$ = 50×0.1 = 5 millimol of $C_2O_4^{2-}$
Using reaction: 5 mol of $C_2O_4^{2-}$ are oxidised by = 2 mol of MnO_4^-

5 millimol of $C_2O_4^{2-}$ are oxidised by = 2 millimol of MnO_4^-

$$0.2\,V = 2$$

$$V = \mathbf{10\ mL}$$

Problem 4. Two acids, A and B are titrated separately. 25 mL of 1 N Na_2CO_3 solution requires 10 mL of A and 40 mL of B for complete neutralisation. What volume of A and B would you mix to produce 1 L of 1 N acid solution?

Plan *Normality of each acid is known by titration. Thus, from total normality and the relation*

$$N_{\text{Total}} = \frac{\Sigma\, NV}{\Sigma\, V}$$

volume of each acid is determined.

Solution 10 mL of N_A acid A = 25 mL of 1 N Na_2CO_3

$$N_A = 2.5$$

40 mL of N_B acid B = 25 mL of 1 N Na_2CO_3

$$N_B = 0.625$$

Let volume of acid $A = x$ L

and that of acid $B = (1-x)$ L

Resultant normality of mixture = 1 N

using relation $N_{\text{mix}} = \dfrac{\Sigma\, NV}{\Sigma\, V} = \dfrac{N_A V_A + N_B V_B}{V_A + V_B}$

$$1 = \frac{2.5x + 0.625(1-x)}{1}$$

$$x = 0.2\ \text{L}$$

Volume of acid A = **0.2 L**

Volume of acid B = **0.8 L**

Problem 5. A sample supposed to be pure $CaCO_3$ is used to standardise a solution of HCl. The substance really was a mixture of $MgCO_3$ and $BaCO_3$, but the standardisation of HCl was accurate. Find the percentage of $BaCO_3$ and $MgCO_3$ in the mixture. (Ca = 40, Mg = 24, Ba = 137)

Plan *$CaCO_3$ and mixture of $MgCO_3$ and $BaCO_3$ give same result in the standardisation of HCl hence,*

Equivalent of $CaCO_3$ = equivalent of $MgCO_3$ and $BaCO_3$

Thus, by assuming a quantity of $CaCO_3$ (which is also equal to mixture of $MgCO_3$ and $BaCO_3$), % of each can be determined.

Solution Equivalent of $CaCO_3$ = equivalent of mixture

$$CaCO_3 \equiv (MgCO_3 + BaCO_3)$$

Let $CaCO_3$ be 100 g, g-equivalent $= \dfrac{100}{50} = 2$

Let $MgCO_3 = x$ g

$BaCO_3 = (100-x)$ g

g-equivalent of $MgCO_3 = \dfrac{x}{42}$

g-equivalent of $BaCO_3 = \dfrac{100-x}{197/2} = \dfrac{100-x}{98.5}$

$$\frac{x}{42} + \frac{100-x}{98.5} = 2$$

This gives $x = 72.11$

$\therefore$ $MgCO_3 = \mathbf{72.11\%}$

$BaCO_3 = \mathbf{27.89\%}$

Problem 6. Determine the equivalent weight in each of the following

(a) the reduction half-reaction for the disproportionation of Br_2 in base,

(b) the oxidation half-reaction which yields bromate ion,

(c) the overall reaction.

Plan *Equivalent weight* $= \dfrac{\textit{formula weight}}{\textit{change in oxidation number}}$

and in a disproportionation reaction $= E_O + E_R$

where, E_O is the equivalent weight in oxidation half-reaction and E_R is the equivalent weight in reduction half-reaction.

Solution (a) $2e^- + Br_2 \longrightarrow 2Br^-$

$$\text{Eq. wt.} = \frac{\text{mol. wt. of } Br_2}{2} = \frac{160}{2}$$

$$= \mathbf{80\ g\ equiv^{-1}}$$

(b) $Br_2 + 12OH^- \longrightarrow 2BrO_3^- + 6H_2O + 10e^-$

$$\text{Eq. wt.} = \frac{\text{mol. wt. of } Br_2}{10} = \frac{160}{10} = \mathbf{16\ g\ equiv^{-1}}$$

(c) $6OH^- + 3Br_2 \longrightarrow 5Br^- + BrO_3^- + 3H_2O$

Since, number of electrons involved in each half-reaction should be identical, hence 10 electrons are involved in oxidation then 10 electrons should also be involved in reduction thus,

$$5 \times (a) + (b)$$

$$12OH^- + 6Br_2 \longrightarrow 10Br^- + 2BrO_3^- + 6H_2O$$

Hence, Eq. wt. $(Br_2) = \dfrac{6 \times 160}{10} = \mathbf{96.0\ g\ equiv^{-1}}$.

or net eq. wt. = eq. wt. of step (a) + eq. wt. of step (b) = **96**

Problem 7. 100.0 mL sample of hard water is passed through a column of the ion exchange resin H_2R. The water coming off the column requires 15.17 mL of 0.0265 M NaOH for its titration. What is the hardness of the water expressed as ppm of Ca^{2+}?

Plan *Ca^{2+} in hard water replaces H^+ of ion-exchange column*

$$Ca^{2+} \equiv 2H^+ \equiv 2OH^-$$

$$\textit{mol of } Ca^{2+} \equiv \frac{1}{2} \times \textit{mol of } OH^-$$

Thus, ppm concentration of Ca^{2+} can be determined.

Solution $Ca^{2+} + H_2R \longrightarrow 2H^+ + CaR$

Every Ca^{2+} exchanged gives two H^+ ion.

$$2H^+ + 2OH^- \longrightarrow 2H_2O$$

Thus, $$2OH^- \equiv 2H^+ \equiv 1\ Ca^{2+}$$

$$\text{moles of } Ca^{2+} = \frac{1}{2} \text{ mol of } OH^-$$

$$= \frac{1 \times 0.0265 \times 15.17}{2 \times 1000} \text{ mol}$$

$$= 2.01 \times 10^{-4} \text{ mol}$$

∴ 100 mL of hard water has

$$Ca^{2+} = 2.01 \times 10^{-4} \text{ mol } Ca^{2+}$$

$$= 2.01 \times 10^{-4} \times 40 \text{ g } Ca^{2+}$$

∴ 10^6 mL hard water has Ca^{2+}

$$= \frac{2.01 \times 10^{-4} \times 40 \times 10^6}{100} \text{ g}$$

$$= 80.4 \text{ g}$$

∴ Hardness of water = **80.4 ppm Ca^{2+}**

Alternate method

Equivalent of Ca^{2+} = equivalent of H^+ or OH^-

$$= \frac{15.17 \times 0.0265}{1000} = 4.02 \times 10^{-4}$$

∴ Ca^{2+} in 100 mL water = $4.02 \times 10^{-4} \times 20$ g

$$Ca^{2+} \text{ in } 10^6 \text{ mL water} = \frac{4.02 \times 10^{-4} \times 20 \times 10^6}{100}$$

$$= \mathbf{80.4\ ppm}$$

Problem 8. 1 g of fuming H_2SO_4 (oleum : it is a mixture of conc. H_2SO_4 saturated with SO_3 and having formula $H_2S_2O_7$) diluted with H_2O. This solution is completely neutralised by 26.7 mL of 0.8 N NaOH. Find the percentage of free SO_3 in the oleum.

Plan *SO_3 and H_2SO_4 in oleum both act as a dibasic acid and are neutralised by* NaOH.

Also, equivalents of SO_3 and H_2SO_4 = equivalents of NaOH

Thus, percentage of SO_3 in the oleum can be determined.

Solution $\underset{\text{oleum}}{H_2S_2O_7} + H_2O \longrightarrow 2H_2SO_4$

SO_3 part of $H_2S_2O_7$ is converted into H_2SO_4, hence, SO_3 acts also as a dibasic acid.

$$\text{Equivalent weight } (SO_3) = \frac{M}{2} = 40$$

Let H_2SO_4 in the fuming $H_2SO_4 = x$ g

and $$SO_3 = (1 - x) \text{ g}$$

$$\text{Equivalent of } H_2SO_4 = \frac{x}{49}$$

and that of $$SO_3 = \frac{1-x}{40}$$

$$\text{Equivalent of NaOH used} = \frac{26.7 \times 0.8}{1000} = 0.02136$$

∴ $$\frac{x}{49} + \frac{1-x}{40} = 0.02136$$

∴ $$x = 0.7927 \text{ g } H_2SO_4 \text{ in 1 g oleum.}$$

∴ percentage of H_2SO_4 = **79.27%**

and SO_3 = **20.73%**

Problem 9. A sample of water has a hardness expressed as 77.5 ppm Ca^{2+}. This sample is passed through an ion exchange column and the Ca^{2+} is replaced by H^+. What is the pH of the water after it has been so treated?

Plan *Ca^{2+} in hard water replaces H^+ of cationic ion exchange column.*

$$Ca^{2+} \equiv 2H^+$$

From ppm concentration of Ca^{2+}, molar concentration of H^+ can be determined. Hence, pH = − log H^+ can be determined.

Solution $Ca^{2+} + \underset{\text{(ion exchange column)}}{RH_2} \longrightarrow RCa + 2H^+$

every Ca^{2+} replaced gives two H^+ ions.

$$10^6 \text{ mL of } H_2O \text{ has} = 77.5 \text{ g } Ca^{2+} = \frac{77.5}{40} \text{ mol } Ca^{2+}$$

$$= 2\left(\frac{77.5}{40}\right) \text{ mol } H^+$$

$$10^3 \text{ mL of } H_2O \text{ has } H^+ = \frac{2 \times 77.5}{40} \times 10^{-3} \text{ mol } H^+$$

$$= 3.875 \times 10^{-3} \text{ M}$$

∴ $$pH = -\log [H_3O^+]$$

$$= \mathbf{2.41}$$

Problem 10. A sample of water has its hardness due only to $CaSO_4$. When this water is passed through an ion exchange resin, SO_4^{2-} ions are replaced by OH^-. A 25.00 mL sample of water so treated requires 21.58 mL of 1.00×10^{-3} M H_2SO_4 for its titration. What is the hardness of the water; expressed in ppm of $CaSO_4$? Assume the density of the water is 1.00 g mL^{-1}.

Plan *SO_4^{2-} replaces OH^- ions (in ion-exchange resin) and OH^- in turn is determined by titration with H_2SO_4.*

$$SO_4^{2-} \equiv 2OH^- \equiv H_2SO_4$$

Thus, hardness in ppm of $CaSO_4$ can be determined.

Solution Let an ion exchange resin be $R(OH)_2$

$$R(OH)_2 + SO_4^{2-} \longrightarrow RSO_4 + 2(OH^-)$$

every SO_4^{2-} exchanged gives two (OH^-)

OH^- is neutralised by H_2SO_4

$$2OH^- + H_2SO_4 \longrightarrow 2H_2O + SO_4^{2-}$$

Thus, $1\ CaSO_4 \text{ (in hard water)} \equiv 2OH^- \equiv 1\ H_2SO_4$

Moles of $CaSO_4 \equiv$ moles of $H_2SO_4 = \dfrac{21.58 \times 10^{-3}}{1000}$

$= 2.158 \times 10^{-5}$ mol

This amount is in 25 mL hard water

25 mL of hard water has hardness

$= 2.158 \times 10^{-5}$ mol $CaSO_4$

$= 2.158 \times 10^{-5} \times 136$ g $CaSO_4$

10^6 mL (per million) of hard water has hardness

$= \dfrac{2.158 \times 10^{-5} \times 136 \times 10^6}{25}$ g $CaSO_4$

= 117.4 ppm $CaSO_4$

Problem 11. What mass of Na_2SO_3 must have been present in a sample that required 26.50 mL of 0.0510 M $KMnO_4$ for its oxidation to Na_2SO_4 in acidic solution?

Plan *Taking the fact that equivalent of* $SO_3^{2-} \equiv$ equivalent of MnO_4^- *amount of* Na_2SO_3 *can be calculated. or on comparison of moles of* MnO_4^- *and* SO_3^{2-}, *amount of* Na_2SO_3 *can be calculated.*

Solution Redox reaction is

$$5SO_3^{2-} + 2MnO_4^- + 6H^+ \longrightarrow 2Mn^{2+} + 5SO_4^{2-} + 3H_2O$$

($MnO_4^- \to Mn^{2+}$: $+5e^-$; $SO_3^{2-} \to SO_4^{2-}$: $-2e^-$)

Equivalent weight of $MnO_4^- = \dfrac{M}{5}$

Equivalent weight of $SO_3^{2-} = \dfrac{M}{2}$

26.50 mL of 0.0510 M MnO_4^-

$\equiv$ 26.50 mL of 0.2550 N MnO_4^-

$\equiv$ 26.50 mL of 0.2550 N SO_3^{2-}

Equivalent of $Na_2SO_3 = \dfrac{0.2550 \times 26.50}{1000}$

$= 6.7575 \times 10^{-3}$

$\therefore$ Weight of $Na_2SO_3 = 6.75 \times 10^{-3} \times 63$ g

= 0.426 g

Alternate method

Moles of $MnO_4^- = \dfrac{26.5 \times 0.051}{1000}$

$= 1.35 \times 10^{-3}$ mol

$\therefore$ Moles of $Na_2SO_3 = \dfrac{5}{2} \times 1.35 \times 10^{-3}$ mol

= 0.426 g

Problem 12. 25 mL of 0.1122 M $AgNO_3$ is added to an excess of $K_2CrO_4(aq)$. What mass of Ag_2CrO_4 will precipitate from the solution? (Ag = 108, Cr = 52, N = 14, O = 16)

$$2AgNO_3 + K_2CrO_4 \longrightarrow Ag_2CrO_4 + 2KNO_3$$

Plan $\dfrac{\textit{Molarity} \times \textit{volume in mL}}{1000}$ = *moles of the substances*

Based on reaction 2 mol $AgNO_3 \equiv$ 1 mol Ag_2CrO_4 *hence, number of moles of* Ag_2CrO_4 *obtained from given mol of* $AgNO_3$ *can be calculated.*

Solution From the reaction:

2 mol of $AgNO_3$ gives = 1 mol Ag_2CrO_4

hence, $\dfrac{25 \times 0.1122}{1000}$ mol of $AgNO_3$ gives

$= \dfrac{25 \times 0.1122}{1000 \times 2}$ mol Ag_2CrO_4

$= 0.0014$ mol Ag_2CrO_4

$= 0.0014 \times 332$ g Ag_2CrO_4

= 0.4656 g Ag_2CrO_4

Problem 13. A sample of pure $KHC_2O_4 \cdot H_2C_2O_4 \cdot 2H_2O$ requires 46.2 mL of 0.1 M NaOH for titration. How many millilitres of 0.1 M $KMnO_4$ will the sample react?

Plan NaOH *neutralises three ionisable* H^+ *and each mole of the sample has two moles of* $C_2O_4^{2-}$ *ions oxidised by* MnO_4^-. *Also*

moles of $C_2O_4^{2-} = \dfrac{2}{3}$ *mol of* H^+

From the number of moles of OH^-, *number of moles of* H^+ *and hence that of* $C_2O_4^{2-}$ *are known. Thus, volume of* MnO_4^- *required to oxidise* $C_2O_4^{2-}$ *can be calculated.*

Solution $H^+ + OH^- \longrightarrow H_2O$

$$5C_2O_4^{2-} + 2MnO_4^- + 16H^+ \longrightarrow 2Mn^{2+} + 10CO_2 + 8H_2O$$

Number of moles of OH^-

$\equiv$ number of moles of H^+ ion $= \dfrac{46.2 \times 0.1}{1000}$

$= 4.62 \times 10^{-3}$ mol

$\therefore$ Number of moles of $C_2O_4^{2-}$ ion

$= \dfrac{2}{3} \times 4.62 \times 10^{-3} = 3.08 \times 10^{-3}$ mol

Since, 5 mol of $C_2O_4^{2-} \equiv$ 2 mol of MnO_4^-

$\therefore$ 3.08×10^{-3} mol of $C_2O_4^{2-} = \dfrac{2}{5} \times 3.08 \times 10^{-3}$

$= 1.232 \times 10^{-3}$ mol MnO_4^-

If V mL of 0.1 M MnO_4^- are required, then

$\dfrac{V \times 0.1}{1000} = 1.232 \times 10^{-3}$

$\therefore$ V = **12.32 mL**

Alternate method

Sample has three H^+ ions

$\therefore$ 46.2 mL of 0.1 M NaOH $\equiv 3H^+$ ions

$\therefore$ Equivalent of salt as acid $= \dfrac{46.2 \times 0.1}{3 \times 1000} = 1.54 \times 10^{-3}$

$\therefore$ Equivalent of $C_2O_4^{2-}$ in two oxalates $= 6.16 \times 10^{-3}$

$$\text{Equivalent of 0.1 M } MnO_4^- = \frac{V \times 0.1 \times 5}{1000}$$

$$\therefore \quad \frac{V \times 0.1 \times 5}{1000} = 6.16 \times 10^{-3}$$

$$\therefore \quad V = \mathbf{12.32\ mL}$$

Problem 14. A 0.8040 g sample of an iron ore is dissolved in acid. The iron is then reduced to Fe^{2+} and titrated with 47.22 mL of 0.02242 M $KMnO_4$ solution. Calculate the results of this analysis in terms of (a) % Fe (56 g mol^{-1}) (b) % Fe_3O_4 (232 g mol^{-1})

Plan *(a)* Fe^{2+} *is oxidised to* Fe^{3+} *by* MnO_4^- *in acidic medium.*

$$MnO_4^- + 5Fe^{2+} + 8H^+ \longrightarrow Mn^{2+} + 5Fe^{3+} + 4H_2O$$

From the stoichiometric ratio, on comparison, moles of Fe hence, its amont (in gram) is calculated. Thus % of Fe is also determined.

(b) $\quad 5Fe^{2+} \equiv 1\,MnO_4^-$

$$\therefore \quad 5Fe_3O_4 \equiv 15\,Fe^{2+} \equiv 3MnO_4^-$$

Knowing number of moles of Fe_3O_4 *which is equivalent to 3 mol* MnO_4^-*, its % can be determined.*

Solution $\quad \text{Moles of } MnO_4^- = \dfrac{\text{molarity} \times \text{volume (in mL)}}{1000}$

$$1 \text{ mol } MnO_4^- \equiv 5 \text{ mol of } Fe^{2+}$$

$$\therefore \quad \frac{47.22 \times 0.02242}{1000} \text{ mol } MnO_4^-$$

$$\equiv \frac{5 \times 47.22 \times 0.02242}{1000} \text{ mol } Fe^{2+}$$

$$= 0.00529 \text{ mol } Fe^{2+}$$

$$= 0.00529 \times 56$$

$$= 0.2964 \text{ g } Fe^{2+}$$

Thus, % of pure Fe(Fe^{2+}) in 0.8040 g sample

$$= \frac{0.2964}{0.8040} \times 100$$

$$= \mathbf{36.87\%}$$

(b) 3 mol MnO_4^- = 5 mol Fe_3O_4

$$\therefore \quad \frac{47.22 \times 0.02242}{1000} \text{ mol } MnO_4^-$$

$$\equiv \frac{5}{3} \times \frac{47.22 \times 0.02242}{1000} \text{ mol}$$

$$= 0.001764 \text{ mol } Fe_3O_4$$

$$= 0.001764 \times 232 \text{ g } Fe_3O_4$$

$$= 0.4094 \text{ g}$$

$$\% \text{ as } Fe_3O_4 = \frac{0.4094}{0.8040} \times 100$$

$$= \mathbf{50.91\%}$$

Problem 15. 25 mL of a solution containing 6.1 g L^{-1} of an oxalate of formula $K_xH_y(C_2O_4)_z \cdot nH_2O$ required 18 mL of 0.1 N NaOH and 24 mL of 0.1 N $KMnO_4$ in two separate titrations. What is the compound?

Plan $K_xH_y(C_2O_4)_z \cdot nH_2O$ *has* y *ionisable* H^+ *ions determined by* NaOH *and* z *oxidisable* $C_2O_4^{2-}$ *ions determined by* MnO_4^-, K^+ *and* H^+ *both are monovalent cations and* $C_2O_4^{2-}$ *is divalent anion hence,* $x + y = 2z$

Thus, from three linear equations, value of x, y and z can be determined. Value of n is determined when molecular weight of the compound is also known.

Solution Concentration = 6.1 g/L

K^+ and H^+ are cations and oxalate $C_2O_4^{2-}$ is an anion

hence, $\quad x + y = 2z \quad$ ...(i)

(a) NaOH will neutralize H^+ part

25 mL of N_1 (acid oxalate) ≡ 18 mL of 0.1 N NaOH

$$\therefore \quad N_1 \text{ (acid oxalate)} = \frac{18 \times 0.1}{25} = 0.072 \text{ N}$$

$$\text{Eq. wt. of an acid salt} = \frac{6.1}{0.072} = 84.72 \text{ g equiv}^{-1}$$

(b) $KMnO_4$ will oxidise $C_2O_4^{2-}$

$$\therefore \quad 25 \text{ mL } N_1'(C_2O_4^{2-}) \equiv 24 \text{ mL of } 0.01 \text{ N } KMnO_4$$

$$N_1'(C_2O_4^{2-}) = 0.096 \text{ N}$$

$$\text{Eq. wt. of oxalate (reducing agent)} = \frac{6.1}{0.096} = 63.54$$

Thus, $\dfrac{\text{mol. wt.}}{y}$ = eq wt. (84.72) when it is an acid salt

$\dfrac{\text{mol. wt.}}{2z}$ = eq. wt. (63.54) when it is a reducing agent

$$\therefore \quad \frac{2z}{y} = \frac{84.72}{63.54}$$

$$\therefore \quad y = 1.5z \quad \text{...(ii)}$$

Using Eqs. (i) and (ii), $x + 1.5z = 2z$

$$x = 0.5z$$

∴ *x*, *y* and *z* are in the ratio of 1 : 3 : 2 (to balance charge)

or $\quad x = 1, \quad y = 3, \quad z = 2$

$$\therefore \quad \text{Mol. wt.} = 84.72 \times 3 = 254.16$$

$$39x + y + 88z + 18n = 254.16$$

$$\therefore \quad n = 2$$

hence, salt is $\mathbf{KH_3(C_2O_4)_2 \cdot 2H_2O}$.

Problem 16. 30 mL of an acidified solution of 1.5 N MnO_4^- ions, 15 mL of 0.5 N oxalic acid and 15 mL of 0.4 N ferrous salt solution are added together. Find the molarities of MnO_4^- and Fe^{3+} ions in the final solution.

Plan MnO_4^- *is reduced to* Mn^{2+}, Fe^{2+} *is oxidised to* Fe^{3+} *and* $C_2O_4^{2-}$ *is oxidised to* CO_2

$$molarity = \frac{normality}{change\ in\ oxidation\ number}$$

$$\therefore \quad M(MnO_4^-) = \frac{N}{5}(MnO_4^-)$$

$$M(Fe^{2+}) = N(Fe^{2+})$$

$$M(C_2O_4^{2-}) = \frac{N}{2}(C_2O_4^{2-})$$

On comparison of milliequivalents of each reactant, final concentration of Fe^{3+} *and* MnO_4^- *can be calculated.*

Solution Acidified MnO_4^- oxidises Fe^{2+} into Fe^{3+} and $C_2O_4^{2-}$ into CO_2.

$$MnO_4^- + 8H^+ + 5Fe^{2+} \longrightarrow Mn^2 + 5Fe^{3+} + 4H_2O$$

$$2MnO_4^- + 5C_2O_4^{2-} + 16H^+ \longrightarrow 2Mn^{2+} + 10CO_2 + 8H_2O$$

Milliequivalents of $MnO_4^- = 30 \times 1.5 = 45$

Milliequivalents of $Fe^{2+} = 15 \times 0.4 = 6.0$

Milliequivalents of oxalic acid $= 15 \times 0.5 = 7.5$

Milliequivalents of MnO_4^- used by 6.0 milliequivalents of

$Fe^{2+} = 6.0$

milliequivalents of MnO_4^- used by 6.5 milliequivalents of oxalic acid = 7.5

thus, total milliequivalents of MnO_4^- used = 13.5

milliequivalents of MnO_4^- left = 45 – 13.5

= 31.5 in 60 mL

$$\text{thus,} \quad [MnO_4^-] = \frac{31.5 \times 10^{-3} \times 1000\,N}{60} = \frac{31.5}{60}N$$

$$= \frac{31.5}{60} \times \frac{1}{5}M = \mathbf{0.105\ M}$$

Milliequivalents of Fe^{2+} consumed

= milliequivalents of Fe^{3+} formed

= 6.0 in 60 mL solution

$$\text{thus,} \quad [Fe^{3+}] = \frac{6.0}{60} \times 10^{-3} \times 1000\,N = 0.1\,N$$

$$= \mathbf{0.1\ M}$$

Problem 17. If 10 g of V_2O_5 is dissolved in acid and is reduced to V^{2+} by zinc metal, how many moles of I_2 could be reduced by the resulting solution if it is further oxidised to VO^{2+} ions? (V = 51, O = 16, I = 127)

Plan From the stoichiometric coefficients, number of moles of I_2 *reduced by given amount of* V_2O_5 *can be determined.*

Solution

$$V_2O_5 + 10H^+ + 3Zn \longrightarrow 3Zn^{2+} + 2V^{2+} + 5H_2O$$

$$H_2O + I_2 + V^{2+} \longrightarrow 2I^- + VO^{2+} + 2H^+] \times 2$$

$$V_2O_5 + 6H^+ + 3Zn + 2I_2 \longrightarrow 3Zn^{2+} + 4I^- + 3H_2O + 2VO^{2+}$$

Thus, 1 mol of V_2O_5 reduces = 2 mol of I_2

$$\text{hence,} \quad 10\ g = \frac{10}{182}\ \text{mol of}\ V_2O_5\ \text{reduces}$$

$$= 2 \times \frac{10}{182}\ \text{mol of}\ I_2$$

$$= \mathbf{0.11\ mol\ of\ I_2}$$

Problem 18. 1.44 g pure FeC_2O_4 was dissolved in dil. HCl and solution diluted to 100 mL. Calculate volume of 0.01 M $KMnO_4$ required to oxidise FeC_2O_4 solution completely.

Plan MnO_4^- *oxidises* Fe^{2+} *as well as* $C_2O_4^{2-}$. *Thus, from the number of moles of* Fe^{2+} *and* $C_2O_4^{2-}$ *number of moles hence, volume of* MnO_4^- *can be determined.*

Solution Fe^{2+} is oxidised to Fe^{3+} and $C_2O_4^{2-}$ is oxidised to CO_2 by MnO_4^- in acidic medium.

$$5Fe^{2+} + MnO_4^- + 8H^+ \longrightarrow 5Fe^{3+} + Mn^{2+} + 4H_2O$$

$$5C_2O_4^{2-} + 2MnO_4^- + 16H^+ \longrightarrow 10CO_2 + 2Mn^{2+} + 8H_2O$$

$$5FeC_2O_4 + 3MnO_4^- + 24H^+ \longrightarrow 5Fe^{3+} + 10CO_2 + 3Mn^{2+} + 12H_2O$$

Thus, 5 mol of FeC_2O_4 are oxidised by = 3 mol of MnO_4^-

1.44 g (= 0.01 mol) FeC_2O_4 is oxidised by

$$= \frac{3}{5} \times 0.01 = 0.006\ \text{mol}\ MnO_4^-$$

If V mL of 0.01 M $KMnO_4$ are required then

$$\frac{0.01 \times V}{1000} = 0.006$$

$$\therefore \quad V = \mathbf{600\ mL}$$

$$\left(\text{moles} = \frac{mV}{1000} = \frac{0.01 \times V}{1000} = 0.006\right)$$

Problem 19. For titrating a certain volume of a reducing substance by 1 M $KMnO_4$, it is found that 20 mL was used in acidic medium, 33.4 mL was used in alkali medium and 100 mL was used in neutral medium. If MnO_4^- is reduced to Mn^{2+} in acidic medium, to what state MnO_4^- is reduced to in alkaline and neutral medium?

Plan From the volume of MnO_4^- *used in acidic medium, normality of* MnO_4^- *in basic medium as well as in neutral medium can be determined. Since*

normality = molarity × change in O.N.

hence, O.N. of Mn *in final state is known.*

Solution In acidic medium MnO_4^- is reduced to Mn^{2+} (by gain of $5e^-$)

$$\underset{+7}{Mn}O_4^- + 8H^+ + 5e^- \longrightarrow \underset{+2}{Mn^{2+}} + 4H_2O$$

$$\text{thus,} \quad 1\,M\,MnO_4^- = 5\,N\,MnO_4^-$$

(normality = molarity × change in oxidation number)

Let normality in alkaline medium be N' and in neutral medium N''

$$20 \text{ mL of } 5\text{ N } MnO_4^- \equiv 33.4 \text{ mL } N'\,MnO_4^-$$

$$\equiv 100 \text{ mL of } N''\,MnO_4^-$$

$$\therefore \quad N'(\text{alkaline}) = 3\text{ N}$$

$$N''(\text{neutral}) = 1\text{ N}$$

Thus, change in oxidation number in alkaline medium = 3

Hence, MnO_4^- is reduced to MnO_2 (+4)

$$\text{Thus,} \quad \underset{+7}{MnO_4^-} + 2H_2O + 3e^- \longrightarrow \underset{+4}{MnO_2} + 4OH^-$$

MnO_4^- is reduced to MnO_4^{2-} (+6) in neutral medium when change in oxidation number is one unit :

$$\text{Thus,} \quad \underset{+7}{MnO_4^-} + e^- \longrightarrow \underset{+6}{MnO_4^{2-}}$$

Problem 20. If 1.00 mL $KMnO_4$ solution will react with 0.125 g Fe^{2+} and if 1.00 mL of $KHC_2O_4 \cdot H_2C_2O_4$ solution will react with 0.175 mL of $KMnO_4$ solution, how many millilitres of 0.20 M NaOH will react with 1.00 mL of $KHC_2O_4 \cdot H_2C_2O_4$ solution in which all the protons (H^+) are ionisable? ($Fe = 56 \text{ g mol}^{-1}$)

Plan *We are to relate oxidisable $C_2O_4^{2-}$ ions with ionisable H^+ ions in $KHC_2O_4 \cdot H_2C_2O_4$. Molar concentration of MnO_4^- is determined by Fe^{2+}. In turn $C_2O_4^{2-}$ is determined by MnO_4^-. Thus, number of ionisable H^+ hence, volume of NaOH can be determined.*

Solution

$$5Fe^{2+} + MnO_4^- + 8H^+ \longrightarrow 5Fe^{3+} + Mn^{2+} + 4H_2O$$

$$5C_2O_4^{2-} + 2MnO_4^- + 16H^+ \longrightarrow 10CO_2 + 2Mn^{2+} + 8H_2O$$

$$H^+ + OH^- \longrightarrow H_2O$$

Let molarity of MnO_4^- solution = M_1

and molarity of $KHC_2O_4 \cdot H_2C_2O_4$ solution = M_2

$$\text{Number of moles of } Fe^{2+} = \frac{0.125}{56} \text{ mol}$$

$$\text{Number of moles of } MnO_4^- = \frac{1 \times M_1}{1000} = 0.001\,M_1$$

$$\text{Also,} \quad 5 \text{ mol of } Fe^{2+} = 1 \text{ mol } MnO_4^-$$

$$\therefore \quad \frac{0.125}{56} \text{ mol of } Fe^{2+} = \frac{1}{5} \times \frac{0.125}{56} \text{ mol } MnO_4^-$$

$$\frac{0.125}{5 \times 56} = 0.001\,M_1$$

$$\therefore \quad M_1\,(MnO_4^-) = 0.446\text{ M}$$

$$\text{Also, } 1.00 \text{ mL of } M_2 KHC_2O_4 \cdot H_2C_2O_4$$

$$\equiv 0.175 \text{ mL of } 0.446\text{ M } MnO_4^-$$

$$1 \times M_2 \text{ millimol } KHC_2O_4 \cdot H_2C_2O_4$$

$$= 0.001\,M_2 \text{ mole of } KHC_2O_4 \cdot H_2C_2O_4$$

$$= 0.002\,M_2 C_2O_4^{2-} \text{ ions}$$

(since, $KHC_2O_4 \cdot H_2C_2O_4$ has two oxidisable $C_2O_4^{2-}$ ions)

$$5 \text{ mol of } C_2O_4^{2-} = 2 \text{ mol of } MnO_4^-$$

$$\therefore \quad 0.002\,M_2\;C_2O_4^{2-} = \frac{2}{5} \times 0.002\,M_2\;MnO_4^-$$

$$\therefore \quad \frac{0.004}{5} M_2 = \frac{0.175 \times 0.446}{1000} \text{ mol } MnO_4^-$$

$$\therefore \quad M_2 = \frac{0.175 \times 0.446 \times 5}{1000 \times 0.004}$$

$$M_2(KHC_2O_4 \cdot HC_2O_4) = 0.0976\text{ M}$$

Problem 21. A mixture containing Na_2CO_3, NaOH and inert matter weighs 0.7500 g. When the aqueous solution is titrated with 0.5000 N HCl, the colour of the phenolphthalein disappears when 21.00 mL of the acid has been added. Methyl orange is then added and 7.00 mL more of the acid is requried to give a red colour to the solution. Find the percentage composition of the sample.

Plan *Phenolphthalein indicates complete reaction of NaOH and 50% reaction of Na_2CO_3. Remaining 50% reaction of Na_2CO_3 is indicated when methyl orange is added at this stage. Thus, from the volume of HCl,% of each is determined.*

$$I: \quad NaOH + HCl \longrightarrow NaCl + H_2O$$

$$II: \quad Na_2CO_3 + HCl \longrightarrow NaHCO_3 + NaCl$$

$$III: \quad NaHCO_3 + HCl \longrightarrow NaCl + H_2O + CO_2$$

Solution Let volume of 0.5000 N HCl for complete neutralisation of NaOH = x mL

and for that of $Na_2CO_3 = y$ mL

Phenolphthalein indicates 100% reaction of NaOH but 50% reaction of Na_2CO_3 upto $NaHCO_3$ stage.

$$NaOH + HCl \longrightarrow NaCl + H_2O$$

$$Na_2CO_3 + HCl \longrightarrow NaHCO_3 + NaCl$$

On the other hand, methyl orange indicates 100% reaction each of NaOH and Na_2CO_3 both.

Thus, first end-point appears when complete NaOH and 50% Na_2CO_3 have reacted (stage I and II)

$$\text{thus,} \quad x + \frac{y}{2} = 21.00 \text{ mL}$$

Second end-point appears when remaining Na_2CO_3 reacts.

$$\text{thus,} \quad \frac{y}{2} = 7 \text{ mL}, \; y = 14 \text{ mL}$$

$$\text{hence,} \quad x = 14 \text{ mL}$$

$$\text{thus,} \quad 14 \text{ mL of } 0.5000\text{ N HCl} \equiv NaOH$$

$$\text{thus, NaOH in mixture} = \frac{14 \times 0.5}{1000} = 7 \times 10^{-3} \text{ equivalent}$$

$$= 7 \times 10^{-3} \times 40 = 0.28 \text{ g NaOH}$$

Similarly, 14 mL of 0.5000 N HCl

$$\equiv Na_2CO_3 = 7 \times 10^{-3} \text{ equivalent}$$

$$= 7 \times 10^{-3} \times 53 \text{ g} = 0.371 \text{ g } Na_2CO_3$$

$$\% \text{ of NaOH in mixture} = \frac{0.28}{0.75} \times 100 = \mathbf{37.33\%}$$

$$\% \text{ of } Na_2CO_3 \text{ in mixture} = \frac{0.371}{0.750} \times 100 = \mathbf{49.47\%}$$

Master Exercises

Exercise 1
(Stage 1 : Learning)

Short Answer Type Questions

1. What volume of 0.125 M NaOH, in mL contains 4.0 g of NaOH?

2. If 1.06 g of Na_2CO_3 is dissolved in enough water to make 250 mL of solution, what is molar concentration of Na_2CO_3? What are molar concentrations of Na^+ and CO_3^{2-} ions?

3. For each solution, identify the ions that exist in aqueous solution, and specify the concentration of each ion
   (a) 0.25 M $(NH_4)_2SO_4$ (b) 0.123 M Na_2CO_3
   (c) 0.056 M HNO_3

4. If you dilute 25.0 mL of 1.50 M HCl to 500 mL, what is molar concentration of the dilute acid?

5. Which of the following methods would you use to prepare 1.00 L of 0.125 M H_2SO_4 ?
   (a) Dilute 20.8 mL of 6.00 M H_2SO_4 to a volume of 1.00 L.
   (b) Add 950 mL of water to 50.0 mL of 3.00 M H_2SO_4.

6. What volume of 0.750 M $Pb(NO_3)_2$ is mL, is required to react completely with 1.00 L of 0.225 M NaCl solution?

$$Pb(NO_3)_2(aq) + 2NaCl(aq) \longrightarrow PbCl_2(s) + 2NaNO_3(aq)$$

7. If 38.55 mL of HCl is required to titrate 2.150 g of Na_2CO_3 as given, what is the concentration (mol L^{-1}) of the HCl solution?

$$Na_2CO_3(aq) + 2HCl(aq) \longrightarrow 2NaCl(aq) + CO_2(g) + H_2O(l)$$

8. You have 0.490 g of unknown acid, H_2A, which reacts with KOH, according to

$$H_2A(aq) + 2KOH(aq) \longrightarrow K_2A(aq) + 2H_2O(l)$$

If 100 mL of 0.1 M KOH is required to titrate the acid to the second equivalence point, what is the molar mass of acid?

9. In which of the following cases, equilibrium molarity and analytical molarity are identical?
   (a) 1 M glucose solution
   (b) 1 M H_3PO_4 solution
   (c) saturated AgI solution
   (d) solution of LiCl in ether.

10. Acetic acid forms 100% dimer in benzene by hydrogen bonding as shown below by --- lines.

$$CH_3{-}C\begin{matrix}\nearrow O{-}{-}{-}H{-}O\searrow \\ \searrow O{-}H{-}{-}{-}O\nearrow\end{matrix}C{-}CH_3$$

If analytical molarity is 1 M, what is its equilibrium molarity?

11. Calculate the percentage of MnO_2 in a mineral specimen if the iodine liberated by a 0.1344 g sample in the net reaction

$$MnO_2(s) + 4H^+ + 2I^- \longrightarrow Mn^{2+} + I_2 + 2H_2O$$

required 32.30 mL of 0.07220 M $Na_2S_2O_3$. (MnO_2 = 87 g mol^{-1})

Analytical Questions

1. A 5.00×10^2 mL sample of 2.00 M HCl is treated with 4.47 g of magnesium. Calculate the concentration of the acid solution after all the metal has reacted. Assume that the volume remains unchanged.

2. A useful application of oxalic acid is the removal of rust (Fe_2O_3) from a cloth:

$$Fe_2O_3(s) + 6H_2C_2O_4 \longrightarrow 2Fe(C_2O_4)_3^{3-} + 3H_2O + 6H^+$$

Calculate the number of grams of rust that can be removed by 5.00×10^2 mL of 1 M oxalic acid.

3. 30 mL of x M HCl neutralise completely 10 mL of 0.2 M $NaHCO_3$ and further 5 mL of 0.2 M Na_2CO_3 to phenolphthalein end-point. What is value of x?

4. 300 mL of H_2SO_4 solution of a certain molarity was mixed with 200 mL of a $Ba(OH)_2$ solution of same molarity. The pH of the mixture solution is 1. Calculate normality of H_2SO_4 and $Ba(OH)_2$.

5. What volume of 6M NaOH would be required to react with the SO_2 produced from 10^3 kg of coal containing 0.05% by mass of FeS_2 (iron pyrites).

$$2FeS_2 + 5.5\,O_2 \longrightarrow Fe_2O_3 + 4SO_2$$
$$SO_2 + 2NaOH \longrightarrow Na_2SO_3 + H_2O$$

6. 25 mL solution containing a mixture of Na_2CO_3 and NaOH required 19.5 mL of 0.995 N HCl for the phenolpthalein end-point and in a separate titration 25 mL of the same mixture required 25.9 mL of same HCl for methyl orange end-point. Calculate concentration of each base in mixture.

7. A drop (0.05 mL) of 12.0 M HCl is spread over a sheet of thin aluminium foil. Assuming that all the acid dissolves through the foil, what will be the area, in cm^2, of the hole produced? (density of Al = 2.70 g cm^{-3}; thickness of the foil = 0.10 mm)

8. Titration of a 0.7439 g sample of impure $Na_2B_4O_7$ (borax) required 31.64 mL of 0.108 M HCl for reaction. Express the result of this analysis in terms of per cent (B = 11)

(a) $Na_2B_4O_7$ (b) $Na_2B_4O_7 \cdot 10H_2O$
(c) B_2O_3 (d) B

$$B_4O_7^{2-} + 2H_3O^+ + 3H_2O \longrightarrow 4H_3BO_3$$

9. Average human blood serum contains 18 mg of K^+ and 365 mg of Cl^- per 100 mL. Calculate

(a) the molar concentration for each of these species; use 1.00 g mL^{-1}, for the density of the serum
(b) p_{K^+} and p_{Cl^-} for human serum.

10. A mixture of HCl and H_3PO_4 is titrated with 0.1 M NaOH. The first end-point (methyl red) occurs at 35.00 mL and the second end-point (bromothymol blue) occurs at a total of 50.00 mL (15.00 mL after the first end-point). Calculate the millimoles of HCl and H_3PO_4 present in the solution.

11. A sample of P_2O_5 contains some H_3PO_4 impurity. A 0.405 g sample is reacted with water ($P_2O_5 + 3H_2O \longrightarrow 2H_3PO_4$), and the resulting solution is titrated with 0.25 M NaOH requiring 42.5 mL (upto Na_2HPO_4). What is the per cent H_3PO_4 impurity?

12. 1.2 g sample of Na_2CO_3 and K_2CO_3 was dissolved in water to form 100 mL of solution. 20 mL of this solution required 40 mL of 0.1 N HCl for complete neutralisation. Calculate the weight of Na_2CO_3 in mixture. If another 20 mL of this solution is treated with excess of $BaCl_2$, what will be the weight of the precipitate?

13. 2.4 g of a mixture of ($KCl + NH_4Cl$) is heated till constant weight is obtained. The solid residue is dissolved in water and the solution is made up to 250 mL. 25 mL of this solution required 15 mL of 0.1 N $AgNO_3$ for precipitation of chloride. Calculate percentage of chloride by weight in the mixture.

14. The concentration of Cu^{2+} ions in the water discharged from a certain industrial plant is determined by adding excess sodium sulphide solution to 0.80 L of water.

$$S^{2-} + Cu^{2+} \longrightarrow CuS(s)$$

What is molar concentration of Cu^{2+} in water sample if 0.0177 g of solid CuS is formed?

15. $Mn^{2+}(aq)$ can be determined by titration with $MnO_4^-(aq)$.

$$Mn^{2+} + MnO_4^- + OH^- \longrightarrow MnO_2 + H_2O \text{ (unbalanced)}$$

A 25.00 mL sample of $Mn^{2+}(aq)$ requires 34.77 mL of 0.05876 M $KMnO_4(aq)$ for its titration. What is the molarity of the $Mn^{2+}(aq)$?

16. Oxalic acid ($H_2C_2O_4$) is present in many plants and vegetables. If 24.0 mL of 0.01 M $KMnO_4$ solution is needed to titrate 1.00 g of $H_2C_2O_4$ to the equivalence-point, what is the per cent by mass of $H_2C_2O_4$ in the sample?

17. The calcium oxalate isolated from blood is dissolved in acid and titrated against a standardised $KMnO_4$ solution. In one test, it is found that the calcium oxalate isolated from a 10.0 mL sample of blood requires 24.2 mL of 9.56×10^{-4} M $KMnO_4$ for titration. Calculate the number of milligrams of calcium per mL of blood.

18. A solution contains 2.52 g of a reductant per litre. 25 mL of this solution of the reductant required 20 mL of 0.01 M $KMnO_4$ in acidic medium for oxidation. Find the molecular weight of the reductant given that each of the two atoms which undergo oxidation per molecule of the reductant, suffers an increase in the oxidation state by one unit.

19. In an experiment 10 L of air (containing O_3) at 1 atm and 27°C were passed through an alkaline KI solution. After neutralisation, free iodine required 1.5 mL of 0.01 N $Na_2S_2O_3$ solution. Calculate mol% of O_3 in air.

20. A polyvalent metal weighing 0.1 g and having atomic weight 51 reacted with dil. H_2SO_4 to give 43.9 mL of H_2 at STP. This solution containing the metal in lower oxidation state is found to require 58.8 mL of 0.1 N $KMnO_4$ for complete oxidation. What are the oxidation states of the metal in two reactions?

21. 0.015 mol of $K_2Cr_2O_7$ oxidises 2.18 g of a mixture of XO and X_2O_3 into XO_4^- in acidic medium. If 0.0187 mol of XO_4^- is formed, calculate atomic weight of X.

$$6XO + 5Cr_2O_7^{2-} + 34H^+ \longrightarrow 6XO_4^- + 10Cr^{3+} + 17H_2O$$
$$3X_2O_3 + 4Cr_2O_7^{2-} + 26H^+ \longrightarrow 6XO_4^- + 8Cr^{3+} + 13H_2O]$$

22. What mass of N_2H_4 can be oxidised to N_2 by 24.0 g of K_2CrO_4 which is reduced to $[Cr(OH)_4]^-$?

$$N_2H_4 \longrightarrow N_2 + 4H^+ + 4e^-$$
$$CrO_4^{2-} + 4H^+ + 3e^- \longrightarrow [Cr(OH)_4]^-$$
$$3N_2H_4 + 4CrO_4^{2-} + 4H^+ \longrightarrow 3N_2 + 4[Cr(OH)_4]^-$$

23. The alcohol content in a 10.0 g sample of blood from a driver required 4.23 mL of 0.07654 M $K_2Cr_2O_7$ for titration. Should the police prosecute the individual for drunken-driving when the current legal limit is 0.1 per cent by mass?

$$3CH_3CH_2OH + 2Cr_2O_7^{2-} + 16H^+ \longrightarrow 3CH_3COOH + 4Cr^{3+} + 11H_2O$$

24. A 1 g sample of H_2O_2 solution containing x per cent H_2O_2 by weight required x mL of $KMnO_4$ solution for complete oxidation in acidic medium. Calculate normality of $KMnO_4$ solution.

25. In basic solution CrO_4^{2-} oxidises $S_2O_3^{2-}$ to form $Cr(OH)_4^-$ and SO_4^{2-}. How many mL of 0.154 M CrO_4^{2-} are required to react with 40.0 mL of 0.246 M $S_2O_3^{2-}$?

[**Hint :** $0.154\ M = 0.154 \times 3N\ CrO_4^{2-}$
$0.246\ M = 0.246 \times 8N\ S_2O_3^{2-}$]

26. 25 mL of a solution containing ferrous sulphate and ferric sulphate acidified with H_2SO_4 is reduced by metallic zinc. The solution required 34.25 mL of 0.1 N solution of $K_2Cr_2O_7$ for oxidation. However, before reduction with zinc, 25 mL of same solution required 22.45 mL of same $K_2Cr_2O_7$. Calculate the weight of $FeSO_4$ and $Fe_2(SO_4)_3$ per litre of the solution.

27. (a) The ion A^{n+} is oxidised to AO_3^- by MnO_4^- changing to Mn^{2+} in acid solution. Given that 2.68×10^{-3} mol of A^{n+} requires 1.61×10^{-3} mol of MnO_4^-, what is the value of n ?

(b) What is the weight of one-gram equivalent of ACl_n for the above reaction if the atomic mass of A is 97.0. ?

28. 0.518 g sample of limestone is dissolved and calcium is precipitated as calcium oxalate CaC_2O_4. After filtration and washing the precipitate, it required 40 mL of 0.05 M $KMnO_4$ solution acidified with sulphuric acid to titrate it. What is the percentage of CaO in the limestone?

29. 20 mL of a solution containing 0.2 g of impure sample of H_2O_2 reacts with 0.316 g of $KMnO_4$ in presence of H_2SO_4.

(a) What is % purity of H_2O_2?

(b) Calculate the volume of dry O_2 formed at 27°C and 750 mmHg pressure.

30. $Na_2C_2O_4$ and $KHC_2O_4 \cdot H_2C_2O_4$ are mixed in such a proportion by weight that the normality of the resulting solution as a reducing agent is 3.62 times the normality as an acid. What is the proportion?

31. A standard solution of potassium dichromate contains 5.442 g L^{-1}. What is the amount in terms of Fe_3O_4 per mL $K_2Cr_2O_7$?

32. A 0.5 g sample containing MnO_2 is treated with HCl, liberating Cl_2. The Cl_2 is passed into a solution of KI and 30.0 mL of 0.1 M $Na_2S_2O_3$ are required to titrate the liberated iodine. Calculate the percentage of MnO_2 in the sample. (Mn = 55)

33. Calculate the mass of oxalic acid, $H_2C_2O_4$, which can be oxidised to CO_2 by 100 mL of an MnO_4^- solution, 10 mL of which is capable of oxidising 50.0 mL of 1.00 N I^- to I_2

$$MnO_4^- + 8H^+ + 5e^- \longrightarrow Mn^{2+} + 4H_2O$$
$$H_2C_2O_4 \longrightarrow 2CO_2 + 2H^+ + 2e^-$$
$$2I^- \longrightarrow I_2 + 2e^-$$

34. 5.0 g of bleaching powder was suspended in water and volume made up to half a litre. 20 mL of this suspension when acidified with acetic acid and treated with excess of potassium iodide solution liberated iodine which required 20 mL of a decinormal hypo solution for titration. Calculate the percentage of available chlorine in bleaching powder.

Exercise 2
(Stage 2 : High Skill Problem Solving)

Only One Option Correct

1. To prepare a solution that is 0.50 M KCl starting with 100 mL of 0.40 M KCl

(a) add 0.75 g KCl
(b) add 20 mL of water
(c) add 0.10 mol KCl
(d) evaporate 10 mL water

2. 10.78 g of H_3PO_4 in 550 mL solution is 0.40 N. Thus this acid

(a) has been neutralised to HPO_4^{2-}
(b) has been neutralised to PO_4^{3-}
(c) has been reduced to $H_2PO_2^-$
(d) has been neutralised to $H_2PO_4^-$

3. How many equivalents are there per mol of H_2S in its oxidation to SO_2?
(a) 2 (b) 4
(c) 6 (d) 8

4. In hot alkaline solution, Br_2 disproportionates to Br^- and BrO_3^-

$$3Br_2 + 6OH^- \longrightarrow 5Br^- + BrO_3^- + 3H_2O$$

hence, equivalent weight of Br_2 is (molecular weight = M)
(a) $\frac{M}{6}$ (b) $\frac{M}{5}$
(c) $\frac{3M}{5}$ (d) $\frac{5M}{3}$

[**Hint** : Each half-reaction, oxidation and reduction should involve equal number of electrons, balance it. $3Br_2 \equiv 5e^-$]

5. M (molarity) and M' (molality) are related to each other by equation:
(m = molecular weight of solute, d = density g/mL]
(a) $M\left(\frac{m}{1000} + \frac{1}{M'}\right) = d$ (b) $M' = \left(\frac{1000\,M}{1000\,d - Mm}\right)$
(c) Both (a) and (b) (d) None of these

6. Maximum number of moles of $PbSO_4$ that can be precipitated by mixing 20.00 mL of 0.1 M $Pb(NO_3)_2$ and 30.00 mL of 0.1 M Na_2SO_4 will be
(a) 0.002 (b) 0.003
(c) 0.005 (d) 0.001

7. $[Na^+]$ in a solution prepared by mixing 30.00 mL of 0.12 M NaCl with 70 mL of 0.15 M Na_2SO_4 is
(a) 0.135 M (b) 0.141 M
(c) 0.210 M (d) 0.246 M

8. When 80 mL of 0.20 M HCl is mixed with 120 mL of 0.15 M KOH, the resultant solution is the same as a solution of
(a) 0.16 M KCl and 0.02 M HCl
(b) 0.08 M KCl
(c) 0.08 M KCl and 0.01 M KOH
(d) 0.08 M KCl and 0.01 M HCl

9. Equal volumes of 0.200 M HCl and 0.400 M KOH are mixed. The concentration of the principal ions in the resulting solution are :
(a) $[K^+] = 0.400$ M, $[Cl^-] = 0.200$ M, $[H^+] = 0.200$ M
(b) $[K^+] = 0.200$ M, $[Cl^-] = 0.200$ M, $[OH^-] = 0.100$ M
(c) $[K^+] = 0.200$ M, $[Cl^-] = 0.100$ M, $[OH^-] = 0.100$ M
(d) $[K^+] = 0.200$ M, $[Cl^-] = 0.100$ M, $[OH^-] = 0.200$ M

10. Mass of KHC_2O_4 (potassium acid oxalate) required to reduce 100 mL of 0.02 M $KMnO_4$ in acidic medium (to Mn^{2+}) is x g, and to neutralise 100 mL of 0.05 M $Ca(OH)_2$ is y g then,
(a) $x = y$ (b) $2x = y$
(c) $x = 2y$ (d) None of these

11. $2H_2O_2(l) \longrightarrow 2H_2O(l) + O_2(g)$
100 mL of X molar H_2O_2 gives 3 L of O_2 gas under the condition when 1 mol occupies 24 L, value of X is
(a) 2.5 (b) 1.0
(c) 0.5 (d) 0.25

12. If equal volumes of 1 M $KMnO_4$ and 1 M $K_2Cr_2O_7$ solutions are allowed to oxidise Fe (II) to Fe (III), then Fe (II) oxidised will be
(a) more by $KMnO_4$ (b) more by $K_2Cr_2O_7$
(c) equal in both cases (d) data is incomplete

13. 100 mL of 1 M $KMnO_4$ oxidised 100 mL of H_2O_2 in acidic medium (when MnO_4^- is reduced to Mn^{2+}); volume of same $KMnO_4$ required to oxidise 100 mL of H_2O_2 in basic medium (when MnO_4^- is reduced to MnO_2) will be
(a) $\frac{100}{3}$ mL (b) $\frac{500}{3}$ mL
(c) $\frac{300}{5}$ mL (d) 100 mL

14. 10 mL of $H_2\,O_2$ solution (volume strength $= x$) required 10 mL of $\frac{N}{0.56}$ MnO_4^- solution in acidic medium. Hence, x is
(a) 0.56 (b) 5.6
(c) 0.1 (d) 10.0

15. Aqueous solution of chloride of an element A containing 1×10^{-3} mol after complete hydrolysis required 30 mL of 0.1 M $AgNO_3$ solution to form AgCl. Element A is
(a) Al (b) P
(c) Si (d) S

16. Molality of 18 M H_2SO_4 ($d = 1.8$ g mL^{-1}) is
(a) 36 mol kg^{-1} (b) 200 mol kg^{-1}
(c) 500 mol kg^{-1} (d) 18 mol kg^{-1}

17. 10 L of hard water required 0.56 g of lime (CaO) for removing hardness. Hence, temporary hardness in ppm of $CaCO_3$ (parts per million *ie* 10^6) is
(a) 100 (b) 200
(c) 10 (d) 20

18. 1 g equivalent of a substance is the weight of that amount of a substance which is equivalent to
(a) 0.25 mol of O_2 (b) 0.50 mol of O_2
(c) 1 mol of O_2 (d) 8 mol of O_2

19. In an experiment, 50 mL of 0·1 M solution of a metallic salt reacted exactly with 25 mL of 0.1 M solution of sodium sulphite. In the reaction SO_3^{2-} is oxidised to SO_4^{2-}. If the original oxidation number of the metal in the salt was 3, what would be the new oxidation number of the metal?
(a) 0 (b) 1
(c) 2 (d) 4

20. 20 mL of x M HCl neutralises completely 10 mL of 0.1 M $NaHCO_3$ and a further 5 mL of 0.2 M Na_2CO_3 solution to methyl orange end-point. The value of x is
(a) 0.167 M (b) 0.133 M
(c) 0.150 M (d) 0.200 M

21. The molality of a H_2SO_4 solution is 9. The weight of the solute in 1 kg H_2SO_4 solution is
(a) 900.0 g (b) 468.65 g
(c) 882.0 g (d) 9.0 g

22. The density of 1 M solution of NaCl is 1.0585 g/mL. The molality of the solution is
(a) 1.0585 (b) 1.00
(c) 0.10 (d) 0.0585

23. Which is not true about H_3PO_2?
(a) It is tribasic acid
(b) One mol is neutralised by 0.5 mol $Ca(OH)_2$
(c) NaH_2PO_2 is normal salt
(d) It disproportionates to H_3PO_3 and PH_3 on heating

24. RH_2 (ion exchange resin) can replace Ca^{2+} in hard water

$$RH_2 + Ca^{2+} \longrightarrow RCa + 2H^+$$

1 L of hard water after passing through RH_2 has pH 2. Hence, hardness in ppm of Ca^{2+} is
(a) 200 (b) 100
(c) 50 (d) 125

25. 1 mol BaF_2 + 2 mol H_2SO_4 $\longrightarrow$ resulting mixture will be neutralised by
(a) 1 mol KOH
(b) 2 mol $Ca(OH)_2$
(c) 3 mol NaOH
(d) 2 mol KOH

26. 100 mL of a mixture of NaOH and Na_2SO_4 is neutralised by 10 mL of 0.5 M H_2SO_4. Hence, amount of NaOH in 100 mL mixture is
(a) 0.2 g (b) 0.4 g
(c) 0.6 g (d) 1.0 g

27. 10 mL of $NaHC_2O_4$ solution is neutralised by 10 mL of 0.1 M NaOH solution. 10 mL of same $NaHC_2O_4$ solution is oxidised by 10 mL of $KMnO_4$ solution in acidic medium. Hence, molarity of $KMnO_4$ is
(a) 0.1 M (b) 0.2 M
(c) 0.04 M (d) 0.02 M

28. For a given mixture of $NaHCO_3$ and Na_2CO_3, volume of a given HCl required is x mL with phenolphthalein indicator and further y mL required with methyl orange indicator. Hence, volume of HCl for complete reaction of $NaHCO_3$ is
(a) $2x$ (b) $\frac{x}{2}$
(c) y (d) $(y - x)$

29. 1 mol of ferric oxalate is oxidised by x mol of MnO_4^- and also 1 mol of ferrous oxalate is oxidised by y mol of MnO_4^- in acidic medium. The ratio $\frac{x}{y}$ is
(a) 2 : 1 (b) 1 : 2
(c) 3 : 1 (d) 1 : 3

30. 40 mL of 0.05 M solution of sesquicarbonate ($Na_2CO_3 \cdot NaHCO_3 \cdot 2H_2O$) is titrated against 0·05 M HCl. x mL of HCl is used when phenolphthalein is the indicator and y mL of HCl is used when methyl orange is the indicator in two separate titrations, hence $(y - x)$ is
(a) 80 mL (b) 30 mL
(c) 120 mL (d) None of these

31. 3 moles of a mixture of $FeSO_4$ and $Fe_2(SO_4)_3$ required 100 mL of 2 M $KMnO_4$ solution in acidic medium. Hence, mole fraction of $FeSO_4$ in the mixture is
(a) $\frac{1}{3}$ (b) $\frac{2}{3}$
(c) $\frac{2}{5}$ (d) $\frac{3}{5}$

32. In the following reaction

$$28NO_3^- + 3As_2S_3 + 4H_2O \longrightarrow 6AsO_4^{3-} + 28NO + 9SO_4^{2-} + 8H^+$$

equivalent weight of As_2S_3 (with molecular weight M) is
(a) $\frac{M}{2}$ (b) $\frac{M}{4}$
(c) $\frac{M}{24}$ (d) $\frac{M}{28}$

33. In the following reaction :

$$Cr(OH)_3 + OH^- + IO_3^- \longrightarrow CrO_4^{2-} + H_2O + I^-$$

(a) IO_3^- is oxidising agent
(b) $Cr(OH)_3$ is oxidised
(c) $6e^-$ are being taken per I atom
(d) All of the above are correct

34. 5.3 g of M_2CO_3 is dissolved in 150 mL of 1 N HCl. Unused acid required 100 mL of 0.5 N NaOH. Hence, equivalent weight of M is
(a) 23 (b) 12
(c) 24 (d) 13

35. Equivalent weight of H_3PO_2 when it disproportionates into PH_3 and H_3PO_3 (mol. wt. = M) is
(a) M (b) $\frac{M}{2}$
(c) $\frac{M}{4}$ (d) $\frac{3M}{4}$

36. I_2 obtained from 0.1 mol of $CuSO_4$ required 100 mL of 1 M hypo solution, hence, mol percentage of pure $CuSO_4$ is

(a) 100 (b) 50
(c) 25 (d) 40

37. 1.2 g of Mg is treated with 100 mL of 1 M H_2SO_4. Molar concentration of the H_2SO_4 solution after complete reaction is
(a) 0.5 M (b) 0.005 M
(c) 0.10 M (d) 0.20 M

38. Hardness in water is expressed in terms of ppm of $CaCO_3$. If hardness is 200 ppm $CaCO_3$, then in terms of molarity it is
(a) 0.004 M (b) 0.002 M
(c) 0.001 M (d) 0.0005 M

39. 10 mL of a blood sample (containing calcium oxalate) is dissolved in acid. It required 20 mL of 0.001 M $KMnO_4$ (which oxidises oxalate to carbon dioxide) hence, Ca^{2+} ion in 10 mL blood is
(a) 0.200 g (b) 0.02 g
(c) 2.00 g (d) 0.002 g

40. Volume of 18.0 M H_2SO_4 required to prepare 1.0 L of a 0.9 M solution of H_2SO_4 is
(a) 50.0 mL (b) 10.0 mL
(c) 500.0 mL (d) 5.0 mL

41. Volume of 0.50 M NaOH solution required to react with 40.0 mL of 0.05 M H_2SO_4 solution is
(a) 40.0 mL (b) 80.0 mL
(c) 20.0 mL (d) 8.0 mL

42. Stock phosphoric acid solution is 85% H_3PO_4 and has a specific gravity of 1.70. Hence, molarity of H_3PO_4 solution is
(a) 8.51 M (b) 1.70 M
(c) 14.74 M (d) 7.37 M

43. 150 mL of 6.00 M H_2SO_4 solution is mixed with 250 mL of 3.00 M H_2SO_4. Resulting molarity is
(a) 4.125 M (b) 8.250 M
(c) 4.500 M (d) 1.650 M

44. If 100 mL of 1.00 M H_2SO_4 is mixed with 200 mL of 1.00 M KOH solution, molarity of the salt formed is
(a) 0.100 M (b) 0.200 M
(c) 0.333 M (d) 0.666 M

45. Commercial "11.2 volume" H_2O_2 solution has a molarity of
(a) 1 (b) 0.5
(c) 11.2 (d) 1.12

46. Equivalent weight of $Cr_2O_7^{2-}$ (mol. wt. = M) in the following reaction is

$$Cr_2O_7^{2-} + 6I^- + 14H^+ \longrightarrow 2Cr^{3+} + 3I_2 + 7H_2O$$

(a) $\frac{M}{6}$ (b) $\frac{M}{3}$
(c) $\frac{M}{2}$ (d) M

47. 0.116 g of $C_4H_4O_4(A)$ is neutralised by 0.074 g of $Ca(OH)_2$. Hence, protonic hydrogen (H^+) in (A) will be
(a) 1 (b) 2
(c) 3 (d) 4

48. 10 mL of $NaHC_2O_4$ is oxidised by 10 mL of 0.02 M MnO_4^-. Hence, 10 mL of $NaHC_2O_4$ is neutralised by
(a) 10 mL of 0.1 M NaOH
(b) 10 mL of 0.02 M NaOH
(c) 10 mL of 0.1 M $Ca(OH)_2$
(d) 10 mL of 0.05 N $Ba(OH)_2$

49. Oxalic acid ($H_2C_2O_4$) forms two series of salt $HC_2O_4^-$ and $C_2O_4^{2-}$. If 0.9 g of oxalic acid is in 100 mL solution, $HC_2O_4^{2-}$ and $C_2O_4^{2-}$ have normality respectively
(a) 0.1 N, 0.1 N (b) 0.1 N, 0.2 N
(c) 0.2 N, 0.2 N (d) 0.2 N, 0.1 N

50. At a given condition, 1 mol O_2 occupies 30 L. Volume occupied by 1 equivalent of O_2 is
(a) 7.5 L (b) 15 L
(c) 30.0 L (d) 22.4 L

51. 10 g of MnO_2 on reaction with conc. HCl liberated 0.1 equivalent of Cl_2 (Mn = 55). Hence, per cent purity of MnO_2 is
(a) 87.0 (b) 21.75
(c) 50.0 (d) 43.5

52. The equivalent mass of a metal is double that of oxygen. How many times is the equivalent mass of its oxide greater than the mass of the metal?
(a) 2 times (b) 1.5 times
(c) 3 times (d) 0.5 time

53. 0.106 g of Na_2CO_3 completely neutralises 40.0 mL of H_2SO_4. Hence, normality of H_2SO_4 solution is
(a) 0.05 N (b) 0.025 N
(c) 0.10 N (d) 0.20 N

54. Volume of 0.02 M MnO_4^- solution required to oxidise 40.0 mL of 0.1 M Fe^{2+} solution is
(a) 200 mL (b) 100 mL
(c) 40 mL (d) 20 mL

55. A 20.0 mL solution of Na_2SO_3 required 30 mL of 0.01 M $K_2Cr_2O_7$ solution for the oxidation to Na_2SO_4. Hence, molarity of Na_2SO_3 solution is
(a) 0.015 M (b) 0.045 M
(c) 0.030 M (d) 0.0225 M

56. Normality of a solution that contains 12.64 g of $KMnO_4$ is 500 mL of solution to be used in the reaction that produces K_2MnO_4 ion as the reducing product is:
(a) 0.16 N (b) 0.32 N
(c) 0.80 N (d) 0.08 N

57. Iodate ion, IO_3^-, oxidises SO_3^{2-} to SO_4^{2-} in acidic medium. If 100 mL sample of solution containing 2.14 g of KIO_3 reacts with 60 mL of 0.5 M Na_2SO_3 solution, then final oxidation state of iodine is
(a) +5 (b) +3
(c) +1 (d) –1

58. A 0.60 g nitrogen containing compound was boiled with NaOH and NH_3, thus formed required 100 mL of 0.2 N H_2SO_4 for neutralisation; percentage of nitrogen in the compound is
(a) 46.67% (b) 23.34%
(c) 60.00% (d) 20.00%

59. 1.00 L of 0.15 NaOH absorbed 11.2 millimol of CO_2 from air. Hence, new molarity of NaOH is
(a) 0.1276 M (b) 0.1500 M
(c) 0.0224 M (d) 0.0112 M

60. Molarity of Na_2CO_3 solution formed in the above question will be
(a) 0.0244 M (b) 0.0112 M
(c) 0.1500 M (d) 0.1276 M

61. A sample of iron ore weighing 0.740 g containing 24% Fe_2O_3 is dissolved and the iron reduced to Fe^{2+} by the addition of 25.0 mL of 0.05 M $SnCl_2$. The excess Sn^{2+} is oxidised to Sn^{4+} using 0.05 M $HgCl_2$. Volume of $HgCl_2$ used is
(a) 5.2 mL (b) 6.2 mL
(c) 5.6 mL (d) 2.6 mL

62. A sample containing only $Na_2C_2O_4$ and KHC_2O_4 required three times the volume of 0.1 N $KMnO_4$ for titration as of 0.1 N base (same size sample in each case). Percentage of KHC_2O_4 in the mixture is
(a) 65.64% (b) 34.36%
(c) 30.28% (d) 69.72%

63. A 0.6234 g sample that might contain NaOH, Na_2CO_3, $NaHCO_3$ or a mixture of NaOH + Na_2CO_3 or Na_2CO_3 + $NaHCO_3$ is titrated with 0.1062 M HCl by two-indicator method. It is found that 40.38 mL of the acid is required to reach the phenolphthalein end-point. Methyl orange is then added to the solution, and the titration continued using an additional 12.83 mL of the acid. Thus, sample contained
(a) NaOH + Na_2CO_3 + $NaHCO_3$
(b) NaOH + Na_2CO_3
(c) Na_2CO_3 + $NaHCO_3$
(d) can't be predicted

64. Percentage of NaOH in the mixture of the above example is
(a) 18.77% (b) 23.17%
(c) 37.54% (d) 46.34%

65. A bottle which contains 200 mL of 0.100 M NaOH absorbs 1.00 mmol of CO_2 from the air. If the solution is then titrated with standard acid using phenolphthalein indicator, what normality of the acid will be found?
(a) 0.190 N (b) 0.380 N
(c) 0.095 N (d) 0.0475 N

One or More Than One Options Correct

1. 1 g equivalent of a substance is the weight of that amount of a substance which is equivalent to
(a) 0.25 mol O_2 (b) 8 g O_2
(c) 16 g O_2 (d) 0.50 mol O_2

2. 2.84 g of P_4O_{10} is dissolved in 1 L aqueous solution. Thus,
(a) resulting solution is 0.01 M
(b) it is neutralised to Na_3PO_4 by 400 mL of 0.3 M NaOH solution
(c) it is neutralised to $CaHPO_4$ by 400 mL of 0.1 M $Ca(OH)_2$ solution
(d) it cannot be neutralised by basic solution

3. A solution of $KHC_2O_4 \cdot H_2C_2O_4 \cdot 2H_2O$ is 0.2 N as an acid. Then it is
(a) 0.267 N as reducing agent
(b) 0.6 M as an acid
(c) 0.067 M as an acid
(d) 0.067 M as reducing agent

4. A sample of water has a hardness expressed as 77.5 ppm Ca^{2+}. This sample is passed through an ion exchange column and the Ca^{2+} is replaced by H^+. Select correct statement(s).
(a) pH of the water after it has been so treated is 2.4
(b) Every Ca^{2+} ion is replaced by one H^+ ion
(c) Every Ca^{2+} ion is replaced by two H^+ ions
(d) pH of the solution remains unchanged

5. 10.78 g of H_3PO_4 in 550 mL solution is 0.40 N. Thus, this acid
(a) has been neutralised to HPO_4^{2-}
(b) has been neutralised to PO_4^{2-}
(c) has been reduced to HPO_3^{2-}
(d) has been neutralised to $H_2PO_4^-$

6. Mass of KHC_2O_4 (potassium acid oxalate) required to reduce 100 mL of 0.02 M $KMnO_4$ in acidic medium (to Mn^{2+}) is x g and to neutralise 100 mL of 0.05 M $Ca(OH)_2$ is y g, then
(a) $x = y$ (b) $2x = y$
(c) $x = 2y$ (d) None of these

7. A mixture of $Na_2C_2O_4$ (A) and $KHC_2O_4 . H_2C_2O_4$ (B) required equal volumes of 0.1 M $KMnO_4$ and 0.1 M NaOH separately. Molar ratio of A and B in the mixture is
(a) 1 : 1 (b) 1 : 5.5
(c) 5.5 : 1 (d) 3.1 : 1

8. $H_2C_2O_4$ and $NaHC_2O_4$ behave as acids as well as reducing agents. Which are correct statements?

(a) Equivalent weight of $H_2C_2O_4$ and $NaHC_2O_4$ are equal to their molecular weights when behaving as reducing agents

(b) 100 mL of 1 M solution of each is neutralised by equal volumes of 1 M $Ca(OH)_2$

(c) 100 mL of 1 N solution of each is neutralised by equal volumes of 1 N $Ca(OH)_2$

(d) 100 mL of 1 M solution of each is oxidised by equal volumes of 1 M $KMnO_4$

9. 100 mL of 1.44% solution of pure FeC_2O_4 in dil. HCl is oxidised by 0.01 M $KMnO_4$. Then volume of $KMnO_4$ required is

(a) 120 mL (b) 600 mL
(c) 200 mL (d) 60 mL

10. 1 mole of MnO_4^{2-} in neutral aqueous solution disproportionates to

(a) $\frac{2}{3}$ mol of MnO_4^- and $\frac{1}{3}$ mol of MnO_2

(b) $\frac{1}{3}$ mol of MnO_4^- and $\frac{2}{3}$ mol of MnO_2

(c) $\frac{1}{3}$ mol of Mn_2O_7 and $\frac{1}{3}$ mol of MnO_2

(d) $\frac{2}{3}$ mol of Mn_2O_7 and $\frac{1}{3}$ mol of MnO_2

Brain Twisters

1. You place 2.56 g $CaCO_3$ (calcium carbonate) in a beaker containing 250 mL of 0.125 M HCl. When the reaction has ceased, does any $CaCO_3$ remain? What mass of $CaCl_2$ can be produced?

$$CaCO_3(s) + 2HCl(aq) \longrightarrow CaCl_2(aq) + CO_2(g) + H_2O(l)$$

2. Gold can be dissolved from gold-bearing rock by treating the rock with sodium cyanide (NaCN) in the presence of oxygen.

$$4Au(s) + 8NaCN(aq) + O_2(g) + 2H_2O(l) \longrightarrow 4NaAu(CN)_2(aq) + 4NaOH(aq)$$

(a) Name the oxidising and reducing agents in this reaction. What has been oxidised and what has been reduced?

(b) If you have exactly one metric ton (1 metric ton = 1000 kg) of gold-bearing rock, what volume of 0.075 M NaCN, in litres, do you need to extract the gold if the rock is 0.0197% gold? (Au = 197)

3. 500 mL of 0.25 M Na_2SO_4 solution is added to an aqueous solution of 15.00 g of $BaCl_2$ solution resulting in the formation of white precipitate of $BaSO_4$. Which is the limiting reactant? Calculate moles of $BaSO_4$ formed. (Ba = 137, S = 32, O = 16, Na = 23)

4. We define *p*-value of a term (say *x*) as

$$p_x = -\log x$$

Calculate *p*-value for each ion in a solution that is 2.00×10^{-3} M in NaCl and 5.4×10^{-4} M HCl.

5. Calculate molar concentration of Ag^+ in a solution that has $p_{Ag} = 6.6990$

6. What mass of Ag_2CO_3 is formed when 25.0 mL of 0.2 M $AgNO_3$ is mixed with 50.0 mL of 0.08 M Na_2CO_3?

7. What is the molar concentration of Na_2CO_3 in the solution when 25.0 mL of 0.2 M $AgNO_3$ are mixed with 50.0 mL of 0.08 M Na_2CO_3 solution?

8. Calculate the *p*-value for each of the indicated ions in the followng

(a) Cu^{2+}, Zn^{2+} and NO_3^- in a solution that is 4.78×10^{-2} M in $Cu(NO_3)_2$ and 0.104 M in $Zn(NO_3)_2$.

(b) H^+, Ba^{2+} and ClO_4^- in a solution that is 3.35×10^{-4} M in $Ba(ClO_4)_4$ and 6.75×10^{-4} M in $HClO_4$.

9. Sea water contains an average of 1.08×10^3 ppm of Na^+ and 270 ppm of SO_4^{2-}. Calculate

(a) the molar concentration of SO_4^{2-} and Na^+ given that average density of sea water is 1.02 g/mL

(b) the p_{Na} and $p_{SO_4^{2-}}$ for sea water

10. A solution was prepared by dissolving 5.76 g of $KCl \cdot MgCl_2 \cdot 6H_2O$ in sufficient water to give 2.00 L. Calculate.

(a) the molar analytical concentration of $KCl \cdot MgCl_2$

(b) the molar concentration of Mg^{2+} and Cl^-

(c) the number of millimoles of Cl^- in 25.0 mL of this solution

(d) ppm K^+

11. What volume of 0.01 M $AgNO_3$ would be required to precipitate all the I^- in a 200.0 mL of a solution that contained 24.32 ppt of KI?

12. Exactly 750.0 mL of a solution that contained 480.4 ppm of $Ba(NO_3)_2$ were mixed 200.0 mL of a solution that 0.03090 M $Al_2(SO_4)_3$

(a) What mass of solid $BaSO_4$ was formed?

(b) What was the molarity of the unreacted reagent [$Al_2(SO_4)_3$ or $Ba(NO_3)_2$]?

13. The phosphorus in a 4.258 g sample of a plant food was converted to PO_4^{3-} and precipitated as Ag_3PO_4 through the addition of 50.0 mL of 0.082 M $AgNO_3$. The excess of $AgNO_3$ was back-titrated with 4.86 mL of 0.0625 M KSCN. Express the results of this analysis in terms of P_2O_5.

$$P_2O_5 + 9H_2O \longrightarrow 2PO_4^{3-} + 6H_3O^+$$
$$2PO_4^{3-} + 6Ag^+ \longrightarrow 2Ag_3PO_4(s)$$
$$Ag^+ + SCN^- \longrightarrow AgSCN(s)$$

14. How many mL of 0.30 M $CaCl_2$ must be added to 400 mL of 0.20 M KCl to produce a solution of 0.25 M Cl^- ?

15. How many kg of wet NaOH containing 12% water is required to prepare 60 L of 0.50 M NaOH solution?

16. Exactly 50.0 mL of Na_2CO_3 solution is equivalent to 56.3 mL of 0.102 N HCl in acid-base neutralisation. How many gram $CaCO_3$ would be precipitated if an excess of $CaCl_2$ solution were added to 100 mL of this Na_2CO_3 solution?

17. What mass of phosphoric acid, H_3PO_4 is required to prepare 550 mL of 0.40 N solution?
(a) assuming complete neutralisation of acid
(b) assuming reduction to HPO_3^{2-}

$$H_3PO_4 + 3OH^- \longrightarrow 3H_2O + PO_4^{3-}$$
$$H_3PO_4 + 2e^- \longrightarrow HPO_3^{2-} + H_2O$$

18. The tetraethyl lead, TEL, $[Pb(C_2H_5)_4]$ in a 25.00 mL sample of aviation gasoline was shaken with 15.00 mL of 0.02095 M I_2

$$Pb(C_2H_5)_4 + I_2 \longrightarrow Pb(C_2H_5)_3I + C_2H_5I$$

After the reaction, unused I_2 required 6.09 mL of 0.03465 M $Na_2S_2O_3$ solution. Calculate molar concentration of TEL.

Passage Comprehension Questions

Passage 1

Chromium (III) ion forms many compounds with ammonia. To find the formula of one of these compounds, you titrate the NH_3 in the compound with standarised acid

$$\underset{(A)}{Cr(NH_3)_xCl_3(aq)} + xHCl \longrightarrow xNH_4^+(aq) + Cr^{3+}(aq) + (x+3)Cl^-(aq)$$

Assume that 40 mL of 1.5 M HCl is used to titrate 2.635 g of (*A*).

Based on the above experiment, answer the following questions.

1. Value of *x* is
(a) 3 (b) 4
(c) 5 (d) 6

2. If 1.5 M H_2SO_4 is used instead of 1.5 M HCl, titre value is
(a) 40 mL (b) 20 mL
(c) 10 mL (d) 15 mL

3. Probable complex (assume that coordination number of chromium is six) can be
(a) $[Cr(NH_3)_6]Cl_3$
(b) $[Cr(NH_3)_5Cl]Cl_2$
(c) $[Cr(NH_3)_4Cl_2]Cl$
(d) $[CrCl_3]\cdot 6NH_3$

Passage 2

Copper containing alloy weighing 0.3175 g is dissolved in an acid, and an excess of KI is added. Estimation of copper in alloy is based on following reaction

$$2Cu^{2+}(aq) + 5I^-(aq) \longrightarrow Cu_2I_2(s) + I_3^-(aq)$$
$$I_3^-(aq) + 2S_2O_3^{2-}(aq) \longrightarrow S_4O_6^{2-}(aq) + 3I^-(aq)$$

Answer the following questions based on the above reaction.

1. Oxidising agent (O) and reducing agent (R) in the above reaction :

	O	R
(a)	$Cu^{2+}, S_2O_3^{2-}$	I_3^-, I^-
(b)	$Cu^{2+}, S_2O_3^{2-}$	I_3^-
(c)	Cu^{2+}, I_3^-	$I^-, S_2O_3^{2-}$
(d)	Cu^{2+}, I^-	$I_3^-, S_2O_3^{2-}$

2. Estimation is based on
(a) iodometry (b) iodimetry
(c) gravimetry (d) colorimetry

3. If 25 mL of 0.10 M $Na_2S_2O_3$ is required, Cu present in alloy is
(a) 20% (b) 40%
(c) 45% (d) 50%

4. I_2 formed is converted to KI_3 in the given reaction. This shows
(a) I_2 is soluble in H_2O (b) I_2 is a Lewis acid
(c) I_2 is ionic compound (d) I^- is a Lewis base

Passage 3

Consider the titration of solution *X* of 1 M HCl using 0.1 M NaOH solution with phenolphthalein indicator, by two students *A* and *B*.

Student	Process	Titre volume of NaOH
A	20 mL HCl + 20 mL H_2O + 1 drop phenolphthalein	V_A
B	20 mL HCl + 60 mL H_2O + 1 drop phenolphthalein	V_B

Answer the following questions

1. What is the relation between V_A and V_B ?
(a) $V_A = V_B$ (b) $2V_A = V_B$
(c) $V_A = 2V_B$ (d) $3V_A = V_B$

2. Before titration, resultant molarity of HCl of the student *A* after addition of H_2O is
(a) same as that of *B* (b) double as that of *B*
(c) half as that of *B* (d) one third as that of *B*

3. End-point is
(a) appearance of blue colour
(b) disappearance of blue colour
(c) appearance of pink colour
(d) disappearance of blue colour

Passage 4

For titrating a certain volume of a reducing substance by 1 M $KMnO_4$, it is found that 20 mL was used in acidic medium, 33.4 mL was used is alkaline medium and 100 mL was used in neutral medium. MnO_4^- is reduced to Mn^{2+} in acidic medium.

Based on this study, answer the following questions.

1. To what state MnO_4^- is reduced to in alkaline medium?
   (a) +4 (b) +1
   (c) +5 (d) +6
2. To what state MnO_4^- is reduced to in neutral medium?
   (a) +4 (b) +5
   (c) +6 (d) +7
3. 1 M $KMnO_4$ expressed in terms of normality is in

	Acidic medium	Alkaline medium	Neutral medium
(a)	1 N	1 N	1 N
(b)	5 N	3 N	1 N
(c)	1 N	3 N	5 N
(d)	0.2 N	0.33 N	1 N

Passage 5

Both $Cr_2O_7^{2-}(aq)$ and $MnO_4^-(aq)$ can be used to titrate $Fe^{2+}(aq)$. You have been provided with 0.1 M solution of each and volumetric estimation is carried in acidic medium. Suppose for the given volume of Fe^{2+} solution :

$$V_1 = \text{volume of } MnO_4^-$$
$$V_2 = \text{volume of } Cr_2O_7^{2-}$$

required in the oxidation to Fe^{3+}.

Based on this experimental study, answer the following questions :

1. Select correct alternate.
   1 Fe^{2+} is equivalent to
   (a) $\frac{1}{5}$ mol of MnO_4^-
   (b) $\frac{1}{6}$ mol of $Cr_2O_7^{2-}$
   (c) 1 equivalent each of MnO_4^- and $Cr_2O_7^{2-}$
   (d) All of the above
2. (a) $V_1 = V_2$
   (b) $V_1 > V_2$
   (c) $V_1 < V_2$
   (d) nothing can be predicted
3. 24.50 mL of 0.1 M $Cr_2O_7^{2-}$ is required in the oxidation of given volume of 0.1 M Fe^{2+}. 0.1 M MnO_4^- required will be
   (a) 29.40 mL (b) 24.50 mL
   (c) 20.42 mL (d) 20.10 mL

Passage 6

Following experiment is designed to determine calcium content in the blood. Read the experiment and answer the questions at the end of it.

Calcium oxalate (CaC_2O_4) is insoluble in water. For this reason it can be used to determine the amount of Ca^{2+} ions in fluids such as blood. The calcium oxalate isolated from blood is dissolved in acid and titrated against a standardized $KMnO_4$ solution.

Answer the following questions

1. In one test it is found that the calcium oxalate isolated from a 10.0 mL sample of blood requires 25 mL of 1×10^{-3} M $KMnO_4$ for titration. Calculate the number of milligrams of calcium per mL of blood
   (a) 0.50 (b) 10.0
   (c) 0.25 (d) 5.0
2. Milliequivalents of $KMnO_4$ used were
   (a) 0.125 (b) 0.025
   (c) 0.250 (d) 0.050
3. Calcium content expressed in ppm is
   (a) 250 (b) 200
   (c) 150 (d) 100
4. Above titration takes place in acidic medium. Acid chosen is
   (a) H_2SO_4 (b) HCl
   (c) CH_3COOH (d) HCOOH
5. Indicator in the above titration is
   (a) phenolphthalein
   (b) methylene blue
   (c) methyl orange
   (d) $KMnO_4$ (self-indicator)

Passage 7

Following titration method is given to determine total content of the species with variable *oxidation states. Answer the questions given at the end of it.*

A quantity of 25.0 mL of solution containing both Fe^{2+} and Fe^{3+} ions is titrated with 25.0 mL of 0.0200 M $KMnO_4$ (in dilute H_2SO_4). As a result, all of the Fe^{2+} ions are oxidised to Fe^{3+} ions. Next 25 mL of the original solution is treated with Zn metal. Finally, the solution requires 40.0 mL of the same $KMnO_4$ solution for oxidation to Fe^{3+}.

$$MnO_4^- + 5Fe^{2+} + 8H^+ \longrightarrow Mn^{2+} + 5Fe^{3+} + 4H_2O$$

Answer the following questions

1. Molar concentration of Fe^{2+} in the original solution is
   (a) 0.01 M (b) 0.02 M
   (c) 0.10 M (d) 0.20 M
2. Molar concentration of Fe^{3+} in the original solution is
   (a) 0.06 M (b) 0.16 M
   (c) 0.032 M (d) 0.012 M
3. Zinc added in the second titration will
   (a) oxidize Fe^{2+} to Fe^{3+} (b) reduce Fe^{3+} to Fe^{2+}
   (c) reduce Fe^{3+} to Fe (d) reduce Fe^{2+} to Fe

4. If 0.02 M $K_2Cr_2O_7$ is used instead of 0.02 M $KMnO_4$, its volume required in these titrations are respectively
(a) 25 mL, 40 mL (b) 25 mL, 15 mL
(c) 20.8 mL, 33.3 mL (d) 10.4 mL, 16.7 mL

Passage 8

Following short write-up describes synthesis of I_2 from natural source chile saltpetre. Answer the questions at the end of it.

Chile saltpetre is a natural source of $NaNO_3$ which also contains $NaIO_3$. The $NaIO_3$ can be used as a source of iodine produced in the following reactions:

$$IO_3^- + 3HSO_3^- \longrightarrow I^- + 3H^+ + 3SO_4^{2-} \quad ...(i)$$

$$5I^- + IO_3^- + 6H^+ \longrightarrow 3I_2(s) + 3H_2O \quad ...(ii)$$

1.00 L of the starting solution which contains 5.80 g $NaIO_3$ / L, is treated with stoichiometric quantity of $NaHSO_3$. Then a further quantity of the starting solution is added to the reaction mixture to bring about the second reaction. [I = 127, Na = 23, S = 32]

Answer the following questions

1. How many grams of $NaHSO_3$ are required in step (i)?
(a) 5.80 g (b) 17.4 g
(c) 9.14 g (d) 3.0 g

2. What additional volume of the starting solution must be added in step (ii)?
(a) 0.2 L (b) 0.4 L
(c) 0.8 L (d) 1.0 L

3. How many equivalents of IO_3^- are present in one mole of $NaIO_3$ in step (i)?
(a) 2 (b) 4
(c) 6 (d) 8

4. How many equivalents of IO_3^- are used in step (ii)?
(a) 6 (b) 5
(c) 0.06 (d) 0.035

Passage 9

Following passage is for the drunken drivers on roads which are becoming fatal by their wrong driving.

Every year in our country about 25,000 people are killed and 500,000 more are injured as a result of drunk driving. In spite of efforts to educate the public about the dangers of driving while intoxicated and stiffer penalties for drunk driving offenses, law enforcement agencies still have to devote a great deal of work to remove drunk drivers from roads.

The police often use a device called a breath analyser to test drivers suspected to being drunk. The chemical basis of this device is a redox reaction. A sample of the driver's breath is drawn into the breath analyser, where it is treated with an acidic solution of potassium dichromate.

$$3CH_3CH_2OH + 2K_2Cr_2O_7 + 8H_2SO_4 \longrightarrow 3CH_3COOH + 2Cr_2(SO_4)_3 + 2K_2SO_4 + 11H_2O$$

The driver's blood alcohol level can be determined readily by measuring the degree of this colour change (read from a calibrated meter on the instrument). The current legal limit of blood alcohol content is 0.1 per cent by mass. Anything higher constitutes intoxication.

Read the passage and answer the questions given below

1. Above test is based on
(a) smell of ethanol and ethanoic acid
(b) change of colour yellow to blue
(c) change of colour orange to green
(d) change of colour orange to colourless

2. Equivalent mass of $K_2Cr_2O_7$ is
(a) $\frac{\text{molar mass}}{6}$ (b) $\frac{\text{molar mass}}{3}$
(c) $\frac{\text{molar mass}}{2}$ (d) $\frac{\text{molar mass}}{1}$

3. Equivalent mass of CH_3CH_2OH is
(a) $\frac{\text{molar mass}}{1}$ (b) $\frac{\text{molar mass}}{2}$
(c) $\frac{\text{molar mass}}{3}$ (d) $\frac{\text{molar mass}}{4}$

4. The alcohol content in a 10.0 g sample of blood from a driver required 4.23 mL of 0.07654 M $K_2Cr_2O_7$ for titration. Should the police prosecute the individual for drunken driving?
(a) Yes (b) No
(c) Nothing can be predicted

Assertion & Reason

Codes :
(a) Both **A** and **R** are true and **R** is the correct explanation of **A**.
(b) Both **A** and **R** are true but **R** is not the correct explanation of **A**.
(c) **A** is true but **R** is false.
(d) **A** is false but **R** is true.

1. **Assertion (A) :** If a solution is 20% by weight of solution, then it is 25% by weight of solvent.

Reason (R) : Solution also includes weight of solute and thus, pure solvent is 80 g which contains 20 g solute.

2. **Assertion (A) :** Molality is independent of temperature.

Reason (R) : Volume of solution does not change with temperature.

3. **Assertion (A) :** In the following half reaction:

$$Cr_2O_7^{2-} \longrightarrow Cr^{3+}$$

$$\text{Equivalent weight of } Cr_2O_7^{2-} = \frac{\text{Molar mass}}{3}$$

Reason (R) : Change in oxidation number of chromium is 3 per atom.

4. **Assertion (A) :** 1.575 g $H_2C_2O_4 \cdot 2H_2O$ in 250 mL solution makes it 0.1 N.
Reason (R) : $H_2C_2O_4 \cdot 2H_2O$ is a dihydrate organic acid.
5. **Assertion (A) :** Following reaction is a disproportionation reaction:
$$Br_2 + 2OH^- \longrightarrow BrO^- + Br^- + H_2O$$
Reason (R) : Br_2 is oxidised to BrO^- as well as reduced to Br^-.

True & False

1. A solution 20% by mass of solvent is 25% by mass of solution.
2. Equivalent mass of $Cr_2O_7^{2-}$ in the following reaction is M/6.
$$Cr_2O_7^{2-} \longrightarrow Cr^{3+}$$
3. 100 mL of 1 M SO_2Cl_2 aqueous solution is neutralised by 100 mL of 4 N NaOH.
4. H_2O_2 can be estimated by MnO_4^- in acidic as well as in basic medium.
5. Iodimetric estimation includes $CuSO_4, Cr_2O_7^{2-}, MnO_4^-$, $CaOCl_2$, MnO_2/HCl which on treatment with KI liberate I_2.
6. In iodometric titration of $Cr_2O_7^{2-}$, there is bluish green colour at the end point.
7. Concentration of 1 ppm means that 1 L solution contains 1 mg solute.
8. "10 volume H_2O_2" means 1 mL of H_2O_2 concentrated solution has been diluted to 10 mL.
9. "1 volume H_2O_2" solution means that 1 mL of H_2O_2 at NTP gives 1 L of O_2.
10. Ferric sulphate is oxidised by MnO_4^- / H^+.
11. In the titration of Fe^{2+} with MnO_4^-, MnO_4^- is self-indicator.
12. Concentration of $H_2C_2O_4$ can be determined by MnO_4^-/H^+ as well as by NaOH.
13. Only 50% of Na_2CO_3 solution is indicated if it is neutralised by HCl using phenolphthalein as the indicator.
14. Estimation of nitrogen is done by Kjeldahl method.
15. SO_2 turns acidified $K_2Cr_2O_7$ solution to colourless solution.

Fill in the Blanks

1. If molecular weight of $KMnO_4$ is 158, then its equivalent weight is .... in acidic medium.
2. Equivalent of $KMnO_4$ is 158 when it changes to ....
3. Pyrolusite can be estimated by ....
4. Concentration of Na_2CO_3 solution can be determined by titration with acid solution using .... as the indicator.
5. End point in the titration of $CuSO_4$ iodometrically (with starch as the indicator) is the ....
6. Concentration of 1 mg per litre is equal to .... ppm.
7. If 900 mL H_2O is added to 100 mL of 1 M H_2SO_4, then it is .... N H_2SO_4.
8. MnO_4^- oxidises Mn^{2+} to MnO_2 and itself reduces to .... in the titration method.
9. Normality of the mixture of 100 mL 1 N HCl and 50 mL 2 N H_2SO_4 is ....
10. If 100 mL of SO_2 solution required 50 mL of 0.01 M $Cr_2O_7^{2-}$ in acidic medium then SO_2 solution is ....
11. Normality of the mixture of 100 mL 1 N HCl and 50 mL 1 N NaOH is .... and the mixture is ....
12. 100 mL of a mixture of $FeSO_4$ and $Fe_2(SO_4)_3$ required 50 mL of 0.1 N $KMnO_4$ solution, then concentration of Fe^{2+} is .... g L^{-1}.
13. Ozone in the atmosphere is determined by ....
14. BrO_3^- oxidises I^- to .... and thus equivalent weight of BrO_3^- is ....
15. I_2 oxidises N_2H_4 to ....

Matrix-Match Type Questions

I. Only One Correct

1. Match the reaction (in List I) with titration method (in List II).

List I	List II
A. $N_2H_4 + 2I_2 \longrightarrow N_2 + 4H^+ + 4I^-$	1. Iodometric
B. $2CuSO_4 + 4KI \longrightarrow Cu_2I_2 + 2K_2SO_4 + I_2$	2. Redox
C. $MnO_4^- + Fe^{2+} \longrightarrow Mn^{2+} + Fe^{3+}$	3. Kjeldahl
D. $NH_4Cl + NaOH \xrightarrow{\Delta} NH_3$; $2NH_3 + H_2SO_4 \longrightarrow (NH_4)_2SO_4$	4. Iodimetric

2. Match the half-reaction (in List I) with equivalent mass (molar mass = *M*) (in List II).

	List I	List II
A.	$Cr_2O_7^{2-} \longrightarrow Cr^{3+}$	1. *M*
B.	$C_2O_4^{2-} \longrightarrow CO_2$	2. *M*/2
C.	$MnO_4^- \longrightarrow MnO_2$	3. *M*/6
D.	$HC_2O_4^- \longrightarrow C_2O_4^{2-}$	4. *M*/3

3. Match the reactants (in List I) with the indicator (in List II).

List I	List II
A. $Fe^{2+}/Cr_2O_7^{2-}$	1. Ferric alum
B. $AgNO_3/CNS^-$	2. MnO_4^- (self)
C. $CuSO_4/KI/S_2O_3^{2-}$	3. $K_3[Fe(CN)_6]$
D. Fe^{2+}/MnO_4^-	4. Starch

4. Match the reaction (in List I) with the colour at the final stage (in List II).

List I	List II
A. $MnO_4^-/H^+/Fe^{2+}$	1. Deep reddish brown
B. MnO_4^-/conc. OH^-/Mn^{2+}	2. Colourless
C. MnO_4^-/H_2O/dil. OH^-	3. Blackish brown
D. $CuSO_4/KI$	4. Green

II. One or More Correct

1. Match the reactant (in Column I) with the type of titration (in Column II) used for estimation.

Column I	Column II
A. KHC_2O_4	1. Acid-Base titration
B. $FeSO_4 \cdot (NH_4)_2SO_4 \cdot 6H_2O$	2. Redox titration
C. $CuSO_4$	3. Iodometric
D. I_2	4. Iodimetric
E. $K_2Cr_2O_7$	

2. Match the half-reaction (oxidation or reduction) (in Column I) with number of electrons exchanged (in Column II).

Column I	Column II
A. $3I^- \longrightarrow I_3^-$	1. 5
B. $Cr_2O_7^{2-} \longrightarrow 2Cr^{3+}$	2. 6
C. $Cl^- \longrightarrow ClO_3^-$	3. 2
D. $N_2 \longrightarrow N_2H_4$	4. 4
E. $MnO_4^- \longrightarrow Mn^{2+}$	
F. $C_2O_4^{2-} \longrightarrow 2CO_2$	

Integer Answer Types

This section contains 8 questions. The answer to each of the questions is a single digit integer, ranging from 0 to 9. The appropriate bubbles below the respectively question numbers in the *ORS* have to be darkened. For example, if the correct answers to question number *X, Y, Z* and *W* (say) are 6, 0, 9 and 2, respectively, then the correct darkening of bubbles will like the following :

X	Y	Z	W
0	0	0	0
1	1	1	1
2	2	2	2
3	3	3	3
4	4	4	4
5	5	5	5
6	6	6	6
7	7	7	7
8	8	8	8
9	9	9	9

Set I

1. 1 M MnO_4^- solution is x equiv. L^{-1} when reduced to Mn^{2+}. Thus, x is ....

2. 1 M $Cr_2O_7^{2-}$ solution is y equiv. L^{-1}. Thus, y is ....

3. 1 M $C_2O_4^{2-}$ solution is z equiv. L^{-1}. Thus, z is ....

4. 6.3 g of an acid (molar mass = 126 g mol^{-1}) reacts completely with 100 mL of 0.5 $Ca(OH)_2$ solution. Thus basicity of acid is ....

5. We want to prepare 2 L of 0.025 M CH_3COOH solution. Amount of CH_3COOH dissolved is ....

6. 100 mL of 0.01 M XO_4^- is reduced to x^{n+} by 100 mL of 0.05 M Fe^{2+} in acidic medium. Thus, oxidation state of X in X^{n+} is ....

7. $Cr(NH_3)_xCl_3(aq)$ solution is titrated with $HCl(aq)$ solution.

$$Cr(NH_3)_xCl_3(aq) + xHCl(aq) \longrightarrow xNH_4^+(aq) + Cr^{3+}(aq) + (x+3)Cl^-(aq)$$

Assume that 24.26 mL of 1.5 M HCl is used to titrate 1.580 g of $Cr(NH_3)_xCl_3$. Thus, x is ....

8. 250 mL of 0.1 M NaOH contains x g of NaOH. Thus, x is ....

Set II

1. Basicity of H_3PO_4 is ....
2. Basicity of H_3PO_3 is ....
3. Basicity of H_3PO_2 is ....
4. 2 moles of MnO_4^- oxidise .... moles of $NaHC_2O_4$.
5. Ionisable H atoms in acetic acid $C_2H_4O_2$ is ....
6. 20 mL of 1 M MnO_4^- are required in acidic medium and 100 mL of 1 M MnO_4^- are required in neutral medium in a redox titration. If MnO_4^- is reduced to Mn^{2+} in acidic medium, oxidation number of Mn in reduced species in neutral medium is ....
7. 22.4 L is the volume of .... equivalent(s) of O_2 gas at NTP.
8. 25 mL of 0.1 M solution of metallic salt (A) oxidises 25 mL of 0.1 M sodium sulphite to sodium sulphate. If oxidation number of the metal in the salt (A) is 3, then new oxidation number of the metal is ....

Set III

1. Complex $Cr(NH_3)_xCl_3$ (A) is titrated with 1.5 M HCl. 40 mL of HCl are required to titrate 2.635 g of A. Thus, x is ....
2. Mixture containing 1 mole BaF_2 and 2 moles H_2SO_4 is to be made neutral by adding $Ca(OH)_2$. Moles of $Ca(OH)_2$ required ....
3. If $[BaCl_2] = 0.05$ M then p_{Cl^-} = ...
4. 2.78 g $FeSO_4 \cdot xH_2O$ in 100 mL solution gives 0.1 M solution. Thus, x = ...
5. 25 mL of a solution containing 6.1 gL^{-1} of an oxalate of formula $K_xH_y(C_2O_4)_z \cdot nH_2O$ required 18 mL of 0.1 N NaOH and 24 mL of 0.1 N $KMnO_4$ in two separate titrations.
   Thus, x = ...
6. Thus, y = ...
7. Thus, z = ...
8. Thus, n = ...

(Refer Problem 15 (P-323) for solution of Question 5-8)

Answers & Solutions
Master Exercises

Exercise 1

Short Answer Type Questions

1. NaOH = 0.125 M

Volume = V (in litres) = ?

Amount = 4.0 g

$$MV = \text{moles} = \text{moles} \times 40 \text{ g}$$
$$0.125 \times V = \text{moles}$$
$$0.125 \times V \times 40 = 4 \text{ g (given)}$$
$$V = 0.8 \text{ L} = \mathbf{800 \text{ mL}}$$

or $$w = \frac{MmV}{1000}$$

$$\text{Amount in grams} = \frac{\text{molarity} \times \text{molar mass} \times \text{volume (in mL)}}{1000}$$

$$\therefore \quad V = \frac{1000w}{Mm} = \frac{1000 \times 4}{0.125 \times 40} = 800 \text{ mL}$$

2. $Na_2CO_3 \rightleftharpoons 2Na^+ + CO_3^{2-}$

$$\text{Moles of } Na_2CO_3 = \frac{1.06}{106} = 0.01 \text{ mol}$$

$$\text{Volume} = 250 \text{ mL} = 0.250 \text{ L}$$

$$\text{Molarity} = \frac{\text{mole}}{\text{litre}} = \frac{0.010}{0.250} = 0.04 \text{ M}$$

$$\therefore \quad [Na^+] = \mathbf{0.08 \text{ M}}$$
$$[CO_3^{2-}] = \mathbf{0.04 \text{ M}}$$

3. (a) $[NH_4^+] = 0.50$ M, $[SO_4^{2-}] = \mathbf{0.25 \text{ M}}$

(b) $[Na^+] = 0.246$ M, $[CO_3^{2-}] = \mathbf{0.123 \text{ M}}$

(c) $[H^+] = [NO_3^-] = \mathbf{0.056 \text{ M}}$

4. For dilution

$$M_1V_1 = M_2V_2$$
before after

$$1.50 \times 25.0 = M_2 \times 500$$
$$M_2 = \frac{1.50 \times 25.0}{500}$$
$$= 0.75 \text{ M}$$

5. $$M_1V_1 = M_2V_2$$
$$6.00 \times 20.8 = M_2 \times 1000$$
$$M_2 = \mathbf{0.125 \text{ M of volume 1 L}}$$

(b) 50.0 mL of 3.00 M is diluted to 1000 mL

$$M_1V_1 = M_2V_2$$
$$3.00 \times 50.0 = M_2 \times 1000$$
$$M_2 = 0.15 \text{ M}$$

Thus, method (a) is employed.

6. Moles of NaCl $= 1.00 \times 0.225$
$= 0.225$ mol

Moles of $PbCl_2 = V \text{ (required)} \times 0.750$

2 mol NaCl react with = 1 mol of $Pb(NO_3)_2$

Thus, 0.225 mol reacts with = 0.1125 mol $Pb(NO_3)_2$

$$\therefore \quad V \times 0.750 = 0.1125$$
$$V = 0.15 \text{ L}$$
$$= \mathbf{150 \text{ mL}}$$

7. Method I : In terms of moles

Volume of HCl = 38.55 mL
= 0.03855 L

Moles of HCl = 0.03855 × molarity (required)

$$\text{Moles of } Na_2CO_3 = \frac{2.150}{106} = 0.0203$$

Based on reaction

1 mol Na_2CO_3 reacts with = 2 mol HCl

∴ 0.0203 mol Na_2CO_3 reacts with = 0.0406 mol HCl

$$\therefore \quad 0.03855 \times M = 0.0406$$
$$M = \mathbf{1.05 \text{ M}}$$

Method II : In terms of equivalent

In any reaction.

Equivalent of reactant I ≡ Equivalent of reactant II

$$\text{Equivalent of } Na_2CO_3 = \frac{2.150}{53 \text{ (equivalent mass)}}$$

$$\text{Equivalent of HCl} = \frac{NV}{1000} = \frac{N \times 38.55}{1000}$$

$$\therefore \quad \frac{N \times 38.55}{1000} = \frac{2.150}{53}$$

$$N(\text{HCl}) = 1.05 \text{ N}$$
$$\therefore \quad M(\text{HCl}) = N(\text{HCl}) = \mathbf{1.05 \text{ N}}$$

8. As above

Equivalent of H_2A = Equivalent of KOH

$$\text{Equivalent of } H_2A = \frac{0.490}{E}$$

$$\text{Equivalent of KOH} = \frac{100 \times 0.1}{1000}$$
$$= 0.01$$

$$\therefore \quad \frac{0.49}{E} = 0.01$$

$$\therefore \quad E = 49 \text{ g equivalent}^{-1}$$

H_2A is a dibasic acid, hence,

$$\text{molar mass of } H_2A = 2 \times \text{equivalent mass}$$
$$= \mathbf{98\ g\ mol^{-1}}$$

9. (a) Glucose is non-electrolyte

(b) $H_3PO_4 \rightleftharpoons H^+ + H_2PO_4^-$

(c) AgI is insoluble in H_2O

(d) LiCl behaves like non-electrolyte in ether.

Hence, in (a), (c) and (d), analytical and equilibrium concentrations (molarity) are identical.

10. $\underset{\text{monomer}}{2CH_3COOH} \rightleftharpoons \underset{\text{dimer}}{(CH_3COOH)_2}$

In benzene, there is 100% dimer formation of acetic acid. Thus, molecular weight of acetic acid in benzene $= 120\ g\ mol^{-1}$

$$\therefore \quad [CH_3COOH]_{eq} = 0.0\ M$$
$$[(CH_3COOH)_2]_{eq} = 0.5\ M$$

Note : If CH_3COOH (without dimer) = 1 M it means 60 g L^{-1} of solution. But when dimer is formed

$$\text{equilibrium molarity} = \frac{60\ gL^{-1}}{120\ g\ mol^{-1}}$$
$$= 0.5\ mol\ L^{-1}$$

11. $MnO_2(s) + 4H^+ + 2I^- \longrightarrow Mn^{2+} + I_2 + 2H_2O$

$$I_2 + 2S_2O_3^{2-} \longrightarrow S_4O_6^{2-} + 2I^-$$

Method I : (Based on number of equilivalents)

$$\text{Equivalent of } MnO_2 \equiv \text{equivalents of } S_2O_3^{2-}$$

Equivalent mass of $MnO_2 = 43.5$

$$0.07220\ M\ S_2O_3^{2-} \equiv 0.07220\ N\ S_2O_3^{2-}$$

$$\text{Equivalents of } S_2O_3^{2-} = \frac{0.07220 \times 32.30}{1000} = 2.332 \times 10^{-3}$$

Equivalents of $MnO_2 = 2.332 \times 10^{-3}$

$\therefore$ Pure MnO_2 in the sample $= 2.332 \times 10^{-3} \times 43.5$ g

$$= 0.1014\ g$$

$$\%\text{ of pure } MnO_2 = \frac{0.1014}{0.1344} \times 100$$
$$= 75.48\%$$

Method II : (Based on stoichiometric coefficients)

$$\text{Numbr of moles } S_2O_3^{2-} = \frac{32.30 \times 0.0722}{1000}$$
$$= 2.332 \times 10^{-3}\ mol$$

$$\text{Number of moles of } MnO_2 = \frac{\text{mol of } S_2O_3^{2-}}{2}$$
$$= 1.166 \times 10^{-3}\ mol$$

Pure MnO_2 in the sample $= 1.166 \times 10^{-3} \times 87$ g

$$\%\ MnO_2 = 75.48\%$$

Analytical Questions

1. $\underset{\substack{1\ mol \\ 24\ g}}{Mg} + \underset{\substack{2\ mol \\ 73\ g}}{2HCl} \longrightarrow MgCl_2 + H_2$

$$HCl = 5 \times 10^2 \times 2 \text{ millimol}$$
$$= 1000 \text{ millimol} = 1 \text{ mol}$$

$$Mg = 4.47\ g = 0.18625\ mol$$
$$\text{HCl required} = 0.3725\ mol$$

$\therefore$ HCl unreacted $= (1 - 0.3725)$ mol in 500 mL solution

$$\text{Molarity of HCl} = \frac{0.6275}{0.5} = \mathbf{1.255\ M}$$

2. $\underset{1\ mol}{Fe_2O_3} + \underset{6\ mol}{6H_2C_2O_4} \longrightarrow 2Fe(C_2O_4)_3^{3-} + 3H_2O + 6H^+$

Oxalic acid taken $= 500 \times 1 = 500$ millimol $= 0.5$ mol

$$\therefore \quad Fe_2O_3 \text{ required} = \frac{0.5}{6}\ mol = \frac{1}{12}\ mol$$
$$= \frac{1}{12} \times 160 = \mathbf{13.33\ g}$$

3. $NaHCO_3$ is completely neutralised and Na_2CO_3 is neutralised to phenolphthalein end point indicating neutralisation upto $NaHCO_3$

$$NaHCO_3 + HCl \longrightarrow NaCl + H_2O + CO_2$$
$$Na_2CO_3 + HCl \longrightarrow NaHCO_3 + NaCl$$

Thus, $\quad 0.2\ M\ NaHCO_3 = 0.2\ N\ NaHCO_3$

and $\quad 0.2\ M\ Na_2CO_3 = 0.2\ N\ Na_2CO_3$

$$\underbrace{N_1V_1}_{\text{acid}} = \underbrace{N_2V_2}_{\text{base } (Na_2CO_3 \text{ and } NaHCO_3)}$$
$$30x = 10 \times 0.2 + 5 \times 0.2$$
$$x = \mathbf{0.1\ M}$$

4. Solution is acidic (pH = 1)

$$[H^+] = 10^{-pH}\ M$$
$$\therefore \quad [H^+] = 10^{-1}\ M$$

Let molarity of H_2SO_4 or $Ba(OH)_2 = 1$ M

$$\therefore \quad \text{Normality} = 2\ M$$
$$\therefore \quad N_1V_1\ (\text{acid}) = 300 \times 2 = 600\ M$$
$$N_2V_2\ (\text{base}) = 200 \times 2 = 400\ M$$
$$N(H^+) = \frac{N_1V_1 - N_2V_2}{V_1 + V_2}$$
$$0.1 = \frac{200\ M}{500}$$
$$M = \frac{0.1 \times 500}{200} = 0.25$$
$$\therefore \quad \text{Normality} = 2\ M = \mathbf{0.5\ N}$$

5. $2FeS_2 + 5.5\ O_2 \longrightarrow Fe_2O_3 + 4SO_2$

$$4SO_2 + 8NaOH \longrightarrow 4Na_2SO_3 + 4H_2O$$

Thus, $\quad 2FeS_2 \equiv 4SO_2 \equiv 8NaOH$

$$\therefore \quad 1\ FeS_2 \equiv 4\ NaOH$$

$$FeS_2 \text{ in } 1000\ kg\ (10^6 \text{ g coal}) = 10^6 \times \frac{0.05}{100} = 500\ g$$
$$= \frac{500}{120} = 4.167\ mol$$

$$\therefore \quad \text{NaOH required} = 16.67\ mol$$

6 M NaOH contains 6 mol NaOH in = 1000 mL

$$\therefore \quad 16.67 \text{ mol are in} = \frac{1000 \times 16.67}{6}\ mL$$
$$= \mathbf{2778\ mL}$$

or $\dfrac{MV}{1000} = \text{mol}$

$\dfrac{6 \times V}{1000} = 16.67$

$\therefore \quad V = \mathbf{2778\ mol}$

6. Phenolphthalein indicates 50% reaction of Na_2CO_3 (upto $NaHCO_3$) and 100% reaction of NaOH while methyl orange indicates 100% reaction of both.

Let volume of 0.995 N HCl required by $Na_2CO_3 = x$ mL and by NaOH = y mL

with **phenolphthalein**

$$\frac{x}{2} + y = 19.5 \text{ mL}$$

with **methyl-orange**

$$x + y = 25.9 \text{ mL}$$

$\therefore \quad x = 12.8$ mL

$\therefore \quad y = 13.1$ mL

$$N_1V_1 = N_2V_2$$
$$Na_2CO_3 \quad HCl$$
$$N_1 \times 25 = 0.995 \times 12.8$$
$$N_1(Na_2CO_3) = 0.50944 \text{ N}$$
$$= 0.50944 \times 53$$
$$= \mathbf{27.0\ gL^{-1}}$$

$$N_1V_1 = N_2V_2$$
$$NaOH \quad HCl$$
$$N_1 \times 25 = 0.995 \times 13.1$$
$$N_1 (NaOH) = 0.5214 \text{ N}$$
$$= 0.5214 \times 40$$
$$= \mathbf{20.86\ gL^{-1}}$$

7. $Al + 3HCl \longrightarrow AlCl_3 + \frac{3}{2}H_2$

$$3 \text{ mol HCl} \equiv 1 \text{ mol Al}$$

$\therefore \quad \dfrac{0.05 \times 12}{1000} \text{ mol HCl} = \dfrac{0.05 \times 12}{1000 \times 3} \text{ mol Al}$

$$= 2 \times 10^{-4} \text{ mol Al}$$
$$= 2 \times 10^{-4} \times 27 \text{ g Al}$$
$$= \frac{2 \times 10^{-4} \times 27}{2.7} \text{ cm}^3 \text{ Al}$$
$$= 2 \times 10^{-3} \text{ cm}^3 \text{ Al}$$

$$\left(\text{volume} = \frac{\text{mass}}{\text{density}}\right)$$

thickness × area = volume

$$\frac{0.1}{10} \times \text{area} = 2 \times 10^{-3}$$

$\therefore \quad \text{area} = 0.2 \text{ cm}^2$

8. $Na_2B_4O_7$ borax behaves like a weak base

0.7439 g sample of borax ≡ 31.64 mL of 0.108 M HCl

Based on reaction

$$\frac{\text{moles of } Na_2B_4O_7}{\text{moles of HCl}} = \frac{1}{2}$$

$$\text{Moles of HCl} = \frac{31.64 \times 0.108}{1000}$$
$$= 3.417 \times 10^{-3} \text{ mol}$$

$\therefore$ Moles of $Na_2B_4O_7 = 1.7085 \times 10^{-3}$ mol

$\therefore \quad Na_2B_4O_7 = 1.7085 \times 10^{-3} \times 202$ g

$= 0.3451$ g in 0.7439 g borax

(a) Per cent $Na_2B_4O_7 = 46.39\%$

(b) $Na_2B_4O_7 \cdot 10H_2O = 1.7085 \times 10^{-3} \times 382$ g

Per cent $Na_2B_4O_7 \cdot 10H_2O = 87.73\%$

(c) $B_2O_3 = 2 \times$ moles of $Na_2B_4O_7$

$= 2 \times 1.7085 \times 10^{-3} \times 70$ g

$= 0.2392$ g

Per cent $B_2O_3 = 32.15\%$

(d) $Na_2B_4O_7 = 4\,B = 4 \times 1.7085 \times 10^{-3}$ mol

$= 4 \times 1.7085 \times 10^{-3} \times 11$ g

$\therefore \quad B = 0.0752$ g

Per cent = 10.11%

9. (a) 100 mL of serum has $K^+ = 18$ mg

$$= 18 \times 10^{-3} \text{ g}$$
$$= \frac{18 \times 10^{-3}}{39} \text{ mol}$$

$\therefore \quad [K^+] = \dfrac{18 \times 10^{-3}}{39} \Big/ 0.1$

$$= \mathbf{4.615 \times 10^{-3}\ M}$$

Similarly, $\quad [Cl^-] = \dfrac{365 \times 10^{-3}}{35.5} \Big/ 0.1$

$$= \mathbf{0.103\ M}$$

(b) $\quad p_{K^+} = -\log [K^+] = \mathbf{2.336}$

$\quad p_{Cl^-} = -\log [Cl^-] = \mathbf{0.988}$

10. $HCl \rightleftharpoons H^+ + Cl^-$: end-point indicated by methyl-red

$H_3PO_4 \rightleftharpoons H^+ + H_2PO_4^-$: end-point indicated by methyl-red

$H_2PO_4^- \rightleftharpoons H^+ + HPO_4^{2-}$: end-point indicated by bromothymol blue

Let HCl = x millimol

$H_3PO_4 = y$ millimol

then $x + y$ (first ionisation step) = 35 × 0.1 = 3.5 millimol

y (second ionisation step) = 15 × 0.1 = 1.5 millimol

$\therefore \quad HCl = 3.5 - 1.5 = \mathbf{2.00}$ **millimol**

and $\quad H_3PO_4 = \mathbf{1.5}$ **millimol**

11. Let $P_2O_5 = x \text{ g} = \dfrac{x}{142} \text{ mol} = \dfrac{2x}{142} \text{ mol } H_3PO_4$

(since one mole of P_2O_5 in aqueous solution forms 2 moles H_3PO_4)

$\therefore \quad H_3PO_4 = (0.405 - x) \text{ g} = \left(\dfrac{0.405 - x}{98}\right) \text{mol}$

$$H_3PO_4 + 2NaOH \longrightarrow Na_2HPO_4 + 2H_2O$$

NaOH required by $P_2O_5 = \dfrac{4x}{142} \text{ mol} = \dfrac{2x}{71}$

and NaOH required by H_3PO_4

$$= 2\left(\frac{0.405 - x}{98}\right) = \left(\frac{0.405 - x}{49}\right)$$

$$\text{Actual NaOH required} = \frac{42.5 \times 0.25}{1000}$$

$$= 0.010625 \text{ mol}$$

$$\therefore \quad \frac{2x}{71} + \left(\frac{0.405 - x}{49}\right) = 0.010625$$

$$\therefore \quad x = 0.3040 \text{ g}$$

$$\therefore \quad H_3PO_4 = (0.4050 - 0.3040) = 0.1010 \text{ g}$$

$$\% \, H_3PO_4 = \mathbf{24.93\%}$$

12. $$Na_2CO_3 + 2HCl \longrightarrow 2NaCl + H_2O + CO_2$$

$$K_2CO_3 + 2HCl \longrightarrow 2KCl + H_2O + CO_2$$

Conc. of mixture = 1.2 g/100 mL = 12 g/L

$$Na_2CO_3 = x \text{ gL}^{-1} = \frac{x}{53} \text{ N}$$

$$K_2CO_3 = \frac{12 - x}{69} \text{ N}$$

$$N_1V_1 = N_2V_2$$

$$N_1 \text{ (mix)} \times 20 = 40 \times 0.1$$

$$\therefore \quad N_1 = 0.2 \text{ N}$$

$$\therefore \quad \frac{x}{53} + \frac{12 - x}{69} = 0.2$$

$$x = 5.9625 \text{ g}$$

$$\therefore \quad Na_2CO_3 = 5.9625 \text{ gL}^{-1} = 0.59625 \text{ g/100 mL}$$

$$K_2CO_3 = 6.0375 \text{ gL}^{-1} = 0.60375 \text{ g/100 mL}$$

$$Na_2CO_3 + BaCl_2 \longrightarrow BaCO_3$$

$$K_2CO_3 + BaCl_2 \longrightarrow BaCO_3$$

$$Na_2CO_3 \text{ in 20 mL solution} = \frac{5.9625}{1000} \times 20 \text{ g}$$

$$= \frac{5.9625 \times 20}{1000 \times 106} \text{ mol}$$

$$= 1.125 \times 10^{-3} \text{ mol}$$

$$\therefore \quad BaCO_3 \text{ formed} = 1.125 \times 10^{-3} \text{ mol}$$

$$K_2CO_3 \text{ in 20 mL solution} = \frac{6.0375}{1000} \times 20 \text{ g}$$

$$= \frac{6.0375 \times 20}{1000 \times 138}$$

$$= 8.75 \times 10^{-4} \text{ mol}$$

$$BaCO_3 \text{ formed} = 8.75 \times 10^{-4} \text{ mol}$$

$$\text{Total } BaCO_3 \text{ formed} = 2 \times 10^{-3} \text{ mol}$$

$$= 2 \times 10^{-3} \times 197 = \mathbf{0.394 \text{ g}}$$

13. $$KCl + NH_4Cl \xrightarrow{\Delta} KCl\downarrow + \underset{\text{sublimates}}{NH_4Cl\uparrow}$$

Only residue left is KCl (dissolved in 250 mL solution)

25 mL of KCl ≡ 15 mL of 0.1 N $AgNO_3$

$$\therefore \quad N\text{ (KCl)} = \frac{15 \times 0.1}{25} = 0.06 \text{ N}$$

$$\therefore \quad \text{KCl in mixture} = 0.06 \times 74.5 = 4.47 \text{ gL}^{-1}$$

$$= 1.1175 \text{ g in 250 mL}$$

$$= \mathbf{0.015 \text{ mol}}$$

$$\therefore \quad NH_4Cl = 2.4 - 1.1175$$

$$= 1.2825 \text{ g}$$

$$= \mathbf{0.024 \text{ mol}}$$

$$\text{Total Cl} = 0.039 \text{ mol}$$

$$= 1.3845 \text{ g}$$

$$\% \text{ of Cl} = \frac{1.3845}{2.4} \times 100 = \mathbf{57.69\%}$$

14. $$\underset{}{S^{2-}} + \underset{\substack{1 \text{ mol} \\ 63.5 \text{ g}}}{Cu^{2+}} \longrightarrow \underset{\substack{1 \text{ mol} \\ 95.5 \text{ g}}}{CuS\downarrow}$$

95.5 g of CuS is from = 63.5 g of Cu^{2+}

$$\therefore \quad 0.0177 \text{ g of CuS is from} = \frac{63.5 \times 0.0177}{95.5} \text{ g } Cu^{2+}$$

$$= 0.01177 \text{ g}$$

$$= 1.8534 \times 10^{-4} \text{ mol } Cu^{2+} \text{ in } 0.80 \text{ L } H_2O$$

$$\therefore \quad [Cu^{2+}] = \mathbf{2.317 \times 10^{-4} \text{ M}}$$

15. $$\underset{+2}{Mn^{2+}} + \underset{+7}{MnO_4^-} + OH^- \longrightarrow \underset{+4}{MnO_2} + H_2O$$

$$\text{Normality of } Mn^{2+} = \text{molarity} \times 2$$

$$\text{Normality of } MnO_4^- = \text{molarity} \times 3$$

$$\text{Let molarity of } Mn^{2+} = x$$

$$\text{Molarity of } MnO_4^- = 0.05876 \text{ M}$$

$$\underset{Mn^{2+}}{N_1V_1} = \underset{MnO_4^-}{N_2V_2}$$

$$2x \times 25 = 34.77 \times 0.05876 \times 3$$

$$x = \mathbf{0.1226 \text{ M}}$$

16. 24 mL of 0.01 M $KMnO_4 = 0.24 \times 10^{-3}$ mol $KMnO_4$

Redox reaction is :

$$2MnO_4^- + 5C_2O_4^{2-} + 16H^+ \longrightarrow 10CO_2 + 2Mn^{2+} + 8H_2O$$

$$\therefore \quad C_2O_4^{2-} \text{ (pure) reacted} = \frac{5}{2} \times 0.24 \times 10^{-3} \text{ mol}$$

$$= 0.60 \times 10^{-3} \text{ mol}$$

Thus, oxalic acid (pure) $= 0.60 \times 10^{-3} \times 90$ g

= 0.054 g in 1.0 g impure oxalic acid and

% purity = **5.4%**

17. Oxalate part of calcium oxalate is oxidised by $KMnO_4$

$$\text{Moles of } MnO_4^- \text{ used} = \frac{24.2 \times 9.56 \times 10^{-4}}{1000}$$

$$= 2.31352 \times 10^{-5}$$

$$\therefore \quad \text{Moles of calcium oxalate} = \frac{5}{2} \times \text{moles of } KMnO_4$$

(Refer Q. 16 above) $= 5.7838 \times 10^{-5}$ mol

$$[Ca^{2+}] = 5.7838 \times 10^{-5} \text{ mol}$$

$$= 5.7838 \times 10^{-5} \times 40 \text{ g}$$

$$= 5.7838 \times 10^{-5} \times 40 \times 1000 \text{ mg in 10 mL blood}$$

= 2.314 mg in 10 mL blood

= **0.2314 mg per mL blood**

18. Let molar mass of the reductant = m g mol^{-1}

$\therefore$ $$\text{Molarity} = \frac{2.52}{m} \text{ mol L}^{-1}$$

$$0.01 \text{ M } MnO_4^- = 0.05 \text{ N } MnO_4^-$$

25 mL of N_1 (reductant) = 20×0.05 N MnO_4^-

$\therefore$ $$N_1 \text{ (reductant)} = \frac{20 \times 0.05}{25}$$

$$= 0.04 \text{ N}$$

Change in oxidation state of reductant = 2 units

$\therefore$ Molarity (reductant) = 0.02 M

$\therefore$ $$0.02 = \frac{2.52}{m}$$

$\therefore$ $$m = \mathbf{126 \text{ g mol}^{-1}}$$

19. $$O_3 + 6I^- + 3H_2O \longrightarrow 3I_2 + 6OH^-$$

$$\text{Eq. wt. of } O_3 = \frac{M}{6} = \frac{48}{6} = 8$$

$$I_2 + 2S_2O_3^{2-} \longrightarrow S_4O_6^{2-} + 2I^-$$

Equivalent of $S_2O_3^{2-}$ $\equiv$ Equivalent of I_2 $\equiv$ Equivalent of O_3

$$= \frac{1.5 \times 0.01}{1000} = 1.5 \times 10^{-5}$$

$\therefore$ $$= \frac{1.5 \times 10^{-5}}{6} \text{ mol}$$

$$\text{Air taken} = \frac{pV}{RT} = \frac{1 \times 10}{0.0821 \times 300} \text{ mol}$$

$$= 0.406 \text{ mol}$$

$\therefore$ Mol % of O_3 in air $= \dfrac{1.5 \times 10^{-5} \times 100}{6 \times 0.406}$

$$= \mathbf{6.16 \times 10^{-4}\%}$$

20.
$$\underset{\substack{2 \times 51 \text{ g} \\ 0.1 \text{ g}}}{2M} + nH_2SO_4 \longrightarrow M_2(SO_4)_n + \underset{\substack{n \times 22400 \text{ mL at STP} \\ 43.9 \text{ mL at STP}}}{nH_2}$$

$$\frac{2 \times 51}{0.1} = \frac{n \times 22400}{43.9}$$

$\therefore$ $$n = \mathbf{2}$$

M is oxidised to M^{2+} by H_2SO_4

$$\underset{+2}{M^{2+}} + \underset{+7}{MnO_4^-} + H^+ \longrightarrow \underset{\substack{(x>2) \\ +x}}{M^{x+}} + \underset{+2}{Mn^{2+}}$$

Also 0.1 N MnO_4^- = 0.02 M MnO_4^-

$\therefore$ Change in oxidation number of $M^{2+} = (x - 2)$

Change in oxidation number of $MnO_4^- = 5$

$\therefore$ $5M^{2+} + (x-2)\,MnO_4^- \longrightarrow 5M^{x+} + (x-2)\,Mn^{2+}$

$\therefore$ $$\frac{\text{moles of } M^{2+}}{\text{moles of } MnO_4^-} = \frac{5}{x-2} = \frac{0.1/51}{58.8 \times 0.02/1000}$$

$\therefore$ $$x = \mathbf{3}$$

[58.8 mL of 0.1 N MnO_4^- $\equiv$ 58.8 mL of 0.02 M MnO_4^-

$$\equiv \frac{58.8 \times 0.02}{1000} \text{ mol]}$$

21. Let $$XO = w \text{ g} = \frac{w}{(X+16)} \text{ mol}$$

and $$X_2O_3 = (2.18 - w) \text{ g} = \frac{2.18 - w}{(2X + 48)} \text{ mol}$$

Number of moles of $Cr_2O_7^{2-}$ required by $XO = \dfrac{5w}{6(X+16)}$

Number of moles of $Cr_2O_7^{2-}$ required by X_2O_3
$$= \frac{4(2.18 - w)}{3(2X + 48)}$$

Number of moles of XO_4^- from $XO = \dfrac{w}{(X+16)}$

Number of moles of XO_4^- from $X_2O_3 = \dfrac{2(2.18 - w)}{(2X + 48)}$

$\therefore$ $$\frac{5w}{6(X+16)} + \frac{4(2.18 - w)}{3(2X+48)} = 0.015$$

and $$\frac{w}{(X+16)} + \frac{2(2.18 - w)}{(2X+48)} = 0.0187$$

$$\frac{2.18 - w}{2X + 48} = 0.0018$$

On solving $w = 1.7372$ g and $\mathbf{X = 99}$

22. $$24 \text{ g } K_2CrO_4 = \frac{24}{194} \text{ mol} = 0.1237 \text{ mol}$$

$$N_2H_4 \text{ used} = 0.1237 \times \frac{3}{4}$$

$$= 0.0928 \text{ mol}$$

$$= 0.0928 \times 32 \text{ g}$$

$$= \mathbf{2.969 \text{ g}}$$

23. $$\text{Moles of } K_2Cr_2O_7 = \frac{4.23 \times 0.07654}{1000} = 3.24 \times 10^{-4} \text{ mol}$$

$\therefore$ CH_3CH_2OH oxidised $= 3.24 \times 10^{-4} \times \dfrac{3}{2}$ mol

$$= 4.86 \times 10^{-4} \text{ mol}$$

$$= 4.86 \times 10^{-4} \times 46 \text{ g}$$

= 0.0223 g in 10 g sample of blood

= 0.223% above 0.1% prescribed limit

Hence, **police will prosecute the person.**

24. $$2MnO_4^- + 5H_2O_2 + 6H^+ \longrightarrow 5O_2 + 2Mn^{2+} + 8H_2O$$

$$\text{Pure } H_2O_2 = \frac{x}{100} \text{ g} = \frac{x}{1700} \text{ equivalent}$$

$$MnO_4^- \text{ used} = \frac{Nx}{1000} \text{ equivalent}$$

$\therefore$ $$\frac{Nx}{1000} = \frac{x}{1700}$$

$$N(MnO_4^-) = \mathbf{\frac{10}{17}}$$

25. $$\underset{+6}{8CrO_4^{2-}} + \underset{+4}{3S_2O_3^{2-}} \longrightarrow \underset{12}{6SO_4^{2-}} + \underset{+3}{8Cr(OH)_4^-}$$

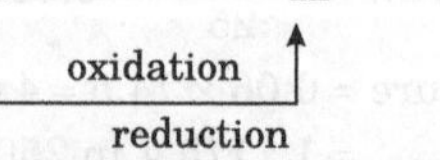

$$0.154 \text{ M } CrO_4^{2-} = 0.154 \times 3 \text{ N } CrO_4^{2-}$$

$$0.246\ M\ S_2O_3^{2-} = 0.246 \times 8\ N\ S_2O_3^{2-}$$

$$\therefore \quad N_1V_1 = N_2V_2$$

$$V \times 0.154 \times 3 = 0.246 \times 8 \times 40$$

$$V = \mathbf{170.4\ mL}$$

Note : Oxidation number of two S-atoms in $S_2O_3^{2-}$ = 4 and that of in SO_4^{2-} = 12

26. $K_2Cr_2O_7$ is required by oxidation of $FeSO_4$ only $Fe_2(SO_4)_3$ is already in oxidised state. However, metallic zinc reduces $Fe_2(SO_4)_3$ to $FeSO_4$ and thus,

$$N(FeSO_4 \text{ before reduction}) = \frac{22.45 \times 0.1}{25} = 0.0898\ N$$

$$N(FeSO_4 \text{ after reduction}) = \frac{34.25 \times 0.1}{25} = 0.137\ N$$

$$\therefore N\,[\text{total } FeSO_4 \text{ and } Fe_2(SO_4)_3] = 0.137\ N$$

$$\therefore \quad FeSO_4 \text{ in mixture} = 0.0898 \times 152 = \mathbf{13.6406\ gL^{-1}}$$

$$N(Fe_2(SO_4)_3 = 0.137 - 0.0898 = 0.0472\ N$$

$$\therefore \quad Fe_2(SO_4)_3 \text{ in mixture} = \mathbf{9.44\ gL^{-1}}$$

(Eq. wt. of $FeSO_4$ = 152 g equiv^{-1},
$Fe_2(SO_4)_3$ = 200 g equiv^{-1})

27.
$$\underset{O.N. = n}{A^{n+}} + \underset{+7}{MnO_4^-} \longrightarrow \underset{+2}{Mn^{2+}} + \underset{+5}{AO_3^-}$$

(Mn: $+5e^-$; A: $(5-n)\,e^-$)

$$\therefore \quad 5A^{n+} + (5-n)MnO_4^- \longrightarrow (5-n)\,Mn^{2+} + 5AO_3^-$$

$$\therefore \quad \frac{5 \text{ mol of } A^{n+}}{(5-n) \text{ mol of } MnO_4^-} = \frac{2.68 \times 10^{-3}}{1.61 \times 10^{-3}}$$

$$5 - n = 3$$

$$\therefore \quad n = 2$$

$\therefore$ Chloride is ACl_2

$$\therefore \text{ Equivalent wt. of } ACl_2 = \frac{\text{molar mass}}{\text{change in O.N.}} = \frac{97 + 71}{3} = \mathbf{56}$$

28. 40 mL of 0.05 M MnO_4^- ≡ 40 mL of 0.05 × 5 N MnO_4^-

= 10 milliequivalent
= 0.01 equivalent

$\therefore$ $C_2O_4^{2-}$ = 0.01 equivalent
CaO = 0.01 equivalent
= 0.005 mol
= 0.005 × 56 g = 0.28 g
in 0.518 g sample of limestone

$\therefore$ % of CaO = **54.05%**

29. (a) $2MnO_4^- + 5H_2O_2 \longrightarrow 5O_2 + 2Mn^{2+}$

$$\text{Moles of } KMnO_4 = \frac{0.316}{158} = 2 \times 10^{-3} \text{ mol}$$

$$\therefore \quad H_2O_2 = \frac{5}{2} \times 2 \times 10^{-3} \text{ mol pure}$$

= 5×10^{-3} mol pure
= 0.17 g pure H_2O_2 in 0.2 g sample

$\therefore$ % of pure H_2O_2 = **85%**

(b) $H_2O_2 \longrightarrow O_2$

O_2 formed from 0.17 g pure H_2O_2

$$= \frac{32}{34} \times 0.17 \text{ g} = 0.16 \text{ g}$$

$$= \frac{0.16}{32} \text{ mol}$$

$$= 5.0 \times 10^{-3} \text{ mol}$$

$$pV = nRT$$

$$\therefore \quad V = \frac{nRT}{p} = \frac{5.0 \times 10^{-3} \times 0.0821 \times 300}{(750/760)}$$

$$= 0.1248\ L = \mathbf{124.8\ mL}$$

30. Let $Na_2C_2O_4 = x \text{ g} = \dfrac{x}{134} \text{ mol} = \dfrac{x}{134} \text{ mol } C_2O_4^{2-}$

$$KHC_2O_4 \cdot H_2C_2O_4 = y \text{ g} = \frac{y}{218} \text{ mol} = \frac{2y}{218} \text{ mol } C_2O_4^{2-}$$

and $\dfrac{3y}{218}$ mol H^+

$$\text{Total } C_2O_4^{2-} = \left(\frac{x}{134} + \frac{2y}{218}\right) \text{mol}$$

$$\text{Total } H^+ = \frac{3y}{218} \text{ mol}$$

If volume of the solution is V litres, then

$$M\,(C_2O_4^{2-}) = \left(\frac{x}{134} + \frac{2y}{218}\right)\Big/ V$$

$$\therefore \quad N(C_2O_4^{2-}) = 2\left(\frac{x}{134} + \frac{2y}{218}\right)\Big/ V$$

$$N(H^+) = \frac{3y}{218}\Big/ V$$

$C_2O_4^{2-}$ is a reducing agent and H^+ is an acid.

Given, $N(C_2O_4^{2-}) = 3.62\ N(H^+)$

$$\therefore \quad \frac{2\left(\dfrac{x}{134} + \dfrac{2y}{218}\right)}{V} = \frac{3.62 \times 3y}{218}\Big/ V$$

$$\frac{x}{134} + \frac{2y}{218} = \frac{3.62 \times 1.5y}{218}$$

$$\frac{x}{y} = 2.1 : 1$$

31. $6Fe^{2+} + Cr_2O_7^{2-} + 14H^+ \longrightarrow 6Fe^{3+} + 2Cr^{3+} + 7H_2O$

$$1000 \text{ mL contains} = 5.442 \text{ g } K_2Cr_2O_7 = \frac{5.442}{294} \text{ mol}$$

$$= 0.01851 \text{ mol } K_2Cr_2O_7$$

$$= 0.01851 \times 6 \text{ mol } Fe^{2+}$$

$$= \frac{0.01851 \times 6}{3} \text{ mol } Fe_3O_4$$

$$= 0.03702 \text{ mol } Fe_3O_4$$

$$= 0.03702 \times 232 \text{ g } Fe_3O_4$$

$\therefore$ Fe_3O_4 per mL = $\mathbf{8.59 \times 10^{-3}\ g}$

32.

$$MnO_2 + 4HCl \longrightarrow MnCl_2 + 2H_2O + Cl_2$$
$$Cl_2 + 2KI \longrightarrow 2KCl + I_2$$
$$2Na_2S_2O_3 + I_2 \longrightarrow 2NaI + Na_2S_4O_6$$
$$2Na_2S_2O_3 \equiv I_2 \equiv Cl_2 \equiv MnO_2$$

$$\text{Moles of } Na_2S_2O_3 \text{ used} = \frac{30 \times 0.1}{1000} = 3 \times 10^{-4} \text{ mol}$$

$$\therefore \text{ Moles of pure } MnO_2 = \frac{30 \times 10^{-4}}{2} \text{ mol}$$

$$= \frac{30 \times 10^{-4}}{2} \times 87 \text{ g pure } MnO_2$$

$$= 0.1305 \text{ g in 0.5 g impure } MnO_2$$

$$\therefore \quad \% \text{ purity} = \frac{0.1305}{0.5} \times 100 = \mathbf{26.1\%}$$

33. 10 mL of N_1 $MnO_4^- \equiv 50$ mL of 1.0 N I^-

$$\therefore \quad 10 \times N_1 = 50 \times 1$$
$$\therefore \quad N_1 = 5 \text{ N}$$

$$\text{Equivalent of } KMnO_4 = \frac{5 \times 100}{1000} = 0.5$$

$\therefore$ Equivalents of oxalic acid $= 0.5$

$\therefore$ Mass of oxalic acid $= 0.5 \times 45$ g $=$ **22.5 g**

34.

$$CaOCl_2 + H_2O \longrightarrow Ca(OH)_2 + Cl_2$$
$$Cl_2 + 2KI \longrightarrow 2KCl + I_2$$
$$I_2 + 2Na_2S_2O_3 = Na_2S_4O_6 + 2NaI$$
$$CaOCl_2 \equiv Cl_2 \equiv I_2 \equiv 2Na_2S_2O_3$$

$$\therefore \text{ Equivalent mass of } Cl_2 = \frac{71}{2} = 35.5$$

$$\therefore \quad \underset{\text{Cl}_2 \text{ in bleaching powder}}{N_1V_1} = \underset{\text{hypo}}{N_2V_2}$$

$$N_1 \times 20 = 0.1 \times 20$$
$$\therefore \quad N_1 = 0.1 \text{ N}$$
$$\therefore \quad CaOCl_2 = 0.1 \text{ N}$$
$$\therefore \quad Cl_2 = 0.1 \times 35.5 \text{ g L}^{-1}$$
$$= 3.55 \text{ gL}^{-1} \text{ (pure)}$$

$$\text{Impure sample} = \frac{5.0 \text{ g}}{0.5 \text{ L}} = 10 \text{ g L}^{-1}$$

$$\therefore \quad \% \text{ of pure } Cl_2 = \frac{3.55}{10} \times 100$$
$$= \mathbf{35.5\%}$$

Exercise 2

Only One Option Correct

1. 100 mL of 0.40 M KCl $= 0.04$ mol KCl

100 mL of 0.50 M KCl $= 0.05$ mol KCl

Thus, KCl required $= 0.05 - 0.04$

$= 0.01$ mol

$= 0.01 \times 74.5$ g

$=$ **0.745 g KCl**

Hence, **(a)**

2. $H_3PO_4 = 10.75$ g in 550 mL

$$= \frac{10.75}{98} \text{ mol in 0.550 L}$$

$$[H_3PO_4] = \frac{10.75/98}{0.550} \text{ mol L}^{-1} = 0.20 \text{ M}$$

But it is 0.40 N

Hence, change in oxidation number $= 2$ units

or H^+ neutralised $= 2$

In (a) H_3PO_4 changes to HPO_4^{2-}

Hence, **(a)**

3.

$$\underset{-2}{H_2S} \longrightarrow \underset{+4}{SO_2}$$

Change in oxidation number $= 6$ unit

$\therefore$ 1 mol H_2S = **6 equivalents H_2S**

Hence, **(c)**

4. In the disproportionation reaction

net equivalent weight $= (\text{Eq. wt.})_O + (\text{Eq. wt.})_R$

$$\frac{1}{2}\underset{0}{Br_2} \longrightarrow \underset{+5}{BrO_3^-} \qquad (\text{Eq. wt.})_O = \frac{M}{10}$$

$$\frac{1}{2}\underset{0}{Br_2} \longrightarrow \underset{-1}{Br^-} \qquad (\text{Eq. wt.})_R = \frac{M}{2}$$

$$\text{Net} = \frac{M}{10} + \frac{M}{2} = \mathbf{0.6\ M}$$

Hence, **(c)**

5. **(c)**

6. 20.00 mL of 0.1 M $Pb^{2+} = 2 \times 10^{-3}$ mol Pb^{2+}

30.00 mL of 0.1 M $SO_4^{2-} = 3 \times 10^{-3}$ mol SO_4^{2-}

$$Pb^{2+} + SO_4^{2-} \longrightarrow PbSO_4$$

Pb^{2+} is in the limiting a quantity and every 1 mol of Pb^{2+} gives equal mole of $PbSO_4$ hence, $PbSO_4$ formed $=$ **0.002 mol.**

Hence, **(a)**

7.

$$NaCl \rightleftharpoons Na^+ + Cl^-$$
$$Na_2SO_4 \rightleftharpoons 2Na^+ + SO_4^{2-}$$

$$[Na^+]_{mix} = \frac{M_1V_1 + M_2V_2}{V_1 + V_2}$$

$$= \frac{0.12 \times 30 + 0.30 \times 70}{30 + 70}$$

$$= \mathbf{0.246}$$

Hence, **(d)**

8. $M_1V_1(\text{HCl, acid}) = 16$

$M_2V_2(\text{KOH, base}) = 18 > M_1V_1(\text{acid})$

Mixture is basic.

$$[\text{KOH}]_{\text{mix}} = \frac{M_2V_2 - M_1V_1}{V_1 + V_2}$$

$$= \frac{18 - 16}{200} = 0.01 \text{ M}$$

$$\text{KOH} + \text{HCl} \longrightarrow \text{KCl} + \text{H}_2\text{O}$$

KCl is from HCl and 80 mL of 0.2 KCl is diluted to 200 mL

$$80 \times 0.2 = \text{M(KCl)} \times 200$$

$$\text{M(KCl)} = 0.08 \text{ M}$$

Thus, resultant mixture has KOH (0.01 M) and KCl (0.08 M)

Hence, **(c)**

9. K^+ and Cl^- remain unchanged but due to dilution (in equal volumes), molar concentration of each is halved.

$$[\text{Cl}^-] = 0.1 \text{ M}$$

$$[\text{K}^+] = 0.2 \text{ M}$$

Also, $[OH^-] > [H^+]$, mixture is basic

$$[\text{OH}^-]_{\text{mix}} = \frac{0.4 - 0.2}{2} = \mathbf{0.1\ M}$$

Hence, **(c)**

10. $$5\text{HC}_2\text{O}_4^- + 2\text{MnO}_4^- \longrightarrow 2\text{Mn}^{2+} + 10\text{CO}_2$$

100 mL of 0.02 M $MnO_4^- = 0.002$ mol MnO_4^-

$= 0.005$ mol of $HC_2O_4^- = x$

100 mL of 0.05 M $Ca(OH)_2$

$$\equiv \frac{100 \times 0.05}{1000} \text{ mol Ca(OH)}_2$$

$$\equiv 0.005 \text{ mol Ca(OH)}_2$$

$$\equiv 0.01 \text{ mol H}^+$$

$$\equiv 0.01 \text{ mol HC}_2\text{O}_4^- = y$$

$$\frac{x}{y} = \frac{0.005}{0.01} = \frac{1}{2}$$

$\therefore$ $y = 2x$

Hence, **(b)**

11. 100 mL of XM $H_2O_2 = 0.1 X$ mol H_2O_2

$$\underset{2\text{ mol}}{2\text{H}_2\text{O}_2} \longrightarrow 2\text{H}_2\text{O} + \underset{1\text{ mol} = 24\text{ L}}{\text{O}_2}$$

$$24 \text{ L O}_2 = 1 \text{ mol O}_2$$

$$3 \text{ L O}_2 = \frac{1}{8} \text{ mol O}_2 = \frac{1}{4} \text{ mol H}_2\text{O}_2$$

$$0.1\,X = \frac{1}{4}$$

$\therefore$ $X = \mathbf{2.5}$

Hence, **(a)**

12. $$\text{MnO}_4^- + 5\text{Fe}^{2+} \longrightarrow 5\text{Fe}^{3+} + \text{MnO}_4^-$$

$$\text{Cr}_2\text{O}_7^{2-} + 6\text{Fe}^{2+} \longrightarrow 6\text{Fe}^{3+} + 2\text{Cr}^{3+}$$

$Cr_2O_7^{2-}$ oxidises more Fe^{2+}

Hence, **(b)**

13. (I) $H_2O_2 \longrightarrow O_2$, $MnO_4^- \longrightarrow Mn^{2+}$ } O.N. change = 5

(II) $H_2O_2 \longrightarrow O_2$, $MnO_4^- \longrightarrow MnO_2$ } O.N. change = 3

In I : (acidic) $\underset{H_2O_2}{N_1V_1} = \underset{MnO_4^-}{N_2V_2}$

$$2M_1 \times 100 = 0.5 \times 100$$

$$M_1\,(\text{H}_2\text{O}_2) = 2.5 \text{ M}$$

In II : (Basic) $N_1V_1 = N_2'\,V_2'$

$$2M_1 \times 100 = 3 \times M_2 \times V_2'$$

$$2 \times 2.5 \times 100 = 3 \times 1 \times V_2'$$

$$V_2' = \frac{\mathbf{500}}{\mathbf{3}} \textbf{ mL}$$

Hence, **(b)**

14. Normality of "x" volume $H_2O_2 = \dfrac{x}{5.6}$

$$10 \times N_1\,(\text{H}_2\text{O}_2) = 10 \times \frac{1}{0.56}\,(\text{MnO}_4^-)$$

$$N_1 = \frac{1}{0.56} = \frac{x}{5.6}$$

$\therefore$ $x = \mathbf{10}$

Hence, **(d)**

15. $$\underset{1 \times 10^{-3}}{M\text{Cl}_n} + 3\text{H}_2\text{O} \longrightarrow M(\text{OH})_n + \underset{n \times 10^{-3}}{n\text{HCl}}$$

30 mL of 0.1 M $Ag^+ = 3 \times 10^{-3}$ mol Ag^+

$= n \times 10^{-3}$

$\therefore$ $n = 3$ thus, trivalent chloride.

Hence, **(a)**

16. Use relation, $M'(\text{molality}) = \dfrac{1000\,M}{(1000\,d - Mm)}$

Hence, **(c)**

17. Temporary hardness is due to HCO_3^- of Ca^{2+} and Mg^{2+}

$$\text{Ca(HCO}_3)_2 + \underset{\substack{56\text{ g}\\0.56\text{ g}}}{\text{CaO}} \longrightarrow \underset{\substack{200\text{ g}\\2\text{ g}}}{2\text{CaCO}_3} + \text{H}_2\text{O}$$

10 L (= 10^4 g) hard water contains = 2 g $CaCO_3$

10^6 g (ppm) hard water contains = 200 g $CaCO_3$

Hence, **(b)**

18. 1 equivalent of O_2 = 8 g = $\dfrac{1}{4}$ mol

Hence, **(a)**

19. SO_3^{2-} is oxidised to SO_4^{2-} (change in O.N. = 2)

25 mL of 0.1 M SO_3^{2-} = 2.5 millimoles

= 5.0 milliequivalents of SO_3^{2-}

= 5.0 milliequivalents of M^{3+}

50 mL of 0.1 M M^{3+} = 5 millimol (given)

Thus, decrease in O.N. of M^{3+} should be 1

So that 5 millimoles = 5 milliequivalents

Thus, new O.N. of $M^+ = 2$

Hence, **(c)**

20. Methyl orange indicates 100% reaction of Na_2CO_3

$$Na_2CO_3 + 2HCl \longrightarrow 2NaCl + H_2O + CO_2$$

$$\therefore \quad 20x = 10 \times 0.1 + 5 \times 0.4 = 3$$

$$x = 0.15\text{ M}$$

Thus, **(c)**

21. Molality = 9

9 mol H_2SO_4 (= 9 × 98 g H_2SO_4) is in 1000 g H_2O

Total mass of 9 molal H_2SO_4 solution = 1000 + 98 × 9 = 1882 g

1882 g solution has H_2SO_4 = 882 g

1000g (= 1 kg) solution has H_2SO_4 = **469 g**

Hence, **(b)**

22. As in (16)

$$M' = \frac{1000 \times 1}{[1000 \times 1.0585 - 58.5 \times 1]} = 1$$

Hence, **(b)**

23. H_3PO_2 is monobasic acid

$$H_3PO_2 \rightleftharpoons H^+ + H_2PO_2^-$$

Hence, **(a)**

24. pH = 2 ∴ $[H^+] = 10^{-2}$ M

$$[Ca^{2+}] = \frac{10^{-2}}{2}\text{ M} = \frac{10^{-2}}{2}\text{ mol/L}$$

$$= \frac{10^{-2}}{2} \times 40\text{ g/L}$$

$$= \frac{10^{-2}}{2} \times \frac{40 \times 10^6}{10^3}\text{ ppm}$$

$$= \mathbf{200\ ppm\ Ca^{2+}}$$

Hence, **(a)**

25. $\underset{1\text{ mol}}{BaF_2} + \underset{2\text{ mol}}{H_2SO_4} \longrightarrow BaSO_4 + H_2F_2 + \underbrace{H_2SO_4}_{\text{unreacted}}$

$$H_2F_2 = 2OH^- \equiv 1\text{ mol } Ca(OH)_2 = 2\text{ mol KOH}$$

$$H_2SO_4 = 2OH^- \equiv 1\text{ mol } Ca(OH)_2 = 2\text{ mol KOH}$$

$$2\text{ mol } Ca(OH)_2 = 4\text{ mol KOH}$$

Hence, **(b)**

26. Na_2SO_4 remains unreacted

$$N_1V_1 = N_2V_2$$

$$N_1 \times 100 = (2 \times 0.5) \times 10$$

$$N_1\text{ (NaOH)} = 0.1$$

100 mL of 0.1 N NaOH = 0.01 mol NaOH

$$= 0.01 \times 40\text{ g} = \mathbf{0.4\ g}$$

Hence, **(b)**

27. 10 mL of x M $NaHC_2O_4$ ≡ 10 mL of 0.1 M NaOH

$$\left\{\begin{array}{l}H^+\text{ ion is neutralised}\\ \therefore x\text{ M} = x\text{N}\end{array}\right\}$$

$$\therefore \quad 10 \times x = 10 \times 0.1$$

$$x = 0.1\text{ M}$$

with $KMnO_4$, $C_2O_4^{2-}$ is oxidised

$$\therefore \quad 0.1\text{ M} = 0.2\text{ N } HC_2O_4^-$$

$$10 \times 0.2 = 10 \times N'(MnO_4^-)$$

$$N' = 0.2\text{ N}$$

MnO_4^- is reduced to Mn^{2+}

Hence, $\quad M(MnO_4^-) \times 5 = N'$

$$M(MnO_4^-) = \mathbf{0.04\ M}$$

Hence, **(c)**

28. With phenolphthalein, 50% reaction of Na_2CO_3 is only indicated. With methyl orange, next 50% reaction of Na_2CO_3 and 100% reaction of $NaHCO_3$ is indicated.

V_1 = volume of HCl for 100% reaction of Na_2CO_3

V_2 = volume of HCl for 100% reaction of $NaHCO_3$

$$\therefore \quad \frac{V_1}{2} = x$$

$$V_2 + \frac{V_1}{2} = y$$

$$\therefore \quad V_2 = (y - x)$$

Hence, **(d)**

29. $\underset{1\text{ mol}}{Fe_2(C_2O_4)_3} \rightleftharpoons 2Fe^{3+} + \underset{3\text{ mol}}{3C_2O_4^{2-}}$

Only $C_2O_4^{2-}$ is oxidised.

$$2MnO_4^- + 5C_2O_4^{2-} \longrightarrow 10CO_2 + 2Mn^{2+}$$

$$5\text{ mol } C_2O_4^{2-} \equiv 2\text{ mol } MnO_4^-$$

$$\therefore \quad 3\text{ mol } C_2O_4^{2-} = 1.2\text{ mol } MnO_4^-$$

$$FeC_2O_4 = Fe^{2+} + C_2O_4^{2-}$$

$$5Fe^{2+} + MnO_4^- \longrightarrow 5Fe^{3+} + Mn^{2+}$$

$$5C_2O_4^{2-} + 2MnO_4^- \longrightarrow 10CO_2 + 2Mn^{2+}$$

$$5\text{ mol } FeC_2O_4 \equiv 3\text{ mol } MnO_4^-$$

$$1\text{ mol } FeC_2O_4 \equiv \frac{3}{5}\text{ mol } MnO_4^-$$

$$1\text{ mol } Fe_2(C_2O_4)_3 \equiv \frac{6}{5}\text{ mol } MnO_4^-$$

Thus, ratio of MnO_4^- required by 1 mol of $Fe_2(C_2O_4)_3$ and FeC_2O_4 = 2 : 1

Hence, **(a)**

30. Na_2CO_3 and $NaHCO_3$ are in 1 : 1 ratio with phenolphthalein 50% reaction of Na_2CO_3 up to $NaHCO_3$ is indicated and with methyl orange complete reaction of Na_2CO_3 and $NaHCO_3$ is indicated.

40 mL of 0.05 M $Na_2CO_3 \cdot NaHCO_3 \cdot 2H_2O$ = 2 millimoles

Na_2CO_3 = 1 millimol

$NaHCO_3$ = 1 millimol

$x \times 0.05$ millimol HCl ≡ 0.5 millimol of Na_2CO_3 when 50% of Na_2CO_3 (1 millimol) is neutralised

$y \times 0.05$ millimol HCl = 2

$$\therefore \quad x = 10$$

$$y = 40$$

$$\therefore \quad (y - x) = 30\text{ mL}$$

Thus, **(b)**

31. MnO_4^- oxidises only $FeSO_4$

$$MnO_4^- + 5Fe^{2+} \longrightarrow 5Fe^{3+} + Mn^{2+}$$

1 mol MnO_4^- oxidises = 5 mol Fe^{2+} ($FeSO_4$)

$\frac{100 \times 2}{1000}$ mol MnO_4^- oxidises = 1 mol $FeSO_4$ in 3 mol mixture

$\therefore$ mole fraction of $FeSO_4 = \frac{1}{3}$

Hence, (a)

32. (b) **33.** (d)

34. V mL of 1 N unused HCl = 100 mL of 0.5 N NaOH

$$V = 50 \text{ mL}$$

Used 1 N HCl = 100 mL

100 mL of 1 N HCl = 0.1 equivalent HCl

= 0.1 equivalent of M_2CO_3

$\therefore$ Equivalent mass of $M_2CO_3 = \frac{2M + 60}{2} = (M + 30)$

$$\therefore \quad \frac{5.3}{M + 30} = 0.1$$

$$\therefore \quad M = \mathbf{23}$$

Hence, (a)

35. Proceed as in **Q. No. 4.** (d)

36.
$$2CuSO_4 + 4KI \longrightarrow Cu_2I_2 + I_2 + 2K_2SO_4$$
$$I_2 + 2Na_2S_2O_3 \longrightarrow 2NaI + Na_2S_4O_6$$
$$2CuSO_4 = 2Na_2S_2O_3$$
$$CuSO_4 = Na_2S_2O_3$$

100 mL of 1 M hypo = 0.1 mol hypo

= 0.1 mol pure $CuSO_4$

Hence, 100% pure

Hence, (a)

37.
$$\underset{24\text{ g}}{Mg} + \underset{(98\text{ g} = 1\text{ mol})}{H_2SO_4} \longrightarrow MgSO_4 + H_2$$

1.2 g Mg reacts with = 0.05 mol H_2SO_4

H_2SO_4 taken = 100 mL of 1 M H_2SO_4

= 0.1 mol

$\therefore$ H_2SO_4 left = 0.1 − 0.05

= 0.05 mol in 100 mL solution

$\therefore$ $[H_2SO_4] = 0.05 \times 10 = \mathbf{0.5\ M}$

Hence, (a)

38. 10^6 mL of water has = 200 g $CaCO_3$

10^3 mL of water has = 0.200 g $CaCO_3$

= 0.002 mol $CaCO_3$

$\therefore$ $[CaCO_3] = \mathbf{0.002\ M}$

Hence, (b)

39.
$$5C_2O_4^{2-} + 2MnO_4^- \longrightarrow 10CO_2 + 2Mn^{2+}$$

Also, $1\,Ca^{2+} = 1\,C_2O_4^{2-}$

20 mL of 0.001 M $MnO_4^- = 2 \times 10^{-5}$ mol MnO_4^-

$= 5 \times 10^{-5}$ mol $C_2O_4^{2-}$

$= 5 \times 10^{-5}$ mol Ca^{2+}

$= 5 \times 10^{-5} \times 40$ g Ca^{2+}

$= \mathbf{0.002\ g\ Ca^{2+}}$

Hence, (d)

40.
$$M_1V_1 = M_2V_2$$
$$V_1 = \frac{M_2V_2}{M_1} = \frac{1000 \times 0.9}{18}$$
$$= \mathbf{50\ mL}$$

Hence, (a)

41. (d)

42. 100 g H_3PO_4 solution has = 85 g H_3PO_4

$\frac{100}{1.7}$ mL H_3PO_4 solution has = 85 g $H_3PO_4 = \frac{85}{98}$ mol

$$\therefore \quad \text{Molarity} = \frac{85/98 \text{ mol}}{\left(\frac{100}{1.7 \times 1000}\text{ L}\right)} = \mathbf{14.74\ M}$$

Hence, (c)

43.
$$M\text{(mixture)} = \frac{M_1V_1 + M_2V_2}{V_1 + V_2}$$
$$= \mathbf{4.125\ M}$$

Hence, (a)

44.
$$\underset{\substack{1\text{ mol}\\0.1\text{ mol}}}{H_2SO_4} + \underset{\substack{2\text{ mol}\\0.2\text{ mol}}}{2KOH} \longrightarrow \underset{\substack{1\text{ mol}\\0.1\text{ mol in 300 mL}}}{K_2SO_4} + 2H_2O$$

$$\therefore \quad [K_2SO_4] = \frac{0.1}{0.3} = \mathbf{0.333\ M}$$

Hence, (c)

45.
$$\text{Molarity} = \frac{\text{“}x\text{” volume}}{11.2}$$
$$= \frac{11.2}{11.2} = 1$$

Hence, (a)

46. (a)

47. 0.116 g $C_4H_4O_4(A) = \frac{0.116}{116} = 0.001$ mol

$= 0.001\,x$ equivalent

0.074 g $Ca(OH)_2 = \frac{0.074}{74} = 0.001$ mol

= 0.002 equivalent

$$\therefore \quad 0.001x = 0.002$$
$$\therefore \quad x = \mathbf{2}$$

Hence, (b)

48. Let $[NaHC_2O_4] = x$ M

$= x$ N as an acid when H^+ is neutralised

$= 2x$ N as a reducing agent when $C_2O_4^{2-}$ is oxidised

10 mL of $2x$ $NaHC_2O_4 \equiv$ 10 mL of 0.02 M MnO_4^-

= 10 mL of 0.1 N MnO_4^-

$$\therefore \quad x = 0.05 \text{ M}$$
$$= 0.05 \text{ N}$$

Hence, 10 mL of 0.05 N $NaHC_2O_4$ is neutralised by 10 mL of 0.05 N of any strong base.

Hence, **(d)**

49. $$\underset{}{H_2C_2O_4} \rightleftharpoons H^+ + \underset{\text{I}}{HC_2O_4^-} \rightleftharpoons H^+ + \underset{\text{II}}{C_2O_4^{2-}}$$

Up to stage I — monobasic acid

II — dibasic acid

Hence, **(b)**

50. $$1 \text{ mol } O_2 = 4 \text{ equivalents } O_2 = 30 \text{ L}$$

$$\therefore \quad 1 \text{ equivalent } O_2 = \mathbf{7.5\ L}$$

Hence, **(a)**

51. $MnO_2 + 4HCl \longrightarrow MnCl_2 + Cl_2 + 2H_2O$

$$0.1 \text{ equivalent } Cl_2 = \frac{0.1}{2} \text{ mol } Cl_2$$
$$= 0.05 \text{ mol } MnO_2$$
$$= 4.35 \text{ g } MnO_2 \text{ in 10 g sample}$$

Thus, **43.5%** pure.

Hence, **(d)**

52. Equivalent mass of O = 8

Equivalent mass of metal = 16

Equivalent mass of metal oxide = 24

= **1.5 times of metal**

Hence, **(b)**

53. $$0.106 \text{ g } Na_2CO_3 = \frac{0.106}{106} \text{ mol}$$
$$= 2 \times 10^{-3} \text{ equivalent}$$
$$H_2SO_4 = \frac{40 \text{ N}}{1000} \text{ equivalents}$$
$$\frac{40 \text{ N}}{1000} = 2 \times 10^{-3}$$
$$\therefore \quad \text{N} = \mathbf{0.05\ N\ H_2SO_4}$$

Hence, **(a)**

54. $$MnO_4^- + 5Fe^{2+} \longrightarrow 5Fe^{3+} + Mn^{2+}$$
$$M_1V_1(MnO_4^-) = 5M_1V_1(MnO_4^-) \text{ milliequivalent}$$
$$M_2V_2(Fe^{2+}) = M_2V_2(Fe^{2+}) \text{ milliequivalent}$$
$$N_1V_1(MnO_4^-) = N_2V_2(Fe^{2+})$$
$$5 \times 0.02 \times V_1 = 40 \times 0.1$$
$$V_1 = \mathbf{40\ mL}$$

Hence, **(c)**

55. $$SO_3^{2-} + Cr_2O_7^{2-} \longrightarrow 2Cr^{3+} + SO_4^{2-}$$

O.N. = 4   12   6   6

(+6e⁻ from 12 to 6; −2e⁻ from 4 to 6)

$$N_1V_1(SO_3^{2-}) = N_2V_2(Cr_2O_7^{2-})$$
$$2M_1 \times 20 = 6 \times 0.01 \times 30$$
$$M_1 = \mathbf{0.045\ M}$$

Hence, **(b)**

56. $$\underset{+7}{MnO_4^-} \longrightarrow \underset{+6}{MnO_4^{2-}}$$

Change in O.N. = 1 unit

$\therefore$ Normality = Molarity

$$12.64 \text{ g } KMnO_4 = 0.08 \text{ mol in 0.5 L}$$
$$= 0.16 \text{ M}$$
$$= 0.16 \text{ N}$$

Hence, **(a)**

57. Oxidation state of I in $KIO_3 = +5$

$$S \text{ in } SO_3^{2-} = +4$$
$$S \text{ in } SO_4^{2-} = +6$$
$$\underset{+5}{IO_3^-} + \underset{+4}{SO_3^{2-}} \longrightarrow \underset{+6}{SO_4^{2-}} + \underset{x<5}{I^x}$$
$$\text{Equivalent of } SO_3^{2-} = \frac{60 \times 0.5 \times 2}{1000} = 0.06$$
$$\text{Equivalent of } IO_3^- = (5 - x) \times \frac{2.14}{214}$$
$$= (5 - x)\,0.01$$
$$(5 - x)\,0.01 = 0.06$$
$$5 - x = 6$$
$$x = \mathbf{-1}$$

Hence, **(d)**

58. $$N\% = \frac{1.4N_1V_1}{w} = \frac{1.4 \times 0.2 \times 100}{0.6}$$
$$= \mathbf{46.67\%}$$

Hence, **(a)**

59. $$\underset{2\text{ mol}}{2NaOH} + \underset{1\text{ mol}}{CO_2} \longrightarrow \underset{1\text{ mol}}{Na_2CO_3} + H_2O$$

$$11.2 \text{ millimol} = 11.2 \times 10^{-3} \text{ mol } CO_2$$

1 mol CO_2 reacts with = 2 mol NaOH

$\therefore$ 11.2×10^{-3} mol CO_2 reacts with = 0.0224 mol NaOH

NaOH at start = 0.15 mol

NaOH unreacted = 0.150 − 0.224

= 0.1276 in 1 L

$\therefore$ [NaOH] = **0.1276 M**

Hence, **(a)**

60. $$[Na_2CO_3] = [CO_2] = 0.0112 \text{ mol per L}$$
$$= \mathbf{0.0112\ M}$$

Hence, **(b)**

61. Mass of iron-ore= 0.740 g

$$\therefore \quad Fe_2O_3 = 0.740 \times \frac{24}{100} = 0.1776 \text{ g} = 1.11 \times 10^{-3} \text{ mol}$$
$$\therefore \quad Fe^{3+} = 2.22 \times 10^{-3} \text{ mol}$$
$$2Fe^{3+} + Sn^{2+} \longrightarrow 2Fe^{2+} + Sn^{4+}$$
$$Sn^{2+} + 2Hg^{2+} \longrightarrow Sn^{4+} + Hg_2^{2+}$$
$$Sn^{2+} \text{ required by } Fe^{3+} = \frac{Fe^{3+}}{2} \text{ mol}$$
$$= \frac{2.22 \times 10^{-3}}{2} = 1.11 \times 10^{-3} \text{ mol}$$

Sn^{2+} taken $= \frac{25 \times 0.05}{1000} = 1.25 \times 10^{-3}$ mol

Sn^{2+} unreacted $= (1.25 \times 10^{-3} - 1.11 \times 10^{-3})$ mol $= 0.14 \times 10^{-3}$ mol

$\therefore$ Hg^{2+} required for oxidising Sn^{2+}

$= 2 \times [Sn^{2+}]$ mol

$= 0.28 \times 10^{-3}$ mol

Let volume of $HgCl_2 = V$ mL

$\therefore \quad \frac{V \times 0.05}{1000} = 0.28 \times 10^{-3}$

$= \mathbf{5.6\ mL}$

Hence, **(c)**

62. $Na_2C_2O_4 \longrightarrow 2Na^+ + C_2O_4^{2-}$

$KHC_2O_4 \longrightarrow K^+ + H^+ + C_2O_4^{2-}$

Let $Na_2C_2O_4 = x$ mol

$KHC_2O_4 = y$ mol

Then $C_2O_4^{2-} = (x + y)$ mol $= 2(x + y)$ equivalents

$H^+ = y$ mol $= y$ equivalents

Let volume of 0.1 N base required $= V$ mL

Then $\frac{0.1 \times V}{1000} = y$

$\therefore \quad V = 10000\, y$

Volume of 0.1 N $KMnO_4 = 3$ V

$\therefore \quad \frac{0.1 \times 3\, V}{1000} = 2(x + y)$

$\frac{0.1 \times 3 \times 10000 y}{1000} = 2(x + y)$

$3y = 2x + 2y$

$\therefore \quad y = 2x$

$\therefore \quad Na_2C_2O_4 = 134\, x$ g

$KHC_2O_4 = y = 2x$ mol $= 2x \times 128$ g $= 256x$ g

$\therefore \quad \%$ of $KHC_2O_4 = \frac{256x \times 100}{256x + 134x}$

$= \mathbf{65.64\%}$

Hence, **(a)**

63. Since, 40.38 mL > 12.83 mL, the sample must contain NaOH and Na_2CO_3.

Hence, **(b)**

64. The volume of titrant used by Na_2CO_3 in the second step $= 12.83$ mL.

Thus, volume of HCl used by NaOH in first step

$= 40.38 - 12.83$

$= 27.55$ mL

$\therefore \quad \%$ of NaOH $= \frac{27.55 \times 0.1062 \times 40 \times 100}{1000 \times 0.6234}$

$= \mathbf{18.77}$

Hence, **(a)**

65. Moles of NaOH initially $= \frac{200 \times 0.100}{1000} = 0.02$ mol

CO_2 absorbed = 1 millimol = 0.001 mol

$2NaOH + CO_2 \longrightarrow Na_2CO_3 + H_2O$

Thus, NaOH reacted $= 0.001 \times 2 = 0.002$ mol

Na_2CO_3 formed = 0.001 mol

NaOH unreacted $= 0.02 - 0.002 = 0.018$ mol

Thus, HCl required by NaOH for phenolphthalein end point

$= 0.018$ mol

Equivalent of $Na_2CO_3 = 0.002$ equivalents

$\therefore$ Moles of HCl required by 0.002 equivalents for phenolphthalein end point $= \frac{0.002}{2} = 0.001$ (50% reaction is indicated upto $NaHCO_3$)

$\therefore$ Total HCl $= (0.018 + 0.001)$ mol in 200 mL

$\therefore \quad N(HCl) = \frac{(0.018 + 0.001)}{0.2}$ mol L^{-1}

$= 0.095$ M

$= \mathbf{0.095\ N}$

Hence, **(c)**

One or More Than One Options Correct

1. $O_2 + 4e^- \longrightarrow 2O^{2-}$

$4e^- \equiv 1$ mol $O_2 = 0.25$ mol $= 8$ g O_2

Thus, (a) and (b) correct

2. 2.84 g $P_4O_{10} = \frac{2.84}{284} = 0.01$ mol L^{-1} P_4O_{10}

Also, $P_4O_{10} + 6H_2O \longrightarrow 4H_3PO_4$

Thus, 0.01 M $P_4O_{10} = 0.01 \times 4$ M H_3PO_4

$= 0.04$ M H_3PO_4

1 L of 0.04 M $H_3PO_4 \equiv 1$ L of 0.12 N H_3PO_4

$\equiv 1$ L of 0.12 N NaOH

$\equiv 400$ mL of 0.3 N NaOH (to Na_3PO_4 stage)

$\equiv 400$ mL of 0.1 M (= 0.2 N) $Ca(OH)_2$ (to $CaHPO_4$)

Thus, (b) and (c) are true

3. $KHC_2O_4 \cdot H_2C_2O_4 \cdot 2H_2O$ has

(i) three ionisable H^+. If it is 0.2 N then it is $= \frac{0.2}{3}$ M $= 0.067$ M

(ii) two $C_2O_4^{2-}$ oxidised to CO_2

$2C_2O_4^{2-} \longrightarrow 4CO_2 + 4e^-$

Thus, 0.067 M $= 0.067 \times 4 = 0.267$ N

Thus, (a) and (c) are true.

4. $Ca^{2+} + RH_2 \longrightarrow CaR + 2H^+$

77.5 ppm = 77.5 g Ca^{2+} per 10^6 mL

$= 77.5 \times 10^{-3}$ g L^{-1} Ca^{2+}

$= \frac{77.5 \times 10^{-3}}{40}$ mol L^{-1} Ca^{2+} ion

$= \frac{77.5 \times 10^{-3} \times 2}{40}$ mol L^{-1} H^+ ion

$\therefore \quad [H^+] = 3.875 \times 10^{-3}$

$\therefore \quad pH = -\log(H^+) = 2.4$

Thus, (a) and (c) are true.

5. 10.78 g H_3PO_4 is 0.550 L $= \dfrac{10.78/98}{0.550} = 0.2$ M

But it is 0.4 N hence

H_3PO_4 has been changed to HPO_4^{2-} (by neutralisation of $2H^+$). Thus, (a) is true.

When H_3PO_4 (O.N. of P = +5) is reduced to HPO_3^{2-} (O.N. of P = +3)

Change in O.N. = 2 units

Thus, 0.2 M = 0.4 N

Then, (c) is also true. Thus, **(a)** and **(c)**

6.
$$KHC_2O_4 \rightleftharpoons K^+ + H^+ + C_2O_4^{2-}$$
$$H^+ \equiv 1\,OH^- \equiv \frac{1}{2}Ca(OH)_2$$
$$5C_2O_4^{2-} + 2MnO_4^- \longrightarrow 2Mn^{2+} + 10CO_2$$
$$5C_2O_4^{2-} \equiv 2MnO_4^-$$
$$\therefore \quad 1\,C_2O_4^{2-} \equiv \frac{2}{5}MnO_4^-$$

100 mL of 0.02 M $MnO_4^- \equiv \dfrac{100 \times 0.02}{1000}$ mol MnO_4^-

$$2 \text{ mol } MnO_4^- \equiv 5 \text{ mol } KHC_2O_4$$
$$\therefore \quad 0.002 \text{ mol } MnO_4^- = 0.005 \text{ mol } KHC_2O_4$$
$$= 128 \times 0.005 \text{ g } KHC_2O_4$$
$$= x$$

100 mL of 0.05 M $Ca(OH)_2 \equiv \dfrac{100 \times 0.05}{1000}$ mol $Ca(OH)_2$

$$\equiv 0.005 \times 2 \text{ mol } (OH^-)$$
$$\equiv 0.01 \text{ mol } H^+$$
$$\equiv 0.01 \text{ mol } KHC_2O_4$$
$$\equiv 0.01 \times 128 \text{ g } KHC_2O_4 = y$$
$$\therefore \quad \frac{x}{y} = \frac{128 \times 0.005}{128 \times 0.01} = \frac{1}{2}$$
$$\therefore \quad 2x = y$$

Thus, **(b)**

7. Let $(A)\,Na_2C_2O_4 = x \text{ mol} \equiv x \text{ mol } C_2O_4^{2-}$

$(B)\,KHC_2O_4 \cdot H_2C_2O_4 = y$ mol

$= 3y =$ mol H^+ and $2y$ mol $C_2O_4^{2-}$

Total oxalate $= (x + 2y)$

Let volume of $KMnO_4$ or NaOH = V mL

$$H^+ + OH^- \longrightarrow H_2O$$
$$5C_2O_4^{2-} + 2MnO_4^- \longrightarrow 2Mn^{2+} + 10CO_2$$
$$1H^+ \equiv 1OH^-$$
$$\therefore \quad 3y \text{ mol } H^+ = 3y \text{ mol } OH^-$$
$$\therefore \quad 3y = \frac{V \times 0.1}{1000} \text{ mol NaOH}$$
$$y = \frac{V}{30000}$$
$$5\,C_2O_4^{2-} \equiv 2\,MnO_4^-$$
$$\therefore \quad (x + 2y) \text{ mol } C_2O_4^{2-} = \frac{2(x + 2y)}{5} \text{ mol } MnO_4^-$$
$$\therefore \quad \frac{2(x + 2y)}{5} = \frac{V \times 0.1}{1000} \text{ mol } KMnO_4$$
$$(x + 2y) = \frac{5V}{20000}$$
$$\frac{x + 2y}{y} = \frac{5 \times 3}{2} = 7.5$$
$$\frac{x}{y} + 2 = 7.5$$
$$\frac{x}{y} = 5.5$$

Thus, $x : y = 5.5 : 1$

Thus, (c)

8. (a) $C_2O_4^{2-} \longrightarrow 2CO_2 + 2e^-$

$$\text{Equivalent weight} = \frac{\text{Molecular weight}}{2}$$

Thus, (a) is not true.

(b) $H_2C_2O_4$ has $2H^+$ ions while $NaHC_2O_4$ has $1\,H^+$ ion.

Thus, (b) is not true

(c) Equivalent of $H_2C_2O_4 \equiv NaHC_2O_4$

Thus, (c) is true

(d) $5C_2O_4^{2-} \equiv 2MnO_4^-$

Thus, (d) is also true.

Thus, (c) and (d)

9. 100 mL has = 1.44 g FeC_2O_4

$$= \frac{1.44}{144} \text{ mol } FeC_2O_4$$
$$= 0.01 \text{ mol } FeC_2O_4 \rightleftharpoons 0.01 \text{ mol } Fe^{2+} + 0.01 \text{ mol } C_2O_4^{2-}$$
$$5Fe^{2+} + MnO_4^- \longrightarrow 5Fe^{3+} + Mn^{2+}$$
$$5C_2O_4^{2-} + 2MnO_4^- \longrightarrow 10CO_2 + 2Mn^{2+}$$
$$5FeC_2O_4 \equiv 3MnO_4^-$$
$$\therefore \quad 0.01 \text{ mol } FeC_2O_4 \equiv \frac{3}{5} \times 0.01$$
$$= 0.006 \text{ mol } MnO_4^-$$

Let volume of $KMnO_4 = V$ mL

$$\therefore \quad \frac{V \times 0.01}{1000} = 0.006$$
$$V = \frac{0.006 \times 1000}{0.01} = 600 \text{ mL}$$

Thus, (b)

10.
$$\underset{+6}{MnO_4^{2-}} \longrightarrow \underset{+4}{MnO_2} + \underset{+7}{MnO_4^-}$$
$$MnO_4^{2-} + 2e^- \longrightarrow MnO_2$$
$$[MnO_4^{2-} \longrightarrow MnO_4^- + e^-] \times 2$$
$$3MnO_4^{2-} \longrightarrow 2MnO_4^- + MnO_2$$
$$MnO_4^{2-} \longrightarrow \frac{2}{3}MnO_4^- + \frac{1}{3}MnO_2$$

Thus, **(a)**

Brain Twisters

1. $CaCO_3 + 2HCl \longrightarrow CaCl_2 + CO_2 + H_2O$

1 mol : 2 mol : 1 mol

$\frac{2.56}{100}$ mol : $\frac{250 \times 0.125}{1000}$ mol

0.0256 mol : 0.03125 mol

$CaCO_3$ reacts with 0.0512 mol of HCl : HCl reacts with 0.015625 mol of $CaCO_3$

Thus, HCl is limiting reactant and $CaCO_3$ is in excess.

$CaCO_3$ left unreacted $= 0.0256 - 0.0156$
$= 0.01$ mol
$= 1.0$ g

$CaCl_2$ formed = moles of $CaCO_3$ reacted
$= 0.015625$ mol $=$ **1.7344 g**

2. (a) Oxidising agent : O_2, oxidises Au

Reducing agent : Au, reduces O_2

(b) Gold = 1000 kg impure

$= 1000 \times 1000 \times \frac{0.0197}{100}$ g pure

$= 197$ g

$= 1$ mol

Moles of NaCN = V (litres required) $\times M$
$= 0.075$ V

4 mol Au require = 8 mol NaCN

3. 1 mol Au requires = 2 mol NaCN

$\therefore \quad 0.075\, V = 2$

$V =$ **26.67 L**

500 mL of 0.25 mol $Na_2SO_4 = 0.125$ mol

$BaCl_2 = 15$ g $= 0.0721$ mol

$BaCl_2 + Na_2SO_4 \longrightarrow BaSO_4 + 2NaCl$

Thus, $BaCl_2$ is limiting reactant

$BaCl_2 = BaSO_4 =$ **0.0721 mol**

4. NaCl and HCl are strong electrolytes thus, ionise completely in aqueous solution.

$p_{Na^+} = -\log 2 \times 10^{-3}$
$= 3 - \log 2$
$= 3 - 0.3010 = 2.6990$

$p_{H^+} = -\log 5.4 \times 10^{-4}$
$= 4 - \log 5.4$
$= 4 - 0.7324$
$= 3.2676$

Total $[Cl^-] = 2 \times 10^{-3} + 5.4 \times 10^{-4}$
$= 2.54 \times 10^{-3}$ M

$p_{(Cl^-)} = 2.595$

5. $p_{Ag^+} = 6.6990$

$[Ag^+] = 10^{-6.6990} = 10^{-7}\, 10^{0.3010}$
$= 2 \times 10^{-7}$

6. $2AgNO_3 + Na_2CO_3 \longrightarrow Ag_2CO_3 + 2NaNO_3$

2 mol : 1 mol : 1 mol

25.0 mol of : 50.0 mol of

0.2 M : 0.08 M

0.005 mol : 0.004 mol

will react with : will react

0.0025 mol of : with 0.008 mol

Na_2CO_3 : of $AgNO_3$

Thus, $AgNO_3$ is limiting reactant

Thus, Ag_2CO_3 formed $= \frac{1}{2}$ mol of $AgNO_3$
$= 0.0025$ mol
$= 0.0025 \times 276 =$ **0.69 g**

7. In the above case Na_2CO_3 left unreacted
$= 0.004 - 0.0025$
$= 0.0015$ mol in total 75 mL
$= (0.075$ L$)$

Thus, molar concentration of $Na_2CO_3 = \frac{0.0015 \text{ mol}}{0.075 \text{ L}}$
$=$ **0.02 M**

8. (a) $[Cu^{2+}] = 4.78 \times 10^{-2}$

$\therefore \quad p_{Cu^{2+}} = -\log [Cu^{2+}]$
$= -\log 4.78 \times 10^{-2}$
$= 2 - \log 4.78$
$= 2 - 0.6794$
$=$ **1.3206**

$[Zn^{2+}] = 0.104 = 1.04 \times 10^{-1}$

$p_{Zn^{2+}} = -\log [Zn^{2+}]$
$= -\log 1.04 \times 10^{-1}$
$= 1 - \log 1.04$
$= 1 - 0.0170$
$= 0.9830$

$[NO_3^-] = 2 \times 4.78 \times 10^{-2}$ [from $Cu(NO_3)_2$]
$+ 2 \times 0.104$ [from $Zn(NO_3)_2$]
$= 2.1756 \times 10^{-1}$

$\therefore \quad p_{NO_3^-} = -\log [NO_3^-]$
$= -\log (2.1756 \times 10^{-1})$
$= 1 - \log 2.1756$
$= 1 - 0.3376$
$= 0.6624$

(b) $[H^+] = 6.75 \times 10^{-4}$

$\therefore \quad p_{H^+} = -\log [H^+]$
$= -\log 6.75 \times 10^{-4}$
$= 3.1707$

$[Ba^{2+}] = 3.35 \times 10^{-4}$

$\therefore \quad p_{Ba^{2+}} = -\log 3.35 \times 10^{-4}$
$=$ **3.4750**

$[ClO_4^-] = 2 \times 3.35 \times 10^{-4}$ from $Ba(ClO_4)_2$
$+ 6.75 \times 10^{-4}$ from $HClO_4$
$= 13.45 \times 10^{-4}$ M

$p_{ClO_4^-} = -\log 13.45 \times 10^{-4}$
$= \mathbf{2.8713}$

9. (a) $Na^+ = 1.08 \times 10^{-3}$ ppm

Thus, 10^6 g of sea water has $= 1.08 \times 10^3$ g Na^+

$\frac{10^6}{1.02}$ mL of sea water has $= 1.08 \times 10^3$ g Na^+

$\therefore$ 10^3 mL of sea water has

$$= \frac{1.08 \times 10^3 \times 1.02 \times 10^3}{10^6} \text{ g Na}^+$$
$$= 1.1016 \text{ gL}^{-1}$$
$$= 4.8 \times 10^{-2} \text{ M}$$

$\therefore$ Similarly $[SO_4^{2-}] = \frac{270 \times 1.02 \times 10^3}{10^6} = 0.2754 \text{ gL}^{-1}$

$= \mathbf{2.87 \times 10^{-3}\ M}$

(b) $p_{Na^+} = 1.32$

$p_{SO_4^{2-}} = 2.54$

10. $KCl \cdot MgCl_2 \cdot 6H_2O = 5.76$ g in 2 L solution

$= \frac{5.76}{277.5}$ mol in 2 L solution

$= 0.01$ M

(a) $[KCl \cdot MgCl_2] = 0.01$ M

(b) $[Mg^{2+}] = 0.01$ M

$[Cl^-] = 0.03$ M

(c) Millimoles of Cl^- in 25.0 mL solution

$= 0.03 \times 25 = 0.75$

(d) $[K^+] = 0.01$ M

$= 0.39 \text{ gL}^{-1}$

$= \frac{0.39}{1000} \times 10^6$ g per million

$= 390$ ppm

11. KI = 24.32 ppt

$= 24.32$ g per thousand mL solution

$= \frac{24.32}{1000} \times 200$ g in 200 mL solution

$= 4.864$ g

$= 0.038$ mol

Volume of $AgNO_3$ required $= V$ mL

Molarity $= 0.01$ M

$\therefore$ $0.01 \text{ M} \times V = 0.01\,V$ millimol

$= 1 \times 10^{-5}\,V$ mol

$1 \times 10^{-5}\,V = 0.038$

$V = \frac{0.038}{1 \times 10^{-5}} = \mathbf{3800\ mL}$

12. $Al_2(SO_4)_3 + 3Ba(NO_3)_2 \longrightarrow 3BaSO_4 + 2Al(NO_3)_3$
1 mol | 3 mol | 3 mol

$Ba(NO_3)_2 = 480.4$ ppm (parts per million)

$= 480.4$ g in 10^6 mL

$= \frac{480.4 \times 750}{10^6}$ g in 750 mL

$= 0.3603$ g

$= \frac{0.3603}{261}$

$= 1.3804 \times 10^{-3}$ mol

$Al_2(SO_4)_3 = 200 \times 0.03090$ millimol

$= 6.18$ millimol

$= 6.18 \times 10^{-3}$ mol

$\frac{Al_2(SO_4)_3}{Ba(NO_3)_2} = \frac{1}{3}$ based on stoichiometry

$= \frac{6.18 \times 10^{-3}}{1.3804 \times 10^{-3}} = 4.5$

Thus, $Ba(NO_3)_2$ is limiting reactant and $Al_2(SO_4)_3$ is in excess.

(a) Thus $BaSO_4$ formed $= 1.3804 \times 10^{-3}$ mol

$= 1.3804 \times 10^{-3} \times 233$ g

$= \mathbf{0.3216\ g}$

(b) Unreacted $Al_2(SO_3)_3 = 6.18 \times 10^{-3} - \frac{1.3804 \times 10^{-3}}{3}$

$= 5.72 \times 10^{-3}$ mol in **950 mL**

$= \mathbf{6.02 \times 10^{-3}\ M}$

13. Let unreacted $AgNO_3$ be $= V$ mL

V mL of 0.082 M $AgNO_3 \equiv 4.86$ mL of 0.0625 M KSCN

$\therefore$ $V = \frac{4.86 \times 0.0625}{0.082}$

$= 3.70$ mL

$\therefore$ used $AgNO_3 = (50.0 - 37.0)$ mL

$= 46.30$ mL

Moles of $AgNO_3$ used $= \frac{46.30 \times 0.082}{1000}$

$= 3.8 \times 10^{-3}$ mol

$P_2O_5 \longrightarrow 2PO_4^{3-}$

$2PO_4^{3-} + 6Ag^+ \longrightarrow 2Ag_3PO_4$

Thus, P_2O_5 content $= \frac{3.8 \times 10^{-3}}{6}$ mol

$= \mathbf{6.33 \times 10^{-4}\ mol}$

$= \frac{3.8 \times 10^{-3}}{6} \times 142$ g

$= 0.0898$ g

Thus, % of P_2O_5 in plant food $= \frac{0.0898}{4.258} \times 100$

$= \mathbf{2.11\%}$

14. 0.30 M $CaCl_2 = 0.60$ M Cl^- (in $CaCl_2$)

Volume of $CaCl_2 = V_1$ mL

$KCl = 0.20 \text{ M} = 0.20 \text{ M Cl}^-$

Volume of KCl solution = 400 mL

Resultant molarity of Cl^- = 0.25 M

$$\text{Resultant molarity} = \frac{M_1V_1 + M_2V_2}{V_1 + V_2}$$

$$0.25 = \frac{0.60V_1 + 0.20 \times 400}{V_1 + 400}$$

$$V_1 = \mathbf{57.14\ mL}$$

15. 60 L of 0.50 NaOH ≡ 60 × 0.5 mol NaOH if pure

= 30 mol NaOH if pure

= 30 × 40 g NaOH if pure

$= \frac{30 \times 40 \times 100}{88}$ g NaOH if wet

= **1363.64 g**

16. 50 mL of N_1 Na_2CO_3 ≡ 56.3 mL of 0.102 N HCl

$\therefore$ $50 \times N_1 = 56.3 \times 0.102$

$N_1 = 0.115$ N Na_2CO_3

= 0.0575 M Na_2CO_3

$\therefore$ 100 mL of 0.0575 M $Na_2CO_3 = 5.75 \times 10^{-3}$ mol

$$\underset{\substack{1\text{ mol}\\5.75\times10^{-3}\text{ mol}}}{CaCl_2 + Na_2CO_3} \longrightarrow \underset{\substack{1\text{ mol}\\5.75\times10^{-3}\text{ mol}}}{CaCO_3} + 2NaCl$$

$CaCO_3$ formed $= 5.75 \times 10^{-3}$ mol

$= 5.75 \times 10^{-3} \times 100$ g

= **0.575 g**

17. (a) If H_3PO_4 is completely neutralised then it behaves as a tribasic acid thus, 0.40 N $H_3PO_4 \equiv \frac{0.40\text{ M}}{3} H_3PO_4$

$$\text{Moles in one litre} = \frac{0.40}{3}\text{ mol}$$

$$\therefore \quad \text{Moles in 550 mL} = \frac{0.40 \times 550}{3 \times 1000}$$

$$\text{Amount of } H_3PO_4 = \frac{0.40 \times 550 \times 98}{3 \times 1000}\text{ g}$$

$$= \mathbf{7.187\ g}$$

(b) $H_3PO_4 + 2e^- \longrightarrow HPO_3^{2-} + H_2O$

Due to reduction as a result of gain of two electrons.

$$0.40\text{ N } H_3PO_4 = \frac{0.40\text{ M}}{2} H_3PO_4$$

Moles of H_3PO_4 in one litre = 0.20 mol

$$\text{Moles in 550 mL} = \frac{0.20 \times 550}{1000}$$

$$\text{Amount of } H_3PO_4 = \frac{0.20 \times 550 \times 98}{1000}\text{ g}$$

$$= \mathbf{10.78\ g}$$

Alternate (short method)

(a) $H_3PO_4 + 3OH^- \longrightarrow 3H_2O + PO_4^{3-}$

Equivalent weight of $H_3PO_4 = \frac{98}{3}$

$$\therefore \quad w\text{ (amount)} = \frac{NEV}{1000}$$

$$= \frac{0.40 \times 98 \times 550}{1000 \times 3}$$

$$= 7.187\text{ g}$$

(b) $H_3PO_4 + 2e^- \longrightarrow HPO_3^{2-} + H_2O$

Equivalent weight of $H_3PO_4 = \frac{98}{2} = 49$

$$\therefore \quad w\text{ (amount)} = \frac{NEV}{1000}$$

$$= \frac{0.40 \times 49 \times 550}{1000}$$

$$= \mathbf{10.78\ g}$$

Passage Comprehension Questions

Passage 1

Molar mass of $Cr(NH_3)_xCl_3 = 52 + 17x + 35.5 \times 3$

$= (158.5 + 17x)$

NH_3 in the complex is neutralised by HCl.

$$\text{Moles of HCl required} = \frac{40 \times 1.5}{1000} = 0.06$$

$$\text{Moles of } A = \frac{2.635}{(158.5 + 17x)}$$

$$\text{Thus,} \quad \frac{0.06}{x} = \frac{2.635}{(158.5 + 17x)}$$

$$\therefore \quad x = 5.89 \approx \mathbf{6}$$

1. Thus, $x = 6$, Thus **(d)**
2. 1.5 M H_2SO_4 being dibasic, volume of H_2SO_4 required will be = 20 mL
   Thus, **(b)**
3. Thus, complex is $(Cr(NH_3)_6)Cl_3$
   Thus, **(a)**

Passage 2

1. (c) **2.** (a) **3.** (d) **4.** (b), (d)

Passage 3

A. Resultant molarity of HCl $= \frac{20 \times 1}{40}$

due to dilution from 20 mL to 40 mL = **0.5 M**

Thus, 40 mL of 0.5 M HCl ≡ V_A mL of 0.1 M NaOH

$$\therefore \quad V_A = 200\text{ mL}$$

B. Resultant molarity of HCl due to dilution from 20 mLto 80 mL $= \frac{20 \times 1}{80}$

= 0.25 M

Thus, 80 mL of 0.25 M HCl ≡ V_B mL of 0.1 M

$$V_B = 200\text{ mL}$$

1. Thus, $V_A = V_B$, Thus, **(a)**
2. (b)

3. At the end-point additional drop of NaOH, imparts pink colour.

Thus, (c)

Passage 4

$MnO_4^- \xrightarrow{H^+} Mn^{2+}$	Change in O.N.	Volume of $KMnO_4$
+7 +2	5 unit	V_A 20 mL
$\xrightarrow{OH^-} Mn^{x+}$	$(7-x)$ unit	V_B 33.4 mL
$\xrightarrow{\text{neutral}} Mn^{y+}$	$(7-y)$ unit	V_C 100 mL

$$1 \text{ M } MnO_4^- = 5 \text{ N in acidic medium}$$
$$= (7-x) \text{ N in basic medium}$$
$$= (7-y) \text{ N in neutral medium}$$

$$20 \text{ mL of } 5 \text{ N } MnO_4^- = 33.4 \text{ mL of } (7-x) \text{ N } MnO_4^-$$
$$\frac{100}{33.4} = 7 - x$$
$$x = 4$$

Also, $20 \text{ mL of } 5 \text{ N } MnO_4^- = 100 \text{ mL of } (7-y) \text{ N } MnO_4^-$

$$\therefore \quad (7-y) = 1$$
$$y = 6$$

1. (a) **2.** (c) **3.** (b)

Passage 5

$$6Fe^{2+} \longrightarrow 6Fe^{3+} + 6e^-$$
$$Cr_2O_7^{2-} + 14H^+ + 6e^- \longrightarrow 2Cr^{3+} + 7H_2O$$
$$N(Cr_2O_7^{2-}) = 6 \times M(Cr_2O_7^{2-})$$
$$= 0.6 \text{ N}$$
$$5Fe^{2+} \longrightarrow 5Fe^{3+} + 5e^-$$
$$MnO_4^- + 8H^+ + 5e^- \longrightarrow Mn^{2+} + 4H_2O$$
$$N(MnO_4^-) = 5 \times M(MnO_4^-)$$
$$= 0.5 \text{ N}$$
$$\text{Volume of } Fe^{2+} = V \text{ mL (say)}$$
$$\text{Molarity of } Fe^{2+} = 1 \text{ M (say)} = 1 \text{ N}$$

1.
$$6Fe^{2+} \equiv Cr_2O_7^{2-}$$
$$1\,Fe^{2+} \equiv \frac{Cr_2O_7^{2-}}{6}$$
$$5\,Fe^{2+} \equiv MnO_4^-$$
$$1\,Fe^{2+} \equiv \frac{MnO_4^-}{5}$$

In all cases

$$1 \text{ equivalent of } X \equiv 1 \text{ eqivalent of } Y$$

Thus, (d)

2.
$$V \times 1 \equiv 0.5 \times V_1 \quad \text{for } MnO_4^-$$
$$V_1 = 2\text{ V}$$
$$V \times 1 \equiv 0.6 \times V_2 \quad \text{for } Cr_2O_7^{2-}$$
$$V_2 = 1.33 \text{ V}$$

Thus, (b)

3. 24.5 of mL 0.1 M $Cr_2O_7^{2-} \equiv$ 24.5 mL of 0.6 N $Cr_2O_7^{2-}$

V mL of 0.1 M $MnO_4^- \equiv V$ mL of 0.5 N MnO_4^-

Thus, $\quad 24.5 \times 0.6 \equiv V \times 0.5$

$$\therefore \quad V = 29.4 \text{ mL}$$

Thus, (a)

Passage 6

1. (c) 25 mL of 1×10^{-3} M MnO_4^-

$$\equiv 25 \text{ mL of } 5 \times 10^{-3} \text{ N } MnO_4^-$$
$$= 0.125 \text{ milliequivalent of } MnO_4^-$$
$$\equiv 0.125 \text{ milliequivalent of } CaC_2O_4$$
$$\equiv 0.125 \text{ milliequivalent of } Ca^{2+}$$
$$= 0.125 \times 20 \text{ mg of } Ca^{2+} \text{ in } 10 \text{ mL}$$
$$= 0.250 \text{ mg/mL } Ca^{2+} \text{ solution}$$

2. (a) Explained above

3. (a) 0.250 mg/mL = 0.250 g/L

$$= 250 \text{ g}/10^6 \text{ mL} = 250 \text{ ppm}$$

4. (c) In H_2SO_4, HCl, it is decomposed.

HCOOH is oxidised by $KMnO_4$, hence cannot be used.

5. (d)

Passage 7

1. (c) Only Fe^{2+} is oxidised and MnO_4^- is reduced to Mn^{2+}

$$25 \times M\,(Fe^{2+}) = 25 \times 0.02 \times 5\,(MnO_4^-)$$
$$M = 0.10 \text{ M}$$

2. (a) Total Fe^{2+} in second part (including that of from Fe^{3+})$^-$

$$25 \times y = 40 \times 0.02 \times 5$$
$$y = 0.16 \text{ M} = 0.16 \text{ N}$$
$$\therefore \quad Fe^{3+} = 0.16 - 0.10\,(Fe^{2+}) = 0.06 \text{ M}$$

3. (b)

4. (c) 25 mL of 0.10 M $Fe^{2+} = V_1$ mL of 0.02 M $K_2Cr_2O_7$

$$= V_1 \times 0.02 \times 6 \text{ N } K_2Cr_2O_7$$
$$V_1 = 20.83 \text{ mL}$$

($Cr_2O_7^{2-}$ is reduced to Cr^{3+})

Similarly for 0.16 (Fe^{2+} to Fe^{3+})

$$V_2 = 33.33 \text{ mL}$$

Passage 8

1. (c) $\quad 5.8 \text{ g } NaIO_3 = \frac{5.8}{198} \text{ mol } NaIO_3$

$$NaHSO_3 = \frac{3 \times 5.8}{198} \text{ mol}$$
$$= \frac{3 \times 5.8 \times 104}{198} \text{ g} = 9.14 \text{ g}$$

2. (a) $\quad I^-$ formed in step I $= \frac{5.8}{198}$ mol

$$IO_3^- \text{ required in step II} = \frac{5.8}{198 \times 5} \text{ mol} = 1.16 \text{ g } (NaIO_3)$$
$$\text{volume} = \frac{1.16}{5.8} = 0.2 \text{ L}$$

3. (c) $\underset{+5}{IO_3^-} \longrightarrow \underset{-1}{I^-}$

change in ON = 6 units

Thus, 1 mol $NaIO_3$ = 6 equivalents

4. (d) $1.16 \text{ g } NaIO_3 = \dfrac{1.16}{198}$ mol

$= \dfrac{6.96}{198} = 0.035$ equivalent

Passage 9

1. (c) $\underset{\text{orange}}{Cr_2O_7^{2-}} \longrightarrow \underset{\text{green}}{Cr^{3+}}$

2. (a)

3. (d) $CH_3CH_2OH \longrightarrow CH_3COOH$

change in oxidation number = 4 units

4. (a) Number of moles of $K_2Cr_2O_7$ required

$= \dfrac{4.23 \times 0.07654}{1000} = 3.24 \times 10^{-4}$

Number of moles of CH_3CH_2OH

$= 3.24 \times 10^{-4} \times 1.5$

$= 4.86 \times 10^{-4}$ mol

$= 0.0224$ g CH_3CH_2OH in 10 g sample

$= 0.224\%$ which is greater than legal limit.

Assertion & Reason

1. (a) **2.** (c) **3.** (d) **4.** (b) **5.** (a)

True & False

1. F **2.** T **3.** T **4.** T **5.** F **6.** T **7.** T
8. F **9.** T **10.** F **11.** T **12.** T **13.** T **14.** T
15. F

Fill in the Blanks

1. 31.6 **2.** K_2MnO_4 **3.** iodometric method
4. methyl red **5.** disappearance of blue colour
6. 1 **7.** 0.2 **8.** MnO_2 **9.** 1.33 N **10.** 0.015 M
11. 0.333 N, acidic **12.** 2.8 **13.** iodometric method
14. I_2, M/6 **15.** N_2

Matrix-Match Type Questions

I. Only One Correct

1. (A)—(4); (B)—(1); (C)—(2); (D)—(3);
2. (A)—(3); (B)—(2); (C)—(4); (D)—(1);
3. (A)—(3); (B)—(1); (C)—(4); (D)—(2);
4. (A)—(2); (B)—(4); (C)—(3); (D)—(1)

II. One or More Correct

1. (A)—(1,2); (B)—(1,2); (C)—(2,3); (D)—(4)
(E)—(2,3)
2. (A)—(3); (B)—(2); (C)—(2); (D)—(4)
(E)—(1); (F)—(3)

Integer Answer Types

Set I

Questions →	1	2	3	4	5	6	7	8
Answers →	5	6	2	2	3	2	6	1
	⓪	⓪	⓪	⓪	⓪	⓪	⓪	⓪
	①	①	①	①	①	①	①	①
	②	②	②	②	②	②	②	②
	③	③	③	③	③	③	③	③
	④	④	④	④	④	④	④	④
	⑤	⑤	⑤	⑤	⑤	⑤	⑤	⑤
	⑥	⑥	⑥	⑥	⑥	⑥	⑥	⑥
	⑦	⑦	⑦	⑦	⑦	⑦	⑦	⑦
	⑧	⑧	⑧	⑧	⑧	⑧	⑧	⑧
	⑨	⑨	⑨	⑨	⑨	⑨	⑨	⑨

Set II

Questions →	1	2	3	4	5	6	7	8
Answers →	3	2	1	5	1	6	4	1
	⓪	⓪	⓪	⓪	⓪	⓪	⓪	⓪
	①	①	①	①	①	①	①	①
	②	②	②	②	②	②	②	②
	③	③	③	③	③	③	③	③
	④	④	④	④	④	④	④	④
	⑤	⑤	⑤	⑤	⑤	⑤	⑤	⑤
	⑥	⑥	⑥	⑥	⑥	⑥	⑥	⑥
	⑦	⑦	⑦	⑦	⑦	⑦	⑦	⑦
	⑧	⑧	⑧	⑧	⑧	⑧	⑧	⑧
	⑨	⑨	⑨	⑨	⑨	⑨	⑨	⑨

Set III

Questions →	1	2	3	4	5	6	7	8
Answers →	6	2	1	7	1	3	2	2
	⓪	⓪	⓪	⓪	⓪	⓪	⓪	⓪
	①	①	①	①	①	①	①	①
	②	②	②	②	②	②	②	②
	③	③	③	③	③	③	③	③
	④	④	④	④	④	④	④	④
	⑤	⑤	⑤	⑤	⑤	⑤	⑤	⑤
	⑥	⑥	⑥	⑥	⑥	⑥	⑥	⑥
	⑦	⑦	⑦	⑦	⑦	⑦	⑦	⑦
	⑧	⑧	⑧	⑧	⑧	⑧	⑧	⑧
	⑨	⑨	⑨	⑨	⑨	⑨	⑨	⑨

> **There is nothing in the universe but alkali and acid, from which nature composes all things.** O. Tachenius

s-Block Elements —The Alkali Metals

Quick Points

- Effective nuclear charge Z^* (due to shielding effect)

$$Z^* = Z - S$$

where Z is atomic number and
S is the shielding constant
Value of Z^* is derived by using Slater's rule;

$S = 0.25$ for $1s$-electron when it is alone (only for He)
$= 0.35$ per electron in n^{th} orbit
$= 0.85$ per electron in $(n-1)^{th}$ orbit
$= 1.00$ per electron in $(n-2), (n-3) \dots$ orbit

Thus, for Li ($Z = 3, 1s^2 2s^1$)

$Z^* = 3 -$ shielding constant of two electrons in $1s$-suborbit $(n-1)^{th}$ orbit
$= 3 - 2 \times 0.85 = 1.30$

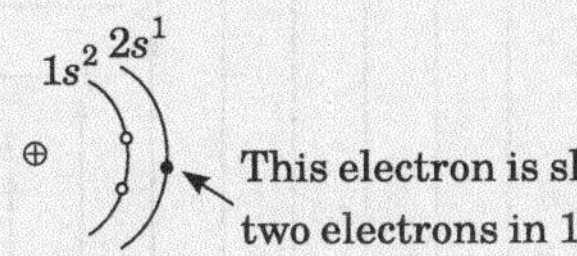

For second period, $Z^* = 0.65\,(n+1)$
where n is the number of valence electrons.
For Na ($Z = 11, 1s^2 2s^2 2p^6 3s^1$)
Electron in $3s$-suborbit is shielded by 8 electrons in 2nd, *i.e.*, $(n-1)^{th}$ orbit and 2 electrons in 1st, *i.e.*, $(n-2)^{th}$ orbit

Thus, $S = 8 \times 0.25 + 2 \times 1 = 8.8$
$\therefore$ $Z^* = 11 - 8.8 = 2.2$

- Species with N-unpaired electrons is paramagnetic and magnetic moment $= \sqrt{N(N+2)}$ BM

Learning Framework of the Chapter ...

- Introduction
- Periodicity at a Glance
- Study of Group 1 (IA) — The Alkali Metals
- Compounds of Sodium
- Compounds of Potassium
- Diagonal Relationship —Anomalous Behaviour of Lithium
- Additional Reactions and Facts of Alkali Metals 1A

Alkali metals 1	Alkaline earths metals 2
$_1$H	
$_3$Li	$_4$Be
$_{11}$Na	$_{12}$Mg
$_{19}$K	$_{20}$Ca
$_{37}$Rb	$_{38}$Sr
$_{55}$Cs	$_{56}$Ba
$_{87}$Fr	$_{88}$Ra

- KO_2 has O_2^- ion (superoxide ion) and based on molecular orbital theory, bond-order is 1.5
- Amphoteric species are acidic (soluble in base) as well as basic (soluble in acid)
- Amphiprotic species is H^+ ion donor as well as H^+ ion acceptor

$$\underset{\text{acid}}{HCO_3^-} \rightleftharpoons H^+ + CO_3^{2-}$$

$$\underset{\text{base}}{HCO_3^-} + H^+ \rightleftharpoons H_2CO_3$$

- Born-Haber cycle is based on Hess's law of constant heat summation and is used to derive Lattice energy of the compound and also other unknown parameter. For the salt $MX(s)$

$$\begin{array}{lcll} & & & \Delta H^\circ \\ M(s) & \longrightarrow & M(g) & q_1 \\ M(g) & \longrightarrow & M^+(g) + e^- & q_2 \\ \frac{1}{2}X_2(g) & \longrightarrow & X(g) & q_3 \\ X(g) + e^- & \longrightarrow & X^-(g) & q_4 \\ M^+(g) + X^-(g) & \longrightarrow & MX(s) & -U \text{ (lattice energy)} \\ \hline M(s) + \frac{1}{2}X_2(g) & \longrightarrow & MX(s) & \Delta H_f^\circ = Q \end{array}$$

then $q_1 + q_2 + q_3 + q_4 - U = Q$

Thus, U is calculated.

Introduction

Shells up to $(n-1)$ are completely filled and differentiating (last) electron enters into *ns* sub-orbit—elements under this class are called **s-block elements.**

General electronic configuration of alkali group 1 (IA) is [inert gas] ns^1 where n signifies group.

Alkali metals form compounds with +1 oxidation state in which stable nearest inert gas configuration is attained.

Alkali metals				EC of M^+ (of IA) M^{2+} (of IIA)
1(IA)	Z	EC	Ion	
Li	3	[He] $2s^1$	Li^+	[He]
Na	11	[Ne] $3s^1$	Na^+	[Ne]
K	19	[Ar] $4s^1$	K^+	[Ar]
Rb	37	[Kr] $5s^1$	Rb^+	[Kr]
Cs	55	[Xe] $6s^1$	Cs^+	[Xe]
Fr	87	[Rn] $7s^1$	Fr^+	[Rn]

- Group IA elements are collectively known as alkali metals after the **Arabic word** *al-quis* meaning plant ashes which are rich in the carbonates of sodium and potassium.

Periodicity at a Glance

Detailed periodicity has been given in chapter 3. Table 8.1 describes periodicity of s-block elements at a glance.

- There is uniform gradation in properties except in case of Li and Be which being smallest in their groups show abnormal behaviour. Li resembles Mg and Be resembles Al that we call **diagonal relationship.**

Table 8.1 **Periodicity in *s*-Block Elements**

A	:	Atomic number
B	:	Effective nuclear charge $(Z^* = Z - \sigma)$ where σ is screening constant
C	:	Number of shells
D	:	Atomic radius
E	:	Ionisation energy
F	:	Hydration and hydration energy
G	:	Size of hydrated ion
H	:	Polarising power (based on Fajan's rule)
I	:	Lattice energy of chloride
J	:	Reactivity
K	:	Ionic conductance

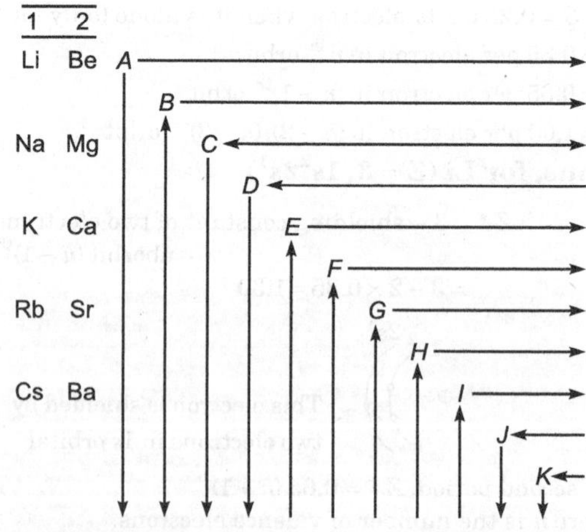

- From the Table 8.1, it is concluded that due to higher (*IE*) of alkaline earth metals, they are less reactive than alkali metals.

Role of *s*-Block Ions in Biological Systems

- The sodium and potassium ions, taken together, constitute **sodium pump** which is also known as '$Na^+ - K^+$ pump'. Operation of this process is biological in each and every cell of all animals in which Na^+ ions from intracellular fluids are transferred to the extracellular fluids with the help of **carrier-proteins'**. Transfer of K^+ ions is however, in a reverse way and thus an electrical potential gradient is set-up across the cell-membrane which is responsible for the nerve signals in animals.

- Some NaCl is essential in the diet of humans and other animals because many biological functions are controlled by the concentration of Na^+ and Cl^- ions. The fact that salt has long been recognised as important is evident in surprising ways. For example, we are paid a **"salary"** for workdone. This word is derived from the Latin salarium which meant "salt-money" because Roman soliders were paid in salt.

Study of Group 1 (IA)—The Alkali Metals

The elements of group 1 (IA) illustrate, more clearly than any other group of elements, the effects of increasing the size of the atoms or ions on the physical and chemical properties which are closely related to their electronic structures (Table 8.2) and sizes.

- The elements are all metals, excellent conductors of electricity, and are typically soft and highly reactive.
- They have one loosely held valence electron (ns^1) in their outer shell, and thus form univalent, ionic and colourless compounds. M^+ has stable nearest noble gas configuration.
- The hydroxides and oxides are very strong bases and the oxo salts are very stable.
- Lithium, the first element in the group, shows considerable difference from the rest of the group. This is due to high charge density of the Li^+ ion compound compared to other alkali metal ions. (charge density = ionic charge/ionic radius)

DEEP Focus

- Ionic size increases in the order :
$$Li^+ < Na^+ < K^+ < Rb^+ < Cs^+$$
and size of hydrated ion increases in the order (smaller the size, larger the charge, greater the hydration) :
$$Cs^+ < Rb^+ < K^+ < Na^+ < Li^+$$
hence ionic conductance is in order :
$$Cs^+ > Rb^+ > K^+ > Na^+ > Li^+$$

Table 8.2 **Electronic Structure**

Element	Symbol	Z	Electronic structure	
Lithium	Li	3	$1s^2 2s^1$	[He] $3s^1$
Sodium	Na	11	$1s^2 2s^2 2p^6 3s^1$	[Ne] $3s^1$
Potassium	K	19	$1s^2 2s^2 2p^6 3s^2 3p^6 4s^1$	[Ar] $4s^1$
Rubidium	Rb	37	$1s^2 2s^2 2p^6 3s^2 3p^6 4s^2 4p^6 5s^1$	[Kr] $5s^1$
Caesium	Cs	55	$1s^2 2s^2 2p^6 3s^2 3p^6 3d^{10} 4s^2 4p^6 4d^{10} 5s^2 5p^6 6s^1$	[Xe] $6s^1$
Francium	Fr	87	.... $6s^2 6p^6 7s^1$	[Rn] $7s^1$

- All the alkali metals give characteristic colours in bunsen flame. On heating in bunsen flame, outermost electrons get excited to higher energy levels (due to very low $(IE)_1$. When these electrons return to the ground state absorbed energy is released in the form of light in the visible region of the spectrum.

Element	Colour	λ/nm
Li	crimson red	670.8
Na	golden yellow	589.2
K	violet	766.8
Rb	red violet	780.0
Cs	blue	455.5

Colour emission by these elements is the basis of analytical methods (like flame photometry, atomic absorption spectroscopy) for the precise estimation of these metals.

Occurrence and Uses of Alkali Metals

Element	Abundance	Main mineral	Uses
Lithium	0.002%	First isolated in 1817 from petalite ($LiAlSi_4O_{10}$) named after the Greek word lithos meaning stone. Main source is spodumene ($LiAlSi_2O_6$)	All-purpose automative grease.
Sodium	2.36% sixth most abundant element in the earth's crust and the most abundant metal, after Al, Fe, Ca and Mg	First prepared in 1807 from caustic soda, NaOH. NaCl (halite or rock salt), $Na_2CO_3 \cdot NaHCO_3 \cdot 2H_2O$. (trona) $NaNO_3$ (saltpeter) are main sources	Glass, rubber, pharmaceutical agents use sodium or its salts in their production.
Potassium	2.09% eighth most abundant element in the earth's crust and next most abundant metal after Na.	First prepared in 1807, name derived from potash (K_2CO_3) KCl (sylvite), $KCl \cdot MgCl_2 \cdot 6H_2O$ (carnallite) are main sources.	As a plant fertilizer.

Although Na and K are almost equally abundant in the crusted rocks of the earth, Na is some 30 times as abundant as K in the oceans. This is partly because K salts with the larger anions tends to be less soluble than the Na salts and like wise, K is more strongly bound to the complex silicates and aluminosilicates in soils (ion exchange in clays)

Also K leached from rocks is preferentially absorbed and used by plants whereas Na can proceed to sea.

The Alkalide Ions

The reaction $2Na(g) \rightleftharpoons Na^+(g) + Na^-(g)$

is endothermic by 438 kJ mol^{-1} which indicates –1 oxidation state nonexisting but under unusual conditions this –1 state exists as in crown ethers (polycyclic ethers).

Solutions of sodium in ethylenediamine or methylamine show electrical conductance of normal 1 : 1 electrolytes— possibly due to Na^+ and Na^-. Alkali metals also dissolve in certain ethers, tetrahydro furan and other solvents forming relatively unstable solutions. The optical spectra of such solutions clearly indicate two distinct absorption bands — one for the solvated electrons and the other for the M^- species called alkalide ion. This is supported by the following facts :

(i) The solutions are diamagnetic, as expected for spin-paired M^- ions in the ground state.

(ii) The molar conductivity of solutions containing sodium decreases slowly as the concentration is raised. This is due to the formation of ion pairs $Na^+ Na^-$.

(iii) Absorption band also confirms the existence ofNa^-

Periodicity of Alkali Metals

Table 8.3 summarises physical properties of the elements of group 1 (IA). Form this we conclude

– they have low $(IE)_1$

– they have large negative electrode potentials

– they are a group of active metals.

Table 8.3 **Properties of the Alkali Metals**

Properties	Li	Na	K	Rb	Cs
Effective nuclear charge	1.3	2.2	2.2	2.2	2.2
Metallic radius (pm)	123	157	203	216	235
Ionic radius (pm)	76	102	138	152	167
Ionic charge density (ionic charge/ionic radius (Å)	1.32	40.98	0.72	0.66	0.60
Radius of hydrated ions (pm)	340	276	232	228	220
(*EN*), electronegativity	1.0	0.9	0.8	0.8	0.7
Melting point (°C)	181	98	63	39	29
Boiling point (°C)	1331	890	766	701	688
Density ($10^3 \times$ kg m^{-3})	0.53	0.97	0.86	1.53	1.87
$(IE)_1, M(g) \rightarrow M^+(g) + e^-$ or $\Delta H^\circ_{ionis.}$ (kJ mol^{-1})	520	496	419	403	376
$(IE)_2$, $M^+(g) \rightarrow M^{2+}(g) + e^-$ kJ mol^{-1}	7297	4561	3069	2650	2420
Sublimation energy $M(s) \rightarrow M(g), \Delta H^\circ_{subl.}$ (kJ mol^{-1})	155	109	90	86	79
Hydration energy $M^+(g) \longrightarrow M^+(aq), \Delta H^\circ_{hydr.}$ (kJ mol^{-1})	–506	–397	–318	–289	–259
$\Delta H^\circ_{subli.} + \Delta H^\circ_{ionis.} + \Delta H^\circ_{hydr.}$ (kJ mol^{-1})	169	208	191	200	196
$E^\circ_{red}(V), M^+(aq) + e^- \longrightarrow M(s)$	–3.045	–2.714	–2.925	–2.925	–2.923
Abundance on Earth%	0.002	2.36	2.09	0.009	0.0001
Ionic mobility	33.5	43.5	64.5	67.5	68.0
Lattice energy (of *M*Cl), ΔH_U kJ mol^{-1}	–845	–770	–703	–674	–644
ΔS°_{hydr} of M^+, JK^{-1} mol^{-1}	–140	–110	–70	–70	–60
ΔG°_{hydr} of M^+, kJ mol^{-1}	–485	–379	–308	–285	–362

1. Size

Alkali metals are the largest in their corresponding periods in the Periodic Table. The size of the atom or its ion increases on descending the group. This is due to the addition of an extra shell of electrons as we move down the group from one element to the next. The addition of the extra shell of electrons overweighs the effect of increased nuclear charge and thus there is an increase in size from Li to Cs. The size is based on

$$r_n = \frac{n^2 a_0}{Z^*}$$

where n = number of shells
a_0 = Bohr's radius
Z^* = Effective nuclear charge due to screening
(Z^* = 1.3 for Li and 2.2 for Na—Fr)

Thus, atomic ionic size is mainly dependent on n.

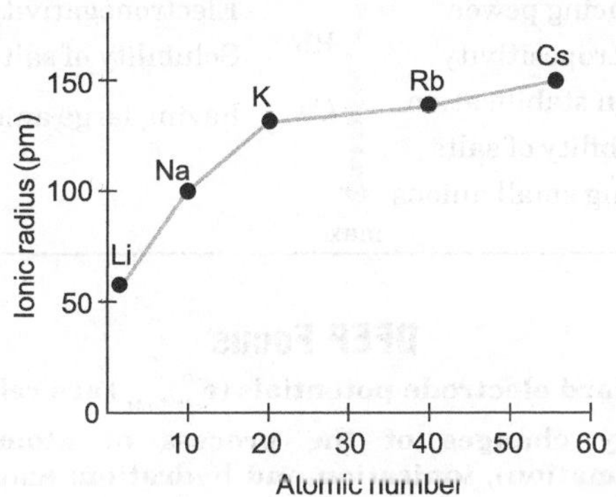

Fig. 8.1 Trend in ionic radii of group one elements.

2. Density

In a group, density increases on going down the group. Since, the elemental structure are often the same within any group, the factors which determine the density are atomic mass and volume. In this group, increase in atomic weight from one element to the next in passing down the group overweighs the effect of increase in the size of the atom. Density of potassium is however less than that of sodium.

3. Melting Point and Boiling Point

These metals are soft and can be cut with a knife. As a result of increase in size and repulsion of the non-bonded electrons, their cohesive energy and tendency for metallic bonding decrease down the group and thus softness increases from Li to Cs. These metals have low melting point and boiling point which also reflect the low values of cohesive energy between the atoms.

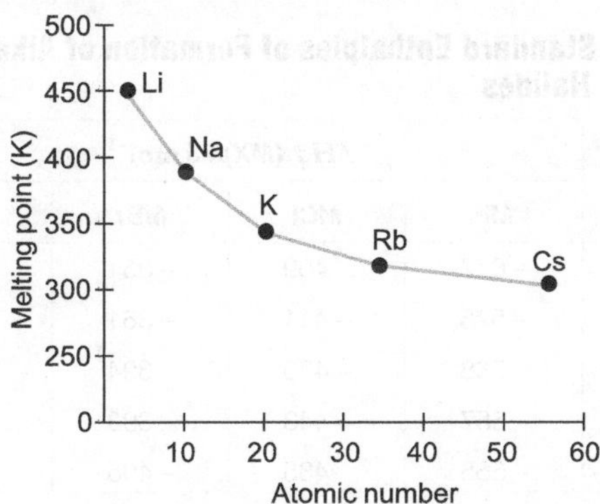

Fig. 8.2 Trend in the melting point of group one elements.

4. Thermal and Electrical Conductivity

In alkali metals, electrons of the noble gas core efficiently shield the lone valence shell electron from the nuclear charge. Thus, effective nuclear charge felt by the electron in the valence shell of an atom of an alkali metal is the least and the sole valence shell electron is very loosely held by the nucleus. It can move freely from one metal to the other in the lattice, thereby the alkali metals are good conductors of heat and electricity. **The loosely bound electron is also responsible for the silvery lusture of the alkali metals when freshly cut.**

5. Ionisation Energy, Electrode Potential and Hydration Energy

Low values of $(IE)_1$ reflect the tendency of the element to oxidise — the electropositive character. Rb and Cs are so highly electropositive that they emit electrons even when exposed to light, that is they exhibit photoelectric effect.

- What is the significance of the large negative electrode potentials of the alkali metals ? They mean that the metal M^+ can be reduced to the free metal M only with great difficulty. On the other hand, the reverse process— oxidation of the metals occurs with great ease.

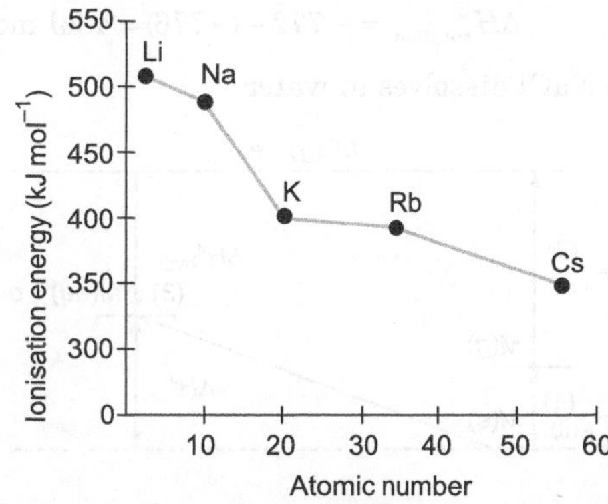

Fig. 8.3 Trend in first ionisation of alkali metals.

DEEP Focus

- If we use the magnitude of the first ionisation energy $(IE)_1$ as an indicator of ease of oxidation, Li is the most difficult of the alkali metals to oxidise. On the other hand, if we use $E^\circ_{\text{redn.}}$ (standard reduction electrode potential) as our criterion, Li is the easiest to oxidise. How can we explain this discrepancy? Actually these two criteria $(IE)_1$ and $E^\circ_{\text{redn.}}$ measure the same thing. $(IE)_1$ measures the tendency for a gaseous atom to ionise, $E^\circ_{\text{redn.}}$ measures the tendency for a metal atom to go from solid state to the ion in aqueous solution which involves:

(1) subliming the solid metal to the gaseous state
(2) ionising the gaseous metal atoms to gaseous metal ions
(3) hydrating the ions by dissolving $M^+(g)$ in water to produce $M^+(aq)$

Thus, $\Delta H^\circ_{ox.} = \Delta H^\circ_1 + \Delta H^\circ_2 + \Delta H^\circ_3$

$= \Delta H^\circ_{subli.} + \Delta H^\circ_{ionis.} + \Delta H^\circ_{hydr.}$

From Table 8.3, it is clear that they ($\Delta H^\circ_{oxdn.}$) are positive values, but that the value is smallest for Li (169 kJ). This is primarily because of the large negative ΔH for the hydration of Li^+. Thus it appears **that Li(*s*) should be easiest to oxidise of the alkali metals. If lithium takes part in reactions that do not involve water, then it does show less reducing power than the other members of the group.**

In terms of lattice energy ΔH°_U, heat of solution is

$$\Delta H^\circ_{solution} = \Delta H^\circ_{hydra.} - \Delta H^\circ_U$$

In general, if $\Delta H_{hydra} > \Delta H_U$, or $\Delta H^\circ_{solution} > 0$, the salt dissolves in water.

For NaCl $\quad \Delta H_{hydra} = -772 \text{ kJ mol}^{-1}$

$\Delta H_U = -776 \text{ kJ mol}^{-1}$

$\Delta H^\circ_{solution} = -772 - (-776) = 4 \text{ kJ mol}^{-1}$

hence NaCl dissolves in water.

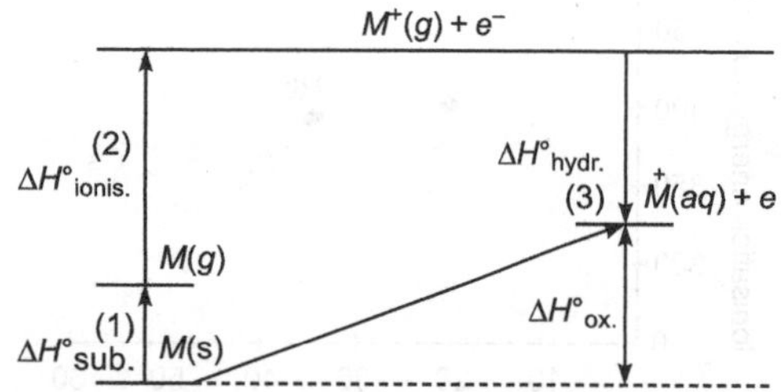

- With increase in atomic number, the atoms become larger and the strength of the metallic bonding decreases.
- Enthalpies of hydration (ΔH°_{hydr}) and entropies of hydration (ΔS°_{hydr}) both become less negative as ionic size increases, but the variation in free energy of hydration (ΔG°_{hydr}) follows fairly closely that in enthalpy of hydration.
- Due to the fact that the due to the largest size of the hydrated Li^+ ion, it is least conducting in solution order being $Li^+ < Na^+ < K^+ < Rb^+ < Cs^+$.

Gradation in the Properties of Alkali Metals

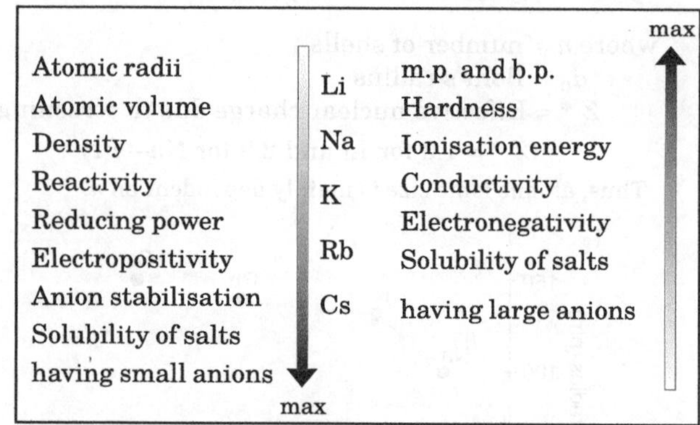

DEEP Focus

- **Standard electrode potentials ($E^\circ_{M^+/M}$) are related to energy changes of the process of atomisation (sublimation), ionisation and hydration; among the alkali metals these changes almost cancel one another and $E^\circ_{M^+/M}$ for the different M^+/M systems is nearly constant.** The lower reactivity of Li towards water is therefore kinetic rather than thermodynamic in origin : as a harder and less easily melted metal, Li is less easily dispersed and hence, reacts more slowly than the other metals.
- The second ionisation energies, $(IE)_2$, of all alkali metals are so high that the formation of M^{2+} ions under chemically realisable conditions is not possible.
- All of the halides (MX, X = F, Cl, Br, I) have strongly negative standard enthalpy of formation, ΔH°_f, but the actual values show two different trends as is evident from Table 8.4.

(a) **ΔH°_f for fluorides becomes less negative as we go down the group, while reverse is true for chlorides, bromides and iodides.**

(b) **For a given metal, ΔH°_f always becomes less negative as we go from the fluoride to iodide.**

Table 8.4 **Standard Enthalpies of Formation of Alkali Metal Halides**

M	ΔH°_f (*MX*)/kJ mol^{-1}			
	MF	*MCl*	*MBr*	*MI*
Li	− 617	− 409	− 351	− 270
Na	− 575	− 411	− 361	− 288
K	− 568	− 436	− 394	− 328
Rb	− 557	− 443	− 393	− 333
Cs	− 555	− 435	− 405	− 348

6. Lattice Energy and Solubility

As shown in Table 8.5, the variation in lattice energy (which is defined as the energy released when cations and anions in gaseous state combine to form one mole of ionic crystals) for fluorides is greater than the variation in lattice energy for chlorides, bromides and iodides. This is because lattice energy is proportional to $\left(\frac{1}{r_+ + r_-}\right)$ and to the charge on the ions hence, the variation in lattice energy is the greatest when r_- is smallest and least when r_- is largest. **Lattice energy is the driving force for the formation of an ionic compound and its stability.**

Lattice energy is measured by **Born-Haber cycle**

$$M(s) + \frac{1}{2}X_2(g) \xrightarrow{\Delta H_f^\circ} MX(s)$$

$$M(s) \xrightarrow{S} M(g) \xrightarrow{I,\ -e^-} M^+(g)$$

$$\frac{1}{2}X_2(g) \xrightarrow{\frac{D}{2}} X(g) \xrightarrow{-E,\ +e^-} X^-(g)$$

$$M^+(g) + X^-(g) \xrightarrow{-U} MX(s)$$

S = heat of sublimation of $M(s)$

I = ionisation energy of $M(g)$

D = dissociation energy of $X_2(g)$

E = electron-affinity of $X(g)$

U = lattice energy of $MX(s)$

ΔH_f° = heat of formation of $MX(s)$

Based on above cycle

$$\Delta H_f^\circ = S + I + \frac{D}{2} - E - U$$

$$\therefore \quad U = -\Delta H_f^\circ + S + I + \frac{D}{2} - E$$

The solubility of most of the salts of group 1 (alkali metals) elements M water **decreases on descending the group.** For a substance to dissolve, the energy evolved when the ions are hydrated (hydration energy) must be larger than the energy required to break the crystal lattice (crystal energy). We can say that if the solid is insoluble.

Hydration Energy < Lattice energy

In case of alkali metals, lattice energy only changes slightly, but the free energy of hydration changes rather more.

	Free energy of hydration $\Delta G^\circ_{\text{hydration}}$	Lattice energy MCl (U)	
Li^+	–485	–845	difference
Na^+	–379	–770	67
K^+	–308	–703	
Rb^+	–285	–674	
Cs^+	–362	–644	

all values in kJ mol^{-1}

Difference in lattice energy between NaCl and KCl is 67 kJ mol^{-1}, and yet the difference in $\Delta G_{(\text{hydration})}$ for Na^+ and K^+ is 71 kJ mol^{-1}. Thus, KCl is less soluble than NaCl.

Table 8.5 **Lattice Energies of Alkali Metal Halides**

M	U (MX)/kJ mol^{-1}			
	MF	MCl	MBr	MI
Li	–1035	–845	–800	–740
Na	–908	–770	–736	–690
K	–803	–703	–674	–636
Rb	–770	–674	–653	–615
Cs	–720	–644	–623	–590

We observe that the lattice energies decrease regularly from Li to Cs and from F^- to I^-. (Table 8.5).

- The higher the charge and the smaller the size of the ion, the more is the hydration energy. In a group, lattice energy and hydration energy decrease as we go down. While the decrease in lattice energy favours the solubility, the decrease in hydration energy makes the compound insoluble.

DEEP Focus

- Another important factor contributing to the solubility of the compound is the comparison of the size of cation and anion. Whenever cation is small and anion is large or *vice-versa*, the solubility will be larger. Thus, solubility of CsF > CsCl > CsBr > CsI and solubility of LiF < LiCl < LiBr < LiI.
- LiCl, LiBr, LiI, and NaI are readily soluble in some oxygen-containing organic solvents such as ether. NaI and LiI both are very soluble and form complexes in NH_3. The unstable complex $[Na(NH_3)_4]I$ has been isolated and shown to contain a tetrahedrally coordinated sodium ion.
- **In the vapour phase, alkali metal halides are present as ion-pairs, but measurement of bond length and dipole moment suggests that under these conditions electron-sharing is also involved to a considerable extent, especially for the lithium halides.**

- For the salts of small anions (like F^-, OH^-, O^{2-}, CO_3^{2-} etc.), lattice energy which is inversely proportional to ($r_+ + r_-$) is very sensitive to the change in the size of the cation, and anion being very small in size has little contribution in the total and decreases sharply along the group, and decrease in lattice energy is greater than decrease in hydration energy and thus solubility of these salts increases as we go down the group. Solubility of the alkali fluorides is thus in order:

$$LiF < NaF < KF < RbF < CsF$$

$$Li_2CO_3 < Na_2CO_3 < K_2CO_3 < Rb_2CO_3 < Cs_2CO_3$$

Thermal Stability of Salts

The stability of the compounds can also be explained by using the concept of polarising power. The simple idea is that, as **the charge density of the metal ions increases, the thermal stability of the salts of large polarisable anions, relaive to some decomposition product decreases.** In general, the least polarising metal ions are those of the most electropositive metal ions and these from the most stable salts with large anions. In other words, **small cations form stable salts with small anions and large cations form stable salts with large anions.** Let us take the example of carbonates of Group I metals. The carbonates of sodium, potassium rubidium and caesium are resistant to the heat of a bunsen burner flame, however, lithium carbonate decomposes to its oxide and carbon dioxide under the same conditions :

DEEP Focus

Small cations and small anions or Large cations and large anions → Stable salts

$$Li_2CO_3 \xrightarrow{\Delta} Li_2O + CO_2$$

$$Na_2CO_3 \xrightarrow{\Delta} \text{No effect}$$

The tendency of Li_2CO_3 **to undergo thermal decomposition may be explained in terms of the gain in electrostatic attraction that occurs when extremely small** Li^+ ion combines with the smaller oxide ion rather than the much **larger carbonate ion.** The other carbonates of Group I (Na–Cs) are more stable because the cations have a lower charge density and considerably larger in size and thus decomposition in less favourable energetically.

DEEP Focus

All the metals except lithium form stable bicarbonates (lithium bicarbonate is formed only aqueous solutions and has not been isolated. It is due to strong polarising power of Li^+ on the HCO_3^- ion of larger size.) When we heat the alkali metal bicarbonates, they are decomposed to carbonates and simultaneously, carbon dioxide and water are liberated.

$$2NaHCO_3 \xrightarrow{\Delta} Na_2CO_3 + CO_2 + H_2O$$

This reaction forms the basis for the use of $NaHCO_3$ in baking powder.

As the electropositive character increases down the group the stability of carbonates and bicarbonates increases. (Na < K < Rb < Cs)

The crystal structures of $NaHCO_3$ and $KHCO_3$ both show hydrogen bonding, but are different. In $NaHCO_3$ the HCO_3^- ions are linked into **infinite chain,** while in $KHCO_3$ a dimeric anion is formed.

$$\left[O{-}C\langle^{O-H\cdots O}_{O\cdots H-O}\rangle C{-}O \right]^{2-}$$

H-bonding in $KHCO_3$ (a-dimer)

H-bonding in $NaHCO_3$ (a-chain)

The thermal stability of Group I hydroxides also follows a similar trend as that of carbonates. Thus, except LiOH, which on heating decomposes to Li_2O, all other Group I hydroxides are stable.

$$2LiOH \xrightarrow{\Delta} Li_2O + H_2O$$

$$\left.\begin{array}{l} 2NaOH \xrightarrow{\Delta} \\ KOH \longrightarrow \\ \cdots \longrightarrow \end{array}\right] \text{no effect, thermally stable.}$$

The ease of thermal decomposition of a salt is related to the **enthalpy of formation** of the salt. The **enthalpy of formation,** ΔH_f°, of a salt *MA* (*M* is the metal, *A* is anion) is given by

$$\Delta H_f^\circ = (\Delta H_{atom} + I)_{metal} + (\Delta H_{atom} - E_A)_{anion} - \Delta H_{latt}$$

where ΔH_{atom} is the **enthalpy of atomisation,** *I* is the ionisation energy and E_A is the electron affinity. Since for any salt in a particular group the terms involving the anion alone remain constant, the value of ΔH_f° for such compounds is dependent upon the sum of the enthalpy terms of the particular metal, $(\Delta H_{atom} + I)_{metal}$ and the lattice energy, ΔH_{latt}. The larger the lattice energy, the more negative the enthalpy of formation and so, more stable is the compound. All these terms become smaller on descending the series from lithium to caesium. The relative stabilities of the salts are, therefore, decided by the parameter which decreases more rapidly—the lattice energy or the sum of the metal enthalpies.

In the salts having **small anions of high charge density, *e.g.,*** F^-, N^{3-}, OH^-, O^{2-} **etc., the change in lattice energy is much dependent on the size of the cation and decreases rapidly on descending the group.** Thus, as the size of the cation increases, lattice energy decreases more than the change in the sum of the metal enthalpies. Therefore, as we go down the group, the stability of these salts having small anions decreases. Thus, in alkali metal fluorides, the stability decreases in the order **LiF > NaF > KF > RbF > CsF.**

DEEP Focus

Similarly, lithium nitrate also decomposes on heating to give Li_2O, NO_2 and O_2 but all other alkali metal nitrates decompose on strong heating to nitrites liberating oxygen : Stability being

$$LiNO_3 < NaNO_3 < KNO_3 < RbNO_3 < CsNO_3$$

$$4LiNO_3 \longrightarrow 2Li_2O + 4NO_2 + O_2$$

$$2NaNO_3 \longrightarrow 2NaNO_2 + O_2$$

- The increasing stability of the peroxides (M_2O_2) or superoxides (MO_2), as the size of the metal ion increases is due to the stabilisation of large anions by larger cations through lattice energy effects.
- The m.p. and b.p. of alkali halides always follow trend $MF > MCl > MBr > MI$.

Hydride

The alkali metals react with H_2 at high temperature to form the hydrides.

$$2Li + H_2 \xrightarrow{1073\text{ K}} 2LiH$$

$$2K + H_2 \xrightarrow{673\text{ K}} 2KH$$

- Metallic hydrides are ionic solids with high melting points. They form complex hydrides when find applications in synthesis of organic compounds.

$$LiH + AlH_3 \longrightarrow LiAlH_4$$

$$NaH + BH_3 \longrightarrow NaBH_4$$

They also react with proton donors such as alcohols, gaseous ammonia and alkynes.

Complex Halide and Polyhalides

Complex halide formation such as represented by the equations :

$$MF + BF_3 \longrightarrow M[BF_4]$$

and $$MI + ICl \longrightarrow M[ICl_2]$$

is favoured by increase in the size of the cation.

- The alkali metal halides react ith the halogen and interhalogen compounds forming ionic polyhalide compounds

$$KI + I_2 \longrightarrow K[I_3]$$

$$KBr + ICl \longrightarrow K[BrICl]$$

$$KF + BrF_2 \longrightarrow K[BrF_4]$$

Alkali and Coinage Metals

Elements of group I except hydrogen are all metals arranged in two sub-groups:

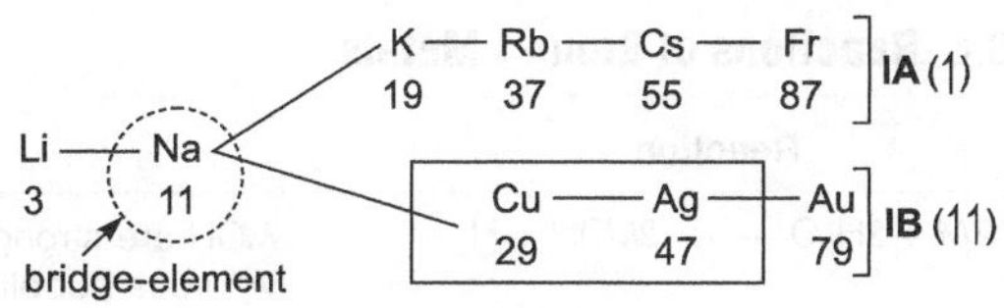

(1) IA — Li, Na, K, Rb, Cs, Fr
(11) IB — Cu, Ag, Au.

Sodium (11) acts as a bridge for sub-groups A and B. Distinction between alkali and coinage metals has been summarised below:

	Properties	IA(1)	IB(11)
1.	Electronic configuration	[Inert gas] ns^1	[Inert gas] $(n-1)\,d^{10}\,ns^1$
2.	Block	*s*-block	*d*-block
3.	(*IE*)	low, form stable M^+ ion	higher than IA, form M^+ ion, Cu also forms Cu^{2+} while Au form Au^{3+} salts
4.	Reactivity	highly reactive even with water	non-reactive with water but tarnished by air
5.	Nature of salts	halides, sulphates, nitrates, carbonates, bicarbonates water soluble (Li^+ is exception, its salts are predominantly covalent due to smaller size)	CuCl, CuI, CuBr, AgCl, AgBr, AgI, CuS, Ag_2S etc., insoluble
6.	Size of ions	larger than coinage metal $Na^+ > Cu^+, K^+ > Ag^+$	Smaller than alkali metal
7.	Electropositive nature	placed above H in electrochemical series, more reactive than coinage metals $Li^+(aq) + e^- \longrightarrow Li(s)$ $E° = -3.045$ V	placed below H in electrochemical series hence, less reactive than alkali metals, do not displace H_2 with acids $Cu^{2+}(aq) + 2e^- \longrightarrow Cu(s)$ $E° = +0.34$ V

Reactions of Group I Metals

Alkali metals, due to low (*IE*), are reactive and are always found in combined state. Table 8.6 summarises reactions of alkali metals. The reactivity of the alkali metals increases from Li to Cs except towards N_2. The slow reactivity of Li is due to kinetic factor.

Table 8.6 **Reactions of Group I Metals**

Reaction	Comment
$2M + 2H_2O \longrightarrow 2MOH + H_2$	MOH are strongest bases. All alkali metals react with H_2O. Lithium undergoes the reaction with vigorous bubbling as H_2 is released, sodium reacts rapidly with evolution of heat and potassium reacts so violently that H_2 produced bursts instantly into flame.
$2Li + \frac{1}{2}O_2 \longrightarrow Li_2O$ $2Na + O_2 \longrightarrow Na_2O_2$	Monoxide is formed by Li and to a small extent by Na (oxidation number (O.N.) of O = – 2) Normal oxides M_2O, (M = Li, Na, K, Rb) have antifluorite structure . This structure is related to the CaF_2 structure but the sites occupied by the cations and anions interchanged so that M replaces F and O replaces Ca in the structure. There is a trend to increasing coloration with increasing atomic number, Li_2O, Na_2O — pure white K_2O — yellowish white Rb_2O — bright yellow Cs_2O — orange Peroxide formed by Na and to a small extent by Li (O.N. of O = – 1)
$K + O_2 \longrightarrow KO_2$	Superoxide (MO_2) formed by K, Rb and Cs. Superoxides are coloured and paramagnetic (ON of O = – 1/2) Under appropriate conditions pure compounds M_2O, M_2O_2 and MO_2 can be prepared for all five metals.
$2M + H_2 \xrightarrow{\Delta} 2MH$	Ionic salt-like hydride, LiH is the most stable.
$MH + H_2O \longrightarrow MOH + H_2$	Hydrides are decomposed by H_2O forming H_2 and thus are reducing agents
$3Li + \frac{1}{2}N_2 \longrightarrow Li_3N$	Only Li forms nitride, slow at room temperature rapid at elevated temperature.
$2Na + C_2H_2 \longrightarrow Na_2C_2 + H_2$	Carbide contains carbide ion $(C{\equiv}C)^{2-}$
$2M + F_2 \longrightarrow 2MF$ $2M + Cl_2 \longrightarrow 2MCl$ $2M + Br_2 \longrightarrow 2MBr$ $2M + I_2 \longrightarrow 2MI$	All metals form fluorides, chlorides, bromides and iodides. The reactivity of the alkali metals increases as their (IE) decreases giving reactivity order: Cs > Rb > K > Na > Li The higher members can form polyhalides, KI_3 (K^+ and I_3^-) and $CsBr_3$ (Cs^+, Br_3^-)
$2M + S \longrightarrow M_2S$	Very vigorous reaction. Polysulphides are also formed. They are *zig-zag* chain structure of sulphur atoms.
$2M + 2NH_3 \xrightarrow{Fe} 2MNH_2 + H_2$ $Na(l) \longrightarrow Na^+(l) + e^-$ $e^- + NH_3 \longrightarrow NH_2^- + \frac{1}{2}H_2$ $Na^+\,(am) + e^-\,(am) + NH_3(l) \longrightarrow NaNH_2(s) + \frac{1}{2}H_2$	All the metals form amides. Dilute solutions of alkali metals in liquid NH_3 are dark blue in colour and the main species present are solvated electrons and the solvated metal ions. If the blue solution is allowed to stand, the colour slowly fades until it disappears owing to the formation of metal amide.

where 'am' denotes solution in ammonia.

Sesquoxide "M_2O_3" have been prepared as dark-coloured paramagnetic powders by careful thermal decomposition of MO_2 (M = K, Rb, Cs) They can also be prepared by oxidation of liquid ammonia solutions of the metals or by controlled oxidation of the peroxides, and are considered to be peroxide disuperoxide $[(M^+)_4(O_2^{2-})(O_2^-)_2]$

- **Ozonides MO_3 have been prepared for Na, K, Rb and Cs by the reaction of O_3 on powdered anhydrous MOH at low temperature and extraction of the red MO_3 by liquid NH_3**

$$3MOH(s) + 2O_3(g) \longrightarrow 2MO_3(s) + MOH \cdot H_2O(s) + \frac{1}{2}O_2(g)$$

The ozonides, on standing, slowly decompose to oxygen and the superoxide MO_2, but on hydrolysis forms MOH

$$MO_3 \longrightarrow MO_2 + \frac{1}{2}O_2$$

$$4MO_3 + 2H_2O \longrightarrow 4MOH + 5O_2$$

Rb and Cs also form suboxides in when the formal oxidation state of the metal is considerably lower. Partial oxidation of Rb at low temperature gives Rb_6O which decompose above –7.3°C to give copper-coloured metallic crystals of Rb_9O_2 :

$$2Rb_6O \xrightarrow{-7.3\,°C} Rb_9O_2 + 3Rb$$

Alkali metals react with H_2O, liberating H_2 and forming hydroxides. The reaction becomes increasingly violent on descending the group. The Li reacts gently, sodium melts on the surface of water and molten metal even catches fire. Similarly for K.... . Thus, Na, K are placed in **kerosene oil.**

- Alkali metals are reactive and tarnish rapidly in dry air. Na, K, Rb and Cs form oxides of various types, but Li forms mixture of the oxide and the nitride, Li_2O and Li_3N. **Li is the only element of the alkali metals that forms nitride.**

Solutions in Liquid Ammonia

The alkali metals dissolve in pure liquid ammonia without evolution of hydrogen; the metals can be recovered from such solution by evaporation. The saturation solubilities are quite high, *e.g.*, at –33°C.

Element	Solubility in kg/kg NH_3
Li	108
Na	250
K	463

1. Colour Dilute solution of alkali metals in liquid NH_3 are blue due to solvated electrons and the solvated metal ions. **But in the concentration of NH_3 above 3*M*, metal ion cluster are formed, the solutions are copper-coloured with metallic lustre.**

(Solvated electron)

Estimated diameter of the cavity = 3 – 3.4Å

2. Conductance and magnetic properties The solutions are good conductors of electricity and are paramagnetic in nature, which decreases as concentration increases.

Explanation

When alkali metal is dissolved in liquid NH_3, there is

Step I. Ionisation of alkali metal

$$M(s) \longrightarrow M^+(g) + e^-$$

Step II. Cation M^+ and e^- both are solvated by NH_3 forming $[M(NH_3)_x]^+$ and $[e(NH_3)_y]^-$.

$$M^+ + (x+y)\,NH_3 \longrightarrow [M(NH_3)_x]^+ + [e(NH_3)_y]$$

It is **the ammoniated electron which is responsible for the blue colour of the solution, and the electrical conductivity is due to the ammoniated cation, $[M(NH_3)_x]^+$ as well as the ammoniated electron, $[e(NH_3)_y]^-$, values of *x* and *y* depend on the extent of solvation (by NH_3). Dilute solutions are paramagnetic due to free electrons. In concentrated solutions ammoniated electrons associate to form electron-pairs and their paramagnetic character decreases.**

The solutions of alkali metals in liquid NH_3 are extremely good conductors of electricity : they have higher conductivity than that of completely ionised salt in any known solvent. But the equivalence conductance decreases as the solution become more concentrated and reaches a minimum around 0.05 M. Beyond this concentration the conductivity increases dramatically sharply and the saturated solutions practically conduct electricity like liquid metals :

Conductivity of Hg $\approx 10^4$ ohm^{-1} of Na in liquid $\approx 0.5 \times 10^4$ ohm^{-1} NH_3

$$[e(NH_3)_y]^- + [e(NH_3)_y]^- \longrightarrow [e(NH_3)_y]_2^{2-}$$

The solutions are themselves unstable with respect to amide formation.

$$Na^+(NH_3) + e^-(NH_3) \longrightarrow NaNH_2 + H_2$$

However, under anhydrous conditions and in the absence of catalytic inpurities such as transition metal ions, solutions can be stored for several days.

3. Reducing Property Ammoniated solutions of alkali metals are reducing agents due to the presence of free or solvated electrons

$$O_2 \xrightarrow{e^-(NH_3)_y} O_2^- \xrightarrow{e^-(NH_3)_y} O_2^{2-}$$

$$K_2[Ni^{II}(CN)_4] + 2K \xrightarrow{e^-(NH_3)_y} K_4[\overset{0}{Ni}(CN)_4]$$

- The alkali metals dissolve readily in mercury forming more reactive species *M*(Hg) called **amalgams**

$$M(s) + Hg(l) \longrightarrow M(Hg), \quad \Delta H = -ve$$

Formation of an amalgam is an exothermic process.

Solvation

When a metal ion is surrounded by the solvent molecules, the phenomenon is called **solvation** of the metal ion. When the solvent is water it is called **hydration.** The alkali metal ions are highly hydrated. The smaller the size of the ion, the greater is its degree of hydration. This is because the smaller the size, the more will be its charge density and the more will be its attraction for the polar solvent molecules. Thus, Li^+ ion, which is the smallest gets more hydrated than the Na^+ ion and so on. The degree of hydration decreases on moving down the group. As a result of differences in their degree of hydration, the hydrated ionic radii of the alkali metal ions decrease as we go down the group from lithium to cesium. Li^+ ion has the largest hydrated radius while Cs^+ has the smallest hydrated radius in the first group. One would agree that the smaller the size of the ion and the lighter it is, the more will be its mobility and thus conductance. In this regard we should expect the highest conductance for Li^+ which is the smallest of the alkali metals but this is not so. **The hydrated radius of Li^+ is the largest of all the alkali metal ions, in solution. Therefore, Li^+ ion is the least conducting in solution. The ionic conductance in solution actually decreases in the order $Cs^+ > Rb^+ > K^+ > Na^+ > Li^+$.**

Target Practice 1

1. Arrange Li, Na, K, Rb, Cs in increasing.
   (a) atomic radii ....
   (b) ionisation energy ....
   (c) electronegativity ....
   (d) electron-affinity ....
2. Arrange alkali metal carbonates in increasing
   (a) thermal stability ....
   (b) covalent nature ....
   (c) melting point ....
   (d) lattice energy ....
3. Compare hydration energy of
   (a) Na^+ and Be^{2+} ....
   (b) Li^+ and Be^{2+} ....
   (c) Li^+ and Mg^{2+} ....
   (d) Be^{2+} and Al^{3+} ....
4. The single-bonded metallic radius of sodium is 157 pm. Assume that the increment between radii of different magnitudes is 60 pm. Calculate the
   (a) covalent radius of Na ....
   (b) ionic radius Na^+ ....
   (c) ionic radius Na^- ....
   (d) van der Waals' radius Na ....
5. Write compounds of low solubility of Fr. ....
6. How could you remove unreacted Na (metal) in liquid ammonia safely? ....

Answers

1. (a) Li < Na < K < Rb < Cs (b) Cs < Rb < K < Na < Li
   (c) Cs < Rb = K < Na < Li (d) Cs < Rb < K < Na < Li
2. (a) $Li_2CO_3 < Na_2CO_3 < K_2CO_3 < Rb_2CO_3 < Cs_2CO_3$
   (b) $Cs_2CO_3 < Rb_2CO_3 < K_2CO_3 < Na_2CO_3 < Li_2CO_3$
   (c) $Li_2CO_3 < Na_2CO_3 < K_2CO_3 < Rb_2CO_3 < Cs_2CO_3$
   (d) $Cs_2CO_3 < Rb_2CO_3 < K_2CO_3 < Na_2CO_3 < Li_2CO_3$
3. (a) $Be^{2+} > Na^+$ (b) $Be^{2+} > Li^+$ (c) $Mg^{2+} > Li^+$
   (d) $Al^{3+} > Be^{2+}$
4. (a) 157 pm (b) 97 pm (c) 217 pm (d) 217 pm
5. $FrClO_4$, Fr_2PtCl_6 (due to larger alkali metal ion)
6. Add NH_4Cl, (a strong acid in NH_3)
   $$2Na + 2NH_4Cl \longrightarrow 2NaCl + NH_3 + H_2$$

Practice Exercise 1

1. Differentiating electron in K(19) goes into *ns* and this is *s*-block element. What happens if Aufbau rule is not followed ?
2. What would be electronic configuration of the following in the absence of Aufbau rule?
   (a) Ca^{2+} (b) Sr^{2+}
   (c) Rb^+
3. Na and Mg are also called bridge element. Why?
4. Which is more covalent in each pair?
   (a) LiCl or NaCl (b) LiCl or $BeCl_2$
5. $(IE)_1$ of the two elements of *s*-block of 2nd period are 520 and 899 kJ mol^{-1}. Identify these elements.
6. Compare Li and Be in the following properties:
   (a) $(IE)_2$
   (b) size of hydrated ion
   (c) polarisation (of cations)
   (d) thermal stability of carbonates
   (e) nature of oxide
   (f) nature of hydride
   (g) oxidation state
7. Compare Li and Mg in the following properties:
   (a) effect of heat on their carbonates
   (b) hydration of chloride
   (c) nature of oxides
   (d) tendency to form nitrides
   (e) electronegativity
8. Alkali metals are good reducing agents. Explain.
9. $(IE)_1$ of Li is highest in first group but it is best reducing agent in aqueous solution. Explain.
10. Arrange He, Li, Be, B, C, N, O, F, Ne in increasing second ionisation energy.
11. In Table 9.3, size of the hydrated ion is in order $Li^+ > Na^+ > K^+ > Rb^+ \geq Cs^+$ which is just reverse of ionic size. Explain.
12. Calculate lattice energy of NaCl (*s*) from the following data :
    heat of sublimation of Na(*s*) $= S$
    $(IE)_1$ of Na (*g*) $= I$
    bond dissociation energy of $Cl_2(g) = D$
    electron affinity of Cl(*g*) $= -E$
    heat of formation of NaCl(*s*) $= -Q$
13. Arrange the following in increasing lattice energy :
    (i) NaF, NaCl, NaBr, NaI
    (ii) LiF, LiCl, LiBr, LiI
    (iii) LiCl, NaCl, KCl, RbCl, CsCl
14. A mixture contains 2 moles of Na_2CO_3 and 1 mole of Li_2CO_3. Calculate volume of CO_2 formed on heating this mixture and the data converted at STP.
15. Identify (*A*), (*B*), (*C*), (*D*) and (*E*) in the following:
    $$\text{Alkali metal } (A) + N_2 \xrightarrow{\Delta} (B) \xrightarrow{H_2O} (C) + (D)$$
    (*C*) and (*D*) both react with HCl, but (*D*) gives white fumes (*E*) with HCl.

16. A binary salt (*X*) of alkali metal on heating with sulphur give (*Y*); (*Y*) with $BaCl_2$ solution gave white ppt. (*Z*) insoluble in conc. HCl. 14.2 g of (*X*) gave 23.3 g of (*Z*). Identify (*X*), (*Y*) and (*Z*) and explain reactions.

17. (a) Which element of the alkali earth metals (group II) will form nitride?

(b) 1 mole of a mixture of Li_2CO_3 and Na_2CO_3 forms 0.75 mole of CO_2 at NTP. What is percentage of each in the mixture?

(c) $Li + H_2 \longrightarrow (A) \xrightarrow{H_2O} (B) + (C)$; $(A) \xrightarrow{AlH_3} (D)$

What are (*A*), (*B*), (*C*) and (*D*)? What is main application of (*D*)?

18. Match the following column (*A*) with column (*B*):

(A)	(B)
1. KO_2	(a) Diamagnetic
2. Na_2O_2	(b) Conductor of electricity
3. Na/NH_3	(c) Paramagnetic

19. $2Na(s) + 2H_2O(l) \longrightarrow 2Na^+(aq) + 2OH^-(aq) + H_2(g)$
Write the ON of all the atoms in various species.

Answers

1. K(19) in the absence of Aufbau rule will have EC as: $1s^2, 2s^22p^6, 3s^23p^63d^1$ and thus differentiating electron enters into 3*d*— a *d*-block element.

2. (a) $Ca^{2+} = 1s^2, 2s^22p^6, 3s^23p^6$
(b) $Sr^{2+} = 1s^2, 2s^22p^6, 3s^23p^63d^{10}, 4s^24p^6$
(c) $Rb^+ = 1s^2, 2s^22p^6, 3s^23p^6\,3d^{10}, 4s^24p^6$

3. Na is called bridge element since it links Group IA and Group IB

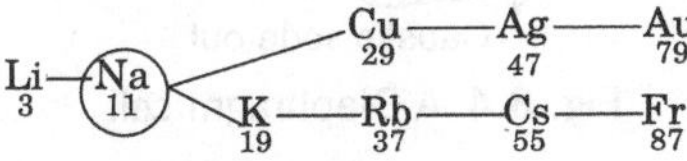

Similarly, Mg links IIA and IIB

Be(4) — Mg(12) — Zn(30) — Cd(48) — Hg(80); Mg(12) — Ca(20) — Sr(38) — Ba(56) — Ra(88)

4. (a) LiCl (b) $BeCl_2$
Smaller cationic size, larger, charge, larger polarising power

5. Li and Be

6. (a) Li > Be (b) $Li^+ < Be^{2+}$ (c) $Li^+ < Be^{2+}$
(d) $Li_2CO_3 > BeCO_3$ (e) Li_2O—basic (f) LiH—ionic
(g) Li = + 1 BeO—amphoteric BeH_2—covalent Be = + 2

7. (a) Both decompose to form CO_2 and oxide
(b) Both hydrated (c) Basic oxides
(d) Both form nitrides (Li_3N, Mg_3N_2)
(e) $\chi_{Li} = 1.0$ $\chi_{Mg} = 1.2$

8. In case of alkali metals, outer *ns*-electron is very well shielded by the inner electrons. Thus effective nuclear charge (= $Z - \sigma$) is very small. As we go down the group, shielding wins over the effect of the increasing number of protons in the nucleus. Thus $(IE)_1$ decreases, hence tendency to lose electrons, which is also a measure of the reducing nature, increases down the group.

9. See text

10. Be < C < B < N < O < F < Ne < He < Li

11. Greater the charge or charge density, smaller the ionic size, greater the hydration hence, greater the size of the hydrated ion.

12. Lattice energy $-U = S + I + \frac{D}{2} - E - Q$

13. (i) NaI < NaBr < NaCl < NaF (ii) LiI < LiBr < LiCl < LiF
(iii) CsCl < RbCl < KCl < NaCl < LiCl

14. 22.4 L (only Li_2CO_3 decomposes)

$$\underset{1\text{ mol}}{Li_2CO_3} \longrightarrow Li_2O + \underset{1\text{ mol}}{CO_2} \quad \text{at STP}$$

15. (A) : Li (only Li forms nitride with N_2) (B) : Li_3N
(C) : LiOH (D) : NH_3 (E) : NH_4Cl

16. Let alkali metal be *M*, since *X* is binary and combines with S to give *Y*. *Y* gives white precipitate with $BaCl_2$ insoluble in conc. HCl. It means *Y* has SO_4^{2-} and thus *X* is an oxide and thus, *X* is an oxide. Since *X* is binary hence, *X* is

$$\underset{\substack{2(M+32)\\14.2}}{2MO_2} + S \longrightarrow M_2SO_4 \xrightarrow{BaCl_2} \underset{\substack{233\\23.3}}{BaSO_4}$$

Thus, $\frac{2(M+32)}{14.2} = \frac{233}{23.3}$

Thus, *M* = 39. Thus *M* is potassium.
(*X*) : KO_2 (potassium superoxide, $K^+O_2^-$ (binary)
(*Y*) : K_2SO_4 (*Z*) : $BaSO_4$

$$\underset{\substack{(X)\\2\times 71\text{ g}\\14.2\text{ g}}}{2KO_2} + S \longrightarrow \underset{(Y)}{K_2SO_4} \xrightarrow{BaCl_2} \underset{\substack{(Z)\\233\text{ g}\\23.3\text{ g}}}{BaSO_4}$$

17. (a) Mg (due to diagonal relationship with Li)

(b) $Li_2CO_3 \xrightarrow{\Delta} Li_2O + CO_2$
$Na_2CO_3 \xrightarrow{\Delta} X$ (no effect)

CO_2 is due to Li_2CO_3 only
hence, 0.75 mol CO_2 = 0.75 mol of Li_2CO_3
75 mol % Li_2CO_3
25 mol % Na_2CO_3

(c) (*A*) : LiH (*B*) : LiOH (*C*) : H_2 (*D*) : $LiAlH_4$
(*D*) is a reducing agent in the preparation of alcohols from carbonyl compounds, acids, esters, etc.

$$>C{=}O \xrightarrow{LiAlH_4} >CHOH$$

$$RCOOH \xrightarrow{LiAlH_4} RCH_2OH$$

$$RCOOR' \xrightarrow{LiAlH_4} RCH_2OH + R'OH$$

18.

(A)	(B)
1.	c
2.	a
3.	b

19. $2Na(s) + 2H_2O(l) \longrightarrow 2Na^+(aq) + 2OH^-(aq) + H_2(g)$

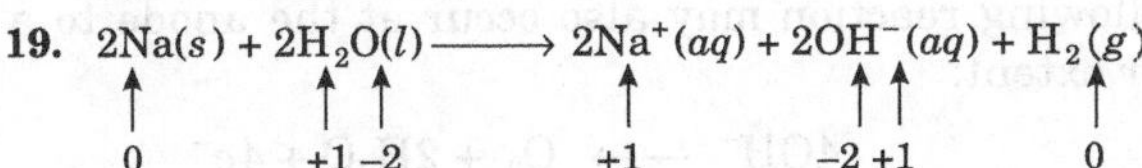

Compounds of Sodium

- The metals of this group are too reactive to be found in the free state. Their compounds are amongst the most stable to heat, so thermal decomposition is impractical.
- The metals are the strongest chemical reducing agents and oxides can't be reduced to metals.
- The metals are generally obtained by electrolysis of a fused salt, usually the fused halide, often with impurity added to lower the melting point.
- Sodium is obtained by electrolysis of molten mixture of:
  (a) 40% NaCl
  (b) 60% $CaCl_2$ in a Downs cell.

$CaCl_2$ lowers the m.p. of NaCl (803° C) to 600° C. Some Ca is formed but it remains insoluble in sodium, which collects at a copper or iron cathode that surrounds a circular graphite anode. Chlorine, a valuable by-product, collects at anode.

Sodium Hydroxide

Until World War II, the **Soda-Lime process** the principal method of producing sodium hydroxide, NaOH, involved conversion of $Ca(OH)_2$ to $CaCO_3$ using Na_2CO_3

$$Ca(OH)_2 + Na_2CO_3 \longrightarrow CaCO_3 + 2NaOH$$

The reaction goes to completion as is evident by measuring pH of the solution after reaction.

The insoluble $CaCO_3$ by-product is filtered off, then heated (calcined) to convert it to lime (CaO), which is recycled into the reaction system.

$$CaCO_3(s) \longrightarrow CaO(s) + CO_2(g)$$

$$CaO(s) + H_2O \longrightarrow Ca(OH)_2$$

- Electrolysis of aqueous NaCl gives NaOH at cathode

$$NaCl\,(aq) \rightleftharpoons Na^+ + Cl^-$$

Anode: $2Cl^- \longrightarrow Cl_2 + 2e^-$

Cathode: $Na^+ + e^- \longrightarrow Na$

$$2Na + 2H_2O \longrightarrow 2NaOH + H_2$$

Side reactions may also occur if the products formed mix with one each other.

$$2NaOH + Cl_2 \longrightarrow NaCl + NaClO + H_2O$$

or $$2OH^- + Cl_2 \longrightarrow 2OCl^- + H_2$$

and following reaction may also occur at the anode to a smaller extent:

$$4OH^- \longrightarrow O_2 + 2H_2O + 4e^-$$

Nelson Cell (Diaphragm cell)

If H_2 and Cl_2 gases mix, they react with explosion. A porous diaphragm of asbestos is used to keep the H_2 and Cl_2 (and thus cathode and anode) gases separated from one another. Product in this cell is a mixture of 11% NaOH and 16% NaCl. Concentration in a steam evaporator gives solution containing 50% NaOH and 1% NaCl. (Fig. 8.4)

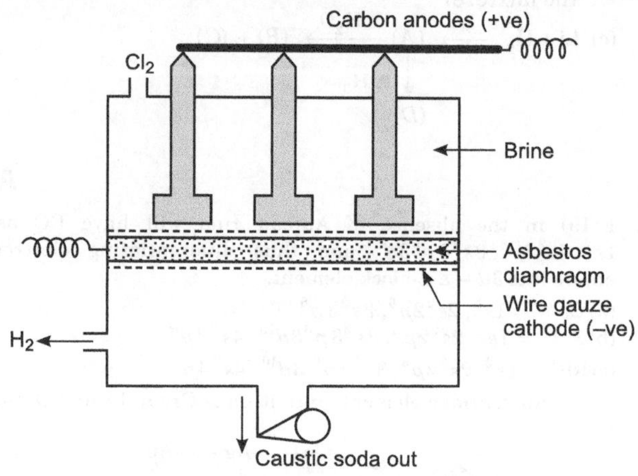

Fig. 8.4 A Diaphragm cell

Mercury Cathode Cell (Castner-Kellner Cell)

During the electrolysis of brine, Na^+ is discharged at cathode

$$Na^+ + e^- \longrightarrow Na$$

If the cathode is made of mercury, the Na atoms produced are dissolved in mercury and form an amalgam. The amalgam is pumped to a different compartment and water reacts to form NaOH.

$$Na\text{ (amalgam)} + H_2O \longrightarrow NaOH + \frac{1}{2}H_2 + Hg$$

Originally the anodes were made of graphite, but because traces of oxygen are produced in a side reaction, they become pitted with corrosion. The anodes are now made of steel coated with titanium (Fig. 8.5).

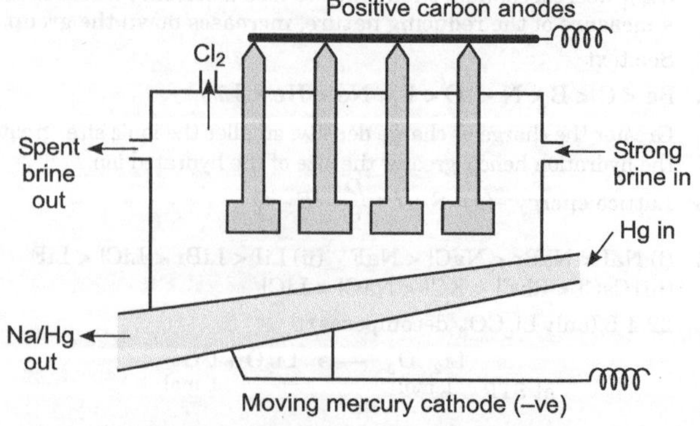

Fig. 8.5 The Castner-Kellner cells

Properties

White crystalline deliquescent solid, highly soluble in water but sparingly in alcohol corrosive to skin. It breaks down the proteins of skin to a pasty mass hence called caustic soda.

- A strong base ionised almost completely

$$NaOH \rightleftharpoons Na^+ + OH^-$$

$$NaOH + H^+ \rightleftharpoons Na^+ + H_2O$$

Equivalent weight of NaOH = molecular weight

$$= 40 \text{ g equiv}^{-1}$$

With Acidic Oxide

It can absorb CO_2, SO_2 NO_2 etc. (acidic oxide) to form salt

$$2NaOH + CO_2 \longrightarrow Na_2CO_3 + H_2O$$

$$2NaOH + SO_2 \longrightarrow Na_2SO_3 + H_2O$$

$$2NaOH + 2NO_2 \longrightarrow NaNO_2 + NaNO_3 + H_2O$$

With P_4

$$\underset{\text{yellow}}{P_4} + 3NaOH + 3H_2O \xrightarrow{\Delta} \underset{\text{sodium hypo phosphite}}{3NaH_2PO_2} + \underset{\text{phosphine}}{PH_3}$$

With S

It is a disproportionation reaction of P_4 to $H_2PO_2^-$ (by oxidation) and PH_3 (by reduction)

- $$4S + 6NaOH \xrightarrow{\Delta} \underset{\text{hypo (sodium thiosulphate)}}{Na_2S_2O_3} + 2Na_2S + 3H_2O$$

With Halogens

$X_2(Cl_2, Br_2, I_2)$ disproportionates to X^- and XO^- in cold and dilute alkaline solution (NaOH)

- $$X_2 + \underset{\text{cold}}{2NaOH} \longrightarrow NaX + \underset{\text{sodium hypohalite}}{NaXO} + H_2O$$
- $X_2(Cl_2, Br_2, I_2)$ also disproportionates to X^- and XO_3^- in hot and concentrated alkaline solution (NaOH)

$$3X_2 + \underset{\text{hot}}{6NaOH} \longrightarrow 5NaX + \underset{\text{sodium haloate}}{NaXO_3} + 3H_2O$$

With Metals

Elements like Zn, Al, Sn, B etc. liberate H_2 with NaOH

$$Zn + 2NaOH \longrightarrow \underset{\text{sodium zincate}}{Na_2ZnO_2} + H_2$$

$$Sn + 2NaOH + H_2O \longrightarrow \underset{\text{sodium stannate}}{Na_2SnO_3} + 2H_2$$

$$2Al + 2NaOH + 2H_2O \longrightarrow \underset{\text{sodium meta aluminate}}{2NaAlO_2} + 3H_2$$

$$2B + 2NaOH + 2H_2O \longrightarrow \underset{\text{sodium meta borate}}{2NaBO_2} + 3H_2$$

With Oxide

Amphoteric oxides (as well as hydroxides) dissolve in NaOH.

$$Al_2O_3 + 2NaOH \longrightarrow 2NaAlO_2 + H_2$$

$$Al(OH)_3 + NaOH \longrightarrow NaAlO_2 + 2H_2O$$

$$ZnO + 2NaOH \longrightarrow Na_2ZnO_2 + H_2O$$

$$Zn(OH)_2 + 2NaOH \longrightarrow Na_2ZnO_2 + H_2O$$

(similarly with SnO, PbO)

With Salts

Salts like $ZnCl_2, SnCl_2, AlCl_3$ etc. give white ppt with NaOH; however precipitate dissolves in excess of it.

$$ZnCl_2 + 2NaOH \longrightarrow \underset{\text{white ppt.}}{Zn(OH)_2} \xrightarrow{NaOH} \underset{\text{soluble}}{Na_2ZnO_2}$$

$$AlCl_3 + 3NaOH \longrightarrow \underset{\text{white ppt.}}{Al(OH)_3} \xrightarrow{NaOH} \underset{\text{soluble}}{NaAlO_2}$$

$$SnCl_2 + 2NaOH \longrightarrow \underset{\text{white ppt.}}{Sn(OH)_2} \xrightarrow{NaOH} Na_2SnO_2$$

- NH_4^+ salts are decomposed by NaOH forming NH_3

$$NH_4^+ + OH^- \xrightarrow{\Delta} NH_3\uparrow + H_2O$$

- Salts of transitional metals give insoluble hydroxide with NaOH.

$$FeCl_3 \xrightarrow{NaOH} \underset{\text{brown ppt.}}{Fe(OH)_3\downarrow}$$

$$CrCl_3 \xrightarrow{NaOH} \underset{\text{green ppt.}}{Cr(OH)_3\downarrow}$$

$$CuSO_4 \xrightarrow{NaOH} \underset{\text{blue ppt.}}{Cu(OH)_2\downarrow}$$

However, AgOH and $Hg(OH)_2$ decompose to oxide on heating.

$$2AgOH \xrightarrow{\Delta} Ag_2O + H_2O$$

$$Hg(OH)_2 \xrightarrow{\Delta} HgO + H_2O$$

With CO

HCOONa is formed when CO is passed into NaOH at high pressure and temperature

$$CO + NaOH \xrightarrow[100\text{–}200^\circ C]{5\text{–}10 \text{ atm}} HCOONa$$

Application

NaOH finds wide application in

(1) paper industry
(2) as a laboratory reagent.
(3) alumina production
(4) soap industry
(5) rayon industry

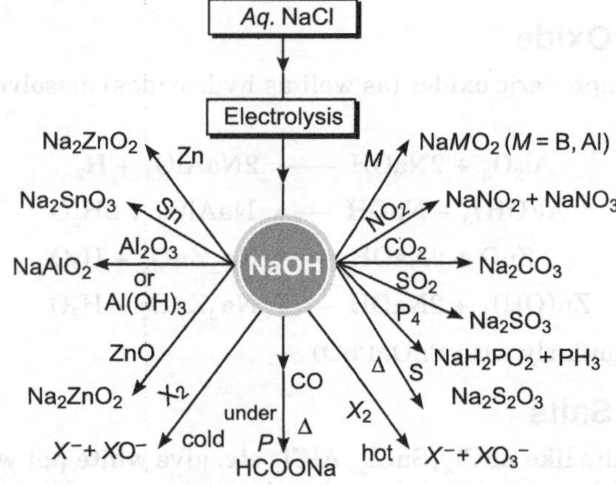

Microcosmic Salt

- Colourless crystals of microcosmic salt are obtained when NH_4Cl and Na_2HPO_4 in 1 : 1 molar ratio are dissolved in hot water and cooled.

$$NH_4Cl + Na_2HPO_4 \longrightarrow Na(NH_4)HPO_4 + NaCl$$

- On heating, $NaPO_3$ is formed, which forms coloured beads of orthophosphates with oxides of transition metal and cloudy bead with SiO_2.

$$Na(NH_4)HPO_4 \xrightarrow{\Delta} NaPO_3 + H_2O\uparrow + NH_3\uparrow$$

$$NaPO_3 + CuO \xrightarrow{\Delta} CuNaPO_4 \text{ (blue bead)}$$

$$NaPO_3 + CoO \xrightarrow{\Delta} CoNaPO_4 \text{ (blue bead)}$$

$$NaPO_3 + MnO \xrightarrow{\Delta} NaMnPO_4 \text{ (violet bead)}$$

$N_2 + H_2 \longrightarrow NH_3 \xrightarrow{HCl} NH_4Cl \xrightarrow{Na_2HPO_4} Na(NH_4)HPO_4 \xrightarrow{\Delta} NaPO_3$

$NaPO_3 \xrightarrow[\Delta]{CuO, O} CuNaPO_4$ blue

$NaPO_3 \xrightarrow[\Delta]{MnO} NaMnPO_4$ violet bead

$NaPO_3 \xrightarrow[\Delta]{CoO} CoNaPO_4$ blue bead

Sodium Oxide and Peroxide

- Sodium peroxide along with sodium oxide is formed when sodium burns in air or with $NaNO_2$.

$$2Na + O_2 \longrightarrow Na_2O_2$$

$$2Na + \frac{1}{2}O_2 \longrightarrow Na_2O$$

$$Na_2O_2 + 2Na \longrightarrow 2Na_2O$$

$$2NaNO_2 + 6Na \longrightarrow 4Na_2O + N_2$$

Oxone is a commercial name for the preparation of Na_2O_2 that is used in breathing apparatus.

- Na_2O is colourless ionic solid and aqueous solution is a strong base.

$$Na_2O + H_2O \longrightarrow 2NaOH$$

DEEP Focus

- Na_2O_2 is diamagnetic (all electrons paired) and is regarded as a salt of dibasic acid H_2O_2. O_2^{2-} is isoelectronic of F_2

$$Na_2O_2 + 2H_2O \longrightarrow 2NaOH + H_2O_2$$

- Na_2O_2 is colourless in pure form but is pale yellow in colour due to presence of sodium superoxide NaO_2 and is an oxidising agent.

– Al is oxidised to Al_2O_3.

$$3Na_2O_2 + 2Al \longrightarrow Al_2O_3 + 3Na_2O$$

– Cr^{3+} salt is oxidised to CrO_4^{2-} (yellow).

$$Na_2O_2 + 2H_2O \longrightarrow 2NaOH + H_2O_2$$

$$2Cr^{3+} + 3H_2O_2 + 10OH^- \longrightarrow \underset{\text{yellow}}{2CrO_4^{2-}} + 8H_2O$$

DEEP Focus

Because it reacts with CO_2 in the air, it has been used to purify air in submarines and confined spaces, as it both removes CO_2 and produces O_2.

$$Na_2O_2 + CO \longrightarrow Na_2CO_3$$

$$2Na_2O_2 + 2CO_2 \longrightarrow 2Na_2CO_3 + O_2$$

- $2Na_2O \xrightarrow{> 400°C} Na_2O_2 + 2Na$ (vapour)

In the absence of oxygen on oxidisable material, the peroxides (except Li_2O_2) are stable towards, thermal decomposition up to quite high temperature ($Na_2O_2 \approx 675°$ C).

- Na_2O_2 gives H_2O_2 with dil acids

$$Na_2O_2 + \underset{\text{dil}}{H_2SO_4} \longrightarrow H_2O_2 + Na_2SO_4$$

- O_2 is formed when Na_2O_2 reacts with hot acids.

$$Na_2O_2 + 4HCl\text{ (hot)} \longrightarrow 4NaCl + 2H_2O + O_2$$

- Na_2O_2 oxides $Mn(OH)_2$ to MnO_4^{2-} (green)

$$Mn(OH)_2 + 2Na_2O_2 \longrightarrow Na_2MnO_4 + 2NaOH$$

When mixture of $Ca(OH)_2$ and Na_2O_2 is compressed, CaO_2 is formed.

$$Na_2O_2 + Ca(OH)_2 \longrightarrow CaO_2 + 2NaOH$$

$Na_2O \xrightarrow{H_2O} NaOH$

$Na_2O \xrightarrow{Al_2O_3} NaAlO_2$

$Na_2O \xrightarrow{\text{above } 400°C} Na_2O_2 + Na$

$Na_2O \xrightarrow{ZnO} Na_2ZnO_2$

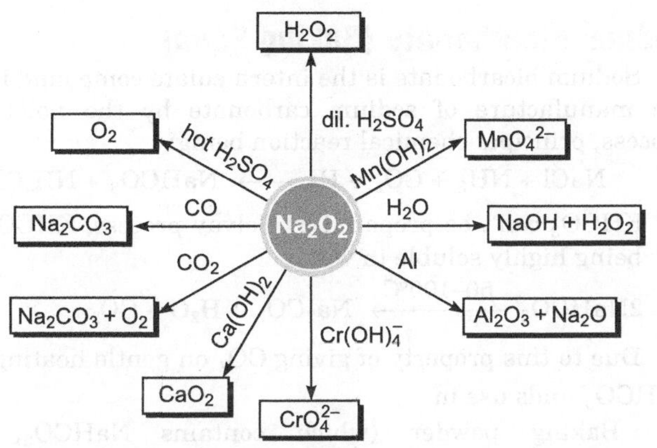

O_2^{2-} has following MO electronic configuration

$$(\sigma 1s)^2(\sigma * 1s)^2(\sigma 2s)^2(\sigma * 2s)^2(\sigma 2p_x)^2$$
$$(\pi 2p_y)^2(\pi 2p_z)^2(\pi * 2p_y)^2(\pi * 2p_z)^2$$

with bond-order one and diamagnetic (all electrons paired)

Sodium Sulphate

Anhydrous sodium sulphate (Na_2SO_4), called salt cake, is the by-product of HCl industry

$$2NaCl + H_2SO_4 \longrightarrow Na_2SO_4 + 2HCl$$

- If aqueous solution of Na_2SO_4 is cooled below 32° C, Glauber's salt $Na_2SO_4 \cdot 10H_2O$ separates. However, if cooling is done below 12° C, $Na_2SO_4 \cdot 7H_2O$ crystals are formed.
- In dry air $Na_2SO_4 \cdot 10H_2O$ effloresces forming anhydrous Na_2SO_4

$$Na_2SO_4 \cdot 10H_2O \longrightarrow Na_2SO_4 + 10H_2O$$

- Na_2SO_4 finds use in paper industry, detergent and glass manufacture.

Sodium Carbonate

Since pre-historic times a natural deposit called **Trona**, $Na_2CO_3 \cdot NaHCO_3 \cdot 2H_2O$ (sesqui-carbonate) has been obtained from dried-up lake beds in Egypt. Trona is sometimes called sodium sesqui-carbonate (sesqui means one and a half) and on heating it, Na_2CO_3 is obtained.

$$2(Na_2CO_3 \cdot NaHCO_3 \cdot 2H_2O) \xrightarrow{\Delta} 3Na_2CO_3 + CO_2 + 5H_2O$$

The Solvay (ammonia-soda) process

- The Solvay (ammonia-soda) process is widely used to produce Na_2CO_3 synthetically after Leblanc process proved non-economical.

Reactions involved in the preparation of Na_2CO_3 by the Solvay process are :

$$NH_3 + H_2O + CO_2 \longrightarrow NH_4HCO_3$$
$$NaCl + NH_4HCO_3 \longrightarrow NaHCO_3 + NH_4Cl$$
$$2NaHCO_3 \xrightarrow{150°C} Na_2CO_3 + H_2O + CO_2$$

CO_2 is obtained by the decomposition of $CaCO_3$

$$CaCO_3 \xrightarrow{\Delta,\ 1100°C} CaO + CO_2$$

CaO (*aq*) can decompose NH_4Cl to form NH_3 which can be reused

$$Ca(OH)_2 + 2NH_4Cl \longrightarrow 2NH_3 + 2H_2O + CaCl_2$$

Thus, NH_3 is required in first instance only; it is recycled in further cycles. It is shown in the (Fig. 8.6 – 8.8).

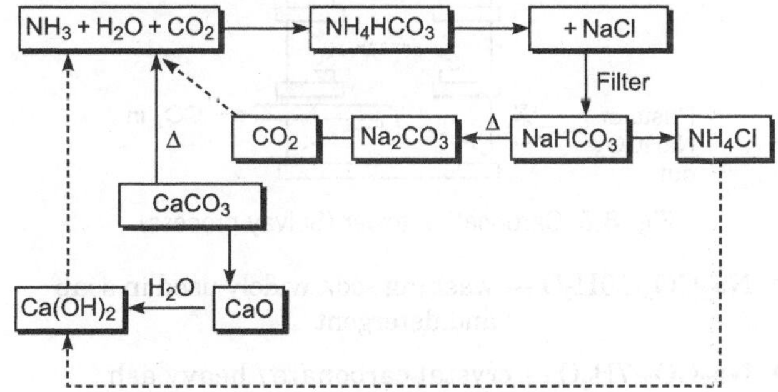

Fig. 8.6 Solvay process for the manufacture of Na_2CO_3 (...) represents recycled process.

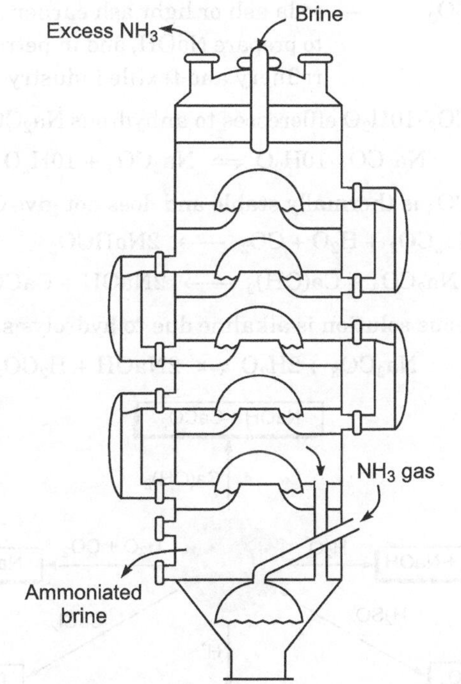

Fig. 8.7 The ammoniation of brine solution (Solvay process)

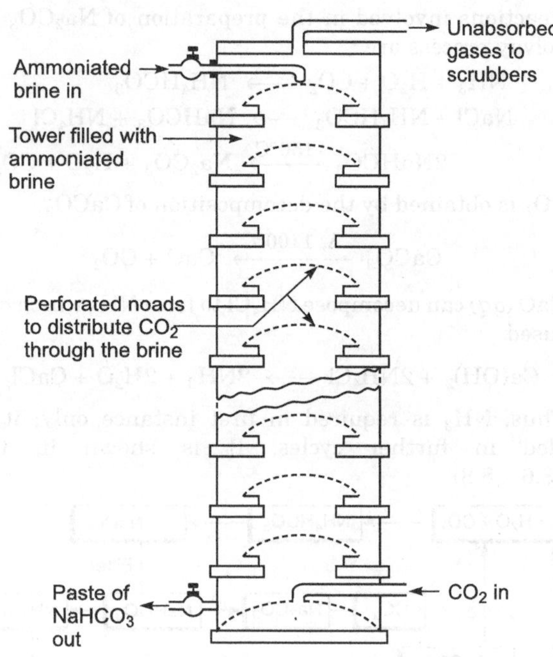

Fig. 8.8 Carbonating tower (Solvay process)

- $Na_2CO_3 \cdot 10H_2O$ — washing soda widely used in soap and detergent
- $Na_2CO_3 \cdot 7H_2O$ — crystal carbonate / heavy ash $Na_2CO_3 \cdot H_2O$ widely used in glass sheets and glass-fibres manufacturing
- Na_2CO_3 — soda ash or light ash earlier used to prepare NaOH, and in petroleum refinery and textile industry
- $Na_2CO_3 \cdot 10H_2O$ effloresces to anhydrous Na_2CO_3.

$$Na_2CO_3 \cdot 10H_2O \rightleftharpoons Na_2CO_3 + 10H_2O$$

- Na_2CO_3 is thermally stable and does not give CO_2.

$$Na_2CO_3 + H_2O + CO_2 \longrightarrow 2NaHCO_3$$
$$Na_2CO_3 + Ca(OH)_2 \longrightarrow 2NaOH + CaCO_3$$

- Aqueous solution is alkaline due to hydrolysis.

$$Na_2CO_3 + 2H_2O \rightleftharpoons 2NaOH + H_2CO_3$$

$NaOH + CaCO_3$
$Ca(OH)_2$
$H_2CO_3 + NaOH$ — H_2O — Na_2CO_3 — $H_2O + CO_2$ — $NaHCO_3$
H_2SO_4 — Na_2SO_4
H^+ — CO_2
$CaCl_2$ — $CaCO_3$

Sodium Bicarbonate (Baking Soda)

Sodium bicarbonate is the intermediate compound in the manufacture of sodium carbonate by the Solvay process, principal chemical reaction being:

$$NaCl + NH_3 + CO_2 + H_2 \longrightarrow NaHCO_3 + NH_4Cl$$

- $KHCO_3$ can't be prepared by Solvay process, $KHCO_3$ being highly soluble in water.
- $2NaHCO_3 \xrightarrow{50\text{–}100°C} Na_2CO_3 + H_2O + CO_2$

Due to this property of giving CO_2 on gentle heating, $NaHCO_3$ finds use in

- Baking powder (which contains $NaHCO_3$, $Ca(H_2PO_4)_2$ and starch). Improved baking powder contains 40% starch, 30% $NaHCO_3$, 20% $NaAl(SO_4)_2$ and 10% $Ca(H_2PO_4)_2$.
- Pharmaceutical products (including antacid preparation for indigestion)
- Fire extinguishers.
- $NaHCO_3$ is also added in small amount to table salt (NaCl). NaCl is often contaminated with small amount of $MgCl_2$ which being hygroscopic make NaCl moist. Adding $NaHCO_3$ converts $MgCl_2$ to $MgCO_3$ which is non-hygroscopic salt.

$$MgCl_2(s) + 2NaHCO_3(s) \longrightarrow MgCO_3(s) + 2NaCl(s) + H_2O(l) + CO_2(g)$$

- It is amphiprotic: $HCO_3^- + H^+ \rightleftharpoons H_2CO_3$

$$HCO_3^- \rightleftharpoons H^+ + CO_3^{2-}$$

Amphiprotic species is H^+ ion donor (acid) and H^+ ion acceptor (base).

Compounds of Potassium

Potassium compounds occur in the earth's crust especially as silicates. All land plants contain considerable amounts of potassium salts due to absorption from the soil. Deficiency of the potassium salts in soil is made up by adding potassium fertilizers.

Potassium can also be made by electrolysis of molten KCl. But molten potassium is soluble in molten KCl and thus separation of metal is difficult. The preferred method for preparation of potassium uses the reaction of sodium vapour with molten KCl, with potassium being continually removed from the equilibrium mixture

$$Na(g) + KCl(l) \rightleftharpoons K(l) + NaCl(l)$$

Potassium Superoxide

- It is formed as a chrome-yellow (orange) powder by burning potassium in excess of moisture-free oxygen or air.

$$K + O_2 \longrightarrow KO_2$$

- KO_2 is better (than K_2O_2) oxidising agent since it produces O_2 and H_2O_2 both in aqueous or acidic solution.

$$2KO_2 + 2H_2O \longrightarrow 2KOH + H_2O_2 + O_2$$
$$2KO_2 + H_2SO_4 \longrightarrow K_2SO_4 + H_2O_2 + O_2$$
$$2KO_2 + CO \longrightarrow K_2CO_3 + O_2$$
$$4KO_2 + 2CO_2 \longrightarrow 2K_2CO_3 + 3O_2$$
$$4KO_2 + 4CO_2 + 2H_2O \longrightarrow 4KHCO_3 + 3O_2$$

KO_2 thus finds applications to be used in space capsules, submarines and breathing masks, because it both produces O_2 and removes CO_2 and CO. Both functions are important in life support system.

- The superoxide contains O_2^- ion, which has an unpaired electron and hence, it is paramagnetic. Bond order is 1.5

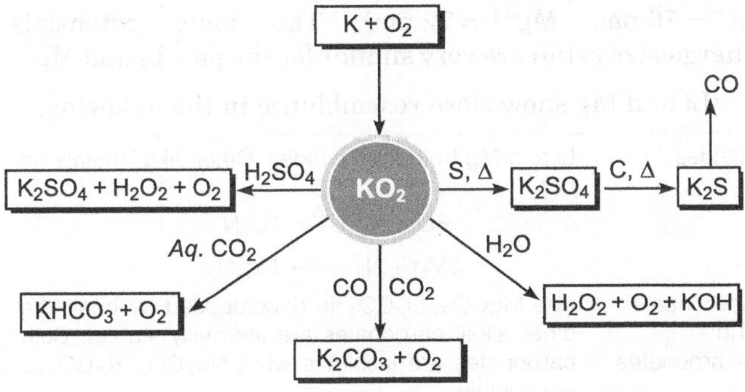

Potassium Carbonate (Potash or Pearl Ash)

- **Solvay process is not adopted to prepare $KHCO_3$ (or K_2CO_3) being highly water soluble.**
- Leblanc process involves conversion of KCl (from carnallite) first into K_2SO_4 and then heating it (K_2SO_4) with carbon and limestone ($CaCO_3$).

$$KCl + H_2SO_4 \longrightarrow KHSO_4 + HCl$$
$$KHSO_4 + KCl \longrightarrow K_2SO_4 + HCl$$
$$K_2SO_4 + 2C + CaCO_3 \xrightarrow{\Delta} K_2CO_3 + CaS + 2CO_2\uparrow$$

- Magnesia process (also called Precht's process) involves passing of CO_2 into aqueous solution of KCl and $MgCO_3{\cdot}3H_2O$ at 17°C, when a double salt $KHCO_3{\cdot}MgCO_3{\cdot}4H_2O$ is precipitated.

$$KCl + 3(MgCO_3{\cdot}3H_2O) + CO_2 \longrightarrow \underset{\text{white ppt. }(A)}{2(KHCO_3{\cdot}MgCO_3{\cdot}4H_2O)} + MgCl_2$$

- Precipitate (*A*) on heating under pressure gives a mixture of K_2CO_3 (soluble) and $MgCO_3$ (insoluble) and thus separated. $MgCO_3$ is reused in further preparation.

$$2(KHCO_3{\cdot}MgCO_3{\cdot}4H_2O) \xrightarrow[\text{under pressure}]{\Delta} K_2CO_3 + 2MgCO_3 + CO_2 + 9H_2O$$

- Colourless deliquescent powder soluble in water (more than Na_2CO_3 and Li_2CO_3).
- Aqueous solution is alkaline due to hydrolysis.

$$K_2CO_3 + 2H_2O \longrightarrow 2KOH + H_2CO_3$$
$$K_2CO_3 + H_2O\text{ (steam)} \xrightarrow{\Delta} 2KOH + CO_2\uparrow$$

K_2CO_3 finds wide use in the manufacture of soft soap, and is washing wool.

Potassium Sulphate

- **A naturally occurring mineral, Schonite — $K_2SO_4{\cdot}MgSO_4{\cdot}6H_2O$ is treated with KCl in aqueous solution and colourless anhydrous crystals of K_2SO_4 separate.**

$$K_2SO_4{\cdot}MgSO_4\cdot 6H_2O + 2KCl \longrightarrow 2K_2SO_4 + MgCl_2 + 6H_2O$$

- It is highly soluble in H_2O and forms double salts like potash alum

$$K_2SO_4 + Al_2(SO_4)_3 + (aq) \longrightarrow \underset{\substack{\text{potash alum}\\\text{(potassium aluminium sulphate)}}}{K_2SO_4\cdot Al_2(SO_4)_3{\cdot}24H_2O}$$

$$K_2SO_4 + 4C \xrightarrow{\Delta} K_2S + 4CO\uparrow$$

K + O2 → KO2 —(S, Δ)→ K2SO4 —(aq Al2(SO4)3)→ Alum

K2SO4 —(BaCl2)→ BaSO4

K2SO4 —(C, Δ)→ K2S

Potassium Nitrate

$NaNO_3$ (chile saltpeter) is converted to KNO_3 by following exchange reaction :

$$NaNO_3(aq) + KCl(aq) \rightleftharpoons KNO_3(aq) + NaCl(s)$$

Above reaction is spontaneous in forward side as, NaCl being less soluble than KNO_3, is removed by crystallisation, and KNO_3 remains in solution, and can be obtained by evaporation of water. KNO_3 is used in gun powder (a mixture of sulphur, carbon and KNO_3)

$$2KNO_3(s) + 4C(s) \longrightarrow K_2CO_3(s) + 3CO(g) + N_2(g)$$
$$2KNO_3(s) + 2S(s) \longrightarrow K_2SO_4(s) + SO_2(g) + N_2(g)$$

Diagonal Relationship
—Anomalous Behaviour of Lithium

The first element of a group in many ways differs from the other heavier members of the group. These differences arise on account of three factors :

(a) the small size
(b) the high electronegativity, and
(c) the unavailability of *d*-orbitals among the first row elements (Li to F).

Its polarising power (*i.e.*, charge/radius ratio) is the greatest of all the alkali metal ions. This results in increased covalent character of lithium compounds which is responsible for their solubility in organic solvents.

Due to the very small size of lithium, the metallic bonding between the atoms in the metallic lattice is very strong giving rise to strong cohesive forces. This is shown in its relatively higher melting and boiling points, hardness and homonuclear bond energy.

The relatively higher attraction of lithium for its outer electron results in its relatively higher electronegativity, ionisation energy, hydration energy, electron affinity and of course smaller atomic radii relative to the other homologues.

Similar anomalies are also found in the chemical properties but the differences appear greater.

- Lithium salts of large polarisable anions are thermally less stable than those of other alkali metals, *e.g.*, lithium carbonate decomposes at 950 K, whereas no decomposition of sodium carbonate take place above 1050 K.
- Lithium forms no solid bicarbonate, triiodide and superoxide as these are unstable at room temperature, whereas those of other alkali metals require a higher temperature to affect their decomposition.
- Lithium salts of anions of high charge density are less soluble than those of other alkali metals. The halides of lithium are more covalent than the other halides and are more soluble in organic solvents.
- Lithium forms stable salts with anions of high charge density owing to their high lattice energy. For example, in air, lithium forms the normal oxide, whereas the others form higher oxides. Lithium reacts with nitrogen to form nitride, Li_3N, the others do not react. Lithium hydride is more stable than the other hydrides and lithium carbide is formed more easily with acetylene.
- Lithium reacts slowly with water.
- Lithium forms more stable covalent bonds than other alkali metals and, therefore, forms more stable complex compounds. For example, lithium cannot be recovered unchanged from its liquid ammonia solution, owing to the formation of $[Li(NH_3)_4]^+$.

The first element of group often shows resemblance to the second element of the neighbouring group on the right. This type of behaviour is known as **diagonal relationship.**

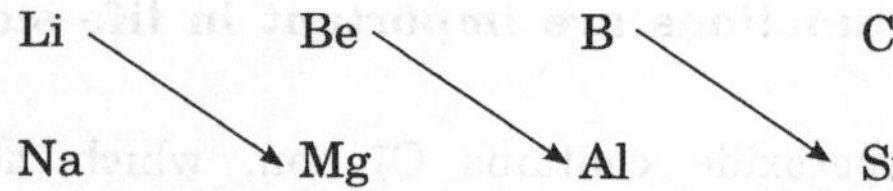

This similarity between Li and Mg is particularly striking and arises because of their similar ionic sizes (Li^+ = 76 pm; Mg^{2+} = 72 pm). The ionic potentials (charge/size ratio) are very similar for the pair Li and Mg.

Li and Mg show close resemblance in the following:

Nitrides	Li and Mg both form nitrides. Other alkali metals do not. $6Li + N_2 \xrightarrow{\Delta} 2Li_3N$ $3Mg + N_2 \xrightarrow{\Delta} Mg_3N_2$
Carbonates and bicarbonates	Like $MgCO_3$, Li_2CO_3 is decomposed by heat (the other alkali carbonates are thermally stable). Both carbonates are insoluble while Na_2CO_3, K_2CO_3,... are soluble. $MgCO_3 \xrightarrow{\Delta} MgO + CO_2$ $Li_2CO_3 \longrightarrow Li_2O + CO_2$ Solid bicarbonates of lithium and magnesium do not exist
Nitrates	$LiNO_3$ decomposes to give Li_2O like $Mg(NO_3)_2$, but other alkali metal nitrates give nitrite. $Mg(NO_3)_2 \xrightarrow{\Delta} MgO + 2NO_2 + \frac{1}{2}O_2$ $2LiNO_3 \xrightarrow{\Delta} Li_2O + 2NO_2 + \frac{1}{2}O_2$ $2NaNO_3 \longrightarrow 2NaNO_2 + O_2$
Oxides	Both give their normal oxides, Li_2O, MgO, when they burn in oxygen; peroxides and superoxides are not formed. Na forms peroxide, Na_2O_2 while K forms superoxide, KO_2.
Chlorides	Both chlorides are hydrated, $LiCl{\cdot}2H_2O$, $MgCl_2{\cdot}6H_2O$, both are deliquescent and anhydrous salts are soluble in ether.
Phosphates	Solubility of both Li and Mg phosphate is low.
Fluorides	LiF and MgF_2 are sparingly soluble in water.
Hydration	Both Li^+ and Mg^{2+} are heavily hydrated.
Hydroxide	LiOH, $Mg(OH)_2$ both are weak bases and very slightly soluble in water, and their hydroxides decompose on heating.
Perchlorates	Lithium perchlorate ($LiClO_4$) and magnesium perchlorate [$Mg(ClO_4)_2$] both are extremely soluble in alcohol and used as water absorber.
Alkyl and aryl	Li alkyls and aryls are important in organic syntheses and Grignard reagent (*R*MgX)

Uses of Alkali Salts

- Lithium stearate is used as a thickener and gelling agent to transform oils into lubricating greases.
- Lithium hydride (LiH) is used to generate hydrogen in military, meterological, and other applications.
- $LiAlH_4$ is a reducing agent in organic synthesis.

$$\left.\begin{array}{l} RCOOR' \\ RCHO \\ RCOOH \\ RCOCl \\ (RCO)_2O \end{array}\right] \xrightarrow{LiAlH_4} RCH_2OH$$

- Li_2CO_3 when taken orally provides an effective treatment for manic-depressive psychoses.
- Li/FeS_x battery systems are emerging as a potentially viable energy storage system and as a non-polluting silent source of power for electric cars.
- NaOH, Na_2CO_3 find applications in soap, paper and glass industry and as reagent in the laboratory.
- K_2CO_3 is used chiefly in high-quality decorative glassware, in optical lenses, colour TV tubes and fluorescent lamps; it is also used in china wares, textile dyes and pigments.
- KNO_3 is a powerful oxidising agent and mainly used in gunpowders and pyrotechnics, and in fertilizers.
- $KMnO_4$ is used as an oxidiser, decolourisers, bleacher and purification agent; its major application is in the manufacture of saccharin.
- KO_2 is used in breathing apparatus as an auxilliary supply of O_2 in mines, submarines and space vehicles.

$$4KO_2 + 2CO_2 \longrightarrow 2K_2CO_3 + 3O_2$$
$$4KO_2 + 4CO_2 + 2H_2O \longrightarrow 4KHCO_3 + 3O_2$$

- $KClO_3$ is used in matches and explosive.
- KBr in photography and as a source of Br_2 in organic synthesis.
- Na_2O_2 finds widespread use as a bleaching agent for fabrics, paper pulp wood etc.
- Li is used in batteries and alloys.
- K is used for analytical reagents because K^+ salts not so highly solvated as those of Na^+, making it easier to obtain anhydrous reagents.
- Cs is used in photocells because of its low ionisation energy.

Target Practice 2

1. Mixture containing one mole Na_2CO_3 and one mole $NaHCO_3$ is heated. What is amount (in moles) of the residue? ....
2. Table salt is always contaminated with $NaHCO_3$. It is to ....
3. NaOH is prepared industrially by electrolysis of brine (NaCl)

$$2NaCl(aq) + 2H_2O(l) \longrightarrow Cl_2(g) + 2NaOH(aq) + H_2(g)$$

What is the name of this process? ....

4. In United States, sodium carbonate is obtained from naturally occurring deposits of the mineral trona. Write formula of trona. ....
5. How can you stabilise solutions containing Na^+ ? ....
6. Name oxides of alkali metals which are
   (i) paramagnetic in nature ....
   (ii) binary salt ....
   (iii) used in life saving devices ....
   (iv) diamagnetic ....

Answers

1. $\underset{1\text{ mol}}{Na_2CO_3} \xrightarrow{\text{heat}}$ no effect. Thus, residue is 1 mol Na_2CO_3

$$\underset{\substack{2\text{ mol}\\1\text{ mol}}}{2NaHCO_3} \xrightarrow{\text{heat}} \underset{\substack{1\text{ mol}\\0.5\text{ mol}}}{Na_2CO_3} + H_2O\uparrow + CO_2\uparrow$$

Residue is 0.5 mol Na_2CO_3. Thus total residue is 1.5 mol Na_2CO_3.

2. Table salt NaCl has also $MgCl_2$ as impurity (from sea). The magnesium salt is hygroscopic; that is it picks water up from the air and thus NaCl becomes wet. If $NaHCO_3$ is added to table salt, it converts $MgCl_2$ to $MgCO_3$ which is non-hygroscopic salt.

$$MgCl_2 + 2NaHCO_3 \longrightarrow MgCO_3 + NaCl + H_2O + CO_2$$

3. Soda-lime process
4. $Na_2CO_3 \cdot NaHCO_3 \cdot 2H_2O$
5. Larger anion as O_2^- (superoxides) can stabilise Na^+.
   NaO_2 is stable.
6. (a) KO_2 (due to O_2^-)
   (b) KO_2
   (c) Na_2O_2
   (d) Na_2O_2

Practice Exercise 2

1. What is the nature of CO when it reacts with NaOH?
2. Gas (A) + NaOH $\longrightarrow$ (B) $\xrightarrow{\Delta}$ (C) $\xrightarrow{H^+}$ (D)

   (B) $\xrightarrow{CaCl_2}$ white ppt (E)

   (C) and (D) both decolourise acidified $KMnO_4$.
   Identify (A), (B), (C) and (D) and explain reactions.
3. (A) gives white ppt with $AgNO_3$ soluble in *aq* NH_3. (A) gives green ppt (B) with NaOH but ppt dissolves to yellow solution (C) when heated with H_2O_2. (C) gives yellow ppt (D) with $BaCl_2$. What are (A), (B), (C) and (D)?
4. (A) gives white ppt (B) with NaOH, but (B) dissolves in excess of NaOH forming (C). (C) gives white ppt (D) when H_2S gas is passed into it. (A) also gives white ppt (E) with $BaCl_2$. (E) is insoluble in conc. HNO_3. Identify (A), (B), (C), (D) and (E).
5. PbO_2 is soluble in NaOH and also in HCl. What does it reflect about the nature of PbO_2?
6. A mixture contains Cr^{3+} and Zn^{2+} salts. How will you separate and analyse them?
7. Colourless salt (A) $\xrightarrow{NaOH,\ \Delta}$ (B) (gas)

   (A) $\xrightarrow{CaCl_2 \text{ solution}}$ white ppt (C)

   (B) gives white fumes with HCl.
   (C) decolorises acidified $KMnO_4$.
   What is (A)? Explain reactions.
8. If NaOH solution is placed in an open vessel, will there be any change in its weight? Explain.
9. What is difference in the chemistry of microcosmic bead test and borax bead test?
10. $H_3PO_4 + 2NaOH \longrightarrow$ (A) $\xrightarrow{NH_4Cl}$ (B) $\longrightarrow$ (C) $\xrightarrow{(E)\ \Delta}$ Violet bead

    (E) is obtained along with gas (F) when (D) is heated. Gas (F) turns lime water milky and $K_2Cr_2O_7/H^+$ solution green.
    What are (A), (B), (C), (D), (E) and (F) and explain reactions.
11. Sodium oxide solution can't be stored in Zn or Al vessel. Explain.
12. There is white deposit in a glass vessel in which NaOH is stored. Explain.
13. What do you consider about Na_2O_2, a salt of monobasic acid or a salt of dibasic acid?
14. How will you make distinction between Na_2O_2 and Na_2O?
15. Sodium forms two oxides (A) and (B) having 25.81% and 41.03% oxygen respectively. What is reaction of (A) and (B) with
    (a) H_2O (b) H_2SO_4 (c) $Cr(OH)_3$ (d) MnO_2
16. Sodium superoxide NaO_2 can't be prepared directly from sodium, but from Na_2O_2. Write balanced chemical equation. How much of Na_2O_2 is required to prepare 11 g of NaO_2?
17. Fill in the blanks :
    1. Na_2O has .... structure.
    2. Bond orders of O_2^{2-}, O_2^-, O_2 are respectively ...., ..., and ....
    3. Na_2O_2 is ....absorber and .... producer and is thus used in ....
18. What happens when equimolar quantity of Na_2SO_4 and conc. H_2SO_4 is cooled?
19. Na_2SO_4 + carbon $\longrightarrow$ (A) $\xrightarrow{Na_2[Fe(CN)_5NO]}$ (B) (purple)

    (A) $\xrightarrow{CdCO_3}$ yellow ppt (C)

    Identify (A), (B) and (C).
20. A colourless salt (X) has 50% Na_2SO_3 and 50% H_2O. What is the formula of (X)? How much of SO_2 at NTP is obtained when 2.52 g of (X) reacts with excess of dil. H_2SO_4?
21. Na_2CO_3 is widely used in softening of hard water. If 1 L of hard water required 0.0106 g of Na_2CO_3, calculate hardness in ppm (parts per million *i.e.*, 10^6 mL) of $CaCO_3$.
22. Silver can be extracted from silver coin and Na_2CO_3 is used in one of the steps. Outline the extraction process.
23. 3 moles of mixture of $NaHCO_3$ and Na_2CO_3 is strongly heated at 200°C, when 1 mole of CO_2 is obtained. What is the weight of residue? What is molar percentage of $NaHCO_3$ and Na_2CO_3 in the mixture?
24. Baking powder is said to contain $NaHCO_3$, $Ca(H_2PO_4)_2$, starch and $NaAl(SO_4)_2$. What is the function of each content?
25. Fire extinguisher contains $NaHCO_3$ and the compound (X). What is the compound (X)? Explain.
26. Arrange the following in increasing solubility in water.
    (i) LiOH, NaOH, KOH
    (ii) Li_2CO_3, Na_2CO_3, K_2CO_3
    (iii) $LiHCO_3$, $NaHCO_3$, $KHCO_3$
    (iv) LiCl, NaCl, KCl
27. What is the formula of the three oxides of potassium with percentage of oxygen being
    (A) : 17.02% (B) : 29.09%
    (C) : 45.07%
    Which of them will show paramagnetism?
28. Which is most thermally stable out of Li_2CO_3, $MgCO_3$ and K_2CO_3?
29. 1 mole of mixture of Li_2CO_3 and K_2CO_3 on strongly heating gave 0.25 mole of CO_2. What is mole % of each in the mixture?
30. Which indicator, phenolphthalein or methyl orange will you choose in the titration of K_2CO_3 solution with HCl?
31. Identify (A) to (C) in the following :

    $KO_2 + S \xrightarrow{\Delta}$ (A) $\xrightarrow{BaCl_2}$ (B)

    (A) $\xrightarrow{Al_2(SO_4)_3 \text{ solution}}$ (C)
32. 10 mL of a solution of K_2SO_4 on reaction with $BaCl_2$ gave 0.233 g of white ppt. What is concentration of K_2SO_4 solution?
    (Ba = 137, S = 32, O = 16, K = 39)

33. A gaseous mixture contains N_2, O_2 and CO_2. It is passed over heated Mg, then over heated Cu and finally into aqueous KOH. What happens?

34. $Li + N_2 \xrightarrow{\Delta} (A) \xrightarrow{H_2O} (B) + (C)$ gas

$(A) \xrightarrow{\Delta} (D) + (E)$ gas

What are (*A*), (*B*), (*C*), (*D*) and (*E*)?

35. Compare Li^+ and Na^+ salts in terms of
(a) formation of ionic carbide
(b) complex formation
(c) solubility of CO_3^{2-}, PO_4^{3-} and F^- in H_2O
(d) nature of Cl^-
(e) nature of bicarbonate

36. Sodium does not form carbide Na_2C_2 with carbon but Na_2C_2 can be formed by other source. What can be it?

37. Which is more stable out of
(i) Li_2O and Li_2CO_3
(ii) Na_2O or Na_2CO_3

Answers

1. $\underset{\text{acid}}{CO} + \underset{\text{base}}{NaOH} \longrightarrow HCOONa$

CO is acidic oxide in this reaction.

2. (*A*) : CO (*B*) : HCOONa (sodium formate)

(*C*) : $\begin{matrix} COONa \\ | \\ COONa \end{matrix}$ (sodium oxalate)

(*D*) : (oxalic acid) (*E*) : CaC_2O_4 (calcium oxalate)

$$\underset{(A)}{CO} + NaOH \longrightarrow \underset{(B)}{HCOONa} \xrightarrow[-H_2]{\Delta} \underset{(C)}{(COONa)_2} \xrightarrow{H^+} \underset{(D)}{(COOH)_2}$$

$$2MnO_4^- + 5C_2O_4^{2-} + 16H^+ \longrightarrow 10CO_2 + 2Mn^{2+} + 8H_2O$$

3. (*A*) : $CrCl_3$ (*B*) : $Cr(OH)_3$ (*C*) : Na_2CrO_4 (*D*) : $BaCrO_4$

4. (*A*) : $ZnSO_4$ (*B*) : $Zn(OH)_2$ (*C*) : Na_2ZnO_2 (*D*) : ZnS (*E*) : $BaSO_4$

5. PbO_2 is an amphoteric oxide

$$PbO_2 + 2NaOH \longrightarrow \underset{\text{sodium plumbate (IV)}}{Na_2PbO_3} + H_2O$$

$$PbO_2 + 4HCl \longrightarrow 2H_2O + PbCl_4$$

6. $Cr^{3+} + Zn^{2+} \xrightarrow[\text{(ii) filter}]{\text{(i) NaOH, excess}}$ green ppt. $Cr(OH)_3\downarrow$

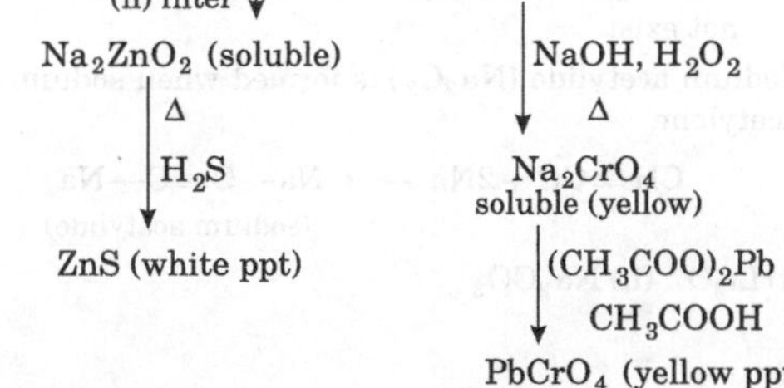

7. (*A*) : $(NH_4)_2C_2O_4$ (*B*) : NH_3 (*C*) : CaC_2O_4

8. NaOH absorbs CO_2 from atmosphere hence weight of the solution would increase.

9. $Na_2B_4O_7 \xrightarrow{\Delta} 2NaBO_2 + B_2O_3$

$B_2O_3 + CuO \longrightarrow Cu(BO_2)_2$

In the case of borax-bead test we take borax when coloured bead of the type $Cu(BO_2)_2$ is formed.

In case of microcosmic bead test, we take microcosmic salt.

10. (*A*) : Na_2HPO_4 (*B*) : $Na(NH_4)HPO_4$ (*C*) : $NaPO_3$ (*D*) : $MnSO_3$ (*E*) : MnO (*F*) : SO_2

$$Ca(OH)_2 + SO_2 \longrightarrow \underset{\text{milky}}{CaSO_3\downarrow}$$

$$Cr_2O_7^{2-}/H^+ + SO_2 \longrightarrow \underset{\text{green}}{Cr^{3+}}$$

11. Zn or Al reacts with Na_2O (*aq*), (actually with NaOH)

$$Zn + Na_2O + H_2O \longrightarrow Na_2ZnO_2 + H_2$$

$$2Al + Na_2O + 3H_2O \longrightarrow 2NaAlO_2 + 3H_2$$

12. Glass has SiO_2. SiO_2 reacts with NaOH.

$$SiO_2 + 2NaOH \longrightarrow Na_2SiO_3 + H_2O$$

Na_2SiO_3 appears as white deposit.

13. Na_2O_2 is a salt of dibasic acid H_2O_2 which is formed on reaction with H_2O.

$$Na_2O_2 + 2H_2O \longrightarrow 2NaOH + H_2O_2$$

14. Na_2O does not show oxidising properties while Na_2O_2 gives

$$Na_2O + H_2O \longrightarrow 2NaOH$$

$$Na_2O_2 + 2H_2O \longrightarrow 2NaOH + H_2O_2$$

Add Cr^{3+} salt into aqueous solution of each

—Na_2O gives green ppt of $Cr(OH)_3$

$$Cr^{3+} + Na_2O + H_2O \longrightarrow \underset{\text{green}}{Cr(OH)_3\downarrow}$$

—Na_2O_2 oxidises Cr^{3+} into CrO_4^{2-} (yellow)

$$Na_2O_2 + H_2O + Cr^{3+} \longrightarrow \underset{\text{yellow}}{CrO_4^{2-}}$$

15.

	A		*B*	
	O	Na	O	Na
	25.81	74.19	41.03	58.97
Mol ratio	1.61	3.22	2.56	2.56
	1 : 2		1 : 1	

A	*B*
Na_2O	Na_2O_2
(a) $Na_2O + H_2O \longrightarrow 2NaOH$	(a) $Na_2O_2 + 2H_2O \longrightarrow 2NaOH + H_2O_2$
(b) $Na_2O + H_2SO_4 \longrightarrow 2Na_2SO_4 + H_2O$	(b) $Na_2O_2 + H_2SO_4 \longrightarrow Na_2SO_4 + H_2O_2$
(c) $Na_2O + Cr(OH)_3 \longrightarrow$ no effect	(c) $Cr(OH)_3 + Na_2O_2 + H_2O \longrightarrow \underset{\text{(yellow)}}{Na_2CrO_4}$
(d) $Na_2O + MnO_2 \longrightarrow$ no effect	(d) $Na_2O_2 + MnO_2 + H_2O \longrightarrow \underset{\text{(green)}}{Na_2MnO_4}$
	Na_2O_2 is oxidant in (c) and (d) due to H_2O_2.

16. $$\underset{\substack{78\text{ g} \\ 7.8\text{ g }NaO_2}}{Na_2O_2} + O_2 \xrightarrow[300\text{ atm}]{450°C} \underset{\substack{2 \times 55\text{ g} \\ 11\text{ g}}}{2NaO_2}$$

17. (1) antifluorite (2) 1, 1.5 and 2 (3) CO_2, O_2, submarines

18. Prism type crystals of $NaHSO_4$ are formed.

$$Na_2SO_4 + H_2SO_4 \longrightarrow 2NaHSO_4$$

19. (A) : Na_2S (B) : $Na_2[Fe(CN)_5NOS]$ (C) : CdS

20. (X) : $Na_2SO_3 \cdot 7H_2O$

$$\underset{252\text{ g}}{Na_2SO_3 \cdot 7H_2O} + H_2SO_4 \longrightarrow Na_2SO_4 + \underset{\substack{22.4\text{ L}\\ \text{at NTP}}}{SO_2\uparrow} + 8H_2O$$

SO_2 : 0.224 L at NTP from 2.52 g $Na_2SO_3 \cdot 7H_2O$.

21. $CaCl_2 + Na_2CO_3 \longrightarrow CaCO_3\downarrow + 2NaCl$

106 g 100 g

0.0106 g 0.010 g in 1 L (1000 mL) water

$$\frac{0.016 \times 10^6}{10^3} \text{ g in } 10^6 \text{ mL } H_2O \text{ (ppm)}$$

10 ppm $CaCO_3$

22. Silver coin $\xrightarrow[HNO_3]{\text{boil with}}$ $AgNO_3 + Cu(NO_3)$ etc. $\xrightarrow{NaCl}$ $AgCl\downarrow$

(other remain soluble)

$\xrightarrow{Na_2CO_3} Ag_2CO_3\downarrow \xrightarrow{\Delta} Ag$

23. $\underset{2\text{ mol}}{2NaHCO_3} \xrightarrow{\Delta} \underset{1\text{ mol}}{Na_2CO_3} + \underset{1\text{ mol}}{H_2O\uparrow} + \underset{1\text{ mol}}{CO_2\uparrow}$

$Na_2CO_3 \xrightarrow{\Delta}$ no effect

CO_2 is due to $NaHCO_3$ only

1 mol $CO_2 \equiv$ 2 mol $NaHCO_3$

$NaHCO_3 = 66.67\%$ $Na_2CO_3 = 33.33\%$

residue = 1 mol Na_2CO_3 (in mixture)

+ 1 mol Na_2CO_3 (from 2 mol $NaHCO_3$)

= 2 mol Na_2CO_3 (H_2O and CO_2 escape at 200°C)

24. $NaHCO_3$] $Ca(H_2PO_4)_2$ is acidic in aqueous solution.

$Ca(H_2PO_4)_2$ acidic solution of $Ca(H_2PO_4)_2$ liberates CO_2 on reaction with $NaHCO_3$

starch] as a filler

$NaAl(SO_4)_2$] retarder so that CO_2 is given off more slowly while preparing cake or bread.

25. Fire extinguisher contains $NaHCO_3$ and compound *X* can be an acid (like CH_3COOH) that can decompose $NaHCO_3$ to give CO_2.

$$NaHCO_3 + \underset{(X)}{H^+} \longrightarrow Na^+ + H_2O + CO_2$$

$NaHCO_3$ and acid (*X*) are in a sealed flask (but placed inside separately). in case of emergency (when there is fire), seal is broken and *X* and $NaHCO_3$ react instantly to give CO_2 which can put off fire.

26. In all cases compound of Li^+, Na^+ and K^+ have solubility in the order :

$$Li^+ < Na^+ < K^+$$

27. (A) : K_2O (B) : K_2O_2 (C) : KO_2

KO_2 is paramagnetic (with one unpaired electron in O_2^-) while K_2O and K_2O_2 are diamagnetic.

28. K_2CO_3

29. Li_2CO_3 : 25% K_2CO_3 : 75% (CO_2 is due to Li_2CO_3 only

$$Li_2CO_3 \xrightarrow{\Delta} Li_2O + CO_2$$

$$K_2CO_3 \xrightarrow{\Delta} \text{no effect)}$$

30. Methyl orange; phenolphthalein ceases to indicate colour change after 50% of K_2CO_3 has reacted with acid.

$$K_2CO_3 + HCl \longrightarrow KHCO_3 + HCl$$

31. (A) : K_2SO_4 (B) : $BaSO_4$ (C) : $K_2SO_4 \cdot Al_2(SO_4)_3 \cdot 24H_2O$

32. 17.4 g L^{-1}

33. (a) Mg reacts with N_2 as well as O_2.

$$3Mg + N_2 \xrightarrow{\Delta} Mg_3N_2$$

(b) O_2 if left unreacted in (a) reacts with Cu

$$Cu + O_2 \xrightarrow{\Delta} 2CuO$$

(c) KOH absorbs CO_2

$$2KOH + CO_2 \longrightarrow K_2CO_3 + H_2O$$

34. (A) : Li_3N (B) : LiOH (C) : NH_3 (D) : Li (E) : N_2

35. (a) Li forms Li_2C_2, Na does not form carbide with carbon

(b) $[Li(NH_3)_4]I$ exists, Na does not form

(c) Li_2CO_3, Li_3PO_4, LiF are insoluble while sodium salts are soluble

(d) LiCl is covalent while NaCl is ionic

(e) $NaHCO_3$ is stable in solid and aqueous solution while $LiHCO_3$ exists in aqueous solution but solid $LiHCO_3$ does not exist.

36. Sodium acetylide (Na_2C_2) is formed when sodium reacts with acetylene

$$CH{\equiv}CH + 2Na \longrightarrow \underset{\text{(sodium acetylide)}}{Na{-}C{\equiv}C{-}Na}$$

37. (i) Li_2O (ii) Na_2CO_3

Additional Reactions and Facts of Alkali Metals IA

Reactions :

1. $2NaNO_3 \xrightarrow{500°C} 2NaNO_2 + O_2$
2. $4NaNO_3 \xrightarrow{800°C} 2Na_2O + 5O_2 + 2N_2$
3. $KNO_3 + Zn \xrightarrow{\Delta} KNO_2 + ZnO$
4. $2KOH + 4NO \longrightarrow 2KNO_2 + N_2O + H_2O$

   $4KOH + 6NO \longrightarrow 4KNO_2 + N_2 + 2H_2O$
5. $KI + I_2 \rightleftharpoons KI_3$
6. $KBr + ICl \rightleftharpoons K[BrICl]$
7. $KF + BrF_3 \rightleftharpoons K[BrF_4]$
8. $2[Co(NH_3)_6]Cl_3 + 6NaOH \longrightarrow 12NH_3 + Co_2O_3 + 3NaCl + 3H_2O$
9. $2Na_2O_2 + 2CO_2 \longrightarrow 2Na_2CO_3 + O_2$
10. $4KO_2 + 4CO_2 + 2H_2O \longrightarrow 4KHCO_3 + 3O_2$
11. $Na_2CO_3 + NO + NO_2 \longrightarrow 2NaNO_2 + CO_2$
12. $\underset{\text{cold}}{2NaOH} + Cl_2 \longrightarrow NaCl + \underset{\text{sodium hypochlorite}}{NaClO} + H_2O$
13. $\underset{\text{hot}}{6NaOH} + Cl_2 \longrightarrow 5NaCl + \underset{\text{sodium chlorate}}{NaClO_3} + 3H_2O$

14. $2CuSO_4 + 4KI \longrightarrow \underset{\text{white ppt}}{Cu_2I_2} + K_2SO_4 + I_2$

$$I_2 + KI \rightleftharpoons \underset{\text{(reddish brown)}}{KI_3}$$

More reactions of KI are given in halogen family of *p*-block elements.

15. $Mg_2C_3 + 4H_2O \longrightarrow 2Mg(OH)_2 + \underset{\text{propyne}}{CH_3—C{\equiv}CH}$

16. $NaF + LiI \xrightarrow{\Delta} NaI + LiF$

17. $>C—Cl + KF \longrightarrow\ >C—F + KCl$

18. $2BF_3 + 6NaH \xrightarrow{200°C} B_2H_6 + 6NaF$

19. $BF_3 + 4NaH \xrightarrow{Et_2O,\ 125°C} NaBH_4 + 3NaF$

20. $AlBr_3 + 4NaH \xrightarrow{Me_2O} NaAlH_4 + 3NaBr$

21. $2SO_2 + 2NaOH \longrightarrow \underset{\text{sodium dithionite}}{2Na_2S_2O_4} + H_2$

22. $CO_2 + NaH \longrightarrow \underset{\text{sodium formate}}{HCO_2Na}$

23. $TiCl_4 + NaH \xrightarrow{400°C} Ti + 4NaCl + H_2$

Important Facts of Alkali Metals

- Francium is radioactive; its longest-lived isotope ^{223}Fr has a half-life of only 21 min.
- Li and Mg, show diagonal relationship.
- Among the alkali metal K, Rb and Cs form superoxides MO_2 ; O_2^- is paramagnetic.
- Increasing stability of the peroxide or superoxide, as the size of the metal ion increases is due to the stabilisation of large anions by larger cation through lattice energy effects.
- Li forms Li_3N (nitrides).
- Li^+ being of smallest size has the highest hydration enthalpy which accounts for its high negative $E°$ value and its reducing power.
- Alkali metals dissolve in liquid NH_3; solutions are coloured due to solvated electrons. In dilute solution, ammoniated solution is paramagnetic but in concentrated solution, unpaired electrons get paired and it becomes diamagnetic.
- **$NaOH \cdot nH_2O$ (n = 1, 2, 3, 4, 5 and 7) has been prepared.**
- Li*X* have partial covalent character due to smaller size of cation.
- Carbonates and in most cases bicarbonates also (except that of Li) are highly stable to heat. Along a group, stability increases. $LiHCO_3$ does not exist in solid state.
- LiCl is deliquescent and crystallises as a hydrate $LiCl \cdot 2H_2O$.
- KO_2 is CO_2 and CO absorber, and O_2 emitter. This finds application in submarines, space capsules and breathing masks.
- NaOH is prepared from electrolysis of aqueous NaCl solution using
  - Castner Kellner cell
  - Diaphragm cell
  - Nafion membrane cell
- Na_2CO_3 is prepared by Solvay process.
- Alkali metals impart colour in flame.
- KNO_3 has been used for centuries as the oxidising agent in gun powder. A mixture of KNO_3, charcoal and sulphur will spontaneously react when ignited.
- Li_2CO_3 has been used for than 40 yr as a treatment for bipolar disorder.

Total Practice Set
(Read, Plan & Solve!)

Problem 1. On the basis of the following values of hydration energies and lattice energies, calculate heat of solution of LiCl, NaCl and KCl.

Hydration energy $\Delta H_{hydr.}$ (kJ mol^{-1})		Lattice energy ΔH_U (kJ mol^{-1})	
Li^+	−499	LiCl	−840
Na^+	−390	NaCl	−776
K^+	−305	KCl	−703
Cl^-	−382		

Solution $\Delta H_{solution} = \Delta H_{hydr.} - \Delta H_U$

A solute M^+X^- dissolves in water if $\Delta H_{solution}$ (heat of solution) is positive.

$$\Delta H_{hydr.}: \quad LiCl = -499 - 382 = -881 \text{ kJ mol}^{-1}$$
$$NaCl = -390 - 382 = -772 \text{ kJ mol}^{-1}$$
$$KCl = -305 - 382 = -687 \text{ kJ mol}^{-1}$$
$$\Delta H_{sol.}: \quad LiCl = -881 - (-840) = -41 \text{ kJ mol}^{-1}$$
$$NaCl = -772 - (-776) = +4 \text{ kJ mol}^{-1}$$
$$KCl = -687 - (-703) = +16 \text{ kJ mol}^{-1}$$

Problem 2. Which of the following methods would you use to extinguish a fire of lithium, sodium or potassium metals ? Explain why some of these are unsuitable, and give the reaction involved?

(a) Water (b) Nitrogen (c) Carbon dioxide
(d) Asbestos blanket

Solution (a) Water can't be used since metal catches fire forming hydroxide and hydrogen.

$$2M + 2H_2O \longrightarrow 2MOH + H_2$$

(b) Nitrogen forms nitride with Li.

$$6Li + N_2 \longrightarrow 2Li_3N$$

(c) If metal has converted to oxide or peroxide, CO_2 may react producing O_2 that would further worsen the situation.

$$2Na_2O_2 + 2CO_2 \longrightarrow 2Na_2CO_3 + O_2$$

(d) Asbestos blanket is best suited.

Problem 3. NaCl in earlier days used to manufacture NaOH and Cl_2 involving following steps. Identify (*A*) to (*I*) in the following :

$$NaCl + \text{conc. } H_2SO_4 \xrightarrow{\Delta} (A) + (B) \text{ (gas)}$$
$$(B)g \xrightarrow{MnO_2} (C) \text{ gas}$$
$$(A) + NaCl \xrightarrow{\Delta} (D) + (B) \text{ gas}$$
$$(D) + \text{carbon} + CaCO_3 \longrightarrow (E) + (F)$$
$$CaCO_3 \xrightarrow{\Delta} (G) + (H) \text{ gas}$$
$$(G) + H_2O \longrightarrow (I)$$
$$(I) + (E) \longrightarrow NaOH + CaCO_3$$

Solution $NaCl + \text{conc. } H_2SO_4 \longrightarrow \underset{(A)}{NaHSO_4} + \underset{(B)}{HCl}$

$$\underset{(B)}{HCl} \xrightarrow{MnO_2} Cl_2$$
$$MnO_2 + \underset{(B)}{4HCl} \xrightarrow{\Delta} MnCl_2 + \underset{(C)}{Cl_2} + 2H_2O$$
$$\underset{(A)}{NaHSO_4} + NaCl \xrightarrow{\Delta} \underset{(D)}{Na_2SO_4} + HCl$$
$$\underset{(D)}{Na_2SO_4} + \text{carbon} + CaCO_3 \xrightarrow{\Delta} \underset{(E)}{Na_2CO_3} + \underset{(F)}{CaSO_4}$$
$$CaCO_3 \xrightarrow{\Delta} \underset{(G)}{CaO} + \underset{(H)}{CO_2}$$
$$\underset{(G)}{CaO} + H_2O \longrightarrow \underset{(I)}{Ca(OH)_2}$$
$$\underset{(I)}{Ca(OH)_2} + \underset{(E)}{Na_2CO_3} \longrightarrow 2NaOH + CaCO_3$$

Problem 4. Propose a method of synthesising Na_2CO_3 from NaCl.

Solution

$$NaCl(s) + H_2SO_4 \text{ (conc.)} \xrightarrow{\Delta} NaHSO_4(s) + HCl(g)$$
$$NaCl(s) + NaHSO_4(s) \longrightarrow Na_2SO_4(s) + (HCl)(g)$$
$$Na_2SO_4(s) + 4C(s) \xrightarrow{\Delta} Na_2S(s) + 4CO(g)\uparrow$$
$$Na_2S(s) + CaCO_3(s) \xrightarrow{\Delta} CaS(s) + Na_2CO_3(s)$$

Problem 5. In some foam-type fire extinguishers the reactants are $Al_2(SO_4)_3$ (*aq*) and $NaHCO_3(aq)$. When the extinguisher is activated, these reactants are allowed to mix producing $Al(OH)_3(s)$ and $CO_2(g)$. The $Al(OH)_3 - CO_2$ foam extinguishes the fire. Write net ionic equation to represent this reaction.

Solution $Al_2(SO_4)_3$ solution is acidic due to hydrolysis.

$$Al_2(SO_4)_3 + 6H_2O \longrightarrow 2Al(OH)_3 + 3H_2SO_4$$
$$H_2SO_4 \longrightarrow 2H^+ + SO_4^{2-}$$
$$NaHCO_3 + H^+ \longrightarrow Na^+ + H_2O + CO_2$$
$$Al^{3+} + 3HCO_3^- \longrightarrow Al(OH)_3 + 3CO_2$$
$$\text{or } [Al(H_2O)_6]^{3+} + 3HCO_3^- \longrightarrow Al(OH)_3(s)\downarrow + 3CO_2(g) + 6H_2O$$

Problem 6. Mention various factors involved in the control of the pH of blood.

Solution Several factors are involved in the control of the pH of blood. A particularly important one is the ratio of dissolved HCO_3^- (from $NaHCO_3$) to H_2CO_3 (*aq*. CO_2, $CO_2 + H_2O$). CO_2 is soluble in water and in aqueous solution it reacts to a limited extent to produce H_2CO_3.

$$H_2O + CO_2 \rightleftharpoons H_2CO_3(aq)$$
$$H_2CO_3 + H_2O \rightleftharpoons H_3O^+ + HCO_3^- \quad K_1 = 4.2\times 10^{-7},\ pK_1 = 6.38$$
$$HCO_3^- + H_2O \rightleftharpoons H_3O^+ + CO_3^{2-} \quad K_2 = 5.6\times 10^{-11},\ pK_2 = 10.25$$

Since $K_2 <<< K_1$, hence K_1 is of importance and involved in the buffer. Then

$$pH = pK_1 + \log\frac{[HCO_3^-]}{[H_2CO_3]}$$

Generally $\frac{[HCO_3^-]}{[H_2CO_3]} = 10:1$ so that pH (blood) = 7.4

If lactic acid is produced due to exercise, HCO_3^- can be converted to H_2CO_3, but due to large ratio of $[HCO_3^-]/[H_2CO_3]$, pH remains almost constant.

To control alkalinity, extra H_2CO_3 is required. CO_2 (in the lungs) can be absorbed to keep H_2CO_3 in the blood constant and thus its pH.

Problem 7. Explain the following:

(a) Why should an alkali never be put into a burette?
(b) Large quantities of rock salt are used on roads in winter (in cold countries).
(c) Why is it that lithium salts have a greater degree of covalent character than other halides of the group?
(d) Why is it that although Li^+ is far smaller than the other metal ions, it moves through a solution less rapidly than the others?
(e) LiF has the lowest solubility of the group 1 metal halides.

Solution (a) It gives a deposit of a carbonate with CO_2 in the air, which can jam the tap.

(b) They lower the freezing point of water, thus preventing ice formation.
(c) The small size of Li^+ gives it a large polarising power hence covalent nature of its compounds increases. It can be viewed in terms of Fajan's rule:
— smaller the cation
— larger the anion
— greater the charge
— thus greater the polarising power
hence, greater the covalent nature.
(d) Smaller the size of cation, larger the charge (hence, greater the charge density) then greater the hydration. The dense charge of Li^+ attracts several layers of water molecules around it. They increase the effective size of the ion, thus slowing it down.
(e) The small size of both the Li^+ and F^- ions leads to a very large lattice energy hence solubility is least.

Problem 8. Give the distinct features of structures of sodium and potassium bicarbonates.

Solution The structures of sodium and potassium bicarbonates provide interesting examples of hydrogen bonding. The potassium salt contains a dimeric anion as shown in the following structure.

In the sodium bicarbonate, the HCO_3^- anions form an infinite chain

Problem 9. Lithium compounds have covalent character. Explain.

Solution High polarising power of Li^+ ion due to its smaller size makes lithium compounds (particularly

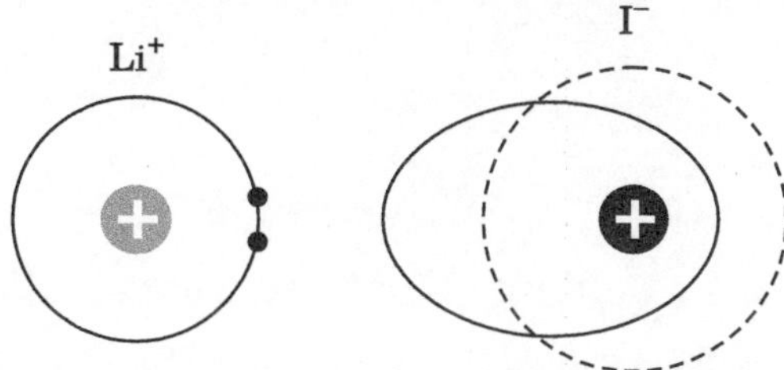

Polarisation of electron charge cloud of iodide by lithium ion

lithium halides) predominantly covalent. If Li^+I^- is considered fully ionic compound, its dipole moment in the gaseous state would be 11.5 debye (D) but its experimental value comes out to be 3.25 D. This is due to distortion of electron charge cloud of the iodide ion by the lithium ion, which results in decrease in ionic character (hence the decrease in dipole moment) and increase in covalent character of LiI.

Explanation

(a) **LiF *is much less soluble than* LiCl, *but* AgF *is much more soluble than* AgCl.**

– LiF has such a high lattice energy (U) that the high solvation energy of F^- is not sufficient to make it more soluble than LiCl. The difference in U is not so great for AgF and AgCl and the high solvation energy of F^- determines relative solubilities.

(b) ***Lithium salts are not involved in alums.***

– In alums, $\overset{I}{M}_2SO_4 \cdot \overset{II}{M}_2(SO_4)_3 \cdot 24H_2O$

(M^{I} = alkali metal, M^{III} = a trivalent metal), each alkali metal is coordinated to six separate water molecules :

$$[M(H_2O)_6]^+[Al(H_2O)_6]^{3+}[SO_4^{2-}]_2$$

Lithium does not form alums, probably because the small $[Li(H_2O)_6]^+$ is too small to form stable lattice required for alum formation.

(c) ***In lithium salts, lithium is surrounded octahedrally by six water molecules.***

– Due to smallest size, Li^+ ion is hydrated to maximum and it is surrounded octahedrally by six water .

Molecules — the octahedral share faces, forming chains.

$$\overset{+}{Li} \langle (OH_2)_3 \rangle \overset{+}{Li} \langle (OH_2)_3 \rangle \overset{+}{Li} —$$

(d) ***Solubility of iodides of alkali metals in water is in codes***

LiI > NaI > KI > RbI > CsI

But that of fluorides is in reverse oxides

LiF < NaF < KF < RbF < CsF

– LiI > NaI > KI > RbI > CsI (solubility)

$$\text{Lattice energy} \frac{1}{r_+ + r_-}$$

Since, $r_- > r_+$, the sum will not change too much as r_+ increases. As a result the lattice energy will not change significantly as r_+ increases. Thus decrease in lattice energy is not as fast as decrease in hydration energy. Thus, more decrease in hydration energy results in decrease in solubility. Thus, solubility decreases as

LiI > NaI > KI > RbI > CsI

With a small anion as (F^-), the lattice energy decreases more rapidly along a series of salts with increasing cation size and thus solubility increases.

LiF < NaF < KF < RbF < CsF

(e) ***Ionic mobility of alkali metals ion is in order***

$$\mathbf{Cs^+ > Rb^+ > K^+ > Na^+ > Li^+}$$

– Size of alkai metal ions is in order

$$Li^+ > Na^+ < K^+ < Rb^+ < Cs^+$$

Smaller the size, larger the hydration. Thus, Li^+ 15 most hydrated and Cs^+ is least hydrated. Hydration results in decrease in ionic mobility. Thus, Li^+ has least ionic mobility and Cs^+ maximum, order being:

Hydration : $Li^+ > Na^+ > K^+ > Rb^+ > Cs^+$

Ionic mobility : $Li^+ < Na^+ < K^+ < Rb^+ < Cs^+$

(f) ***On being heated in excess supply of air,* K, Rb *and* Cs *form superoxide rather than oxides and peroxides.***

– As a general rule, to the large cations stabilise larger anions. K^+, Rb^+ and Cs^+ are larger cations and O_2^- (superoxide) is also larger anion. Thus, when MO_2 is formed, it is stabilised, hence its formation is preferred to oxides and peroxides (which are relatively smaller ions).

Master Exercises

Exercise 1
(Stage 1 : Learning)

Short Answer Type Questions

1. Name the bicarbonate of alkali metal which does not exist in solid state?

2. Name the binary salt of potassium which shows paramagnetic character and which is used is life-saving devices?

3. Compare the following :
(a) Li^+ and Be^{2+} in ionic conductance
(b) Li^+ and K^+ in hydrated ionic size
(c) LiCl and NaCl in lattice energy
(d) Na_2CO_3 and Li_2CO_3 in thermal stability
(e) Li^+ and Na^+ in polarising power
(f) LiI and CsI in solubility in water.

4. Arrange $Na^+, Mg^{2+}, O^{2-}, F^-, Al^{3+}, N^{3-}$ (each having 10 electrons) in increasing ionic radii.

5. Arrange LiCl, NaCl, KCl, RbCl and CsCl in increasing lattice energy.

6. Arrange Li_2CO_3, Na_2CO_3, K_2CO_3 and Rb_2CO_3 in increasing melting point.

7. Identify A to in the following sequence of reaction :

$$Na + O_2 \longrightarrow (A) \xrightarrow{HCl} (B) \xrightarrow[\text{ether}]{H^+/K_2Cr_2O_7} \underset{\text{blue in ethereal layer}}{(C)}$$

8. Arrange $LiClO_4$, $NaClO_4$, $KClO_4$ and $RbClO_4$ in increasing solubility.

9. Arrange fluorides of alkali metals in increasing solubility.

10. Potassium cobaltinitrite is used in qualitative analysis. Explain.

11. Na^+ (cation) is smaller than Na (atom)

12. You have a mixture of LiCl and NaCl. Suggest a suitable solvent to separate the mixture.

13. LiCl has lower melting point than NaCl. Explain

14. Li_2CO_3 decomposes into CO_2 and Li_2O but other alkali carbonates are thermally stable. Explain.

15. What is oxone?

16. Why sodium is less reactive than potassium?

17. Cs metal is used in photoelectric cells, Li is not used. Explain.

18. Lithium is the only metal in alkali metals that from nitride. Explain.

19. (a) What is formula of potassium superoxide?
(b) What is oxidation state of potassium?
(c) What is bond-order of superoxide ion?
(d) What is magnetic moment of superoxide ion?

20. Aqueous Na_2CO_3 is alkaline in nature. Explain

21. In the laboratory, you would have noticed that NaOH solution has some white deposit inside reagent bottle. Explain.

22. Pallets of KOH become wet when exposed to air. Explain.

23. Name the metal which floats on water without any apparent reaction with water.

24. Li resembles to element X which is not the member of alkali metals.
(a) Name X
(b) What is group of X?
(c) Mention two properties which resembles to that of Li

25. Lithium on heating in air forms mainly Li_2O (oxide) and not peroxide (Li_2O_2). Explain.

26. Give an example to verify that larger anion stabilise larger cation.

27. Name the salt of alkali metal that does not form double salts such as alums.

28. Baking soda or baking powder, they are different or same.

29. $KHCO_3$ or K_2CO_3 cannot be obtained Solvay process. Explain.

30. Give two oxidising properties given by Na_2O_2.

Analytical Questions

1. The ionisation energies of group 1 elements (alkali metals) suggest that caesium should be most reactive but standard electrode potentials suggest that lithium is the most reactive. Reconcile these two observations.

2. In the following sequence of reactions (unbalanced)

$$Li + N_2 \longrightarrow Li_3N$$
$$Li_3N + H_2O \longrightarrow NH_3$$
$$Li_3N + H_2O + CO_2 \longrightarrow NH_4HCO_3$$
$$NH_4HCO_3 + NaCl \longrightarrow NaHCO_3$$
$$NaHCO_3 \longrightarrow Na_2CO_3$$

Calculate volume of N_2 required at NTP to form 10.6 kg of Na_2CO_3.

3. Comment on each of the following observations:

(a) The mobilities of the alkali metal ions in aqueous solutions are:

$$Li^+ < Na^+ < K^+ < Rb^+ , Cs^+$$

(b) Lithium is the only alkali metal to form a stable nitride.

(c) The solubility of sodium sulphate in water increases up to 32°C and thereafter decreases.

(d) $E°$ for $M^+(aq) + e^- \rightleftharpoons M(s)$ is nearly constant for the alkali metals.

(e) For the replacement of chlorine in organic compounds by fluorine by the autoclave reaction

$$\rangle C—Cl + MF \longrightarrow \rangle C—F + MCl$$

KF is a better reagent than NaF.

4. One mole each of C_2H_5OH and $CH{\equiv}CH$ is treated with excess of Na metal.

(a) Write reactions

(b) In which case, yield of H_2 is better (in terms of volume of H_2 at STP)

(c) Compare % yield (H_2) in terms of mass of reactant.

5. Explain polarisation by Li^+ ion.

6. Na and K are strong reducing agents. Give examples.

7. Identify *A* to *C* in the following :

$$Mn(OH)_2 \xrightarrow{Na_2O_2} \underset{\text{green}}{(A)} \xrightarrow{Cl_2} \underset{\text{violet}}{(B)} \xrightarrow{Na_2O_2/H^+} \underset{\text{colourless}}{(C)}$$

8. Identify *A* to *C* in the following

$$CrCl_3 \xrightarrow{NaOH} A \xrightarrow{Na_2O_2} \underset{\text{yellow}}{B} \xrightarrow{(CH_3COO)_2Pb} \underset{\text{yellow precipitate}}{C}$$

9. (a) A white solid is either $\mathbf{Na_2O}$ or $\mathbf{Na_2O_2}$. A piece of red litmus paper turns white when it is dipped into a freshly made aqueous solution of the white solid. Identify the substance and explain with balanced equation.

(b) Explain what would happen to red litmus if the white solid were the other compound?

10. (*A*) is binary salt of alkali metal. 0.71 g of *A* on heating with excess of sulphur forms (*B*) and which in aqueous solution gives 2.33 g white precipitate (insoluble in conc. HCl) with aqueous $BaCl_2$ solution. A crystalline double salt (*C*) formed from $Al_2(SO_4)_3$ and (*B*) (Ba = 137, Al = 27, S = 32)

(a) What is *A*? (b) What is *B*?

(c) What is *C*? (d) What is amount of *C*?

11. Calculate the energy change for the alternating reaction that yields chlorine sodium (Cl^+Na^-) and then explain that why sodium chloride is prepared

$$2Na(s) + Cl_2(g) \longrightarrow 2Cl^+Na^-(s)$$

Lattice energy of Cl^+Na^- = – 787 kJ mol^{-1}

Electron affinity of Na = – 52.9 kJ mol^{-1}

Ionisation energy of Cl = + 1251 kJ mol^{-1}

BE of Cl_2 = 244 kJ mol^{-1}

Heat of sublimation of Na(*s*) = 107.3 kJ mol^{-1}

ΔH_f(NaCl) = – 411 kJ mol^{-1}

Also make Born-Haber cycle for $Cl^+Na^-(s)$

NCERT Problems

1. The alkali metals follow the noble gases in their atomic structure. What properties of these metals can be predicted from this information?

2. What happens when

(i) sodium metal is dropped in water?

(ii) sodium metal is heated in free supply of air?

(iii) sodium peroxide dissolves in water?

3. Name the alkali metals which form superoxides when heated in excess of air.

4. Name the metal which floats on water without any apparent reaction with it.

5. List three properties of lithium in which it differs from the rest of the alkali metals.

6. Name the chief factor responsible for the anomalous behaviour of lithium.

7. When an alkali metal dissolves in liquid ammonia the solution acquires different colours. Explain the reasons for this type of colour change.

8. What is meant by 'diagonal relationship' in the Periodic Table? What is it due to?

9. Describe in detail the manufacture of sodium carbonate by the Solvay process. State the principles involved in this process.

10. Describe three industrial uses of caustic soda. Describe one method of manufacture of sodium hydroxide. What happens when sodium hydroxide reacts with (i) aluminium metal (ii) CO_2 (iii) SiO_2 ?

11. Why is that the *s*-block elements never occur free in nature? What are their usual modes of occurrence and how are they generally prepared?

12. Compare the solubility and thermal stability of the following compounds of the alkali metals with those of the alkaline earth metals
(a) nitrates (b) carbonates
(c) sulphates

13. Explain what happens when
(i) sodium hydrogen carbonate is heated?
(ii) sodium amalgam reacts with water?
(iii) fused sodium metal reacts with ammonia?

14. Mention the general trends in Group 1 and in Group 2 with increasing atomic number with respect to
(a) density (b) melting point
(c) atomic size (d) ionization enthalpy

15. Starting with sodium chloride how would you proceed to prepare (state the steps only)?
(i) Sodium metal (ii) Sodium hydroxide
(iii) Sodium peroxide (iv) Sodium carbonate

16. State as to why
(i) lithium on being heated in air mainly forms the monoxide and not the peroxide?
(ii) an aqueous solution of sodium carbonate gives alkaline tests?
(iii) sodium is prepared by electrolytic method and not by chemical method?

17. How do the following properties change on moving from group 1 to group 2 in the Periodic Table?
(i) Atomic size (ii) Ionisation enthalpy
(iii) Density (iv) Melting points

18. Like lithium in group 1, beryllium shows anomalous behaviour in group 2. Write three such properties of beryllium which make it anomalous in the group.

19. Comment on each of the following observations :
(a) The mobilities of the alkali metal ions in aqueous solution are $Li^+ < Na^+ < K^+ < Rb^+ < Cs^+$
(b) Lithium is the only alkali metal to form a nitride directly
(c) $E°$ for $M^{2+}(aq) + 2e^- \longrightarrow M(s)$ (where M = Ca, Sr or Ba) is nearly constant
(d) LiF is least soluble among the fluorides of alkali metals

20. Why does the following reaction

$$>C—Cl + MF \longrightarrow >C—F + MCl$$

proceed better with KF with than NaF?

21. Compare and contrast the chemistry of group 1 metals with that of group 2 metals with respect to
(i) nature of oxides
(ii) solubility and thermal stability of carbonates
(iii) polarizing power of cations
(iv) reativity and reducing power

22. Describe two important uses of each of the following.
(i) caustic soda (ii) sodium carbonate
(iii) quicklime

23. State as to why
(a) A solution of Na_2CO_3 is alkaline?
(b) Alkali metals are prepared by electrolysis of their fused chlorides?
(c) Sodium is found more useful than potassium?

24. Write balanced equations for reactions between
(a) Na_2O_2 and water
(b) KO_2 and water
(c) Na_2O and CO_2

Exercise 2
(Stage 2 : High Skill Problem Solving)

Only One Option Correct

1. Which is/are not correct configuration of *s*-block elements?
(a) (Ar) $3d^{10} 4s^2$ (b) (Ar) $3d^{10} 4s^1$
(c) Both (a) and (b) (d) None of these

2. Select correct statement(s).
(a) Li_2CO_3 decomposes into oxides while other alkali carbonates are thermally stable
(b) LiCl is predominantly covalent
(c) Li_3N is stable
(d) all of the above

3. Alkali metals resemble IB (copper family) in following respects
(a) +1 valency
(b) sulphates are water soluble
(c) oxides are strong base
(d) oxides are strong acids

4. Select incorrect statement.
(a) Na_2CO_3 is thermally stable while Ag_2CO_3 decomposes into Ag, CO_2 and O_2
(b) Ag^+ forms complexes, Na^+ does not
(c) NaCl is water soluble, AgCl is insoluble
(d) NaCl and AgCl both give colour in flame when ignited

5. A colourless solid (*X*) on heating evolved CO_2 and also gave a white residue, soluble in water. Residue also gave CO_2 when treated with dilute acid. (*X*) is
(a) Na_2CO_3 (b) $CaCO_3$
(c) $Ca(HCO_3)_2$ (d) $NaHCO_3$

6. Which of the following is preferentially formed on reaction of oxygen and potassium?
(a) K_2O_2 (b) KO_2
(c) K_2O (d) K_2O_3

7. CO_2 gas along with solid (*Y*) is obtained when sodium salt (*X*) is heated. (*X*) is again obtained when CO_2 gas is passed into aqueous solution (*Y*). (*X*) and (*Y*) are
(a) Na_2CO_3, Na_2O (b) Na_2CO_3, NaOH
(c) $NaHCO_3$, Na_2CO_3 (d) Na_2CO_3, $NaHCO_3$

8. Which salt can be used to identify coloured cation?
(a) Borax (b) Microcosmic salt
(c) Both (a) and (b) (d) None of these

9. Intermediate formed by heating microcosmic salt and which forms coloured bead with coloured cation is
(a) NH_3 (b) H_3PO_3
(c) $NaPO_3$ (d) H_2O

10. Molten sodium chloride conducts electricity due to the presence of
(a) free electrons (b) ions
(c) Na atom (d) Cl atom

11. Which is the major constituent of gun powder?
(a) Nitre (b) Sulphur
(c) Charcoal (d) Chile salt petre

12. When SO_2 gas is passed into aqueous Na_2CO_3, product formed is
(a) $NaHSO_4$ (b) Na_2SO_4
(c) $NaHSO_3$ (d) Na_2SO_3

13. CO_2 cannot be obtained by heating
(a) Na_2CO_3 (b) $CaCO_3$
(c) Li_2CO_3 (d) $Ca(HCO_3)_2$

14. Some of alkali metal salts are coloured. Na_2CrO_4—yellow, $KMnO_4$—pink, K_2MnO_4 —green. It is due to
(a) cations are coloured ions
(b) anions are coloured ions
(c) both (a) & (b) are correct
(d) None of the above is correct

15. Na and Li are placed in dry air. We get
(a) NaOH, Na_2O, Li_2O
(b) Na_2O, Li_2O
(c) Na_2O, Li_2O, Li_3N, NH_3
(d) Na_2O, Li_3N, Li_2O

16. All alkali metals form following compound except
(a) amide, MNH_2
(b) superoxide like KO_2
(c) ionic 'salt-like' hydride *MH*
(d) basic oxides

17. Match the compounds (given in *X*) with their colours (given in *Y*)

X	Y
A. Li_3N	I Pale yellow
B. K_2O	II Orange
C. Cs_2O, KO_2	III Bright yellow
D. Rb_2O	IV Ruby red

	A	B	C	D
(a)	IV	I	II	III
(b)	II	III	I	IV
(c)	I	II	III	IV
(d)	None of these			

18. Which of the following is best CO_2 absorber as well as source of O_2 in submarines?
(a) KO_2 (b) NaOH
(c) KOH (d) LiOH

19. Following are the ionisation potential values of (I_1) 899 kJ mol^{-1}, (I_2) 1757 kJ mol^{-1}, (I_3) 15000 kJ mol^{-1}
(a) Na (b) K
(c) Be (d) Ne

20. A compound which can be used in space vehicles both to absorb CO_2 and liberate O_2 is
(a) NaOH (b) Na_2O
(c) Na_2O_2 (d) CaO + NaOH

21. Which of the following changes is not realised in the laboratory ?
(a) Absorption of NO by alkaline sodium sulphite to form a compound
(b) Combustion of metallic Mg in CO_2
(c) Heating hydrated magnesium chloride to get the anhydrous salt
(d) 'Displacement' of chlorine from $KClO_3$ by iodine to form KIO_3

22. Which is not *s*-block element?
(a) $[Ar]4s^2\,3d^{10}\,4p^6\,5s^1$ (b) $1s^2\,2s^2\,2p^1$
(c) $1s^2\,2s^1\,2p^1$ (d) $[Ar]4s^1\,4p^1$

23. Which will give N_2O on heating?
(a) NH_4NO_2 (b) NH_4NO_3
(c) $NaNO_3$ (d) $LiNO_3$

24. Nitrate can be converted into metal oxide on heating in case of
(a) Li (b) Na
(c) both (a) and (b) (d) None of these

25. Automobile grease is obtained from
(a) CH_3COONa (b) $C_{17}H_{35}COONa$
(c) CH_3CH_2COOLi (d) $C_{17}H_{35}COOLi$

26. Li_2CO_3 is
(a) added to bauxite in the electrolytic production of aluminium
(b) used to toughen glass
(c) used as medicine as it affects the balance between Na^+ and K^+, and Mg^{2+} and Ca^{2+}
(d) all of the above are correct

27. Which cannot be used to generate H_2?
(a) Al + NaOH (b) Zn + NaOH
(c) Mg + NaOH (d) $LiH + H_2O$

28. There is loss in weight when mixture of Li_2CO_3 and $Na_2CO_3{\cdot}10H_2O$ is heated strongly. This loss is due to
(a) Li_2CO_3 (b) $Na_2CO_3{\cdot}10H_2O$
(c) both (a) and (b) (d) None of these

29. Which disproportionates into M_2O_2 (peroxide) and M (metal) on heating?
(a) Li_2O (b) Na_2O
(c) Both (a) and (b) (d) None of these

30. Reagent used as fixer in photography
(a) AgBr (b) $AgNO_3$
(c) $Na_2S_2O_3{\cdot}5H_2O$ (d) both (a) and (c)

31. Which can dissolve I_2?
(a) KI (b) NaI
(c) Both (a) and (b) (d) None of these

32. The deep colour produced when iodine is dissolved in a solution of potassium iodide is caused by the presence of
(a) I_2 (b) I^-
(c) I_3^- (d) I_2^-

33. Alkali metals dissolve in liquid NH_3, then which of the following observation are true?
(a) H_2 gas is liberated
(b) Solution is blue due to the presence of solvated electrons
(c) Solution is conducting
(d) All of the above are correct

34. Which is incorrect statement?
(a) The heats of hydration of the dipositive alkaline earth metal ions decrease with an increase in their ionic size.
(b) $NaNO_3$ forms Na_2O on heating
(c) Hydration of alkali metal ion is less than that of II A.
(d) Alkaline earth metal ions, because of their much larger charge to size ratio, exert a much stronger electrostatic attraction on the oxygen of water molecule surrounding them.

35. NaOH is not used in
(a) paper industry (b) soap industry
(c) rayon industry (d) plastic industry

36. Match compounds given in (X) with their uses in (Y)

X	Y
A. Na_2CO_3	I glass
B. Na_2SO_3	II bleach
C. NaOH	III SO_2 absorber
D. NaOCl	IV detergent

hence, correct order is

	A	B	C	D
(a)	IV	I	III	II
(b)	I	III	IV	II
(c)	II	IV	I	III
(d)	III	II	IV	I

37. Match the compounds/metal in (X) with their uses in (Y).

X	Y
A. Liquid sodium metal	I. Breathing apparatus submarine
B. Potassium stearate	II. Explosive
C. Potassium nitrate	III. Coolant in nuclear reaction
D. Potassium superoxide	IV. Soft soap

hence correct order is

	A	B	C	D
(a)	I	III	II	IV
(b)	III	IV	II	I
(c)	II	I	III	IV
(d)	IV	II	III	I

38. Lattice energy (numerical value) of chloride of alkali metals is in order
(a) LiCl > NaCl > KCl > RbCl > CsCl
(b) LiCl < NaCl < KCl < RbCl < CsCl
(c) NaCl < KCl < LiCl < RbCl < CsCl
(d) NaCl < KCl < RbCl < CsCl < LiCl

39. Which is most stable halide of alkali metal?
(a) Sodium fluoride (b) Sodium bromide
(c) Sodium chloride (d) Sodium iodide

40. Which is a pair of paramagnetic species?
(a) KO_2, NO_2 (b) K_2O_2, KO_2
(c) K_2O, NO_2 (d) NO_2, N_2O_2

41. Baking powder contains
(a) $NaHCO_3$, $Ca(H_2PO_2)_2$ and starch
(b) $NaHCO_3$, $Ca(H_2PO_2)_2$
(c) $NaHCO_3$, starch
(d) $NaHCO_3$

42. Which does not exist in solid state?
(a) $NaHCO_3$ (b) $NaHSO_3$
(c) $LiHCO_3$ (d) $CaCO_3$

43. Following compounds are used in fire-works
(a) $LiNO_3$ (b) $BaCl_2$
(c) $(NH_4)_2Cr_2O_7$ (d) all of these

44. When NO is passed into KOH solution, products are
(a) N_2 (b) N_2O
(c) both (a) and (b) (d) None of these

45. Which has least molar solubility in H_2O?
(a) LiCl (b) NaCl
(c) KCl (d) CsCl

Q. 46 to 49 are based on following reactions:

$$(A) \xrightarrow{\Delta} (B)\,(\text{oxide}) + CO_2$$
$$(B) + H_2O \longrightarrow (C)$$
$$(C) + CO_2 \longrightarrow (A)\,(\text{milky})$$
$$(C) + NH_4Cl \xrightarrow{\Delta} (D)\,\text{gas}$$
$$(D) + H_2O + CO_2 \longrightarrow (E)$$
$$(E) + NaCl \longrightarrow (F)$$
$$(F) \xrightarrow{\Delta} Na_2CO_3 + CO_2 + H_2O$$

46. Name of the process is
(a) Solvay (b) ammonia-soda
(c) both (a) & (b) correct (d) None is correct

47. (*A*) is
(a) $Ca(HCO_3)_2$ (b) $CaCO_3$
(c) CaO (d) Na_2CO_3

48. (*B*) and (*C*) are
(a) CaO, $Ca(OH)_2$ (b) $Ca(OH)_2$, $CaCO_3$
(c) $CaCO_3$, $Ca(OH)_2$ (d) $Ca(OH)_2$, CaO

49. (*D*), (*E*) and (*F*) are
(a) NH_3, NH_4Cl, NH_4HCO_3
(b) NH_3, NH_4HCO_3, $NaHCO_3$
(c) NH_4HCO_3, Na_2CO_3, $NaHCO_3$
(d) None of the above

50. Out of AgF, CaF_2, BeF_2, MgF_2
(a) AgF and BeF_2 are soluble, and CaF_2 and MgF_2 are insoluble in water
(b) AgF is soluble, others insoluble
(c) AgF is insoluble, others are soluble
(d) all are insoluble

51. To an acidified dichromate solution, a pinch of Na_2O_2 and ether is added and shaken. What is observed?
(a) Deep violet colour
(b) Red colour changing to green
(c) Copious evolution of oxygen
(d) Bluish-green precipitate

52. Nitrolim is a
(a) mixture of calcium carbide and nitrogen
(b) mixture of calcium cyanamide and carbon
(c) mixture of calcium cyanide and carbon
(d) mixture of NH_4CN and CaCN

53. The common name, "cream of tartar" refers to
(a) $KHC_4H_4O_6$
(b) $K(SbO)C_4H_4O_6$
(c) $KNaC_4H_4O_6$
(d) $Mg_2(OH)_2(C_4H_4O_6)H_2O$

54. Which of the following mixtures cannot be prepared?
(a) $NaHCO_3 + Na_2CO_3$
(b) $Na_2CO_3 + NaOH$
(c) $NaHCO_3 + NaOH$
(d) $H_2CO_3 + NaHCO_3$

55. Noble gases (like He, Ne, Ar, Kr etc.) are isolated from air. One of the steps is/are
(a) heating air with Li or Mg
(b) bubbling air into NaOH solution
(c) both (a) and (b) are correct
(d) None of the above is correct

56. Which is used to treat acid indigestion?
(a) $Be(OH)_2$ (b) KOH
(c) $Mg(OH)_2$ (d) $Ca(OH)_2$

57. Which of the following is not used as food preservatives?
(a) NaCl (b) CH_3COOH
(c) C_6H_5COONa (d) HCHO

58. Which are involved in various physiological functions in animals?
(a) Na^+ (b) K^+
(c) Both (a) and (b) (d) None of these

59. Which is used in the treatment of manic-depressive disorders?
(a) Na_2CO_3 (b) Li_2CO_3
(c) K_2CO_3 (d) $MgCO_3$

60. Select correct statement(s).
(a) Li_2CO_3 is only sparingly soluble in water and no $LiHCO_3$ has been isolated.
(b) K_2CO_3 cannot be made by a method similar to the ammonia-soda process.
(c) Li_2CO_3 and $MgCO_3$ both are thermally stable.
(d) $Na_2CO_3 \cdot NaHCO_3 \cdot 2H_2O$ is a mineral called trona.

61. Select correct statement(s).
(a) $CaCO_3$ is more soluble in a solution of CO_2 than in H_2O
(b) Na_2CO_3 is converted to Na_2O and CO_2 on heating
(c) Li_2CO_3 is thermally stable
(d) Presence of $CaCl_2$ or $CaSO_4$ in water causes temporary hardness

62. The alkalide ion is
(a) diamagnetic (b) represented as $\overset{-}{C}H_3$
(c) paramagnetic (d) represented as $\overset{+}{C}H_3$

63. Select correct statement(s).
(a) Mg^{2+} ions are necessary for the activation of phosphate transfer enzymes.

(b) Mg is present in chlorophyll used in photosynthesis in green plants.
(c) Operation of $Na^+ - K^+$ pumps is biological.
(d) All the above are correct statements

64. The alkali metals dissolve in liquid NH_3, it is found that
(a) the dilute solutions are blue but the colour changes to bronze with increasing concentration
(b) the blue colour is due to the presence of solvated electrons
(c) the blue solutions are paramagnetic but the bronze-coloured solutions are diamagnetic
(d) all the facts given above are found

65. Select correct statement(s).
(a) Stability of peroxides and superoxides of alkali metals increases with increase in size of the metal ion
(b) NaOH does not form hydrated salt
(c) Increase in stability in (a) is due to stabilisation of large anions by larger cations through lattice energy effects
(d) The low solubility of LiF is due to its high lattice energy whereas low solubility of CsI is due to smaller hydration energy

66. Which one of the following is correct sequence followed by molar ionic conductance of the ions?
(a) $Li^+ < Na^+ < K^+ < Rb^+$
(b) $Rb^+ < K^+ < Na^+ < Li^+$
(c) $Sr^{2+} < Ca^{2+} < Mg^{2+} < Be^{2+}$
(d) $Na^+ < K^+ < Li^+ < Rb^+$

67. Which one among the following is the most soluble in water?
(a) LiI (b) LiBr
(c) LiCl (d) LiF

68. The stability of K_2O, K_2O_2 and KO_2 is in order $K_2O < K_2O_2 < KO_2$. This increasing stability at the size of metal ion increases is due to stabilisation of
(a) larger cation by smaller anions through lattice energy effects
(b) larger cation by larger anions through lattice energy effects
(c) smaller cations by smaller anions through m.p.
(d) smaller cations by larger anions through m.p.

69. M_2O_3 (sesquoxides) is paramagnetic in nature. It is due to presence of
(a) peroxide ion (b) superoxide ion
(c) oxide ion (d) all of these

70. Which is/are coloured and paramagnetic?
(a) KO_2 (b) K_2O_3
(c) Both (a) and (b) (d) None of these

71. Suboxide of Rb has the formula
(a) RbO_2 (b) RbO_3
(c) Rb_2O_2 (d) Rb_6O

72. MO_3 an hydrolysis forms
(a) MOH and H_2 (b) MOH and O_2
(c) MOH and M_2O_2 (d) MOH and H_2O_2

73. Which is used as a treatment for bipolar disorder (an illness that involves alternating periods of depression)?
(a) Li_2CO_3 (b) K_2CO_3
(c) LiCl (d) $LiCH_3$

74. $NaNO_3$ is converted to KNO_3 by following reaction :

$$NaNO_3(aq) + KCl(aq) \rightleftharpoons KNO_3(aq) + NaCl(s)$$

In the above equilibrium, which is least soluble in hot water?
(a) $NaNO_3$ (b) KCl
(c) KNO_3 (d) NaCl

75. Gun powder spontaneously react when ignited. Reaction can be
(a) $2KNO_3(s) + 4C(s) \longrightarrow K_2CO_3(s) + 3CO(g) + N_2(g)$
(b) $2KNO_3(s) + 2S(s) \longrightarrow K_2SO_4(s) + SO_2(g) + N_2(g)$
(c) both (a) & (b)
(d) None of the above

One or More Than One Options Correct

1. Select correct statement(s).
(a) MO_2 contains the paramagnetic ion O_2^-
(b) O_2^- is stable only in the presence of large cations as K, Rb and Cs
(c) KO_2 is paramagnetic and has orange colour
(d) K_2O has antifluorite structure

2. Recently sodium naphthenide ($C_{10}H_8Na$) has been used as reductant in complex formation. $C_{10}H_8Na$ is
(a) paramagnetic (b) deep-green in colour
(c) diamagnetic (d) colourless

3. Sesquoxides (M_2O_3) of alkali metals
(a) have been prepared by careful thermal decomposition of MO_2
(b) are dark-coloured paramagnetic substances
(c) are colourless diamagnetic substances
(d) are coloured diamagnetic substance

4. Select correct statement(s).
(a) Alkali metals ozonides are coloured and paramagnetic
(b) Alkali metal-ozonides are colourless and diamagnetic
(c) Sesquoxides of alkali metals are peroxides disuperoxides $(M^+)_4(O_2^{2-})(O_2^-)_2$
(d) Sesquoxides are paramagnetic due to presence of superoxide ion O_2^-.

5. Select correct alternate(s).
(a) In $NaHCO_3$, the HCO_3^- ions are linked into infinite chain
(b) In $KHCO_3$, a dimer is formed by H-bonding.
(c) In $NaHCO_3$, a dimer is formed by H-bonding
(d) In $KHCO_3$, the HCO_3^- ions are linked into infinite chain

6. Select correct alternate(s).
(a) All alkali metals form solid bicarbonates
(b) Except $LiHCO_3$, all alkali metals bicarbonates are solid
(c) Li_2CO_3 decomposes into CO_2 and Li_2O
(d) Na_2CO_3 decomposes into CO_2 and Na_2O

7. Select corrrect alternate(s).
(a) Solid $LiNO_3$ and $NaNO_3$ and deliquescent
(b) KNO_3 is used in preference to $NaNO_3$ in gun powder
(c) $NaNO_3$ is used in preference to $NaNO_3$ in gun powder
(d) NH_4HCO_3 exists in solid state

8. $LiAlH_4$ can reduce
(a) carbonyl compounds into alcohols
(b) alkenes into alkanes
(c) —CHO group into —CH_3
(d) either into alcohols

9. Select correct statement(s).
(a) Radius of hydrated Li^+ ion is smaller than that of hydrated Cs^+ ion
(b) Ionic mobility of hydrated Li^+ ion is smaller than that of hydrated Cs^+ ion
(c) Hydrated Cs^+ is more conducting that hydrated Na^+ ion
(d) Ionic mobility of hydrated
$Cs^+ > Rb^+ > K^+ > Na^+ > Li^+$

10. Li does not resemble other alkali metals in following properties
(a) Li_2CO_3 decomposes into oxides while other alkali carbonates are thermally stable
(b) LiCl is predominantly covalent
(c) Li_3N is stable
(d) LiCl is non-conducting in molten state

11. Alkali metals resemble 1B (copper family) in following respects
(a) +1 valency
(b) sulphates are water soluble
(c) oxides are strong bases
(d) oxides are strong acids

12. Na^+ and Ag^+ differ in
(a) Na_2CO_3 is thermally stable while Ag_2CO_3 decomposes into Ag, CO_2 and O_2
(b) Ag^+ forms complexes, Na^+ does not
(c) NaCl is water soluble, AgCl is insoluble
(d) NaCl and AgCl both give colour in flame when ignited

13. Which is/are true statement(s)?
(a) The heats of hydration of the dipositive alkaline earth metal ions decrease with an increase in their ionic size
(b) Hydration of alkali metal ion is less than that of IIA.
(c) Alkaline earth metal ions, because of their much larger charge to size ratio exert a much stronger electrostatic attraction on the oxygen of water molecule surrounding them
(d) None of the above statements is correct

14. Which are true statements about *s*-block elements?
(a) Metals are obtained by the electrolysis of fused chlorides
(b) Oxides are basic except BeO
(c) +1 valency by alkali metals and +2 valency by alkaline earth metals is shown
(d) Carbonates are thermally stable

15. Li_2CO_3 is
(a) added to bauxite in the electrolytic production of aluminium
(b) used to toughen glass
(c) used as medicine as it affects the balance between Na^+ and K^+, and Mg^{2+} and Ca^{2+}
(d) thermally stable

Passage Comprehension Questions

Passage 1

Read the following passage and answer the questions at the end of it.

Dilute solutions of alkali metals in liquid NH_3 are blue. It is the ammoniated electron which is responsible for the blue colour of the solution, and the electrical conductivity is due to the ammoniated cation, $[M(NH_3)_x]^+$ as well as the ammoniated electron, $[e(NH_3)_y]^-$, values of x and y depend on the extent of solvation (by NH_3). Dilute solutions are paramagnetic due to free electrons.

1. What happens if alkali metal is allowed to react with concentrated ammonia solution?
(a) Paramagnetic character of solvated electrons is retained
(b) Solvated electrons associate to form electron-pairs and paramagnetic character decreases
(c) Reducing character is increased
(d) Reducing character is not affected

2. Ammoniated solutions of alkali metals are reducing agents due to the presence of free or solvated electrons and can reduce
(a) O_2 to O_2^{2-}
(b) $K_2(Ni(CN)_4]$ to $K_4[Ni(CN)_4]$
(c) both (a) & (b)
(d) none of the above

Passage 2

Questions given below are based on the following values of hydration energy and lattice energy.

Hydration energy $\Delta H_{hydr.}$ (kJ mol^{-1})		Lattice energy ΔH_U (kJ mol^{-1})	
Li^+	−499	LiCl	−840
Na^+	−390	NaCl	−776
K^+	−305	KCl	−703
Cl^-	−382		

1. Which salt has maximum heat of hydration?
(a) LiCl (b) NaCl
(c) KCl (d) LiCl and KCl equally

2. Heat of hydration (numerical value) of LiCl, NaCl and KCl in the increasing order is
(a) LiCl < KCl < NaCl (b) LiCl < NaCl < KCl
(c) LiCl = KCl < NaCl (d) KCl < NaCl < LiCl

3. Maximum heat is absorbed in the dissolution of one mole of
(a) LiCl (b) NaCl
(c) KCl (d) equally

4. Which salt can be used to control humidity?
(a) LiCl (b) NaCl
(c) KCl (d) None of these

Passage 3

Answer the questions based on following flow-sheet.

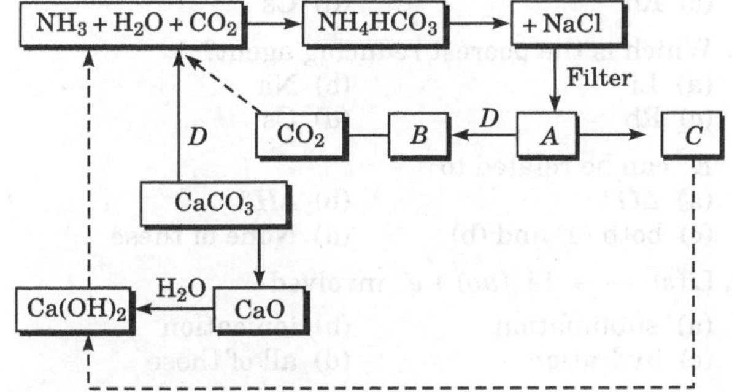

1. This flow-sheet represents
(a) Solvay process of NaOH
(b) Solvay process of Na_2CO_3
(c) Dow process of Na_2CO_3
(d) none of the above is correct

2. Filtrate *A* contains
(a) NH_4Cl (b) Na_2CO_3
(c) $NaHCO_3$ (d) $CaCO_3$

3. Species *B* formed is
(a) NH_3 (b) CO_2
(c) Na_2CO_3 (d) $NaHCO_3$

4. Following are some statements about species *C*. Select correct statement(s).
(a) *C* is NH_4Cl
(b) *C* is $NaHCO_3$
(c) *C* is recycled into $Ca(OH)_2$ chamber to form NH_3
(d) *C* is Na_2CO_3

Passage 4

Read the following passage and answer the questions given at the end.

One of the cheif uses of sodium carbonate has been in the conversion of $Ca(OH)_2(s)$ to concentrated aqueous solution of NaOH which in turn was used in the manufacture of soap

$$Ca(OH)_2(s) + 2Na^+(aq) + CO_3^{2-}(aq) \rightleftharpoons CaCO_3(s) + 2Na^+(aq) + 2OH^-(aq)$$

In 1775, in order to gain independence from the importation of natural Na_2CO_3, the French government offered a prize to any one who could devise a process to prepare Na_2CO_3 from NaCl.

1. Who first devised a process of preparing Na_2CO_3 from NaCl?
(a) Solvay (b) Nicolas Leblanc
(c) Peterson (d) Hofmann

2. A more economical process but based on same principle was given by
(a) Solvay (b) Leblanc
(c) Williamson (d) Haber

3. What is natural source of Na_2CO_3?
(a) Nitre
(b) Chile salt peter
(c) Dolomite
(d) Trona

4. Reaction given in the passage is spontaneous because
(a) NaOH formed is ionic
(b) $CaCO_3$ is removed from the reaction mixture
(c) reaction is exothermic
(d) reaction is endothermic

5. NaOH can also be prepared by electrolysis of aqueous NaCl. Amount of NaOH formed when 0.445 L of NaCl (*aq*) is electrolysed for 137 s with a current of 1.08 A, is
(a) 0.09 g
(b) 0.12 g
(c) 0.06 g
(d) amount of NaCl has not been given

6. pH of the above solution (in Q. 5) is
(a) 11.70 (b) 11.53
(c) 11.83 (d) data is incomplete

Passage 5

$NaHCO_3$ can be used on its own to make cakes or bread "rise". Baking powder is more commonly used, and contains $NaHCO_3$, $Ca(H_2PO_4)_2$ and starch. An improved combination baking powder contains about 40% starch, 30% $NaHCO_3$, 20% $NaAl(SO_4)_2$ and 10% $Ca(H_2PO_4)_2$.

Answer the following questions based on above study.

1. Rise in bread and cakes is due to .... when it decomposes between 50°C and 100°C
(a) CO_2 (b) H_2O
(c) both (a) and (b) (d) None of these

2. $Ca(H_2PO_4)_2$ in baking powder
(a) behaves as an acid in moist condition and decomposes $NaHCO_3$ to give CO_2
(b) behaves as a base in moist condition and gives H_3PO_4
(c) behaves as a binder
(d) behaves as a filler

3. Filler in baking powder is
(a) $NaHCO_3$ (b) $Ca(H_2PO_4)_2$
(c) starch (d) $NaAl(SO_4)_2$

4. In improved baking powder, $NaAl(SO_4)_2$
(a) makes the decomposition of $NaHCO_3$ faster
(b) slows the decomposition of $NaHCO_3$
(c) acts as filler
(d) makes the bread rise

Passage 6

The crystal structures of $NaHCO_3$ and $KHCO_3$ both show hydrogen bonding, but are different. Answer the following questions based on this

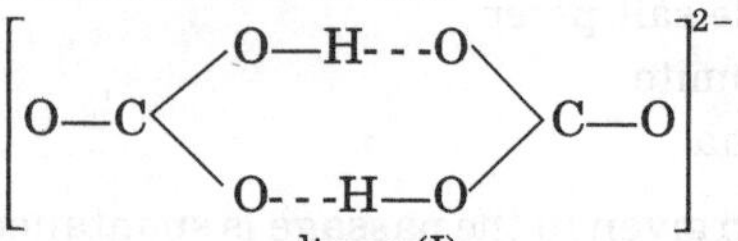

dimer (I)

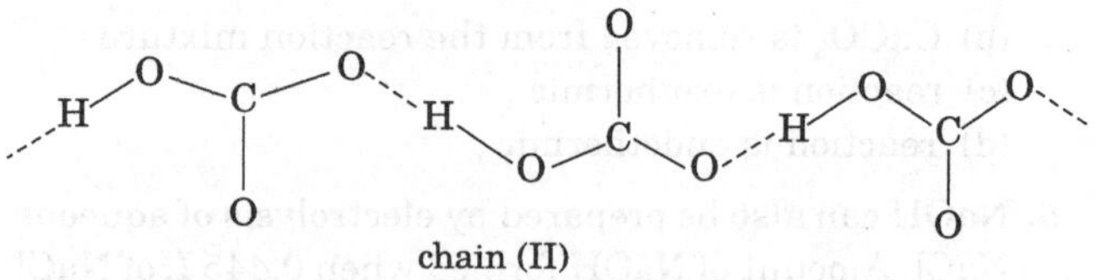

chain (II)

1. Which structure is adopted by $NaHCO_3$?
(a) I (b) II
(c) Both of these (d) None of these

2. Which structure is adopted by $KHCO_3$?
(a) I (b) II
(c) Both of these (d) None of these

3. Which is used in baking powder?
(a) $LiHCO_3$ (b) $NaHCO_3$
(c) $KHCO_3$ (d) Na_2CO_3

Passage 7

Sesquoxides "M_2O_3" have been prepared for K, Rb and Cs. Ozonides MO_3 have been prepared for Na, K, Rb and Cs.

Answer the following questions.

1. Which are paramagnetic in nature?
(a) M_2O_3, MO_3, MO_2 (b) M_2O_2, M_2O
(c) Both (a) & (b) (d) None of these cases

2. Sesquides is said to be double oxide of
(a) oxide, peroxide (b) oxide, superoxide
(c) peroxide, disuperoxide (d) oxide, disuperoxide

3. A typical sesquoxide of Rb/Rb_4O_6 can be represented as
(a) $[Rb_4(O_2^{2-})(O_2^-)_2]$ (b) $Rb_4(O^{2-})_6$
(c) $RbO_2, (Rb_2O_2)_2$ (d) $Rb_2O\cdot(Rb_2O_2)_2$

Passage 8

Reduction potentials of group 1 (alkali metals) are given below :

Element	$E°(U)$
$Li^+ + e^- \longrightarrow Li$	–3.045
$Na^+ + e^- \longrightarrow Na$	–2.714
$K^+ + e^- \longrightarrow K$	–2.925
$Rb^+ + e^- \longrightarrow Rb$	–2.925
$Cs^+ + e^- \longrightarrow Cs$	–2.923

Answer the following questions.

1. Which is the best reducing agent?
(a) Li (b) Na
(c) Rb (d) Cs

2. Which is the poorest reducing agent?
(a) Li (b) Na
(c) Rb (d) Cs

3. $E°$ can be related to
(a) $\Delta G°$ (b) $\Delta H°$
(c) both (a) and (b) (d) None of these

4. $Li(s) \longrightarrow Li^+(aq) + e^-$ involved
(a) sublimation (b) ionisation
(c) hydration (d) all of these

Assertion & Reason

Codes :
(a) Both **A** and **R** are true and **R** is the correct explanation of **A**.
(b) Both **A** and **R** are true but **R** is not the correct explanation of **A**.
(c) **A** is true but **R** is false.
(d) **A** is false but **R** is true.

1. Assertion (A) : Li^+ forms series of trihydrates $LiX\cdot 3H_2O\,[X = Cl, Br, I, ClO_2, ClO_4, MnO_4, NO_3, BF_4]$
Reason (R) : This is due to its smaller size (Li^+) and highest ion dipole interaction.

2. Assertion (A) : Lithium does not form alums.

Reason (R) : Small $[Li(H_2O)_6]^+$ is too small to form a stable lattice required for alum formation.

3. Assertion (A) : Alkali metals form solid bicarbonates.

Reason (R) : $LiHCO_3$ exists in aqueous solution.

4. Assertion (A) : NH_4HCO_3 exists in solid state.

Reason (R) : Alkali metals form solid bicarbonates.

5. Assertion (A) : $NaHCO_3$ and $KHCO_3$ exist in solid state.

Reason (R) : Alkali metal ions are basic in nature.

6. Assertion (A) : K_2CO_3 cannot be prepared by solvay process.

Reason (R) : $KHCO_3$ being fairly soluble in water does not precipitate in carbonation tower.

7. Assertion (A) : CsF_2 could not be formed from its elements.

Reason (R) : Enthalpy of formation of CsF is much larger which makes CsF_2 unstable.

8. Assertion (A) : The superoxide MO_2 contains the paramagnetic ion O_2^-.

Reason (R) : O_2^- is stable in presence of large cations as K, Rb, Cs.

9. Assertion (A) : FrCl has CsCl structure with coordination number 8.

Reason (R) : CN 8 is favoured by the large cation size.

10. Assertion (A) : Fr burns in air forming FrO_2.

Reason (R) : The larger alkali metal ions stabilise O_2^- (superoxide).

11. Assertion (A) : $FrClO_4$ and Fr_2PtCl_6 are the salts of low solubility.

Reason (R) : All alkali metal ions form insoluble salts.

12. Assertion (A) : Li is used in photocells.

Reason (R) : Cs has lower ionisation energy than Li.

13. Assertion (A) : The mobility of sodium ion is lower than that of potassium ion.

Reason (R) : The ionic mobilities depend on the effective radius of the ion.

14. Assertion (A) : Among the alkali metals, lithium salts exhibit the least electrical condutance in aqueous solutions.

Reason (R) : Smaller the radius of the hydrated cation, lower is the electrical conductance in aqueous solutions.

15. Assertion (A) : Ether can extract LiCl from the aqueous solution of a mixture of LiCl, NaCl and KCl.

Reason (R) : Hydration of Li^+ ion is greater than that of Na^+ and K^+ ions.

16. Assertion (A) : NaOH cannot be stored in a vessel made of Al or Zn.

Reason (R) : A protective layer of oxide is formed on the surface of the metal.

True & False

1. K_2O_2 is paramagnetic but KO_2 is diamagnetic.

2. When aqueous KO_2 solution reacts with CO_2, O_2 is formed thus KO_2 is used in submarines.

3. Borax or microcosmic salt can be used to identify coloured cations.

4. Li_2CO_3 and Na_2CO_3 are thermally stable.

5. When 1 mole of Na_2CO_3 is heated strongly, 1 mole of CO_2 is formed.

6. Solution of Na in liquid NH_3 is of blue colour due to solvated electron.

7. Lattice energy of alkali metal chlorides is in order:

$$LiCl > NaCl > KCl > RbCl > CsCl$$

8. $LiHCO_3$ does not exist in solid state.

9. Polarising power of Li^+ ion is nearly the same as that of Mg^{2+} ion.

10. Alkali metals form covalent hydrides.

11. The enthalpy of hydration decreases in going from Li^+ to Cs^+.

12. Potassium burns in air forming potassium superoxide.

13. Aqueous solution of brine on electrolysis forms sodium.

14. Carnallite is $KCl \cdot MgCl_2 \cdot 6H_2O$.

15. Li_2SO_4 forms double salts with $Al_2(SO_4)_3$.

16. $K_3[Co(NO_2)_6]$ is insoluble in water.

17. $KHCO_3$ is less soluble than $NaHCO_3$ and is thus prepared easily by following reaction.

$$NH_4HCO_3 + KCl \longrightarrow NH_4Cl + KHCO_3$$

18. Alkali metal hydrides are ionic in nature.

19. The solubility of carbonates and bicarbonates of alkali metals increases down the group.

20. Zinc displaces H_2 on reaction with aqueous NaOH as well as with dil. HCl.

Fill in the Blanks

1. .... of IA and .... of IIA show diagonal relationship.
2. .... is binary oxide of alkali metal.
3. NaOH can't be stored in the vessel made of ....
4. .... can react with NaOH as well as with HCl. This salt gives CO_2 on heating.
5. The fact that Li has maximum reducing nature is due to its maximum negative .... value which is actually in aqueous solution $M^+(aq) + e^- \longrightarrow M(s)$.
6. The first element of a group in many ways differs from the other heavier members of the group. These differences arise on account of three factors :
   (i) .... (ii) .... (iii) ....
7. Alkali metals dissolve in liquid ammonia to give .... coloured solutions which upon standing slowly liberate H_2. The .... colour of metal-ammonia solutions is due to the presence of .... electrons.
8. Of the alkali metals only .... forms nitrides.
9. Of the alkali metal carbonates, only .... decomposes to give CO_2 and metal oxide.
10. Alkali metal and alkaline earth metal chlorides can be detected by .... test.
11. KO_2 is .... absorber and .... producer and is thus used in ....
12. A solute dissolves in water if $\Delta H_{\text{hydration}}$ is .... than ΔH_U.
13. Of the O_2^-, O_2^{2-} and O_2, ... has odd order bond and, ... and .... are paramagnetic.
14. .... is used to control humidity.
15. Li_3N is decomposed by water forming .... gas.
16. Lithium ion has the .... hydration enthalpy of all the alkali metals.
17. Alkali metals are good .... agents.
18. The phenomenon of crumbling to powder of sodium carbonate crytals on exposure to air is called ....
19. Potassium salts impart .... colour to the flame.
20. Aqueous solution of Na_2CO_3 is .... due to .... of ....
21. *R*COOH decomposes $NaHCO_3$ to liberate CO_2 which comes from ....
22. Sodium metals when dissolved in ammonia gives a .... solution.
23. Sodium metal when dissolved in ammonia conducts electricity because of ....
24. Sodium silicate dissolved in water, and its concentrated aqueous solution is called ....
25. The Castner's process uses molten .... for the extraction of sodium.

Matrix-Match Type Questions

I. Only One Correct

1. Match the commercial names (in List I) with the chemical formulae (in List II).

List I	List II
A. Caustic potash	1. Na_2CO_3
B. Common salt	2. $Na_2SO_4 \cdot 10H_2O$
C. Soda ash	3. NaCl
D. Lithia	4. KOH
E. Caustic soda	5. $NaHCO_3$
F. Baking soda	6. KNO_3
G. Glaubers' salt	7. Li_2O
H. Salt petre	8. NaOH

2. Select the metals (in List I) with the specific properties of the compounds (in List II).

List I	List II
A. Li	1. Super oxide
B. Na	2. Lewis acid—a chloride
C. K	3. Most negative value of E°_{redn}
D. Be	4. Thermally stable carbonate
E. Mg	5. No flame colouration of the chloride

3. Match List I (Compounds) with List II (Associated Uses).

List I (Compounds)	List II (Associated Uses)
A. KO_2	1. Baking
B. $NaHCO_3$	2. Photovoltaic cell
C. LiCl	3. Submarine
D. Cs	4. Humidity control

II. One or More Correct

1. Match the species in Column I with given property (ies) in Column II.

Column I	Column II
A. Alkalide of halonium	1. Paramagnetic
B. Superoxide of alkali metal	2. Diamagnetic
C. Peroxide of alkali metal	3. Life saving devices
D. Oxide of alkali metal	4. Oxidant
	5. Bleaching agent
	6. Dissolves Zn, Al

2. Match the species in Column I with given property (ies) in Column II.

Column I	Column II
A. Li_3N	1. Paramagnetic
B. $LiCl$	2. Ether soluble
C. KO_2	3. Humidity control
D. Rb_2O_3	4. Automobiles air bags
	5. Coloured compounds

Integer Answer Types

This section contains 8 questions. The answer to each of the questions is a single digit integer, ranging from 0 to 9. The appropriate bubbles below the respectively question numbers in the *ORS* have to be darkened. For example, if the correct answers to question number *X*, *Y*, *Z* and *W* (say) are 6, 0, 9 and 2, respectively, then the correct darkening of bubbles will like the following :

X	*Y*	*Z*	*W*
0	0	0	0
1	1	1	1
2	2	2	2
3	3	3	3
4	4	4	4
5	5	5	5
6	6	6	6
7	7	7	7
8	8	8	8
9	9	9	9

Set I

1. Unpaired electron(s) in superoxide ion ....

2. Potassium sesquoxide has actually .... O atoms

3. When one mole of Na_2CO_3 is heated, CO_2 obtained is .... mole.

4. Number of HCO_3^- ions joined by H-bonding in $KHCO_3$ is ....

5. When $LiNO_3$ is heated, change in oxidation number of N is ....

6. When $NaNO_3$ is heated, change in oxidation number of N is ....

7. Number of *s*-electrons in Na^- is ....

8. Number of elements in second period showing diagonal relationship is ....

Set II

1. Number of radioactive elements in *s*-block is ....

2. Hydrated sodium sulphite has 50% H_2O. Number of molecules of hydration is ....

3. One mole of lithium nitride is decomposed by H_2O and resultant solution is neutralised by HCl. Number of moles of HCl required is ....

4. Cl_2 disproportionates to NaCl and $NaClO_3$ on passing into hot NaOH solution. In this process each mole of Cl_2 consumes .... mole(s) of NaOH.

5. KO_2 is CO_2 absorber and O_2 producer as shown

$$4KO_2 + CO_2 \longrightarrow K_2CO_3 + xO_2$$

Value of x is ....

6. Na_2CO_3 removes permanent hardness :

$$CaSO_4 + Na_2CO_3 \longrightarrow Na_2SO_4 + CaCO_3$$

10 L of hard water required 0.0848 g Na_2CO_3. Thus, hardness in ppm of $CaCO_3$ is ....

7. Potassium superoxide has .... unpaired electron(s).

8. Bond order of oxide in sodium peroxide is ....

Test Yourself on Periodic Table

1. Mark the element which is invariably bivalent and whose oxide is soluble in excess of NaOH and its dipositive ion has a noble gas core :

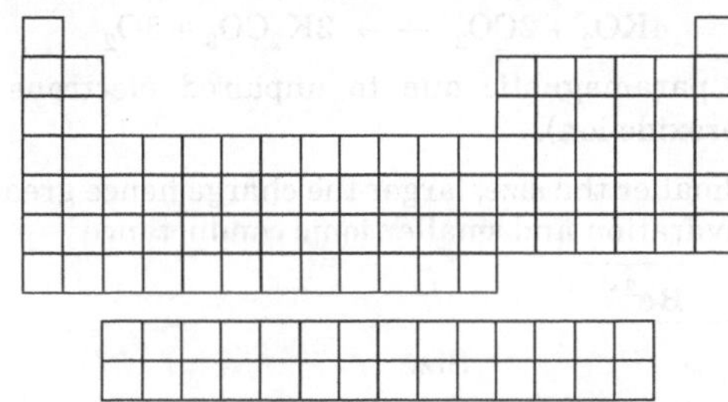

2. Identify the elements in the Periodic Table:

A : which shows diagonal relationship to Be

B : which shows diagonal relationship to Li

C : whose alkali metal carbonate is decomposed on heating

D : coinage metals

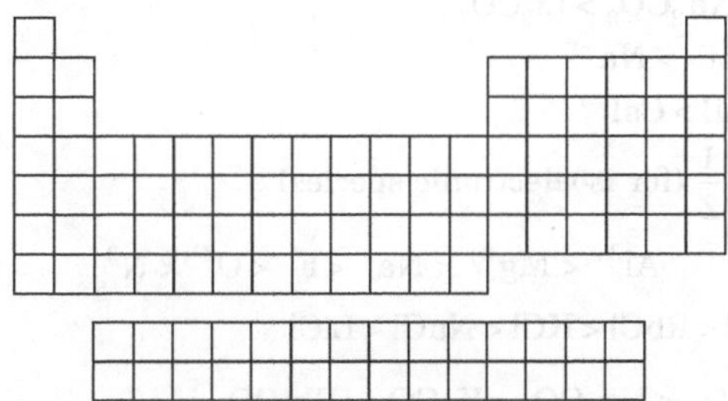

3. Out of the elements marked A, B, C, D, E, F, G, and H

(a) which form superoxide?

(b) which form thermally stable carbonate?

(c) which forms strongest base?

(d) which show diagonal relationship?

(e) which forms amphoteric oxide?

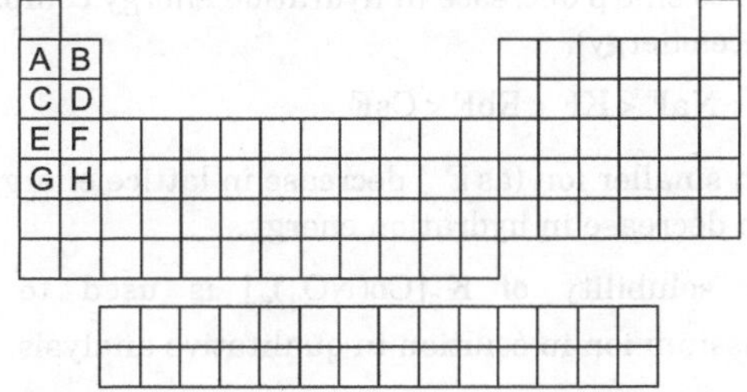

Answers & Solutions
Master Exercises
Exercise 1

Short Answer Type Questions

1. $LiHCO_3$

2. KO_2 : It absorbs CO_2 and releases O_2 thus is used in life-saving devices.

$$4KO_2 + 2CO_2 \longrightarrow 2K_2CO_3 + 3O_2$$

It is paramagnetic due to unpaired electrons in O_2^- (superoxide ion).

3. (a) Smaller the size, larger the charge hence greater the hydration and smaller ionic conductance.

Li^+	Be^{2+}	
>		Size
<		Charge
<		Hydration
>		Ionic conductance

(b) $Li^+ > K^+$

(c) $LiCl > NaCl$ (smaller cation Li^+ polarises to a greater extent than Na^+)

(d) $Na_2CO_3 > Li_2CO_3$

(e) $Li^+ > Na^+$

(f) $LiI > CsI$

4. $r_n \propto \frac{1}{Z}$ (for isoelectronic species)

$$Al^{3+} < Mg^{2+} < Na^+ < F^- < O^{2-} < N^{3-}$$

5. $CsCl < RbCl < KCl < NaCl < LiCl$

6. $Li_2CO_3 < Na_2CO_3 < K_2CO_3 < Rb_2CO_3$

7. A : Na_2O_2

B : H_2O_2

C : CrO_3 (chromic anhydride blue in ethereal layer)

8. $RbClO_4 < KClO_4 < NaClO_4 < LiClO_4$

(Due to sharp decrease in hydration energy compared to lattice energy).

9. $LiF < NaF < KF < RbF < CsF$

With smaller ion (as F^-) decrease in lattice energy more than decrease in hydration energy.

10. Low solubility of $K_3[Co(NO_2)_6]$ is used to detect potassium ion in solution in qualitative analysis.

11. $Na \longrightarrow Na^+ + e^-$

When cation is formed, outermost shell is removed (in case of alkali metal) and also due to increase in force of attraction, outermost shell is pulled inward. Hence, ionic six decreases.

12. In aqueous solution of LiCl and NaCl, add ether. LiCl being covalent is extracted into ether. Evaporate ether when LiCl is left as residue.

13. $Li^+ < Na^+$ (size)

Thus, polarisation by Li^+ is larger than that of Na^+. Thus, LiCl is more covalent than NaCl. Hence, m.p. of LiCl (being covalent) is lower than that of NaCl.

14. Refer text

15. Oxone is the commercial name for the preparation of Na_2O_2 (sodium peroxide)

$$2Na + O_2 \longrightarrow Na_2O_2$$

Na_2O_2 is used in life-saving devices in submarines. Na_2O_2 absorbs CO_2 and releases O_2

$$2Na_2O_2 + 2CO_2 \longrightarrow 2Na_2CO_3 + O_2$$

16. Along a group, ionisation energy decreases and thus tendency to oxides increases.

$$Na \longrightarrow Na^+ + e^- \qquad I_1$$
$$K \longrightarrow K^+ + e^- \qquad I_1' < I_1$$

Thus, Na is less reactive than K.

17. Lithium has highest ionisation energy while caesium has the lowest among the alkali metals. Photoelectric cells are based on emission of electrons. Hence, Cs with least ionisation energy has better tendency (as compared to Li) to emit electrons hence thus finds use in photoelectric cells.

18. N^{3-} is very small in ionic size and thus can be set into crystalline lattice with smaller Li^+ ion. Thus, Li_3N is formed

$$6Li + N_2 \longrightarrow 2Li_3N$$

19. (a) KO_2 (with superoxide as –1)

(b) $K^+O_2^-$ thus $K = +1$

(c) $O_2^-(17e^-)$ has M.O. electronic configuration as

$\sigma 1s^2 \ \sigma^* 1s^2 \ \sigma 2s^2 \ \sigma^* 2s^2 \ \sigma 2p_x^2 \ \pi 2p_y^2 \ \pi 2p_z^2 \ \pi^* 2p_y^2 \ \pi^* 2p_z^1$

N_B = Electrons in bonding MO = 10

N_A = Electrons in anti-bonding MO = 7

Number of bonds $= \frac{N_B - N_A}{2} = \frac{10-7}{2} = 1.5$

Number of unpaired electron $= N = 1$

Thus, magnetic moment $= \sqrt{N(N+2)}$ BM

$= \sqrt{3}$ BM

It is paramagnetic due to unpaired electron.

20. Na_2CO_3 is salt of weak acid (H_2CO_3) and strong base (NaOH). CO_3^{2-} is hydrolysed forming free OH^- thus it is alkaline in nature

$$Na_2CO_3 + 2H_2O \rightleftharpoons 2NaOH + \underset{\text{weak acid}}{H_2CO_3}$$

$$2Na^+ + CO_3^{2-} + 2H_2O \rightleftharpoons 2Na^+ + 2OH^- + H_2CO_3$$

Thus, $\boxed{CO_3^{2-} + 2H_2O \rightleftharpoons 2OH^- + H_2CO_3}$

21. NaOH absorbs atmospheric CO_2 and thus while deposit of Na_2CO_3 is formed. Also SiO_2 of glass reacts with NaOH forming insoluble Na_2SiO_3 (white deposit)

$$2NaOH + CO_2 \longrightarrow Na_2CO_3 + H_2O$$
$$2NaOH + SiO_2 \longrightarrow Na_2SiO_3 + H_2O$$

22. KOH is hygroscopic and thus absorbs H_2O. Thus, KOH pallets become wet.

23. Lithium (density $= 0.53$ g mL^{-1}) is lighter than water hence it floats on water. Being less reactive (due to high IE) it does not react with H_2O to a greater extent.

24. (a) Li resembles Mg

(b) Mg belongs to group II (alkaline earth metals)

(c) (i) $MgCO_3 \xrightarrow{\Delta} MgO + CO_2$

$Li_2CO_3 \xrightarrow{\Delta} Li_2O + CO_2$

(ii) $3Mg + N_2 \longrightarrow Mg_3N_2$

$6Li + N_2 \longrightarrow 2Li_3N$

25. Li^+ being smaller in size cannot stabilise larger anion as O_2^{2-} (peroxide). Thus, Li_2O_2 is not formed.

26. $LiI + KF \longrightarrow LiF + KI$

This reaction is spontaneous in forward side indicating that Li^+ (smaller cation) is attached to F^- (smaller anion) and K^+ (larger cation) is attached to I^- (larger anion).

27. Li_2SO_4 is the only alkali metal salt that does not form double salt (as alum) as it cannot set cavity of alum.

28. Baking soda and baking powder, they are different.

Baking soda is $NaHCO_3$ and baking powder is a mixture of $NaHCO_3$, $Ca(HPO_4)_2$ starch and potassium hydrogen tartarate.

29. In Solvay process, intermediate reactions are

$$NH_4HCO_3 + KCl \longrightarrow KHCO_3 + NH_4Cl$$

$KHCO_3$ is highly soluble in water and cannot be separated by crystallisation.

30. (a) $2Na_2O_2 + Mn(OH)_2 \longrightarrow \underset{\text{green}}{Na_2MnO_4} + 2NaOH$

(b) $3Na_2O_2 + 3H_2O \longrightarrow 6NaOH + 3[O]$

$$2Cr(OH)_3 + 4NaOH + 3[O] \longrightarrow 2Na_2CrO_4 + 5H_2O$$

$$\underset{\text{Green ppt}}{2Cr(OH)_3} + 3Na_2O_2 \longrightarrow \underset{\text{Yellow}}{2Na_2CrO_4} + 2NaOH + 2H_2O$$

Analytical Questions

1. The oxidation of a metal (M) to form $M^+(aq)$ takes place in following stages :

Stage I : Sublimation

$M(s) \longrightarrow M(g)$ ΔH°_{sub} Endothermic

Stage II : Ionisation

$M(g) \longrightarrow M^+(g) + e^-$

ΔH°_{ioni} Endothermic

Stage III : Hydration

$M^+(g) + H_2O\,(\text{excess}) \longrightarrow M^+(aq)$

$\Delta H^\circ_{Hydration}$ Exothermic

Overall : $M(s) + H_2O\,(\text{excess}) \longrightarrow M^+(aq) + e^-$

It is feasible only if more energy is released in stage III than that absorbed in stages I and II. Li^+ being smallest in size, its hydration is maximum hence energy released in stage III (exothermic) is much larger than that absorbed overall in stages I and II together. Thus, Li acts as strongest reducing agent and is the best reducing agent.

2. $Li + N_2 \longrightarrow 2Li_3N] \times 6$

$Li_3N + 3H_2O \longrightarrow 3LiOH + NH_3] \times 2$

$NH_3 + H_2O + CO_2 \longrightarrow NH_4HCO_3] \times 2$

$NH_4HCO_3 + NaCl \longrightarrow NaHCO_3 + NH_4Cl] \times 2$

$2NaHCO_3 \longrightarrow Na_2CO_3 + H_2O + CO_2$

Overall

$$6Li + N_2 + 7H_2O + CO_2 + 2NaCl \longrightarrow 6LiOH + Na_2CO_3 + 2NH_4Cl$$

Thus, every one mole of N_2 gives enough NH_3 that can be used to absorb CO_2 and thus gives one mole $\mathbf{Na_2CO_3}$

$N_2 \equiv Na_2CO_3$

1 mole $\equiv$ 1 mole

22.4 L at NTP $\equiv$ 106 g

22.4 kL at NTP $\equiv$ 10.6 kg

Thus, $\mathbf{22.4 \times 10^3}$ **L** N_2 is required.

3. (a) Greater the charge, smaller the radius, larger the hydration and thus heavier (hydrated) ion more slower.

$Li^+ < Na^+ < K^+ < Rb^+ < Cs^+$

ionic size ⟶ max

hydration ⟵ max

Ionic mobility ⟶ max

(b) Smaller cation (Li^+) can accommodate smaller anion (N_3^-) hence, Li^+ forms nitride LiN_3.

(c) When Na_2SO_4 dissolves in water, there is hydration also

$$Na_2SO_4 + xH_2O \rightleftharpoons Na_2SO_4 \cdot xH_2O, \quad \Delta H = -\text{ve}$$

This hydration process is exothermic, hence increase in temperature shifts above equilibrium in the direction in which heat is absorbed (*i.e.*, backward side) hence, solubility is decreased. This is in accordance with Le-Chatelier's principle.

(d) Refer text

(e) $\rangle C^+ + Cl^- + M^+ + F^- \longrightarrow \rangle C—F + MCl$

only variable is $U(MCl) - U(MF)$

This opposes reaction but less when M^+ is larger.

4. (a) $C_2H_5OH + Na \longrightarrow C_2H_5ONa + \frac{1}{2}H_2$

1 mol, 46 g; 0.5 mol = 11.2 L at NTP 1 g H_2

$CH{\equiv}CH + 2Na \longrightarrow NaC{\equiv}CNa + H_2$

1 mol, 26 g; 1 mole = 22.4 L at NTP 2 g H_2

(b) $CH{\equiv}CH$ gives better yield in terms of volume of H_2

(c) % yield from C_2H_5OH in terms of mass

$$= \frac{1}{46} \times 100 = \mathbf{2.174\%}$$

$$\% \text{ yield from } C_2H_2 \text{ in terms of mass} = \frac{2 \times 100}{26} = \mathbf{7.692\%}$$

5. The distortion of the electron cloud of a neighbouring anion by a cation is called polarisation. Polarisation leads to lowering of the charges on the both the ions due to partial neutralisation of charges and thus compounds assume covalent nature factors that favour polarisation.

(i) Smaller cation
(ii) Larger anion
(iii) Larger charge on ion (cation/anion)
(iv) Presence of *d*- and *f*-electrons

Li^+ being smallest in ionic size has maximum polarising tendency, and it can easily polarise larger anion as I^-, thus LiI shows covalent nature compound to NaI, KI etc.

6. $Al_2O_3 + 6Na \longrightarrow 3Na_2O + 2Al$

$BeCl_2 + 2Na \longrightarrow Be + 2NaCl$

$3CO_2 + 4Na \longrightarrow C + 2Na_2CO_3$

$SiO_2 + 4Na \longrightarrow Si + 4Na_2O$

Similarly with K.

7. $2Na_2O_2 + Mn(OH)_2 \longrightarrow \underset{(A)\text{ green}}{Na_2MnO_4} + 2NaOH$

$2Na_2MnO_4 + Cl_2 \longrightarrow 2NaCl + \underset{(B)\text{ violet}}{2NaMnO_4}$

$Na_2O_2 + 2HCl \longrightarrow 2NaCl + H_2O_2$

MnO_4^- is decolourised by H_2O_2 in acidic medium.

$H_2O_2 \longrightarrow 2H^+ + O_2 + 2e^-$

$MnO_4^- + 8H^+ + 5e^- \longrightarrow \underset{\text{colourless}}{Mn^{2+}} + 4H_2O$

8. $CrCl_3 + 3NaOH \longrightarrow \underset{(A)}{Cr(OH)_3} + 3NaCl$

$Cr(OH)_3 + Na_2O_2 + H_2O \longrightarrow \underset{(B)\text{ yellow}}{Na_2CrO_4}$

$Na_2CrO_4 + (CH_3COO)_2Pb \longrightarrow \underset{(C)\text{ yellow ppt.}}{PbCrO_4\downarrow}$

9. (a) $Na_2O_2 + 2H_2O \longrightarrow H_2O_2 + 2NaOH$

H_2O_2 bleaches red litmus

Thus, white powder is Na_2O_2.

(b) If it would be Na_2O, then red litmus turned blue by alkaline (NaOH) solution

$Na_2O + H_2O \longrightarrow 2NaOH$

10. *A* contain oxygen since *B* gives white precipitate with $BaCl_2$ indicating that *B* has SO_4^{2-}.

A is a binary salt with M^+ cation and oxygen containing anion.

Thus, *A* is MO_2

$$\underset{(M+32)\text{ g}}{MO_2} + S \longrightarrow \frac{1}{2}(M_2SO_4) \quad \left(\frac{2M + SO_4}{2}\right) = (M + 48)$$

$$M_2SO_4 + BaCl_2 \longrightarrow \underset{(233\text{ g})}{BaSO_4} + 2MCl$$

$$\frac{MO_2}{BaSO_4} = \frac{0.71}{2.33}$$

$$\frac{M+32}{233} = \frac{0.71}{2.33} = \frac{71}{233}$$

$\therefore \quad M = 39$

(a) Thus, *M* is K (potassium)

(b) *B* is K_2SO_4

(c) C is $K_2SO_4 \cdot Al_2(SO_4)_3 \cdot 24H_2O$

(d) K_2SO_4 obtained from 0.71 g KO_2 = 0.87 g and double salt (alum) obtained = 4.74 g

11. $Cl(g) \longrightarrow Cl^-(g) + e^-$ $\Delta H = 1251$ kJ mol^{-1} (IE)

$Na(g) + e^- \longrightarrow Na^-(g)$ $\Delta H = -52.9$ kJ mol^{-1} (EA)

$Na^-(g) + Cl^+(g) \longrightarrow Cl^+Na^-(s)$ $\Delta H = -787$ kJ mol^{-1} (*U*)

$\frac{1}{2}Cl_2(g) \longrightarrow Cl(g)$ $\Delta H = 122$ kJ mol^{-1}

$Na(s) \longrightarrow Na(g)$ $\Delta H = 107.3$ kJ mol^{-1}

$Na(s) + \frac{1}{2}Cl_2(g) \longrightarrow Cl^+Na^-(s)$ $\Delta H = 640.4$ kJ mol^{-1} by addition

$$\Delta H_f(Na^+Cl^-) = -411 \text{ kJ mol}^{-1}$$

$$\Delta H_f(Cl^+Na^-) = +640 \text{ kJ mol}^{-1}$$

Thus, formation of Cl^-Na^- is not favoured.

NCERT Problems

Refer Text

Exercise 2

Only One Option Correct

1. (a) [Ar] $3d^{10}4s^2$ — Last filling electron is in $3d$ thus it is *d*-block element

(b) [Ar] $3d^{10}4s^1$ — same as (a)

Thus, both are *d*-block element

Thus, **(c)**

2.

(a) $Li_2CO_3 \xrightarrow{\Delta} Li_2O + CO_2$ Thus, (a) is true	$Na_2CO_3 \xrightarrow{\Delta} X$
(b) LiCl soluble in ether thus covalent Thus, (b) is true	NaCl insoluble in ether but soluble in H_2O thus ionic
(c) Li forms nitride Thus, (c) is true	Na does not form nitride

Thus, **(d)**

3.

IA (Alkali)	IB (Coinage)
(a) NaCl	CuCl +1 valency thus true
(b) Na_2SO_4 soluble	$CuSO_4$, Ag_2SO_4 less soluble
(c) $Na_2O + H_2O \longrightarrow \underset{\text{strong base}}{2NaOH}$	$Cu_2O + H_2O \longrightarrow \underset{\text{weak base}}{Cu(OH)_2}$
	Thus false
(d) also false	

Thus, (a)

4. NaCl — golden yellow

AgCl — no flame

Thus, **(d)** is incorrect

5. (a) $Na_2CO_3 \xrightarrow{\Delta}$ no CO_2

(b) $CaCO_3 \longrightarrow CaO + CO_2$

$CaO \xrightarrow{H_2O} Ca(OH)_2$

$CaO \xrightarrow{HCl}$ no gas

(c) $Ca(HCO_3)_2 \longrightarrow \underset{\text{insoluble in } H_2O}{CaCO_3} + H_2O + CO_2$

(d) $2NaHCO_3 \longrightarrow Na_2CO_3 + H_2O + CO_2$

$Na_2CO_3 \xrightarrow{HCl} CO_2 \uparrow$

Na_2CO_3 is water soluble

Thus, **(d)**

6. KO_2 is stable due to larger anion O_2^- which polarises K^+

Thus, **(b)**

7. $\underset{(X)}{NaHCO_3} \longrightarrow \underset{(Y)}{Na_2CO_3} + H_2O + CO_2$

$Na_2CO_3 \xrightarrow{CO_2} NaHCO_3$

Thus, **(c)**

8. (c) **9.** (c)

10. In molten state $NaCl(l) \rightleftharpoons Na^+ + Cl^-$

Thus, **(b)**

11. (d)

12. $Na_2CO_3 + SO_2 \longrightarrow Na_2SO_3 + CO_2$

Thus, **(d)**

13. Na_2CO_3 is thermally stable

Thus, **(a)**

14. Alkali salts are colourless

NaCl colourless Na_2CrO_4 yellow since CrO_4^{2-} is yellow

KCl colourless K_2CrO_4 — yellow

$KMnO_4$ — pink since MnO_4^- is pink

K_2MnO_4 — green since MnO_4^{2-} is green

Thus, **(b)**

15. $Na + O_2 \longrightarrow Na_2O$

$Li + O_2 \longrightarrow Li_2O$

$Li + Ni \longrightarrow Li_3N$

Thus, **(d)**

16. Li and Na do not form superoxide MO_2

Thus, **(b)**

17. (a)

18. $4KO_2 + 2H_2O + 4CO_2 \longrightarrow 4KHCO_3 + 3O_2$

Thus, **(a)**

19. Third IE (I_3) is very high, thus it represents alkali earth metal (Be)

Thus, **(c)**

20. $2Na_2O_2 + 2CO_2 \longrightarrow 2Na_2CO_3 + O_2$

Thus, **(c)**

21. (c)

22. (a) (*s*-block)

(b) (*p*-block)

(c) $1s^22s^2$ in ground state — (*s*-block) and $1s^22s^12p^1$ in excited state.

(d) [Ar] $4s^2$ — ground state

[Ar] $4s^14p^1$ — excited state

Thus, *s*-block.

Thus, **(b)**

23. $NH_4NO_3 \longrightarrow N_2O + 2H_2O$
Thus, **(b)**

24. $2LiNO_3 \longrightarrow Li_2O + 2NO_2 + \frac{1}{2}O_2$

$Mg(NO_3)_2 \longrightarrow MgO + 2NO_2 + \frac{1}{2}O_2$

Thus, **(c)**

25. (d) **26.** (d)

27. Mg does not react with NaOH.
Thus, **(c)**

28. $Li_2CO_3 \longrightarrow Li_2O + CO_2\uparrow$

$Na_2CO_3 \cdot 10H_2O \longrightarrow Na_2CO_3 + 10H_2O\uparrow$

Thus, **(c)**

29. $2Na_2O \longrightarrow Na_2O_2 + 2Na$
Thus, **(b)**

30. $Na_2S_2O_3$ (as fixer) dissolves unreacted AgBr as
$2Na_2S_2O_3 + AgBr \longrightarrow Na_3[Ag(S_2O_3)_2] + NaBr$
Thus, **(c)**

31. $KI + I_2 \rightleftharpoons KI_3$
Thus, **(a)**

32. Refer Q. 31
Thus, **(c)**

33. (d)

34. $2NaNO_3 \longrightarrow 2NaNO_2 + O_2$

Thus, (b) is incorrect.

35. (d) **36.** (a) **37.** (b)

38. Smaller cation polarises anion to a greater extent hence larger covalent nature and thus lattice energy.
Thus, **(a)**

39. (a)

40. KO_2 (O_2^- is paramagnetic due to one unpaired electron)

NO_2 (is paramagnetic due to one unpaired electron on N)
Thus, **(a)**

41. (a) **42.** (c)

43. $LiNO_3$ on ignition imparts crimson red,

$BaCl_2 \longrightarrow$ green

$(NH_4)_2Cr_2O_7 \longrightarrow$ green

Thus, **(d)**

44. (c)

45. Molar solubility is in **mol L^{-1}** (Reference concise inorganic chemistry J.D. Lee)

LiCl = 19.6
NaCl = 6.2
KCl = 4.8
RbCl = 7.5
CsCl = 11.0

Thus, **(c)**

Solution 46-50

$CaCO_3 \longrightarrow CaO + CO_2$
(*A*) (*B*)

$CaO + H_2O \longrightarrow Ca(OH)_2$
(*C*)

$Ca(OH)_2 + CO_2 \longrightarrow CaCO_3$
(*A*)

$Ca(OH)_2 + NH_4Cl \longrightarrow NH_3$
(*D*)

$NH_3 + CO_2 + H_2O \longrightarrow NH_4HCO_3$
(*E*)

$NH_4HCO_3 + NaCl \longrightarrow NaHCO_3 + NH_4Cl$
(*F*)

46. (a) **47.** (b) **48.** (a) **49.** (b) **50.** (a)

51. $Na_2O_2 + 2H^+Cl^- \longrightarrow H_2O_2 + 2NaCl$

$H_2O_2 + Cr_2O_7^{2-} + 2H^+ \longrightarrow CrO_5$
deep violet soluble in ether

Thus, **(a)**

52. $(CaCN_2 + C)$
Thus, **(b)**

53. Potassium antimonyl tartrate
Thus, **(b)**

54. $NaHCO_3$ is acidic and decomposed by NaOH forming Na_2CO_3

$NaHCO_3$ (acid) + NaOH (base) $\longrightarrow Na_2CO_3 + H_2O$

Thus, **(c)**

55. (a) $Li + N_2 \longrightarrow Li_3N$

$Mg + O_2 \longrightarrow MgO$

O_2 and N_2 are removed

(b) $NaOH + CO_2 \longrightarrow Na_2CO_3$

CO_2 is removed

Thus, **(c)**

56. (c)

57. HCHO is used as preservatives of dead biological specimen.
Thus, **(d)**

58. (c) **59.** (b)

60. Li_2CO_3 and $MgCO_3$ are decomposed by heating

$Li_2CO_3 \longrightarrow Li_2O + CO_2$

$MgCO_3 \longrightarrow MgO + CO_2$

Thus, **(c)**

61. $CaCO_3 + CO_2 + H_2O \longrightarrow Ca(HCO_3)_2$

$Na_2CO_3 \xrightarrow{\Delta} X$

$Li_2CO_3 \xrightarrow{\Delta} Li_2O + CO_2$

$CaCl_2/CaSO_4$ make permanent hardness.
Thus, **(a)**

62. Na^- is alkalide ion
$Na\,(1s^2 2s^2 2p^6 4s^1)$ ← Paramagnetic due to unpaired electrons
$Na^-(1s^2 2s^2 2p^6 3s^2)$ ← diamagnetic due to paired electrons
Thus, **(a)**

63. (d) **64.** (d) **65.** (b)

66. Smaller cation, larger hydration thus smaller molar ionic conductance
Ionic size $Li^+ < Na^+ < K^+ < Rb^+$
Hydration $Li^+ > Na^+ > K^+ > Rb^+$
Size of hydration ion $Li^+ > Na^+ > K^+ > Rb^+$
Molar ionic conductance $Li^+ < Na^+ < K^+ < Rb^+$
Thus, **(a)**

67. (a) **68.** (b)

69. M_2O_3 actually exists as $M_4O_6\,[M_4(O_2^{2-})(O_2^-)_2]$
O_2^- is paramagnetic.
Thus, **(b)**

70. KO_2 is paramagnetic due to O_2^-
K_2O_3 is actually K_4O_6

$$K_4(O_2^{2-})(O_2^-)_2$$

↑ peroxide ↑ superoxide

Thus, it also paramagnetic.
Thus, **(c)**

71. (d) **72.** (b) **73.** (a) **74.** (d) **75.** (c)

One or More Than One Options Correct

1. (a), (b), (c), (d) **2.** (a), (b) **3.** (a), (b)
4. (a), (c), (d) **5.** (a), (b) **6.** (b), (c) **7.** (a), (b), (d)
8. (a), (d) **9.** (b), (c), (d) **10.** (a), (b), (c) **11.** (a)
12. (a), (b), (c) **13.** (a), (b), (c) **14.** (a), (b), (c)
15. (a), (b), (c)

Passage Comprehension Questions

Passage 1

1. (b) **2.** (c)

Passage 2

1. (a) $\Delta H_{(Hydration)} = \Delta H_{hydration}(\text{cation}) + \Delta H_{hydration}(\text{anion})$

2. (d)

3. (c) $\Delta H_{solution} = \Delta H_{hydration} - \Delta H_U$

$$\Delta H_{solution}\,(LiCl) = -499 - 382 + 840 = -41\ \text{kJ mol}^{-1}$$
$$NaCl = -4\text{kJ mol}^{-1}$$
$$KCl = +16\ \text{kJ mol}^{-1}$$

4. (a)

Passage 3

1. (b) **2.** (c) **3.** (c) **4.** (a), (c)

Passage 4

1. (b) **2.** (a) **3.** (d) **4.** (b)

5. (c) $w = zit = \dfrac{40\times 1.08\times 137}{96500\times 1} = 0.06$ g

6. (b) 0.06 g NaOH per 0.0445 L,

$$\text{Conc. in mol L}^{-1} = \frac{0.06}{40\times 0.0445} = 3.37\times 10^{-3}\ \text{M}$$
$$\therefore\quad [OH^-] = 3.37\times 10^{-3}\ \text{M}$$
$$p_{OH} = 2.47$$
$$p_H = 11.53$$

Passage 5

1. (a) **2.** (a) **3.** (c) **4.** (b)

Passage 6

1. (b) **2.** (a) **3.** (b)

Passage 7

1. (a) **2.** (c) **3.** (a)

Passage 8

1. (a) **2.** (b) **3.** (c) **4.** (d)

Assertion & Reason

1. (a) **2.** (a) **3.** (b) **4.** (b) **5.** (a) **6.** (a)
7. (a) **8.** (b) **9.** (a) **10.** (a) **11.** (c) **12.** (d)
13. (a) **14.** (c) **15.** (b) **16.** (c)

True & False

1. F, (KO_2 is paramagnetic) **2.** T **3.** T

4. F : Li_2CO_3 is decomposes into CO_2 and Li_2O, Na_2CO_3 does not

5. F, Na_2CO_3 is not decomposed on heating.

6. T **7.** T **8.** T **9.** T

10. F : Ionic hydride **11.** T **12.** T

13. F : Na formed reacts with H_2O giving NaOH and H_2.

14. T

15. F : Smaller size of Li^+ cannot set into double salt.

16. T **17.** F : $KHCO_3$ is more soluble.

18. T **19.** T **20.** T

Fill in the Blanks

1. Li, Mg **2.** KO_2 **3.** Zn/Al/Be **4.** $NaHCO_3$

5. standard reduction electrode potential

6. smaller size, greater charge density, lack of *d*-orbital

7. blue, blue, solvated **8.** Li **9.** Li_2CO_3

10. flame **11.** CO_2, O_2, submarines **12.** greater
13. O_2^-; O_2, O_2^- **14.** LiCl **15.** NH_3 **16.** highest
17. reducing **18.** efflorescence **19.** violet
20. alkaline, hydrolysis, CO_3^{2-} **21.** $NaHCO_3$
22. conducting **23.** solvated electrons
24. water glass **25.** NaCl

Matrix-Match Type Questions

I. Only One Correct

1. (A)—(4); (B)—(3); (C)—(1); (D)—(7); (E)—(8); (F)—(5); (G)—(2); (H)—(6);
2. (A)—(3); (B)—(4); (C)—(1); (D)—(2) (E)—(5)
3. (A)—(3); (B)—(1); (C)—(4); (D)—(2)

II. One or More Correct

1. (A)—(2); (B)—(1,3,4); (C)—(2,4,5); (D)—(2,6)
2. (A)—(4,5); (B)—(2,3); (C)—(1,5); (D)—(1,5)

Integer Answer Types

Set I

Questions →	1	2	3	4	5	6	7	8
Answers →	1	6	0	2	1	2	6	3
	0	0	0	0	0	0	0	0
	1	1	1	1	1	1	1	1
	2	2	2	2	2	2	2	2
	3	3	3	3	3	3	3	3
	4	4	4	4	4	4	4	4
	5	5	5	5	5	5	5	5
	6	6	6	6	6	6	6	6
	7	7	7	7	7	7	7	7
	8	8	8	8	8	8	8	8
	9	9	9	9	9	9	9	9

Set II

Questions →	1	2	3	4	5	6	7	8
Answers →	2	7	4	2	3	8	1	1
	0	0	0	0	0	0	0	0
	1	1	1	1	1	1	1	1
	2	2	2	2	2	2	2	2
	3	3	3	3	3	3	3	3
	4	4	4	4	4	4	4	4
	5	5	5	5	5	5	5	5
	6	6	6	6	6	6	6	6
	7	7	7	7	7	7	7	7
	8	8	8	8	8	8	8	8
	9	9	9	9	9	9	9	9

Test Yourself on Periodic Table

1.

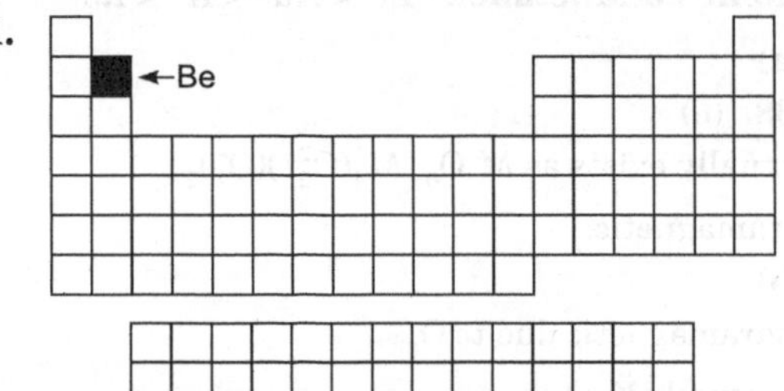

2.

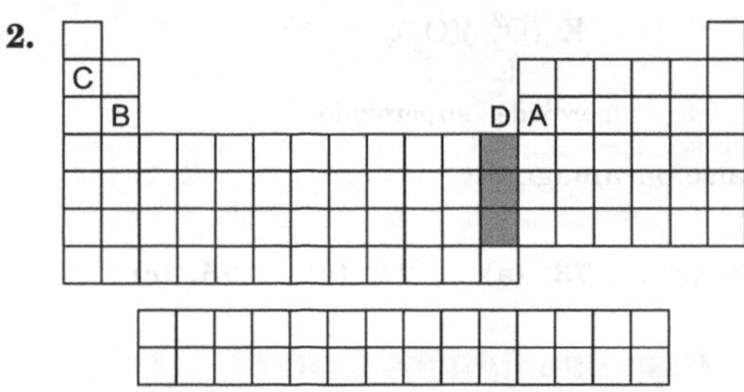

3. (a) E and G (b) C, E, and G (c) G (d) A and D (e) B

9

s-Block Elements —The Alkaline Earth Metals

Group 2 family members (alkaline-earth elements) are not free from tension. The youngest of the family (Be) and the younger one (Mg) does not have patch with other members. Be has tied up with Al (Group 13) and Mg with Li (Group 1) — a diagonal relationship. *"Ghar ka Bhedi Lanka Dhaye."*

- General electronic configuration is [Inert gas] ns^2 — Alkaline earth metals; Group 2(IIA) n signifies period and number of valence electrons in ns, its group.

Alkaline Earth Metals			
Iron	**EC**	**Z**	**2(IIA)**
Be^{2+}	[He] $2s^2$	4	Be
Mg^{2+}	[Ne] $3s^2$	12	Mg
Ca^{2+}	[Ar] $4s^2$	20	Ca
Sr^{2+}	[Kr] $5s^2$	38	Sr
Ba^{2+}	[Xe] $6s^2$	56	Ba
Ra^{2+}	[Ra] $7s^2$	88	Ra

Alkali earth metals form compounds with +2 oxidation state ions, having nearest stable noble gas configuration.

- Magnesium is an important constituent of chorophyll which initiates the process of photosynthesis in green plants.
- Magnesium ion are concentrated more in intracellular than in extracellular fluids in animal bodies. Their presence is also necessary for the activation of phosphate-transfer enzymes. These enzymes take part in the biochemical process (exothermic) occurring in animal body. Mg^{2+} ions are also involved in carbohydrate metabolism.
- Ca^{2+} ions are essential for the formation of bones and teeth. The enamel on teeth is a double salt $3Ca_3(PO_4)_2 \cdot CaF_2$.
- Mg^{2+} and Ca^{2+} ions are also involved in the transmission of electrical impulses along the nerve fibre and for the contraction of muscles.

Learning Framework of the Chapter ...

^1_1H	
$_3Li$	$_4Be$
$_{11}Na$	$_{12}Mg$
$_{19}K$	$_{20}Ca$
$_{37}Rb$	$_{38}Sr$
$_{55}Cs$	$_{38}Ba$
$_{87}Fr$	$_{88}Ra$

Periodicity at a Glance

Detailed periodicity has been given in chapter 3. Periodicity of s-block elements is summarised below.

There is uniform gradation in properties except in case of Li and Be which being smallest in their groups show abnormal behaviour. Li resembles Mg and Be resembles Al that we call **diagonal relationship.**

Periodicity in *s*-Block Elements

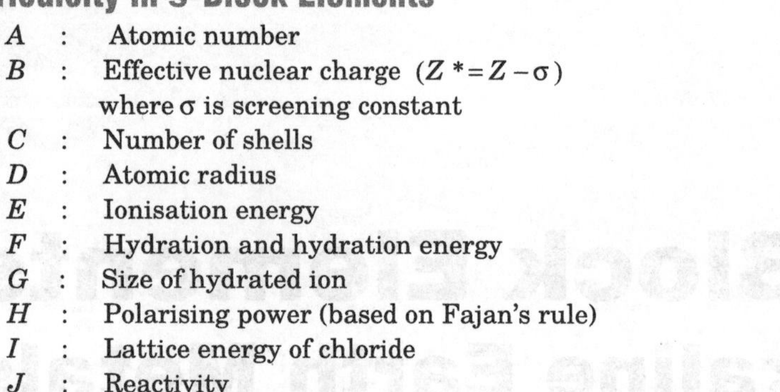

A : Atomic number
B : Effective nuclear charge $(Z^* = Z - \sigma)$ where σ is screening constant
C : Number of shells
D : Atomic radius
E : Ionisation energy
F : Hydration and hydration energy
G : Size of hydrated ion
H : Polarising power (based on Fajan's rule)
I : Lattice energy of chloride
J : Reactivity
K : Ionic conductance

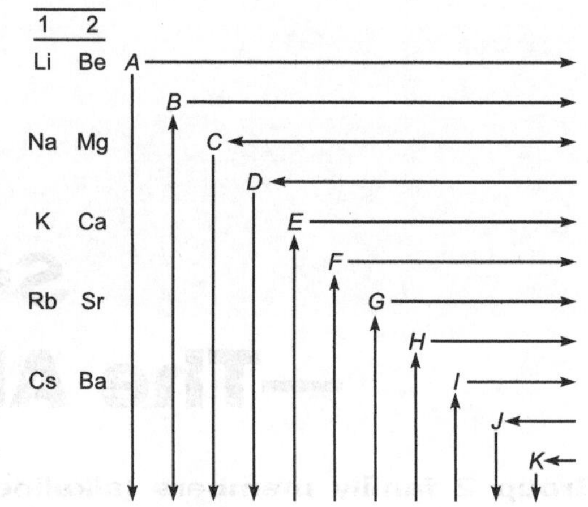

- From the above table, it is concluded that due to higher (*IE*) of alkaline earth metals, they are less reactive than alkali metals.

Table 9.1 Summarises some of the properties of alkaline earth metals [Group 2(IIA)]

Table 9.1 **Properties of the Alkaline Earth Metals**

Properties	Be	Mg	Ca	Sr	Ba	Ra
Atomic number	4	12	20	38	56	88
Atomic mass	9.012	24.3050	40.078	87.62	137.327	226.0254
Effective nuclear charge	1.95	2.85	2.85	2.85	2.85	2.85
Ground state electron configuration	[He] $2s^2$	[Ne] $3s^2$	[Ar] $4s^2$	[Kr] $5s^2$	[Xe] $6s^2$	[Rn] $7s^2$
Metallic radius, (pm)	112	160	197	215	222	—
Ionic radius, (pm)	31	72	100	118	135	—
Ionic charge density (ionic charge/radius (Å))	6.45	2.78	2.00	1.69	1.48	—
$(IE)_1$, kJ mol^{-1}	899	737	590	549	503	508
$(IE)_2$, kJ mol^{-1}	1757	1450	1145	1064	965	975
$[(IE)_1 + (IE)_2]$ kJ mol^{-1}	2656	2187	1735	1613	1468	1483
$(IE)_3$, (kJ mol^{-1})	14850	7730	4940	4150	3440	—
Pauling's electronegativity	1.5	1.2	1.0	1.0	0.9	—
$\Delta H°_{hydr}$, Hydration energy, kJ mol^{-1} $[M^{2+}(g) + aq \longrightarrow M^{2+}(aq)]$	−2494	−1921	−1577	−1443	−1305	—
$\Delta S°_{hydra}$ of M^{2+} /JK^{-1} mol^{-1}	−300	−320	−230	−220	−200	—
$\Delta G°_{hydra}$ of M^{2+}/kJ mol^{-1}	−2410	−1845	−1525	−1400	−1265	—
$\Delta H°_{atomis}$ / kJ mol^{-1}	326	149	177	164	178	130
Density at 25°C (g cm^{-3})	1.848	1.738	1.55	2.54	3.51	5
Colour	Grey	Silver	Silver	Silver	Silver-yellow	Silver
Enthalpy of fusion $\Delta H°_{fusion}$/kJ mol^{-1}	11.6	8,95	8.95	9.62	7.66	7.25
Bond enthalpy of M_2 (kJ mol^{-1})	9.46					
Electrode potential, $E°$, (V) $[M^{2+}(aq) + 2e^- \longrightarrow M(s)]$	−1.85	−2.37	−2.87	−2.89	−2.90	−2.92
Melting point/°C	1287	649	839	768	727	700
Boiling point/°C	2500	1105	1494	1381	1850	1140
Abundance on earth %	2.8×10^{-3}	2.33	4.15	0.038	0.042	trace

- Like the alkali metals they also form predominantly ionic compounds but tendency towards covalency is greater, particularly with Be and Mg because of their smaller atomic and ionic radii.
- Be forms compounds which are essentially covalent.
- Alkaline earth metal ions, because of their larger charge to size ratio, exert a much stronger electrostatic attraction on the oxygen of water molecules surrounding them.
- Since the alkaline earth metals (except Be) tend to lose their valence electrons readily, they act as strong reducing agents as indicated by E°_{red} values. (Table 9.1). The particularly less negative value for Be arises from the large hydration energy associated with the small size of Be^{2+} and the relatively large value of heat of sublimation.
- Basic nature of oxides increases down the group but solubilities of sulphates and carbonates decrease as ionic size increases.

Thermal Stability of Oxy Salts

All the group 2 elements form oxysalts. The thermal stability of the oxysalts increases with the increase in electropositivity of the metals. Thus it increases down the group. The thermal stability of the salts of group 2 is less than those of the group 1 metals.

- The sulphates are stable to heat whereas the carbonates decompose to give MO and CO_2, the temperature of decomposition increasing from Mg to Ba. **$BeCO_3$ is kept in the atmosphere of CO_2 to prevent its decomposition.**

$$MCO_3 \longrightarrow CO_2 + MO$$
$$MSO_4 \longrightarrow SO_3 + MO$$

$BeCO_3$	$MgCO_3$	$CaCO_3$	$SrCO_3$	$BaCO_3$
< 100°C	540°C	900°C	1290°C	1360°C
$BeSO_4$	$MgSO_4$	$CaSO_4$	$SrSO_4$	—
575°C	875°C	1100°C	1375°C	—

Alkali metal nitrates decomposes into nitrites on heating whereas alkaline earth metal nitrates decompose on heating to metal oxide, nitrogen dioxide and oxygen.

$$2Ca(NO_3)_2 \longrightarrow 2CaO + 4NO_2 + O_2$$

DEEP Focus

Atomic and Ionic Radii

The atoms of the alkaline earth metals are smaller than those of the corresponding group 1 elements. This is because of the increase in effective nuclear charge with increase in atomic number. Due to increase in effective nuclear charge, valence shell electrons are pulled in more firmly by the nucleus, thereby reducing the size of the atoms. Similarly, their ionic radii are also smaller than those of group 1 elements, because the removal of two orbital electrons increase the effective nuclear charge even further.

The atomic and ionic radii increase from Be to Ra due to the effect of extra shells of electrons being added. This outweighs the effect of increased nuclear charge.

These elements are denser than group 1 metals because they have two valence electrons per atom for bonding the atoms into a metallic lattice and as a result more mass can be packed into a smaller volume. The density decreases slightly on moving down the group from Be to Ca but increases considerably there after up to Ra.

Melting Point

Group 2 metals have higher melting points when compared to the group 1 metals. The reason being +2 charge on the cations in the metallic lattice, causing them to be more strongly attracted to the 'sea of electrons' and making it difficult to pull them apart.

Ionisation Energy

The first ionisation energy (Table 9.1) of alkaline earth metals is more than that of corresponding alkali metals. This is because the alkaline earth metals have higher effective nuclear charge and are smaller in size, thereby the electrons are more tightly held to the nucleus. The second ionisation energy of these elements is almost twice first ionisation energy. This is because once one electron has been removed, the effective nuclear charge felt by orbital electrons is increased, so that the remaining electrons are more tightly held and hence, much more energy is needed to remove the second electron. However, their second ionisation energy is less than that of the corresponding alkali metals because of stability of a closed shell configuration of the univalent cations that are formed in the latter cases. The ionisation energy of alkaline earth metals also decreases on moving down the group.

Lower second ionisation energies of the alkaline earth metals as compared to those of alkali metals can be examplified below :

$$\underset{(2,\,8,\,2)}{Mg} \longrightarrow \underset{(2,\,8,\,1)}{Mg^+} + e^-$$

$$\underset{(2,\,8,\,1)}{Mg^+} \longrightarrow \underset{\substack{(2,\,8)\\ \text{stable closed}\\ \text{shell}}}{Mg^{2+}} + e^-$$

(Electronic configruration of the metal and the ion is given in brackets)

$$\underset{(2,\,8,\,1)}{Na^+} \longrightarrow \underset{\substack{(2,\,8)\\ \text{stable}}}{Na^+} + e^-$$

$$\underset{\substack{(2,\,8)\\ \text{stable closed}\\ \text{shell}}}{Na^+} \longrightarrow \underset{(2,\,7)}{Na^{2+}} + e^-$$

Electropositive character and the reducing property (tendency to lose electrons) increase on moving down the group.

As the alkaline earth metals may lose electrons quite easily, they form divalent cations which have a noble gas structure with no unpaired electrons. Therefore, their compounds are diamagnetic and colourless, unless the anion is coloured. Ca, Sr and Ba compounds give characteristic flame colourations which are used to identify them —

Ion	Colour
Ca^{2+}	Brick-red
Sr^{2+}	Crimson
Ba^{2+}	Apple-green
Ra^{2+}	Carmine-red

Solubility, Lattice Energy and Hydration Energy

The metal ions are easily hydrated, *e.g.*, $MgCl_2 \cdot 6H_2O$, $CaCl_2 \cdot 6H_2O$, $BaCl_2 \cdot 2H_2O$. The hydration energies of these ions are much greater than those of alkali metal ions (Table 9.2), because of their smaller size and increased cationic charge. The lattice energies (Table 9.2) of alkaline earth metal salts are also much higher than those of alkali metal salts.

DEEP Focus

Hydration and lattice energies decrease with increase in size of metal ions. Decreasing lattice energy favours increased solubility, whilst decreasing hydration energy favours decreased solubility. If on moving down a group the hydration energy decreases more rapidly than the lattice energy, the compound becomes less soluble.

This occurs with most of the compounds except for fluorides and hydroxides for example, solubility of sulphates decreases from $BeSO_4$ to $BaSO_4$. Due to their small ionic radii, Be^{2+} and Mg^{2+} have high hydration energies. Thereby $BeSO_4$ and $MgSO_4$ are soluble in water. $CaSO_4$ is only slightly soluble in water, whereas $SrSO_4$ and $BaSO_4$ are almost insoluble in water. This solubility of sulphates is in order :

$$BeSO_4 > MgSO_4 > CaSO_4 > SrSO_4 > BaSO_4$$

DEEP Focus

In case of florides and hydroxides, the lattice energy decreases more rapidly than their hydrration energy. This causes a reverse trend, *i.e.,* the fluorides and hydroxides increases in solubility on moving down the group. Thus, order of solubility is

$$Ba(OH)_2 > Sr(OH)_2 > Ca(OH)_2 > Mg(OH)_2 > Be(OH)_2$$

Table 9.2 **Enthalpies of Hydration, ΔH_{hyd}, of Alkaline Earth Metal Ions, M^{2+} and Lattice Energies, ΔH_{latt}, of their Oxides, Carbonates Fluorides and Iodides in kJ mol^{-1}**

Element	ΔH_{hyd}	ΔH_{latt}			
	M^{2+}	MO	MCO_3	MF_2	MI_2
Be	−2494	—	—	—	—
Mg	−1921	−3923	−3178	−2906	−2292
Ca	−1577	−3517	−2986	−2610	−2058
Sr	−1443	−3312	−2718	−2459	—
Ba	−1305	−3120	−2614	−2367	—

Table 9.3 **Occurrence and Uses of Alkaline Earth Metals**

Element	Abundance	Main mineral	Uses
Beryllium	2.8×10^{-3}%	First detected in 1798 in the gemstone beryl and emerald ($Be_3Al_2Si_6O_{18}$).	Used in corrosion resistant alloys.
Magnesium	2.33%, seventh most abundant element in the earth's crust	Pure Mg first prepared in 1808, named after the Magnesia district in Thessaly, Greece where large deposits of the mineral are found.	When alloyed with Al, Mg is widely used as structural material because of its high strength, low density and ease in machining.
Calcium	4.15%, fifth most abundant element in the earth's crust	$CaCO_3$, $CaSO_4 \cdot 2H_2O$ (gypsum), obtained in pure form in 1808, calcium is derived from the Latin word calx, meaning "lime".	As an alloying agent to harden aluminium, calcium is the primary constituent of teeth and bones.
Strontium	0.038%	Discovered in 1787 and named after the small town of Strontian (Scotland)	$SrCO_3$ is used for the manufacture of glass for colour TV picture tubes.
Barium	0.042%	Found in the minerals witherite ($BaCO_3$) and barite ($BaSO_4$) after which it is named.	$BaSO_4$ is used in medicine as a contrast medium for stomach and intestinal X-rays.
Radium	traces	Isolated as chloride in 1898 from the mineral pitchblende	Used in cancer-radiotherapy

Group IIA (Alkaline Earth Metals) and Group IIB (Zn, Cd, Hg)

Mg acts as a bridge element between IIA (2) and IIB(12):

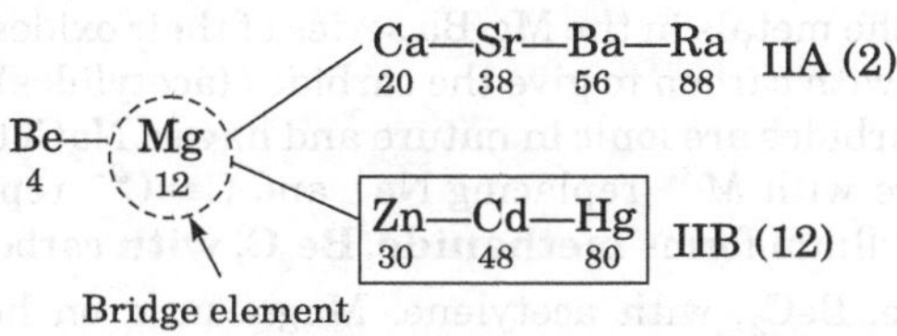

Table 9.4 **Distinction Between IIA(2) and IIB(12) Elements**

	Properties	IIA (Be, Mg, Ca, Sr, Ba, Ra)	IIB (Zn, Cd, Hg)
1.	Electronic configuration	[Inert gas] ns^2	[Inert gas] $(n-1)d^{10}\ ns^2$
2.	Block	*s*-block	*d*-block
3.	Oxidation state	+2	+2, mercury also forms dimeric Hg_2^{2+}
4.	Nature of oxide	BeO is amphoteric, other oxides are basic MgO < CaO < SrO < BaO < RaO	ZnO is amphoteric, CdO and HgO are basic.
5.	Nature of halides	Electron-deficient BeX_2 covalent, others (MX_2) are ionic : $MgCl_2 < CaCl_2 < SrCl_2 < BaCl_2$	$ZnCl_2$, $CdCl_2$ are ionic but less than IIA, $HgCl_2$ is covalent.
6.	Nature of sulphates	Less soluble in water and solubility decreases down the group $BeSO_4 > MgSO_4 > CaSO_4 > SrSO_4 > BaSO_4$	more soluble than IIA.
7.	Nature of hydroxides	Solubility of hydroxides increases as we move down the group	Solubility of hydroxides decreases as we move down the group.
8.	Nature of sulphides	Soluble	ZnS, CdS, HgS insoluble and precipitate in salt analysis.
9.	Reactivity	Increases as we move down the group Be < Mg < Ca < Sr < Ba	Decreases as we move down the group Zn > Cd> Hg

Reactions of Alkaline Earth Metals

Table 9.5 summarises reactions of group 2 (IIA). From this, it is evident that as we go down the group, reactivity increases.

Table 9.5 **Reactions of Group 2 (IIA)**

Reaction	Comment
$M + 2H_2O \longrightarrow M(OH)_2 + H_2$	Be probably reacts with steam, Mg with hot water, and Ca, Sr and Ba react rapidly with cold water.
$M + 2HCl \longrightarrow MCl_2 + H_2$	All the metals react with acids liberating H_2. With HNO_3, Be becomes passive due to formation of oxide layer.
$3M + N_2 \longrightarrow M_3N_2$ $M_3N_2 + 6H_2O \longrightarrow 2NH_3 + 3M(OH)_2$	All form nitrides at high temperatures. Stability Be > Mg > Ca (hydrolysis to NH_3).
$M + X_2 \longrightarrow MX_2$ ($X_2 = F_2, Cl_2, Br_2, I_2$)	All the metals form MX_2. No polyhalides are formed. (except BeX_2)
$M + 2NH_3 \longrightarrow M(NH_2)_2 + H_2$	All the metals dissolve in liquid NH_3 to give deep blue black solutions from which ammoniates $[M(NH_3)_6]^{2+}$ can be recovered and form amides at high temperatures.
$2M + O_2 \longrightarrow 2MO$ $Ba + O_2 \longrightarrow BaO_2$	All the metals form normal oxide. Ba also forms peroxide (BaO_2).
$M + H_2 \longrightarrow MH_2$	Ionic salt-like hydrides formed at high temperatures by Ca, Sr and Ba. Be forms electron-deficient hydrides.
$Be + 2H_2O + 2OH^- \longrightarrow [Be(OH)_4]^{2-} + H_2(g)$	Not with other alkaline earth metals.
$M(s) + 2C(s) \longrightarrow MC_2(s)$	At high temperature Be forms Be_2C, ionic compounds.
$M + S \longrightarrow MS(s)$	The sulphides are soluble, but hydrolyse if heated in water.

With Water

- **Ca, Sr and Ba have reduction potentials (E°) similar to those of the corresponding alkali metal and are quite high in the electrochemical series. They react with cold water quite readily.**

$$Ca + 2H_2O \longrightarrow Ca(OH)_2 + H_2$$

- **Mg has an intermediate value and it does not react with cold water but it decomposes hot water.**

$$Mg + 2H_2O \longrightarrow Mg(OH)_2 + H_2$$
or
$$Mg + H_2O \longrightarrow MgO + H_2$$

Mg *forms a protective layer of oxide, so despite its favourable reduction potential, it does not react readily unless the oxide layer is removed by amalgamating with mercury. In the formation of the oxide film, it resembles aluminium.*

- **Be is amphoteric, as it reacts with acid as well as with base.**

$$\underset{}{Be + 2NaOH + 2H_2O} \longrightarrow \underset{\text{sodium beryllate}}{Na_2[Be(OH)_4]} + H_2$$

Mg, Ca, Sr and Ba do not react with NaOH and thus are purely basic.

With Ammonia

- All the metals dissolve in liquid ammonia as do the alkali metals. Dilute solutions are **bright blue** in colour due to the spectrum from the solvated electrons. These solutions decompose very slowly, forming amide and H_2, but the reaction is accelerated by many transition metals and their compounds.

$$2NH_3 + 2e^- \longrightarrow 2NH_2^- + H_2$$

Evaporation of the ammonia from solutions of alkali metals yields the metal, but in case of alkaline earth metals, evaporation of ammonia gives hexammoniates of the metals, which slowly decompose to give amides:

$$M(NH_3)_6 \longrightarrow M(NH_2)_2 + 4NH_3 + H_2$$

Concentrated solutions of the metals in ammonia are bronze coloured, due to the formation of metal clusters.

DEEP Focus

Formation of Hydrides

All the group 2 element except beryllium form hydrides, MH_2, by direct combination with hydrogen. Beryllium hydride can be formed by reducing beryllium chloride with lithium aluminium hydride, $LiAlH_4$. All these hydrides are reducing agents which react with water and liberate hydrogen. Calcium, stronium and barium hydrides are ionic and contain the hydride ion, H^+. Beryllium and magnesium hydrides are covalent and polymeric. $(BeH_2)_n$ has an interesting structure. The polymeric solid contains hydrogen bridges between beryllium atoms.

Beryllium hydride polymer

Each beryllium atom is bonded to four hydrogen atoms and each hydrogen atoms forms two bonds as it is bridging two Be atoms. Since Be has two valence electrons and H only one, it is apparent that there are not enough electrons to form the usual electron pair bonds in which two electrons are shared between two atoms. Instead of this, **three-centre bonds are formed in which a 'banana-shaped' molecular orbital covers three atoms Be...H...Be, and contains two electrons.** The monomeric molecule BeH_2, if formed with normal bonds, would have only four electrons in the outer shell of the beryllium atom and would be electron deficient.

This would make the molecule very unstable; that is why BeH_2 exists as a polymer. This is an example of a cluster compound in which each atom shares its electrons with several neighbours and receives a share in their electrons to acquire a stable configuration.

Formation of Carbides

All the metals in the Mg-Ba series of their oxides react directly with carbon to give the carbides (acetylides), MC_2. These carbides are ionic in nature and have a NaCl type of structure with M^{2+} replacing Na^+ and $C{\equiv}C^{2-}$ replacing Cl^-. Beryllium forms **methanide, Be_2C, with** carbon and acetylide, BeC_2, with acetylene. Magnesium on heating with carbon forms Mg_2C_3 also, which is an allylide since with water it liberates **allylene (methylacetylene).**

The carbides are named depending upon the hydrocarbon they liberate on reaction with water. If it liberates acetylene, it will be named acetylide, and if methane is liberated, it will be called as **'methanide'**, etc.

$$\underset{\text{carbide}}{CaC_2} + 2H_2O \longrightarrow Ca(OH)_2 + \underset{\text{acetylene}}{C_2H_2}$$

$$\underset{\text{methanide}}{Be_2C} + 4H_2O \longrightarrow 2Be(OH)_2 + \underset{\text{methane}}{CH_4}$$

$$\underset{\text{allylide}}{Mg_2C_3} + 4H_2O \longrightarrow 2Mg(OH)_2 + \underset{\text{allylene}}{CH_3{-}C{\equiv}CH}$$

Formation of Nitrides

The alkaline earth metals burn **in nitrogen and form nitrides, M_3N_2. It requires a lot of energy** to convert the stable N_2 molecule into nitride ion, N^{3-}, and this is recovered from the very high lattice energies of the alkaline earth metal nitrides. The beryllium compound is rather volatile while others are not. They are all colourless crystalline solids which decompose on heating and react with water to liberate ammonia and form either the metal oxide or hydroxide, *e.g.*,

$$3Mg(s) + N_2(g) \longrightarrow Mg_3^{2-}N_2^{3-}(s)$$

$$Mg_3^{2+}N_2^{3-}(s) + 6H_2O(l) \longrightarrow 3Mg^{2+}(OH^-)_2 + 2NH_3(g)$$

Complexation Behaviour

Complex formation is favoured by small, highly charged cations with suitable empty orbitals of approximately the right energy with which the ligand orbitals can combine.

Alkaline earth metals form more complexes as compared to alkali metals. The tendency to form complexes (mostly with O and N donors) decreases with increasing atomic number. Thus, of the heavier ions only Ca^{2+} forms a complex wth ethanol. Beryllium having the smallest ion in the group tends to form complexes most readily. It mostly form complexes with tetrahedral arrangement because of the available orbitals as shown below :

	1s	2s	2p	
$_4$Be (ground state)	⇅	⇅		No unpaired electrons hence no covalent bond formation
Be (excited state)	⇅	↑	↑	Two unpaired electron can form two covalent bonds in BeF_2.
Be in $(BeF_4)^{2-}$	⇅	⇅	⇅ ⇅ ⇅	Electron donated by two fluoride ions forming coordinate bonds.

However, once these are formed, all the Be—F bonds tend to become similar.

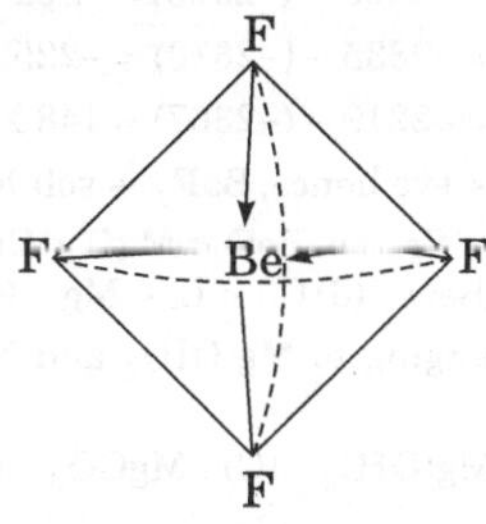

In the hydrated salts, *e.g.,* $BeCO_3 \cdot 4H_2O$ and $BeSO_4 \cdot 4H_2O$, beryllium ions exist in the form $[Be(H_2O)_4]^{2+}$, where they show a coordination number of four.

Magnesium is known to form a very important complex occurring in nature, *viz*; chlorophyll; a green pigment of the plants which produces sugar for the plants in presence of sunlight, CO_2 and H_2O in a process called photosynthesis.

$$6CO_2 + 6H_2O \xrightarrow[\text{Sunlight}]{\text{Chlorophyll}} C_6H_{12}O_6 + 6O_2$$

Magnesium, in chlorophyll, is coordinated to four nitrogen atom in the heterocyclic prophyrin ring system as shown.

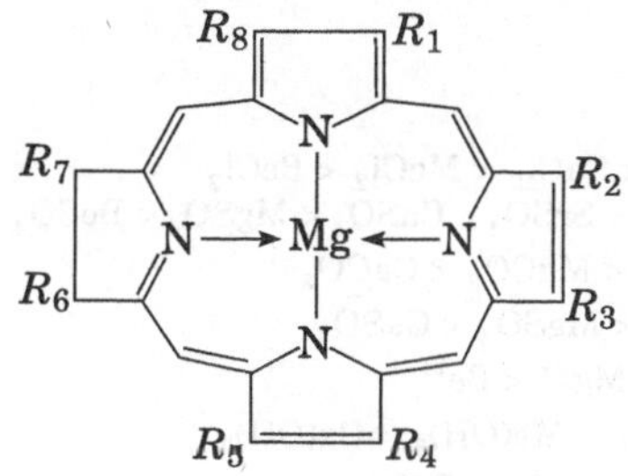

Skeleton of chlorophyll molecule

The rest of the element from calcium to barium form complexes only with strong complexation agents such as acetylacetone, ethylenediamineteraacetic acid (EDTA), etc. In fact, titrations are performed using EDTA in buffer solutions to estimate the amounts of Ca^{2+} and Mg^{2+} present in water, to determine the hardness of water.

Target Practice 1

1. Out of Be^{2+}, Mg^{+} and Ca
   (a) Which has maximum IE? ....
   (b) Which is maximum hydrated? ....
   (c) Which has least size? ....
   (d) Which has maximum size? ....
   (e) Which is most stable? ....
   (f) Which is least stable? ....
2. $BeCO_3$ is decomposes easily. Hence, it is kept in the atmosphere of .... to prevent its decomposition.
3. In Group 2 (IIA)
   (a) Maximum basic oxide is ....
   (b) Maximum soluble sulphate is ....
   (c) Maximum soluble carbonate is ....
   (d) Maximum soluble fluoride is ....
   (e) Maximum hydrated ion is ....

Answers

1. (a) Be^{2+} (b) Be^{2+} (c) Be^{2+} (d) Ca (e) Ca (f) Mg^{+}
2. CO_2
3. (a) RaO (b) $BeSO_4$ (c) $BeCO_3$ (d) RaF_2 (e) Be^{2+}

Practice Exercise 1

1. Arrange the following properties, as indicated, in the increasing order
   (a) Lattice energy of $BeCl_2$, $MgCl_2$, $CaCl_2$, $SrCl_2$
   (b) Solubility of $BeSO_4$, $MgSO_4$, $CaSO_4$, $SrSO_4$, $BaSO_4$
   (c) Stability of $BeCO_3$, $MgCO_3$, $CaCO_3$
   (d) Stability of $BeSO_4$, $MgSO_4$, $CaSO_4$
   (e) Size of the hydrated ions of Be^{2+}, Mg^{2+}, Ca^{2+}
   (f) Solubility of $Be(OH)_2$, $Mg(OH)_2$, $Ca(OH)_2$
   (g) Covalent nature of $BeCl_2$, $MgCl_2$, $CaCl_2$
   (h) Hydrolysis of $BeCl_2$, $MgCl_2$, $CaCl_2$

2. Lattice energies of BeF_2, MgF_2 and BaF_2 are given in Table 9.2.
Hydration energies of Be^{2+}, Mg^{2+}, Ba^{2+} and F^- are – 2494, –1921, –1305 and – 457 kJ mol^{-1}. Which of the BeF_2, MgF_2 and BaF_2 is soluble in water?
3. Arrange the following in increasing order:
(a) Basic nature of Be, Mg, Ca, Sr
(b) Reactivity of BeO, MgO, CaO with H_2O
(c) Refractory properties of BeO, MgO, CaO
(d) Reaction of Li, Mg and Ca with N_2
(e) Reaction of Be, Mg, Ca with H_2O
4. What are the possible products when Mg is burnt in air and they are decomposed by H_2O?
5. Identify (*A*) to (*E*) in the following:

$$Mg + air \xrightarrow{\Delta} (A) + (E)$$
$$(A) + H_2O \longrightarrow (B) \xrightarrow{CO_2} (C)$$
$$(E) + H_2O \longrightarrow (D) + (B)$$
$$(D) \xrightarrow{HCl} \text{white fumes}$$

1 mole of (*B*) neutralises 2 moles of HCl.

Answers

1. (a) $SrCl_2 < CaCl_2 < MgCl_2 < BeCl_2$
(b) $BaSO_4 < SrSO_4 < CaSO_4 < MgSO_4 < BeSO_4$
(c) $BeCO_3 < MgCO_3 < CaCO_3$
(d) $BeSO_4 < MgSO_4 < CaSO_4$
(e) $Ca^{2+} < Mg^{2+} < Be^{2+}$
(f) $Be(OH)_2 < Mg(OH)_2 < Ca(OH)_2$
(g) $CaCl_2 < MgCl_2 < BeCl_2$
(h) $CaCl_2 < MgCl_2 < BeCl_2$
2. Hydration energy of

$$BeF_2 = -2494 + 2\times(-457) = -3408 \text{ kJ mol}^{-1}$$
$$MgF_2 = -1921 + 2\times(-457) = -2835 \text{ kJ mol}^{-1}$$
$$BaF_2 = -1305 + 2\times(-457) = -2219 \text{ kJ mol}^{-1}$$

$$\Delta H_{Solution} = \Delta H_{Hydration} - \text{Lattice energy}$$
$$BeF_2 = -3408 - (-2906) = -502 \text{ kJ mol}^{-1}$$
$$MgF_2 = -2835 - (-2610) = -225 \text{ kJ mol}^{-1}$$
$$BaF_2 = -2219 - (-2367) = 148 \text{ kJ mol}^{-1}$$

In BaF_2, $\Delta H_{soln.}$ is +ve hence, BaF_2 is soluble in water.

3. (a) Be < Mg < Ca < Sr (b) BeO < MgO < CaO
(c) CaO < MgO < BeO (d) Ca < Li < Mg (e) Be < Mg < Ca
4. MgO, Mg_3N_2 changing to $Mg(OH)_2$ and NH_3. (also refer Q.5 below)
5. (A) : MgO (B) : $Mg(OH)_2$ (C) : $MgCO_3$ (D) : NH_3
(E) : Mg_3N_2

Compounds of Alkaline Earth Metals

Oxides and Hydroxides

- Oxides are formed either by heating the metals in oxygen:

$$2Be + O_2 \longrightarrow 2BeO$$
$$2Ca + O_2 \longrightarrow 2CaO$$

or by heating their carbonates at high temperatures:

$$CaCO_3 \xrightarrow{\Delta} CaO + CO_2$$
$$MgCO_3 \xrightarrow{\Delta} MgO + CO_2$$

- The oxides are highly stable due to large ionic crystal lattice energies. The values are so high in the case of oxides of Be and Mg that these compounds are almost insoluble in water. BeO is amphoteric and other oxides are basic and yield hydroxides in aqueous solution.

$$MgO + H_2O \longrightarrow Mg(OH)_2$$
$$CaO + H_2O \longrightarrow Ca(OH)_2$$

- Basic strength of the hydroxides increases going down the group

$$Be(OH)_2 < Mg(OH)_2 < Ca(OH)_2 < Sr(OH)_2 < Ba(OH)_2$$

BeO is insoluble in H_2O but dissolves in acid and alkali and is thus amphoteric

$$BeO + 2HCl \longrightarrow BeCl_2 + H_2O$$
$$BeO + 2NaOH \longrightarrow \underset{\text{sodium beryllate}}{Na_2BeO_2} + H_2O$$

- CaO combines with solid acidic oxides at high temperatures

$$CaO + SiO_2 \longrightarrow CaSiO_3$$
$$6CaO + P_4O_{10} \longrightarrow 2Ca_3(PO_4)_2$$

- MgO is not very reactive, especially if it has been ignited at high temperature and for this reason it is used as a refractory. BeO is also used as a refractory.
- Solubility of hydroxides increases while going down the group

$$Be(OH)_2 < Mg(OH)_2 < Ca(OH)_2 < Sr(OH)_2 < Ba(OH)_2$$

- $Ca(OH)_2$ is called **lime water** or **slaked lime** and $Ba(OH)_2$ is called **baryta water**; lime water or baryta turns milky by CO_2

$$Ca(OH)_2 + CO_2 \longrightarrow \underset{\text{milky}}{CaCO_3} + H_2O$$
$$Ba(OH)_2 + CO_2 \longrightarrow \underset{\text{milky}}{BaCO_3} + H_2O$$

$Ba(OH)_2$ gives turbidity (milkyness) even on exhaling breath on it.

Milkyness disappears when excess of CO_2 is passed, due to formation of soluble bicarbonates.

$$CaCO_3 + H_2O + \underset{\text{excess}}{CO_2} \longrightarrow \underset{\text{soluble}}{Ca(HCO_3)_2}$$

Temporary hardness in water is due to dissolved $Ca(HCO_3)_2$ and $Mg(HCO_3)_2$

- $Ca(OH)_2 + Na_2CO_3 \longrightarrow CaCO_3 \downarrow + 2NaOH$
  This reaction was used in earlier days to manufacture NaOH from $Ca(OH)_2$.
- $2Ca(OH)_2 + 2Cl_2 \longrightarrow CaCl_2 + \underset{\text{calcium hypochlorite}}{Ca(ClO)_2} + 2H_2O$
- Solutions of beryllium salts are acidic and dissolve appreciable quantities of $Be(OH)_2$. In alkali solution $[Be(OH)_4]^{2-}$ is formed

$$[Be(H_2O)_4]^{2+} \rightleftharpoons [Be(H_2O)_3(OH)]^+ + H^+$$
$$[Be(H_2O)_3(OH)]^+ \rightleftharpoons [Be(H_2O)_2(OH)_2] + H^+$$
$$[Be(H_2O)_2(OH)_2] \rightleftharpoons [Be(H_2O)(OH)_3]^- + H^+$$
$$[Be(H_2O)(OH)_3]^- \rightleftharpoons [Be(OH)_4]^{2-} + H^+$$

Bicarbonates and Carbonates

- The carbonates are all ionic, but due to excessive hydration of Be^{2+}, $BeCO_3$ is unusual because it contains the hydrated ion $[Be(H_2O)_4]^{2+}$.
- $BeCO_3$ decomposes at low temperature and is placed in the atmosphere of CO_2

$$BeCO_3 \longrightarrow BeO + CO_2$$

Thermal stability increases with increase in atomic weight down the group or with increasing cationic size.

$$BeCO_3 < MgCO_3 < CaCO_3 < SrCO_3 < BaCO_3$$

- $Ca(HCO_3)_2$ and $Mg(HCO_3)_2$ exist in water causing temporary hardness which can be removed as insoluble carbonates by boiling:

$$Ca(HCO_3)_2 \xrightarrow{\Delta} CaCO_3 \downarrow + H_2O + CO_2$$
$$Mg(HCO_3)_2 \xrightarrow{\Delta} MgCO_3 \downarrow + H_2O + CO_2$$

- $Ca(HCO_3)_2$ and $Mg(HCO_3)_2$ are amphoteric thus neutralised by acid as well as base:

$$Ca(HCO_3)_2 + H_2SO_4 \longrightarrow CaSO_4 + 2H_2O + 2CO_2$$
$$Ca(HCO_3)_2 + Ca(OH)_2 \longrightarrow 2CaCO_3 \downarrow + 2H_2O$$

- All of these carbonates are much soluble in a solution of CO_2 than in water due to formation of bicarbonates

$$CaCO_3 + H_2O + CO_2 \longrightarrow Ca(HCO_3)_2$$

- $SrCO_3$ is used in the manufacture of glass for colour TV picture tube.

Calcium and barium bicarbonates are stable only in solution and decompose on removal of water to give carbonates. This property of bicarbonates is the reason for **stalactite** and **stalagmite** formation.

Stalactite The downward growth of $CaCO_3$ formed on the roof of a cave by the trickling of water containing calcium compounds.

Stalagmite The upward growth from the floor of a cave; of the same nature and origin as a stalactite.

Sulphates

- The solubility of the sulphates in water decreases down the group.

$$\underbrace{BeSO_4 > MgSO_4}_{\text{soluble}} > \underbrace{CaSO_4}_{\text{sparingly soluble}} > \underbrace{SrSO_4 > BaSO_4}_{\text{insoluble}}$$

High solubilities of $BeSO_4$ and $MgSO_4$ are due to the high enthalpy of hydration of the smaller Be^{2+} and Mg^{2+} ions, which overcomes the lattice energy factor.

- The sulphates decompose into oxides and SO_3.

$$MSO_4 \xrightarrow{\Delta} MO + SO_3$$

Thermal stability of the sulphates increases going down the group as shown by decomposition temperatures :

$BeSO_4$	$MgSO_4$	$CaSO_4$	$SrSO_4$
500°C	845°C	1149°C	1374°C

- $MgSO_4 \cdot 7H_2O$ is called Epsom salt and is used as mild laxative. Correct formulation is $[Mg(H_2O)_6]SO_4 \cdot H_2O$ and is isomorphous with zinc sulphate $ZnSO_4 \cdot 7H_2O$. $MgSO_4$ is deliquescent and readily soluble in water. It forms double salts with alkali metal sulphates. $K_2SO_4 \cdot MgSO_4 \cdot 6H_2O$ is sold as a fertilizer in the name of **'potash magnesis'**.
- $CaSO_4 \cdot 2H_2O$ is called gypsum. Heating converts it into variety of compounds depending on the temperature.

$$\underset{\text{gypsum}}{CaSO_4 \cdot 2H_2O} \xrightarrow{150°C} \underset{\text{Plaster of Paris}}{CaSO_4 \cdot \tfrac{1}{2}H_2O} \xrightarrow{200°C} \underset{\text{anhydrite}}{CaSO_4}$$
$$\xrightarrow{1100°C} CaO + SO_3$$

when powdered Plaster of Paris, $CaSO_4 \cdot \frac{1}{2}H_2O$ ($2CaSO_4 \cdot H_2O$ or $(CaSO_4)_2 \cdot H_2O$) is mixed with stoichiometric amount of water, it sets into a solid mass $CaSO_4 \cdot 2H_2O$ — gypsum. Plaster of Paris finds use

- for plastering walls
- for making plaster casts (moulds)
- as industrial material
- as sculptural material and
- for encasing limbs so that broken bones are set straight.

- $CaSO_4 \cdot H_2O$ is called Alabaster — a shiny-like marble and is used to make ornaments.
- $MSO_4 + 4C \xrightarrow{\Delta} MS + 4CO \uparrow$

- **$BaSO_4$ is used in medicine as a contrast medium for stomach and intestinal X-rays.**

Halides

The tendency for the smaller cations to get hydrated is so strong that it is not possible to prepare anhydrous salts by simply heating the hydrated ones. Thus, anhydrous $BaCl_2$ may be prepared by heating the hydrated salts beyond 127°C.

$$BaCl_2 \cdot 2H_2O \xrightarrow{\text{heat}} BaCl_2 + 2H_2O$$

However analogous reaction cannot be carried out for magnesium chloride.

$$MgCl_2 \cdot 6H_2O \xrightarrow{\text{heat}} MgCl(OH) + HCl + 5H_2O$$
$$MgCl(OH) \longrightarrow MgO + HCl$$

If anyhydrous $MgCl_2$ is required, it may be obtained by mixing the hydrated salt with thionyl chloride $SOCl_2$.

$$MgCl_2 \cdot 6H_2O + 6SOCl_2 \longrightarrow MgCl_2 + 6SO_2 + 12HCl$$

- Halides MX_2 of alkaline earth metals can be prepared by heating the metals with the halogen acid on either metal or its carbonate :

$$M + 2HCl \longrightarrow MCl_2 + H_2$$
$$MCO_3 + 2HCl \longrightarrow MCl_2 + H_2O + CO_2$$

- **The beryllium halides are covalent, hygroscopic and fumes in air due to hydrolysis:**

$$BeCl_2 + 2H_2O \longrightarrow Be(OH)_2 + 2HCl \quad \text{(fumes)}$$

- BeX_2 sublimes on heating and are bad conductor of electricity. In aqueous solution it is difficult to crystallise anhydrous BeX_2 since dehydration results in hydrolysis:

$$[Be(H_2O)_4]Cl_2 \xrightarrow{\Delta} Be(OH)_2 + 2HCl + 2H_2O$$

- Anhydrous BeX_2 can be obtained by :

$$BeO + C + Cl_2 \xrightarrow{700^\circ C} BeCl_2 + CO$$
$$2BeO + CCl_4 \xrightarrow{800^\circ C} 2BeCl_2 + CO_2$$
$$BeO \xrightarrow[\Delta]{NH_4F} (NH_4)_2BeF_4 \xrightarrow{\Delta} BeF_2 + 2NH_4F$$

Anhydrous $BeCl_2$ is polymeric. Vapours exist as monomer $BeCl_2$ and dimer $(BeCl_2)_2$. Bond length data show that the polymer has three centre two-electron bonds.

Cl—Be(μ-Cl)₂Be—Cl

dimer

polymer $(BeCl_2)_n$

- $BeCl_2$ forms complexes $M_2(BeCl_4)$ with alkali metal chlorides but are decomposed by water.
- BeF_2 is very soluble in water owing to the high solvation energy of Be^{2+} in forming $[Be(H_2O)_4]^{2+}$. The other fluorides MF_2 ($M^{2+} = Mg^{2+}, Ca^{2+}, Sr^{2+}, Ba^{2+}$) are all almost insoluble.
- The chlorides, bromides and iodides of Mg, Ca, Sr and Ba are ionic and are readily soluble in water.
- The halides form hydrates and they are hygroscopic.
- $CaCl_2$ is widely used for treating ice on roads, particularly in very cold countries when it prevents ice formation even up to −55° C.
- Mixed chloride hydrides of formula $MClH$ which have layer lattices (the different anions occupying different layers) are formed when the hydrides and chlorides of alkaline earth metals are heated together:

$$CaH_2 + CaCl_2 \longrightarrow 2CaHCl$$

Anomalous Behaviour of Beryllium

- Because of high (*IE*) and small atomic size it forms compounds which are largely covalent and its salts are easily hydrolysed.
- Be ($1s^2\ 2s^2\ 2p^0$) can use only one 2*s* and three 2*p* orbital in coordination thus maximum coordination number (CN) of Be is 4 while other elements can show CN of 6 in their compounds by use of *d*-orbitals in addition to *s*- and *p*-orbitals.
- BeO and $Be(OH)_2$ are amphoteric while other oxides are basic.
- The cohesive properties of beryllium are much greater than those of magnesium and other elements in the group. As a result, it is much harder and has higher melting and boiling points.
- Its higher polarising power leads to all its compounds being largely covalent, with lower melting and boiling points and enthalpies of formation. (MP of BeF_2 is 800°C that of other fluorides about 1300°C). All the compounds of Be are more soluble in organic solvents than the corresponding magnesium compounds. They hydrolyse in water, and in this respect they rather resemble aluminium (diagonal relationship)

Diagonal Relationship

- Beryllium shows a diagonal relationship to aluminium :
  - The chlorides of both Be and Al have bridged structures in vapour phase

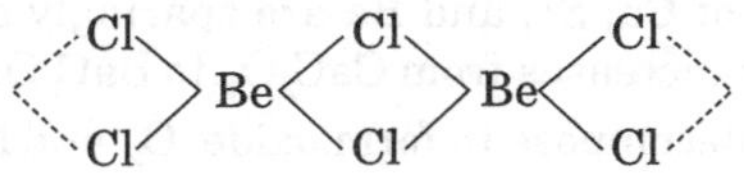

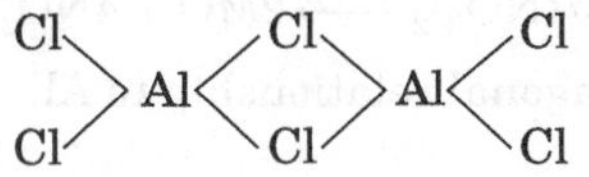

$$\cdots Cl \diagdown Be \diagup Cl \diagdown Be \diagup Cl \cdots$$
(polymeric chain: $BeCl_2$ units bridged by Cl)

$$Cl_2Al(\mu\text{-}Cl)_2AlCl_2$$
(Cl, Cl / Al / Cl, Cl / Al / Cl, Cl — bridged dimer)

- Two metals have similar standard electrode potential:

$$Be^{2+}(aq) + 2e^- \longrightarrow Be(s), \quad E^\circ = -1.85\text{ V}$$
$$Al^{3+}(aq) + 3e^- \longrightarrow Al(s), \quad E^\circ = -1.66\text{ V}$$

- Both are rendered passive by nitric oxide due to formation of oxide layer

$$2HNO_3 \longrightarrow H_2O + 2NO_2 + O$$
$$Be + O \longrightarrow \underset{\text{passive}}{BeO} \xrightarrow{HNO_3} \text{no reaction}$$
$$Al + O \longrightarrow \underset{\text{passive}}{Al_2O_3} \xrightarrow{HNO_3} \text{no reaction}$$

Thus, HNO_3 can be stored in a vessel made of Al or Be.

- Both form electron-deficient hydrides and volatile chlorides. BeH_2, AlH_3, etc, are some electron-deficient hydrides.
- Both form carbides which on hydrolysis give methane.

$$Al_4C_3 + 12H_2O \longrightarrow 3CH_4 + 4Al(OH)_3$$
$$Be_2C + 2H_2O \longrightarrow 2BeO + CH_4$$
$$BeO + H_2O \longrightarrow Be(OH)_2$$

- Both metals dissolve in aqueous alkali evolving H_2

$$Be + 2NaOH \longrightarrow Na_2BeO_2 + H_2$$
$$2Al + 2NaOH + 2H_2O \longrightarrow 2NaAlO_2 + 3H_2$$

- Oxides and hydroxides are amphoteric

In dil acid (showing basic nature of oxide or hydroxide)	In alkali solution (showing acidic nature of oxide or hydroxide)
$BeO + 2HCl \longrightarrow BeCl_2 + H_2O$	$BeO + 2NaOH \longrightarrow Na_2BeO_2 + H_2O$
$Be(OH)_2 + 2HCl \longrightarrow BeCl_2 + 2H_2O$	$Be(OH)_2 + 2NaOH \longrightarrow \underset{\text{sodium berrylate}}{Na_2BeO_2} + 2H_2O$
$Al_2O_3 + 6HCl \longrightarrow 2AlCl_3 + 3H_2O$	$Al_2O_3 + 2NaOH \longrightarrow$
$Al(OH)_3 + 3HCl \longrightarrow AlCl_3 + 3H_2O$	$\underset{\text{sodium meta-aluminate}}{2NaAlO_2} + H_2O$

- The two elements have same electronegativity (Be = 1.5 and Al = 1.5) and their charge/radius ratios (0.064 and 0.060 respectively) are very similar indicating similar field strengths.
- They have also strong tendency to form complexes:

Be forms tetrahedral complexes as BeF_4^{2-} and $[Be(C_2O_4)_2]^{2-}$

Al forms octahedral complexes like AlF_6^{3-} and $[Al(C_2O_4)_3]^{3-}$.

Additional Reactions and Facts of Alkaline Earth Metal Compounds (IIA)

Reactions

1. $2Ba(NO_3)_2 \xrightarrow{\Delta} 2BaO + 4NO_2(g) + O_2(g)$
2. $CaH_2 + 2H_2O \longrightarrow Ca(OH)_2 + 2H_2$
3. $Ca_3N_2 + 6H_2O \longrightarrow 3Ca(OH)_2 + 2NH_3$
4. $Be_2C + 4H_2O \longrightarrow 2Be(OH)_2 + \underset{\text{methane}}{CH_4}$
5. $CaC_2 + 2H_2O \longrightarrow Ca(OH)_2 + \underset{\text{ethyne}}{C_2H_2}$
6. $CaC_2 + N_2 \longrightarrow \underset{\text{nitrolim}}{CaCN_2 + C}$
7. $CaCN_2 + 5H_2O \longrightarrow CaCO_3 + 2NH_4OH$
8. $BaC_2 + N_2 \longrightarrow Ba(CN)_2$
9. $Ca(OH)_2 + Cl_2 \longrightarrow \underset{\text{bleaching powder}}{CaCl\cdot OCl} + H_2O$

(Bleaching powder contains Cl^- and ClO^- ions)

10. $Mg_2C_3 + 4H_2O \longrightarrow 2Mg(OH)_2 + \underset{\text{propyne}}{CH_3—C{\equiv}CH}$

Important Facts of Alkaline Earth Metals

- Be and Mg are kinetically inert to oxygen and water because of the formation of a surface film of oxide.
- Be and Mg burn in air forming oxides and nitrides (BeO, MgO/Be_3N_2, Mg_3N_2). Ca, Sr and Ba are readily attacked by air forming oxides and nitrides.
- Alkaline earth metals dissolve in liquid NH_3 to give deep blue black solutions from which ammoniates $[M(NH_3)_6]^{2+}$ can be recovered.
- Metals are electropositive and strong reducing agents. The particularly less negative value for Be arises from the large hydration energy associated with the small size of Be^{2+} and the relatively large value of heat of sublimation.
- The solubility, thermal stability and the basic character of hydroxides increase with atomic number from $Mg(OH)_2$ to $Ba(OH)_2 \cdot Be(OH)_2$ / BeO is amphoteric.
- Except BeX_2 (X = F, Cl, Br, I), all other hallides MX_2 are ionic. BeX_2 exists in gas phase only but in condensed phase it exists as polymer. Solubility in water increases with increase in size of cation.

- Carbonates and bicarbonates decompose on heating. $BeCO_3$ is least stable and is kept in the atmosphere of CO_2.
- Sulphates are stable to heat. $BeSO_4$ and $MgSO_4$ are soluble in water but solubility decreases from $CaSO_4$ to $BaSO_4$. The greater hydration energies of Be^{2+} and Mg^{2+} ions overcome the lattice energy factor, and hence sulphates are soluble.
- Oxalates of Ca, Sr, and Ba are sparingly soluble and solubility increases from CaC_2O_4 to BaC_2O_4.
- Nitrates decompose to form oxide, O_2 and NO_2

$$2M(NO_3)_2 \xrightarrow{\Delta} 2MO + 4NO_2 + O_2$$

- Be shows diagonal relationship to Al.

Target Practice 2

1. Mixture containing one mole each of $CaCO_3$ and $Ca(HCO_3)_2$ is strongly heated. What is the amount (in moles) of the residue left? ....
2. For several thousand years, lime has been used in mortar (a lime, sand and water paste) in building houses, etc. Explain its reaction ....
3. (a) What is hybridisation of Be in BeF_2? ....
   (b) When BeF_4^{2-} is formed hybridisation of Be changes from .... to ....
4. BeO is .... based on following reactions

$$BeO + 2NaOH \longrightarrow Na_2BeO_2 + H_2O$$
$$BeO + 2HCl \longrightarrow BeCl_2 + H_2O$$

5. CaO is .... based on following reaction

$$6CaO + P_4O_{10} \longrightarrow 2Ca_3(PO_4)_2$$

6. Give the reaction used to prepare NaOH from $Ca(OH)_2$ is earlier days ....
7. $MgSO_4 \cdot 7H_2O$ is correctly represented as ....
8. Name the salt which is used in medicine as a contrast medium for stomach and intestinal X-rays. ....

Answers

1. $\underset{1\text{ mol}}{CaCO_3} \longrightarrow \underset{1\text{ mol}}{CaO} + CO_2\uparrow$

   $\underset{1\text{ mol}}{Ca(HCO_3)_2} \longrightarrow \underset{1\text{ mol}}{CaO} + H_2O\uparrow + 2CO_2\uparrow$

   Residue is **2 moles** CaO
2. $CaO + H_2O \longrightarrow Ca(OH)_2$
3. (a) *sp* (b) *sp*, *sp*^{3}
4. amphoteric
5. basic
6. $Na_2CO_3 + Ca(OH)_2 \longrightarrow 2NaOH + CaCO_3\downarrow$
7. $[Mg(H_2O)_6]SO_4 \cdot H_2O$
8. $BaSO_4$

Practice Exercise 2

1. Gas (*X*) turns lime water milky. Gas (*X*) also turns acidified $K_2Cr_2O_7$ green. What is gas (*X*)?
2. Beryllium turns passive in presence of HNO_3. Explain.
3. Identify (*A*) to (*C*) and the metal ion M^{2+}.

$$\underset{(X)}{MCO_3} \xrightarrow{\Delta} \underset{\text{(solid)}}{(A)} + \underset{\text{(gas)}}{(B)}$$
$$(A) + H_2O \longrightarrow (C)$$
$$(C) + (B) \longrightarrow (X)$$

   Paste of the salt (*X*) in conc. HCl given green flame on ignition.
4. Following reaction is used to remove temporary hardness which is due to dissolved bicarbonates of Ca^{2+} and Mg^{2+}

$$Ca(HCO_3)_2 + Ca(OH)_2 \longrightarrow 2CaCO_3\downarrow + 2H_2O$$

   Actually lime (CaO) is added which in H_2O forms $Ca(OH)_2$ and precipitates $CaCO_3$

$$CaO + H_2O \longrightarrow Ca(OH)_2$$

   If 10 L of hard water requires 0.56 g of CaO, what is temporary hardness in ppm of $CaCO_3$ (parts per million—ppm.)?
5. $Ca(HCO_3)_2$ and $Ca(OH)_2$ can't be stored in the same vessel. Explain.
6. $Ca(HCO_3)_2$ is determined volumetrically by titration with acid. Which of the following indicators is better?
   (a) Phenolphthalein
   (b) Methyl red
7. Ca^{2+} can be determined by precipitating it as CaC_2O_4 gravimetrically. Calculate percentage of Ca^{2+} in a sample, if 0.5 g of Ca^{2+} salt gave 1.28 g of CaC_2O_4.
   (Ca = 40, Ca = 12, O = 16)
8. Identify (*A*) to (*D*) in the following:

$$KO_2 + S \xrightarrow{\Delta} (A) \xrightarrow{BaCl_2} (B)\text{ (white ppt)} \xrightarrow[\Delta]{\text{Carbon}} (C) \xrightarrow{HCl} D\text{ (gas)}$$

   (*D*) gas turns lead acetate paper black.
9. Presence of $CaSO_4$ causes 'permanent hardness' in water. It can be removed by precipitating it as $CaCO_3$ by adding Na_2CO_3. Calculate amount of Na_2CO_3 to soften 100 L of hard water which is 200 ppm $CaCO_3$.

10. While moving down the group (in alkaline earth metals) following property decreases
(a) radius
(b) solubility of hydroxides
(c) solubility of sulphates
(d) thermal stability of carbonates
Choose the correct alternate.

11. (*A*) is binary oxide of alkaline earth metal. On reaction with HCl, (*A*) forms solid (*B*) and liquid (*C*). (*B*) gives apple green flame on ignition and liquid (*C*) decolourises $KMnO_4$ in acidic medium. What are (*A*), (*B*) and (*C*)?

12. $BeCl_2$ in aqueous solution exists as $[Be(H_2O)_4]^{2+}$ rather than Be^{2+}. This is acidic in nature. Explain.

13. Arrange $BeCl_2$, $MgCl_2$, $CaCl_2$, $SrCl_2$ and $BaCl_2$ in increasing
(i) tendency of hydrolysis
(ii) lattice energy
(iii) solubility in H_2O
(iv) covalent nature.

14. Match the following chlorides with the colour they impart on ignition in flame:

Salt	Colour in flame
1. $CaCl_2$	(A) Apple green
2. $SrCl_2$	(B) Brick red
3. $BaCl_2$	(C) Crimson red

Answers

1. Gas (*X*) is SO_2

$$Ca(OH)_2 + SO_2 \longrightarrow \underset{\text{milky}}{CaSO_3} + H_2O$$

$$SO_2 + Cr_2O_7^{2-} + H^+ \longrightarrow \underset{\text{green}}{Cr^{3+}} + SO_4^{2-} + H_2O$$

2. HNO_3 is an oxidising agent. It oxidises Be to BeO. This, oxide layer prevents attack of HNO_3 on it. Hence, Be turns passive.

3. (*X*) : $BaCO_3$, M^{2+} : Ba^{2+} (*A*) : BaO (*B*) : CO_2 (*C*) : $Ba(OH)_2$

4. $$\underset{\substack{56\text{ g}\\0.56\text{ g}}}{CaO} \longrightarrow Ca(OH)_2 \xrightarrow{Ca(HCO_3)_2} \underset{200\text{ g}}{2CaCO_3}$$

2.00 g in 10^4 mL H_2O

10^4 mL H_2O hardness = 2 g

10^6 mL water has hardness = **200 ppm**

5. $Ca(HCO_3)_2$ behaves as an acid with $Ca(OH)_2$

$$Ca(HCO_3)_2 + Ca(OH)_2 \longrightarrow 2CaCO_3\downarrow + H_2O$$

Hence, neutralisation takes place.

6. Methyl red is only used. Phenolphthalein does not indicate any colour change.

7. 80% pure.

8. (*A*) : K_2SO_4 (*B*) : $BaSO_4$ (*C*) : BaS (*D*) : H_2S

9. $$\underset{106\text{ g}}{Na_2CO_3} + CaSO_4 \longrightarrow \underset{100\text{ g}}{CaCO_3}\downarrow + Na_2SO_4$$

(1 million) = 10^6 mL of hard water has = 200 g $CaCO_3$

100 L = 10^5 mL of hard water has = 20 g $CaCO_3$

If hardness is 100 g, Na_2CO_3 required = 106 g

If hardness is 20 g, Na_2CO_3 required = 21.2 g

10. (c)

11. (*A*) : $BaO_2 \rightleftharpoons Ba^{2+} + O_2^{2-}$ (*B*) : $BaCl_2$ (*C*) : H_2O_2

12. $[Be(H_2O)_4]^{2+}$ is acidic because of hydrolysis.

$$[Be(H_2O)_4]^{2+} + H_2O \rightleftharpoons [Be(H_2O)_3(OH)]^+ + H_3O^+$$

This happens because the Be—O bond is very strong, and this weakens the O—H bond and thus, proton is lost.

$$H—O—H \rightleftharpoons H^+ + OH^-$$

$$[Be(H_2O)_4]^{2+} + OH^- \rightleftharpoons [Be(H_2O)_3(OH)]^+ + H_2O$$

$$H_2O + H^+ \rightleftharpoons H_3O^+$$

13. (i) $BaCl_2 < SrCl_2 < CaCl_2 < MgCl_2 < BeCl_2$
(ii) $BaCl_2 < SrCl_2 < CaCl_2 < MgCl_2 < BeCl_2$
(iii) $BeCl_2 < MgCl_2 < CaCl_2 < SrCl_2 < BaCl_2$
(iv) $BaCl_2 < SrCl_2 < CaCl_2 < MgCl_2 < BeCl_2$

14. 1—(B) 2—(C) 3—(A)

Total Practice Set

(Read, Plan & Solve!)

Problem 1. What is biological role of Ca^{2+} and Mg^{2+} ions?

Solution Mg^{2+} ions are concentrated in animal cells, and Ca^{2+} are concentrated in the body fluids outside the cell. Mg^{2+} is important in chlorophyll, in the green parts of plants.

Ca^{2+} is important in bones and teeth as apatite $Ca_3(PO_4)_2$ and the enamel on teeth as fluoroapatite $[3(Ca_3(PO_4)_2) \cdot CaF_2]$.

Ca^{2+} ions are important in blood clotting, and are required to trigger the contraction of muscles and to maintain the regular beating of heart.

Problem 2. Beryllium gives a compound X with the following percentage composition : Be–6.1%; N –37.8% Cl –48%, H —8.1%. Molecular weight of (X) is 148 g mol^{-1} and that of Be is 9 g mol^{-1}.

(i) What is the molecular formula of the compound?

(ii) In aqueous solution, 1 mole of the compound reacts with 2 moles of $AgNO_3$ giving white ppt. What is structural formula of (X)?

Solution

Atom	Atomic weight (A)	%	%/A	Simplest ratio
Be	9	6.1	0.68	1
N	14	37.8	2.70	4
Cl	35.5	48.0	1.35	2
H	1	8.1	8.10	12

(i) Molecular formula is $BeN_4Cl_2H_{12}$.

(ii) Since, 1 mole of X reacts with 2 moles of $AgNO_3$, hence two Cl are outside the complex. Since, there are four N and twelve H, this indicates four NH_3 molecules attached to Be. This gives structural formula as:

$$[Be(NH_3)_4]\,Cl_2$$

Problem 3. A student suggested that calcium should be made if calcium oxide is reacted with aluminium powder. Was the student correct ?

$$\Delta G^\circ_f\ (CaO) = -604.2\ kJ\ mol^{-1},$$
$$\Delta G^\circ_f\ (Al_2O_3) = -1582.4\ kJ\ mol^{-1}$$

Solution A given reaction is spontaneous, if ΔG°_f is negative.

$$3CaO + 2Al \longrightarrow Al_2O_3 + 3Ca$$
$$\Delta G^\circ_f = \Delta G^\circ_f\,(Al_2O_3) - 3\Delta G^\circ_f\,(CaO)$$
$$= -1582.4 - 3(-604.2)$$
$$= -1582.4 + 1812.6$$
$$= +230.2\ kJ\ mol^{-1}$$

The positive sign tells us that the reaction cannot occur under standard conditions.

Note : Under non-standard conditions *e.g.*, under high temperatures, ΔG is negative and reaction takes place. ΔG°_f (element) = 0.

Problem 4. 0.2 g of magnesium ribbon was placed in a crucible and heated with the lid on until the magnesium began to burn rapidly. At the end of the experiment there was 0.3 g of a white powder left. Show that this result does not agree with the equation:

$$2Mg(s) + O_2(g) \longrightarrow 2MgO(s)$$

What might have gone wrong?

Solution

$$\underset{2\times 24\ g}{2Mg(s)} + O_2(g) \longrightarrow \underset{2\times 40\ g}{2MgO(s)}$$

48 g of Mg gives = 80 g of MgO

$$0.2\ g\ of\ Mg\ gives = \frac{80}{48} \times 0.2$$
$$= 0.33\ g\ of\ MgO$$

But residue left = 0.30 g (given)

Hence, the results are not in accordance with this equation.

It can be due to

(1) some MgO may have escaped as vapours

(2) some Mg may not have reacted

(3) some Mg_3N_2 can be formed.

Problem 5. Give the trends in solubilities of chlorides, carbonates, sulphates and hydroxides of alkaline earth metals.

Solution

	Chlorides	Carbonates	Sulphates	Hydroxides
Be	least soluble ↓	most soluble ↑	most soluble ↑	least soluble ↓
Mg				
Ca				
Sr				
Ba	most soluble	least soluble	least soluble	most soluble

Problem 6. What is the importance of calcium and magnesium carbonates and hydrogen carbonates in making water hard or soft?

Solution When slightly acidic rain water percolates through carbonate rocks, some of the rocks dissolve. Ca^{2+} and Mg^{2+} ions are dissolved in the water. It is these ions which give scum with soap or make scale (white deposit) when hard water is boiled.

Problem 7. Explain anomalous behaviour of beryllium.

Solution

- It is extremely small in size and as determined by Fajan's rule, smaller cations tend to polarise larger anion and thus, Be^{2+} forms covalent compounds.
- Be has high electronegativity and compounds like BeF_2 and BeO are covalent.
- BeX_2 exists as polymeric $(BeX_2)_4$ in solid state and coordination number = 4 is shown in which each halogen forms

Cl—Be(μ-Cl)₂Be—Cl (bridging Cl atoms with coordinate bonds)

one normal covalent bond, and uses a lone pair to form a coordinate bond.

- In water, beryllium salts are extensively hydrolysed to give a series of hydroxo complexes of the type :

$$\left[(HO)_2Be(\mu\text{-}OH)_2Be(OH)_2\right]^{2-}$$

If alkali is added to these solutions, the polymers break down to form $[Be(OH)_4]^{2-}$ — a tetrahedral structure.

- Be^{2+} salts are extensively hydrated due to larger (charge/radius) ratio (called charge density). $[Be(H_2O)_4]SO_4$, $[Be(H_2O)_4](NO_3)_2$ and $[Be(H_2O)_4]Cl_2$ are known. These salts are acidic due to following equilibrium :

$$[Be(H_2O)_4]^{2+} + H_2O \rightleftharpoons [Be(H_2O)_3OH]^+ + H_3O^+$$

Problem 8. Give a comparison of the elements Be and Al.

Solution **Be [2nd period, group 2 (IIA)] and Al [3rd period, group 13 (IIIA)] show diagonal relationship. Be differs in most of the properties of its group but resembles Al (diagonally placed to Be) in following respects:**

Properties	Explanations
With HNO_3	Both metals are rendered passive and do not dissolve due to formation of oxide layer.
With alkali	Both dissolve in alkali giving off H_2. $2Al + 2NaOH + 2H_2O \longrightarrow 2NaAlO_2 + 3H_2$; $Be + 2NaOH \longrightarrow Na_2BeO_2 + H_2$
Amphoteric nature	Both metals react with acid and base and thus are amphoteric.
Oxides	Both BeO and Al_2O_3 are amphoteric.
Chlorides	$BeCl_2$ and $AlCl_3$ both are electron deficient and thus are Lewis acids.

Problem 9. **The standard enthalpy of formation ΔH_f°, of hypothetical CaCl(*s*) theoretically found to be -188 kJ mol^{-1} and that of $CaCl_2(s)$ -795 kJ mol^{-1}. Calculate the value of standard heat of reaction for the following disproportionation reaction:**

$$2CaCl(s) \longrightarrow CaCl_2(s) + Ca(s)$$

Solution **A disproportionation reaction is that in which the same species (in this case Ca^+) is oxidised as well as reduced.**

$$2\,CaCl(s) \longrightarrow CaCl_2(s) + Ca(s)$$

+1 → +2 (oxidation); +1 → 0 (reduction)

To determine standard heat of reaction, we use

$$\Delta H^\circ = \Delta H_f^\circ\ \text{(products)} - \Delta H_f^\circ\ \text{(reactants)}$$

and ΔH_f° of the element = 0

$$\text{Thus, } \Delta H^\circ = \Delta H_f^\circ(CaCl_2) + \Delta H_f^\circ(Ca) - 2\Delta H_f^\circ(CaCl)$$

$$= (-795 + 0) - (-2 \times 188)$$

$$= -419 \text{ kJ mol}^{-1}$$

This shows that CaCl(*s*) does not exist and any synthetic route to get *MX* in case of the alkaline earth metals will result in the formation of MX_2.

Master Exercises

Exercise 1
(Stage 1 : Learning)

Short Answer Type Questions

1. Arrange Be, Ca, Ba, Ra in increasing order of
(a) size (b) ionisation energy
(c) electronegativity (d) electron-affinity

2. Arrange alkaline-earth metal chlorides in increasing
(a) covalent nature (b) melting point
(c) lattice energy (d) thermal stability

3. Compare thermal stability of
(a) $NaHCO_3$ and $Ca(HCO_3)_2$
(b) $CaCO_3$ and Na_2CO_3
(c) $BeCO_3$ and Li_2CO_3
(d) $NaHCO_3$ and Na_2CO_3

4. When magnesium burns in air, it forms both an oxide and nitride. Write balanced equations.

5. In the preparation of barium oxide by thermal decomposition of barium carbonate, why is carbon added to the reaction mixture?

6. What is sorel cement?

Analytical Questions

1. Do you know of a biologically important molecule in which magnesium atoms are held in a position by organic groups?

2. Beryllium chloride exists as $BeCl_2 \cdot 4H_2O$ due to solvation (hydration). What is the hydrated cation ? Does it behave as an acid or base, explain?

3. $Ca(OH)_2 + Na_2CO_3 \longrightarrow CaCO_3 \downarrow + 2NaOH$

This reaction made the basis of manufacturing of NaOH in earlier days and pH of the solution was found to be greater than 14. What do you think about the completeness of reaction ? Explain.

4. When $BeCl_2$ is hydrolysed, there is formation of fumes, explain.

5. (a) Describe how the following properties vary on descending group IA and IIA (alkali and alkaline earth metals) and state how these properties differ between the two groups ?
(i) $(IE)_1$ and $(IE)_2$
(ii) cationic radii
(iii) rates of reaction with cold water
(b) For the same s-block elements, discuss :
(i) action of heat on their nitrates
(ii) the crystal structures of their chlorides, *XY* type only.

6. Compare the chemistries of Mg and Ca with reference to the following :
(i) solubilities of their sulphates in water
(ii) the thermal stabilities of their carbonates
(iii) the reactions of their oxides with H_2O.

7. (*A*) is binary electrolyte of a divalent cation. (*A*) on reaction with H_2SO_4 gave a white ppt. (*B*) and a colourless liquid (*C*) that can decolourise $KMnO_4$ in acidic medium. 1.69 g of (*A*) gave 2.33 g of (*B*). What are (*A*), (*B*) and (*C*) and explain reactions?

8. "The chemistry of beryllium is not essentially ionic." Justify the statement by making reference to the nature of oxide, chloride and fluoride of beryllium.

9. Comment on each of the following observations
(a) $E°$ for the reaction, $M^{2+}(aq) + 2e^- \longrightarrow M(s)$ is nearly constant (M =Ca, Sr, Ba).
(b) LiF is least soluble among the fluorides of alkali metals.

10. How would you explain?
(a) BeO is insoluble but $BeSO_4$ is soluble in water.
(b) BaO is soluble but $BeSO_4$ is insoluble in water.
(c) LiI is more soluble than KI in ethanol.
(d) $NaHCO_3$ is isolated in solid state but $Ca(HCO_3)_2$ is not isolated in solid state.

11. Make distinction between in terms of action of heat
(a) Na_2CO_3 and $CaCO_3$
(b) $MgCl_2 \cdot 6H_2O$ and $CaCl_2 \cdot 6H_2O$
(c) $Ca(NO_3)_2$ and $NaNO_3$

NCERT Problems

1. Compare the solublity and thermal stability of the following compounds of the alkali metals with those of the alkaline earth metals.
(a) nitrates (b) carbonates
(c) sulphates

2. Mention the general trends in group 1 and in group 2 with increasing atomic number with respect to
(a) density (b) melting point
(c) atomic size (d) ionisation enthalpy

3. How do the following properties change on moving from group I to group 2 in the Periodic Table?
(i) Atomic size (ii) Ionisation enthalpy
(iii) Density (iv) Melting point

4. Like lithium in group I, beryllium shows anomalous behaviour in group 2. Write three such properties of beryllium which make it anomalous in the group.

5. Boryllium exhibits some similarities with aluminium. Point out three such properties.

6. The chemistry of beryllium is not essentially ionic. Justify the statement by making a reference to the nature of oxide, chloride and fluoride of beryllium.

7. Name the chief forms of occurrence of magnesium in nature. How is magnesium extracted from one of its ores?

8. Commercial aluminium always contains some magnesium. Name two such alloys of aluminium. What properties are imparted by the addition of magnesium to these alloys?

9. The enthalpy of formation of hypothetical $CaCl(s)$ theoretically found to be -118 kJ mol^{-1} and the $\Delta_f H^\ominus$ for $CaCl_2(s)$ is -795 kJ mol^{-1}. Calculate the $\Delta_f H^\ominus$ for the disproportionation reaction.

$$2CaCl(s) \longrightarrow CaCl_2(s) + Ca(s)$$

[Hint. $\Delta_f H^\circ(CaCl_2) - 2\Delta_f H^\circ(CaCl)$
$= -795 + 2 \times 188$
$= -419 \text{ kJ mol}^{-1}$]

10. Compare and contrast the chemistry of group 1 metals with that of group 2 metals with respect to
(i) nature of oxides
(ii) solubility and thermal stability of carbonates
(iii) polarizing power of cations
(iv) reactivity and reducing power

11. Describe the importance of the following
(i) limestone (ii) cement
(iii) Plaster of Paris

12. Mention the main constitutes of Portland cement.

13. What happens when
(i) magnesium is burnt in air?
(ii) quicklime is heated with silica?
(iii) chlorine reacts with slaked lime?
(iv) calcium nitrate is heated?

14. Draw the structure of
(i) $BeCl_2$ (vapour) (ii) $BeCl_2(s)$

15. List the raw materials required in the manufacture of Portland cement. What is the role of gypsum in it?

16. Differentiate between (a) quick-lime (b) lime-water (c) slaked-lime.

17. How is plaster of Paris prepared? Describe its chief property due to which it is widely used.

18. How would you explain?
(i) BeO is insoluble but $BeSO_4$ is soluble in water
(ii) BaO in soluble but $BaSO_4$ is insoluble in water
(iii) LiI is more soluble than KI in ethanol
(iv) $NaHCO_3$ is known in solid state but $Ca(HCO_3)_2$ is not isolated in solid state

19. Contrast the action of heat on the following and explain your answer
(i) Na_2CO_3 and $CaCO_3$
(ii) $MgCl_2 \cdot 6H_2O$ and $CaCl_2 \cdot 6H_2O$
(iii) $Ca(NO_3)_2$ and $NaNO_3$

20. Complete the following equations for the reaction between
(i) $Ca + H_2O$ (ii) $Ca(OH)_2 + Cl_2$
(iii) $BeO + NaOH$ (iv) $BaO_2 + H_2SO_4$

Exercise 2
(Stage 2 : High Skill Problem Solving)

Only One Option Correct

1. Mg and Zn does not resemble in following properties
(a) oxides are amphoteric
(b) carbonates on heating form metal oxides
(c) widely used as electrodes
(d) used to prevent corrosion

2. When Cl_2 is passed into moist slaked lime, compound formed is
(a) CaO_2Cl_2 (b) CaO_2Cl
(c) $CaOCl_2$ (d) $CaCl_2O_4$

3. A colourless solid (*X*) on heating evolved CO_2 and also gave a white residue, soluble in water. Residue also gave CO_2 when treated with dilute acid. (*X*) is

(a) Na_2CO_3 (b) $CaCO_3$
(c) $Ca(HCO_3)_2$ (d) $NaHCO_3$

4. II A (alkaline earth metals) and II B (zinc family) resemble
(a) $MgSO_4 \cdot 7H_2O$ is isomorphous with $ZnSO_4 \cdot 7H_2O$
(b) II A and II B cations are not precipitated by H_2S in acidic medium
(c) both (a) and (b)
(d) None of the above

5. Be and Al show diagonal relationship hence, both have
(a) same degree of electronegativity
(b) amphoteric nature of oxides
(c) approximately same charge/radius ratio
(d) all the properties above

6. Be and Al do not resemble in
(a) both become passive on reaction with HNO_3 due to formation of oxide layer
(b) their chlorides are Lewis bases
(c) chlorides exist in polymeric form
(d) hydroxides are soluble in alkali as well as in acid

7. Going down to II A group, following properties increase except
(a) solubility of hydroxides in H_2O
(b) hydration energy
(c) thermal stability of carbonates
(d) ionic radius

8. Identify the correct statement.
(a) Gypsum is obtained by heating plaster of Paris.
(b) Plaster of Paris can be obtained by hydration of gypsum.
(c) Plaster of Paris contains higher percentage of calcium than does gypsum.
(d) Plaster of Paris is obtained from gypsum by oxidation.

9. Following are the ionisation potential values of (I_1) 899 kJ mol^{-1}, (I_2) 1757 kJ mol^{-1}, (I_3) 15000 kJ mol^{-1}
(a) Na (b) K
(c) Be (d) Ne

10. Which is most stable out of the following?
(a) $[Be(H_2O)_4]^{2+}$ (b) $[Mg(H_2O)_4]^{2+}$
(c) $[Ca(H_2O)_4]^{2+}$ (d) $[Sr(H_2O)_4]^{2+}$

11. Lattice energy of II A group compounds (oxides, carbonates, flourides)
(a) decreases as size of the ion increases
(b) increases as size of the ion increases
(c) constant for a given type of anion
(d) all the above are incorrect

12. Which of the following statements are true about II A group elements?
(a) All form nitrides in air
(b) Be is amphoteric
(c) MH_2 is ionic 'salt-like' hydride
(d) All the above are correct statements

13. Choose the incorrect statement.
(a) $BeCO_3$ is kept in the atmosphere of CO_2 since it is least thermally stable
(b) Be dissolves in alkali forming $[Be(OH)_4]^{2-}$
(c) BeF_2 forms complex ion with NaF in when Be goes with cation
(d) BeF_2 forms complex ion with NaF in which Be goes with anion

14. Which are true statements about *s*-block elements?
(a) Metals are obtained by the electrolysis of fused chlorides
(b) Only one type of valency, +1 for I A and +2 for II A, is shown
(c) Oxides are basic except BeO
(d) All the above are correct statements

15. Which cannot be used to generate H_2?
(a) Al + NaOH (b) Zn + NaOH
(c) Mg + NaOH (d) LiH + H_2O

16. Portland cement does not contain
(a) $CaSiO_4$ (b) $CaSiO_3$
(c) $Ca_3Al_2O_6$ (d) $Ca_3(PO_4)_2$

17. Setting of plaster of Paris is
(a) oxidation with atmospheric oxygen
(b) combination with atmospheric CO_2
(c) dehydration
(d) hydration to yield another hydrate

18. Mixture of $MgCl_2$ and MgO is called
(a) Portland cement (b) Sorel's cement
(c) double salt (d) None of these

19. Setting of cement is
(a) exothermic reaction
(b) endothermic reaction
(c) neither endothermic nor exothermic
(d) None of the above

20. Which is main constituent of egg-shell?
(a) $CaCO_3$ (b) $CaSiO_3$
(c) $CaSO_4 \cdot \frac{1}{2} H_2O$ (d) $CaSO_4 \cdot 2H_2O$

21. The deliquescent among the following is
(a) $CaCl_2$ (b) $FeSO_4 \cdot 7H_2O$
(c) $CuSO_4 \cdot 5H_2O$ (d) $BaCl_2 \cdot 2H_2O$

22. Which is incorrect statement?
(a) The heats of hydration of the dipositive alkaline earth metal ions decrease with an increase in their ionic size.
(b) $NaNO_3$ forms Na_2O on heating
(c) Hydration of alkali metal ion is less than that of II A.

(d) Alkaline earth metal ions, because of their much larger charge to size ratio, exert a much stronger electrostatic attraction on the oxygen of water molecule surrounding them.

23. Which is used for fixing atmospheric nitrogen?
(a) $CaCN_2$ (nitrolim) (b) Li_3N
(c) Mg_3N_2 (d) All of these

24. $BaC_2 + N_2 \xrightarrow{\Delta} (A)$
$CaC_2 + N_2 \xrightarrow{\Delta} (B)$
(A) and (B) are
(a) $BaCN_2$, $CaCN_2$ (b) $Ba(CN)_2$, $Ca(CN)_2$
(c) $Ba(CN)_2$, $CaCN_2$ (d) None is correct

25. Estimation of calcium and magnesium is done by
(a) EDTA (b) oxalate
(c) phosphate (d) None of these

26. Which of the following is/are correct statement(s)?
(a) $Ca_3(PO_4)_2$ in part of bones
(b) $3Ca_3(PO_4)\cdot CaF_2$ is part of enamel on teeth
(c) Ca^{2+} ions are important in blood clotting
(d) All of the above are correct

27. 1 mole of a substance (X) was treated with an excess of water. 2 moles of readily combustible gas were produced along with solution which when reacted with CO_2 gas produced a white turbidity. The substance (X) could be
(a) Ca (b) CaH_2
(c) $Ca(OH)_2$ (d) $Ca(NO_3)_2$

28. A basic refractory material among the following is
(a) Al_2O_3 (b) SiO_2
(c) Fe_2O_3 (d) CaO

29. In water
(a) temporary hardness is due to the bicarbonates of Ca^{2+} and Mg^{2+}
(b) permanent hardness is due to chlorides and sulphates of Ca^{2+} and Mg^{2+}
(c) hardness can be removed by adding phosphates
(d) all of the above properties are true

30. $Na_2[Be(OH)_4]$ is formed when
(a) BeO reacts with NaOH solution
(b) Be reacts with NaOH solution
(c) both (a) and (b) correct
(d) None of the above is correct

31. Which is used to treat acid indigestion?
(a) $Be(OH)_2$ (b) KOH
(c) $Mg(OH)_2$ (d) $Ca(OH)_2$

32. Which is not obtained when metal carbides react with H_2O?
(a) $Al_4C_3 + H_2O \longrightarrow CH{\equiv}CH$
(b) $CaC_2 + H_2O \longrightarrow CH{\equiv}CH$
(c) $Mg_4C_3 + H_2O \longrightarrow CH_3C{\equiv}CH$
(d) $Be_2C + H_2O \longrightarrow CH_4$

33. Select correct statement(s).
(a) Beryllium and magnesium hydride are covalent and polymeric
(b) CaH_2, SrH_2 and BaH_2 are ionic
(c) BeH_2 contains three-centre two-electron bond
(d) All the above are correct statements

34. Select correct statement(s).
(a) $CaCO_3$ is more soluble in a solution of CO_2 than in H_2O
(b) $NaCO_3$ is converted to Na_2O and CO_2 on heating
(c) Li_2CO_3 is thermally stable
(d) Presence of $CaCl_2$ or $CaSO_4$ in water causes temporary hardness

35. Select correct statement(s).
(a) Mg^{2+} ions are necessary for the activation of phosphate transfer enzymes.
(b) Mg is present in chlorophyll used in photosynthesis in green plants.
(c) Operation of $Na^+ - K^+$ pumps is biological.
(d) All the above are correct statements

36. A metal M readily forms water soluble MSO_4. It also forms oxide MO which becomes inert on heating. Hydroxide $M(OH)_2$ is insoluble in water but soluble in NaOH solution. What is M?
(a) Mg (b) Ba
(c) Ca (d) Be

37. Which one of the following has magnesium?
(a) Vitamin B_{12} (b) Chlorophyll
(c) Haemocyanin (d) Carbonic anhydrase

38. In which of the following the hydration energy is higher than the lattice energy?
(a) $BaSO_4$ (b) $MgSO_4$
(c) $RaSO_4$ (d) $SrSO_4$

39. Epsom salt is
(a) $BaSO_4 \cdot 2H_2O$ (b) $CaSO_4 \cdot H_2O$
(c) $MgSO_4 \cdot 2H_2O$ (d) $MgSO_4 \cdot 7H_2O$

40. Slaked lime and chlorine reacts to produce
(a) quick lime
(b) $CaCl_2$
(c) $CaOCl_2$
(d) mixture of $CaCl_2$ and $Ca(OCl_2)$

41. Mg and Li and similar in their properties due to
(a) same e/m ratio (b) same electron affinity
(c) same group (d) same radius

42. Which one of the following hydroxide is insoluble in water?
(a) $Ca(OH)_2$ (b) $Ba(OH)_2$
(c) $Be(OH)_2$ (d) $Mg(OH)_2$

43. The pair of amphoteric hydroxide is
(a) $Al(OH)_3$, LiOH (b) $Be(OH)_2$, $Mg(OH)_2$
(c) $B(OH)_3$, $Be(OH)_2$ (d) $Be(OH)_2$, $Zn(OH)_2$

44. The pair whose both species are used in antacid medicinal preparation is
(a) $NaHCO_3$ and $Mg(OH)_2$
(b) Na_2CO_3 and $Ca(HCO_3)_2$
(c) $Ca(HCO_3)_2$ and $Mg(OH)_2$
(d) $Ca(OH)_2$ and $NaHCO_3$

45. Lithophone is mixture of
(a) $ZnCO_3, BaCO_3$ (b) ZnS, Na_2SO_4
(c) $ZnSO_4, BaSO_4$ (d) $ZnS, BaSO_4$

46. Gypsum is
(a) $CaSO_4 \cdot 2H_2O$ (b) $CaCO_3 \cdot 2H_2O$
(c) $CaSO_4 \cdot \frac{1}{2} H_2O$ (d) SiO_2

47. Alkaline earth metals are
(a) reducing agent (b) dehydrating agent
(c) amphoteric (d) oxidising agent

48. A covalent chloride is
(a) $BeCl_2$ (b) NaCl
(c) $MgCl_2$ (d) $CaCl_2$

49. Plaster of Paris is hardened by
(a) giving out water
(b) uniting with water
(c) changing into $CaCO_3$
(d) liberating CO_2

50. Which of the following halide of calcium is insoluble in water?
(a) $CaCl_2$ (b) CaI_2
(c) CaF_2 (d) $CaBr_2$

51. The substance used as pigment in paint is
(a) borax (b) alumina
(c) lithophone (d) None of these

52. Lightest alkaline earth metal is
(a) Be (b) Mg
(c) Ca (d) Sr

53. Which of the following is not an ore of magnesium?
(a) Carnallite (b) Magnesite
(c) Dolomite (d) Gypsum

54. The activity of alkaline earth metals as reducing agent
(a) decreases from Be to Ba
(b) increases from Be to Ba
(c) increase form Be to Ca and decreases from Ca to Ba
(d) decreases from Be to Ca and increases from Ca to Ba

55. Magnalium contains
(a) aluminium + magnesium
(b) magnesium + copper
(c) magnesium + iron
(d) magnesium + silver

56. Be in $BeCl_2$ undergoes
(a) linear hybridisation
(b) trigonal hybridisation
(c) tetrahedral hybridisation
(d) no hybridisation

57. $CaCl_2$ is used as
(a) disinfectant (b) desiccant agent
(c) medicine (d) None of these

58. The product obtained on fusion of $BaSO_4$ and Na_2CO_3 is
(a) $BaCO_3$ (b) BaO
(c) $Ba(OH)_2$ (d) $BaHSO_4$

59. Which of the following is known as dead burnt?
(a) Gypsum (b) Plaster of Paris
(c) Anhydrite (d) None of these

60. The decomposition temperature is maximum for
(a) $MgCO_3$ (b) $CaCO_3$
(c) $BaCO_3$ (d) $SrCO_3$

61. Bleaching powder loses its power on keeping for a long time because
(a) it changes into calcium hypochlorate
(b) it changes into calcium chloride and calcium hydroxide
(c) it absorbs moisture
(d) it changes into calcium chloride and calcium chlorate

62. Among Na^+, Na, Mg and Mg^{2+} the largest particle is
(a) Mg^{2+} (b) Mg
(c) Na (d) Na^+

63. The electron affinity of Be is similar to
(a) He (b) B
(c) Li (d) Na

64. Bleaching action of $CaOCl_2$ is due to
(a) nascent oxygen (b) chlorine
(c) HClO (d) HCl

65. In alkaline earth metal sulphates, the value of hydration energy decreases down the group because of the
(a) decrease in size
(b) increase in size
(c) greater lattice energy
(d) None of the above

66. The correct order of increasing ionic character is
(a) $BeCl_2 < CaCl_2 < MgCl_2 < BaCl_2$
(b) $BeCl_2 < MgCl_2 < BaCl_2 < CaCl_2$
(c) $BeCl_2 < BaCl_2 < MgCl_2 < CaCl_2$
(d) $BaCl_2 < CaCl_2 < MgCl_2 < BeCl_2$

67. Ripening of fruits can be caried out in presence of
(a) Na_2SO_4 (b) NaCl
(c) $CaCl_2$ (d) CaC_2

68. The right order of the solubility of sulphates of alkaline earth metals in water
(a) Be > Ca > Mg > Ba > Sr
(b) Mg > Be > Ba > Ca > Sr
(c) Be > Mg > Ca > Sr > Ba
(d) Mg > Ca > Ba > Be > Sr

69. Burning of Mg is extinguished by
(a) throwing N_2 liquid (b) throwing sand
(c) throwing ice (d) throwing water

70. Flame test is not given by
(a) Ca (b) Ba
(c) Mg (d) Li

71. Lithium shows similarities with magnesium in its chemical behaviour because
(a) similar size, greater electronegativity and lower polarising power
(b) similar size, same electronegativity and lower polarising power
(c) similar size, same electronegativity and similar high polarising power
(d) None of the above

72. The element whose electronic configuration is $1s^2 2s^2 2p^6 3s^2$ is
(a) non-metal (b) noble gas
(c) metalloid (d) metal

73. On strong heating $MgCl_2 \cdot 6H_2O$ the product obtained is
(a) $MgCl_2$ (b) MgO
(c) $MgCl_2 \cdot 2H_2O$ (d) $MgCl_2 \cdot 4H_2O$

74. Which of the following carbonate decompose most easily on heating?
(a) Rb_2CO_3 (b) K_2CO_3
(c) Na_2CO_3 (d) $MgCO_3$

75. The solubilities of carbonates decrease down the magnesium group due to a decrease in
(a) lattice energies of solids
(b) hydration energies of cations
(c) interionic attractions
(d) entropy of solution formation

76. The charge/size ratio of a cation determines its polarising power. Which one of the following sequences represents the increasing order of the polarising power of cationic species; K^+, Ca^{2+}, Mg^{2+}, Be^{2+}?
(a) $Mg^{2+} < Be^{2+} < K^+ < Ca^{2+}$
(b) $Be^{2+} < K^+ < Ca^{2+} < Mg^{2+}$
(c) $K^+ < Ca^{2+} < Mg^{2+} < Be^{2+}$
(d) $Ca^{2+} < Mg^{2+} < Be^{2+} < K^+$

77. What is impurity (as a salt) associated with table salt obtained from sea-water?
(a) $NaHCO_3$ (b) $MgCO_3$
(c) $MgCl_2$ (d) NaI

78. Select correct statement(s).
(a) Presence of $MgCl_2$ in table salt causes it to clump
(b) Addition of $NaHCO_3$ to table salt converts $MgCl_2$ to non-hygroscopic salt
(c) Both (a) and (b)
(d) None of the above

79. In polymeric $(BeCl_2)_n$, there are
(a) three centre two-electron bonds
(b) three centre three-electron bonds
(c) two centre three-electrons bonds
(d) two centre two-electron bonds

80. Be and Al resemble in the following but not in
(a) both form electron deficient hydrides
(b) both are rendered passive by HNO_3
(c) both form amphoteric oxides
(d) both have *sp*-hybridisation in their compounds

81. Select incorrect statement.
(a) NaOH can be stored in a vessel made of Al
(b) HNO_3 can be stored in a vessel made of Be/Al alloy
(c) HF can be stored in a vessel made of wax
(d) HF attacks glass

82. Metal carbides on reaction with H_2O from CH_4. Carbide can be
(a) CaC_2 (b) Mg_3C_2
(c) Be_2C (d) SiC

83. Select correct statement.
(a) Be and Al show diagonal relationship
(b) Be form tetrahedral complexes $[Be(C_2O_4)_2]^{2-}$
(c) Al forms AlF_6^{3-}, an octahedral complex
(d) All the above are correct statements

Brain Twisters

1. Metals reacts with hydrogen halides (such as HCl) to give metal halide and hydrogen :

$$M(s) + nHX(g) \longrightarrow MX_n(s) + \frac{1}{2}nH_2(g)$$

$$\Delta G^\circ = \Delta G^\circ(MX_n) - n\Delta G^\circ(HX(g))$$

$$\Delta G^\circ(HCl(g)) = -95.1 \text{ kJ mol}^{-1}.$$

What is relative value of ΔG° for MX_n for the reaction to be product-favoured?

2. A mineral $Mg_xBa_y(CO_3)_z$ was analysed as given below:

A sample of the mineral was dissolved in excess HCl and the solution made up to 100 mL with water. During the process 48 mL of CO_2 measured a 25°C and 1 atmosphere pressure were evolved.

A 25.0 mL portion of the resulting solution required 25.0 mL of EDTA solution of 0.02 M to reach end point. A further 25.0 mL portion gave a precipitate of $BaSO_4$ of mass 0.058 g on treatment with excess of dil. H_2SO_4. You may assume that group 2 (alkaline earth metals) metal ions form 1 : 1 complexes with EDTA.

(Given molar volume of any gas at 25°C and 1 atmosphere pressure = 24 L)

3. Use the following data to calculate second electron affinity of oxygen, *i.e.*, for the process

$$O^-(g) + e^-(g) \longrightarrow O^{2-}(g)$$

Is the O^{2-} ion stable in the gas phase? Why is it stable in solid MgO?

Heat of sublimation of Mg(*s*) = + 147.7 kJ mol^{-1}

Ionisation energy of Mg(*g*) to form $Mg^{2+}(g)$ = + 2189.0 kJ mol^{-1}

Bond dissociation energy for O_2 = + 498.4 kJ mol^{-1}

First electron affinity of O(*g*) = – 141.0 kJ mol^{-1}

Heat formation of MgO(*s*) = – 601.7 kJ mol^{-1}

Lattice energy of MgO = – 3791.0 kJ mol^{-1}

4. A 1.005 g sample of an unknown alkaline earth metal was allowed to react with a volume of chlorine gas that contains 1.91×10^{22} Cl_2 molecules. The resulting metal chloride was analysed for chlorine by dissolving 0.436 g sample in water and adding an excess of $AgNO_3(aq)$. The analysis yielding 1.9126 g of AgCl.

(a) What is the identity of the alkaline earth metal?
(b) What is the % of Cl in the alkaline earth chloride.

Passage Comprehension Questions

Passage 1

Based on the following analytical data, answer the questions.

A mineral, which can be represented by the formula $Mg_xBa_y(CO_3)_z$, was analysed as described below:

A sample of the mineral was dissolved in excess hydrochloric acid and the solution made up to 100 cm^3 with water. During the process 48 cm^3 of carbon dioxide, measured at 25°C and 1 atmosphere pressure, were evolved.

A 25.0 cm^3 portion of the resulting solution required 25.0 cm^3 of EDTA solution of concentration 0.02 mole dm^3 to reach an end-point. A further 25.0 cm^3 portion gave a precipitate of barium sulphate of mass 0.058 g on treatment with excess dilute sulphuric acid. You may assume that group-2 metal ions form 1 : 1 complexes with EDTA.

Molar volume of any gas at 25°C and 1 atmosphere pressure = 24 dm^3).

1. Formula of the mineral is
(a) $MgBa(CO_3)_4$ (b) $MgBa(CO_3)_2$
(c) $MgBa(CO_3)_3$ (d) $Mg_2Ba(CO_3)_4$

2. In the mineral barium ion can be identified by
(a) green colour in flame
(b) yellow precipitate with CrO_4^{2-} solution
(c) white precipitate with SO_4^{2-}
(d) all of the above

3. Identical mineral of calcium and magnesium ions is
(a) dolomite (b) magnesite
(c) epsomite (d) talc

Passage 2

Questions given below are based on the following enthalpy values

The standard enthalpy of formation ΔH_f°, of hypothetical CaCl(*s*) theoretically found to be –188 kJ mol^{-1} and that of $CaCl_2(s)$ –795 kJ mol^{-1}.

1. Which of the following compounds is more stable?
(a) CaCl(*s*)
(b) $CaCl_2(s)$
(c) Both (a) & (b) are equally stable
(d) Nothing can be said

2. Select most spontaneous reaction
(a) $2CaCl(s) \longrightarrow CaCl_2(s) + Ca(s)$
(b) $CaCl_2(s) + Ca(s) \longrightarrow 2CaCl(s)$
(c) $CaCl(s) \longrightarrow Ca(s) + Cl(g)$
(d) $CaCl_2(s) \longrightarrow Ca(s) + Cl_2(g)$

Passage 3

ΔG° values (kJ mol^{-1}) for the following cases can given

	MCO_3	$\longrightarrow$ MO	+ CO_2
			–394.4
Mg	–1028.2	–568.9	64.9
Ca	–1129.2	603.4	165.9
Ba	–1134.4	–520.4	219.6

Answer the following questions

1. Maximum value of ΔG° is for the reaction when *M* is
(a) Mg (b) Ca
(c) Ba (d) equal for all cases

2. Thermal stability of the carbonates is in order
(a) $MgCO_3 < CaCO_3 < BaCO_3$
(b) $BaCO_3 < CaCO_3 < MgCO_3$
(c) $MgCO_3 < BaCO_3 < CaCO_3$
(d) $CaCO_3 < BaCO_3 < MgCO_3$

3. Which is decomposed most easily?
(a) $CaCO_3$ (b) $MgCO_3$
(c) $BaCO_3$ (d) Equally

4. $BeCO_3$ decomposes at low temperature. To avoid its decomposition, it is placed in the atmosphere of
(a) N_2 (b) He
(c) Ne (d) CO_2

Passage 4

"Calcium oxide, CaO, is used to remove SO_2 from power plant exhaust."

Based on the above statement, answer the following questions

1. In what form CaO removes SO_2 ?
   (a) $CaSO_3$ (b) $CaSO_4$
   (c) $Ca(HSO_3)_2$ (d) $Ca(HSO_4)_2$
2. What mass of SO_2 can be removed using 1.4×10^3 kg CaO?
   (a) 1.4×10^3 kg (b) 1.6×10^3 kg
   (c) 2.2×10^3 kg (d) 2.4×10^3 kg
3. Presence of SO_2 from power plant exhaust can be detected by
   (a) $Cr_2O_7^{2-}$ which changes to green
   (b) MnO_4^- which changes to colourless
   (c) both (a) & (b) above
   (d) None of the above

Assertion & Reason

Codes :

(a) Both **A** and **R** are true and **R** is the correct explanation of **A.**

(b) Both **A** and **R** are true but **R** is not the correct explanation of **A.**

(c) **A** is true but **R** is false.

(d) **A** is false but **R** is true.

1. **Assertion (A) :** $BaSO_4$ is likely insoluble in water but readily dissolves in solution of sodium salt of ethylene diamine tetra acetic acid (EDTA).

   Reason (R) : Ba^{2+} forms a very stable complex with the anion of the sodium salt EDTA which is water soluble.
2. **Assertion (A) :** Temporary hardness can be removed by addition of lime.

   Reason (R): $Ca(HCO_3)_2$ in hard water is converted to insoluble $CaCO_3$ on moderate heating.
3. **Assertion (A) :** Group 2 metals have higher melting points when compared to the group 1 metals.

   Reason (R) : +2 charge on the cations in the metallic lattice causes them to be more strongly attracted to the 'sea of electrons' and making it difficult to pull them apart.
4. **Assertion (A) :** The metals of group 2 (except Be) from ionic compounds.

   Reason (R) : Electropositive character and the reducing property increase on moving down the group.
5. **Assertion (A) :** Divalent compounds of alkaline earth metals are diamagnetic.

   Reason (R) : M^{2+} ions of group 2 have noble gas configuration.

True & False

1. When CO_2 is passed into lime water, white turbidity appears which dissolves in excess of CO_2.
2. SO_2 turns lime water milky.
3. $[Be(H_2O)_4]^{2+}$ is an acid.
4. Solubility of sulphates of alkaline earth metals increases going down the group.
5. Mg_2C_3 on reaction with water forms propane.
6. $CaCO_3$ is the main constituent of egg shell.
7. EDTA is used to estimate Ca and Mg.
8. Permanent hardness is due to dissolved bicarbonates of Ca^{2+} and Mg^{2+}.
9. $BeCl_2$ is a nucleophile.
10. BeO and Al_2O_3 are amphoteric.
11. Be and Al are made passive by HNO_3.
12. Chlorophyll contains Mg and chlorine.
13. Chlorophyll produces sugars for the plants in presence of sunlight, CO_2 and H_2O by photosynthesis.
14. Chlorophyll is an organometallic compound.
15. The lattice energies of alkaline earth metal salts are much higher than these of alkali metal salts.
16. Decreasing lattice energy favours increased solubility, while decreasing hydration energy favours decreased solubility.
17. BeO has hexagonal ZnS structure.
18. Beryllium forms hydrate peroxide $BeO_2 \cdot H_2O$.
19. Calcium and barium bicarbonates are stable only in solution and decompose on removal of water to give carbonates.
20. Formation of stalactite and stalagmite is the property of calcium and barium bicarbonates which change to carbonates.

Fill in the Blanks

1. .... of IA and .... of IIA show diagonal relationship.
2. $(IE)_1$ of Be is .... than that of B and $(IE)_2$ of Be is .... than that of B.
3. Of the alkaline earth metals only .... carbide gives CH_4 on reaction with H_2O.
4. Alkali metal and alkaline earth metal chlorides can be detected by .... test.
5. $BeCl_2$ is an .... and has .... structure in solid state.

6. $Be(OH)_2$ is ....

7. Solubility of sulphates and carbonates of alkaline earth metals .... as the atomic number of the metal increases (down the group).

8. $BeCO_3$ is placed in an atmosphere of .... since, it is ....

9. Be and .... show diagonal relationship.

10. Be is not readily attacked by acids due to formation of .... layer.

11. Be dissolves in alkali to give ....

12. HCO_3^- is ....

13. Be forms complexes with coordination number ....

14. The thermal stability of group 2 oxysalts is .... than those of the group 1 oxy salts.

15. $Ca(NO_3)_2$ decomposes into gases .... and ....

16. .... is the primary constituent of teeth and bones

17. Amphoteric oxide of group 2 is ....

18. MgO is used as ....

19. $Ca(OH)_2$ is called .... and $Ba(OH)_2$ is called ....

20. Most soluble sulphate of group 2 is ....

21. Potash magnesis is ....

22. $CaSO_4 \cdot H_2O$ is called .... and is used to make ....

23. $BeCl_2$ fumes in air due to formation of ....

24. $BeCl_2$ exists as monomer in .... state.

25. Polymeric $BeCl_2$ has .... centre .... electron bonds.

Matrix-Match Type Questions

I. Only One Correct

1. Match the commercial names (in List I) with the chemical formulae (in List II).

List I	List II
A. Magnesia	1. BeO
B. Epsom salt	2. $Ca(OH)_2$
C. Anhydrone	3. $MgSO_4 \cdot 7H_2O$
D. Hydrolith	4. CaH_2
E. Beryllia	5. MgO
F. Slaked lime	6. $Mg(ClO_4)_2$

2. Match the mineral names (in List I) with the chemical formulae (in List II).

List I	List II
A. Magnesite	1. $KCl \cdot MgCl_2 \cdot 6H_2O$
B. Dolomite	2. $MgSO_4 \cdot H_2O$
C. Kieserite	3. $MgCO_3$
D. Carnallite	4. $MgCO_3 \cdot CaCO_3$

3. Match compounds (in List I) with the specific properties of the compounds (in List II).

List I	List II
A. $BeCO_3$	1. Temporary hardness
B. $BaCl_2$	2. Permanent hardness
C. $MgSO_4$	3. Decomposes readily
D. $Ca(HCO_3)_2$	4. Gives green edged flame

II. One or More Correct

1. Match the species in Column I with their corresponding property (ies) in Column II.

Column I	Column II
A. BeO	1. Amphoteric
B. Al_2O_3	2. Diagonal relationship
C. MgO	3. Lime water
D. CaO	4. Bartya water
E. BaO	5. Water insoluble
	6. Maximum soluble in water
	7. Estimation by EDTA

2. Match the species in Column I with their corresponding property (ies) in Column II.

Column I	Column II
A. Li	1. Diagonal relationship
B. Be	2. Maximum standard reduction electrode potential ($E°$)
C. Mg	3. Cation is maximum hydrated
D. Al	4. Cation has maximum ionic mobility in aqueous solution
E. Ca	5. Fluoride and carbonate least soluble in water
	6. Coordination number six in the complexes
	7. Ions are involved in the transmission of electrical impulses in human body.

Integer Answer Types

This section contains 8 questions. The answer to each of the questions is a single digit integer, ranging from 0 to 9. The appropriate bubbles below the respectively question numbers in the *ORS* have to be darkened. For example, if the correct answers to question number *X, Y, Z* and *W* (say) are 6, 0, 9 and 2, respectively, then the correct darkening of bubbles will like the following :

X	Y	Z	W
0	0	0	0
1	1	1	1
2	2	2	2
3	3	3	3
4	4	4	4
5	5	5	5
6	6	6	6
7	7	7	7
8	8	8	8
9	9	9	9

Set I

1. Coordination number of complexes of beryllium is ....
2. Number of water molecules in gypsum is ....
3. Epsom salt contains .... water molecules (of hydration)
4. Magnesium allyide is Mg_2C_x where x is ....
5. Magnesium nitride is Mg_xN_y where $(x + y)$ is ....
6. Magnesium is coordinated to .... N atoms in chlorophyll.
7. Bleaching powder has two types of chlorine one with oxidation number −1 and other with oxidation number ....
8. One mole of $Ca(HCO_3)_2$ is neutralised by .... mole(s) of KOH.

Set II

1. Number of electrons in magnesium ion formed by three steps ionisation of magnesium is ....
2. Number of neutrons in ${}^{9}_{4}Be^{3+}$ is ....
3. Magnesium is coordinated to N-atoms in chlorophyll. Number of rings in chlorophyll is ....
4. When Cl_2 is passed into hot alkali solution, products formed have two types of chlorine. Difference in oxidation number of two chlorine atoms is ....
5. 96 g of Mg is burnt in air in a closed chamber. On analysis 25% of Mg is converted into oxide and remaining Mg into other possible product. Residue is dissolving in H_2O and neutralised by H_2SO_4. Number of moles of H_2SO_4 required is ....
6. $E^\circ_{M^{2+}/M}$ values of some metals are

Be	:	− 1.85 V
Mg	:	−2.37 V
Ca	:	−2.87 V
Sr	:	−2.89 V
Cu	:	+0.34 V
Hg	:	0.80 V

Number of metals which can displace H_2 gas from acid is ....

7. $Mg_xBa_y(CO_3)_z$ has 8.54% Mg, 48.75% Ba and remaining carbonate. 0.281 kg is decomposed by H_2SO_4. CO_2 formed is .... mole(s).
8. In question (7), $x + y + z =$ ....

Test Yourself on the Periodic Table

1. Identify the elements in the Periodic Table
   A : which shows diagonal relationship to Be
   B : which shows diagonal relationship to Li
   C : whose alkali metal carbonate is decomposed on heating
   D : coinage metals

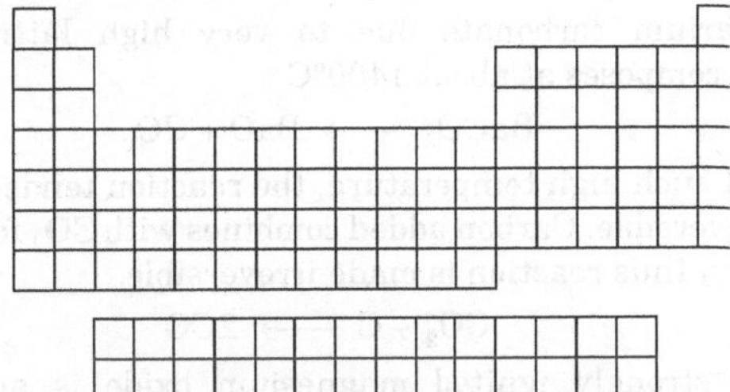

2. Out of the elements marked A, B, C, D, E, F, G, and H
   (a) which form superoxide?
   (b) which form thermally stable carbonate?
   (c) which forms strongest base?
   (d) which show diagonal relationship?
   (e) which forms amphoteric oxide?

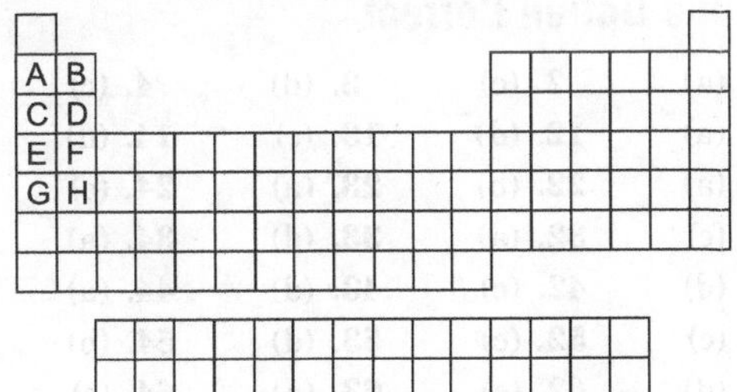

Answers & Solutions
Master Exercises

Exercise 1

Short Answer Type Questions

1. (a) $Be < Ca < Ba < Ra$

   (b) $Ba < Ra < Ca < Be$

   (c) $Ba < Ca < Be$

   (d) $Ba < Ca < Be (\approx 0)$

2. (a) $RaCl_2 < BaCl_2 < SrCl_2 < CaCl_2 < MgCl_2 < BeCl_2$

   (b) $BeCl_2 < MgCl_2 < CaCl_2 < SrCl_2 < BaCl_2 < RaCl_2$

   (c) $RaCl_2 < BaCl_2 < SrCl_2 < CaCl_2 < MgCl_2 < BeCl_2$ (numerical value)

   (d) $RaCl_2 < BaCl_2 < SrCl_2 < CaCl_2 < MgCl_2 < BeCl_2$

3. (a) $Ca(HCO_3)_2 < NaHCO_3$

   (b) $CaCO_3 < Na_2CO_3$

   (c) $BeCO_3 < Li_2CO_3$

   (d) $NaHCO_3 < Na_2CO_3$

4. $Mg + \frac{1}{2}O_2 \longrightarrow MgO$

   $3Mg + N_2 \longrightarrow Mg_3N_2$

5. Barium carbonate due to very high lattice energy decomposes at about 1400°C

   $$BaCO_3 \longrightarrow BaO + CO_2$$

   At such high temperature, the reaction tends to become reversible. Carbon added combines with CO_2 forming CO and thus reaction is made irreversible.

   $$CO_2 + C \longrightarrow 2CO$$

6. If strongly ignited magnesium oxide is added to a concentrated solution magnesium chloride, the slurry so obtained solidifies in the course of a few hours, to a solid mass. This is called **sorel** cement or magnesia cement.

   When mixed with saw-dust or cork waste, sorel cement gives a weather proof composition.

Analytical Questions

1. Chlorophyll

2. Hydrated cation is $[Be(H_2O)_4]^{2+}$.

   $[Be(H_2O)_4]^{2+} + H_2O \longrightarrow [Be(H_2O)_3(OH)]^+ + H_3O^+$

   hydrated cation is H^+ ion donor, hence, it behaves as an acid (Lowry-Bronsted protonic concept).

3. This indicates that concentration of NaOH is very high hence, pH is greater than 14.

4. $BeCl_2 + 2H_2O \longrightarrow Be(OH)_2 + 2HCl$

   HCl causes fumes formation.

5. (a) See text (b) See text **6.** See text

7. (A) : BaO_2 (B) : $BaSO_4$ (C) : H_2O_2 **8.** Refer text

9. (a) Refer text (b) Refer text **10.** (a) to (d) Refer text

11. (a) $Na_2CO_3 \xrightarrow{\Delta}$ no effect

    $CaCO_3 \xrightarrow{\Delta} CaO + CO_2$

    (b) $MgCl_2 \cdot 6H_2O \xrightarrow{\Delta} Mg(OH)Cl \cdot 5H_2O + HCl$

    $CaCl_2 \cdot 6H_2O \xrightarrow{\Delta} CaCl_2 + 6H_2O$

    (c) $2Ca(NO_3)_2 \xrightarrow{\Delta} 2CaO + 2NO_2 + 3O_2$

    $2NaNO_3 \xrightarrow{\Delta} 2NaNO_2 + O_2$

NCERT Problems

Refer Text

Exercise 2

Only One Option Correct

1. (a)	**2.** (c)	**3.** (d)	**4.** (c)	**5.** (d)	**6.** (b)	**7.** (b)	**8.** (c)	**9.** (c)	**10.** (a)
11. (a)	**12.** (d)	**13.** (c)	**14.** (d)	**15.** (c)	**16.** (d)	**17.** (d)	**18.** (b)	**19.** (a)	**20.** (a)
21. (a)	**22.** (b)	**23.** (a)	**24.** (c)	**25.** (a)	**26.** (d)	**27.** (b)	**28.** (d)	**29.** (d)	**30.** (c)
31. (c)	**32.** (a)	**33.** (d)	**34.** (a)	**35.** (d)	**36.** (d)	**37.** (b)	**38.** (b)	**39.** (d)	**40.** (d)
41. (d)	**42.** (c)	**43.** (d)	**44.** (a)	**45.** (d)	**46.** (a)	**47.** (a)	**48.** (a)	**49.** (b)	**50.** (c)
51. (c)	**52.** (c)	**53.** (d)	**54.** (b)	**55.** (a)	**56.** (a)	**57.** (b)	**58.** (b)	**59.** (c)	**60.** (c)
61. (d)	**62.** (c)	**63.** (a)	**64.** (c)	**65.** (b)	**66.** (a)	**67.** (d)	**68.** (a)	**69.** (c)	**70.** (c)
71. (c)	**72.** (d)	**73.** (b)	**74.** (d)	**75.** (a)	**76.** (c)	**77.** (c)	**78.** (c)	**79.** (a)	**80.** (d)
81. (a)	**82.** (c)	**83.** (d)							

Brain Twisters

1. If the reaction is to be product-favoured, *i.e.*, it is spontaneous in forward side then $\Delta G°$ should be negative and $\Delta G°$ for MX_n should be greater than and more negative than $(95.1) \times n$ kJ.

2. 24 L of CO_2 at 298 K and 1 atm = 1 mol CO_2

$$48 \text{ mL} (= 0.048 \text{ L}) = \frac{1}{24} \times 0.048$$
$$= 0.002 \text{ mol } CO_2 \text{ in 100 mL solution}$$
$$= \mathbf{0.02\ M}$$

EDTA complexes with Mg^{2+} and Ba^{2+}

$$M_1V_1 = M_2V_2$$
$$M_1(Mg^{2+} \text{ and } Ba^{2+}) \times 25 = 25 \times 0.02$$
$$= 0.02 \text{ M} = 0.02 \text{ mol L}^{-1}$$
$$M_1(Mg^{2+}) + M_1(Ba^{2+}) = 0.02 \text{ M}$$
$$Ba^{2+} + SO_4^{2-} \longrightarrow BaSO_4$$
$$137 \text{ g} = 1 \text{ mol} \qquad 233 \text{ g} = 1 \text{ mol}$$

233 g $BaSO_4$ is from = 137 g Ba^{2+} = 1 mol Ba^{2+}

$\therefore$ 0.058 g $BaSO_4$ is from $= \frac{0.058}{233} = 2.5 \times 10^{-4}$ mol Ba^{2+} in 25 mL

$$\therefore \quad [Ba^{2+}] = 0.01 \text{ mol L}^{-1}$$
$$\therefore \quad [Mg^{2+}] = 0.01 \text{ mol L}^{-1}$$

Thus, Mg^{2+}, Ba^{2+} and CO_3^{2-} are in molar ratio of 1 : 1 : 2

$\therefore$ Salt is $MgBa(CO_3)_2$

3. Stepwise formation of MgO involves

Step	Reaction	ΔH kJ mol^{-1}
I	$Mg(s) \longrightarrow Mg(g)$	147.7
II	$Mg(g) \longrightarrow Mg^{2+}(g) + 2e^-$	2189.0
III	$\frac{1}{2} O_2(g) \longrightarrow O(g)$	$\frac{498.4}{2}$
IV	$O(g) + e^- \longrightarrow O^-(g)$	−141.0
V	$O^-(g) + e^- \longrightarrow O^{2-}(g)$	q
VI	$Mg^{2+}(g) + O^{2-}(g) \longrightarrow MgO(s)$	−3791.0
	$Mg(s) + \frac{1}{2} O_2(g) \longrightarrow MgO(s)$	

$$\Delta H = -1346.1 + q$$
$$Mg(s) + \frac{1}{2} O_2(g) \longrightarrow MgO(s)$$
$$\Delta H_2 = -601.7$$

By **Born-Haber Cycle** (based on Hess law)

$$\Delta H_1 = \Delta H_2$$
$$-1346.1 + q = -601.7$$
$$q = \mathbf{744.4\ kJ\ mol^{-1}}$$

4. Let molar mass of the element $(M) = A$

$$\text{Moles of element} = \frac{1.005}{A}$$

$$\underset{\substack{(A+71)\\0.436}}{MCl_2} + 2AgNO_3 \longrightarrow \underset{\substack{245.3 \times 2\\1.126}}{\mathbf{2AgCl}} + M(NO_3)_2$$

$$\frac{A+71}{0.436} = \frac{243.5 \times 2}{1.9126}$$
$$\therefore \quad A = 40 \text{ g mol}^{-1}$$

(a) Thus element is **Ca**

(b) Molar mass of $CaCl_2$ = 111 g mol^{-1}

$$\% \text{ of Cl} = \frac{71}{111} \times 100$$
$$= \mathbf{63.96\%}$$

Passage Comprehension Questions

Passage 1

1. (b) **2.** (d) **3.** (a)

Passage 2

1. (b) **2.** (a)

Passage 3

1. (c) **2.** (a) **3.** (b) **4.** (d)

Passage 4

1. (a) **2.** (b) **3.** (c)

Assertion & Reason

1. (a) **2.** (b) **3.** (a) **4.** (a) **5.** (b)

True & False

1. T **2.** T **3.** T **4.** F (decreases)

5. F (propyne) **6.** T **7.** T

8. F (due to Cl^- and SO_4^{2-})

9. F ($BeCl_2$ is an electrophile)

10. T **11.** T **12.** F **13.** T

14. F (Mg is coordinated to N)

15. T **16.** T **17.** T **18.** F **19.** T **20.** T

Fill in the Blanks

1. Li, Mg **2.** higher, smaller **3.** beryllium

4. flame **5.** electrophile, polymeric **6.** amphoteric

7. decreases **8.** CO_2, thermally unstable **9.** Al

10. oxide **11.** Na_2BeO_2 and H_2 **12.** amphoteric

13. four **14.** less **15.** NO_2, O_2 **16.** $Ca_3(PO_4)_2$

17. BeO **18.** refractory **19.** lime water, baryta water

20. $BeSO_4$ **21.** $K_2SO_4 \cdot MgSO_4 \cdot 6H_2O$

22. Alabaster, ornaments **23.** HCl **24.** vapour

25. three, two

Matrix-Match Type Questions

I. Only One Correct

1. (A)—(5); (B)—(3); (C)—(6); (D)—(4); (E)—(1); (F)—(2);
2. (A)—(3); (B)—(4); (C)—(2); (D)—(1)
3. (A)—(3); (B)—(4); (C)—(2); (D)—(1)

II. One or More Correct

1. (A)—(1,2,5); (B)—(1,2,5); (C)—(2,7); (D)—(3,7); (E)—(4,6)
2. (A)—(1,2,); (B)—(1,3); (C)—(1,6,7); (D)—(1,6); (E)—(4,5,6,7)

Integer Answer Types

Set I

Questions →	1	2	3	4	5	6	7	8
Answers →	4	2	7	3	5	4	1	2
	0	0	0	0	0	0	0	0
	1	1	1	1	1	1	1	1
	2	2	2	2	2	2	2	2
	3	3	3	3	3	3	3	3
	4	4	4	4	4	4	4	4
	5	5	5	5	5	5	5	5
	6	6	6	6	6	6	6	6
	7	7	7	7	7	7	7	7
	8	8	8	8	8	8	8	8
	9	9	9	9	9	9	9	9

Set II

Questions →	1	2	3	4	5	6	7	8
Answers →	9	5	8	8	5	4	2	4
	0	0	0	0	0	0	0	0
	1	1	1	1	1	1	1	1
	2	2	2	2	2	2	2	2
	3	3	3	3	3	3	3	3
	4	4	4	4	4	4	4	4
	5	5	5	5	5	5	5	5
	6	6	6	6	6	6	6	6
	7	7	7	7	7	7	7	7
	8	8	8	8	8	8	8	8
	9	9	9	9	9	9	9	9

Test Yourself on the Periodic Table

1.

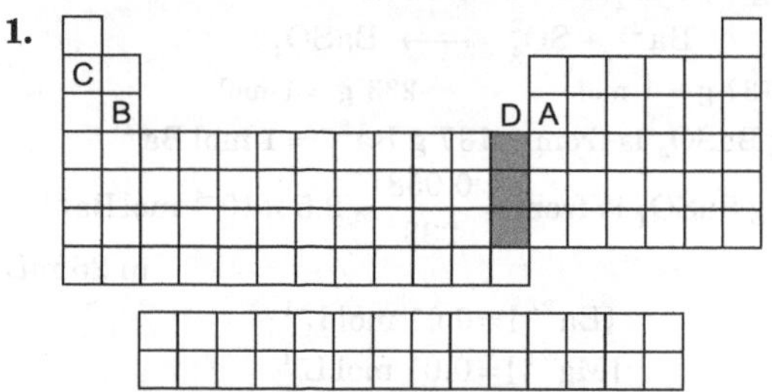

2. (a) E and G (b) C, E, and G (c) G (d) A and D (e) B

10

Hydrogen and Its Compounds

The science fiction writer Jules Verne had an uncanny knack of predicting the future. In 1874, he wrote "I believe that water will one day be employed as fuel, that hydrogen and oxygen which constitute it, used singly or together, will furnish an inexhaustible source of heat and light" Many now believe that this prophecy has come true in the twenty-first century when a major research work is being carried out with hydrogen and it may become a universal fuel.

Learning Framework of the Chapter ...

- Hydrogen
- Hydrogen Peroxide

Quick Points

- Ionisation Energy is the energy required to convert gaseous atom to gaseous ion, *i.e.*,
$$M(g) \longrightarrow M^+(g) \qquad \Delta H = IE$$
- Electronegativity is measured in Pauling scale (w.r.t. H)
- Entropy is the measure of the disorder of the system
$$\Delta S = \frac{q}{T}$$
- Ionic product of H_2O, $K_w = [H_3O^+][OH^-] = 1 \times 10^{-14}$ at 298 K

  Ionisation constant of a weak acid HA (with ionisation step $HA + H_2O \longrightarrow H_3O^+ + A^-$) is
$$K_a = \frac{[H_3O^+][A^-]}{[HA]}$$
- In an electrochemical cell, oxidation takes place at anode and reduction takes place at cathode.
- MnO_4^- is used to determine concentration of H_2O_2 volumetrically
$$MnO_4^- + 8H^+ + 5e^- \longrightarrow Mn^{2+} + 4H_2O$$
$$H_2O_2 \longrightarrow 2H^+ + 2e^- + O_2$$
- 1 mol of every gas at STP occupies volume 22.4 L or 22400 mL or 0.0224 m^3
- Molarity is the concentration in mol L^{-1}
- Normality is the concentration in equivalent L^{-1}

In many ways, hydrogen is unique and although it is hardly found on earth as the free element, life here would be impossible without it (Fig. 10.1). It accounts for virtually all of our energy, either directly in the form of sunlight, or indirectly as fossil fuels. Without hydrogen there would be no water to drink and DNA molecules would not form the double-helix structure that allows our genetic code to be copied and passed on to future generations.

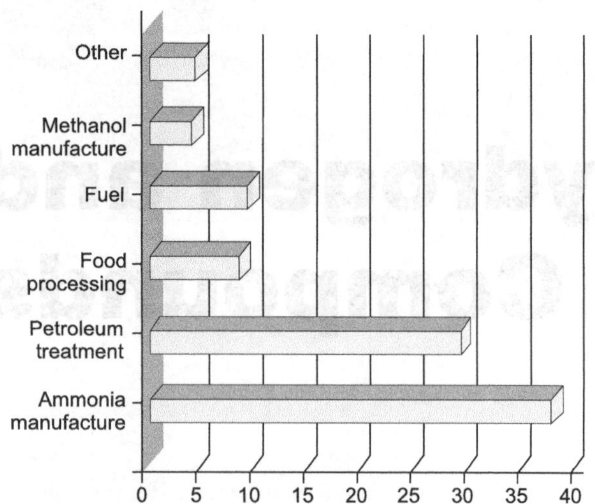

Fig. 10.1 Some of the most important industrial uses of hydrogen

DEEP Focus

Hydrogen is the most abundant lightest element in the universe and oxygen is the most abundant element on the earth's surface. In the earth's crust and oceans, **hydrogen is the ninth most abundant element on a mass basis (0.9%) and the third most abundant on an atom basis (15.4 atom %).** Hydrogen combines with every element in the Periodic Table except the noble gases and forms more compounds than any other element.

Henry Cavendish, an English chemist and physicist, was the first person to isolate hydrogen in pure form (but did not name it) by action of dilute acids such as HCl, H_2SO_4 on metals such as Zn, Sn, etc.

$$Zn(s) + 2H_3O^+(aq) \longrightarrow Zn^{2+}(aq) + H_2(g) + 2H_2O(l)$$

The French chemist Lavoisier called the gas **"hydrogen"** meaning **"water former"** because it combines with oxygen to produce water.

Hydrogen

Electronic Structure ($1s^1$)

Hydrogen has the simplest atomic structure of all the elements, and consists of a nucleus containing one proton with a charge +1 and one orbital electron ($1s^1$). Atoms of hydrogen may attain stability in three different ways :

- by forming an electron pair (covalent) bond with another atom as in H_2, H_2O, NH_3, HCl, CH_4
- by losing an electron to form H^+; a proton is of approximately 1.5×10^{-5} Å radius compared to that of H which is 0.7414 Å. Due to its very small size, H^+ has a very high polarising power, and therefore distorts the electron cloud on other atoms. H^+ (protons) is always associated with other atoms or molecules, as H_3O^+, $H_9O_4^+$ or $H(H_2O)_n^+$ ions.

$$H_2O + H^+ \longrightarrow H_3O^+$$
$$(H_2O)_n + H^+ \longrightarrow [H(H_2O)_n]^+$$

Free protons do not exist under normal conditions though they are found in low pressure gaseous beams (as in spectrometer)

- by gaining an electron to form H^-; crystalline solid such as LiH contains the H^- ion and are formed by highly electropositive metals (as groups I, II).

Position in the Periodic Table

- Hydrogen is the first element in the Periodic Table. Hydrogen is placed in no specific group due to its property of gaining electron (when H^- is formed) and also losing electron (when H^+ is formed).
- Hydrogen is placed in Group I (alkali metals) as :
  - it has one electron in its (outer) shell—$1s^1$ like other alkali metals which have **[inert gas]** $\boldsymbol{ns^1}$ configuration
  - it forms monovalent H^+ ion like Li^+, Na^+,....
  - its valency is also 1
  - its oxide (H_2O) is stable as Li_2O, Na_2O
  - it is a good reducing agent (in atomic as well as molecular state) like Na, Li, ...
- Hydrogen also resembles halogens (Group 17) as :
  - it is also diatomic (H_2) like F_2, Cl_2,....
  - it also forms anion H^- like F^-, Cl^-,.... by gain of one electron
  - H^- has stable inert gas [He] configuration as F^-, Cl^-, Br^-, and I^- have
  - it forms covalent compounds with carbon as CH_4, C_2H_6 like halogens CCl_4, CF_2Cl_2 etc.
  - H is one electron short of duplet (stable configuration) like F, Cl, ... which are also one electron deficient than octet

$$F : 2s^2 2p^5 \qquad Cl : 3s^2 3p^5$$

  - (*IE*) of H(1312 kJ mol^{-1}) is of the same order as that of halogens, as :

Atom	(*IE*), kJ mol^{-1}
H	1312
F	1681
Cl	1256
Br	1143
I	1009

- (*IE*) of H is very high in comparison with alkali metals. Also size of H^+ is very small compared to that of alkali metal ions. H forms stable hydride only with strongly electropositive metals due to smaller value of its electron affinity (72.8 kJ mol^{-1}).
- In view of the anomalous behaviour of hydrogen, it is difficult to assign any definite position to it in the Periodic Table. Hence, it is customary to place it in Group I (along with alkali metals) as well as in Group (17) VIIA (along with halogens).

1(IA)	17(VIIA)
H	H
Li	F
Na	Cl
K	Br
Rb	I
Cs	At
Fr	

Table 10.1 **A Comparison of Some Properties of Hydrogen, Lithium, Carbon and Fluorine**

Property	Hydrogen	Lithium (Group 1)	Carbon (Group 14)	Fluorine (Group 17)
Electron configuration	$1s^1$	[He] $2s^1$	[He] $2s^2 2p^2$	[He] $2s^2 2p^5$
First ionization energy/kJ mol^{-1}	+1312	+519	+1090	+1681
Electron affinity/kJ mol^{-1}	–72	–52	–120	–348
Electronegativity (Pauling)	2.1	1.0	2.5	4.0
Structure	diatomic molecules	metallic lattice	covalent macromolecule	diatomic molecule
Atomization energy /kJ mol^{-1}	+218	+161	+715	+79

Manufacture of Hydrogen

- **Water is the natural abundant source for the manufacture of hydrogen.**

Reduction of water of hydrogen can be carried out.

(1) Chemically (2) Electrolytically common reducing agents are (1) Coke (2) Carbon monoxide (3) Hydrocarbons

(1) Hydrogen is manufactured by allowing steam to react with red hot coke at about 1250 K.

$$C + H_2O \xrightarrow{1250\text{ K}} CO + H_2$$

This mixture of **CO and H_2** is also known **as water** gas and the reaction is referred to as water gas reaction. As this mixture ($CO + H_2$) is used for the synthesis of methanol an a number of hydrocarbons., this is also called **synthesis gas.**

(2) Hydrogen is now produced by the reaction of natural gas (chiefly methane) with steam in presence of a nickel catalyst.

$$CH_4 + H_2O \xrightarrow[920\text{-}1170\text{ K}]{\text{Ni catalyst}} CO + 3H_2$$

Similar reaction can occur with other hydrocarbons also. In both the above processes CO is converted to CO_2 and more hydrogen generated by passing the gases ($CO + H_2$) and steam over iron oxide or cobalt oxide catalyst.

$$CO + H_2O + H_2 \xrightarrow[450\text{–}550\text{ K}]{\text{catalyst (iron oxide)}} CO_2 + 2H_2$$

This reaction is called water gas shift reaction

DEEP Focus

Hydrogen and carbon dioxide can be easily separated from each other by bubbling the gas mixture through water in which CO_2 is fairly soluble and H_2 virtually insoluble.

$$H_2O + CO_2 \longrightarrow H_2CO_3$$

(3) At higher temperature, in the presence of catalysts (silica, alumina) hydrocarbons decompose and rearrange in what are called **cracking reactions.** These reactions, which are used in refining of petroleum, produce hydrogen as a by-product.

$$\underset{\text{propane}}{2CH_3{-}CH_2{-}CH_3} \xrightarrow[\text{catalyst}]{875\text{ K}} CH_4 + CH_2{=}CH_2 + CH_3{-}CH{=}CH_2 + H_2\uparrow$$

(4) **Electrolysis of acidified water using** platinum electrodes. On a large scale, very pure hydrogen (> 99.95%) can be obtained from the electrolysis of aqueous solutions of barium hydroxide between nickel electrodes.

$$2H_2O \xrightarrow{\text{current}} \underset{\text{at cathode}}{2H_2(g)} + \underset{\text{at anode}}{O_2(g)}$$

Hydrogen is economically obtained as a by-product in the electrolysis of brine solution in the manufacture of sodium hydroxide.

At anode :

During electrolysis there is a **competition** at the anode between the oxidation of chloride ion and the oxidation of water.

$$2Cl^-(aq) \longrightarrow Cl_2(g) + 2e^-$$
$$2H_2O(l) \longrightarrow O_2(g) + 4H^+(aq) + 4e^-$$

When a concentrated salt solution (brine) is used, the first reaction is the one that is observed.

At cathode :

The reaction is the reduction of water because it is more easily reduced than Na^+

$$2H_2O(l) + 2e^- \longrightarrow H_2(g) + 2OH^-(aq)$$

Overall reaction is

$$2Cl^- + 2H_2O \longrightarrow Cl_2(g) + H_2(g) + 2OH^-$$

Water is very stable to thermal decomposition, it does not decompose to hydrogen and oxygen to any appreciable extent at practically viable temperatures. Great efforts are being made to make electrolysis of water more energy-efficient so that the use of hydrogen as a fuel can become economically viable.

(5) **Non-electrolytic methods of preparation of hydrogen,** *e.g.,* thermochemical and photochemical. In thermochemical hydrogen generation, water is heated with certain chemicals to liberate hydrogen at much lower temperatures. Two potentially useful cycles are given below :

Cycle 1

(i) $3FeCl_2 + 4H_2O \xrightarrow{775\,K} Fe_3O_4 + 6HCl + H_2$

(ii) $Fe_3O_4 + \frac{3}{2}Cl_2 + 6HCl \xrightarrow{375\,K} 3FeCl_3 + 3H_2O + \frac{1}{2}O_2$

(iii) $3FeCl_3 \xrightarrow{575\,K} 3FeCl_2 + \frac{3}{2}Cl_2$

Cycle 2

(i) $CaBr_2 + H_2O \xrightarrow{1025\,K} CaO + 2HBr$

(ii) $Hg + 2HBr \xrightarrow{375\,K} HgBr_2 + H_2$

(iii) $HgBr_2 + CaO \xrightarrow{280\,K} HgO + CaBr_2$

(iv) $HgO \xrightarrow{775\,K} Hg + \frac{1}{2}O_2$

Thus, by repeating reactions (i)—(iii) in cycle 1 or reactions (i)—(iv) in cycle 2, water can be splitting to H_2 and O_2.

Net reaction is $\quad H_2O \longrightarrow H_2 + \frac{1}{2}O_2$

Decomposition of water by solar energy in presence of catalyst is known as **photochemial decomposition** of water. Using special catalysts, scientists in France have been able to achieve the efficient decomposition of water under visible and ultraviolet light.

Properties

- Some of the physical properties of hydrogen atoms are summarised in Table 10.2.

Table 10.2 **Physical Properties of Hydrogen**

Property	Value
Relative atomic mass	1.008
Electron structure	$1s^1$
(*IE*), H(*g*) $\longrightarrow$ $H^+(g) + e^-$	1312 kJ mol^{-1}
Electronegativity (Pauling scale)	2.1
Electron affinity	–72 kJ mol^{-1}
Molecular formula	H_2
Melting point	–259°C
Boiling point	–253°C
Density at STP	0.09 g L^{-1}
B.E. (H—H)	436 kJ mol^{-1}
Bond length (H—H)	74 pm
Covalent radius	37 pm
van der Waals' radius	120 pm
Ionic radius (H^-)	208 pm
Standard entropy	131 JK^{-1} mol^{-1}
Heat capacity (25°C)	297 JK^{-1} mol^{-1}

- At ordinary temperatures and pressures, hydrogen is a colourless, odourless and tasteless gas composed of diatomic H_2 molecules. Because H_2 is a non-polar molecule that contains only two electrons, **intermolecular forces are very weak hence,** it (in the solid form it has hexagonal close-packed structure) has very low melting point (–259°C) and low boiling point (–253°C). The bonding forces within the H_2 molecule are exceptionally strong and H_2 molecule is thus, thermally stable. Even at 2000 K, only 1 of every 2500 H_2 molecules is dissociated into H-atoms.

Atomic Hydrogen

Hydrogen is the lightest element known. It is a colourless, odourless and tasteless gas. Hydrogen molecule is thermally stable and has little tendency to dissociate at normal temperature, since the reaction.

$$H_2(g) \longrightarrow 2H(g), \qquad \Delta H = 436\ \text{kJ mol}^{-1}$$

is highly endothermic. However, at high temperatures in an electric arc or under ultraviolet light, it does dissociate. The **atomic hydrogen** produced, exists for less than half a second after which it recombines to give molecular hydrogen and liberates a large amount of energy (436 kJ mol^{-1}) in the form of heat. Most of the transition metals catalyse the recombination reaction of atomic hydrogen.

Atomic hydrogen is a powerful **reducing agent** and reduces copper, silver and mercury salts to the metallic state. It combines with alkali metals to form **hydrides,** reduces sulphur to hydrogen sulphide and carbon monoxide to formaldehyde. It also reacts with oxygen at room temperature to form hydrogen peroxide. The reactions can be written as :

$$HgO + 2H \longrightarrow Hg + H_2O$$
$$CuO + 2H \longrightarrow Cu + H_2O$$

$$AgNO_3 + H \longrightarrow Ag + HNO_3$$
$$Na + H \longrightarrow NaH$$
$$CO + 2H \longrightarrow HCHO$$
$$O_2 + 2H \longrightarrow H_2O_2$$
$$S + 2H \longrightarrow H_2S$$

Atomic hydrogen is generally produced by passing ordinary hydrogen through an electric arc maintained between two tungsten electordes. The atoms set free are carried away by a stream of incoming hydrogen gas. These free atoms recombine at once on coming in contact with a metallic surface liberating a large amount of heat and thus raising the temperature of the metal very high, say to 4000–5000 K. **This principle is utilised in making of the 'atomic hydrogen welding torch' (Fig. 10.2). It provides an opportunity of welding at a very high temperature but in a reducing atmosphere.**

Despite fairly high bond dissociation energy of hydrogen molecule, it is moderately reactive because hydrogen forms strong bonds with many other elements. It reacts with almost all elements except the noble gases.

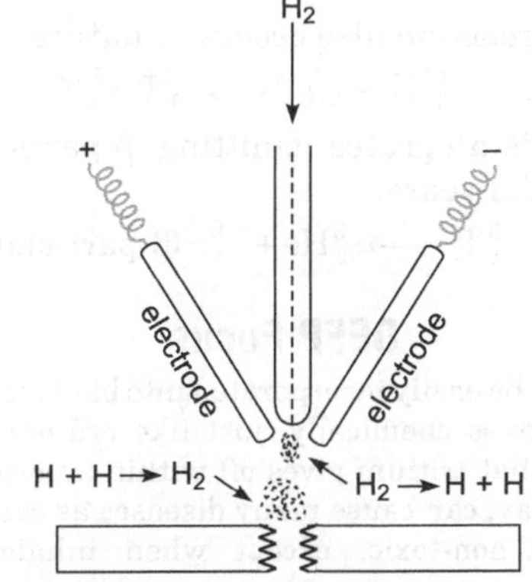

Fig. 10.2 Atomic hydrogen welding torch

Reducing Agent

Hydrogen is easily oxidised to water and; therefore, it acts as a very good **reducing agent in a variety** of situations. Hydrogen is used in metallurgy to reduce metal oxides to metals in cases where carbon cannot be used because the metal, *e.g.*, Mo and W can form a carbide :

$$WO_3 + 3H_2 \longrightarrow W + 3H_2O$$
$$(NH_4)_2MoO_4 + 3H_2 \longrightarrow Mo + 2NH_3 + 4H_2O$$

- Hydrogen adds on the **multiple bonds** in organic compounds. In the presence of catalysts such as finely divided nickel, palladium or mixtures of metal oxides, unsaturated organic compounds are thus reduced to saturated compounds.

$$\underset{\text{2-butene}}{CH_3CH{=}CHCH_3} + H_2 \xrightarrow[\text{Pd, 1 atm}]{278\text{ K}} \underset{n\text{-butane}}{CH_3CH_2CH_2CH_3}$$

$$\underset{\text{ethyl cyanide}}{CH_3C{\equiv}N} + 2H_2 \xrightarrow[ZnO/Cr_2O_3]{673\text{ K}} \underset{\text{ethyl amine}}{CH_3CH_2NH_2}$$

- **Catalytic hydrogenation of unsaturated liquid vegetable oils to solid edible fats.**

$$\underset{\text{(oil) oleate}}{CH_3(CH_2)_7CH{=}CH(CH_2)_7COOR} + H_2 \longrightarrow \underset{\text{(fat) stearate}}{CH_3(CH_2)_{16}COOR}$$

- Hydrogen reacts with carbon monoxide in the presence of catalysts to form methanol :

$$CO + 2H_2 \xrightarrow{\text{cobalt}} CH_3OH$$

This reaction is known as **hydroformylation** reaction and is used in industrial preparation of methanol.

Isotopes of Hydrogen

Hydrogen has three isotopes :

Name	Symbol	Atomic mass	P	N	Percentage	Nuclear stability	Nuclear spin quantum number	Relative atomic mass
Protium	$_1H^1$	1	1	0	99.986	Stable	$\frac{1}{2}$	1.007825
Deuterium	$_1H^2$ (D)	2	1	1	0.014	Stable	1	2.014102
Tritium	$_1H^3$ (T)	3	1	2	7×10^{-16}	Radioactive	$\frac{1}{2}$	3.106049

- Physical properties of H_2, D_2 and T_2 are summarised in Table 10.3. These isotopes have same electronic configuration and therefore their chemical properties are almost the same. The only difference is in the rate of reaction. Hydrogen has lower energy of activation than deuterium in its reaction with halogen hence, reacts faster.

Table 10.3 **Physical Properties of Isotopes of Hydrogen**

Physical contents	H_2	D_2	T_2
Freezing point	– 259° C	– 254.3°C	– 252.4°C
Boiling point	– 252.6° C	– 249.3°C	– 248.0°C
T_C (critical temperature)	–239.8°C	– 234.6°C	– 232.4°C
Bond length (H—H)	74 pm	74 pm	74 pm
BE (H—H)	436.0 kJ mol^{-1}	443.3 kJ mol^{-1}	446.9 kJ mol^{-1}
Latent heat of vaporisation	0.904 kJ mol^{-1}	1.226 kJ mol^{-1}	1.393 kJ mol^{-1}
Latent heat of fusion	0.117 kJ mol^{-1}	0.197 kJ mol^{-1}	0.250 kJ mol^{-1}
Water	H_2O	D_2O	T_2O
Melting point	0.00°C	3.81°C	4.48°C
Boiling point	100.0°C	101.42°C	101.51°C
Density at 25°C	0.997 g/cm^3	1.104 g/cm^3	1.214 g/cm^3
Dissociation constant (K_w)	1.01× 10^{-14}	0.195× 10^{-14}	0.06× 10^{-14}
Temperature of maximum density	4°C	11.6°C	—

DEEP Focus

Although, D_2O chemically resembles H_2O in many respects, it is toxic substance. The reason is that deuterium is heavier than hydrogen; thus, its compounds often react more slowly than those of the lighter isotope. Regular drinking of D_2O instead of H_2O could prove fatal because of the slower rate of transfer of D^+ compared with that of H^+ in the acid-base reactions involved in enzyme catalysis. This **kinetic isotopic effect** is also observed in the ionisation of an acid. Ionisation constant of acetic acid.

$$CH_3COOH(aq) \rightleftharpoons CH_3COO^-(aq) + H^+(aq), \quad K_a = 1.8\times 10^{-5}$$

is about three times as large as that of deuterated acetic acid:

$$CH_3COOD(aq) \rightleftharpoons CH_3COO^-(aq) + D^+(aq), \quad K_a = 6\times 10^{-6}$$

Deuterium undergoes exchange reactions, replacing protium.

$$H_2 + D_2 \longrightarrow 2HD$$
$$CH_4 + 2D_2 \longrightarrow CD_4 + 2H_2$$
$$2NH_3 + 3D_2 \longrightarrow 2ND_3 + 3H_2$$

To prepare deuterium rich compounds, heavy water (D_2O) is used

$$Mg_3N_2 + 3D_2O \longrightarrow 2ND_3 + 3MgO$$
$$SO_3 + D_2O \longrightarrow D_2SO_4$$
$$P_4O_{10} + 6D_2O \longrightarrow 4D_3PO_4$$

(If T_2O is used, tritium rich compounds are formed)

$$Mg_3N_2 + 3T_2O \longrightarrow 3NT_3 + 3MgO$$

Tritium differs from the other two isotopes of hydrogen in being radioactive. It was first obtained synthetically by the bombardment of deuterium compounds such as $(ND_4)_2SO_4$ with fast deuterons

$${}^2_1D + {}^2_1D \longrightarrow {}^3_1T + {}^1_1H$$

It is now prepared on a large scale by attack of slow neutrons on isotopes of lithium-6

$${}^6_3Li + {}^1_0n \longrightarrow {}^3_1T + {}^4_2He$$

Following reaction also occurs in nature

$${}^{14}_7N + {}^1_0n \longrightarrow {}^3_1T + {}^{12}_6C$$

Tritium disintegrates emitting β-particle, half-life period being 12.3 years.

$${}^3_1T \longrightarrow {}^3_2He + {}^0_{-1}e \text{ (β-particle)}$$

DEEP Focus

Tritium can be easily incorporated into biological molecules because it behaves chemically, just like ordinary hydrogen. The radiation that tritium gives off within an organism, as a result of its decay, can cause many diseases as cancer.

Tritium is non-toxic, except when inhaled in large quantities.

Ortho and *Para* Hydrogen

- In the H_2 molecule, the two nuclei may be spinning in either the same direction or in opposite direction. This gives rise to **spin isomerism**, that is, two different forms of hydrogen may exist :
  - if two nuclei have same spin then it is called ***ortho*** H_2
  - if two nuclei have different spins then it is called ***para*** H_2.

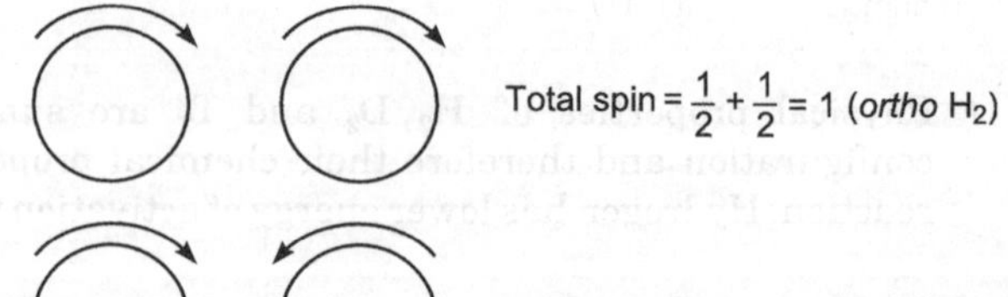

- There are considerable differences in physical properties of the *ortho* and *para* H_2 because of differences in their internal energy. *Para* H_2 has a lower internal energy than the *ortho* H_2.

Hydrogen gas is an equilibrium mixture of *ortho* and *para* hydrogen. The ratio of *ortho* to *para* hydrogen varies with temperature as shown in Fig. 10.3

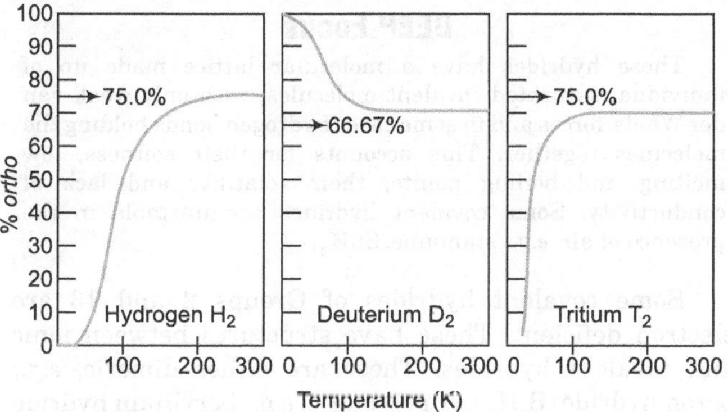

Fig. 10.3 *Ortho-para* equilibria for H_2, D_2 and T_2.

Evidently, this ratio increases with the rise in temperature upto a temperature of about 300 K (27°C) and remains constant thereafter. The percentage of *ortho* hydrogen at 300 K and above is 75%. This means that it is not possible to get 100% *ortho* hydrogen at any temperature. The equilibrium mixture of *para* and *ortho* hydrogen changes to almost 100% *para* hydrogen when cooled nearly to absolute zero. *Para* hydrogen is stable for weeks in the absence of catalysts like activated charcoal, Fe, Ni, Pt, O_2, NO_2, etc., which catalyse the conversion of *para* to *ortho* hydrogen. **Similarly, deuterium and tritium also exhibit spin isomerism and exist in *ortho* and *para* forms. However, the ratio of *ortho* and *para* forms in deuterium and tritium is different from that in hydrogen.** The variation of *ortho*/*para* ratio at different temperatures is also different. From Fig. 10.3 it is evident that tritium resembles hydrogen more closely than deuterium in this respect.

DEEP Focus

- In spacecrafts, H_2 gas is used in fuel cells for generating electrical energy and for providing clean drinking water to the astronauts. In fuel cell, electrical energy is generated by the reaction of H_2 and O_2 without evolution of any heat : called **"cold combustion"**, A H_2—O_2 fuel cell may have an alkaline or acidic electrolyte.

With acidic electrolyte, the half-cell reactions are:

at anode: $H_2(g) \longrightarrow 2H^+ + 2e^-, \quad E° = 0\text{ V}$

at cathode: $2H^+ + \frac{1}{2}O_2(g) \longrightarrow H_2O(l), \quad E° = 1.23\text{ V}$

$$H_2(g) + \frac{1}{2}O_2(g) \longrightarrow H_2O(l), \quad E° = 1.23\text{ V}$$

With alkaline electrolyte, the half-cell reactions are:

at anode: $H_2(g) + 2OH^- \longrightarrow 2H_2O(l) + 2e^-, \quad E° = 0.83\text{ V}$

at cathode $H_2O(l) + \frac{1}{2}O_2(g) + 2e^- \longrightarrow 2OH^-, \quad E° = 0.40\text{ V}$

$$H_2(g) + \frac{1}{2}O_2(g) \longrightarrow H_2O(l) \quad E° = 1.23\text{ V}$$

Thus, $E°$ is 1.23 V irrespective of the electrolyte because we are using the same reactants (H_2 and O_2) at the electrodes. Fuel-cells have greater efficiency (about 75%) than other rechargeable secondary cells (about 40%)

Combustion of hydrogen is a highly exothermic reaction and produces no pollutants :

$$2H_2(g) + O_2(g) \longrightarrow 2H_2O(l), \quad \Delta H = -572\text{ kJ}$$

Liquid hydrogen is, therefore, used as a **rocket fuel.**

It has been recognised during the past two decades that the world reserves of fossil fuels like coal, oil and gas are finite, so they cannot last for ever. Similarly, the nuclear and hydroelectric power also cannot meet all our energy requirements. Moreover, these sources pose a danger of environmental degradation. This has prompted an active search for alternative sources of energy. In addition to solar power, hydrogen is being considered a potential fuel for the future. Hydrogen-run vehicles are being tried. Hydrogen as a fuel has many advantages over the conventional fossil fuels and electric power.

Hydrides

Binary compounds of the elements (metal or non-metal) with hydrogen are called hydrides. The type of hydride which an element forms depends on its electronegativity and hence the type of bond formed. While there is no sharp division between ionic, covalent and metallic bonding, they are classified as:

(a) ionic or salt like hydride

(b) covalent or molecular hydride

(c) metallic or interstitial hydride

They have been summarised in Fig. 10.4.

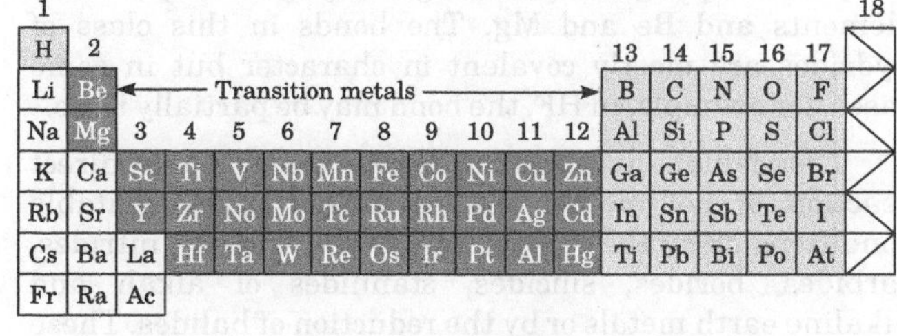

Fig. 10.4 Hydrides of the Elements

Ionic Hydrides

- They are formed when molecular hydrogen combines directly with any alkali metal or with the alkaline earth metals Ca, Sr or Ba (electropositive metal).

$$2Li(s) + H_2(g) \xrightarrow{973\text{ K}} 2LiH(s)$$

$$Ca(s) + H_2(g) \xrightarrow{973\text{ K}} CaH_2(s)$$

All ionic hydrides are solids that have the high melting point characteristic of ionic compounds and conduct electricity in fused state liberating H_2 at the anode. Anion of these hydrides is H^- which is a very strong **Bronsted** base (H^+ ion acceptor) and accepts proton from a proton donor such as water:

$$\underset{\text{Base}_1}{H^-(aq)} + \underset{\text{Acid}_2}{H_2O(l)} \longrightarrow \underset{\text{Base}_2}{OH^-(aq)} + \underset{\text{Acid}_1}{H_2(g)}$$

Due to their high reactivity with water, ionic hydrides are used to remove traces of water from organic solvents.

They are powerful reducing agent specially at higher temperature, *e.g.,*

$$2CO + NaH \longrightarrow HCOONa + C$$
$$SiCl_4 + 4NaH \longrightarrow SiH_4 + 4NaCl$$
$$PbSO_4 + 2CaH_2 \longrightarrow PbS + 2Ca(OH)_2$$

Li^+H^- and Na^+H^- are used in making valuable reducing agents like lithium aluminium hydride ($LiAlH_4$) and sodium borohydride ($NaBH_4$). These complex hydrides are frequently used in the reduction of aldehydes, ketones, acids and their derivatives of give alcohols.

$$\underset{\text{organic acid}}{R\text{—COOH}} \xrightarrow{LiAlH_4/NaBH_4} \underset{\text{primary alcohol}}{R\text{—CH}_2\text{OH}}$$

$$\underset{\text{ketone/aldehyde}}{>C{=}O} \xrightarrow{LiAlH_4/NaBH_4} \underset{\text{secondary alcohol}}{>CH\text{—OH}}$$

Covalent Hydride

Covalent hydrides are formed by elements of comparatively higher electronegativity, generally *p*-block elements and Be and Mg. The bonds in this class of hydrides are mostly covalent in character but in some cases, for example, in HF, the bond may be partially ionic.

The covalent hydrides can be prepared either by direct reaction of non-metals with hydrogen under suitable conditions or by the reaction of H_2O or acids on nitrides, carbides, borides, silicides, stannides of alkali and alkaline earth metals or by the reduction of halides. These are illustrated by the following reactions :

$$N_3 + 3H_2 \xrightarrow{\text{catalyst, 750 K}} 2NH_3$$

$$Mg_3N_2 + 6H_2O \xrightarrow{600\text{–}1000\text{ atm}} Mg(OH)_2 + 2NH_3$$
$$CaC_2 + 2H_2O \longrightarrow Ca(OH)_2 + C_2H_2$$
$$AsCl_3 + 6H \xrightarrow{Zn + HCl} 4AsH_3 + 3HCl$$
$$4PCl_3 + 3LiAlH_4 \xrightarrow{\text{ether}} 4PH_3 + 3LiCl + 3AlCl_3$$

DEEP Focus

These hydrides have a molecular lattice made up of individual saturated covalent molecules, with only weak van der Waals' forces and in some cases hydrogen bonds holding the molecules together. This accounts for their softness, low melting and boiling points, their volatility and lack of conductivity. Some covalent hydrides are unstable in the presence of air, *e.g.,* stannane, SnH_4.

Some covalent hydrides of Groups 2 and 13 are electron deficient. These have structures between ionic and covalent hydrides. These are either dimeric, *e.g.,* boron hydride (B_2H_6). or polymeric, *e.g.,* beryllium hydride $(BeH_2)_n$, magnesium hydride $(MgH_2)_n$ and aluminium hydride $(AlH_3)_n$.

Metallic Hydrides

On heating, hydrogen reacts with many transition metals, lanthanides and actinides to form metallic hydrides. Most of these have metallic appearance and are less dense than the parent metal. They all conduct heat and electricity though not as well as the parent metal (of ionic hydrides). Unlike the saline hydrides, they are almost always non-stoichiometric, being deficient in hydrogen. For example, $TiH_{1.5-1.8}$, $VH_{0.56}$, $CH_{1.7}$, $NiH_{0.6-0.7}$, $PdH_{0.6-0.8}$, $TaH_{2.76}$, $LaH_{2.87}$, $YbH_{2.55}$, etc., approach the stoichiometric formulae given in following Table. Most of these hydrides are stable to water up to 375 K but are quantitatively decomposed by acids and show reducing properties.

Stoichiometric Formulae of Some Metallic Hydrides

MH MH (M = V, Cr, Ni, Cu, Ta, Pd)

MH_2 (M = Ti, Zr, Hf, V, Nb, Cr, Zn, Y; La-Lu except Pm; Th, Np, Pu, Am, Cm)

MH_3 (M = Y, La, LA-Lu except Pm and Eu; Pa, U, Np, Pu, Am)

Formerly, these hydrides were termed as interstitial compounds in which hydrogen was thought to be accommodated in the interstices in the metal lattice producing distortion but no change in its type. But recent studies have shown that except for the hydrides of nickel, palladium, cerium and actinium, other hydrides of this class have lattice of a type different from that of the parent metal.

DEEP Focus

For example, the hexagonal close packed lattice of some lanthanides is transformed to a face-centred cubic lattice in their dihydrides. These interstitial hydrides are poorer conductors of electricity, exhibit less paramagnetism and are more brittle than the parent metal. These characteristics suggest that hydrogen is present in the metal lattice as hydrogen atoms rather than as hydrogen molecules. The single electron of hydrogen is paired with an electron of the metal, thereby reducing the extent of metallic bonding. Breaking of the H—H bond is in agreement with the fact that these metals catalyse reactions of hydrogen.

The physical and chemical properties of these compounds change from **ionic to covalent across a given period** (moving left to right).

LiH → ionic high m.p. (950 K)

BeH_2 → covalent polymeric in solid state

B_2H_6 CH_4 → non polar

NH_3 H_2O HF → polar with H atom as positive end of the polar bond; HF: acidic in nature

- As we move down any group, the compounds change from covalent to ionic.

BeH_2
MgH_2
CaH_2
SrH_2
BaH_2

(Covalent ↑ ; Ionic ↓)

The ΔH_f° values (Table 10.4) show that the energetic stability of the covalent hydrides tends to increase on passing from left to right across a period, but to decrease on passing down any group. The main reason for this is the change in the strength of the X—H bond as shown by the bond enthalpy figures in Table 10.4. The attraction of an atom for a shared pair of electrons increases across a period as nuclear charge increases and atomic radius falls, but decreases down a group as both atomic size and the number of inner screening electrons increase.

Although all of the hydrides shown in Table 10.4 are described as covalent, the bonds are polar to some extent. **The degree of this polarity depends upon the electronegativity difference between the two elements concerned.** In the case of the Group 14 hydrides, however, this bond polarity does not result in a polar molecule. In the tetrahedral XH_4 molecule, there is no overall dipole moment as the centre of the δ^+ charges coincides with the centre of the δ^- charge, for example in methane.

As they have no overall dipole, **the only forces of attraction between XH_4 molecules are relatively weak van der Waals' forces due to temporary dipoles.** Since, these forces result from mutual repulsion of the electron clouds of neighbouring molecules, they increase as the size of the molecule increases, thus, accounting for the steady increase in the boiling point of the hydrides on passing down group 14.

In the case of the hydrides of Group 15, 16 and 17, bond polarity does result in a permanent molecular dipole.

Permanent dipoles in period 2 hydrides

Table 10.4 **Some Common Covalent Hydrides of Periods 2-5. Electronegativity Differences in Brackets Indicate that Hydrogen is more Electronegative Than the *p*-Block Element, that is, the Polarity of the Bond is $X^{\delta+}$—$H^{\delta-}$**

	Group 14	Group 15	Group 16	Group 17
Period 2	**CH_4**	**NH_3**	**H_2O**	**HF**
ΔH/kJ mol^{-1}	−74	−46	−286	−271
Bond enthalpy/kJ mol^{-1}	+436	+431	+494	+574
Boiling point/K	112	238	373	293
Electronegativity difference (Pauling)	0.35	0.84	1.24	1.78
Period 3	**SiH_4**	**PH_3**	**H_2S**	**HCl**
ΔH/kJ mol^{-1}	34	−10	+20	−92
Bond enthalpy/kJ mol^{-1}	+323	+323	+364	+432
Boiling point/K	161	185	213	188
Electronegativity difference (Pauling)	(0.3)	(0.01)	0.38	0.96
Period 4	**GeH_4**	**AsH_3**	**H_2Se**	**HBr**
ΔH/kJ mol^{-1}	+91	+66	+73	−36
Bond enthalpy/kJ mol^{-1}	289	—	—	+363
Boiling point/K	185	211	232	206
Electronegativity difference (Pauling)	(0.19)	(0.02)	0.35	0.76
Period 5	**SnH_4**	**SbH_3**	**H_2Te**	**HI**
ΔH/kJ mol^{-1}	+163	+145	+100	+26
Bond enthalpy/kJ mol^{-1}	253	—	—	+295
Boiling point/K	221	256	269	238
Electronegativity difference (Pauling)	(0.24)	(0.2)	(0.1)	0.46

In these cases, as well as attraction due to temporary dipoles, there is an extra attraction due to these permanent molecular dipoles. The positively charged hydrogen atom will attract a lone pair of electrons on the *p*-block atom of a neighbouring molecule.

DEEP Focus

As a result of this extra permanent dipole attraction, in any period the boiling points of the Group 15, 16 and 17 hydrides are all higher than that of the non-polar Group 14 hydride. This effect is much greater in period 2 than in the lower periods (Fig. 10.5). Since, the attraction resulting from the presence of N—H, O—H or F—H bonds in much stronger than other permanent dipoles, in these cases only the effect is referred to as *hydrogen bonding*. Typically such 'hydrogen bonds' 5-10% of the strength of a covalent and in the case of water this is enough to make it liquid at room temperature.

Hydrogen bonding in hydrogen fluoride

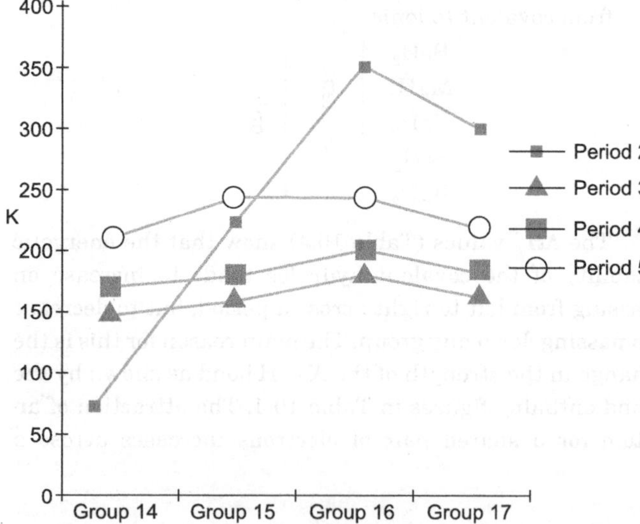

Fig. 10.5 Boiling point of covalent hydrides plotted by period

At first glance it might seem surprising that hydrogen fluoride has a lower boiling point than water, even though it contains the more polar H—F bond. Whilst the individual hydrogen bonds in HF are indeed stronger than those in water, the later can form more hydrogen bonds. **Each water molecule contains two $H^{\delta+}$ atoms and each $O^{\delta-}$ has two lone pairs and so may form up to four hydrogen bonds with its neighbours.**

$\underset{\delta+}{H}$—$O^{\delta-}$ ⋯ $H_{\delta+}$ ⋯ $O^{\delta-}$ (water hydrogen bonding diagram)

Water can hydrogen bond to several other water molecules.

On average both HF and NH_3 can form two hydrogen bonds with neighbouring molecules and, as expected, hydrogen bonding between the less polar NH_3 molecules results in a lower boiling point for ammonia than hydrogen fluoride.

Atomic Hydrogen

Hydrogen gas dissociates into atoms when it (H_2) is subjected to an electric discharge under low pressure.

$$H_2 \xrightarrow{\text{electric discharge}} 2H$$

- **Atomic hydrogen is very reactive and can be used as a reducing agent as well as oxidising agent**
- **As a reducing agent**

$$CuO + 2H \longrightarrow Cu + H_2O$$
$$NaNO_3 + H \longrightarrow Na + HNO_3$$
$$AgCl + H \longrightarrow Ag + HCl$$
$$AuCl_3 + 3H \longrightarrow Au + 3HCl$$

- **As an oxidising agent** (when hydrides are formed)

$$Li + H \longrightarrow LiH$$
$$Ca + 2H \longrightarrow CaH_2$$
$$Mg + 2H \longrightarrow MgH_2$$
$$Al + 3H \longrightarrow AlH_3$$

- **Reaction with non-metals**

$$S + 2H \longrightarrow H_2S$$
$$P + 3H \longrightarrow PH_3$$

Target Practice 1

1. Why does hydrogen resemble the alkali metals?

....

....

2. Chemical properties of isotopes of hydrogen are similar. Explain.

....

....

3. Select True/False
   (i) *Ortho* and *para* hydrogen forms are different due to difference in their nuclear spins.
   (ii) *Ortho* and *para* hydrogen forms are different due to difference in their electron spins.
   (iii) Physical properties of *ortho* and *para* hydrogen are similar.
   (iv) *Para* hydrogen is more stable at lower temperatures.
   (v) At 100 K, percentage of *ortho* hydrogen in the mixture is 70%.
4. Complete the following chemical reactions which take place during the formation of hydrogen
   (i) $C + H_2O \longrightarrow$ .... + ...
   (ii) $CH_4 + H_2O \longrightarrow$ .... + ...
   (iii) $CO + H_2O \longrightarrow$ .... + ...
   (iv) $CaH_2 + H_2O \longrightarrow$ .... + ...
5. Hydroformylation reaction is the industrial preparation of methanol. Write this reaction .... $\longrightarrow CH_3OH$.

Answers

1. Hydrogen resembles alkali metals because like alkali metals it also has one electron in *s*-orbital
   $H : 1s^2$
   $Li : 1s^2 2s^1$
   $Na : [Ne]\, 3s^1$
2. Most of the chemical properties are governed by the electronic configuration of the element. As the isotopes of hydrogen have the same electronic configuration, their chemical properties are closely similar.
3. (i) T (ii) F (iii) F (iv) T (v) F
4. (i) $\longrightarrow CO + H_2$
   (ii) $\longrightarrow CO + 3H_2$
   (iii) $\longrightarrow CO_2 + H_2$
   (iv) $\longrightarrow Ca(OH)_2 + H_2$
5. $CO + 2H_2 \longrightarrow CH_3OH$

Practice Exercise 1

1. What is the equation for the nuclear reaction that takes place when tritium undergoes beta decay? What is the product in the reaction ?
2. Before natural gas can be used to make hydrogen, it has to be purified. Especially sulphur compounds must not be allowed through; why ?
3. If Zn pieces are added into aqueous $FeCl_3$ solution, deep yellow colour of $FeCl_3$ fades and changes to light green. But if H_2 gas is passed into $FeCl_3$ solution, there is no change. Explain.
4. For H_2O, $K_W = 1\times 10^{-14}$; D_2O, $K_W = 3.0\times 10^{-15}$ Compare $[H^+]$ and $[D^+]$ in H_2O and D_2O. If both are electrolysed, which is liberated faster, H_2 or D_2?
5. Ionic hydrides are the **ready-made** source of hydrogen. Explain.
6. What is the nature of hydride in the following reaction,

$$LiH + H_2O \longrightarrow LiOH + H_2$$

Answers

1. ${}^3_1H \longrightarrow {}^3_2He + {}^{0}_{-1}e$
2. Sulphur is very effective poison of catalysts, so it has to be removed before it reaches the nickel.
3. $FeCl_3$ solution is acidic in nature due to hydrolysis :

$$Fe^{3+} + 3H_2O \rightleftharpoons Fe(OH)_3 + 3H^+$$

On adding Zn, nascent hydrogen is formed in situ, that being very reactive causes reduction of Fe^{3+} (deep yellow) to Fe^{2+} (light green). Molecular H_2 is not effective in this case :

$$Zn + 2H^+ \rightleftharpoons Zn^{2+} + 2H$$

$$FeCl_3 + H \longrightarrow FeCl_2 + HCl$$

4. $$H_2O \rightleftharpoons H^+ + OH^-, \quad K_w = 1\times 10^{-14}$$

$$\therefore \quad [H^+] = \sqrt{1\times 10^{-14}} = 1\times 10^{-7}\ M$$

$$D_2O \rightleftharpoons D^+ + OD^-, \quad K_w = 3.0\times 10^{-15}$$

$$[D^+] = \sqrt{3.0\times 10^{-15}}\ M$$

Thus, $[H^+] > [D^+]$

H_2 gas is liberated much faster than D_2 at cathode.

5. $CaH_2 + 2H_2O \longrightarrow Ca(OH)_2 + 2H_2$
6. LiH (H^-) is H^+ acceptor from H_2O and changes to H_2. Hence, LiH is a Lowry-Bronsted base

Hydrogen Peroxide

- It is the hydride of oxygen.
- BaO_2 is decomposed by H_2SO_4; Na_2O_2 and KO_2 by H_2O forming H_2O_2:

$$\underset{\text{barium peroxide}}{BaO_2} + H_2SO_4 \longrightarrow BaSO_4 + H_2O_2$$

$$\underset{\text{sodium peroxide}}{Na_2O_2} + 2H_2O \longrightarrow 2NaOH + H_2O_2$$

$$\underset{\text{potassium superoxide}}{2KO_2} + 2H_2O \longrightarrow 2KOH + H_2O_2 + O_2$$

- Electrolysis of H_2SO_4 or NH_4HSO_4 at high current density using platinum electrodes forms peroxosulphuric acid which on subsequent hydrolysis gives H_2O_2 :

$$H_2SO_4 \xrightleftharpoons{\text{current}} H^+ + HSO_4^-$$

At anode : $$2HSO_4^- \longrightarrow H_2S_2O_8 + 2e^-$$

$$H_2S_2O_8 + H_2O \xrightarrow{\text{rapid}} H_2SO_5 + H_2SO_4$$

$$H_2SO_5 + H_2O \xrightarrow{\text{slow}} H_2SO_4 + H_2O_2$$

At cathode : $$2H^+ + 2e^- \longrightarrow H_2$$

- $$2NH_4HSO_4 \longrightarrow H_2 + (NH_4)_2S_2O_8$$

$$(NH_4)_2S_2O_8 + 2H_2O \longrightarrow 2NH_4HSO_4 + H_2O_2$$

This solution is concentrated to 98 per cent by low pressure fráctionation.

- Industrially it is obtained by a **Cyclic Process.** 2-ethyl anthraquionol (*A*) is oxidised by air to 2-ethyl anthraquinone (*B*) and H_2O_2 : (*B*) is reduced back to (*A*) with H_2 at a moderate temperature using Pt, Pd or Raney Ni as a catalyst. The cycle is repeated. The reaction is carried out in a mixture of organic solvents which must :
  - dissolve both (*A*) and (*B*)

(A) 2-ethyl anthraquinol (OH, OH, C_2H_5) $\underset{H_2/Pd}{\overset{O_2}{\rightleftharpoons}}$ (B) 2-ethyl anthraquinone (O, O, C_2H_5) $+ H_2O_2$

  - resist oxidation
  - be immiscible with water.

Solvent is generally a mixture of ester and hydrocarbon or octanol and methyl naphthalene.

H_2O_2 is extracted with water as a 1% solution. This is concentrated under reduced pressure. H_2O_2 solutions are stored in plastic or wax coated glass vessels, often with negative catalysts **such as urea or sodium stannate added as stabilisers.**

- An alternative process is based on the partial oxidation of 2-propanol ;

$$Me_2CHOH + O_2 \longrightarrow Me_2C\begin{matrix} OH \\ OOH \end{matrix}$$

$$Me_2C\begin{matrix} OH \\ OOH \end{matrix} \longrightarrow Me_2CO + H_2O_2$$

- Pure H_2O_2 is a colourless liquid which resembles water quite closely. It is more hydrogen bonded than is water so has a higher boiling point and is more dense than H_2O.

Property	H_2O_2	H_2O
boiling point	152°C (decomposes)	100°C
melting point	–0.4°C	0°C
density	1.4 g cm^{-3}	1.0 g cm^{-3}

- At 68 mmHg pressure, it boils at 84°C.
- H_2O_2 is unstable and the rate at which it disproportionates, depends on the temperature and concentration. Many impurities catalyse the decomposition, which may become very violent, especially with concentrated solutions.

$$2H_2O_2 \longrightarrow 2H_2O + O_2$$

(reduction: $H_2O_2 \to H_2O$; oxidation: $H_2O_2 \to O_2$)

Catalysts include metal ions Fe^{2+}, Fe^{3+}, Cu^{2+}, Ni^{2+}, metal surfaces such as Pt or Ag, MnO_2, charcoal or alkali. They are best stored in polythene bottles in a refrigerator. Traces of complexing agents (*e.g.*, 8-hydroquinoline)or adsorbing material (*e.g.*, sodium stannate) are often added as stabilisers.

- Mixtures of hydrogen peroxides and organic or other easily oxidised materials are dangerously explosive. **Mixture of hydrogen peroxide and hydrazine has been used for rocket propulsion.**
- The ionisation constant of hydrogen peroxide in aqueous solution at 20°C for the acid equilibrium is

$$H_2O_2 \rightleftharpoons H^+ + HO_2^-$$

$$K_a = \frac{[H^+][HO_2^-]}{[H_2O_2]} = 2 \times 10^{-12}$$

In alkaline medium, therefore, it is present as the HO_2^- ion.

DEEP Focus

- Many reactions of H_2O_2 are radical reactions. The oxidation of Fe(II) at low hydrogen ion concentrations proceeds as:

$$Fe^{2+} + H_2O_2 \longrightarrow FeOH^{2+} + OH$$
$$Fe^{2+} + OH \longrightarrow FeOH^{2+}$$

and a mixture of H_2O_2 and Fe(II) is thus a source of hydroxyl radicals for organic reactions.

- **H_2O_2 as reducing agent :** (when itself oxidised to O_2)

$$H_2O_2 \longrightarrow 2H^+ + O_2 + 2e^-$$

* $2MnO_4^- + 5H_2O_2 + 6H^+ \longrightarrow 2Mn^{2+} + 5O_2 + 8H_2O$

* $IO_4^- + H_2O_2 \longrightarrow IO_3^- + O_2 + H_2O$

* $2Ce^{4+} + H_2O_2 \longrightarrow 2Ce^{3+} + O_2 + 2H^+$

* $Ag_2O + H_2O_2 \longrightarrow 2Ag + H_2O + O_2$

* $PbO_2 + H_2O_2 \longrightarrow PbO + H_2O + O_2$

* $O_3 + H_2O_2 \longrightarrow 2O_2 + H_2O$

* $Au_2O + H_2O_2 \longrightarrow 2Au + H_2O + O_2$

* $MnO_2 + H_2O_2 + 2H^+ \longrightarrow Mn^{2+} + O_2 + 2H_2O$

* $[Fe^{III}(CN)_6]^{3-} + H_2O_2 + 2OH^- \longrightarrow [Fe^{II}(CN)_6]^{4-} + 2H_2O + O_2$

* $2MnO_4^- + 3H_2O_2 \longrightarrow 2MnO_2\downarrow + 3O_2 + 2H_2O + 2OH^-$

DEEP Focus

Traces studies using ^{18}O show that in reactions all the oxygen that is formed comes from the H_2O_2, none from the solvent H_2O, so the O—O bond is not broken.

- **H_2O_2 as an oxidising agent :** (when itself reduced to H_2O)

$$H_2O_2 \longrightarrow H_2O + O$$
$$H_2O_2 + 2H^+ + 2e^- \longrightarrow 2H_2O$$

and at $[H^+] = 1$ M, $E° = +1.77$ V

hence, it is a strong oxidising agent.

- Its bleaching action is also due to its oxidising property.

Oil paintings contain Pb^{2+} salts which are blackened by atmospheric H_2S (due to PbS). PbS (black), on reaction with H_2O_2, changes to $PbSO_4$ (white), hence H_2O_2 is used to clean oil paintings :

$$PbS + 4H_2O_2 \longrightarrow PbSO_4 + 4H_2O$$

* $2Fe^{2+} + 2H^+ + H_2O_2 \longrightarrow 2Fe^{3+} + 2H_2O$

* $SO_3^{2-} + H_2O_2 \longrightarrow SO_4^{2-} + H_2O$

* $NO_2^- + H_2O_2 \longrightarrow NO_3^- + H_2O$

* $2I^- + H_2O_2 + 2H^+ \longrightarrow I_2 + 2H_2O$

This reaction can be used to estimate H_2O_2 iodometrically. I_2 formed is titrated against $Na_2S_2O_3$ solution :

$$I_2 + 2S_2O_3^{2-} \longrightarrow 2I^- + S_4O_6^{2-}$$
$$H_2O_2 \equiv I_2 \equiv 2S_2O_3^{2-}$$
$$I_2 \equiv 2I \equiv H_2O_2 \equiv 2S_2O_3^{2-}$$

$$\therefore \quad \text{Eq. wt. } (H_2O_2) = \left(\frac{\text{mol. wt.}}{2}\right)$$

$$\text{Eq. wt. } (S_2O_3^{2-}) = \text{mol. wt.}$$

* $AsO_3^{3-} + H_2O_2 \longrightarrow AsO_4^{3-} + H_2O$
* $2[Fe^{II}(CN)_6]^{4-} + H_2O_2 + 2H^+ \longrightarrow 2[Fe^{III}(CN)_6]^{3-} + 2H_2O$
* Mn^{2+} (colourless when dilute) changes to MnO_2 (black) and Cr^{3+} (green) changes to CrO_4^{2-} (yellow) in alkaline H_2O_2.

$$Mn^{2+} + H_2O_2 + 2OH^- \longrightarrow \underset{\text{black}}{MnO_2\downarrow} + 2H_2O_2$$

$$\underset{\text{green}}{2Cr^{3+}} + 3H_2O_2 + 10OH^- \longrightarrow \underset{\text{yellow}}{2CrO_4^{2-}} + 8H_2O$$

DEEP Focus

- HCHO (formaldehyde) is oxidised to HCOOH (formic acid) in presence of pyrogallol and is itself reduced to hydrogen.

$$2HCHO + H_2O_2 \longrightarrow 2HCOOH + H_2$$

The reaction in the dark is accompanied by emission of light (chemiluminescence).

- **As an acid :**

H_2O_2 decomposes CO_2 from carbonates and thus behaves as an acid. It turns blue litmus red but aqueous solutions has no effect.

$$Na_2CO_3 + H_2O_2 \longrightarrow Na_2O_2 + H_2O + CO_2$$

A 30% solution of H_2O_2 is sold in the market under the trade name **perhydrol**. This is also known as **"100 volume" peroxide.**

- **Concentration Determination** :

Concentration of H_2O_2 is expressed in terms of **"volume of O_2"**

"10 volume" H_2O_2 means that 1 mL of H_2O_2 at NTP gives 10 mL O_2 gas. These values can be changed into other concentration terms like density, g L^{-1}, normality and molarity.

$$\underset{68\text{ g}}{2H_2O} \longrightarrow 2H_2O + \underset{1\text{ mol} = 22400\text{ mL at NTP}}{O_2}$$

$$\underset{?}{} \qquad\qquad \underset{x\text{ mL}}{}$$

22400 mL O_2 is from = 68 g H_2O_2 at NTP

x mL O_2 gas is from = $\dfrac{68x}{22400}$ g H_2O_2 NTP

This amount is in **1 mL H_2O_2** (by definition)

Thus, $\quad 1\text{ mL } H_2O_2 = \dfrac{68\,x}{22400}$ g

$\therefore \quad \text{density} = \dfrac{68\,x}{22400}$ g/mL

$\text{conc. in g/L} = \dfrac{68\,x}{22400} \times 1000$

$\text{g/L conc.} = \dfrac{680\,x}{224}$ g/L

$\%\text{ conc.} = \dfrac{68\,x}{224}$

$$\text{Normality }(N) = \frac{\text{conc in g/L}}{\text{equiv. wt.}} = \frac{680\,x}{224} \times \frac{1}{17}$$

$$N = \frac{x}{5.6}$$

$$\text{Molarity }(M) = \frac{\text{conc. in g/L}}{\text{mol. wt.}} = \frac{680\,x}{224} \times \frac{1}{34}$$

$$M = \frac{x}{11.2}$$

$$\begin{matrix} CH_2 \\ \| \\ CH_2 \end{matrix} + H_2O_2 \longrightarrow \underset{\text{glycol}}{\begin{matrix} CH_2OH \\ | \\ CH_2OH \end{matrix}}$$

This reaction confirms the presence of two —OH groups in H_2O_2

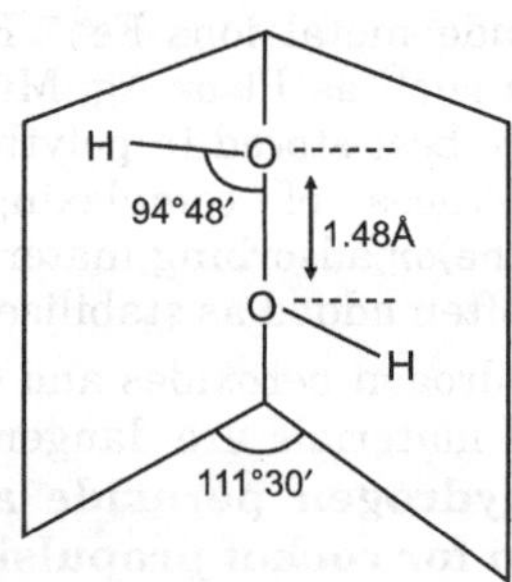

- H_2O_2 finds use
  - as a bleaching agent
  - as an antiseptic in surgery.
  - as a fuel for rockets (85–90% solution).

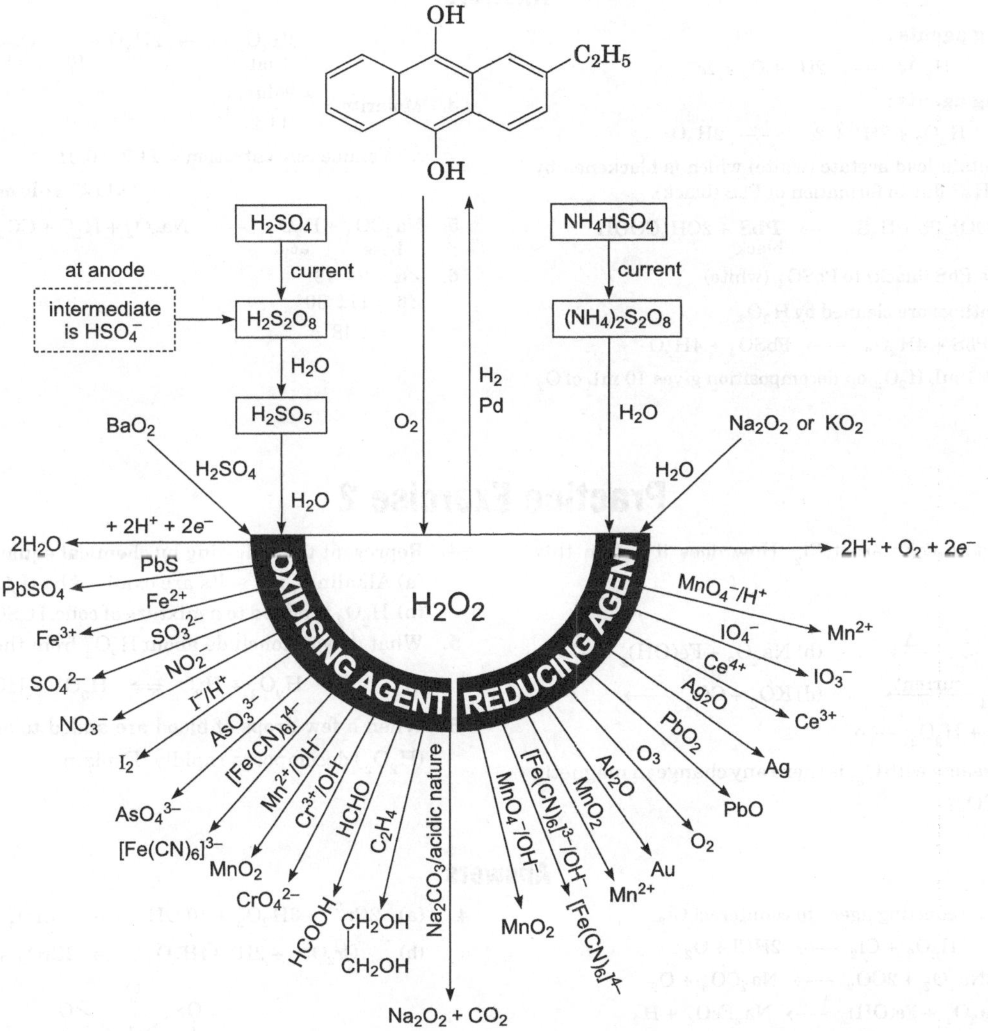

OH

C_2H_5

OH

H_2SO_4 → current → $H_2S_2O_8$ → H_2O → H_2SO_5 → H_2O

at anode

intermediate is HSO_4^-

NH_4HSO_4 → current → $(NH_4)_2S_2O_8$ → H_2O

O_2

H_2 Pd

BaO_2 H_2SO_4

Na_2O_2 or KO_2 H_2O

H_2O_2

OXIDISING AGENT

REDUCING AGENT

$2H_2O$ ← $+ 2H^+ + 2e^-$

→ $2H^+ + O_2 + 2e^-$

PbS → $PbSO_4$

Fe^{2+} → Fe^{3+}

SO_3^{2-} → SO_4^{2-}

NO_2^- → NO_3^-

I^-/H^+ → I_2

AsO_3^{3-} → AsO_4^{3-}

$[Fe(CN)_6]^{4-}$ → $[Fe(CN)_6]^{3-}$

Mn^{2+}/OH^- → MnO_2

Cr^{3+}/OH^- → CrO_4^{2-}

HCHO → HCOOH

C_2H_4 → CH_2OH—CH_2OH

Na_2CO_3/acidic nature → $Na_2O_2 + CO_2$

MnO_4^-/OH^- → MnO_2

$[Fe(CN)_6]^{3-}/OH^-$ → $[Fe(CN)_6]^{4-}$

MnO_2 → Mn^{2+}

Au_2O → Au

O_3 → O_2

PbO_2 → PbO

Ag_2O → Ag

Ce^{4+} → Ce^{3+}

IO_4^- → IO_3^-

MnO_4^-/H^+ → Mn^{2+}

Target Practice 2

1. Write ionic reactions showing H_2O_2 as
   (a) **Reducing agent :** ....
   (b) **Oxidising agent :** ....
2. H_2O_2 is used to clean oil-paintings. Write reactions making paintings black and then cleaning with H_2O_2.
   ....
   ....
3. Specify the meaning of "10 volume" H_2O_2.
   ....
   ....
4. Molarity of H_2O_2 solution is 0.01 mol L^{-1}. Thus, its "volume" concentration is ....
5. H_2O_2 behaves as an acid. Write reaction
   ....
   ....
6. Structure of H_2O_2 is represented as
   Angle $\angle\alpha$ = ...
   Angle $\angle\beta$ = ...
   Bond-angle length x = ...

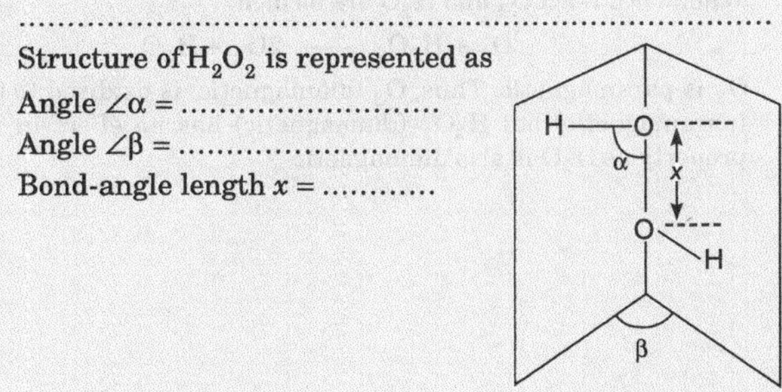

Answers

1. (a) **Reducing agents :**

$$H_2O_2 \longrightarrow 2H^+ + O_2 + 2e^-$$

(b) **Oxidising agents :**

$$H_2O_2 + 2H^+ + 2e^- \longrightarrow 2H_2O$$

2. Oil paints contain lead acetate (white) which is blackened by atmospheric H_2S due to formation of PbS (black).

$$(CH_3COO)_2Pb + H_2S \longrightarrow \underset{\text{black}}{\mathbf{PbS}} + 2CH_3COOH$$

H_2O_2 oxidises PbS (black) to $PbSO_4$ (white)

Thus, oil-paintings are cleaned by H_2O_2

$$PbS + 4H_2O_2 \longrightarrow PbSO_4 + 4H_2O$$

3. It means that 1 mL H_2O_2 on decomposition gives 10 mL of O_2 gas at NTP

$$\underset{\text{1 mL}}{2H_2O_2} \longrightarrow 2H_2O + \underset{\text{10 mL at NTP}}{O_2}$$

4. $\text{Molarity} = \dfrac{x \text{ volume}}{11.2}$

$\therefore$ Volume concentration $= 11.2 \times 0.01$

$=$ **"0.112" volume**

5. $\underset{\text{base}}{Na_2CO_3} + \underset{\text{acid}}{H_2O_2} \longrightarrow Na_2O_2 + H_2O + CO_2$

6. $\angle\alpha = 94°48'$

$\angle\beta = 111°30'$

$x = 1.48$ Å

Practice Exercise 2

1. H_2O_2 is used to counteract Cl_2. How does it act in this process?
2. Complete :

(a) $Na_2O_2 + CO_2 \xrightarrow{\Delta}$ (b) $Na_2O_2 + Fe(OH)_2 \xrightarrow{\Delta}$

(c) $NH_4HSO_4 \xrightarrow{\text{current}}$ (d) $KO_2 + CO_2 \longrightarrow$

(e) $Ba(OH)_2 + H_2O_2 \longrightarrow$

3. When H_2O_2 reacts with O_3, is there any change in magnetic behaviour of O_3?
4. Represent the following by chemical equations :

(a) Alkaline Cr^{3+} salts are oxidised by H_2O_2.

(b) H_2O_2 is added to a mixture of conc. H_2SO_4 and $K_2Cr_2O_7$.

5. What do you conclude about H_2O_2 from the reaction

$$H_2O_2 + H_2O \rightleftharpoons H_3O^+ + HO_2^-?$$

6. When a few drops of blood are added to aqueous H_2O_2, it (H_2O_2) decomposes rapidly. Explain.

Answers

1. H_2O_2 acts as a reducting agent to counteract Cl_2

$$H_2O_2 + Cl_2 \longrightarrow 2HCl + O_2$$

2. (a) $2Na_2O_2 + 2CO_2 \longrightarrow Na_2CO_3 + O_2$

(b) $Na_2O_2 + Fe(OH)_2 \longrightarrow Na_2FeO_4 + H_2$

(c) $2NH_4HSO_4 \xrightarrow{\text{current}} \underset{\text{anode}}{(NH_4)_2S_2O_8} + \underset{\text{cathode}}{H_2}$

$(NH_4)_2S_2O_8 + 2H_2O \longrightarrow 2NH_4HSO_4 + H_2O_2$

(d) $4KO_2 + 2CO_2 \longrightarrow 2K_2CO_3 + 3O_2$

(e) $Ba(OH)_2 + H_2O_2 \longrightarrow BaO_2 + 2H_2O$

3. Peroxide (O_2^{2-}) is diamagnetic and O_3 is also diamagnetic. When both react O_2 and H_2O are formed:

$$O_3 + H_2O_2 \longrightarrow 2O_2 + H_2O$$

O_2 is paramagnetic. Thus, O_3 (diamagnetic) is oxidised to O_2 (paramagnetic) but H_2O_2 (diamagnetic) has no effect in its property as H_2O is also diamagnetic.

4. (a) $2Cr^{3+} + 3H_2O_2 + 10\,OH^- \longrightarrow 2CrO_4^{2-} + 8H_2O$

(b) $Cr_2O_7^{2-} + 2H^+ + 4H_2O_2 \longrightarrow 2CrO_5 + 5H_2O$

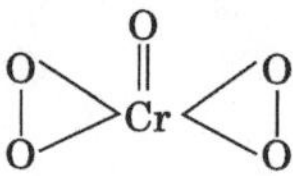

(CrO_5 gives blue colour in ethereal layer; it is diperoxide)

5. This indicates that H_2O_2 behaves as acid (though weak)

$$\underset{\text{acid}}{H_2O_2} + \underset{\text{base}}{H_2O} \rightleftharpoons \underset{\text{acid}}{H_3O^+} + \underset{\text{base}}{HO_2^-}$$

6. The decomposition of H_2O_2 is catalysed by **catalase,** an enzyme present in blood. There is formation of bubbles due to O_2.

Total Practice Set

(Read, Plan & Solve!)

Problem 1. Compare bond lengths, bond order and magnetic behaviour of

$$O_2^{2-}, O_2, O_2^-$$

Solution Based on molecular orbital theory :

$$\text{bond order of } O_2^{2-} = 1$$
$$O_2^- = 1.5$$
$$O_2 = 2.0$$

Greater the bond order, smaller the bond length, thus bond length of $O_2 < O_2^- < O_2^{2-}$

O_2 has two unpaired electrons hence, paramagnetic,

O_2^- has one unpaired electron hence, paramagnetic,

O_2^{2-} is without unpaired electron hence diamagnetic.

Problem 2. When blood is added to a solution of H_2O_2, the solution bubbles furiously. This reaction is also observed when H_2O_2 is poured on a skin wound to cleanse it. What is the overall reaction that occurs?

Solution An enzyme or biological catalyst (called catalase), present in tissue and in blood, decomposes the H_2O_2 and produces bubbles of O_2. The natural function of this enzyme is to prevent the accumulation of H_2O_2 in the body, since such a build-up of H_2O_2 could damage tissue.

Problem 3. To 25 mL H_2O_2, excess of acidified solution of KI was added. The iodine liberated required 20 mL of 0.3 N $Na_2S_2O_3$ solution. Calculate volume strength of H_2O_2 solution.

Solution $H_2O_2 + 2I^- + 2H^+ \longrightarrow 2H_2O + I_2$

$$I_2 + 2S_2O_3^{2-} \longrightarrow S_4O_6^{2-} + 2I^-$$

$$N_1V_1 = N_2V_2$$
$$H_2O_2 \qquad Na_2S_2O_3$$
$$N_1 \times 25 = 0.3 \times 20$$
$$N_1 = 0.24\text{ N}$$

Normality of H_2O_2 is related to "x volume" strength by relation

$$N = \frac{x}{5.6}$$

$$\therefore \quad x = N_1 \times 5.6$$
$$= 0.24 \times 5.6$$
$$= \textbf{“1.344 volume”}$$

Problem 4. (a) Give structures of O^{2-}, O_2^{2-} and O_2^-.
(b) What is their behaviour towards H_2O ?
(c) Give applications of H_2O_2 and KO_2.

Solution (a) $\left[:\ddot{\underset{\cdot\cdot}{O}}:\right]^{2-} \quad \left[:\ddot{\underset{\cdot\cdot}{O}}:\ddot{\underset{\cdot\cdot}{O}}:\right]^{2-} \quad \left[:\ddot{\underset{\cdot\cdot}{O}}:\ddot{\underset{\cdot}{O}}:\right]^{-}$

(b)

$$\underset{\text{oxide}}{O^{2-}} + H_2O \longrightarrow 2OH^-$$

$$\underset{\text{peroxide}}{2O_2^{2-}} + 2H_2O \longrightarrow O_2 + 4OH^-$$

$$\underset{\text{superoxide}}{4O_2^-} + 2H_2O \longrightarrow 3O_2 + 4OH^-$$

All the three ions are unstable towards H_2O and give alkaline solution. The decomposition of O_2^{2-} and O_2^- involve oxidation and reduction. Both of these ions, especially O_2^-, are good oxidising agents.

(c) H_2O_2 is used as :

- bleaching agent
- a substitute for Cl_2 in water and sewage treatment
- an antiseptic in curing wounds.

KO_2 is used in **life supports** in spacecrafts, submarines and emergency breathing apparatus since it absorbs CO_2 and releases O_2

$$4KO_2 + 2CO_2 \longrightarrow 2K_2CO_3 + 3O_2$$

Problem 5. Starting with deuterium oxide (D_2O), describe how you would prepare?
(a) NaOD, (b) DCl, (c) ND_3, (d) C_2D_2,
(e) CD_4, (f) D_2SO_4

Solution (a) $2Na + 2D_2O \longrightarrow \mathbf{2NaOD} + D_2$

(b) $D_2O + Cl_2 \longrightarrow \mathbf{DCl} + DOCl$

$\xrightarrow{\text{Electrolysis}} D_2 \xrightarrow{Cl_2} 2DCl$ (at cathode)

(c) $D_2O \longrightarrow D_2 \xrightarrow{N_2} \mathbf{ND_3}$

(d) $CaC_2 + 2D_2O \longrightarrow Ca(OD)_2 + \mathbf{C_2D_2}$

(e) $Al_4C_3 + 12D_2O \longrightarrow 4Al(OD)_3 + \mathbf{3CD_4}$

(f) $SO_3 + D_2O \longrightarrow \mathbf{D_2SO_4}$

Problem 6. A thermochemical cycle for hydrogen production involves at least one element (say Cu) that can exist in two different oxidation states; in a typical cycle *A* (ox) represents oxidised form and *A* (red) represents reduced form in a photolytic production of H_2 as given below:

$$H_2O + h\nu + 2A\,(ox) \longrightarrow 2A\,(red) + \frac{1}{2}O_2 + 2H^+ + 2e^-$$

$$2A\,(red) + 2H^+ + 2e^- \longrightarrow 2A\,(ox) + \mathbf{H_2}$$

Starting with $FeCl_2$, H_2O and Cl_2, explain formation of H_2. You may consider oxidised species as $FeCl_3$.

Solution $6FeCl_2 + 8H_2O \xrightarrow{600°C} 2Fe_3O_4 + 12HCl + 2H_2$

$$2\,Fe_3O_4 + 3Cl_2 + 12HCl \xrightarrow{150°C} 6FeCl_3 + 6H_2O + O_2$$

$$6FeCl_3 \xrightarrow{400°C} 6FeCl_2 + 3Cl_2$$

Net $\quad 2H_2O \longrightarrow 2H_2 + O_2$

Various devices may be employed to get round the difficulty created by the simultaneous production of the two gases (H_2 and O_2)

Problem 7. Hydrogen peroxide can also be obtained by partial oxidation of 2-propanol. Outline the scheme of this oxidation.

Solution

Step I $(H_3C)_2CHOH + O_2 \longrightarrow (H_3C)_2C(OH)(O{-}OH)$

2-propanol $\quad$ intermediate

Step II $(H_3C)_2C(OH)(O{-}OH) \longrightarrow (H_3C)_2C{=}O + H_2O_2$

Net $(H_3C)_2CHOH + O_2 \longrightarrow (H_3C)_2C{=}O + H_2O_2$

Problem 8. Oxygen forms three types of oxides : O^{2-}, O_2^{2-} and O_2^-. Classify them as base/acid when reacted with H_2O. Also classify the reaction

Solution $:\ddot{\underset{..}{O}}:^{2-}$ (oxide) $\quad :\ddot{\underset{..}{O}}:\ddot{\underset{..}{O}}:^{2-}$ (peroxide) $\quad :\ddot{\underset{..}{O}}:\ddot{\underset{..}{O}}\cdot^{-}$ (superoxide)

The ions are all strong **Bronsted-Lowry bases** and react with water as follows :

Oxide ion $\underset{-2}{O^{2-}}(aq) + \underset{+1,\,-2}{H_2O}\,(l) \longrightarrow \underset{-2,\,+1}{2OH^-}(aq)$

There is no change in oxidation number of O or H, hence, it is simply **hydrolysis** of O^{2-}.

Peroxide ion $\underset{-1}{O_2^{2-}}(aq) + 2H_2O(l) \longrightarrow \underset{0}{O_2}(g) + \underset{-2\,+1}{4OH^-}(aq)$

There is change in oxidation number of O, thus it is a **redox reaction** (disproportionation of O_2^{2-})

Superoxide ion

$$\underset{-0.5}{4O_2}(aq) + 2H_2O(l) \longrightarrow \underset{0}{3O_2}(g) + \underset{-2\,+1}{4OH^-}(aq)$$

This is also a redox reaction (disproportionation of O_2^-)

Master Exercises

Exercise 1
(Stage 1 : Learning)

Short Answer Type Questions

1. Which of the following compounds would be expected to form intermolecular hydrogen bonds in the liquid state?

(a) CH_3OCH_3 (b) CH_4
(c) HF (d) CH_3CO_2H
(e) Br (f) CH_3OH

2. What is meant by metallisation of hydrogen?

3. Name the binary compounds of hydrogen.

4. Hydrides are reducing agent. Explain

5. (a) Hydrides of boron are called ....
(b) Hydrides of silicon are called ....
(c) Hydrides of carbon are called ....
(d) Hydrides of that can reduce aldehydes to alcohol ....

6. 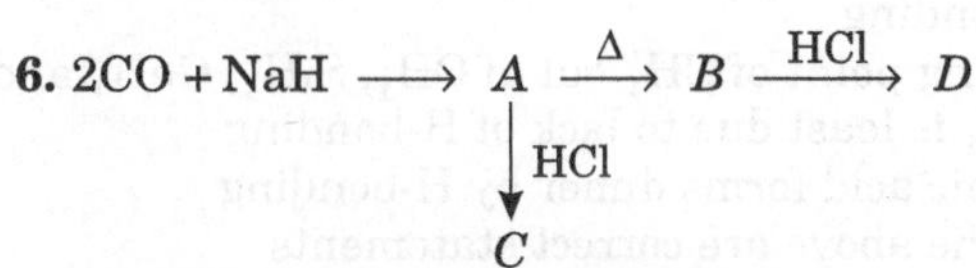

B and *D* decolourise $KMnO_4$/acidic
C reduces Tollen's reagent
What are *A* to *D*?

7. Complete the following reactions :

(a) $SiCl_4 + Li[AlH_4] \longrightarrow$
(b) $AgNO_3 + SiH_4 \longrightarrow$
(c) $CuSO_4 + PH_3 \longrightarrow$
(d) $SiH_4 + Cl_2 \longrightarrow$
(e) $Al_4C_3 + H_2O \longrightarrow$

8. Name the hydrides used in nuclear reaction technology.

Analytical Questions

1. A 1 g sample of H_2O_2 solution containing x per cent H_2O_2 by weight required x mL of $KMnO_4$ for complete oxidation in acidic medium. Calculate normality of $KMnO_4$ solution.

2. A 5.0 mL of H_2O_2 solution liberates 0.508 g of I_2 from an acidified KI solution. Calculate the strength of H_2O_2 solution in terms of volume strength at STP.

3. You have been given two liquids (*A*) and (*B*) made of same atoms. Both are diamagnetic. Liquid (*A*) + KI/H^+ + starch $\longrightarrow$ blue colour, liquid (*B*) is neutral to litmus and also does not give above test with starch. Identify (*A*) and (*B*).

4. $Na[Cr(OH)_4]$ green solution changes to yellow colour solution when boiled with H_2O_2. Yellow solution changes to orange on acidification. Write reactions of these chemical changes.

5. H_2O_2 ionises as:

$$H_2O_2 \rightleftharpoons H^+ + HO_2^-$$

If pH of H_2O_2 at 20°C is found to be 5.91, calculate ionic product (auto-protolysis constant) of H_2O_2.

6. To prepare H_2 from water, would you allow water to react with an oxidising agent or a reducing agent? Which of the following metals could be used in the reaction?

(a) Ag (b) Al
(c) Au (d) Ca

7. Which of the following molecules have similar properties and which have quite different properties?

(a) $H_2\overset{16}{O}$ (b) $D_2\overset{17}{O}$
(c) $H_2\overset{16}{O}_2$ (d) $H_2\overset{17}{O}$
(e) $D_2\overset{17}{O}_2$

8. Ionic metal hydrides react with water to give hydrogen gas and an aqueous solution of the metal hydroxide.

(a) On reaction of equal masses of LiH and CaH_2 with water, which compound gives more H_2?

(b) How many kilograms of CaH_2 is needed to fill a 100 L tank with compressed H_2 gas at 150 atm pressure and at 25°C? (Li = 9, H = 1, Ca = 40)

$$LiH + H_2O \longrightarrow LiOH + \frac{1}{2} H_2$$

$$CaH_2 + 2H_2O \longrightarrow Ca(OH)_2 + 2H_2$$

9. Which of the following are oxidised by water? Which are reduced by water? Which undergo a disproportionation reaction when treated with water?

(a) Cl_2 (b) F_2
(c) K (d) Br_2

10. How many kilograms of water must be processed to obtain 2.0 L of D_2 at 25°C and 0.90 atm? Assume that deuterium abundance is 0.015 per cent and that recovery is 80 per cent.

11. Identify *A* to *E* in the following

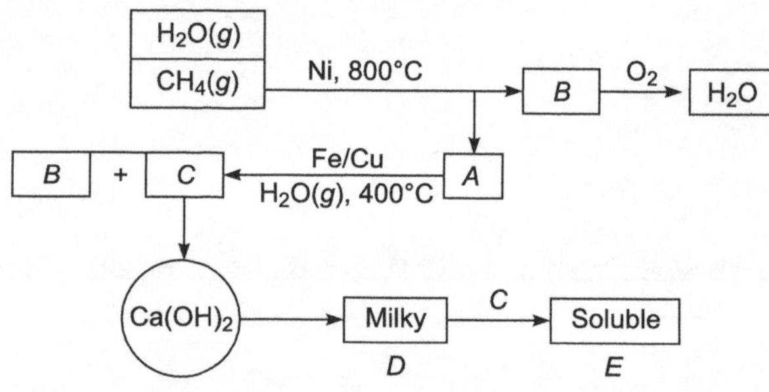

12. What conclusion can you draw about the structures of *A*, and *B* based on following reactions?

$$\underset{(A)}{H_2SO_5} + H_2O \longrightarrow H_2SO_4 + H_2O_2$$

$$\underset{(B)}{H_2S_2O_8} + 2H_2O \longrightarrow 2H_2SO_4 + H_2O_2$$

Exercise 2
(Stage 2 : High Skill Problem Solving)

Only One Option Correct

1. H_2 gas is liberated at cathode and anode both by electrolysis of the following aqueous solution except in

(a) NaH (b) HCOONa
(c) NaCl (d) LiH

2. Statement S : Trimethyl ammonium hydroxide (*A*) is a weaker base than tetramethyl ammonium hydroxide (*B*).

Reason R : O of OH group in (*A*), is hydrogen bonded to the Me_3NH—group while in (*B*), hydrogen bonding can't occur.

(a) S and R both are correct
(b) S is correct but R is incorrect
(c) S is incorrect but R is correct
(d) S and R both are incorrect

3. In the absence of hydrogen bonding, boiling point of water would have been

(a) 100°C (b) 0°C
(c) −100°C (d) 373°C

4. Which is most volatile out of *o*-, *m*- and *p*-nitrophenol?

(a) *o*- (b) *m*-
(c) *p*- (d) All are equally volatile

5. Which is correct order of boiling point of VA, VIA and VIIA hydride?

(a) $NH_3 > SbH_3 > AsH_3 > PH_3$
(b) $H_2O > H_2Te > H_2Se > H_2S$
(c) HF > HI > HBr > HCl
(d) All of the above are correct

6. Which statements is/are correct?

(a) Boiling point of H_2O, NH_3, HF are maximum in their respective group due to intermolecular H-bonding
(b) Boiling point of CH_4 out of CH_4, SiH_4, GeH_4 and SnH_4 is least due to lack of H-bonding
(c) Formic acid forms dimer by H-bonding
(d) All the above are correct statements

7. H_2O_2 can be obtained when following reacts with H_2SO_4 except with

(a) PbO_2 (b) BaO_2
(c) Na_2O_2 (d) SrO_2

8. The set of elements which could form stable, covalent hydrogen bonded hydrides are

(a) nitrogen, oxygen and fluorine
(b) lithium, sodium and potassium
(c) sulphur, selenium and tellurium
(d) chlorine, bromine and iodine

9. The fuel gas obtained by blowing steam over incandescent coal is known as

(a) coal gas (b) water gas
(c) producer gas (d) natural gas

10. Steam is passed over red hot carbon and the gaseous products cooled and passed first through a solution of alkali, then through ammoniacal cuprous chloride and then through water. What is finally collected?
(a) CO_2
(b) A mixture of hydrocarbons
(c) $CO + H_2$
(d) H_2

11. Which among the following is a hydride?
(a) Rongalite (b) Nitrolim
(c) Hydrolith (d) Minium

12. Which of the following easily reacts with water producing hydrogen?
(a) PH_3 (b) B_2H_6
(c) CH_4 (d) H_2S

13. How does H_2O_2 differ from O_3 in its chemical action?
(a) In oxidising PbS to $PbSO_4$
(b) In liberating I_2 from KI
(c) In decolorising acidified $KMnO_4$
(d) In oxidising $K_4[Fe(CN)_6]$ to $K_3[Fe(CN)_6]$

14. Which of the following statements about H_2O_2 is not true?
(a) H_2O_2 is used to clean oil paintings
(b) H_2O_2 acts as oxidising as well as reducing agent
(c) Two hydroxyl groups in H_2O_2 lie in the same plane
(d) It retains same structure in liquid and solid form

15. H_2O_2 is used as but not as
(a) oxidant, reductant (b) bleaching agent
(c) antiseptic (d) catalyst

16. Which is accepted structure of H_2O_2 in gas phase?
(a) 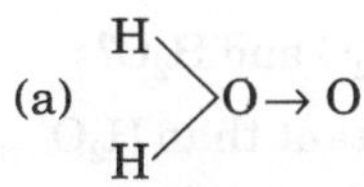 (b) 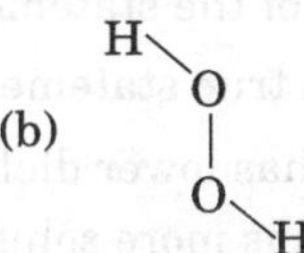
(c) Both (a) and (b) (d) None of these

17. "10 volume" H_2O_2 means
(a) 1 mL H_2O_2 gives 10 mL O_2 at NTP
(b) 1 g H_2O_2 gives 10 mL O_2 at NTP
(c) 1 mol H_2O_2 gives 10 mL O_2 at NTP
(d) none of the above is correct

18. Industrially H_2O_2 is obtained from
(a) 2-ethyl anthraquinol by oxidation and then reduction in a cyclic process
(b) H_2SO_5
(c) $H_2S_2O_8$
(d) BaO_2

19. Density of H_2O is maximum at
(a) 0°C (b) 100°C
(c) –273°C (d) 4°C

20. Temporary hardness and permanent hardness in water can be removed respectively by addition of
(a) CaO, $CaCO_3$ (b) CaO, Na_2CO_3
(c) Na_2CO_3, CaO (d) $NaHCO_3$, $CaCl_2$

21. Decomposition of H_2O_2 can be prevented by addition of
(a) ferrous sulphate (b) oxalic acid
(c) ozone (d) glycerol

22. Which is true statement?
(a) The layer of ice on the surface of river in the winter acts as a thermal insulator between the water below and the air above
(b) The fish and other marine organisms are enabled to survive long periods of freezing weather due to the fact that ice is lighter than water
(c) Both (a) and (b) are correct
(d) None of the above is correct

23. Hydrides of metals are named like
(a) alkane (b) alkene
(c) alkyne (d) none is correct

24. In water, or in aqueous solutions of HCl or H_2SO_4, proton exists as
(a) H_3O^+ (b) $H(H_2O)_4^+$
(c) $H(H_2O)_n^+$ (d) all are correct

25. Hydrogen
(a) is placed in I A since it forms monovalent cation H^+
(b) is placed in VII A (halogen family) since it forms monovalent anion H^-
(c) is placed in IV A (carbon family) since both have a half filled shell of electrons
(d) follows all of the above facts

26. Which is true statement about H_2O?
(a) Hardness can be removed by passing through ion exchange resin
(b) Its presence can be detected by anhydrous $CuSO_4$
(c) It is amphiprotic
(d) All the above are correct statement

27. H_2O_2 is "5.6 volume", then
(a) it is 1.7% weight by volume
(b) it is 1 N
(c) both (a) and (b) are true
(d) none of the above is true

28. Bond angles H—O—H and H—O—O— in water and hydrogen peroxide respectively are
(a) 104.5° in both (b) 94.8° in both
(c) 104.5°, 94.8° (d) 94.8° and 104.5°

29. $H_2O_2 \longrightarrow H_2O + O_2$
This represents
(a) oxidation of H_2O_2
(b) reduction of H_2O_2
(c) disproportionation of H_2O_2
(d) acidic nature of H_2O_2

30. Which is a source of nascent hydrogen?
I : Zn + dil. HCl,
II : CH_3OH + Na
III : Electrolysis of H_2O
IV : Silent electric discharge
(a) I, II (b) II, III
(c) I, II, III, IV (d) IV

31. Select correct statement(s) abut lime light.
(a) Oxy-hydrogen flame
(b) Used in welding
(c) Temperature of lime light is sufficient to melt even platinum
(d) All the above are correct

32. H_2 is adsorbed on palladium surface. It is a case of
(a) occlusion (b) diffusion
(c) effusion (d) electro-osmosis

33. D_2O (heavy water) and H_2O differ in following except in
(a) freezing point
(b) density
(c) ionic product of water
(d) its reaction with sodium

34. (*A*) $FeCl_3$ solution + Zn $\longrightarrow$ product *X*
(*B*) $FeCl_3$ solution + H_2 gas $\longrightarrow$ product *Y*
$FeCl_3$ solution gives blue colour with $K_4[Fe(CN)_6]$ hence
(a) *X* also gives blue colour with $K_4[Fe(CN)_6]$
(b) *Y* also gives blue colour with $K_4[Fe(CN)_6]$
(c) both *X* and *Y* give blue colour with $K_4[Fe(CN)_6]$
(d) none gives colour with $K_4[Fe(CN)_6]$

35. H^+, D^+ and T^+ differ in all except in
(a) number of electrons (b) number of neutrons
(c) ionic mass (d) all of these

36. Which is radioactive isotope of hydrogen?
(a) ${}^1_1H$ (b) ${}^2_1H$
(c) ${}^3_1H$ (d) All of these

37. Which is true about different forms of hydrogen?
(a) *Ortho* hydrogen has same spins of two nuclei clockwise or anticlockwise
(b) *Para* hydrogen has different spins of two nuclei
(c) At absolute zero, there is 100% *para* form and at high temperature, there is 75% *ortho* form
(d) All the above are correct

38. Which is not correct statement?
(a) *s*-block elements, except Be and Mg, form ionic hydride
(b) BeH_2, MgH_2, CuH_2, ZnH_2, CaH_2 and HgH_2 are intermediate hydride
(c) *p*-block elements form covalent hydride
(d) *d*- and *f*-block elements form ionic hydride

39. H_2 can be obtained from
(a) water gas (CO + H_2) by liquefaction of CO at low temperature under pressure
(b) water gas by oxidation of CO into CO_2 (by steam) which can be easily removed by dissolving in H_2O
(c) electrolysis of water
(d) all the above methods

40. Out of the following metals which will give H_2 on reaction with NaOH
I : Zn, II : Mg, III : Al, IV : Be
(a) I, II, III, IV (b) I, III, IV
(c) II, IV (d) I, III

41. Choose correct statement(s).
(a) H_2 is more rapidly adsorbed on to surfaces than D_2
(b) H_2 reacts over 13 times faster with Cl_2 than D_2 because H_2 has a lower energy of activation
(c) Both (a) and (b) are true
(d) None of the above is true

42. Abundance of H_2 in the earth's atmosphere is very small. This is because
(a) the earth's gravitation field is too small to hold so light an element
(b) H_2 exists in *ortho* and *para* form
(c) H_2 is diatomic gas
(d) H_2 is not the metal

43. H_2 reacts much faster with Cl_2 than D_2, because
(a) rate of diffusion of H_2 is greater than D_2
(b) H_2 has lower energy of activation than D_2
(c) both (a) and (b) are correct
(d) none of the statements is correct

44. Which is true statement about D_2O and H_2O?
(a) D_2O has lower dielectric constant than H_2O
(b) NaCl is more soluble in D_2O than in H_2O
(c) Both (a) and (b) are correct
(d) None of the above is correct

45. Out of LiH, MgH_2 and CuH
(a) all are ionic hydrides
(b) LiH, MgH_2 are ionic and CuH covalent hydride
(c) all are covalent hydrides
(d) LiH is ionic, MgH_2 and CuH are intermediate hydrides

46. There are three isotopes of hydrogen and three naturally occurring isotopes of oxygen (${}^{16}O$, ${}^{17}O$ and ${}^{18}O$). How many kinds of water are possible?
(a) 18 (b) 16
(c) 8 (d) 9

47. In the following compounds H is covalent bonded in case of
(a) BaH_2 (b) CaH_2
(c) SiH_4 (d) NaH

48. Which of the following disproportionates when treated with water?
(a) SO_3 (b) F_2
(c) N_2 (d) Cl_2

49. Select the correct statement.
(a) Ammonia is more soluble in aqueous ammonium chloride than in pure water.
(b) Solid ammonium fluoride and ice are miscible in all proportions.
(c) Both (a) and (b) are correct
(d) None of the above statements is correct

50. H, D and T (isotopes of hydrogen) have nuclear spin quantum number respectively as
(a) $\frac{1}{2}, 1, \frac{1}{2}$ (b) $\frac{1}{2}, \frac{1}{2}, 1$
(c) $\frac{1}{2}, 1, \frac{3}{2}$ (d) $\frac{1}{2}, \frac{1}{2}, \frac{1}{2}$

51. D_2O has maximum density at
(a) 4°C (b) 11.6°C
(c) 0°C (d) 0 K

52. H_2O_2 is the hydride of
(a) O_2 (b) H_2O
(c) both (a) and (b) (d) none of these

53. H_2O_2 can also be obtained by the partial oxidation of 2-propanol.

$$(CH_3)_2CHOH + O_2 \longrightarrow (CH_3)_2CO + H_2O_2$$

Intermediate in this reaction is
(a) $(CH_3)_2C(OH)(OH)$
(b) $(CH_3)_3\overset{+}{C}$
(c) $(CH_3)_2C(OH)(O—O—H)$
(d) none of the above

54. H_2O_2 exists as .... in alkaline medium.
(a) HO_2^- (b) $HO_2^{\oplus}$
(c) O_2^{2-} (d) $H_2O^{\oplus}$

55. In the following reaction using $H_2\overset{18}{O}_2$,

$$2MnO_4^- + 5H_2\overset{18}{O}_2 + 6H^+ \longrightarrow 2Mn^{2+} + 8H_2O + 5O_2$$

^{18}O goes with
(a) H_2O (b) O_2
(c) both (a) and (b) (d) none of these

56. Water gas is a mixture of
(a) CO and H_2O (b) CO and N_2
(c) CO and H_2 (d) CO and CH_4

57. H^- is a
(a) Lewis base (b) Lowry-Bronsted base
(c) both (a) and (b) (d) None of these

58. Consider following reactions

I : $AlH_3 + H^- \longrightarrow AlH_4^-$

II : $H_2O + H^- \longrightarrow H_2 + OH^-$

Select correct statements based on these reactions.
(a) H^- is a Lewis acid in I and Lewis base in II
(b) H^- is a Lewis base in I and Bronsted base in II
(c) H^- is a Lewis acid in I and Bronsted acid in II
(d) H^- is a Lewis base in I and II

59. Consider following statements :
I : Rate of transfer of D^+ from D_2O is slower than that of H^- from H_2O
II : K_a for $CH_3COOH \rightleftharpoons CH_3COO^- + H^+$ is smaller than that of K_a for
$$CH_3COOD \rightleftharpoons CH_3COO^- + D^+$$
III : Tritium is a radioactive isotope.
Select correct statements
(a) I, II (b) II, III
(c) I, III (d) I, II, III

60. Which is hydrolysis reaction?
(a) $O^{2-}(aq) + H_2O(l) \longrightarrow 2OH^-(aq)$
(b) $O_2^{2-}(aq) + 2H_2O(l) \longrightarrow O_2(g) + 4OH^-(aq)$
(c) $4O_2^-(aq) + 2H_2O(l) \longrightarrow 3O_2(g) + 4OH^-(aq)$
(d) All of the above

61. The hydride ion H^- is stronger base than its hydroxide ion OH^-. Which of the following reactions will occur if sodium hydride (NaH) is dissolved in water?
(a) $2H^+(aq) \longrightarrow H_2 + 2e^-$
(b) $H^+(aq) + H_2O(l) \longrightarrow OH^- + H_2$
(c) $H^- + H_2O(l) \longrightarrow$ No reaction
(d) None of the above

62. Permanent hardness of water can be removed by adding
(a) $NaHCO_3$ (b) Na_2CO_3
(c) $CaOCl_2$ (d) Cl_2

63. The correct decreasing order of basic strength is
(a) $AsH_3 > SbH_3 > PH_3 > NH_3$
(b) $SbH_3 > AsH_3 > PH_3 > NH_3$
(c) $NH_3 > PH_3 > AsH_3 > SbH_3$
(d) $PH_3 > AsH_3 > SbH_3 > NH_3$

64. Which statement is not correct for hydrogen?
(a) It has a very high ionisation potential
(b) It is always collected at cathode
(c) It can form bonds in +1 as well as –1 oxidation state
(d) It has same electronegativity as halogens

65. The most stable is
(a) NaH (b) RbH
(c) KH (d) LiH

66. H_2O_2 is a reducing agent in the reaction
(a) $Ag_2O + H_2O_2 \longrightarrow 2Ag + H_2O + O_2$
(b) $2KI + H_2O_2 \longrightarrow 2KOH + I_2$
(c) $H_2O_2 + SO_2 \longrightarrow H_2SO_4$
(d) $PbS + 4H_2O_2 \longrightarrow PbSO_4 + 4H_2O$

67. Which cannot be oxidised by H_2O_2 ?
(a) Na_2SO_3 (b) PbS
(c) KI (d) O_3

68. Dil. H_2SO_4 and oxide react to produce hydrogen peroxide. The oxide is
(a) MnO_2 (b) PbO_2
(c) TiO_2 (d) Na_2O_2

69. Heavy water is
(a) water at 4°C
(b) water containing heavy ions
(c) D_2O
(d) water containing sulphate of Ca

70. Adsorbed hydrogen by palladium is known as
(a) nascent (b) atomic
(c) heavy (d) occluded

71. Pure water can be obtained from sea water by
(a) centrifugation (b) plasmolysis
(c) reverse osmosis (d) sedimentation

72. Zn gives H_2 gas with H_2SO_4 and HCl but not with HNO_3 because
(a) Zn acts as an oxidising agent when reacts with HNO_3
(b) HNO_3 is weaker acid than H_2SO_4 and HCl
(c) in electrochemical series Zn is above hydrogen
(d) NO_3^- is reduced in preference to hydronium ion

73. Mass of one atom is 6.66×10^{-23} g. Its percentage in an hydride is 95.24. Thus, hydride is
(a) MH (b) MH_2
(c) MH_3 (d) MH_4

74. On burning hydrogen in air, the colour of flame is
(a) green (b) light bluish
(c) yellow (d) none of these

75. Triple point of water is
(a) 273.16 K (b) 373.15 K
(c) 203.12 K (d) 193.16 K

76. Hydrogen can be fused to form helium at
(a) high temperature and high pressure
(b) high temperature and low pressure
(c) low temperature and high pressure
(d) low temperature and low pressure

77. Water gas is produced by
(a) passing steam over red hot coke
(b) passing steam and air over red hot coke
(c) burning coke in excess of air
(d) burning coke in limited supply of air

78. Hydrogen is produced by the reaction
(a) $Na_2O_2 + 2HCl$ (b) $Mg + H_2O$
(c) $BaO_2 + HCl$ (d) $H_2S_4O_8 + H_2O$

79. Which metal gives hydrogen gas on heating with hot concentrated alkali?
(a) Ag (b) Ni
(c) Zn (d) Cu

80. Hydrogen peroxide is prepared in the laboratory by
(a) passing CO_2 into BaO_2
(b) adding MnO_2 to dil. H_2SO_4
(c) adding Na_2O_2 to cold water
(d) adding PbO_2 into $KMnO_4$

81. The hardness of water is estimated by
(a) EDTA method (b) titrimetric method
(c) conductivity method (d) distillation method

82. Hydrogen peroxide is reduced by
(a) ozone
(b) barium peroxide
(c) acidic solution of $KMnO_4$
(d) lead sulphide suspension

83. Hard water is not fit for washing clothes because
(a) it contains impurities
(b) it is acidic in nature
(c) it gives precipitate with soap
(d) it contains Na_2SO_4 and KCl

84. The acid, not suitable for the preparation of hydrogen by the action of metals, is
(a) HCl (b) CH_3COOH
(c) HNO_3 (d) H_2SO_4

85. Maximum number of hydrogen bonding in H_2O is
(a) 1 (b) 2
(c) 3 (d) 4

86. $Ag_2O + H_2O_2 \longrightarrow 2Ag + H_2O + O_2$
In an above reaction H_2O_2 acts as a/an
(a) oxidising agent (b) reducing agent
(c) bleaching agent (d) None of these

87. Ionic hydride reacts with water to give
(a) hydride ions (b) acidic solution
(c) protons (d) basic solutions

88. H_2O_2 is manufactured these days by
(a) the action of H_2SO_4 on BaO_2
(b) the action of H_2SO_4 on Na_2O_2
(c) electrolysis of 50% Na_2O_2
(d) burning hydrogen in excess oxygen

89. Which of the following is heavy water?
(a) H_2O^{18} (b) H_2O^{16}
(c) H_2O_3 (d) D_2O

90. The volume of oxygen liberated at NTP from 15 mL of 20 volume H_2O_2 is
(a) 250 mL (b) 300 mL
(c) 150 mL (d) 200 mL

91. Spin isomerism is shown by
(a) dichloro benzene (b) hydrogen
(c) dibasic acid (d) *n*-butane

92. Select incorrect statement.
(a) *Ortho* and *para* hydrogen are different due to difference in their nuclear spins
(b) *Ortho* and *para* hydrogen are different due to difference in their electron spins
(c) *Para* hydrogen has a lower internal energy than that of *ortho* hydrogen
(d) *Para* hydrogen is more stable at lower temperature

93. SiH_4 is an example of which of the following type of hydrides
(a) ionic (b) interstitial
(c) metallic (d) covalent

94. Molar ratio of H_2O_2 oxidised by one mole of MnO_4^- in basic and acidic medium is
(a) 1:1 (b) 1:2
(c) 3:5 (d) 5:3

95. Electrolysis of aqueous NaCl and NaH differs in
(a) formation of basic solution at the cathode by NaCl only
(b) formation of basic solution at the cathode by NaH only
(c) formation of H_2 gas at cathode and anode both by NaH only
(d) formation of H_2 gas at cathode and anode both by NaCl only

96. Select correct statement(s).
(a) Hydride ion is larger than any of the halide ions except iodide
(b) Hydrides ions are reducing agents
(c) Boranes are electron deficient hydrides
(d) All of the above are correct statements

97. Metallic hydrides
(a) are also called interstitial hydrides
(b) are non-stoichiometric, being deficient in hydrogen
(c) are poor conductors of electricity, exhibit less paramagnetism and have hydrogen as atom and not as molecule
(d) have all the properties given above

98. Hydride ion is a
(a) strong conjugate acid of H_2
(b) strong conjugate base of H_2
(c) strong conjugate acid of H^+
(d) strong conjugate base of H^-

99. Select incorrect statement.
(a) H^+ can exist as $H_9O_4^+$ in water
(b) H_2 is thermally stable
(c) Ionisation of CH_3COOH is slower than that of CH_3COOD
(d) Kinetic isotopic effect is observed when there is retardation in the rate water H_2O is replaced by D_2O.

100. H_2, D_2 and T_2 do not differ in
(a) freezing point (b) boiling point
(c) critical temperature (d) bond energy

101. H_2O_2 is present as ...... in acidic medium.
(a) HO_2^- ion (b) $H_3O_2^+$ ion
(c) H^+ ion (d) O_2^{2-} ion

102. Hydroxyl radicals are used in organic reactions. These are obtained from
(a) glycol (b) glycerol
(c) H_2O_2/Fe^{2+} (d) H_2O_2/Fe^{3+}

103. H_2O_2 is a reducing agent as shown by
(a) $H_2O_2 \longrightarrow 2H^+ + O_2 + 2e^-$
(b) $H_2O_2 + 2H^+ + 2e^- \longrightarrow 2H_2O$
(c) $H_2O_2 \longrightarrow H^+ + HO_2^-$
(d) $H_2O_2 + H^+ \longrightarrow H_3O_2^+$

104. Industrial source of H_2O_2 is
(a) by oxidation of 2-ethyl anthraquinol in a cyclic process
(b) by partial oxidation of 2-propanol
(c) by both (a) and (b)
(d) none of the above

105. Fuel used for rocket propulsion is a mixture of
(a) hydrazine and hydrogen peroxide
(b) hydrazine and TNT
(c) hydroxyl amine and TNT
(d) hydroxyl amine and hydrogen peroxide

106. Select correct statement(s).
(a) H_2O_2 reduces MnO_4^- to Mn^{2+} in acidic medium
(b) H_2O_2 reduces MnO_4^- to MnO_2 in basic medium
(c) H_2O_2 can be used to bleach blackened oil paintings
(d) All the above are correct statement

107. Electron-deficient hydride is/are
(a) BH_3 (b) AlH_3
(c) BeH_2 (d) all of these

108. Perhydrol is
(a) 30% solution of H_2O_2
(b) "100 volume" H_2O_2
(c) both (a) and (b) are correct
(d) none of the above is correct

109. 10 mL of H_2O_2 solution on treatment with KI and titration of liberated I_2, required 10 mL of 1 N hypo. Thus H_2O_2 is
(a) 1N (b) 5.6 volume
(c) 17 gL^{-1} (d) all are correct

110. In alkaline H_2O_2
(a) Mn^{2+} changes to MnO_2
(b) Cr^{3+} changes to CrO_4^{2-}
(c) both (a) and (b) are correct
(d) none of the above is correct

One or More Than One Options Correct

1. Select correct statements.
(a) Boiling point of H_2O, NH_3, HF are maximum in their respective group due to intermolecular H-bonding
(b) Boiling point of CH_4 out of CH_4, SiH_4, GeH_4 and SnH_4 is least due to lack of H-bonding
(c) Benzoic acid forms dimer by H-bonding
(d) H_2O exists as $H_9O_4^+$ in acidic medium

2. H_2O_2 can be obtained when following reacts with H_2SO_4 except with
(a) PbO_2 (b) BaO_2
(c) Na_2O_2 (d) SrO_2

3. How does H_2O_2 differ from O_3 in its chemical action?
(a) In oxidising PbS to $PbSO_4$
(b) In liberating I_2 from KI
(c) In decolourising acidified $KMnO_4$
(d) In oxidising $K_4[Fe(CN)_6]$ to $K_3[Fe(CN)_6]$

4. Which of the following statements about H_2O_2 are true?
(a) H_2O_2 is used to clean oil paintings
(b) H_2O_2 acts as oxidising as well as reducing agent
(c) Two hydroxyl groups in H_2O_2 lie in the same plane
(d) It retains same structure in liquid and solid form

5. H^+, D^+ and T^+ differ in all except in
(a) number of electrons
(b) number of neutrons
(c) ionic mass
(d) electronic configuration

6. Which is/are radioactive isotope(s) of hydrogen?
(a) ${}_1^1H$ (b) ${}_1^2H$
(c) ${}_1^3H$ (d) ${}_1^4H$

7. Which is a source of nascent hydrogen?
(a) Zn + dil. HCl,
(b) CH_3OH + Na
(c) Electrolysis of H_2O
(d) Silent electric discharge

8. Select correct statement(s) about lime light
(a) Oxy-hydrogen flame
(b) Used in welding
(c) Temperature of lime light is sufficient to melt even platinum
(d) Produced in an endothermic reaction

9. "10 volume" H_2O_2 means
(a) 1 mL H_2O_2 gives 10 mL O_2 at NTP
(b) 1 g H_2O_2 gives 10 mL O_2 at NTP
(c) 1 mol H_2O_2 gives 10 mL O_2 at NTP
(d) 10 mL gives 1 mol H_2 and 1 mol O_2 at NTP

10. Industrially H_2O_2 is obtained from
(a) 2-ethyl anthraquinol by oxidation and then reduction in a cyclic process
(b) H_2SO_5
(c) $H_2S_2O_8$
(d) BaO_2

11. Density of H_2O is maximum at
(a) 0°C (b) 100°C
(c) –273°C (d) 4°C

12. Temporary hardness and permanent hardness can be removed respectively by addition of
(a) CaO, $CaCO_3$ (b) CaO, Na_2CO_3
(c) Na_2CO_3, CaO (d) $NaHCO_3$, $CaCl_2$

13. Decomposition of H_2O_2 can be prevented by addition of
(a) ferrous sulphate (b) oxalic acid
(c) ozone (d) glycerol

14. Which is/are true statement(s)?
(a) The layer of ice on the surface of river in the winter acts as a thermal insulator between the water below and the air above
(b) The fish and other marine organisms are enabled to survive long periods of freezing weather due to the fact that ice is lighter than water
(c) When ice is formed volume decreases
(d) Density of ice is maximum at 0°C

15. Hydrides of metals are named like
(a) alkane (b) alkene
(c) alkyne (d) none is correct

16. In water, or in aqueous solutions of HCl or H_2SO_4, proton exists as
(a) H_3O^+ (b) $H(H_2O)_4^+$
(c) $H(H_2O)_n^+$ (d) $[H_3O^+][OH^-]$

17. H_2O_2 is "5.6 volume", then
(a) it is 1.7% weight by volume
(b) it is 1 N
(c) it is 1 M
(d) it is 5.6 M

18. H_2 can be obtained from
(a) water gas ($CO + H_2$) by liquefication of CO at low temperature under pressure
(b) water gas by oxidation of CO into CO_2 (by steam) which can be easily removed by dissolving in H_2O
(c) electrolysis of water or NaOH
(d) reaction of NaOH on Zn

19. Out of the following metals which will give H_2 on reaction with NaOH
I : Zn, II : Mg, III : Al, IV : Be
(a) I, II, III, IV (b) I, III, IV
(c) II, IV (d) I, III

20. Which is hydrolysis reaction?

(a) $O^{2-}(aq) + H_2O(l) \longrightarrow 2OH^-(aq)$

(b) $O_2^{2-}(aq) + 2H_2O(l) \longrightarrow O_2(g) + 4OH^-(aq)$

(c) $4O_2^-(aq) + 2H_2O(l) \longrightarrow 3O_2(g) + 4OH^-(aq)$

(d) $Fe^{3+} + 3H_2O \longrightarrow Fe(OH)_3 + 3H^+$

Brain Twisters

1. How many grams of barium hydride must be treated with water to obtain 4.36 L of hydrogen at 20°C and 0.975 atm pressure? (Ba = 137).

2. An element X reacts with H_2 gas at 200°C to form compound Y. When Y is heated to a higher temperature, it decomposes to the element X and H_2 gas in the ratio of 559 mL of H_2 (measured at STP) for 1.00 g of X reacted. X also combines with Cl_2 to form a compound Z, which contains 63.89 per cent by mass of chlorine. Deduce the identify of X, Y and Z.

3. H_2O_2 is reduced rapidly by Sn^{2+}. H_2O_2 decomposes slowly at room temperature to yield O_2 and H_2O. Calculate the volume of O_2 gas produced at 20°C and 1 atm when 200 g H_2O_2 having 10% by mass H_2O_2 is treated with 100 mL of 2.00 M Sn^{2+} and then the mixture is allowed to stand until no further reaction occurs.

4. H_2O_2 is marked "22.4 volume". How much of it (in mL) are required to oxidise 3.4 g H_2S gas?

5. A 'hair-bleach' solution contains H_2O_2. The amount of H_2O_2 in a 12.5 g sample was determined by titration with $KMnO_4$ in acidic solution. If 39.3 mL of 0.5045 N $KMnO_4$ was required, how many equivalents of H_2O_2 are contained in the solution ? What is the mass percentage of H_2O_2 in the bleach solution?

6. A 1.84 g sample of an alkaline-earth metal hydride was treated with an excess of dil. HCl and the resulting gas was collected in 1.0 L container at 20°C. The measured pressure of the gas was 750.0 mmHg. Identify the alkaline earth-metal.

7. A 5.0 mL of H_2O_2 solution liberates 0.508 g I_2 from an acidified KI solution. Calculate volume strength of H_2O_2 solution.

Passage Comprehension Questions

Passage 1

Read the following experiments and answer the questions at the end of it.

"Research scholar A added zinc pieces into aqueous $FeCl_3$ solution and performed some experiments with resultant solution." Research scholar B passed H_2 gas into aqueous $FeCl_3$ solution and performed some experiments with resultant solution."

1. Yellow coloured $FeCl_3$ solution changed to light green (appeared as colourless) in the experiment of

(a) A (b) B

(c) both of these (d) none of these

2. Select correct statements.

(a) Zn pieces liberate nascent hydrogen on reaction with acidic solution of $FeCl_3$

(b) $FeCl_3$ solution is reduced to $FeCl_2$ in the experiments of A and B both

(c) Blue colour complex is formed in both the experiments on reaction with $K_4[Fe(CN)_6]$

(d) A blood red colour is formed on reaction with KCNS in experiment B

Passage 2

Depletion of resources of fossil fuels will at some future time make hydrogen, either for use directly by combustion and electrochemically in fuel cells or indirectly *via* hydrogenation of coal, the major alternative to nuclear energy; hence, arises the current interest in the so-called **hydrogen economy.**

A thermochemical cycle for hydrogen production involves at least one element that can exist in two different oxidation states.

$$2H_2O \longrightarrow 2H_2 + O_2$$

Read the above passage and answer the following questions.

1. Production of H_2 from H_2O requires a net input of energy that would come from

(a) nuclear source (b) solar source

(c) both of these (d) None of these

2. In the above thermochemical cycle, $FeCl_2$ is one of the catalyst involved and thus regenerated at the end of the reaction, after formed into

(a) $FeCl_3$, FeO, Fe_2O_3 (b) $FeCl_3$, Fe_2O_3, Fe_3O_4

(c) $FeCl_3$, FeO (d) $FeCl_3$, Fe_3O_4

3. H_2 is produced in the following thermochemical reaction

(a) $FeCl_2 + H_2O \longrightarrow$

(b) $Fe_3O_4 + Cl_2 + HCl \longrightarrow$

(c) $FeO + H_2O \longrightarrow$

(d) $FeCl_3 + H_2O \longrightarrow$

Assertion & Reason

Codes :

(a) Both **A** and **R** are true and **R** is the correct explanation of **A.**

(b) Both **A** and **R** are true but **R** is not the correct explanation of **A.**

(c) **A** is true but **R** is false.

(d) **A** is false but **R** is true.

(e) **A** and **R** both are false.

1. **Assertion (A) :** On adding zinc pieces to $FeCl_3(aq)$ solution, colour changes from deep yellow to light green.

   Reason (R) : Aqueous $FeCl_3$ solution is acidic and on adding zinc, nascent hydrogen is produced which reduces deep yellow $FeCl_3$ solution to light green $FeCl_2$ solution.

2. **Assertion (A) :** On passing H_2 gas into aqueous Fe^{3+} solution, resulting solution gives test of Fe^{3+}.

   Reason (R) : $H_2(g)$ reduces Fe^{3+} to Fe^{2+} but Fe^{2+} is oxidised to Fe^{3+} by atmospheric oxygen.

3. **Assertion (A) :** Drinking of heavy water (D_2O) mixed water (H_2O) could prove fatal.

   Reason (R) : There is slower rate of transfer of D^+ compared with that of H^+ ion in acid-base reactions involved in enzyme catalysis.

4. **Assertion (A) :** Ionisation constant of CH_3COOH and CH_3COOD are very equal.

   Reason (R) : The kinetic isotopic effect is not observed in the ionisation of weak acid.

5. **Assertion (A) :** In space crafts H_2 gas is used in fuel cells for generating electrical energy and for providing clean drinking water to the astronauts.

   Reason (R) : A fuel cell may have an alkaline or acidic electrolyte.

6. **Assertion (A) :** Hydride (H^-) is a conjugate base of hydrogen (H_2).

   Reason (R) : Every negative ion is an electron-pair donor and is thus, a Lewis base.

7. **Assertion (A) :** Ionic hydrides are the ready-made source of H_2.

   Reason (R) : LiH is an ionic hydride.

8. **Assertion (A) :** H_2O_2 decomposes Na_2CO_3 to give CO_2.

   Reason (R) : H_2O_2 has two protons and is thus a strong acid.

9. **Assertion (A) :** When blood is added to a solution of H_2O_2, the solution bubbles furiously.

   Reason (R) : Catalase (an enzyme) present in blood decomposes H_2O_2 and produces bubbles of O_2.

10. **Assertion (A) :** KO_2 is used as life supports in space crafts.

    Reason (R) : KO_2 absorbs CO_2 and releases O_2.

True & False

1. KO_2 is used as life supports in space crafts, submarines and emergency breathing apparatus since it absorbs CO_2 and releases O_2.
2. O_2 is diamagnetic and KO_2 is paramagnetic.
3. Peroxides and oxides are decomposed by H_2O forming O_2 and their behaviour becomes alkaline.
4. Due to very small size of H^+, it has very high polarising power and therefore, distorts the electron-cloud on other atoms.
5. H^+ exists as $H_9O_4^+$.
6. Ionic size of H, H^+ and H^- are equal.
7. Ionisation constant of CH_3COOH and CH_3COOD are same.
8. Spin isomerism is observed in H_2.
9. On electrolysis of $HCOONa(aq)$, H_2 is obtained at cathode and anode both.
10. Ionic hydrides conduct electricity in fused state.
11. Hydride is a conjugate acid of H_2.
12. Hydride is a Lewis base.
13. H_2O_2 is "10 volume" if 1 mL gives 10 mol O_2.
14. H_2O_2 behaves as an acid, oxidising agent and reducing agent.
15. H_2O_2 is used as fuel for rockets.
16. H_2O_2 oxidises Cr^{3+} to CrO_4^{2-} in basic medium.
17. True structure of H_2O_2 is $\begin{matrix} H \\ H \end{matrix}\!\!>\!O \longrightarrow O$.
18. Catalase present in blood interacts with H_2O_2 forming O_2.
19. Density of H_2O is not affected by change in temperature.
20. Tritium is the radioactive isotope of hydrogen.

Fill in the Blanks

1. Aqueous solution of .... on electrolysis gives H_2 at anode and cathode both.
2. .... and .... are two spin isomers of H_2.
3. At absolute zero there is only .... form of H_2.
4. Different atomic masses of three isotopes of hydrogen is due to the difference in the number of ....

5. Hydrogen resembles .... metals in the formation of .... cation and .... elements in the formation of .... anion.

6. Hydrogen is the .... most abundant element in the earth's crust.

7. Water gas contains CO and ....

8. Very pure H_2 is obtained by electrolysis of ....

9. H_2 is .... rapidly adsorbed on to the surfaces than D_2.

10. H_2 reacts over 13 times faster with Cl_2 than D_2, because H_2 has a .... energy of activation.

11. Auto-protolysis constant of H_2O is .... times more than that of D_2O.

12. Due to lower value of dielectric constant of D_2O than H_2O, ionic compounds are .... soluble in D_2O than in H_2O.

13. B.E. of O—D bond in D_2O is .... than that of O—H bond in H_2O.

14. ${}_3^6Li + {}_0^1n \longrightarrow {}_2^4He +$ ....

15. When water is electrolysed, H_2 is liberated much .... than D_2 and thus remaining water becomes enriched in ....

16. Tritium is widely used as radioactive ....

17. Considerable differences between the physical properties of *ortho* and *para* forms of H_2 is because of differences in their .... energy.

18. LiH, NaH, KH are some examples of .... hydrides.

19. Interstitial hydrides are made by .... metals.

20. $H_9O_4^+$ is .... H^+ ion and can be written as ....

Matrix-Match Type Questions

I. One or More Correct

1. Match the species in Column I with corresponding properties in Column II.

Column I	Column II
A. H_2O	1. Reducing agent
B. H_2O_2	2. Oxidising agent
C. Aqueous Na_2O	3. Paramagnetic
D. Aqueous KO_2	4. Diamagnetic
E. Aqueous CaH_2	5. Turns red litmus blue
	6. Turns blue litmus red

2. Match the species in Column I with the corresponding properties in Column II.

Column I	Column II
A. LiH	1. Reducing agent
B. H_2O	2. Lewis base
C. H_3O^+	3. Conjugate base
D. BH_3	4. Lewis acid
E. H	5. Lowry Bronsted acid
	6. Lowry Bronsted base

Integer Answer Types

This section contains 8 questions. The answer to each of the questions is a single digit integer, ranging from 0 to 9. The appropriate bubbles below the respectively question numbers in the *ORS* have to be darkened. For example, if the correct answers to question number *X, Y, Z* and *W* (say) are 6, 0, 9 and 2, respectively, then the correct darkening of bubbles will like the following

X	Y	Z	W
⓪	⓪	⓪	⓪
①	①	①	①
②	②	②	②
③	③	③	③
④	④	④	④
⑤	⑤	⑤	⑤
⑥	⑥	⑥	⑥
⑦	⑦	⑦	⑦
⑧	⑧	⑧	⑧
⑨	⑨	⑨	⑨

Set I

1. Number of neutrons in the heaviest isotopes of hydrogen is ....

2. Number of peroxy linkage (s) in $H_2S_2O_8$ is ....

3. Two moles of MnO_4^- reduces *x* mole(s) of H_2O_2 in basic medium. *x* is ....

4. 68 g H_2O_2 will make .... equivalent(s).

5. H_2O_2 is "33.6" volume. Thus, is normality is ....

6. CrO_5 has peroxy linkage(s). *x* is ....

7. How many moles of H_2O_2 are oxidised by 2 moles of $KMnO_4$ in acidic medium?

8. 5.0 mL of H_2O_2 solution liberates 1.016 g of I_2 from an acidified KI solution. Thus, volume strength (at STP) is approximately ....

Set II

1. How many of the following make alkaline solution in water?

   O^{2-}, O_2^{2-}, O_2^-

2. How many of the following are oxidant?

   $H_2O, H_2O_2, KO_2, Na_2O_2, Na_2O, O_3$

3. Ionic product, $[H^+][HO_2^-]$ of H_2O_2 was found to be 1×10^{-12} at a given temperature. Thus, pH of H_2O_2 is ....

4. How many types of water can be formed out of $^1_1H, ^2_1H, ^3_1H$ and $^{16}_8O$?

5. H_2 is formed by electrolysis of aqueous brine solution. In an experiment 0.112 mL of H_2 gas is formed at NTP. What is the pH after electrolysis of brine solution?

6. How many of the following undergo disproportionation when dissolved in water?

   $Cl_2, F_2, Na_2O_2, KO_2, K_2O$

7. $^{14}_7N$ undergoes nuclear reaction

   $$^{14}_7N \xrightarrow{(n,\ T)}$$

   What is atomic number of the atom formed?

8. 5.6 g of a unsaturated hydrocarbon occupies 2.24 L at NTP. It also required 4.48 L of H_2 gas for its reduction. How many π bonds does it have?

Test Yourself on the Periodic Table

1. Look at the location of elements *A, B, C* and *D* in the following Periodic Table

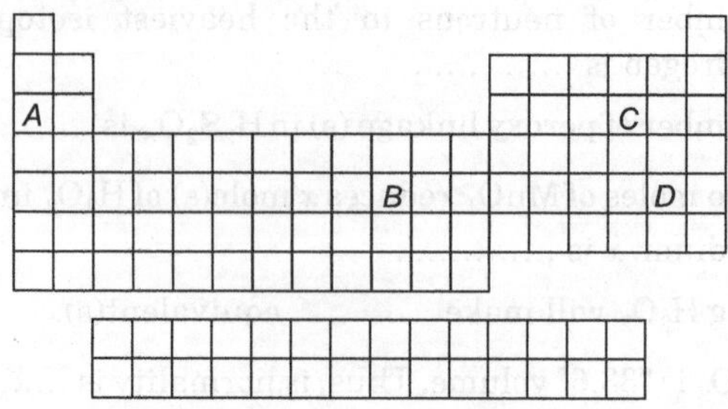

   (a) Write the formula of the simplest binary hydride of each element.
   (b) Classify each binary hydride as ionic, covalent or interstitial.
   (c) Which of these hydrides are molecular?
   (d) What is the oxidation state of the element in hydrides *A, C* and *D*?

2. Arrange the compounds of hydrogen with elements

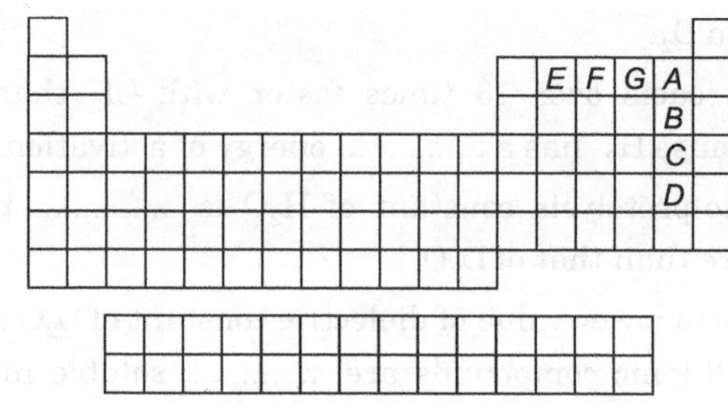

   (a) *A, B, C, D* in increasing acid strength
   (b) *A, B, C, D* in increasing boiling point
   (c) *E, F, G, A* in increasing H-bonding strength

3. In the Periodic Table given below

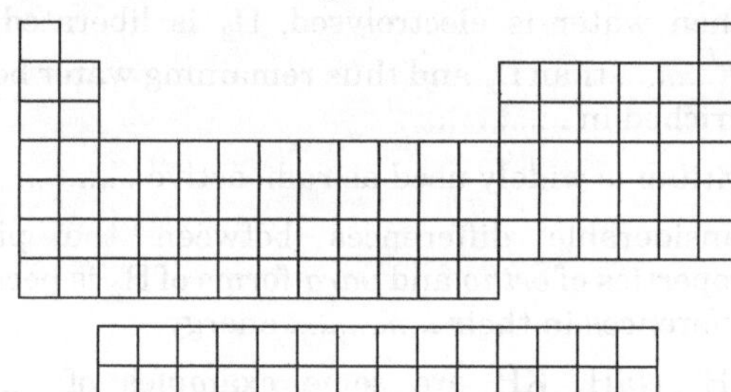

   trace the element(s)
   (a) that has maximum dissociation energy in diatomic state
   (b) that forms double bond in diatomic state
   (c) that forms triple bond in diatomic state
   (d) that forms compounds with +1 oxidation state with hydrogen

Answers & Solutions
Master Exercises

Exercise 1

Short Answer Type Questions

1. (c) $\overset{\delta+}{H}-\overset{\delta-}{F}\cdots\overset{\delta+}{H}-\overset{\delta-}{F}\cdots$

(d) $CH_3-C\begin{matrix}\overset{\delta-}{O}\cdots\overset{\delta+}{H}-O \\ O-H\cdots\overset{\delta-}{O}\end{matrix}C-CH_3$ (dimer)

(f) $CH_3-\overset{\delta-}{O}-\overset{\delta+}{H}\cdots\overset{\delta-}{O}(CH_3)-\overset{\delta+}{H}$

2. Hydrogen is a non-metal in its natural gaseous state as well as in liquid state. If a very thin layer of liquid hydrogen (about 0.5 mm) placed in between two single-crystal alumina is subjected to shock compression under a very high pressure (about 2 million atmosphere), then liquid hydrogen behaves like metal and its conducting power is very comparable to Rb and Cs. This transition of a non-metal to a metal under very high pressure is called metallisation of hydrogen.

This phenomenon is taking place at the planet Jupiter.

3. Hydrides : LiH, BeH_2, BH_3

4. In aqueous solution, hydrides are decomposed by H_2O forming which show reducing nature.

$$CaH_2 + 2H_2O \longrightarrow Ca(OH)_2 + H_2$$

5. (a) boranes (b) silanes (c) alkanes (d) $LiAlH_4$

6. A: HCOONa

B: $\begin{matrix}COONa \\ | \\ COONa\end{matrix}$ oxalate decolourises MnO_4^-/H^+

$$C_2O_4^{2-} \longrightarrow 2CO_2 + 2e^-$$
$$MnO_4^- + 8H^+ + 5e^- \longrightarrow Mn^{2+} + 4H_2O$$

C: HCOOH HCOOH reduces Tollen's reagent

D: $\begin{matrix}COOH \\ | \\ COOH\end{matrix}$

$$HCOOH + \underset{\text{Tollen's reagent}}{Ag_2O} \xrightarrow{\Delta} 2Ag + H_2O + CO_2$$

7. (a) $SiCl_4 + Li[AlH_4] \longrightarrow SiH_4 + AlCl_3 + HCl$

(b) $4AgNO_3 + SiH_4 \longrightarrow 4Ag + 4HNO_3 + Si$

(c) $3CuSO_4 + 2PH_3 \longrightarrow Cu_3P_2 + 3H_2SO_4$

(d) $SiH_4 + 4Cl_2 \longrightarrow SiCl_4 + 4HCl$

(e) $Al_4C_3 + 12H_2O \longrightarrow 4Al(OH)_3 + 3CH_4$

8.

S. no.	Metallic hydride	Property	Used as
1.	ZrH_2	Low thermal neutron capture cross-section area	Moderation in nuclear reaction
2.	TiH_2	High neutron-capture cross-section area	Shielding material in nuclear reaction

Analytical Questions

1. $H_2O_2 = 1$ g (impure) $= \dfrac{x}{100}$ g pure

$$\text{Equivalent of } H_2O_2 = \frac{x}{100} \times \frac{1}{17}$$

Let normality of $KMnO_4 = N$

$\therefore$ Equivalent of $KMnO_4 = \dfrac{Nx}{1000}$

$\therefore$ $\dfrac{Nx}{1000} = \dfrac{x}{1700}$

$\therefore$ $N = \dfrac{10}{17}$

2. Let normality of $H_2O_2 = N$

$$\text{Equivalent of } H_2O_2 = \frac{5N}{1000} = \frac{N}{200}$$

$$\text{Equivalent of } I_2 = \frac{0.508}{127}$$

$$\text{Equivalent of } H_2O_2 = \frac{0.508}{127}$$

$\therefore$ $\dfrac{N}{200} = \dfrac{0.508}{127} \Rightarrow N = 0.8\text{ N } H_2O_2$

If volume strength $= x$

then $N = \dfrac{x}{5.6}$

$\therefore$ $x = N \times 5.6$
$= 0.8 \times 5.6 = 4.48$ volume

3. (A): H_2O_2; (B): H_2O

4. $2Cr(OH)_4^- + 3H_2O_2 + 2OH^- \longrightarrow \underset{\text{yellow}}{2CrO_4^{2-}} + 8H_2O$

$$2CrO_4^{2-} + 2H^+ \longrightarrow \underset{\text{orange}}{Cr_2O_7^{2-}} + H_2O$$

5. pH = 5.91

$\therefore$ $[H^+] = 10^{-5.91} = 10^{-6} \times 10^{0.09} = 1.23 \times 10^{-6}$ M

$$[HO_2^-] = 1.23 \times 10^{-6}\ M$$

$$\therefore \quad K(H_2O_2) = [H^+][HO_2^-] = (1.23 \times 10^{-6})^2 = \mathbf{1.51 \times 10^{-12}\ M^2}$$

6. React H_2O with a reducing agent to produce H_2

$$xH_2O + 2M \longrightarrow M_2O_x + xH_2$$

(b) Al and (d) Ca could be used

7. (a), (b), (d) are different kinds of water and have similar properties.

(c) and (e) are different kinds of hydrogen peroxides and have similar properties.

Properties of (a), (b) and (d) are quite different from those of (c) and (e).

8. (a)

$$\underset{\substack{8g\\1\ mol}}{LiH} + H_2O \longrightarrow Li(OH) + \underset{\substack{2g\\1\ mol}}{H_2}$$

$$\underset{\substack{42g\\1\ mol}}{CaH_2} + 2H_2O \longrightarrow Ca(OH)_2 + \underset{\substack{4\ g\\2\ mol}}{2H_2}$$

8 g LiH gives = 2 g H_2

1 g LiH gives = 0.25 g H_2

42 g CaH_2 gives = 4 g H_2

1 g CaH_2 = 0.095 g.

Thus, LiH

(b)

$$pV = nRT$$

$$n(H_2) = \frac{pV}{RT} = \frac{150 \times 100}{0.0821 \times 298} = 613.1\ mol$$

2 mol H_2 is obtained from = 1 mol CaH_2 = 42 g CaH_2

$\therefore$ 613.1 mol H_2 is obtained from $= \frac{42}{2} \times 613.1$ g CaH_2

$= 12875.1$ g

$= 12.875$ kg

9. K is oxidised by water :

$$2K + 2H_2O \longrightarrow 2KOH + H_2$$

F_2 is reduced by water :

$$2F_2 + 2H_2O \longrightarrow 4HF + O_2$$

Cl_2 and Br_2 disproportionate when treated with H_2O

$$Cl_2 + H_2O \longrightarrow HCl + HOCl$$

10. Moles of $D_2 = \frac{pV}{RT} = \frac{2 \times 0.9}{0.0821 \times 298} = 0.0736$

$\therefore \quad H_2O = 0.0736$ mol $= 0.0736 \times 18$ g]

Actual water required with 0.015% D_2 and 80% recovery

$$= \frac{0.0736 \times 18}{\frac{0.015}{100} \times \frac{80}{100}}$$

$= 11040$ g

$= 11.040$ kg

11. A : CO; B : H_2; C : CO_2; D : $CaCO_3$; E : $Ca(HCO_3)_2$

12. A and B both are peroxo acids since, H_2O_2 is formed due to their reactions with H_2O.

$$HO-\overset{\overset{O}{\uparrow}}{\underset{\underset{O}{\downarrow}}{S}}-O-O-H + H_2O \longrightarrow HO-\overset{\overset{O}{\uparrow}}{\underset{\underset{O}{\downarrow}}{S}}-OH + H_2O_2$$

peroxy linkage

$$HO-\overset{\overset{O}{\uparrow}}{\underset{\underset{O}{\downarrow}}{S}}-O-O-\overset{\overset{O}{\uparrow}}{\underset{\underset{O}{\downarrow}}{S}}-OH + 2H_2O \longrightarrow 2HO-\overset{\overset{O}{\uparrow}}{\underset{\underset{O}{\downarrow}}{S}}-OH + 2H_2O_2$$

Exercise 2

Only One Option Correct

1. (c)

Salt	Ionisation	At cathode	At anode
(a) NaH	$NaH \rightleftharpoons Na^+ + H^-$	$Na^+ + e^- \longrightarrow Na$; $Na + H_2O \longrightarrow NaOH + \frac{1}{2}H_2$	$H^+ + e^- \longrightarrow \frac{1}{2}H_2 + e^-$
(b) HCOONa	$HCOONa \rightleftharpoons HCOO^- + Na^+$	same	$HCOO^- \longrightarrow \frac{1}{2}H_2 + CO_2 + e^-$
(c) NaCl	$NaCl \rightleftharpoons Na^+ + Cl^-$	same	$Cl^- \longrightarrow \frac{1}{2}Cl_2 + e^-$
(d) LiH	$LiH \rightleftharpoons Li^+ + H^-$	$Li^+ + e^- \longrightarrow Li$; $Li + H_2O \longrightarrow LiOH + \frac{1}{2}H_2$	$H^- \longrightarrow \frac{1}{2}H_2 + e^-$

Thus, **(c)**

2. (a)

3. (c) Force of attraction is minimum thus most volatile

4. (a) *o*-nitrophenol has intramolecular H-bonding thus most volatile.

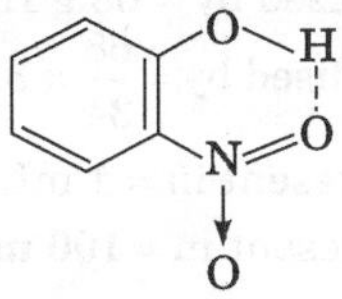

Thus, (a)

5. (d) **6.** (d)

7. (a) PbO_2 does not have peroxy linkage

Thus, (a)

8. (a) **9.** (b) **10.** (d) **11.** (c) **12.** (b)

13. (c) O_3 does not decolourise $KMnO_4$

14. (c) **15.** (d) **16.** (b) **17.** (a) **18.** (a)

19. (d)

20. (b) Temporary hardness is due to dissolved HCO_3^- of Ca^{2+} and Mg^{2+}

$$Ca(HCO_3)_2 + CaO \longrightarrow 2CaCO_3 + H_2O$$

Permanent hardness is due to dissolved Cl^- and SO_4^{2-} of Ca^{2+} and Mg^{2+}

$$CaCl_2 + Na_2CO_3 \longrightarrow CaCO_3\downarrow + 2NaCl$$

$$CaSO_4 + Na_2CO_3 \longrightarrow CaCO_3\downarrow + Na_2SO_4$$

21. (d) **22.** (c) **23.** (a) **24.** (d) **25.** (d)

26. (d) **27.** (c) **28.** (c)

29. (c)

$$\underset{-1}{H_2O_2} \longrightarrow \underset{-2}{H_2O} + \underset{0}{O_2}$$

reduction

oxidation

30. (a) **31.** (d) **32.** (a) **33.** (d)

34. (b) $FeCl_3$ solution is acidic due to hydrolysis when Zn is added, nascent hydrogen is formed which reduces $FeCl_3$ to $FeCl_2$ and thus no reaction with $K_4[Fe(CN)_6]$

$$FeCl_3 + K_4[Fe(CN)_6] \longrightarrow \underset{\text{blue}}{KFe[Fe(CN)_6]} + 3KCl$$

$\downarrow$ H(Zn + HCl)

$$FeCl_2 + K_4[Fe(CN)_6] \longrightarrow \text{No colour}$$

$$FeCl_3 + 3H_2O \longrightarrow Fe(OH)_3 + 3HCl$$

$$Zn + 2HCl \longrightarrow ZnCl_2 + 2H$$

H_2 gas is not effective in reducing $FeCl_3$ to $FeCl_2$

Thus, (b)

35. (b) **36.** (c) **37.** (d) **38.** (d) **39.** (d)

40. (b) **41.** (c) **42.** (a) **43.** (b) **44.** (a)

45. (d) **46.** (a) **47.** (c) **48.** (d) **49.** (c)

50. (a) **51.** (b) **52.** (a) **53.** (c) **54.** (a)

55. (b) **56.** (c) **57.** (c) **58.** (b) **59.** (d)

60. (a) **61.** (b) **62.** (b) **63.** (c) **64.** (b)

65. (d) **66.** (a) **67.** (d) **68.** (d) **69.** (c)

70. (d) **71.** (c) **72.** (d)

73. (b) Mass of one atom $= 6.66 \times 10^{-23}$ g

Mass of N_0 atoms $= 6.66 \times 10^{-23} \times 6.02 \times 10^{23}$ g

$= 40$ g

Thus, atomic weight of the element = 40

Element	%	%/at wt.	Ratio
M	95.24	2.381	1
H	4.76	4.76	2

Thus, hydride is MH_2

Thus, **(b)**

74. (b) **75.** (a) **76.** (a) **77.** (a)

78. (b) $Mg + H_2O \xrightarrow{\Delta} Mg(OH)_2 + H_2$, other are for H_2O_2

79. (c)

80. (a) $BaO_2 + CO_2 + H_2O \longrightarrow BaCO_3 + H_2O_2$

81. (a) **82.** (d) **83.** (c) **84.** (c) **85.** (d)

86. (b) **87.** (d) **88.** (c) **89.** (d)

90. (b) "20 volume" means

1 mL H_2O_2 gives = 20 mL O_2 at NTP

thus, 15 mL H_2O_2 gives = 300 mL O_2 at NTP

91. (b) **92.** (b) **93.** (d) **94.** (c) **95.** (c)

96. (d) **97.** (d) **98.** (b) **99.** (c) **100.** (d)

101. (b) **102.** (b) **103.** (d) **104.** (b) **105.** (c)

106. (d) **107.** (d) **108.** (b) **109.** (c) **110.** (d)

One or More Than One Options Correct

1. (a, b, c, d) **2.** (a) **3.** (c) **4.** (a,b,d) **5.** (a,d)

6. (c) **7.** (a, b) **8.** (a, b, c) **9.** (a) **10.** (a)

11. (d) **12.** (b) **13.** (d) **14.** (a, b) **15.** (a)

16. (a, b, c) **17.** (a, b) **18.** (a, b, c, d) **19.** (b)

20. (a, d)

Brain Twisters

1. $\underset{139\text{ g}}{BaH_2} + 2H_2O \longrightarrow Ba(OH)_2 + \underset{2g\,=\,1\text{ mol}}{H_2}$

$$pV = nRT$$

$$n = \frac{pV}{RT} = \frac{0.975 \times 4.36}{0.0821 \times 293}$$

$$= 0.1767 \text{ mol}$$

1 mol H_2 is obtained from = 139 g BaH_2

Hence, 0.1767 mol H_2 is obtained from $= 139 \times 0.1767$

$= 24.56$ g BaH_2

2. $X + H_2 \longrightarrow Y$

$$\underset{}{Y} \longrightarrow \underset{1.00\text{ g}}{X} + \underset{559\text{ mL at STP}}{H_2}$$

$$22400 \text{ mL } H_2 \text{ at STP} = 2\text{g } H_2$$

$$\therefore \quad 559 \text{ mL } H_2 \text{ at STP} = \frac{2}{22400} \times 559 = 0.05 \text{ g } H_2$$

Also 559 mL H_2 combines with = 1 g X

$\therefore$ 22400 mL H_2 combines with = 40 g X

Thus, atomic mass of X = 40 g mol^{-1}

Thus, X is **Ca**

Molecular formula of Y

Element	Amount	%	Molar ratio	Simplest ratio
Ca	1 g	95.24%	2.38	1
H	0.05 g	4.76%	4.76	2

Thus, Y is **CaH_2**

Molecular formula of Z

Element	%	Molar ratio	Simplest ratio
Cl	63.89%	1.80	2
Ca	36.11%	0.90	1

Thus, Z is **$CaCl_2$**.

3. $2H^+ + H_2O_2 + Sn^{2+} \longrightarrow Sn^{4+} + 2H_2O$ ...(i)

$2H_2O_2 \longrightarrow 2H_2O + O_2$ ...(ii)

$$200 \text{ g } H_2O_2 \text{ sample} = 200 \times \frac{10}{100}$$

$= 20$ g pure H_2O_2

$= 0.588$ mol pure H_2O_2

and 100 mL of 2M Sn^{2+} = 0.2 mol Sn^{2+}

By reaction (i) :

0.2 mol Sn^{2+} reduces = 0.2 mol H_2O_2

hence unreacted H_2O_2 = 0.588 – 0.200

= 0.388 mol

By reaction (ii) :

$$O_2 \text{ obtained from } 0.388 \text{ mol } H_2O_2 = \frac{0.388}{2} = 0.194 \text{ mol}$$

$$\therefore \text{Volume of } O_2 \text{ gas} = \frac{nRT}{p} = \frac{0.194 \times 0.0821 \times 293}{1}$$

$= \mathbf{4.67\ L}$

4. H_2O_2 is marked "22.4 volume" which means 1 mL H_2O_2 at NTP gives 22.4 mL of O_2 at NTP.

$$\underset{(2\text{ mol} = 68\text{ g})}{2H_2O_2} \longrightarrow 2H_2O + \underset{22400\text{ mL at NTP}}{O_2}$$

22400 mL of O_2 is from = 68 g H_2O_2

$$22.4 \text{ mL of } O_2 \text{ is from} = \frac{68}{22400} \times 22.4 \text{ g } H_2O_2 = 0.068 \text{ g}$$

This amount is present in 1 mL H_2O_2 solution.

$$\underset{34\text{ g}}{H_2S} + \underset{68\text{ g}}{H_2O_2} \longrightarrow 2H_2O + S$$

34 g H_2S is oxidised by = 68 g H_2O_2

$$3.4 \text{ g } H_2S \text{ is oxidised by} = \frac{68}{34} \times 3.4 = \mathbf{6.8\ g\ H_2O_2}$$

0.068 g H_2O_2 is present in = 1 mL

$\therefore$ 6.8 g H_2O_2 is present in = **100 mL**

$$\text{Or} \quad \text{density of "}x\text{ volume" } H_2O_2 = \frac{68\,x}{22400} \text{ g/mL}$$

$$\text{density of "22.4 volume" } H_2O_2 = \frac{68 \times 22.4}{22400} = 0.068 \text{ g/mL}$$

$\therefore$ 6.8 g H_2O_2 is in = **100 mL**

5. $2MnO_4^- + 5H_2O_2 + 6H^+ \longrightarrow 2Mn^{2+} + 8H_2O + 5O_2$

1000 mL of 1N MnO_4^- ≡ 1000 mL of 1N H_2O_2

Or equivalent of H_2O_2 = equivalent of MnO_4^-

$$= \frac{39.3 \times 0.5045}{1000} = 0.0198$$

$\therefore$ Amount of pure H_2O_2 = 0.0198 × 17 = 0.337 g

$$\% \text{ of } H_2O_2 \text{ in bleach solution} = \frac{0.337}{12.5} \times 100 = \mathbf{2.69\%}$$

6. Alkaline-earth metal hydride has M^{2+} (divalent cation) and H^-, hence, it can be represented as MH_2.

$$MH_2 + 2H_2O \xrightarrow{HCl} M(OH)_2 + 2H_2$$

Number of moles of H_2 collected

$$= \frac{pV}{RT} = \frac{\frac{750}{760} \times 1.0}{0.0821 \times 293}$$

= 0.041 mol

$$\text{Thus, number of moles of } MH_2 = \frac{\text{moles of } H_2}{2} = \frac{0.041}{2}$$

= 0.0205 mol

Let molar mass of M be = M

then molar mass of MH_2 = $(M + 2)$

$$\text{given number of moles } MH_2 = \frac{1.84}{M+2}$$

$$\therefore \quad \frac{1.84}{M+2} = 0.0205$$

$$M + 2 = \frac{1.84}{0.0205} = 89.7$$

$\therefore$ $M = 87.7$

Thus, M is **Sr** and hydride is **SrH_2**.

7. $$0.508 \text{ g } I_2 = \frac{0.508}{127} \text{ equivalent } I_2$$

$$= \frac{0.508}{127} \text{ equivalent } H_2O_2 \text{ in 5 mL solution}$$

$$\therefore \quad N(H_2O_2) = \frac{0.508}{127} \times \frac{1000}{5} = 0.8 \text{ N}$$

Volume strength = $N \times 5.6$

$= 0.8 \times 5.6$

$= 4.48$ volume

Passage Comprehension Questions

Passage 1

1. (a) **2.** (a,d)

Passage 2

1. (c) **2.** (d) **3.** (a)

Assertion & Reason

1. (a) **2.** (c) **3.** (a) **4.** (e) **5.** (b) **6.** (b)
7. (b) **8.** (c) **9.** (a) **10.** (a)

True & False

1. T **2.** F **3.** T **4.** T **5.** T **6.** F
7. F **8.** T **9.** T **10.** T **11.** F **12.** T
13. F **14.** T **15.** T **16.** T **17.** F **18.** T
19. F **20.** T

Fill in the Blanks

1. NaH or HCOONa **2.** *Ortho* and *para* **3.** *para*
4. neutrons **5.** alkali, monovalent, Group VII, monovalent
6. tenth **7.** H_2 **8.** H_2O(acidified) or *aq.* KOH **9.** more
10. lower
11. 3.33 $[K_w(H_2O) = 1 \times 10^{-14};\ K_w(D_2O) = 3 \times 10^{-15}]$ **12.** less
13. greater **14.** ${}^3_1T$ **15.** faster, D_2O **16.** tracer
17. internal **18.** ionic **19.** transition
20. hydrated, $H(H_2O)_4^+$

Matrix-Match Type Questions

I. One or More Correct

1. (A) — (4); (B) — (1,2,4,5); (C) — (4,5); (D) — (2,3,5) (E) — (1,4,5)

2. (A) — (1,2,3); (B) — (2,5,6); (C) — (3,6); (D) — (1,4); (E) — (1)

Integer Answer Types

Set I

Questions →	1	2	3	4	5	6	7	8
Answers →	2	1	3	4	6	2	5	9
	0	0	0	0	0	0	0	0
	1	1	1	1	1	1	1	1
	2	2	2	2	2	2	2	2
	3	3	3	3	3	3	3	3
	4	4	4	4	4	4	4	4
	5	5	5	5	5	5	5	5
	6	6	6	6	6	6	6	6
	7	7	7	7	7	7	7	7
	8	8	8	8	8	8	8	8
	9	9	9	9	9	9	9	9

Set II

Questions →	1	2	3	4	5	6	7	8
Answers →	3	4	6	6	9	4	6	2
	0	0	0	0	0	0	0	0
	1	1	1	1	1	1	1	1
	2	2	2	2	2	2	2	2
	3	3	3	3	3	3	3	3
	4	4	4	4	4	4	4	4
	5	5	5	5	5	5	5	5
	6	6	6	6	6	6	6	6
	7	7	7	7	7	7	7	7
	8	8	8	8	8	8	8	8
	9	9	9	9	9	9	9	9

Test Yourself on the Periodic Table

1. (a) **A** : NaH;
B : PdH_x;
C : H_2S;
D : HI
(b) NaH — Ionic; PdH_x — interstitial
H_2S and HI—covalent
(c) H_2S and HI—molecular
NaH and PdH_x—three dimensional crystals.
(d) Na = +1; S = –2; I = –1

2. (a) hydrogen compounds with *A*, *B*, *C* and *D* are

$$\xrightarrow[\text{increasing acid strength}]{\text{HF} < \text{HCl} < \text{HBr} < \text{HI}}$$

(b) HCl < HBr < HI < HF (increasing boiling point)
(c) C--- H < N---H < O---H < F---H
(A = F, B = Cl, C = Br, D = I, E = C, F = N, G = O)

3.

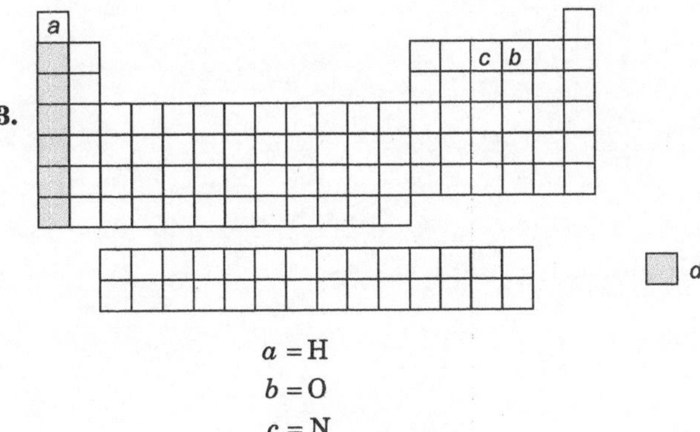

a = H
b = O
c = N
d = alkali metals

With Group 3 (boron family), we see the first evidence of a change from metallic behaviour of the elements at the left side of the Periodic Table to non-metal behaviour on the right side of the Table. Boron is a metalloid, whereas all the other elements of this group are metals.

11

p-Block Elements —Boron Family (Group 13)

Learning Framework of the Chapter ...

- Group 13(IIIA) (The Boron Family)
- Compounds of Boron
- Compounds of Aluminium

Shells up to $(n-1)$ are completely filled and differentiating electron (last filling electron) enters into *np*-suborbits — elements constitute what we call *p*-block elements.

Group → Period ↓	13 (IIIA)	14 (IVA)	15 (VA)	16 (VIA)	17 (VIIA)	18 (VIIIA)
2	B	C	N	O	F	Ne
3	Al	Si	P	S	Cl	Ar
4	Ga	Ge	As	Se	Br	Kr
5	In	Sn	Sb	Te	I	Xe
6	Tl	Pb	Bi	Po	At	Rn
	ns^2np^1	ns^2np^2	ns^2np^3	ns^2np^4	ns^2np^5	ns^2np^6

- The *p*-block elements are characterised by the ns^2np^x valence shell electronic configuration where x varies from 1 to 6 and n from 2 to 6 but inner core for different elements may be different.
- The highest oxidation state = (group number − 10) Stability of this oxidation state (OS) decreases and that of [(OS) − 2] state increases as we move down the group—a **case of inert pair effect.** This is shown below in Table 11.1
- Trends in properties of *p*-block elements have been represented in Fig. 11.1

Table 11.1 **Common Oxidation States of *p*-Block Elements**

Group	13	14	15	16	17	18
Oxidation state	B +3	C +4, −4	N +5 to −3	O −1, −2	F −1	
	Al +3	Si +4	P, As +3, +5, −3	S, Se, Te −2, +2, +4, +6	Cl, Br, I, −1, +1, +3 +5, +7	Xe +2, +4, +6, +8
	Ga, In, Tl +3, +1	Ge, Sn, Pb +4, +2	Sb, Bi +3, +5			

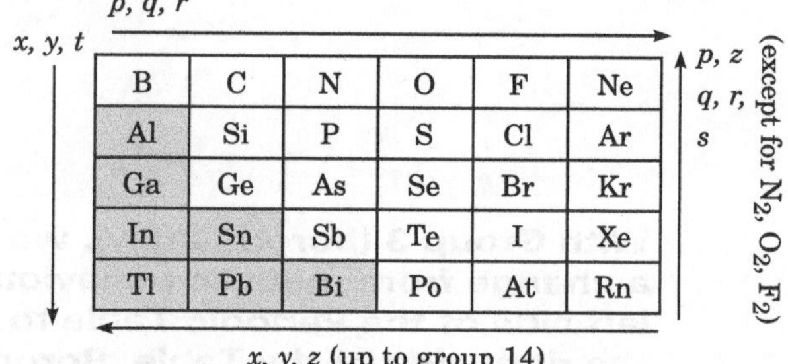

where ... represents
x : covalent/van der Waals' radius
y : metallic character
z : heat of sublimation
p : electronegativity
q : ionisation energy (N > O, P > S)
r : oxidising power
s : stability of higher oxidation state within a group
t : stability of lower oxidation state within a group

Fig. 11.1 Trends in properties of *p*-Block elements in the direction of arrow

The van Arkel-Ketelaar Triangle

Before looking at the compounds of the *p*-block elements in detail, it will be useful to consider the general types of compound we will expect and this, as we shall see, can often be rationalized on the basis of electronegativities, χ, and electronegativity differences, $\Delta\chi$. **Thus, compounds with a large $\Delta\chi$ will tend to be ionic and those with medium $\Delta\chi$ are likely to form polar covalent bonds and therefore, have polymeric or macromolecular structures.** For compounds with a small $\Delta\chi$, the type of compound will depend on whether the elements have a low or high electronegativity. In the former case, the compounds are likely to be metallic for much the same reason that elements with low electronegativity are metals. In the latter case, we shall expect relatively non-polar molecular species.

DEEP Focus

A useful way to visualize these ideas is with the so called **van Arkel-Ketelaar or element triangle, shown in Fig. 11.2 in which the vertices are labelled metallic,** covalent and ionic in such a way that the vertical and horizontal axes represent electronegativity, difference, $\Delta\chi$, and absolute electronegativity, $\Sigma\chi$, respectively and are defined such that, for a molecule *AB*, $\Delta\chi = \chi_B - \chi_A$ and $\Sigma\chi = (\chi_A + \chi_B)/2$.

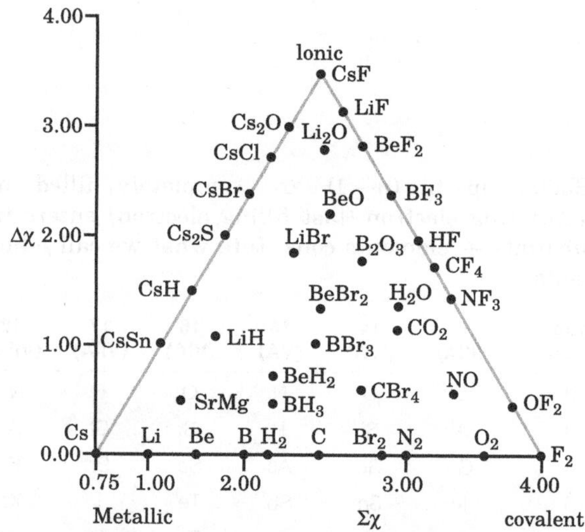

Fig. 11.2 A van Arkel-Ketalaar triangle showing the position of selected compounds and elements

The metallic-covalent edge of the van Arkel-Ketelaar triangle comprises the elements themselves, since $\Delta\chi$ is zero, and ranges from metals on the left to covalent species such as F_2 on the right.

The ionic-covelent edge is characterized by decreasing $\Delta\chi$ which is well illustrated with the fluorides of the elements. Thus, CsF and LiF are ionic with large $\Delta\chi$ whereas NF_3 and CF_4 both have small $\Delta\chi$ and are covalent; compounds with intermediate $\Delta\chi$ such as BeF_2 are polymeric or macromlecular solids.

Group 13 (IIIA) (The Boron Family)

This group marks the beginning of the *p*-block elements. Table 11.2 summarises the properties of this group, and trends shown in Fig. 11.3.

DEEP Focus

Elements of group 13 are less metallic than those of groups 1 and 2. Within the group, there is a variation in metallic character. Boron, the smallest element in this group is a non-metal. The other elements in this group are **fairly reactive metals.**

The elements of group 13 have smaller atomic radii and higher electronegativities as compared to *s*-block elements of the same period. Since, the size an the ionisation energies of atoms depend on effective nuclear charge of the atom, insertion of transition elements result in higher effective nuclear charge of the fourth row elements Ga, Ge, etc., than expected by simple extrapolation from the second and third row elements. In other words, **the nuclei of these fourth row elements attract electrons more strongly than expected and this affects their properties. Thus, the atomic size of gallium is smaller; its electronegativity and ionisation energies higher than expected.** Atoms with d^{10} inner shell, in general, are smaller and have higher ionisation energies. In a similar way, the inclusion of fourteen electrons in $4f$-orbital further affects the size and ionisation energy of Tl. As a result of this, irregularities in atomic radii, electronegativity and ionisation energies are seen from B to Tl.

Table 11.2 shows that densities of these elements show a gradual increase while melting and boiling points fall in value. **The exceptionally low melting point of gallium (303 K), however, has, so far, no simple explanation.**

From Table 11.2 we observe that these elements form small ions of high charge density and the value of the sum of their first, second and third ionisation energies is very high. These properties lead us to the conclusion that these elements will prefer to form covalent rather than ionic compounds. **Boron is always covalent and does not form B^{3+} ions, because the energy required to remove three electrons is very high. Many simple compounds of Al and Ga like $AlCl_3$ and $GaCl_3$ are also covalent when anhydrous. However, in solution, the large amount of hydration energy evolved compensates the high ionisation energy and all the metal ions exist in a hydrated state.**

This can be explained with the help of **Boron-Haber cycle as** given below.

	ΔH (kJ)
$AlCl_3(s) \longrightarrow Al(s) + \frac{3}{2}Cl_2(g)$	704
$Al(s) \longrightarrow Al(g)$	326
$\frac{3}{2}Cl_2(g) \longrightarrow 3Cl(g)$	366
$Al(g) \longrightarrow Al^{3+}(g) + 3e^-$	5137
$3Cl(g) + 3e^- \longrightarrow 3Cl^-(g)$	–1044
$Al^{3+}(g) + aq \longrightarrow Al^{3+}(aq)$	–4630
$3Cl^-(g) + aq \longrightarrow 3Cl^-(aq)$	–1155
$AlCl_3(s) + aq \longrightarrow Al^{3+}(aq) + 3Cl^-(aq)$	

$\Delta H_{soln} = -$ **296 kJ mol^{-1}**

Born-Haber cycle for formation of Al^{3+} (*aq*) ion

Thus, despite the large ionisation energies of aluminium, the enthalpy of solution of $AlCl_3$ has a large negative value. Therefore, in aqueous solution, $AlCl_3$ exsits as $Al^{3+}(aq)$ and $Cl^-(aq)$. BCl_3 hydrolyses in aqueous solution to give boric acid and not $B^{3+}(aq)$

$$BCl_3 + 3H_2O \longrightarrow H_3BO_3 + 3HCl$$

This group is the most metallic in nature of the *p*-block elements and only boron is non-metallic.

Boron and aluminium show an oxidation state of +3 in their compounds while gallium, indium and thallium also show +1 oxidation state (due to inert pair effect).

- Both boron and aluminium are amphoteric.

$$2B(s) + 2OH^-(aq) + 2H_2O(l) \longrightarrow \underset{\text{metaborate ion}}{2BO_2^-(aq)} + 3H_2(g)$$

$$2Al(s) + 2OH^-(aq) + 2H_2O(l) \longrightarrow \underset{\text{metaaluminate ion}}{2AlO_2^-(aq)} + 3H_2(g)$$

but the ions are better represented as complex ions.

$$[Al(OH)_4(H_2O)_2]^- \text{ or } [Al(OH)_6]^{3-}$$

- B and Al show +3 oxidation state. Ga, In and Tl show +3 and +1 oxidation state but stability order is

$$Ga^{3+} > In^{3+} > Tl^{3+}$$
$$Ga^+ < In^+ < Tl^+$$

due to **inert-pair effect.**

Thus, Ga^+ salts tend to stabilise by changing to Ga^{3+} ($Ga^{3+} > Ga^+$) and thus are better reducing agents.

$$Ga^+ \longrightarrow Ga^{3+} + 2e^-$$

On the other hand, Tl^{3+} salts tend to stabilise by changing to Tl^+($Tl^+ > Tl^{3+}$) and thus are better oxidising agents.

$$Tl^{3+} + 2e^- \longrightarrow Tl^+$$

Ga*X*, In*X* and Tl*X* (*X* = Cl, Br, I) exist. Ga*X* and In*X* disproportionate in water

$$3GaX(s) \longrightarrow Ga(s) + Ga^{3+}(aq) + 3X^-(aq)$$

All the above reaction are spontaneous and $\Delta G° < 0$

Table 11.2 **Physical Properties of the Elements of Group 13 (IIIA)**

Property	B	Al	Ga	In	Tl
Electronic structure	[He] $2s^2\ p^1$	[Ne] $3s^2\ 3p^1$	[Ar] $3d^{10}\ 4s^2\ 4p^1$	... $5s^2 5p^1$	... $6s^2 6p^1$
Electronegativity (Pauling)	2.0	1.5	1.6	1.7	1.8
$(IE)_1$ kJ mol^{-1}	801	577	579	558	589
$(IE)_2$ kJ mol^{-1}	2427	1816	1979	1820	1971
$(IE)_3$ kJ mol^{-1}	3659	2744	2962	2704	2877
m.p./°C	2180	660	30	157	303
b.p./°C	3650	2467	2403	2080	1457
Density ($10^3 \times$ kg m^{-3})	2.34	2.7	5.91	7.3	11.8
Metallic radius/pm	88.5	143	141	166	170
Ionic radius/pm M^{3+}	27.0	53.5	62	80	88.5
M^+	—	—	120	140	150
Principal coordination number	3, 4	3, 4, 6	3, 6	3, 6	3, 6
$E^\circ_{M^{3+}/M}$/V *	—	–1.66	–0.56	–0.38	+2.18
$E^\circ_{M^+/M}$/V	—	0.55	—	–0.18	–0.34

* For Al and Ga in acidic solutions, others in basic solution.

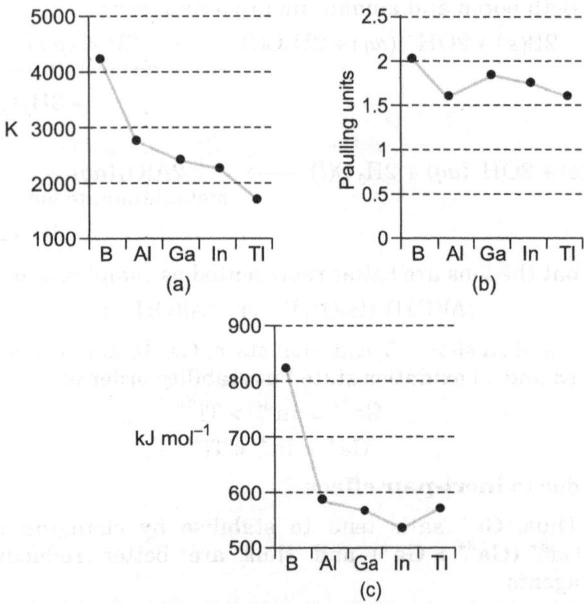

Fig. 11.3 Some group trends of the group 13 elements : (a) Boiling point, (b) Pauling electronegativity, and (c) The first ionisation energy.

The term **'inert pair effect'** is somewhat misleading. The decrease stability of the higher oxidation state (+3) with increasing atomic number arises because of the decrease in bond energy with size from Al to Tl. As a result, the energy required to unpair ns^2 electrons is not compensated by the energy released in forming the two additional bonds.

- Except boron, the other elements of group 13 (IIIA) show increasing metallic character from Al to Tl. Small size (atomic radius = 88.5 pm) and relatively high nuclear charge provide boron with non-metallic properties.
- Except boron, other exist in aqueous solution in tripositive (M^{3+}) state. $[M(H_2O)_6]^{3+}$, $[M(H_2O)_2(OH)_4]^-$ and $[M(OH)_4]^-$ exist in aqueous solution.
- Stability of hydrides (MH_3) decreases as we move down the group. AlH_3 is a solid and is polymerised *via* Al—H—Al bridging units. Hydrides are electron-deficients and thus act as Lewis acids (lone pair acceptor) and form adducts with strong Lewis bases.

$$\underset{\text{Lewis acid}}{AlH_3} + \underset{\text{Lewis base}}{H^-} \longrightarrow \underset{\text{adduct}}{AlH_4^-}$$

- They form halides of the type MX_3, all of which are known except Tl(III) iodide.
  - fluorides are ionic with high melting point
  - chlorides, bromides and iodides are covalent with low melting point
  - halides have halogen bridged dimeric structures as shown.
  - MX_3 are strong Lewis acids. Anhydrous $AlCl_3$ is used as a catalyst is Friedel-Craft's reaction

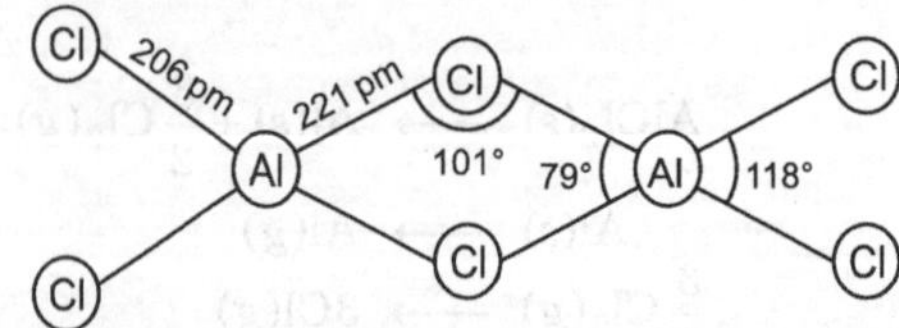

- They form oxides and hydroxides of the type M_2O_3 and $M(OH)_3$ and their basic nature increases as we move down the group. **However, Al_2O_3 and Ga_2O_3 (and their hydroxides) show amphoteric nature.**

$$\underset{\text{base}}{2Al(OH)_3(s)} + \underset{\text{acid}}{6HCl(aq)} \longrightarrow 2AlCl_3(aq) + 3H_2O(l)$$

$$\underset{\text{acid}}{Al(OH)_3(s)} + \underset{\text{base}}{NaOH(aq)} \longrightarrow NaAl(OH)_4(aq)$$

- They form octahedral aqua complexes of the type $[M(H_2O)_6]^{3+}$ (M = Al, Ga, In, Tl)
- Aluminium sulphate forms double salt with alkali sulphates of the type $MAl(SO_4)_2 \cdot 12H_2O$ ($M^+ = Na^+$ or K^+). These are called **alums.**
- $(IE)_1$ of boron is relatively high and $(IE)_2$ and $(IE)_3$ are still higher. Thus, the total energy required to give B^{3+} ions is far more than that which would be compensated by lattice energies of ionic compounds or hydration of such ions in solution. As a result, boron does not easily form a cation; however their covalent compounds are known.
- The melting point of B is very high which suggests a strong bonding between individual atoms in solid state and is used as a **semi-conductor.**

Reactions of Elements

Pure crystalline boron is very unreactive. However, it is attacked at high temperatures by strong oxidising agents such as a mixture of hot concentrated H_2SO_4 and HNO_3 or Na_2O_2.

$$2B(s) + 6HNO_3(aq) \longrightarrow 2H_3BO_3(aq) + 6NO_2(g)$$

In contrast, finely divided amorphous boron (containing about 2–5% impurities) is more reactive. Reactions are summarised in Table 11.3.

Table 11.3 **Reactions of Group 13 (IIIA) Elements**

Reaction	Comment
• $4M + 3O_2 \longrightarrow 2M_2O_3$	Reaction at high temperature; with Al—exothermic, a protective oxide layer is formed and Tl also forms Tl_2O.
• $2M + N_2 \longrightarrow 2MN$	Only Al and (B) form nitrides. BN, a slippery white solid with layer structure similar to graphite.
• $2M + 3X_2 \longrightarrow 2MX_3$	All form MX_3. Tl also form TlX. Iodide of Tl^{3+} (TlI_3) is not formed; but TlI reacts with I_2 forming TlI_3 ($Tl^+ I_3^-$).
• $2M + 2NaOH + 2H_2O \longrightarrow 2NaMO_2 + 3H_2$	B, Al and Ga only form $NaMO_2$.
• $2M + 6NH_3 \longrightarrow 2M(NH_2)_3 + 3H_2$ $2B + 2NH_3 \longrightarrow 2BN + 3H_2$	All form amides except B which forms BN.
• $2M + 6H^+ \longrightarrow 2M^{3+} + 3H_2$	Metals liberate H_2, but HNO_3 makes Al passive.

- Boron reacts directly with elements forming borides which are hard and refractory.

$$3Mg + 2B \longrightarrow Mg_3B_2$$

Borides are decomposed by acids forming boranes.

$$Mg_3B_2 + 6HCl \longrightarrow 3MgCl_2 + \underset{\text{diborane}}{B_2H_6}$$

DEEP Focus

- Aluminium is far more reactive than boron but in air and water it is stable. It is due to the formation of aluminium oxide, Al_2O_3 layer, that makes it passive for further attack. HNO_3 also makes Al passive by oxide layer formation.
  The layer is so useful that in industry it is purposely increased by an electrolytic process called **anodising**. This is done by electrolysing dilute H_2SO_4 with Al as anode. This produces much thicker layer of oxide on the surface (0.01 mm). *This layer can take up pigments; thus colouring the aluminium.*

- Reaction of Al with O_2 is exothermic and is called **thermite reaction.**

$$2Al(s) + \frac{3}{2}O_2(g) \longrightarrow Al_2O_3(s), \Delta H = -1670 \text{ kJ}$$

The thermite reaction involves so much energy that it can be dangerous. The Al becomes white hot and often causes fires. For this precise reason, mixture of Al and an oxide such as Fe_2O_3 or SiO_2 (a source of oxygen) were used to make incendiary bombs during World War II. Warships are sometimes made of aluminium alloys to reduce their weight. A thermite reaction can be started if the ship is hit by a missile. Thermite reaction also finds applications in the metallurgical extraction of other metals from their oxides :

$$8Al + 3Mn_3O_4 \longrightarrow 4Al_2O_3 + 9Mn$$
$$2Al + Cr_2O_3 \longrightarrow Al_2O_3 + 2Cr$$

In recent years, Al has been under suspicion as a possible course of **Alzheimer's** disease. This disease induces senility in relatively young men and women. The disease causes them to lose their memory.

- Most of the reactions of boron are at high temperatures.

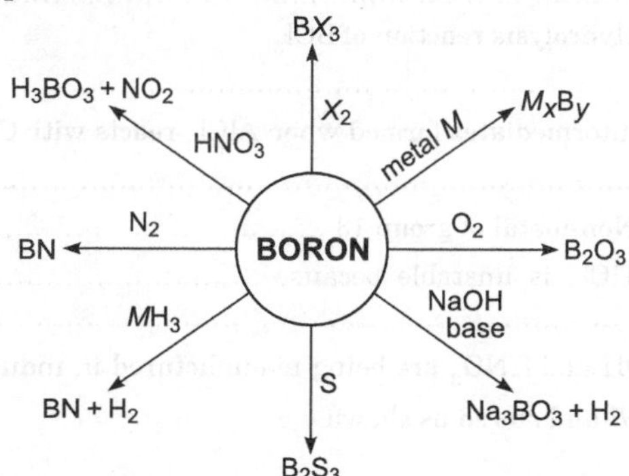

Differences between Boron and the other Elements

Boron because of its small size has high ionisation energy and electronegativity. It is always covalent and non-metallic. It differs other elements of its group in the following respects :

Boron	Other elements
• B_2O_3 is an acidic oxide (like SiO_2)	• Al_2O_3 is amphoteric
• H_3BO_3 is acidic	• $Al(OH)_3$ is amphoteric
• Simple borate and silicate ions can polymerise forming isopolyacids by sharing oxygen atoms	• Aluminates are formed but do not polymerise
• Boranes are gaseous compounds readily hydrolysed and spontaneously inflammable	• Aluminium hydride $(AlH_3)_n$ is a polymeric solid
• Halides are easily hydrolysed	• Halides are easily hydrolysed

Boron : Extraction

Amorphous boron (called Moisasan boron) is obtained by reduction of B_2O_3 with Na or Mg at a high temperature. B_2O_3 in turn is obtained by decomposing borax with acid and subsequent heating.

$$Na_2B_4O_7 \cdot 10H_2O \xrightarrow{acid} \underset{\text{orthoboric acid}}{H_3BO_3} \xrightarrow{\Delta} H_2O + B_2O_3$$

$$B_2O_3 + 6Na \xrightarrow{\Delta} 2B + 3Na_2O$$

$$B_2O_3 + 3Mg \xrightarrow{\Delta} 2B + 3MgO$$

Boron obtained is 95–98% pure and black in colour. Pure crystalline boron is obtained by

- reducing BCl_3 with H_2:

$$2BCl_3 + 3H_2 \xrightarrow[\text{filament}]{\text{red hot W or Ta}} 2B + 6HCl$$

- pyrolysis of BI_3

$$2BI_3 \xrightarrow[\text{(van Arkel method)}]{\text{red hot W or Ta filament}} 2B + 3I_2$$

- thermal decomposition of B_2H_6

$$B_2H_6 \xrightarrow{\Delta} 2B + 3H_2$$

Target Practice 1

1. *A* and *B* with valence electrons are represented below

$$:\dot{\underset{\bullet}{A}}\bullet \quad \times B \underset{\times}{\times}$$

What can be possible formula of the compound of *A* and *B*?

....

....

2. Write structure of Al_2Cl_6.

....

3. Do as directed

(a) Increasing order of stability of M^{3+} of Group 13 in 4th, 5th and 6th period.

....

(b) Hydrolysis reaction of BCl_3.

....

(c) Intermediates formed when $AlCl_3$ reacts with CH_3Cl.

....

(d) Non-metal of group 13 ....

(e) $TlCl_3$ is unstable because ....

....

4. NaOH and HNO_3 are being manufactured in industries *A* and *B* and stored as shown

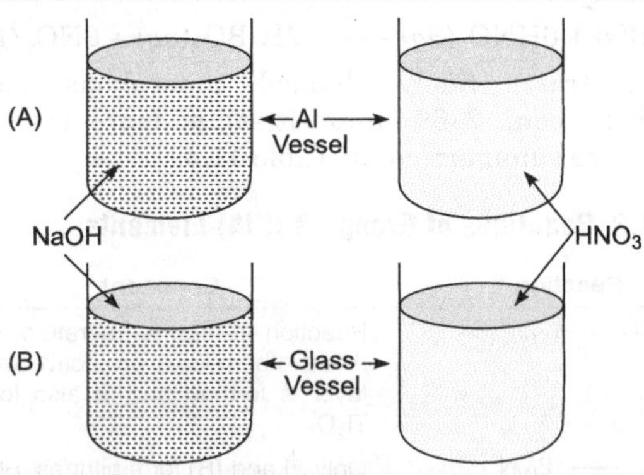

Comment ....

....

....

5. Disease due to Al metal is ....
This diseases causes ....

6. Write the nature of the following

(a) B_2O_3 ....

(b) Al_2O_3 ....

(c) $Al(OH)_3$ ....

(d) $B(OH)_3$ ....

Answers

1. Based on number of valence electrons, *A* is more electronegative than *B*. Thus, *BA*

2. 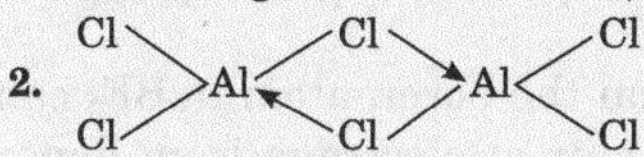

3. (a) $Tl^{3+} < In^{3+} < Ga^{3+}$

(b) $BCl_3 + 3H_2O \longrightarrow H_3BO_3 + 3HCl$

(c) $AlCl_3 + CH_3Cl \longrightarrow AlCl_4^{\ominus} + \underset{\text{carbocation}}{CH_3^{\oplus}}$

(d) B

(e) $TlCl_3$ change to TlCl due to greater stability of Tl^+.

$$Tl^{3+} + 2e^- \longrightarrow Tl^+$$

4. Storage method in industry *A* is not correct for NaOH since, NaOH reacts with Al.

$$2NaOH + 2Al \longrightarrow 2NaAlO_2 + 3H_2 + 2H_2O$$

However, HNO_3 can be stored in Al vessel since, with HNO_3, there is formation of oxide layer which makes Al passive for further reaction. Storage method in industry *B* is correct.

5. Alzheimer; it induces senility and causes loss of memory.

6. (a) B_2O_3 : Acidic
(b) Al_2O_3 : Amphoteric
(c) $Al(OH)_3$: Amphoteric
(d) $B(OH)_3$: Acidic

Practice Exercise 1

1. Third ionisation energy, $(IE)_3$ of group 13 (IIIA) is generally high. Explain.
2. Boron shows amphoteric nature. Illustrate with chemical reactions.
3. $TlCl_3$ is unstable while TlCl is stable. Explain.
4. Al^{3+} ion is hydrated and exists as $[Al(H_2O)_6]^{3+}$. It is said to behave as an acid. Explain by chemical equation.
5. *X* and *Y* have respectively 3 and 5 electrons in the valence shell. Write the formula of the compound formed from *X* and *Y*.
6. Identify (*A*), (*B*), (*C*) and (*D*) :

$$\text{Metal } (A) + N_2 \xrightarrow{\Delta} (B) \xrightarrow{H_2O} (C) + (D)$$

(*C*) (white ppt) dissolves in NaOH solution while (*D*) (gas) gives white fumes in HCl. (*A*) has atomic number 13.

7. Al is amphoteric, give chemical reactions to illustrate it.
8. The following reaction is exothermic :

$$2Al(s) + \frac{3}{2}O_2(g) \longrightarrow Al_2O_3(s), \qquad \Delta H = -1670 \text{ kJ}$$

What is this reaction called? Give two main applications.

9. HNO_3 can be stored in Al vessel but NaOH can't be. Explain.
10. Why can the thermite reaction not be stopped by water like an ordinary fire?

Answers

1. Due to increase in effective nuclear charge; this increases force of attractions hence *IE* increases; in general

$$(IE)_3 > (IE)_2 > (IE)_1$$

2. $2B + 2NaOH + 2H_2O \longrightarrow \underset{\text{meta borate}}{2NaBO_2} + 3H_2$

$$2B + 6HCl \longrightarrow BCl_3 + 3H_2$$

B reacts with NaOH (base) as well as with HCl (acid) hence B is amphoteric.

3. Due to 'inert pair' effect, stability of higher oxidation state decreases and that of lower oxidation state increases. Hence

$$Tl^{3+} < Tl^+ \text{ (stable)}$$

4. $\underset{\text{acid}}{[Al(H_2O)_6]^{3+}} + \underset{\text{base}}{H_2O} \longrightarrow \underset{\text{base}}{[Al(H_2O)_5(OH)]^{2+}} + \underset{\text{acid}}{H_3O^+}$

5. *XY*

6. (*A*) : Al (*B*) : AlN (*C*) : $Al(OH)_3$ (*D*) : NH_3

$$2Al + N_2 \longrightarrow 2AlN \quad (A) \; (B)$$

$$AlN + 3H_2O \longrightarrow \underset{(C)}{Al(OH)_3} + \underset{(D)}{NH_3}$$

$$\underset{\text{white ppt}}{Al(OH)_3} + NaOH \longrightarrow \underset{\text{soluble}}{NaAlO_2} + 2H_2O$$

$$NH_3 + HCl \longrightarrow \underset{\text{white fumes}}{NH_4Cl}$$

7. Refer Q. (2)
8. This is called thermite reaction. It is used (i) to reduce oxides into metals :

$$Cr_2O_3 + 2Al \longrightarrow Al_2O_3 + 2Cr$$
$$Fe_2O_3 + 2Al \longrightarrow Al_2O_3 + 2Fe$$

(ii) in welding.

9. Al turns passive with HNO_3 due to formation of protective oxide layer hence, reaction ceases. This makes Al-vessel suitable for storing HNO_3. Al dissolves in NaOH forming $NaAlO_2$ hence it can't be stored in Al-vessel:

$$2Al + 2NaOH + 2H_2O \longrightarrow 2NaAlO_2 + 3H_2$$

10. In an ordinary fire the reactions are kept going by the material reacting with oxygen in the air. Water not only cools the fire, but also keeps oxygen out. In the thermite reaction, the oxygen involved is in the metal oxide, so keeping atmospheric oxygen out has no effect. Also high temperature (about 1000°C) will allow reaction between Al and H_2O to give off H_2.

Compounds of Boron

Boron is the first member of group 13 (IIIA) of the Periodic Table and is the only non-metal of this group, having three valence electrons. Because of its small size, high ionisation energy and low electronegativity, boron forms covalent compounds and in this respect resembles carbon and silicon. Since, compounds formed are electron-deficients, they behave as Lewis acids (electron-pair acceptors)

Boron Halides

- Boron halides are covalent. It is due to the extremely small size of B^{3+} ion (27 pm) and can polarise the anion easily (smaller the size of cation, larger the size of anion, greater the charge ⇒ greater the polarising power hence, greater the covalent nature —Fajan's rule)
  - BF_3 is a gas
  - BCl_3 is a liquid
  - BI_3 is a solid
- When borates are treated with HF (or CaF_2 and concentrated H_2SO_4), the volatile compound BF_3 is formed. If the BF_3 gas produced is introduced into a bunsen flame, a characteristic green colouration is formed :

$$CaF_2 + \underset{\text{Conc}}{H_2SO_4} \longrightarrow CaSO_4 + 2HF$$

$$3HF + H_3BO_3 \longrightarrow BF_3 + 3H_2O$$

This can be used to test borates.

- BX_3 is electron-deficient (octet of B incomplete) and thus behaves as a Lewis acid. The shape of the BF_3 molecule is a planar triangle with bond angles of 120°.
- The bond lengths in BF_3 are 1.30Å and are shorter than the sum of the covalent radii (B = 0.80 Å and F = 0.72Å). The bond energy is very high (646 kJ mol^{-1}) and is higher than for any single bond. The shortness and strength of (B—X) bond is interpreted in terms of a $p\pi-p\pi$ interaction. The empty $2p_z$ atomic orbital on B (which is not involved in hybridisation) is perpendicular to the triangle containing the sp^2 hybrid orbitals. This $2p_z$ orbital may accept an electron pair from a full p_z orbital on any one of the three fluorine atoms. Thus a dative bond is formed and the B atom attains an octet of electrons. If one localised double bond existed, there would be one short bond and two longer ones. **But all B—F bonds are equivalent and based on modern explanation, the double bond is delocalised.**

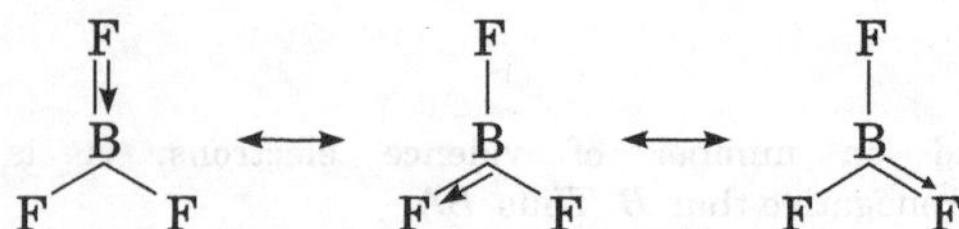

- The empty $2p_z$ orbitals on the boron atom in BF_3 can also be filled by a lone pair of electrons from donor molecules such as ether (Et_2O), NH_3, $(Me)_3N$ or F^- and tetrahedral molecule or ion is formed.

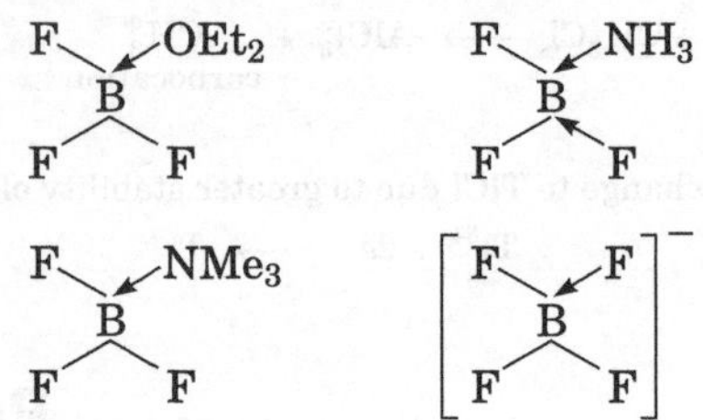

- Boron also forms halides of the formula B_2X_4. These decompose slowly at room temperature. B_2Cl_4 can be made as follows :

$$2BCl_3 + 2Hg \xrightarrow[\text{low pressure}]{\text{electric discharge}} B_2Cl_4 + Hg_2Cl_2$$

There is free rotation about the B—B bond, and in the gaseous and liquid states the molecule adopts a non-eclipsed conformation. In the solid state the molecule is planar, because of crystal forces and ease of packing.

$Cl_2B—BCl_2$ (non-eclipsed) $Cl_2B—BCl_2$ (planar)

non-eclipsed planar

- BF_3 hydrolyses incompletely and forms fluoroborates. This is because the HF first formed reacts with the H_3BO_3 :

$$4BF_3 + 12H_2O \longrightarrow 4H_3BO_3 + 12HF$$

$$12HF + 3H_3BO_3 \longrightarrow 3H^+ + 3[BF_4]^- + 9H_2O$$

$$\text{Net}: \; 4BF_3 + 3H_2O \longrightarrow H_3BO_3 + 3H^+ + 3[BF_4]^-$$

The other halides hydrolyse completely, giving boric acid.

$$BCl_3 + 3H_2O \longrightarrow H_3BO_3 + 3HCl$$

$CaF_2 + H_2SO_4 + H_3BO_3$ (or B_2O_3)
↓ Δ
BF_3 $\xrightarrow{H_2O}$ $H_3BO_3 + BF_4^-$
↓ 180°C NaH
B_2H_6

Boron Hydrides

- The boron hydrides are sometimes called **boranes** by analogy with the alkanes (hydrocarbons).
- There are two series :
  - B_nH_{n+4} called **nido-boranes,** (B_2H_6)
  - B_nH_{n+6} called **arachno-boranes,** (less stable), (B_4H_{10});
  - of all these, diborane is the simplest borane.

Source

$$Mg_3B_2 + H_3PO_4 \longrightarrow \underset{(\text{mainly } B_4H_{10})}{\text{mixture of boranes}} \xrightarrow{\Delta} \underset{\text{diborane}}{B_2H_6}$$

$$B_2O_3 + 3H_2 + 2Al \xrightarrow[750\text{ atm}]{150°C} B_2H_6 + Al_2O_3$$

$$4BF_3 + 3LiAlH_4 \longrightarrow 2B_2H_6 + 2LiF + 3AlF_3$$

(industrial method)

$$\underset{\text{gas}}{2BF_3} + 6NaH \xrightarrow{180°C} B_2H_6 + 6NaF$$

(industrial method)

$$2NaBH_4 + I_2 \xrightarrow{\text{ether}} B_2H_6 + 2NaI + H_2$$

Properties

- Diborane is a colourless gas, highly reactive (b.p. 180 K). It catches fire spontaneously in air and explodes with oxygen :

$$B_2H_6 + 3O_2 \longrightarrow B_2O_3 + 3H_2O, \quad \Delta H° = -2165 \text{ kJ mol}^{-1}$$

- It is hydrolysed by water :

$$B_2H_6 + 6H_2O \longrightarrow \underset{\text{boric acid}}{2H_3BO_3} + 3H_2$$

- In alkaline solution, hypoborates are formed.

$$B_4H_{10} + 4KOH \longrightarrow \underset{\text{hypoborates}}{4KOBH_3} + H_2$$

$$B_2H_6 \xrightarrow{\text{red heat}} 2B + 3H_2$$

- $$B_2H_6 + 2LiH \longrightarrow 2Li[BH_4]$$

$$2B_2H_6 + 2Na \longrightarrow Na[BH_4] + Na[B_3H_8]$$

Both $LiBH_4$ and $NaBH_4$ **are used as reducing agents in organic synthesis.**

$$B_2H_6 + HCl \longrightarrow B_2H_5Cl + H_2$$

$$B_2H_6 + 3Cl_2 \longrightarrow 2BCl_3 + 6HCl$$

DEEP Focus

- B_2H_6 selectively changes alkenes to alcohol with alkaline H_2O_2 against Markownikoff's rule :

$$CH_3CH{=}CH_2 \xrightarrow[\text{(ii) } H_2O_2,\ OH^-]{\text{(i) } B_2H_6} CH_3CH_2CH_2OH$$

- All the boranes act as Lewis acids and can accept electron pairs. Thus they react with amines, forming simple adducts. They also react with ammonia to give variety of products depending on experimental conditions:

$$B_2H_6 + 2Me_3N \longrightarrow 2[Me_3N \cdot BH_3]$$

$$B_2H_6 + 2CO \longrightarrow 2[BH_3 \cdot CO]$$

A complex reaction occurs when B_2H_6 is heated in a sealed tubes. The B_2H_6 molecule probably breaks into the reactive intermediate BH_3 which combines with B_2H_6 forming another intermediate B_3H_9 which can how further reactions :

$$B_2H_6 \longrightarrow 2BH_3$$

$$BH_3 + B_2H_6 \longrightarrow B_3H_9$$

$$B_3H_9 \longrightarrow B_3H_7 + H_2$$

$$B_3H_7 + BH_3 \longrightarrow B_4H_{10}$$

B_2H_6 splits into two symmetrical fragments (BH_3) in presence of Lewis acid (L) to give borane-adducts $BH_3 \cdot L$. **This is called symmetric or homolytic cleavage.**

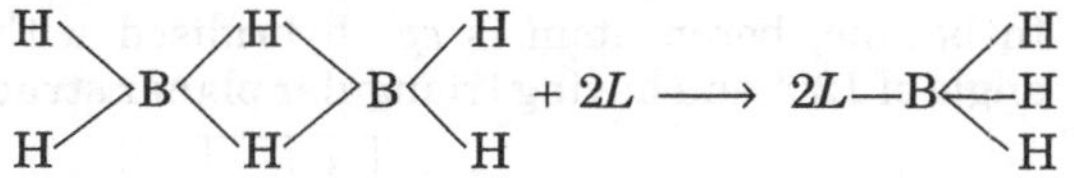

$$H_2B(\mu\text{-}H)_2BH_2 + 2L \longrightarrow 2L\text{—}BH_3$$

- $$B_2H_6 + NH_3 \xrightarrow[\text{low temperature}]{\text{excess } NH_3} B_2H_6 \cdot 2NH_3$$

$$\xrightarrow[\text{high temperature}]{\text{excess } NH_3} \underset{\text{boron nitride}}{(BN)_x}$$

$$\xrightarrow[\text{high temperature}]{\text{ratio : } 2NH_3\text{:}1B_2H_6} \underset{\text{borazine}}{B_3N_3H_6}$$

$B_2H_6 \cdot 2NH_3$ is ionic and **forms borazine on heating :**

$$B_2H_6{\cdot}2NH_3 \longrightarrow [H_3N \longrightarrow BH_2 \longleftarrow NH_3]^+ + [BH_4]^-$$

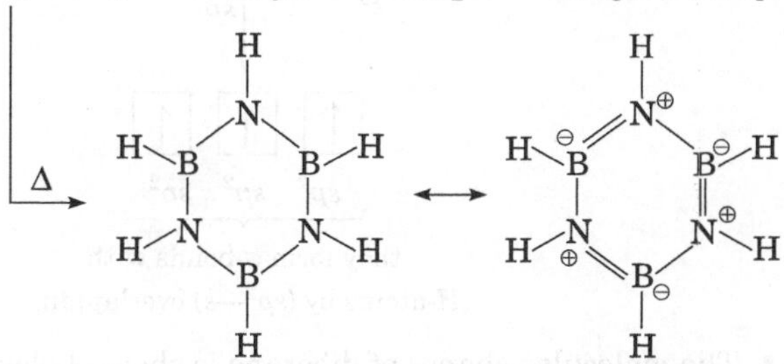

$\xrightarrow{\Delta}$ (I) $\longleftrightarrow$ (II)

Borazine

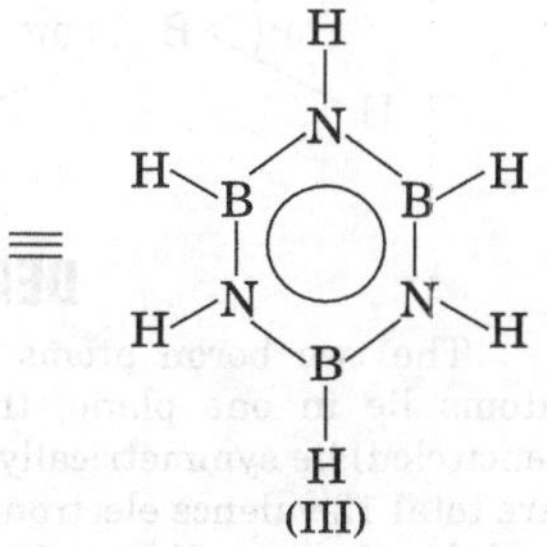

(III)

DEEP Focus

Borazine, $B_3N_3H_6$ (I) is called 'inorganic benzene' because its structure shows some formal similarity with benzene, with delocalised electrons (II) and aromatic character (III).

- Boron nitride is a white slippery solid. One B atom and one N atom together have the same number of valence electrons as two C atoms. Thus boron nitride has almost the same structure as graphite, with sheets made up of hexagonal rings of alternate B and N atoms joined together (inorganic graphite).

boron-nitride   graphite

- In borane, boron atom is sp^2 hybridised with bond angle of 120° and having triangular planar structure.

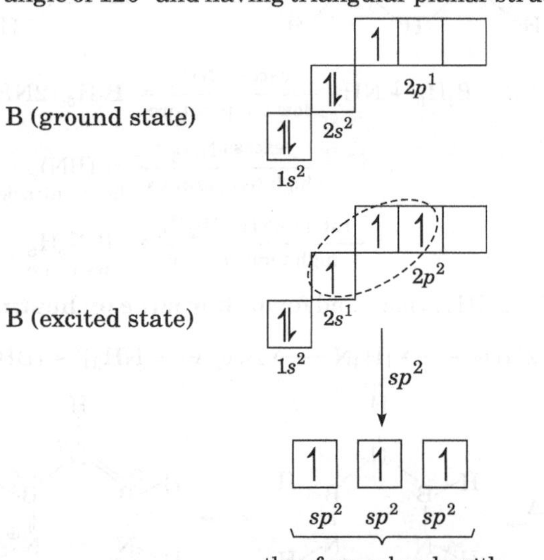

- The molecular shapes of diborane is shown below :

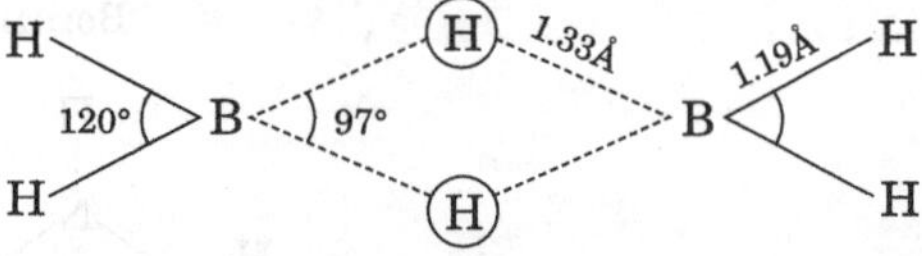

DEEP Focus

The two boron atoms and the four terminal hydrogen atoms lie in one plane; the two bridging hydrogen atoms (encircled) lie symmetrically above and below the plane. There are total 12 valence electrons—6 in two B and 6 in six H atoms and they form in all eight B—H bonds. Evidently, there are not enough electrons to fill all the available orbitals to form eight normal—two-centre two-electron (2c–2*e*) or electron pair bonds. The two bridging hydrogen atoms lie symmetrically above and below this plane (3c – 2*e* bonding). This has actually multicentre bonds or odd-electron bonds and B—H bond lengths (of terminal sides) is less than that of central B—H bonds.

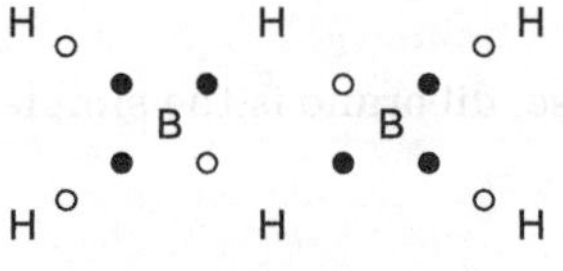

(Bridging groups are electron deficient).

2c – 2*e*   B—H   terminal bond : B—H

3c – 2*e*   B—H—B   bridging bond : B⟨H⟩B

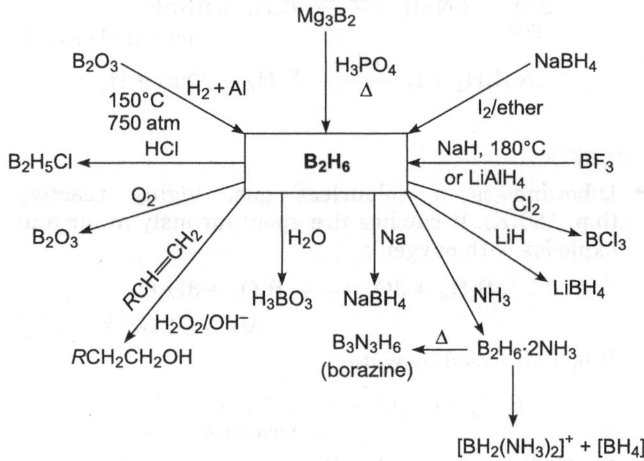

Similarity between Boron and Silicon (Diagonal Relationship)

- Both form volatile hydrides (called boranes and silanes) which spontaneously catch fire on exposure to air and are easily hydrolysed.
- Halides of both (except BF_3) are easily hydrolysed :

$$BCl_3 + 3H_2O \longrightarrow H_3BO_3 + 3HCl$$
$$SiCl_4 + 4H_2O \longrightarrow H_4SiO_4 + 4HCl$$

- Boron forms binary compounds with several metals—these are called borides. Silicon form silicides. Borides and silicides are hydrolysed by H_2O forming boranes and silanes :

$$Mg_3B_2 + 6H_2O \longrightarrow 3Mg(OH)_2 + B_2H_6$$
$$Al_4Si_3 + 12H_2O \longrightarrow 4Al(OH)_3 + 3SiH_4$$

- B_2O_3 and SiO_2 are acidic in nature and are important constituents of glass. Borate and silicates have tetrahedral BO_4 and SiO_4 structural units, respectively.

- Both are semi-conductors.
- Boric acid (H_3BO_3) and silicic acids (H_4SiO_4) are weak acids.
- Both exist in two allotropic forms—amorphous and crystalline.
- They do not dissolve in cold dilute acids but do so in alkalies—boron only in fused alkalies and silicon in both fused and aqueous alkalies.

$$2B + 2NaOH + 2H_2O \longrightarrow 2NaBO_2 + 3H_2$$

$$Si + 2NaOH + H_2O \longrightarrow Na_2SiO_3 + 2H_2$$

Borax : ($Na_2B_4O_7 \cdot 10H_2O$)

Source

- Tincal (*suhaga*) occurs naturally and contains about 45% borax. Tincal is dissolved in water and after removing impurities (by filtration), aqueous solution is evaporated to get crystals of borax.
- Orthoboric acid, H_3BO_3 on neutralisation with Na_2CO_3 gives borax.

$$Na_2CO_3 + 4H_3BO_3 \longrightarrow Na_2B_4O_7 + 6H_2O + CO_2$$

- Colemanite, $Ca_2B_6O_{11}$ is converted into borax by boiling it with concentrated solution of Na_2CO_3.

$$Ca_2B_6O_{11} + 2Na_2CO_3 \longrightarrow \underset{\text{borax}}{Na_2B_4O_7} + 2CaCO_3 + \underset{\substack{\text{sodium metaborate}\\\text{(in mother liquor)}}}{2NaBO_2}$$

CO_2 is passed into mother liquor containing $NaBO_2$ when borax is formed.

$$4NaBO_2 + CO_2 \longrightarrow Na_2B_4O_7 + Na_2CO_3$$

Properties

- Borax is
  - prismatic, $Na_2B_4O_7 \cdot 10H_2O$ obtained by crystallising salt at ordinary temperature.
  - octahedral, $Na_2B_4O_7 \cdot 5H_2O$ obtained by crystallising salt above 370 K. It is also called **jeweller's borax.**
  - borax glass, $Na_2B_4O_7$, obtained by heating ordinary borax above its melting point. It is colourless glassy substance which absorbs moisture from air and is converted into $Na_2B_4O_7 \cdot 10H_2O$.
- Borax dissolves in water to give alkaline solutions :

$$Na_2B_4O_7 + 2H_2O \longrightarrow 2NaOH + H_2B_4O_7$$

Borax is therefore, used as a water softner and cleansing agent.

DEEP Focus

- One mole of borax reacts with two moles of HCl (or one mole H_2SO_4). This is because when borax is dissolved in water both $B(OH)_3$ and $[B(OH)_4]^-$ are formed, but only the $[B(OH)_4]^-$ reacts with HCl. Each $[B(OH)_4]^-$ has one neutralisable OH^- ion.

$$[B_4O_5(OH)_4]^{2-} + 5H_2O \rightleftharpoons 2B(OH)_3 + 2[B(OH)_4]^-$$

$$2[B(OH)_4]^- + 2H_3O^+ \longrightarrow 2B(OH)_3 + 4H_2O$$

- Borax on strongly heating forms B_2O_3 which form coloured glassy bead with coloured oxide (generally of transition metals). It is called **borax bead test.**

$$Na_2B_4O_7 \xrightarrow{\Delta, 740°C} \underset{\text{(transparent bead)}}{2NaBO_2} + B_2O_3$$

$$B_2O_3(s) + CoO(s) \longrightarrow \underset{\text{blue bead}}{Co(BO_2)_2}$$

$$B_2O_3(s) + CuO(s) \longrightarrow \underset{\text{red bead}}{Cu(BO_2)_2}$$

DEEP Focus

- Borax is used as a primary standard in titrations against acids.

$$Na_2B_4O_7 \cdot 10H_2O + 2HCl \longrightarrow 2NaCl + 4H_3BO_3 + 5H_2O$$

One of the products H_3BO_3 is itself a weak acid. Thus, the indicator used to detect end point of this reaction must be one that is unaffected by H_3BO_3. **Methyl orange (pH range 3.1–4.4) is normally used.**

- Borax is actually made of two tetrahedra and two triangular units joined as shown and should be written as $Na_2[B_4O_5(OH)_4] \cdot 8H_2O$.

```
            OH
            |
            B⊖
         O/ | \O
  HO—B<     O    >B—OH
         O\ | /O
            B⊖
            |
            OH
```

- It is used as a flux in welding and brazing metals for dissolving any oxide which may form on the surfaces to be joined; in the manufacture of high quality heat resistant glass (borosil-mark) Jena and pyrex glasses contain boric oxide.

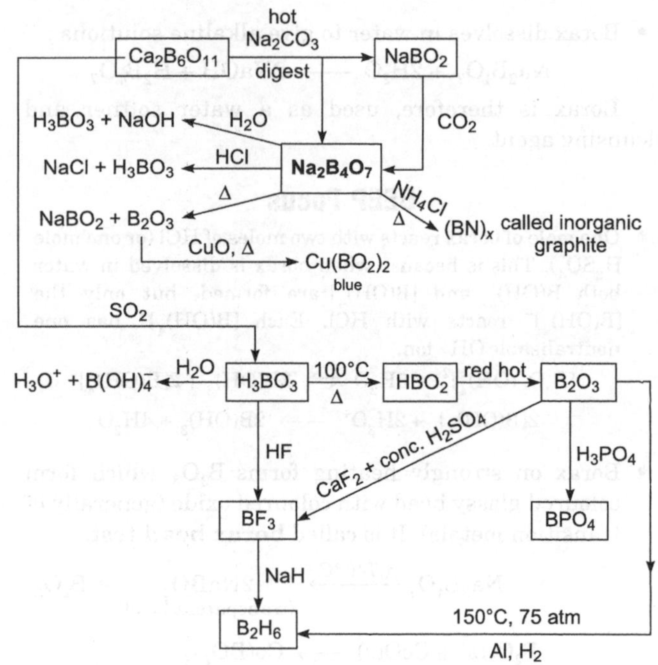

Boric acids and oxides

(i) H_3BO_3 or $B(OH)_3$ orthoboric acid.

(ii) HBO_2 or BO(OH) metaboric acid.

(iii) $H_2B_4O_7$ pyroboric acid

Boric acid is the trivial name for ortho boric acid. In aqueous solution *meta* and pyroboric acid are changed into *ortho* boric acid.

Source

- $Na_2B_4O_7 + H_2SO_4 + 5H_2O \longrightarrow Na_2SO_4 + 4H_3BO_3$

$$H_3BO_3 \xrightarrow{100°C} HBO_2 + H_2O$$

- On a commercial scale, boric acid is prepared from colemanite, $Ca_2B_6O_{11}\cdot 5H_2O$. Mineral (colemanite) is powdered finely and SO_2 is passed into hot aqueous suspension. On cooling crystals of boric acid are obtained.

$$Ca_2B_6O_{11} + 9H_2O + 2SO_2 \longrightarrow 2CaSO_3 + 6H_3BO_3$$

Excess of SO_2 dissolves $CaSO_3$ as $Ca(HSO_3)_2$

$$\underset{\text{insoluble}}{CaSO_3} + H_2O + SO_2 \longrightarrow \underset{\text{soluble}}{Ca(HSO_3)_2}$$

Properties

- Both form B_2O_3 (boron sesquioxide — sesqui means one and a half, so the oxide should have $BO_{1\frac{1}{2}}$ or B_2O_3)

$$H_3BO_3 \xrightarrow{100°C} HBO_2 \xrightarrow{140°C} H_2B_4O_7 \xrightarrow{\text{red hot}} B_2O_3$$

DEEP Focus

- Ortho boric acid, H_3BO_3 is soluble in water and behaves as a weak monobasic acid. It does not donate protons like most acids, but rather it accepts OH^-. It is therefore a Lewis acid, and is better written as $B(OH)_3$.

$$B(OH)_3 + 2H_2O \rightleftharpoons H_3O^+ + [B(OH)_4]^-; \qquad pK_a = 9.25$$

plane triangle — tetrahedral metaborate ion

- Polymeric metaborate species are formed at higher concentration, for example :

$$3B(OH)_3 \rightleftharpoons H_3O^+ + [B_3O_3(OH)_4]^- + H_2O \qquad pK_a = 6.84$$

- Orthoboric acid contains triangular BO_3^{3-} units. In the solid the $B(OH)_3$ units are hydrogen bonded together into two-dimensional sheets with almost hexagonal symmetry (Fig. 11.4). The layers are quite a larger distance apart (3.18Å) and thus the crystal breaks quite easily into very fine particles.

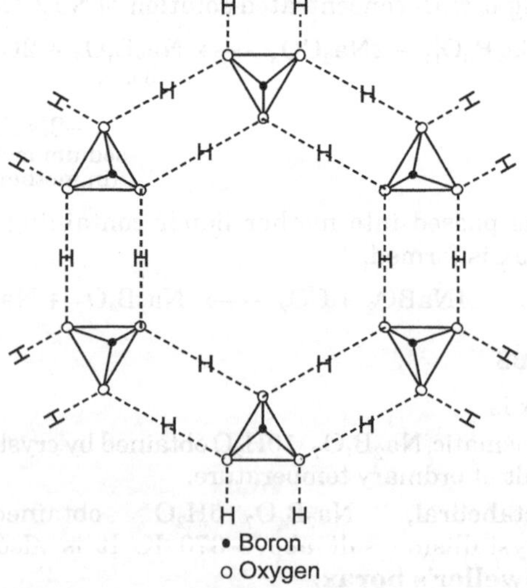

Fig.11.4 Hydrogen bonded structure of orthoboric acid

- It is used in the manufacture of enamels and glazes for pottery and as disinfectant in medicine.

Boron trioxide (B_2O_3)

- This (B_2O_3) is obtained when boron is burnt in oxygen

$$4B + 3O_2 \longrightarrow 2B_2O_3$$

- It is also obtained as a colourless glass when boric acid is heated to redness

$$2H_3BO_3 \longrightarrow B_2O_3 + 3H_2O$$

- B_2O_3 is a white non-volatile glassy solid.
- It is an acidic oxide giving rise to boric acid, when dissolved in water

$$B_2O_3 + 3H_2O \rightleftharpoons 2H_3BO_3$$

- Boron trioxide reacts with various metallic salts at high temperatures giving characteristic coloured beads. This is called **borax-bead test** of the identification of coloured salts

$$CuSO_4 + B_2O_3 \xrightarrow{\Delta} \underset{\text{blue}}{Cu(BO_2)_2} + SO_3$$

- B_2O_3 also shows weakly basic properties as indicated by its reaction with ortho phosphoric acid

$$B_2O_3 + 2H_3PO_4 \longrightarrow 2BPO_4 + 3H_2O$$

Compounds of Aluminium

Aluminium Halides

$AlX_3 (X = Cl, Br)$ exists as dimer, thus attaining an octet of electrons. The dimer formula is retained when the halides dissolve in non-polar solvents (like ether, benzene).

$$\begin{array}{c} Cl \quad\; Cl \quad\; Cl \\ >Al< \; >Al< \\ Cl \quad\; Cl \quad\; Cl \end{array}$$

DEEP Focus

However, when the halides dissolve in H_2O, the high enthalpy of hydration is sufficient to break the covalent dimer into $[M(H_2O)_6]^{3+}$ and $3X^-$ ions.

Aqueous solution of $AlCl_3$ is acidic due to hydrolysis

$$AlCl_3 + 3H_2O \longrightarrow Al(OH)_3 + 3HCl$$

$AlCl_3$, being electron deficient, finds use as Lewis acid catalyst in Friedel-Craft's reactions of aromatic class.

Aluminium Oxide and Hydroxide

Source

- Alumina, Al_2O_3 can be made by dehydrating $Al(OH)_3$, or from elements. Al_2O_3 is

$$\alpha\text{-}Al_2O_3 \text{ or corundum}$$

$$\gamma\text{-}Al_2O_3$$

$$\gamma\text{-}Al_2O_3 \xrightarrow{\Delta, 1000^\circ C} \alpha\text{-}Al_2O_3$$

Properties

- Corundum is extremely hard and is used as 'jewellers rouge' to polish glass. Its impure form containing Fe_2O_3 and silica is called **emery** and is used to make emery paper (sand paper) used to polish metals. It is also unaffected by acids and is used as refractory to line furnaces.
- $Al(OH)_3 \xrightarrow[\Delta]{450^\circ C} \gamma\text{-}Al_2O_3$

$\gamma\text{-}Al_2O_3$ dissolves in acids, absorbs moisture and is used in chromatography.

- Alumina is white in colour and addition of Cr_2O_3 or Fe_2O_3 makes it coloured.

DEEP Focus

- $Al(OH)_3$ is amphoteric.

(a) $\underset{\text{acid}}{Al(OH)_3} + \underset{\text{base}}{3H^+} \longrightarrow Al^{3+} + 3H_2O$

Al^{3+} exists as hydrated $[(Al(H_2O)_6]^{3+}$ ion.

(b) $\underset{\text{base}}{Al(OH)_3} + \underset{\text{acid}}{NaOH} \longrightarrow Na[Al(OH)_4]$

or $\underset{\text{sodium meta aluminate}}{NaAlO_2 \cdot 2H_2O}$

In (a) it behaves as a base while in (b) it behaves as an acid—thus amphoteric. *However, Al^{3+} is precipitated as $Al(OH)_3$ on passing CO_2, showing that the acidic properties are very weak.*

$$CO_2 + H_2O \longrightarrow H_2CO_3 \rightleftharpoons CO_3^{2-} + 2H^+$$

$$2Al^{3+} + 3CO_3^{2-} + 3H_2O \longrightarrow 2Al(OH)_3 + 3CO_2$$

- Structure of aluminate ion changes with both pH and concentration :
  - between pH 8 and 12, the ions polymerise using OH bridges and each Al is octahedrally coordinated.
  - in dilute solution above pH 13, a tetrahedral $Al(OH)_4^-$ exists.
  - in concentrated solutions above 1.5 M and pH greater than 13, it exists a dimer $[(OH)_3Al—O—Al(OH)_3]^{2-}$.

Alums

An aqueous solution containing equimolar amounts of $Al_2(SO_4)_3$ and K_2SO_4 crystallises as potassium aluminium sulphate, $KAl(SO_4)_2 \cdot 12H_2O$. This is generally called **alums** with general formula

$$M(\text{I})\, M'(\text{III})\, (SO_4)_2 \cdot 12H_2O$$

where M(I) is almost any unipositive cation [except Li^+, because this ion is too small to meet the structural requirements of the crystal) and M'(III) is a tripositive cation— $Al^{3+}, Ti^{3+}, V^{3+}, Cr^{3+}, Fe^{3+}, Mn^{3+}, Co^{3+}, Ga^{3+}, In^{3+}, Re^{3+}, Ir^{3+}$]. **Alums contain the ions $[M(H_2O)_6]^+, [M'(H_2O)_6]^{3+}$ and SO_4^{2-} in the ratio 1 : 1: 2.** Most common alums have $M(\text{I}) = K^+$ or NH_4^+ and $M'(\text{III}) = Al^{3+}$.

Alums are used as mordant in dyeing. The fabric to be dyed is dipped in a solution of the alum and heated with steam. Hydrolysis of $[Al(H_2O)_6]^{3+}$ deposits $Al(OH)_3$ into the fibres of the material and the dye is adsorbed on the $Al(OH)_3$.

Cement

- $Ca_3Al_2O_6$ (tricalcium aluminate) is important constituent of Portland and high alumina cements.
- General formula is $Ca_9[Al_6O_{18}]$ because it contains 12-membered rings of Si—O—Si—O made by joining six $\mathbf{AlO_4}$ tetrahedra.
- Portland cement is made by heating mixture of $CaCO_3$, SiO_2 and clay (aluminosilicate) at a temperature of 1450 – 1600° C in a rotary kiln. When mixed with sand and water, Portland cement sets to give a concrete, a hard whitish solid (with appearance similar to Portland cement). About 2-5% of gypsum $[CaSO_4 \cdot 2H_2O]$ is added to slow down the setting process to increase strength.

  Portland cement has:

CaO	70%	Fe_2O_3	3%
SiO_2	20%	$CaSO_4 \cdot 2H_2O$	2%
Al_2O_3	5%		

- High alumina cement is made by fusing limestone and bauxite with small amount of SiO_2 and TiO_2 at 1400–1500°C. It is better than Portland cement in terms of :
  - it sets much quicker,
  - it develops high strength within one day,
  - it has good resistance to sea water and dilute mineral acids.

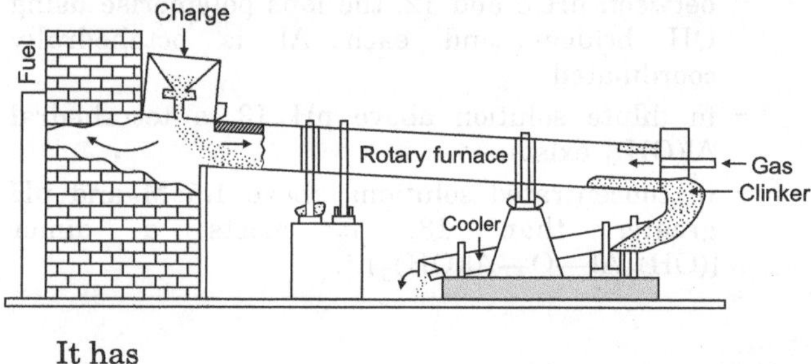

It has

CaO	40%	SiO_2	10%
Al_2O_3	40%	Fe_2O_3	10%

DEEP Focus

- When mixed with water, the setting of cement takes place. Chemically, it is hydration of the molecules of constituents and their rearrangement. The adhesion of other particles to each other and to the embedded aggregates is responsible for the strength of the cement which is due, ultimately, to the formation of Si—O—Si—O bonds.

Uses

Boron is used to increase the hardness of steels. Crystalline boron is used in transistors. Boron is a good neutron absorber and is used as shields and control rods in **nuclear reactors.** Boranes are **used as high energy fuels,** for example, in rockets. **Boric acid is used as an antiseptic.** Borax is used to make **heat resistant borosilicate glass.** It is also used for **vitreous enamelling** of baths, domestic appliances and for glazing tiles and pottery. Another use of borax is to make peroxyborates, *e.g.,* $NaBO_2 \cdot H_2O_2 \cdot 3H_2O$, which are **useful cleaning and bleaching** agents. In the laboratory, borax is used for standardising acids and in the borax bead test in qualitative analysis.

Aluminium exhibits useful properties of low density, high thermal and electrical conductivity, good corrosion resistance with non-toxic nature of the metal and its compounds. Due to these properties, it is the most widely used non-ferrous metal. Aluminium is used for making electrical conductors, cooking utensils and wrapping materials. Aluminium is extensively used for structural purposes, either alone or alloyed, in aircraft, ship and building industries. Large amounts of aluminium are converted into alloys, such as, duralumin and magnalium containing a few per cent of copper or magnesium. These alloys are harder and stronger than pure aluminium but possess almost the same properties of lightness and corrosion resistance which can be further increased by anodising process. Aluminium beryllium alloys are harder and lighter than other alloys of aluminium and are extensively used in space-craft. However, toxic nature of beryllium makes their handling difficult. Due to its strong affinity for oxygen, aluminium is a good reducing agent and is used in aluminothermic process for extraction of metals and in welding. Suspension of finely powdered aluminium in oil is used as paint.

Anhydrous $AlCl_3$ is an important catalyst, used in organic synthesis and in the cracking of petroleum. $Al_2(SO_4)_3$ as such or as potash alum, $K_2SO_4 \cdot Al_2(SO_4)_3 \cdot 24H_2O$ is used for sizing paper, for tanning leather, for waterproofing cloth and as a mordant for dying cotton. It is extensively used for purification of water and in sewage treatment.

Because of its extreme hardness, high melting point, non-volatility, chemical inertness and good electrical insulating properties, **corundum (Al_2O_3) finds many applications in abrasives, refractories and ceramics.** Large crystals of α-Al_2O_3 when coloured with metal-ion impurities are prized as gemstones, *e.g.,* ruby (Cr^{3+}, red), sapphire ($Fe^{2+/3+}Tl^{4+}$, blue), oriental emerald (Cr^{3+}/V^{3+}, green), oriental topaz (Fe^{3+}, yellow), etc. Aluminates are imporant constituents of Portland cement. Gallium has the longest liquid range (303–2343 K) of any known substance and so finds use as a high temperature thermometer liquid.

Some Gemsones Based on Al_2O_3

Gem	Impurity
White sapphire	none
Blue sapphire	Fe, Ti
Green sapphire	Co
Yellow sapphire	Ni, Mg
Star sapphire	Ti
Ruby	Cr

Target Practice 2

1. Boron finds application as semi-conductor. How is pure crystalline boron obtained?
(a) ....
(b) ....
(c) ....
2. Borax is made of two tetrahedra and two triangular units. Write this structure
....
3. How is Cu^{2+} identified by borax bead test?
....
4. Aqueous solution containing one mole of borax is treated with H_2SO_4. Write reaction ....
Thus, H_2SO_4 used is ....
5. Write the names of following boric acids?
(a) H_3BO_3 ....
(b) HBO_2 ....
(c) $H_2B_4O_7$ ....
(a) change to (b) by ....
(b) changes to (c) by ....
6. Fill in the blanks
$B(OH)_3$ is .... acceptor. Thus, it is a ....
7. Write the structure of the species formed when H_3BO_3 reacts with H_2O.
....
8. Compare covalent nature of
(a) B_2F_4 ☐ TlF
(b) TlF_3 ☐ TlF
9. (a) TlF dissolves in H_2O forming ....
(b) TlF_3 hydrolyses in H_2O forming ....
10. Chemist '*X*' tried the synthesis of 1-propanol by following method but each time yield of 1-propanol was very low

$$CH_3CH{=}CH_2 \xrightarrow{H_3O^+} CH_3CH_2CH_2OH$$

(a) Suggest a suitable device to get better yield of 1-propanol.
....
(b) What was the main product of the method used by '*X*'?
....
11. Write two resonance structures of inorganic benzene
(i) ....
(ii) ....
12. Write the common names of hybrides of
(a) Si .... (b) B .... (c) C ....
13. $AlCl_3$ dimerises to Al_2Cl_6. This dimer exists as .... in ether and as .... in water.
14. What is the role of $AlCl_3$ in Friedel-Craft's reaction?
....
15. Al_2O_3 dissolves in NaOH as well as in HCl. This property is called
....
16. Write the percentage of each component in Portland cement
....
17. Explain **"Setting of Cement".**
....

Answers

1. Refer text.
2. Refer text.
3. Refer text.
4. $$\underset{1\text{ mol}}{Na_2B_4O_7} + 2H_2O \longrightarrow \underset{2\text{ mol}}{2NaOH} + \underset{\text{weak acid}}{H_2B_4O_7}$$
$$2NaOH + H_2SO_4 \longrightarrow Na_2SO_4 + 2H_2O$$
Thus, H_2SO_4 used = 1 mol
5. (a) Orthoboric acid, (b) Metaboric acid, (c) Pyrobaric acid.
$$H_2B_4O_7 + H_2O \rightleftharpoons 3H_3BO_3$$
$$\underset{(a)}{H_3BO_3} \xrightarrow{100°C} \underset{(b)}{HBO_2} \xrightarrow{140°C} \underset{(c)}{H_2B_4O_7}$$
6. $B(OH)_3$ is OH^- acceptor. Thus, it is a Lewis acid.
7. $H_3BO_3 + 2H_2O \rightleftharpoons H_3O^+ + [B(OH)_4]^-$
8. (a) B_2F_4 > TlF (b) TlF_3 > TlF
9. (a) $TlF + H_2O \longrightarrow [Tl(H_2O)]^+ F^-$
Ionic nature
(b) $\underset{\text{Covalent nature}}{TlF_3} + 3H_2O \longrightarrow Tl(OH)_3 + 3HF$
10. (a) Oxidation of $CH_3CH{=}CH_2$ by $B_2H_6 \cdot THF/H_2O_2, OH^-$ gives 1-propanol.
$$CH_3CH{=}CH_2 \xrightarrow[\text{(ii) } H_2O_2,\, OH^-]{\text{(i) } B_2H_6 \cdot THF} CH_3CH_2CH_2OH$$
(b) $CH_3\underset{\substack{|\\ OH}}{CH}CH_3$
11. Refer text.
12. (a) Silane, (b) Borane, (c) Alkane.
13. Al_2Cl_6, $[Al(H_2O)_6]^{3+}$ and $3Cl^-$.
14. As Lewis acid, it helps in the formation of electrophile, without which electrophilic substitution at benzene nucleus does not take place.
$$CH_3Cl + AlCl_3 \longrightarrow \underset{\text{electrophile}}{\overset{\oplus}{C}H_3} + AlCl_4^{\ominus}$$
15. Amphoteric nature.
16. Refer text.
17. Refer text.

Practice Exercise 2

1. The end-point of a borax–acid titration is at about pH = 4. What indicator would you use in the titration?
2. $Na_2B_4O_7 + \text{conc. } H_2SO_4 \longrightarrow (A) \xrightarrow[\text{(ii) ignite}]{\text{(i) } C_2H_5OH} (B)$

   (B) is identified by the characteristic colour of the flame. Identify *(A)* and *(B)*.
3. You should know that the oxides of non-metals are acidic and the oxides of the metals are basic. Would you expect boron to be more acidic or more basic than aluminium in its reactions?
4. Identify *(A)*, *(B)* and *(C)*.

   $$Na_2B_4O_7 + NaOH \longrightarrow (A) \xrightarrow{HCl} (B) \xrightarrow{H_2O} (C)$$
5. Borates are identified by **"Flame Test"**. Give the chemistry of this test.
6. What happens when glycol reacts with orthoboric acid?
7. B_2O_3 is an acidic oxide. Give examples.
8. Identify *(A)*, *(B)* and *(C)*.

   $$Na_2B_4O_7 \xrightarrow{\Delta} (A) + (B)$$
   $$(A) + NaOH \longrightarrow (B)$$
   $$(A) + HF \xrightarrow{H_2SO_4} (C)$$
9. BCl_3 or BI_3 can be converted into pure B. Write chemical reactions.
10. BF_3 finds use as Lewis acid catalyst in Friedel-Crafts reaction. Give example.
11. Identify *(A)*, and *(B)* in the following:

    Methylenecyclohexane ($=CH_2$) $\xrightarrow[\text{(ii) } H_2O_2,\, OH^-]{\text{(i) } B_2H_6} (A)$

    Methylenecyclohexane $\xrightarrow{H_3O^+} (B)$
12. *R*CN, *R*NO_2 and *R*CHO can be reduced by B_2H_6. What are the products due to reduction?
13. Benzene $+ CH_3Cl \xrightarrow{AlCl_3}$ Toluene ($C_6H_5CH_3$)

    What is the function of $AlCl_3$ in this reaction ? Also name the reaction.
14. Aluminium salts turn blue litmus red. Explain.
15. A mixture contains ZnO and Al_2O_3. Give a scheme to separate them.

Answers

1. Methyl orange.
2. *(A)* : H_3BO_3 *(B)* : $B(OC_2H_5)_3$: gives green-edged flame.
3. Being nearer the top of the group, we would expect boron to show more non-metallic nature than aluminium; so it should be more acidic in nature than aluminium.
4. *(A)* : $NaBO_2$ *(B)* : HBO_2 *(C)* : H_3BO_3

   $$Na_2B_4O_7 + 2NaOH \longrightarrow \underset{(A)}{4NaBO_2} + H_2O$$
   $$NaBO_2 + HCl \longrightarrow \underset{(B)}{HBO_2} + NaCl$$
   $$HBO_2 + H_2O \longrightarrow \underset{(C)}{H_3BO_3} \text{ or } B(OH)_3$$
5. $$\underset{\text{borax}}{Na_2B_4O_7} + H_2SO_4 \xrightarrow{\Delta} Na_2SO_4 + 4H_3BO_3 + 5H_2O$$
   $$H_3BO_3 + 3C_2H_5OH \xrightarrow{\Delta} \underset{\text{triethyl borate}}{B(OC_2H_5)_3} + 3H_2O$$

   Triethyl borate burns with green-edged flame when ignited.
6. *Cis*-di-ol form very stable complex with $[B(OH)_4]^-$ which is formed when H_3BO_3 reacts with H_2O :

   $$H_3BO_3 + 2H_2O \longrightarrow [B(OH)_4]^- + H_3O^+$$
   $$[B(OH)_4]^- + \text{glycol} \longrightarrow$$ cyclic complex: H_2C–O and H_2C–O bonded to B, which also carries two OH groups.
7. $$B_2O_3 + 6HCl \longrightarrow 2BCl_3 + 3H_2O$$
   $$B_2O_3 + 2NaOH \longrightarrow 2NaBO_2 + H_2O$$
8. *(A)* : B_2O_3 *(B)* : $NaBO_2$ *(C)* : BF_3
9. $$2BCl_3 + 3H_2 \xrightarrow{\text{red hot W}} 2B + 6HCl$$
   $$2BI_3 \xrightarrow{\text{red hot W}} 2B + 3I_2$$
10. Benzene $+ CH_3Cl \xrightarrow{BF_3}$ Toluene ($C_6H_5CH_3$) $+ HCl$
11. *(A)* : Cyclohexyl–CH_2OH *(B)* : 1-methylcyclohexanol (ring carbon bearing CH_3 and OH)

12. $RCN \longrightarrow RCH_2NH_2$

$RNO_2 \longrightarrow RNH_2$

$RCHO \longrightarrow RCH_2OH$

13. It is Friedel-Craft's reaction. $AlCl_3$ acts as Lewis acid that helps in the formation of carbocation ion (in this case $\overset{\oplus}{C}H_3$ without which S_E (Electrophilic substitution) reaction is not possible:

$$CH_3Cl + AlCl_3 \longrightarrow \overset{\oplus}{C}H_3 + AlCl_4^-$$

$$H^+ + AlCl_4^- \longrightarrow HCl + AlCl_3$$

$$C_6H_6 + \overset{\oplus}{C}H_3 \longrightarrow C_6H_5CH_3 + H^+$$

14. Aluminium salts are acidic due to hydrolysis of Al^{3+}

$$Al^{3+} + 3H_2O \longrightarrow Al(OH)_3 + 3H^+$$

Aluminum salts also exist as $[Al(H_2O)_6]^{3+}$.

In the following reaction it behaves like an acid (H^+ ion donor)

$$\underset{\text{acid}}{[Al(H_2O)_6]^{3+}} + \underset{\text{base}}{H_2O} \longrightarrow [Al(H_2O)_5(OH)]^{2+} + H_3O^+$$

15. ZnO and Al_2O_3 both are amphoteric. They can be separated by:

Step I : Dissolve in dil. HCl when $ZnCl_2$ and $AlCl_3$ are formed.

Step II : Add NH_4Cl and then NH_4OH dropwise.

Al^{3+} is precipated as $Al(OH)_3$ and Zn^{2+} is soluble.

$$AlCl_3 + 3NH_4OH \longrightarrow Al(OH)_3 \downarrow + 3NH_4Cl$$

$$ZnCl_2 + 4NH_4OH \longrightarrow \underset{\text{soluble}}{[Zn(NH_3)_4]Cl_2} + 4H_2O$$

Total Practice Set
(Read, Plan & Solve!)

Problem 1. Whereas $AlCl_3$ is always covalent, it ionises in solution inspite of high ionisation energy. Explain.

Given $Al(g) \longrightarrow Al^{3+}(g) + 3e^-, \Delta H = 5114 \text{ kJ mol}^{-1}$

$Al^{3+}(g) + aq \longrightarrow Al^{3+}(aq), \quad \Delta H = -4665 \text{ kJ mol}^{-1}$

$Cl^-(g) + aq \longrightarrow Cl^-(aq), \quad \Delta H = -381 \text{ kJ mol}^{-1}$

Solution Al^{3+} will be formed from $AlCl_3$ in solution if heat of hydration of $AlCl_3$ is greater than total ionisation energy of Al(*g*).

$$Al^{3+}(g) + 3Cl^-(g) + aq \longrightarrow Al^{3+}(aq) + 3Cl^-(aq)$$

$$\Delta H = -4665 + 3(-381) = -5808 \text{ kJ mol}^{-1}$$

$$Al(g) \longrightarrow Al^{3+}(g) + 3e^-,$$

$$\Delta H = 5114 \text{ kJ mol}^{-1}$$

Thus, $\Delta H(\text{hydration}) > \Delta H(\text{ionisation})$

Thus, $AlCl_3$ ionises in solution.

Problem 2. $E^\circ_{Al^{3+}/Al} = -1.66$ V; $E^\circ_{SO_4^{2-}/SO_2} = 1.17$ V.

What, if anything, would happen if you reacted aluminium with sulphuric acid? If you think a reaction would take place, write down the equation.

Solution The given reaction (say dissolution of Al into H_2SO_4) is possible if E°_{cell} is positive.

$$Al \longrightarrow Al^{3+} + 3e^-$$

$$E^\circ_{Al/Al^{3+}} = -E^\circ_{Al^{3+}/Al} = 1.66 \text{ V}$$

$$SO_4^{2-} + 2e^- + 4H^+ \longrightarrow SO_2 + 2H_2$$

$$E^\circ_{SO_4^{2-}/SO_2} = 1.17 \text{ V.}$$

Net reaction is

$$\mathbf{2Al + 3SO_4^{2-} + 12H^+ \longrightarrow 2Al^{3+} + 3SO_2 + 6H_2O}$$

Thus, E°_{cell} is 2.83 V hence Al can reduce SO_4^{2-} to SO_2.

Problem 3. The standard free energies of formation of B_2O_3 and MgO are -1194 kJ mol^{-1} and -569 kJ mol^{-1} respectively. Should it be possible to prepare boron by reacting B_2O_3 with magnesium.

Solution $B_2O_3 + Mg \longrightarrow 2B + 3MgO$

$$\Delta G^\circ = \Delta G^\circ_{product} - \Delta G^\circ_{reactant}$$

$$= 3\Delta G^\circ_f(MgO) - \Delta G^\circ_f(B_2O_3)$$

$$= 3(-569) - (-1194)$$

$$= -513 \text{ kJ mol}^{-1} \quad [\Delta G^\circ_f(\text{element}) = 0]$$

The **negative sign of ΔG° shows that the reaction should be spontaneous.** *(However, the reaction is initially endothermic when two reactants are heated.)*

Problem 4. Suggest synthesis (starting from deuterium oxide as the only source of deuterium) for:

(a) B_2D_6 (b) $B(OD)_3$ (c) $B_3N_3D_6$ (d) $B(CH_2CH_2D)_3$

Solution (a) $P_2O_5 + 3D_2O \longrightarrow 2D_3PO_4$

$$Mg_3B_2 + 2D_3PO_4 \longrightarrow Mg_3(PO_4)_2 + \mathbf{B_2D_6}$$

(b) $B_2O_3 + 3D_2O \longrightarrow \mathbf{2B(OD)_3}$

(c) $Mg_3N_2 + D_2O \longrightarrow ND_3 + Mg(OD)_2$

$$2ND_3 + B_2D_6 \longrightarrow [D_2B(ND_3)_2][BD_4] \xrightarrow{\Delta} \mathbf{B_3N_3D_6}$$

(d) $6CH_2{=}CH_2 + B_2D_6 \longrightarrow \mathbf{2B(CH_2CH_2D)_3}$

Explanation

(a) ***Boron does not usually form a cation :***

B has

$(IE)_1 = 801$ kJ mol^{-1}, $(IE)_2 = 2427$ kJ mol^{-1}, $(IE)_3 = 3659$ kJ mol^{-1}

– Thus, the total energy required to give B^{3+} ions is far more than that which would be compensated by lattice energies of ionic compounds or hydration of such ions in solution. Thus formation of cation (like B^{3+}) is not possible.

(b) ***The polarity of* B—*X bonds is in the order :* B—F > B—Cl > B—Br *but Lewis acidicity shows the sequence :* $BF_3 < BCl_3 < BBr_3$.**

– With increase in polarity of B—*X* bond, acidity also increases and should be thus in the order :

$$BF_3 > BCl_3 > BBr_3$$

But Lewis acidity is in reverse order :

$$BF_3 < BCl_3 < BBr_3$$

There is lateral overlap of the vacant 2*p*-orbital of B with one completely filled orbital of F leading to $p\pi - p\pi$ bonds between B and F. This B—F bond thus acquires double bond character. This also leads to compensate electron deficiency of boron and thus Lewis acid character of BF_3 is reduced. This $p\pi$-$p\pi$ bonding decreases going from BF_3 to BBr_3 and thus Lewis acidic nature increases in the order :

$$BF_3 < BCl_3 < BBr_3$$

B ground state: $2s^1$ ⇅, $2p^1$ ↿ —excite→ excited state: $2s^1$ ↿, $2p^2$ ↿ ↿

F, F, B, F (B—F with $p\pi$-$p\pi$ overlap) → F, F > B ---- F

(c) ***Al and Ga are trivalent in their compounds but in case of Tl, monovalent compounds are the most stable. Why ?***

– It is due to '**inert pair**' effect due to which stability of +3 state decreases and that of +1 state increases while going down the group (in this case group 13).

$$\underset{\text{(most stable)}}{Al^{3+}} > Ga^{3+} > In^{3+} > Tl^{3+}$$

$$Al^{+} < Ga^{+} < In^{+} < \underset{\text{(most stable)}}{Tl^{+}}$$

(d) BF_3 ***reacts with*** F^- ***to give*** BF_4^-, ***but*** AlF_3 ***reacts with*** F^- ***to give*** AlF_6^{3-}.

– Boron has only 2*s* and 2*p* valence orbitals and can form maximum of four bonds. Aluminium has available 3*d*-orbitals and can use octahedral sp^3d^2-hybrid orbitals to bond to six F^- ions.

(e) ***Aqueous solutions of aluminium salts are acidic***

$$[Al(H_2O)_6]^{3+} + H_2O \rightleftharpoons H_3O^+ + [Al(H_2O)_5OH^-]^{2+}$$

– Due to attractive power of the small, highly charged Al^{3+} ion for electron, an O—H bond in a ligand H_2O molecule breaks thus releasing a proton to an H_2O molecule outside the coordination sphere. Thus, H_2O ligand is converted to OH^-, and the complex ion to $[Al(H_2O)_5OH^-]^{2+}$

$$[Al(H_2O)_5(HOH)]^{3+} \longrightarrow [Al(H_2O)_5(OH^-)]^{2+} + H^+$$

One of H_2O ligands releases H^+ (which goes to H_2O solvent) and OH^- (which remains are ligand in complex)

$$H^+ + H_2O \rightleftharpoons H_3O^+$$

Thus, net reaction is

$$[Al(H_2O)_6]^{3+} + H_2O \rightleftharpoons [Al(H_2O)_5(OH^-)]^{3+} + H_3O^+$$

Thus aqueous Al^{3+} salts are acidic.

Master Exercises

Exercise 1

(Stage 1 : Learning)

Short Answer Type Questions

1. Rewrite the equation :

$$B_4O_7^{2-}(aq) + 2H^+(aq) + 5H_2O(l) \longrightarrow 4H_3BO_3(aq)$$

showing the reaction as one that involves the $B_4O_5(OH)_4^{2-}$ ion instead of $B_4O_7^{2-}$. Does it matter very much which equation is used?

2. Aqueous solution of borax containing one mole of it reacts with two moles of acid. Explain.

3. Borax is also used as a buffer. Explain.

4. H_3BO_3 can't be titrated satisfactorily with NaOH, why?

5. (a) The formula of boric acid does not represent its structure. What is its formula?

(b) The formula could be written in a different way. What is it?

(c) How might you classify 'boric acid' if this formula were used?

6. Complete the following reactions :

(a) $BCl_3 + NH_4Cl \xrightarrow[C_6H_5Cl]{140°C} (A) \xrightarrow{NaBH_4} (B)$

(b) $Na_2B_4O_7 + NH_4Cl \xrightarrow{\text{red heat}} (C)$

(c) $BCl_3 + LiAlH_4 \longrightarrow (D)$

(d) $BCl_3 + (C_2H_5)_4N^+Cl^- \longrightarrow (E)$

(e) $BCl_3 + H_2 + C_{\text{fibre}} \xrightarrow{1700\text{–}1800°C} (F)$

7. What are bond order of B—H bonds (*X*) and (*Y*) in diborane?

H (X) and H (Y) terminal on left B; two bridging H between the B atoms; two terminal H on right B:

$$\begin{matrix} H \overset{X}{\diagdown} & & \overset{Y}{\diagup} H \diagdown & & \diagup H \\ & B & & B & \\ H \diagup & & \diagdown H \diagup & & \diagdown H \end{matrix}$$

8. Boron-nitride is also called 'inorganic graphite and borazine is also called inorganic benzene'. Why?

9. $Al(OH)_3$ finds uses in 'fire extinguisher' and in 'dyeing'. Explain.

10. Hot Al_2O_3 will catalyse the decomposition of ethanol. A colourless gas is given off, which decolorises Br_2 water. What are the products of the reaction?

11. $AlCl_3$ can be converted into AlCl. What are the necessary conditions?

12. $(A) \xrightarrow{NaOH,\ \Delta}$ gas (*B*) giving white fumes with HCl alongwith white ppt (*D*) which dissolves in excess of NaOH to give (*E*)

$(A) \xrightarrow{aq\ BaCl_2}$ white ppt (*C*)

0.453g of (*A*) gave 0.466g of (*C*). H_2O content in (*A*) is 47.68%.

Identify (*A*), (*B*), (*C*), (*D*) and (*E*), and explain reactions.

13. Which type of cations are capable of replacing aluminium in alums? Give examples.

Analytical Questions

1. Explain :

(i) BCl_3 exists as a monomer whereas $AlCl_3$ is a dimer.

(ii) Al forms $[AlF_6]^{3-}$ but $[BF_6]^{3-}$ is not formed.

(iii) Boric acid is a weak acid but it can be titrated as a strong acid in presence of ethylene glycol.

2. Some baking powders contain the solids $NaHCO_3$ and $NaAl(SO_4)_2$. When water is added to this mixture of compounds, $CO_2(g)$ and $Al(OH)_3(s)$ are two of the products. Write the chemical equations to represent this change.

3. In the purification of bauxite ore as a preliminary step in the production of aluminium, $[Al(OH)_4]^-(aq)$ can be converted to $Al(OH)_3(s)$ by passing $CO_2(g)$ through it. Write an equation for the same. Can HCl be used instead of CO_2?

4. Which of the following compounds are molecular, and which have an extended three-dimensional structure?

(a) B_2H_6 (b) $KAlSi_3O_6$
(c) SO_3 (d) $GeCl_4$

5. Which oxide in each of the following pairs is more acidic?
(a) Al_2O_3 or P_4O_{10} (b) B_2O_3 or Ga_2O_3
(c) SO_2 or SnO_2

6. Identify the group 13 (IIIA) element that best fits each of the following descriptions:
(a) has an unusually low melting point,
(b) is the most electronegative,
(c) is the extremely toxic,
(d) forms an acidic oxide,
(e) is the most abundant element of the group,
(f) is the most stable in +1 oxidation state,
(g) is a semiconductor,
(h) forms a molecular fluoride.

NCERT Problems

1. Write balanced equations for the reaction of elemental boron with elemental chlorine, oxygen and nitrogen at high temperature.

2. Why are boron halides are diborane referred to as "electron deficient compounds?

3. Write the structure of diborane and explain the nature of bonding in it.

4. What happens when a borax solution is acidified? Write a balaned equation for the reaction.

5. Describe what happens when boric acid is heated?

6. Describe briefly how elemental boron can be prepared?

7. Describe the shapes of BF_3 and BH_4^-. What type of hybridisation can we assign to boron in each of these compounds?

8. Why do boron halides form "addition compounds" with amines?

9. By means of a balanced chemical equation show how $B(OH)_3$ behaves as an acid in water.

Exercise 2
(Stage 2 : High Skill Problem Solving)

Only One Option Correct

1. Select incorrect statements.
(a) B_2H_6 has all B—H bond equal
(b) AlH_3 is colourless solid and is polymeric containing Al—H—Al bridges
(c) $LiAlH_4$ is ionic compound
(d) AlX_3 (X= Cl, Br, I) have low melting point, are covalent and have the halogen-bridged dimeric structure

2. Select correct statements.
(a) Hydrides of B and Si are volatile and catches fire on exposure to air
(b) Oxides of B and Si(B_2O_3 and SiO_2) are acidic in nature
(c) Borates and silicates have tetrahedral BO_4 and SiO_4 structural units
(d) All the above are correct statements

3. NaOH can't be stored in
(a) Al vessel (b) Zn vessel
(c) Both (a) & (b) (d) None of these

4. The dipole moments of the given molecules are such that
(a) $BF_3 > NF_3 > NH_3$ (b) $NF_3 > BF_3 > NH_3$
(c) $NH_3 > NF_3 > BF_3$ (d) $NH_3 > BF_3 > NF_3$

5. Consider the following boron halides
1. BF_3 2. BCl_3
3. BBr_3 4. BI_3
The Lewis acid characters of these halides are such that
(a) $1 < 2 < 3 < 4$ (b) $1 < 3 < 2 < 4$
(c) $4 < 3 < 2 < 1$ (d) $4 < 2 < 3 < 1$

6. The correct structural representation of diborane is
(a) $[BH_2]^+ + [BH_4]^-$
(b) H·B—B·H (each B also bonded to one H above and one H below)
(c) H, H bonded to B; B bridged by two H to B; B bonded to H, H
(d) H_2 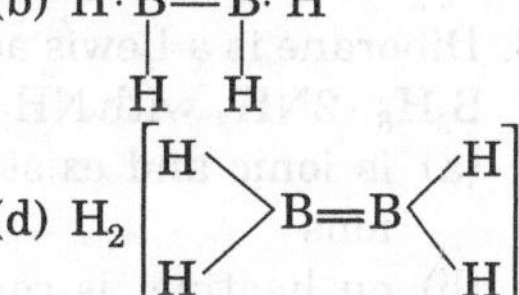

7. Consider the following statements for diborane
1. Boron is approximately sp^3 hybridized.
2. B—H—B angle is 180°
3. There are two terminal B—H bonds for each boron atom.
4. There are only 12 bonding electrons available.
Of these statements:
(a) 1,2 and 4 are correct (b) 1,2 and 3 are correct
(c) 2,3 and 4 are correct (d) 1,3 and 4 are correct

8. Molecule(s) possessing three-centre-two electron bonds and three-centre-four electron bonds would include
(a) B_2H_6 and SiF_4 (b) XeF_6 and BF_6
(c) B_2H_6 and XeF_6 (d) B_2H_6 alone

9. Which of the following has the regular tetrahedral structure?
(a) BF_4^- (b) SF_4
(c) XeF_4 (d) $[Ni(CN)_4]^{2-}$

10. H_3BO_3 and HBO_2 do not differ in
(a) oxidation number (b) basicity
(c) melting point (d) structure

11. Light-emitting diodes (LED), laser-diodes and memory chips of computers are made of
(a) gallium (b) aluminium
(c) arsenic (d) gallium-arsenide

12. Indium is used
(a) to dope crystals to make *p-n-p* transistors
(b) in thermistors
(c) in low melting point solder
(d) all the above are correct

13. When boron is fused with NaOH, products formed are
(a) $Na_2B_4O_7$ and H_2 (b) $NaBO_2$ and H_2
(c) Na_3BO_3 and H_2 (d) B_2O_3 and H_2

14. Thermite welding uses Al because of
(a) its low melting point
(b) its lightness
(c) its greater affinity for oxygen
(d) all the properties given above

15. Consider following reactions
I : $ZnO + C \longrightarrow Zn + CO$
II : $Cr_2O_3 + 2Al \longrightarrow Al_2O_3 + 2Cr$
III : $Al_2O_3 + 2Cr \longrightarrow Cr_2O_3 + 2Al$
In this, possible reactions are
(a) I, III (b) II, III
(c) I, II (d) I, II, III

16. Diborane is a Lewis acid forming addition compound $B_2H_6 \cdot 2NH_3$ with NH_3, a Lewis base. This
(a) is ionic and exists as $[BH_2(NH_3)_2]^+$ and $[BH_4]^-$ ions
(b) on heating, is converted into borazine, $B_3N_3H_6$ (called inorganic benzene)
(c) both (a) and (b) are correct
(d) none of the above is correct

17. $Al_2(SO_4)_3$ is used in the following but not
(a) as a coagulant and precipitant in treating drinking water and sewage
(b) in paper industry
(c) as a mordant in dyeing
(d) in plastic industry

18. General formula of aluminium alums is
(M^I is monovalent cation)
(a) $[M^I(H_2O)_6][Al(H_2O)_6](SO_4)_2$
(b) $[M^I(H_2O)_4][Al(H_2O)_6](SO_4)_2$
(c) $[M^I(H_2O)_4][Al(H_2O)_4](SO_4)_4$
(d) $[M^I(H_2O)_6][Al(H_2O)_6](SO_4)_4$

19. Aqueous solution of borax reacts with 2 moles of acids. This is because of
(a) formation of 2 moles of $B(OH)_3$ only
(b) formation of 2 moles of $[B(OH)_4]^-$ only
(c) formation of 1 mole each of $B(OH)_3$ and $[B(OH)_4]^-$
(d) formation of 2 moles each of $[B(OH)_4]^-$ and $B(OH)_3$, of which only $[B(OH)_4]^-$ reacts with acid

20. Borax is used as a buffer since
(a) its aqueous solution contains equal amount of weak acid and its salt
(b) it is easily available
(c) its aqueous solution contains equal amount of strong acid and its salt
(d) statement that borax is a buffer, is wrong

21. While testing BO_3^{3-}, there is green-edged flame on heating the salt with conc. H_2SO_4 and CH_3OH. Green colour is of
(a) $(CH_3)_3B$ (b) $(CH_3O)_3B$
(c) B_2O_3 (d) H_3BO_3

22. Which is insoluble in excess of NH_4OH?
(a) $Al(OH)_3$ (b) $Cu(OH)_2$
(c) $Zn(OH)_2$ (d) $Cd(OH)_2$

23. Borax is converted into B by steps
$$\text{Borax} \xrightarrow{\text{I}} H_3BO_3 \xrightarrow{\Delta} B_2O_3 \xrightarrow{\text{II}} B$$
I and II reagents are
(a) acid, Al (b) acid, C
(c) acid, Fe (d) acid, Mg

24. Boron carbide is used
(a) in nuclear reactor to absorb neutrons
(b) as an abrasive
(c) as both (a) & (b)
(d) as none of the above

25. Borax is used
(a) as a flux in brazing and in silver soldering
(b) in making enamel
(c) in leather tanning
(d) in all given above

26. Soda free glass fibre are made of
(a) H_3BO_3, B_2O_3 and $Ca_3(BO_3)_2$
(b) H_3BO_3, B_2O_3
(c) H_3BO_3
(d) B_2O_3

27. Na_3AlF_6 is added to Al_2O_3 to
(a) improve the electrical conductivity of the cell

(b) increases rate of production
(c) increase the melting point
(d) decrease the electrical conductivity

28. $Na_2B_4O_7 \cdot 10H_2O \xrightarrow{\Delta} NaBO_2 + (A) + H_2O$
$(A) + MnO \xrightarrow{\Delta} (B)$
(A) and (B) are
(a) $Na_3BO_3, Mn_3(BO_3)_2$
(b) $Na_2(BO_2)_2, Mn(BO_2)_2$
(c) $B_2O_3, Mn(BO_2)_2$
(d) none of the above is correct

29. Mixture of $Al(OH)_3$ and $Fe(OH)_3$ can be separated by
(a) HCl (b) NH_4OH
(c) NaOH (d) HNO_3

30. Boric acid is the trivial name for
(a) orthoboric acid (b) metaboric acid
(c) pyroboric acid (d) none of these

31. In H_2O, Ga^+ changes to
(a) Ga^{3+} (b) Ga
(c) Ga and Ga^{3+} (d) Ga^{2+} and Ga^{3+}

32. Al^{3+} of $KAl(SO_4)_2 \cdot 12H_2O$ (alum) can be replaced by
(a) Cr^{3+} (b) Co^{3+}
(c) Fe^{3+} (d) all of these

33. Borax $Na_2B_4O_7 \cdot 10H_2O$ is actually
(a) $Na_2[B_4O_5(OH)_4] \cdot 8H_2O$
(b) $Na_2[B_4O_4(OH)_6] \cdot 7H_2O$
(c) $Na_2[B_4O_3(OH)_8] \cdot 6H_2O$
(d) $Na_2[B_4O_2(OH)_{10}] \cdot 5H_2O$

34. Which reactions can be used to prepare diborane
I. $NaBH_4 + BF_3$ (in ether) $\longrightarrow$
II. $NaBH_4 + I_2 \longrightarrow$
III. $BF_3 + NaH \longrightarrow$
(a) I, III (b) I, II
(c) II, III (d) I, II and III

35. Which of the following statements about H_3BO_3 is not correct?
(a) It is strong tribasic acid
(b) It is prepared by acidifying an aqueous solution of borax
(c) It has a layer structure in which planar BO_3 units are joined by hydrogen bonds
(d) It does not acts as a Lewis acid by accepting hydroxyl ion

36. The type of hybridisation of boron in diborane is
(a) *sp* (b) sp^2
(c) sp^3 (d) sp^3d^2

37. Reactivity of borazole is greater than that of benzene because
(a) borazole is polar compound
(b) borazole is non-polar compound
(c) borazole is electron deficient compound
(d) of localised electrons in it

38. Arsenic drugs are mainly used for the treatment of
(a) cholera (b) typhoid
(c) jaundice (d) syphilis

39. $B(OH)_3 + NaOH \rightleftharpoons NaBO_2 + Na[B(OH)_4] + H_2O$
How can this reaction is made to proceed in forward direction?
(a) Addition of *cis*-1,2-diol
(b) Addition of borax
(c) Addition of *trans*-1,2-diol
(d) Addition of Na_2HPO_4

40. H_3BO_3 is
(a) monobasic acid and weak Lewis acid
(b) monobasic and weak Bronsted acid
(c) monobasic and strong Lewis acid
(d) tribasic and weak Bronsted acid

41. $AlCl_3$ fumes in moist air because it
(a) is covalent (b) is volatile
(c) is hygroscopic (d) forms HCl in moist air

42. The solution of sodium meta aluminate on boiling with ammonium chloride gives a white precipitate of
(a) Al_2O_3 (b) $AlCl_3$
(c) $Al(OH)_3$ (d) $(NH_4)_2SO_4 \cdot Al_2(SO_4)_3$

43. The incorrect statement is
(a) Anhydrous $AlCl_3$ is a covalent substance
(b) Ionic compounds generally have low melting point and boiling point.
(c) A molecule shows a more stable forms as compared to individual atom
(d) CCl_4 is a non-polar solvent

44. Which of the following has the highest first ionisation energy?
(a) Lithium (b) Beryllium
(c) Boron (d) Carbon

45. Fusion of borane with NaOH produce
(a) $B_2O_3 + H_2O$ (b) $NaBO_2 + H_2O$
(c) $Na_2B_4O_7 + H_2O$ (d) $Na_3BO_3 + H_2O$

46. Radius of Ga is less than Al because of
(a) lanthanoid contraction
(b) greater screening effect
(c) inert pair effect
(d) None of the above

47. In the reaction $B_2O_3 + C + Cl_2 \longrightarrow A + CO$
A is
(a) BCl_2 (b) B_2Cl_2
(c) CCl_2 (d) BCl_3

48. Which of the following compounds is/are possible?
(a) B_2Cl_4, BCl_2, Cl_2B (b) $GaCl_2$, GaS
(c) $Ga[GaCl_4] \cdot Ga_2S_2$ (d) All of these

49. Diamagnetic species can be
(a) B_2Cl_4 (b) $GaCl_2$
(c) GaS (d) all of these

50. Which of the following are used as transition materials and light-emitting devices (LED)?
I : Gallium-phosphides
II : Indium-phosphides
III : Gallium-arsenides
IV : Indium-arsenides
(a) All (b) Except I all
(c) Except I, II all (d) Only IV

51. Select correct statemet(s).
(a) Aluminium dissolves in dilute mineral acids but is made passive by concentrated nitric acid
(b) Aluminium vessel can be used as storage of NaOH
(c) Both of the above
(d) None of the above

52. Thallium and lead are classified as chemially soft because
(a) they have higher affinities for soft anions as I^- and S^{2-} ions
(b) they are soft in reactivity
(c) they are stable in lower oxidation state (Tl^+, Pb^{2+})
(d) they are stable in higher oxidation state (Tl^{3+}, Pb^{4+})

53. The relative stability of the different oxidation state

$$Tl^+ > Tl^{3+}, > Ga^{3+} > Ga^+$$
$$Tl^+ > In^+ > Ga^+$$

is an example of
(a) redox potential (b) disproportionation
(c) inert pair effect (d) electron-affinity

54. Thermal stability of boron compounds is in order
(a) $B_2F_4 < B_2Cl_4 < B_2(OR)_4 < B_2(NR_2)_4$
(b) $B_2Cl_4 < B_2F_4 < Br(OR)_4 < B_2(NR_2)_4$
(c) $B_2(NR_2)_4 < B_2(OR)_4 < B_2Cl_4 < B_2F_4$
(d) $B_2(NR_2)_4 < B_2(OR)_4 < B_2F_4 < B_2Cl_4$

55. Anion containing both three-and four-coordinated boron is
(a) $[B(OH)_4]^-$ (b) $B_3O_6^{3-}$
(c) $[B_4O_5(OH)_4]^{2-}$ (d) $BO_3BO_4^{4-}$

56. Which is spontaneous in forward side?
(a) $Al^{3+} + 2e^- \longrightarrow Al^+$ (b) $Ga^{3+} + 2e^- \longrightarrow Ga^+$
(c) $In^{3+} + 2e^- \longrightarrow In^+$ (d) $Tl^{3+} + 2e^- \longrightarrow Tl^+$

57. Light elements B, Al, C and Si are
(a) oxophiles (high affinity for oxygen)
(b) fluorophiles (high affinity for fluorine)
(c) both type
(d) none of the types given

58. Stabilities of adducts formed with boron-halides is in order
(a) $BF_3 < BCl_3 < BBr_3$ (b) $BBr_3 < BCl_3 < BF_3$
(c) $BCl_3 < BBr_3 < BF_3$ (d) $BF_3 < BBr_3 < BCl_3$

59. The thermal disproportionation of B_2Cl_4 gives
(a) BCl_3 (b) B_4Cl_4
(c) B_8Cl_8 (d) all of these

60. Boric acid is a weak acid. Thus, it behaves as
(a) proton donor as in

$$H_3BO_3 + H_2O \rightleftharpoons H_3O^+ + H_2BO_3^-$$

(b) an electron acceptor as in

$$B(OH)_3 + H_2O \rightleftharpoons [B(OH)_4]^- + H^+$$

(c) proton acceptor as in

$$H_3BO_3 + H_2O \rightleftharpoons H_4BO_3^+ + OH^-$$

(d) an electron donor as in

$$H_3BO_3 \rightleftharpoons H_3BO_3^+ + e^-$$

61. The decrease stability of higher oxidation state in *p*-block with increasing atomic number is due to
(a) decrease in bond energy as going down the group
(b) energy required to unpair ns^2 electrons is not compensated by the energy released in forming the two additional bonds
(c) both (a) and (b) are correct
(d) none of the above is correct

62. Select correct statement(s) about group 13 (IIIA) elements.
(a) Stability of hydrides (MH_3) decreases as we move down the group
(b) MX_3 is a strong Lewis acid
(c) $[Al(H_2O)_6]^{3+}$ exists in aqueous solution
(d) All the above are correct statements

63. AlO_2^- is known as
(a) aluminium dioxo ion (b) meta aluminate ion
(c) dioxo aluminium ion (d) aluminium oxide ion

64. In aqueous solution GaCl disproportionates to
(a) $GaCl_2$ and $GaCl_3$ (b) Ga and $GaCl_3$
(c) $GaCl_2$ and Ga (d) $GaCl_3$ and $GaCl_5$

65. Boron does not form B^{3+} cation easily. It is due to
(a) energy required to form B^{3+} ion is for more than that which would be compensated by lattice energies or hydration energies of such ion
(b) boron is non-metal
(c) boron is semiconductor
(d) none of the above

66. An alloy of boron and aluminium is allowed to react with NH_3. Products formed are
(a) BN, AlN and H_2
(b) $B(NH_2)_3$, $Al(NH_2)_3$ and H_2
(c) BN, $Al(NH_2)_3$ and H_2
(d) $B(NH_2)_3$, AlN and H_2

67. Element $M + N_2 \xrightarrow{\Delta} \xrightarrow{H_2O} NH_3$
Element *M* belonging to group 13 can be
(a) B or Al (b) Ga or Al
(c) Mg or Al or B (d) Mg or B

68. N_2 gas is passed or heated Mg and B. Products formed are
(a) Mg_3N_2, BN (b) Mg_3N_2, BN and Mg_3B_2
(c) MgN, BN (d) no reaction

69. Anodising can be done by electrolysing dilute H_2SO_4 with Al as anode. This results in
(a) the formation of protective oxide layer (Al_2O_3)
(b) the formation of $Al_2(SO_4)_3$ and SO_2 gas
(c) the formation of AlH_3 and SO_2 gas
(d) the formation of $Al(HSO_3)_3$ and H_2 gas

70. Alzheimer's disease is caused due to Al interaction with internal organs of the body if food is contaminated with Al. This diseases
(a) induces senility in young persons
(b) causes memory loss
(c) both (a) and (b) are correct
(d) none of the above is correct

71. Pure boron is best prepared by
(a) heating B_2O_3 with H_2
(b) heating B_2O_3 with Na or K
(c) heating KBF_4 with Na or K
(d) heating BBr_3 with H_2 in presence of a catalyst

72. The borax bead test can be used to detect the presence of
(a) Al^{3+} (b) Mg^{2+}
(c) Fe^{3+} (d) Na^+

73. Water softener is
(a) borax (b) zeolite
(c) both (a) and (b) (d) none of these

74. $Na_2B_4O_7 \cdot 10H_2O$ is correctly represented as
(a) $Na_2[B_4O_5(OH)_4] \cdot 8H_2O$
(b) $2NaBO_2 \cdot Na_2B_2O_3 \cdot 10H_2O$
(c) $Na_2[B_4(H_2O)_4O_7] \cdot 6H_2O$
(d) all of the above

75. B can be obtained from halide by van Arkel method. This involves reaction
(a) $2BCl_3 + 3H_2 \xrightarrow[\text{filament}]{\text{red hot W or Ta}} 2B + 6HCl$
(b) $2BI_3 \xrightarrow{\text{red hot W or Ta}} 2B + 3I_2$
(c) both (a) and (b)
(d) none of the above

76. Which of the following statements regarding ortho boric acid (H_3BO_3) is false?
(a) It acts as a monobasic acid
(b) It is soluble in hot water
(c) It has a planar structure
(d) It acts as a tribasic acid

77. When orthoboric acid is heated to red heat the residue is
(a) boron (b) boric oxide
(c) metaboric acid (d) pyroboric acid

78. Orthoboric acid H_3BO_3 behaves as weak monobasic acid giving H_3O^+ and
(a) $H_2BO_2^+$ (b) $H_2BO_2^-$
(c) $[B(OH)_4]^-$ (d) $[B(OH)_4]^+$

79. Boron sesquioxide is
(a) B_2O_3 (b) BO_3
(c) BO_3^{3-} (d) BO_2^-

80. Select correct statement(s) about H_3BO_3
(a) It has triangular BO_3^{3-} units
(b) In solid states, molecules are hydrogen bonded
(c) Both (a) & (b) are correct
(d) None of the statements is correct

81. The structure of BF_3 is
(a) tetrahedral (b) pyramidal
(c) planar triangular (d) *T*-shaped

82. In BF_3
(a) all the B—F bond are single covalents in nature
(b) boron-fluorine bond has some double bond character and this bond is delocalised
(c) bond energy and bond length of B—F bond indicate its single bond character
(d) all the bonds are ionic

83. BF_3 behaves as a
(a) Lewis acid and lone pair from the Lewis base is filled into empty $2p_z$ orbital
(b) Lewis base and lone pair on B is donated to empty orbital of the Lewis acid
(c) amphoteric species due to interaction with acid and base both
(d) polymer substance

84. B_2Cl_4
(a) can be formed by passing BCl_3 over Hg at low pressure in an electric discharge
(b) can exist on non-eclipsed and planar structure
(c) decomposes at low temperature
(d) can show all the facts given above

85. BF_3 on hydrolysis forms
(a) H_3BO_3 (b) HBF_4
(c) both (a) and (b) (d) None of these

86. Which of the following is false statement?
(a) BH_3 is a Lewis acid
(b) All the B—H distance in diborane (B_2H_6) are equal
(c) Boranes are easily hydrolysed
(d) $LiAlH_4$ reduces BCl_3 to borane

87. In B_2H_6
(a) there is direct boron-boron bond
(b) the B—H bonds are ionic
(c) it is isostructural to C_2H_6
(d) boron atoms are linked through hydrogen bridges

88. From B_2H_6 all the following can be prepared except
(a) H_3BO_3 (b) $NaBH_4$
(c) $B_2(CH_3)_6$ (d) B_2O_3

89. Boron nitride has the structure of the type
(a) both diamond and graphite
(b) graphite
(c) diamond
(d) NaCl

90. Which of the following is the false statement?
(a) Reducing action of $NaBH_4$ and $LiAlH_4$ are specific
(b) B_2H_6 reduces alkene to primary alcohol
(c) Both (a) and (b) are the false statements
(d) None of the above is false statement.

91. Beryllium and aluminium have similar properties because
(a) they belong to same group
(b) they have different electronegativity
(c) they have same electronegativity
(d) they have same ionisation energy

92. The solubility of $Al(OH)_3$ in strong NaOH solution is best explained by the formation of
(a) AlO_3^- ion
(b) double salt
(c) a peptised colloidal sol
(d) a complex hydroxyl aluminate ion

93. On passing CO_2 into aqueous solution containing Al^{3+}
(a) $Al_2(CO_3)_3$ is formed
(b) $Al(OH)_3$ is precipitated
(c) $[Al(OH)_4]^-$ is formed
(d) colloidal $Al(OH)_3$ is formed

94. Addition of sodium carbonate to a solution of an oxide in water produces CO_2. This experiment indicates that
(a) the oxide is that of a non-metal
(b) the oxide is basic
(c) the oxide is amphoteric
(d) the oxide is neutral

95. Dimer formation can take place with
(a) BH_3 (b) $AlCl_3$
(c) NO_2 (d) all of these

96. The major constituent in cement is
(a) silica (b) magnesium oxide
(c) calcium carbonate (d) iron oxide

97. Hardening of cement is due to
(a) dehydration
(b) hydrolysis
(c) hydration and hydrolysis
(d) polymerisation

98. Alums are used as mordant is dyeing because
(a) dye is adsorbed on $Al(OH)_3$ which is deposited on fibre in the hydrolysis process
(b) dye is adsorbed on KOH formed due to hydrolysis
(c) of both (a) and (b)
(c) of none of the above

99. Alum is found to contain hydrated monovalent cation $[M(H_2O)_6]^+$, trivalent cation $[M'(H_2O)_6]^{3+}$ and SO_4^{2-} in the ratio of
(a) 1 : 1 : 1 (b) 1 : 2 : 3
(c) 1 : 3 : 2 (d) 1 : 1 : 2

100. About 2 to 5% of gypsum ($CaSO_4 \cdot 2H_2O$) is added to Portland cement. It is just to
(a) increase the setting rate of the cement during hydration
(b) slow down the setting rate to increase strength during hydration
(c) both (a) and (b)are correct
(d) none of the above is correct

One or More Than One Options Correct

1. Which fumes in moist air?
(a) $AlCl_3$ (b) BF_3
(c) $CuSO_4$ (d) NH_3

2. Select correct statement(s).
(a) BF_3 fumes strongly in moist air and is partially hydrolysed by excess of water
(b) BF_3 is converted into the adducts $BF_3 \cdot H_2O$ and $BF_3 \cdot 2H_2O$ with small amounts of water at low temperature
(c) H_3BO_3 is a weak acid but HBF_4 is a very strong acid
(d) KBF_4 is sparingly soluble in water

3. Select correct statement(s).
(a) BF_3 forms adducts with NH_3, ROH and RNH_2
(b) BCl_3 form adducts with NH_3, H_2O and RNH_2
(c) BCl_3 reacts with liquid NH_3 and ethanol forming $B(NH_2)_3$ and $B(OEt)_3$
(d) $[BCl_4]^-$, $[BBr_4]^-$ and $[BI_4]^-$ ions can be made only in non-aqueous media

4. Select correct statement(s).
(a) Boron trimethyl is a weaker Lewis acid than the boron trihalides or monoborane
(b) Lewis acid property when Me_3N is a donor is in order :

$$BBr_3 > BCl_3 > BF_3 \sim BH_3 > BMe_3$$

(c) Lewis acid property when Me_3P is a donor is in order :

$$BBr_3 > BCl_3 \sim BH_3 > BF_3 \sim BMe_3$$

(d) Lewis acid property when CO is a donor : no adduct is formed with halides and Lewis acid property of

$$BH_3 >> BF_3 \cdot BMe_3$$

5. Select correct statement(s).
(a) A saturated solution of boric acid in water is neutral to the indicator bromocresol green
(b) AlF_3 is insoluble in anhydrous HF
(c) Aqueous solution of KHF_2 is basic to the indicator bromocresol green
(d) Aqueous solution of boric acid and potassium hydrogen difluoride is alkaline to bromocresol indicator

6. Select correct statement(s).
(a) Diborane is an electron-deficient compound
(b) Diborane is stoichiometrically analogous to ethane (C_2H_6)
(c) Total electrons involved in bonding are 12 in diborane and ethane
(d) Total electrons involved in bonding are 12 in diboroane and 14 in ethane

7. Boric acid is a weak Lewis acid proton comes from
(a) by its ionisation
(b) when it forms complex with H_2O
(c) when it forms borate esters with alcohol
(d) all of the above

8. Select correct statement(s).
(a) On the basis of Lewis acidity, an oxygen atom is more effective than an fluorine atom as a π-donor towards boron
(b) 1,2-diols have a strong tendency to form borate esters on account of chelate effect
(c) Borate esters are stable due to chelate effect
(d) None of the above is correct statement

9. Actual source of protons in boric acid is based on the following
(a) $B(OH)_3(aq) + 2H_2O(l) \rightleftharpoons H_3O^+(aq) + [B(OH)_4]^-(aq)$
(b) $3B(OH)_3(aq) \rightleftharpoons [B_3O_3(OH)_4]^-(aq) + H^+(aq) + 2H_2O(l)$
in neutral or basic solution
(c) $B(OH)_3 \rightleftharpoons H_2BO_3^- + H^+$
(d) all of the above

10. Select correct statement(s).
(a) Aluminium monohalides are formed by the reactions of the trihalides with aluminium at 1300K followed by rapid cooling
(b) Aluminium monochloride (red) is formed when aluminium reacts with HCl at 1200 K
(c) Aluminium monohalides disproportionates at room temperature
(d) Ga_2Cl_4 exists as $Ga^+[GaCl_4]^-$ and is diamagnetic and has crystalline structure.

11. Consider following statements
I : Boric acid is a mild antiseptic and is used as a food preservative.
II : Borax and other borates are used in water-treatment, timber preservation, glass manufacture.
III :Sodium peroxoborate is an important constituent of washing powder.
Select correct statement.
(a) I, II (b) II, III
(c) I, III (d) I, II, III

12. Select correct alternate(s).
(a) The group BN is isoelectronic with C_2
(b) No species with B≡N group is formed
(c) Common form of boron nitride has an ordered layer structure closely resembling that of graphite
(d) AlN has the wurtzite structure

Passage Comprehension Questions

Passage 1

The tetrafluoroborate anion BF_4^-, is a Lewis base. The other tetrahaloborate anions, BCl_4^- and BBr_4^-, can be prepared in non-aqueous solvents.

Based on the above short write-up, answer the following questions :

1. BF_4^- is
(a) F^- donor
(b) a lone-pair acceptor
(c) a lone-pair donor
(d) solvated in aqueous solution

2. BCl_4^- and BBr_4^- are stable in
(a) H_2O (b) alcohol
(c) both (a) and (b) (d) None of these

Passage 2

Certain aspects of the aqueous chemistry of aluminium compounds derive from the nature of the hydrated aluminium ion-$[Al(H_2O)_6]^{3+}$.

$$[Al(H_2O)_6]^{3+} + H_2O \rightleftharpoons [Al(H_2O)_5(OH^-)]^{2+} + H_3O^+$$

Read the above write-up and answer the following questions.

1. Above equilibrium indicates that
(a) H_2O is a Lowry-Bronsted acid
(b) H_2O is a Lowry-Bronsted base
(c) H_2O is a Lowry-Bronsted acid as well as a base
(d) H_2O is neither acid nor base

2. Above equilibrium is repressed in
(a) strongly acidic solution
(b) alkaline solution
(c) aqueous solution
(d) any of the above

3. Ionisation of the above equilibrium is favoured in
(a) strongly acidic solution
(b) alkaline solution

(c) aqueous solution
(d) any of the above

4. Aluminate ion can be written as
(a) AlO_2^- (b) $Al(OH)_4^-$
(c) $[Al(H_2O)_2(OH)_4]^-$ (d) all of these

Passage 3

Diborane is an electron deficient compound. Stoichiometrically, it is analogous to ethane, C_2H_6. However, the total number of electrons involved in bonding and less in diborane.

Read the above short note and answer the following questions.

1. Bonding electrons involved
(a) in B_2H_6 is 14 and in C_2H_6 16
(b) in B_2H_6 is 12 and in C_2H_6 14
(c) in B_2H_6 is 10 and in C_2H_6 12
(d) in B_2H_6 is 16 and in C_2H_6 18

2. Octet is complete in
(a) boron in diborane
(b) carbon in ethane
(c) boron in diborane and carbon in ethane
(d) None of the above

3. In diborane there is
(a) two-electron-three-centre bonds
(b) three-electron-three-centre bonds
(c) two-electron-two-centre bonds
(d) three-electron-two-centre bonds

4. Consider the following statements for diborane
I. Boron is approximately sp^3 hybridised
II. B—H—B angle is 180°
III. There are two terminal B—H bonds for each boron atom
IV. There are only 12 bonding electrons available
Of these statements
(a) I, III and IV are correct
(b) I, II and III are correct
(c) II, III and IV are correct
(d) I, II and IV are correct

Passage 4

Study the following sequence of reactions :

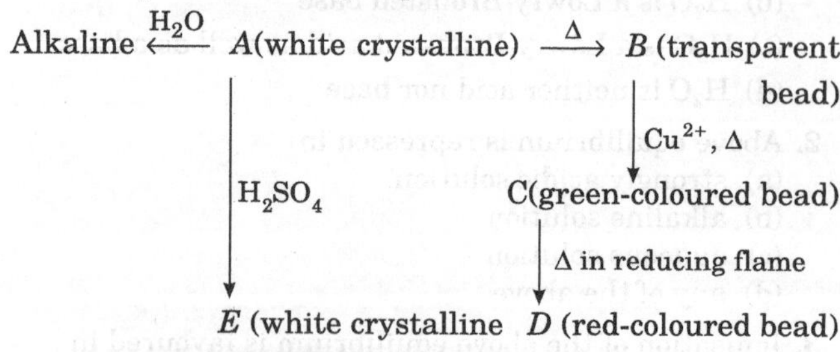

1. *A* is
(a) $NaBO_2$ (b) $Na_2B_4O_7$
(c) H_3BO_3 (d) HBO_3

2. *B* is
(a) $NaBO_2$ (b) B_2O_3
(c) Na_3BO_3 (d) H_3BO_3

3. *C* and *D* respectively are
(a) Cu_2BO_2, CuO (b) $Cu(BO_2)_2$, $CuBO_2$
(c) $Cu_2B_2O_4$, CuO (d) $Cu(BO_2)_2$, $Cu_2B_2O_4$

4. Select correct statement(s) about *E*
(a) *E* is a weak acid (H_3BO_3)
(b) It is not a protonic acid but electron-pair acceptor
(c) *E* shows following equilibrium
$$B(OH)_3 + H_2O \rightleftharpoons [B(OH)_4]^- + H_3O^+$$
(d) All the above are correct statements

Passage 5

The chemical properties of boron, carbon, silicon and germanium are distinctly those of non-metals. The light elements, boron, aluminium, carbon and silicon are strong oxophiles and fluorophiles, in the sense that they have high affinities for oxygen and fluorine, respectively.

Based on the above short write-up, answer the following questions.

1. Oxophilic character is observed in
(a) borates, aluminates, carbonates, silicates
(b) boranes, alkanes, silanes, alanes
(c) both of the above
(d) none of the above

2. The similar oxophilicity and fluorophilicity concludes that diagonal relationship is between
(a) boron and aluminium
(b) boron and silicon
(c) boron and carbon
(d) aluminium and silicon

Assertion & Reason

Codes :
(a) Both **A** and **R** are true and **R** is the correct explanation of **A**.
(b) Both **A** and **R** are true but **R** is not the correct explanation of **A**.
(c) **A** is true but **R** is false.
(d) **A** is false but **R** is true.

1. Assertion (A) : Boron forms BF_4^-.
Reason (R) : BF_3 is an electron deficient species.

2. Assertion (A) : B_2Cl_4, $GaCl_2$, GaS are diamagnetic.
Reason (R) : They show +2 oxidation state.

3. Assertion (A) : $NaBH_4$ has NaCl structure.
Reason (R) : $NaBH_4$ is an ionic salt.

4. **Assertion (A) :** HBF_4 is a very strong acid.

Reason (R) : $B(OH)_3$ behaves as an acid in the following reaction :

$$B(OH)_3 + 4HF \longrightarrow H_3O^+ + [BF_4]^- + 2H_2O$$

5. **Assertion (A) :** Cryolite is added to bauxite in metallurgical extraction of aluminium.

Reason (R) : Cryolite, $Na_3[AlF_6]$ contains an octahedrally coordinated aluminium atom.

6. **Assertion (A) :** $AlCl_3$ exists as $[Al(H_2O)_6]Cl_3$ in dilute aqueous solution.

Reason (R) : When water is dropped to a solid $AlCl_3$, vigrous hydrolysis ensues.

7. **Assertion (A) :** $B(OH)_3$ is a good lubricant and slippery.

Reason (R) : $B(OH)_3$ has a layer structure.

8. **Assertion (A) :** Boric acid does not act as a proton donor but an olectron-acceptor.

$$B(OH)_3 + H_2O \rightleftharpoons [B(OH)_4]^- + H^+$$

Reason (R) : Boric acid is a very weak acid.

9. **Assertion (A) :** The group BN is isoelectronic with C_2.

Reason (R) : No species containing a B≡N group has yet been discovered.

10. **Assertion (A) :** $B_3N_3H_{12}$ (similar to cyclohexane) exists as chair conformation.

Reason (R) : Borazine ($B_3N_3H_6$) is much more reactive than benzene.

11. **Assertion (A) :** Aluminium trimethyl is a dimer having structure like that of $AlCl_3$.

Reason (R) : There is three-centre two-electron bonding in Al—C—Al bridge.

12. **Assertion (A) :** Borazine is called inorganic benzene.

Reason (R) : Borazine undergoes addition and substitution reactions with HCl and Br_2 etc.

13. **Assertion (A) :** The bond lengths in BF_3 are 1.30 Å and are shorter than the sum of the covalent radii.

Reason (R) : $p\pi$-$p\pi$ interaction exists and all B—F bonds are equivalents are based on modern explanation, the double bond is delocalised.

14. **Assertion (A) :** $AlCl_3$ forms dimer Al_2Cl_6 but is dissolves in H_2O forming $[Al(H_2O)_6]^{3+}$ and Cl^- ions.

Reason (R) : Aqueous solution of $AlCl_3$ is acidic due to hydrolysis.

15. **Assertion (A) :** In aqueous solution $AlCl_3$ forms dimer Al_2Cl_6.

Reason (R) : The high enthalpy of hydration is sufficient to break the covalent dimer into $[M(H_2O)_6]^{3+}$ and Cl^- ions.

16. **Assertion (A) :** In water, orthoboric acid behaves as a weak monobasic acid.

Reason (R) : In water orthoboric acid acts as a proton donor.

17. **Assertion (A) :** Boron always forms covalent bond.

Reason (R) : The small size of B^{3+} favours formation of covalent bond.

True & False

1. Borosilicate glass has a lower coefficient of thermal expansion.
2. H_3BO_3, B_2O_3 and $Ca_3(BO_3)_2$ are used to make soda-free glass fibre.
3. Al_2O_3 is a good conductor.
4. When boron is fused with sodium hydroxide, sodium metaborate ($NaBO_2$) and hydrogen are formed.
5. H_3PO_2 is a weak monobasic acid.
6. Borax is also a good buffer.
7. Metaborate beads of many transition metals have characteristic colours.
8. H_3BO_3 is not a Bronsted acid but Lewis acid and accepts OH^-.
9. NaH_2PO_2 and $NaHPO_3$ are acid salts.
10. Borazine $B_3N_3H_6$ is called inorganic benzene.
11. In diborane all the B—H bonds are equivalents
12. Structure of aluminate ion changes with both pH and concentration.
13. $Al(OH)_3$ is used as mordant in dyeing.
14. In diborane (B_2H_6) there are (*2c-2e*) (*3c-2e*) type of bonds.
15. Aqueous solution of $AlCl_3$ is acidic due to hydrolysis.

Fill in the Blanks

1. Correct representation of the formula of borax is ....
2. Conversion of BI_3 into B by pyrolysis is called ....
3. Cryolite .... the electrical conductivity of Al_2O_3 in the extraction of Al.
4. Stability order of $Ga^+ < In^+ < Tl^+$ is described by ....

5. Concentrated HNO_3 renders Al passive due to formation of ....
6. Al_2O_3 is an .... oxide.
7. H_3BO_3 behaves as ....
8. H_3BO_3 does not donate protons but accepts .... ion.
9. In diborane, boron atom is .... hybridised.
10. S_2 is .... since, it has .... electrons.
11. AlO is formed when aluminium powder is heated in oxygen at 3260°C and is called ....
12. When ethereal solutions of $AlCl_3$ and $LiAlH_4$ are mixed, ... is formed.
13. $AlCl_3$ fumes in moisture due to ....
14. When AlI_3 is heated with CCl_4, ... and .... are formed.
15. Aluminium chloride forms dimer Al_2Cl_6 but in aqueous solution $[Al(H_2O)_6]^{3+}$ and Cl^- ions are formed since, covalent bonds of dimer are broken by ....
16. *p*-block elements are characterised by valence shell electronic configuration as ....
17. Boron and aluminium both are amphoteric and form .... with NaOH.
18. Halides of group 13(IIIA) have halogen bridged .... structures.
19. Reaction of Al with O_2 is .... and is called .... reaction.
20. Use of Al causes disease called ....
21. Borax is used as a .... standard in titrations against acids.
22. B_2H_6 changes $CH_3CH{=}CH_2$ with alkaline H_2O_2 into ....
23. Diborane (B_2H_6) has .... electron bonds.
24. Cu^{2+} salts form coloured bead with borax of the formula ....
25. Corundum is ....
26. Gypsum is added to cement to slow down the .... to increase strength.
27. Be_2C and Al_4C_3 are called .... because they react with H_2O yielding methane.
28. Si_3N_4, B_4C and SiC are .... ceramics.

Matrix-Match Type Questions

I. Only One Correct

1. Match the compounds (in List I) with their use (in List II).

Compound (List I)	Use (List II)
A. Boric acid	1. In Friedel-Craft acylation
B. Borax	2. Mild antiseptic
C. Aluminium chloride	3. Washing powder
D. Sodium peroxoborate	4. Buffer

2. Match the types of borates (in List I) with their formula (in List II).

Types of borates (List I)	Formula (List II)
A. Metaborate	1. $NaBO_3 \cdot 4H_2O$
B. Orthoborates	2. $Mg_2B_2O_5$
C. Pyroborates	3. $K_3(BO_2)_3$
D. Perborate	4. $Ca_3(BO_3)_2$

3. Match the items (in List I) with their formula (in List II).

Items (List I)	Formula (List II)
A. Tetraboric acid	1. $B_3N_3H_6$
B. Borax	2. $Ga(OH)_2$
C. Borazine	3. $H_2B_4O_7$
D. Amphoteric	4. $Na_2B_4O_7 \cdot 10H_2O$

II. One or More Correct

1. Match the compounds in Column I with their corresponding property (ies) in Column II.

Column I	Column II
A. Borax	1. Electron deficient
B. Ortho boric acid	2. sp^2 hybridisation
C. Borane	3. Coloured beads with CuO
D. Alums	4. Mordant in dyeing
E. Boron nitride	5. Triangular planar
	6. Changes $CH_3CH{=}CH_2$ into alcohol by anti-Markownikoff's addition

2. Match the species in Column I with their characteristics in Column II.

Column I	Column II
A. Benzene	1. Six-membered ring
B. Borazine	2. Delocalisation of π-electrons
C. Inorganic graphite	3. sp^2-hybridised
D. Borane	4. Mole fraction of hydrogen 0.5
	5. Mole fraction of nitrogen and boron equal
	6. Mole fraction of boron 0.25

Integer Answer Types

This section contains 8 questions. The answer to each of the questions is a single digit integer, ranging from 0 to 9. The appropriate bubbles below the respectively question numbers in the *ORS* have to be darkened. For example, if the correct answers to question number *X, Y, Z* and *W* (say) are 6, 0, 9 and 2, respectively, then the correct darkening of bubbles will like the following :

X	Y	Z	W
0	0	0	0
1	1	1	1
2	2	2	2
3	3	3	3
4	4	4	4
5	5	5	5
6	6	6	6
7	7	7	7
8	8	8	8
9	9	9	9

Set I

1. Maximum covalency of aluminium is

2. In the hydrolysis of borax

$$Na_2B_4O_7 + 7H_2O \longrightarrow 2Na[B(OH)_4] + B(OH)_3$$

products formed are to be neutralised by HCl. Number of moles of HCl required per mole of borax hydrolysed is

3. Total number of B and N atom in borazine is ....

4. $B(OH)_3$ can accept .... hydroxide ion(s) from H_2O.

5. Aluminium methanide contains .... carbon atom(s).

6. ΔG is ngative for the following reaction

$$Tl^{3+} \longrightarrow Tl^{x+}$$

when x = ...

7. Out of B_2Cl_4, $GaCl_2$, GaS and S_2 number of diamagnetic substances are ....

8. Number of tribasic acid out of the following is ....

$H_3BO_3, H_3PO_3, H_3PO_2, H_3PO_4$

Set II

1. Alum is found to contain hydrated monovalent cation, trivalent cation and sulphate in the simplest ratio $x : y : z$ where $x + y + z$ =

2. Borax is found to have .... tetrahedral unit(s).

3. B_2H_6 has B—H terminal bonds and B—H bridging bonds which is equal to ....

4. Total electrons involved in all the B—H bridging bond is ....

5. Total electrons involved in all the B—H terminal bonds is ....

6. A x—membered ring is formed in $(BN)_n$ where x is ....

7. Al(*s*) is to be converted into $Al^{3+}(aq)$ ion. How many steps are involved? (Assume ionisation in one step)

8. $Cl_2(g)$ is to be converted into $Cl^-(aq)$ ion. How many steps are involved?

Answers & Solutions

Master Exercises

Exercise 1

Short Answer Type Questions

1. $B_4O_5(OH)_4^{2-}(aq) + 2H^+(aq) + 4H_2O(l) \longrightarrow 4H_3BO_3(aq)$

No, it does not; the mole ratio of the ion to the H^+ ions is the same in both cases.

2. When borax is dissolved in water, both $B(OH)_3$ and $B(OH)_4^-$ are formed, but only $[B(OH)_4]^-$ reacts with HCl. $[B(OH)_3$, called orthoboric acid, does not react with HCl]

$$\underset{1\text{ mol}}{Na_2B_4O_7} + 7H_2O \longrightarrow \underset{2\text{ mol}}{2Na[B(OH)_4]} + \underset{2\text{ mol}}{2H_3BO_3}$$

$$2Na[B(OH)_4] + 2HCl \longrightarrow 2NaCl + 2H_3BO_3 + 2H_2O$$

or $$Na_2B_4O_7 + 2H_2O \longrightarrow 2NaCl + \underset{\text{weak acid}}{H_2B_4O_7}$$

$$2NaOH + 2HCl \longrightarrow 2NaCl + 2H_2O$$

3. Borax can be used as a buffer since, its aqueous solution contains equal moles of weak acid (H_3BO_3) and its salt $Na[B(OH)_4]$.

$$B_4O_7^{2-} + 7H_2O \longrightarrow \underset{\text{salt}}{2[B(OH)_4]^-} + \underset{\text{weak acid}}{2H_3BO_3}$$

4. Since, $B(OH)_3$ only partially reacts with H_2O to form H_3O^+ and $[B(OH)_4]^-$, it behaves as a weak acid. Thus, sharp end point is not obtained when it is titrated with NaOH :

$$B(OH)_3 + 2H_2O \longrightarrow [B(OH)_4]^- + H_3O^+$$

5. (a) H_3BO_3 (b) $B(OH)_3$ (c) it looks like a hydroxide

6. (a) $$3BCl_3 + 3NH_4Cl \xrightarrow[C_6H_5Cl]{140°C} \underset{(A)}{B_3N_3H_3Cl_3} \xrightarrow{NaBH_4} \underset{\substack{\text{Borazine}\\\text{(inorganic benzene)}\\(B)}}{B_3N_3H_6}$$

```
        H
        |
 H    B    H
  \ /   \ /
   N     N
   |     |
   B     B
  / \   / \
 H    N    H
      |
      H
```

(b) $$Na_2B_4O_7 + 2NH_4Cl \xrightarrow{\text{red heat}} \underset{\substack{(C)\\\text{Boron nitride}\\\text{(inorganic graphite)}}}{2BN} + B_2O_3 + 2NaCl + 4H_2O$$

(c) $$4BCl_3 + 3LiAlH_4 \longrightarrow 3LiCl + 3AlCl_3 + \underset{\substack{\text{Diborane}\\(D)}}{B_2H_6}$$

(d) $$4BCl_3 + 6H_2 + C_{(\text{fibre})} \xrightarrow{1700-1800°C} \underset{(F)}{B_4C_{(\text{fibre})}} + 12HCl$$

7. Bond order : bond $X = 1$ (formed by two electrons)
$Y = 0.5$ (formed by one electron)
(Also see text)

8. Refer text.

9. Fire extinguishers can be made that contain aluminium sulphate together with the usual carbonate and acid. When the extinguisher is used, $Al(OH)_3$ is produced :

$$2Al^{3+}(aq) + 3CO_3^{2-}(aq) + 3H_2O \longrightarrow 2Al(OH)_3(s) + 3CO_2(g)$$

The hydroxide has the virtue of stabilising the foam, which covers the fire.

Particles of $Al(OH)_3$ are able to attach themselves to the fibres in cloth. The hydroxide is able to adsorb dye molecules much more efficiently than the fibres alone. A substance like $Al(OH)_3$ that assists in dyeing of a fabric is called a **mordant.**

10. $$C_2H_5OH \xrightarrow{\Delta,\ Al_2O_3} \underset{\text{ethene}}{CH_2{=}CH_2} + H_2O$$

11. $AlCl_3$ forms AlCl in the gas phase at elevated temperatures:

$$AlCl_3 + 2Al \xrightarrow[\text{temperature}]{\text{high}} 3AlCl$$

12. $(A): NH_4Al(SO_4)_2 \cdot 12H_2O$ $(B): NH_3$ $(C): BaSO_4$ $(D): Al(OH)_3$ $(E): NaAlO_2$

$$\underset{(A)}{NH_4Al(SO_4)_2 \cdot 12H_2O} \xrightarrow{NaOH} \underset{(B)}{NH_3\uparrow} + \underset{(D)}{Al(OH)_3}$$

$$\underset{(A)}{NH_4Al(SO_4)_2 \cdot 12H_2O} \xrightarrow{BaCl_2} \underset{(C)}{BaSO_4} \qquad \underset{(D)}{Al(OH)_3} \xrightarrow{NaOH} \underset{(E)}{NaAlO_2}$$

13. Ti^{3+}, Cr^{3+}, Mn^{3+}, Fe^{3+}, Co^{3+}, etc. (of approximately same size as that of Al^{3+}) can replace Al^{3+} in alums.

Analytical Questions

1. (i) BCl_3 is stable due to stability of boron-chlorine π-bonding. When there is dimer formation, this bonding is disrupted as there would be no vacant orbital for π-bonding. It is also likely that the boron atom is too small to form the stable 4-membered ring system in such dimers.

   $AlCl_3$ is electron-deficient in Al (octet-incomplete) thus behaves as Lewis acid and Al completes its octet by taking electron-pair from Cl-atom

   (Structure: Al_2Cl_6 dimer — two Al atoms each bonded to two terminal Cl, bridged by two Cl atoms donating lone pairs to Al)

   The crystalline solid has a layer-lattice with six-coordinated Al. On melting it changes to the halogen-bridged dimer Al_2Cl_6 in which coordination number drops to four. This is accompanied by a dramatic increase in volume-density of the melt is nearly 45% lower than that of the solid at the melting point. The electrical conductivity also approaches zero. The dimers also dominate in gas phase.

   (ii) Boron has only 2*s*- and 2*p*-valence orbitals and can form maximum of four bonds. Aluminium has available 3*d*-orbitals and can use octahedral sp^3d^2-hybrid orbitals to bond to six F^- ions.

   (iii) Orthoboric acid, H_3BO_3, is soluble in water and behaves as a weak monobasic acid. It does not donate protons like most acids, but rather it accepts OH^-.

$$B(OH)_3 + 2H_2O \rightleftharpoons H_3O^+ + [B(OH)_4]^-;\ pK_a = 9.2$$

   Above equilibrium can be shifted largely towards right by forming a **stable chelate** with *cis*-dihydroxy compounds (as glycol)

$$[B(OH)_4]^- + 2\ \begin{matrix} >C-OH \\ | \\ >C-OH \end{matrix} \longrightarrow \left[\begin{matrix} -C-O \\ | \\ -C-O \end{matrix} B \begin{matrix} O-C- \\ | \\ O-C- \end{matrix}\right]^- + 4H_2O$$

   The proton can now be titrated against strong alkali solution using phenolphthalein indicator.

2. $$NaAl(SO_4)_2 \rightleftharpoons Na^+ + Al^{3+} + 2SO_4^{2-}$$
$$Na^+ + SO_4^{2-} + Al^{3+} + H_2O \rightleftharpoons Al(OH)_3 + NaOH + H_2SO_4$$
$$NaHCO_3 + H^+ \rightleftharpoons Na^+ + H_2O + CO_2$$

3. $$[Al(OH)_4]^- + H_2O + CO_2 \longrightarrow Al(OH)_3 + H_2O + HCO_3^-$$
$$[Al(OH)_4]^- + H^+ \longrightarrow Al(OH)_3 + H_2O \xrightarrow{HCl} AlCl_3$$
   HCl converts it into $AlCl_3$.

4. (a), (c) and (b) : molecular
   (b) : three dimensional

5. (a) P_4O_{10} (b) B_2O_3 (c) SO_2

6. (a) Ga (b) B (c) Tl (d) B (e) Al (f) Tl (g) B (h) B

NCERT Problems

Refer Text

Exercise 2

Only One Option Correct

1. (a)	**2.** (d)	**3.** (c)	**4.** (c)	**5.** (c)	**6.** (c)	**7.** (d)	**8.** (d)	**9.** (a)	**10.** (a)
11. (d)	**12.** (d)	**13.** (c)	**14.** (c)	**15.** (c)	**16.** (c)	**17.** (d)	**18.** (a)	**19.** (d)	**20.** (a)
21. (b)	**22.** (a)	**23.** (d)	**24.** (c)	**25.** (d)	**26.** (a)	**27.** (a)	**28.** (c)	**29.** (c)	**30.** (a)
31. (c)	**32.** (d)	**33.** (a)	**34.** (d)	**35.** (a)	**36.** (c)	**37.** (d)	**38.** (d)	**39.** (a)	**40.** (a)
41. (c)	**42.** (c)	**43.** (b)	**44.** (d)	**45.** (d)	**46.** (b)	**47.** (d)	**48.** (d)	**49.** (d)	**50.** (a)
51. (a)	**52.** (a)	**53.** (c)	**54.** (b)	**55.** (c)	**56.** (d)	**57.** (c)	**58.** (a)	**59.** (d)	**60.** (b)
61. (c)	**62.** (d)	**63.** (b)	**64.** (b)	**65.** (a)	**66.** (c)	**67.** (a)	**68.** (b)	**69.** (a)	**70.** (c)
71. (b)	**72.** (c)	**73.** (c)	**74.** (a)	**75.** (b)	**76.** (d)	**77.** (b)	**78.** (c)	**79.** (a)	**80.** (c)
81. (c)	**82.** (b)	**83.** (a)	**84.** (d)	**85.** (c)	**86.** (b)	**87.** (d)	**88.** (c)	**89.** (b)	**90.** (d)
91. (c)	**92.** (d)	**93.** (b)	**94.** (a)	**95.** (d)	**96.** (c)	**97.** (c)	**98.** (a)	**99.** (d)	**100.** (b)

One or More Than One Options Correct

1. (a,b) **2.** (a,b,c,d) **3.** (a,c,d) **4.** (a,b,c,d)
5. (a,b,d) **6.** (a,b,d) **7.** (b) **8.** (a,b,c) **9.** (a)
10. (a,b,c,d) **11.** (d) **12.** (a,b,c,d)

Passage Comprehension Questions

Passage 1

1. (c) **2.** (d)

Passage 2

1. (c) 2. (a) 3. (b) 4. (d)

Passage 3

1. (b) 2. (b) 3. (a) 4. (a)

Passage 4

1. (b) 2. (b) 3. (b) 4. (d)

Passage 5

1. (a) 2. (b)

Assertion & Reason

1. (a) 2. (b) 3. (b) 4. (c) 5. (b) 6. (b)
7. (a) 8. (b) 9. (b) 10. (b) 11. (b) 12. (a)
13. (a) 14. (b) 15. (d) 16. (c) 17. (a)

True & False

1. T 2. T 3. F 4. F 5. T 6. T
7. T 8. T 9. F 10. T 11. F 12. T
13. T 14. T 15. T

Fill in the Blanks

1. $Na_2[B_4O_5(OH)_4]\cdot 8H_2O$ 2. van Arkel method
3. increases 4. inert-pair effect
5. protective layer of oxide on the surface 6. amphoteric
7. weak monobasic acid 8. OH^- 9. sp^3
10. paramagnetic, two unpaired 11. aluminium suboxide
12. AlH_3 13. hydrolysis 14. $AlCl_3$ and Cl_4
15. high enthalpy of hydration 16. ns^2np^{1-6}
17. sodium metaborate and meta-aluminate respectively
18. dimeric 19. exothermic, thermite 20. alzheimer
21. primary 22. $CH_3CH_2CH_2OH$ 23. odd
24. $Cu(BO_2)_2$ 25. Al_2O_3 26. setting process
27. methanides 28. non-oxide

Matrix-Match Type Questions

I. Only One Correct

1. (A)—(2); (B)—(4); (C)—(1); (D)—(3)
2. (A)—(3); (B)—(4); (C)—(2); (D)—(2)
3. (A)—(3); (B)—(4); (C)—(1); (D)—(2)

II. One or More Correct

1. (A)—(3); (B)—(1,2,3,5); (C)—(2,5,6); (D)—(4) (E)—(2)
2. (A)—(1,2,3,4); (B)—(1,2,3,4,5,6); (C)—(1,2,3,5); (D)—(3,6)

Integer Answer Types

Set I

Questions →	1	2	3	4	5	6	7	8
Answers →	6	2	6	1	3	1	3	1

Set II

Questions →	1	2	3	4	5	6	7	8
Answers →	4	2	8	4	8	6	3	3

12

Group 14 (IVA—The Carbon Family)

"Carbon is a non-metal; silicon and germanium are classified as metalloids; tin and lead are metals. As a result, the elements of group 14 (Carbon family) have a broad range of chemical behaviour."

Learning Framework of the Chapter ...

- Group 14 (IVA—The Carbon Family) Study of Properties
- Compounds of Carbon
- Compounds of Graphite
- The Carbon Cycle and Green-House Effect
- Silicon
- Compounds of Silicon
- Compounds of Tin
- Compounds of Lead

Group 14 (IVA—The Carbon Family) Study of Properties

Table 12.1

Element		Z	Electronic configuration	Oxidation state
Carbon	C	6	[He] $2s^2 2p^2$	+2, +4
Silicon	Si	14	[Ne] $3s^2 3p^2$	+4
Germanium	Ge	32	[Ar] $3d^{10} 4s^2 4p^2$	+2, +4
Tin	Sn	50	[Kr] $4d^{10} 5s^2 5p^2$	+2, +4
Lead	Pb	82	[Xe] $4f^{14} 5d^{10} 6s^2 6p^2$	+2, +4

Group 14 (IVA, carbon family) has ns^2np^2 configuration of the valence shell, their most common oxidation state is +4, as in CCl_4, $SiCl_4$, $GeCl_4$, $SnCl_4$ and $PbCl_4$. These are volatile, molecular liquids in which atom uses tetrahedral sp^3 hybrid orbitals to form covalent bonds to the Cl atoms.

Carbon is a non-metal. Silicon and germanium are metalloids and are used in semiconductors. Tin and lead are metals.

Carbon and silicon have tetravalent nature in most of the compounds when ns^2 pair is also involved in bonding. Germanium, tin and lead show +2 and +4 oxidation states depending on whether ns^2 pair is inert (and is not involved in bonding) or ns^2 pair is involved in bonding—a case of **inert pair effect**. As we go down the group, inertness of ns^2 pair increases hence tendency to exhibit +2 oxidation state increases and that of +4 oxidation state decreases. Thus, stability of

$Ge^{4+} > Sn^{4+} > Pb^{4+}$ and $Ge^{2+} < Sn^{2+} < Pb^{2+}$

Also $Sn^{2+} < Sn^{4+}$ and $Pb^{2+} > Pb^{4+}$

Thus, $\Delta G° < 0$ for the reactions :

$$Sn^{2+} \longrightarrow Sn^{4+} + 2e^-$$
$$Pb^{4+} + 2e^- \longrightarrow Pb^{2+}$$
$$Ge^{2+} \longrightarrow Ge^{4+} + 2e^-$$

Thus, Sn^{2+} salts are better reducing agents and Pb^{4+} salts are better oxidising agents. Physical properties of group 14 (IVA) are summarised in Table 12.2 and Fig. 12.1.

The +2 oxidation state occurs for tin and lead and is the most stable oxidation state for lead. Both $Sn^{2+}(aq)$ and $Pb^{2+}(aq)$ are common solution species, but there are no simple $M^{4+}(aq)$ ions for any of the group 14 (IVA) elements. Instead M^{4+} species exist in solution as covalently bonded complex ions, *e.g.*, SiF_6^{2-}, $GeCl_6^{2-}$, $Sn(OH)_6^{2-}$ and $Pb(OH)_6^{2-}$. In general, the +4 oxidation state compounds are covalent, and the compounds with the Sn and Pb in the +2 oxidation state are largely ionic.

Table 12.2 **Physical Properties of Carbon Family [Group 14, (IVA)]**

Property	C	Si	Ge	Sn	Pb
Covalent radius/(pm)	77	117	122	140	146
Ionic radius *M* (IV)/(pm)	—	40	53	69	78
M (II)/(pm)	—	—	73	118	119
$(IE)_1$ kJ mol	1086	786	760	707	715
$(IE)_2$ kJ mol^{-1}	2354	1573	1534	1409	1447
$(IE)_3$ kJ mol^{-1}	4622	3232	3300	2943	3087
$(IE)_4$ kJ mol^{-1}	6223	4351	4409	3821	4081
Electronegativity (Pauling)	2.5	1.8	1.8	1.8	1.8
Melting point/°C	4100	1420	945	232	327
Boiling point/°C	—	3280	2850	2623	1751
Electrode potential $E°$				–	
$M^{2+} + 2e^- \longrightarrow M(s)$	—	—	—	– 0.14V	– 0.13V
$M^{4+} + 2e^- \longrightarrow M^{2+}$	—	—	—	0.15 V	1.46V
Abundance in earth's crust (mass%)	0.020	28.2	0.0015	0.0002	0.0013
Density (g/cm^3)	3.51 (diamond)	2.33	5.35	7.28 (white)	11.3
Occurrence	Free state and combined state	Found in rocks as silicon or silicate	Rare element	Found as tinstone SnO_2	Found as galena PbS
Allotropic forms	Crystalline (diamond and graphite) and amorphous	Crystalline and amorphous	Two crystalline forms	White grey rhombic	None
M–M bond energy (kJ mol^{-1})	348.0	297.0	260.0	240	—
Nature	Non-metal	Metalloids	Metalloids	Metal	Metal

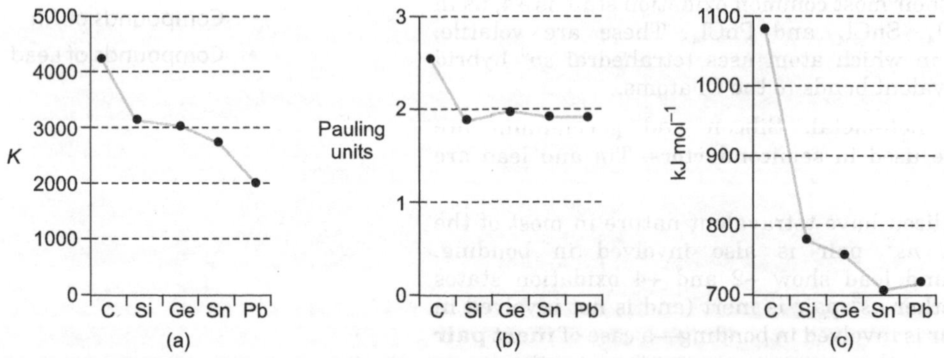

Fig. 12.1 Some group trends of the group 14 elements; (a) Boiling point, (b) Pauling electronegativity and (c) First ionisation energy.

Occurrence

- Carbon is the **seventeenth,** (0.8%) and silicon the **second most** (27.6%) abundant element by mass in the earth's crust. Crude oil and coal are the most common forms of combined and free carbon respectively. Flakes of graphite occur in metamorphased sedimentary rocks like quartz. Diamond occurs embedded in other rocks (as Kimberlite) Germanium is about 1-2 ppm in tin about 2.1 ppm and lead about 13-16 ppm.

Carbon	0.8%
Silicon	27.6%
Germanium	1-2 ppm
Tin	2.1 ppm
Lead	13-16 ppm

Some stable isotopes

Carbon ^{12}C, ^{13}C, ^{14}C

Silicon ^{28}Si, ^{29}Si, ^{30}Si

Germanium ^{70}Ge, ^{72}Ge, ^{73}Ge, ^{74}Ge, ^{76}Ge

Tin Ten stable isotopes of mass numbers : 112, 114, 115, 116, 118, 119, 120, 122, 124

Lead ^{204}Pb, ^{206}Pb, ^{207}Pb, ^{208}Pb

Ionisation Energy

The ionisation energies (Table 12.2 indicate that a very large amount of energy is required to form a M^{4+} ion, therefore, not many ionic compounds are formed by group 14 elements in the +4 oxidation state. On the other hand, the promotion of one of the *ns* electrons to the *np* vacant orbital, as shown below, can allow a tetracovalency. In fact, most of the tetravalent compounds formed by elements of this group are covalent in nature. Electronic configuration of carbon in the ground and excited state is given as an example.

	1s	2s	2p		
Electronic configuration of C in ground state	⇅	⇅	1	1	
C in excited state	⇅	1	1	1	1

Four unpaired electrons can form four equivalent covalent bonds — a case of sp^3 hybridisation.

DEEP Focus

Since, the sum of the first two ionisation energies also decreases down the group, it becomes increasingly favourable thermodynamically, for **Sn and Pb to form compounds in oxidation state +2 which are more ionic in character.**

If we go through the data for ionisation energies given in Table 12.2, we find that the first and second ionisation energies decrease down the group from C to Sn as expected, but they increase slightly when we move to Pb. On the other hand, both the 3rd and the 4th ionisation energies vary irregularly, their sequence is $C > Ge > Si > Pb > Sn$.

The decrease in the 1st and 2nd ionisation energies from C to Sn is because of the increase in atomic size down the group. The 3rd and 4th ionisation energies increases from Si to Ge because of poor shielding of 4*s*-electrons offered by the 3*d*-electrons against the increased nuclear charge in Ge. All the ionisation energies are slightly higher in case of Pb. This increase can again be attributed to the poor shielding effect of the 4*f* and 5*d* orbital electrons. Further, the *s*-electrons being more penetrating, experience higher nuclear charge which makes it difficult to remove them.

- On account of high ionisation energies, simple M^{4+} ions of the group are not known.
- The tendency to form negative ions by gaining electrons is only in carbon. For C and Si, stable divalent compounds MX_2 are very rare.

Size

The covalent radii of the elements, as expected, increase with increase in atomic number. This increase, however, is not uniform. When we move from carbon to silicon the covalent radius increases sharply, and then increases in gradual.

Explanation

A large increase in the size of silicon is due to the introduction of the 3rd shell. When we go down to germanium, the 4th shell is introduced, no doubt, but the effective nuclear charge resulting from addition of eighteen protons to the nucleus out balances the effect of the additional shell leading to only a marginal increase in size. Similarly, in case of Pb, the increase in covalent radius is very small in comparison to that in Sn. This is further reflected in the ionisation energies and electron affinities of these elements.

Electronegativity

Carbon has higher electronegativity as compared to other elements of the group. **As a result,** it can accept (gain) **electrons and can form negative ions of the kind C_2^{2-} in acetylides and C^{4-} in methanides, while the other elements in the group are unable to do so.** When we see the electronegativity trends in the group, we find that the value decreases down the group as expected, with Ge as an exception whose electronegativity is slightly higher than that of Si. As explained above, this could be due to the higher effective nuclear charge in case of Ge.

Catenation

Catenation is a property by virtue of which elements form long chain compounds by single or multiple bond formation between atoms of the same element. Carbon shows this tendency to a remarkable extent because of its small size and high bond energy (Table 12.2), Si and Ge to a lesser extent and Sn and Pb to a very small extent. In case of carbon we have alkanes, *e.g.*, ethane propane butane, pentane, etc. These are chains of carbon hydrides having the general formula C_nH_{2n+2}, where n can be a large number, alkanes with $n = 100$ are known. Carbon also forms a number of catenated halides, *e.g.*, polyvinyl chloride (PVC). Silicon forms silanes having the general formula Si_nH_{2n+2}, but n does not exceed 8. Its halids also shows catenation and have a general formula Si_nX_{2n+2}, where if $X = F$, $n = 1$ to 14; if $X = Cl$, Br or I, $n = 1$ to 10. Germanium forms germanes, with general formula Ge_nH_{2n+2}, where $n = 1$ to 9. Germanium halides form dimers, *e.g.*, Ge_2Cl_6. Hydrides of Sn and Pb are much less stable, while SnH_4 and Sn_2H_6 have been prepared, analogous methods fail for the preparation of PbH_4. No catenated halides are known for Sn and Pb.

- There is steady decrease in the *M—M* bond energy as we go down the group.

Multiple Bonding

In group 14, carbon is the only element capable of forming comparatively stable multiple bonds with another C atom and also with N, O, etc. Examples are alkenes ($>C{=}C<$), alkynes ($-C{\equiv}C-$), imines ($>C{=}N-$), nitriles ($-C{\equiv}N$), ketones ($>C{=}O$), etc. None of the other elements of the group forms stable compounds containing multiple bonds. Recently, some success has been achieved in synthesising compounds with $>Si{=}Si<$ and $>Si{=}C<$ double bonds. Stoichiometric analogues like SiO_2, SnO_2 are $(Me_2SiO)_n$ are known which are giant macromolecules containing *M—O—M* linkages.

The tendency of formation of multiple bonds easily, in case of carbon, is because of the high bond energy (Table 12.3) and a small atomic radius of carbon as compared to other elements of the group. So its *p*-orbitals can approach the *p*-orbitals of the other C, O or N atom for an effective overlap essential for π-bond formation.

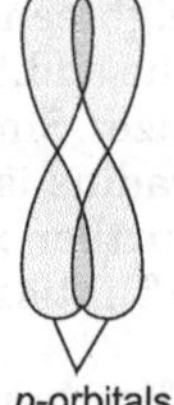
p-orbitals

Table 12.3 **Bond Energies of Group 14 Elements (kJ mol^{-1})**

Bond	C	Si	Ge	Sn	Pb
E–*E*	347	226	118	150	—
E=*E*	611	318	—	—	—
E≡*E*	837	—	—	—	—
E–H	414	318	285	251	—
E–O	360	464	360	—	—
E=O	736	640	—	—	—

*E** = Element C, Si, Ge, Sn or Pb.

Allotropes

- Carbon and graphite are two allotropes of carbon. In diamond (Fig. 12.2(a)) each carbon atom is linked tetrahedrally to four, other carbon atoms by sp^3 bonds and a giant covalent network is formed. It is, therefore, hard and non-conducting. Graphite (Fig. 12.2(b) and (c)) is more thermodynamically stable allotrope.

DEEP Focus

Graphite consists of planar hexagonal layers of carbon atoms held together by relatively weak van der Waals' forces. Within each layer, a carbon atom is bonded to three other carbon atoms by sp^2 hybrid bonds and the additional *p* electron from each carbon forms an extended delocalised π-bonding system encompassing the entire layer. This makes graphite a good conductor of electricity in directions along the planes of carbon atoms. The covalent bonds in graphite account for its hardness; however, because the layers can slide over one another, graphite is slippery to touch and is effective as a lubricant.

Silicon exists only in the diamond structure due to tendency to form $p\pi$–$p\pi$ bonds to itself. Fullerene, [Fig. 12.2(d)] a carbon allotrope discovered in 1990 as a constituent of soot, consists of spherical C_{60} molecules with the extraordinary shape of a soccer ball. The C_{60} ball has 12 pentagonal and 20 hexagonal faces with each atom sp^2-hybridised and bonded to three other atoms.

C_{60} was first prepared in 1985, by laser vaporisation of graphite (at about 10000°C) followed by sweeping of the carbon atoms by sweeping of the carbon atoms by a blast of high pressure helium gas and rapid cooling in a vacuum chamber (Richard Smaller, Rice university Texas and Harry kroto, University of Sussex, U.K. Noble Prize Winners (1990).)

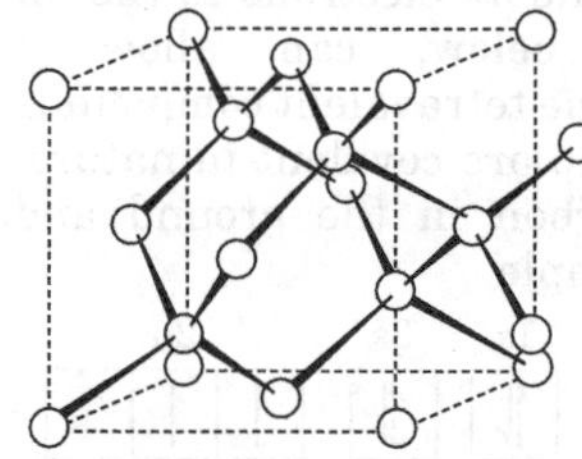
(a) The crystal structure of diamond

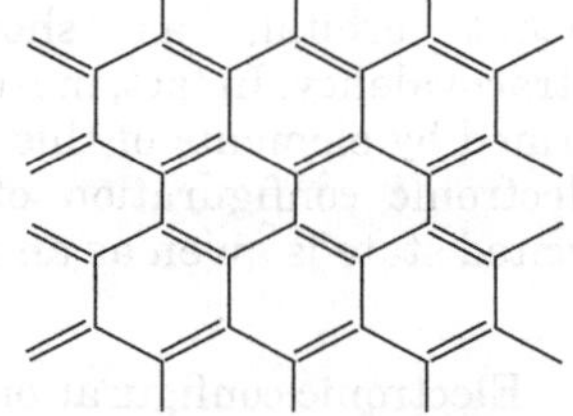
(b) The structure of a graphite sheet

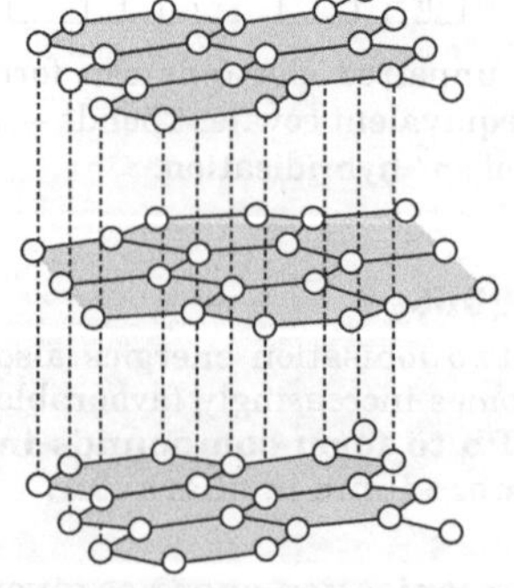
(c) Structure of α-graphite

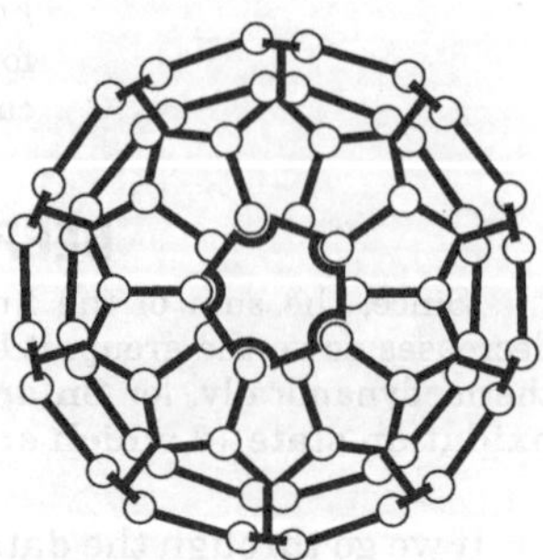
(d) Structure of Buckminster fullerene

Fig. 12.2 Allotropes carbon

DEEP Focus

- Silicon also uses d-orbitals in multiple bonding, **In case of $N(CH_3)_3$, geometry is pyramidal, but in case of $N(SiH_3)_3$ it is planar. It is due to the fact that in the latter, the lone pair of N-atom is transferred to the empty d-orbitals of silicon ($p\pi$–$d\pi$ overlapping) :**

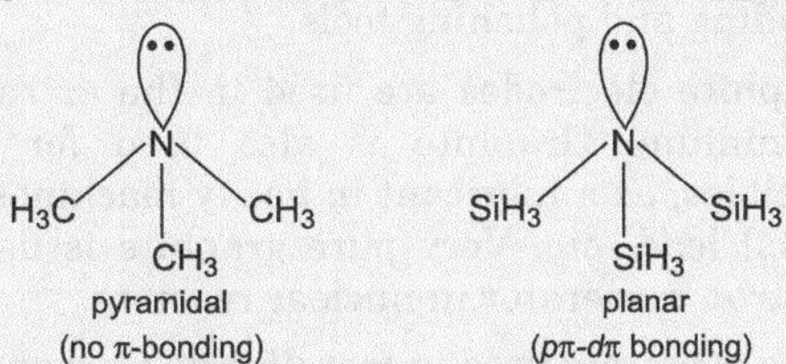

- The normal valency of the elements is four, but apart from carbon, the elements can make more than four bonds. This is because they make use of a set of d-orbitals in bonding. Thus availability of d-orbitals (empty but electrons of ns^2 and np^2 can be excited into this) is responsible for the ability of the elements, except carbon, to make complex ions such as SiF_6^{2-} (Fig. 12.3).

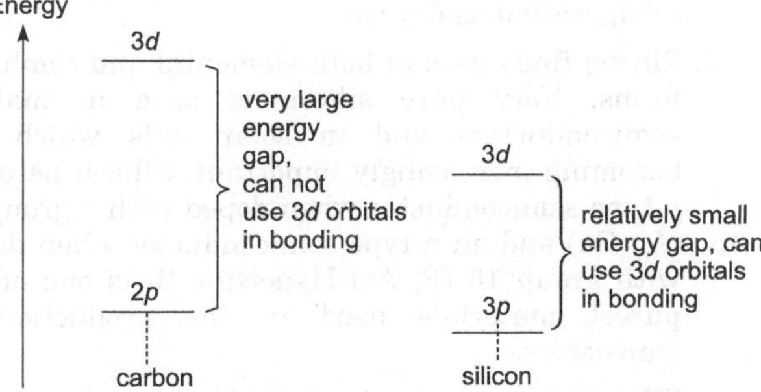

Fig. 12.3 Energy difference between 2p and 3d energy levels in carbon and between 3p and 3d energy levels in silicon

- The main chemical differences between silicon and carbon are
  - Si does not form stable double bonds,
  - Si does not form very stable Si—Si bonds,
  - Si has vacant 3d-orbitals in its valence shell into which it can accept electrons from donor atoms.
- Some carbon compounds are less reactive than the corresponding compounds of the other members of the group. CCl_4 has no interaction with H_2O (hydrolysis) whereas $SiCl_4$ does by use of empty 3d-orbitals.
- All the elements give hydrides. Carbon, of course gives an immense number, but silicon and germanium also show a wide variety.

CH_4	methane (alkane)
SiH_4	stannate
GeH_4	germane
SnH_4	stannane
PbH_4	plumbane

Carbon hydrides will not ignite in air unless a flame is put to them. Apart from silane, SiH_4, the silicon hydrides are less well behaved. Si_3H_8 is spontaneously flammable in air :

$$Si_3H_8(l) + 5O_2(g) \longrightarrow 3SiO_2(s) + 4H_2O(l)$$

Alkane as well as silane are not hydrolysed however, traces of alkali converts silane into hydrated silica $SiO_2 \cdot nH_2O$.

- As we descend the group, acidic nature of the oxides decreases and basic nature of the oxides increases.

Oxides	Nature	Oxides	Nature
CO, CO_2	acidic	SnO, SnO_2	amphoteric
SiO_2	acidic	PbO, PbO_2	basic
GeO, GeO_2	amphoteric	Pb_3O_4	amphoteric

Si, Ge, Sn and Pb form oxides $MO_2 \cdot SiO_2$ is an infinite three-dimensional network solid of silicon and oxygen atoms connected by single covalent bonds. GeO_2, SnO_2 and PbO_2 are all solids and exist in several modifications.

Illustration 1 *Why is graphite a better electrical conductor than diamond and why does the conductivity of graphite depend on direction?*

Solution Graphite has a two-dimensional sheet like structure in which each carbon atom uses sp^2 hybrid orbitals to form trigonal planar σ bonds to three neighbouring carbon atoms. In addition, each carbon atom uses its remaining p-orbital, perpendicular to the plane of the sheet, to form a π bond. Because each carbon atom must share its π bond with its three neighbours, the π electrons are delocalised and are free to move in the plane of the sheet. As a result, the electrical conductivity of graphite in a direction parallel to the sheet is about 10^{20} times greater than the conductivity of diamond. The conductivity of graphite perpendicular to the sheets of carbon atoms is lower because electrons must hop from one sheet to the next. The substance or crystal that show different properties in different directions is called **Anisotropic.**

Chemical Reactivity

Elements in this group are relatively unreactive, but reactivity increases down the group. **Lead often appears more noble (unreactive) than expected. It is due to a surface coating of oxide and partly due to high overpotential for the reduction of H^+ to H_2 at a lead surface.** The production of H_2 from H^+ at a lead electrode is kinetically unfavourable, hence a much larger potential is required than the standard reduction potential. Table 12.4 summarises reactions of the elements with different reagents :

Table 12.4 **Reactivity of the Elements of Group 14 (IVA)**

Reagent	Reactivity
H_2O	• C, S, Ge unaffected by H_2O. • Sn reacts with steam forming SnO_2 and H_2: $Sn + 2H_2O \text{ (steam)} \longrightarrow SnO_2 + 2H_2$ • Pb is unaffected by H_2O probably because of protective oxide film.
Dilute acids	• C, Si, Ge unaffected by dilute acid. • Sn dissolves in dilute HNO_3 forming $Sn(NO_3)_2$. • Pb dissolves slowly in dilute HCl forming $PbCl_2$ (sparingly soluble) and quite readily in dilute HNO_3 forming $Pb(NO_3)_2$ and oxides of nitrogen. • Pb does not dissolve in dilute H_2SO_4 due to formation of $PbSO_4$ coating.
Concentrated acids	• Diamond is unaffected by concentrated acids, but graphite changes to mellitic acid ($C_{12}H_6O_{12}$) also called benzene hexa-carboxylic acid (A) A: benzene ring bearing six COOH groups (HOOC, COOH at each position) with hot concentrated HNO_3 and to graphite oxide with hot concentrated HF/HNO_3. • Si is oxidised and changes to SiF_4 by hot concentrated HNO_3/HF. • Ge dissolves slowly in hot concentrated H_2SO_4 and HNO_3. • Sn also dissolves in concentrated acids. • Pb does not dissolve in concentrated HCl due to formation of $PbCl_2$ coating.
Alkalies	• Carbon is unaffected by alkalies. • Si reacts slowly with cold aqueous NaOH solution, and readily with hot solution forming SiO_4^{4-}. • Sn and Pb are slowly attacked by cold alkali, and rapidly by hot alkali, giving stannates $Na_2[Sn(OH)_6]$ and plumbates $Na_2[Pb(OH)_6]$.
Halogens	• Diamond is unreactive, but graphite reacts forming $(CF)_n$. • Si and Ge form volatile SiX_4 and GeX_4. • Sn and Pb are less reactive. Sn reacts with Cl_2 and Br_2 in cold, and with F_2 and I_2 on warming. Lead reacts with F_2 in cold and with Cl_2 on heating forming PbX_2.
Air	• Graphite burns above 700°C (CO) Diamond burn above 900°C (CO). Si, Ge, Sn, burn on strong heating (At 1400°C, Si also forms SiN and Si_3N_4) Finely divided lead powder is pyrophoric (combines highly exothermically with oxygen) but the surface is practically inactivated by a thin protective layer of oxide, chloride, sulphate. PbO is formed above 600°C.

Uses

1. Diamonds are used in jewellery on account of their high refractive and dispersive power. Due to their exceptionally hard nature, inferior quality diamonds are used for industrial purposes mainly in the form of grit and powder in cutting, drilling, grinding and polishing tools.

   Graphite electrodes are used in the extraction of aluminium. Graphite is also used for making crucibles, as a lubricant in heavy machines and as pencil lead, etc. Very pure graphite is used as a neutron moderator in nuclear reactors.

   Coke is mainly used in metallurgy. Carbon black is used to strengthen rubber, as a pigment in inks, paints, paper and plastics. Activated carbon has enormous surface area and is used extensively in sugar industry as a decolourising agent. It is a very efficient adsorbent it absorbs organic pollutants from drinking water, noxious gases from the air. It is also used in the treatment of sewage water and as a catalyst. Amongst the compounds of carbon, CO_2 finds an important use as a fire extinguisher and in refrigeration as dry ice.

2. Silicon finds uses in both elemental and combined forms. Very pure silicon is used in making semiconductors and in solar cells which are becoming increasingly important. Silicon becomes *p*-type semiconductor when doped with a group 13 (Al, Ga) and an *n*-type semiconductor when doped with group 15 (P, As) Hyperfine Si is one of the purest materials used in the production of transistors.

   Silicates are used in glass and cement manufacture.

   Silicones are also widely used in lubricants. Their inert nature makes them ideal for use in gas chromatography, *e.g.*, methylphenyl silicones. Germanium is used largely in transistor technology. However, its use in optics is growing; germanium is transparent in infrared region and is, therefore, used in infrared windows, prisms and lenses.

   • Silicon is extensively used in steel making as a deoxidiser. Ferrosilicon is the usual form. Steels with high Si content are corrosion resistant.

3. Due to its resistance to corrosion, tin is used to coat mild steel plate. Tinning of copper and brass utensils has been an age old practice in India. Tin is also used for making various alloys (Refer Appendix) like bronze, pewter, gun metal, etc. Tin-lead alloys are fusible and low melting. Solders, which are low melting alloys, are used to join pieces of metal together. Soldering is a very common practice for joining components of an integrated circuit. Sn—Pb alloys are also used as bearing metal or type metal.

Lead is used as an inert material, *e.g.*, for building roofs, and for cable sheathing. Lead gets covered with a layer of $PbSO_4$, which is insoluble and does not get attacked by common reagents. Lead pipes were used in municipal water supply at one time. Soft water, however, tends to dissolve lead and may cause lead poisoning. Therefore, galvanised iron (GI) pipes are now used for water supply.

Due to their low chemical reactivity, lead coated chambers are used in many industrial preparations like that of sulphuric acid. In view of its high density, lead is also used in making shielding blocks for radioactivity experiments. Organolead compounds, *e.g.*, tetraethyllead, Et_4Pb and tetramethyllead, Me_4Pb, are used as anti-knock additives to petrol for internal combustion engines. However, in recent times, because of their polluting nature, many countries have reduced or altogether stopped their use.

One of the major uses of lead still is in the storage batteries. Storage batteries also called lead-acid batteries have a supporting grid of lead-tin alloy (91 : 9) with PbO_2 as the oxidising agent and spongy Pb as the reducing agent. The electrolyte is dilute H_2SO_4.

Following table summarises uses of elements of group 14 (IVA):

Elements	Uses
Carbon	Diamond in jewellery and glass cutter, graphite as lubricant and electrode
Silicon	As a semiconductor, in silicone waxes and polymers, as silicates in the ceramic and glass industries.
Germanium	As a semiconductor.
Tin	In alloys, as a protective coating for iron and steel.
Lead	In alloys, in car batteries.

Compounds of Carbon

Carbon is unique in the variety and complexity of compounds that it can form. The molecules that are responsible for the growth and development of living organisms, such as vitamins, proteins, enzymes, hormones and DNA, are structures built from chains of carbon.

Oxides of Carbon

- In oxides of carbon there are pπ–pπ multiple bonds between carbon and oxygen.

  stable oxides : CO and CO_2

  less stable : C_3O_2 (carbon suboxide), C_5O_2, $C_{12}O_9$

  least stable : C_2O (graphite oxide), C_2O_3

- CO is a colourless, odourless, poisonous gas. It is formed when carbon is burnt in limited supply of air.

$$2C + O_2 \longrightarrow 2CO$$

Dehydration of formic acid with conc. H_2SO_4 gives CO (laboratory method).

$$H-\overset{\overset{\displaystyle O}{\|}}{C}-OH \xrightarrow{\text{Conc. } H_2SO_4,\ \Delta} H_2O + CO$$

- CO can be detected because it burns with a blue flame.
- CO reduces I_2O_5 to I_2

$$5CO + I_2O_5 \longrightarrow I_2 + 5CO_2$$

I_2 thus liberated is determined by titration with $Na_2S_2O_3$ (hypo)

$$2Na_2S_2O_3 + I_2 \longrightarrow 2NaI + Na_2S_4O_6$$

This is a quantitative method of estimation of CO. $PdCl_2$ is also reduced to Pd

$$PdCl_2 + CO + H_2O \longrightarrow Pd + 2HCl + CO_2$$

Following reduction takes place in blast furnace :

$$CuO + CO \longrightarrow Cu + CO_2$$

$$Fe_2O_3 + 3CO \longrightarrow 2Fe + 3CO_2$$

- **CO is toxic because it forms a complex with haemoglobin in the blood, and this complex is more stable than oxy-haemoglobin. This prevents the haemoglobin in the red blood corpuscles from carrying oxygen round the body. This causes an oxygen deficiency, leading to unconsciousness and then death. The blood of victims of carbon monoxide poisoning is a brilliant red-pink colour rather than the dull red of normal blood.**

$$\underset{\text{oxyhaemoglobin}}{Hb-O_2} + CO \rightleftharpoons \underset{\text{carboxyhaemoglobin}}{Hb-CO} + O_2$$

- Although CO has no acidic or basic properties (in terms of proton H^+), it can behave as a **Lewis base—electron-pair donor**, and thus forms complexes with transition metals (mostly with zero oxidation state) :

$$Ni(CO)_4,\ Fe(CO)_5,\ Cr(CO)_6$$

These are called carbonyl compounds and can be made by :

$$Ni + 4CO \xrightarrow{28°C} Ni(CO)_4 \xrightarrow{230°C} Ni + 4CO\uparrow$$

This is **Mond** process of the purification of Ni metal.

$$Fe + 5CO \xrightarrow[\text{under pressure}]{200°C} Fe(CO)_5$$

$$2Fe(CO)_5 \xrightarrow{\text{photolysis}} Fe_2(CO)_9 + CO$$

- Bond order between C and O is three and metal carbonyl bond is dative

$$M \leftarrow C{\equiv}O$$

- CO is sparingly soluble in water and is neutral oxide.
- CO is an important fuel

$$CO + \frac{1}{2}O_2 \longrightarrow CO_2, \qquad \Delta H° = -282.25 \text{ kJ mol}^{-1}$$

It is associated with following fuels :

Water gas : Equimolecular mixture of CO and H_2

$$C + H_2O \xrightarrow{\text{red hot}} CO + H_2, \quad \Delta H^\circ = +131 \text{ kJ}$$

Producer gas : A mixture of CO and N_2

$$\underset{\substack{\text{red hot}\\ \text{coke}}}{C} + \underbrace{O_2 + 4N_2}_{\text{air}} \longrightarrow CO_2 + 4N_2 \xrightarrow{+C} \underset{\text{(producer gas)}}{2CO + 4N_2}$$

Producer gas is less efficient than water gas as only part of the gas will burn. The overall reaction is exothermic :

$$2C + O_2 \longrightarrow 2CO, \quad \Delta H^\circ = -221 \text{ kJ}$$

Coal gas : A mixture of CO, H_2 CH_4 and CO_2.

Water gas	Co, H_2
Producer gas	Co, N_2
Coal gas	Co, H_2, CH_4, CO_2

- $CO + S \longrightarrow COS$ (carbonyl sulphide)

  $CO + Cl_2 \longrightarrow COCl_2$ (carbonyl chloride)
- CO_2 is a colourless and odourless gas and is generally recovered as by-product from

  — the **manufacture of H_2**

  $$CO + H_2O \rightleftharpoons CO_2 + H_2$$
  $$CH_4 + 2H_2O \longrightarrow CO_2 + 4H_2$$

  — the **fermentation processes in breweries**

  $$C_6H_{12}O_6 \xrightarrow{\text{yeast}} 2CO_2 + 2C_2H_5OH$$
- CO_2 gas can be liquefied under pressure between –57° C and +31° C. Solid CO_2 called **'dry ice'** sublimes directly to the vapour state (without going through the liquid state) at –78° C under atmospheric pressure. Dry ice (also called **cardice**) is used to freeze meat, frozen foods and ice cream, and in the laboratory as a coolant.
- CO_2 is used in many fire extinguishers:
  - Water based extinguishers have a sealed bottle of acid immersed in a solution of HCO_3^- (say $NaHCO_3$). When the extinguisher is struck, the bottle bursts, and reaction between acid and HCO_3^- produces CO_2 instantly in the form of watery spray (or in a foam if other chemicals have been added).

    $$HCO_3^- + H^+ \longrightarrow H_2O + CO_2$$
  - Other extinguishers use pure CO_2 under pressure. The gas given out covers the fire and prevents oxygen in the air from reacting with hot materials.
- CO_2 is an acidic oxide and reacts with bases, forming salts:

  $$Ca(OH)_2 + CO_2 \longrightarrow CaCO_3 + H_2O$$

  Solution of CO_2 in water behaves as weak dibasic acid, the solubility in water increases remarkably under pressure. This property is used to carbonate drinks (Coca-Cola etc.)

  $$CO_2 + H_2O \longrightarrow H_2CO_3$$
  $$H_2CO_3 \rightleftharpoons H^+ + HCO_3^- \quad \text{(salt : NaHCO}_3)$$
  $$HCO_3^- \rightleftharpoons H^+ + CO_3^{2-} \quad \text{(salt : Na}_2\text{CO}_3)$$

  and thus forms two series of salt.

$NaHCO_3$	:	an acid salt
Na_2CO_3	:	normal salt

  A hydrate $CO_2 \cdot 8H_2O$ can be crystallised at 0° C under a pressure of 50 atm CO_2.

DEEP Focus

- Free carbonic acid has not been isolated but its acid salt ($NaHCO_3$) and normal salt (Na_2CO_3) have been isolated. An etherate (*A*) of probable composition $OC(OH)_2 \cdot O(CH_3)_2$ is obtained by reacting Na_2CO_3 with HCl in presence of dimethyl ether at very low temperature (–50°C). The white crystalline solid decomposes at 5°C. H_2CO_3 is believed to be present in the gaseous phase by thermal decomposition of NH_4HCO_3

$$Na_2CO_3 + 2HCl \longrightarrow 2NaCl + \underset{(A)}{H_2CO_3}$$

$$O{=}C(OH)_2 + (CH_3)_2O \longrightarrow \underset{(A)}{O{=}C(OH)_2 \cdot O(CH_3)_2}$$

$$NH_4HCO_3 \xrightarrow{\Delta} NH_3(g) + H_2CO_3(g)$$

- Biologically, CO_2 is important in photosynthesis where the green parts of plants manufacture glucose sugar

  $$6CO_2 + 6H_2O \xrightarrow{\text{sunlight}} 6O_2 + \underset{\text{glucose}}{C_6H_{12}O_6}$$

  The reverse reaction occurs during the process of respiration, where animals and plants release energy :

  $$C_6H_{12}O_6 + 6O_2 \longrightarrow 6CO_2 + 6H_2O + \text{energy}$$
- CO_2 turns lime water milky which changes to colourless on passing excess of CO_2 in it

  $$\underset{\text{lime water}}{Ca(OH)_2} + CO_2 \longrightarrow \underset{\substack{\text{milky}\\ \text{(white ppt)}}}{CaCO_3 \downarrow} + H_2O$$
  $$\underset{\text{milky}}{CaCO_3} + H_2O + CO_2 \longrightarrow \underset{\text{soluble}}{Ca(HCO_3)_2}$$
- Carbon suboxide C_3O_2 is a foul smelling gas with boiling point 6° C.

  $$\underset{\text{malonic acid}}{CH_2(COOH)_2} \xrightarrow[150^\circ C]{P_4O_{10}} O{=}C{=}C{=}C{=}O + 2H_2O$$

  linear molecule stable at –78°C.

  $$C_3O_2 + 2H_2O \longrightarrow \underset{\text{malonic acid}}{CH_2(COOH)_2}$$
  $$C_3O_2 + 2HCl \longrightarrow \underset{\text{malonyl chloride}}{CH_2(COCl)_2}$$
  $$C_3O_2 + 2NH_3 \longrightarrow \underset{\text{malonic amide}}{CH_2(CONH_2)_2}$$

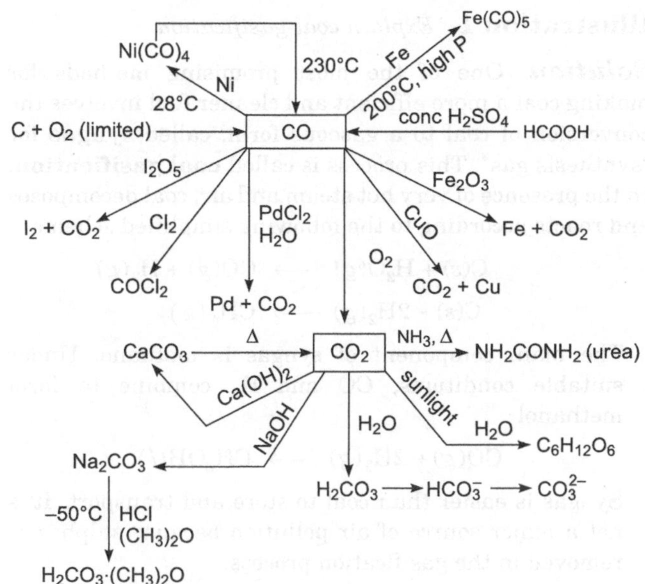

Compounds of Graphite

Type I : Electrically Conducting Compounds

Compounds in which the π-electron system of graphite is retained and thus electrical conductivity is retained due to mobile electrons (or even many increase). In such compounds separation between layers increases by as much as 10Å. Such compounds are called **Lamellar** compounds.

Preparation

(i) Such Lamellar compounds are formed by insertion of various atoms (as alkali metals), molecules or ions (as $FeCl_3$) between the graphite layers.

$$\underset{\text{(graphite)}}{C + K} \longrightarrow \underset{\text{deep blue}}{C_{60}K} \xrightarrow{K} \underset{\text{steel blue}}{C_{24}K} \xrightarrow{K} \underset{\text{bronze}}{C_8K}$$

(ii) **By electrochemical method**

Reduction : $nC + M^+ + e^- \rightleftharpoons M^+ + C_n^-$

Oxidation : $nC + X^- \rightleftharpoons C_n^+ + X^-$

The high electrical conductivity may arise from transfer of electron(s) from the metal to the π-orbitals of the carbon plane, where they are free to move under an electric field.

Type II : Electrically non-conducting Compounds

Graphite changes to $(CF)_n$ — a white compound when it is treated with F_2 at 400–600°C. It is found by X-ray study that in $(CF)_n$ layers separation is about 6.6 Å. Also carbon atoms in the layers have sp^3 hybridisation with boat or chair conformation and (C—F) covalent bond exists. Due to loss of π-electron, compound $(CF)_n$ is non-conducting.

Carbides

- Solid compounds of carbon with elements other than hydrogen are generally called **carbides**.
- **Ionic or salt-like carbides** are formed by the most electropositive metals such as alkali and alkaline earth metals and aluminium:

$$2BeO + 2C \xrightarrow{2000°C} Be_2C + 2CO$$

$$4Al + 3C \xrightarrow{\text{electric furnace}} Al_4Cl_3$$

Both Be_2C and Al_4C_3 are called **methanides** because they react with H_2O yielding methane:

$$Be_2C + 4H_2O \longrightarrow 2Be(OH)_2 + CH_4$$

$$Al_4C_3 + 12H_2O \longrightarrow 4Al(OH)_3 + 3CH_4$$

CaC_2 (calcium acetylide) is obtained by strongly heating lime and coke:

$$CaO + 3C \xrightarrow{2000°C} CaC_2 + CO$$

It is used to manufacture acetylene and calcium cyanamide:

$$CaC_2 + 2H_2O \longrightarrow Ca(OH)_2 + CH{\equiv}CH$$

$$CaC_2 + N_2 \xrightarrow{1100°C} CaCN_2 + C$$

Calcium cyanamide, Ca(NCN), is used as nitrogenous fertiliser, and to make urea:

$$CaCN_2 + H_2O + CO_2 \longrightarrow CaCO_3 + NH_2CN$$

$$NH_2CN + H_2O \xrightarrow[\text{or pH > 12}]{\text{pH < 2}} \underset{\text{urea}}{NH_2CONH_2}$$

The cyanamide ion $[N{=}C{=}N]^{2-}$ is isoelectronic with CO_2 and has the same linear structure.

C^{2-} (acetylide) has multiple bonding of the type

$$^-C{\equiv}C^-$$

while CaC_2 has NaCl type of lattice with Ca^{2+} replacing Na^+ and C_2^{2-} replacing Cl^-.

Mg_2C_3 has ($^-C{\equiv}C{-}C^{3-}$) ion and gives propyne on reaction with H_2O.

$$Mg_2C_3 + 4H_2O \longrightarrow 2Mg(OH)_2 + CH_3C{\equiv}CH$$

- **Interstitial carbides** are formed by transition metals in which carbon atoms occupy tetrahedral holes in the close-packed arrays of metal atoms. These are generally very hard, electrically conducting and have very high melting points (tungsten carbide (WC)). Carbides of Cr, Mn, Fe, Co and Ni are hydrolysed by water or dilute acids.
- **Covalent carbides** are formed by metalloids like Si and B. SiC and B_4C (and more correctly as $B_{13}C_2$) are extremely hard, infusible and chemically inert. SiC has a diamond like structure (hence also called **artificial diamond**) in which C and Si atoms are each tetrahedrally surrounded by four of the other kind of atoms under the name **carborundum**, it is used in cutting tools and abrasives. It is produced by

heating quartz or sand with an excess of coke in an electric arc furnace at 2000—2500° C.

$$SiO_2 + C \longrightarrow Si + 2CO$$
$$Si + C \longrightarrow SiC$$

SiC reacts with NaOH in presence of air and with Cl_2 at 100° C.

$$SiC + 2NaOH + 2O_2 \longrightarrow Na_2SiO_3 + CO_2 + H_2O$$
$$SiC + 2Cl_2 \longrightarrow SiCl_4 + C$$

Boron carbide is even harder than silicon carbide and is used both as an abrasive and as a shield from radiation.

- Be_2C — red solid
  Al_4C_3 — pale yellow solid
  CaC_2 — colourless
  SiC — pale yellow (pure)

Cyanides

- The **cyanides** contains the anion group $:\!C\equiv N\!:^-$. The cyanide ions are extremely toxic because they bind almost irreversibly to the Fe(III) ion in cytochrome oxidase, a key enzyme in metabolic processes involved in the oxidation of food molecules. With CN^- attached to the iron, the enzyme is unable to function, cellular energy production comes to a halt, and rapid death follows. HCN being volatile (b.p. 26°C is more dangerous and can cause death within minutes.
- Cyanide ions are used to extract gold and silver:

$$\underset{\text{(in mineral)}}{4Au(s)} + 8CN^-(aq) + O_2(g) + 2H_2O(l) \longrightarrow 4[Au(CN)_2]^-(aq) + 4OH^-(aq)$$
$$2[Au(CN)_2]^- + Zn(s) \longrightarrow [Zn(CN)_4]^{2-} + 2Au(s)$$

- $Fe^{2+}(aq)$ combines with CN^- forming ferrocyanide (*aq*) ion, $[Fe(CN)_6]^{4-}$.

$$Fe^{2+}(aq) + 6CN^-(aq) \longrightarrow [Fe(CN)_6]^{4-}(aq)$$

DEEP Focus

- Mixture of Cu^{2+} (*aq*) and Cd^{2+} (*aq*) can be separated by using $CN^-(aq)$. Cu^{2+} form stable $[Cu^I(CN)_4]^{3-}$ and unstable $[Cd(CN)_4]^{2-}$ complexes. On passing H_2S gas yellow ppt of CdS are obtained without affecting $[Cu(CN)_4]^{3-}$ complex.

$$Cu^{2+} + 2CN^- \longrightarrow \underset{\text{yellow ppt}}{Cu(CN)_2\downarrow}$$
$$2Cu(CN)_2 \longrightarrow \underset{\text{white ppt}}{2CuCN\downarrow} + \underset{\text{cyanogen (poisonous gas)}}{(CN)_2\uparrow}$$
$$CuCN + \underset{\text{excess}}{3CN^-} \longrightarrow \underset{\text{stable}}{[Cu(CN)_4]^{3-}}$$
$$Cd^{2+} + 4CN^- \longrightarrow \underset{\text{unstable}}{[Cd(CN)_4]^{2-}}$$
$$[Cd(CN)_4]^{2-} \rightleftharpoons Cd^{2+} + 4CN^-$$
$$Cd^{2+} + S^{2-} \longrightarrow \underset{\text{yellow ppt}}{CdS\downarrow}$$

Illustration 1 *Explain coal-gasification.*

Solution One of the most promising methods for making coal a more efficient and cleaner fuel involves the conversion of coal to a gaseous form, called **syngas** for "synthesis gas". This process is called **coal gasification.** In the presence of very hot steam and air, coal decomposes and reacts according to the following simplified scheme

$$C(s) + H_2O(g) \longrightarrow CO(g) + H_2(g)$$
$$C(s) + 2H_2(g) \longrightarrow CH_4(g)$$

The main component of syngas is methane. Under suitable conditions, CO and H_2 combine to form methanol

$$CO(g) + 2H_2(g) \longrightarrow CH_3OH(l)$$

Syngas is easier than coal to store and transport. It is not a major source of air pollution because sulphur is removed in the gasification process.

The Carbon Cycle and Green-House Effect

- Carbon cycle is shown by the following scheme (Fig. 12.4)

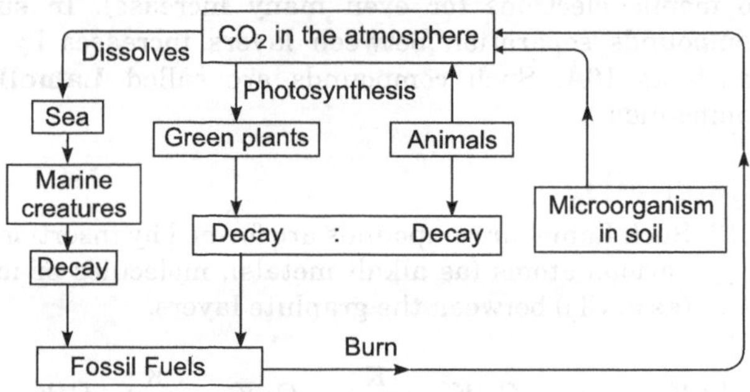

Fig. 12.4 Carbon cycle

Energy from the sun reaches the earth in the form of light. Neither CO_2 nor H_2O vapour absorbs the visible light so they do not prevent it from reaching the surface of the earth.

- The CO_2 as well as H_2O absorbs strongly in the infrared region, and its presence in the atmosphere decreases the loss of heat from the earth by radiation. This global warming is called the **'green house effect'** (other gases, including the oxides of nitrogen from car exhausts, freons from aerosols and refrigerators and methane from bacteria in the soil and in the rumen of cows, also add to the green house effect). The concentration of atmospheric CO_2 has increased by 10%. This is resulting in the increase in the mean temperature of the earth by 2.5° C, varying from 2° C at the equator to 4°C at the poles **(Global**

Warming). This could have dramatic effects on the climate. The increased temperature would cause more evaporation of water and hence more rain, flooding and tropical storms in certain parts of the world. CO_2 may dissolve in water and pH of the soil is affected.

$$CO_3^{2-} + CO_2 + H_2O \longrightarrow 2HCO_3^-$$

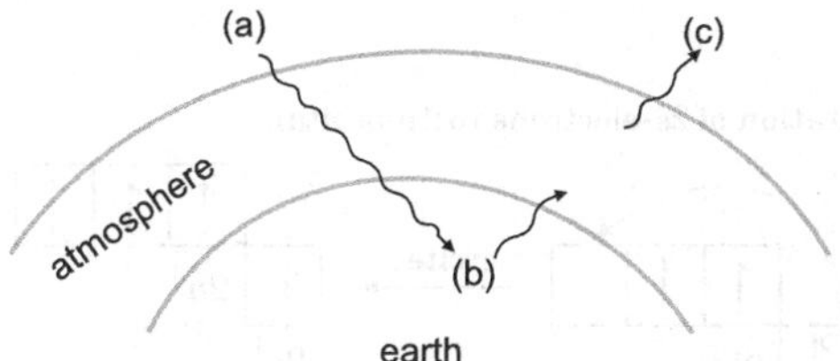

Fig. 12.5 The 'Green-House Effect' (a) Sunlight received by earth, (b) Infrared radiation reradiated by the surface of earth, (c) Infrared radiation emitted to space, following absorption of some of this

Target Practice 1

1. **Tetra covalent nature of the element with following valence shell electronic configuration is explained on the basis of**

....

$2s^2$: ⇅ ; $2p^2$: ↿ ↿ □

2. **Covalent radii of**

C	Si	Ge
77 pm	117 pm	122 pm

C → Si: 40 pm; Si → Ge: 5 pm

Show that there is difference of 40 pm while going from C to Si but there is difference of only 5 pm while going from Si to Ge. This is due to

....

....

3. **(a) Write the element with maximum catenation property.**

....

(b) Write the formulae of compounds/ions in which carbon forms .... with nitrogen.

(i) single bond ....

(ii) double bond ....

(iii) triple bond ....

4. **You have seen electrican joining two points in a circuit by using a wire which is low melting and immediately solidifies.**

(i) What is its name? ....

(ii) What is the composition? ....

5. **Write name/formulae and use of the following**

	Name/Formula	Use
(i) Dry ice	....	....
(ii) Semi conductor	....	....
(iii) Solder	....	....
(iv) TEL	....	....
(v) CNG	....	....
(vi) LPG	....	....

6. **In the laboratory where you perform chemistry practical, you would have noticed NaOH and $Ca(OH)_2$ containing reagent bottles with a white coating inside. This coating is of .... and formed by reaction ....**

7. **Ni is metal but dissolved in CO gas due to formation of *A*. *A* decomposes into metal**

(a) *A* is ....

(b) *A* is formed at a temperature ....

(c) *A* decomposes into metal at a temperature ....

8. **What are the components in water based fire-extinguisher (a)........(b)........ and reaction releasing CO_2 is ....**

9. **Mention two biological/medical uses of CO_2**

(a) ....

(b) ....

10. **Carbide of magnesium yields propyne on reaction with H_2O. Write bonding structure of this carbide ion.**

....

11. **Silver containing ore can be converted into silver in cyanide process, with following steps. Write the species formed in**

$Ag + CN^- \longrightarrow$ (a)

(a) + Zn $\longrightarrow$ Ag^+ ....(b)

12. **In the complex** $[Fe(CN)_6]^{3-}$ and $[Fe(CN)_6]^{4-}$

	$[Fe(CN)_6]^{3-}$	$[Fe(CN)_6]^{4-}$
(a) Write number of unpaired electrons	....	....
(b) What is magnetic moment	....	....

13. **When KCN (excess) is added to a Cu^{2+} final species formed is .... and on passing H_2S gas into it ....**

14. **pH of neutral water is considered as 7. When green house gases dissolve in water, pH will ....**

15. **There is strong warning due to global warming "New York City will submerge". Comment in three or four lines**

....

....

....

....

Answers

1. Excitation of 2*s*-electrons to 2*p*-orbital.

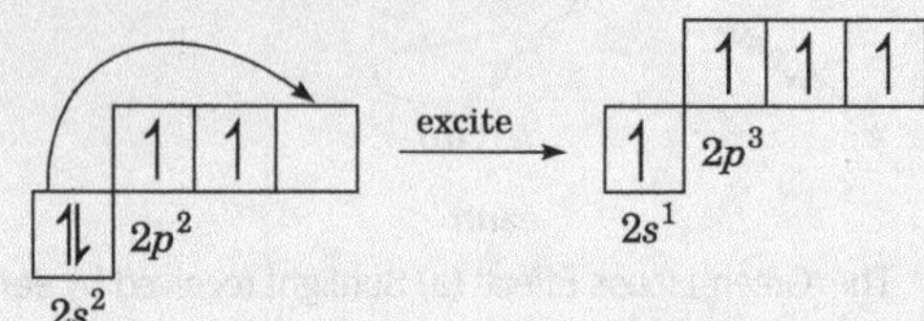

2. Refer text

3. (a) Carbon

(b) (i) $-\overset{|}{\underset{|}{C}}-\underset{\underset{H}{|}}{N}-H$

(ii) $>C=N-H$

(iii) $-C\equiv N$

4. (i) Solder (ii) Sn and Pb

5.

	Name/ Formula	Use
(i) Dry ice	CO_2	Referigerant for preserving food
(ii) Semi conductor	Ge	Electronic diode triode in computer
(iii) Solder	Sn/Pb	joining circuits
(iv) TEL	Tetra ethyl lead	Anti-knocking compound for petroleum products
(v) CNG	Compressed natural gas	Eco-friendly fuel for automobiles
(vi) LPG	Liquefied petroleum gas	Cooking gas

6. $Na_2CO_3, CaCO_3$

$$2NaOH + CO_2 \longrightarrow CO_2 \longrightarrow Na_2CO_3 + H_2O$$

$$Ca(OH)_2 + CO_2 \longrightarrow CaCO_3 + H_2O$$

There is also reaction between silicates (of glass) and alkali.

7. (a) $Ni(CO)_4$ (b) 28°C (c) 230°C

8. (a) Acid (like CH_3COOH)

(b) $NaHCO_3$

$$HCO_3^- + H^+ \longrightarrow H_2O + CO_2$$

9. (a) Photosynthesis

(b) CO_2 (as HCO_3^-) is an important component of buffer of blood which controls pH.

$$CO_2 + H_2O \rightleftharpoons H_2CO_3$$

$$H_2CO_3 \rightleftharpoons H^+ + HCO_3^-$$

10. $^-C\equiv C-C^{3-}$

11. $$Ag^+ + 2CN^- \longrightarrow [Ag(CN)_2]^-$$

$$2[Ag(CN)_2]^- + Zn \longrightarrow \underset{(a)}{Ag} + \underset{(b)}{[Zn(CN)_4]^{2-}}$$

12.

	$[Fe(CN)_6]^{2-}$	$[Fe(CN)_6]^{4-}$
(a)	1	0
(b)	$\sqrt{3}$BM	0

13. $[Cu(CN)_4]^{3-}$

no precipitation takes place being a stable complex.

14. Decrease.

15. Due to global warming, glaciers are melting and due to this sea-level will be certainly increasing which may cause flooding of cities situated at the bank of sea. New York is one of them.

Practice Exercise 1

1. (a) Which theory other than valence bond theory could be used to explain the bonding in SiF_6^{2-}?

(b) Predict the shape of this ion.

2. Use $E°$ values given in Table 12.2 to decide whether following reactions are feasible or not?

(a) $Pb^{4+} + Sn^{2+} \longrightarrow Sn^{4+} + Pb^{2+}$

(b) $Sn^{4+} + Pb^{2+} \longrightarrow Pb^{4+} + Sn^{2+}$

3. Explain :

(a) $(CH_3)_3N$ is pyramidal in shape while $(SiH_3)_3N$ is planar.

(b) Graphite is a good conductor and diamond is non-conductor.

4. Sn and Pb are amphoteric. Explain.

5. Sn reacts with HNO_3 under different conditions. What are various products?

6. Sn and Pb dissolve in hot NaOH solution. Write chemical reactions.

7. Every year some people die by being in a room containing a faulty heater that uses coal, gas or oil. How might the death occur?

8. Identify (*A*) to (*F*) in the following:

$$CaCO_3 \longrightarrow \underset{(g)}{(A)} + \underset{(solid)}{(B)}$$

$$(B) + H_2O \longrightarrow (C) \xrightarrow{(A)} (D) \xrightarrow{(A)\ excess} (E)$$

$$(A) + NH_3 \xrightarrow{\Delta} (F)$$

9. $$\underset{\text{oxalic acid}}{H_2C_2O_4} \xrightarrow{\Delta} \text{gas } (A) + \text{gas } (B) + \text{liquid } (C)$$

Gas (*A*) burns with blue flame and is oxidised to gas (*B*). Gas (*B*) turns lime water milky.

$$\text{Gas } (A) + Cl_2 \longrightarrow (D) \xrightarrow{NH_3, \Delta} (E) \xleftarrow{NH_3, \Delta} (B)$$

Identity (*A*) to (*E*) and explain reactions.

10. A binary salt (*A*) on reaction with H_2O gives *B*(*aq*) and *C*(*g*). (*C*) gas on passing into ammoniacal $AgNO_3$ gives white ppt *D*(*aq*). CO_2 gas turns *B* (*aq*) milky. Identify (*A*), (*B*), (*C*) and (*D*).

11. $$CaCO_3(s) \xrightarrow{\Delta} A(s) + B(g)$$

$$A(s) \xrightarrow{\text{carbon}, \Delta} C(s) + D(g)$$

$$(C) + N_2 \longrightarrow E(s)$$

$$(E) + H_2O + (B) \longrightarrow (F) + CaCO_3$$

$$(F) + H_2O \longrightarrow (G)$$

Identify (*A*) to (*G*).

12. Give the number of σ and π bonds in (i) C_2^{2-} (ii) C_3^{4-} and (iii) CN_2^{2-}.

13. (i) $BaC_2 + N_2 \xrightarrow{\Delta} (A)$

(ii) $CaC_2 + N_2 \xrightarrow{\Delta} (B)$

Identify (*A*) and (*B*).

14. Is CO_2 the only atmospheric gas responsible for the greenhouse effect.

15. 10 L of a mixture of CO and CO_2 gas is passed over heated coke when volume increases by 2L. What is % composition of the mixture ?

16. CO_2 goes to the atmosphere (greenhouse effect) and thus gets dissolved in H_2O. pH of the water increases or decreases and what is the effect on the fertility of the soil?

17. Which of the following gases qualify as a greenhouse gas : CO, NO, NO_2, Cl_2, H_2, Ne?

Answers

1. (a) Molecular orbital theory (b) Octahedral

2. Only (a) feasible due to the fact that stability of $Pb^{2+} > Pb^{4+}$ and $Sn^{4+} > Sn^{2+}$

$$\begin{array}{llr} & & E^\circ \\ \text{Also,} & Sn^{2+} \longrightarrow Sn^{4+} + 2e^- & -0.15\text{ V} \\ & Pb^{4+} + 2e^- \longrightarrow Pb^{2+} & 1.46\text{ V} \\ \hline & Pb^{4+} + Sn^{2+} \longrightarrow Sn^{4+} + Pb^{2+} & E^\circ_{cell} = +1.31\text{ V} \end{array}$$

Thus, spontaneous.

3. See text.

4. Both metals react with NaOH (base) as well as with HCl (acid), hence are amphoteric :

$$Sn + 2NaOH + H_2O \longrightarrow Na_2SnO_3 + 2H_2$$
$$Sn + 2HCl \longrightarrow SnCl_2 + H_2$$

Similarly with lead.

5. $$\underset{\text{dilute}}{4Sn + 10HNO_3} \longrightarrow 4Sn(NO_3)_2 + NH_4NO_3 + 3H_2O$$
$$Sn + \underset{\text{conc}}{4HNO_3} \longrightarrow \underset{\textit{meta}\text{ stannic acid}}{H_2SnO_3} + H_2O + 4NO_2$$

6. $Sn + 2NaOH + H_2O \longrightarrow Na_2SnO_3 + 2H_2$

Or $Sn + 2NaOH + 4H_2O \longrightarrow Na_2[Sn(OH)_6] + 2H_2$

Similarly with Pb.

7. If the flue (chimney) to the heater becomes blocked, or there is insufficient ventilation, the fumes of CO and CO_2 build up in the room. Unless the person lies on the floor, it is the CO that kills.

8. (*A*) : CO_2 (*B*) : CaO (*C*) : $Ca(OH)_2$ (*D*) : $CaCO_3$ (*E*) : $Ca(HCO_3)_2$ (*F*) : NH_2CONH_2

9. (*A*) : CO (*B*) : CO_2 (*C*) : H_2O (*D*) : $COCl_2$ (*E*) : NH_2CONH_2

10. (*A*) : CaC_2 (Ca^{2+}, C_2^{2-} binary) (*B*) : $Ca(OH)_2$ (*C*) : C_2H_2 (*D*) : C_2Ag_2

11. *A*: CaO *B*: CO_2 *C*: CaC_2 *D*: CO

E : $CaCN_2$ (calcium cyanamide nitrolim)

F : NH_2CN (cyanamide)

G : NH_2CONH_2 (urea)

12. (i) $C{\equiv}C$ one σ, two π (ii) $^-C{\equiv}C{-}C^{3-}$ two σ, two π

(iii) $^-N{=}C{=}N^-$ two σ, two π

13. *A* : Ca (N=C=N) Calcium cyanamide

Ba $(C{\equiv}N)_2$ barium cyanide.

14. There are several gases that are even stronger infrared absorbers than CO_2 [for example CH_4, O_3, N_2O, chlorofluoro carbons (CFC)]. About one-half of the anticipated greenhouse effect is due to CO_2 and the remainder of these other gases. Furthermore, the atmospheric concentration of some of these other gases (especially CFC) have been growing at a faster rate than CO_2.

15. $CO_2 + C \xrightarrow{\Delta} 2CO$

let $CO_2 = x$ L

$CO = (10 - x)$ L

CO due to passage over heated coke = $2x$ (from CO_2, no effect on CO, already in mixture)

Thus, $10x + 2x = 12$

$x = 2$L

Thus, $CO_2 = 2L, 20\%$

$CO = 8L, 80\%$

16. $$CO_2 + H_2O \rightleftharpoons H_2CO_3$$
$$H_2CO_3 \longrightarrow H^+ + HCO_3^-$$

$[H^+]$ increases hence pH decreases. Fertility of the soil also decreases.

17. CO, NO and NO_2

Silicon

Isolation of Silicon

Silicon is a hard grey, semiconducting solid that melts at 1410°C. It crystallises in a diamond like structure but does not form a graphite allotrope because of the relatively poor overlap of silicon π orbitals. In nature, silicon is generally found combined with oxygen in SiO_2 and in various silicate minerals.

It is obtained in elemental form by reduction of silica sand (SiO_2) with coke (C) in an electric furnace :

$$SiO_2(l) + 2C(s) \xrightarrow{\text{heat}} Si(s) + 2CO(g)$$

The silicon used for making solid-state semiconductor devices such as transistors, computer chips, and solar cells must be ultrapure, with impurities at a level of less than $10^{-7}\%$ (1 ppb). Silicon is purified by converting it to $SiCl_4$, a volatile liquid (b.p. 58°C) that can be separated from impurities by fractional distillation and then converted back to elemental silicon by reduction with hydrogen.

$$Si(s) + 2Cl_2(g) \longrightarrow SiCl_4(l)$$
$$SiCl_4(g) + 2H_2(g) \xrightarrow{\text{heat}} Si(s) + 4HCl(g)$$

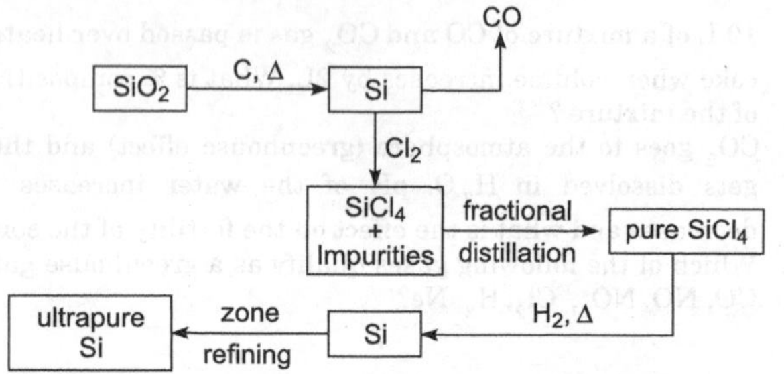

The silicon is purified further by a process called **zone refining** (Fig. 12.6). In this process, a heater melts a narrow zone of a silicon rod. Impurities concentrate in the molten zone and as the heater sweeps slowly down the rod, ultrapure silicon crystallises at the trailing zone of the molten zone.

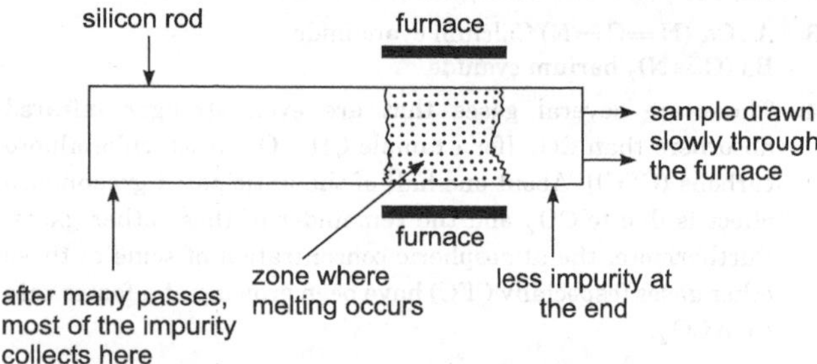

Fig. 12.6 Zone-refining. The sample is passed through the furnace many times. The impurities collect at one end of the

Compounds of Silicon

- A silicon atom, like a carbon atom, can form four bonds simultaneously. It uses all of its valence electrons ($3s^2 3p^2$) in an ***sp*^{3}** hybridisation scheme, and crystallises in an **fcc** structure similar to diamond. There is no allotrope of silicon equivalent to the graphite. **Diamond is an electrical insulator, graphite is good electrical conductor but silicon is a semiconductor.**

Diamond	Insulator
Graphite	Conductor
Silicon	Semiconductor

- Silicon is one of the most abundant of elements, being the essential ingredient of a large number of minerals that make up the earth's surface.

Silicates

- The majority of the compounds consist of silicates, in which silicon and oxygen make **giant** three dimensional structures. They are all based on the same base building block made from one silicon atom and four oxygen atoms at the corners of a tetrahedron. These SiO_4 tetrahedra may be :
  - separate units,
  - joined into chains or rings in groups of 2, 3, 4 or 6,
  - joined together into long single or double chains,
  - arranged in sheets,
  - linked into three dimensional network.
- The name **'silica'** covers an entire group of minerals, which have the general formula SiO_2, the most common of which is **quartz**. Quartz is a framework silicate with SiO_4 tetrahedra arranged in spirals. *The spirals can turn in a clockwise or anti-clockwise direction—a feature that results in there being two mirror image, optically active, varieties of quartz*. Quartz is used as a piezoelectric material (crystal oscillators and transducers). Silica get and fumed silica are amorphous forms of silica. Silica gel is formed when sodium silicate is acidified, and its dehydrated form is a drying agent in chromatography and catalyst support. Some silicate minerals are summarised in Table 12.5.

Table 12.5 **Some Representative Silicate Minerals**

Type	Anion type	Mineral	Composition
1. Ortho silicate (neso-silicates)	SiO_4^{4-}	zircon olivine	$ZrSiO_4, Mg_2SiO_4$
2. Pyrosilicate (soro-silicates)	$Si_2O_7^{6-}$	thortveitite	$Sc_2Si_2O_7$
	$Si_6O_{18}^{12-}$	beryl	$Be_3Al_2Si_6O_{18}$
3. Pyroxene	Si—O chains	spodumene	$LiAl(SiO_3)_2$
4. Amphibole	Si—O double chains	hornblende	$(Ca, Na, K)_{2-3}$ $(Mg, Fe, Al)_5$ $(Si, Al)_2$ $Si_6O_{22}(OH)_2$
5. Mica	Si—O sheets	muscovite	$KAl_2(AlSi_3O_{10})(OH)_2$
6. Zeolite	Si—O three-D network	natrolite	$Na_2Al_2Si_3O_{12} \cdot 2H_2O$
7. Single-chain silicate	SiO_4^{4-}	diopside	$[CaMg(SiO_3)_2]_n$
8. Double-chain silicate	SiO_4^{4-}	asbestos	$[Ca_2Mg_5(Si_4O_{11})_2(OH)_2]_n$
9. Sheet-like silicate	SiO_4^{2-}	talc	$[Mg_3Si_4O_{10}(OH)_2]_n$

- The vast majority of silicates are insoluble in water; only the simple alkali metal silicates are soluble.
- The extreme stability of the silicates is due to the donation of extra electrons from O atoms into vacant $3d$ orbitals of Si.
- In many common minerals, called **aluminosilicates,** Al atoms replace some Si atoms with very little structural change. In such cases univalent cations such as K^+ or Na^+ are also joined to balance positive charge of Si.

DEEP Focus

Following points are helpful in understanding the structural principles in silicates :

(a) The strong Si—O bond (ΔH = 466 kJ mol^{-1}) may be considered 50% ionic and 50% covalent from electronegativity difference (which is 3.5 – 1.8 = 1.7).

(b) The radii of Si^{4+} ion and O^{2-} ion are 0.40 and 1.40 Å respectively. The radius ratio 0.29 concludes coordination number 4 around Si which is in agreement with sp^3 hybridized tetrahedral arrangement around Si.

(c) The radius of Al^{3+} ion (0.53 Å) is close to that of the Si^{4+} ion and thus Al^{3+} can replace Si^{4+} in many silicates. Monovalent cation as Na^+ or K^+ maintains electrical neutrality. A pair of ions like (Si^{4+} + Na^+, total charge +5) may also be replaced by isoelectrical ion pairs like (Al^{3+} + Mg^{2+}) or (Al^{3+} + Ca^{2+}).

(d) The SiO_4 tetrahedral units may occur singly or may share one, two, three or four O-atoms through corners giving rise to cyclic groups, chains, infinite layers or infinite three-dimensional frameworks.

(e) SiO_4^{4-} unit is represented as

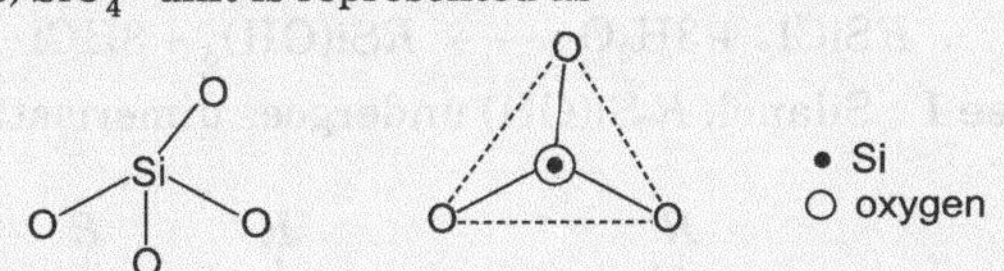

Taking above facts we consider following cases (also refer Table 12.5).

Case I: No oxygen atom shared (orthosilicate)

Unit : SiO_4^{4-}

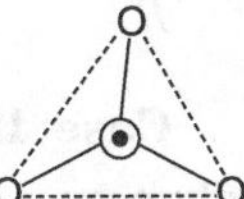

Case II: One oxygen atom shared (pyrosilicate)

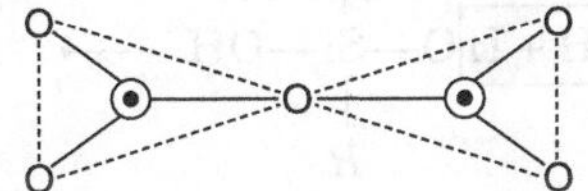

Unit : $Si_2O_7^{6-}$ ion, two SiO_4 units joined by sharing an oxygen atom.

Case III: Two oxygen atoms shared (closed ring : cyclosilicates and continuous chains : inosilicates)

Each SiO_4 tetrahedron shares an O-atom with each of two neighbouring tetrahedra resulting **ring** or **chain** anions

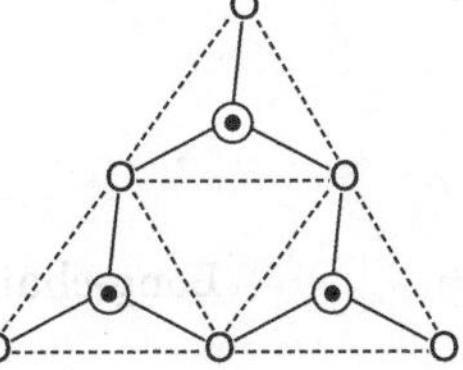

Ring anions : (1) $Si_3O_9^{6-}$—three SiO_4 tetrahedra linked together as in benitoite ($BaTiSi_3O_9$) and in wollastonite ($Ca_3Si_3O_9$).

(2) $Si_6O_{18}^{12-}$—six SiO_4 tetrahedra linked together as in beryl ($Be_3Al_2Si_6O_{18}$)

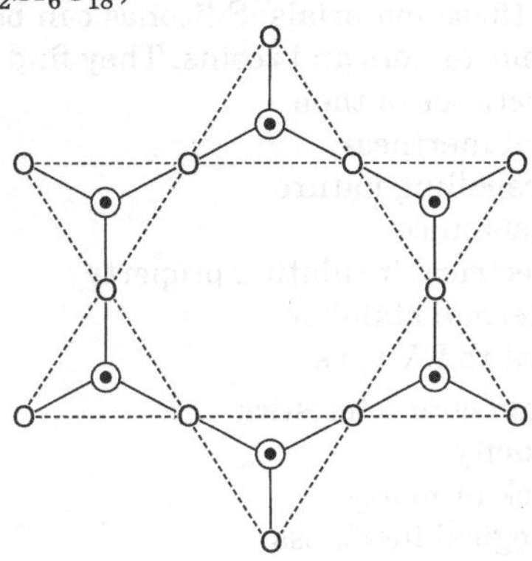

Chain-Silicate Anions

The minerals having silicate chains are collectively called pyroxenes : $MgSiO_3$, $CaMg(SiO_3)_2$.

Double chains result by joining two simple chains by sharing oxygens. Examples are :

Amphiboles having anions like $Si_4O_{11}^{6-}$

Asbestos mineral as **tremolite** $Ca_2Mg_5(Si_4O_{11})_2(OH)_2$

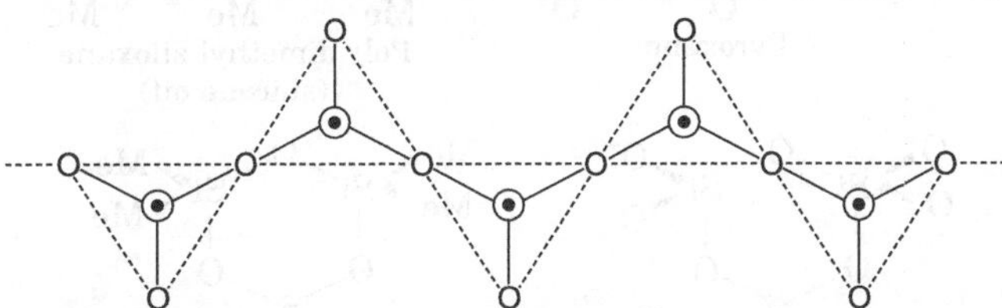

Illustration 1 *The silicate anion in the mineral kinoite is a chain of three* SiO_4 *tetrahedra that share corners with adjacent tetrahedra. The mineral also contains* Ca^{2+} *ions,* Cu^{2+} *ions, and water molecules in a 1:1:1 ratio.*

(a) *Give the formula and charge of the silicate anion*

(b) *Give the complete formula for the mineral*

Solution (a) The silicate anion has three SiO_4 tetrahedra that share corners with adjacent tetrahedra thus silicate is Si_3O_{10}, hence it can be represented as with charge as = 3 × 4 + 10 × (–2) = – 8

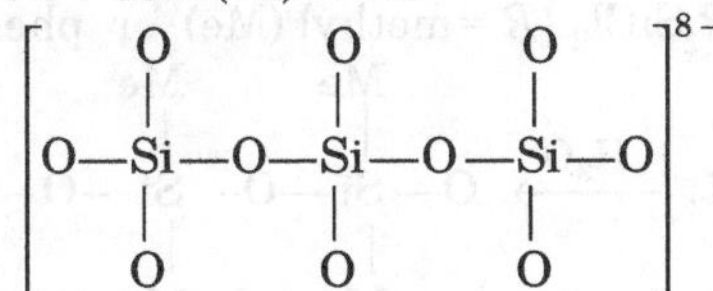

(b) Ca^{2+}, Cu^{2+} and H_2O are in the ratio of 1 : 1 : 1 and to balance (– 8) charge of silicate anion, (+8) charge is required thus there are two units each of Ca^{2+}, Cu^{2+} and H_2O thus, kinoite has formula $\mathbf{Ca_2Cu_2Si_3O_{10} \cdot 2H_2O}$.

Silicones

Silicones are synthetic organo-silicon polymers containing repeated R_2SiO units. Since the empirical formula is that of a ketone (R_2CO), the name silicone has been given to these materials. Silicones can be made into oils, rubbery elastomers and resins. They find a variety of applications because of their

(i) chemical inertness
(ii) water-repelling nature
(iii) heat-resistance
(iv) good electrical insulating property
(v) high thermal stability
(vi) resistant to UV rays
(vii) anti-foam characteristics
(viii) non-toxicity
(ix) non-stick property
(x) physiological inertness

The wide range of chain, cyclic and cross-linked silicone polymers are based on structural patterns similar to those in silicates. For example, the O^- in silicates have been replaced by the formally isoelectronic CH_3(Me) groups in the dimethyl silicones

$$-\overset{O^-}{\underset{O^-}{Si}}-O-\overset{O^-}{\underset{O^-}{Si}}-O-\overset{O^-}{\underset{O^-}{Si}}- \qquad -\overset{Me}{\underset{Me}{Si}}-O-\overset{Me}{\underset{Me}{Si}}-O-\overset{Me}{\underset{Me}{Si}}-$$

Pyroxene — Poly dimethyl siloxane (silicone oil)

$$(O^-)_2Si(-O-)_3 \text{ ring: } [(O^-)_2SiO]_3 \qquad [(Me)_2SiO]_3$$

Cyclic metasilicate — Cyclic dimethyl siloxane

Both silicones and silicates are formed due to the reluctance of silicon to form silicon to oxygen double bonds. But the silicates are all solids while the silicones may be oils; greases, rubbers or resins depending upon the extent of polymerisation.

Commercial silicon polymers are usually methyl derivatives and to a lesser extent phenyl derivatives and are synthesised by the hydrolysis of chlorosilanes.

R_2SiCl_2 [R = methyl (Me) or phenyl (ϕ)]

$$Me_2SiCl_2 \xrightarrow{H_2O} O-\overset{Me}{\underset{Me}{Si}}-O-\overset{Me}{\underset{Me}{Si}}-O-\overset{Me}{\underset{Me}{Si}}-O-$$

Thus, synthesis of silicones involves two steps

Step I Preparation of chlorosilanes

Step II Hydrolysis of chlorosilanes to silanols and then to silicones

Step I Chlorosilanes may be prepared by the

(i) reaction of Grignard reagent with $SiCl_4$

$$RMgCl + SiCl_4 \longrightarrow RSiCl_3 + MgCl_2$$

(R = alkyl, aryl)

$$RSiCl_3 + RMgCl \longrightarrow R_2SiCl_2 + MgCl_2$$
$$R_2SiCl_2 + RMgCl \longrightarrow R_3SiCl + MgCl_2$$

(ii) reactions of alkyl halides with silicon (Rochow direct method)

$$RCl + Si \xrightarrow[573\text{ K}]{Cu} R_2SiCl_2$$

Nature of chlorosilanes may be changed by changing molar concentration of RCl.

Above reaction is exothermic hence in the course of the reaction, removal of heat is necessary to make better yield.

Free radicals are involved as the intermediates.

$$RCl + 2Cu \longrightarrow [CuR] + [CuCl]$$
$$[CuR] \longrightarrow Cu + \underset{\text{alkyl free radical}}{[R^\bullet]}$$
$$[CuCl] + Si \longrightarrow Cu + [SiCl]$$
$$[R^\bullet] + [SiCl] \longrightarrow RSiCl$$
$$[CuR] \longrightarrow Cu + [R^\bullet]$$
$$[CuCl] \longrightarrow Cu + Cl^\bullet$$
$$[R^\bullet] + [Cl^\bullet] + [RSiCl] \longrightarrow \mathbf{R_2SiCl_2}$$

Step II **Hydrolysis of chlorosilanes to silanol**

$$R_3SiCl + H_2O \longrightarrow R_3Si(OH) + HCl$$
$$R_2SiCl_2 + 2H_2O \longrightarrow R_2Si(OH)_2 + 2HCl$$
$$RSiCl_3 + 3H_2O \longrightarrow RSi(OH)_3 + 3HCl$$

Case I Silanol, $R_3Si(OH)$ undergoes dimerisation

$$R-\overset{R}{\underset{R}{Si}}-\boxed{OH + H}O-\overset{R}{\underset{R}{Si}}-R \longrightarrow \underset{\text{Dimer}}{R-\overset{R}{\underset{R}{Si}}-O-\overset{R}{\underset{R}{Si}}-R}$$

Case II Disilanol, $R_2Si(OH)_2$ produces long-chain polymers

$$HO-\overset{R}{\underset{R}{Si}}-\boxed{OH + H}O-\overset{R}{\underset{R}{Si}}-OH \longrightarrow$$

$$HO-\overset{R}{\underset{R}{Si}}-O-\overset{R}{\underset{R}{Si}}-OH$$

$$\text{Long-chain polymer}\left[\cdots O-\overset{R}{\underset{R}{Si}}-O-\overset{R}{\underset{R}{Si}}-O\cdots\right]$$

Case III Trisilanol, $RSi(OH)_3$ produces cross-linked polymers

```
        R             R
        |             |
  HO—Si—[OH + H]O—Si—OH  ⟶
        |             |
        OH            OH

     R      R
     |      |
---O—Si—O—Si—O---                   R     R
     |      |                       |     |
     O      O      ⟵          HO—Si—O—Si—OH
     |      |                       |     |
  R—Si—O—Si—R                       OH    OH
     |      |
     O      O
     :      :
```

Cross-linked polymer

As the degree of polymerisation increases, silicones may be changed to solid state from liquid state.

- Liquid (oils)
- Semi-solids (greases)
- Elastomers (rubbers)
- Solids (resins)

Some special adhesives are made by mixing two silicones-based materials, which react together when mixed to give a cross-linked polymer.

The hydrolysis is usually initiated by a small quantity of concentrated H_2SO_4 but KOH may also be used to prepare certain silicone rubbers. Hydrolysis of $PhSiCl_3$ (Ph = phenyl) and $PhSiCl_2$ in toluene produces silicone resins. The polymerisation is allowed to proceed to a stage when the products is still soluble in toluene. It is shaped or moulded in this form and heated to 200°C in presence of a catalyst (a quaternary ammonium compound) when cross-linking develops the final 3-D network by condensing the remaining OH groups.

Given properties of the silicones may be related to their silica like (Si—O—Si—O) skleton surrounded by alkyl groups which make them water-repelling.

Silicone fluids containing striaght chain polymers (20-50 units) are used in making cement work and fibres water-repellant, as dielectric insulating media (in high voltage transformer), hydraulic oils and light duty lubricating oil. Mixed with lithium stearate soaps, they make quality grease for use in high vacuum apparatus. Silicone oils are also useful broad spectrum anti-foaming agents.

Silicone rubbers are made by reinforcing straight chain polymers (having 6×10^3 to 6×10^5 Si units) with fillers like SiO_2 or graphite. They can be **'vulcanised'** by increasing cross links through oxidations with benzoyl peroxide. These rubbers retain their quality from −100°C to 250°C. Such rubbers find extensive uses as industrial sealants and adhesives, space suits, medical tubing, etc.

Silicone resins are used in electrical insulations, as laminates for printed circuit boards or in encapsulation resistors, integrated circuit, etc. They are used in coating cooking ware and industrial tyre moulds.

Glass

Solids are most stable in crystalline form. However, if a solid is formed rapidly (for example, when a liquid is cooled quickly), its atoms or molecules do not have time to align themselves and may become locked in positions other than those of a regular crystal.

If we mix sodium carbonate and calcium carbonate with sand and fuse the mixture at about 1500°C, we get a liquid mixture of sodium and calcium silicates. When this mixture is cooled, it becomes viscous and eventually ceases to flow; it becomes solid—called glass. **Glass is thus super-cooled liquid** (solid solution). Amorphous solids, such as glass, lack regular three dimensional arrangement of atoms. **Glass thus refers to an optically transparent fusion product of inorganic materials that has cooled to a rigid state without crystallising.**

$$Na_2CO_3 + SiO_2 \xrightarrow{\Delta} Na_2SiO_3 + CO_2$$

$$CaCO_3 + SiO_2 \longrightarrow CaSiO_3 + CO_2$$

By varying the proportions of the three basic ingredients, and by adding other substances, we can alter the properties of the glass as given :

- Calcium-alkali silicate glass — normal domestic glass for windows

  Sodium silicate — soda glass used in cheap laboratory glassware

  Potassium silicate — potash glass

  PbO (if CaO in glass is replaced by PbO) — lead glass has a higher refractive index used in making optical parts and glass ornaments.

  B^{3+} replaces some Si^{4+} in the tetrahedral — borosilicate glass—used in heat resistant glasswares (Pyrex)

- Fining agents such as $NaNO_3$ or As_2O_3 are added to remove bubbles.
- Decolourising agents may be added to eliminate impurities and to obtain colourless glass. CaF_2 is sometimes added to make **opal** glass.
- The colour of glass is due largely to the presence of metal ions (as oxides)

Colour of glass	Metal oxide
Green	Fe_2O_3, CuO
Yellow	UO_2
Blue	CuO, CoO
Red	Colloidal Cu, Au

Following table shows the composition and properties of quartz, Pyrex, and soda-lime glass.

Name	Composition	Properties and uses
Pure quartz glass	100% SiO_2	Low thermal expansion, transparent to wide range of wavelengths; used in optical research.
Pyrex glass	SiO_2 : 60–80% B_2O_3 : 10–25% Al_2O_3 : small%	Low thermal expansion; transparent to visible and infrared, but not to UV radiation; used mainly in laboratory and household cooking glassware.
Soda-lime glass	SiO_2 : 75% Na_2O : 15% CaO : 10%	Easily attacked by chemicals and sensitive thermal shocks. Transmits visible light, but absorbs UV radiation; used mainly in windows and bottles.

Glass is attacked by HF and this property is used to make marking on the glass **(etching)**

$$\underset{\text{(in glass)}}{Na_2SiO_3} + 6HF \longrightarrow Na_2SiF_6 + 3H_2O$$

$$CaSiO_3 + 6HF \longrightarrow CaSiF_6 + 3H_2O$$

HF, however does not attack wax. The portion of the glass on which **etching** is to be done is covered with wax layer. Markings are carved on the wax layer with a pin and HF is poured on the carved parts and left for sometime. After washing with water, wax is removed and permanent marking is obtained.

Ceramics

Ceramics are inorganic, non-metallic, non-molecular solids, including both crystalline materials such as quartz (SiO_2) and amorphous materials such as glasses. Traditional silicate ceramics, such as pottery and porcelain are made by heating aluminosilicate clays to high temperatures. Modern, so-called **advanced ceramics** materials that have hightech engineering, electronic and biomedical applications include oxide ceramics, such as alumina (Al_2O_3), and non-oxide ceramics, such as alumina carbide (SiC) and silicon nitride. Additional examples are given in Table 12.6 which compares properties of ceramics with those of aluminium and steel.

Table 12.6 **Properties of Some Ceramic and Metallic Materials**

Material	M.P/°C	Density/g cm^{-3}
Oxide ceramics		
Alumina, Al_2O_3	2072	4.0
Beryllia, BeO	2530	3.0
Zirconia, ZrO_2	2700	5.9
Non-oxide ceramics		
Boron carbide, B_4C	2350	2.5
Silicon carbide, SiC	2700	3.2
Silicon nitride, Si_3N_4	1900	3.4
Metals		
Aluminium	660	2.7
Plain carbon steel	1515	7.9

Compounds of Tin

Tin oxides

Two types of oxides are known :

SnO and SnO_2

Both are amphoteric. As we know, basic character of oxides increases as we go down the group.

$$\underbrace{CO,\ CO_2,\ SiO_2}_{\text{acidic}}\quad \underbrace{GeO,\ GeO_2}_{\text{amphoteric}}\quad \underbrace{SnO,\ SnO_2}_{\text{amphoteric}}\quad \underbrace{PbO,\ PbO_2}_{\text{basic}}$$

(A) $\underset{\text{basic}}{SnO} + 2H^+ \longrightarrow Sn^{2+} + H_2O$

(B) $\underset{\text{acidic}}{SnO} + 4OH^- + H_2O \longrightarrow Sn(OH)_6^{4-}$

In (A), SnO reacts with H^+, hence SnO is basic oxide.

In (B), SnO reacts with OH^-, hence SnO is an acidic oxide.

These two reactions confirm that SnO is an amphoteric oxide.

$[Sn(OH)_6]^{4-}$ or $[Sn(OH)_3]^-$ is called stannite ion and is also written as $HSnO_2^-$ and SnO_2^{2-}. $[Sn(OH)_6]^{2-}$ is called stannate ion and is also written as SnO_3^{2-} and SnO_4^{4-}.

Tin Chlorides

$SnCl_2$ [tin (II) chloride] is obtained by dissolving tin in concentrated HCl; on cooling crystals of $SnCl_2 \cdot 2H_2O$ separate out. Anhydrous $SnCl_2$ is formed by heating Sn in a current of HCl vapour.

$SnCl_2$ and $SnCl_4$ are common chlorides but due to 'inert-pair effect', Sn(II) is less stable than Sn(IV), hence the reaction

$$Sn^{2+} \longrightarrow Sn^{4+} + 2e^- \qquad \Delta G° = -\text{ve}$$

is spontaneous. Thus Sn^{2+} salts are reducing agents.

- $SnCl_2$ reduces $HgCl_2$ first to Hg_2Cl_2 and then to Hg.

$$SnCl_2 + 2HgCl_2 \longrightarrow \underset{\text{white ppt}}{Hg_2Cl_2\downarrow} + SnCl_4$$

$$Hg_2Cl_2 + SnCl_2 \longrightarrow \underset{\text{grey}}{2Hg} + SnCl_4$$

- $SnCl_2$ reduces $FeCl_3$ to $FeCl_2$

$$2FeCl_3 + SnCl_2 \longrightarrow 2FeCl_2 + SnCl_4$$

- It also reduces $CuCl_2$ to CuCl

$$SnCl_2 + 2CuCl_2 \longrightarrow SnCl_4 + \underset{\text{white ppt}}{2CuCl\downarrow}$$

- $SnCl_2$ (and also SnF_2) dissolve in solutions containing Cl^- (or F^-) ions

$$SnCl_2 + Cl^- \longrightarrow [SnCl_3]^-$$

- $SnCl_2$ partly hydrolyses in water forming the basic chloride Sn(OH)Cl

$$SnCl_2 + H_2O \longrightarrow Sn(OH)Cl$$

- $SnCl_4$ also hydrolyses in dilute solution but it is incomplete and can be repressed in presence of halogen acid :

$$SnCl_4 + 4H_2O \longrightarrow Sn(OH)_4 + 4HCl$$
$$Sn(OH)_4 + 4HCl \longrightarrow SnCl_4 + 4H_2O$$
$$SnCl_4 + 2HCl \longrightarrow H_2SnCl_6$$

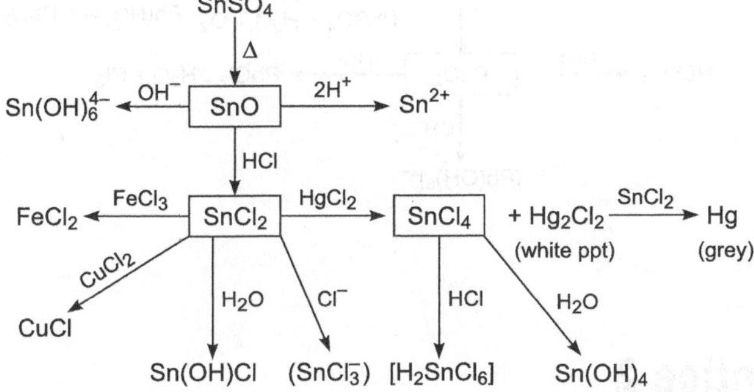

Compounds of Lead

Lead Chlorides

- Lead reacts with dil. HCl producing insoluble $PbCl_2$ which protects the metal from further attack. But due to the formation of the soluble complex ion $[PbCl_3]^-$, $PbCl_2$ dissolves in conc. HCl. Thus lead dissolves in conc. HCl.

$$Pb + 3HCl\,(\text{conc.}) \longrightarrow [PbCl_3]^- + H^+ + H_2(g)$$

- Pb can also be converted into $PbCl_2$ on reaction with Cl_2:

$$Pb + Cl_2 \longrightarrow PbCl_2$$

- $PbCl_2$ is partially hydrolysed by H_2O :

$$PbCl_2 + H_2O \longrightarrow Pb(OH)Cl + HCl$$

or $$Pb^{2+} + 2H_2O \longrightarrow [Pb(OH)]^+ + H_3O^+$$

- Aqueous hot solution of $PbCl_2$ gives black ppt of PbS on passing H_2S gas:

$$PbCl_2 + H_2S \longrightarrow \underset{\text{black ppt}}{PbS\downarrow} + 2HCl$$

- $PbCl_4$ decomposes very readily into $PbCl_2$ and Cl_2 due to greater stability of Pb^{2+} than Pb^{4+} (inert-pair effect)

$$PbCl_4(s) \longrightarrow PbCl_2(s) + Cl_2(g)$$
$$Pb^{4+} + 2e^- \longrightarrow Pb^{2+}$$
$$2Cl^- \longrightarrow Cl_2 + 2e^-$$

Thus, Pb^{4+} salts are better oxidising agents.

- Lead shows the ability to make a complex with chloride ions. If ice cold conc. HCl is added to lead dioxide [lead (IV) oxide], PbO_2, yellow hexachloro plumbate (IV) ions, $PbCl_6^{2-}$ are made. $PbCl_4^{2-}$ are made from PbO

$$PbO_2(s) + 6Cl^-(aq) + 4H^+(aq) \longrightarrow PbCl_6^{2-}(aq) + 2H_2O(l)$$
$$PbO(s) + 4Cl^-(aq) + 2H^+(aq) \longrightarrow PbCl_4^{2-}(aq) + H_2O(l)$$

Lead oxides

Three oxides of lead are known :

(a) PbO — lead (II) oxide called **litharge** (a yellow-orange solid)

(b) PbO_2 — lead (IV) oxide also called **lead dioxide** (a dark brown solid)

(c) Pb_3O_4 — dilead (II) lead (IV) oxide. It is a combination of (a) and (b) : $2PbO \cdot PbO_2$. It is also called **red lead.**

- PbO is obtained by roasting of PbS (galena):

$$2PbS + 3O_2 \xrightarrow{\Delta} 2PbO + 2SO_2$$

PbO is reduced to Pb by heating with carbon:

$$PbO + C \xrightarrow{\Delta} Pb + CO$$
$$PbO + CO \longrightarrow Pb + CO_2$$

PbO is amphoteric, being soluble both in acid and alkali:

$$PbO + 2H^+ \longrightarrow Pb^{2+} + H_2O$$
$$PbO + 2OH^- + H_2O \longrightarrow [Pb(OH)_4]^{2-}$$

Crystal structure shows that it is predominantly covalent character.

- PbO_2 is an oxidising agent and reduced to PbO since stability of Pb (II) > Pb (IV) based on 'inert pair' effect:

$$Pb^{4+} + 2e^- \longrightarrow Pb^{2+}$$
$$PbO_2 + 4HCl \longrightarrow PbCl_2 + 2H_2O + Cl_2$$
$$PbO_2 + SO_2 \longrightarrow PbSO_4$$
$$PbO_2 + 2H_2SO_4 \longrightarrow 2PbSO_4 + 2H_2O + O_2$$
$$2PbO_2 \xrightarrow{\Delta} 2PbO + O_2$$

It is insoluble in an acid but soluble in concentrated alkali thus indicating its acidic nature:

$$PbO_2 + 2OH^- + 2H_2O \longrightarrow [Pb(OH)_6]^{2-}$$

- Pb_3O_4, a double oxide, is obtained by heating lead (II) oxide in air:

$$3PbO + \frac{1}{2} O_2 \xrightarrow{\Delta} Pb_3O_4$$

Its crystal structure shows that it is predominantly covalent and composed of a combination **of two types of lead atom. One type is bonded to six oxygen atoms, the other to three oxygen atoms. But in the reactions, it behaves like a mixture of PbO and PbO_2 :**

$$2Pb_3O_4 \xrightarrow{\Delta} 6PbO + O_2$$

Pb_3O_4 can be written as :

$$2(2PbO \cdot PbO_2) \equiv 4PbO + 2PbO_2$$

$$2PbO_2 \longrightarrow 2PbO + O_2$$

and out of two oxides only PbO_2 is affected by heat :

When it reacts with H_2SO_4, $PbSO_4$ is formed along with O_2 only :

$$Pb_3O_4 + 3H_2SO_4 \longrightarrow 3PbSO_4 + 3H_2O + \frac{1}{2} O_2$$

DEEP Focus

With HNO_3, PbO_2 is precipitated and soluble $Pb(NO_3)_2$ is also formed :

$$Pb_3O_4 + 4HNO_3 \longrightarrow 2Pb(NO_3)_2 + PbO_2\downarrow + 2H_2O$$

This reaction clearly indicates that $Pb(NO_3)_2$ *is from* PbO *while* PbO_2 *remains insoluble in* HNO_3 *and thus gets precipitated :*

$$(2PbO \cdot PbO_2) + 4HNO_3 \longrightarrow 2Pb(NO_3)_2 + PbO_2\downarrow + 2H_2O$$

Lead oxides are used in the manufacture of lead-acid (storage) batteries, glass, ceramic glazes, cements (PbO), metal protecting paints (Pb_3O_4), matches (PbO_2), and explosives (PbO_2).

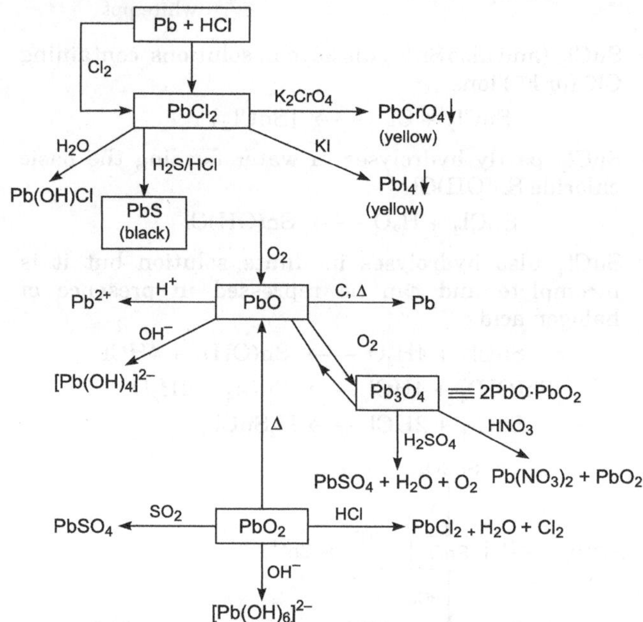

Target Practice 2

1. SiO_4^{4-} unit is represented as

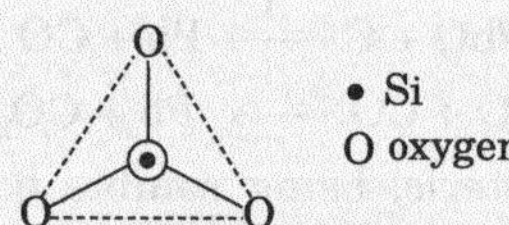

Represent $Si_2O_7^{6-}$ similarly such that one oxygen atom is shared between two units of SiO_4.

....

2. (a) Silicones are ....

(b) They are formed when .... reacts with ....

(c) They are used as ....

(d) Write structure of simple silicone.

....

....

3. A lab assistant was employed in a chemical industry manufacturing NaOH, HF and HNO_3. He was given three vessels for storing these chemicals. Can you help in exact matching without any loss to the vessel.

Chemicals : NaOH, HF, HNO_3.

Vessles made of : Al, glass, wax

Matching is : (i) NaOH ....

(ii) HF ....

(iii) HNO_3 ....

4. $SnCl_2$ can be volumetrically determined by titration with $FeCl_3$

$$2Fe^{3+} + Sn^{2+} \longrightarrow 2Fe^{2+} + Sn^{4+}$$

Can you write the suitable indicator based on previous sections.

....

5. SnO is **called an amphoteric oxide.** You are asked to prove this statement then chemicals required are

(a) .... (b) ....

6. I. $[Sn(OH)_6]^{4-}$ can be represented as .... or ....
(a) (b)

and is called ....
(c)

II. $[Sn(OH)_6]^{2-}$ can be written as .... or ....
(a) (b)

and is called ....
(c)

7. Write the products when Pb_3O_4 reacts with

(a) HCl $\longrightarrow$ ....

(b) H_2SO_4 $\longrightarrow$ ....

(c) HNO_3 $\longrightarrow$ ....

8. Following reactions are spontaneous not.

(a) $PbO_2 \longrightarrow PbO + O_2$ Yes/No

(b) $PbCl_4 \longrightarrow PbCl_2 + Cl_2$ Yes/No

(c) $SnCl_4 + PbCl_2 \longrightarrow PbCl_4 + SnCl_2$ Yes/No

(d) $2FeCl_3 + SnCl_2 \longrightarrow SnCl_4 + 2FeCl_2$ Yes/No

9. An oxide of lead changes to another insoluble oxide and soluble salt on reaction with HNO_3

(a) oxide is ....

(b) insoluble oxide is ....

(c) soluble salt is ....

Answers

1. $Si_2O_7^{6-}$ ion with two (SiO_4) units joined by sharing as oxygen atom.

2. (a) Synthetic polymers containing repeated R_2SiO.

(b) R_2SiCl_2, H_2O

(c) Refer text

(d) $—O—Si(R)_2—O—Si(R)_2—O—Si(R)_2—O—$

3. (i) NaOH : Glass

(ii) HF : Wax

(iii) HNO_3 : Al

4. KCNS, its gives red colour with Fe^{3+} ion.

5. (a) HCl (b) NaOH

6. I : (a) $HSnO_2^-$ (b) SnO_2^{2-} (c) stannite ion

II : (a) SnO_3^{2-} (b) SnO_4^{4-} (c) stannate ion

7. (a) $Pb_3O_4 + 4HCl \longrightarrow 2PbCl_2 + PbO_2 + 2H_2O$

$PbO_2 + 4HCl \longrightarrow PbCl_2 + 2H_2O + Cl_2$

(b) $Pb_3O_4 + 3H_2SO_4 \longrightarrow 3PbSO_4 + 3H_2O + \frac{1}{2}O_2$

(c) $Pb_3O_4 + 4HNO_3 \longrightarrow 2Pb(NO_3)_2 + PbO_2 + 2H_2O$

8. (a) Yes (b) Yes (c) No (d) Yes

9. (a) Pb_3O_4 (b) PbO_2 (c) $Pb(NO_3)_2$

Practice Exercise 2

1. $SnCl_2 + 2NaOH \longrightarrow (A)$ (white ppt) $\xrightarrow{\text{NaOH (excess)}} (B)$

Identify (*A*) and (*B*).

2. $FeCl_3$ can be reduced to $FeCl_2$ by $SnCl_2$ as well as by the addition of Sn into it. Explain.

3. $FeCl_3$ gives blood red colour with NH_4CNS. But after $FeCl_3$ has been treated with $SnCl_2$, there is no colour with NH_4CNS. Explain.

4. Identify (*A*) and (*B*) based on following results:

Colourless salt (*A*) + dil. HCl $\longrightarrow$ white ppt (*B*) soluble in hot water.

In this reaction vinegar smell is also observed.

5. You are given a colourless solution and was told to add a little NaCl solution to it. A cloudy white precipitate was made. What conclusion should you draw about the nature of the colourless solution?

6. If you were to attempt to dissolve lead (II) oxide in acid, which acid would you choose?

7. Explain why the reactions with HCl and SO_2 show the oxidising nature of PbO_2?

8. Predict the result of reacting Pb_3O_4 with conc. HCl.

Answers

1. (*A*) : $Sn(OH)_2$ (*B*) : Na_2SnO_2 (sodium stannite)

2. $2FeCl_3 + SnCl_2 \longrightarrow SnCl_4 + 2FeCl_2$

This reaction is possible since Sn^{4+} is more stable than Sn^{2+} (inert-pair effect)

Also $FeCl_3$ solution is acidic due to hydrolysis :

$$FeCl_3 + 3H_2O \rightleftharpoons Fe(OH)_3 + 3HCl$$

Sn added reacts with HCl forming nascent hydrogen that can reduce $FeCl_3$ to $FeCl_2$:

$$2HCl + Sn \longrightarrow SnCl_2 + \underset{\text{(nascent)}}{2H}$$

$$FeCl_3 + H \longrightarrow FeCl_2 + HCl$$

3. $SnCl_2$ reduces $FeCl_3$ to $FeCl_2$ hence no colour with NH_4CNS.

4. (*A*) : $Pb(CH_3COO)_2$ (*B*) : $PbCl_2$

5. White ppt can be of AgCl or $PbCl_2$. To make further distinction, warm the solution. $PbCl_2$ dissolves in hot water while AgCl remains insoluble. AgCl dissolves in *aq* NH_3.

6. HNO_3 (since $PbCl_2$ and $PbSO_4$ formed by using HCl or H_2SO_4, are insoluble).

7. The lead changes from Pb (IV) to Pb(II); this represents gain of electrons, *i.e.*, reduction hence PbO_2 is an oxidising agent. Also stability of $Pb^{2+} > Pb^{4+}$, hence the reaction

$$Pb^{4+} + 2e^- \longrightarrow Pb^{2+}$$

is spontaneous in forward side.

8. $Pb_3O_4 + 8HCl \longrightarrow PbCl_2 + 4H_2O + Cl_2$

Total Practice Set
(Read, Plan & Solve!)

Problem 1. What would you expect to happen when:

(a) Tin is heated with a concentrated aqueous solution of sodium hydroxide.

(b) Sulphur dioxide is passed over lead dioxide.

(c) Carbon disulphide is shaken with aqueous sodium sulphide.

(d) Dichlorosilane is hydrolysed by water.

(e) 4 moles of $ClCH_2SiCl_3$ react with 3 moles of $LiAlH_4$ in diethyl ether solution?

Solution (a) $Sn + 2OH^- + 4H_2O \longrightarrow [Sn(OH)_6]^{2-} + 2H_2$

(b) $PbO_2 + SO_2 \longrightarrow PbSO_4$

(c) $CS_2 + Na_2S \longrightarrow \underset{\text{sodium thiocarbonate}}{Na_2CS_3}$

(d) $SiH_2Cl_2 + 2H_2O \longrightarrow H_2Si(OH)_2 + 2HCl$

$$H_2Si(OH)_2 \longrightarrow H_2Si{=}O + H_2O$$

(e) $4ClCH_2SiCl_3 + 3LiAlH_4 \longrightarrow 4ClCH_2SiH_3 + 3LiCl + 3AlCl_3$

Problem 2. If you had a mixture of CO and CO_2, how would you find out relative proportions of the two gases in the mixture ?

Solution **Method I :** Mixture of CO and CO_2 is run into $Ca(OH)_2$ solution. CO_2 is absorbed by $Ca(OH)_2$ and thus residual volume will be that of CO.

$$Ca(OH)_2 + CO_2 \longrightarrow CaCO_3 + H_2O$$

Method II : Mixture of CO and CO_2 is passed over heated charcoal :

$$CO_2 + C \longrightarrow 2CO$$

Increase in volume will be due to CO_2. By measuring total volume after passing over heated charcoal, composition of mixture of CO and CO_2 can be determined.

Problem 3. A mixture of CS_2 vapour and nitrogen monoxide, NO, gives a vivid blue flame when ignited. A yellow solid is left afterwards. Predict the products of the reaction, and write the equations.

Solution You should have predicted that a yellow solid is sulphur formed by oxidation of CS_2 and thus NO is reduced to N_2 which appears as a vivid blue flame :

$$2NO + CS_2 \longrightarrow \underset{\text{(yellow solid)}}{2S} + N_2 + CO_2$$

Problem 4. CO and CN^- ions are highly poisonous. A sign that someone has been poisoned by either of these two chemicals is that the person's blood changes colour. They both have the effect of preventing oxygen being taken up by the blood stream. What might happen to haemoglobin to stop oxygen being taken up?

Solution CO and CN^- ions are able to bond more strongly to the iron atom in haemoglobin than can oxygen. If you were to breathe in CO for any length of time, your haemoglobin would become saturated with CO molecules rather than oxygen. As a result, all the body processes that rely on a supply of oxygen gradually cease. Death is the result.

Problem 5. Comment on each of the following observations:

(a) When an aqueous solution of potassium cyanide is added to aluminium sulphate, $Al(OH)_3$ is precipitated.

(b) The carbide Mg_2C_3 liberated propyne ($CH_3C{\equiv}CH$) on treatment with water.

(c) Magnesium silicide reacts with ammonium bromide in liquid ammonia to form silane.

(d) $Me_2Si\langle(SiMe_2)(SiMe_2)\rangle Si\langle(Me)(H)$ is hydrolysed by aqueous alkali at the same rate as the corresponding Si-D compound.

Solution (a) Aqueous solution of KCN is alkaline due to hydrolysis of CN^- ion.

$$CN^- + H_2O \longrightarrow \underset{\text{weak acid}}{HCN} + OH^-$$

OH^- formed causes precipitation of $Al_2(SO_4)_3$ as $Al(OH)_3$

$$Al^{3+} + 3OH^- \longrightarrow Al(OH)_3 \downarrow$$

(b) $$Mg_2C_3 \longrightarrow 2Mg^{2+} + C_3^{4-}$$

C_3^{4-} ion is decomposed by H_2O giving $CH_3C{\equiv}CH$

$$C^{3-}{-}C{\equiv}C^- + 4H_2O \longrightarrow CH_3{-}C{\equiv}CH + 4OH^-$$

$$2Mg^{2+} + 4OH^- \longrightarrow 2Mg(OH)_2$$

Net : $\mathbf{Mg_2C_3 + 4H_2O \longrightarrow 2Mg(OH)_2 + CH_3{-}C{\equiv}CH}$

(c) Solutions of NH_4Br (in general ammonium halide) in NH_3 may be used as acids and SiH_4 (silane) is formed when NH_4Br in NH_3 reacts with Mg_2Si (magnesium silicide)

$$\underset{\text{magnesium silicide}}{Mg_2Si} + 4NH_4Br \longrightarrow \underset{\text{silane}}{SiH_4} + 2MgBr_2 + 4NH_3$$

(d) Si—H or Si—D bond is not broken in rate determining step, it is probable that there is attack of OH^- on Si.

Explanation

(a) CO_2 *is a gas but* SiO_2 *is a solid*

– Carbon and silicon (Group 14, IVA) have four valence electrons. We might expect carbon and silicon to form oxides with similar properties. In CO_2 the ability of C and O atoms to form π bonds through the side-wise overlap of their $2p$ orbitals is strong. The result is strong C— to —O double bonds and a very stable triatomic molecule.

$$O{=}C{=}O$$

Silicon, being in third period, would have to use $3p$ orbitals to form double bonds with oxygen. The side-wise overlap of these orbitals with the $2p$ orbitals of oxygen is too limited for π bond formation. From an energy stand point, a stronger bonding arrangement results if the Si atoms form four single bonds with O atoms (bond energy : 464 kJ/mol) rather than two double bonds (bond energy : 640 kJ/mol) since each O atom must be simultaneously bonded to two Si atoms, the result is a network of —Si—O—Si— bonds and thus a hard (giant) solid :

(b) $(SiH_3)N$ *is planar while* $(CH_3)_3N$ *is pyramidal*

– In case of $(CH_3)_3N$, lone pair on N and three covalent bonds with CH_3 groups results in sp^3 hybridisation but lone pair-bond pair repulsion causes distortion hence gives pyramidal structure :

But in case of $(SiH_3)_3N$, lone pair on the nitrogen is transferred to the empty d-orbitals of silicon ($p\pi$-$d\pi$ overlap) thereby causing planarity of unit.

(c) $SiCl_4$ *is hydrolysed but* CCl_4 *is inert to hydrolysis*

– $SiCl_4$ and CCl_4 are covalent compounds. Carbon has maximum covalency of four while silicon has six.

Thus CCl_4 is inert towards water and is not hydrolysed but $SiCl_4$ can coordinate with 2 moles of H_2O molecules and is thus readily hydrolysed :

$$SiCl_4 + 2H_2O \longrightarrow Si(OH)_4 + 2HCl$$

Silicon with covalency six

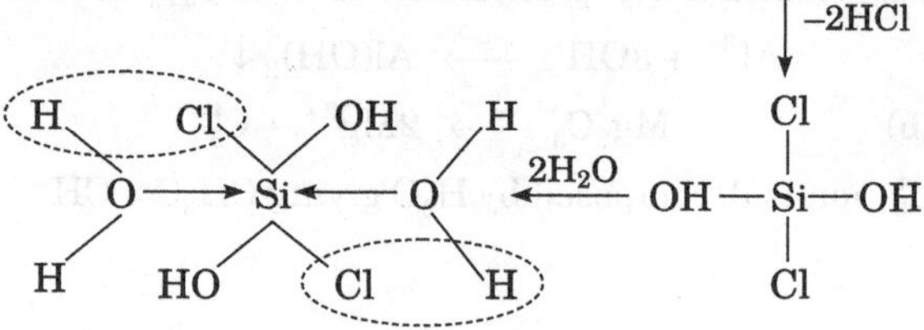

Silicon with covalency six

−2HCl

(d) *Glass made of* SiO_2 *is attacked by* HF *with the formation of* SiF_6^{2-} *anion. The analogous* CF_6^{2-} *does not exist :*

– Silicon has $3d$-orbitals and can use octahedral sp^3d^2-hybrid orbitals to bond to six F^- ions. Carbon lacks $2d$-orbitals ($2d$ does not exist) and with just $2s$ and $2p$ valence orbitals, carbon can form maximum of only four bonds.

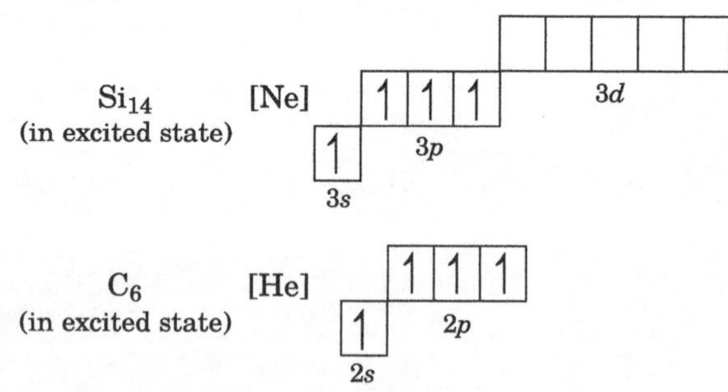

(e) $SnCl_2 \cdot 2H_2O$ *readily loses* 1 *molecule of* H_2O *at* 80°C.

– Pyramidal $SnCl_2 \cdot H_2O$ molecule has a lone pair on Sn. The other H_2O, hydrogen-bonded to the coordinated H_2O is easily lost.

Master Exercises

Exercise 1
(Stage 1 : Learning)

Short Answer Type Questions

1. Carbon has coordination number (C.N.) of four while other elements of this group have C.N. of six. Explain.

2. Identify the group 14A element that best fits each of the following descriptions:
(a) prefers the +2 oxidation state
(b) forms the strongest π bonds
(c) is the second most abundant element in the earth's crust
(d) forms the most acidic oxide

3. Give the formula of a compound in which carbon exhibits an oxidation state of:
(a) +4 (b) +2
(c) −4

4. List three properties of diamond, and account for them in terms of structure and bonding.

5. Identify (*A*), (*B*), (C), (*D*) and (*E*) :

(a) $Si \xrightarrow[\text{conc, } \Delta]{HNO_3/HF} (A) \xrightarrow{HF} (B)$; $(A) \xrightarrow{H_2O} (B) + (C)$

(b) $\text{Graphite} \xrightarrow{\text{hot conc } HNO_3} (D)$

(c) $\underset{\text{fused}}{Na_2CO_3} + Si \longrightarrow (E)$

6. Solution of CO_3^{2-} (say Na_2CO_3) is alkaline. Explain.

7. Explain why a solution containing $Ca^{2+}(aq)$ and $HCO_3^-(aq)$ ions is said to be hard?

8. What happens if solution of Na_2CO_3 is heated with Al?

9. A mixture of CS_2 vapour and nitrogen monoxide, NO, gives a vivid blue flame when ignited. A yellow solid is left afterwards. Predict the products of reaction, and write the equation.

10. A student placed a pellet of NaOH on a watch glass. A few days later, she noticed that the pellet was covered with a white solid. What is the identity of this solid?

11. List three commercial uses for CO_2, and relate each use to one of carbon dioxide's properties.

12. Is it wise to write

$$Al_2C_3 \longrightarrow 4Al^{3+} + 3C^{4-} ?$$

13. Draw a Lewis structure for the C_2^{2-} and C_3^{3-} ions.

14. Why are CO and CN^- harmful?

15. Give an example of an ionic carbide. What is the oxidation state of carbon in this compound?

16. Which of the following is a more effective greenhouse gas CO or H_2O?

17. Instead of monitoring carbon dioxide, suggest another gas that scientists could study to substantiate the fact that CO_2 concentration is steadily increasing in the atmosphere.

18. The equilibrium constant (K_p) for the reaction

$$2CO(g) + O_2(g) \rightleftharpoons 2CO_2(g)$$

is 1.4×10^{90} at 25°C. Given this enormous value, why doesn't CO convert totally to CO_2 in the troposphere?

19. What is glanzkohlenstoff?

20. Talc, $Mg_3Si_4O_{10}(OH)_2$ is the softest material. It has a smooth, greasy touch. It is used in talcum powder. Muscovite, $KAl_2(Si_3Al)O_{10}(OH)_2$, is one of the micas, which split into thin layers extremely easily. Both minerals have layer structures. What type of force holds the layers together?

21. Silica will react in a similar way with metal sulphates (and phosphates) as it does with carbonates, giving a solid residue and a gas. 1 mole of silica reacts with 1 mole of sodium sulphate, Na_2SO_4. Write the equation.

22. What would be the structure of the silicone molecule made from $Si(CH_3)_3Cl$?

23. One type of silicone adhesive uses a polymer that has two OH groups left on the ends of the chain :

$$HO{-}Si(CH_3)_2[O{-}Si(CH_3)_2]_n{-}O{-}Si(CH_3)_2{-}OH$$

It has been found that this kind of polymer reacts easily with an organo-silicon molecule that contains acetate (CH_3COO) group. Which of these three molecules

$CH_3COOSi(CH_3)_3$, $(CH_3COO)_2Si(CH_3)_2$, $(CH_3COO)_3SiCH_3$

would you choose to add to the hydroxy polymer in order to make a strong adhesive?

24. The following pictures represent various silicate anions. Give the formula and charge of each anions.

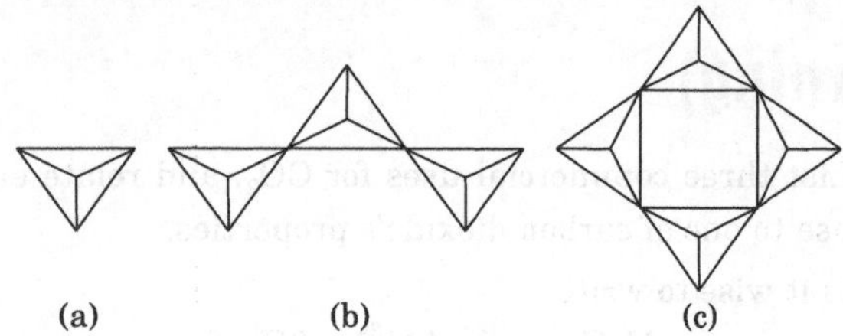

25. Draw the structure of the silicate anion in :

(a) K_4SiO_4

(b) $Ag_{10}Si_4O_{13}$

26. A colourless solution (*X*) gives black ppt (*Y*) on passing H_2S gas. (*X*) also gives white ppt (*Z*) with $SnCl_2$. (*Z*) changes to grey. Identify (*X*), (*Y*) and (*Z*).

27. One problem with making crystals of tin (II) chloride is that they are easily hydrolysed by warm water. Instead of dissolving tin in HCl(*aq*) how else might the anhydrous chloride be made?

28. HF can't be stored in a glass vessel. Explain.

29. Fill in the blanks :

(a) Glass is a .... liquid.

(b) Markings on the glass plate are made by use of ....

(c) Glass with high refractive index contains ....

(d) Pyrex glass can withstand heat and contains ....

(e) Opal glass contains ....

Analytical Questions

1. Explain :

(i) Why silicon is not expected to form an allotrope with graphite like structure?

(ii) Contrary to $(CH_3)_2O$, $(SiH_3)_2O$ does not form addition compound with BF_3.

2. Given below the phase diagram of carbon, answer the following questions:

(a) How many triple points are there and what are the phases that can coexist at each triple point?

(b) Which has a higher density, graphite or diamond?

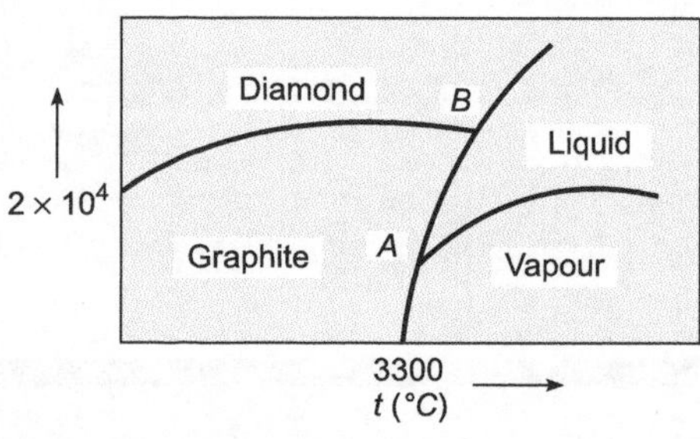

(c) Synthetic diamond can be made from graphite. Using the phase diagram, how would you go about making diamond?

3. Consider the following list of elements: C, Se, B, Sn, Cl. Identify the element on this list that

(a) has the largest atomic radius,

(b) is the most electronegative,

(c) is the best electrical conductor,

(d) has a maximum oxidation state of +6,

(e) forms a hydride with the empirical formula XH_3

4. Carbon is an essential element in the molecule on which life is based. Would silicon be equally satisfactory? Explain.

5. The name dilead(II) lead(IV) oxide and plumbous orthoplumbate have both been used to describe red lead, Pb_3O_4. Show that these names are consistent with this mixed oxide.

6. Write chemical equations to represent the following reactions:

(a) The dissolving of PbO in NaOH (*aq*).

(b) The disproportionation of SnO to Sn and SnO_2.

(c) The oxidation of HCl(*aq*) to $Cl_2(g)$ by $PbO_2(s)$.

(d) The production of basic lead carbonate (white lead) $2PbCO_3 \cdot Pb(OH)_2$ by the action of H_2O and CO_2 on PbO.

7. One natural source of atomospheric CO_2 is precipitation reaction such as the precipitation of silicates in the oceans :

$$Mg^{2+}(aq) + SiO_2\ (\text{dispersed}) + 2HCO_3^-(aq) \longrightarrow MgSiO_3(s) + 2CO_2(g) + H_2O(l)$$

How many grams of magnesium silicate would be precipitated during the formation of 100 L of CO_2 at 30°C and 775 Torr?

(Mg = 24, Si = 28, C = 12, O = 16)

8. The organic solvent acetone has the molecular formula $(CH_3)_2CO$. The silicon analogue, a thermally stable lubricant, is a polymer, $[(CH_3)_2SiO]_n$. Account for the difference in structure.

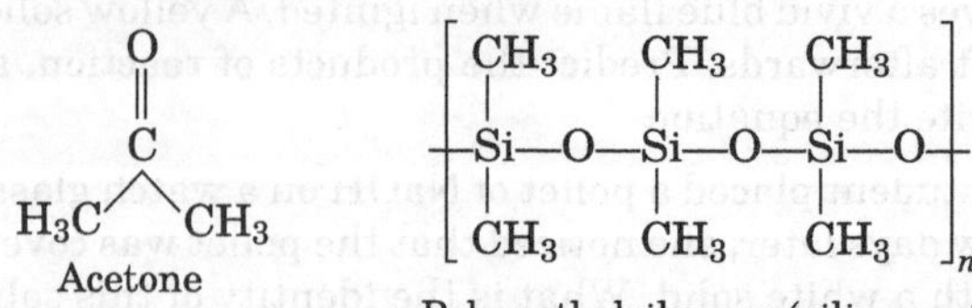

Acetone

Polymethylsiloxane (silicone oil)

NCERT Problems

1. Write balanced equations for the following reactions.
   (a) Combustion of C_4H_{10} in a limited supply of oxygen to form carbon monoxide and water.
   (b) The reaction of calcium carbide with water to form acetylene.
   (c) The preparation of calcium cyanamide from calcium carbide and nitrogen.
   (d) Formation of hydrogen cyanide from methane and ammonia.
2. Write resonance structures for CO_3^{2-} and HCO_3^-.
3. What is the oxidation state of carbon in each of the following compounds?
   (a) CO (b) HCN
   (c) H_2CO_3 (d) CaC_2
4. What is dry ice? Why is it so called?
5. What is the state of hybridisation of carbon in
   (a) CO_3^{2-} (b) HCN
   (c) diamond (d) graphite
6. Explain the differences in the properties of diamond and graphite on the basis of their structures.
7. What are fullerenes? How are they prepared?

Exercise 2
(Stage 2 : High Skill Problem Solving)

Only One Option Correct

1. Stability of hydrides of carbon family is in order
   (a) $CH_4 > SiH_4 > GeH_4 > SnH_4 > PbH_4$
   (b) $CH_4 < SiH_4 < GeH_4 < SnH_4 < PbH_4$
   (c) $CH_4 > SnH_4 > GeH_4 > SiH_4 > PbH_4$
   (d) None of the above
2. Select correct statement.
   (a) oxides of carbon family (MO_2) are all network solids with octahedral coordination
   (b) Silicon dioxide (silica) is a network solid with tetrahedral coordination and is a giant molecule
   (c) GeO_2, SnO_2 and PbO_2 are all acidic oxides
   (d) None of the above appears correct
3. Dry ice is
   (a) $H_2O(s)$ (b) $NH_3(g)$
   (c) $CO_2(s)$ (d) $PH_3(g)$
4. Select correct statements.
   (a) Hydrides of B and Si are volatile and catch fire on exposure to air
   (b) Oxides of B and Si(B_2O_3 and SiO_2) are acidic in nature
   (c) Borates and silicates have tetrahedral BO_4 and SiO_4 structural units
   (d) All the above are correct statements
5. Select correct statements.
   (a) Catenation is maximum in carbon
   (b) Carbon has pronounced ability to form $p\pi$-$p\pi$ multiple bonds to itself and to other elements like O and N
   (c) Both (a) and (b) are correct
   (d) None of the above is correct
6. Consider following statements,
   I : In diamond, each carbon atom is linked tetrahedrally to four other carbon atoms by sp^3 bonds.
   II : Graphite has planar hexagonal layers of carbon atoms held together by weak van der Waals' forces.
   III : Silicon exists only in diamond structure due to its tendency to form $p\pi$-$p\pi$ bonds to itself.
   In this
   (a) Only I and II are correct
   (b) Only I is correct
   (c) Only II and III are correct
   (d) All the above are correct statement
7. Out of CO_2, SiO_2, GeO_2, SnO_2 and PbO_2
   (a) CO_2 and SiO_2 are acidic, SnO_2 is amphoteric and PbO_2 is an oxidising agent
   (b) PbO_2 is converted to $Pb(NO_3)_2$ on reaction with HNO_3
   (c) both (a) and (b) are correct
   (d) none of the above is correct
8. Which is incorrect statement about silicones?
   (a) They are repeating units (SiO_4) in silicates
   (b) They are synthetic polymers containing repeated R_2SiO_2 units
   (c) They are formed by hydrolysis of R_2SiCl_2
   (d) All the above are incorrect statements
9. Which one of the following bonds has the highest bond energy?
   (a) C—C (b) Si—Si
   (c) Ge—Ge (d) Sn—Sn
10. Which of the following species have undistorted octahedral structures?
    1. SF_6 2. PF_6^-
    3. SiF_6^{2-} 4. SeF_6
    Select the correct answer using the codes given below
    (a) 2, 3 and 4 (b) 1, 3 and 4
    (c) 1, 2 and 3 (d) 1, 2 and 4

11. Which of the following statements are true for zeolites?
1. They are formed by the replacement of some of the silicon atoms of the SiO_2 lattice, by say, aluminium.
2. They have a more closed structure than feldspar.
3. They can absorb CO_2, NH_3 and EtOH.
4. They can separate straight chain hydrocarbons from a mixture containing both straight, chain and branched chain hydrocarbons.

Of these statements
(a) 1, 2 and 3 are correct (b) 1, 3 and 4 are correct
(c) 1, 2 and 4 are correct (d) 2, 3 and 4 are correct

12. Artificial gem used for cutting glass is
(a) graphite (b) diamond
(c) SiC (d) $CaCN_2$

13. Red lead used as primer for iron to prevent it from rusting is
(a) Pb_3O_4 (b) PbO
(c) PbO_2 (d) Pb_2O

14. Lead as impurity in the extraction of silver is removed by
(a) Parke's process (b) Solvay process
(c) Cyanide process (d) None of these

15. Which involves breaking of covalent bonds?
(a) Boiling of H_2S (b) Melting of SiO_2
(c) Melting of KCN (d) Boiling of CCl_4

16. Which is silane?
(a) SiO_2 (b) SiO_3
(c) SiH_4 (d) Si

17. Holme's signals can be given by using
(a) $CaC_2 + CaCN_2$ (b) $CaC_2 + Ca_3P_2$
(c) $CaC_2 + CaCO_3$ (d) $Ca_3P_2 + CaCN_2$

18. Which is/are true statements?
(a) Diamond is unaffected by conc acids but graphite reacts with hot conc. HNO_3 forming mellitic acid, $C_6(COOH)_6$
(b) CO is toxic because it forms a complex with haemoglobin in the blood
(c) C_3O_2, carbon suboxide, is a foul-smelling gas
(d) All the above are true statements

19. Carborundum is
(a) CaC_2 (b) Fe_3C
(c) $CaCO_3$ (d) SiC

20. Gas that strikes in thundering of light is
(a) CO (b) NO
(c) CO_2 (d) H_2

21. NO_2 is not obtained when following is heated
(a) $Pb(NO_3)_2$ (b) $AgNO_3$
(c) $LiNO_3$ (d) KNO_3

22. In which case hydrolysis is faster

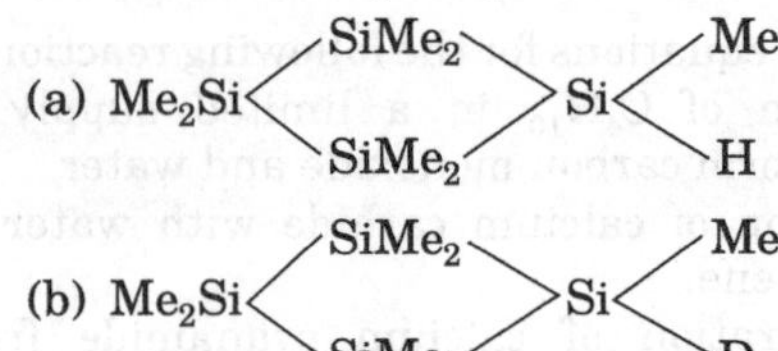

(c) at equal rate
(d) no hydrolysis

23. Which of the following is the ionic carbide?
(a) Fe_3C (b) SiC
(c) CaC_2 (d) Cu_2C

24. The silicate anion in the mineral kinoite is a chain of three SiO_4 tetrahedra that share corners with adjacent tetrahedra. The mineral also contains Ca^{2+} ions, Cu^{2+} ions, and water molecules in a 1 : 1 : 1 ratio. The mineral is
(a) $Ca_2Cu_2Si_3O_{10} \cdot 2H_2O$
(b) $CaCuSi_3O_{10} \cdot H_2O$
(c) $Ca_2Cu_2Si_3O_8 \cdot 2H_2O$
(d) $Ca_2Cu_2Si_3O_6 \cdot 2H_2O$

25. The best oxidising agent of the following oxides is
(a) SnO_2 (b) PbO_2
(c) HgO (d) MgO

26. Dipole moment of CO is very small (0.1 D) (inspite of the larger difference in electronegativity between C and O) and negative end of the dipole lies near the less electronegative C-atom. It is due to
(a) occupation of anti-bonding molecular orbitals
(b) occupation of empty *d*-orbitals
(c) occupation of lone-pairs
(d) occupation of the lone pairs as well as empty *d*-orbitals

27. Carbon and silicon belong to (IV) group. The maximum coordination of carbon is commonly occurring compounds is 4, whereas that of silicon is 6. This is due to
(a) large size of silicon
(b) more electropositive nature of silicon
(c) availability of *d*-orbital in silicon
(d) both (a) and (b)

28. In graphite, additional electrons are
(a) localised on each C-atom
(b) localised on every third C-atom
(c) present in anti-bonding orbital
(d) delocalised forming extended π-bonding system

29. $PbO_2 \longrightarrow PbO \quad \Delta G_{298} < 0$
$SnO_2 \longrightarrow SnO \quad \Delta G_{298} > 0$
Most probable oxidation state of Pb and Sn will be
(a) Pb^{4+}, Sn^{4+} (b) Pb^{4+}, Sn^{2+}
(c) Pb^{2+}, Sn^{2+} (d) Pb^{2+}, Sn^{4+}

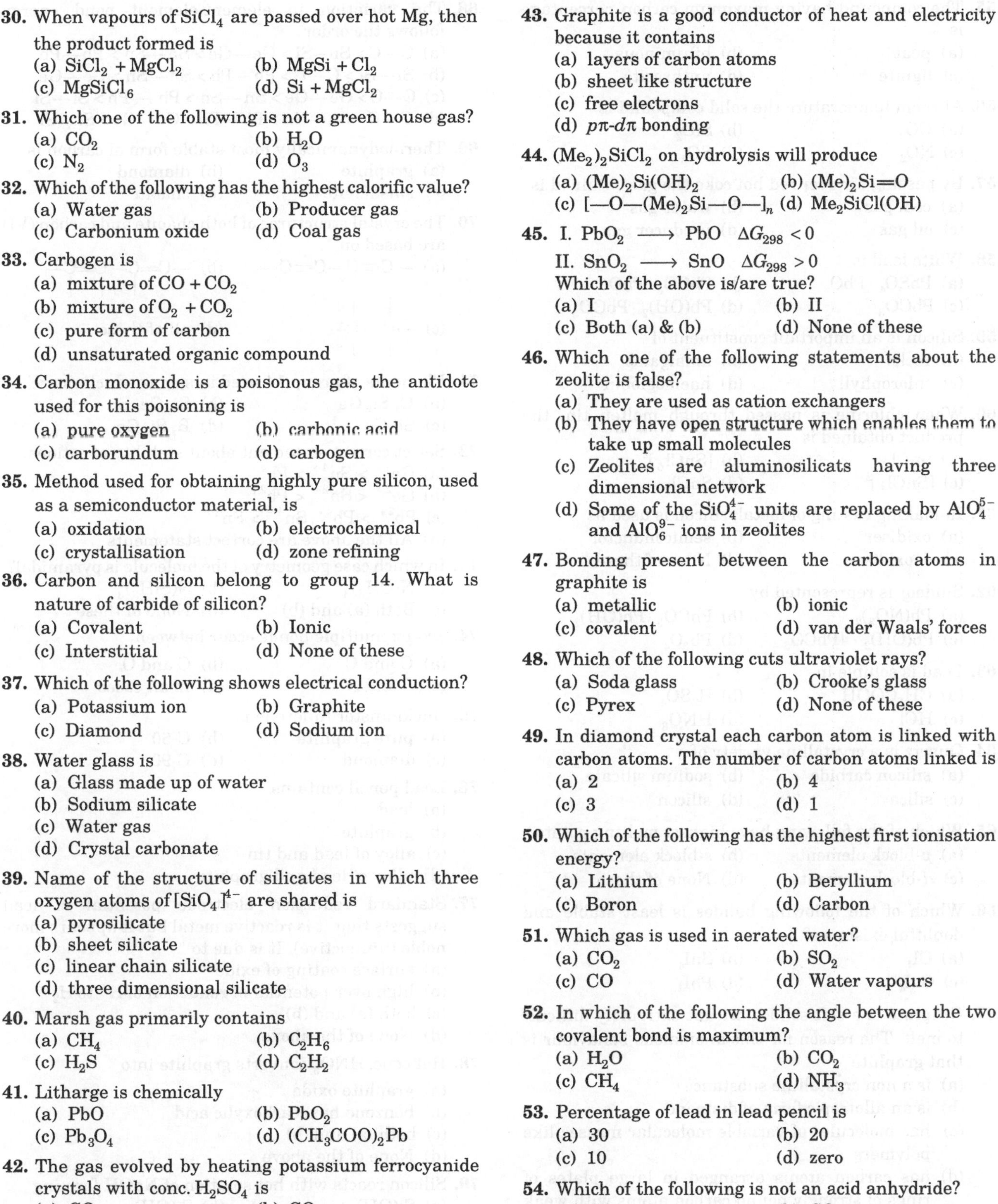

30. When vapours of $SiCl_4$ are passed over hot Mg, then the product formed is
(a) $SiCl_2 + MgCl_2$ (b) $MgSi + Cl_2$
(c) $MgSiCl_6$ (d) $Si + MgCl_2$

31. Which one of the following is not a green house gas?
(a) CO_2 (b) H_2O
(c) N_2 (d) O_3

32. Which of the following has the highest calorific value?
(a) Water gas (b) Producer gas
(c) Carbonium oxide (d) Coal gas

33. Carbogen is
(a) mixture of $CO + CO_2$
(b) mixture of $O_2 + CO_2$
(c) pure form of carbon
(d) unsaturated organic compound

34. Carbon monoxide is a poisonous gas, the antidote used for this poisoning is
(a) pure oxygen (b) carbonic acid
(c) carborundum (d) carbogen

35. Method used for obtaining highly pure silicon, used as a semiconductor material, is
(a) oxidation (b) electrochemical
(c) crystallisation (d) zone refining

36. Carbon and silicon belong to group 14. What is nature of carbide of silicon?
(a) Covalent (b) Ionic
(c) Interstitial (d) None of these

37. Which of the following shows electrical conduction?
(a) Potassium ion (b) Graphite
(c) Diamond (d) Sodium ion

38. Water glass is
(a) Glass made up of water
(b) Sodium silicate
(c) Water gas
(d) Crystal carbonate

39. Name of the structure of silicates in which three oxygen atoms of $[SiO_4]^{4-}$ are shared is
(a) pyrosilicate
(b) sheet silicate
(c) linear chain silicate
(d) three dimensional silicate

40. Marsh gas primarily contains
(a) CH_4 (b) C_2H_6
(c) H_2S (d) C_2H_2

41. Litharge is chemically
(a) PbO (b) PbO_2
(c) Pb_3O_4 (d) $(CH_3COO)_2Pb$

42. The gas evolved by heating potassium ferrocyanide crystals with conc. H_2SO_4 is
(a) CO (b) CO_2
(c) SO_2 (d) SO_3

43. Graphite is a good conductor of heat and electricity because it contains
(a) layers of carbon atoms
(b) sheet like structure
(c) free electrons
(d) $p\pi$-$d\pi$ bonding

44. $(Me_2)_2SiCl_2$ on hydrolysis will produce
(a) $(Me)_2Si(OH)_2$ (b) $(Me)_2Si{=}O$
(c) $[-O-(Me)_2Si-O-]_n$ (d) $Me_2SiCl(OH)$

45. I. $PbO_2 \longrightarrow PbO \quad \Delta G_{298} < 0$
II. $SnO_2 \longrightarrow SnO \quad \Delta G_{298} > 0$
Which of the above is/are true?
(a) I (b) II
(c) Both (a) & (b) (d) None of these

46. Which one of the following statements about the zeolite is false?
(a) They are used as cation exchangers
(b) They have open structure which enables them to take up small molecules
(c) Zeolites are aluminosilicats having three dimensional network
(d) Some of the SiO_4^{4-} units are replaced by AlO_4^{5-} and AlO_6^{9-} ions in zeolites

47. Bonding present between the carbon atoms in graphite is
(a) metallic (b) ionic
(c) covalent (d) van der Waals' forces

48. Which of the following cuts ultraviolet rays?
(a) Soda glass (b) Crooke's glass
(c) Pyrex (d) None of these

49. In diamond crystal each carbon atom is linked with carbon atoms. The number of carbon atoms linked is
(a) 2 (b) 4
(c) 3 (d) 1

50. Which of the following has the highest first ionisation energy?
(a) Lithium (b) Beryllium
(c) Boron (d) Carbon

51. Which gas is used in aerated water?
(a) CO_2 (b) SO_2
(c) CO (d) Water vapours

52. In which of the following the angle between the two covalent bond is maximum?
(a) H_2O (b) CO_2
(c) CH_4 (d) NH_3

53. Percentage of lead in lead pencil is
(a) 30 (b) 20
(c) 10 (d) zero

54. Which of the following is not an acid anhydride?
(a) CO_2 (b) CO
(c) SO_2 (d) SO_3

55. The compound having maximum carbon percentage is
(a) peat (b) bituminous
(c) lignite (d) anthracite

56. At room temperature the solid compound is
(a) CO_2 (b) SiO_2
(c) NO_2 (d) SO_2

57. By passing air over red hot coke the gas obtained is
(a) coal gas (b) water gas
(c) oil gas (d) producer gas

58. White lead is
(a) $PbSO_4 \cdot PbO$ (b) $PbCO_3 \cdot PbO$
(c) $PbCO_3$ (d) $Pb(OH)_2 \cdot PbCO_3$

59. Silicon is an important constituent of
(a) rocks (b) amalgams
(c) chlorophyll (d) haemoglobin

60. When chlorine is passed through molten tin, the product obtained is
(a) $SnCl_4$ (b) $[SnCl_6]^{2-}$
(c) $[SnCl_4]^{2-}$ (d) $SnCl_2$

61. In making casting of metal, silicon is used as
(a) oxidiser (b) semiconductor
(c) deoxidiser (d) None of these

62. Sindoor is represented by
(a) $Pb(NO_3)_2$ (b) $PbCO_3 \cdot Pb(OH)_2$
(c) $Pb(OH)_2 \cdot 4PbCO_3$ (d) Pb_3O_4

63. Lead is soluble in
(a) CH_3COOH (b) H_2SO_4
(c) HCl (d) HNO_3

64. Quartz is a crystalline variety of
(a) silicon carbide (b) sodium silicate
(c) silica (d) silicon

65. Which of the following have highest melting points?
(a) *p*-block elements (b) *s*-block elements
(c) *d*-block elements (d) None of these

66. Which of the following halides is least stable and doubtful existence?
(a) CI_4 (b) SnI_4
(c) GeI_4 (d) PbI_4

67. Graphite is a soft solid lubricant extremely difficult to melt. The reason for this anomalous behaviour is that graphite
(a) is a non crystalline substance
(b) is an allotropic form of diamond
(c) has molecules of variable molecular masses like polymers
(d) has carbon atoms arranged in large plates of rings of strongly bound carbon atoms with weak interplate bonds

68. The variation in element-element bond energy follows the order
(a) C—C > Si—Si > Ge—Ge > Sn—Sn > Pb—Pb
(b) Si—Si > C—C > Pb—Pb > Sn—Sn > Ge—Ge
(c) C—C > Ge—Ge > Sn—Sn > Pb — Pb > Si—Si
(d) C—C > Sn—Sn > Pb—Pb > Ge—Ge > Si—Si

69. Thermodynamically most stable form of carbon is
(a) graphite (b) diamond
(c) ionsdaleite (d) chaoite

70. The crystal structures of both chaoite and carbon (VI) are based on
(a) —C≡C—C≡C— (b) —C=C—C=C— (with single bonds projecting downward from the three right-hand carbons)
(c) —C—C— (each carbon with bonds projecting up and down) (d) all of these

71. Elements of group 14 used in semiconductors are
(a) C, Si, Ge (b) Si, Ge, Sn
(c) Si, Ge (d) B, Si, Ge

72. Select correct statement about stability of cations.
(a) $Ge^{4+} > Sn^{4+} > Pb^{4+}$
(b) $Ge^{2+} < Sn^{2+} < Pb^{2+}$
(c) $Pb^{2+} > Pb^{4+}, Sn^{4+} > Sn^{2+}$
(d) All the above are correct statements

73. In which case geometry of the molecule is pyramidal?
(a) $N(CH_3)_3$ (b) $N(SiH_3)_3$
(c) Both (a) and (b) (d) None of these

74. $p\pi$-$p\pi$ multiple bonds occur between
(a) C and C (b) C and O
(c) C and N (d) in all cases

75. Buckminster fullerene is
(a) pure graphite (b) C-60
(c) diamond (d) C-90

76. Lead pencil contains
(a) lead
(b) graphite
(c) alloy of lead and tin
(d) alloy of lead and graphite

77. Standard reduction electrode potential of lead suggests that it is reactive metal but it appears more noble (unreactive). It is due to
(a) surface coating of oxide
(b) high over potential of reduction of H^+ to H_2
(c) both (a) and (b)
(d) None of the above

78. Hot conc. HNO_3 converts graphite into
(a) graphite oxide
(b) benzene hexacarboxylic acid
(c) both (a) and (b)
(d) None of the above

79. Silicon reacts with hot solution of NaOH forming
(a) $Si(OH)_4$ (b) $Si(OH)_2$
(c) SiO_2 (d) Na_4SiO_4

80. Select incorrect statement.
(a) Mellitic acid is benzene hexa-carboxylic acid
(b) Pb dissolves in conc. HCl forming $PbCl_2$
(c) Pb dissolves in hot NaOH forming $Na_2[Pb(OH)_6]$
(d) Diamond is unreactive but graphite forms $(CF)_n$ with F_2

81. When a mixture of carbon monoxide and chlorine is exposed to sunlight the product formed is
(a) thionyl chloride
(b) phosgene
(c) phosphine
(d) carbon tetrachloride

82. Stable oxides of carbon are
(a) CO, CO_2, C_3O_2 (b) CO_2, C_3O_2, C_2O
(c) CO, CO_2 (d) C_2O, C_3O_2

83. CO behaves as
(a) Lewis acid (b) Lewis base
(c) amphoteric oxide (d) None of these

84. Which reaction takes place during respiration?
(a) $6CO_2 + 6H_2O \longrightarrow 6O_2 + C_6H_{12}O_6$
(b) $C + O_2 \longrightarrow CO_2$
(c) $C + \frac{1}{2}O_2 \longrightarrow CO$
(d) $C_6H_{12}O_6 + 6O_2 \longrightarrow 6CO_2 + 6H_2O$

85. CO_2 in water behaves as
(a) weak dibasic acid H_2CO_3
(b) weak monobasic acid $HO{-}CO_2H$
(c) weak diacid base $CO(OH)_2$
(d) weak monoacid base $HO{-}CO_2H$

86. A certain compound on burning in air forms three oxides. One of the oxides turned lime water milky, the other turned anhydrous $CuSO_4$ blue and third formed a solution of pH = 9. Compound is formed of
(a) S, N and H (b) S, N and C
(c) S, C and H (d) S, H and Na

87. Concentrations of the atmospheric CO_2 have been rising because of
(a) use of fossil fuels (b) acid rain
(c) photochemical smog (d) ozone depletion

88. Principal cause of ozone depletion is the
(a) presence of fluorocarbons
(b) acid rain
(c) photochemical smog
(d) green-house effect

89. Global warming is called
(a) photochemical smog (b) green-house effect
(c) acid rain (d) respiration

90. Green house effect is due to
(a) disturbance of the earth's delicate thermal balance
(b) absorption of heat radiation by atmospheric gases particularly water vapours
(c) both (a) and (b)
(d) None of the above

91. Methanides are
(a) Mg_2C_3, Be_2C, Al_4S_3 and CaC_2
(b) Mg_2C_3, Be_2C and Al_4C_3
(c) Be_2C, Al_4C_3 and CaC_2
(d) Be_2C and Al_4C_3

92. Select correct statements(s).
(a) Cyanamide ion (CN_2^{2-}) is isoelectronic with CO_2 and has the same linear structure
(b) Mg_2C_3 reacts with water to form propyne
(c) CaC_2 has NaCl type of lattice
(d) all the above are correct statements

93. Select incorrect statement(s).
(a) Interstitial carbides are formed by metalloids like Si and B
(b) Covalent carbides are formed by metalloids
(c) CO and CN^- both are fatal due to complex formation with Fe(III) present in blood
(d) SiC is called carborundum

94. Which is/are used as ligand in complexes?
(a) CN^- (b) CO
(c) Both (a) and (b) (d) None of these

95. Sulphide mineral + $CN^- \xrightarrow{O_2} [M(CN)_2]^-$

$2[M(CN)_2]^- + Zn \longrightarrow [Zn(CN)_4]^{2-} + 2M$

M can be
(a) Mg, Ag, Au (b) Ag, Au
(c) Pt, Au (d) Pt, Ag

96. When zeolite is treated with hard water, there is exchange reaction between calcium/magnesium ion/and ......
(a) aluminium ion (b) sodium ion
(c) water of hydration (d) sulphate ion

97. Select correct statement(s).
(a) Si uses all of its valence electrons in an sp^3-hybridisation and crystallises in an fcc structure similar to diamond.
(b) Diamond is an electrical insulator, graphite is an electrical conductor but silicon is a semiconductor.
(c) Most common silica is quartz.
(d) All of the above are correct statements.

98. Silicones are
(a) synthetic polymers containing repeated R_2SiO units
(b) silicates with common SiO_4 unit
(c) ketones with silyl group (SiH_3) similar to alkyl $(SiH_3)_2CO$
(d) Zircon (neso silicates)

99. Formula of the following silicate anion is

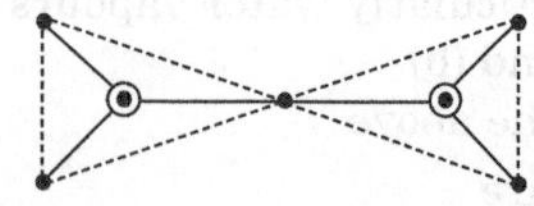

(a) SiO_4^{4-} (b) $Si_2O_7^{6-}$
(c) $Si_2O_8^{6-}$ (d) $Si_3O_9^{6-}$

100. The minerals having silicate chains are collectively called
(a) olivine (b) zircon
(c) pyroxenes (d) natrolite

101. Glass is best described as a
(a) solid (b) liquid
(c) super cooled liquid (d) colloidal sol

102. Which among the following statements is false regarding glass?
(a) it absorbs ultra-violet light
(b) it consists of silicates of sodium and alkali
(c) HF causes marking on glass
(d) all the halogen acids (HF, HCl, HBr and HI) are stored in glass bottles

103. The colour of blue glass is due to the presence of oxide of
(a) chromium (b) cobalt
(c) gold (d) silver

104. Non-oxide ceramics can be
(a) B_4C (b) SiC
(c) Si_3N_4 (d) all of these

105. Glass having higher refreactive index is prepared using
(a) NiO (b) CoO
(c) PbO (d) CaO

106. Amphoteric oxides are
(a) SnO, SnO_2 (b) GeO, GeO_2
(c) both (a) and (b) (d) None of these

107. Compound used on thermal insulator is
(a) silica gel (b) zeolites
(c) silicones (d) asbestos

108. $SnCl_4$ can be used as
(a) reducing agent
(b) oxidising agent
(c) catalyst in Friedel Craft's reaction
(d) a base

109. Aqueous solution of SnO liberates CO_2 with $NaHCO_3$. This shows that SnO is
(a) an acidic oxide
(b) a basic oxide
(c) an amphoteric oxide
(d) a Lewis acid

110. Which silicon compound is used as lubricant?
(a) Asbestos (b) Silicones
(c) Zeolite (d) Mica

111. Out of Si, Ge, Sn and Pb which forms stable ionic compounds in divalent state?
(a) Pb (b) Sn
(c) Ge (d) Si

112. Select correct statement about PbO.
(a) Red oxide is called litharge
(b) Yellow oxide is called massicot
(c) Both (a) and (b) are correct
(d) None of the above is correct

113. Pb_3O_4 reacts with HNO_3 forming nitrate and oxide which are
(a) $Pb(NO_3)_4$ from PbO_2, and PbO remains unreacted
(b) $Pb(NO_3)_2$ from PbO, and PbO_2 remains unreacted
(c) $Pb(NO_3)_2$ and $Pb(NO_3)_4$, and PbO remains unreacted
(d) Pb_3O_4 remains insoluble in HNO_3

114. Unstable lead compounds are
(a) $PbCl_4, PbBr_4$ and PbI_4 (b) $PbCl_2, PbBr_2$ and PbI_2
(c) PbO, PbO_2 and Pb_3O_4 (d) $PbCl_4^{2-}, PbCl_6^{2-}$

115. $PbCl_2$ is soluble in
(a) dil. HCl (b) CH_3COOH
(c) hot water (d) dil. H_2SO_4

One or More Than One Options Correct

1. Select correct statement(s).
(a) Graphite is composed of planar, two dimensional sheets of sp^2 hybridised carbon atoms.
(b) Each sheet is a network of fused, hexagonal rings of carbons.
(c) The layers in graphite are held together by relatively weak van der Waals forces of attraction.
(d) Graphite is thermodynamically the most stable allotropic form of carbon.

2. Select correct statement(s).
(a) In graphite, only three electrons (out of four) are involved in σ bonding.
(b) π-bonding electrons in graphite are delocalised over the whole structure.
(c) Conduction of electricity in graphite is due to mobile electrons.
(d) Diamond is insulator due to absence of mobile electrons.

3. Conc. H_2SO_4 reacts with carbon forming
(a) CO_2 (b) SO_2
(c) H_2CO_3 (d) H_2SO_3

4. Carbon suboxide (C_3O_2) is formed when
(a) carbon reacts with conc. H_2SO_4
(b) malonic acid reacts with P_4O_{10}
(c) carbon reacts with strong alkali solution
(d) oxalic acid is heated strongly

5. $SiCl_4$ is hydrolysed but CCl_4 is not hydrolysed because

(a) Si has vacant *d*-orbitals and can accommodate lone-pair of electrons from oxygen of water

(b) Si has relatively large size and can increase its coordination number from four to five

(c) Si has relatively smaller size and thus interaction with H_2O is spontaneous

(d) Si has oxidation number of six while carbon has oxidation number of four in all its compounds.

6. Select correct statement(s).

(a) In α-graphite, the layers are arranged in the sequence *ABAB* ......

(b) In β-graphite, the layers are arranged in the sequence *ABCABC* ......

(c) In α-graphite, the layers are arranged in the sequence *ABCABC* ......

(d) In β-graphite, the layers are arranged in the sequence *ABAB* ......

7. Select corrrect statement(s).

(a) The enthalpy difference between α- and β-graphite is very large and thus two forms are not interconvertible

(b) The enthalpy difference between α- and β-graphite is very small and two forms are interconvertible

(c) α-graphite changes to β-graphite at about 1300 K.

(d) β-graphite changes to α-graphite at about 1300 K

8. Select ionic carbides

(a) Al_4C_3, CaC_2, Mg_2C_3 (b) Al_4C_3, SiC, B_4C

(c) CaC_2, Mg_2C_3 (d) CaC_2, SiC, Mg_2C_3

9. Silly putty

(a) is a silicone polymer

(b) has a composition intermediate between silicone oils and silicone rubbers

(c) is an asbestos

(d) is not related to any chemical.

10. The great thermal and chemical stability of silicones is attributed to high strength of

(a) silicon-carbon bond (b) silicon-oxygen bond

(c) silicon-silicon bond (d) carbon-carbon bond

11. Correct statement(s) out of the following is

(a) $N(CH_3)_3$ has pyramidal structure

(b) $N(SiH_3)_3$ shows planar arrangement

(c) SiC is highly volatile

(d) SiO_2 is called silane

12. Group 14 (carbon family) elements have the following properties

(a) They are all network solids with octahedral network

(b) Only SiO_2 is a network solid with tetrahedral coordination and is a giant molecule

(c) SnO_2 and PbO_2 are amphoteric oxides.

(d) All oxides are acidic

13. Silicones

(a) are synthetic polymer containing repeating (R_2SiO_2) units

(b) are formed by hydrolysis of R_2SiCl_2

(c) are natural occurring repeating (SiO_4) units

(d) single R_2SiO_2 unit

14. Select correct statement(s).

(a) $COCl_2$ is called phosphine gas

(b) C_3O_2 (carbon suboxide) has sp^2 hybridised carbons

(c) CO is toxic because it forms a complex with haemoglobin in the blood

(d) Diamond is unaffected by concentrated acid but graphite is oxidised to mellitic acid with concentrated HNO_3.

15. Gases responsible for "green-house effect" are

(a) CO_2 (b) H_2O

(c) O_3 (d) N_2

Passage Comprehension Questions

Passage 1

Questions are based on following medical report.

In the first biological application of buckyball, chemists at the University of California at San Francisco and Santa Barbara made a discovery in 1993 that could help in designing drugs to treat AIDS. The human immunodeficiency virus (HIV) that causes AIDS reproduces by synthesising a long protein chain, which is cut into smaller segments by an enzyme called HIV-protease. One way to stop AIDS, then. might be to inactivate the enzyme. When the chemists reacted a water-soluble derivative of buckyball with HIV-protease, they found that it binds to the portion of the enzyme that would ordinarily cleave the reproductive protein, thereby preventing the HIV virus from reproducing. Consequently, the virus could no longer infect the human cells they had grown in the laboratory. The buckyball compound itself is not a suitable drug for use against AIDS because of potential side effects and delivery difficulties, but it does provide a model for the development of such drugs.

1. Buckyball is the allotrope of

(a) phosphorus (b) sulphur

(c) carbon (d) titanium

2. What is the formula of the buckyball?

(a) P_4 (b) S_8

(c) Ti_3 (d) C_{60}

3. In buckyball each atom is

(a) sp^2 hybridized element with extensive delocalised molecular orbital

(b) sp^2 hybridised element with localised molecular orbital

(c) sp^3 hybridised element with delocalised molecular orbital

(d) sp^3 hybridised element with localised molecular orbital

4. Consider following statements about buckyball

A : It is also called fullerence

B : It is also called buckminster fullerence

C : Bucky tubes (made of fullerenes) are several times stronger than steel wires

D : Buckyball is a plastic polymer

Select correct statement(s).

(a) *A, C, D* (b) *A, B, C*

(d) *A, B, D* (d) *B, C, D*

Passage 2

Read the following discovery and answer the questions at the end.

"Elemental carbon appears in many structural forms or allotropes. Three of these forms are crystalline—diamond, graphite and the recently discovered fullerence (buckyball)—while more than 40 others, including coke and carbon black, are amorphous. Now there seems to be set a fourth crystalline allotrope of carbon, reported in 1995 by Richard Lagow at the University of Texas."

1. Newly discovered allotrope of carbon has the form

(a) polyyne (b) fullerene

(c) buckyball (d) none of these

2. Structures of different allotropes of carbon have been compared. Which represents incorrect comparison?

(a) Allotrope discovered in 1995 — *sp*–hybridised carbon

(b) Buckyball — *sp*–hybridised carbon

(c) Graphite — sp^2–hybridised carbon

(d) Diamond — sp^3–hybridised carbon

Passage 3

Read the following short write-up and answer the questions at the end of it.

The name '**silica**' covers an entire group of minerals, which have the general formula SiO_2, the most common of which is **quartz**. Quartz is a framework silicate with SiO_4 tetrahedra arranged in spirals. *The spirals can turn in a clockwise or anti-clockwise direction—a feature that results in there being two mirror images, optically active, varieties of quartz.*

1. The following pictures represent various silicate anions. Their formulae are respectively:

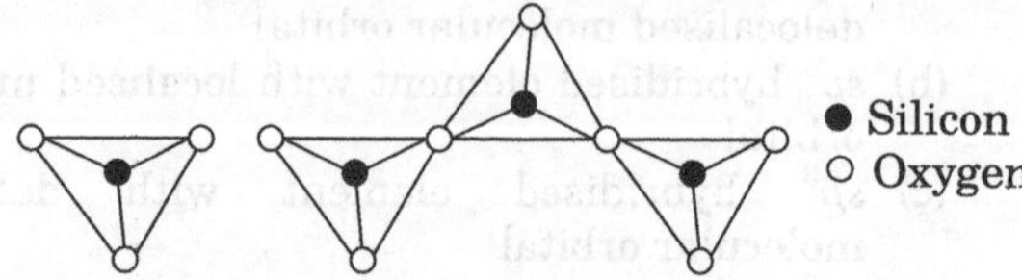

(a) SiO_3^{2-} $Si_3O_7^{2-}$

(b) SiO_4^{4-} $Si_3O_{10}^{8-}$

(c) SiO_4^{2-} $Si_3O_9^{2-}$

(d) SiO_3^{4-} $Si_3O_7^{8-}$

2. $Si_3O_9^{6-}$ (having three tetrahedra) is represented as

(a)

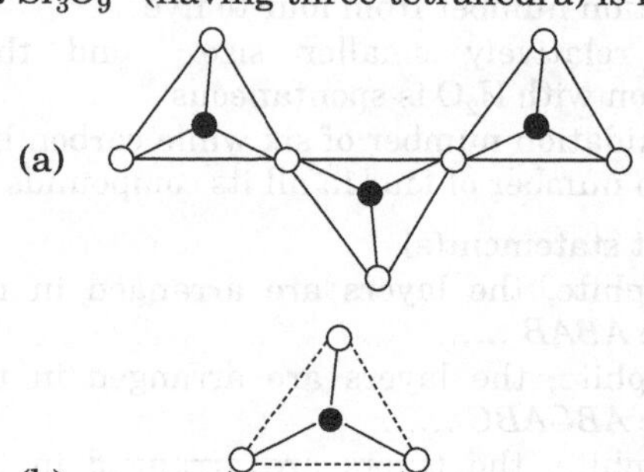

(b)

(c) both (a) and (b)

(d) none of the above

3. The silicate anion in the mineral kinoite is a chain of three SiO_4 tetrahedra that share corners with adjacent tetrahedra. The mineral also contains Ca^{2+} ions, Cu^{2+} ions, and water molecules in a 1 : 1 : 1 ratio mineral is represented as

(a) $CaCuSi_3O_{10} \cdot H_2O$ (b) $CaCuSi_3O_{10} \cdot 2H_2O$

(c) $Ca_2Cu_2Si_3O_{10} \cdot 2H_2O$ (d) None of these

Passage 4

Based on following short report, answer the questions given at the end.

In some foam-type fire extinguishers, the reactants are $Al_2(SO_4)_3(aq)$ and $NaHCO_3(aq)$. When the extinguisher is activated, these reactants are allowed to mix producing $Al(OH)_3(s)$ and $CO_2(g)$. The $Al(OH)_3$—CO_2 foam extinguishes the fire.

1. CO_2 is formed as a result of

(a) reaction between Al^{3+} and HCO_3^-

(b) reaction between hydrolysis product of Al^{3+} and HCO_3^-

(c) reaction between hydrolysis product of $NaHCO_3$ and $Al_2(SO_4)_3$

(d) direct reaction between $Al_2(SO_4)_3$ and $NaHCO_3$

2. Net ionic reaction of the above chemical change is

(a) $Al^{3+} + HCO_3^- + H_2O \longrightarrow Al(OH)_3 + CO_2$

(b) $Al^{3+} + HCO_3^- + OH^- \longrightarrow Al(OH)_3 + CO_2 + H_2O$

(c) $[Al(H_2O)_6]^{3+} + HCO_3^- \longrightarrow Al(OH)_3 + CO_2 + H_2O$

(d) $Al_2(SO_4)_3 + NaHCO_3 \longrightarrow Na_2SO_4 + Al(OH)_3 + H_2O$

3. Addition of Na_2CO_3 to a solution of an oxide in water produces CO_2. This experiment indicates that
(a) the oxide is that of a non-metal
(b) the oxide is basic
(c) the oxide is amphoteric
(d) the oxide is neutral

Passage 5

Questions given below are based on following technical report.

The CO_2 as well as H_2O absorbs strongly in the infrared region, and its presence in the atmosphere decreases the loss of heat from the earth by radiation. This global warming is called the **'green house effect'** (other gases, including the oxides of nitrogen from car exhausts, freons from aerosols and refrigerators and methane from bacteria in the soil and in the rumen of cows, also add to the green house effect). The concentration of atmospheric CO_2 has increased by 10%. This is resulting in the increase in the mean temperature of the earth by 2.5°C, varying from 2°C at the equator to 4°C at the poles. This could have dramatic effects on the climate.

1. As a result of green house effect, there can be
A : increase in rate of evaporation of water, thus, untimely more rain, flooding, etc.
B : tropical storms in certain parts of the world
C : decrease in pH of the soil
D : increase in pH of the soil

Select correct alternate
(a) B, C, D (b) A, C, D
(c) A, B, D (d) A, B, C

2. Which of the following is growing at a faster rate than CO_2 thus responsible for the green house effect?
(a) CFC (b) N_2O
(c) O_3 (d) CH_4

3. Instead of monitoring carbon dioxide, suggest another gas that scientists could study to substantiate the fact that CO_2 concentration is steadily increasing in the atmosphere
(a) N_2O (b) O_2
(c) CH_4 (d) O_3

4. The equilibrium constant (K_p) for the reaction

$$2CO(g) + O_2(g) \rightleftharpoons 2CO_2(g)$$

is 1.4×10^{90} at 25°C. Given this enormous value, why doesn't CO convert totally to CO_2 in the troposphere?
(a) CO forms complex with haemoglobin
(b) CO has low solubility in H_2O
(c) CO has high activation energy
(d) CO is toxic in nature

Passage 6

Read the following write-ups and answer the questions at the end of it.

Silicones are synthetic polymers containing repeated R_2SiO units. Since, the empirical formula is that of a ketone (R_2CO), the name silicone has been given to these materials. Silicones can be made into oils, rubbery elastomers and resins. They find a variety of applications because of their chemical inertness, water repelling nature, heat-resistance and good electrical insulating property.

Commercial silicon polymers are usually methyl derivatives and to a lesser extent phenyl derivatives and are synthesised by the hydrolysis of

R_2SiCl_2[R = methyl (Me) or phenyl (ϕ)]

$$Me_2SiCl_2 \xrightarrow{H_2O} —O—Si(Me)_2—O—Si(Me)_2—O—Si(Me)_2—O—$$

1. If we mix $SiMe_3Cl$ with $SiMe_2Cl_2$, we get silicones of the type
(a) $Me—Si(Me)_2—O—Si(Me)_2—O—Si(Me)_2—O—Si(Me)_2—Me$
(b) Cross-linked sheet: two chains $—O—Si(Me)—O—Si(Me)—O—$, each Si joined to the Si of the other chain through —O—
(c) both (a) and (b)
(d) None of the above

2. If we start with $SiMeCl_3$ as the starting material, silicones formed is
(a) $Me—Si(Me)_2—O—Si(Me)_2—O—Si(Me)_2—O—Si(Me)_2—Me$
(b) Cross-linked sheet: two chains $—O—Si(Me)—O—Si(Me)—O—$, each Si joined to the Si of the other chain through —O—
(c) both (a) and (b)
(d) None of the above

Assertion & Reason

Codes :

(a) Both **A** and **R** are true and **R** is the correct explanation of **A.**

(b) Both **A** and **R** are true but **R** is not the correct explanation of **A.**

(c) **A** is true but **R** is false.

(d) **A** is false but **R** is true.

1. Assertion (A) : SiF_6^{2-} is known but $SiCl_6^{2-}$ is not known.

Reason (R) : Interaction of F lone pair electrons with Si is stronger than that of chlorine and due to smaller size of F steric repulsion will be less.

2. Assertion (A) : If a metal exhibits more than one oxidation state, the halide in the higher oxidation state will be more covalent than the one in lower oxidation state.

Reason (R) : $SnCl_4, PbCl_4, SbCl_5$ and UF_6 are more covalent than $SnCl_2$, $PbCl_2, SbCl_3$ and UF_4.

3. Assertion (A) : CO is toxic.

Reason (R) : CO forms complex with haemoglobin in the blood, and this complex is more stable than oxy-haemoglobin.

4. Assertion (A) : Lime water turns milky when the gas liberated from an acidified carbonate solution is passed into it.

Reason (R) : The liberated CO_2 reacts with lime water to give calcium bicarbonate as the precipitate.

5. Assertion (A) : $PbCl_2$ is more stable than $PbCl_4$.

Reason (R) : $PbCl_4$ is powerful oxidising agent.

6. Assertion (A) : $SiCl_4$ is easily hydrolysed as compared to CCl_4.

Reason (R) : Silicon can extend its coordination number beyond four.

7. Assertion (A) : $N(CH_3)_3$ is pyamidal but $N(SiH_3)_3$ is planar.

Reason (R) : In $N(SiH_3)_3$ lone-pair on N-atom is transferred to the empty *d*-orbitals of silicon ($p\pi$-$d\pi$ overlapping).

8. Assertion (A) : Between $SiCl_4$ and CCl_4, only $SiCl_4$ reacts with water.

Reason (R) : $SiCl_4$ is ionic and CCl_4 is covalent.

9. Assertion (A) : Large number of organic compounds are formed with long carbon-chain.

Reason (R) : Maximum catenation occurs in carbon atoms.

10. Assertion (A) : Graphite cannot be converted into diamond.

Reason (R) : Graphite is conducting in nature.

11. Assertion (A) : Diamond is insulator and graphite is conducting.

Reason (R) : C-atom is sp^3 hybridised in both.

12. Assertion (A) : Carbon dioxide is a gas but silicon dioxide is an infusible solid.

Reason (R) : Carbon dioxide has a linear π-bonded structure.

True & False

1. $N(CH_3)_3$ and $N(SiH_3)_3$ both have sp^3-hybridised N-atom and pyramidal structure.

2. Cu^{2+} and Cd^{2+} both form soluble complex with CN^- but are decomposed by H_2S.

3. Quartz is framework silicate with SiO_4 tetrahedra arranged in spirals.

4. Silicones are synthetic polymers containing repeated R_2SiO units.

5. Fe_2O_3 and CuO impart green colour to the glass.

6. PbO_2 is called red-lead.

7. Pb_3O_4 on reaction with HNO_3 forms $Pb(NO_3)_2$ and PbO_2.

8. Ultra pure silicon is prepared by reduction of SiO_2 with carbon.

9. Si becomes a *p*-type semiconductor when doped with group 13 element (Al, Ga).

10. Si becomes an *n*-type semiconductor when doped with group 15 element (P, As).

11. Graphite is the least stable form of carbon.

12. Pb(II) salts are easily oxidised to Pb(IV) salts.

13. $SnCl_2$ is better reducing agent than $HgCl_2$.

14. Finely divided lead powder is pyrophoric.

15. Pb does not dissolve in hot concentrated HCl or H_2SO_4.

Fill in the Blanks

1. Silicon is the .... abundant element by mass in the earth's crust.

2. Graphite consists of planar hexagonal layers of carbon atoms held together by ....

3. Silicon exists only in the .... structure due to tendency to form $p\pi$-$p\pi$ bonds to itself.

4. In case of $N(CH_3)_3$, geometry is .... and in case of $N(SiH_3)_3$ geometry is ....

5. Dry ice is ....

6. Be_2C and Al_4C_3 are called .... because they react with H_2O yielding methane.

7. The cyanide ions are toxic because they bind almost irreversibly to the Fe(III) ion in ....

8. The CO_2 as well as H_2O absorbs strongly in the infrared region and its presence in the atmosphere decreases the loss of heat from the earth by radiation. This global warming is called ....

9. Si_3N_4, B_4C and SiC are .... ceramics.

10. SiO_2 is ...., GeO_2 is .... SnO is .... and PbO is ....

11. Concentrated HNO_3 converts graphite into ....

12. CCl_4 is kinetically stable towards ....

13. $SnCl_2 \cdot 2H_2O$ changes to .... on heating at 80°C.

14. $C_{60}K$ is an ....

15. Graphite fluoride and graphite oxide are ....

Matrix-Match Type Questions

I. One Or More Correct

1. Match the oxide of Column I with their corresponding property (ies) in Column II.

Column I	Column II
A. Pb_3O_4	1. oxidation state +2
B. SnO_2	2. oxidation state +4
C. PbO_2	3. amphoteric
D. CO_2	4. basic
E. GeO	5. acidic

2. Match the species in Column I with their corresponding property (ies) in Column II.

Column I	Column II
A. CO_2	1. Dry ice
B. $Al_2(SO_4)_3$	2. Fire extinguisher
C. KCN	3. Acidic
D. CaC_2	4. Ligand
E. Mg_2C_3	5. *sp*-hybridisation
	6. Green house effect

Integer Answer Types

This section contains 8 questions. The answer to each of the questions is a single digit integer, ranging from 0 to 9. The appropriate bubbles below the respectively question numbers in the *ORS* have to be darkened. For example, if the correct answers to question number *X*, *Y*, *Z* and *W* (say) are 6, 0, 9 and 2, respectively, then the correct darkening of bubbles will like the following :

X	*Y*	*Z*	*W*
⓪	⓪	⓪	⓪
①	①	①	①
②	②	②	②
③	③	③	③
④	④	④	④
⑤	⑤	⑤	⑤
⑥	⑥	⑥	⑥
⑦	⑦	⑦	⑦
⑧	⑧	⑧	⑧
⑨	⑨	⑨	⑨

Set I

1. Total number of atoms of carbon and oxygen in graphite oxide ....

2. When malonic acid is dehydrated using P_4O_{10} at 150°C, species formed has total number of atoms ....

3. Number of isotopes of carbon ....

4. How many of the following oxides are basic oxides?

$CO_2, SnO_2, PbO, PbO_2, Pb_3O_4, Al_2O_3$

5. For the half-cell reaction,

$$M^{2+} + 2e^- \longrightarrow M$$

$E°$ (in volts) are given for different metals

Cu	Zn	Fe	Pb	Sn
0.34	–0.76	–0.48	–0.13	–0.14

How many of these metals can reduce Cu^{2+} to Cu?

6. Out of the elements C, S, Ge, Sn and Pb only +4 oxidation state is shown by how many element(s)?

7. More than four bonds are made with how many elements in carbon family?

8. How many elements in group 14 are used as semiconductor?

Set II

1. How many of the following have two σ and two π bonds?

$C_3^{4-}, CN_2^{2-}, CO_2, C_2^{2-}, C_2H_4, CN^-$

2. How many of the fuel gases have CO and H_2 both as combustible gases?

LPG, water gas, Producer gas, Coal gas, CNG

3. A silicate mineral has three SiO_4 tetrahedra that share corners with adjacent tetrahedra. Mineral also contains Ca^{2+}, Cu^{2+} and H_2O in 1 : 1 : 1 molar ratio. Thus, total number of atoms of Cu, Si and Ca per unit of mineral is ....

4. A mixture contains 3 moles of CO and CO_2. On passing the mixture over heated charcoal, volume increases to 5 moles. Thus, CO_2 in the mixture is .... mole(s).

5. Total number of σ and π bonds in carbon suboxide (C_3O_2) is ....

6. What is total negative charge on the silicate $[Si_4O_{11}]$?

7. $CaCO_3(s) \rightleftharpoons CaO(s) + CO_2(g)$, $K = 10^{-9}$ at 300 K. What is p_{CO_2} ?

8. Inert pair effect is shown by how many elements? Ga, Al, Tl, Pb, Sn, Ge

Answers & Solutions
Master Exercises
Exercise 1

Short Answer Type Questions

1. Due to available *d*-orbital in which $(n-1)p$ electrons. can be excited to make C.N. of six. Carbon lacks *d*-orbital.

2. (a) Pb (b) C (c) Si (d) C

3. (a) CCl_4, CO_2 (b) CO (c) CH_4

4. Diamond is very hard, high melting solid. It is an electrical insulator. Diamond has a covalent network structure in which each C-atom uses sp^3-hybrid orbitals to form a tetrahedral array of σ-bonds. The interlocking, three-dimensional network of strong bonds makes diamond the hardest known substance with highest m.p. for an element. Because the valence electrons are localised in the σ-bonds, diamond is an electrical insulator.

5. (a) (*A*) A : SiF_4

(*B*) : H_2SiF_6

(*C*) : H_4SiO_4

$$3SiF_4 + 4H_2O \longrightarrow H_4SiO_4 + 2H_2SiF_6$$

(b) (*D*) : $C_6(COOH)_6$, mellitic acid (benzene hexacarboxylic acid)

(c) (*E*) : $(Na_2SiO_3 + C)$

6. $CO_3^{2-} + 2H_2O \rightleftharpoons H_2CO_3 + 2OH^-$

CO_3^{2-} solution is alkaline due to hydrolysis.

7. The Ca^{2+} ions give precipitates (scum) with soaps. On heating the HCO_3 ions revert to CO_3^{2-} which gives a precipitate with Ca^{2+} ions :

$$Ca(HCO_3)_2 \longrightarrow CaCO_3 + H_2O + CO_2$$

8. Al reacts with NaOH forming H_2 gas. NaOH is formed due to hydrolysis of Na_2CO_3 (Q. 6 above)

$$2Al + 2NaOH + 2H_2O \longrightarrow 2NaAlO_2 + 3H_2$$

9. $2NO(g) + CS_2 \longrightarrow \underset{\text{yellow solid}}{2S(s)} + N_2(g) + CO_2(g)$

10. Na_2CO_3 formed when atmospheric CO_2 reacts with NaOH.

11. (a) To provide a "bite" in soft drinks.

$$CO_2(g) + H_2O(l) \rightleftharpoons H_2CO_3(aq)$$

(b) As fire extinguisher; CO_2 is non-flammable and 1.5 times more dense than air.

(c) As refrigerant; dry ice sublimes at −78°C.

12. No, as such a high charge separation is unlikely.

13. $^{\ominus}{:}C{\vdots\vdots}C{:}^{\ominus}$ $\quad$ $^{\ominus}{:}\overset{\ominus}{\underset{\ominus}{\ddot{C}}}{\cdot\cdot}C{\vdots\vdots}C{:}^{\ominus}$

14. Refer text.

15. Be_2C and Al_4C_3 are ionic carbides; oxidation state of carbon = −4

16. H_2O **17.** O_2

18. Due to high activation energy.

19. Highly lustrous form (in layer) of carbon is called glanzkohlenstoff. It consists of very small crystals of graphite. It is found to have greater adsorption power and greater hardness than graphite. This crystalline form is obtained when the flame of burning methane is impinged on glazed porcelain placed at 900 K.

20. van der Waals' forces

21. $SiO_2(s) + Na_2SO_4(s) \longrightarrow Na_2SiO_3(s) + SO_3(g)$

22.

```
        CH3     CH3
         |       |
    H3C—Si—O—Si—CH3
         |       |
        CH3     CH3
```

23. $(CH_3COO)_3SiCH_3$. This would give the cross-linking for necessary strength.

24. (a) SiO_4^{4-} (b) $Si_3O_{10}^{8-}$ (c) $Si_4O_{12}^{8-}$

25. (a)

SiO_4^{4-}

(b)

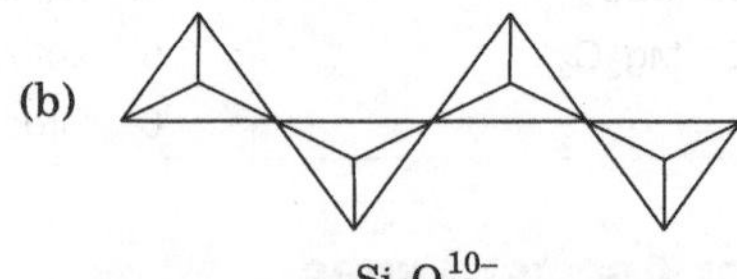

$Si_4O_{13}^{10-}$

26. (*X*) : $HgCl_2$

(*Y*) : HgS

(*Z*) : Hg_2Cl_2

27. Sn is treated with HCl gas to make anhydrous $SnCl_2$.

28. HF attacks silicates of glass

$$\left[\begin{array}{l} CaSiO_3 + 6HF \longrightarrow CaSiF_6 + 3H_2O \\ Na_2SiO_3 + 6HF \longrightarrow Na_2SiF_6 + 3H_2O \end{array}\right.$$

29. (a) supercooled (b) HF (c) PbO (d) B_2O_3 (e) CaF_2

Analytical Questions

1. (i) There is feasibility of p-p π bond in silicon.
   (ii) π-bonding between oxygen and silicon involving p- and d-orbitals decreases availability of lone-pairs of electrons on oxygen.
2. (a) **Triple points : Two Phase**
   Point A : Graphite $\rightleftharpoons$ Liquid $\rightleftharpoons$ Vapour
   Point B : Diamond $\rightleftharpoons$ Graphite $\rightleftharpoons$ Liquid
   (b) Diamond
   (c) Apply high pressure at high temperature
3. (a) Sn (b) Cl (c) Sn (d) Se (e) B
4. Carbon forms various stable compounds with itself and other elements such as N. O, H, S. Bimolecules contain chains and rings with many C—C bonds. Si—Si bonds are much less stable and chain of Si atoms are uncommon.
5. Pb_3O_4 on reaction with HNO_3 forms $Pb(NO_3)_2$ and PbO_2.
   $$Pb_3O_4 + 4HNO_3 \longrightarrow 2Pb(NO_3)_2 + 2H_2O + PbO_2$$
   2 moles of $Pb(NO_3)_2$ are from dilead(II) oxide while PbO_2 is from lead(IV) oxide. Thus it is $2PbO \cdot PbO_2$.
6. (a) $PbO + NaOH + H_2O \longrightarrow Na[Pb(OH)_3]$
   (b) $2SnO \longrightarrow Sn + SnO_2$
   (c) $PbO_2 + 4HCl \longrightarrow PbCl_2 + 2H_2O + Cl_2$
   (d) $3PbO + H_2O + CO_2 \longrightarrow 2PbCO_3 \cdot Pb(OH)_2$
7. Moles of CO_2 formed $= n = \dfrac{pV}{RT}$
   $$= \frac{\frac{775}{760} \times \text{atm} \times 100\text{ L}}{0.0821\text{ L atm mol}^{-1}\text{ K}^{-1} \times 303\text{ K}}$$
   $= 4.0992$
   If 2 moles CO_2 are formed then $MgSiO_3$ precipitated is = 1 mol and when 4.0992 moles CO_2 and formed $MgSiO_3$ would be $= \dfrac{4.0992}{2}$
   $= 2.0496$ mol
   (molar mass of $MgSiO_3 = 100\text{ g mol}^{-1}$)
   $= 2.0496 \times 100$ g
   $= 204.96$ g
8. Refer text.

NCERT Problems

Refer Text

Exercise 2

Only One Option Correct

1. (a)	**2.** (b)	**3.** (c)	**4.** (d)	**5.** (c)	**6.** (d)	**7.** (c)	**8.** (a)	**9.** (a)	**10.** (d)
11. (b)	**12.** (c)	**13.** (a)	**14.** (a)	**15.** (b)	**16.** (c)	**17.** (b)	**18.** (d)	**19.** (d)	**20.** (b)
21. (d)	**22.** (c)	**23.** (c)	**24.** (a)	**25.** (b)	**26.** (a)	**27.** (c)	**28.** (d)	**29.** (d)	**30.** (d)
31. (d)	**32.** (d)	**33.** (b)	**34.** (d)	**35.** (d)	**36.** (a)	**37.** (b)	**38.** (b)	**39.** (b)	**40.** (a)
41. (a)	**42.** (a)	**43.** (c)	**44.** (b)	**45.** (c)	**46.** (d)	**47.** (c)	**48.** (b)	**49.** (b)	**50.** (d)
51. (a)	**52.** (b)	**53.** (d)	**54.** (b)	**55.** (d)	**56.** (b)	**57.** (d)	**58.** (d)	**59.** (a)	**60.** (a)
61. (c)	**62.** (d)	**63.** (d)	**64.** (c)	**65.** (a)	**66.** (d)	**67.** (d)	**68.** (a)	**69.** (a)	**70.** (a)
71. (c)	**72.** (d)	**73.** (a)	**74.** (d)	**75.** (b)	**76.** (b)	**77.** (c)	**78.** (b)	**79.** (d)	**80.** (b)
81. (b)	**82.** (c)	**83.** (b)	**84.** (d)	**85.** (a)	**86.** (d)	**87.** (d)	**88.** (a)	**89.** (b)	**90.** (b)
91. (d)	**92.** (d)	**93.** (a)	**94.** (c)	**95.** (b)	**96.** (a)	**97.** (d)	**98.** (a)	**99.** (b)	**100.** (c)
101. (c)	**102.** (d)	**103.** (b)	**104.** (d)	**105.** (c)	**106.** (c)	**107.** (d)	**108.** (a)	**109.** (a)	**110.** (b)
111. (a)	**112.** (c)	**113.** (b)	**114.** (a)	**115.** (c)					

One or More Than One Options Correct

1. (a,b,c,d) **2.** (a,b,c,d) **3.** (a,b,c,d) **4.** (b)
5. (a,b) **6.** (a,b) **7.** (b,d) **8.** (a) **9.** (a,b)
10. (a,b) **11.** (a,b) **12.** (b,c) **13.** (a,b)
14. (c,d) **15.** (a,b)

Passage Comprehension Questions

Passage 1

1. (c) **2.** (d) **3.** (a) **4.** (b)

Passage 2

1. (a) **2.** (b)

Passage 3

1. (b) 2. (b) 3. (c)

Passage 4

1. (b) 2. (c) 3. (a)

Passage 5

1. (d) 2. (a) 3. (b) 4. (c)

Passage 6

1. (a) 2. (b)

Assertion & Reason

1. (a) 2. (b) 3. (a) 4. (c) 5. (b) 6. (a)
7. (a) 8. (c) 9. (a) 10. (b) 11. (c) 12. (b)

True & False

1. F 2. F 3. T 4. T 5. T 6. F 7. T
8. F 9. T 10. T 11. F 12. F 13. T
14. T 15. T

Fill in the Blanks

1. Second most 2. van der Waals' forces 3. diamond
4. pyramidal, planar 5. solid CO_2 6. methanides
7. cytochrome oxidase 8. green house effect
9. non-oxide 10. acidic, amphoteric, amphoteric, basic
11. mellitic acid $C_6(COOH)_6$ 12. hydrolysis
13. $SnCl_4 \cdot H_2O$ 14. electrically conducting compound
15. electrically non-conducting compounds

Matrix-Match Type Questions

I. One or More Correct

1. (A)—(1,2,3); (B)—(2,3); (C)—(2,4); (D)—(2,5); (E)—(1,3);
2. (A)—(1,2,3,5,6); (B)—(2,3); (C)—(4,5); (D)—(5); (E)—(5)

Integer Answer Types

Set I

Questions →	1	2	3	4	5	6	7	8
Answers →	3	5	3	2	4	1	4	2
	0	0	0	0	0	0	0	0
	1	1	1	1	1	1	1	1
	2	2	2	2	2	2	2	2
	3	3	3	3	3	3	3	3
	4	4	4	4	4	4	4	4
	5	5	5	5	5	5	5	5
	6	6	6	6	6	6	6	6
	7	7	7	7	7	7	7	7
	8	8	8	8	8	8	8	8
	9	9	9	9	9	9	9	9

Set II

Questions →	1	2	3	4	5	6	7	8
Answers →	3	2	7	2	8	6	9	5
	0	0	0	0	0	0	0	0
	1	1	1	1	1	1	1	1
	2	2	2	2	2	2	2	2
	3	3	3	3	3	3	3	3
	4	4	4	4	4	4	4	4
	5	5	5	5	5	5	5	5
	6	6	6	6	6	6	6	6
	7	7	7	7	7	7	7	7
	8	8	8	8	8	8	8	8
	9	9	9	9	9	9	9	9

13

> **"Whether it may be laughing gas (which does not make you laugh) or Holme's signal or flickering light (called *'will-o-the-wisp'*) are clues from Nitrogen family"**

Group 15 (VA—The Nitrogen Family)

Learning Framework of the Chapter ...

- Group 15 —Periodicity
- Study of Nitrogen and its Compounds
- Study of Phosphorus and its Compounds

$_7N$
$_{15}P$
$_{33}As$
$_{51}Sb$
$_{83}Bi$

The elements in this group (N, P, As, Sb, Bi) are also called Pnictogen (pnicogen) or pnictides from the Greek synonym for choking (as nitrogen does not support combustion). The elements N and P are essential constituents of the living system. Heavier members of the group, particularly arsenic are extremely toxic.

Nitrogen is highly abundant (about 78% by volume and 75% by mass) in air as N_2. However abundance of nitrogen in rocks and soils is low (19 ppm) as KNO_3 and $NaNO_3$.

Naturally occurring nitrogen consists of ^{14}N stable isotope, ^{15}N also occurs slightly, relative abundance being $^{14}N/^{15}N$ ~ 272 : 1

Phosphorus is highly abundant in crystal rocks (≈ 1120 ppm) most as orthophosphate. It also occurs in plants and animals where it is an essential constituent of life.

^{31}P is the only stable isotope of phosphorus. Among six radioactive isotopes, ^{32}P (β-emitter) is most important.

The elements As, Sb, Bi have very low abundances

As < 2 ppm

Sb 0-2 ppm

Bi 0-01 ppm

They occur as sulphides (As_2S_3, Sb_2S_3, Bi_2S_3) and are thus chalcophilic.

Group 15 — Periodicity

Physical properties of the elements in nitrogen family [Group 15 (VA)] are summarised in Table 13.1 and Fig. 13.1.

Table 13.1 **Physical Properties of the Elements of Group 15 (VA)**

	N	P	As	Sb	Bi
Electronic configuration	[He] $2s^2 2p^3$	[Ne] $3s^2 3p^3$	[Ar] $3d^{10} 4s^2 4p^3$	[Kr] $4d^{10} 5s^2 5p^3$	[Xe] $4f^{14} 5d^{10} 6s^2 6p^3$
Electronegativity (Pauling) IE kJ mol^{-1}	3.0	2.1	2.0	1.9	1.9
$(IE)_1$	1403	1012	947	834	703
$(IE)_2$	2857	1897	1950	1590	1610
$(IE)_3$	4578	2910	2732	2440	2467
Melting point/° C	–210	44 (white) 597 (red)	816	631	271
Boiling point/° C	–195.8	281 (white) 431 (red)	615 (sublimes)	1587	1564
Atomic radius/ pm	74	110	121	141	152
Principal oxidation number	+3, 0, +5, –3, –2, –1 +1, +2, +4	+3, + 5	+3, + 5	+3, + 5	+3
Abundance			0.0002%	0.00002%	0.00002%
Physical state (1 atm, 25°C)	gas	solid	solid	solid	solid

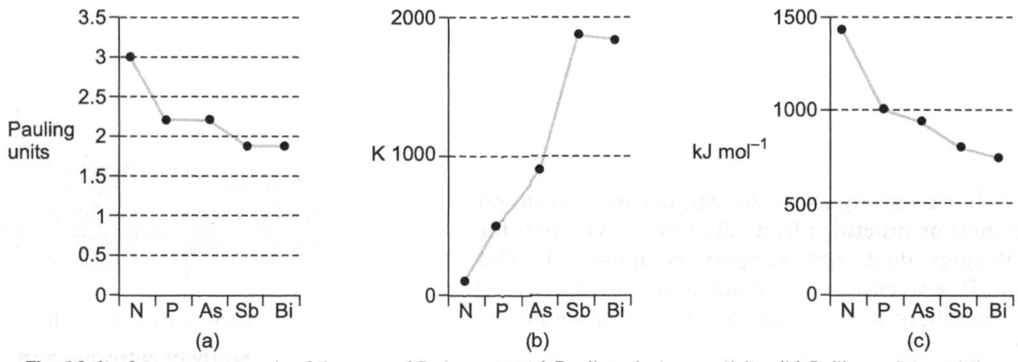

Fig. 13.1 Some group trends of the group 15 elements: (a) Pauling electronegativity, (b) Boiling point, and the (c) First ionisation energy

Physical State

Group 15 elements are sometimes referred to as **pnictogens.** All the elements are important and bismuth is notable for being the heaviest element with non-radioactive isotope.

Elements of this group show a systematic gradation of properties (Table 13.1) Nitrogen is gaseous in nature and exists as a diatomic molecule N≡N. This is due to the very high bond energy of the N≡N bond, 946 kJ mol^{-1}, which is nearly six times the N—N bond energy, 160 kJ mol^{-1}. On the contrary P≡P bond is quite weak, its energy of 490 kJ mol^{-1} is only ~ 2.5 times the P—P bond energy, 209 kJ mol^{-1}. Therefore, phosphorus exists as tetra-atomic P_4 molecules in the gaseous state. In the solid state, it exists in different allotropic forms.

The change from non-metallic to metallic character with increase in the atomic number is well illustrated in this group. The first two elements, nitrogen and phosphorus, are non-metals, arsenic and antimony are metalloids whereas bismuth is predominantly metallic in nature. In keeping with this trend, the electronegativity of these elements decreases gradually on moving down the group. The density, melting and boiling points of the elements increase with increase in atomic number.

DEEP Focus

Radii, IE, Oxidation State

Atomic radii, as expected, increase with increase in atomic number. As a consequence, ionisation energies decrease on descending the group. Nitrogen shows an exceptionally high ionisation energy which is due to its small atomic size. As discussed above, all the elements of this group have five electrons in their valence shell, they exhibit a highest oxidation state of +5 by using all five electrons in forming bonds. As one goes down the group, the tendency of the pair of *s* electrons to remain inert increases and only *p*-electrons are used in bonding, resulting in an oxidation state of +3. The stability of +3 state relative to the +5 state thus increases on moving down the group. In the case of nitrogen, avery wide range of oxidation state exists :

Compound	Oxidation state of N
NH_3	–3
N_2H_4	–2
NH_4OH	–1
N_2	0
N_2O	+1
NO	+2
NO_2	+4
HNO_3	+5

The negative oxidation state of nitrogen arise because of its higher electronegativity (3.05) than that of hydrogen (2.1). As you can see from the Table 13.1 the energy required to produce M^{5+} is so immense that it is nevery formed. The compounds formed by these elements in +5 oxidation state are thus predominantly covalent. However, in case of Sb and Bi, the sum of the first three ionisation energies is just low enough to form M^{3+} ions. But fluorine is the only element which gives an electronegativity difference enough to permit the formation of ionic bonds. Thus, SbF_3 and BiF_3 exist as ionic solids.

DEEP Focus

Since, it is not easy from energy considerations to gain three electrons to achieve stable noble gas configuration, the formation of the tri-negative ion (M^{3-}) is rare. But, nitrogen being the smallest and the most electronegative of all the members of the group, forms trivalent anion with reactive metals which have low ionisation energies. Thus, nitrogen forms ionic nitrides such as $Li_3N, Be_3N_2, Mg_3N_2, Ca_3N_2$. As the electronegativity of the other elements of the group, *i.e.*, P, As, Sb and Bi, is low, most of their compounds **are covalent** in nature.

Nitrogen is not able to extend its coordination number beyond four, the other elements have coordination numbers of five as well as six. The extension of coordination number is due to the availability of *d*-orbitals in P, As, Sb and Bi, using which these elements can attain sp^3d and sp^3d^2 hybridisation, *e.g.*, in PCl_5 and PCl_6^-, respectively. **Thus, nitrogen does not function as a Lewis acid whereas phosphorus, arsenic, antimony and bismuth do so. The hydrides of this group and their alkyl derivatives act as Lewis bases, formming a number of** addition complexes, *e.g.,* $(NH_3 \cdot BF_3)$, $(CaCl_2 \cdot 4NH_3)$, etc.

- N and P show the typical properties of non-metals :
  - are poor conductors of heat and electricity,
  - form acidic oxides,
  - form covalent compounds predominantly.
- Metallic nature increases as we go down the group, N and P are typical non-metals, As and Sb are metalloids and Bi is a metal.
- Phosphorus and the other members of the group can make use of *d*-orbitals in their bonding, thus compounds MX_3 as well as MX_5 are formed (PCl_3, PCl_5 etc.), but NCl_5 is not formed; N does not have *d*-orbital.

DEEP Focus

- Stability of +3 state increases and that of +5 state decreases going down the group due to "inert pair effect", thus

$$Bi^{3+} > Sb^{3+} > As^{3+}$$
$$Bi^{5+} < Sb^{5+} < As^{5+}$$

Bi^{5+} salts are very rare and are good oxidising agents :

$$Bi^{5+} + 2e^- \longrightarrow Bi^{3+}$$

As^{3+} salts are good reducing agents :

$$As^{3+} \longrightarrow As^{5+} + 2e^-$$

Nature of Compounds

- As we go down the group, basic nature of oxides increases and acidic nature decreases. Oxides of N and P are acidic, those of As and Sb are amphoteric and that of Bi are basic.

Element	Oxides	Trend
N	$N_2O, NO, N_2O_3, N_2O_4, NO_2, N_2O_5$	acidic
P	P_4O_6, P_4O_{10}	
As	As_4O_6, As_4O_{10}	amphoteric
Sb	Sb_4O_6, Sb_4O_{10}	
Bi	Bi_2O_3	basic

DEEP Focus

- They form trihalide and pentahalide (except N, which does not form NCl_5 etc). As we go down the group, covalent nature decreases and ionic nature increases; thus NCl_3 and PCl_3 are covalent while $BiCl_3$ is ionic. This is also in accordance with **Fajan's rule :** *Smaller the size of cation, larger the size of anion, greater the charge then greater the polarising power and hence greater the covalent nature.*

Halides are hydrolysed in water forming oxyacids or oxychlorides :

$$PCl_3 + 3H_2O \longrightarrow H_3PO_3 + 3HCl$$

$$SbCl_3 + H_2O \longrightarrow \underset{\text{(orange ppt)}}{SbOCl \downarrow} + 2HCl$$

$$BiCl_3 + H_2O \longrightarrow \underset{\text{(white ppt)}}{BiOCl} + 2HCl$$

- Trihalides (PF_3) have a pyramidal structure in the gaseous state.
- Nitrogen is relatively unreactive, which is why it has accumulated in such large amounts in the atmosphere.

White phosphorus catches fire when exposed to air, burning to form P_4O_{10}. It is **stored under water to prevent this**. Red phosphorus is stable in air at room temperature, though it reacts on heating.

Arsenic is stable in dry air, but tarnishes in moist air, giving first a brown and then a black tarnish. White arsenic is As_2O_3.

Antimony is less reactive, and is stable towards water and to air at room temperature.

Hydrides

- Elements of group 15 (VA), all form volatile hydrides of formula MH_3, which are all poisonous, foul smelling gases. On descending the group from NH_3 to BiH_3 :
  - the hydrides become increasingly difficult to prepare
  - their stability decreases
  - their reducing power decreases
  - the ease of replacing hydrogen atoms by other groups such as Cl or Me (methyl) decreases
  - Lewis-base nature (tendency to donate lone pair of electrons) decreases
  - PH_3 is a much weaker base than NH_3; AsH_3; SbH_3 and BiH_3 do not show any basic properties.
  - hydrides have a pyramidal structure
  - **boiling point increases but NH_3 is the exception due to intermolecular H-bonding**

$$\underset{-87°C}{PH_3} < \underset{-55°C}{AsH_3} < \underset{-33.5°C}{NH_3} < \underset{-18°C}{SbH_3}$$

  (BiH_3 is very unstable)

  — **bond angle decreases**

$$NH_3 > PH_3 > AsH_3 > SbH_3$$

- The ionisation energy to produce M^{5+} is so large for N that it does not show +5 oxidation state. Nitrogen also differs from other elements in that it can form strong $p\pi$-$p\pi$ multiple bonds and hence it produces a number of compounds which have no counter parts in other elements. Thus, NO_3^-, NO_2^-, N_3^- (azide), $C{\equiv}N^-$, N_2O, NO, NO_2, N_2O_4, N_2O_3, $N{\equiv}N$, azo and diazo compounds are formed. Oxides N_2O_3 and N_2O_5 are monomeric while trioxides and pentoxides of other elements are dimeric as in P_4O_6 and P_4O_{10}. Phosphorus and the heavier elements of this group show $d\pi$-$p\pi$ bonding.

Illustration 1 *Could the strain in the P_4 molecule be reduced by using sp^3 hybrid orbitals in bonding instead of pure p-orbitals? Explain.*

Solution The angle required by P_4 is 60°. The strain would not be reduced by using sp^3 hybrid orbitals because their angle is ~ 109°.

Study of Nitrogen and its Compounds

- N_2 makes up almost 78% of the atmosphere, and is therefore the most common gas. N_2 is unreactive (inert), however it reacts with some metals forming metal nitrides which are ionic with N^{3-} ion :

$$3Mg + N_2 \longrightarrow Mg_3N_2 \quad \text{(magnesium nitride)}$$

- The most important compounds of nitrogen are in living things, *e.g.*, amino acids and proteins. Among other things, the high electronegativity of nitrogen allows it to take part in H-bonding and thus helps to keep the strands of DNA together.
- Some microorganisms capture nitrogen from the atmosphere and convert it into useful compounds—a case of **nitrogen-fixation**. It involves a complicated system of biochemical reactions involving an enzyme, **nitrogenase** and **adenosine triphosphate** (ATP). Some plants fix nitrogen by use of *Rhizobium* bacteria, which are often found in the root system. Nitrogen is also fixed by chemical reactions in the atmosphere, particularly in thunderstorms.

$$N_2(g) + O_2(g) \xrightarrow{\text{lightning}} 2NO(g)$$

$$2NO(g) + O_2(g) \longrightarrow 2NO_2(g)$$

$$2NO_2(g) + H_2O(l) \longrightarrow HNO_2(aq) + HNO_3(aq)$$

DEEP Focus

- Hence, the energy released by lightning converts nitrogen into nitrogen oxides and nitric acid, which falls to earth in the rain. However, the proportion of nitrogen in the atmosphere is not continually decreasing because that lost by fixation is replaced by the decomposition of nitrogenous compounds by bacteria. This process is called **denitrification.** ***Currently, the total rate of fixation exceeds the rate of return of nitrogen to the atmosphere which means that nitrogen compounds are accumulating in soil, ground water, surface water and the oceans with subsequent environmental effects,*** *e.g.*, NO_3^- occurs in ground water in higher concentration as a result of fertilisers runoff and drainage from cattle feedlots. When this water is drunk, nitrate reducing bacteria in the gastro-intestinal tract can convert NO_3^- into NO_2^-. NO_2^- interferes with the oxygen transport ability of blood, which can lead to a condition known as **methemoglobinemia**. This order can be fatal in infants.

Nitrogen cycle is represented in Fig. 13.2.

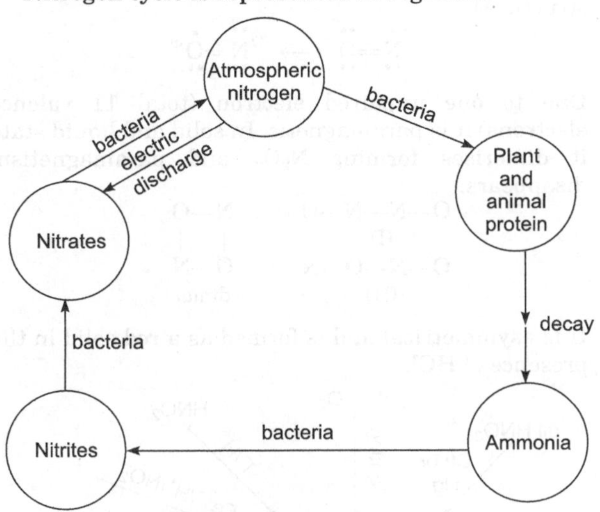

Fig. 13.2 Nitrogen cycle in nature

- **On a commercial scale, N_2 is obtained by the fractional evaporation of liquid air. Liquid air usually contains 50–60 per cent oxygen instead of 23 per cent in gaseous air. Since N_2 and O_2 differ widely in their b.p. ($N_2 = -195°C$, $O_2 = -183°C$) hence separation of N_2 and O_2 can be easily affected by fractional distillation. N_2 thus obtained has some traces of rare gases like He and Ne.**

Nitrogen Oxides

- **The oxides show the wide range of oxidation numbers of N, from +1 to +5.**
- **All the oxides are covalent.**
- **All are gases at room temperature (except N_2O_5 which is unstable solid at room temperature).**
- **The lower oxides are neutral and the higher ones are acidic. Oxides are summarised in Table 13.2.**

Table 13.2 **Oxides of Nitrogen**

Name	Formula	Oxyacid	Oxidation no.	Preparation	Magnetic properties
1. Dinitrogen oxide (nitrous oxide)	N_2O	$H_2N_2O_2$ (hyponitrous acid)	+1	$NH_4NO_3 \xrightarrow{\Delta} N_2O + 2H_2O$	Diamagnetic
2. Nitrogen monoxide (nitric oxide)	NO	—	+2	$\underset{\text{dil.}}{3Cu + 8HNO_3} \xrightarrow{\Delta} 3Cu(NO_3)_2 + 2NO + 4H_2O$	Paramagnetic
3. Nitrogen dioxide	NO_2	—	+4	$2Pb(NO_3)_2 \xrightarrow{\Delta} 2PbO + 4NO_2 + O_2$	Paramagnetic
4. Dinitrogen tetraoxide	N_2O_4	—	+4	Cool NO_2, $N_2O_4 \underset{\text{cool}}{\overset{\Delta}{\rightleftharpoons}} 2NO_2$	Diamagnetic
5. Dinitrogen trioxide (nitrogen sesquioxide)	N_2O_3	HNO_2	+3	$NO + NO_2 \longrightarrow N_2O_3$ $2HNO_2 \xrightarrow{\Delta} N_2O_3 + H_2O$	Diamagnetic
6. Dinitrogen pentoxide	N_2O_5	HNO_3	+5	$2HNO_3 \xrightarrow{P_2O_5} N_2O_5 + H_2O$	Diamagnetic

Nitrous oxide (N_2O)

- $\underset{\text{(molten)}}{NH_4NO_3} \xrightarrow{\Delta, 250°C} N_2O + 2H_2O$
- $\underset{\text{hydroxyl amine}}{NH_2OH} + HNO_2 \longrightarrow N_2O + 2H_2O$
- **It is stable, relativity unreactive colourless gas, it is a neutral oxide.**
- **It is soluble in water (1 : 1 by volume), but solution is neutral probably the following equilibrium lies in left-hand side**

$$N_2O + H_2O \rightleftharpoons \underset{\text{hyponitrous acid}}{H_2N_2O_2}$$

It may be taken as anhydride of $H_2N_2O_2$

- **It forms sodium azide on reaction with $NaNH_2$:**

$$NaNH_2 + N_2O \longrightarrow NaN_3 + H_2O$$
$$NaNH_2 + H_2O \longrightarrow NaOH + NH_3$$
$$\overline{2NaNH_2 + N_2O \longrightarrow NaN_3 + NaOH + NH_3}$$

- **N_2O resembles molecular oxygen (O_2) in that in supports combustion. It does so because it decomposes to molecular nitrogen and molecular oxygen when heated**

$$2N_2O \longrightarrow 2N_2 + O_2$$

- **The molecule is linear as would be expected for a triatomic molecule with 16 outer shell electrons (as for CO_2).**

$:N{\equiv}N{-}\ddot{\underset{..}{O}}: \longleftrightarrow \ddot{:N}{=}N{=}\ddot{O}:$

$N \xrightarrow{1.126\,Å} N \xrightarrow{1.86\,Å} O$

- **The largest use of N_2O is as a propellant for whipped ice-cream. Because it has no taste, and is non-toxic, it meets the strict food and health regulations.**
- **N_2O is also used as an anaesthetic, particularly by dentists. It is also called laughing gas, because small amounts cause euphoria.**

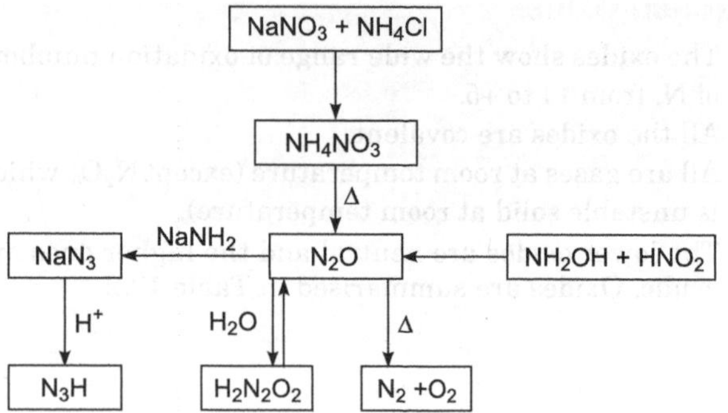

Nitric Oxide (NO)

- $N_2 + O_2 \longrightarrow 2NO$
- $3Cu + \underset{dil}{8HNO_3} \longrightarrow 3Cu(NO_3)_2 + 2NO + 4H_2O$
- $2HNO_2 + 2I^- + 2H^+ \longrightarrow 2NO + I_2 + 2H_2O$
- NO can be prepared in the laboratory by reacting sodium nitrite ($NaNO_2$) with a reducing agent such as Fe^{2+} in an acidic medium

$$NO_2^-(aq) + Fe^{2+}(aq) + 2H^+(aq) \longrightarrow NO(g) + Fe^{3+}(aq) + H_2O(l)$$

- $2NO_3^- + 6Hg + 8H^+ \longrightarrow 3Hg_2^{2+} + 2NO + 4H_2O$
- Lighting also contributes to the atmospheric concentration of NO. Exposed to air, nitric oxide quickly forms brown nitrogen dioxide gas:

$$2NO(g) + O_2(g) \longrightarrow \underset{\text{(brown fumes)}}{2NO_2(g)}$$

Nitrogen dioxide is a major component of smog.

DEEP Focus

- NO is an odd-electron molecule; alternately we can say it is a free-radical. Hence it is highly reactive :

$$2NO + Cl_2 \longrightarrow 2NOCl$$

$$3NO \xrightarrow[\text{high pressure}]{30\text{–}50°C} N_2O + NO_2$$

NO readily forms coordination complexes with transition metal ions. These complexes are called **nitrosyls.**

$[Fe(H_2O)_5NO]^{2+}$ (brown ring)

- Bond-order is 2.5 (as described by M.O. theory) and bond length is 1.15Å which is intermediate between a double bond and a triple bond.

- It can be represented by the following resonance structures:

$$\dot{N}{=}\ddot{O} \longleftrightarrow {}^{\ominus}\ddot{N}{=}\ddot{O}^{\oplus}$$

Due to one unpaired electron (total 11 valence electrons) it is paramagnetic. In solid and liquid state it dimerises forming N_2O_2 and paramagnetism disappears.

O—N—N—O (I)

O—N—O—N (II)

N—O / O—N (dimer)

II is asymmetrical and is formed as a red solid in the presence of HCl.

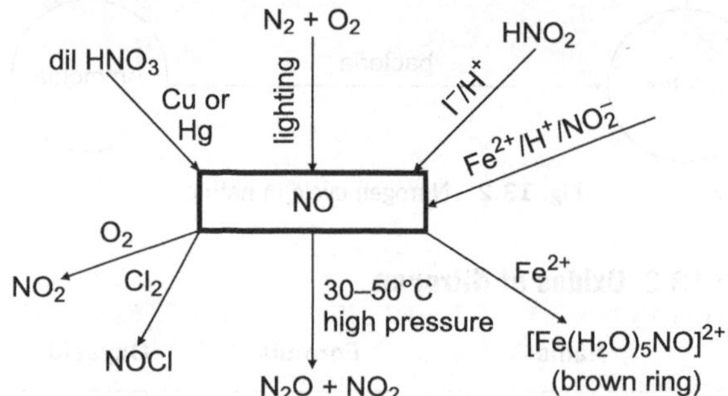

Nitrogen dioxide (NO_2), dinitrogen tetraoxide (N_2O_4)

$$Cu + 4H^+ + 2NO_3^- \xrightarrow{\Delta} Cu^{2+} + 2H_2O + 2NO_2$$

$$2Pb(NO_3)_2 \xrightarrow{\Delta} 2PbO + 4NO_2 + O_2$$

On condensation, NO_2 is obtained as a pale liquid and ultimately a colourless solid (N_2O_4 due to dimerisation).

- NO_2 is paramagnetic due to unpaired electron but in dimer formation paramagnetism disappears due to pairing of unpaired electrons

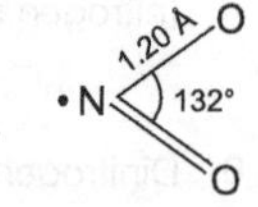

$$2NO_2 \rightleftharpoons N_2O_4$$

O₂N—NO₂ structure: $\begin{matrix} O & & & & O \\ & \rangle N — N \langle & \\ O & & & & O \end{matrix}$

This reaction occurs in both the gas phase and the liquid phase.

- It is represented by resonance structures:

$$\ddot{O}{=}\dot{N}{-}\ddot{O}: \longleftrightarrow :\ddot{O}{-}\dot{N}{=}\ddot{O}$$

- NO_2 molecule is angular with an O—N—O angle of 132° and N—O bond length of 1.20Å (which is intermediate between a single and a double bond).

DEEP Focus

- N_2O_4 has planar structure studied by X-ray

$$\begin{matrix} O & & & & O \\ & \rangle N & \underline{1.64\text{ Å}} & N\langle & \\ O & & & & O \end{matrix}$$

N—N bond length (1.64 Å) in N_2O_4 is about 20% larger than the usual N—N single bond. *We can think of the positive charge formed on the adjacent* N *atoms which causes repulsion and thus increases the bond length.*

- N_2O_4 is a mixed anhydride as indicated by its reaction with H_2O when HNO_2 (nitrous acid) and HNO_3 (nitric acid) are formed

$$\underset{\substack{\uparrow\downarrow \\ 2NO_2}}{N_2O_4} + H_2O \longrightarrow HNO_2 + HNO_3$$

The HNO_2 formed decomposes to give NO and NO_2

$$2HNO_2 \longrightarrow NO_2 + NO + H_2O$$

Thus moist NO_2 and N_2O_4 gases are strongly acidic.

- Liquid N_2O_4 is useful as a non-aqueous solvent and self-ionises as

$$N_2O_4 \rightleftharpoons \underset{\text{acid}}{NO^+} + \underset{\text{base}}{NO_3^-}$$

Thus, $ZnCl_2 + 2N_2O_4 \longrightarrow 2NOCl + Zn(NO_3)_2$

- NO_2—N_2O_4 system is a strong oxidising agent

$$2NO_2 \xrightarrow{150°C} 2NO + O_2$$

$$2NO_2 + F_2 \longrightarrow \underset{\text{nitryl fluoride}}{2NO_2F}$$

$$2NO_2 + 4HCl \longrightarrow \underset{\text{nitrosyl chloride}}{2NOCl} + Cl_2 + 2H_2O$$

$$NO_2 + CO \longrightarrow CO_2 + NO$$

$$2NO_2 \rightleftharpoons N_2O_4$$

N_2O_4 solidifies at −11.2° C and is colourless.

As the temperature increases, proportion of brown NO_2 increases and the colour of the liquid darkens. At around 140° C, the mixture is almost entirely NO_2.

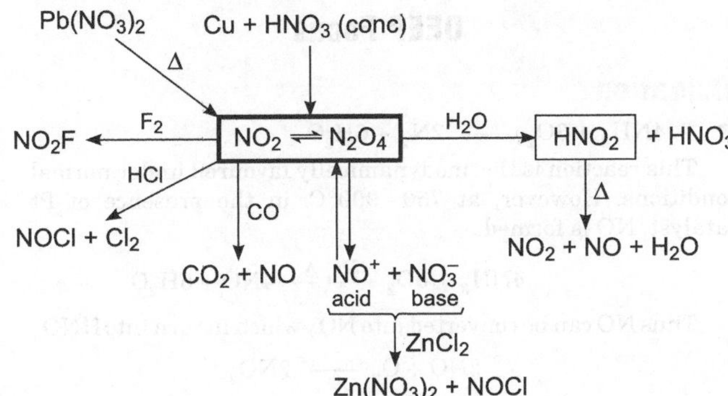

Dinitrogen Trioxide (N_2O_3)

- It is obtained by condensing equimolar amount of NO and NO_2 together, or by reacting NO with O_2 in a fixed ratio.

$$NO + NO_2 \longrightarrow N_2O_3$$

$$4NO + O_2 \longrightarrow 2N_2O_3$$

- It is an acidic oxide and is the anhydride of HNO_2

$$N_2O_3 + H_2O \longrightarrow 2HNO_2$$

$$N_2O_3 + 2NaOH \longrightarrow \underset{\text{sodium nitrite}}{2NaNO_2} + H_2O$$

- It is an intensely blue liquid or a pale blue solid. It exists only at low temperature and readily dissociates to give NO and NO_2

$$N_2O_3 \rightleftharpoons NO + NO_2$$

- The N_2O_3 molecule has an O_2N—NO structure in the gas phase and at low temperature, with an extremely long (1.86Å) N—N bond which is consistent with its easy dissociation.

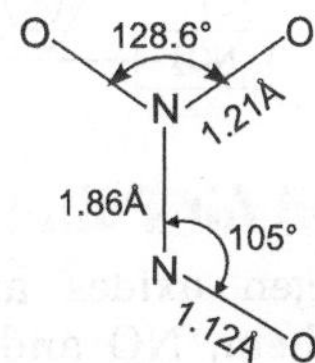

Two interconvertible structures (in presence of light radiations of a particular wavelength) are assigned to it.

$$O_2N—NO \rightleftharpoons O=N—O—N=O$$

NO + NO_2 ⇅ N_2O_3
NH_4NO_2 $\xleftarrow{NH_4OH}$ N_2O_3 $\xrightarrow{H_2O}$ HNO_2
NH_4NO_2 $\xrightarrow{\Delta}$ $N_2 + H_2O$
N_2O_3 $\xrightarrow{NaOH}$ $NaNO_2$

Dinitrogen Pentoxide (N_2O_5)

- $$2HNO_3 \xrightarrow[\text{low temp.}]{P_2O_5} N_2O_5 + H_2O$$

  Thus N_2O_5 is the anhydride of HNO_3.

- It is colourless deliquescent solid and highly reactive.
- In the gaseous phase :

$$N_2O_5 \xrightarrow{\Delta} NO_2 + NO + O_2$$

$$N_2O_5 + O_3 \longrightarrow 2NO_3 + O_2$$

- In solid state :

$$N_2O_5 + H_2O \longrightarrow 2HNO_3$$

$$N_2O_5 + Na \longrightarrow NaNO_3 + NO_2$$

$$N_2O_5 + NaCl \longrightarrow NaNO_3 + NO_2Cl$$

- With conc. H_2SO_4, it forms NO_2^+ (nitronium ion) which can attack benzene nucleus forming nitrobenzene

$$N_2O_5 + 3H_2SO_4 \longrightarrow 2NO_2^+ + 3HSO_4^- + H_3O^+$$

$$C_6H_6 + NO_2^+ \longrightarrow C_6H_5NO_2 + H^+$$

- X-ray diffraction shows that solid N_2O_5 is ionic $NO_2^+ NO_3^-$ (nitronium nitrate). In the solution and gaseous phase, it is covalent with structure :

$$O_2N—O—NO_2$$

Each N atom is sp^2-hybridised.

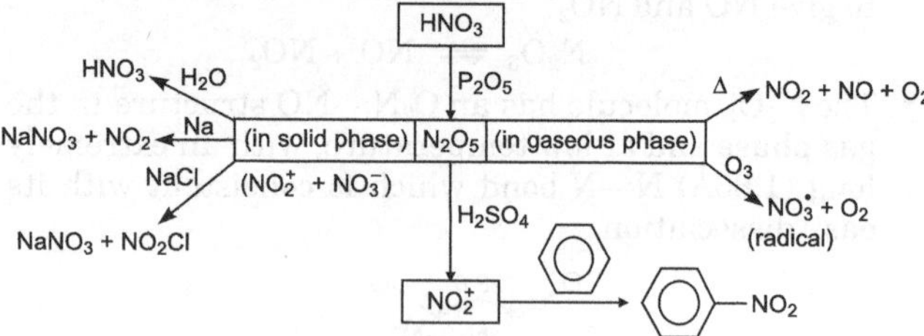

Nitrogen Oxides and Air Pollution

Collectively, nitrogen oxides are often given the symbol NO_x. Two of them, NO and NO_2 are of special importance in environmental chemistry, causing considerable amount of air pollution. They are given off in car exhaust fumes and when fossil fuels are burnt, as well as being made during thunderstorms. In each case NO is made and then NO_2.

$$N_2 + O_2 \longrightarrow 2NO$$

DEEP Focus

NO_2 is more unpleasant and gives rise to acid rain and also takes part in photochemical reactions forming smog. The photons making up sunlight can cause the NO_2 molecules to fall apart, releasing oxygen atoms:

$$NO_2 \xrightarrow{h\nu} NO + O$$

Oxygen atoms are extremely reactive and will attack other molecules, especially those in car exhaust fumes. For examples hydrocarbons that escape combustion can be changed into aldehydes. The oxygen atoms can also react with O_2 to form O_3.

NO_2 is also involved in the production of one of the most unpleasant of all chemicals present in photochemical smog. It is known as PAN (peroxyacetyl nitrate):

(PAN) $CH_3—C(=O)—O—O—NO_2$

Its presence in the atmosphere irritates eyes and lungs and generally makes breathing difficult.

Illustration 1 *What is difference in the nature of the products obtained when $CO_2(g)$ and $NO_2(g)$ are dissolved separately in water?*

Solution NO_2 is an acidic oxide; it reacts rapidly with cold water to form both nitrous acid, HNO_2, and nitric acid, HNO_3

$$\underset{+4}{2NO_2(g)} + H_2O(l) \longrightarrow \underset{+3}{HNO_2(g)} + \underset{+5}{HNO_3(g)}$$

Oxidation number

This is a disproportionation reaction in which the oxidation number of nitrogen changes from +4 (in NO_2) to +3 (in HNO_2) and +5 (in HNO_3). This reaction is quite different from that between CO_2 and H_2O in which only one acid (H_2CO_3, carbonic acid) is formed without change in oxidation number of carbon

$$CO_2 + H_2O \rightleftharpoons H_2CO_3$$

Ammonia (NH_3)

- NH_3 is formed by decomposing NH_4^+ salts with a base

$$NH_4X + OH^- \xrightarrow{\Delta} NH_3 + H_2O + X^-$$

$$N_2 + 3H_2 \underset{}{\overset{\text{catalyst}}{\rightleftharpoons}} 2NH_3 \quad \Delta H = -40 \text{ kJ mol}^{-1}$$

$$K_{298} = 10^3 \text{ atm}^{-2}$$

This is the basis of the Haber process and is carried out at 400–500°C and at a pressure of 10^2–10^3 atm in the presence of a catalyst.

$$\underset{\text{calcium cyanamide}}{CaNCN} + 3H_2O \longrightarrow 2NH_3 + CaCO_3$$

- Ammonia is extremely soluble in water. Although aqueous solution is generally referred to as solution of weak base NH_4OH, undissociated NH_4OH does not exist. The solutions are best described as $NH_3(aq)$ with equilibrium :

$$NH_3(aq) + H_2O \rightleftharpoons NH_4^+ + OH^-$$

$$K_b = \frac{[NH_4^+][OH^-]}{[NH_3]} = 1.8 \times 10^{-5} \text{ mol L}^{-1}$$

- NH_3 is colourless, pungent and quite poisonous gas (b.p −33.35° C)

DEEP Focus

Oxidation

- $4NH_3 + 3O_2 \longrightarrow 2N_2 + 6H_2O$

This reaction is thermodynamically favoured under normal conditions. However, at 750 – 900°C, in the presence of Pt catalyst, NO is formed.

$$4NH_3 + 5O_2 \xrightarrow{Pt, \Delta} 4NO + 6H_2O$$

Thus NO can be converted into NO_2 which in turn into HNO_3

$$2NO + O_2 \longrightarrow 2NO_2$$

$$3NO_2 + H_2O \longrightarrow 2HNO_3 + NO$$

Thus the sequence in industrial utilisation of atmospheric nitrogen is as follows:

$$N_2 \xrightarrow[\text{Haber process}]{H_2} NH_3 \xrightarrow[\text{Ostwald process}]{O_2} NO \xrightarrow{O_2 + H_2O} HNO_3$$

- Liquid NH_3 resembles water in its physical behaviour, being highly associated *via* strong hydrogen bonding. Like H_2O it also self-ionises:

$$H_2O + H_2O \rightleftharpoons H_3O^+ + OH^-$$

$$NH_3 + NH_3 \rightleftharpoons NH_4^+ + NH_2^-$$

or $$NH_3 \rightleftharpoons H^+ + NH_2^-$$

As a Ligand

- Because $NH_3(l)$ has a much lower dielectric constant than water, it is a better solvent for organic compounds but generally a poorer one for ionic inorganic compounds. But complex formation by NH_3 (as ligand) is better than that by H_2O. AgCl is insoluble in H_2O but soluble in NH_3 due to complex formation :

$$AgCl + 2NH_3 \longrightarrow [Ag(NH_3)_2]Cl$$

Reducing Nature

- NH_3 can reduce oxides into metal :

$$3CuO + 2NH_3 \xrightarrow{\Delta} 3Cu + N_2 + 3H_2O$$

$$3PbO + 2NH_3 \xrightarrow{\Delta} 3Pb + N_2 + 3H_2O$$

- $$\underset{\text{excess}}{8NH_3} + 3Cl_2 \longrightarrow N_2 + 6NH_4Cl$$

$$NH_3 + \underset{\text{excess}}{3Cl_2} \longrightarrow NCl_3 + 3HCl$$

- $$NH_3 + 3H_2O_2 \longrightarrow HNO_2 + 4H_2O$$

- $$2NH_3 + \underset{\substack{\text{sodium hypochlorite}\\ \text{excess}}}{3NaClO} \longrightarrow N_2 + 3NaCl + 3H_2O$$

$$\underset{\text{excess}}{2NH_3} + NaClO \longrightarrow \underset{\text{hydrazine}}{NH_2—NH_2} + NaCl + H_2O$$

$$2NH_3 + \underset{\text{bleaching powder}}{3CaOCl_2} \longrightarrow 3CaCl_2 + N_2 + 3H_2O$$

Solvated Electrons

- $$2NH_3 + 2Na \longrightarrow \underset{\text{sodamide}}{2NaNH_2} + H_2$$

Dilute solutions of alkali metals in liquid NH_3 are dark blue in colour and the main species are solvated metal ions and solvated electrons.

Test

- $$\underset{\substack{\text{Nessler's}\\ \text{reagent}}}{2K_2HgI_4} + NH_3 + 3KOH \longrightarrow \underset{\substack{\text{iodide of Millon's base}\\ \text{(brown ppt)}}}{H_2NHgOHgI\downarrow} + 7KI + 2H_2O$$

- $$CaSO_4 + 2NH_3 + CO_2 + H_2O \longrightarrow CaCO_3 + (NH_4)_2SO_4$$

- $$2NH_3 + CO_2 \rightleftharpoons \underset{\substack{\text{ammonium}\\ \text{carbamate}}}{NH_2COONH_4} \overset{\Delta}{\rightleftharpoons} \underset{\text{urea}}{NH_2CONH_2} + H_2O$$

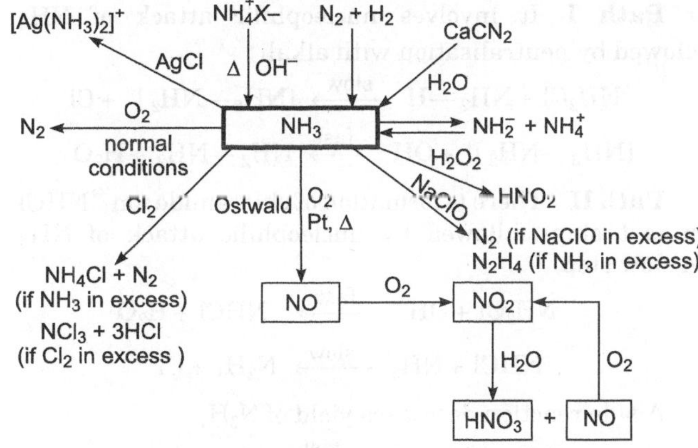

DEEP Focus

- Liquid ammonia functions as a good solvent for many substances and many types of reactions. **All those substances which dissolve in liquid NH_3 to give NH_4^+ ions are acids** and those which give amides, NH_2^- ions are bases. NH_4Cl, NH_4NO_3 are acids in liquid NH_3 due to existence of NH_4^+ ions.

$$NH_4Cl \rightleftharpoons NH_4^+ + Cl^-$$

$$NH_4NO_3 \rightleftharpoons NH_4^+ + NO_3^-$$

$NaNH_2$ is a base in liquid NH_3 due to existence of NH_2^- ion:

$$NaNH_2 \rightleftharpoons Na^+ + NH_2^-$$

Acid-base neutralisation reaction in liquid NH_3 can be thus represented as,

$$\underset{\text{acid}}{2NH_4Cl} + \underset{\text{base}}{PbNH} \xrightarrow{NH_3} PbCl_2 + 3NH_3$$

$$3NH_4Cl + AlN \xrightarrow{NH_3} AlCl_3 + 4NH_3$$

- Acids which are weak in water will be highly acidic in liquid NH_3:

$$\underset{\text{weak acid}}{CH_3COOH} + H_2O \rightleftharpoons CH_3COO^- + H_3O^+$$

$$\underset{\text{strong acid}}{CH_3COOH} + NH_3 \rightleftharpoons CH_3COO^- + NH_4^+$$

Hydrazine (N_2H_4)

Preparation

Raschig Process

It is obtained by controlled oxidation of NH_3 (in excess) by alkaline solution of sodium hypochlorite (NaOCl) :

$$2NH_3 + NaOCl \longrightarrow N_2H_4 + NaCl + H_2O$$

Reaction involves rapid formation of chloramine (NH_2Cl)

$$NH_3 + OCl^- \longrightarrow NH_2Cl + OH^-$$

NH_2Cl changes to N_2H_4 by path I or path II

Path I It involves nucleophilic attack of NH_3 followed by neutralisation with alkali

$$NH_2Cl + NH_2—H \xrightarrow{\text{slow}} [NH_3—NH_2]^+ + Cl^-$$

$$[NH_3—NH_2]^+ + OH^- \xrightarrow{\text{fast}} NH_2—NH_2 + H_2O$$

Path II There is formation of chloramide ion $^{\ominus}NHCl$ (in fast step) followed by nucleophilic attack of NH_3 (in slow step)

$$NH_2Cl + OH^- \xrightarrow{\text{fast}} {}^-NHCl + H_2O$$

$${}^-NHCl + NH_3 \xrightarrow{\text{slow}} N_2H_4 + Cl^-$$

A side reaction decreases yield of N_2H_4

$$N_2H_4 + 2NH_2Cl \xrightarrow{\text{fast}} N_2 + 2NH_4Cl$$

The reaction is catalysed by traces of heavy metal ions (*e.g.*, Cu^{2+}). Addition of gelatin or glue suppresses the side reaction by complexing the metal ions and yield of hydrazine is increased.

Industrially and laboratory method involve the same type of reactants. Urea may replace ammonia.

Properties

- Colourless fuming liquid with faint ammoniacal odour.

Melting point	: 20°C
Boiling p[oint	: 113.5°C
Density	: 1 g/cm^3 at 25°C
Dielectric constant	: 51.7 at 25°C
ΔH_f°	: 50.6 kJ mol^{-1} (an endothermic compound)
ΔG_f°	: 149.0 kJ mol^{-1}

- Pure hydrazine and its aqueous solution are kinetically stable inspite of its endothermic nature.
- **Basic Nature** It is a weaker base than NH_3 but its behaviour is **bifunctional.**

$$N_2H_4(aq) + H_2O \rightleftharpoons N_2H_5^+(aq) + OH^- \quad K_b = 8.5 \times 10^{-7}$$

$$N_2H_5^+(aq) + H_2O \rightleftharpoons N_2H_6^{2+} + OH^- \quad K_b = 8.9 \times 10^{-16}$$

$$NH_3 + H_2O \rightleftharpoons NH_4^+ + OH^- \quad K_b = 1.8 \times 10^{-5}$$

Hydrazinium salts of $N_2H_5^+$ are stable in water but those of $N_2H_6^{2+}$ are hydrolysed in aqueous solution.

Such salts as $N_2H_6^{2+}$ are less soluble than those of $N_2H_5^+$ and may be crystallised in presence of acid.

These series of salts are formed with H_2SO_4

$(N_2H_5)_2SO_4$ with $N_2H_5^+$ cation

$(N_2H_6)SO_4$ with $N_2H_6^{2+}$ cation

$(N_2H_6)(HSO_4)_2$ with $N_2H_6^{2+}$ cation

- **Reducing Nature : N_2H_4 is oxidised to N_2** and thus it is a better reducing agent.

In **acidic medium**

$$N_2H_5^+ \longrightarrow N_2 + 5H^+ + 4e^-$$

In **basic medium**

$$N_2H_4 + 4OH^- \longrightarrow N_2 + 4H_2O + 4e^-$$

$$\text{Equivalent mass} = \frac{\text{molar mass}}{4} = \frac{32}{4} = 8.0 \text{ g equivalent}^{-1}$$

– $AuCl_3$ is reduced to N_2

$$4AuCl_3 + 3N_2H_4 \longrightarrow 4Au + 3N_2 + 12HCl$$

– CuO is reduced to Cu_2O (red ppt)

$$4CuO + N_2H_4 \xrightarrow{\text{Cold}} 2Cu_2O + N_2 + 2H_2O$$

– IO_3^- is reduced to I^+

$$N_2H_4 + \underset{(O.N. = +5)}{KIO_3} + 2HCl \longrightarrow N_2 + KCl + \underset{(O.N. = +1)}{ICl} + 3H_2O$$

– $AgNO_3$ is reduced to Ag

$$4AgNO_3 + N_2H_4 \longrightarrow 4Ag + 4HNO_3 + N_2$$

– $PtCl_4$ is reduced to Pt

$$PtCl_4 + N_2H_4 \longrightarrow Pt + 4HCl + N_2$$

N_2H_4 is oxidised to NH_3 and N_2 or N_3^-

Medium

Acidic $\quad N_2H_5^+ \longrightarrow NH_4^+ + \frac{1}{2}N_2 + H^+ + e^-$

Basic $\quad N_2H_4 + OH^- \longrightarrow NH_3 + \frac{1}{2}N_2 + H_2O + e^-$

Equivalent mass = molar mass

Acidic $\quad N_2H_5^+ \longrightarrow \frac{1}{2}NH_4^+ + \frac{1}{2}N_3H + \frac{5}{2}H^+ + 2e^-$

Basic $\quad N_2H_5^+ + \frac{5}{2}OH^- \longrightarrow \frac{1}{2}NH_3 + \frac{1}{2}N_3^- + \frac{5}{2}H_2O + 2e^-$

$$\text{Equivalent mass} = \frac{\text{molar mass}}{2}$$

Fe^{3+} is reduced to Fe^{2+}	$Fe^{3+} + e^- \longrightarrow Fe^{2+}$
MnO_4^- is reduced to Mn^{2+}	$MnO_4^- + 8H^+ + 5e^- \longrightarrow Mn^{2+} + 4H_2O$

Oxidising Nature

In presence of a powerful reducing agent (as Zn/HCl), it is an oxidising agent

$$\underset{-4}{N_2H_4} + Zn + 2HCl \longrightarrow ZnCl_2 + \underset{-6}{2NH_3}$$

$$\text{Equivalent mass of } N_2H_4 = \frac{\text{molar mass}}{2}$$

As a Rocket Fuel Hydrazine being an endothermic compound burns rapidly and completely in air with evolution of heat

$$N_2H_4 + \underset{\text{air}}{O_2} \longrightarrow N_2 + 2H_2O \quad \Delta H = -620 \text{ kJ mol}^{-1}$$

Thus, N_2H_4 is used as a rocket fuel in guided missile, space shuttles and lunar missions.

Structure

Each N-atom is sp^3-hybridised and tetrahedrally surrounded by one nitrogen, two hydrogen and one lone pair of electrons. The molecule has a *gauche* conformation with an angle of rotation of 95° from a purely eclipsed position.

Uses

- As a rocket fuel; methyl derivatives $MeNHNH_2$ and Me_2NNH_2 are also used in combination with it. Oxidants are O_2 (liquid), N_2O_4 or H_2O_2.
- As a reducing agent to prepare copper and silver mirror; to prepare platinum metal.
- Isonicotinic hydrazine ($C_5H_4NCONHNH_2$) is an antitubercular drug.

Similarity Between Hydrazine and Hydrogen Peroxide

	N_2H_4	H_2O_2
1. Structures	H_2N-NH_2	H–O–O–H
2. Association	Form liquid and solid through hydrogen-bonding	
3. Hybridisation	N-atom is sp^3-hybridised	O-atom is sp^3-hybridised
4. Decomposition	$3N_2H_4 \longrightarrow N_2 + 4NH_3$	$2H_2O_2 \longrightarrow 2H_2O + O_2$
5. Oxidising nature	$N_2H_5^+ + 3H^+ + 2e^- \longrightarrow 2NH_4^+$	$H_2O_2 + 2H^+ + 2e^- \longrightarrow 2H_2O$
6. Reducing nature	$N_2H_5^+ \longrightarrow N_2 + 5H^+ + 4e^-$	$H_2O_2 \longrightarrow O_2 + 2H^+ + 2e^-$

Hydrogen Azide (Hydrazoic acid) N_3H

Preparation

1. By oxidation of aqueous hydrazine with nitrous acid
$$N_2H_4 + HNO_2 \longrightarrow N_3H + 2H_2O$$
Oxidising agent can be H_2O_2, HNO_3, ClO_3^- etc.
2. By careful distillation of sodium azide (NaN_3) with dilute H_2SO_4
$$NaN_3 + H_2SO_4 \longrightarrow N_3H + NaHSO_4$$

NaN_3 is formed by

$$N_2O + 2NaNH_2 \longrightarrow NaN_3 + NaOH + NH_3$$
$$NaNO_3 + 3NaNH_2 \longrightarrow NaN_3 + 3NaOH + NH_3$$
$$3N_2O + 4Na + NH_3 \longrightarrow NaN_3 + 3NaOH + 2N_2$$

Properties

(i) Colourless liquid, boiling point 37°C. The liquid and vapour are deadly poisonous and explode on heating
$$2N_3H \longrightarrow H_2 + 3N_2$$

(ii) Weak acid, stronger than acetic acid
$$N_3H + H_2O \rightleftharpoons H_3O^+ + N_3^- \quad K_a = 2.8 \times 10^{-5}$$
$$CH_3COOH + H_2O \rightleftharpoons H_3O^+ + CH_3COO^- \quad K_a' = 1.8 \times 10^{-5}$$
$$(K_a > K_a')$$

(iii) N_3H dissolves Mg forming H_2
$$Mg + 2N_3H \longrightarrow Mg(N_3)_2 + H_2$$

(iv) Metals like Cu, Zn, Fe form NH_3 and N_2 with N_3H
$$Cu + 3N_3H \longrightarrow Cu(N_3)_2 + N_2 + NH_3$$

(v) Reducing nature

(a) $$\underset{+3}{HNO_2} + \underset{-\frac{1}{3}}{N_3H} \longrightarrow \underset{0}{N_2} + \underset{+1}{N_2O} + H_2O$$
oxidation; reduction

N_3H may be estimated by titration with $NaNO_2$ solution using $FeCl_3$ as an indicator N_3H gives red colour with $FeCl_3$.

(b) $$\underset{+1}{HOCl} + \underset{-\frac{1}{3}}{2N_3H} \longrightarrow \underset{0}{3N_2} + \underset{-1}{HCl} + H_2O$$
oxidation; reduction

Structure The azide ion has very much similarity with halide ions and cyanide ion and is termed as **pseudo halide.**

H–N (1.01 Å), N–N 1.24 Å, N–N 1.13 Å, angle 120°

$$H-\ddot{N}-\overset{+}{N}\equiv N: \longleftrightarrow H-\ddot{N}=\overset{+}{N}=\overset{-}{N}:$$

Hydroxyl Amine (NH_2OH)

Preparation

It is stable in the form of acid salt as $[NH_3OH]^+Cl^-$ or $[NH_3OH]^+HSO_4^-$

1. NH_2OH in the form of salt is formed when ammonium nitrite (NH_4NO_2) reacts with ammonium hydrogen sulphite (NH_4HSO_3) in presence of sulphur dioxide (SO_2).

$$NH_4NO_2 + NH_4HSO_3 + SO_2 + 2H_2O \longrightarrow [NH_3OH]^+HSO_4^- + (NH_4)_2SO_4$$

2. Methyl nitrite (CH_3NO_2) changes to $[NH_3OH]^+HSO_4^-$ on reaction with H_2SO_4

$$CH_3NO_2 + H_2SO_4 \longrightarrow [NH_3OH]^+HSO_4^- + CO$$

Properties

1. Pure NH_2OH is colourless, odourless deliquescent crystals — melting point = 33° C

boiling point = 58° C at 72 mmHg

2. Dielectric constant = 78

highly soluble in water and alcohol but slightly in ether or benzene.

3. Its aqueous solution is less basic than NH_3 and N_2H_4

$$NH_2OH + H_2O \rightleftharpoons NH_3OH^+ + OH^- \quad K_b = 6.6 \times 10^{-9}$$

4. In absence of H_2O, it also behaves as a weak acid

$$NH_2OH \rightleftharpoons NH_2O^- + H^+$$

$$\underset{\text{acid}}{NH_2OH} + \underset{\text{base}}{Ca(OH)_2} \longrightarrow \underset{\text{salt}}{Ca(OH)(ONH_2)}$$

5. **NH_2OH as an oxidising agent**

NH_2OH can be reduced to NH_3 hence it can be used as an oxidising agent

Medium

Acidic $NH_2OH + 2H^+ + 2e^- \longrightarrow NH_3 + H_2O$

Basic $NH_2OH + H_2O + 2e^- \longrightarrow NH_3 + OH^-$

$$\text{Equivalent mass} = \frac{\text{molar mass}}{2}$$

(a) $SnCl_2$ is oxidised to $SnCl_4$ in acidic medium

$$SnCl_2 + NH_2OH + 2HCl \longrightarrow SnCl_4 + NH_3 + H_2O$$

(b) SO_2 is oxidised to SO_4^{2-} and $(NH_4)_2SO_4$ is the end product.

$$NH_2OH + 2H^+ + 2e^- \longrightarrow NH_3 + H_2O$$

$$SO_2 + 2H_2O \longrightarrow SO_4^{2-} + 4H^+ + 2e^-$$

$$2NH_3 + 2H^+ + SO_4^{2-} \longrightarrow (NH_4)_2SO_4$$

(c) HI is oxidised to I_2

$$NH_2OH + 2HI \longrightarrow NH_3 + H_2O + I_2$$

(d) $Fe(OH)_2$ is oxidised to $Fe(OH)_3$ in basic medium

$$2Fe(OH)_2 + NH_2OH + H_2O \longrightarrow 2Fe(OH)_3 + NH_3$$

(e) Na_3AsO_3 (sodium argenite) is oxidised to Na_3AsO_4 (sodium argenate) in basic medium

$$Na_3AsO_3 + NH_2OH \longrightarrow Na_3AsO_4 + NH_3$$

6. **NH_2OH as a reducing agent**

Medium

Acidic $2NH_2OH \longrightarrow N_2O + H_2O + 4H^+ + 4e^-$

Basic $2NH_2OH + 4OH^- \longrightarrow N_2O + 5H_2O + 4e^-$

$$\text{Equivalent mass} = \frac{\text{molar mass}}{4}$$

(a) CuO(II) is reduced to Cu_2O(I)

$$4CuO + 2NH_2OH \longrightarrow \underset{\text{red ppt}}{Cu_2O\downarrow} + N_2O + 3H_2O$$

(b) Fe^{3+}(III) is reduced to Fe^{2+}(II)

$$2NH_2OH + 4FeCl_3 \longrightarrow N_2O + 4FeCl_3 + H_2O + 4HCl$$

Uses

(i) As a reducing agent-antioxidant in photographic developers and reduction of Cu(II) in the dyeing of acrylic fibres.

(ii) In the preparation of caprolactam (as intermediate in the manufacture of nylon-6)

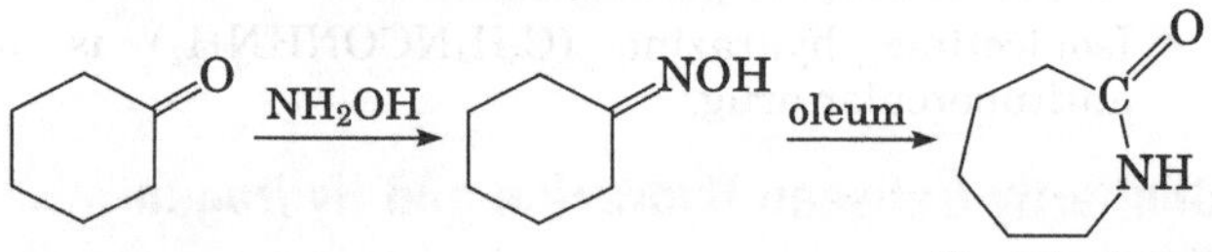

Cyclohexanone $\xrightarrow{NH_2OH}$ cyclohexanone oxime (=NOH) $\xrightarrow{\text{oleum}}$ Caprolactam (cyclic C=O / NH)

Caprolactam

Structure

N-atom is sp^3-hybridised with one lone pair of electrons. The N—O bond distance is `1.47Ã which corresponds to single bond.

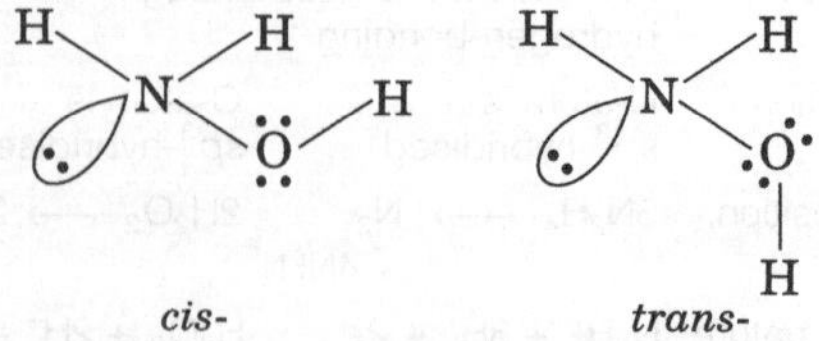

cis- trans-

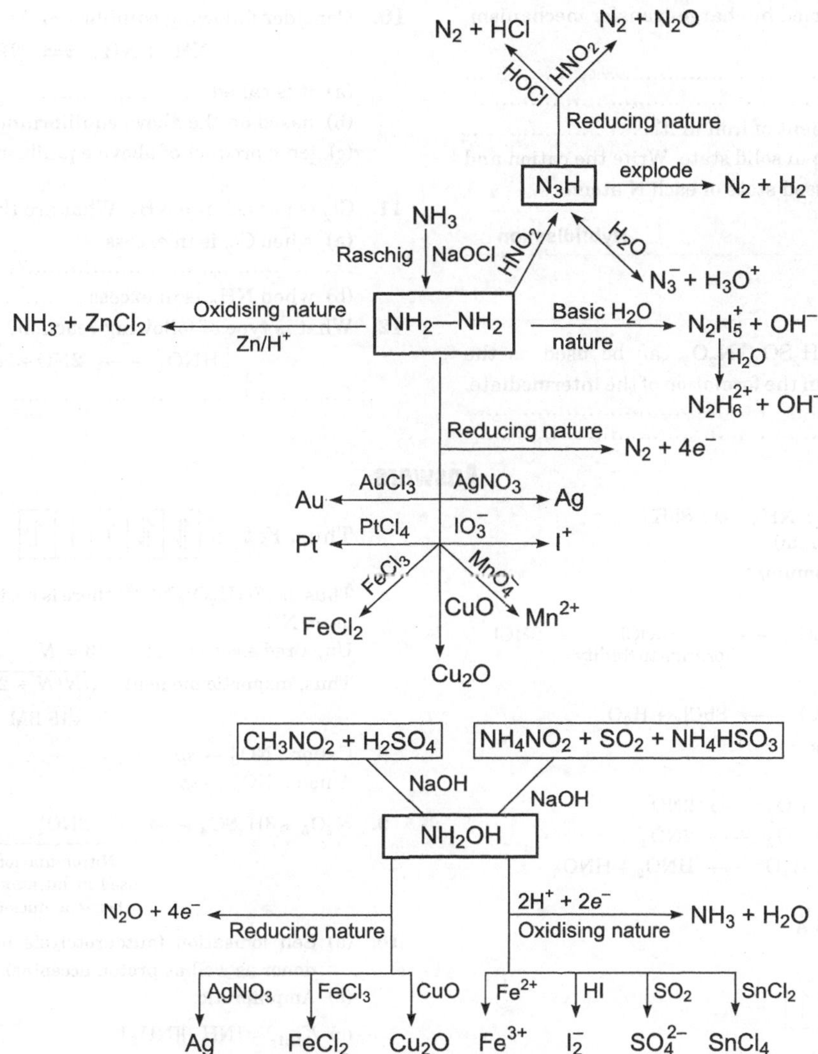

Target Practice 1

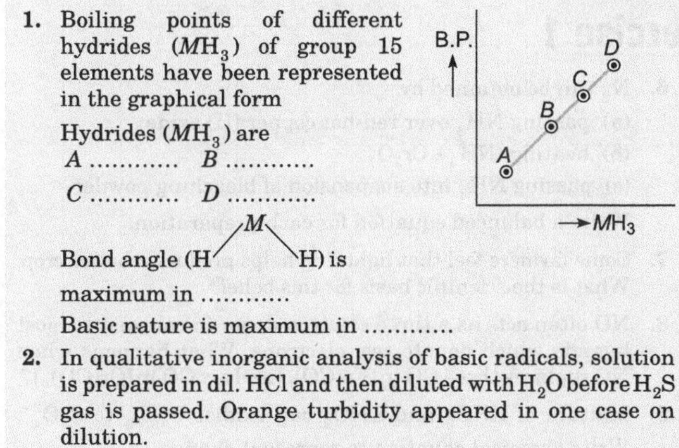

1. Boiling points of different hydrides (MH_3) of group 15 elements have been represented in the graphical form

Hydrides (MH_3) are

A .... *B* ....

C .... *D* ....

Bond angle (H–M–H) is maximum in ....

Basic nature is maximum in ....

2. In qualitative inorganic analysis of basic radicals, solution is prepared in dil. HCl and then diluted with H_2O before H_2S gas is passed. Orange turbidity appeared in one case on dilution.

(a) What do you conclude about basic radical? ....

(b) Turbidity dissolves when we add ....

3. Write full form of ATP ....

4. When there is thunderstorm accompanying lighting and rain, some oxoacids are formed.

Oxoacids are ....

and chemical reactions taking place are

....

....

5. Nitrite interferes in the oxygen transport ability of blood causing a disease known as ....

6. NO is a free radical. What is

(a) bond order? ....

(b) number of unpaired electrons? ....

(c) bond length between N and O? ....

7. $[Fe(H_2O)_5NO]^{2+}$ is formed by charge-transfer mechanism. Explain it.

....

....

What is magnetic moment of iron in it? ....

8. N_2O_5 is said to be ionic in solid state. Write the cation and anion in it and hybridised state of each N atom.

Species		Hybridisation
Cation	...	...
Anion	...	...

9. In presence of conc. H_2SO_4, N_2O_5 can be used as the nitrating agent. Explain the formation of the intermediate.

....

....

10. Consider following equilibrium

$$NH_3 + NH_3 \rightleftharpoons NH_4^+ + NH_2^-$$

(a) it is called ....

(b) based on the above equilibrium, NH_3 is ....

(c) ionic product of above equilibrium is written as ....

11. Cl_2 is passed into NH_3. What are the products

(a) when Cl_2 is in excess

....?

(b) when NH_3 is in excess ....?

12. What is type of following reaction?

$$3HNO_2 \longrightarrow 2NO + HNO_3 + H_2O$$

....

Answers

1. $A: PH_3$ $B: AsH_3$ $C: NH_3$ $D: SbH_3$
   NH_3 (bond angle maximum)
   NH_3 (basic nature maximum)
2. (a) Sb^{3+}

$$SbCl_3 + H_2O \longrightarrow \underset{\text{orange turbidity}}{SbOCl} + 2HCl$$

(b) HCl

$$SbOCl + 2HCl \longrightarrow SbCl_3 + H_2O$$

3. Adenosine tri-phosphate
4. HNO_2, HNO_3

$$N_2 + O_2 \longrightarrow 2NO$$
$$2NO + O_2 \longrightarrow 2NO_2$$
$$2NO_2 + H_2O \longrightarrow HNO_2 + HNO_3$$

5. Methemoglobinemia.
6. (a) 2.5 (b) 1 (c) 1.15 Å
7. Fe : [Ar] $4s^2 3d^6$

Fe^{2+} : | ⇅ | 1 | 1 | 1 | 1 | ($3d^6$)

NO donates its electrons to Fe^{2+} (charge transfer)

$$NO \longrightarrow NO^+ + e^-$$
$$Fe^{2+} + e^- \longrightarrow Fe^+$$

Thus, Fe^+ : | ⇅ | ⇅ | 1 | 1 | 1 |

Thus, in $[Fe(H_2O)_5NO]^{2+}$ there is positive charge on Fe as well as on NO.

Unpaired electron in $Fe^+ = 3 = N$

Thus, magnetic moment $= \sqrt{N(N+2)}$ BM

$= \sqrt{15}$ BM

8. Cation : NO_2^+ — sp
   Anion : NO_3^- — sp^2
9. $N_2O_5 + 3H_2SO_4 \longrightarrow \underbrace{2NO_2^+}_{\text{Nitronium ion used in nitration of benzene nucleus}} + 3HSO_4^- + H_3O^+$
10. (a) Self ionisation (autoprotolysis in which NH_3 is proton donor as well as proton acceptor).
    (b) Amphiprotic.
    (c) $K_{NH_3} = [NH_4^+][NH_2^-]$
11. (a) NCl_3, HCl ($NH_3 + 3Cl_2 \longrightarrow NCl_3 + 3HCl$)
    (b) NH_4Cl, N_2 ($8NH_3 + 3Cl_2 \longrightarrow 6NH_4Cl + N_2$)
12. It is a disproportionation reaction.

Practice Exercise 1

1. Give examples of compounds of N with oxidation number –3 to +5.
2. Nitrogen does not form NCl_5 or NF_5 but phosphorus can. Explain.
3. P_4 exists but N_4 does not; N_2 exists but P_2 does not. Explain.
4. What happens when PH_3 reacts with
   (a) Aqueous Ag^+ (b) Aqueous Cu^{2+} (c) O_2
5. There are two colourless solutions (*A*) and (*B*) both give white ppt with $AgNO_3$; ppt dissolves in *aq* NH_3. Also (*A*) on reaction with H_2O gives orange turbidity while (*B*) gives white turbidity. Identify (*A*) and (*B*). Suggest the reagent that can dissolve turbidity.
6. N_2 can be obtained by
   (a) passing NH_3 over red-hot copper (II) oxide
   (b) heating $(NH_4)_2Cr_2O_7$.
   (c) passing NH_3 into suspension of bleaching powder
   Write a balanced equation for each preparation.
7. Some farmers feel that lightning helps produce a better crop. What is the scientific basis for this belief?
8. NO often acts as a three electron donor in contrast to most ligands which donate two electrons. What happens when NO replaces three CO in $[Fe(CO)_5]$ and six CO in $[Cr(CO)_6]$?
9. Mixture of KNO_3 and KNO_2 can oxidise Cr_2O_3 to CrO_4^{2-}. Write chemical equation to represent change.

10. Complete the following:
(a) $MnO_4^- + NO + H^+ \longrightarrow$ (b) $SO_2 + H^+ + NO \longrightarrow$
(c) $Cr^{2+} + H^+ + NO \longrightarrow$ (d) $NaNO_3 + NH_4Cl \xrightarrow{\Delta}$

11. NH_3 gas is dried by CaO and not by P_2O_5 and H_2SO_4. Explain.

12. Identify (*A*), (*B*), (*C*) and (*D*) in the following:

$$Al + N_2 \xrightarrow{\Delta} (A) \xrightarrow{H_2O} (B) \xrightarrow{(C)} (D) \text{ (brown ppt).}$$

13. $NH_3 + O_2 \xrightarrow{Pt} (A)$
$(A) + O_2 \longrightarrow (B)$ (brown fumes)
$(B) + H_2O \longrightarrow (C) + (D)$ (both oxy acid)
$(C) + I^- \longrightarrow (E)$ (violet vapours)
Identify (*A*), (*B*), (*C*), (*D*) and (*E*).

Answers

1. NH_3 : −3; N_2H_4 : −2; NH_2OH : −1
N O
N_2 : 0 (zero); $H_2N_2O_2$: +1; NO : +2
$\left[\begin{matrix}N_2O_3 \\ HNO_2\end{matrix}\right.$: +3; $\left[\begin{matrix}NO_2 \\ N_2O_4\end{matrix}\right.$: +4; $\left[\begin{matrix}N_2O_5 \\ HNO_3\end{matrix}\right.$: +5

2. Nitrogen ($1s^2 2s^2 2p^3$) lacks 2*d*-orbitals in which electrons of 2*s* can be excited to make five orbitals unpaired. Only three electrons remain unpaired hence only NX_3 is formed. In case of P($1s^2 2s^2 2p^6 3s^2 3p^3$), 3*s* electron can be excited into 3*d* vacant orbitals thus making five unpaired electrons, thus PX_5 is possible.

3. 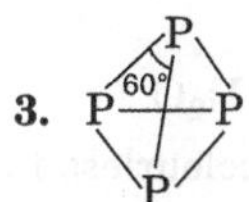

(white phosphorus : P_4)

N≡N 946 kJ mol^{-1}, P≡P 481 kJ mol^{-1}
N—N 160 kJ mol^{-1}, P—P 215 kJ mol^{-1}
Greater the B.E., greater the stability [compare BE of N≡N and P≡P and N—N (in N_4) and P—P (in P_4)]

4. (a) $PH_3 + 6Ag^+ + 3H_2O \longrightarrow H_3PO_3 + 6Ag + 6H^+$
(b) $PH_3 + 4Cu^{2+} + 4H_2O \longrightarrow H_3PO_4 + 6Cu + 8H^+$
(c) $4PH_3 + 8O_2 \longrightarrow P_4O_{10} + 6H_2O$

5. (*A*) : $SbCl_3$ (*B*) : $BiCl_3$

$$SbCl_3 \text{ or } BiCl_3 + AgNO_3 \longrightarrow \underset{\text{white ppt}}{AgCl} \downarrow \text{ (due to } Cl^-)$$

$$AgCl + 2NH_3\,(aq) \longrightarrow \underset{\text{soluble}}{[Ag(NH_3)_2]\,Cl}$$

$$SbCl_3 + H_2O \longrightarrow \underset{\text{orange}}{SbOCl} \downarrow + 2HCl$$

$$SbOCl + 2HCl \longrightarrow \underset{\text{soluble}}{SbCl_3} + H_2O$$

$$BiCl_3 + H_2O \longrightarrow \underset{\text{white}}{BiOCl} \downarrow + 2HCl$$

$$BiOCl + 2HCl \longrightarrow \underset{\text{soluble}}{BiCl_3} + H_2O$$

Turbidity can be dissolved in dil. HCl.

6. (a) $3CuO + 2NH_3 \longrightarrow 3Cu + N_2 + 3H_2O$
(b) $(NH_4)_2Cr_2O_7 \longrightarrow N_2 + 4H_2O + Cr_2O_3$
(c) $3CaOCl_2 + 2NH_3 \longrightarrow 3CaCl_2 + 3H_2O + N_2$

7. Lightning forms NO which is converted to NO_2, HNO_2 and HNO_3 under atmospheric conditions. Thus, NO_3^- is a source of nutrients (fertilisers) to the plant. Thus, there are chances of better crops when lightning occurs.

$$N_2 + O_2 \xrightarrow{\text{lightening}} 2NO$$

$$2NO + O_2 \longrightarrow 2NO_2$$

$$2NO_2 + H_2O \longrightarrow HNO_2 + HNO_3$$

8. $Fe(CO)_5 + 2NO \longrightarrow [Fe(CO)_2(NO)_2] + 3CO$
CO donates two electrons while NO three electrons, hence electrons lost (due to CO) = six
electrons gained (due to NO) = six
Similarly, $Cr(CO)_6 + 4NO \longrightarrow [Cr(NO)_4] + 6CO$

9. $3KNO_2 + KNO_3 + Cr_2O_3 \xrightarrow{\Delta} 2K_2CrO_4 + 4NO$
(but KNO_2 and KNO_3 are in molten states)

10. (a) $3MnO_4^- + 5NO + 4H^+ \longrightarrow 5NO_3^- + 3Mn^{2+} + 2H_2O$
(b) $SO_2 + 2NO + H_2O \longrightarrow N_2O + SO_4^{2-} + 2H^+$
(c) $3Cr^{2+} + NO + 3H^+ \longrightarrow NH_2OH + 3Cr^{3+}$
(d) $NaNO_3 + NH_4Cl \longrightarrow NH_4NO_3 + NaCl$
$\xrightarrow{\Delta} N_2O + 2H_2O$

11. Moist NH_3 can react with P_2O_5 forming phosphate and with conc. H_2SO_4 forming sulphates :

$$6NH_3 + P_2O_5 + 3H_2O \longrightarrow 2(NH_4)_3PO_4$$

$$2NH_3 + H_2SO_4 \longrightarrow (NH_4)_2SO_4$$

hence can't be used for drying NH_3. CaO being basic would only cause drying of NH_3.

12. (*A*) : AlN (*B*) : NH_3 (*C*) : K_2HgI_4/OH^- (Nessler's reagent)
(D) : H_2N—Hg—O—Hg—I (Iodide of Millon's base)

or $O\left\langle\begin{matrix}Hg \\ Hg\end{matrix}\right\rangle NH_2I$

13. (*A*) : NO (*B*) : NO_2 (*C*) : HNO_2 (*D*) : HNO_3 (*E*) : I_2

Oxyacids of Nitrogen

Acid	Formula	Ox. no.	Anhydride
Hyponitrous acid	$H_2N_2O_2$	+1	N_2O*
Nitrous acid	HNO_2	+3	N_2O_3
Nitric acid	HNO_3	+5	N_2O_5

* $$H_2N_2O_2 \xrightarrow{\Delta} H_2O + N_2O$$

$$N_2O + H_2O \longrightarrow \text{(no acid formation)}$$

$H_2N_2O_2$ gives N_2O (anhydride) only on heating but N_2O when dissolved in H_2O will not form $H_2N_2O_2$.

Hyponitrous Acid ($H_2N_2O_2$)

- It is obtained starting from $NaNO_2$:

$$NaNO_2 \xrightarrow[\text{reduction}]{\text{Na—Hg}} Na_2N_2O_2 \xrightarrow{AgNO_3} AgN_2O_2 \xrightarrow[-AgCl]{HCl} H_2N_2O_2$$

- Colourless leaflets hygroscopic crystals.
- In aqueous solution it decomposes :

$$H_2N_2O_2 \longrightarrow H_2O + N_2O$$

- Very weak dibasic acid but a strong reducing agent :
  - I_2 is reduced to HI

$$H_2N_2O_2 + 3I_2 + 3H_2O \longrightarrow HNO_3 + HNO_2 + 6HI$$

  - MnO_4^- is reduced to Mn^{2+}

$$8MnO_4^- + 5H_2N_2O_2 + 24H^+ \longrightarrow 10HNO_3 + 8Mn^{2+} + 12H_2O$$

- It has zero dipole moment which is based on its *trans* structure :

$$HO{-}N{=}N{-}OH \text{ (trans)}$$

Nitrous Acid (HNO_2)

- Nitrous acid is unstable except in dilute solution. It is easily made by acidifying a solution of a nitrite :

$$Ba(NO_2)_2 + H_2SO_4 \longrightarrow BaSO_4 + 2HNO_2$$

$$NaNO_2 + HCl \longrightarrow NaCl + HNO_2$$

It appears as a pale blue solution (due to N_2O_3) when nitrite is dissolved in acid. To prevent its decomposition, it must be kept below 5° C.

- $$N_2O_3 + H_2O \longrightarrow 2HNO_2$$
- $$NH_3 + 3H_2O_2 \longrightarrow HNO_2 + 4H_2O$$
- It is a weak acid ($K_a = 4.5 \times 10^{-5}$) and

$$HNO_2 + NaOH \longrightarrow NaNO_2 + H_2O$$

- It disproportionates to HNO_3 (oxidation) and NO (reduction) even in cold solution

$$\underset{+3}{3HNO_2} \longrightarrow \underset{+2}{2NO} + \underset{+5}{HNO_3} + H_2O$$

(+3 → +2: redn.; +3 → +5: oxidn.)

- **As an oxidising agent :**

$$HNO_2 + H^+ + e^- \longrightarrow NO + H_2O \quad ...(I)$$

  - it oxidises I^- to I_2

$$2I^- \longrightarrow I_2 + 2e^-$$

  - it oxidises H_2S to S

$$H_2S \longrightarrow 2H^+ + S + 2e^-$$

  - it oxidises SO_2 to SO_4^{2-}

$$SO_2 + 2H_2O \longrightarrow SO_4^{2-} + 4H^+ + 4e^-$$

  - it oxidises Fe^{2+} to Fe^{3+}

$$Fe^{2+} \longrightarrow Fe^{3+} + e^-$$

Note : Oxidation half-reaction may be added to reduction half reaction (I) to cancel electron in final reaction.

- **As a reducing agent :**

$$HNO_2 + H_2O \longrightarrow NO_3^- + 3H^+ + 2e^- \quad ...(II)$$

  - It reduces acidified MnO_4^- (purple) to Mn^{2+} (colourless)

$$MnO_4^- + 8H^+ + 5e^- \longrightarrow Mn^{2+} + 4H_2O$$

Note : Mn^{2+} salt are pink-coloured but appears colourless in aqueous solution.

  - it reduces acidified $Cr_2O_7^{2-}$ (orange) to Cr^{3+} (green)

$$Cr_2O_7^{2-} + 14H^+ + 6e^- \longrightarrow 2Cr^{3+} + 7H_2O$$

  - it reduces H_2O_2 to H_2O

$$H_2O_2 + 2H^+ + 2e^- \longrightarrow 2H_2O$$

- Aliphatic amines and amides on reaction with HNO_2 give N_2 gas

$$RNH_2 + HNO_2 \longrightarrow ROH + N_2 + H_2O$$

$$\underset{\text{urea}}{NH_2CONH_2} + 2HNO_2 \longrightarrow 2N_2 + CO_2 + 3H_2O$$

This provides the basis of Duma's method of estimation of nitrogen quantitatively in organic compounds.

- However, aromatic amines form diazonium salts at low temperature (diazotisation)

$$C_6H_5{-}NH_2{\cdot}HCl + HNO_2 \xrightarrow{0°C} C_6H_5{-}N_2Cl + 2H_2O$$

- It behaves as a monobasic acid and has structure

$$H{-}O{-}\ddot{N}{=}O$$

Microwave spectroscopic studies have shown that nitrous acid in the gaseous state has *trans* configuration :

H—O: 0.95 Å; O—N: 1.43 Å; N=O: 1.18 Å

$\angle\alpha = 110.7°$

$\angle\beta = 102.1°$

NO_2^- has planar triangular structure, with N at the centre, two corners occupied by O atoms, and the third corner occupied by the lone pair.

Nitrite ion can exist in the following resonance structures :

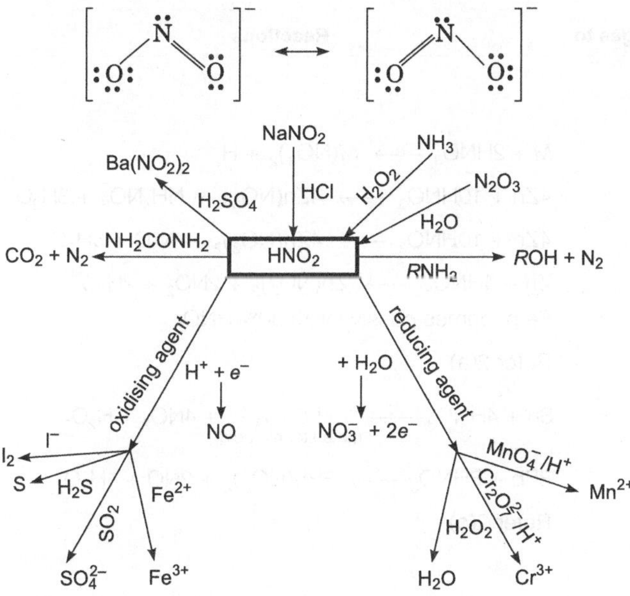

Nitric Acid (HNO_3)

- $$KNO_3 + \underset{\text{conc}}{H_2SO_4} \xrightarrow{\Delta} KHSO_4 + HNO_3$$

On condensation, vapours of HNO_3 changes to liquid state (a laboratory method).

- **Birkeland-Eyde process** involves electric sparking of N_2 and O_2 and passing of the gas into H_2O.

$$N_2 + O_2 \overset{\text{spark}}{\rightleftharpoons} 2NO, \quad \Delta H = 181 \text{ kcal mol}^{-1}$$

$$2NO + O_2 \longrightarrow 2NO_2$$

$$2NO_2 + H_2O \longrightarrow HNO_2 + HNO_3$$

$$3HNO_2 \longrightarrow HNO_3 + H_2O + 2NO$$

- **Ostwald process** depends on the catalytic oxidation of NH_3 to NO, followed by oxidation of NO to NO_2, and conversion of NO_2 with H_2O to HNO_3.

$$4NH_3 + 5O_2 \xrightarrow[850°C]{Pt/Rh, 5 \text{ atm}} 4NO + 6H_2O$$

The NO and air are cooled and the mixture of gases is absorbed in a counter current of water.

$$2NO + O_2 \rightleftharpoons 2NO_2$$

$$2NO_2 + H_2O \longrightarrow HNO_3 + HNO_2$$

$$2HNO_2 \longrightarrow H_2O + NO_2 + NO$$

$$3NO_2 + H_2O \longrightarrow 2HNO_3 + NO$$

Overall reaction being

$$\mathbf{NH_3 + 2O_2 \longrightarrow HNO_3 + H_2O}$$

Acid is 60% by weight and concentrated HNO_3 (98% by weight) is obtained by dehydration with conc. H_2SO_4. Fuming HNO_3 contains dissolved NO_2 in concentrated nitric acid.

- Anhydrous HNO_3 is colourless fuming and pungent smelling liquid, soluble in water; boiling point = 84.1° C and freezing point = 41.55° C. On account of photochemical decomposition (in sunlight), it is often of yellow colour.

$$2HNO_3 \xrightarrow{h\nu} 2NO_2 + H_2O + \frac{1}{2}O_2$$

- Its auto-ionisation gives NO_2^+ (nitronium ion) and NO_3^-

$$2HNO_3 \longrightarrow NO_2^+ + NO_3^- + H_2O$$

- It is a strong monobasic acid and forms nitrates with carbonates, oxides and hydroxides.

- HNO_3 is a powerful **oxidising agent :**

$$\underset{\text{concentrated}}{2NO_3^-} + 4H^+ + 2e^- \longrightarrow 2NO_2 + 2H_2O$$

$$\underset{\text{dilute}}{NO_3^-} + 4H^+ + 3e^- \longrightarrow NO + 2H_2O$$

$$\underset{\text{very dilute}}{NO_3^-} + 10H^+ + 8e^- \longrightarrow NH_4^+ + 3H_2O$$

Thus oxidising power of HNO_3 depends on the extent of dilution.

$$\underset{+5}{NO_3^-} \longrightarrow \underset{+4}{NO_2} \longrightarrow \underset{+2}{NO} \longrightarrow \underset{-3}{NH_3}$$

- Non-metals are oxidised to their oxyacids by conc. HNO_3.

$$C \longrightarrow H_2CO_3 \text{ (carbonic acid)}$$

$$S \longrightarrow H_2SO_4 \text{ (sulphuric acid)}$$

$$P \longrightarrow H_3PO_4 \text{ (o-phosphoric acid)}$$

$$I_2 \longrightarrow HIO_3 \text{ (iodic acid)}$$

One reaction is given to explain, others may be written similarly.

$$[2NO_3^- + 4H^+ + 2e^- \longrightarrow 2NO_2 + 2H_2O] \times 3$$

$$S + 4H_2O \longrightarrow SO_4^{2-} + 8H^+ + 6e^-$$

$$S + 6NO_3^- + 4H^+ \longrightarrow 6NO_2 + SO_4^{2-} + 2H_2O$$

Metals are oxidised forming nitrates and NO_2, NO, N_2O, N_2 or NH_3 depending on the nature of the metal, the temperature and the concentration of the acid. Reactions of metals with HNO_3 are summarised in Table 13.3.

DEEP Focus

- **Noble metals like Au, Pt, Rh and Ir are not attacked by nitric acid. Aqua regia, which is a mixture of 75% conc HCl and 25% conc. HNO_3 can dissolve Au and Pt. The enhanced ability to dissolve metals shown by aqua-regia arises from the oxidising power of HNO_3 coupled with the ability of Cl^- to form complexes with the metal ions.**

$$\underbrace{HNO_3 + 3HCl}_{\text{aqua-regia}} \longrightarrow 2H_2O + 2Cl + NOCl$$

$$Au + 3Cl \longrightarrow AuCl_3$$

$$AuCl_3 + HCl \longrightarrow \underset{\text{chloro auric acid}}{HAuCl_4}$$

$$Pt + 4Cl \longrightarrow PtCl_4$$

$$PtCl_4 + 2HCl \longrightarrow \underset{\text{chloro platinic acid}}{H_2PtCl_6}$$

Table 13.3 **Reactions of Elements (Metals/Metalloids with HNO_3)**

Element	Nature of HNO_3	HNO_3 changes to	Reactions
(A) Metals placed above H in electrochemical series (ECS)			
1. Mg, Mn	dilute	$M(NO_3)_2$	$M + 2HNO_3 \longrightarrow M(NO_3)_2 + H_2$
2. Zn, Fe	(a) very dilute	NH_4NO_3	$4Zn + 10HNO_3 \longrightarrow 4Zn(NO_3)_2 + NH_4NO_3 + 3H_2O$
	(b) dilute	N_2O	$4Zn + 10HNO_3 \longrightarrow 4Zn(NO_3)_2 + N_2O + 5H_2O$
	(c) conc	NO_2	$Zn + 4HNO_3 \longrightarrow Zn(NO_3)_2 + 2NO_2 + 2H_2O$ Fe becomes passive with 80% HNO_3
3. Sn	(a) dilute	NH_4NO_3	Refer 2(a).
	(b) conc	NO_2	$Sn + 4HNO_3 \longrightarrow \underset{\text{meta stannic acid}}{H_2SnO_3} + 4NO_2 + H_2O$
4. Pb	(a) dilute	NO	$3Pb + 8HNO_3 \longrightarrow 3Pb(NO_3)_2 + 2NO + 4H_2O$
	(b) conc	NO_2	Refer 2(c).
(B) Metals below H in ECS			
5. Cu, Ag	(a) dilute	NO	Refer 4(a). Hg forms $Hg_2(NO_3)_2$
Hg	(b) conc	NO_2	Refer 4(c).
(C) Metalloids			
Sb, As	conc	NO_2	$Sb + 5HNO_3 \longrightarrow \underset{\text{antimonic acid}}{H_3SbO_4} + 5NO_2 + H_2O$

- **HNO_3 can also be used to oxidise compounds :**

$$2HNO_3 \longrightarrow H_2O + 2NO_2 + O$$

$$H_2S \longrightarrow S + H_2O$$

$$SO_2 \longrightarrow SO_4^{2-}$$

$$Fe^{2+} \longrightarrow Fe^{3+}$$

$$FeS \longrightarrow FeSO_4$$

$$\underset{\text{cane sugar}}{C_{12}H_{22}O_{11}} \longrightarrow \underset{\text{oxalic acid}}{(COOH)_2}$$

- **Structure can be represented as:**

$$H{-}\ddot{O}{-}N(=O)(-O^-) \;\longleftrightarrow\; H{-}\ddot{O}{-}N(-O^-)(=O)$$

(N–O 1.22 Å; O–N 1.41 Å)

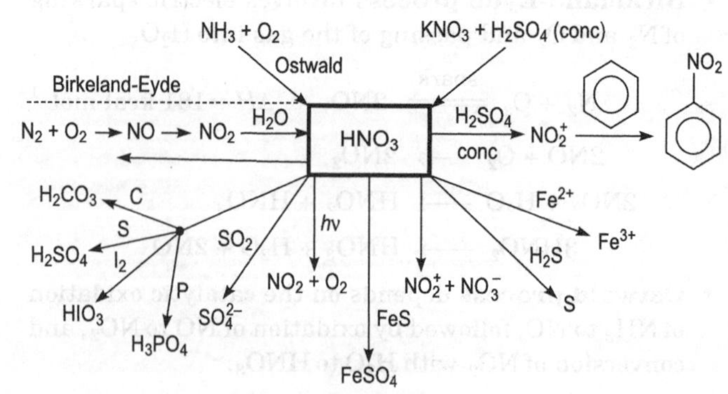

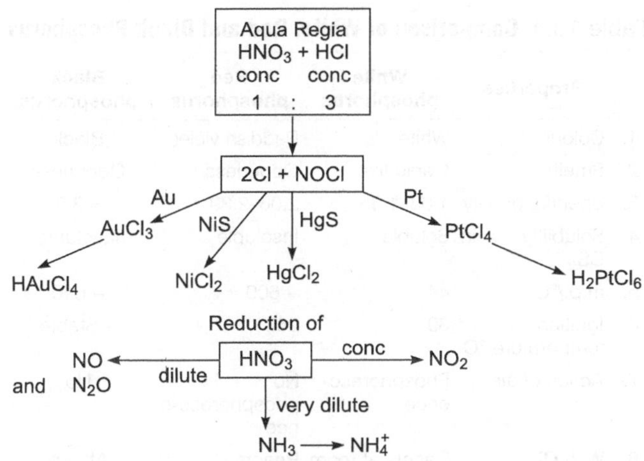

$NaNO_2$ and $NaNO_3$ as Food Additives

The brown colour of "old" meat is the result of oxidation of blood and is objectionable to many consumers. Nitrites and nitrates are added to food to retard this oxidation and also to prevent growth of botulism bacteria. NO_3^- ions are reduced to NO_2^- ions, which are then converted to NO. This in turn reacts with the brown oxidised form of the heme in blood. This reaction keeps meat red longer. However, there are doubts that nitrites may combine with amines under the acidic conditions in the stomach to give carcinogenic nitrosoamines.

$$HNO_2 + R_2NH \longrightarrow \underset{\text{nitrosoamine}}{R_2N{-}N{=}O} + H_2O$$

Study of Phosphorus and its Compounds

Isolation of Phosphorus

Phosphorus is the most abundant element of group 15(VA), accounting for **0.1%** of the mass of the earth's crust. Phosphorus is also important in living systems and is the **sixth** most abundant element in the human body. Our bones are mostly $Ca_3(PO_4)_2$, and tooth enamel is almost pure $Ca_5(PO_4)_3OH$. Phosphate groups are an integral part of the nucleic acid DNA and RNA, the molecules that pass genetic information from generation to generation.

Minerals

Phosphorite	:	$Ca_3(PO_4)_2$
Fluorapatite	:	$3Ca_3(PO_4)_2 \cdot CaF_2$ or $Ca_5(PO_4)_3F$
Chlorapatite	:	$3Ca_3(PO_4)_2 \cdot CaCl_2$
Aluminium phosphate	:	$AlPO_4$
Ferric phosphate	:	$FePO_4$

Phosphorus exists in three common **allotropic** forms
(i) white phosphorus
(ii) red phosphorus
(iii) black phosphorus

(i) White phosphorus In earlier days, white phosphorus was obtained by the distillation of metaphosphoric acid (HPO_3) with carbon at high temperature. HPO_3, in turn was obtained by decomposing bone-ash [$Ca_3(PO_4)_2$] with 60% H_2SO_4.

$$Ca_3(PO_4)_2 + H_2SO_4 \longrightarrow CaSO_4 \downarrow + H_3PO_4$$

$$H_3PO_4 \xrightarrow{\text{Evaporate}} HPO_3 + H_2O$$

$$4HPO_3 + 10C \longrightarrow P_4 + 10CO + 2H_2O$$

$Ca_3(PO_4)_2$ —60% H_2SO_4→ Filter → H_3PO_4
Filter → $CaSO_4$
H_3PO_4 → Evaporate → H_2O
Evaporate → HPO_3 —C, Δ→ P_4 → CO

In modern electrochemical process, white phosphorus is manufactured by heating a mixture of $Ca_3(PO_4)_2$, sand and coke at 1500°C in a furnace by means of an electric arc (Fig. 13.3). The reactions are:

$$2Ca_3(PO_4)_2 + 6SiO_2 \longrightarrow 6CaSiO_3 + P_4O_{10}$$

$$P_4O_{10} + 10C \longrightarrow P_4 + 10CO \uparrow$$

$Ca_3(PO_4)_2 + SiO_2 + C$ —In electric arc, 1500°C→ P_4 + CO
→ $CaSiO_3$
P_4 + CO → H_2O → CO
H_2O → P_4 in H_2O —250°C, Δ, air absent→ Red phosphorus

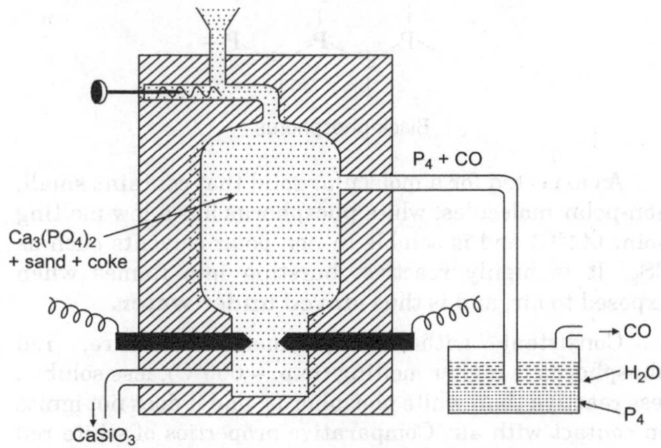

Fig. 13.3 Manufacture of phosphorus by electrochemical process

The calcium silicate forms a slag and is run off through an outlet at the bottom of the furnace. Phosphorus vaporus issuing through an opening at the top of the furnace is condensed under water.

(ii) Red phosphorus Red phosphorus is prepared by heating white phosphorus in the absence of air at 250°C for a few hours with a trace of iodine at catalyst in a closed iron vessel fitted with a safety valve. At the end of conversion, the product is ground under water and treated with hot caustic soda solution to eliminate any unchanged white phosphorus. It is finally washed with hot water and dried by heating with steam.

(iii) Black phosphorus It is obtained by heating white phosphorus at 470 K under high pressure. A series of phases of black phosphorus are formed and one of these phases consists of an extended layer structure in which each phosphorus is bonded to three neighbours by single covalent bonds. Thermodynamically, black phosphorus is the most stable form.

White phosphorus, the form produced in the industrial synthesis, is a toxic, waxy, white solid that contains discrete tetrahedral P_4 molecules. Red phosphorus, by contrast, is essentially nontoxic and has a polymeric structure:

White phosphorus

Red phosphorus

Black phosphorus

As expected for a molecular solid that contains small, non-polar molecules, white phosphorus has a low melting point (44°C) and is soluble in non-polar solvents such as CS_2. It is highly reactive, bursting into flames when exposed to air, and is thus **stored under water.**

Consistent with its polymeric structure, red phosphorus is higher melting (m.p. ~ 600°C), less soluble, less reactive than white phosphorus, and it does not ignite on contact with air. Comparative properties of white red and black phosphorus have been summarised, in Table 13.4.

Table 13.4 **Comparison of White, Red and Black Phosphorus**

Properties	White phosphorus	Red phosphorus	Black phosphorus
1. Colour	White	Reddish-violet	Black
2. Smell	Garlic-like	Odourless	Odourless
3. Specific gravity	1.83–1.85	2.06–2.39	≈ 3.9
4. Solubility in CS_2	Soluble	Insoluble	Insoluble
5. m.p./°C	44	~ 600	~ 610
6. Ignition temperature/°C	30	260	~ Stable
7. Action of air	Phosphoresc-ence	No phosphoresc-ence	No
8. With Cl_2	Reacts at room temperature	Reacts only when heated	At m.p
9. With NaOH	PH_3 is formed	No reaction	No reaction
10. Physiological effect	Poisonous	Non-poisonous	Non-poisonous

Illustration 1 *Explain the high reactivity of white phosphorus.*

Solution The high reactivity of white phosphorus is due to an unusual bonding that produces considerable strain in the P_4 molecules. If each P atom uses 3*p* orbitals to form three P—P bonds, all the bond angles should be 90°. The geometry of P_4, however, requires that all the bonds have 60° angles, which means that the *p*-orbitals cannot overlap in a head-on-fashion. As a result, the P—P bonds are "bent", relatively weak, and highly reactive.

Note : *White phosphorus was once used in matches, but because of its toxicity it has been replaced by* P_4S_3.

Compounds of Phosphorus

DEEP Focus

- Phosphorus and the other members of the group can make use of *d*-orbitals in their bonding, and thus +5 oxidation state is shown beside +3 which is common in all the compounds. Thus, PCl_5 is formed while NCl_5 is not formed.

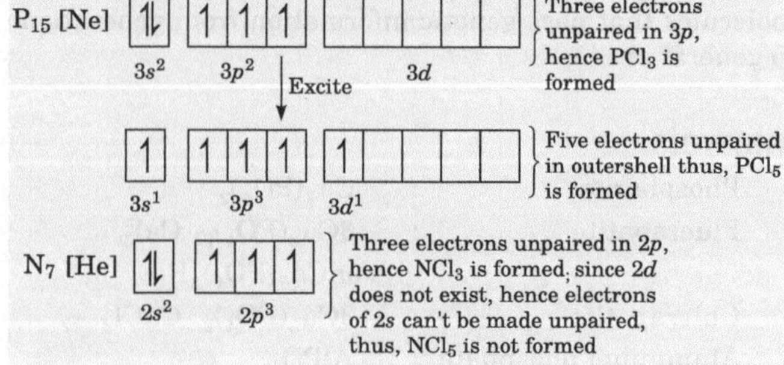

Phosphine (PH_3)

- Phosphine can be formed by :

$$\underset{\text{calcium phosphide}}{Ca_3P_2} + 6H_2O \longrightarrow 2PH_3 + 3Ca(OH)_2$$

$$P_4 + 3NaOH + 3H_2O \xrightarrow{\text{warm}} PH_3 + 3NaH_2PO_2$$

It is necessary to exclude air from the apparatus because, although pure PH_3 will not burn unless ignited, it is often accompanied with P_2H_4 and H_2 which are spontaneously flammable.

$$2P + 2NaOH + 2H_2O \longrightarrow 2NaH_2PO_2 + H_2\uparrow$$

$$6P + NaOH + 4H_2O \longrightarrow 4NaH_2PO_2 + P_2H_4$$

- $$2AlP + 3\underset{\text{dil.}}{H_2SO_4} \longrightarrow 2PH_3 + Al_2(SO_4)_3$$
- It is a colourless, highly reactive and extremely toxic gas, which smells slightly of garlic or bad fish.
- **It is better reducing agent than NH_3**
  - Ag^+ is reduced to Ag metal

$$6Ag^+ + PH_3 + 3H_2O \longrightarrow H_3PO_3 + 6Ag\downarrow + 6H^+$$

  - Cu^{2+} is reduced to Cu

$$4Cu^{2+} + PH_3 + 4H_2O \longrightarrow H_3PO_4 + 4Cu\downarrow + 8H^+$$

- PH_3, unlike NH_3, is not very soluble in water. Its aqueous solution is neutral.
- It is more soluble in CS_2 and other organic solvents.
- Anhydrous PH_3 and anhydrous HCl (not in aqueous solution) combine to form $PH_4^+Cl^-$
- Pure PH_3 is stable in air, but it catches fire when heated to about 150°C.

$$PH_3 + 2O_2 \longrightarrow H_3PO_4$$

- In liquid NH_3, phosphine dissolves to give $NH_4^+PH_2^-$

$$\underset{\text{base}}{NH_3} + \underset{\text{acid}}{PH_3} \longrightarrow NH_4^+PH_2^-$$

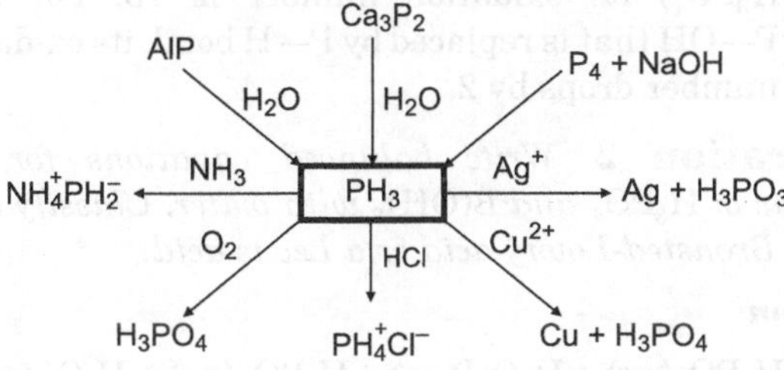

Oxides of Phosphorus

Two important oxides of phosphorus are known :

Oxides	Oxoacids	Oxidation number
P_4O_6	H_3PO_3	+3
P_4O_{10}	H_3PO_4	+5

The oxides of phosphorus react vigorously with H_2O to give acidic solution :

$$P_4O_6 + 6H_2O \longrightarrow \underset{o\text{-phosphorous acid}}{4H_3PO_3}$$

$$P_4O_{10} + 6H_2O \longrightarrow \underset{o\text{-phosphoric acid}}{4H_3PO_4}$$

This takes place in stages :

$$P_4O_{10} + 2H_2O \longrightarrow 4HPO_3$$
$$4HPO_3 + 2H_2O \longrightarrow 2H_4P_2O_7$$
$$2H_4P_2O_7 + 2H_2O \longrightarrow 4H_3PO_4$$

DEEP Focus

- **P_4O_{10} is a white flocculent powder that has greater affinity for water hence, is used as dehydrating agent in:**

$$4HNO_3 + P_4O_{10} \longrightarrow 4HPO_3 + 2N_2O_5$$
$$2CH_3CONH_2 + P_4O_{10} \longrightarrow 2CH_3CN + 4HPO_3$$
$$4CH_3COOH + P_4O_{10} \longrightarrow 2(CH_3CO)_2O + 4HPO_3$$
$$2H_2SO_4 + P_4O_{10} \longrightarrow 2SO_3 + 4HPO_3$$

- $$6CaO + P_4O_{10} \longrightarrow 2Ca_3(PO_4)_2$$
- $$10C + P_4O_{10} \longrightarrow 4P + 10CO\uparrow$$
- P_4O_{10} reacts with alcohols and ethers forming phosphate esters:

$$P_4O_{10} + 6EtOH \longrightarrow 2O{=}P(OH)(OH)(OEt) + 2O{=}P(OEt)(OH)(OEt)$$

$$P_4O_{10} + 6Et_2O \longrightarrow 2O{=}P(OEt)_3$$

(Formation of esters confirms that H_3PO_4 has structure $O{=}P(OH)_3$)

- The basic building block of both of them is a tetrahedron of P-atoms. In P_4O_6; oxygen atoms act as a bridge between the corners of the tetrahedron. In P_4O_{10}, each P-atom is also bonded to an extra oxygen atom. **Each oxide (P_4O_6, P_4O_{10}) has six POP bonds.**

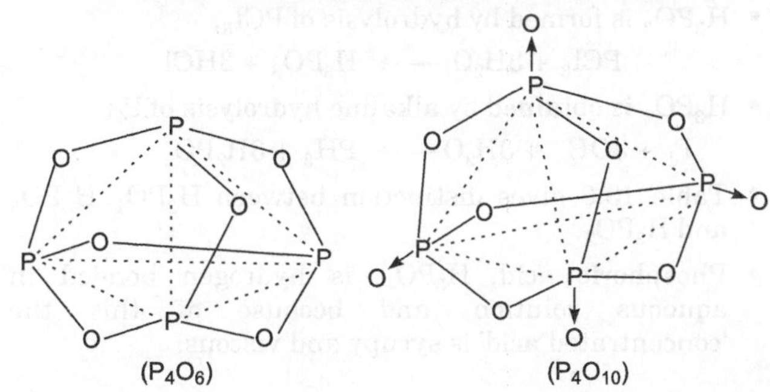

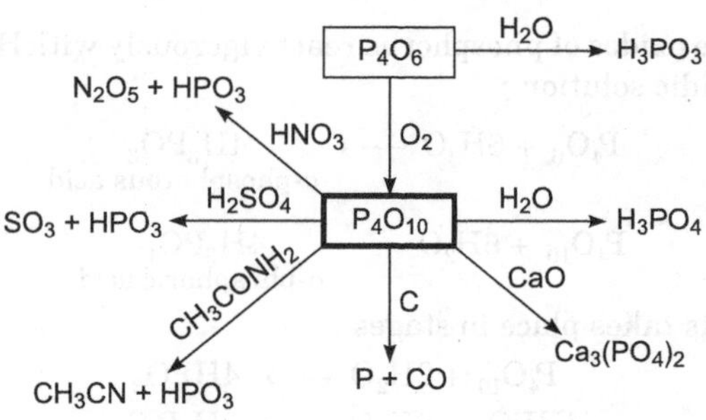

Oxoacids of Phosphorus

- Following oxoacids of phosphorus are commonly known :

Table 13.5

Acid	Basicity	Name	Oxidation no.
H_3PO_2	1	Phosphinic acid (hypophosphorous acid)	+1
H_3PO_3	2	Phosphonic acid (*o*-phosphorous acid)	+3
H_3PO_4	3	Phosphoric acid (*o*-phosphoric acid)	+5
HPO_3	1	Metaphosphoric acid	+5
$H_4P_2O_7$	4	Pyrophosphoric acid	+5
$H_5P_3O_{10}$	5	Triphosphoric acid	+5

Wet Process

- H_3PO_4 is obtained when phosphate rock reacts with H_2SO_4.

$$Ca_3(PO_4)_2 + 3H_2SO_4 \longrightarrow 3CaSO_4 + 2H_3PO_4$$

$$[3Ca_3(PO_4)_2CaF_2] + 10H_2SO_4 \longrightarrow 6H_3PO_4 + 10CaSO_4 + 2HF$$

Furnace Process

- $$P_4 + 5O_2 \longrightarrow P_4O_{10}$$
$$P_4O_{10} + 6H_2O \longrightarrow 4H_3PO_4$$

- H_3PO_2 and H_3PO_3 disproportionates to PH_3 and H_3PO_4 on heating :

$$3H_3PO_2 \longrightarrow 2H_3PO_3 + PH_3$$

$$4H_3PO_3 \longrightarrow 3H_3PO_4 + PH_3$$

- $P_4 + 20HNO_3 \longrightarrow 4H_3PO_4 + 20NO_2 + 4H_2O$
- H_3PO_3 is formed by hydrolysis of PCl_3:

$$PCl_3 + 3H_2O \longrightarrow H_3PO_3 + 3HCl$$

- H_3PO_2 is obtained by alkaline hydrolysis of P_4:

$$P_4 + 3OH^- + 3H_2O \longrightarrow PH_3 + 3H_2PO_2^-$$

- Table 13.6 gives distinction between H_3PO_4, H_3PO_3 and H_3PO_2.
- Phosphoric acid, H_3PO_4, is hydrogen bonded in aqueous solution, and because of this the 'concentrated acid' is syrupy and viscous.

DEEP Focus

- In all of these acids, P is four-coordinate and tetrahedrally surrounded wherever possible $p\pi - p\pi$ back bonding gives rise to P=O bonds. **The hydrogen atoms in O—H groups are ionisable and are acidic, but P—H bonds found in phosphorous acids have reducing and not the acidic properties.**
- Also basicity = number of ionisable H (protonic) = H attached to O-atoms (H encircled are acidic and H squared indicates reducing properties)

tribasic

dibasic

monobasic

- Thus :
  - all P atoms in the oxoacids and oxoanions (conjugate base) are 4-coordinate and contain at least one P—O unit,
  - all P atoms in the oxoacids have at least one P—OH group. In every case, the H atom is ionisable as H^+.
  - some oxoacids have one or more P—H bonds. This H atom is not ionisable as H^+.
  - Polymerisation can occur by P—O—P bond formation to give both linear and cyclic species. Two P atoms are never joined by more than one P—O—P bridge.
  - When P atom is surrounded by O atoms (as in H_3PO_4) its oxidation number is +5. For each P—OH that is replaced by P—H bond, its oxidation number drops by 2.

Illustration 2 *Write balanced equations for the reactions of* H_3PO_4 *and* $B(OH)_3$ *with water. Classify each acid as Bronsted-Lowry acid or a Lewis acid.*

Solution

$$H_3PO_4(aq) + H_2O(l) \rightleftharpoons H_2PO_4^-(aq) + H_3O^+(aq)$$

Since, H_3PO_4 is proton-donor hence, H_3PO_4 is a Bronsted-Lowry acid and H_2O which is a proton-acceptor is a Bronsted-Lowry base.

$$B(OH)_3(aq) + 2H_2O(l) \rightleftharpoons B(OH)_4^- + H_3O^+$$

$B(OH)_3$ has electron-deficient B (octet incomplete) hence, it accepts electron-pair from H_2O. Thus, $B(OH)_3$ is a Lewis acid.

Table 13.6 **Distinctive Properties of Oxoacids of Phosphorus**

Properties	H_3PO_2	H_3PO_3	H_3PO_4
1. Basicity	Monobasic $H_3PO_2 \rightleftharpoons H^+ + H_2PO_2^-$ hence only one type of salt (like NaH_2PO_2) is formed. NaH_2PO_2 is normal salt (which does not further react with base or which is without protonic H^+).	Dibasic $H_3PO_3 \rightleftharpoons H^+ + H_2PO_3^-$ $H_2PO_3^- \rightleftharpoons H^+ + HPO_3^{2-}$ Two types of salts, NaH_2PO_3 and Na_2HPO_3 are formed. NaH_2PO_3 is an acid salt but Na_2HPO_3 is normal salt.	Tribasic $H_3PO_4 \rightleftharpoons H^+ + H_2PO_4^-$ $H_2PO_4^- \rightleftharpoons H^+ + HPO_4^{2-}$ $HPO_4^{2-} \rightleftharpoons H + PO_4^{3-}$ Three series of salts, NaH_2PO_4, Na_2HPO_4 and Na_3PO_4 are formed. NaH_2PO_4 and Na_2HPO_4 are acid salts but Na_3PO_4 is normal salt.
2. Effect of heat	$2H_3PO_2 \xrightarrow{\Delta} H_3PO_4 + PH_3$ disproportionation	$4H_3PO_3 \xrightarrow[200°C]{\Delta} 3H_3PO_4 + PH_3$ disproportionation	$H_3PO_4 \xrightarrow[600°C]{\Delta} HPO_3 + H_2O$ dehydration
3. Reducing properties	Most reducing agent $AgNO_3 \longrightarrow Ag$ $HgCl_2 \longrightarrow Hg$ $CuSO_4 \longrightarrow Cu_2H_2$ $Cl_2 \longrightarrow HCl$	$AgNO_3 \longrightarrow Ag$ $HgCl_2 \longrightarrow Hg_2Cl_2$ $CuSO_4 \longrightarrow Cu$ $MnO_4^- \longrightarrow Mn^{2+}$	Does not reduce salts into metals.

Phosphorus Halides

Two series of halides are known PX_3 and PX_5, where X = halogen. Among these PI_5 is doubtful, whereas iodine forms a stable di-iodide P_2I_4.

$$P_4 + 6X_2 \longrightarrow 4PX_3 \quad \text{(in the limited supply of } X_2\text{)}$$
$$P_4 + 10X_2 \longrightarrow 4PX_5 \quad \text{(in excess amount of } X_2\text{)}$$

All these halides are gases, volatile liquids or low-melting solids.

- PCl_3 is a colourless liquid (sp. grav. = 1.6, b.p. = 76° C). PCl_5 is an off-white solid m.p. = 167° C).

It fumes strongly in moist air and reacts strongly with H_2O giving phosphorous acid.

$$PCl_3 + Cl_2 \longrightarrow PCl_5$$
$$PCl_3 + H_2O \longrightarrow H_3PO_3 + 3H_2O$$

- $SO_3 + PCl_3 \longrightarrow POCl_3 + SO_2$
- $SO_3 + PCl_5 \longrightarrow SO_2Cl_2 + POCl_3$
- $S_2Cl_2 + 3PCl_3 \longrightarrow 2PSCl_3 + PCl_5$

$$PCl_3 + 3NH_3 \longrightarrow P(NH_2)_3 + 3HCl$$
$$PCl_3 + Br_2 \longrightarrow PCl_3Br_2$$
$$PCl_3 + 3HI \longrightarrow PI_3 + 3HCl$$
$$PCl_3 + S \longrightarrow PSCl_3$$
$$2PCl_3 + O_2 \longrightarrow 2POCl_3$$
$$PCl_3 + 3AgNCO \longrightarrow 3AgCl + P(NCO)_3$$

- $3RCOOH + PCl_3 \longrightarrow 3RCOCl + H_3PO_3$
- $RCOOH + PCl_5 \longrightarrow RCOCl + POCl_3 + HCl$
- $3ROH + PCl_3 \longrightarrow 3RCl + H_3PO_3$
- $ROH + PCl_5 \longrightarrow RCl + POCl_3 + HCl$
- $SO_2(OH)_2 + PCl_5 \longrightarrow SO_2Cl_2 + 2POCl_3 + 2HCl$

In all the oxo-acids and hydroxy compounds, —OH group is replaced by Cl.

$$HNO_3(HO—NO_2) + PCl_5 \longrightarrow NO_2Cl + POCl_3 + HCl$$

- In carbonyl compounds, O is replaced by two Cl atoms.

$$RCHO + PCl_5 \longrightarrow RCHCl_2 + POCl_3$$
$$RCOR + PCl_5 \longrightarrow RCCl_2R + POCl_3$$

$POCl_3$ is the chloroanhydride of phosphoric acid.

- In the liquid and vapour states, PCl_5 has the trigonal bipyramidal structure, P atom being sp^3d-hybridised.

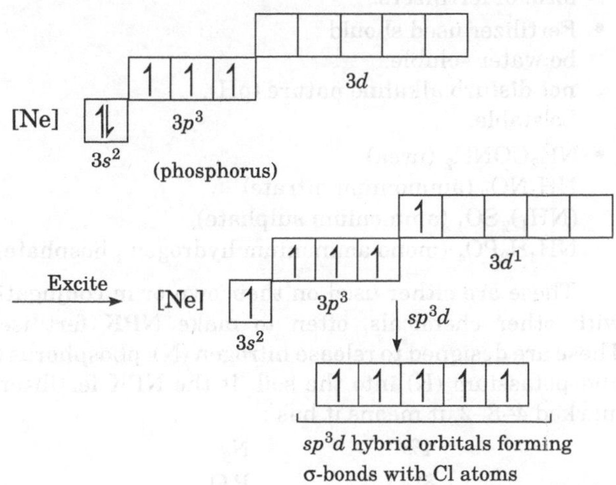

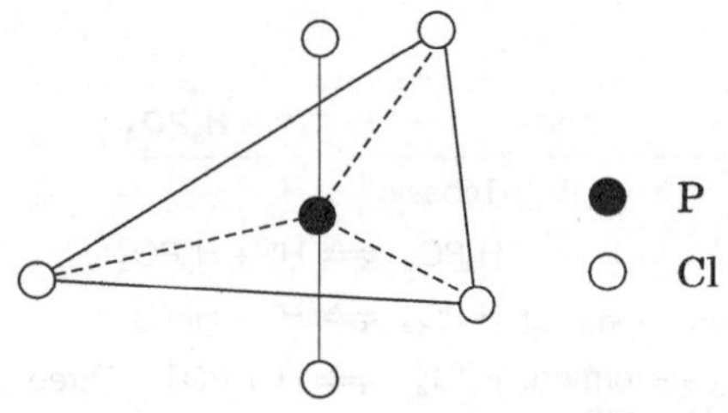

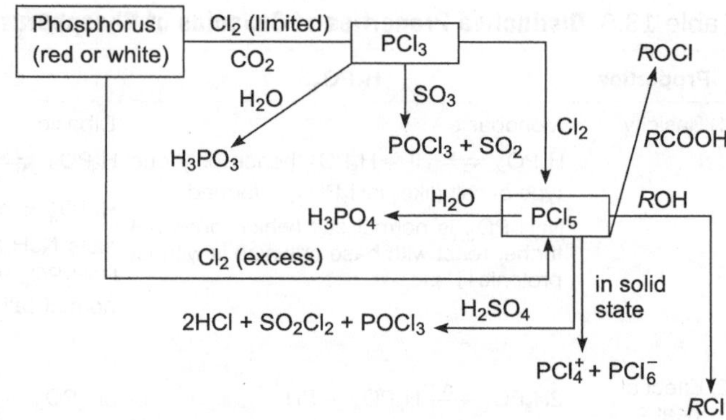

Table 13.7 **Comparison between Nitrogen and Phosphorus**

Properties	Nitrogen	Phosphorus
1. Physical state	N_2 is a gas which can be condensed to a liquid at very low temperature.	P_4 is a solid at room temperature.
2. Bond energy	N≡N (multiple bonds) are present in N_2 N≡N B.E. : 941.4 kJ mol^{-1} Due to high value of bond energy (N≡N), it is less reactive and more stable $N_2 + O_2 \xrightarrow{3300°C} 2NO$	Multiple bonds are absent in P P—P B.E. : 79.08 kJ mol^{-1} Due to low value of (P—P) bond energy P_4 is less stable and more reactive $P_4 + 5O_2 \longrightarrow P_4O_{10}$ This reaction takes place at room temperature.
3. Formation of compounds	NCl_3 is formed but NCl_5 is not formed since N cannot expand its octet due to lack of *d*-orbital.	Both PCl_3 and PCl_5 are formed due to expansion of octet.
4. Hydride	NH_3 is more stable than PH_3. NH_3 is highly soluble in water due to H-bonding with water, giving alkaline solution.	Also PH_4^+ is less stable than NH_4^+. PH_3 is sparingly soluble in water and the solution is neutral-the hydrogen bond in latter is very weak.
5. Oxyacids	Oxyacids of nitrogen are stronger than those of phosphorus. HNO_3 is a strong oxidising agent.	H_3PO_4 is weaker acid than HNO_3. H_3PO_4 is not oxidising agent.

Fertilizers

- Nitrogen, phosphorus and potassium are three most essential plant nutrients consumed by them. Fertility of soil is maintained by using these nutrients in the form of fertilizers.
- Fertilizer used should :
  be water soluble.
  not disturb alkaline nature (pH).
  be stable.
- NH_2CONH_2 (urea)
  NH_4NO_3 (ammonium nitrate)
  $(NH_4)_2SO_4$ (ammonium sulphate)
  $NH_4H_2PO_4$ (mono ammonium hydrogen phosphate)

These are either used on their own or in conjugation with other chemicals, often to make **NPK** fertilisers. These are designed to release nitrogen (N), phosphorus (P) and potassium (K) into the soil. If the NPK fertiliser is marked 4–8–2, it means it has :

4%	N_2
8%	P_2O_5
2%	K_2O

- Phosphate rocks such as fluoroapatite $(Ca_3(PO_4)_2 \cdot CaF_2)$ are very insoluble, and thus are not used as fertilisers. **Superphosphate** is made by treating phosphate rock with conc. H_2SO_4. The acid salt $Ca(H_2PO_4)_2$ is more soluble, and over a period of weeks this would dissolve in the soil water. $CaSO_4$ is waste product of this fertiliser :

$$3[Ca_3(PO_4)_2 \cdot CaF_2] + 7H_2SO_4 \longrightarrow \underbrace{3Ca(H_2PO_4)_2 + 7CaSO_4}_{\text{superphosphate}} + 2HF$$

- **'Triple superphosphate'** is made using H_3PO_4 and thus waste product $CaSO_4$ is not formed.

$$[Ca_3(PO_4)_2 \cdot CaF_2] + 6H_3PO_4 \longrightarrow \underset{\text{triple superphosphate}}{4Ca(H_2PO_4)_2} + 2HF$$

Uses

A major use of phosphoric acid is to impart corrosion resistance to metal objects such as nuts, bolts, tools and car engine parts by plunging the object into hot acid bath. Car bodies are similarly treated with phosphoric acid containing metal ions such as Zn^{2+}, and aluminium trim is "polished" by treating it with acid.

Red phosphorus is used in matches, fire works and as a component of phosphor-bronze alloys used for making bearings. The main use of elemental arsenic, antimony and bismuth is in the production of various alloys.

Besides the use of elemental nitrogen, compounds of nitrogen have extensive applications in various fields. Ammonia is used in various forms as a fertiliser, *e.g.*, as NH_4NO_3, $(NH_4)_2SO_4$, $(NH_4)_2HPO_4$, etc. Its applications also include refrigeration, as a pH controller in food and beverage industry, pharmaceuticals and in water purification. Nitric acid is another important compound of nitrogen. Its major uses include manufacture of NH_4NO_3, explosives like nitroglycerine, nitrocellulose an TNT. **Hydrazine, N_2H_4 and its methyl derivatives, $MeNHNH_2$ and Me_2NNH_2, are used as rocket fuels. Nitrous oxide, N_2O, is used as an anaesthetic.**

Sulphides of phosphorus, P_4S_3 and P_4S_{10} are used in match industry. Phosphoric acid is used is fertilisers, cleaning of metals, rust proofing, pharmaceuticals, etc.

Disodium hydrogen phosphate, Na_2HPO_4, is widely used as an emulsifier in the manufacture of pasteurised cheese. Sodium dihydrogen phosphate, NaH_2PO_4, is used as a constituent in the undercoat for metal paints. Calcium phosphates are used in baking powders and tooth-pastes. **Synthetic $Ca_5(PO_4)_3OH$ (1-2%) is added to table salt to impart free flowing properties. Sodium tripolyphosphate, $Na_5P_3O_{10}$, is used in detergents and in softening water.** Phosphorus compounds like melathion, parathion, etc., are used as pesticides.

Arsenic compounds find extensive use in agriculture as herbicides for weed and pest control, *e.g.*, monosodium methyl arsenate, **MSMA** and disodium methyl arsenate, **DSMA.** Arsenic acid is used as a wood preservative. Sodium arsenite is used for aquatic weed control. Arsenic oxide, As_2O_5, is used to decolourise bottle glass. Antimony compound, antimonial lead, is used in storage batteries. **Compounds of antimony with aluminium, gallium and indium, AlSb, GaSb and InSb, find applications in infrared devices and in diodes.**

Target Practice 2

1. Nitrite ion (NO_2^-) is said to interfere in the ring-test of nitrate ion (NO_3^-). Name the chemical that can decompose NO_2^-.

....

....

2. $CH_3CH_2CH_2NH_2$ on reaction with HNO_2 changes to 2-propanol as the main product. Explain the formation of 2-propanol.

....

....

3. Write the species of aqua-regia.

....

What is the product when

(a) gold is dissolved in it? ....

(b) platinum is dissolved in it? ....

4. Nitrites and nitrates are added to food to keep meat red for longer time. But there are harmful effects.

Harmful effects are ....

....

5. Anhydrous HNO_3 is colourless but it is often yellow in colour due to ....

....

6. White phosphorus is manufactured by modern electrochemical process represented in the flow-sheet. Fill in the blank space *X*, *Y* and *Z*.

$Ca_3(PO_4)_2$ + (X) + C —1500°C→ P_4 + CO —H_2O→ (Z)

(first step gives $CaSiO_3$; second step gives (Y))

7. Compare white and red phosphorus in the following properties.

	White	Red
1. Smell	....	....
2. Solubility in CS_2	....	....
3. With NaOH	....	....

8. Write formula of phosgene and phosphine

....

Of these

(a) Which is poisonous? ....

(b) Which is reducing agent ....

Write reactions of their formation.

Phosgene

Phosphine

9. NH_3 is more basic than PH_3. It can be proved by ....

....

10. Many hundred years ago, a ship was transporting calcium carbide and calcium phosphide (stored in different container). By chance sea-water entered sea and luminous flame appeared. Passengers travelling in ship ran here to there assuming a ghost would have come. However, a chemist travelling in it explained it based on reactions. These reactions are ....

....

....

11. 1 mole each of H_3PO_2, H_3PO_3 and H_3PO_4 were given in three test tubes. In a volumetric titration, standard 1 M $Ca(OH)_2$ solution was provided. Volume of $Ca(OH)_2$ used by

H_3PO_2 ....

H_3PO_3 ....

H_3PO_4 ....

12. H_3PO_2 is represented by

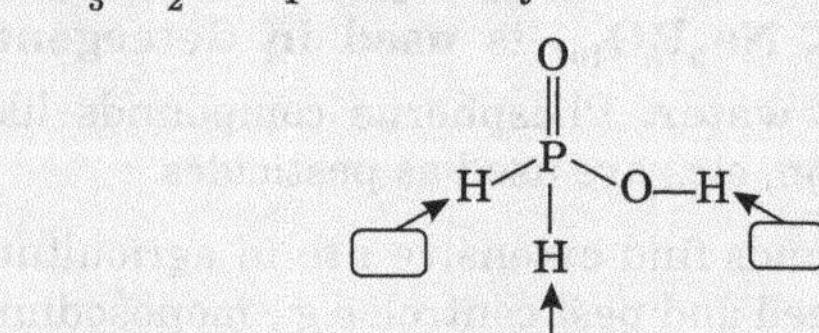

Specify the nature of different H (acidic/basic/reducing)

13. Write full form of
(a) DSMA ....
(b) MSMA ....
(c) NPK ....

14. Write formulae of
(a) Superphosphate ....
(b) Triple superphosphate ....

15. Write applications of the following
(a) N_2H_4 ....
(b) N_2O ....
(c) P_4S_3 ....
(d) $Ca_5(PO_4)_3OH$ ....
(e) $Na_5P_3O_{10}$ ....
(f) MSMA, DSMA ....

Answers

1. Urea

$$2HNO_2 + NH_2CONH_2 \longrightarrow 2N_2 + CO_2 + 3H_2O$$

2. $CH_3CH_2NH_2 + HNO_2 \longrightarrow \underset{\text{intermediate}}{CH_3CH_2CH_2N_2^{\oplus}}$

$$CH_3CH_2CH_2N_2^{\oplus} \longrightarrow \underset{1^\circ \text{ carbocation}}{CH_3CH_2CH_2^{\oplus}} + N_2$$

$$CH_3CH_2CH_2^{\oplus} \xrightarrow{\text{1,2-hydride shift}} \underset{2^\circ \text{ carbocation}}{CH_3\overset{\oplus}{C}HCH_3}$$

$$CH_3\underset{\oplus}{C}HCH_3 \longrightarrow \underset{\text{2-propanol}}{CH_3\underset{OH}{CH}CH_3}$$

(2° carbocation is more stable than 1° carbocation hence, 1,2-hydride shift takes place)

3. 75% conc. HCl and 25% conc. HNO_3

$$HNO_3 + 3HCl \longrightarrow 2H_2O + 2Cl + 2NOCl$$

(a) $HAuCl_4$ (chloro auric acid)

(b) H_2PtCl_6 (chloro platinic acid)

4. They cause formation of carcinogenic nitrosoamines.

5. Photochemical decomposition (in sunlight)

$$2HNO_3 \xrightarrow{h\nu} H_2O + 2NO_2 + \frac{1}{2}O_2$$

6. $X: SiO_2$ $Y: P_4$ $Z: CO$ **7.** Refer text

8. Phosgene $COCl_2$
Phosphine PH_3
(a) both (b) PH_3

Formation Phosgene is formed by air oxidation of $CHCl_3$

$$CHCl_3 + \frac{1}{2}O_2 \longrightarrow COCl_2 + HCl$$

Phosphine is formed by action of NaOH on

$$P_4 + 3NaOH + 3H_2O \longrightarrow PH_3 + 3NaH_2PO_2$$

9. $NH_3 + PH_3 \longrightarrow NH_4^+PH_2^-$
NH_3 is H^+ ion acceptor thus, NH_3 is more basic than PH_3 which is H^+ ion donor.

10. $\underset{\text{Calcium carbide}}{CaC_2} + 6H_2O \longrightarrow Ca(OH)_2 + \underset{\text{Acetylene}}{C_2H_2}$

$$\underset{\text{Calcium phosphide}}{Ca_3P_2} + 6H_2O \longrightarrow 3Ca(OH)_2 + \underset{\text{Phosphine}}{2PH_3}$$

CaC_2 and Ca_3P_2 both were stored in containers but when water reacted with them PH_3 and C_2H_2 were formed. Some highly spontaneously flammable gases (P_2H_4 and H_2) are also formed and as a result of thus, C_2H_2 gas also burns (which people assumed as ghost)

11. $\underset{2\text{ mol}}{2H_3PO_2} + \underset{1\text{ mol}}{Ca(OH_2)} \longrightarrow Ca(H_2PO_2)_2 + 2H_2O$

1 mol 0.5 mol ≡ 500 mL of 1 M $Ca(OH)_2$

$$\underset{1\text{ mol}}{H_3PO_3} + \underset{1\text{ mol}}{Ca(OH)_2} \longrightarrow CaHPO_3 + 2H_2O$$

≡ 1000 mL of 1M $Ca(OH)_2$ solution

$$\underset{2\text{ mol}}{2H_3PO_4} + \underset{3\text{ mol}}{3Ca(OH)_2} \longrightarrow Ca_3(PO_4)_2 + 3H_2O$$

1 mol 1.5 mol ≡ 1500 mL of 1 M $Ca(OH)_2$

Thus, H_3PO_2 ≡ 500 mL 1 M $Ca(OH)_2$
H_3PO_3 ≡ 1000 mL of 1 M $Ca(OH)_2$
H_3PO_4 ≡ 1500 mL of 1 M $Ca(OH)_2$

12.

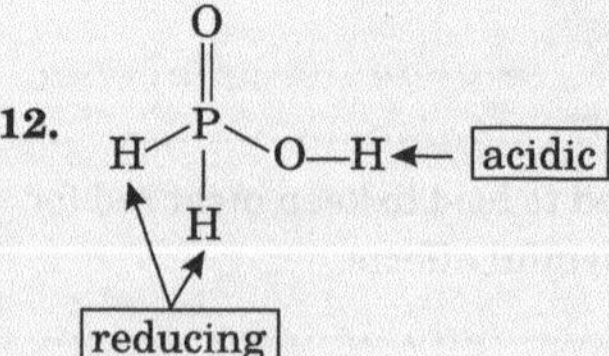

13. Refer text
14. Refer text
15. Refer text

Practice Exercise 2

1. HNO_2 can oxidise I^- to I_2; suggest a suitable reagent that can estimate liberated I_2. Also determine equivalent weight of HNO_2 in this estimation.
2. Fe^{2+} as well as HNO_2 can be used to estimate MnO_4^- in acidic medium. If 100 mL of 1 M Fe^{2+} can be oxidised by V mL of 1 M MnO_4^-/H^+ solution, calculate volume of 1 M HNO_2 that can be oxidises by V mL of 1 M MnO_4^-/H^+ solution.
3. Vessels made of like Al, Fe, Cr, Ni etc., can be used to store conc. HNO_3. Explain.
4. Make distinction between NO_3^- and NO_2^-.
5. What is effect of heat on NH_4NO_3 and NH_4NO_2?
6. NO_2^- interferes in the 'ring test' of NO_3^-. How is it removed from the mixture?
7. What is hybridisation of N-atom in HNO_2 and HNO_3?
8. Complete the following in acidic medium :
   (a) $Fe^{2+} + NO_3^- \longrightarrow$
   (b) $SO_3^{2-} + NO_3^- \longrightarrow$
   (c) $SO_3^{2-} + NO_2^- \longrightarrow$
9. Arrange NH_3, PH_3, AsH_3, SbH_3 in increasing:
   (a) covalent nature (b) reducing nature
   (c) bond angle (d) thermal stability
10. What is the origin of the flickering light in marshes ?
11. What is the chemistry of Holme's signal?
12. What is the type of the following reaction?
$$P_4 + 3NaOH + 3H_2O \xrightarrow{\Delta} PH_3 + 3NaH_2PO_2$$
13. Explain reducing action of H_3PO_2 and H_3PO_3 with some examples.
14. Classify acids and normal salts out of :
   (a) NaH_2PO_3 (b) NaH_2PO_4
   (c) NaH_2PO_2 (d) Na_2HPO_3
   (e) Na_2HPO_4
15. How is phosphorus estimated?
16. (a) $4\underline{H_3PO_3} \longrightarrow 3H_3PO_4 + PH_3$
   (b) $2\,\underline{H_3PO_2} \longrightarrow H_3PO_4 + PH_3$
   Determine equivalent weights of the underlined.

Answers

1. $HNO_2 + H^+ + e^- \longrightarrow NO + H_2O$
$$2I^- \longrightarrow I_2 + 2e^-$$
$Na_2S_2O_3$ (hypo sodium thiosulphate) is used to estimate liberated I_2.
$$2Na_2S_2O_3 + I_2 \longrightarrow 2NaI + Na_2S_4O_6$$
$$HNO_2 \equiv NO \quad \text{or} \quad HNO_2 \equiv 1e^-$$
Equivalent weight of HNO_2 = molecular weight

2. $MnO_4^- + 8H^+ + 5e^- \longrightarrow Mn^{2+} + 4H_2O$ Eq. wt. $= M/5$
$Fe^{2+} \longrightarrow Fe^{3+} + e^-$ $= M/1$
$HNO_2 + H_2O \longrightarrow NO_3^- + 3H^+ + 2e^-$ $= M/2$
V mL of 1 $MnO_4^- \equiv$ 100 mL of 1 M Fe^{2+}
or V mL of 5N $MnO_4^- \equiv$ 100 mL of 1N Fe^{2+} ...(i)
Also V mL of 1 M $MnO_4^- \equiv V'$ mL of 1M HNO_2
V mL of 5 N $MnO_4^- \equiv V'$ mL of 2N HNO_2 ...(ii)
Thus, from (i) and (ii) $V' = 50$ mL

3. There is formation of a protective oxide layer on the surface of the metal when it reacts with HNO_3. Hence, vessels of such metals can be used to store HNO_3.

4. Add KI solution. NO_2^- would oxidise I^- to I_2. NO_3^- would give 'ring-test'.

5. $NH_4NO_3 \xrightarrow{\Delta} N_2O + 2H_2O$
$NH_4NO_2 \xrightarrow{\Delta} N_2 + 2H_2O$

6. If NH_4Cl or urea is added to a solution containing NO_2^- and NO_3^-, NO_2^- is decomposed as N_2 and thus removed :
$$NaNO_2 + NH_4Cl \longrightarrow NaCl + N_2 + 2H_2O$$
$$2NaNO_2 + NH_2CONH_2 \longrightarrow 2N_2 + Na_2CO_3 + 2H_2O$$

7. H—O—N̈=O   H—O—N(=O)(→O)

N-atom is sp^2 hybridised in both.

8. (a) $3Fe^{2+} + 4H^+ + NO_3^- \longrightarrow 3Fe^{3+} + 2H_2O + NO$
(b) $SO_3^{2-} + 2NO_3^- + 2H^+ \longrightarrow SO_4^{2-} + 2NO_2 + H_2O$
(c) $SO_3^{2-} + 2NO_2^- + 2H^+ \longrightarrow SO_4^{2-} + 2NO + H_2O$

9. (a) Covalent $NH_3 > PH_3 > AsH_3 > SbH_3$
(b) Reducing nature $NH_3 < PH_3 < AsH_3 < SbH_3$
(c) Bond angle $NH_3 > PH_3 > AsH_3 > SbH_3$
107° 48′ 93°36′ 91°48′ 91°18′

10. PH_3 frequently contains traces of diphosphine (P_2H_4) which causes it to catch fire spontaneously. This is the origin of the flickering light called **'will-o-the-wisp'**, which is sometimes seen in marshes.

11. Mixture of CaC_2 and Ca_3P_2 placed in the container with holes is thrown into the sea. Both react with H_2O :
$$CaC_2 + 2H_2O \longrightarrow Ca(OH)_2 + C_2H_2$$
$$Ca_3P_2 + 6H_2O \longrightarrow 3Ca(OH)_2 + 2PH_3$$
Some P_2H_4 is also produced, which catches fire instantly, which in turn ignites C_2H_2 (acetylene) and then luminous flame is produced. This guides the approaching ship. This signal is called **Holme's** signal.

12. In this reaction P_4 (ox. no. = 0) has been reduced to PH_3 (ox. no. = – 3) as well oxidised to $H_2PO_2^-$ (ox. no. = + 1). Hence, this is a disproportionation reaction.

13. $$H_3PO_2 + H_2O \longrightarrow H_3PO_3 + 2H$$
$$H_3PO_3 + H_2O \longrightarrow H_3PO_4 + 2H$$
H produced causes reduction of salts into metals :
$$AgNO_3 + H \longrightarrow HNO_3 + Ag$$
$$CuSO_4 + H \longrightarrow H_2SO_4 + Cu$$
$$2HgCl_2 + 2H \longrightarrow Hg_2Cl_2 + 2HCl$$
$$Hg_2Cl_2 + 2H \longrightarrow 2Hg + 2HCl$$

14. Salts with ionisable H (attached to O-atom) are acid salts, others are normal salts :

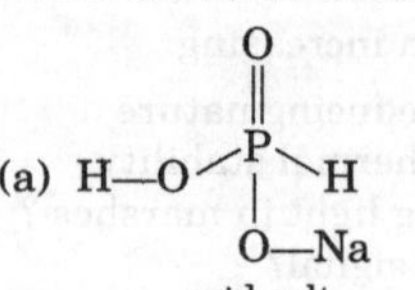

(a) H—O—P(=O)(—H)—O—Na
acid salt

(b) H—O—P(=O)(—O—H)—O—Na
acid salt

(c) H—P(=O)(—H)—O—Na
normal salt

(d) Na—O—P(=O)(—H)—O—Na
normal salt

(e) Na—O—P(=O)(—O—H)—O—Na
acid salt

15. Phosphorus can be estimated by converting it into phosphates and then precipitating as $MgNH_4PO_4$ (by adding NH_4OH and Mg^{2+}).

$$P \xrightarrow{HNO_3,\ \Delta} H_3PO_4$$

$$H_3PO_4 + NH_4^+ + Mg^{2+} \longrightarrow MgNH_4PO_4 + 3H^+$$

On ignition, white ppt of $Mg_2P_2O_7$ is obtained :

$$2MgNH_4PO_4 \xrightarrow{\Delta} Mg_2P_2O_7 + 2NH_3 + H_2O$$

16. (a) $\underset{+3}{H_3PO_3} \longrightarrow \underset{+5}{H_3PO_4}$ (oxidn)

Equivalent weights

$$E_1 = \frac{\text{mol. wt.}}{2}$$

$\underset{+3}{H_3PO_3} \longrightarrow \underset{-3}{PH_3}$ (redn)

$$E_2 = \frac{\text{mol. wt.}}{6}$$

$\therefore$ **Eq. wt. of H_3PO_3 in disproportionation reaction**

$$= E_1 + E_2$$

$$= \frac{2}{3}\text{ mol. wt.} = \frac{2}{3} \times 82$$

$$= \mathbf{54.67}$$

(b) $2\underset{+1}{H_3PO_2} \longrightarrow \underset{+5}{H_3PO_4} + \underset{-3}{PH_3}$ (oxidation; reduction)

$$E_1 \text{ (oxidation)} = \frac{M}{4}$$

$$E_2 \text{ (reduction)} = \frac{M}{4}$$

$$\text{Eq. wt. } (H_3PO_2) = E_1 + E_2 = \frac{M}{2} = 33$$

Eq. wt. of (X) when it undergoes disproportionation

$$= E_1 \text{ (oxidation)} + E_2 \text{ (reduction)}$$

where, E_1 (oxidation) is the eq. wt. in oxidation reaction and E_2 (reduction) that of reduction reaction.

Total Practice Set
(Read, Plan & Solve!)

Problem 1. Given the following bond dissociation enthalpies (kJ mol^{-1}) :

P≡P	490
P—P	209
N≡N	946
N—N	160

Compare the enthalpy changes for the process
(i) $2P_2(g) \longrightarrow P_4(g)$
(ii) $2N_2(g) \longrightarrow N_4(g)$

Solution The process involves replacement of $2(X{\equiv}X)$ units by 6 X—X bonds in the X_4 tetrahedron. Energy is released when bonds are formed and thus the species formed is stable.

(i) $2P_2(g) \longrightarrow P_4(g)$

$$2(P{\equiv}P) \longrightarrow P_4 \text{ (tetrahedron)}$$

$$\Delta H_f = -6 \times BE\,(P - P) + 2BE\,(P{\equiv}P)$$
$$= -6 \times 209 + 2 \times 490$$
$$= -274 \text{ kJ mol}^{-1}$$

Formation of P_4 is exothermic thus possible.

(ii) $$2N_2(g) \longrightarrow N_4(g)$$

$$\Delta H_f = -6BE\,(N - N) + 2BE\,(N{\equiv}N)$$
$$= -6 \times 10 + 2 \times 946$$
$$= +932 \text{ kJ mol}^{-1}$$

Formation of N_4 is endothermic, thus is not feasible.

Problem 2. The action of concentrated sulphuric acid on urea (NH_2CONH_2) results in the production of a white crystalline solid X of formula H_3NO_3S. This is a monobasic acid. On treatment with sodium nitrate and dilute hydrochloride acid at 0°C it liberates 1 mole of N_2 per mol of X, and on addition of aqueous barium chloride the resulting solution yields 1 mole of $BaSO_4$ per mol of X taken.

Deduce the structure of X.

Solution NH_2CONH_2 (Urea) reacts with conc. H_2SO_4 forming $X(H_3NO_3S)$. X contains one ionisation H (since, X is a monobasic acid). Thus,

$$\underset{\text{Urea}}{NH_2—\overset{O}{\overset{\|}{C}}—NH_2} + \underset{\text{Sulphuric acid}}{H—O—\overset{O}{\overset{\|}{S}}—O—H} \longrightarrow \underset{X}{NH_2—SO_3H} + NH_3 + CO$$

X contains free —NH_2 as well as free —SO_3H. ($NaNO_2$ + HCl) converts free —NH_2 into free —OH group liberating N_2.

$$NaNO_2 + HCl \longrightarrow HNO_2 + NaCl$$

$$\underset{1\text{ mol }X}{NH_2—SO_3H} + HNO_2 \longrightarrow \underset{1\text{ mol }X}{N_2\uparrow} + H_2O + \underset{1\text{ mol}}{H_2SO_4}$$

$$\underset{1\text{ mol}}{H_2SO_4} + BaCl_2 \longrightarrow \underset{1\text{ mol}}{BaSO_4} + 2HCl$$

Thus, X is **NH_2—SO_3H** (amino sulphonic acid), 1 mole of which liberated 1 mol of N_2 on reaction with ($NaNO_2$ + HCl) at 0°C and in turn resulting solution gives 1 mole of $BaSO_4$ on reaction with $BaCl_2$.

Problem 3. Suggest synthesise for,
(a) $Na^{15}NH_2$, $^{15}N_2$ from $K^{15}NO_3$.
(b) $^{32}PH_3$, $H_3{}^{32}PO_3$ from $Ca_3(^{32}PO_4)_2$.

Solution (a) Isotopic ^{15}N compounds are to be synthesised from $K^{15}NO_3$.

$$K\overset{15}{N}O_3 + NaOH + Zn \longrightarrow \overset{15}{N}H_3 \text{ (unbalanced)}$$

$$\overset{15}{N}H_3 \xrightarrow{CuO,\ \Delta} {}^{15}N_2$$

$$\overset{15}{N}H_3 \xrightarrow{Na} Na\overset{15}{N}H_2$$

(b) $$Ca_3(\overset{32}{P}O_4)_2 + SiO_2 + C \xrightarrow{\Delta} \overset{32}{P}_4$$

$$\overset{32}{P}_4 \xrightarrow{NaOH} \overset{32}{P}H_3$$

$$\overset{32}{P}_4 \xrightarrow{I_2} \overset{32}{P}I_3 \xrightarrow{H_2O} H_3\overset{32}{P}O_3$$

Problem 4. Deduce what you can about the nature of the following reactions:

(a) 1 mole of NH_2OH reacts with 2 moles of Ti(III) in the presence of excess alkali, and the Ti(III) is converted into Ti(IV).

(b) Silver phosphite, Ag_2HPO_3 is warmed with water, and the silver is precipitated as the metal.

(c) When 1 mole of hypophosphorous acid is treated with excess of iodine in acidic solution, 1 mole of iodine is reduced; on making the solution alkaline, another mole of iodine is consumed.

Solution (a) Based on change in oxidation number of two moles Ti(III) from 3 to 4, NH_2OH is reduced to NH_3.

$$\underset{+3}{2Ti(III)} \xrightarrow{\text{oxidation}} \underset{+4}{2Ti(IV)}$$

two units change

$$\underset{-1}{NH_2OH} \xrightarrow{\text{reduction}} \underset{-3}{NH_3}$$

two units change

Thus, NH_2OH is reduced to NH_3

$$2Ti^{3+} + NH_2OH + H_2O \longrightarrow 2Ti^{4+} + NH_3 + 2OH^-$$

(b) Ag^+ (in Ag_2HPO_3) is reduced to Ag

$$2Ag^+ \xrightarrow[\text{two units change}]{\text{reduction}} 2Ag$$

Hence, HPO_3^{2-}(oxidation number of P = +3) should be oxidised to PO_4^{3-} (oxidation number of P = +5) giving a change of oxidation number of two units.

$$2Ag^+ + HPO_3^{2-} + H_2O \longrightarrow PO_4^{3-} + 2Ag + 3H^+$$

oxidation

reduction

(c) $$H_3PO_2 + I_2 + H_2O \longrightarrow H_3PO_3 + 2I^- + 2H^+$$

$$H_3PO_3 + I_2 + 2OH^- \longrightarrow H_3PO_4 + 2I^- + H_2O$$

Problem 5. 25.0 cm^3 of 0.0497 M solution of NH_2OH was boiled with excess of ferric alum solution in H_2SO_4. The Fe(II) formed required 24.7 cm^3 of 0.1007 N $K_2Cr_2O_7$ solution. Identify the product formed by oxidation of NH_2OH.

Solution Fe^{3+} is reduced to Fe^{2+} which in turn is oxidised to Fe^{3+} by $Cr_2O_7^{2-}$ in acidic medium.

$$NH_2OH + Fe^{3+} \longrightarrow Fe^{2+} + \text{(oxidation product)}$$

$$6Fe^{2+} + Cr_2O_7^{2-} + 14H^+ \longrightarrow 6Fe^{3+} + 2Cr^{3+} + 7H_2O$$

$$\text{Equivalent of } Cr_2O_7^{2-} = \frac{24.7 \times 0.1007}{1000} = 0.0025$$

$$\text{Equivalent of } Fe^{3+} \text{ or } Fe^{2+} = 0.0025$$

$$\text{Equivalent of } NH_2OH = 0.0025$$

$$\text{Moles of } NH_2OH \text{ (taken)} = \frac{25 \times 0.0497}{1000} = 0.00125$$

Moles (NH_2OH) = Change in oxidation number × Equivalent of NH_2OH

$$\therefore \text{ Change in oxidation number of } NH_2OH = \frac{0.0025}{0.00125} = 2$$

Oxidation number of N is $NH_2OH = -1$

Oxidation number of N is oxidised produced

$$= -1 + 2 = +1$$

Thus, compound formed = $\mathbf{N_2O}$

Problem 6. Give a scheme of reactions that do not involve disproportionation of the oxides of nitrogen.

Solution

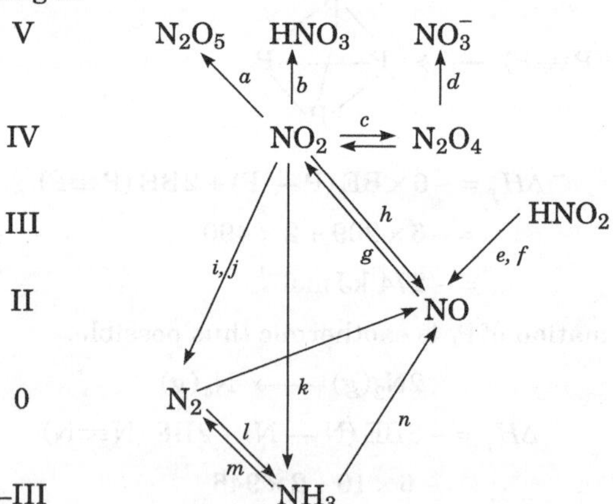

Reactions :

(a) $$2NO_2 + O_3 \longrightarrow N_2O_5 + O_2$$

(b) $$2NO_2 + H_2O_2 \longrightarrow 2HNO_3$$

(c) $$2NO_2 \rightleftharpoons N_2O_4$$

(d) $$N_2O_4 + Cu \longrightarrow Cu(NO_2)_2$$

(e) $2HNO_2 + 2HI \longrightarrow I_2 + 2NO + 2H_2O$

(f) $Fe^{2+} + HNO_2 + H^+ \longrightarrow Fe^{3+} + NO + H_2O$

(g) $2NO + O_2 \longrightarrow 2NO_2$

(h) $2Cu + NO_2 \longrightarrow Cu_2O + NO$

(i) $C + NO_2 \longrightarrow CO_2 + \frac{1}{2}N_2$

(j) $NO_2 + 2H_2 \longrightarrow \frac{1}{2}N_2 + 2H_2O$

(k) $2NO_2 + 7H_2 \longrightarrow 2NH_3 + 4H_2O$

(l) $N_2 + 3H_2 \longrightarrow 2NH_3$

(m) $4NH_3 + 3O_2 \longrightarrow 2N_2 + 6H_2O$

(n) $4NH_3 + 5O_2 \longrightarrow 4NO + 6H_2O$

(o) $N_2 + O_2 \longrightarrow 2NO$

Problem 7. Give a scheme of reactions which involve disproportionation of the oxides of nitrogen.

Solution

Oxidation state of nitrogen		
V	HNO_3	NO_3^- NO_2^-
IV	NO_2	N_2O_4
III	HNO_2	NO^+
II	NO	
I	N_2O	

Arrows: $NO_2 \xrightarrow{a} HNO_3$, $NO_2 \xrightarrow{a} HNO_2$; $HNO_2 \xrightarrow{d} HNO_3$, $HNO_2 \xrightarrow{d} NO$; $NO \xrightarrow{e} NO_2$, $NO \xrightarrow{e} N_2O$; $N_2O_4 \xrightarrow{b} NO_3^-$, $N_2O_4 \xrightarrow{b} NO^+$; $N_2O_4 \xrightarrow{c} NO_2^-$, $N_2O_4 \xrightarrow{c} NO^+$

Reactions :

(a) $2NO_2 + H_2O \longrightarrow HNO_3 + HNO_2$

(b) $N_2O_4 \longrightarrow NO^+ + NO_3^-$

(c)
$N_2O_4 + 3H_2SO_4 \longrightarrow NO^+ + NO_2^+ + 3HSO_4^- + H_3O^+$

(d) $3HNO_2 \longrightarrow HNO_3 + 2NO + H_2O$

(e) $3NO \longrightarrow N_2O + NO_2$

Explanation

(a) $(SiH_3)N$ *is planar while* $(CH_3)_3N$ *is pyramidal*

- In case of $(CH_3)_3N$, lone pair on N and three covalent bonds with CH_3 groups results in sp^3 hybridisation but lone pair-bond pair repulsion causes distortion hence gives pyramidal structure :

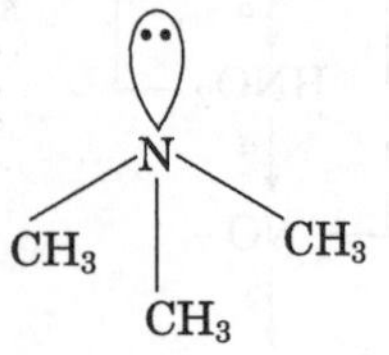

[Structure: N with lone pair above, bonded to three CH_3 groups]

But in case of $(SiH_3)_3N$, lone pair on the nitrogen is transferred to the empty *d*-orbitals of silicon ($p\pi$-$d\pi$ overlap) thereby causing planarity of unit.

(b) NF_3 *is stable while* NCl_3 *and* NI_3 *are explosive*

- In NF_3 the electronegativity difference (N= 3.0; F= 4.0) permits strong $p\pi$-$p\pi$ bonding and the molecule is stable. But in NCl_3 and NI_3, the electronegativity difference is nil or small and the molecules are unstable with $p\pi$-$p\pi$ bonding weak. Hence, NF_3 is a stable gas, NCl_3 an explosive liquid and NI_3 an explosive solid.

(c) *Hydrogen azide is explosive, while sodium azide is not.*

- $2N_3H \xrightarrow{\text{slightly heat}} 3N_2 + H_2$

Due to increase in volume, there is explosion. Thus, N_3H is used as detonators and explosives.

However, ionic halides are quite stable. NaN_3 decomposes smoothly (but not explosively) at about 600 K

$$2NaN_3 \xrightarrow{600\text{ K}} 2Na + 3N_2$$

It is due to linear structure of N_3^- which is having resonating structures is stable.

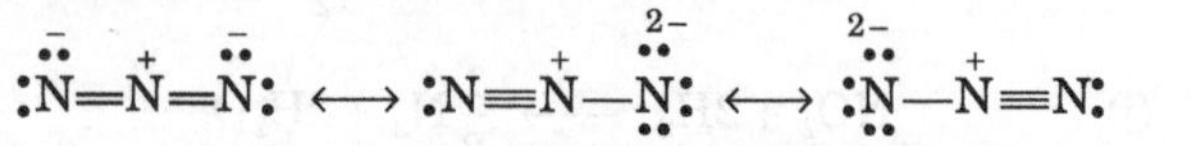

$$:\overset{..}{\overset{-}{N}}=\overset{+}{N}=\overset{..}{\overset{-}{N}}: \longleftrightarrow :N\equiv\overset{+}{N}-\underset{..}{\overset{..}{\overset{2-}{N}}}: \longleftrightarrow :\underset{..}{\overset{..}{\overset{2-}{N}}}-\overset{+}{N}\equiv N:$$

(d) H_3PO_2 *is a powerful reducing agent but* H_3PO_4 *is not.*

	H_3PO_4	H_3PO_2
oxidation number of P	+5	+1

H_3PO_2 is a monobasic acid with oxidation number of P as +1 which can be further increased to **+3** and to **+5**. Thus, H_3PO_2 can be used as a reducing agent. H_3PO_4 is already in oxidised state (oxidation number of P = + 5) Thus, it is not a reducing agent H_3PO_4 with all the ionisable to hydrogens is a tribasic acid and H_3PO_2 with only one ionisable hydrogen is a monobasic acid. Two other hydrogen directly attached to P provide reducing nature to H_3PO_2.

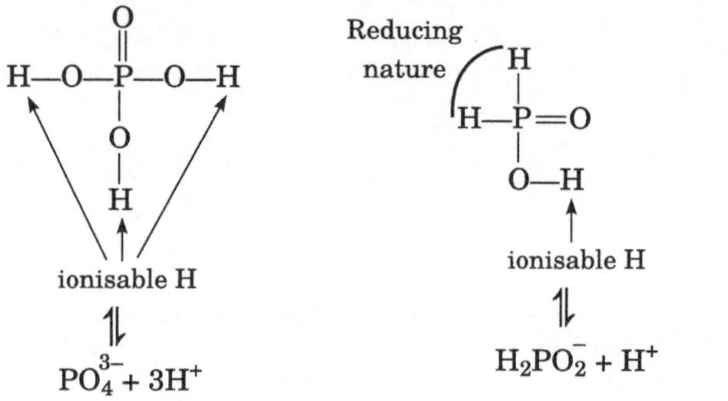

ionisable H

$$\rightleftharpoons PO_4^{3-} + 3H^+$$

Reducing nature

ionisable H

$$\rightleftharpoons H_2PO_2^- + H^+$$

Master Exercises

Exercise 1
(Stage 1 : Learning)

Short Answer Type Questions

1. Identify the group 15 (VA) elements (s) that best fits each of the following descriptions:
(a) makes up part of bones and teeth
(b) forms stable salts containing M^{3+} ions
(c) is the most abundant element in the atmosphere
(d) forms a basic oxide.

2. Draw an electron-dot structure for N_2, and explain why this molecule is so unreactive?

3. In the 'ring test' of NO_3^- ion, there is formation of brown colour ring. What is its formula?

4. What is the nature of the nitride ion?

5. Apart from N_2O, which other colourless gas supports combustion?

6. Write an equation for the reaction of burning Mg with NO_2.

7. Automobiles are now provided with catalytic converters. What are their role in reducing air-pollution?

8. Colourless salt (*A*) + NaOH $\xrightarrow{\Delta}$ gas (*B*) giving white fumes with HCl + alkaline solution (*C*)

$(C) + \text{Zn} \longrightarrow \text{gas } (B) \quad (A) \xrightarrow{\Delta} \underbrace{\text{gas } (D) + \text{liquid } (E)}_{\text{both triatomic}}$

Identify (*A*), (*B*), (*C*), (*D*) and (*E*).

9. A few drops of concentrated ammonia solution added to a calcium bicarbonate solution causes a white precipitate to form. Write a balanced equation for the reaction.

10. What is the type of the following reaction?

$$NH_3 + NH_3 \rightleftharpoons NH_4^+ + NH_2^-$$

11. Identify salts (*A*) and (*B*) based on following reactions:

$(A) \text{ or } (B) \xrightarrow{\text{NaOH, }\Delta} NH_3 \xleftarrow{H_2O}$ ┐

$(A) \xrightarrow{\Delta} (C) \text{ (gas) } \xrightarrow{\text{Al, }\Delta} (D)$

$(B) \xrightarrow{\Delta} (E) \text{(gas)}$

(*E*) is used as an anaesthetic.

12. Write balanced equation to account for each of the following observations:
(a) HNO_3 turns brown when exposed to air
(b) HNO_3 turns yellow-brown on standing
(c) Ag dissolves in dilute HNO_3, yielding a colourless gas
(d) N_2H_4 (hydrazine) reduces iodine to I^- and in the process is oxidised to N_2 gas

13. What substances form when concentrated HNO_3 is reacted with As_2S_3?

14. Arrange oxides (M_2O_3) of nitrogen family in increasing acidic nature.

15. $POCl_3$, like P_2O_5, is also used as dehydrating agent. Write reaction.

16. How many (P—O) bonds are there in P_4O_6 and P_4O_{10}? (Coordinate bonds may also be included.)

17. One of the oxoacid salts on heating with conc. HNO_3 and ammonium molybdate gives yellow ppt. What is this ion? Give chemical reactions.

18. What are structures of metaphosphoric acid and trimetaphosphoric acid?

19. $H_4P_2P_7 + H_2O \longrightarrow 2H_3PO_4$

What conclusion can you make from this reaction regarding structure of $H_4P_2O_7$?

20. The structures of various phosphate and silicate anions have many similarities. Give examples.

21. Starting with elemental phosphorus, P_4, show how you would prepare phosphoric acid.

22. The product $Ca(H_2PO_4)_2$ is known as triple phosphate but it is said to be richer in phosphorus than other superphosphates. Explain.

23. What happens when urea slowly hydrolyses in the soil?

24. Which has maximum% of (P) out of
(a) superphosphate
(b) triple superphosphate
(c) fluoroapatite rock

Analytical Questions

1. Explain :

(i) PCl_5 is common but NCl_5 is unknown.

(ii) The N—O$^+$ ion has a shorter bond length than does NO, even though the latter has an extra electron

(**Hint :** compare bond order)

2. Predict the structures of the following species:

(a) NF_4^+ (b) $N_2F_3^+$

(c) NH_2OH (d) $SPCl_3$

(e) PF_3Cl_2.

3. It has been claimed that NH_4NO_3 fertiliser can be rendered unexplodable by a process that involves additives such as diammonium hydrogen phosphate $(NH_4)_2HPO_4$. Analysis of such a "densensitised" sample of NH_4NO_3 showed the mass per cent nitrogen to be 33.81%.

Assuming that the mixture contains only NH_4NO_3 and $(NH_4)_2HPO_4$, what is the mass per cent of each of these two components?

4. A mixture of the sulphides of As_2S_3, Sb_2S_3 and Bi_2S_3 was treated with a Na_2S solution. Which sulphide remained undissolved? Write the equations of the sulphide dissolving reactions.

5. Hyponitrous acid and nitramide both have the formula $H_2N_2O_2$. Hyponitrous acid is a weak diprotic acid; nitramide contains the amide group (—NH_2). Write structures of both of these compounds.

6. Peroxonitrous acid is an unstable intermediate formed in the oxidation of HNO_2 by H_2O_2. If has the same formula as nitric acid, HNO_3. Show how you would expect these two acids to differ in structure?

NCERT Problems

1. Give examples of compounds in which nitrogen exhibits oxidation states of –3, +3 and +5.

2. Discuss the conditions required in the Haber process for the manufacture of ammonia.

3. How is dinitrogen prepared in the laboratory?

4. Write the products of the following reactions (give balanced equations)

$$Mg_3N_2 + H_2O \longrightarrow$$
$$I_2 + HNO_3 \text{ (concentrated)} \longrightarrow$$
$$Cu + HNO_3 \text{ (concentrated)} \longrightarrow$$
$$Li + N_2 \longrightarrow$$
$$HNO_3 + P_4O_{10} \longrightarrow$$

5. How would you prepare a sample of deuterated ammonia, ND_3 ?

6. Illustrate how nitrogen compounds provide good examples of multiple bonding and resonance?

7. Determine the oxidation number of nitrogen in

(a) N_2O (b) NO_2

(c) HNO_3 (d) NH_3

8. Write the resonance Lewis structure for N_2O, NO_2 and NO_3^-.

9. Illustrate how a metal such as copper can give different products with nitric acid? Give balanced equations.

10. Write balanced equation(s) for the manufacture of nitric acid by the oxidation of ammonia.

11. Describe the uses of ammonia and nitric acid.

12. Classify the following compounds into acidic, basic and amphoteric oxides

SrO, SiO_2, Al_2O_3, N_2O_5, SO_2, P_4O_6, Cl_2O_7. Write balanced equations for the reaction of each of these compounds with water, a base or an acid as the case may be.

13. How far do you agree with the phenomenon concept of 'inert pair effect'? Justify your answer.

14. Write balanced equations for the following reactions

(a) $Ca_3P_2 + H_2O \longrightarrow$

(b) $P_4O_{10} + H_2O \longrightarrow$

(c) $As_4 + Cl_2$ (excess) $\longrightarrow$

(d) $P_4 + KOH + H_2O \longrightarrow$

15. Describe the method for the isolation of elemental phosphorus from $Ca_3(PO_4)_2$.

16. What is the oxidation state of phosphorus in the following?

(a) H_3PO_3 (b) PCl_3

(c) Ca_3P_2 (d) Na_3PO_4

(e) POF_3 ?

17. Calculate the volume of 0.1 M NaOH solution required to neutralise the solution produced by dissolving 1.1 g of P_4O_6 in water.

Hint : $P_4O_6 + 6H_2O \longrightarrow 4H_3PO_3$

$$H_3PO_3 + 2NaOH \longrightarrow Na_2HPO_3 + 2H_2O] \times 4$$

$$P_4O_6 \equiv 4H_3PO_3 \equiv 8NaOH \text{ (}H_3PO_3\text{ is a dibasic acid)}$$

$$1.1 \text{ g } P_4O_6 = \frac{1.1}{220} = 5 \times 10^{-3} \text{ mol } P_4O_6$$

$$= 4 \times 10^{-2} \text{ mol NaOH}$$

Let $\quad NaOH = V \text{ mL}$

$$\therefore \text{ Moles of NaOH} = \frac{V \times 0.1}{1000}$$

$$\frac{V \times 0.1}{1000} = 4 \times 10^{-2}$$

$$\therefore \quad V = 400 \text{ mL}$$

$$= \mathbf{400 \text{ mL of } 0.1 \text{ M NaOH}}$$

18. Compare the structures of white phosphorus, P_4O_6 and P_4O_{10}.

Exercise 2
(Stage 2 : High Skill Problem Solving)

Only One Option Correct

1. Of the following acids
I : hypo phosphorous acid
II : hydrofluoric acid,
III : oxalic acid
IV : glycine
(a) I, II are monobasic; III dibasic acid and IV amphoteric
(b) II monobasic; I, III dibasic acid; IV amphoteric
(c) I monobasic, II, III dibasic, IV amphoteric
(d) I, II, III dibasic; IV amphoteric

2. Following metals are soluble in aqua-regia
(a) Pt (b) Au
(c) Ag (d) all of these

3. Which forms $p\pi-p\pi$ multiple bonds with itself and with C and O?
(a) P, As (b) N, As
(c) N, P (d) N

4. In P_4 (tetrahedral)
(a) each P is joined to four P
(b) each P is joined to three P
(c) each P is joined to two P
(d) P_4 does not exist

5. Maximum number of covalent bonds formed by N and P respectively are
(a) 3, 5 (b) 3, 6
(c) 4, 5 (d) 4, 6

6. For the hydrides of nitrogen family going down the group
(a) stability decreases
(b) reducing activity increases
(c) bond angle HMH decreases
(d) all the above variations followed

7. Which of the following compounds do not exist?
N_4, P_4, PCl_5, NCl_5, NCl_3, P_2O_5, NO_2, PO_2
(a) N_4, NCl_5, NO_2 (b) N_4, NCl_5, PO_2
(c) PCl_5, NCl_5, PO_2 (d) All of these

8. Which of the following is/are paramagnetic?
NO_2, NO, N_2O_4, N_2O_2, N_2O_5
(a) Only NO_2 (b) NO_2, NO
(c) NO, NO_2, N_2O_5 (d) All are paramagnetic

9. Which of the following dibasic acids shows geometrical isomerism?
(a) Hyponitrous acid $H_2N_2O_2$
(b) Maleic acid $C_4H_4O_4$
(c) Both (a) and (b)
(d) None of the above

10. Which is/are correct statements about P_4O_6 and P_4O_{10}?
(a) Both form oxyacids H_3PO_3 and H_3PO_4 respectively
(b) In P_4O_6 each P is joined to four O and in P_4O_{10} each P is joined to six O atoms
(c) Both (a) and (b)
(d) None of the above

11. 1 mole each of H_3PO_2, H_3PO_3 and H_3PO_4 will neutralise x moles of NaOH, y moles of $Ca(OH)_2$ and z moles of $Al(OH)_3$ (assuming all as strong electrolytes). x, y, z are in the ratio of
(a) 3 : 1.5 : 1 (b) 1 : 2 : 3
(c) 3 : 2 : 1 (d) 1 : 1 : 1

12. The dipole moments of the given molecules are such that
(a) $BF_3 > NF_3 > NH_3$ (b) $NF_3 > BF_3 > NH_3$
(c) $NH_3 > NF_3 > BF_3$ (d) $NH_3 > BF_3 > NF_3$

13. The electronegativities of N, C, Si and P are such that
(a) P < Si < C < N (b) Si < P < N < C
(c) Si < P < C < N (d) P < Si < N < C

14. Glacial phosphoric acid is
(a) HPO_3 (b) H_3PO_3
(c) H_3PO_4 (d) $H_4P_2O_7$

15. Phosgene can be obtained when
(a) white phosphorus reacts with alkali
(b) calcium phosphide reacts with water
(c) chloroform reacts with air
(d) bone comes in contact with water

16. Which are acid salts?
(a) NaH_2PO_2, Na_2HPO_3 (b) Na_2HPO_3, Na_2HPO_4
(c) $NaHCO_3$, Na_2HPO_4 (d) All of these

17. Which of the following statements are true?
(a) Cold and very dilute HNO_3 forms NH_4NO_3 with Zn or Sn
(b) Concentrated HNO_3 forms H_2SnO_3 with Sn
(c) Cold and more concentrated HNO_3 forms NO_2 with Cu
(d) All the above are correct

18. NO_2 is not obtained when following is heated
(a) $Pb(NO_3)_2$ (b) $AgNO_3$
(c) $LiNO_3$ (d) KNO_3

19. Which is least basic?
(a) NF_3 (b) NCl_3
(c) NBr_3 (d) NI_3

20. Which is/are used as an anaesthetic?
(a) N_2O (b) C_2H_4
(c) $CHCl_3$ (d) All of these

21. Which is a set of acid salts and can react with base?
(a) NaH_2PO_2, Na_2HPO_3, NaH_2PO_4
(b) Na_2HPO_3, NaH_2PO_3, Na_2HPO_4
(c) $NaHCO_3$, NaH_2PO_3, Na_2HPO_4
(d) All of the above

22. Which is used as a rocket fuel?
(a) N_2H_4 (b) Polybudiene
(c) Both (a) and (b) (d) None of these

23. NH_3 can't be obtained by
(a) heating of NH_4NO_3 or NH_4NO_2
(b) heating of NH_4Cl or $(NH_4)_2CO_3$
(c) heating of NH_4NO_3 with NaOH
(d) reaction of AlN or Mg_3N_2 or $CaCN_2$ with H_2O

24. Red and white phosphorus will differ but not in
(a) smell
(b) solubility in $CHCl_3$
(c) exhibiting phosphorescence
(d) reactions with HNO_3

25. Choose correct statements.
(a) Superphosphate is $[3Ca(H_2PO_4)_2 + 7CaSO_4]$
(b) triple superphosphate is $4Ca(H_2PO_4)_2$
(c) Both (a) and (b) are correct
(d) None of the above is correct

26. N_2O (laughing gas) finds use in the following except
(a) as a propellant for whipped ice-cream
(b) as an anaesthetic
(c) for the preparation of N_3H
(d) as fuel for rockets

27. Following are neutral oxides except
(a) NO (b) N_2O
(c) CO (d) NO_2

28. Select incorrect statement about N_2O_4.
(a) It self ionises as NO^+, NO_3^-
(b) Substance containing NO^+ is said to be acid and that containing NO_3^- is said to be base
(c) N_2O_4 is paramagnetic
(d) NO_2 dimerises to N_2O_4 with disappearance in paramagnetism

29. Acid rain may cause
(a) rusting easier
(b) stone-cancer in Taj Mahal
(c) non-fertility of soil
(d) all of the above

30. Select correct statement.
(a) Mixture of NH_4Cl and $NaNO_3$ on heating gives N_2 gas
(b) CFC is used as refrigerating fluid and as propellant in aerosols
(c) Phosgene is formed when P_4 reacts with NaOH
(d) Phosgene dissolves in water forming P_2O_5

31. 'Anhydrone' is a very effective desiccant (water absorber) used in 'dry batteries'. It is
(a) conc. H_2SO_4 (b) P_2O_5
(c) $CaCl_2$ (d) $Mg(ClO_4)_2$

32. N_2O is formed when
(a) moist Fe reacts with NO
(b) Sn^{2+} reacts with conc. HNO_3 in presence of HCl
(c) cold dil. HNO_3 reacts with Cu and Zn
(d) by all the reactions

33. Which of the following compounds has the greatest ionic character?
(a) PCl_5 (b) $SiCl_4$
(c) CCl_4 (d) BCl_3

34. Fe^{3+} oxidises NH_2OH to
(a) NO_2 (b) N_2O
(c) N_2 (d) NO

35. Fe^{2+} reduces NH_2OH to
(a) NH_3 (b) N_3H
(c) N_2H_4 (d) N_2

36. All of the following are bases except
(a) N_2H_4 (b) NH_2OH
(c) NH_3 (d) N_3H

37. An example of an *"ortho"* acid is
(a) $HAsO_2$ (b) H_3AsO_4
(c) $(HPO_3)_n$ (d) $H_4As_2O_7$

38. Compound used in safety matches is
(a) P_4S_3 (b) P_4
(c) P_2O_5 (d) PCl_3

39. Repeated use of which one of the following fertilizers would increase the acidity of the soil?
(a) Urea
(b) Superphosphate of lime
(c) Ammonium sulphate
(d) Potassium nitrate

40. Arsenic drugs are mainly used for the treatment of
(a) cholera (b) typhoid
(c) jaundice (d) syphilis

41. The wrong statement about N_2O is
(a) it is nitrous oxide
(b) it is least reactive oxide of nitrogen
(c) it is not a linear molecule
(d) it is known as laughing gas

42. The true statement for the acids of phosphorus. H_3PO_2, H_3PO_3 and H_3PO_4 is
(a) their acidic nature is : $H_3PO_4 < H_3PO_3 < H_3PO_2$
(b) all of them are reducing in nature
(c) all of them are tribasic acids
(d) the geometry of phosphorus is tetrahedral in all the three

43. The oxide which is not a reducing agent is
(a) CO_2 (b) NO_2
(c) SO_2 (d) ClO_2

44. Nitrogen is obtained by the thermal decomposition of
(a) NH_4NO_2 (b) NH_4NO_3
(c) $AgNO_3$ (d) $Pb(NO_3)_3$

45. The substance used as a fast drying agent in the laboratory is
(a) Na_2PO_4
(b) P_2O_5
(c) charcoal
(d) anhydrous calcium chloride

46. Laughing gas is prepared by heating
(a) $NH_4Cl + NaNO_3$ (b) NH_4Cl
(c) $(NH_4)_2SO_2$ (d) NH_4NO_2

47. The sides of safety matches contain
(a) red phosphorus + sand powder
(b) P_4S_3
(c) $Ca_3(PO_4)$ + glass pieces
(d) $KClO_3, KNO_3$, sulphur + antimony

48. Cold solution of barium nitrite on mixing with sulphuric acid produces
(a) $BaSO_4 + HNO_2$ (b) $BaSO_4 + HNO_3$
(c) $BaSO_4 + NO_2$ (d) $BaSO_4 + N_2 + O_2$

49. $4HNO_3 + P_4O_{10} \longrightarrow 4HPO_3 + X$
In the above reaction the product X is
(a) NO_2 (b) N_2O_3
(c) N_2O_4 (d) N_2O_5

50. Which of the following oxides of nitrogen is paramagnetic?
(a) N_2O (b) N_2O_5
(c) NO_2 (d) N_2O_3

51. The structural formula of hypophosphorus acid is

(a) 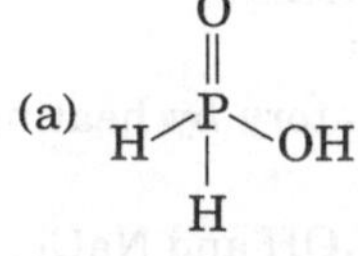

(b) 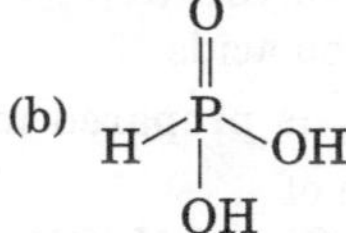

(c) 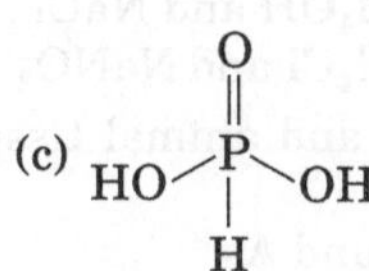

(d) O
HO P OH
OH

52. Among the following, the paramagnetic compound is
(a) Na_2O_2 (b) P_2O_5
(c) N_2O (d) KO_2

53. Which of the following is most acidic?
(a) N_2O_5 (b) P_2O_5
(c) As_2O_5 (d) Sb_2O_5

54. Which of the following phosphorus is the most reactive?
(a) Red phosphorus (b) White phosphorus
(c) Scarlet phosphorus (d) Violet phosphorus

55. The percentage of p-character in the orbitals forming P—P bonds in P_4 is
(a) 25 (b) 33
(c) 50 (d) 75

56. Which blue-liquid is obtained on reacting equimolar amounts of two gases at –30°C?
(a) N_2O (b) N_2O_3
(c) N_2O_4 (d) N_2O_5

57. Calcium cyanamide on treatment with steam produces
(a) $NH_3 + CaO$ (b) $NH_3 + CaHCO_3$
(c) $NH_3 + CaCO_3$ (d) $NH_3 + Ca(OH)_2$

58. Which gas cannot be collected over water?
(a) O_2 (b) PH_3
(c) N_2 (d) SO_2

59. NaOCl reacts with NH_3 to produce
(a) NH_2OH (b) NH_2NH_2
(c) N_2 (d) NO

60. Which one of the following acid possesses oxidising, reducing and complex forming properties?
(a) HNO_3 (b) HCl
(c) H_2SO_4 (d) HNO_2

61. On hydrolysis, PCl_3 gives
(a) H_3PO_3 (b) H_3PO_4
(c) $POCl_3$ (d) HPO_3

62. On photochemical decomposition of HNO_3 the products formed are
(a) NO and O_2 (b) NO_2 and O_2
(c) N_2O_4 and NO_2 (d) N_2O_5 and NO_2

63. Nitric oxide is formed by the reaction of
(a) $C + N_2O$ (b) $Cu + HNO_3$
(c) $Cu + N_2O$ (d) $Cu + NO_2$

64. On heating lead nitrate the product formed is
(a) NO (b) NO_2
(c) N_2O (d) N_2O_5

65. The number of P—O—P bonds in cyclic metaphosphoric acid is
(a) zero (b) two
(c) three (d) four

66. Polyphosphates are used as water softening agents because they
(a) form soluble complexes with anionic species
(b) precipitate anionic species
(c) form soluble complexes with cationic species
(d) precipitate cationic species

67. Solid crystalline PCl_5 has structure which of the following?
(a) Bi-pyramidal moieties
(b) Octahedral and tetrahedral ions
(c) Square-pyramidal moieties
(d) Pentagonal moieties

68. Hydrazine is not
(a) a reducing agent (b) an oxidising agent
(c) an acid (d) a base

69. Which of the following oxides of nitrogen is solid?
(a) NO_2 (b) N_2O
(c) N_2O_3 (d) N_2O_5

70. The element which forms oxides in all oxidation states +1 to +5 is
(a) N (b) P
(c) As (d) Sb

71. Which of the following is arranged in the increasing order of enthalpy of vaporisation?
(a) NH_3, PH_3, AsH_3 (b) AsH_3, PH_3, NH_3
(c) NH_3, AsH_3, PH_3 (d) PH_3, AsH_3, NH_3

72. The statement true for N_3^- is
(a) It has non-linear structure
(b) It is called pseudohalogen
(c) the formal oxidation state of nitrogen in this anion is +1
(d) It is isoelectronic with N_2O

73. Metallic tin and conc. HNO_3 reacts to form
(a) metastannic acid (b) stannic nitrate
(c) stannous nitrite (d) stannous nitrate

74. PI_3 upon hydrolysis give
(a) monobasic acid and dibasic acid
(b) monobasic acid and tribasic acid
(c) monbasic acid and a salt
(d) dibasic acid and tribasic acid

75. Common name of phosgene is
(a) carbon tetrachloride (b) phosphoryl chloride
(c) carbonyl chloride (d) phosphorus trichloride

76. The bond angle in AsH_3 is greater than that in
(a) NH_3 (b) H_2O
(c) BCl_3 (d) None of these

77. When PCl_3 reacts with sulphuric acid sulphuryl chloride is formed as the final product. This shows that sulphuric acid
(a) has two hydroxyl groups in its structure
(b) is a derivative of sulphur dioxide
(c) is a dibasic acid
(d) has greater affinity for water

78. Pentavalence is phosphorus is more stable when compared to that of nitrogen even through they belong to same group is due to
(a) reactivity of phosphorus
(b) inert nature of nitrogen
(c) presence of *d*-orbital
(d) large size of phosphorus atom

79. Bond angle of 109°28′ is found in
(a) NH_3 (b) H_2O
(c) $\overset{\oplus}{C}H_5$ (d) $\overset{\oplus}{N}H_4$

80. What is role of phosphate ion in a detergent?
(a) It reduces pH of the water
(b) It increase pH of the water
(c) It removes Ca^{2+} and Mg^{2+} ions from water that causes hardness
(d) It increases its solubility in water

81. Among the VA(15) group elements one of them forms a compound sold under the name **'Pearl White'**. The element used is
(a) N (b) P
(c) Sb (d) Bi

82. White arsenic has the chemical composition
(a) As_2 (b) As_4
(c) As_2O_5 (d) As_2O_3

83. Both NF_3 and NCl_3 are covalent but they differ in the extent of hydrolysis because
(a) NF_3 is more stable than NCl_3 and hydrolysis product of NF_3, HOF, does not exist
(b) dipole moment of NF_3 is greater than that of NCl_3
(c) electronegativity of F is greater than that of Cl
(d) Cl can expand its octet by using *d*-orbitals

84. Nitrogen differs from P, As, Sb and Bi in the following properties
(a) It forms diatomic molecule
(b) It is not able to extent its coordination number beyond four
(c) Nitrogen does not function as a Lewis acid whereas P, As, Sb and Bi do so
(d) In all the properties given above

85. $d\pi$- $p\pi$ bonding is shown in
(a) $NO_3^-, NO_2^-, N^{3-}, CN^-$ (b) $P_2O_3, P_2O_5, PO_4^{3-}$
(c) NH_3, PH_3, BiH_3 (d) CO, NO, CO_2, NO_2

86. In plants, atmospheric nitrogen is converted into
(a) carbohydrate (b) proteins
(c) amino acids (d) fats

87. Pure N_2 is prepared in the laboratory by heating a mixture of
(a) NH_4Cl and NaOH (b) NH_4OH and NaCl
(c) NH_4Cl and $NaNO_2$ (d) NH_4Cl and $NaNO_3$

88. Essential constituents of plants and animal tissues are
(a) N and P (b) N and As
(c) Cu and Mg (d) Ca and Mg

89. Major minerals containing nitrogen are
(a) DNA (b) RNA
(c) KNO_3, $NaNO_3$ (d) all of these

90. Most abundant uncombined element present in atmosphere is
(a) Ca (b) Mg
(c) Al (d) N

91. Substances burn more readily in N_2O than in air because N_2O
(a) is reative at high temperature
(b) dissociates to give O_2 that supports combustion

(c) the activation energy is increased on increasing temperature
(d) acts as a catalyst

92. Nitric oxide is paramagnetic in
(a) gaseous state (b) liquid state
(c) solid state (d) polymeric state

93. Nitrosonium ion (NO^+) is isoelectronic with
(a) carbon dioxide (b) carbon monoxide
(c) nitrogen dioxide (d) nitric oxide

94. $p\pi$-$p\pi$ multiple bonding is present in
(a) oxides of nitrogen (b) oxides of phosphorus
(c) halides of nitrogen (d) halides of phosphorus

95. Paramagnetic species are
(a) NO, NO_2, O_2 (b) N_2O_2, N_2O_4, O_3
(c) NO^+, NO_2^+, O_2^+ (d) O_2, CO, CO_2

96. With excess of Cl_2 ammonia forms
(a) ammonium chloride (b) nitrogen trichloride
(c) nitrosyl chloride (d) nitrogen

97. The dipole moment of NF_3 is less than that of NH_3 because
(a) F is more reactive than H
(b) NH_3 forms associated molecules
(c) the resultant of the individual bond polarities is less
(d) the resultant of the individual bond polarities is opposed by the polarity of lone pair

98. $NH_4Cl(s)$ is heated in a test tube. Vapours are brought in contact with red litmus paper, which changes to blue and then to red. It is because of
(a) formation of NH_4OH and HCl
(b) formation of NH_3 and HCl
(c) greater diffusion of NH_3 than HCl
(d) greater diffusion of HCl than NH_3

99. Select incorrect statement about hydrides of group 15 elements.
(a) The central atom in the hydride is sp^2 hybridised
(b) NH_3 readily form NH_4^+ salts with H^+; PH_4^+ salts are formed with H^+ under anhydrous condition
(c) The tetrahedron is distorted due to repulsion between the lone pair of electrons and the bond pairs
(d) The bond energy of the M—H bond decreases from NH_3 to BiH_3 because of increase in the size of the element

100. In liquid NH_3
(a) NH_4Cl is an acid
(b) $NaNH_2$ is a base
(c) CH_3COOH behaves as strong acid
(d) all of the above facts are true

101. A gaseous substance dissolves in water giving a pale blue solution which decolorises $KMnO_4$ and oxidises KI to I_2. Gaseous substances is
(a) N_2O_5 (b) NH_3
(c) N_2O_3 (d) HNO_3

102. $p\pi$-$p\pi$ multiple bonding between nitrogen atoms is present in
(a) hyponitrous acid (b) nitrous acid
(c) nitric acid (d) in all of these

103. In which case geometrical isomerism is not exhibited?
(a) Hyponitrous acid (b) 2-butene
(c) 1-butene (d) 2-pentene

104. Ambidentate ligands are
(a) NO_2^- (b) CN^-
(c) CNS^- (d) all of these

105. MnO_4^-/H^+ oxidises NO_2^- to
(a) N_2 (b) NH_3
(c) NO_3^- (d) N_2O

106. A colourless salt gives a white ppt (soluble in ammonium acetate) and a brown coloured pungent gas on reaction with conc. H_2SO_4. Salt is
(a) $Ba(NO_3)_2$ (b) $Pb(NO_3)_2$
(c) $NaNO_3$ (d) NH_4NO_3

107. Nitration of aromatic compounds is done using a mixture of conc. H_2SO_4 and conc. HNO_3. Intermediate formed in the nitration process is
(a) NO_2^+ (nitronium ion)
(b) NO_3^- (nitrate ion)
(c) NO^+ (nitrosonium ion)
(d) NO_2^- (nitrite ion)

108. $NaNO_2$ and $NaNO_3$ are used as food additives and generally added to meat to
(a) retard oxidation of meat to brown coloured material
(b) prevent growth of botulism bacteria
(c) impart red colour to oxidised meat by NO by reaction of heme in blood
(d) make all of the functions given above effective

109. Nitrites in water may react with amines under acidic conditions in the stomach to
(a) give carcinogenic compounds
(b) give proteins
(c) keep stomach acidity free
(d) produce free amino acids

110. HNO_3 in pure state is colourless but it is often of yellow colour. It is due to
(a) unstable structure of HNO_3 which immediately changes to NO_2^-
(b) photochemical decomposition in presence of sunlight
(c) interaction of atmospheric gases with HNO_3
(d) conversion of HNO_3 into NO_2^+

111. Among these ores the highest phosphorus content is in
(a) chlorapatite (b) phosphorite
(c) fluorspar (d) equal

112. A person working with phosphorus suffers from a disease in which bones decay. It is known as
(a) arthrities (b) phossy jaw
(c) rickets (d) cancer

113. In the preparation of red phosphorus from white phosphorus
(a) MnO_2 is used as a catalyst
(b) the white phosphorus is treated in an electric furnace
(c) a little iodine is used as catalyst
(d) the gas P_4 is released

114. Compound A undergoes hydrolysis to produce a colourless gas with rotten fish smell. The gas gives a vortex ring. The gas is
(a) PH_3 (b) P_2O_3
(c) P_2O_5 (d) P_2S_5

115. One of the acid listed below is formed only from P_2O_3; the rest are formed from P_2O_5. Acid formed from P_2O_3 is
(a) HPO_3 (b) $H_4P_2O_7$
(c) H_3PO_4 (d) H_3PO_3

116. Consider the reaction

$$(NH_4)_2HAsO_3 + CuSO_4 \longrightarrow \underbrace{CuHAsO_3}_{\text{green ppt}} + (NH_4)_2SO_4$$

The green precipitate is also known as
(a) Pair's green (b) Scheele's green
(c) Verdigris' green (d) Rinmann's green

117. A substance which gives an yellow precipitate when boiled with an excess of nitric acid and ammonium molybdate, and red precipitate with $AgNO_3$ is
(a) orthophosphate (b) pyrophosphate
(c) metaphosphate (d) hypophosphate

118. This salt of oxoacids of phosphorus is used as washing detergent
(a) wavellite (b) microcosmic salt
(c) calgon (d) chlorapatite

119. Calgon (sodium metaphosphate) finds application as
(a) artificial jewel (b) paint
(c) a resin (d) washing powder

120. With $FeCl_3$, a neutral solution of orthophosphate gives an yellow precipitate insoluble in
(a) CH_3COOH (b) HCl
(c) H_2SO_4 (d) NaOH

121. Phosphorus trichloride, PCl_3, undergoes hydrolysis to produce an oxoacid. It has the formula
(a) HPO_3 (b) H_3PO_3
(c) H_3PO_4 (d) H_3PO_2

122. The solid PCl_5 exists as
(a) PCl_3 (b) PCl_4^+
(c) PCl_6^- (d) PCl_4^+ and PCl_6^-

123. $\underset{X}{PCl_5} + \underset{Y}{CH_3}\underset{Z}{CH_2OH} \longrightarrow$

In the above reaction Cl replaces one of the H-atoms in CH_3CH_2OH. This H is of
(a) C (in X) (b) C (in Y)
(c) O (in Z) (d) any of X, Y and Z

124. The high reactivity and low volatility of white phosphorus is due to
(a) tetrahedrally arranged P_4 units
(b) bond angle of 60°
(c) weak van der Waals' force of attraction
(d) increased steric (strain) factor

125. Stability of pentahalides is in order
(a) P > Sb > As > Bi for given X
(b) F > Cl > Br for given element
(c) both (a) and (b)
(d) None of the above

One or More Than One Options Correct

1. N_2H_4 and H_2O_2 show similarity in
(a) hybridisation of central atoms
(b) oxidising nature
(c) reducing nature
(d) molar mass

2. In acidic medium N_2H_4
(a) is oxidised to N_2
(b) loses four electrons
(c) has equivalent mass 8
(d) reduces $AuCl_3$ to AuCl

3. N_2H_4 reduces IO_3^-/H^+
(a) to I^+
(b) with I_2 as an intermediate indicated by violet colour in CCl_4 layer
(d) indicated by blue colour with starch
(d) to I^-

4. N_2H_4
(a) is an endothermic compound
(b) burns in air with evolution of heat in an exothermic compound
(c) is kinetically stable
(d) reduces $FeCl_3$ to $FeCl_2$ in acidic medium

5. Thermodynamic tendency of $PH_3(g)$ to react with $HX(g)$ would be favoured by
(a) low H—X bond energy
(b) high electron affinity of X
(c) high lattice energy of $PH_4X(s)$
(d) high electronegativity of P

6. The following side reaction in the production of N_2H_4

$$N_2H_4 + 2NH_2Cl \longrightarrow N_2 + 2NH_4Cl$$

(a) is catalysed by traces of heavy metals as Cu^{2+}
(b) is suppressed by addition of gelatin or glue

(c) is made reversible by removing N_2
(d) is made reversible by adding NaOH

7. NH_2OH can be used as
(a) on oxidising agent
(b) a reducing agent
(c) an autioxidant in photographic developers
(d) oxime forming reagent (with carbonyl compound)

8. Select correct statement(s).
NH_2OH forms *syn* and *anti* isomer with
(a) benzaldehyde (b) acetone
(c) acetaldehyde (d) benzophenone

9. Nylon-6 is formed from caprolactam which is formed from
(a) cyclohexanone and hydroxyl amine
(b) cyclohexanone and hydrazine
(c) benzophenone and hydrazine
(d) benzophenone and hydroxylamine

10. Select correct statement(s).
(a) $SbCl_3$ is hydrolysed to orange SbOCl soluble in dil. HCl
(b) $BiCl_3$ is hydrolysed to white BiOCl soluble in dil. HCl
(c) PCl_5 can change to PCl_6^- and PCl_4^+ with no change in hybridisation of P.
(d) NCl_5 and PCl_5 are stable compounds.

Passage Comprehension Questions

Passage 1

Consider the six second and third-row elements in group 15–17 of the Periodic Table. Possible molecular structures for common allotropes of these elements are shown below:

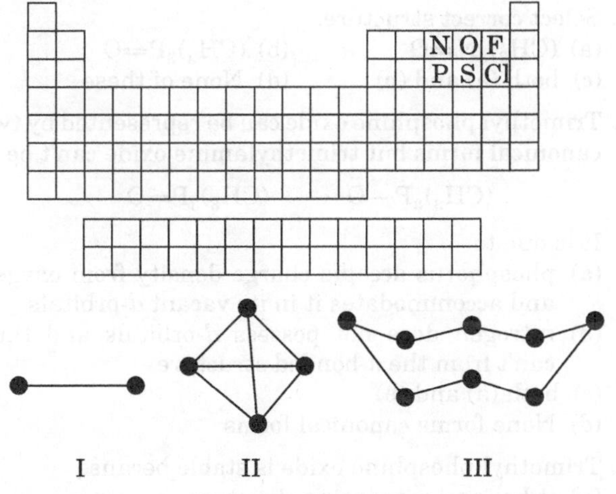

1. Following is the matching of the structure of molecular form. Which is not the correct matching?
(a) Nitrogen I (b) Phosphorus II
(c) Sulphur III (d) Ozone I

2. Bond energy of
$(N{\equiv}N) = 946 \text{ kJ mol}^{-1}$
$(P{\equiv}P) = 481 \text{ kJ mol}^{-1}$
$(N{-}N) = 160 \text{ kJ mol}^{-1}$
$(P{-}P) = 215 \text{ kJ mol}^{-1}$
which molecules do not exist?
(a) P_4 (b) N_4
(c) N_2 (d) P_2

3. Following table compares properties of red phosphorus and white phosphorus. Which is incorrect comparison?

	Red	White
(a)	Nontoxic and has polymeric structure	Toxic, waxy discrete tetrahedral P_4 molecule
(b)	Discrete tetrahedral P_4 molecule	Polymeric structure
(c)	High m.p., less reactive	Low m.p., more reactive
(d)	Shows no phosphorescence in air	Shows phosphorescence in air

4. In P_4 (tetrahedral)
(a) each P is joined to four P
(b) each P is joined to three P
(c) each P is joined to two P
(d) P_4 does not exist

5. High reactivity of white phosphorus is due to
(a) unusual bonding that produces considerable strain
(b) high solubility in water
(c) oxide layer made on it
(d) all of the above

6. White phosphorus reacts with NaOH forming PH_3 and $H_2PO_2^-$. This is an example of
(a) neutralisation
(b) thermal decomposition
(c) redox reaction
(d) disproportionation

Passage 2

Question given below is based on the following structures:

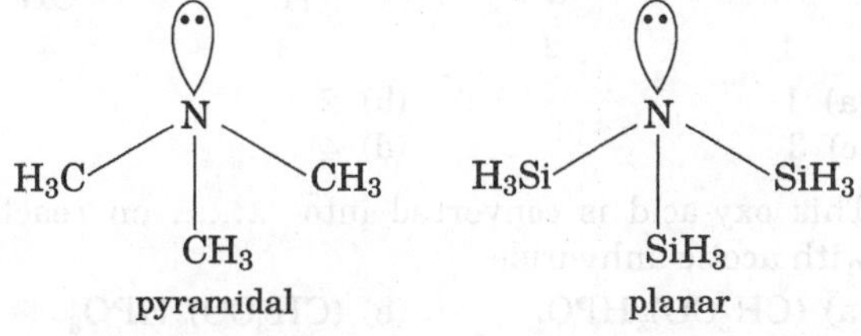

1. Structures of $N(CH_3)_3$ and $N(SiH_3)_3$ are different. It is due to the fact that
(a) silicon also uses *d*-orbitals in multiple bonding

(b) in case of $N(SiH_3)_3$, lone pair of N-atom is transferred to the empty d-orbitals of silicon ($p\pi$-$d\pi$ overlapping)
(c) both of the above
(d) none of the above

Passage 3

Consider following scheme involving oxides and oxy-acids of nitrogen. Based on this answer the questions given at the end.

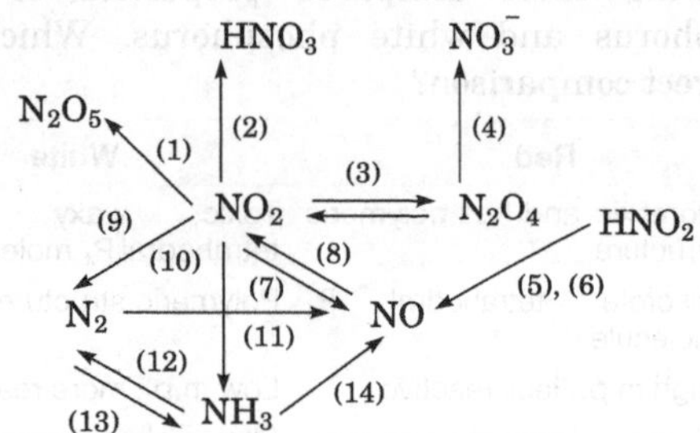

1. Out of the following which reactions are disproportionations?
(a) 2, 10 (b) 2, 3
(c) 1, 11 (d) 13, 14

2. In which case change in oxidation number is maximum?
(a) 1 (b) 10
(c) 5 (d) 11

Passage 4

Questions given below are based on the following experiment.

An oxyacid of phosphorus has the following properties

Complete neutralisation of the acid with sodium hydroxide solution gives an aqueous solution of sodium ions and oxy-acid anions in the ratio 2 : 1. When a solution of the acid is warmed with silver nitrate solution metallic silver is deposited.

1. What is the structure of the oxy-acid?

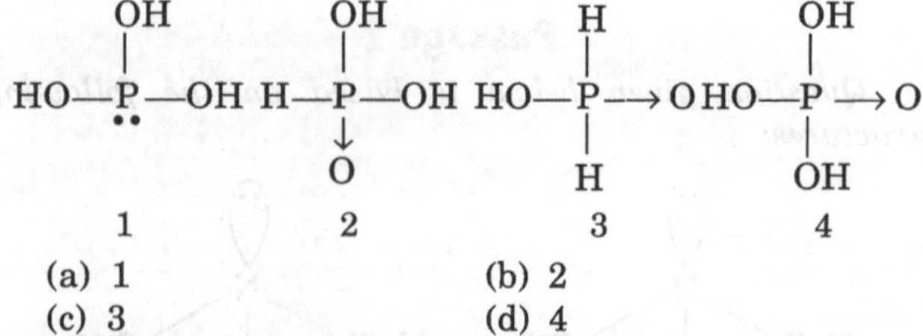

(a) 1 (b) 2
(c) 3 (d) 4

2. This oxy-acid is converted into .... on reaction with acetic anhydride
(a) $(CH_3CO)_2HPO_4$ (b) $(CH_3CO)_2HPO_3$
(c) P_2O_3 (d) P_2O_5

3. pH of 0.01 M solution of the sodium salt of this acid formed (assume $pK_{a_1} = 2x$) is
(a) $6 + 2x$ (b) $6 - 2x$
(c) $6 - x$ (d) $6 + x$

Passage 5

The following flow diagram represents the industrial preparation of nitric acid from ammonia:

$$NH_3 + O_2 \text{ (excess air)} \xrightarrow[900°C]{(A)} NO \xrightarrow[\text{air}]{(B)} (C) \xrightarrow[\text{water}]{} HNO_3 + NO$$

(NO from the last step is returned to the NO stage.)

Answer the questions given below

1. Which line of entry describes the undefined reagents, products and reaction conditions?

	A	B	C
(a)	Catalyst	cool (– 25°C)	NO_2
(b)	Catalyst	cool (– 25°C)	N_2O
(c)	Catalyst	high pressure	NO_2
(d)	High pressure	catalyst	N_2O_3

2. Formation of HNO_3 when (C) is dissolved in H_2O takes place through various reactions. Select the reaction not observed in this step.
(a) $NO_2 + H_2O \longrightarrow HNO_3 + HNO_2$
(b) $HNO_2 \longrightarrow H_2O + NO + NO_2$
(c) $NO_2 + H_2O \longrightarrow HNO_3 + NO$
(d) None of the above

Passage 6

Trimethylamine oxide and trimethyl phosphine oxide can be represented as

$$\underset{\text{(I)}}{(CH_3)_3\overset{+}{N}—O^-} \qquad \underset{\text{(II)}}{(CH_3)_3\overset{+}{P}—\overset{-}{O}}$$

Answer the following questions based on these

1. Select correct structure.
(a) $(CH_3)_3N{=}O$ (b) $(CH_3)_3P{=}O$
(c) both (a) and (b) (d) None of these

2. Trimethyl phosphine oxide can be represented by two canonical forms but trimethylamine oxide can't be

$$(CH_3)_3\overset{+}{P}—\overset{-}{O} \longleftrightarrow (CH_3)_3P{=}O$$

It is due to
(a) phosphorus accepts charge density from oxygen and accommodates it in its vacant d-orbitals
(b) nitrogen does not possess d-orbitals and thus can't from the π-bonded structure
(c) both (a) and (b)
(d) None forms canonical forms

3. Trimethyl phosphine oxide is stable because
(a) it has two resonating structure
(b) phosphorus-oxygen double bond has high dissociation energy
(c) both (a) and (b)
(d) None of the above

Passage 7

N_2H_4 reduces IO_3^- to I^+ in acidic medium

$$N_2H_4 + IO_3^- + H^+ \longrightarrow N_2 + I^+ + H_2O$$

Answer the following questions

1. Equialent mass of N_2H_4 (molar mass = 42 g mol^{-1}) is

(a) 42 (b) 21
(c) 10.5 (d) 14

2. Equivalent mass of KIO_3 (molar mass = 214 g mol^{-1}) is

(a) 71.33 (b) 107.0
(c) 214.0 (d) 53.5

3. Intermediate in the above reaction is I_2 which can be detected by

(a) taking solvent CCl_4 in which I_2 is soluble and thus layer becomes violet
(b) starch as indicator which turns blue
(c) both (a) and (b)
(d) Nonc of thc abovc

4. N_2H_4 is oxidised to N_2. Both N-atom in N_2 are from N_2H_4. It can be verified by

(a) taking ^{15}N enriched N_2H_4,
(b) taking ^{2}D enriched N_2H_4
(c) taking both (a) & (b)
(d) taking none of the above

Assertion & Reason

Codes :

(a) Both **A** and **R** are true and **R** is the correct explanation of **A.**
(b) Both **A** and **R** are true but **R** is not the correct explanation of **A.**
(c) **A** is true but **R** is false.
(d) **A** is false but **R** is true.

1. Assertion (A) : On cooling, the brown colour of NO_2 disappears.

Reason (R) : On cooling NO_2 forms dimer (N_2O_4) resulting in the pairing of the odd electrons in NO_2.

2. Assertion (A) : Both H_3PO_3 and H_3PO_4 have the same number of hydrogen atoms but H_3PO_4 is a tribasic acid and H_3PO_3 is a dibasic acid.

Reason (R) : 1 mole of H_3PO_3 is neutralised by 2 moles of NaOH while 1 mole of H_3PO_4 is neutralised by 3 moles of NaOH.

3. Assertion (A) : N_2H_4 is kinetically stable compound.

Reason (R) : Two N-atoms are bonded by triple bonds.

4. Assertion (A) : N_3H explodes violently on slight heating.

Reason (R) : N_3^- has linear structure and is stable due to resonance.

5. Assertion (A) : N_2H_4 is a weaker base than NH_3.

Reason (R) : Its behaviour is bifunctional.

6. Assertion (A) : Equivalent mass of N_2H_4 is 8 in acidic medium or basic medium.

Reason (R) : It is oxidised to N_2 with loss of four electron per N_2H_4 unit.

7. Assertion (A) : IO_3^- is reduced to I^+ but intermediate gives violet colour in CCl_4 (taken as medium).

Reason (R) : I_2 is formed as the intermediate.

8. Assertion (A) : Each N-atom is sp^3-hybridised in N_2H_4.

Reason (R) : It has *gauche* conformation.

9. Assertion (A) : H_2O_2 and N_2H_4 have oxidant and reductant nature.

Reason (R) : Both have equivalent mass of 8.

10. Assertion (A) : N_3H is a weak acid and NH_3 is a weak base.

Reason (R) : There is proton-donor acceptor equilibrium

$$N_3H + NH_3 \rightleftharpoons N_3^- + NH_4^+$$

True & False

1. H_3PO_2 is a weak monobasic acid.

2. N_3H is a stronger acid than acetic acid.

3. N_3H is thermodynamically stable.

4. N_3^- has linear structure.

5. N_2H_4 is thermodynamically endothermic compound but is kinetically stable.

6. P_4O_{10} changes malonic acid into CO_2.

7. NaH_2PO_2 and $NaHPO_3$ are acid salts.

8. Na_3PO_4 and $NaHPO_3$ are normal salts.

9. NH_3, PH_3 and AsH_3 have sp^3-hybridised central atom and thus each has bond angle of 109°28′.

10. NH_4Cl, NH_4NO_3 and $(NH_4)_2CO_3$ give NH_3 gas on heating.

11. N_2O resembles molecular O_2 since, it supports combustion.

12. Structure of N_2O is similar to that of CO_2.

13. N—O bond length in N_2O_4 is greater than expected due to charge repulsion.

14. $H_2N_2O_2$ exists as *cis* and *trans* geometrical isomer.

15. $NaNO_2$ and $NaNO_3$ are used as food additives and generally added to meat.

Fill in the Blanks

1. When mixture of NH_4Cl and $NaNO_2$ is heated, gaseous product formed is ....

2. Formation of N_2H_4 is an .... reaction but combustion is an .... reaction.

3. Out of NO, NO_2, N_2O_4, N_2O_2, paramagnetic species are ....

4. N_2H_4 is .... stable compound.

5. N_2H_4 is a .... base than NH_3 and N_3H is a .... acid than CH_3COOH.

6. N_2H_4 is oxidised to .... in acidic medium and to .... in basic medium.

7. N_3^- is stable due to ....

8. H_3PO_4 is a .... acid,
H_3PO_3 is a .... acid and
H_3PO_2 is a .... acid

9. Reducing nature of H_3PO_4, H_3PO_3 and H_3PO_2 is in order ....

10. N_2O, NO are .... As_4O_6, As_4O_{10} are .... Bi_2O_3 is a ....

11. Out of PCl_3, PCl_5, NCl_5 and NCl_3, least stable is ....

12. As^{3+} salts are .... white Bi^{5+} salts are ....

13. If MCl_3 is hydrolysed and there is an orange turbidity then MCl_3 is .... and when there is white turbidity then MCl_3 is ....

14. N_3^- resembles .... and is also called ....

15. N_2H_4 can be prepared by ....

Matrix-Match Type Questions

I. Only One Correct

1. Match the acids (in List I) with number of acid salts formed by them (in List II).

List I		List II
A. Hypophosphorus acid	1.	Two
B. Orthophosphorus acid	2.	Zero
C. Orthophosphoric acid	3.	One
D. Mellitic acid	4.	Five

2. Match the reactions of metals with dilute HNO_3 (in List I) with the nitrogen compounds (obtained by oxidation/reduction (in List II).

List I	List II
A. Mg + dil. HNO_3	1. NO
B. Zn + dil. HNO_3	2. H_2
C. Sn + dil. HNO_3	3. N_2O
D. Pb + dil. HNO_3	4. NH_4NO_3

3. Match the compounds (in List I) with their oxidation number of N (in List II).

List I	List II
A. N_2H_4	1. –1
B. NH_2OH	2. 3
C. HNO_2	3. 4
D. NO_2	4. 5
E. HNO_3	5. –2

II. One or More Correct

1. Match the reaction in Column I with their corresponding characteristic(s) in Column II.

Column I	Column II
A. $3H_3PO_2 \longrightarrow 2H_3PO_3 + PH_3$	1. Disproportion-ation
B. $2NO_2 + H_2O \longrightarrow HNO_3 + HNO_2$	2. Reducing nature of any reactant
C. $H_3PO_4 + Ca(OH)_2 \longrightarrow CaHPO_4 + 2H_2O$	3. Dibasic acid
D. $2NH_2OH + 4CuO \longrightarrow Cu_2O\downarrow + N_2O + 3H_2O$	4. Equivalent weight (reactant) $= \frac{\text{mol. wt}}{4}$
	5. Reactant is paramagnetic
	6. Oxidising nature of any reactant

2. Match the half-reactions in Column I with the corresponding characteristic(s) in Column II.

Column I	Column II
A. $N_2H_5^+ \longrightarrow N_2 + 5H^+$	1. Oxidation
B. $N_2H_4 + OH^- \longrightarrow N_2 + H_2O$	2. Reduction
C. $2N_3^- \longrightarrow 3N_2$	3. Equivalent weight $= \frac{\text{molar mass}}{4}$
D. $NH_2OH \longrightarrow NH_3$	4. Equivalent weight $= \frac{\text{molar mass}}{2}$
E. $NH_2OH \longrightarrow N_2O$	5. Exchange of 4 electrons
	6. Exchange of 2 electrons

3. Match the species in Column I with their characteristic(s) in Column II.

Column I	Column II
A. NH_2OH	1. Reducing agent
B. N_3H	2. Oxidising agent
C. $NH_2—NH_2$	3. Weak base
D. NH_3	4. Molecule or anion with linear structure
E. N_2O	5. Isoelectronic of CO_2

Integer Answer Types

This section contains 8 questions. The answer to each of the questions is a single digit integer, ranging from 0 to 9. The appropriate bubbles below the respectively question numbers in the *ORS* have to be darkened. For example, if the correct answers to question number *X, Y, Z* and *W* (say) are 6, 0, 9 and 2, respectively, then the correct darkening of bubbles will like the following :

X	Y	Z	W
0	0	0	0
1	1	1	1
2	2	2	2
3	3	3	3
4	4	4	4
5	5	5	5
6	6	6	6
7	7	7	7
8	8	8	8
9	9	9	9

Set I

1. How many of the species are isoelectronic?
$N_2O, CO_2, CN_2^{2-}, N_3^-, NH_2OH, N_3H$

2. How many of the species are paramagnetic?
$N_2O, NO, NO_2, O_2, NO_2^+, NO^+, CO^+$

3. In the following, equilibrium partial pressure of NH_3, N_2 and H_2 gases are 4, 1 and 2 atm respectively at 300 K. What is value of K_p ?

$$N_2(g) + 3H_2(g) \rightleftharpoons 2NH_3(g)$$

4. In the oxidation of N_2H_4 to N_2, equivalent weight of N_2H_4 would be ....

5. Total number of σ bonds in N_3H is ....

6. NO_3^- is reduced to NH_4^+. What is change in oxidation number?

7. A phosphate mineral has formula $Ca_xP_3O_{12}F$. Thus x is ....

8. How many species out of the following have reducing properties?
$NH_3, PH_3, H_3PO_2, H_3PO_3, H_3PO_4, LiAlH_4, BH_3$

Set II

1. How many POP bonds are there in P_4O_{10} ?

2. How many P—H bonds are there in H_3PO_2 ?

3. In PCl_5, how many (P—Cl) are inclined at 120° with one each.

4. 25.0 mL of 0.05 M solution of NH_2OH was boiled with excess of Fe^{3+} in acidic medium. The Fe^{2+} formed required 25.0 mL of 0.1 N $Cr_2O_7^{2-}$ in acidic medium. What is oxidation number of the N in new product?

5. What is pH of 0.01 M Na_2HPO_4 if pK_a is 6?

6. Consider following reactions :

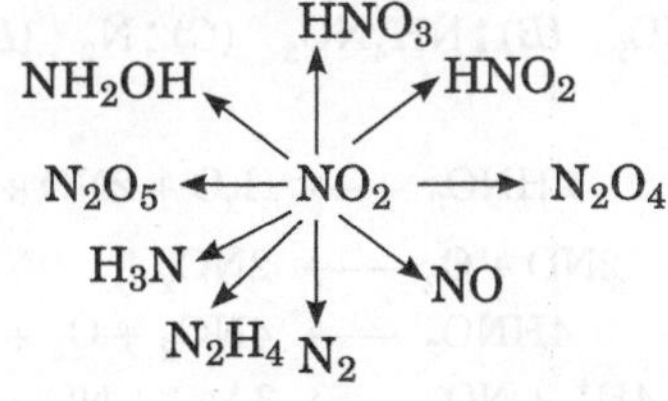

In how many reactions NO_2 has been oxidised?

7. In how many reactions NO_2 has been reduced in Q. 7?

8. Maximum change in oxidation number of nitrogen is ....

Answers & Solutions
Master Exercises

Exercise 1

Short Answer Type Questions

1. (a) P (b) Sb, Bi (c) N (d) Bi

2. :N⋮⋮N:, N_2 is unreactive since, bond energy is very high.

3. $[Fe(H_2O)_5NO]^{2+}$

4. $N_3^-(aq) + 3H_2O(l) \longrightarrow NH_3(g) + 3OH^-(aq)$

   N_3^- accepts proton hence, it is a strong Bronsted Lowry base and reacts with water to produce NH_3 and OH^- ions.

5. O_2 6. $4Mg + 2NO_2 \longrightarrow 4MgO + N_2$

7. Hydrocarbons are oxidised into CO_2 and H_2O in the presence of an oxidation catalyst (Pt or Pd) and NO is reduced to N_2 by reduction catalyst. Catalytic converter has dual catalyst system.

$$C_xH_y + \left(x + \frac{y}{4}\right)O_2 \longrightarrow xCO_2 + \frac{y}{2}H_2O$$

$$2NO + 2CO \longrightarrow N_2 + 2CO_2$$

8. $(A): NH_4NO_3$ $(B): NH_3$ $(C): (NaNO_3 + NaOH)$
   $(D): N_2O$ $(E): H_2O$

9. $Ca(HCO_3)_2 + 2NH_3 \longrightarrow \underset{\text{white ppt}}{CaCO_3\downarrow} + (NH_4)_2CO_3$

10. It is an acid-base reaction in terms of Bronsted Lowry protonic concept

$$\underset{\text{Acid (I)}}{NH_3} + \underset{\text{Base (II)}}{NH_3} \rightleftharpoons \underset{\text{Base (I)}}{NH_2^-} + \underset{\text{Acid (II)}}{NH_4^+}$$

11. $(A): NH_4NO_2$ $(B): NH_4NO_3$ $(C): N_2$ $(D): AlN$
    $(E): N_2O$

12. (a) $2HNO_3 \longrightarrow H_2O + 2NO + 3(O)$
    $2NO + O_2 \longrightarrow 2NO_2$
    (b) $4HNO_3 \longrightarrow 4NO_2 + O_2 + 2H_2O$
    (c) $3Ag + 4H^+ + NO_3^- \longrightarrow 3Ag^+ + NO + 2H_2O$
    (d) $N_2H_4 + 2I_2 \longrightarrow N_2 + 4H^+ + 4I^-$

13. $As_2S_3 + 10HNO_3 \longrightarrow 2H_3AsO_4 + 10NO_2 + 3S + 2H_2O$

14. $Bi_2O_3 < Sb_2O_3 < As_2O_3 < P_2O_3 < N_2O_3$

15. $CH_3—\underset{\underset{O}{\|}}{C}—NH_2 \longrightarrow CH_3CN + H_2O$

$$POCl_3 + 3H_2O \longrightarrow H_3PO_4 + 3HCl$$

16. In P_4O_6 : 12 (P—O) bonds; In P_4O_{10} : 16 (P—O) bonds — Refer structure

17. It is PO_4^{3-} (phosphate) ion.

$$H_3PO_4 + 21HNO_3 + 12(NH_4)_2MoO_4 \longrightarrow (NH_4)_3PO_4 \cdot 12MoO_3 + 21NH_4NO_3 + 12H_2O$$

18. $HO—P(=O)_2$ metaphosphoric acid

(cyclic P_3O_9 ring structure: three P atoms joined by bridging O atoms, each P bearing OH and =O) trimetaphosphoric acid

19. $H_4P_2O_7$ has two units of H_3PO_4 joined by (—O—) formed due to loss of water :

$$HO—\overset{\overset{O}{\|}}{\underset{\underset{OH}{|}}{P}}—OH + HO—\overset{\overset{O}{\|}}{\underset{\underset{OH}{|}}{P}}—OH \longrightarrow HO—\overset{\overset{O}{\|}}{\underset{\underset{OH}{|}}{P}}—O—\overset{\overset{O}{\|}}{\underset{\underset{OH}{|}}{P}}—OH$$

20. In silicate and phosphate anions both Si and P are surrounded by tetrahedra of O-atoms, which can link together to form chains and rings.

$$[PO_4]^{3-} \qquad [SiO_4]^{4-}$$

21. $P_4(s) + 5O_2(g) \longrightarrow P_4O_{10}(s) \xrightarrow{H_2O} H_3PO_4$

22. It is without $CaSO_4$ hence richer in P-content.

23. $NH_2CONH_2 + 2H_2O \longrightarrow (NH_4)_2CO_3$

24. (b) maximum % of P

Analytical Questions

1. Refer text

2. (a) Tetrahedral (b) Planar
   (c) Pyramidal at N and bent at O (d) Tetrahedral
   (e) Trigonal bipyramidal with apical Cl atom.

3. Let $(NH_4)_2HPO_4 = x\%$

$$NH_4NO_3 = (100 - x)\%$$

$$\text{Nitrogen in } (NH_4)_2HPO_4 = \frac{28x}{132}$$

$$\text{Nitrogen in } NH_4NO_3 = \frac{28(100-x)}{80}$$

$$\therefore \quad \frac{28x}{132} + \frac{28(100-x)}{80} = 33.81$$

$$\frac{x}{132} + \frac{100-x}{80} = 1.2075$$

$$x = 8.63\%$$

$$\therefore \quad (NH_4)_2HPO_4 = 8.63\%$$

$$\therefore \quad NH_4NO_3 = 91.37\%$$

4. Bi_2S_3 remains undissolved

$$As_2S_3 + 3Na_2S \longrightarrow \underset{\text{soluble}}{2Na_3[AsS_3]}$$

$$Sb_2S_3 + 3Na_2S \longrightarrow \underset{\text{soluble}}{2Na_3[SbS_3]}$$

5. N—O—H ‖ N—O—H *cis* hyponitrous acid (*cis* and *trans*); $H_2N—NO_2$ (H, H on N—N, with =O and →O) nitramide

6. H—O—N(=O)(→O) nitric acid; H—O—O—N=O peroxo nitrous acid

NCERT Problems

Refer text

Exercise 2

Only One Option Correct

1. (c)	**2.** (d)	**3.** (d)	**4.** (b)	**5.** (a)	**6.** (d)	**7.** (b)	**8.** (b)	**9.** (c)	**10.** (c)
11. (d)	**12.** (c)	**13.** (c)	**14.** (c)	**15.** (c)	**16.** (c)	**17.** (d)	**18.** (d)	**19.** (a)	**20.** (d)
21. (c)	**22.** (c)	**23.** (a)	**24.** (d)	**25.** (c)	**26.** (d)	**27.** (d)	**28.** (c)	**29.** (d)	**30.** (a)
31. (d)	**32.** (d)	**33.** (b)	**34.** (b)	**35.** (a)	**36.** (d)	**37.** (b)	**38.** (a)	**39.** (c)	**40.** (d)
41. (c)	**42.** (d)	**43.** (a)	**44.** (a)	**45.** (d)	**46.** (a)	**47.** (b)	**48.** (a)	**49.** (d)	**50.** (c)
51. (a)	**52.** (d)	**53.** (a)	**54.** (b)	**55.** (d)	**56.** (b)	**57.** (c)	**58.** (d)	**59.** (b)	**60.** (d)
61. (a)	**62.** (b)	**63.** (b)	**64.** (b)	**65.** (c)	**66.** (c)	**67.** (b)	**68.** (c)	**69.** (c)	**70.** (a)
71. (d)	**72.** (d)	**73.** (a)	**74.** (a)	**75.** (c)	**76.** (d)	**77.** (a)	**78.** (c)	**79.** (d)	**80.** (d)
81. (c)	**82.** (d)	**83.** (a)	**84.** (d)	**85.** (b)	**86.** (b)	**87.** (c)	**88.** (a)	**89.** (c)	**90.** (d)
91. (b)	**92.** (a)	**93.** (b)	**94.** (a)	**95.** (a)	**96.** (b)	**97.** (d)	**98.** (c)	**99.** (a)	**100.** (d)
101. (c)	**102.** (a)	**103.** (c)	**104.** (d)	**105.** (c)	**106.** (b)	**107.** (a)	**108.** (d)	**109.** (a)	**110.** (b)
111. (b)	**112.** (b)	**113.** (c)	**114.** (a)	**115.** (d)	**116.** (b)	**117.** (a)	**118.** (c)	**119.** (d)	**120.** (a)
121. (b)	**122.** (d)	**123.** (c)	**124.** (d)	**125.** (c)					

One or More Than One Options Correct

1. (a,b,c) **2.** (a,b,c) **3.** (a,b,c) **4.** (a,b,c,d) **5.** (a,b,c) **6.** (a,b) **7.** (a,b,c,d) **8.** (a,c) **9.** (a) **10.** (a,b)

Passage Comprehension Questions

Passage 1

1. (d) **2.** (b,d) **3.** (b) **4.** (b) **5.** (a) **6.** (d)

Passage 2

1. (c)

Passage 3

1. (a,c) **2.** (d)

Passage 4

1. (b) **2.** (c)

3. (d)

$$pH = 7 + \frac{pK_a}{2} + \frac{\log C}{2}$$

$$= 7 + \frac{2x}{2} - 1 = (6 + x)$$

Passage 5

1. (a) **2.** (d)

Passage 6

1. (b) **2.** (a) **3.** (c)

Passage 7

1. (c) **2.** (d) **3.** (c) **4.** (a)

Assertion & Reason

1. (b) **2.** (b) **3.** (c) **4.** (b) **5.** (b) **6.** (a)
7. (a) **8.** (b) **9.** (c) **10.** (a)

True & False

1. T **2.** T **3.** F **4.** T **5.** T **6.** F **7.** F
8. T **9.** F **10.** F **11.** T **12.** T **13.** F **14.** T
15. T

Fill in the Blanks

1. N_2 **2.** endothermic, exothermic **3.** NO, NO_2
4. kinetically **5.** weaker, stronger **6.** N_2, N_2
7. resonance **8.** tribasic, dibasic, monobasic
9. $H_3PO_4 < H_3PO_3 < H_3PO_2$
10. acidic, amphoteric, basic
11. NCl_5 **12.** reducing agents, oxidising agents
13. $SbCl_3$, $BiCl_3$ **14.** halide, pseudo halide
15. Raschig method

Matrix-Match Type Questions

I. Only One Correct

1. (A) — (2); (B) — (3); (C) — (1); (D) — (4)
2. (A) — (2); (B) — (3); (C) — (4); (D) — (1);
3. (A) — (5); (B) — (1); (C) — (2);
(D) — (3); (E) — (4);

II. One or More Correct

1. (A) — (1,2,6); (B) — (1,2,5,6); (C) — (3);
(D) — (2,4,6)
2. (A) — (1,3,5); (B) — (1,3,5); (C) — (1,4,6);
(D) — (2,4,6); (E) — (1,4,6)
3. (A) — (1,2,3); (B) — (1,4,5); (C) — (1,2);
(D) — (1,3); (E) — (1,2,4,5);

Integer Answer Types

Set I

Questions →	1	2	3	4	5	6	7	8
Answers →	5	4	2	8	3	8	5	6
	0	0	0	0	0	0	0	0
	1	1	1	1	1	1	1	1
	2	2	2	2	2	2	2	2
	3	3	3	3	3	3	3	3
	4	4	4	4	4	4	4	4
	5	5	5	5	5	5	5	5
	6	6	6	6	6	6	6	6
	7	7	7	7	7	7	7	7
	8	8	8	8	8	8	8	8
	9	9	9	9	9	9	9	9

Set II

Questions →	1	2	3	4	5	6	7	8
Answers →	6	2	3	1	9	2	6	7
	0	0	0	0	0	0	0	0
	1	1	1	1	1	1	1	1
	2	2	2	2	2	2	2	2
	3	3	3	3	3	3	3	3
	4	4	4	4	4	4	4	4
	5	5	5	5	5	5	5	5
	6	6	6	6	6	6	6	6
	7	7	7	7	7	7	7	7
	8	8	8	8	8	8	8	8
	9	9	9	9	9	9	9	9

14

Group 16 (VIA—The Chalcogens)

"I am king" of acids. I can dehydrate every thing even skin. I can cause boiling of water with violent splashing of added into my concentrated form" ***Guess me.***

Learning Framework of the Chapter ...

- Periodicity in Group 16 (VIA—The Chalogens)
- Ozone (O_3)
- Allotropy and Polymorphism of Sulphur
- Study of Compounds of Group 16 (VIA)

Oxygen is the most abundant element in the earth's crust, representing slightly less than 50% of it by weight. It is present as elemental oxygen in the atmosphere and is combined with other elements in water and in many minerals. Sulphur, seventeenth in abundance in the earth's crust occurs as sulphide (HgS, PbS etc) in natural gas and oil, as disulphide ion (FeS_2 —"fools' gold") as sulphate ($CaSO_4$)

Selenium and tellurium are comparatively rare on earth's crust, having abundance as selenide/telluride of Cu, Ag, Fe, As.

Periodicity in Group 16 (VIA—The Chalcogens)

- The first four elements are called chalcogens meaning ore forming (oxides, sulphides, selenides, tellurides).
- Some representative properties of group 16 (VI A) elements are given in Table 14.1 and Fig. 14.1.

Similar to the groups discussed earlier, the metallic character in group 16 increases with increasing atomic number. Oxygen and sulphur are entirely non-metallic in their chemical behaviour. Selenium and tellurium, though essentially non-metallic, assume increasing metallic character and are termed as metalloids. Polonium is most metallic in the group.

Density, melting and boiling points covalent and ionic radii, show a regular increase from oxygen to polonium. The large difference in melting and boiling points between oxygen and sulphur can be explained on the basis of their structure. Oxygen exists mostly as a diatomic molecule held together by weak van der Waals' forces while others exist as polyatomic molecules, *e.g.*, S_8, Se_8, etc., where the atoms are bonded by covalent bonds associated with high dissociation energy. Their existence as diatomic and polyatomic molecules can be explained as follows: The O=O bond more than three times as strong as the O—O bond. By comparison the S=S bond is less than twice as strong as the S—S single bond. This results in catenated —O—O—O— chains begining unstable relative to O=O; but catenated —S—S—S— chains being stable relative to the molecule S=S.

Bond	Bond energy kJ mol^{-1}
O=O	498
O—O	142
S=S	431
S—S	265

The elements of group 16 are characterised by high ionisation energies. (Fig 14.1) decreasing gradually from oxygen to polonium. The high values indicate reluctance of these elements to from cations. **Their electronegativities decrease with increasing atomic number. This, is view of the fall in electronegativity, metallic character within the group increases with increasing atomic size.**

Oxygen, the second most electronegative element, fluorine being the first, has a strong tendency to accept two electrons and give O^{2-} ion. Thus, almost all metal oxide are ionic and contain O^{2-} ions.

DEEP Focus

Usually oxygen exhibits as oxidation state –2 in its other compounds also. **It exhibits positive oxidation state only in a few compounds formed with fluorine, *i.e.*, OF_2 and O_2F_2.** The tendency for the formation of divalent anions decreases from sulphur downwards because of the increasing size and decreasing electronegativity of the elements. Sulphur, selenium an tellurium show a tendency for covalency with formal oxidation states +2, +4 and +6 in compounds in which they are combined with more electronegative elements such as oxygen and halogens. In the higher oxidation state of +4 and +6 of these elements electrons are unpaired and promoted to vacant *d*-orbitals.

Table 14.1 **Physical Properties of the Element in Group 16 (VIA)**

	O_8	S_{16}	Se_{34}	Te_{52}	Po_{84}
Electronic configuration	[He] $2s^2 2p^4$	[Ne] $3s^2 3p^4$	[Ar] $3d^{10} 4s^2 4p^4$	[Kr] $4d^{10} 5s^2 5p^4$	[Xe]$4f^{14} 5d^{10} 6s^2 6p^4$
Covalent radius (pm)	74	104	114	137	168
Ionic radius (M^{2-}, pm)	140	184	198	221	—
$(IE)_1$/kJ mol^{-1}	1314	999	941	869	813
Electronegativity	3.5	2.5	2.4	2.1	2.0
Melting point (°C)	–229	114*	221$^{\alpha}$	452	254
Boiling point (°C)	–183	445	685	1087	962
Density ($10^3 \times$ kg m^{-3})	1.14	2.07	4.79	6.25	9.4
Oxidation states	–2, –1, +1, +2	–2, +2, +4, +6	–2, +2, +4, +6,	–2, +2, +4, +6	+2, +4
Electron affinity (kJ mol^{-1})	–141	–200	–195	–190	–183
$E°/V$ for $X + 2H^+ + 2e^- \rightarrow H_2X$	1.23	0.14	– 0.40	– 0.79	—

*Rhombic sulphur, α-Grey Se

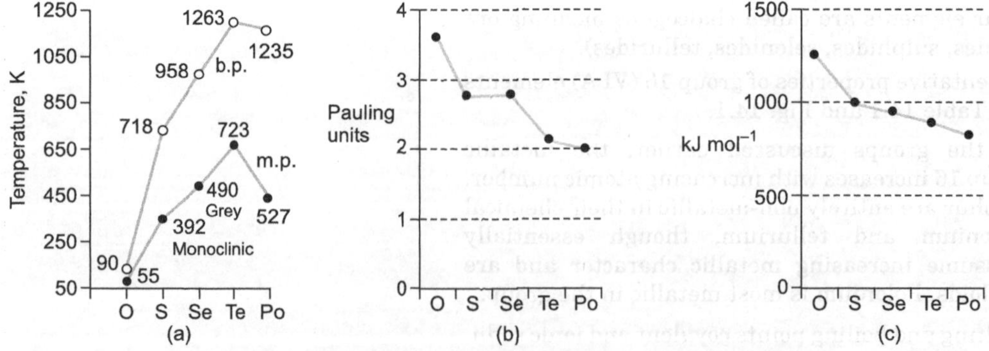

Fig. 14.1 Some group trends of the group 16 elements : (a) Boiling point and melting point, (b) Pauling electronegativity, and (c) The first ionisation energy

DEEP Focus

- One of the points about oxygen is that as the first member of a group it differs in some important ways from other group members due to:
  - small size,
  - high electronegativity, and
  - non-availability of *d*-orbitals for bonding.

Table 14.2 summarises similarities and differences between oxygen and sulphur.

Table 14.2 **Comparison of Oxygen and Sulphur**

Oxygen	Sulphur
• Exists as $O_2(g)$ at 298 K and 1 atm	• Exist as $S_8(s)$ rings at 298 K and 1 atm. At higher temperature S_2 predominates.
• Two allotropes, $O_2(g)$ and $O_3(g)$	• Two solid polymorphic forms and many different molecular species in liquid and gaseous states.
• $\dot{O}_2(g)$ and $\dot{O}_3(g)$ are very good oxidising agents.	• $\dot{S}(s)$ is a poor oxidising agent.
• Metallic oxides are generally ionic.	• Metallic sulphides are ionic (Na_2S, K_2S ...) as well as covalent insoluble in water (CuS, NiS, CdS ... generally of *d*-block)
• O_2 paramagnetic	• S_2 paramagnetic
• Can form only two and three atom chains as in H_2O_2 and O_3. Compounds with O—O bonds decompose readily.	• Can form molecules with up to six S atoms per chain in compounds such as H_2S_n, Na_2S_n, $H_2S_nO_6$.
• Maximum coordination number of oxygen is four, as in water : (H—O—H, O, H, H, O—H)	• Maximum coordination number of sulphur is six, as in SF_6 : (F, F, F—S—F, F, F)
• H_2O : – is a liquid at 298 K and 1 atm – is extensively hydrogen bonded – has a large dipole moment – is a good solvent – is oxidised with difficulty – forms hydrated and aqua complexes.	• H_2S – is a (poisonous) gas at 298 K and 1 atm – is not hydrogen bonded – has a small dipole moment – is a poor solvent – forms no complexes – is easily oxidised.

- Selenium has unique property of photoconductivity and is used in photocopying (xerox) machines and also a decoloriser of glass. Te and Po are highly toxic, Po is higher due to its intense radioactivity. Se and Te are found as selenides and tellurides in sulphide ores (Cu_2Se, Cu_2Te). **The colour of the red glass in the traffic signals is due to CdSe.**

Oxidation States

- With valence electronic configuration ns^2np^4, the group 16 (VI A) elements are just two electron short of an octet configuration, and –2 oxidation state is therefore, a common one. The stability of –2 state decreases with increasing metallic character as indicated by $E°$ (redox) values. Oxygen is powerful oxidising agent and H_2Se and H_2Te are reducing agents ($E°$ values of Se and Te are negative). Because S, Se, and Te are much less electronegative than oxygen, they are commonly found in positive oxidation states as +4 in SF_4, SO_2 and H_2SO_3 and +6 in SF_6, SO_3, H_2SO_4.
- Unlike oxygen, the maximum coordination number of sulphur and other elements can exceed **four** and valency is not limited to **two** because of available *d*-orbitals for bonding. Stability of +6 state decreases and that of +4 state increases going down the group due to inert-pair effect. It is:

$$Se^{4+} < Te^{4+} < Po^{4+}$$
$$Se^{6+} > Te^{6+} > Po^{6+}$$

Hydrides

- H_2Se and H_2Te are unpleasant gaseous hydrides like H_2S but less stable than H_2S, order of thermal stability being $H_2O > H_2S > H_2Se > H_2Te > H_2Po$.
- Oxyacids of S, Se and Te exist and H_2SO_4 and H_2SeO_4 are considered as strong acids.

Catenation

- The tendency for catenation decreases markedly as we go down the group. This property is prominently displayed by sulphur (S_8). The S—S bond is important in biological systems and is found in compounds such as cysteine, some proteins and enzymes. Stable polyoxides and polysulphides are known. Examples are:

H—O—O—H, H—S—S—H, H—S—S—S—H, H—S—S—S—S—H.

Metallic Nature

- The metallic character increases as we descend the group. S is non-metal and insulator. Se and Te are metalloids an are semiconductors. Po is a metal.

Multiple Bonding

- The tendency to form multiple bonds to C, N and O decreases as we descend the group from Se to Te. Thus S=C=S is moderately stable; Se=C=Se decomposes readily whereas Te=C=Te does not exist.
- Acidic nature of hydride increases as we go down the group:

$$H_2O < H_2S < H_2Se < H_2Te < H_2Po$$

Halides

- The form halides of different oxidation number +1, +2, +4 and +6.

Halides	X	Oxidation numbers
S_2X_2	F, Cl, Br, I	+1
SX_2	F, Cl	+2
SX_4	F	+4
SX_6	F	+6
Se_2X_2	Cl, Br	+1
SeX_2, TeX_2	Cl, Br	+2
SeX_4, TeX_4	F, Cl, Br	+4
TeX_4	I	+4
SeX_6, TeX_6	F	+6

The stability of the halides decreases in the order F > Cl > Br > I. +6 oxidation is observed in fluoride, while iodide of Te(TeI_4) is formed.

SF_6 is thermodynamically inert, non-toxic gas. Interness of SF_6 is due to the presence of sterically protected sulphur atom. SF_6 is used as a gaseous insulator in high voltage generators. SF_4 an SeF_4 are good fluorinating agents and —COOH group can be converted into CF_3 and >C=O and —P=O into CF_2 and $—PF_2$ groups.

- Oxides of sulphur are more stable than the corresponding oxides of other elements.

Reactions

Reactions of the group 16 (VIA) elements are summarised below:

General equations	Remarks
$xE + yM \longrightarrow M_yE_x$	With many metals
$zE + M_xE_y \longrightarrow M_xE_{y+z}$	Especially with S and Se
$E + H_2 \longrightarrow H_2E$	Decreasing in the series O_2, S, Se, Te
$E + 3F_2 \longrightarrow EF_6$	With S, Se, Te and excess F_2
$2E + Cl_2 \longrightarrow E_2Cl_2$	With S, Se (Te gives $TeCl_2$); also with Br_2
$E_2Cl_2 + Cl_2 \longrightarrow 2ECl_2$	With S, Se, also with Br_2
$E + 2Cl_2 \longrightarrow ECl_4$	With S, Se, Te and excess Cl_2; also with Br_2
$E + O_2 \longrightarrow EO_2$	With S (with Se, use O_2 and NO_2)

(*E* = O, S, Se, Te, Po and *M* metal).

Ozone (O_3)

- In the laboratory, O_3 can be made by passing oxygen through a strong electric field. An equilibrium is set up :

$$3O_2(s) \rightleftharpoons 2O_3(g)$$

It is only a metastable allotrope, always having a tendency to convert back into oxygen.

- Some 30 km above the earth's surface oxygen molecules can be split apart by UV light from the sun. Some of the atoms join with other oxygen molecules to make ozone :

$$O_2(g) + O(g) \longrightarrow O_3(g)$$

- O_3 is an unstable, dark blue **diamagnetic** gas, boiling point – 112°C. The colour is due to intense absorption of red light. It also absorbs strongly in the UV region. *This is particularly important since there is a layer of O_3 in the upper atmosphere which absorbs harmful UV radiation from the sun, thus protecting people on the earth.* The use of chlorofluorocarbons (CFC) in aerosols and refrigerators, and their subsequent escape into the atmosphere, is blamed for making holes in the ozone layer over the Antarctic and Arctic. It is feared that this will allow an excessive amount of UV light to reach the earth which will cause **melanoma** (skin cancer) in humans. Oxides of nitrogen (from car exhausts) and the halogens can also damage the O_3 layer.

Ozone destroying reactions:

— based on oxides of nitrogen

$$NO + O_3 \longrightarrow NO_2 + O_2$$
$$O_3 + h\nu \longrightarrow O_2 + O$$
$$NO_2 + O \longrightarrow NO + O_2$$

Net reaction : $2O_3 \xrightarrow{h\nu} 3O_2$

– based on reactive chlorine species from CFC

$$Cl + O_3 \longrightarrow ClO + O_2$$
$$O_3 + h\nu \longrightarrow O + O_2$$
$$ClO + O \longrightarrow Cl + O_2$$

Net reaction : $2O_3 + h\nu \longrightarrow 3O_2$

- O_3 is thermodynamically unstable, and decomposes to O_2:

$$2O_3 \longrightarrow 3O_2, \quad \Delta G = -163 \text{ kJ mol}^{-1}$$

- O_3 is an extremely powerful **oxidising agent:**

$$O_3 + 2H^+ + 2e^- \rightleftharpoons O_2 + H_2O$$
$$O_3 + H_2O + 2e^- \rightleftharpoons O_2 + 2OH^-$$
$$3PbS + 4O_3 \longrightarrow 3PbSO_4$$
$$6NO_2 + O_3 \longrightarrow 3N_2O_5$$
$$S + H_2O + O_3 \longrightarrow H_2SO_4$$
$$2KOH + 5O_3 \longrightarrow 2KO_3 + 5O_2 + H_2O$$
$$O_3 + 2KI + H_2O \longrightarrow I_2 + 2KOH + O_2$$
$$3SnCl_2 + 6HCl + O_2 \longrightarrow 3SnCl_4 + 3H_2O$$
$$2[Fe(CN)_6]^{4-} + H_2O + O_3 \longrightarrow 2[Fe(CN)_6]^{3-} + O_2 + 2OH^-$$

- **Potassium ozonide KO_3 is an orange coloured solid and contains the paramagnetic O_3^- ion.**

DEEP Focus

- The amount of O_3 in a gas mixture can be determined by passing the gas into a KI solution (at a constant pH 9.2 using borate buffer). The iodine that is liberated is titrated with sodium thiosulphate solution

- O_3 can also be decomposed catalytically and change in volume measured :

$$\underset{(2\ \text{volume})}{2O_3} \longrightarrow \underset{(3\ \text{volume})}{3O_2}$$

- O_3 also adds to unsaturated organic compounds at room temperature forming ozonides which can be cleaved to aldehydes and ketones in solution :

$$CH_2{=}CH_2 + O_3 \longrightarrow \text{(ozonide: } CH_2-O-CH_2 \text{ with } O-O \text{ bridge)} \xrightarrow{H_2O/Zn} 2HCHO$$

$$CH_3{-}CH{=}CH{-}CH_3 + O_3 \xrightarrow{H_2O/Zn} 2CH_3CHO$$

DEEP Focus

To derive products of ozonolysis, break (C=C) bond, place O at breaking point,

$$CH_3CH{\vdots}{=}CHCH_3 \xrightarrow[\text{Place O}]{\text{break}} CH_3CHO + CH_3CHO$$

- O_3 is used as a disinfectant; it is used to purify drinking water since, it can destroy bacteria and viruses. It is better than Cl_2 since, it (O_3) avoids the unpleasant smell and taste of Cl_2 and any excess O_3 soon decomposes to O_2.

$$O_3 + 2I^- + H_2O \longrightarrow I_2 + 2OH^- + O_2$$

$$I_2 + 2S_2O_3^{2-} \longrightarrow 2I^- + S_4O_6^{2-}$$

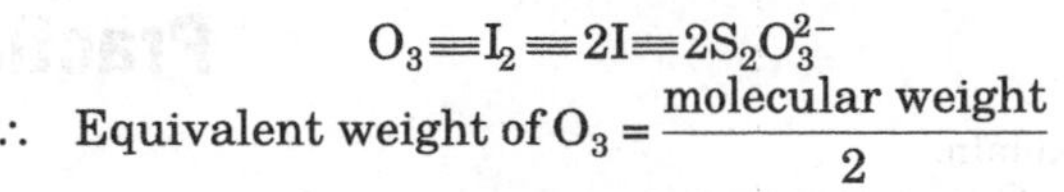

$$O_3 \equiv I_2 \equiv 2I \equiv 2S_2O_3^{2-}$$

$$\therefore \ \text{Equivalent weight of } O_3 = \frac{\text{molecular weight}}{2}$$

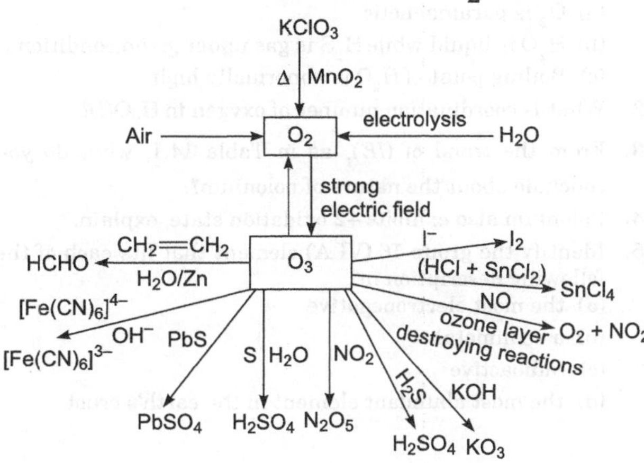

The molecules O_3 is bent with an O—O—O angle 116.8° and equal O—O distance of 128 pm and may be described 90° canonical forms.

(Structure: O at apex, 116.8°, 218 pm between terminal O atoms; resonance forms O^+ with O^- ⟷ O^+ with O^-)

Target Practice 1

1. Write the formulae of the compounds with given oxidation number of the given elements.

	Element	O.N.	Compounds
(a)	O	+2	....
(b)	O	–1	....
(c)	O	–0.5	....
(d)	S	–2	....
(e)	S	+4	....
(f)	S	+6	....

2. Based on following reaction identify alkene A

$$A \xrightarrow{O_3/H_2O/Zn} 2HCHO + \begin{array}{c} CHO \\ | \\ CHO \end{array}$$

3. What is bond length (O—O) in O_3 ....

and bond angle $O{\overset{O}{\wedge}}O$ in O_3 ....?

4. $2O_3 \longrightarrow 3O_2; \Delta G° = -163 \text{ kJ mol}^{-1}$

Is this reaction spontaneous. Yes/No

....

5. Write reaction representing O_3 as an oxidising agent

(a) in acidic medium ....

(b) in basic medium ....

Equivalent weight in (a) ....

Equivalent weight in (b) ....

6. Compare oxygen and fluorine in following properties

	Property			
(a)	Electronegativity	O	☐	F
(b)	Electron affinity	O	☐	F
(c)	Reactivity	O_2	☐	F_2
(d)	Bond energy to form gaseous atoms	O_2	☐	F_2
(e)	Covalent radius	O	☐	F
(f)	Ionic radius	O^{2-}	☐	F^-

Answers

1. (a) OF_2 (b) H_2O_2 (c) KO_2 (d) H_2S (e) SO_2 (f) SO_3
2. $CH_2{=}CH{-}CH{=}CH_2$
3. Bond length = 128 pm, Bond-angle = 116.8°
4. Yes (since $\Delta G° < 0$)
5. (a) $O_3 + 2H^+ + 2e^- \rightleftharpoons O_2 + H_2O$

   (b) $O_3 + H_2O + 2e^- \rightleftharpoons O_2 + 2OH^-$

   Equivalent weight in (a) $= \frac{48}{2} = 24$ g equivalent^{-1}

   Equivalent weight in (b) $= \frac{48}{2} = 24$ g equivalent^{-1}.

6. (a) O < F (b) O < F (c) $O_2 > F_2$ (d) $O_2 > F_2$ (e) O > F (f) $O^{2-} > F^-$

Practice Exercise 1

1. Explain.
   (a) O_2 is paramagnetic
   (b) H_2O is liquid while H_2S is gas under given conditions
   (c) Boiling point of H_2O is abnormally high
2. What is coordination number of oxygen in $H_2O(l)$?
3. From the trend of $(IE)_1$ as in Table 14.1, what do you conclude about the nature of polonium?
4. Polonium also exhibits +2 oxidation state, explain.
5. Identify the group 16 (VI A) element that fits each of the following description in:
   (a) the most electronegative
   (b) a semimetal
   (c) radioactive
   (d) the most abundant element in the earth's crust.
6. O_3 is diamagnetic and O to O bond lengths are equal. Explain.
7. O_3 is a powerful oxidising agent. Write equations to represent oxidation of
   (a) I^- to I_2 in acidic solutions,
   (b) sulphur to sulphuric acid in the presence of moisture,
   (c) $[Fe(CN)_6]^{4-}$ to $[Fe(CN)_6]^{3-}$ in basic solution.
8. It has been estimated that if all the O_3 in the atmosphere were brought to sea level at STP, the gas would form a layer 0.3 cm thick. Estimate the number of O_3 molecules in the earth's atmosphere.
   (radius of the earth is 4000 mile, 1 km = 0.6214 mile)
9. Alkene $(C_6H_{12}) \xrightarrow[Zn]{O_3/H_2O} 2CH_3COCH_3$
   Identify alkene.

Answers

1. (a) It is due to presence of two unpaired electrons in anti-bonding molecular orbitals
   $$(\sigma 1s)^2(\sigma^* 1s)^2(\sigma 2s)^2(\sigma^* 2s)^2(\sigma 2p_x)^2(\pi 2p_y)^2(\pi 2p_z)^2 (\pi^* 2p_y)^1(\pi^* 2p_z)^1.$$
   (b) H_2O molecules have intermolecular H-bonding due to greater electronegativity of oxygen hence, H_2O is liquid; H_2S lacks H-bonding hence, gas.
   (c) Due to intermolecular H-bonding.
2. Four. **3.** It is least, hence metallic.
4. It is due to **inert-pair effect of *ns*-pair** as well as ***np*-pair**

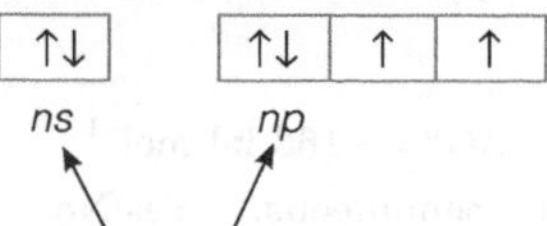

(both are inert due to shielding effect hence difficult to unpair—hence +2 oxidation state)

5. (a) O (b) Te (c) Po (d) O
6. 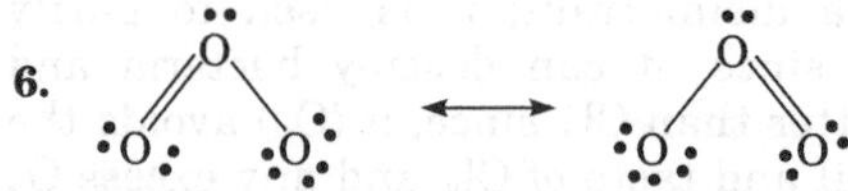

   Due to resonance, two O—O bonds are equivalent and have a bond order of 1.5 (bond length = 128 pm, angle = 116.5°)
   All electrons appear paired hence diamagnetic.
7. (a) $O_3 + 2I^- + 2H^+ \longrightarrow O_2 + I_2 + H_2O$
   (b) $3O_3 + S + H_2O \longrightarrow H_2SO_4 + 3O_2$
   (c) $O_3 + 2[Fe(CN)_4]^{4-} + H_2O \longrightarrow 2(Fe(CN)_6]^{3-} + 2OH^- + O_2$
8. 4×10^{37} ozone molecules.
9. $(H_3C)_2C{=}C(CH_3)_2$

Allotropy and Polymorphism of Sulphur

Many different molecular species are possible for the elemental sulphur, and this accounts for the existence of the large number of physical forms in which the element may appear as shown in Fig. 14.2. Allotropy is more complex for the element sulphur than for any other element of group 16(VI). Sulphur may exist as:

- **rhombic sulphur (S_α)** which has sixteen S_8 rings in a unit cell and at 95.5°C converts to:
- **monoclinic sulphur (S_β)** which has six S_8 rings in its unit cell. It melts at 119°C, yielding :
- **liquid sulphur (S_λ)** comprised of S_8 molecules. This is yellow, transparent, mobile liquid. At 160°C, a remarkable transformation occurs. The S_8 rings open up and join together into long spiral-chain molecules, resulting in :
- **liquid sulphur (S_μ)** which is a dark in colour and very thick and viscous. The chain length and viscosity of the liquid reach a maximum at about 180°C. At higher temperatures the chains break up and viscosity decreases again. At 445°C, this liquid boils, producing :

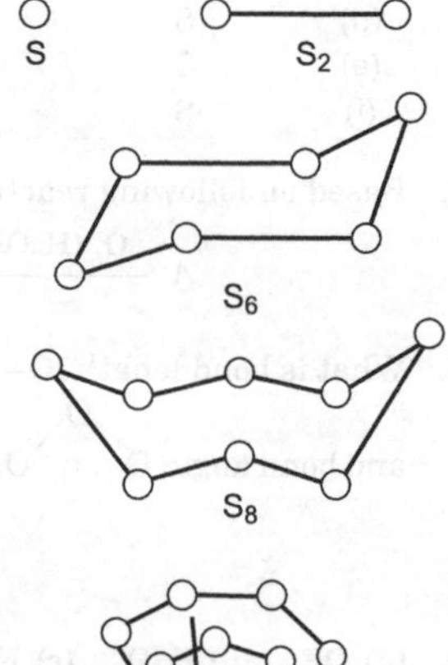

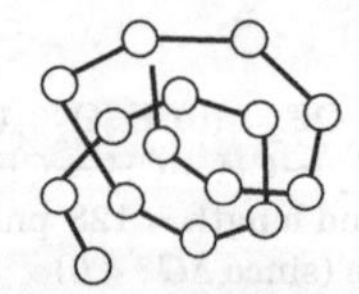

Fig. 14.2 Different molecular forms of sulphur

- **Sulphur vapours,** which consists of molecules ranging from S_2 to S_{10}, but predominantly S_8. **At higher temperatures, S_2 predominates. S_2 is paramagnetic.**
- Plastic sulphur forms if liquid sulphur (μ) is poured into cold water. It consists of chain-like molecules and has rubber-like properties when first formed. On standing, it becomes brittle and finally converts to rhombic sulphur.

  Above transformations can be written as:

$$S_\alpha \xrightarrow{95.5°C} S_\beta \xrightarrow{119°C} S_\lambda \xrightarrow{160°C} S_\mu \xrightarrow{445°C} S_8(g)$$

$$S \xleftarrow{2000°C} S_2 \xleftarrow{1000°C} S_4 \longleftarrow S_6$$

- If rhombic sulphur is heated rapidly, it fails to convert into monoclinic sulphur and melts at 113°C.
- In **cyclo-S_6,** the ring adopts the chain form chain polymers, **catena-S_n are also known.**

Production of Sulphur

Frasch Process

Sulphur occurs abundantly in the earth's crust as elemental sulphur, as mineral sulphides and sulphates, as H_2S in natural gas, and as organic sulphur compounds in oil and coal. The sulphur is mined in an unusual way, known on the **Frasch process** (Fig. 14.3).

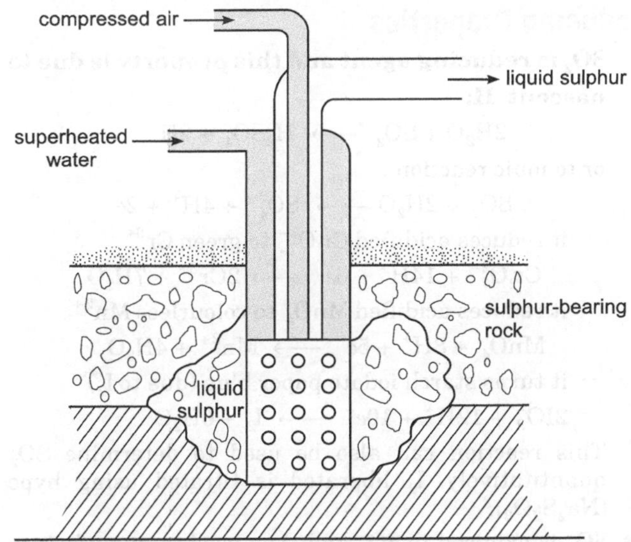

Fig. 14.3 Frasch process for production of sulphur

A mixture of superheated water and steam (at about 160°C and 16 atm) is forced down the outermost of three concentric pipes into an underground bed of sulphur containing rock. The sulphur melts and forms a liquid pool. Compressed air (at 20–25 atm) is pumped down the innermost pipe and forces liquid sulphur up the remaining pipe. The liquid sulphur is either stored and shipped as a liquid (by tank car, ship, or pipeline) or allowed to solidify to a solid.

Although the Frasch process was once the principal source of elemental sulphur this is no longer the case. The change has been brought, about by the need to control sulphur emissions from industrial operations. In one process to remove H_2S from a gaseous mixture, the gas stream is split into two parts. One part (consisting of about one-third of the stream) is burnt to convert H_2S to SO_2. The streams are rejoined in a catalytic converter at 200 to 300°C, where this reaction occurs.

$$2H_2S(g) + SO_2(g) \longrightarrow 3S(g) + 2H_2O(g)$$

Study of Compounds of Group 16 (VIA)

Hydrides (H_2S)

- All form **covalent** hydrides. Water (H_2O) is liquid at room temperature but others are all colourless, foul smelling toxic gases.
- A comparative study of hydrides of this group are given in Table 14.3.
- H_2S is prepared by reaction of FeS on dil. H_2SO_4:

$$FeS + H_2SO_4 \longrightarrow FeSO_4 + H_2S$$

DEEP Focus

- Now a days, H_2S used in qualitative analysis is prepared by the hydrolysis of thioacetamide:

$$CH_3{-}C(=S)NH_2 + 2H_2O + H^+ \longrightarrow CH_3{-}C(=O)OH + NH_4^+ + H_2S$$

thioacetamide

Use of thioacetamide is useful as it produces H_2S *in situ*, and thus causes no pollution.

- H_2S is soluble in water and burns in air with a blue flame :

$$H_2O + H_2S \rightleftharpoons H_3O^+ + HS^-$$

$$HS^- + H_2O \rightleftharpoons H_3O^+ + S^{2-}$$

$$2H_2S + 3O_2 \longrightarrow 2H_2O + 2SO_2$$

- A saturated solution of H_2S is used as a laboratory reagent for qualitative analysis of basic radicals (cations) :
  - in acidic medium (HCl), Cu^{2+}, Pb^{2+} .... of II group are precipitated as sulphides

$$CuSO_4 + H_2S \longrightarrow \underset{\text{black}}{CuS\downarrow} + H_2SO_4$$

$$CdCl_2 + H_2S \longrightarrow \underset{\text{yellow}}{CdS\downarrow} + 2HCl$$

  - in basic (ammoniacal) medium, Zn^{2+}, Ni^{2+} .... of IV group are precipitated as sulphides :

$$ZnCl_2 + H_2S \xrightarrow{NH_4OH} \underset{\text{white}}{ZnS\downarrow} + 2HCl$$

- H_2S is a weak dibasic acid and forms two series of salts :

$$H_2S + NaOH \longrightarrow NaHS + H_2O$$
$$NaHS + NaOH \longrightarrow Na_2S + H_2O$$

- O_3 can oxidise H_2S to H_2SO_4:

$$H_2S + 2O_3 \longrightarrow H_2SO_4 + O_2$$

- H_2S decolourises acidified $KMnO_4$:

$$5H_2S + \underset{\text{purple}}{2MnO_4^-} + 6H^+ \longrightarrow \underset{\text{yellow}}{5S\downarrow} + \underset{\text{colourless}}{2Mn^{2+}} + 8H_2O$$

- In alkaline medium MnO_4^- is reduced to MnO_2 (black).

$$2MnO_4^- + 3H_2S \longrightarrow 2S\downarrow + 2MnO_2\downarrow + 2H_2O + 2OH^-$$

- H_2S also turns acidified $K_2Cr_2O_7$ to green :

$$3H_2S + \underset{\text{orange}}{Cr_2O_7^{2-}} + 8H^+ \longrightarrow \underset{\text{yellow}}{3S\downarrow} + \underset{\text{green}}{2Cr^{3+}} + 7H_2O$$

- H_2S is also oxidised by Cl_2 :

$$H_2S + Cl_2 \longrightarrow 2HCl + S\downarrow$$

- $FeCl_3$ (yellow in aqueous solution) changes to $FeCl_2$ (light green)

$$2FeCl_3 + H_2S \longrightarrow 2FeCl_2 + 2HCl + S$$

Table 14.3 **Some Physical Properties of Hydrides of Group 16 (VIA)**

Property	H_2O	H_2S	H_2Se	H_2Te
Melting point/K	273	188	208	222
Boiling point/K	373	213	232	269
M—H Bond length (pm)	96	134	146	169
H*M*H angle (°)	104.5	92	91	90
Bond energy (kJ mol^{-1})	463	347	276	238
K_a (dissociation constant) in *aq* solution.	1.8×10^{-16}	1.3×10^{-7}	1.3×10^{-4}	2.3×10^{-3}

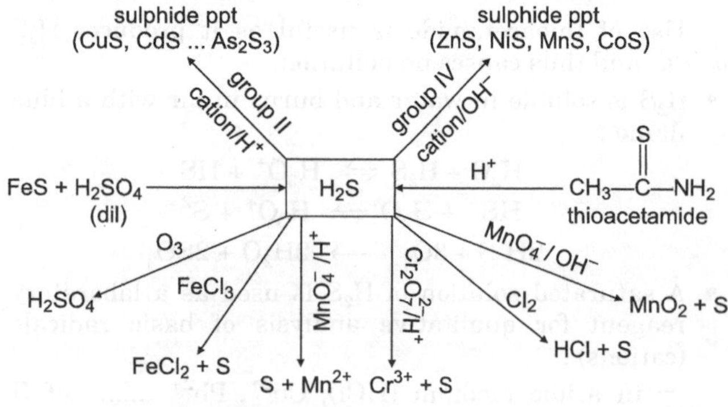

Oxides of Sulphur

- Sulphur forms **two main gaseous** oxides :

SO_2 (anhydride of H_2SO_3)

SO_3 (anhydride of H_2SO_4)

- Oxides of other elements (SeO_2, TeO_2, PoO_2, SeO_3, TeO_3, TeO and PoO) are solids; oxides of Se are volatile.

Source

- SO_2 is produced commercially on a vast scale :
  - by burning sulphur and H_2S in air

$$S + O_2 \longrightarrow SO_2$$
$$2H_2S + 3O_2 \longrightarrow 2H_2O + 2SO_2$$

  - by roasting metal sulphide ores with air

$$2ZnS + 3O_2 \longrightarrow 2ZnO + 2SO_2$$

- In the laboratory, it can be prepared by the action of dilute acid on a sulphite ion,

$$Na_2SO_3 + 2HCl \longrightarrow 2NaCl + H_2O + SO_2$$

Or by the action of concentrated H_2SO_4 on Cu.

$$Cu + 2H_2SO_4 \longrightarrow CuSO_4 + 2H_2O + SO_2$$

Properties

- SO_2 is a colourless gas (b.p. –10°C, m.p. –75.5°C) which has a choking smell, and is very soluble in water. SO_2 in solution is almost completely present as $SO_2 \cdot 6H_2O$ and only traces of H_2SO_3 (sulphurous acid) may be present. **SO_2 levels above 5 ppm are poisonous to man, but plants are harmed at appreciably lower levels.**
- An acidic oxide, it reacts with water as follows:

$$SO_2 + H_2O \rightleftharpoons HSO_3^- + H^+$$

Reducing Properties

- **SO_2 is reducing agent and this property is due to nascent H:**

$$2H_2O + SO_2 \longrightarrow H_2SO_4 + 2H$$

or to ionic reaction :

$$SO_2 + 2H_2O \longrightarrow SO_4^{2-} + 4H^+ + 2e^-$$

  - it reduces acidified $Cr_2O_7^{2-}$ to green Cr^{3+}

$$Cr_2O_7^{2-} + 14H^+ + 6e^- \longrightarrow 2Cr^{3+} + 7H_2O$$

  - it reduces acidified MnO_4^- to colourless Mn^{2+}

$$MnO_4^- + 8H^+ + 5e^- \longrightarrow Mn^{2+} + 4H_2O$$

  - it turns starch iodate paper blue (due to I_2)

$$2IO_3^- + 12H^+ + 10e^- \longrightarrow I_2 + 6H_2O$$

This reaction can also be used to determine SO_2 quantitatively. I_2 liberated is titrated using hypo ($Na_2S_2O_3$).

- SO_2 is oxidised to SO_3 and thus used to manufacture H_2SO_4

$$2SO_2 + O_2 \rightleftharpoons 2SO_3$$
$$SO_3 + H_2O \longrightarrow H_2SO_4$$

Oxidising Properties

H_2S is oxidised to colloidal sulphur

$$2H_2S + SO_2 \longrightarrow 2H_2O + 3S$$

- SO_2 is used to make other products :

$$2SO_2 + Na_2CO_3 + H_2O \longrightarrow 2NaHSO_3 + CO_2$$
$$2NaHSO_3 + Na_2CO_3 \longrightarrow 2Na_2SO_3 + H_2O + CO_2$$
$$Na_2SO_3 + S \longrightarrow Na_2S_2O_3 \text{ (hypo)}$$

DEEP Focus

- Liquid SO_2 has also been used as a non-aqueous solvent and is a useful reaction medium for organic and inorganic compounds :

$$2SO_2 \rightleftharpoons SO^{2+} + SO_3^{2-}$$

Radioactive isotopic study shows that $SOCl_2$ ionises as

$$SOCl_2 \rightleftharpoons SOCl^+ + Cl^-$$

and not as $\quad SOCl_2 \rightleftharpoons SO^{2+} + 2Cl^-$

Thus, formation of SO^{2+} in liquid SO_2 is not favourable.

- The SO_3 is not usually isolated, and practically all of it is converted to H_2SO_4.
- **H_2SO_4 saturated with SO_3 is called oleum (pyrosulphuric acid)**

$$H_2SO_4 + SO_3 \longrightarrow H_2S_2O_7$$

DEEP Focus

- SO_3 is used for the sulphonation of long chain alkyl benzene compounds (like dodecyl benzene). The sodium salt of these alkyl benzene sulphonic acid are anionic surface-active agents, and are the active ingredients of detergents :

$R-C_6H_4-SO_3Na$ (ABS)

$R = C_{12}H_{25}$

ABS (Alkyl Benzene Sulphonates)

- $SO_3 + H_2SO_4 + NH_2CONH_2 \longrightarrow \underset{\text{sulphamic acid}}{2NH_2SO_3H} + CO_2$

Structure

- SO_2 gas forms discrete V-shaped molecule and this structure is retained in solid state.
- The sulphur atom is sp^2 hybridised but bond angle $\angle OSO$ is 119°30′. It is due to VSEPR (valence shell electron pair repulsion) :

$$:\ddot{O}=\ddot{S}\rightarrow\ddot{O}: \longleftrightarrow :\ddot{O}\leftarrow\ddot{S}=\ddot{O}:$$

- In the gas phase SO_3 has a plane triangular structure. The bonding is best described as S forming three σ bonds, giving rise to plane triangle, and three delocalised π-bonds :

$$O=S(=O)=O$$

At room temperature, SO_3 is solid and exists as :

$$-O-\underset{O}{\overset{O}{\underset{\|}{\overset{\|}{S}}}}-O-\underset{O}{\overset{O}{\underset{\|}{\overset{\|}{S}}}}-O-\underset{O}{\overset{O}{\underset{\|}{\overset{\|}{S}}}}-O-$$

α-SO_3 (chains)

γ-SO_3 (cyclic, ice-like)

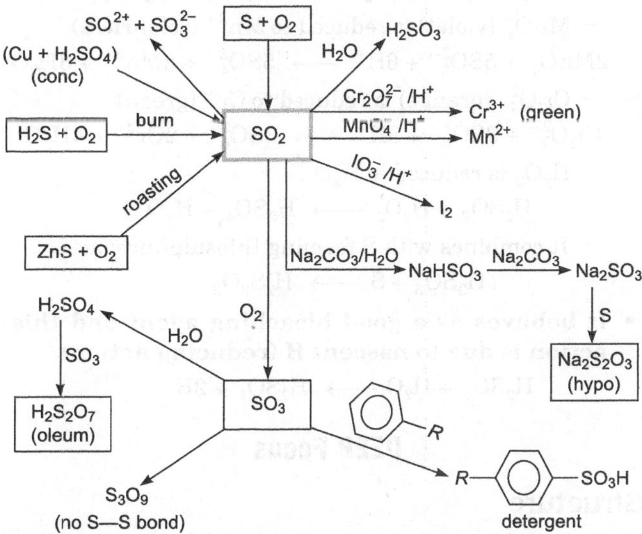

Oxoacids of Sulphur

- The oxoacids of sulphur are more numerous and are summarised in Table 14.4.
- Acids ending in **-ous** have S in the **oxidation state (+IV)**, and form salts ending in **-ite**. Acids ending in **-ic** have S in the **oxidation state (+VI)** and form salt ending in **-ate.**
- In peroxy acids, there is a peroxy linkage (with oxidation number of each oxygen in peroxide –1) hence oxidation number of sulphur is (+VI).
- In $H_2S_2O_3$ one sulphur is attached to central S, hence, its oxidation state is (–II) as in oxygen.
- H_2SO_4 (oxidation number of S = VI) is stronger than H_2SO_4 (oxidation number of S = IV).

Sulphurous Acid (H_2SO_3)

- The acid is obtained by dissolving SO_2 gas in H_2O :

$$SO_2 + H_2O \rightleftharpoons H_2SO_3$$

It actually exists as **$SO_2 \cdot H_2O$** and this acid is known only in solution.

- A strong diprotic acid:

$$H_2SO_3 \rightleftharpoons H^+ + HSO_3^-$$
$$HSO_3^- \rightleftharpoons H^+ + SO_3^{2-}$$

$NaHSO_3$ is thus an acid salt and Na_2SO_3 is normal salt and both liberate SO_2 gas on reaction with acid :

$$NaHSO_3 + HCl \longrightarrow NaCl + H_2O + SO_2$$
$$Na_2SO_3 + 2HCl \longrightarrow 2NaCl + H_2O + SO_2$$

- The acid is a good **reducing agent :**

$$H_2SO_3 + H_2O \longrightarrow H_2SO_4 + 2H^+ + 2e^-$$

– $X_2(Cl_2, Br_2, I_2)$ is reduced to HX

$$H_2SO_3 + I_2 + H_2O \longrightarrow H_2SO_4 + 2HI$$

– Fe^{3+} salts are reduced to Fe^{2+} salts

$$2FeCl_3 + H_2SO_3 + H_2O \longrightarrow 2FeCl_2 + H_2SO_4 + 2HCl$$

– MnO_4^- (violet) is reduced to Mn^{2+} (colourless)

$$2MnO_4^- + 5SO_3^{2-} + 6H^+ \longrightarrow 5SO_4^{2-} + 2Mn^{2+} + 3H_2O$$

– $Cr_2O_7^{2-}$ (orange) is reduced to Cr^{3+} (green)

$$Cr_2O_7^{2-} + 3SO_3^{2-} + 8H^+ \longrightarrow 3SO_4^{2-} + 2Cr^{3+} + 4H_2O$$

– H_2O_2 is reduced to H_2O

$$H_2SO_3 + H_2O_2 \longrightarrow H_2SO_4 + H_2O$$

– It combines with S forming thiosulphuric acid

$$H_2SO_3 + S \longrightarrow H_2S_2O_3$$

- **It behaves as a good bleaching agent and this action is due to nascent H (reducing action)**

$$H_2SO_3 + H_2O \longrightarrow H_2SO_4 + 2H$$

DEEP Focus

Structure

- The sulphite ion exists in crystal and has a pyramidal structure, with one position occupied by lone pair. S-atom is sp^3 hybridised but bond angle O—S—O is 106° (slightly distorted due to lone pair) and the bond lengths are 1.51Å. The $(d\pi - p\pi)$ bond between S and O is delocalised, and hence, the S—O bonds have a bond order of 1.33

$$O{=}\ddot{S}(O^-)_2 \longleftrightarrow \bar{O}{-}\ddot{S}(O^-){=}O \longleftrightarrow O{-}\ddot{S}(=O)(O^-)$$

Fe^{2+}, Mn^{2+}, HCl, Cr^{3+}

Cl_2, Fe^{3+}, MnO_4^-/H^+, $Cr_2O_7^{2-}/H^+$

$H_2SO_4 + 2H^+ + 2e^-$ —H_2O_2→ H_2O

reducing action of H_2SO_3

H_2O

$H_2SO_4 + 2H$ ←H_2O— bleaching action — H_2SO_3 —S→ $H_2S_2O_3$

NaOH

$NaHSO_3$ —NaOH→ Na_2SO_3

Table 14.4 **Oxoacids of Sulphur**

Formula	Oxidation state	Name	Structure
1. Sulphurous acid series			
(a) H_2SO_3	S (IV)	sulphurous acid	(HO)$_2$S=O
(b) $H_2S_2O_5$	S (+V) S (+III)	disulphurous acid or pyrosulphurous acid	H—O—S(=O)$_2$—S(=O)—OH
(c) $H_2S_2O_4$	S (+III)	dithionous acid	HO—S(=O)—S(=O)—OH
2. Sulphuric acid series			
(a) H_2SO_4	S (VI)	sulphuric acid	HO—S(=O)$_2$—OH
(b) $H_2S_2O_3$	S (VI) S(–II)	thiosulphuric acid	OH—S(=O)(=S)—OH
(c) $H_2S_2O_7$	S (VI)	pyrosulphuric acid	HO—S(=O)$_2$—O—S(=O)$_2$—OH
3. Peroxo acid series			
(a) H_2SO_5	S (VI)	peroxomono-sulphuric acid (Caro's acid)	HO—S(=O)$_2$—O—OH
(b) $H_2S_2O_8$	S (VI)	peroxodisulphuric acid (Marshall's acid)	HO—S(=O)$_2$—O—O—S(=O)$_2$—OH

Sulphuric Acid

- **Contact process** (a commercial method)
  Involves four stages:

Stage I : $S + O_2 \xrightarrow{\Delta} SO_2$

Stage II : $2SO_2 + O_2 \xrightarrow{V_2O_5} 2SO_3$

Stage III : $SO_3 + H_2O \longrightarrow H_2SO_4$

This is a violent reaction; instead SO_3 is passed into 98% H_2SO_4

$$H_2SO_4 + SO_3 \longrightarrow \underset{\text{oleum (pyrosulphuric acid)}}{H_2S_2O_7}$$

Some trisulphuric acid $H_2S_3O_{10}$ is also formed .

Stage IV : $H_2S_2O_7 + H_2O \longrightarrow 2H_2SO_4$

$$H_2S_3O_{10} + 2H_2O \longrightarrow 3H_2SO_4$$

Lead chamber process (which is no longer in use) Involves three stages:

Stage I : $2NO\ (catalyst) + O_2 \longrightarrow 2NO_2$

Stage II : $2SO_2 + 2NO_2 \longrightarrow 2SO_3 + 2NO$

$$2SO_2 + O_2 \xrightarrow{NO} 2SO_3$$

Stage III : $SO_3 + H_2O \longrightarrow H_2SO_4$

$$NO_2 + SO_2 + H_2O \longrightarrow H_2SO_4 + NO\ (recycled)$$

$$2NO + O_2 \longrightarrow 2NO_2$$

- **Pure H_2SO_4 is strongly hydrogen bonded and in the absence of water it does not react with metals to produce H_2**

---HO, O---HO, O---HO, O--- (S) ---HO, O---HO, O---HO, O

- If pure H_2SO_4 is heated, a little SO_3 is evolved, and an azeotropic mixture of 98.3% H_2SO_4 and 1.7% water is produced.

DEEP Focus

- Anhydrous H_2SO_4 and concentrated H_2SO_4 mix with water in all proportions, and evolve a great deal of heat (880 kJ mol^{-1}). If water is poured into concentrated sulphuric acid, the heat evolved leads to boiling of the drops of water and causes violent splashing. **The safe way to dilute strong acid is to carefully pour the acid into the water with stirring.**

- It is a diprotic acid. It is colourless viscous liquid (m.p. 10.4°C). The concentrated sulphuric acid we use in the laboratory is 98 per cent H_2SO_4 by mass (density 1.84 g/cm^3) which corresponds to a concentration of 18 M.

As a strong acid :

In aqueous solution, it is regarded as a strong acid ionising in two stages :

$$H_2SO_4 + H_2O \longrightarrow HSO_4^- + H_3O^+$$

$$HSO_4^- + H_2O \rightleftharpoons SO_4^{2-} + H_3O^+$$

Only the first dissociation is complete; the second is partial.

When it is dilute, it shows all the usual properties of acids.

$$Zn + H_2SO_4 \longrightarrow ZnSO_4 + H_2$$

$$Zn + 2H_3O^+ \longrightarrow Zn^{2+} + H_2 + 2H_2O$$

- **As an oxidising agent :**

The oxidising strength of sulphuric acid depends on its temperature and concentration. A cold, dilute sulphuric acid solution reacts with metals above hydrogen in the activity series (as shown above with Zn), thereby liberating molecular hydrogen (H_2) in a displacement reaction; Mg reacts in the same way

$$Mg + 2H_3O^+ \longrightarrow Mg^{2+} + H_2 + 2H_2O$$

This is a typical reaction of an active metal with an acid.

DEEP Focus

The strength of sulphuric acid as an oxidising agent is greatly enhanced when it is **both hot and concentrated.** In such a solution, **the oxidising agent is actually the sulphate ion** rather than the hydrated proton, $H^+(aq)$. Thus,

$$SO_4^{2-} + 4H^+ + 2e^- \longrightarrow SO_2 + 2H_2O$$

$$Cu + 2H_2SO_4 \longrightarrow CuSO_4 + 2H_2O + SO_2$$

$$S + 2H_2SO_4 \longrightarrow 2H_2O + 3SO_2$$

$$C + 2H_2SO_4 \longrightarrow CO_2 + 2H_2O + 2SO_2$$

$$2HBr + H_2SO_4 \longrightarrow Br_2 + SO_2 + 2H_2O$$

$$K_4[Fe(CN)_6] + 6H_2SO_4 + 6H_2O \longrightarrow 2K_2SO_4 + FeSO_4 + 3(NH_4)_2SO_4 + 6CO$$

Depending on the nature of the reducing agents, the sulphate ion may be further reduced to elemental sulphur or the sulphide ion

$$8HI + H_2SO_4 \longrightarrow H_2S + 4I_2 + 4H_2O$$

- **As a dehydrating agent :**

Concentrated H_2SO_4 is also a dehydrating agent :

$$C_6H_{12}O_6 \xrightarrow{conc.\ H_2SO_4} 6C + 6H_2O$$

$$CH_3CH_2OH \xrightarrow[160°C]{conc.\ H_2SO_4} CH_2{=}CH_2 + H_2O$$

$$2CH_3CH_2OH \xrightarrow[140°C]{conc\ H_2SO_4} CH_3CH_2{-}O{-}CH_2CH_3 + H_2O$$

- A mixture of conc. HNO_3 and conc. H_2SO_4 is used as a nitrating mixture in the nitration of aromatic compounds :

$$HNO_3 + 2H_2SO_4 \longrightarrow \underset{\text{nitronium ion}}{NO_2^+} + H_3O^+ + 2HSO_4^-$$

Nitronium ion is a sufficiently good electrophile to attack a benzene ring :

$$C_6H_6 + NO_2^+ \longrightarrow \underset{\text{nitrobenzene}}{C_6H_5{-}NO_2} + H^+$$

- Conc. H_2SO_4 is also used for sulphonation of organic compounds :

$$C_6H_6 + H_2SO_4 \longrightarrow \underset{\text{benzene sulphonic acid}}{C_6H_5{-}SO_3H} + H_2O$$

- $$HO{-}SO_2{-}OH + 2PCl_5 \longrightarrow \underset{\text{sulphuryl chloride}}{SO_2Cl_2} + 2POCl_3 + 2HCl$$

- When conc. H_2SO_4 is subjected to electrolysis there is formation of peroxy disulphuric acid and H_2O_2 at the anode

$$H_2SO_4 \rightleftharpoons H^+ + HSO_4^-$$

at anode

$$2HSO_4^- \longrightarrow H_2S_2O_8 + 2e^-$$

$$H_2S_2O_8 + 2H_2O \longrightarrow 2H_2O + 2H_2SO_4$$

at cathode

$$2H^+ + 2e^- \longrightarrow H_2$$

Structure

S-atom is sp^3 hybridised. SO_4^{2-} ion is tetrahedral. The bond lengths are all equal (1.49Å) and are all rather short. The bond order of the S—O bonds is approximately 1.5. There are four σ-bonds and two π bonds between S and O atoms (with two π-bonds delocalised over the S and the four O atoms).

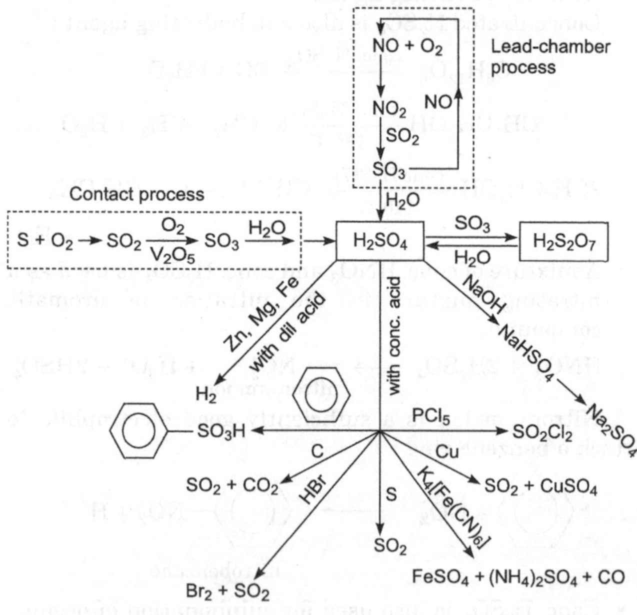

Sodium Thiosulphate ($Na_2S_2O_3 \cdot 5H_2O$ —Hypo)

- $H_2S_2O_3$, thiosulphuric acid, because of its low stability, can't be prepared. However, its salts called thiosulphates can be prepared.

- $$\underset{\text{neutral or alkaline}}{Na_2SO_3} + S \xrightarrow{\text{boiling}} \underset{\text{sodium thiosulphate}}{Na_2S_2O_3}$$

- $$2Na_2S_3 + 3O_2 \xrightarrow{\text{heat in air}} 2Na_2S_2O_3 + 2S$$

- Following reactions called **Spring's reactions,** are also used to convert Na_2S or S into $Na_2S_2O_3$:

$$Na_2S + Na_2SO_3 + I_2 \longrightarrow Na_2S_2O_3 + 2NaI$$

$$2Na_2S + 4SO_2 + Na_2CO_3 \longrightarrow 3Na_2S_2O_3 + CO_2$$

$$6NaOH + 4S \xrightarrow{\Delta} Na_2S_2O_3 + 2Na_2S + 3H_2O$$

- Hydrated sodium thiosulphate, $Na_2S_2O_3 \cdot 5H_2O$, is called **'hypo'**. It forms very large colourless hexagonal crystals, melting point 48°C.
- It is highly soluble in water.
- It is an efflorescent substance and at 215°C, it is completely dehydrated to $Na_2S_2O_3$.
- Dilute acids like H_2SO_4, HCl decompose it into SO_2 and colloidal sulphur (yellowish-white turbidity):

$$Na_2S_2O_3 + 2HCl \longrightarrow 2NaCl + SO_2 + S + H_2O$$

This is a disproportionation reaction.

DEEP Focus

- I_2 oxidises $Na_2S_2O_3$ into $Na_2S_4O_6$:

$$2Na_2S_2O_3 + I_2 \longrightarrow \underset{\text{sodium tetrathionate}}{Na_2S_4O_6} + 2NaI$$

This reaction is used for the iodometric estimation of oxidising agents like $CuSO_4$, $CaOCl_2$, MnO_2, $K_2Cr_2O_7$, $KMnO_4$ etc., which on reaction with KI liberate I_2 which in turn is determined by titration using $Na_2S_2O_3$.

$$2CuSO_4 + 4KI \longrightarrow 2K_2SO_4 + \underset{\text{White ppt}}{Cu_2I_2\downarrow} + I_2$$

- $Na_2S_2O_3$ is used in the bleaching industry to destroy any excess Cl_2 on fabrics :

$$Na_2S_2O_3 + 4Cl_2 + 5H_2O \longrightarrow 2NaHSO_4 + 8HCl$$

$$2FeCl_3 + 2Na_2S_2O_3 \longrightarrow Na_2S_4O_6 + 2FeCl_2 + 2NaCl$$

- $AgNO_3$ reacts with $Na_2S_2O_3$ giving white ppt changing to yellow, then to brown and finally black :

$$2AgNO_3 + Na_2S_2O_3 \longrightarrow \underset{\text{white ppt}}{Ag_2S_2O_3} + 2NaNO_3$$

$$Ag_2S_2O_3 + H_2O \longrightarrow \underset{\text{black ppt}}{Ag_2S} + H_2SO_4$$

- Hypo is used in photography for **'fixing'** films and prints. It dissolves unreacted AgBr from the photographic emulsion by complexation :

$$AgBr + 2Na_2S_2O_3 \longrightarrow \underset{\text{soluble}}{Na_3[Ag(S_2O_3)_2]} + NaBr$$

The thiosulphate ion is structurally similar to the sulphate ion; one oxygen atom has been replaced by sulphur atom. Based on Lewis structure oxidation number **+6** has been assigned to central sulphur and **–2** for the terminal sulphur. Thus, average oxidation state is **+2**.

In tetrathionate ion $S_4O_6^{2-}$, two of the sulphur atoms present in the ion are in oxidation state 0 (zero) and two are in oxidation state +5. Alternately one can say that $S_4O_6^{2-}$ is the adduct formed by binding of the Lewis base. S_2^{2-} to SO_3

$$\underset{\text{Lewis base}}{S_2^{2-}} + \underset{\text{Lewis acid}}{2SO_3} \longrightarrow \underset{\text{adduct}}{S_4O_6^{2-}}$$

Isotopic tracer technique has been used to verify the existence of two types of sulphur in $S_2O_3^{2-}$. Sulphite ion (SO_3^{2-}) is treated with radioactive isotope (^{35}S) and the thiosulphate ion ($S_2O_3^{2-}$) is decomposed by acid. ^{35}S is found to be ony in precipitated sulphur. This indicates that sulphur (^{35}S) is not distributed with other sulphur is $S_2O_3^{2-}$

$$SO_3^{2-} + S*(^{35}S) \longrightarrow S*SO_3^{2-}$$
$$S*SO_3^{2-} + 2H^+ \longrightarrow H_2SO_3 + S*$$

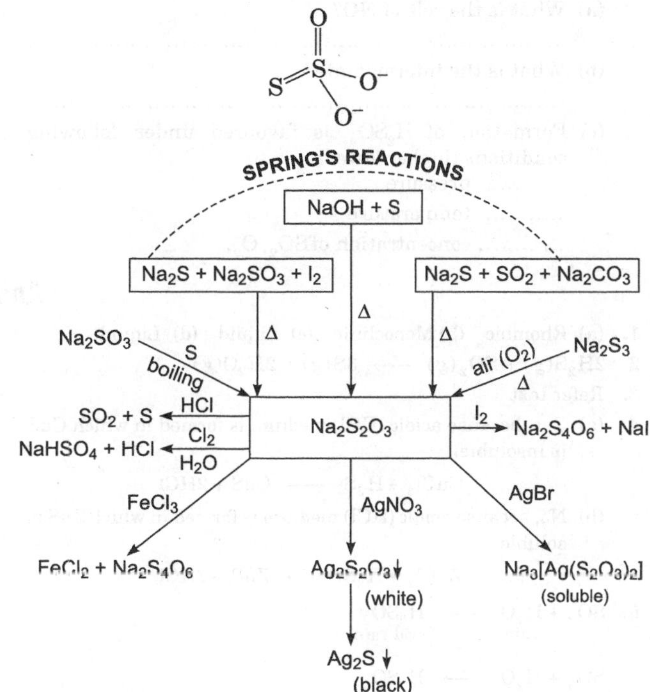

Target Practice 2

1. What are following sulphur allotropes called?
(a) S_α .... (b) S_β ....
(c) S_γ .... (d) S_μ ....

2. H_2S changes to sulphur in Frasch process by reaction ....

3. Differentiate H_2O and H_2S in following properties (smaller/larger)

	H_2O	H_2S
(a) Boiling point	☐	☐
(b) Bond angle	☐	☐
(c) Acid-strength	☐	☐

4. H_2S gas is passed into aqueous solution
(a) $CuCl_2$ solution ⟶ (H_2S) CuS is precipitated [Yes/No]
(b) $ZnCl_2$ solution ⟶ (H_2S) ZnS is precipitated [Yes/No]

5. Oxide of sulphur are part of acid rain. Represent phenomenon of acid rain by chemical reaction
....
....

6. $2SO_2 \rightleftharpoons [SO^{2+}] + [SO_3^{2-}]$
....
Name the cation and anion.

7. Write the formulae of
(a) Pyrosulphuric acid ....
(b) Peroxosulphuric ....
(c) Peroxodisulphuric acid ....

8. Equivalent weight of SO_2 is found to be 32 in a reaction taking place in acidic medium. This reaction is represented as ....

9. A gaseous mixture of CO_2 and SO_2 was to be identified in a laboratory. Mr *X* and Mr *Y*. performed following experiments with common reagents but at different stages.

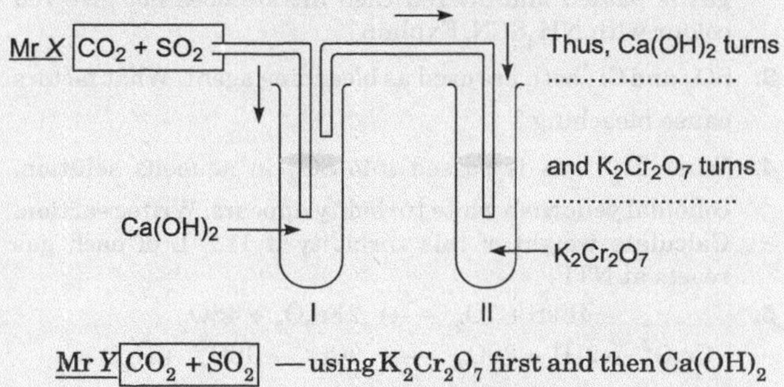

Mr *Y* $CO_2 + SO_2$ —using $K_2Cr_2O_7$ first and then $Ca(OH)_2$
Thus, K_2CrO_7 turns ....
and $Ca(OH)_2$ turns ....

10. Sulphuric acid is produced by lead chamber process with following exothermic reaction

$$2SO_2 + O_2 \underset{}{\overset{NO}{\rightleftharpoons}} 2SO_3 \xrightarrow{H_2O} H_2SO_4$$

(a) What is the role of NO?

....

(b) What is the intermediate?

....

(c) Formation of H_2SO_4 is favoured under following conditions (higher/lower)

.... pressure

.... temperature

.... concentration of SO_2, O_2.

11. Hypo dissolves unreacted AgBr of the photographic film by reaction

....

12. In the following

$$2Cu^{2+} + 4I^- \longrightarrow Cu_2I_2 + I_2$$
$$2S_2O_3^{2-} + I_2 \longrightarrow S_4O_6^{2-} + 2I^-$$

(a) equivalent mass of Cu^{2+} is ....

(b) equivalent mass of $S_2O_3^{2-}$ is ....

Answers

1. (a) Rhombic (b) Monoclinic (c) Liquid (d) Liquid

2. $2H_2S(g) + SO_2(g) \longrightarrow 3S(g) + 2H_2O(g)$

3. Refer text

4. (a) Yes, because acidic (HCl) medium is formed in which CuS is insoluble.

$$CuCl_2 + H_2S \longrightarrow CuS + 2HCl$$

(b) No, because acidic (HCl) medium is formed in which ZnS is soluble

$$ZnCl_2 + H_2S \longrightarrow ZnS + 2HCl$$

5. $\underset{\text{rain}}{SO_2 + H_2O} \longrightarrow \underset{\text{acid rain}}{H_2SO_3}$

$\underset{\text{rain}}{SO_3 + H_2O} \longrightarrow \underset{\text{acid rain}}{H_2SO_4}$

6. SO^{2+} : Sulpharyl

SO_3^{2-} : Sulphite

7. (a) $H_2S_2O_7$ (b) H_2SO_5 (c) $H_2S_2O_8$

8. $SO_2 + 2H_2O \longrightarrow SO_4^{2-} + 4H^+ + 2e^-$

9. Mr *X* : Both CO_2 and SO_2 react with $Ca(OH)_2$ and lime water turns milky. No effect on $K_2Cr_2O_7$.

Mr *Y* : SO_2 turns $K_2Cr_2O_7$ green and CO_2 does not react at this stage. CO_2 reacts at stage II with $Ca(OH)_2$ and it turns milky.

10. (a) Catalyst

(b) NO_2

(c) Higher pressure

Lower temperature (theoretically)

Higher concentration of SO_2 and O_2

11. $AgBr + 2S_2O_3^{2-} \longrightarrow [Ag(S_2O_3)_2]^{3-} + Br^-$

12. (a) $2Cu^{2+} \equiv I_2 \equiv 2I$

$\therefore\ Cu^{2+} \equiv 1$

equivalent mass of Cu^{3+} ≡ ionic weight

(b) $2S_2O_3^{2-} \equiv I_2 \equiv 2I$

$\therefore\ S_2O_3^{2-} \equiv I$

$\therefore$ Equivalent mass of $S_2O_3^{2-}$ = ionic weight.

Practice Exercise 2

1. You have been given an inorganic mixture with a knowledge that group II is absent, but when pass H_2S gas into the mixture solution acidified with dil. HCl, yellowish turbidity appears. Explain.

2. $FeCl_3$ solution gives blood red colour with NH_4SCN. If H_2S gas is passed and filtered then filtrate does not give red colour with NH_4SCN. Explain.

3. SO_2 and Cl_2 both are used as bleaching agent. What factors cause bleaching ?

4. When H_2S gas is passed into SO_2 in aqueous solution, colloidal yellowish white turbidity appears. Write reaction. Calculate weight of this turbidity if 11.2 L of each gas reacts at NTP.

5.
$$4FeS + 7O_2 \longrightarrow 2Fe_2O_3 + 4SO_2$$
$$Cr_2O_7^{2-} + 2H^+ + 3SO_2 \longrightarrow 2Cr^{3+} + 3SO_4^{2-} + H_2O$$

1 g sample of FeS was roasted and SO_2 thus formed required 10 mL of 0.05 M $K_2Cr_2O_7$ solution in acidic medium. Calculate percentage of FeS in the given sample. (Fe = 56, S = 32)

6. Give examples of reactions for preparing SO_2 that

(a) are attended by a change in oxidation state of sulphur

(b) are not attended by a change in its oxidation state

7. Sodium salt (*A*) of a dibasic acid $\xrightarrow{HCl}$ gas (*B*)

Gas (*B*) turns $K_2Cr_2O_7$ green and also lime water milky. Identify (*A*) and (*B*).

8. How will you estimate $NaHSO_3$ volumetrically?

9. What is structure of SO_3^{2-} ion?

10. Cl_2 also behaves as bleaching agent like H_2SO_3 (*aq* SO_2). Explain.

11. How will you identify SO_3^{2-} in presence of CO_3^{2-}?

12. Which of nitric acid and sulphuric acid is the stronger Bronsted acid, *i.e.*, which donates proton to the other ?

13. Zinc reacts with dil H_2SO_4 to give H_2. It also reacts with conc H_2SO_4 to give SO_2. Make difference between these two reactions.

Answers

1. It is due to presence of some oxidising agents (acidic radicals like NO_2^-, NO_3^-) which oxidise H_2S to colloidal sulphur (yellowish turbidity)

$$H_2S + O \longrightarrow H_2O + S \text{ (colloidal yellowish turbidity)}$$

2. $FeCl_3 + NH_4SCN \longrightarrow \underset{\text{red}}{[Fe(SCN)]Cl_2} + NH_4Cl$

$FeCl_3$ changes to $FeCl_2$ on passing H_2S hence, no colour with NH_4SCN.

3. $SO_2 + 2H_2O \longrightarrow H_2SO_4 + 2H$

$Cl_2 + H_2O \longrightarrow 2HCl + O$

Bleaching action of SO_2 is due to H (that causes reduction) and that of Cl_2 is due to O (that causes oxidation).

4. $\underset{1\text{ mol}}{SO_2} + \underset{2\text{ mol}}{2H_2S} \longrightarrow 2H_2O + \underset{3\text{ mol (colloidal turbidity)}}{3S}$

$\frac{1}{2}$ mol $\quad \frac{1}{2}$ mol $\quad \frac{3}{4}$ mol = 24 g

(11.2 L of each at NTP = $\frac{1}{2}$ mol)

H_2S is limiting reactant, hence S formed = $\frac{3}{4}$ mol

5. 10 mL of 0.05 M $K_2Cr_2O_7 = \frac{10 \times 0.05}{1000}$ mol $= 5 \times 10^{-4}$ mol

$$SO_2 = 3 \times 5 \times 10^{-4} \text{ mol}$$

$$\text{pure FeS} = 3 \times 5 \times 10^{-4} \text{ mol}$$
$$= 15 \times 10^{-4} \times 88 \text{ g}$$
$$= 0.132 \text{ g in 10g impure FeS}$$
$$= 1.32\%$$

6. (a) $\underset{0}{S} + O_2 \longrightarrow \underset{+4}{SO_2}$] change in oxidation state of sulphur

(b) $Na_2\underset{+4}{SO_3} + 2HCl \longrightarrow 2NaCl + H_2O + \underset{+4}{SO_2}$] no change

7. (*A*) : Na_2SO_3 (*B*) : SO_2

8. $NaHSO_3$ is treated with excess (known) amount of I_2 and unreacted I_2 is determined by back titration using $Na_2S_2O_3$ (hypo) iodometrically.

$$I_2 + NaHSO_3 + H_2O \longrightarrow NaHSO_4 + 2HI$$
$$I_2 + 2Na_2S_2O_3 \longrightarrow Na_2S_4O_6 + 2NaI$$

9. See text

10. $Cl_2 + H_2O \longrightarrow 2HCl + (O)$

It is due to nascent oxygen.

11. $(CO_3^{2-} + SO_3^{2-}) + \text{dil. HCl} \longrightarrow \underbrace{CO_2 + SO_2}$

Both gases turn lime water milky. Hence, proceed :

(a) First pass the gaseous mixture into acidified $K_2Cr_2O_7$ solution. Only SO_2 reacts and turns it green :

$$Cr_2O_7^{2-} + 3SO_2 + 2H^+ \longrightarrow \underset{\text{green}}{2Cr^{3+}} + 3SO_4^{2-} + H_2O$$

(b) Gas unreacted is CO_2. Pass it into lime water. It turns milky.

$$Ca(OH)_2 + CO_2 \longrightarrow \underset{\text{milky}}{CaCO_3} + H_2O$$

12. $H_2SO_4 + HNO_3 \longrightarrow HSO_4^- + H_2NO_3^+$

H_2SO_4 is proton donor.

13. (a) $Zn + \underset{\text{dil}}{H_2SO_4} \longrightarrow ZnSO_4 + H_2$

In this reaction H^+ is reduced by Zn

$$Zn + 2H^+ \longrightarrow H_2 + Zn^{2+}$$

(b) $Zn + \underset{\text{conc.}}{2H_2SO_4} \longrightarrow ZnSO_4 + SO_2 + 2H_2O$

Zn reduces SO_4^{2-} to SO_2

$$SO_4^{2-} + 4H^+ + Zn \longrightarrow Zn^{2+} + SO_2 + 2H_2O$$

Total Practice Set
(Read, Plan & Solve!)

Problem 1. Zinc dissolves in dil HCl to liberate $H_2(g)$, but from conc H_2SO_4, SO_2 is evolved. Explain.

Solution (A) $Zn + \underset{\text{dil}}{2H^+} \longrightarrow Zn^{2+} + H_2 \quad E° = 0.76\ V$

(B) $Zn + SO_4^{2-} + 4H^+ \longrightarrow 2H_2O + SO_2 + Zn^{2+}$

$$E° = 0.93\ V$$

$E°$ of reaction (B) is large than that of (A).

In conc. H_2SO_4, SO_4^{2-} is a better oxidising agent than H^+. Hence, with conc. H_2SO_4, reaction (B) predominates.

Problem 2. $E°_{Al^{3+}/Al} = -1.66\ V; E°_{SO_4^{2-}/SO_2} = 1.17\ V.$

What, if anything, would happen if you reacted aluminium with sulphuric acid ? If you think a reaction would take place, write down the equation.

Solution The given reaction (say dissolution of Al into H_2SO_4) is possible if $E°_{cell}$ is positive.

$$Al \longrightarrow Al^{3+} + 3e^-$$

$$E°_{Al/Al^{3+}} = -E°_{Al^{3+}/Al} = 1.66\ V$$

$$SO_4^{2-} + 2e^- + 4H^+ \longrightarrow SO_2 + 2H_2$$

$$E°_{SO_4^{2-}/SO_2} = 1.17\ V.$$

Net reaction is

$$\mathbf{2Al + 3SO_4^{2-} + 12H^+ \longrightarrow 2Al^{3+} + 3SO_2 + 6H_2O}$$

Thus, $E°_{cell}$ is 2.83 V hence, Al can reduce SO_4^{2-} to SO_2.

Problem 3. What mass of H_2SO_4 could be produced in the process given below if 1.00 kg of FeS_2 is used? The unbalanced equations for the process are:

$$FeS_2(s) + O_2(g) \longrightarrow Fe_2O_3(s) + SO_2(g)$$

$$SO_2(g) + O_2(g) \longrightarrow SO_3(g)$$

$$SO_3(g) + H_2SO_4(l) \longrightarrow H_2S_2O_7(l)$$

$$H_2S_2O_7(l) + H_2O(l) \longrightarrow H_2SO_4(aq) \text{(Fe = 56, S = 32)}$$

Solution Sulphuric acid contains one sulphur atom per molecule, and each FeS_2 molecule gives two atoms of sulphur. Thus,

$$FeS_2 \longrightarrow 2SO_2 \longrightarrow 2SO_3 \longrightarrow H_2S_2O_7 \longrightarrow 2H_2SO_4$$

Thus, 1 mol of FeS_2 gives = 2 mol H_2SO_4

$$= 2 \times 98\ g\ H_2SO_4$$

Hence, $1.00\ kg\ FeS_2 = 1000\ g\ FeS_2 = \frac{1000}{120}\ mol\ FeS_2$

$$H_2SO_4 \text{ from } \frac{1000}{120}\ mol\ FeS_2 = \frac{2 \times 98 \times 1000}{120}\ g$$

$$= \mathbf{1.63\ kg\ H_2SO_4}$$

Problem 4. The action of concentrated sulphuric acid on urea (NH_2CONH_2) results in the production of a white crystalline solid X of formula H_3NO_3S. This is a monobasic acid. On treatment with sodium nitrate and dilute hydrochloride acid at 0° C it liberates 1 mol of N_2 per mol of X, and on addition of aqueous barium chloride the resulting solution yields 1 mol of $BaSO_4$ per mol of X taken.

Deduce the structure of X.

Solution NH_2CONH_2 (Urea) reacts with conc. H_2SO_4 forming $X(H_3NO_3S)$. X contains one ionisation H (since, X is a monobasic acid). Thus,

$$\underset{\text{Urea}}{NH_2—\overset{\overset{O}{\|}}{C}—NH_2} + \underset{\text{Sulphuric acid}}{H—O—\overset{\overset{O}{\|}}{S}—O—H} \longrightarrow \underset{X}{NH_2—SO_3H} + NH_3 + CO$$

X contains free $—NH_2$ as well as free $—SO_3H$. ($NaNO_2 + HCl$) converts free $—NH_2$ into free $—OH$ group liberating N_2.

$$NaNO_2 + HCl \longrightarrow HNO_2 + NaCl$$

$$\underset{1\ mol\ X}{NH_2—SO_3H} + HNO_2 \longrightarrow \underset{1\ mol\ X}{N_2 \uparrow} + H_2O + \underset{1\ mol}{H_2SO_4}$$

$$\underset{1\ mol}{H_2SO_4} + BaCl_2 \longrightarrow \underset{1\ mol}{BaSO_4} + 2HCl$$

Thus X is $\mathbf{NH_2—SO_3H}$ (amino sulphonic acid), 1 mol of which liberated 1 mol of N_2 on reaction with ($NaNO_2 + HCl$) at 0°C and in turn resulting solution gives 1 mol of $BaSO_4$ on reaction with $BaCl_2$.

Explanation

(a) ***When metallic copper is heated with conc. sulphuric acid, in addition to copper(II) sulphate, $CuSO_4$, and sulphur dioxide SO_2, some copper(II) sulphide, CuS, is also formed.***

- When conc H_2SO_4 is used, Cu is oxidised to Cu^{2+} and SO_4^{2-} is reduced to SO_2.

$$Cu \longrightarrow Cu^{2+} + 2e^-$$

$$SO_4^{2-} + 4H^+ + 2e^- \longrightarrow SO_2 + 2H_2O$$

There is also side reaction assisted by very high [H^+] and very low solubility of CuS.

$$SO_4^{2-} + 8H^+ + 8e^- \longrightarrow S^{2-} + 4H_2O$$

(b) ***The S—OH and S—O distance in the tetrahedral species are as follows:***

	S—OH (Å)	S—O (Å)
$H_2SO_4(g)$	1.54	1.42
HSO_4^-	1.56	1.47
SO_4^{2-}	—	1.49

Comment on the increase in S—O bond length.

- As the π-bonding is distributed over a larger number of atoms, bond order decreases resulting in an increase in bond length.

Master Exercises

Exercise 1
(Stage 1 : Learning)

Short Answer Type Questions

1. Give the names and formulae of the compounds in which sulphur exhibits an oxidation state of:
(a) –2 (b) +4 (c) +6

2. Explain why SCl_6, SBr_6 and SI_6 cannot be prepared?

3. Give applications of O_3.

4. What is the change in oxidation number for the following reaction?

$$3O_2 \longrightarrow 2O_3$$

5. One of the steps involved in the depletion of ozone in the stratosphere by nitric oxide (NO) may be represented as:

$$NO(g) + O_3(g) \longrightarrow NO_2(g) + O_2(g)$$

What is change in oxidation number of N and O?

6. After the ozonisation of a certain volume of oxygen, the volume of the oxygen was found to diminish by 500 mL. What volume of ozone was formed? What amount of heat was absorbed in its formation if ΔH_f° is 144.2 kJ mol^{-1} for ozone? Assume values at STP.

7. Black $(A) + H_2SO_4 \longrightarrow$ gas $(B) + (C)$

$(B) + (CH_3COO)_2Pb \longrightarrow$ black ppt (D)

$(C) + K_3[Fe(CN)_6] \longrightarrow$ blue (E)

(C) also decolorises acidified $KMnO_4$. Identify (A) to (E).

8. What properties does H_2S show when it reacts with aqueous solution of $KMnO_4$, H_2O_2 and NaOH?

9. Acid salt (A) $\xrightarrow{\text{heat}}$ (B) $\xrightarrow{HCl}$ Gas (C) $\xrightarrow{Cr_2O_7^{2-}/H^+}$ Green (G)
Molar mass 104

(B) $\xrightarrow{BaCl_2}$ White ppt (D) soluble in HCl $\longrightarrow$ White ppt (H)

Gas (C) $\xrightarrow{Ca(OH)_2}$ Milky $\xrightarrow{\text{Gas } (C)}$ Soluble (F)

Identify (A) to (H) in the above sequence of reactions.

10. Which of the following desiccants can be used to remove moisture from SO_2: conc. H_2SO_4; KOH, P_2O_5, K_2CO_3?

11. Conc. H_2SO_4 can't be used for drying up H_2S and NH_3. Explain.

Analytical Questions

1. $O_3(g)$ is a powerful oxidising agent. Write equations to represent oxidation of :
(a) I^- to I_2 in acidic solution.
(b) $[Fe(CN)_6]^{4-}$ to $[Fe(CN)_6]^{3-}$ in basic solution.
(c) S in the presence of moisture to H_2SO_4.

2. It has been estimated that if all the O_3 in the atmosphere were brought to sea level at STP, the gas would form a layer 0.3 cm thick. Estimate the number of O_3 molecules in the earth's atmosphere. (radius of the earth = 4000 miles; 1 km = 0.6214 mile)

3. The redox reaction involving the reducing power of hydrogen sulphide is :

$$S + 2H^+ + 2e^- \longrightarrow H_2S, \qquad E^\circ_{S/H_2S} = +0.14V$$

Two other half equations are :

$$Fe^{3+} + e^- \longrightarrow Fe^{2+} \qquad E^\circ_{Fe^{3+}/Fe^{2+}} = +0.77\ V$$

$$Br_2 + 2e^- \longrightarrow 2Br^- \qquad E^\circ_{Br_2/Br^-} = 1.07\ V$$

Under standard conditions :

(i) will hydrogen sulphide react with iron (III) ions?

(ii) will it react with bromine?

If you decide a reaction should take place, write the equation.

4. Explain :
(i) Oxygen is never more than divalent, but S, Se and Te can form four or six bonds.
(ii) SO_2 is gas but SeO_2 is a solid at room temperature.
(iii) SF_6 is unreactive but TeF_6 is readily hydrolysed by water.

(iv) The melting point and boiling point show the sequence :

$H_2O >> H_2S < H_2Se < H_2Te$

(v) Acidity increases in the sequence :

$H_2O < H_2S < H_2Se \leq H_2Te$

(vi) The bond angle decreases in the order :

$H_2O > H_2S > H_2Se$

(vii) Liquid oxygen sticks to the poles of a magnet but liquid nitrogen does not.

[**Hint** : By MO theory, O_2 is paramagnetic)

(viii) The O—O bond in O_2F_2 is much shorter than that in H_2O_2.

(ix) The two sulphur atoms in thiosulphate ion are not identical.

5. Account for the following facts:

(a) H_2SO_4 is a stronger acid than H_2SO_3

(b) SF_4 exists, but OF_4 does not.

(c) The S_8 ring is non-planar.

6. How many litres of SO_2 taken in standard conditions have to be passed through a solution of $HClO_4$ to reduce 16.9 g of it to HCl?

7. Which of the compounds listed below enter into a reaction with $Na_2S_2O_3$?

(a) HCl (b) NaCl

(c) NaI (d) I_2

(e) $KMnO_4$ if:

(1) the initial substance becomes decolorised,

(2) a precipitate forms,

(3) a gas evolved.

8. Oxides and peroxides are not the only types of compounds that metals can make with oxygen, superoxides are also possible. 1.15 g of sodium on reaction with oxygen at high pressure gives 2.75 g of a white powder, sodium superoxide.

(i) What mass of oxygen was combined with the sodium?

(ii) How many moles of oxygen atoms in this?

(iii) How many moles of sodium were used?

(iv) What is the formula of the oxide and superoxide ion?

Exercise 2
(Stage 2 : High Skill Problem Solving)

Only One Option Correct

1. Oil of vitriol is

(a) HNO_3 (b) HCl

(c) H_2SO_4 (d) H_3PO_4

2. Which of the following has peroxy linkage?

(a) $H_2S_2O_3$ (b) H_2SO_5

(c) $H_2S_2O_7$ (d) $H_2S_4O_6$

3. Which of the following can be used as dehydrating agents?

(a) Conc. H_2SO_4 (b) $POCl_3$

(c) P_2O_5 (d) All of these

4. The thermal stability of hydrides of oxygen family is in order

(a) $H_2Po < H_2Te < H_2Se < H_2S < H_2O$

(b) $H_2Po < H_2O < H_2Te < H_2Se < H_2S$

(c) $H_2S < H_2O < H_2Te < H_2Se < H_2Po$

(d) $H_2O < H_2S < H_2Te < H_2Se < H_2Po$

5. In case of oxygen family (Group 16)

(a) the tendency for catenation decreases markedly as we go down the group

(b) maximum coordination of oxygen is four due to lack of *d*-orbital but that of other elements is six due to presence of *d*-orbital

(c) the tendency to form multiple bonds with C, N and O decreases as going down the group from S to Te.

(d) all of the above are correct statements

6. Which does not have (S—S) bond?

(a) $H_2S_2O_3$ (b) $H_2S_2O_5$

(c) $H_2S_2O_4$ (d) S_3O_9

7. Which has maximum pK_a value?

(a) H_2O (b) H_2S

(c) H_2Se (d) H_2Te

8. Consider the following species :

1. $[O_2]^{2-}$ 2. $[CO]^+$

3. $[O_2]^+$

Among these species sigma bond alone is present in

(a) 1, 2, and 3 (b) 2 alone

(c) 2 and 3 (d) 1 alone

9. Which of the following species have undistorted octahedral structures?

1. SF_6 2. PF_6^-

3. SiF_6^{2-} 4. SeF_6

Select the correct answer using the codes given below

(a) 2, 3 and 4 (b) 1, 3 and 4

(c) 1, 2 and 3 (d) 1, 2 and 4

10. Consider the following compounds
1. Sulphur dioxide
2. Hydrogen peroxide 3. Ozone

Among these compounds, those which can act as bleaching agents would include
(a) 1 and 3 (b) 2 and 3
(c) 1 and 2 (d) 1, 2 and 3

11. A considerable part of the harmful UV rays of the sun does not reach the surface of the earth. This is because high above the earth's atmosphere, there is a layer of
(a) O_3 (b) CO_2
(c) SO_2 (d) NO

12. Hypervalent ion is/are
(a) SO_3^{2-} (b) PO_4^{3-}
(c) SO_4^{2-} (d) ClO_4^-

13. Which is/are true statements?
(a) Sulphur trioxide exists as cyclic trimer in solid state, S_3O_9
(b) Selenium trioxide solid is a cyclic tetramer, Se_4O_{12}
(c) TeO_3 is a solid with a network structure in which TeO_6 octahedra share all vertices
(d) All of the above are correct

14. Which is thermocromic?
(a) S_4N_4 (b) $B_3N_3H_6$
(c) S_8 (d) None of these

15. The best reducing agent of the following is
(a) H_2S (b) Cl^-
(c) SO_4^{2-} (d) O_3

16. All of the following have a tetrahedral shape except
(a) SO_4^{2-} (b) XeF_4
(c) ClO_4^- (d) XeO_4

17. The term 'thio' is used in the names of all of the following compounds except
(a) $Na_2S_2O_3$ (b) $NaCS_3$
(c) NaSCN (d) Na_2SO_3

18. Cold solution of barium nitrite on mixing with sulphuric acid produces
(a) $BaSO_4 + HNO_2$ (b) $BaSO_4 + HNO_3$
(c) $BaSO_4 + NO_2$ (d) $BaSO_{4+} N_2 + O_2$

19. Which of the following product is formed by the reaction of sulphur dioxide with chlorine in presence of sunlight?
(a) SO_2Cl (b) SO_2Cl_2
(c) $SOCl_2$ (d) SO_3Cl

20. Which of the following bonds has the least energy?
(a) Se—Se (b) Te—Te
(c) S—S (d) O—O

21. Oxalic acid on heating with conc. H_2SO_4 produce
(a) H_2O and CO_2 (b) CO, CO_2 and H_2O
(c) CO_2 and H_2S (d) CO and CO_2

22. Which of the following is not a chalcogen?
(a) Se (b) O
(c) S (d) Na

23. Which of the following gas is insoluble in water?
(a) SO_2 (b) H_2
(c) NH_3 (d) CO_2

24. Which one of the following acid possesses oxidising, reducing and complex forming properties?
(a) HNO_3 (b) HCl
(c) H_2SO_4 (d) HNO_2

25. Na_2CO_3 reacts with SO_2 in alkaline medium to produce
(a) Na_2SO_4 (b) Na_2SO_3
(c) $NaHSO_4$ (d) $NaHSO_3$

26. The number of S—S bonds in sulphur trioxide trimer, (S_3O_9) is
(a) three (b) two
(c) one (d) zero

27. Sulphur dioxide levels in the atmosphere can be reduced by using
(a) catalytic converters in industry
(b) static eleectricity to attract it in factory chimneys
(c) more efficient car-engines
(d) low-sulphur fuels

28. What does Green Chemisry in terms of environment mean?
(a) Greenhouse effect
(b) Reactions related to depletion of ozone layer
(c) Photosynthetic reactions in plants
(d) Reduction in the use and production of hazardous chemicals

29. Which one of the following compounds is not a protonic acid?
(a) $SO(OH)_2$ (b) $SO_2(OH)_2$
(c) $B(OH)_3$ (d) $Po(OH)_3$

30. Which of the following gas mixture is used by the divers inside the sea?
(a) O_2 + He (b) O_2 + Xe
(c) O_2 + Ar (d) O_2 + N_2

31. A group 16 element exists in the monoatomic state in metallic lattice. It also exists in two crystalline forms. The metal is
(a) sulphur (b) polonium
(c) selenium (d) tellurium

32. To a piece of charcoal sulphuric acid is added. Then
(a) there is no reaction
(b) water gas is formed

(c) SO_2 and CO_2 are evolved
(d) CO and SO_2 are evolved

33. The correct order of O—O bond length in O_2, H_2O_2 and O_3 is
(a) $H_2O_2 > O_3 > O_2$ (b) $O_2 > H_2O_2 > O_3$
(c) $O_3 > H_2O_2 > O_2$ (d) $O_2 > O_3 > H_2O_2$

34. Which of the following is used in vulcanization of rubber?
(a) SF_6 (b) CF_4
(c) Cl_2F_2 (d) C_2F_2

35. Aqueous solutions of hydrogen sulphide and sulphur dioxide when mixed together yield
(a) sulphur trioxide and water
(b) presence of *d*-orbital
(c) sulphur and water
(d) hydrogen preroxide and sulphur

36. Which of the following have highest melting points?
(a) *p*-block elements (b) *s*-block elements
(c) *d*-block elements (d) None of these

37. Electron affinity is positive, when
(a) O changes into O^-
(b) O^- changes into O^{2-}
(c) O changes into O^+
(d) electron affinity is always negative

38. Consider the following compounds
1. $Na_2S_2O_3$
2. $Na_2S_4O_6$
3. $CaOCl_2$

Among these compounds identify those that have different oxidation states for same type of atoms
(a) 1 and 3 (b) 1 and 3
(c) 1 and 2 (d) 1, 2 and 3

39. Select correct statement(s).
(a) O_3 molecule is resonance stabilised
(b) There is 3-centre, 4-electron π-bond system in O_3
(c) Ozone layer is being depleted by fluorocarbons
(d) All the above are correct statements

40. Alkali ozonides are
(a) paramagnetic in nature
(b) hydrolytically unstable and decompose to superoxides
(c) decreasing in stability as size of alkali metal decreases
(d) having all the above statements as correct

41. Among CH_4, NH_3 and H_2O, acidicity of H_2O is maximum because
(a) oxygen contains two lone pairs of electrons
(b) bond angle in H_2O is less than 109°28′ due to VSEPR
(c) oxygen is more electronegative than C or N
(d) H_2O is an associated liquid

42. An experiment involving absorption of oxygen and its quantitative estimation would involve use of
(a) caustic soda (b) pyrogallol
(c) conc sulphuric acid (d) tupentine oil

43. Maximum coordination number of oxygen in liquid water is
(a) two (b) three
(c) four (d) five

44. The colour of the red glass is traffic signals is due to
(a) Sb_2S_3 (b) HgI_2
(c) CdS (d) CdSe

45. There is $p\pi$-$d\pi$ multiple bonding in
(a) CS_2 (b) CO_2
(c) NO_2 (d) NO

46. Ozone reacts with all the following reagents except one of the following
(a) K_2MnO_4
(b) moist iodine
(c) potassium ferrocyanide
(d) mercury

47. The absorption of UV radiation by O_3
(a) protects the inhabitants of our planet from injurious radiation
(b) maintains an equilibrium between the concentrations of O_2 and O_3
(c) makes both the function effective
(d) makes no function effective

48. Ozone layer is being depleted. This is due to
(a) No emission from supersonic jets
(b) chlorofluorocarbon used as aerosols
(c) both (a) and (b)
(d) None of the above

49. Estimation of ozone can be made quantitatively by
(a) decomposition into O_2 and absorption of O_2 into pyrogallol
(b) volumetric method using KI and titration of the liberated iodine using hypo solution
(c) oxidative ozonolysis method
(d) all methods given above

50. Oxidative ozonolysis of 2-butene in presence of Zn or dimethyl sulphide forms
(a) CH_3CHO (b) CH_3COOH
(c) CO_2 (d) $(CH_3)_2CO$

51. The process of obtaining sulphur by the borehole method is called
(a) the Frasch process
(b) the Lablanc process
(c) the Calcaroni process
(d) the Mannheium process

52. When molten sulphur is suddenly cooled by pouring into water, it takes the form of
(a) milk of sulphur (b) colloidal sulphur
(c) flower of sulphur (d) plastic sulphur

53. $FeS + HCl \longrightarrow A + B(g)$
Gas B is passed into aqueous solution of C to form A
Select correct statements based on the above facts
(a) Colour of C changes from yellow to very light green
(b) Gas B turns lead acetate paper black
(c) A form deep blue colour with $K_3[Fe(CN)_6]$
(d) All the above facts are true

54. Yellow oils of sulphur is/are
(a) H_2S (b) H_2S_2, H_2S_3
(c) H_2SO_4 (d) CS_2, NH_2CSNH_2

55. H_2S precipitates aqueous $ZnCl_2$ solution in
(a) acidic medium (b) neutral medium
(c) both (a) and (b) (d) None of these

56. In presence of H_2O, SO_2 acts as
(a) an oxidising agent (b) a reducing agent
(c) a hydrolysing agent (d) a redox reagent

57. SO_2 behaves as a reducing agent when
(a) passed over hot CuO
(b) mixed with moist H_2S
(c) passed through acidified $KMnO_4$ solution
(d) passed through $FeSO_4$ solution

58. When an article is bleached by SO_2 it loses its colour. The colour can be restored by
(a) exposure to air (b) heating
(c) dilution (d) None of these

59. The compound of sulphur used as a solvent in rubber industry is
(a) $SO_2(OH)Cl$ (b) SO_2
(c) SO_3 (d) S_2Cl_2

60. S—S bond is present in
(a) $\alpha\text{-}(SO_3)_n$ (b) $\gamma\text{-}(S_3O_9)$
(c) $H_2S_2O_3$ (d) $H_2S_2O_8$

61. The compound which has no reaction with $KMnO_4$ is
(a) perdisulphuric acid (b) sulphurous acid
(c) hydrogen sulphide (d) hydrogen peroxide

62. A bottle completely fillled with conc. H_2SO_4 is left unstoppered for several days and we observe spontaneous overflow of acid. It is due to
(a) change in temperature
(b) hygroscopic nature of H_2SO_4
(c) absorption of air by H_2SO_4
(d) dehydration of H_2SO_4

63. Out of $H_2S_2O_3$, $H_2S_4O_6$, H_2SO_5 and $H_2S_2O_8$ peroxy acids are
(a) $H_2S_2O_3, H_2S_4O_6$ (b) $H_2S_4O_6, H_2SO_5$
(c) $H_2SO_5, H_2S_2O_8$ (d) $H_2S_2O_3$ and $H_2S_2O_8$

64. Which of the following can convert acidified $Cr_2O_7^{2-}$ to green?
(a) $SO_2/H_2SO_3/H_2SO_4$ (b) $SO_3/H_2SO_3/H_2S$
(c) $SO_3^{2-}/H_2S/Fe^{2+}$ (d) $S_2O_3^{2-}/SO_3/Fe^{3+}$

65. In SO_3^{2-}
(a) $d\pi\text{-}p\pi$ bond between S and O is delocalised
(b) bonds between S and O are equivalents
(c) there is sp^3 hybridised sulphur atom
(d) all of the facts given above are true

66. Liver of sulphur is a mixture of
(a) potassium pentasulphide (A) and potassium thiosulphate (B)
(b) potassium carbonate (C) and (B) above
(c) potassium disulphide (D) and (B) above
(d) (B) and (D) above

67. Bleaching of a fabric cloth is done using A and excess of chlorine is removed using B is
(a) $CaOCl_2, Na_2SO_3$ (b) $Na_2S_2O_3, CaOCl_2$
(c) $CaCl_2, Na_2S_2O_3$ (d) $CaOCl_2, Na_2S_2O_3$

68. $Na_2S_2O_3$ is used in photography for fixing the negative. It removes the AgBr by
(a) complex formation $Na_3[Ag(S_2O_3)_2]$
(b) oxidation of AgBr to Br_2
(c) reduction of AgBr to Ag
(d) formation of double salt

69. Aqueous hyposolution on reaction with aqueous $AgNO_3$ gives
(a) yellow ppt changing to black
(b) white ppt changing to black
(c) orange ppt changing to blue
(d) no ppt

70. $S_2O_3^{2-}$ has
(a) S—S linkage (b) S—O linkage
(c) both (a) and (b) (d) None of these

Brain Twisters

1. Is the same amount of sulphuric acid needed to dissolve 40 g of nickel if we take concentrated acid in one case and dilute acid in another? How many grams of sulphuric acid will go to oxidise the nickel in each case? (Ni = 59)

2. What mass of H_2SO_4 could be produced in the process given below if 1.00 kg of FeS_2 is used? The unbalanced equations for the process are:

$$FeS_2 + O_2 \longrightarrow Fe_2O_3 + SO_2$$
$$SO_2 + O_2 \longrightarrow SO_3$$
$$SO_3 + H_2SO_4 \longrightarrow H_2S_2O_7$$
$$H_2S_2O_7 + H_2O \longrightarrow H_2SO_4$$

3. What is equivalent weight of $Na_2S_2O_3 \cdot 5H_2O$?

4. Calculate number of moles of $Na_2S_2O_3$ that will react with I_2 obtained when 1 mole of $K_2Cr_2O_7$ reacts with excess of KI in acidic medium.

5. Colourless salt (*A*) decolorises I_2 solution and gives white ppt (changing to black) with $AgNO_3$ solution. (*A*) also changes $FeCl_3$ (yellow) solution to $FeCl_2$ (green). Identify (*A*) and explain reactions.

6. A 1.1 g sample of copper ore is dissolved and $Cu^{2+}(aq)$ is titrated with excess KI. The liberated I_2 requires 12.12 mL of 0.1 M $Na_2S_2O_3$ for its titration. What is the % copper by mass in the ore?

Passage Comprehension Questions

Passage 1

Peroxydisulphuric acid and its salts are powerful oxidising agents, thus can be used to estimate reducing agents and also to study the kinetics of the reaction.

Answer the following questions.

1. Peroxydisulphuric acid
(a) is also called Marshall's acid
(b) has two peroxy linkages
(c) liberates I_2 with KI in a fast reaction
(d) has oxidation number of sulphur as +7

2. As compared to H_2SO_5, $H_2S_2O_8$
(a) has two peroxy linkage
(b) liberates I_2 from KI solwly and thus kinetics of oxidation of KI by $S_2O_8^{2-}$ can be studied
(c) has different oxidation number
(d) reacts with $KMnO_4$

Passage 2

Two acids have been derived from H_2O_2 by replacing H by (SO_2OH) group. Both the acids have one peroxy linkage

H—O—O—H H—O—O—SO_2—OH
H_2SO_5

HO—SO_2—O—O—SO_2—OH
$H_2S_2O_8$

Based on the above study, answer the following questions.

1. Which is called Marshall's acid and which is called Caro's acid?
(a) H_2SO_5, $H_2S_2O_8$ (b) $H_2S_2O_8$, H_2SO_5
(c) H_2SO_5, $H_2S_2O_7$ (d) $H_2S_2O_3$, H_2SO_5

2. $H_2S_2O_8$ is obtained at anode by the electrolysis of ......
(a) H_2SO_3 (b) $H_2S_2O_4$
(c) H_2SO_4 (d) H_2O_2 and H_2SO_4

3. Which reacts with $KMnO_4$?
(a) H_2SO_3 (b) H_2SO_5
(c) H_2SO_4 (d) $H_2S_2O_8$

Passage 3

Sodium thiosulphate is the laboratory reagent used in iodometric and iodimetric titration. This also finds used in photography.

Based on the above passage, answer the following questions.

1. In the following reaction

$$I_2 + 2S_2O_3^{2-} \longrightarrow S_4O_6^{2-} + 2I^-$$

equivalent mass of $S_2O_3^{2-}$ is (molar mass = M)
(a) M (b) $\frac{M}{2}$
(c) $\frac{M}{3}$ (d) $\frac{M}{4}$

2. If I_2 is the titre in a titration *A* and I_2 is formed *in situ* in a titration *B*, then these are respectively
(a) iodometric, iodimetric
(b) iodimetric, iodometric
(c) both iodimetric
(d) both iodometric

3. Sodium thiosulphate is called ......(I) and is used as ......(II) and removes unreacted light sensitive AgBr by forming the complex ......(III)

	I	II	III
(a)	hypo	developer	$[AgS_2O_3]^-$
(b)	hypo	fixer	$[Ag(S_2O_3)_2]^{3-}$
(c)	hypo	fixer	$[Ag_2(S_2O_3)]$

(d) none of the above is correct

Passage 4

When metallic copper is heated with conc. sulphuric acid, various types of products are formed depending on sttandard reduction potential.

Answer following questions based on above study.

1. Which reaction is possible?
(a) $Cu + 2H^+ \longrightarrow Cu^{2+} + H_2$
(b) $Cu^{2+} + H_2 \longrightarrow Cu + 2H^+$
(c) Both (a) & (b)
(d) None of the above

2. When Cu reacts with conc. H_2SO_4
(a) Cu is oxidised to Cu^{2+}
(b) SO_4^{2-} is reduced to SO_2
(c) both of the above reactions take place
(d) none of the above reaction take place

3. There is also formation of CuS which is formed by
(a) $SO_4^{2-} + 8H^+ + 8e^- \longrightarrow S^{2-} + H_2O$
(b) $SO_2 + 6H^+ + 6e^- \longrightarrow H_2S + 2H_2O$
(c) both (a) & (b)
(d) None of the above

Assertion & Reason

Codes :

(a) Both **A** and **R** are true and **R** is the correct explanation of **A**.

(b) Both **A** and **R** are true but **R** is not the correct explanation of **A**.

(c) **A** is true but **R** is false.

(d) **A** is false but **R** is true.

(e) **A** and **R** both are false.

1. Assertion (A) : Oxygen is the most abundant element on the surface of earth.

Reason (R) : Oxygen is paramagnetic with double bond character.

2. Assertion (A) : Of all the elements of the Group-16, sulphur has the largest tendency to catenation.

Reason (R) : Sulphur-sulphur bond energy is very high.

3. Assertion (A) : $S_3O_6^{2-}$ (trithionate) has S—S bond.

Reason (R) : It is a conjugate base of $HS_3O_6^-$.

4. Assertion (A) : O_2 gas is effected in magnetic field.

Reason (R) : It has north and south pole.

5. Assertion (A) : SO_3 is a nucleophile.

Reason (R) : SO_3 has lone-pair of electrons.

6. Assertion (A) : SO_2 turns acidified $K_2Cr_2O_7$ green.

Reason (R) : $K_2Cr_2O_7$ is reduced to Cr^{3+} (green)salt by SO_2.

7. Assertion (A) : SO_2 is a bleaching agent.

Reason (R) : Nascent oxygen causes bleaching.

8. Assertion (A) : One mole of SO_2Cl_2 in aqueous solution is neutralised by 3 moles of NaOH.

Reason (R) : 3 moles of H^+ are formed due to reaction with water.

9. Assertion (A) : Anhydrous sulphuric acid has very high viscosity.

Reason (R) : H-bonding makes it viscous.

10. Assertion (A) : Anhydrous H_2SO_4 undergoes autoprotolysis

$$H_2SO_4 + H_2SO_4 \rightleftharpoons H_3SO_4^+ + HSO_4^-$$

Reason (R) : Anhydrous H_2SO_4 has very high electrical conductivity.

11. Assertion (A) : In aqueous solution, SO_2 reacts with H_2S forming sulphur.

Reason (R) : SO_2 is an effective reducing agent.

12. Assertion (A) : Structure of SO_4^{2-} is identical to $S_2O_3^{2-}$.

Reason (R) : Sulphur atoms in $S_2O_3^{2-}$ are in different oxidation states.

True & False

1. Sulphur is an insulator while selenium and tellurium are semiconductor.
2. Maximum catenation among non-metals is in sulphur.
3. O_2, O_2^- are paramagnetic while O_3, O_2^{2-} are diamagnetic.
4. There is one (S—S) bond in $H_2S_2O_8$.
5. S_4N_4 is thermochromic.
6. S_4N_4 is kinetically stable but endothermic compound.
7. Peroxy linkage is present in oleum $H_2S_2O_7$.
8. Caro's is an oxidising agent.
9. SO_3 is an electrophile in the sulphonation of benzene.
10. Highest fluorides of oxygen is OF_2 whereas sulphur forms SF_6.
11. KO_2 is paramagnetic.
12. Bond order of O_2^- is 1.5.
13. KO_2, BaO_2, Na_2O_2 are superoxides.
14. SO_2 and CO_2 both turns lime water milky.
15. Oxygen is by far the most abundant metal on the earth's crust.

Fill in the Blanks

1. Paramagnetic behaviour is observed in .... out of O_3, O_2, NO and NO_2.
2. Highly electrical conducting behaviour is present in anhydrous H_2SO_4 due to ....
3. Anhydrous H_2SO_4 has high viscosity due to ....
4. Peroxy linkage is present in acid ....
5. When Marshall's acid is hydrolysed there is formation of .... and ....
6. (S—S) bond is present in the acid ....
7. Tetrasulphur tetranitride is ....
8. In group 16 maximum catenation is in ....
9. Common oxidaton states of group 16 elements are ....
10. Unlike sulphur which is an .... selenium and tellerium are ....

11. Superoxide ion is .... and is .... in character.

12. Bond orders of O_2, O_2^-, O_2^{2-}, O_2^+ and O_2^{2+} in increasing order are ....

Matrix-Match Type Questions

I. One or More Correct

1. Match the species in Column I with their corresponding properties in Column II.

	Column I		Column II
A.	SO_2	1.	Oxidant
B.	SO_3	2.	Reductant
C.	O_3	3.	Resonance stabilised
D.	H_2S	4.	sp^2 hybridisation
E.	H_2O	5.	sp^3 hybridisation
		6.	Maximum boiling point
		7.	Minimum boiling point

2. Match the acids in Column I with their corresponding properties in Column II.

	Column I		Column II
A.	H_2SO_5	1.	Hydrolysed to give H_2O_2
B.	H_2SO_4	2.	Hydrolysed to give H_2SO_4
C.	$H_2S_2O_7$	3.	S—S bond present
D.	$H_2S_2O_3$	4.	Peroxy linkage present
E.	$H_2S_2O_8$	5.	S in different oxidation state
		6.	O in different oxidation state
		7.	Hygroscopic

Integer Answer Types

This section contains 8 questions. The answer to each of the questions is a single digit integer, ranging from 0 to 9. The appropriate bubbles below the respectively question numbers in the ORS have to be darkened. For example, if the correct answers to question number *X*, *Y*, *Z* and *W* (say) are 6, 0, 9 and 2, respectively, then the correct darkening of bubbles will like the following

X	*Y*	*Z*	*W*
0	0	0	0
1	1	1	1
2	2	2	2
3	3	3	3
4	4	4	4
5	5	5	5
6	6	6	6
7	7	7	7
8	8	8	8
9	9	9	9

Set I

1. O_2 is paramagnetic due to .... unpaired electrons.

2. SO_2 reduces $Cr_2O_7^{2-}$ to Cr^{3+}; change in oxidation number of Cr is ....

3. Maximum oxidation state shown by sulphur in its compounds is ....

4. There are equal number of sulphur atoms in thiosulphuric acid, Marshall's acid and oleum. This number is ....

5. SO_2 reduces MnO_4^- to Mn^{2+}. Number of electrons involved in the reduction of MnO_4^- to Mn^{2+} is ....

6. SO_2 turns lime water milky, 3.2 g of SO_2 gives milkyness .... g.

7. One mole of SO_2Cl_2 on hydrolysis gives acids which can be neutralised by .... moles of NaOH.

8. Ozonolysis of 2,5-dimethyl-2,4-hex-diene gives total number of carbonyl compounds ....

Set II

1. 4.35 g of pyrolusite (MnO_2 mineral) is digested with conc. HCl and KI added. Mixture required 5 mL of 1 M $Na_2S_2O_3$ solution for complete reaction of KI_3 formed. What is % of pure MnO_2 in pyrolusite? (Mn = 55)

2. How many S—S bonds are there in S_3O_9?

3. There are two types of sulphur with different oxidation states in thiosulphate ion. Difference in oxidation states is ....

4. Acid rain is due to dissolved non-metallic oxides (SO_2, NO_2, P_2O_5, etc). 10 mL of acid rain sample required 5 mL of 0.01 M $Ca(OH)_2$ for making it neutral. What is pH of acid rain?

5. How many $\widehat{SOS}$ angles are there in S_3O_9?

6. How many of the following have peroxy linkages?

$H_2S_2O_3, H_2S_2O_8, H_2SO_5, CrO_5, BaO_2, KO_2$

7. There is also formation of insoluble CuS (black) when copper reacts with conc. H_2SO_4. What is change in oxidation number of sulphur?

8. How many of the following can be the oxidising agent?

$H_2S, H_2SO_4, H_2SO_3, H_2S_2O_3$

Answers & Solutions
Master Exercises
Exercise 1

Short Answer Type Questions

1. (a) H_2S (b) SO_2, H_2SO_3
(c) $H_2SO_4, H_2SO_5, H_2S_2O_8, SO_3$

2. Cl, Br and I atoms are two bulky around the sulphur atom.

3. It absorbs UV light. It is also used as a disinfectant to purify drinking water, since it destroys bacteria and viruses. *Its advantage over Cl_2 for this purpose is that it avoids the unpleasant smell and taste of chlorine since any excess O_3 soon decompose to O_2.*

4. No change

5. $\underset{+2\ -2}{NO} + \underset{0}{O_3} \longrightarrow \underset{+4\ -2}{NO_2} + \underset{0}{O_2}$

6. 1000 mL O_3, 6.44 kJ

$$3xO_2 \rightleftharpoons 2xO_3$$
$$(3x - 2x) = x = 500 \text{ mL}$$
$$\text{Volume of formed } O_3 = 1000 \text{ mL} = \frac{1000}{22400} \text{ mol}$$
$$= 0.0446 \text{ mol}$$
$$\text{Heat absorbed} = 144.2 \times 0.0446 \text{ kJ} = 6.44 \text{ kJ}$$

7. (A) : FeS (B) : H_2S (C) : $FeSO_4$ (D) : PbS
(E) : $K\,Fe^{II}[Fe^{III}(CN)_6]$ **Turnbull's blue**

$$MnO_4^- + 5Fe^{2+} + 8H^+ \longrightarrow 5Fe^{3+} + \underset{\text{colourless}}{Mn^{2+}} + 4H_2O$$

8. $MnO_4^- + H^+ + H_2S \longrightarrow S + Mn^{2+}$, H_2S is a reducing agent

$H_2S + H_2O_2 \longrightarrow H_2O + H_2SO_4$, H_2S is a reducing agent

$H_2S + NaOH \longrightarrow NaHS \longrightarrow Na_2S$, H_2S is a diprotic acid.

9. (A) : $NaHSO_3$ (B) : Na_2SO_3 (C) : SO_2 (D) : $BaSO_3$
(E) : $CaSO_3$ (F) : $Ca(HSO_3)_2$ (G) : Cr^{3+} (H) : $BaSO_4$

10. P_2O_5

11. H_2S is oxidised to S by H_2SO_4 (which is reduced to SO_2)

$$H_2S + H_2SO_4 \longrightarrow 2H_2O + SO_2 + S$$

while NH_3 forms salt with H_2SO_4

$$2NH_3 + H_2SO_4 \longrightarrow (NH_4)_2SO_4$$

hence, H_2SO_4 can't be used for drying up these gases.

Analytical Questions

1. (a) $O_3 + 2I^- + 2H^+ \longrightarrow O_2 + I_2 + H_2O$
(b) $2[Fe(CN)_6]^{4-} + O_3 + H_2O \longrightarrow 2[Fe(CN)_6]^{3-} + 2OH^- + O_2$
(c) $S + 3O_3 + H_2O \longrightarrow H_2SO_4 + 3O_2$

2.
$$\text{radius} = 4000 \text{ miles}$$
$$4000 \text{ miles} = \frac{4000}{0.6214} \text{ km} = \frac{4 \times 10^6}{0.6214} \text{ m}$$
$$\text{thickness} = 0.3 \text{ cm} = 0.003 \text{ m}$$
$$\text{Volume of } O_3 \text{ molecules at the surface} = 4\pi r^2 \text{ thickness}$$
$$= 4 \times \frac{22}{7} \times \left(\frac{4 \times 10^6}{0.6214}\right)^2 \times 0.003 \text{ m}^3$$
$$V = 1.56273 \times 10^{12} \text{ m}^3$$
$$P = 1 \text{ atm} = 1.01325 \times 10^5 \text{ Pascal}$$
$$\text{Moles} = n = \frac{pV}{RT}$$
$$= \frac{1.01325 \times 10^5 \times 1.56273 \times 10^{12}}{8.3143 \times 273}$$
$$= 6.976 \times 10^{13}$$
$$\therefore \quad \text{Molecules} = nN_0 = 6.976 \times 10^{13} \times 6.02 \times 10^{23}$$
$$= \mathbf{4.2 \times 10^{37}}$$

3. (i) Yes, Fe^{3+} ions will oxidise H_2S to S

$$2Fe^{3+} + H_2S \longrightarrow 2Fe^{2+} + 2H^+ + S$$

Since, $E° = 0.77 - 0.14 = 0.63$ V

and thus $\Delta G° = -nFE°_{cell} = -$ ve

(ii) Yes, Br_2 also oxidises H_2S to S

$$Br_2 + H_2S \longrightarrow 2Br + 2H^+ + S$$

Since, $E° = 1.07 - 0.14 = 0.93$V

and $\Delta G° = -$ ve

Note : A cell reaction is spontaneous if $E°_{cell} = +$ ve and $\Delta G° = -$ ve

$$E°_{cell} = E°_{ox} + E°_{red}$$

(Take values as per half-reactions)

4. Refer Text.

5. (a) Acid strength increases as the number of O-atoms increases.

(b) In comparison with S, O is much too electronegative to form compounds of O in +4 oxidation state. Also uses sp^3d^2-hybrid orbitals for bonding in SF_4 but O does not have valence *d*-orbitals and it cannot form four bonds to F.

(c) Each S is sp^3-hybridised with two lone pair of electrons. The bond angles are 109.5°. A planar ring would require bond angles of

$$= \left(180 - \frac{360}{8}\right) = 135°$$

6. $\underset{+7}{ClO_4^-} \longrightarrow \underset{-1}{Cl^-}$ Change in oxidation number 8 units

$\underset{+4}{SO_4} \longrightarrow \underset{+6}{SO_4^{2-}}$ 2 units

Thus, $ClO_4^- + 4SO_2 \longrightarrow 4SO_4^{2-} + Cl^-$

Since, 1 mol $HClO_4$ is reduced by = 4 mol SO_2

hence, 16.9 g $\left(= \frac{16.9}{100.5} \text{ mole}\right)$ $HClO_4$ is reduced by

$$= \frac{4 \times 16.9}{100.5} \text{ mol } SO_2$$

$$= 0.673 \text{ mol}$$

1 mol of STP = 22.4 L SO_2

Thus, 0.673 mol at STP = 22.4 × 0.673

$$= \mathbf{15.067\ L}$$

7. (a) $Na_2S_2O_3 + 2HCl \longrightarrow 2NaCl + SO_2 + S + H_2O$

If $Na_2S_2O_3$ reacts with HCl, there is formation of precipitate (S) — thus (2) and formation of gas (SO_2) — thus (3)

(a) — 2, 3

(b) None

(c) None

(d) $Na_2S_2O_3 + \underset{\text{violet}}{I_2} \longrightarrow \underset{\text{colourless}}{2NaI} + Na_2S_4O_6$

Thus, initial substance is decolourised (1)

Thus, (d) — 1

(e) $KMnO_4$ (pink-violet) is also decolourised.

Thus, (e) — 1

8. $Na + O_2 \longrightarrow NaO_2$ (superoxide)

Na	O_2	NaO_2
1 mol	1 mol	1 mol
23 g	32 g	55 g

1.15 g = 0.05 mol 0.05 mol

$$2.75 \text{ g} = \frac{2.75}{55} = 0.05 \text{ mol}$$

(i) O_2 combined = 0.05 mol = 1.6 g

(ii) O_2 = 0.05 mol = 0.10 mol O-atoms

(iii) sodium used = 0.05 mol

(iv) superoxide = NaO_2

Exercise 2

Only One Option Correct

1. (c)	**2.** (b)	**3.** (d)	**4.** (a)	**5.** (d)	**6.** (d)	**7.** (a)	**8.** (d)	**9.** (d)	**10.** (c)
11. (a)	**12.** (d)	**13.** (d)	**14.** (a)	**15.** (a)	**16.** (b)	**17.** (d)	**18.** (a)	**19.** (b)	**20.** (b)
21. (b)	**22.** (d)	**23.** (b)	**24.** (d)	**25.** (b)	**26.** (d)	**27.** (d)	**28.** (d)	**29.** (c)	**30.** (a)
31. (b)	**32.** (c)	**33.** (a)	**34.** (a)	**35.** (c)	**36.** (a)	**37.** (b)	**38.** (d)	**39.** (d)	**40.** (d)
41. (c)	**42.** (b)	**43.** (c)	**44.** (d)	**45.** (a)	**46.** (b)	**47.** (c)	**48.** (c)	**49.** (b)	**50.** (a)
51. (a)	**52.** (d)	**53.** (d)	**54.** (b)	**55.** (b)	**56.** (b)	**57.** (c)	**58.** (a)	**59.** (d)	**60.** (c)
61. (a)	**62.** (b)	**63.** (c)	**64.** (c)	**65.** (d)	**66.** (d)	**67.** (d)	**68.** (a)	**69.** (b)	**70.** (c)

Brain Twisters

1. $Ni + \underset{\text{(dil.)}}{H_2SO_4} \longrightarrow NiSO_4 + H_2$

59 g 98 g

40 g = **66.44 g H_2SO_4**

$Ni + \underset{\text{(conc.)}}{2H_2SO_4} \longrightarrow NiSO_4 + 2H_2O + SO_2$

59 g 2 × 98 g

40 g = **132.88 g H_2SO_4**

Thus, amount of dil. H_2SO_4 and conc. H_2SO_4 to dissolve 40 g Ni is different.

2. $2FeS_2 + O_2 \longrightarrow Fe_2O_3 + 4SO_2$

$4SO_2 + O_2 \longrightarrow 4SO_3$

$4[SO_3 + H_2SO_4 \longrightarrow H_2S_2O_7]$

$4[H_2S_2O_7 + H_2O \longrightarrow 2H_2SO_4]$

Thus, $2FeS_2 \equiv 4H_2SO_4$

$$\underset{FeS_2}{2[56 + 64] \text{ g}} \equiv 4 \times 98 \text{ g}$$

$$1000 \text{ g } FeS_2 \text{ give} = \frac{4 \times 98}{240} \times 1000 \text{ g } H_2SO_4$$

$$= 1633.33 \text{ g}$$

$$= \mathbf{1.6333\ kg}$$

3. $2Na_2S_2O_3 + I_2 \longrightarrow Na_2S_4O_6 + 2NaI$

$2Na_2S_2O_3 \equiv I_2 \equiv 2I$

Thus, $I \equiv Na_2S_2O_3$

Equivalent weight of $Na_2S_2O_3 \cdot 5H_2O \equiv$ molar mass $= 248$ g

3. $2Na_2S_2O_3 + I_2 \longrightarrow 2NaI + Na_2S_4O_6$

$$2Na_2S_2O_3 \equiv I_2 \equiv 2I$$

Or $$Na_2S_2O_3 \equiv I$$

Thus equivalent weight of $(Na_2S_2O_3 \cdot 5H_2O)$ = mol. wt. $= 248.0$

Or in terms of electrons : $2S_2O_3^{2-} \longrightarrow S_4O_6^{2-} + 2e^-$

$$S_2O_3^{2-} \equiv 1e^-$$

$$\therefore \text{ equivalent weight} = \frac{\text{molecular weight}}{\text{electrons gained/lost}}$$

$$= \text{molecular weight}$$

4. $Cr_2O_7^{2-} + 14H^+ + 6I^- \longrightarrow 2Cr^{3+} + 7H_2O + 3I_2$

$$3I_2 + 6Na_2S_2O_3 \longrightarrow 3Na_2S_4O_6 + 6NaI$$

$$1 \text{ mol } K_2Cr_2O_7 \xrightarrow{KI} 3 \text{ mol } I_2 \equiv 6 \text{ mol } Na_2S_2O_3$$

5. $(A) : Na_2S_2O_3$

$$Na_2S_2O_3 + 2AgNO_3 \longrightarrow \underset{\text{white ppt}}{Ag_2S_2O_3\downarrow} + 2NaNO_3$$

$$Ag_2S_2O_3 + H_2O \longrightarrow H_2SO_4 + \underset{\text{black}}{Ag_2S\downarrow}$$

$$\underset{\text{yellow}}{2FeCl_3} + 2Na_2S_2O_3 \longrightarrow \underset{\text{green}}{2FeCl_2} + Na_2S_4O_6 + 2NaCl$$

6. $2Cu^{2+} + 4I^- \longrightarrow Cu_2I_2 + I_2$

$$I_2 + 2S_2O_3^{2-} \longrightarrow S_4O_6^{2-} + 2I^-$$

Equivalent mass of Cu^{2+} = molar mass

Equivalent mass of $S_2O_3^{2-}$ = molar mass

Thus, millimoles of Cu^{2+} = millimoles of $S_2O_3^{2-}$

$= 12.12 \times 0.1$

$= 1.212$ millmol

$= 1.212 \times 10^{-3}$ mol

$= 1.212 \times 10^{-3} \times 63.5$ g pure Cu

$= 0.076962$ g in 1.1 g Cu – ore

Thus, % of copper $= \dfrac{0.076962}{1.1} \times 100$

$= \mathbf{7.0\%}$

Passage Comprehension Questions

Passage 1

1. (a) **2.** (b)

Passage 2

1. (b) **2.** (c) **3.** (a)

Passage 3

1. (a) **2.** (b) **3.** (b)

Passage 4

1. (b) **2.** (c) **3.** (a)

Assertion & Reason

1. (b) **2.** (a) **3.** (b) **4.** (c) **5.** (e) **6.** (a)
7. (c) **8.** (e) **9.** (a) **10.** (b) **11.** (b) **12.** (b)

True & False

1. T **2.** Carbon has maximum catenation **3.** T

4. there is (O—O) bond in $H_2S_2O_8$

5. T **6.** T

7. $H_2S_2O_7$ (oleum) is SO_3 saturated H_2SO_4 (no peroxy linkage)

8. T **9.** T **10.** T **11.** T **12.** T

13. Only KO_2 is superoxide BaO_2, Na_2O_2 are peroxides

14. T

15. it is not metal, it is non-metal.

Fill in the Blanks

1. O_2, NO and NO_2

2. Autoprotolysis $(H_2SO_4 + H_2SO_4 \longrightarrow H_3SO_4^+ + HSO_4^-)$

3. Strong hydrogen bonding

4. H_2SO_5 and $H_2S_2O_8$

5. H_2SO_4, H_2O_2

6. $H_2S_2O_3$ (thiosulphuric acid)

7. thermochromic

8. sulphur

9. –2, +4, +6

10. insulator, semiconductors

11. O_2^-, paramagnetic

12. $O_2^{2-}(1) < O_2^-(1.5) < O_2(2) < O_2^+(2.5) < O_2^{2+}(3)$

Matrix-Match Type Questions

I. One or More Correct

1. (A) — (1,2,3,4); (B) — (1,3,4); (C) — (1,3,4);
(D) — (2,5,7) (E) — (5,6)

2. (A) — (1,2,4,6); (B) — (7); (C) — (2,7);
(D) — (3,5); (E) — (1,2,4,6)

Integer Answer Types

Set I

Questions →	1	2	3	4	5	6	7	8
Answers →	2	6	6	2	5	6	4	3
	0	0	0	0	0	0	0	0
	1	1	1	1	1	1	1	1
	2	2	2	2	2	2	2	2
	3	3	3	3	3	3	3	3
	4	4	4	4	4	4	4	4
	5	5	5	5	5	5	5	5
	6	6	6	6	6	6	6	6
	7	7	7	7	7	7	7	7
	8	8	8	8	8	8	8	8
	9	9	9	9	9	9	9	9

Hints

2. $Cr_2O_7^{2-} \longrightarrow 2Cr^{3+}$

$\uparrow$ 12 $\qquad \uparrow$ 6 $\qquad$ charge 6 units

6. $SO_2 + Ca(OH)_2 \longrightarrow CaSO_3$

64g $\qquad$ 120 g

3.2 g $\qquad$ 6 g

7. $SO_2Cl_2 + 2H_2O \longrightarrow \underbrace{2HCl + H_2SO_4}_{\text{Total } 4H^+}$

$\equiv 4NaOH$

8. C—C$\neq$C—C$\neq$C—C $\longrightarrow$ (with CH_3 groups on the second and fifth carbons)

these bonds break

hence, **three** products.

Set II

Questions →	1	2	3	4	5	6	7	8
Answers →	5	0	8	2	3	4	8	2
	0	0	0	0	0	0	0	0
	1	1	1	1	1	1	1	1
	2	2	2	2	2	2	2	2
	3	3	3	3	3	3	3	3
	4	4	4	4	4	4	4	4
	5	5	5	5	5	5	5	5
	6	6	6	6	6	6	6	6
	7	7	7	7	7	7	7	7
	8	8	8	8	8	8	8	8
	9	9	9	9	9	9	9	9

Hints

1. $MnO_2 + 4HCl \longrightarrow MnCl_2 + Cl_2$

$$Cl_2 + 2KI \longrightarrow I_2$$

$$I_2 + 2Na_2S_2O_3 \longrightarrow 2NaI + Na_2S_4O_6$$

$$MnO_2 \equiv Cl_2 \equiv I_2 \equiv 2Na_2S_2O_3$$

$$\text{Moles of } Na_2S_2O_3 = \frac{5 \times 1}{1000} = 5 \times 10^{-3}$$

$$\text{Moles of } MnO_2 = 2.5 \times 10^{-3}$$

$$\text{Pure } MnO_2 = 2.5 \times 10^{-3} \text{ mol}$$

$$= 2.5 \times 10^{-3} \times 87 \text{ g}$$

$$= 0.2175 \text{ g in } 4.35 \text{ g impure sample}$$

$$\% \text{ of pure } MnO_2 = \frac{0.2175}{4.35} \times 100$$

$$= 5\%$$

4. 10 mL of N_1 $(H^+) \equiv$ 5 mL of 0.01 M $Ca(OH)_2$

$= 5$ mL of 0.02 N $Ca(OH)_2$

$\therefore \quad N_1(H^+) = 10^{-2}$ N

$\therefore \quad pH = -\log [H^+] = 2$

$= -\log 10^{-2} = 2$

"A unique group of non-metals in different molecular forms. F_2, Cl_2 (gas), Br_2 (liquid) and I_2 (solid)."

15

Group 17 (VIIA—The Halogen Family)

Learning Framework of the Chapter ...

- Gradation of Physical Properties
- Study of Halogens (X_2)
- Interhalogen Compounds
- Pseudohalogens and Pseudohalides
- Anomalous Behaviour of Fluorine
- Halogen Acids
- Oxoacids of Halogen
- Oxides of Halogen

$_9F$

$_{17}Cl$

$_{35}Br$

$_{53}I$

$_{85}At$

Gradation of Physical Properties

- The name halogen comes from the Greek, and means salt former. All the elements react with metals to form salts, and they are also very reactive with non-metals. Fluorine is the most reactive element known.

Stable isotope

Fluorine	^{18}F, ^{19}F (tracer)
Chlorine	^{35}Cl (75%), ^{37}Cl (25%)
Bromine	^{79}Br (50.7%), ^{81}Br (49.3%)
Iodine	^{127}I

- All the elements have seven electrons in their outer shell and thus complete their octet by gaining one electron (in ionic compounds as in NaCl) or by sharing electrons (thus forming covalent bonds as in CCl_4). Physical properties of the elements are summarised in Table 15.1 and Fig. 15.1

Physical properties like melting and boiling points are related to the size and mass of the molecules as well as intermolecular attraction. The effect of size and mass, which gradually increase as we go down the group, is easy to understand. As for the intermolecular attraction in non-polar homonuclear diatomic molecules like halogens, which do not have any permanent polarity, the only forces of attraction are the weak van der Waals' forces. The polarisability of halogens increases as we go down the group, being maximum in iodine and least in fluorine. So, van der Waals' forces of attraction are maximum in iodine and least in fluorine, with bromine and chlorine coming in between. As a consequence of this, fluorine and chlorine are gases at ordinary temperature, bromine is a liquid and iodine a solid. This is also reflected in the trends observed in their enthalpies of fusion and vaporisation.

As we move along a period, the effective nuclear charge increases reaching a maximum at the noble gases. Halogens which immediately precede the noble gases have a very high effective nuclear charge coupled with small size and thus have the highest ionisation energies in the respective periods, next only to the noble gases. Like the trend in other groups, ionisation energy of halogens also decreases in going down the group from fluorine to iodine.

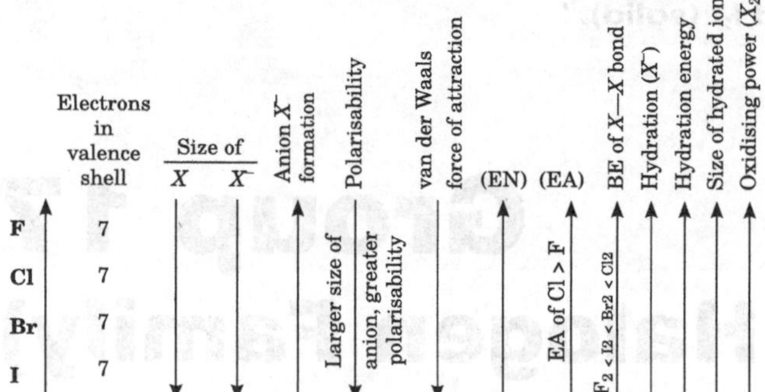

Halogens have seven electrons in their valence shell, they have a very strong tendency of gaining an electron to acquire a stable noble gas configuration. Therefore, they have very high electron affinities. In fact, their electron affinities are the highest in their respective periods. **Their electron affinity follows the order Cl > F > Br > I. As explained earlier, the smaller electron affinity of fluorine than that of chlorine is a consequence of its small size.**

As we go across the *p*-block elements in a period, the electronegativity increases reaching a maximum at the halogen group. Thus, halogens are the most electronegative elements in their respective periods. Electronegativity decreases on moving down a group, making **fluorine the most and iodine the least electronegative among the halogens.**

DEEP Focus

In going down the group from chlorine to iodine, the *X*—*X* bond dissociation energy gradually decreases. This is easily explained by considering once again the size factor. In the chlorine molecule, which is the smallest of the three, *viz.* Cl_2, Br_2, I_2, the two bonding electrons are nearer to both the nuclei and are held strongly, while in bromine and iodine, the distance of the bonding electrons from the nuclei gradually increases resulting in lesser attraction and consequent weakening of the bond. Furthermore, as the size of the atom increases, it results in a less effective overlap of the orbitals and therefore, progressively weaker bonds are formed as we go down the group. The bond dissociation energy increases in the order $I_2 < Br_2 < Cl_2$ and if this trend was to continue, **we should expect the F—F bond dissociation energy to be greater than the bond energy of chlorine, 210 kJ mol^{-1}. But this is not so. The actual bond dissociation energy of fluorine molecule is, however, surprisingly low and has the value of 126 kJ mol^{-1} only. The anomalously low bond dissociation energy of fluorine molecule is attributed to the fact that fluorine atom is very small and the non-bonding electrons on fluorine are nearer to each other, resulting in a much greater lone pair-lone pair repulsion, which weakens the covalent bond and lowers its dissociation energy. This repulsion is not so great in relatively larger halogen molecules like chlorine, bromine and iodine where the lone pairs are at a greater average distance from each other.**

- The halogens show very close group similarities. Fluorine differs in several ways from the rest of the group due to :
  - smaller in size,
  - non availability of *d*-orbitals which may be used for bonding.

Fluorine **is the most reactive of all the halogens.** The difference in reactivity between fluorine and chlorine is greater than between chlorine and bromine.

- **HF has a high boiling point (19.5°C) as a result of strong intermolecular hydrogen bonding, whereas all** other hydrogen halides have much lower boiling point (Fig. 15.2)

Table 15.1 **Physical Properties of the Halogens**

Properties	F	Cl	Br	I
Atomic number	9	17	35	53
Electronic configuration	[He] $2s^2 2p^5$	[Ne] $3s^2 3p^5$	[Ar] $3d^{10} 4s^2 4p^5$	[Kr] $4d^{10} 5s^2 5p^5$
Electronegativity (Pauling)	4.0	3.0	2.8	2.5
(*IE*)/kJ mol^{-}	1681	1256	1143	1009
EA/kJ mol^{-}	– 333	– 349	– 325	– 296
Hydration energy of X^-/kJ mol^{-}	– 515	– 381	– 347	– 305
Atomic radius/pm	72	99	114	133
Ionic radius(X^-)/pm	133	184	196	220
Enthalpy of atomisation (kJ mol^{-1})	79.1	122	111	106
$X_2 + 2e^- \rightarrow 2X^-\ E°/V$	2.87	1.40	1.09	0.62
Bond energy (*X*—*X*)/kJ mol^{-}	126	210	158	118
Melting point (*X*)/°C	– 219	– 101	– 7	114
Boiling point (*X*)/°C	– 188	– 34	60	185
Density of solid X_2 (g/cc)	1.3	1.9	3.4	4.93
Oxidising power (*X*)	most	–	–	least
Common oxidation state	– 1	–1, +1, +3, +5, +7	–1, +1, +3, +5, +7	–1, +1, +3, +5, +7
Physical form (X_2) at room temperature	pale yellow gas	yellow green gas	dark red liquid	violet-black solid

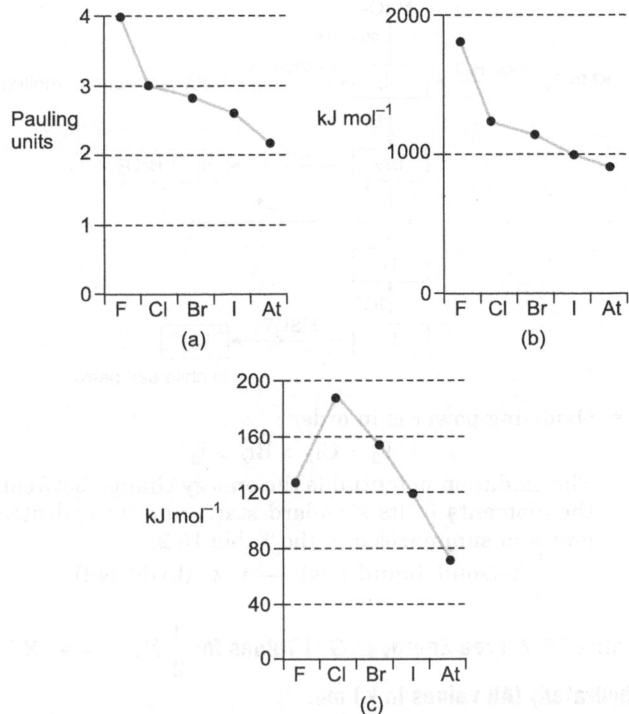

Fig. 15.1 Some group trends of the group 17 elements (a) Pauling electronegativity, (b) The first ionisation energy, and (c) Bond enthalpy in the halogens X_2.

- HF is a weak acid, whereas all other hydrohalic acids (HCl, HBr and HI) are strong acids.
- F_2 reacts with cold NaOH solution to produce OF_2 (oxygen difluoride) gas

$$2F_2 + 2NaOH \longrightarrow 2NaF + H_2O + OF_2$$

The same reaction with chlorine or bromine produces a halides (X^-) and hypohalite (XO^-)

$$X_2 + 2NaOH \longrightarrow NaX + NaXO + H_2O$$

X stands for Cl or Br. Iodine (I_2) does not react under the same condition.

- AgF is soluble in water while other silver halides (AgCl, AgBr and AgI) are insoluble.
- Chlorine, bromine and iodine occur as halides in sea-water, and fluorine occur in the mineral fluorite (CaF_2) and cryolite (Na_3AlF_6)
- Fluorine is the most electronegative element and has no basic properties. Chlorine shows a slight tendency to from cations. ClF ionises to form Cl^+ and F^- (due to higher electronegativity of fluorine)

$$ClF \longrightarrow Cl^+ + F^-$$

- Electrolysis of ICN in pyridine solution gives iodine at the cathode. This indicates that ICN ionises to give I^+ and CN^-.
- Iodine dissolves in oleum giving a bright blue solution which has been shown to have I_2^+ and I_3^+.

$$2I_2 + \underbrace{6H_2SO_4 \cdot SO_3}_{\text{oleum}} \longrightarrow 2I_2^+ + 2HS_3O_{10}^- + 5H_2SO_4 + SO_2$$

$$3I_2 + \underbrace{6H_2SO_4 \cdot SO_3}_{\text{oleum}} \longrightarrow 2I_3^+ + 2HS_3O_{10}^- + 5H_2SO_4 + SO_2$$

- Electrical conductivity of molten iodine is due to the presence of ($I_3^+ + I_3^-$) species produced by self ionisation of iodine:

$$3I_2 \rightleftharpoons I_3^+ + I_3^-$$

Study of Halogens (X_2)

- F_2 is extremely reactive, and this causes great difficulties in the preparation and handling of the element.
  - Moissan (1886) prepared it from CaF_2 :

$$CaF_2 + H_2SO_4 \longrightarrow CaSO_4 + 2HF$$

$$KF + HF \longrightarrow KHF_2$$

$$KHF_2 \xrightarrow{\text{electrolysis}} \underset{\text{at cathode}}{H_2} + \underset{\text{at anode}}{F_2}$$

Electrodes : Pt–Ir alloy

Working temperature : –40° C to –23° C (using boiling CH_3Cl, –23° C)

  - Dennis and others (1931) also prepared F_2 by using fused KHF_2 with graphite electrodes in a copper vessel.
  - Modern method uses graphite anode and steel cathode in a steam heated steel tank.
  - If F_2 were formed by the electrolysis of aqueous fluoride solution, it would immediately oxidise water to O_2

$$F_2 + 2H_2O \longrightarrow 4HF + O_2$$

Most recent method of the preparation of F_2 involves reaction

$$K_2MnF_6 + 2SbF_5 \longrightarrow 2KSbF_6 + MnF_3 + \frac{1}{2}F_2$$

- The commercial production of Cl_2 is by electrolysis, either of concentrated aqueous NaCl or of molten salt such as NaCl(l) or $MgCl_2$(l):

$$2NaCl + 2H_2O \xrightarrow{\text{electrolysis}} \underbrace{2NaOH + H_2}_{\text{at cathode}} + \underbrace{Cl_2}_{\text{at anode}}$$

This process is called **chlor-alkali process**

DEEP Focus

This reaction yields two useful by-products, NaOH and H_2. The cells are designed to separate the molecular chlorine from the sodium hydroxide solution and the molecular hydrogen to prevent side reactions such as

$$2NaOH + Cl_2 \longrightarrow NaOCl + NaCl + H_2O$$

$$H_2 + Cl_2 \longrightarrow 2HCl$$

The reaction between H_2 and Cl_2 is explosive.

$$\underset{\text{(molten)}}{2NaCl(l)} \xrightarrow{\text{electrolysis}} \underset{\text{at cathode}}{2Na} + \underset{\text{at anode}}{Cl_2}$$

- Cl_2 is also prepared (in laboratory) by heating conc. HCl with $KMnO_4$ or MnO_2 (both oxidising agent) :

$$2MnO_4^- + 10Cl^- + 16H^+ \longrightarrow 2Mn^{2+} + 5Cl_2 + 8H_2O$$

$$MnO_2 + 4H^+ + 2Cl^- \longrightarrow Mn^{2+} + 2H_2O + Cl_2$$

- Br_2 is obtained from sea water and brine lakes. Sea water contains about 65 ppm Br^-. Br_2 is isolated in following steps :

  - First sea water is adjusted to pH 3.5 and treated with Cl_2 which oxidises Br^- to Br_2

$$Cl_2 + 2Br^- \longrightarrow Br_2 + 2Cl^-$$

  - Br_2 is quite volatile hence is removed by a stream of air.

DEEP Focus

- Br_2 is absorbed into aqueous Na_2CO_3 solution when NaBr and $NaBrO_3$ are formed :

$$3Na_2CO_3 + 3Br_2 \longrightarrow 5NaBr + NaBrO_3 + 3CO_2$$

Or $3Br_2 + 6OH^- \longrightarrow 5Br^- + BrO_3^- + 3H_2O$

- Finally the solution is acidified and distilled to give pure Br_2 :

$$5NaBr + NaBrO_3 + 3H_2SO_4 \longrightarrow 5HBr + HBrO_3 + 3Na_2SO_4$$

$$5HBr + HBrO_3 \longrightarrow 3Br_2 + 3H_2O$$

or $5Br^- + BrO_3^- + 6H^+ \longrightarrow 3Br_2 + 3H_2O$

- I_2 is obtainable in small quantities from dried seaweed, since certain marine plants absorb and concentrate I^- selectively in presence of Cl^- and Br^-. Low concentration of I^- are also found in some natural brines (salt solutions) associated with oil fields.

Chile saltpetre is mainly $NaNO_3$ but it contains traces of sodium iodate $NaIO_3$ and sodium periodate $NaIO_4$. I_2 is recovered by reducing IO_3^- to I^- by $NaHSO_3$ and then oxidation of I^- to I_2 by IO_3^- itself :

$$IO_3^- + 3HSO_3^- \longrightarrow I^- + 3SO_4^{2-} + 3H^+$$

$$5I^- + IO_3^- + 6H^+ \longrightarrow 3I_2 + 3H_2O$$

Natural brine is treated with Cl_2 water to oxidise I^- to I_2 and the solution is passed through an ion-exchange resin which adsorbs I_2 as I_3^-. I_3^- is removed from the resin by treatment with alkali.

- Reactivity of halogens is based on bond-energy (BE) value:

$$F_2 > Cl_2 > Br_2 > I_2$$

Greater the BE, smaller the reactivity.

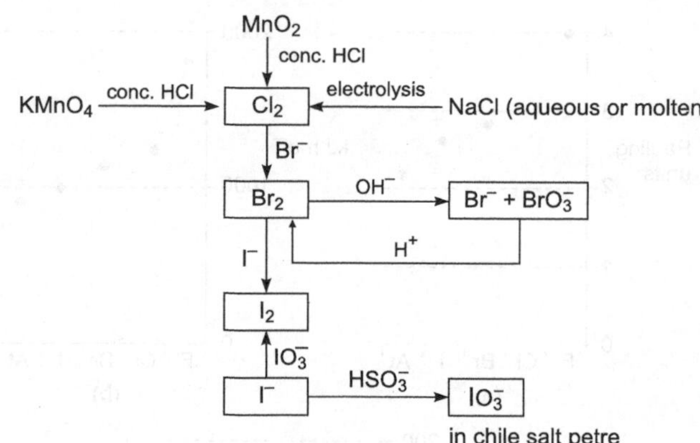

- Oxidising power is in order :

$$F_2 > Cl_2 > Br_2 > I_2$$

The oxidation potential is the energy change between the elements in its standard state, and its hydrated ions and summarised in the Table 15.2.

$$\frac{1}{2}X_2(\text{solid, liquid, gas}) \longrightarrow X^-(\text{hydrated})$$

Table 15.2 **Free Energy (ΔG°) Values for $\frac{1}{2}X_2 \longrightarrow X^-$ (hydrated) (All values in kJ mol^{-1})**

	1/2 Free energy of fusion	1/2 Free energy of evaporation	1/2 Free energy of dissociation	Electron affinity	Free energy of hydration	Sum of ΔG°
F_2	—	—	126/2	–333	–460	–730
Cl_2	—	—	210/2	–349	–348	–592
Br_2	—	31/2	158/2	–325	–318	–548.5
I_2	15/2	44/2	118/2	–296	–279	–486.5

DEEP Focus

F_2 has the most negative ΔG° value, hence is strongest oxidising agent and can oxidise Cl^-, Br^- and I^- to respectively Cl_2, Br_2 and I_2. F_2 can also oxidise H_2O to O_2

$$F_2 + H_2O \longrightarrow 2HF + \frac{1}{2}O_2$$

Similarly Cl_2 will oxidise Br^- and I^-, and Br_2 will oxidise I^-. **In general any halogen (X_2) of lower atomic number will oxidise halide ions (X^-) of higher atomic number.**

Some reactions of the halogens are summarised in Table 15.3.

DEEP Focus

- Cl_2 is used as a **bleach or as a disinfectant** (as in public water supplies). It reacts slowly with H_2O to form HCl and HOCl. The hypochlorous acid then decomposes into HCl and O radicals, which kill bacteria.

$$Cl_2 + H_2O \rightleftharpoons HCl + HOCl$$

$$HOCl \longrightarrow HCl + :\ddot{O}\cdot$$

These oxygen radicals are very strong oxidising as well as effective bleaching and disinfecting agent in aqueous solution of Cl_2 or hypochlorite salts.

- **Iodine is necessary for the normal functioning of the thyroid gland. Insufficient iodine in the diet leads to Goitre (enlargement of the thyroid gland) 'Iodised salt' contains KI/KIO_3.**
- **For** $I_2 + H_2O \longrightarrow 2H^+ + 2I^- + \frac{1}{2}O_2$, $\Delta G^\circ = 105$ kJ mol^{-1}

and thus for reverse reaction $\Delta G^\circ = -105$ kJ mol^{-1}, hence reverse reaction occurs spontaneously and even atmospheric O_2 oxidises I^- to I_2.

$$2HI + \frac{1}{2}O_2 \longrightarrow H_2O + I_2$$

Table 15.3 **Some Reactions of Halogen**

Reaction	Comment
• With H_2O	
$2F_2 + 2H_2O \longrightarrow 4H^+ + 4F^- + O_2$ $2X_2 + 2H_2O \longrightarrow 4H^+ + 4X^- + O_2$ $X_2 + H_2O \xrightarrow{\text{disproportionate}} X^- + HOX + H^+$	Vigorous reaction with F_2; atmospheric O_2 can oxidise I^- to I_2 hence, reverse reaction $Cl_2 > Br_2 > I_2$ (F_2 does not disproportionate).
• With H_2	
$H_2 + X_2 \longrightarrow 2HX$	All the halogens, with Br_2 photochemical reaction, with I_2 very slow even at high temperature.
• With metals	
$2M + nX_2 \longrightarrow 2MX_n$	Most metals form halides.
• With CO	
$CO + X_2 \longrightarrow COX_2$	Only Cl_2, Br_2 form carbonyl halide
• With P	
$2P + 3X_2 \longrightarrow 2PX_3$ $2P + 5X_2 \longrightarrow 2PX_5$	For F, Cl, Br, I For F, Cl, Br
• With S	
$2S + X_2 \longrightarrow S_2X_2$ $S + 2Cl_2 \longrightarrow SCl_4$ $S + 3F_2 \longrightarrow SF_6$	Cl, Br
• With H_2S	
$H_2S + X_2 \longrightarrow 2HX + S$	All the halogens oxidise H_2S (S^{2-}) to S
• With SO_2	
$SO_2 + X_2 \longrightarrow SO_2X_2$	F and Cl
• With NH_3	
$8NH_3 + 3X_2 \longrightarrow N_2 + 6NH_4X$	F, Cl, Br
• With halogens	
$X_2 + X'_2 \longrightarrow 2XX'$	Interhalogen compounds
• With cold NaOH	
$2NaOH + 2F_2 \longrightarrow 2NaF + OF_2 + H_2O$ $2NaOH + X_2 \longrightarrow NaX + NaOX + H_2O$	(X = Cl, Br) I_2 does not react under the same condition.
Cl_2 reacts with dry slaked lime forming bleaching powder $Ca(OH)_2 + Cl_2 \longrightarrow CaCl(OCl) + H_2O$	
• With hot NaOH	
$6NaOH + 3X_2 \longrightarrow 5NaX + NaXO_3 + 3H_2O$	(X = Cl, Br, I)

Colour of Halogens

In gaseous phase, colours of the halogens vary remarkably :

F_2 — pale yellow
Cl_2 — pale greenish yellow
Br_2 — deep red-brown
I_2 — deep violet

All these colours may be assigned to transition of an electron from the highest occupied π^* molecular orbital to the lowest occupied σ^* molecular orbital. As one moves from F_2 to I_2, the difference between the two energy levels decreases and probability of this transition also increases at the same time. Hence, the absorption maximum gradually shifts from the near UV to the red region of the

visible spectrum; the intensity of absorption also increases side by side. Hence, we observe the complimentary colour with increasing intensity.

Interhalogen Compounds

- The halogens react with each other to form **interhalogen compounds** of the type AX, AX_3, AX_5 and AX_7 of which A is of higher atomic number as shown :

AX	*AX*$_3$	*AX*$_5$	*AX*$_7$
ClF	ClF$_3$	ClF$_5$	
BrF	BrF$_3$	BrF$_5$	
ICl	ICl$_3$ (I$_2$Cl$_6$)	IF$_5$	IF$_7$
IBr			
BrCl			

- There are never more than two different halogens in a molecule. The bonds are covalent because of the small electronegativity difference. Ternary combinations occur in polyhalide ions (*e.g.*, $IBrCl^-$)
  - Neutral interhalogen compounds are of the type AX_n where n is an odd number and X is always lighter halogen when $n > 1$. Because n is odd the compounds are diamagnetic, their valence electrons are present as bonding pairs or as unshared pairs.
  - Interhalogen ions can be either cations or anions. Halogen fluorides react with fluoride ion acceptors,

$$2ClF + AsF_5 \longrightarrow FCl_2^+AsF_6^-$$

or with fluoride ion donors,

$$IF_5 + CsF \longrightarrow Cs^+IF_6^-$$

  - The pale yellow triiodide ion (I_3^-) is formed by dissolving I_2 in aqueous KI.

$$I_2 + I^- \rightleftharpoons I_3^-$$

Other ions are not usually stable in aqueous solution although they can be obtained in CH_3OH or CH_3CN and as crystalline salts of large cations as Cs^+ or $R_4\overset{+}{N}$. Cl_3^- is formed only in concentrated solution

$$Cl_2 + Cl^- \rightleftharpoons Cl_3^-$$

  - The electrical conductance of molten I_2 is ascribed to self-ionisation

$$3I_2 \rightleftharpoons I_3^+ + I_3^-$$

  - The interhalogens are generally more reactive than the halogens (except F_2). This is because the A—X bond in interhalogens is weaker than the X—X bond in the halogens.
  - Hydrolysis of interhalogens gives halide and oxohalide. **Oxohalide is always formed from larger halogens present :**

$$BrF_5 \xrightarrow{H_2O} 5F^- + \underset{\text{bromate ion}}{BrO_3^-}$$

$$ICl \xrightarrow{H_2O} Cl^- + \underset{\text{iodate ion}}{IO_3^-}$$

$$ICl \xrightarrow{H_2O} Cl^- + \underset{\text{hypoiodite ion}}{IO^-}$$

  - Interhalogen compounds can fluorinate many metal oxides, metal halides and metals :

$$3UO_2 + 4BrF_3 \longrightarrow 3UF_4 + 2Br_2 + 3O_2$$

$$UF_4 + ClF_3 \longrightarrow UF_6 + ClF$$

Some interhalogen compounds are summarised below in Table 15.4

Table 15.4 **Interhalogen Compounds and their Properties**

Type of inter-halogen compound	Formulae and names of compounds	Physical state and colour at ordinary temperatures	Melting point (K)	Boiling point (K)
XY	ClF Chlorine fluoride	Gas, colourless	119	163
	BrF Bromine fluoride	Liquid below 293 K, Pale-brown gas above 293 K	240	293
	BrCl Bromine chloride	Liquid, reddish yellow	207	283
	ICl Iodine chloride	Solid, rubby red	—	310
	IBr Iodine bromide	Solid, brown like iodine	232	
XY$_3$	ClF$_3$ Chlorine trifluoride	Liquid below 261 K, colourless gas above 261 K	197	261
	BrF$_3$ Bromine trifluoride	Liquid, colourless	282	400
	BrCl$_3$ Bromine trichloride	Solid, orange	374 (decomposes)	—
XY$_5$	ClF$_5$ Chlorine pentalfluoride	Gas, colourless	170	259
	BrF$_5$ Bromine pentafluoride	Liquid colourless	213	314
	IF$_5$ Bromine pentafluoride	Liquid colourless	283	314
XY$_7$	IF$_7$ Iodine heptafluoride	Gas, colourless	279.5	

Structure of Interhalogen Compounds

The interhalogen compounds are generally covalent in which the larger atom lies in the centre. Their shapes can be predicted by VSEPR theory (Chapter 7) Shapes of some of the interhalogen compounds are given below

Molecule	Total number of electron pairs around X	Bond pairs	Lone pairs	Shape
XY	4	1	3	Linear
XY_3	5	3	2	T-shaped
XY_5	6	5	1	Square pyramidal
XY_7	7	7	0	Pentagonal bipramidal

Structure of interhalogens can be explained with the help of Valence Bond Theory also in the following manner :

(i) Type *XY* As expected, the compounds of the type *XY* are linear. Thus, ClF, BrF, BrCl, IBr and ICl all have a linear sturcture. In these compounds both the halogen atoms have an unpaired electron in the p_z orbital. Axial overlap of the p_z orbitals of the two halogen atoms results in a **linear interhalogen** molecule.

(ii) Type XY_3 Structure of the interhalogens of this type can be explained on the basis of sp^3d hybridisation of the central halogen atom *X*. Three of the hybrid orbitals having an electron each are used in making three covalent bonds with three atoms of the other halogen atom. The remaining two orbitals accommodate the two lone pairs of the central atom. In order of minimise lone pair-lone pair and lone pair-bond pair repulsions, the lone pairs occupy the two equatorial sites around the central halogen atom, thereby giving a T-shape to the molecule. **Thus, ClF_3, BrF_3 and ICl_3 have a T-shaped structure.**

X atom in excited state 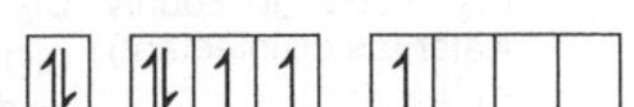

sp^3d hybridisation-trigonal bipyramidal structure with two equatorial sites occupied by lone pairs

T-shaped structure of interhalogen molecules of the types XY_3

(iii) Type XY_5 This type of compounds have a square pyramidal structure (instead of octahedral due to VSEPR). The central atom *X* uses five of its seven valence electrons in forming five bonds with *Y* and two electrons remains as a lone pair. Thus, *X*-atom in XY_5 molecule is sp^3d^2 hybridised. The lone pair occupies the axial hybrid orbital.

X atom in excited state

sp^3d^2 hybridisation-octahedral structure with one of the axial sites occupied by lone pair

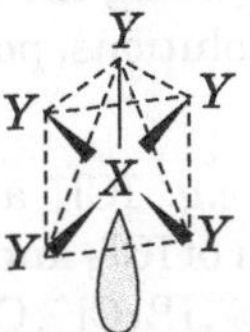

Square pyramidal structure of interhalogen molecules of the type XY_5

(iv) Type XY_7 In these molecules, the central *X* atom uses all its seven valence electrons in forming seven *X*—*Y* bonds. Formation of this type of compounds can be explained on the basis of sp^3d^3 hybridisation of the *X* atom. These compounds have a pentagonal bipyramidal structure with two axial *X*—*Y* bonds and five equatorial *X*—*Y* bonds. Bond angle is 72° for five sp^3d^3 orbitals and other two are at 90° with first the sp^3d^3 orbitals.

X atom in excited state

sp^3d^3 hybridisation-pentagonal bipyramidal structure

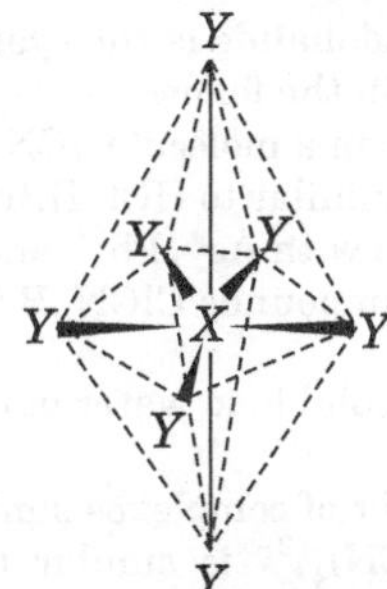

Pentagonal bipyramidal structure of interhalogen molecules of the type XY_7

Polyhalides and Polyhalonium Ions

Halide ions often associate with molecules of halogens or with interhalogen compounds to form polyhalide ions. For example, on addition of potassium iodide, the solubility of iodine in water is greatly enhanced due to the formation of the triiodide ion, I_3^- :

$$KI + I_2 \longrightarrow KI_3$$

More complex ions of iodine, such as pentaiodide, I_5^-, heptaiodide, I_7^- and noniodide, I_9^- are also known.

These are generally found in the form of salts of large metal ions, as $Rb^+I_9^- \cdot 2C_6H_6$ or large complex cations, as $NH_4^+I_5^-$, $(C_2H_5)_4N^+I_7^-)$. A number of mixed polyhalide ions containing two or three different halogens are also known, as ICl_2^-, ICl_4^- and $IBrF^-$ and $IBrCl^-$.

All the polyhalides are hydrolysed to some extent in aqueous solution. The ease of hydrolysis increases with the introduction of the more electronegative halogen, into the ion. Thus, the triiodide being the most stable and Cl_3^- the least. In non-aqueous solutions, polyhalides behave as mild halogenating agents.

Polyhalonium cations, *e.g.*, ICl_4^+ and BrF_2^+ are formed as a result of autoionisation of ICl_3 and BrF_3. Other cations like $Br_2^+, I_2^+, I_3^+, ClF_2^+, IF_2^+, IBr_2^+, IBrCl^+, ClF_4^+, BrF_4^+, IF_4^+, ClF_6^+$, BrF_6^+, IF_6^+, etc. are also known. These are stabilised by large anions like $BF_4^-, PF_6^-, AsF_6^-, AlCl_4^-, SbCl_6^-$, etc.

Pseudohalogens and Pseudohalides

Some molecules like cyanogen, $(CN)_2$, thiocyanogen, $(SCN)_2$ and selenocyanogen, $(SeCN)_2$, have properties similar to those of the halogens. Therefore, these are called pseudohalogens. Pseudohalogens form hydro acids similar to the hydrohalic acids, *e.g.*, H—CN. They also form anions such as cyanide, CN^-, cyanate, OCN^-, thiocynate, SCN^-, selenocyanate, $SeCN^-$, tellurocyanate. $TeCN^-$ and azide, N_3^-, which resemble the halide ions to some extent.

DEEP Focus

The best known pseudohalide is the cyanide, CN^-, which resembles Cl^-, Br^- and I^- in the following respects :

- It can be oxidised to form a molecule, $(CN)_2$.
- It forms an acid HCN similar to HCl, HBr, etc.
- It forms insoluble salts with Ag^+, Pb^{2+} and Hg^+.
- Interpseudohalogen compounds ClCN, BrCN and ICN can also be formed.
- Like AgCl, AgCN is insoluble in water but soluble in liquor ammonia.
- It forms a large number of complexes similar to the halide complexes, *e.g.*, $[Cu(CN)_4]^{2-}$ is similar to $[CuCl_4]^{2-}$ and $[Co(CN)_6]^{3-}$ is similar to $[CoCl_6]^{3-}$.

Anomalous Behaviour of Fluorine

Fluorine differs from other elements of the group because of its exceptionally small atomic and ionic size and low fluorine—fluorine bond dissociation energy. The result of these differences is that fluorine is the most electronegative element in the Periodic Table and is a powerful oxidant. Some difference between fluorine and other halogens are :

- It is more reactive than other halogens because of low F—F bond energy, also due to its high electronegativity the bond between fluorine and other elements is very strong so its compounds are more stable. Some of them are inert, *e.g.*, SF_6.
- Fluorine is almost invariably monocoordinate (coordination number = 1) and is never more than monocovalent.
- Fluorides are more ionic.
- Fluorine forms strong hydrogen bonds resulting in the properties of hydrogen fluoride being anomalous.
- Fluorine is the strongest oxidising agent and oxidises the elements to their highest oxidation state, for example in IF_7, iodine has oxidation number seven.
- The reactions of fluorine are also different from other halogens. Fluorine fumes in moist air and decomposes water to give oxygen, whereas other halogens are sparingly soluble in water and react partly to give hydrohalic and hypohalous acids.

$$2F_2 + 2H_2O \longrightarrow 2HF + O_2$$
$$X_2 + H_2O \rightleftharpoons HOX + HX$$

Fluorine reacts with alkalies to yield the oxide, OF_2 :

$$2F_2 + 2NaOH \longrightarrow OF_2 + 2NaF + H_2O$$

The other halogens react with alkalies to yield a solution of hypohalite ions, which may disproportionate. The oxoacid of fluorine, HOF, prepared recently is very unstable. The oxides of fluorine are not acidic.

$$Cl_2 + 2NaOH \longrightarrow NaClO + NaCl + H_2O$$

Uses of Halogens

Halogen	Uses	Remark
F_2	In the preparation of UF_6	UF_6 used in the gaseous diffusion process for separating isotopes of uranium (U-235 and U-238)
	In the preparation of teflon $-(CF_2-CF_2)_n-$	Teflon is used in electrical insulators, high temperature plastic and cooking utensils
Cl_2	Cl^- ion is the principal anion in intracellular and extracellular fluids	—
	Cl_2 as bleaching agent	Household laundry bleach contains the active ingredient NaOCl (sodium hypochlorite about 5% by mass)
	Cl_2 used to purify water (as disinfectant)	$Cl_2 + H_2O \longrightarrow HCl + HOCl$ ClO^- ions destroy bacteria by oxidising life-sustaining compounds within them
	As a chlorinating agent to prepare various organic compounds	$CHCl_3$, CCl_4, $CH_2{=}CHCl$, BHC, DDT
Br_2	To prepare ethylene dibromide $BrCH_2CH_2Br$	$BrCH_2CH_2Br$ is used as an insecticide and as a scavenger for lead in gasoline to keep lead deposits from clogging engines.
	To prepare AgBr	AgBr is used in photographic plate
I_2	To prepare tincture of iodine —an antiseptic	50% (by mass) alcohol solution of iodine is called tincture of iodine
	In iodised salt	Iodised salt contains 0.01 per cent KI or NaI which is more than sufficient to satisfy 1 mg iodine per week required for the formation of thyroxine in the human body.
	To prepare AgI	Used in cloud seeding—a process for inducing rainfall.

Target Practice 1

1. Cl_2 gas is passed aqueous NaOH solution under given conditions

(a) $Cl_2 + NaOH(cold) \longrightarrow$ ....

(b) $Cl_2 + NaOH(hot) \longrightarrow$ ....

Write balanced ionic reactions

Equivalent mass of Cl_2 in (a) is ....

Equivalent mass of Cl_2 in (b) is ....

2. (a) $(CN)_2$ cyanogen is also called ....

(b) CN^- cyanide ion is also called ....

(c) Bond energy of F_2, Cl_2, Br_2 and I_2 in increasing order is ....

(d) Electron-affinity is maximum (out of F, Cl Br and I) for ....

(e) OF_2 is formed by the reaction ....

3. ΔG_f° (standard free energy of formation) of

$$HF(g) = -273.20 \text{ kJ mol}^{-1}$$
$$HI(g) = +1.72 \text{ kJ mol}^{-1}$$

Based on the above data, select which reaction is most favourable in each case

(a) $H_2 + I_2 \longrightarrow 2HI$ or $2HI \longrightarrow H_2 + I_2$

(b) $H_2 + F_2 \longrightarrow 2HF$ or $2HF \longrightarrow H_2 + F_2$

4. Out of HF, HCl, HBr and HI, select

(a) strongest acid ....

(b) weaker acid ....

(c) acid with maximum boiling point ....

(d) acid that attacks glass ....

(e) acid with maximum dipole moment ....

(f) acid which has been used as non-aqueous solvent ......

Answers

1. (a) $2NaOH(cold) + Cl_2 \longrightarrow NaCl + NaOCl + H_2O$ (Cl_2 undergoes reduction to NaCl and oxidation to NaOCl)

(b) $6NaOH(hot) + 3Cl_2 \longrightarrow 5NaCl + NaClO_3 + 3H_2O$ (reduction to NaCl, oxidation to $NaClO_3$)

Equivalent mass of Cl_2 in (a)

$$\underset{0}{\frac{1}{2}Cl_2} \xrightarrow{\text{reduction}} \underset{-1}{Cl^-} \qquad \text{Equivalent mass} = \frac{M}{2}$$

$$\underset{0}{\frac{1}{2}Cl_2} \xrightarrow{\text{oxidation}} \underset{+1}{OCl^-} \qquad \text{Equivalent mass} = \frac{M}{2}$$

Net equivalent mass = in reduction part + in oxidation part

$$= \frac{M}{2} + \frac{M}{2} = M$$

$$= 71.0 \text{ g equivalent}^{-1}$$

Equivalent mass of Cl_2 in (b)

$$\underset{0}{\frac{1}{2}Cl_2} \xrightarrow{\text{reduction}} \underset{-1}{Cl^-} \qquad \text{Equivalent mass} = \frac{71}{2}$$

$$\underset{0}{\frac{1}{2}Cl_2} \longrightarrow \underset{+5}{ClO_3^-} \qquad \text{Equivalent mass} = \frac{71}{10}$$

$$\therefore \text{ Net equivalent mass} = \frac{71}{2} + \frac{71}{10} = 42.6 \text{ g equivalent}^{-1}$$

2. (a) Pseudohalogen (b) Pseudohalide

(c) $I_2 < F_2 < Br_2 < Cl_2$ (d) Cl

(e) $2F_2 + 2NaOH \longrightarrow 2NaF + H_2O + OF_2$

3. (a) $H_2 + I_2 \longrightarrow 2HI \quad \Delta G_f^\circ = +\text{ve}$

Decomposition of HI is kinetically controlled. (rather than thermodynamically controlled)

(b) $H_2 + F_2 \longrightarrow 2HF$

Very fast since, $\Delta G_f^\circ = -\text{ve}$

4. (a) HI (b) HF (c) HF (d) HF (e) HF (f) HF

Practice Exercise 1

1. (a) ICl is found to be partially ionised in fused state. Give reaction.

(b) ICl is also found to form addition compound with NaBr and KI. Write the formulae of the addition compounds.

2. Identify (*A*) and (*B*).

$$Br_2 + OH^- \xrightarrow[\text{hot}]{} (A) + (B)$$

$$(A) + (B) + H^+ \longrightarrow Br_2$$

(*A*) gives yellow ppt with $AgNO_3$.

3. When I_2 is estimated by $Na_2S_2O_3$ iodometrically, blue colour (due to starch) changes to colourless when I_2 has completely reacted

$$2S_2O_3^{2-} + I_2 \longrightarrow S_4O_6^{2-} + 2I^-$$

but after sometimes blue colour reappears. Explain

4. SbF_5, AsF_5 and BF_3 accept F^- from HF as :

$$AsF_5 + HF \rightleftharpoons H^+ + [AsF_6]^-$$

$$BF_3 + HF \rightleftharpoons H^+ + [BF_4]^-$$

Identify nature of HF and fluorides.

5. CN^- resembles Cl^-, Br^- and I^- in many of its properties. What is it called ? Give some other examples which resemble halides.

6. Select the species which are soluble in H_2O and which are soluble in $NH_3(aq)$ AgF, AgCN, AgCl, AgI, CaF_2, $CaCl_2$.

7. Identify the products (which can be in the form of ions) in the following :

(a) $KF + IF_5 \longrightarrow$ (b) $CsF + IF_5 \longrightarrow$

(c) $KCl + ICl \longrightarrow$ (d) $CsI_3 \xrightarrow{\Delta}$

(e) $Rb[ICl_2] \xrightarrow{\Delta}$ (f) $K[BrICl] \xrightarrow{\Delta}$

(g) $I_2 + ICl + AlCl_3 \xrightarrow{\Delta}$

8. H_2SO_4 is a weaker acid than HCl, yet HCl is evolved when conc. H_2SO_4 is added to NaCl. Explain

9. HF can be prepared by the action of H_2SO_4 on NaF. Explain why HBr cannot be prepared by the action of the same acid on NaBr.

Answers

1. (a) $2ICl \longrightarrow I^+ + [ICl_2]^-$

(b) $NaBr + ICl \longrightarrow Na^+[BrICl]^-$

$KI + ICl \longrightarrow K^+[I_2Cl]^-$

2. Method is for enriching Br_2

$$3Br_2 + 6OH^- \longrightarrow \underset{(A)}{5Br^-} + \underset{(B)}{BrO_3^-}$$

$$5Br^- + BrO_3^- + 6H^+ \longrightarrow 3Br_2 + 3H_2O$$

3. This is because some atmospheric oxidation has taken place, forming I_2, which reacts with starch to give the blue colour again :

$$2I^- + \frac{1}{2}O_2 + 2H^+ \longrightarrow I_2 + H_2O$$

4. SbF_5, AsF_5, BF_3 etc., are F^- acceptors thus they are behaving as an acids and HF behaves as a base :

$$\underset{\text{acid}}{SbF_5} + \underset{\text{base}}{HF} \longrightarrow [SbF_6]^- + H^+$$

5. CN^- is called **pseudohalide.** Like X^-:

– it forms HCN

– it can be oxidised to $(CN)_2$

– it forms insoluble salts with Ag^+, Pb^{2+} and Hg_2^{2+}.

Other **pseudohalides** are :

SCN^- (thiocyanate), OCN^- (cyanate),

NCN^{2-} (cyanamide), $SeCN^-$ (selenocyanate)

ONC^- (fulminate), N_3^- (azide)

6. H_2O **soluble :** AgF, $CaCl_2$

NH_3 ***(aq)*** **soluble :** AgCN, AgCl

AgI and CaF_2 are insoluble in H_2O and NH_3 *(aq)* both.

7. (a) $K^+[IF_6]^-$ (b) $Cs^+[IF_6]^-$ (c) $K^+[ICl_2]^-$ (d) $CsI + I_2$ (e) $RbCl + ICl$ (f) $KCl + IBr$ (g) $[I_3]^+[AlCl_4]^-$

8. $NaCl + H_2SO_4 \longrightarrow NaHSO_4 + HCl$

The HCl gas escapes taking the equilibrium to the right

9. H_2SO_4 oxidises HBr formed to Br_2.

Halogen Acids

- All the halogen acids (HF, HCl, HBr, HI) are known. In aqueous solution, they are typical mineral acids whose acid strengths decrease in the direction of increasing bond energy, that is

$$\underbrace{HI > HBr > HCl}_{\text{strong acids}} > \underset{\text{weak acid}}{HF}$$

- HX can be prepared by direct combination of H_2 and X_2

$$H_2(g) + X_2(g) \longrightarrow 2HX(g)$$

– reaction between H_2 and F_2 is very fast,

– reaction between H_2 and Cl_2 is also rapid,

– but reaction between H_2 and Br_2 or I_2 is very slow.

Free energies of formation of HX are given :

HX	ΔG_f°, kJ mol^{-1}
HF(g)	–273.20
HCl(g)	–95.30
HBr(g)	–53.43
HI(g)	+1.72

– thus we see that ΔG_f° of HF, HCl and HBr are large and negative suggesting that for them reaction

$$H_2 + X_2 \longrightarrow 2HX$$

goes to completion.

DEEP Focus

For HI(g), ΔG_f° is small and positive suggesting that HI should dissociate

$$2HI(g) \longrightarrow H_2(g) + I_2(g)$$

to some extent. **However, because of the high activation energy the dissociation occurs only very slowly in the absence of a catalyst. As a result, HI(g) is quite stable at room temperature. Thus, at room temperature, decomposition of HI(g) is kinetically controlled (rather than thermodynamically controlled).**

- $CaF_2 + H_2SO_4 \xrightarrow{\Delta} CaSO_4 + 2HF$
- CaF_2 has also SiO_2 as impurity otherwise HF formed would react with SiO_2 forming SiF_4 and H_2SiF_6:

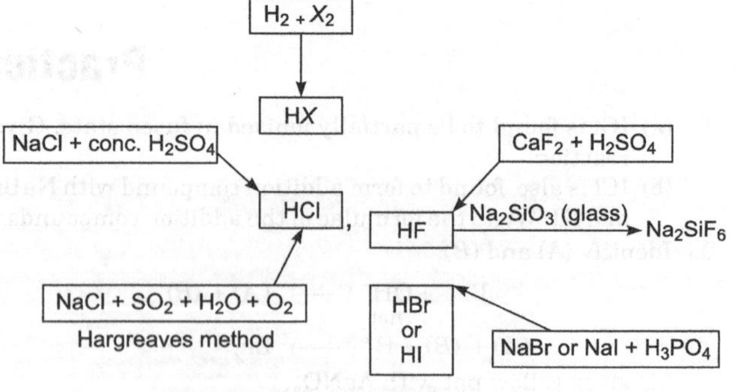

$$SiO_2 + 4HF \longrightarrow SiF_4 + 2H_2O$$

$$SiF_4 + 2HF \longrightarrow H_2[SiF_6]$$

- $$\underset{\text{conc.}}{NaCl + H_2SO_4} \xrightarrow{150\,°C} HCl + \underset{\text{salt cake}}{NaHSO_4}$$

$$NaHSO_4 + NaCl \xrightarrow{550\,°C} HCl + Na_2SO_4$$

This method called **'salt-cake'** method involves formation of $NaHSO_4$ (salt-cake).

NaCl is now replaced by NH_4Cl since NH_4HSO_4 (intermediate) is water soluble and reaction goes to completion:

$$2NH_4Cl + H_2SO_4 \longrightarrow (NH_4)_2SO_4 + 2HCl$$

The **Hargreaves** process is a variant of the salt-cake process in which NaCl is reacted with a gaseous mixture of SO_2 and H_2O :

$$2NaCl + SO_2 + H_2O + \frac{1}{2}O_2 \longrightarrow Na_2SO_4 + 2HCl.$$

- HBr and HI are made by the reaction of concentrated phosphoric acid on metal bromides and iodides :

$$NaBr + H_3PO_4 \longrightarrow NaH_2PO_4 + HBr$$
$$NaI + H_3PO_4 \longrightarrow NaH_2PO_4 + HI$$

ALERT

Concentrated H_2SO_4 (a strong oxidising agent) can't be used since it would oxidise HBr to Br_2 and HI to I_2.

These are also formed from red phosphorus and $X_2(Br_2, I_2)$ which give PX_3 and its hydrolysis gives HX :

$$\underset{\text{red}}{2P} + 3Br_2 \longrightarrow 2PBr_3$$
$$2PBr_3 + 6H_2O \longrightarrow 6HBr + 2H_3PO_3$$
$$2P + 3Br_2 + 6H_2O \longrightarrow 6HBr + 2H_3PO_3$$

- HF is only just liquid at room temperature and HCl, HBr and HI are gases. HF has abnormally high boiling point due to intermolecular H-bonding which is not possible in other halogen acids due to their smaller electronegativity values.

Boiling point HCl < HBr < HI < HF

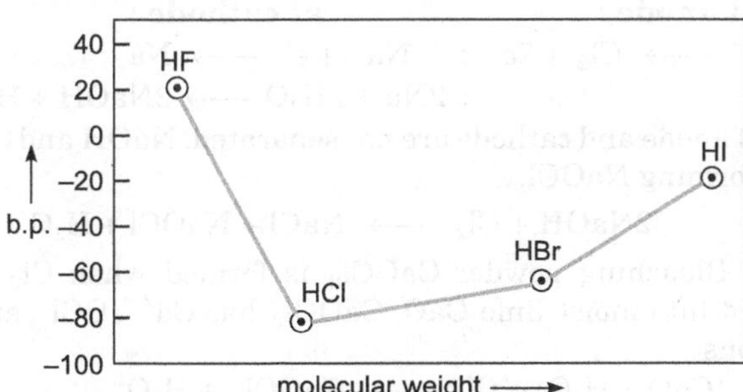

Fig. 15.2 Boiling point of HX

Some H-bonding also occurs in the gas, which consists of a mixture of cyclic $(HF)_6$ polymers, dimeric $(HF)_2$ and the monomeric HF.

(Hydrogen-bonded chain in solid HF)

DEEP Focus

Because of the tendency to form hydrogen bond, metal fluorides are solvated by HF giving species of the type HF_2^- etc. KHF_2 (KF·HF), KH_2F_3 (KF·2HF) and KH_3F_4 (KF·3HF) are known examples.

- HF attacks glass, other halogen acids do not. Glass is silicates of sodium and calcium and are converted into complex fluorides :

$$Na_2SiO_3 + 6HF \longrightarrow Na_2SiF_6 + 3H_2O$$
$$CaSiO_3 + 6HF \longrightarrow CaSiF_6 + 3H_2O$$

Hence, **HF can't be stored in a glass vessel. Instead, it is stored in a wax-vessel which does not react with HF. This unique property of HF of attacking glass is used for making permanent marking on glass-wares.**

- HF also reacts with CCl_4 to form **freons** (refrigerant)

$$CCl_4 + HF \longrightarrow CFCl_3 + HCl$$
$$CFCl_3 + HF \longrightarrow CF_2Cl_2 + HCl$$

- In the gaseous state, HX are essentially covalent. In aqueous solution, they ionise to form H_3O^+ and X^- but not H^+ and X^-

$$\underset{\text{base}}{H_2O} + \underset{\text{acid}}{HX^-} \rightleftharpoons \underset{\text{acid}}{H_3O^+} + \underset{\text{base}}{X^-}$$

HF has more ionic character (due to greater electronegativity difference) than HCl, HBr and HI but acid strength varies as :

$$\underset{\text{weak}}{HF} < \underbrace{HCl < HBr < HI}_{\text{strong}}$$

DEEP Focus

Acid strength is dependent not on the electronegativity difference of H and X **but on the tendency of the hydrated molecule** to form hydrogen ions :

HX (hydrated) $\longrightarrow$ H^+ (hydrated) + X^-(hydrated)

The acid strength is equal to the sum of all the energy terms in :

	ΔH
= free energy of dehydration	x_1
+ free energy of dissociation	x_2
+ ionisation energy of H(g)	x_3
+ electron affinity of $X(g)$	x_4
+ free energy of hydration of H^+ and X^-	x_5

HX (hydrated) $\longrightarrow HX(g) + aq$ $\quad \Delta H = x_1$

$HX(g) \longrightarrow H(g) + X(g)$ $\quad = x_2$

$H(g) \longrightarrow H^+(g) + e^-$ $\quad = x_3$

$X(g) + e^- \longrightarrow X^-(g)$ $\quad = x_4$

$H^+(g) + X^-(g) + aq \longrightarrow$ H^+ (hydrated) + X^- (hydrated) $\quad = x_5$

HX (hydrated) $\longrightarrow$ H^+ (hydrated) + X^-(hydrated)

$$\Delta H = x_1 + x_2 + x_3 + x_4 + x_5$$

Due to maximum value of x_1 and x_2 for HF due to H-bonding, net ΔG (free energy change) for HF comes out to be positive which disfavours dissociation; (–ve sign of ΔG favours spontaneous dissociation, +ve sign disfavours dissociation).

$$\Delta G_f^\circ \; \underset{+ve}{HI} < HBr < HCl < \underset{-ve}{HF}$$

- Dipole moment of HF is maximum and is in order

$$\underset{1.91D}{HF} > \underset{1.05D}{HCl} > \underset{0.80D}{HBr} > \underset{0.42D}{HI}$$

- Liquid HF has been used as a non-aqueous solvent. It undergoes self ionisation.

$$2HF \rightleftharpoons [H_2F]^+ + F^-$$

DEEP Focus

Acid-base reactions occur in this solvent system. **However, the solvent itself has a very strong tendency to donate protons. Mineral acids like HNO_3, H_2SO_4 and HCl, when dissolved in HF, behave as bases in HF as they accept proton from HF :**

$$\underset{base}{HCl} + \underset{acid}{HF} \rightleftharpoons \underset{acid}{H_2Cl^+} + \underset{base}{F^-}$$

$$\underset{base}{H_2SO_4} + \underset{acid}{HF} \rightleftharpoons \underset{acid}{H_3SO_4^+} + \underset{base}{F^-}$$

$$\underset{base}{HNO_3} + \underset{acid}{HF} \rightleftharpoons \underset{acid}{H_2NO_3^+} + \underset{base}{F^-}$$

$HClO_4$ (perchloric acid) however behaves as an acid in liquid HF which acts like a base :

$$\underset{acid}{HClO_4} + \underset{base}{HF} \rightleftharpoons \underset{acid}{H_2F^+} + \underset{base}{ClO_4^-}$$

Oxoacids of Halogen

Five series of oxoacids are known and are summarised in Table 15.5.

Table 15.5 **Oxoacids of Halogen**

	HOX hypohalous acid	HXO_2 halous acid	HXO_3 halic acid	HXO_4 perhalic acid	H_5XO_6 para perhalic acid
	HOF	—	—	—	—
	HOCl	$HClO_2$	$HClO_3$	$HClO_4$	—
	HOBr	—	$HBrO_3$	$HBrO_4$	—
	HOI	—	HIO_3	HIO_4	H_5IO_6
Anhydride	X_2O	X_2O_3	X_2O_5	X_2O_7	I_2O_7
Oxidation state of X	+I	+III	+V	+VII	+VII
Oxidising power	strong ←			weak	
Acid strength*	weak		→	strong	
Thermal stability	weak		→	good	
Hydrolysis of anion	max ←			min	
Structure of anion	(OCl^-) linear	(ClO_2^-) angular	(ClO_3^-) trigonal pyramidal	(ClO_4^-) tetrahedral	

*Acid strength of one type oxoacid increases as size of halogen atom increases

HOX series

- $+I$ oxidation state in HOF appears very doubtful but this has been prepared by trapping F_2 and H_2O in an unreactive matrix of solid N_2 (at very low temperature) and photolysing the gases :

$$F_2 + H_2O \xrightarrow{h\nu} HOF + HF$$

It has been recently prepared by passing F_2 over ice at 0°C and removing the product into a cold trap:

$$H_2O + F_2 \xrightleftharpoons{-40°C} HOF + HF$$

HOCl, HOBr and HOI are not very stable and are known only in aqueous solution.

– HOCl is formed when Cl_2 is dissolved in H_2O. Cl_2, Br_2 and I_2 all dissolve to some extent in water forming hydrated X_2 molecules, and X^- and OX^- ions :

$$H_2O + Cl_2 \longrightarrow HOCl + HCl$$

(Bleaching action of chlorine water is due to formation of HOCl)

– HOX are also formed by shaking X_2 with freshly precipitated HgO in water :

$$2HgO + 2Cl_2 + H_2O \longrightarrow HgO \cdot HgCl_2 + 2HOCl$$

The solution, after filtration, is distilled whereby the dilute acid passes over. The maximum concentration which can be obtained is only 25 per cent. In the process of distillation, the acid decomposes into its anhydride, Cl_2O.

$$2HOCl \longrightarrow H_2O + Cl_2O$$

- Salts of HOX are called **hypohalites** and are bleaching agent. NaOCl is formed due to electrolysis of cold NaCl (brine) solution :

$$NaCl(aq) \rightleftharpoons Na^+ + Cl$$

at anode :   **at cathode :**

$$2Cl^- \longrightarrow Cl_2 + 2e^- \;:\; Na^+ + e^- \longrightarrow Na$$

$$: 2Na + 2H_2O \longrightarrow 2NaOH + H_2$$

If anode and cathode are not separated, NaOH and Cl_2 mix forming NaOCl.

$$2NaOH + Cl_2 \longrightarrow NaCl + NaOCl + H_2O$$

* Bleaching powder $CaOCl_2$ is formed when Cl_2 is passed into moist lime CaO. $CaOCl_2$ has Ca^{2+}, OCl^- and Cl^- ions

$$(CaO + H_2O) + Cl_2 \longrightarrow CaOCl_2 + H_2O$$

DEEP Focus

* OCl^- disproportionates in hot solution

$$3OCl^- \xrightarrow{hot} 2Cl^- + ClO_3^-$$

ON of Cl: +1 → −1 (red.), +1 → +5 (oxid.)

* Fe^{2+} is oxidised to Fe^{3+}

$$HOCl + 2Fe^{2+} + H^+ \longrightarrow Cl^- + 2Fe^{3+} + H_2O$$

* OCl^- also oxidises I^- to I_2 in acidic medium.

$$OCl^- + 2I^- + 2H^+ \longrightarrow H_2O + Cl^- + I_2$$

HXO$_2$ series

- Only stable acid known is $HClO_2$ and exists in solution :

$$Ba(ClO_2)_2 + H_2SO_4 \longrightarrow BaSO_4 + 2HClO_2$$

- Salts are called **chlorites** and are made from :

$$2ClO_2 + 2NaOH \longrightarrow NaClO_2 + NaClO_3 + H_2O$$

$$2ClO_2 + Na_2O_2 \longrightarrow 2NaClO_2 + O_2$$

- It is stable in alkaline solution but disproportionates in acid solution :

$$\underset{+3}{5HClO_2} \longrightarrow \underset{+4}{4ClO_2} + \underset{-1}{HCl} + 2H_2O$$

and $$4HClO_2 \longrightarrow 2ClO_2 + HClO_3 + HCl + H_2O$$

HXO$_3$ series

- $HClO_3$, $HBrO_3$, HIO_3 are known of which most stable is HIO_3. Salts like $NaClO_3$, $NaBrO_3$, $NaIO_3$ are stable :

$$Ba(ClO_3)_2 + H_2SO_4 \longrightarrow BaSO_4 + 2HClO_3$$

After reaction, $BaSO_4$ is removed by filtration, and the filtrate is evaporated in vacuum till 40 per cent solution is obtained. However, further concentration by evaporation leads to decomposition :

$$3HClO_3 \longrightarrow HClO_4 + Cl_2 + 2O_2 + H_2O$$

Similarly for $HBrO_3$.

HIO_3 is formed by oxidation of I_2 with concentrated HNO_3 or O_3.

$$8H^+ + 10NO_3^- + I_2 \longrightarrow 2IO_3^- + 10NO_2 + 4H_2O$$

Chlorates, etc. may be made :

– by passing Cl_2 into hot alkali solution.

Thus, $$2NaOH + Cl_2 \xrightarrow{\text{cold}} NaCl + NaOCl + H_2O$$

$$6NaOH + 3Cl_2 \xrightarrow{\text{hot}} 5NaCl + NaClO_3 + 3H_2O$$

$$2OH^- + Cl_2 \xrightarrow{\text{cold}} Cl^- + OCl^- + H_2O$$

$$6OH^- + 3Cl_2 \xrightarrow{\text{hot}} 5Cl^- + ClO_3^- + 3H_2O$$

In this Cl_2 undergoes disproportionation.

– by electrolysing hot halide solutions (with vigorous stirring)

$$2Cl^- + 2H_2O \xrightarrow{\text{electrolysis}} \underset{\text{at anode}}{Cl_2} + \underbrace{2OH^- + H_2}_{\text{at cathode}}$$

$$6NaOH + 3Cl_2 \longrightarrow 5NaCl + NaClO_3 + 3H_2O$$

- $HClO_3$ on evaporation gives ClO_2 :

$$4HClO_3 \longrightarrow 4ClO_2 + 2H_2O + O_2$$

- $HClO_3$ oxidises SO_2 to SO_3

$$HClO_3 + 3SO_2 \longrightarrow HCl + 3SO_3$$

- IO_3^- oxidises I^- to I_2 :

$$IO_3^- + 5I^- + 6H^+ \longrightarrow 3I_2 + 3H_2O$$

- ClO_3^- oxidises SO_3^{2-} to SO_4^{2-} :

$$ClO_3^- + 3SO_3^{2-} \longrightarrow Cl^- + 3SO_4^{2-}$$

- $$2KClO_3 \xrightarrow[MnO_2]{400\text{–}500^\circ C} 2KCl + 3O_2$$

Following disproportionation of ClO_3^- also takes place as side reaction at low temperature and in the absence of catalyst :

$$\underset{+5}{4KClO_3} \longrightarrow \underset{-1}{KCl} + \underset{+7}{3KClO_4}$$

- Chlorates are used to make fireworks and matches.

HXO$_4$ series

- $HClO_4$ and HIO_4 and their salts are known.

$$NH_4ClO_4 + HNO_3 \longrightarrow HClO_4 + NH_4NO_3$$

$$4HClO_3 \xrightarrow{\Delta} 3HClO_4 + HCl$$

$HClO_4$ is commercially available as 70% $HClO_4$ and is dihydrate **$HClO_4 \cdot 2H_2O$** which can be dehydrated to $HClO_4$ by oleum :

$$HClO_4 \cdot 2H_2O + \underset{\text{oleum}}{H_2S_2O_7} \longrightarrow HClO_4 + 2H_2SO_4 + H_2O$$

$$NaClO_3 + H_2O \xrightarrow{\text{Electrolysis}} NaClO_4 + H_2$$

Oxidation of ClO_3^- occurs at a Pt anode through the half-reaction :

$$ClO_3^- + H_2O \longrightarrow ClO_4^- + 2H^+ + 2e^-$$

- $HClO_4$ is the strongest of all other oxoacids of Cl :

$$HClO < HClO_2 < HClO_3 < HClO_4$$

DEEP Focus

- The dissociation of an oxoacid involves two energy terms : breaking an O—H bond to produce an anion and H^+ ion

$$HOX \rightleftharpoons H^+ + OX^- \text{ and}$$

– hydrating both ions.

– ClO_4^- ion is larger than OCl^- so the hydration energy of OCl^- is larger than ClO_4^-. Also electronegativity of oxygen is greater than that of chlorine. In HOCl, $HClO_2$, $HClO_3$, $HClO_4$, an increasing number of oxygen atoms are bonded to the chlorine atom. **The more oxygen atoms that are bonded, the more the electrons will be pulled away from the O—H bond, and the more this bond is weakened. Thus, greater the number of oxygen bonds attached to halogen, least is the energy required to break O—H bond, hence greater the acid strength**

$$HClO < HClO_2 < HClO_3 < HClO_4$$

- Metals like Zn, Mg and Fe dissolve in dilute solutions of the acid forming their perchlorates:

$$Zn + 2HClO_4 \longrightarrow Zn(ClO_4)_2 + H_2\uparrow.$$

Hot concentrated acid (73%) behaves as a remarkable oxidising agent:

$$4HClO_4 \longrightarrow 2Cl_2 + 7O_2 + 2H_2O$$

- $Mg(ClO_4)_2$ is used in **dry batteries** and is also a effective desicant called **anhydrone**. $KClO_4$ is used **in fireworks and flares.**

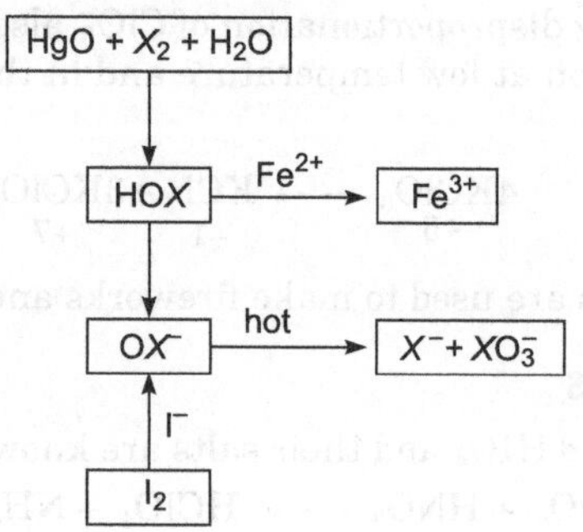

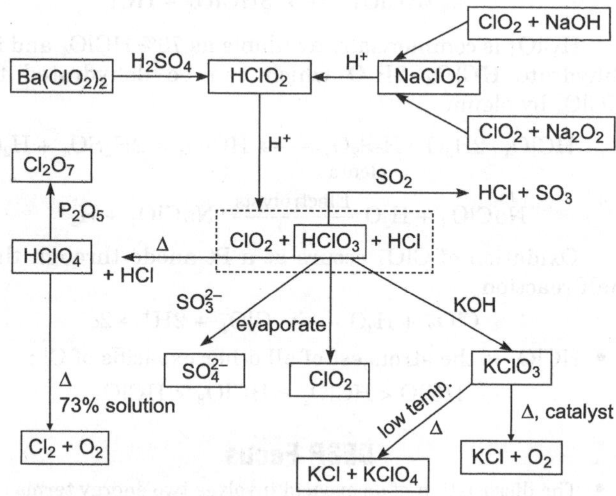

Bleaching Powder

- Commonly it is written as $CaOCl_2$ with one Cl^- and one OCl^- (hypochlorite) ions **(Cl with two different oxidation number)** and is thus calcium chloro hypochlorite

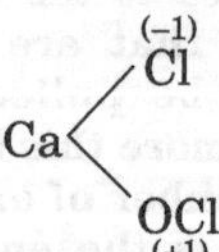

- Its actual composition is $Ca(OCl)_2 \cdot CaCl_2 \cdot Ca(OH)_2 \cdot 2H_2O$ containing 35% **available** chlorine. It is obtained by action of Cl_2 gas on slaked lime :

$$3Ca(OH)_2 + 2Cl_2 \longrightarrow Ca(OCl)_2 \cdot CaCl_2 \cdot Ca(OH)_2 \cdot 2H_2O$$

DEEP Focus

"Available chlorine" content is defined as the weight of Cl_2 which liberates the same amount of I_2 from HI as does the given weight of the compound; it is expressed as a percentage.

Available chlorine can be determined by treating aqueous solution of bleaching powder with KI and titrating liberated I_2 using $Na_2S_2O_3$ solution (hypo). CH_3COOH is also added to neutralise $Ca(OH)_2$ formed.

$$CaOCl_2 + H_2O \longrightarrow Cl_2 + Ca(OH)_2$$

$$Cl_2 + 2KI \longrightarrow 2KCl + I_2$$

$$I_2 + 2Na_2S_2O_3 \longrightarrow 2NaI + Na_2S_4O_6$$

Thus, $CaOCl_2 \equiv Cl_2 \equiv I_2 \equiv 2Na_2S_2O_3$

$$\text{Eq. wt. of } CaOCl_2 = \left(\frac{\text{molecular weight}}{2}\right)$$

- Pale yellow solid with odour of Cl_2; in aqueous solution it gives Cl^-, OCl^- and Ca^{2+} ions :

$$CaOCl_2 \xrightleftharpoons{H_2O} Ca^{2+} + Cl^- + OCl^-$$

- It undergoes disproportionation on standing:

$$6CaOCl_2 \longrightarrow Ca(ClO_3)_2 + 5CaCl_2$$

- When CO_2 is passed into suspension of bleaching powder in water, HOCl is formed

$$2CaOCl_2 + H_2O + CO_2 \longrightarrow CaCl_2 + CaCO_3 + 2HOCl$$

- It is a **bleaching agent** as well as **oxidising agent** due to oxygen formed in acidic medium/neutral medium :

$$CaOCl_2 + H_2O \longrightarrow Ca(OH)_2 + Cl_2$$

$$Cl_2 + H_2O \longrightarrow 2HCl + O$$

$$CaOCl_2 + HCl \longrightarrow CaCl_2 + HOCl$$

$$HOCl \longrightarrow HCl + O$$

- $CaOCl_2$ oxidises
  - **H_2S into S**

$$CaOCl_2 + H_2S \longrightarrow CaCl_2 + H_2O + S$$

  - **$FeSO_4$ into $Fe_2(SO_4)_3$**

$$CaOCl_2 + 2FeSO_4 + H_2SO_4 \longrightarrow Fe_2(SO_4)_3 + CaCl_2 + H_2O$$

  - **NH_3 into N_2**

$$3CaOCl_2 + 2NH_3 \longrightarrow 3CaCl_2 + 3H_2O + N_2$$

  - **KNO_2 into KNO_3 and Na_3AsO_3 into Na_3AsO_4**

$$KNO_2 + CaOCl_2 \longrightarrow KNO_3 + CaCl_2$$

  - **Carbonyl compounds ($CH_3—\overset{O}{\overset{\|}{C}}—H$, $—\overset{O}{\overset{\|}{C}}—CH_3$) and alcohols ($—\underset{OH}{\underset{|}{CH}}—CH_3$) are oxidised to chloroform.**

$$CaOCl_2 + H_2O \longrightarrow Ca(OH)_2 + Cl_2$$
$$Cl_2 + H_2O \longrightarrow 2HCl + O$$
$$CH_3CH_2OH \xrightarrow{O} CH_3CHO + H_2O$$
$$CH_3CHO + 3Cl_2 \longrightarrow CCl_3CHO + 3HCl$$
$$2CCl_3CHO + Ca(OH)_2 \longrightarrow (HCOO)_2Ca + 2CHCl_3$$

Similarly for CH_3CHO, $RCOCH_3$. This test, called **'Halform test'**, is given by

- CH_3CHO
- CH_3CH_2OH
- all 2° alcohols (at C_2) $-CH(OH)-CH_3$
- all ketones (with carbonyl group at C_2 as $CH_3-C(=O)-$)

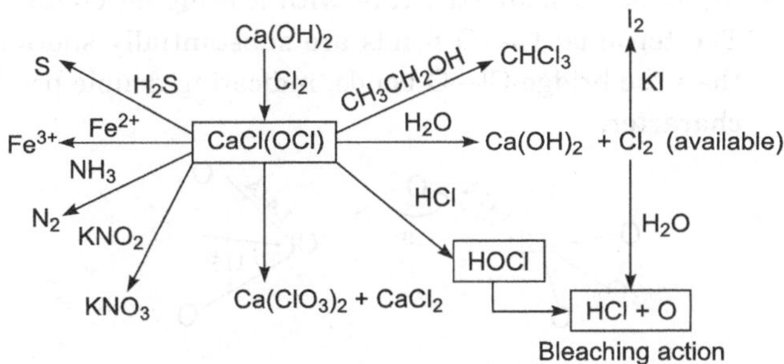

Oxides of Halogen

Following oxides of chlorine are known.

Formula	Name	Oxyacid	Oxidation state of chlorine
Chlorine monoxide	Cl_2O	HOCl	+1
Chlorine dioxide	ClO_2	$HClO_2$	+3
Chlorine hexoxide	Cl_2O_6	$HClO_3$	+5
Chlorine heptoxide	Cl_2O_7	$HClO_4$	+7

Cl_2O (Chlorine monoxide)

- When Cl_2 gas is passed over HgO at 300–500°C, Cl_2O is formed.

$$2Cl_2 + 2HgO \longrightarrow HgO \cdot HgCl_2 + Cl_2O\uparrow$$

Reaction may also be carried out with chlorine dissolved in CCl_4.

- Cl_2O is a brownish-yellow gas which condenses to an orange liquid (b.p. 2°C)
- It is thermally unstable and explodes on heating

$$2Cl_2O \longrightarrow 2Cl_2 + O_2$$

- It is soluble in water giving orange yellow coloured solution of HOCl

$$Cl_2O + H_2O \longrightarrow 2HOCl$$

- It forms salt with alkali solution indicating its acidic character.

$$Cl_2O + 2KOH \longrightarrow 2KOCl + H_2O$$

- It is an oxidising agent

$$3Cl_2O + 10NH_3 \longrightarrow 2N_2 + 6NH_4Cl + 3H_2O$$

- Structure is angular with sp^3 hybridised O-atom and is related to a tetrahedron with two positions occupied by lone pairs of electrons.

Cl–O–Cl: O–Cl 1.71Å, angle 111°

DEEP Focus

- The bond angle varies in the order FOF < ClOCl < BrOBr. This is because electrons in the case of OF_2 are nearer to fluorine due to high electronegativity of F compared to Cl to Br. The bonded electron pairs in Cl_2O and Br_2O are closer to oxygen making the repulsion between them more and thereby reducing the lone pair-lone pair repulsion on oxygen to some extent. Also due to the bulkiness of Cl and Br, the angles ClOCl and BrOBr increase to such an extent that 109°28′, (the tetrahedral angle), is approached.

ClO_2 (Chlorine dioxide)

- ClO_2 (as yellow-coloured gas with unpleasant smell) is obtained when $KClO_3$ reacts with conc. H_2SO_4 or oxalic acid

$$3KClO_3 + 3H_2SO_4 \longrightarrow HClO_4 + 3KHSO_4 + H_2O + 2ClO_2\uparrow$$

$$2KClO_3 + \underset{\text{oxalic acid}}{2H_2C_2O_4} \xrightarrow{60°C} K_2C_2O_4 + 2H_2O + 2CO_2 + 2ClO_2\uparrow$$

- $$2AgClO_3 + Cl_2 \xrightarrow{90°C} 2AgCl\downarrow + 2ClO_2\uparrow + O_2\uparrow$$
- Commercially it is prepared by passing SO_2 gas into a mixture of $NaClO_3$ and H_2SO_4 having NaCl in traces.

$$2NaClO_3 + SO_2 + H_2SO_4 \xrightarrow[\text{NaCl}]{\text{trace}} 2ClO_2 + 2NaHSO_4$$

- Aqueous solution of gas is neutral and is a strong oxidising agent :

$$10HI + 2ClO_2 \longrightarrow 2HCl + 4H_2O + 5I_2$$

DEEP Focus

- ClO_2 reacts with KOH forming $KClO_3$ and $KClO_2$.

$$2KOH + 2ClO_2 \longrightarrow KClO_2 + KClO_3 + H_2O$$

This reaction indicates that ClO_2 is a mixed anhydride of $HClO_2$ and $HClO_3$.

- It is strong oxidising agent and chlorinating agent :

$$Mg + 2ClO_2 \longrightarrow Mg(ClO_2)_2$$
$$Al + 3ClO_2 \longrightarrow Al(ClO_2)_3$$

These metals in powdered form in aqueous suspension are oxidised.

- Ozone oxidises ClO_2 to Cl_2O_6

$$2ClO_2 + 2O_3 \longrightarrow Cl_2O_6 + 2O_2$$

- Alkaline H_2O_2 is oxidised to O_2

$$2ClO_2 + 2NaOH + H_2O_2 \longrightarrow 2NaClO_2 + O_2 + 2H_2O$$

- Cl atom is sp^3 hybridised in the angular molecule with bond angle of 118° and Cl—O bond length of 1.47Å. It has odd number of electrons and is **thus paramagnetic** and the stability is due to the presence of 3-electron bonds stabilised by resonance. Molecules containing odd number of electrons are reactive and generally dimerise in order to pair the electrons, but ClO_2 is the exception.

Cl_2O_6 or ClO_3 [Chlorine hexoxide (trioxide)]

- When ClO_2 reacts with ozonised O_2 at 0°C, Cl_2O_6 is obtained.

$$2ClO_2 + 2O_3 \longrightarrow Cl_2O_6 + 2O_2$$

- Cl_2O_6 is unstable red oil (m.p. 3.5°C) and explodes on heating.
- It is a strong oxidising agent

$$Cl_2O_6 + 2NaOH + H_2O_2 \longrightarrow 2NaClO_3 + O_2 + 2H_2O$$

- Anhydrous HF reversibly gives $FClO_2$ and $HClO_4$

$$Cl_2O_6 + HF \rightleftharpoons FClO_2 + HClO_4$$

- $$NO + Cl_2O_6 \longrightarrow ClO_2 + [NO]^+[ClO_4]^-$$
$$NO_2 + Cl_2O_6 \longrightarrow ClO_2 + [NO_2]^+[ClO_4]^-$$

DEEP Focus

- When Cl_2O_6 (or ClO_3) reacts with KOH, $KClO_3$ and $KClO_4$ are formed indicating that Cl_2O_6 is a mixed anhydride of $HClO_3$ and $HClO_4$

$$2KOH + Cl_2O_6 \longrightarrow KClO_3 + KClO_4 + H_2O$$

- In the liquid state and in aqueous solution, there is paramagnetism indicating existence of ClO_3 formed due to dissociation of Cl_2O_6

$$Cl_2O_6 \rightleftharpoons 2ClO_3$$

Paramagnetism is due to unpaired electrons in ClO_3, Cl_2O_6 is diamagnetic in solid state which is due to paired electrons.

Solid has the ionic composition $ClO_2^+ \; ClO_4^-$ confirmed by X-ray study. The liquid may have either of two structures :

Cl_2O_7 (Chlorine heptoxide)

- When $HClO_4$ is added to P_2O_5 at –10°C, Cl_2O_7 is obtained which separates on distilling the mixture at 85°C.

$$2HClO_4 + P_2O_5 \longrightarrow Cl_2O_7 + 2HPO_3$$

- Cl_2O_7 is colourless, oily liquid (b.p. 82°C)
- Cl_2O_7 is extremely explosive.
- $$Cl_2O_7 + H_2O \rightleftharpoons 2HClO_4$$
- Cl_2O_7 reacts with alkali forming perchlorates.

$$Cl_2O_7 + 2KOH \longrightarrow 2KClO_4 + H_2O$$

- Cl_2O_7 has a bent structure with a bridging O-atom. The terminal Cl—O bonds are substantially shorter than the bridge Cl—O bonds, indicating double bond character.

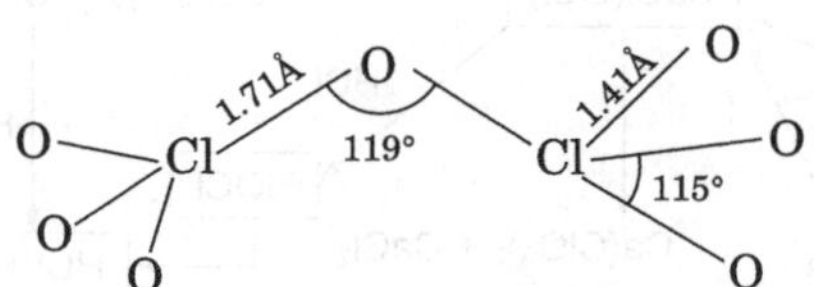

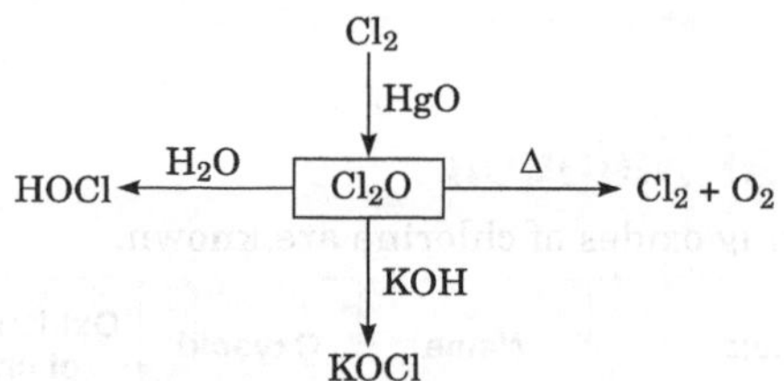

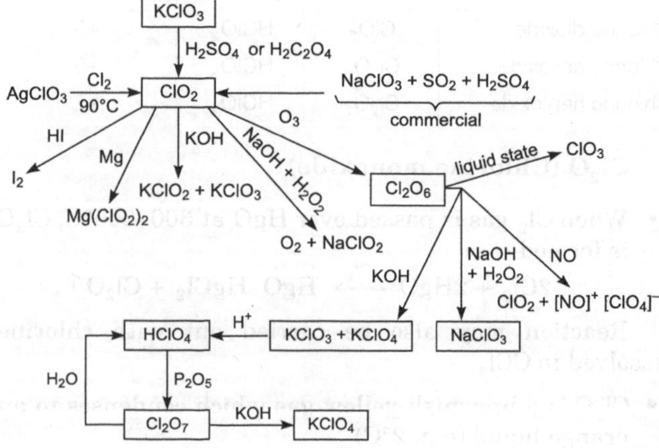

Target Practice 2

1. Arrange HOCl, $HClO_2$, $HClO_3$ and $HClO_4$ in increasing
   (a) oxidation state of halogen atom ....
   (b) oxidising power ....
   (c) acid strength ....
   (d) thermal stability
   (e) *s*-character in hybridised state ....
2. What are the anhydrides of
   (a) HOCl ....
   (b) $HClO_2$ ....
   (c) $HClO_3$ ....
   (d) $HClO_4$ ....
3. Mixture of KBr and $KBrO_3$ in acidic medium can be used as brominating agent. Explain by chemical reaction
   ....
   ....
4. $KClO_3$ is used ....
   $KClO_4$ is used as ....
   $Mg(ClO_4)_2$ is used as ....
5. In aqueous solution bleaching powder reacts as
   $$CaOCl_2 + H_2O \longrightarrow Cl_2 + Ca(OH)_2$$
   (a) Cl_2 thus, obtained is called ....
   (b) Equivalent weight of $CaOCl_2$ is ....
   (c) What is the type of above reaction? ....
6. Which of the following pairs can be distinguished by **iodoform test?**
   (a) CH_3CHO and CH_3COCH_3
   (b) CH_3CHO and CH_3CH_2OH
   (c) CH_3OH and CH_3CH_2OH
   (d) HCHO and CH_3CHO
7. Calculate magnetic moment of Cl_2O_6 and ClO_3.
   (a) Cl_2O_6 .... BM
   (b) ClO_3 .... BM
8. Consider bond lengths x and y and bond angles α and β in the following and compare these values

   O, Cl, O, O, x, α, Cl, y O, O, β, O

   Bond lengths : $x \square y$
   Bond angles : $\alpha \square \beta$

Answers

1. (a) HOCl < $HClO_2$ < $HClO_3$ < $HClO_4$
   (b) HOCl < $HClO_2$ < $HClO_3$ < $HClO_4$
   (c) HOCl < $HClO_2$ < $HClO_3$ < $HClO_4$
   (d) HOCl< $HClO_2$ < $HClO_3$ < $HClO_4$
   (e) Equal (25%) each Cl-atom is sp^3 hybridised.
2. (a) Cl_2O (b) ClO_2 (c) Cl_2O_6 (d) Cl_2O_7
3. $5Br^- + BrO_3^- + 6H^+ \longrightarrow 3Br_2 + 3H_2O$
4. (a) an explosive, (b) fireworks and flares
   (c) dry batteries
5. available chlorine
   (b) $CaOCl_2 \equiv 2Cl_2 \equiv 2Cl$
   $CaOCl_2 \equiv 2Cl$
   $$\therefore \text{ Equivalent weight} = \frac{\text{Molecular weight}}{2}$$
   (c) There are two types of Cl in $CaOCl_2$
   Cl^- (oxidation number = – 1) oxidised to Cl_2
   ClO^- (oxidation number = + 1) reduced to Cl_2
   Thus, it is a intramolecular redox reaction
6. (a) No (b) No (c) Yes (d) Yes
7. (a) Zero BM (no unpaired electron)
   (b) $\sqrt{3}$ BM (with one unpaired electron)
8. (a) $x > y$ (b) $\alpha > \beta$

Practice Exercise 2

1. Complete following reactions :
   (a) $ClO_3^- + I^- + 6H^+ \longrightarrow$ (b) $ClO_3^- + Fe^{2+} + H^+ \longrightarrow$
   (c) $Zn(ClO_3)_2 \xrightarrow{\Delta}$ (d) $O_3 + I_2 + H_2O \longrightarrow$
   (e) $KBrO_3 + F_2 + KOH \longrightarrow$
2. Which is **stronger acid** in each pair ?
   (a) HClO $HClO_2$ (b) H_2SO_4 H_2SO_3
   (c) H_3PO_2 H_3PO_3 (d) HNO_3 HNO_2
   (e) HF HCl (f) H_3PO_3 H_3PO_4
3. C_4H_8O gives iodoform test. Identify it.
4. $C_8H_8O\,(A) \xrightarrow[\Delta]{CaOCl_2/H_2O} CHCl_3 + (B)$
   $(B) \xrightarrow{H^+}$ C_6H_5–COOH
   What are (*A*) and (*B*)?
5. 10 g of bleaching powder is dissolved in water containing some CH_3COOH and diluted to 1 L. 20 mL of this solution on reaction with KI required 20 mL of 0.1 N $Na_2S_2O_3$ solution. Calculate percentage of available chlorine in it.
6. $CaOCl_2$ in aqueous solution changes to Cl_2. What is the type of this change ?
7. Give structure of oxoacids of halogens.
8. What are the corresponding anhydrides of the following oxyacids,
   $HClO_4$, $HClO_3$, $HClO_2$ and HOCl
9. Identify *A*, *B* and *C* in the following

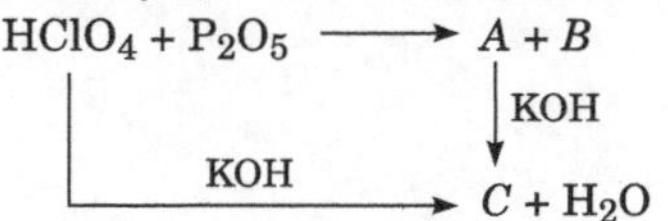

Answers

1. (a) $ClO_3^- + 6I^- + 6H^+ \longrightarrow Cl^- + 3I_2 + 3H_2O$
(b) $ClO_3^- + 6Fe^{2+} + 6H^+ \longrightarrow Cl^- + 6Fe^{3+} + 3H_2O$
(c) $2Zn(ClO_3)_2 \xrightarrow{\Delta} 2ZnO + 2Cl_2 + 5O_2$
(d) $3O_3 + I_2 + H_2O \longrightarrow 2IO_3^- + 2O_2 + 2H^+$
(e) $KBrO_3 + F_2 + 2KOH \longrightarrow KBrO_4 + 2KF + H_2O$

2. (a) $HClO_2$ (b) H_2SO_4 (c) almost same (d) HNO_3
(e) HCl (f) almost same

In H_3PO_2, H_3PO_3 and H_3PO_4, terminal O is one, and electronegativity values of P and H are almost same, hence, acid strengths of these acids are same.

3. $CH_3—\overset{O}{\overset{\|}{C}}—CH_2—CH_3$ (it should have $CH_3—\overset{O}{\overset{\|}{C}}—$ group of the given carbonyl formula)

4. (*A*): $C_6H_5—\overset{O}{\overset{\|}{C}}—CH_3$ (*B*): $C_6H_5—\overset{O}{\overset{\|}{C}}—O^-$

5. $CaOCl_2 \equiv Cl_2 \equiv I_2 \equiv 2Na_2S_2O_3$
$CaOCl_2 \equiv Cl_2 \equiv I_2 \equiv 0.1N\ (N_1V_1 = N_2V_2)$
$Cl_2 = 3.55\ g\ L^{-1}$ in $10\ g\ L^{-1}$ bleaching powder.
Available $Cl_2 = 35.5\%$

6. $Ca\begin{cases}\overset{(-1)}{Cl}\\ \underset{(+1)}{OCl}\end{cases} \xrightarrow{H_2O} \overset{0}{Cl_2} + Ca(OH)_2$

Cl^- (alone) is oxidised to Cl_2 and Cl (of OCl) is reduced to Cl_2. Hence, it is a redox reaction.

7. (a) $H—O—\ddot{\underset{..}{Cl}}:$ (b) $H—O—\ddot{\underset{..}{Cl}}: \longrightarrow O$

(c) $H—O—\ddot{Cl} \rightarrow O$ with $\uparrow O$ above and $\downarrow O$ below (d) $H—O—Cl \longrightarrow O$ with $\uparrow O$ above and $\downarrow O$ below

In all cases Cl atom is sp^3 hybridised.

8. $HClO_4 : Cl_2O_7$; $HClO_3 : Cl_2O_6$; $HClO_2 : ClO_2$;
$HOCl : Cl_2O$

9. (*A*) : Cl_2O_7 (B) : HPO_3 (C) : $KClO_4$.

Total Practice Set

(Read, Plan & Solve!)

Problem 1. An important oxidation-reduction reaction used to determine trace amounts of O_3 and other oxidants in polluted air involves oxidation of iodide ion :

$$2I^- + O_3 + H_2O \longrightarrow 2OH^- + I_2 + O_2$$

The solution is acidified and the I_2 is titrated with sodium thiosulphate solution using starch indicator. One litre of a mixture of O_3 and O_2 at STP was allowed to react with an excess of acidified solution of KI and I_2 liberated required 40 mL of 0.1 M sodium thiosulphate solution. What is the percentage of ozone in the mixture?

Solution $2I^- + O_3 + H_2O \longrightarrow 2OH^- + I_2 + O_2$

$$I_2 + 2Na_2S_2O_3 \longrightarrow Na_2S_4O_6 + 2NaI$$

From the above reaction :

$$2Na_2S_2O_3 \equiv I_2 \equiv O_3$$

$$\text{moles of } Na_2S_2O_3 = \frac{40 \times 0.1}{1000} = 0.004$$

$$\text{moles of } I_2 = \text{moles of } O_3 = 0.002$$

$$\text{volume of } O_3 \text{ at STP} = 0.002 \times 22.4 \text{ L}$$

$$= 0.0448 \text{ L present in 1 L mixture}$$

$\therefore$ Percentage of O_3 = **4.48% (by volume)**

Problem 2. Although fluorite, CaF_2, is its chief mineral source, fluorine can also be obtained as a by-product of the production of phosphate fertilisers. These fertilisers are derived from phosphate rock $[3Ca_3(PO_4)_2 \cdot CaF_2]$. What is the maximum mass of fluorine that could be extracted as by-product from 1000 kg of phosphate rock? (Ca = 40, P = 31, F = 19)

Solution $3\,Ca_3(PO_4)_2 \cdot CaF_2 \equiv 1\,F_2$

$$3(120 + 190) + (40 + 38) \equiv 1 \times 38$$

1008 g phosphate rock gives = 38 g F_2

1000 g phosphate rock gives = 37.698 g F_2

1000 kg phosphate rock gives = **37.698 kg F_2**

Problem 3. In neutral solution, one mole of periodate ion (IO_4^-) reacts with excess of iodide ion to produce one mole of iodine; on acidification of the resulting solution, a further three moles of iodine are liberated. Derive equations for the reactions which occur under these conditions.

Solution $IO_4^- + 2I^- + H_2O \longrightarrow IO_3^- + I_2 + 2OH^-$

$$IO_3^- + 5I^- + 6H^+ \longrightarrow 3I_2 + 3H_2O$$

Problem 4. Describe in outline how you would attempt to determine the equilibrium constant and standard enthalpy change for the reaction in aqueous solution.

$$Cl_2 + H_2O \rightleftharpoons HCl + HOCl$$

Solution This question is based on the study of physical aspects of the equilibrium; equilibrium constant is determined by using equation,

$$K = \frac{[HCl][HOCl]}{[Cl_2]}$$

(H_2O as reactant is not considered in equilibrium constant expression).

When equilibrium has been attained [HCl] = [HOCl]

Concentration of acid is determined by pH measurement:

$$pH = -\log [H^+]$$

To determine concentration of Cl_2, reactions mixture is treated with KI (adding some $NaHCO_3$ to neutralise acid) and titrating liberated I_2 against standard $Na_2S_2O_3$ (sodium thiosulphate) using starch indicator.

$$Cl_2 + 2KI \longrightarrow 2KCl + I_2$$

$$\underset{\text{sodium thiosulphate}}{I_2 + 2Na_2S_2O_3} \longrightarrow 2NaI + \underset{\text{sodium tetrathionate}}{Na_2S_4O_6}$$

Based on the above reactions,

$$2Na_2S_2O_3 \equiv I_2 \equiv Cl_2$$

Hence, form the concentration of HCl, HOCl and Cl_2 equilibrium constant can be determined.

To determine ΔH (standard enthalpy change) we measure value of equilibrium constant K at different temperature and using van't Hoff equation,

$$\Delta H = \frac{2.303RT_1T_2}{(T_2 - T_1)} \log \frac{K_2}{K_1}$$

Problem 5. Describe in outline how you would attempt to show that oxide I_4O_9, reported to be formed by the action of ozone on iodine, reacts with water according to the equation,

$$5I_4O_9 + 9H_2O \longrightarrow 18HIO_3 + I_2$$

Solution Neutralise the solution of a weighted amount of the oxide with $NaHCO_3$ and titrate the I_2 against $Na_2S_2O_3$ (sodium thiosulphate) solution,

$$I_2 + 2Na_2S_2O_3 \longrightarrow 2NaI + Na_2S_4O_6$$

There is consumption of $Na_2S_2O_3$ due to formation of I_2, hence, I_4O_9 reacts with H_2O to give I_2.

Problem 6. In a strongly alkaline solution containing excess of barium ions, a solution containing 0.01587 g of I^- was treated with 0.1 M MnO_4^- until a pink colour persisted in the solution. 10.0 mL of the MnO_4^- solution were required. Under these conditions MnO_4^- is converted into the sparingly soluble $BaMnO_4$. What is the product of oxidation of the iodide? (I = 127)

Solution Let oxidation product of I^- by $MnO_4^- = IO_x^-$

Oxidation number of I in $IO_x^- = n = (2x - 1)$

MnO_4^- is reduced to MnO_4^{2-} in the oxidation of I^-. Thus,

$$\underset{-1}{I^-} + \underset{+7}{MnO_4^-} \longrightarrow \underset{+6}{MnO_4^{2-}} + \underset{(2x-1)}{IO_x^-}$$

reduction: change in ON = 1

oxidation: change in ON = $2x$

Thus, $I^- + 2x\, MnO_4^- \longrightarrow 2x\, MnO_4^{2-} + IO_x^-$

$$1 \text{ mol } I^- = 2x \text{ mol of } MnO_4^-$$

$$\frac{0.01587}{127} \text{ mol } I^- = \frac{2x \times 0.01587}{127} \text{ mol } MnO_4^- = \frac{10 \times 0.1}{1000}$$

$$x = 4$$

Thus, oxidation product of I^- is IO_4^- and

$$\mathbf{I^- + 8MnO_4^- + 8OH^- \longrightarrow 8MnO_4^{2-} + IO_4^- + 4H_2O}$$

Explanation

(a) ***The polarity of B—X bonds is in the order :* B—F > B—Cl > B—Br *but Lewis acidicity shows the sequence :* $BF_3 < BCl_3 < BBr_3$.**

– With increase in polarity of B—X bond, acidity also increases and should be thus in the order :

$$BF_3 > BCl_3 > BBr_3$$

But Lewis acidity is in reverse order :

$$BF_3 < BCl_3 < BBr_3$$

There is lateral overlap of the vacant 2*p* orbital of B with one completely filled orbital of F leading to $p\pi-p\pi$ bonds between B and F. This B—F bond thus acquires double bond character. This also leads to compensate electron deficiency of boron and thus Lewis acid character of BF_3 is reduced. This $p\pi-p\pi$ bonding decreases going from BF_3 to BBr_3 and thus Lewis acidic nature increases in the order :

$$BF_3 < BCl_3 < BBr_3$$

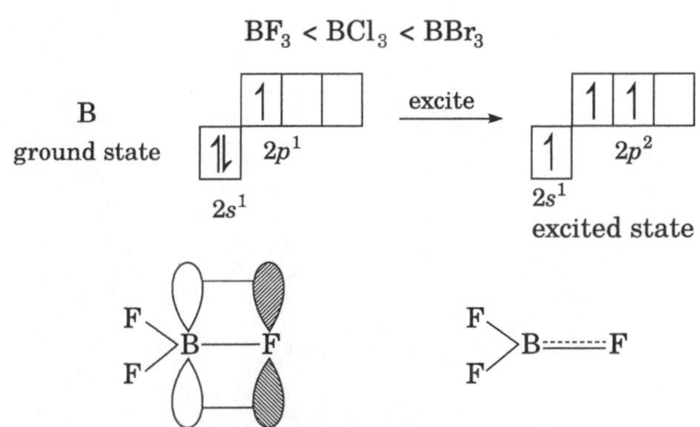

(b) **NF_3 *is stable while* NCl_3 *and* NI_3 *are explosive.***

– In NF_3 the electronegativity difference (N = 3.0; F = 4.0) permits strong $p_\pi - p_\pi$ bonding and the molecule is stable. But in NCl_3 and NI_3, the electronegativity difference is nil or small and the molecules are unstable with $p_\pi - p_\pi$ bonding weak. Hence, NF_3 is a stable gas, NCl_3 an explosive liquid and NI_3 an explosive solid.

(c) ***In the series* HXO*, the order of stability is :* HOCl > HOBr > HOI**

– HOX and OX^- are unstable and undergo disproportionation :

$$\underset{+1}{OCl^-} \longrightarrow \underset{-1}{Cl^-} + \underset{+5}{ClO_3^-}$$

The instability is due to decrease in bond energy from chlorine to iodine.

(d) ***Silver nitrate gives a white precipitate with aqueous sodium thiosulphate; the precipitate dissolves in excess of thiosulphate. If the precipitate is heated with water it turns black, and the supernatant liquid then gives a white precipitate with acidified barium nitrate solution.***

– When sodium thiosulphate ($Na_2S_2O_3$) solution is added to silver nitrate ($AgNO_3$) solution, there is a white precipitate due to the formation silver thiosulphate ($Ag_2S_2O_3$); these precipitate dissolve in excess of sodium thiosulphate due to the formation of the complex $[Ag(S_2O_3)_2]^{3-}$.

$$2AgNO_3 + Na_2S_2O_3 \longrightarrow \underset{\substack{\text{silver thiosulphate} \\ \text{(white precipitate)}}}{Ag_2S_2O_3\downarrow} + 2NaNO_3$$

$$Ag_2S_2O_3 + 3Na_2S_2O_3 \longrightarrow \underset{\text{sodium argentothiosulphate}}{2Na_3[Ag(S_2O_3)_2]}$$

If the precipitate of $Ag_2S_2O_3$ is heated, there is disproportionation giving black precipitate of Ag_2S and soluble H_2SO_4 which remains in the upper (supernatant) layer which gives white precipitate (of $BaSO_4$) with $Ba(NO_3)_2$ solution.

$$\underset{\text{white precipitate}}{Ag_2S_2O_3} + H_2O \longrightarrow \underset{\text{black precipitate}}{AgS\downarrow} + H_2SO_4$$

$$H_2SO_4 + Ba(NO_3)_2 \longrightarrow \underset{\text{white precipitate}}{BaSO_4\downarrow} + 2HNO_3$$

(e) ***Silver chloride and silver iodide are soluble in saturated aqueous potassium iodide, but insoluble in saturated aqueous potassium chloride.***

– Iodide complex of silver must be more stable than chloride complex.

(f) ***Although the hydrogen bonding in hydrogen fluoride is stronger than that in water, water has much the higher boiling-point.***

– Hydrogen fluoride vapour is polymeric, hence, the hydrogen bonds are not broken on volatisation; those in water are broken thus requiring higher enthalpy. Hence, boiling point of H_2O is higher than that of HF.

(g) ***Fluorine combines with other elements in the higher oxidation states.***

– Following factors contribute the unique ability of fluorine to combine with other elements in their higher oxidation states.

(i) the highest oxidising power

(ii) high lattice energy of fluoride

(iii) high enthalpy of hydration of fluoride

(iv) lower F—F bond energy of F—F bond

(h) ***The bond dissociation energy increases from I_2 to Cl_2. Thus, bond dissociation energy of F_2 should be higher than that of Cl_2. But it is not so.***

- Fluorine has the smallest size of all the halogens. There is considerable repulsion in the non-bonding electrons which weakers the F—F bond. Thus, bond dissociation energy of F_2 is less than that of Cl_2.

(i) HF ***is the weakest acid***

- HF has the highest bond dissociation enthalpy amongst all the halogen acids. The hydration enthalpy of the fluoride ion is also highest, the former predominates making HF the weakest of the halogen acids.

(j) ***Salts of hypohalous acids are alkaline in nature due to hydrolysis.***

- NaClO (sodium hypochlorite) is a salt of HClO (hypochlorous acid) and NaOH. HClO is a fairly weak acid. On hydrolysis ClO^- forms HClO (weak acid) and free OH^- ions which makes the solution alkaline.

$$ClO^- + H_2O \longrightarrow \underset{\text{(weak acid)}}{HClO} + \underset{\text{(free)}}{OH^-}$$

(k) ***Tap water requires fluoridation.***

- The main constituent of tooth enamel is $Ca_5(OH)(PO_4)_3$ called hydroxyapatite, which in presence of fluoride ions is converted into harder, acid-resistant fluorapatite, $Ca_5F(PO_4)_3$

$$Ca_5(OH)(PO_4)_3 + F^- \longrightarrow Ca_5F(PO_4)_3 + OH^-$$

Thus, addition of fluoride makes the teeth harder and prevents their decay. If we use tap water deficient in fluoride, it can cause formation of cavity.

Imp Paradoxically, consuming 2 ppm or more fluoride also causes mottling of teeth.

Master Exercises

Exercise 1
(Stage 1 : Learning)

Short Answer Type Questions

1. Sea water contains around 65×10^{-3} g of bromide ions in 1 dm^3. If all the bromide ions are converted into Br_2, how many dm^3 of sea-water are needed to produce 1 kg of bromine Br_2 ?

2. Complete the following reactions:

(a) $CH_3CH{=}CH_2 + ICl \longrightarrow$

(b) $Cl_2 + \underset{\text{hot}}{NaOH} \longrightarrow$

(c) $Cl_2 + S^{2-} \longrightarrow$

(d) $SO_2 + I_2 + H_2O \longrightarrow$

(e) $SnCl_2 + HCl + I_2 \longrightarrow$

3. I_3^- is formed when I_2 combines with I^-. Which of them is a Lewis base?

4. Which is **stronger base** in each pair?

(a) ClO_2^-	ClO_3^-	(b) OH^-	CN^-
(c) I^-	Cl^-	(d) $CH_3CH_2^-$	$CH_2{=}CH^-$
(e) NO_3^-	NO_2^-	(f) $H_2PO_2^-$	$H_2PO_3^-$

5. Iodine pentoxide, I_2O_5, is sometimes used to remove CO from the air by forming CO_2 and I_2. Write a balanced equation for this reaction and identify species that are oxidised and reduced.

6. HOCl can also be prepared by passing CO_2 through a suspension of bleaching powder in water. Write balanced equation for this reaction.

7. Cl_2O_6 is diamagnetic but in liquid state, it shows paramagnetism. Explain.

8. Identify A to F in the following :

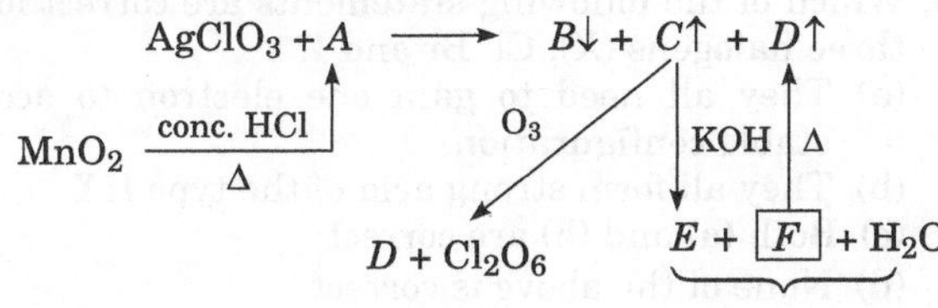

9. Name the halogen

(a) has two complete electron shells below the valence shells

(b) is solid at room temperature

(c) has highest electronegativity value

(d) lowest ionisation energy

(e) lowest formation of $X(g)$

10. Complete the following reactions

(a) $I_2 + IF_3 \longrightarrow$ (b) $Cl_2 + H_2O \longrightarrow$

(c) $Cl_2 + H_2S \longrightarrow$

11. Write the names of the following oxoacids and deduce the oxidation number of halogen atom in each

(a) HOCl (b) HOClO

(c) HOBr (d) $HOBrO_2$

(e) $HOIO_3$ (f) $HOClO_3$

Analytical Questions

1. The redox reaction involving the reducing power of hydrogen sulphide is :

$$S + 2H^+ + 2e^- \longrightarrow H_2S, \quad E^\circ_{S/H_2S} = +0.14V$$

Two other half equations are :

$$Fe^{3+} + e^- \longrightarrow Fe^{2+} \quad E^\circ_{Fe^{3+}/Fe^{2+}} = +0.77\ V$$

$$Br_2 + 2e^- \longrightarrow 2Br^- \quad E^\circ_{Br_2/Br^-} = 1.07\ V$$

Under standard conditions :

(i) will hydrogen sulphide react with iron (III) ions?

(ii) will it react with bromine?

If you decide a reaction should take place, write the equation.

2. Explain :

(i) Electron affinity of Cl is the highest among the halogens; yet, F_2 is the strongest oxidising agent.

(ii) ClO_2 is a free radical with one unpaired electron but has no tendency to dimerise like NO_2.

(iii) HCl(g) and HCl (aq) have different behaviour towards litmus.

3. 0.01 M solutions of iodine in *n*-hexane, benzene, ethanol and pyridine are violet, purple, brown and yellow respectively. When 0.001 mol of pyridine is added to 100 mL of each of the solutions of iodine in *n*-hexane, benzene and ethanol, all of these become yellow, Explain.

Exercise 2
(Stage 2 : High Skill Problem Solving)

Only One Option Correct

1. Of the following acids
I : Hypophosphorus acid
II : Oxalic acid
III : Glycine
(a) I, II are monobasic; III dibasic and amphoteric
(b) II monobasic; I, dibasic acid; III amphoteric
(c) I monobasic, II, dibasic, III amphoteric
(d) I, II dibasic; III amphoteric

2. Which incorrect statement?
(a) All halogens form oxyacids
(b) All halogens show –1, +1, +3, +5, +7 oxidation states
(c) Hydrofluoric acid forms KHF_2 and K_2F_2 and attacks glass
(d) Oxidising power is in order $F_2 > Cl_2 > Br_2 > I_2$

3. $HClO_4$, HNO_3 and HCl are all very strong acids in aqueous solution. In glacial acetic acid medium, their acid strength is such that
(a) $HClO_4 > HNO_3 > HCl$
(b) $HNO_3 > HClO_4 > HCl$
(c) $HCl > HClO_4 > HNO_3$
(d) $HCl > HClO_4 \sim HNO_3$

4. Tincture of iodine is
(a) CHI_3 in alcohol (b) I_2 in alcohol
(c) I_2 in KI (d) CHI_3 in KI

5. "Chlorine-type" laundry bleaches are in reality aqueous solution of
(a) HOCl (b) $HClO_3$
(c) $NaClO_3$ (d) NaOCl

6. The active constituent of bleaching powder is
(a) $Ca(OCl)_2$ (b) Ca(OCl)Cl
(c) $Ca(ClO_2)_2$ (d) $Ca(ClO_2)Cl$

7. Which one of the following is the weakest base as per the Bronsted concept?
(a) $[ClO_4]^-$ (b) $[ClO_3]^-$
(c) $[ClO_2]^-$ (d) $[ClO]^-$

8. Which one of the following is not a true "per acid"?
(a) $HClO_4$ (b) H_2SO_5
(c) $H_2S_2O_8$ (d) H_3PO_5

9. Which is/are true statement(s)?
(a) Basic nature of X^- is in order $F^- > Cl^- > Br^- > I^-$
(b) HI is strongest acid of HF, HCl, HBr and HI
(c) The ionic character of *M*—*X* bond decreases in the order *M*—F > *M*—Cl > *M*—Br > *M*—I
(d) All the above are correct statements

10. Select incorrect statement.
(a) ClO_2 and Cl_2O are used as bleaching agents for paper pulp and textiles
(b) OCl^- (hypohalites) salts are used as detergent
(c) OCl^- disproportionates in alkaline medium
(d) BrO_3^- is oxidised to Br_2 by Br^- in acidic medium

11. Consider the following boron halides:
I. BF_3 II. BCl_3
II. BBr_3 IV. BI_3
The Lewis acid characters of these halides are such that
(a) I < II < III < IV (b) I < III < II < IV
(c) IV < III < II < I (d) IV < II < III < I

12. Which of the following compounds possesses Lewis acid character?
I. BF_3 II. SiF_4
III. PF_5
Select the correct answer using the codes given below
Codes
(a) I alone (b) I, II and III
(c) II and III (d) I and III

13. Inter-halogen compounds can be
(a) ICl_3 (b) BrF_5
(c) IF_7 (d) all of these

14. Oxidation of hydrogen halide, H*X* affords a method for the industrial and laboratory preparation of the halogen, X_2, in the free state in respect of all of the following except
(a) fluorine (b) chlorine
(c) bromine (d) iodine

15. Which of the following statements are correct for all three halogens (*X*), Cl, Br and I?
(a) They all need to gain one electron to acquire stable configuration
(b) They all form strong acid of the type H*X*
(c) Both (a) and (b) are correct
(d) None of the above is correct

16. Which of the underlined atoms in oxyacids have sp^3 hybridised atoms?
   (a) $H\underline{Cl}O_4, H_2\underline{S}O_4, H\underline{N}O_2$
   (b) $H_2\underline{S}O_4, H_3\underline{P}O_4, H\underline{N}O_3$
   (c) $H\underline{Cl}O_4, H_2\underline{S}O_4, H_2\underline{S}O_5$
   (d) $H\underline{Cl}O_4, H\underline{N}O_3, H\underline{Cl}O_3$

17. In the following
   (a) CsI_3 will ionise to give Cs^+ and I_3^-
   (b) $RbICl_2$ will give RbCl and ICl
   (c) KBrICl will give KCl and IBr
   (d) all of the above are correct

18. Which of the following atoms has the highest ionisation energy?
   (a) F (b) Cl
   (c) Br (d) I

19. Which one of the following arrangements does not truly represent the property indicated against it?
   (a) $Br_2 < Cl_2 < F_2$: Oxidising power
   (b) $Br < Cl < F$: Electronegativity
   (c) $Br < F < Cl$: Electron affinity
   (d) $Br_2 < Cl_2 < F_2$: Bond energy

20. The mixture of concentrated HCl and HNO_3 made in 3 : 1 ratio contains
   (a) ClO_2 (b) NOCl
   (c) NCl_3 (d) N_2O_4

21. The formula for calcium chlorite is
   (a) $Ca(ClO_4)_2$ (b) $Ca(ClO_3)_2$
   (c) $CaClO_2$ (d) $Ca(ClO_2)_2$

22. Which one of the following oxoacids of chlorine is the least oxidising in nature?
   (a) HClO (b) $HClO_2$
   (c) $HClO_3$ (d) $HClO_4$

23. Least stable oxide of chlorine is
   (a) Cl_2O_4 (b) ClO_2
   (c) Cl_2O_7 (d) Cl_2O_6

24. Which reaction is not feasible?
   (a) $2KI + Br_2 \longrightarrow 2KBr + I_2$
   (b) $2KBr + I_2 \longrightarrow 2KI + Br_2$
   (c) $2KBr + Cl_2 \longrightarrow 2KCl + Br_2$
   (d) $2H_2O + 2F_2 \longrightarrow 4HF + O_2$

25. Shape of O_2F_2 is similar to that of
   (a) C_2F_2 (b) H_2O_2
   (c) H_2F_2 (d) C_2H_2

26. Iodine test is shown by
   (a) glucose (b) starch
   (c) glycogen (d) polypeptide

27. Which one will liberate Br_2 from KBr?
   (a) I_2 (b) SO_2
   (c) HI (d) Cl_2

28. An important product in the ozone depletion by chlorofluorocarbons is
   (a) Cl_2 (b) OCl
   (c) F_2 (d) O_2F_2

29. Which one is the strongest bond?
   (a) Br—F (b) F—F
   (c) Cl—F (d) Br—Cl

30. The interhalogen compound not obtained is
   (a) ICl (b) IF_5
   (c) BrF_5 (d) $BrCl_7$

31. Which of the following will displace the halogen from the solution of the halide?
   (a) Br_2 added to NaI (b) Br_2 added to NaCl
   (c) Cl_2 added to KCl (d) Cl_2 added to NaF

32. Which is the strongest oxidising agent out of the following?
   (a) I_2 (b) Cl_2
   (c) Br_2 (d) F_2

33. The high oxidising power of fluorine is due to
   (a) high electron affinity
   (b) high heat of dissociation and low heat of hydration
   (c) low heat of dissociation and high heat of hydration
   (d) high heat of dissociation and high heat of hydration

34. When Cl_2 is passed through cold dil. NaOH, the products are
   (a) NaCl, NaOCl and H_2O
   (b) NaCl, $NaClO_2$ and H_2O
   (c) NaCl, $NaClO_3$ and H_2O
   (d) NaCl, $NaClO_4$ and H_2O

35. The greater reactivity of F_2 is due to
   (a) lower electron affinity of F
   (b) lower bond energy of F—F bond
   (c) higher electronegativity of F
   (d) gaseous state of F_2

36. Dissociation of HI into H_2 and I_2 is
   (a) kinetically controlled
   (b) thermodynamically controlled
   (c) kinetically as well as thermodynamically controlled
   (d) neither kinetically nor thermodynamically controlled

37. In the preparation of HBr or HI, NaX (X = Br, I) is treated with H_3PO_4 and not by conc. H_2SO_4 and since
   (a) H_2SO_4 makes the reaction reversible
   (b) H_2SO_4 oxidises HX to X_2 (Br, I)
   (c) Na_3PO_4 is water soluble
   (d) Na_2SO_4 is water soluble

38. Acid strength of $HX (X = F, Cl, Br, I)$ is dependent on
(a) the electronegativity differences of H and X
(b) the tendency of the hydrated molecule to form H^+
(c) both (a) and (b)
(d) None of the above

39. Select correct statement(s) regarding behaviour of HF as non-aqueous solvent.
(a) HCl behaves as an acid and HF as a base
(b) $HClO_4$ behaves as a base and HF as an acid
(c) HNO_3 and H_2SO_4 behave as base and HF as an acid
(d) All of the above are correct statements

40. Dipole moment and ionisation constant are maximum in case of
(a) HF, HF (b) HF, HI
(c) HI, HF (d) HI, HI

41. The tetrahedral arrangement of perchlorate ion is due to
(a) presence of a lone pair of electrons
(b) trigonal bipyramidal shape of the ion
(c) sp^3 hybridisation
(d) sp^2 hybridisation

42. Acid strength of oxoacids of chlorine is in order
(a) $HOCl < HClO_2 < HClO_3 < HClO_4$
(b) $HClO_4 < HClO_3 < HClO_2 < HOCl$
(c) $HClO_3 < HClO_4 < HClO_2 < HOCl$
(d) $HClO_2 < HClO_3 < HOCl < HClO_4$

43. Acid strength of oxoacids of halogen is in order
(a) HOF > HOCl > HOBr > HOI
(b) HOI > HOBr > HOCl > HOF
(c) HOI > HOCl > HOBr > HOF
(d) HOCl > HOBr > HOI > HOF

44. Which has peroxy linkage?
(a) Perchloric acid (b) Hypochloro acid
(c) Para perchloric acid (d) None of these

45. SO_2 can reduce
(a) $HClO_3$ to HCl (b) $Cr_2O_7^{2-}/H^+$ to Cr^{3+}
(c) MnO_4^-/H^+ to Mn^{2+} (d) all of these

46. Bleaching powder contains
(a) CaO and Cl_2 molecules
(b) Ca^{2+} and OCl_2^{2-} ions
(c) Ca^{2+}, Cl^- and OCl^- ions
(d) Ca^{2+}, O^{2-} ions and Cl_2 molecule

47. Bleaching powder is disinfectant for purification of water when water born germs are killed. But disinfectant activity is destroyed. It is due to its disproportionation into
(a) $CaCl_2$ and Cl_2
(b) $CaCl_2$ and $Ca(ClO_3)_2$
(c) CaO and Cl_2
(d) CaO, Cl_2 and $CaCl_2$

48. 10 g of bleaching powder on reaction with KI required 100 mL of 1 N hypo. Thus, % of pure bleaching powder is
(a) 100% (b) 80%
(c) 63.5% (d) 35.5%

49. $CaOCl_2 + H_2O + CO_2 \longrightarrow CaCO_3 + CaCl_2 + HOCl$
This reaction is simply
(a) oxidation
(b) reduction
(c) redox
(d) hydrolysis in presence of CO_2

50. CH_3CH_2OH on reaction with bleaching powder forms
(a) $CHCl_3$ (b) CCl_3CHO
(c) CH_2ClCHO (d) CCl_4

51. Cl_2O, Br_2O, I_2O have positive value of ΔG (free energy) indicating that
(a) these oxides are stable
(b) these oxides are unstable and changes to X_2 and O_2
(c) these disproportionate into X^- and XO^-
(d) these oxides can form interhalogen compounds

52. Select correct statement(s).
(a) Cl_2O and ClO_2 are used as bleaching agents and as germicides
(b) I_2O_5 is used in the quantitative estimation of CO
(c) Bond angle XOX varies in the order
FOF < ClOCl < BrOBr
(d) All of the above are correct statements

53. The bond angle in OF_2 out of OF_2, Cl_2O, Br_2O is minimum. It is because in case of OF_2
(a) electrons are nearer to fluorine due to high electronegativity of F compared to Cl to Br
(b) lone pair-lone pair repulsion decreases bond angle
(c) both (a) and (b) are correct
(d) None of the above is correct

54. ClO_2 is the anhydride of
(a) HOCl (b) $HClO_2$
(c) $HClO_3$ (d) $HClO_2$ and $HClO_3$

55. ClO_3 is the mixed anhydride of
(a) $HClO_2$ and $HClO_3$ (b) $HClO_3$ and $HClO_4$
(c) $HClO_2$ and $HClO_4$ (d) $HClO_2$ and $HClO_3$

56. Chlorine dioxide
(a) is paramagnetic in nature
(b) has odd-electron bond
(c) is stable
(d) has all the above properties true

57. Select incorrect statement(s).
(a) Cl_2O_6 is diamagnetic in solid state
(b) ClO_3 is paramagnetic and exists in liquid state
(c) Cl_2O_6 is anhydride of $HClO_3$
(d) ClO_3 forms dimer due to unpaired electron

58. Cl_2 gas is passed into aqueous solution of KBr and KI and $CHCl_3$ added. It is observed that there is
(a) violet colour in $CHCl_3$ layer
(b) yellow colour in $CHCl_3$ layer
(c) yellow colour in aqueous layer
(d) violet colour in $CHCl_3$ layer

59. Consider following reactions in which atoms have been labelled with isotopes (indicated by *)
(I) $S^*SO_3^{2-} + 2H^+ \longrightarrow H_2SO_4 + S$
(II) $Na_2HgI_2I_2^* + 2AgNO_3 \longrightarrow 2AgI + HgI_2 + 2NaNO_3$
In which case isotopes are equivalent in the products?
(a) I (b) II
(c) Both I and II (d) None of these

60. In very recent Sunami in Japan, food materials were banned as they were contaminated with radioactivity. Which of the following was found to have radioactive beyond prescribed limit?
(a) F (b) Cl
(c) I (d) At

One or More Than One Options Correct

1. Which are correct statements?
(a) All halogens form oxoacids.
(b) All halogens show –1, +1, +3, +5, +7 oxidation states.
(c) Hydrofluoric acid forms KHF_2 and K_2F_2 and attacks glass.
(d) Oxidising power is in order $F_2 > Cl_2 > Br_2 > I_2$

2. Select correct statement(s).
(a) ClO_2 and Cl_2O are used as bleaching agents for paper pulp and textiles
(b) OCl^- salts are used as detergents
(c) OCl^- disproportionates in alkaline medium
(d) BrO_3^- is oxidised to Br_2 by Br^- in acidic medium

3. In the structure of
(a) IF_2^+, bond angle is smaller than the tetrahedral angle because of lone pair-lone pair repulsion
(b) IF_4^+, there is sp^3 hybridisation with bond angle of 109°28′
(c) IF_6^+, there are six bonding electrons and no lone pairs.
(d) IF_4^+, two F-atoms one at equatorial positions and two are inclined at an angle less than 120° due to lone pair-bond pair repulsion

4. In the conversion of Br^- to BrO_3^-
(a) there is anodic oxidation is basic solution
(b) equivalent mass of Br^- is 20 (atomic weight of Br = 80)
(c) there is anodic oxidation is acidic medium
(d) there is cathodic reduction in basic solution

5. In the conversion of BrO_3^- to BrO_4^-
(a) there is no change in oxidation number
(b) reduction takes place in basic solution
(c) reaction also takes place by XeF_2
(d) equivalent mass of BrO_3^- is one-half of ionic mass

6. Select correct statement(s).
(a) Chile saltpetre ($NaNO_3$) is a main source of I_2
(b) IO_3^- is reduced to I^- and which in turn oxidised to I_2 by IO_3^- in acidic medium
(c) Anhydrous HF is a non-conductor of electricity
(d) F_2 is obtained by electrolysis of molten KHF_2

7. Select ocrrect statement(s).
(a) Colours of the halogens (F_2, Cl_2, Br_2, I_2) in the gas phase are complimentary colours.
(b) Colour is due to transition of an electron from the highest occupied π^* molecular orbital to the lowest unoccupied σ^* molecular orbital
(c) ClO_2 is paramagnetic due to unpaired electrons
(d) ClO_2 dimerises to Cl_2O_4 as NO_2 dimerises to N_2O_4

8. Base on $E°$ values gives

$$\frac{1}{2}F_2 + e^- \longrightarrow F^- \qquad E° = 2.87\text{ V}$$
$$\frac{1}{2}Cl_2 + e^- \longrightarrow Cl^- \qquad E° = 1.40\text{ V}$$
$$\frac{1}{2}Br_2 + e^- \longrightarrow Br^- \qquad E° = 1.09\text{ V}$$
$$\frac{1}{2}I_2 + e^- \longrightarrow I^- \qquad E° = 0.62\text{ V}$$

(a) Oxidising power of F_2 is maximum
(b) Oxidising power of I_2 is maximum
(c) Oxidising power is in order $F_2 > Cl_2 > Br_2 > I_2$
(d) Oxidising power is in order $F_2 < Cl_2 < Br_2 < I_2$

9. $CHCl_3$ and H_2O form two separate immiscible layers Br_2 (orange) and I_2 (violet) are soluble in organic layer due to a greater extent. In test tube *A* ($KI + CHCl_3 + H_2O$) and test tube *B* ($KI + KBr + CHCl_3 + H_2O$) Cl_2 gas is passed then (based on $E°$ values in Q. 8)
(a) organic layer in *A* appear violet
(b) organic layer in *B* appear orange
(c) organic layer in *B* appear violet
(d) organic layer in *B* first appears violet then orange

10. Select correct statement(s).
(a) pK_a value of HI (strongest halogen acid) is most negative
(b) high H—F bond strength makes H—F a weak acid in dilute aqueous solution
(c) Halogen forms clatharates by freezing solution in water
(d) K_a values of HX is in order HF < HCl < HBr < HI

Passage Comprehension Questions

Passage 1

In the gas phase, colours of the halogens very remarkably :

Fluorine — Pale yellow
Chlorine — Pale greenish–yellow
Bromine — Deep red brown
Iodine — Deep violet

Answer the following questions

1. This colour is due to
   (a) transition of an electron from the highest occupies π^* molecular orbital to the lowest unoccupied σ^* molecular orbital
   (b) occupied σ^* MO to the lowest unoccupied π^* MO
   (c) Occupied σ MO to the lowest unoccupied π MO
   (d) None of the above
2. As one moves from F_2 to I_2, difference between the two energy levels
   (a) decreases and probability of transition increases
   (b) increases and probability of transition increases
   (c) remains unchanged and probability of transition remains unchanged
   (d) depends on the temperature
3. Due to the above transition, absorption maximum gradually shifts from
   (a) red region of the visible spectrum to near UV region
   (b) the near UV region to the red region of the visible spectrum
   (c) the near UV region to the far IR region
   (d) the near IR region to visible region

Passage 2

Consider following change,

$$BrO_2 \longrightarrow Br_2O + A$$

Br_2O and A are formed by the change of equal number oxidation number. Based on this change, answer the following questions.

1. Oxide A could be
   (a) Br_2O_3 (b) Br_2O_5
   (c) Br_2O_7 (d) BrO
2. Balanced equation is
   (a) $4BrO_2 \longrightarrow Br_2O + Br_2O_7$
   (b) $8BrO_2 \longrightarrow Br_2O + 3Br_2O_5$
   (c) $6BrO_2 \longrightarrow Br_2O + 2Br_2O_3$
   (d) None of the above
3. Br_2O has structure
   (a) Br—Br=O (b) Br—O—Br
   (c) Br=O=Br (d) None of these
4. Equivalent mass of Br_2 is
   (a) $\frac{M}{2}$ (b) $\frac{M}{3}$
   (c) $\frac{2M}{3}$ (d) $\frac{M}{6}$

Passage 3

This question concerns the elements of Group 17 — fluorine, chlorine, bromine and iodine.

When sodium chloride is treated with concentrated sulphuric acid, a colourless gas, X, which fumes in moist air, is formed. When sodium iodide is treated in the same way a coloured vapour, Y, is product.

If 90% phosphoric(v) acid is used instead of sulphuric acid, a colourless gas is produced in each reaction.

1. Gases X and Y are respectively
   (a) Cl_2, I_2 (b) HCl, I_2
   (c) HCl, HI (d) $HClO, I_2$
2. With 90% phosphoric acid (H_3PO_4), colourless gases formed are
   (a) HCl, HI (b) Cl_2, I_2
   (c) HCl, I_2 (d) Cl_2, HI
3. Behaviour of H_2SO_4 and H_3PO_4 is different towards NaI because
   (a) H_2SO_4 is stronger acid than H_3PO_4
   (b) H_2SO_4 is a strong oxidising agent and oxidises colourless gas to coloured vapour Y
   (c) H_3PO_4 is tribasic acid
   (d) H_2SO_4 is a dibasic acid

Passage 4

Water pollution can be by dissolved gases. Bad smell of water is due to H_2S and can be removed by treatment with bleaching powder.

$$\underset{\text{bleaching powder}}{CaOCl_2} + H_2O \longrightarrow Ca(OH)_2(aq) + Cl_2(aq)$$

$$H_2S(aq) + Cl_2(aq) \longrightarrow 2HCl(aq) + S(s)$$

Based on the above treatment answer the following questions:

1. If H_2S content of contaminated water is 22 ppm by mass, Cl_2 required to remove all the H_2S from 200 gallons of water (1 gallon = 3.785 L) is
   (a) 30.00 g (b) 71.00 g
   (c) 17.5 g (d) 35.00 g
2. 25.0 L of a natural gas measured at 25°C and 740 mm Hg is bubbled through $Pb^{2+}(aq)$, a precipitate weighing 0.3938 g is obtained. ‰(ppt—parts per thousand) of H_2S by volume in the natural gas is, (Pb = 206)
   (a) 0.225‰ (b) 2.25‰
   (c) 5.25‰ (d) 0.525‰

Assertion & Reason

Codes :

(a) Both **A** and **R** are true and **R** is the correct explanation of **A**.

(b) Both **A** and **R** are true but **R** is not the correct explanation of **A**.

(c) **A** is true but **R** is false.

(d) **A** is false but **R** is true.

1. Assertion (A) : CaF_2 has been given the name fluorspar.

Reason (R) : Solid CaF_2 emits light when heated.

2. Assertion (A) : HF is stored in the vessel made of wax.

Reason (R) : HF is a weak acid.

3. Assertion (A) : Fluorine has only –1 oxidation state in compounds.

Reason (R) : Fluorine has maximum electronegativity and electron-affinity.

4. Assertion (A) : Cl_2 or Br_2 changes KI into I_2 (violet in CCl_4 layer), Cl_2 changes KBr into Br_2 (orange-yellow in CCl_4 layer). Thus, Cl_2 would change mixture of KI and KBr into violet layer in CCl_4.

Reason (R) : Oxidising power is in order $I_2 < Br_2 < Cl_2 < F_2$.

5. Assertion (A) : Fluorine is not super halogen.

Reason (R) : It is most reactive of all the halogen.

6. Assertion (A) : Molten I_2 is conducting in nature.

Reason (R) : Bond dissociation energy increases in the order

$$Br_2 < Cl_2 < I_2$$

7. Assertion (A) : Hydrofluoric acid forms two series of salts KHF_2, K_2F_2.

Reason (R) : It is a dibasic acid due to hydrogen bonding.

8. Assertion (A) : CN^- can be oxidised to $(CN)_2$.

Reason (R) : CN^- is pseudohalide.

9. Assertion (A) : Bond dissociation energy of F_2 is smaller than that of Cl_2.

Reason (R) : Bond dissociation energy is in order $F_2 < Cl_2 < Br_2 < I_2$.

10. Assertion (A) : Aqueous solution of Na*OX* is alkaline.

Reason (R) : It is a salt of weak acid (HO*X*) are strong base (NaOH).

True & False

1. $FClO_3$ (perchloryl fluoride) has the highest dielectric constant.

2. HF is a weak acid but H (FHF) is a strong acid.

3. Solid CaF_2 could emit light when heated.

4. When NaCl(*aq*) is electrolysed, sodium metal is formed at the cathode and chlorine gas is formed at the anode.

5. It very recent Japan Sunami, there has been increase in radioactive iodine content is salt.

6. When Cl_2 gas is dissolved in water, there is disproportion to HClO and HCl.

7. Solid propellent in the rockets is aluminium powder with ammonium perchlorate as reducing agent.

8. Br_2 oxidised KCl into Cl_2 and KI into I_2.

9. ClO_3^+ has structure with sp^2 hybridisation having bond angle of 120°.

10. IF_6^+ has regular octahedron structure.

11. Interhalogen compounds are strong oxidising agents.

12. Interhalogen compounds are strong reducing agents.

13. X—Y bond is more polar than X—X and Y—Y bonds.

14. Interhalogen compounds are ionic.

15. XY_3 types of molecules have square pyramidal structure.

Fill in the Blanks

1. $FClO_3$ is an inert and non-corrosive gas with .... structure.

2. Perchloryl fluoride ($FClO_3$) is used as a rocket .... as well as an .... in high-voltage system.

3. BrF_3 has .... shaped structure.

4. ClF_3 is a .... in the reaction

$$ClF_3 + AsF_3 \longrightarrow [ClF_2]^+[AsF_6]^-$$

and is a .... in the reaction

$$ClF_3 + NOF \ldots\rightarrow [NO]^+[ClF_4]^-$$

5. IF_7 has .... structure.

6. Decreasing order of reactivity of F_2, ClF and ClF_3 is ....

7. CN, CNO^-, CNS^- are called ....

8. Out of chloric acid ($HClO_3$) and perchloric acid ($HClO_4$) more stable is ....

9. Chlorous acid is unstable and decomposes into .... and ....

10. Ammonium perchlorate is used as the .... in the solid booster rockets for the space shuttle.

11. .... displaces three halogens from their compounds.

12. Cl^- can be oxidised by ....

13. Br_2 can be reduced by ....

14. Fluorine and .... can oxidise water to oxygen.

15. Oxidation of .... and .... by Cl_2 will give Br_2 and I_2 respectively.

Matrix-Match Type Questions

I. One or More Correct

1. Match the oxides in Column I with their corresponding properties in Column II.

Column I	Column II
A. Cl_2O	1. Anhydride of HOCl
B. ClO_2	2. Anhydride of $HClO_2$
C. Cl_2O_6	3. Anhdride of $HClO_3$
D. Cl_2O_7	4. Anhydride of $HClO_4$
	5. sp^3 hybridization
	6. Paramagnetic in gaseous state
	7. Paramagnetic in liquid state

2. Match the species in Column I with the corresponding properties in Column II.

Column I	Column II
A. SO_2	1. Turns lime water milky
B. CO_2	2. Turns $Cr_2O_7^{2-}/H^+$ solution green
C. Cl_2	3. Turns KI/starch blue
D. Br_2	4. Disinfectant for drinking water
E. O_3	5. Green house effect
	6. Acid rain
	7. Controls pH of blood

Integer Answer Types

This section contains 8 questions. The answer to each of the questions is a single digit integer, ranging from 0 to 9. The appropriate bubbles below the respectively question numbers in the *ORS* have to be darkened. For example, if the correct answers to question number *X, Y, Z* and *W* (say) are 6, 0, 9 and 2, respectively, then the correct darkening of bubbles will like the following

X	Y	Z	W
0	0	0	0
1	1	1	1
2	2	2	2
3	3	3	3
4	4	4	4
5	5	5	5
6	6	6	6
7	7	7	7
8	8	8	8
9	9	9	9

Set I

1. ${}^{35}_{17}Cl$ and ${}^{37}_{17}Cl$ are two isotopes of chlorine in mass ratio of $3 : x$ where $x =$ ....

2. $CaOCl_2$ (bleaching powder) has two types of chlorine with different oxidation numbers. Sum of oxidation number is ....

3. Number of series of salts formed by hydrofluoric acid is ....

4. Cl_2 disproportionates to Cl^- and species *A* such that equivalent mass of Cl_2 is 71. Thus, oxidation number of chlorine in species *A* is ....

5. Number of electrons forming bond between Cl and O is ClO_2 is ....

6. Maximum acidic character is shown by the oxyacids of halogens with oxidation number of halogens as ....

7. Br^- and BrO_x^- in acidic medium forms Br_2 where $x =$

8. $3Cl_2 + xNaOH \longrightarrow NaCl + NaClO_3$. x is ....

Set II

1. How many lone-pairs and bond-pairs are present in IF_7 ?

2. How many lone-pairs are present in ClI_3 ?

3. ClO_2 has odd-electron bonds. How many electrons are involved in (Cl—O) bond?

4. ClO_2 is paramagnetic due to .... unpaired electron(s).

5. How many of the following have underlined atoms in different oxidation states?

$K_2Hg\underline{I}_4, CaO\underline{Cl}_2, Na_2\underline{S}_2O_3, Na_2\underline{S}_4O_6, H_2\underline{S}O_5, H_2S_2\underline{O}_8, Cr\underline{O}_5, \underline{I}_3^-$

6. How many of the following have Cl atom in sp^3 hybridised state?

$ClO_2, Cl_2O, Cl_2O_7, Cl_2O_6, ClO_4^-$

7. How many of the are water insoluble?

AgF, AgCl, AgBr, AgI, CaF_2, $BaSO_4$

8. 6.35 of impure bleaching powder paste is mixed with KI and I_3^- formed required 50 mL of 0.1 M hypo in neutral medium. Thus % purity of bleaching powder is ....

Answers & Solutions

Master Exercises

Exercise 1

Short Answer Type Questions

1. $$Br^- \longrightarrow \frac{1}{2}Br_2 + e^-$$

65×10^{-3} g Br^- = 32.5×10^{-3} g Br_2 is in 1 dm^3

$\therefore$ 1000 g Br_2 is in $= \frac{1000}{32.5 \times 10^{-3}} = 30769$ dm^3

2. (a) $CH_3—CH(Cl)—CH_2(I)$ (ICl $\longrightarrow$ $I^+ + Cl^-$)

(b) $3Cl_2 + 6NaOH \longrightarrow 3NaCl + NaClO_3 + 3H_2O$

(c) $Cl_2 + S^{2-} \longrightarrow 2Cl^- + S$

(d) $SO_2 + I_2 + 2H_2O \longrightarrow H_2SO_4 + 2HI$

(e) $SnCl_2 + 2HCl + I_2 \longrightarrow SnCl_4 + 2HI$

3. $I^- \rightarrow I_2$; I^- behaves as a Lewis base.

4. If acid is strong, its conjugate base is weak and *vice-versa*.

(a) ClO_2^- (b) OH^- (c) Cl^- (d) $CH_3CH_2^-$ (e) NO_2^-
(f) almost same

5. $$I_2O_5 + 5CO \longrightarrow I_2 + 5CO_2$$

I_2O_5 is reduced to I_2, CO is oxidised to CO_2

6. $2Ca(OCl)Cl + H_2O + CO_2 \longrightarrow CaCl_2 + CaCO_3 + 2HOCl$

7. Cl_2O_6 has paired electrons. In liquid state or in aqueous solution Cl—Cl bond breaks giving one unpaired electron hence, diamagnetic nature of Cl_2O_6 changes to paramagnetic nature in ClO_3.

$O_3Cl—ClO_3$ (O—Cl—Cl—O with two O above and two O below each Cl) $\quad$ $O_3Cl\bullet$

8. (A) : Cl_2 (B) : AgCl (C) : ClO_2 (D) : O_2 (E) : $KClO_2$
(F) : $KClO_3$.

9. (a) chlorine (b) iodine (c) fluorine (d) iodine
(e) chlorine

10. $\longrightarrow$ BIF
$\longrightarrow$ HCl + HOCl
$\longrightarrow$ 2HCl + S

11. (a) hypochlorous acid +1
(b) chlorous acid +3
(c) hypobromous acid +1
(d) bromic acid +5
(e) periodic acid +7
(f) perchloric acid +7

Analytical Questions

1. (i) Oxidation of H_2S by Fe^{3+} can be written as

$$2Fe^{3+} + H_2S \longrightarrow 2Fe^{2+} + 2H^+ + S$$

(Oxidation: $H_2S \rightarrow S$; Reduction: $Fe^{3+} \rightarrow Fe^{2+}$)

$$H_2S \longrightarrow S + 2H^+ + 2e^- \quad E^\circ_{ox} = -0.14 \text{ V}$$

$$2Fe^{3+} + 2e^- \longrightarrow 2Fe^{2+} \quad E^\circ_{red} = 0.77 \text{ V}$$

Thus, $E^\circ_{cell} = 0.63$ V

Thus, $\Delta G^\circ = -nFE^\circ_{cell} = -$ ve.

Thus, oxidation of H_2S by Fe^{3+} is spontaneous.

(ii) Oxidation of H_2S by Br_2 can be written as

$$Br_2 + H_2S \longrightarrow 2H^+ + S + 2Br^-$$

(Oxidation: $H_2S \rightarrow S$; Reduction: $Br_2 \rightarrow Br^-$)

$$H_2S \longrightarrow 2H^+ + S + 2e^- \quad E^\circ_{ox} = -0.14 \text{ V}$$

$$Br_2 + 2e^- \longrightarrow 2Br^- \quad E^\circ_{red} = 1.07 \text{ V}$$

Thus, $E^\circ_{cell} = 0.93$ V

$\Delta G^\circ = -nFE^\circ_{cell} = -$ ve

Hence, oxidation of H_2S by Br_2 is spontaneous.

2. Refer text

3. Solution in *n*-hexane contains uncomplexed I_2 molecule; charge-transfer complex formed with benzene; probably O— and N—donor complexes with ethanol and pyridine; pyridine complex being the most stable.

Exercise 2

Only One Option Correct

1. (c)	**2.** (b)	**3.** (a)	**4.** (b)	**5.** (a)	**6.** (b)	**7.** (a)	**8.** (a)	**9.** (d)	**10.** (b)
11. (a)	**12.** (a)	**13.** (d)	**14.** (a)	**15.** (c)	**16.** (c)	**17.** (d)	**18.** (a)	**19.** (d)	**20.** (b)
21. (d)	**22.** (d)	**23.** (a)	**24.** (b)	**25.** (b)	**26.** (b)	**27.** (d)	**28.** (b)	**29.** (c)	**30.** (d)
31. (a)	**32.** (d)	**33.** (c)	**34.** (a)	**35.** (b)	**36.** (a)	**37.** (b)	**38.** (c)	**39.** (c)	**40.** (b)
41. (c)	**42.** (a)	**43.** (b)	**44.** (d)	**45.** (d)	**46.** (c)	**47.** (b)	**48.** (c)	**49.** (d)	**50.** (a)
51. (b)	**52.** (d)	**53.** (a)	**54.** (d)	**55.** (b)	**56.** (d)	**57.** (c)	**58.** (a)	**59.** (b)	**60.** (c)

One or More Than One Options Correct

1. (c,d) **2.** (a,c,d) **3.** (a,c,d) **4.** (a,b)

5. (a,b,c,d) **6.** (a,b,c,d) **7.** (a,b,c) **8.** (a,c)

9. (a,c) **10.** (a,b,c,d)

Passage Comprehension Questions

Passage 1

1. (a) **2.** (a) **3.** (b)

Passage 2

1. (c) **2.** (a) **3.** (b) **4.** (c)

Passage 3

1. (b) **2.** (a) **3.** (b)

Passage 4

1. (d) **2.** (b)

Assertion & Reason

1. (a) fluorspar name is given after fluorescence (meaning emitting light)

2. (b)

3. (c) Cl has maximum electron-affinity

4. (a) Cl_2 oxidises Br^- to Br_2, and Br_2 oxidises I^- to I_2 (violet)

5. (a) I^+ and I_3^- are formed, order is $Cl_2 < Br_2 < I_2$

6. (c) **7.** (a) **8.** (b) **9.** (c) **10.** (b)

True & False

1. T **2.** T **3.** T **4.** F **5.** T **6.** T

7. F **8.** F **9.** T **10.** T **11.** T **12.** F

13. T **14.** F **15.** F

Fill in the Blanks

1. tetahedral **2.** propellant, insulator **3.** T

4. Lewis base, Lewis acid

5. pentagonal bipyramidal structure

6. $ClF_3 > ClF > F_2$ **7.** pseudohalides **8.** $HClO_4$

9. HCl, ClO_2 **10.** oxidises **11.** F_2 **12.** F_2

13. I^- **14.** chlorine **15.** Br^-, I^-

Matrix-Match Type Questions

I. One or More Correct

1. (A) — (1,5); (B) — (2,3,5,6); (C) — (3,4,5,7); (D) — (4,5)

2. (A) — (1,2,6); (B) — (1,5,7); (C) — (3,4); (D) — (3); (E) — (3,4)

Integer Answer Types

Set I

Questions →	1	2	3	4	5	6	7	8
Answers →	1	0	2	1	3	7	3	6

Set II

Questions →	1	2	3	4	5	6	7	8
Answers →	7	2	3	1	6	5	5	5

16

The Transition and Inner Transition Elements

"If elements of *s*- and *p*-blocks make two hands of the Periodic Table, then elements of *d*- and *f*-blocks are its heart and kidney"

Learning Framework of the Chapter ...

- Transition Elements (*d*-Block)
- Physical Properties
- Compounds
- Inner-Transition Elements (*f*-Block)

Quick Points

- **Electrons in various subshells of a given shell are filled based on Aufbau rule (ascending energy rule) and last filling electron decides block.**
- **Chemical reactivity is based on the position of the element in electrochemical series (ECS). A particular electrochemical reaction is possible of reducing agent in ECS is above oxidising agent, *i.e.*, more negative value of standard reduction potential ($E^{\circ}_{M^{n+}/M}$) indicates the better reducing nature of the metal.**
- **Equivalent mass in a redox reaction**

$$= \frac{\text{molar mass}}{\text{change in oxidation number}}$$

$$= \frac{\text{molar mass}}{\text{electrons exchanged in a redox reaction}}$$

Thus, for $(MnO_4^- \xrightarrow{+5e^-} Mn^{2+})$

$$\text{Equivalent mass of } MnO_4^- = \frac{\text{molar mass}}{5}$$

and for, $(Cr_2O_7^{2-} \xrightarrow{+6e^-} 2Cr^{3+})$

$$\text{Equivalent mass of } Cr_2O_7^{2-} = \frac{\text{molar mass}}{6}$$

Transition Elements (*d*-Block)

Group	III B	IV B	V B	VI B	VII B	VIII B			I B	II B
	3	4	5	6	7	8	9	10	11	12
Period										
4th	Sc	Ti	V	Cr	Mn	Fe	Co	Ni	Cu	Zn
5th	Y	Zr	Nb	Mo	Tc	Ru	Rh	Pd	Ag	Cd
6th	La	Hf	Ta	W	Re	Os	Ir	Pt	Au	Hg
7th	Ac	Rf	Db	Sg	Bh	Hs	Mt	Ds		

On August 16, 2003 IUPAC approved the name for the element of atomic number 110, as **DARMSTADTIUM** with symbol Ds. It is a super heavy element which was first produced in **1994**

Physical Properties

As discussed in the **Periodicity**, three series of elements are formed by filling the 3*d*, 4*d* and 5*d*-shells of electrons. Together these comprise the *d*-block elements. They are often called **transition elements** because their position in the Periodic Table is between the *s*-block and *p*-block elements. The electrons are added to the *penultimate shell,* expanding it from 8 to 18 electrons.

DEEP Focus

Thus, transition metals typically have incompletely filled *d*-subshells or readily give rise to ions with incompletely *d*-subshells. The group 12 (II B) metals—Zn, Cd and Hg do not have this characteristic electron configuration, and so although they are some times called transition metals, they really do not belong in this category.

- The elements with a partially filled *f*-subshell in common oxidation states are known as ***f*-block transition elements** or **inner-transition elements**. They are the two rows elements at the bottom of the Periodic Table. The elements in the first row are called the **lanthanides or rare earths. The elements in the second row are called actinides.** All of the actinides are radioactive, and from neptunium ($Z = 93$) onwards, are synthetic. They have been produced in nuclear reactor or by using particle accelerators.
- The lowest-energy configuration of the lutetium atom is [Xe] $4f^{14}\ 5d^1\ 6s^2$, and in its only common oxidation state (+3), the configuration is [Xe] $4f^{14}$. Although both configurations have a filled *f*-subshell, lutetium is usually considered a lanthanide element.

DEEP Focus

- Nuclear charge increases from Sc to Cu, but electrons are being added to the inner 3*d*-subshell. These 3*d* electrons shield the 4*s*-electrons from the increasing nuclear charge somewhat more effectively than outer-shell electrons can shield one another, hence the atomic radii decrease less rapidly. For the same reason, electronegativities and ionisation energies increase only slightly from Sc across to Cu compared with the increase from Na to Ar.

Physical properties of 3*d*-series [Sc(21)—Zn(30)] have been summarised in Table 16.1.

Comparison of the First Row Transition Metals Through the *d*-Electron Configuration

The d^0 configuration Of the simple ions, only Sc^{3+} is known to have this configuration. This configuration then occurs for those metals in which the formal oxidation states equal the total number of 3*d* and 4*s*-electron. This is true for Ti(IV), V(V), Cr(VI), and Mn(VII) but for Fe(VIII) is unknown.

The d^1 configuration Except vanadium(IV), all others with this configuration are either reducing or undergo disproportionation. For example, disproportionation occurs for Cr(V) an Mn(VI) as follows :

$$3CrO_4^{3-} + 8H^+ \longrightarrow 2CrO_4^{2-} + Cr^{3+} + 4H_2O$$

$$3MnO_4^{2-} + 4H^+ \longrightarrow 2MnO_4^- + MnO_2 + 2H_2O$$

The d^2 configuration This configuration ranges from Ti^{II} which is very strongly reducing, to Fe^{VI}, which is very strongly oxidising. Vanadium (III) is also reducing.

The d^3 configuration Chromium (III) is the important species with this configuration which is stable and well known for complex formation. In other cases, this configuration is relatively unimportant.

The d^4 configuration There are really no stable species with this configuration. The chromium (II) is strongly reducing and manganese(III) disproportionates.

The d^5 configuration The two important species with this configuration are Mn^{2+} and Fe^{3+}, the latter may, however, be reduced to Fe^{2+}.

The d^6 configuration Iron(II) and cobalt(III) are important species with this configuration. Iron(II) is quite stable although a mild reducing agent and cobalt(III) is stable in the presence of strong complexing reagents.

The d^7 configuration The species with this configuration is cobalt(II) which is stable in aqueous solutions but gets oxidised to form Co(III) complexes in the presence of strong ligands.

The d^8 configuration Nickel(II) is the most important species with this configuation.

The d^9 configuration The configuration is found in Cu^{2+} compounds and is by far the most important in the chemistry of copper. Otherwise, this configuration is of little importance.

The d^{10} configuration The two species Cu^+ and Zn^{2+} are important with this configuration. Whereas, copper(I) is easily oxidised to copper(II), zinc(II) is the only state known for zinc.

Metallic Character

Nearly all the transition metals have the simple **hcp, ccp** or **bcc** lattice characteristics of true metals. They have relatively high densities. They are malleable, ductile and have high tensile strength, thermal and electrical conductivity and lustre. Melting points of 3*d*-series (shown in Fig. 16.1) rise to a maximum and fall as the atomic number increases except in Mn where melting point is minimum. **High melting point and high heat of atomisation indicate that the atoms in these elements are held together by strong metallic bonds that are present in the molten states as well. Greater the number of valence shell, stronger is the resultant bonding.**

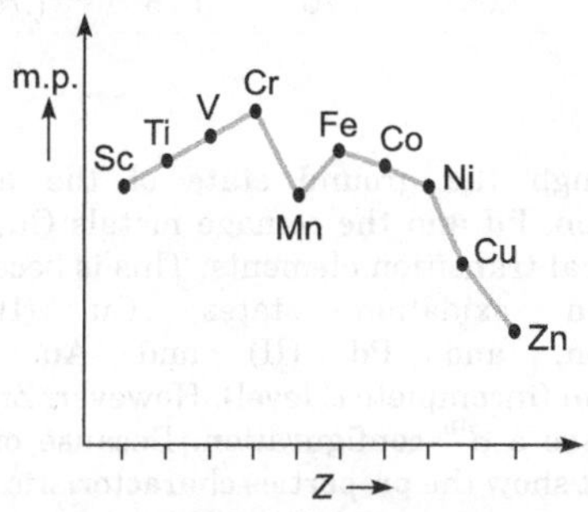

Fig. 16.1 Melting point of 3*d*-series elements (not according to scale)

The periodic trends in the boiling points are similar to those in the melting points. As the process of boiling requries almost complete breaking of bonds and such metallic bonding exists in the liquid state to some extent, high temperatures are necessary. Therefore, the boiling points of metals are much higher than their melting points.

Oxidation States

A characteristic property of the *d*-transition metals is their ability to exhibit several oxidation states differing usually by units of one. **This is due to the fact that (*n* – 1)*d* electrons may get involved along with *ns*-electrons in bonding, as electrons in (*n* – 1)*d* orbitals are in an energy state comparable to *ns*-electrons.** The stability of a given oxidation state depends on the nature of the element with which the transition metal is combined. **The highest oxidation states are found in the compounds of fluorides and oxides because of most electronegative nature of F and O.** The different oxidation states of 3*d*-series have been given below (very rare oxidation states are provided in parenthesis).

Sc	Ti	V	Cr	Mn	Fe	Co	Ni	Cu	Zn
			I					I	
(II)	(II)	II	II	II	II	II	II	II	II
III	III	III	III	(III)	III	III	(III)	(III)	
	IV	IV	IV	IV	(IV)	(IV)	(IV)		
		V	(V)	(V)	(V)	(V)			
			VI	VI	VI				
				VII					

- The lesser number of oxidation states in the beginning of the series can be due to the presence of too few electrons to loose or share, towards the end of the series it can be ascribed to the presence of too many electrons and thus fewer empty orbitals to share electrons with the ligands.
- Except Sc, the most common oxidation state of the first row transition elements is II which arises from the loss of two 4*s*-electron. This means that after Sc, *d*-orbitals become more stable than the *s*-orbital.
- In the II and III oxidation states, bonds formed are mostly ionic. In compounds of the higher oxidation states, the bonds are formed by the removal or sharing of *d*-electron hence, the bonds formed are essentially covalent (as in MnO_4^-).
- Within a group, the maximum oxidation state increases with atomic numbers.

Fe	II, III
Ru, Os	II, III, IV, VI, VIII

- Transition metals also form compounds in low oxidation states (1,0). Such compounds are expected to be unstable except in cases where vacant *d*-orbitals are used for accepting lone-pairs from π-bonding ligands *e.g.*, $Ni(CO)_4$, $[Ag(CN)_2]^-$, $[Ag(NH_3)_2]^+$.

DEEP Focus

- **When an element exists in more than one oxidation state, standard electrode potential data provide a clue to the relative stabilities of different states w.r.t. oxidation. Cr^{2+} is unstable in water w.r.t. oxidation whereas Cr^{3+} is stable in water; Fe^{2+} is unstable in aerated water (and changes to Fe^{3+}) w.r.t. oxidation.**

- Another feature is the reduced tendency of higher oxidation states towards the end of the series. This could be due to steady increase in the effective nuclear charge along the series thus pulling the *d*-orbitals into the electron core and not making them readily available for bonding. Thus, only oxidation of zinc is Zn(II) where no *d*-orbital is involved. Also, it is difficult to form species that do not utilise the *d*-electrons, *i.e.*, Sc(II) is virtually unknown and Ti (IV) is more stable than Ti (II).

Table 16.1 **Selected Properties of Elements of the First Transition Series**

Properties	Sc (21)	Ti (22)	V (23)	Cr (24)	Mn (25)	Fe (26)	Co (27)	Ni (28)	Cu (29)	Zn (30)
Electron configuration	[Ar] $3d^1 4s^2$	$3d^2 4s^2$	$3d^3 4s^2$	$3d^5 4s^1$	$3d^5 4s^2$	$3d^6 4s^2$	$3d^7 4s^2$	$3d^8 4s^2$	$3d^{10} 4s^1$	$3d^{10} 4s^2$
Radius/pm	161	145	132	127	124	124	125	125	128	133
M^+	–	90	88	85	80	77	75	69	72	–
M^{2+}	81	77	74	64	66	60	64	–	–	74
Covalent radius/pm	144	132	122	117	117	117	116	115	117	125
$(IE)_1$	631	656	650	652	717	762	758	736	775	906
$(IE)_2$	1235	1309	1413	1591	1509	1561	1645	1751	1958	1733
$(IE)_3$	2389	2650	2828	2986	3250	2956	3231	3393	3578	3833
Abundance ppm by weight	25	6320	136	122	1060	60000	29	99	68	76
Melting point/°C	1397	1672	1710	1900	1244	1530	1495	1455	1083	420
$E^\circ_{M^{2+}/M}$/V	–2.08 ($Sc^{3+} + 3e^- \rightarrow Sc$)	–1.63	–1.18	–0.91	–1.19	–0.44	–0.28	–0.23	–0.34	–0.76
Density g/cm^3	2.99	4.49	5.96	7.20	7.20	7.86	8.90	8.91	8.92	7.14
Electrical conductivity [w.r.t. Ag which is 100 (say)]	—	2	3	10	2	17	24	24	97	100
Common oxidation state	+3	+4	+3	+3	+2	+2	+2	+2	+1	+2
			+4	+6	+4	+3	+3		+2	
			+5		+7					
Sublimation energy/kJ mol^{-1}	305	429	459	349	220	351	382	372	305	115
Electronegativity (A/R)	1.20	1.30	1.45	1.55	1.60	1.65	1.70	1.75	1.75	1.65

Table 16.2 **Reactivity of Some Oxidation States of 3*d*-series Elements in Aqueous Solution**

Reducing agent	Most stable	Oxidising agents
—	Sc^{3+}	—
Ti^{2+}, Ti^{3+}	Ti^{3+}	—
V^{2+}, V^{3+}	V^{4+}	V^{5+}
Cr^{2+}	Cr^{3+}	$Cr^{6+}(CrO_4^{2-})$
—	Mn^{2+}	Mn^{3+}, $Mn^{4+}(MnO_2)$ $Mn^{7+}(MnO_4^-)$
Fe^{2+}	Fe^{3+}	—
—	Co^{2+}	Co^{3+}
—	Ni^{2+}	—
Cu^+	Cu^{2+}	—

DEEP Focus

- The electronic structure of the atoms in the second and third rows do not always follow the pattern of the first row. The structures of the nickel group are :

Ni	$3d^8\ 4s^2$
Pd	$4d^{10} 5s^0$
Pt	$5d^9\ 6s^1$

Since a full shell of electrons is a stable arrangement, the place where this occurs is of importance.

The *d*-levels are complete in :

Ni		Cu	$3d^{10}\ 4s^1$	Zn	$3d^{10}\ 4s^2$
Pd	$4d^{10}\ 5s^0$	Ag	$4d^{10}\ 5s^1$	Cd	$4d^{10}\ 5s^2$
Pt		Au	$5d^{10}\ 6s^1$	Hg	$5d^{10}\ 6s^2$

Even though the ground state of the atom has a d^{10}-configuration, Pd and the coinage metals Cu, Ag and Au behave as typical transition elements. This is because in their most common oxidation states, Cu (II) has a d^9-configuration, and Pd (II) and Au (III) have d^8-configuration (incomplete *d* level). However, Zn (II), Cd (II) and Hg (II) have a d^{10} configuration. Because of this, these elements do not show the properties characteristic of transition elements.

Ionisation Energies

- The magnitude of ionisation energies provide an indication of the energy needed to raise the metal to a particular oxidation state in a compound.
- **$(IE)_1$ of 5*d*-elements are higher than those of the 3*d* and 4*d*-elements. This is due to greater effective nuclear charge acting on outer valence electrons because of the weak shielding of the nucleus by 4*f*-electrons.** The ionisation energies of the 3*d* and 4*d*-elements are irregular.

DEEP Focus

- It is important to note that though the 4*s*-orbitals are occupied before 3*d*-orbitals, we cannot say that they are always more stable. In fact, the ionisation of transition elements takes place by the loss of *ns*-electrons first. What happens actually is that when the electron is ionised from any transition element, say from iron ($Z = 26$) of 3*d*-series, the effective nuclear charge experienced by the 3*d*-electrons is greatly enhanced over that of any 4*s*-electron as a direct consequence of the greater stability attained by the 3*d*-orbitals. Thus, 3*d*-orbitals are expected to drop in energy below 4*s*-orbitals. Thus, ionisation of two or more electrons from an atom of a transition element will take place with the removal of *s*-electrons in preference to the *d*-electrons.

- Common oxidation states for each element include +2 and +3 or both. The +3 oxidation states are more stable at the beginning of the series, whereas towards the end of +2 oxidation states are more stable. This can be explained by study of variation of (*IE*) plots. *IE* increases gradually from left to right. However, the third (*IE*), when an electron is removed from 3*d*-orbital, increases more rapidly than the second and third (*IE*). Because it takes more energy to remove the third electron from the metals near the end of the row than from those near the beginning, the metals near the end tend to form M^{2+} ions rather than M^{3+} ions. (Fig. 16.2)

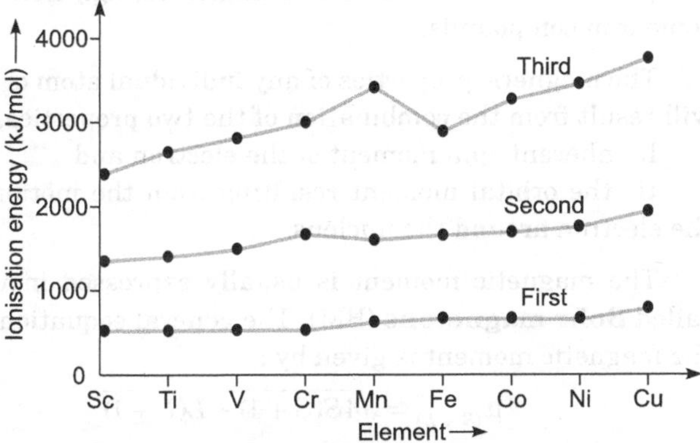

Fig. 16.2 Variation of the first, second, and third ionisation energies for the first transition metals.

DEEP Focus

- It is possible to determine thermodynamic stability of the transition metal compounds based on the values of (*IE*) of the metals. $(IE)_1$, $(IE)_2$, $(IE)_3$ and $(IE)_4$ of Ni and Pt in mJ mol^{-1} are compared.

	$(IE)_1 + (IE)_2$	$(IE)_3 + (IE)_4$	Total
Ni	2.49	8.80	11.29
Pt	2.66	6.70	9.36

Thus, nickel (II) compounds tend to be thermodynamically more stable than platinum (II) while platinum (IV) compounds are relatively more stable than nickel (IV). $K_2\overset{IV}{Pt}Cl_6$ is well known but such type of nickel compound is not formed. Following factors also decide the stability of the compounds :

– the energy of the sublimation,
– the lattice energy,
– the solvation energy.

Electrode Potential

When a metal is placed in a solution of its ions a potential difference is set up between the metal and the solution. There is a tendency for the metal ions to leave the metal lattice and go into the solution thus leaving an excess of electrons and hence a negative charge on the metal; there is also a reverse tendency for the metal ions from the solution to deposit on the metal leading to a positive charge on the metal. In practice one of these effects is greater than the other, bringing about a potential difference between the metal and the solution. The value of this potential difference for a particular metal depends upon the nature of metal, the concentration of the metal ions in solution an the temperature. **By convention, the potential difference set up in a 1 M solution of metal ions at 298 K is called the standard electrode potential.** (measured w.r.t. to reference half-cell)

In general, we can say that more negative the value of the electrode potential for the couple M^{n+}/M, more is the reducing power of the element. Similarly, more positive is the value of electrode potential for the couple M^{n+}/M, more is the oxidising power of the element. The values of some standard electrode potentials for the elements of first transition series are given in Table 16.1. Electrode potential is a measure of the electropositive character and the reactivity of the metals. In general along a period, there is a decrease in electropositive character. The reactivity of metals also decreases along a period and down a group. All the elements of the first transition series, except copper, have negative values and can react with acids (H^+) producing hydrogen. A plot of variation of the electrode potential of the transition elements of 3*d*-series is shwon in Fig. 16.3.

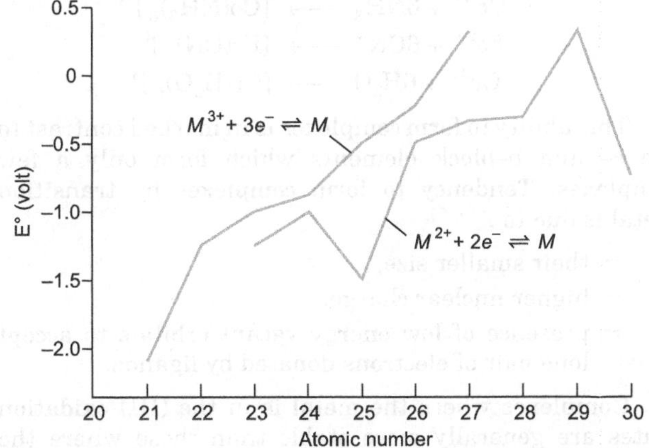

Fig. 16.3 Trends in electrode potentials of transition metals of 3*d*-series

A qualitative treatment of the stability of transition metal ion in different oxidation states in aqueous medium is obtained from the electrode potential data. The value of $E°$ for M^{2+}/M denotes the emf of the cell in which the reaction

$$2H^+(aq) + M(s) \longrightarrow M^{2+}(aq) + H_2(g) + 2e^-$$

occurs under reversible conditions (values given in Table 16.1). The total energy ΔH of process

$$M(s) \longrightarrow M^{2+}(aq) + 2e^-$$

can be represented as equal to sum of the energy of the following steps :

$$M(s) \longrightarrow M(g); \quad \Delta H_{sub} \text{ (enthalpy of sublimation)}$$
$$M(g) \longrightarrow M^{2+}(g) + 2e^-; \quad (IE)_1 + (IE)_2$$
$$M^{2+}(g) \longrightarrow M^{2+}(aq); \Delta H_{hyd} \text{ (enthalpy of hydration)}$$
$$M(s) \longrightarrow M^{2+}(aq) + 2e^-$$
$$\Delta H = \Delta H_{sub} + (IE)_1 + (IE)_2 + \Delta H_{hyd}.$$

DEEP Focus

- Although the transition metals are less electropositive (or more electronegative) than the alkali and alkaline-earth metals, their standard reduction potentials suggest that all of them except copper should react with strong acids such as dil. HCl to produce H_2 gas. However, most transition metals are inert towards acids or react slowly with them because of a protective layer of oxide. In case of Cr, despite a negative standard reduction potential, it is quite inert chemically because of the formation of chromium (III) oxide, Cr_2O_3 on its surfaces. This fact find application of Cr to use as a protective and noncorrosive plating on other metals.

Complexes

The transition elements have an unparalleled tendency to form coordination compounds with Lewis bases that is with groups which are able to donate an electron pair (called ligands). A ligand may be a neutral molecule such as NH_3, or an ion such as Cl^- or CN^-.

$$Co^{3+} + 6NH_3 \longrightarrow [Co(NH_3)_6]^{3+}$$
$$Fe^{2+} + 6CN^- \longrightarrow [Fe(CN)_6]^{4-}$$
$$Cr^{3+} + 6H_2O \longrightarrow [Cr(H_2O)_6]^{3+}$$

This ability to form complexes is in marked contrast to the *s*– and *p*–block elements which form only a few complexes. Tendency to form complexes by transition metal is due to :

- their smaller size,
- higher nuclear charge,
- presence of low energy vacant orbitals to accept lone pair of electrons donated by ligands.

Complexes where the metal is in the (III) oxidation states are generally more stable than those where the metal is in the (II) state. Thus, $[Fe^{III}(CN)_6]^{3-}$ is more stable than $[Fe^{II}(CN)_6]^{4-}$ (Detailed study of the complexes is given in Chapter 18 of the Text.

Electronegativity

Transition elements have fairly low values of electronegativity. It increases from Sc to Cu with a fall at Mn and Zn. However, this increase in electronegativity is much slower because the additional electron is being added to an inner shell which provides relatively good shielding to the outer electrons from the nucleus. The increasing electronegativity from Sc to Cu means that the elements become slightly less metallic and this is reflected in the increasing positive electrode potentials of their ions M^{2+} and M^{3+} (Table 16.1)

Magnetic Properties

When you place an iron piece near a magnet, you will see that it is immediately drawn towards the magnet. However, some elements are repelled by the magnets. The property of an element to be attracted or repelled by a magnet differs from element to element. Substances which are weakly repelled by the magnetic field are called **diamagnetic** (with no unpaired electrons) while the substances which are weakly attracted by the magnetic field and lose their magnetism when removed from the field are called **paramagnetic.** If the force of attraction is very large and the permanent magnetisation is retained, the substance is said to be **ferromagnetic,** *e.g.,* iron and some iron compounds.

The magnetic properties of any individual atom or ion will result from the combination of the two properties,

I. inherent spin moment of the electron and

II. the orbital moment resulting from the motion of the electron around the nucleus.

The magnetic moment is usually expressed in unit called **Bohr magnetons** (BM). The general eequation for the magnetic moment is given by :

$$\mu_{(S+L)} = \sqrt{4S(S+1) + L(L+1)}$$

In above expression, S is the sum of the spin quantum numbers and L is the sum of orbital angular momentum quantum numbers of all the electrons. In many compounds including those of the first row transition elements, the orbital contribution is compensated by the electric fields of the surrounding atoms and as an approximation, the observed magnetic moment may be considered to arise **only from unpaired spins.** Putting $L = 0$ in the above expression, you can get the spin-only magnetic moment μ_S.

Thus, $$\mu_S = \sqrt{4S(S+1)}$$

The spin-only magnetic moment, μ_S can also be related to the number of unpaired electrons, N, in any species, as the total spin quantum number $S = \frac{N}{2}$

Hence, $$\mu_S = \sqrt{4S(S+1)}$$
$$= \sqrt{\frac{4N}{2}\left(\frac{N}{2}+1\right)}$$
$$\mu_S = \sqrt{N(N+2)}$$

Above expression gives the values of magnetic moment in Bohr magnetons which can be converted into SI unit of Ampere square meter (A m^2) by the following relationship :

$$1 \text{ BM} = 9.274 \times 10^{-24} \text{ A m}^2$$

Magnetic moment increases as number of unpaired electrons increases. For 3*d*-series elements, variation of magnetic moment is given in Fig. 16.4. This clearly shows magnetic moment is maximum for chromium (with six unpaired electrons).

	Sc	Ti	V	Cr	Mn	Fe	Co	Ni	Cu	Zn
Z	21	22	23	24	25	26	27	28	29	30
N	1	2	3	6	5	4	3	2	1	0
μ	$\sqrt{3}$	$\sqrt{8}$	$\sqrt{15}$	$\sqrt{48}$	$\sqrt{35}$	$\sqrt{24}$	$\sqrt{15}$	$\sqrt{8}$	$\sqrt{3}$	0

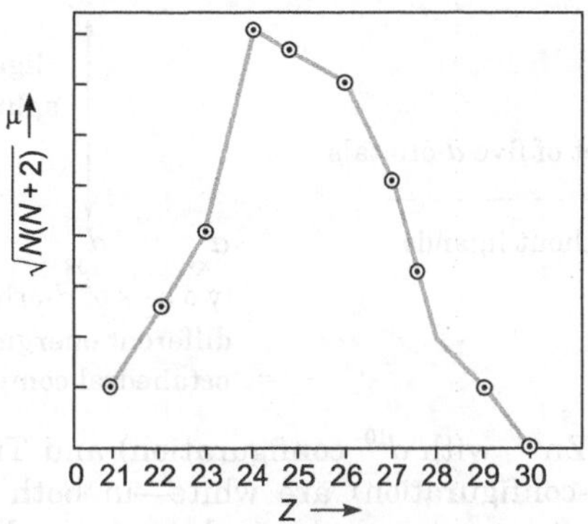

Fig. 16.4 Magnetic moment of 3*d*-series elements (on arbitrary scale)

In some cases (*e.g.*, Mn^{2+}, or Fe^{3+}, in which all the *d*-orbitals are occupied singly by electrons for which $m_l = 2, 1, 0, -1$ and -2, giving $L = 0$) the observed magnetic moment values agree very well with the spin-only value as given in Table 16.3. But generally, experimental values differ from the spin-only values. This is because the orbital motion of the electron also makes some contribution to the moment.

Table 16.3 **Predicted and Observed Magnetic Moment Values of Some Transition Metal Hydrated Ions**

Ion	Electronic configuration	Unpaired electrons	Magnetic moment (BM) [Calculated]	Magnetic moment (BM) [Experimental (μ_S)]
$[Ti(H_2O)_6]^{3+}$	$3d^1$ ↑	1	1.73	1.75
$[V(H_2O)_6]^{3+}$	$3d^2$ ↑↑	2	2.84	2.75
$[V(H_2O)_6]^{2+}$	$3d^3$ ↑↑↑	3	3.87	3.86
$[Cr(H_2O)_6]^{2+}$	$3d^4$ ↑↑↑↑	4	4.90	4.80
$[Mn(H_2O)_6]^{2+}$	$3d^5$ ↑↑↑↑↑	5	5.92	5.96
$[Fe(H_2O)_6]^{2+}$	$3d^6$ ↑↓↑↑↑↓	6	4.90	5.00
$[Co(H_2O)_6]^{2+}$	$3d^7$ ↑↓↑↓↑↑↑	7	3.87	4.40
$[Ni(H_2O)_6]^{2+}$	$3d^8$ ↑↓↑↓↑↓↑↑	8	2.84	2.90
$[Cu(H_2O)_6]^{2+}$	$3d^9$ ↑↓↑↓↑↓↑↓↑	9	1.73	1.80

Colour

Many ionic and covalent compounds of transition elements (and also inner transition elements) are coloured. In contrast, compounds of the *s*- and *p*- block elements are almost always white. Colour may arise from an entirely different cause in ions with **incomplete *d* or *f*-shells**. In a free isolated gaseous ions, the five *d*-orbitals are degenerate, that is, they are identical in energy. The surrounding groups, which can be solvent molecules in solution or ligands in a complex or other ions in a crystal lattice, affect the energy of some *d*-orbitals more than others. Thus, the *d*-orbitals no longer degenerate and at their simplest they form two groups of orbitals of different energy. Thus in transition element with a partly filled *d*- shell, it is possible to promote electrons from one *d*-level to another *d** level of higher energy (called *d*–*d** transition). This corresponds to a fairly small energy difference, and so light is absorbed in the visible region. **The colour of a transition metal complex is dependent on how big the energy difference is between the two *d*-levels. This in turn depends on the nature of the ligand, and the type of the complex formed. The colour which we see is the colour of the transmitted wavelength and is the complimentary colour absorbed by the compounds**

$[Ni(NH_3)_6]^{2+}$	blue
$[Ni(H_2O)_6]^{2+}$	green
$[Ni(NO_2)_6]^{2-}$	brown-red

Energy d_{z^2} $d_{x^2-y^2}$

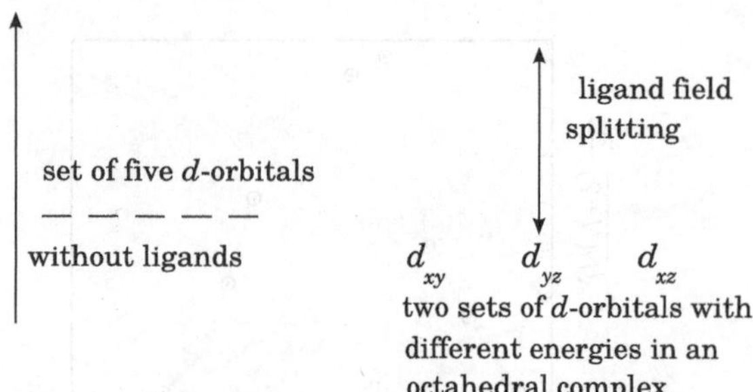

- $ZnSO_4$ (Zn^{2+} with d^{10} configuration) and TiO_2 (Ti^{4+} with d^0-configuration) are white—in both of them *d*–*d** spectra are impossible and they are colourless.

DEEP Focus

- In the series Sc (+III), Ti (+IV), V (+V) Cr (+VI) and Mn (+VII), all have empty *d*-orbitals, hence *d*–*d** spectra are impossible and they should be colourless. But following ions in aqueous solution are:

VO_2^{2+} (oxidation state +V) — pale yellow
CrO_4^{2-} (oxidation state +VI) — deep yellow
MnO_4^- (oxidation state +VII) — intense purple

The **colours arise by charge transfer,** for example, in MnO_4^- an electron is momentarily transferred from O to the metal and thus oxygen changes from O^{2-} to O^- and manganese from Mn(+VII) to Mn(+VI). (Also refer Chapter 18 of this Text). Table 16.4 gives the colour and oxidation states of the metal ions present in some hydrated ions of transition elements.

Table 16.4 **Oxidation States and Observed Colours for Some Aqua Species**

Element	+2	+3	+6	+7
Sc		Colourless $[Sc(H_2O)_6]^{3+}$		
Ti		Violet $[Ti(H_2O)_6]^{3+}$		
V	Violet $[V(H_2O)_6]^{2+}$	Green $[V(H_2O)_6]^{3+}$		
Cr	Blue $[Cr(H_2O)_6]^{2+}$	Violet/green $[Cr(H_2O)_6]^{3+}$	Orange yellow $Cr_2O_7^{3-}$, CrO_4^{2-}	
Mn	Pink $[Mn(H_2O)_6]^{2+}$	Red $[Mn(H_2O)_6]^{3+}$	Green MnO_4^{2-}	Purple MnO_4^-
Fe	Pale green $[Fe(H_2O)_6]^{2+}$	Yellow/ brown $[Fe(H_2O)_6]^{3+}$		
Co	Pink $[Co(H_2O)_6]^{2+}$	Blue $[Co(H_2O)_6]^{3+}$		
Ni	Green $[Ni(H_2O)_6]^{2+}$			
Cu	Blue $[Cu(H_2O)_6]^{2+}$			
Zn	Colourless $[Zn(H_2O)_6]^{2+}$			

Size of Atoms and Ions

- The covalent radii of the elements (summarised in Table 16.5) decrease from left to right across a row in the transition series, until near the end when the size increases slightly. On passing from left to right, extra protons are placed in the nucleus and extra orbital electrons are added. The orbital electrons shield the nucleus charge incompletely (order being $d < p < s$). Thus the nuclear charge attracts all the electrons more strongly, hence, a contraction in size occurs.

DEEP Focus

- **The radii of the elements from chromium to copper, however, are very close to one another. This is due to the fact that successive addition of *d*-electrons screen the outer electrons (4*s*) from the inward pull of the nucleus. As a result of this, the size of the atom does not alter much in moving from chromium to copper.**

- The relative constancy in covalent radii for the later elements is partly responsible for the similarity in properties of the Groups 8, 9 and 10 (VIII B) elements (which are Fe, Co, and Ni).
- Atoms of the transition elements are smaller than those of the Group I or II elements (*s*-block) in the same horizontal period. This is partly because of the usual contraction in size across a horizontal period and partly because the orbital electrons are added to the penultimate *d*-shell rather than to the outer shell of the atom.

However, it is important to emphasise here that shielding of the outer *ns*-electron(s) by $(n-1)$ *d*-electron(s) is more efficient than the shielding of a *ns*-electron by another *ns* electron (or that of a *np*-electon by another *np*-electron). **This is why the decrease in atomic radius from sodium to chlorine is greater than that from scandium to copper.** The elements which occur immediately after the transition elements are smaller than expected from simple.

The rate of decrease in size along the lanthanide series is even less than that in the transition series since in the lanthanides the electrons are added to the penultimate $[(n-2)f]$ shell and these shield the outer electrons much more effectively. The presence of 4*f*-electrons in the lanthanides affects the atomic size and therefore, the chemistry of the elements following the lanthanides. The atomic radii of the elements of third transition series are much smaller than expected. This is due to the effect of the greater than expected effective nuclear charge felt by the electrons of the elements of the third row transition series, hafnium to gold, owing to the insertion of lanthanides.

Metallic radii (pm) of some elements of Group 1-13

1	2	3	4	5	6	7	8	9	10	11	12	13
K	Ca	Sc	Ti	V	Cr	Mn	Fe	Co	Ni	Cu	Zn	Ga
235	197	164	147	135	130	135	126	125	125	128	137	141
Rb	Sr	Y	Zr	Nb	Mo	Tc	Ru	Rh	Pd	Ag	Cd	In
248	215	178	160	146	139	136	134	134	137	144	154	166
Cs	Ba	La	Hf	Ta	W	Re	Os	Ir	Pt	Au	Hg	Tl
267	222	188	160	149	141	137	135	136	139	146	157	171

Data in Table 16.1 and Fig. 16.5 show that the general trend of decreasing size is reversed towards the end of the series. This could be due to an increase in inter-electronic repulsion after the addition of sufficient number of electrons in the *d*-orbitals leading to the gradual increase in size.

Table 16.5 **Covalent Radii of the Transition Elements (pm)**

K	Ca	Sc	Ti	V	Cr	Mn	Fe	Co	Ni	Cu	Zn
203	174	144	132	122	117	117	117	116	115	117	125
			Zr	Nb							
			145	134							
			Hf	Ta							
			144	134							

There is a gradual decrease in size of the 14 lanthanide elements from cerium (58) to lutetium (71). This is called **Lanthanide Contraction.** ***The lanthanide contraction cancels almost exactly the normal size increase on descending a group of transition elements. Thus covalent and ionic radii of* Nb *are the same as the values for* Ta. *Same is the case for Zr and* Hf** (also refer Table 16.6 and Fig. 16.5).

Table 16.6 **Effect of Lanthanide Contraction on Ionic Radii (pm)**

Ca^{2+} 100	Sc^{3+} 74.5	Ti^{4+} 60.5	V^{3+} 64.0
Sr^{2+} 118	Y^{3+} 90.0	Zr^{4+} 72.0	Nb^{3+} 72.0
Ba^{2+} 135	La^{3+} 103.2	Hf^{4+} 71.0	Ta^{3+} 72.0

The effects of lanthanide contraction are less pronounced towards the right of the d-block.

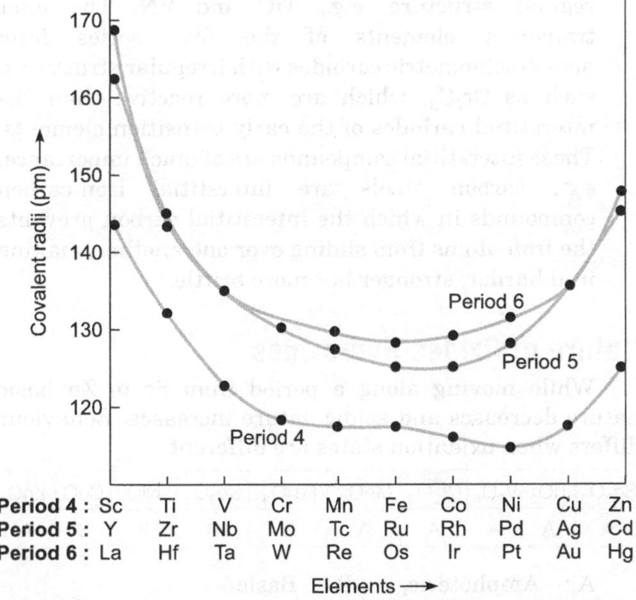

Fig. 16.5 Comparison of the covalent radii for transition elements of different periods. Note that elements in the same column of the fifth and sixth periods have approximately the same radii.

Atomic volume of an element is directly related to its size and therefore, atomic volumes follow the same trend as the atomic size. Similarly density is also related to the size of the element. The smaller the size, the higher the density of the element as shown in Fig. 16.6. For 4*d* and 5*d* elements, this increase is not as regular as the increase in densities for 3*d*-elements.

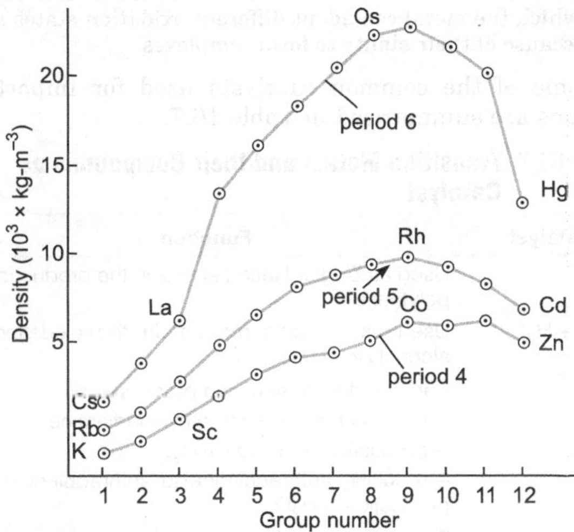

Fig. 16.6 Trend in densities of the alkali, alkaline earth and transition metals of the fourth, fifth and sixth periods.

Non-stoichiometry

Transition elements have the property of forming compounds of indefinite structures and proportions. These are called **Non-stoichiometric** compounds. **Iron (II) oxide FeO is written as $\overline{FeO}$ to indicate** that ratio of Fe and O atoms is not exactly 1 : 1. Analysis shows that the formula varies between $Fe_{0.94}O$ and $Fe_{0.84}O$. Vanadium and selenium form a series of compounds ranging from $VSe_{0.98}$ to VSe_2 :

$$\overline{VSe}\ (VSe_{0.98} \longrightarrow VS_{1.2})$$
$$\overline{V_2Se_3}\ (VSe_{1.2} \longrightarrow VSe_{1.6})$$
$$\overline{V_2Se_4}\ (VSe_{1.6} \longrightarrow VSe_2)$$

CuS (insoluble) is actually a mixture of CuS and Cu_2S. Non-stoichiometry is shown particularly among transition metal compounds of the Group VI elements (O, S, Se, Te) and is mostly due to **variable valency** of transition elements, and also sometimes due to **defects** in solid structures.

Catalytic Properties

Many transition metals and their compounds have catalytic properties.

DEEP Focus

These metals can function as catalysts because they can utilise both *d* and *s*-electrons for the formation of bonds between reactant molecules and the surface catalyst atoms. This increases the concentration of the reactants at the catalyst surface and weakens the bonds in the reactant molecules with the result that the activation energy is lowered. Compounds of transition metals are able to act as catalysts because of the ease with which the metal can adopt different oxidation states and also because of their ability to form complexes.

Some of the common catalysts used for important reactions are summarised in Table 16.7.

Table 16.7 **Transition Metals and their Compounds as Catalyst**

Catalyst	Function
$TiCl_4$	Used as Ziegler Natta catalyst in the production of polythene
$FeSO_4 + H_2O_2$	Used as Fenton's reagent in the oxidation of alcohols to aldehydes.
Ni	Raney nickel in reduction processes : —manufacture of hexamethylenediamine, —production of H_2 from NH_3, —reducing anthraquinone to anthraquinol in the production of H_2O_2
Ni complexes	Reppe synthesis (polymerisation of alkynes to give benzene, cyclooctatetraene).
Cu	Used in manufacturing of $(CH_3)_2SiCl_2$ being used to make silicones.
Cu/V	Used in the oxidation of cyclohexanol/ cyclohexanone mixture to acetic acid which is used to prepare Nylon–66.
$CuCl_2$	Deacon process of making Cl_2 from HCl.
MnO_2	Used in the decomposition of $KClO_3$ to KCl and O_2.
V_2O_5	Used in contact process of H_2SO_4.
Fe/Mo	Promoted Fe used in the Haber-Bosch process of NH_3
$FeCl_3$	Used in the production of CCl_4 from CS_2 and Cl_2
$PdCl_2$	Wacker process for converting C_2H_4 into CH_3CHO.
Pt/PtO	Adams catalyst in reduction.
Pt	In three stage convertors for cleaning car exhaust fumes.
Pt/Rh	In Ostwald process for HNO_3 from NH_3
ZnO/CuO	In synthesis of CH_3OH from CO and H_2
Co/Fe	In synthesis of hydrocarbons by CO and H_2 in Fischer-Tropsch process.
Pt/PtO	Adians Catalyst for reduction

Enzymes are catalysts that enhance the rates of specific reactions. Some enzymes require the presence of metal ions as cofactors called as **metalloenzymes** which generally contain transition metal, *e.g.*, Mo in xanthine oxidase (metabolism of purines).

A cobalt atom lies at the centre of the vitamin B_{12} coenzyme. Iron atoms are importantly involved in haemoglobin of blood and in the ferredoxins of photosynthetic process. Fe and Mo are present in nitrogen fixing enzyme.

Alloy Formation

An alloy is a blend of metals prepared by mixing the components. Alloys (refer Appendix VI) may be homogeneous solid solutions in which the atoms of one metal are distributed randomly among the atoms of the other. Such alloys are formed by atoms with metallic radii that are within about 15 per cent of each other. Because of similar radii and other characteristics of transition metals, alloys are readily formed by these metals. The alloys so formed are hard and have often high melting points. The best known are ferrous alloys; chromium, vanadium, tungsten, molybdenum, and manganese are used for the production of variety of steels and stainless steel. Alloys of transition metals with non-transition metals as brass (copper-zinc), and bronze (copper-tin) are also of considerable industrial importance.

Interstitial Compounds

- Transition metals can trap some small atoms like hydrogen, boron carbon, nitrogen, etc., in vacant spaces in their crystal lattice forming interstitial compounds. Carbon and nitrogen always occupy octahedral holes; hydrogen is smaller and always occupies tetrahedral holes. As only transition metals form such compounds, the *d*-electrons are, therefore, presumably involved in the bonding. The structure of the metal often changes during the formation of such compounds. The composition of these compounds is generally non-stoichiometric, *e.g.*, $TiH_{1.73}$, $PdH_{0.56}$, $VH_{0.56}$, but may approach regular stoichiometry and a regular structure, *e.g.*, TiC and VN. The later transition elements of the first series form non-stoichiometric carbides with irregular structures, such as Cr_7C_3, which are more reactive than the interstitial carbides of the early transition elements. These interstitial compounds are of much importance, *e.g.*, carbon steels are interstitial iron-carbon compounds in which the interstitial carbon prevents the iron atoms from sliding over one another, making iron harder, stronger but more brittle.

Nature of Oxides/Hydroxides

While moving along a period from Sc to Zn basic nature decreases and acidic nature increases. Behaviour differs when oxidation states are different

Sc_2O_3	TiO_2	V_2O_5	Cr_2O_3	MnO_2	Fe_2O_3	Co_2O_3	NiO	CuO	ZnO
A	A	A	A	A	B	B	B	B	A

A : Amphoteric, B : Basic

With increase in the oxidation state of a given transition metal, the covalent character of its compound increases and thus acidic character also increases.

	MnO	Mn_3O_4	Mn_2O_3	MnO_2	MnO_3	Mn_2O_7
Oxidation state of Mn	+2	+8/3	+3	+4	+6	+7
Nature of oxide	Basic	Amphoteric			Weakly acidic	Strongly acidic

Characteristics of oxides and ions of vanadium and chromium have been summarised in Table 16.8.

The basic and amphoteric oxides dissolve in non-oxidising acids and form hexaaquo ions $[M(H_2O)_6]^{n+}$. The acidic oxides like CrO_3 and Mn_2O_3 dissolve in water to form the oxyacids These oxides also react with bases to form oxosalts.

$$CrO_3 + H_2O \longrightarrow \underset{\text{(chromic acid)}}{H_2CrO_4}$$

$$CrO_3 + 2KOH \longrightarrow K_2CrO_4 + H_2O$$

$$2CrO_3 + H_2O \longrightarrow \underset{\text{(dichromic acid)}}{H_2Cr_2O_7}$$

$$2CrO_3 + 2KOH \longrightarrow \underset{\text{potassium dichromate}}{K_2Cr_2O_7} + H_2O$$

$$Mn_2O_7 + H_2O \longrightarrow \underset{\text{(permanganic acid)}}{2HMnO_4}$$

$$Mn_2O_7 + 2KOH \longrightarrow 2KMnO_4 + H_2O$$

As we move along a given transition series, +2 oxidation state of the metals becomes more and more stable and hence the tendency of these metals to lose electrons increases hence, reducing nature of the metal oxide increases. VO and CrO are strong reducing agents.

Table 16.8 **Chracteristics of Oxides and Some Ions of V and Cr**

O.S.	Oxide/ Hydroxide	Behaviour	Ion	Name of ion	Colour of ion
+2	VO	basic	V^{2+}	vanadium (II) (vanadous)	violet
+3	V_2O_5	basic	V^{3+}	vanadium (III) (vanadic)	green
+4	VO_2	amphoteric	VO^{2+}	oxovanadium (IV) (vanadyl)	blue
			$V_4O_9^{2-}$	hypovanadate (vanadite)	brown
+5	V_2O_5	amphoteric	VO_2^+	dioxovanadium (V) orthovanadate	yellow
			VO_4^{3-}		colourless
+2	CrO, $Cr(OH)_2$	basic	Cr^{2+}	chromium (II) (chromous)	light blue
+3	Cr_2O_3, $Cr(OH)_3$	amphoteric	Cr^{3+}	chromium (III) (chromic)	violet
			$Cr(OH)_4^-$	chromite	green
+6	CrO_3, $CrO_2(OH)_2$, $H_2Cr_2O_7$	acidic	CrO_4^{2-}	chromate	yellow
			$Cr_2O_7^{2-}$	dichromate	orange

Table 16.9 **Fourth Period Transition Elements Essential to Human Nutrition**

Element	Some biochemical substances	Functions
Chromium	Glucose tolerance factor	Utilisation of glucose
Cobalt	Cobalamin (vitamin B_{12})	Development of red blood cells
Copper	Ceruloplasmin	Synthesis of haemoglobin
	Cytochrome oxidase	Cell energetics
Iron	Haemoglobin and myoglobin	Transport and storage of oxygen
	Cytochrome C	Cell energetics
	Catalase	Decomposition of hydrogen peroxide
Manganese	Isocitrate dehydrogenase	Cell energetics
Zinc	Carbonic anhydrase	Elimination of CO_2
	Carboxypeptidase A (Pancreatic juice)	Protein digestion
	Alcohol dehydrogenase	Oxidation of ethanol

Target Practice 1

1. What are the types of the following reactions?
   (a) $3CrO_4^- + 8H^+ \longrightarrow 2CrO_4^{2-} + Cr^{3+} + 4H_2O$
   ....
   (b) $3MnO_4^{2-} + 4H^+ \longrightarrow 2MnO_4^- + MnO_2 + 2H_2O$
   ....
2. Write EC of
   (a) Cu(II) ....
   (b) Cr(0) ....
   (c) Co(III) ....
3. Name the element with maximum oxidation state in *d*-block elements
   ....
4. (a) Cr^{2+} is unstable in ....
   (b) Fe^{2+} is oxidised to Fe^{3+} in ....
5. Why a transition series contains ten elements?
   ....
   ....
6. $(IE)_1$ of 5*d*-elements are higher than those of the 3*d* and 4*d* elements. It is due to ....
7. Stepwise $(IE)_s$ of Ni and Pt are given below

	$(IE)_1 + (IE)_2$	$(IE)_3 + (IE)_4$
Ni	2.49	8.80
Pt	2.66	6.70

   Compare thermodynamic stability of Ni and Pt salts in (II) and (IV) oxidation state
   Ni(II) ☐ Pt(II)
   Ni(IV) ☐ Pt(IV)
8. $E°$ values of reactions
   $Fe^{2+} + 2e^- \longrightarrow Fe;$ $E° = -0.44$ V
   $Cu^{2+} + 2e^- \longrightarrow Cu;$ $E° = +0.34$ V
   The $E°$ of $Fe + Cu^{2+} \rightleftharpoons Fe^{2+} + Cu$ is .... and thus, reaction is spontaneous/non-spontaneous.
9. Select ions which can form complexes out of $Fe^{2+}, Ni^{2+}, Ca^{2+}, K^+, Mg^{2+}, Na^+, Cu^{2+}, Zn^{2+}$
   ....
10. Which complex is more stable in each pair?
    (a) $[Fe(CN)_6]^{4-}$ ☐ $[Fe(CN)_6]^{3-}$
    (b) $[Ni(CN)_4]^{2-}$ ☐ $[Pt(CN)_4]^{2-}$
    (c) $[NiCl_6]^{2-}$ ☐ $[PtCl_6]^{2-}$
11. Mention three factors which favour formation of complexes by *d*-block elements
    (a) ....
    (b) ....
    (c) ....
12. State whether the following complexes are coloured or not
    (a) $[Sc(H_2O)_6]^{3+}$ Yes/No
    (b) $[Cu(CN)_4]^{3-}$ Yes/No
    (c) $[Cu(H_2O)_4]^{2+}$ Yes/No
    (d) $[Zn(H_2O)_4]^{2+}$ Yes/No
13. Consider following complexes
    I : $[Fe(H_2O)_6]^{2+}$ II : $[Fe(CN)_6]^{4-}$
    III : $[Ni(CO)_4]$ IV : $[Ni(H_2O)_4]^{2+}$
    V : $[Ni(CN)_4]^{2-}$
    Out of these select the complex(es) with.
    (a) Maximum number of unpaired electrons ....
    (b) Maximum number of magnetic moment ....
    (c) Equal number of unpaired electrons ....
    (d) Equal magnetic moment ....
    (e) Square planar geometry ....
    (f) Octahedral geometry ....
    (g) Tetrahedral geometry ....
14. Convert 3.00 BM into SI unit (A m^2)
    ....
15. Calculate magnetic moment of V^{3+} ($Z = 23$)
    .... BM
16. What is oxidation state of iron in $[Fe(H_2O)_5NO]^{2+}$?
    ....
17. Give one example in which ion is coloured due to charge transfer.
    ....
    ....
18. Give oxidation of the underlined and give the colour of the ion

	Oxidation state	Coloured/colourless
VO_2^+		
CrO_4^{2-}	....	....

19. Give examples of coloured ions with following oxidation state
    (a) Cr(VI) ....
    (b) Cu (II) ....
    (c) Mn(VI) ....
    (d) Ni(II) ....
    (e) V(III) ....
    (f) Co(III) ....

Answers

1. (a) CrO_4^-(V) is oxidised to CrO_4^{2-} (VI) and reduced to Cr^{3+}(III)
   (b) MnO_4^{2-}(VI) is oxidised to MnO_4^-(VII) and reduced to MnO_2(IV)
   Thus, (a) and (b) both disproportionation reactions.
2. (a) [Ar] $3d^9$ (b) [Ar] $3d^5 4s^1$ (c) [Ar] $3d^6$
3. Os, Ru(VIII)
4. (a) water (b) aerated water (CO_2 mixed water)
5. In transition series, the electrons enter the *d*-orbitals which can accommodate ten electrons and therefore, there are ten elements in one transition series.
6. This is due to greater effective nuclear charge acting on outer valence electrons because of the weak shielding of the nucleus by 4*f*-electrons.

7. Ni(II) > Pt(II)
Ni(IV) < Pt(IV)

8. $E^\circ = E^\circ_{Fe/Fe^{2+}} + E^\circ_{Cu^{2+}/Cu}$
$= -(-0.44) + 0.34 = 0.78$ V

Spontaneous.

9. $Fe^{2+}, Ni^{2+}, Ca^{2+}, Cu^{2+}, Zn^{2+}, Mg^{2+}$
(Ca^{2+} and Mg^{2+} form complexes with EDTA)

10. (a) $[Fe(CN)_6]^{4-}$ < $[Fe(CN)_6]^{3-}$
(b) $[Ni(CN)_4]^{2-}$ > $[Pt(CN)_4]^{2-}$
(c) $[NiCl_6]^{2-}$ < $[PtCl_6]^{2-}$

11. (a) Smaller size,
(b) Higher nuclear charge,
(c) Presence of low energy vacant orbitals to accept lone pair of electrons donated by ligands.

12. (a) No (b) No (c) Yes (d) No
(Complex with unpaired electrons in *d*-orbital in the atom/ion is coloured)

13. (a) I (b) I (c) None (d) None
(e) V (f) I, II (g) III, IV

14. $3 \times 9.274 \times 10^{-24}$ A m^2

15. $V(Z=23): [Ar]\ 4s^2 3d^3$
$V^{3+}: [Ar]\ 4s^0 3d^2$
Magnetic moment $= \sqrt{N(N+2)}$ BM $= \sqrt{8}$ BM

16. This is a typical case of complex formed by electron exchange.

NO change to NO^+ by loss of electron and this electron is gained by Fe^{2+} which change to Fe^+

$$NO \longrightarrow NO^+ + e^-$$
$$Fe^{2+} + e^- \longrightarrow Fe^+$$

Thus, oxidation state of iron = +1

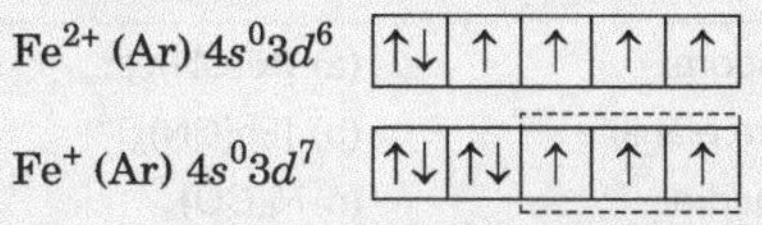

Three unpaired electrons

+1 oxidation state is confirmed by magnetic moment of iron $= \sqrt{15} = 3.87$ BM and also by diamagnetic character of NO. (which is only possible when it has no unpaired electron)

17. MnO_4^-
Mn is in +7 oxidation state,
Mn : [Ar] $4s^2 3d^5$
Mn^{2+} : [Ar] with no unpaired electrons; oxygen atom transfers electrons to Mn (in its *d*-orbital) and thus MnO_4^- is coloured.

Oxidation state

18. VO_2^+ +5 coloured
CrO_4^{2-} +6 coloured

19. (a) CrO_4^{2-} (b) $[Cu(NH_3)_4]^{2+}$ (c) MnO_4^{2-}
(d) $[Ni(H_2O)_4]^{2+}$ (e) $[V(H_2O)_6]^{3+}$ (f) $[Co(H_2O)_6]^{3+}$

Practice Exercise 1

1. What is the oxidation state of the underlined atoms in
(a) $\underline{Ni}(CO)_4$ (b) MnO_4^{2-}
(c) $\underline{Cr}O_2Cl_2$ (d) $\underline{Cr}O_5$
(e) $[\underline{Cr}(CN)_6]^{3-}$ (f) $[\underline{Fe}(CN)_5\overset{\oplus}{NO}]^{2-}$

2. Give the electron configuration of the metal in each of the following atoms or ions:
(a) Ni (b) Cr^{3+}
(c) FeO_4^{2-} (ferrate ion) (d) MnO_2

3. Arrange the following hydroxy compounds in order of increasing acid strength, and account for the trend
(a) $CrO_2(OH)_2$ (b) $Cr(OH)_2$
(c) $Cr(OH)_3$

4. Explain how $Cr(OH)_3$ can act both as an acid and as a base.

5. $E^\circ_{Cu^{2+}/Cu} = 0.34$ V, $E^\circ_{Zn^{2+}/Zn} = -0.76$ V, $E^\circ_{Ag^+/Ag} = 0.80$ V
Calculate E° values of following reactions :
(a) $2Ag^+ + Cu \longrightarrow 2Ag + Cu^{2+}$
(b) $Cu + Zn^{2+} \longrightarrow Zn + Cu^{2+}$
(c) $2Ag + Zn^{2+} \longrightarrow 2Ag^+ + Zn$
Are they spontaneous.

6. Following arbitrary values are given for the conversion of $M(s)$ into $M^{2+}(aq)$

$$M(s) \longrightarrow M(g); \quad \Delta H = 1000 \text{ kJ mol}^{-1}$$
$$M(g) \longrightarrow M(g) + e^-; \quad \Delta H = 950 \text{ kJ mol}^{-1}$$
$$M^+(g) \longrightarrow M^{2+}(g) + e^-; \quad \Delta H = 1100 \text{ kJ mol}^{-1}$$
$$M^{2+}(g) + aq \longrightarrow M^{2+}(aq); \quad \Delta H = -750 \text{ kJ mol}^{-1}$$

Calculate ΔH for

$$M(s) \longrightarrow M^{2+}(aq) + 2e^-$$

7. Which of the following compounds in each pair is more stable?
(a) $K_2[PtCl_6]$ $K_2[NiCl_6]$
(b) $[Ni(CN)_4]^{2-}$ $[Pt(CN)_4]^{2-}$
(c) $PtCl_4$ $NiCl_4$
(d) $PtCl_2$ $NiCl_2$

8. Why does chromium seem to be less reactive than its standard reduction potential suggests?

9. Write a balanced net ionic equation for reaction of each of the following metals with HCl in the absence of air. If no reaction occurs, indicate N.R.
(a) Cr (b) Zn
(c) Cu (d) Fe

10. $[Fe^{III}(CN)_6]^{3-}$ ion has magnetic moment of 1.73 B.M., while $[Fe^{III}(H_2O)_6]^{3+}$ has a magnetic moment of 5.92 B.M. Explain.

11. Which of the complexes use *d*-orbital of the metal/metal ion in hybridisation ?
(a) MnO_4^- (b) $\underline{Ni}(CO)_4$
(c) $[\underline{Ni}(CN)_4]^{2-}$ (d) $[\underline{Fe}(H_2O)_6]^{3+}$
(e) $[\underline{Fe}(CN)_6]^{3-}$

12. What is the EAN (effective atomic number) of the underlined atoms in the following complexes?
(a) $[\underline{Fe}(CN)_6]^{4-}$ (b) $[\underline{Cr}(en)_3]^{3+}$,
(c) $\underline{Ni}(CO)_4$ (d) $[Fe(H_2O)_6]^{2+}$

[**Hint :** EAN = *Z*(atomic no.) – oxid no. + 2 × (CN), CN is coordination number]

13. Match the geometry (given in column A) with the complexes (given in column B) in :

A	B
(I) octahedral	(a) $[Ni(CN)_4]^{2-}$
(II) square planar	(b) $[Fe(CN)_6]^{4-}$
(III) tetrahedral	(c) $Ni(CO)_4$

14. Calculate magnetic moment [$=\sqrt{N(N+2)}$ BM where N = number of unpaired electrons] of the underlined atoms in

(a) $\underline{Ni}(CO)_4$ (b) $[\underline{Fe}(CN)_6]^{4-}$
(c) $[\underline{Fe}(H_2O)_6]^{3+}$ (d) $[Fe(H_2O)_6]^{2+}$

15. Explain the term **'Lanthanide contraction.'**

16. Which is larger in each pair ?
(a) Fe^{2+} and Fe^{3+} (b) Ni and Ni^{2+}
(c) Nb^{3+} and Ta^{3+} (d) Y^{3+} and La^{3+}
(e) Cu and Zn (f) Cu and Ni
(g) I^- and I^+

17. Due to **'lanthanide contraction'**, second and third row transition elements (called 4d and 5d series respectively) have similar radii. What are the other properties do you think would be similar?

Answers

1. (a) 0 (b) +6 (c) +6 (d) +6 (e) +3 (f) +2

2. (a) $[Ar]3d^8\ 4s^2$ (b) $[Ar]\ 3d^3$ (c) $[Ar]3d^2$ (d) $[Ar]\ 3d^3$

3. $Cr(OH)_2 < Cr(OH)_3 < CrO_2(OH)_2$
Acid strength increases with polarity of the bond which increases with increase in oxidation state.

4. $\underset{\text{base}}{Cr(OH)_3} + 3HCl \longrightarrow CrCl_3 + 3H_2O$

$\underset{\text{acid}}{Cr(OH)_3} + NaOH \longrightarrow Na[Cr(OH)_4]$

5. (i) $E° = 0.46$ V spontaneous
(ii) $E° = -1.10$ V non-spontaneous
(iii) $E° = -1.56$ V non-spontaneous

6. $\Delta H = 2300$ kJ mol^{-1} (Add all the reactions to get the desired reaction)

7. (a) K_2PtCl_6 (b) $[Ni(CN)_4]^{2-}$ (c) $PtCl_4$ (d) $NiCl_2$

8. because of formation of oxide layer

9. (a) $2Cr + 6H^+ \longrightarrow Cr^{3+} + 3H_2$
(b) $Zn + 2H^+ \longrightarrow Zn^{2+} + H_2$
(c) $Cu + 2H^+ \longrightarrow$ no reaction
(d) $Fe + 2H^+ \longrightarrow Fe^{2+} + H_2$

10. $[Fe(CN)_6]^{3-}$ has one unpaired electron (d^2sp^3 hybridised Fe^{3+})

$$\text{Magnetic moment} = \sqrt{N(N+2)}\ \text{BM} = \sqrt{1\ (1+2)}\ \text{BM} = \sqrt{3} = 1.73\ \text{BM}$$

while $[Fe(H_2O)_6]^{3+}$ has five unpaired electrons (sp^3d^2 hybridised Fe^{3+}).

$$\text{Magnetic moment} = \sqrt{5(5+2)} = 35 = 5.92\ \text{BM}$$

Note : Magnetic moment = $\sqrt{N(N+2)}$ BM, where N = number of unpaired electrons

11. (a) MnO_4^- sp^3
(b) $Ni(CO)_4$ sp^3
(c) $[Ni(CN)_4]^{2-}$ dsp^2
(d) $[Fe(H_2O)_6]^{3+}$ sp^3d^2
(e) $[Fe(CN)_6]^{3-}$ d^2sp^3
hence (c), (d), (e) use *d*-orbital in bonding.

12. (a) 36 (b) 33 (c) 36 (d) 36

13.

A	B
I	(b)
II	(a)
III	(c)

14.

Complex	Unpaired electrons (N)	Magnetic moment $=\sqrt{N(N+2)}$ BM
(a)	0	0
(b)	0	0
(c)	5	$\sqrt{35}$ BM
(d)	1	$\sqrt{3}$ BM

15. There is a steady decrease in the radii as the atomic number of the lanthanide element increases. It is because, for every additional proton in a nucleus, the corresponding electron goes into 4*f*-shell which is too diffused to screen the nucleus as effectively as more localised inner shell. Hence, the attraction of the nucleus for the outermost electrons increases steadily with the atomic number of the lanthanides. This contraction in size is quite regular. This is called **Lanthanide contraction.**

16. (a) $Fe^{2+} > Fe^{3+}$ (b) $Ni > Ni^{2+}$ (c) $Nb^{3+} = Ta^{3+}$
(d) $Y^{3+} < La^{3+}$ (e) Cu < Zn (f) Cu > Ni (g) $I^- > I^+$

17. Lattice energies, solvation energies and ionisation energies would be similar.

Compounds

Potassium Dichromate ($K_2Cr_2O_7$)

Preparation

$K_2Cr_2O_7$ is obtained from chromite ore ($FeCr_2O_4$) in following steps:

Step I :	Chromite ore is fused with molten Na_2CO_3 in the presence of air when Na_2CrO_4 and Fe_2O_3 are formed. Fe_2O_3 is water insoluble while Na_2CrO_4 is water soluble hence, separation is done.	$4FeCr_2O_4 + 8Na_2CO_3 + 7O_2 \xrightarrow{1100°C} \underset{\text{water soluble}}{8Na_2CrO_4} + \underset{\text{water insoluble}}{Fe_2O_3} + 8CO_2\uparrow$
Step II :	Aqueous solution of Na_2CrO_4 is acidified when $Na_2Cr_2O_7$ is formed.	$2Na_2CrO_4 + \underset{\text{dil.}}{H_2SO_4} \longrightarrow Na_2Cr_2O_7 + Na_2SO_4 + H_2O$
Step III :	Sodium dichromate ($Na_2Cr_2O_7$) is more soluble and less stable than $K_2Cr_2O_7$ hence, $Na_2Cr_2O_7$ changes to $K_2Cr_2O_7$ on reaction with KCl. On cooling concentrated solution of $K_2Cr_2O_7$, orange crystals are obtained.	$Na_2Cr_2O_7 + 2KCl \longrightarrow \underset{\text{orange crystals}}{K_2Cr_2O_7}\downarrow + 2NaCl$

Properties

Solubility

- It exists as orange-red crystals (m.p. 398° C) and is moderately soluble in cold water but freely soluble in hot water.

Behaviour in acid and base

- In alkaline solution (pH > 7), orange colour of $Cr_2O_7^{2-}$ (dichromate ion) changes to yellow colour due to formation of CrO_4^{2-} (chromate ion) and again yellow colour changes to orange colour in acid medium (pH < 7) :

$$\underset{\text{orange-red}}{Cr_2O_7^{2-}} + 2OH^- \longrightarrow \underset{\text{yellow}}{2CrO_4^{2-}} + H_2O$$

$$2CrO_4^{2-} + 2H^+ \longrightarrow Cr_2O_7^{2-} + H_2O$$

Thus CrO_4^{2-} and $Cr_2O_7^{2-}$ exist in equilibrium and are interconvertible by altering the pH of the solution :

$$2CrO_4^{2-} + 2H^+ \rightleftharpoons 2HCrO_4^- \rightleftharpoons Cr_2O_7^{2-} + H_2O$$

Effect of heat

- $4K_2Cr_2O_7 \xrightarrow{\Delta} K_2CrO_4 + 2Cr_2O_3 + 3O_2$
- $(NH_4)_2Cr_2O_7 \xrightarrow{\Delta} N_2 + \underset{\text{green}}{Cr_2O_3} + 4H_2O$

Once this reaction starts, it will keep going at a reasonable rate. Also a small amount of it gives a much larger volume of $N_2(g)$ along with Cr_2O_3 (green). Both features of the reaction lead to it being used in indoor fireworks.

Chromyl chloride test

- $K_2Cr_2O_7$ when heated with conc. H_2SO_4 and soluble chloride (like NaCl, KCl), gives deep red vapours of CrO_2Cl_2 (chromyl chloride).

$$K_2Cr_2O_7 + \underset{\text{conc.}}{2H_2SO_4} \xrightarrow{\Delta} 2KHSO_4 + 2CrO_3 + H_2O$$

Chromium (VI) oxide (also called chromium trioxide) is red, crystalline compound. It precipitates when conc. H_2SO_4 is added to conc solution of $K_2Cr_2O_7$.

$$NaCl + \underset{\text{conc.}}{H_2SO_4} \xrightarrow{\Delta} NaHSO_4 + HCl$$

$$CrO_3 + 2HCl \xrightarrow{\Delta} CrO_2Cl_2 + H_2O$$

CrO_2Cl_2 (chromyl chloride) when passed into aqueous NaOH solution, yellow colour solution of Na_2CrO_4 is obtained. This on reaction with $(CH_3COO)_2Pb/CH_3COOH$ gives yellow ppt of $PbCrO_4$:

$$CrO_2Cl_2 + 2NaOH \longrightarrow Na_2CrO_4 + 2HCl$$

$$Na_2CrO_4 + (CH_3COO)_2Pb \longrightarrow PbCrO_4\downarrow + 2CH_3COONa$$

This test called **'Chromyl-Chloride Test'** is used to identify Cl^- in inorganic salt-analysis.

CrO_2Cl_2 is an acid chloride and is analogous to SO_2Cl_2 (sulphuryl chloride)

Oxidising agent

- $Cr_2O_7^{2-}$ is a good oxidising agent in acidic medium

$$\underset{\text{orange}}{Cr_2O_7^{2-}} + 14H^+ + 6e^- \longrightarrow \underset{\text{green}}{2Cr^{3+}} + 7H_2O$$

$$E° = 1.33 \text{ V (A)}$$

The reaction is spontaneous ($E° > 0$) and $Cr_2O_7^{2-}$ is used as a primary standard in volumetric (titrimetric) analysis.

$$Cr_2O_7^{2-} \equiv 6e^-; \quad \therefore \text{ eq. wt. } (Cr_2O_7^{2-}) = \frac{\text{mol. wt.}}{6}$$

Following reactions are affected by $Cr_2O_7^{2-}/H^+$ and solution turns green due to formation of Cr^{3+} (green) :

(a) I^- is oxidised to I_2

$$6I^- \longrightarrow 3I_2 + 6e^-$$

(b) H_2S is oxidised to S

$$3H_2S \longrightarrow 6H^+ + 3S + 6e^-$$

(c) Fe^{2+} is oxidised to Fe^{3+}

$$6Fe^{2+} \longrightarrow 6Fe^{3+} + 6e^-$$

(d) SO_2 is oxidised to SO_4^{2-}

$$3SO_2 + 6H_2O \longrightarrow 3SO_4^{2-} + 12H^+ + 6e^-$$

(Use equation (A) to write complete reaction)

(e) When H_2O_2 is added to an acidified solution of a dichromate [or any other Cr(VI) species], a complicated reaction occurs. The products depend on the pH and the concentration of Cr(VI) ion :

$$Cr_2O_7^{2-} + 2H^+ + 4H_2O_2 \longrightarrow 2CrO(O_2)_2 + 5H_2O$$

A deep blue-violet coloured peroxo compound $CrO(O_2)_2$, called chromic peroxide is formed. This decomposes rapidly in aqueous solution into Cr^{3+} and oxygen. The peroxo compound can be extracted into ether, where it is stable; it reacts with pyridine forming the adduct : Py. $CrO(O_2)_2$

(chromic peroxide)

DEEP Focus

In **less acidic solution,** $K_2Cr_2O_7$ and H_2O_2 give salts which are violet coloured and diamagnetic due to formation of $[CrO(O_2)(OH)]^-$.

In **alkaline** solution with 30% H_2O_2, a red-brown compound $\mathbf{K_3CrO_8}$ is formed. It is tetraperoxo species $[Cr(O_2)_4]^{3-}$ and thus contains Cr (V) ion.

In **ammoniacal solution** $(NH_3)_3CrO_4$, a dark red-brown compound with Cr (IV) ion is formed.

$K_2SO_4 \cdot Cr_2(SO_4)_3 \cdot 24H_2O$ — Chrome alum

It is a deep purple compound used for leather tanning.

- Chrome alum is obtained when acidified $K_2Cr_2O_7$ solution is saturated with SO_2:

$$K_2Cr_2O_7 + H_2SO_4 + 3SO_2 \xrightarrow{\Delta} \underbrace{K_2SO_4 + Cr_2(SO_4)_3}_{\text{crystallises as}}$$

$$\downarrow + H_2O$$

$$KCr(SO_4)_2 \cdot 12H_2O$$

Estimation of $Cr_2O_7^{2-}$

$K_2Cr_2O_7$ can be estimated volumetrically by a iodometric titration.

Acidified $K_2Cr_2O_7$ is treated with KI and liberated I_2 is titrated using hypo ($Na_2S_2O_3$) solution with starch indicator.

$$Cr_2O_7^{2-} + 14H^+ + 6e^- \longrightarrow 2Cr^{3+} + 7H_2O$$

$$6I^- \longrightarrow 3I_2 + 6e^-$$

$$Cr_2O_7^{2-} + 14H^+ + 6I^- \longrightarrow 2Cr^{3+} + 7H_2O + 3I_2$$

$$I_2 + 2S_2O_3^{2-} \longrightarrow S_4O_6^{2-} + 2I^-$$

Thus, $Cr_2O_7^{2-} \equiv 3I_2 \equiv 6I^- \equiv 6e^-$

$\therefore$ $1\,e^- \equiv 1\,I = \dfrac{Cr_2O_7^{2-}}{6}$

Equivalent weight of $K_2Cr_2O_7 = \dfrac{\text{molecular weight}}{6}$

$$= \frac{294}{6} = 49$$

$$2S_2O_3^{2-} \equiv I_2 \equiv 2I \equiv 2e^-$$

$\therefore$ $S_2O_3^{2-} \equiv I \equiv 1\,e^-$

Equivalent weight of $Na_2S_2O_3 \cdot 5H_2O = \dfrac{248}{1}$

• $PbCrO_4$	—	Chrome yellow
$PbCrO_4 \cdot PbO$	—	Chrome red
$PbCrO_4 + [FeFe(CN)_6]^-$ Prussian blue	—	Chrome green
$Zn(OH)_2 \cdot ZnCrO_4 \cdot H_2O$	—	Yellow
$Cr_2O_3 \cdot 2H_2O$	—	Guigrete's green
CrO_3	—	Orange (called chromic acid)
$K_3[CrO_8]$	—	Red-brown
$[Cr(H_2O)_6]^{2+}$	—	Bright blue
$[Cr(H_2O)_6]^{3+}$	—	Violet

- In $Cr_2O_7^{2-}$ two CrO_3 units have been joined by O

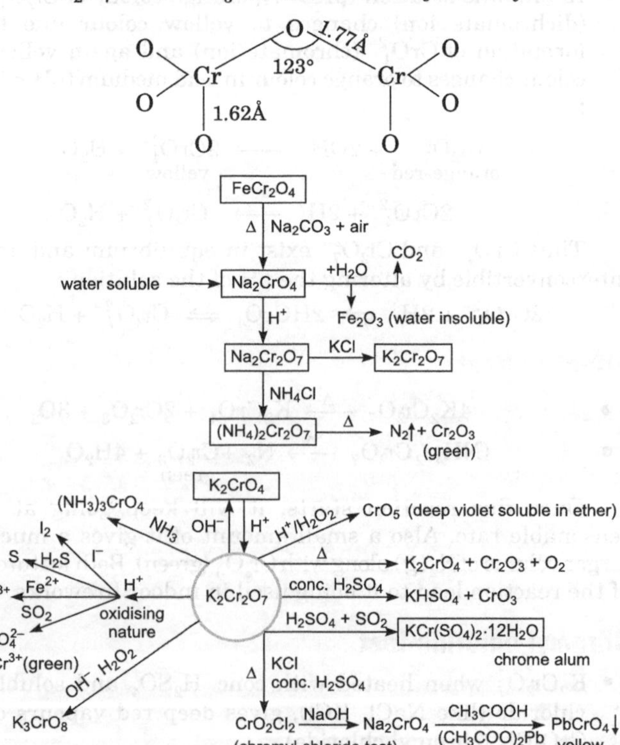

Potassium Permanganate ($KMnO_4$)

Preparation

$KMnO_4$ is prepared from MnO_2 (pyrolusite) in following steps :

Step I :	MnO_2 is fused with KOH and $KClO_3$ or KNO_3 when potassium manganate (K_2MnO_4) is formed as indicated by its green colour. $KClO_3/KNO_3$ acts as oxidising agent.	$3MnO_2 + 6KOH + KClO_3 \xrightarrow{\Delta} \underset{\text{green}}{3K_2MnO_4} + KCl + 3H_2O$
Step II :	The fused mass is extracted with water and solution is green (MnO_4^{2-})	
	(a) when Cl_2 gas is passed, K_2MnO_4 is oxidised to $KMnO_4$ (purple).	$\underset{\text{green}}{2K_2MnO_4} + Cl_2 \longrightarrow 2KCl + \underset{\text{purple}}{2KMnO_4}$
	(b) MnO_4^{2-} can be oxidised to MnO_4^- electrochemically at anode.	**at anode :** $MnO_4^{2-} \longrightarrow MnO_4^- + e^-$ **at cathode :** $2H^+ + 2e^- \longrightarrow H_2$

- Strong oxidising agents such as PbO_2 or sodium bismuthate ($NaBiO_3$) can also oxidise Mn^{2+} to MnO_4^-.
- MnO_4^{2-} is unstable in acidic medium and disproportionates:

$$3MnO_4^{2-} + 4H^+ \longrightarrow MnO_2 + 2MnO_4^- + 2H_2O$$

- $KMnO_4$ crystallises as deep purple rhombic prisms.

DEEP Focus

MnO_4^- (VII) has $3d^0$ electronic configuration in Mn and thus MnO_4^- would be expected to be colourless, but colour arises due to charge transfer by oxygen to Mn and thus one of the oxygen is O^- and Mn changes from +VII to +VI (with one unpaired electron in 3d) hence, coloured.

- It is isomorphous with $KClO_4$, potassium perchlorate; highly soluble in water :

Effect of heat

- $2KMnO_4 \xrightarrow{\Delta} K_2MnO_4 + MnO_2 + O_2$

Oxidising agent

I. $KMnO_4$ is a good oxidising agent; it finds use in volumetric analysis as secondary standard in the estimation of Fe^{2+}, $C_2O_4^{2-}$, H_2O_2 etc. in acidic medium:

$$\underset{\text{(purple)}}{MnO_4^-} + 8H^+ + 5e^- \longrightarrow \underset{\text{(colourless very-very light pink)}}{Mn^{2+}} + 4H_2O$$

The purple colour of MnO_4^- acts as its own indicator (self) in titrations

$$MnO_4^- \equiv 5e^-$$

Thus, equivalent weight of MnO_4^-

$$= \frac{\text{mol. wt.}}{5} \text{ in acidic medium.}$$

Alternately,

$$2KMnO_4 + 3H_2SO_4 \longrightarrow K_2SO_4 + 2MnSO_4 + 3H_2O + 5(O)$$

$2KMnO_4 \equiv 5(O)$ taking weight as 16.

$\equiv 10\ (O)$ taking weight as 8.

$$\frac{KMnO_4}{5} \equiv 1(O)$$

$\therefore$ Eq. wt. of $KMnO_4$ in acidic medium $= \left(\frac{\text{mol. wt.}}{5}\right)$

Some oxidation reactions (using MnO_4^-/H^+) are :

(a) Fe^{2+} is oxidised to Fe^{3+}

$$Fe^{2+} \longrightarrow Fe^{3+} + e^-$$

(b) Oxalate is oxidised to CO_2,

$$C_2O_4^{2-} \longrightarrow 2CO_2 + 2e^-$$

(c) I^- is oxidised to I_2

$$2I^- \longrightarrow I_2 + 2e^-$$

(d) Cl^- is oxidised to Cl_2

$$2Cl^- \longrightarrow Cl_2 + 2e^-$$

(e) H_2O_2 is oxidised to O_2

$$H_2O_2 \longrightarrow O_2 + 2H^+ + 2e^-$$

(f) SO_2 (SO_3^{2-}) is oxidised to SO_4^{2-}

$$H_2O + SO_3^{2-} \longrightarrow SO_4^{2-} + 2H^+ + 2e^-$$

(g) H_2S is oxidised to S

$$H_2S \longrightarrow S + 2H^+ + 2e^-$$

(h) CH_3CH_2OH is oxidised to CH_3CHO

$$CH_3CH_2OH \xrightarrow{O} CH_3CHO + H_2O$$

II. In dilute alkaline solution, MnO_2 is formed.

$$MnO_4^- + 2H_2O + 3e^- \longrightarrow MnO_2 + 4OH^-$$

$$\text{Eq. wt. (in basic medium)} = \left(\frac{\text{mol. wt.}}{3}\right)$$

H_2O_2 is oxidise to O_2

$$2MnO_4^- + 3H_2O_2 \longrightarrow 2MnO_2 + 3O_2 + 3OH^- + 2H_2O$$

III. In very strong base, and with an excess of MnO_4^-, however manganate ion (MnO_4^{2-}) is formed.

$$4MnO_4^- + 4OH^- \longrightarrow 4MnO_4^{2-} + O_2 + 2H_2O$$

$$MnO_4^- + e^- \longrightarrow MnO_4^{2-}$$

and $\quad \text{eq. wt.} = \frac{\text{mol wt.}}{1}$

Thus, equivalent weight of $KMnO_4$ varies with medium.

Alkaline $KMnO_4$ is called **Baeyer's reagent** and is used to test **unsaturation** in an organic compound, when it decolorises $KMnO_4$.

$$CH_2{=}CH_2 + \underbrace{H_2O + O}_{\text{alkaline } KMnO_4} \longrightarrow \underset{OH}{CH_2}-\underset{OH}{CH_2}$$

In dilute alkali, water or acidic solutions, MnO_4^{2-} again disproportionates:

$$3MnO_4^{2-} + 4H^+ \longrightarrow 2MnO_4^- + MnO_2 + 2H_2O$$

$$3MnO_4^{2-} + 2H_2O \longrightarrow 2MnO_4^- + MnO_2 + 4OH^-$$

Stability in acidic medium

- Permanganate solutions are intrinsically unstable is acidic solution, and decompose slowly. **Decomposition is catalysed by sunlight, so $KMnO_4$ solutions should be stored in dark bottles and they must be standarised at the moment of its use in volumetric analysis.**

$$4MnO_4^- + 4H^+ \longrightarrow 4MnO_2 + 3O_2 + 2H_2O$$

With conc. H_2SO_4

- If a small amount of $KMnO_4$ is added to concentrated H_2SO_4, a green solution containing MnO_3^+ ions is formed:

$$KMnO_4 + 3H_2SO_4 \longrightarrow K^+ + \underset{\text{green}}{MnO_3^+} + 3HSO_4^- + H_3O^+$$

MnO_3^+ changes to MnO_3 with Na_2CO_3

$$2MnO_3^+ + Na_2CO_3 \longrightarrow 2MnO_3 + CO_2 + \frac{1}{2}O_2 + 2Na^+$$

If $KMnO_4$ is in excess, an explosive oil Mn_2O_7 is formed:

$$2KMnO_4 + \underset{\text{conc.}}{H_2SO_4} \xrightarrow{\Delta} Mn_2O_7 + K_2SO_4 + H_2O$$

MnO_4^- has following structure

$$\left[\begin{array}{c} O \\ | \\ O - Mn - O \\ | \\ O \end{array}\right]^-$$

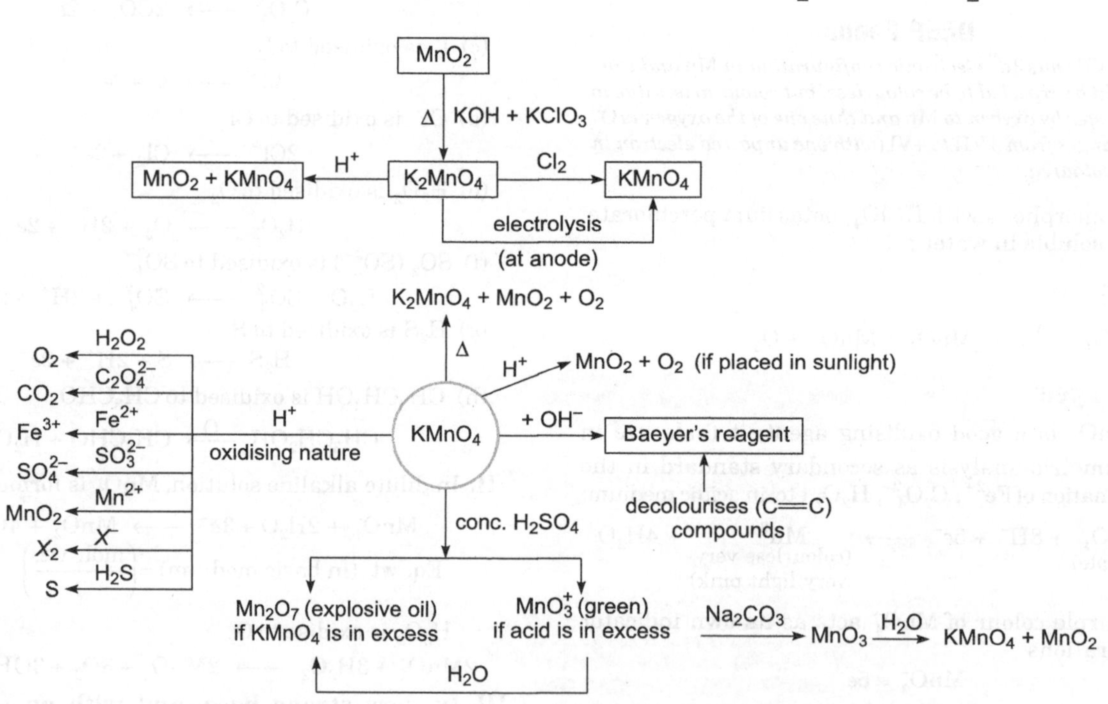

Zinc Oxide (ZnO) Zinc white

Preparation

- ZnO is formed when ZnS is oxidised :

$$2ZnS + 3O_2 \longrightarrow 2ZnO + 2SO_2$$

$Zn(OH)_2$ on strongly heating gives ZnO :

$$Zn(OH)_2 \xrightarrow{\Delta} ZnO + H_2O$$

Zinc on burning in air gives ZnO (commercial method)

$$2Zn + O_2 \longrightarrow 2ZnO$$

Properties

- ZnO is white when it is cold, a property that has given it a use as a pigment in paints. However, **it changes colour, when hot, to a pale yellow. This is due to change in the structure of lattice.**

- ZnO is soluble both in acid and alkali and is thus amphoteric in nature :

$$\underset{\text{base}}{ZnO} + \underset{\text{acid}}{2H^+} \longrightarrow Zn^{2+} + H_2O$$

$$\underset{\text{acid}}{ZnO} + \underset{\text{base}}{2OH^-} + H_2O \longrightarrow \underset{\text{zincate ion}}{[Zn(OH)_4]^{2-} \text{ or } ZnO_2^{2-}}$$

Or $$ZnO + 2HCl \longrightarrow ZnCl_2 + H_2O$$

$$ZnO + 2NaOH \longrightarrow \underset{\text{sodium zincate}}{Na_2ZnO_2} + H_2O$$

- $$ZnO + C \xrightarrow[>1000°C]{\Delta} Zn + CO$$

$$ZnO + CO \xrightarrow{\Delta} Zn + CO_2$$

- It is preferred to white lead as it is not blackened by H_2S. It is also used in medicine and in the preparation of **Rinmann's green ($ZnCo_2O_4$).**

Zinc Sulphate ($ZnSO_4$)

Preparation

- $ZnSO_4 \cdot 7H_2O$ (also called white vitriol) is formed:
  - by decomposing $ZnCO_3$ with dil. H_2SO_4

$$ZnCO_3 + H_2SO_4 \longrightarrow ZnSO_4 + H_2O + CO_2$$

  - by heating ZnS (zinc blende) in air at lower temperature and dissolving the product in dil. H_2SO_4

$$2ZnS + 3.5\ O_2 \xrightarrow{\Delta} ZnO + ZnSO_4 + SO_2$$

$$ZnO + H_2SO_4 \longrightarrow ZnSO_4 + H_2O$$

Properties

- Highly soluble in water and solution is acidic in nature due to hydrolysis

$$ZnSO_4 + 2H_2O \longrightarrow Zn(OH)_2 + H_2SO_4$$

- $$ZnSO_4 \cdot 7H_2O \xrightarrow[\Delta]{100°C} ZnSO_4 \cdot 6H_2O \xrightarrow[\Delta]{280°C} ZnSO_4 \xrightarrow{T > 760°} ZnO + SO_3$$

It slowly effloresces when exposed to air.

- It is isomorphous with Epsom salt and used in the manufacture of **lithophone** (which is a mixture of $BaS + ZnSO_4$ and is used as white pigment).

Silver Nitrate (Lunar Caustic) $AgNO_3$

Preparation

- When Ag is heated with dil. HNO_3, $AgNO_3$ is formed. Crystals separate out on cooling the concentrated solution of $AgNO_3$.

$$3Ag + 4HNO_3 \xrightarrow{\Delta} 3AgNO_3 + NO + 2H_2O$$

Properties

- Colourless crystalline compound soluble in H_2O and alcohol; m.p. 212°C
- When exposed to light, it decomposes hence, stored in a brown coloured bottle :

$$2Ag + 2NO_2 + O_2 \xleftarrow{\Delta, \text{red hot}} 2AgNO_3 \xrightarrow{\Delta, T > 212°C} 2AgNO_2 + O_2$$

- It is reduced to metallic Ag by more electropositive metals like Cu, Zn, Mg and also by PH_3.

$$2AgNO_3 + Cu \longrightarrow Cu(NO_3)_2 + 2Ag$$

$$6AgNO_3 + PH_3 + 3H_2O \longrightarrow 6Ag + 6HNO_3 + H_3PO_3$$

- It dissolves in excess of KCN :

$$AgNO_3 \xrightarrow{KCN} \underset{\substack{\text{white}\\\text{ppt}}}{AgCN} \xrightarrow{KCN} \underset{\substack{\text{soluble}\\\text{potassium argentocyanide}}}{K[Ag(CN)_2]}$$

- $AgNO_3$ gives white precipitate with $Na_2S_2O_3$; white precipitate changes to black.

$$2AgNO_3 + Na_2S_2O_3 \longrightarrow \underset{\text{white ppt}}{Ag_2S_2O_3} + 2NaNO_3$$

$$Ag_2S_2O_3 + H_2O \longrightarrow \underset{\text{black}}{Ag_2S} + H_2SO_4$$

DEEP Focus

- It possesses powerful corrosive action on organic tissues, which it turns black especially in presence of light. The blackening is due to finely divided metallic silver, reduced by the organic matter.
- Ammoniacal $AgNO_3$ is called **Tollen's reagent** and is used to identify reducing sugars (including aldehydes) :

$$RCHO + 2Ag^+ + 3OH^- \xrightarrow{\Delta} RCOO^- + 2Ag \downarrow + 2H_2O$$

It is called **'silver mirror test'** of aldehydes and reducing sugar (like glucose, fructose).

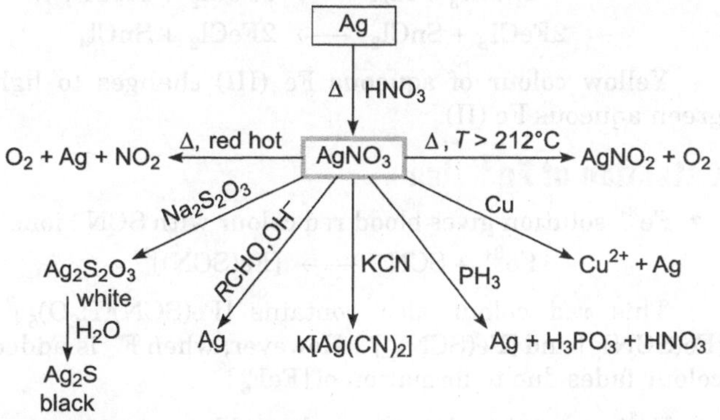

Ag
Δ HNO₃ ↓
$O_2 + Ag + NO_2 \xleftarrow{\Delta, \text{red hot}} AgNO_3 \xrightarrow{\Delta, T > 212°C} AgNO_2 + O_2$
$AgNO_3 \xrightarrow{Na_2S_2O_3} Ag_2S_2O_3$ (white) $\xrightarrow{H_2O} Ag_2S$ (black)
$AgNO_3 \xrightarrow{RCHO, OH^-} Ag$
$AgNO_3 \xrightarrow{KCN} K[Ag(CN)_2]$
$AgNO_3 \xrightarrow{PH_3} Ag + H_3PO_3 + HNO_3$
$AgNO_3 \xrightarrow{Cu} Cu^{2+} + Ag$

Ferric Chloride ($FeCl_3$)

Preparation

- $$2Fe + 3Cl_2 \xrightarrow{\Delta} 2FeCl_3 \text{ (anhydrous)}$$

- When Fe is dissolved in **aqua-regia** or Fe_2O_3 is dissolved in HCl, hydrate $FeCl_3 \cdot 6H_2O$ is formed :

$$HNO_3 + 3HCl \longrightarrow NOCl + 2H_2O + 2Cl$$
$$Fe + 4HCl + 2Cl \longrightarrow 2FeCl_3 + 2H_2$$
$$Fe_2O_3 + 6HCl \longrightarrow FeCl_3 + 3H_2O$$

Properties

- Anhydrous $FeCl_3$ is black but hydrate $FeCl_3 \cdot 6H_2O$ is yellowish brown, highly deliquescent crystalline solid.
- It sublimes at 300°C giving a dimeric gas :

$$2FeCl_3 \xrightarrow{300°C} \underset{(FeCl_3)_2 \text{ gas dimer}}{Cl_2Fe(\mu\text{-}Cl)_2FeCl_2}$$

- $FeCl_3$ dissolves in both ether and water giving solvated monomeric species :

$$\underset{\text{solvated } FeCl_3}{(C_2H_5)_2O \longrightarrow FeCl_3,} \quad \underset{\text{hydrate } FeCl_3}{[Fe(H_2O)_4Cl_2]\,Cl \cdot 2H_2O}$$

- It normally consists of *trans* $[Fe(H_2O)_4Cl_2]Cl \cdot 2H_2O$ but in strong HCl it forms tetrahedral $(FeCl_4)^-$ ions.

Hydrolysis

- Aqueous solution is **acidic** due to hydrolysis :

$$FeCl_3 + 3H_2O \rightleftharpoons Fe(OH)_3 + 3HCl$$

or $$\underset{\text{acid}}{[Fe(H_2O)_6]^{3+}} + \underset{\text{base}}{H_2O} \rightleftharpoons \underset{\text{base}}{[Fe(H_2O)_5(OH)]^{2+}} + \underset{\text{acid}}{H_3O^+}$$

Oxidising agent

- $FeCl_3$ is also an **oxidising agent :**

$$2FeCl_3 + H_2S \longrightarrow 2FeCl_2 + 2HCl + S$$
$$2FeCl_3 + SnCl_2 \longrightarrow 2FeCl_2 + SnCl_4$$

Yellow colour of aqueous Fe (III) changes to light green aqueous Fe (II).

Detection of Fe^{3+} ion

- Fe^{3+} solution gives blood red colour with SCN^- ions :

$$Fe^{3+} + SCN^- \longrightarrow [Fe(SCN)]^{2+}$$

This red colour also contains $[Fe(SCN)(H_2O)_5]^{2+}$, $[Fe(SCN)_3]$ and $[Fe(SCN)_4]^-$. However, when F^- is added, colour fades due to formation of $[FeF_6]^{3-}$.

- Fe^{3+} solution also gives deep blue precipitate of **Prussian blue** with $K_4[Fe(CN)_6]$, potassium ferrocyanide :

$$4\overset{III}{Fe}Cl_3 + 3K_4[\overset{II}{Fe}(CN)_6] \longrightarrow \underset{\text{Prussian blue}}{\overset{III}{Fe}_4[\overset{II}{Fe}(CN)_6]_3} + 12KCl$$

or $$FeCl_3 + K_4[Fe(CN)_6] \longrightarrow \underset{\text{Prussian blue}}{K\overset{III}{Fe}[\overset{II}{Fe}(CN)_6]} + 3KCl$$

- A deep blue colour is also produced when Fe^{2+} reacts with $K_3[Fe(CN)_6]$, potassium ferricyanide :

$$\overset{II}{Fe}SO_4 + K_3[\overset{III}{Fe}(CN)_6] \longrightarrow \underset{\text{Turnbull's blue}}{K\overset{II}{Fe}[\overset{III}{Fe}(CN)_6]}$$

DEEP Focus

Recent X-ray work, IR and other spectroscopic methods have proved that Turnbull's blue is identical to Prussian blue. The intense colour arises from electron transfer between Fe(+II) and Fe(+III).

Fe^{3+} partly oxidises $[\overset{II}{Fe}(CN)_6]^{4-}$ to $[\overset{III}{Fe}(CN)_6]^{3-}$ and itself is reduced to Fe^{2+}, and also Fe^{2+} partly reduces $[\overset{III}{Fe}(CN)_6]^{3-}$ to $[\overset{II}{Fe}(CN)_6]^{4-}$ and itself is oxidised to Fe^{3+}. Thus, $\overset{II}{Fe}[\overset{III}{Fe}(CN)_6]^-$ and $\overset{III}{Fe}[\overset{II}{Fe}(CN)_6]^-$ are identical.

Ferric Oxide (Fe_2O_3)

Preparation

Hydrolysis of $FeCl_3$ actually gives a red-brown gelatinous precipitate of the hydrous oxide $Fe_2O_3(H_2O)_n$ which on heating at 200°C gives red-brown α-Fe_2O_3 (which occurs as the mineral haematite).

- On oxidation of Fe_3O_4, γ-Fe_2O_3 is formed.

Properties

- α-Fe_2O_3 has hexagonally close-packed lattice of O^{2-} ions with Fe^{3+} ion in two thirds of the octahedral holes while γ-Fe_2O_3 has cubic close-packed arrangement of O^{2-} ions with Fe^{3+} ions randomly distributed in both the octahedral and tetrahedral sites.

Effect of heat

- $$6Fe_2O_3 \xrightarrow{1400°C, \Delta} 4Fe_3O_4 + O_2$$

Fe_3O_4 is a mixed oxide $\overset{II}{Fe}O \cdot \overset{III}{Fe}_2O_3$.

FeO, Fe_2O_3 and Fe_3O_4 all tend to be non-stoichiometric. Fe_3O_4 occurs in nature as magnetite.

Solubility in water

- Fe_2O_3 is predominantly basic. Freshly precipitated $Fe_2O_3 \cdot (H_2O)_n$ dissolves in acid giving pale violet $[Fe(H_2O)_6]^{3+}$ ion. This $[Fe_2O_3 \cdot (H_2O)_n]$ also dissolves in concentrated NaOH forming $[Fe(OH)_6]^{3-}$. This shows that Fe_2O_3 is slightly amphoteric.

Fusion with Na_2CO_3

- Fusion of Fe_2O_3 with Na_2CO_3 gives $NaFeO_2$ (sodium ferrite) which is hydrolysed to Fe_2O_3 and NaOH :

$$Na_2CO_3 + Fe_2O_3 \longrightarrow 2NaFeO_2 + CO_2$$
$$2NaFeO_2 + H_2O \longrightarrow 2NaOH + Fe_2O_3$$

(In earlier days this method, called **Lowig Process, was used to manufacture NaOH).**

With Cl_2

- If Cl_2 gas is passed into an alkaline solution of hydrated ferric oxide, a red purple solution is formed containing the ferrate ion $[\overset{VI}{Fe}O_4]^{2-}$.

$$Fe_2O_3 + 2NaOH \longrightarrow 2NaFeO_2 + H_2O$$
$$2NaFeO_2 + Cl_2 + 4NaOH \longrightarrow 2Na_2FeO_4 + 2NaCl + 2H_2$$

DEEP Focus

Na_2FeO_4 can also be obtained by oxidation of Fe_2O_3 with NaOCl (sodium hypochlorite). Na_2FeO_4 has Fe(+VI) and is a strong oxidising agent (like $KMnO_4$).

Ferrous Sulphate ($FeSO_4 \cdot 7H_2O$ – Green Vitriol)

Preparation

- $$FeS + H_2SO_4 \longrightarrow FeSO_4 + H_2S \uparrow$$
$$Fe + \underset{\text{dil.}}{H_2SO_4} \xrightarrow{\Delta} FeSO_4 + H_2 \uparrow$$

$FeSO_4$ solution is concentrated when pale green crystals are obtained.

Effect of heat

- $$2FeSO_4 \xrightarrow{\Delta} Fe_2O_3 + SO_2 + SO_3$$

Atmospheric oxidation

- Anhydrous $FeSO_4$ is colourless/white. On exposure to atmosphere it turns brownish- yellow due to formation of basic ferric sulphate :

$$4FeSO_4 + 2H_2O + O_2 \longrightarrow 4Fe(OH)SO_4$$

- $FeSO_4 \cdot 7H_2O$ is isomorphous with $ZnSO_4 \cdot 7H_2O$ (white vitriol) and $MgSO_4 \cdot 7H_2O$ (Epsom salt).

Hydrolysis

- Aqueous solution of $FeSO_4$ is acidic due to hydrolysis of Fe^{2+} :

$$Fe^{2+} + 2H_2O \rightleftharpoons Fe(OH)_2 + 2H^+$$

Oxidation

- Acidified MnO_4^- and $Cr_2O_7^{2-}$ oxidise Fe^{2+} to Fe^{3+} :

$$MnO_4^- + 8H^+ + 5Fe^{2+} \longrightarrow 5Fe^{3+} + Mn^{2+} + 4H_2O$$
$$Cr_2O_7^{2-} + 14H^+ + 6Fe^{2+} \longrightarrow 6Fe^{3+} + 2Cr^{3+} + 7H_2O$$

Thus, MnO_4^- and $Cr_2O_7^{2-}$ can be used to estimate Fe^{2+} salts by volumetric method in acidic medium.

$$MnO_4^- \xrightarrow{5e^-} Mn^{2+}$$
$$Fe^{2+} \longrightarrow Fe^{3+} + e^-$$
$$MnO_4^- \equiv 5Fe^{2+}$$

Similarly, $Cr_2O_7^{2-} \equiv 6Fe^{2+}$

Detection

- Fe^{2+} salts form blue precipitate of Turnbull's blue with $K_3[Fe(CN)_6]$. Turnbull's blue is used as pigment in ink and paint (also refer $FeCl_3$)

$$Fe^{2+} + [Fe^{III}(CN)_6]^{4-} \longrightarrow \underset{\text{Turnbull's blue}}{Fe^{II}[Fe^{III}(CN)_6]^-}$$

Ring test (for NO_3^-)

- The complex $[Fe(H_2O)_5NO]^{2+}$ is formed in the brown ring test for nitrates when freshly prepared $FeSO_4$ solution is added to aqueous solution of NO_3^- followed by addition of conc. H_2SO_4. The colour is due to charge transfer. Its magnetic moment is approximately 3.873 BM confirming the presence of three unpaired electrons in iron. Thus iron exists as Fe(+I) and nitrosyl as NO^+.

$$NO \longrightarrow NO^+ + e^-$$
$$Fe^{2+} + e^- \longrightarrow Fe^+$$

↑↓	↑	↑	↑	↑

↑↓	↑↓	↑	↑	↑

gained electron from NO — three unpaired electrons in Fe^+

Thus, magnetic moment $= \sqrt{N(N+2)}$ BM
$= \sqrt{15}$
$= 3.873$ BM

- Mixture of $FeSO_4$ and H_2O_2 is used as **Fenton's** reagent for producing hydroxyl radicals and for oxidising alcohols to aldehydes.
- Saturated solution of $FeSO_4$ and $(NH_4)_2SO_4$ on concentration and cooling gives crystals of Mohr's salt, $FeSO_4 \cdot (NH_4)_2SO_4 \cdot 6H_2O$ which is widely used as **primary standard** for standarisation of $KMnO_4$ and $K_2Cr_2O_7$ solutions.

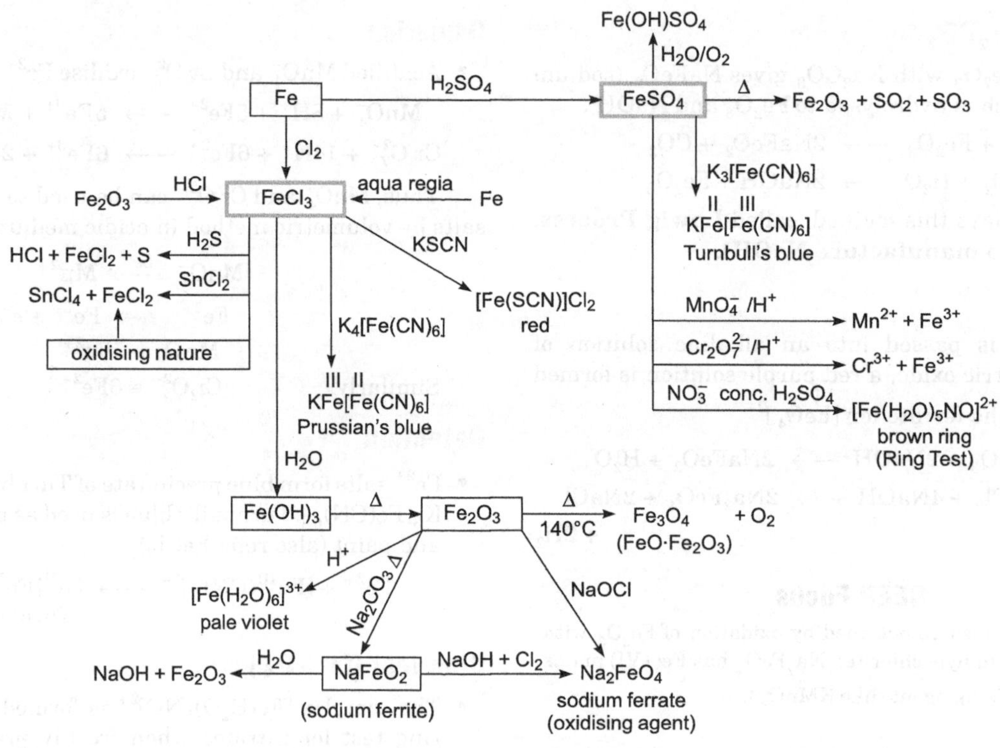

Target Practice 2

1. Iron(II) oxide is written as $\overline{FeO}$. What does it mean?

....

....

2. Mention catalysts in the following

(a) $KClO_3 \longrightarrow KCl + O_2$

(b) $CS_2 + Cl_2 \longrightarrow CCl_4$

(c) $CH_3C{\equiv}CCH_3 \longrightarrow$ (H)(H$_3$C)C=C(CH$_3$)(H)

(d) $C_2H_4 \xrightarrow{O_2} CH_3CHO$

(e) $CO + H_2 \longrightarrow CH_3OH$

3. When CO_2 is passed into aqueous K_2CrO_4 (yellow) solution, this changes to orange. This is explained by following reactions

....

....

4. Which of the following pairs is more stable?

(a) CrO_3	☐	WO_3
(b) MnO_4^-	☐	ReO_4^-
(c) Cr^{2+}	☐	Cr^{3+}
(d) Mn^{2+}	☐	MnO_4^-
(e) CrO_4^{2-}	☐	MoO_4^{2-}
(f) Mn_2O_7	☐	Re_2O_7
(g) V^{2+}	☐	VO_4^{3-}

5. Cr_2O_3 is an amphoteric oxide. Thus, it dissolves in acid forming .... and dissolves in base forming ....

6. $Cr_2O_7^{2-} \xrightarrow{pH > 7} CrO_4^{2-} \xrightarrow{pH < 7} Cr_2O_7^{2-}$

Explain the change based on reactions

....

....

7. Name the compound which imparts odour in fire-works

....

8. Write the formulae of

(a) Chromyl chloride ....

(b) Chrome yellow ....

(c) Chrome green ....

(d) Chrome alum ....

9. What is oxidation state of Cr in

(a) CrO_5 ....

(b) $[CrO_3(OH)]^-$ ....

(c) $[Cr(O_2)_4]^{3-}$ ....

(d) CrO_2Cl_2 ....

10. Consider structure of Cr_2O_7

$$O_2Cr(=O)\overset{x}{-}O\overset{\alpha}{-}Cr(=O)O_2$$

State (a) Angle $\angle\alpha$ = ...

(b) Bond length x = ...

(c) Bond length y = ...

11. Alcohol content in a drunkard person can be determined by $K_4Cr_2O_7$. Explain in their lines.

....

....

....

12. (a) Lithophone is ....
(b) Rinmann's green is ....
(c) Zinc blende is ....
(d) White vitriol is ....

13. ZnO is converted into .... in basic medium and into .... in acidic medium.

14. ZnO is amphoteric. Name two other oxides of *d*-block elements which are amphoteric

(a) ....
(b) ....

15. (a) ZnO is white when cold and is yellow when hot. This colour change is due to ....
(b) $ZnSO_4 \cdot 7H_2O$ loses H_2O on exposure to air. This property is called ....

16. Calculate equivalents in one mole each of .... in following reactions

one mole =

(a) $Cr_2O_7^{2-} \longrightarrow 2Cr^{3+}$ .... equivalents
(b) $MnO_4^- \longrightarrow MnO_4^{2-}$ .... equivalents
(c) $MnO_4^- \longrightarrow Mn^{2+}$ .... equivalents
(d) $MnO_4^- \longrightarrow MnO_2$ .... equivalents

Answers

1. It is to indicate that ratio of Fe and O atoms is not exactly 1 : 1. Analysis shows that formula varies between $Fe_{0.94}O$ and $Fe_{0.84}O$.

2. (a) MnO_2 (b) $FeCl_3$ (c) Li/NH_3 (d) $PdCl_2$ (e) ZnO/CuO

3. $CO_2 + H_2O \rightleftharpoons H_2CO_3$

$H_2CO_3 \rightleftharpoons H^+ + HCO_3^-$

H^+ change K_2CrO_4 into $K_2Cr_2O_7$

$$\underset{\text{yellow}}{2CrO_4^{2-}} + 2H^+ \longrightarrow \underset{\text{orange}}{Cr_2O_7^{2-}} + H_2O$$

4. (a) WO_3 (b) ReO_4^- (c) Cr^{3+} (d) Mn^{2+} (e) MoO_4^{2-} (f) Re_2O_7 (g) VO_4^{3-}

5. Cr^{3+} (acid medium), $Cr(OH)_4^-$ (basic medium)

6.

pH > 7 (basic medium)	pH < 7 (acidic medium)
$Cr_2O_7^{2-}$ changes to CrO_4^{2-}	CrO_4^{2-} changes to $Cr_2O_7^{2-}$

$$Cr_2O_7^{2-} + 2OH^- \longrightarrow 2CrO_4^{2-} + H_2O$$
$$2CrO_4^{2-} + 2H^+ \longrightarrow Cr_2O_7^{2-} + H_2O$$

7. $(NH_4)_2Cr_2O_7$

8. (a) CrO_2Cl_2 (b) $PbCrO_4$ (c) $PbCrO_4 + Fe^{III}[Fe^{II}(CN)_6]^-$ (d) $[K_2SO_4 \cdot Cr_2(SO_4)_3 \cdot 24H_2O]$

9. (a) Cr = + 6 (b) Cr = + 4 (c) Cr = + 5 (d) Cr = + 6

10. (a) $\angle\alpha = 123°$ (b) $x = 1.77$ Å (c) $y = 1.62$ Å

11. Drunkard person takes a breath, alcohol content is oxidised to CH_3CHO by $K_2Cr_2O_7/H^+$ which in turn is reduced to Cr^{3+} (green). Formation of green colour indicates alcohol content.

12. (a) Mixture of BaS and $ZnSO_4$ (b) $ZnCo_2O_4$ (c) ZnS (d) $ZnSO_4 \cdot 7H_2O$

13. $[Zn(OH)_4]^{2-}$ or ZnO_2^{2-} (basic medium); Zn^{2+} (acidic medium)

14. (a) TiO_2 (b) V_2O_5

15. (a) Change in structure of lattice (b) efflorescence

16. (a) 6 (b) 1 (c) 5 (d) 3

Change oxidation number per mole

Practice Exercise 2

1. VO^{2+} is oxidised to VO_2^+ by MnO_4^- in acidic medium, which in turn is reduced to Mn^{2+}. Write balanced equation for this reaction.

2. One of the following oxide is also called chromic acid. Select that one : CrO_2, Cr_2O_3, CrO_3, CrO

3. The red oxide CrO_3 dissolves in water to produce a strongly acidic solution to form orange coloured solution. Solution changes to yellow on making it alkaline. Explain these observations in terms of chemical equations.

4. What is oxidation number of Cr in
(a) CrO_2Cl_2
(b) CrO_5 or $CrO(O_2)_2$
(c) $(NH_4)_2Cr_2O_7$
(d) $(NH_3)_3CrO_4$ or $(NH_3)_3Cr(O_2)_2$
(e) $[Cr(O_2)_4]^{3-}$

5. Identify *A, B, C* ... *G*.

$$K_2Cr_2O_7 \xrightarrow{KOH} (A) \xrightarrow[CH_3COOH]{Pb(CH_3COO)_2} (B)$$

$\downarrow NH_4Cl$

$$(C) \xrightarrow{\Delta} \underset{\text{gas}}{(D)} + \underset{\text{green}}{(E)},\ (E) \xrightarrow{Al,\ \Delta} (F),\ (\text{metal}) + (G),\ (\text{oxide})$$

6. Green coloured compound (*A*)

$$(A) \xrightarrow{\Delta,\ NaOH + H_2O_2} (C) \text{ yellow coloured solution}$$

$$(A) \xrightarrow{\Delta\ K_2Cr_2O_7 + \text{conc. } H_2SO_4} \text{deep-red coloured gas } (B) \xrightarrow{NaOH} \text{yellow coloured solution } (C)$$

$$\xrightarrow{(CH_3COO)_2Pb + CH_3COOH} (D) \text{ yellow ppt}$$

Identify *A, B, C* and *D*.

7. Explain by reactions.
(a) Mn^{3+} disproportionates to Mn^{2+} and MnO_2 in acidic medium.

(b) MnO_4^{2-} also disproportionates to MnO_2 and MnO_4^- in alkaline medium.

(c) MnO_4^- oxidises Mn^{2+} into MnO_2 in acidic medium.

8. MnO_4^{2-} changes to MnO_2 and MnO_4^- in acidic medium. Calculate equivalent weight of MnO_4^{2-}.

9. (a) A qualitative analysis test for Mn^{2+} involves its oxidation to MnO_4^- in acidic solution by sodium bismuthate, $NaBiO_3$. The bisulphate ion is reduced to Bi^{3+}. Write a balanced equation for this test.

(b) What happens if PbO_2 is used instead of $NaBiO_3$?

10. Give structures of MnO_4^- and CrO_4^{2-} ions.

11. H_2O_2 can be oxidised to O_2 by MnO_4^- in acidic as well as in basic medium.

(a) Write reactions.

(b) If 100 mL of 1M MnO_4^- is used in acidic medium, calculate volume of 1M MnO_4^- to be used in basic medium by same H_2O_2.

12. What happens when CO_2 gas is passed into MnO_4^{2-} solution?

13. You have been given $AlCl_3$ and $ZnCl_2$ solution. To each NaOH and NH_4OH are added in excess. What happens?

14. ΔG_f° (standard free energy of formation) of ZnO and CO are -318.2 kJ mol^{-1} and -137.3 kJ mol^{-1}. Is reduction of ZnO by carbon possible ?

15. Colourless salt (*A*) gives white ppt (*B*) with NaOH; ppt (*B*) dissolves in excess of NaOH to form (*C*). On passing H_2S gas into (*C*) white ppt (*D*) appears. (*A*) also gives white ppt (*E*) with $AgNO_3$ solution. Identify (*A*) to (*E*).

16. Colourless salt (*A*) $\xrightarrow{\Delta}$ (*B*) + (*C*) gas. (*B*) dissolves both in acid and alkali solution. Gas (*C*) turns lime water milky and acidified $K_2Cr_2O_7$ solution green. Ammoniacal solution of (*A*) gives white ppt (*D*) with H_2S gas. Identify (*A*), (*B*), (*C*) and (*D*).

17. What is the advantage of using ZnO in paints instead of Pb^{2+} salts?

18. Compare thermal stability of ZnO, CdO and HgO.

Answers

1. $5VO^{2+} + MnO_4^- + H_2O \longrightarrow 5VO_2^+ + Mn^{2+} + 2H^+$

2. CrO_3

3. $2CrO_3 + H_2O \longrightarrow Cr_2O_7^{2-} + 2H^+$

$Cr_2O_7^{2-} + 2OH^- \longrightarrow 2CrO_4^{2-}$

4. (a) VI (b) VI (O = – 2, diperoxo = – 4) (c) VI (d) IV (e) V

(oxidation number of O in peroxide = – 1 and in oxide = – 2)

5. (*A*) : K_2CrO_4 (*B*) : $PbCrO_4$ (*C*) : $(NH_4)_2Cr_2O_7$ (*D*) : N_2

(*E*) : Cr_2O_3 (F) : Cr (G) : Al_2O_3

6. (*A*) : $CrCl_3$ (*B*) : CrO_2Cl_2 (by chromyl chloride test of Cl^-)

(*C*) : Na_2CrO_4 (due to oxidation of Cr^{3+} by $NaOH/H_2O_2$ and due to dissolution of CrO_2Cl_2 in NaOH)

(*D*) : $PbCrO_4$

7. (a) $\underset{+3}{2Mn^{3+}} + 2H_2O \longrightarrow \underset{+2}{Mn^{2+}} + \underset{+4}{MnO_2} + 4H^+$

(b) $\underset{+6}{3MnO_4^{2-}} + 2H_2O \longrightarrow \underset{+4}{MnO_2} + \underset{+7}{2MnO_4^-} + 4OH^-$

(c) $2MnO_4^- + 3Mn^{2+} + 2H_2O \longrightarrow 5MnO_2 + 4H^+$

8. $\underset{+6}{3MnO_4^{2-}} + 2H^+ \longrightarrow \underset{+4}{MnO_2} + \underset{+7}{2MnO_4^-} + 2H_2O$

It is a disproportionation reaction. In disproportionation reaction, equivalent weight of the species undergoing disproportionation = Eq. wt. (in oxidation) + Eq. wt. (in reduction)

$\underset{+6}{MnO_4^{2-}} \longrightarrow \underset{+4}{MnO_2}$ Eq. wt. (redn.) $= \dfrac{\text{mol. wt.}}{2}$

$\underset{+6}{MnO_4^{2-}} \longrightarrow \underset{+7}{MnO_4^-}$ Eq. wt. (oxidn.) $= \dfrac{\text{mol. wt.}}{1}$

$\therefore$ Eq. wt. of $MnO_4^{2-} = \dfrac{M}{2} + \dfrac{M}{1} = \dfrac{3}{2}M$

9. (a) $5BiO_3^- + 2Mn^{2+} + 14H^+ \longrightarrow 2MnO_4^- + 5Bi^{3+} + 7H_2O$

(b) $5PbO_2 + 2Mn^{2+} + 4H^+ \longrightarrow 2MnO_4^- + 5Pb^{2+} + 2H_2O$

10. $\left[\text{tetrahedral } CrO_4: \text{Cr bonded to four O}\right]^{2-}$, $\left[\text{tetrahedral } MnO_4: \text{Mn bonded to four O}\right]^{-}$

11. (a) $2MnO_4^- + 5H_2O_2 + 6H^+ \longrightarrow 2Mn^{2+} + 5O_2 + 8H_2O$

$2MnO_4^- + 3H_2O_2 \longrightarrow 2MnO_2 + 3O_2 + 2OH^- + 2H_2O$

(b) In acidic medium $MnO_4^- \longrightarrow Mn^{2+}$

$$1M = 5N$$

$\therefore$ $H_2O_2 \equiv$ 100 mL of 5N $KMnO_4$

In basic medium $MnO_4^- \longrightarrow MnO_2$

$$1M = 3N$$

$\therefore$ $H_2O_2 \equiv$ V mL of 3N $KMnO_4$

$\therefore$ $NV = 3V$

$\therefore$ $3V = 500$

$$V = \frac{500}{3} = 166.67 \text{ mL}$$

12. Aqueous CO_2 releases H^+ which can convert MnO_4^{2-} into MnO_2 and MnO_4^- in a disproportionation reaction.

$$CO_2 + H_2O \rightleftharpoons H^+ + HCO_3^-$$

$$3MnO_4^{2-} + 4H^+ \longrightarrow MnO_2 + 2MnO_4^- + H_2O$$

(reduction: $MnO_4^{2-} \to MnO_2$; oxidation: $MnO_4^{2-} \to MnO_4^-$)

13.

	$AlCl_3$	$ZnCl_2$
(a) **NaOH added:**	$Al(OH)_3$ white precipitate which dissolves in excess of it. $AlCl_3 + 3NaOH \longrightarrow Al(OH)_3 \downarrow + 3NaCl$; $Al(OH)_3 + NaOH \longrightarrow \underset{\text{sodium meta aluminate (soluble)}}{NaAlO_2} + 2H_2O$	$Zn(OH)_2$ white ppt dissolved in excess of it. $ZnCl_2 + 2NaOH \longrightarrow Zn(OH)_2 \downarrow + 2NaCl$; $Zn(OH)_2 + 2NaOH \longrightarrow \underset{\text{sodium zincate (soluble)}}{Na_2ZnO_2} + 2H_2O$
(b) **NH_4OH added:**	$Al(OH)_3$ white ppt; precipitate remains insoluble in excess of it	$Zn(OH)_2$ white ppt. but precipitate dissolves in excess of it. $Zn(OH)_2 + 4NH_4OH \longrightarrow \underset{\text{soluble}}{[Zn(NH_3)_4]^{2+}} + 4H_2O + 2OH^-$

14. $ZnO + C \longrightarrow Zn + CO$

$$\Delta G^\circ = \Delta G^\circ_f\ (CO) - \Delta G^\circ_f\ (ZnO)$$
$$= -137.3 - (-318.2) = 180.9 \text{ kJ mol}^{-1}$$

Positive value of ΔG° indicates that reaction is non-spontaneous. (However, it is spontaneous above 1000°C).

15. (*A*) : $ZnCl_2$ (*B*) : $Zn(OH)_2$ (*C*) : Na_2ZnO_2 (*D*) : ZnS (*E*) : AgCl

16. (*A*) : $ZnSO_3$ (*B*) : ZnO (C) : SO_2 (*D*) : ZnS

17. ZnO in white paint by atmospheric H_2S gives ZnS (white) but Pb^{2+} salts in white paint turns black by atmospheric H_2S due to formation of PbS.

18. ZnO > CdO > HgO

Copper Sulphate ($CuSO_4 \cdot 5H_2O$ – Blue Vitriol)

Preparation

- $$CuO + H_2SO_4 \longrightarrow CuSO_4 + H_2O$$
$$2Cu + 2H_2SO_4 + O_2 \xrightarrow{\Delta} 2CuSO_4 + 2H_2O$$

On concentration and cooling, blue crystals of $CuSO_4 \cdot 5H_2O$ separate.

Properties

- The Cu(II) ion or cupric ion is written more precisely as $[Cu(H_2O)_4]^{2+}$ when in aqueous solution. It has a bright blue colour. In $CuSO_4 \cdot 5H_2O$ **four of the water molecules are associated with Cu^{2+}, and the fifth is hydrogen – bonded to the sulphate ion as well as to the water molecules on the copper ion,** hence, this it is written as $[Cu(H_2O)_4]SO_4 \cdot H_2O$.

Effect of heat

- When this is heated, it loses its water of hydration in stages:

$$\underset{\text{blue}}{CuSO_4 \cdot 5H_2O} \xrightarrow{100^\circ C} \underset{\text{bluish-white}}{CuSO_4 \cdot H_2O} \xrightarrow{230^\circ C} \underset{\text{white}}{CuSO_4} \xrightarrow{750^\circ C} CuO + 5O_2 + O_2$$

DEEP Focus

It is the complexing of water molecules to the copper ion that is responsible for the blue colour of the pentahydrate.

Complex formation with KCN

- Addition of KCN to $CuSO_4$ solution first causes reduction and then precipitates CuCN (cuprous cyanide). This reacts with excess CN^-, forming a soluble four coordinate complex $[Cu(CN)_4]^{3-}$ which is tetrahedral in shape :

$$2Cu^{2+} + 4CN^- \longrightarrow 2Cu^ICN + \underset{\text{cyanogen}}{(CN)_2}$$
$$CuCN + 3CN^- \longrightarrow [Cu^I(CN)_4]^{3-}$$

$[Cu(CN)_4]^{3-}$ is a stable complex. Cd^{2+} also forms complex $[Cd(CN)_4]^{2-}$ but unstable, which ionises to form Cd^{2+} and CN^-. Thus if Cd^{2+} and Cu^{2+} both are present, Cu^{2+} and Cd^{2+} both are complexed by CN^-. If H_2S gas is passed, Cd^{2+} (in unstable $[Cd(CN_4)^{2-}]$ gives yellow ppt of CdS.

Complex formation with NH_3

- Addition of aqueous $NH_3(aq)$ to Cu^{2+} salts give blue ppt of $Cu(OH)_2$ which dissolves in excess of $NH_3(aq)$ due to the formation of the deep blue complex $[Cu(NH_3)_4]^{2+}$:

$$CuSO_4 + 2NH_4OH \longrightarrow Cu(OH)_2 \downarrow + (NH_4)_2SO_4$$
$$Cu(OH)_2 + 4NH_3 \longrightarrow \underset{\text{deep blue (soluble)}}{[Cu(NH_3)_4]^{2+}} + 2OH^-$$

With hypo

- $CuSO_4$ solution on reaction with $Na_2S_2O_3$ gives colourless (white) $NaCuS_2O_3$ due to reduction of Cu(II) to Cu(I).

$$CuSO_4 + Na_2S_2O_3 \longrightarrow CuS_2O_3 + Na_2SO_4$$
$$2CuS_2O_3 + 2Na_2S_2O_3 \longrightarrow 2NaCu^IS_2O_3 + Na_2S_4O_6$$

On boiling or in acid solution, $NaCuS_2O_3$ changes to Cu_2S.

$$2NaCuS_2O_3 + H_2O \longrightarrow Na_2S_2O_3 + Cu_2S + H_2SO_4$$

- When the copper (II) ion (complexed with tartrate ion) is heated in basic solution with a reducing sugar, such as glucose), the ion is reduced to Cu_2O (brick red precipitate). Mixture of alkaline $CuSO_4$ and sodium potassium tartrate is called Fehling's solution and this test called **Fehling's test** of aldehydes and reducing sugars:

$$RCHO + 2Cu^{2+} + 5OH^- \longrightarrow RCOO^- + Cu_2O\downarrow + 3H_2O$$

- Benedict's solution contains Cu(II) ion complexed to citrate ion. When it is heated with reducing sugar (as glucose) Cu(II) is reduced to Cu(I) oxide (Cu_2O) as in the Fehling's test.

This test called **Benedict's test** is used to test glucose in urine sample.

Estimation

- If excess of KI is added to an acidified solution of $CuSO_4(Cu^{2+}$ ions), I_2 is produced by oxidation of I^- which reduces Cu^{2+} to CuI (or Cu_2I_2) :

$$2CuSO_4 + 4KI \longrightarrow \underset{\text{white ppt}}{2CuI\downarrow} + I_2$$

$$I_2 + KI \rightleftharpoons KI_3$$

I_2 produced can be determined by titration with hypo solution ($Na_2S_2O_3$)

$$I_2 + 2Na_2S_2O_3 \longrightarrow 2NaI + Na_2S_4O_6$$

Thus $CuSO_4$ (Cu^{2+} salts) can be determined iodometrically:

$$2CuSO_4 \equiv I_2 \equiv 2Na_2S_2O_3$$

$$\therefore \quad CuSO_4 \equiv I \equiv Na_2S_2O_3$$

Eq. wt. of $CuSO_4$ = mol. wt.

Eq. wt. of $Na_2S_2O_3$ = mol. wt.

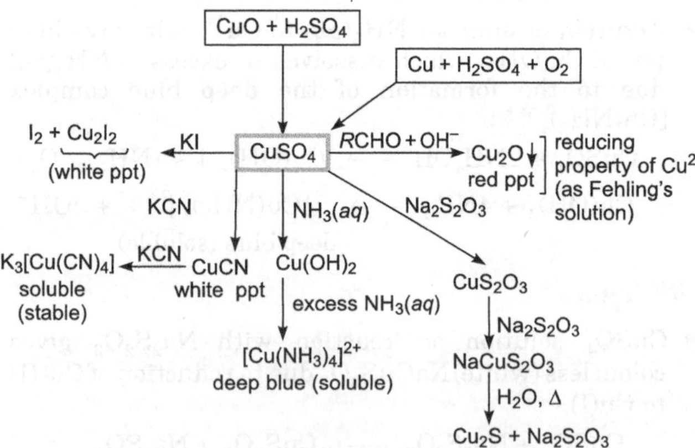

Mercurous Chloride (Hg_2Cl_2–Calomel)

Preparation

- Hg_2Cl_2 (calomel) is formed as white precipitate when soluble chloride (say NaCl) is added to soluble mercurous salt (say $Hg_2(NO_3)_2$:

$$Hg_2(NO_3)_2 + 2NaCl \longrightarrow Hg_2Cl_2\downarrow + 2NaNO_3$$

Hg_2Cl_2 is also formed when aqueous $HgCl_2$ and aqueous $SnCl_2$ are mixed in 1 : 1 molar ratio, or $HgCl_2$ is reduced by Hg.

$$HgCl_2 + SnCl_2 \longrightarrow Hg_2Cl_2 + SnCl_4$$

Excess of $SnCl_2$ would reduce Hg_2Cl_2 to Hg :

$$Hg_2Cl_2 + SnCl_2 \longrightarrow 2Hg + SnCl_4$$

$$HgCl_2 + Hg \xrightarrow{\Delta} Hg_2Cl_2$$

Properties

- If treated with aqueous NH_3, it turns black due to formation of $Hg(NH_2)Cl \cdot Hg$

$$Hg_2Cl_2 + 2NH_3 \longrightarrow \underbrace{HgNH_2Cl + Hg}_{\text{black}} + NH_4Cl$$

- In Hg_2Cl_2 two Hg atoms are bonded together using the 6s orbitals.

DEEP Focus

- Mercurous ion is **diamagnetic;** this indicates that there is no unpaired electron, hence, mercurous ion is not Hg^+ but Hg_2^{2+} :

$$Hg_2Cl_2 + Cl_2 \longrightarrow 2HgCl_2$$

$$Hg_2Cl_2 \xrightarrow{\Delta} HgCl_2 + Hg$$

- It dissolves in aqua-regia forming $HgCl_2$

$$Hg_2Cl_2 + 2Cl \longrightarrow 2HgCl_2$$

It is used in medicine as purgative.

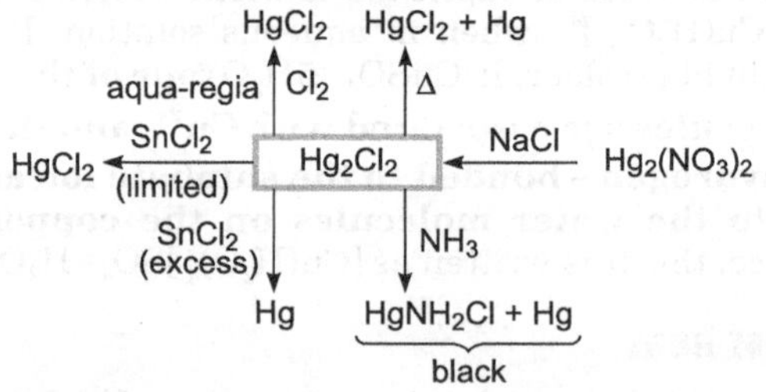

Mercuric Chloride ($HgCl_2$–Corrosive sublimate)

Preparation

- $Hg + Cl_2 \xrightarrow{\Delta} HgCl_2$
- $Hg + 2Cl \text{ (aqua regia)} \longrightarrow HgCl_2$
- $Hg_2Cl_2 + Cl_2 \longrightarrow 2HgCl_2$
- $HgSO_4 + 2NaCl \xrightarrow{MnO_2, \Delta} HgCl_2 + Na_2SO_4$

Properties

MnO_2 prevents formation of Hg_2Cl_2.

- On account of corrosive property, it is called **corrosive sublimate.**
- Highly soluble in hot water, ether and alcohol.

- With aqueous NH_3, white ppt of $HgNH_2Cl$ is formed :

$$HgCl_2 + 2NH_3 \longrightarrow HgNH_2Cl + NH_4Cl$$

- $SnCl_2$ reduces it to Hg_2Cl_2 and then to Hg :

$$SnCl_2 + 2HgCl_2 \longrightarrow \underset{\text{white ppt}}{Hg_2Cl_2 \downarrow} + SnCl_4$$

$$Hg_2Cl_2 + SnCl_2 \longrightarrow 2Hg + SnCl_4$$

- When treated with KI, it forms orange ppt of HgI_2, precipitate dissolves in excess of KI forming soluble K_2HgI_4 **(Nessler's reagent) :**

$$HgCl_2 + 2KI \longrightarrow HgI_2 \downarrow + 2KCl$$

$$HgI_2 + 2KI \longrightarrow K_2HgI_4 \text{ (soluble)}$$

Radioactive tracer technique indicates that in HgI_4^{2-} all the iodine atoms are identical.

- It dissolves AgCl,

$$HgCl_2 \rightleftharpoons HgCl^+ + Cl^-$$

$$HgCl^+ + Cl^- \text{ (from AgCl)} \rightleftharpoons HgCl_2$$

- $HgCl_2 + 2NaOH \longrightarrow HgO + 2NaCl + H_2O$
- It is used in medicine as an antiseptic and dressing skins, furs etc.

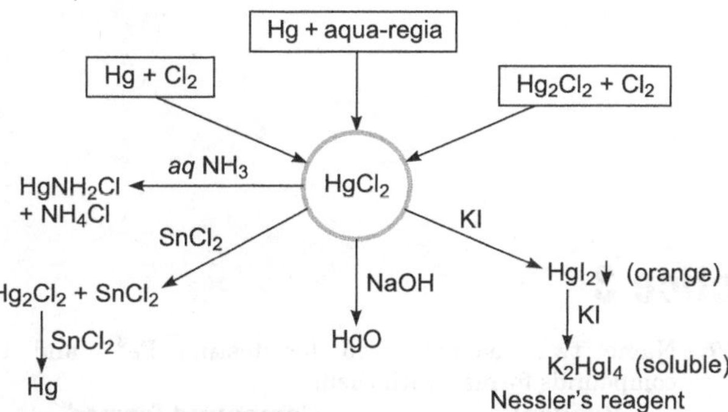

Titanium (IV) oxide

Preparation

Rutile (TiO_2) is natural occurring titanium (IV) oxide and is invariably coloured by impurities. Pure TiO_2 is obtained commercially by **chloride process**

$$\underset{\text{rutile}}{TiO_2} + 2C + Cl_2 \xrightarrow{900°C} TiCl_4 + 2CO \uparrow$$

$TiCl_4$ is volatile and can be separated from impurities. $TiCl_4$ is heated with O_2 at about 1200°C forming TiO_2 and Cl_2.Cl_2 is recycled for further conversion of TiO_2 into $TiCl_4$

$$TiCl_4 + O_2 \longrightarrow TiO_2 + 2Cl_2$$

TiO_2 —(Cl_2 + C, Δ)→ $TiCl_4$ —(O_2)→ TiO_2 (with Cl_2 recycled)

Properties

- TiO_2 is amphoteric and thus dissolves in both bases and acids

$$\underset{\text{titanyl sulphate}}{TiO \cdot SO_4} \xleftarrow{\text{conc. } H_2SO_4} TiO_2 \cdot (H_2O)_n \xrightarrow{\text{conc. NaOH}} \underset{\text{sodium titanate}}{Na_2TiO_3 \cdot (H_2O)_n}$$

- TiO_2 is used as white pigment in paint. It has three advantages over lead paints
  - it covers better
  - it is non-toxic and
  - it is not blackened by H_2S
- TiO_2 is also used for whitening paper and as a filler in plastics and rubber.

Titanium (IV) Chloride ($TiCl_4$)

Preparation

- It is prepared by passing Cl_2 gas over heated TiO_2 and C.

$$TiO_2 + Cl_2 + 2C \xrightarrow{900°C} TiCl_4 + 2CO \uparrow$$

Properties

- It is colourless, diamagnetic, covalent fuming liquid.
- It is hydrolysed vigorously by water, giving fumes in moist air

$$TiCl_4 + 4H_2O \longrightarrow Ti(OH)_2 + 4HCl$$
$$\downarrow$$
$$TiO_2$$

- If hydrolysed in presence of aqueous HCl, $TiOCl_2$ is formed

$$TiCl_4 + H_2O \xrightarrow{HCl} TiOCl_2 + 2HCl$$

- It acts as a Lewis acid (electron-pair acceptor) forming octahedral complexes:

$$TiCl_4 + \text{conc. HCl} \longrightarrow \underset{\text{(very unstable)}}{[TiCl_6]^{2-}}$$

$$TiCl_4 \xrightarrow{650°C} \underset{\text{violet}}{TiCl_3} \xrightarrow{H_2O} [Ti(H_2O)_6]Cl_3 \text{ , } [Ti(H_2O)_5Cl]Cl_2 \xrightarrow[\text{disproportionates}]{\Delta} TiCl_4 + TiCl_2$$

$$TiCl_4 \xrightarrow{Ti} TiCl_2$$

DEEP Focus

- Solution of $AlEt_3$ (triethyl aluminium) and $TiCl_4$ in a hydrocarbon solvent (as cyclohexane) react exothermically to form a brown solid called **Ziegler-Natta** catalyst. This catalyst is used for polymerising ethene into polythene.

Photography

Ordinary black-and-white photographic film consists of a celluloid strip that has been coated with a gelatin emulsion containing silver halides, usually AgBr. Photography involves a series of redox reactions (in stages) as given :

Exposure

When exposed to light, the surfaces of the AgBr grains turn dark because of light-induced redox reaction in which Br^- transfers an electron to Ag^+, producing atoms of elemental Ag. (The Br_2 produced in the reaction reacts with the gelatin emulsion)

$$2AgBr(s) \xrightarrow{\text{light}} 2Ag(s) + Br_2(l)$$

Developing

The exposed film is subjected to developing by adding an organic reducing agent (probably hydroquinone in basic medium) when more AgBr is reduced

$$2AgBr(s) + 2OH^-(aq) + \underset{\text{hydroquinone}}{C_6H_4(OH)_2} \longrightarrow 2Ag(s) + 2H_2O(l) + 2Br^-(aq) + \underset{\text{quinone}}{C_6H_4O_2}$$

By carefully monitoring the time period allowed for reduction of the AgBr with hydroquinone it is possible to amplify the latent image on the exposed film and make it visible. At the desired point, further developing is stopped by transferring the film from the basic hydroquinone solution to a weakly acidic one.

Fixing

Once the image is fully formed, the film is fixed by washing away the remaining unreacted AgBr with a solution of sodium thiosulphate (hypo)

$$AgBr(s) + 2S_2O_3^{2-}(aq) \longrightarrow [Ag(S_2O_3)_2]^{3-}(aq) + Br^-(aq)$$

The negative is now ready and can be processed in day-light. Negative is converted to positive by allowing the light to pass through it on to AgBr containing photographic paper which is then treated in a similar way as the negative.

Target Practice 3

1. Name two substances that can reduce $AgNO_3$ to Ag
   (a) .... (b) ....
2. Silver coin contains mainly silver. Pure silver is obtained by first dissolving it in HNO_3 and then other steps
   Step I : $Ag + HNO_3 \xrightarrow{\Delta} AgNO_3$
   Step II : $AgNO_3 \xrightarrow{+A} B$
   Step III : $B \xrightarrow{\Delta} Ag + CO_2 + O_2$
   Name A ....
   Name B ....
3. $AgNO_3$ is a common reagent of testing halides.
   (a) AgCl is of .... ppt soluble in ....
   (b) AgBr is of .... ppt partially soluble in ....
   (c) AgI is of .... ppt insoluble in ....
4. Silver-mirror test is for testing aldehydes reagent used is called .... and contains ....
5. Identify conjugate acid-base pair in the following acid-base reaction
   $$[Fe(H_2O)_6]^{2+} + H_2O \rightleftharpoons [Fe(H_2O)_5OH]^{2+} + H_3O^+$$
   A : B : C : D :
6. It is said that $FeCl_3 \cdot 6H_2O$ consists of *trans*-complex. This *trans*-complex is ....
   ....
7. Name two reagent used for testing Fe^{3+} and the compounds formed with them

	Reagent	Compound formed
I :	....	....
II :	....	....

8. Following tests are performed with *aq* $FeCl_3$ solution. What is the observation and conclusion?

Test	Observation	Conclusion
(a) $NaHCO_3$ is added	....	....
(b) Blue litmus is added	....	....
(c) Red litmus is added	....	....

9. Write reaction of manufacturing NaOH in earlier days by Lowig process.
   ....
   ....
10. Name the compound of iron in the +6 oxidation state. How is it prepared.
   ....
   ....
11. Name three non-stoichiometric oxides of iron
   ....
   ....
12. $Fe_2O_3 \cdot (H_2O)_4$ (freshly precipitated) is dissolved in acid as well as in base in two different test tubes. Write the compounds formed

(a) in acid ....

(b) in base ....

13. Name the compounds with which green vitriol is isomorphous

(I) .... (II) ....

14. $FeSO_4$ is generally used freshly prepared. It is due to ....

15. $FeSO_4$ or $FeSO_4(NH_4)_2SO_4 \cdot 6H_2O$ is used an volumetric reagent for the estimation of oxidising agent like

(I) .... (II) ....

16. Mohr's salt can be estimated by MnO_4^-/H^+, with MnO_4^- as self-indicator. How does it work? ....

17. Mohr's salt can be estimated by $Cr_2O_7^{2-}/H^+$. Name the indicator used ....

18. Name the contents of

(a) Fehling solution ....

(b) Benedict's solution ....

(c) Fenton's reagent ....

19. Compare stability of

(a) $[Cu(CN)_4]^{2-}$ ☐ $[Cu(CN)_4]^{3-}$

(b) $[Cd(CN)_4]^{2-}$ ☐ $[Cu(CN)_4]^{3-}$

20. Specify the temperature for the following changes

$$CuSO_4 \cdot 5H_2O \xrightarrow{A:\ K} CuSO_4 \cdot H_2O \xrightarrow{B:\ K} CuSO_4$$

$$CuO \xleftarrow{C:\ K} CuSO_4$$

21. Give two chemical tests to identify Cu^{2+} in laboratory

(a) ....

(b) ....

22. Calculate moles of $Na_2S_2O_3$ to react with I_2 obtaiend when 0.5 mol of $CuSO_4$ reacts with excess of KI. Write balanced equations.

....

....

(For more problems refer **Target Practice 4**)

Answers

1. (a) Cu (b) HCHO

2. A : Na_2CO_3 B : Ag_2CO_3

3. (a) White, aqueous NH_3 (b) Yellow, aqueous NH_3 (c) Yellow, NH_3

4. Tollen's reagent, ammoniacal $AgNO_3$

5. A : Acid B : Base C : Conjugate base of A D : Conjugate acid of B

6. $[Fe(H_2O)_4Cl_2]Cl \cdot 2H_2O$

$[\ldots]^+$: Cl (top), H_2O, OH_2, Fe (centre), H_2O, OH_2, Cl (bottom)

7.

Reagent	Compounds formed
I. $K_4[Fe(CN)_6]$	$Fe^{III}[Fe^{II}(CN)_6]^-$ Purssian blue
II. KCNS	$[Fe(SCN)]^{2+}$ with some $[Fe(SCN)(H_2O)_5]^{2+}$ blood red

8.

	Observation	Conclusion
(a)	CO_2 is evolved	$FeCl_3$ is acidic
(b)	turns red	$FeCl_3$ is acidic
(c)	no effect	......

9.

$$Na_2CO_3 + Fe_2O_3 \longrightarrow 2NaFeO_2 + CO_2$$
$$2NaFeO_2 + H_2O \longrightarrow 2NaOH + Fe_2O_3$$
$$\text{Net}: Na_2CO_3 + H_2O \longrightarrow 2NaOH + CO_2$$

10.

$$Fe_2O_3 + 2NaOH \longrightarrow 2NaFeO_2 + H_2O$$
$$2NaFeO_2 + Cl_2 + 4NaOH \longrightarrow 2Na_2FeO_4 + 2NaCl + 2H_2$$
$$\text{Net}: Fe_2O_3 + 6NaOH + Cl_2 \longrightarrow \underset{\text{red purple}}{2Na_2FeO_4} + 2NaCl + 2H_2 + H_2O$$

Na_2FeO_4 (a strong oxidising agent like $KMnO_4$) has Fe in +6 state.

11. FeO, Fe_2O_3 and Fe_3O_4

12. (a) $[Fe(H_2O)_6]^{3+}$ pale-violet (b) $[Fe(OH)_6]^{3-}$

13. I : $ZnSO_4 \cdot 7H_2O$ (white vitriol) II : $MgSO_4 \cdot 7H_2O$ (epsom salt)

14. The fact that on exposure to atmosphere it turns brownish-yellow due to the formation of basic ferric sulphate :

$$4FeSO_4 + 2H_2O + O_2 \longrightarrow 4Fe(OH)SO_4$$

15. (i) $KMnO_4$ (ii) $K_2Cr_2O_7$

16. MnO_4^- changes to Mn^{2+} (very light pink but appears colourless in aqueous solution) when Fe^{2+} is completely oxidised. One additional drop of MnO_4^- imparts its own purple-violet colour at the end point.

17. $K_3[Fe(CN)_6]$ as external indicator. Fe^{2+}solution is tested at the intermediate stages. $K_3[Fe(CN)_6]$ gives blue colour with Fe^{2+} but when it is completely oxidised there is no blue colour

$$Fe^{2+} + [Fe(CN)_6]^{3-} \longrightarrow \text{blue}$$
$$Fe^{3+} + [Fe(CN)_6]^{3-} \longrightarrow \text{no colour}$$

18. (a) Mixture of alkaline $CuSO_4$ and sodium potassium tartrate.

(b) Cu(II) complexed to citrate ion.,

(c) $FeSO_4$ and H_2O_2

19. (a) $[Cu(CN)_4]^{2-} < [Cu(CN)_4]^{3-}$

(b) $[Cd(CN)_4^{2-}] < [Cu(CN)_4]^{3-}$

20. A : 373 K B : 503 K C : 1023 K

21. (a) Add aqueous NH_3 in excess. Formation of deep blue coloured complex indicates Cu^{2+}

$$CuSO_4 + 4NH_3 \longrightarrow \underset{\text{deep blue}}{[Cu(NH_3)_4]SO_4}$$

(b) Add KI solution. Deep brown colour which change to white precipitate on addition of hypo ($Na_2S_2O_3$)

$$2CuSO_4 + 4KI \longrightarrow \underset{\text{white ppt}}{Cu_2I_2} + I_2 + 2K_2SO_4$$

$$I_2 + KI \rightleftharpoons KI_3$$

$$I_2 + 2Na_2S_2O_3 \longrightarrow 2NaI + Na_2S_4O_6$$

Deep brown colour is due to KI_3 but when hypo is added, I_2 disappears and Cu_2I_2 appears in the form of white precipitate.

22. Based on chemical reaction in 21(b)

$$2CuSO_4 \equiv 2Na_2S_2O_3$$

$$\therefore \quad Na_2S_3O_3 = 0.5 \text{ mol}$$

Practice Exercise 3

1. $AgNO_3$ is soluble in excess of hypo. Give reaction.

2. $AgNO_3$ solution can be used to identify following radicals by the colour of the ppt obtained :
(a) Cl^- (b) Br^-
(c) I^- (d) CrO_4^{2-}
(e) S^{2-}
Specify the colour of the precipitate

3. Unreacted AgBr in photography is removed by hypo used as **'fixer'**. What is the reaction ?

4. $Ag_2S \xrightarrow{aq\ NaCN} (A) \xrightarrow{Zn} (B)$
Identify (*A*) and (*B*).

5. Give the scheme of extraction of silver from silver coins.

6. How is $AgNO_3$ determined volumetrically?

7. (a) When H_2S gas is passed into $FeCl_3$ solution, yellow colour of $FeCl_3$ changes to light green. Explain
(b) Same behaviour is also observed when $SnCl_2$ solution is added. Explain.

8. $K_3[Fe(CN)_6]$, potassium ferricyanide, is used as external indicator in the volumetric estimation of Fe^{2+} salts by $K_2Cr_2O_7/H^+$ solution How does it indicate end-point?

9. Aqueous $FeCl_3$ forms acidic solution due to hydrolysis and there is evidence of $[Fe(H_2O)_5OH]^{2+}$ ion. This forms dimer at pH 4 – 5. What is structure of dimer ?

10. Identify (*A*) to (*F*) in the following:

$$FeSO_4\ (A) \xrightarrow{\Delta} (B) + (C) + (D)$$

(*C*) and (*D*) are gases, (*B*) is red-brown solid, (*C*) can be oxidised to (*D*). (*C*) also turns $Cr_2O_7^{2-}/H^+$ solution green. (*B*) dissolves in HCl forming deep yellow solution (*E*). (*E*) gives blue colour (*F*) with $K_4[Fe(CN)_6]$. (*D*) dissolves in H_2O forming a well known mineral acid called "'king of acids'".

11. In earlier days Fe_2O_3 was used to prepare NaOH in which Fe_2O_3 was recycled. Give scheme of this method.

12. Complete
(a) $Fe_2O_3 \cdot H_2O + NaOCl \longrightarrow$
(b) $Fe_2O_3 + LiOH \xrightarrow{\Delta}$ (c) $\underset{\text{(rust)}}{Fe_2O_3} + H_2C_2O_4 \longrightarrow$
(d) $Fe_2O_3 + CO \xrightarrow{\Delta}$ (e) $Fe_2O_3 + H_2 \xrightarrow{\Delta}$

13. What happens when :
(a) $Fe_2(SO_4)_3$ reacts with Fe.
(b) $FeCO_3$ reacts with dil H_2SO_4.
(c) $FeSO_4$ reacts with $K_2Cr_2O_7/H^+$ and then $K_4[Fe(CN)_6]$ is added.
(d) $FeSO_4$ is strongly heated.

14. $FeSO_4$ can be determined volumetrically by titration with $K_2Cr_2O_7$ using $K_3[Fe(CN)_6]$ as an external indicator. What is the end-point ?

15. 2.78 g of $FeSO_4 \cdot xH_2O$ is treated with excess of $BaCl_2$ when 2.33 g of white ppt are obtained. Calculate the value of *x*. (Ba = 137, S = 32, Fe = 56, H = 1, O = 16)

16. Give chemical tests for making distinction between Fe^{2+} and Fe^{3+} ions.

17. Aqueous $FeSO_4$ gives test for Fe^{2+} and SO_4^{2-} but after excess of KCN is added, solution does not give test for Fe^{2+}. Explain.

18. Copper (II) ion in basic solution is reduced by formaldehyde, HCHO to copper (I) oxide. Formaldehyde is oxidised to formate, $HCOO^-$. Write the balanced equations for this reaction.

19. You are supposed to be 'pathologist' in a medical lab. You are given a sample of urine of a diabetic patient. How will you detect the presence of sugar in it ?

20. A blue coloured mixture is supposed to contain Cd^{2+} ion also along with Cu^{2+} ion. How is the presence of Cd^{2+} detected ?

21. Cu^{2+} ion gives deep violet colour with glycine $CH_2(NH_2)COOH$. Give its equation.

(For more problem refer **Practice Exercise 4**)

Answers

1. $$2AgNO_3 + Na_2S_2O_3 \longrightarrow \underset{\text{white ppt.}}{Ag_2S_2O_3} + 2NaNO_3$$

$$Ag_2S_2O_3 + 3Na_2S_2O_3 \longrightarrow \underset{\substack{\text{sodium argento thiosulphate} \\ \text{(soluble)}}}{2Na_3[Ag(S_2O_3)_2]}$$

2. (a) $$Ag^+ + Cl^- \longrightarrow AgCl\downarrow$$
white ppt. soluble in NH_3 (*aq*)

(b) $$Ag^+ + Br^- \longrightarrow AgBr\downarrow$$
yellow ppt. partially soluble in NH_3 (*aq*)

(c) $$Ag^+ + I^- \longrightarrow AgI\downarrow$$
yellow ppt. insoluble in NH_3 (*aq*)

(d) $2Ag^+ + CrO_4^{2-} \longrightarrow Ag_2CrO_4 \downarrow$ red ppt.

(e) $2Ag^+ + S^{2-} \longrightarrow Ag_2S \downarrow$ black ppt.

3. $AgBr + 2Na_2S_2O_3 \longrightarrow Na_3[Ag(S_2O_3)_2] + NaBr$

4. $Ag_2S + 4NaCN \longrightarrow \underset{(A)}{2Na[Ag(CN)_2]} + Na_2S$

$\downarrow$ Zn

$\underset{(B)}{Ag} + Na_2[Zn(CN)_4]$

5. Silver coin has silver and copper metals (in the form of alloy).

$$\text{Coin} \xrightarrow{HNO_3, \Delta} Cu(NO_3)_2 + AgNO_3$$

(i) NaCl, (ii) Filter: Filtrate Cu^{2+} ⟵ ; ↓ white ppt (AgCl)

$K[Ag(CN)_2] \xleftarrow{KCN}$ white ppt (AgCl)

$\downarrow$ Zn

Ag (also refer Q.4)

6. $AgNO_3$ can be determined volumetrically by titration using NH_4SCN (ammonium sulphocyanide) with ferric alum as indicator:

$$Ag^+ + SCN^- \longrightarrow \underset{\text{white ppt.}}{AgSCN \downarrow}$$

When precipitation is complete, additional drop of SCN^- gives red colour with Fe^{3+} of ferric alum:

$$Fe^{3+} + SCN^- \longrightarrow \underset{\text{red colour}}{[Fe(SCN)]^{2+}}$$

7. In both cases Fe^{2+} ions are formed due to reduction. Hence, solution appears light green:

(a) $2FeCl_3 + H_2S \longrightarrow 2FeCl_2 + 2HCl + S$

(b) $2FeCl_3 + SnCl_2 \longrightarrow 2FeCl_2 + SnCl_4$

8. $Fe^{2+} + [Fe^{III}(CN)_6]^{3-} \longrightarrow$ blue colour

$Fe^{3+} + [Fe^{III}(CN)_6]^{3-} \longrightarrow$ no blue colour

Initially when Fe^{2+} is present, indicator $K_3[Fe(CN)_6]$ gives blue colour. After Fe^{2+} is completely oxidised to Fe^{3+}, there is no colour with indicator.

9. $[Fe(H_2O)_6]^{3+} + H_2O \underset{}{\overset{\text{hydrolysis}}{\rightleftharpoons}} [Fe(H_2O)_5OH]^{2+} + H_3O^+$

$$2[Fe(H_2O)_5OH]^{2+} \longrightarrow \underset{\text{dimer}}{\left[(H_2O)_5Fe\langle{}^{OH}_{OH}\rangle Fe(H_2O)_5\right]^{4+}}$$

10. $(A): FeSO_4$ $(B): Fe_2O_3$ $(C): SO_2$ $(D): SO_3$ $(E): FeCl_3$ $(F): KFe^{III}[Fe^{II}(CN)_6]$, Prussian blue

11. $Fe_2O_3 + Na_2CO_3 \longrightarrow 2NaFeO_2 + CO_2$

$2NaFeO_2 + H_2O \longrightarrow 2NaOH + Fe_2O_3$

(Lowig method) — NaOH recycled

12. (a) $Fe_2O_3 \cdot H_2O + 4NaOCl \longrightarrow 2Na_2Fe^{IV}O_4 + 2HCl + Cl_2$

(b) $Fe_2O_3 + 2LiOH \xrightarrow{\Delta} 2LiFeO_2 + H_2O$

(c) $Fe_2O_3 + 6H_2C_2O_4 \longrightarrow \underset{\text{water soluble complex}}{2[Fe(C_2O_4)_3]^{3-}} + 3H_2O + 6H^+$

(d) $Fe_2O_3 + 3CO \xrightarrow{\Delta} 2Fe + 3CO_2$

(e) $Fe_2O_3 + 3H_2 \xrightarrow{900°C} 2Fe + 3H_2O$

Reaction 12(c) has wide application in removing iron rust stains (Fe_2O_3) from fabric. If oxalic acid is used, then Fe(+III) forms complex $[Fe(C_2O_4)]^{3-}$ which can be washed with water.

13. (a) $Fe_2(SO_4)_3 + Fe \longrightarrow 3FeSO_4$

(b) $FeCO_3 + H_2SO_4 \longrightarrow FeSO_4 + H_2O + CO_2$

(c) $6Fe^{2+} + Cr_2O_7^{2-} + 14H^+ \longrightarrow 6Fe^{3+} + 2Cr^{3+} + 7H_2O$

$$Fe^{3+} + [Fe^{II}(CN)_6]^{4-} \longrightarrow \underset{\text{Prussian blue}}{Fe^{III}[Fe^{II}(CN)_6]^-}$$

(d) $2FeSO_4 \longrightarrow Fe_2O_3 + SO_3 + SO_2$

14. Fe^{2+} salts give blue colour with $K_3[Fe^{III}(CN)_6]$ being used as an external indicator. After Fe^{2+} has been completely oxidised to Fe^{3+}, there is no colour change when tested with $K_3[Fe(CN)_6]$.

15. $FeSO_4 \cdot xH_2O + BaCl_2 \longrightarrow BaSO_4$

$152 + 18x$ 233

2.78   2.33

$\Rightarrow \quad x = 7$

16.

Reagent	Fe^{2+}	Fe^{3+}
NaOH	green ppt	red-brown ppt
$K_4[Fe(CN)_6]$	white ppt turning blue rapidly	Prussian blue ppt
$K_3[Fe(CN)_6]$	Turnbull's blue ppt	red brown colour (no ppt.)
KSCN	no colour	deep red colour

17. It is due to formation of the complex $K_4[Fe(CN)_6]$ in which CN^- has been coordinated to Fe^{2+}

$$FeSO_4 + 6KCN \longrightarrow K_2SO_4 + K_4(Fe(CN)_6]$$

18. $2Cu^{2+} + 5OH^- + HCHO \xrightarrow{\Delta} HCOO^- + Cu_2O \downarrow + 3H_2O$

19. Urine sample of the diabetic patient is supposed to contain glucose. Hence, when it is heated with Fehling's solution, (see text) a brick red ppt. appears.

$$2Cu^{2+} + 5OH^- + \underset{\text{(in urine sample)}}{RCHO} \xrightarrow{\Delta} RCOO^- + \underset{\text{brick red}}{Cu_2O \downarrow} + 3H_2O$$

20. If KCN in excess is added to the mixture, Cu^{2+} forms a stable complex $[Cu^{I}(CN)_4]^{3-}$ while Cd^{2+} forms unstable complex $[Cd(CN)_4]^{2-}$. This complex ionises :

$$[Cd(CN)_4]^{2-} \rightleftharpoons Cd^{2+} + 4CN^-$$

When H_2S gas is passed, yellow ppt. of CdS appears if Cd^{2+} is present.

21. $Cu^{2+} + 2CH_2(NH_2)COOH \longrightarrow [Cu(CH_2(NH_2)COO^-]_2] + 2H^+$

or

$$\begin{array}{ccccc} CH_2NH_2 & & & & NH_2-CH_2 \\ | & \searrow & Cu & \swarrow & | \\ O=C-O & \nearrow & & \nwarrow & O-C=O \end{array}$$

Inner-Transition Elements (*f*-Block)

As we have specified earlier, the elements with a partially filled *f*-subshell in common oxidation states are known as ***f*-block transition elements** or **inner-transition elements.** They are the **two rows** elements at the bottom of the Periodic Table. (4*f* and 5*f*-series).

- Fourteen elements of the first row (4*f*-series) from cerium ($Z = 58$) to lutetium ($Z = 71$) which follow lanthanum ($Z = 57$) in the Periodic Table are called **Lanthanides** or **Lanthanoids** or **Lanthanones.** These elements are characterised by successive filling of 4*f*-orbitals in their atoms. Lanthanides along with yttrium ($Z = 39$) and lanthanum ($Z = 57$) were originally called as rare earths since their oxides (earths) were rare.

Fourteen elements of the second row (5*f*-series) from thorium ($Z = 90$) to lawrencium ($Z = 103$) following actinium are known as **Actinides** or **Actinoids** or **Actinones.** These are characterised by successive filling of 5*f*-orbital. Actinides from $Z = 93$ (after uranium $Z = 92$) have been produced artificially and are known as **transuranium elements.**

The Lanthanoids

Electronic Configuration

- Shells $n, (n-1)$ and $(n-2)$ are incomplete and $(n-2)f$, *i.e.,* 4*f*-orbital is being successively filled.
- Lanthanum ($Z = 57$) has electronic configuration [Xe] $5d^1 6s^2$.
- There is an electron in 5*d*-orbital only in Ce($Z = 58$) Gd($Z = 64$) and in Lu($Z = 71$). This is just to explain extra stability of half-filled or completely filled 4*f*-orbitals.
- Lanthanides mostly from ionic compounds in +3 oxidation state, the electronic configuration of M^{3+} ions from $Z = 58$ to $Z = 71$ varies in a regular manner from [Xe] $4f^1$ to [Xe] $4f^{14}$. Electronic configuration have been summarised in Table 16.10.

Atomic and Ionic Sizes : (The Lanthanide Contraction)

The observed values of the atomic radii and ionic radii of the trivalent lanthanide ion (Ln^{3+}) have been summarised in Table 16.10. Like the ions of the transition elements, the ionic radii show a progressive decrease with increasing atomic number. The decrease in size of Ln^{3+} ion in 4*f*-series is known as **lanthanide contraction.** (Fig. 16.7)

- In lanthanides, electrons in 4*f*-orbitals are being filled progressively. Since addition of an extra electron in 4*f* (screening constant of $(n-2)f$ electron = 1) orbital exactly balances the additional unit of nuclear charge, and thus effective nuclear charge (Z_{eff}) remains unchanged. Hence, it would be predicted that the atomic radii should also remain constant. Table 16.10. shows that there is slight decrease in the value of atomic radius along the series (with exception that of Eu and Yb. These two metals crystallise with cubic lattices and attain oxidation state of +2. It is to be noted that the 4*f*-orbitals are half-filled ($4f^7$) and completely filled ($4f^{14}$).

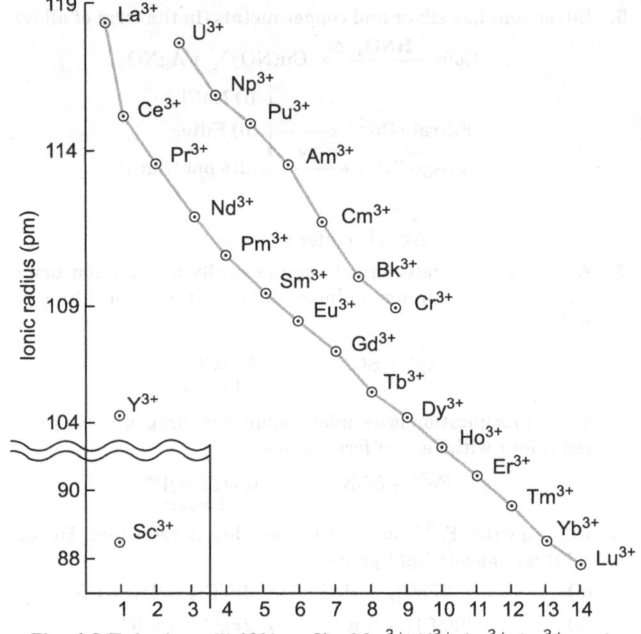

Fig. 16.7 Ionic radii (CN = 6) of Sc^{3+}, Y^{3+}, La^{3+}, Ln^{3+} and An^{3+} ions

When the trivalent ion Ln^{3+} is formed, then screening power of 4*f*-electron is no longer equal to unity (but 0.85 since 4*f* is now [$(n-1)$, penultimate shell], and each time an electron is added to 4*f*-orbitals, the effective nuclear charge (Z_{eff}) is increased by 0.15 and thus ionic size $\left(r_n \propto \dfrac{1}{Z_{eff}}\right)$ of Ln^{3+} decreases continuously from Ce^{3+} to Lu^{3+}.

DEEP Focus

- As a result of lanthanide contraction, the normal increase in size from Sc→ Y→ La disappears after the lanthanide and the pairs of elements such as **Zr and Hf, Nb and Ta, Mo and W,** etc. have nearly same size. The properties of these elements are very similar.
- The similarities in properties within these pairs make their separation very difficult.
- Due to lanthanide contraction, the elements of 5*d* and 4*d*-transition series resemble each other much more closely than do the elements of 4*d* and 3*d* series.

Table 16.10 **Electronic Configuration and Other Physical Constants of Lanthanum and Lanthanides**

Name	Atomic number	Symbol	Oxidation state	Electronic configuration outside [Xe] Core				Radius/pm		$E°/V$ M^{3+}/M	$I_1 + I_2 + I_3$ kJ mol^{-1}
				Ln	Ln^{2+}	Ln^{3+}	Ln^{4+}	Ln	Ln^{3+}		
Lanthanum	57	La	III	$5d^1 6s^2$	$5d^1$	$4f^0$		187	106	–2.52	3455
Cerium	58	Ce	III, IV	$4f^1 5d^1 6s^2$	$4f^2$	$4f^1$	$4f^0$	183	103	–2.48	3524
Praseodymium	59	Pr	III (IV)	$4f^3 6s^2$	$4f^3$	$4f^2$	$4f^1$	182	101	–2.46	3631
Neodymium	60	Nd	III	$4f^4 6s^2$	$4f^4$	$4f^3$	$4f^2$	181	100	–2.43	3692
Promethium	61	Pm	III	$4f^5 6s^2$	$4f^5$	$4f^4$		181	98	–2.42	3728
Samarium	62	Sm	III(II)	$4f^6 6s^2$	$4f^6$	$4f^5$		179	96	–2.41	3871
Europium	63	Eu	III(II)	$4f^7 6s^2$	$4f^7$	$4f^6$		204	95	–2.41	4057
Gadolinium	**64**	**Gd**	**III**	$\mathbf{4f^7 5d^1 6s^2}$	$\mathbf{4f^7 5d^1}$	$\mathbf{4f^7}$		**180**	**94**	**–2.40**	**3763**
Terbium	65	Tb	III(IV)	$4f^9 6s^2$	$4f^9$	$4f^8$	$4f^7$	178	92	–2.39	3798
Dysprosium	66	Dy	III(IV)	$4f^{10} 6s^2$	$4f^{10}$	$4f^9$	$4f^8$	177	91	–2.35	3918
Holmium	67	Ho	III	$4f^{11} 6s^2$	$4f^{11}$	$4f^{10}$		176	89	–2.32	3920
Erbium	68	Er	III	$4f^{12} 6s^2$	$4f^{12}$	$4f^{11}$		175	88	–2.30	3930
Thulium	69	Tm	III(II)	$4f^{13} 6s^2$	$4f^{13}$	$4f^{12}$		174	87	–2.28	4044
Ytterbium	70	Yb	III(II)	$4f^{14} 6s^2$	$4f^{14}$	$4f^{13}$		194	86	–2.27	4193
Lutetium	**71**	**Lu**	**III**	$\mathbf{4f^{14} 5d^1 6s^2}$	$\mathbf{4f^{14} 5d^1}$	$\mathbf{4f^{14}}$		**174**	**85**	**–2.26**	**3887**

Oxidation States

The sum of the first three ionisation energies of the lanthanides is comparatively low, so the elements are highly electropositive. They readily form Ln^{3+} ions by losing both the 6*s* and one 5*d*-electrons. In the absence of 5*d* electron, one of the electrons present in the $(n-2)f$ orbitals is lost. Besides the +3 state, some of the lanthanides show other oxidation states. Following stable ions also exist as:

$4f^0$ — La^{3+}
$4f^7$ — Eu^{2+}, Gd^{3+}, Tb^{4+}
$4f^{14}$ — Yb^{2+}, Lu^{3+}

Stability of these ions depend on
— ionisation energy
— sublimation energy
— lattice energy.

Colour of Ions

Ions of lanthanides are coloured in the solid state as well as in aqueous solution (except La^{3+}, Ce^{3+}, Gd^{3+}, Yb^{3+} and Lu^{3+}). Colour of these ions arise due to electronic transition in 4*f*-orbitals $(f - f^*)$. Colours of Ln^{3+} ions have been summarised in Table 16.11.

Magnetic Properties

Paramagnetism is associated with the presence of unpaired electrons in a substance. Ln^{3+} ions other than f^0 type (La^{3+} and Ce^{4+}) and the f^{14} type (Yb^{2+} and Lu^{3+}) are all paramagnetic. The values of magnetic moment have been given in Table 16.11

Table 16.11 **The Colours and Observed and Calculated Magnetic Moment of La^{3+} and Ln^{3+} Ions**

Ion	Colour (aquo complex)	Electronic configuration	*N* (unpaired electrons)	Magnetic moment of *N* unpaired electrons/μ BM	
				Calculated (only total spin $\sqrt{N(N+2)}$	Observed
La^{3+}	colourless	$4f^0$	0	0.0	0
Ce^{3+}	colourless	$4f^1$	1	1.73	2.3–2.5
Pr^{3+}	green	$4f^2$	2	2.83	3.4–3.6
Nd^{3+}	lilac	$4f^3$	3	3.87	3.5–3.6
Pm^{3+}	yellow	$4f^4$	4	4.90	2.70
Sm^{3+}	yellow	$4f^5$	5	5.92	1.5–1.6
Eu^{3+}	pale pink	$4f^6$	6	6.93	3.4–3.6
Gd^{3+}	colourless	$4f^7$	7	7.94	7.8–8.0
Tb^{3+}	pale pink	$4f^8$	6	6.93	9.4–9.6
Dy^{3+}	yellow	$4f^9$	5	5.92	10.4–10.5
Ho^{3+}	pale yellow	$4f^{10}$	4	4.90	10.3–10.5
Er^{3+}	rose-red (pink)	$4f^{11}$	3	3.87	9.4–9.6
Tm^{3+}	pale-green	$4f^{12}$	2	2.83	7.1–7.4
Yb^{3+}	colourless	$4f^{13}$	1	1.73	4.4–4.9
Lu^{3+}	colourless	$4f^{14}$	0	0.00	0

Colours of Ln^{4+}, Ln^{2+} ions are given below

Ce^{4+} — orange-red $4f^0$ isoelelctronic of La^{3+}

Sm^{2+} — blood-red $4f^6$ isoelectronic of Eu^{3+}

Eu^{2+} — pale greenish yellow $4f^7$ isoelectronic of Gd^{3+}

Yb^{2+} — yellow $4f^{14}$ isoelectronic of Lu^{3+}

Chemical Reactivity

- The lanthanide elements (as well as La and Y) are all silvery white, soft and malleable with high electrical conductivity. Eu and Yb are pale yellow.
- They are highly electropositive and reactive. All of them burn in air to form oxides Ln_2O_3 and nitrides LnN (cerium gives CeO_2). Powdered cerium is even pyrophoric. Yttrium usually forms a protective oxide coating on it and remains unattacked by air even at 1273 K.
- The hydrides, formed in stable in H_2 and LnH_3 phases, usually **have a defect lattice and are non-stoichiometric. They react with H_2O liberating H_2 gas.**

$$LnH_3 + 3H_2O \longrightarrow Ln(OH)_3 + 3H_2$$

The hydrides are remarkable stable to heat up to 1200 K

- They form trihalides with halogens but Ce and Tb give CeF_4 and TbF_4 respectively. Complex fluorides of Pr(IV), Nd(IV) and Dy(IV) are also formed.
- The hydroxides are ionic and basic. They are less basic than $Ca(OH)_2$ but more basic than amphoteric $Al(OH)_3$. **The base strength decreases from $Ce(OH)_3$ to $Lu(OH)_3$.**
- Non-availability of *d*-electrons for π-bonding does not favour the formation of carbonyls by most lanthanides. However, an unstable $Nd(CO)_6$ has been prepared.
- Since $E°$ for $Ln^{3+} + 3e^- \longrightarrow Ln$ is negative, they react directly with water, rapidly on heating, to liberate H_2 Reactions with acids are more vigorous, but the metals do not dissolve in alkalies. They are also stable in HF and H_3PO_4 owing to the formation of a protective film of insoluble salts.

DEEP Focus

Europium and ytterbium dissolve in liquid ammonia at 195 K forming blue solutions which give spectra characteristic of M^{2+} and ammoniated electrons. The solution decolourises slowly but the concentrated solution appear golden.

- An alloy of lanthanide metal is **mischmetal.** It has lanthanide metal (95%) and iron about 5% with traces of S, C, Ca and Al and is used in the production of bullets, shells and lighter flint. Mixed oxides of lanthanides are used as catalysts in petroleum cracking. Oxides are also used as **phosphorus** in television screens.

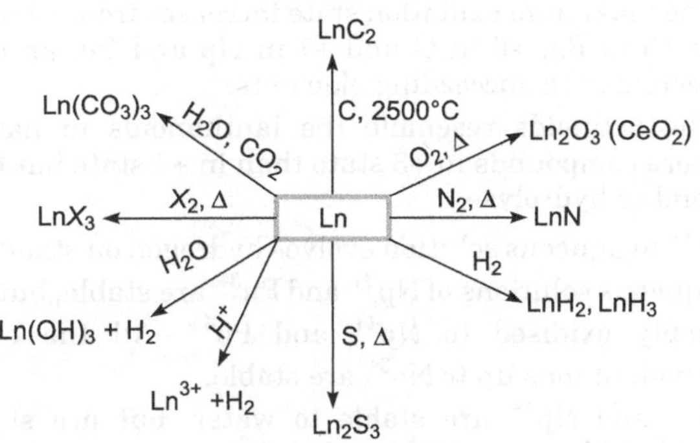

The Actinoids

The actinoids are the fourteen elements from Th to Lr following actinium in the seventh period of the Periodic Table. They are radioactive, and therefore of the greatest importance in the study of nuclear chemistry and of the principles which underline the conversion of mass into energy.

Electronic Configuration

Three shells (*n*, (*n* – 1) and (*n* – 2)) are incompletely filled and differentiating electron enters into 5*f*-orbital [*i.e.*, (*n* – 2)*f*]. Electronic configurations (EC) of actinoids and their ions have been given in Table 17.13. The variation in the (EC) of the actinoids like those in lanthanoids is due to extra stabilities of f^0, f^7 and f^{14} EC of the 5*f*-orbitals. Thus, EC of **Am and Cm** are with $5f^7$ arrangements.

$$Am(Z = 95) \quad : \quad [Rn]5f^7 7s^2$$
$$Cm(Z = 96) \quad : \quad [Rn]5f^7 6d^1 7s^2$$

Similarly Th(*Z* = 90) has $5f^0$ and No (*Z* = 102) and Lr(*Z* = 103) have $5f^{14}$ arrangements

$$Th(Z = 90) \quad : \quad [Rn]5f^0 6d^2 7s^2$$
$$No(Z = 102) \quad : \quad [Rn]5f^{14} 7s^2$$
$$Lr(Z = 103) \quad : \quad [Rn]5f^{14} 6d^1 7s^2$$

DEEP Focus

The allocation of electrons to the 5*f* or 6*d*-orbitals in the actinide atoms is more uncertain than the allocation to the 4*f* or 5*d*-orbitals in the lanthanide atoms, because the external shielding of the 5*f*-electrons is less than that of the 4*f*-electrons.

Transuranic Elements

Elements with atomic numbers greater than 92 (atomic number of uranium) are called the transuranium elements. All these elements are man-made, *i.e.*, they are produced in the laboratory through artificial nuclear reactions.

$$^{238}_{92}U + ^{1}_{0}n \longrightarrow ^{239}_{92}U + \gamma$$

$$^{239}_{92}U \xrightarrow{T_{50} = 23.5 \text{ min}} ^{239}_{93}Np + ^{0}_{-1}e$$

$$^{239}_{93}U \xrightarrow{T_{50} = 2.3 \text{ day}} ^{239}_{94}Pu + ^{0}_{-1}e$$

- **At present, transuranium elements with atomic number upto 110 have been reported. Recently, on August 16, 2003, IUPAC approved the name for the element with atomic number 110 as DARMSTADTIUM (Ds)**

The IUPAC proposed a system for naming elements with *Z* > 100.

Digits	Name	Abbrevation
0	nil	n
1	un	u
2	bi	b
3	tri	t
4	quad	q
5	pont	p
6	hex	h
7	sept	s
8	oct	o
9	enn	e

Table 16.12 **IUPAC Nomenclature of Elements with Atomic Number above 100**

Atomic number	Name	Symbol	IUPAC official name	IUPAC symbol
101	Un-nil-unnium	Unu	Mendelevium	Md
102	Un-nil-bium	Unb	Nobelium	No
103	Un-nil-trium	Unt	Lawrencium	Lr
104	Un-nil-quadium	Unq	Rutherfordium	Rf
105	Un-nil-pentium	Unp	Dubnium	Db
106	Un-nil-hexium	Unh	Seaborgium	Sg
107	Un-nil-septium	Uns	Bohrium	Bh
108	Un-nil-octium	Uno	Hassnium	Hs
109	Un-nil-ennium	Une	Meitnerium	Mt
110	Un-un-nillium	Unn	Darmstaditium***	Ds
111	Un-un-unnium	Uuu	**	
112	Un-un-bium	Uub	**	
113	Un-un-trium	Uut	*	
114	Un-un-quadium	Uuq	**	
115	Un-un-pentium	Uup		
116	Un-un-hexium	Uuh	*	
117	Un-un-septium	Uus	*	
118	Un-un-octium	Uuo	*	

***Elements yet to be discovered,**
****IUPAC names yet to be announced.**
*****IUPAC approved the name on August 16, 2003.**

Atomic and Ionic Sizes

The radii of the trivalent (M^{3+}) and quadrivalent (M^{4+}) ions have been given in Table 16.13. The ionic radii for the M^{3+} and M^{4+} ions contract slightly with increasing atomic number. There is thus an **actinide contraction** corresponding to the lanthanide contraction. (Fig. 16.7). There are two observed differences between the two series of ions :

(1) although the actinide contraction initially parallels that of the lanthanides, the elements from curium on are smaller than might be expected, probably resulting from poorer shielding by 5*f*-electrons in these elements;

(2) the lanthanide curve consists of two very shallow arcs with a discontinuity at the spherically symmetrical Gd^{3+} ($4f^7$) ion. A similar discontinuity is not clearly seen at Cm^{3+}.

Oxidation States

The actinoids in general show +3 oxidation state. There is also a greater range of oxidation states due to comparable energies of 5*f*, 6*d* and 7*s* levels, and have been shown in Table 16.14.

- The maximum oxidation state increases from +4 in Th to +5 in Pa, +6 in U and +7 in Np and Pu, an then decreases in succeeding elements.
- The actinoids resemble the lanthanoids in having more compounds in +3 state than in +4 state but they tend to hydrolyse.
- U^{3+} in aqueous solution evolves hydrogen on standing; aqueous solutions of Np^{3+} and Pu^{3+} are stable, but are easily oxidised to Np^{4+} and Pu^{4+}. All the other trivalent ions up to No^{3+} are stable.
- U^{4+} and Np^{4+} are stable to water, but are slowly oxidised by air to UO_2^{2+} and NpO_2^{2+}.

DEEP Focus

- Pu^{4+} disproportionates to Pu^{3+} and PuO_2^{2+} when the solution is strongly acidic

$$\underset{+4}{Pu^{4+}} \longrightarrow \underset{+3}{Pu^{3+}} + \underset{+6}{PuO_2^{2+}}$$

(+4 → +3: reduction; +4 → +6: oxidation)

UO_2^{+} disproportionates to U^{4+} and UO_2^{2+}.

UO_2^{2+} is stable and difficult to reduce.

Table 16.13 **Some Properties of Actinium and Actinoides**

Atomic number	Name	Symbol	Electronic configuration outside [Rn] core			Ionic radius/Pm	
			An	An^{3+}	An^{4+}	An^{3+}	An^{4+}
89	Actinium	Ac	$6d^17s^2$	$5f^0$	—	111	
90	Thorium	Th	$6d^27s^2$	$5f^1$	$5f^0$	—	99
91	Protactinium	Pa	$5f^26d^17s^2$	$5f^2$	$5f^1$	—	96
92	Uranium	U	$5f^36d^17s^2$	$5f^3$	$5f^2$	103	93
93	Neptunium	Np	$5f^46d^17s^2$	$5f^4$	$5f^3$	101	92
94	Plutonium	Pu	$5f^67s^2$	$5f^5$	$5f^4$	100	90
95	Americium	Am	$5f^77s^2$	$5f^6$	$5f^5$	99	89
96	Curium	Cm	$5f^76d^17s^2$	$5f^7$	$5f^6$	99	88
97	Berkelium	Bk	$5f^97s^2$	$5f^8$	$5f^7$	98	87
98	Californium	Cf	$5f^{10}7s^2$	$5f^9$	$5f^8$	98	86
99	Einsteinium	Es	$5f^{11}7s^2$	$5f^{10}$	$5f^9$	—	—
100	Fermium	Fm	$5f^{12}7s^2$	$5f^{11}$	$5f^{10}$	—	—
101	Mendelevium	Md	$5f^{13}7s^2$	$5f^{12}$	$5f^{11}$	—	—
102	Nobelium	No	$5f^{14}7s^2$	$5f^{13}$	$5f^{12}$	—	—
103	Lawrencium	Lr	$5f^{14}6d^17s^2$	$5f^{14}$	$5f^{13}$	—	—
110	Darmstadium	Ds	$5f^{14}6d^87s^2$	$5f^{14}6d^7$	$5f^{14}6d^6$	—	—

Table 16.14 **Oxidation States of Actinium and Actinoids**

Ac	Th	Pa	U	Np	Pu	Am	Cm	Bk	Cf	Es	Fm	Md	No	Lr	Ds
3		3	3	3	3	3	3	3	3	3		3	3	3	3
	4	4	4	4	4	4	4	4							
		5	5	5	5	5									
			6	6	6	6									
				7	7										

Magnetic Properties

The magnetic properties of the actinoids is again dependent on unpaired electrons (N) and are given by

$$\mu \text{ (magnetic moment)} = \sqrt{N(N+2)} \text{ BM}$$

but observed values are more complex than those of lanthanoids and have been summarised in Table 16.15.

Table 16.15 **Magnetic Moment of Actinoids**

Element	Z	N	$\mu = \sqrt{N(N+2)}$ BM (only total spin)
Th	90	2	2.83
Pa	91	3	3.87
U	92	4	4.90
Np	93	5	5.92
Pu	94	6	6.93
Am	95	7	7.94
Cm	96	8	8.94
Bk	97	5	5.92
Cf	98	4	4.90
Es	99	3	3.87
Fm	100	2	2.83
Md	101	1	1.73
No	102	0	0.00
Lr	103	1	1.73

Ionisation Energy

- Ionisation energy of the early actinoids are smaller than those of early lanthanoids. When electrons are being filled into 5f-orbitals they will penetrate less into the inner core of the electrons and thus these occupied 5f-electrons will be more effectively shielded from the nuclear charge than are the 4f-electrons of the corresponding lanthanoids.

Physical and Chemical Properties

- The actinoids metals are all silvery in appearance with a variety of structure.
- The dominant feature of the actinoids is their nuclear instability as they have tendency to undergo spontaneous fission and decay becomes more pronounced with the heavier elements. The radioactivity of Th and U is probably responsible for much of the earth's internal heat.
- The actinoids metals are more electropositive and are reactive, reactivity increase with increase in atomic number.
- They tarnish rapidly in air, forming an oxide coating which is protective in case of Th but less for the other elements.
- The metals react with most non-metals especially on heating, but resist alkali attack and are less reactive towards acids than might be expected. Concentrated HCl react rapidly but concentrated HNO_3 makes Th, U and Pu passive, which can be prevented in presence of F^-.
- With boiling water or steam, H_2 is obtained and metal is converted into oxide.
- Metals react rapidly with H_2 forming hydrides which are decomposed by H_2O.

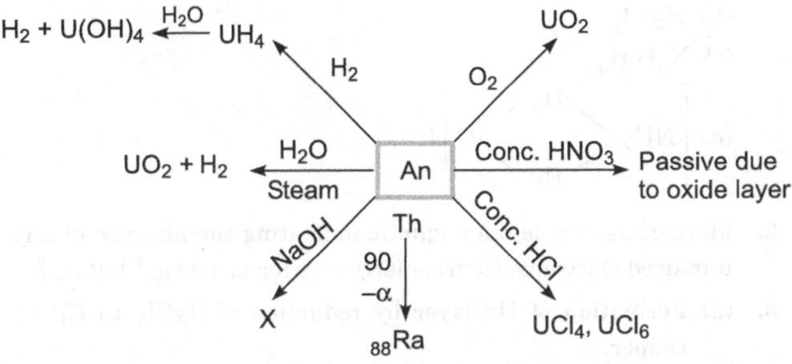

Target Practice 4

Note : This also includes problems from d-block elements.

1. Write formulae of
(a) Calomel ....
(b) Corrosive sublimate ....
(c) Nessler's reagent ....
(d) Iodide of Millon's base ....

2. Mercury (I) chloride is written as Hg_2Cl_2 and not as HgCl because ....
....

3. A magician showed following performances in a magic show. But you immediately disclosed their secrets by chemical reactions.
(a) He rubbed a white substance on a copper coin which was covered with a silvery layer. It is due to
....
(b) He took two colourless solutions. On adding one into another dropwise, orange precipitate appeared which dissolved on adding excess of that. It was explained as
....
....
(c) He placed some white substance with 10 paise coin on the palm of a child and asked to close it tightly. After sometime, palm got hot and child cried. It was due to
....
....

4. TiO_2 can be converted into $TiCl_4$ and *vice-versa*. Explain based on reactions
....
....

5. What are the contents of the Ziegler-Natta catalyst?

....

6. Select species which are amphoteric or amphiprotic?

Species		Amphoteric	Amphiprotic
I :	TiO_2		
II :	H_2O		
III :	Al_2O_3		
IV :	ZnO		
V :	HSO_4^-		
VI :	$H_2PO_2^-$		
VII :	$H_2PO_3^-$		
VIII :	HPO_4^{2-}		

7. What are the contents of mischmetal?

....

....

8. Mention ion of following electronic configuration in valence shell

(a) $4f^0$ ....

(b) $4f^7$ ....

(c) $4f^{14}$ ....

9. Lanthanides does not form carbonyl because ....

What happens when europium or ytterbium is dissolved in liquid ammonia?

....

....

10. Write maximum oxidation states of the following

(a) U .... (b) Np .... (c) Bk ....

11. (a) In acidic solution Pu^{4+} disproportionates. Write chemical reation

$$Pu^{4+} \longrightarrow \underset{\text{Oxidation}}{.........} + \underset{\text{Reduction}}{.........}$$

(b) UO_2^+ disproportionates to .... and ....

12. Balance following nuclear reactions

(a) ${}^{238}_{92}U \longrightarrow A + {}^{4}_{2}He$

(b) ${}^{238}_{92}U \longrightarrow {}^{206}_{82}Pb + B + C$

13. Write IUPAC symbol for the elements with atomic number

(a) 100 (b) 101 (c) 110 (d) 112 (e) 114

Answers

1. (a) Hg_2Cl_2

(b) $HgCl_2$

(c) K_2HgI_4

(d) $\left[NH_2\langle{}^{Hg}_{Hg}\rangle O\right]I$

2. Mercurous ions is diamagnetic indicating the absence of any unpaired electrons. Hence, mercurous ion is not Hg^+ but Hg_2^{2+}.

3. (a) Formation of Hg layer by reduction of $HgCl_2$ to Hg by copper.

$$HgCl_2 + Cu \longrightarrow CuCl_2 + \underset{\text{silvery}}{Hg}$$

(b) $HgCl_2 + 2KI \longrightarrow \underset{\text{orange}}{HgI_2} + 2KCl$

(c) Exothermic reaction

$$3HgCl_2 + \underset{\text{(coin)}}{2Al} \longrightarrow 2AlCl_3 + 3Hg + \text{Heat}$$

4. $TiO_2 + 2Cl_2 + 2C \xrightarrow{1173\text{ K}} TiCl_4 + 2CO\uparrow$

$$TiCl_4 + 4H_2O \longrightarrow Ti(OH)_4 + 4HCl \xrightarrow{\Delta} TiO_2$$

5. Triethyl aluminium and titanium (IV) chloride.

6.

Species		Amphoteric	Amphiprotic
I	TiO_2	✔	×
II	H_2O	×	✔
III	Al_2O_3	✔	×
IV	ZnO	✔	×
V	HSO_4^-	✔	✔
VI	$H_2PO_2^-$	×	×
VII	$H_2PO_3^-$	✔	✔
VIII	HPO_4^{2-}	✔	✔

7. 95% lanthanide, 5% iron with traces of sulphur, carbon, calcium and aluminium.

8. (a) $4f^0 : La^{3+}$

(b) $4f^7 : Eu^{2+}, Gd^{3+}, Tb^{4+}$

(c) $4f^{14} : Yb^{2+}, Lu^{2+}$

9. of non-availability of *d*-electrons for π-bonding.

10. (a) U = + 6 (b) Np = + 7 (c) Bk = + 4

11. (a) $\underset{+4}{Pu^{4+}} \longrightarrow \underset{+6}{PuO_2^{2+}} + \underset{+3}{Pu^{3+}}$

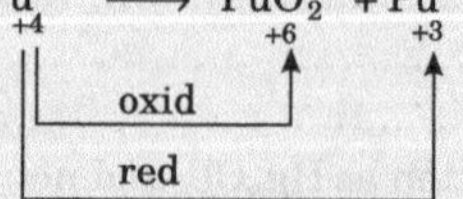

(b) U^{4+} and UO_2^{2+}

12. (a) $A : {}^{234}_{90}Th$ (b) $B : 8\alpha$ $C : 6\beta$

13. (a) Unn (b) Unu (c) Uun (d) Uub (e) Uuq

Practice Exercise 4

1. Mercurous ion has zero magnetic moment. It is represented as Hg^+, Hg^{2+} or Hg_2^{2+}, explain.
2. What is van't Hoff factor of $Hg_2(NO)_2$ if it is 100% ionised in aqueous solution?
3. Give the scheme of separation of mixture of insoluble AgCl, Hg_2Cl_2 and $PbCl_2$.
4. When K_2HgI_4 reacts with NH_3, brown ppt is formed. Explain the formation of brown ppt
5. Make distinction between Hg_2^{2+} and Hg^{2+} salts.
6. (a) What happens when a solution of $HgCl_2$ is boiled with a solution of HgO?
   (b) What happens when $HgCl_2$ is dissolved in conc. HCl?
7. Ilmenite ($FeTiO_3$) can also be converted into $TiCl_4$ by heating with carbon and chlorine. Write a balanced equation for this reaction.
8. $TiCl_4$ can be converted to pure Ti by reduction with Na or Mg. Write chemical equation.
9. Explain lanthanide contraction
10. Write EC of (a) Gd, (b) Gd^{3+} (c) Lu^{3+}
    Also determine magnetic moment (assuming total spin only).
11. Compare base strength of $Ln(OH)_3$, $Ca(OH)_2$ and $Al(OH)_3$
12. Compare lanthanoids and actinoids in terms of:
    (a) EC (b) ionic size
    (c) radioactivity (d) oxidation state
13. Complete the following reactions:
    (a) $Pu^{4+} \xrightarrow{H^+}$ (b) $UO_2^{2+} \xrightarrow{H^+}$
    (c) $U^{4+} \xrightarrow{H_2O/O_2}$

Answers

1. Hg^{2+} has no electron unpaired hence, its magnetic moment is zero, but it is called mercuric [Hg(II)] ion. Hg^+ has one unpaired electron, it has magnetic moment of $\sqrt{3}$ BM, but mercurous ion has zero magnetic moment, it is only possible if it has all paired electrons. Thus it is Hg_2^{2+}.
2. $Hg_2(NO_3)_2 \rightleftharpoons Hg_2^{2+} + 2NO_3^-$
   Total ions = $y = 3$
   i (van't Hoff factor) = $(1 + (y-1)x]$
   $= 1 + 2x$ (x = degree of ionisation = 1)
   $= 1 + 2 = 3$
3. (a) $PbCl_2$ is soluble in hot water; Hg_2Cl_2 and AgCl are insoluble in hot water.
   (b) AgCl is soluble in *aq* NH_3.
   (c) Hg_2Cl_2 is blackened by *aq* NH_3

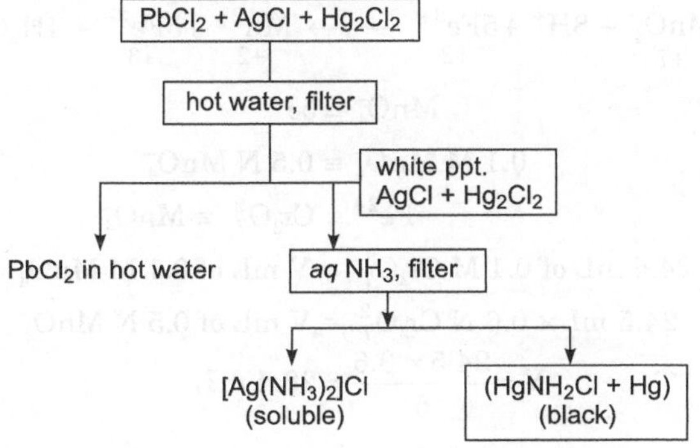

4. $2K_2HgI_4 + NH_3 + 3NaOH \longrightarrow \left[O\langle^{Hg}_{Hg}\rangle NH_2\right]I + 4KI + 3NaI + 2H_2O$
   (iodide of Millon's base)
   brown ppt.
5.

Reagent	Hg_2^{2+}	Hg^{2+}
• KI	green ppt. which dissolves in excess of KI	orange ppt. which dissolves in excess of KI (K_2HgI_4)
• HCl	white ppt (Hg_2Cl_2)	no ppt.
• H_2S gas	no ppt.	black ppt (HgS).

6. (a) Mercury oxychlorides are formed:
   $2HgCl_2 \cdot HgO$ (red)
   $HgCl_2 \cdot 2HgO$ (black)
   $HgCl_2 \cdot 3HgO$ (yellow)
   $HgCl_2 \cdot 4HgO$ (black)
   (b) H_2HgCl_4 is formed which crystallises on cooling.
7. $2FeTiO_3 + 6C + 7Cl_2 \longrightarrow 2TiCl_4 + 2FeCl_3 + 6CO$
8. $TiCl_4 + 2Mg \longrightarrow Ti + 2MgCl_2$
   $TiCl_4 + 4Na \longrightarrow Ti + 4NaCl$
9. Refer text 10. Refer text
11. Refer text 12. Refer text
13. Refer text

Total Practice Set
(Read, Plan & Solve!)

Problem 1. In an atmosphere with industrial smog, Cu corrodes to a basic sulphate $Cu_2(OH)_2SO_4$ and basic carbonate $Cu_2(OH)_2CO_3$. Propose a series of chemical reactions to describe this corrosion.

Solution $2Cu + H_2O + SO_3 + O_2 \longrightarrow Cu_2(OH)_2SO_4$

$$2Cu + H_2O + CO_2 + O_2 \longrightarrow Cu_2(OH)_2CO_3$$

Fortunately this corrosion product forms a tough adherent coating that protects the underlying metal. Overall reaction is a combination of oxidation-reduction, acid-base and precipitation reaction.

Problem 2. When Zn is added to an acidified solution of $K_2Cr_2O_7$, the colour of the solution changes from orange to green, then to blue and over a period of time, back to green. Write equations for this series of reactions.

Solution Zn reduces $Cr_2O_7^{2-}$(orange)/H^+ to Cr^{3+} (green) and then to Cr^{2+} (blue), but atmospheric O_2 oxides Cr^{2+} (blue) back to Cr^{3+} (green) :

$$Zn \longrightarrow Zn^{2+} + 2e^-$$

$$Cr_2O_7^{2-} + 14H^+ + 6e^- \longrightarrow 2Cr^{3+} + 7H_2O$$

$$Cr^{3+} + e^- \longrightarrow Cr^{2+}$$

$$4Cr^{2+} + \underset{(air)}{O_2} + 4H^+ \longrightarrow 4Cr^{3+} + 2H_2O$$

Problem 3. In qualitative analysis, a confirmatory test for chromium involves the formation of a blue transient peroxo species, CrO_5, which is better represented as $CrO(O_2)_2$. It is formed when H_2O_2 is added to on acidic solution containing $Cr_2O_7^{2-}$. What is oxidation state of Cr is $CrO(O)_2$? Write the equation of formation of CrO_5.

Solution $Cr_2O_7^{2-} + 2H^+ + 4H_2O_2 \longrightarrow 2CrO(O_2)_2 + 5H_2O$

Oxidation state of oxygen in peroxide = – 1

Oxidation state of oxygen in oxide = – 2

Hence, in Cr O $\boxed{(O_2)_2}$ ⟵ four oxygen of peroxide.

$$x - 2 - 4 = 0$$

$$x = +6$$

Problem 4. Both $Cr_2O_7^{2-}$ and MnO_4^- solutions can be used to titrate Fe^{2+} in acidic medium.

Suppose you have available 0.1 M solutions of each. For a given sample of Fe^{2+} solution, if a given titration requires 24.5 mL of 0.1M $Cr_2O_7^{2-}$ solution, how many mL of 0.1 M MnO_4^- solution would have been required if it had been used instead ?

Solution $\underset{+12}{Cr_2O_7^{2-}} + 14H^+ + \underset{+2}{6Fe^{2+}} \longrightarrow \underset{+6}{2Cr^{3+}} + \underset{+3}{6Fe^{3+}} + 7H_2O$

$Cr_2O_7^{2-} \equiv 6e^-$] gain of $6e^-$ by two Cr atoms.

$\therefore$ 0.1M $Cr_2O_7^{2-}$ = 0.6 N $Cr_2O_7^{2-}$

$$\underset{+7}{MnO_4^-} + 8H^+ + \underset{+2}{5Fe^{2+}} \longrightarrow \underset{+2}{Mn^{2+}} + \underset{+3}{5Fe^{3+}} + 4H_2O$$

$$MnO_4^- \equiv 5e^-$$

$\therefore$ 0.1 M $MnO_4^- \equiv$ 0.5 N MnO_4^-

$$Fe^{2+} \equiv Cr_2O_7^{2-} \equiv MnO_4^-$$

24.5 mL of 0.1 M $Cr_2O_7^{2-} \equiv$ V mL of 0.1 M MnO_4^-

24.5 mL × 0.6 N $Cr_2O_7^{2-} \equiv$ V mL of 0.5 N MnO_4^-

$$V = \frac{24.5 \times 0.6}{5} = \mathbf{29.4\ mL}$$

Problem 5. Nickel can be determined by the precipitation of nickel dimethyl glyoximate. What is the reaction?

Solution (a) Ni^{2+} salts (in basic medium) on reaction with dimethyl glyoxime give cherry red ppt of nickel dimethyl glyoximate :

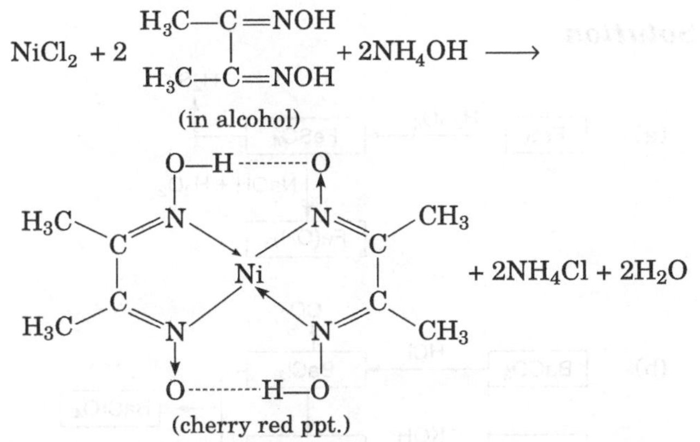

$$NiCl_2 + 2\begin{array}{l} H_3C\text{—}C{=}NOH \\ \quad\quad\; | \\ H_3C\text{—}C{=}NOH \end{array} + 2NH_4OH \longrightarrow$$

(in alcohol)

$$[Ni(dmg)_2] + 2NH_4Cl + 2H_2O$$

(cherry red ppt.)

Problem 6. (a) When $Mn(OH)_2$ is made by adding an alkali to a solution containing Mn^{2+} ions, the precipitate quickly darkens, and eventually goes black. What might be the chemical giving the black colour, and how is it madc?

(b) Dimercury (I) iodide, Hg_2I_2 is a greenish colour and is precipitated if iodide ions are added to a solution of dimercury (I) sulphate, Hg_2SO_4. Likewise the red mercury (II) iodide, HgI_2, is precipitated from a solution of mercury (II) sulphate, $HgSO_4$. However, both precipitates dissolve in excess iodide solution. What might be the reason for this?

Solution (a) The black colour is due to the manganese (IV) oxide, MnO_2. It is made by the $Mn(OH)_2$ being oxidised by oxygen in the air:

$$Mn(OH)_2 \longrightarrow MnO + H_2O$$

$$MnO + \underset{\text{air}}{\frac{1}{2}O_2} \longrightarrow \underset{\text{black}}{MnO_2}$$

(b) It is due to formation of HgI_4^{2-} (a soluble complex) in both the cases with HgI_2 :

$$HgI_2 + 2I^- \longrightarrow HgI_4^{2-}$$

But in Hg_2I_2, first there is oxidation of Hg(I) to Hg(II) and then complex formation takes place; it is by following disproportionation reaction :

$$\underset{+1}{Hg_2^{2+}} + 4I^- \longrightarrow \underset{+2}{HgI_4^{2-}} + \underset{0}{Hg}$$

Problem 7. (a) CO and CN^- are toxic. Explain by a suitable example.

(b) $S_2O_8^{2-}$ (peroxy disulphate ion) oxidises Ag^+ in the presence of excess pyridine. Write chemical reaction.

Solution (a) CO is toxic because it forms a complex with iron of haemoglobin in the blood, and this complex is more stable than oxy-haemoglobin. This prevents the haemoglobin in the red blood corpuscles from carrying oxygen round the body. This causes an oxygen deficiency, leading to unconsciousness and then death.

The extreme toxicity of cyanides is due to CN^- complexing with metals in cytochrome-enzymes and haemoglobin in the body, thus preventing normal metabolism.

(b) $$2Ag^+ + S_2O_8^{2-} \longrightarrow 2Ag^{2+} + 2SO_4^{2-}$$

$$Ag^{2+} + 4Py \longrightarrow \underset{\text{red}}{[Ag(Py)_4]^{2+}}$$

Problem 8. Explain the terms:

(a) Rusting, (b) Passivity.

Solution (a) **Rusting :** When exposed to air and moisture, iron gets rapidly oxidised to a hydrated oxide, $Fe_2O_3 \cdot xH_2O$, known as rust of iron. Actual composition is not known, but it consists mainly of $2Fe_2O_3 \cdot 3H_2O$ together with a small quantity of $FeCO_3$.

Rusting is an electrochemical phenomenon. When impure iron comes in contact with water containing CO_2, an internal voltaic cell is set up. The iron and the other metals present as impurities constitute the electrode while aerated water (having CO_2) acts as an electrolyte. As iron is more electropositive than many other metals, it passes into solution as iron (II) ion more readily than the rest of the metals.

Anode (Fe)	**Cathode (Impurities/Noble metals)**
$Fe \xrightarrow{(H^+)} Fe^{2+} + 2e^-$ (Fe dissolves)	$H_2O + CO_2 \rightleftharpoons H_2CO_3 + H^+ + HCO_3^-$
$Fe^{2+} \xrightarrow[H_2O]{O_2} FeO \cdot OH$	$2H^+ + 2e^- \longrightarrow H_2$
Fe_2O_3	$Fe(OH)_2 + CO_2 \longrightarrow FeCO_3 + H_2O$
Fe_3O_4	$Fe_2(CO_3)_3 + H_2O \longrightarrow Fe_2O_3 \cdot H_2O + 3CO_2$

Iron is protected from rusting by:

- galvanisation (by a coating of Zn)
- electroplating with Ni or Cr
- heating the iron to redness in steam when a protective coating of Fe_3O_4 (ferrosoferric oxide) is formed – Barf Process.
- painting with Pb_3O_4, $PbCrO_4$, $ZnCrO_4$ etc.

(b) **Passivity :** When a piece of iron is dipped in conc. HNO_3, a slight reaction takes place which then ceases completely. There is no change in the appearance of the iron, but this does not displace Cu from $CuSO_4$ solution which is however a common property of Fe. This behaviour is also observed with Cr and Al when in contact with HNO_3. These metals in this form are said to be **passive**.

Passive iron returns to normal state when scratched or heated in a reducing atmosphere of H_2 or CO or immersed in a conc. alkaline solution.

Passivity of iron is believed to be due to the formation of an extremely thin layer of Fe_3O_4 on the surface. **Passivity of iron prevents it from rusting.**

Problem 9. In acidic solution silver (II) oxide first dissolves to produce Ag^{2+}. This is followed by oxidation of H_2O to O_2 and the reduction of Ag^{2+} to Ag^+. Write chemical equations for the dissolution and redox reactions.

Solution

Dissolution : $AgO + 2H^+ \longrightarrow Ag^{2+} + H_2O$

Redox : $4Ag^{2+} + 2H_2O \longrightarrow 4Ag^+ + 4H^+ + O_2$

Problem 10. How are the following prepared?

(a) Ferrochrome (b) Chrome alum
(c) Mohr's salt (d) Ferromanganese

Solution (a) $\underset{\text{chromite}}{FeCr_2O_4} + 4C \xrightarrow{\Delta} \underbrace{Fe + 2Cr}_{\text{ferrochrome}} + 4CO\uparrow$

Ferrochrome is an alloy (of Fe and Cr) and can be added directly to iron, together with other metals, to produce steel.

(b) $K_2Cr_2O_7 + H_2SO_4 + 3SO_2 \longrightarrow \underbrace{K_2SO_4 + Cr_2(SO_4)_3}_{} + H_2O$

crystallises as ↓

$KCr(SO_4)_2 \cdot 12H_2O$
(chrome alum)

It is used in tanning hides, dyeing textiles and fixing baths in photography.

(c) $FeSO_4 + (NH_4)_2SO_4 + H_2O \xrightarrow[\text{(ii) cool}]{\text{(i) concentrate}} \underbrace{FeSO_4 \cdot (NH_4)_2SO_4 \cdot 6H_2O}_{\text{Mohr's salt}}$

(d) $MnO_2 + Fe_2O_3 + 5C \xrightarrow{\Delta} \underbrace{Mn + 2Fe}_{\text{ferromanganese}} + 5CO\uparrow$

Problem 11. Suggest a series of reactions, using common chemicals, by which each of the following syntheses can be performed?

(a) $Fe(OH)_3$ from FeS
(b) $BaCrO_4$ from $BaCO_3$ and $K_2Cr_2O_7$
(c) $CrCl_3$ from $(NH_4)_2Cr_2O_7$
(d) $MnCO_3$ from MnO_2

Solution

(a) FeS $\xrightarrow{H_2SO_4}$ $FeSO_4$ ($\uparrow H_2S$) $\xrightarrow{NaOH + H_2O_2}$ $Fe(OH)_3$

(b)

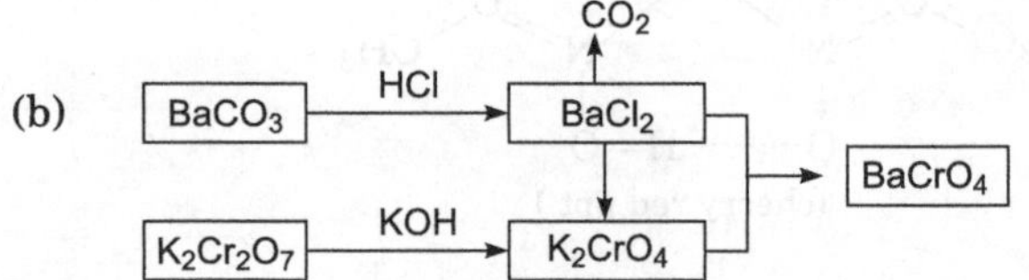

(c)

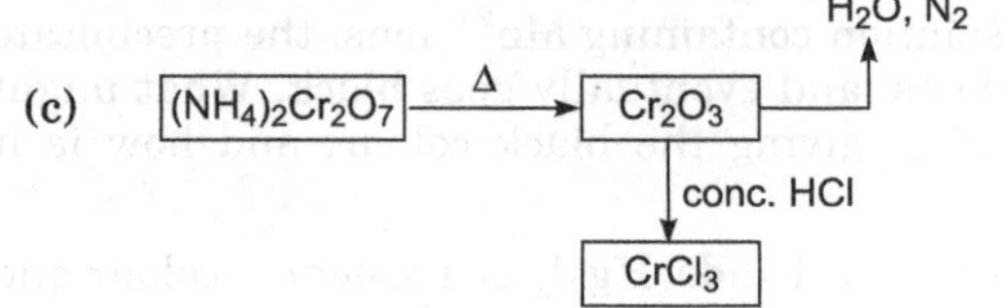

(d)

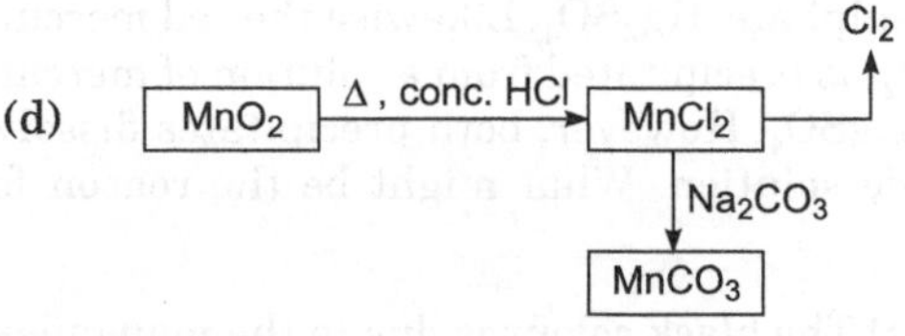

Problem 12. Write chemical reactions for the following :

(a) Mn^{3+} disproportionates to Mn^{2+} and MnO_2 in acidic medium
(b) MnO_4^{2-} disproportionates to MnO_2 and MnO_4^- in acidic medium
(c) MnO_4^{2-} is obtained if MnO_2 reacts with MnO_4^- in basic medium.
(d) Zn reduces VO^{2+} to V^{3+} in acidic medium

Solution

(a) $2Mn^{3+} + H_2O \longrightarrow Mn^{2+} + MnO_2 + 4H^+$

(b) $3MnO_4^{2-} + 4H^+ \longrightarrow MnO_2 + 2MnO_4^- + 2H_2O$

(c) $MnO_2 + 2MnO_4^- + 4OH^- \longrightarrow 3MnO_4^{2-} + 2H_2O$

(d) $Zn + 2VO^{2+} + 4H^+ \longrightarrow Zn^{2+} + 2V^{3+} + 2H_2O$

Problem 13. Calculate the sum of the first two ionisation energies for the first-series transition elements, from Table 16.1 and account for the general trend.

Solution (From Table 16.1)

Element	Z	$(IE)_1$	$(IE)_2$	$(IE)_2 + (IE)_2$
Sc	21	631	1235	1866
Ti	22	656	1309	1965
V	23	650	1413	2063
Cr	24	652	1591	2243
Mn	25	717	1509	2226
Fe	26	762	1561	2323
Co	27	758	1645	2403
Ni	28	736	1751	2487
Cu	29	775	1958	2733
Zn	30	906	1733	2639

Across the first transition element series, Z_{eff} increases and there is an almost linear increase in the sum of first two ionisation energies. This is what is expected if the two electrons are removed from the 4*s*-orbital. Higher than the expected values for the sum of first two ionisation energies are observed for Cr and Cu, because of their anomalous electron configurations (Cr : $3d^5 4s^1$ and Cu : $3d^{10} 4s^1$). An increasing Z_{eff} affects 3*d*-orbitals more than the 4*s*-orbital and the second ionisation energy for an electron from the 3*d*-orbital is higher than expected.

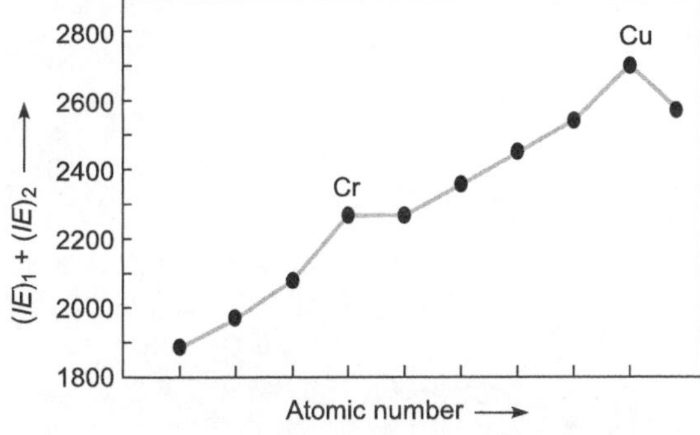

Problem 14. A solution may contain any of the following ions : $Fe^{2+}, Ni^{2+}, Cr^{3+}, Zn^{2+}, Mn^{2+}$. The solution is treated as outlined below, with the results indicated. Which of the ions is (are) probably present? Are there some ions about which you have insufficient information to reach a decision? Explain.

(1) The original unknown solution is treated with $(NH_4)_2S(aq)$ in a buffered basic solution. A dark precipitate is obtained.

(2) With the exception of a small residue of sulphur, the precipitate from (1) is completely soluble in aqua-regia (HNO_3 + HCl)

(3) The solution from (2) is treated with NaOH(*aq*) and $H_2O_2(aq)$. A dark precipitate separates from the solution. The solution is colourless.

(4) The precipitate from (3) dissolves in dil. HCl to produce a coloured solution.

(5) The solution from (4) is treated with $NH_3(aq)$. A dark precipitate forms.

(6) The precipitate from (5) is soluble in HCl(*aq*) and the solution develops an intense red colour when treated with $SCN^-(aq)$.

Solution

Scheme	Conclusion
(1) and (2)	A dark precipitate may be due to NiS, Cr_2S_3, Fe_2S_3, MnS. (ZnS is colourless)
(3)	Cr^{3+} is not possible since this treatment makes Cr^{3+} as Na_2CrO_4 soluble. Possibility of Ni^{2+} and Mn^{2+} is rejected since solution (in upper layer) is colourless. A dark precipitate is due to $Fe(OH)_3$.
(4)	$Fe(OH)_3$ is soluble in dil. HCl forming $FeCl_3$ (coloured)
(5)	$FeCl_3$ is precipitated as $Fe(OH)_3$ on addition of NH_3 (*aq*).
(6)	$Fe(OH)_3$ is again soluble in dil. HCl, and $FeCl_3$ formed gives intense rod colour with SCN^- $Fe^{3+} + SCN^- \longrightarrow \underset{\text{red}}{[Fe(SCN)]^{2+}}$

Probably present : $\mathbf{Fe^{3+}}$

Probably absent : $\mathbf{Cr^{3+}}$

Uncertain : $\mathbf{Ni^{2+}, Zn^{2+}, Mn^{2+}}$

Problem 15. Predict the products of the following reactions :

(a) Zn is added to a solution of $Cr_2(SO_4)_3$ in acid.

(b) $TiCl_3$ is added to a solution containing an excess of CrO_4^{2-} ion.

(c) Cr^{2+} is added to CrO_4^{2-} is acid solution.

(d) Mn is heated with CrO_3

(e) CrO is added to HNO_3 solution

(f) $CrCl_3$ is added to NaOH solution

(g) Fe is heated in an atmosphere of steam

(h) NaOH is added to solution of $Fe(NO_3)_3$

(i) $Fe(NO_3)_2$ is added to an acidic solution of K_2CrO_4

(j) A solution of $Fe(NO_3)_2$ and HNO_3 is allowed to stand in air

(k) $FeCO_3$ is added to a solution of $HClO_4$

(l) Fe is heated in air.

Solution (a) $2Zn(s) + 2Cr^{3+}(aq) + 2H_3O^+(aq) \longrightarrow 2Zn^{2+}(aq) + H_2(g) + 2H_2O(l) + 2Cr^{2+}(aq)$

(b) $3TiCl_3(s) + CrO_4^{2-}(aq) + 5H^+(aq) \longrightarrow 3Ti^{4+}(aq) + Cr(OH)_3(s) + 9Cl^-(aq) + H_2O(l)$

(c) $3Cr^{2+}(aq) + CrO_4^{2-}(aq) + 8H_3O^+(aq) \longrightarrow 4Cr^{3+}(aq) + 12H_2O(l)$

(d) $8CrO_3(s) + 9Mn(s) \longrightarrow 4Cr_2O_3(s) + 3Mn_3O_4(s)$

(e) $CrO(s) + 2H_3O^+(aq) + 2NO_3^-(aq) \longrightarrow Cr^{3+}(aq) + 2NO_3^-(aq) + 3H_2O(l)$

(f) $CrCl_3(s) + 3NaOH(aq) \longrightarrow Cr(OH)_3(s) + 3NaCl(aq)$

(g) $3Fe(s) + 4H_2O(g) \longrightarrow Fe_3O_4(s) + 4H_2(g)$

(h) $Fe(NO_3)_3(aq) + 3NaOH(aq) \longrightarrow Fe(OH)_3(s) + 3NaNO_3(aq)$

(i) $2CrO_4^{2-}(aq) + 2H^+(aq) \longrightarrow Cr_2O_7^{2-}(aq) + H_2O(l)$

$6Fe^{2+}(aq) + Cr_2O_7^{2-}(aq) + 14H^+(aq) \longrightarrow 6Fe^{3+}(aq) + 2Cr^{3+}(aq) + 7H_2O(l)$

(In presence of H^+, CrO_4^{2-} changes to $Cr_2O_7^{2-}$ which oxidises Fe^{2+} to Fe^{3+})

(j) $4Fe^{2+}(aq) + O_2(g) + 4HNO_3(aq) \longrightarrow 4Fe^{3+}(aq) + 2H_2O(l) + 4NO_3^-(aq)$

(k) $FeCO_3(s) + 2HClO_4(aq) \longrightarrow Fe(ClO_4)_2(aq) + H_2O(l) + CO_2(g)$

(l) $3Fe(s) + 2O_2(g) \longrightarrow Fe_3O_4(s)$

Explanation

(a) *Explain why a green solution of potassium manganate (VI) K_2MnO_4 turns purple and a brown solid is precipitated when CO_2 is bubbled into the solution.*

– CO_2 in aqueous solution is acidic

$$CO_2 + H_2O \rightleftharpoons H_2CO_3 \rightleftharpoons H^+ + HCO_3^-$$

In presence of H^+, MnO_4^{2-} disproportionates to MnO_4^- (purple, by oxidation) and MnO_2 (brown solid, by reduction)

$$\underset{+VI}{3MnO_4^{2-}} + 4H^+ \longrightarrow \underset{+VII}{2MnO_4^-} + \underset{+IV}{MnO_2} + 2H_2O$$

(b) *When mercury is oxidised with a limited amount of oxidising agent (i.e., an excess of Hg) then Hg^I compounds are formed. If there is an excess of oxidising agent, then Hg^{II} compounds are formed.*

$$Hg^{2+} + 2e^- \longrightarrow Hg \quad E° = 0.85\ V$$
$$Hg_2^{2+} + 2e^- \longrightarrow 2Hg \quad E° = 0.79\ V$$

– The standard reduction potentials are so close that oxidising agents like HNO_3 will convert Hg to Hg^{2+} rather than Hg(I) if the oxidising agent is present in excess.

The reduction potential diagram shows that Hg_2^{2+} is stable to disproportionation by a small margin under standard conditions :

$$Hg^{2+} + Hg \rightleftharpoons Hg_2^{2+} \qquad E° = 0.06\ V$$

(c) *The mercurous ion is written as Hg_2^{2+} while the cuprous ion is written as Cu^+.*

– Hg_{80} : $[Xe]_{54}\, 4f^{14} 5d^{10} 6s^2$

Hg^+ : $[Xe]\, 4f^{14} 5d^{10} 6s^1$

In Hg^I, there is one unpaired electron and we expect Hg^I (mercurous) salt to be paramagnetic. But magnetic moment of mercurous salt is zero indicating that it is diamagnetic which is only possible when 6s electron has been used by two Hg atoms in bonding. Hence, mercurous ion is Hg_2^{2+}.

Cu_{29} : $[Ar] 3d^{10} 4s^1$

Cu^+ : $[Ar]\, 3d^{10}$

Cu^I (cuprous) salt is diamagnetic by experiment and thus it is not dimeric and is thus Cu^+.

(d) *ZnO is yellow when hot, but white when cold.*

– Zn^{2+} has $3d^{10}$ EC hence all electrons paired thus form white salt. The yellow colour of hot ZnO is due to defects in the solid structure which increases with temperature.

(e) *Cu^+ is a d^{10} ion and colourless but Cu_2O is red and Cu_2S is black.*

– Cu^+ is a d^{10} ion so that there is no d—d^* transition and thus the ion is colourless. But Cu_2O and Cu_2S are coloured because of charge transfer of electrons from O^{2-} or S^{2-} to the vacant orbital of Cu^+.

(f) *Prussian blue and Turnbull's blue have different chemical formula, yet they are identical.*

– Prussian blue is formed as an intense blue precipitate when a solution of ferric salt is treated with $K_4[Fe^{II}(CN)_6]$.

(I) $4FeCl_3 + 3K_4[Fe(CN)_6] \longrightarrow Fe_4[Fe(CN)_6]_3 \downarrow + 12KCl$

Turnbull's blue is formed as a deep blue precipitate when a solution of ferrous salt is treated with $K_3[Fe^{III}(CN)_6]$.

(II) $3FeCl_2 + 2K_3[Fe(CN)_6] \longrightarrow Fe_3[Fe(CN)_6]_2 \downarrow + 6KCl$

X-ray studies have established that Prussian blue and Turnbull's blue are identical. When either of the reactions (I) and (II) is performed, a redox reaction also takes place partially :

$$Fe^{3+} + [Fe^{II}(CN)_6]^{4-} \longrightarrow Fe^{2+} + [Fe^{III}(CN)_6]^{3-}$$
$$Fe^{2+} + [Fe^{III}(CN)_6]^{3-} \longrightarrow Fe^{3+} + [Fe^{II}(CN)_6]^{4-}$$

Thus, both of these compounds have a mixture of

$$Fe^{2+}, Fe^{3+}, [Fe(CN)_6]^{4-} \text{ and } [Fe(CN)_6]^{3-}$$

hence we write Prussian blue and Turnbull's blue as $Fe[Fe(CN)_6]^-$ that can be due to Fe^{2+} and $[Fe^{III}(CN)_6]^{3-}$ shown as $Fe^{II}(Fe^{III}(CN)_6]$ or due to Fe^{3+} and $[Fe^{II}(CN)_6]^{4-}$ shown as $Fe^{III}[Fe^{II}(CN)_6]^-$.

(g) *$[Co(NH_3)_6]^{3+}$ is diamagnetic and $[CoF_6]^{3-}$ is strongly paramagnetic.*

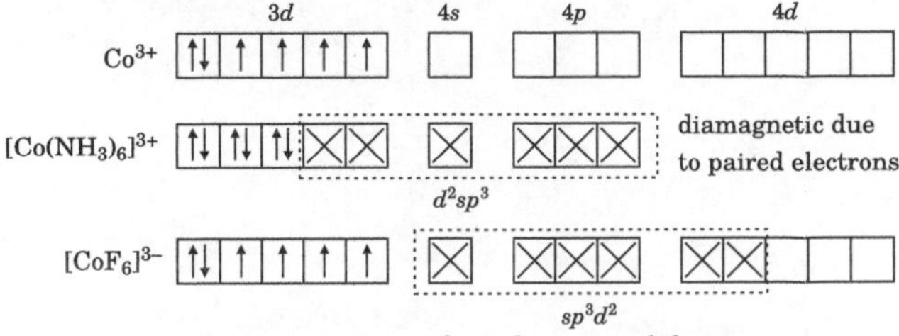

Co^{3+} has $3d^6$-configuration with four unpaired electrons in ground state. In presence of NH_3 (strong ligand) all the unpaired electrons in Co^{3+} get **paired** and thus $[Co(NH_3)_6]^{3+}$ has d^2sp^3 hybridisation (octahedral structure), thus it is diamagnetic (no electron unpaired). F^- is a weak ligand hence, six lone pairs of six F^- are filled in outer d-orbitals of Co^{3+} which has now four electrons unpaired. Thus CoF_6^{3-} has sp^3d^2 hybridisation in Co^{3+} and is thus paramagnetic due to unpaired electrons.

(h) ***For the first row transition metal the E° values are***

E°	V	Cr	Mn	Fe	Co	Ni	Cu
(M^{2+}/M)	−1.18	−0.91	−1.18	−0.44	−0.28	−0.25	+0.34

Explain the irregularity in the above values.

The $E°(M^{2+}/M)$ values are not regular which can be explained from the irregular variation of ionization energies $(IE_1 + IE_2)$ and also the sublimation energies which are relatively much less for manganese and vanadium. (240 kJ mol^{-1} for manganese and 470 kJ mol^{-1} for vanadium).

(i) ***Explain as to why the E° value for the Mn^{3+}/Mn^{2+} couple is much more positive than that for Cr^{3+}/Cr^{2+} or Fe^{3+}/Fe^{2+}.***

– Much larger third ionization energy to Mn (where the required change is d^5 to d^4) is mainly responsible for this. This also explains why the +3 state of Mn is of little importance.

(j) ***Although Mn^{2+} ion has five unpaired electrons, $[Mn(H_2O)_6]^{2+}$ is not intensely coloured.***

– Spin multiplicity is defined as = (2s + 1)

where, s = sum of spin.

$$Mn(25) = [Ar]\,4s^3 3d^5$$

$$Mn^{2+} = [Ar]\,3d^5 = [Ar]\ \boxed{\uparrow}\boxed{\uparrow}\boxed{\uparrow}\boxed{\uparrow}\boxed{\uparrow}$$

Thus, $s = 5 \times \frac{1}{2} = \frac{5}{2}$

Hence, multiplicity = (2s + 1) = 6

Multiplity of excited state can't be six.

Thus, electronic transition in Mn^{2+} are **spin forbidden,** hence $[Mn(H_2O)_6]^{2+}$ salts appear colourless.

Also, $[Mn(H_2O)_6]^{2+}$ has centre of symmetry and in such cases electronic transitions are expected to be **laporate forbidden**.

Master Exercises

Exercise 1
(Stage 1 : Learning)

Short Answer Type Questions

1. How many ions are obtained in the following complexes due to ionisation in aqueous solution ?

(a) $PtCl_4 \cdot 6H_2O$ (b) $CuCl_2 \cdot 2H_2O$
(c) $CrCl_3 \cdot 5H_2O$ (d) $CuCl_2 \cdot 4NH_3$

2. Calculate magnetic moments of M^{2+} ions (divalent cations) of $3d$-series and arrange them in the increasing order.

3. Magnetic moment of $[Fe(CN)_6]^{4-}$ is zero but that of $[Fe(H_2O)_6]^{2+}$ is $\sqrt{24}$ BM Explain.

4. Arrange $[Fe(CN)_6]^{4-}$, $[Fe(CN)_6]^{3-}$, $[Ni(CN)_4]^{2-}$ and $[Ni(H_2O)_4]^{2+}$ in increasing order of magnetic moment.

5. Magnetic of Fe^{x+} ion is $\sqrt{35}$ BM. Determine number of unpaired electrons. Also write its electronic configuration.

6. s- and p-block elements form colourless ions. Explain.

7. Which of the following ions are coloured/colourless ?

(a) Zn^{2+} (b) Cu^+
(c) Cu^{2+} (d) Sc^{3+}
(e) Ti^{2+} (f) V^{3+}
(g) Mn^{2+} (h) Ni^{2+}
(i) Na^+ (j) Ca^{2+}

8. Assuming Aufbau rule is not known then which of the following ions would be coloured?

(a) K^+ (b) Cu^+
(c) Ca^+ (d) Sc^{2+}

9. Hg^{2+} **salts are colourless. Explain.**

10. Which of the following are paramagnetic, coloured, and coloured and paramagnetic both?

$NO_2, Cu^+, O_2, O_3, Hg_2^{2+}, Cd^{2+}$, Al, C (graphite)

11. Compounds, containing the Sc^{3+} ion are colourless, whereas those containing the Ti^{3+} ion are coloured. Explain.

12. [A] Compare covalent size of the following

(a) Nb	☐	Ta
(b) Cr	☐	Mn
(c) Cu	☐	Zn
(d) Cu	☐	Ni
(e) Zr	☐	Hf
(f) Ca	☐	Sc
(g) Cr	☐	Mo

[B] Calculate effective nuclear charge of $3d$-series elements (Sc—Zn)

	Screening constant(s)	Effective nuclear charge $Z^* = Z - S$
(a) Sc(21)	....	....
(b) Ti(22)	....	....
(c) V(23)	....	....
(d) Cr(24)	....	....
(e) Mn(25)	....	....
(f) Fe(26)	....	....
(g) Co(27)	....	....
(h) Ni(28)	....	....
(i) Cu(29)	....	....
(j) Zn(30)	....	....

13. **(a)** CrO_5 **forms addition compound with pyridine in ethereal layer. What is structure of the addition product?**

(b) In ammoniacal solution, dichromate in presence of H_2O_2 **forms** $(NH_3)_3CrO_4$ **in which Cr is in +IV state. What is structure?**

(c) CrO_3 **is used as cleansing agent for glassware. How is it prepared?**

(d) Give structure of $Cr_2O_7^{2-}$ **ion.**

14. What happens when $(NH_4)_2Cr_2O_7$ is heated? What is the type of reaction?

15. Gas (A) and gas (B) both turns $K_2Cr_2O_7/H^+$ green. Gas (A) also turns lead acetate paper black. When gas (A) is passed into gas (B) in aqueous solution, yellowish white turbidity appears. Identify gas (A) and (B) and explain reactions.

16. If $CO_2(g)$ under pressure is passed into $Na_2CrO_4(aq)$, yellow colour solution changes to orange-red solution. What is it due to ? What is the function of $CO_2(g)$?

17. Arrange the following ions in order of increasing strength as a reducing agent, and account for the trend:
(a) Cr^{2+} (b) Cr^{3+}
(c) $Cr_2O_7^{2-}$

18. (a) Zn^{2+} salts appear acidic due to hydrolysis. Identify acid and base in the following :
$$[Zn(H_2O)_4]^{2+} + H_2O \rightleftharpoons [Zn(H_2O)_3OH]^+ + H_3O^+$$
(b) Which other metal ion (other than Zn^{2+}) behaves like $[Zn(H_2O)_4]^{2+}$ in giving an acidic solution?

19. What happens if $NaHCO_3$ is added to aqueous $ZnSO_4$ solution?

20. Calculate magnetic moment of Mn in MnO_4^-
.... unpaired electrons on Mn in MnO_4^-
.... BM

21. Determine number of moles of MnO_4^- which are required to oxidise one mole of ferrous oxalate. Write reactions.
....
....

22. Rust stains contain Fe^{3+} salts ($Fe_2O_3 \cdot H_2O$). These stains from fabric can be removed by the addition of oxalic acid. Explain.

23. $FeCl_3$ solution gives CO_2 with $NaHCO_3$ solution. Explain.

24. How might you separate the following pairs of ions by addition of a single reagent?
(a) Fe^{3+} and Na^+ (b) Cr^{3+} and Fe^{3+}
(d) Fe^{3+} and Cu^{2+}

25. (*A*) is a blue coloured hydrated salt giving white ppt. (*B*) with $BaCl_2$. 2.495 g of (*A*) gave 2.33 g of white ppt. (*B*). (*A*) gives deep reddish-brown solution on reaction with KI but after addition of hypo, reddish-brown solution appears in the form of white ppt. (*C*). (*A*) also gives chocolate coloured ppt. (*D*) with $K_4[Fe(CN)_6]$. Identify (*A*) to (*D*) and explain reactions.

26. When aqueous KCN is added to aqueous $CuSO_4$ solution, a white precipitate soluble in an excess of KCN, is formed. No precipitate is formed when H_2S gas is passed through the solution this point. Explain.

27. ΔG_f° (standard free energy of formation) of ZnO and CO are -318.2 kJ mol^{-1} and -137.3 kJ mol^{-1}. Is reduction of ZnO by carbon possible.

28. Colourless salt (*A*) gives white ppt. (*B*) with NaOH; ppt. (*B*) dissolves in excess of NaOH to form (*C*). On passing H_2S gas into (*C*) white ppt. (*D*) appears. (*A*) also gives white ppt (*E*) with $AgNO_3$ solution. Identify (*A*) to (*E*).

29. Colourless salt (*A*) $\xrightarrow{\Delta}$ (*B*) + (*C*) gas. (*B*) dissolves both in acid and alkali solution. Gas (*C*) turns lime water milky and acidified $K_2Cr_2O_7$ solution green. Ammoniacal solution of (*A*) gives white ppt. (*D*) with H_2S gas. Identify (*A*), (*B*), (*C*) and (*D*).

30. What is the advantage of using ZnO in paints instead of Pb^{2+} salts?

31. Compare thermal stability of ZnO, CdO and HgO.

32. Match the compounds in column (*A*) with their uses in column (*B*) :

Column A	Column B
(I) ZnO	(a) Metallurgical extraction of Ag.
(II) ZnO	(b) Zerox printing.
(III) $ZnSO_4$	(c) Luminous paints.
(IV) Zn	(d) Lithophone, eye lotion.

33. An impure metal is allowed to react with carbon monoxide at 50°C and the volatile gas thus formed is collected and heated further to about 200°C. This procedure gives the metal of 99.99% purity. What is the metal?

Analytical Questions

1. (a) Calculate number of moles of MnO_4^- required in acidic medium to oxidise :
(i) 1 mol of ferrous oxalate,
(ii) 1 mol of ferric oxalate.
(b) Calculate number of moles of $Cr_2O_7^{2-}$ required in acidic medium to oxidise :
(i) 1 mol of ferrous oxalate,
(ii) 1 mol of ferric oxalate.

2. Black coloured solid (*A*) $\xrightarrow[\Delta]{KNO_3 + KOH}$ green coloured solution (*B*) $\xrightarrow{CO_2}$ (*C*) pink + (*A*)

(*C*) is decolorised by Fe^{2+}. Identify (*A*), (*B*) and (*C*) and explain reaction.

3. Mineral (*A*) + dil. H_2SO_4 $\longrightarrow$ $\underbrace{(B) + (C)}_{\text{soluble}}$ + (*D*), (gas)

(*B*) or (*C*) + $BaCl_2$ solution $\longrightarrow$ white ppt. (*E*)

(*B*) + $K_3[Fe(CN)_6]$ $\longrightarrow$ blue ppt (*F*)

(*C*) + *aq* NH_3 (excess) $\longrightarrow$ deep blue solution (*G*)

(*C*) + (*D*) gas $\longrightarrow$ black ppt. (*H*) soluble in dil. HNO_3

What are (*A*) to (*H*) and explain reactions?

4. A student prepared three coordination compounds containing chromium with the following properties:

Formula	Colour	Cl^- ions in solution per formula unit
(a) $CrCl_3 \cdot 6H_2O$	Violet	3
(b) $CrCl_3 \cdot 6H_2O$	Light green	2
(c) $CrCl_3 \cdot 6H_2O$	Dark green	1

Write modern formulae for these compounds and suggest a method for confirming the number of Cl^- ions present in solution in each case.

5. Titanium, used in making jet aircraft engines, is much harder than potassium or calcium. Explain.

6. The highest oxidation state for the early transition metals Sc, Ti, V, Cr and Mn is the periodic group number. The highest oxidation state for the later transition elements Fe, Co, and Ni is less than the periodic group number. Explain.

7. (a) Which is stronger oxidising agent, Cr^{2+} or Cu^{2+}?
(b) Which is stronger reducing agent, V^{2+} or Co^{2+}?
(c) Which is more easily oxidised, Cr^{2+} or Ni^{2+}?
(d) Which is more easily reduced, Ti^{3+} or Co^{3+}?

8. Arrange the following substances in order of increasing strength as an oxidising agent, and account for the trend:
(a) Mn^{2+} (b) MnO_2
(c) MnO_4^-

9. What product is formed when dilute H_2SO_4, a non oxidising acid is added to each of the following ions or compounds?
(a) CrO_4^{2-} (b) $Cr(OH)_3$
(c) $Cr(OH)_2$ (d) $Fe(OH)_2$
(e) $Cu(OH)_2$

10. Write a balanced net ionic equation for each of the following reactions:
(a) A CrO_4^{2-} solution turns from yellow to orange upon addition of acid.
(b) $Fe^{3+}(aq)$ reacts with aqueous KSCN to give a deep red solution.
(c) Copper metal reacts with HNO_3 to give NO gas and a blue solution.
(d) A deep green solution of $Cr(OH)_3$ in excess base turns yellow on addition of H_2O_2.
(e) Dissolution of CuS in hot HNO_3 give NO gas, a blue solution and a yellow solid.

11. For each of the following reactions in acidic solution, predict the products and write a balanced net ionic equation:
(a) $Co^{3+}(aq) + H_2O(l) \longrightarrow$
(b) $Cr^{2+}(aq) + O_2(g) \longrightarrow$
(c) $Cu(s) + Cr_2O_7^{2-}(aq) \longrightarrow$

12. Identify *A, B, C, D* and *E* in the following :

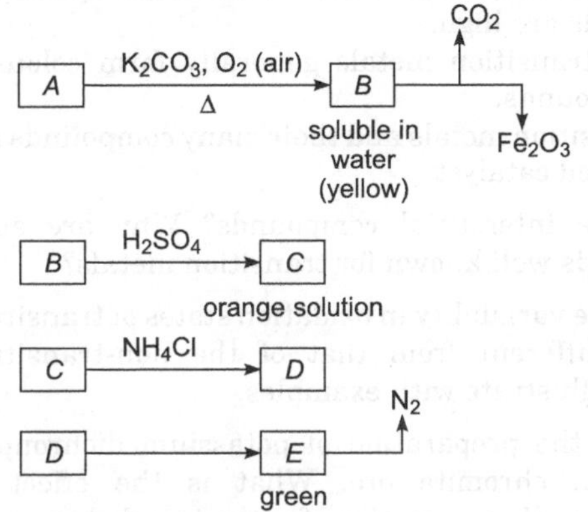

NCERT Problems

1. Write down the electronic configuration of
(a) Cr^{3+} (b) Cu^+
(c) Co^{2+} (d) Mn^{2+}
(e) Pm^{3+} (f) Ce^{4+}
(g) Lu^{2+} (h) Th^{4+}

2. Why are Mn^{2+} compounds more stable than Fe^{3+} towards oxidation to their +3 state?

3. Explain briefly how +2 state becomes more and more stable in the first half of the first row transition element with increasing atomic number?

4. To what extent do the electronic configuration decide the stability of oxidation states in the first series of the transition elements? Illustrate your answer with example.

5. What may be the stable oxidation state of the transition element with the following *d*-electron configurations in the ground state of their atoms : $3d^3, 3d^5, 3d^8$, and $3d^4$?

6. Name the oxometal anions of the first series of the transition metals in which the metal exhibits the oxidation state equal to its group number.

7. What is lanthanoid contraction? What are the consequences of lanthanoid contraction?

8. What are the characteristics of the transition elements and why are they called transition elements? Which of the *d*-block elements may not be regarded as the transition elements?

9. In what way is the electronic configuration of the transition elements different from that of the non-transition elements?

10. What are the different oxidation states exhibited by the lanthanoids?

11. Explain giving reason
(a) Transition metals and many of their compounds show paramagnetic behaviour.

(b) The enthalpies of atomisation of the transition metals are high.
(c) The transition metals generally form coloured compounds.
(d) Transition metals and their many compounds act as good catalyst.

12. What are interstitial compounds? Why are such compounds well known for transition metals?

13. How is the variability in oxidation states of transition metals different from that of the non-transition metals? Illustrate with examples.

14. Describe the preparation of potassium dichromate from iron chromite ore. What is the effect of increasing pH on a solution of potassium dichromate?

15. Describe the oxidising action of potassium dichromate and write the ionic equations for its reaction with
(a) iodide (b) iron(II) solution
(c) H_2S

16. Describe the preparation of potassium permanganate. How does the acidified permanganate solution reacts with (a) iron(II) (b) SO_2 (c) oxalic acid? Write the ionic equations for the reactions.

17. For M^{2+}/M and M^{3+}/M^{2+} systems that $E^{\ominus}$ values for some metals are as follows :

Cr^{2+}/Cr	−0.9 V	Cr^{3+}/Cr^{2+}	−0.4 V
Mn^{2+}/Mn	−1.2 V	Mn^{3+}/Mn^{2+}	+1.5 V
Fe^{2+}/Fe	−0.4 V	Fe^{3+}/Fe^{2+}	+0.8 V

Use this data to comment upon :
(a) The stability of Fe^{3+} in acid solution as compared to that of Cr^{3+} or Mn^{3+}
(b) The ease with which iron can be oxidised as compared to the similar process for either chromium or manganese metal.

18. Predict which of the following will be coloured in aqueous solution?
Ti^{3+}, V^{3+}, Cu^{+}, Sc^{3+}, Mn^{2+}, Fe^{3+}, Co^{2+} and MnO_4^-.
Give reason for each.

19. Compare the stability of +2 oxidation state for the elements of the first transition series.

20. Describe the chemistry of the three stages of photography, *i.e.*, exposure, developing, and fixing.

21. Compare the chemistry of actinoids with that of the lanthanoids with special reference to
(a) electronic configuration
(b) oxidation state
(c) atomic and ionic sizes and
(d) chemical reactivity

22. How would you account for the following
(a) Of the d^4 species, Cr^{2+} is strongly reducing while manganese(III) is strongly oxidising.
(b) Cobalt(II) is stable in aqueous solution but in the presence of complexing reagents it is easily oxidised.
(c) The d^1 configuration is very unstable in ions.

23. What is meant by 'disproportionation'? Give two examples of disproportion reaction in aqueous solution.

24. Which metal in the first series of transition metals exhibits +1 oxidation state most frequently and why?

25. Calculate the number of unpaired electrons in following gaseous ions : Mn^{3+}, Cr^{3+}, V^{3+}, and Ti^{3+}. Which one of these is the most stable in aqueous solution?

26. Give example and suggest reason for the following features of the transition metal chemistry :
(a) The lowest oxide of transition metal is basic, the highest is acidic.
(b) A transition metal exhibits higher oxidation states in oxides and fluorides.
(c) The highest oxidation state is exhibited in oxoanions of a metal.

27. Indicate the steps in the preparation of
(a) $K_2Cr_2O_7$ from chromite ore
(b) $KMnO_4$ from pyrolusite ore
(c) Copper sulphate from metallic copper
(d) Calomel from corrosive sublimate

28. What happens when aqueous ammonia reacts with
(a) silver chloride
(b) mercury (I) chloride
(c) mercury (II) chloride?

29. Describe the two uses of each of the following
(a) Copper sulphate (b) Silver nitrate
(c) Silver bromide

30. What are alloys? Name an important alloy which contains some of the lanthanoid metals. Mention its uses.

31. What are inner transition elements? Decide which of the following atomic numbers are the atomic numbers of the inner transition elements : 29, 59, 74, 95, 102, 104.

32. The chemistry of the actinoid elements is not so smooth as that of the lanthanoids. Justify this statement by giving some examples from the oxidation state of these elements.

33. Which is the last element in the series of the actionoids? Write the electronic configuration of these elements. Comment on the possible oxidation state of this element.

34. Use Hund's rule to derive the electronic configuration of Ce^{3+} ion, and calculate the magnetic moment on the basis of 'spin-only' formula.

35. Name the members of the lanthanoid series which exhibit +4 oxidation states and those which exhibit

+2 oxidation states. Try to correlate this type of behaviour with the electronic configuration of these elements.

36. Compare the chemistry of the actinoids with that of lanthanoids with reference to :
(i) electronic configuration
(ii) oxidation states, and
(iii) chemical reactivity.

37. What is lanthanoid contraction? What is its effect on the chemistry of the elements which follow the lanthanoids?

38. Write the electronic configurations of the elements with the atomic number 61, 91, 101, and 109.

Exercise 2
(Stage 2 : High Skill Problem Solving)

Only One Option Correct

1. When KCN is added to $CuSO_4$ solution
(a) KCN acts a reducing agent
(b) KCN acts as complexing agent
(c) $K_3[Cu(CN)_4]$ is formed
(d) all of the above are correct

2. Out of $AgNO_3$, AgF and $AgClO_4$, water soluble salts are
(a) AgF
(b) AgF, $AgNO_3$
(c) AgF, $AgNO_3$, $AgClO_4$
(d) None of the above

3. Stability of Cu^+ and Ag^+ halide complexes are in order
(a) I > Br > Cl > F (b) F > Cl > Br > I
(c) Cl > F > I > Br (d) Br > I > Cl > F

4. Which of the following exists as white salt in anhydrous state?
(a) CuF_2 (b) $CuSO_4$
(c) Both (a) and (b) (d) None of these

5. Which oxides will not give metal on heating?
(a) HgO (b) ZnO
(c) Ag_2O (c) All of these

6. When KI (excess) is added to
I : $CuSO_4$ II : $HgCl_2$
III : $Pb(NO_3)_2$
(a) a white ppt. of CuI in I, a orange ppt. HgI_2 in II and a yellow ppt. PbI_2 in III
(b) a white ppt. of CuI in I, an orange ppt. dissolving to HgI_4^{2-} in II, and a yellow ppt. of PbI_2 in III
(c) a white ppt. of CuI, HgI_2 and PbI_2 in each case
(d) None of the above is correct

7. Most common oxidation state of lanthanides is
(a) +2 (b) +3
(c) +4 (d) +5

8. The purest form of Fe is
(a) stainless steel (b) steel
(c) cast iron (d) wrought iron

9. In $Na_2[Fe(CN)_5NO]$, sodium nitroprusside,
(a) oxidation state of Fe is +2
(b) this has NO^+ as ligand
(c) both (a) and (b) correct
(d) none of the above is correct

10. $FeSO_4$ solution gives brown colour ring in testing nitrates or nitrites. This is
(a) $[Fe(H_2O)_5NO]^{2+}$ (b) $[Fe(H_2O)_5NO_2]^{2+}$
(c) $[Fe(H_2O)_4(NO)_2]^{2+}$ (d) $[Fe(H_2O)_4NO]^{2+}$

11. $FeCl_3 \cdot 6H_2O$ is actually
(a) $[Fe(H_2O)_6]Cl_3$
(b) $[Fe(H_2O_5Cl]Cl_2 \cdot H_2O$
(c) $[Fe(H_2O)_4Cl_2]Cl \cdot 2H_2O$
(d) $[Fe(H_2O)_3Cl_3] \cdot 3H_2O$

12. Ni^{2+}, in traces, can be tested using
(a) sodium nitroprusside
(b) dimethyl glyoxime
(c) ammonium sulphocyanide
(d) potassium ferrocyanide

13. $CuSO_4$ can be estimated volumetrically
(a) by reaction with KI followed by reaction with $Na_2S_2O_3$
(b) by reaction with $BaCl_2$
(c) by reaction with $K_4Fe(CN)_6$
(d) none of the above is correct

14. Which is correct statement?
(a) Ammoniacal CuCl is used to measure the amount of CO in gas samples
(b) Ammoniacal CuCl gives red ppt. with $CH \equiv CH$
(c) Both (a) and (b) are correct
(d) None of the above is correct

15. Which is not true statement about FeO?
(a) It is non-stoichiometric and is metal deficient
(b) It is basic oxide
(d) Its aqueous solution changes to $Fe(OH)_3$ and then to $Fe_2O_3 \cdot (H_2O)_n$ by atmospheric oxygen
(d) It gives red colour with KCNS

16. Match List I with List II and select the correct answer using the codes given below the lists:

List I (Metals)	List II (Ores)
A. Zinc	1. Azurite
B. Tin	2. Carnallite
C. Copper	3. Calamine
D. Magnesium	4. Cassiterite

Codes

	A	B	C	D
(a)	3	4	2	1
(b)	3	4	1	2
(c)	4	1	3	2
(d)	4	3	2	1

17. Haemoglobin and chlorophyll contain respectively
(a) Fe, Co (b) Fe, Mn
(c) Mg, Fe (d) Fe, Mg

18. Due to lanthanide contraction
(a) Fe, Co, Ni have equal size
(b) Zr and Hf have equal size
(c) all *f*-block ions have equal size
(d) all isoelectronic ions have equal size

19. In $Cr_2O_7^{2-}$ every Cr is linked to
(a) two O atoms (b) three O atoms
(c) four O atoms (d) five O atoms

20. For Ni and Pt different IP in MJ mol^{-1} are given below

	$(IP)_1 + (IP)_2$	$(IP)_3 + (IP)_4$
Ni	2.49	8.80
Pt	2.60	6.70

hence:
(a) nickel (II) compounds tend to be thermodynamically more stable than platinum (II)
(b) platinum (IV) compounds tend to be more stable than nickel (IV)
(c) both (a) and (b) are correct
(d) none of the above is correct

21. Paramagnetism is given by the relation $\mu = 2\sqrt{s(s+1)}$ magnetons where '*s*' is the total spin. On this basis, the paramagnetism of Cu^+ ion is
(a) 3.88 magnetons (b) 2.83 magnetons
(c) 1.41 magnetons (d) zero

22. The oxygen carrying pigment, oxy-haemocyanin, containing two copper ions is diamagnetic, because
(a) the two copper ions are in +1 oxidation state
(b) one of the copper ions is in +1 oxidation state and the other is in +2 oxidation state
(c) there are strong anti-ferromagnetic interactions between the two copper ions
(d) there are ferromagnetic interactions between the two copper ions

23. Match List I with List II and select the correct answer using the codes given below the lists:

List I (Alloys)	List II (Constituents)
A. Gun metal	1. Lead + tin
B. German silver	2. Copper + tin + zinc
C. Brass	3. Copper + zinc
D. Solder	4. Copper + zinc + nickel

Codes

	A	B	C	D
(a)	1	3	4	2
(b)	4	2	1	3
(c)	2	4	3	1
(d)	3	1	2	4

24. An acidic solution contains Cu^{2+}, Pb^{2+} and Zn^{2+}. If hydrogen sulphide gas is passed through this solution, the precipitate will contain
(a) CuS and ZnS (b) PbS and ZnS
(c) CuS and PbS (d) CuS, PbS and ZnS

25. Ti^{2+} is purple while Ti^{4+} is colourless, because
(a) there is no crystal field effect in Ti^{4+}
(b) Ti^{2+} has $3d^2$ configuration
(c) Ti^{4+} has $3d^2$ configuration
(d) Ti^{4+} is a very small cation when compared to Ti^{2+} and hence, does not absorb any radiation

26. Silver ornaments turn black by atmospheric
(a) O_2 (b) N_2
(c) Cl_2 (d) H_2S

27. $Cr_2O_7^{2-} \xrightarrow{pH = x} CrO_4^{2-} \xrightarrow{pH = y} Cr_2O_7^{2-}$
pH values *x* and *y* can be
(a) 4 and 5 (b) 4 and 8
(c) 8 and 4 (d) 8 and 9

28. In dilute alkaline solution MnO_4^- changes to
(a) MnO_4^{2-} (b) MnO_2
(c) Mn_2O_3 (d) MnO

29. Cl_2 gas is obtained by various reactions but not by
(a) $KMnO_4$ + conc. HCl $\xrightarrow{\Delta}$
(b) KCl + $K_2Cr_2O_7$ + conc. H_2SO_4 $\xrightarrow{\Delta}$
(c) MnO_2 + conc. HCl $\xrightarrow{\Delta}$
(d) KCl + F_2 $\longrightarrow$

30. Which is not true statement about $KMnO_4$?
(a) Its solution is unstable in acidic medium
(b) Its small quantity added to conc. H_2SO_4, a green coloured solution containing MnO_3^+ ions is formed
(c) MnO_4^- changes to Mn^{2+} in basic solution
(d) It is self-indicator in Fe^{2+} or $C_2O_4^{2-}$ titration

31. Pyrolusite in MnO_2 is used to prepare $KMnO_4$. Steps are :

$$MnO_2 \xrightarrow{I} MnO_4^{2-} \xrightarrow{II} MnO_4^-$$

I and II are
(a) fuse with KOH/air, electrolytic oxidation
(b) fuse with KOH/air, electrolytic reduction
(c) fuse with conc HNO_3/air, electrolytic reduction
(d) all the above are correct

32. Fe is made passive by
(a) dil. H_2SO_4 (b) dil. HCl
(c) aqua-regia (d) conc. H_2SO_4

33. Maximum magnetic moment is shown by
(a) d^5 (b) d^6
(c) d^7 (d) d^8

34. Magnetic moment of Cr ($Z = 24$), Mn^+ ($Z = 25$) and Fe^{2+} ($Z = 26$) are x, y, z. They are in order
(a) $x < y < z$ (b) $x = y < z$
(c) $z < x = y$ (d) $x = y = z$

35. Interstitial compound is formed by
(a) Fe, Co (b) Co, Ni
(c) Fe, Ni (d) all of these

36. A jeweller is selling 22-carat gold articles with 95% purity, it is approximately
(a) exact (b) 3.5% higher
(c) 3.5% lower (d) 5% lower

37. Which is not true statement?
(a) Ions of d-block elements are coloured due to d—d transition
(b) Ions of f-block elements are coloured due to f—f transition
(c) $[Sc(H_2O)_6]^{3+}$, $[Ti(H_2O)_6]^{4+}$ are coloured complexes
(d) Cu^+ is colourless ion

38. Match the catalysts in column X with their uses in column Y.

Column X	Column Y
A. $TiCl_4$	1. Adams catalyst in reduction
B. $PdCl_2$	2. in preparation of $(CH_3)_2SiCl_2$
C. Pt/PtO	3. Reppe synthesis
D. Cu	4. used as the ziegler-Natta catalyst in polythene production
E. Ni-complexes	5. Wacker process for converting C_2H_4 to CH_3CHO

	A	B	C	D	E
(a)	4	5	1	2	3
(b)	4	5	2	1	3
(c)	5	4	1	3	2
(d)	2	1	3	5	4

39. When (A) NH_4VO_3 is heated.
(B) $(NH_4)_2Cr_2O_7$ is heated
(a) in both cases N_2 is formed
(b) in both cases NH_3 is formed
(c) in (A) NH_3 and in (B) N_2 are formed
(d) in (A) N_2 and in (B) NH_3 are formed

40. $FeCr_2O_4$ (chromite) is converted to Cr by following steps:

$$\text{Chromite} \xrightarrow{\text{I}} Na_2CrO_4 \xrightarrow{\text{II}} Cr_2O_3 \xrightarrow{\text{III}} Cr$$

I, II and III are

	I	II	III
(a)	Na_2CO_3/air, Δ	C	C
(b)	NaOH/air, Δ	C, Δ	Al, Δ
(c)	NaOH/air, Δ	C, Δ	Mg, Δ
(d)	conc. H_2SO_4, Δ	NH_4Cl, Δ	C, Δ

41. When H_2O_2 is added to an acidified solution of $K_2Cr_2O_7$
(a) solution turns green due to formation of Cr_2O_3
(b) solution turns yellow due to formation of K_2CrO_4
(c) a deep blue-violet coloured compound $CrO(O_2)_2$ is formed
(d) solution gives green ppt. of $Cr(OH)_3$

42. Out of $SiCl_4$, $TiCl_4$, PO_4^{3-}, SO_4^{2-}, CrO_4^{2-}, CCl_4 isostructural are
(a) $SiCl_4$, $TiCl_4$ (b) SO_4^{2-}, CrO_4^{2-}
(c) both (a) and (b) (d) None of these

43. Maximum oxidation state is shown by
(a) Os (b) Mn
(c) Cr (d) Co

44. Following elements do not show the properties characteristic of d-block elements
(a) Cu, Ag, Au (b) Zn, Hg, Cd
(c) Sc, Ti, V (d) Fe, Co, Ni

45. The ability of d-block elements to form complexes is due to
(a) small and highly charged ions
(b) vacant low energy orbitals to accept lone pair of electrons from ligands
(c) both (a) and (b) are correct
(d) None of the above is correct

46. AgCl and NaCl are colourless. NaBr and NaI are also colourless but AgBr and AgI are coloured. This is due to
(a) Ag^+ polarises Br^- and I^-
(b) Ag^+ has unpaired d-orbital
(c) Ag^+ depolarises Br^- and I^-
(d) None of above is correct

47. Which is correct statement?
(a) In less acidic solution $K_2Cr_2O_7$ and H_2O_2 give violet coloured diamagnetic $[CrO(O_2)(OH)]^-$ ion
(b) In alkaline H_2O_2, K_3CrO_8 (with tetraperoxo species) $[Cr(O_2)_4]^{3-}$ is formed
(c) In ammoniacal solution, $K_2Cr_2O_7$ gives $(NH_3)_3CrO_4$
(d) All of the above are correct statements

48. Match the compounds of column X with oxidation state of column Y.

	Column X	Column Y
A.	$[Cr(H_2O)_6]Cl_3$	5
B.	CrO_5	4
C.	K_3CrO_8	6
D.	$(NH_3)_3CrO_4$	3

	A	B	C	D
(a)	3	6	5	4
(b)	3	4	5	6
(c)	4	5	6	3
(d)	6	5	4	3

49. For CrO_3 following is not true statement
(a) it is called chromic acid
(b) it is colourless due to $3d^{10}$ configuration
(c) it is bright orange solid and colour arises due to charge transfer
(d) it is toxic and corrosive

50. Only +2, +3, +4 oxidation states are shown by
(a) Sc, Ti (b) Fe, Ni
(c) Ti, Ni (d) Zn, Ni

51. Coagulation of blood takes place by
(a) ferric alum (b) potash alum
(c) both (a) and (b) (d) None of these

52. When KCN comes in contact with blood, one dies immediately, it is due to
(a) CN^- forms stable complex with iron of haemoglobin of blood
(b) CN^- combines with H_2O of blood causing it poison
(c) both (a) and (b) are correct
(d) None of the above is correct

53. MnO_4^- is of intense pink colour, though Mn is in (+7) oxidation state. It is due to
(a) oxygen gives colour to it
(b) charge transfer when Mn gives its electron to oxygen
(c) charge transfer when oxygen gives its electron to Mn making it Mn (+VI) hence, coloured
(d) None of the above is correct

54. Out of $[Fe(CN)_6]^{4-}$, $[Ni(CN)_4]^{2-}$ and $[Ni(CO)_4]$
(a) all have identical geometry
(b) all are paramagnetic
(c) all are diamagnetic
(d) $[Fe(CN)_6]^{4-}$ is diamagnetic but $[Ni(CN)_4]^{2-}$ and $[Ni(CO)_4]$ are paramagnetic

55. In $[Fe(CN)_5(NO^+)]^{2-}$, Fe has +3 state. It can be decided by
(a) magnetic measurement
(b) colligative property
(c) colour
(d) hybridisation

56. In $3d$ transition series, if nuclear charge increases, the screening effect
(a) increases
(b) decreases
(c) first decreases and then increases
(d) first increases and then decreases

57. Maximum oxidation state of Os is
(a) +6 (b) +7
(c) +8 (d) +5

58. Which forms interstitial compound?
(a) Fe (b) Co
(c) Ni (d) All of these

59. Which shows maximum magnetic moment among the bivalent ions of the first transition series?
(a) Fe^{2+} (b) Co^{2+}
(c) Ni^{2+} (d) Mn^{2+}

60. Select incorrect statement(s).
(a) Ionisation energies of $5d$-elements are greater than those of $3d$ and $4d$ elements
(b) Cu(I) is diamagnetic while Cu(II) is paramagnetic
(c) $[Ti(H_2O)_6]^{3+}$ is coloured while $[Sc(H_2O)_6]^{3+}$ is colourless
(d) Transition elements cannot form complexes

61. Guigret's green is
(a) $NiO \cdot 2H_2O$ (b) $Cr_2O_3 \cdot 2H_2O$
(c) $CuSO_4 \cdot 2H_2O$ (d) $CrO \cdot 2H_2O$

62. Which of the following metals has more than one oxidation state?
(a) Zn (b) Mn
(c) Sr (d) Ca

63. Which of the following metals have only one oxidation state?
(a) Co (b) Fe
(c) Sc (d) Al

64. Which of the following compounds is amphoteric?
(a) $Cr(OH)_2$ (b) $Fe(OH)_2$
(c) $Cr(OH)_3$ (d) $Fe(OH)_3$

65. Increasing basic properties of TiO_2, ZrO_2 and HfO_2 are in order
(a) $TiO_2 < ZrO_2 < HfO_2$ (b) $HfO_2 < ZrO_2 < TiO_2$
(c) $HfO_2 < TiO_2 < ZrO_2$ (d) $ZrO_2 < TiO_2 < HfO_2$

66. Which is called chromic acid?
(a) CrO (b) Cr_2O_3
(c) CrO_3 (d) CrO_2

67. What are the species A and B in the following

$$CrO_3 + H_2O \longrightarrow A \xrightarrow{OH^-} B?$$

(a) $H_2CrO_4, H_2Cr_2O_7$ (b) $H_2Cr_2O_7, Cr_2O_3$
(c) $CrO_4^{2-}, Cr_2O_7^{2-}$ (d) $H_2Cr_2O_7, CrO_4^{2-}$

68. Very recently (in Aug 2003), IUPAC has approved the name of the element of atomic number 110. Name of the element is
(a) Darmstadtium (b) Mountanium
(c) Rhenium (d) Bhorium

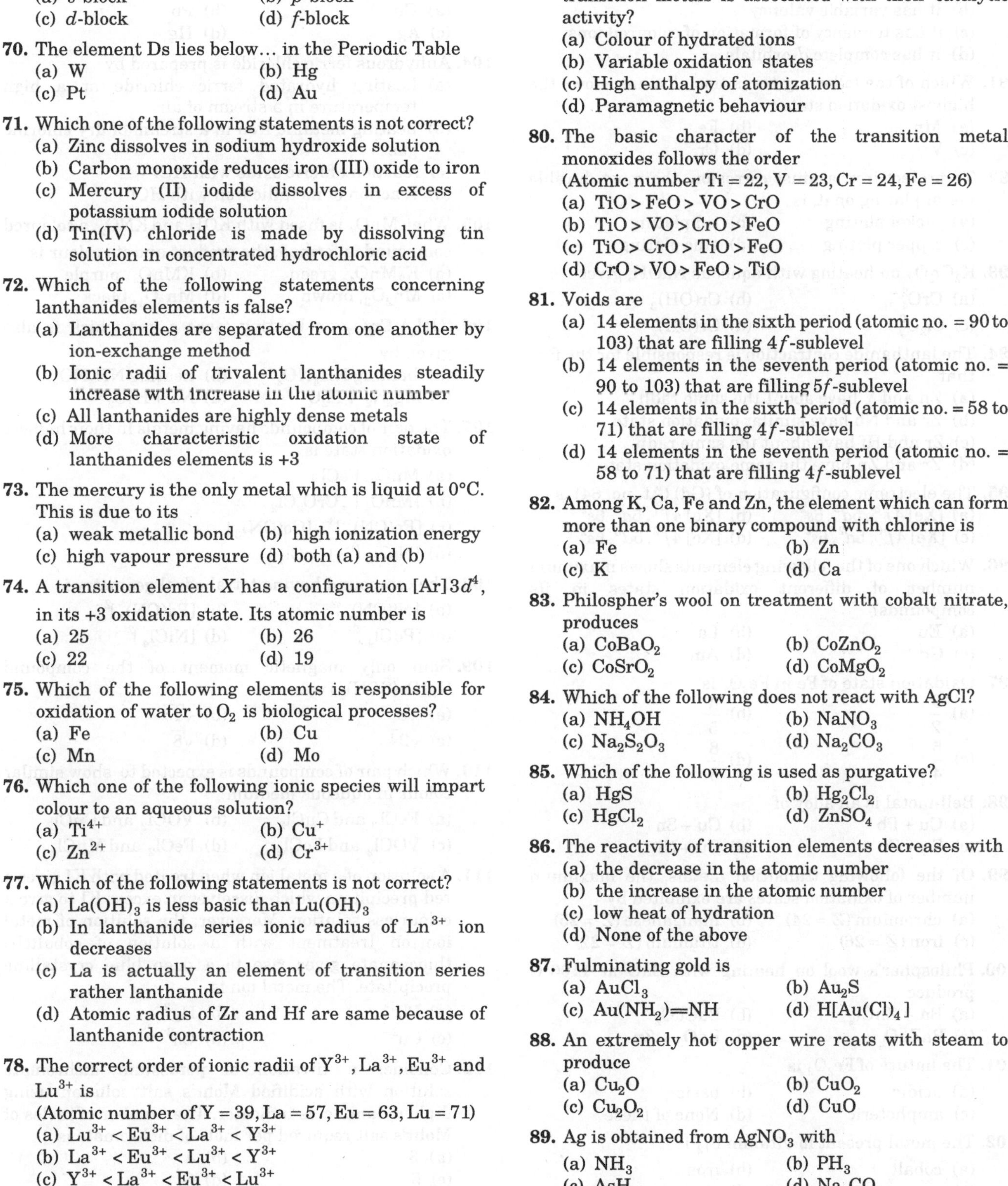

69. The element Ds lies in
(a) *s*-block (b) *p*-block
(c) *d*-block (d) *f*-block

70. The element Ds lies below... in the Periodic Table
(a) W (b) Hg
(c) Pt (d) Au

71. Which one of the following statements is not correct?
(a) Zinc dissolves in sodium hydroxide solution
(b) Carbon monoxide reduces iron (III) oxide to iron
(c) Mercury (II) iodide dissolves in excess of potassium iodide solution
(d) Tin(IV) chloride is made by dissolving tin solution in concentrated hydrochloric acid

72. Which of the following statements concerning lanthanides elements is false?
(a) Lanthanides are separated from one another by ion-exchange method
(b) Ionic radii of trivalent lanthanides steadily increase with increase in the atomic number
(c) All lanthanides are highly dense metals
(d) More characteristic oxidation state of lanthanides elements is +3

73. The mercury is the only metal which is liquid at 0°C. This is due to its
(a) weak metallic bond (b) high ionization energy
(c) high vapour pressure (d) both (a) and (b)

74. A transition element *X* has a configuration $[Ar]3d^4$, in its +3 oxidation state. Its atomic number is
(a) 25 (b) 26
(c) 22 (d) 19

75. Which of the following elements is responsible for oxidation of water to O_2 is biological processes?
(a) Fe (b) Cu
(c) Mn (d) Mo

76. Which one of the following ionic species will impart colour to an aqueous solution?
(a) Ti^{4+} (b) Cu^{+}
(c) Zn^{2+} (d) Cr^{3+}

77. Which of the following statements is not correct?
(a) $La(OH)_3$ is less basic than $Lu(OH)_3$
(b) In lanthanide series ionic radius of Ln^{3+} ion decreases
(c) La is actually an element of transition series rather lanthanide
(d) Atomic radius of Zr and Hf are same because of lanthanide contraction

78. The correct order of ionic radii of Y^{3+}, La^{3+}, Eu^{3+} and Lu^{3+} is
(Atomic number of Y = 39, La = 57, Eu = 63, Lu = 71)
(a) $Lu^{3+} < Eu^{3+} < La^{3+} < Y^{3+}$
(b) $La^{3+} < Eu^{3+} < Lu^{3+} < Y^{3+}$
(c) $Y^{3+} < La^{3+} < Eu^{3+} < Lu^{3+}$
(d) $Y^{3+} < Lu^{3+} < Eu^{3+} < La^{3+}$

79. Which one of the following characteristics of the transition metals is associated with their catalytic activity?
(a) Colour of hydrated ions
(b) Variable oxidation states
(c) High enthalpy of atomization
(d) Paramagnetic behaviour

80. The basic character of the transition metal monoxides follows the order
(Atomic number Ti = 22, V = 23, Cr = 24, Fe = 26)
(a) TiO > FeO > VO > CrO
(b) TiO > VO > CrO > FeO
(c) TiO > CrO > TiO > FeO
(d) CrO > VO > FeO > TiO

81. Voids are
(a) 14 elements in the sixth period (atomic no. = 90 to 103) that are filling 4*f*-sublevel
(b) 14 elements in the seventh period (atomic no. = 90 to 103) that are filling 5*f*-sublevel
(c) 14 elements in the sixth period (atomic no. = 58 to 71) that are filling 4*f*-sublevel
(d) 14 elements in the seventh period (atomic no. = 58 to 71) that are filling 4*f*-sublevel

82. Among K, Ca, Fe and Zn, the element which can form more than one binary compound with chlorine is
(a) Fe (b) Zn
(c) K (d) Ca

83. Philospher's wool on treatment with cobalt nitrate, produces
(a) $CoBaO_2$ (b) $CoZnO_2$
(c) $CoSrO_2$ (d) $CoMgO_2$

84. Which of the following does not react with AgCl?
(a) NH_4OH (b) $NaNO_3$
(c) $Na_2S_2O_3$ (d) Na_2CO_3

85. Which of the following is used as purgative?
(a) HgS (b) Hg_2Cl_2
(c) $HgCl_2$ (d) $ZnSO_4$

86. The reactivity of transition elements decreases with
(a) the decrease in the atomic number
(b) the increase in the atomic number
(c) low heat of hydration
(d) None of the above

87. Fulminating gold is
(a) $AuCl_3$ (b) Au_2S
(c) $Au(NH_2)=NH$ (d) $H[Au(Cl)_4]$

88. An extremely hot copper wire reats with steam to produce
(a) Cu_2O (b) CuO_2
(c) Cu_2O_2 (d) CuO

89. Ag is obtained from $AgNO_3$ with
(a) NH_3 (b) PH_3
(c) AsH_3 (d) Na_2CO_3

90. The incorrect statement for d-block element is
(a) it shows magnetic property
(b) it has variable valency
(c) it has tendency of formation of coloured ions
(d) it has complete d-orbitals

91. Which of the following transition element shows the highest oxidation state?
(a) Mn (b) Fe
(c) V (d) Cr

92. To protect iron against corrosion, the most durable metal plating on it, is
(a) nickel plating (b) tin plating
(c) copper plating (d) zinc plating

93. $K_2Cr_2O_7$ on heating with aqueous NaOH, gives
(a) CrO_4^{2-} (b) $Cr(OH)_3$
(c) $Cr_2O_7^{2-}$ (d) $Cr(OH)_2$

94. The lanthanide contraction is responsible for the fact that
(a) Zn and Y have about the same radii
(b) Zr and Nb have similar oxidation state
(c) Zr ahd Hf have about the same radii
(d) Zr and Zn have the same oxidation state

95. The electronic configuration of (Gd) (At. no. 64) is
(a) $[Xe]4f^8, 5d^9, 6s^2$ (b) $[Xe]4f^7, 5d^1, 6s^2$
(c) $[Xe]4f^6, 5d^2, 6s^2$ (d) $[Xe]4f^3, 5d^3, 6s^2$

96. Which one of the following elements shows maximum number of different oxidation states in its compounds?
(a) Eu (b) La
(c) Gd (d) Am

97. Oxidation state of Fe in Fe_3O_4 is
(a) $\frac{3}{2}$ (b) $\frac{4}{5}$
(c) $\frac{5}{4}$ (d) $\frac{8}{3}$

98. Bell-metal is an alloy of
(a) Cu + Pb (b) Cu + Sn
(c) Cu + Zn (d) Cu + Ni

99. Of the following transition metals, the maximum number of oxidation states are exhibited by
(a) chromium ($Z = 24$) (b) manganese ($Z = 25$)
(c) iron ($Z = 26$) (d) titanium ($Z = 22$)

100. Philospher's wool on heating with BaO at 1100°C produce
(a) Ba + $ZnCl_2$ (b) $BaCdO_2$
(c) $BaZnO_2$ (d) BaO_2 + Zn

101. The nature of Fe_2O_3 is
(a) acidic (b) basic
(c) amphoteric (d) None of these

102. The metal present in vitamin B_{12} is
(a) cobalt (b) iron
(c) manganese (d) magnesium

103. Which among the following is consumed by humans in the elemental form?
(a) Cu (b) Pb
(c) Ag (d) Hg

104. Anhydrous ferric chloride is prepared by
(a) heating hydrated ferric chloride at a high temperature in a stream of air
(b) heating metallic iron in a stream of dry chlorine gas
(c) reaction of feric oxide with HCl
(d) reaction of metallic iron with HCl

105. When MnO_2 is fused with KOH and KNO_2, a coloured compound is formed, the product and its colour is
(a) K_2MnO_4, green (b) $KMnO_4$, purple
(c) Mn_2O_3, brown (d) Mn_3O_4, black

106. $(NH_4)_2Cr_2O_7$ on heating gives a gas which is also given by
(a) heating NH_4NO_2 (b) heating NH_4NO_3
(c) $Mg_3N_2 + H_2O$ (d) Na + H_2O_2

107. The pair of compounds having metals in their highest oxidation state is
(a) $MnO_2, FeCl_3$
(b) $[MnO_4]^-, CrO_2Cl_2$
(c) $[Fe(CN)_6]^{3-}, [Co(CN)_3]$
(d) $[NiCl_4]^{2-}, [CoCl_4]^-$

108. The compound having tetrahedral geometry is
(a) $[Ni(CN)_4]^{2-}$ (b) $[Pd(CN)_4]^{2-}$
(c) $[PdCl_4]^{2-}$ (d) $[NiCl_4]^{2-}$

109. Spin only magnetic moment of the compound $Hg[Co(SCN)_4]$ is
(a) $\sqrt{3}$ (b) $\sqrt{15}$
(c) $\sqrt{24}$ (d) $\sqrt{8}$

110. Which pair of compounds is expected to show similar colour in aqueous medium?
(a) $FeCl_3$ and $CuCl_2$ (b) $VOCl_2$ and $CuCl_2$
(c) $VOCl_2$ and $FeCl_2$ (d) $FeCl_2$ and $MnCl_2$

111. A solution of a metal ion when treated with KI gives a red precipitate which dissolves in excess KI to give a colourless solution. Moreover, the solution of metal ion on treatment with a solution of cobalt(II) thiocyanate gives rise to a deep blue crystalline precipitate. The metal ion is
(a) Pb^{2+} (b) Hg^{2+}
(c) Cu^{2+} (d) Co^{2+}

112. Consider a titration of potassium dichromate solution with acidified Mohr's salt solution using diphenylamine as indicator. The number of moles of Mohr's salt required per mole of dichromate is
(a) 3 (b) 4
(c) 5 (d) 6

113. Consider following statements,

I. The sizse of the lanthanide M^{3+} ions decreases as the atomic number of M increases.

II. Electronic spectra of lanthanides show very broad bands.

III. As with transition metals, coordination number six is very common in lanthanide complexes.

Which of the statements given above is/are correct?

(a) I only (b) I and II
(c) I and III (d) III only

114. A red solid is insoluble in water. However, it becomes soluble if some KI is added to water. Heating the red solid in a test tube results in liberation of some violet coloured fumes and droplets of a metal appear on the cooler parts of the test tube. The red solid.

(a) $(NH_4)_2Cr_2O_7$ (b) HgI_2
(c) HgO (d) Pb_3O_4

115. What would happen when as solution of potassium chromate is treated with an excess of dilute nitric acid?

(a) Cr^{3+} and $Cr_2O_7^{2-}$ are formed
(b) $Cr_2O_7^{2-}$ and H_2O are formed
(c) CrO_4^{2-} is reduced to +3 state of Cr
(d) None of the above

116. Which one of the following nitrates will leave behind a metal on strong heating?

(a) Ferric nitrate (b) Copper nitrate
(c) Manganese nitrate (d) Silver nitrate

117. Cerium ($Z = 58$) is an important member of lanthanides. Which of the following statements about cerium is incorrect?

(a) The common oxidation states of cerium are +3 and +4
(b) The +3 oxidation states of cerium is more stable than the +4 state
(c) The +4 oxidation state of cerium is not known in solutions
(d) Cerium (IV) acts as an oxidising agent

118. Calomel (Hg_2Cl_2) on reaction with ammonium hydroxide gives

(a) HgO (b) Hg_2O
(c) $NH_2—Hg—Hg—Cl$ (d) $HgNH_2Cl$

119. Which of the following factors may be regarded as the main cause of lanthanide contraction?

(a) Greater shielding of $5d$-electron by $4f$-electrons
(b) Poorer shielding of $5d$-electron by $4f$-electrons
(c) Effective shielding of one $4f$-electrons by another in the subshell
(d) Poor shielding of one of $4f$-electrons by another in the subshell

120. Amongst the following, identify the species with an atom in +6 oxidation state

(a) MnO_4^- (b) $Cr(CN)_6^{3-}$
(c) NiF_6^{2-} (d) CrO_2Cl_2

121. When MnO_2 is fused with KOH and $KClO_3$, a coloured compound is formed, the product and its colour is

(a) K_2MnO_4, green (b) $KMnO_4$, purple
(c) Mn_2O_3, brown (d) Mn_3O_4 black

122. CrO_2Cl_2 is formed while testing

(a) NO_3^- (b) Cl^-
(c) Cr^{3+} (d) Fe^{3+}

123. The most abundant element is

(a) Cu (b) Hg
(c) Cd (d) Fe

124. Which represents correct comparison of the stability of ions?

(a) $MnO_4^- < Mn^{2+}$ (b) $Cr^{2+} < Cr^{3+}$
(c) $CrO_4^{2-} < Cr^{3+}$ (d) All of these

125. On August 2003, IUPAC approved the name of the element with atomic number 110 as

(a) Ds (b) Mt
(c) Uun (d) Nnu

126. Different (variable) oxidation state is shown by transition elements. It is due to the fact

(a) $(n-1)d$ electrons may be excited to ns orbital
(b) ns electrons may be excited to $(n-1)d$ orbitals
(c) $(n-1)d$-electron may get involved along with ns electrons in bonding
(d) none of the above facts is correct

127. Select correct statement for the lesser number of oxidations states in $3d$-series.

(a) In the beginning of the series it can be due to the presence of too few electrons to loose or share
(b) Towards the end of the series, it can be ascribed to the presence of too many electrons and thus fewer empty orbitals to share electrons with the ligands
(c) Both (a) and (b) are correct
(d) None of the above is correct

128. In which case(s) there is change in oxidation number

(a) Aqueous solution of CrO_4^{2-} is acidified
(b) SO_2 gas is passed into $Cr_2O_7^{2-}/H^+$
(c) $Cr_2O_7^{2-}$ solution is made alkaline
(d) CrO_2Cl_2 is dissolved in NaOH

129. Factors which affect the stability of the compounds are

(a) the energy of sublimation
(b) the lattice energy
(c) the solvation energy
(d) all of the above

130. Which forms protective and non-corrosive oxide layer?

(a) Cr (b) Ni
(c) Zn (d) Cu

131. First *IE* of 5*d*-elements are higher than those of 3*d* and 4*d*-elements. This is due to
(a) greater effective nuclear charge acting on outer valence electrons
(b) greater effective nuclear charge is experienced because of the weak shielding of the nucleus by 4*f*-electrons
(c) both (a) and (b)
(d) None of the above

132. NH_4Cl is used to clear metal surfaces because
(a) it dissociates into NH_3 and HCl on heating
(b) NH_3 forms soluble complexes with the metal
(c) HCl forms a volatile
(d) None of the above

133. Most transition metals
I : forms sets of compounds which display different oxidation states of the metal.
II : form coloured ions in solution
III : burn vigorously in oxygen
IV : replace H_2 from dilute acids
of these
(a) I, II, III are correct
(b) II, III, IV and correct
(c) I, II are correct
(d) all are correct

134. A blue solution of copper sulphate becomes darker when treated with excess of ammonia. This is because
(a) ammonia molecules replace water molecules in the solution
(b) ammonia is stronger ligand than water
(c) ammonia forms a stable complex ion $[Cu(NH_3)_4]^{2+}$ with Cu^{2+} ions.
(d) all of the above are correct

135. Effective atomic number (EAN) of Fe in brown ring complex $[Fe(H_2O)_5NO]^{2+}$
(a) 36 (b) 37
(c) 38 (d) 39

136. Select the correct relative stability.
(a) $[Cu(CN)_4]^{3-} > [Cd(CN)_4]^{3-}$
(b) $PtCl_4^{2-} > NiCl_4^{2-}$
(c) $[NiCl_6]^{2-} > [PtCl_6]^{2-}$
(d) $[Cd(CN)_4]^{2-} > [Cu(CN)_4]^{3-}$

137. Magnetic moment of $[Ni(CN)_4]^{2-}$ is zero but that of $[Ni(H_2O)_4]^{2+}$ is 2.83 BM.
It is because of
(a) different oxidation state of Ni in two complexes
(b) CN^- is a strong ligand making two unpaired electrons in Ni^{2+} paired while in $[Ni(H_2O)_4]^{2+}$, two electrons remain unpaired H_2O being weak ligand
(c) both (a) and (b)
(d) none of the above

138. There are three electrons unpaired in $[Co(H_2O)_6]^{2+}$ and calculated value of magnetic moment is 3.87 BM which is quite different from the experimental value of 4.40 BM. This is because of
(a) increase in number of unpaired electrons
(b) some contribution of the orbital motion of the electron to the magnetic moment
(c) change in orbital spin of the electron
(d) *d-d** transition

139. Cu^{2+} has a stronger polarising power than that of Ca^{2+} because
(a) Cu^{2+} ion is smaller than Ca^{2+} ion
(b) Ca^{2+} has inert gas configuration whereas Cu^{2+} ion does not
(c) copper shows variable valency, calcium does not
(d) Cu^{2+} is smaller than Ca^{2+} ion and the *d*-electrons in Cu^{2+} ion shield the nucleus poorly

140. Atoms of the transition elements are smaller than those of the *s*-block elements. This is because of
(a) usual contraction in size across a horizontal period
(b) orbital electrons added to the penultimate *d*-shell rather than to the outer shell of the atom
(c) both (a) and (b)
(d) none of the above

141. Select correct order of sizes of different species.
(a) Zr = Hf; Nb = Ta; Fe = Co = Ni
(b) Zr^{4+} < Zr; $Nb^{3+} < Ta^{3+}$, $Fe^{3+} < Fe^{2+}$ < Fe
(c) $Zr^{4+} = Hf^{4+}$; $Nb^{3+} = Ta^{3+}$; Fe < Co < Ni
(d) $Zr^{4+} < Hf^{4+}$ $Nb^{3+} = Ta^{3+}$, Ni < Cu < Co

142. The radii of the elements from chromium ($Z = 24$) to copper ($Z = 29$) are very close to one another. This is due to
(a) lanthanide contraction
(b) the fact that successive addition of *d*-electrons screen the outer electrons (4*s*) from the inward pull of the nucleus
(c) increase in radii due to increase in *n* is compensated by decrease in radii due to increase in *Z*
(d) atomic radii do not remain constant but decrease in a normal gradation

143. The lanthanoid contraction is due to
(a) filling of 4*f* before 4*d*
(b) filling of 4*d* before 4*f*
(c) filling of 4*f* before 5*d*
(d) filling of 5*d* before 4*f*

144. Magnetic moment of $[Ti(H_2O)_6]^{3+}$ is
(a) 1.73 BM (b) 2.83 BM
(c) 3.87 BM (d) 4.90 BM

145. Magnetic moment of Fe is similar to that of
(a) $[Fe(H_2O)_6]^{2+}$ (b) $[Cr(H_2O)_6]^{2+}$
(c) both (a) and (b) (d) None of these

146. The atomic numbers of vanadium (V), chromium (Cr), manganese (Mn) and iron are respectively 23, 24, 25 and 26. Which one of these may be expected to have the highest second ionization enthalpy?
(a) V (b) Cr
(c) Mn (d) Fe

147. There are three unpaired electrons in $[Co(H_2O)_6]^{2+}$ and calculated value of magnetic moment is 3.87 BM which is quite different from the experimental value of 4.40 BM. This is because of
(a) increase in number of unpaired electrons
(b) some contribution of the orbital motion of the electron to the magnetic moment
(c) change in orbital spin of the electron
(d) *d-d** transition

148. Increasing value of magnetic moments of (I) $Ni(CO)_4$, II : $[Ti(H_2O)_6]^{2+}$, III : $[V(H_2O)_6]^{3+}$, IV: $[V(H_2O)_6]^{2+}$ is
(a) I < II < III < IV (b) IV < III < II < I
(c) II < III < I < IV (d) II < I < III < IV

149. Which of the following is colourless?
(a) $[Zn(H_2O)_6]^{2+}$ (b) $[V(H_2O)_6]^{2+}$
(c) $[Mn(H_2O)_6]^{3+}$ (d) $[Fe(H_2O)_6]^{2+}$

150. Which is the coloured ion?
(a) $[Cu(CN)_4]^{3-}$ (b) $[Sc(H_2O)_6]^{3+}$
(c) $[Mn(H_2O)_6]^{3+}$ (d) All of these

151. Oxidation state of Mn in MnO_4^- is +7 indicating all electrons paired in Mn but MnO_4^- is coloured. It is due to
(a) charge transfer
(b) presence of unpaired electron in *d*-orbital in oxygen
(c) both (a) and (b)
(d) none of the above

152. Which of the following is paramagnetic as well as coloured ion?
(a) Cu^+ (b) Cu^{2+}
(c) Sc^{3+} (d) Ti^{4+}

153. Select correct statement(s).
(a) Coloured the ion arises due to *d-d** transition
(b) Colour we observe is the complimentary colour absorbed by the compound
(c) Both (a) and (b) are correct
(d) None of the above is correct

154. Nonstoichiometry is shown
(a) due to variable valency of transition elements
(b) due to defects in solid structures
(c) due to both (a) and (b)
(d) due to none of the above

155. Cementite is
(a) interstitial compound of iron and carbon
(b) an alloy of Fe and Cr
(c) a compound resembling cement
(d) an ore of iron

156. Which catalyst is matched according to its name and function?

Catalyst	Name	Function
(a) $TiCl_4$	Fenton's reagent	Oxidation of alcohols
(b) $FeSO_4/H_2O_2$	Ziegler-Natta	Polythene preparation
(c) Pd/C	Lindlar	Hydrogen to give *cis*-alkene
(d) Pt/PtO	Adam	Synthesis of CH_3OH

157. Select correct statement.
(a) Fe and Mo atoms are present in nitrogen fixing enzymes
(b) A cobalt atom lies at the centre of the vitamin B_{12} coenzyme
(c) Fe atoms are involved in the ferrodoxins of photosynthetic process
(d) All of the above are correct statements

158. Which is the set of nonstoichiometric compounds?
(a) NaCl, FeO, $MgCl_2$
(b) FeO, CuS, VSc
(c) Fe_3O_4, NaCl, CuS
(d) CuCl, CuS, MgO

159. Which one of the following elements is the main metallic constituent of haemoglobin?
(a) Al (b) Cu
(c) Fe (d) Mn

160. $KMnO_4$ is the oxo salt of
(a) MnO_2 (b) Mn_2O_7
(c) MnO_3 (d) Mn_2O_3

161. CrO_3 liberates CO_2 with Na_2CO_3 solution. Thus, CrO_3 is
(a) an acidic oxide
(b) a basic oxide
(c) an amphoteric oxide
(d) a nonstoichiometric oxide

162. VO_2 is an amphoteric oxide and in acidic medium it forms
(a) VO^{2+} (b) VO_2^+
(c) V^{3+} (d) VO_2^{2+}

163. CrO_2^{2+} is the cation from
(a) CrO (b) Cr_2O_3
(c) CrO_5 (d) CrO_3

164. The oxide Cr_2O_3 is
(a) acidic (b) amphoteric
(c) basic (d) ozonide

165. At pH = 4, $Cr_2O_7^{2-}$ exists as
(a) CrO_4^{2-} (b) CrO_3
(c) CrO_2^{2+} (d) $Cr_2O_7^{2-}$

166. At pH = 12, $Cr_2O_7^{2-}$ changes to
(a) CrO_3 (b) CrO_2^{2+}
(c) CrO_4^{2-} (d) no change

167. In acidic medium H_2O_2 changes $K_2Cr_2O_7$ to CrO_5 (deep violet solution in ether) having two peroxy linkage. Oxidation number of Cr in CrO_5 is

(a) +10 (b) −10
(c) +4 (d) +6

168. In alkaline H_2O_2, $Cr_2O_7^{2-}$ changes to tetraperoxo species ... having oxidation number of Cr as ...

(a) CrO_4^{2-} 6
(b) CrO_5 6
(c) CrO_8^{3-} 5
(d) CrO_8^{3-} 11

169. MnO_4^{2-} can be converted to MnO_4^-

(a) by oxidation with Cl_2
(b) by electrochemical oxidation at anode
(c) by both (a) and (b) methods
(d) by none of the above methods

170. In acidic medium MnO_4^{2-}

(a) disproportionates to MnO_2 and MnO_4^-
(b) is oxidised to MnO_4^-
(c) is reduced to MnO_2
(d) is reduced to Mn^{2+}

171. Alkaline $KMnO_4$ (Baeyer's reagent) can be used to test unsaturation in (*A*).

[Structure: benzene ring with —CH=CH$_2$ side chain] (*A*)

In this case

(a) unsaturation in side-chain is affected
(b) unsaturation in benzene nucleus is affected
(c) unsaturation in both is affected
(d) Baeyer's reagent cannot be used

172. $KMnO_4$ spot can be bleached by

(a) H_2O_2/H^+ (b) SO_2/H^+
(c) $C_2O_4^{2-}/H^+$ (d) all of these

173. Traces of MnO_4^- in conc. H_2SO_4 may change to

(a) Mn_2O_7 (b) MnO_3^+
(c) MnO_2 (d) MnO_4^{2-}

174. A white solid *Y*, on heating gives off a gas which turns lime water milky; the residue is yellow when hot; white when cold. The solid *Y* is probably

(a) $ZnCO_3$ (b) $PbCO_3$
(c) $ZnSO_4$ (d) $Zn(NO_3)_2$

175. Aqueous ZnO can neutralise HCl as well as NaOH solution. Thus, ZnO is an

(a) acidic oxide (b) basic oxide
(c) amphoteric oxide (d) amphiprotic oxide

176. Which is not blackened by atmosphere H_2S?

(a) $Pb(CH_3COO)_2$ (b) $Zn(CH_3COO)_2$
(c) $Cu(CH_3COO)_2$ (d) $Hg(CH_3COO)_2$

177. ZnO in CH_3COOH gives bluish white ppt. with *X*. *X* is

(a) $K_4[Fe(CN)_6]$ (b) $K_3[Fe(CN)_6]$
(c) $Na_2[Fe(CN)_5NO]$ (d) $Fe(CO)_5$

178. Select correct statement.

(a) $ZnSO_4 \cdot 7H_2O$ is called Epsom salt
(b) $ZnSO_4 \cdot 7H_2O$ is isomorphous with Glauber's salt
(c) $ZnSO_4 \cdot 7H_2O$ with BaS is called lithophone
(d) All the above are correct statements

179. Silver nitrate solution is kept in brown bottles in the laboratory because

(a) it reacts with ordinary bottles
(b) ordinary bottles catalyst the decomposition
(c) the materials of brown bottles do not react with it
(d) brown bottles cut the passage of light through it

180. On addition of $AgNO_3$ to four different test tubes containing different solutions, one of them gave a white precipitate. It may be

(a) $CHCl_3$ (b) $CaCl_2$
(c) KNO_3 (d) CCl_4

181. Select correct statement.

(a) PH_3 reduces $AgNO_3$ to metallic Ag
(b) Organic tissues turn $AgNO_3$ black by reducing it to Ag
(c) AgCN is soluble in KCN
(d) All of the above are correct statements

182. Hypo ($Na_2S_2O_3$)

(a) dissolves AgBr in photographic plate
(b) gives white precipitate with $AgNO_3$; white precipitate changes to black on dilution
(c) gives both reactions
(d) gives none of the above reactions

183. Sugar in urine sample can be detected by

I : Fehling's solution
II : Benedict's solution
III : Tollen's solution.

Select correct alternate

(a) I, II, III (b) I, III
(c) I, II (d) II, III

184. If a person is asked to prepare the blue print of a building plan, he can use

(a) $FeCl_3 + K_4[Fe(CN)_6]$
(b) $FeCl_2 + K_3[Fe(CN)_6]$
(c) $FeCl_2 + K_4[Fe(CN)_6]$
(d) $Fe(Ct) + K_3[Fe(CN)_6]$ Ct is citrate

185. If $FeCl_3 \cdot 6H_2O$ would exist as a complex, one mole of it in aqueous solution on reaction with excess of $AgNO_3$ forms

(a) 1 mole of AgCl (white precipitate)
(b) 2 moles of AgCl
(c) 3 moles of AgCl
(d) no reaction

186. $K_3[Fe(CN)_6]$ is used as an external indicator in the dichromate estimation of Fe^{2+}. Following change is observed
(a) colourless to blue (b) blue to red
(c) colourless to red (d) blue to colourless

187. If zinc pieces are added yellow $FeCl_3$ solution, it turns to very light green. It is because
(a) $FeCl_3$ solution is acidic due to hydrolysis and zinc added produces nascent hydrogen causing reduction of $FeCl_3$ to $FeCl_2$
(b) $FeCl_3$ solution is basic and zinc added produces H_2 which reduces $FeCl_3$ to $FeCl_2$
(c) Zinc reduces $FeCl_3$ to $FeCl_2$
(d) none of the above is correct

188. At 300°C, $FeCl_3$
(a) decomposes into $FeCl_2$ and Cl_2
(b) decomposes into Fe and Cl_2
(c) sublimes to give liquid $FeCl_3$
(d) sublimes to give gaseous dimer $(FeCl_3)_2$

189. The compound that get oxidised even on exposure to air is
(a) Fe_2O_3 (b) $FeCl_2$
(c) $FeCl_3$ (d) $Fe_2(SO_4)_3$

190. Rust contains $Fe_2O_3 \cdot H_2O$. Rust spots can be removed by
(a) $Na_2S_2O_3$ (hypo) (b) SO_2
(c) $H_2C_2O_4$ (oxalic acid) (d) $KMnO_4$

191. Select correct statement(s).
(a) α-Fe_2O_3 has hexagonally close-packed lattice of O^{2-} ions with Fe^{3+} ions in two-thirds of the octahedra holes.
(b) γ-Fe_2O_3 has cubic close-packed arrangement of O^{2-} ion with Fe^{3+} ions randomly distributed in both the octahedral and tetrahedral sites.
(c) Fe_3O_4, Fe_2O_3 and FeO all and tend to be non-stoichiometric.
(d) All the above are correct statements

192. On passing Cl_2 gas into alkaline hydrated Fe_2O_3 solution
(a) red purple solution of Na_2FeO_4 is formed
(b) colourless $NaFeO_2$ is formed due to dissolution of Fe_2O_3 into NaOH
(c) $Fe(OH)_3$ is formed
(d) no reaction takes place

193. Fe_3O_4 is called
(a) haematite (b) magnetite
(c) casserite (d) dolomite

194. Isomorphous salts are
(a) green vitriol, blue vitriol and Epsom salt
(b) green, vitriol, white vitriol and blue vitriol
(c) green vitriol, white vitriol and Epsom salt
(d) blue vitriol, white vitriol and Epsom salt

195. $[Fe(H_2O)_5NO]^{2+}$ is brown-ring complex in NO_3^-. In this complex
(a) NO transfers its electron to Fe^{2+} so that there are three electrons unpaired making iron as Fe(I) and nitrosyl as NO^+
(b) magnetic moment of Fe is 3.87 BM
(c) the colour is due to charge transfer
(d) all of the above are correct statements

196. $K_2Cr_2O_7/H^+$ changes to green by
(a) Fe^{2+} (b) SO_3^{2-}
(c) both (a) and (b) (d) none of these

197. Some of the following reagents are used as **primary standard.**
I : $KMnO_4$; II : NaOH; III : $K_2Cr_2O_7$;
IV : $FeSO_4 \cdot (NH_4)_2SO_4 \cdot 6H_2O$; V : $H_2C_2O_4 \cdot 2H_2O$
Select primary standard
(a) all except II, IV (b) all except I, II
(c) all except I, II, III (d) only IV

198. One mole of FeC_2O_4 is oxidised by $KMnO_4$ in acidic medium. Number of moles of $KMnO_4$ used are
(a) 0.6 mol (b) 1.2 mol
(c) 0.4 mol (d) 1 mol

199. Fehling's solution consists of two separate alkaline solutions. One solution contains $CuSO_4$. The other solution contains.
(a) $NaHCO_3$ (b) $KNaC_4H_4O_6$
(c) $KHCO_3$ (d) K_2CO_3

200. If H_2S gas is passed into a solution of Cu^{2+}, Cd^{2+} having excess of KCN
(a) CuS and CdS both are precipitated
(b) soluble complex $[Cu(CN)_4]^{3-}$ and $[Cd(CN)_4]^{2-}$ are formed and no effect of passing H_2S gas
(c) soluble complex $[Cu(CN)_4]^{3-}$ and $[Cd(CN)_4]^{2-}$ as are formed, of which CdS is precipitated as yellow ppt.
(d) soluble complex $[Cu(CN)_4]^{3-}$ and $[Cd(CN)_4]^{2-}$ are formed of which CuS is precipitated as black ppt.

201. When K_2CrO_4 is added to $CuSO_4$ solution, there is formation of $CuCrO_4$ as well as $CuCr_2O_7$. Formation of $CuCr_2O_7$ is due to
(a) basic nature of $CuSO_4$ solution which converts CrO_4^{2-} to $Cr_2O_7^{2-}$
(b) acidic nature of $CuSO_4$ solution which converts CrO_4^{2-} to $Cr_2O_7^{2-}$
(c) $CuSO_4$ has the typical property of converting $CuCrO_4$ is formed to $CuCr_2O_7$
(d) no $CuCr_2O_7$ is formed

202. In an atmosphere with industrial smog, Cu corrodes to
(a) basic sulphate $Cu_2(OH)_2SO_4$
(b) basic carbonate $Cu_2(OH)_2CO_3$
(c) both (a) and (b)
(d) none of the above

203. $Na_2S_2O_3$
(a) reduces Cu^{2+} to Cu^+
(b) reduces I_2 to I^-
(c) complexes AgBr as $[Ag(S_2O_3)_2]^{3-}$
(d) undergoes all the above are correct

204. The element which forms ions in dimeric state is
(a) iron (b) mercury
(c) cadmium (d) nickel

205. Mixture is supposed to contain nitrates of Pb^{2+} and Hg_2^{2+}. On adding dil. HCl both are precipitated as white precipitate
Select correct statement(s).
(a) $PbCl_2$ (white precipitate) is soluble in hot water and formation of yellow precipitate on adding KI to hot solution confirms Pb^{2+}
(b) Hg_2Cl_2 is blackened by NH_3, and changes to $HgCl_2$ by aqua-regia. $HgCl_2$ can be detected by KI when orange precipitate soluble in excess of KI is formed.
(c) both tests are conclusive
(d) none of the tests is conclusive

206. Hg_2Cl_2 ionises as .... and cation has .... unpaired electron(s)
(a) $2Hg^+$ and $2Cl^-$, two (b) Hg_2^{2+} and $2Cl^-$, two
(c) Hg_2^{2+} and $2Cl^-$, one (d) Hg_2^{2+} and $2Cl^-$, no

207. Alkaline NH_4Cl solution reacts with K_2HgI_4 (Nessler's reagent) forming
(a) $(HgNH_2Cl + Hg)$ — black
(b) $\left[O\langle^{Hg}_{Hg}\rangle N\langle^{H}_{H}\right] I$ — brown iodide of Millon's base
(c) $K_2[O—Hg—NH_2Cl]\,I$ — brown iodide of Millon's base
(d) none of the above

208. $HgCl_2$ is soluble in
(a) cold water
(b) NaCl solution due to formation of $HgCl_4^{2-}$
(c) both (a) and (d)
(d) none of the above

209. In the following reaction

$$nCH_2{=}CH_2 \longrightarrow [—CH_2—CH_2—]_n$$

Catalyst is
(a) Adam (b) Ziegler-Natta
(c) Ni/Pd (d) Fe/Mo

210. $TiCl_4$ is a
(a) Bronsted-Lowry acid (b) Bronsted-Lowry base
(c) Lewis acid (d) Lewis base

211. Which of the following are sets of diamagnetic species?
(a) $TiCl_4, O_2, [Ni(CN)_4]^{2-}, [Ni(H_2O)_4]^{2+}$
(b) $TiCl_4, [Ni(CN)_4]^{2-}, CO, [Ni(H_2O)_4]^{2+}$
(c) $TiCl_4, [Ni(CN)_4^{2-}], CO$
(d) $TiCl_4, [Ni(CN)_4]^{2-}, O_2$

212. Which are not blackend by atmospheric H_2S?
(a) TiO_2 (b) ZnO
(c) $ZnSO_4$ + BaS (d) All of these

213. Elements after atomic number 92 are called
(a) lanthanoids
(b) actinoids
(c) inner-transition elements
(d) transuranium elements

214. Ln^{3+} (trivalent lanthanides ions) have EC
(a) $[Xe]\,4f^1$ to $[Xe]\,4f^{14}$
(b) $[Xe]\,4d^1 4f^1$ to $[Xe]\,4d^1 4f^{14}$
(c) $[Xe]\,4d^2 4f^0$ to $[Xe]\,4d^1 4f^{14}$
(d) $[Xe]\,4f^0$ to $[Xe]\,4f^{14}$

215. A certain metal will liberate hydrogen from dilute acids. It will react with water to form hydrogen only when the metal is heated and the water is in the form of steam. The metal is probably
(a) iron (b) potassium
(c) copper (d) mercury

216. The magnetic moment μ, of transition metals is related to the number of unpaired electrons, *n* as
(a) $\mu = n(n+2)^2$ (b) $\mu = n^2(n+2)$
(c) $\mu = \dfrac{n}{(n+2)}$ (d) $\mu = \sqrt{n(n+2)}$

217. Misch metal is
(a) an alloy of lathanide and copper
(b) an alloy of lanthanide and nickel
(c) an alloy of lanthanide, iron and carbon
(d) an alloy of calcium and copper

218. Which of the following have completely filled 5*f* orbitals?
(a) No (b) Lr
(c) Both (a) and (b) (d) None of these

219. An^{3+} (trivalent actinoid ions) have configuration
(a) $5f^0$ to $5f^{14}$ (b) $5f^1$ to $5f^{14}$
(c) $5f^2$ to $5f^{14}$ (d) $5f^0$ to $5f^{13}$

220. Consider following radioactive decay of

$$\text{I}: {}_{92}U \xrightarrow{-\alpha} {}_{90}Th$$

$$\text{II}: {}_{90}Th \xrightarrow{-\alpha} {}_{88}Ra$$

In which case group of the parent and daughter element remains unchanged?
(a) in I (b) in II
(c) in both (a) and (b) (d) in none of these

221. Select correct statement(s).
(a) Pu^{4+} disproportionates to Pu^{3+} and PuO_2^{2+} in strongly acidic solution
(b) Maximum oxidation state of Np is +7

(c) UO_2^{2+} is stable
(d) All of the above are correct statement

222. Uranium reacts with conc. HNO_3
(a) to give $U(NO_3)_4$ and H_2
(b) to give $U(NO_3)_2$ and NO_2
(c) $UO_2(NO_3)_2$ and H_2
(d) to make metal passive

223. Which of the following statements is not correct?
(a) $La(OH)_3$ is less basic than $Lu(OH)_3$
(b) In lanthanide series ionic radius of Ln^{3+} ion decreases
(c) La is actually an element of transition series rather lanthanide
(d) Atomic radius of Zr and Hf are same because of lanthanide contraction

224. Lanthanoids are
(a) 14 elements in the sixth period (atomic number = 90 to 103) that are filling $4f$-sublevel
(b) 14 elements in the seventh period (atomic no. = 90 to 103) that are filling $5f$-sublevel
(c) 14 elements in the sixth period (atomic no. = 58 to 71) that are filling $4f$-sublevel
(d) 14 elements in the seventh period (atomic no. = 58 to 71) that are filling $4f$-sublevel

225. Four successive members of the first row transition elements are listed below with their atomic numbers. Which one of them is expected to have the highest third ionization enthalpy?
(a) Vanadium ($Z = 23$) (b) Chromium ($Z = 24$)
(c) Iron ($Z = 26$) (d) Manganese ($Z = 25$)

226. The number of moles of $KMnO_4$ reduced by one mole of KI in alkaline medium is
(a) one fifth (b) five
(c) one (d) two

227. The main reason for larger number of oxidation state exhibited by the actinides than that corresponding lanthanides, is
(a) lesser energy difference between $5f$ and $6d$ orbitals than between $4f$ and $5d$-orbitals
(b) larger atomic size of actinides than the lanthanides
(c) more energy difference between $5f$ and $6d$ orbitals than between $4f$ and $5d$-orbitals
(d) greater reactive nature of the actinides than the lanthanides

228. The common oxidation state of the elements of lanthanide series is
(a) +1 (b) +3
(c) +4 (d) +6

229. The metal that does not displace hydrogen from an acid is
(a) Al (b) Ca
(c) Zn (d) Hg

230. Ferric sulphate is represented by which formula?
(a) $FeSO_4$ (b) $FeSO_3$
(c) $Fe(SO_4)_2$ (d) $Fe_2(SO_4)_3$

231. The transition metal with least atomic number
(a) Os (b) Zr
(c) Pt (d) Ru

232. Lanthanide and actinides resemble in
(a) electronic configuration
(b) oxidation state
(c) ionisation energy
(d) formation of complexes

233. Stainless steel is an alloy of
(a) Cu (b) Ni and Cr
(c) Mn (d) Zn

234. Solder is an alloy of
(a) 70% lead 30% tin (b) 30% lead, 70% tin
(c) 80% lead, 20% tin (d) 90% copper, 10% tin

235. Which of the following is a highly corrosive salt?
(a) $FeCl_2$ (b) $PbCl_2$
(c) Hg_2Cl_2 (d) $HgCl_2$

236. Which of the following is potassium ferricyanide?
(a) $K_4[Fe(CN)_6)]$ (b) $K_3[Fe(CN)_6]$
(c) $K_3[Fe(CN)_3]$ (d) $K_3[Fe(CN)_4]$

237. German silver is an alloy of
(a) Fe, Cr, Ni (b) Ag, Cu, Au
(c) Cu, Zn, Ni (d) Cu, Zn, Sn

238. The Mohr's salt is shown by
(a) $FeSO_4(NH_4)_2SO_4 \cdot 6H_2O$
(b) $FeSO_4(NH_3)_2SO_4 \cdot 6H_2O$
(c) $K_2SO_4Al_2(SO_4)_3 \cdot 24H_2O$
(d) $FeSO_2(NH_2)_4SO_4 \cdot 6H_2O$

239. The composition of duralumin is
(a) Al 94%, Mg 6%
(b) Cu 56%, Zn 24%, Ni 20%
(c) Cu 95, Al 5%
(d) Al 95%, Cu 4%, Mn 0.5%, Mg 0.5%

240. Lanthanide for which +II and +III oxidation states are common is
(a) La (b) Nd
(c) Ce (d) Eu

241. The colourless species is
(a) VCl_3 (b) $VOSO_4$
(c) Na_3VO_4 (d) $[V(H_2O)_6]SO_4 \cdot H_2O$

242. MnO_4^{2-} (1 mol) in neutral aqueous medium disproportionates to
(a) $\frac{2}{3}$ mol of MnO_4^- and $\frac{1}{3}$ mol of MnO_2
(b) $\frac{1}{3}$ mol of MnO_4^- and $\frac{2}{3}$ mol of MnO_2

(c) $\frac{1}{3}$ mol of Mn_2O_7 and $\frac{1}{7}$ mol of MnO_2

(d) $\frac{2}{3}$ mol of Mn_2O_7 and $\frac{1}{3}$ mol of MnO_2

243. An aqueous solution of $CoCl_2$ on addition of excess of concentrated HCl turn blue due to formation of

(a) $[Co(H_2O)_4Cl_2]$ (b) $[Co(H_2O)_2Cl_4]^{2-}$
(c) $[CoCl_4]^{2-}$ (d) $[Co(H_2O)_2Cl_2]$

244. The diamagnetic species is

(a) $[Ni(CN)_4]^{2-}$ (b) $[NiCl_4]^{2-}$
(c) $[CoCl_4]^{2-}$ (d) $[CoF_6]^{2-}$

245. Among the following pairs of ions, the lower oxidation state in aqueous solutions is more stable than the other in

(a) Ti^+, Ti^{3+} (b) Cu^+, Cu^{2+}
(c) Cr^{2+}, Cr^{3+} (d) V^{2+}, VO^{2+}

246. The colour imparted by Co(II) compounds to glass is

(a) green (b) deep blue
(c) yellow (d) red

247. Given below, catalyst and corresponding process/reaction are matched. The mismatch is

(a) $[RhCl(PPh_3)_2]$: Hydrogenation
(b) $TiCl_4 + Al(C_2H_5)_3$: Polymerisation
(c) V_2O_5 : Haber-Bosch process
(d) Nickel : Hydrogenation

248. KI and $CuSO_4$ solutions on mixing produce

(a) $Cu_2I_2 + K_2SO_4$ (b) $Cu_2I_2 + KI_3 + K_2SO_4$
(c) $CuI_2 + K_2SO_4$ (d) $CuI_2 + KI_3 + K_2SO_4$

249. The reactivity of transition elements decreases with

(a) the decrease in the atomic number
(b) the increase in the atomic number
(c) low heat of hydration
(d) None of the above

250. Catalyst used in making H_2SO_4 in contact process is

(a) V_2O_5 (b) Fe_2O_3
(c) Cr_2O_3 (d) CrO_3

251. The incorrect configuration is

(a) $K = [Ar]4s^1$ (b) $Cr = [Ar]3d^5, 4s^1$
(c) $Cr = [Ar]3d^4, 4s^2$ (d) $Cu = [Ar]3d^{10}, 4s^1$

252. By annealing, steel

(a) becomes soft
(b) becomes liquid
(c) becomes hard and brittle
(d) is covered with a thin film of Fe_3O_4

253. Transition metals show paramagnetism due to

(a) characteristic configuration
(b) high lattice energy
(c) variable oxidation states
(d) unpaired electrons

254. The oxides, CrO_3, MoO_3 and WO_3 are strongly

(a) neutral (b) acidic
(c) basic (d) None of these

255. The outer electronic configuration of transitional elements is

(a) $(n-1)s^2nd^{1-2}$
(b) $(n+1)s^2nd^{1-5}$
(c) $(n-1)s^2p^6(n-1)d^{1-10}, ns^{1,2}$
(d) $ns^2(n+1)d^{1-10}$

256. The highest magnetic moment will be shown by

(a) Ni (b) Co
(c) Fe (d) Sc

257. Which of the following is not a member of 3*d*-transition series?

(a) Fe (b) Co
(c) Au (d) Cu

258. Which is most soluble in water?

(a) AgBr (b) AgCl
(c) AgF (d) AgI

259. Addition of $SnCl_2$ to $HgCl_2$ give precipitate

(a) white turning to red
(b) white turning to grey
(c) black turning to white
(d) None of the above

260. The stability of ferric ion is due to

(a) half filled *f*-orbitals
(b) half filled *d*-orbitals
(c) completely filled *f*-orbitals
(d) completely filled *d*-orbitals

261. Which of the following lanthanide is commonly used?

(a) Lanthanium (b) Nobelium
(c) Thorium (d) Cerium

262. $4K_2Cr_2O_7 \xrightarrow{\text{Heat}} 4K_2CrO_4 + 3O_2 + X$; in the above reaction *X* is

(a) CrO_3 (b) Cr_2O_7
(c) Cr_2O_3 (d) CrO_5

263. The colour of $FeSO_4 \cdot (NH_4)_2SO_4 \cdot 6H_2O$ is

(a) blue (b) green
(c) white (d) red

264. Which of the following is most stable among Cu^+, Fe^+, Fe^{2+} and Fe^{3+} ?

(a) Cu^+ (b) Fe^+
(c) Fe^{2+} (d) Fe^{3+}

265. Brown glass and cement have, which element common in them?

(a) Fe (b) Al
(c) Na (d) All of these

266. The atomic size of cerium and promethium is quite close, due to

(a) they are in same period in Periodic Table
(b) they electronic configuration is same
(c) *f*-electrons have poor shielding effect
(d) nuclear charge is higher on cerium than promethium

267. The trivalent ion having largest size is
(a) Ti (b) Zr
(c) Hf (d) La

268. The transition metal used as a catalyst is
(a) nickel (b) platinum
(c) cobalt (d) all of these

269. Transition elements does not show
(a) paramagnetism (b) colour
(c) fixed valency (d) all of these

270. Aluminium appears like gold when mixed with
(a) 90% Cu (b) 75% Ni
(c) 80% Sn (d) 80% Co

271. Iron is dropped in very dil. HNO_3, it gives
(a) ferric nitrate
(b) ferric nitrate and NO_2
(c) ferrous nitrate and ammonium nitrate
(d) ferrous nitrate and nitric oxide

272. Stainless steel contains
(a) Fe + Cr + Cu (b) Fe + C + Ni
(c) Fe + Cr + Ni (d) Fe + Ni + Cu

273. Which of the following alloys is used for making magnets for hearing aids?
(a) Alnico (b) German silver
(c) Invar (d) Monel metal

274. Cuprous ion is colourless while cupric ion is coloured because
(a) both have half-filled *p*- and *d*-orbitals
(b) both have unpaired electrons in *d*-orbitals
(c) cuprous ion has incomplete *d*-orbital and cupric ion has a complete *d*-orbital
(d) cuprous ion as a complete *d*-orbital and cupric ion has an incomplete *d*-orbital

275. Stainless steel does not rust because
(a) chromium and nickel combine with iron
(b) chromium forms an oxide layer and protects iron from rusting
(c) nickel present in it, does not rust
(d) iron forms a hard chemical compound with chromium present in it

276. Which of the following pairs of element cannot form an alloy?
(a) Zn, Cu (b) Fe, Hg
(c) Fe, C (d) Hg, Na

277. Which of the following types of metals form the most efficient catalysts?
(a) Alkali metals (b) Alkaline earth metals
(c) Transition metals (d) All of these

278. Impure metal forms volatile compound (X) with CO and then (X) gives pure metal on heating. Metal is
(a) Cu (b) Fe
(c) Ni (d) Pt

279. Which of the following chloride is water insoluble?
(a) HCl (b) AgCl
(c) Both (a) and (b) (d) None of these

280. Which of the following dissolves in hot conc. NaOH solution?
(a) Fe (b) Zn
(c) Cr (d) Ag

281. Reason of passivity of iron is
(a) Fe_2O_3 (b) Fe_3O_4
(c) FeO (d) $Fe_2O_3 \cdot 3H_2O$

282. Which is wrongly matched?
(a) Duralumin — Al + Cu + Mg + Mn
(b) Alnico — Fe + Al + Ni + Cu
(c) German silver — Cu + Zn + Ni
(d) Monel metal — Cu + Zn + Sn

283. Formula of green vitriol oil is
(a) $FeSO_4 \cdot 7H_2O$ (b) $MgSO_4 \cdot 7H_2O$
(c) $ZnSO_4 \cdot 7H_2O$ (d) $CuSO_4 \cdot 5H_2O$

284. Which of the following is called white vitriol?
(a) $ZnCl_2$ (b) $MgSO_4 \cdot 7H_2O$
(c) $ZnSO_4 \cdot 7H_2O$ (d) $Al_2(SO_4)_3$

285. Which of the following is not an actinide?
(a) Curium (b) Californium
(c) Uranium (d) Terbium

286. Europium is
(a) *s*-block element (b) *p*-block element
(c) *d*-block element (d) *f*-block element

287. Which of the following is not an actinoid?
(a) Am (b) Cm
(c) Fm (d) Tm

288. Most oxidising agent is
(a) $[WO_4]^{2-}$ (b) $[CrO_4]^{2-}$
(c) $[CeO_4]^{2-}$ (d) $[MnO_4]^{2-}$

289. Which of the following mixture is chromium acid?
(a) $K_2Cr_2O_7$ + HCl (b) K_2SO_4 + conc. H_2SO_4
(c) $K_2Cr_2O_7$ + conc. H_2SO_4 (d) H_2SO_4 + HCl

290. In reduction of dichromate by Fe(II), the number of electrons involved per chromium atom is
(a) 2 (b) 3
(c) 4 (d) 1

291. Which alloy contains Cu, Sn and Zn?
(a) Gun metal (b) Solder
(c) Type metal (d) Bronze

292. Finely divided iron combines with CO to give
(a) $Fe(CO)_5$ (b) $Fe_2(CO)_9$
(c) $Fe_2(CO)_{12}$ (d) $Fe(CO)_6$

293. Green vitriol is formed by
(a) $FeS_2 + H_2O + O_2$ (b) $FeS_2 + H_2O + CO_2$
(c) $FeS_2 + CO + CO_2$ (d) $FeS_2 + CO$

294. Thermite is a mixture of
(a) $MgO + Al$ (b) $Fe_2O_3 + Al$
(c) $Zn + CaCO_3$ (d) $Zn + P_2O_5$

295. Which plays a major role in the formation of complex compound?
(a) Transition metal
(b) Lanthanides and actinides
(c) Representative elements
(d) *p*-block element

One or More Than One Options Correct

1. When KCN is added to $CuSO_4$ solution
(a) KCN acts as a reducing agent
(b) KCN acts as a complexing agent
(c) complex $K_3[Cu(CN)_4]$ is formed
(d) complex $K_2[Cu(CN)_4]$ is formed

2. Which will not give metal on heating?
(a) Ag_2CO_3 (b) $ZnCO_3$
(c) HgO (d) CuO

3. Electron transfer from $Fe(H_2O)_6^{2+}$ to $Fe(H_2O)_6^{3+}$ is likely to occur *via*
(a) *d-d* transition
(b) inner sphere electron transfer
(c) S_N1 mechanism
(d) outer sphere electron transfer

4. Pyrolusite is MnO_2 used to prepare $KMnO_4$. Steps are

$$MnO_2 \xrightarrow{I} MnO_4^{2-} \xrightarrow{II} MnO_4^-$$

Steps I and II are
(a) fuse with KOH/air, electrolytic oxidation
(b) fuse with KOH/air, electrolytic reduction
(c) fuse with conc. HNO_3/air, electrolytic reduction
(d) dissolve in H_2O, oxidation

5. Fe is made passive by
(a) dil. H_2SO_4 (b) dil. HCl
(c) aqua-regia (d) conc. H_2SO_4

6. N_2 gas can be obtained by heating
(a) $(NH_4)_2Cr_2O_7$ (b) NH_4NO_2
(c) NH_4VO_3 (d) NH_4NO_3

7. Which are correct statements about $KMnO_4$?
(a) Its solution is unstable in acidic medium
(b) Its small quantity added to conc. H_2SO_4, a green coloured solution containing MnO_3^+ ions is formed
(c) MnO_4^- changes to Mn^{2+} in basic solution
(d) It is self-indicator in Fe^{2+} or $C_2O_4^{2-}$ titration

8. Which are correct statements?
(a) In less acidic solution $K_2Cr_2O_7$ and H_2O_2 gives violet coloured diamagnetic $[CrO(O_2)(OH)]^-$ ion
(b) In alkaline H_2O_2, K_2CrO_8 (with tetraperoxo species $[Cr(O_2)_4]^{3-}$) is formed
(c) In ammoniacal solution, $(NH_3)_3CrO_4$ is formed
(d) CrO_4^{2-} changes to $Cr_2O_7^{2-}$ by oxidation

9. For CrO_3 following is not true statement.
(a) It is called chromic acid
(b) It is colourless due to $3d^0$ configuration
(c) It is bright orange solid and colour arises due to charge transfer
(d) It is toxic and corrosive

10. Lanthanoids are
(a) 14 elements in the sixth period (atomic no. = 90 to 103) that are filling 4*f*-sublevel
(b) 14 elements in the seventh period (atomic no. = 90 to 103) that are filling 5*f*-sublevel
(c) 14 elements in the sixth period (atomic no. = 58 to 71) that are filling 4*f*-sublevel
(d) 14 elements in the seventh period (atomic no. = 58 to 71) that are filling 4*f*-sublevel

11. When CO_2 is passed into aqueous
(a) Na_2CrO_4 solution, its yellow colour changes to orange
(b) K_2MnO_4 solution, it disproportionates to $KMnO_4$ and MnO_2
(c) $Na_2Cr_2O_7$ solution, its orange colour changes to green
(d) $KMnO_4$ solution, its pink colour changes to green

12. Select incorrect statement(s).
(a) Nickel (II) compounds tend to be thermodynamically more stable than platinum (II)
(b) Platinum (IV) compounds are relatively more stable than nickel (IV) compounds
(c) $K_2Pt^{IV}Cl_6$ exists
(d) $K_2Ni^{IV}Cl_6$ exists

13. Standard reduction electrode potential at Zn^{2+}/Zn is –0.76 V. This means
(a) ZnO is reduced to Zn by H_2
(b) Zn liberates H_2 with conc. acids
(c) Zn is generally the anode in an electrochemical cell
(d) Zn is generally the cathode in an electrochemical cell

14. Select correct statement(s) about $[CoCl_6]^{4-}$ complex.
(a) It is paramagnetic
(b) It is a low-spin complex
(c) Oxidation number of cobalt is –4
(d) The coordination number of cobalt is 6

15. Why does aqueous Fe(III) ion develop intense red colour when it reacts with SCN^- ion while Fe(II) ion does not?
(a) Fe(III) ion forms a charge transfer complex with SCN^- ions

(b) Fe(III) is reduced to Fe(I) which is deep red in colour

(c) SCN^- ion is oxidised to CN^- which forms red coloured complex with Fe(III) ion

(d) SCN^- does not form any complex with Fe(III) ion

16. Select correct statement(s).

(a) Stabilities of variable oxidation states can be explained by standard electrode potential

(b) Cr^{2+} is unstable and Cr^{3+} is stable in aqueous solution

(c) Fe^{2+} is stable and Fe^{3+} is unstable is aerated water

(d) All of the above are correct statements

17. Select correct statement(s).

(a) When $FeCl_3$ solution is added to $K_4[Fe(CN)_6]$ solution, in addition to $Fe^{III}[Fe^{II}(CN)_6]^-$, $Fe^{II}[Fe^{III}(CN)_6]^-$ is also formed due to side redox reaction

(b) When $FeCl_2$ solution is added to $K_3[Fe(CN)_6]$ solution, in addition to $Fe^{II}[Fe^{II}(CN)_6]^-$ $Fe^{III}[Fe^{II}(CN)_6]^-$ is also formed due to side redox reaction

(c) $Fe^{III}[Fe^{II}(CN)_6]^-$ is paramagnetic while $Fe^{II}[Fe^{III}(CN)_6]^-$ is diamagnetic

(d) $Fe^{III}[Fe^{II}(CN)_6]^-$ is diamagnetic while $Fe^{II}[Fe^{III}(CN)_6]^-$ is paramagnetic

Brain Twisters

1. A 0.81 g sample of pyrolusite ore (impure MnO_2) is treated with 1.651 g of oxalic acid ($H_2C_2O_4 \cdot H_2O$) in an acidic medium. Following this reaction the excess oxalic acid is titrated with 0.1 M $KMnO_4$, 30.6 mL being required. What is the % of MnO_2 in the ore? (Mn = 55)

$$H_2C_2O_4 + MnO_2 + 2H^+ \longrightarrow Mn^{2+} + 2H_2O + 2CO_2$$

$$5H_2C_2O_4 + 2MnO_4^- + 6H^+ \longrightarrow 2Mn^{2+} + 8H_2O + 10CO_2$$

2. A steel sample is to be analysed for Cr and Mn simultaneously. By suitable treatment the Cr is oxidised to $Cr_2O_7^{2-}$ and Mn to MnO_4^-. A 10.0 g of sample of steel is used to produce 250.0 mL of solution containing $Cr_2O_7^{2-}$ and MnO_4^-. A 10.0 mL portion of this solution is added to a $BaCl_2$ solution, and by proper adjusting the acidity, the Cr is precipitated as $BaCrO_4$; 0.0549 g is obtained. A second 10 mL of the solution requires exactly 15.95 mL of 0.0750 M standard Fe^{2+} solution for its titration is acid solution. Calculate the % Mn and % Cr in the steel sample. (Ba = 137, Cr = 52, Mn = 55)

3. When a black powder (*A*) was heated strongly, it gave off a colourless non-inflammable gas. When (*A*) was boiled with excess dil. H_2SO_4 and a little $H_2C_2O_4$, a non-inflammable odourless gas was evolved. The solution formed gave no precipitate when saturated with H_2S, but this saturated solution gave a light coloured precipitate when NH_4OH solution was added to it. A small sample of (*A*) was warmed with excess of conc. HCl and the gas liberated was passed into a conc. KI solution. This solution was made up to 200 mL; 20 mL of this solution was found to required 180 mL of 0.1 N $Na_2S_2O_3$ solution iodometrically. Identify (*A*) and calculate the weight of (*A*) taken. Also write necessary reactions.

4. A monomeric compound of cobalt gives the following data on quantitative analysis :
Co^{3+} : 21.24% Cl^- : 12.81% NH_3 : 24.77%
SO_4^{2-} : 34.65% H_2O : 6.53%
What is empirical formula of the complex?

5. One mole of the aqueous solution of the complex $CoCl_3 \cdot 5H_2O$ when treated with excess of aqueous $AgNO_3$ solution gave one mole of white precipitate of AgCl. What is the complex? (For this you should know that how many Cl^- ions are not coordinated to Co^{3+}— naturally one which is precipitated as AgCl)

6. For the following cell

Hg	0.005 M mercurous nitrate in 0.1 M HNO_3	0.05 M mecurous nitrate in 0.1 M HNO_3	Hg

EMF is 0.0295 V. What is charge on mercurous ion?

7. A solution made by dissolving 0.875 g of $Co(NH_3)_4Cl_3$ in 25.0 g of water freezes at –0.56°C. Calculate the number of moles of ions produced when 1 mole of $Co(NH_3)_4Cl_3$ is dissolved in water, and suggest a structure for the complex ion present in the compound. Molal depression constant (K_f) for water is 1.86 K kg mol^{-1} (molar mass of solute = 235.5 g mol^{-1})

Passage Comprehension Questions

Passage 1

Read the following report from a science journal and answer the questions at the end of it.

CNIC (Commission on Nomenclature of Inorganic Chemistry) appointed by IUPAC in 1994, approved official names for elements with atomic number 104 to 109. With $Z > 109$, IUPAC recommended a nomenclature to be followed for naming these elements until their names are officially recognised. This nomenclature is based on the Latin words for the atomic numbers of the elements.

Very recently, on August 16, 2003, IUPAC approved the name for the superheavy element which was first produced in 1994.

1. Atomic number of the recently discovered element is

(a) 107 (b) 108

(c) 109 (d) 110

2. Name and symbol given to this new element is

(a) Darmstadtium Ds

(b) Bohrium Bh

(c) Hassnium Hs

(d) Meitnerium Mt

3. IUPAC nomenclature of element with atomic number 110 is

(a) Unu (b) Uun

(c) Unn (d) Uum

Passage 2

From the data given in the table, answer the following questions:

Covalent Radii of the Transition Elements (pm)

K	Ca	Sc	Ti	V	Cr	Mn	Fe	Co	Ni	Cu	Zn
203	174	144	132	122	117	117	117	116	115	117	125
			Zr	Nb							
			145	134							
			Hf	Ta							
			144	134							

1. Atoms of the transition elements are smaller than those of the *s*-block elements. This is because of

(a) usual contraction in size across a horizontal period

(b) orbital electrons added to the penultimate *d*-shell rather than to the outer shell of the atom

(c) both of the above

(d) none of the above

2. The radii of the elements from chromium ($Z = 24$) to copper ($Z = 29$) are very close to one another. This is due to

(a) lanthanide contraction

(b) the fact that successive addition of *d*-electrons screen the outer electrons (4*s*) from the inward pull of the nucleus

(c) increase in radii due to increase in *n* is compensated by decrease in radii due to increase in *Z*

(d) all of the above

3. From the table, atomic radii of Zr and Hf and those of Nb and Ta are same. It is due to

(a) lanthanide contraction

(b) the fact that increase in size due to increase in *n* is compensated by decrease in size due to increase in *Z* based on equation $r_n = \frac{n^2 a_0}{Z}$

(c) both of the above

(d) none of the above

Passage 3

Observe the following data and answer the questions at the end of it:

A characteristic property of the *d*-transition metals is their ability to exhibit several oxidation states. The stability of a given oxidation state depends on the nature of the element with which the transition metal is combined. The highest oxidation states are found in the compounds of fluorides and oxides because of most electronegative nature of F and O. The different oxidation states of 3*d*-series have been given below (very rare oxidation states are provided in parenthesis).

Sc	Ti	V	Cr	Mn	Fe	Co	Ni	Cu	Zn
			I					I	
(II)	(II)	II	II	II	II	II	II	II	II
III	III	III	III	(III)	III	III	(III)	(III)	
	IV	IV	IV	IV	(IV)	(IV)	(IV)		
		V	(V)	(V)	(V)	(V)			
			VI	VI	VI				
				VII					

1. Nickel also shows (0) oxidation state. This oxidation state is in

(a) nitriles (b) carbonyls

(c) cyclopentadienyl (d) in all of these

2. These elements do not show the properties characteristic of transition elements

(a) Fe, Co, Ni (b) Sc, Y, La

(c) Cu, Ag, Au (d) Zn, Cd, Hg

3. In which case(s) there is change in oxidation number?

(a) aqueous solution of CrO_4^{2-} is acidified

(b) SO_2 gas is passed into $Cr_2O_7^{2-}/H^+$

(c) $Cr_2O_7^{2-}$ solution is made alkaline

(d) CrO_2Cl_2 is dissolved in NaOH

4. Most transition metals are inert towards acids or react slowly with them because of

(a) negative standard reduction potential

(b) protective layer of oxide

(c) both of the above

(d) none of the above

5. For Ni and Pt, different I.P. in MJ mol^{-1} are given below:

	$(IP)_1 + (IP)_2$	$(IP)_3 + (IP)_4$
Ni	2.49	8.80
Pt	2.60	6.70

hence

(a) nickle (II) compounds tend to be thermodynamically more stable than platinum (II)
(b) platinum (IV) compounds tend to be more stable than nickel (IV)
(c) both (a) and (b)
(d) none of the above is correct

6. Match the compounds of column X with oxidation state of chromium in column Y.

Column X	Column Y
I $[Cr(H_2O)_6]Cl_3$	5
II CrO_5	4
III K_3CrO_8	6
IV $(NH_3)_3CrO_4$	3

	I	II	III	IV
(a)	3	6	5	4
(b)	3	4	5	6
(c)	4	5	6	3
(b)	6	5	4	3

Passage 4

Read the following short write-up and answer the questions given:

"The transition elements have an unparalleled tendency to form coordination compounds with Lewis bases that is with groups which are able to donate an electron pair (called ligands)."

1. The tendency to form complexes by transition metal compared to s and p-block elements is due to
(a) their smaller size
(b) higher nuclear charge
(c) presence of low energy vacant orbitals to accept lone pair of electrons donated by ligands
(d) all of the above

2. Out of $[Fe(CN)_6]^{4-}$, $[Ni(CN)_4]^{2-}$ and $[Ni(CO)_4]$
(a) all have identical geometry
(b) all are paramagnetic
(c) all are diamagnetic
(d) $[Fe(CN)_6]^{4-}$ is diamagnetic but $[Ni(CN)_4]^{2-}$ and $[Ni(CO)_4]$ are paramagnetic

Passage 5

Following diagram represents variation of magnetic moment (μ) with atomic number Z for $3d$-series elements:

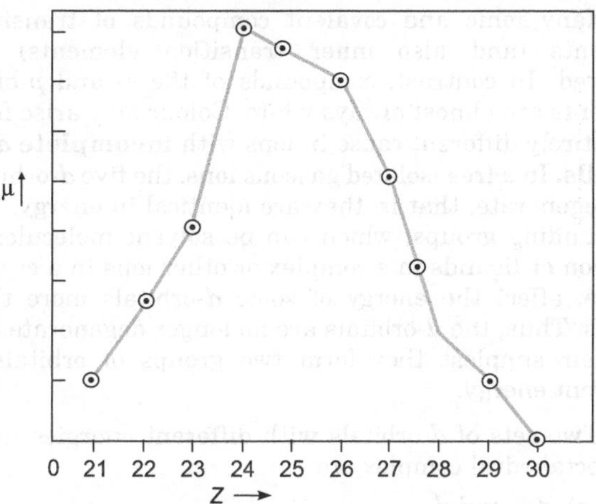

Magnetic moment of $3d$-series elements (on arbitrary scale)

Based on this, answer the following questions.

1. Consider following statements:
A : The magnetic moment arises from the spin and orbital motions. It is measured in the unit of BM.
B : Every coloured ion is paramagnetic and every paramagnetic ion is coloured.
C : Fe, Co and Ni exhibit ferromagnetism.
D : Magnetic moment of N unpaired electron is $N(N+2)$

Select correct statements.
(a) A, B (b) B, C
(c) C, D (d) A, C

2. Maximum magnetic moment is 6.93 for chromium. This means number of unpaired electrons is
(a) 4 (b) 5
(c) 6 (d) 7

3. Magnetic moment of Fe^{x+} ion is $\sqrt{35}$. Thus, x, number of unpaired (N) electrons and its electronic configuration are

	x	N	E-C
(a)	+3	5	$[Ar]3d^5$
(b)	+2	4	$[Ar]3d^6$
(c)	+1	3	$[Ar]3d^7$
(d)	0	4	$[Ar]3d^5 4s^1$

4. The oxygen carrying pigment, oxy-haemocyanin, containing two copper ions is diamagnetic, because
(a) the two copper ions are in +1 oxidation state
(b) one of the copper ions is in +1 oxidation state and the other is in +2 oxidation state
(c) there are strong anti-ferromagnetic interactions between the two copper ions
(d) there are ferromagnetic interactions between the two copper ions

Passage 6

Read the following passage and answer the questions given at the end of it:

Many ionic and covalent compounds of transition elements (and also inner transition elements) are coloured. In contrast, compounds of the *s*- and *p*-block elements are almost always white. Colour may arise from an entirely different cause in ions with **incomplete *d* or *f*-shells.** In a free isolated gaseous ions, the five *d*-orbitals are degenerate, that is, they are identical in energy. The surrounding groups, which can be solvent molecules in solution or ligands in a complex or other ions in a crystal lattice, affect the energy of some *d*-orbitals more than others. Thus, the *d*-orbitals are no longer degenerate and at their simplest they form two groups of orbitals of different energy.

1. Two sets of *d*-orbitals with different energies in an octahedral complex are
(a) d_{z^2} and $d_{x^2-y^2}$
(b) $d_{x^2-y^2}$ and d_{xy}
(c) $d_{z^2}, d_{x^2-y^2}$ and d_{xy}, d_{yz}, d_{zx}
(d) d_{xy}, d_{yz} and d_{xz}

2. Consider following statements
A : Colour of a transition metal complex is dependent on energy difference between two *d*-levels
B : Colour of the complex is dependent on nature of the ligand and the type of complex formed
C : $ZnSO_4$ and TiO_2 are white-in both *d*—*d** spectra are impossible
Select correct statements.
(a) A, B, C (b) A, B
(c) B, C (d) A, C

3. Out of the following, select colourless complexes
$Ni(CO)_4$ (*A*), $[Ni(CN)_4^{2-}]$ (*B*), $[Ni(NH_3)_4]^{2+}$ (*C*),
$[Ni(H_2O)_4]^{2+}$ (*D*), VO_2^{2+} (*E*), CrO_4^{2-} (*F*), MnO_4^- (*G*)
(a) *D, E* (b) *A, B, E*
(c) *A, E* (d) *A, B*

4. MnO_4^- is of intense pink colour, though Mn is in (+ 7) oxidation state. It is due to
(a) oxygen gives colour to it
(b) charge transfer when oxygen gives its electron to Mn making it Mn (+ VI) hence, coloured
(c) charge transfer when Mn gives its electron to oxygen
(d) none of the above is correct

5. Which is not true statement?
(a) Ions of *d*-block elements are coloured due to *d*—*d* transition
(b) Ions of *f*-block elements are coloured due to *f*—*f* transition
(c) $[Sc(H_2O)_6]^{3+}$, $[Ti(H_2O)_6]^{4+}$ are coloured complexes
(d) Cu^+ is colourless ion

Passage 7

Taking into account the following experimental fact answer the questions given at the end.

"When metallic copper is heated with concentrated sulphuric acid, in addition to copper(II) sulphate and sulphur dioxide, some copper (II) sulphide is also formed."

1. In which reaction, SO_2 is formed?
(a) $SO_4^{2-} + 2e^- \longrightarrow$ (b) $SO_4^{2-} + 8e^- \longrightarrow$
(c) $SO_4^{2-} + 6e^- \longrightarrow$ (d) in all of these

2. Copper (II) sulphide is formed because following side reaction is favourable due to its low solubility in acidic medium
(a) $SO_4^{2-} + 4H^+ + 2e^- \longrightarrow$
(b) $SO_4^{2-} + 8H^+ + 8e^- \longrightarrow$
(c) both of the above
(d) none of the above

Passage 8

Read the following short write-up and answer the questions given at the end.

Recent X-ray work, IR and other spectroscopic methods have proved that Turnbull's blue is identical to Prussian blue.

1. What is the common formula of Turnbull's blue and Prussian blue?
(a) $Fe_3[Fe(CN)_6]_2$ (b) $Fe_4[Fe(CN)_6]_3$
(c) $KFe[Fe(CN)_6]$ (d) $KFe_2[Fe(CN)_6]$

2. Intense blue colour arises as a result of
(a) electron transfer between Fe(II) and Fe(I)
(b) electron transfer between Fe(II) and Fe(III)
(c) *d*—*d** transition
(d) spin magnetic moment

Passage 9

Based on the following statements answer the questions given at the end.

Statement : The mercurous ion is written as Hg_2^{2+} while the cuprous ion is written as Cu^+.

Statement : Cu^+ is d^{10} ion and colourless but Cu_2O is red and Cu_2S is black

1. Mercurous ion is written as Hg_2^{2+} because
(a) magnetic moment of mercurous ion is zero and thus, 6*s* unpaired electron has been used in bonding to make Hg_2^{2+}
(b) in aqueous solution two Hg^+ ions are solvated forming Hg_2^{2+}
(c) $HgCl_2$ disproportionates to Hg_2Cl_2 and Hg
(d) none of the above is correct

2. There is no *d*—*d** transition in Cu^+(I) ion, but Cu_2O(I) and Cu_2S(I) are coloured because of
(a) presence of electronegative ions O^{2-}, S^{2-}

(b) presence of divalent anions
(c) charge transfer of electrons from O^{2-} or S^{2-} to the vacant orbitals of Cu^+
(d) all of the above

Passage 10

Based on the following experimental facts answer the questions given at the end:

"Green solution of potassium manganate (VI), turns purple and a brown solid is precipitated when CO_2 is bubbled into the solution".

1. Purple colour is due to formation of
(a) manganese dioxide (IV)
(b) potassium permanganate (VII)
(c) manganese (II) ion
(d) none of the above

2. Brown solid that is precipitated is of
(a) manganese dioxide (IV)
(b) potassium permanganate (VII)
(c) manganese oxide (II)
(d) manganese trioxide (III)

3. Above change is observed when CO_2 is passed into aqueous solution of the given ion. Function of CO_2 is
(a) it makes solution acidic due to formation of H_2CO_3
(b) it makes the solution basic due to formation of CO_3^{2-}
(c) it is simply the medium of the reaction
(d) none of the above is correct

4. Given experiment involves
(a) acid-base reaction
(b) a redox reaction
(c) disproportionation reaction
(d) decomposition reaction

5. What happens when CO_2 is passed into aqueous chromate (VI) solution
(a) CrO_3 is formed
(b) CrO_5 is formed
(c) Cr^{3+} is formed
(d) $Cr_2O_7^{2-}$ is formed

Passage 11

Read the following passage and answer the questions at the end of it.

When a piece of iron is dipped in conc. HNO_3, a slight reaction takes place which then ceases completely. There is no change in the appearance of the iron, but this does not displace Cu from $CuSO_4$ solution which is however a common property of Fe. This behaviour is also observed with Cr and Al when in contact with HNO_3.

1. Above phenomenon is called
(a) corrosion (b) rusting
(c) passivity (d) reduction

2. When exposed to air and moisture, iron gets rapidly oxidised to
(a) FeO (b) $Fe_2O_3 \cdot xH_2O$
(c) Fe_3O_4 (d) $FeO \cdot Fe_2O_3$

3. Select some facts about passivity and corrosion (rusting) of iron:
A : Rusting is an electrochemical phenomenon
B : Passivity of iron prevents it from rusting
C : Passivity and rusting are reversible processes
Select correct statements.
(a) A, B (b) B, C
(c) A, C (d) A, B, C

Passage 12

Based on the following qualitative test of chromium, answer the questions given at the end of it.

In qualitative analysis, a confirmatory test for chromium involves the formation of a blue transient poroxo species, CrO_5. It is formed when H_2O_2 is added to on acidic solution containing $Cr_2O_7^{2-}$.

1. Select correct statement about CrO_5.
(a) It has two types of oxygen-peroxide and oxide
(b) Oxidation number of Cr is + 10
(c) It is formed by oxidation of $Cr_2O_7^{2-}$
(d) All of the above are correct statements

2. CrO_5 can be represented as

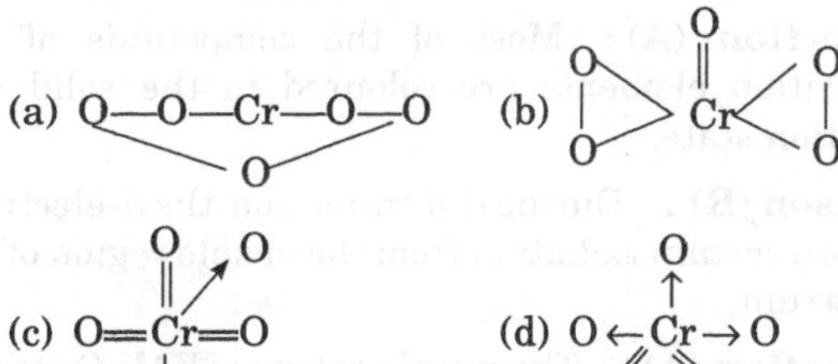

3. When $Cr_2O_7^{2-}$ is heated with Cl^- and conc. H_2SO_4 deep red vapours of chromyl chloride are formed. Deep red vapours have formula
(a) $CrOCl_2$ (b) $(CrO)_2Cl_2$
(c) CrO_2Cl_2 (d) CrO_3Cl

Passage 13

Read the following observation and answer the questions given at the end.

In an atmosphere with industrial smog, Cu corrodes to a basic sulphate $Cu_2(OH)_2SO_4$ and basic carbonate $Cu_2(OH)_2CO_3$.

1. Basic sulphate is generally formed during acid rain with a reaction
(a) $2Cu + 2H_2O + SO_3 \longrightarrow Cu_2(OH)_2SO_4$
(b) $2Cu + H_2O + SO_3 + O_2 \longrightarrow Cu_2(OH)_2SO_4$
(c) $2Cu + H_2O + SO_2 + O_2 \longrightarrow Cu_2(OH)_2SO_4$
(d) all of the above

2. Formation of basic carbonate and basic sulphate involves some of the steps
A : a redox reaction
B : acid-base neutralisation reaction
C : precipitation reaction
D : decomposition
Select the steps actually taking place
(a) A, C, D (b) B, C, D
(c) A, B, D (d) A, B, C

3. This type of corrosion
(a) forms a tough adherent coating that protects the underlying metal
(b) causes loss of copper and thus the articles made of copper
(c) causes poisoning of the atmosphere
(d) gives a direct method of preparing basic copper carbonate and basic copper sulphate

Assertion & Reason

Codes :
(a) Both **A** and **R** are true and **R** is the correct explanation of **A.**
(b) Both **A** and **R** are true but **R** is not the correct explanation of **A.**
(c) **A** is true but **R** is false.
(d) **A** is false but **R** is true.
(e) **A** and **R** both are false.

1. **Assertion (A) :** Most of the compounds of the transition elements are coloured in the solid and solution state.
Reason (R) : During *d-d* transition the *d*-electrons absorb certain radiation from the visible region of the spectrum.

2. **Assertion (A) :** The purple colour of $KMnO_4$ is due to the charge transfer transition.
Reason (R) : The intense colour, in most of the transition metal complexes is due to *d-d* transition.

3. **Assertion (A) :** Nickel (II) compounds tend to be thermodynamically more stable than platinum (II).
Reason (R) : Platinum (IV) compounds are relatively more stable than nickel (IV) compounds.

4. **Assertion (A) :** Chromium can be used as a protective and non-corrosive plating on other metals.
Reason (R) : Due to atmospheric oxidation, chemically a inert layer of Cr_2O_3 is formed.

5. **Assertion (A) :** $[Fe(CN)_6]^{3-}$ is more stable than $[Fe(CN)_6]^{4-}$.
Reason (R) : Complexes where the iron is in the (III) oxidation state are more stable than those in (II) oxidation state.

6. **Assertion (A) :** Ionic radii of Ta and Nb are same.
Reason (R) : The lanthanide contraction cancels almost exactly the normal size increase on descending a group of transition elements.

7. **Assertion (A) :** Aqueous solution of ZnO is purely basic in nature.
Reason (R) : ZnO cannot be dissolved in dil. acid.

8. **Assertion (A) :** CrO_3 is an acidic oxide.
Reason (R) : CrO_3 liberates CO_2 with Na_2CO_3.

9. **Assertion (A) :** When CrO_3 is dissolved in NaOH, CrO_3 goes in the cationic part.
Reason (R) : Na_2CrO_4 (yellow) is formed.

10. **Assertion (A) :** $Cu^{2+}(aq)$ and $Cd^{2+}(aq)$ both form stable complexes with $CN^-(aq)$.
Reason (R) : $[Cu(CN)_4]^{3-}$ is not decomposed by H_2S.

11. **Assertion (A) :** Green solution of potassium manganate (VI), K_2MnO_4 turns purple and brown solid is precipitated when CO_2 is bubbled into it.
Reason (R) : K_2MnO_4 changes to K_3MnO_8 by CO_2.

12. **Assertion (A) :** Prussian blue and Turnbull's blue are formed by different reagents, yet they are identical.
Reason (R) : They have formula as $Fe[Fe(CN)_6]^-$.

13. **Assertion (A) :** Copper(I) chloride is white, white Cu(I) oxide is red.
Reason (R) : In Cu(I) oxide, crystal structure has excess copper (I) ions making it defective.

True & False

1. On heating MnO_2 with Fe_2O_3 and carbon, ferromanganese is formed.
2. Zn reduces VO^{2+} to V^{3+} in acidic medium.
3. Cu^+ and Hg_2^{2+} both are paramagnetic.
4. ZnO is yellow when hot, but white when cold.
5. When KCN is added to $CuSO_4$ solution, a stable complex $K_3[Cu(CN)_4]$ is formed.
6. Cast iron is the purest form of iron.
7. When excess of $AgNO_3$ solution is added to 1 mol of $aq. FeCl_3 \cdot 6H_2O$, 2 mol of AgCl is formed.
8. Aqueous solution of CO_2 changes CrO_4^{2-} (yellow) into $Cr_2O_7^{2-}$ (orange).
9. Chromium and iron and made passive (inert) by HNO_3.

10. Due to lanthanide contraction, Zr and Hf have equal size.

11. Oxidation state of Cr in CrO_5 is +10.

12. Maximum oxidation state of osmonium is +6.

13. In $[Fe(H_2O)_5NO]^{2+}$, magnetic moment of Fe is $\sqrt{15}$ BM.

14. Equivalent weight of MnO_4^- in acidic medium is one-fifth of the molecular weight.

15. Stains of blue ink on the cloth can be removed by oxalic acid.

16. When H_2S gas is passed into $K_2[Cd(CN)_4]$, no ppt. is found since Cd^{2+} is in the complex form.

17. When H_2S gas is passed into aqueous $ZnCl_2$ solution, white ppt. of ZnS are formed.

18. MnO_4^- has Mn in +7 state thus it is colourless due to no unpaired electron in *d* orbital.

19. Nickel(II) compounds tend to be thermodynamically more stable than platinum(II) while platinum (IV) compounds are relatively more stable than nickel(IV).

20. $[Fe(H_2O)_6]^{3+}$ and $[Fe(CN)_6]^{3-}$ have same magnetic moment and identical structure.

21. The colour of the *d*-block ions is due to *d-d** transition.

22. $ZnSO_4$ and TiO_2 are colourless (white) due to absence of *d-d** transition.

23. CrO_3 is also called chromic acid.

24. $KMnO_4$ solution is used as a primary standard solution in volumetric titration.

25. Mohr's salt is isomorphous with white vitriol and epsom salt.

26. Correct formula of blue vitriol is $[Cu(H_2O)_5]SO_4$.

27. Hg_2Cl_2 is blackened by addition of *aq.* NH_3 due to formation of $[Hg_2(NH_3)_2]Cl_2$.

28. $HgCl_2$ is soluble in excess of KI due to formation of K_2HgI_4.

29. $HgCl_2$ is called calomel and Hg_2Cl_2 is called corrosive sublimate.

30. Solution of $Al(C_2H_5)_3$ and $TiCl_4$ in cyclohexane react to form Ziegler-Natta catalyst.

31. AgBr is insoluble in $Na_2S_2O_3$ due to formation of white precipitate of $Ag_2S_2O_3$.

32. The element (Unn) has atomic number of 100.

33. The actinoids are more electropositive and are reactive, reactivity increases with increase in atomic number.

34. Maximum oxidation state of Np is +6.

35. When CrO_2Cl_2 is dissolved in NaOH, yellow colour is due to the anion.

Fill in the Blanks

1. Coagulation of blood takes place by use of .... solution.

2. Rust stains can be removed by using .... which forms water soluble complex with .... ion of rust.

3. Mercurous ion is .... and cuprous ion is .... (magnetic behaviour).

4. 'Brown ring' in ring test of nitrate is due to the formation of $[Fe(H_2O)_5NO]^{2+}$ in which magnetic moment is found to be .... BM due to presence of .... unpaired electrons.

5. In 'ring test', NO transfers its electron to Fe^{2+} (charge transfer) thus there are .... and .... ions in $[Fe(H_2O)_5NO]^{2+}$.

6. Measurement of magnetic moment ($\sqrt{8}$ BM) of $Na_2[Fe(CN)_5NO]$ indicates that Fe is .... state and NO exists as ....

7. $[Fe(CN)_6]^{4-}$ and $[Fe(CN)_6]^{3-}$ differ in their .... while their structures are similar.

8. $[Ni(CN)_4]^{2-}$ and $[Ni(CO)_4]$ are .... but $[Ni(CN)_4]^{2-}$ is .... and $[Ni(CO)_4]$ is tetrahedral in structure.

9. Titanium is called the .... metal.

10. Rusting of iron is a special case of ....

11. In case of Cu^+, all the d^{10} electrons are paired, hence it is .... and it does not give .... transition.

12. In MnO_4^-, Mn is in +7 state with d^0 electronic configuration but it is coloured ion due to .... from .... to ....

13. Due to charge transfer from O^{2-} to Mn(VII), O is .... and Mn(VII) changes to .... hence MnO_4^- is ....

14. Magnetic moment of Fe^{2+} (24 electrons) is .... and is thus .... than Cr that of (24 electrons).

15. FeO is a .... compound and is .... deficient.

16. Fe, Co and Ni have approximately ....size.

17. Due to .... Zr and Hf, and also Nb and Ta have equal size.

18. V_2O_5 is used as catalyst in .... process for the manufacture of ....

19. $[Fe(H_2O)_6]^{3+}$ reacts with H_2O forming ....and H_3O^+ and is thus .... in nature.

20. .... is used to make many ferrous alloys including stainless and hard chromium steel.

21. Fe^{2+} salts can be determined volumetrically using .... and .... in acidic medium.

22. $Cr_2O_7^{2-}/H^+$ and H_2O_2 gives a deep blue-violet compound .... in which oxidation number of Cr is ....

23. Oxidation number of Cr in K_3CrO_8 with $[Cr(O_2)_4]^{3-}$ ion is ....

24. In $(NH_3)_3CrO_4$, oxidation number of Cr is +4 hence, it is ....

25. Chromyl ion is ....

26. $(NH_4)_2Cr_2O_7$ (orange) changes to green due to formation of .... by .... and .... gas is obtained by ....

27. MoO_3 and CrO_3 are strongly .... hence dissolve in ....

28. CrO_4^{2-} changes to $Cr_2O_7^{2-}$ in .... solution and $Cr_2O_7^{2-}$ changes to CrO_4^{2-} in .... solution.

29. Mn^{3+} disproportionates in H_2O/H^+ to .... and ....

30. In acidic medium, equivalent weight of MnO_4^- is .... of molecular weight and in dilute alkaline medium it is .... of molecular weight.

31. When CO_2 gas is passed into K_2MnO_4 solution, there is formation of .... and .... by ....

32. CO is toxic since when inhaled, it forms complex with .... of .... present in blood.

33. When a small amount of $KMnO_4$ is added to conc. H_2SO_4, a green solution containing .... ion is formed.

34. When excess of $KMnO_4$ is added to conc. H_2SO_4, an explosive oil .... is formed.

35. $Fe(OH)_2$ dissolves in concentrated solution of NaOH forming a blue-green complex ....

36. Turnbull's blue and Prussian blue are identical with formula ....

37. Fe_3O_4 is a .... oxide ....

38. $NaFeO_2$ on hydrolysis gives .... and ....

39. Fehling solution contains .... and sodium potassium tartrate (Rochelle salt).

40. $CuSO_4$ is determined iodometrically by its reactions with .... and then titration with .... solution.

Matrix-Match Type Questions

I. Only One Correct

1. Match the alloys (in List I) with the constituents metal (in List II).

List I	List II
A. Gun metal	1. Pb, Sn
B. German silver	2. Cu, Sn, Zn
C. Brass	3. Cu, Zn
D. Solder	4. Cu, Zn, Ni

2. Match the natural coordination compounds (in List I) with metals therein (in List II).

List I	List II
A. Nitrogenases	1. Cu
B. Cytochrome oxidase	2. Mo
C. Cytochrome C	3. Zn
D. Carboxy peptidase	4. Fe

3. Match the lanthanide ions (Ln^{3+}, in List I) with their calculated magnetic moments (in List II).

List I	List II
A. Ce^{3+}	1. 7.94 BM
B. Nd^{3+}	2. 4.90 BM
C. Gd^{3+}	3. 1.73 BM
D. Pm^{3+}	4. 3.87 BM

4. Match the catalysts in List I with their uses in List II.

List I	List II
A. $TiCl_4$	1. Adams catalyst in reduction
B. $PdCl_2$	2. In preparation of $(CH_3)_2SiCl_2$
C. Pt/PtO	3. Reppe synthesis
D. Cu	4. Used as the Natta catalyst in polythene production
E. Ni	5. Wake process for converting C_2H_4 to CH_3CHO

5. Match the alloys (in List I) with their constituents (in List II).

List I	List II
A. Invar	1. 70% Cu + 30% Zn
B. Brass	2. 66% Ni + 33% Cu
C. Monel	3. 36% Ni
D. Coinage metal	4. 60% Ni + 20% Fe + 20% Cr
E. Nichrome	5. 75% Cu + 25% Ni

6. Match the compounds of List I with oxidation state of List II.

List I	List II
A. $[Cr(H_2O)_6]Cl_3$	1. 5
B. CrO_5	2. 4
C. K_3CrO_8	3. 6
D. $(NH_3)_3CrO_4$	4. 3

7. Match the compounds in List I with their uses in List II.

List I	List II
A. Hg_2Cl_2	1. Metallurgial extraction of Ag
B. ZnO	2. Electrode
C. $ZnSO_4$	3. Luminous paints
D. Zn	4. Lithophone

8. Match the underlined atoms in List I with oxidation number in List II.

List I	List II
A. $\underline{Mn}O_4^-$	1. 1
B. $\underline{Cr}O_4^{2-}$	2. 2
C. $\underline{Fe}\,[Fe^{II}(CN)_6]^-$	3. 3
D. $\underline{Zn}O_2^{2-}$	4. 6
E. $[\underline{Ag}(CN)_2]^-$	5. 7

9. Match the alloys (in List I) with their constituents (in List II).

List I	List II
A. Invar	1. 70% Cu + 30% Zn
B. Brass	2. 66% Ni + 33% Cu
C. Monel	3. 36% Ni
D. Coinage metal	4. 60% Ni + 20% Fe + 20% Cr.
E. Nicrome	5. 75% Cu + 25% Ni

10. Match catalyst is (List I) according to its name and funtion.

List I Catalyst	List II Name	List III Function
A. $TiCl_4$	1. Fenton's reagent	P. Oxidation of alcohol
B. $FeSO_4/H_2O_2$	2. Ziegler-Natta	Q. Polythene preparation
C. Pd/C	3. Lindlar	R. Hydrogenation of alkynes to give *cis*-alkene
D. Pt/PtO	4. Adam	S. Synthesis of CH_3OH

11. Match the complex ion (List I) with its spin-only spin-only magnetic moment (List II).

List I	List II
A. $[Co(NH_3)_6]^{3+}$	1. 1.73 BM
B. $[Fe(CN)_6]^{3-}$	2. 5.92 BM
C. $[Mn(H_2O)_6]^{2+}$	3. 0.0 BM
D. $[Ni(H_2O)_6]^{2+}$	4. 2.83 BM

II. One or More Correct

1. Match the species in Column I that can react (oxidise, reduce, give ppt) species in Column II.

Column I	Column II
A. Fe^{2+}	1. $Cr_2O_7^{2-}$
B. $C_2O_4^{2-}$	2. CrO_4^{2-}
C. Pb^{2+}	3. $[Fe(CN)_6]^{3-}$
D. Ag^+	4. MnO_4^-
E. SO_2	5. $S_2O_3^{2-}$

2. Match the species in Column I with their characteristics in Column II.

Column I	Column II
A. Cu^+	1. Paramagnetic
B. Cu^{2+}	2. Diamagnetic
C. Hg^{2+}	3. Corrosive sublimate
D. Hg_2^{2+}	4. Calomel
E. Zn^{2+}	5. Colour with SCN^-
	6. d^{10}
	7. Reacts with KI

Integer Answer Types

This section contains 9 questions. The answer to each of the questions is a single digit integer, ranging from 0 to 9. The appropriate bubbles below the respectively question numbers in the *ORS* have to be darkened. For example, if the correct answers to question number *X, Y, Z* and *W* (say) are 6, 0, 9 and 2, respectively, then the correct darkening of bubbles will like the following.

X	Y	Z	W
⓪	⓪	⓪	⓪
①	①	①	①
②	②	②	②
③	③	③	③
④	④	④	④
⑤	⑤	⑤	⑤
⑥	⑥	⑥	⑥
⑦	⑦	⑦	⑦
⑧	⑧	⑧	⑧
⑨	⑨	⑨	⑨

Answer of this integer type question are in the form of colours given in number.

1. Violet
2. Indigo
3. Blue
4. Green
5. Yellow
6. Orange
7. Red
8. Prussian blue
9. Cherry Red
0. Colourless (white)

Colour of the

Set I

1. Complex when KCNS is added to $FeCl_3$ solution.
2. Complex when NH_3 is added to $CuSO_4$ solution.
3. Compounds when $K_2Cr_2O_7$ solution is made alkaline.
4. Compound when K_2CrO_4 solution is made acidic.
5. Compound when Cl_2 gas is passed into K_2MnO_4 solution.
6. Compound when SO_2 gas is passed into acidified $K_2Cr_2O_7$ solution.
7. Compound when $CuSO_4$ is treated with KI and then titrated with $Na_2S_2O_3$ solution.
8. Complex when dimethyl-glyoxime (in alcohol) added to ammoniacal nickel chloride solution.
9. Complex when $K_4[Fe(CN)_6]$ is added to $FeCl_3$ solution.

Set II

This section contains 8 questions

1. Fe^{2+} (24 electrons) has .... electrons in 3*d*-orbitals.
2. Cr^+ (23 electrons) has .... electrons in $(n-1)\,d$ orbitals.
3. Maximum oxidation state shown by manganese is ....
4. Magnetic moment of Mn in $MnSO_4$ is 5.91 BM. Thus it has .... electrons in 3*d*-orbitals.
5. CrO_5 has .... peroxy linkage(s).
6. 2.674 g of $CoCl_3 \cdot 6NH_3$ (molar mass = 267.4 g mol^{-1}) in aqueous solution gave 4.305 g of white precipitate of AgCl (molar mass = 143.5 g mol^{-1}). On reaction with excess of $AgNO_3$ solution. Thus Cl atoms in outer sphere of the complex is ....
7. Number of oxygen atom(s) between two chromium atoms in $Cr_2O_7^{2-}$ is ....
8. When CrO_5 is treated with H_2O_2 and KOH, a red-brown compound $K_3[Cr(O)_x]$ is formed. *x* is ....

Set III

This section contains 8 questions

1. 2.78 g of $FeSO_4 \cdot xH_2O$ on treatment with excess of $BaCl_2$ solution gives 2.33 g of white precipitate of $BaSO_4$. What is value of *x*? (Ba = 137, S = 32, Fe = 56, O = 16)
2. How many of the following also forms dimer?
   $NO_2, AlCl_3, FeCl_3, N_2O_5, CuSO_4, PCl_3$
3. How many of the following are paramagnetic as well as coloured species?
   O_2, NO_2, Cu^{2+}, Hg_2^{2+}, Fe^{2+}, Fe^{3+}, $[Fe(CN)_6]^{4-}$, $[Fe(CN)_6]^{3-}$, $[Ni(H_2O)_6]^{2+}$, $[Ni(CN)_4]^{2-}$
4. What is van't Hoff factor of $Hg_2(NO_2)_2$ if it is 100% ionised in aqueous solution?
5. How many of the following have underlined atoms in different oxidation states?
   $Hg\underline{I_4}^{2-}, \underline{S_2}O_3^{2-}, \underline{S_4}O_6^{2-}, Cr\underline{O_5}, \underline{Fe_3}O_4, \underline{Pb_3}O_4$
6. How many of the following pairs have approximately equal radii
   (Zr, Hf), (Nb, Ta), (Cr, Mn), (Mn, Fe), (Nb^{3+}, Ta^{3+}), (Zr^{4+}, Hf^{4+})
7. How many of the following are amphoteric?
   $ZnO, Al_2O_3, VO_2, V_2O_5, P_2O_5, Cr_2O_3, CrO_3$
8. How many H_2O molecules are in coordination sphere in $FeCl_3 \cdot 6H_2O$

Test Yourself on the Periodic Table

1. Locate with each of the following electron configuration on the Periodic Table given the transition elements. Identify each element.
   (a) $[Ar]3d^7 4s^2$ (b) $[Ar]3d^5 4s^1$
   (c) $[Kr]4d^2 5s^2$ (d) $[Xe]4f^3 6s^2$

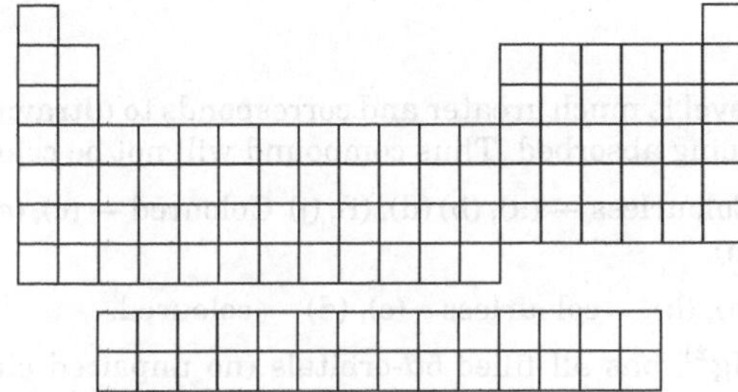

2. Locate the transition elements, on the Periodic Table, with following specifications :
   (a) Element in $3d$ series giving +7 oxidation state
   (b) Element in 5th period whose divalent cation has d^{10} configuration
   (c) Element which has been recently named as Ds
   (d) Elements which forms colourless ion in M^+ state

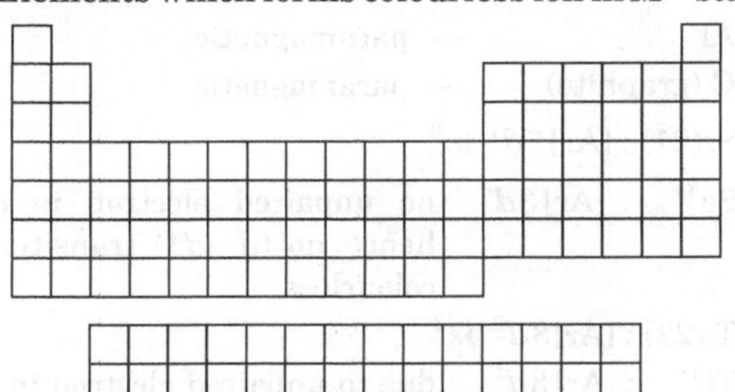

3. Locate the period, group and block of the element Ds recently named by IUPAC. Also write its electronic configuration.

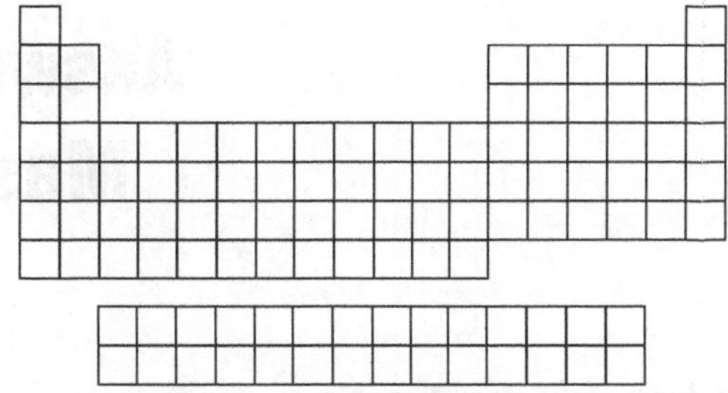

4. On the Periodic Table locate
   (1) Coinage metals (2) Lanthanides
   (3) Actinides (4) Alkaline earth elements
   (5) Chalcogens

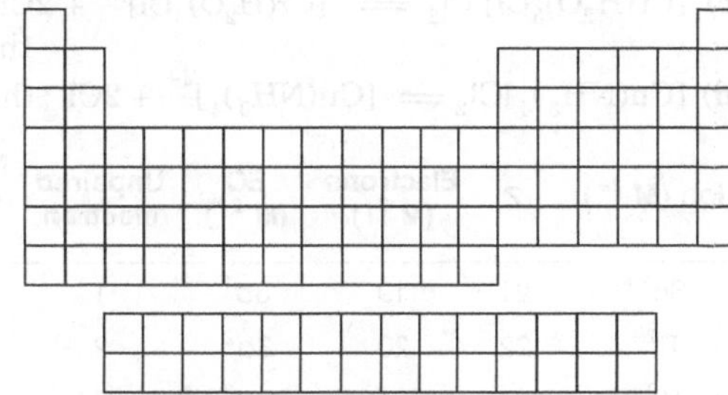

5. On the Periodic Table, locate the elements
   (a) in $3d$ series which form insoluble sulphides
   (b) in $4d$ series which form insoluble sulphides
   (c) with $s^2 p^6$ configuration of valence shell
   (d) with [Ar] configuration in +3 state

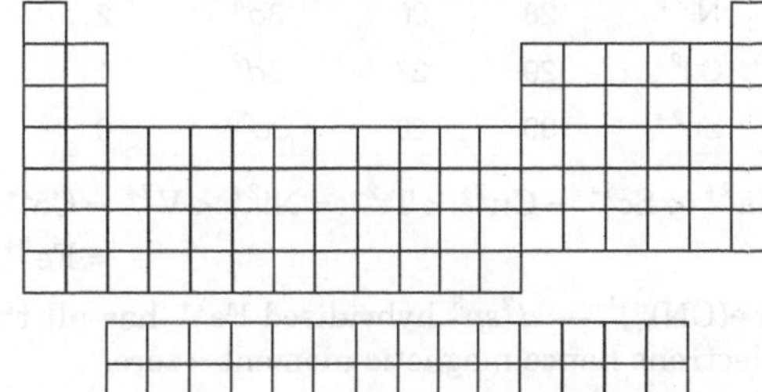

Answers & Solutions
Master Exercises

Exercise 1

Short Answer Type Questions

1. In all cases CN of the metal [Cu = 4, Cr = 6, Pt = 6] is to be satisfied.

(a) $[Pt(H_2O)_6]Cl_4 \rightleftharpoons [Pt(H_2O)_6]^{4+} + 4Cl^-$; five ions.

(b) $[Cu(H_2O)_2Cl_2] \rightleftharpoons$ no ionisation; exists as molecule.

(c) $[Cr(H_2O)_5Cl]\,Cl_2 \rightleftharpoons [Cr(H_2O)_2Cl]^{2+} + 2Cl^-$; three ions.

(d) $[Cu(NH_3)_4]Cl_2 \rightleftharpoons [Cu(NH_3)_4]^{2+} + 2Cl^-$; three ions.

2.

Ion (M^{2+})	Z	Electrons (M^{2+})	EC (M^{2+})	Unpaired electron	Magnetic moment (BM)
Sc^{2+}	21	19	$3d^1$	1	$\sqrt{3}$
Ti^{2+}	22	20	$3d^2$	2	$\sqrt{8}$
V^{2+}	23	21	$3d^3$	3	$\sqrt{15}$
Cr^{2+}	24	22	$3d^4$	4	$\sqrt{24}$
Mn^{2+}	25	23	$3d^5$	5	$\sqrt{35}$
Fe^{2+}	26	24	$3d^6$	4	$\sqrt{24}$
Co^{2+}	27	25	$3d^7$	3	$\sqrt{15}$
Ni^{2+}	28	26	$3d^8$	2	$\sqrt{8}$
Cu^{2+}	29	27	$3d^9$	1	$\sqrt{3}$
Zn^{2+}	30	28	$3d^{10}$	0	0

$Zn^{2+} < Sc^{2+} = Cu^{2+} < Ti^{2+} = Ni^{2+} < V^{2+} = Co^{2+} < Cr^{2+} = Fe^{2+} < Mn^{2+}$

3. $[Fe(CN)_6]^{4-}$— d^2sp^3 hybridised Fe^{2+} has all the paired electrons hence magnetic moment = zero.

$[Fe(H_2O)_6]^{2+}$ — sp^3d^2 hybridised Fe^{2+} has four unpaired electrons, hence magnetic moment = $\sqrt{24}$ BM

4. $[Fe(CN)_6]^{4-}$ unpaired electron (N) = 0,

$$\mu = \sqrt{N(N+2)}\text{ BM} = 0$$

$[Fe(CN)_6]^{3-}$ $\quad N = 1 \quad = \sqrt{3}$

$[Ni(CN)_4]^{2-}$ $\quad N = 0 \quad = 0$

$[Ni(H_2O)_4]^{2+}$ $\quad N = 2 \quad = \sqrt{8}$

5. $\sqrt{N(N+2)} = \sqrt{35} \Rightarrow N = 5$ (unpaired electrons)

$\Rightarrow \quad x = 3\ (Fe^{3+})$

$Fe^{3+} = [Ar]3d^5$

6. The s- and p- block elements do not have a partially filled d-shell so there cannot be any $d-d^*$ transition. The energy to promote an s- or p- electron to a higher energy level is much greater and corresponds to ultraviolet light being absorbed. Thus compound will not be coloured.

7. Colourless — (a), (b) (d), (i), (j) Coloured — (c), (e), (f), (g), (h).

8. (a), (b) — colourless (c), (d) — coloured.

9. Hg^{2+} has all filled $5d$-orbitals (no unpaired electron is $5d$), hence no d—d^* transition, hence colourless.

10.

NO_2	— coloured and paramagnetic
Cu^+	— colourless
O_2	— paramagnetic
O_3	— diamagnetic
Hg_2^{2+}	— colourless
Cd^{2+}	— colourless
Al	— paramagnetic
C (graphite)	— paramagnetic.

11. Sc(21) : $[Ar]\,3d^1\,4s^2$

Sc^{3+} : $[Ar]\,3d^0$ no unpaired electron in d-orbital, hence no $(d-d^*)$ transition hence, colourless

Ti(22) : $[Ar]3d^2\,4s^2$

Ti^{3+} : $[Ar]\,3d^1$ due to unpaired electron in d-orbital $(d-d^*)$ transition possible hence, coloured.

12. A.

(a) Nb	=	Ta		(b) Cr	=	Mn
(c) Cu	<	Zn		(d) Cu	>	Ni
(e) Zr	≈	Hf		(f) Ca	>	Sc
(g) Cr	<	Mo				

B.

	Element (Z)	S	$Z^* = Z - S$
(a)	Sc(21)	18.0	3.0
(b)	Ti(22)	18.85	3.15
(c)	V(23)	19.70	3.30
(d)	Cr(24)	20.55	3.45
(e)	Mn(25)	21.40	3.60
(f)	Fe(26)	22.25	3.75
(g)	Co(27)	23.10	3.90
(h)	Ni(28)	23.95	4.05
(i)	Cu(29)	24.80	4.20
(j)	Zn(30)	25.65	4.35

13. (a) [Structure: Cr with O above, Py, and four O in a pentagon]
Pentagonal pyramid

(b) [Structure: Cr with NH_3 above and below, NH_3 and four O in a pentagon]
Pentagonal bipyramid

(c) $\underset{\text{saturated } aq \text{ solution}}{Na_2Cr_2O_7} + \underset{\text{conc.}}{H_2SO_4} \longrightarrow Na_2SO_4 + 2CrO_3 + H_2O$

(d) $^-O-Cr(=O)_2-O-Cr(=O)_2-O^-$ [dichromate structure]

14. $(NH_4)_2Cr_2O_7 \xrightarrow{\Delta} N_2 + Cr_2O_3 + 4H_2O$

It is a redox reaction

— NH_4^+ ion is oxidised to N_2

— $Cr_2O_7^{2-}$ ion is reduced to Cr^{3+}

15. (A) : H_2S, oxidised to S by $Cr_2O_7^{2-}/H^+$

(B) : SO_2, oxidised to SO_4^{2-} by $Cr_2O_7^{2-}/H^+$

$$2H_2S + SO_2 \longrightarrow \underset{\text{yellowish white}}{3S} + 2H_2O$$

16. $CO_2 + H_2O \rightleftharpoons H_2CO_3 \rightleftharpoons H^+ + HCO_3^-$

$$\underset{\text{yellow}}{2CrO_4^{2-}} + 2H^+ \longrightarrow \underset{\text{orange}}{Cr_2O_7^{2-}} + H_2O$$

Aqueous CO_2 is acidic and changes yellow CrO_4^{2-} into orange $Cr_2O_7^{2-}$

17. $Cr_2O_7^{2-} < Cr^{3+} < Cr^{2+}$

Smaller the oxidation state of the element in the ion, greater the reducing nature.

18. (a) $[Zn(H_2O)_4]^{2+}$ (acid) $+ H_2O$ (base) $\rightleftharpoons [Zn(H_2O)_3OH]^+$ (base) $+ H_3O^+$ (acid)

(b) $[Al(H_2O)_6]^{3+}$

19. $ZnSO_4$ solution is acidic due to hydrolysis. When $NaHCO_3$ is added to it, CO_2 is given off and $ZnCO_3$ is also precipitated:

$$[Zn(H_2O)_4]^{2+} + 2HCO_3^- \longrightarrow \underset{\text{white ppt.}}{ZnCO_3\downarrow} + 5H_2O + CO_2$$

($[Zn(H_2O)_4]^{2+}$ formed from $ZnSO_4$ with H_2O)

20. Oxidation number of Mn in $MnO_4^- = +7$

It is expected that (Mn in) MnO_4^- should have [Ar] configuration and should be colourless. But one electron is transferred to Mn from oxygen and thus Mn in MnO_4^- has one unpaired electron making it paramagnetic.

Magnetic moment $= \sqrt{N(N+2)} = \sqrt{3}$ BM

21. $\underset{\text{ferrous oxalate}}{FeC_2O_4} \rightleftharpoons Fe^{2+} + C_2O_4^{2-}$

$$5Fe^{2+} + MnO_4^- + 8H^+ \longrightarrow 5Fe^{3+} + Mn^{2+} + 4H_2O$$
$$5C_2O_4^{2-} + 2MnO_4^- + 16H^+ \longrightarrow 10CO_2 + 2Mn^{2+} + 8H_2O$$
$$5FeC_2O_4 + 3MnO_4^- + 24H^+ \longrightarrow 5Fe^{3+} + 10CO_2 + 3Mn^{2+} + 12H_2O$$

$$5FeC_2O_4 \equiv 3MnO_4^-$$
$$\therefore \quad 1FeC_2O_4 \equiv 0.6\,MnO_4^-$$

(**Note** Both Fe^{2+} and $C_2O_4^{2-}$ are oxidised by MnO_4^-. Ask your teacher if one takes one mole of ferric oxalate)

22. Oxalic acid forms a dark-coloured complex with rust stains (Fe^{3+}). This complex is water soluble. The rust stains can be removed by treating the fabric with oxalic acid and then washing with water.

$$Fe^{3+} + 3C_2O_4^{2-} \longrightarrow \underset{\text{soluble}}{[Fe(C_2O_4)]^{3-}}$$

23. $FeCl_3 + 3H_2O \longrightarrow Fe(OH)_3 + 3HCl$

$NaHCO_3 + HCl \longrightarrow NaCl + H_2O + CO_2$

24. (a) Excess of KOH separates Fe^{3+} as $Fe(OH)_3$

$$Fe^{3+} + 3OH^- \longrightarrow Fe(OH)_3\downarrow$$

(b) Excess of KOH precipitates Fe^{3+} as $Fe(OH)_3$ while Cr^{3+} remains soluble $KCr(OH)_4$

(c) Excess of $NH_3(aq)$ precipitates Fe^{3+} as $Fe(OH)_3$ while Cu^{2+} remains soluble as $[Cu(NH_3)_4]^{2+}$

White ppt. with $BaCl_2 \Rightarrow SO_4^{2-}$

Choclate coloured ppt with $K_4[Fe(CN)_6] \Rightarrow Cu^{2+}$

Thus, hydrated salt is : $CuSO_4 \cdot xH_2O$

$CuSO_4 \cdot xH_2O$	$\longrightarrow$	$BaSO_4$
$159.5 + 18x$		233
2.495		2.33

$\therefore \quad x = 5$

25. (*A*) : $CuSO_4 \cdot 5H_2O$ (*B*) : $BaSO_4$ (*C*) : Cu_2I_2

(*D*) : $Cu_2[Fe(CN)_6]$

Reddish brown solution is due to formation of $KI_3 (KI + I_2 \rightleftharpoons KI_3)$. On adding hypo, I_2 is reduced to I^- and white ppt. of Cu_2I_2 appears.

26. White precipitate is due to CuCN

$$2Cu^{2+} + 4CN^- \longrightarrow 2CuCN + \underset{\text{cyanogen}}{C_2N_2}$$

White precipitate of CuCN dissolves in excess of KCN due to formation of a soluble and stable complex $[Cu(CN)_4]^{3-}$ and even in the presence of H_2S, Cu^{2+} is not precipitated as CuS.

$$\overset{\text{I}}{Cu}CN + 3KCN \longrightarrow \underset{\text{stable}}{K_3[\overset{\text{I}}{Cu}(CN)_4]} \rightleftharpoons 3K^+ + [Cu(CN)_4]^{3-}$$

27.
$$ZnO + C \longrightarrow Zn + CO$$
$$\Delta G^\circ = \Delta G^\circ_f(CO) - \Delta G^\circ_f(ZnO)$$
$$= -137.3 - (-318.2)$$
$$= 180.9 \text{ kJ mol}^{-1}$$

Positive value of ΔG° indicates that reaction is non-spontaneous. (However, it is spontaneous above 1000°C).

28. (*A*) : $ZnCl_2$ (*B*) : $Zn(OH)_2$ (*C*) : Na_2ZnO_2 (*D*) : ZnS (*E*) : AgCl

$$\underset{(C)}{Na_2ZnO_2} + H_2S \longrightarrow \underset{(D)}{ZnS} + 2NaOH$$

29. (*A*) : $ZnSO_3$ (*B*) : ZnO (*C*) : SO_2 (*D*) : ZnS

ZnO is amphoteric soluble in NaOH and HCl

30. ZnO in white paint by atmospheric H_2S gives ZnS (white) but Pb^{2+} salts in white paint turns black by atmospheric H_2S due to formation of PbS.

31. ZnO > CdO > HgO

32.

A	B
I	(b)
II	(c)
III	(d)
IV	(a)

33. Ni (obtained by Mond's process)

Analytical Questions

1. (a) MnO_4^- oxidises Fe^{2+} to Fe^{3+} and $C_2O_4^{2-}$ to CO_2 in ferrous oxalate (i)and only $C_2O_4^{2-}$ to CO_2 in (ii)

$$5Fe^{2+} + MnO_4^- + 8H^+ \longrightarrow 5Fe^{3+} + Mn^{2+} + 4H_2O$$

$$5C_2O_4^{2-} + 2MnO_4^- + 16H^+ \longrightarrow 10CO_2 + 8H_2O$$

$$5(Fe^{2+} + C_2O_4^{2-}) + 3MnO_4^- + \ldots \longrightarrow \ldots$$

Thus, $5Fe^{2+} \equiv MnO_4^-$

$5C_2O_4^{2-} \equiv 2MnO_4^-$

$5FeC_2O_4 \equiv 3MnO_4^-$

$1\,FeC_2O_4 \equiv 0.6\,MnO_4^-$

$$\underset{\text{Ferric oxalate}}{Fe_2(C_2O_4)_3} \longrightarrow 2Fe^{3+} + 3C_2O_4^{2-}$$

1 mol of ferric oxalate $\equiv$ 3 mol of $C_2O_4^{2-} \equiv \frac{2}{5} \times 3 = 1.2$ mol of MnO_4^-

(b) As above

$$6Fe^{2+} + Cr_2O_7^{2-} + 14H^+ \longrightarrow 6Fe^{2+} + 7H_2O + 2Cr^{3+}$$

$$6C_2O_4^{2-} + 2Cr_2O_7^{2-} + 28H^+ \longrightarrow 12CO_2 + 4Cr^{3+} + 14H_2O$$

Thus, $6(Fe^{2+} + C_2O_4^{2-}) \equiv 3Cr_2O_7^{2-}$

$\therefore$ 1 mol $FeC_2O_4 \equiv$ 0.5 mol $Cr_2O_7^{2-}$

Thus, 1 mol ferric oxalate $\equiv$ 3 mol oxalate $\equiv$ 1.0 mol of $Cr_2O_7^{2-}$

2. (*A*) : MnO_2 (*B*) : K_2MnO_4 (*C*) : $KMnO_4$

$$\underset{(A)}{MnO_2} + KOH + O_2(KNO_3) \xrightarrow{\Delta} \underset{\text{green}}{K_2MnO_4} \; (B)$$

$$CO_2 + H_2O \rightleftharpoons H_2CO_3 \rightleftharpoons H^+ + HCO_3^-$$

$$3MnO_4^{2-} + 4H^+ \longrightarrow MnO_2 + 2MnO_4^- + 2H_2O$$

3. (*A*) : $CuFeS_2$ (*B*) : $FeSO_4$ (*C*) : $CuSO_4$ (*D*) : H_2S (*E*) : $BaSO_4$ (*F*) : $KFe[Fe(CN)_6]$, Turnbull's blue. (*G*) : $[Cu(NH_3)_4]^{2+}$ (*H*) : CuS.

4. (a) $[Cr(H_2O)_6]Cl_3$ (b) $[Cr(H_2O)_5Cl]Cl_2 \cdot H_2O$ (c) $[Cr(H_2O)_4Cl_2]Cl \cdot 2H_2O$

1 mol of each is treated with $AgNO_3$.

(a) gives 3 mol AgCl (b) given 2 mol AgCl

(c) gives 1 mol AgCl

Thus, nature of the complex can be decided.

5. Due to sharing of *s*-as well as *d*-electrons there is formation of stronger metallic bonding.

6. Refer text (oxidation state)

7. (a) Cu^{2+} (b) V^{2+} (c) Cr^{2+} (d) Co^{3+}

8. $Mn^{2+} < MnO_2 < MnO_4^-$

9. (a) $\underset{\text{yellow}}{2CrO_4^{2-}} + 2H^+ \longrightarrow \underset{\text{orange}}{Cr_2O_7^{2-}} + H_2O$

(b) $Cr(OH)_3 + 3H^+ \longrightarrow Cr^{3+} + 3H_2O$

(c) $Cr(OH)_2 + 2H^+ \longrightarrow Cr^{2+} + 2H_2O$

(d) $Fe(OH)_2 + 2H^+ \longrightarrow Fe^{2+} + 2H_2O$

(e) $Cu(OH)_2 + 2H^+ \longrightarrow Cu^{2+} + 2H_2O$

In cases (b) to (e), metal hydroxides (insoluble) dissolve in dil. H_2SO_4.

10. (a) $\underset{\text{yellow}}{2CrO_4^{2-}} + 2H^+ \longrightarrow \underset{\text{orange}}{Cr_2O_7^{2-}} + H_2O$

(b) $Fe^{3+} + SCN^- \longrightarrow \underset{\text{deep red}}{[Fe(SCN)]^{2+}}$

(c) $3Cu + 2NO_3^- + 8H^+ \longrightarrow \underset{\text{blue}}{3Cu^{2+}} + 2NO + 4H_2O$

(d) $Cr(OH)_3 + OH^- \longrightarrow [Cr(OH)_4]^-$

$2[Cr(OH)_4]^- + 3H_2O_2 \longrightarrow \underset{\text{yellow}}{2CrO_4^{2-}} + 6H_2O + 2H^+$

(e) $3CuS + 8H^+ + 2NO_3^- \longrightarrow \underset{\text{blue}}{3Cu^{2+}} + \underset{\text{yellow}}{3S} + 2NO + 4H_2O$

11. (a) $4Co^{3+}(aq) + 2H_2O(l) \longrightarrow 4Co^{2+}(aq) + O_2(g) + 4H^+(aq)$

(b) $4Cr^{2+}(aq) + O_2(g) + 4H^+(aq) \longrightarrow 4Cr^{3+}(aq) + 2H_2O(l)$

(c) $3Cu(s) + Cr_2O_7^{2-}(aq) + 14H^+(aq) \longrightarrow 3Cu^{2+}(aq) + 2Cr^{3+}(aq) + 7H_2O(l)$

12. *A* : $FeO \cdot Cr_2O_3$ *B* : K_2CrO_4 *C* : $K_2Cr_2O_7$ *D* : $(NH_4)_2Cr_2O_7$ *E* : Cr_2O_3

NCERT Problem

Refer Text.

Exercise 2

Only One Option Correct

1. (d) Cu^{2+} is reduced to Cu^+ by CN^-
Thus, (a) is true
Cu^+ undergoes complexion with CN^- forming $K_3[Cu(CN)_4]$. Thus, (b) and (c) are true
Thus, (d)

2. (c)

3. Larger anion stabilises to a greater extent.
Thus, (a)

4. (c)

5. (b) Metals lying above H in electrochemical series are not obtained from their oxides by heating

$$HgO \longrightarrow Hg + O_2$$
$$ZnO \not\longrightarrow Zn + O_2$$
$$Ag_2O \longrightarrow Ag + O_2$$

6. I : $2CuSO_4 + 4KI \longrightarrow \underset{\text{white ppt.}}{Cu_2I_2} + I_2 + K_2SO_4$

II : $HgCl_2 + 2KI \longrightarrow 2KCl + \underset{\text{orange ppt.}}{HgI_2\downarrow}$

$HgI_2 + 2KI \longrightarrow \underset{\text{soluble}}{K_2HgI_4}$

III : $Pb(NO_3)_2 + 2KI \longrightarrow \underset{\text{yellow ppt}}{PbI_2\downarrow} + 2KNO_3$

Thus, (b)

7. (b) 8. (d)

9. (c) $NO \longrightarrow NO^+ + e^-$
$Fe^{3+} + e^- \longrightarrow Fe^{2+}$
Oxidation state of Fe is +2 and NO^+ is ligand
Thus, (c)

10. (a) Ring is formed by charge transfer
$NO \longrightarrow NO^+ + e^-$
$Fe^{2+} + e^- \longrightarrow Fe^{2+}$
Thus, (a)

11. (c)

12. (b) $Ni^{2+} + DMG \longrightarrow \underset{\text{cherry red ppt.}}{[Ni(DMG)_2]^{2+}}$

13. (a) $2CuSO_4 + 4KI \longrightarrow K_2SO_4 + Cu_2I_2 + I_2$
$I_2 + 2Na_2S_2O_3 \longrightarrow 2NaI + Na_2S_4O_6$
Thus, (a)

14. (c)

15. (d) Fe(III) gives red colours with KCNS hence, (d) is incorrect.

16. (b) 17. (d) 18. (b) 19. (c) 20. (c)

21. (d) Cu^+ : $[Ar]3d^{10}$
No unpaired electron, thus $s = 0$
Thus, (d)

22. (a) 23. (c)

24. (c) II group (of cation-analysis) contains Pb^{2+} and Cu^{2+} and are precipitated as sulphides on passing H_2S in acidic medium. ZnS is precipitated in presence of NH_4OH and NH_4Cl.
Thus, (c)

25. (b) Ti(22) : $[Ar]4s^2 3d^2$ [↑|↑| | |]
Ti^{2+} : $[Ar]3d^2$
Ti^{4+} : $[Ar]3d^0$
Ti^{2+} has two unpaired electrons in $3d$ and thus d-d* transition is possible due to absorption of light in visible region.

26. (d) $2Ag^+ + H_2S \longrightarrow Ag_2S$ (black)

27. (c) $Cr_2O_7^{2-} + 2OH^- \longrightarrow 2CrO_4^{2-} + H_2O$
$pH > 7 \ (x > 7)$
$2CrO_4^{2-} + 2H^+ \longrightarrow Cr_2O_7^{2-} + H_2O$
$pH < 7 \ (y < 7)$
Thus, (c)

28. (a) $MnO_4^- \longrightarrow MnO_4^{2-} + e^-$

29. (b) In this case CrO_2Cl_2 is formed

30. (c) MnO_4^- charges to MnO_4^{2-} or MnO_2 in basic medium.

31. (a) 32. (c)

33. (a) Maximum unpaired electrons [↑|↑|↑|↑|↑] d^5

34. (c) Magnetic moment $\mu = \sqrt{N(N+2)}$ BM.
where, N = unpaired electrons

Cr(24)	$[Ar]3d^5 4s^1$	$N = 6$, $\mu = \sqrt{48}$ BM $= x$
Mn^+	$[Ar]3d^5 4s^1$	$N = 6$, $\mu = \sqrt{48}$ BM $= y$
Fe^{2+}	$[Ar]3d^6$	$N = 4$, $\mu = \sqrt{24}$ BM $= z$

[↑↓|↑|↑|↑|↑]
$N = 4$

Thus, $z < x = y$
Thus, (c)

35. (d)

36. (b) 24 carat gold is 100%
Thus, 22-carat gold is 91.67%
Thus, 3.5% higher
Thus, (b)

37. (c) 38. (a) 39. (c) 40. (b) 41. (c)

42. (c) 43. (a)

44. (b) generally salts are colourless due to d^{10}-configuration.

45. (c) **46.** (a) **47.** (d) **48.** (a) **49.** (b)

50. (c) **51.** (c) **52.** (a) **53.** (c) **54.** (c)

55. (a) **56.** (a) **57.** (c) **58.** (a)

59. (d) Mn^{2+} (d^5) ↑ ↑ ↑ ↑ ↑

60. (d) **61.** (b) **62.** (b) **63.** (d) **64.** (c)

65. (a) **66.** (c)

67. (d) $CrO_3 + H_2O \longrightarrow \underset{(A)}{H_2CrO_4} \xrightarrow{OH^-} \underset{(B)}{CrO_4^{2-}}$

Thus, (d)

68. (a) **69.** (c) **70.** (c) **71.** (d) **72.** (b)

73. (a) **74.** (a) **75.** (c) **76.** (d) **77.** (a)

78. (d) **79.** (b) **80.** (b) **81.** (c) **82.** (a)

83. (b) **84.** (b) **85.** (b) **86.** (b) **87.** (c)

88. (d) **89.** (b) **90.** (d) **91.** (a) **92.** (d)

93. (a) **94.** (c) **95.** (b) **96.** (d) **97.** (d)

98. (b) **99.** (b) **100.** (c) **101.** (b) **102.** (a)

103. (c) **104.** (b) **105.** (a)

106. (a) $(NH_4)_2Cr_2O_7 \xrightarrow{\Delta} Cr_2O_3 + N_2 + 4H_2O$

$NH_4NO_2 \xrightarrow{\Delta} N_2 + 2H_2O$

107. (b) **108.** (d)

109. (b) $[Co(SCN)_4]^{2-}$, cobalt is +2 with three unpaired electrons in $3d$. Thus, $\mu = \sqrt{15}$ BM.

110. (b) **111.** (b)

112. (d) $6Fe^{2+} + Cr_2O_7^{2-} \longrightarrow 6Fe^{3+} + 2Cr^{3+}$

113. (b)

114. (b) $HgI_2 + 2KI \longrightarrow K_2HgI_4$

$HgI_2 \longrightarrow Hg + I_2$ (violet)

115. (b) $CrO_4^{2-} + 2H^+ \longrightarrow Cr_2O_7^{2-} + H_2O$

116. (d) $AgNO_3 \xrightarrow{\Delta} Ag_2O \xrightarrow{\Delta} 2Ag + O_2$

117. (c)

118. (d) $Hg_2Cl_2 + 2NH_3 \longrightarrow \underset{\text{white}}{HgNH_2Cl} + \underset{\text{black}}{Hg} + NH_4Cl$

119. (d) **120.** (d) **121.** (a) **122.** (a) **123.** (b)

124. (d) **125.** (a) **126.** (c) **127.** (c)

128. (b) $\underset{+6}{Cr_2O_7^{2-}} + \underset{+4}{SO_2} \longrightarrow \underset{+6}{SO_4^{2-}} + \underset{+3}{Cr^{3+}}$

129. (d) **130.** (a) **131.** (b) **132.** (b) **133.** (c)

134. (d)

135. (b) $NO \longrightarrow NO^+ + e^-$ $\quad 25 + 12 \longrightarrow 37$

$Fe^{2+} + e^- \longrightarrow Fe^+ (25e)$

136. (a) **137.** (b) **138.** (b) **139.** (d) **140.** (c)

141. (a) **142.** (b) **143.** (c) **144.** (a) **145.** (c)

146. (b) Cr^{3+} : [Ar] $3d^5$ stable

147. (a) **148.** (a) **149.** (a) **150.** (c)

151. (a) **152.** (b) **153.** (c) **154.** (c) **155.** (a)

156. (c) **157.** (d) **158.** (b) **159.** (c) **160.** (b)

161. (a)

162. (a) $VO_2 + 2H^+ \longrightarrow VO^{2+} + H_2O$

163. (d) $CrO_3 + 2H^+ \longrightarrow CrO_2^{2+} + H_2O$

164. (b)

165. (d) $2CrO_4^{2-} + 2H^+ \longrightarrow Cr_2O_7^{2-} + H_2O$

166. (c) $Cr_2O_7^{2-} + 2OH^- \longrightarrow 2CrO_4^- + H_2O$

167. (d) **168.** (c) **169.** (c)

170. (a) $2MnO_4^{2-} \xrightarrow{H^+} MnO_2 + MnO_4^-$

171. (a) **172.** (d) **173.** (b) **174.** (a) **175.** (c)

176. (b) ZnS is white.

177. (a) **178.** (c) **179.** (d) **180.** (b) **181.** (d)

182. (c) **183.** (a) **184.** (a)

185. (a) $[Fe(H_2O)_4Cl_2]Cl \cdot 2H_2O$

186. (d) $Fe^{2+} + [Fe(CN)_6]^{3-} \longrightarrow$ blue

$\xrightarrow[\text{(by oxidation)}]{Fe^{3+}}$ no colour

187. (a) **188.** (d) **189.** (b)

190. (c) Fe^{3+} forms soluble complex with oxalic acid.

191. (d)

192. (a) $Fe_2O_3 + OH^- + Cl_2 \longrightarrow [FeO_4]^{2-}$

193. (b) **194.** (c)

195. (d) $NO \longrightarrow NO^+ + e^-$,

$\underset{3d^6}{Fe^{2+}} + e^- \longrightarrow \underset{3d^7}{Fe^+}$ (three unpaired electrons)

↑↓ ↑ ↑ ↑ ↑ $\quad$ ↑↓ ↑↓ ↑ ↑ ↑

Magnetic moment $= \sqrt{N(N+2)} = \sqrt{15}$ BM

196. (c) **197.** (b)

198. (a) Fe^{2+} as well $C_2O_4^{2-}$ is oxidised by MnO_4^- in acidic medium

$$FeC_2O_4 \rightleftharpoons Fe^{2+} + C_2O_4^{2-}$$

$$5Fe^{2+} + MnO_4^- \longrightarrow Mn^{2+} + 5Fe^{3+}$$

$$5C_2O_4^{2-} + 2MnO_4^- \longrightarrow Mn^{2+} + 10CO_2$$

$$5FeC_2O_4 \equiv 3MnO_4^-$$

$$\therefore \quad 1\,FeC_2O_4 \equiv \frac{3}{5}\,MnO_4^-$$

$$= 0.6 \text{ mol}$$

Thus, (a)

199. (b) **200.** (c)

201. (b) $CuSO_4 + 2H_2O \rightleftharpoons Cu(OH)_2 + H_2SO_4$

$$2CrO_4^{2-} + 2H^+ \rightleftharpoons Cr_2O_7^{2-} + H_2O$$

Thus, (b)

202. (c) **203.** (d) **204.** (b) **205.** (c) **206.** (d)

207. (b) **208.** (b) **209.** (b) **210.** (c) **211.** (c)

212. (d) **213.** (d) **214.** (a) **215.** (a) **216.** (d)

217. (c) **218.** (c) **219.** (b)

220. (a) Element (90-103) are in IIIB

221. (d) **222.** (d) **223.** (a) **224.** (c)

225. (d) Third electron which is removed in third ionisation potential enthalpy belongs to $3d^3$ subshell in V.

$$\text{In } _{23}V = 1s^2 2s^2 2p^6\, 3s^2 3p^6 3d^3 4s^2$$

Third electron which is removed in third ionisation potential enthalpy belongs to $3d^4$ subshell.

$$_{24}Cr = 1s^2 2s^2 2p^6 3s^2 3p^6 3d^5 4s^1$$

Third electron which is removed in third ionisation potential enthalpy belongs to $3d^6$ subshell.

$$_{26}Fe = 1s^2 2s^2 2p^6 3s^2 3p^6 3d^6 4s^2$$

Third electron which is removed in third ionisation potential belongs to $3d^5$ subshell.

In all elements shell and subshells are same. Required amount of energy is based upon the stability of d-subshell. Hence, Mn shows highest third ionisation potential or enthalpy.

$$_{25}Mn = 1s^2 2s^2 2p^6 3s^2 3p^6 3d^5 4s^2$$

226. (d) In alkaline solution, $KMnO_4$ is reduced to MnO_2 (black).

$$2KMnO_4 + 2H_2O \longrightarrow 2MnO_2 + 2KOH + 3[O]$$

$$KI + 3[O] \longrightarrow KIO_3$$

$$\overline{2KMnO_4 + 2H_2O + KI \longrightarrow 2MnO_2 + 2KOH + KIO_3}$$

Hence, 2 mol of $KMnO_4$ are reduced to MnO_2 by 1 mol of KI.

227. (a) **228.** (b) **229.** (d) **230.** (d) **231.** (b)

232. (b) **233.** (b) **234.** (b) **235.** (d) **236.** (b)

237. (c) **238.** (a) **239.** (d) **240.** (d) **241.** (c)

242. (a) $3MnO_4^{2-} \longrightarrow MnO_2 + 2MnO_4^-$

243. (c) **244.** (a)

245. (b) $Cu^+ : [Ar]\,3d^{10}$ all electrons paired.

246. (b) **247.** (c)

248. (b) $2CuSO_4 + 4KI \longrightarrow Cu_2I_2 + K_2SO_4 + I_2$

$$I_2 + KI \longrightarrow KI_3$$

249. (b) **250.** (a) **251.** (c) $[Ar]\,4s^1 3d^5$

252. (a) **253.** (d)

254. (b) $CrO_3 + H_2O \longrightarrow H_2CrO_4$ (acidic)

255. (c)

256. (c) Fe has 4 unpaired electrons

257. (c) **258.** (c)

259. (b) $SnCl_2 + HgCl_2 \longrightarrow \underset{\text{white}}{Hg_2Cl_2} \xrightarrow{SnCl_2} \underset{\text{grey}}{Hg}$

260. (b) **261.** (d) **262.** (c) **263.** (b)

264. (a) $Fe^{3+} : [Ar]\,3d^5$; $Cu^+ : [Ar]\,3d^{10}$

265. (a) **266.** (c) **267.** (d) **268.** (d) **269.** (c)

270. (a)

271. (c) Nascent H reduces NO to NH_3 which changes to NH_4NO_3

272. (c) **273.** (a) **274.** (d) **275.** (b) **276.** (b)

277. (c) **278.** (c) **279.** (b)

280. (b) $Zn + 2NaOH \longrightarrow Na_2ZnO_2 + H_2$

281. (b) **282.** (d) **283.** (a) **284.** (c) **285.** (d)

286. (d) **287.** (d) **288.** (b) **289.** (c)

290. (b) $Cr_2O_7^{2-} + 6e^- + 14H^+ \longrightarrow 2Cr^{3+} + 7H_2O$

electron per Cr atom = 3

291. (a) **292.** (a) **293.** (a) **294.** (b) **295.** (a)

One or More Than One Options Correct

1. (a,b,c) **2.** (b,d) **3.** (b) **4.** (a) **5.** (c)

6. (a,b) **7.** (a,b,d) **8.** (a,b,c) **9.** (b) **10.** (c)

11. (a,b) **12.** (a,b,c) **13.** (c) **14.** (a,d)

15. (a) **16.** (a,b) **17.** (a,b,d)

Brain Twisters

1. MnO_4^- used $= 30.6 \times 0.1$ millimol

$$= 3.06 \times 10^{-3} \text{ mol}$$

$$2 \text{ mol } MnO_4^- = 5 \text{ mol } (H_2C_2O_4 \cdot H_2O)$$

thus, 3.06×10^{-3} mol $MnO_4^- = 7.65 \times 10^{-3}$ mol unreacted $(H_2C_2O_4 \cdot H_2O)$

$$H_2C_2O_4 \cdot H_2O \text{ taken } \frac{1.651}{108} = 1.53 \times 10^{-2} \text{ mol}$$

Used $H_2C_2O_4 \cdot H_2O$ by $MnO_2 = 7.650 \times 10^{-3}$ mol

1 mol $H_2C_2O_4 \cdot H_2O \equiv 1$ mol MnO_2

$\therefore\ 7.64 \times 10^{-3}$ mol $H_2C_2O_4 \cdot H_2O$

$$= 7.64 \times 10^{-3} \text{ mol } MnO_2$$

$$= 7.64 \times 10^{-3} \times 87 \text{ g pure } MnO_2$$

$$= 0.6644 \text{ g in } 0.81 \text{ g impure } MnO_2$$

$$\therefore \text{ Percentage of pure } MnO_2 = \frac{0.6644}{0.81} \times 100$$

$$= 82.02\%$$

2. Cr is in the form of $Cr_2O_7^{2-}$ and Mn is in the form MnO_4^-

Steel content (as MnO_4^- and $Cr_2O_7^{2-}$) = 10 g in 250 mL

$$= \frac{100}{250} \text{ g in 10 mL}$$
$$= 0.4 \text{ g}$$

Cr	$\longrightarrow \frac{1}{2} Cr_2O_7^{2-}$	$\longrightarrow BaCrO_4$
1 mol	0.5 mol	1 mol
2.17×10^{-4} mol	1.085×10^{-4} mol	0.0549 g
= 0.011284 g	in 10 mL	$= 2.17 \times 10^{-4}$ mol
in 0.4 g steel	= 0.01085 M (mol L^{-1})	

Thus, percentage of Cr $= \frac{0.011284}{0.4} \times 100$
$$= 2.821\%$$

$$\underset{1 \text{ mol}}{Mn} \longrightarrow \underset{1 \text{ mol}}{MnO_4^-}$$

$$MnO_4^- + 5Fe^{2+} \longrightarrow 5Fe^{3+} + Mn^{2+}$$
$$Cr_2O_7^{2-} + 6Fe^{2+} \longrightarrow 6Fe^{3+} + 2Cr^{3+}$$

MnO_4^- and $Cr_2O_7^{2-}$ both oxidises Fe^{2+} to Fe^{3+}

$$N(Cr_2O_7^{2-} + MnO_4^-) = \frac{15.95 \times 0.075}{10}$$
$$= 0.119625 \text{ N}$$

since, $[Cr_2O_7^{2-}] = 0.01085$ M
$$= 0.0651 \text{ N}$$

($N = 6 \times M$ as $Cr_2O_7^{2-}$ is reduced to Cr^{3+})

$\therefore$ $N(MnO_4^-) = 0.119625 - 0.0651$
$$= 0.054525 \text{ N}$$
$$= 0.010905 \text{ M (mol L}^{-1})$$

$\therefore$ ($N = 5 \times M$ as MnO_4^- is reduced to Mn^{2+})

$\therefore$ Mn = 0.010905 mol L^{-1}
$= 1.0905 \times 10^{-4}$ mol in 10 mL solution
= 0.006 g Mn in 10 mL solution
containing 0.4 g steel

$\therefore$ Percentage of Mn $= \frac{0.006}{0.4} \times 100$
$$= 1.50\%$$

3. (*A*) : MnO_2

$$MnO_2 + C_2O_4^{2-} + H^+ \longrightarrow Mn^{2+} + H_2O + CO_2$$
$$Mn^{2+} + H_2S \longrightarrow MnS\downarrow$$
$$MnO_2 + 4HCl \longrightarrow MnCl_2 + 2H_2O + Cl_2$$
$$Cl_2 + 2KI \longrightarrow 2KCl + I_2$$
$$I_2 + 2Na_2S_2O_3 \longrightarrow 2NaI + Na_2S_4O_6$$
$$MnO_2 \equiv Cl_2 \equiv I_2 \equiv 2Na_2S_2O_3$$

(Mn = 55)

20 mL of I_2 containing solution ≡ 180 mL of 0.1 N $Na_2S_2O_3$

$$N(I_2) = \frac{180 \times 0.1}{20}$$
$$= 0.9 \text{ N}$$

Equivalent weight of $MnO_2 = \frac{\text{mol. wt.}}{2} = \frac{87}{2} = 43.5$

Thus, $MnO_2 = 0.9$ N

$\therefore$ Conc in g/L $= 0.9 \times 43.5$ g/L
$$= 39.15 \text{ g/L}$$

Thus, MnO_2 in 200 mL $= \frac{39.15}{5} = 7.83$ g

4.

Species	Ionic/molar mass	%	Mole ratio	Simplest ratio
Co^{3+}	59.0	21.24	0.36	1
Cl^-	35.5	12.81	0.36	1
SO_4^{2-}	96	34.65	0.36	1
NH_3	17	24.77	1.45	4
H_2O	18	6.53	0.36	1

Thus, empirical formula is : $[Co(NH_3)_4(SO_4)(Cl)(H_2O)]$

5. Coordination number of cobalt = 6

Since, one mole of the complex gives one mole of AgCl on reaction with $AgNO_3$, only one **Cl** is outside coordination sphere, and two **Cl** satisfy primary and secondary valency. Out of fine water molecules, four are as ligand and remaining one water unit makes the complex hydrated. Thus comples is $[Co(H_2O)_4Cl_2]Cl \cdot H_2O$

6. Let charge on the mercurous ion = *n*
= number of electrons exchanged

Reactions $E°$

At anode $2Hg \longrightarrow \underset{0.005}{Hg_2^{n+}} + ne^-$ $E°_{ox} = x$

At cathode $\underset{0.05}{Hg_2^{n+}} + ne^- \longrightarrow 2Hg$ $E°_{red} = -x$

$$\underset{0.05 \text{ M}}{Hg_2^{n+}} \longrightarrow \underset{0.005 \text{ M}}{Hg_2^{n+}} \qquad E°_{cell} = 0.00 \text{ V}$$

$$K = \frac{0.005}{0.05} = \frac{1}{10}$$

$\therefore$ $E_{cell} = E°_{cell} - \frac{0.0501}{n} \log \frac{1}{10}$

$$0.0295 = 0 + \frac{0.0591}{n} \log 10$$

$\therefore$ $n = 2$

7. Using equation

$$\Delta T_f \text{ (depression in freezing point)} = \frac{1000 K_f w_1 (i)}{m_1 w_2}$$

where, K_f = molar depression constant = 1.86° mol^{-1} kg
w_1 = mass of solute = 0.875 g
w_2 = mass of solvent = 25.0 g
$\Delta T_f = 0.56°$

i = van't Hoff factor = Number of ions from one mole complex (solute)

m_1 = molar mass of solute

$$0.56° = \frac{1000 \times 1.86 \times 0.875(i)}{233.5 \times 25}$$

$$i = 2$$

Thus, only one Cl^- is outside coordination sphere (NH_3 is not ionisable).

$$[Co(NH_3)_4Cl_2]Cl \rightleftharpoons \underbrace{[Co(NH_3)_4Cl_2]^+ + Cl^-}_{\text{Two ions}}$$

Passage Comprehension Questions

Passage 1

1. (d) **2.** (a) **3.** (b)

Passage 2

1. (c) **2.** (b) **3.** (a)

Passage 3

1. (b) **2.** (d) **3.** (b) **4.** (b) **5.** (c) **6.** (a)

Passage 4

1. (d) **2.** (c)

Passage 5

1. (d) **2.** (c) **3.** (a) **4.** (c)

Passage 6

1. (c) **2.** (a) **3.** (d) **4.** (b) **5.** (d)

Passage 7

1. (a) **2.** (b)

Passage 8

1. (c) **2.** (b)

Passage 9

1. (a) **2.** (c)

Passage 10

1. (b) **2.** (a) **3.** (a) **4.** (c) **5.** (d)

Passage 11

1. (c) **2.** (b) **3.** (a)

Passage 12

1. (a) **2.** (b) **3.** (c)

Passage 13

1. (b) **2.** (d) **3.** (a)

Assertion & Reason

1. (a) **2.** (b) **3.** (b) **4.** (b) **5.** (b) **6.** (a)
7. (e) **8.** (b) **9.** (d) **10.** (d) **11.** (c) **12.** (b)
13. (a)

True & False

1. T **2.** T **3.** F **4.** T **5.** T **6.** F
7. F **8.** T **9.** T **10.** T **11.** F **12.** F
13. T **14.** T **15.** T **16.** F **17.** F **18.** F
19. T **20.** F **21.** T **22.** T **23.** T **24.** F
25. T **26.** F **27.** F **28.** T **29.** F **30.** T
31. F **32.** T **33.** T **34.** F **35.** T

Fill in the Blanks

1. $FeCl_3$ **2.** oxalic acid, Fe^{3+}
3. diamagnetic, diamagnetic
4. $\sqrt{15}$, three **5.** Fe^+ and NO^+ **6.** +2, NO^+
7. magnetic moment **8.** diamagnetic, square planar
9. 'wonder' **10.** corrosion.
11. colourless (white), d–d*
12. charge transfer, O^{2-}, Mn (+VII)
13. O^-, Mn(VI), coloured.
14. $\sqrt{24}$ BM, smaller (Fe^{2+} has $3d^6 4s^0$, Cr has $3d^5 4s^1$ EC)
15. nonstoichiometric, metal
16. equal **17.** lanthanide contraction
18. contact, H_2SO_4 **19.** $[Fe(H_2O)_5OH]^{2+}$, acidic
20. Ferrochrome **21.** MnO_4^-, $Cr_2O_7^{2-}$
22. $CrO(O_2)_2$ (a diperoxo compound), +6. **23.** +5
24. $(NH_3)_3Cr(O_2)_2$ (with two peroxide) **25.** CrO_2^{2+}
26. Cr_2O_3, reduction, N_2, oxidation.
27. acidic, alkali (say NaOH) **28.** acidic, alkaline
29. Mn^{2+}, MnO_2 **30.** $\frac{1}{5}$th, $\frac{1}{3}$rd
31. MnO_2, MnO_4^- disproportionation
32. Fe, haemoglobin
33. MnO_3^+ **34.** Mn_2O_7 **35.** $Na_4[Fe(OH)_6]$
36. $Fe[Fe(CN)_6]^-$ **37.** mixed, $\overset{II}{Fe}O\overset{III}{Fe_2}O_3$
38. Fe_2O_3, NaOH **39.** $CuSO_4$ **40.** KI, hypo

Matrix-Match Type Questions

I. Only One Correct

1. (A)—(2); (B)—(4); (C)—(3); (D)—(1);
2. (A)—(2); (B)—(1); (C)—(4); (D)—(3);
3. (A)—(3); (B)—(4); (C)—(1); (D)—(2);
4. (A)—(4); (B)—(5); (C)—(1); (D)—(2); (E)—(3)
5. (A)—(3); (B)—(1); (C)—(2); (D)—(5); (E)—(4)
6. (A)—(4); (B)—(3); (C)—(1); (D)—(2);
7. (A)—(2); (B)—(3); (C)—(4); (D)—(1);
8. (A)—(5); (B)—(4); (C)—(3); (D)—(2); (E)—(1)
9. (A)—(3); (B)—(1); (C)—(2); (D)—(5); (E)—(4)
10. (A)—(2-Q); (B)—(1-P); (C)—(3-R); (D)—(4-S);
11. (A)—(3); (B)—(1); (C)—(2); (D)—(4)

II. One or More Correct

1. (A)—(1,3,4); (B)—(1,4); (C)—(2); (D)—(2,5) (E)—(1,4)
2. (A)—(2,6); (B)—(1,5,7); (C)—(2,3,6,7); (D)—(2,4,6,7) (E)—(2,6)

Integer Answer Types

Set I

Questions →	1	2	3	4	5	6	7	8	9
Answers →	7	2	5	6	1	4	0	9	8
	0	0	0	0	0	0	0	0	0
	1	1	1	1	1	1	1	1	1
	2	2	2	2	2	2	2	2	2
	3	3	3	3	3	3	3	3	3
	4	4	4	4	4	4	4	4	4
	5	5	5	5	5	5	5	5	5
	6	6	6	6	6	6	6	6	6
	7	7	7	7	7	7	7	7	7
	8	8	8	8	8	8	8	8	8
	9	9	9	9	9	9	9	9	9

Set II

Questions →	1	2	3	4	5	6	7	8
Answers →	6	0	7	5	2	3	1	8
	0	0	0	0	0	0	0	0
	1	1	1	1	1	1	1	1
	2	2	2	2	2	2	2	2
	3	3	3	3	3	3	3	3
	4	4	4	4	4	4	4	4
	5	5	5	5	5	5	5	5
	6	6	6	6	6	6	6	6
	7	7	7	7	7	7	7	7
	8	8	8	8	8	8	8	8
	9	9	9	9	9	9	9	9

Set III

Questions →	1	2	3	4	5	6	7	8
Answers →	7	3	7	3	5	6	5	4
	0	0	0	0	0	0	0	0
	1	1	1	1	1	1	1	1
	2	2	2	2	2	2	2	2
	3	3	3	3	3	3	3	3
	4	4	4	4	4	4	4	4
	5	5	5	5	5	5	5	5
	6	6	6	6	6	6	6	6
	7	7	7	7	7	7	7	7
	8	8	8	8	8	8	8	8
	9	9	9	9	9	9	9	9

Test Yourself on the Periodic Table

1.

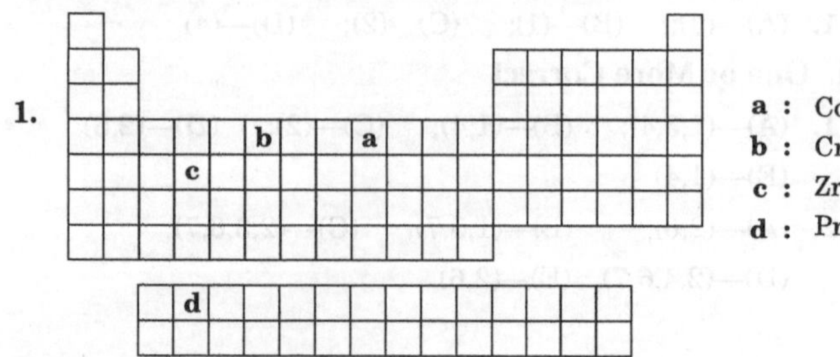

2.

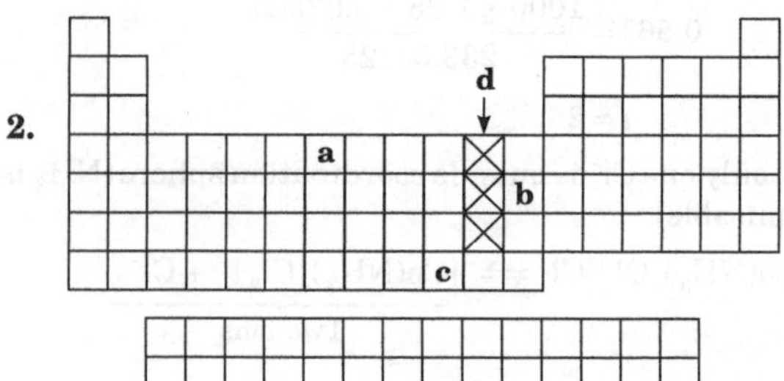

3.

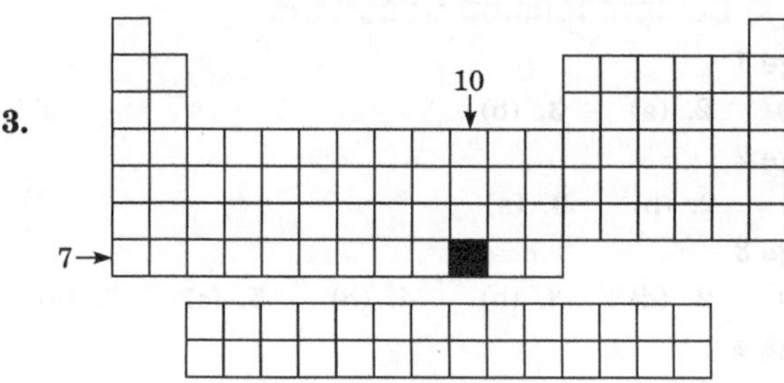

Period : 7th

Group : 10 (VIIIB)

Atomic no. : 110

(It lies below Pt)

6*d*-series element of *d*-block

Electronic configuration:

$[Rn]\,5f^{14}6d^{9}7s^{1}$

4.

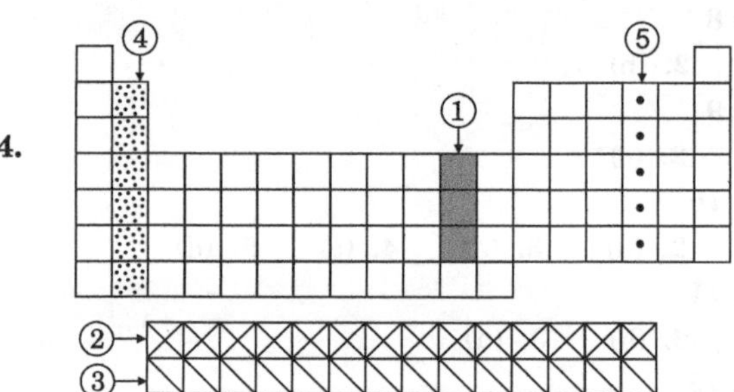

5.

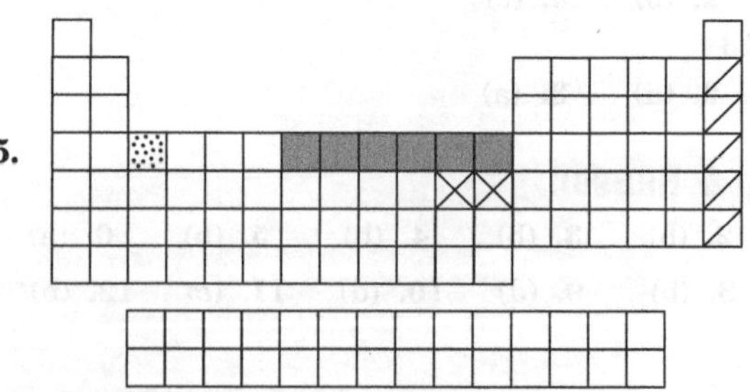

(a) insoluble sulphides of 3*d*-series

(b) insoluble sulphides of 4*d*-series

(c) s^2p^6 configuration

(d) M^{3+} with [Ar] configuration

17

"Gold, silver, copper, tin and iron were probably the metals discovered by our ancestors"

Metallurgical Extraction

Learning Framework of the Chapter ...

- The Occurrence and Isolation of the Elements
- Extraction of Copper
- Extraction of Magnesium
- Extraction of Lead
- Extraction of Tin
- Extraction of Silver
- Extraction of Aluminium
- Extraction of Iron
- Extraction of Zinc

Metallurgy is the science and technology of separating metals from their ores and of compounds alloys. An **alloy** is a solid solution either of two or more metals, or of a metal or metals with one or more non-metals.

Recovery of a metal from its ores involves three principal steps (three P's)

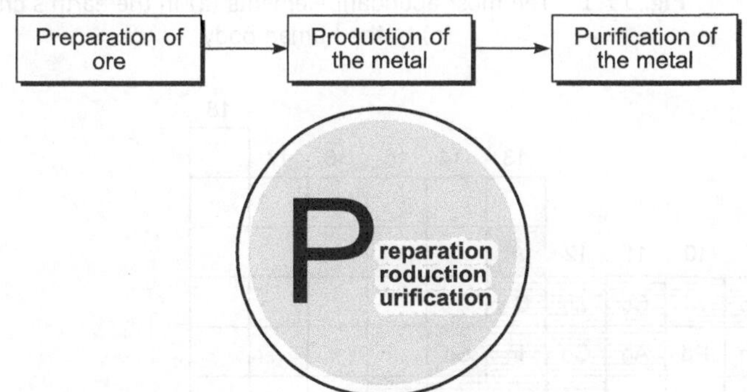

The Occurrence and Isolation of the Elements

- Most metals occur in nature as **minerals**; silicates and aluminates are the most abundant minerals (Table 17.1), but they are difficult to concentrate and reduce and therefore generally unimportant for commercial source of metals. Other minerals are oxides as haematite (Fe_2O_3) and rutile (TiO_2) and sulphides as cinnabar (HgS) and galena (PbS). Principal ores of some important metals have been summarised in Table 17.2.

DEEP Focus

- The early transition metals on the left hand side of the *d*-block generally occur as oxides, and the more electronegative, late transition metals on the right hand side of the *d*-block occurs as sulphides. **Less electronegative (more electropositive) metals can form ionic compounds by losing electrons to oxygen (and thus oxides are possible). More electronegative (less electropositive) metals tend to form covalent sulphide minerals with less electronegative sulphur.**

- We might also expect to find oxide ores for the *s*-block metals and sulphide ores for the more electronegative *p*-block metals. In fact, sulphide ores are common for the *p*-block metals, except for Al and Sn.
- Oxides of the *s*-block metals are strongly basic hence, form carbonates and silicates with acidic oxides as CO_2 and SiO_2

$$\underset{\text{basic}}{CaO} + \underset{\text{acidic}}{CO_2} \longrightarrow CaCO_3$$

$$\underset{\text{basic}}{MgO} + \underset{\text{acidic}}{SiO_2} \longrightarrow MgSiO_3$$

- Only gold and platinum-group metals (Ru, Os, Rh, Ir, Pd and Pt) are sufficiently unreactive to occur as free metals (in natural state).
- The most abundant elements in the earth's crust (by weight) are shown in Table 17.1 and Fig. 17.1. This indicates :
  - oxygen is the most abundant element,
  - aluminium is the most abundant metal,
  - first five elements comprise almost 92% by weight, and
  - first ten elements make up over 99.5% of the earth's crust.
- Other very abundant elements are nitrogen (78% of the atmosphere) and hydrogen, which occurs as water in the ocean.

Table 17.1 **The Most Abundant Elements**

	Element	% of Earth's crust
1.	Oxygen	46.4
2.	Silicon	28.2
3.	Aluminium	8.32
4.	Iron	5.63
5.	Calcium	4.15
6.	Sodium	2.36
7.	Magnesium	2.14
8.	Potassium	2.09
9.	Titanium	0.57
10.	Hydrogen	0.14
11.	Phosphorus	0.112
12.	Manganese	0.106

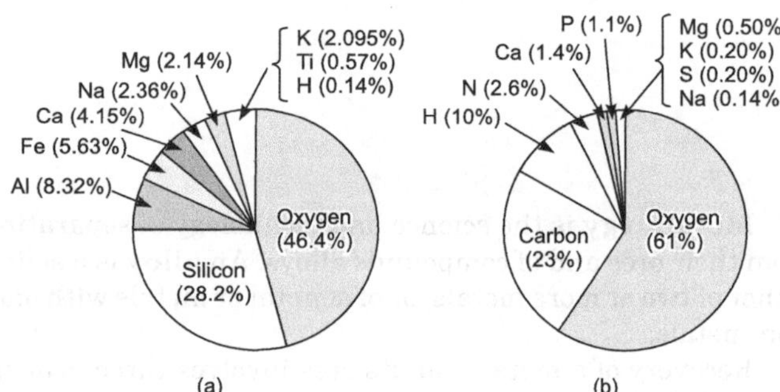

Fig. 17.1 Ten most abundant elements (a) In the earth's crust (b) In the human body

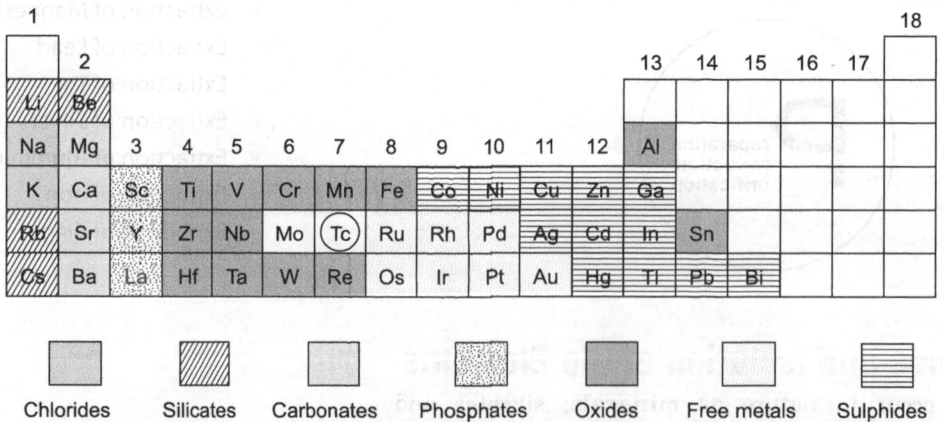

Fig. 17.2 Primary mineral sources of metals. The *s*-block metals occur as chlorides, silicates, and carbonates. The *d*- and *p*-block metals are found as oxides and sulphides, except for the group 3 metals, which occur as phosphates, and the platinum-group metals and gold, which occur in uncombined form. There is no mineral source of technitium (Tc in group 7), a radioactive element that is made in nuclear reactors.

Table 17.2 **Principal Ores of Some Important Metals**

Metal	Ore	Formula
Aluminium	Bauxite	$Al_2O_3 \cdot x\,H_2O$
Chromium	Chromite	$FeCr_2O_4$
Copper	Chalcopyrite	$CuFeS_2$
Iron	Haematite	Fe_2O_3
Lead	Galena	PbS
Manganese	Pyrolusite	MnO_2
Mercury	Cinnabar	HgS
Tin	Cassiterite	SnO_2
Titanium	Rutile	TiO_2
	Ilmenite	$FeTiO_3$
Zinc	Sphalerite	ZnS

Table 17.2 and Fig. 17.2 summarise principal type of minerals as anion. Other minerals have been given in Appendix I

Table 17.3 **Principal Types of Minerals Representing as Anions**

Type	
Uncombined metals	Ag, Au, Bi, Cu, Pd, Pt (Native ores)
Carbonates	$CaCO_3$ (calcite, limestone), $MgCO_3$ (magnesite), $CaCO_3 \cdot MgCO_3$ (dolomite) $PbCO_3$ (cerussite), $ZnCO_3$ (smithsonite), $Cu(OH)_2 \cdot CuCO_3$ (malachite), $FeCO_3$ (siderite)
Halides	CaF_2 (fluorite), NaCl (halite or rock salt), KCl (sylvite), Na_3AlF_6 (cryolite), $KCl \cdot MgCl_2 \cdot 6H_2O$ (carnallite), AgCl (horn siliver)
Oxides	$Al_2O_3 \cdot 2H_2O$ (bauxite), Cu_2O (cuprite), MnO_2 (pyrolusite), SnO_2 (cassiterite) Al_2O_3 (corundum), Fe_2O_3 (haematite), TiO_2 (rutile), ZnO (zincite), Fe_3O_4 (magnetite)
Phosphates	$Ca_3(PO_4)_2$ (phosphate rock), $Ca_5(PO_4)_3OH$ (hydroxyapatite), $3Pb_3(PO_4)_2 \cdot PbCl_2$ (pyromorphite)
Silicates	$NaAlSi_3O_8$ (albite), $Mg_3(Si_4O_{10})(OH)_2$ (talc) $CaMg_3Si_4O_{12}$ (asbestos) $K_2O \cdot Al_2O_3 \cdot 6SiO_2$ (felspar) $2ZnO \cdot SiO_2 \cdot H_2O$ (hemimorphite) $KAl_3H_2Si_3O_{12}$ (mica)
Sulphides	Ag_2S (argentite), CdS (greenockite), Cu_2S (chalcocite), FeS_2 (pyrite), HgS (cinnabar), PbS (galena), ZnS (sphalerite) $CuFeS_2$ (chalcopyrite), NiS (nickel blende)
Sulphates	$BaSO_4$ (barite), $CaSO_4$ (anhydrite), $MgSO_4 \cdot 7H_2O$ (epsomite), $PbSO_4$ (anglesite) $CaSO_4 \cdot 2H_2O$ (gypsum)

Benefication of Ores

Most of the ores available in nature contain large amount of impurities, *i.e.,* gangue. The pretreatment of ores, based on physical properties and without bringing out any major chemical change in the ore, is known as beneficatìon or concentration of the ore or ore-dressing.

Table (17.4) summarises the various methods adpoted for the ore-dressing

Table 17.4 **Methods of Concentration of the Ore**

Method	Ores specification	Process
1. **Gravity separation (Tabling) by hydraulic washing**	Based on the difference in the specific gravities of the ore and gangue. Cassiterite or tin-stone, chromite and pitchblende are concentrated	Crushed ore is kept on top of a slope table, which is made to vibrate. A stream of water is passed in the direction perpendicular to the slope. The lighter particles are thrown up by vibration and are removed by water stream
2. **Magnetic separation**	Based on the difference in magnetic properties of minerals. If the ore but not the gangue is attracted by a magnetic field, it can be concentrated to yield a sample which is rich in metal. Magnetite (Fe_3O_4), haematite (Fe_2O_3), wulframite ($FeWO_4$), chromite ($FeO \cdot Cr_2O_3$) and ilmenite ($FeO \cdot TiO_2$) separated from non-magnetic impurities	The pulverised mineral is passed over a rubber belt which moves on a pulley in a magnetic field. The non-magnetic gangue particles fall off in a vertical position when the belt passess over a pulley, but the magnetic ore clings to the belt. When the belt is out of magnetic field, the ore separates. (Fig. 17.3)

Fig. 17.3 Magnetic separation of ores

Method	Ores specification	Process
3. **Froth flotation process**	Used to concentrate sulphide ores; based on the difference in wettability of different minerals	Refer Fig. 17.5. Sometimes depressants are used to prevent certain type of particles from forming the froth with the bubbles. NaCN is used as a depressants in the separation of CuS and PbS. NaCN acts a depressant for ZnS but does not prevent PbS from the formation of the froth, due to water soluble complex formation. $ZnS + 4NaCN \longrightarrow Na_2[Zn(CN)_4] + Na_2S$ Some additives increase the efficiency of this process. Cresols (froth stabiliser) increase the life time of froth.
4. **Leaching**	Useful in case the ore is soluble in a suitable solvent. Pure Al_2O_3 is obtained from the bauxite ore in the Baeyer's process by leaching, also employed in concentration of silver and gold ores when leaching is done using NaCN in presence of air.	Leaching of bauxite $Al_2O_3(s) + 2OH^-(aq) + 3H_2O \longrightarrow 2Al(OH)_4^-$ $\xrightarrow[CO_2]{\text{dilution}} Al(OH)_3 + OH^- \xrightarrow{\Delta} Al_2O_3$ Fe_2O_3 if present as impurity is not affected. Leaching of silver and gold $4M + 8CN^- + 2H_2O + O_2 \longrightarrow \underset{\text{soluble}}{[M(CN)_2]^-}$ (M = Ag or Au) $Ag_2S + 4CN^- \longrightarrow 2[Ag(CN)_2]^- + S^{2-}$
5. **Calcination**	Used when concentrated ore is in the form of hydroxide or carbonate; volatile matter is burnt away. It is done in a reverberatory furnace.	$2Al(OH)_3 \xrightarrow{\Delta} Al_2O_3 + 3H_2O$ $CaCO_3 \xrightarrow{\Delta} CaO + CO_2$ It also drives out any organic matter that remains with the ore after concentration. It also removes moisture or water of crystallisation $2Fe_2O_3 \cdot 3H_2O \longrightarrow 2Fe_2O_3 + 3H_2O$
6. **Roasting**	Sulphide ore is heated in a regular supply of air at a temperature below the melting point of the metal. It converts sulphide into oxide or sulphate; part of sulphide may also act as a reducing agent.	$2ZnS + 3O_2 \xrightarrow{\Delta} 2ZnO + 2SO_2$ $2Cu_2S + 3O_2 \xrightarrow{\Delta} 2Cu_2O + 2SO_2$ $2Cu_2O + Cu_2S \xrightarrow{\Delta} 6Cu + SO_2$ Process is carried out in a reverberatory furnace. The impurities which form volatile oxides are also driven off during roasting. $P_4 + 5O_2 \longrightarrow P_4O_{10}$ $As_4 + 5O_2 \longrightarrow As_4O_{10}$ $C + O_2 \longrightarrow CO_2\uparrow$ $S + O_2 \longrightarrow SO_2\uparrow$

Separation and Extraction of Elements

The different methods for separating and extracting elements may be divided into five classes :

Mechanical Separation : A surprisingly large number of elements occur in the free elemental state. They have remained in the native form because they are unreactive as Cu, Ag, Au and Pt.

- Au is found in the native form, as grains in quartz, as nuggets and in the silt of river beds because of very high density (19.3 g cm^{-3}). Ag and Cu are sometimes found in the native form as nuggets. All the three metals lie below hydrogen and other reactive metals in electrochemical series and thus are easily displaced to the metallic form.
- Palladium and platinum are also found as native metals.
- Liquid droplets of Hg are found associated with cinnabar, HgS.
- Diamonds are found in the earth and are obtained by mechanical separation of large amount of earth and rock.

Thermal Decomposition Method : A few compounds will decompose into their constituent elements simply by heating :

$$2AsH_3 \xrightarrow{\text{red hot tube}} 2As + 3H_2$$

$$\underset{\text{sodium azide}}{2NaN_3} \xrightarrow{\Delta} 2Na + 3N_2$$

In Mond's process : $Ni(CO)_4 \xrightarrow{230°C} Ni + 4CO$

$$ZrI_4 \xrightarrow{W,\ \Delta} Zr + 2I_2$$

- $2BI_3 \xrightarrow{W,\ \Delta} 2B + 3I_2$
- In van Arkel-de-Boer process : $2HgO \xrightarrow{\Delta} 2Hg + O_2$
- $Ag_2CO_3 \xrightarrow{\Delta} Ag_2O + CO_2 \xrightarrow{\Delta} 2Ag + \frac{1}{2}O_2$

DEEP Focus

Oxides of the metals lying below Hg (and that of Hg) in electrochemical series decompose on heating giving metal.

A reaction is spontaneous if reducing agent is above oxidising agent in ECS.

Zn and Mg can reduce Fe^{2+} to Fe (in prevention of rust) since both are above Fe in ECS.

Displacement of Element by Another : In principle any element may be displaced from solution by another element which is higher in electrochemical series (ECS).

- Zn displaces Cu from $CuSO_4$ solution since Zn is above Cu in ECS :

$$Cu^{2+} + Zn \longrightarrow Zn^{2+} + Cu$$

- Cu displaces Ag from $AgNO_3$ solution due to the same fact:

$$Cu + 2Ag^{+} \longrightarrow Cu^{2+} + 2Ag$$

- Sea water has Br^-. Cl_2 can oxidise Br^- to Br_2

$$2Br^- + Cl_2 \longrightarrow 2Cl^- + Br_2$$

This method is quite applicable when displaced metal is costlier than the other metal which is oxidised. ECS is in order:

Li > K > Ca > Na > Mg > Al > Mn > Zn > Fe > Cu > Hg > Ag > Au

High Temperature Chemical Reduction Methods : Once an ore has been concentrated, it is reduced to the free metal, either by **chemical reduction** or by **electrolysis** as given in Tables 17.5 and 17.6. The method used depends on the activity of the metal as measured by its standard reduction potential. The most active metals have the most negative standard reduction potentials and are the most difficult to reduce; the least active metals have the most positive standard reduction potentials and are the easiest to reduce.

- ***Reduction by carbon:***

$$Fe_2O_3 + 3C \xrightarrow{\text{blast furnace}} 2Fe + 3CO$$

$$ZnO + C \xrightarrow{1200°C} Zn + CO$$

$$2Ca_3(PO_4)_2 + 10\,C + 6SiO_2 \xrightarrow{\text{electric furnace}} P_4 + 6CaSiO_3 + 10CO$$

$$MgO + C \xrightarrow[\text{electric furnace}]{2000°C} Mg + CO$$

$$PbO + C \longrightarrow Pb + CO$$

- ***Reduction by another metal:***

If the temperature needed for carbon to reduce an oxide is too high for economic or practical purpose, the reduction may be affected by another highly electropositive metal such as aluminium, which liberates a large amount of energy (1675 kJ mol^{-1}) on oxidation to Al_2O_3. This is the basis of the **thermite process :**

$$3Mn_3O_4 + 8Al \longrightarrow 9Mn + 4Al_2O_3$$

$$B_2O_3 + 2Al \longrightarrow 2B + Al_2O_3$$

$$Cr_2O_3 + 2Al \longrightarrow 2Cr + Al_2O_3$$

Magnesium is used in a similar way to reduce oxides in certain cases where the oxide is too stable to reduce, electropositive metals are used to reduce halides :

$$TiO_2 + 2Mg \longrightarrow Ti + 2MgO$$

$$TiO_2 + 4Na \longrightarrow Ti + 2Na_2O$$

$$TiCl_4 + 2Mg \xrightarrow[\text{Kroll process}]{1000\text{-}1150°C} Ti + 2MgCl_2$$

$$TiCl_4 + 4Na \xrightarrow{\text{IMI process}} Ti + 4NaCl$$

(IMI—Imperial Metal Industries)

- ***Reduction of oxides with*** H_2**:**

$$CuO + H_2 \longrightarrow Cu + H_2O$$

$$Co_3O_4 + 4H_2 \longrightarrow 3Co + 4H_2O$$

This method is not widely used, because many metals react with H_2 at elevated temperatures forming hydrides. Also there are chances of explosion of H_2 with O_2.

- ***Self reduction:***

Sulphides of Cu, Pb and Sb when roasted in air are converted partially into oxides. On further roasting in the absence of air, self reduction takes place:

$$CuS \xrightarrow{\text{roast in air}} CuO \xrightarrow[\text{absence of air}]{\text{CuS, roast in the}} Cu + SO_2$$

It is carried out in a **Bessemer converter** or **Pierce–Smith converter.**

Electrolytic Reduction

To get metal of high purity, electrochemical refining is done making impure metal as anode.

- ***In an aqueous solution***

It is used in cases when the electrolyte is soluble in water and products do not react with it. Copper, Zn, Ag and other metal are obtained by the electrolysis of aqueous solution of their salts ($ZnSO_4$, $AgNO_3$, $CuSO_4$). Impure metal is oxidised at anode and is deposited at the cathode (of pure metal).

- ***In fused melts:***

Na, K, Ca, Mg etc., react with H_2O hence, these metals are extracted by electrolysis of fused halides.

  - Na and Cl_2 are obtained by electrolysis of fused NaCl.
  - Al is obtained by the electrolysis of fused mixture of Al_2O_3 and $Na_3[AlF_6]$ (cryolite).

- ***In other solvents:***

Electrolysis can be carried out in solvents other than water. F_2 reacts violently with H_2O hence, it is produced by electrolysis of KHF_2 in HF.

Table 17.5 **Reduction Methods for Producing Some Common Metals**

	Metal	Reduction method
Least active	Au, Pt	None; found in nature as the free metal
↓	Cu, Ag, Hg	Roasting of the metal sulphide
	V, Cr, Mn, Fe	Reduction of the metal oxide with carbon, hydrogen or a more active metal
	Al	Electrolysis of molten Al_2O_3 in cryolite
Most active	Li, Na, Mg	Electrolysis of the molten metal chloride

Table 17.6 **Some Specific Reduction Processes**

Metal	Compound (ore)	Reduction process	Comments
Mercury	HgS (cinnabar)	Roast reduction; heating of ore in air $HgS + O_2 \xrightarrow{\Delta} Hg + SO_2$	
Copper	Sulphides such as Cu_2S (chalcocite)	Blowing of oxygen through purified molten Cu_2S $Cu_2S + O_2 \xrightarrow{\Delta} 2Cu + SO_2$	Preliminary ore concentration and purification steps required to remove FeS impurities
Zinc	ZnS (sphalerite)	Conversion to oxide and reduction with carbon $2ZnS + 3O_2 \xrightarrow{\Delta} 2ZnO + 2SO_2$ $ZnO + C \xrightarrow{\Delta} Zn + CO$	Process also used for the production of lead from galena PbS
Iron	Fe_2O_3 (haematite)	Reduction with carbon monoxide $2C\text{ (coke)} + O_2 \xrightarrow{\Delta} 2CO$ $Fe_2O_3 + 3CO \xrightarrow{\Delta} 2Fe + 3CO_2$	
Titanium	TiO_2 (rutile)	Conversion of oxide to halide salt and reduction with an active metal $TiO_2 + 2Cl_2 + 2C \xrightarrow{\Delta} TiCl_4 + 2CO$ $TiCl_4 + 2Mg \xrightarrow{\Delta} Ti + 2MgCl_2$	Also used for the reduction of UF_4 obtained from UO_2, pitchblende
Tungsten	$FeWO_4$ (wolframite)	Reduction with hydrogen $WO_3 + 3H_2 \xrightarrow{\Delta} W + 3H_2O$	Used also for molybdenum
Aluminium	$Al_2O_3 \cdot nH_2O$ (bauxite)	Electrolytic reduction (electrolysis) in molten cryolite. $Na_3[AlF_6]$ at 1000°C. $2Al_2O_3 \xrightarrow{\Delta} 4Al + 3O_2$	
Sodium	NaCl (sea water)	Electrolysis of molten chlorides $2NaCl \xrightarrow{\Delta} 2Na + Cl_2$	Also for calcium, magnesium, and other active metals in groups IA(1) and IIA(2)

Table 17.7 **Refining of Crude Metals**

	Methods	Metals purified
1.	Distillation	Zinc, mercury and cadmium. This methods is used in cases where metal has lower boiling point than impurities. Impure metal is heated to the boiling point in a closed retort when vapours formed are condensed to the pure metal. Non-volatile impurities are left unreacted.
2.	Electrolysis	Copper, gold, silver, lead, zinc aluminium. Impure metal is made anode and the pure metal is cathode. The net result is the transfer of pure metal from anode to cathode. At anode : $\underset{\text{(impure)}}{Cu} \longrightarrow Cu^{2+} + 2e^-$ At cathode : $Cu^{2+} + 2e^- \longrightarrow \underset{\text{(pure)}}{Cu}$; Cu (impure) $\longrightarrow$ Cu (pure)
3.	Liquation	A low melting metal like tin can be made to flow on a sloping surface and thus separated from higher melting impurities. 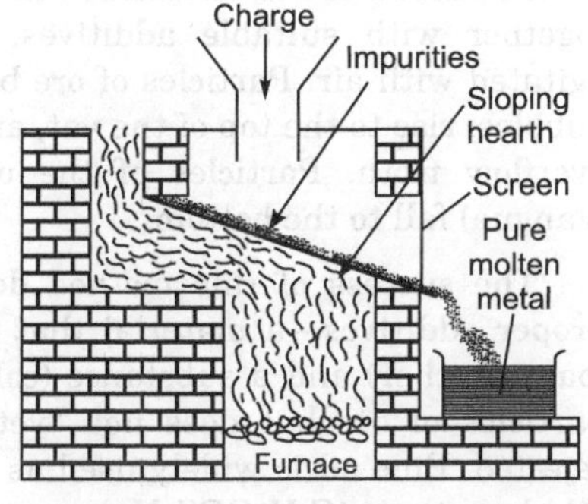
4.	Zone refining	Metal of high purity are obtained. Silicon, germanium, boron, gallium, indium are purified. (which are used in semiconductors). It is based on the fact that impurities are more soluble in the melt than in the pure metal. Impure metal (in the form of rod) is moved through a circular heater. When melting point is reached, pure metal melts and impurities are concentrated in a particular zone and pure metal in the other zone and crystallises out. Process is repeated several times.
5.	Vapour phase refining	In Mond process for the refining of nickel $\underset{\text{impure}}{Ni} + 4CO \xrightarrow{80°C} Ni(CO)_4 \xrightarrow{200°C} \underset{\text{pure}}{Ni} + 4CO\uparrow$ In van Arkel method for zirconium $\underset{\text{impure}}{Zr} + 2I_2 \xrightarrow{600°C} ZrI_4 \xrightarrow{1800°C} \underset{\text{pure}}{Zr} + 2I_2$

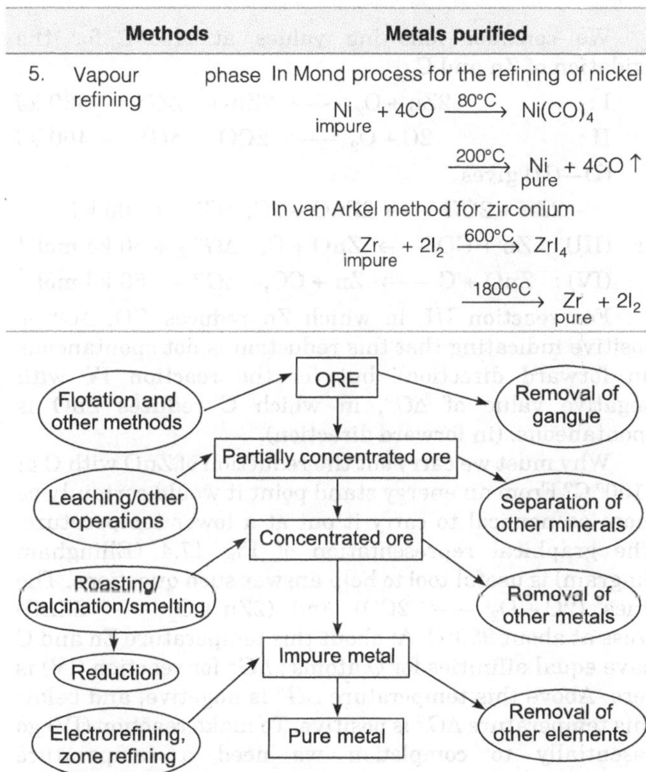

Factors Influencing the Choice of Extraction Process

The type of process used commercially for any particular element depends on a number of factors :

- Is the element unreactive enough to exist in the free state?
- Are any of the compounds unstable to heat?
- Does the element exists as an ionic compound, and is the element stable in H_2O?
- Is there a cheap element above it in the ECS which can be sacrificed to displace it from solution?
- Does the element occur as sulphide ores which can be roasted, or oxide ores which can be reduced by carbon (which can be the cheapest) or by Al, Mg and Na (which can be still costlier)?
- If all other methods fail, electrolysis is preferred (in aqueous solution which is the best to get metal of highest purity or in fused state which is again costlier).

Thermodynamics of Extractive Metallurgy

It is interesting to think of the reduction of ZnO with C as involving a competition between Zn and C for O atoms.

We consider following values at 1100° C for the oxidation of Zn and C :

I : $2Zn + O_2 \longrightarrow 2ZnO \quad \Delta G° = -360$ kJ

II : $2C + O_2 \longrightarrow 2CO \quad \Delta G° = -460$ kJ

(I)—(II) gives

$$2Zn + 2CO \longrightarrow 2ZnO + 2C, \Delta G° = +100 \text{ kJ}$$

or (III) : $Zn + CO \longrightarrow ZnO + C, \quad \Delta G° = +50 \text{ kJ mol}^{-1}$

(IV) : $ZnO + C \longrightarrow Zn + CO, \quad \Delta G° = -50 \text{ kJ mol}^{-1}$

For reaction III, in which Zn reduces CO, $\Delta G°$ is positive indicating that this reduction is not spontaneous (in forward direction), but for the reaction IV with negative value of $\Delta G°$, in which C reduces ZnO is spontaneous. (in forward direction).

Why must we carry out the reduction of ZnO with C at 1100° C? From an energy stand point it would certainly be more economical to carry it out at a lower temperature. The graphical representation of Fig 17.4 (Ellingham diagram) is useful tool to help answer such questions. The lines ($2C + O_2 \longrightarrow 2CO$) and ($2Zn + O_2 \longrightarrow 2ZnO$) cross at about 950°C. At about this temperature Zn and C have equal affinities for O atoms : $\Delta G°$ for reaction (IV) is zero. Above this temperature $\Delta G°$ is negative, and below this temperature $\Delta G°$ is positive. To make reaction (IV) go essentially to completion we need a temperature somewhat in excess of 950°C.

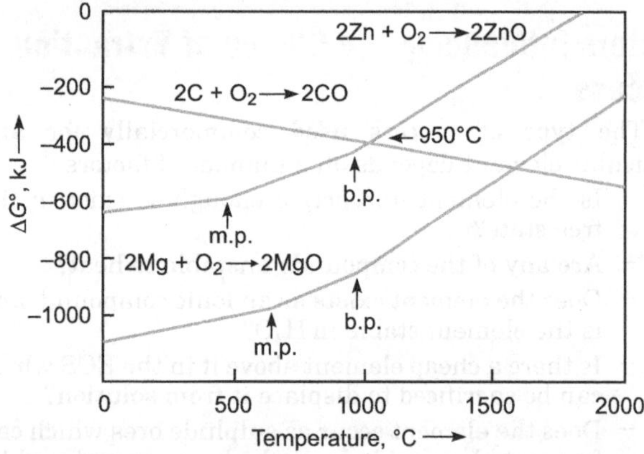

Fig. 17.4 The points noted by arrows are the melting and boiling points of the metals zinc and magnesium. $\Delta G°$ as a function of temperature for some reactions of extractive metallurgy.

Extraction of Copper

Occurrence

- Cuprite Cu_2O
- Malachite $CuCO_3 \cdot Cu(OH)_2$
- Azurite $2CuCO_3 \cdot Cu(OH)_2$
- Copper pyrites or Chalcopyrites $CuFeS_2$

Main source is copper pyrites. Some copper is also found in native state.

Pyrometallurgical Method (Dry Method)

Sulphide ore is converted into copper in following steps :

Step I **: Concentration by Froth-Flotation process (Fig. 17.5)**

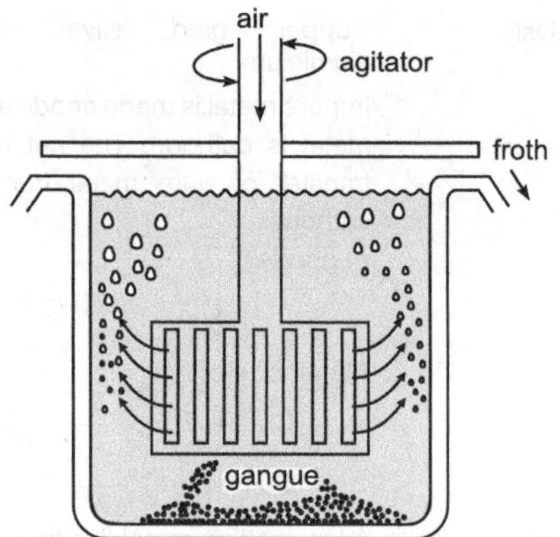

Fig. 17.5 Concentration of an ore by flotation

Powdered ore is suspended in water in a large vat, together with suitable additives, and the mixture is agitated with air. Particles of ore become attached to air bubbles, rise to the top of the vat, and are collected in the overflow froth. Particles of the undesired waste rock (gangue) fall to the bottom.

The success of this method depends on the use of proper additives—a material that will produce a stable foam (frother) and a substance (collector) that coats the particles of ore but does not "wet" the particles to be rejected. Pine oil is widely used as a frother and sodium ethyl xanthate ($C_2H_5OCS_2Na$) as a collector.

Step II **: Roasting** is done in multiple hearth furnace. Volatile impurities are removed as thin oxide (As_2O_3, Sb_2O_3, P_2O_5). It also converts iron sulphide to iron oxide. The copper remains as the sulphide if the temperature is kept below 800°C.

$$2FeS + 3O_2 \xrightarrow{\Delta} 2FeO + 2SO_2 \uparrow$$

Step III **: Smelting** of the roasted ore is done in a furnace at 1400°C. In this furnace charge melts and separates into two layers. The bottom layer is **copper-matte,** consisting chiefly of the molten sulphides of copper and iron. The top layer is a silicate slag formed by the reaction of oxides of Fe, Ca, and Al with SiO_2 (which either is present in the ore or is added)and is removed.

$$FeO + SiO_2 \longrightarrow FeSiO_3$$

***Step IV* : Bessemerisation** (or **Converting**) occurs in another furnace where air is blown through the molten copper matte. First, the remaining iron sulphide is converted to oxide (as in step II) followed by slag formation (as in step III). The slag is poured off and air is again blown through the furnace. Now the following reactions occur, yielding a product that is about 98 to 99% Cu.

$$2Cu_2S + 3O_2 \longrightarrow 2Cu_2O + 2SO_2$$
$$2Cu_2O + Cu_2S \longrightarrow 6Cu + SO_2$$

The product of this step is called **Blister copper** because of frozen bubbles of SO_2. It can be used where high purity is not required.

***Step V* : Electrorefining :**

Anode	:	impure copper
Cathode	:	pure copper
Electrolyte	:	3.5% $CuSO_4$ + 15% H_2SO_4
Voltage	:	0.3 to 0.5 V
Temperature	:	50.55° C

Cu from impure copper anode goes into solution and is deposited at the cathode :

$$\text{Cu} \xrightarrow{\text{Anode}} Cu^{2+} + 2e^- \qquad Cu^{2+} + 2e^- \xrightarrow{\text{Cathode}} \text{Cu}$$

Anode-mud (anode-sludge) contains Au, Pt, Ag and other precious metals (noble metals).

Hydrometallurgical Process (Wet Process)

***Step I* :** Leaching with H_2SO_4 converts salts into sulphates:

$$CuCO_3 + H_2SO_4 \xrightarrow{\Delta} CuSO_4 + H_2O + CO_2$$
$$Cu(OH)_2 + H_2SO_4 \xrightarrow{\Delta} CuSO_4 + 2H_2O$$
$$CuFeS_2 + 2H_2SO_4 \xrightarrow{\Delta} CuSO_4 + FeSO_4 + 2H_2S$$

***Step II* : Electrochemical Displacement :** Fe is above Cu in ECS, hence if Fe scrap is added to $CuSO_4$ solution, Cu is displaced :

$$CuSO_4 + Fe \longrightarrow FeSO_4 + Cu\downarrow$$

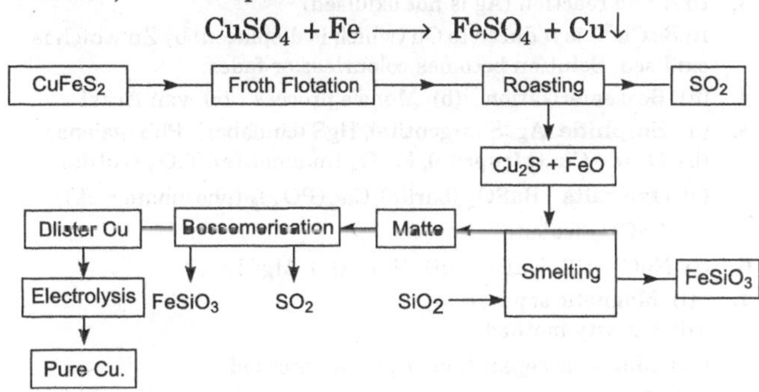

Target Practice 1

1. NaCl solution is neutral. If it undergoes electrolysis, pH increases/decreases/remains constant ....
2. When Cu turnings are added to $AgNO_3$ solution, it turns blue. It is due to formation of ....
3. We have $CuSO_4$ solutions (blue) in two beakers *A* and *B*. In beaker *A* we add Ag pieces and in *B* we add Zn pieces. What will happen?
   In *A* ....
   In *B* ....
4. Name the process (unbalanced reactions)
   (a) $2Cu_2S + 3O_2 \longrightarrow Cu_2O + SO_2$
   $Cu_2O + Cu_2S \longrightarrow Cu + SO_2$ ....
   (b) $Ni + 4CO \xrightarrow{80°C} Ni(CO)_4 \xrightarrow{200°C} Ni + 4CO\uparrow$ ....
   (c) $Zn + 2I_2 \xrightarrow{600°C} ZrI_4 \xrightarrow{1800°C} Zn + 2I_2$
5. Write three minerals (in each case) which occur as
   (a) Sulphide ....
   (b) Oxide ....
   (c) Oxosalts ....
6. You are living very near to sea-coast. What types of minerals can be present in direct supply of tap water?
   (i) .... (ii) ....
   (iii) .... (iv) ....
7. What method can be used for benefication of the following?
   (i) Haematite ....
   (ii) Cassiterite ....
   (iii) Chromite ....
   (iv) Copper pyrites ....
8. A mineral is rich in ZnS and PbS. Give scheme to concentrate it by froth flotation process.
   ....
   ....
9. What are the main steps of pyrometallurgy?
   ....
   ....
10. Explain the term hydrometallurgy with example.
    ....
    ....
11. What are the criteria of selection of a good reducing agent for extraction of metals?
    ....
    ....
12. Why is carbon reduction not used to obtain certain metals from their ores?
13. From the Ellingham diagram (Fig. 17.4) compute the temperature of reduction of MgO, CaO, TiO_2 and Al_2O_3 by carbon
    ....
    ....
14. Define the terms
    (a) **Copper matte** ....
    (b) **Blister copper** ....
    (c) **Anode-mud** ....

15. Minerals are rich in the following as given
Mineral I : Rich in AgCl, CuS, FeS
Mineral II : Rich in ZnS, CuS, FeS
Mineral III : Rich in Ag_2S, ZnS, FeS
If concentration of the above minerals are done separately without knowng the presence of their contents, metals obtained in the native state from
(I) ....
(II) ....
(III) ....
16. Chemical reactions of the hydrometallurgical extracton of copper are
....

Answers

1. Increases (NaCl in neutral; due to electrolysis NaOH is formed which being basic has pH > 7).
2. $Cu^{2+}(aq)$
3. In *A* : no reaction (Ag is not oxidised)
In *B* : Cu^{2+} is reduced to Cu (which is displaced) by Zn which is oxidised. Solution becomes colourless or fades.
4. (a) Bessemerisation (b) Mond's process (c) van Arkel
5. (a) **Sulphide** Ag_2S (argentite), HgS (cinnabar). PbS (galena)
(b) **Oxide** Cu_2O (cuprite), Fe_3O_4 (magnetite), TiO_2 (rutile)
(c) **Oxo salts** $BaSO_4$ (barite), $Ca_3(PO_4)_2$ (phosphate rock). $CaCO_3$ (calcite)
6. (i) NaCl (ii) NaBr (iii) NaI (iv) $MgCl_2$
7. (i) Magnetic separation
(ii) Gravity method
(iii) Magnetic separation or gravity method
(iv) Froth flotation process
8. NaCN is added benefication by froth flotation process (FFP). ZnS is converted to soluble complex $Na_2[Zn(CN)_4]$ and PbS goes into froth by FFP.
9. In pyrometallurgy, the concentration ore is heated to a high temperature and reduction is done with a suitable reducing agent. Steps involved are
(a) Calcination, (b) Roasting, (c) Smelting
10. Hydrometallurgy is generally used in the case of low grade ores, which cannot be concentrated economically. The powdered ore is first treated with an aqueous solution of a suitable chemical whereby the metal is obtained in the form of its soluble salts leaving behind gangue particles. (leaching)

$$CuO + H_2SO_4 \longrightarrow CuSO_4 + H_2O$$
$$AgCl + 2NaCN \longrightarrow Na[Ag(CN)_2] + NaCl$$
$$Ag_2S + 4NaCN \rightleftharpoons 2Na[Ag(CN)_2] + Na_2S$$

11. A good reducing agent for extraction of metals should possess the following properties
(i) It should be cheaper than the metal being produced.
(ii) It should be able to reduce the metal compounds to free metal in a pure form at a suitable temperature.
(iii) It should not combine with the metal being produced.
12. Carbon reduction is not used to obtain certain metals from their ores because it can combine with them to produce metal carbides.
13. MgO : 1935 K
CaO : 2445 K
TiO_2 : 1865 K
Al_2O_3 : 2570 K
14. Refer text
15. I : Ag II : Cu III : Ag
16.
$$CuO + H_2SO_4 \longrightarrow CuSO_4 + H_2O$$
$$CuCO_3 + H_2SO_4 \longrightarrow CuSO_4 + H_2O + CO_2$$
$$Cu_2S + 2H_2SO_4 + 2O_2 \longrightarrow 2CuSO_4 + SO_2 + 2H_2O$$

Practice Exercise 1

1. Look at the location of elements *A*, *B*, *C* and *D* in the following periodic table :

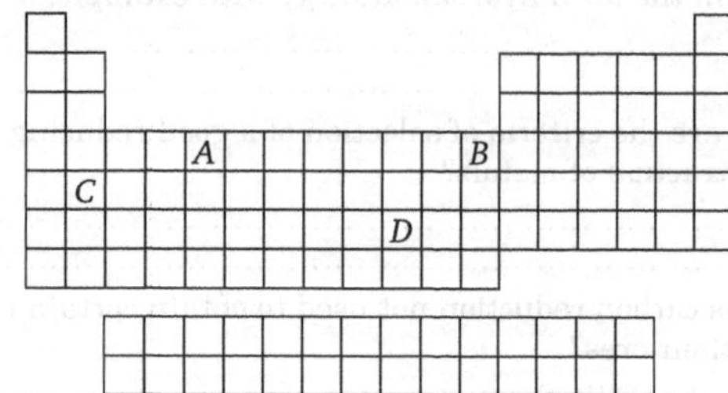

Predict whether these elements are likely to be found in nature as carbonates, oxides, sulphides or in uncombined state.
2. List three metals that are found in nature as oxide ores.
3. Explain, why early transition metals occur in nature as oxides but the late transition metals occur as sulphides.
4. Imagine a planet with an atmosphere that contains O_2 and SO_2 but no CO_2. Give the chemical composition of the minerals you would expect to find for the alkaline earth metals on such a planet.
5. Na metal can't be obtained by electrolysis of aqueous NaCl solution. Explain.
6. (A) Al_2O_3 can't be reduced by carbon to get Al metal. Explain
(B) NaCN is sometimes added in the froth flotation process as a depressant when ZnS and PbS minerals are expected because
(a) $Pb(CN)_2$ is precipitated while no effect on ZnS
(b) ZnS forms soluble complex $Na_2[Zn(CN)_4]$ while PbS forms froth
(c) PbS forms soluble complex $Na_2[Pb(CN)_4]$ while ZnS forms froth
(d) NaCN is never added in froth flotation process
7. Cu can reduce Ag^+ to metallic Ag but Ag can't reduce Cu^{2+} to metallic Cu. Explain.

8. Among the methods for extracting metals from their ores are (i) roasting a metal sulphide, (ii) chemical reduction of a metal oxide, and (iii) electrolysis. The preferred method depends on $E°$ value for the reduction half-reaction:

$$M^{n+}(aq) + ne^- \longrightarrow M(s)$$

(a) Which method is preferred for the metals with the most negative $E°$ values?

(b) Which method is preferred for the metals with the most positive $E°$ values?

9. The roasting of an ore of a metal usually results in conversion of the metal to the oxide. Why does the roasting of cinnabar, HgS, produce metallic mercury rather than an oxide of mercury?

10. A metal is in combined state as sulphide. Identify the steps (*A*), (*B*), (*C*)

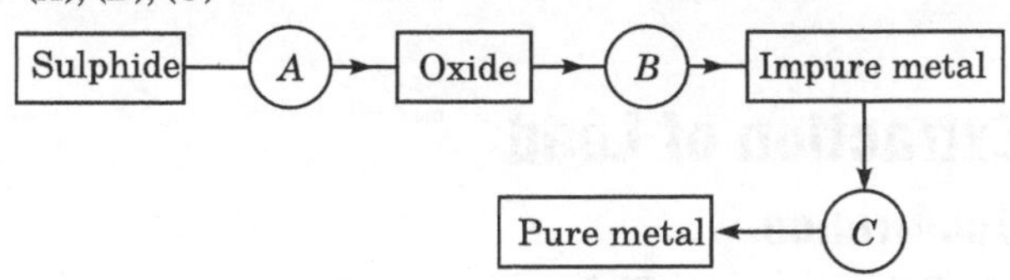

11. Write chemical equations for metallurgical processes to represent :

(a) roasting of galena (PbS),

(b) reduction of Cu_2O using charcoal as a reducing agent,

(c) deposition of pure silver from an aqueous solution of Ag^+,

12. Write chemical equations to represent the most probable outcome in each of the following. If no reaction is likely to occur, state so:

(a) $CdCO_3 \xrightarrow{\Delta}$ (b) $MgO \xrightarrow{\Delta}$

(c) $SnO_2 + CO \xrightarrow{\Delta}$ (d) $CdSO_4(aq) \xrightarrow{\text{electrolysis}}$

(e) $HgO \xrightarrow{\Delta}$ (f) $MgO + Zn \xrightarrow{\Delta}$

13. Using data in Fig. 17.4, predict whether the reduction of MgO with C is spontaneous or not at 1500°C?

14. Name two metals and two non-metals obtained by electrolysis.

15. The following reactions take place during the extraction of copper from copper ore:

(a) $2Cu_2S(l) + 3O_2(g) \longrightarrow 2Cu_2O(l) + 2SO_2(g)$

(b) $2Cu_2O(l) + Cu_2S(l) \longrightarrow 6Cu(l) + SO_2(g)$

Identify the oxidising and reducing agents.

Answers

1. *A* : Oxide *B* : Sulphide *C* : Carbonate *D* : Free metal

2. Fe as Fe_3O_4, Ti as TiO_2, Sn as SnO_2 **3.** Refer text

4. MSO_4, MSO_3

5. When aqueous NaCl solution is electrolysed, Cl_2 is formed at the anode and Na at the cathode. But Na is very reactive with H_2O forming NaOH and H_2. Thus, Na can't be prepared by electrolysis of *aq* NaCl

$$NaCl(aq) \rightleftharpoons Na^+ + Cl^-$$

At anode : $2Cl^- \longrightarrow Cl_2 + 2e^-$

At cathode : $2Na^+ + 2e^- \longrightarrow 2Na$

$$2Na + 2H_2O \longrightarrow 2NaOH + H_2$$

6. A. Al has greater affinity for O_2 that carbon has, hence C can't be used to reduce Al_2O_3 to Al. Thermodynamically for this reaction :

$$Al_2O_3 + 3C \longrightarrow 2Al + 3CO$$

$\Delta G°_f$ is positive hence, reaction is non-spontaneous.

B. (b)

7. $Cu + 2Ag^+ \longrightarrow 2Ag + Cu^{2+}$

$2Ag + Cu^{2+} \longrightarrow X$ (no reaction)

In ECS, Cu is above Ag hence Cu can reduce Ag^+ to metallic Ag.

8. (a) Electrolysis (b) Roasting a metal sulphide

9. $HgS \longrightarrow HgO \longrightarrow Hg.$

Since, Hg lies much below H in electrochemical series hence, HgO is further reduced to Hg.

10. (A) : roasting

(B) : reduction by carbon or more electropositive metal.

(C) : electrolysis

11. (a) $2PbS + 3O_2 \longrightarrow 2PbO + 2SO_2$

(b) $Cu_2O + C \xrightarrow{\Delta} 2Cu + CO$

(c) $Ag^+ + e^- \xrightarrow{\Delta} Ag$ (at cathode)

12. (a) $CdCO_3 \xrightarrow{\Delta} CdO + CO_2$

(b) No reaction.

(c) $SnO_2 + 2CO \xrightarrow{\Delta} Sn + 2CO_2$

(d) $CdSO_4(aq) \xrightarrow{\text{current}} \underbrace{H_2SO_4 + O_2}_{\text{at anode}} + \underbrace{Cd}_{\text{at cathode}}$

(e) $2HgO \xrightarrow{\Delta} 2Hg + O_2$

(f) no reaction

13. From Fig. 17.4 at 1500°C

$2C + O_2 \longrightarrow 2CO,$ $\Delta G° \approx -530$ kJ

$2MgO \longrightarrow 2Mg + O_2,$ $\Delta G° \approx +730$ kJ

$2MgO + 2C \longrightarrow 2Mg + 2CO,$ $\Delta G° = +80$ kJ

Or $MgO + C \longrightarrow Mg + CO,$ $\Delta G° = +40$ kJ mol^{-1}

non-spontaneous $\Delta G°$ being positive.

Remember $\Delta G° = \Delta G°_f$ (products) — $\Delta G°_f$ (reactants)

$\Delta G° = -$ ve forward reaction is favoured

$= +$ ve backward reaction is favoured

The positive value of $\Delta G°$ indicates that the reduction of MgO with C does not occur to a significant extent at 1500°C.

14. Metals : Na (from fused NaCl), Al (from fused $Al_2O_3 + Na_3AlF_6$)

Non-metals : Cl_2 (from fused NaCl), F_2 (from fused KHF_2 in HF)

15.

Oxidising agent	Reducing agent
(a) O_2 (itself reduced to O^{2-})	S^{2-} (in Cu_2S) oxidised to SO_2
(b) Cu_2O oxidises S^{2-} of Cu_2S to SO_2	Cu_2S, reduces Cu_2O to Cu, and itself also reduced to Cu.

Extraction of Magnesium

Occurrence

Magnesite	:	$MgCO_3$
Dolomite	:	$MgCO_3 \cdot CaCO_3$
Kieserite	:	$MgSO_4 \cdot H_2O$
Epsomite	:	$MgSO_4 \cdot 7H_2O$
Sea water	:	Mg^{2+} as SO_4^{2-} and Cl^-
Carnallite	:	$MgCl_2 \cdot KCl \cdot 6H_2O$

Main sources are **dolomite and sea water.**

Dow Sea-Water Process

Sea water contains about 0.13% Mg^{2+} ions, and the extraction of Mg depends on the fact that **$Mg(OH)_2$ is very much less soluble than $Ca(OH)_2$.** Slaked lime, $Ca(OH)_2$ is added to sea water, and calcium ions dissolve and $Mg(OH)_2$ is precipitated. This is filtered off, treated with HCl to produce $MgCl_2$, and electrolysed.

$$Ca(OH)_2 + MgCl_2 \longrightarrow Mg(OH)_2 \downarrow + CaCl_2$$

$$Mg(OH)_2 + 2HCl \xrightarrow{\Delta} MgCl_2 + 2H_2O$$

The process has been summarised in Fig. 17.6. and 17.7.

DEEP Focus

The actual electrolyte is a mixture of molten NaCl, $CaCl_2$ and $MgCl_2$. This mixture has a lower melting point and higher electrical conductivity than does $MgCl_2$ alone, but only Mg^{2+} is reduced under the conditions employed in the electrolysis.

Oxidation : $2Cl^- \longrightarrow Cl_2(g) + 2e^-$

Reduction : $Mg^{2+} + 2e^- \longrightarrow Mg(l)$

Ca(OH)2
HCl
HCl
Sea water containing Mg^{2+}
Precipitation: $Mg^{2+} + 2OH^- \longrightarrow Mg(OH)_2(s)$
$Mg(OH)_2$
Dissolving: $Mg(OH)_2(s) + 2H^+ + 2Cl^- \longrightarrow Mg^{2+} + 2Cl^- + 2H_2O$
$MgCl_2(aq)$
Evaporation: $Mg^{2+}(aq) + 2Cl^-(aq) \xrightarrow{\Delta} MgCl_2(s)$
$MgCl_2$ molten
Electrolysis: $Mg^{2+} + 2Cl^- \xrightarrow{\Delta} Mg(l) + Cl_2(g)$
$Cl_2(g)$
HCl Production: $2Cl_2(g) + 2H_2O(g) \xrightarrow{\Delta} 4HCl(g) + O_2(g)$
Product Mg

Fig. 17.6 The Dow process for the production of Mg. The main reaction sequence is traced by solid arrows. Recycling of $Cl_2(g)$ is shown by broken arrows.

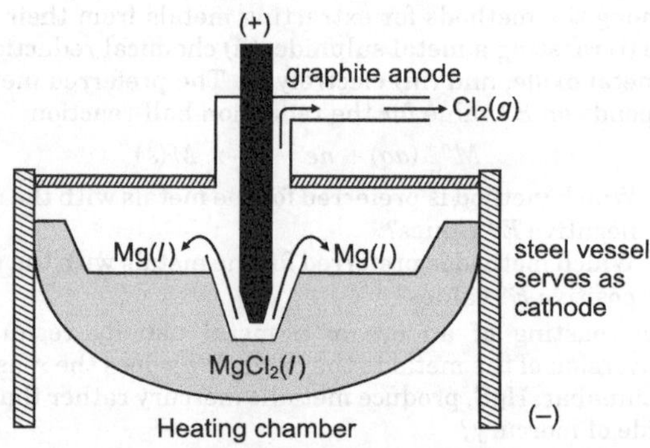

Fig. 17.7 The electrolysis of molten $MgCl_2$

Extraction of Lead

Occurrence

- Galena : PbS
- Anglesite : $PbSO_4$
- Cerussite : $PbCO_3$
- Lanakite : $PbO \cdot PbSO_4$
- Matlockite : $PbCl_2 \cdot PbO$

Main source is PbS.

Self Reduction Method (Air Reduction Process)

Step I : Concentration by **froth flotation process. (Fig 17.5)**

Step II : **Roasting** in a limited supply of air :

$$PbS + 3O_2 \longrightarrow 2PbO + 2SO_2$$

$$PbS + 2O_2 \longrightarrow PbSO_4$$

Step III : **Smelting** is done along with step II. When 50% of the PbS has been oxidised, the air supply is cut off and the temperature is increased. PbS reduces PbO and $PbSO_4$ to Pb :

$$PbS + 2PbO \longrightarrow 3Pb + SO_2$$

$$PbS + PbSO_4 \longrightarrow 2Pb + 2SO_2$$

The metal contains a number of metallic impurities : Cu, Ag, Sn, As, Sb, Bi and Zn. These are removed by cooling to near the freezing point of Pb, when first Cu and then Zn containing most of the Ag and Au solidify (Parke's method). Preferential oxidation converts As, Sb and Sn to As_2O_3, Sb_2O_3 and SnO_2 which float on the surface of the molten metal and may be skimmed off.

Further refining is done by electrolysis.

Carbon Reduction Process

Roasted ore as obtained in step II above is mixed with coke and melted.

$$PbO + C \xrightarrow{\Delta} Pb + CO$$

Extraction of Tin

Occurrence

- Cassiterite (tin stone) : SnO_2

It contains $FeWO_4 \cdot MnWO_4$ (Wolframite) as a magnetic impurity.

***Step I* : Concentration** is done in stages :

- **Electromagnetic separation** : To remove magnetic impurities ($FeWO_4 \cdot MnWO_4$)
- **Gravity separation** : To remove gangue, ore is washed with current of water.
- **Roasting** : To remove volatile impurities as oxides:

$$S + O_2 \longrightarrow SO_2$$

$$2As_2S_3 + 9O_2 \longrightarrow 2As_2O_3 + 6SO_2$$

***Step II* : Reduction (Smelting) :** Concentrated ore from step I is mixed with coal, lime stone and sand and heated in a reverberatory furnace at 1200°–1300° C

$$SnO_2 + C \longrightarrow Sn + 2CO$$

***Step III* : Refining :** Sn obtained above has some SnO_2. It is refined by :

(i) **Poling** process (molten tin is stirred with poles of green wood which reduces SnO_2 to Sn).

(ii) **Liquation** (bars of tin are heated on the hearth of a reverberatory furnace when the readily fusible tin flows leaving the impurities like Fe, Cu, W, As etc., (behind).

***Step IV* : Electro-refining**

Cathode	:	sheet of pure tin
Anode	:	blocks of impure tin
Electrolyte	:	tin sulphate ($SnSO_4$) in dil. H_2SO_4 and some phenol sulphonic acid

Pure tin from the blocks (anode) is deposited at the cathode.

Extraction of Silver

Occurrence

- Argentite (silver glance) : Ag_2S
- Horn silver (chlorargyrite) : AgCl
- Argentiferrous galena : $PbS \cdot Ag_2S$
- Pyragyrite or Ruby silver : $3Ag_2S \cdot Sb_2S_3$

Main source is **argentite**.

Cyanide Process (Based on Hydrometallurgy) Mac Arthur —Forrest Method

***Step I* : Complexation of Ag/Ag₂S /AgCl**

The powdered ore is treated with NaCN solution and O_2 is bubbled through the solution. Ag if any present in native state is oxidised to Ag^+ · Ag^+ is complexed by CN^- forming soluble $[Ag(CN)_2]^-$

$$4Ag + 8CN^- + 2H_2O + O_2 \longrightarrow 4[Ag(CN)_2]^- + 4OH^-$$

$$AgCl + 2CN^- \longrightarrow [Ag(CN)_2]^- + Cl^-$$

$$Ag_2S + 4CN^- \longrightarrow 2[Ag(CN)_2]^- + S^{2-}$$

Last reaction is reversible due to very low solubility of Ag_2S in water. Air oxidises S^{2-} to SO_4^{2-}, S and $S_2O_3^{2-}$

$$S^{2-} + 2O_2 \longrightarrow SO_4^{2-}$$

***Step II* : Precipitation of Ag**

Zn is added to soluble $[Ag(CN)_2]^-$ when Ag is displaced.

$$Zn + 2[Ag(CN)_2]^- \longrightarrow 2Ag \downarrow + [Zn(CN)_4]^{2-}$$

***Step III* : Refining by electrolysis**

Cathode	:	pure Ag
Anode	:	impure Ag
Electrolyte	:	$AgNO_3(aq)$

When electric current is passed, impure Ag (anode) is oxidised to Ag^+ and is deposited at pure Ag (cathode).

At anode : $Ag \longrightarrow Ag^+ + e^-$

At cathode: $Ag^+ + e^- \longrightarrow Ag$

In case of gold, $O_2(g)$ oxidises free metal to Au^+, which complexes with CN^-

$$4Au(s) + 8CN^- + O_2(g) + 2H_2O \longrightarrow 4[Au(CN)_2]^- + 4OH^-$$

The pure metal is then displaced from the solution by an active metal (as Zn).

$$2[Au(CN)_2]^- + Zn \longrightarrow 2Au(s) \downarrow + [Zn(CN)_4]^{2-}$$

Lead Process

The process depends upon the fact that when ores of silver are smelted down with lead, an alloy of lead and silver is formed; from this alloy lead is removed by oxidation.

$$\text{ores of silver} \xrightarrow[\text{lead}]{\text{smelted with}} (Pb-Ag)\text{ alloy} \xrightarrow{\text{cupellation}} \text{silver}$$

- (Pb–Ag) alloy rich in silver : **Cupellation process is used**
- (Pb–Ag) alloy poor in silver : **Parke's or Pattinson process is used**

Mexican Amalgamation Process

Mercury and magistral (burnt pyrites sulphates and oxides of copper and iron) are added to the powdered mineral containing H_2O and NaCl. The mixture is kept for several days. Silver is formed in the amalgam form. On washing, drying and subsequent distillation, silver is obtained.

$$CuSO_4 + 2NaCl \longrightarrow CuCl_2 + Na_2SO_4$$
$$CuCl_2 + Ag_2S \longrightarrow Cu_2Cl_2 + 2AgCl + S$$
$$Cu_2Cl_2 + Ag_2S \longrightarrow Cu_2S + 2AgCl$$
$$2AgCl + 2Hg \longrightarrow Hg_2Cl_2 + 2Ag$$
$$Ag + Hg \longrightarrow Ag(Hg) \xrightarrow{\text{distill}} Ag\downarrow + Hg\uparrow$$

Desilverisation of Lead : When lead-silver alloy is poor in silver (such as argentiferrous lead or galena), desilverisation of lead is affected by **Parke's process**. It depends upon the fact that:

(i) when zinc is added to a melted alloy of silver and lead, zinc takes away silver from the alloy and itself forms an alloy with silver and not with lead

(ii) alloy of zinc and silver melts at a higher temperature than lead

(iii) the alloy of silver and zinc is lighter than lead and forms two separate layers

(iv) on distillation of silver and zinc alloy, zinc separates

Pattinson's Process

It depends upon the fact that the alloy of lead and silver containing less than 1.8 per cent of silver, melts at a lower temperature than pure lead. If this type of alloy is melted, and the melted alloy is allowed to cool, lead first separates and repeating the process, one becomes richer in silver. Further recovery is done by cupellation process.

Extraction of Aluminium

Occurrence

Bauxite	:	$Al_2O_3 \cdot 2H_2O$
Kaolin	:	$Al_2O_3 \cdot 2SiO_2 \cdot 2H_2O$
Cryolite	:	Na_3AlF_6
Felspar	:	$K_2O \cdot Al_2O_3 \cdot 6SiO_2$
Corundum	:	Al_2O_3
Orthoclase	:	$KAlSi_3O_8$
Beryl	:	$Be_3Al_2Si_6O_{18}$

Main source is bauxite.

Felspar or orthoclase on weathering gives rise to porcelain-clay, China-clay or Kaolin.

Purification of Bauxite

Baeyer's Process

Step I : Bauxite contains Fe_2O_3 and TiO_2 as impurities. Al_2O_3 is amphoteric and dissolves in NaOH. The other oxides are basic oxides and therefore, insoluble in NaOH.

$$Al_2O_3 + 2OH^- + 3H_2O \longrightarrow \underset{\text{soluble}}{2[Al(OH)_4]^-}$$

Insoluble oxides (as Fe_2O_3) are removed by filtration

Step II : Soluble $[Al(OH)_4]^-$ is diluted and then acidified when $Al(OH)_3$ is precipitated which on ignition at 1250°C gives pure alumina :

$$Al(OH)_4^- + H_3O^+ \longrightarrow Al(OH)_3\downarrow + 2H_2O$$
$$2Al(OH)_3 \xrightarrow{1250°C} Al_2O_3 + 3H_2O$$

The complex is decomposed into $Al(OH)_3$ by passing CO_2 through it

$$2[Al(OH)_4]^- + CO_2 \longrightarrow 2Al(OH)_3 + CO_3^{2-} + H_2O$$

Serpeck's Process

$$Al_2O_3 + 3C + N_2 \xrightarrow{\Delta} 2AlN + 3CO\uparrow$$
$$AlN + 3H_2O \longrightarrow Al(OH)_3 + NH_3$$
$$2Al(OH)_3 \xrightarrow{\Delta} Al_2O_3 + 3H_2O$$

Electrolysis of Al_2O_3 (Hall-Heroult Process)

Al is usually extracted by this process. Al_2O_3 is melted with cryolite $Na_3[AlF_6]$ and electrolysed in a graphite lined steel tank, which serves as cathode. The anodes are also made of graphite. The cell runs continuously and at intervals molten Al (m.p. 660°C) is drained from the bottom of the cell and more bauxite is added (Fig. 19.8). Cryolite improves the electrical conductivity of the cell as Al_2O_3 is poor conductor. This also serves as an added impurity and lowers the m.p. of the mixture to about 950°C.

Al of 99.6 to 99.8% purity is obtained. Probable net reactions are.

Oxidation at anode:

$$C + 2O^{2-} \longrightarrow CO_2 + 4e^-] \times 3$$

Reduction at cathode:

$$Al^{3+} + 3e^- \longrightarrow Al] \times 4$$

$$\overline{3C + 4Al^{3+} + 6O^{2-} \longrightarrow 4Al + 3CO_2}$$

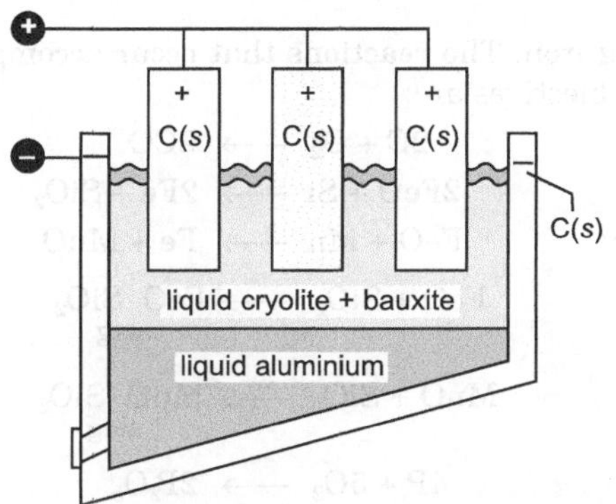

Fig. 17.8 Electrolysis cell for aluminium production. The cathode is a carbon lining in a steel tank. The anodes are also made of carbon. Liquid aluminium is more dense than the electrolyte medium and collects at the bottom of the tank. A crust of frozen electrolyte forms at the top of the cell.

Reaction can also be written as

Anodo (oxidation)

$$2O^{2-} \longrightarrow O_2 + 4e^-] \times 3$$

Cathode (reduction)

$$Al^{3+} + 3e^- \longrightarrow Al] \times 4$$

Overall $\quad 2Al_2O_3 \longrightarrow 4Al + 3O_2$

Oxygen gas reacts with carbon anodes (at elevated temperature) to form CO

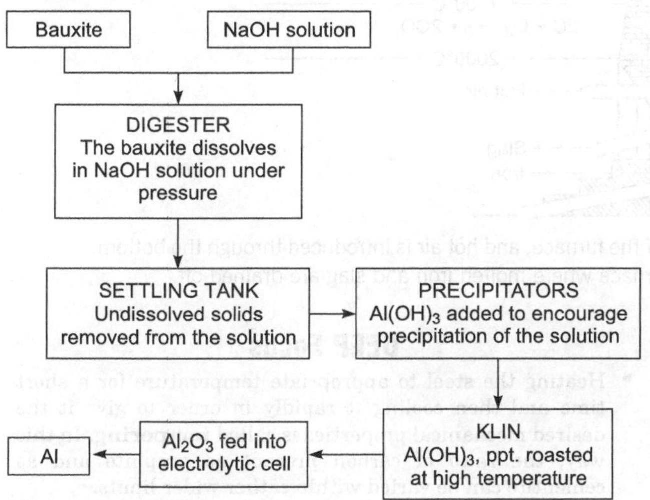

Extraction of Iron

Occurrence

Haematite	:	Fe_2O_3
Magnetite	:	Fe_3O_4
Limonite	:	$FeO(OH)$ or $Fe_2O_3 \cdot 3H_2O$
Siderite	:	$FeCO_3$
Pyrites (fool's gold)	:	FeS_2
Pyrrhotite	:	FeS
Ilmenite	:	$FeTiO_3$

Oxide-ores are main sources of iron.

Carbon Reduction Method

***Step I* : Concentration** by

- gravity separation method (which removes gangue)
- magnetic separator (which enriches it by removing non-magnetic impurities)

***Step II* : Roasting** is done

- to decompose carbonates to Fe_2O_3 and also to remove moisture :

$$4FeCO_3 + O_2 \longrightarrow 2Fe_2O_3 + 4CO_2$$

- to oxidise volatile impurities :

$$4P + 5O_2 \longrightarrow 2P_2O_5$$
$$4As + 3O_2 \longrightarrow 2As_2O_3$$
$$S + O_2 \longrightarrow SO_2$$

***Step III* : Reduction** or **Smelting**

It is done in a blast furnace (Fig. 17.9), the reactions taking place in a blast furnace (along with working temperature) are :

- Formation of reducing agents, principally CO(*g*) and $H_2(g)$:

$$2C + O_2 \longrightarrow 2CO \quad (1700°C)$$
$$C + CO_2 \longrightarrow 2CO \quad (> 1000°C)$$
$$C + H_2O \longrightarrow CO + H_2 \quad (> 600°C)$$

- Reduction of iron oxide :

$$3CO + Fe_2O_3 \longrightarrow 2Fe + 3CO_2 \quad (900°C)$$
$$3H_2 + Fe_2O_3 \longrightarrow 2Fe + 3H_2O \quad (900°C)$$

- Slag formation to remove impurities from ore :

$$CaCO_3 \longrightarrow CaO + CO_2 \quad (800-900°C)$$
$$CaO + SiO_2 \longrightarrow CaSiO_3(l) \quad (1200°C)$$
$$3CaO + P_2O_5 \longrightarrow Ca_3(PO_4)_2(l) \quad (1200°C)$$

- Impurity formation in iron :

$$MnO + C \longrightarrow Mn + CO \quad (1400°C)$$
$$SiO_2 + 2C \longrightarrow Si + 2CO \quad (1400°C)$$
$$P_2O_5 + 5C \longrightarrow 2P + 5CO \quad (1400°C)$$

- Iron obtained from a blast furnace is called **pig iron**. It contains about 95% Fe, 4% C and varying quantities of other impurities. **Cast iron** can be obtained by pouring pig iron directly into molds of the desired shape. Cast iron is very hard and brittle and can be used where it will not be subjected to mechanical or thermal shock.
- Wrought iron is prepared by **"Pudding process"** in which cast iron is fused in a reverberatory furnace, the hearth of which is lined with haematite. Reactions taking place are:

$$3C + Fe_2O_3 \longrightarrow 2Fe + 3CO$$
$$3Mn + Fe_2O_3 \longrightarrow 2Fe + 3MnO$$
$$3Si + 2Fe_2O_3 \longrightarrow 4Fe + 3SiO_2$$
$$MnO + SiO_2 \longrightarrow MnSiO_3 \text{ (slag).}$$

- **Pig iron** is converted into steel by following changes :
  1. Reduce the carbon content from 3–4% in pig iron to 0–1.5% in steel.
  2. Remove, through slag formation, Si, Mn, and P (each present in pig iron to the extent of 1% or so), together with other minor impurities.
  3. Add alloying elements (such as Cr, Ni, Mn, V, Mo, and W) to give the steel its desired end properties.

The most important method of steel making today is the **basic oxygen process (BOP).** The process is carried out in a steel vessel with a refractory lining (usually made of dolomite, $CaCO_3 \cdot MgCO_3$). Oxygen gas at about 10 atm pressure and a stream of powdered limestone are fed through a water-cooled lance and discharged above the molten pig iron. The reactions that occur accomplish the first two objectives are:

$$2C + O_2 \longrightarrow 2CO$$
$$2FeO + Si \longrightarrow 2Fe + SiO_2$$
$$FeO + Mn \longrightarrow Fe + MnO$$
$$FeO + SiO_2 \longrightarrow \underset{\text{slag}}{FeO \cdot SiO_2}$$
$$MnO + SiO_2 \longrightarrow \underset{\text{slag}}{MnO \cdot SiO_2}$$
$$4P + 5O_2 \longrightarrow 2P_2O_5$$
$$3CaO + P_2O_5 \longrightarrow \underset{\text{slag}}{Ca_3(PO_4)_2}$$

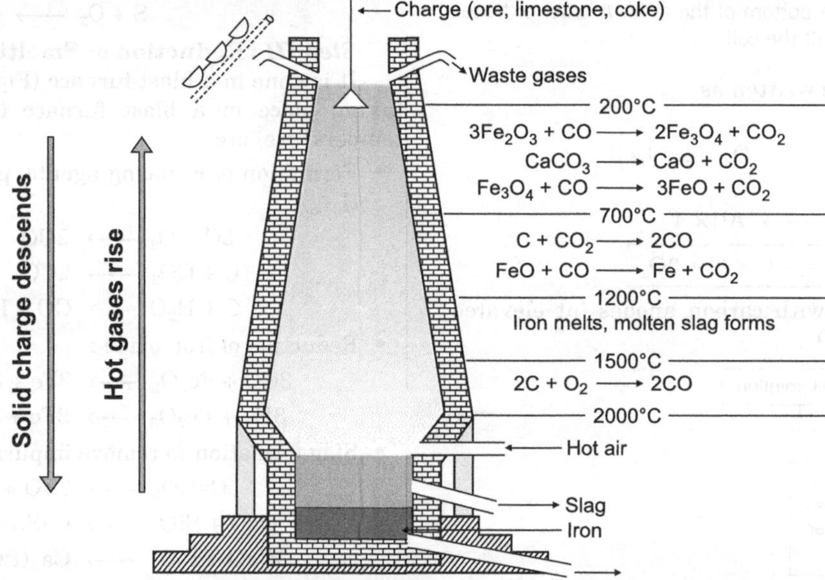

Fig. 17.9 Iron ore, coke, and limestone are added at the top of the furnace, and hot air is introduced through the bottom. Maximum temperatures are attained near the bottom of the furnace where molten iron and slag are drained off.

The properties of steel depend on its chemical composition but also on heat treatment. At high temperatures, iron and carbon in steel combine to form iron carbide, Fe_3C, called cementite:

$$3Fe(s) + C(s) \rightleftharpoons Fe_3C(s)$$

The forward reaction is endothermic, so that the formation of cementite is favoured at high temperatures. When steel containing cementite is cooled slowly, the above equilibrium shifts to the left, and the carbon separates as small particles of graphite, which give the steel a gray colour. If the steel is cooled rapidly, equilibrium is not attained and the carbon remains large in the form of cementite, Fe_3C. **Steel containing cementite is light in colour, and is harder and more brittle than that containing graphite.**

DEEP Focus

- Heating the steel to appropriate temperature for a short time and then cooling it rapidly in order to give it the desired mechanical properties is called **tempering.** In this way, the ratio of carbon present as graphite and as cementite can be varied within rather wider limits.

Based on % of carbon, steels are named

Name	%C
Mild	0.15–0.3
Medium	0.3–0.6
High-Carbon	0.6–0.8
Tool	0.5–1.4

Table 17.8 shows the composition, properties, and uses of various types of steel.

Table 17.8 **Types of Steel**

Composition	Type-mass%		
	Plain	High strength	Stainless
C	1.35	0.25	0.03–1.2
Mn	1.65	1.65	1.0–10
P	0.04	0.04	0.04–0.06
S	0.05	0.05	0.03
Si	0.06	0.15–0.90	1–3
Ni	—	0.4 –1.0	1.22
Cr	—	0.3 –1.3	4.0–27
Others	Cu (0.2 – 0.6)	Cu (0.01–0.08)	—
Uses	sheet products, tools	construction, steel turbines	kitchen utensils, razor blades

Extraction of Zinc

Zinc occurs in the earth's crust to the extent of 132 ppm by weight. It is the twenty-fourth most abundant element.

Occurrence

- Zinc blende (Sphaelerite) : $(ZnFe)S$
- Smithsonite (Calamine) : $ZnCO_3$
- Hemimorphite : $Zn_4(OH)_2(Si_2O_7)\cdot H_2O$

Main source is zinc blende and is extracted in following steps :

Step I : Concentration by froth-flotation process

Step II : **Roasting**

$$ZnS \longrightarrow ZnO + SO_2$$

SO_2 is the important side product and is used to manufacture H_2SO_4.

$$2SO_2 + O_2 \longrightarrow 2SO_3 \xrightarrow{2H_2O} 2H_2SO_4$$

Step III : **Reduction of ZnO in a smelter**

$$ZnO + CO \xrightleftharpoons{1200°C} Zn + CO_2$$

Above reaction is reversible and thus high temperature is required to shift the equilibrium towards formation of Zn. Thermodynamically :

$$\Delta G° = \Delta H° - T\Delta S°$$

To make $\Delta G°$ negative (for spontaneous reaction)

$$T\Delta S > \Delta H \quad \text{and} \quad T > \frac{\Delta H°}{\Delta S°}$$

If we use standard thermodynamic data, T should exceed 1200°C. Zn and CO_2 both are in gaseous state at this temperature. Reversible nature of the above reaction can be minimised :

(a) by having excess of carbon so that CO_2 formed is converted to CO

$$C + CO_2 \longrightarrow 2CO$$

(b) by cooling the gaseous products so that equilibrium is not attained immediately.

Zinc obtained in this way is of 99% purity.

Step IV : **Refinement by Electrolysis**

ZnO is dissolved in dil. H_2SO_4 and electrolysed.

Anode : Impure zinc plate

Cathode : Pure zinc plate

Electrolyte : $ZnSO_4/H_2SO_4$

Zinc from impure zinc plate is oxidised at anode, goes into solution and reduced to pure zinc at cathode.

Anode : $Zn \longrightarrow Zn^{2+} + 2e^-$

Cathode : $Zn^{2+} + 2e^- \longrightarrow Zn$

Target Practice 2

1. Sulphide ore is concentrated by the method ....
2. Name the following process

$$SnO_2 + C \xrightarrow{1500\ K} Sn + 2CO \quad ....$$

3. Ag is precipitated when Zn is added to $[Ag(CN)_2]^-$. Write the reaction............
4. In the extraction of silver, some lead is also present as impurity in final step. This lead can be removed by (i) .... (ii) ....
5. Carbon can't be used to reduce Al_2O_3 to Al while ZnO can be reduced by carbon. It is due to ....
6. What is the name of the process involving following reaction :
$CaO\cdot MgO(s) + Fe/Si \xrightarrow{\Delta} Mg(l) + Fe + Ca_2SiO_4$

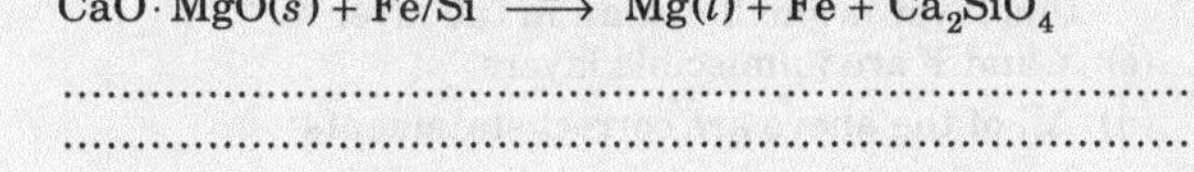

7. Based on following Ellingham diagram, indicate the temperature favourable for the reduction of metal oxide to metal.

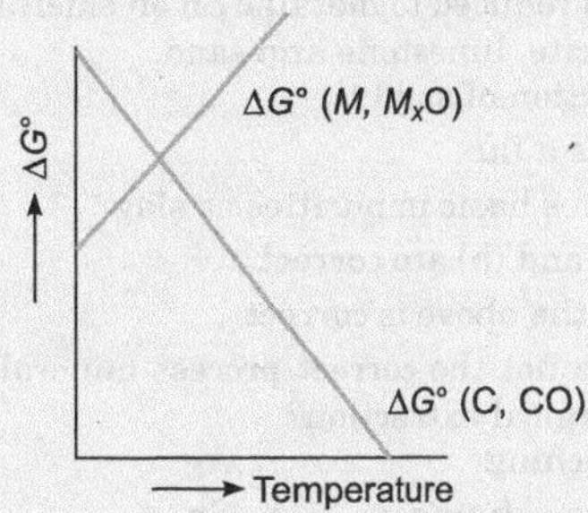

8. What is the principle of the following reaction?

$$Ca(OH)_2 + MgCl_2 \longrightarrow Mg(OH)_2 + CaCl_2$$

....

....

9. Alumina (Al_2O_3) and silica (SiO_2) get dissolved in a concentrated solution of NaOH under pressure.

(a) In what forms are these soluble? Write reactions.

....

....

(b) $Al(OH)_3$ is precipitated when CO_2 is passed or by seeding with some freshly precipitated $Al(OH)_3$. Explain.

....

....

(c) Is silica precipitated by treatment in (b).

10. Match the following properties of Al with the uses to which the metal or its compounds are put.

(i) Good thermal conductivity	(a) Building ships and aircrafts
(ii) Good electrical conductivity	(b) Utensils
(iii) Low density and resistance to corrosion	(c) Electric wires
(iv) Non-toxicity	(d) Adsorbent
(v) Gel nature of $Al(OH)_3$	(e) Food-packaging

11. In the production of iron, intermediate are oxides. Name the oxides of iron

(a) .... (b) ....

12. Basic impurities are removed as slag in the production of iron. Write various formulae of slags.

(a) .... (b) ....

(c) ....

Answers

1. Froth flotation process
2. Smelting (reduction)
3. $2[Ag(CN)_2]^- + Zn \longrightarrow [Zn(CN)_4]^{2-} + 2Ag$
4. (i) Cupellation process if rich in silver
   (ii) Parke's or Pattinson process is used.
5. Al has greater reactivity with oxygen hence, can't be reduced by carbon
6. Pidgeon process.
7. The formation of carbon monoxide from carbon can reduce the metal oxide to metal at a temperature higher than the point of intersection of the two lines.
   (To make $\Delta G = -$ ve, $T\Delta S > \Delta H$)
   Sine $\Delta G = \Delta H - T\Delta G$
8. Solubility of $Mg(OH)_2$ is lower than that of $Ca(OH)_2$.
9. (a) $Al_2O_3 + 2NaOH + 3H_2O \longrightarrow 2NaAl(OH)_4$
   $SiO_2 + 2NaOH \longrightarrow Na_2SiO_3 + H_2O$
   (b) $2NaAl(OH)_4 + CO_2 \longrightarrow 2Al(OH)_3 \downarrow + Na_2CO_3 + H_2O$
   (c) Silica is more acidic oxide than alumina hence, silicate remains in solution.
10. (i) (b) (ii) (c) (iii) (a) (iv) (e) (v) (d)
11. (a) FeO (b) Fe_2O_3 (c) Fe_3O_4
12. (a) $CaSiO_3$ (b) $MgSiO_3$ (c) $MnSiO_3$

Practice Exercise 2

1. Pb can also be obtained by reduction of roasted ore with coke. Outline the process.

2. Considering 100% pure, which of the following has maximum percentage of Pb?

(a) PbS (b) $PbSO_4$

(c) $PbCO_3$ (d) PbO : $PbSO_4$

3. SnO_2 can be precipitated as $Sn(OH)_4$. Can this be used to concentrate tin stone.

4. (A) Tin stone is amphoteric. Explain.

(B) SnO_2 is reduced to metallic Sn on smelting oxide with anthracite, limestone and sand.

In this, function of sand is:

(a) it acts as a flux

(b) it removes basic impurities as slag

(c) both (a) and (b) are correct

(d) none of the above is correct

(C) Which is not the correct process-mineral matching in metallurgical extraction?

(a) Leaching : Ag

(b) Zone-refining : Sn

(c) Liquation : Sn

(d) van Arkel : Zr

5. Scrap tin plate can be recycled by dissolving the tin in NaOH solution, followed by electrolysis. Write equations for the reactions involved.

6. To prevent the air oxidation of aqueous solutions of Sn^{2+} to Sn^{4+}, sometimes metallic tin is kept in contact with the Sn^{2+}. Suggest how this helps to prevent the oxidation ?

7. Gold is also extracted by cyanide process as in case of silver. Outline the reactions.

8. When 1 mol of pure Ag_2CO_3 is strongly heated, calculate weight of residue obtained.

9. How is Ag extracted from silver coin?

10. $(Ag + Pb) \text{ alloy} \xrightarrow{\text{melt and add zinc}} (Ag + Pb + Zn) \text{ melt} \xrightarrow{\text{Cool}}$ Layer *X* / Layer *Y*

Select correct statement based on above scheme:

(a) Layer *X* contains zinc and silver

(b) Layer *Y* contains lead and silver but amount of silver in this layer is smaller than in the layer *X*.

(c) *X* and *Y* are immiscible layers

(d) All of the above are correct statements

11. In the purification of bauxite ore as preliminary step in the production of Al, $[Al(OH)_4]^-$ can be converted to $Al(OH)_3$ by passing CO_2 through it. Write an equation for the reaction that occurs.

12. Addition of NH_4Cl can also precipitate $Al(OH)_3$ from soluble $Na[Al(OH)_4]$. Explain.

13. Coke does not cause reduction of Al_2O_3. Explain.

ΔG_f° (in kJ mol^{-1}) for

Al_2O_3 : –1582

CO : –137.2

14. What happens when $Al(NO_3)_3$ is heated?

15. In the **Baeyer's process**:
(a) Al_2O_3 goes into solution as soluble $Al(OH)_4^-$ while other basic oxides as TiO_2 and Fe_2O_3 remain insoluble.
(b) Al_2O_3 changes to AlN which in turn decomposed by H_2O
(c) Al_2O_3 changes to $Al_2(CO_3)_3$ which changes to $AlCl_3$
(d) none of the above is correct

16. In electrolysis of Al_2O_3 by Hall-Heroult process.
(a) Cryolite $Na_3[AlF_6]$ lowers the melting point of Al_2O_3 and increases its electrical conductivity.
(b) Al is obtained at cathode and probably CO_2 at anode
(c) both (a) and (b) are correct
(d) none of the above is correct

17. $$\boxed{\text{Bauxite}} \xrightarrow[\uparrow N_2]{+C} \boxed{\text{AlN}} (\uparrow CO) \xrightarrow{H_2O} Al(OH)_3 \xrightarrow{\Delta} Al_2O_3$$

This flow-sheet is for
(a) Baeyer's process (b) Serpeck's process
(c) Hall's process (d) Kroll's process

18. Which mineral has been named incorrectly?
(a) Bauxite : $Al_2O_3 \cdot 2H_2O$
(b) Corundum : Al_2O_3
(c) Cryolite : $3NaF \cdot AlF_3$
(d) Felspar : $Be_3Al_2Si_6O_{18}$

19. Identify (*A*) to (*F*) in the following :

$$FeSO_4 \xrightarrow{\Delta} \underset{\text{solid}}{(A)} + \underbrace{(B) + (C)}_{\text{gas}}$$

$$(B) \xrightarrow{O_2} (C)$$

$$(A) \xrightarrow{H_2O} (D) \text{ brown ppt}$$

$$(D) \xrightarrow{\Delta} (A)$$

$$(A) \xrightarrow{HCl} \text{deep yellow solution } (E)$$

$$(E) \xrightarrow{K_4[Fe(CN)_6]} \text{blue ppt. } (F)$$

20. It is possible to identify an **"Iron cycle"** of the same kind as a water or nitrogen cycle. Suggest some stages in the cycle.

21. Al is costlier than Fe but Al is sacrificed to convert Fe_2O_3 into Al. Why this change is possible? Name this process.

$$\Delta G^\circ_f (Fe_2O_3) = -742.2 \text{ kJ mol}^{-1}$$
$$\Delta G^\circ_f (Al_2O_3) = -1582 \text{ kJ mol}^{-1}$$

22. When iron ore is reduced in a blast furnace, some of the SiO_2 impurity is also reduced by reaction with carbon to give elemental silicon and carbon monoxide. The silicon is subsequently reoxidised in basic oxygen process, and the resulting SiO_2 reacts with CaO, yielding slag, which is then separated from the molten steel. Write balanced equations for the three reactions involving SiO_2.

23. Although iron is only about two-third as abundant as aluminium in Earth's crust, mass for mass it costs only about one-quarter as much to produce. Why?

24. Dolomite ($MgCO_3 \cdot CaCO_3$) can also be treated to get $MgCl_2$ which in turn is electrolysed to get Mg. (Dow natural brine process). Give reactions of this process

25. $$[CaCO_3 \cdot MgCO_3] \xrightarrow{\Delta} CaO \cdot MgO \xrightarrow[1150°C]{Fe/Si} Mg + Fe + Ca_2SiO_4$$

Name this process.

Answers

1. $PbS \xrightarrow[O_2]{\text{roasting}} PbO, PbSO_4$. This mixture is smelted with coke, SiO_2 and $CaCO_3$

$$(PbO + PbSO_4) \xrightarrow[\text{silica}]{SiO_2} PbSiO_3$$

$$PbO + C \longrightarrow Pb + CO \uparrow$$

$$PbO + CO \longrightarrow Pb + CO_2 \uparrow$$

$$CaCO_3 \longrightarrow CaO + CO_2$$

$$CaO + PbSiO_3 \longrightarrow \underset{\text{slag}}{CaSiO_3} + PbO$$

2.

Ore	% of Pb
(a) PbS	86.55 (maximum)
(b) $PbSO_4$	68.21
(c) $PbCO_3$	77.44
(d) $PbO \cdot PbSO_4$	78.62

3. $SnO_2 \xrightarrow[\Delta]{H_2O} Sn(OH)_4 \xrightarrow{\Delta} SnO_2$

This process will not precipitate impurities hence, **precipitation** of $Sn(OH)_4$ and **subsequent heating** will enrich SnO_2.

4. (A) SnO_2 is soluble in acid and alkali both hence, amphoteric.

$$\underset{\text{basic}}{SnO_2} + 4HCl \longrightarrow SnCl_4 + 2H_2O$$

$$\underset{\text{acidic}}{SnO_2} + 2NaOH \longrightarrow Na_2SnO_3 + H_2O$$

(B) (c) (C) (b)

5. $Sn + 2NaOH \longrightarrow Na_2SnO_2 + H_2$

$Na_2SnO_2 \xrightarrow{\text{electrolysis}} Sn$ (at cathode)

6. $Sn^{4+} + Sn \longrightarrow 2Sn^{2+}$
Sn reduces Sn^{4+} to Sn^{2+}.

7. In the cyanide process, crushed rock containing traces of gold is treated with 0.1–0.2% solution of NaCN and aerated. O_2 (air) oxidises free metal to Au^+ which complexes with CN^-

$$4Au + 8CN^- + 2H_2O + O_2 \longrightarrow 4[Au(CN)_2]^- + 4OH^-$$

The pure metal is displaced from solution by an active metal.

$$2[Au(CN)_2]^- + Zn \longrightarrow [Zn(CN)_4]^{2-} + 2Au\downarrow$$

8. $$Ag_2CO_3 \xrightarrow{\Delta} Ag_2O + CO_2\uparrow$$

$$Ag_2O \xrightarrow{\Delta} 2Ag + \frac{1}{2}O_2\uparrow$$

Residue is metallic Ag.

$$1 \text{ mol } Ag_2CO_3 = 2 \text{ mol Ag}$$
$$= 216 \text{ g Ag}$$

9. $$\text{Silver coin } (Cu + Ag) \xrightarrow{\text{conc. } HNO_3} (Cu^{2+} + Ag^+)NO_3^- \xrightarrow{NaCl} AgCl\downarrow \text{ (white ppt)}$$

$$2AgCl + Na_2CO_3 \longrightarrow Ag_2CO_3 + 2NaCl$$

$$Ag_2CO_3 \xrightarrow{\Delta} 2Ag\downarrow + CO_2\uparrow + \frac{1}{2}O_2\uparrow$$

Or use cyanide process with **AgCl.**

10. (d)

11. Aqueous solution of CO_2 is acidic. Hence, CO_2 when passed into $[Al(OH)_4]^-$ solution makes it acidic hence, $Al(OH)_3$ is precipitated.

$$CO_2 + 2H_2O \longrightarrow H_3O^+ + HCO_3^-$$
$$[Al(OH)_4]^- + H_3O^+ \longrightarrow Al(OH)_3\downarrow + 2H_2O$$

12. NH_4Cl solution is acidic due to hydrolysis :

$$NH_4Cl + H_2O \rightleftharpoons NH_4OH + HCl.$$

hence addition of NH_4Cl solution precipitates $Al(OH)_3$ when added to $NaAl(OH)_4$

$$Na[Al(OH)_4] + HCl \longrightarrow NaCl + Al(OH)_3\downarrow + H_2O$$

13. $$Al_2O_3 + C \longrightarrow 2Al + 3CO$$

$$\Delta G^\circ = 3\Delta G^\circ_f(CO) - \Delta G^\circ_f(Al_2O_3)$$
$$= -3(137.2) + 1582$$
$$= +ve$$

hence, reaction is not spontaneous.

14. $$2Al(NO_3)_3 \xrightarrow{\Delta} Al_2O_3 + 6NO_2 + 1.5O_2$$

15. (a) **16.** (c) **17.** (b) **18.** (d)

19. $(A): Fe_2O_3$ $(B): SO_2$ $(C): SO_3$ $(D): Fe(OH)_3$ $(E): FeCl_3$ $(F): KFe^{III}[Fe^{II}(CN)_6]$, Prussian blue.

20. **The cycle might be :**

Iron oxide $\longrightarrow$ Fe $\longrightarrow$ iron and steel products $\longrightarrow$ rusting $\longrightarrow$ iron oxide.

21. $$2Al + Fe_2O_3 \longrightarrow 2Fe + Al_2O_3$$

$$\Delta G^\circ = \Delta G^\circ_f(Al_2O_3) - \Delta G^\circ_f(Fe_2O_3)$$
$$= -1582 + 742.2$$
$$= -ve$$

hence, this reduction of Fe_2O_3 by Al is possible. This is called **Thermite welding.**

DEEP Focus

Remember:

* $\Delta G^\circ = \Delta G^\circ_f$ *(products)* — ΔG°_f *(reactants)*
* $\Delta G^\circ = +ve$, *then backward reaction*
$= -ve$, *then forward reaction*

22. $$SiO_2(s) + 2C(s) \longrightarrow Si(s) + 2CO(g)$$
$$Si(s) + O_2(g) \longrightarrow SiO_2(s)$$
$$CaO(s) + SiO_2(s) \longrightarrow CaSiO_3(l) \text{ (slag)}$$

23. Iron production does not require electrolysis.

24. $$MgCO_3 \cdot CaCO_3 \xrightarrow[\Delta]{\text{calcination}} MgO + CaO + 2CO_2\uparrow$$

$$MgO + CaO \xrightarrow{HCl} MgCl_2 + CaCl_2 \xrightarrow{CO_2} MgCl_2 + CaCO_3\downarrow$$

[On passing CO_2 into *aq* $MgCl_2$ and $CaCl_2$ solution, $CaCO_3$ is precipitated leaving $MgCl_2$ soluble].

25. It is called Pidgeon process.

Total Practice Set
(Read, Plan & Solve!)

Problem 1. Following method of extracting Zn is based on thermodynamics:

(A) $2ZnS + 3O_2 \longrightarrow 2ZnO + 2SO_2$

(B) $ZnO + C \longrightarrow Zn + CO$

ΔG°_f (standard free energies of formation, in kJ mol^{-1}) of

$ZnS = -205.4$; $ZnO = -318.2$

$SO_2 = -300.4$; $CO = -137.3$

Calculate free energy changes of the above reactions and comment on the result.

Solution **For reaction (*A*)**

$$\Delta G^\circ = 2\Delta G^\circ_f(ZnO) + 2\Delta G^\circ_f(SO_2) - 2\Delta G^\circ_f(ZnS)$$
$$= 2[-318.2 - 300.4 + 205.4]$$
$$= -826.4 \text{ kJ} \qquad (\Delta G^\circ_f(\text{element}) = 0)$$

For reaction (*B*)

$$\Delta G^\circ = \Delta G^\circ_f(CO) - \Delta G^\circ_f(ZnO)$$
$$= -137.3 + 328.2$$
$$= +180.9 \text{ kJ}$$

For reaction (*A*), ΔG° is negative hence, reaction is spontaneous; reaction (*B*) is non-spontaneous ΔG° being positive for it.

DEEP Focus

Again remember :

* $\Delta G^\circ = \Delta G^\circ_f$ (products) $- \Delta G^\circ_f$ (reactants)
* $\Delta G^\circ = +ve$ indicates non-spontaneous reaction (in forward direction)
* $\Delta G^\circ = -ve$ indicates spontaneous reaction (in forward direction)

Problem 2. Explain the difference between **hydro-metallurgy** and **pyro-metallurgy**.

Solution The metallurgical procedure based on the roasting of an ore, followed by its reduction (usually with carbon), is known as **pyro-metallurgy.** A key consideration in pyro-metallurgy is the temperature at which reaction becomes spontaneous. As the term **"pyro"** suggests, often this temperature is quite high. Some of the characteristics of this process are :

- large quantities of waste material generated in concentrating low grade ores;
- high energy consumption to maintain high temperatures necessary for roasting and reduction of ores;
- gaseous emissions, such as SO_2 in roasting, that must be controlled.

In **hydro-metallurgy,** the materials handled are water and aqueous solutions at moderate temperatures rather than dry materials at high temperatures. It involves :

- leaching with acid/water;
- purification and/or concentration;
- precipitation.

Low grade oxide, carbonate and sulpohide ores of copper are treated with dilute sulphuric acid in the presence of oxygen :

$$CuO + H_2SO_4 \longrightarrow CuSO_4 + H_2O$$
$$CuCO_3 + H_2SO_4 \longrightarrow CuSO_4 + CO_2 + H_2O$$
$$Cu_2S + 2H_2SO_4 + 2O_2 \longrightarrow 2CuSO_4 + SO_2 + 2H_2O$$

When the silver ore, AgCl, is treated with an aqueous solution of sodium cyanide, AgCl dissolves in it due to the formation of $Na[Ag(CN)_2]$:

$$AgCl + 2NaCN \longrightarrow Na[Ag(CN)_2] + NaCl$$

Sulphide ore, Ag_2S, dissolves only slowly as the reaction is reversible :

$$Ag_2S + 4NaCN \rightleftharpoons 2Na[Ag(CN)_2] + Na_2S$$

If air is passed through this solution, sodium sulphide is oxidised to sodium sulphate and the forward reaction goes to completion dissolving all the sulphide ore. In the presence of air, native silver is also leached out in the form of $Na[Ag(CN)_2]$

$$4Ag + 8NaCN + 2H_2O + O_2 \longrightarrow 4Na[Ag(CN)_2] + 4NaOH$$

The leached out metals are recovered from the solution either by precipitation on treatment with a more electropositive metal or by electrolysis. For example, copper can be recovered from its solution by adding metals like Fe, Al, etc. Silver is obtained from its solution by treatment with Zn or Al :

$$CuSO_4 + Fe \longrightarrow Cu + FeSO_4$$
$$2Na[Ag(CN)_2] + Zn \longrightarrow 2Ag + Na_2[Zn(CN)_4]$$

Problem 3. In some foam-type fire extinguishers, the reactants are $Al_2(SO_4)_3$ and $NaHCO_3$ solutions. When the extinguisher is activated, these reactants are allowed to mix, producing $Al(OH)_3$ and CO_2. The $Al(OH)_3$—CO_2 foam extinguishes the fire. Write the net ionic equation to represent this reaction.

Solution $Al_2(SO_4)_3$ solution is acidic due to hydrolysis :

$$Al^{3+} + 4H_2O \longrightarrow [Al(OH)_4]^- + 4H^+$$

$[Al(OH)_4]^-$ reacts with HCO_3^- forming $Al(OH)_3$—CO_2 foam :

$$[Al(OH)_4]^- + HCO_3^- \longrightarrow Al(OH)_3 + H_2O + CO_2$$

Problem 4. Use the relationship $\Delta G° = -nFE°_{cell}$ to estimate the minimum voltage required to electrolyse Al_2O_3 in the Hall–Heroult process.

$$\Delta G°_f\ (Al_2O_3) = -1520\ \text{kJ mol}^{-1}$$

$$\Delta G°_f\ (CO_2) = -394\ \text{kJ mol}^{-1}$$

Show that the oxidation of the graphite anode to CO_2 permits the electrolysis to occur at a lower voltage than if the electrolysis reactions were

$$Al_2O_3 \longrightarrow 2Al + 3O_2$$

Solution Net reaction in Hall-Heroult process is:

$$3C + 2Al_2O_3 \longrightarrow 4Al + 3CO_2$$

Or $\quad 4Al^{3+} + 12e^- \longrightarrow 4Al,$

Number of electrons $(n) = 12$, ΔG_f° **(element)** $= 0$

$$\Delta G° = 3\Delta G°_f(CO_2) - 2\Delta G°_f(Al_2O_3)$$

$$= -3 \times 394 - 2(-1520)$$

$$= 1858\ \text{kJ}$$

$$\Delta G° = -nFE°_{cell}$$

$$-E°_{cell} = \frac{\Delta G°}{nF} = \frac{1858 \times 1000}{12 \times 96500}$$

$$= \mathbf{1.60\ V}$$

For the reaction $Al_2O_3 \longrightarrow 2Al + 3O_2$

$$\Delta G° = 1520\ \text{kJ}$$

$$2Al^{3+} + 6e^- \longrightarrow 2Al,\ n = 6$$

$$\therefore\quad -E°_{cell} = \frac{\Delta G}{nF} = \frac{1520 \times 1000}{6 \times 96500}$$

$$= \mathbf{2.62\ V}$$

Thus Hall-Heroult process takes place at lower voltage.

Problem 5. The names dilead (II) lead (IV) oxide and plumbous orthoplumbate have been used to describe red lead Pb_3O_4. Show that these names are true.

Solution When Pb_3O_4 is treated with HNO_3, $Pb(NO_3)_2$ is formed and PbO_2 is precipitated :

$$Pb_3O_4 + 2HNO_3 \longrightarrow 2Pb^{II}(NO_3)_2 + Pb^{IV}O_2 \downarrow + 2H_2O$$

This shows that red lead has PbO and PbO_2 in 2 : 1 molar ratio and thus it is $2PbO \cdot PbO_2$ and is thus dilead (II) lead (IV) oxide. Only PbO reacts with HNO_3 forming $Pb(NO_3)_2$ while PbO_2 remains unreacted and is precipitated.

Problem 6. The following are two reaction schemes involving Mg.

Scheme I Mg $\xrightarrow{\text{air}\ \Delta}$ colourless solid A $\xrightarrow{\text{HCl}}$ colourless solution B $\xrightarrow{Na_2CO_3}$ white ppt C $\xrightarrow{\Delta}$ colourless gas E + D; colourless gas E $\xrightarrow{Ca(OH)_2}$ white ppt F

Scheme II Mg $\xrightarrow{\text{dil. } H_2SO_4}$ colourless solution G $\xrightarrow{NaOH}$ white ppt H $\xrightarrow{HNO_3}$ colourless solution $\xrightarrow{\text{evaporate}}$ I $\xrightarrow{\Delta}$ $D + O_2 +$ gas J

Identify A to J

Solution A : MgO; $\quad B$: $MgCl_2$; $\quad C$: $MgCO_3$;
D : MgO; $\quad E$: CO_2; $\quad F$: $CaCO_3$;
G : $MgSO_4$; $\quad H$: $Mg(OH)_2$;
I : $Mg(NO_3)_2$ $\quad J$: NO_2

Problem 7. Explain following terms

(a) Electrometallurgy
(b) Zone refining
(c) van Arkel de-Boer process
(d) Mond's process

Solution

(a) Electrometallurgy

The two metallurgical processes, namely pyrometallurgy and hydrometallurgy can be used in the extraction of a fairly large number of metals. These methods, however, cannot be used in cases:

Where the metal is highly reactive, *e.g.,* Na, Li, etc. There are no any chemical-reducing agent strong enough to prepare these metals.

Where the oxide gets reduced only at very high temperature at which formation of carbides also takes place, *e.g.,* Al, Mg, etc.

In these cases, metals can be extracted by electrolysis of their salts in molten state. Thus, sodium and magnesium are prepared by electrolysis of fused chlorides, where the metals are liberated at the cathode and chlorine gas is evolved at the anode. Following reactions take place during electrolysis :

$$\text{Anode :} \quad 2Cl^- \longrightarrow Cl_2(g) + 2e^-$$
$$\text{Cathode :} \quad 2Na^+ + 2e^- \longrightarrow 2Na(l)$$
$$Mg^{2+} + 2e^- \longrightarrow Mg(l)$$

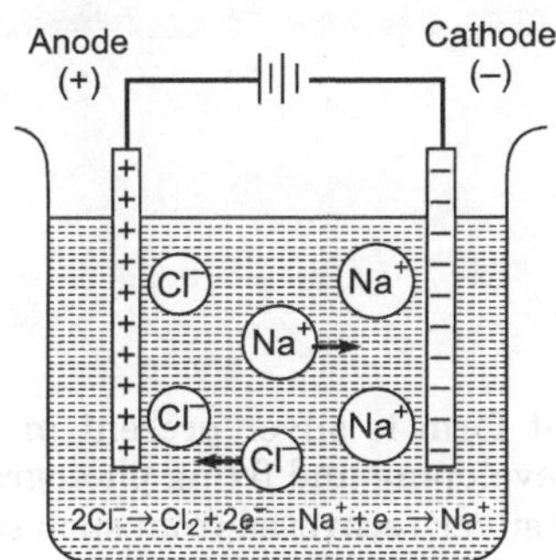

Electrolysis of molten sodium chloride

In theory, aluminium metal could be made the same way. But, aluminium trichloride is covalent and it does not conduct electricity. Aluminium is obtained by electrolytic reduction of alumina in fused cryolite at 1100–1300 K using carbon anode and iron cathode. Electrolysis yields aluminium at cathode and O_2 at anode which reacts with carbon to produce CO_2. The reactions at electrodes are :

$$\text{Anode :} \quad 2O^{2-} \longrightarrow O_2(g) + 4e^-$$
$$C(s) + O_2(g) \longrightarrow CO_2(g)$$
$$\text{Cathode :} \quad Al^{3+} + 3e^- \longrightarrow Al(s)$$

(b) Zone Refining

This method is used to obtain metals of very high purity. The basic principle involved in this process is similar to fractional crystallisation. A small heater is used to heat a bar of the impure metal. The heater melts a small band of metal as it is slowly moved along the rod. As small bands of metal are thus melted sequentially, the pure metal crystallises out of the melt, while impurities pass into the adjacent molten zone. The impurities thus collect at the end of the bar. This end can be cut off and removed. High grade germanium and the silicon are obtained by purifying them by zone refining.

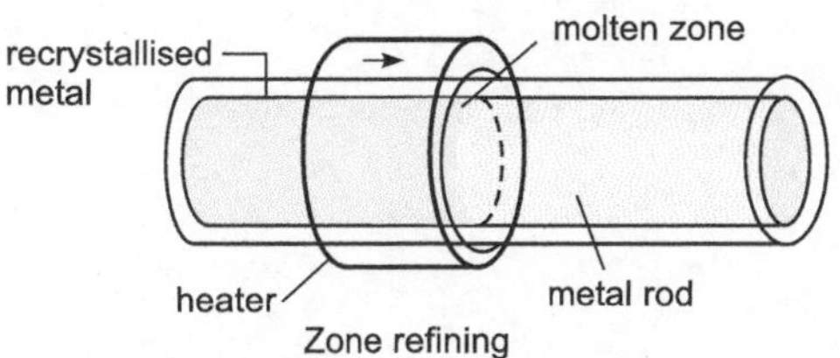

Zone refining

(c) van Arkel de-Boer Process

This method is based on the thermal decomposition of a volatile metal compound like an iodide. In this method, first a metal iodide is formed by direct reaction of iodine and the metal to be purified at a temperature of 475–675 K in an evacuated vessel. The vapours of metal iodide, thus formed are heated strongly on a tungsten or tantalum filament at 1300-1800 K. The metal iodide decomposes to yield the pure metal, as in the case of zirconium.

$$Zr(s) + 2I_2(s) \xrightarrow{473\text{-}673\text{ K}} ZrI_4(g) \xrightarrow[\text{Tungsten filament}]{1300\text{-}1800\text{ K}} Zr(s) + 2I_2(g)$$

Titanium is also purified by this method. The impure metal is heated with iodine and TiI_4 thus formed is decomposed by heating at 1700K over tungsten filament.

$$Ti(s) + 2I_2(g) \longrightarrow TiI_4(g) \xrightarrow[\text{Tungsten filament}]{1700\text{ K}} Ti(s) + 2I_2(g)$$

The regenerated iodine is used over and over again. This process is very expensive and is employed for the preparation of limited amounts of very pure metals for special uses.

(d) Mond Process

Some metals are purified by obtaining their volatile carbonyl compounds which on heating strongly decompose to yield pure metal. Purification of nickel is done by this method. Impure nickel is reacted with carbon monoxide at 325 K to give volatile nickel carbonyl leaving solid impurities behind. Pure nickel is obtained by heating nickel carbonyl at 450-475 K.

$$Ni(s) + 4CO(g) \xrightarrow{325\text{ K}} Ni(CO)_4(g) \xrightarrow{450\text{-}475\text{ K}} Ni(s) + 4CO(g)$$

Master Exercises

Exercise 1
(Stage 1 : Learning)

Short Answer Type Questions

1. Is it possible to reduce a sulphide ore directly with carbon.

2. Consider following reactions

(a) $2ZnS + 3O_2 \longrightarrow 2ZnO + 2SO_2$

$ZnO + C \longrightarrow Zn + CO$

(b) $2ZnS + C \longrightarrow 2Zn + CS_2$

(i) Which reaction is more spontaneous?

(ii) Which reaction has more negative value of ΔG ?

(iii) Which reaction has more positive value of ΔG ?

3. Which reactions are favourable?

(a) $MgO + Si \longrightarrow$

(b) $SiO_2 + Mg \longrightarrow$

4. Specify magnetic impurity present in tin stone ....

5. Pure tin is obtained by electro-refining. Specify the indicated (by arrow) contents.

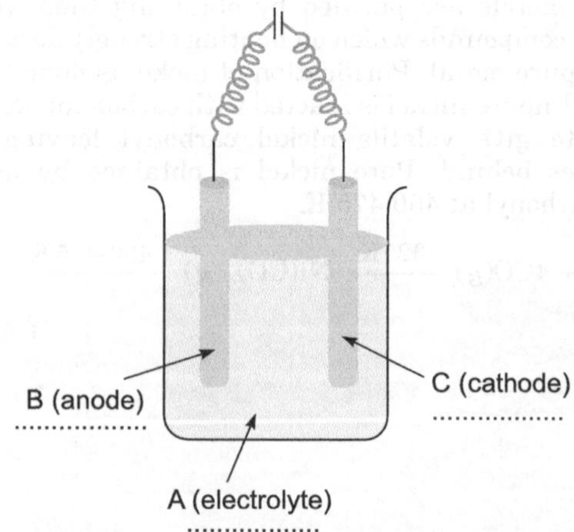

6. The recovery of silver from the solutions used in developing and fixing photographic film and prints presents just such a problem. Silver is an expensive metal, and the recovery is profitable. Moreover, if not recovered, the silver ions would constitute and undesirable pollutant in water supplies.

(a) In what form is silver present in the solutions after developing and fixing photographic film?

(b) How is silver recovered from such solution?

(c) What is undesirable pollutant?

7. Following reaction is reversible

$$Ag_2S + 4NaCN \rightleftharpoons 2Na[Ag(CN)_2] + Na_2S$$

Suggest a method to make it spontaneous.

8. ZnS is always contaminated with PbS. Outline a method that can enrich PbS by froth flotation process.

9. Consider following facts

I : Zinc and lead are immiscible in molten state; zinc is lighter than lead.

II : Solubility of silver in molten zinc is higher than that in molten lead.

III : Zinc is volatile.

(a) Outline method of separation of silver from lead based on above facts.

(b) Name the method

10. (a) Write full from of **BOP** process of making steel.

(b) Write various forms of iron.

(c) Which is the purest form?

11. Consider following equilibrium

$$3Fe(s) + C(s) \rightleftharpoons Fe_3C(s), \qquad \Delta H^\circ > 0$$

What is the effect of temperature on this equilibrium?

Analytical Questions

1. (a) Calculate the free energy changes for these reactions

I : $2CuS + C \longrightarrow 2Cu + CS_2$

II : $2CaS + C \longrightarrow 2Ca + CS_2$

III : $2CuS + 3O_2 \longrightarrow 2CuO + 2SO_2$

Given ΔG°_f (in kJ mol^{-1}) for

$CuS = -49$, $CS_2 = 63.6$

$CaS = -1320$; $CuO = -127.2$

$SO_2 = -300.4$

(b) Which of them are spontaneous under standard conditions?

(c) What do you notice about the value of $\Delta G^\circ_f(CS_2)$ compared to $\Delta G^\circ_f(SO_2)$?

(d) Why are the reactions not carried out at standard conditions?

2. A student suggested that calcium should be made if calcium oxide is reacted with aluminium powder. Was the student correct?

$$\Delta G^\circ_f(CaO) = -604.2 \text{ kJ mol}^{-1}$$
$$\Delta G^\circ_f(Al_2O_3) = -1582 \text{ kJ mol}^{-1}$$

3. Although Au is soluble in aqua regia, Ag is not. What is the likely reason for this difference ?

4. $FeCr_2O_4$ (chromite) can be used to extract Cr. Outline scheme of extraction.

5. Lead metal is purified by electrolysis in a manner similar to that for copper; the electrolyte is lead (II) hexafluorosilicate $PbSiF_6$. Describe the process.

6. (a) Pure iron is prepared for special purposes by precipitating iron (III) oxide and reducing the dry oxide with H_2 gas. Write the balanced equation.

(b) HCl can't used to precipitate $Al(OH)_3$ from soluble $Na[Al(OH)_4]$ but addition of NH_4Cl can cause precipitation. Explain by reactions.

(c) AgCl (Horn silver) is converted into Ag by pyrometallurgical method. Describe reactions.

7. Zinc, cadmium and mercury are found in nature as sulphides. Values of standard reduction potentials for the half-reactions are :

$$M^{2+}(aq) + 2e^- \longrightarrow M(s)$$

M	E°
Zn	– 0.76 V
Cd	– 0.40 V
Hg	+ 0.79 V

Explain, why roasting of ZnS gives ZnO but roasting of HgS gives elemental Hg. What product would you expect from roasting CdS?

8. When an inert atmosphere is needed for a metallurgical process, nitrogen is frequently used. However, in the reduction of $TiCl_4$ by magnesium, helium is used. Explain, why nitrogen is not suitable for this process?

9. The following equations represent reactions used in some important metallurgical processes:

(a) $Fe_3O_4 + CO(g) \longrightarrow Fe(l) + CO_2(g)$

(b) $MgCO_3(s) + SiO_2(s) \longrightarrow MgSiO_3(l) + CO_2(g)$

(c) $Au(s) + CN^-(aq) + H_2O(l) + O_2(g) \longrightarrow [Au(CN)_2]^-(aq) + OH^-(aq)$

Balance the equations. Which one (s) represent reduction to a free metal?

10. Assuming complete recovery of metal, which of the following ores would yield the greater quantity of copper on a mass basis?

(a) an ore containing 3.30% (by mass) azurite $Cu(OH)_2 \cdot 2CuCO_3$ or

(b) an ore containing 4.95% (by mass) chalcopyrite $CuFeS_2$.

11. When 1.164 g of a certain metal sulphide was roasted in air, 0.972 g of the metal oxide was formed. If the oxidation number of the metal is +2, calculate the molar mass of the metal.

12. How much aluminium can be produced by a current of 94.0. A flowing through a Hall-Heroult cell for 1.00h?

13. Sea-water contains 1272 g of Mg^{2+} per metric ton (1 mega gram). How much of slaked lime must be added to 1.0 metric ton of sea water to precipitate all of the Mg^{2+} ion? (Mg = 24, Ca = 40)

Exercise 2
(Stage 2 : High Skill Problem Solving)

Only One Option Correct

1. Si of high purity to be used in semiconductor can be prepared by following methods:

I. $SiO_2 + 2C \longrightarrow Si + 2CO$

II. $Si + 2Cl_2 \longrightarrow SiCl_4$

$SiCl_4 + 2Mg \longrightarrow Si + 2MgCl_2$

Better method is

(a) I (b) II
(c) both (a) and (b) (d) None of these

2. Main source of lead is PbS. It is converted to Pb by:

I. $PbS \xrightarrow[\Delta]{air} PbO + SO_2$, then $PbO \xrightarrow{C} Pb + CO_2$

II. $PbS \xrightarrow[\Delta]{air} PbO + PbS$, then $PbO \xrightarrow{PbS} Pb + SO_2$

Self-reduction process is

(a) I (b) II
(c) both (a) and (b) (d) None of these

3. Match the compounds (A) with their applications/ colour in (B)

A	B
I. Pb_3O_4	W. white lead
II. $(PbCO_3)_2 \cdot Pb(OH)_2$	X. rust proofing sheets
III. Ca_2PbO_4	Y. road signs and markings
IV. $PbCrO_4$	Z. red lead

Choose the correct alternate

	I	II	III	I
(a)	Y	Z	X	W
(b)	X	W	Y	Z
(c)	Z	W	X	Y
(d)	X	Y	W	Z

4. P_4O_{10} is converted to white phosphorus by
(a) Al (b) C
(c) Na (d) all are correct

5. Phosphine, PH_3, is prepared by
(a) $Ca_3P_2 + H_2O \longrightarrow$ (b) $P_4 + NaOH \longrightarrow$
(c) both are correct (d) None is correct

6. Formation of $Ni(CO)_4$ and subsequent its decomposition into Ni and CO (recycled) makes the basis of Mond's process

$$Ni + 4CO \xrightarrow{T_1} Ni(CO)_4 \xrightarrow{T_2} Ni + 4CO$$

T_1 and T_2 are
(a) 100° C, 50° C (b) 50° C, 100° C
(c) 50° C, 230° C (d) 230° C, 50° C

7. Oxides of the various metals are converted into metal by heating but not in
(a) Ag_2O (b) CuO
(c) HgO (d) all are correct

8. Of the following reduction processes
I. $Fe_2O_3 + C \longrightarrow Fe$
II. $ZnO + C \longrightarrow Zn$
III. $Ca_3(PO_4)_2 + C \longrightarrow P$
IV. $PbO + C \longrightarrow Pb$
Correct processes are
(a) all of these (b) all but III
(c) all but IV (d) II and IV

9. Following reaction is not involved in *Thermite* process
(a) $3Mn_3O_4 + 8Al \longrightarrow 9Mn + 4Al_2O_3$
(b) $Cr_2O_3 + 2Al \longrightarrow Al_2O_3 + 2Cr$
(c) $2Fe + Al_2O_3 \longrightarrow 2Al + Fe_2O_3$
(d) $B_2O_3 + 2Al \longrightarrow 2B + Al_2O_3$

10. Consider the following steps

$$CuS \xrightarrow{\text{roast in air}} (A) \xrightarrow{\text{roast without air}} (B)$$

Which is not the correct statement?
(a) It is self-reduction
(b) It involves disproportionation
$Cu_2S \longrightarrow Cu + CuS$
(c) (*A*) is a mixture of CuO and CuS and (*B*) is a mixture of Cu and SO_2
(d) All are incorrect statements

11. When $FeCr_2O_4$ (chromite) is reduced with C in an electric-arc furnace
(a) Cr and Fe_2O_3 are formed
(b) Fe and Cr_2O_3 are formed
(c) Fe and Cr (ferrochrome) are formed
(d) $FeCrO_4$ is formed

- $FeCr_2O_4$ (chromite) is a good source of Cr and its compounds like Na_2CrO_4, $Na_2Cr_2O_7$. Questions 12 to 14 are based on following reactions:

I. $FeCr_2O_4 + NaOH + \text{air} \longrightarrow (A) + Fe_2O_3$
II. $(A) + (B) \longrightarrow Na_2Cr_2O_7$
III. $Na_2Cr_2O_7 + X \xrightarrow{\Delta} Cr_2O_3$
IV. $Cr_2O_3 + Y \xrightarrow{\Delta} Cr$

12. Compounds (*A*) and (*B*) are
(a) Na_2CrO_4, H_2SO_4
(b) $Na_2Cr_2O_7$, HCl
(c) Na_2CrO_5, H_2SO_4
(d) $Na_4[Fe(OH)_6]$, H_2SO_4

13. (*X*) and (*Y*) are
(a) C and Al (b) Al and C
(c) C in both (d) Al in both

14. Na_2CrO_4 and Fe_2O_3 are separated by dissolving in
(a) conc. H_2SO_4 (b) NH_3
(c) H_2O (d) dil. HCl

15. Match the reactions taking place in blast furnace with temperature-range of operations:

Reactions	Temperature range
(A) : $Fe_2O_3 + CO \longrightarrow CO_2 + Fe$	(I) 1200° C
(B) : $C + CO_2 \longrightarrow CO$	(II) 900° C
(C) : $C + H_2O \longrightarrow CO + H_2$	(III) 600° C
(D) : $CaO + SiO_2 \longrightarrow CaSiO_3$	(IV) 1000° C

Choose the correct alternate

	A	B	C	D
(a)	IV	III	II	I
(b)	II	IV	III	I
(c)	I	II	IV	III
(d)	I	II	III	IV

16. Match column (*A*) (process) with column (*B*) (electrolyte).

A (process)	B (electrolyte)
(I) Downs cell	(W) Fused $MgCl_2$
(II) Dow sea water process	(X) Fused (Al_2O_3 + Na_3AlF_6)
(III) Hall-Heroult	(Y) Fused KHF_2
(IV) Moissan	(Z) Fused (40% NaCl + 60% $CaCl_2$)

Choose the correct alternate

	I	II	III	IV		I	II	III	IV
(a)	Z	W	X	Y	(b)	X	Y	Z	W
(c)	W	Z	X	Y	(d)	X	Z	W	Y

17. In the Pidgeon process, Mg is produced by
(a) electrolysis of fused $MgCl_2$
(b) reducing calcined dolomite with ferrosilicon at high temperature under pressure
(c) both (a) and (b)
(d) none of the above

18. Boron can be obtained by various methods but not by
(a) thermal decomposition of B_2H_6
(b) pyrolysis of BI_3 (van Arkel)
(c) reducing BCl_3 with H_2
(d) electrolysis of fused BCl_3

19. Following method is not used for extraction of Al
(a) van Arkel (b) Serpeck
(c) Bayer (d) Hall-Heroult

20. B_4C (boron carbide) is used except
(a) to extract boron
(b) as an abrasive for polishing
(c) for making bullet-proof clothing
(d) for making diborane

21. In the extraction of nickel by Mond's process, the metal is obtained by
(a) electrochemical reduction
(b) thermal decomposition
(c) chemical reduction by aluminium
(d) reduction by carbon

22. Which one of the following statements is not correct?
(a) Nickel forms $Ni(CO)_4$
(b) All the transition metals form monometallic carbonyls
(c) $Ni(CO)_4$ is volatile and decomposes into Ni and CO
(d) Transition metals form complexes

23. Match List I with List II and select the correct answer using the codes given below the lists:

List I	List II
A. Carnallite	1. Zinc
B. Calamine	2. Titanium
C. Ilmentite	3. Magnesium
D. Chalocopyrite	4. Copper

	A	B	C	D
(a)	1	3	2	4
(b)	1	3	4	2
(c)	3	1	4	2
(d)	3	1	2	4

24. Which of the following is not an ore?
(a) Malachite (b) Calamine
(c) Stellite (d) Cerussite

25. In the metallurgy of iron, when $CaCO_3$ is added to blast furnace, calcium ion appears as
(a) CaO (b) metallic Ca
(c) gangue (d) slag

26. Iron obtained from blast furnace is
(a) wrought iron (b) cast iron
(c) pig iron (d) steel

27. Four metals and their methods of refinement are given:
(i) Ni, Cu, Zr, Ga
(ii) Electrolysis, van Arkel process, Zone refining, Mond's process
Choose the right method for each:
(a) Ni : Electrolysis, Cu : van Arkel process, Zr : Zone refining, Ga : Mond's process
(b) Ni : Mond's process, Cu : Electrolysis, Zr : van Arkel process, Ga : Zone refining
(c) Ni : Mond's process, Cu : van Arkel process, Zr : Zone refining, Ga : Electrolysis
(d) Ni : Electrolysis, Cu : Zone refining, Zr : van Arkel process, Ga : Mond's process

28. On igniting Fe_2O_3 at 1400° C, the product obtained is
(a) Fe_2O_3 melt (b) FeO
(c) Fe_3O_4 (d) metallic iron

29. Consider the following metallurgical processes
I. Heating impure metal with CO and distilling the resulting volatile carbonyl (b.p. 43° C) and finally decomposing at 150° to 200° C to get the pure metal
II. Heating the sulphide ore in air until a part is converted to oxide and then further heating in the absence of air to let the oxide react with unchanged sulphide
3. Electrolysing the molten electrolyte containing approximately equal amounts of the metal chloride and $CaCl_2$ to obtain the metal
The processes used for obtaining sodium, nickel and copper are respectively
(a) I, II and III (b) II, III and I
(c) III, I and II (d) II, I and III

30. Extraction of aluminium from aluminium oxide (Al_2O_3) is best done by
(a) electrolytic reduction of Al_2O_3
(b) reduction of Al_2O_3 with carbon
(c) reduction of Al_2O_3 with sodium
(d) reduction of Al_2O_3 with CO

31. Consider the following statements:
Roasting is carried out to
I. convert sulphide to oxide and sulphate
II. remove water of hydration
III. melt the ore
4. remove arsenic and sulphur impurities
Of these statements
(a) I, II and III are correct

(b) I and IV are correct
(c) I, II and IV are correct
(d) II, III and IV are correct

32. The element which could be extracted by electrolytic reduction of its oxide dissolved in a high temperature melt is
(a) sodium (b) magnesium
(c) fluorine (d) aluminium

33. Chemical leaching is useful in the concentration of
(a) copper pyrites (b) bauxite
(c) galena (d) cassiterite

34. Carbon cannot be used in the reduction of Al_2O_3 because
(a) it is an expensive proposition
(b) the enthalpy of formation of CO_2 is more than that of Al_2O_3
(c) pure carbon is not easily available
(d) the enthalpy of formation of Al_2O_3 is too high

35. In which of the following isolations no reducing agent is required?
(a) Iron from haematite
(b) Aluminium from bauxite
(c) Mercury from cinnabar
(d) Zinc from zinc blende

36. Which one of the following is not a method of concentration of metals?
(a) Gravity separation
(b) Froth flotation process
(c) Electromagnetic separation
(d) Smelting

37. Blister Cu is about
(a) 60% Cu (b) 90% Cu
(c) 98% Cu (d) 100% Cu

38. There are following extraction process of silver but not
(a) as a side product in electrolytic refining of copper
(b) Parke's process in which Zn is used to extract silver by solvent extraction from molten lead
(c) by reaction of silver sulphide with KCN and then reaction of soluble complex with Zn.
(d) by heating $Na[Ag(CN)_2]$

39. $Ag_2S + NaCN \longrightarrow (A)$
$(A) + Zn \longrightarrow (B)$
(B) is a metal. Hence, *(A)* and *(B)* are
(a) $Na_2[Zn(CN)_4]$, Zn (b) $Na[Ag(CN)_2]$, Ag
(c) $Na_2[Ag(CN)_4]$, Ag (d) $Na_3[Ag(CN)_4]$, Ag

40. Refining of silver is done by
(a) liquation (b) Poling
(c) cupellation (d) van Arkel method

41. Bessemerisation is carried out for
I : Fe, II : Cu, III : Al, IV : silver
(a) I, II (b) II, III
(c) III, IV (d) I, III

42. Match the method of concentration of the ore in column I with the ore in column II and select the correct alternate:

Column I	Column II
X : Magnetic separation	A : Ag_2S
Y : Froth flotation	B : $FeCr_2O_4$
Z : Gravity separation	C : $Al_2(SiO_3)_3$

	X	Y	Z
(a)	A	B	C
(b)	B	A	C
(c)	C	A	B
(d)	B	C	A

43. Which of the following does not contain Mg?
(a) Magnetite (b) Magnesite
(c) Asbestos (d) Carnallite

44. Calcination is the process of heating the ore
(a) in inert gas
(b) in the presence of air
(c) in the absence of air
(d) in the presence of CaO and MgO

45. $Ag_2S + NaCN + Zn \longrightarrow Ag$

This method of extraction of Ag by complex formation and then its displacement is called
(a) Parke's method
(b) Mac Arthur-Forrest method
(c) Serpeck method
(d) Hall's method

46. The slag obtained during the extraction of copper from copper pyrites is composed of
(a) Cu_2S (b) $CuSiO_3$
(c) $FeSiO_3$ (d) SiO_2

47. In zone-refining method the molten zone
(a) consists of impurities only
(b) contains more impurity than the original metal
(c) contains the purified metal only
(d) moves to either side

48. Reduction of a metal oxide by excess carbon at high temperature is a method for the commercial preparation of some metals. This method can be successfully applied in the case of
(a) BeO and Al_2O_3 (b) ZnO and Fe_2O_3
(c) CaO and Cr_2O_3 (d) BaO and U_3O_8

49. Match List I with List II and select the correct answer using the codes gives below the lists :

List I	List II
A. van Arkel method	1. Manufacture of caustic soda
B. Solvay process	2. Purification of titanium
C. Cupellation	3. Manufacture of Na_2CO_3
D. Poling	4. Purification of copper
	5. Refining of silver

Codes

	A	B	C	D
(a)	2	1	3	4
(b)	4	3	2	5
(c)	2	3	5	4
(d)	5	1	3	4

50. Formation of metallic copper from the sulphide ore in the normal thermo-metallurgical process essentially involves which one of the following reaction?

(a) $CuS + \frac{3}{2}O_2 \longrightarrow CuO + SO_2;$

(b) $CuS + \frac{3}{2}O_2 \longrightarrow CuO + SO_2;$

$2CuO + CuS \longrightarrow 3Cu + SO_2$

(c) $CuS + 2O_2 \longrightarrow CuSO_4;$

(d) $CuS + \frac{3}{2}O_2 \longrightarrow CuO + SO_2;$

$CuSO_4 + CuS \longrightarrow 2Cu + 2SO_2$

$CuO + CO \longrightarrow Cu + CO_2$

51. Annealing of steel is the process of heating steel
(a) to a bright red hot and then cooling it slowly
(b) to a bright red hot and then cooling it suddenly
(c) to a temperature much below redness and cooling it slowly
(d) none of the above

52. Out of Cu_2S, HgS, Ag_2S and ZnS, roasting will convert the minerals into metal in case of
(a) Cu_2S, ZnS (b) HgS, ZnS
(c) Cu_2S, Ag_2S (d) HgS

53. The method of zone refining of metals is based on the principle of
(a) greater noble character of the solid metal than that of the impurity
(b) greater solubility of the impurity in the molten state than in the solid
(c) greate mobility of the pure metal than that of impurity
(d) higher melting point of the impurity than that of the pure metal

54. Splitting of silver is
(a) reduction of ammoniacal silver nitrate solution by tartarate
(b) extraction of silver from its ore Ag_2S, by hydrometallurgy
(c) making of silver amalgam during filling of teeth
(d) cooling of molten silver with the evolution of oxygen causing violent spurting

55. In the extraction of Cu, the reaction takes place in bessemer converter is
(a) $2Cu_2O + Cu_2S \longrightarrow 6Cu + SO_2$
(b) $2CuFeS_2 + O_2 \longrightarrow Cu_2S + FeS + SO_2$
(c) $2Cu_2S + 3O_2 \longrightarrow 2Cu_2O + 2SO_2$
(d) $2FeS + 3O_2 \longrightarrow 2FeO + 2SO_2$

56. Sodium is made by the electrolysis of molten mixture of about 40% NaCl and 60% $CaCl_2$ because
(a) $CaCl_2$ helps in conduction of electricity
(b) Ca^{2+} can reduce NaCl to Na
(c) Ca^{2+} can displace Na from NaCl
(d) this mixture has a lower melting point than NaCl

57. Calcium is obtained by the
(a) roasting of limestone
(b) electrolysis of solution of calcium chloride in H_2O
(c) electrolysis of molten anhydrous calcium chloride
(d) reduction of calcium chloride with carbon

58. Which one of the following elements constitutes a major impurity in pig iron?
(a) Silicon (b) Oxygen
(c) Sulphur (d) Graphite

59. Purification of aluminium by electrolytic refining, is known as
(a) Hall's process (b) Baeyer's process
(c) Hoope's process (d) Serpeck's process

60. Cassiterite is an ore of
(a) Mn (b) Ni
(c) Sb (d) Sn

61. The process used for purification $Al_2O_3 \cdot 2H_2O$ is
(a) magnetic separation (b) froth flotation
(c) leaching (d) liquation

62. Froth flotation is a process of
(a) oxidation (b) reduction
(c) refining (d) concentration

63. Malachite is an ore of
(a) silver (b) mercury
(c) magnesium (d) copper

64. Flux is used to remove
(a) basic impurities
(b) acidic impurities
(c) all types of impurities
(d) acidic and basic both impurities

65. Mac Arthur process is used for the extraction of
(a) Al (b) Ag
(c) Cu (d) Zn

66. The chemical process in the production of steel from haematite ore involves
(a) reduction
(b) oxidation
(c) reduction followed by oxidation
(d) oxidation followed by reduction

67. Electrolytic reduction of alumina to aluminium by Hall-Heroult process is carried out
(a) in the presence of NaCl
(b) in the presence of fluorite

(c) in the presence of cryolite which forms a melt with lower melting temperature
(d) in the presence of cryolite which forms a melt with higher melting temperature

68. Which of the following ore is concentrated by both, magnetic and gravimetric separation?
(a) Dolomite (b) Tinstone
(c) Galena (d) Bauxite

69. Which of the following processes is used in extractive metallurgy of magnesium?
(a) Fused salt electrolysis
(b) Self reduction
(c) Aqueous solution electrolysis
(d) Thermite reduction

70. In the process of extraction of gold,

$$\text{Roasted gold ore} + CN^- + H_2O \xrightarrow{O_2} [X] + OH^-$$

$$[X] + Zn \longrightarrow [Y] + Au$$

Identify the complexes $[X]$ and $[Y]$.
(a) $X = [Au(CN)_2^-]$, $Y = [Zn(CN)_4]^{2-}$
(b) $X = [Au(CN)_4]^{3-}$, $Y = [Zn(CN)_4]^{2-}$
(c) $X = [Au(CN)_2]^-$, $Y = [Zn(CN)_6]^{4-}$
(d) $X = [Au(CN)_4]^-$, $Y = [Zn(CN)_4]^{2-}$

71. The methods chiefly used for the extraction of lead and tin from their ores are respectively
(a) self reduction and carbon reduction
(b) self reduction and electrolytic reduction
(c) carbon reduction and self reduction
(d) cyanide process and carbon reduction

72. Extraction of zinc from zinc blende is achieved by
(a) electrolytic reduction
(b) roasting followed by reduction with carbon
(c) roasting followed by reduction with another metal
(d) roasting followed by self-reduction

73. Which one of the following statements is correct? In the metallurgy of sodium by electrolysis, excess of calcium chloride is mixed with sodium chloride to
(a) make the latter a good conductor
(b) make the latter soft
(c) generate more energy for the electrolytic cell
(d) assist liquefication of the latter at a much lower temperature

74. Consider the following statements
I : Pig iron is malleable and ductile.
II : Pig iron can be melted to produce cast iron of desired shapes.
III : Wrought iron is obtained by heating pig iron with iron oxide in a furnace.

Which of the statement(s) given above is/are correct?
(a) I only (b) I and II
(c) II and III (d) III only

75. Main process for extracting nickel in the pure form uses which one of the following?
(a) Vapour phase refining
(b) Zone refining
(c) Electrolysis
(d) Solvent extraction

76. Which represents correct matching of metals with their minerals?

	Mg	Na	Cu	Al
(a)	chloride	chloride	sulphide	silicate
(b)	carbonate	chloride	sulphide	oxide
(c)	carbonate	carbonate	oxide	phosphate
(d)	oxide	chloride	sulphide	oxide

77. Which of the following metals are found in native state?

Ag, Pt, C, Si, N, O, Mg, Na, Pb.

(a) Ag, Pt, C, N, O (b) Ag, Pt, Mg
(c) Ag, Pt, Pb, Mg (d) Ag, Pt

78. Most abundant metal in earth crust is
(a) Al (b) O
(c) Fe (d) Si

79. Followed by oxygen which element is the most abundant in human body
(a) C (b) H
(c) Ca (d) P

80. Select correct statement.
(a) Based on reactivity series, occurrence of certain elements takes place in native state
(b) Due to basic nature of oxides of alkaline earth elements, they combine with atmospheric acidic oxides giving salts
(c) Both (a) and (b) are correct
(d) None of the above is correct

81. Which is not the correct statement?
(a) Cassiterite, chromite and pitchblende are concentrated by hydraulic washing (Tabling).
(b) Pure Al_2O_3 is obtained from the bauxite ore by leaching in the Baeyer's process.
(c) Sulphide ore is concentrated by calcination method.
(d) Roasting can convert sulphide into oxide or sulphate and part of sulphide may also act as a reducing agent.

82. NaCN is sometimes added in the froth flotation process as a depressant when ZnS and PbS minerals are expected because
(a) $Pb(CN)_2$ is precipitated while no effect as ZnS
(b) ZnS forms soluble complex $Na_2[Zn(CN)_4]$ while PbS forms froth
(c) PbS forms soluble complex $Na_2[Pb(CN)_4]$ while ZnS forms froth
(d) NaCN is never added in froth flotation process

83. Which method is not correct given for refining of crude metals?
(a) **Distillation :** zinc and mercury
(b) **Liquation :** tin
(c) **van Arkel :** Zirconium
(d) **Mond process :** lead

84. Self-reduction of CuS to Cu can be carried out in:
(a) Bessemer converter (b) Pierce-Smith converter
(c) both (a) and (b) (d) None of these

85. Select incorrect reduction process
(a) $2[Ag(CN)_2]^- + Zn \longrightarrow [Zn(CN)_4]^{2-} + 2Ag$
(b) $CuO + H_2 \longrightarrow Cu + H_2O$
(c) $ZnO + H_2 \longrightarrow Zn + H_2O$
(d) $MgO + C \longrightarrow Mg + CO$

86. Ellingham diagram represents
(a) change of ΔG with temperature
(b) change of ΔH with temperature
(c) change of ΔG with pressure
(d) change of $(\Delta G - T\Delta S)$ with temperature

87. Selection of temperature to carry out a reduction process depends so as to make
(a) ΔG negative (b) ΔG positive
(c) ΔH negative (d) ΔH positive

88. The reduction of an oxide by aluminium is called
(a) Ellingham process
(b) Goldschmidt's aluminothermite process
(c) Kroll's process
(d) van Arkel process

89. Select correct statement.
(a) In the decomposition of an oxide, into oxygen and gaseous metal, entropy increases.
(b) Decomposition of an oxide is an endothermic change.
(c) To make $\Delta G°$ negative, temperature should be high enough so that $T\,\Delta S° > \Delta H°$
(d) All of the above are correct statements

90. Which does not represent correct method?
(a) $TiCl_2 + 2Mg \longrightarrow Ti + 2MgCl_2$: Kroll
(b) $Ni(CO)_4 \longrightarrow Ni + 4CO$: Mond
(c) $Ag_2CO_3 \longrightarrow 2Ag + CO_2 + \frac{1}{2}O_2$: van Arkel
(d) $ZrI_4 \longrightarrow Zr + 2I_2$: van Arkel

91. Pyrometallurgical method used for the extraction of copper from sulphide ore
(a) is a dry method
(b) involves concentration by leaching the sulphide ore with dil. H_2SO_4
(c) involves concentration of the sulphide ore by froth flotation process
(d) statement given in (b) is incorrect

92. Blister copper is
(a) impure copper
(b) obtained in self-reduction process during bessemerisation
(c) both (a) and (b) are correct
(d) none of the above is correct

93. Froth flotation process used for the concentration of sulphide ore
(a) is based on the difference in wettability of different minerals
(b) uses sodium ethyl xanthate, $C_2H_5OCS_2Na$ as collector
(c) uses NaCN as depressant in the mixture of ZnS and PbS when ZnS forms soluble complex and PbS forms froth
(d) all of the above are correct

94. Select correct statement(s) about differentiation flotation
(a) is used to concentrate mixture of sulphides at different stages
(b) zinc sulphate is used to depress zinc sulphide and PbS forms froth
(c) copper sulphate is used to activate depressed lead sulphide and O_2 is blown to make froth
(d) all of the above are correct statements

95. Which step is not involved in hydrometallurgical process?
(a) $Cu_2S + 2Cu_2O \longrightarrow 6Cu + SO_2$
(b) $CuFeS_2 + 2H_2SO_4 \longrightarrow CuSO_4 + FeSO_4 + 2H_2S$
(c) $CuSO_4 + Fe \longrightarrow FeSO_4 + Cu$
(d) $CuCO_3 + H_2SO_4 \longrightarrow CuSO_4 + H_2O + CO_2$

96. When alloy of silver and lead is rich in silver,
(a) cupellation process is used
(b) Parke's method is used
(c) Pattinson's method is used
(d) any of the above methods can be used

97. Magistral is the burnt pyrites containing
(a) sulphates of iron and copper
(b) sulphates and oxides of iron and copper
(c) oxides of iron and copper
(d) sulphides of silver and lead

98. In the leaching of Ag_2S with NaCN, a stream of air is also passed. It is because of
(a) reversible nature of reaction between Ag_2S and NaCN
(b) to oxide Na_2S formed into $Na_2S_2O_3$ and sulphur
(c) both (a) and (b)
(d) none of the above

99. $(Ag + Pb)$ alloy $\xrightarrow[\text{is added}]{\text{melt and zinc}}$ $(Ag + Pb + Zn)$ melt $\xrightarrow{\text{cool}}$ [Layer X / Layer Y]

Select correct statement(s) based on above scheme
(a) layer X contains zinc and silver
(b) layer Y contains lead and silver but amount of silver in this layer is smaller than in the layer X.
(c) X and Y are immiscible layers
(d) all of the above are correct statement

100. $AgCl + Na_2CO_3 \longrightarrow Ag_2CO_3 \xrightarrow{\Delta} X$, X is
(a) Ag_2O and CO_2 (b) Ag, O_2 and CO_2
(c) Ag_2O_2 and CO_2 (d) no effect

101. Purest form of iron is
(a) cast iron (b) wrought iron
(c) pig iron (d) none of these

102. The most important method of making steel is
(a) BOP process (b) BOD process
(c) pudding process (d) smelting

103. Cementite is
(a) predominant content of cement
(b) a mineral of calcium
(c) a carbide of iron in steel
(d) none of the above

104. Phosphorus separates, in the extraction of iron, as
(a) slag, $Ca_3(PO_4)_2$ (b) volatile, P_2O_5
(c) slag, $Mg_3(PO_4)_2$ (d) Ca_3P_2

105. Tempering of steel
(a) is the heating the steel to appropriate temperature and then cooling it rapidly
(b) increases mechanical strength
(c) changes ratio of carbon in cementite
(d) all of the above

106. Mg can be obtaineds
(a) by heating $MgCl_2$ (anhydrous) with Na in the atmosphere of coal gas
(b) by electrolysis of fused carnallite
(c) by both the methods
(d) by none of the above methods

107. During the electrolysis of carnallite, $MgCl_2$ is decomposed and not KCl. This is because of
(a) lower decomposition voltage of $MgCl_2$ than that of KCl
(b) reverse reaction $MgCl_2 + 2K \longrightarrow Mg + 2KCl$ if KCl is decomposed under other experimental condition
(c) both (a) and (b)
(d) none of the above

108. NaCl and $CaCl_2$ are also added to fused $MgCl_2$ in the electrolysis of $MgCl_2$ since
(a) melting point is decreased and conductivity is increased
(b) melting point is increased and conductivity is decreased
(c) melting point and conductivity both are decreased
(d) melting point and conductivity both are increased

109. Which is the sulphate ore of Mg?
(a) Dolomite (b) Carnallite
(c) Magnesite (d) Kieserite

110. Which is the chloride ore of Mg?
(a) Kieserite (b) Carnallite
(c) Epsomite (d) Dolomite

111. Black tin is
(a) an alloy of Sn (b) an allotrope of Sn
(c) 60–70 per cent SnO_2 (d) 100 per cent SnO_2

112. SnO_2 is reduced to metallic Sn on smelting oxide with anthracite, limestone and sand.
In this, function of sand is
(a) it acts as a flux
(b) it removes basic impurities as slag
(c) both as in (a) and (b)
(d) none of the above

113. Poling process
(a) reduced SnO_2 to Sn
(b) oxidises impurities like iron and removes as scum
(c) uses green poles
(d) involves all of the above

114. Refining of tin cannot be done by
(a) cupellation (b) liquation
(c) poling (d) electrorefining

115. Which is not the correct process-mineral matching in metallurgical extraction?
(a) Leaching : Ag
(b) Zone-refining : Sn
(c) Liquation : Sn
(d) van Arkel : Zr

116. Softening of lead means:
(a) melting pure lead at high temperature
(b) removal of impurities, except silver, present in commercial lead
(c) formation of lead alloy
(d) formation of 100% pure lead

117. Softened lead is desilverised by
(a) Parke's method (b) Pattinson' method
(c) both (a) and (b) (d) none of these

118. Self reduction of the sulphide ore takes place during
(a) roasting (b) smelting
(c) calcination (d) cupellation

119. Lead pencil contains
(a) graphite (b) softened lead
(c) galena (d) anglesite

120. Select incorrect statement.
(a) When the lead-silver alloy is rich in silver, lead is removed by the cupellation process
(b) When the lead-silver alloy is rich in lead, lead is removed by Parke's or Pattinson's process
(c) Zinc forms an alloy with lead from which lead is separated
(d) Zinc forms an alloy with silver, from which zinc is separated by distillation

121. In the **Baeyer's process**:
(a) Al_2O_3 goes into solution as soluble $Al(OH)_4^-$ while other basic oxides as TiO_2 and Fe_2O_3 remain insoluble.

(b) Al_2O_3 changes to AlN which in turn decomposed by H_2O
(c) Al_2O_3 changes to $Al_2(CO_3)_3$ which changes to $AlCl_3$
(d) none of the above is correct

122. In electrolysis of Al_2O_3 by Hall-Heroult process
(a) Cryolite $Na_3[AlF_6]$ lowers the m.p. of Al_2O_3 and increases its electrical conductivity.
(b) Al is obtained at cathode and probably CO_2 at anode
(c) both (a) and (b) are correct
(d) none of the above is correct

123. $$\boxed{\text{Bauxite}} \xrightarrow[N_2]{+C} \boxed{\text{AlN}} (\uparrow CO) \xrightarrow{H_2O} Al(OH)_3 \xrightarrow{\Delta} Al_2O_3$$
This flow-sheet is for
(a) Baeyer's process (b) Serpeck's process
(c) Hall's process (d) Kroll's process

124. Corundum is .... mineral of Al.
(a) silicate (b) oxide
(c) double salt (d) sulphate

125. Which mineral has been named incorrectly?
(a) Bauxite : $Al_2O_3 \cdot 2H_2O$
(b) Corundum : Al_2O_3
(c) Cryolite : $3NaF \cdot AlF_3$
(d) Feldspar : $Be_3Al_2Si_6O_{18}$

126. The chemical process in the production of steel from haematite ore involves
(a) reduction
(b) oxidation
(c) reduction followed by oxidation
(d) oxidation followed by reduction

127. Electrolytic reduction of alumina to aluminium by Half-Heroult process is carried out
(a) in the presence of NaCl
(b) in the presence of fluorite
(c) in the presence of cryolite which forms a melt with lower melting temperature
(d) in the presence of cryolite which forms a melt with higher melting temperature

128. The chemical composition of 'slag' formed during the smelting process in the extraction of copper is
(a) $Cu_2O + FeS$ (b) $FeSiO_3$
(c) $CuFeS_2$ (d) $Cu_2S + FeO$

129. Which of the following processes is used in extractive metallurgy of magnesium?
(a) Fused salt electrolysis
(b) Self reduction
(c) Aqueous solution electrolysis
(d) Thermite reduction

130. Calamine is
(a) ZnS (b) $PbCO_3$
(c) $ZnCO_3$ (d) $MgCO_3$

131. The methods chiefly used for the extraction of lead and tin from their ores are respectively
(a) self reduction and carbon reduction
(b) self reduction and electrolytic reduction
(c) carbon reduction and self reduction
(d) cyanide process and carbon reduction

132. Which ore contains both iron and copper?
(a) Cuprite (b) Chalcocite
(c) Chalcopyrite (d) Malachite

133. Extraction of zinc from zinc blende is achieved by
(a) electrolytic reduction
(b) roasting followed by reduction with carbon
(c) roasting followed by reduction with another metal
(d) roasting followed by self-reduction

134. The method of zone refining of metals is based on the principle of
(a) greater noble character of the solid metal than that of the impurity
(b) greater solubility of the impurity in the molten state than in the solid
(c) greater mobility of the pure metal than that of impurity
(d) higher melting point of the impurity than that of the pure metal

135. Al_2O_3 can be converted to anhydrous $AlCl_3$ by heating
(a) Al_2O_3 with HCl gas
(b) Al_2O_3 with NaCl in solid state
(c) a mixture of Al_2O_3 and carbon in dry Cl_2 gas
(d) Al_2O_3 with Cl_2 gas

136. Flux is used to remove
(a) silica (b) metal oxide
(c) silica and metal oxide (d) impurities from ore

137. The process of heating and suddenly cooling of steel is known as
(a) tempering (b) annealing
(c) hardening (d) quenching

138. The metal which cannot be extracted by smelting process
(a) Zn (b) Al
(c) Pb (d) Fe

139. Which of the following processes is related with the removal of sulphur by heating in the air?
(a) Smelting (b) Calcination
(c) Annealing (d) Roasting

140. In electrorefining of copper, some gold is deposited as
(a) cathode (b) electrode
(c) cathode mud (d) anode mud

141. Bessemer converter is used for preparation of
(a) steel (b) wrought iron
(c) pig iron (d) cast iron

142. Colemanite is
(a) $Ca[B_3O_4(OH_2)] \cdot 2H_2O$
(b) $Ca_2B_6O_{11} \cdot 5H_2O$
(c) $Ca(OH)_2$
(d) $Na_2B_4O_7 \cdot 2H_2O$

143. Mac Arthur process is used for the extraction of
(a) Au (b) Ag
(c) Cu (d) Zn

144. Which of the following is not an iron ore?
(a) Caniterite (b) Magnetite
(c) Limonite (d) Copper pyrite

145. Thermite is a mixture of iron oxide and
(a) aluminium powder (b) zinc powder
(c) potassium metal (d) sodium metal

146. In the extraction of copper from its sulphide ore, the metal is formed by reduction of Cu_2O with
(a) FeS (b) CO
(c) Cu_2S (d) SO_2

147. Which of the following is a carbonate ore?
(a) Pyrolusite (b) Malachite
(c) Diaspore (d) Cassiterite

148. Purification of alumina takes place by
(a) Bosch process (b) Hall's process
(c) Hoope's process (d) Quartation process

149. The hardest naturally occurring substance is
(a) iron (b) graphite
(c) diamond (d) astatine

150. Wolframite ore is separated from tinstone ore by the process of
(b) roasting (b) electromagnetic
(c) smelting (d) calcination

151. By annealing, steel
(a) becomes soft
(b) becomes liquid
(c) becomes hard and brittle
(d) is covered with a thin film of Fe_3O_4

152. The process of purification of metals is represented by the following scheme

$$Ti + 2I_2 \xrightarrow{250°C} TiI_4 \xrightarrow{1400°C} \underset{\text{pure}}{Ti} + I_2$$

(a) cupellation (b) poling
(c) electrolytic refining (d) van-Arkel method

153. Among the following statements, the correct one is
(a) zinc blende and iron pyrites are sulphides
(b) calamine and siderite are carbonates
(c) malachite and chalcocite are the copper ore
(d) argentite and cuprite are oxides

154. The silver is extracted by Parke's process. The basis of this method is
(a) silver is immiscible in molten Zn
(b) Ag is miscible in NaCN
(c) Ag is more miscible in molten zinc than in molten Pb
(d) Ag is more miscible in molten Pb in comparison to molten zinc

155. Railway wagon axles are made by heating iron rods embedded in charcoal powder. This process is known as
(a) tempering (b) annealing
(c) sherar dising (d) case hardening

156. Electric refining is used for refining of
(a) Cu (b) Fe
(c) Pb (d) Al

157. To dissolve argentite ore which of the following is used
(a) $Na[Ag(CN)_2]$ (b) NaCN
(c) NaCl (d) HCl

158. In aluminothermic process, Al acts as a
(a) flux (b) oxidising agent
(c) reducing agent (d) solder

159. Which of the following is not an ore of magnesium?
(a) Carnallite (b) Magnesium
(c) Dolomite (d) Gypsum

160. The formula of malachite ore is
(a) Fe_3O_4 (b) Mn_3O_4
(c) NiAs (d) $Cu(OH)_2CuCO_3$

161. Which of the following has maximum percentage of copper?
(a) Cu_2S (b) Cu_2O
(c) $CuFeS_2$ (d) $CuCO_3 \cdot Cu(OH)_2$

162. The chief impurity present in red bauxite is
(a) SiO_2 (b) Fe_2O_3
(c) K_2SO_4 (d) NaF

163. The extraction of which of the following metals involves bessemerisation?
(a) Fe (b) Ag
(c) Al (d) Cu

164. Which of the following metal is correctly matched with its ore?

	Metal	Ore
(a)	Zinc	Calamine
(b)	Silver	Illmenite
(c)	Magnesium	Cassiterite
(d)	Tin	Azurite

165. A cuprous ore among the following is
(a) cuprite (b) malachite
(c) chalcopyrites (d) azurite

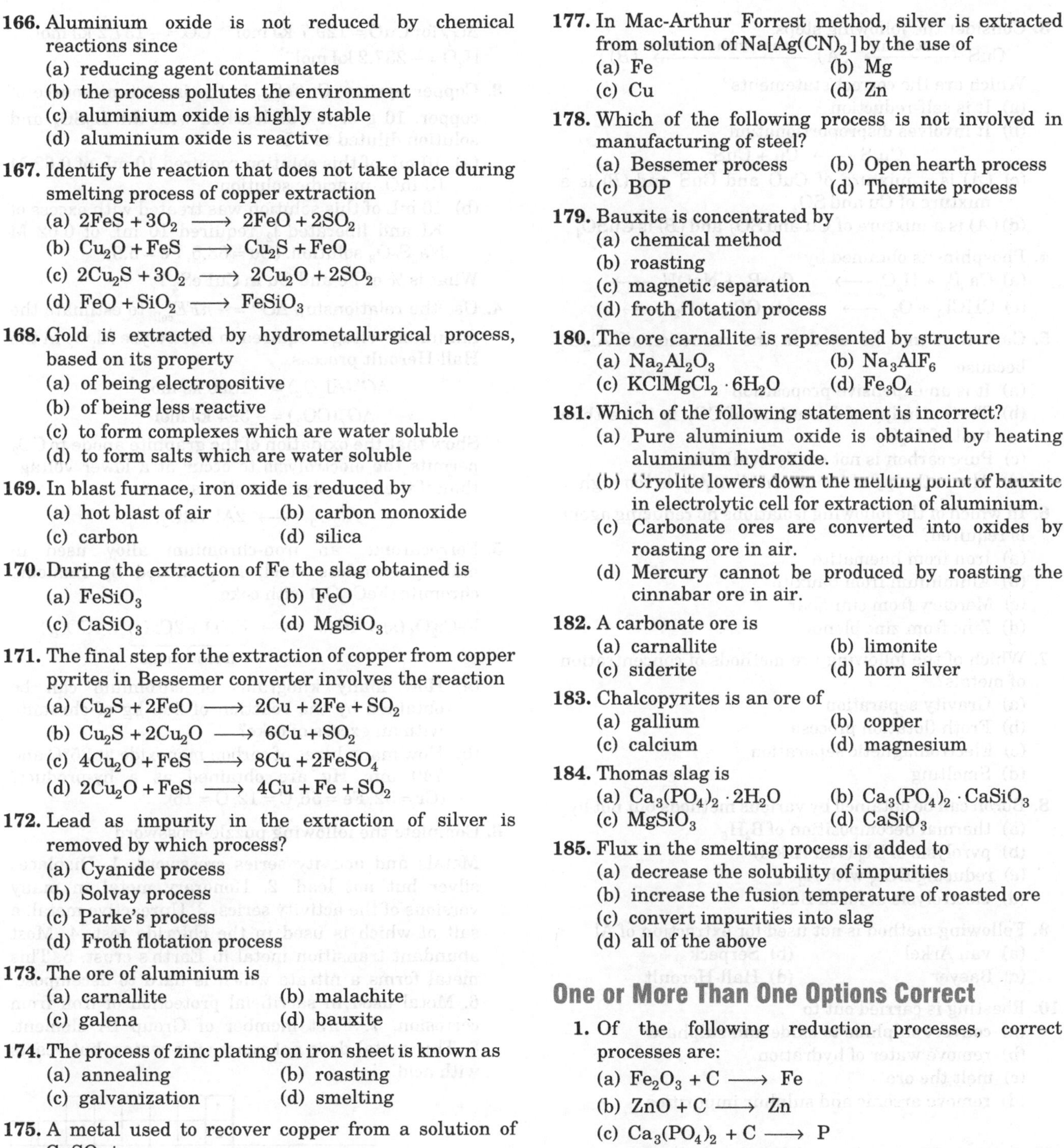

166. Aluminium oxide is not reduced by chemical reactions since
(a) reducing agent contaminates
(b) the process pollutes the environment
(c) aluminium oxide is highly stable
(d) aluminium oxide is reactive

167. Identify the reaction that does not take place during smelting process of copper extraction
(a) $2FeS + 3O_2 \longrightarrow 2FeO + 2SO_2$
(b) $Cu_2O + FeS \longrightarrow Cu_2S + FeO$
(c) $2Cu_2S + 3O_2 \longrightarrow 2Cu_2O + 2SO_2$
(d) $FeO + SiO_2 \longrightarrow FeSiO_3$

168. Gold is extracted by hydrometallurgical process, based on its property
(a) of being electropositive
(b) of being less reactive
(c) to form complexes which are water soluble
(d) to form salts which are water soluble

169. In blast furnace, iron oxide is reduced by
(a) hot blast of air (b) carbon monoxide
(c) carbon (d) silica

170. During the extraction of Fe the slag obtained is
(a) $FeSiO_3$ (b) FeO
(c) $CaSiO_3$ (d) $MgSiO_3$

171. The final step for the extraction of copper from copper pyrites in Bessemer converter involves the reaction
(a) $Cu_2S + 2FeO \longrightarrow 2Cu + 2Fe + SO_2$
(b) $Cu_2S + 2Cu_2O \longrightarrow 6Cu + SO_2$
(c) $4Cu_2O + FeS \longrightarrow 8Cu + 2FeSO_4$
(d) $2Cu_2O + FeS \longrightarrow 4Cu + Fe + SO_2$

172. Lead as impurity in the extraction of silver is removed by which process?
(a) Cyanide process
(b) Solvay process
(c) Parke's process
(d) Froth flotation process

173. The ore of aluminium is
(a) carnallite (b) malachite
(c) galena (d) bauxite

174. The process of zinc plating on iron sheet is known as
(a) annealing (b) roasting
(c) galvanization (d) smelting

175. A metal used to recover copper from a solution of $CuSO_4$ is
(a) Fe (b) Hg
(c) Na (d) Ag

176. Gravity separation process is used for the concentration of
(a) calamine (b) haematite
(c) chalcopyrites (d) bauxite

177. In Mac-Arthur Forrest method, silver is extracted from solution of $Na[Ag(CN)_2]$ by the use of
(a) Fe (b) Mg
(c) Cu (d) Zn

178. Which of the following process is not involved in manufacturing of steel?
(a) Bessemer process (b) Open hearth process
(c) BOP (d) Thermite process

179. Bauxite is concentrated by
(a) chemical method
(b) roasting
(c) magnetic separation
(d) froth flotation process

180. The ore carnallite is represented by structure
(a) $Na_2Al_2O_3$ (b) Na_3AlF_6
(c) $KClMgCl_2 \cdot 6H_2O$ (d) Fe_3O_4

181. Which of the following statement is incorrect?
(a) Pure aluminium oxide is obtained by heating aluminium hydroxide.
(b) Cryolite lowers down the melting point of bauxite in electrolytic cell for extraction of aluminium.
(c) Carbonate ores are converted into oxides by roasting ore in air.
(d) Mercury cannot be produced by roasting the cinnabar ore in air.

182. A carbonate ore is
(a) carnallite (b) limonite
(c) siderite (d) horn silver

183. Chalcopyrites is an ore of
(a) gallium (b) copper
(c) calcium (d) magnesium

184. Thomas slag is
(a) $Ca_3(PO_4)_2 \cdot 2H_2O$ (b) $Ca_3(PO_4)_2 \cdot CaSiO_3$
(c) $MgSiO_3$ (d) $CaSiO_3$

185. Flux in the smelting process is added to
(a) decrease the solubility of impurities
(b) increase the fusion temperature of roasted ore
(c) convert impurities into slag
(d) all of the above

One or More Than One Options Correct

1. Of the following reduction processes, correct processes are:
(a) $Fe_2O_3 + C \longrightarrow Fe$
(b) $ZnO + C \longrightarrow Zn$
(c) $Ca_3(PO_4)_2 + C \longrightarrow P$
(d) $PbO + C \longrightarrow Pb$

2. Following reaction(s) is/are not involved in Thermite process
(a) $3Mn_3O_4 + 8Al \longrightarrow 9Mn + 4Al_2O_3$
(b) $Cr_2O_3 + 2Al \longrightarrow Al_2O_3 + 2Cr$
(c) $2Fe + Al_2O_3 \longrightarrow 2Al + Fe_2O_3$
(d) $B_2O_3 + 2Al \longrightarrow 2B + Al_2O_3$

3. Consider the following steps:

$$CuS \xrightarrow{\text{roast in air}} (A) \xrightarrow{\text{roast without air}} (B)$$

Which are the correct statements?
(a) It is self-reduction
(b) It involves disproportionation

$$Cu_2S \longrightarrow Cu + CuS$$

(c) (*A*) is a mixture of CuO and CuS and (*B*) is a mixture of Cu and SO_2
(d) (*A*) is a mixture of Cu and SO_2 and (*B*) is $CuSO_4$

4. Phosphine is obtained by
(a) $Ca_3P_2 + H_2O \longrightarrow$ (b) $P_4 + NaOH \longrightarrow$
(c) $CHCl_3 + O_2 \longrightarrow$ (d) $CCl_4 + O_2 \longrightarrow$

5. Carbon cannot be used in the reduction of Al_2O_3 because
(a) It is an expensive proposition
(b) The enthalpy of formation of CO_2 is more than that of Al_2O_3
(c) Pure carbon is not easily available
(d) The enthalpy of formation of Al_2O_3 is too high

6. In which of the following isolations no reducing agent is required?
(a) Iron from haematite
(b) Aluminium from bauxite
(c) Mercury from cinnabar
(d) Zinc from zinc blende

7. Which of the following are methods of concentration of metals?
(a) Gravity separation
(b) Froth flotation process
(c) Electromagnetic separation
(d) Smelting

8. Boron can be obtained by various methods but not by
(a) thermal decomposition of B_2H_6
(b) pyrolysis of BI_3 (van Arkel)
(c) reducing BCl_3 with H_2
(d) electrolysis of fused BCl_3

9. Following method is not used for extraction of Al
(a) van Arkel (b) Serpeck
(c) Baeyer (d) Hall-Heroult

10. Roasting is carried out to
(a) convert sulphide to oxide and sulphate
(b) remove water of hydration
(c) melt the ore
(d) remove arsenic and sulphur impurities

Brain Twisters

1. $E^\circ_{Fe^{2+}/Fe} = 0.44$ V, $E^\circ_{Cu^{2+}/Cu} = 0.34$ V, $E^\circ_{Ag^+/Ag} = 0.80$ V
Will Fe displace Cu from $CuSO_4$ solution?
Will Ag displace Cu from $CuSO_4$ solution?

2. CuO can't be reduced by carbon but can be reduced by H_2. Explain in terms of thermodynamics, given :
ΔG°_f for CuO = 129.7 kJ mol^{-1}, CO = – 137.2 kJ mol^{-1}, H_2O = – 237.2 kJ mol^{-1}

3. Copper pyrites, $CuFeS_2$, is the important source of copper. 10 g of it was leached with dil. H_2SO_4 and solution diluted to 1 L.
(a) 10 mL of this solution required 10 mL of 0.02 M $KMnO_4$ in acidic solution
(b) 10 mL of this solution was treated with excess of KI and liberated I_2 required 10 mL of 0.02 M $Na_2S_2O_3$ solution. (Cu = 63.5, Fe = 5.6)
What is % of Fe and Cu in $CuFeS_2$?

4. Use the relationship $\Delta G^\circ = -nFE^\circ_{cell}$ to estimate the minimum voltage required to electrolyse Al_2O_3 in the Hall-Heroult process.

$$\Delta G^\circ_f(Al_2O_3) = -1520 \text{ kJ mol}^{-1}$$
$$\Delta G^\circ_f(CO_2) = -394 \text{ kJ mol}^{-1}$$

Show that the oxidation of the graphite anode to CO_2 permits the electrolysis to occur at a lower voltage than if the electrolysis reactions were

$$Al_2O_3 \longrightarrow 2Al + 3O_2.$$

5. Ferrochrome, an iron-chromium alloy used in making stainless steel, is produced by reducing chromite ($FeCr_2O_4$) with coke

$$FeCr_2O_4(s) + 4C(s) \longrightarrow \underbrace{Fe(s) + 2Cr(s)}_{\text{ferrochrome}} + 4CO(g)$$

(a) How many kilograms of chromium can be obtained by the reaction of 236 kg of chromite with an excess of coke?
(b) How many litres of carbon monoxide at 25°C and 740 mm Hg are obtained as a by-product? (Cr = 52, Fe = 56, C = 12, O = 16)

6. Complete the following puzzle-crossword
Metals and activity series crossword. 1. Displaces silver but not lead. 2. Honorary metal in many versions of the activity series. 3. Unreactive metal, a salt of which is used in the chloride test. 4. Most abundant transition metal in Earth's crust. 5. This metal forms a nitrate which is hard to decompose. 6. Metal used in sacrificial protection of iron from corrosion. 7. First member of Group IA element. 8. This metal does not react with water, but reacts with acid.

1
2
3
4
5
6
7
8

7. 18.4 g of a mixture of $CaCO_3$ and $MgCO_3$ on strongly heating gave 8.8 g of CO_2 gas. Calculate mole% of each in the mixture.

Passage Comprehension Questions

Passage 1

Read the following short write-up about recovery of magnesium and answer the questions at the end of it.

Magnesium is a valuable, light weight metal used as a structural material as well as in alloys, in batteries, and in chemical synthesis. Although magnesium is plentiful in Earth's crust, it is cheaper to mine the metal from sea-water. Magnesium forms the second most abundant cation in the sea (after sodium); there are about 1.3 g of magnesium in a kilogram of sea-water. The process for obtaining magnesium from sea-water employs all three types of reactions precipitation, acid-base, and redox reactions.

1. Precipitation reaction involves formation of
(a) insoluble $MgCO_3$ by adding Na_2CO_3
(b) insoluble $Mg(OH)_2$ by adding $Ca(OH)_2$
(c) insoluble $MgSO_4$ by adding Na_2SO_4
(d) insoluble $MgCl_2$ by adding NaCl

2. Acid-base reaction involves reaction between
(a) $MgCO_3$ and HCl (b) $Mg(OH)_2$ and H_2SO_4
(c) $Mg(OH)_2$ and HCl (d) $MgCO_3$ and H_2SO_4

3. Redox reaction takes place (in the extraction of Mg)
(a) in the electrolytic cell when fused $MgCl_2$ is subjected to electrolysis
(b) when fused $MgCl_2$ is heated
(c) when fused $MgCO_3$ is strongly heated
(d) none of the above

4. Instead of calcium oxide, why don't we simply add sodium hydroxide to sea-water to precipitate magnesium hydroxide?
(a) Solubility of $Ca(OH)_2$ is smaller than that of NaOH so that $Mg(OH)_2$ is precipitated
(b) NaOH may dissolve $Mg(OH)_2$ formed
(c) NaOH may also precipitate other species, being a strong electrolyte
(d) NaOH, being a weak electrolyte will not coagulate $Mg(OH)_2$

5. Which is the best source of the given process in the metallurgical process?
(a) Magnesite (b) Kieserite
(c) Epsomite (d) Dolomite

Passage 2

Following passage describes extraction of zinc from zinc sulphide. Answer the questions at the end of it.

The chief ore of zinc is the sulphide, ZnS. The ore is concentrated by flotation process and then heated in air, which converts the ZnS to ZnO.

$$2ZnS + 3O_2 \xrightarrow{90.6\%} 2ZnO + 2SO_2 \quad ...(i)$$

The ZnO is then treated with dilute H_2SO_4

$$ZnO + H_2SO_4 \xrightarrow{100\%} ZnSO_4 + H_2O \quad ...(ii)$$

to produce $ZnSO_4(aq)$ which produces Zn metal on electrolysis.

$$2ZnSO_4 + 2H_2O \xrightarrow{98.2\%} 2Zn + 2H_2SO_4 + O_2 \quad ...(iii)$$

1. What mass of Zn will be obtained from an ore containing 225 kg of ZnS? Efficiencies of the process have been indicated above the arrow mark. (Zn = 65, S = 32, O = 16, H = 1)
(a) 134 kg (b) 112 kg
(c) 102 kg (d) 130 kg

2. What amount of current is required (with 100% efficiency) in step (iii) if it takes one month?
(a) 10.2 A (b) 15.4 A
(c) 17.0 A (d) 154.0 A

3. ZnO in step (ii) can also be dissolved in NaOH forming
(a) $Zn(OH)_2$ (b) Na_2ZnO_2
(c) Zn_2O_3 (d) $NaZn(OH)_4$

4. How many kilomoles of NaOH are required to dissolve all the ZnO of step (i) assuming 100% yield?
(a) 2.32 (b) 1.16
(c) 4.64 (d) 9.28

5. What volume of 98% H_2SO_4 (by weight, density 1.8 g/mL) is required in step (ii) assuming 100% yield of steps (i) and (ii)
(a) 130 L (b) 140 L
(c) 120 L (d) 150 L

Passage 3

Questions given below are based on the following sequence of reactions:

At high temperature carbon reacts with water to produce a mixture of carbon monoxide (CO) and hydrogen (H_2).

$$C + H_2O \xrightarrow{\text{red heat}} CO + H_2$$

CO is separated from H_2, and then used to separate nickel from cobalt by forming a volatile compound, nickel tetracarbonyl, $Ni(CO)_4$.

$$Ni + 4CO \longrightarrow Ni(CO)_4 \quad (Ni = 59)$$

1. How many moles of $Ni(CO)_4$ could be obtained from the CO produced by the reaction of 75.0 g of carbon? Assume 100% reaction and 100% recovery in both steps.
(a) 6.25 (b) 1.563
(c) 3.125 (d) 25.0

2. Formation of volatile $Ni(CO)_4$ and its subsequent heating gives pure Ni. This process is called
(a) Hall (b) Dow
(c) Serpeck (d) Mond

Passage 4

$FeCr_2O_4$ (chromite) is a good source of chromium and its compounds like Na_2CrO_4, $Na_2Cr_2O_7$. Questions are based on the following reactions.

P : $FeCr_2O_4 + NaOH + \text{air} \longrightarrow A + Fe_2O_3$

Q : $A + B \longrightarrow Na_2Cr_2O_7$

R : $Na_2Cr_2O_7 + X \xrightarrow{\Delta} Cr_2O_3$

S : $Cr_2O_3 + Y \longrightarrow Cr$

1. Compounds (A) and (B) are
(a) Na_2CrO_4, H_2SO_4 (b) $Na_2Cr_2O_7$, HCl
(c) Na_2CrO_5, H_2SO_4 (d) $Na_4[Fe(OH)_6], H_2SO_4$

2. (X) and (Y) are
(a) Al and C (b) C and Al
(c) C in both (d) Al in both

3. Na_2CrO_4 and Fe_2O_3 are separated by
(a) dissolving in conc. H_2SO_4
(b) dissolving in NH_3
(c) dissolving in H_2O
(d) dissolving in dil. HCl

Passage 5

Questions given below are based on the Periodic Table showing different ores / minerals.

1	2											13	14	15	16	17	18
Li	Be																
Na	Mg	3	4	5	6	7	8	9	10	11	12	Al					
K	Ca	Sc	Ti	V	Cr	Mn	Fe	Co	Ni	Cu	Zn	Ga					
Rb	Sr	Y	Zr	Nb	Mo	Tc	Ru	Rh	Pd	Ag	Cd	In	Sn				
Cs	Ba	La	Hf	Ta	W	Re	Os	Ir	Pt	Au	Hg	Tl	Pb	Bi			

Chlorides Silicates Carbonates Phosphates Oxides Free metals Sulphides

Primary mineral sources of metals. The *s*-block metals occur as chloride, silicates, and carbonates. The *d*- and *p*-block metals are found as oxides and sulphides, except for the group 3, metals, which occur as phoshates, and the platinum group metals and gold, which occur in uncombined form. There is no mineral source of technitium (Tc in group 7), a radioactive element that is made in nuclear reactors.

1. The early transition metals on the left hand side of the *d*-block generally occur as oxides because
(a) they are above hydrogen in electrochemical series
(b) they are more electropositive and thus form ionic compounds (in the form of oxides) by more electronegative oxygen by exchange of electrons
(c) these oxides are soluble in water
(d) these oxides remain insoluble in water

2. Transition metals on the right hand side of the *d*-block occurs as sulphides because of
(a) their high electronegativity values as a result of which covalent sulphides are formed
(b) their much low electronegativity values as a result of which ionic sulphides are formed
(c) high melting point of sulphides
(d) low melting point of sulphides

3. *s*-block elements generally occurs as carbonates and silicates (as $MgCO_3$, $CaSiO_3$) because
(a) these elements react with CO_2 and SiO_2
(b) these elements are decomposed by acid reacts
(c) their oxides are basic and oxides react with acidic oxides in an acid-base reactions
(d) their oxides are acidic and oxides react with basic oxides in an acid-base reaction

4. Elements found in native state are
(a) Pb and Bi as they are in *p*-block
(b) Hg and Os as they are in liquid state
(c) C and Si as they are non-metals
(d) Pt and Au as they are at the bottom of the electrochemical series

Passage 6

Questions given below are based on the natural occurring minerals / ores.

Look at the location of elements *A, B, C* and *D* in the following Periodic Table and answer the questions given below

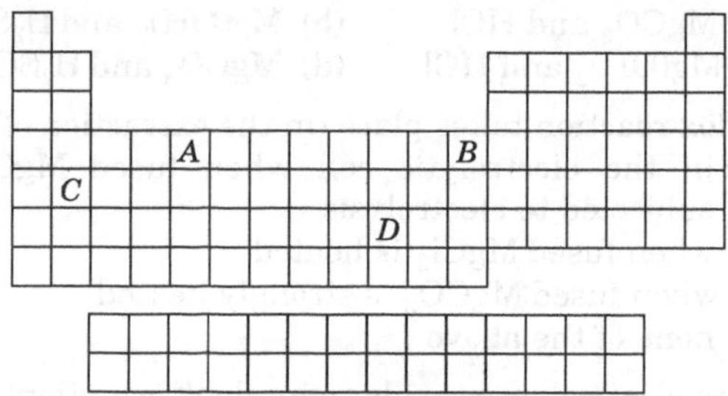

1. Which of the elements indicated by *A, B, C* and *D* is expected to be found in native state?
(a) *A* (b) *B*
(c) *C* (d) *D*

2. Which is found as its sulphide?
(a) *A* (b) *B*
(c) *C* (d) *D*

3. Which is found as its carbonate?
(a) *A* (b) *B*
(c) *C* (d) *D*

4. Which is found as its oxide?
(a) *A* (b) *B*
(c) *C* (d) *D*

5. Imagine a planet with an atmosphere that contains O_2 and SO_2 but no CO_2. What is the chemical composition of the mineral you would expect to find for the alkaline earth metals (M) on such a planet?
(a) MO_2, M_2O_2 (b) MSO_3, MSO_4
(c) $M(HSO_3)_2, M(HSO_4)_2$ (d) all of these

Passage 7

Read the following cyclic process of recovery of copper from a copper wire and answer the questions at the end of it.

The following cycle of copper experiment is performed in some general chemistry laboratories. The series of reactions starts with copper and ends with metallic copper. The steps are as follows: (1) A piece of copper wire of known mass is allowed to react with concentrated nitric acid [the products are copper(II) nitrate, nitrogen dioxide, and water]. (2) The copper (II) nitrate is treated with a sodium hydroxide solution to form copper (II) hydroxide precipitate. (3) On heating copper (II) hydroxide decomposes to yield copper (II) oxide. (4) The copper (II) oxide is reacted with concentrated sulphuric acid to yield copper (II) sulphate. (5) Copper(II) sulphate is treated with an excess of zinc metal to form metallic copper. (6) The remaining zinc metal is removed by treatment with hydrochloric acid, and metallic copper is filtered, dried, and weighted. (Cu = 63.5)

1. Assuming that a student started with 65.6 g of copper, calculate the theoretical yield of copper sulphate
   (a) 165 g (b) 82.4 g
   (c) 90 g (d) 100.2 g
2. Copper obtained at the end of the cycle is .... amount originally taken
   (a) less than
   (b) greater than
   (c) nearly equal
   (d) only copper salts are obtained
3. $CuSO_4$ solution thus formed is subjected to electrolysis. To deposit all the copper formed at the end, a current of 50 A will have to be used for approximately
   (a) 10 h (b) 1 h
   (c) 5 h (d) 0.5 h

Passage 8

Given flow-sheet is for the extraction of copper. Based on this, answer the questions given at the end

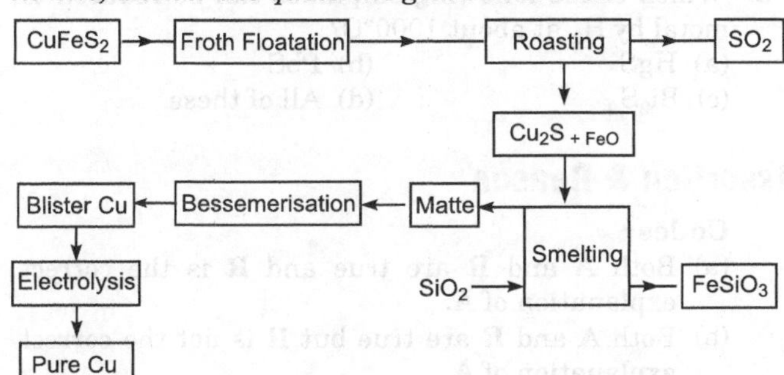

1. Froth-flotation process can be made more effective by
   (a) use of pine oil which produces stable froth
   (b) use of sodium ethyl xanthate as a collector that coats the particles of ore
   (c) both (a) and (b)
   (d) none of the above
2. Smelting process is carried at about 1400°C. Two layers are formed
   (a) Bottom layer contains copper-matte
   (b) Upper layer contains copper-matte
   (c) Copper-matte remains in both the layers
   (d) Copper-matte is not present in any of the layers
3. Some of the reactions taking place during bessemerisation are

   A : $FeO + SiO_2 \longrightarrow FeSiO_3$

   B : $2FeS + 3O_2 \longrightarrow 2FeO + 2SO_2$

   C : $2Cu_2S + 3O_2 \longrightarrow 2Cu_2O + 2SO_2$

   D : $2Cu_2O + Cu_2S \longrightarrow 6Cu + SO_2$

   Select reactions actually taking place
   (a) *A, B* (b) *B, C*
   (c) *A, C* (d) *C, D*
4. In the electrorefining method
   (a) anode is pure copper
   (b) anode is impure copper
   (c) anode mud contains noble metals
   (d) cathode mud contains noble metals
5. This method is called
   (a) pyrometallurgical (b) hydrometallurgical
   (c) parametallurgical (d) semimetallurgical

Passage 9

Questions given below are based on the given diagram for extractive metallurgy.

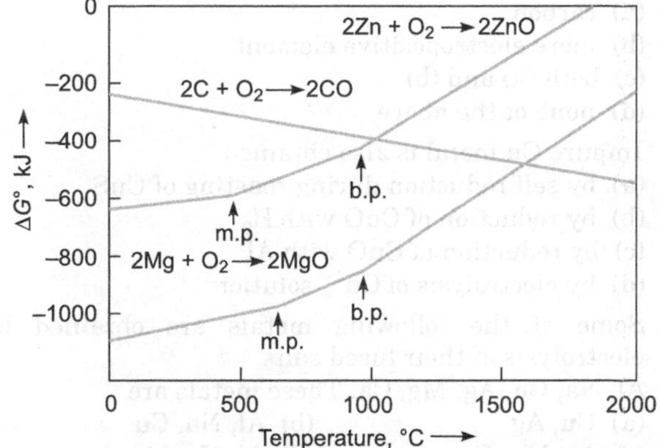

The points noted by arrows are the melting and boiling points of the metals zinc and magnesium. $\Delta G°$ as a function of temperature for some reactions of extractive metallurgy.

1. At what approximate temperature, zinc and carbon have equal affinity for oxygen?
   (a) 1000°C (b) 1500°C
   (c) 500°C (d) 1200°C

2. At this temperature ΔG° of the reaction is

$$ZnO + C \longrightarrow Zn + CO$$

(a) –ve (b) +ve
(c) zero (d) nothing can be said

3. To make the following reduction process spontaneous, temperature should be

$$ZnO + C \longrightarrow Zn + CO$$

(a) < 1000°C
(b) > 1100°C
(c) < 500°C
(d) > 500°C but < 1000°C

4. At 1100°C, which reaction is spontaneous to a maximum extent?
(a) $MgO + C \longrightarrow Mg + CO$
(b) $ZnO + C \longrightarrow Zn + CO$
(c) $MgO + Zn \longrightarrow Mg + ZnO$
(d) $ZnO + Mg \longrightarrow MgO + Zn$

Passage 10

From the following flow-sheet for the extraction of pure metal, answer the questions at the end of it.

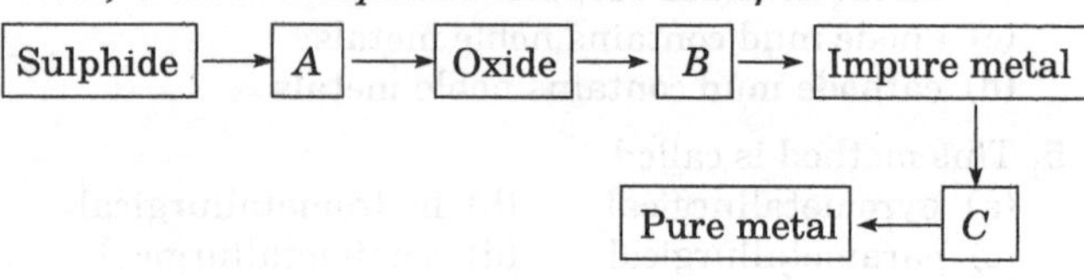

1. Step *A* is
(a) roasting (b) smelting
(c) calcination (d) bessemerisation

2. Step *B* (reduction) can be carried out using
(a) carbon
(b) more electropositive element
(c) both (a) and (b)
(d) none of the above

3. Impure Cu metal is also obtained
(a) by self reduction during roasting of CuS
(b) by reduction of CuO with H_2
(c) by reduction of CuO with Al
(d) by electrolysis of Cu^{2+} solution

4. Some of the following metals are obtained by electrolysis of their fused salts
Al, Na, Cu, Ag, Mg, Ca. These metals are
(a) Cu, Ag (b) Al, Na, Cu
(c) Ag, Mg, Ca (d) Al, Na, Mg, Ca

5. Reduction of oxides to elements with carbon generally takes place at high temperature, hence Al is used in aluminothermite process. It is because
(a) Al is more electropositive than the other metals (to be formed by reduction)
(b) Al has maximum affinity for oxygen
(c) reduction is higly exothermic and heat liberated makes the process spontaneous
(d) reduction is highly endothermic and heat absorbed makes the process spontaneous.

6. Following reaction is not involved in Thermite process
(a) $3Mn_3O_4 + 8Al \longrightarrow 9Mn + 4Al_2O_3$
(b) $Cr_2O_3 + 2Al \longrightarrow Al_2O_3 + 2Cr$
(c) $2Fe + Al_2O_3 \longrightarrow 2Al + Fe_2O_3$
(d) $B_2O_3 + 2Al \longrightarrow 2B + Al_2O_3$

Passage 11

The Ellingham diagram for a number of metallic sulphides is reproduced below.

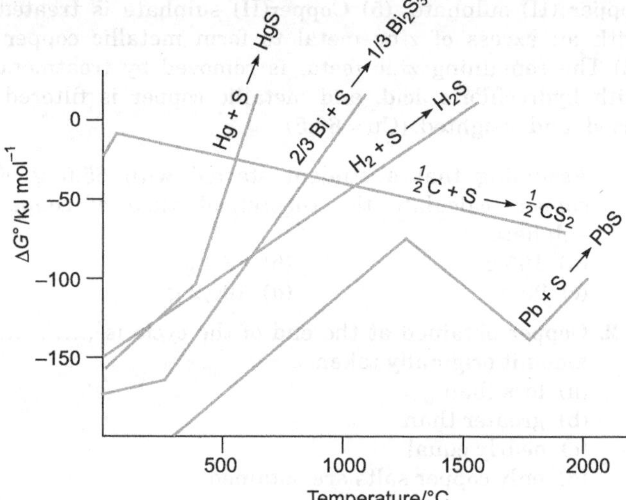

Answer the questions given below

1. Formation of which of the sulphides is most spontaneous?
(a) HgS (b) Bi_2S_3
(c) PbS (d) CS_2

2. Which occurs to minimum extent in nature?
(a) HgS (b) H_2S
(c) Bi_2S_3 (d) CS_2

3. Which of the following sulphides can be reduced to metal by H_2 at about 1000°C?
(a) HgS (b) PbS
(c) Bi_2S_3 (d) All of these

Assertion & Reason

Codes :
(a) Both **A** and **R** are true and **R** is the correct explanation of **A**.
(b) Both **A** and **R** are true but **R** is not the correct explanation of **A**.
(c) **A** is true but **R** is false.
(d) **A** is false but **R** is true.
(e) **A** and **R** both are false.

1. **Assertion (A) :** Gold occurs in native state.

   Reason (R) : Gold dissolves in aqua-regia.

2. **Assertion (A) :** Froth-flotation process is used to concentrate sulphide ores.

   Reason (R) : There is difference in the wettability of different minerals.

3. **Assertion (A) :** Al_2O_3 is converted into Al by reduction with carbon.

   Reason (R) : Carbon (graphite) has greater affinity for oxygen than aluminium has.

4. **Assertion (A) :** Metals of high impurity are obtained by zone-refining.

   Reason (R) : Impurities in the ore are more soluble in the melt than in the pure metal.

5. **Assertion (A) :** Reduction of ZnO with carbon is done at 1100°C.

   Reason (R) : At this temperature, $\Delta G°$ is negative thus process is spontaneous.

6. **Assertion (A) :** Desilverisation of lead is done by Parke's method.

   Reason (R) : When lead-silver alloy is poor in silver, zinc is added to molten ore.

True & False

1. Lunar caustic is KOH.
2. KCN forms soluble complex with silver glance.
3. Zinc can displace silver from $[Ag(CN)_2]^-$.
4. Main source of aluminium is corundum.
5. Wrought iron is prepared by Pudding process.
6. Pig iron is the purest form of iron.
7. Steel is mostly made by BOP method.
8. Heating the steel to appropriate temperature for a short time and then cooling it rapidly in order to give it the desired mechanical properties is called the tempering.
9. Oxides of the *s*-block metals are strongly acidic and form salts with basic oxides.
10. Oxygen is the most abundant metal on the earth's crust.
11. Mond's process is used for the extraction of Mg.
12. In the bessemerisation of sulphide ore of copper, there is self-reduction giving copper metal.
13. Na is obtained by the electrolysis of aqueous NaCl solution.
14. Al is obtained by the electrolysis of fused mixture of Al_2O_3 and Na_3AlF_6.
15. Metals of high purity (to be used in semiconductors) are obtained by zone-refining.
16. Concentration of sulphide ore is done by calcination method.
17. Sulphide ore is converted into pure metal (say Cu) by following steps.

    Bessemerisation $\longrightarrow$ Roasting $\longrightarrow$ Electrolysis
18. Dow sea-water process is used for the extraction of Mg.
19. $FeWO_4$ and $MnWO_4$ are magnetic impurities.
20. Calcium silicate is called cementite.

Fill in the Blanks

1. When $(NH_4)_2Cr_2O_7$ is heated, green coloured residue is .... along with gas ....
2. The reducing agent used in thermite process is ....
3. In cyanide process, Zn is added to soluble ....to get Ag.
4. Temperature required in pyrometallurgical process is .... than that required in hydrometallurgical process.
5. .... iron is the purest form of it.
6. .... iron is brittle.
7. Most abundant element in the earth's crust is ....
8. Most abundant metal in the earth's crust is ....
9. Sulphur is extracted by ....
10. ZrI_4 is converted to Zr by the ....
11. Mg can be used to reduce $TiCl_4$ into Ti in ....
12. Self-reduction process is used in the extraction of .... and lead.
13. Dow sea-water process is used in extraction of ....
14. In Serpeck process, bauxite is converted into .... which is hydrolysed to precipitate ....
15. Cryolite, Na_3AlF_6 is added to Al_2O_3 in .... process to .... m.p. and to .... electrical conductivity.
16. Parke's process is used to separate .... from argentiferrous lead.
17. Be_2C and Al_4C_3 are also called .... because they react with H_2O yielding ....
18. CO is toxic because it forms complex with .... in the blood.
19. When $FeCr_2O_4$ is heated with carbon an alloy called .... is obtained.
20. FeS_2 is also called .... because it has been mistaken for gold.

21. '18 carrat' gold is .... pure.

22. 'Puddling' process was used originally to make ....

23. Puddling process involved mixing molten iron with haematite and burning off all the C and other impurities to give .... iron.

24. BOP is the modern process to prepare ....

25. Siemen open-hearth process is used in making ....

26. Mond's process is used in the extraction of ....

27. Cu_2S is called ....

28. Malachite (green) is ....

29. Zinc is used to extract silver by solvent extraction from molten lead in .... process

30. Paris green is .... made from basic copper acetate, arsenious oxide and acetic acid.

Matrix-Match Type Questions

I. Only One Correct

1. Match the compounds (in List I) with their applications/colours (in List II).

List I	List II
A. Pb_3O_4	1. White lead
B. $(PbCO_3)_2 \cdot Pb(OH)_2$	2. Rust proofing sheets
C. Ca_2PbO_4	3. Road signs and markings
D. $PbCrO_4$	4. Red lead

2. Match the reactions taking place in blast furnace (in List I) with temperature-range of operations (in List II).

List I	List II
A. $Fe_2O_3 + CO \longrightarrow CO_2 + Fe$	1. 1200°C
B. $C + CO_2 \longrightarrow CO$	2. 900°C
C. $C + H_2O \longrightarrow CO + H_2$	3. 600°C
D. $CaO + SiO_2 \longrightarrow CaSiO_3$	4. 1100°C

3. Match process (in List I) with electrolytes (in List II).

List I	List II
A. Downs cell	1. Fused $MgCl_2$
B. Dow sea-water process	2. Fused ($Al_2O_3 + Na_3AlF_6$)
C. Hall-Heroult	3. Fused KHF_2
D. Moissan	4. Fused (40% NaCl + 60% $CaCl_2$)

4. Match ores in List I with metals in List II.

List I	List II
A. Carnallite	1. Zinc
B. Calamine	2. Titanium
C. Ilmenite	3. Magnesium
D. Chalcopyrite	4. Copper

5. Match the extraction process listed in List I with metals listed in List II.

List I	List II
A. Self reduction	1. Zirconium
B. Carbon reduction	2. Silver
C. Complex formation and displacement by metal	3. Copper
D. Decomposition of iodide	4. Boron

6. Match the method in List I with related reaction in List II.

List I	List II
A. Mond's process	1. $Cr_2O_3 + 2Al \longrightarrow Al_2O_3 + 2Cr$
B. van Arkel-de-Boer process	2. $TiCl_4 + Mg \xrightarrow{1000-1150°C} Ti + 2MgCl_2$
C. Thermite process	3. $Ni(CO)_4 \xrightarrow{230°C} Ni + 4CO$
D. Kroll process	4. $2CuO + CuS \longrightarrow 3Cu + SO_2$
E. Self reduction	5. $ZrI_4 \xrightarrow{\Delta} Zr + 2I_2$

II. One or More Correct

1. Match the species in Column I with then corresponding methods of extraction/concentration in Column II.

List I	List II
A. Copper pyrites	1. Electrolysis
B. Galena	2. Mac-Arthur Forrest method
C. Silver glance	3. Extraction by complex formation
D. Corundum	4. Magnetic separation
E. Ilmenite	5. Concentration by gravity separation
	6. Concentration by froth-flotation process

2. Match the elements in Column I with their corresponding mineral formation in Column II.

List I	List II
A. Calcium	1. Oxide
B. Copper	2. Sulphide
C. Magnesium	3. Phosphate
D. Silver	4. Carbonate
E. Aluminium	5. Chloride
	6. Fluoride
	7. Sulphate

Integer Answer Types

This section contains 9 questions. The answer to each of the questions is a single digit integer, ranging from 0 to 9. The appropriate bubbles below the respectively question numbers in the *ORS* have to be darkened. For example, if the correct answers to question number *X, Y, Z* and *W* (say) are 6, 0, 9 and 2, respectively, then the correct darkening of bubbles will like the following

X	Y	Z	W
⓪	⓪	⓪	⓪
①	①	①	①
②	②	②	②
③	③	③	③
④	④	④	④
⑤	⑤	⑤	⑤
⑥	⑥	⑥	⑥
⑦	⑦	⑦	⑦
⑧	⑧	⑧	⑧
⑨	⑨	⑨	⑨

Set I

1. Bauxite is $Al_2O_3 \cdot xH_2O$, where x = ...
2. Epsomite is $MgSO_4 \cdot xH_2O$, where y = ...
3. When one mole of Ag_2CO_3 is strongly heated, residue is .... mole(s).
4. In the following reaction slag $CaSiO_3$ is formed
$$Ca_3(PO_4)_2 + C + SiO_2 \longrightarrow P_4 + CaSiO_3 + CO$$
If one mole of P_4 is formed, then $CaSiO_3$ formed is .... moles.
5. One unit formula of azurite contains .... copper atoms
6. Total number of atoms in one unit of chalcopyrites are ....
7. One unit of corundum has .... oxygen atoms.
8. Mineral $CuFeS_2$ is 50% pure.
9. 5 moles of $CuFeS_2$ is dissolved in H_2SO_4. Mixture will require .... mole(s) $KMnO_4$ to oxidise Fe^{2+} to Fe^{3+}.

Set II

This section contains 8 questions.

1. E°_{red} values are :

	E°
$Ag^+ + e^- \longrightarrow Ag$	+ 0.80 V
$Zn^{2+} + 2e^- \longrightarrow Zn$	− 0.76 V
$Fe^{2+} + 2e^- \longrightarrow Fe$	− 0.44 V
$Cu^{2+} + 2e^- \longrightarrow Cu$	+ 0.34 V
$Hg^{2+} + 2e^- \longrightarrow Hg$	+ 0.79 V
$Mg^{2+} + 2e^- \longrightarrow Mg$	− 2.37 V
$Al^{3+} + 3e^- \longrightarrow Al$	− 1.66 V

How many of the metals can displace Cu from $CuSO_4$ solution?

2. If reduction of Cu^{2+} by Al is possible then E°_{cell} will be ...... volt. (based on Q. 1)
$$3Cu^{2+} + 2Al \longrightarrow 2Al^{3+} + 3Cu$$
3. How many of the following ores can be enriched by FFP?

dolomite, cerussite, argentite, pyrites, anhydrite, cinnabar, chalcopyrite, galena, greenockite, chalcocite

4. Phosphate rock is represented as $Pb_{10}(PO_4)_xCl_2$ what is value of x?
5. In the electrolysis of $AgNO_3$ solution, 97.2 g of Ag is deposited on passing current for 9650 s. What is the value of current in amperes?
6. Al_2O_3 is converted to AlN on heating with carbon in the atmosphere of N_2. How much carbon is required to convert 23 g of Al_2O_3 into AlN?
7. $Al(OH)_3$ is soluble in alkaline solution of pH 8.3010. Thus $[NaOH] = y \times 10^{-6}$ what is value of y?
8. Gold is found to be 37.5% pure. What is its grading in terms of carrat?

Test Yourself On The Periodic Table

1. Locate the following metals in the Periodic Table and predict whether they are likely to be found in nature as oxides, sulphides, carbonates or in uncombined state

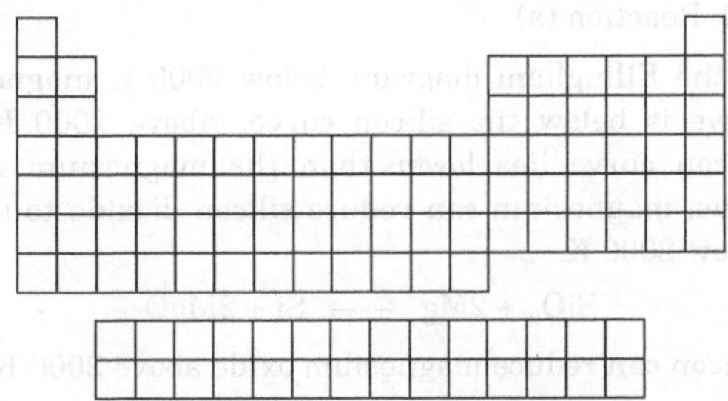

A : Copper B : Calcium
C : Sodium D : Chromium
E : Mercury F : Gold

2. Name the final process to obtain the metals indicated in the Periodic Table below:

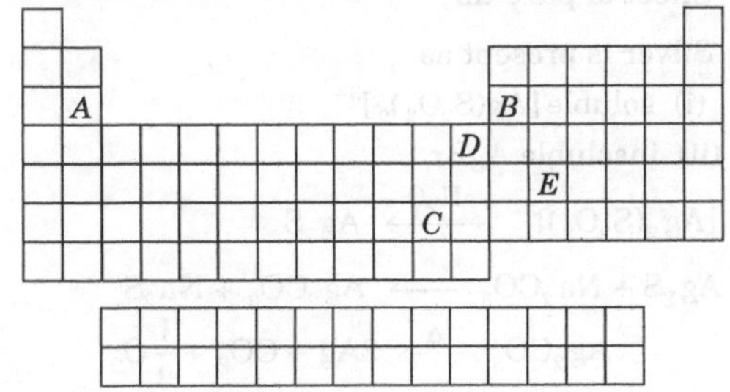

Answers & Solutions
Master Exercises
Exercise 1

Short Answer Type Questions

1. In practice, sulphide ore is first roasted in air to the corresponding oxide which in turn is reduced to metal with carbon.

$$2ZnS + 3O_2 \longrightarrow 2ZnO + 2SO_2\uparrow$$
$$ZnO + C \longrightarrow Zn + CO\uparrow$$
$$2ZnO + C \longrightarrow 2Zn + CO_2\uparrow$$

CO, CO_2 and SO_2 being gases escape from the reaction and thus, entropy change is positive and reaction is thermodynamically favourable.

If ZnS is directly reduced with carbon, CS_2 (liquid) is formed which remain in reaction mixture. Entropy change is less positive and reaction is thermodynamically less favourable.

$$2ZnS + C \longrightarrow 2Zn + CS_2 \text{ (liquid)}$$

2. (i) Reaction (a) since gaseous species (CO, SO_2) escape from the mixture
   (ii) Reaction (a)
   (iii) Reaction (a)

3. In the Ellingham diagram, below 2000 K magnesium curve is below the silicon curve. Above 2000 K, the silicon curve lies lower than the magnesium curve. Thus, magnesium can reduce silicon dioxide to silicon below 2000 K.

$$SiO_2 + 2Mg \longrightarrow Si + 2MgO$$

Silicon can reduce magnesium oxide above 2000 K

$$2MgO + Si \longrightarrow SiO_2 + 2Mg$$

4. (a) $FeWO_4$ (b) $MnWO_4$

5. A : $SnSO_4$ in dil. H_2SO_4
   B : Blocks of impure tin
   C : Sheet of pure tin

6. (a) Silver is present as
   (i) soluble $[Ag(S_2O_3)_2]^{3-}$
   (ii) insoluble AgBr

(b) $[Ag_2(S_2O_3)]^{3-} \xrightarrow{H_2O} Ag_2S$

$$Ag_2S + Na_2CO_3 \longrightarrow Ag_2CO_3 + Na_2S$$
$$Ag_2CO_3 \xrightarrow{\Delta} 2Ag + CO_2 + \frac{1}{2}O_2$$

(c) AgBr and Ag_2S

7. Air is passed when S^{2-} is oxidised to SO_4^{2-}

$$S^{2-} + 2O_2 \longrightarrow SO_4^{2-}$$

8. Add NaCN. ZnS changes to soluble complex

$$ZnS + 4NaCN \longrightarrow Na_2[Zn(CN)_4] + Na_2S$$

PbS can now be enriched by froth flotation process without affecting ZnS.

9. (a) Mineral containing silver and lead is melted, and molten zinc is added. Ag is distributed between zinc and lead layers (zinc being lighter forms upper layer). Since, solubility of Ag in molten zinc is higher that in molten lead, hence, most of the silver goes from lead layer to zinc layer. Two layers are separated. Ag containing zinc layer is distilled when zinc separates leaving behind silver.
   (b) Parke's method.

10. (a) Basic oxygen process
    (b) (i) pig iron (ii) cast iron (iii) wrought iron
    (c) wrought iron

11. Forward reaction is endothermic so that formation of cementite (Fe_3C) is favoured at high temperatures. When steel containing Fe_3C is cooled slowly, equilibrium is displaced in left hand side and the carbon separates. If the steel is cooled rapidly, equilibrium is not attained and the carbon remains in the form of Fe_3C.

Analytical Questions

1. (a) $\Delta G° = \Delta G_f°$ (products) $- \Delta G_f°$ (reactants)

(I) $\Delta G°$ (element) $= 0$

$$\Delta G° = \Delta G°(CS_2) - 2\Delta G°(CuS)$$
$$= 63.6 - 2(-49)$$
$$= 161.6 \text{ kJ mol}^{-1}$$

(II) $\Delta G° = 2703.6 \text{ kJ mol}^{-1}$

(b) If $\Delta G° = -$ ve then spontaneous $(\rightarrow)$
$= +$ ve then non-spontaneous $(\leftarrow)$

(III) $\Delta G_f° = -757.2 \text{ kJ mol}^{-1}$

(c) Formation of SO_2 is more spontaneous

(d) $\Delta G = \Delta H - T\Delta S$

If T is low, ΔS is also low such that $T\Delta S < \Delta H$ making ΔG positive hence, non-spontaneous.

At higher temperature, reactions are spontaneous.

2. $$3CaO + 2Al \longrightarrow Al_2O_3 + 3Ca$$

$$\Delta G^\circ = \Delta G^\circ_f(Al_2O_3) - 3\Delta G^\circ_f(CaO)$$
$$= -1582 + 1812.6$$
$$= +ve$$

hence non-spontaneous.
Student was wrong.

3. Au dissolves in aqua-regia (1 part conc. HNO_3 and 3 parts conc. HCl) forming soluble $HAuCl_4$, but Ag forms insoluble AgCl.

4. $$FeCr_2O_4 \xrightarrow[\Delta]{\text{air, } Na_2CO_3} Na_2CrO_4 \xrightarrow{H_2SO_4} Na_2Cr_2O_7$$
$$Cr \xleftarrow[\Delta]{Al} Cr_2O_3 \xleftarrow{\text{heat}} (NH_4)_2Cr_2O_7 \xleftarrow{NH_4Cl} Na_2Cr_2O_7$$

5. The impure lead metal serves as the anode and pure lead serves as the cathode. Impurities either remain at the anode or go into solution.

Anode : $Pb(s) \longrightarrow Pb^{2+}(aq) + 2e^-$

Cathode : $PbSiF_6(aq) + 2e^- \longrightarrow Pb(s) + SiF_6^{2-}(aq)$

6. (a) $$Fe_2O_3 + 3H_2 \longrightarrow 2Fe + 3H_2O$$

(b) $Na[Al(OH)_4] + HCl \longrightarrow NaCl + Al(OH)_3 + H_2O$

but strong acid like HCl will further dissolve $Al(OH)_3$ which is amphoteric. Aqueous solution of NH_4Cl is slightly acidic and this acidic solution will cause precipitation of $Al(OH)_3$ from $Na[Al(OH)_4]$

$$NH_4^+ + 2H_2O \rightleftharpoons NH_4OH + H_3O^+$$
$$[Al(OH)_4]^- + H_3O^+ \rightleftharpoons Al(OH)_3 \downarrow + 2H_2O$$

(c) $AgCl + Na_2CO_3 \xrightarrow{\Delta} Ag_2CO_3 \xrightarrow{\Delta} Ag_2O + CO_2$

$$Ag_2O \xrightarrow{\Delta} 2Ag + \frac{1}{2}O_2$$

7. Since, $E^\circ < 0$ for Zn^{2+}; the reduction of Zn^{2+} is not favoured. Since, $E^\circ > 0$ for Hg^{2+}, the reduction of Hg^{2+} is favoured. The roasting of CdS should give CdO since $E^\circ < 0$ for Cd^{2+}.

8. Since, N_2 can also combine with Mg giving magnesium nitride, and inert atmosphere no longer remains

$$3Mg + N_2 \longrightarrow Mg_3N_2$$

9. (a)
$$Fe_3O_4(s) + 4CO(g) + \frac{1}{2}O_2(g) \longrightarrow 3Fe(l) + 4CO_2(g)$$

(b) balanced

(c) $2Au(s) + 4CN^-(aq) + H_2O(l) \longrightarrow 2[Au(CN)_2]^-(aq) + 2OH^-(aq)$

(c) is used to convert mineral to a free metal

10. (a) $$\underbrace{Cu(OH)_2 \cdot 2CuCO_3}_{} \equiv 3Cu$$
$$344.5 \text{ g} \equiv 3 \times 63.5 \text{ g Cu}$$
$$0.033 \text{ g} = \frac{3 \times 63.5 \times 0.033 \text{ g}}{344.5} \text{ Cu} = 0.0182 \text{ g}$$

(b) $CuFeS_2 \equiv Cu$
$$183.5 \text{ g} \equiv 63.5 \text{ g}$$
$$0.0495 \text{ g} = \frac{63.5 \times 0.0495}{183.5} \text{ g Cu} = 0.0171 \text{ g}$$

(a) gives greater quantity of ore

11. Sulphide is = MS (oxidation number of metal ion being +2)

$$\underset{\substack{(M+32)\\1.164}}{MS} + \frac{3}{2}O_2 \longrightarrow \underset{\substack{(M+16)\\0.972}}{MO} + SO_2$$

$$\therefore \quad \frac{M+32}{1.164} = \frac{M+16}{0.972}$$

$$\therefore \quad M = 64$$

12. $W = zit$, $Al^{3+} + 3e^- \longrightarrow Al$

where, z = electrochemical equivalent
$$= \frac{\text{atomic mass}}{\text{electrons} \times 96500} = \frac{27}{3 \times 96500}$$
$$i = \text{current } (A) = 94.0 \text{ A}$$
$$t = \text{time (seconds)} = 1 \text{ h} = 3600 \text{ s}$$
$$\therefore \quad w = \frac{9 \times 94 \times 3600}{96500}$$
$$= \mathbf{31.56 \text{ g}}$$

13. $$\underset{\substack{24\text{ g}\\1272\text{ g}}}{Mg^{2+}} + \underset{\substack{74\text{ g}\\?}}{Ca(OH)_2} \longrightarrow Mg(OH)_2 + Ca^{2+}$$

24 g of Mg^{2+} is precipitated by = 74 g $Ca(OH)_2$

$\therefore$ 1272 g of Mg^{2+} is precipitated by $= \frac{74 \times 1272}{24}$ g
$$= 3922 \text{ g}$$
$$= \mathbf{3.922 \text{ kg}}$$

Exercise 2

Only One Option Correct

1. (b) **2.** (b) **3.** (c) **4.** (b) **5.** (c) **6.** (c) **7.** (b) **8.** (a) **9.** (c) **10.** (b)
11. (c) **12.** (a) **13.** (a) **14.** (a) **15.** (b) **16.** (a) **17.** (b) **18.** (d) **19.** (a) **20.** (d)
21. (b) **22.** (b) **23.** (d) **24.** (c) **25.** (d) **26.** (c) **27.** (b) **28.** (d) **29.** (c) **30.** (a)
31. (c) **32.** (d) **33.** (b) **34.** (d) **35.** (c) **36.** (d) **37.** (c) **38.** (d) **39.** (b) **40.** (c)
41. (a) **42.** (b) **43.** (a) **44.** (c) **45.** (b) **46.** (c) **47.** (b) **48.** (b) **49.** (c) **50.** (b)
51. (a) **52.** (d) **53.** (b) **54.** (d) **55.** (a) **56.** (d) **57.** (c) **58.** (d) **59.** (a) **60.** (d)
61. (c) **62.** (d) **63.** (d) **64.** (d) **65.** (b) **66.** (d) **67.** (c) **68.** (b) **69.** (a) **70.** (a)
71. (a) **72.** (b) **73.** (a) **74.** (b) **75.** (a) **76.** (b) **77.** (d) **78.** (a) **79.** (a) **80.** (c)
81. (c) **82.** (b) **83.** (d) **84.** (c) **85.** (c) **86.** (a) **87.** (a) **88.** (b) **89.** (d) **90.** (c)
91. (d) **92.** (c) **93.** (d) **94.** (d) **95.** (a) **96.** (a) **97.** (b) **98.** (c) **99.** (d) **100.** (b)
101. (b) **102.** (c) **103.** (c) **104.** (a) **105.** (b) **106.** (c) **107.** (c) **108.** (a) **109.** (d) **110.** (b)
111. (c) **112.** (c) **113.** (d) **114.** (a) **115.** (b) **116.** (b) **117.** (c) **118.** (b) **119.** (a) **120.** (c)
121. (a) **122.** (c) **123.** (b) **124.** (b) **125.** (d) **126.** (d) **127.** (c) **128.** (b) **129.** (a) **130.** (c)
131. (a) **132.** (c) **133.** (b) **134.** (b) **135.** (c) **136.** (d) **137.** (d) **138.** (b) **139.** (d) **140.** (d)
141. (a) **142.** (b) **143.** (b) **144.** (a) **145.** (a) **146.** (c) **147.** (b) **148.** (b) **149.** (c) **150.** (b)
151. (a) **152.** (d) **153.** (d) **154.** (c) **155.** (d) **156.** (a) **157.** (b) **158.** (c) **159.** (d) **160.** (d)
161. (b) **162.** (b) **163.** (d) **164.** (a) **165.** (a) **166.** (c) **167.** (a) **168.** (c) **169.** (b) **170.** (c)
171. (b) **172.** (c) **173.** (d) **174.** (c) **175.** (a) **176.** (b) **177.** (d) **178.** (d) **179.** (a) **180.** (c)
181. (c) **182.** (c) **183.** (b) **184.** (b) **185.** (d)

One or More Than One Options Correct

1. (a,b,c,d) **2.** (c) **3.** (a,c,d) **4.** (a,b) **5.** (d)
6. (c) **7.** (a,b,c) **8.** (d) **9.** (a) **10.** (a,c,d)

Brain Twisters

1.
$$Fe + Cu^{2+} \longrightarrow Fe^{2+} + Cu$$
$$E^\circ = E^\circ_{Fe/Fe^{2+}} + E^\circ_{Cu^{2+}/Cu}$$
$$= 0.44 + 0.34 = 0.78 \text{ V}$$

This reaction (a) is spontaneous

$$2Ag + Cu^{2+} \longrightarrow 2Ag^+ + Cu$$
$$E^\circ = E_{Ag/Ag^+} + E_{Cu^{2+}/Cu}$$
$$= -0.80 + 0.34$$
$$= -0.46 \text{ V}$$

This reaction (b) is not spontaneous.

2.
$$CuO + C \longrightarrow Cu + CO$$
$$\Delta G^\circ = \Delta G_f^\circ(CO) - \Delta G_f^\circ(CuO)$$
$$= -137.2 + 129.7$$
$$= -7.5 \text{ kJ}$$
$$CuO + H_2 \longrightarrow Cu + H_2O$$
$$\Delta G^\circ = \Delta G_f^\circ(H_2O) - \Delta G_f^\circ(CuO)$$
$$= -237.2 + 129.7$$
$$= -107.5 \text{ kJ}$$

More negative value of ΔG° for reaction (II) indicates that H_2 is better reducing agent.

3.
$$CuFeS_2 + 2H_2SO_4 \longrightarrow CuSO_4 + FeSO_4 + 2H_2S$$

(a)
$$Fe^{2+} \equiv MnO_4^-$$
$$10 \text{ mL of } Fe^{2+} \equiv 10 \text{ mL of } 0.02 \text{ M } MnO_4^-$$
$$\equiv 10 \text{ mL of } 0.1 \text{ N } MnO_4^-$$
$$\therefore \quad Fe^{2+} = 0.1 \text{ N}$$
$$= 5.6 \text{ g L}^{-1} \text{ in } 10 \text{ gL}^{-1} \text{ mineral}$$
$$= \mathbf{56\% \ Fe}$$

(b)
$$2CuSO_4 + 4KI \longrightarrow Cu_2I_2\downarrow + 2K_2SO_4 + I_2$$
$$I_2 \equiv Na_2S_2O_3$$
$$\therefore \quad Cu^{2+} \equiv Na_2S_2O_3$$
$$10 \text{ mL of } Cu^{2+} \equiv 10 \text{ mL of } 0.02 \text{ M } Na_2S_2O_3$$
$$\equiv 10 \text{ mL of } 0.02 \text{ N } Na_2S_2O_3$$
$$\therefore \quad Cu^{2+} = 0.02 \text{ N}$$
$$= 1.27 \text{ gL}^{-1} \text{ in } 10 \text{ gL}^{-1} \text{ mineral}$$
$$= \mathbf{12.7\% \ Cu}$$

4. Net reaction in Hall-Heroult process is
$$3C + 2Al_2O_3 \longrightarrow 4Al + 3CO_2$$
Or $\quad 4Al^{3+} + 12e^- \longrightarrow 4Al,$

Number of electrons $(n) = 12$
$$\Delta G^\circ = 3\Delta G_f^\circ(CO_2) - 2\Delta G_f^\circ(Al_2O_3)$$
$$= -3 \times 394 - 2(-1520)$$
$$= 1858 \text{ kJ}$$

$$\Delta G^\circ = -nFE^\circ_{cell}$$

$$-E^\circ_{cell} = \frac{\Delta G^\circ}{nF} = \frac{1858 \times 1000}{12 \times 96500}$$

$$= \mathbf{1.60\ V}$$

For the reaction $Al_2O_3 \longrightarrow 2Al + 3O_2$

$$\Delta G^\circ = 1520\ kJ$$

$$2Al^{3+} + 6e^- \longrightarrow 2Al, \quad n = 6$$

$$\therefore \quad -E^\circ_{cell} = \frac{\Delta G}{nF} = \frac{1520 \times 1000}{6 \times 96500}$$

$$= \mathbf{2.62\ V}$$

Thus, Hall-Heroult process takes place at lower voltage.

5. (a) $FeCr_2O_4(s) + 4C(s) \longrightarrow Fe(s) + 2Cr(s) + 4CO(g)$

1 mol (224 g) ... 2 mol (104 g) ... 4 mol (112 g)

224 kg of chromite gives = 104 kg of Cr

Thus, 236 kg of chromite gives $= \frac{104 \times 236}{224}$ kg of Cr

$= \mathbf{109.57\ kg\ Cr}$

(b) 224 kg chromite gives = 112 kg CO

Thus, 236 kg chromite gives $= \frac{112 \times 236}{224}$ kg CO

= 118 kg CO

= 4214.2857 mol

Using equation $pV = nRT$; $T = 298$ K, $p = \frac{740}{760}$ atm

$$V = \frac{nRT}{p} = \frac{4214.2857 \times 0.0821 \times 298}{(740/760)}$$

$$= \mathbf{105.892 \times 10^3\ L\ CO}$$

6. **(1) Copper (2) Platinum (3) Silver (4) Iron (5) Barium (6) Zinc (7) Lithium (8) Magnesium**

7. Let $CaCO_3 = x$ g

then, $MgCO_3 = (18.4 - x)$ g

$CaCO_3 \longrightarrow CaO + CO_2$

1 mol ... 1 mol ... 1 mol

100 g ... 56 g ... 44 g

x g ... $\left(\frac{44x}{100}\right)$ g

$MgCO_3 \longrightarrow MgO + CO_2$

1 mol ... 1 mol ... 1 mol

84 g ... 40 g ... 44 g

$\therefore$ $(18.4 - x)$ g ... $\frac{44}{84}(18.4 - x)$ g

Total CO_2 $\frac{44x}{100} + \frac{44(18.4 - x)}{84} = 8.8$ (given)

This gives $x = 10$ g

$\therefore$ $CaCO_3 = 10\ g = 0.1$ mol

$MgCO_3 = 18.4 - 10 = 8.4\ g = 0.1$ mol

Total moles of $CaCO_3$ and $MgCO_3 = 0.2$ mol

$\therefore$ $CaCO_3 = 50$ mol%

$MgCO_3 = 50$ mol%

Passage Comprehension Questions

Passage 1

1. (b) **2.** (c) **3.** (a) **4.** (c) **5.** (d)

Passage 2

1. (a) $\mathbf{ZnS} = 90.6\% = ZnO \xrightarrow{100\%} ZnSO_4 \xrightarrow{98.2\%} Zn$

1 mol **ZnS** gives 1 mol **Zn** but based on % yield, 1 mol ZnS gives $= 0.906 \times 1.0 \times 0.982$

= 0.8897 mol Zn

225 kg ZnS $= \frac{225}{97}$ kmol ZnS = 2.32 kmol ZnS

= 2.06 kilomol Zn

= **134 kg Zn**

2. (d) $w = zit$

$$i = \frac{w}{zt}$$

$$= \frac{134 \times 10^3\ g}{\left(\frac{65}{2 \times 96500}\right)(30 \times 24 \times 60 \times 60)}$$

$$= 154\ A$$

where, z = electrochemical equivalent

$$= \left(\frac{\text{atomic mass}}{\text{electrons} \times 96500}\right)$$

t = time in seconds

3. (b)

4. (c) $ZnO + 2NaOH \longrightarrow Na_2ZnO_2 + H_2O$

5. (a) 98% H_2SO_4 by weight H_2SO_4 ($d = 1.8$ g/mL) is **18 M**

100 g H_2SO_4 solution has = 98 g H_2SO_4

$\frac{100}{1.8}$ mL H_2SO_4 solution has = 98 g H_2SO_4

$$\text{molarity} = \frac{\text{mol}}{\text{litre}} = \frac{(98/98)}{\frac{100}{1.8 \times 1000}}$$

$$= 18\ M$$

ZnO = 2.32 kilomol in step (i) assuming 100% yield

$$H_2SO_4 = 2.32 \times 10^3\ \text{mol}$$

18 mol H_2SO_4 are in = 1 L

$\therefore$ 2.32×10^3 mol are in = 129 mol

≈ 130

Passage 3

1. (b) **2.** (d)

Passage 4

1. (a) **2.** (b) **3.** (a)

Passage 5

1. (b) **2.** (a) **3.** (c) **4.** (d)

Passage 6

1. (d) **2.** (b) **3.** (c) **4.** (a) **5.** (b)

Passage 7

1. (a) **2.** (c)

3. (b) $w = zit$ $\quad Cu^{2+} + 2e^- \longrightarrow Cu$

$$\text{Equivalent weight} = \frac{63.5}{2}$$

where z = electrochemical equivalent

$$= \frac{\text{equivalent weight}}{96500}$$

i = current (A)

t = time (s)

$$65.6 = \frac{63.5 \times 50 \times t}{2 \times 96500}$$

$$t = \frac{2 \times 96500 \times 65.6}{63.5 \times 50}\ \text{s}$$

$$= 1.1\ \text{h}$$

Passage 8

1. (c) **2.** (a) **3.** (d) **4.** (b,c) **5.** (a)

Passage 9

1. (a)

2. (c)

$$ZnO \longrightarrow Zn + \frac{1}{2}O_2 \qquad \Delta G^\circ_1 = x$$

$$C + \frac{1}{2}O_2 \longrightarrow CO \qquad \Delta G^\circ_2 = -x$$

$$ZnO + C \longrightarrow Zn + CO, \qquad \Delta G^\circ_3 = 0$$

System is at equilibrium;

$\Delta G^\circ_1 = -\Delta G^\circ_2$ since zinc and carbon have equal affinity for oxygen.

3. (b) This will be above a temperature when two curves intersect.

4. (d) ΔG° = most negative.

Passage 10

1. (a) **2.** (c) **3.** (a,b) **4.** (d) **5.** (a,c)

6. (c)

Passage 11

1. (c), ΔG° being more negative **2.** (d) **3.** (a,c)

Assertion & Reason

1. (b) **2.** (a) **3.** (e) **4.** (b) **5.** (a) **6.** (b)

True & False

1. F **2.** T **3.** T **4.** F **5.** T **6.** F
7. T **8.** T **9.** F **10.** F **11.** F **12.** T
13. F **14.** T **15.** T **16.** F **17.** F **18.** T
19. T **20.** F

Fill in the Blanks

1. Cr_2O_3, N_2 **2.** Al **3.** $[Ag(CN)_2]^-$ **4.** higher
5. Wrought **6.** Pig **7.** oxygen **8.** aluminium
9. Frasch process **10.** van Arkel-de-Boer process
11. Kroll process **12.** copper **13.** magnesium
14. AlN, $Al(OH)_3$ **15.** Hall-Heroult process, lower, increase
16. silver **17.** methanides, methane **18.** haemoglobin
19. ferrochrome **20.** fool's gold **21.** 75% **22.** steel
23. pig, wrought **24.** steel **25.** steel **26.** nickel
27. chalcocite **28.** $CuCO_3 \cdot Cu(OH)_2$ **29.** Parke's
30. an insecticide

Matrix-Match Type Questions

I. Only One Correct

1. (A)—(4); (B)—(1); (C)—(2); (D)—(3)
2. (A)—(2); (B)—(4); (C)—(3); (D)—(1)
3. (A)—(4); (B)—(1); (C)—(2); (D)—(3)
4. (A)—(3); (B)—(1); (C)—(2); (D)—(4)
5. (A)—(3); (B)—(4); (C)—(2); (D)—(1)
6. (A)—(3); (B)—(5); (C)—(1); (D)—(2)
(E)—(4)

II. One or More Correct

1. (A)—(1,6); (B)—(6); (C)—(1,2,3,6);
(D)—(1,5) (E)—(4,5)
2. (A)—(3,4,6,7); (B)—(1,2,4); (C)—(4,5,7);
(D)—(2,5) (E)—(1,6)

Integer Answer Types

Set I

Questions →	1	2	3	4	5	6	7	8
Answers →	2	7	2	6	3	4	3	1
	0	0	0	0	0	0	0	0
	1	1	1	1	1	1	1	1
	2	2	2	2	2	2	2	2
	3	3	3	3	3	3	3	3
	4	4	4	4	4	4	4	4
	5	5	5	5	5	5	5	5
	6	6	6	6	6	6	6	6
	7	7	7	7	7	7	7	7
	8	8	8	8	8	8	8	8
	9	9	9	9	9	9	9	9

Set II

Questions →	1	2	3	4	5	6	7	8
Answers →	4	2	7	6	9	9	2	9
	⓪	⓪	⓪	⓪	⓪	⓪	⓪	⓪
	①	①	①	①	①	①	①	①
	②	2	②	②	②	②	2	②
	③	③	③	③	③	③	③	③
	4	④	④	④	④	④	④	④
	⑤	⑤	⑤	⑤	⑤	⑤	⑤	⑤
	⑥	⑥	⑥	6	⑥	⑥	⑥	⑥
	⑦	⑦	7	⑦	⑦	⑦	⑦	⑦
	⑧	⑧	⑧	⑧	⑧	⑧	⑧	⑧
	⑨	⑨	⑨	⑨	9	9	⑨	9

Test Yourself on the Periodic Table

1. Minerals are

A : Sulphide B : Carbonate C : Chloride D : Oxide

E : Sulphide F : Uncombined form

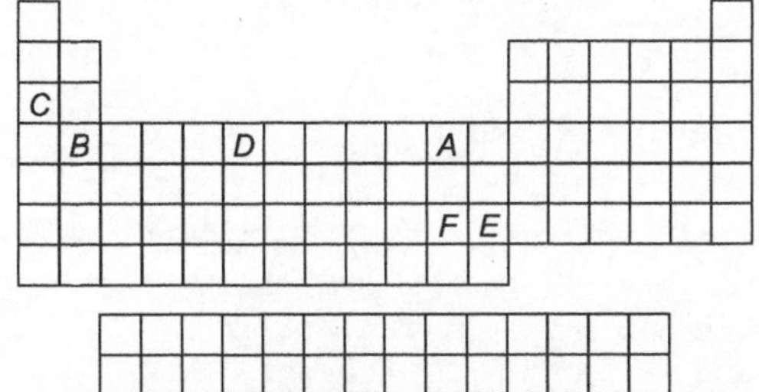

2. A (Mg) : Electrolysis of molten $MgCl_2$

B (Al) : Electrolytic reduction of $(Al_2O_3 + Na_3AlF_6)$

C (Au) : Cyanide reduction process

$$2\,[Au(CN)_2]^- + Zn \longrightarrow [Zn(CN)_4]^{2-} + 2Au\downarrow$$

D (Cu) : Bessemerisation followed by electrolysis of *aq*. $CuSO_4$ solution using impure Cu as anode

E (Sn) : Reduction of SnO_2 by charcoal

18

Coordination Compounds

"Even when, to judge by the valence number, the combining power of certain atoms is exhausted, they still possess in most cases, the power of participating further in the construction of complex molecules with the formation of very definite linkages. The possibility of this action is to be traced back to the fact that, besides the affinity bonds designated as principal valences, still other bonds called auxiliary valences, may be called into action."

Alfred Werner

Quick Points

- For the stepwise formation of ML_2 from M and L

$$M + L \rightleftharpoons ML \quad K_1$$
$$ML + L \rightleftharpoons ML_2 \quad K_2$$
$$M + 2L \rightleftharpoons ML_2 \quad K_3$$

$$K_3 = K_1K_2$$
$$\log K_3 = \log K_1 + \log K_2$$
$$pK_3 = pK_1 + pK_2$$

- O_2^- is called superoxide ion.
- Anions outside coordination sphere ionise, greater the number of ions larger the conductance

$$[Co(H_2O)_6]Cl_3 \rightleftharpoons [Co(H_2O)_6]^{3+} + 3Cl^-$$

- van't Hoff factor i

$$i = [1 + (y - 1)\, x]$$

where, y = number of ions per unit of the complex/species

x = degree of ionisation

$i = (1 + 3x)$ in the above case since, $y = 4$

- Electron-pair donor is a Lewis base and electron-pair acceptor is a Lewis acid.

Learning Framework of the Chapter ...

- Double Salts and Coordination Compounds
- Ligands
- Coordination Number
- Nomenclature
- Effective Atomic Number
- Isomerism in Complexes
- Werner Theory of Coordination Compounds
- Bonding in Complexes
- Stability Constants and Stability of Coordination Compounds
- Applications of Coordination Chemistry

Double Salts and Coordination Compounds

- When $Fe(CN)_2$ solution is treated with KCN solution, species formed, no longer, gives tests of Fe^{2+} and CN^-.
- When K_2SO_4 solution is treated with $Al_2(SO_4)_3$ solution, species formed gives tests of K^+, Al^{3+} and SO_4^{2-}.

$$Fe(CN)_2 + 4KCN \longrightarrow \underset{(A)}{Fe(CN)_2 \cdot 4KCN}$$

$$K_2SO_4 + Al_2(SO_4)_3 + 24H_2O \longrightarrow \underset{(B)}{K_2SO_4 \cdot Al_2(SO_4)_3 \cdot 24H_2O}$$

We call them addition compounds of

Type I : those which lose their identity in solution (double salt) as alum (*B*).

Type II : those which retain their identity in solution (complexes) as potassium ferrocyanide (*A*).

Most *d*-transition metal ions have vacant *d*-orbitals that can accommodate electron pairs. Many act as Lewis acids by forming coordinate covalent bonds in **coordination compounds (coordination complexes** or **complex ions).**

A complex ion is a metal ion with Lewis bases (ligand) attached to it through coordinate covalent bonds (also called coordination). A complex (or coordination compound) is a compound consisting either of complex ions with other ions of opposite charge.

$K_4[Fe(CN)_6]$ — **K^+ is other ion**
complex anion

$[Pt(en)_3]Cl_4$ — **Cl^- is other ion**
complex cation

$[Pt(en)_2Cl_2]$
neutral complex

Many important biological substances are coordination compounds. Haemoglobin and chlorophyll are two such examples. Haemoglobin is a protein that carries O_2 in blood. It contains iron (II) ions bound to large porphyrin rings. Chlorophyll is necessary for photosynthesis in plants. It contains magnesium ions bound to porphyrin rings. Vitamin B-12 is a large complex of cobalt.

Ligands

A ligand is a species that is capable of donating an electron pair(s) to a central metal or metal ion. It is a **Lewis base.** In accepting electron pairs, the central ion acts as a **Lewis acid.**

- Ligand is said to be **unidentate** if it has only one pair of electrons that it can donate (Table 18.1).
- Ligand is said to be **bidentate, tridentate, tetradentate** etc. depending on the number of electron-pairs that it can donate. (Table 18.2).

Table 18.1 **Some Common Unidentate Ligands**

Neutral		Anions (replacing ide by o)		Anions (replacing e by o)	
Formula	**Name as ligand**	**Formula**	**Name as ligand**	**Formula**	**Name as ligand**
$\underline{O}H_2$	aqua	$\underline{F}^-$ (fluoride)	fluoro	$\underline{S}O_4^{2-}$	sulphato
$\underline{N}H_3$	ammine	$\underline{Cl}^-$	chloro	$\underline{S}_2O_3^{2-}$	thiosulphato
$\underline{C}O$	carbonyl	$\underline{Br}^-$	bromo	$\underline{C}O_3^{2-}$	carbonato
$\underline{N}O$	nitrosyl	$\underline{I}^-$	iodo	$\underline{N}O_2$	nitro*
$\underline{N}H_2CH_3$	methylamine	$\underline{O}^{2-}$ (oxide)	oxo	$\underline{O}NO^-$	nitrito*
$\underline{N}C_5H_5$ or Py	pyridine	$\underline{O}H^-$ (hydroxide)	hydroxo	$\underline{S}CN^-$	thiocyanato*
$\underline{N}(C_2H_5)_3$	triethylamine	$\underline{C}N^-$ (cyanide)	cyano*	$\underline{N}CS^-$	isothiocyanato*
$\underline{N}(CH_3)_3$	trimethylamine	$\underline{N}C$	isocyano*		
$\underline{P}(C_6H_5)_3$	triphenyl phosphine	$\underline{S}^{2-}$	sulphido		
$\underline{P}H_3$	phosphine	$\underline{H}^-$	hydrido		

* underlined atoms are electron-donating, ambidentate.

Table 18.2 **Some Common Multidentate Ligands (Chelating Agents)**

Abbreviation	Multi	Name	Formula	Charge
en	bidentate	ethylenediamine	CH_2NH_2 | CH_2NH_2	0
ox	bidentate	oxalate ion	COO^- | COO^-	–2
gly	bidentate	glycinate ion	CH_2NH_2 | COO^-	–1
DMG	bidentate	dimethylglyoximate ion	$CH_3—C{=}NO^-$ | $CH_3—C{=}NOH$	–1
dien	tridentate	diethylene-triamine	$H_2N(CH_2)_2—NH—(CH_2)_2NH_2$	0
EDTA	hexadentate	ethylenediamine tetraacetate ion	$[(^-OOCCH_2)_2N—CH_2—CH_2—N(CH_2COO^-)_2]$	–4

(a) $\ddot{N}H_3$, one electron-pair available for donation–unidentate.

(b) $CH_2\ddot{N}H_2$ | $CH_2\ddot{N}H_2$, two electron-pairs available for donation– bidentate.

- Some ligands can be coordinated to the metal or metal ion through either of the sides. They are called **"ambidentate"** ligands.

  If the nitrite ion is attached through N atom ($—NO_2$), it is nitro; if attached through O atom (—ONO), then nitrito.

- Ligands can be
  — neutral molecules such as H_2O
  — anions such as CN^-, $C_2O_4^{2-}$
  — cations [but in rare cases since an electron pair on a cation is held firmly by the positive charge and would not be involved in coordinate bonding.]
- Ligands forming complex with metal do not ionise.
- In some cases it is not possible to identify donor atoms, because the bonding electrons are not localised on specific atoms. Ethylene ($CH_2{=}CH_2$) bonds to transition metal through the electrons in their double bonds.

Chelates

Some ligands are capable of donating more than a single electron pair, from different atoms in the ligand and to different sites in the geometric structure of a complex.

DEEP Focus

These are called **multidentate** ligands (Table 18.2). When the bonding of a multidentate ligand to a metal ion produces a ring (usually five– or six–membered), we refer to the complex as a **chelate** (pronounced-keylate). The multidentate ligand is called a **chelating agent** and the process of chelate formation is called **chelation.** The number of such ligating groups indicate the denticity (unidentate, bidentate ....) of the ligand.

nickel (II) dimethyl glyoximate (chelate)
[bis-(dimethylglyoximato) nickel (II)]

copper (II) glycinate (chelate)
[*trans*-bisglycinato copper (II)]

Coordination Number

The coordination number (C.N.) of a metal atom in a complex is the total number of bonds the metal atom forms with ligands. Some common coordination numbers of metal ions are summarised in Table 18.3.

Table 18.3 **Coordination Number (C.N.) of Metal Ions**

Metal ion	C.N.
Ag^+	2
Cu^+	2, 4
Cu^{2+}	4, 6
Au^+	2, 6
Ca^{2+}	6
Fe^{2+}, Fe^{3+}	6
Co^{2+}	4, 6
Co^{3+}	6
Ni^{2+}	4, 6
Zn^{2+}	4
Al^{3+}	4, 6
Sc^{3+}	6
Cr^{3+}	6
Pt^{2+}	4
Pt^{4+}	6

Examples of complexes of various C.N. are given in Table 18.4.

Table 18.4 **Examples of Complexes of Various Coordination Numbers**

Complex	C.N.
$[Ag(NH_3)_2]^+$	2
$[HgI_3]^-$	3
$PtCl_4^{2-}$, $Ni(CO)_4$	4
$Fe(CO)_5$, $[Co(CN)_5]^{3-}$	5
$[Co(NH_3)_6]^{3+}$, $W(CO)_6$	6
$[Mo(CN)_7]^{3-}$	7
$[W(CN)_8]^{4-}$	8

Nomenclature

The definite source for the naming of inorganic compounds is **Nomenclature of Inorganic Compounds 2005 (N.G. Connelly and T. Damhus, Sr. Eds. Royal Society of Chemistry 2005)**

Following rules are adopted for naming a complex ion;

(a) Cations are named before anions

(b) Oxidation state (O.S.) of the central metal ion is denoted by Roman numeral.

	Cation	O.S. anion
CuCl	Copper (I)	chloride
$CuCl_2$	Copper (II)	chloride
$FeCl_2$	Iron (II)	chloride
$FeCl_3$	Iron (III)	chloride

(c) The names of ligands are given first followed by the name of the central metal ion.

(d) The names of ligands that are anions and ending with

'**ide**' are changed to '**o**'
'**ite**' are changed to '**ito**'
'**ate**' are changed to '**ato**' (Table 18.1 and 18.2)

(e) Many ligands that are molecules carry the unmodified name (Table 18.1 and 18.2)

(f) Positive groups end in —ium

NH_2—$NH_3^{\oplus}$ hydrazinium.

DEEP Focus

(g) When there are several ligands of the same kind, we normally use the prefixes **di, tri, tetra, penta** and **hexa** to show the number of ligands of that type. An exception occurs when the name of the ligands includes a number, for example as in ethylenediamine (en). To avoid confusion in such cases, **bis, tris** and **tetrakis** are used instead of di, tri, and tetra, and the name of the ligand is placed in brackets.

bis (ethylenediamine)

(h) If anion is a complex, then metal ends with '**ate**'

$[Ni(CN)_4]^{2-}$
tetracyanonickelate (II) ion

lead — **plumbate**
gold — **aurate**
zinc — **zincate**
tin — **stannate**
silver — **argentate**
cobalt — **cobaltate**
iron — **ferrate**
aluminium — **aluminate**
manganese — **manganate**
copper — **cuprate**
chromium — **chromate**
platinum — **platinate**

(i) If the complex contains two or more metal atoms, it is termed polynuclear. The bridging ligands which link the two metal atoms together are indicated by the prefix μ —.

(j) Ambidentate ligands may be attached through different atoms

M—NO_2 (NO_2 joined to metal *M* through N; it is **nitro**)

M—ONO (NO_2 joined to metal *M* through O; it is **nitrito**)

Similarly, the SCN group may bond *M*—SCN (thiocyanato) or *M*—NCS (isothiocyanato). They may be named

thiocyanato–S [to indicate S coordinated to metal *M*]

thiocyanato–N [to indicate N coordinated to metal *M*]

(k) When writing (not naming) the formula of the complex :

- complex ion should be enclosed by square brackets and

- ligands are placed after metal in the alphabetical order but first negative ligands, then neutral, then positive.

$[Pt(H_2O)_4(NH_3)_2]^{4+}$ — wrong
aqua ammine

$[Pt(NH_3)_2(H_2O)_4]^{4+}$ — right

$[Cr(CN)_4(NH_3)_2]^-$ — right
negative neutral

Examples

(1) metal ligands
K_4 $[Fe(CN)_6]$
cation anion
potassium hexacyanoferrate (II)
cation anion

- anion is a complex hence its metal ends with **ate** (ferrate),
- ligands are named before metal ion of the complex.

(2) $[Co(NH_3)_6]Cl_3$
cation anion
hexaamminecobalt (III) chloride

- cation is a complex
- anion will not change its name (since it is not the ligand)

(3) $[Co(en)_3]Cl_3$
tris (ethylenediamine)cobalt (III) chloride

- tris is used to indicate three ligands **en** (ethylenediamine) since it also carries **di**.

(4) $[PtCl_2(NH_3)_4]Cl_2$
tetraamminedichloroplatinum (IV) chloride

- NH_3 (ammine) and Cl (chloro) are two types of ligands **named in alphabetical order:**

(5) $[(NH_3)_5Co—NH_2—Co(NH_3)_5](NO_3)_5$
bridging ligand will start with μ
μ-amido bis [pentaamminecobalt (III)] nitrate

1. In the following examples, **complex ion** exists as **cation:**

(a) $[Fe(NH_3)_6]Cl_3$
Hexaammineiron (III) chloride

(b) $[CoCl(NH_3)_5]^{2+}$
Pentaamminechlorocobalt (III) ion

(c) $[Co(SO_4(NH_3)_4)]NO_3$
Tetraamminesulphatocobalt (III) nitrate

(d) $[Co(Cl)(NO_2)(NH_3)_4]^+$
Tetraamminechloronitrocobalt (III) ion

2. In the following examples, **complex ion** exists as **anion**

(a) $[ZnCl_4]^{2-}$
Tetrachlorozincate (II) ion

(b) $[AlH_4]^-$
Tetrahydridoaluminate (III) ion

(c) $K_4[Fe(CN)_6]$
Potassium hexacyanoferrate (II)

(d) $Na_3[Ag(S_2O_3)_2]$
Sodium dithiosulphatoargenate (I)

(e) $Na_2[OsCl_5N]$
Sodium pentachloronitridoosmate (VI)

(f) $K_2[PtCl_6]$
Potassium hexachloroplatinate (IV)

(g) $K_2[Cr(CN)_2(O_2)_2(O_2)NH_3]$
Potassium amminedicyanodioxoperoxo chromate (VI)

(h) $[Zn(NCS)_4]^{2-}$
Tetrathiocyanato-N-zincate (II) ion
(N indicates that ligand joined to metal ion through N-side)

3. In the following examples, **organic ligands** have been used :

(a) $[Pt(Py)_2][PtCl_4]$
Tetrapyridineplatinum (II) tetrachloroplatinate (II)

(b) $[Cr(en)_3]Cl_3$
d-or-*l* tris (ethylenediamine) chromium (III) chloride

(c) $[Pt(en)_2]^{2+}$
Bis (ethylenediamine) Platinum (II) ion

(d) $[Cr(C_6H_6)_2]$
Bis (benzene) chromium (0)

(e) $[Fe(C_5H_5)_2]$
Bis (cyclopentadienyl) iron (II)

4. In the following examples, **bridging groups** have been used:

(a) $\left[(en)_2Co\begin{matrix}NH\\OH\end{matrix}Co(en)_2\right]^{4+}$
Tetrakis(ethylenediamine)-μ-hydroxo-μ-imidodicobalt (III) ion.

(b) $[(CO)_3Fe(CO)_3Fe(CO)_3]$
Tri-μ-carbonyl-bis (tricarbonyliron (0))

(c) $(C_6H_5)_3P$, Cl, Pd, Cl, Pd, Cl, Cl, $P(C_6H_5)_3$
(.... line used to indicate symmetry)
Trans-bis(chlorotriphenylphosphine-μ-chloropalladium(I))
or
Chlorotriphenylphosphine palladium (II) μ-dichlorochlorotriphenylphosphine palladium (II)

Effective Atomic Number

Each ligand donates an **electron pair to the metal ion, thus forming a coordinate bond.** Transition metals form coordination compounds very readily because they have vacant *d*-orbitals which can accommodate these electron pairs. Metal ion in the complex tends to attain nearest stable inert gas configuration by gaining electrons from ligands. **Effective atomic number** (EAN) of the metal in a complex is given by

$$\text{EAN} = Z - (\text{O.N.}) + 2(\text{C.N.})$$

(where O.N. is oxidation number, C.N. the coordination number and Z the atomic number).

EAN can also be given by

$$\text{EAN} = Z - (\text{O.N.}) + 2(\text{Ligands})$$

assuming that each ligand is unidentate.

EAN of some metals in complexes is summarised in Table 18.5.

Table 18.5 **Effective Atomic Numbers of Some Metal in Complexes**

Atom	Atomic number (Z)	Complex	C.N.	Electrons lost in ion formation (O.N.)	Electrons gained by coordination	EAN
Cr	24	$[Cr(CO)_6]$	6	0	12	36
Ni	28	$Ni(CO)_4$	4	0	8	36
Fe	26	$[Fe(CN)_6]^{4-}$	6	2	12	36
Fe	26	$[Fe(NH_3)_6]^{2+}$	6	2	12	36 } [Kr]
Cu	29	$[Cu(CN)_4]^{3-}$	4	1	8	36
Co	27	$[Co(en)_3]^{3+}$, (en) is bidentate	6	3	12	36
Pd	46	$[Pd(H_2O)_6]^{4+}$	6	4	12	54[Xe]
Pt	78	$[PtCl_6]^{2-}$	6	4	12	86[Rn]
Fe	26	$[Fe(CN)_6]^{3-}$	6	3	12	35
Ag	47	$[Ag(CN)_2]^-$	2	1	4	50
Al	13	$[Al(C_2O_4)_3]^{3-}$, ox is bidentate	6	3	12	22

18-Electrons System

If there are total of 18-electrons attained due to gain of coordination electrons of ligands including $(n-1)d^x ns^y$ electrons of metal/metal ion, the complex is said to attain 18-electrons configuration.

$In[Fe(CN)_6]^{4-}$

$Fe : [Ar]3d^6 4s^2$ Electron gained from $6CN^- = 12$

$Fe^{2+} : [Ar]3d^6$

Thus, $3d^6 + 12e^- = 18e^-$

Isomerism in Complexes

Compounds that have the same chemical formula but different structural arrangement are called **isomers.**

A. Structural Isomerism

Structural isomers are isomers that differ in how the atoms are joined together—that is, in the order in which the atoms are bonded to one another.

(H—N=C=O and N≡C—O—H are structural isomers)

These are of the following types:

(a) Ionisation Isomerism

This type of isomerism occurs when there is an interchange of groups between the coordination sphere of the metal ion and the ions outside this sphere. Examples are:

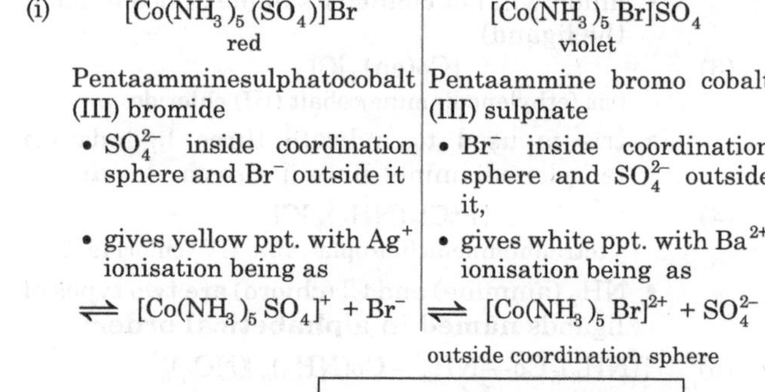

(ii) $[Pt(NH_3)_4Cl_2]Br_2$ $[Pt(NH_3)_4Br_2]Cl_2$

inside coordination sphere

Other examples are:

$[Pt(NH_3)_4SO_4](OH)_2$ and $[Pt(NH_3)_4(OH)_2]SO_4$

$[Co(NH_3)_5NO_2]SO_4$ and $[Co(NH_3)_5SO_4]NO_2$

$[Cr(NH_3)_5SO_4]Br$ and $[Cr(NH_3)_5Br]SO_4$

(b) Coordination Isomerism

When both the positive and negative ions are complex ions, isomerism may be caused by the interchange of ligands between the anion and cation. This arises only if cationic and anionic metals have same C.N. and charge. Examples are:

same charge

(i) $[Co(NH_3)_6]^{3+}$ $[Cr(CN)_6]^{3-}$

same C.N. = 6

Hexaamminecobalt (III) hexacyanochromate (III)

$[Cr(NH_3)_6]^{3+}$ $[Co(CN)_6]^{3-}$

Hexamminechromium (III) hexacyanocobaltate (III)

(ii) $[Co(en)_3][Cr(CN)_6]$

$[Cr(en)_3][Co(CN)_6]$

(iii) $[Pt(NH_3)_4Cl_2][PtCl_4]$

$[Pt(NH_3)_4][PtCl_6]$

(c) Linkage Isomerism

This type of isomerism arises when the ligand may attach to the central metal ion of a complex in different ways. (Such ligands are called **ambidentate ligands**). Nitrite ion has electron pairs available for coordination both on N and O atoms. (refer Table 18.1)

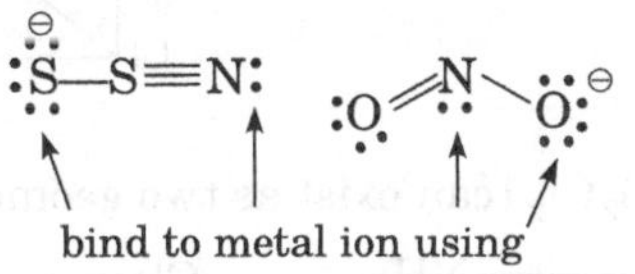

$:\overset{\ominus}{S}-S\equiv N:$ $\quad :O=N-O:^{\ominus}$

bind to metal ion using either lone pair

Examples :

(i) $[Co(NH_3)_5ONO]Cl_2$ and $[Co(NH_3)_5NO_2]Cl_2$

pentaamminenitritocobalt (III) chloride (red), decomposes in acids

pentaamminenitrocobalt (III) chloride (yellow), stable in acids

(ii) $[Mn(CO)_5(SCN)]^+$ and $[Mn(CO)_5(NCS)]^+$

pentacarbonylthiocyanato–S–manganese (II) ion

pentacarbonylthiocyanato–N-manganese (II) ion

(d) Coordination Position Isomerism

In polynuclear complexes an interchange of ligands between the different metal nuclei gives rise to position isomerism.

Example :

$$\left[(NH_3)_4Co\left<\begin{matrix}NH_2\\O_2\end{matrix}\right>Co(NH_3)_2Cl_2\right]Cl_2$$

$$\left[Cl(NH_3)_3Co\left<\begin{matrix}NH_2\\O_2\end{matrix}\right>Co(NH_3)_3Cl\right]Cl_2$$

(e) Hydrate (Solvate) Isomerism

Hydrate isomers are isomers of a complex that differ in the placement of water molecules in the complex. $CrCl_3 \cdot 6H_2O$ have following hydrate isomers:

$[Cr(H_2O)_6]Cl_3$ — violet (anhydrous)

$[Cr(H_2O)_5Cl]Cl_2 \cdot H_2O$ — light green (monohydrate)

$[Cr(H_2O)_4Cl_2]Cl \cdot 2H_2O$ — dark green (dihydrate)

Illustration 1 *One mole of the complex* $CrCl_3 \cdot 6H_2O$ *on reaction with excess of aq.* $AgNO_3$ *gives two moles of white precipitate of* AgCl. *What is the complex?*

Solution $\underset{1\text{ mol}}{CrCl_3 \cdot 6H_2O} + \underset{\text{excess}}{AgNO_3} \longrightarrow \underset{2\text{ mol}}{AgCl}$

This reaction indicates that two Cl are outside the coordinate sphere and one Cl inside it. Coordination number of six of Cr ion is satisfied by five H_2O and one Cl as ligands. Thus, this is monohydrate complex $[Cr(H_2O)_5Cl]Cl_2 \cdot H_2O$

$$\underset{1\text{ mol}}{[Cr(H_2O)_5Cl]Cl_2} \rightleftharpoons [Cr(H_2O)_5Cl]^{2+} + \underset{2\text{ mol}}{2Cl^-}$$

$$2Cl^- + 2Ag^+ \longrightarrow \underset{2\text{ mol}}{2AgCl\downarrow}$$

(f) Electronic Isomerism

The complex $[Co(NH_3)_5NO]Cl_2$ exists in two forms. One is black paramagnetic while the other is pink and diamagnetic. Strucutral studies show that the black isomer is a Co(II) complex containing neutral NO group whereas the pink one is a Co(III) complex with NO^- ligand. This type of isomerism is called **electronic isomerism.**

(g) Ligand Isomerism

Isomerism may also arise due to isomerism in the ligand set. The effect of changing an alkyl chain on a ligand from a linear to a branched structure is sufficient to induce quite profound strucutral changes in the complex.

Consider reaction of **A**, *bis*-(salicylaldehyde) nickel (II) with propyl amine to give **B**, *bis*-(N-propylasalicylidene aminato) nickel (II).

The use of *n*-propyl amine or isopropylamine gives two isomers in this reaction. The more stearically hindered isopropyl amine results in a complex distorted away from the square planar geometry displayed by the *n*-propyl complex towards a tetrahedral geometry.

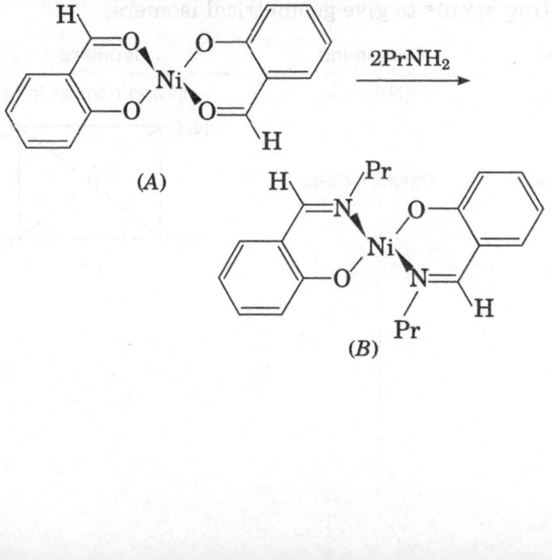

H, O, O, Ni, O, O, H $\xrightarrow{2PrNH_2}$

(*A*)

H, Pr, N, O, Ni, O, N, H, Pr

(*B*)

(h) Polymerisation Isomerism

This term is used to describe compounds haivng same empirical formula but different molecular compositions. The difference from normal polymerisation is that here we have differences in arrangements of group as multiplication of molecular weight. Thus, value of n in the empirical formula $[ML_m]_n$ varies.

$[Pt(NH_3)_2Cl_2]_n$ gives
(A) $[Pt(NH_3)_2Cl_2]$ if $n = 1$ and
(B) $[Pt(NH_3)_4][PtCl_4]$ if $n = 2$

and thus (A) and (B) are polymerisation isomers, $[Co(NH_3)_3(NO_2)_3]$ gives following isomers

Formula	Molecular weight
$[Co(NH_3)_6][Co(NO_2)_6]$	double
$[Co(NH_3)_4(NO_2)_2][Co(NH_3)_2(NO_2)_4]$	double
$[Co(NH_3)_5(NO_2)][Co(NH_3)_2(NO_2)_4]_2$	triple
$[Co(NH_3)_6][Co(NH_3)_2(NO_2)_4]_3$	quadruple
$[Co(NH_3)_4(NO_2)_2]_3[Co(NO_2)_6]$	quadruple
$[Co(NH_3)_5(NO_2)]_3[Co(NO_2)_6]_2$	quintuple

B. Stereo Isomers

These are the isomers in which the atoms are bonded to each other in the same order but that differ in the precise arrangement of these atoms in space. These are of the following types:

(a) Geometric Isomerism

Geometric isomers are isomers in which the atoms are joined to one another in the same way but that differ because some atoms occupy different relative positions in space.

Geometric isomerism of metals of C.N. 4

For square planar complex Ma_4, Ma_3b or Mab_3 where a and b etc., are monodenate ligands, geometrical isomerism is not possible. However, square planar complexes of the type Ma_2b_2, Ma_2bc, $Mabcd$ and $M(AA)_2$, $M(AB)_2$—when AA and AB represent symmetrical and unsymmetrical chelating agents to give geometrical isomers.

Type	Compound	Isomers
Ma_2b_2	$Pt(NH_3)_2Cl_2$	*cis* and *trans* as in (i)
Ma_2bc	$Pt(NH_3)_2ClBr$	NH_3, Cl, Pt, H_3N, Br — *cis*

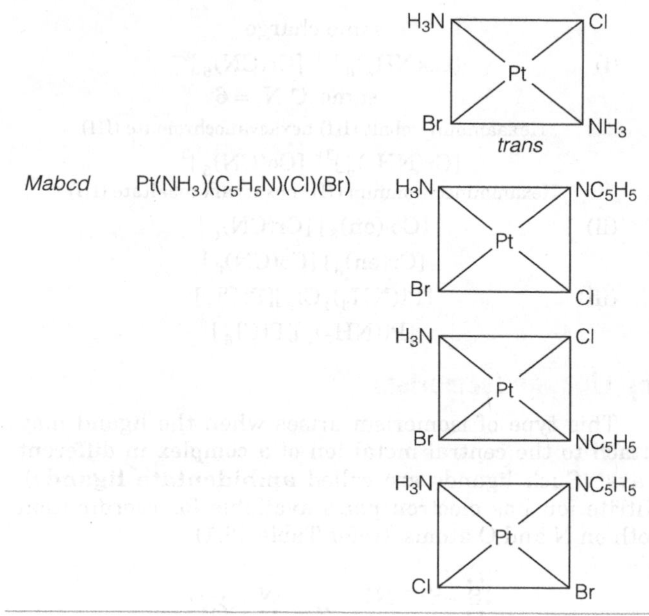

(i) $[Pt(NH_3)_2Cl_2]$ can exist as two geometric isomers:

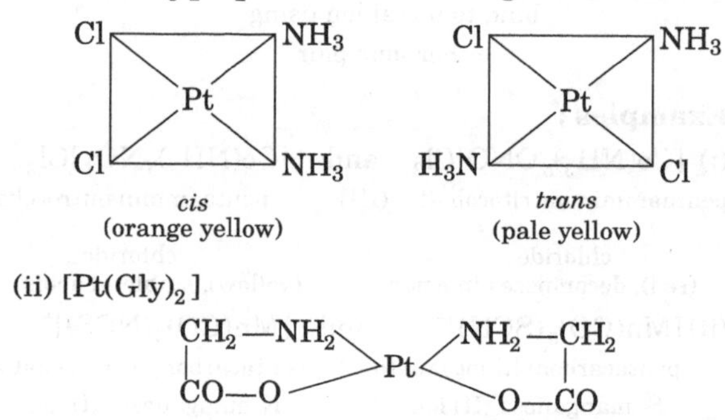

(ii) $[Pt(Gly)_2]$

CH_2—NH_2 > Pt < NH_2—CH_2; CO—O, O—CO

cis

CO—O > Pt < NH_2—CH_2; CH_2—NH_2, O—CO

trans

Geometric isomerism of metals of C.N. 6

(i) $[Co(NH_3)_4Cl_2]^+$ can exist as

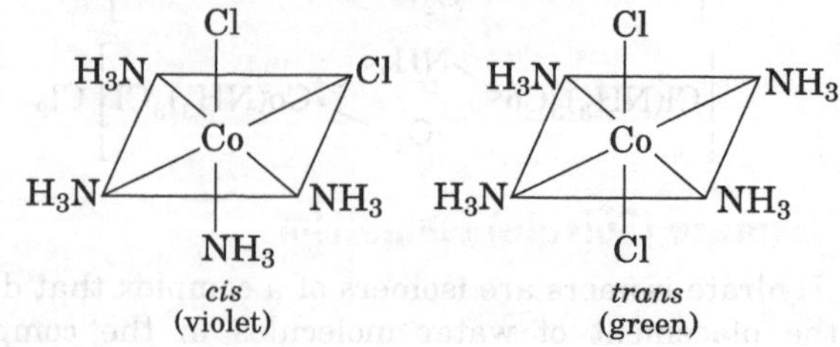

(ii) $[Pt(NH_3)_2Cl_2Br_2]$ can exist as

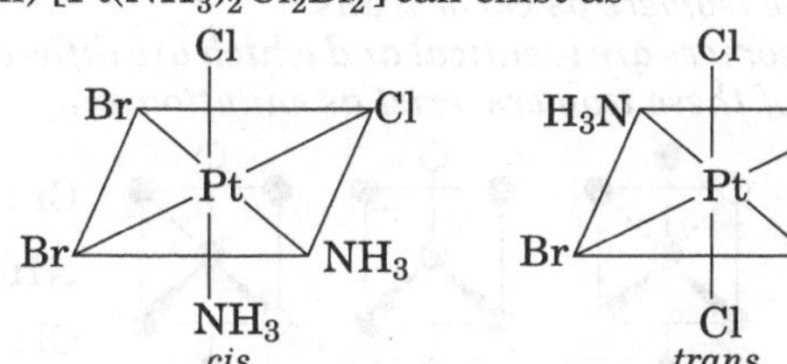

There are so many other arrangements also.

Another type of geometrical isomerism occurs in octahedral coordination entities of the type $[MX_3Y_3]$ like $[Co(NO_2)_3(NH_3)_3]$. If each trio of donor atoms occupy adjacent positions at the corners of an octahedral face, it is called **facial (*fac*-) isomer.** When the positions are around the meridian of the octahedron the isomer is called **meridional (*mer*-) isomer.**

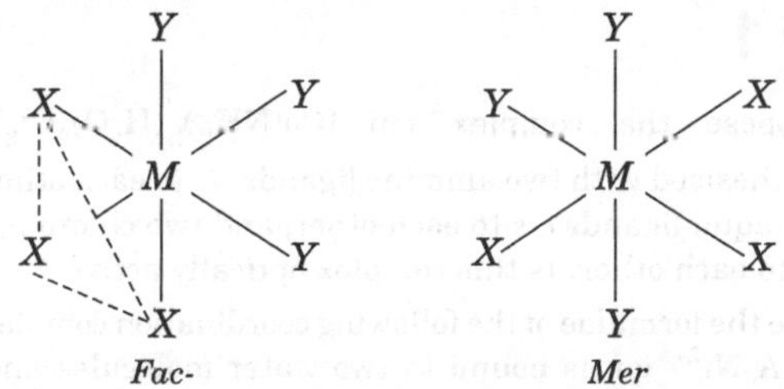

Bridged binuclear complexes like $[Pt(PEt_3)Cl_2]_2$ may exist in three isomers forms

Et_3P >Pt< Cl Cl >Pt< Cl PEt_3 — *trans*

Et_3P >Pt< Cl Cl >Pt< PEt_3 Cl Cl — *cis*

Et_3P Et_3P >Pt< Cl Cl >Pt< Cl Cl — *unsymmetrical*

As we increase the number of different ligands, the possible number of isomers increases.

(b) Optical Isomerism

If a molecule is asymmetric, it can't be superimposed on its mirror image. The two forms have the type of symmetry shown by the left and right hands and are called **enantiomorphic pair.** The two forms are called **optical isomer.** They are called either *dextro* or *laevo* (*d* or *l*). This depends on the direction they rotate the plane of polarised light in a polarimeter.

- Optical isomerism is common in octahedral complexes involving bidentate groups.
- $[Co(en)_2Cl_2]^+$ exists as *cis*- and *trans*-isomers. But *cis*-isomer can have *d* and *l*-optical isomers.

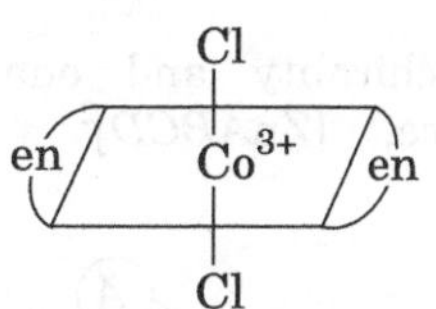

trans-(no optical isomer possible)

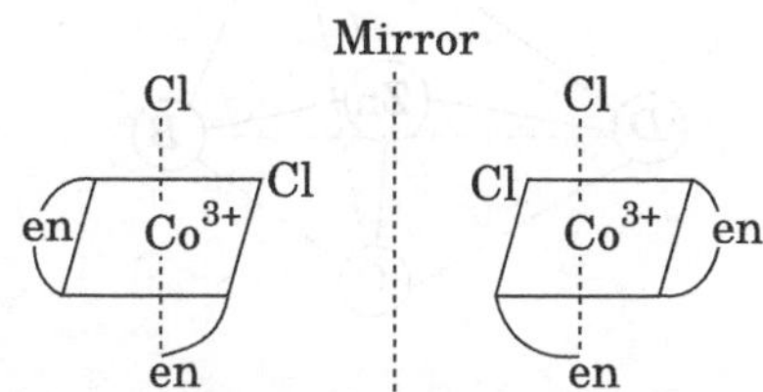

enantiomorphic pair *d* and *l*

cis-dichlorobis (ethylenediammine) cobalt (III)ion

- Optical isomers of $[Co(en)_3]^{3+}$ are:

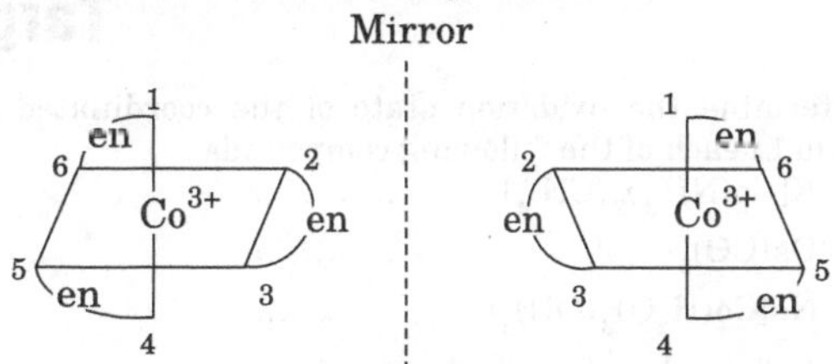

- In $[Fe(C_2O_4)_3]^{3-}$, $C_2O_4^{2-}$ is a bidentate ligand and that three $C_2O_4^{2-}$ ligands mean that the metal has a coordination number of 6. Mirror images of complex $[Fe(C_2O_4)_3]^{3-}$ are not superimposable; therefore two optical isomers are possible.

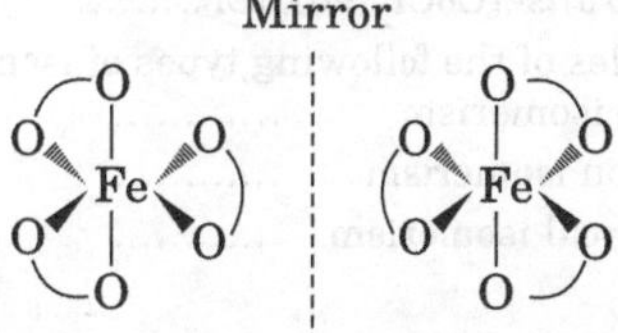

Non-superimposable mirror images of $[Fe(ox)_3]^{3-}$

Illustration 2 *If A, B, C, and D are four different ligands,*

(a) how many geometric isomers will be found for square planar $[PtABCD]^{2+}$*?*

(b) will tetrahedral $[ZnABCD]^{2+}$ *display optical isomerism?*

Solution (a) Following three different isomers are possible for the square planar $[PtABCD]^{2+}$.

(You may write others but they will be identical)

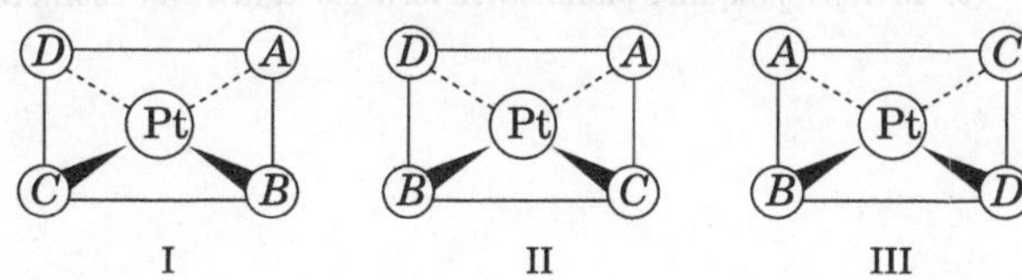

(b) There is chirality and enantiomers do not superimpose hence, $[ZnABCD]^{2+}$ displays optical isomerism.

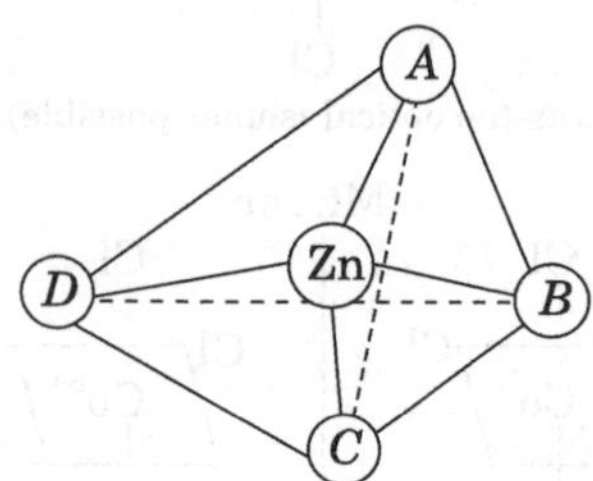

Illustration 3 *Consider the given isomers of* $[Cr(NH_3)_2Cl_4]^-$

(a) Label the isomers as cis or trans.
(b) Which isomers are identical and which are different.
(c) Do any of these isomers exist as enantiomers.

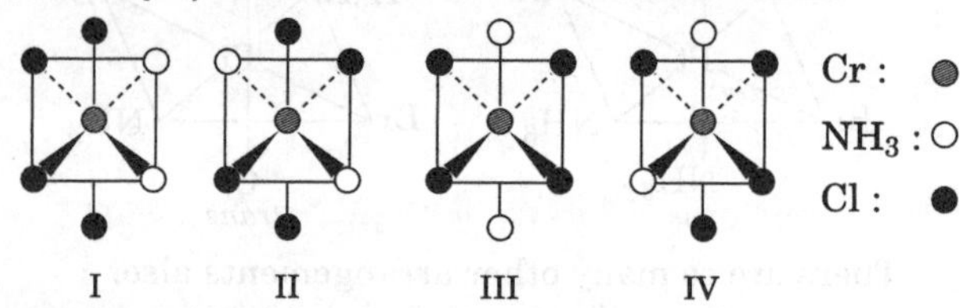

Solution (a) I, IV *cis*; II, III *trans*
(b) I and IV are the same; II and III are the same
(c) None of the isomers exists as enantiomers because their mirror images are identical.

Target Practice 1

1. Determine the oxidation state of the coordinated metal atom in each of the following compounds :
   (a) $K[Co(NH_3)_2(CN)_4]$ ....
   (b) $Os(CO)_5$ ....
   (c) $Na[Co(H_2O)_3(OH)_3]$ ....
2. Write formulas of
   (a) Sodium tricarbonatocobaltate (III) ....
   (b) Diamminediaquadichloroplatinum (IV) bromide ....
   (c) Sodium tetranitratoborate (III) ....
3. How might one distinguish between the following isomers?
   (a) $[Co(NH_3)_5Br]SO_4$ and $[Co(NH_3)_5SO_4]Br$ ....
   I II
   (b) *cis* and *trans*- $[CoCl_2(en)_2]Cl$ ....
4. Give examples of the following types of isomerisation.
   (a) Hydrate isomerism ....
   (b) Ionisation isomerism ....
   (c) Geometrical isomerism ....
5. Suppose the complex ion $[Co(NH_3)_2(H_2O)_2Cl_2]^+$ is synthesised with two ammine ligands *cis* to each other, the two aqua ligands *cis* to each other, and two chloro ligands *cis* to each other. Is this complex optically active.
6. Give the formulae of the following coordination complexes :
   (a) A Ni^{2+} ion is bound to two water molecules and two bidentate oxalate ions. ....
   (b) A Co^{3+} ion is bound to one Cl^- ion, one ammonia molecule, and two bidentate ethylenediamine (en) molecules ....
7. Identify the type of isomerism shown by
   (a) $[Co(NH_3)_4Cl_2]^+$ ....
   (b) $[Pt(NH_3)_2(CN)_2]$ (square planar) ....
   (c) $Zn(NH_3)_2Cl_2$ (tetrahedral) ....
   (d) $[Fe(C_2O_4)_3]^{3-}$ ....
   (e) $[Co(NH_3)_5(SCN)]^{2+}$ ....

Answers

1. (a) +3 (b) zero (c) +2
2. (a) $Na_3[Co(CO_3)_3]$
   (b) $[Pt(NH_3)_2(H_2O)_2Cl_2]Br_2$
   (c) $Na[B(NO_3)_4]$
3. (a) I gives white precipitate with $BaCl_2(aq)$ while II gives yellow precipitate with $AgNO_3(aq)$.
   (b) Dipole moment of *cis-* > *trans-*
4. (a) $[Co(H_2O)_3Cl_3]\cdot 3H_2O$ and $[Co(H_2O)_6]Cl_3$
   (b) $[Co(NH_3)_5Br]SO_4$ and $[Co(NH_3)_5SO_4]Br$
   (c) MCl_2Br_2 (square planar) will form *cis-* and *trans-* isomers
5. $[Co(NH_3)_2(OH_2)_2Cl_2]^+$ structures: left — H_3N, H_3N, OH_2, Cl, Cl, OH_2 around Co (*cis*); right (mirror image) — OH_2, NH_3, Cl, H_2O, Cl, NH_3 around Co (*cis*)

   Mirror

   Mirror images are not superimposable hence the complex is optically active (being chiral)
6. (a) $[Ni(H_2O)_2(C_2O_4)_2]^{2-}$ (b) $[Co(NH_3)(en)_2Cl]^{2+}$
7. (a) geometrical *cis, trans-* (b) geometrical *cis-, trans-*
   (c) only single structure is possible (d) optical and
   (e) linkage S—bonded SCN^- N—bonded NCS^-

Practice Exercise 1

1. Classify double salts and coordination compounds out of
   (a) $3NaF \cdot AlF_3$ which does not give tests of F^- and Al^{3+} ions
   (b) $4KCN \cdot Fe(CN)_2$ which does not give tests of CN^- and Fe^{2+}
   (c) Alum which gives tests of Al^{3+} and SO_4^{2-} ions
   (d) Mohr's salt which gives tests of NH_4^+, Fe^{2+} and SO_4^{2-} ions.
2. What is the type of complex formed, cationic or anionic, in the following?
   Also designate the coordinate ion in each case.
   (a) $FeSO_4 + KCN \longrightarrow$ (b) $CoCl_2 + KNO_2 \longrightarrow$
   (c) $PtCl_4 + KCl \longrightarrow$ (d) $AgCl + NH_3 \longrightarrow$
   (e) $AgBr + Na_2S_2O_3 \longrightarrow$ (f) $HgI_2 + KI \longrightarrow$
3. Following complexes ionise in aqueous solution. What are the ions in each case?
   (a) $KFe^{II}[Fe^{III}(CN)_6]$ (b) $NaAlO_2$
   (c) $Ni(CO)_4$ (d) Na_3AlF_6
   (e) $(NH_4)_2PtCl_6$ (f) $Na_2[Ni(CN)_4]$
4. When aqueous KCN is added to a solution of $CuSO_4$, a white precipitate, soluble in excess of KCN, is formed. No precipitate is formed when H_2S gas is bubbled through the solution at this point. Explain.
5. (a) Formation of $[Ni(en)_3]^{2+}$ from Ni^{2+} and ethylenediamine is a Lewis acid-base reaction. Explain.
   (b) Identify the Lewis acid and Lewis base in the reaction of oxalate ions ($C_2O_4^{2-}$) with Fe^{3+} to give $[Fe(C_2O_4)_3]^{3-}$.
6. Classify the following ligands as unidentate, bidentate, ... polydentate:
   (a) CH_3COO^- (b) $\ddot{:}OCN^-$
   (c) $:NCO^-$ (d) CO_3^{2-}
   (e) $\ddot{N}H_2$—$CH_2CH(CH_3)$—$\ddot{N}H_2$
   (propylene diamine written as **pn**),
   (f) CH_3—C(=$\ddot{O}$:)—CH=C(—:$\ddot{O}$:)—CH_3
   (acetylacetonato written as **acac**)
   (g) 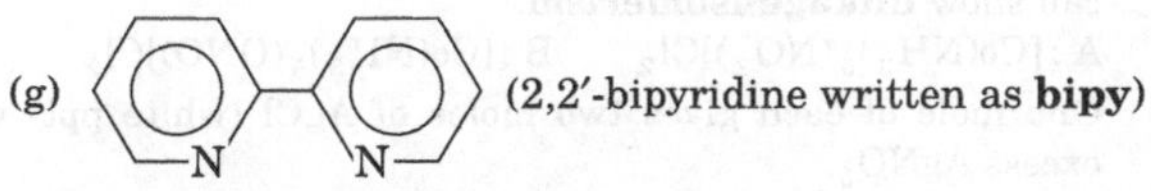

   (2,2′-bipyridine written as **bipy**)
   (h) Ph—$\ddot{P}$—Ph (triphenyl phosphine)
   Ph
7. Write the formula of
   (a) a complex ion having Co^{2+} as the central ion and two NH_3 molecules and four Cl^- ions as ligands,
   (b) a complex ion of manganese (III) having coordination number of 6 and CN^- as ligands,
   (c) a coordination compound comprised of two types of complex ions; one is a complex of Cr(III) with ethylenediamine (en), having a coordination number of 6 and the other is a complex of Ni(II) with CN^- and having a coordination number of 5.
8. What are the coordination numbers of the underlined atoms in the following? Also specify the charge on the complex, oxidation number of the metal ion has been given :
   (a) $[\underline{Fe}^{II}(CN)_6]$, (b) $[\underline{Ni}^0(CO)_4]$,
   (c) $[\underline{Ca}^{II}(EDTA)]$, (d) $[\underline{Pt}^{IV}(en)Cl_2Br_2]$
   (e) $[\underline{Al}^{III}(C_2O_4)_3]$, (f) $[\underline{Ag}^{I}(S_2O_3)_2]$
   (g) $\left[(en)_2Fe^{III}\begin{matrix} O_2 \\ NH_2 \end{matrix}Fe^{III}(H_2O)_4\right]Cl_4$
9. One mole of $CoCl_3 \cdot 4NH_3$ on reaction with excess of $AgNO_3$ gives one mole of white precipitate. What is the complex? What is the coordination number of cobalt?
10. Give the oxidation number of the metals in the following species:
    (a) $K_3[Fe(CN)_6]$ (b) $K_3[Cr(C_2O_4)_3]$
    (c) $[Ni(CN)_4]^{2-}$ (d) Na_2MoO_4
    (e) $Fe(CO)_5$ (f) $MgWO_4$
11. Complete the following statements for the complex ion: $[Cr(C_2O_4)_2(H_2O)_2]^-$
    (a) The oxidation number of Cr is ....
    (b) The coordination number of Cr is ....
    (c) .... is a bidentate ligand
12. Name the following compounds:
    (a) $[Co(NH_3)_4(NO_2)_2]NO_3$ (b) $[Pt(en)Cl_4]$
    (c) $Na_3[Ag(S_2O_3)_2]$ (d) $K_3[Co(ONO)_6]$
    (e) $Na_2[Zn(EDTA)]$ (f) $[Ru(en)_2Br_2]Br$
    (g) $[Pt(Py)_4][PtF_4]$
13. Write the formulae of each of the following compounds :
    (a) cesium trichlorofluoro iodate (III)
    (b) tris (ethylenediamine)nickel (II) nitrate
    (c) potassium dibromodioxalatocobaltate (III)
    (d) pentamminechlorochromium (III) sulphate
    (e) ethylenediaminetetraiodoplatinum (IV)
    (f) potassium tetrafluoroargentate (III)
    (g) rubidium ethylenediaminetetraacetatocobaltate (III) dihydrate
14. Calculate EAN of the metal in the following complexes, and compare it with the atomic number of the nearest rare gas:
    (a) $AuCl_2^-$ (b) $Fe(C_2O_4)_3^{3-}$
    (c) CdI_4^{2-} (d) $[Cr(H_2O)_4(NH_3)_2]^{2+}$
    (e) $Co(NO_2)_6^{3-}$ (f) $Fe(CO)_5$
    (g) $[Ca(EDTA)]^{2-}$ (h) $[Ni(Py)(en)(NH_3)_3]^{2+}$
    (i) $[Zn(OH)_4]^{2-}$ (j) HgI_4^{2-}
15. Determine EAN of iron in
    (i) $Fe(CO)_5$ (ii) $Fe(C_5H_5)_2$
16. Predict value of x in
    (a) $H_xCr(CO)_5$ (b) $H_xCo(CO)_4$
17. Make hydrate isomers of:
    (a) $[Fe(H_2O)_6]Cl_3$ (b) $[Cu(H_2O)_4]Cl_2$
    Also arrange them in the increasing conducting power.
18. One cationic complex has two isomers *A* and *B*. Each has one Co^{3+}, five NH_3, one Br^- and one SO_4^{2-} stoichiometrically. *A* gives white ppt. with $BaCl_2$ while *B* give yellow ppt. with $AgNO_3$. What are *A* and *B*? What is the type of isomerism shown?
19. Two isomers of $[Co(NH_3)_5(NO_2)]Cl_2$ differ in colour. What is the cause? If each reacts with excess $AgNO_3$, what is precipitated and how much per mole of the complex?
20. Which of the following can show coordination isomerism?
    (a) $[Cu(NH_3)_4][PtCl_4]$ (b) $[Fe(NH_3)_6]_2[Pt(CN)_6]_3$
    (c) $[Co(NH_3)_6][Cr(C_2O_4)_3]$ (d) $[Pt(en)_3](SO_4)_2$

21. Write the formula of ionisation isomer of each of the following compounds:
(a) $[Ni(en)_2(NO_2)_2]Cl_2$ (b) $[Fe(NH_3)_5CN]SO_4$

22. Draw the structures of all the isomers of the square-planar complexes:
(a) $Pd(NH_3)_2BrI$ (b) $Ni(en)Cl_2$

23. For which of the following complexes are optical isomers possible? Explain your answers and draw structure of enantiomers:
(a) $[Cr(NH_3)_4(C_2O_4)]^+$ (b) $[Cr(NH_3)_2(C_2O_4)_2]^-$
(c) $[FeCl_5F]^{3-}$ (d) $[Co(en)_2NH_3Cl]^{2+}$
(e) $[Cr(C_2O_4)_3]^{3-}$

24. A complex has a composition corresponding to the formula $CoBr_2Cl \cdot 4NH_3$. What is the structural formula if conductance measurements show two ions per formula unit? Silver nitrate solution gives ppt. of AgCl but no AgBr. Write the structural formula of an isomer.

25. How many geometric isomers are in the following species?
(a) $[Co(NH_3)_2Cl_4]^-$ (b) $[Co(NH_3)_3Cl_3]$

26. Draw structures of all the geometric and optical isomers of each of the following cobalt complexes:
(a) $[Co(NH_3)_4Cl_2]^+$ (b) $[Co(en)_3]^{3+}$

27. Which of the following complexes are chiral?
(a) $[Pt(en)Cl_2]$ (b) *cis*-$[Co(NH_3)_4Br_2]^+$
(c) *cis*-$[Cr(en)_2(H_2O)_2]^{3+}$ (d) $[Cr(C_2O_4)_3]^{3-}$

Answers

1. (a) Na_3AlF_6 ; (b) $K_4[Fe(CN)_6]$;
(c) $K_2SO_4 \cdot Al_2(SO_4)_3 \cdot 24H_2O$ (d) $FeSO_4 \cdot (NH_4)_2SO_4 \cdot 6H_2O$

2. (a) $K_4[Fe(CN)_6]$ — anionic complex;
(b) $K_3[Co(NO_2)_6]$ — anionic complex;
(c) $K_2[PtCl_6]$ — anionic complex;
(d) $[Ag(NH_3)_2]Cl$ — cationic complex;
(e) $Na_3[Ag(S_2O_3)_2]$ — anionic complex;
(f) K_2HgI_4 — anionic complex;

3. (a) K^+, Fe^{2+}, $[Fe(CN)_6]^{3-}$; (b) Na^+, AlO_2^- (c) no ionisation
(d) Na^+, AlF_6^{3-}; (e) NH_4^+, $PtCl_6^{2-}$; (f) Na^+, $[Ni(CN)_4]^{2-}$

4.
$$CuSO_4 + 2KCN \longrightarrow \underset{\text{white ppt.}}{Cu(CN)_2} + K_2SO_4$$
$$Cu(CN)_2 + KCN \longrightarrow \underset{\text{stable}}{K_3[Cu(CN)_4]}$$
$$K_3[Cu(CN)_4] \rightleftharpoons 3K^+ + [Cu(CN)_4]^{3-}$$
Stable $[Cu(CN)_4]^{3-}$ does not ionise to give Cu^+ hence no precipitate with H_2S gas.

5. (a) Ni^{2+} is electron-pair acceptor (a Lewis acid) and en (ethylene diamine) is electron-pair donor (a Lewis base). Hence, complex formation is a Lewis acid-base reaction.
$$\underset{\text{Lewis acid}}{Ni^{2+}} + 3\underset{\text{Lewis base}}{\begin{matrix} CH_2NH_2 \\ | \\ CH_2NH_2 \end{matrix}} \longrightarrow [Ni(en)_3]^{2+}$$
(b) Fe^{3+} :Lewis acid; $C_2O_4^{2-}$:Lewis base

6. (a), (b), (c), (d) and (h) — unidentate; (e), (f), (g) — bidentate

7. (a) $[Co(NH_3)_4Cl_2]$ (b) $[Mn(CN)_6]^{3-}$
(c) $[Cr(en)_3][Ni(CN)_5]$

8.

	(a)	(b)	(c)	(d)	(e)	(f)	(g)
C.N.	6	4	6	6	6	2	6
Charge	–4	0	–2	0	–3	–3	+4

9. Complex $[Co(NH_3)_4Cl_2]Cl$; C.N. of cobalt is six.

10. (a) Fe = +3 (b) Cr = +3 (c) Ni = +2 (d) Mo = +6
(e) Fe = 0 (f) W = +6

11. (a) +3 (b) 6 (c) $C_2O_4^{2-}$

12. (a) tetramminedinitrocobalt (III) nitrate
(b) tetrachloroethylenediamine platinum (IV)
(c) sodium dithiosulphatoargentate (I)
(d) potassium hexanitritocobaltate (III)
(e) sodium ethylenediaminetetraacetatozincate (II)
(f) dibromo bis-(ethylenediamine)ruthenium (III) bromide
(g) tetrapyridineplatinum (II) tetrafluoroplatinate (II)

13. (a) $Cs[ICl_3F]$ (b) $[Ni(en)_3](NO_3)_2$ (c) $K_3[Co(C_2O_4)_2Br_2]$
(d) $[Cr(NH_3)_5Cl]SO_4$ (e) $Pt(en)I_4$ (f) $K[AgF_4]$
(g) $Rb[Co(EDTA)] \cdot 2H_2O$

14. (a) Au = 82 (b) Fe = 35 (c) Cd = 54 [Xe] (d) Cr = 34
(e) Co = 36 [Kr] (f) Fe = 36 [Kr] (g) Ca = 30 (h) Ni = 38
(i) Zn = 36 [Kr] (j) Hg = 86 [Rn]

15. (i) EAN of Fe = 26 + 10 (from five CO) = 36 = Kr
(ii) EAN of Fe = 24 + 10 (from two $C_5H_5^-$ ions) = 34

16. (a) $H_2Cr(CO)_4$
(b) $HCo(CO)_4$ (in each case EAN of metal is 36]

17. (a) (I) $[Fe(H_2O)_5Cl]Cl_2 \cdot H_2O$ (II) $[Fe(H_2O)_4Cl_2]Cl \cdot 2H_2O$
(III) $[Fe(H_2O)_3Cl_3] \cdot 3H_2O$
Conductance : III < II < I
(b) (I) $[Cu(H_2O)_3Cl]Cl \cdot H_2O$ (II) $[Cu(H_2O)_2Cl_2] \cdot 2H_2O$
Conductance : II < I

18. *A* : $[Co(NH_3)_5Br]SO_4$ *B* : $[Co(NH_3)_5SO_4]Br$
A and *B* are ionisation isomers.

19. Difference in colour is due to difference in linkage of NO_2 that can show **linkage-isomerism.**
A : $[Co(NH_3)_5(NO_2)]Cl_2$ B : $[Co(NH_3)_5(ONO)]Cl_2$
One mole of each gives two moles of AgCl (white ppt) with excess $AgNO_3$.

20. (a) and (c).

21. (a) $[Ni(en)_2Cl_2](NO_2)_2$ (b) $[Fe(NH_3)_5SO_4]CN$

22. (a)

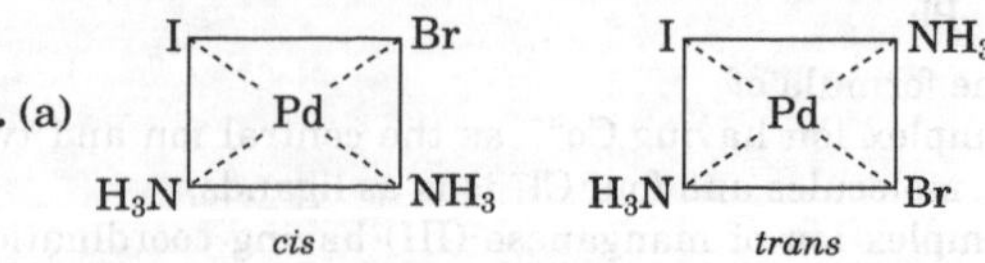

(b) Only one isomer is possible as ethylenediamine cannot span the *trans*-position.
(chelating agent like $CH_2NH_2—CH_2NH_2$ with at least two coordination points is symbolised by a curved line giving the symbol for the ligand in the middle —en—)

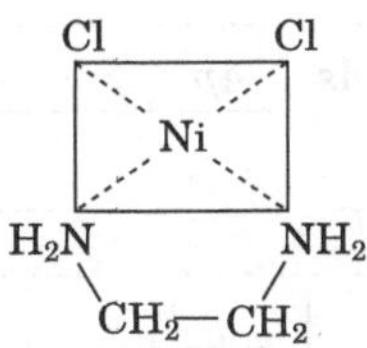

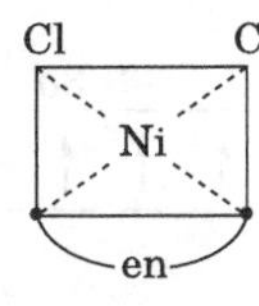

23. (b) and (e) for structure see text.
24. $[Co(NH_3)_4BrCl]Cl$, isomer $[Co(NH_3)_4Cl_2]Br$
25. (a) 2 (b) 2
26. (a) Two geometical isomers
(b) Two optical isomer. For structures see text
27. (c) and (d)

Werner Theory of Coordination Compounds

- Beginning of the coordination chemistry is not based on the first formation of $CoCl_3 \cdot 6NH_3$ by Tassqert but is actually considered based on the outstanding work of the Alsatian-Swiss chemist Alfred Werner in 1892.
- Werner prepared and isolated many different coordination compounds from $CoCl_3$ and NH_3 like
(a) $CoCl_3 \cdot 6NH_3$ $[Co(NH_3)_6]Cl_3$ orange-yellow
(b) $CoCl_3 \cdot 5NH_3$ $[Co(NH_3)_5Cl]Cl_2$ violet (purple)
(c) $CoCl_3 \cdot 4NH_3$ $[Co(NH_3)_4Cl_2]Cl$ violet
(d) $CoCl_3 \cdot 3NH_3$ $[Co(NH_3)_3Cl_3]$ green

Treatment of these compounds with aqueous hydrochloric acid did not remove the ammonia, which suggested that the ammonia was somehow closely bound with the cobalt ions. Treatment with aqueous silver nitrate at 273 K, however gave interesting results. One mole each of the complexes (a), (b) and (c) gave respectively thus, two and one mole of white precipitates of AgCl, with excess of $AgNO_3$. Complex (d) did not give the precipitate at all.

Postulates

- There are two types of valencies shown by central metal/metal ion in a compound :
(a) primary (b) secondary
Primary valency corresponds to oxidation number and secondary valency corresponds to coordination number.
- In the complex $[Co(NH_3)_6]Cl_3$, primary valency of Co is **three** and secondary valency **six.**
- Primary valency is satisfied by negative ions (anions) while secondary valency is satisfied by ligands which can be neutral, and as well as negative (positive in rare cases)
  - In $[Co^{III}(NH_3)_6]Cl_3$, six NH_3 ligands satisfy secondary valencies while three Cl^- satisfy primary valencies.
  - In $[Co^{II}(NH_3)_4(H_2O)_2]SO_4$, four NH_3 and two H_2O (ligands) satisfy secondary and one SO_4^{2-} satisfy primary valencies.
- Ligands/ions attached by secondary valency do not ionise while ions attached by primary valency ionise.

$$K_4[Fe(CN)_6] \rightleftharpoons 4K^+ + [Fe(CN)_6]^{4-}$$

$$[Pt(NH_3)_6]Cl_4 \rightleftharpoons [Pt(NH_3)_6]^{4+} + 4Cl^-$$

- The secondary valencies are directional and are responsible for the isomerism in complexes.
- Primary valency in the complex is represented by .... while secondary valency by —— .
- In all cases, metal or metal ion should satisfy primary and secondary valencies both. Some negative ions may satisfy primary as well as secondary valencies **(dual nature).** In the complexes, they are represented as ....

Werner's Representation

Consider the case of $CoCl_3 \cdot xNH_3$
where maximum value of x = C.N. of Co(III) = 6
and minimum value of x = C.N. – O.N. = 3
Various structures are summarised in Table 18.6.

Table 18.6 **Werner's Coordination Compounds**

Werner complex	Modern notation	Ionisation	Secondary valency satisfied by	Primary valency satisfied by
(A) $CoCl_3 \cdot 6NH_3$	$[Co(NH_3)_6]Cl_3$	$[Co(NH_3)_6]^{3+} + 3Cl^-$	six (NH_3)	three (Cl^-)
(B) $CoCl_3 \cdot 5NH_3$	$[Co(NH_3)_5Cl]Cl_2$	$[Co(NH_3)_5Cl]^{2+} + 2Cl^-$	five (NH_3) and one (Cl^-)	three (Cl^-) including one (Cl^-) with dual nature
(C) $CoCl_3 \cdot 4NH_3$	$[Co(NH_3)_4Cl_2]Cl$	$[Co(NH_3)_4Cl_2]^+ + Cl^-$	four (NH_3) and two (Cl^-)	three (Cl^-) including two (Cl^-) with dual nature
(D) $CoCl_3 \cdot 3NH_3$	$[Co(NH_3)_3Cl_3]$	$[Co(NH_3)_3Cl_3]$ (single species)	three (NH_3) and three (Cl^-)	three (Cl^-) all with dual nature

- **From the Table 18.6, it is clear that conduction of the complexes which depends on the number of ions will be in order:**

$$D < C < B < A$$

- **They are represented as :**

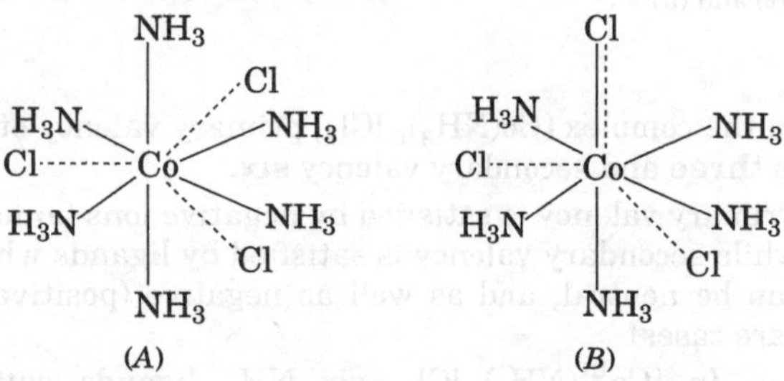

[only Cl^- ions joined by (........) will ionise].

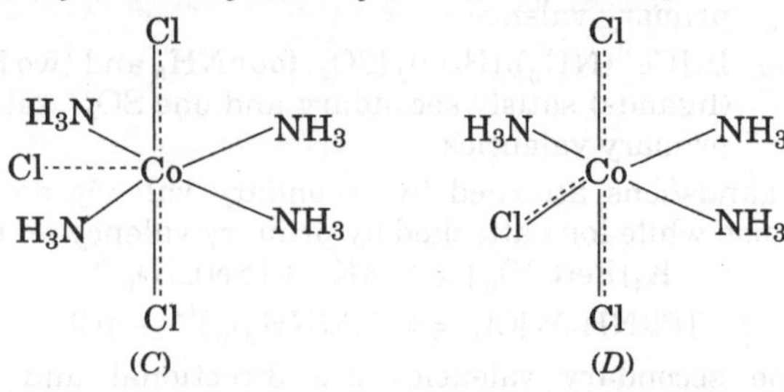

Bonding in Complexes

Valence Bond Theory (VBT) One way to think about bonds between a metal ion and its ligands is based on coordinate covalency. The central ion furnishes the orbitals and the ligands furnish electron pairs (refer Chapter-7). Although this view is now considered inadequate, it does account for the observed coordination number of the central ion, and it also describes the geometric structure of the complex ion.

Let us study some complexes to see how the valence bond theory explains their properties. Take example of $[Fe(H_2O)_6]^{3+}$. The complex is known to be paramagnetic, with a magnetic moment corresponding to five unpaired electrons. This can be described if we follow:

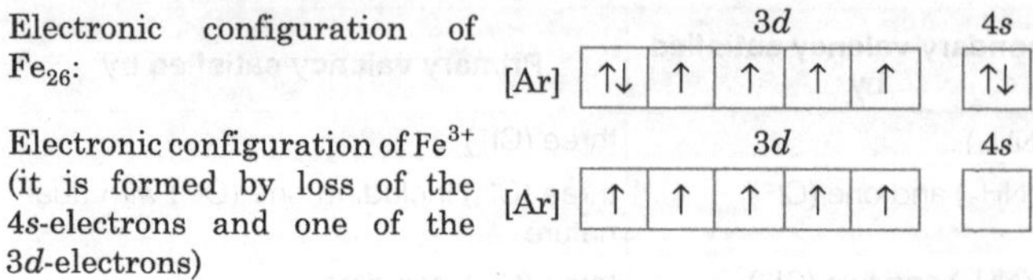

To account for the experimentally observed fact that $[Fe(H_2O)_6]^{3+}$ has five unpaired electrons, each 3*d*-orbitals is assumed to have one unpaired electrons in the complex. The vacant $4d_{x^2-y^2}$ and $4d_{z^2}$ orbitals are hybridised with the vacant 4*s*-and 4 *p*-orbitals.

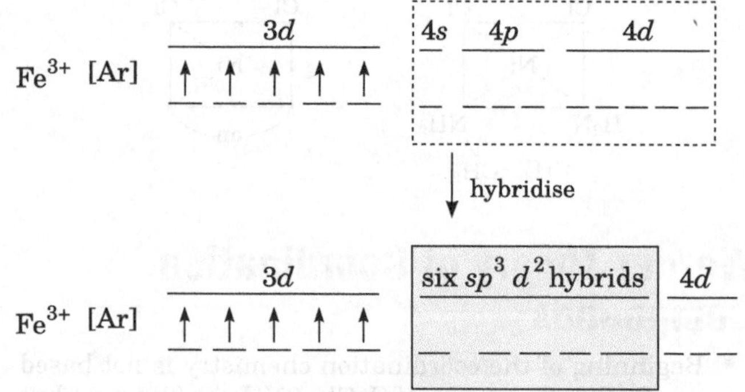

Each of the six H_2O ligands donates one electron pair into one of the six sp^3d^2-orbitals, forming six coordinate-covalent bonds. These donated electrons (from O of H_2O ligands) have been represented as "×" rather than "↑". These are the only bonding electrons; none of the 3*d*-electron of iron is involved in bonding.

A set of outer *d* orbitals is used in hybridisation, and hence $[Fe(H_2O)_6]^{3+}$ is called an outer orbital complex.

	3d	sp^3d^2	4d
$[Fe(H_2O)_6]^{3+}$ [Ar]	↑ ↑ ↑ ↑ ↑	xx xx xx xx xx xx	— — —

In deciding whether the hybridisation at the central metal ion of octahedral complexes is sp^3d^2 (outer orbital) or d^2sp^3 (inner orbital) we must know the results of magnetic measurements. These indicate the number of unpaired electrons (*N*). Magnetic moment is given by

$$\mu = \sqrt{N(N+2)} \text{ BM}$$

(BM is Bohr Magneton)

In $[Fe(CN)_6]^{3-}$, Fe^{3+} is involved (a d^5 ion). But magnetic measurements indicate only one unpaired electron per iron atom. The $3d_{x^2-y^2}$ and $3d_{z^2}$ orbitals (rather than those in 4*d*-shell) are involved in d^2sp^3 hybridisation. All but one of the non-bonding 3*d*-electrons of Fe^{3+} are paired in this complex.

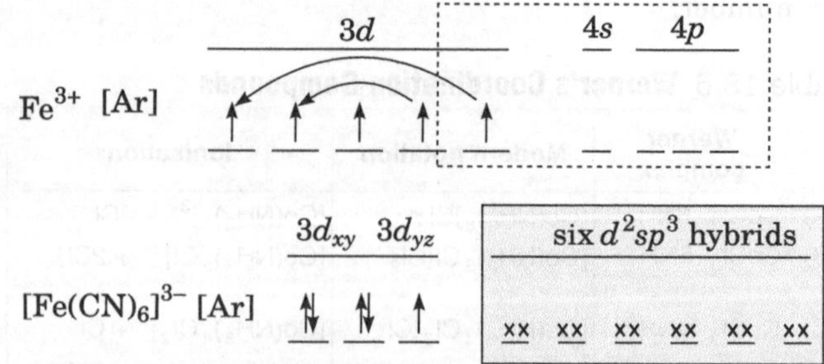

Only inner *d*-orbitals are used in hybridisation, hence $[Fe(CN)_6]^{3-}$ is called an inner orbital complex. Additional illustrations have been summarized in Table 18.7 (refer Chapter-7)

Illustration 1 *Explain the paramagnetic behaviour of* $[Co(NH_3)_6]^{2+}$.

Solution Oxidation state of Co = + 2

		3d	4s	4p	5s
Co (ground state)	[Ar]	↑↓ ↑↓ ↑ ↑ ↑	↑↓	_ _ _	
Co^{2+}	[Ar]	↑↓ ↑↓ ↑ ↑ ↑ (one 3d electron excited to 5s)	_	_ _ _	_

↓ hybridise

		$3d_{xy}$ $3d_{xz}$ $3d_{yz}$	d^2sp^3	5s
$[Co(NH_3)_6]^{2+}$	[Ar]	↑↓ ↑↓ ↑↓	xx xx xx xx xx xx	↑

When hybridisation occurs, one of the 3*d*-electrons is excited to 5*s*-orbital.

The suggested bonding for $[Co(NH_3)_6]^{2+}$ is consistent with the fact that this complex is easily oxidised to $[Co(NH_3)_6]^{3+}$, an extremely stable complex ion.

Since there exists one unpaired electron, this complex ion is paramagnetic. Magnetic moment is given by:

$$\mu = \sqrt{N(N+2)}\ \text{B.M.} = \sqrt{3}\ \text{BM}$$

Table 18.7 **Bonding and Hybridisation in Some Octahedral Complexes**

Metal ion	Outer electron configuration (3d / 4s / 4p / 4d)	Complex ion	Type	Outer electron configuration (3d / hybrid / 4d)
Cr^{3+} (d^3)	3d: ↑ ↑ ↑ _ _; 4s: _; 4p: _ _ _; 4d: _ _ _ _ _	$[Cr(NH_3)_6]^{3+}$	inner orbital (paramagnetic)	3d: ↑ ↑ ↑; d^2sp^3: xxxxxxxxxxxx; 4d: _ _ _ _ _
Mn^{2+} (d^5)	3d: ↑ ↑ ↑ ↑ ↑; 4s: _; 4p: _ _ _; 4d: _ _ _ _ _	$[Mn(CN)_6]^{2-}$	inner orbital (paramagnetic)	3d: ↑↓ ↑↓ ↑; d^2sp^3: xxxxxxxxxxxx; 4d: _ _ _ _ _
Fe^{2+} (d^6)	3d: ↑↓ ↑ ↑ ↑ ↑; 4s: _; 4p: _ _ _; 4d: _ _ _ _ _	$[Fe(CN)_6]^{4-}$	inner orbital (diamagnetic)	3d: ↑↓ ↑↓ ↑↓; d^2sp^3: xxxxxxxxxxxx; 4d: _ _ _ _ _
Co^{3+} (d^6)	3d: ↑↓ ↑ ↑ ↑ ↑; 4s: _; 4p: _ _ _; 4d: _ _ _ _ _	$[Co(NH_3)_6]^{3+}$	inner orbital (diamagnetic)	3d: ↑↓ ↑↓ ↑↓; d^2sp^3: xxxxxxxxxxxx; 4d: _ _ _ _ _
		$[CoF_6]^{3-}$	outer orbital (paramagnetic)	3d: ↑↓ ↑ ↑ ↑ ↑; sp^3d^2: xxxxxxxxxxxx; 4d: _ _ _
Co^{2+} (d^7)	3d: ↑↓ ↑↓ ↑ ↑ ↑; 4s: _; 4p: _ _ _; 4d: _ _ _ _ _	$[Co(OH_2)_6]^{2+}$	outer orbital (diamagnetic)	3d: ↑↓ ↑↓ ↑ ↑ ↑; sp^3d^2: xxxxxxxxxxxx; 4d: _ _ _
Ni^{2+} (d^8)	3d: ↑↓ ↑↓ ↑↓ ↑ ↑; 4s: _; 4p: _ _ _; 4d: _ _ _ _ _	$[Ni(OH_2)_6]^{2+}$	outer orbital (paramagnetic)	3d: ↑↓ ↑↓ ↑↓ ↑ ↑; sp^3d^2: xxxxxxxxxxxx; 4d: _ _ _

Crystal Field Theory

The valence-bond theory has some shortcomings:

- It does not provide insight into the origin of the characteristic colour of the complex ions.
- It does not distinguish between weak and strong ligands.
- It does not give quantitative interpretation of the thermodynamic or kinetic stabilities of coordination compounds.
- There is no quantitative interpretation of magnetic data.
- It does not explain why some complexes are inner-orbital complex and some complexes are outer-orbital.

Most of the shortcomings of VBT of coordination compounds were removed in **Crystal Field Theory** (CFT). It attributes partial covalent character and partial ionic character to bonds.

Complexes with C.N. 6 (Octahedral Crystal Field)

- Five *d*-orbitals (d_{xy}, d_{yz}, d_{xz}, $d_{x^2-y^2}$ and d_{z^2}) are of equal energy and thus are degenerate.
- The $d_{x^2-y^2}$ and d_{z^2} orbitals are directed along a set of mutually perpendicular *x*, *y* and *z*-axes. As a group, these orbitals are called $\boldsymbol{e_g}$ **orbitals.**
- The d_{xy}, d_{yz} and d_{xz} orbitals, lie between the axes and collectively called $\boldsymbol{t_{2g}}$ **orbitals.**
- The ligand donor atoms approach the metal ion along the axes to form octahedral complexes. Crystal field theory proposes that the approach of the six donor atoms along the axes sets up an electric field (crystal field). Electrons on the ligand repel electrons in $\boldsymbol{e_g}$ orbitals on the metal ion more strongly than they repel those in $\boldsymbol{t_{2g}}$ orbitals. This removes the degeneracy of the set of *d*-orbitals and splits them into two sets, the $\boldsymbol{e_g}$ set at higher energy and the $\boldsymbol{t_{2g}}$ set at lower energy.

The energy separation between the two sets is called $\Delta_{octahedral}$ or Δ_{oct}. It is proportional to the crystal field of the strength of the ligands.

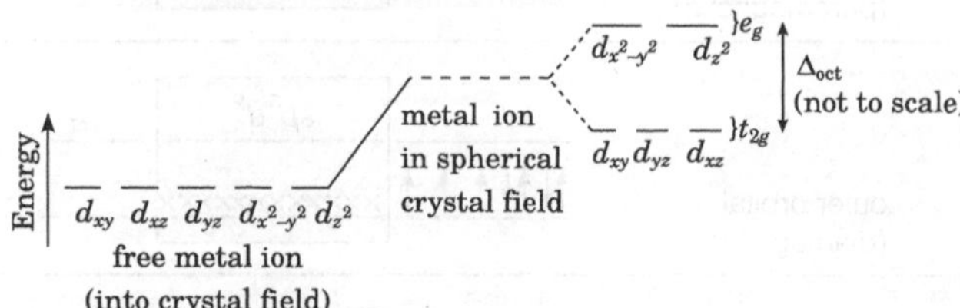

- The *d*-electrons on a metal ion occupy the t_{2g} set in preference to the higher energy, e_g set. Electrons that occupy the e_g orbitals are strongly repelled by the relatively close approach of ligands. Electron in e_g set tends to destabilise octahedral complexes.
- Based on valence bond theory, $[CoF_6]^{3-}$ is a paramagnetic outer orbital complex while $[Co(NH_3)_6]^{3+}$ is a diamagnetic inner orbital complex. Both contain d^6 Co^{3+} ions (four unpaired electron) in its 3*d*-orbitals.

	3d				
Co^{3+} (Ar)	↑↓	↑	↑	↑	↑

Magnetic measurements indicate that $[CoF_6]^{3-}$ also has four unpaired electrons per ion. Thus, there must be four electrons in t_{2g} orbitals and two in e_g orbitals.

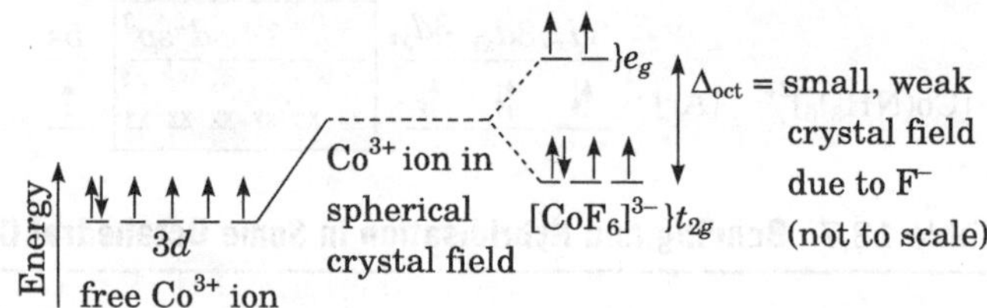

On the other hand, $[Co(NH_3)_6]^{3+}$ is diamagnetic, thus, all the six *d*-electrons must be paired in the t_{2g} orbitals.

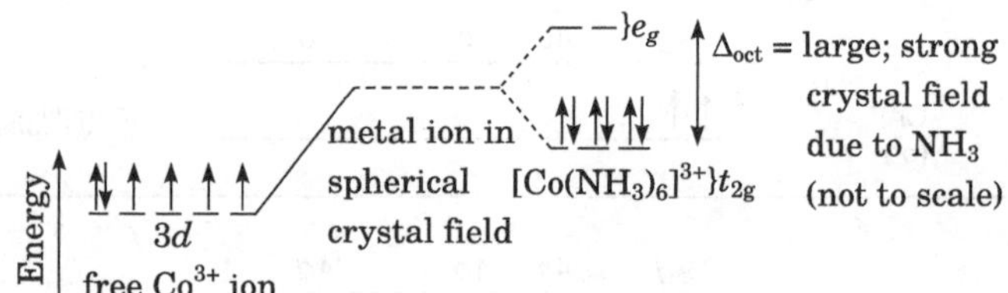

The difference in configuration between $[CoF_6]^{3-}$ and $[Co(NH_3)_6]^{3+}$ is due to the relative magnitude of the crystal field splitting, $\Delta_{oct,}$ caused by the different crystal field strengths of F^- and NH_3. The NH_3 molecule interacts with vacant metal orbitals more strongly than the F^- ion does.

$$\Delta_{oct} \text{ for } [Co(NH_3)_6]^{3+} > \Delta_{oct} \text{ for } [CoF_6]^{3-}$$

The crystal field splitting for $[CoF_6]^{3-}$ complex ion is very small and thus, an energetically more favourable condition may arise if two electrons remain unpaired in the anti-bonding, e_g orbitals. After all *d*-orbitals are half-filled, additional electrons will pair with electrons in t_{2g} set. This is the case for $[CoF_6]^{3-}$, which is called a **high spin complex.**

For $[CoF_6]^{3-}$: F^- is a weak field ligand, so $\Delta_{oct} < P$ (electron pairing energy), thus, **high spin complex.**

A high spin complex in crystal field terminology corresponds to an outer orbital complex in VBT.

For $[Co(NH_3)_6]^{3+}$: NH_3 is a strong field ligand so that $\Delta_{oct} > P$, thus **low spin complex.**

A low spin complex corresponds to an inner orbital complex.

- Low spin configurations exist only for octahedral complexes having metal ions with d^4, d^5, d^6 and d^7 configurations. For $d^1 - d^3$ and $d^8 - d^{10}$ ions configuration is designated as high spin. All d^n possibilities have been shown in Table 18.8

Table 18.8 **Electron Configurations and Crystal Field Stabilisation Energies for High and Low Spin Octahedral Complexes**

Configuration		d^1	d^2	d^3	d^4	d^5	d^6	d^7	d^8	d^9	d^{10}
Examples		Ti^{3+}	Ti^{2+}, V^{3+}	V^{2+}, Cr^{3+}	Mn^{3+}, Re^{3+}	Mn^{2+}, Fe^{3+}	Fe^{2+}, Pd^{4+}	Co^{2+}, Rh^{2+}	Ni^{2+}, Pt^{2+}	Cu^{2+}	Zn^{2+}, Ag^{+}
High spin	e_g	— —	— —	— —	↑ —	↑ ↑	↑ ↑	↑ ↑	↑ ↑	↑↓ ↑	↑↓ ↑↓
	t_{2g}	↑ — —	↑ ↑ —	↑ ↑ ↑	↑ ↑ ↑	↑ ↑ ↑	↑↓ ↑ ↑	↑↓ ↑↓ ↑	↑↓ ↑↓ ↑↓	↑↓ ↑↓ ↑↓	↑↓ ↑↓ ↑↓
	CFSE	$-\frac{2}{5}\Delta_o$	$-\frac{4}{5}\Delta_o$	$-\frac{6}{5}\Delta_o$	$-\frac{3}{5}\Delta_o$	0	$-\frac{2}{5}\Delta_o$	$-\frac{4}{5}\Delta_o$	$-\frac{6}{5}\Delta_o$	$-\frac{3}{5}\Delta_o$	0
Low spin	e_g				— —	— —	— —	↑ —			
	t_{2g}				↑↓ ↑ ↑	↑↓ ↑↓ ↑	↑↓ ↑↓ ↑↓	↑↓ ↑↓ ↑↓			
	CFSE	Same as high spin			$-\frac{8}{5}\Delta_o$	$-\frac{10}{5}\Delta_o$	$-\frac{12}{5}\Delta_o$	$-\frac{9}{5}\Delta_o$	Same as high spin		

Complexes with C.N. 4 (Tetrahedral Crystal Field)

In tetrahedral complex formation, the d-orbital splitting is inverted and smaller as compared to the octahedral field splitting. If Δ_t is the magnitude of energy separation in tetrahedral crystal field, then for the same metal, the same ligands and metal-ligand bond distance,

$$\Delta_t = -\frac{4}{9}\Delta_{oct}$$

In such cases, low spin configuration are rarely observed.

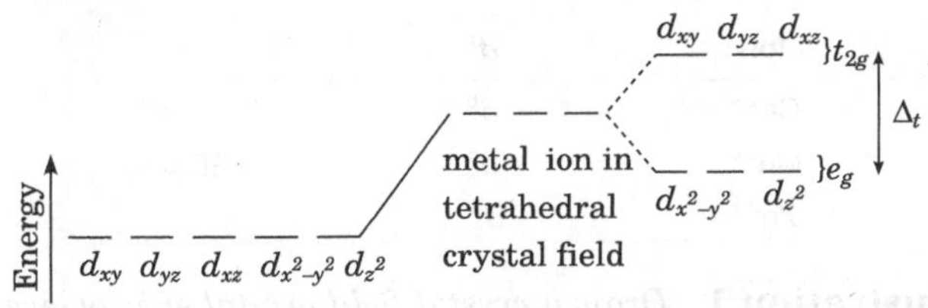

- The magnitude of Δ_{oct} or Δ_t, the orbitals splitting energy is decided by
  - — **nature of the ligand** : which have been arranged in increasing field strength as:

    $I^- < Br^- < Cl^- < F^- < OH^- < H_2O < C_2O_4^{2-} < NH_3 < en < NO_2^- < CN^-$
  - — **geometry of the complex:** Δ_t value (of tetrahedral complex) is about 50% compared to Δ_{oct} value (of octahedral complexes)
  - — **nature of the metal ion:** going from Cr to W (through Mo), Δ_{oct} increases 50% in each series change (*i.e.*, 3d to 4d to 5d). Thus, there is greater tendency of the second (4d) and third (5d) transition series to be low spin as compared to first (3d) transition series.
  - — **oxidation state of the metal ion**

    $\Delta(Fe^{3+}) > \Delta(Fe^{2+})$

    $\Delta(Cu^{2+}) > \Delta(Cu^{+})$

Thus, greater the ionic charge on the central metal ion, the greater the value of Δ.

Colour and the Spectrochemical Series

A substance appears coloured because it absorbs light that corresponds to one or more of the wavelengths in the visible region of the electromagnetic spectrum (4000 to 7000Å) and transmits or reflects the other wavelengths. A combination of all wavelengths in the visible region is called **"white light"**, sunlight is an example. The absence of all wavelengths in the visible region is blackness.

Most transition metal compounds are coloured. In transition metal compounds, the d-orbitals in any one energy level of the metal are not degenerate. They are often split into two sets of orbitals separated by energies, Δ_{oct}, that correspond to wavelength of light in the visible region. The absorption of visible light causes electronic transition between orbitals in these sets. Colour of some transition metal nitrates in aqueous solutions are given below:

Transition metal ion	Colour of aqueous solution
Cr^{3+}	Deep blue
Mn^{2+}	Pale pink
Fe^{2+}	Pale green
Fe^{3+}	Orchid
Co^{2+}	Pink
Ni^{2+}	Green
Cu^{2+}	Blue

One transition of a high spin octahedral Co(III) complex is represented below :

Group state		**Excited state**	**Energy of light Absorbed**
e_g ↑ ↑ t_{2g} ↑↓ ↑ ↑	Absorption of light ⟶	↑↓ ↑ ↑ ↑ ↑	$\Delta E = h\nu$ depend on Δ_{oct}

The frequency (ν), and thus the wavelength and colour, of the light absorbed are related to Δ_{oct}. This, in turn, depends on the crystal-field strength of the ligands. Hence, the colour and the visible absorption spectra of transition metal complexes, as well as their magnetic properties, provide information about the strengths of the ligand-metal interactions.

Table 18.9 gives the relationship among colours absorbed and colours transmitted or reflected in the visible region. The **spectral colour** is the colour associated with the wavelengths of the light absorbed by the sample. The **complimentary colour** is the colour associated with the wavelengths that are not absorbed by the sample, and is seen when the spectral colour is removed from white light. Complimentary colour of orange is spectral colour blue and *vice-versa* and is represented in the form of a wheel (to remember)

Table 18.9 **Complimentary Colours**

Wavelength absorbed (Å)	Spectral colour (colour absorbed)	Complimentary colour (colour observed)
4100	violet	lemon yellow
4300	indigo	yellow
4800	blue	orange
5000	blue-green	red
5300	green	purple
5600	lemon yellow	violet
5800	yellow	indigo
6100	orange	blue
6800	red	blue-green

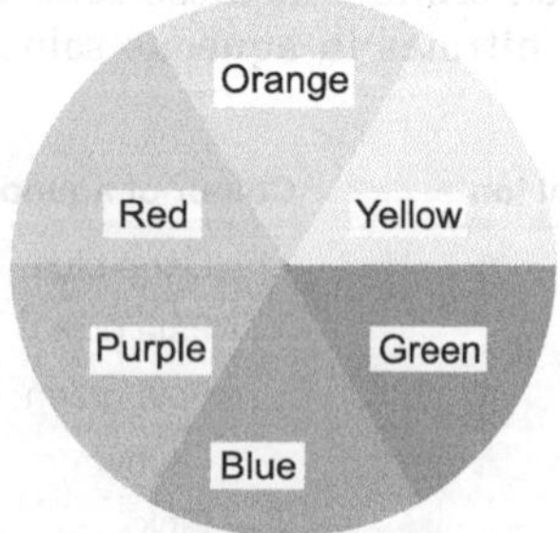

Fig. 18.1 Spectral colour and complimentary colour are opposite to each other

DEEP Focus

- By study of visible spectra of many complexes, it is possible to arrange common ligands in order of increasing crystal field strength:

$$I^- < Br^- < Cl^-, F^- < OH^- < H_2O < C_2O_4^{2-} < NH_3 < en < NO_2^- < CN^-, CO$$

This arrangement is called the **spectrochemical series.**

- Strong field ligands, such as CN^-, usually produce **low spin** complexes, where possible, and **large crystal field-splittings.**
- Weak field ligands, such as Cl^-, usually produce **high spin** complexes and **small crystal field-splittings**. Low spin complexes usually absorb higher-energy (shorter-wavelength) light than do high spin complexes.
- In $[Cr(NH_3)_6]^{3+}$, the Cr(III) is bonded to six NH_3 ligands, which produce a relatively high value of Δ_{oct}. This absorbs light in the blue and violet regions of visible light and thus orange-yellow complimentary colour is observed.
- H_2O is a weaker field ligand than NH_3, hence,

$$\Delta_{oct}\ ([Cr(H_2O)_6]^{3+}) < \Delta_{oct}\ ([Cr(NH_3)_6]^{3+})$$

and thus $[Cr(H_2O)_6]^{3+}$ absorbs lower-energy light and appears blue (complimentary colour)

Crystal Field Stabilisation Energy

Electrons in the t_{2g} orbitals of an octahedral complex are lower in energy, while electrons in e_g orbitals are higher in energy, than they would be in a spherical field. **The lower the total energy of a system, the more stable it is.**

The **crystal field stabilisation energy** (CFSE) of a complex is a measure of the net energy of stabilisation of a metal ions electron's. CFSE is zero for weak field (high spin) octahedral d^0, d^5 and d^{10} complexes.

Ion	d^n	
Ca^{2+}	d^0	
Mn^{2+}	d^5	CFSE = 0
Zn^{2+}	d^{10}	

Illustration 1 *Draw a crystal field orbital energy level diagram, and predict the number of unpaired electrons for each of the following complexes:*

(*a*) $[Cr(en)_3]^{3+}$ (*b*) $[Mn(CN)_6]^{3-}$ (*c*) $[Co(H_2O)_6]^{2+}$

Solution (a) Cr^{3+} ($[Ar]3d^3$) has three unpaired electrons. In the complex, they occupy the lower-energy set of *d*-orbitals.

(b) Mn^{3+} ($[Ar]3d^4$) can have a high-spin or low-spin configuration. Because CN^- is a strong-field ligand, all four *d*-electrons go into the lower-energy *d*-orbitals. The complex is low-spin with two unpaired electrons.

(c) Co^{2+}($[Ar]3d^7$) has a high-spin configuration with three unpaired electrons because H_2O is a weak-field ligand.

In the following orbital energy level diagrams, the relative values of the crystal field splitting Δ are in accordance with the position of the ligands in the spectrochemical series (H_2O < en < CN^-) :

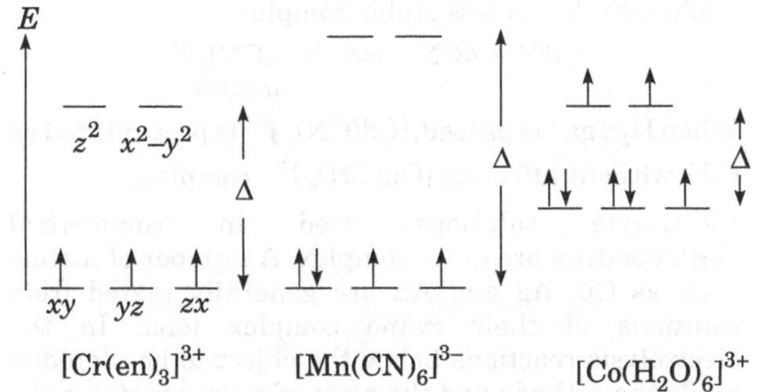

Stability Constants and Stability of Coordination Compounds

Consider the stepwise formation of the complex ML_4 from the metal ion M^{4+} (Lewis acid) and ligand L^-(Lewis base). Let k_1, k_2, k_3 and k_4 be stepwise formation constants and $\beta_4(= k_f)$ be the overall formation constant. Thus,

$$M^{4+} + L^- \rightleftharpoons ML^{3+}, \qquad k_1 = \frac{[ML^3]}{[M^{4+}][L^-]}$$

$$ML^{3+} + L^- \rightleftharpoons ML_2^{2+}, \qquad k_2 = \frac{[ML_2^{2+}]}{[ML^{3+}][L^-]}$$

$$ML_2^{2+} + L^- \rightleftharpoons ML_3^{+}, \qquad k_3 = \frac{[ML_3^{+}]}{[ML_2^{2+}][L^-]}$$

$$ML_3^{+} + L^- \rightleftharpoons ML_4, \qquad k_4 = \frac{[ML_3^{+}]}{[ML_3^{+}][L^-]}$$

$$\therefore \quad M^{4+} + 4L^- \rightleftharpoons ML_4, \qquad \beta_4 = \frac{[ML_4]}{[M^{4+}][L^-]^4}$$

$$\therefore \quad \beta_4 = k_1k_2k_3k_4$$

$$\therefore \quad \log\beta_4 = \log k_1 + \log k_2 + \log k_3 + \log k_4$$

If the complex is ML_n, and β_n the overall formation constant, then

$$M + nL \rightleftharpoons ML_n \text{(charges omitted)}$$

$$\beta_n = \frac{[ML_n]}{[M][L]^n} = k_1k_2k_3k_4 \ldots k_n$$

$$\log\beta_n = \log k_1 + \log k_2 + \log k_3 + \ldots + \log k_n$$

- β_n, the stability constant, is related to thermodynamic stability when the system has reached equilibrium. Most of the measurements have been made from aqueous solutions, which means that when complex is formed, stronger ligand L (say CN^-, en) displaces weaker ligand H_2O of the aqua complex of the metal ion.

$$M(H_2O)_n + nL \rightleftharpoons ML_n + nH_2O$$

DEEP Focus

- **For a given ion and ligand, greater the charge on the metal ion, greater the stability. Thus, stability of**

$$[Fe^{III}(CN)_6]^{3-} > [Fe^{II}(CN)_6]^{4-}$$

- **For the divalent ions of the first row transition elements (3*d*-series), the stabilities vary in the Irving-Williams order:**

$$Mn^{II} < Fe^{II} < Co^{II} < Ni^{II} < Cu^{II} < Zn^{II}$$

- **Ligands containing N, O or F donor atoms form most stable coordination compounds when metal ions are:**
  - **groups 1 and 2 elements, the inner transition elements and transition elements of groups 3 to 6 (Sc, T, V, Cr) called 'class a' acceptors.**
  - **Transition elements—Rh, Pd, Ag, Ir, Au and Hg having relatively completely filled *d*-orbitals called 'class b' acceptors.**
  - **The stability also depends on the formation of chelate rings and stabilisation due to chelation is called the chelation effect. This effect is maximum for the five and six-membered rings.**
- **If a multidentate ligand happens to be cyclic and there are no unfavourable steric effects, a further increase in stability occurs. This is termed the macrocyclic effect.**

Kinetic Consideration of Ligand Displacement

When we add $NH_3(aq)$ to a solution containing Cu^{2+}, we observe a change in colour from pale blue to deep blue. The reaction involves NH_3 molecules displacing H_2O molecules as ligands.

$$\underset{\text{(pale blue)}}{[Cu(H_2O)_4]^{2+}} + 4NH_3 \longrightarrow \underset{\text{(very deep blue)}}{[Cu(NH_3)_4]^{2+}} + 4H_2O$$

This reaction occurs very rapidly as rapidly as the two reactions can be brought together. The addition of $HCl(aq)$ to an aqueous solution of Cu^{2+} produces an immediate colour change from pale blue to green, or even yellow if the $HCl(aq)$ is sufficiently concentrated.

$$\underset{\text{pale blue}}{[Cu(H_2O)_4]^{2+}} + 4Cl^- \longrightarrow \underset{\text{yellow}}{[CuCl_4]^{2-}} + 4H_2O$$

Complex ions in which ligands can be interchanged rapidly are said to be **labile.** $[Cu(H_2O)_4]^{2+}$, $[Cu(NH_3)_4]^{2+}$, and $[CuCl_4]^{2-}$ are all labile.

In freshly prepared $CrCl_3(aq)$, $[Cr(H_2O)_4Cl_2]^+$ produces a green colour, but colour gradually turns to violet. This colour change results from the very slow exchange of H_2O for Cl^- as ligands. A complex ion that exchanges ligands slowly is said to be **non-labile or inert.** In general, complex ions of the first transition series, except for those Cr(III) and Co(III) are kinetically labile. Those of the second and third transition series are generally kinetically inert.

These terms (labile/inert) are used to describe the different kinetics of the reactions that are thermodynamically allowed. Large activation energy barriers for ligand substitution reactions of inert complexes make those reactions slow even though there may be thermodynamic tendency to proceed. In the substitution reaction

$$[Co(NH_3)_6]^{3+}(aq) + 6H_3O^+(aq) \rightleftharpoons [Co(H_2O)_6]^{3+}(aq) + 6NH_4^+(aq)$$

The products are favoured by an enormous amount, yet the inert $[Co(NH_3)_6]^{3+}$ complex ion last for weeks in acidic solution because there is no low-energy pathway for the reaction. The cobalt (III) ion, $[Co(NH_3)_6]^{3+}$, is thermodynamically unstable relative to $[Co(H_2O)_6]^{3+}$ but kinetically stable (inert) The closely related cobalt (II) complex, $[Co(NH_3)_6]^{2+}$, reacts with water in a matter of seconds.

$$[Co(NH_3)_6]^{2+}(aq) + 6H_3O^+(aq) \longrightarrow [Co(H_2O)_6]^{2+}(aq) + 6NH_4^+(aq)$$

The complex $[Co(NH_3)_6]^{2+}$ is thermodynamically unstable and also kinetically labile.

Illustration 1 *Formation constant K_f for following complexes have been given* :

$$[Ni(en)_3]^{2+}, \quad K_f = 1.1 \times 10^{18}$$
$$[Ni(NH_3)_6]^{2+}, \quad K_f = 5.3 \times 10^{8}$$

Explain the relative stability of these complexes.

Solution Ethylenediamine (en) is a bidentate ligand and ammine (NH_3) is an unidentate ligand. Complex ion $[Ni(en)_3]^{2+}$ with multidentate ligand has much larger formation constant than do those with unidentate ligands. Thus, $[Ni(en)_3]^{2+}$ is more stable than $[Ni(NH_3)_6]^{2+}$.

This additional stability of complex ions associated with chelate formation by multidentate ligands is known as the **chelation effect.**

Applications of Coordination Chemistry

- The photographic "fixing" process removes unreacted AgBr which forms soluble complex with sodium thiosulphate:

$$AgBr(s) + 2S_2O_3^{2-} \longrightarrow \underset{\text{soluble}}{[Ag(S_2O_3)_2]^{3-}} + Br^-$$

- In qualitative analysis of group I, there is formation of white precipitate of AgCl, Hg_2Cl_2, and $PbCl_2$. $PbCl_2$ is soluble in hot water. AgCl can be removed as soluble $[Ag(NH_3)_2]Cl$ on adding $NH_3(aq)$:

$$AgCl(s) + 2NH_3(aq) \rightleftharpoons \underset{\text{soluble}}{Ag[NH_3)_2]Cl}$$

In group II, Cd^{2+} can be precipitated as CdS in presence of Cu^{2+}. By treating the solution of Cu^{2+} and Cd^{2+} with an excess of CN^-, Cu^{2+}forms a stable complex in a redox reaction :

$$2Cu^{2+} + 10CN^- \longrightarrow \underset{\text{stable}}{2[Cu(CN)_4]^{3-}} + C_2N_2(g)$$

while Cd^{2+} forms less stable complex:

$$Cd^{2+} + 4CN^- \rightleftharpoons \underset{\text{unstable}}{[Cd(CN)_4]^{2-}}$$

When H_2S gas is passed, $[Cd(CN)_4]^{2-}$ is precipitated as CdS without affecting $[Cu(CN)_4]^{3-}$ complex.

- Electrolyte solutions used in commercial electroplating are quite complex. A number of metals such as Cu, Ag and Au are generally plated from solutions of their cyano complex ions. In the electrolysis reactions below the object to be plated is made the cathode and the piece of copper metal is the anode.

Anode: $Cu + 4CN^- \longrightarrow [Cu(CN)_4]^{3-} + e^-$

Cathode : $[Cu(CN)_4]^{3-} + e \longrightarrow Cu + 4CN^-$

The net change simply involves the transfer of Cu metal from the anode to the cathode through the formation, migration and decomposition of complex ion $[Cu(CN)_4]^{3-}$.

- Purification of water is also through complex formation. One method of water treatment involves **chelation.** Among the chelating agents widely employed are the salts of EDTA, usually the sodium salt. The formation constants of $[Ca(EDTA)]^{2-}$ and $[Mg(EDTA)]^{2-}$ are very large. Hence, on treating water with sodium salt of EDTA, concentration of Ca^{2+} and Mg^{2+} can be reduced to the point that these ions will not precipitate with soap or detergent.

$$Ca^{2+} + (EDTA)^{4-} \rightleftharpoons [Ca(EDTA)]^{2-}$$
$$Mg^{2+} + (EDTA)^{4-} \rightleftharpoons [Mg(EDTA)]^{2-} \quad K_f = 4 \times 10^8$$

- Chelation with EDTA can also be used in the treatment of metal poisoning. If a person suffering from lead-poisoning is fed $[Ca(EDTA)]^{2-}$, the following exchange occurs:

$$Pb^{2+} + [Ca(EDTA)]^{2-} \longrightarrow [Pb(EDTA)]^{2-} + Ca^{2+}$$

because K_f of $[Pb(EDTA)]^{2-}$ is 1×10^{18} and is thus more stable than $[Ca(EDTA)]^{2-}$. The lead complex is excreted by the body and the Ca^{2+} remains as the body nutrient.

- The extraction of silver and gold involves the formation of cyanide complex:

$$4Au(s) + 8CN^-(aq) + O_2(g) + 2H_2O(l) \longrightarrow 4OH^-(aq) + 4[Au(CN)_2](aq)$$

$$2[Au(CN)_2]^-(aq) + Zn(s) \longrightarrow [Zn(CN)_4]^{2-}(aq) + 2Au(s)$$

Mond's process involves formation of carbonyl complex of nickel:

$$Ni(s) + 4CO(g) \longrightarrow Ni(CO)_4(g)$$

- Traces of Ni^{2+} salt can be detected by dimenthyl glyoxime when cherry red precipitate of nickel-dimethyl glyoximate (chelate) is formed:

Ni^{2+} + 2 HO—C=N—CH_3 / HO—C=N—CH_3 ⟶

O—H---O
H_3C—C=N   N=C—CH_3
>Ni<
H_3C—C=N   N=C—CH_3
O---H—O

- *Cis*-diammine dichloroplatinum (II), $[Pt(NH_3)_2Cl_2]$, is

H_3N, H_3N >Pt< Cl, Cl

one of a number of platinum coordination compounds used in the treatment of cancer. Commonly known as ***cis*-platin,** this compound has the ability to block the uncontrolled division of cancerous cells that results in the growth of tumours. Recent studies show that *cis*-platin can cause serious side effects, including severe kidney damage. It is being replaced by dinuclear platinum complex.

$[$Cl, H_3N >Pt< NH_3, $NH_2(CH_2)_nH_2N$ >Pt< Cl, $NH_3]^{2+}$

trans-isomer (n = 2 to 6)

- Organometallic compound must contain, at least one metal-carbon bond. $Ni(CO)_4$ is also considered organometallic compound, through CO is not considered to be an organic compound. The suffix metallic often includes metalloids like boron, silicon, arsenic and tellurium as well as true metals. Metal-cyanides, though possessing an M—C bond are not included in organometallics.
- **Main group organometallics**

E.C. Frankland was the first chemist to synthesise an organometallic compound.

The *s*- and *p*-block organometallics are named according to substituent names used in organic chemistry.

CH_3Li	methyl lithium
$B(CH_3)_3$	trimethyl boron
$Si(CH_3)_4$	tetramethyl silane
$As(CH_3)_3$	trimethyl arsane

The oxidation number of the metallic element in an organometallic compound is based on the organic part being considered to be anionic. Thus, CH_3 group in $Zn(CH_3)_2$ has (–1) charge.

- ***d*- and *f*-block organometallics**

W.C. Zeise was the first to prepare organometallic compound of *d*-block. Examples are

– trichloroethene platinum (II) : $[PtCl_3(C_2H_4)]^-$
– tetracarbonyl nickel (0) : $Ni(CO)_4$
– bis(cyclopentadienyl) iron (II) : $Fe(C_5H_5)_2$ also called ferrocene.

Ferrocene is stable in air, forms orange crystals and is diamagnetic. It is soluble in organic solvents (alcohol, ether and benzene) and insoluble in water, NaOH solution and conc. HCl solution.

It is thermally stable up to 500°C. The X-ray structure shows that this has a sandwich structure in which the metal atom is sandwiched between two planar cyclopentadienyl rings.

In metal carbonyls, CO as a ligand binds itself to metal atom through the carbon atom. It is a weak donor (a weak Lewis base) and forms a weak σ-bond to the central atom. CO is also and acceptor (Lewis acid) ligand and forms π bond to the metal. This characteristic property of back bonding stabilises the metal-ligand interaction.

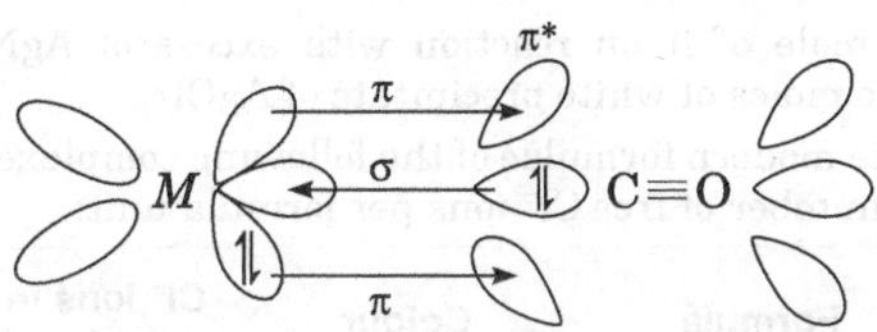

Target Practice 2

1. The complex $CoCl_4^{2-}$ has a tetrahedral structure.
   (a) How many *d*- electrons are on the cobalt? ....
   (b) What is its electronic configuration? ....
2. Make complexes with varying NH_3 ligands to satisfy primary and secondary valencies of platinum.
   $PtCl_4 \cdot xNH_3$ ....
   ....
   ....
3. Arrange the complexes in (2) in increasing electrical conductivity ....
4. What is CFSE of high-spin d^5 complex of Fe^{3+}? ....
5. What is pK value of the complex $[PtCl_4]$ if

$$Pt^{4+} + Cl^- \longrightarrow [PtCl_3]^{3+} \qquad K_1 = 10^{-2}$$
$$[PtCl_3]^{3+} + Cl^- \longrightarrow [PtCl_2]^{2+} \qquad K_2 = 10^{-3}$$
$$[PtCl_2]^{2+} + Cl^- \longrightarrow [PtCl_3]^{+} \qquad K_3 = 10^{-4}$$
$$[PtCl_3]^{+} + Cl^- \longrightarrow [PtCl_4] \qquad K_4 = 10^{-5}$$

6. Dilute equimolar solutions of each of the following compounds are prepared. Arrange the compounds in order of increasing molar conductivity (which is proportional to the number of ions) of these solutions :
   (a) $Na_3[Co(NO_2)_6]$ (b) $K[Co(EDTA)]$
   (c) $[Co(Py)_3(CN)_3]$ (d) $[Cr(NH_3)_5Cl]Cl_2$
   (e) $[Pt(NH_3)_6]Br_4$
7. While practice in the laboratory, you have some yellow spot of $FeCl_3$ on your white shirt. Suggest a suitable reagent that can remove this yellow spot ....

Answers

1. (a) 7

(b) t_{2g}(5) $d_{xy}^2\ d_{yz}^2\ d_{zx}^2$ ↑↓ ↑↓ ↑

e_g (2) $d_{z^2}^1\ d_{x^2-y^2}^1$ ↑ ↑

2. I. $[Pt(NH_3)_6]Cl_4$
II. $[Pt(NH_3)_5Cl]Cl_3$
III. $[Pt(NH_3)_4Cl_2]Cl_2$
IV. $[Pt(NH_3)_3Cl_3]Cl$
V. $[Pt(NH_3)_2Cl_4]$

3. V < IV < III < II < I

4. Zero

5. On adding

$$Pt^{4+} + 4Cl^- \longrightarrow PtCl_4$$

∴ Overall stability constant

$$K = K_1K_2K_3K_4 = 10^{-14}$$

$$\log K = \log K_1 + \log K_2 + \log K_3 + \log K_4 = -14$$

$$-\log K = 14$$

$$pK = 14$$

6. (c) < (b) < (d) < (a) < (e), ionisation being as:
(c) does not ionise
(b) $\rightleftharpoons K^+ + [Co(EDTA)]^-$ (two ions)
(d) $\rightleftharpoons [Cr(NH_3)_5Cl]^{2+} + 2Cl^-$ (three ions)
(a) $\rightleftharpoons 3Na^+ + [Co(NO_2)_6]^{3-}$ (four ions)
(e) $\rightleftharpoons [Pt(NH_3)_6]^{4+} + 4Br^-$ (five ions)

7. Yellow spots can be removed by complexing Fe^{3+} by oxalate ion and washing it with water.

$$Fe^{3+} + 3C_2O_4^{2-} \longrightarrow \underset{\text{colourless (soluble)}}{[Fe(C_2O_4)_3]^{3-}}$$

Practice Exercise 2

1. $PtCl_4 \cdot 6H_2O$ on reaction with $AgNO_3$ forms 2 moles of AgCl (white ppt.) per mol of the $PtCl_4 \cdot 6H_2O$. What is complex? Give its Werner's representation.
2. $CuCl_2 \cdot xH_2O$ (x = 2, 3, 4) can be represented by Werner's formula. What are those? Arrange them in increasing conducting power.
3. What are types of ions joined by primary valency and secondary valency of the metal in the following complexes?
(a) $FeCl_3 \cdot 6H_2O$ (b) $CuSO_4 \cdot 5H_2O$
(c) $PtCl_4 \cdot 6NH_3$ (d) $PtCl_2 \cdot 2H_2O$
4. What is the true formula of the complex $PtCl_4 \cdot 5NH_3$ if one mole of it on reaction with excess of $AgNO_3$ yields three moles of white precipitate of AgCl?
5. Write modern formulae of the following complexes based on the number of free Cl^- ions per formula unit:

Formula	Colour	Cl^- ions in solution per formula unit
(a) $CrCl_3 \cdot 6H_2O$	violet	3
(b) $CrCl_3 \cdot 6H_2O$	light green	2
(c) $CrCl_3 \cdot 6H_2O$	dark green	1

6. Transition metal complexes containing, CN^- ligands are often yellow in colour, whereas those containing H_2O ligands are often green or blue. Explain.
7. The complex $[FeCl_6]^{3-}$ is more paramagnetic than $[Fe(CN)_6]^{3-}$. Explain.
8. Give a valence bond description of the bonding in each of the following complexes. Include orbital diagram for the free metal ion and the metal ion in the complex. Specify type of hybridisation and number of unpaired electrons.
(a) $[Ti(H_2O)_6]^{3+}$ (b) $[NiBr_4]^{2-}$
(c) $[Fe(CN)_6]^{3-}$ (d) $[MnCl_6]^{3-}$
9. The absorption maximum for the complex ion $[Co(NH_3)_6]^{3+}$ occurs at 470 nm.
(a) Predict the colour of the complex and
(b) Calculate the crystal field splitting in kJ/mol.
10. Draw a crystal field orbital energy level diagram, and predict the number of unpaired electrons for each of the following complexes:
(a) $[Fe(H_2O)_6]^{2+}$ (b) $[Fe(CN)_6]^{4-}$
(c) $(VF_6)^{3-}$
11. Explain why $[CoCl_4]^{2-}$ (blue) and $[Co(H_2O)_6]^{2+}$ (pink) have different colours? Which complex has its absorption bands at longer wavelength?
12. What is chelation effect?
13. Give the Irving —Williams order of stability of metal ions.
14. Stability constants of $[Cd(CN)_4]^{2-}$ and $[Ni(CN)_4]^{2-}$ are 1.28×10^{17} and 1.0×10^{31} respectively. Which is more stable?
15. Which of the following complexes is expected to be inert to the ligand substitution?
(a) $[Ni(en)_3]^{2+}$ (b) $[Mg(EDTA)]^{2-}$
(c) $[Sc(H_2O)_6]^{3+}$ (d) $[Co(NO_2)_6]^{3-}$
(e) $[Cr(H_2O)_6]^{2+}$ (f) $[Al(OH)_4]^-$
(g) $[FeF_6]^{3-}$
16. Give the classification of organometallic compounds.
17. Discuss the nature of bonding in metal carbonyls.
18. Draw the structure of iron oxalate complex $[Fe(C_2O_4)_3]^{3-}$. Describe the coordination geometry and identify any chelate ring. What is the oxidation number and the coordination number of iron?
19. What happens when $[Ni(H_2O)_6]^{2+}$ is mixed with ethylenediamine in excess?

Answers

1. $[PtCl_2(H_2O)_4]Cl_2 \cdot 2H_2O$

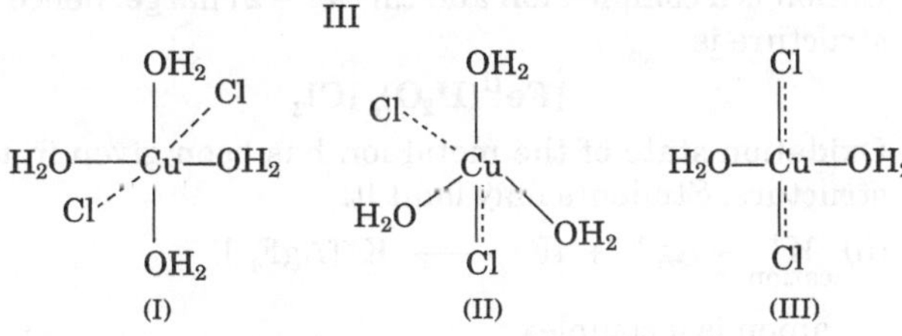

Since two moles of AgCl are obtained per mole of the compound hence, two Cl are joined by primary valency and two by secondary valency.

2. $[Cu(H_2O)_4]Cl_2$ (I), $[Cu(H_2O)_3Cl]Cl$ (II), $[Cu(H_2O)_2Cl_2]$ (III)

(I) (II) (III)

Conductance III < II < I

3. (a) $[Fe(H_2O)_6]Cl_3$: Three Cl joined by primary valency (PV) and six H_2O joined by secondary valency (SV)

(b) $[Cu(H_2O)_4]SO_4 \cdot H_2O$: SO_4^{2-} joined by (PV) and four H_2O by (SV)

(c) $[Pt(NH_3)_6]Cl_4$: Four Cl by (PV) and six NH_3 by (SV)

(d) $[Pt(H_2O)_2Cl_2]$: Two Cl have dual nature joined by (PV) as well as (SV) while two H_2O by (SV).

4. $[Pt(NH_3)_5Cl]Cl_3$

5. (a) $[Cr(H_2O)_6]Cl_3$ (b) $[Cr(H_2O)_5Cl]Cl_2 \cdot H_2O$
(c) $[Cr(H_2O)_4Cl_2]Cl \cdot 2H_2O$

6. CN^- is a strong ligand. Absorption near UV (blue), hence appear yellow.

7. Based on VBT, $[FeCl_6]^{3-}$ has Cl^- as weak ligands, Fe is sp^3d^2 hybridised (outer orbital complex) having five unpaired electrons. $[Fe(CN)_6]^{3-}$ has CN^- as strong ligand. Fe is d^2sp^3 hybridised (inner orbital complex) having only one unpaired electron, hence, $(FeCl_6)^{3-}$ is more paramagnetic than $[Fe(CN)_6]^{3-}$. Also refer text.

8. Refer text

9. (a) orange (b) 254.38 kJ mol^{-1}

10.

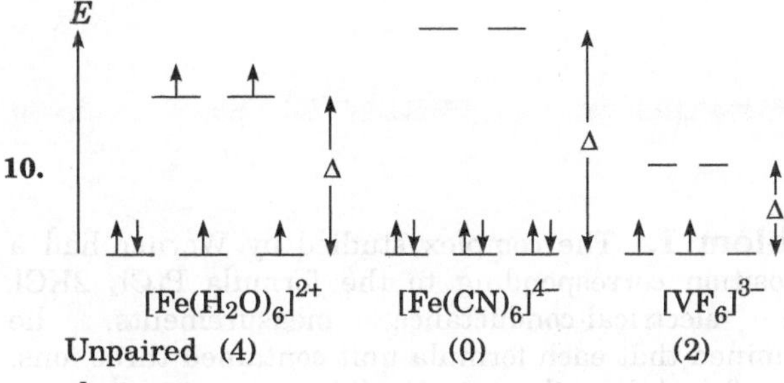

$[Fe(H_2O)_6]^{2+}$ $[Fe(CN)_6]^{4-}$ $[VF_6]^{3-}$

Unpaired electrons : (4) (0) (2)

11. $\Delta_{tet} < \Delta_{oct}$ hence, different colour $[CoCl_4]^{2-}$ has absorption peak at higher wavelength.

12. Increase in stability due to presence of multidenate ligands (which causes ring formation) is called chelation effect.

13. $Mn^{2+} < Fe^{2+} < Co^{2+} < Ni^{2+} < Cu^{2+} < Zn^{2+}$

14. $[Ni(CN)_4]^{2-}$

15. Only (d), $[Co(NO_2)_6]^{3-}$, due to very high value of formation constant.

16. Refer text

17. Refer text

18. Oxidation number of Fe = + 3

Coordination number of Fe = 6

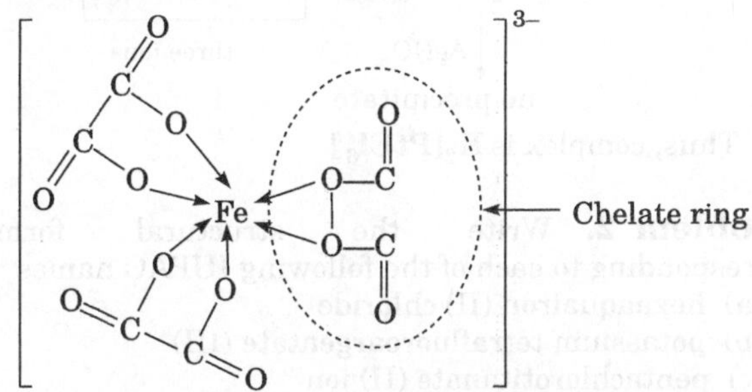

$$[Ni(H_2O)_6]^{2+} + 3en \longrightarrow [Ni(en)_3]^{2+} + 6H_2O$$

19. Ethylenediamine (en) displaces H_2O forming a chelate.

Total Practice Set

(Read, Plan & Solve!)

Problem 1. The complex studied by Werner had a composition corresponding to the formula $PtCl_4 \cdot 2KCl$. From electrical-conductance measurements, he determined that each formula unit contained three ions. He also found that silver nitrate did not give a precipitate of AgCl with this complex. Write a formula for this complex that agrees with this information.

Plan *Since Ag^+ does not give precipitate with the complex, it means all the Cl are in coordinate sphere. Electrical conductance shows three ions per mol of complex, thus, formula can be derived.*

Solution Given complex does not give precipitate with $AgNO_3$. It shows the absence of Cl^- ions. Thus, all the Cl ions are **ligands** and are in coordination sphere.

$$2KCl + PtCl_4 \rightleftharpoons K_2[PtCl_6]$$

$$K_2[PtCl_6] \rightleftharpoons \underbrace{2K^+ + [PtCl_6]^{2-}}_{\text{three ions}}$$

$$K_2[PtCl_6] \xrightarrow{AgNO_3} \text{no precipitate}$$

Thus, complex is $\mathbf{K_2[PtCl_6]}$

Problem 2. Write the structural formula corresponding to each of the following IUPAC names :

(a) hexaaquairon (II) chloride
(b) potassium tetrafluoroargentate (III)
(c) pentachlorotitanate (II) ion
(d) tetraamminedichlorocobalt (III) chloride
(e) dicyanoargentate (I) ion
(f) diamminetetrachloronickelate (II) ion
(g) tris (ethylenediamine) copper (II) sulphate
(h) sodium diaquatetrahydroxoaluminate (III)
(i) amminechlorobis (ethylene diamine) chromium (III) sulphate
(j) potassium tetracyanozincate (II)
(k) hexaamminechromium (III) tetrachlorocuprate (II)
(l) diamminedichloroplatinum (II)
(m) tetracarbonyl nickel (0)
(n) tetraammineplatinum (II) amminetrichloroplatinate (II)
(o) sodium bis(thiosulphato) argentate (I)
(p) potassium tetracyanonickelate (0)
(q) bis (cyclopentadienyl) iron (II)
(r) tetrathiocyanato–N–zincate (II) ion
(s) potassium tetraoxomanganate (VII)
(t) potassium trisoxalatoaluminate (III)
(u) tetrapyridineplatinum (II) tetrachloroplatinate (II)

Solution (a) $Fe^{2+} + 6H_2O + Cl^- \longrightarrow [Fe(H_2O)_6]^{2+}Cl^-$

Cation is a complex ion and carries + 2 charge, hence its structure is

$$\mathbf{[Fe^{II}(H_2O)_6]Cl_2}$$

Oxidation state of the metal ion has been given in the structure. Students may omit it.

(b) $\underset{\text{cation}}{K^+} + Ag^{3+} + 4F^- \longrightarrow K^+[AgF_4]^-$

anion is a complex

$$K^+[Ag^{III}F_4]^-$$

(c) $[Ti^{II}Cl_5]^{3-}$
(d) $[Co^{III}Cl_2(NH_3)_4]Cl$
(e) $[Ag^{I}(CN)_2]^-$
(f) $[Ni^{II}Cl_4(NH_3)_2]^{2-}$
(g) $[Cu^{II}(en)_3]SO_4$
(h) $Na[Al^{III}(OH)_4(H_2O)_2]$
(i) $[Cr^{III}Cl\,(NH_3)(en)_2]SO_4$
(j) $K_2[Zn^{II}(CN)_4]$
(k) $[Cr^{III}(NH_3)_6]_2[Cu^{II}Cl_4]_3$
(l) $[Pt^{II}Cl_2(NH_3)_2]$
(m) $[Ni^{0}(CO)_4]$
(n) $[Pt^{II}(NH_3)_4][Pt^{II}Cl_3NH_3]_2$
(o) $Na_3[Ag^{I}(S_2O_3)_2]$
(p) $K_4[Ni^{0}(CN)_4]$
(q) $Fe[\eta^5-C_5H_5]_2$

η^5 means that all the five carbon atoms of cyclopentadienyl anion are coordinated to the metal ion.

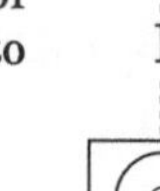

(r) $[Zn^{II}(NCS)_4]^{2-}$
(s) $K[Mn^{VII}O_4]$
(t) $K_3[Al^{III}(C_2O_4)_3]$
(u) $[Pt^{II}(Py)_4][Pt^{II}Cl_4]$

Problem 3. Write IUPAC names of the following:

(a) $Na_2[Fe(CN)_5NO]$ (b) $[Ni(DMG)_2]$
(c) $NH_4[Cr(SCN)_4(NO)_2]$ (d) $[Pd(H_2O)_2(ONO)_2\ I_2]$

(e) $[Co(NH_3)_4(NCS)\,Cl]^+$ (f) $K_4[FeO_4]$
(g) $K_2[Co(N_3)_4]$ (h) $[Ni(PPh_3)_2Cl_2]$
(i) $[Cr(acac)_3]$ (j) $[Pt(C_5H_5N)_4][PtCl_2]$
(k) $[Co(NH_3)_6][CuCl_5]$ (l) $[V(H_2O)_6]Cl_3$
(m) $(NH_4)_3[Co(C_2O_4)_3]$
(n) $\left[(NH_3)_4Co\langle{}^{OH}_{OH}\rangle Co(en)_2\right]Cl_4$
(o) $\left[(en)_2Co\langle{}^{OH}_{NH}\rangle Co(en)_2\right]^{4+}$
(p) $Na_2[SiF_6]$
(q) $K_2[CrO_4]$
(r) $[(NH_3)_5Cr—OH—Cr(NH_3)_5]Cl_5$
(s) $[Fe(en)_3][Fe(CO)_4]$
(t) $TiCl_4(Et_2O)_2$
(u) $Mn_2(CO)_{10}$
(v) $[VO(acac)_2]$
(w) $Fe[Fe(CN)_6]^+$

Solution (a) Sodium pentacyanonitrosoferrate (II)
(b) Bis (dimethylglyoximato) nickel (II)
(c) Ammonium tetrathiocyanatodinitrosochromate (III)
(d) Diaquadiiododinitritopalladium (IV)
(e) Tetraamminechloroisothiocyanatocobalt (III) ion
(f) Potassium tetraoxoferrate (IV)
(g) Potassium tetraazidocobalt (II) [N_3^- is azide]
(h) Dichlorobis(triphenyl phosphine) nickel (II)
(i) Tris (acetylacetonato)chromate (III) [$(acac)^-$ is $CH_3COCH_2CH_2^-$)]
(j) Tetrakis (pyridine)platinum (II) tetrachloroplatinate (II)
(k) Hexaammine cobalt (III) pentachlorocuprate (II)
(l) Hexaaquavanadium (III) chloride
(m) Ammonium tris(oxalato)cobaltate (III)
(n) Tetraamminebis (ethylenediammine)-μ-dihydroxo dicobalt (III) ion
(o) Tetrakis (ethylenediamine)-μ-hydroxo-μ-imidodicobalt (III) ion
(p) Sodium hexafluorosilicate (IV)
(q) Potassium tetraoxochromate (VI)
(r) μ–hydroxodecaamminedichromium (III) chloride
(s) Tris(ethylenediamine)iron (III) tetracarbonyl iron (III) (metal in this complex can also be iron (II))
(t) Tetrachlorobis (diethyl ether) titanium (IV)
(u) Decacarbonyldiamanganese (0)
(v) Bis (acetylacetonato) oxovanadium (IV)
(w) Iron (III) hexacyanoferrate (II) ion (also called Prussian blue)

Problem 4. A compound $Co(en)_2(NO_2)_2Cl$ has been prepared in a number of isomeric forms. One form undergoes no reaction with $AgNO_3$ or (en) and is optically inactive. A second form reacts with $AgNO_3$ but not with (en) and is optically inactive. A third form is optically active and reacts with both $AgNO_3$ and (en). Identify each of these isomeric forms. Name and sketch each of the structures.

Plan Based on reactions with $AgNO_3$ *and (en), and optical activity, isomers can be identified.*

Solution **First form :** There is no reaction with $AgNO_3$, hence no Cl^- ions outside coordination sphere. Also there is no reaction with bidentate (en), hence these ligands are *trans* to each other. Optical inactivity is also due to *trans* structure. Thus, it may have structure :

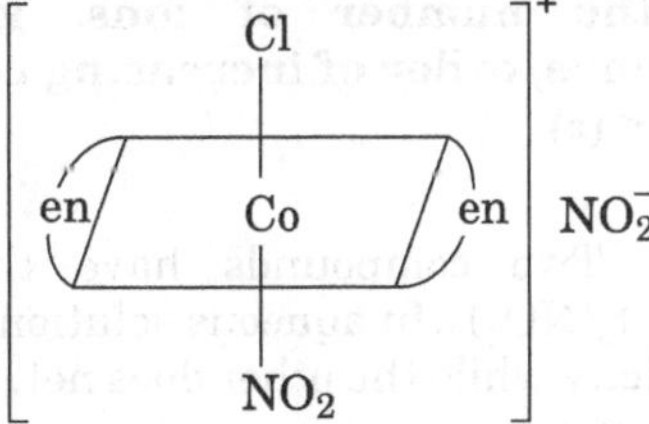

trans-chloronitrobis(ethylenediamine)cobalt (III) nitrite.

Second form : In this Cl^- is outside coordination sphere since it reacts with $AgNO_3$. As in the first form NO_2^- ligands are *trans* to each other being optically inactive. This is represented as,

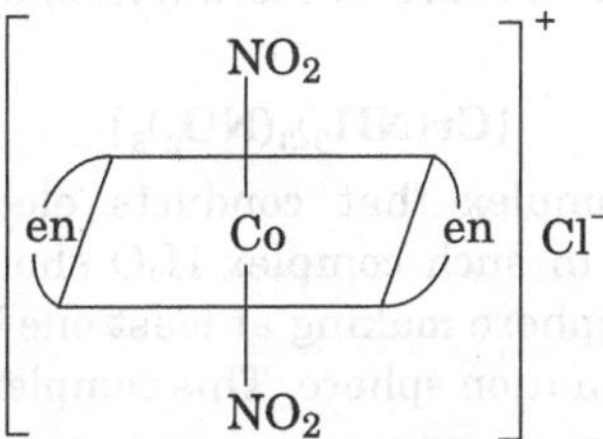

trans-bis (ethylenediamine) dinitrocobalt (III) chloride

Third form : In this case also, Cl^- is outside coordination sphere. Also it shows reaction with (en) hence monodentate ligands are *cis* to each other. Being optically active, mirror image should not superimpose. Thus, it can have structure:

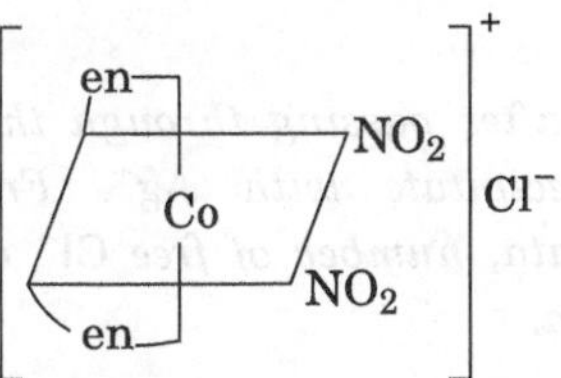

cis-bis (ethylenediamine) dinitrocobalt (III) chloride

Problem 5. Each of the compounds (a) $Pt(NH_3)_6Cl_4$, (b) $Cr(NH_3)_6Cl_3$, (c) $Co(NH_3)_4Cl_3$ and (d) K_2PtCl_6 has been dissolved in water to make its 0.001 M solution. Rank them in order of their increasing conductivity in solution (assume 100% ionisation in each case).

Solution

	Complex	C.N. of metal ion	Actual presentation	Ionisation of complex	Total ions
(a)	$Pt(NH_3)_6Cl_4$	6	$[Pt(NH_3)_6]Cl_4$	$[Pt(NH_3)_6]^{4+} + 4Cl^-$	5
(b)	$Cr(NH_3)_6Cl_3$	6	$[Cr(NH_3)_6]Cl_3$	$[Cr(NH_3)_6]^{3+} + 3Cl^-$	4
(c)	$Co(NH_3)_4Cl_3$	6	$[Co(NH_3)_4Cl_2]Cl$	$[Cr(NH_3)_4Cl_2]^+ + Cl^-$	2
(d)	K_2PtCl_6	6	$K_2[PtCl_6]$	$2K^+ + [PtCl_6]^{2-}$	3

Plan *Based on C.N., their correct formulae are written and number of ions per mol of complex are determined. Conductivity is proportional to the number of ions.*

Greater the number of ions, greater the conduction hence, order of increasing conductivity is (c) < (d) < (b) < (a)

Problem 6. Two compounds have the empirical formula $Cr(NH_3)_3(NO_2)_3$. In aqueous solution, one of these conducts electricity while the other does not. Deduce their probable structures.

Plan *Complex responsible for conductance ionises in aqueous solution. Hence, formula can be derived.*

Solution We know C.N. of Cr is six, hence the complex that does not conduct electricity, would not ionise. Hence, NO_2^- and NH_3 ligands are in coordination sphere and the complex is

$$[Cr(NH_3)_3(NO_2)_3]$$

The other complex that conducts electricity would ionise. Thus, in such complex H_2O should enter into coordination sphere making at least one NO_2^- available outside coordination sphere. This complex can be

$$[Cr(H_2O)(NH_3)_3(NO_2)_2]NO_2$$

Problem 7. A solution containing 2.665 g of $CrCl_3 \cdot 6H_2O$ is passed through a cation exchanger. The chloride ions obtained in solution react with $AgNO_3$ and give 2.87 g of AgCl. Determine the structure of the compound (Cr = 52, Cl = 35.5, H = 1, O = 16, Ag = 108, N = 14)

Plan *Free* Cl^- *after passing through the ion exchange gives white precipitate with* Ag^+. *From the given stoichiometric data, number of free* Cl^- *can be derived, hence the formula.*

Solution AgCl is obtained only with free (ionisable) ions

$$CrCl_3 \cdot 6H_2O + AgNO_3 \longrightarrow AgCl$$

$$\text{mole of AgCl obtained} = \frac{2.87}{143.5} = 0.02$$

$$\text{mole of } CrCl_3 \cdot 6H_2O = \frac{2.665}{266.5} = 0.01$$

0.01 mole of complex gives = 0.02 mol AgCl

1 mole of complex gives = 2 mol AgCl

Thus, numbers of free Cl^- = 2

Also C.N. of Cr is six, hence complex is

$$[CrCl(H_2O)_5]Cl_2 \cdot H_2O$$

$$[CrCl(H_2O)_5]Cl_2 \cdot H_2O \rightleftharpoons [CrCl(H_2O)_5]^{2+} \cdot H_2O + 2Cl^-$$

Problem 8. A solution containing 0.319 g of $CrCl_3 \cdot 6H_2O$ was passed through a cation exchange resin and acid coming out of the cation exchange resin required 28.5 mL of 0.125 M NaOH. Determine correct formula of the complex.

(mol. wt. of the complex = 266.5)

Plan $Cl^- \equiv HCl \equiv NaOH$

From the number of moles of NaOH, *number of moles of* Cl^- *are determined. Hence, complex is known.*

Solution Let the number of Cl^- ions outside the coordination sphere or number of chloride ions which can be ionised be n. When the solution of the complex is passed through cation exchanger, nCl^- ions will combine with H^+ (of the cation exchanger) to form HCl

$$nCl^- + nH^+ \longrightarrow nHCl$$

Thus, 1 mole of the complex will form n moles of HCl

1 mole of complex $\equiv n$ mol HCl $\equiv n$ moles NaOH

$$\text{mole of the complex} = \frac{0.319}{266.5} = 0.0012 \text{ mol}$$

$$\text{mole of NaOH used} = \frac{28.5 \times 0.125}{1000} = 0.0036 \text{ mol}$$

0.0012 mol of complex ≡ 0.0036 mol NaOH

≡ 0.0036 mol HCl

$$1 \text{ mol of complex} \equiv \frac{0.0036}{0.0012} = 3 \text{ mol HCl}$$

$\therefore \quad n = 3$

Thus, all the Cl^- ions are outside coordination sphere. Hence, complex is $\mathbf{[Cr(H_2O)_6]Cl_3}$.

Problem 9. Metal carbonyls can have formula $M(CO)_x$ where x is the number of carbonyl units coordinated to metal M. If EAN of each metal in the carbonyl is 36, what are formulae of the carbonyl of Fe (26), Cr (24) and Ni (28) ?

Plan Based on EAN and atomic number, value of x in each complex is determined. Thus, their formulae are known.

Solution Carbonyl is neutral ligand, hence, oxidation number of M = 0. EAN of metal M (neutral) in the complex = 36. Each CO donates 2 electrons.

(i) $\mathbf{Fe(CO)_x}$ $\quad$ EAN = Z − (O.N.) + 2x

$36 = 26 - 0 + 2x$

$x = 5$

hence, complex is $\mathbf{Fe(CO)_5}$

(ii) $\mathbf{Cr(CO)_x}$ $\quad Z = 24$

EAN = 36 = 24 − 0 + 2x

$x = 6$

hence, complex is $Cr(CO)_6$

(iii) $\mathbf{Ni(CO)_x}$ $\quad Z = 28$

EAN = 36 = 28 − 0 + 2x

$x = 4$

hence, complex is $\mathbf{Ni(CO)_4}$

Problem 10. The EAN of each Mn (Z = 25) in its carbonyl is 36. What is the structure of the carbonyl with molecular formula $Mn_2(CO)_{10}$?

Plan Based on EAN and Z, structure can be derived.

Solution Electrons from each Mn = 25

Electrons from five CO ligands = 2 × 5 = 10

Electron from (Mn—Mn) bond = 1

Thus, $\quad$ EAN = 25 + 10 + 1 = 36

Thus, in the complex five CO (ligands) are coordinated to each Mn atom and sixth coordination number is attained by other Mn atom of (Mn–Mn) bond. Thus, it can have structure :

CO CO CO CO
OC—(Mn)——(Mn)—CO
CO CO CO CO

Problem 11. Give the name describing the type of structural isomerism displayed by each of the following pairs :

(a) $[Co(en)_3][Cr(CN)_6]$ and $[Cr(en)_3][Co(CN)_6]$
(b) $[Mn(CO)_5(SCN)]$ and $[Mn(CO)_5(NCS)]$
(c) $[Co(NH_3)_5(NO_3)]SO_4$ and $[Co(NH_3)_5(SO_4)]NO_3$
(d) $[Co(NH_3)_4(H_2O)Cl]Cl_2$ and $[Co(NH_3)_4Cl_2]Cl \cdot H_2O$

Solution (a) Coordination isomerism
(b) Linkage isomerism
(c) Ionisation isomerism
(d) Hydrate isomerism

Problem 12. Draw/write possible isomers of
(a) $[Pt(NH_3)_2Cl_2]$ (b) $CrCl_3 \cdot 6H_2O$
(c) $[Co(NH_3)_5(SO_4)]Cl$ (d) $[Co(en)_2NH_3Br]SO_4$
(e) $[Pt(NH_3)(H_2O)(C_5H_5N)(NO_2)]Cl$

Solution (a)

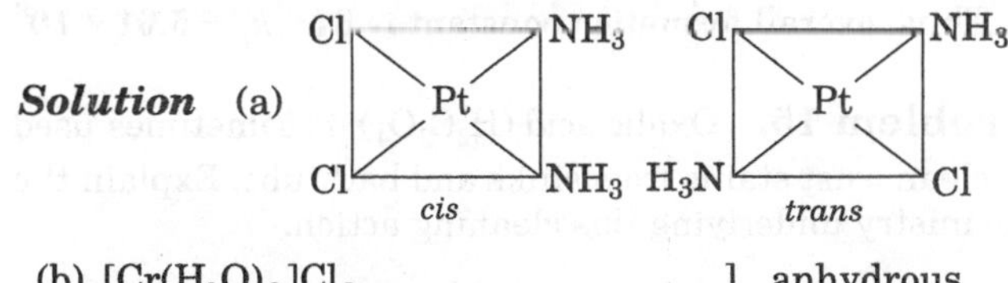

(b) $[Cr(H_2O)_6]Cl_3$	anhydrous
$[Cr(H_2O)_5Cl]Cl_2 \cdot H_2O$	hydrate
$[Cr(H_2O)_4Cl_2]Cl \cdot 2H_2O$	hydrate
$[Cr(H_2O)_3Cl_3] \cdot 3H_2O$	hydrate
(c) $[Co(NH_3)_5Cl]SO_4$	ionisation
(d) $[Co(en)_2(NH_3)SO_4]Br$	ionisation
(e) (i) $[Pt(NH_3)(H_2O)(C_5H_5N)Cl]NO_2$	ionisation
(ii) $[Pt(NH_3)(NO_2)(C_5H_5N)Cl]H_2O$	hydrate
(iii) $[Pt(NH_3)(ONO)(C_5H_5N)(H_2O)]Cl$	linkage

Problem 13. Write equations to represent the following observation :

When NaOH is added to $CuSO_4(aq)$, a pale blue precipitate forms. If $NH_3(aq)$ is added, the precipitate redissolves, producing a solution with an intense deep blue colour. If this deep blue solution is made acidic with $HNO_3(aq)$, the colour is converted back to pale blue.

Solution $Cu^{2+} + 2OH^- \longrightarrow \underset{\text{pale blue}}{Cu(OH)_2(s)}$

$$Cu(OH)_2(s) + 4NH_3 \longrightarrow \underset{\text{deep blue}}{[Cu(NH_3)_4]^{2+}} + 2OH^-$$

$$[Cu(NH_3)_4]^{2+} + 4H_3O^+ \longrightarrow [Cu(H_2O)_4]^{2+} + 4NH_4^+$$

Problem 14. Write a series of equations to show the stepwise displacement of H_2O ligands in $[Fe(H_2O)_6]^{3+}$ by ethylenediamine (en), for which $\log k_1 = 4.34$; $\log k_2 = 3.31$ and $\log k_3 = 2.05$. What is the overall formation constant, $\beta_3 (= k_f)$ for $[Fe(en)_3]^{3+}$?

Plan *If k_1, k_2 and k_3 are stepwise formation constants and β_3 is the overall formation constant, then*

$$\beta_3 = k_1 k_2 k_3$$

$$\log \beta_3 = \log k_1 + \log k_2 + \log k_3$$

Thus, β_3 can be calculated.

Solution Each ethylenediamine (en) being a bidentate ligand replaces two H_2O ligands (unidentate)

$$[Fe(H_2O)_6]^{3+} + en \longrightarrow [Fe(en)(H_2O)_4]^{3+} + 2H_2O, \quad k_1$$

$$[Fe(en)(H_2O)_4]^{3+} + en \longrightarrow [Fe(en)_2(H_2O)_2]^{3+} + 2H_2O, \quad k_2$$

$$[Fe(en)_2(H_2O)_2]^{3+} + en \longrightarrow [Fe(en)_3]^{3+} + 2H_2O, \quad k_3$$

$$[Fe(H_2O)_6]^{3+} + 3\ en \longrightarrow [Fe(en)_3]^{3+} + 6H_2O, \quad \beta_3 = k_f$$

Thus, $\beta_3 = k_f = k_1 k_2 k_3$

$\therefore$ $\log \beta_3 = \log k_1 + \log k_2 + \log k_3 = 4.34 + 3.31 + 2.05 = 9.70$

$\therefore$ $\beta_3 = 5.01 \times 10^9$

Thus, overall formation constant is $\beta_3 = k_f = \mathbf{5.01 \times 10^9}$

Problem 15. Oxalic acid ($H_2C_2O_4$) is sometimes used to clean rust stains from sinks and bathtubs. Explain the chemistry underlying this cleaning action.

Plan *Rust is $Fe_2O_3 \cdot xH_2O$. We are to study what happens to the rust when oxalic acid is used for cleaning.*

Solution Oxalic acid forms a dark green coloured complex with rust stains (Fe^{3+} ions).

$$Fe^{3+} + 3C_2O_4^{2-} \longrightarrow [Fe(C_2O_4)_3]^{3-}$$

This complex is water soluble. Thus, rust stains can be removed by treating the fabric with oxalic acid and then washing with water.

Problem 16. Carbon monoxide (CO) is toxic when inhaled. Explain.

Plan *We are to study the effect of CO on the internal body system.*

Solution CO is toxic because it forms a complex with haemoglobin in the blood and this complex is more stable than oxy-haemoglobin. This prevents the haemoglobin in the red blood corpuscles from carrying oxygen round the body. This causes an oxygen deficiency, leading to unconsciousness and then death. The blood of victims of carbon monoxide poisoning is a brilliant red-pink colour rather than the dull red of normal blood.

Master Exercises

Exercise 1
(Stage 1 : Learning)

Short Answer Type Questions

1. Which of the following complexes obey the rule of 18 (EAN rule)?

(a) $[Cu(NH_3)_4]^{2+}, [Cu(en)_3]^{2+}, [Cu(CN)_4]^{3-}$

(b) $[Fe(CN)_6]^{4-}, [Fe(CN)_6]^{3-}, Fe(CO)_5$

(c) $[Ni(NH_3)_6]^{2+}, [Ni(CN)_4]^{2-}, [Ni(CO)_4]$

(d) $[Cr(NH_3)_6]^{3+}, [Cr(CO)_6]$

(e) $[Co(NH_3)_6]^{3+}, [CoCl_4]^{2-}$

2. Suppose 0.010 mole of each of the following compounds is dissolved (separately) in 1.0 L water : $KNO_3, [Co(NH_3)_6Cl_3], [Cu(NH_3)_2Cl_2], [Na_2(PtCl_6)]$

Rank the resulting four solutions in order of conductivity, from lowest to highest.

3. Predict which of the following octahedral complexes has the shortest λ_{max} : $[FeF_6]^{3-}$, $[Fe(CN)_6]^{3-}$, $[Fe(H_2O)_6]^{3+}$?

4. Is the coordination compound $[Co(NH_3)_6]Cl_2$ diamagnetic or paramagnetic.

5. Will methylamine $[CH_3NH_2]$ be a monodentate or a bidentate ligand. With which of its atoms will it bind to a metal ion?

6. Consider following octahedral complexes

I : $[Fe(CN)_6]^{3-}$

II : $[Fe(H_2O)_6]^{3+}$

(a) Which has fewer unpaired electrons?

(b) How many unpaired electrons are present in each?

(c) What is CFSE in terms of Δ_o?

7. The chromium (III) ion in aqueous solution is blue-violet.

What is the complimentary colour to blue-violet?

8. Which of the following molecules/ions satisfies the EAN rule?

(a) [cyclopentadienyl–Fe(OC)(CO)]$^-$ ....

(b) $Mn(CO)_5^-$ ....

(c) cyclopentadienyl–Co with two $H_2C{=}CH_2$ ligands (H_2C, CH_2, CH_2, CH_2) ....

9. The anion $[NiCl_4]^{2-}$ is paramagnetic, but when CN^- ions are added, the product $[Ni(CN)_4]^{2-}$ is diamagnetic. Explain the observations.

$$[NiCl_4]^{2-}(aq) + 4CN^-(aq) \longrightarrow [Ni(CN)_4]^{2-}(aq) + 4Cl^-(aq)$$

10. In water, the titanium (III) ion $[Ti(H_2O)_6]^{3+}$ has a broad absorption bond at about 500 nm. What colour light is absorbed by the ion?

11. How many unpaired electrons expected for high spin and low-spin complexes of Fe^{2+}?

Analytical Questions

Coordination Number, Effective Atomic Number and Oxidation Number

1. Determine EAN of the metal ion in the following complexes:

(i) $[Cu(NH_3)_4]^{2+}$ (ii) $[Cr(NH_3)_6]Cl_3$

(iii) $[Cr(en)_3]^{3+}$

2. What is coordination number and the oxidation state of the central metal ion in each of the following complexes?

(a) $[Ni(NH_3)_6]^{2+}$ (b) $[AlF_6]^{3-}$
(c) $[Cu(CN)_4]^{2-}$ (d) $[Cr(NH_3)_3Br_3]$
(e) $[Ag(S_2O_3)_2]^{3-}$

3. Identify the ligands and give the coordination number for the central atom or ion in each of the following
(a) $[Co(NH_3)_2(NO_2)_4]^-$ (b) $K_4[Fe(CN)_6]$
(c) $[Cr(NH_3)_5Cl]Cl_2$ (d) $[Pd(NH_3)_4]^{2+}$
(e) $Na[Au(CN)_2]$ (f) $[Co(en)_3]^{3+}$

4. Only two-third of the chlorine contained in the solution of the complex salt $CoCl_3 \cdot 5NH_3$ were precipitated from it by silver nitrate. No cobalt or free ammonia were detected in the solution of the salt. The electrical conductivity of the solution shows that the salt dissociates into three ions. What is the coordination structure of the complex salt?

5. Silver nitrate precipitates the entire chlorine as silver chloride from a solution of the complex salt $PtCl_4 \cdot 6NH_3$ and only one-fourth of the chlorine from a solution of the salt $PtCl_4 \cdot 3NH_3$. Write the coordination formulae of these salts, and determine the coordination number of the platinum in each case.

6. A sufficient amount of $AgNO_3$ solution was added to a solution containing 23.35 g of the complex salt $CoCl_3 \cdot 4NH_3$. The mass of the precipitated AgCl was 14.35 g. Derive the coordination formula of the salt. (Co = 59 g mol^{-1})

7. Manganese carbonyl is a dimer, $(CO)_nMn—Mn(CO)_n$. If each Mn atom has an EAN of 36, deduce the value of n, assuming single covalent bond between two Mn atoms.

Nomenclature

8. Name the following:
(a) $[Ag(NH_3)_2]^+$ (b) $[Fe(H_2O)_5OH]^{2+}$
(c) $[ZnCl_4]^{2-}$ (d) $[Pt(en)_2]^{2+}$
(e) $[Co(NH_3)_4(NO_2)Cl]^+$ (f) $[Co(NH_3)_5Br]SO_4$
(g) $[Co(NH_3)_5SO_4]Br$ (h) $[Cr(NH_3)_6][Co(CN)_6]$
(i) $Na_3[Co(NO_2)_6]$ (j) $[Co(en)_3]Cl_3$

9. Write IUPAC name of
(a) $[Cr(NH_3)_5CO_3]Cl$ (b) $[Co(NH_3)_5ONO]Cl_2$
(c) $K_3[Cr(CN)_6]$ (d) $[Co(NH_3)_6]Cl_3$

10. Write the formulae of the following complexes:
(a) Pentamminechlorocobalt (III) ion
(b) Lithiumtetrahydroaluminate (III)

Isomerism

11. Draw the possible geometrical isomers for the following :
(i) $[Pt(gly)_2]$ (ii) $[Pt(P(C_2H_5)_3)Cl_2]_2$

12. The complex ion; $[Pt(NH_3)(NH_2OH)(NO_2)(Py)]^+$ has been found to be optically active. What is the probable geometry of the complex?

13. How many different structures are possible for each of the following complex ions?
(a) $[Co(NH_3)_5H_2O]^{3+}$ (b) $[Co(NH_3)_4(H_2O)_2]^{3+}$
(c) $[Co(NH_3)_3(H_2O)_3]^{3+}$ (d) $[Co(NH_3)_2(H_2O)_4]^{3+}$

14. Indicate what type of isomerism may be found in each of the following cases?
(a) $[Zn(NH_3)_4][CuCl_4]$ (b) $[Fe(CN)_5SCN]^{4-}$
(c) $[Ni(NH_3)_5Cl]^+$ (d) $[Pt(Py)Cl_3]^-$
(e) $[Cr(NH_3)_3(OH)_3]^-$

15. The structure of four complex ions are given. Determine which, if any, of these complex ions are isomers (geometric or optical); which, if any, are identical and which, if any, are distinctly different.

(a)
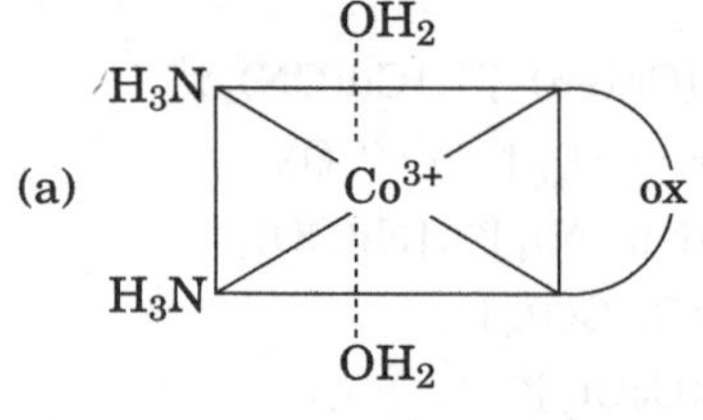

(b)
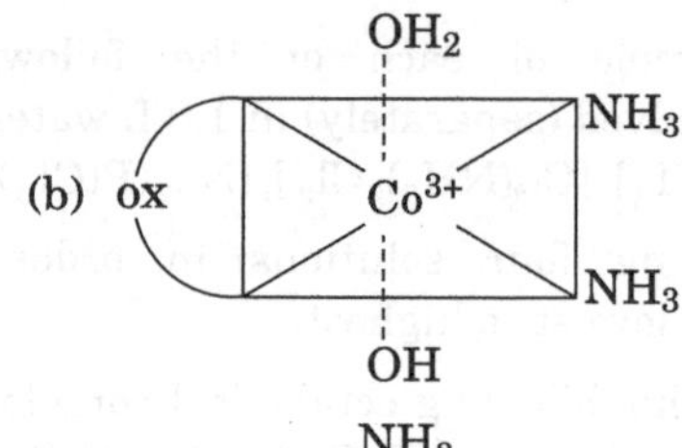

(c)
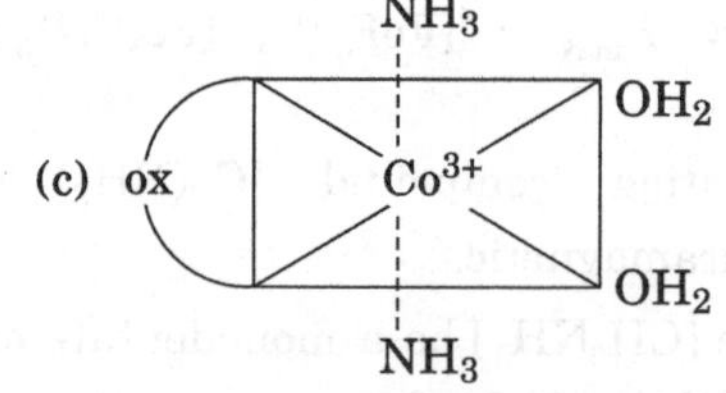

(d)
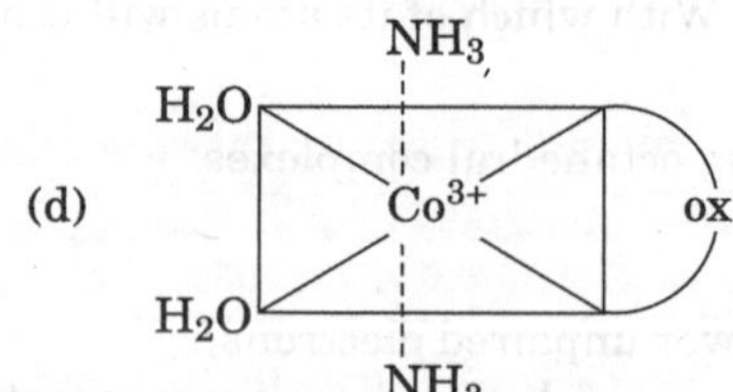

16. Consider the following ethylenediamine complexes

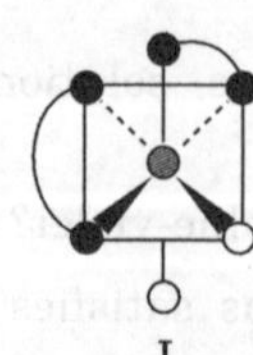
I

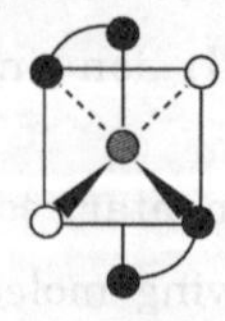
II

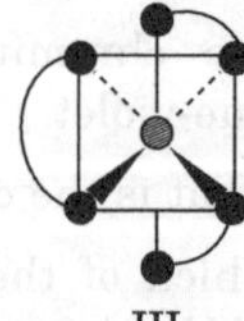
III

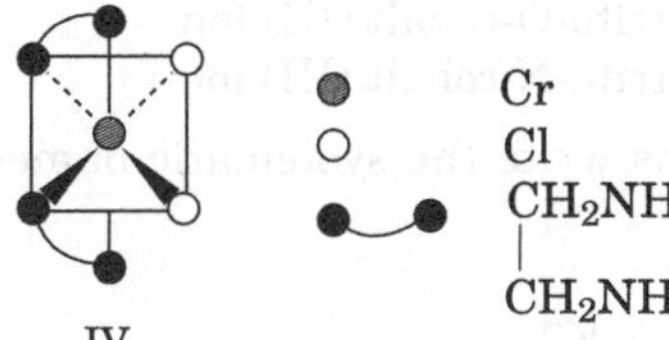

(a) Which complexes are chiral, and which are achiral?
(b) Draw the enantiomer of each chiral complex.
(c) Which, if any, of the chiral complexes are enantiomers of one another?

17. Which of the following complexes can exist as diastereoisomers?
(a) $[Cr(NH_3)_2Cl_4]^-$
(b) $[Co(NH_3)_5Br]^{2+}$
(c) $[FeCl_2(NCS)_2]^{2-}$ (tetrahedral)
(d) $[PtCl_2Br_2]^{2-}$ (square planar)

18. Two different compounds of Cr(III) have the same empirical formula, $CrCl_3 \cdot 6H_2O$. One (A) is green and the other (B) is violet. (A) on reaction with excess $AgNO_3$ gives 1 mole of AgCl per mol of (A) while (B) on reaction with excess $AgNO_3$ gives 3 moles AgCl per mol of (B). Deduce formula of (A) and (B). What is affect of heating on each? Also compare their conductance.

19. A monomeric compound of cobalt gives the following data on quantitative analysis :
Co^{3+} : 21.24%; NH_3 : 24.77% , Cl^- : 12.81% ;
SO_4^{2-} : 34.65% ; H_2O : 6.53%.
What is empirical formula of the complex ? Write possible isomers.

20. A , B and C are three complexes of chromium (III) with the empirical formula $H_{12}O_6Cl_3Cr$. All the three complexes have water and chloride ions as ligands. Complex A does not react with concentrated H_2SO_4, whereas complexes B and C lose 6.75% and 13.5% of their original weight, respectively, on treatment with concentrated H_2SO_4. Identify A , B and C.

Bonding

21. A solution containing 1 g of the complex $[Cr(H_2O)_5Cl]Cl_2 \cdot H_2O$ was passed through a cation exchanger. The acid liberated was made up to 1 L. Calculate the strength of the acid formed.

22. The $[Ti(NCS)_6]^{3-}$ ion exhibits a single absorption band at 544 nm. Calculate the crystal field splitting energy Δ (in kJ mol^{-1}). Is NCS^- a stronger or weaker field ligand than water? Predict the colour of $[Ti(NCS)_6]^{3-}$.

23. Predict
(a) the number of unpaired electrons expected for the tetrahedral complex ion $[CoCl_4]^{2-}$;
(b) whether the square planar complex ion $[Cu(py)_4]^{2+}$ is diamagnetic or paramagnetic,
(c) whether the octahedral complex ion $[Mn(CN)_6]^{3-}$ or tetrahedral $[FeCl_4]^-$ has the greater number of unpaired electrons.

24. Of the complex ions, $[Co(H_2O)_6]^{3+}$ and $[Co(en)_3]^{3+}$, one has a yellow colour in aqueous solution and the other blue. Match each ion with a expected colour and state your reason for doing so.

25. On the basis of the spectrochemical series, determine whether each of the following complexes is inner orbital or outer orbital and diamagnetic or paramagnetic.
(a) $[Cu(H_2O)_6]^{2+}$ (b) $[MnF_6]^{3-}$
(c) $[Co(CN)_6]^{3-}$ (d) $[Cr(NH_3)_6]^{3+}$

26. Draw a crystal field energy level diagram, for each of the following complex ions and predict the number of unpaired electrons.
(a) $[CrF_6]^{3-}$ (b) $[V(H_2O)_6]^{3+}$
(c) $[Fe(CN)_6]^{3-}$

27. The $[Fe(CN)_6]^{3-}$ ion has magnetic moment of 1.73 BM (Bohr Magneton) while $[Fe(H_2O)_6]^{3+}$ has a magnetic moment of 5.92 BM. Explain.

28. Among the following ions which has the highest paramagnetism?
(a) $[Cr(H_2O)_6]^{3+}$ (b) $[Fe(H_2O)_6]^{2+}$
(c) $[Cu(H_2O)_6]^{2+}$ (d) $[Zn(H_2O)_6]^{2+}$

29. Identify the complexes which are expected to be coloured, explain.
(a) $Ti(NO_3)_4$ (b) $[Cu(NCCH_3)]^+BF_4^-$
(c) $[Cr(NH_3)_6]^{3+}3Cl^-$ (d) $K_3[VF_6]$

30. Select correct answer.
(I) KF combines with HF to form KHF_2. The compound contains the species
(a) K^+, F^- and H^+ (b) K^+ , F and HF
(c) K^+ and $[HF_2]^-$ (d) $[KHF]^+$ and F^-
(II) Which compound is formed when excess of KCN is added to aqueous solution of copper sulphate?
(a) $Cu(CN)_2$ (b) $K_2[Cu(CN)_4]$
(c) $K[Cu(CN)_2]$ (d) $K_3[Cu(CN)_4]$

31. In nitroprusside ion the iron and NO exist as Fe^{II} and NO^+ rather than Fe^{III} and NO. These forms can be differentiated by
(a) estimating the concentration of iron
(b) measuring the concentration of CN^-
(c) measuring the solid state magnetic moment
(d) thermally decomposing the compound

32. Draw the structures of $[Co(NH_3)_6]^{2+}$, $[Ni(CN)_4]^{2-}$ and $[Ni(CO)_4]$.

33. A metal complex having composition $Cr(NH_3)_4Cl_2Br$ has been isolated in two forms (A) and (B). The form

(*A*) reacts with $AgNO_3$ to give a white precipitate readily soluble in dilute aqueous ammonia, whereas (*B*) gives a pale yellow precipitate soluble in concentrated ammonia. Write the formula of (*A*) and (*B*) and state the hybridisation of chromium in each. Calculate their magnetic moments (spin–only value).

34. Deduce the structures of $[NiCl_4]^{2-}$ and $[Ni(CN)_4]^{2-}$ considering hybridisation of the metal ion., Calculate the magnetic moment (spin only) of the species.

Applications

35. Ethylenediamine tetraacetic acid, EDTA, in the form of its calcium salt, is administered as an antidote for lead poisoning. Explain why this reagent might be an effective medicine? Why is the calcium salt administered rather than the free acid?

36. Explain the following observations in terms of complex ion formation:
(a) $Al(OH)_3$ is soluble in $NaOH(aq)$ but insoluble in $NH_3(aq)$;
(b) $ZnCO_3(s)$ is soluble in $NH_3(aq)$ but $ZnS(s)$ is not;
(c) $CoCl_3$ is unstable in water solution, being reduced to $CoCl_2$ and liberating $O_2(g)$. On the other hand, $[Co(NH_3)_6]Cl_3$ can be easily maintained in aqueous solution.

37. Potassium alum, $KAl(SO_4)_2 \cdot 12H_2O$ is obtained in the form of octahedral crystals when a solution of K_2SO_4 and $Al_2(SO_4)_3$ is concentrated by evaporation. Suggest specific experiments to specify whether it is a complex or double salt.

NCERT Problems

1. Complete the following statements for the coordination entity (complex ion) $[CrCl_2(OX)_2]^{3-}$
(a) OX is abbreviation for ....
(b) The oxidation number of chromium is ....
(c) .... is a bidentate ligand.

2. Specify the oxidation numbers of the metals in the following coordination entities
(a) $[Co(CN)(H_2O)(en)_2]^{2+}$ (b) $[PtCl_4]^{2-}$
(c) $[CrCl_3(NH_3)_3]$ (d) $[CoBr_2(en)_2]^+$
(e) $K_3[Fe(CN)_6]$

3. Using IUPAC norms write the formulae for the following
(a) Tetrahydroxozincate (II) ion
(b) Hexaammine cobalt (III) sulphate
(c) Potassiumtetrachloropalladate (II)
(d) Potassium tris (oxalato) chromate (III)
(e) Diamminedichloroplatinum (II)
(f) Hexaammineplatinum (IV) ion
(g) Potassium tetracyanonickelate (II)
(h) Tetrabromocuprate (II) ion
(i) Pentaamminenitrito-O-cobalt (III) ion
(j) Pentaamminenitrito-N-cobalt (III) ion

4. Using IUPAC norms write the systematic names of the following
(a) $[Co(NH_3)_6]Cl_3$
(b) $[CoCl(NO_2)(NH_3)_4]Cl$
(c) $[Ni(NH_3)_6]Cl_2$
(d) $[PtCl(NH_2CH_3)(NH_3)_2]Cl$
(e) $[Mn(H_2O)_6]^{2+}$
(f) $[Co(en)_3]^{3+}$
(g) $[Ti(H_2O)_6]^{3+}$
(h) $[NiCl_4]^{2-}$
(i) $[Ni(CO)_4]$

5. How many geometric isomers are possible in the following coordination entities?
(a) $[Cr(OX)_3]^{3-}$ (b) $[CoCl_3(NH_3)_3]$

6. Draw the structures of optical isomers of
(a) $[Cr(OX)_3]^{3-}$
(b) $[PtCl_2(en)_2]^{2+}$
(c) $[CrCl_2(en)(NH_3)_2]^+$

7. Draw all the isomers (geometric and optical) of
(a) $[CoCl_2(en)_2]^+$
(b) $[CoCl(en)_2(NH_3)]^{2+}$
(c) $[CoCl_2(en)(NH_3)_2]^+$

8. Draw the structure of
(a) *Cis*-dichlorotetracyano-chromate (III)
(b) *Mer*-triamminetrichlorocobalt (III)
(c) *Fac*-triaquatrinitro-N-cobalt (III)

9. Write the correct formulae for the following coordination compounds
(a) $CrCl_3 \cdot 6H_2O$ (violet, with 3 chloride ions/unit formula)
(b) $CrCl_3 \cdot 6H_2O$ (light green colour, with 2 chloride ions/unit formula)
(c) $CrCl_3 \cdot 6H_2O$ (dark green colour, with 1 chloride ion/unit formula)

[**Hints :** Some of these compounds may exist as hydrates].

10. Aqueous copper sulphate solution (blue in colour) gives
(a) a green precipitate with aqueous potassium fluoride, and
(b) a bright green solution with aqueous potassium chloride.
Explain these experimental results.

11. Discuss the nature of bonding in the following coordination entities on the basis of valence bond theory
(a) $[Fe(CN)_6]^{4-}$ (b) $[FeF_6]^{3-}$
(c) $[Co(OX)_3]^{3-}$ (d) $[CoF_6]^{3-}$

12. Write the valence bond description of
(a) $[Ni(CN)_4]^{3-}$
(b) $[NiCl_4]^{2-}$

13. Write the formulae of the following
(a) methyllithium
(b) tetramethylsilane
(c) trimethylbismuth
(d) trimethylarsane
(e) hexamethyldialuminium
(f) trimethylboron

14. Give IUPAC name and draw the structure of
(a) $Ni(CO)_4$ (b) $Fe(CO)_5$
(c) $[PtCl_3(C_2H_2)]^-$ (d) $[Cr(CO)_6]$

Exercise 2
(Stage 2 : High Skill Problem Solving)

Only One Option Correct

1. A compound has the empirical formula $CoCl_3 \cdot 5NH_3$. When an aqueous solution of this compound is mixed with excess silver nitrate, 2 moles of AgCl precipitate per mol of compound. On reaction with excess HCl, no NH_4^+ is detected. Hence, it is
(a) $[Co(NH_3)_5Cl_2]Cl$ (b) $[Co(NH_3)_5Cl]Cl_2$
(c) $[Co(NH_3)_5Cl_3]$ (d) $[Co(NH_3)_4Cl_2]Cl \cdot NH_3$

2. In the above compound in (Q. 1),
(a) all the Cl show primary valency (PV)
(b) two Cl show (PV) and one Cl secondary valency (SV)
(c) two Cl show (PV) and one Cl (PV) as well as (SV)
(d) all the Cl show (SV)

3. If there is 100% ionisation of this complex in (Q. 1) above, its 1 M solution at 300 K will have osmotic pressure equal to
(a) 8.21 atm (b) 24.63 atm
(c) 49.28 atm (d) 73.89 atm

4. Why is $[Ni(en)_3]^{2+}$, nearly 10^{10} times more stable than $[Ni(NH_3)_6]^{2+}$?
(a) NH_3 evaporates easily and causes instability to $[Ni(NH_3)_6]^{2+}$ complex
(b) Six NH_3 ligands cause steric hindrance around the Ni^{2+} centre
(c) 'en' is a chelating ligand and forms thermodynamically more stable complexes
(d) NH_3 is the weakest ligand known

5. Which of the following complexes is diamagnetic?
(a) $[Fe(CN)_6]^{4-}$ (b) $[Cu(NH_3)_4]^{2+}$
(c) $[Ti(H_2O)_6]^{3+}$ (d) $[Ni(en)_2]^{2+}$

6. Which of the following cations is not paramagnetic?
(a) $Sc^{3+}(aq)$ (b) $Ti^{3+}(aq)$
(c) $V^{3+}(aq)$ (d) $Cr^{3+}(aq)$

7. Which of the following complexes is not a chelate?
(a) bis (dimethylglyoximato) nickel (II)
(b) potassium ethylenediamine tetrathiocyanato chromate (III)
(c) tetrammine carbonatocobalt (III) nitrate
(d) *trans*-diglycinatoplatinum (II)

8. Which of the following complexes exhibit optical isomerism?
(a) *trans*-tetramminedithiocyanatochromium(III) ion
(b) *cis*-diamminedicarbonatocobalt (III) ion
(c) *trans*-diamminedicarbonatocobalt (III) ion
(d) *cis*-glycinatoplatinum (II)

9. What is the ratio of uncomplexed to complexed Zn^{2+} ion in a solution that is 10 M in NH_3, if the stability constant of $[Zn(NH_3)_4]^{2+}$ is 3×10^9 ?
(a) 3.3×10^{-9} (b) 3.3×10^{-11}
(c) 3.3×10^{-14} (d) 3×10^{-13}

10. The formula of a carbonyl complex of cobalt, $(CO)_nCo—Co(CO)_n$, in which there a single covalent Co—Co bond is
(a) $Co_2(CO)_4$ (b) $Co_2(CO)_6$
(c) $Co_2(CO)_8$ (d) $Co_2(CO)_{10}$

11. If excess of $AgNO_3$ solution is added to 100 mL of a 0.024 M solution of dichlorobis (ethylene diamine) cobalt (III) chloride. How many moles of AgCl be precipitated?
(a) 0.0012 (b) 0.0016
(c) 0.0024 (d) 0.0048

12. Among $Ni(CO)_4$, $[Ni(CN)_4]^{2-}$ and $NiCl_4^{2-}$
(a) $Ni(CO)_4$ and $[Ni(CN)_4]^{2-}$ are diamagnetic and $NiCl_4^{2-}$ is paramagnetic
(b) $Ni(CO)_4$ and $NiCl_4^{2-}$ are diamagnetic and $[Ni(CN)_4]^{2-}$ is paramagnetic
(c) $Ni(CO)_4$ is diamagnetic and $[Ni(CN)_4]^{2-}$ and $NiCl_4^{2-}$ are paramagnetic
(d) $NiCl_4^{2-}$ and $[Ni(CN)_4]^{2-}$ are diamagnetic and $Ni(CO)_4$ is paramagnetic

13. $[Cr(NH_3)_5Br]Cl$ and $[Cr(NH_3)_5Cl]Br$ can be distinguished by/and isomerism shown is
(a) $BaCl_2$, ionisation (b) $AgNO_3$, ionisation
(c) $AgNO_3$, coordinate (d) $BaCl_2$, linkage

14. Which is true for $[Ni(en)_2]^{2+}$, $Z(Ni) = 28$?
(a) Paramagnetism, dsp^2, square planar, C.N. of Ni = 2
(b) Diamagnetism, dsp^2, square planar, C.N. of Ni = 4

(c) Diamagnetism, sp^3, tetrahedral, C.N. of Ni = 4
(d) Paramagnetism, sp^3, square planar, C.N. of Ni = 4

15. Which is Prussian blue?
(a) $KFe^{III}[Fe^{II}(CN)_6]$ (b) $KFe^{III}[Fe^{III}(CN)_6]$
(c) $K_2[Fe^{II}(CN)_6]$ (d) $K_3[Fe(CN)_6]$

16. $[Ni(CN)_4]^{2-}$ and $[NiCl_4]^{2-}$ have similarity but not in
(a) magnetic moment (b) C.N. and O.N.
(c) structure (d) both (a) and (c)

17. Which has maximum EAN of the underlined atoms?
(Cr = 24 , Co = 27 , Fe = 26 , Ni = 28)
(a) $[\underline{Cr}(EDTA)]^-$ (b) $[\underline{Co}(en)_3]^{3+}$
(c) $[\underline{Fe}(C_2O_4)_3]^{3-}$ (d) $[\underline{Ni}(CN)_4]^{2-}$

18. Oxidation number of Cr in the following complex is

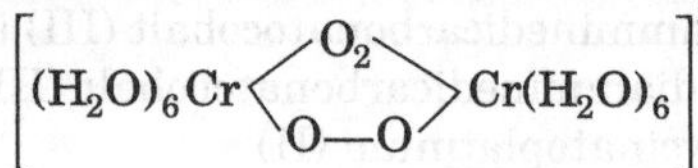

(a) 3 (b) 6
(c) 4 (d) 5

19. Which has maximum coordinating (donor) points?
(a) DMG (b) EDTA
(c) en (d) py

20. Following ligands can show linkage isomerism
(a) CNS (b) NO_2
(c) CN (d) all of these

21. $[Fe^{II}(O_2)(CN)_4Cl]^{4-}$ is named as
(a) chlorotetracyanodioxoferrate (II) ion
(b) chlorotetracyanoperoxoferrate (II) ion
(c) chlorotetracyanosuperoxoferrate (II) ion
(d) none of the above is correct

22. Which has maximum conductance?
(a) $PtCl_4 \cdot 6NH_3$ (b) $PtCl_4 \cdot 5NH_3$
(c) $PtCl_4 \cdot 4NH_3$ (d) Equal

23. Primary and secondary valency of Pt in $[Pt(en)_2Cl_2]Cl_2$ **are**
(a) 4, 4 (b) 4, 6
(c) 6, 6 (d) 4, 4

24. Which is used in cancer-chemotherapy?
(a) *Cis*-platin (b) Zeisse's salt
(c) Both (a) and (b) (d) None of these

25. Which one of the following complex species does not obey the EAN rule?
(a) $[Cu(CN)_4]^{3-}$ (b) $[Cr(NH_3)_6]^{3+}$
(c) $[Fe(CN)_6]^{4-}$ (d) $[Ni(CO)_4]$

26. Which has aromatic ring in complex?
(a) DMG in dimethyl glyoximate
(b) Cyclopenta-dienyl anion in ferrocene
(c) Both (a) and (b)
(d) None of the above

27. Extraction of Ag from sulphide ore and removal of unreacted silver from photographic plate involve complexes
(a) $[Ag(S_2O_3)_2]^{3-}$ in both
(b) $[Ag(CN)_2]^-$ in both
(c) $[Ag(S_2O_3)_2]^{3-}$, $[Ag(CN)_2]^-$
(d) $[Ag(CN)_2]^-$, $[Ag(S_2O_3)_2]^{3-}$

28. $[Co(NH_3)_5(NO_2)]^{2+}$ and $[Co(NH_3)_5(ONO)]^{2+}$ will differ in
(a) colour (b) structure
(c) hybridisation (d) magnetic moment

29. In which case racemic mixture is obtained on mixing its mirror images in 1 : 1 molar ratio?
(a) $[Cr(en)_3]^{3+}$ (b) $[Ni(DMG)_2]$
(c) *Cis*-$[Cu(Gly)_2]$ (d) In all cases

30. EAN of the elements (*) are equal in
(a) $\overset{*}{Ni}(CO_4)_4$, $[\overset{*}{Fe}(CN)_6]^{4-}$
(b) $[\overset{*}{Ni}(en)_2]^{2+}$, $[\overset{*}{Fe}(H_2O)_6]^{2+}$
(c) $[\overset{*}{Co}(CN)_6]^{3-}$, $[\overset{*}{Fe}(CN)_6]^{3-}$
(d) $[\overset{*}{Ni}(en)_2]^{2+}$, $[\overset{*}{Sc}(H_2O)_6]^{3+}$

31. Consider the following complexes:
(1) K_2PtCl_6 (2) $PtCl_4 \cdot 2NH_3$
(3) $PtCl_4 \cdot 3NH_3$ (4) $PtCl_4 \cdot 5NH_3$
Their electrical conductances in aqueous solutions are
(a) 256, 0, 97, 404 (b) 404, 0, 97, 256
(c) 256, 97, 0, 404 (d) 404, 97, 256, 0

32. Which one of the following is an example of coordination isomerism?
(a) $[Co(NH_3)_5Br]SO_4$ and $[Co(NH_3)_5SO_4]Br$
(b) $[Co(NH_3)_5NO_2]Cl_2$ and $[Co(NH_3)_5ONO]Cl_2$
(c) $[Cr(H_2O)_6]Cl_3$ and $[Cr(H_2O)_5Cl]Cl_2 \cdot H_2O$
(d) $[Cr(NH_3)_6][Co(CN)_6]$ and $[Co(NH_3)_6][Cr(CN)_6]$

33. Which of the following compounds show optical isomerism?
I. *Cis*-$[Co(NH_3)_4Cl_2]^+$ II. *Trans*-$[Co(en)_2Cl_2]^+$
III. *Cis*-$[Co(en)_2Cl_2]^+$ IV. $[Co(en)_3]^{3+}$
Select the correct answer using the codes given below
(a) I and II (b) II and III
(c) III and IV (d) I, III and IV

34. Lead poisoning in the body can be removed by
(a) EDTA in the form of calcium dihydrogen salt
(b) *cis*-platin
(c) Zeisse's salt
(d) DMG

35. $[Pt(NH_3)(NO_2)Py(NH_2OH)]^+$ will form geometrical isomers
(a) 2 (b) 3
(c) 4 (d) 5

36. The coordination number of Pt in the complex ion $[Pt(en)_2Cl_2]^{2+}$ is
(a) 3 (b) 4
(c) 5 (d) 6

37. Of the following complex ions one exhibits optical isomerism. That one is
(a) *cis*- $[Co(en)_2Cl_2]^+$ (b) $[Co(NH_3)_4Cl_2]^+$
(c) $[Co(NH_3)_2Cl_4]^-$ (d) *trans*-$[Co(en)_2Cl_2]^+$

38. Of the following complex ions, one exhibits isomerism. That is
(a) $[Ag(NH_3)_2]^+$ (b) $[Co(NH_3)_5NO_2]^{2+}$
(c) $[Pt(en)Cl_2]$ (d) $[Co(NH_3)_5Cl]^{2+}$

39. The number of unpaired electrons expected for the complex ion $[Cr(NH_3)_6]^{2+}$ is
(a) 2 (b) 3
(c) 4 (d) 5

40. Of the following complex ions, one is a Bronsted-Lowry acid. That one is
(a) $[Cu(NH_3)_4]^{2+}$ (b) $[FeCl_4]^-$
(c) $[Fe(H_2O)_6]^{3+}$ (d) $[Zn(OH)_4]^{2-}$

41. Of the following complex ions, the one that probably has the largest overall formation constant, K_f, is
(a) $[Co(NH_3)_6]^{3+}$ (b) $[Co(H_2O)_6]^{3+}$
(c) $[Co(H_2O)_4(NH_3)_2]^{3+}$ (d) $[Co(en)_3]^{3+}$

42. When concentrated HCl is added to a, solution of $[Co(H_2O)_6]^{2+}$ ion, an intense blue colour develops due to the formation which one of the following?
(a) $[CoCl_6]^{4-}$ (b) $[CoCl_4]^{2-}$
(c) $[CoCl_2(H_2O)_4]$ (d) $[CoCl(H_2O)_5]^+$

43. Select the correct statement.
(a) Complex ion $[MoCl_6]^{3-}$ is paramagnetic
(b) Complex ion $[Co(en)_3]^{3+}$ is diamagnetic
(c) Both (a) and (b) are correct
(d) None of the above is correct

44. Match the geometry (given in column A) with the complexes (given in column B) in

Geometry : A	Complex : B
I : Octahedral	P : $[Ni(CN)_4]^{2-}$
II : Square planar	Q : $Ni(CO)_4$
III : Tetrahedral	R : $[Fe(CN)_6]^{4-}$

	I	II	III		I	II	III
(a)	P	Q	R	(b)	R	P	Q
(c)	R	Q	P	(d)	Q	P	R

45. The complex $[Fe(H_2O)_5NO]^{2+}$ is formed in the brown ring test for nitrates when freshly prepared $FeSO_4$ solution is added to aqueous solution of NO_3^- followed by addition of conc. H_2SO_4. Select correct statement about this complex.
(a) Colour change is due to charge transfer
(b) It has iron in +1 oxidate state and nitrosyl as NO^+
(c) It has magnetic moment of 3.87 BM confirming three unpaired electrons in Fe
(d) All the above are correct statements

46. While Ti^{3+}, V^{3+}, Fe^{3+} and Co^{2+} afford a large number of tetrahedral complexes, Cr^{3+} never does this, the reason being
(a) Cr^{3+} forces high crystal field splitting with a varieties of ligands
(b) crystal field stabilisation energy in octahedral *vis-a-vis* tetrahedral Cr^{3+} system plays the deciding role
(c) the ionic radius of Cr^{3+} is the largest among the other M^{3+} ions mentioned
(d) electronegativity of Cr^{3+} is the largest among these trivalent 3*d*-metals and so chromium prefers to be associated with as many ligands as its ionic radius permits

47. Isomerisms exhibited by $[Cr(NH_3)_2(H_2O)_2Cl_2]^+$ are
(a) ionisation, optical
(b) hydrate, optical
(c) geometrical, optical
(d) coordinate, geometrical

48. The crystal field-splitting for Cr^{3+} ion in octahedral field increases for ligands I^-, H_2O, NH_3, CN^- and the order is
(a) $I^- < H_2O < NH_3 < CN^-$
(b) $CN^- < I^- < H_2O < NH_3$
(c) $CN^- < NH_3 < H_2O < I^-$
(d) $NH_3 < H_2O < I^- < CN^-$

49. Which of the following types of bonds are present in $CuSO_4 \cdot 5H_2O$?
I : Electrovalnt II : Covalent III : Coordinate
Select correct answer out of
(a) I, II (b) II, III
(c) I, III (d) I, II, III

50. Arrange the following in order of decreasing number of unpaired electrons:
I : $[Fe(H_2O)_6]^{2+}$ II : $[Fe(CN)_6]^{3-}$
III : $[Fe(CN)_6]^{4-}$ IV : $[Fe(H_2O)_6]^{3+}$
(a) IV, I, II, III (b) I, II, III, IV
(c) III, II, I, IV (d) II, III, I, IV

51. In an octahedral complex, if ligands on one axis are displaced little away from their ideal positions, the crystal field splitting of *d*-orbitals for this complex is as given below
$d_{x^2-y^2}$ d_{z^2} d_{xy} d_{yz} d_{xz}
The ligands are displaced on which axis/ axes
(a) *x*- and *y* (b) *y*-only
(c) *x*-only (d) *z*-only

52. Which one of the following compounds has tetrahedral geometry?
(a) $[Ni(CN)_4]^{2-}$ (b) $[NiCl_4]^{2-}$
(c) $[PdCl_4]^{2-}$ (d) $[Pd(CN)_4]^{2-}$

53. Aqueous Fe(III) ion develops intense red colour with SCN^- while Fe(II) does not. It is due to
(a) Fe(III) ion forms a charge transfer complex with SCN^- ions
(b) Fe(III) is reduced to Fe(I) which is deep red in colour
(c) SCN^- ion oxidises to CN^- ion that form red complex with Fe(III) ion
(d) SCN^- ion does not form any complex with Fe(II) ion

54. The intense blue colour of Prussian blue salts arises from which one of the following?
(a) *d-d* transition
(b) inter valence electron transfer
(c) ligand to metal charge transfer
(d) metal to ligand charge transfer

55. Which one of the following is the correct order of the wavelengths of absorption for complexes?
I : $[Ni(H_2O)_6]^{2+}$ II : $[Ni(NH_3)_6]^{2+}$
III : $[Ni(NO_2)_6]^{4-}$
(a) I > II > III (b) II > I > III
(c) III > II > I (d) III > I > II

56. What are the spin-only magnetic moments (in BM) for Ni(II) ion in square-planar and octahedral geometry, respectively?
(a) 0, 2.83 (b) 2.83, 2.83
(c) 2.83, 0 (d) 0,0

57. What is CFSE of a free Co(II) ion on forming the tetrahedral chloro complex $[CoCl_4]^{2-}$ (in the units of Δ_o)?
(a) 0.6 (b) 1.2
(c) 1.8 (d) 2.4

58. Which one of the following complexes is expected to have lowest Δ_o values?
(a) $[Co(NH_3)_6]^{3+}$ (b) $[CoF_6]^{3-}$
(c) $[Rh(NH_3)_5]^{3+}$ (d) $[Ir(NH_3)_6]^{3+}$

59. If Δ_o and Δ_t represent crystal field splitting energies for *d*-orbitals for octahedral and tetrahedral geometries respectively, then for d^6 (high spin in both cases), what are CFSE (ignore the pairing energy) respectively?
(a) 0.6 Δ_o and 0.6 Δ_t (b) 0.4 Δ_o and 0.4 Δ_t
(c) 0.4 Δ_o and 0.6 Δ_t (d) 0.6 Δ_o and 0.4 Δ_t

60. The complex $[Co(NH_3)_5(NO_2)]^{2+}$ and $[Co(NH_3)_5(ONO)]^{2+}$ are called
(a) ionisation isomer (b) linkage isomers
(c) coordination isomer (d) geometrical isomer

61. The complex

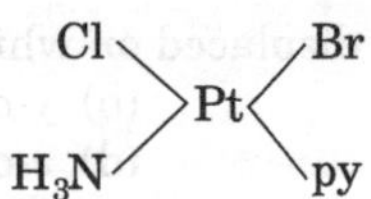

can be formally formed from $K_2[PtCl_4]$ by which one of the following sequences of substitution of the chloride ions?
(a) Py, Br^-, NH_3 (b) Br^-, Py, NH_3
(c) Br^-, NH_3, Py (d) NH_3, Br^-, Py

62. Consider the following statements in respect of $[CoCl_6]^{4-}$ complex ion
I : It is paramagnate
II : It is low-spin complex
III : Oxidation number of cobalt is-4
IV : The coordination number of cobalt is 6
Select correct statements.
(a) all of the above (b) III and IV only
(c) I and IV only (d) I and II only

63. The reaction between metallic silver and aqueous NaCN forming a soluble complex occurs in the presence of
(a) nitrogen (b) helium
(c) argon (d) oxygen

64. What type of isomerism is present in the pair of complexes $[Co(NH_3)_5Br]SO_4$ and $[Co(NH_3)_5SO_4]Br$?
(a) Linkage isomerism (b) Ligand isomerism
(c) Ionisation isomerism (d) Coordination isomerism

One or More Than One Options Correct

1. Which are correct statements?
(a) $[Ag(NH_3)_2]^+$ is linear with *sp* hybridised Ag^+ ion
(b) $NiCl_4^{2-}$, VO_4^{3-} and MnO_4^- have tetrahedral geometry
(c) $[Cu(NH_3)_4]^{2+}$, $[Pt(NH_3)_4]^{2+}$ and $[Ni(CN)_4]^{2-}$ have dsp^2 hybridisation of the metal ion
(d) $Fe(CO)_5$ have bipyramidal structure with dsp^3 hybridised iron

2. Which of the following are paramagnetic?
(a) $[Ni(CN)_4]^{2-}$ (b) $[NiCl_4]^{2-}$
(c) $[CoF_6]^{3-}$ (d) $[Co(NH_3)_6]^{3+}$

3. Which of the following can show coordination isomerism?
(a) $[Cu(NH_3)_4][PtCl_4]$
(b) $[Fe(NH_3)_6]_2[Pt(CN)_6]_3$
(c) $[Co(NH_3)_6][Cr(C_2O_4)_3]$
(d) $[Pt(en)_3](SO_4)_2$

4. 0.0012 mole of $CrCl_3 \cdot 6H_2O$ was passed through a cation exchange resin and acid coming out of it required 28.5 mL of 0.125 M NaOH. Hence, complex is
(a) $[Cr(H_2O)_5Cl]Cl_2 \cdot H_2O$ (b) $[Cr(H_2O)_4Cl_2] \cdot 2H_2O$
(c) $[Cr(H_2O)_6]Cl_3$ (d) $[Cr(H_2O)_3Cl_3] \cdot 3H_2O$

5. Which is/are correct statement(s)?
(a) $[Co(en)_3][Cr(CN)_6]$ will display coordination isomerism
(b) $[Mn(CO)_5(SCN)]$ will display linkage isomerism
(c) $[Co(NH_3)_5(NO_3)]SO_4$ will display ionisation isomerism
(d) None of the above is correct

6. Identify the complexes which is/are expected to be coloured
(a) $Ti(NO_3)_4$ (b) $[Cu(NCCH_3)]^+BF_4^-$
(c) $[Cr(NH_3)_6]Cl_3$ (d) $K_3[VF_6]$

7. Nitroprusside ion is
A : $[Fe^{II}(CN)_5NO^+]^{2+}$ and not
B : $[Fe^{III}(CO)_5NO]^{2+}$. *A* and *B* can be differentiated by
(a) estimating the concentration of iron
(b) measuring the concentration of CN^-
(c) measuring the magnetic moment
(d) thermally decomposing the compound

8. The effective atomic number of $Co(CO)_4$ is 35 and hence, is less stable. It attains stability by
(a) oxidation of Co (b) reduction of Co
(c) dimerization (d) tetramerization

9. Bidentate ligands are
(a) $C_2O_4^{2-}$ (oxalate)
(b) en (ethylenediamine)
(c) DMG (dimethyl glyoxime)
(d) Gly (glycine)

10. Which of the following complexes are chelates?
(a) *bis*-(dimethyl glyoximato) nickel (II)
(b) Potassium ethylenediaminetetracyanato chromate (III)
(c) Tetraamminedicyanocobalt (III) nitrate
(d) *trans*-diglycinatopalladium (II)

11. CN^- is a strong field ligand. This is due to the fact that
(a) it carries negative charge
(b) it is a conjugate base of weak acid
(c) it can accept electrons from metal species
(d) it forms high spin complexes with metal species

12. Considering H_2O as a weak field ligand, the number of unpaired electrons in $[Mn(H_2O)_6]^{2+}$ will be (At. no. of Mn = 25)
(a) three (b) five
(c) two (d) four

13. Which of the following coordination compounds would exhibit optical isomerism?
(a) Pentaamminenitrocobalt(III) iodide
(b) Diamminedichloroplatinum(II)
(c) *Trans*-dicyanobis (ethylenediamine) chromium(III) chloride
(d) *Tris*-(ethylenediamine) cobalt(III) bromide

14. In which of the following compounds transition metal may have zero oxidation state?
(a) $[Fe(CO)_5]$ (b) $[Ni(CN)_4]^{2-}$
(c) Fe_2O_3 (d) CrO_5

15. Which one of the following complexes will have four different isomers?
(a) $[Co(en)_2Cl_2]Cl$
(b) $[Co(en)(NH_3)_2Cl_2]Cl$
(c) $[Co(PPh_3)_2(NH_3)Cl_2]Cl$
(d) $[Co(en)_3]Cl_3$

Passage Comprehension Questions

Passage 1

Three different compounds (*A*, *B* and *C*) are known to have the empirical formula $CrCl_3 \cdot 6H_2O(X)$

When exposed to a dehydrating agent

A loses 2 moles of water per mole of the compound (*X*).

B loses 1 mole of water and compounds *C* loses no water. Answer the following questions.

1. Thus, *A*, *B* and *C* are

	A	*B*	*C*
(a)	$[Cr(H_2O)_4Cl_2]Cl \cdot 2H_2O$	$[Cr(H_2O)_5Cl]Cl_2 \cdot H_2O$	$[Cr(H_2O)_6]Cl_3$
(b)	$[Cr(H_2O)_2Cl_3] \cdot 4H_2O$	$[Cr(H_2O)Cl_3] \cdot 5H_2O$	$[CrCl_3] \cdot 6H_2O$
(c)	$[Cr(H_2O)_3Cl_3] \cdot 3H_2O$	$[Cr(H_2O)_4Cl_2] \cdot 2H_2O$	$[Cr(H_2O)_6]Cl_3$
(d)	$[Cr(H_2O)_6]Cl_3$	$[Cr(H_2O)_5Cl]Cl_2 \cdot H_2O$	$[Cr(H_2O)_4Cl_2]Cl \cdot 2H_2O$

2. Which is most conducting in aqueous solution?
(a) *A* (b) *B*
(c) *C* (d) Equal

3. One mole each of *A*, *B* and *C* is treated with excess of $AgNO_3$. There is formation of white precipitate of AgCl.
Moles of white precipitates are

	A	*B*	*C*
(a)	1	2	3
(b)	3	2	1
(c)	1	3	2
(d)	3	1	2

Passage 2

The octahedral complex ions $[FeCl_6]^{3-}$ and $[Fe(CN)_6]^{3-}$ are both paramagnetic but the former is high spin and the latter is low spin.

Based on the above study, answer the following questions

1. High spin complex $[FeCl_6]^{3-}$ has the *d*-configuration as
(a) $t_{2g}^3 e_g^2$ (b) t_{2g}^5
(c) $t_{2g}^2 e_g^3$ (d) e_g^5

2. Low spin complex $[Fe(CN)_6]^{3-}$ has the *d*-configuration as
(a) $t_{2g}^3 e_g^2$ (b) t_{2g}^5
(c) $t_{2g}^2 e_g^3$ (d) e_g^5

3. Unpaired electrons in $[FeCl_6]^{3-}$ and $[Fe(CN)_6]^{3-}$ are respectively
(a) 5, 2 (b) 5, 1
(c) 3, 2 (d) 2, 3

4. CFSE for the complexes are

	$[FeCl_6]^{3-}$	$[Fe(CN)_6]^{3-}$
(a)	0	0
(b)	$-2\Delta_o$	0
(c)	0	$-2\Delta_o$
(d)	$-2\Delta_o$	$-2\Delta_o$

Passage 3

The Alsatian-Swiss chemist Alfred Werner pioneered the field of coordination chemistry in the late nineenth century. At that time, a number of compounds of cobalt (III) chloride with ammonia were known, with general formulae

$$CoCl_3 \cdot xNH_3 \qquad x = 3 \text{ to } 6$$

Treatment of these compounds with aqueous HCl acid did not remove the ammonia. However treatment of the compounds with excess of $AgNO_3$ at 273 K gave white precipitate of AgCl in varying ratio depending on the structure. Also each complex had electrical conductivity of varying nature. Based on the above study, answer the following questions.

1. Which complex has the maximum electrical conductivity?
   (a) $CoCl_3 \cdot 3NH_3$ (b) $CoCl_3 \cdot 4NH_3$
   (c) $CoCl_3 \cdot 5NH_3$ (d) $CoCl_3 \cdot 6NH_3$
2. There is no reaction of the compound with aqueous HCl indicating that there is no free NH_3. Thus, NH_3 ligand is bonded to cobalt (III) by
   (a) primary valency (b) secondary valency
   (c) both of these (d) None of these
3. One mole of the compound gave three moles of white precipitate of AgCl, on reaction with aqueous $AgNO_3$. Thus, complex is
   (a) $CoCl_3 \cdot 4NH_3$ (b) $CoCl_3 \cdot 5NH_3$
   (c) $CoCl_3 \cdot 6NH_3$ (d) all of these

Passage 4

Consider following reactions :

I. $[Co(NH_3)_6]^{3+}(aq) + 6H_3O^+(aq) \rightleftharpoons [Co(H_2O)_6]^{3+}(aq) + 6NH_4^+(aq)$

The products are favoured in the above reaction

II. $[Co(NH_3)_6]^{2+}(aq) + 6H_3O^+(aq) \longrightarrow [Co(H_2O)_6]^{2+}(aq) + 6NH_4^+(aq)$

This, reactions (II) takes place with H_2O in a matter of seconds

Based on the above reactions, answer the following questions.

1. Select correct statement.
   (a) $[Co(NH_3)_6]^{3+}$ is thermodynamically and kinetically stable relative to $[Co(H_2O)_6]^{3+}$
   (b) $[Co(NH_3)_6]^{3+}$ is thermodynamically unstable but kinetically stable relative to $[Co(H_2O)_6]^{3+}$
   (c) $[Co(NH_3)_6]^{3+}$ is thermodynamically and kinetically unstable relative to $[Co(H_2O)_6]^{3+}$
   (d) $[Co(NH_3)_6]^{3+}$ is thermodynamically stable but kinetically unstable relative to $[Co(H_2O)_6]^{3+}$
2. Which is true in terms of $\Delta G°$?
   (a) ΔG_I° is more negative than that of II
   (b) ΔG_I° is less negative than that of II
   (c) $\Delta G_I^\circ = \Delta G_{II}^\circ < 0$
   (d) $\Delta G_I^\circ = \Delta G_{II}^\circ > 0$
3. Which is thermodynamically unstable and also kinetically labile?
   (a) $[Co(H_2O)_6]^{3+}$ (b) $[Co(H_2O)_6]^{2+}$
   (c) $[Co(NH_3)_6]^{3+}$ (d) $[Co(NH_3)_6]^{2+}$

Passage 5

The octahedral complex ion $[FeCl_6]^{3-}$ and $[Fe(CN)_6]^{3-}$ are both paramagnetic. Answer the following questions.

1. Select correct statements.
   (a) $[FeCl_6]^{3-}$ and $[Fe(CN)_6]^{3-}$ both are high-spin complex
   (b) $[FeCl_6]^{3-}$ is high-spin and $[Fe(CN)_6]^{3-}$ is low-spin complex
   (c) $[FeCl_6]^{3-}$ is low-spin and $[Fe(CN)_6]^{3-}$ is high-spin complex
   (d) $[FeCl_6]^{3-}$ and $[Fe(CN)_6]^{3-}$ both are low-spin complexes
2. $[FeCl_6]^{3-}$ has
   (a) five unpaired electrons ($t_{2g}^3 e_g^2$)
   (b) one unpaired electron (t_{2g}^1)
   (c) three unpaired electrons ($t_{2g}^1 e_g^2$)
   (d) all the above electrons paired
3. Select correct statement.
   The CFSE for
   (a) $[FeCl_6]^{3-}$ is zero (b) $[Fe(CN)_6]^{3-}$ is $-2\Delta_o$
   (c) Both are correct (d) None is correct

Passage 6

Consider following exchange reactions :

	k (rate constant)
$[Fe(H_2O)_6]^{2+} + Cl^- \longrightarrow [Fe(H_2O)_5Cl]^+ + H_2O$	$10^6\ M^{-1}s^{-1}$
$[Ru(H_2O)_6]^{2+} + Cl^- \longrightarrow [Ru(H_2O)_5Cl]^+ + H_2O$	$10^{-2}\ M^{-1}s^{-1}$

Now answer the following questions

1. Rate constant of the exchange reaction with iron complex is very high. It is attributed to that Fe^{2+} complex involving a first-row transition metal is
   (a) high-spin d^6 and labile
   (b) low-spin d^6 and kinetically inert
   (c) paramagnetic and kinetically inert
   (d) diamagnetic and kinetically inert

2. Ru^{2+}, like other second- and third-row transition metal ions
(a) forms low-spin complexes
(b) low-spin d^6 species are kinetically inert
(c) both of the above
(d) none of the above

Passage 7

Read the following experiment and answer the questions at the end of it.

"A student in 1895 prepared three coordination compounds containing chromium with the following properties:

	Formula	Colour	Cl^- ions in solution per formula units
(*A*)	$CrCl_3 \cdot 6H_2O$	Violet	3
(*B*)	$CrCl_3 \cdot 6H_2O$	Light green	2
(*C*)	$CrCl_3 \cdot 6H_2O$	Dark green	1

1. Which complex has maximum molecules of hydration?
(a) *A* (b) *B*
(c) *C* (d) Equally

2. IUPAC name of the complex *A* is
(a) hexaaquotrichlorochromium (III)
(b) hexaaquochromium(III)chloride
(c) pentaaquochlorochromium(III)chloride
(d) tetraaquodichlorochromium(III)chloride

3. Complexes *A*, *B*, *C* are represented as
(a) A : $[Cr(H_2O)_6]Cl_3$
B : $[Cr(H_2O)_5Cl]Cl_2 \cdot H_2O$
C : $[Cr(H_2O)_4Cl_2]Cl \cdot 2H_2O$
(b) A : $[Cr(H_2O)_3Cl_3] \cdot 3H_2O$
B : $[Cr(H_2O)_4Cl_2]Cl \cdot 2H_2O$
C : $[Cr(H_2O)_5Cl]Cl_2 \cdot H_2O$
(c) A : $[Cr(H_2O)_6]Cl_3$
B : $[Cr(H_2O)_6]Cl_3$
C : $[Cr(H_2O)_6]Cl_3$
(d) none of the above

4. Structural formula of *B* indicating primary and secondary valencies both is

(a) Cr bonded to H_2O, H_2O, H_2O, OH_2, OH_2, OH_2 (primary valencies shown dotted to Cl, Cl, Cl)

(b) Cr bonded to H_2O, H_2O, OH_2, $OH_2 \cdot H_2O$ and Cl (primary valencies shown dotted to Cl, Cl)

(c) Cr bonded to H_2O, H_2O, OH_2, $OH_2 \cdot 2H_2O$ and Cl (primary valencies shown dotted to Cl, Cl)

(d) None of the above

5. Conducting nature of these complexes in the increasing order is
(a) $A < B < C$ (b) $C < A < B$
(c) $B < A < C$ (d) $C < B < A$

Passage 8

Read the following experimental method and answer the questions at the end of it.

A rose-coloured compound has the empirical formula $CoCl_3 \cdot 5NH_3 \cdot H_2O$. Two moles of this compound react with concentrated sulphuric acid to form $HCl(g)$ and one mole of a new compound with empirical formula $Co_2(SO_4)_3 \cdot 10NH_3 \cdot 5H_2O$. When this new compound is dried at room temperature, it loses three moles of water per mol of $Co_2(SO_4)_3 \cdot 10NH_3 \cdot 5H_2O$.

1. What is the formula of the rose-coloured compound?
(a) $[Co(NH_3)_5Cl]Cl_2 \cdot H_2O$
(b) $[Co(NH_3)_5H_2O]Cl_2$
(c) $[Co(NH_3)_3(H_2O) \cdot Cl_2]Cl$
(d) None of the above

2. What is the structure of the new complex formed?
(a) $[Co—(NH_3)_5—Co—(NH_3)_5](SO_4)_3 \cdot 5H_2O$
(b) $[Co—(NH_3)_5—Co—(NH_3)_5(SO_4)](SO_4)_2 \cdot 5H_2O$
(c) 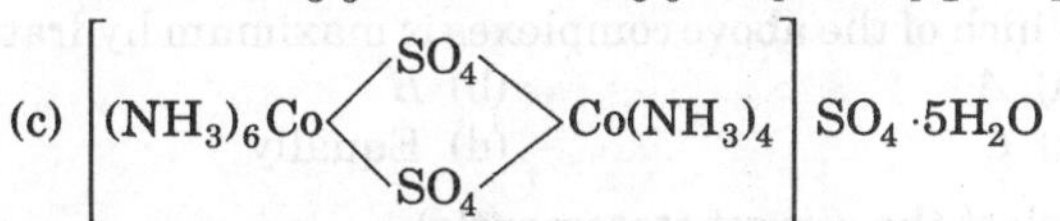

(d) None of the above

Passage 9

Mr. William isolated two complexes and in Chemical Absract, he reported

"A metal complex having composition $Cr(NH_3)_4Cl_2Br$ has been isolated in two forms (*A*) and (*B*). The form (*A*) reacts with $AgNO_3$ to give a white precipitate readily soluble in dilute aqueous ammonia, whereas (*B*) gives a pale yellow precipitate soluble in concentrated ammonia.

Read the above report from Chemical Abstract and answer the questions at the end of it.

1. Complex *A* and *B* are respectively
(a) $[Cr(NH_3)_4Cl_2]Br, [Cr(NH_3)_4Br]Cl_2$
(b) $[Cr(NH_3)_4Br]Cl_2, [Cr(NH_3)_4Cl_2]Br$
(c) $[Cr(NH_3)_4ClBr]Cl, [Cr(NH_3)_4Cl_2]Br$
(d) $[Cr(NH_3)_4Cl_2]Br, [Cr(NH_3)_4ClBr]Cl$

2. Cr atom is .... hybridised
(a) sp^3d^2 in both complexes
(b) d^2sp^3 in both complexes
(c) sp^3d^2 in *A* and d^2sp^3 in *B*
(d) d^2sp^3 in *A* and sp^3d^2 in *B*

3. EAN and magnetic moment of Cr in *A* and *B* are
(a) 33, $\sqrt{15}$ BM (b) 36, $\sqrt{15}$ BM
(c) 33, $\sqrt{24}$ BM (d) 36, $\sqrt{24}$ BM

Passage 10

Study the following short write up and answer the questions at the end of it.

A, *B* and *C* are three complexes of chromium (III) with the empirical formula $H_{12}O_6Cl_3Cr$. All the three complexes have water and chloride ions as ligands. Complex *A* does not react with concentrated H_2SO_4, whereas complexes *B* and *C* lose 6.75% and 13.5% of their original weight respectively, on treatment with concentrated H_2SO_4.

1. Complex *A* is
(a) $[Cr(H_2O)_6]Cl_3$
(b) $[Cr(H_2O)_5Cl]Cl_2 \cdot H_2O$
(c) $[Cr(H_2O)_4Cl_2]Cl \cdot 2H_2O$
(d) None of the above

2. Loss in weight on treatment with concentrated H_2SO_4 is due to
(a) Cl^- is lost as HCl
(b) Cr^{3+} is lost as $Cr_2(SO_4)_3$
(c) H_2O is lost due to absorption
(d) Cl^- is lost as Cl_2

3. Which of the above complexes is maximum hydrated?
(a) *A* (b) *B*
(c) *C* (d) Equally

4. Select the correct statement(s).
(a) Conductance in these complexes is in order $C < B < A$
(b) EAN and nuclear spin of chromium are identical in each complex
(c) Non-reactivity of *A* with conc. H_2SO_4 is due to absence of water molecules outside coordinated sphere
(d) All of the above are correct

Passage 11

Consider following experiments and answer the questions at the end of it.

(*A*) When $Fe(CN)_2$ solution is treated with KCN solution, species formed, no longer, gives tests of Fe^{2+} and CN^-.

(*B*) When K_2SO_4 solution is treated with $Al_2(SO_4)_3$ solution, species formed gives tests of K^+, Al^{3+} and SO_4^{2-}.

1. Species formed in experiment *A* does not give test of Fe^{2+} and CN^-. It is due to formation of
(a) $K_2[Fe(CN)_4]$ (b) $K_3[Fe(CN)_5]$
(c) $K_3[Fe(CN)_6]$ (d) $K_4[Fe(CN)_6]$

2. Species formed in experiment (*B*) is
(a) complex (b) double salt
(c) liquid crystal (d) None of these

3. EAN of iron formed in (*A*) is
(a) 26 (b) 24
(c) 36 (d) 38

4. When the species formed in (*A*) is treated with $FeCl_3$, a blue colour is obtained. It is due to the formation of
(a) $Fe^{II}[Fe^{III}(CN)_6]^-$ (b) $Fe^{III}[Fe^{II}(CN)_6]^-$
(c) both (a) and (b) (d) None of these

Passage 12

Consider the following isomers of $[Cr(NH_3)_2Cl_4]^-$ *and answer the questions*

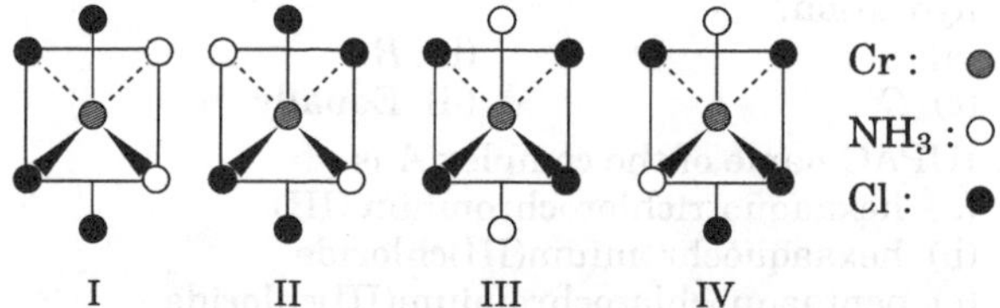

1. Out of these isomers
(a) I and IV are *trans* and II and III are *cis*
(b) I and IV are *cis* and II and III are *trans*
(c) I and II are *cis* and III and IV are *trans*
(d) I and II are *trans* and III and IV are *cis*

2. Select the correct statement.
(a) Pair of I and IV are same and pair of II and III are also same
(b) All have chiral centres
(c) I and III are enantiomers
(d) II and IV are enantiomers

Passage 13

Cis-diamminedichloroplatinum (II), $[Pt(NH_3)_2Cl_2]$, is

$$\begin{matrix} H_3N & & Cl \\ & Pt & \\ H_3N & & Cl \end{matrix}$$

One of the number of platinum coordination compounds is used in the treatment of cancer. Commonly known as ***cis*-platin,** this compound has the ability to block the uncontrolled division of cancerous cells that results in the growth of tumours. Recent studies show that *cis*-platin can cause serious side effects, including severe kidney damage.

Read the above medical report and answer the following questions.

1. *Cis*-platin is being replaced by

(a) $\left[\begin{matrix} Cl & & NH_3 & H_3N & & Cl \\ & Pt & & & Pt & \\ H_3N & & NH_2(CH_2)_n & H_2N & & NH_3 \end{matrix} \right]^{2+}$

trans-isomer ($n = 2$ to 6)

(b) $\left[\begin{matrix} H_3N & & Cl & Cl & & NH_3 \\ & Pt & & & Pt & \\ H_3N & & NH_2(CH_2)_n & H_2N & & NH_3 \end{matrix} \right]$

($n = 2$ to 6)

(c) $\left[\begin{matrix} H_3N & & Cl & H_3N & & Cl \\ & Pt & & & Pt & \\ H_3N & & NH_2(CH_2)_n & H_2N & & NH_3 \end{matrix} \right]$

(d) None of the above

Passage 14

Following experiment has been given to identify isomer. Read the experiment and answer the questions.

A compound $Co(en)_2(NO_2)_2Cl$ has been prepared in a number of isomeric forms. One form undergoes no reaction with $AgNO_3$ or (en) and is optically inactive. A second form reacts with $AgNO_3$ but not with (en) and is optically inactive. A third form is optically active and reacts with both $AgNO_3$ and (en), complexes are:

(A) $[Co(en)_2(NO_2)_2]^+ Cl^-$

cis-bis (ethylenediamine)dinitrocobalt (III) chloride

(B) 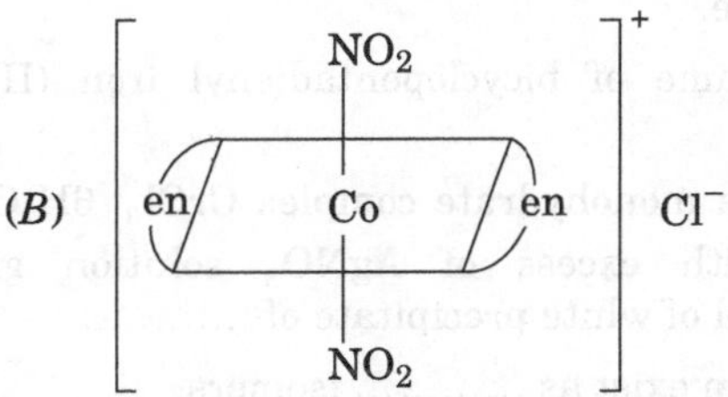

trans-bis (ethylenediamine)dinitrocobalt (III) chloride

(C) 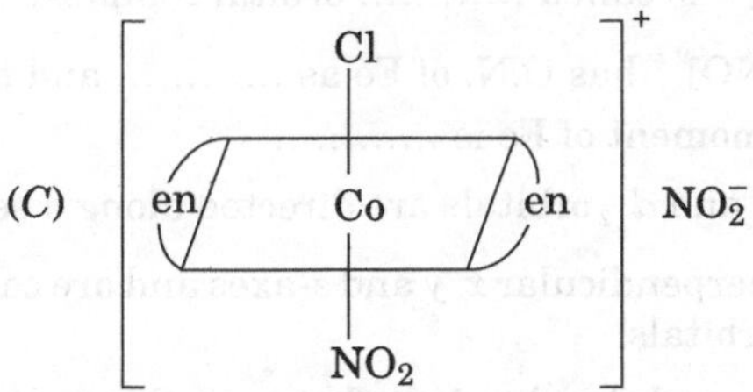

trans-chloronitrobis(ethylenediamine)cobalt (III) nitrite

1. First form of the complex is

(a) *A* (b) *B*
(c) *C* (d) None of these

2. One of the complexes turns KI-starch paper blue in acidic medium. This complex is

(a) *C* (b) *B*
(c) *A* (d) all of these

Passage 15

A research-guide instructed his two students to synthesize complex

$$[Co(NH_3)_5(NO_2)]Cl_2$$

They synthesised the complexes with identical molecular formula, molar mass, geometry conductance and spin, but they differed in colour. Based on the above facts answer the following questions.

1. The difference in colour is due to

(a) optical isomerism
(b) geometrical isomerism
(c) linkage isomerism
(d) nuclear isomerism

2. Which of the ligands can show ambident property?

(a) NO_2^- (b) NH_3
(c) H_2O (d) CO_3^{2-}

3. Complexes synthesized can be

(a) $[Co(NH_3)_5(NO_2)]Cl_2$ (b) $[Co(NH_3)_5(ONO)]Cl_2$
(c) $[Co(NH_3)_5Cl_2]NO_2$ (d) all of these

Passage 16

Read the following short write-up and answer the questions at the end of it.

One cationic complex has two isomers *A* and *B*. Each has one Co^{3+}, five NH_3, one Br^- and one SO_4^{2-} stoichiometrically. *A* gives white ppt with $BaCl_2$ while *B* give yellow ppt with $AgNO_3$.

1. *A* can be

(a) $[Co(NH_3)_5]BrSO_4$
(b) $[Co(NH_3)_5SO_4]Br$
(c) $[Co(NH_3)_5Br]SO_4$
(d) $[Co(NH_3)_4SO_4]Br \cdot NH_3$

2. *B* can be

(a) $[Co(NH_3)_5SO_4]Br$
(b) $[Co(NH_3)_5Br]SO_4$
(c) $[Co(NH_3)_3Br(SO_4)] \cdot 2NH_3$
(d) None of the above

3. Complexes *A* and *B* have similarity in the following but not in

(a) molar conductance (b) van't Hoff factor
(c) EAN (d) colour

Assertion & Reason

Codes :

(a) Both **A** and **R** are true and **R** is the correct explanation of **A.**

(b) Both **A** and **R** are true but **R** is not the correct explanation of **A.**

(c) **A** is true but **R** is false.

(d) **A** is false but **R** is true.

1. Assertion (A) : All the iron-carbon bond distances in ferrocene are equal.

Reason (R) : The π-electrons in the cyclopentadienyl group of ferrocene are delocalised.

2. Assertion (A) : The electronic absorption spectrum of the tetrahedral complex, $[CoCl_4]^{2-}$ shows bands which are more intense compared to the absorption bands for the octahedral complex $[Co(H_2O)_6]^{2+}$.

Reason (R) : The octahedral complexes possess a centre of symmetry.

3. Assertion (A) : $Ni(CO)_4$ is a diamagnetic complex.

Reason (R) : All the electrons in the complex are paired.

4. Assertion (A) : NO_2^- is an ambident ligand.

Reason (R) : NO_2^- is a Lewis base.

5. Assertion (A) : A stable complex is formed when ethylene diamine (en) is added to $[Ni(H_2O)_4]^{2+}$.

Reason (R) : (en) being a strong ligand replaces weak ligand H_2O and thus, chelate is formed.

6. Assertion (A) : $CoCl_3 \cdot 4NH_3$ gives two ions when ionised.

Reason (R) : Complex is $[Co(NH_3)_4]Cl_3$.

7. Assertion (A) : Spot of iron-rusting can be removed by addition of oxalic acid.

Reason (R) : Water soluble complex is formed with oxalic acid.

True & False

1. Use of *cis*-platin in cancer treatment also causes severe kidney damage.

2. Lead poisoning can be removed by use of EDTA.

3. Out of $[Cu(CN)_4]^{3-}$ and $[Cd(CN)_4]^{2-}$, $[Cu(CN)_4]^{3-}$ is unstable and gives black precipitate of CuS on passing H_2S gas.

4. AgBr is soluble in *aq*. NH_3 as well as in hypo.

5. $[Cu(NH_3)_4]^{2+}$ and $[Cu(CN)_4]^{2-}$ are labile complexes.

6. $[Cu(H_2O)_4]^{2+} + 4NH_3 \longrightarrow [Cu(NH_3)_4]^{2+} + 4H_2O$ is spontaneous in forward side, and $\Delta G = -ve$

7. $[Fe(CN)_6]^{3-}$ is more stable than $[Fe(CN)_6]^{4-}$

8. A cyclic multidentate ligand increases the stability of the complex.

9. $[Fe(H_2O)_6]^{2+}$ is an inner *d*-complex.

10. H_2O is a weaker field ligand than NH_3 and $\Delta_{oct}\ ([Cr(NH_3)_6]^{3+}) < \Delta_{oct}\ ([Cr(H_2O)_6]^{3+})$

11. Orange is the complimentary colour of blue spectral colour.

12. $\Delta_t = \frac{4}{9}\Delta_{oct}$

13. The d_{xy}, d_{yz} and d_{xz} orbitals lie between the axes and collectively called t_{2g} orbitals.

14. Geometry of $[Ni(CN)_4]^{2-}$ is tetrahedral.

15. In the complex $CoCl_3 \cdot 5NH_3$ one Cl^- ion satisfies primary and secondary valencies of cobalt(III).

16. *cis*$[Co(en)_2Cl_2]^+$ can have *d* and *l*-optical isomers.

17. $[Co(NH_3)_5(SO_4)]Br$ and $[Co(NH_3)_5Br]SO_4$ are linkage isomers.

18. CN of Fe in $[Fe(en)_3]^{3+}$ is three.

19. Ferrocene is stable since π-electrons in cyclopentadienyl ion are delocalised.

20. EDTA is hexadentate ligand and is used to remove hardness in water.

Fill in the Blanks

1. Ligands which can be coordinated to a metal or metal ion through either of the sides are called ....

2. Vitamin B_{12} is a complex of ....

3. Haemoglobin is a complex of Fe(II) bound to .... rings.

4. .... is the number of ligating groups in a multidentate.

5. Common name of bicyclopentadienyl iron (II) is ....

6. 1 mol of the monohydrate complex $CrCl_3 \cdot 6H_2O$ on reaction with excess of $AgNO_3$ solution gives .... mol of white precipitate of ....

7. $[Pt(Gly)_2]$ can exist as .... isomers.

8. $[Fe(H_2O)_6]^{3+}$ is called .... orbital complex.

9. $[Fe(H_2O)_5NO]^{2+}$ has O.N. of Fe as .... and thus magnetic moment of Fe is ....

10. The $d_{x^2-y^2}$ and d_{z^2} orbitals are directed along a set of mutually perpendicular *x*, *y* and *z*-axes and are called .... orbitals.

11. The d_{xy}, d_{yz} and d_{xz} orbitals lie between the axes and collectively called .... orbitals.

12. $\Delta_t = -\frac{4}{9}$ ....

13. Blue-green is the complimentary colour of .... spectral colour.

14. Strong field ligands, such as CN^-, usually produce .... spin complexes and .... crystal field splitting.

15. Reaction between NH_3 and $[Cu(H_2O)_4]^{2+}$ is ....

16. $[Cu(H_2O)_4]^{2+}$, $[Cu(NH_3)_4]^{2+}$ and $[CuCl_4]^{2-}$ are called .... complexes.

17. Stability of a complex with multidentate ligand is increased due to ....

18. Lead poisoning can be decreased by addition of ....

19. Gold and silver are extracted by complex formation with ....

20. Unreacted silver bromide in photographic plate is removed by the addition of ....

Matrix-Match Type Questions

I. Only One Correct

1. Match the complex (in List I) with the hybridisation of the central ion (in List II).

List I	List II
A. $Ni(CO)_4$	1. sp^3d^2
B. $[Ni(CN)_4]^{2-}$	2. sp^3
C. $[Fe(CN)_6]^{4-}$	3. d^2sp^3
D. $[MnF_6]^{4-}$	4. dsp^2

2. Match the complex (in List I) with the geometry (in List II).

List I	List II
A. $[Ni(CN)_4]^{2-}$	1. Tetrahedral
B. $[ZnCl_4]^{2-}$	2. Trigonal bipyramidal
C. $[Fe(CO)_5]$	3. Square planar
D. $[Co(NO_2)_6]^{4-}$	4. Square pyramid
	5. Octahedral

3. Match the complex (in List I) with the equivalent conductance (in List II).

List I	List II
A. $[Pt(NH_3)_5Cl]Cl_3$	1. 229
B. $[Pt(NH_3)_4Cl_2]Cl_2$	2. 97
C. $[Pt(NH_3)_3Cl_3]Cl$	3. 404
D. $[Pt(NH_3)_6]Cl_4$	4. 523

4. Match the complex (in List I) with the oxidation number of Co (in List II).

List I	List II
A. $[Co(NCS)(NH_3)_5]SO_3$	1. −1
B. $Na[Co(CO)_4]$	2. 0
C. $Na_4[Co(S_2O_3)_3]$	3. +1
D. $Co_2(CO)_5$	4. +2
	5. +3

5. Match the complex (in List I) with the type of isomerism (in List II).

List I	List II
A. $[Co(NH_3)_4Cl_2]$	1. Optical
B. *cis*$[Co(en)_2Cl_2]$	2. Ionisation
C. $[Co(en)_2(NO_2)Cl]SCN$	3. Coordination
D. $[Co(NH_3)_6][Cr(CN)_6]$	4. Geometrical

6. Match complex ion (in List I) with number of unpaired electrons (in List II).

List I	List II
A. $[Fe(H_2O)_6]^{2+}$	1. 0
B. $[Fe(H_2O)_6]^{3+}$	2. 1
C. $[Fe(CN)_6]^{4-}$	3. 2
D. $[Fe(CN)_6]^{3-}$	4. 3
E. $[Ni(H_2O)_4]^{2+}$	5. 4
	6. 5

II. One or More Correct

1. Match the complex (in Column I) with its corresponding property (ies) (given in Column II).

Column I	Column II
A. $[Ni(DMG)_2]$	1. Chelation
B. $Na_3[Fe(C_2O_4)_3]$	2. Paramagnetic
C. $[Ni(en)_2]Cl_2$	3. Diamagnetic
D. $Ni(Co)_4$	4. Conducting
	5. H-bonding

2. Match the complex (in Column I) with its corresponding property (ies) (in Column II).

Column I	Column II
A. $[Cr(NO_2)_2(NH_3)_2Cl_2]$	1. Geometrical isomerism
B. $[Cr(NH_3)_4Cl_2]Br$	2. Ionisation isomerism
C. $[Cr(en)_3]^{3+}$	3. Optical isomerism
D. $[Pt(NH_3)_2Cl_2][Cu(CN)_4]$	4. Linkage isomerism
	5. Coordinate isomerism

Integer Answer Types

This section contains 8 questions. The answer to each of the questions is a single digit integer, ranging from 0 to 9. The appropriate bubbles below the respectively question numbers in the *ORS* have to be darkened. For example, if the correct answers to question number *X, Y, Z* and *W* (say) are 6, 0, 9 and 2, respectively

X	Y	Z	W
0	0	0	0
1	1	1	1
2	2	2	2
3	3	3	3
4	4	4	4
5	5	5	5
6	6	6	6
7	7	7	7
8	8	8	8
9	9	9	9

Set I

1. $[Co(NH_3)_6]^{3+}$ has .... unpaired electron(s) in t_{2g} orbitals.

2. There are .... electron in e_g orbitals is $[CoF_6]^{3-}$

3. There are .... electron in t_{2g} orbitals in $[CoF_6]^{3-}$

4. In the high spin complexes of Cr^{3+} number of unpaired electron in t_{2g} orbitals is ....

5. Total number of ions due to ionisation of the complex $CoCl_3 \cdot 6H_2O$ is ....

6. Total number of orbitals used in hybridisation of octahedral complex $[Fe(CN)_6]^{3-}$ is ....

7. Square-planar $[Ni(CN)_4]^{2-}$ uses .... *d*-orbital(s) in hybridisation.

8. CFSE of high-spin d^5-Mn^{2+} complex is ....

Set II

This section contains 10 questions.

1. Ionisable chloride ions in $CoCl_5 \cdot 4NH_3$ are ....

2. Magnetic moment of a complex is 4.9 B.M. Thus, unpaired electron(s) may be ....

3. Nickel dimethylglyoximate is stable due to chelation. Chelate ring(s) is/are ....

4. Ethylenediamine displaces H_2O in the complex $[Fe(H_2O)_6]^{3+}$ in three steps

$$[Fe(H_2O)_6]^{3+} + 3\,en \longrightarrow [Fe(en)_3]^{3+} + 6H_2O$$

Stepwise formation constants are

$$\log k_1 = 4.34$$
$$\log k_2 = 3.31$$
$$\log k_3 = 2.05$$

Overall formation constant is $A \times 10^B$
A is ....

5. *B* in question (4) is ....

6. Ambidentate ligands out of

Py, en, CNS, CN, CNO, $C_2O_4^{2-}$, NO_2, are

7. EDTA is a multidentate ligand. Its deniticity (multicity) is ....

8. $CrCl_3H_{12}O_6$ loses 13.5% water on treatment with conc. H_2SO_4. Number of water molecules lost is ....

9. Number of aromatic rings in ferrocene is ....

10. Effective atomic number of Ni in $Ni(CO)_x$ is 36. Thus, $x =$ ....

Answers & Solutions
Master Exercises

Exercise 1

Short Answer Type Questions

1. To obey the rule of 18 (EAN), the sum of the number of electrons in the metal valence shell [$(n-1)\,dsp$] and those donated by the ligands should be 18, filling the low-energy metal orbitals.

$Cu[Ar]\,3d^{10}4s^1$ $Ni[Ar]\,3d^8\,4s^2$

$Cr(Ar)\,3d^5\,4s^2$ $Co[Ar]\,3d^7 4s^2$

$Fe(Ar)\,3d^6 4s^2$

$Cu^+[Ar]\,3d^{10}$ $Ni^{2+}[Ar]\,3d^8$

$Fe^{2+}[Ar]\,3d^6$ $Cr^{3+}[Ar]\,3d^3$

$Fe^{3+}[Ar]\,3d^5$ $Co^{3+}[Ar]\,3d^6$

(a) $[Cu(NH_3)_4]^{2+}$ $9+8=17e^-$

$[Cu(en)_3]^{2+}$ $9+12=21e^-$

$[Cu(CN)_4]^{3-}$ $10+8=18e^-$

Thus, $[Cu(CN)_4]^{3-}$

(b) $[Fe(CN)_6]^{4-}$ $6+12=18e^-$

$[Fe(CN)_6]^{3-}$ $5+12=17e^-$

$[Fe(CO)_5]$ $8+10=18e^-$

Thus, $[Fe(CN)_6]^{4-}$ and $[Fe(CO)_5]$

(c) $[Ni(NH_3)_6]^{2+}$ $8+12=20e^-$

$[Ni(CN)_4]^{2-}$ $8+8=16e^-$

$[Ni(CO)_4]$ $10+8=18e^-$

Thus, $[Ni(CO)_4]$

(d) $[Cr(NH_3)_6]^{3+}$ $3+12=15e^-$

$[Cr(CO)_6]$ $6+12=18e^-$

Thus, $[Cr(CO)_6]$

(e) $[Co(NH_3)_6]^{3+}$ $6+12=18e^-$

$[CoCl_4]^{2-}$ $7+8=15e^-$

Thus, $[Co(NH_3)_6]^{3+}$

2. $\underset{\text{neutral}}{[Cu(NH_3)_2Cl_2]} < \underset{\text{two ions}}{KNO_3} < \underset{\text{three ions}}{Na_2[PtCl_6]} < \underset{\text{four ions}}{[Co(NH_3)_6]Cl_3}$

Larger the number of ions, greater the electrical conductivity.

3. $[Fe(CN)_6]^{3-}$ has the strongest field ligands of the three complexes; thus, its energy levels are split by the greatest amount. The frequency of the light absorbed should be greatest, and λ_{max} should be the shortest for this ion.

$[Fe(CN)_6]^{4-}$: red colour — absorb blue-violet light

$[Fe(H_2O)_6]^{3+}$: pale violet — absorb red light

$[FeF_6]^{3-}$: colourless — absorb light in visible spectrum

4. $[Co(NH_3)_6]^{2+}$ is paramagnetic due to presence of one electron unpaired in presence of strong field ligand (low-spin complex).

5. Monodentate

N-atom is bonded to metal ion

6.

	Ligand		Unpaired electron (b)	CFSE (c)
I.	$[Fe(CN)_6]^{3-}$	(d^5) strong (low spin complex)	1	$-\frac{10}{5}\Delta_o$
II.	$[Fe(H_2O)_6]^{3+}$	(d^5) weak (high spin complex)	5	zero

(a) $[Fe(CN)_6]^{3-}$ has fewer unpaired electron

7. Yellow

8. (a) The overall charge is −1, which is equal to the charge on the $C_5H_5^-$ group. Thus, the Fe atom must have no charge. Thus,

$$Fe=[Ar]\,3d^6 4s^2$$

(6 electrons from $C_5H_5^-$) + (8 electrons from Fe) + (4 electrons from 2 CO groups) $=18e^-$

(b) $Mn=[Ar]\,3d^5 4s^2$

$Mn^-=[Ar]\,3d^6 4s^2$

Valence electrons in $Mn^-=8$

Electrons donated of $5CO=10$

Thus, $8+10=18e^-$

(c) $Co=[Ar]\,3d^7 4s^2$

$Co^+=[Ar]\,3d^7 4s^1$

Electrons of $Co^+=8$

Electrons donated by $C_5H_5^-=6$

Electrons donated by two $C_2H_4=4$

Thus, $8+6+4=18e^-$

9. In $[NiCl_4]^{2-}$ Cl^- is a weak ligand thus electrons are unpaired in Ni^{2+} making it paramagnetic sp^3–tetrahedral structure

When strong ligand CN^- is added, substitution takes place and electrons are paired in $[Ni(CN)_4]^{2-}$ dsp^2 — square planar.

10. Blue green

11. Fe^{2+} —$[Ar]3d^6$ high spin — four unpaired electrons

low spin — none unpaired electrons

Analytical Questions

1. (i) 35 (ii) 33 (iii) 33

2.

	(a)	(b)	(c)	(d)	(e)
C.N.	6	6	4	6	2
O.N.	2	3	2	3	1

3.

	Ligands	C.N.	O.N.
(a)	NH_3, NO_2	6	+3
(b)	CN	6	+2
(c)	NH_3, Cl	6	+3
(d)	NH_3	4	+2
(e)	CN	2	+1
(f)	en	6	+3

4. Only two-third of the chlorine is precipitated indicating that two chlorine atoms are outside coordinate sphere. Thus, complex can be $[Co(NH_3)_5Cl]Cl_2$. It also ionises as

$$[Co(NH_3)_5Cl]Cl_2 \rightleftharpoons [Co(NH_3)_5Cl]^{2+} + 2Cl^-$$

5. $[Pt(NH_3)_6]Cl_4$; C.N. of Pt = 6

$[Pt(NH_3)_3Cl_3]Cl$; C.N. of Pt = 6

6. Moles of the complex $= \dfrac{23.25}{233.5} = 0.1$

Moles of AgCl ppt $= \dfrac{14.35}{143.5} = 0.1$

Thus, 1 mole of the complex gives 1 mole of AgCl indicating only one chlorine is outside coordinate sphere. Thus, complex is $[Co(NH_3)_4Cl_2]Cl$.

7. $n = 5$ **8.** See text

9. (a) Pentamminecarbonatechromium (III) chloride

(b) Pentamminenitritocobalt (III) chloride

(c) Potassium hexacyanochromate (III)

(d) Hexammine cobalt (III) chloride

10. (a) $[Co(NH_3)_5Cl]^{2+}$ (b) $Li[AlH_4]$

11. See text **12.** See text **13.** See text **14.** See text

15. (a) and (b) are identical

(a) and (d) are geometrical isomers

(c) is distinct

16. (a) I, II, IV — Chiral

II — Achiral

(b) Refer text

(c) I, IV are enantiomers

17. Refer text

18. $\underset{1\text{ mol}}{(A)} + AgNO_3 \longrightarrow \underset{1\text{ mol}}{AgCl}$

Thus, (A) has only one ionisable Cl atom. Thus, A is

$$[Cr(H_2O)_4Cl_2]Cl\cdot 2H_2O$$

$$\underset{1\text{ mol}}{(B)} + AgNO_3 \longrightarrow \underset{3\text{ mol}}{AgCl}$$

Thus, (B) has three ionisable Cl-atoms. Thus, B is $[Cr(H_2O)_6]Cl_3$ — an anhydrous complex.

$A : [Cr(H_2O)_4Cl_2]Cl\cdot 2H_2O$,

$B : [Cr(H_2O)_6]Cl_3$

A will change to $[Cr(H_2O)_4\cdot Cl_2]Cl$ on heating being hydrated complex. B has no effect of heating, all H_2O molecules being ligands. A will ionise as $[Cr(H_2O)_4Cl_2]^+$ and Cl^-

B will ionise as $[Cr(H_2O)_6]^{3+}$ and $3Cl^-$ hence conductance of B is greater than that of A.

19. $A : [Co(NH_3)_4Cl(H_2O)]SO_4$ $B : [Co(NH_3)_4(SO_4)(H_2O)]Cl$

$C : [Co(NH_3)_4(SO_4)Cl]\cdot H_2O$

A and B are ionisation isomers, C is hydrate isomer each of A and B.

(You can also make geometrical isomers of each)

20. A does not lose water with conc. H_2SO_4 indicating that H_2O are in coordinate sphere.

Thus, A is $[Cr(H_2O)_6]Cl_3$

Molar mass of the complex $= 52 + 108 + 106.5$

$= 266.5\text{ g mol}^{-1}$

After loss with conc. H_2SO_4 in $(B) = \dfrac{266.5 \times 93.25}{100}$

$= 249.18$

Net loss $= 266.5 - 249.18$

$= 17.32\text{ g} = 1\text{ mol } H_2O$

Thus, one water molecule is lost per molecule of the complex. Thus, complex (B) is

$$\mathbf{[Cr(H_2O)_5Cl_2]Cl_2\cdot H_2O}$$

After loss with conc. H_2SO_4, molar mass of (C) is

$= \dfrac{266.5 \times 86.5}{100}$

$= 230.52\text{ g mol}^{-1}$

Net loss $= 35.98$ g

Number of water molecules lost $= \dfrac{35.98}{18} = 2$. Thus, two H_2O molecules are as hydrate molecules.

Thus, (C) is $[Cr(H_2O)_4Cl_2]Cl\cdot 2H_2O$

21. While passing through the cation exchanger there is formation of HCl. Molar mass of the complex $Cr(H_2O)_6Cl_2 = 52 + 108 + 106.5$

$= 266.5\text{ g mol}^{-1}$

$$[Cr(H_2O)_5Cl]Cl_2 \cdot H_2O \longrightarrow 2HCl$$
$$\underset{1\text{ mol}}{} \qquad \underset{2\text{ mol}}{}$$
$$\frac{1}{266.5} \qquad \frac{2}{266.5}\text{ mol}$$
$$[HCl] = \mathbf{7.5 \times 10^{-3}\ M}$$
$$= 7.50 \times 10^{-3} \times 36.5\ gL^{-1}$$
$$= \mathbf{0.24\ gL^{-1}}$$

22.
$$\Delta E = \frac{hc}{\lambda}$$
$$= \frac{6.62 \times 10^{-34} \times 3 \times 10^{6}}{544 \times 10^{-9}}$$
$$= 3.65 \times 10^{-19}\ J\ mol^{-1}$$
$$= 3.65 \times 10^{-19} \times 6.02 \times 10^{23}\ J\ mol^{-1}$$
$$= \mathbf{219.79\ kJ\ mol^{-1}}$$

23. (a) 3 unpaired electrons; (b) paramagnetic;

(c) $[FeCl_4]^-$ has the greater number of unpaired electrons than $[Mn(CN)_6]^{3-}$.
Cl^- — a weak ligand
CN^- — a strong ligand

24. $[Co(en)_3]^{3+}$ --- yellow
$[Co(H_2O)_6]^{3+}$ --- blue
Since (en) produces greater *d*-level splitting than does H_2O.

25. (a) inner, paramagnetic (b) outer, paramagnetic
(c) inner, diamagnetic (d) inner, paramagnetic

26.

$[CrF_6]^{3-}$	$[V(H_2O)_6]^{3+}$	$[Fe(CN)_6]^{3-}$
e_g: — — ; Δ ; t_{2g}: ↑ ↑ ↑	e_g: — — ; Δ ; t_{2g}: ↑ ↑ —	e_g: — — ; Δ ; t_{2g}: ↑↓ ↑↓ ↑
Three unpaired electrons	Two unpaired electrons	One unpaired electron

27. $[Fe(CN)_6]^{3-}$ has d^2sp^3 (inner *d*-complex) hybridisation with one electron unpaired while $[Fe(H_2O)_6]^{3+}$ has sp^3d^2 (outer *d*–complex) hybridisation with five unpaired electrons

magnetic moment $= \sqrt{N(N+2)}$ BM

where, N is the number of unpaired electrons.

28. (b) **29.** Only (c) is correct **30.** (I) (c) (II) (d)

31. (c) They differ in number of unpaired electrons.

32. Refer text

33. A : $[Cr(NH_3)_4ClBr]Cl$

B : $[Cr(NH_3)_4Cl_2]Br$

Cr is d^2sp^3-hybridised

unpaired electrons $N = 3$

magnetic moment $= \sqrt{N(N+2)}$ BM $= \sqrt{15}$ BM

34. Also text for structures.

	Hybridisation	Unpaired electron	Magnetic moment
$[NiCl_4]^{2-}$	sp^3, tetrahedral	2	$\sqrt{8}$ BM
$[Ni(CN)_4]^{2-}$	dsp^2, square planar	0	0

35. EDTA can form complex with lead present in the body. Calcium salt is preferred otherwise calcium of the body will also be removed as calcium-EDTA complex (that is harmful).

36. (a) Al^{3+} forms a hydroxo compound such as $[Al(H_2O)_2(OH)_4]^-$ but not an ammine complex;

(b) Solubility product (K_{sp}) for ZnS is much smaller than for $ZnCO_3$, and a saturated solution of ZnS does not produce a large enough $[Zn^{2+}]$ for the complex ion to form;

(c) $[Co^{3+}]$ is kept sufficiently low in a solution of the stable complex ion, $[Co(NH_3)_6]^{3+}$, that it is unable to oxidise water.

37. (a) One can determine conductance;

(b) One can determine freezing point;

(c) One can detect Al^{3+}, SO_4^{2-}, K^+.

NCERT Problems

1. (a) oxalate (b) six (c) oxalate

2. (a) +3 (b) +2 (c) +3 (d) +3 (e) +3

3. (a) $[Zn(OH)_4]^{2-}$ (b) $[Co(NH_3)_6](SO_4)_3$ (c) $K_2[PdCl_4]$
(d) $K_3[Cr(OX)_3]$ (e) $[Pt(NH_3)_2Cl_2]$
(f) $[Pt(NH_3)_6]^{4+}$ (g) $K_2[Ni(CN)_4]$ (h) $[CuBe_4]^{2-}$
(i) $[Co(NH_3)_5ONO]^{2+}$ (j) $[Co(NH_3)_5NO_2]^{2+}$

4. Refer text.

5. Refer text.

6. Refer text.

7. Refer text.

8. Refer text.

9. (a) $[Cr(H_2O)_6]Cl_3$ (b) $[Cr(H_2O)_5Cl]Cl_2 \cdot H_2O$
(c) $[Cr(H_2O)_4Cl_2]Cl \cdot 2H_2O$

10. Aqueous $CuSO_4$ is $[Cu(H_2O)_4]SO_4 \cdot H_2O$

(a) $[Cu(H_2O)_4]^{2+} + KF \longrightarrow \underset{\text{green ppt}}{[CuF_4]^{2-}}$

(b) $[Cu(H_2O)_4]^{2+} + KCl \longrightarrow \underset{\text{light green}}{[CuCl_4]^{2-}}$

11. Refer text.

12. Refer text.

13. (a) CH_3—Li (b) $Si(CH_3)_4$ (c) $Bi(CH_3)_3$
(d) $As(CH_3)_3$ (e) $Al(CH_3)_6$ (f) $B(CH_3)_3$

14. Refer text.

Exercise 2

Only One Option Correct

1. (b) **2.** (c)

3. (d) Complex is $[Co(NH_3)_5Cl]Cl_2$

$$\rightleftharpoons$$

$[[Co(NH_3)_5Cl]^{2+} + 2Cl^-]$ $y = 3$ ions per mol

i = van't Hoff factor $= 1 + (y - 1)x = 3$

$$\pi = MSTi = 1 \times 0.0821 \times 300 \times 3$$
$$= 73.89 \text{ atm.}$$

Thus, **(d)**

4. (c) **5.** (a) **6.** (a) **7.** (c) **8.** (b)

9. (c) $Zn^{2+} + 4NH_3 \rightleftharpoons [Zn(NH_3)_4]^{2+}$

$$K_f = \frac{[Zn(NH_3)_4]^{2+}}{[Zn^{2+}][NH_3]^4}$$
$$= 3 \times 10^9$$

$$\therefore \quad \frac{[Zn(NH_3)_4]^{2+}}{[Zn^{2+}]} = [NH_3]^4 \times 3 \times 10^9$$
$$= 10^4 \times 3 \times 10^9$$
$$= 3 \times 10^{13}$$

$$\therefore \quad \frac{[Zn^{2+}]}{[Zn(NH_3)_4]^{2+}} = 3.3 \times 10^{-14}$$

Hence, **(c)**

10. (c)

11. (c) 100 mL of 0.024 M = 0.0024 mol of the complex

complex is $[Co(en)_2Cl_2]Cl$ and thus, one Cl^- is formed per mol of the complex which gives 1 mole of AgCl.

Hence, AgCl formed = 0.0024 mol

Hence, **(c)**

12. (a) **13.** (b) **14.** (b) **15.** (a) **16.** (d)

17. (b) **18.** (a) **19.** (b) **20.** (d) **21.** (c)

22. (a) **23.** (b) **24.** (a) **25.** (b) **26.** (b)

27. (d) **28.** (a) **29.** (a) **30.** (a)

31. (a) Greater the number of ions, greater the conductance.

1. $K_2[PtCl_6] \rightleftharpoons \underbrace{2K^+ + [PtCl_6]^{2-}}_{\text{three ion}}$

2. $PtCl_4 \cdot 2NH_3 \rightleftharpoons [Pt(NH_3)_2Cl_4] \longrightarrow$ no ion (least)

3. $[Pt(NH_3)_3Cl_3]Cl \rightleftharpoons$ two ions

4. $[Pt(NH_3)_5Cl]Cl_3 \rightleftharpoons$ four ions (maximum)

Hence, **(a)**

32. (d) **33.** (d) **34.** (a) **35.** (b) **36.** (d)

37. (a) **38.** (b) **39.** (a) **40.** (c) **41.** (d)

42. (a) **43.** (c) **44.** (b)

45. Fe^{2+}

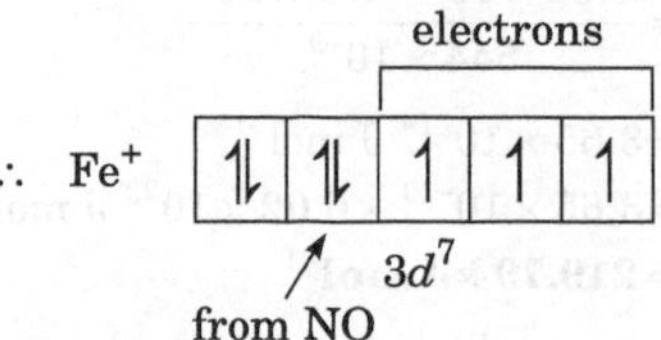

$3d^6$

$NO \longrightarrow NO^+ + e^-$
$Fe^{2+} + e^- \longrightarrow Fe^+$] Charge transfer

three unpaired electrons

$\therefore$ Fe^+ ⇅ ⇅ ↿ ↿ ↿ $3d^7$ (from NO)

Thus, (a), (b), (c) are true.

Hence, **(d)**

46. (b)	**47.** (c)	**48.** (a)	**49.** (d)	**50.** (a)
51. (a)	**52.** (b)	**53.** (a)	**54.** (b)	**55.** (a)
56. (a)	**57.** (b)	**58.** (d)	**59.** (c)	**60.** (b)
61. (d)	**62.** (c)	**63.** (d)	**64.** (c)	

One or More Than One Options Correct

1. (a), (b), (c), (d) **2.** (b), (c) **3.** (a), (c) **4.** (c)

5. (a), (b), (c) **6.** (c) **7.** (c) **8.** (c)

9. (a), (b), (c), (d) **10.** (a), (b), (d) **11.** (b) **12.** (b)

13. (d) **14.** (a) **15.** (b)

Passage Comprehension Questions

Passage 1

1. (a) **2.** (c) **3.** (a)

Passage 2

1. (a) **2.** (b) **3.** (b) **4.** (c)

Passage 3

1. (d) **2.** (b) **3.** (c)

Passage 4

1. (b) **2.** (b) **3.** (d)

Passage 5

1. (b) **2.** (a) **3.** (c)

Passage 6

1. (a) **2.** (c)

Passage 7

1. (c) **2.** (b) **3.** (a) **4.** (b) **5.** (d)

Passage 8

1. (b) **2.** (d)

Passage 9

1. (c) **2.** (b) **3.** (a)

Passage 10

1. (a) **2.** (c) **3.** (a) **4.** (d)

Passage 11

1. (d) **2.** (b) **3.** (c) **4.** (c)

Passage 12

1. (b) **2.** (a)

Passage 13

1. (a)

Passage 14

1. (c) **2.** (a)

Passage 15

1. (c) **2.** (a) **3.** (a), (b)

Passage 16

1. (c) **2.** (a) **3.** (d)

Assertion & Reason

1. (b) **2.** (b) **3.** (a) **4.** (b) **5.** (a) **6.** (c)
7. (a)

True & False

1. T **2.** T **3.** F **4.** F **5.** T **6.** T **7.** T
8. T **9.** F **10.** F **11.** T **12.** F **13.** T **14.** F
15. T **16.** T **17.** F **18.** F **19.** T **20.** T

Fill in the Blanks

1. ambidentate **2.** cobalt **3.** porphyrin **4.** Denticity
5. ferrocene **7.** two, AgCl **8.** *cis*-, and *trans*
8. outer **9.** +1, $\sqrt{15}$ B.M. **10.** e_g **11.** t_{2g}
12. Δ_{oct} **13.** red **14.** low, large **15.** spontaneous
16. labile **17.** chelation **18.** EDTA **19.** CN^-
20. hypo ($Na_2S_2O_3$)

Matrix-Match Type Questions

I. Only One Correct

1. (A)—(2); (B)—(4); (C)—(3); (D)—(1);
2. (A)—(3); (B)—(1); (C)—(2); (D)—(5);
3. (A)—(3); (B)—(1); (C)—(2); (D)—(4);
4. (A)—(5); (B)—(1); (C)—(4); (D)—(2);
5. (A)—(4); (B)—(1); (C)—(2); (D)—(3);
6. (A)—(5); (B)—(6); (C)—(1); (D)—(2); (E)—(3)

II. One or More Correct

1. (A)—(1,3,5); (B)—(1,2,4); (C)—(1,3,4); (D)—(3);
2. (A)—(1,4); (B)—(1,2); (C)—(3); (D)—(1,4,5);

Integer Answer Types

Set I

Questions →	1	2	3	4	5	6	7	8
Answers →	0	2	2	3	4	6	1	0
	⓪	⓪	⓪	⓪	⓪	⓪	⓪	⓪
	①	①	①	①	①	①	①	①
	②	②	②	②	②	②	②	②
	③	③	③	③	③	③	③	③
	④	④	④	④	④	④	④	④
	⑤	⑤	⑤	⑤	⑤	⑤	⑤	⑤
	⑥	⑥	⑥	⑥	⑥	⑥	⑥	⑥
	⑦	⑦	⑦	⑦	⑦	⑦	⑦	⑦
	⑧	⑧	⑧	⑧	⑧	⑧	⑧	⑧
	⑨	⑨	⑨	⑨	⑨	⑨	⑨	⑨

Set II

Questions →	1	2	3	4	5	6	7	8	9	10
Answers →	3	4	4	5	9	4	6	2	2	4
	⓪	⓪	⓪	⓪	⓪	⓪	⓪	⓪	⓪	⓪
	①	①	①	①	①	①	①	①	①	①
	②	②	②	②	②	②	②	②	②	②
	③	③	③	③	③	③	③	③	③	③
	④	④	④	④	④	④	④	④	④	④
	⑤	⑤	⑤	⑤	⑤	⑤	⑤	⑤	⑤	⑤
	⑥	⑥	⑥	⑥	⑥	⑥	⑥	⑥	⑥	⑥
	⑦	⑦	⑦	⑦	⑦	⑦	⑦	⑦	⑦	⑦
	⑧	⑧	⑧	⑧	⑧	⑧	⑧	⑧	⑧	⑧
	⑨	⑨	⑨	⑨	⑨	⑨	⑨	⑨	⑨	⑨

19

Salt Analysis

"Salt-analysis is the "Magic-show" of chemists playing with colours."

Quick Points

- **Common-ion Effect** Ionisation of weak electrolyte or sparingly soluble salt is decreased in presence of common ion. It is called **common-ion effect.**
  - Ionisation of weak acid (say CH_3COOH) is decreased in presence of CH_3COONa (with CH_3COO^- as common ion)
  - Ionisation of weak base (say NH_4OH) is decreased in presence of NH_4Cl (with NH_4^+ as common ion)
  - Solubility of sparingly soluble salt (say AgCl) is decreased in presence of $AgNO_3$ (Ag^+ as common ion) or KCl (Cl^- as common ion)
- Solubility product of solute A_xB_y is represented as
$$K_{sp} = [A^{y+}]^x [B^{x-}]^y = x^x y^y s^{x+y}$$
where, s is the solubility of solute A_xB_y (in mol L^{-1})
For AgCl, $K_{sp} = s^2$
$Mg(OH)_2$, $K_{sp} = 4s^3$
$Al(OH)_3$, $K_{sp} = 27s^4$
$Fe_2(SO_4)_3$, $K_{sp} = 108s^5$
- Any solute AB is precipitated if
$$[A^+][B^-] > K_{sp}$$
- Salts have been divided into groups based on common-ion effect and solubility product.
- $$pH = -\log [H_3O^+]$$
$$pH + pOH = 14 = pK_w$$
$$K_w = [H_3O^+][OH^-] = 1 \times 10^{-14} \text{ at } 298 \text{ K}$$
- Species is said to be amphoteric if it acts as an acid as well as a base. It is amphiprotic if can donate H^+ ion as well as can accept H^+
$$\underset{\text{acid}}{H_2O} \rightleftharpoons H^+ + OH^- \; ; \quad \underset{\text{base}}{H_2O} + H^+ \rightleftharpoons H_3O^+$$

Learning Framework of the Chapter ...

- Physical Examination
- Analysis of Acid (Anion) Radicals
- Analysis of Basic (Cation) Radicals
- Dry Tests
- One Reagent for Testing Different Ions

Physical Examination

Solubility Rules for Common Compounds

The following solubility rules are not comprehensive, but they cover most of the compounds you encountered in general chemistry. Compounds that dissolve in water to the extent of approximately 0.02 mol per litre (0.02 M) are usually classified as "soluble" compounds, while those that are less soluble are classified as "insoluble" compounds. No gaseous or solid substances are infinitely soluble in water.

1. All common inorganic acids are soluble in water. Low molecular weight organic acids are soluble.
2. All common compounds of the group IA metal (Na, K, etc.) and the ammonium ion, NH_4^+, are soluble in water.
3. All common nitrates NO_3^-, acetates, CH_3COO^-, and perchlorates, ClO_4^-, are soluble in water.
4. (a) All common chlorides, Cl^-, are soluble in water except AgCl, Hg_2Cl_2, and $PbCl_2$.
   (b) The common bromides, Br^-, and iodides, I^-, show approximately the same solubility behaviour as chlorides, but there are some exceptions. As the halide ions (Cl^-, Br^-, I^-), increase in size, the solubilities of their slightly soluble compounds decrease. Although $HgCl_2$ is readily soluble in water, $HgBr_2$ is only slightly soluble and HgI_2 is even less soluble.
   (c) The solubilities of the pseudo-halide ions, CN^- (cyanide) and SCN^- (thiocyanate), are quite similar to those of the corresponding iodides. Additionally, both CN^- and SCN^- show strong tendencies to form soluble complex compounds.
5. All common sulphates, SO_4^{2-}, are soluble in water except $PbSO_4$, Hg_2SO_4, and $BaSO_4$; $CaSO_4$, and Ag_2SO_4 are sparingly soluble.
6. All common metal hydroxides are insoluble in water except those of the group IA metals and the lower members of the group IIA metals, beginning with $Ca(OH)_2$.
7. All common carbonates, CO_3^{2-}, phosphates, PO_4^{3-}, and arsenates, AsO_4^{3-}, are insoluble in water except those of the group IA metals and $NH_4^+ \cdot MgCO_3$ is fairly soluble.
8. All common sulphides, S^{2-} are insoluble in water except those of the group IA and group IIA metals and the ammonium ion. Solubility and colour of different salts are given in Table 19.1

Table 19.1 **Solubility and Colour of Different Salts**

Salt/Complex	Formula	Colour	(Soluble / Insoluble) in water
Ammonium carbonates	$(NH_4)_2CO_3$	white	soluble
Ammonium chloride	NH_4Cl	white	soluble
Ammonium nitrate	NH_4NO_3	white	soluble
Ammonium phosphate	$(NH_4)_3PO_4$	white	soluble
Ammonium oxalate	$(NH_4)_2C_2O_4$	white	soluble
Ammonium sulphate	$(NH_4)_2SO_4$	white	soluble
Ammonium dichromate	$(NH_4)_2Cr_2O_7$	orange (due to $Cr_2O_7^{2-}$ ion)	soluble
Ammonium molybdate	$(NH_4)_2MoO_4$	white	soluble in presence of NH_4OH
Arsenic(III) sulphide	As_2S_3	yellow	insoluble
Arsenic(V) sulphide	As_2S_5	yellow	insoluble
Arsenic(III) iodide	AsI_3	red	insoluble
Antimony sulphide	Sb_2S_3	orange	insoluble
Antimonyl chloride	SbOCl	orange	insoluble
Antimony iodide	SbI_3	red-yellow	insoluble
Bismuth sulphide	Bi_2S_3	black	insoluble
Bismuth iodide	BiI_3	black	insoluble
Bismuth hydroxide	$Bi(OH)_3$	white	insoluble
Bismuthyl chloride	BiOCl	white	insoluble
Bismuth(V) oxide	Bi_2O_5	brown	insoluble
Bismuthyl iodide	BiOI	red-orange	insoluble
Barium chloride	$BaCl_2$	white	soluble

Salt/Complex	Formula	Colour	$\left(\frac{\text{Soluble}}{\text{Insoluble}}\right)$ in water
Barium nitrate	$Ba(NO_3)_2$	white	soluble
Barium sulphate	$BaSO_4$	white	insoluble
Barium chromate	$BaCrO_4$	yellow (due to CrO_4^{2-})	insoluble (also in CH_3COOH) but soluble in dil. HCl
Chromium(III) oxide	Cr_2O_3	green	insoluble
Chromium sulphate	$Cr_2(SO_4)_3$	green	partially soluble
Chromium chloride	$CrCl_3$	green	soluble
Chromium chloride (mineral)	$CrCl_3 \cdot 6H_2O$	violet	soluble
Chromyl chloride	CrO_2Cl_2	orange red gas	soluble giving H_2CrO_4
Chromium hydroxide	$Cr(OH)_3$	dark green	insoluble
Chromate ion	CrO_4^{2-}	yellow	soluble (salts of alkali and ammonium ion)
Copper(I) sulphide (cupric sulphide)	Cu_2S	black	insoluble
Copper(II) sulphide (cupric sulphide)	CuS	black	insoluble
Hydrated copper (II) sulphate	$CuSO_4 \cdot 5H_2O$	blue	soluble
Anhydrous copper(II) sulphate	$CuSO_4$	white	soluble (colour changes to blue)
Copper(II) hydroxide	$Cu(OH)_2$	sky blue	insoluble
Copper(II) chloride (cupric chloride)	$CuCl_2$	bluish green	soluble
Copper(I) oxide (cuprous oxide)	Cu_2O	red	insoluble
Cuprous iodide	Cu_2I_2	white	insoluble
Cuprous thiosulphate	$Cu_2S_2O_3$	white	soluble
Tetrammine copper(II) sulphate	$[Cu(NH_3)_4]SO_4$	deep blue	soluble (insoluble in ethanol)
Cupric ferrocyanide	$Cu_2[Fe(CN)_6]$	brown	insoluble
Cupric oxide, (copper(II) oxide)	CuO	black	insoluble
Copper(II) cyanide	$Cu(CN)_2$	yellow	insoluble
Cpper(I) cyanide	$CuCN$	white	insoluble
Copper(II) thiocyanate	$Cu(SCN)_2$	black	insoluble
Copper(I) thiocyanate	$CuSCN$	white	insoluble
Cobalt sulphide	CoS	black	insoluble
Potassium cobalti nitrite	$K_3[Co(NO_2)_6]$	yellowish orange	insoluble
Cobalt(II) chloride dihydrate	$CoCl_2 \cdot 2H_2O$	brown	soluble
Cobalt(II) chloride tetrahydrate	$CoCl_2 \cdot 4H_2O$	brown	soluble
Cobalt(II) chloride hexahydrate	$CoCl_2 \cdot 6H_2O$	red	soluble
Cobalt(II) cyanide	$Co(CN)_2$	buff-coloured	insoluble
Cadmium oxide	CdO	brown	insoluble
Cadmium sulphide	CdS	yellow	insoluble
Cadmium carbonate	$CdCO_3$	white	partly soluble
Cadmium hydroxide	$Cd(OH)_2$	white	insoluble
Dichromate ion	$Cr_2O_7^{2-}$	orange	soluble
Zinc sulphide	ZnS	white	insoluble
Zinc hydroxide	$Zn(OH)_2$	white	insoluble, soluble in acid as Zn^{2+}, soluble in alkali as ZnO_2^{2-}
Zinc oxide	ZnO	white	(-do-)
Zinc carbonate	$ZnCO_3$	white	insoluble
Lead sulphite	$PbSO_3$	white	insoluble
Lead sulphate	$PbSO_4$	white	insoluble
Lead chloride	$PbCl_2$	white	soluble in hot water

Salt/Complex	Formula	Colour	(Soluble / Insoluble) in water
Lead bromide	$PbBr_2$	white	soluble in hot water
Lead oxide (litharge)	PbO	orange-red	insoluble
Lead double oxide	$2PbO \cdot PbO_2 (Pb_3O_4)$	red	insoluble
Lead dioxide	PbO_2	blackish brown	insoluble
Lead carbonate	$PbCO_3$	white	insoluble
Lead chromate (basic)	$PbO \cdot PbCrO_4$	red	insoluble
Lead chromate	$PbCrO_4$	yellow	insoluble (soluble in HNO_3 as $PbCr_2O_7$ and in NaOH as $[Pb(OH)_4^{2-}]$
Lead iodide	PbI_2	yellow	insoluble
Lead sulphide	PbS	black	insoluble
Lead hydroxide	$Pb(OH)_2$	white	insoluble (soluble in NaOH as $[Pb(OH)_4]^{2-}$
Lead cyanide	$Pb(CN)_2$	white	insoluble
Lead carbonate (basic)	$PbO \cdot PbCO_3$	white	insoluble
Lead phosphate	$Pb_3(PO_4)_2$	white	insoluble
Manganese dioxide	MnO_2	brownish-red	insoluble
Manganese sulphide	MnS	buff coloured	insoluble
Manganese chloride	$MnCl_2$	light pink	soluble
Manganese hydroxide	$Mn(OH)_2$	light pink	insoluble
Permanganate (potassium)	$KMnO_4$	violet (purple)	soluble
Manganate (potassium)	K_2MnO_4	green	soluble
Manganese oxide	MnO	green	insoluble
Manganese carbonate	$MnCO_3$	brown	insoluble
Calcium chloride	$CaCl_2$	white	soluble
Calcium carbonate	$CaCO_3$	white	insoluble
Calcium oxalate	CaC_2O_4	white	insoluble
Calcium nitrate	$Ca(NO_3)_2$	white	soluble
Calcium bicarbonate	$Ca(HCO_3)_2$	white	soluble
Calcium sulphate	$CaSO_4$	white	insoluble
Magnesium chloride	$MgCl_2$	white	soluble
Magnesium ammonium phosphate	$Mg(NH_4)PO_4$	white	insoluble
Iron(III) oxide	Fe_2O_3	red	insoluble
Iron(II) carbonate	$FeCO_3$	green	soluble
Iron(II) chloride	$FeCl_2$	green	soluble
Iron(II) sulphide	FeS	black	insoluble
I	FeS_2	yellow	insoluble
Iron(II) phosphate	$Fe_3(PO_4)_2$	brown	insoluble
Iron(III) hexacyanoferrate(II) (Prussian blue) ion	$Fe^{III}[Fe^{II}(CN)_6]^-$	blue	soluble, $Fe_4[Fe(CN)_6]_3$ insoluble
Iron(II) hexacyanoferrate(III) ion (Turnbull's blue)	$Fe^{II}[Fe(CN)_6]^-$	blue	soluble
Complex of S^{2-} with sodium nitroprusside	$Na_2[Fe(CN)_5NOS]$	purple	soluble
Magnetite	Fe_3O_4	red	insoluble
	$[Fe(CNS)]^{2+}$	blood red	soluble
Iron(III) carbonate	$Fe_2(CO_3)_3$	red	insoluble
Iron(III) fluoride	FeF_3	white	insoluble
Iron(II) fluoride	FeF_2	green	insoluble

Salt/Complex	Formula	Colour	(Soluble / Insoluble) in water
Iron(III) oxalate	$Fe_2(C_2O_4)_3$	blackish-red	insoluble
Iron(III) hydroxide	$Fe(OH)_3$	reddish-brown	insoluble
Mercury(II) sulphide	HgS	black	insoluble
Mercury(I) oxide	Hg_2O	black	insoluble
Mercury(II) chloride (corrosive sublimate)	$HgCl_2$	white	soluble
Mercury(I) chloride (calomel)	Hg_2Cl_2	white	insoluble
Mercury(II) bromide	$HgBr_2$	white	
	$HgNH_2Cl \cdot HgO$	black	insoluble
Mercury(II) oxide	HgO	yellow	insoluble
Mercury(II) iodide	HgI_2	reddish orange	insoluble
Iodide of Millon's base	$H_2N{-}Hg{-}O{-}Hg{-}I$	brown	insoluble
Mercury(I) chromate	Hg_2CrO_4	red	insoluble
Mercury(I) carbonate	Hg_2CO_3	yellowish white	insoluble
Mercury(I) hydrogen phosphate	$HgHPO_4$	white	insoluble
Mercury(II) cyanide	$Hg(CN)_2$	white	soluble
Nickel sulphide	NiS	black	insoluble
Nickel chloride	$NiCl_2$	green	soluble
Nickel hydroxide	$Ni(OH)_2$	green	insoluble
Nickel(III) oxide	Ni_2O_3	black	insoluble
Silver chloride	$AgCl$	white	insoluble (soluble in *aq.* NH_3 as $[Ag(NH_3)_2]Cl$)
Silver bromide	$AgBr$	yellow	insoluble
Silver iodide	AgI	yellow	insoluble
Silver sulphide	Ag_2S	black	insoluble
Silver oxide	Ag_2O	black/brown	insoluble
Silver thiosulphate	$Ag_2S_2O_3$	white	insoluble
Silver nitrate	$AgNO_3$	white crystals	soluble
Silver phosphate	Ag_3PO_4	yellow	insoluble
Silver carbonate	Ag_2CO_3	yellowish white	insoluble
Silver arsenate	Ag_3AsO_4	red	insoluble
Silver chromate	Ag_2CrO_4	red	insoluble
Silver sulphite	Ag_2SO_3	white	insoluble
Silver arsenite	Ag_3AsO_3	yellow	insoluble
Silver arsenate	Ag_3AsO_4	brownish red	insoluble
Tin(IV) sulphide	SnS_2	yellow	insoluble
Tin(II) chloride	$SnCl_2$	white	soluble
Tin(II) sulphide	SnS	brown	insoluble
Tin(II) iodide	SnI_2	red	insoluble
Strontium carbonate	$SrCO_3$	white	insoluble
Strontium sulphate	$SrSO_4$	white	(insoluble in dilute mineral acid)
Strontium sulphite	$SrSO_3$	white	insoluble
Strontium oxalate	SrC_2O_4	white	insoluble in CH_3COOH but soluble in mineral acids

A Convenient Classification of Bases

1. Strong soluble bases	Bases that are soluble in water and are completely, or nearly completely, ionised (dissociated) in dilute aqueous solutions. The only *common* strong soluble bases are the hydroxides of the group IA metals and the hydroxides of the lower members of the group IIA metals. $Ca(OH)_2$ and $Sr(OH)_2$ have limited solubilities but are classified as strong soluble bases. Most metal hydroxides are "insoluble" in water.
2. Weak soluble bases	Bases that are soluble in water but which ionise only slightly in dilute aqueous solutions. Aqueous ammonia and its low molecular weight organic derivatives, the amines, are common weak soluble bases.
3. Insoluble bases	Bases that are very slightly soluble in water. As a matter of common usage, very slightly soluble bases are called "insoluble" bases. With the exception of those listed as strong soluble bases, all common low oxidation state metal hydroxide are "insoluble".
4. Amphoteric hydroxides (Bases)	Insoluble bases that react with both acids and strong soluble bases to complex ions of the general formula $[M(OH)_n]^{x-}$ where "*n*" varies from 3 to 6 and "*x*" is either 1 or 2 and metalloids which form amphoteric hydroxides are listed together with the formula for the amphoteric hydroxides and the complex ions formed in an excess of strong soluble bases.

Metal ion or metalloid	Insoluble amphoteric hydroxide	Complex ion formed in an excess of a strong soluble base
Be^{2+}	$Be(OH)_2$	$[Be(OH)_4]^{2-}$
Al^{3+}	$Al(OH)_3$	$[Al(OH)_4]^-$
Cr^{3+}	$Cr(OH)_3$	$[Cr(OH)_4]^-$
Zn^{2+}	$Zn(OH)_2$	$[Zn(OH)_4]^{2-}$
Sn^{2+}	$Sn(OH)_2$	$[Sn(OH)_3]^-$
Sn^{4+}	$Sn(OH)_4$	$[Sn(OH)_6]^{2-}$
Pb^{2+}	$Pb(OH)_2$	$[Pb(OH)_4]^{2-}$
As^{3+}	$As(OH)_3$	$[As(OH)_4]^-$
Sb^{3+}	$Sb(OH)_3$	$[Sb(OH)_4]^-$
Si^{4+}	$Si(OH)_4$	SiO_4^{4-} and SiO_3^{2-}
Co^{2+}	$Co(OH)_2$	$[Co(OH)_4]^{2-}$
Cu^{2+}	$Cu(OH)_2$	$[Cu(OH)_4]^{2-}$

Metal Hydroxides that are Soluble in Excess of Aqueous Ammonia

Some metal hydroxides dissolve in an excess of aqueous ammonia to form soluble complex compounds. The common examples are listed in the Table 19.2.

Table 19.2 **Common Metal Ions that Forms Ammine Complexes with an Excess of Aqueous Ammonia**

Metal ions	Insolubles hydroxides formed by limiting amount *aq.* NH_3	Complex ion formed by excess *aq.* NH_3
Co^{2+}(*)	$Co(OH)_2$	$[Co(NH_3)_6]^{2+}$
Co^{3+}	$Co(OH)_3$	$[Co(NH_3)_6]^{3+}$
Ni^{2+}	$Ni(OH)_2$	$[Ni(NH_3)_6]^{2+}$
Cu^+(**)	$CuOH \longrightarrow \frac{1}{2} Cu_2O$	$[Cu(NH_3)_2]^+$
Cu^{2+}(*)	$Cu(OH)_2$	$[Cu(NH_3)_4]^{2+}$
Ag^+(**)	$AgOH \longrightarrow \frac{1}{2} Ag_2O$	$[Ag(NH_3)_2]^+$
Zn^{2+}	$Zn(OH)_2$	$[Zn(NH_3)_4]^{2+}$
Cd^{2+}	$Cd(OH)_2$	$[Cd(NH_3)_4]^{2+}$
Hg^{2+}	$Hg(OH)_2$	$[Hg(NH_3)_4]^{2+}$

*$Co(OH)_2$ and $Cu(OH)_2$ are only slightly amphoteric, *i.e.*, a very large excess of strong soluble base is required to dissolve small amounts of these insoluble hydroxides.

**CuOH and AgOH are unstable and decompose to the corresponding oxides.

The common ions of Rh, Ir, Pd, Pt, and Au show similar behaviour, but since these ions are seldom encountered in elementary courses, they are not included here. Note that these metals and those in the table are all members of the Co, Ni, Cu, and Zn families of the Periodic Table.

Colour Change on Heating

Certain oxides change colour on heating and this fact can be used to identify salt.

Oxides	Colour	
	in cold	on heating
ZnO	white	yellow
SnO_2, Bi_2O_3	yellow	yellowish brown
Fe_2O_3	brown	black/red
PbO	yellow	

Observation	Inference
(i) Sublimate formed with smell of NH_3	NH_4^+ salts
(ii) Sublimation formed without smell of NH_3	Hg^{2+} salts
(iii) Brown fumes	Some nitrates

Analysis of Acid (Anion) Radicals

Salt or mixture is treated with dil. H_2SO_4 and also with conc. H_2SO_4 separately and by observing the types of gases evolved, confirmatory tests of anions are performed.

Observation with Dilute H_2SO_4

S. no.	Observation	Acid radical	Confirmatory test*
1.	Brisk effervescence with evolution of colourless and odourless gas	CO_3^{2-} (carbonate)	Gas turns lime water (or baryta water) milky but milkyness disappears on passing gas in excess (A)*
2.	Brown fumes	NO_2^- (nitrite)	Add KI and starch solution—blue colour (B)*
3.	Smell of rotten eggs (H_2S smell) on heating	S^{2-} (sulphide)	Gas turns lead acetate paper black (C) Sodium carbonate extract (SE)† + sodium nitroprusside solution — purple colour (D)
4.	Colourless gas with pungent smell of burning sulphur	SO_3^{2-} (sulphite)	Gas turns acidified $K_2Cr_2O_7$ solution green [(E), different from CO_3^{2-}] since gas also turns lime water milky (F)
5.	Solution gives smell of vinegar	CH_3COO^- (acetate)	Aqueous solution + neutral $FeCl_3$ → blood red colour (G) changing to brownish red ppt on boiling
6.	White or yellowish white turbidity on warming due to colloidal sulphur	$S_2O_3^{2-}$ (thiosulphate)	Aqueous solution + $AgNO_3$ → white ppt changing to black (H) on warming

*(A), (B) .... indicates reaction number given separately.

† S E is prepared by boiling mixture and Na_2CO_3 (1 : 3) in H_2O and filtering. Filtrate has soluble sodium salt of anions.

Carbonate

$$\text{(A)}\ Na_2CO_3 + H_2SO_4 \longrightarrow Na_2SO_4 + H_2O + CO_2$$

$$\underset{\text{lime water}}{Ca(OH)_2} + CO_2 \longrightarrow \underset{\text{milky}}{CaCO_3\downarrow} + H_2O$$

$$\underset{\text{baryta water}}{Ba(OH)_2} + CO_2 \longrightarrow \underset{\text{milky}}{BaCO_3\downarrow} + H_2O$$

$$CaCO_3 + H_2O + CO_2 \longrightarrow \underset{\text{soluble}}{Ca(HCO_3)_2}$$

- Soluble carbonate gives white precipitate of silver carobnate with silver nitrate

$$2Ag^+ + CO_3^{2-} \longrightarrow \underset{\text{white ppt}}{Ag_2CO_3\downarrow}$$

The precipitate is soluble in nitric acid and in ammonia. The precipitate becomes yellow or brown upon addition of excess reagent owing to the forming of silver oxide; the same happens if the mixture is boiled:

$$Ag_2CO_3\downarrow \xrightarrow{\Delta} Ag_2O\downarrow + CO_2\uparrow$$

DEEP Focus

- Aqueous carbonate solution is alkaline due to hydrolysis (with H_2O)

$$CO_3^{2-} + 2H_2O \longrightarrow H_2CO_3 + 2OH^-$$

Alkaline solution turns pink when a drop of phenolphthalein is added to it.

If CO_2 (liberated from the reaction of given carbonate with dilute H_2SO_4) is passed into above pink-coloured solution, then it is decolourised by HCO_3^- formed with CO_2.

$$CO_2 + CO_3^{2-} + H_2O \longrightarrow 2HCO_3^-$$

Nitrite

$$\text{(B)}\ 2NaNO_2 + H_2SO_4 \longrightarrow Na_2SO_4 + 2HNO_2$$

$$HNO_2 \longrightarrow NO\ (\text{colourless})$$

$$2NO + O_2(\text{air}) \longrightarrow 2NO_2\ (\text{brown})$$

$$2KI + H_2SO_4 + 2HNO_2 \longrightarrow K_2SO_4 + 2H_2O + 2NO + I_2$$

$$I_2 + \text{starch} \longrightarrow \text{blue colour}$$

When the nitrite solution is added carefully to a saturated solution of iron(II) sulphate acidified with acetic acid or with dilute sulphuric acid, a browk ring, due to the compound $[Fe \cdot NO]SO_4$, is formed at the junction of two liquids.

$$NO_2^- + CH_3COOH \longrightarrow HNO_2 + CH_3COO^-$$

$$3HNO_2 \longrightarrow H_2O + HNO_3 + 2NO$$

$$Fe^{2+} + SO_4^{2-} + NO \longrightarrow [Fe \cdot NO]SO_4$$

Iodides, bromides, coloured ions, and anions that give coloured compounds with iron(II) ions must be absent.

NO_2^- solution also decolourises MnO_4^-/H^+

$$5NO_2^- + 2MnO_4^- + 6H^+ \longrightarrow 5NO_3^- + 2Mn^{2+} + 3H_2O$$

NO_2^- in CH_3COOH solution on treatment with thiourea forms SCN^- which gives red colour with $FeCl_3$ and HCl.

$$\underset{\text{thiourea}}{(NH_2)_2CS} + HNO_2 \longrightarrow N_2 + H^+ + SCN^- + 2H_2O$$

$$FeCl_3 + SCN^- \longrightarrow \underset{\text{blood red}}{[Fe(SCN)]Cl_2} + Cl^-$$

Sulphide

(C) $$Na_2S + H_2SO_4 \longrightarrow H_2S + Na_2SO_4$$

$$H_2S + (CH_3COO)_2Pb \longrightarrow \underset{\text{(black)}}{PbS} + 2CH_3COOH$$

(D) $$Na_2S + \underset{\text{sodium nitroprusside}}{Na_2[Fe(CN)_5NO]} \longrightarrow \underset{\text{(purple)}}{Na_4[Fe(CN)_5NOS]}$$

Sulphite

(E) $$Na_2SO_3 + H_2SO_4 \xrightarrow{\Delta} Na_2SO_4 + H_2O + SO_2$$

$$Cr_2O_7^{2-} + 3SO_2 + 2H^+ \longrightarrow \underset{\text{(green)}}{2Cr^{3+}} + 3SO_4^{2-} + H_2O$$

Acidified $KMnO_4$ solution is also decolourised by SO_3^{2-}

$$5SO_3^{2-} + \underset{\text{dark pink}}{2MnO_4^-} + 6H^+ \longrightarrow \underset{\text{colourless}}{2Mn^{2+}} + 5SO_4^{2-} + 3H_2O$$

(F) $$Ca(OH)_2 + SO_2 \longrightarrow \underset{\text{(milky)}}{CaSO_3}$$

Acetate

(G) $$6CH_3COO^- + 3Fe^{3+} + 2H_2O \longrightarrow \underset{\text{blood red}}{[Fe_3(OH)_2(CH_3COO)_6]^+} + 2H^+$$

$$[Fe_3(OH)_2(CH_3COO)_6]^+ + 4H_2O \xrightarrow{\text{boil}} \underset{\text{brownish-red ppt}}{3[Fe(OH)_2CH_3COO]\downarrow} + 3CH_3COOH + H^+$$

Thiosulphate

$$S_2O_3^{2-} + 2H^+ \longrightarrow S\downarrow + SO_2\uparrow + H_2O$$

SO_2 gas turns $Cr_2O_7^{2-}/H^+$ solution green.

(H) $$Na_2S_2O_3 + 2AgNO_3 \longrightarrow \underset{\text{white}}{Ag_2S_2O_3\downarrow} + 2NaNO_3$$

$$Ag_2S_2O_3 + H_2O \xrightarrow{\text{warm}} \underset{\text{black}}{Ag_2S\downarrow} + H_2SO_4$$

I_2 solution (in KI) is decolourised by thiosulphate solution

$$I_3^- + 2S_2O_3^{2-} \longrightarrow S_4O_6^{2-} + 3I^-$$

$S_2O_3^{2-}$ gives white precipitate with $Pb(NO_3)_2$ or $(CH_3COO)_2Pb$; white precipitate changes to black on boiling :

$$S_2O_3^{2-} + Pb^{2+} \longrightarrow \underset{\text{white}}{PbS_2O_3\downarrow}$$

$$PbS_2O_3 + H_2O \longrightarrow \underset{\text{black}}{PbS\downarrow} + 2H^+ + SO_4^{2-}$$

When $FeCl_3$ solution is added to $S_2O_3^{2-}$ solution, a dark-violet colour appears due to formation of the complex

$$2S_2O_3^{2-} + Fe^{3+} \longrightarrow \underset{\text{dark-violet (dithiosulphatoiron (III) complex)}}{[Fe(S_2O_3)_2]^-}$$

But on standing colour disappears rapidly due to the following change

$$[Fe(S_2O_3)_2]^- + Fe^{3+} \longrightarrow \underset{\text{colourless}}{2Fe^{2+}} + \underset{\text{tetrathionate ion}}{S_4O_6^{2-}}$$

Observation with Concentrated H_2SO_4

S. no.	Observation	Acid radical	Confirmatory test
1.	Colourless pungent gas giving white fumes with *aq* NH_4OH	Cl^- (chloride)	Add MnO_2 in the same test tube and heat — pale green Cl_2 gas (*A*), SE + HNO_3 + $AgNO_3$ solution $\longrightarrow$ white ppt soluble in *aq* NH_3 (*B*) Chromyl chloride test (*C*)
2.	Reddish brown fumes	Br^- (bromide)	Add MnO_2 and heat — yellowish brown Br_2 gas (*D*). SE + HNO_3 + $AgNO_3$ solution — pale yellow ppt partially soluble *aq* NH_3 (*E*). Layer test (*F*)
3.	Violet pungent vapours turning starch paper blue	I^- (iodide)	SE + HNO_3 + $AgNO_3$ $\longrightarrow$ yellow ppt insoluble in *aq* NH_3 (*G*) Layer test (*H*)
4.	Brown pungent fumes intensified by the addition of Cu-turnings	NO_3^- (nitrate)	Ring test (I)
5.	Colourless gases turning lime water milky and burning with blue flame	$C_2O_4^{2-}$ (oxalate)	Acidified $KMnO_4$ solution is decolorised (*J*) SE + CH_3COOH + $CaCl_2$ solution — white ppt decolorising acidified $KMnO_4$ solution (*K*)

Chloride

(A) $$KCl + \underset{\text{conc.}}{H_2SO_4} \longrightarrow KHSO_4 + HCl$$

$$HCl + NH_3(aq) \longrightarrow \underset{\text{(white fumes)}}{NH_4Cl}$$

$$4HCl + MnO_2 \xrightarrow{\Delta} MnCl_2 + Cl_2\uparrow + 2H_2O$$

(B) $$KCl + AgNO_3 \longrightarrow \underset{\text{white ppt}}{AgCl\downarrow} + KNO_3$$

$$AgCl + 2NH_3(aq) \longrightarrow \underset{\text{soluble}}{[Ag(NH_3)_2]Cl}$$

(C) Chromyl-Chloride Test

Chloride + $K_2Cr_2O_7$ (solid) + conc. $H_2SO_4 \xrightarrow{\text{heat}}$ reddish brown vapours of chromyl-chloride (CrO_2Cl_2). Pass these vapours into NaOH solution, when yellow Na_2CrO_4 solution is formed. On adding CH_3COOH and $(CH_3COO)_2Pb$, yellow ppt of lead chromate ($PbCrO_4$) is formed.

$$KCl + \underset{\text{conc.}}{H_2SO_4} \xrightarrow{\Delta} KHSO_4 + HCl$$

$$K_2Cr_2O_7 + 2H_2SO_4 \xrightarrow{\Delta} 2KHSO_4 + 2CrO_3 + H_2O$$
conc.

$$CrO_3 + 2HCl \longrightarrow CrO_2Cl_2 + H_2O$$

$$CrO_2Cl_2 + 4NaOH \longrightarrow Na_2CrO_4 + 2NaCl + 2H_2O$$

$$Na_2CrO_4 + (CH_3COO)_2Pb \longrightarrow \underset{\text{yellow ppt}}{PbCrO_4\downarrow} + 2CH_3COONa$$

Bromide

(D) $KBr + H_2SO_4 \xrightarrow{\Delta} KHSO_4 + HBr$ (conc.)

$$4HBr + MnO_2 \xrightarrow{\Delta} Br_2 + 2H_2O + MnBr_2$$

(E) $NaBr + AgNO_3 \longrightarrow \underset{\text{pale yellow ppt}}{AgBr\downarrow} + NaNO_3$

$$AgBr + 2NH_3(aq) \longrightarrow \underset{\text{partially soluble}}{[Ag(NH_3)_2]Br}$$

(F) Layer Test

SE + Cl_2 water + $CHCl_3 \xrightarrow{\text{shake}}$ yellowish orange colour in $CHCl_3$ layer (CS_2 or CCl_4 can be taken instead of $CHCl_3$)

$$2NaBr + Cl_2 \longrightarrow 2NaCl + \underset{\substack{\text{orange yellow}\\ \text{soluble in } CHCl_3}}{Br_2}$$

In case of I^-, violet colour of I_2 in $CHCl_3$ layer (H)

$$2NaI + Cl_2 \longrightarrow 2NaCl + I_2 \text{ (violet)}$$

Iodide

(G) $KI + H_2SO_4 \xrightarrow{\Delta} KHSO_4 + HI$ (conc.)

$$2HI + H_2SO_4 \longrightarrow I_2 \text{ (violet)} + 2H_2O + SO_2$$

(H) Layer Test See (F) above.

Nitrate

$NaNO_3 + H_2SO_4 \longrightarrow NaHSO_4 + HNO_3$ (conc.)

$$4HNO_3 \longrightarrow \underset{\text{brown fumes}}{4NO_2} + O_2 + 2H_2O$$

$$Cu + 4HNO_3 \longrightarrow Cu(NO_3)_2 + 2NO_2 + 2H_2O$$

(I) Ring Test To water extract (all NO_3^- are water soluble) add freshly prepared $FeSO_4$ solution and then conc. H_2SO_4 carefully by the side of the test-tube. A dark brown ring of $[Fe(H_2O)_5NO]^{2+}SO_4^{2-}$ at the interface between the two liquids is formed.

$$2NaNO_3 + H_2SO_4 \longrightarrow 2NaHSO_4 + 2HNO_3$$

$$2HNO_3 + 6FeSO_4 + 3H_2SO_4 \longrightarrow 3Fe_2(SO_4)_3 + 2NO + 4H_2O$$

$$[Fe(H_2O)_6]SO_4 + NO \longrightarrow [Fe(H_2O)_5NO]^{2+}SO_4^{2-} + H_2O$$

DEEP Focus

The colour is due to charge transfer. This complex formally contains Fe (+I) and NO^+ since Fe(II) gains one electron from NO which changes to NO^+. Its magnetic moment is 3.87 B.M. confirming the presence of three unpaired electrons. Nitrites give the brown colour before H_2SO_4 is added.

$$NO \longrightarrow NO^+ + e^-$$

$$Fe^{2+} + e^- \longrightarrow Fe^+$$

⥮	1	1	1	1

⥮	⥮	1	1	1

The ring test is unreliable in the presence of bromide, iodide, nitrite, chlorate and chromate, hence must be removed before ring-test of NO_3^- is performed.

Oxalate

$$Na_2C_2O_4 + H_2SO_4 \longrightarrow Na_2SO_4 + H_2O + CO + CO_2$$

CO burns with blue flame and CO_2 turns lime water milky.

(J) $5C_2O_4^{2-} + \underset{\text{violet}}{2MnO_4^-} + 16H^+ \longrightarrow 10CO_2 + \underset{\text{colourless}}{2Mn^{2+}} + 8H_2O$

(K) $CaCl_2 + Na_2C_2O_4 \longrightarrow \underset{\text{white ppt}}{CaC_2O_4\downarrow} + 2NaCl$

CaC_2O_4 decolourises acidified $KMnO_4$.

Specific Tests in Solution

(A) Sulphate

To SE add dil. HCl (to decompose CO_3^{2-} until reaction ceases). Add $BaCl_2$ solution. White ppt insoluble in conc. HNO_3.

$$BaCl_2 + Na_2SO_4 \longrightarrow \underset{\text{white ppt.}}{BaSO_4\downarrow} + 2NaCl$$

$Hg(NO_3)_2$ gives yellow precipitate of basic mercury (II) sulphate

$$SO_4^{2-} + 3Hg^{2+} + 2H_2O \longrightarrow \underset{\text{yellow}}{HgSO_4 \cdot 2HgO} + 4H^+$$

(B) Borate

Ignite the mixture containing borate, conc. H_2SO_4 and ethanol in a china-dish with a burning splinter—green edged flame of ethyl borate.

$2Na_3BO_3 + 3H_2SO_4 \longrightarrow 2H_3BO_3 + 3Na_2SO_4$ (conc.)

$$H_3BO_3 + \underset{\text{ethanol}}{3C_2H_5OH} \xrightarrow{\Delta} \underset{\substack{\text{ethyl borate}\\ \text{burns with green flame}\\ \text{(volatile)}}}{(C_2H_5O)_3B} + 3H_2O$$

In presence of Cu^{2+}, perform this test in a test-tube since Cu^{2+} salts are not volatile.

$AgNO_3$ gives white precipitate of silver metaborate $AgBO_2$ from fairly concentrated borax solution, soluble in dilute *aq.* NH_3 and in CH_3COOH. On boiling the precipitate with H_2O, a brown precipitate of Ag_2O is obtained due to hydrolysis.

$$B_4O_7^{2-} + 4Ag^+ + H_2O \longrightarrow \underset{\text{white}}{4AgBO_2\downarrow} + 2H^+$$

$$2AgBO_2 + 3H_2O \longrightarrow Ag_2O\downarrow + 2H_3BO_3$$

Barium salts also give white precipitate of $Ba(BO_2)_2^-$ barium metaborate.

(C) Phosphate

SE + HNO_3 + ammonium molybdate solution. Heat; yellow crystalline ppt confirms PO_4^{3-}.

$$Na_3PO_4 + 12(NH_4)_2MoO_4 + 24HNO_3 \xrightarrow{\Delta} \underset{\text{yellow ppt}}{(NH_4)_3PO_4\cdot12MoO_3} + 21NH_4NO_3 + 3NaNO_3 + 12H_2O$$

Arsenate ion also gives this test. Hence, presence of phosphate should also be checked after group II.

(D) Fluoride

Sand + salt (F^-) + conc. H_2SO_4; heat and bring a water-wetted rod in contact with vapours at the mouth of the test-tube. A white deposit on the rod shows the presence of F^-.

$$NaF + H_2SO_4 \xrightarrow{\Delta} NaHSO_4 + HF$$

$$SiO_2 + 4HF \xrightarrow{\Delta} SiF_4 + 2H_2O$$

$$3SiF_4 + 4H_2O \longrightarrow 2H_2SiF_6 + \underset{\text{white (silicic acid)}}{H_4SiO_4}$$

Target Practice 1

1. It is experimental facts that all common inorganic acids dissolve in water. What are the ions formed if acid is HCl?

....

....

2. Specify the colour and soluble/insoluble character in water of the following

	Salt	Colour	Soluble/Insoluble
(i)	CuCl	....	....
(ii)	$CuCl_2$	....	....
(iii)	KI	....	....
(iv)	I_2	....	....
(v)	I_2 in KI	....	....
(vi)	KNO_3	....	....

3. Specify the cases in which soluble nature of the salt changes?

(a) KCN is added to AgCl (insoluble) ....

(b) NH_3 is added to AgCl (insoluble) ....

(c) KI is added to HgI_2 (insoluble) ....

(d) H_2O is added $BiCl_3$ (soluble) in HCl ....

(e) H_2O is added to $SbCl_3$ (soluble) in HCl ....

4. Consider following activity with salt X

$X + H_2O \longrightarrow$ Aqueous solution turns red litmus blue

(+ phenolphthalein)

$X \xrightarrow{+\text{ dil. }H_2SO_4}$ gas $\longrightarrow$ pink $\longrightarrow$ colourless

X contains (A) .... as acid radicals.

Solution of X is (B) ....

Formation of pink colour with phenolphthalein indicates that aqueous solution of X is (C) ....

Colourless solution is formed due to (D) ....

5. MnO_4^-/H^+ is decolourised by radicals

(i) ....

(ii) ....

(iii) ....

6. Two different test tubes contain CH_3COO^- and $S_2O_3^{2-}$ (thiosulphate). Specify one reagent which can detect both.

Reagent ....

	Product	Colour
I. $S_2O_3^{2-}$	....	....
II. CH_3COO^-	....	....

7. Write the formulae of

(a) Chromyl chloride

....

(b) Brown-coloured complex (ring) in nitrate test

....

(c) AgCl in the form of complex with aqueous NH_3

....

8. Brown-coloured ring is formed in nitrate test

(a) Write EC. of iron in the complex

....

(b) What is EAN of iron?

....

(c) What is magnetic moment of iron?

....

(d) What is charge on NO?

....

9. Take solution oxalate in a test-tube and heat it strongly. Take a burning match-stick at the mouth of the test-tube. You observe that a gas is buring at mouth of the test tube but on further heating gas is put-off itself.

(a) gas burns due to ....

(b) gas is put-off due to ....

Answers

1. H_3O^+ and Cl^-
2. (i) White insoluble
   (ii) Blue soluble
   (iii) White soluble
   (iv) Violet insoluble
   (v) Brown (I_3^-) soluble
   (vi) White soluble
3. (a) AgCl changes to $K[Ag(CN)_2]$ (soluble)
   (b) AgCl change to $[Ag(NH_3)_2]Cl$ (soluble)
   (c) HgI_2 changes to K_2HgI_4 (soluble)
   (d) $BiCl_3$ is hydrolysed to BiOCl (insoluble)
   (e) $SbCl_3$ is hydrolysed to SbOCl (insoluble)
4. (A) CO_3^{2-} (B) Basic (C) Alkaline
   (D) If CO_2 (CO_3^{2-} + dil. H_2SO_4) is passed into above pink-coloured solution, HCO_3^- formed decolourises it
   $$CO_3^{2-} + CO_2 + H_2O \longrightarrow HCO_3^-$$
5. (i) Fe^{2+} (ii) $C_2O_4^{2-}$ (iii) NO_2^-
   There are so many others.
6. Reagent $FeCl_3$

		Product	Colour
I.	$S_2O_3^{2-}$	$[Fe(S_2O_3)_2]^-$	dark-violet
II.	CH_3COO^-	$[Fe_2(OH)_2(CH_3COO)_6]^+$	blood dark

7. (a) CrO_2Cl_2 (b) $[Fe(H_2O)_5NO]^{2+}$ (c) $[Ag(NH_3)_2]Cl$
8. Fe is in (+1) state
   (a) $[Ar]\,4s^0 3d^7$ ↑↓ | ↑↓ | ↑ | ↑ | ↑
   (b) $$\text{EAN} = Z - (\text{O.N.}) + 2 \times (\text{ligands})$$
   $$= 26 - 1 + 12 = 37$$
   (c) $= \sqrt{N(N+2)} = \sqrt{15} = 3.87$ BM due to three unpaired electrons.
   (d) +1
9. (a) CO (b) CO_2
   Note : CO (molar mass = 28) diffuses more rapidly that CO_2 (molar mass = 44) hence, (CO) gas burns. After further heating CO_2 goes upwards which puts-off burning CO.

Practice Exercise 1

1. Oil paintings turn blackish after sometime. What is the salt formed? Assume oil paint contains Pb^{2+}.
2. A metallic statue under 'acid-rain' attack is turned bluish-green. What can be the probable metal and the ion formed?
3. Potassium permanganate is purple in colour. On adding with KOH, it turns green. What is the compound formed ?
4. Aqueous solution of $K_2Cr_2O_7$ (orange) changes to yellow on adding alkali. Can you explain.
5. $FeCl_3$ is yellow in aqueous solution but on passing H_2S gas, solution turns green. Explain.
6. CO_2 and SO_2 both turn lime water milky. How will you detect the presence if both are present in a mixture?
7. Identify (*A*), (*B*) and (*C*).
   (*A*) + dil. H_2SO_4 ⟶ brown colour vapours turning (KI + starch) paper blue.
   (*A*) + NaOH $\xrightarrow{\Delta}$ $NH_3(g)$
   (*A*) $\xrightarrow{\Delta}$ (*B*) (*g*) + (*C*) (*g*), (but (*C*) liquid at room temperature)
8. (*A*), (black) + dil. H_2SO_4 ⟶ (*B*) (*g*) + (*C*) (light green colour solution). Gas (*B*) turns lead acetate paper black. What are (*A*), (*B*) and (*C*)?
9. NO_2^- interferes in the 'Ring Test' of NO_3^-. Suggest a chemical method of removal of NO_2^-.
10. I^- also interferes in the 'Ring Test' of NO_3^-. Suggest a chemical reagent that can remove I^-.
11. Arrange AgF, AgCl, AgBr and AgI in the increasing order of solubility in water.
12. While testing oxalate, gas obtained burns with a blue flame initially but is put off instantly even as gas appears coming. Explain.
13. Can you detect Br^- and I^- by 'Layer Test' if present together.
14. Colourless solid (*A*) $\xrightarrow{\Delta}$ (*B*)(*g*) + (*C*)(*g*). Aqueous solutions of (*A*) and (*B*) turn blue litmus red while that of (*C*) turns red litmus blue. Aqueous solutions of (*A*) and (*B*) also give white ppt with $AgNO_3$ solution soluble in aqueous solution of (*C*). Identify (*A*), (*B*) and (*C*).

Answers

1. PbS
2. Probable metal : Cu
   Salt formed : Cu^{2+}
3. $2KMnO_4 + 2KOH \longrightarrow \underset{\text{green}}{2K_2MnO_4} + H_2O + \frac{1}{2}O_2$
4. $\underset{\text{orange}}{Cr_2O_7^{2-}} + 2OH^- \longrightarrow \underset{\text{yellow}}{2CrO_4^{2-}} + H_2O$
5. $2FeCl_3 + H_2S \longrightarrow \underset{\text{green}}{2FeCl_2} + 2HCl + S$
6. Pass the mixture of gases first into acidified $K_2Cr_2O_7$ solution then into lime water.
   * If $K_2Cr_2O_7$ solution turns green —SO_2
   * If lime water turns milky —CO_2
7. (*A*) : NH_4NO_2 (*B*) : N_2 (*C*) : H_2O
8. (*A*) : FeS (*B*) : H_2S (*C*) : $FeSO_4$
9. Boil the mixture with NH_4Cl, NO_2^- is decomposed as N_2:
   $$NaNO_2 + NH_4Cl \xrightarrow{\Delta} NaCl + N_2\uparrow + 2H_2O$$

10. $HgCl_2$ removes I^- as HgI_2

$$HgCl_2 + 2I^- \longrightarrow \underset{\text{orange ppt}}{HgI_2 \downarrow} + 2Cl^-$$

11. $AgI < AgBr < AgCl < AgF$

12. $H_2C_2O_4 \xrightarrow{\Delta} H_2O(l) + CO(g) + CO_2(g)$

CO burns with blue flame. It diffuses faster than CO_2. Once CO burns with blue flame, it is put off by CO_2 which diffuses later.

13. Reducing power of $I^- > Br^- > Cl^- > F^-$. Add Cl_2 water and $CHCl_3$ into mixture of I^- and Br^-.

$$2I^- + Cl_2 \longrightarrow \underset{\text{violet}}{I_2} + 2Cl^-$$

$$2Br^- + Cl_2 \longrightarrow \underset{\text{orange-yellow}}{Br_2} + 2Cl^-$$

Br_2 will also oxidise I^- to I_2. Thus, first there will be violet colour in $CHCl_3$ layer. Extract aqueous layer (containing Br^-) and repeat adding Cl_2 water and $CHCl_3$. Orange-yellow colour in $CHCl_3$ layer confirms Br^-.

14. (*A*) : NH_4Cl (*B*) : HCl (*C*) : NH_3

$$NH_4^+ + H_2O \rightleftharpoons \underset{\text{acidic}}{NH_4OH + H^+}$$

Analysis of Basic (cation) Radicals

Classification into Groups

Based on the values of solubility products (K_{sp}), basic radicals are classified into groups summarised in Table 19.3

Group I

Ag^+, Pb^{2+}, Hg_2^{2+} are precipitated as insoluble chlorides on adding dil. HCl to aqueous solution. If no precipitate, this group is absent.

$$Ag^+ + Cl^- \longrightarrow AgCl \downarrow$$

$$Pb^{2+} + 2Cl^- \longrightarrow PbCl_2 \downarrow$$

$$Hg_2^{2+} + 2Cl^- \longrightarrow Hg_2Cl_2 \downarrow$$

Precipitates are separated. Filtrate is for group II.

DEEP Focus

- $PbCl_2$ (*A*) is soluble in hot water and thus separated.
- AgCl is soluble in aqueous NH_3 as $[Ag(NH_3)_2]Cl$

$$AgCl + 2NH_3 \longrightarrow \underset{\text{diamine silver chloride (soluble), (B)}}{[Ag(NH_3)_2]Cl}$$

- Hg_2Cl_2 is blackened by NH_3

$$Hg_2Cl_2 + 2NH_3 \longrightarrow \underset{\text{black}}{HgNH_2Cl} + Hg + NH_4Cl$$

Table 19.3 **Classification of Basic Radicals into Groups Based on K_{sp} Values**

Group	Basic radicals	Group reagent	Precipitate as	Explanation
I	Ag^+, Hg_2^{2+}(I), Pb^{2+}	dil. HCl	chloride (AgCl, Hg_2Cl_2, $PbCl_2$)	K_{sp} values of chlorides are low, hence precipitated. Others have higher K_{sp} values hence not precipitated.
II	Cu^{2+}, Cd^{2+}, Pb^{2+}, Hg^{2+}(II), Bi^{3+}, Sb^{3+}, Sn^{2+}, As^{3+}	H_2S gas in presence of dil. HCl	sulphides (CuS, As_2S_3 etc)	K_{sp} values of sulphides are low hence precipitated by low $[S^{2-}]$ ion. HCl (with common H^+ ion) decreases ionisation of H_2S which gives low $[S^{2-}]$. Hence, II group is precipitated. Others with higher K_{sp} values are not precipitated.
III	Al^{3+}, Cr^{3+}, Fe^{3+}	NH_4OH in presence of NH_4Cl	hydroxide, $Al(OH)_3$ etc.	K_{sp} values of $Al(OH)_3$ etc are low. NH_4Cl (with common NH_4^+ ion) decrease ionisation of NH_4OH giving low $[OH^-]$. Hence, group III is precipitated.
IV	Zn^{2+}, Ni^{2+}, Mn^{2+}, Co^{2+}	H_2S in ammoniacal medium	sulphides (ZnS etc.)	K_{sp} values of sulphides of group IV are high hence precipitation takes place in higher $[S^{2-}]$ Basic medium increases ionisation of H_2S, thus increasing $[S^{2-}]$ hence precipitation of group IV.
V	Ca^{2+}, Ba^{2+}, Sr^{2+}	$(NH_4)_2CO_3 + NH_4Cl$	carbonates ($CaCO_3$ etc.)	K_{sp} values of carbonate are less than that of group VI (Mg^{2+}) hence precipitation before Mg^{2+}.
VI	Mg^{2+}, (Na^+, K^+) also included	$NH_4OH + Na_2HPO_4$ (only for Mg^{2+})	white ppt ($MgHPO_4$)	—
0 (zero)	NH_4^+	—	—	tested independently from original solution.

Pb^{2+} : *A* : soluble $PbCl_2$ (in hot water) solution.

$$(A) + KI \longrightarrow \text{yellow ppt}$$

$$PbCl_2 + 2KI \longrightarrow \underset{\text{yellow ppt}}{PbI_2 \downarrow} + 2KCl$$

$$(A) + K_2CrO_4 \longrightarrow \text{yellow ppt}$$

$$PbCl_2 + K_2CrO_4 \longrightarrow \underset{\text{yellow ppt}}{PbCrO_4 \downarrow} + 2KCl$$

HNO_3 are NaOH dissolve the precipitate of $PbCrO_4$

$$2PbCrO_4 + 2H^+ \rightleftharpoons 2Pb^{2+} + Cr_2O_7^{2-} + H_2O$$

$$PbCrO_4 + 4OH^- \rightleftharpoons [Pb(OH)_4]^{2-} + CrO_4^{2-}$$

Ag^+ : *B* : soluble $[Ag(NH_3)_2]Cl$

- $(B) + KI \longrightarrow$ yellow ppt

$$[Ag(NH_3)_2]Cl + KI \longrightarrow \underset{\text{yellow ppt}}{AgI\downarrow} + KCl + 2NH_3$$

- $(B) + HNO_3 \longrightarrow$ white ppt

$$[Ag(NH_3)_2]Cl + 2HNO_3 \longrightarrow \underset{\text{white ppt}}{AgCl\downarrow} + 2NH_4NO_3$$

$$(B) + K_2CrO_4 \longrightarrow \text{brick red ppt}$$

$$2[Ag(NH_3)_2]Cl + K_2CrO_4 \longrightarrow \underset{\text{brick red ppt}}{Ag_2CrO_4\downarrow} + 2KCl + 2NH_3$$

Hg_2^{2+} : Black $[Hg(NH_2)Cl + Hg]$ + aqua-regia, heat to dissolve

- Add $SnCl_2 \longrightarrow$ white ppt changing to grey

$$\underbrace{\underset{\text{conc}}{HNO_3} + \underset{\text{conc}}{3HCl}}_{\text{aqua-regia}} \xrightarrow{\Delta} NOCl + 2H_2O + 2Cl$$

$$Hg(NH_2)Cl + Hg + 2HCl + 2Cl \xrightarrow{\Delta} 2HgCl_2 + NH_4Cl$$

$$2HgCl_2 + SnCl_2 \longrightarrow \underset{\text{white ppt}}{Hg_2Cl_2\downarrow} + SnCl_4$$

$$Hg_2Cl_2 + SnCl_2 \longrightarrow \underset{\text{(grey)}}{2Hg\downarrow} + SnCl_4$$

- Distinction: Salt + $K_2CrO_4 \longrightarrow A \xrightarrow{NH_3} B$

	Pb^{2+}	Hg_2^{2+}	Ag^+
A :	Yellow $PbCrO_4\downarrow$	Red $Hg_2CrO_4\downarrow$	Red $Ag_2CrO_4\downarrow$
B :	no change	black $Hg\downarrow$	dissolves as $[Ag(NH_3)_2]^+$

Group II

Filtrate of group I + dil. HCl + H_2O. Pass H_2S gas $\longrightarrow$

- yellow ppt — CdS
- black ppt — PbS, HgS, Bi_2S_3, CuS (salt is blue)
- bright yellow ppt — As_2S_3
- dark brown ppt — SnS
- orange ppt — Sb_2S_3
- As_2S_3, SnS, Sb_2S_3 (II B—Arsenic group) are soluble in yellow ammonium sulphide (YAS) or NaOH.

$$As_2S_3 + \underset{\text{YAS}}{(NH_4)_2S_2} \longrightarrow \underset{\substack{\text{ammonium}\\\text{thioarsenate}}}{(NH_4)_3AsS_4} + S$$

$$Sb_2S_3 + \underset{\text{YAS}}{(NH_4)_2S_2} \longrightarrow \underset{\substack{\text{ammonium}\\\text{thioantimonate}}}{(NH_4)_3SbS_4} + S$$

$$SnS_2 + \underset{\text{YAS}}{(NH_4)_2S_2} \longrightarrow \underset{\substack{\text{ammonium}\\\text{thiostannate}}}{(NH_4)_2SnS_3} + S$$

$$As_2S_3 + 6NaOH \longrightarrow Na_3AsO_3 + Na_3AsS_3 + 3H_2O$$

- PbS, CdS, HgS, Bi_2S_3, CuS (II A—Copper group) are insoluble in YAS.
- If on dilution with H_2O (before passing H_2S), there is white turbidity—Bi^{3+} confirmed, or orange turbidity—Sb^{3+} confirmed. Turbidity disappears on adding dil. HCl.

$$BiCl_3 + H_2O \longrightarrow \underset{\text{white turbidity}}{BiOCl} + 2HCl$$

$$SbCl_3 + H_2O \longrightarrow \underset{\text{orange turbidity}}{SbOCl} + 2HCl$$

$$BiOCl + 2HCl \longrightarrow \underset{\text{soluble}}{BiCl_3} + H_2O$$

$$SbOCl + 2HCl \longrightarrow \underset{\text{soluble}}{SbCl_3} + H_2O$$

- PbS, CdS, CuS, Bi_2S_3 are soluble in HNO_3 (1 : 1) while HgS is insoluble; HgS, however, is soluble in aqua regia.

Pb^{2+} :

- PbS (black) is soluble in HNO_3 (1 : 1)
- $3PbS + 8HNO_3 \longrightarrow \underset{\text{soluble }(A)}{3Pb(NO_3)_2} + 2NO + 3S + 4H_2O$
- $(A) + KI \longrightarrow$ yellow ppt (PbI_2)
- $(A) + K_2CrO_4 \xrightarrow{CH_3COOH}$ yellow ppt ($PbCrO_4$)

Cd^{2+} :

- $3CdS\text{ (yellow)} + 8HNO_3 \xrightarrow{\Delta} \underset{(B)}{3Cd(NO_3)_2} + 2NO + 4H_2O + 3S$
- $(B) + CH_3COOH + K_4[Fe(CN)_6] \longrightarrow$ bluish white ppt

$$Cd(NO_3)_2 + 2CH_3COOH \longrightarrow (CH_3COO)_2Cd + 2HNO_3$$

$$2(CH_3COO)_2Cd + K_4[Fe(CN)_6] \longrightarrow \underset{\text{bluish white ppt}}{Cd_2[Fe(CN)_6]} + 4CH_3COOH$$

Cu^{2+} (salt is blue)

- $3CuS\text{ (black)} + 8HNO_3 \longrightarrow \underset{\substack{\text{soluble (blue)}\\(C)}}{3Cu(NO_3)_2} + 2NO + 3S + 4H_2O$

$$(C) + CH_3COOH + K_4[Fe(CN)_6] \longrightarrow \text{reddish brown ppt}$$

$$Cu(NO_3)_2 + 2CH_3COOH \longrightarrow (CH_3COO)_2Cu + 2HNO_3$$

$$2(CH_3COO)_2Cu + 3K_4[Fe(CN)_6] \longrightarrow \underset{\text{reddish brown ppt}}{Cu_2[Fe(CN)_6]} + 4CH_3COOH$$

$$(C) + \text{aqueous } NH_3\text{(excess)} \longrightarrow \text{deep blue}$$

$$Cu^{2+} + 4NH_3(aq) \longrightarrow \underset{\text{deep blue}}{[Cu(NH_3)_4]^{2+}}$$

$$C + KSCN \longrightarrow \text{black ppt}$$

$$Cu^{2+} + 2SCN^- \longrightarrow \underset{\text{black ppt}}{Cu(SCN)_2\downarrow}$$

Bi^{3+} :

- $Bi_2S_3\text{ (black)} + 8HNO_3 \longrightarrow \underset{(D)}{2Bi(NO_3)_3} + 2NO + 4H_2O + 3S$

- $(D) + NH_4OH \longrightarrow$ white ppt of $Bi(OH)_3$

$$Bi(NO_3)_3 + 3NH_4OH \longrightarrow Bi(OH)_3 + 3NH_4NO_3$$

$$Bi(OH)_3 + \text{dil. } HCl \longrightarrow \text{ppt dissolves forming } BiCl_3$$

$$BiCl_3 + \underset{\text{sodium stannite}}{Na_2SnO_2} \xrightarrow{NaOH} \text{black residue}$$

$$BiCl_3 + 6NaOH + 3Na_2SnO_2 \longrightarrow 3Na_2SnO_3 + \underset{\text{black}}{2Bi\downarrow} + 3H_2O + 6NaCl$$

Hg^{2+} : HgS (black) is soluble in aqua regia.

$$3HgS + \underbrace{2(HNO_3 + 3HCl)}_{\text{aqua regia}} \xrightarrow{\Delta} \underset{(E)}{3HgCl_2} + 2NO + 3S + 4H_2O$$

$$(E) + SnCl_2 \longrightarrow \text{white ppt changing to grey}$$

$$2HgCl_2 + SnCl_2 \longrightarrow \underset{\text{white ppt}}{2Hg_2Cl_2\downarrow} + SnCl_4$$

$$Hg_2Cl_2 + SnCl_2 \longrightarrow \underset{\text{grey}}{2Hg} + SnCl_4$$

$$(E) + KI \longrightarrow \text{orange ppt soluble in excess of KI}$$

$$HgCl_2 + 2KI \longrightarrow \underset{\text{orange}}{HgI_2\downarrow} + 2KCl$$

$$HgI_2 + 2KI \longrightarrow \underset{\substack{\text{soluble}\\\text{(Nessler's reagent)}}}{K_2HgI_4}$$

- (E) + NaOH (in small amount) $\longrightarrow$ brownish red ppt which changes to yellow on adding NaOH in stoichiometric amount.

$$Hg^{2+} + 2OH^- \longrightarrow \underset{\text{yellow}}{HgO\downarrow} + H_2O$$

Group III

Boil off H_2S gas from the filtrate of group II. Add NH_4Cl and one drop of dil. HNO_3; heat, cool and add $NH_4OH \longrightarrow$ ppt

- Reddish brown ppt of $Fe(OH)_3$ if Fe^{3+} is present (salt is brown).
- Dirty green ppt of $Cr(OH)_3$ if Cr^{3+} is present (salt is green).
- White ppt of $Al(OH)_3$ if Al^{3+} is present (salt is colourless).

DEEP Focus

- **dil. HNO_3 is added to convert Fe^{2+} (light green) into Fe^{3+} (deep yellow).**
- NH_4Cl by common-ion effect decreases ionisation of NH_4OH.

Al^{3+} :

- White ppt of $Al(OH)_3$ (salt is colourless) + NaOH solution (excess) $\longrightarrow$ ppt dissolves forming $NaAlO_2$.

$$Al(OH)_3 + NaOH \longrightarrow \underset{\substack{\text{sodium meta-aluminate}\\(A)\text{ (soluble)}}}{NaAlO_2} + 2H_2O$$

White ppt reappears if NH_4Cl is added into soluble meta-aluminate and boiled.

$$NH_4Cl + \underset{(A)}{NaAlO_2} + H_2O \xrightarrow{\Delta} NaCl + \underset{\text{white ppt}}{Al(OH)_3\downarrow} + NH_3$$

$$Al^{3+} + CH_3COONa \longrightarrow \text{no ppt in cold, neutral solution}$$

$$\xrightarrow[\text{excess } CH_3COONa]{\text{blue}} \text{white ppt}$$

$$Al^{3+} + 3CH_3COO^- + 2H_2O \longrightarrow Al(OH)_2CH_3COO\downarrow + 2CH_3COOH$$

Fe^{3+} :

$Fe(OH)_3$ ppt are insoluble in NaOH but soluble in conc. HCl.

$$Fe(OH)_3 + 3HCl \longrightarrow \underset{\text{soluble }(B)}{FeCl_3} + 3H_2O$$

- $$(B) + CH_3COOH + \underset{\substack{\text{potassium}\\\text{ferrocyanide}}}{K_4[Fe(CN)_6]} \longrightarrow \text{blue ppt}$$

A large excess of the reagent dissolves it (Prussian blue) partly or entirely when an intense blue colour is obtained.

$$4FeCl_3 + 3K_4[Fe(CN)_6] \longrightarrow \underset{\substack{\text{ferric ferrocyanide}\\\text{(Prussian blue)}}}{Fe_4[Fe(CN)_6]_3} + 12KCl$$

$$Fe_4[Fe(CN)_6]_3 + K_4[Fe(CN)_6] \longrightarrow \underset{\text{soluble}}{4KFe[Fe(CN)_6]}$$

$$(B) + KCNS \longrightarrow \text{red colour}$$

$$FeCl_3 + \underset{\substack{\text{potassium}\\\text{sulphocyanide}}}{KCNS} \longrightarrow \underset{\substack{\text{blood red}\\\text{colour}}}{[Fe(CNS)]Cl_2} + KCl$$

Actual composition of blood red colour is $[Fe(CNS)(H_2O)_5]^{2+}$

- B + sodium acetate solution $\longrightarrow$ a reddish brown colour

$$3Fe^{3+} + 6CH_3COO^- + 2H_2O \longrightarrow [Fe_3(OH)_2(CH_3COO)_6]^+ + 2H^+$$

The reaction becomes complete only if the strong acid, which is formed, is removed by the addition of an excess of reagent, which acts as a buffer.

Fe^{2+} *does not give these tests, hence at the start of this group,* HNO_3 *is added to convert* Fe^{2+} *into* Fe^{3+}.

Cr^{3+} :

$Cr(OH)_3$ (green ppt) is insoluble in NaOH but dissolves in presence of Br_2 water (oxidising agent) giving yellow coloured Na_2CrO_4 solution.

$$Br_2 + H_2O \longrightarrow 2HBr + O$$

$$4NaOH + 2Cr(OH)_3 + 3(O) \longrightarrow \underset{\text{soluble }(C)}{2Na_2CrO_4} + 5H_2O$$

$$C + CH_3COOH + (CH_3COO)_2Pb \longrightarrow \text{yellow ppt}$$

$$Na_2CrO_4 + (CH_3COO)_2Pb \longrightarrow \underset{\text{yellow ppt}}{PbCrO_4\downarrow} + 2CH_3COONa.$$

Group IV

Filtrate of group III + $NH_4OH + NH_4Cl \xrightarrow{\Delta}$ pass H_2S gas $\longrightarrow$

- white ppt (salt colourless) of ZnS
- buff coloured ppt (salt light pink) of MnS
- black ppt (salt green) of NiS
- black ppt (salt dark blackish brown) of CoS
- Ammoniacal medium increases ionisation of H_2S, hence higher $[S^{2-}]$ will precipitate Zn^{2+}, Co^{2+}, Ni^{2+} and Mn^{2+} as sulphides (having higher K_{sp} values)

Zn^{2+} : ZnS ppt (white) are soluble in dil HCl.

$$ZnS + 2HCl \longrightarrow \underset{(A)}{ZnCl_2} + H_2S$$

- (*A*) + NaOH $\longrightarrow$ white ppt soluble in excess of NaOH.

$$\underset{(A)}{ZnCl_2} + 2NaOH \longrightarrow \underset{\text{white ppt}}{Zn(OH)_2\downarrow} + 2NaCl$$

$$Zn(OH)_2 + 2NaOH \longrightarrow \underset{\text{sodium zincate}}{Na_2ZnO_2} + 2H_2O$$

White ppt of ZnS reappears on passing H_2S into soluble Na_2ZnO_2 solution.

$$Na_2ZnO_2 + H_2S \longrightarrow 2NaOH + \underset{\text{(white ppt)}}{ZnS\downarrow}$$

$$(A) + CH_3COOH + K_4[Fe(CN)_6] \longrightarrow \text{bluish white ppt}$$

$$2ZnCl_2 + K_4[Fe(CN)_6] \longrightarrow \underset{\text{bluish white ppt}}{Zn_2[Fe(CN)_6]\downarrow} + 2KCl$$

Zn^{2+} salts form white ppt of $ZnNH_4PO_4$ on reaction with Na_2HPO_4 in basic buffer ($NH_4Cl + NH_4OH$)

$$ZnSO_4 + Na_2HPO_4 + NH_4Cl + NH_4OH \longrightarrow \underset{\text{(white ppt)}}{ZnNH_4PO_4\downarrow} + Na_2SO_4 + NH_4Cl + H_2O$$

Mn^{2+} : MnS (buff coloured ppt) are soluble in dil. HCl but addition of excess of NaOH gives $Mn(OH)_2$ ppt which changes to brown/black by atmospheric O_2.

$$MnS + 2HCl \longrightarrow MnCl_2 + H_2S$$

$$MnCl_2 + 2NaOH \longrightarrow Mn(OH)_2\downarrow + 2NaCl$$

$$Mn(OH)_2 + \underset{\text{air}}{O} \longrightarrow \underset{\text{brown or black}}{MnO_2\downarrow} + H_2O$$

Dissolve MnO_2 or $Mn(OH)_2$ ppt into HNO_3 and then add oxidising agent $KClO_3$ or PbO_2; heat to boil—purple solution (*B*).

$$MnO_2 + 2HNO_3 \longrightarrow Mn(NO_3)_2 + H_2O + \frac{1}{2}O_2$$

$$2Mn(NO_3)_2 + 5PbO_2 + 6HNO_3 \longrightarrow \underset{\substack{\text{permanganic acid } (B)\\ \text{(purple)}}}{2HMnO_4} + 2H_2O + 5Pb(NO_3)_2$$

$$5PbO_2 + 2Mn^{2+} + 6H^+ \longrightarrow 5Pb^{2+} + 2HMnO_4 + 2H_2O$$

Ni^{2+} : NiS (black ppt) is soluble in conc. HCl in presence of oxidising agent like $KClO_3$.

$$NiS + \underset{\text{conc.}}{2HCl} + O \longrightarrow \underset{\substack{\text{soluble } (C)\\ \text{(green)}}}{NiCl_2} + H_2O + S$$

- (*C*) + $NaHCO_3$ + Br_2 water + NaOH $\xrightarrow{\Delta}$ black residue

$$NiCl_2 + 2NaHCO_3 \xrightarrow{\Delta} NiCO_3 + 2NaCl + H_2O + CO_2$$

$$2NiCO_3 + 4NaOH + O\ (Br_2\ \text{water}) \xrightarrow{\Delta} \underset{\text{black}}{Ni_2O_3\downarrow} + 2Na_2CO_3 + H_2O$$

(*C*) + NH_4OH + dimethyl glyoxime $\longrightarrow$ cherry red ppt

$NiCl_2$ (*C*) in NH_4OH + $2\,CH_3—C{=}NOH$ / $CH_3—C{=}NOH$ (dimethyl glyoxime) $\longrightarrow$

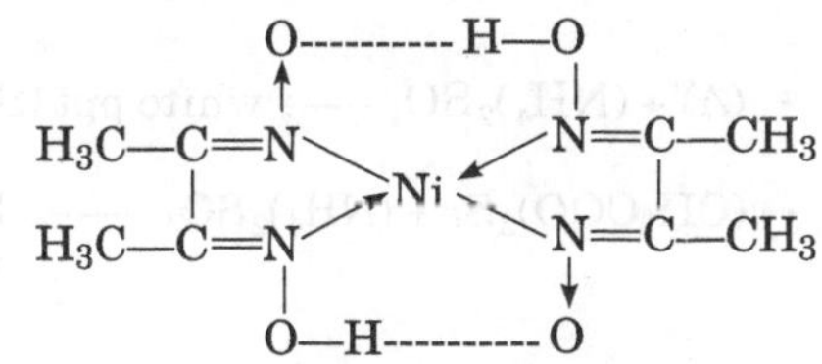

cherry red ppt
(dimethyl glyoximate)

Co^{2+} : Like NiS, CoS (black ppt) is also soluble in conc. HCl in presence of oxidising agent $KClO_3$

$$CoS + \underset{\text{conc.}}{2HCl} + O \xrightarrow{\Delta} \underset{\substack{(D)\\ \text{soluble}}}{CoCl_2} + H_2O + S$$

- (*D*) + NH_4CNS crystal + ether $\longrightarrow$ blue colour in ethereal layer

$$CoCl_2 + 4NH_4CNS \longrightarrow \underset{\text{blue in ethereal layer}}{(NH_4)_2[Co(CNS)_4]} + 2NH_4Cl$$

- (*D*) + CH_3COOH + KNO_2 $\longrightarrow$ yellowish orange ppt

$$CoCl_2 + 2KNO_2 \longrightarrow 2KCl + Co(NO_2)_2$$

$$CH_3COOH + KNO_2 \longrightarrow CH_3COOK + HNO_2$$

$$3KNO_3 + Co^{II}(NO_2)_2 + 2HNO_2 \longrightarrow \underset{\substack{\text{potassium cobaltinitrite}\\ \text{(yellowish orange ppt)}}}{K_3[Co^{III}(NO_2)_6]\downarrow} + H_2O + NO\uparrow$$

- (*D*) + NaOH + $NaHCO_3$ + Br_2 water $\xrightarrow{\Delta}$ green solution

$$CoCl_2 + NaHCO_3 + NaOH \xrightarrow{\Delta} 2NaCl + CoCO_3 + H_2O$$

$$2NaHCO_3 \xrightarrow{\Delta} Na_2CO_3 + H_2O + CO_2$$

$$CoCO_3 + 2Na_2CO_3 \xrightarrow{\Delta} Na_4[Co^{II}(CO_3)_3]$$

$$2Na_4[Co^{II}(CO_3)_3] + 2NaHCO_3 + O \xrightarrow{\Delta} \underset{\substack{\text{(green)}\\ \text{sodium cobalticarbonate}}}{2Na_3[Co^{III}(CO_3)_3]} + 2Na_2CO_3 + H_2O$$

Group V

$$\text{Filtrate of group IV} \xrightarrow[\text{(ii) } NH_4OH,\ (NH_4)_2CO_3]{\text{(i) boil off } H_2S} \text{white ppt}$$

- $BaCO_3, CaCO_3, SrCO_3$ appear as white ppt
- **Flame test** Perform flame test with these group ppt
  Green flame — Ba^{2+}
  Crimson red — Sr^{2+}
  Brick red — Ca^{2+}
- Dissolve the ppt in CH_3COOH.

$$\underset{(M = Ca, Ba, Sr)}{MCO_3} + 2CH_3COOH \xrightarrow{\Delta} \underset{(A)}{(CH_3COO)_2M} + H_2O + CO_2$$

- $(A) + K_2CrO_4 \longrightarrow$ yellow ppt $(BaCrO_4)$

$$(CH_3COO)_2Ba + K_2CrO_4 \longrightarrow \underset{\text{yellow ppt}}{BaCrO_4\downarrow} + 2CH_3COOK$$

- $(A) + (NH_4)_2SO_4 \longrightarrow$ white ppt $(SrSO_4)$

$$(CH_3COO)_2Sr + (NH_4)_2SO_4 \longrightarrow \underset{\text{white ppt}}{SrSO_4\downarrow} + 2CH_3COONH_4$$

- $(A) + (NH_4)_2C_2O_4 \longrightarrow$ white ppt (CaC_2O_4)

$$(CH_3COO)_2Ca + (NH_4)_2C_2O_4 \longrightarrow \underset{\text{white ppt}}{CaC_2O_4\downarrow} + 2CH_3COONH_4$$

$$\text{white ppt} + H^+ + \underset{\text{pink}}{MnO_4^-} \longrightarrow \text{pink colour is discharged}$$

$$CaC_2O_4 + 2H^+ \longrightarrow Ca^{2+} + H_2C_2O_4$$

$$5C_2O_4^{2-} + \underset{\text{pink}}{2MnO_4^-} + 16H^+ \longrightarrow \underset{\text{colourless}}{2Mn^{2+}} + 10CO_2 + 8H_2O$$

Group VI (Mg^{2+})

Filtrate of group V + $NH_4OH + Na_2HPO_4 \longrightarrow$ a fine crystalline ppt on scratching the side of the test tube.

$$MgCl_2 + NH_4OH + Na_2HPO_4 \longrightarrow \underset{\text{white ppt}}{Mg(NH_4)PO_4\downarrow} + 2NaCl + H_2O$$

Group Zero (NH_4^+)

- Salt + NaOH $\xrightarrow{\Delta}$ gas giving white fumes with HCl

$$NH_4Cl + NaOH \xrightarrow{\Delta} NH_3(g) + NaCl + H_2O$$

$$NH_3(g) + HCl \longrightarrow \underset{\text{white fumes}}{NH_4Cl}$$

- Salt + NaOH $\xrightarrow{\Delta}$ gas $\xrightarrow{\text{add Nessler's reagent}}$ brown ppt or brown or yellow colouration (oxydimercuri ammonium iodide) called iodide of Millon's base (A). *A* is also called basic mercury (II) amido iodine.

$$NH_4Cl + NaOH \xrightarrow{\Delta} NH_3$$

$$NH_3 + 3NaOH + \underset{\text{Nessler's reagent}}{2K_2HgI_4} \longrightarrow \underset{(A)}{O\langle^{Hg}_{Hg}\rangle NH_2I} + 4KI + 3NaI + 2H_2O$$

Above reaction can also be written as :

$$NH_4^+ + 2[HgI_4]^{2-} + 4OH^- \longrightarrow HgO\cdot Hg(NH_2)I\downarrow + 7I^- + 3H_2O$$

Dry Tests

Borax Bead Test

If borax, $Na_2B_4O_7 \cdot 10H_2O$ is heated on the platinum loop, a transparent colourless glass like bead of sodium metaborate ($NaBO_2$) and boric anhydride (B_2O_3) is formed.

$$Na_2B_4O_7 \xrightarrow{\Delta} 2NaBO_2 + B_2O_3$$

Characteristics coloured beads are produced with salts of copper, iron, chromium, manganese, cobalt and nickel.

$$CuO + B_2O_3 \longrightarrow \underset{\text{copper (II) metaborate}}{Cu(BO_2)_2}$$

Results have been summarised in Table 19.4.

Table 19.4 **Colour in oxidising and Reducing Flames in Borax-Bead Test**

Oxidising flame		Reducing flame		Metal
Hot	Cold	Hot	Cold	
Green	Blue	Colourless	Opaque red-brown	Copper
Yellowish-brown	Yellow	Green	Green	Iron
Yellow	Green	Green	Green	Chromium
Violet (amethyst)	Amethyst	Colourless	Colourless	Manganese
Blue	Blue	Blue	Blue	Cobalt
Violet	Reddish -brown	Grey	Grey	Nickel
Yellow	Colourless	Brown	Brown	Molybdenum
Rose-violet	Rose-violet	Red	Violet	Gold
Yellow	Colourless	Yellow	Yellowish-brown	Tungsten
Yellow	Pale yellow	Green	Bottle-green	Uranium
Yellow	Greenish -yellow	Brownish	Emerald -green	Vanadium
Yellow	Colourless	Grey	Pale violet	Titanium
Orange-red	Colourless	Colourless	Colourless	Cerium

Microcosmic Salt Bead Test

A test similar to borax bead test is used for identification of coloured cations if microcosmic salt, $Na(NH_4)HPO_4 \cdot 4H_2O$, is used.

$$Na(NH_4)HPO_4 \xrightarrow{\Delta} \underset{\text{transparent bead}}{NaPO_3} + H_2O\uparrow + NH_3\uparrow$$

$$NaPO_3 + CoO \longrightarrow \underset{\text{(blue bead)}}{NaCoPO_4}$$

Results have been summarised in Table 19.5.

Table 19.5 **Microcosmic Salt Bead Test**

Oxidising flame	Reducing flame	Metal
Green when hot, blue when cold	Colourless when hot, red when cold	Copper
Yellowish-or reddish-brown when hot, yellow when cold	Yellow when hot, colourless to green when cold	Iron
Green, hot and cold	Green, hot and cold	Chromium
Violet, hot and cold	Colourless, hot and cold	Manganese
Blue, hot and cold	Blue, hot and cold	Cobalt
Brown, hot and cold	Grey when cold	Nickel
Yellow, hot and cold	Green when cold	Vanadium
Yellow when hot, yellow-green when cold	Green, hot and cold	Uranium
Pale yellow when hot, colourless when cold	Green when hot, blue when cold	Tungsten
Colourless, hot and cold	Yellow when hot, violet when cold	Titanium

Sodium Carbonate Bead Test

The sodium carbonate bead is prepared by fusing a small quantity of sodium carbonate on a platinum wire, loop in the Bunsen flame; a white, opaque bead is produced. If this is moistened, dipped into a little KNO_3 and then into a small quantity of a manganese salt (for example) and the whole heated in the oxidising flame, a green bead of sodium manganate (Na_2MnO_4) is formed.

$$MnO + Na_2CO_3 + O_2 \xrightarrow{\Delta} Na_2MnO_4 + CO_2$$

A yellow bead is obtained with chromium salt due to formation of sodium chromate (Na_2CrO_4)

$$2Cr_2O_3 + 4Na_2CO_3 + 3O_2 \longrightarrow 4Na_2CrO_4 + 4CO_2$$

Flame Test

Paste of the salt and conc. HCl is taken into the lower oxidising zone and colour imparted to the flame by salts is observed; salts, particularly of group V (Ba^{2+}, Ca^{2+}, Sr^{2+}), are identified by colours of the flame and summarised in Table 19.6.

Table 19.6 **Flame Tests**

Colour	Cation
Golden yellow	Na^+
Violet	K^+
Carmine-red	Li^+
Brick-red	Ca^{2+}
Apple-green	Ba^{2+}, Mo^{2+}
Green	Cu^{2+}, (BO_3^{3-}), Tl^{3+}
Crimson-red	Sr^{2+}

The yellow colouration due to sodium masks that of potassium. In such cases view the flame through cobalt glass (Table 19.7), the yellow sodium colour is absorbed and the potassium flame appears crimson.

Target Practice 2

1. One reagent can be used to detect more than one ion. Give colour of the precipitate/solution formed.

Reagent	Colour of the precipitate with Ag^+	Pb^{2+}	Hg^{2+}
(a) KI	....	....	....
(b) K_2CrO_4	....	....	....
(c) HCl	....	....	....

2. Thioacetamide (CH_3CSNH_2) serves as a source of H_2S. Write reactions.

....

....

3. Consider following experiments

(i) Aqueous solution of

$$ZnCl_2 + H_2S \text{ gas} \longrightarrow \text{no precipitate}$$

(ii) Aqueous solution of

$$(CH_3COO)_2Zn + H_2S \text{ gas} \longrightarrow \text{white precipitate}$$

(iii) Aqueous solution of

$$(CH_3COO)_2Zn + \text{dil. HCl} + H_2S \text{ gas} \longrightarrow \text{no precipitate}$$

Observation in case of

(i) is due to ....

(ii) is due to ....

(iii) is due to ....

4. Write the compounds formed/ separated from a mixture containing Ag^+, Hg_2^{2+} and Pb^{2+} in the following scheme.

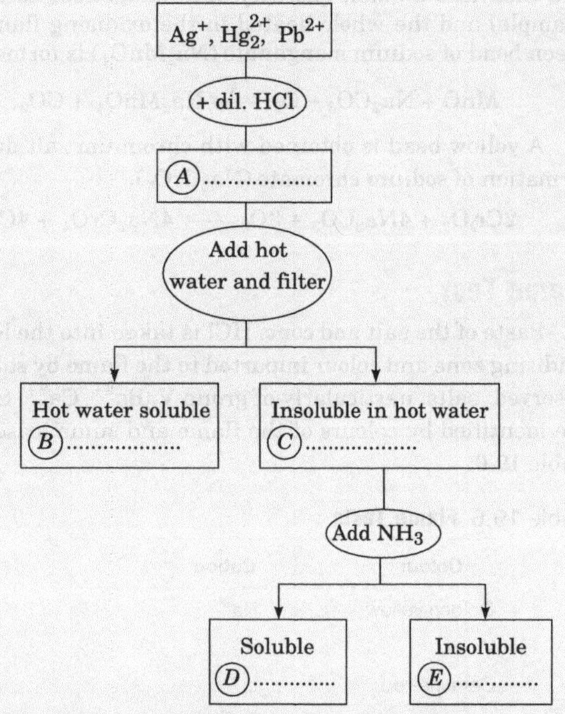

5. Write following reactions

(a) Dissolution of CuS precipitate in dil. HNO_3

....

(b) Dissolution of HgS precipitate in aqua regia

....

(c) Dissolution of Ag_2S_3 in aqueous NaOH

....

(d) Stannite $[Sn(OH)_3]^-$ ion disproportionates on long exposure to air

....

6. Filtrate of group II has been given to you for analysis of group III (which may contain Fe^{2+}, Cr^{3+} and Al^{3+}). Mention systematic steps you would follow

Step I : ....

Step II : ....

Step III : ....

7. Two students (*X* and *Y*) have salts with common cation Fe^{2+} but different anion. Each of them dissolved the salt in water, added dil. HCl, boiled and added NH_4Cl and NH_4OH.

Student *X* : Light green precipitate.

Student *Y* : Brown precipitate.

Anion in *X* can be ....

Anion in *Y* can be ....

Light-green precipitate in ***X*** is due to ....

Brown precipitate in ***Y*** is due to ....

8. Give the reagents to identify Fe^{3+}

(I) ....

(II) ....

(III) ....

9. A reagent can identify Fe^{3+}, Zn^{2+}, Cd^{2+} and Cu^{2+}; reagent is .... and coloured formed with

(a) Fe^{3+} ....

(b) Zn^{2+} ....

(c) Cd^{2+} ....

(d) Cu^{2+} ....

10. Write the names of reagents to identify

(a) Co^{2+} ....

(b) Ni^{2+} ....

11. Mn^{2+} is identified by its oxidation to a salt giving pink colour in solution.

(a) Pink colour is due to formation of ....

(b) Oxidising agent used is ....

(c) Oxidised Mn^{2+} changes to .... by KOH.

12. KCNS can identify following basic radicals

(I) ....

(II) ....

(III) ....

Answers

1.

		Ag^+	Pb^{2+}	Hg^{2+}
(a)	KI	AgI (yellow)	PbI_2 (yellow)	HgI_2 orange soluble in excess of KI forming K_2HgI_4
(b)	K_2CrO_4	Ag_2CrO_4 black red	$PbCrO_4$ (yellow)	—
(c)	HCl	AgCl (white)	$PbCl_2$ (white)	$HgCl_2$ (soluble)

2. $CH_3\overset{\overset{\displaystyle S}{\|}}{C}NH_2 + 2H_2O \longrightarrow CH_3\overset{\overset{\displaystyle O}{\|}}{C}O^- + NH_4^+ + H_2S$

3. (i) ZnS is not precipitated in presence of HCl which is formed in following reaction :

$$ZnCl_2 + H_2S \longrightarrow ZnS + 2HCl$$

(ii) CH_3COOH does not dissolve ZnS which remains insoluble.

(iii) Medium is HCl and ZnS is dissolved.

4. (*A*) : $AgCl, PbCl_2, Hg_2Cl_2$ (*B*) : $PbCl_2$

(*C*) : $AgCl, Hg_2Cl_2$ (*D*) : $[Ag(NH_3)_2]Cl$

(*E*) : $HgNH_2Cl + Hg$

5. (a) $3CuS + 8HNO_3 \longrightarrow 3Cu(NO_3)_2 + 2NO + 4H_2O + 3S$

(b) $3HgS + \underset{\text{aqua regia}}{2(HNO_3 + 3HCl)} \xrightarrow{\Delta} 3HgCl_2 + 2NO + 3S + 4H_2O$

(c) $As_2S_3 + 6NaOH \longrightarrow Na_3AsO_3 + Na_3AsS_3 + 3H_2O$

(d) $2[Sn(OH)_3]^- \longrightarrow Sn(s) + [Sn(OH)_6]^{2-}$ (redn. → Sn(s); oxidn. → $[Sn(OH)_6]^{2-}$)

6. Step I : Boil off H_2S

Step II : Add a drop of HNO_3 (to oxidise Fe^{2+} to Fe^{3+})

Step III : Add NH_4Cl boil, cool and add NH_4OH.

7. Anion in X can be all soluble anions except NO_3^-.
Anion in Y can be NO_3^-
Light green precipitate in X due to $Fe(OH)_2$,
Light green precipitate in Y is due to $Fe(OH)_3$.
(HNO_3 formed in Y oxidises Fe^{2+} to Fe^{3+} hence, $Fe(OH)_3$ is formed)

8. I : $K_4Fe(CN)_6$ II : KCNS III : CH_3COO^-

9. Reagent : $K_4[Fe(CN)_6]$
(a) $Fe_4(Fe(CN)_6)_3$ Prussian blue
(b) $Zn_2[Fe(CN)_6]$ bluish white
(c) $Cd_2[Fe(CN)_6]$ bluish white
(d) $Cu_2[Fe(CN)_6]$ reddish brown

10. (a) NH_4CNS
(b) dimethyl glyoxime (DMG) in NH_4OH/alcoholic solution.

11. (a) MnO_4^-
(b) Br_2 water or PbO_2/H^+
(c) K_2MnO_4 (green)

12. I : Fe^{3+} II : Co^{2+} III : Cu^{2+}

Practice Exercise 2

1. Identify (*A*), (*B*), (*C*) and (*D*) in the following :
Colourless salt (*A*) + conc. $H_2SO_4 \longrightarrow$ brown fumes intensified on adding Cu turnings.
Aqueous solution (*A*) + Cu $\longrightarrow$ blue coloured solution (*B*) and metal (*C*).
Aqueous solution (*A*) + HCl $\longrightarrow$ white ppt soluble in aqueous NH_3 forming (*D*)

2. Colourless salt (*A*) soluble in H_2O $\xrightarrow{\text{dil. HCl}}$ white ppt (*B*) soluble in hot water

(*A*) $\xrightarrow{\text{(i) } FeSO_4 \text{ (ii) conc. } H_2SO_4}$ brown colour ring (*D*)

(*B*) $\xrightarrow{\text{KI}}$ yellow ppt (*C*)

Identify (*A*), (*B*), (*C*) and (*D*).

3. Salt (*A*) makes part of electrode and is insoluble in water. (*A*) is blackened by NH_3 forming (*B*). (*B*) is soluble is aqua regia forming (*C*). (*C*) gives orange ppt with KI but ppt dissolves in excess of KI forming (*D*). Identify (*A*), (*B*), (*C*) and (*D*).

4. Identify (*A*), (*B*), (*C*) and (*D*) and explain reactions.
(*A*) (green coloured salt) + $K_2Cr_2O_7$ + conc. $H_2SO_4 \xrightarrow{\Delta}$ (*B*)
(*B*) (reddish brown gas) + NaOH $\longrightarrow$ (*C*) (yellow coloured solution)
(*C*) + $(CH_3COO)_2Pb \xrightarrow{CH_3COOH}$ (*D*) (yellow ppt)
(*A*) + NaOH + Br_2 water $\xrightarrow{\Delta}$ (*C*)
(*C*) + $(CH_3COO)_2Pb \xrightarrow{CH_3COOH}$ (*D*)

5. Light green solution of (*A*) does not give blue coloured ppt with $K_4[Fe(CN)_6]$ but on adding a drop of HNO_3, blue ppt (*B*) appears. However, (*A*) gives blue colour (*C*) with $K_3[Fe(CN)_6]$. Explain the formation of (*B*) and (*C*). Identify (*A*) if it, also gives white ppt with $AgNO_3$ solution.

6. (*A*) (yellow coloured solution) changes to light green coloured solution (*B*) on passing H_2S gas. (*A*) and (*B*) both give white ppt with $BaCl_2$ solution, insoluble in conc. HCl. A gives blue coloured ppt (*C*) with $K_4[Fe(CN)_6]$, (*B*) does not. What are (*A*), (*B*) and (*C*)?

7. (*A*) (colourless) $\xrightarrow{\Delta}$ (*B*) (residue) + (*C*) (gas) + (*D*) (gas)

(*B*) $\xrightarrow{H_2O}$ solution of (*B*) $\xrightarrow{(D)}$ milky

A gives brick red colour in flame and decolourises MnO_4^-/H^+. Gas (*C*) burns with blue flame. Identify (*A*), (*B*), (*C*) and (*D*).

8. $MCl_2 + K_2CrO_4 \longrightarrow$ Yellow ppt
What can be MCl_2
(a) if it is soluble in hot water?
(b) if it gives green colour in flame?

9. $CaSO_4$ is insoluble but is not precipitated when excess of $(NH_4)_2SO_4$ is added to $CaCl_2$; explain.

10. Identify (*A*), (*B*), (*C*), (*D*) and (*E*).

(*A*) + NaOH $\longrightarrow$ *B*(*g*) + *C*(*aq*) (slightly alkaline)

(*A*) $\xrightarrow{\Delta}$ H_2O + *D*(*g*)

B(*g*) $\xrightarrow{\text{HCl}}$ white fumes

C(*aq*) $\xrightarrow{\text{Zn}}$ *B*(*g*)

D(*g*) $\xrightarrow{\text{Al, }\Delta}$ *E* $\xrightarrow{H_2O}$ *B*(*g*)

11. *A* (orange) $\xrightarrow{\Delta}$ *B*(*g*) + residue (*C*) (green) + H_2O

A $\xrightarrow{\text{NaOH, }\Delta}$ *D*(*g*)

D(*g*) $\xrightarrow{\text{HCl}}$ white fumes

B(*g*) $\xrightarrow[\Delta]{\text{Al}}$ (*E*) $\xrightarrow{H_2O}$ gas (*B*)

residue (*C*) $\xrightarrow{\text{conc. HCl}}$ (*F*) $\xrightarrow[\Delta]{\text{NaOH, } H_2O_2}$ *G* (yellow) $\xrightarrow[(CH_3COO)_2Pb]{CH_3COOH}$ *H* (yellow)

Identify (*A*) to (*H*) and explain reactions.

12. Colourless salt (*A*) gives apple green flame with conc. HCl. (*A*) on reaction with dil. H_2SO_4 gives light brown fumes (*D*) turning KI-starch paper blue.
(*A*) + $CH_3COOH + K_2CrO_4 \longrightarrow$ yellow ppt (*B*)
(*A*) + $H_2SO_4 \longrightarrow$ white ppt (*C*) + (*D*)
(*D*) + $CH_3NH_2 \longrightarrow CH_3OH + H_2O$ + gas (*E*)
(*E*) + Mg $\xrightarrow{\Delta}$ (*F*)
(*F*) + $H_2O \longrightarrow NH_3$
Identify (*A*) to (*F*) and explain reactions.

Answers

1. (A) : $AgNO_3$ (B) : $Cu(NO_3)_2$ (C) : Ag (D) : $[Ag(NH_3)_2]Cl$
2. (A) : $Pb(NO_3)_2$ (B) : $PbCl_2$ (C) : PbI_2 (D) : $[Fe(H_2O)_5NO]^{2+}$
3. (A) : Hg_2Cl_2 (calomel) (B) : $(HgNH_2Cl + Hg)$ (C) : $HgCl_2$ (D) : K_2HgI_4
4. (A) : $CrCl_3$ (B) : CrO_2Cl_2 (by chromyl-chloride test of Cl^-) (C) : Na_2CrO_4 (D) : $PbCrO_4$
5. (A) : $FeCl_2$ (B) : $KFe^{III}[Fe^{II}(CN)_6]$ Prussian blue, (C) : $KFe^{II}[Fe^{III}(CN)_6]$, Turnbull's blue
6. (A) : $FeCl_3$ (B) : $FeCl_2$ (C) : $KFe^{III}[Fe^{II}(CN)_6]$, Prussian blue
7. (A) : CaC_2O_4 (B) : CaO (C) : CO (D) : CO_2
8. (a) $PbCl_2$ (b) $BaCl_2$
9. $CaSO_4 + (NH_4)_2SO_4 \longrightarrow \underset{\text{soluble}}{(NH_4)_2[Ca(SO_4)_2]}$
10. (A) : NH_4NO_2 (B) : NH_3 (C) : $NaNO_2$ (D) : N_2 (E) : AlN
11. (A) : $(NH_4)_2Cr_2O_7$ (B) : N_2 (C) : Cr_2O_3 (D) : NH_3 (E) : AlN (F) : $CrCl_3$ (G) : Na_2CrO_4, (H) : $PbCrO_4$
12. (A) : $Ba(NO_2)_2$ (B) : $BaCrO_4$ (C) : $BaSO_4$ (D) : HNO_2 (E) : N_2 (F) : Mg_3N_2

One Reagent For Testing Different Ions

1. KCN

(a) Pb^{2+} : Pb^{2+} salts form white precipitate of $Pb(CN)_2$ with KCN which is insoluble in excess test of KCN.

$$Pb^{2+} + 2CN^- \longrightarrow Pb(CN)_2\downarrow$$

(b) Hg_2^{2+} : Hg_2^{2+} salts form soluble $Hg(CN)_2$ and a precipitate of Hg

$$Hg_2^{2+} + 2CN^- \longrightarrow Hg + Hg(CN)_2$$

Above reaction is an example of disproportionation reaction.

(c) Ag^+ : CN^- dissolves insoluble AgCl forming soluble $[Ag(CN)_2]^-$

$$AgCl + 2CN^- \longrightarrow \underset{\substack{\text{dicyano argentate(1) ion}\\ \text{(soluble)}}}{[Ag(CN)_2]^-} + Cl^-$$

(d) Cu^{2+} : KCN dissolves the precipitate of CuS, when colourless tetracyanocuprate(I), $[Cu(CN)_4]^{3-}$, and disulphide S_2^{2-} ions, are formed.

$$2CuS + 8CN^- \longrightarrow 2[Cu(CN)_4]^{3-} + S_2^{2-}$$

This is an oxidation-reduction process. Cu^{2+} is reduced to Cu^+ and sulphur is oxidised coupled with formation of a complex.

Soluble Cu^{2+} salts react with KCN forming yellow precipitate of copper(II) cyanide, $Cu(CN)_2$ which quickly decomposes into white precipitate of copper(I) cyanide, CuCN, and a highly poisonous, $(CN)_2$, cyanogen gas

$$Cu^{2+} + 2CN^- \longrightarrow \underset{\text{yellow}}{Cu(CN)_2\downarrow}$$

$$Cu(CN)_2 \longrightarrow \underset{\text{white}}{2CuCN} + \underset{\text{cyanogen}}{\begin{matrix} CN \\ | \\ CN \end{matrix}}\uparrow$$

Excess of KCN dissolves the precipitate of CuCN forming colourless $[Cu(CN)_4]^{3-}$

$$CuCN\downarrow + 3CN^- \longrightarrow [Cu(CN)_4]^{3-}$$

The complex is so stable that no precipitate of CuS is formed on passing H_2S gas

$$[Cu(CN)_4]^{3-} \rightleftharpoons \text{no ionisation}$$

(e) Cd^{2+} : Cd^{2+} salts form white precipitate of $Cd(CN)_2$ white dissolves in excess of KCN forming soluble tetracyanocadimate(II) ions

$$Cd^{2+} + 2CN^- \longrightarrow \underset{\text{white}}{Cd(CN)_2\downarrow}$$

$$Cd(CN)_2 + 2CN^- \longrightarrow \underset{\text{soluble}}{[Cd(CN)_4]^{2-}}$$

$[Cd(CN)_4]^{2-}$ is unstable forming free Cd^{2+} which is precipitate as CdS (yellow) on passing H_2S gas

$$[Cd(CN)_4]^{2-} + H_2S \longrightarrow \underset{\text{yellow}}{CdS\downarrow} + 2H^+ + 4CN^-$$

Thus, mixture of Cu^{2+} and Cd^{2+} can be identified by complex formation with CN^- and then passing H_2S gas.

$$Cu^{2+} + Cd^{2+} \xrightarrow{\text{excess } CN^-} [Cu(CN)_4]^{3-} + [Cd(CN)_4]^{2-}$$

Yellow precipitate ... Yellow ppt $\xleftarrow{H_2S}$ (from $[Cd(CN)_4]^{2-}$)

(a) confirms Cd^{2+} if original solution is colourless

(b) confirms Cd^{2+} and Cu^{2+} both if original solution is blue.

(f) Fe^{2+} : KCN forms yellowish-brown precipitate of $Fe(CN)_2$ with Fe^{2+}, which dissolves in excess of KCN forming $K_4[Fe(CN)_6]$ a pale yellow coloured solution.

$$Fe^{2+} + 2CN^- \longrightarrow \underset{\text{yellowish-brown}}{Fe(CN)_2\downarrow}$$

$$Fe(CN)_2 + 4CN^- \longrightarrow \underset{\substack{\text{hexacyanoferrate(II) ion}\\ \text{(pale yellow solution)}}}{[Fe(CN)_6]^{4-}}$$

(g) Fe^{3+} : KCN forms reddish-brown precipitate of $Fe(CN)_3$ with Fe^{3+}, which dissolves in excess of KCN due to formation of $K_3[Fe(CN)_6]$ — a yellow coloured solution

$$Fe^{3+} + 3CN^- \longrightarrow \underset{\text{reddish-brown}}{Fe(CN)_3\downarrow}$$

$$Fe(CN)_3 + 3CN^- \longrightarrow \underset{\text{hexacyanoferrate(III) ion}}{[Fe(CN)_6]^{3-}}$$

(h) Co^{2+} : KCN forms reddish-brown precipitate of $Co(CN)_2$ with Co^{2+} ions; precipitate dissolves in excess of KCN forming a brown solution of $K_4[Co(CN)_6]$

$$Co^{2+} + 2CN^- \longrightarrow \underset{\text{reddish-brown}}{Co(CN)_2\downarrow}$$

$$Co(CN)_2 + 4CN^- \longrightarrow \underset{\text{brown (soluble)}}{[Co(CN)_6]^{4-}}$$

If dilute/cold HCl is added $Co(CN)_2$ reappears as reddish-brown precipitate

$$[Co(CN)_6]^{4-} + 4H^+ \longrightarrow Co(CN)_2\downarrow + 4HCN$$

Brown solution of $K_4[Co(CN)_6]$ changes to yellow on boiling in air or by addition of H_2O_2.

$$4[Co(CN)_4]^{4-} + O_2 + 2H_2O \longrightarrow \underset{\text{yellow}}{4[Co(CN)_6]^{3-}} + 4OH^-$$

$$2[Co(CN)_6]^{4-} + H_2O_2 \longrightarrow 2[Co(CN)_6]^{3-} + 2OH^-$$

(i) Ni^{2+} KCN forms green precipitate of $Ni(CN)_2$ with Ni^{2+} salts; precipitate dissolves in excess of KCN forming a yellow coloured solution of $[Ni(CN)_4]^{2-}$.

$$Ni^{2+} + 2CN^- \longrightarrow \underset{\text{green}}{Ni(CN)_2\downarrow}$$

$$Ni(CN)_2 + 2CN^- \longrightarrow \underset{\text{yellow}}{[Ni(CN)_4]^{2-}}$$

In presence of dil. HCl, $Ni(CN)_2$ is again precipitated from soluble $[Ni(CN)_4]^{2-}$.

$$[Ni(CN)_4]^{2-} + 2H^+ \longrightarrow Ni(CN)_2\downarrow + 2HCN$$

A black precipitate of $Ni(OH)_3$ is formed if soluble yellow solution of $[Ni(CN)_4]^{2-}$ is heated with sodium hypobromite solution

$$2[Ni(CN)_4]^{2-} + \underset{\text{hypobromite}}{OBr^-} + 4OH^- + H_2O \longrightarrow \underset{\text{black}}{2Ni(OH)_3\downarrow} + 8CN^- + Br^-$$

2. KI

(a) Pb^{2+} : Yellow precipitates of lead iodide are formed.

$$Pb^{2+} + 2I^- \longrightarrow \underset{\text{yellow}}{PbI_2\downarrow}$$

The precipitate is moderately soluble in boiling water to yield a colourless solution, from which it separates as golden yellow plates on cooling.

An excess of a more concentrated solution (6 M) of the reagent dissolves the pecipitate and tetraiodoplumbate(II) ions are formed.

$$PbI_2 + 2I^- \rightleftharpoons [PbI_4]^{2-}$$

The action is reversible, on diluting with water the precipitate reappears.

(b) Hg_2^{2+} (Mercurous ion) : Potassium iodide gives green precipitate of Hg_2I_2 (mercury(I) iodide or mercurous iodide)

$$Hg_2^{2+} + 2I^- \longrightarrow \underset{\text{green}}{Hg_2I_2\downarrow}$$

If excess of KI is added, a disproportionation reaction takes place, soluble tetraiodo mercurate(II) ions and a black precipitate of finely divided mercury being formed

$$Hg_2I_2\downarrow + 2I^- \longrightarrow [HgI_4]^{2-} + Hg\downarrow$$

Disproportionation of Hg_2I_2 also takes place to HgI_2 and Hg on boiling it with H_2O

$$Hg_2I_2 \xrightarrow{\Delta,\, H_2O} \underset{\text{orange}}{HgI_2\downarrow} + Hg\downarrow$$

(c) Ag^+ : KI gives yellow precipitate AgI with Ag^+ ions.

$$Ag^+ + I^- \longrightarrow \underset{\text{yellow}}{AgI\downarrow}$$

The precipitate is soluble in dilute or concentrated ammonia, butdissolves readily in KCN and in $Na_2S_2O_3$

$$AgI + 2CN^- \longrightarrow \underset{\substack{\text{soluble}\\ \text{dicyanoargentate(I) ion}}}{[Ag(CN)_2]^-} + I^-$$

$$AgI + \underset{\text{thiosulphate}}{2S_2O_3^{2-}} \longrightarrow \underset{\substack{\text{soluble}\\ \text{dithiosulphato argentate(I)}\\ \text{ion}}}{[Ag(S_2O_3)_2]^{3-}} + I^-$$

(d) Hg^{2+} (Mercuric ion) : Hg^{2+} gives orange precipitate of HgI_2 with KI. Precipitate dissolves in excess of KI forming HgI_4^{2-}

$$HgCl_2 + 2KI \longrightarrow \underset{\text{orange}}{HgI_2\downarrow} + 2KCl$$

$$HgI_2 + 2KI \longrightarrow \underset{\text{soluble}}{K_2HgI_4}$$

(e) KI is oxidised by Cl_2 or Br_2 forming I_2 which appears as violet colour in CCl_4 or $CHCl_3$ layer

$$2I^- + Cl_2 \longrightarrow \underset{\substack{\text{violet in}\\ CCl_4}}{I_2} + 2Cl^-$$

$$2I^- + Br_2 \longrightarrow \underset{\substack{\text{violet in}\\ CCl_4}}{I_2} + 2Br^-$$

(f) Cu^{2+} : Cu^{2+} $(CuSO_4)$ gives white precipitates of Cu_2I_2 along with I_2 which gives dark brown coloured solution

$$CuSO_4 + 2KI \longrightarrow CuI_2 + K_2SO_4$$

$$2CuI_2 \longrightarrow \underset{\text{white ppt}}{Cu_2I_2\downarrow} + I_2$$

$$I_2 + KI \rightleftharpoons \underset{\text{dark brown}}{KI_3}$$

If $Na_2S_2O_3$ (sodium thiosulphate-hypo) is added, I_2 disappears and white precipitate of Cu_2I_2 appears

$$I_2(I_3^-) + 2Na_2S_2O_3 \longrightarrow Na_2S_4O_6 + 2NaI$$

(g) Bi^{3+} : KI when added dropwise gives black precipitate of BiI_3

$$Bi^{3+} + 3I^- \longrightarrow \underset{\text{black}}{BiI_3\downarrow}$$

The precipitate dissolves readily in excess reagent forming $[BiI_4]^-$ — orange coloured

$$BiI_3 + I^- \rightleftharpoons \underset{\text{tetraiodobismuthate(III) ions}}{[BiI_4]^-}$$

When diluted with H_2O, the above reaction is reversible forming BiI_3 (black). Heating the precipitate with H_2O, it turns orange due to the formation of bismuthyl iodide.

$$BiI_3 + H_2O \longrightarrow \underset{\text{orange}}{BiOI\downarrow} + 2HI$$

(h) Arsenate(V) AsO_4^{3-} : Arsenate(V) ions oxidise I^- to I_2 in acidic medium which appears as brown coloured solution due to formation of I_3^-

$$AsO_4^{3-} + 2H^+ + 2I^- \longrightarrow AsO_3^{3-} + I_2 + H_2O$$

$$I_2 + KI \rightleftharpoons \underset{\text{brown}}{KI_3}$$

(i) Antimony (Sb^{3+}): KI form yellows coloured solution with Sb^{3+} due to formation of $[SbI_6]^{3-}$

$$Sb^{3+} + 6I^- \longrightarrow \underset{\text{yellow}}{[SbI_6]^{3-}}$$

3. K_2CrO_4

DEEP Focus

The metallic chromates are usually coloured solids, yielding yellow solution when soluble in water. In the presence of dilute mineral acids, chromates are converted into dichromates which yield orange-red aqueous solution. In alkali solution, dichromates change to chromate.

$$\underset{\text{yellow}}{2CrO_4^{2-}} + 2H^+ \rightleftharpoons Cr_2O_7^{2-} + H_2O$$

$$\underset{\text{orange-red}}{Cr_2O_7^{2-}} + 2OH^- \rightleftharpoons \underset{\text{yellow}}{2CrO_4^{2-}} + H_2O$$

The reaction may also be expressed as :

$$2CrO_4^{2-} + 2H^+ \rightleftharpoons 2HCrO_4^- \rightleftharpoons Cr_2O_7^{2-} + H_2O$$

(a) Pb^{2+} : K_2CrO_4 gives yellow precipitate of $PbCrO_4$ on reaction of Pb^{2+} salts after acidification with CH_3COOH

$$Pb^{2+} + CrO_4^{2-} \longrightarrow \underset{\text{yellow}}{PbCrO_4\downarrow}$$

HNO_3 or NaOH dissolves the precipitate

$$2PbCrO_4\downarrow + 2H^+ \rightleftharpoons 2Pb^{2+} + Cr_2O_7^{2-} + H_2O$$

$$2PbCrO_4\downarrow + 4OH^- \rightleftharpoons [Pb(OH)_4]^{2-} + CrO_4^{2-}$$

Both reactions are reversible; by buffering the solution with $NH_3(aq)$ or CH_3COOH respectively, $PbCrO_4$ reprecipitates.

(b) Hg_2^{2+} (Mercurous ion) : Hg_2^{2+} salts form red crystalline precipitate of Hg_2CrO_4 (mercury(I) chromatte)

$$Hg_2^{2+} + CrO_4^{2-} \longrightarrow \underset{\text{red}}{Hg_2CrO_4\downarrow}$$

NaOH turns the precipitate into black mercury(I) oxide :

$$Hg_2CrO_4\downarrow + 2OH^- \longrightarrow \underset{\text{black}}{Hg_2O\downarrow} + CrO_4^{2-} + H_2O$$

(c) Ag^+ : Ag^+ salts form red precipitate of Ag_2CrO_4 (silver(I) chromate)

$$2Ag^+ + CrO_4^{2-} \longrightarrow \underset{\text{red}}{Ag_2CrO_4\downarrow}$$

The precipitate is soluble in dilute HNO_3 and in $NH_3(aq)$ solution.

$$2Ag_2CrO_4\downarrow + 2H^+ \rightleftharpoons 2Ag^+ + \underset{\text{orange}}{Cr_2O_7^{2-}} + H_2O$$

$$Ag_2CrO_4\downarrow + 4NH_3 \longrightarrow 2[Ag(NH_3)_2]^+ + CrO_4^{2-}$$

The acidified solution turns to oragne because of the formation of $Cr_2O_7^{2-}$.

HCl converts the precipitate into AgCl (white)

$$Ag_2CrO_4\downarrow + 2Cl^- \longrightarrow 2AgCl\downarrow + CrO_4^{2-}$$

(d) Ba^{2+} : K_2CrO_4 gives yellow precipitate of $BaCrO_4$ with Ba^{2+} salts in presence of CH_3COOH

$$Ba^{2+} + CrO_4^{2-} \longrightarrow \underset{\text{yellow}}{BaCrO_4\downarrow}$$

(e) H_2S : H_2S reduces acidified chromate solution to green chromium(III) ion accompanied by the separation of sulphur.

$$2CrO_4^{2-} + 3H_2S + 10H^+ \longrightarrow \underset{\text{green}}{2Cr^{3+}} + 3S\downarrow + 8H_2O$$

(f) SO_3^{2-} : SO_3^{2-}/H^+ or SO_2 reduces CrO_4^{2-} to Cr^{3+} (green)

$$2CrO_4^{2-} + 3SO_2 + 4H^+ \longrightarrow \underset{\text{green}}{2Cr^{3+}} + 3SO_4^{2-} + 2H_2O$$

(f) Fe^{2+} : Fe^{2+} in presence of mineral acids reduces chromates to Cr^{3+} (green)

$$CrO_4^{2-} + 3Fe^{2+} + 8H^+ \longrightarrow \underset{\text{green}}{Cr^{3+}} + 3Fe^{3+} + 4H_2O$$

(g) I^- (Iodides) : Iodide (KI) reduces CrO_4^{2-} to Cr^{3+} (green), and itself oxidised to I_2

$$2CrO_4^{2-} + 6I^- + 16H^+ \longrightarrow 2Cr^{3+} + 3I_2 + 8H_2O$$

I_2 goes into solution as I_3^- (deep brown or bluish brown)

$$I_2 + I^- \rightleftharpoons I_3^-$$

(h) C_2H_5OH : Ethanol (C_2H_5OH) in presence of mineral acid reduces chromates (or dichromates) slowly in cold but rapidly if the solution is heated

$$2CrO_4^{2-} + 3C_2H_5OH + 10H^+ \longrightarrow \underset{\text{green}}{2Cr^{3+}} + 3CH_3CHO\uparrow + 8H_2O$$

Smell of acetaldehyde (CH_3CHO) is observed on heating the mixture.

(4) $K_4[Fe(CN)_6]$ Potassium hexacyanoferrate(II) (Potassium ferrocyanide)

(a) Cu^{2+} : Cu^{2+} salts on reaction with CH_3COOH and $K_4[Fe(CN)_6]$ give reddish brown precipitate of $Cu_2[Fe(CN)_6]$

$$Cu^{2+} + [Fe(CN)_6]^{4-} \longrightarrow \underset{\text{reddish brown}}{Cu_2[Fe(CN)_6]\downarrow}$$

(b) Cd^{2+} : Cd^{2+} salts give bluish white precipitate of $Cd_2[Fe(CN)_6]$ under on reaction with $K_4[Fe(CN)_6]$ and CH_3COOH

$$Cd^{2+} + [Fe(CN)_6]^{4-} \longrightarrow \underset{\text{bluish white}}{Cd_2[Fe(CN)_6]\downarrow}$$

(c) Zn^{2+} : (Similar to Cd^{2+})

(d) Fe^{3+} : Fe^{3+} salts forms blue coloured precipitate of $Fe_4[Fe(CN)_6]_3$ called **Prussian blue**

$$4FeCl_3 + 3K_4[Fe(CN)_6] \longrightarrow \underset{\text{Prussian blue}}{Fe_4[Fe(CN)_6]_3} + 12KCl$$

A large excess of the reagent dissolves Prussian blue partly or entirely when an intense blue colour is obtained.

$$Fe_4[Fe(CN)_6]_3 + K_4[Fe(CN)_6] \longrightarrow \underset{\text{soluble}}{4KFe[Fe(CN)_6]}$$

(e) Fe^{2+} : When $K_3[Fe(CN)_6]$ is added to Fe^{2+} salts, first $K_3[Fe(CN)_6]$ oxidises Fe^{2+} to Fe^{3+} and itself reduced to $K_4[Fe(CN)_6]$

$$[Fe^{III}(CN)_6]^{3-} + Fe^{2+} \longrightarrow Fe^{3+} + [Fe^{II}(CN)_6]^{4-}$$

Fe^{3+} and $[Fe(CN)_6]^{4-}$ combine to form a blue precipitate of **Turnbull's blue**

$$4Fe^{3+} + 3[Fe(CN)_6]^{4-} \longrightarrow \underset{\text{Turbull's blue}}{Fe_4[Fe(CN)_6]_3\downarrow}$$

DEEP Focus

We can observe that the composition of this precipitate is identical to that of Prussian blue. Earlier it was suggested that its composition was $Fe_3[Fe(CN)_6]_2$ hence the different name. The identical composition and structure of Turnbull's blue and Prussian blue has recently been proved by **Mossbauer spectroscopy.** The precipitate is decomposed by NaOH or KOH forming $Fe(OH)_3$ — brown coloured precipitate.

(f) Ag^+ : Silver salts ($AgNO_3$) give white precipitate of $Ag_4[Fe(CN)_6]$ with $K_4[Fe(CN)_6]$

$$4Ag^+ + [Fe(CN)_6]^{4-} \longrightarrow \underset{\text{white}}{Ag_4[Fe(CN)_6]\downarrow}$$

Precipitate dissolves in KCN and $Na_2S_2O_3$ but remains insoluble in $NH_3(aq)$

$$Ag_4[Fe(CN)_6]\downarrow + 8CN^- \longrightarrow \underset{\text{soluble}}{4[Ag(CN)_2]^-} + [Fe(CN)_6]^{4-}$$

$$Ag_4[Fe(CN)_6]\downarrow + 8S_2O_3^{2-} \longrightarrow \underset{\text{soluble}}{4[Ag(S_2O_3)_2]^{3-}} + [Fe(CN)_6]^{4-}$$

5. KSCN (Potassium thiocyanate solution)

Note : Instead NH_4SCN can also be used

(a) Hg^{2+} (Mercuric ion) When NH_4SCN and cobalt acetate is added to Hg^{2+} solution, a blue colour precipitate is produced

$$Hg^{2+} + Co^{2+} + 4SCN^- \longrightarrow \underset{\substack{\text{blue}\\ \text{cobalt tetrathiocyanato-}\\ \text{mercurate(II)}}}{Co[Hg(SCN)_4]\downarrow}$$

White precipitate of $Hg(SCN)_2$ readily soluble in excess of NH_4SCN is obtained

$$Hg^{2+} + 2SCN^- \longrightarrow Hg(SCN)_2$$

$$Hg(SCN)_2 + 2SCN^- \longrightarrow [Hg(SCN)_4]^{2-}$$

(b) Cu^{2+} : NH_4SCN on reaction with Cu^{2+} salts gives black precipitate of $Cu(SCN)_2$

$$Cu^{2+} + 2SCN^- \longrightarrow \underset{\text{black}}{Cu(SCN)_2\downarrow}$$

The precipitate decomposes slowly to form white precipitate of CuSCN along with thiocyanogen gas

$$\underset{\text{black}}{2Cu(SCN)_2\downarrow} \longrightarrow \underset{\text{white}}{2CuSCN\downarrow} + \underset{\text{thiocyanogen}}{(SCN)_2\uparrow}$$

$Cu(SCN)_2$ can be converted to CuSCN immediately by suitable reducing agent like SO_3^{2-}/H^+

$$SO_3^{2-} + 2H^+ \longrightarrow SO_2 + H_2O$$

$$\underset{\text{black}}{2Cu(SCN)_2} + SO_2 + 2H_2O \longrightarrow \underset{\text{white}}{2CuSCN\downarrow} + 2SCN^- + SO_4^{2-} + 4H^+$$

(c) Fe^{3+} : NH_4SCN reacts with Fe^{2+} salts in slightly acidic medium giving red colour solution of $Fe(SCN)_3$

$$Fe^{3+} + 3SCN^- \longrightarrow \underset{\text{red}}{Fe(SCN)_3}$$

Sometime $[Fe(SCN)]^{2+}$ is also formed. $Fe(SCN)_3$ can be extracted by ether or amyl alcohol. Fluorides, oxalates and Hg(II) ions bleach the red colour of $Fe(SCN)_3$.

$$\underset{\text{red}}{Fe(SCN)_3} + 6F^- \longrightarrow \underset{\substack{\text{more stable}\\ \text{(colourless)}}}{[FeF_6]^{3-}} + 3SCN^-$$

$$2Fe(SCN)_3 + 3Hg^{2+} \longrightarrow 2Fe^{3+} + 3Hg(SCN)_2$$

$$Fe(SCN)_3 + 3\underset{\text{oxalate}}{(COO)_2^{2-}} \longrightarrow [Fe\{(COO)_2\}_3]^{3-} + 3SCN^-$$

(d) Co^{2+} : On adding a few crystals of NH_4SCN to a neutral or acid solution of cobalt(II), a blue colour appears owing to the formation of $[Co(SCN)_4]^{2-}$

$$Co^{2+} + 4SCN^- \longrightarrow \underset{\substack{\text{tetrathiocyanato-cobalt(II) ion} \\ \text{(blue)}}}{[Co(SCN)_4]^{2-}}$$

6. Na_2HPO_4

(a) Pb^{2+} : Na_2HPO_4 gives white precipitate with Pb^{2+} salts

$$3Pb^{2+} + 2HPO_4^{2-} \rightleftharpoons \underset{\text{white}}{Pb_3(PO_4)_2\downarrow} + 2H^+$$

The reaction is reversible; strong acids (HNO_3) dissolve the precipitate. The precipitate is also soluble in NaOH.

(b) Hg_2^{2+} (Mercurous) : Na_2HPO_4 gives white precipitate with Hg_2^{2+}

$$Hg_2^{2+} + HPO_4^{2-} \longrightarrow \underset{\text{white}}{Hg_2HPO_4\downarrow}$$

(c) Ag^+ : Na_2HPO_4 in neutral solution gives yellow precipitate of silver phsophate with Ag^+

$$3Ag^+ + HPO_4^{2-} \longrightarrow \underset{\text{yellow}}{Ag_3PO_4\downarrow} + H^+$$

(d) Fe^{3+} : NaH_2PO_4 gives yellowish-white precipitate with Fe^{3+} salts

$$Fe^{3+} + HPO_4^{2-} \longrightarrow \underset{\text{yellowish white}}{FePO_4\downarrow} + H^+$$

The reaction is reversible due to formation of strong acid (H^+) which dissolves $FePO_4$. Hence, $FePO_4$ is precipitated in presence of acidic buffer of CH_3COOH and CH_3COONa.

(e) Al^{3+} : White gelatinous precipitate of $AlPO_4$ is obtained when Al^{3+} salts react with Na_2HPO_4.

$$Al^{3+} + HPO_4^{2-} \rightleftharpoons AlPO_4\downarrow + H^+$$

The reaction is reversible; strong acids dissolve the precipitate of $AlPO_4$. Precipitates are also soluble in NaOH but insoluble in CH_3COOH.

(f) Mn^{2+} : Mn^{2+} salts are precipitated as manganese ammonium phosphate, $Mn(NH_4)PO_4 \cdot 7H_2O$, (pink) on the addition of Na_2HPO_4 in presence of $NH_3(aq)$

$$Mn^{2+} + NH_3 + HPO_4^{2-} \longrightarrow \underset{\text{pink}}{Mn(NH_4)PO_4\downarrow}$$

If $NH_3(aq)$ is absent, $Mn_3(PO_4)_2$ is precipitated

$$3Mn^{2+} + 2HPO_4^{2-} \longrightarrow Mn_3(PO_4)_2 + 2H^+$$

Both precipitates are soluble in acids.

(g) Zn^{2+} : Zn^{2+} salts are precipitated as $Zn_3(PO_4)_2$ on the addition of Na_2HPO_4.

$$3Zn^{2+} + 2HPO_4^{2-} \longrightarrow \underset{\text{white}}{Zn_3(PO_4)_2\downarrow} + 2H^+$$

However, in presence of $NH_3(aq)$, $Zn(NH_4)PO_4$ is precipitated

$$Zn^{2+} + NH_3(aq) + HPO_4^{2-} \longrightarrow \underset{\text{white}}{Zn(NH_4)PO_4\downarrow}$$

Both the precipitates are soluble in dilute acids. Excess of $NH_3(aq)$ dissolves the precipitates

$$Zn_3(PO_4)_2 + 12NH_3 \longrightarrow \underset{\text{soluble}}{3[Zn(NH_3)_4]^{2+}} + 2PO_4^{3-}$$

$$Zn(NH_4)PO_4 + 3NH_3 \longrightarrow \underset{\text{soluble}}{[Zn(NH_3)_4]^{2+}} + HPO_4^{2-}$$

(h) Mg^{2+} : Mg^{2+} salts are precipitated as $Mg(NH_4)PO_4 \cdot 6H_2O$ when Na_2HPO_4 is added in presence of basic buffer ($NH_4Cl + NH_4OH$).

$$Mg^{2+} + NH_3 + HPO_4^{2-} \longrightarrow \underset{\text{white}}{Mg(NH_4)PO_4\downarrow}$$

The precipitate is sparingly soluble in water; soluble in acetic acid and in mineral acids. The precipitate separates slowly from dilute solutions because of its tendency to form supersaturated solutions; this may be overcome by cooling and by rubbing the test-tube with a glass rod.

In neutral solution, a white flocculant precipitate of $MgHPO_4$ is produced

$$Mg^{2+} + HPO_4^{2-} \longrightarrow MgHPO_4\downarrow$$

7. MnO_4^-

All permanganates are soluble in water forming purple (reddish-violet) solutions

(a) $C_2O_4^{2-}$ (Oxalate) : $C_2O_4^{2-}$ decolourises $KMnO_4$ in acidic medium.

$$5C_2O_4^{2-} + 2MnO_4^- + 16H^+ \longrightarrow 10CO_2 + 2Mn^{2+} + 8H_2O$$

DEEP Focus

The reaction is slow at room temperature but is rapid at 60°C. Manganese(II), Mn^{2+} ions, catalyse the reaction; thus reaction is autocatalytic; once Mn^{2+} ions are formed, it becomes faster.

(b) SO_3^{2-} : SO_3^{2-}/H^+ or SO_2 also decolourises MnO_4^-

$$SO_3^{2-} + 2H^+ \longrightarrow SO_2 + H_2O$$

$$2MnO_4^- + 5SO_2 + 2H_2O \longrightarrow 2Mn^{2+} + 5SO_4^{2-} + 4H^+$$

Some dithionate ($S_2O_6^{2-}$) may also be formed in this reaction, the quantity being dependent upon the experimental conditions :

$$2MnO_4 + 6SO_2 + 2H_2O \longrightarrow 2Mn^{2+} + S_2O_6^{2-} + 4SO_4^{2-} + 4H^+$$

(c) Fe^{2+} : Fe^{2+} reduces MnO_4^- to Mn^{2+} and solution becomes yellow due to formation of Fe^{3+}

$$MnO_4^- + 5Fe^{2+} + 8H^+ \longrightarrow 5Fe^{2+} + Mn^{2+} + 4H_2O$$

The yellow colour disappears if phosphoric acid or potassium fluoride is added due to formation of colourless complexes with Fe^{3+}.

(d) I^- : KI reduces MnO_4^-/H^+ and solution turn dark brown due to formation of I_3^-

$$2MnO_4^- + 10I^- + 16H^+ \longrightarrow 2Mn^{2+} + 8H_2O + 5I_2$$

$$5I_2 + 5I^- \rightleftharpoons \underset{\text{dark brown}}{5I_3^-}$$

(e) NO_2^- : NO_2^- also decolourises MnO_4^-/H^+

$$2MnO_4^- + 5NO_2^- + 6H^+ \longrightarrow 2Mn^{2+} + 5NO_3^- + 3H_2O$$

(f) Upon warming a concentrated solution of $KMnO_4$ with concentrated NaOH solution, a green solution of K_2MnO_4 (potassium manganate) is produced and O_2 is evolved. When K_2MnO_4 solution is diluted or acidified with dil. H_2SO_4, the purple colour of $KMnO_4$ is restored and MnO_2 (manganese dioxide) is precipitated (as $MnO(OH)_2$).

$$4MnO_4^- + 4OH^- \longrightarrow 4MnO_4^{2-} + O_2\uparrow + 2H_2O$$

$$\underset{\substack{\text{green}\\+6}}{3MnO_4^{2-}} + 3H_2O \longrightarrow \underset{\substack{\text{purple}\\+7}}{2MnO_4^-} + \underset{+4}{MnO(OH)_2} + 4OH^-$$

The MnO_4^{2-} disproportionates to MnO_4^- (oxidation) and MnO_2 (reduction).

8. Thioacetamide

(Analysis of cations without H_2S)

H_2S in the laboratory is produced by reaction of H_2SO_4 with FeS

$$FeS + H_2SO_4 \longrightarrow FeSO_4 + H_2S$$

This causes lot of pollution and hence formation of H_2S by the above reaction should be discouraged. Thioacetamide serves as a source of H_2S. On hydrolysis in hot acidic solution, it forms H_2S (*in situ*) and CH_3COONH_4.

$$\underset{\text{thioacetamide}}{CH_3\overset{\overset{S}{\|}}{C}NH_2} + 2H_2O \xrightarrow{\Delta} CH_3\overset{\overset{O}{\|}}{C}ONH_4 + H_2S$$

Thus, simply addition of thioacetamide in hot solution is the instant source of H_2S in the test-tube itself.

$$Pb^{2+} + H_2S \longrightarrow \underset{\text{black}}{PbS\downarrow} + 2H^+$$

$$Cu^{2+} + H_2S \longrightarrow \underset{\text{black}}{CuS\downarrow} + 2H^+$$

Total Practice Set
(Read, Plan & Solve!)

Problem 1. (*A*) is yellow coloured solid partially soluble in aqueous NH_3. (*A*) is soluble in $Na_2S_2O_3$ (hypo) solution forming a complex (*B*) which on heating is converted into (*C*) (black). (*C*) is converted into white ppt (*D*) on reaction with HCl and HNO_3. (*D*) is soluble in aqueous NH_3 forming (*E*). Identify (*A*), (*B*), (*C*), (*D*) and (*E*) and explain reactions.

Solution (*A*) is **AgBr** (present in photographic plate)

$$\underset{}{AgBr} + \underset{\text{hypo}}{2Na_2S_2O_3} \longrightarrow \underset{\substack{\text{soluble}\\(B)}}{Na_3[Ag(S_2O_3)_2]} + NaBr$$

$$\underset{(B)}{} \xrightarrow{\Delta} \underset{(C)}{Ag_2S}$$

$$\underset{(E)}{[Ag(NH_3)_2Cl]} \xleftarrow{NH_3} \underset{(D)}{AgCl\downarrow} \xleftarrow{HCl+HNO_3} \underset{(C)}{Ag_2S}$$

Problem 2. Identify (*A*) and (*B*) in the following and explain reactions given VD of (*A*) is 17.
A decolourises acidified $KMnO_4$. (*A*) also gives brown ppt (*B*) with alkaline $KMnO_4$. (*A*) also liberates I_2 from acidified KI solution. (*A*) also removes black stains from old oil paintings.

Solution Molecular weight of $(A) = 2 \times VD = 34 \text{ g mol}^{-1}$. Also (*A*) removes stains from old oil paintings.

$\Rightarrow$ (*A*) **is H_2O_2** (mol. wt. = 34)

$$2MnO_4^- + 5H_2O_2 + 6H^+ \longrightarrow 2Mn^{2+} + 8H_2O + 5O_2$$

$$2MnO_4^- + 3H_2O_2 \longrightarrow \underset{(B)\text{ (brown)}}{2MnO_2\downarrow} + 3O_2 + 2OH^- + 2H_2O$$

$$4H_2O_2 + \underset{\text{black (of oil paintings)}}{PbS} \longrightarrow PbSO_4 + 4H_2O$$

Problem 3. (*A*) + tap water $\longrightarrow$ white turbidity soluble in *aq* NH_3

$$(A) \xrightarrow{\Delta} \text{residue } (B) + (C) \text{ (oxides of N)} + O_2$$

aqueous (*A*) gives brown ring on adding $FeSO_4$ and conc. H_2SO_4. (*C*) is paramagnetic and forms dimer of VD 46. Identify (*A*), (*B*) and (*C*) and explain reactions.

Solution (*A*) + tap water $\longrightarrow$ white turbidity soluble in *aq* NH_3. Tap water has Cl^- and turbidity is soluble in *aq* NH_3.

$\Rightarrow$ turbidity is of AgCl and thus (*A*) has Ag^+

(*A*) also gives **ring test** of NO_3^-

$\Rightarrow$ (*A*) has NO_3^- and thus (*A*) is **$AgNO_3$**

Explanation : $2AgNO_3 \xrightarrow{\Delta} \underset{(B)}{2Ag} + \underset{(C)}{2NO_2} + O_2$

NO_2 is paramagnetic due to one unpaired electron as shown and thus forms dimer by using unpaired electron.

$$2NO_2 \rightleftharpoons \underset{\text{dimer (VD = 46)}}{N_2O_4}$$

$\cdot N(\to O)(=O)$

Problem 4. (*A*), an important laboratory reagent, turns red litmus blue, imparts golden yellow colour in flame and is a good precipitating agent. (*A*) reacts with Zn or Al forming H_2 gas. (*A*) gives white ppt with $ZnCl_2$ or $AlCl_3$ but ppt dissolves in excess of (*A*). What is (*A*) and explain reactions?

Solution (*A*) turns blue litmus red $\Rightarrow$ (*A*) is basic in nature.

(*A*) imparts golden yellow colour in flame $\Rightarrow$ (*A*) has Na^+.

(*A*) gives H_2 gas with Zn or Al $\Rightarrow$ (*A*) is **NaOH.**

Explanation :

$$2NaOH + Zn \longrightarrow Na_2ZnO_2 + H_2\uparrow$$

$$2NaOH + 2H_2O + 2Al \longrightarrow 2NaAlO_2 + 3H_2\uparrow$$

$$ZnCl_2 + 2NaOH \longrightarrow \underset{\text{white ppt}}{Zn(OH)_2\downarrow} + 2NaCl$$

$$Zn(OH)_2 + 2NaOH \longrightarrow Na_2[Zn(OH)_4] \text{ or } \underset{\text{sodium zincate}}{Na_2ZnO_2}$$

$$AlCl_3 + 3NaOH \longrightarrow \underset{\text{white ppt}}{Al(OH)_3\downarrow} + 3NaCl$$

$$Al(OH)_3 + NaOH \longrightarrow Na[Al(OH)_4] \text{ or } \underset{\text{sodium meta-aluminate}}{NaAlO_2}$$

Problem 5. Identify (*A*) and (*B*) based on following reactions :

(i) $(A) \xrightarrow[\Delta]{NaOH} (B)(g) \xrightarrow{HCl}$ white fumes

(ii) After (*B*) is expelled completely, resultant alkaline solution again gives gas (*B*) on heating with zinc.

(iii) $(A) \xrightarrow{\Delta} N_2O + H_2O$

Solution By (iii) (*A*) gives N_2O, H_2O on heating, hence (A) is **NH_4NO_3**

$$NH_4NO_3 \xrightarrow{\Delta} N_2O + 2H_2O$$

(i) $$NH_4NO_3 + NaOH \xrightarrow{\Delta} NaNO_3 + \underset{(B)}{NH_3} + H_2O$$

$$\underset{(B)}{NH_3} + HCl \longrightarrow \underset{\text{white fumes}}{NH_4Cl} + H_2O$$

(ii) $$2NaOH + Zn \xrightarrow{\Delta} Na_2ZnO_2 + 2H$$

$$NaNO_3 + 8H \xrightarrow{\Delta} NaOH + \underset{(B)}{NH_3} + 2H_2O$$

(NH_3 gas is obtained due to reduction of NO_3^-)

Problem 6. Identify (*A*), (*B*), (*C*) and (*D*) based on following observations :

(i) $(A) \xrightarrow{\Delta}$ glassy transparent bead (*B*) on platinum wire

$(B) + CuSO_4 \longrightarrow$ coloured bead (*C*)

(ii) $(A) + \text{conc. } H_2SO_4 + CH_3CH_2OH \xrightarrow{\text{ignite}}$ green flame (*D*)

(iii) Aqueous solution of (*A*) is alkaline.

Solution (i) (*A*) forms glassy transparent bead which is characteristic property of **borax.**

⇒ (*A*) is **borax ($Na_2B_4O_7 \cdot 10H_2O$)**

$$Na_2B_4O_7 \cdot 10H_2O \xrightarrow{\Delta} \underbrace{NaBO_2 + B_2O_3}_{(B)\text{ glassy bead}} + 10H_2O$$

$$B_2O_3 + CuSO_4 \xrightarrow{\Delta} \underset{\text{coloured bead }(C)}{Cu(BO_2)_2} + SO_3 \uparrow$$

(ii) $$Na_2B_4O_7 + H_2SO_4 + 5H_2O \longrightarrow Na_2SO_4 + 4H_3BO_3$$

$$H_3BO_3 + 3C_2H_5OH \longrightarrow \underset{\substack{\text{green flame}\\ \text{(on ignition)}\\ (D)}}{(C_2H_5)_3BO_3} + 3H_2O$$

(iii) $$Na_2B_4O_7 + 5H_2O \longrightarrow \underset{\text{weak acid}}{2H_3BO_3} + 2Na[B(OH)_4]$$

$Na[B(OH)_4]$ reacts with acid (HCl) hence aqueous solution of (*A*) is alkaline.

Problem 7. A colourless mixture of two salts (*A*) and (*B*) (excess) is soluble in H_2O. (*A*) turns blue litmus red and (*B*) turns red litmus blue. (*A*) gives white ppt with (*B*), which dissolves in excess of (*B*) forming (*C*). (*A*) when placed in atmosphere gives fumes and can form dimer. (*A*) gives white ppt with NH_4Cl and NH_4OH soluble in (*B*). Identify (*A*), (*B*) and (*C*) and explain reactions.

Solution (*A*) gives white ppt with NH_4Cl and NH_4OH

⇒ (*A*) is Al^{3+}

⇒ ppt is of $Al(OH)_3$

$Al(OH)_3$ is soluble in (*B*) ⇒ (*B*) is NaOH

⇒ (*C*) is $NaAlO_2$ (sodium meta-aluminate)

(*A*) forms dimer indicating (*A*) is electron-deficient that is the possible case of $AlCl_3$.

(*A*) is thus **$AlCl_3$** (acidic due to hydrolysis, turns blue litmus red)

(*B*) is **NaOH** (alkaline, turns red litmus blue)

$$\underset{(A)}{AlCl_3} + \underset{(B)}{3NaOH} \longrightarrow \underset{\text{white ppt}}{Al(OH)_3} + 3NaCl$$

$$Al(OH)_3 + \underset{\text{(excess)}}{NaOH} \longrightarrow \underset{\text{soluble }(C)}{NaAlO_2} + H_2O$$

$$AlCl_3 + 3H_2O \longrightarrow Al(OH)_3 + 3HCl \uparrow \text{ (fumes)}$$

$$AlCl_3 + 3NH_4OH \xrightarrow{NH_4Cl} Al(OH)_3 \downarrow + 3NH_4Cl$$

$$2AlCl_3 \longrightarrow Al_2Cl_6$$

(Structure of Al_2Cl_6: two Al atoms each bonded to two terminal Cl, bridged by two Cl atoms.)

Problem 8. Identify (*A*) based on following facts:

(a) (*A*) reduces $HgCl_2$ solution to white ppt changing to grey.

(b) (*A*) turns $FeCl_3$ yellow coloured solution to green.

(c) (*A*) gives white ppt with NaOH soluble in excess of NaOH.

(d) (*A*) gives yellow dirty ppt on passing H_2S gas, soluble in yellow ammonium sulphide (YAS).

(e) (*A*) gives chromyl chloride test.

Solution (*A*) gives yellow dirty ppt with H_2S soluble in YAS

⇒ (*A*) has Sn^{2+}

(*A*) gives chromyl chloride test (of Cl^-)

⇒ (*A*) has Cl^- ion and thus (*A*) is **$SnCl_2$**

Explanation :

(a) $$SnCl_2 + 2HgCl_2 \longrightarrow \underset{\text{white}}{Hg_2Cl_2 \downarrow} + SnCl_4$$

$$SnCl_2 + Hg_2Cl_2 \longrightarrow \underset{\text{grey}}{2Hg \downarrow} + SnCl_4$$

(b) $2FeCl_3 + SnCl_2 \longrightarrow 2FeCl_2 + SnCl_4$
yellow green

(c) $SnCl_2 + 2NaOH \longrightarrow Sn(OH)_2 \downarrow + 2NaCl$
white

$Sn(OH)_2 + 2NaOH \longrightarrow Na_2SnO_2 + 2H_2O$
soluble
(sodium stannite)

(d) $SnCl_2 + H_2S \longrightarrow 2HCl + SnS \downarrow$
yellow

$SnS + (NH_4)_2S_2 \longrightarrow (NH_4)_2SnS_3$
YAS soluble

Problem 9. You have a mixture of CO_2, SO_2 and O_2 gas. You have been provided following reagents for testing
(a) lime water
(b) acidified potassium dichromate solution
(c) pyrogallol
Arrange these reagents in order such that only one gas is tested at one time.

Solution The correct order is (b), (a), (c)

CO_2, SO_2, O_2 → $K_2Cr_2O_7$ —turns green due to SO_2
↓ CO_2, O_2
$Ca(OH)_2$ —turns milky due to CO_2
↓ O_2
pyrogallol —turns black due to O_2

Note:
- SO_2 and CO_2 both turn lime water milky hence $Ca(OH)_2$ can't be used before $K_2Cr_2O_7$.
- O_2 is soluble in pyrogallol.

Problem 10. Identify gas (*X*) based on following facts:
(a) (*X*) occupies 0.35 L per g at NTP
(b) (*X*) turns acidified $K_2Cr_2O_7$ solution green.
(c) (*X*) decolourises acidified $KMnO_4$ solution.
(d) (*X*) gives white turbidity when H_2S gas is passed into its aqueous solution.

Solution (a) (*X*) occupies 0.35 L per g at NTP
⇒ molecular weight of (*X*) is 64 g mol^{-1}
(b) (*X*) turns $K_2Cr_2O_7/H^+$ green, (*X*) is a reducing agent and green colour is due to formation of Cr^{3+}.

$$Cr_2O_7^{2-} + 14H^+ + 6e^- \longrightarrow 2Cr^{3+} + 7H_2O$$

(c) (*X*) also reduces MnO_4^-/H^+

$$MnO_4^- + 8H^+ + 5e^- \longrightarrow Mn^{2+} + 4H_2O$$

⇒ probably (*X*) is $\mathbf{SO_2}$.

$$SO_2 + 2H_2O \longrightarrow SO_4^{2-} + 4H^+ + 2e^-$$

(d) SO_2 in aqueous solution gives white turbidity due to formation of colloidal sulphur.

$$SO_2 + 2H_2S \longrightarrow 2H_2O + \underset{\text{(colloidal)}}{3S}$$

Problem 11. (*A*) is a colourless salt with divalent cation. (*A*) gives yellow ppt with K_2CrO_4 as well as with $AgNO_3$. (*A*) gives apple green colour in flame. (*A*) is precipitated by H_2S neither in acidic nor ammoniacal medium, but addition of $(NH_4)_2CO_3$ in NH_4OH gave white ppt (*B*). 0.297 g of (*A*) gave 0.197 g of (*B*). Identify (*A*) and explain reactions.

Solution (*A*) gives yellow ppt with KI as well as with $AgNO_3$.
⇒ (*A*) can have Pb^{2+} or Ba^{2+} (both divalent cations) and Br^-. (*A*) is not precipitated by H_2S in acidic and ammoniacal medium.
⇒ (*A*) does not have Pb^{2+} (precipitated as PbS in acidic medium by H_2S gas)
⇒ (*A*) contains Ba^{2+} (apple green flame)
⇒ (*A*) is $\mathbf{BaBr_2}$

$$BaBr_2 + 2AgNO_3 \longrightarrow \underset{\text{yellow}}{2AgBr \downarrow} + Ba(NO_3)_2$$

$$BaBr_2 + K_2CrO_4 \longrightarrow \underset{\text{yellow}}{BaCrO_4 \downarrow} + 2KBr$$

$$\underset{(A)}{BaBr_2} + (NH_4)_2CO_3 \longrightarrow \underset{(B)}{BaCO_3} + 2NH_4Br$$

$$\underset{\substack{?\\0.297}}{BaBr_2} + (NH_4)_2CO_3 \longrightarrow \underset{\substack{197\text{ g}\\0.197\text{ g}}}{BaCO_3} + 2NH_4Br$$

$$\frac{M}{0.297} = \frac{197}{0.197}$$

$$\therefore \quad M(BaBr_2) = 297 \text{ g mol}^{-1}$$

Thus, *A* is $\mathbf{BaBr_2}$
(Ba = 137, Br = 80).

Problem 12. A colourless salt (*A*), soluble in water, gives a mixture of three gases (*B*), (*C*) and (*D*) along with water vapours. (*B*) is blue, (*C*) is red and gas (*D*) is neutral towards litmus paper. Gas (*B*) is also obtained when (*A*) is heated with NaOH and gives brown ppt with K_2HgI_4. Solution thus obtained gives white ppt (*E*) with $CaCl_2$ solution in presence of CH_3COOH. (*E*) decolorises MnO_4^-/H^+. Gas (*C*) turns lime water milky while gas (*D*) burns with blue flame and is fatal when inhaled.
Identify (*A*) to (*D*) and explain reactions.

Solution Gas (*B*) gives brown ppt with $\mathbf{K_2HgI_4}$
⇒ gas (*B*) is $\mathbf{NH_3}$
⇒ (*A*) has NH_4^+
Gas (*C*) turns lime water milky
⇒ gas (*C*) can be SO_2 or CO_2
Gas (*D*) is also obtained along with (*C*). Gas (*D*) burns with blue flame and is fatal when inhaled
⇒ gas (*D*) is **CO**
⇒ gas (*C*) is $\mathbf{CO_2}$
⇒ (*A*) has $C_2O_4^{2-}$ ion

It is confirmed by the fact that $CaCl_2$ gives white ppt CaC_2O_4 (*E*) which decolourises MnO_4^- / H^+

$\Rightarrow$ (*A*) is $(NH_4)_2C_2O_4$

Explanation :

$$\underset{(A)}{(NH_4)_2C_2O_4} \xrightarrow{\Delta} \underset{(B)}{2NH_3} + \underset{(C)}{CO_2} + \underset{(D)}{CO} + H_2O$$

(*B*) is blue towards litmus (basic)
(*C*) is red towards litmus (acidic)
(*D*) is neutral

$$(NH_4)_2C_2O_4 + 2NaOH \xrightarrow{\Delta} Na_2C_2O_4 + \underset{(B)}{2NH_3} + 2H_2O$$

$$Na_2C_2O_4 + CaCl_2 \longrightarrow \underset{\substack{\text{white ppt}\\(E)}}{CaC_2O_4\downarrow} + 2NaCl$$

$$3NaOH + NH_3 + 2K_2HgI_4 \longrightarrow \underset{\substack{\text{brown ppt}\\\text{(Iodide of Millon's base)}}}{O\langle^{Hg}_{Hg}\rangle NH_2I\downarrow} + 4KI + 3NaI + 2H_2O$$

$$\underset{\text{violet}}{2MnO_4^-} + 16H^+ + 5C_2O_4^{2-} \longrightarrow 10\,CO_2 + \underset{\text{colourless}}{2Mn^{2+}} + 8H_2O$$

Problem 13. Match the following using three columns with (*X*) containing radicals, (*Y*) the reagents used and (*Z*) the compound formed when (*X*) and (*Y*) react.

	(*X*)	(*Y*)	(*Z*)
(I)	Fe^{2+}	(a) NH_4SCN	(*A*) Cherry red ppt
(II)	Fe^{3+}	(b) DMG	(*B*) Turnbull's blue
(III)	Ni^{2+}	(c) $K_4[Fe(CN)_6]$	(*C*) Prussian blue
(IV)	Co^{2+}	(d) $K_3[Fe(CN)_6]$	(*D*) Blue

Solution

(*X*)	(*Y*)	(*Z*)
(I)	(d)	(*B*)
(II)	(c)	(*C*)
(III)	(b)	(*A*)
(IV)	(a)	(*D*)

Problem 14. Name one common reagent that can precipitate or react and differentiate following pairs

(a) Ag^+ and Ba^{2+} (b) Cu^{2+} and Pb^{2+}
(c) I^- and Cl^- (d) I^- and Br^-
(e) SO_3^{2-} and SO_4^{2-} (f) Fe^{3+} and Cu^{2+}
(g) Co^{2+} and Cu^{2+}

Solution (a) K_2CrO_4:

$$2Ag^+ + CrO_4^{2-} \longrightarrow \underset{\text{red}}{Ag_2CrO_4\downarrow}$$

$$Ba^{2+} + CrO_4^{2-} \longrightarrow \underset{\text{yellow}}{BaCrO_4\downarrow}$$

(b) K_2CrO_4:

$$Cu^{2+} + CrO_4^{2-} \longrightarrow \underset{\text{chocolate}}{CuCrO_4\downarrow}$$

$$Pb^{2+} + CrO_4^{2-} \longrightarrow \underset{\text{yellow}}{PbCrO_4\downarrow}$$

(c) $AgNO_3$:

$$Ag^+ + I^- \longrightarrow \underset{\text{yellow}}{AgI\downarrow}$$

$$Ag^+ + Cl^- \longrightarrow \underset{\text{white}}{AgCl\downarrow}$$

(d) Cl_2 water + $CHCl_3$:

$$2I^- + Cl_2 \longrightarrow \underset{\substack{\text{(violet in}\\CHCl_3\text{ layer)}}}{I_2} + 2Cl^-$$

$$2Br^- + Cl_2 \longrightarrow \underset{\substack{\text{(orange in}\\CHCl_3\text{ layer)}}}{Br_2} + 2Cl^-$$

(e) $BaCl_2$:

$$BaCl_2 + SO_3^{2-} \longrightarrow \underset{\text{white ppt soluble in conc. HCl}}{BaSO_3\downarrow}$$

$$BaCl_2 + SO_4^{2-} \longrightarrow \underset{\text{white ppt insoluble in conc. HCl}}{BaSO_4\downarrow}$$

(f) NH_4SCN :

$$Fe^{3+} + SCN^- \longrightarrow \underset{\text{red colour}}{[Fe(SCN)]^{2+}}$$

$$Cu^{2+} + 2SCN^- \longrightarrow \underset{\text{black}}{Cu(SCN)_2\downarrow}$$

(g) NH_4SCN :

$$Co^{2+} + 4SCN^- \longrightarrow \underset{\text{blue colour}}{[Co(SCN)_4]^{2-}}$$

$$Cu^{2+} + 2SCN^- \longrightarrow \underset{\text{black}}{Cu(SCN)_2\downarrow}$$

Problem 15. What single reagent solution (including H_2O) could be used to affect the separation of the following pairs of solids ?

(a) NaOH and $Fe(OH)_3$ (b) $Ni(OH)_2$ and $Fe(OH)_3$
(c) Cr_2O_3 and $Fe(OH)_3$ (d) MnS and CoS
(e) AgCl and AgI

Solution (a) H_2O can dissolve NaOH, $Fe(OH)_3$ remains insoluble in water.

(b) NH_4OH dissolves $Ni(OH)_2$, $Fe(OH)_3$ is insoluble

$$Ni(OH)_2 + 4NH_4OH \longrightarrow \underset{\text{soluble}}{[Ni(NH_3)_4]^{2+}}$$

(c) Excess of NaOH can dissolve Cr_2O_3 but $Fe(OH)_3$ remains insoluble

$$Cr_2O_3 + 3H_2O \longrightarrow 2Cr(OH)_3$$

$$Cr(OH)_3 + NaOH \longrightarrow \underset{\text{soluble}}{Na[Cr(OH)_4]}$$

(d) Dil. HCl dissolves MnS while CoS remains insoluble.

$$MnS + 2HCl \longrightarrow MnCl_2 + H_2S\uparrow$$

(e) Aqueous NH_3 would dissolve AgCl while AgI remains insoluble

$$AgCl + 2NH_3(aq) \longrightarrow \underset{\text{soluble}}{[Ag(NH_3)_2]Cl}$$

Problem 16. (a) If $CO_2(g)$ under pressure is passed into $Na_2CrO_4(aq)$, $Na_2Cr_2O_7(aq)$ is formed. What is the function of the $CO_2(g)$?

(b) When Zn is added to acidic solution of $Na_2Cr_2O_7$, the colour of the solution changes from orange to green then to blue, and, over a period of time, back to green. Explain.

(c) In acidic solution silver (II) oxide first dissolves to produce Ag^{2+} *(aq)* and then reduces to Ag^+ by H_2O which is oxidised to O_2. Write reactions.

Solution (a) CO_2 makes the solution acidic that can convert Na_2CrO_4 to $Na_2Cr_2O_7$

$$CO_2 + H_2O \longrightarrow HCO_3^- + H^+$$

$$2CrO_4^{2-} + 2H^+ \longrightarrow Cr_2O_7^{2-} + H_2O$$

(b) Zn reduces $Cr_2O_7^{2-}$ to Cr^{3+} (green) and then to Cr^{2+}(blue) over a long time, Cr^{2+} is oxidised to Cr^{3+} by atmospheric O_2 and thus blue colour changes to green.

$$\underset{\text{orange}}{Cr_2O_7^{2-}} + 14H^+ + 3Zn \longrightarrow \underset{\text{green}}{2Cr^{3+}} + 7H_2O + 3Zn^{2+}$$

$$\underset{\text{green}}{Cr^{3+}} + e^- \longrightarrow \underset{\text{blue}}{Cr^{2+}}$$

$$Cr^{2+} \longrightarrow \underset{\text{green}}{Cr^{3+}} + e^-$$

(c)

$$AgO + 2H^+ \longrightarrow Ag^{2+} + H_2O$$

$$4Ag^{2+} + 2H_2O \longrightarrow 4Ag^+ + 4H^+ + O_2$$

Problem 17. Colourless salt *(A)*, on heating with NaOH, gave gas *(B)* that can also be obtained when Mg_3N_2 reacts with H_2O. When reaction of *(A)* with NaOH was complete, solution obtained on reaction with $FeSO_4$ and conc. H_2SO_4 gave a brown coloured ring *(C)* between two layers. *(A)* on heating strongly forms *(D)* and *(E)*. *(E)* is a neutral oxide of a dibasic acid *(F)* that exists as *cis* and *trans* isomer. Identify *(A)* to *(F)* and explain reactions.

Solution $(A) \xrightarrow{NaOH,\ \Delta} (B) \xleftarrow{H_2O} Mg_3N_2$

⇒ *(A)* has NH_4^+ ion and *(B)* is NH_3

$$Mg_3N_2 + 6H_2O \longrightarrow 3Mg(OH)_2 + 2NH_3$$

$$NH_4^+ + NaOH \longrightarrow Na^+ + NH_3 + H_2O$$

Resultant solution after NH_3 has escaped completely contains Na^+ and anion of *(A)* which also gives ring test of NO_3^-.

Thus, *(A)* is $\mathbf{NH_4NO_3}$.

Explanation:

$$\underset{(A)}{NH_4NO_3} + NaOH \xrightarrow{\Delta} \underset{(B)}{NH_3\uparrow} + NaNO_3 + H_2O$$

$$NaNO_3 + H_2SO_4 \longrightarrow NaHSO_4 + HNO_3$$

$$3H_2SO_4 + 6FeSO_4 + 2HNO_3 \longrightarrow 3Fe_2(SO_4)_3 + 2NO + 4H_2O$$

$$[Fe(H_2O)_6]SO_4 + NO \longrightarrow \underset{\text{brown coloured ring }(C)}{[Fe(H_2O)_5NO]SO_4} + H_2O$$

$$\underset{(A)}{NH_4NO_3} \xrightarrow{\Delta} \underset{(D)}{N_2O} + \underset{(E)}{2H_2O}$$

$$N_2O + H_2O \longrightarrow \underset{(F)}{H_2N_2O_2}$$

trans: HO–N=N–OH (OH groups on opposite sides); cis: HO–N=N–OH (OH groups on same side)

trans *cis*

F (hyponitrous acid)

Problem 18. Identify *(A)* to *(G)* in the following scheme and name the process.

$$CaCO_3 \xrightarrow{\Delta} (A) + (B)\text{ gas}$$

$$(A) + H_2O \longrightarrow (C)$$

$$(C) + (B) \longrightarrow CaCO_3 + H_2O$$

$$(D) + (C) \xrightarrow{\Delta} (E)\text{ gas}$$

$$(E) + H_2O + (B) \longrightarrow (F)$$

$$NaCl + (F) \longrightarrow (G) + (D)$$

$$(G) \xrightarrow{\Delta} Na_2CO_3 + H_2O + (B)$$

Solution Scheme represents the Solvay process of manufacture of Na_2CO_3.

$$CaCO_3 \xrightarrow{\Delta} \underset{(A)}{CaO} + \underset{(B)}{CO_2(g)}$$

$$\underset{(A)}{CaO} + H_2O \longrightarrow \underset{(C)}{Ca(OH)_2}$$

$$Ca(OH)_2 + CO_2 \longrightarrow CaCO_3 + H_2O$$

$$\underset{(D)}{2NH_4Cl} + \underset{(C)}{Ca(OH)_2} \xrightarrow{\Delta} \underset{(E)}{2NH_3\uparrow} + CaCl_2 + 2H_2O$$

$$\underset{(E)}{NH_3} + H_2O + \underset{(B)}{CO_2} \longrightarrow \underset{(F)}{NH_4HCO_3}$$

$$NaCl + \underset{(F)}{NH_4HCO_3} \longrightarrow \underset{(G)}{NaHCO_3} + \underset{(D)}{NH_4Cl}$$

$$\underset{(G)}{2NaHCO_3} \xrightarrow{\Delta} Na_2CO_3 + H_2O + \underset{(B)}{CO_2}$$

Problem 19. How might you separate the following pairs of ions by addition of a single reagent? Include formulae for the major products of the reactions.

(a) Fe^{3+} and Na^+

(b) Cr^{3+} and Fe^{3+}

(c) Fe^{3+} and Cu^{2+}

Solution (a) Fe^{3+} **and** Na^+:

Add excess of *aq* KOH. Fe^{3+} is precipitated as $Fe(OH)_3$. $Na^+(aq)$ remains in solution.

$$Fe^{3+}(aq) + 3OH^-(aq) \longrightarrow \underset{\text{brown ppt}}{Fe(OH)_3(s)\downarrow}$$

(b) Cr^{3+} **and** Fe^{3+} :

Add excess of *aq* NaOH. Fe^{3+} is precipitated as $Fe(OH)_3$ while $Cr(OH)_3$ first formed dissolves in excess of *aq* NaOH.

$$Fe^{3+}(aq) + 3OH^-(aq) \longrightarrow Fe(OH)_3(s) \xrightarrow{OH^-} \text{ppt does not dissolve}$$

$$Cr^{3+}(aq) + 3OH^-(aq) \longrightarrow \underset{\text{soluble in excess of NaOH (aq)}}{Cr(OH)_3(s)} \xrightarrow{OH^-} \underset{\text{soluble (deep green)}}{[Cr(OH)_4]^-(aq)}$$

(c) Fe^{3+} **and** Cu^{2+}:

Add excess of *aq* $NH_3(NH_4OH)$. Fe^{3+} is precipitated as $Fe(OH)_3$ while Cu^{2+} forms deep blue coloured soluble complex $[Cu(NH_3)_4]^{2+}$.

$$Fe^{3+}(aq) + 3NH_4OH(aq) \longrightarrow Fe(OH)_3(s) + 3NH_4^+$$

$$Cu^{2+} + 4NH_4OH \longrightarrow [Cu(NH_3)_4]^{2+} + 4H_2O$$

Problem 20. Write a balanced net ionic equation for each of the following reactions :

(a) A CrO_4^{2-} solution turns from yellow to orange upon addition of acid.

(b) $Fe^{3+}(aq)$ reacts with KSCN(*aq*) to give deep red solution.

(c) Copper metal reacts with nitric acid to give NO gas and a blue solution.

(d) A deep green solution of $Cr(OH)_3$ in excess base turns yellow on addition of hydrogen peroxide.

Solution (a) $\underset{\text{yellow}}{2CrO_4^{2-}(aq)} + 2H_3O^+(aq) \longrightarrow \underset{\text{orange}}{Cr_2O_7^{2-}(aq)} + 3H_2O(l)$

(b) $[Fe(H_2O)_6]^{3+}(aq) + SCN^-(aq) \longrightarrow [Fe(H_2O)_5(SCN)]^{2+}(aq) + H_2O(l)$

(c) $3Cu(s) + 2NO_3^-(+H_2O(l)) + 8H^+(aq) \longrightarrow 3Cu^{2+}(aq) + 2NO(g) + 4H_2O(l)$

(d) $Cr(OH)_3(s) + OH^-(aq) \longrightarrow \underset{\text{deep green}}{Cr(OH)_4^-(aq)}$

$$2Cr(OH)_4^-(aq) + 3H_2O_2 + 2OH^-(aq) \longrightarrow \underset{\text{yellow}}{2CrO_4^{2-}(aq)} + 8H_2O(l)$$

Problem 21. The following solid substances are in separate but unlabelled test tubes :

$Al_2(SO_4)_3 \cdot 18H_2O$, $BaCl_2 \cdot 2H_2O$, KOH

Describe how could you identify the compounds by chemical tests using only these substances and water?

Solution All the above substances are unlabelled. Label the test-tubes I, II and III.

Mix I and II in aqueous solution. If there is formation of white ppt, then these are $Al_2(SO_4)_3 \cdot 18H_2O$ and $BaCl_2 \cdot 2H_2O$

$$Ba^{2+} + SO_4^{2-} \longrightarrow \underset{\text{white ppt}}{BaSO_4\downarrow}$$

III is certainly KOH.

- Add III into I, if there is white ppt and ppt dissolves in excess of III then I is $Al_2(SO_4)_3 \cdot 18H_2O$

$$Al_2(SO_4)_3 + 6KOH \longrightarrow \underset{\text{white ppt}}{2Al(OH)_3\downarrow} + 3K_2SO_4$$

$$Al(OH)_3 + KOH \longrightarrow \underset{\text{soluble}}{KAl(OH)_4}$$

- Add III into II and there is no ppt formation then II is certainly $BaCl_2 \cdot 2H_2O$ thus

I is $Al_2(SO_4)_3 \cdot 18H_2O$, II is $BaCl_2 \cdot 2H_2O$ and III is KOH.

Problem 22. Baking powders contain sodium (or potassium) hydrogen carbonate and an acidic substance. When water is added to a baking powder, carbon dioxide is released. One kind of baking powder contains $NaHCO_3$ and sodium aluminium sulphate, $NaAl(SO_4)_2$. Write the net ionic equation for the reaction that occurs in water solution.

Solution $NaHCO_3(aq) \rightleftharpoons Na^+(aq) + HCO_3^-(aq)$

$$NaAl(SO_4)_2(aq) \rightleftharpoons Na^+(aq) + Al^{3+}(aq) + 2SO_4^{2-}(aq)$$

Al^{3+} solution is acidic due to hydrolysis

$$Al^{3+}(aq) + 3H_2O(l) \rightleftharpoons Al(OH)_3 + 3H^+(aq)$$

$H^+(aq)$ decomposes $HCO_3^-(aq)$ forming $CO_2(g)$

$$H^+(aq) + HCO_3^-(aq) \longrightarrow H_2O(l) + CO_2(g)$$

or $Al(H_2O)_6^{3+}(aq) + HCO_3^-(aq) \longrightarrow [Al(H_2O)_5(OH)]^{2+}(aq) + H_2O(l) + CO_2(g)$

Problem 23. A light coloured crystalline solid *A* has 27.55% H_2O. A gives following reactions :

I: $A \xrightarrow{BaCl_2 \text{ solution}}$ white ppt (*B*) insoluble in conc. HNO_3

II : $A \xrightarrow{K_3[Fe(CN)_6] \text{ solution}}$ a dark-blue ppt (*C*)

III : $A \xrightarrow{K_2HgI_6/NaOH \text{ solution}}$ brown ppt (*D*)

Identify *A*

Solution

Observation	Inference
A gives white ppt insoluble in conc. HNO_3	A contains SO_4^{2-} as anion
A gives dark-blue ppt with $K_3[Fe(CN)_6]$	A contains Fe^{2+} as cation
A also gives brown ppt with K_2HgI_4/NaOH (Nessler's reagent)	A also contains NH_4^+ as cation.

Thus, **A** is a double salt.

$$FeSO_4 \cdot (NH_4)_2SO_4 \cdot xH_2O$$

H_2O is 27.55%

$$\text{Molar mass of A is} = (284 + 18x)$$

$$\therefore \quad \% \text{ of } H_2O = \frac{18x}{284 + 18x} \times 100 = 27.55$$

$$\therefore \quad x = 6$$

Thus, A is $\mathbf{FeSO_4 \cdot (NH_4)_2SO_4 \cdot 6H_2O}$

Reactions :

I : $$SO_4^{2-} + Ba^{2+} \longrightarrow BaSO_4\downarrow$$

II : $[Fe(CN)_6]^{3-}$ first oxidises Fe^{2+} to Fe^{3+} and itself reduced to $[Fe(CN)_6]^{4-}$

$$Fe^{2+} + [Fe(CN)_6]^{3-} \longrightarrow Fe^{3+} + [Fe(CN)_6]^{4-}$$

Products thus formed combined to form a blue coloured ppt called Turnbull's blue

$$4Fe^{3+} + 3[Fe(CN)_6]^{4-} \longrightarrow \underset{\text{Turnbull's blue}}{Fe_4[Fe(CN)_6]_3}$$

(However, if $[Fe(CN)_6]^{3-}$ is in excess, ppt dissolves partially or completely forming a blue coloured solution.

$$\underbrace{Fe^{3+} + [Fe(CN)_6]^{4-}}_{\text{final product of redox reaction}} \longrightarrow \underset{\text{Turnbull's blue}}{Fe^{III}[Fe^{II}(CN)_6]^-}$$

DEEP Focus

Note : Composition of this Turnbull's blue is identical to that of Prussian blue and has been confirmed by spectroscopy.

III : $$NH_4^+ + NaOH \longrightarrow NH_3 + H_2O + Na^+$$

$$3NaOH + NH_3 + \underset{\text{Nessler's reagent}}{2K_2HgI_4} \longrightarrow \underset{\text{Iodide of Millon's base (brown ppt)}}{O\langle^{Hg}_{Hg}\rangle NH_2I} + 3NaI + 4KI + 2H_2O$$

Problem 24. A sensitive test for arsenic that has been popularised through mystery novels is the Marsh test. Write the chemical equations.

Solution The suspected arsenic-containing material is treated with zinc in an acidic solution:

$$Zn(s) + 2H^+(aq) \longrightarrow Zn^{2+}(aq) + 2H(g)$$

The freshly prepared (nascent) H(*g*) reduces the arsenic compounds to arsine, AsH_3.

$$\underset{\text{(arsenic sample)}}{As_4O_6} + 24H_2(g) \longrightarrow 4AsH_3(g) + 6H_2O$$

Finally, the AsH_3 is decomposed by heating it in a glass tube; the As deposits as a thin film (a mirror).

$$2AsH_3(g) \longrightarrow \underset{\text{mirror}}{2As(s)} + 3H_2(g)$$

Even trace quantities of arsenic can be detected by this method.

Problem 25. Identification of cations can also be done using a dry test called sodium carbonate-bead test which is similar to borax-bead test. Identify *A*, *B*, *C* and *D* in the following:

$$Na_2CO_3 \xrightarrow[\text{oxidising flame with MnO}]{KNO_3 \text{ fuse and heat in}} A \xrightarrow{Cl_2} B$$

$$\xrightarrow{\text{repeat with } Cr_2O_3} C \xrightarrow{H_2SO_4} D$$

Solution $Na_2CO_3 + O_2$ (from KNO_3) + MnO $\longrightarrow$

$$\underset{(A \text{ green})}{Na_2MnO_4} + CO_2$$

$$\underset{A}{2Na_2MnO_4} + Cl_2 \longrightarrow \underset{B \text{ (purple)}}{2NaMnO_4} + 2NaCl$$

$$2Cr_2O_3 + 4Na_2CO_3 + 3O_2 \longrightarrow \underset{C \text{ (yellow)}}{4Na_2CrO_4} + 4CO_2$$

$$\underset{C}{2Na_2CrO_4} + H_2SO_4 \longrightarrow \underset{D \text{ (orange)}}{Na_2Cr_2O_7} + Na_2SO_4 + H_2O$$

Problem 26. When metallic copper is heated with concentrated sulphuric acid, in addition to copper (II) sulphate and sulphur dioxide, some copper (II) sulphide is also formed. Explain.

Solution When Cu reacts with conc. H_2SO_4 there is formation of $CuSO_4$ (by oxidation of Cu to Cu^{2+}) and SO_2 (by reduction of SO_4^{2-} to SO_2)

$$Cu \longrightarrow Cu^{2+} + 2e^-$$

$$SO_4^{2-} + 2e^- + 4H^+ \longrightarrow SO_2 + 2H_2O$$

There is also side reaction :

$$SO_4^{2-} + 8H^+ + 8e^- \longrightarrow S^{2-} + 4H_2O$$

$$Cu^{2+} + S^{2-} \longrightarrow CuS$$

This reaction is favourable due to low solubility of CuS and high $[H^+]$.

Problem 27. Silver nitrate gives a white precipitate (*A*) with aqueous sodium thiosulphate (*B*); the precipitate dissolves in excess of thiosulphate. If the precipitate is heated with water, it turns black (*C*), and the supernant liquid then gives a white precipitate (*D*) with acidified barium nitrate solution. Explain.

Solution $2AgNO_3 + \underset{(B)}{Na_2S_2O_3} \longrightarrow \underset{\text{white ppt }(A)}{Ag_2S_2O_3\downarrow} + 2NaNO_3$

White ppt dissolves in excess of B forming a complex.

$$Ag_2S_2O_3 + 3Na_2S_2O_3 \longrightarrow \underset{\text{soluble}}{2Na_3[Ag(S_2O_3)_2]}$$

When A is heated with water, there is disproportionation of $S_2O_3^{2-}$ to SO_4^{2-} and S^{2-}

$$Ag_2S_2O_3 + H_2O \longrightarrow \underset{\text{(c) black ppt}}{Ag_2S\downarrow} + \underset{\text{soluble}}{H_2SO_4}$$

H_2SO_4 remains in supernant (upper) layer and gives white ppt (D) with $Ba(NO_3)_2$ solution

$$H_2SO_4 + Ba(NO_3)_2 \longrightarrow \underset{D}{BaSO_4\downarrow} + 2HNO_3$$

Problem 28. Identify (A), (B), (C) ... (H) in the following and explain reactions.

$$(A) + \text{black} + \text{dil. } H_2SO_4 \longrightarrow B(aq) + C(aq) + D(g)$$

$$(B) + KI \longrightarrow \text{brown coloured solution} \xrightarrow{\text{hypo}} \text{white ppt }(E)$$

$$(C) + MnO_4^- + H^+ \longrightarrow \text{colourless solution}$$

$$(C) + K_3[Fe(CN)_6] \longrightarrow \text{blue }(F)$$

$$(D) + (CH_3COO)_2Pb \longrightarrow \text{black }(G)$$

$$(B) \text{ or } (C) + BaCl_2 \longrightarrow \text{white ppt }(H)$$

$$(B) + (D) \longrightarrow \text{black ppt }(I)$$

Solution $(D) + (CH_3COO)_2Pb \longrightarrow \text{black }(G)$

$\Rightarrow (D)$ is H_2S gas and (G) in PbS

$\Rightarrow (A)$ also has S^{2-}

$$(A) + \text{dil. } H_2SO_4 \longrightarrow (B), (C)$$

(B) and (C) both are sulphates as confirmed by white ppt (H) with $BaCl_2$

(B) + KI $\longrightarrow$ brown coloured solution changing to white ppt (E)

Thus, (B) is $CuSO_4$.

$$2CuSO_4 + 4KI \longrightarrow \underset{(E)}{Cu_2I_2} + 2K_2SO_4 + I_2$$

$$I_2 + KI \rightleftharpoons KI_3$$

Due to KI_3, solution appears reddish brown. On adding hypo, I_2 disappears and white ppt of Cu_2I_2 appears.

$$(I_3^-)I_2 + \underset{\text{hypo}}{2Na_2S_2O_3} \longrightarrow 2NaI + Na_2S_4O_6$$

(C) decolourises MnO_4^-/H^+. (C) also gives blue colour with $K_3[Fe(CN)_6]$ thus (C) is $\mathbf{FeSO_4}$.

$$FeSO_4 + K_3[Fe(CN)_6] \longrightarrow \underset{\text{Turnbull's blue }(F)}{KFe^{II}[Fe^{III}(CN)_6]} + K_2SO_4$$

(A) : $CuS + FeS\ (CuFeS_2)$	**(B) :** $CuSO_4$
(C) : $FeSO_4$	**(D) :** H_2S
(E) : Cu_2I_2	**(F) :** $KFe^{II}[Fe^{III}(CN)_6]$
(G) : PbS	**(H) :** $BaSO_4$
(I) : CuS	

Problem 29. $(A)(\text{black}) + \text{dil. } H_2SO_4 \longrightarrow (B)(aq) + (C)(\text{gas})$

$$(C) + (CH_3COO)_2Pb \longrightarrow \text{black}$$

(B) + NaOH $\longrightarrow$ green ppt (D) changing to brown (E) by atmospheric O_2

$$(B) + K_3[Fe(CN)_6] \longrightarrow \text{blue }(F)$$

$$(E) + HCl \xrightarrow{\Delta} \text{deep yellow coloured solution }(G).$$

$$(F) + KCNS \longrightarrow \text{red coloured solution }(H).$$

What are (A) to (H) and explain reactions?

Solution $(C) + (CH_3COO)_2Pb \longrightarrow \text{black}$

$\Rightarrow (C)$ is H_2S and (A) has S^{2-}

$$(B) + K_3[Fe(CN)_6] \longrightarrow \text{blue}$$

$\Rightarrow (B)$ has Fe^{2+} ion and thus (A) is FeS.

Explanation :

$$\underset{(A)}{FeS} + H_2SO_4 \longrightarrow \underset{(B)}{FeSO_4} + \underset{(C)}{H_2S}$$

$$(CH_3COO)_2Pb + H_2S \longrightarrow PbS(\text{black})\downarrow + 2CH_3COOH$$

$$FeSO_4 + 2NaOH \longrightarrow \underset{\text{green ppt }(D)}{Fe(OH)_2\downarrow} + Na_2SO_4$$

$$4Fe(OH)_2 + \underset{\text{air}}{O_2} \longrightarrow \underset{\text{brown }(E)}{2Fe_2O_3 \cdot H_2O} + 2H_2O$$

$$Fe_2O_3 + 6HCl \longrightarrow \underset{\text{deep yellow }(G)}{2FeCl_3} + 3H_2O$$

$$\underset{(B)}{FeSO_4} + K_3[Fe(CN)_6] \longrightarrow \underset{\text{Turnbull's blue }(F)}{KFe[Fe(CN)]_6} + K_2SO_4$$

$$\underset{(G)}{FeCl_3} + KCNS \longrightarrow \underset{\text{red }(H)}{[Fe(CNS)]Cl_2} + KCl$$

Problem 30. Black coloured (insoluble in H_2O) solid (A) does not dissolve in dil. HNO_3. Aqua regia can dissolve (A) forming (B). (B) gives yellow ppt. (C) with NaOH. (B) also gives orange ppt (D) with KI; (D) dissolves in excess of KI forming (E). (E) gives brown ppt with NH_4^+ salt in presence of NaOH. (A) is precipitated if H_2S gas is passed into solution of (B) in dil. HCl. Identify (A) to (E) and explain reactions.

Solution (A) is insoluble in dil. HNO_3.

$\Rightarrow (A)$ is of group II (copper subgroup) and is HgS [only]

HgS (black) is insoluble in dil. HNO_3]

$$\underbrace{(A) + 3HCl + HNO_3}_{\text{aqua regia}} \longrightarrow \underset{(B)}{HgCl_2}$$

$$HgCl_2 + 2NaOH \longrightarrow \underset{\text{yellow ppt }(C)}{HgO\downarrow} + 2NaCl + H_2O$$

$$HgCl_2 + 2KI \longrightarrow \underset{(D)\text{ (orange)}}{HgI_2\downarrow} + 2KCl$$

$$HgI_2 + 2KI \longrightarrow \underset{(E)\text{ (soluble)}}{K_2HgI_4}$$

$$NH_4^+ + NaOH \xrightarrow{\Delta} NH_3 + H_2O + Na^+$$

$$3NaOH + NH_3 + 2K_2HgI_4 \longrightarrow O\langle^{Hg}_{Hg}\rangle NH_2I\downarrow + 4KI + 3NaI + 2H_2O$$

brown ppt (iodide of Millon's base)

$$\underset{(B)}{HgCl_2} + H_2S \longrightarrow \underset{\text{black}}{HgS\downarrow} + 2HCl$$

Problem 31. Identify (A) to (C) in the following :

$$A(aq) + KI \longrightarrow I_2$$

$$A(aq) + CO_2 + CH_3COOH \longrightarrow \text{milky }(B)$$

$$A(aq)\text{(paste)} + CH_3CH_2OH \xrightarrow{\Delta} \underset{\text{(an anaesthetic)}}{\text{product }(C)}$$

Solution $CH_3CH_2OH + A(aq)\text{ (paste)} \longrightarrow \text{product }(C)$

(C) is an anaesthetic from CH_3CH_2OH

$\Rightarrow (C)$ is $CHCl_3$

$\Rightarrow (A)$ contains Cl

(A) with CO_2 gives milkyness

$\Rightarrow (A)$ contains $Ca(OH)_2$ in aqueous solution.

(A) is $CaOCl_2$ (bleaching powder)

Explanation :

$$\underset{(A)}{CaOCl_2} + H_2O \longrightarrow Ca(OH)_2 + Cl_2\text{(available)}$$

$$Cl_2 + 2KI \longrightarrow 2KCl + I_2$$

$$Ca(OH)_2 + Cl_2 + CO_2 \longrightarrow \underset{\text{milky }(B)}{CaCO_3} + H_2O + Cl_2$$

$$CH_3CH_2OH + Cl_2(CaOCl_2) \xrightarrow{\Delta} \underset{(C)}{CHCl_3}$$

Formation of C is explained by following reactions :

$$CaOCl_2 + H_2O \longrightarrow Ca(OH)_2 + Cl_2$$

$$Cl_2 + H_2O \longrightarrow 2HCl + O$$

$$CH_3CH_2OH + O \longrightarrow CH_3CHO + H_2O$$

$$CH_3CHO + 3Cl_2 \longrightarrow CCl_3CHO + 3HCl$$

$$2CCl_3CHO + Ca(OH)_2 \longrightarrow (HCOO)_2Ca + 2CHCl_3$$

Problem 32. Aqueous solution of (A) is deep yellow but (A) (solid) is brown. (A) is aqueous solution gives white ppt with $AgNO_3$ soluble in NH_3. Yellow solution of (A) changes to green solution (B) on passing H_2S. (A) is precipitated as brown ppt (C) on adding NH_4OH in presence of NH_4Cl. Precipitate (C) is insoluble in NaOH even in presence of oxidising agents but (C) dissolves in HCl forming (A) again. Aqueous (A) gives blue colouration (D) with $K_4[Fe(CN)_6]$. Aqueous (B) also gives blue colouration (E) with $K_3[Fe(CN)_6]$. Identify (A) to (E) and explain differece between (D) and (E).

Solution A gives white ppt with $AgNO_3$ soluble in NH_3.

$\Rightarrow (A)$ has Cl^-

$$Cl^- + Ag^+ \longrightarrow \underset{\text{white ppt}}{AgCl\downarrow}$$

$$AgCl + 2NH_3(aq) \longrightarrow \underset{\text{soluble}}{[Ag(NH_3)_2]Cl}$$

(A) is precipitated as brown ppt. (C) by NH_4OH in presence of NH_4Cl

$\Rightarrow (A)$ has Fe^{3+} (of III group) which is confirmed by formation of blue colouration with $K_4[Fe(CN)_6]$

$\Rightarrow (A)$ is $FeCl_3$

Explanation :

$$\underset{(A)\text{ (yellow)}}{2FeCl_3} + H_2S \longrightarrow \underset{(B)\text{ (green)}}{2FeCl_2} + S + 2HCl$$

$$FeCl_3 + 3NH_4OH \longrightarrow \underset{(C)\text{ (brown)}}{Fe(OH)_3\downarrow} + 3NH_4Cl$$

$$\underset{(C)}{Fe(OH)_3} + 3HCl \longrightarrow \underset{(A)}{FeCl_3} + 3H_2O$$

$$\underset{(A)}{FeCl_3} + K_4[Fe(CN)_6] \longrightarrow \underset{(D)\text{ (Prussian blue)}}{KFe^{III}[Fe^{II}(CN)_6]} + 3KCl$$

$$\underset{(B)}{FeCl_2} + K_3[Fe(CN)_6] \longrightarrow \underset{(E)\text{ (Turnbull's blue)}}{KFe^{II}[Fe^{III}(CN)_6]} + 2KCl$$

DEEP Focus

X-ray studies have established that Prussian blue and Turnbull's blue are identical. In (D), Fe^{3+} partially oxidises $[Fe^{II}(CN)_6]^{4-}$ forming some $[Fe^{III}(CN)_6]^{3-}$ and in (E), Fe^{2+} partially reduces $[Fe^{III}(CN)_6]^{3-}$ forming some $[Fe^{II}(CN)_6]^{4-}$. Thus, both (D) and (E) have Fe^{2+}, Fe^{3+}, $[Fe^{II}(CN)_6]^{4-}$ and $[Fe^{III}(CN)_6]^{3-}$.

Problem 33. (*A*) is scarlet (reddish-orange) inorganic salt insoluble in H_2O. (*A*) on reaction with HNO_3 gives blackish-brown residue (*B*) and colourless solution of (*C*). (*C*) gives yellow ppt with KI solution as well as with K_2CrO_4. (*C*) also gives black ppt (*D*) with H_2S in HCl solution, ppt (*D*) dissolves in dil. HNO_3. (*B*) on heating with conc. HNO_3 and $Mn(NO_3)_2$ forms pink coloured solution (*E*). Identify (*A*), (*B*), (*C*), (*D*) and (*E*).

Solution (*C*) is colourless soluble solution.

(*C*) give black ppt with H_2S/HCl

$\Rightarrow$ (*C*) is of group II

(*C*) gives yellow ppt with KI as well as with K_2CrO_4

$\Rightarrow$ (*C*) has Pb^{2+} ion

Thus, (*C*) is $Pb(NO_3)_2$ obtained from (*A*), (*B*) on heating with Mn^{2+} and HNO_3 gives purple colour which is due to oxidaton of Mn^{2+} of $HMnO_4$ (purple)

$\Rightarrow$ (*B*) is oxidising agent

$\Rightarrow$ (*B*) is PbO_2 (blackish-brown)

$\Rightarrow$ (*A*) is double oxide of lead.

(*A*) is Pb_3O_4 (red lead) $(2PbO \cdot PbO_2)$

$$\underset{(A)}{Pb_3O_4} + 4HNO_3 \longrightarrow \underset{(B)}{PbO_2} + \underset{(C)}{2Pb(NO_3)_2} + 2H_2O$$

$$\underset{(C)}{Pb(NO_3)_2} + 2KI \longrightarrow \underset{\text{yellow}}{PbI_2\downarrow} + 2KNO_3$$

$$Pb(NO_3)_2 + K_2CrO_4 \longrightarrow \underset{\text{yellow}}{PbCrO_4\downarrow} + 2KNO_3$$

$$Pb(NO_3)_2 + H_2S \longrightarrow \underset{(D)\text{ black}}{PbS\downarrow} + 2HNO_3$$

$$5PbO_2 + 2Mn^{2+} + 4H^+ \longrightarrow \underset{(E)\text{ pink}}{2MnO_4^-} + 5Pb^{2+} + 2H_2O$$

Problem 34. Chromite ore (*A*) $(FeCr_2O_4)$ is fused with NaOH in presence of H_2O_2 when yellow coloured solution (*B*) and residue (*C*) are obtained. (*C*) is separated by filtration and dissolved in conc. HCl forming a yellow coloured solution (*D*). (*D*) gives red colour with NH_4CNS and blue colour with $K_4[Fe(CN)_6]$. (*D*) changes to green (*E*) when Zn is added into it, however, H_2 gas if passed into (*D*) has no effect. (*B*) changes to orange (*F*) on reaction with dil. H_2SO_4 and again (*F*) changes to (*B*) on adding NaOH. (*F*) on reaction with NH_4Cl and subsequent heating gives *G* (green) and *H* (gas). Identify (*B*) to (*H*) and explain reactions.

Solution $A(FeCr_2O_4)$ is an ore of chromium.

$$\underset{(A)}{2FeCr_2O_4} + 8NaOH + 7H_2O \longrightarrow \underset{(B)}{4Na_2CrO_4} + Fe_2O_3 + 11H_2O$$

$$Fe_2O_3 + \underset{\text{conc.}}{6HCl} \longrightarrow \underset{\text{yellow }(D)}{3FeCl_3} + 3H_2O$$

$$\underset{(D)}{FeCl_3} + NH_4CNS \longrightarrow \underset{\text{red}}{[Fe(CNS)]Cl_2} + NH_4Cl$$

$$\underset{(D)}{Fe^{III}Cl_3} + K_4[Fe^{II}(CN)_6] \longrightarrow \underset{\text{(Prussian blue)}}{KFe^{III}[Fe^{II}(CN)_6]}$$

$$FeCl_3 + 3H_2O \longrightarrow Fe(OH)_3 + 3HCl$$

$$2HCl + Zn \longrightarrow ZnCl_2 + 2H \text{ (nascent hydrogen)}$$

$$FeCl_3 + H \longrightarrow \underset{\text{green }(E)}{FeCl_2} + HCl$$

DEEP Focus

[*Nascent hydrogen is more reactive than molecular* H_2 *hence* H_2 *gas passed into* $FeCl_3$ *solution has no effect.* $FeCl_3$ *solution is acidic due to hydrolysis, hence Zn added into it gives nascent hydrogen (in situ) that causes reduction of* $FeCl_3$ *into* $FeCl_2$]

$$\underset{(B)}{2Na_2CrO_4} + H_2SO_4 \longrightarrow Na_2SO_4 + \underset{\text{orange }(F)}{Na_2Cr_2O_7} + H_2O$$

$$\underset{(F)}{Na_2Cr_2O_7} + 2NaOH \longrightarrow \underset{(B)}{2Na_2CrO_4} + H_2O$$

$$\underset{(F)}{Na_2Cr_2O_7} + 2NH_4Cl \longrightarrow (NH_4)_2Cr_2O_7 + 2NaCl$$

$$(NH_4)_2Cr_2O_7 \xrightarrow{\Delta} \underset{(H)}{N_2} + \underset{\text{green }(G)}{Cr_2O_3} + 4H_2O$$

Note : (Cr_2O_3 can be reduced to Cr by Al.)

Problem 35. Colourless salt, (*A*) (insoluble in water)

$$+ \text{dil. } H_2SO_4 \longrightarrow (B) + C(g)$$

$$(B) \xrightarrow{+\text{NaOH}} (D) \text{ (white ppt)} \xrightarrow[\text{(excess)}]{\text{NaOH}} (E) \text{ (soluble)}$$

(*B*) $\xrightarrow{NH_4Cl + NH_4OH}$ no ppt $\xrightarrow{(C)}$ (*A*)

(*E*) $\xrightarrow{(C)}$ (*A*) (white ppt)

$(C) + (F)$ (yellow coloured solution) $\longrightarrow$ light green (*G*)

- $\xrightarrow{K_4[Fe(CN)_6]}$ blue (*H*)
- $\xrightarrow{AgNO_2}$ white ppt (*I*) (soluble in NH_3)
- $\xrightarrow{+O_2,\ \Delta}$ $H_2O + (J)$ $\xrightarrow{+C}$ colloidal white turbidity
  - (*J*) $\xrightarrow{Cr_2O_7^{2-}/H^+}$ green
  - (*J*) $\xrightarrow{Ca(OH)_2}$ milky

Identify (*A*) to (J) and explain reactions.

Solution Gas (*C*) is vital compound in this scheme.

$$\text{Gas }(C) \xrightarrow[\Delta]{O_2} H_2O + (J)$$

$$\text{milky} \xleftarrow{Ca(OH)_2} \text{Gas }(J) \xrightarrow{Cr_2O_7^{2-}/H^+} \text{green}$$

$\Rightarrow (J)$ is SO_2

$\Rightarrow (C)$ is H_2S (giving H_2O and SO_2 by combustion in O_2).

(*B*) gives white ppt only if (*C*) is passed into its ammoniacal solution containing $NH_4Cl(NH_4OH + NH_4Cl)$

$\Rightarrow (B)$ has Zn^{2+} and (*B*) is $ZnSO_4$

$\Rightarrow$ (*A*) has Zn^{2+} ion

$\Rightarrow (A)$ is **ZnS**

(*F*) gives blue colour with $K_4[Fe(CN)_6]$

$\Rightarrow (F)$ has Fe^{3+} ion

(*F*) gives white ppt with $AgNO_3$

$\Rightarrow (F)$ has Cl^- ion. Thus, **F** is **$FeCl_3$.**

Explanation :

$$\underset{(A)}{ZnS} + H_2SO_4 \longrightarrow \underset{(B)}{ZnSO_4} + \underset{(C)}{H_2S}$$

$$\underset{(B)}{ZnSO_4} + 2NaOH \longrightarrow \underset{(D)}{Zn(OH)_2} + Na_2SO_4$$

$$\underset{(D)}{Zn(OH)_2} + 2NaOH \longrightarrow \underset{\text{soluble }(E)}{Na_2ZnO_2} + 2H_2O$$

$$\underset{(E)}{Na_2ZnO_2} + \underset{(C)}{H_2S} \longrightarrow \underset{(A)}{ZnS} + 2NaOH$$

$$\underset{(C)}{H_2S} + \underset{\substack{(F)\\ \text{yellow}}}{2FeCl_3} \longrightarrow \underset{(G)\text{ green}}{2FeCl_2} + 2HCl + S$$

$$\underset{(F)}{FeCl_3} + K_4[Fe(CN)_6] \longrightarrow \underset{(H),\text{ Prussian blue}}{KFe^{III}[Fe^{II}(CN)_6]} + 3KCl$$

$$FeCl_3 + 3AgNO_3 \longrightarrow Fe(NO_3)_3 + \underset{\text{white ppt }(I)}{3AgCl\downarrow}$$

$$\underset{(I)}{AgCl} + 2NH_3 \longrightarrow \underset{\text{soluble}}{[Ag(NH_3)_2]Cl}$$

$$2H_2S + 3O_2 \xrightarrow{\Delta} 2H_2O + \underset{(J)}{2SO_2}$$

$$\underset{(J)}{SO_2} + \underset{(C)}{2H_2S} \longrightarrow 2H_2O + \underset{\substack{\text{colloidal}\\ \text{white turbidity}}}{3S\downarrow}$$

$$3SO_2 + Cr_2O_7^{2-} + 2H^+ \longrightarrow \underset{\text{green}}{2Cr^{3+}} + 3SO_4^{2-} + H_2O$$

$$SO_2 + Ca(OH)_2 \longrightarrow \underset{\text{milky}}{CaSO_3} + H_2O$$

Problem 36. A solution may contain any of the following ions : Fe^{3+}, Ni^{2+}, Cr^{3+}, Zn^{2+}, Mn^{2+}. Based on the following experiment and results therein, indicate which of the ions would be present? Indicate any wrong information, if any.

(a) The original solution is treated with $(NH_4)_2S$ (a substitute of H_2S) in a buffered basic solution. A dark precipitate is obtained.
(b) The ppt for (a) dissolves in aqua-regia
(c) The filtrate after separating ppt in (a) is treated with NaOH and H_2O_2. A dark ppt is separated. Filtrate is colourless.
(d) The ppt from (c) dissolves in HCl (*aq*) giving a coloured solution.
(e) The solution from (d) is treated with aqueous NH_3. A dark ppt forms.
(f) The ppt from (e) is soluble in HCl (*aq*) and solution develops an intense red colour when treated with $SCN^-(aq)$.

Solution

(a) Precipitate can be of NiS
(b) NiS dissolves in aqua regia
(c) Precipitate can be of $Fe(OH)_3$
(d) $Fe(OH)_3$ is soluble in HCl (*aq*) giving (yellow) coloured solution of $FeCl_3$.
(e) This step is incorrect since filtrate of step (c) is already colourless indicating absence of Cr^{3+}.
(f) This confirms Fe^{3+} by forming intense red colour

$$Fe^{3+} + SCN^- \longrightarrow [Fe(SCN)]^{2+}$$

Thus, Fe^{3+} — **confirmed**

Ni^{2+} — **probable**

$Zn^{2+}, Mn^{2+}, Cr^{3+}$ — **absent**

Problem 37. On treatment with cold water, an element (*A*) reacted quietly, liberating a colourless, odourless gas (*B*), and a solution (*C*). Lithium reacted with (*B*) yielding a solid product (*D*) which effervesced with water to give a strongly basic solution (*E*) and gas (*F*). When CO_2 was bubbled through solution (*C*) and initial white ppt (*G*) was formed, but this re-dissolved, forming solution (*H*) when more CO_2 was added. ppt (*G*) effervesced when moistened with conc. HCl, and gave a deep brick red colouration to a Bunsen burner flame. When (*G*) was heated with carbon at 1000°C a caustic white compound (*I*) was formed, which when heated with N_2 at 1000°C gave a solid (*J*) of some commerical importance. Name the compounds (*A*) to (*J*) and explain reactions.

Solution When CO_2 is passed into (*C*), white ppt (*G*) was formed which re-dissolved in excess of CO_2

$\Rightarrow (C)$ is $Ca(OH)_2$

$\Rightarrow (A)$ is Ca

Explanation :

$$\underset{(A)}{Ca} + 2H_2O \longrightarrow \underset{(C)}{Ca(OH)_2} + \underset{(B)}{H_2}$$

$$2Li + \underset{(B)}{H_2} \longrightarrow \underset{(D)}{2LiH}$$

$$LiH + H_2O \longrightarrow \underset{(E)}{LiOH} + \underset{(F)}{H_2}$$

$$\underset{(C)}{Ca(OH)_2} + CO_2 \longrightarrow \underset{(G)}{CaCO_3} + H_2O$$

$$\underset{(G)}{CaCO_3} + H_2O + CO_2 \longrightarrow \underset{(H)}{Ca(HCO_3)_2}$$

$$CaCO_3 + 2HCl \longrightarrow CaCl_2 + H_2O + CO_2$$

Brick red colour in flame is due to $CaCl_2$

$$CaCO_3 + 4C \xrightarrow{1000^\circ C} \underset{(I)}{CaC_2} + 3CO$$

$$\underset{(I)}{CaC_2} + N_2 \longrightarrow \underset{J\ (\text{nitrolim})}{CaCN_2} + C$$

DEEP Focus

$CaCN_2$ is of industrial importance. It is used to make urea, thiourea etc.

$$CaCN_2 + H_2SO_4 \longrightarrow CaSO_4 + NH_2CN$$
$$NH_2CN + H_2O \longrightarrow NH_2CONH_2\ (\text{urea})$$
$$NH_2CN + H_2S \longrightarrow NH_2CSNH_2\ (\text{thiourea})$$

Problem 38. Identify colourless salt (A) based on the following reactions :

Aqueous solution of (A) $\xrightarrow{NaOH}$ white ppt soluble in excess of NaOH (B)

- $\downarrow H_2SO_4$: white ppt soluble in CH_3COONH_4 (D)
- $\downarrow H_2O_2$: black ppt (C)

Aqueous solution of A $\xrightarrow{AgNO_3}$ white ppt soluble is aqueous $NH_3(E)$

- $\xrightarrow{K_2CrO_4}$ yellow ppt (F)
- $\xrightarrow{HCl,\ H_2S}$ black ppt (G) insoluble in NaOH

Solution $A \xrightarrow{AgNO_3}$ white ppt soluble in $aq\ NH_3$

Thus, A contains Cl^- as anion.

$$Cl^- + Ag^+ \longrightarrow \underset{\text{white ppt}}{AgCl\downarrow}$$

$$AgCl + 2NH_3 \longrightarrow \underset{\text{soluble }(E)}{[Ag(NH_3)_2]Cl}$$

$$A \xrightarrow{HCl,\ H_2S} \text{black ppt insoluble in NaOH.}$$

Thus, A (colourless salt) can contain Hg^{2+} or Pb^{2+} or Bi^{3+} as cation.

Since, A gives yellow ppt with K_2CrO_4, hence A contains Pb^{2+} as cation. Thus, A is $PbCl_2$.

Reactions :

$$\underset{(A)}{PbCl_2} + 2NaOH \longrightarrow \underset{\text{white ppt}}{Pb(OH)_2} + 2NaCl$$

$$Pb(OH)_2 + 2NaOH \longrightarrow \underset{\text{soluble }(B)}{Na_2[Pb(OH)_4]}$$

$$\underset{(B)}{Na_2[Pb(OH)_4]} + H_2O_2 \longrightarrow \underset{\text{black ppt }(C)}{PbO_2} + 2H_2O + 2NaOH$$

$$PbCl_2 + H_2SO_4 \longrightarrow \underset{\text{white ppt}}{PbSO_4\downarrow} + 2HCl$$

$$PbSO_4 + 2CH_3COONH_4 \longrightarrow \underset{\text{soluble }(D)}{(CH_3COO)_2Pb} + (NH_4)_2SO_4$$

$$PbCl_2 + K_2CrO_4 \longrightarrow \underset{\text{yellow ppt }(F)}{PbCrO_4\downarrow} + 2KCl$$

$$PbCl_2 + H_2S \longrightarrow \underset{\text{black ppt }(G)}{PbS\downarrow} + 2HCl$$

Problem 39. Identify (A) to (F) based on the following reactions:

Brownish black oxide (A) + dil. HCl $\longrightarrow$ green coloured solution (B) + gas (C)

(B) + $aq\ NH_3$ + H_2S gas $\longrightarrow$ black ppt (D)

(B) + $aq\ NH_3$ $\longrightarrow$ green ppt (E) soluble in excess of $aq\ NH_3$

(C) + aq KI $\longrightarrow$ reddish brown solution (F)

Solution Ammoniacal solution of B (green) gives black ppt with H_2S hence (B) has Ni^{2+} ion and exists as $NiCl_2$ due to dissolution of (A) in dil HCl.

(A) also contains nickel ion but in +3 oxidation state which is reduced +2 by dil HCl

Thus, A is $\mathbf{Ni_2O_3}$

Reactions :

$$\underset{A}{Ni_2O_3} + 6HCl \longrightarrow \underset{B}{2NiCl_2} + 3H_2O + \underset{C}{Cl_2}$$

$$NiCl_2 + H_2S \longrightarrow \underset{D}{NiS\downarrow} + 2HCl$$

$$NiCl_2 + 2NH_4OH \longrightarrow \underset{\text{soluble in excess } NH_3}{\underset{E}{Ni(OH)_2}} + 2NH_4Cl$$

$$\underset{(C)}{Cl_2} + 2KI \longrightarrow 2KCl + I_2$$

$$I_2 + KI \rightleftharpoons \underset{(F)}{KI_3}$$

Problem 40. Identify (*A*) to (*J*) based on following observations :

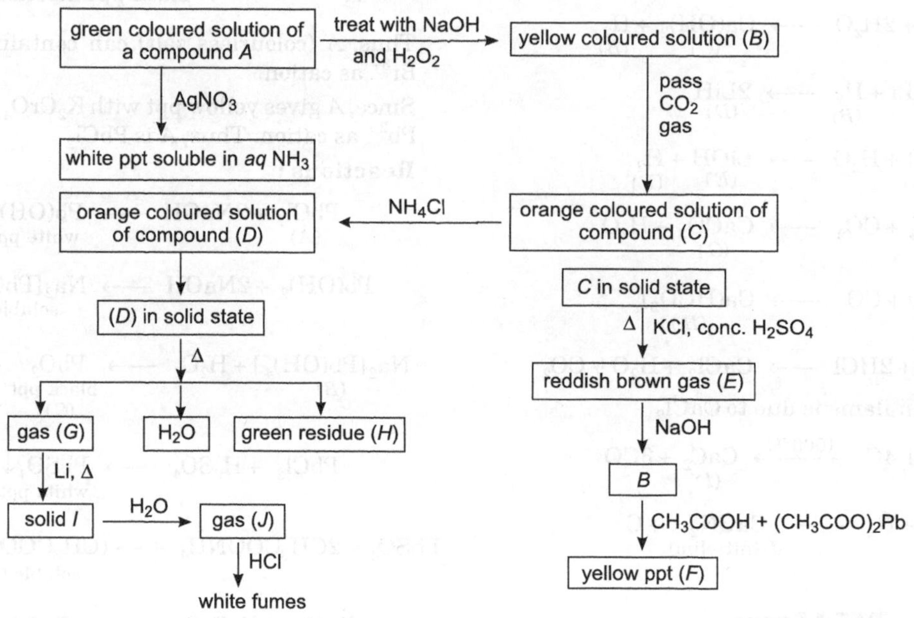

Solution

Observations	Inference
A gives white ppt. soluble in *aq*. NH_3, with $AgNO_3$	A contains Cl^-
C + conc. H_2SO_4 + KCl $\xrightarrow{\Delta}$ $\xrightarrow{NaOH}$ *B*	This is chromyl-chloride test thus *C* is $Na_2Cr_2O_7$.
$B + (CH_3COO)_2Pb \xrightarrow{CH_3COOH}$ yellow ppt	Thus, *B* is CrO_4^- (Na_2CrO_4).

Other compounds are thus :

(*A*) : $CrCl_3$ (*B*) : Na_2CrO_4 (*C*) : $Na_2Cr_2O_7$ (D) : $(NH_4)_2Cr_2O_7$ (*E*) : CrO_2Cl_2

(*F*) : $PbCrO_4$ (*G*) : N_2 (*H*) : Cr_2O_3 (I) : Li_3N (*J*) : NH_3

Reactions (unbalanced/poartial) :

$$\underset{(A)}{CrCl_3} + NaOH + H_2O_2 \xrightarrow{\Delta} \underset{(B)}{Na_2CrO_4} + NaCl + H_2O$$

$$CO_2 + H_2O \longrightarrow HCO_3^- + H^+$$

$$2CrO_4^{2-} + 2H^+ \longrightarrow \underset{(C)}{Cr_2O_7^{2-}} + H_2O$$

$$Na_2Cr_2O_7 + 2NH_4Cl \longrightarrow \underset{(D)}{(NH_4)_2Cr_2O_7} + 2NaCl$$

$$\underset{(D)}{(NH_4)_2Cr_2O_7} \xrightarrow{\Delta} \underset{(G)}{N_2} + \underset{(H)}{Cr_2O_3} + 4H_2O$$

$$4KCl + Na_2Cr_2O_7 + \underset{\text{conc.}}{3H_2SO_4} \longrightarrow \underset{\substack{(E)\\ \text{chromyl chloride}}}{2CrO_2Cl_2} + 2K_2SO_4 + 3H_2O + Na_2SO_4$$

$$\underset{(E)}{CrO_2Cl_2} + 2NaOH \longrightarrow \underset{(B)}{Na_2CrO_4} + 2HCl$$

$$Na_2CrO_4 + (CH_3COO)_2Pb \longrightarrow \underset{(F)}{PbCrO_4} + 2CH_3COONa$$

$$\underset{(G)}{6Li + N_2} \longrightarrow \underset{(I)}{2Li_3N}$$

$$Li_3N + 3H_2O \longrightarrow 3LiOH + \underset{(J)}{NH_3}$$

Problem 41. Identify (A), (B) and (C) in the following scheme :

$$A \text{ (black)} \xrightarrow[\Delta]{KOH, H_2O_2} (B) \text{ (green)}$$

$$(B) \xrightarrow{Cl_2} (C) \text{ (pink-violet)}$$

$$(A) + \text{conc. } HCl \longrightarrow Cl_2(g)$$

$$(C) + \text{conc. } HCl \longrightarrow Cl_2(g)$$

(C) is good laboratory reagent in acidic, basic and neutral medium.

(C) can be decolourised by Fe^{2+}, $C_2O_4^{2-}$ in acidic medium and by C_2H_4 (ethene) in basic medium (in which brown ppt occurs)

$$(C) + KOH \longrightarrow (B)$$

Solution Reactions indicate that (C) is $KMnO_4$ and (A) is thus MnO_2.

$KMnO_4$ can oxidise Fe^{2+}, $C_2O_4^{2-}$ in acidic medium and used as reagent (Baeyer) in basic medium for testing unsaturation of alkenes and alkynes.

Thus, (A) is MnO_2.

Explanation :

$$\underset{(A)\text{ black}}{MnO_2} + 2KOH + H_2O_2 \longrightarrow \underset{(B)\text{ (green)}}{K_2MnO_4} + 2H_2O$$

$$2K_2MnO_4 + Cl_2 \longrightarrow \underset{(C)\text{ (pink)}}{2KMnO_4} + 2KCl$$

$$\underset{(A)}{MnO_2} + \underset{\text{conc.}}{4HCl} \xrightarrow{\Delta} MnCl_2 + Cl_2 + 2H_2O$$

$$\underset{(C)}{2KMnO_4} + \underset{\text{conc.}}{16HCl} \longrightarrow 2KCl + 2MnCl_2 + 8H_2O + 5Cl_2$$

$$MnO_4^- + 8H^+ + 5e^- \xrightarrow{\text{acidic}} Mn^{2+} + 4H_2O$$

$$MnO_4^- + e^- \xrightarrow{\text{neutral}} MnO_4^{2-}$$

$$MnO_4^- + 2H_2O + 3e^- \xrightarrow{\text{basic}} MnO_2 + 4OH^-$$

$$MnO_4^- + 8H^+ + 5Fe^{2+} \longrightarrow Mn^{2+} + 5Fe^{3+} + 4H_2O$$

$$2MnO_4^- + 16H^+ + 5C_2O_4^{2-} \longrightarrow 2Mn^{2+} + 10CO_2 + 8H_2O$$

$$\underset{(C)}{2KMnO_4} + 2KOH \longrightarrow \underset{(B)}{2K_2MnO_4} + H_2O + O$$

(Also refer $KMnO_4$ in chapter 16)

Chapter Proficiency Exercises

Exercise 1 (Stage 1 : Learning)

Short Answer Type Questions

1. $BaCl_2$ gives white precipitate with an aqueous solution of a salt. Identify acid radical.
2. PO_4^{3-} and AsO_4^{3-} can be identified by a reagent. Name the reagent. Write reaction.
3. A mixture contains, CO_3^{2-}, SO_3^{2-} and SO_4^{2-}. Give scheme to identify them.
4. SO_3^{2-} and SO_4^{2-} both give white ppt with $BaCl_2$ solution. How is SO_3^{2-} detected in presence of SO_4^{2-} ?
5. $Na_2B_4O_7 \cdot 10H_2O + \text{conc.}\, H_2SO_4 \xrightarrow{\Delta} (A) \xrightarrow{CH_3OH,\, \Delta} (B)$.
   Identify (*A*) and (*B*).
6. SO_4^{2-} also gives white ppt with lead acetate but ppt dissolves in ammonium acetate solution. Explain.
7. $(A) + KBr \longrightarrow$ yellow ppt (*B*)
   $(A) + \text{conc. } H_2SO_4 \xrightarrow{\Delta}$ brown vapours intensified with Cu-turnings.
   (*B*) dissolves in hypo forming a soluble complex (*C*). What are (*A*), (*B*) and (*C*) and explain their reactions?
8. $(A) + \text{dil. } H_2SO_4 \xrightarrow{\Delta}$ gas (*B*). Gas (*B*) turns $K_2Cr_2O_7/H^+$ solution green.
   Aqueous solution of $(A) + BaCl_2 \longrightarrow$ white ppt (*C*).
   Filtrate after removing $(C) + Br_2$ water $\longrightarrow$ white ppt (*C*). Identify (*A*), (*B*) and (*C*).
9. In the top space write the formula for each substance formed when the following ion pairs are mixed. Use the bottom space to describe any visible change, *i.e.,* white precipitate formed, solution turns red, etc. If no visible change occurs, leave the space blank.

Cations

Reagent added	Pb^{2+}	Cu^{2+}	Fe^{3+}	Co^{2+}	Zn^{2+}	Mg^{2+}	Ba^{2+}	K^+
$(NH_4)_2SO_4$	$PbSO_4$							
	white ppt							
$(NH_4)_2S$								
NaI (Petroleum ether)								
NH_4SCN								
NaOH (Limited amount)								
NaOH (Excess amount)								
NH_3 (*aq*) (Limited amount)								
NH_3 (*aq*) (Excess amount)								

10. If Cu^{2+} and Cd^{2+} both are present, it is difficult to analyse. Outline a scheme to analyse in a mixture.

11. HgS is soluble in aqua regia forming $HgCl_2$. What happens if Cu turnings are added to $HgCl_2$?

12. Sometimes it happens that when H_2S gas is passed into solution in dil. HCl, yellowish white turbidity appears. What do you conclude? What precautions are taken to check this turbidity?

13. II B (arsenic group) sulphides are soluble in YAS. If conc. HCl is added to this soluble portion, coloured ppt are formed. Write reactions.

14. Identify (*A*), (*B*), (*C*), (*D*) and (*E*). (*A*) (black) + dil. HCl $\xrightarrow{\Delta}$ (*B*) (*l*) + (*C*)(*g*). Gas (*C*) turns lead acetate paper black. (*B*) gives orange ppt (*D*) soluble in excess of KI forming (*E*).

15. Test tube (*A*) contains aqueous zinc acetate solution while test tube (*B*) contains aqueous zinc chloride solution. What happens if H_2S gas is passed into each solution?

16. (*A*) (colourless solution) gives white ppt (*B*) with NaOH solution but ppt dissolves in excess of NaOH forming (*C*). (*C*) does not give ppt with H_2S but on boiling with NH_4Cl, white ppt (*B*) appears. (*A*) also gives yellow ppt with $AgNO_3$. Identify (*A*), (*B*) and (*C*).

17. Colourless solution of (*A*) gives white ppt (*B*) with $AgNO_3$ solution, soluble in *aq* NH_3. (*A*) also gives white ppt (*C*) with NaOH soluble in excess of it forming (*D*). (*D*) gives white ppt (*E*) with H_2S. Identify (*A*), (*B*), (*C*), (*D*) and (*E*).

18. A reagent can be used to identify Ba^{2+} and Pb^{2+} both. Name the reagent is.

19. Without going into systematic mixture analysis specify the test that can identify $Ba^{2+}/Ca^{2+}/Sr^{2+}$

20. Teacher demonstrated that ($C_2O_4^{2-}$) oxalate ion decolourised MnO_4^-/H^+. He also demonstrated that Ca^{2+} can be precipitated as CaC_2O_4 (white) and these precipitate also decolourised MnO_4^-/H^+. While performing salt analysis quickly, a student performed this experiment without precipitation of $C_2O_4^{2-}$ but salt given to him confirmed MnO_4^-/H^+ test. He reported $C_2O_4^{2-}$ but teacher awarded him zero. Can you explain positive MnO_4^-/H^+ test?

21. Identify (*A*), (*B*), (*C*), (*D*) and (*E*).

$$(A) + NaOH \xrightarrow{\Delta} B(g) + C(aq) \text{ (slightly alkaline)}$$

$$(A) \xrightarrow{\Delta} H_2O + D(g)$$

$$B(g) \xrightarrow{HCl} \text{white fumes}; \quad C(aq) \xrightarrow{Zn} \text{white fumes } B(g)$$

$$D(g) \xrightarrow{Al,\ \Delta} E \xrightarrow{H_2O} B(g)$$

22. $A(\text{orange}) \xrightarrow{\Delta} D(g) + \text{residue } (C) \text{ (green)} + H_2O$

$$A \xrightarrow[\Delta]{NaOH} B(g)$$

$$B(g) \xrightarrow{HCl} \text{white fumes}$$

$$D(g) \xrightarrow[\Delta]{Al} (E) \xrightarrow{H_2O} \text{gas } (B)$$

$$\text{residue } (C) \xrightarrow{\text{conc. HCl}} (F) \xrightarrow[\Delta]{NaOH,\ H_2O_2} (G) \text{ (yellow)}$$

$$(G) \xrightarrow[(CH_3COO)_2Pb]{CH_3COOH} (H) \text{ (yellow ppt)}$$

Identify (*A*) to (*H*) and explain reactions.

23. Solution containing Mg^{2+} and NH_4^+ ions is called **magnesia mixture** and can be used to identify PO_4^{3-} ions. What can be other anion (acidic radical) which can be identified by this magnesia mixture?

24. You have following salts on your table in a lab. KI, HgI_2, NaOH, NH_4OH, $SnCl_2$.
(a) Select salts to prepare Nassler's reagent.
(b) Above reagent is used to identify which ion?
(c) Write reaction

Analytical Questions

1. A monomeric complex of cobalt gave the following results on analysis :

	Co^{3+}	NH_3	Cl^-	SO_4^{2-}
%	21.24	24.77	12.81	34.65

The compound is diamagnetic and contains no other groups or elements, except water which might be present. What is probable formula of the compound? (Co = 59)

2. Identify (*A*) to (*H*) giving reactions taking place :

$$\text{Mineral } (A) \xrightarrow{\text{dil. } H_2SO_4} (B) + (C) + (D) \text{ (gas)}$$

$$(D) + (CH_3COO)_2Pb \xrightarrow{\Delta} \text{black ppt}$$

$$(B) + (C) \xrightarrow{(D)} E \text{ (black ppt) separated from } (C) \text{ by filtration}$$

$$(E) + \underset{\text{dil.}}{HNO_3} \xrightarrow{\Delta} (F) \text{ blue coloured solution}$$

$$(F) + K_4[Fe(CN)_6] \longrightarrow (C) \text{ chocolate coloured ppt}$$

$$(C) + K_3[Fe(CN)_6] \longrightarrow (H) \text{ blue}$$

3. An aqueous solution of salt (*A*) gives a white crystalline precipitate (*B*) with NaCl solution. The filtrate gives a black precipitate (*C*) when H_2S is passed through it. Compound (*B*) dissolves in hot water and the solution gives yellow precipitate (*D*) on treatment with potassium iodide and on cooling. The compound (*A*) does not give any gas with dilute HCl but liberates a reddish brown gas on heating. Identify the compounds (*A*) to (*D*) giving the involved equations.

4. A white amorphous powder (*A*) when heated gives a colourless gas (*B*), which turns lime water milky and the residue (*C*) which is yellow when hot but white when cold. The residue (*C*) dissolves in dilute HCl and the resulting solution gives a white precipitate on addition of potassium ferrocyanide solution. (*A*) dissolves in dilute HCl with the evolution of a gas which is identical in all respects with (*B*). The solution of (*A*) as obtained above gives a white precipitate (*D*) on addition of excess of NH_4OH and on passing H_2S. Another portion of this solution gives initially a white precipitate (*E*) on addition of NaOH which dissolves in excess of it. Identify (*A*) to (*E*) and explain reactions.

5. Compound (*A*) is a light green crystalline solid. It gives the following tests :

(i) It dissolves in dilute sulpuric acid. No gas is produced.

(ii) A drop of $KMnO_4$ is added to the above solution. The pink colour disappears.

(iii) Compound (*A*) is heated strongly. Gases (*B*) and (*C*) with pungent smell came out. A brown residue (*D*) is left behind.

(iv) The gas mixture [(*B*) and (*C*)] is passed into a dichromate solution. The solution turns green.

(v) The green solution from step (iv) gives a white precipitate (*E*) with a solution of barium nitrate.

(vi) Residue (*D*) from (v) is heated on charcoal in reducing flame. It gives a magnetic substance.

Identify the compound (*A*) to (*E*).

6. When 20.02 g of white solid (*X*) was heated, 4.4 g of acid gas (*A*) that turned lime water milky was driven off together with 1.8 g of a gas (*B*) which condenses to a colourless liquid. The solid (*Y*) that remained is dissolved in water to give an alkaline solution, with which excess barium chloride solution gave a white precipitate (*Z*). The precipitate effervesced with acid giving off carbon dioxide. Identify (*A*), (*B*) and (*Y*) and write down the equation for the thermal decomposition of (*X*).

7. A hydrated metallic salt (*A*), light green in colour, gives a white anhydrous residue (*B*) after being heated gradually. (*B*) is soluble in water and its aqueous solution reacts with NO to give a dark brown compound (*C*). (*B*) on strong heating gives a brown residue (*D*) and a mixture of two gases (*E*) and (*F*). The gaseous mixture, when passed through acidified permanganate, discharged the pink colour and when passed through acidified $BaCl_2$ solution, gives a white precipitate. Identify (*A*), (*B*), (*C*), (*D*), (*E*) and (*F*).

8. Identify (*A*) to (*G*) in the following scheme :

$$\text{Colourless }(A) \xrightarrow{\text{NaOH}} (B)\text{ gas} + (C)\,aq$$

(alkaline due to excess NaOH added)

$$(B) + \text{Na} \longrightarrow (D) + H_2$$

$$C(aq) + \text{Zn} \xrightarrow{\Delta} (B)$$

$$(A) \xrightarrow{\Delta} (E) + H_2O$$

$$(E) + (D) \longrightarrow (F)$$

$$(F) + H_2SO_4 \longrightarrow (G)$$

(*G*) contains 97.67% N, rest hydrogen.

9. 0.98 g of an acid (*A*) of phosphorus in 100 mL solution is 0.1 M and neutralises 300 mL of 0.1 N NaOH.

$$(A) \xrightarrow{MgCl_2} (B) \xrightarrow{\Delta} (C)$$

If 1.11 of (*C*) is obtained from given (*A*), identify (*A*), (*B*) and (*C*).

10. An acid (*A*) can oxidise I^- to I_2. One mole of (*A*) can neutralise one mole of $Ca(OH)_2$ in *aq* solution. One mole of (*A*) in aqueous solution, on hydrolysis, gives two moles of another acid (*B*) and one mole of liquid (*C*). 0.98 g of acid (*B*) gives 2.33 g of white ppt (*D*) on reaction with excess of $BaCl_2$. Liquid (*C*) has oxidising as well as reducing property. (*C*) can decolourise $KMnO_4/H^+$, blackened oil paintings. Identify (*A*) to (*D*) and explain reactions.

11. A certain metal (*A*) is boiled in dilute nitric acid to give a salt (*B*) and an oxide of nitrogen (*C*). An aqueous solution of (*B*) with brine gives a precipitate (*D*) which is soluble in ammonium hydroxide. On adding aqueous solution of (*B*) to hypo solution, a white precipitate (*E*) is obtained. (*E*) on standing turns to a black compound (*F*). Identify (*A*) to (*F*).

12. An aqueous solution of a gas (*X*) gives the following reactions :

(i) It turns an acidified $K_2Cr_2O_7$ solution green.

(ii) On boiling it with H_2O_2, cooling it and then adding an aqueous solution of $BaCl_2$, a precipitate insoluble in dilute hydrochloric acid is obtained.

(iii) On passing H_2S in the solution a white turbidity is obtained.

Identify (*X*) and give equations for the reactions at steps (i), (ii) and (iii).

13. (a) A compound (*X*) imparts a golden yellow flame and shows the following reactions :

(i) Zinc powder when boiled with a concentrated aqueous solution of (*X*) dissolves and hydrogen is evolved.

(ii) When an aqueous solution of (*X*) is added to an aqueous solution of stannous chloride, a white precipitate is obtained first which dissolves in excess of solution of (*X*).

Identify (*X*) and write equations for reactions at steps (i) and (ii).

(b) Write equations for the action of strong heat on borax and green vitriol.

14. A metal chloride (*X*) shows the following reactions :

(i) When H_2S is passed in an acidified aqueous solution of (*X*), a black precipitate is obtained.

(ii) The precipitate obtained at step (i) is not soluble in yellow ammonium sulphide.

(iii) When a solution of stannous chloride is added to an aqueous solution of (*X*), a white precipitate is obtained which turns grey on addition of more of stannous chloride.

(iv) When an aqueous solution of KI is added to an aqueous solution of (*X*), a red precipitate is obtained which dissolves on addition of excess of KI.

Identify (*X*) and write down the equations for the reactions at steps (i), (iii) and (iv).

15. (a) An aqueous solution of a gas (*X*) shows the following reactions :

(i) It turns red litmus blue.

(ii) When added in excess to a copper sulphate solution, a deep blue colour is obtained.

(iii) On addition to a ferric chloride solution, a brown precipitate, soluble in dilute nitric acid, is obtained. Identify (X) and give equations for the reactions at steps (ii) and (iii).

(b) A certain compound (*X*) is used in laboratory for analysis. Its aqueous solution gives the following reactions :

(i) On addition to a copper sulphate solution, a brown precipitate is obtained which turns white on addition of excess of $Na_2S_2O_3$ solution.

(ii) On addition to Ag^+ ion solution, a yellow curdy precipitate is obtained which is insoluble in ammonium hydroxide.

Identify (*X*) and give equations for the reactions at step (i) and (ii).

16. A certain salt (*X*) gives the following tests :

(i) Its aqueous solution is alkaline to litmus.

(ii) On strongly heating it swells to give a glassy material.

(iii) When concentrated sulphuric acid is added to a hot concentrated solution of (*X*), white crystals of a weak acid separate out.

Identify (*X*) and write down the chemical equations for reactions at steps (i), (ii) and (iii).

17. A certain compound (*X*) shows the following reactions :

(i) When KI is added to an aqueous suspension of (*X*) containing acetic acid, iodine is liberated.

(ii) When CO_2 is passed through an aqueous suspension of (*X*), the turbidity transforms to a precipitate.

(iii) When a paste of (*X*) in water is heated with ethyl alcohol, a product of anaesthetic use is obtained.

Identify (*X*) and write down chemical equations for reactions at steps (i), (ii) and (iii).

18. An aqueous solution of an inorganic compound (*X*) shows the following reactions :

(i) It decolourises an acidified $KMnO_4$ solution accompanied by the evolution of oxygen.

(ii) It liberates iodine from an acidified potassium iodide solution.

(iii) It gives a brown precipitate with alkaline $KMnO_4$ solution with the evolution of oxygen.

(iv) It removes black stains from old oil paintings.

Identify (*X*) and give chemical equations for the reactions at steps (i) to (iv).

19. Two solid laboratory reagents (*A*) and (*B*) give the following reactions :

Compound (*A*) :

(i) On strongly heating it gives two oxides of sulphur.

(ii) On adding aqueous NaOH solution to its aqueous solution, a dirty green precipitate is obtained which starts turning brown on exposure to air.

Compound (*B*) :

(iii) It imparts green colour to the flame.

(iv) Its solution does not give a precipitate on passing H_2S.

(v) When it is heated with solid $K_2Cr_2O_7$ and concentrated H_2SO_4, a red gas is evolved. The gas when passed in an *aq* NaOH solution turns it yellow.

Identify (*A*) and (*B*) and give chemical equations for reactions at step (i), (ii)and (v).

20. Identify (*A*) to (*D*) in following steps and give chemical equations :

(i) A white amorphous powder (*A*) on strongly heating gives a colourless non-combustible gas (*B*) and solid (*C*).

(ii) The gas (*B*) turns lime water milky and turbidity disappears with the passage of excess of gas.

(iii) The solution (*C*) in dilute HCl gives a white precipitate with an aqueous solution of $K_4[Fe(CN)_6]$.

(iv) The solution of (*A*) gives a white precipitate (*D*) on passing H_2S in presence of excess of NH_4OH.

21. (i) A black mineral (*A*) on heating in presence of air gives a gas (*B*).

(ii) The mineral (*A*) on reaction with dilute H_2SO_4 gives a gas (*C*) and solution of a compound (*D*).

(iii) On passing gas (*C*) into an aqueous solution of (*B*), a white turbidity is obtained.

(iv) The aqueous solution of compound (*D*) on reaction with potassium ferricyanide gives a blue compound (*E*). Identify (*A*) to (*E*) and give chemical equations for reactions at steps (i) to (iv).

22. An inorganic Lewis acid (*X*) shows the following reactions :

(i) It fumes in moist air.

(ii) The intensity of fumes increases when a rod dipped in NH_4OH is brought near it.

(iii) An acidic solution of (*X*) on addition of NH_4Cl and NH_4OH gives a precipitate which dissolves in NaOH solution.

(iv) An acidic solution of (*X*) does not give precipitate with H_2S. Identify (*X*) and give chemical equations for reactions steps (i) to (iii).

23. The certain inorganic compound (*X*) shows the following reactions :

(i) On passing H_2S through an acidified solution of (*X*), a brown precipitate is obtained.

(ii) The precipitate obtained at step (i) dissolves in excess of yellow ammonium sulphide.

(iii) On adding an aqueous solution of NaOH to a solution of (*X*), first a white precipitate is obtained which dissolves in excess of NaOH.

(iv) The aqueous solution of (*X*) reduces ferric chloride. Idenitfy the cation of (*X*) and give chemical equations for reactions at steps (i), (ii) and (iv).

24. (i) An inorganic iodide (*A*) on heating with a solution of KOH gives a gas (*B*) and the solution of a compound (*C*).

(ii) The gas (*B*) on ignition in air gives a compound (*D*) and water.

(iii) Copper sulphate is finally reduced to the metal on passing (*B*) thorugh its solution.

(iv) A precipitate of compound (*E*) is formed on reaction of (*C*) with $CuSO_4$ solution. Identify (*A*) to (*E*) and explain reactions (i) to (iv).

25. A black mineral (*A*) is soluble in NaCN solution forming (*B*). On adding Zn into (*B*), metal (*C*) is formed. Metal (*C*) on boiling with HNO_3 forms (*D*) which gives white ppt (*E*) with NaCl soluble in NH_3 solution. (*A*) can be obtained when (*D*) reacts with hypo. Identify (*A*) to (*E*).

26. (i) An inorganic compound (*A*) is formed on passing a gas (*B*) through a concentrated liquor containing sodium sulphide and sodium sulphite.

(ii) On adding (*A*) into a dilute solution of silver nitrate, a white precipitate appears which quickly changes into a black coloured compound (*C*).

(iii) On adding two or three drops of ferric chloride into the excess of solution of (*A*), a violet coloured compound (*D*) is formed. This colour disappears quickly.

(iv) On adding a solution of (*A*) into the solution of cupric chloride, a white precipitate is first formed which dissolves on adding excess of (*A*) forming a compound (*E*).

Identify (*A*) to (*E*) and give chemical equations for the reactions at steps (i) to (iv).

27. (i) A black coloured compound (*B*) is formed on passing hydrogen sulphide through the solution of a compound (*A*) in NH_4OH.

(ii) (*B*) on treatment with hydrochloric acid and potassium chlorate gives (*A*).

(iii) (*A*) on treatment with potassium cyanide gives a buff coloured precipitate which dissolves in excess of this reagent forming a compound (*C*).

(iv) The compound (*C*) is changed into a compound (*D*) when its aqueous solution is boiled.

(v) The solution of (*A*) is treated with excess of sodium bicarbonate and then with bromine water. On cooling and shaking for sometime, a green colour of compound (*E*) is formed. No change is observed on heating.

Identify (*A*) to (*E*) and give chemical equations for the reactions at steps (i) to (v).

28. (i) An aqueous solution of a white coloured compound (*A*) on reaction with HCl gives a white precipitate of compound (*B*).

(ii) (*B*) becomes soluble in chlorine water with the formation of (*C*).

(iii) (*C*) reacts with KI to give a precipitate which becomes soluble in excess of if forming a compound (*D*). The compound (*D*) is used for detecting ammonium salts.

(iv) (*B*) and (*C*) both, on treatment with $SnCl_2$ give a grey precipitate of (*E*).

(v) When conc. H_2SO_4 is added slowly into a mixture of cold solutions of (*A*) and $FeSO_4$, a brown ring of compound (*F*) is formed. Identify (*A*) to (*F*) and give chemical equations for the reactions at steps (i) to (v).

29. (i) A blue coloured compound (*A*) on heating gives two products, (*B*) and (*C*).

(ii) A metal (*D*) is deposited on passing hydrogen through heated (*B*).

(iii) The solution of (*B*) in HCl on treatment with $K_4[Fe(CN)_6]$ gives a chocolate brown coloured precipitate of compound (*E*).

(iv) (*C*) turns lime water milky which disappears on continuous passage of (*C*) forming a compound (*F*).

Identify (*A*) to (*F*) and give chemical equations for the reactions at steps (i) to (iv).

30. (i) The yellow coloured precipitate of compound (*A*) is formed on passing H_2S through a neutral solution of a salt (*B*).

(ii) (*A*) is soluble in hot dilute HNO_3, but insoluble in yellow ammonium sulphide.

(iii) The solution of (*B*) on treatment with small quantity of NH_3 gives white precipitate which becomes soluble in excess of it forming a compound (*C*).

(iv) The solution of (*B*) gives white precipitate with small concentration of KCN which becomes soluble in excess of this reagent forming a compound (*D*).

(v) The solution of (*D*) on treatment with H_2S gives (*A*).

(vi) The solution of (*B*) in dilute HCl on treatment with a solution of $BaCl_2$ gives white precipitate of compound (*E*) which is insoluble in conc. HNO_3.

Idenitfy (*A*) to (*E*) and give chemical equations for the reactions at steps (i) and (iii) to (vi).

31. (i) An aqueous solution of a compound (*A*) is acidic towards litmus and (*A*) is sublimed at about 300°C.

(ii) (*A*) on treatment with an excess of NH_4SCN gives a red coloured compound (*B*) and on treatment with a solution of $K_4[Fe(CN)_6]$ gives a blue coloured compound (*C*).

(iii) (*A*) on heating with excess of $K_2Cr_2O_7$ in presence of concentrated H_2SO_4 evolves deep red vapours of (*D*).

(iv) On passing the vapours of (*D*) into a solution of NaOH and then adding the solution of acetic acid and lead acetate, a yellow precipitate of compound (*E*) is obtained.

Identify (*A*) to (*E*) and give chemical equations for the reactions at steps (ii) to (iv).

32. (a) Pyrolusite on heating with KOH in the presence of air gives a dark green compound (*A*). The solution of (*A*) on treatment with H_2SO_4 gives a purple coloured compound (*B*), which gives following reactions :

(i) KI on reaction with alkaline solution of (*B*) changes into a compound (*C*).

(ii) The colour of the compound (*B*) disappears on treatment with the acidic solution of $FeSO_4$.

(iii) With conc. H_2SO_4 compound (*B*) gives (*D*) which can decompose to yield (*E*) and oxygen.

Identify (*A*) to (*E*) and write balanced chemical equations for the formation of (*A*) and (*B*) and for the steps (i) to (iii).

(b) Complete and balance the following chemical equations.

(i) $Au + HCl + HNO_3 \longrightarrow ...... + ... + H_2O$

(ii) $C + HNO_3\,(conc.) \longrightarrow CO_2 + ... + H_2O$

(iii) $Sn + KOH\,(hot) + H_2O \longrightarrow ...... + ...$

(iv) $Cu(OH)_2 + NH_4NO_3 + NH_4OH \longrightarrow ... + H_2O$

33. Unlabeled test tube contains solid $AlCl_3 \cdot 6H_2O$ in one, $Ba(OH)_2 \cdot 8H_2O$ in another, and $MgSO_4 \cdot 7H_2O$ in the other. How could you find what is in each test tube using chemical tests that involve only these compounds plus water?

34. You are given three unlabeled test tubes. One test tube contains a solution of Na_2SO_4, the second contains a solution of $NaHSO_4$ and the third contains a solution of $NaHSO_3$. Describe how you could identify the solutions.

35. Identify each of the following substances from the description.

(a) A yellow solid that burns with a blue flame, giving off a gas with a choking odour.

(b) A white solid that reacts with water to give phosphoric acid.

(c) A colourless gas that poisons by attaching strongly to iron atoms in the haemoglobin of red blood cells.

(d) A hard, lustrous grey solid used as the base for solid state semiconductor devices.

36. Addition of HCl (*aq*) to a solution containing several different cations produces a white precipitate. The filtrate is removed and treated with $H_2S(aq)$ in 0.3 M HCl. No precipitate forms. Which of the following conclusions is/are valid? Explain

(a) Ag^+ and/or Hg_2^{2+} probably present

(b) Mg^{2+} probably not present

(c) Pb^{2+} probably not present

(d) Fe^{2+} probably not present

37. Suppose you did a cation group I analysis and treated the chloride precipitate with $NH_3(aq)$, without first treating it with hot water. What might you observe and what vaild conclusions could you reach about cations present, cations absent, and cations in doubt?

38. Write net ionic equations for the following qualitative analysis procedures :

(a) Precipitation of $PbCl_2(s)$ from a solution containing $Pb^{2+}(aq)$

(b) Dissolving of $Zn(OH)_2(s)$ in $NaOH(aq)$.

(c) Dissolving of $Fe(OH)_3(s)$ in $HCl(aq)$.

(d) Precipitation of antimony sulphide from a solution containing H_2S and the complex ion $SbCl_4^-$.

39. Identify *A* to *E* based on following observations and write balanced reactions therein :

(*A*) —neutral $FeCl_3$ solution→ blood red colour (*B*)

(*A*) —dil. HCl, H_2S gas→ yellow ppt (*C*), (*C* dissolves in hot dil. HNO_3 but insoluble in YAS)

(*A*) —*aq* NH_3→ white ppt soluble in excess of *aq* NH_3 forming (*D*)

(*A*) —*aq* KCN→ white ppt soluble in excess of *aq* KCN forming (*E*) —H_2S gas→ yellow ppt (*C*)

40. Identify *A* to *G* based on following reactions :

$$\text{solution of the compound } (A) \xrightarrow{+\text{ gas } B} (C) \xrightarrow{\Delta} (A) + (D)$$

$$\text{conc. solution of } (C) \xrightarrow{\text{conc. } H_2SO_4} (E) \text{ (weak acid)}$$

$$(D) + \text{cobalt oxide} \xrightarrow{\Delta} (F) \text{ (coloured bead)}$$

$$(E) + C_2H_5OH \xrightarrow{\Delta} \text{green edged flame } (G)$$

41. Colourless substance

$$(A) \xrightarrow{\text{dil } H_2SO_4} \text{gas } (B) + \text{colourless solution } (C)$$

$$\text{gas } (B) \xrightarrow{(CH_3COO)_2\,Pb} \text{black ppt } (D)$$

$$(C) \xrightarrow{\text{NaOH solution}} \text{white ppt } (E) \xrightarrow{\text{NaOH (excess)}} (F) \text{ (soluble)} \xrightarrow{\text{gas } B} (A)$$

$$(E) \xrightarrow[\text{(ii) gas } (B)]{\text{(i) } NH_4Cl,\ \Delta,\ NH_4OH} (A)$$

What are (*A*), (*B*), (*C*), (*D*), (*E*) and (*F*) and explain reactions.

42. Element *M* is a shiny and highly reactive metal and element *X* is a highly reactive non-metal. They react to form a compound *MX*, a colourless brittle white solid. *MX* conducts electricity in aqueous solution and in molten state. When chlorine gas is bubbled through an aqueous solution, containing *MX*, a reddish-brown liquid appears and Cl^- ions are formed.

$$2MX + Cl_2 \longrightarrow 2MCl + X_2$$

X_2 is decolourised by unsaturated hydrocarbon and *aq* solution of *MX* containing 1.19 g *MX* in 100 g H_2O freezes at –3.72°C. Molar depression constant of H_2O is 1.86° mol^{-1} kg. Deduce the identity of the elements *M* and *X*.

43. An element *X* of group 2 reacts with H_2 gas at 200°C to form a compound *Y*. When *Y* is heated to a higher temperature, it decomposes to the element *X* and hydrogen gas in the ratio of 559 mL H_2 (measured at STP) for 1.00 g *X* reacted. *X* also combines with chlorine to form a compound *Z*, which contains 63.89 per cent by mass of chlorine. Deduce the idenity of the element *X*, compound *Y* and *Z*.

44. A student is given samples of three elements *X*, *Y*, and *Z*, which could be an alkali metal, a member of group IVA, and a member of group VA. He makes the following observations :

Element *X* has a metallic lusture and conducts electricity. It reacts slowly with HCl to produce H_2 gas. Element *Y* is a light-yellow solid that does not conduct electricity. Element *Z* has a metallic lusture and conducts electricity. When exposed to air, it slowly forms a white powder giving a basic solution in water. What can you conclude about the elements from these observations?

45. Deduce what you can about the nature of the following reactions :

(a) One mole of NH_2OH reacts with two moles of Ti(III) in the presence of excess of alkali, and the Ti(III) is converted into Ti(IV).

(b) Silver phosphite, Ag_2HPO_3, is warmed with water, and dilute the silver is precipitated as the metal.

(c) When one mole of hypophosphorous acid, H_3PO_2, is warmed with excess of iodine in acidic solution, one mole of iodine is reduced; on making the solution alkaline, another mole of iodine is consumed.

46. The action of concentrated sulphuric acid on urea results in the production of a white crystalline solid *X* of formula H_3NO_3S, a monobasic acid. On treatment with sodium nitrite and dilute hydrochloric acid at 0°C, it liberates one mole of N_2 per mol of *X*, and on addition of aqueous barium chloride, the resulting solution yields one mole of $BaSO_4$ per mol of *X* taken. Deduce the structure of *X*.

47. Identify, *A*, *B*, *C*, *D*, *E* and *F* in the following :

$$(A) + AgNO_3 \longrightarrow \text{white ppt } (B) \xrightarrow[\text{excess}]{(A)} \text{soluble } (C)$$

$$\text{white ppt } (B) \xrightarrow{H_2O,\ \Delta} \text{black ppt } (D) + (E) \text{ (soluble)}$$

$$(E) \xrightarrow{BaCl_2} \text{white ppt } (F)$$

48. In strongly alkaline solution containing excess of barium ions, a solution containing 0.01587 g of I^- was treated with 0.1 molar MnO_4^- until a pink colour presisted in the solution : 10.0 mL of the MnO_4^- solution were required. Under these conditions MnO_4^- is converted into the sparingly soluble $BaMnO_4$. What is the product of the oxidation of iodide?

49. In neutral solution, one mole periodate ion, IO_4^-, reacts with excess of iodide to produce one mole of iodine; on acidification of the resulting solution, a further three moles of iodine are liberated.

Derive equations of the reactions which occur under these conditions.

50. To prevent the formation of oxides, peroxides and superoxides, alkali metals are sometimes stored in an inert atmosphere. Which of the following gases should not be used for lithium? Why? Ne, Ar, N_2, Kr

51. The following are two reaction-schemes involving magnesium. Identify *A* to *J* based on the observations given :

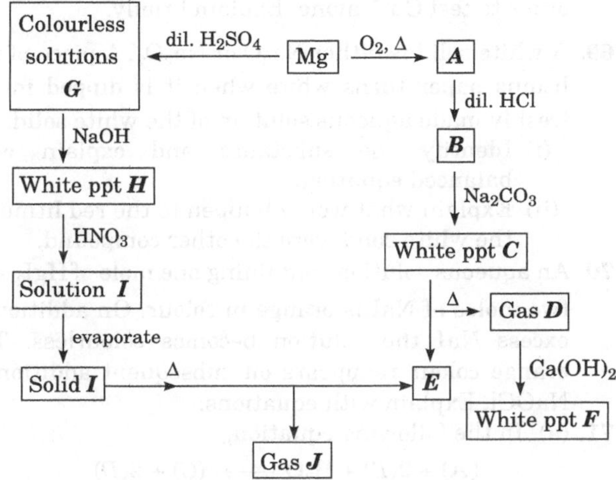

52. When aqueous KCN is added to a solution of copper(II) sulphate, a white precipitate, soluble in an excess of KCN is formed. No precipitate is formed when H_2S is passed into it. Explain.

53. The reaction

$$4MO_2(s) + 2CO_2(g) \longrightarrow 2M_2CO_3(s) + 3O_2(g)$$

is to be used to regenerate oxygen in space-craft. Discuss the choice of *M* among the alkali metals.

54. Write balanced chemical equations for the following reactions :

(i) A mixture of potassium dichromate and sodium chloride is heated with concentrated H_2SO_4.

(ii) Aqueous solution of sodium nitrate is heated with zinc dust and caustic soda solution.

(iii) Sodium iodate is added to a solution of sodium bisulphite.

(iv) Iron reacts with cold dilute nitric acid.

(v) Potassium permanganate is added to a hot solution of manganous sulphate.

55. The gas liberated on heating a mixture of two salts with NaOH, gives a reddish brown precipitate with an alkaline solution of K_2HgI_4. The aqueous solution of the mixture on treatment with $BaCl_2$ gives a white precipitate which is sparingly soluble in conc. HCl. On heating the mixture with $K_2Cr_2O_7$ an conc. H_2SO_4, red vapours (*A*) are produced. The aqueous solution of the mixture gives a deep blue colouration (*B*) with potassium ferricyanide solution. Identify the radicals in the given mixture and write the balanced equations for the formation of (*A*) and (*B*).

56. Write balanced chemical equations for the following :

(i) Sodium nitrite is produced by absorbing the oxides of nitrogen in aqueous solution of washing soda.

(ii) Nitrogen is obtained in the reaction of aqueous ammonia with potassium permanganate.

(iii) Elemental phosphorus reacts with conc. HNO_3 to give phosphoric acid.

(iv) Sulphur is precipitated in the reaction of hydrogen sulphide with sodium bisulphite solution.

(v) Carbon dioxide is passed through a suspension of limestone in water.

57. Give reasons in one or two sentences for the following :

(i) The hydroxides of aluminium and iron are insoluble in water. However, NaOH is used to separate one from the other.

(ii) Ammonium chloride is acidic in liquid ammonia solvent.

58. A light bluish green crystalline compound responds to the following tests :

(i) Its aqueous solution gives a brown precipitate or colouration with alkaline $K_2[HgI_4]$ solution.

(ii) Its aqueous solution gives a blue colour with $K_3[Fe(CN)_6]$ solution.

(iii) Its solution in hydrochloric acid gives a white precipitate with $BaCl_2$ solution.

Identify the ions present and suggest the formula of the compound.

59. Complete and balance the following :

(i) $NH_3 + NaOCl \longrightarrow$ .... + ... + ...

(ii) $AgBr + Na_2S_2O_3 \longrightarrow$ .... + ...

(iii) $(NH_4)_2S_2O_8 + H_2O + MnSO_4 \longrightarrow$ .... + ... + ...

60. The acidic, aqueous solution of ferrous ion forms a brown complex in the presence of NO_3^- by the following two steps. Complete and balance the equations :

$$[Fe(H_2O)_6]^{2+} + NO_3^- + H^+ \longrightarrow (a) + [Fe(H_2O)_6]^{3+} + H_2O$$

$$[Fe(H_2O)_6]^{2+} + (b) \longrightarrow (c) + H_2O$$

61. An orange solid (*A*) on heating gave a green residue (*B*), a colourless gas (*C*) and water vapours. The dry gas (*C*) on passing over heated Mg gave a white solid (*D*). (*D*) on reaction with water gave a gas (*E*) which formed dense white fumes with HCl. Identify (*A*) to (*E*) and give the reactions involved.

62. (*A*) is a binary compound of a univalent metal. 1.422 g of (*A*) reacts completely with 0.321 g of sulphur in an evacuated and sealed tube to give 1.743 g of a white crystalline solid (*B*), that forms a hydrated double salt (*C*) with $Al_2(SO_4)_3$. Identify (*A*), (*B*) and (*C*).

63. When gas (*A*) is passed through dry KOH at low temperature, a deep red coloured compound (*B*) and a gas (*C*) are obtained. The gas *A*, on reaction with but-2-ene, followed by treatment with Zn/H_2O yields acetaldehyde. Identify (*A*), (*B*) and (*C*).

64. A scarlet compound (*A*) is treated with conc. HNO_3 to give a chocolate brown precipitate (*B*). The precipitate is filtered and the filtrate is neutralised with NaOH. Addition of KI to the resulting solution gives a yellow precipitate (*C*). The precipitate (*B*) on warming with conc. HNO_3 in the presence of $Mn(NO_3)_2$ produces a pink-coloured solution due to the formation of (*D*). Identify (*A*), (*B*), (*C*) and (*D*). Write the reactions in sequence.

65. (i) A white precipitate (*B*) is formed when a mineral (*A*) is boiled with Na_2CO_3 solution.

(ii) The precipitate is filtered an the filtrate contains two compounds (*C*) and (*D*). The compound (*C*) is removed by crystallisation and when CO_2 is passed through the mother liquor left, (*D*) changes to (*C*).

(iii) The compound (*C*) on strong heating gives two compound (*D*) and (*E*).

(iv) *E* an heating with CoO forms blue colour bead (*F*). Identify (*A*) to (*F*).

66. A colourless inorganic salt (*A*) decomposes completely at about 250°C to give only two products (*B*) and (*C*) leaving no residue. The oxide (*C*) is a liquid at room temperature and neutral to moist litmus paper while the gas (*B*) is a neutral oxide. White phosphorus burns in excess of (*B*) to produce a strong while dehydrating agent (*D*). Write balanced equations for the reactions involved in the above process.

67. (a) Element (*A*) burns in nitrogen to given an ionic compound (*B*). Compound (*B*) reacts with water to give (*C*) and (*D*). A solution of (*C*) becomes 'milky' on bubbling carbon dioxide. Identify (*A*), (*B*), (*C*) and (*D*).

(b) Aluminium sulphide gives a foul odour when it becomes damp. Write a balanced chemical equation for the reaction.

68. During the qualitative analysis of a mixture containing Cu^{2+} and Zn^{2+} ions, H_2S gas is passed through an acidified solution containing these ions in order to test Cu^{2+} alone. Explain briefly.

69. A white solid is either Na_2O or Na_2O_2. A piece of red litmus paper turns white when it is dipped into a freshly made aqueous solution of the white solid.

(i) Identify the substance and explain with balanced equation.

(ii) Explain what would happen to the red litmus if the white solid were the other compound.

70. An aqueous solution containing one mole of HgI_2 and two moles of NaI is orange in colour. On addition of excess NaI the solution becomes colourless. The orange colour reappears on subsequent addition of NaOCl. Explain with equations.

71. (a) In the following equation,

$$(A) + 2(B) + H_2O \longrightarrow (C) + 2(D)$$

$$[(A) = HNO_2,\ (B) = H_2SO_3,\ (C) = NH_2OH].$$

Identify (*D*). Draw the structures of (*A*), (*B*), (*C*) and (*D*).

(b) The Haber process can be represented by the following scheme :

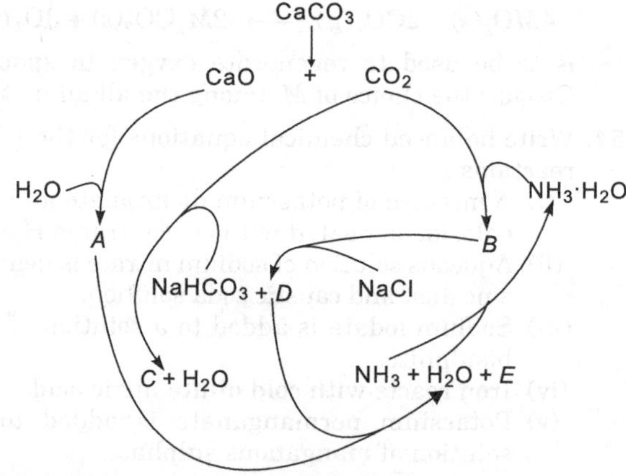

Identify (*A*), (*B*), (*C*), (*D*) and (*E*).

Exercise 2
(Stage 2 : High Skill Problem Solving)

Only One Option Correct

1. Yellow coloured solution of $FeCl_3$ changes to light green when

(a) $SnCl_2$ is added (b) Zn is added
(c) H_2S gas is passed (d) all true

2. Fe^{2+} does not give blue colour with $K_4[Fe(CN)_6]$ but on its reaction with (*X*), blue colour appears. (*X*) can be

(a) MnO_4^-/H^+ (b) H_2SO_4
(c) NH_3 (d) HCl

3. $Fe(OH)_3$ and $Cr(OH)_3$ ppt are separated by

(a) *Aq.* NH_3 (b) HCl
(c) $NaOH/H_2O_2$ (d) H_2SO_4

4. Turnbull's blue and Prussian's blue respectively are

I. $Fe^{II}[Fe^{II}(CN)_6]^{2-}$ II. $Fe^{III}[Fe^{III}(CN)_6]$
III. $Fe^{II}[Fe^{III}(CN)_6]^-$ IV. $Fe^{III}[Fe^{II}(CN)_6]^-$

(a) I, II (b) I, III
(c) III, IV (d) IV, III

5. Which of the following are soluble in excess of NaOH (*X*) : As_2S_3; (*Y*) : CuS; (*Z*): $AlCl_3$?

(a) *X*, *Y*, *Z* (b) *Y*, *Z*
(c) *X*, *Z* (d) *X*, *Y*

6. A mixture on heating gave a gas used as an anaesthetic, 1.1 g of gas occupies 0.56 L at NTP. Mixture contains

(a) $NaNO_3 + NH_4Cl$ (b) $NaNO_2 + NH_4Cl$
(c) $CaCO_3 + MgCO_3$ (d) $NH_4Cl + Na_2SO_4$

7. Aqueous solution contains $Zn(CH_3COO)_2$, $Cd(CH_3COO)_2$ and $Cu(CH_3COO)_2$. On passing H_2S gas, there is a precipitation of ...... as sulphide

(a) Zn^{2+}, Cd^{2+} (b) Cu^{2+}, Cd^{2+}
(c) Zn^{2+}, Cu^{2+} (d) Zn^{2+}, Cu^{2+}, Cd^{2+}

8. Ferric alum gives red colour with NH_4SCN due to formation of

(a) $Al(SCN)_3$ (b) $[Fe(SCN)_3]^-$
(c) $Fe(SCN)_3$ (d) $[Fe(SCN)]^{2+}$

9. Colourless salt (*X*) $\xrightarrow{\Delta}$ (*Y*) $\xrightarrow{Cu^{2+}, \Delta}$ coloured bead (*Z*). (*X*) can be

(a) borax (b) microcosmic salt
(c) both (a) and (b) (d) None of these

10. KCl + conc H_2SO_4 + $K_2Cr_2O_7$ $\xrightarrow{\Delta}$ (*X*) $\xrightarrow{NaOH}$ (*Y*). (*X*) is reddish brown coloured gas soluble in NaOH forming (*Y*). (*X*) and (*Y*) are

(a) Cr_2OCl_2, Na_2CrO_3 (b) $Cr_2O_2Cl_2$, Na_2CrO_3
(c) CrO_2Cl, Na_2CrO_4 (d) CrO_2Cl_2, Na_2CrO_4

11. Aqueous solution of $BaBr_2$ gives yellow ppt with

(a) K_2CrO_4 (b) $AgNO_3$
(c) both (a) and (b) (d) None of these

12. $\underset{\text{orange}}{Cr_2O_7^{2-}} \underset{pH = y}{\overset{pH = x}{\rightleftharpoons}} \underset{\text{green}}{CrO_4^{2-}}$. This change is based on change in pH. Probable values of *x* and *y* can be

(a) 8, 6
(b) 8, 10
(c) 4, 6
(d) change is independent of pH

13. H_2S would separate the following at pH < 7

(a) Zn^{2+}, Co^{2+} (b) Cu^{2+}, Cd^{2+}
(c) Cu^{2+}, Cr^{3+} (d) Cu^{2+}, As^{3+}

14. Solution of (*X*) in dil. HCl + H_2O $\longrightarrow$ white turbidity. (*X*) $\xrightarrow{H_2S/HCl}$ black ppt (*Y*). (*Y*) is soluble in

(a) H_2SO_4 (b) YAS
(c) HNO_3 (d) HCl

15. $K_2Cr_2O_7$ + conc. H_2SO_4 + H_2O_2 + ether $\longrightarrow$ blue perchromic anhydride (in ethereal layer). Blue colour is due to

(a) CrO_3 (b) H_2CrO_4
(c) $H_2Cr_2O_3$ (d) CrO_5

16. There is foul smell in presence of moisture with

(a) $AlCl_3$ (b) $Al_2(SO_4)_3$
(c) FeS (d) $FeSO_4$

17. $AgNO_3$ gives white ppt with hypo changing to black after sometime. Black ppt is of

(a) $Ag_2S_2O_3$ (b) Ag_2SO_4
(c) $Ag_2S_4O_6$ (d) Ag_2S

18. SO_2 and CO_2 both turn lime water (*A*) milky, SO_2 also turns $K_2Cr_2O_7/H^+$ (*B*) green while O_2 is soluble in pyrogallol (*C*) turning it black. These gases are to be detected in order by using these reagents. The order is

(a) (*A*), (*B*), (*C*) (b) (*B*), (*C*), (*A*)
(c) (*B*), (*A*), (*C*) (d) (*A*), (*C*), (*B*)

19. Aluminium sulphate (*X*) is slightly insoluble in water. It is converted into soluble sodium sulphate by using Na_2CO_3 in the preparation of sodium carbonate extract. Moles of (*Y*), required for complete conversion of 1 mole of (*X*) into soluble sulphate, is

(a) 1 (b) 2
(c) 3 (d) 4

20. $CoCl_2$ gives blue colour with NH_4SCN in ethereal layer due to formation of

(a) $(NH_4)_2[Co(SCN)_4]$ (b) $(NH_4)_4[Co(SCN)_6]$
(c) $(NH_4)_3[Co(SCN)_6]$ (d) $(NH_4)[Co(SCN)_4]$

21. $HgCl_2$ + excess of KI $\longrightarrow$ (*A*) $\xrightarrow{NH_3/NaOH}$ (*B*). (*A*) and (*B*) respectively are

(a) K_2HgI_4 (Nessler's reagent), $O\langle{}^{Hg}_{Hg}\rangle NH_2I$
(*X*) (Iodide of Millon's base) (*Y*)

(b) (*Y*), (*X*)
(c) both (*X*)
(d) both (*Y*)

22. NH_4SCN can be used to test one or more out of Fe^{3+}, Co^{2+}, Cu^{2+}

(a) Fe^{3+} only (b) Co^{2+}, Cu^{2+}
(c) Fe^{3+}, Cu^{2+} (d) all of these

23. $K_4[Fe(CN)_6]$ can be used to detect one or more out of $Fe^{2+}, Fe^{3+}, Zn^{2+}, Cu^{2+}, Cd^{2+}$

(a) Fe^{2+}, Fe^{3+} (b) $Fe^{3+}, Zn^{2+}, Cu^{2+}$
(c) all but Fe^{3+} (d) all but Fe^{2+}

24. Aqueous solution containing one mole borax reacts with two moles of acids. This is because of

(a) formation of 2 moles of $B(OH)_3$ only
(b) formation of 2 moles of $[B(OH)_4]^-$ only
(c) formation of 1 mole each of $B(OH)_3$ and $[B(OH)_4]^-$
(d) formation of 2 mol each of $[B(OH)_4]^-$ and $B(OH)_3$, of which only $[B(OH)_4]^-$ reacts with acid.

25. Ag_2S is soluble in NaCN due to formation of

(a) $Na[Ag(CN)_2]$ (b) $Ag(CN)_2$
(c) $Na_2[Ag(CN)_3]$ (d) $Na_2[Ag(CN)_2]$

26. Match the substances given in ***B*** based on reactions given in ***A*** and select correct answer from the alternate

	A (Reactions)		*B* (Substances)
I :	a white, waxy solid, normally stored under water because it spontaneously inflames in air	*P* :	HNO_3
II :	a viscous liquid that reacts with table sugar, giving a charred mass	*Q* :	Cl_2
III :	an acid that reacts with copper metal, releasing brown fumes	*R* :	P
IV :	a pale greenish yellow gas that dissolves in aqueous NaOH to give a solution used as a bleach.	*S* :	H_2SO_4

	I	II	III	IV		I	II	III	IV
(a)	*P*	*Q*	*R*	*S*	(b)	*S*	*R*	*Q*	*P*
(c)	*R*	*S*	*P*	*Q*	(d)	*S*	*R*	*P*	*Q*

27. Cu^{2+} and Ag^+ are both present in the same solution. To precipitate one of the ions and leave the other in solution, add

(a) $H_2S\ (aq)$ (b) $HCl(aq)$
(c) $HNO_3(aq)$ (d) $NH_4NO_3(aq)$

28. The addition of $K_2CO_3(aq)$ to the following solutions is expected to produce a precipitate in every case but one. That one is

(a) $BaCl_2(aq)$ (b) $CaBr_2(aq)$
(c) $Na_2SO_4(aq)$ (d) $Pb(NO_3)_2(aq)$

29. A large excess of $MgF_2(s)$ is maintained in contact with 1.00 L of pure water to produce a saturated solution. When an additional 1.00 L of pure water is added to the solid-liquid mixture and equilibrium re-established, compared to its value in the original saturated solution $[Mg^{2+}]$ will be

(a) the same (b) twice as large
(c) three as large (d) half as large

30. To increase the molar solubility of $CaCO_3(s)$ in a saturated aqueous solution add

(a) more water (b) Na_2CO_3
(c) NaOH (d) $NaHSO_4$

31. $Cu(OH)_2$ is highly soluble in all of the following except one. The exception is

(a) H_2O (b) $NH_3(aq)$
(c) $HCl(aq)$ (d) $HNO_3(aq)$

32. To increase significantly the concentration of free Zn^{2+} ion is a solution of the complex ion $[Zn(NH_3)_4]^{2+}$,

$$Zn^{2+}(aq) + 4NH_3(aq) \rightleftharpoons [Zn(NH_3)_4]^{2+}(aq)$$

add to the solution some

(a) H_2O (b) $HCl(aq)$
(c) $NH_3(aq)$ (d) $NH_4Cl(aq)$

33. The best way to ensure complete precipitation from saturated $H_2S(aq)$ of a metal ion M^{2+}, as its sulphide, $MS(s)$, is to

(a) add an acid
(b) increase $[H_2S]$ in solution
(c) raise the pH
(d) heat the solution

34. Of the following solutions the one that is acidic is

(a) $ZnSO_4(aq)$ (b) $NaAl(OH)_4(aq)$
(c) $NaHCO_3(aq)$ (d) $KNO_3(aq)$

35. Of the following oxides, all are soluble in NaOH(*aq*) except

(a) ZnO (b) Al_2O_3
(c) Fe_2O_3 (d) SnO_2

36. Marsh test is used for detecting

(a) As (b) Zn
(c) Al (d) CH_4

37. With MnO, colour of the bead in sodium carbonate-bead test is

(a) pink (b) black
(c) yellow (d) green

38. With Cr_2O_3, colour of the bead in sodium carbonate-bead test is

(a) red (b) blue
(c) yellow (d) green

39. Colourless salt (*A*) + dil. H_2SO_4 + KI $\longrightarrow$ blue colour with solution (*A*) can be

(a) $K_2Cr_2O_7$ (b) MnO_2
(c) NH_4NO_2 (d) NH_4Cl

40. Three test tubes *A, B, C* contain Pb^{2+}, Hg_2^{2+} and Ag^+ (but unknown). To each aqueous solution NaOH is added in excess. Following changes occur.

A : Black ppt

B : Brown ppt
C : White ppt but dissolves in excess of NaOH
A, *B* and *C* contain respectively
(a) Pb^{2+}, Hg_2^{2+} and Ag^+ (b) Hg_2^{2+}, Ag^+, Pb^{2+}
(c) Ag^+, Pb^{2+}, Hg_2^{2+} (d) Ag^+, Hg_2^{2+}, Pb^{2+}

41. Consider the following equilibrium

$$AgCl\downarrow + 2NH_3 \rightleftharpoons \underset{\text{soluble}}{[Ag(NH_3)_2]^+} + Cl^-$$

White ppt of AgCl appears an adding
(a) NH_3 (b) *aq* NaCl
(c) *aq* HNO_3 (d) *aq* NH_4Cl

42. Ag_2CrO_4 ppt is soluble in
(a) dil. HNO_3 (b) *aq* NH_3
(c) both (a) and (b) (d) None of these

43. A mixture contains Cu^{2+}, Al^{3+} and Ni^{2+}. Following steps have been adopted but written in disorder
I : Filter, boil off H_2S gas and add NH_4Cl, heat and add NH_4OH.
II : Filter, add NH_4OH and pass H_2S gas.
III : Pass H_2S gas into acidified solution of mixture.
Steps will be used in the following order
(a) I, II, III (b) III, I, II
(c) III, II, I (d) I, III, II

44. KI gives precipitate with all the cations given
(a) Ag^+, Hg_2^{2+}, Pb^{2+} (b) Cu^{2+}, Zn^{2+}, Ni^{2+}
(c) Na^+, Ca^{2+}, Mg^{2+} (d) Ag^+, Ca^{2+}, Sr^{2+}

45. Aqueous solution of salt

$$(A) + K_2CrO_4 \longrightarrow \underset{\text{(red ppt)}}{(B)} \xrightarrow{aq\ NH_3} \underset{\text{(black)}}{(C)} \quad A \text{ is}$$

(a) $AgNO_3$ (b) $Pb(NO_3)_2$
(c) $Hg_2(NO_3)_2$ (d) $Ca(NO_3)_2$

46. *A* is a colourless substances. Aqueous solution of *A* gives reddish-orange ppt with KI; ppt dissolves in excess of KI forming a colourless solution. If NH_4Cl and NaOH solution is added to this colourless solution reddish brown ppt is formed. Substance *A* is
(a) Epsom salt (b) Mohr's salt
(c) Calomel (d) Corrosive sublimate

47. CoS (black) obtained in group IV of salt analysis is dissolved in aqua regia, treated with an excess of $NaHCO_3$ and then Br_2 water is added. An apple green coloured stable complex is formed. It is
(a) sodium cobaltocarbonate
(b) sodium cobaltibromide
(c) sodium cobalticarbonate
(d) sodium cobaltobromide

48. Which of the following salt does not give positive test for nitrate ion?
(a) KNO_3 (b) $NaNO_3$
(c) $Pb(NO_3)_2$ (d) $Mg(NO_3)_2$

49. An aqueous solution of MCl_2 containing excess KCN, when treated with H_2S will precipitate *M*S only if '*M*' is
(a) Ni (b) Co
(c) Cu (d) Cd

50. Borax on heating strongly above its melting point melts to a liquid, which then solidifies to a transparent mass commonly known as borax-bead. The transparent glassy mass consists of
(a) sodium pyroborate
(b) boric anhydride
(c) sodium metaborate
(d) boric anhydride and sodium metaborate

51. If CO_2 gas is passed into *aq* Na_2CrO_4 yellow solution
(a) *aq* $Na_2Cr_2O_7$ (orange) solution is formed
(b) *aq* $Cr_2(CO_3)_2$ is formed
(c) $Cr(OH)_3$ is precipitated
(d) no action

52. A mixture of two colourless substances was dissolved in water. When gaseous Cl_2 was passed through the solution, a deep brown colour developed. Addition of $BaCl_2$ or NaOH to the original solution give a white precipitate. The mixture contained
(a) $NaNO_3$ and $AlCl_3$
(b) NaBr and KCl
(c) MgI_2 and $MgSO_4$
(d) $BaSO_4$ and $PbCl_2$

53. Four test tubes containing dil. HCl, $BaCl_2$, $CdCl_2$ and KNO_3 solution. Which of the following will identify $BaCl_2$?
(a) Dil. H_2SO_4 (b) K_2CrO_4
(c) $FeSO_4/H_2SO_4$ (d) $AgNO_3$

54. The ion most difficult to remove as a precipitate is
(a) Ag^+ (b) NH_4^+
(c) Fe^{3+} (d) Cu^{2+}

55. A colourless gas is dissolved in water and the resulting solution turns red litmus blue, the gas may have been which one of the following?
(a) HCl (b) H_2S
(c) SO_2 (d) NH_3

56. A substance that will deliquesce in the normal laboratory atmosphere is:
(a) $CuSO_4$ (b) NaOH
(c) SiO_2 (d) all of these

57. A salt which will turn dark on exposure to sunlight has the formula
(a) $K_2C_2O_4$ (b) $CaCl_2$
(c) AgCl (d) $Fe_4[Fe(CN)_6]$

58. What would you observe if you add with shaking, excess dil. NaOH solution to $ZnCl_2$ solution?
(a) A white ppt
(b) A white ppt which later dissolves
(c) A green ppt
(d) A green ppt which later dissolves

59. A mixture is known to contain NO_3^- and NO_2^-. Before performing **Ring Test** for NO_3^-, the aqueous solution should be made free of NO_2^-. This is done by heating aqueous extract with
(a) conc. HNO_3 (b) dil. HNO_3
(c) urea (d) zinc dust

60. A suspension containing insoluble substances ZnS, CuS, HgS, Ag_2S, and FeS, is treated with 2N HCl. On filtering, the filtrate contained appreciable amounts of which one of the following groups?
(a) Zinc and mercury (b) Silver and iron
(c) Copper and mercury (d) Zinc, copper and iron

61. The compound thioacetamide has been into analytical chemistry to replace which one of the following reagents?
(a) H_2S (b) DMG
(c) YAS (d) $(NH_4)_2C_2O_4$

62. Some white colourless crystals are heated. A cracking sound is heard and brown fumes are given off. The residue is yellow-brown in colour. When a glowing splinter is held in the fumes, it is relighted. The fumes consists of
(a) O_2 (b) NO_2
(c) Cl_2 (d) NO_2 and O_2

63. A yellow solid known to be a single compound is completely insoluble in hot water but dissolves in hot dilute HCl to give an orange solution. When this solution is cooled, a white crystalline ppt is formed. This white ppt redissolves on heating the solution. The compound is
(a) $Fe(OH)_3$ (b) $PbCrO_4$
(c) K_2CrO_4 (d) $Co(OH)_2$

64. '*X*' is a colourless salt giving following reactions:

X ← NaOH — *X* — $AgNO_3/HNO_3$ → White ppt. soluble in aq. NH_3
X: white ppt but dissolves in excess of NaOH
X — NH_4OH/H_2S → White ppt.

X can be
(a) $AlCl_3$ (b) $ZnCl_2$
(c) $Zn(CH_3COO)_2$ (d) $ZnBr_2$

65. A mixture upon adding conc. H_2SO_4 gives orange red fumes. It may contain the anion pair
(a) $CrO_4^{2-} + Cl^-$ (b) $Br^- + Cl^-$
(c) $NO_3^- + Cl^-$ (d) $CrO_4^{2-} + NO_3^-$

66. When calomel reacts with NH_4OH, we get
(a) $Hg(NH_2)Cl + Hg$ (b) $NH_2—Hg—Hg—Cl$
(c) Hg_2O (d) HgO

67. When a substance *A* reacts with water it produces a combustible gas *B* and a solution of substance *C* in water. When another substance *D* reacts with this solution of *C,* it also produces the same gas *B* on warming but *D* can produce gas *B* on reaction with dilute sulphuric acid at room temperature. *A* imparts a deep golden yellow colour to a smokeless flame of Bunsen burner. *A, B, C* and *D,* respectively are
(a) Na, H_2, NaOH, Zn
(b) K, H_2, KOH, Al
(c) CaH_2, $Ca(OH)_2$, Sn
(d) CaC_2, C_2H_2, $Ca(OH)_2$, Fe

68. In the separation of Cu^{2+} and Cd^{2+} in 2nd group of qualitative analysis of cations tetrammine copper(II) sulphate and tetraammine cadmium(II) sulphate react with KCN to form the corresponding cyanide complexes, which one of the following pairs of the complexes and their relative stabilities enables the separation of Cu^{2+} and Cd^{2+}?
(a) $K_3[Cu(CN)_4]$: less stable and
$K_2[Cd(CN)_4]$: more stable
(b) $K_3[Cu(CN)_4]$: more stable and
$K_2[Cd(CN)_4]$: less stable
(c) $K_2[Cu(CN)_4]$: less stable and
$K_2[Cd(CN)_4]$: more stable
(d) $K_2[Cu(CN)_4]$: more stable and
$K_2[Cd(CN)_4]$: less stable

69. Chemical *A* is used for softening of water to remove temporary hardness. *A* reacts with sodium carbonates to produce caustic soda. When CO_2 is bubbled through '*A*' it turns cloudy. Chemically '*A*' is
(a) CaO (b) $CaCO_3$
(c) $Ca(HCO_3)_2$ (d) $Ca(OH)_2$

70. The solution of sodium meta aluminate on boiling with ammonium chloride gives a white precipitate of
(a) Al_2O_3 (b) $AlCl_3$
(c) $Al(OH)_3$ (d) $NaAl(OH)_4$

71. Nitrite (NO_2^-) interferes in the **'ring-test'** of nitrate (NO_3^-). Some of the following reagents can be used for the removal of nitrite.
I : NH_4Cl
II : $(NH_2)_2CS$ (thiourea)
III : NH_2SO_3H (sulphamic acid)
IV : Sulphanilic acid
Correct choice is
(a) I, II (b) I, II, IV
(c) I, II, III (d) I, III, IV

72. Consider following reaction

Nitrite + acetic acid + thiourea

$$\longrightarrow N_2 + SCN^- + 2H_2O$$

Formation of the product in the above reaction can be identified by

(a) $FeCl_3$/dil. HCl when blood red colour appears

(b) $FeCl_3$/dil. HCl when blue colour appears

(c) $K_2Cr_2O_7$/HCl when green colour appears

(d) $KMnO_4$/HCl when colourless solution is formed

73. Zinc pieces are added to acidified solution of SO_3^{2-}. Gas liberated can

(a) turn lead acetate paper black

(b) turn lime water milky

(c) give both of the above tests

(d) give none of the above tests

74. Sodium nitroprusside turns purple when it is exposed in the atmosphere of

(a) CO (b) H_2S

(c) SO_2 (d) CO_2

75. Select correct statement(s).

(a) $Ag_2S_2O_3$ appears as white precipitate when $Na_2S_2O_3$ reacts with $AgNO_3$

(b) $Ag_2S_2O_3$ is unstable turning black on standing due to formation of Ag_2S

(c) $S_2O_3^{2-}$ can form soluble complex $[Ag(S_2O_3)_2]^{3-}$ with Ag^+

(d) All the above are correct statements

76. A gas *X* is passed through water to form a saturated solution. The aqueous solution on treatment with silver nitrate gives a white precipitate. The saturated aqueous solution also dissolves magnesium ribbon with evolution of a colourless gas *Y*. Identify *X* and *Y*

(a) $X = CO_2, Y = Cl_2$ (b) $X = Cl_2, Y = CO_2$

(c) $X = Cl_2, Y = H_2$ (d) $X = H_2, Y = Cl_2$

77. An aqueous solution of a substance gives a white precipitate on treatment with dilute hydrochloric acid, which dissolves on heating. When hydrogen sulphide is passed through the hot acidic solution, a black precipitate is obtained. The substance is a

(a) Hg^{2+} salt (b) Cr^{2+} salt

(c) Ag^+ salt (d) Pb^{2+} salt

78. $[X] + H_2SO_4 \longrightarrow [Y]$ a colourless gas with irritating smell

$$[Y] + K_2Cr_2O_7 + H_2SO_4 \longrightarrow \text{green solution}$$

$[X]$ and $[Y]$ are

(a) SO_3^{2-}, SO_2 (b) Cl^-, HCl

(c) S^{2-}, H_2S (d) CO_3^{2-}, CO_2

79. For H_3PO_3, the correct choice is

(a) H_3PO_3 is dibasic and reducing

(b) H_3PO_3 is dibasic and non-reducing

(c) H_3PO_3 is tribasic and reducing

(d) H_3PO_3 is tribasic and non-reducing

80. A sodium salt of an unknown anion when treated with $MgCl_2$ gives white precipitate only boiling. The anion is

(a) SO_4^{2-} (b) HCO_3^-

(c) CO_3^{2-} (d) NO_3^-

81. A metal nitrate reacts with KI to give a black precipitate which on addition of exces of KI converts into orange colour solution. The cation of metal nitrate is

(a) Hg^{2+} (b) Bi^{3+}

(c) Pb^{2+} (d) Cu^+

82. A solution when diluted with H_2O and boiled, gives a white precipitate. On addition of excess NH_4Cl/NH_4OH, the volume of precipitate decreases leaving behind a white gelatinous precipitate. Identify the precipitate which dissolves in NH_4OH/NH_4Cl

(a) $Zn(OH)_2$ (b) $Al(OH)_3$

(c) $Mg(OH)_2$ (d) $Ca(OH)_2$

83. $CuSO_4$ is decolourised on addition of KCN, the product is

(a) $[Cu(CN)_4]^{2-}$ (b) $[Cu(CN)_4]^{3-}$

(c) $Cu(CN)_2$ (d) CuCN

84. A solution of a metal ion when treated with KI gives a red precipitate which dissolves in excess KI to give a colourless solution. Moreover, the solution of metal ion on treatment with a solution of cobalt(III) thiocyanate gives rise to a deep blue crystalline precipitate. The metal ion is

(a) Pb^{2+} (b) Hg^{2+}

(c) Cu^{2+} (d) Co^{2+}

85. Sodium fusion extract, obtained from aniline, on treatment with iron (II) sulphate and H_2SO_4, in presence of air gives a Prussian blue precipitate. The blue colour is due to the formation of

(a) $Fe_4[Fe(CN)_6]_3$ (b) $Fe_3[Fe(CN)_6]_2$

(c) $Fe_4[Fe(CN)_6]_2$ (d) $Fe[Fe(CN)_6]_3$

86. A dark violet colour mixture in presence of dil. HCl changes to pale yellow solution. Mixture may contain

(a) MnO_4^-, Fe^{2+} (b) $MnO_4^-, C_2O_4^{2-}$

(c) $I^-, Cr_2O_7^{2-}$ (d) none is correct

87. Mixture is initially orange in colour. When solution is prepared in dil. acid, it changes to dark brown colour. Mixture contains

(a) $HgI_2, Cr_2O_7^{2-}$ (b) $I^-, Cr_2O_7^{2-}$

(c) I^-, Cu^{2+} (d) I^-, SO_3^{2-}

88. An orange colour mixture changes to green on acidification. Mixture may contains

(a) HgI_2, CrO_4^{2-} (b) $Fe^{2+}, Cr_2O_7^{2-}$

(c) SO_3^{2-}, MnO_4^- (d) Fe^{2+}, CrO_4^{2-}

89. A colourless salt changes to yellow on heating. Salt is also soluble in NaOH as well as in dil. HCl. Salt can be
(a) FeO (b) PbO
(c) ZnO (d) CdO

90. A salt contains cation A^{2+} and anion B^{2-}. Both decolourise MnO_4^- is acidic medium. Salt is
(a) ferrous oxalate (b) ferric oxalate
(c) both (a) and (b) (d) None of these

91. *A* on reaction with dil. H_2SO_4 gives a colourless pungent gas that can turn $Cr_2O_7^{2-}$ / H^+ green. Green colour is due to formation of
(a) CrO_4^{2-} (b) Cr^{3+}
(c) CrO_3 (d) CrO_2Cl_2

92. Aqueous solution of *A* can dissolve AgBr forming a soluble complex *B*. *A* also reacts with aqueous $AgNO_3$ solution giving white ppt. *C* changing to black ppt *D*. *A, B, C* and *D* are

	A	*B*	*C*	*D*
(a)	$Na_2S_2O_3$	$Na_3[Ag(S_2O_3)_2]$	$Ag_2S_2O_3$	Ag_2S
(b)	$Na_2S_2O_3$	$Ag_2S_2O_3$	Ag_2S	$Na_3[Ag(S_2O_3)_2]$
(c)	NH_3	$[Ag(NH_3)_2]Br$	$Ag(OH)$	Ag_2O
(d)	NH_3	$Na[Ag(OH)_2]$	Ag_2O	$[Ag(NH_3)_2]Br$

93. S^{2-} and SO_3^{2-} can be distinguished using
(a) $(CH_3COO)_2Pb$ (b) $Na_2[Fe(CN)_5NO]$
(c) both (a) and (b) (d) None of these

94. A mixture contains NO_2^- and SO_3^{2-}. On adding dilute H_2SO_4 and KI and then starch, blue colour appears. This is due to
(a) NO_2^- (b) SO_3^{2-}
(c) both (a) and (b) (d) None of these

95. SO_2 can
(a) decolourise MnO_4^-/H^+ solution
(b) turn $Cr_2O_7^{2-}/H^+$ solution green
(c) both (a) and (b) are correct
(d) none of the above is correct

96. Cationic part of chromyl chloride is:
(a) Cr^{3+} (b) CrO_2^+
(c) CrO_2^{2+} (d) CrO^{2+}

97. Aqueous solution of a mixture contains Br^- and I^-. On passing Cl_2 gas and adding $CHCl_3$, then organic layer will be
(a) violet (b) reddish brown
(c) colourless (d) blue

98. $[Fe(H_2O)_5NO]^{2+}$ is formed as brown ring in NO_3^- test. Fe in this complex has ...... unpaired electrons.
(a) one (b) two
(c) three (d) four

99. A colourless salt gives white ppt with $CaCl_2$ solution and can also decolourise MnO_4^-/H^+. Salt is decomposed by conc. H_2SO_4 forming gases
(a) CO_2, SO_2 (b) CO, CO_2
(c) CO, SO_2 (d) N_2, CO, CO_2

100. Which is soluble in H_2O?
(a) AgI (b) AgBr
(c) AgCl (d) AgF

101. Which is the volatile compound burning with green flame?
(a) $(C_2H_5)_3B$ (b) $(C_2H_5)_3BO_3$
(c) $Ba(NO_3)_2$ (d) $BaCl_2$

102. Test of this anion is not performed in glass test tubes. This anion is
(a) F^- (b) AsO_4^{3-}
(c) PO_4^{3-} (d) BO_3^{3-}

103. HF attacks glass (Na_2SiO_3) forming
(a) Na_2SiF_4 (b) Na_2SiF_3
(c) Na_2SiF_6 (d) H_4SiO_4

104. SiF_4 is hydrolysed by H_2O forming
(a) H_2SiF_6, H_4SiO_4 (b) H_2SiF_4, H_2SiF_6
(c) H_2SiF_6, H_2SiO_3 (d) H_2SiF_6, SiO_2

105. Water soluble mixture $\xrightarrow[\text{(ii) Filter}]{\text{(i) } BaCl_2}$ White ppt.

Filtrate + Br_2 water + $BaCl_2$ $\longrightarrow$ White ppt.

Mixture contains
(a) SO_3^{2-} only (b) SO_4^{2-} only
(c) both (a) and (b) (d) None of these

106. White ppt of AgCl and $PbCl_2$ can be separated using
(a) *aq* NH_3 (b) hot water
(c) both (a) and (b) (d) None of these

107. White ppt of *A* on reaction with *aq* NH_3 are blackened. Select correct statement about *A*.
(a) *A* is also called calomel
(b) *A* reacts with *aq* NH_3 forming $HgNH_2Cl$
(c) *A* changes to grey on reaction with $SnCl_2$
(d) all of the above are correct statements

108. Aqueous solution of *A* gives yellow ppt with *aq* K_2CrO_4 *A* may contain
(a) Pb^{2+} or Ag^+ or Ba^{2+} (b) Pb^{2+} or Ba^{2+}
(c) Ag^+ or Ba^{2+} (d) Pb^{2+} or Ag^+

109. Ag_2CrO_4 (brick red ppt) is soluble in
(a) dil. HNO_3 or *aq*. NH_3 (b) dil. HCl or *aq* NH_3
(c) dil. HNO_3 or dil. HCl (d) H_2O or dil. HNO_3

110. $Ag_2CrO_4 + H^+NO_3^- \longrightarrow$ Soluble *A*, *A* is
(a) Ag_2CrO_4 (b) $AgNO_3$
(c) $Ag_2Cr_2O_7$ (d) Ag_2O

111. $K_4[Fe(CN)_6]$ can be used to detect some ions out of Cd^{2+}, Cu^{2+}, Zn^{2+}, Fe^{3+}, Pb^{2+} and Bi^{3+}. Excluded ions are
(a) Fe^{3+}, Bi^{3+}, Pb^{2+} (b) Cu^{2+}, Zn^{2+}, Bi^{3+}
(c) Bi^{3+}, Pb^{2+} (d) Fe^{3+}, Pb^{2+}

112. Which gives ppt with K_2CrO_4?
(a) Hg_2^{2+}, Pb^{2+}, Ag^+, Ba^{2+} (b) Pb^{2+}, Ag^+, Ba^{2+}
(c) Ag^+, Ba^{2+} (d) Pb^{2+}, Ba^{2+}

113. Select correct statement.
(a) $[Cu(CN)_4]^{3-}$ is more stable than $[Cd(CN)_4]^{3-}$
(b) If H_2S gas is passed into the aqueous solution of mixture of $[Cu(CN)_4]^{3-}$ and $[Cd(CN)_4]^{3-}$, formation of yellow ppt indicates presence of Cd^{2+}
(c) HgI_2 and BiI_3 both dissolve in excess of KI forming soluble HgI_4^{2-} and BiI_4^-.
(d) All of the above are correct statements

114. Precipitates of IIA and IIB can be separated by
(a) NaOH (b) yellow $(NH_4)_2S$
(c) both (a) and (b) (d) None of these

115. Eco-friendly reagent that can be used instead of H_2S is
(a) Na_2S (b) $(NH_4)_2CS_3$
(c) $(NH_4)_2S_2$ (d) S_8

116. Fe^{3+} gives different colour with
(a) CNS^- (b) $[Fe(CN)_6]^{4-}$
(c) CH_3COO^- (d) all of these

117. $Fe(OH)_3$ and $Cr(OH)_3$ can be separated using
(a) NaOH (b) $NaOH + H_2O_2$
(c) both (a) and (b) (d) None of these

118. $Fe(OH)_3$ and $Al(OH)_3$ can be separated using
(a) NaOH (b) $KOH + Br_2$ water
(c) both (a) and (b) (d) None of these

119. Select correct statement.
(a) Prussian blue and Turnbull's blue are identical in structure
(b) NH_4Cl decreases ionisation of NH_4OH by common ion effect so as to precipitate only Al^{3+}, Cr^{3+} and Fe^{3+} as hydroxides.
(c) Both (a) and (b) are correct
(d) None of the above is correct

120. By H_2O_2/OH^-, $Cr(OH)_3$ changes to
(a) CrO_3 (b) $Cr_2O_7^{2-}$
(c) CrO_4^{2-} (d) $[Cr(OH)_4]^-$

121. On adding KNO_2 and CH_3COOH solution to the neutral solution of $CoCl_2$, there is formation of yellowish orange precipitate of
(a) $K_3[Co(NO_2)_6]$
(b) $K_4[Co(NO_2)_6]$
(c) $K_3[Co(NO_2)_4(CH_3COO)_2]$
(d) $K_2[Co(NO_2)_4]$

122. Which set does not contain amphoteric species?
(a) $Al(OH)_3$, $Zn(OH)_2$ (b) SnO_2, Al_2O_3
(c) $Fe(OH)_3$, $Ni(OH)_2$ (d) All of these

123. PbO_2 oxidises MnO_2 (black) to
(a) MnO_4^{2-} (b) MnO_4^-
(c) Mn^{2+} (d) Mn_2O_7

124. Test tube *A* contains $ZnCl_2$ aqueous solution while test tube *B* contains *aq* $CdCl_2$ solution. On passing H_2S gas
(a) ZnS is precipitated
(b) CdS is precipitated
(c) both (a) and (b) are precipitated
(d) None of the above is precipitated

125. Dimethyl glyoxime and $NaHSO_3/NH_4CNS$ are used to distinguish and separate Cu^{2+} and Ni^{2+}. These are used in order
(a) HSO_3^-/CNS^- then DMG
(b) DMG then HSO_3^-/CNS^-
(c) at random
(d) given reagents are not suitable

126. Which cation cannot be identified by flame test?
(a) Na^+ (b) K^+
(c) Ba^{2+} (d) Mg^{2+}

127. When microcosmic salt is heated, transparent bead is of
(a) $NaPO_3$ (b) $Na(NH_4)HPO_4$
(c) $NaBO_2$ (d) B_2O_3

128. In borax-bead test there is formation of transparent bead of
(a) $NaPO_3$ and $NaBO_2$
(b) $NaBO_2$ and $Na_2B_4O_7$
(c) $NaBO_2$ and B_2O_3
(d) $NaPO_3$ and B_2O_3

129. In sodium carbonate bead test, chromium salts (green) change to coloured bead
(a) yellow, $PbCrO_4$ (b) yellow, $BaCrO_4$
(c) yellow, Na_2CrO_4 (d) orange, $Na_2Cr_2O_7$

130. Microcosmic salt and borax are used in the identification of cations by dry tests. They are respectively
(a) $NaBO_2$ and $NaPO_3$
(b) $NaNH_4HPO_4 \cdot 4H_2O$ and $Na_2B_4O_7 \cdot 10H_2O$
(c) $NaPO_3$ and $NaBO_2$
(d) $Na_2B_4O_7$ and $NaNH_4HPO_4$

131. Which is not soluble in CH_3COOH?
(a) CaC_2O_4 (b) $Na_2C_2O_4$
(c) $K_2C_2O_4$ (d) $CaCO_3$

132. Acidified MnO_4^- can be decolourised by
(a) CaC_2O_4 (b) H_2O_2
(c) $FeSO_4$ (d) all of these

133. $BaCO_3 + CH_3COOH \longrightarrow A \xrightarrow{Cr_2O_7^{2-} + OH^-} B, B$

(a) $BaCr_2O_7$ (b) $BaCrO_4$
(c) $BaCrO_3$ (d) $Ba_2Cr_2O_4$

134. A (colourless salt) $\xrightarrow{\Delta} \underbrace{B + C}_{gas} + D$; $D \xrightarrow{H_2O} E$.

Gas C turns solution E milky. B burns with blue flame. A also decolourises MnO_4^-/H^+. Thus, A, B, C, D and E are

	A	B	C	D	E
(a)	CaC_2O_4	CO	CO_2	CaO	$Ca(OH)_2$
(b)	CaC_2O_4	CO_2	CO	CaO	$Ca(OH)_2$
(c)	$CaCO_3$	CaO	CO	CO_2	$Ca(OH)_2$
(d)	$CaOCl_2$	Cl_2	O_2	CaO	$Ca(OH)_2$

135. Metal chloride A is soluble in hot water but insoluble in cold water. Select correct statement about A. Thus

(a) A can give yellow ppt. with K_2CrO_4
(b) A can give white ppt with K_2SO_4
(c) A can give yellow ppt with KI
(d) all the above are correct statements

136. $HgCl_2$ + excess KI $\longrightarrow X \xrightarrow{NH_4Cl + NaOH} Y$; X and Y are:

	X	Y
(a)	HgI_2	NH_4I
(b)	K_2HgI_4	$(NH_4)_2HgI_4$
(c)	K_2HgI_4	$O\langle^{Hg}_{Hg}\rangle NH_2I$
(d)	HgI_2	$O\langle^{Hg}_{Hg}\rangle NH_2I$

137. $NaNO_2 + A \longrightarrow B \xrightarrow{\Delta} N_2$

$A + AgNO_3 \longrightarrow$ White ppt soluble in C

$A \xrightarrow{NaOH, \Delta} C$

Hence, A is

(a) NaCl (b) NH_4Cl
(c) NH_4NO_2 (d) NH_4NO_3

138. $A \longrightarrow B$(gas) used by dentist. Hence, A is

(a) NH_4Cl (b) NH_4NO_2
(c) NH_4NO_3 (d) NH_4OH

139. Reagent that can detect presence of NH_3 is called

(a) Fehling's solution
(b) Nessler's reagent
(c) Benedict's solution
(d) Lucas reagent

140. Ammonium salts can be decomposed by

(a) *aq* Na_2O (b) *aq* CaO
(c) both (a) and (b) (d) None of these

One or More Than One Options Correct

1. $K_4[Fe(CN)_6]$ can be used to test

(a) Fe^{3+} (b) Fe^{2+}
(c) Zn^{2+} (d) Cd^{2+}

2. NaOH can dissolve

(a) $Cr(OH)_3$ (b) $Al(OH)_3$
(c) $Zn(OH)_2$ (d) $Fe(OH)_3$

3. Ag^+ and Pb^{2+} ions are precipitated together by addition of

(a) *aq.* NH_3 (b) $K_2C_2O_4$ solution
(c) dil. HNO_3 (d) KI solution

4. An inorganic mixture gives yellow precipitate on boiling with conc. HNO_3 and ammonium molybdate $((NH_4)_2MoO_4)$. Thus, inorganic mixture can be

(a) $(NH_4)_3PO_4$ (b) $(NH_4)_3AsO_4$
(c) $Pb(NO_3)_2$ (d) $(NH_4)_2C_2O_4$

5. KCN is used in the

(a) separation of Cu^{2+} and Cd^{2+}
(b) metallurgical extraction of Ag^+ and Au^+
(c) identification of Fe^{3+}
(d) treatment of cancer

6. Precipitation of cations as sulphide can be done by

(a) H_2S (b) CdS
(c) K_2CS_3 (d) $(NH_4)_2S_2$

7. Borax-bead test can be used to identify following ions

(a) Ca^{2+} (b) Cr^{3+}
(c) Mn^{2+} (d) Cu^{2+}

8. Zinc oxalate can be tested by

(a) NaOH when white precipitate is formed which dissolves in excess of NaOH
(b) $KMnO_4$ which is decolourised
(c) $CaCl_2$ which gives white precipitate
(d) $K_4[Fe(CN)_6]$ which gives yellowish white precipitate

9. Select correct statement(s).

(a) CdS and As_2S_3 can be separated using NaOH or yellow ammonium sulphide
(b) $FeCl_3$ gives blue precipitate with $K_4[Fe(CN)_6]$
(c) $FeCl_3$ gives blood red colour with KCNS
(d) *Aq* $FeCl_3$ is reduced to $FeCl_2$ when Zn pieces are added

10. Acidified $K_2Cr_2O_7$ solution changes to green

(a) when SO_2 gas is passed into it
(b) when $FeCl_3$ is added to it
(c) when $FeSO_4$ is added to it
(d) when H_2O_2 is added to it

Brain Twisters

1. (*A*) is hydrated salt. 2.48 g of (*A*) on heating gives 1.58 g of anhydrous salt (*B*).

(i) Aqueous (*A*) + $AgNO_3 \longrightarrow$ white turbidity changing to black.

(ii) Aqueous (*A*) decolourises I_2 in KI

(iii) Aqueous (*A*) + dil. HCl $\longrightarrow$ white turbidity.

(iv) Aqueous (*A*) dissolves unreacted AgBr of photographic plate.

Identify (*A*) and explain reactions.

2. When a substance (*A*) was treated with dil. HCl, a colourless gas (*B*) was evolved, which turned moist litmus paper red. On bubbling (*B*) thorugh lime water, a precipitate (*C*) was formed, but passage of further gas resulted in a clear solution (*D*). A small sample of (*A*) was moistened with conc. HCl and placed on a paltinum wire, and introduced in to a Bunsen burner flame where it caused a green flame colouration. On strong heating. (*A*) decomposed, giving a white solid (*E*) which turned red litmus paper blue. 1.9735 g of (*A*) was heated strongly and gave 1.5334 g of (*E*). The sample (*E*) was dissolved in water and made upto 250 mL in a standard flask. 25 mL aliquots were titrated with acid and required 20.30 mL of 0.0985 M HCl. Name the compounds (*A*) to (*E*), and give equations for all the reactions. Calculate the gram molecular weight of (*A*).

3. Identify (*A*) to (*O*) in the following (*A*) (hydrated bluish brown coloured double salt) gives white ppt (*B*) with $BaCl_2$ insoluble in conc. HCl. 0.9915 g of (*A*) gave 0.932 g of white ppt.

(*A*) (in dil. HCl) $\xrightarrow{H_2S}$ black ppt (*C*) (insoluble in yellow ammonium sulphide)

black ppt (*C*) $\xrightarrow{\Delta,\ HNO_3}$ blue coloured solution (*D*)

blue coloured solution (*D*) $\xrightarrow{aq\ NH_3}$ blue ppt (*E*) $\xrightarrow{\text{excess } aq.\ NH_3}$ deep blue solution (*F*)

blue coloured solution (*D*) $\xrightarrow{KI}$ white ppt (*G*) (appeared after adding hypo)

blue coloured solution (*D*) $\xrightarrow{NH_4SCN}$ black ppt (*H*)

Filtrate after separating (*C*) $\xrightarrow{NH_4Cl + NH_4OH}$ brown ppt (*I*)

brown ppt (*I*) $\xrightarrow{HCl}$ yellow coloured solution (*J*) $\xrightarrow{\text{Gas } (K)}$ light green coloured solution (*L*)

yellow coloured solution (*J*) $\xrightarrow{SnCl_2}$ green (*L*)

yellow coloured solution (*J*) $\xrightarrow{K_4[Fe(CN)_6]}$ blue (*M*)

yellow coloured solution (*J*) $\xrightarrow{NH_4SCN}$ intense red (*N*)

light green coloured solution (*L*) $\xrightarrow{K_3[Fe(CN)_6]}$ blue (*O*)

($Cu = 63.5, Fe = 56, SO_4^{2-} = 96, H_2O = 18, Ba^{2+} = 137$)

4. (*A*) is binary orange coloured paramnetic compound of a univalent cation. 1.422 g of (*A*) reacts completely with 0.321 g of sulphur in an evacuated and sealed tube to give 1.743 g of a while crystalline solid (*B*) that forms hydrated double salt (*C*) with $Al_2(SO_4)_3$; what are (*A*), (*B*) and (*C*) and explain reactions?

5. Hydrated salt (*A*) + $BaCl_2 \longrightarrow$ white ppt (*B*) insoluble in HNO_3.

(*A*) (in dil. HCl) + $H_2S \longrightarrow$ black ppt soluble in 50% HNO_3

(*A*) + *aq* NH_3 (excess) $\longrightarrow$ deep blue solution (*C*)

(*A*) + microscomic salt $\xrightarrow{\Delta}$ green (*D*) when hot

(*A*) + KI solution $\longrightarrow$ reddish brown changing to white ppt after hypo is added.

2.495 g of (*A*) gave 2.33 g of (*B*). Identify (*A*), (*B*), (*C*) and (*D*) and explain reactions.

6. 0.347 g of a metal (*A*) was dissolved in dil. HNO_3. This solution gave a red colouration to a non-luminous Bunsen burner flame, and on evaporation gave 0.747 g of metal oxide (*B*). (*A*) also reacted with nitrogen, forming a compound (*C*), and with hydrogen, forming (*D*). On reacting 0.1590 g of (*D*) with H_2O, a gas (*E*) was evolved and a sparingly soluble compound (*F*) formed, which gave a strongly basic reaction and required 200 mL of 0.1 M HCl to neutralise it. Identify the substances (*A*) to (*F*) and explain the reactions involved.

7. Identify inorganic salt (*A*) which is hydrated and light green in colour and gives following reactions

(i) (*A*) decolourises acidified MnO_4^- and turns acidified $Cr_2O_7^{2-}$ green.

(ii) (*A*) in aqueous solution gives light green turbidity which dissolves in dil. acid.

(iii) (*A*) gives white ppt with $BaCl_2$ solution insoluble in conc. HCl. 2.78 g of (*A*) gave 2.33 g of white ppt.

(iv) (*A*) $\xrightarrow{\Delta}$ (*B*) (blackish brown) + $\underbrace{(C) + (D)}_{\text{oxides of sulphur}}$

(*C*) $\xrightarrow{\text{oxidation}}$ (*D*)

(*C*) turns acidified $K_2Cr_2O_7$ solution green

(*B*) + HCl $\xrightarrow{\Delta}$ (*E*) deep yellow

(*E*) + KCNS $\longrightarrow$ red colour

Passage Comprehension Questions

Passage 1

Following flow sheet has been given to analyse a salt. Based on the various stages, answer the questions at the end of it.

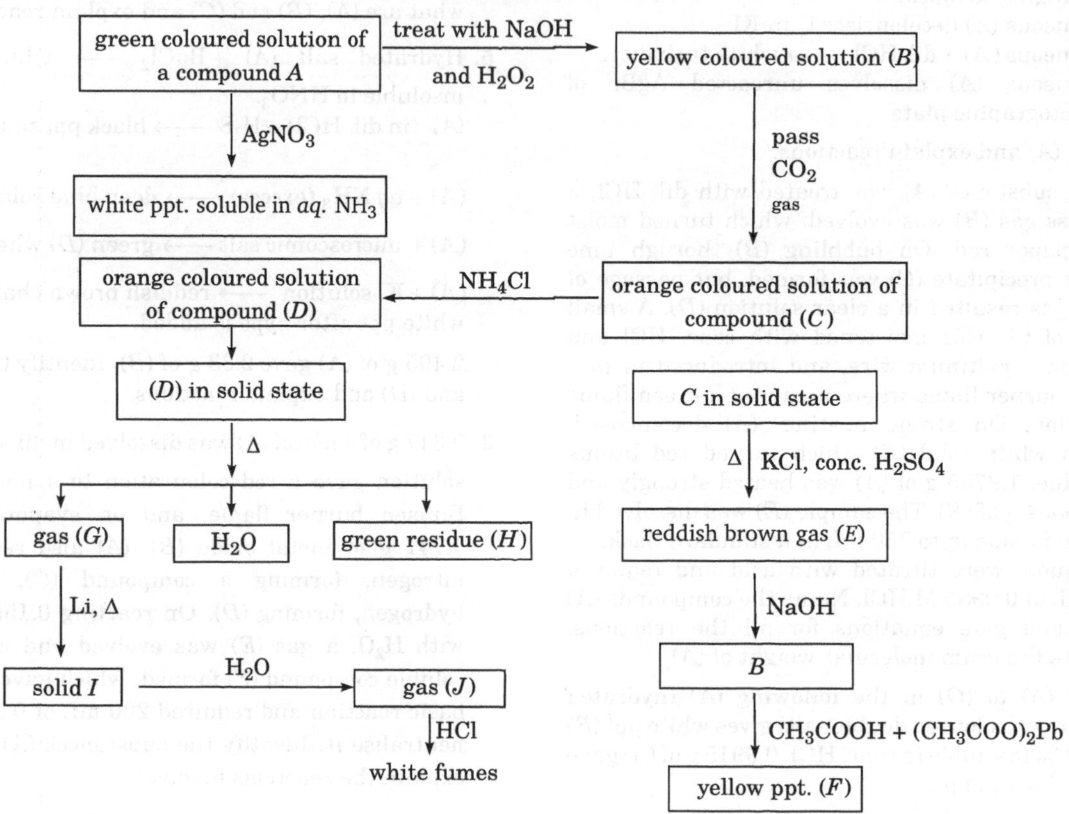

1. Compound (*A*) is
(a) $CuCl_2$ (b) $FeCl_3$
(c) $FeCl_2$ (d) $CrCl_3$

2. Yellow coloured solution (*B*) is of
(a) Na_2CrO_4 formed by oxidation of Cr^{3+} by H_2O_2 in alkaline medium
(b) $FeCl_2$ formed by reduction of $FeCl_2$ formed by reduction of $FeCl_3$ by H_2O_2
(c) $FeCl_3$ formed by oxidation of $FeCl_2$ by H_2O_2
(d) $Cu(OH)_2$ formed by precipitation of $CuCl_2$ by H_2O_2/NaOH

3. Select correct statement(s).
(a) CO_2 makes the solution of (*B*) acidic and converts (*B*) into $FeCl_2$(*C*)
(b) CO_2 makes the solution of (*B*) basic and converts it into $Na_2Cr_2O_7$
(c) CO_2 makes the solution of (*B*) acidic and converts it into $Na_2Cr_2O_7$
(d) None of the above is correct

4. Formation of (*F*) using (*G*), KCl and conc. H_2SO_4 is called
(a) redox test of $Cr_2O_7^{2-}$
(b) chromyl-chloride test of Cl^-
(c) chromyl-chloride test of Cr^{3+}
(d) none of the above is correct

5. Compounds *D*, *E* and *F* respectively are

	D	*E*	*F*
(a)	$(NH_4)_2CrO_4$	CrO_2Cl_2	$PbCrO_4$
(b)	$(NH_4)_2Cr_2O_7$	CrO_3	$PbCrO_4$
(c)	$(NH_4)_2Cr_2O_7$	CrO_2Cl_2	$PbCrO_4$
(d)	$(NH_4)_2CrO_4$	CrO_3	$PbCrO_4$

6. Compounds *G*, *H*, *I* and *J* are respectively

	G	*H*	*I*	*J*
(a)	N_2	CrO_3	Li_3N	NH_3
(b)	O_2	CrO_3	Li_2O	H_2
(c)	O_2	Cr_2O_3	Li_2O	O_2
(d)	N_2	Cr_2O_3	Li_3N	NH_3

Passage 2

Silver nitrate gives a white precipitate (*A*) with aqueous sodium thiosulphate (*B*); the precipitate dissolves in excess of thiosulphate. If the precipitate is heated with water, it turns black (*C*), and the supernant liquid then gives a white precipitate (*D*) with acidified barium nitrate solution.

Read the above analysis-report and answer the questions given below:

1. Species *A* and *B* are respectively

	A	***B***
(a)	$Ag_2S_2O_3$	$Ag(S_2O_3)_2^{3-}$
(b)	AgS_2O_3	$Ag(S_2O_3)_2^{2-}$
(c)	AgS_2O_3	$Ag(S_2O_3)_2^{3-}$
(d)	$Ag(S_2O_3)_2^{3-}$	$Ag_2S_2O_3$

2. Select correct explanation for the formation of *C* and *D*

(a) When (*B*) is heated with H_2O, H_2S is formed which gives black ppt of Ag_2S.
(b) Ag_2S (black *C*) is precipitated alongwith H_2SO_4 (in supernant layer) which gives white ppt with $Ba(NO_3)_2$ solution
(c) Ag_2SO_4 (black *C*) is precipitated alongwith H_2S which gives white ppt with $Ba(NO_3)_2$ solution
(d) None of the above is correct explanation

Passage 3

Questions given below are based on the following analysis.

A coloured solution known to contain two metal ions, was treated with excess cold sodium hydroxide solution. When filtered a whitish solid, slowly changing to brown, was retained on the filter paper and a colourless solution collected as the filtrate. Dropwise addition of hydrochloric acid to the filtrate produced a white precipitate which dissolved in excess acid. Treatment of the residue from the filter paper with a solution of a strong oxidiser produced a reddish-violet solution.

1. Indicate any pairs of ions which on testing as above leads to the observed changes
(a) Zn^{2+} and Mn^{2+} ions (b) Mg^{2+} and Zn^{2+} ions
(c) Mn^{2+} and Mg^{2+} ions (d) Fe^{2+} and Zn^{2+} ions

2. Filtrate obtained after separation of white solid contains
(a) ZnO (b) Na_2ZnO_2
(c) MnO (d) Na_2MnO_2

3. White solid changing to brown is due to formation of
(a) $Mn(OH)_2$ (b) MgO
(c) $Zn(OH)_2$ (d) MnO_2

4. Reddish-violet solution obtained by oxidation is of
(a) ZnO_2^{2-} (b) MnO_4^{2-}
(c) MnO_4^- (d) MnO_2

5. Reddish-violet solution is decolourised by
(a) SO_3^{2-}, $C_2O_4^{2-}$, Fe^{2+} (b) SO_4^{2-}, HCO_3^-, Fe^{3+}
(c) NO_2^-, SO_4^{2-}, Fe^{2+} (d) H_2O_2, $C_2O_4^{2-}$, CO_3^{2-}

Passage 4

You have a mixture of CO_2, SO_2 and O_2 gas. You have been provided following reagents for testing
(A) lime water
(B) acidified potassium dichromate solution
(C) pyrogallol

1. Arrange these reagents in order such that only one gas is tested at one time
(a) *A*, *B*, *C* (b) *B*, *C*, *A*
(c) *B*, *A*, *C* (d) *A*, *C*, *B*

2. If order is *A*, *B*, *C*, then which gas remains unchecked
(a) CO_2 (b) SO_2
(c) O_2 (d) none of these

3. Acidified $K_2Cr_2O_7$ solution has been used to test
(a) SO_2
(b) CO_2
(c) O_2
(d) mixture of SO_2 and CO_2

4. Gas responsible for green-house effect is
(a) O_2 (b) SO_2
(c) CO_2 (d) CO_2 and O_2

Passage 5

A colourless mixture of two salts (*A*) and (*B*) (excess) is soluble in H_2O. Separately (*A*) turns blue litmus red and (*B*) turns red litmus blue. (*A*) gives white ppt. with (*B*), which dissolves in excess of (*B*) forming (*C*). (*A*) when placed in moist air gives fumes and can form dimer. (*A*) gives white ppt. with NH_4Cl and NH_4OH soluble in (*B*). A also gives white ppt. with $AgNO_3$ soluble in NH_4OH.

Read the above analysis report and answer the following questions:

1. Colourless mixture contains
(a) $ZnCl_2$, NaOH (b) $AlCl_3$, NaCl
(c) $AlCl_3$, NaOH (d) $CrCl_3$, NaOH

2. *A* remains soluble in H_2O in presence of excess of *B*. It is due to formation of
(a) $Na_2Zn(OH)_2$ (b) $NaAl(OH)_4$
(c) $Na_2Zn(OH)_4$ (d) Al_2O_3

3. *A* turns blue litmus red. It is due to acidic solution because of
(a) hydrolysis of cation of *A*
(b) hydrolysis of anion of *A*
(c) hydrolysis of both cation and anion
(d) none of the above is correct

4. White fumes are formed in moist air. It is due to formation of
(a) $Al(OH)_3$ (b) HCl
(c) $Zn(OH)_2$ (d) $NaAlO_2$

5. Dimer formed has bonding
(a) ionic, covalent (b) ionic, coordinate
(c) covalent (d) covalent, coordinate

6. NH_4Cl and NH_4OH acts as a buffer for precipitation of *A* as white ppt. If pK_b of $NH_3(aq)$ is 4.74, pH of equimolar $[NH_4^+]$ and $[NH_3]$ will be
(a) 4.74 (b) 9.26
(c) 7.00 (d) 14.00

7. If ppt. formed by addition of NH_4Cl and NH_4OH is of the form $M(OH)_3$ then pH at which 2×10^{-12} *M* M^{3+} ion is precipitated is (K_{sp} of $M(OH)_3 = 2 \times 10^{-33}$)
(a) 7 (b) 5
(c) 6 (d) 8

Passage 6

Presence of Fe^{3+} in an aqueous solution can be made by addition of sodium acetate (in excess) when a reddish-brown colour is formed".

Read the above short **write-up** and answer the following questions:

1. Reddish-brown colour is due to formation of
(a) $[Fe(H_2O)_4(CH_3COO)_2]^+$
(b) $[Fe(CH_3CO_4)_6]^{3-}$
(c) $[Fe_3(OH)_2(CH_3COO)_6]^+$
(d) $[Fe_2(H_2O)_4(CH_3COO)_4]^{2+}$

2. Excess of sodium acetate is added in the above test. It is because
(a) it acts as a buffer to control pH of the reaction otherwise reaction is made reversible
(b) it makes the solution alkaline which destroys acidic effect if any
(c) sodium acetate has low K_{sp} value
(d) sodium acetate has high K_{sp} values

3. Presence of Fe^{3+} can also be detected by .... when .... is formed.
(a) KCNS, $Fe(CNS)_3$ ... blood red
(b) KCNS, $[Fe(CNS)]Cl_2$ ... blood red colour
(c) KCNS, $[Fe(CNS)_2Cl$ ... blood red colour
(d) KCN, $K_3[Fe(CN)_6]$ ... blue precipitate

4. A blue colour solution is formed when potassium ferrocyanide in excess is added to aqueous ferric chloride solution. This blue colour is of
(a) $K\overset{II}{Fe}[\overset{III}{Fe}(CN)_6]$ or $K\overset{III}{Fe}[\overset{II}{Fe}(CN)_6]$
(b) $Fe_4[Fe(CN)_6]_3$
(c) $Fe_3[Fe(CN)_6]_2$
(d) None of the above

Passage 7

"Sometimes it happens that when H_2S gas is passed into a solution in dil. HCl (in qualitative analysis of cations), yellowish white turbidity appears."

Read the above short write-up and answer the following questions:

1. Yellowish white turbidity is of
(a) SO_3^{2-} formed by oxidation of H_2S by HCl
(b) colloidal sulphur formed by oxidation of H_2S by certain oxidising agent
(c) SO_4^{2-} formed by oxidation of H_2S by H_2O
(d) None of the above is correct

2. A colourless salt on reaction with dil. HCl gives a colourless pungent gas that turns $K_2Cr_2O_7$ solution green. A student passed H_2S gas immediately after adding dil. HCl and got colloidal yellowish suspension before any precipitation occurs. This is due to:
(a) SO_3^{2-} which is decomposed to SO_2 forming S with H_2S
(b) NO_2^- which oxidises H_2S to colloidal sulphur
(c) both of the above
(d) none of the above

Passage 8

When a KI solution was added to a solution of mercury (II) chloride, a precipitate [mercury (II) iodide] formed. A student plotted the mass of the precipitate *versus* the volume of the KI solution added and obtained the following graph.

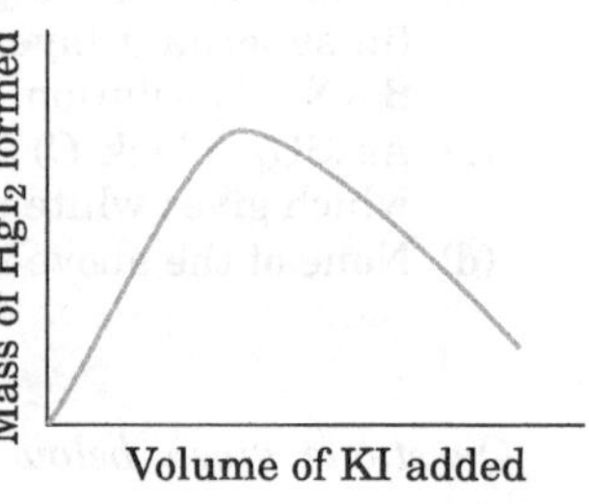

Based on the above graph, answer the following questions:

1. As KI is added to $HgCl_2$ solution, there is increase in the mass of HgI_2 but after a maximum point is reached, mass of HgI_2 decreases. It is due to
(a) formation of soluble K_3HgI_5
(b) formation of soluble K_2HgI_4
(c) decomposition of HgI_2 into Hg_2I_2 and I_2
(d) decomposition of HgI_2 into Hg and I_2

2. Species formed after maximum is attained is called
(a) Nessler's reagent
(b) Tollen's reagent
(c) iodide of Millon's base
(d) none of the above is correct

3. (Refer Q. 2). Species formed in Q. 2 is added to hot NH_4Cl solution containing some NaOH, then there is formation of
(a) brown precipitate of iodide Millon's base
(b) yellow colouration of K_2HgI_4
(c) blue precipitate of $(NH_4)_2HgI_2Cl_2$
(d) none of the above

Assertion & Reason

Codes

(a) Both **A** and **R** are true and **R** is the correct explanation of **A**.

(b) Both **A** and **R** are true but **R** is not the correct explanation of **A**.

(c) **A** is true but **R** is false.

(d) **A** is false but **R** is true.

(e) **A** and **R** both are false.

1. Assertion (A) : When $H_2S(g)$ is passed into aqueous $ZnCl_2$, solution, ZnS is precipitated.
Reason (R) : ZnS is insoluble in dil. HCl.

2. Assertion (A) : When $H_2S(g)$ is passed into aqueous $Zn(CH_3COO)_2$ solution, zinc is precipitated as ZnS.
Reason (R) : CH_3COOH is produced in the reaction which cannot dissolve ZnS formed.

3. Assertion (A) : $PbCl_2$ and Hg_2Cl_2 precipitates can be separated by hot water.
Reason (R) : Hg_2Cl_2 is blackened by *aq*. NH_3.

4. Assertion (A) : $PbCl_2$ and AgCl precipitates can be separated by *aq*. NH_3.
Reason (R) : $PbCl_2$ precipitates change to soluble complex $[Pb(NH_3)_4]Cl_2$.

5. Assertion (A) : Basic radicals (cations) have been divided into groups based on K_{sp} values.
Reason (R) : Mostly alkali salts are water soluble.

6. Assertion (A) : NO_2^- is decomposed by urea in the mixture of NO_3^- and NO_2^-.
Reason (R) : NO_3^- interferes in the ring test of NO_2^-.

7. Assertion (A) : Cu^{2+} and Cd^{2+} solution is complexed by addition of KCNS.
Reason (R) : On passing H_2S gas stable complex $[Cu(CN)_4]^{3-}$ is not affected.

8. Assertion (A) : CO_2 containing H_2O changes yellow CrO_4^{2-} into $Cr_2O_7^{2-}$.
Reason (R) : Aqueous CO_2 is acidic ($H_2CO_3 \rightleftharpoons H^+ + HCO_3^-$) which converts CrO_4^{2-} into $Cr_2O_7^{2-}$.

9. Assertion (A) : Unreacted $AgNO_3$ of photographic plate is removed by $Na_2S_2O_3$ (hypo).
Reason (R) : $Na_3[Ag(S_2O_3)_2]$ is soluble complex.

10. Assertion (A) : Coloured cations can be identified by borax-bead test.
Reason (R) : Transparent bead ($NaBO_2 + B_2O_3$) forms coloured bead with coloured cation.

True & False

1. NH_4Cl can be replaced by $(NH_4)_2SO_4$ in group III.

2. Dilute H_2SO_4 can be used instead of dil. HCl in group II.

3. When H_2S gas is passed into *aq*. $ZnCl_2$ solution, white ppt of ZnS is obtained.

4. When Cl_2 gas is passed into a mixture containing Br^- and I^- and $CHCl_3$, I_2 (violet) first appears in $CHCl_3$ layer.

5. There is precipitation of solute *AB* if its ionic product is greater than K_{sp} value, *i.e.*, $[A][B] > K_{sp}$.

6. If K_{sp} of $M(OH)_3$ is 1×10^{-12} then 0.001 M. M^{3+} is precipitated at a pH > 9.

7. Turnbull's blue and Prussian's blue respectively are $KFe^{II}[Fe^{III}(CN)_6]$, $KFe^{III}[Fe^{II}(CN)_6]$

8. $BaBr_2$ gives yellow ppt. with $AgNO_3$ as well as with K_2CrO_4.

9. $AlCl_3$ is soluble in excess of NaOH forming sodium meta-aluminate $Na[Al(OH)_4]$.

10. NaOH can be used to separate $Al(OH)_3$ and $Fe(OH)_3$.

11. Yellow ammonium sulphide (YAS) can be used to separate SnS and As_2S_3.

12. NH_4SCN can be used to make distinction between Cu^{2+} and Co^{2+}.

13. NaOH can be used to separate $Al(OH)_3$ and $Zn(OH)_2$.

14. In II group, formation of whitish turbidity on dilution with H_2O indicates Sb^{3+}.

15. If acidified solution of $K_2Cr_2O_7$ turns green on addition of a salt then salt may contain Fe^{2+}.

16. White ppt. of $PbCl_2$ is soluble in *aq* NH_3.

17. Hg_2Cl_2 is blackened by NH_3 due to formation of iodide of Millon's base.

18. $K_4[Fe(CN)_6]$ is used to test Cu^{2+}, Fe^{3+}, Zn^{2+} and Cd^{2+} ions.

19. When KNO_2 and CH_3COOH is added to $CoCl_2$ solution, yellow ppt. of $K_4[Co(NO_2)_6]$ is formed.

20. Alkaline solution of NH_4Cl gives brown ppt. with K_2HgI_4.

Fill in the Blanks

1. Separation of basic radicals is based on ....(a)....... and ....(b)....... .

2. NH_4Cl is added alongwith NH_4OH in group ....(a)....... to ....(b)....... concentration of ....(c)....... ion.

3. Precipitation of Cd^{2+} and Cu^{2+} takes place in presence of ....(a)....... by ....(b)....... .

4. Cd^{2+} and Cu^{2+} are separated by ....(a)....... formation using ....(b)....... in which ....(c)....... is more stable than ....(d)....... ; on passing H_2S gas ....(e)....... is precipitated.

5. $PbCl_2$ is soluble in ....(a)....... , AgCl is soluble in ....(b)....... while Hg_2Cl_2 is ....(c)....... by NH_3.

6. Copper subgroup ppt. and arsenic subgroup ppt are separated using.......(a)....... .

7. If orange turbidity appears on dilution with H_2O of the solution in dil. HCl, it is due to ....(a)....... and ....(b)....... ion is assumed confirmed.

8. $Fe(OH)_3$ and $Al(OH)_3$ ppt. can be separated by ....(a)....... when ....(b)....... becomes soluble due to the formation of ......(c)......, and .....(d)..... remains insoluble.

9. $Cr(OH)_3$ is made soluble in NaOH in presence of ....(a)....... when ....(b)....... of ....(c)....... colour is formed and gives yellow ppt. of ....(d)....... when ....(e)....... is added.

10. Reagent used to test Ni^{2+} ion is ....(a)....... .

11. Acidified $KMnO_4$ can be decolourised by ....(a)....... .

12. FeC_2O_4 can decolourise acidified $KMnO_4$ due to the oxidation of ....(a)....... and ....(b)....... .

13. Ferric alum is the indicator in the titration of Ag^+ with SCN^- when ....(a)....... colour appears at the end point.

14. Fe^{2+} gives blue colour, called ....(a)....... with ....(b)....... while Fe^{3+} gives blue colour, called ....(c)....... with ....(d)....... .

15. A reagent that can detect any of Cu^{2+}, Fe^{3+}, Zn^{2+} and Cd^{2+} is ....(a)....... .

16. $PbSO_4$ is soluble in ....(a)....... due to formation of ....(b)....... .

17. $Al(OH)_3$ is precipitated if its ionic product ......(a)...... is ......(b)...... than K_{sp}.

18. $HgCl_2$ gives orange ppt with ....(a)....... , which dissolves in excess of it forming ....(b)....... called ....(c)....... .

19. NH_4SCN gives red colour with ......(a)...... due to the formation of .....(b)..... .

20. Gas that turns lime water milky and acidified $K_2Cr_2O_7$ green is ....(a)....... .

21. Iron (II) sulphide gives foul smell of ....(a)....... in presence of moisture.

22. AgBr is soluble in hypo forming ....(a)....... .

23. NO_3^- is detected by ....(a)....... when ....(b)....... is formed on the addition of $FeSO_4$ and conc. H_2SO_4.

24. Cl^- is confirmed by ....(a)....... .

25. Addition of one drop of HNO_3 in group (III) analysis is to ....(a)....... .

26. Reagent that can detect any of Fe^{3+}, Co^{2+} and Cu^{2+} is ....(a)....... .

27. While testing borate, green edged flame is due to formation of ....(a)....... .

28. When SO_2 is passed into suspension of $CaSO_3$ in water, ...(a)....... is formed.

29. Alkaline NH_4^+ salt gives brown ppt. with K_2HgI_4; brown ppt is called ......(a)....... .

30. Even in the absence of group II, colloidal yellowish ppt appears on passing H_2S gas; it may be due to the presence of ....(a)....... .

31. Yellow ppt of ....(a)....... is formed when $CoCl_2$ reacts with excess of KNO_2 in presence of CH_3COOH.

32. $BaBr_2$ in aqueous solution gives yellow ppt with ....(a)....... as well as with ....(b)....... .

33. Sodium carbonate and mixture are taken in ....(a)....... ratio while preparing sodium carbonate extract.

34. Hg_2Cl_2 is precipitated if ....(a)....... is greater than $K_{sp}(Hg_2Cl_2)$.

35. AgCl is soluble in ....(a)....... and Ag^+ is present in ....(b)....... .

36. Cu^{2+} gives white ppt of ....(a)....... with ....(b)....... and deep blue colour of ....(c)....... with ....(d)....... .

37. On heating the salt with HNO_3 and ammonium molybdate, formation of yellow ppt indicates the presence of ....(a)....... or ....(b)....... .

38. As_2S_3 is soluble in $(NH_4)_2S_2$ (yellow ammonium sulphide) due to the formation of ......(a)....... .

39. Reddish brown colouration when neutral $FeCl_3$ is added to the CH_3COO^- aqueous solution is due to the formation of ....(a)....... .

40. Hypo gives ....(a)....... ppt with $AgNO_3$ which changes to ......(b)....... .

Matrix-Match Type Questions

Only One Correct

1. Match the set of ion and reagent (in List I) with colour obtained (in list II).

List I	List II
A. $Fe^{3+} + [Fe(CN)_6]^{4-}$	1. Cherry red
B. $Fe^{3+} + CNS^-$	2. Pink
C. Ni^{2+} + DMG	3. Blood red
D. $CrO_4^{2-} + Pb^{2+}$	4. Blue
E. $Mn^{2+} + PbO_2 + H^+$	5. Yellow ppt

One or More Correct

1. Match the reagents (in List I) with the cations (in List II) which can be identified.

List I	List II
A. $K_4[Fe(CN)_6]$	1. Zn^{2+}
B. NH_4CNS	2. Fe^{3+}
C. CH_3COONa	3. Cu^{2+}
D. H_2S/H^+	4. Cd^{2+}
E. H_2S/NH_2OH-NH_4Cl	5. Ni^{2+}

2. Match the reagents (in List I) with the cations (in List II) which can be identified.

List I	List II
A. $AgNO_3$	1. Cl^-
B. $FeCl_3$	2. Br^-
C. $BaCl_2$	3. I^-
D. $(CH_3COO)_2Pb$	4. CH_3COO^-
E. Cl_2 water	5. CrO_4^{2-}

3. Match the set of basic radicals (in List I) with the group reagent (in List II) and analytical group (in List III).

List I	List II	List III
A. Zn^{2+}, Ni^{2+}	1. H_2S/H^+	P. I
B. Ag^+, Hg_2^{2+}	2. $(NH_4)_2CO_3/NH_4OH$	Q. II
C. Ca^{2+}, Ba^{2+}	3. $H_2S/NH_4OH, NH_4Cl$	R. III
D. Cu^{2+}, Cd^{2+}	4. dil. HCl	S. IV
E. Al^{3+}, Cr^{3+}	5. NH_4Cl/NH_4OH	T. V

4. Match the reagents (in List I) with cations (in List II) which can be identified.

List I	List II
A. $KMnO_4/H^+$	1. Ca^{2+}
B. $K_2Cr_2O_7/H^+$	2. Ag^+
C. $(NH_4)_2C_2O_4$	3. Pb^{2+}
D. KI	4. Fe^{3+}
E. K_2CrO_4	5. Ba^{2+}

Integer Answer Types

This section contains 8 questions. The answer to each of the questions is a single digit integer, ranging from 0 to 9. The appropriate bubbles below the respectively question numbers in the *ORS* have to be darkened. For example, if the correct answers to question number *X, Y, Z* and *W* (say) are 6, 0, 9 and 2, respectively, then the correct darkening of bubbles will like the following

X	Y	Z	W
0	0	0	0
1	1	1	1
2	2	2	2
3	3	3	3
4	4	4	4
5	5	5	5
6	6	6	6
7	7	7	7
8	8	8	8
9	9	9	9

Set I

1. SO_2 turns acidified $K_2Cr_2O_7$ solution green. Oxidation number of the green-coloured ion is ....

2. Brown coloured ring formed in the nitrate test has formula $[Fe(H_2O)_5NO]SO_4$. Unpaired electrons in Fe is ....

3. In question (2) magnetic moment is $\sqrt{x}$ where x is

4. Aqueous neutral solution of $FeCl_3$ reacts and acetate to form a reddish brown colour compound $[Fe_3(OH)_2(CH_3COO)_x]^+$, where x is ....

5. $FeCl_3$ gives blood red colour of formula $[Fe(CNS)(H_2O)_y]^{2+}$ with KCNS. Here y = ...

6. Copper metaborate is formed in the test of copper ion in borax-bead test. Number of boron atoms in copper metaborate is ....

7. AgBr dissolved in $Na_2S_2O_3$. 1 mol of AgBr is dissolved in x mol of $Na_2S_2O_3$. Where x is ....

8. Blue coloured precipitate of Prussian blue has total of iron atom(s) ....

Set II

This section contains 10 questions

Each number in the following represents a colour give the answer in the form of number related to that colour formed in cation/anion analysis.

1. White	6. Black
2. Blue	7. Grey
3. Milky	8. Green
4. Orange	9. Buff coloured
5. Yellow	0. Reddish brown

1. H_2S gas is passed into aqueous $CdCl_2$ solution

2. H_2S gas is passed into aqueous zinc acetate solution

3. SO_2 gas is passed into lime water

4. $SbCl_3$ solution is diluted with H_2O

5. H_2S gas is passed into lead-acetate solution

6. NH_4OH is added to $CrCl_3$ solution in presence of NH_4Cl

7. H_2S gas is passed into $MnCl_2$ solution in presence of NH_4Cl and NH_4OH

8. KI is added to acidified $K_2Cr_2O_7$ solution

9. $SnCl_2$ is added to $HgCl_2$ solution

10. $K_4[Fe(CN)_6]$ is added to $FeCl_3$ solution

Answers & Solutions
Master Exercises

Exercise 1

Short Answer Type Questions

1. (i) SO_3^{2-} or (ii) SO_4^{2-}

2. $(NH_4)_3PO_4 \cdot 12MoO_3$

For reaction refer text.

3. **Step I :** Add dil. HCl, brisk reaction with pungent odour gas indicates SO_3^{2-} (and also CO_3^{2-}, pungent gas is SO_2). Add $K_2Cr_2O_7$ solution. If mixture turns green then SO_3^{2-} is confirmed. Pass the evolved gas into lime water, if it turns milky, CO_3^{2-} is confirmed.

Explanation : CO_2 and SO_2 both turn lime water milky. If $K_2Cr_2O_7$ is added, SO_2 reduces it to Cr^{3+} (green) and evolved gas is only CO_2 which turns lime water milky

$$CO_3^{2-} + 2H^+ \longrightarrow CO_2 + H_2O$$

$$Ca(OH)_2 + CO_2 \longrightarrow \underset{\text{milky}}{CaCO_3} + H_2O$$

$$SO_3^{2-} + 2H^+ \longrightarrow H_2O + SO_2$$

$$3SO_2 + Cr_2O_7^{2-} + 2H^+ \longrightarrow \underset{\text{green}}{2Cr^{3+}} + 3SO_4^{2-} + H_2O$$

Step II : To fresh mixture solution add dil. **HCl** and **$BaCl_2$** solution. **CO_3^{2-}** and **SO_3^{2-}** are decomposed by HCl without affecting **SO_4^{2-}**. Formation of white precipitate with $BaCl_2$ confirms **SO_4^{2-}**.

4. $BaSO_4$ is insoluble in conc. HCl, while $BaSO_3$ remains soluble. Separate $BaSO_4$ and add Br_2 waer into filtrate. SO_3^{2-} is oxidised to SO_4^{2-} and gives white ppt of $BaSO_4$.

$$Br_2 + H_2O \longrightarrow 2HBr + O$$

$$BaSO_3 + O \longrightarrow BaSO_4\downarrow$$

5. (*A*) : H_3BO_3

(*B*) : $(CH_3O)_3B$, burns with green edged flame

6. $$2CH_3COONH_4 + \underset{\text{white ppt}}{PbSO_4} \longrightarrow \underset{\text{soluble}}{(CH_3COO)_2Pb} + (NH_4)_2SO_4$$

7. (*A*) : $AgNO_3$ (*B*) : $AgBr$ (*C*) : $Na_3[As(S_2O_3)_2]$

8. (*A*) : $SO_3^{2-} + SO_4^{2-}$ (*B*) : SO_2 (*C*) : $BaSO_4$

9. Refer text

10. KCN forms complex with Cu^{2+} and Cd^{2+}

$$Cu^{2+} + 2CN^- \longrightarrow Cu(CN)_2$$

$$Cu(CN)_2 \longrightarrow 2CuCN\downarrow + (CN)_2$$

$$CuCN + 3KCN \longrightarrow \underset{\text{stable}}{K_3[Cu(CN)_4]}$$

$$Cd^{2+} + KCN \longrightarrow \underset{\text{unstable}}{K_2[Cd(CN)_4]}$$

when H_2S gas is passed, unstable complex of Cd^{2+} gives yellow ppt

$$[Cd(CN)_4]^{2-} \rightleftharpoons Cd^{2+} + 4CN^-$$

$$Cd^{2+} + S^{2-} \longrightarrow \underset{\text{yellow}}{CdS\downarrow}$$

11. $HgCl_2 + Cu \longrightarrow CuCl_2 + Hg$. Hg is deposited on Cu.

12. This yellowish turbidity is due to colloidal sulphur, formed by oxidation of H_2S by the presence of NO_3^- or NO_2^- or SO_3^{2-}

$$H_2S + HNO_3 \longrightarrow S\,(\text{colloidal})$$

Before passing H_2S gas, solution is boiled so as to decompose these radicals.

13. $$2\underset{\text{soluble}}{(NH_4)_3AsS_4} + 6HCl \longrightarrow As_2S_5\downarrow + 6NH_4Cl + 3H_2S$$

$$(NH_4)_2SnS_3 + 2HCl \longrightarrow SnS_2\downarrow + 2NH_4Cl + H_2S$$

14. (*A*) : HgS, (*B*) : $HgCl_2$, (*C*) : H_2S, (*D*) : HgI_2,

(*E*) : K_2HgI_4

15. (*A*) $Zn(CH_3COO)_2 + H_2S \longrightarrow ZnS + CH_3COOH$

(*B*) $ZnCl_3 + H_2S \longrightarrow ZnS + 2HCl$

ZnS is precipitated in (*A*). In (*B*) HCl formed dissolved ZnS formed.

16. (*A*) : $AlBr_3$ (*B*) : $Al(OH)_3$ (*C*) : $NaAlO_2$

17. (*A*) : $ZnCl_2$ (*B*) : AgCl (*C*) : $Zn(OH)_2$

(*D*) : Na_2ZnO_2 (*E*) : ZnS

18. $K_2CrO_4 \longrightarrow$ giving yellow precipitate

Or $Na_2SO_4 \longrightarrow$ giving white precipitate

19. **Flame Test :**

Ba^{2+} : Apple green flame

Ca^{2+} : Brick red flame

Sr^{2+} : Crimson red flame

20. Presence of Fe^{2+} in mixture also gave positive MnO_4^-/H^+ test.

21. (*A*) : NH_4NO_2 (*B*) : NH_3 (*C*) : $NaNO_3$ (*D*) : N_2

(*E*) : AlN

22. (*A*) : $(NH_4)_2Cr_2O_7$ (*B*) : NH_3 (*C*) : Cr_2O_3 (*D*) : N_2
(*E*) : AlN (*F*) : $CrCl_3$ (*G*) : Na_2CrO_4, (*H*) : $PbCrO_4$

23. AsO_4^{2-}; $Mg^{2+} + NH_4^+ + AsO_4^{2-} \longrightarrow \underset{\text{white ppt}}{MgNH_4AsO_4}$

24. (a) $KI + HgI_2 + NaOH$
or KI (excess) + $HgCl_2$ + NaOH
(b) NH_4^+ salts (c) Refer text for reaction

Analytical Questions

1.

	%	Mole	Ratio
Co^{3+}	21.24	0.36	1
NH_3	24.77	1.46	4
Cl^-	12.81	0.36	1
SO_4^{2-}	34.65	0.36	1
H_2O	6.53	0.36	1

Thus, empirical formula is $[Co(NH_3)_4ClSO_4]\cdot H_2O$

2. (*A*) : $CuFeS_2$, (*B*) : $CuSO_4$, (*C*) : $FeSO_4$, (*D*) : H_2S
(*E*) : CuS, (*F*) : $Cu(NO_3)_2$, (*G*) : $Cu_2[Fe(CN)_6]$
(*H*) : $KFe^{II}[Fe^{III}(CN)_6]$, Turnbull's blue

3. (*A*) : $Pb(NO_3)_3$, (*B*) : $PbCl_2$, (*C*) : PbS, (*D*) : PbI_2

4. (*A*) : $ZnCO_3$, (*B*) : CO_2, (*C*) : ZnO, (*D*) : ZnS,
(*E*) : $Zn(OH)_3$

5. (*A*) : $FeSO_4\cdot 7H_2O$, (*B*) : SO_2, (*C*) : SO_3,
(*D*) : Fe_2O_3, (*E*) : $BaSO_4$

6. (*X*) is $\underset{2(m+61)\text{ g}}{2MHCO_3} \longrightarrow M_2CO_3 + H_2O + \underset{44\text{ g}}{CO_2}$

$$\frac{2(m+61)}{20.02} = \frac{44}{4.4}$$

Thus, $M = 39.1$

(*X*) : $KHCO_3$, (*Y*) : K_2CO_3, (*Z*) : $BaCO_3$, (*A*) : CO_2,
(*B*) : H_2O

7. (*A*) : $FeSO_4\cdot 7H_2O$, (*B*) : $FeSO_4$,
(*C*) : $FeSO_4\cdot NO$ or $[Fe(H_2O)_5NO]SO_4$
(*D*) : Fe_2O_3, (*E*) : SO_2, (*F*) : SO_3

8. (*A*) : NH_4NO_2, (*B*) : NH_3, (*C*) : $NaNO_3$,
(*D*) : $NaNH_2$, (*E*) : N_2O, (*F*) : N_3Na (sodium azide)
(*G*) : N_3H (hydroazoic acid)

9. Molar mass of acid of phosphorus × molarity = g/L

$$\text{Molar mass} \times 0.1 = 9.8\text{ gL}^{-1}$$

$\therefore$ Molar mass of acid = 98 g mol^{-1}

Let basicity of acid be = x,

then $0.1\text{ M acid} = 0.1 \times x\text{ N}$

$$100 \times 0.1 \times x = 300 \times 0.1\text{ N NaOH}$$

$$x = 3$$

Thus, acid is tribasic which is H_3PO_4.

$A : H_3PO_4$
$B : MgHPO_4$
$C : Mg_2P_2O_7$

Identify if *C* is confirmed by the following reaction

$$\underset{\substack{2\times 98\text{ g}\\0.98\text{ g}}}{2H_3PO_4} \longrightarrow \underset{\substack{222\\1.11\text{ g}}}{Mg_2P_2O_7}$$

10. (*A*) : $H_2S_2O_8$ (Marshall's acid), (*B*) : H_2SO_4,
(*C*) : H_2O_2, (*D*) : $BaSO_4$

11. (*A*) : Ag, (*B*) : $AgNO_3$, (*C*) : NO_2, (*D*) : AgCl,
(*E*) : $Ag_2S_2O_3$, (*F*) : Ag_2S

12. (*X*) : SO_2

(i) $3SO_2 + Cr_2O_7^{2-} + 2H^+ \longrightarrow 2Cr^{3+} + H_2O + 3SO_4^{2-}$

(ii) $SO_2 + H_2O_2 \longrightarrow SO_4^{2-} + 2H^+$

$SO_4^{2-} + Ba^{2+} \longrightarrow BaSO_4\downarrow$

(iii) $2H_2S + SO_2 \longrightarrow 3S + 2H_2O$

13. (a) (*X*) : NaOH

(i) $Zn + 2NaOH \longrightarrow Na_2ZnO_2 + H_2$

(ii) $SnCl_2 + 2NaOH \longrightarrow \underset{\text{white}}{Sn(OH)_2\downarrow}$

$Sn(OH)_2 + 2NaOH \longrightarrow \underset{\text{soluble}}{Na_2SnO_2} + 2H_2O$

(b) $Na_2B_4O_7 \xrightarrow{\Delta} 2NaBO_2 + B_2O_3$

$2FeSO_4\cdot 7H_2O \xrightarrow{\Delta} Fe_2O_3 + SO_2 + SO_3 + 14H_2O$

$Fe_2O_3 + 3B_2O_3 \longrightarrow \underset{\text{yellow}}{2Fe(BO_2)_3}$

14. (*X*) : $HgCl_2$

(i) $HgCl_2 + H_2S \longrightarrow HgS + 2HCl$

(iii) $2HgCl_2 + SnCl_2 \longrightarrow \underset{\text{white}}{Hg_2Cl_2\downarrow} + SnCl_4$

$Hg_2Cl_2 + SnCl_2 \longrightarrow \underset{\text{grey}}{2Hg\downarrow} + SnCl_4$

(iv) $HgCl_2 + 2KI \longrightarrow \underset{\text{orange}}{HgI_2\downarrow}$

$HgI_2 + 2KI \longrightarrow \underset{\text{soluble}}{K_2HgI_4}$

15. (a) (*X*) : NH_3

(i) NH_3 turns red litmus blue

(ii) $CuSO_4 + 4NH_3 \longrightarrow \underset{\text{deep blue}}{[Cu(NH_3)_4]SO_4}$

(iii) $FeCl_3 + 3NH_4OH(aq) \longrightarrow \underset{\text{soluble}}{Fe(OH)_3} + 3NH_4Cl$

$Fe(OH)_3 + 3HNO_3 \longrightarrow \underset{\text{soluble}}{Fe(NO_3)_3} + 3H_2O$

(b) (X) : KI

(i) $2CuSO_4 + 4KI \longrightarrow \underset{\text{white}}{Cu_2I_2\downarrow} + I_2 + 2K_2SO_4$

$I_2 + 2Na_2S_2O_3 \longrightarrow 2NaI + Na_2S_4O_6$

(ii) $Ag^+ + I^- \longrightarrow AgI\downarrow$ (yellow)

16. (*X*) : $Na_2B_4O_7 \cdot 10H_2O$

(i) $Na_2B_4O_7 + 7H_2O \longrightarrow 4H_3BO_3 + 2NaOH$ or $Na[B(OH)_4]$

(ii) $Na_2B_4O_7 \xrightarrow{\Delta} \underbrace{2NaBO_2 + B_2O_3}_{\text{glassy}}$

(iii) $Na_2B_4O_7 \cdot 10H_2O + H_2SO_4 \longrightarrow Na_2SO_4 + 4H_3BO_3 + 5H_2O$

17. (*X*) : $CaOCl_2$

(i) $CaOCl_2 + 2CH_3COOH \longrightarrow (CH_3COO)_2Ca + Cl_2 + H_2O$

$Cl_2 + 2KI \longrightarrow 2KCl + I_2$

(ii) $CaOCl_2 + CO_2 \longrightarrow CaCO_3 + Cl_2$

(iii) $CaOCl_2 + H_2O \longrightarrow Ca(OH)_2 + Cl_2$

$CH_3CH_2OH + Cl_2 + Ca(OH)_2 \xrightarrow{\Delta} CHCl_3 + (HCOO)_2Ca + CaCl_2 + H_2O$

$CHCl_3$ is an anaesthetic.

18. (*X*) : H_2O_2

(i) $2MnO_4^- + 5H_2O_2 + 6H^+ \longrightarrow 2Mn^{2+} + 8H_2O + 5O_2$

(ii) $H_2O_2 + 2I^- + 2H^+ \longrightarrow I_2 + 2H_2O$

(iii) $2MnO_4^- + 3H_2O_2 \longrightarrow 2MnO_2 + 3O_2 + 2H_2O + 2OH^-$

(iv) $PbS + 4H_2O_2 \longrightarrow PbSO_4 + 4H_2O$

19. (*A*) : $FeSO_4$ (*B*) : $BaCl_2$

(i) $2FeSO_4 \longrightarrow Fe_2O_3 + SO_2 + SO_3$

$FeSO_4 + 2NaOH \longrightarrow Fe(OH)_2\downarrow + Na_2SO_4$ (green)

(ii) $2Fe(OH)_2 + H_2O + O_2 \text{ (air)} \longrightarrow 2Fe(OH)_3\downarrow$

(v) Chromyl chloride test (see text)

20. (*A*) : $ZnCO_3$ (*B*) : CO_2 (*C*) : ZnO (*D*) : ZnS

$ZnO + 2HCl \longrightarrow ZnCl_2 + H_2O$

$2ZnCl_2 + K_4Fe(CN)_6 \longrightarrow Zn_2[Fe(CN)_6]\downarrow + 4KCl$ (white)

21. (*A*) : FeS (*B*) : SO_2 (*C*) : H_2S (*D*) : $FeSO_4$

(*E*) : $KFe^{II}[Fe^{III}(CN)_6]$

(i) $2FeS + 3.5O_2 \longrightarrow Fe_2O_3 + 2SO_2$

(ii) $FeS + H_2SO_4 \longrightarrow FeSO_4 + H_2S$

(iii) $SO_2 + 2H_2S \longrightarrow 2H_2O + 3S$

(iv) $FeSO_4 + K_3[Fe(CN)_6] \longrightarrow K\overset{II}{Fe}[\overset{III}{Fe}(CN)_6] + K_2SO_4$

22. (*X*) : $AlCl_3$

(i) $AlCl_3 + H_2O \longrightarrow Al(OH)_3 + 3HCl$

(ii) $NH_3 + HCl \longrightarrow NH_4Cl\downarrow$ (white fumes)

(iii) $AlCl_3 + 3NH_4OH \longrightarrow Al(OH)_3\downarrow + 3NH_4Cl$ (white)

$Al(OH)_3 + NaOH \longrightarrow NaAlO_2 + 2H_2O$ (soluble)

23. (*X*) : $SnCl_2$

(i) $SnCl_2 + H_2S \longrightarrow SnS\downarrow + 2HCl$ (brown)

(ii) $SnS + (NH_4)_2S_2 \text{ (YAS)} \longrightarrow (NH_4)_2SnS_3$ (soluble)

(iii) $SnCl_2 + 2NaOH \longrightarrow Sn(OH)_2\downarrow + 2NaCl$ (white)

$Sn(OH)_2 + 2NaOH \longrightarrow Na_2SnO_3 + H_2O$

(iv) $2FeCl_3 + SnCl_2 \longrightarrow 2FeCl_2 + SnCl_4$

24. (*A*) : PH_4I (*B*) : PH_3 (*C*) : KI (*D*) : P_2O_5 (*E*) : Cu_2I_2

(i) $PH_4I + KOH \xrightarrow{\Delta} PH_3 + H_2O + KI$

(ii) $2PH_3 + 4O_2 \longrightarrow P_2O_5 + 3H_2O$

(iii) $3CuSO_4 + 2PH_3 \longrightarrow Cu_3P_2 + 3H_2SO_4$

$Cu_3P_2 + 5CuSO_4 + 8H_2O \longrightarrow 8Cu + 2H_3PO_4 + 5H_2SO_4$

(iv) $2CuSO_4 + 4KI \longrightarrow Cu_2I_2\downarrow + 2K_2SO_4 + I_2$

25. (*A*) : Ag_2S (*B*) : $Na[Ag(CN)_2]$ (*C*) : Ag (*D*) : $AgNO_3$

(*E*) : $AgCl$

26. (*A*) : $Na_2S_2O_3$ (*B*) : SO_2 (*C*) : Ag_2S (*D*) : $[Fe(S_2O_3)_2]^-$ (violet, immediately changes to colourless)

(E) : $Na_4[Cu_6^I(S_2O_3)_5]$ sodium cupro (I) thiosulphate complex.

(i) $Na_2SO_3 + 2Na_2S + 3SO_2 \longrightarrow 3\,Na_2S_2O_3$ (*A*)

(ii) $2AgNO_3 + Na_2S_2O_3 \longrightarrow Ag_2S_2O_3\downarrow + 2NaNO_3$ (white)

$Ag_2S_2O_3 + H_2O \longrightarrow Ag_2S\downarrow + H_2SO_4$ (black)

(iii) A dark violet coloration is due to formation of dithiosulphato iron (III) complex (D)

$2S_2O_3^{2-} + Fe^{3+} \longrightarrow [Fe(S_2O_3)_2]^-$

This colour disappears rapidly due to formation of $S_4O_6^{2-}$ and Fe^{2+}

$[Fe(S_2O_3)_2]^- + Fe^{3+} \longrightarrow 2Fe^{2+} + S_4O_6^{2-}$

(iv) $2CuCl_2 + 3Na_2S_2O_3 \longrightarrow Cu_2S_2O_3\downarrow + Na_2S_4O_6 + 4NaCl$ (white)

$3Cu_2S_2O_3 + 2Na_2S_2O_3 \longrightarrow Na_4[Cu_6(S_2O_3)_5]$ (soluble)

27. (*A*) : $CoCl_2$ (*B*) : CoS (*C*) : $K_4[Co(CN)_6]$

(*D*) : $K_3[Co(CN)_6]$ (*E*) : $Na_3[Co(CO_3)_3]$

(i) $CoCl_2 + H_2S + 2NH_4OH \longrightarrow CoS\downarrow + 2NH_4Cl + 2H_2O$ (black)

(ii) $CoS + O \text{ (from } KClO_3) + 2HCl \longrightarrow CoCl_2 + S + H_2O$

(iii) $CoCl_2 + 2KCN \longrightarrow Co(CN)_2 + 2KCl$ (buff coloured)

$Co(CN)_2 + 4KCN \longrightarrow K_4[Co(CN)_6]$ (soluble)

(iv) $2K_4[Co(CN)_6] + H_2O + O \longrightarrow 2K_3[Co(CN)_6] + 2KOH$

(v) $2NaHCO_3 \longrightarrow Na_2CO_3 + H_2O + CO_2\uparrow$

$Na_2CO_3 + CoCl_2 \longrightarrow CoCO_3 + 2NaCl$

$CoCO_3 + 2Na_2CO_3 \longrightarrow Na_4[Co(CO_3)_3]$

$Br_2 + H_2O \longrightarrow 2HBr + O$

$2Na_4[Co(CO_3)_3] + 2NaHCO_3 + O \longrightarrow \underset{\text{green, stable}}{2Na_3[Co(CO_3)_3]} + 2Na_2CO_3 + H_2O$

28. (A) : $Hg_2(NO_3)_2$ (B) : Hg_2Cl_2 (C) : $HgCl_2$ (D) : K_2HgI_4 (E) : Hg (F) : $FeSO_4 \cdot NO$

(i) $Hg_2(NO_3)_2 + 2HCl \longrightarrow \underset{\text{white}}{Hg_2Cl_2\downarrow} + 2HNO_3$

(ii) $Hg_2Cl_2 + Cl_2 \longrightarrow 2HgCl_2$

(iii) $HgCl_2 + 2KI \longrightarrow \underset{\text{orange}}{HgI_2} + 2KCl$

$HgI_2 + 2KI \longrightarrow \underset{\text{soluble}}{K_2HgI_4}$

(iv) $SnCl_2 + Hg_2Cl_2 \longrightarrow \underset{\text{grey}}{2Hg\downarrow} + SnCl_4$

(v) $Hg_2(NO_3)_2 + H_2SO_4 \longrightarrow Hg_2SO_4 + 2HNO_3$

$2HNO_3 \longrightarrow H_2O + 2NO + 1.5O_2$

$[Fe(H_2O)_6]^{2+} + NO \longrightarrow [Fe(H_2O)_5NO]^{2+} + H_2O$

29. (A) : $CuSO_4 \cdot 5H_2O$ (B) : CuO (C) : SO_2 (D) : Cu (E) : $Cu_2[Fe(CN)_6]$ (F) : $Ca(HSO_3)_2$

(i) $CuSO_4 \cdot 5H_2O \longrightarrow CuO + SO_2 + \left(5H_2O + \frac{1}{2}O_2\right)$

(ii) $CuO + H_2 \longrightarrow Cu + H_2O$

(iii) $CuO + 2HCl \longrightarrow CuCl_2 + H_2O$

$2CuCl_2 + K_4[Fe(CN)_6] \longrightarrow Cu_2[Fe(CN)_6] + 4KCl$

(iv) $Ca(OH)_2 + SO_2 \longrightarrow \underset{\text{milky}}{CaSO_3} + H_2O$

$CaSO_3 + H_2O + SO_2 \longrightarrow Ca(HSO_3)_2$

30. (A) : CdS (B) : $CdSO_4$ (C) : $[Cd(NH_3)_4]^{2+}$ (D) : $K_2[Cd(CN)_4]$ (E) : $BaSO_4$

31. (A) : $FeCl_3$ (B) : $[Fe(SCN)]Cl_2$ (C) : $\underset{\text{Prussian blue}}{KFe^{III}[Fe^{II}(CN)_6]}$ (D) : CrO_2Cl_2 (E) : $BaCrO_4$

(similar reactions given earlier.)

Note : Actual red colour of (B) is a mixture of $[Fe(SCN)(H_2O)_5]^{2+}$ and also some $Fe(SCN)_3$ and $[Fe(SCN)_4]^-$.

32. (a) (A) : K_2MnO_4 (B) : $KMnO_4$ (C) : KIO_3 (D) : Mn_2O_7 (E) : MnO_2

$MnO_2 + 2KOH + \underset{\text{air}}{\frac{1}{2}O_2} \longrightarrow K_2MnO_4 + H_2O$

$3MnO_4^{2-} + 4H^+ \longrightarrow 2MnO_4^- + MnO_2 + 2H_2O$

(i) $2MnO_4^- + I^- + H_2O \xrightarrow{\Delta} IO_3^- + 2MnO_2 + 2OH^-$

(ii) $MnO_4^- + 8H^+ + 5Fe^{2+} \longrightarrow 5Fe^{3+} + Mn^{2+} + 4H_2O$

(iii) $2MnO_4^- + 2H^+ \longrightarrow Mn_2O_7 + H_2O$

$2Mn_2O_7 \xrightarrow{\Delta} 4MnO_2 + 3O_2$

(b) (i) $Au + 4HCl + HNO_3 \longrightarrow HAuCl_4 + NO + 2H_2O$

(ii) $C + 4HNO_3 \longrightarrow CO_2 + 4NO_2 + 2H_2O$

(iii) $Sn + 2KOH + H_2O \longrightarrow K_2SnO_3 + 2H_2$

(iv) $Cu(OH)_2 + 2NH_4NO_3 + 2NH_4OH \longrightarrow [Cu[(NH_3)_4]_4](NO_3)_2 + 4H_2O$

33. Label them yourself as (A) for $AlCl_3 \cdot 6H_2O$ (B) for $Ba(OH)_2 \cdot 8H_2O$ and (C) for $MgSO_4 \cdot H_2O$.

[(A), (B), and (C) are unknown till confirmed]; mix (B) and (C) white ppt confirms that one of them is $Ba(OH)_2$ and the other $MgSO_4$. Add (B) and (C) each to (A) (which is certainly $AlCl_3 \cdot 6H_2O$) of white ppt [of $Al(OH)_3$] then that compound is (B).

34. By measurement of pH, which would be in order

$NaHSO_4 < NaHSO_3 < Na_2SO_4$

35. (a) sulphur (b) P_2O_5 (c) CO (d) Si

36. (a)

37. If Hg_2Cl_2, AgCl and $PbCl_2$ are precipitated then,

(a) $PbCl_2$ remains insoluble

(b) Hg_2Cl_2 is blackened by addition of NH_3 due to formation of ($Hg + HgNH_2Cl$)

(c) AgCl goes into solution as soluble complex $[Ag(NH_3)_2]Cl$

38. (a) $Pb^{2+}(aq) + 2Cl^-(aq) \longrightarrow PbCl_2(s)\downarrow$

(b) $Zn(OH)_2(s) + 2OH^-(aq) \longrightarrow ZnO_2^{2-}(aq) + 2H_2O(l)$

(c) $Fe(OH)_3(s) + 3H^+(aq) \longrightarrow Fe^{3+}(aq) + 3H_2O(l)$

(d) $2SbCl_4^-(aq) + 3H_2S(aq) \longrightarrow Sb_2S_3(s)\downarrow + 6H^+(aq) + 8Cl^-(aq)$

39. A : $Cd(CH_3COO)_2$, B : $[Fe(CH_3COO)]^{2+}$, C : CdS, D : $[Cd(NH_3)_4]^{3+}$, E : $[Cd(CN)_4]^{2-}$

40. A : $NaBO_2$, B : CO_2, C : $Na_2B_4O_7$, D : B_2O_3, E : H_3BO_3, F : $Co(BO_2)_2$, G : $(C_2H_5O)_3B$

41. A : ZnS, B : H_2S, C : $ZnSO_4$, D : PbS, E : $Zn(OH)_2$, F : Na_2ZnO_2 or $Na_2[Zn(OH)_4]$

42. X_2 is decolourised by unsaturated hydrocarbon thus, X_2 is Br_2 and MX is MBr.

Depression in freezing point

$$(\Delta T_f) = 0 - (-3.72°) = 3.72°$$

$$\Delta T_f = \frac{1000\, K_f w_1(i)}{m_1 w_1}$$

m_1 (molar mass of MBr) = ?

K_f (molar depression constant) = 1.86° mol^{-1} kg

w_1 (mass of MX) = 11.9 g

w_2 (mass of H_2O) = 100 g

i (van't Hoff factor) $= (1 + x) = 1 + 1 = 2$

x = degree of ionisation = 1

(assuming MX is conducting and 100% ionised.)

$$\therefore \quad m_1 = \frac{1000 \times 1.86 \times 11.9 \times 2}{3.72 \times 100}$$

$$= 119 \text{ g mol}^{-1}$$

$\therefore$ Thus, $M + \text{Br} = 119 \text{ g mol}^{-1}$

$$\therefore \quad M + 80 = 119$$

$$M = 39$$

Thus, M is **potassium.**

43. Element X belong to group 2, thus it forms divalent cation X^{2+}

$$X + H_2 \longrightarrow \underset{(Y)}{XH_2}$$

$$XH_2 \longrightarrow X + H_2$$

559 g mL of H_2 combines with = 1 g X

$\therefore$ 22400 mL of H_2 combines with = 40 g X

Thus, atomic mass of $X = 40 \text{ g mol}^{-1}$

Thus, X is **Ca,** Y is CaH_2 and Z is $CaCl_2$

It is confirmed by the fact that

$$\% \text{ of chlorine in } CaCl_2 = \frac{71 \times 100}{111}$$

$$= 63.96 \approx 63.89 \text{ (given)}$$

44. X : Sn or Pb, Y : P, Z : alkali metal

45. (a) NH_2OH is reduced to NH_3

(b) HPO_3^{2-} is oxidised to HPO_4^{2-}

(c) Initially H_3PO_2 is oxidised to H_3PO_3 and finally to H_3PO_4

In the reactions above, total change in oxidation number in oxidation reaction is equal to that in reduction reaction.

46. X : NH_2SO_3H

$$\underset{\text{urea}}{NH_2CONH_2} + \underset{\text{conc.}}{H_2SO_4} \longrightarrow \underset{X}{NH_2SO_3H} + NH_3 + CO_2$$

$$NaNO_2 + HCl \longrightarrow HNO_2 + NaCl$$

$$NH_2SO_3H + HNO_2 \longrightarrow N_2 \uparrow + H_2SO_4 + H_2O$$

$$H_2SO_4 + BaCl_2 \longrightarrow BaSO_4 \downarrow + 2HCl$$

47. A : $Na_2S_2O_3$; B : $Ag_2S_2O_3$; C : $Na_3[Ag(S_2O_3)_2]$

D : Ag_2S; E : H_2SO_4; F : $BaSO_4$

48. $$BaMnO_4 \equiv Ba^{2+} + MnO_4^{2-}$$

Thus, $\underset{(+7)}{MnO_4^-}$ is reduced to $\underset{(+6)}{MnO_4^{2-}}$

Change in oxidation number = 7 − 6 = 1 unit

$$\text{Moles of } MnO_4^- \text{ used} = \frac{10 \times 0.1}{1000} = 0.001$$

$\therefore$ Equivalent of MnO_4^- used = 0.001

Let I^- be oxidised to species with oxidation number = x

$$\therefore \quad \underset{-1}{I^-} \longrightarrow \underset{x}{(I)^{x+}}$$

Change in oxidation number = $x - (-1)$

$$\text{Equivalent of } I^- = \frac{0.01587}{127} = (x + 1)$$

$$\therefore \quad \frac{0.01587\,(x + 1)}{127} = 0.001$$

$$\therefore \quad x = 7$$

Thus, iodide is oxidised to $\mathbf{IO_4^-}$.

49. $$IO_4^- + 2I^- + H_2O \longrightarrow IO_3^- + I_2 + 2OH^-$$

$$IO_3^- + 5I^- + 6H^+ \longrightarrow 3I_2 + 3H_2O$$

50. N_2 can't be used with lithium since it reacts giving nitride.

$$6Li + N_2 \longrightarrow 2Li_3N$$

51. A : MgO; B : $MgCl_2$; C : $MgCO_3$; D : CO_2;

E : MgO; F : $CaCO_3$; G : $MgSO_4$; H : $Mg(OH)_2$;

I : $Mg(NO_3)_2$; J : NO_2

52. It is due to formation of a stable complex which does not ionise to give free Cu^{2+} that can give precipitate with H_2S.

$$\underset{\text{free}}{Cu^{2+}} + S^{2-} \longrightarrow CuS \downarrow$$

$$CuSO_4 + 2KCN \longrightarrow \underset{\text{white ppt}}{Cu(CN)_2} \downarrow + K_2SO_4$$

$$2Cu(CN)_2 \longrightarrow 2\overset{\text{I}}{Cu}CN + \underset{\text{cyanogen}}{(CN)_2}$$

$$CuCN + 3KCN \longrightarrow \underset{\text{stable}}{K_3[Cu(CN)_4]}$$

$$K_3[Cu(CN)_4] \rightleftharpoons 3K^+ + [Cu(CN)_4]^{3-}$$

$$[Cu(CN)_4]^{3-} \longrightarrow \text{no ionisation}$$

53. M is potassium; KO_2 is the most common superoxide.

54. (i) $K_2Cr_2O_7 + 4NaCl + 6H_2SO_4 \longrightarrow 2KHSO_4 + 4NaHSO_4 + 2CrO_2Cl_2 + 3H_2O$

(ii) $NaNO_3 + 4Zn + 7NaOH \longrightarrow 4Na_2ZnO_2 + NH_3 + 2H_2O$

(iii) $2NaIO_3 + 5NaHSO_3 \longrightarrow 2Na_2SO_4 + 3NaHSO_4 + I_2 + H_2O$

(iv) $4Fe + 10HNO_3 \longrightarrow 4Fe(NO_3)_2 + NH_4NO_3 + 3H_2O$

(v) $2KMnO_4 + 3MnSO_4 + 2H_2O \longrightarrow 5MnO_2 + K_2SO_4 + 2H_2SO_4$

55. Radicals : $Fe^{2+}, NH_4^+, SO_4^{2-}, Cl^-$ (A) : CrO_2Cl_2

(B) : $KFe^{II}[Fe^{III}(CN)_6]$, Turnbull's blue

56. (i) $Na_2CO_3 + NO + NO_2 \longrightarrow 2NaNO_2 + CO_2$

(ii) $2NH_3 + 2KMnO_4 \longrightarrow 2KOH + 2MnO_2 + 2H_2O + N_2$

(iii) $P_4 + 20HNO_3 \longrightarrow 4H_3PO_4 + 20NO_2 + 4H_2O$

(iv) $NaHSO_3 + 2H_2S \longrightarrow NaOH + 2H_2O + 3S$

(v) $CaCO_3 + H_2O + CO_2 \longrightarrow Ca(HCO_3)_2$

57. (i) $Al(OH)_3$ is soluble in NaOH forming $NaAlO_2$.

$Fe(OH)_3$ is insoluble in NaOH

(ii) $$NH_4Cl \rightleftharpoons NH_4^+ + Cl^-$$

$$2NH_3 \longrightarrow NH_2^- + NH_4^+$$

In NH_3, NH_4^+ is formed which is one of the ion of $NH_4Cl \cdot NH_4^+$ is an acid.

58. $FeSO_4 \cdot (NH_4)_2SO_4 \cdot 6H_2O$ (Mohr's salt)

59. (i) $$2NH_3 + NaOCl \longrightarrow NH_2 \cdot NH_2 + NaCl + H_2O$$

(ii) $AgBr + 2Na_2S_2O_3 \longrightarrow Na_3[Ag(S_2O_3)_2] + NaBr$

(iii) $(NH_4)_2S_2O_8 + 2H_2O + MnSO_4 \longrightarrow MnO_2 + 2H_2SO_4 + (NH_4)_2SO_4$

60. (a) NO (b) NO (c) $[Fe(H_2O)_5NO]^{2+}$

61. *A* is a binary salt of univalent metal *A* converts S to sulphate that *A* contains oxygen. Since, *A* is a binary salt thus *A* can be MO_2 (superoxide) with O_2^- anion.

$$\underset{(M+32)}{MO_2} + \underset{32}{S} \longrightarrow \underset{(2M+96)}{M_2SO_4}$$

$$\frac{2M+96}{32} = \frac{1.743}{0.321}$$

$\therefore$ $$M = 38.87 \approx 39$$

Thus, binary compound

A is $\mathbf{KO_2}$, *B* is K_2SO_4, *C* is $\underset{\text{(alum)}}{K_2SO_4 \cdot Al_2(SO_4)_2 \cdot 24H_2O}$

62. (*A*): KO_2 (*B*): K_2SO_4 (*C*): $K_2SO_4 \cdot Al_2(SO_4)_3 \cdot 24H_2O$

63. (*A*) : O_3 (*B*) : KO_3 (*C*) : O_2

64. (*A*) : Pb_3O_4 (*B*) : PbO_2 (*C*) : PbI_2 (*D*) : $HMnO_4$ (pink)

65. (*A*) : $Ca_2B_6O_{11}$ (*B*) : $CaCO_3$ (*C*) : $Na_2B_4O_7$

(*D*) : $NaBO_2$ (*E*) : B_2O_3 (*F*) : $Co(BO_2)_2$

$$\underset{(A)}{Ca_2B_6O_{11}} + 2Na_2CO_3 \longrightarrow \underset{(B)}{2CaCO_3} + \underset{(C)}{Na_2B_4O_7} + \underset{(D)}{2NaBO_2}$$

$$\underset{(D)}{4NaBO_2} + CO_2 \longrightarrow \underset{(C)}{Na_2B_4O_7} + Na_2CO_3$$

$$Na_2B_4O_7 \xrightarrow{\Delta} \underset{(D)}{2NaBO_2} + \underset{(E)}{B_2O_3}$$

$$CoO + B_2O_3 \longrightarrow \underset{(F)}{Co(BO_2)_2}$$

66. (*A*) : NH_4NO_3 (*B*) : N_2O (*C*) : H_2O (*D*) : P_4O_{10}

67. (a) (*A*) : Ca (*B*) : Ca_3N_2 (*C*) : $Ca(OH)_2$ (*D*) : NH_3

(b) $Al_2S_2 + 6H_2O \longrightarrow 2Al(OH)_3 \downarrow + \underset{\text{(foul odour)}}{3H_2S \uparrow}$

68. Refer Table 20.3

69. White solid is Na_2O_2

$$Na_2O_2 + 2H_2O \longrightarrow 2NaOH + H_2O_2$$

Bleaching of litmus paper is due to H_2O_2.

If it would have been Na_2O, then its aqueous solution turned red litmus blue due to NaOH.

$$Na_2O + H_2O \longrightarrow 2NaOH \quad \text{(blue towards litmus)}$$

70. $$2NaI + \underset{\text{orange}}{HgI_2} \longrightarrow \underset{\text{colourless}}{Na_2HgI_4}$$

$$Na_2HgI_4 + NaOCl + H_2O \longrightarrow 2NaOH + NaCl + HgI_2 + I_2$$

71. (a) (*D*) : H_2SO_4

(b) (*A*) : $Ca(OH)_2$ (*B*) : NH_4HCO_3 (*C*) : Na_2CO_3

(*D*) : NH_4Cl (*E*) : $CaCl_2$

Exercise 2

Only One Option Correct

1. (d)	**2.** (a)	**3.** (c)	**4.** (c)	**5.** (c)	**6.** (a)	**7.** (d)	**8.** (d)	**9.** (c)	**10.** (d)
11. (c)	**12.** (a)	**13.** (c)	**14.** (c)	**15.** (d)	**16.** (c)	**17.** (d)	**18.** (c)	**19.** (c)	**20.** (a)
21. (a)	**22.** (d)	**23.** (d)	**24.** (d)	**25.** (a)	**26.** (c)	**27.** (b)	**28.** (c)	**29.** (a)	**30.** (d)
31. (a)	**32.** (b)	**33.** (c)	**34.** (a)	**35.** (c)	**36.** (a)	**37.** (d)	**38.** (c)	**39.** (c)	**40.** (b)
41. (c)	**42.** (c)	**43.** (b)	**44.** (a)	**45.** (c)	**46.** (d)	**47.** (c)	**48.** (c)	**49.** (d)	**50.** (d)
51. (a)	**52.** (c)	**53.** (b)	**54.** (b)	**55.** (d)	**56.** (a)	**57.** (c)	**58.** (b)	**59.** (c)	**60.** (d)
61. (a)	**62.** (d)	**63.** (b)	**64.** (b)	**65.** (a)	**66.** (a)	**67.** (a)	**68.** (b)	**69.** (d)	**70.** (c)
71. (c)	**72.** (a)	**73.** (a)	**74.** (b)	**75.** (d)	**76.** (c)	**77.** (d)	**78.** (a)	**79.** (a)	**80.** (b)
81. (b)	**82.** (a)	**83.** (b)	**84.** (b)	**85.** (a)	**86.** (a)	**87.** (b)	**88.** (b)	**89.** (c)	**90.** (a)
91. (b)	**92.** (a)	**93.** (c)	**94.** (a)	**95.** (c)	**96.** (c)	**97.** (a)	**98.** (c)	**99.** (b)	**100.** (d)
101. (b)	**102.** (a)	**103.** (c)	**104.** (a)	**105.** (c)	**106.** (c)	**107.** (d)	**108.** (b)	**109.** (a)	**110.** (c)
111. (c)	**112.** (a)	**113.** (d)	**114.** (c)	**115.** (b)	**116.** (d)	**117.** (b)	**118.** (a)	**119.** (c)	**120.** (c)
121. (a)	**122.** (c)	**123.** (b)	**124.** (b)	**125.** (a)	**126.** (d)	**127.** (a)	**128.** (c)	**129.** (c)	**130.** (b)
131. (a)	**132.** (d)	**133.** (b)	**134.** (a)	**135.** (d)	**136.** (c)	**137.** (b)	**138.** (c)	**139.** (b)	**140.** (c)

One or More Than One Options Correct

1. (a,c,d) **2.** (b,c) **3.** (b,d) **4.** (a,b) **5.** (a,b)
6. (a,c) **7.** (b,c,d) **8.** (a,b,c,d) **9.** (a,b,c)
10. (a,c,d)

Brain Twisters

1. Reactions (i) to (iv) indicate that (*A*) is hypo $\underset{\text{hydrated}}{Na_2S_2O_3 \cdot xH_2O}$

Mol. wt. of hypo (*A*) ($Na_2S_2O_3 \cdot xH_2O$) = 158 + 18*x*
and that of anhydrous hypo (*B*) ($Na_2S_2O_3$) = 158

$$\frac{(A)}{(B)} = \frac{158 + 18x}{158} = \frac{2.48}{1.58}$$

$$\Rightarrow \quad x = 5$$

⇒ (*A*) is $\mathbf{Na_2S_2O_3 \cdot 5H_2O}$

Explanation :

(i) $2AgNO_3 + Na_2S_2O_3 \longrightarrow \underset{\text{white}}{Ag_2S_2O_3\downarrow} + 2NaNO_3$

$$As_2S_2O_3 + H_2O \longrightarrow H_2SO_4 + \underset{\text{black}}{Ag_2S\downarrow}$$

(ii) $I_2 + 2Na_2S_2O_3 \longrightarrow 2NaI + Na_2S_4O_6$
(I_2 is soluble in KI forming KI_3; I_2 is insoluble in H_2O)

(iii) $Na_2S_2O_3 + 2HCl \longrightarrow 2NaCl + SO_2 + \underset{\text{(turbidity)}}{S} + H_2O$

(iv) $AgBr + 2Na_2S_2O_3 \longrightarrow \underset{\text{soluble}}{Na_3[Ag(S_2O_3)_2]} + NaBr$

2. Paste of (*A*) is conc. HCl gave green colouration in flame.
⇒ (*A*) has Ba^{2+}

$$(A) + \text{dil. HCl} \longrightarrow \text{gas } (B) \xrightarrow{\text{lime water}} \text{milky}$$

⇒ gas (*B*) can be either of SO_2 or CO_2
⇒ (*A*) can be $BaSO_3$ or $BaCO_3$ ($BaXO_3$, *X* = S or C)

$$\underset{(A)}{BaXO_3} \longrightarrow \underset{(E)}{BaO} + \underset{(B)}{XO_2}$$

Conformation of *X* : (Ba = 137, *X* = *X*, O = 16)

$$\frac{Ba + X + 3(O)}{Ba + O} = \frac{137 + X + 48}{137 + 16} = \frac{1.9735}{1.5334}$$

$$\Rightarrow \quad X = 11.91 \approx 12$$

Thus, (*X*) is **carbon.**

Confirmation of Ba : 1.5334 g of BaO [suppose it is unknown *MO* of molecular weight (*M* + 16)] where dissolved in 250 mL solution.

$$25 \text{ mL of } MO \equiv 20.30 \text{ mL of } 0.0985 \text{ M HCl}$$
$$\equiv 20.30 \text{ mL of } 0.0985 \text{ N HCl}$$
$$N(MO) = 0.08 \text{ N}$$

***MO* :**

$$\text{Milliequivalent} = 0.08 \times 250$$
$$\text{equivalent} = \frac{0.08 \times 250}{1000} = 0.02$$
$$\text{mol} = 0.01$$

(since *MO* has divalent cation.)

$$\therefore \quad \frac{\text{Amount}}{\text{molecular weight of } MO} = 0.01 \text{ mol}$$

$$\frac{1.5334 \text{ g}}{m} = 0.01$$

$$\therefore \quad m = 153.34 \text{ g mol}^{-1} \text{ for } MO$$

Thus, atomic weight of *M* = 153.34 − 16 = 137.34
⇒ *M* is **Ba**
⇒ *MO* is **BaO**
⇒ *A* is $\mathbf{BaCO_3}$
⇒ Molecular mass of $BaCO_3$ = **197.4 g mol^{-1}**

$$\underset{(A)}{BaCO_3} + 2HCl \longrightarrow BaCl_2 + H_2O + \underset{(B)}{CO_2}$$

$$CaCO_3 + CO_2 \longrightarrow \underset{(C)}{CaCO_3} + H_2O$$

$$CaCO_3 + H_2O + CO_2 \longrightarrow \underset{(D)}{Ca(HCO_3)_2}$$

$$BaCO_3 \longrightarrow \underset{(E)}{BaO} + CO_2$$

3. (*A*) gives white ppt (*B*) with $BaCl_2$ insoluble in conc. HCl
⇒ (*A*) has SO_4^{2-} ion

$$(A) \text{ (in dil. HCl)} \xrightarrow{H_2S} \text{black ppt. } (C)$$

[insoluble in YAS (copper group)]

soluble in HNO_3

$\Delta \downarrow HNO_3$

blue coloured solution (*D*)

⇒ (*A*) contains Cu^{2+}

(*A*) gives brown ppt with ($NH_4Cl + NH_4OH$) buffer indicating presence of Fe^{3+}
Thus, (*A*) contains $Cu^{2+}, Fe^{3+}, SO_4^{2-}, H_2O$

Since it is a double salt, it is a **pseudo alum** with probable formula

$$CuSO_4 \cdot Fe_2(SO_4)_3 \cdot xH_2O \text{ (mol. wt.} = 559.5 + 18x)$$

$$CuSO_4 \cdot Fe_2(SO_4)_3 \cdot xH_2O \xrightarrow{BaCl_2} 4BaSO_4$$

$$\frac{559.5 + 18x}{4 \times 233} = \frac{0.9915}{0.932}$$

$$\Rightarrow \quad x = 24$$

⇒ *A* is $\mathbf{CuSO_4 \cdot Fe_2(SO_4)_3 \cdot 24H_2O}$

Explanation : $\underset{\text{in } (A)}{SO_4^{2-}} + Ba^{2+} \longrightarrow \underset{(B)}{BaSO_4\downarrow}$

$$\underset{\text{in } (A)}{CuSO_4} + H_2S \longrightarrow \underset{(C)}{CuS\downarrow} + H_2SO_4$$

$$3CuS + 8HNO_3 \longrightarrow \underset{(D)}{3Cu(NO_3)_2} + 2NO + 3S + 4H_2O$$

$$\underset{(D)}{Cu(NO_3)_2} + 2NH_4OH \longrightarrow \underset{(E)}{Cu(OH)_2 \downarrow} + 2NH_4NO_3$$

$$\underset{(E)}{Cu(OH)_2} + 4NH_4OH \longrightarrow \underset{(F)}{[Cu(NH_3)_4]^{2+}}$$

$$\underset{(D)}{Cu^{2+}} + 4I^- \longrightarrow \underset{(G)}{Cu_2I_2 \downarrow} + I_2$$

$$I_2 + \underset{\text{hypo}}{2Na_2S_2O_3} \longrightarrow 2NaI + Na_2S_4O_6$$

I_2 (before addition of hypo) remains dissolved in KI as KI_3 (reddish brown). After hypo is added, I_2 disappears and white ppt (*G*) appears.

$$\underset{(D)}{Cu^{2+}} + 2SCN^- \longrightarrow \underset{(H)}{Cu(SCN)_2 \downarrow}$$

Precipitate after separation of (*C*) contains Fe^{3+} ion

$$Fe^{3+} \xrightarrow[\text{(basic buffer)}]{OH^-} \underset{(I)}{Fe(OH)_3 \downarrow}$$

$$Fe(OH)_3 + 3HCl \longrightarrow \underset{(J)}{FeCl_3} + 3H_2O$$

$$\underset{(J)}{2FeCl_3} + SnCl_2 \longrightarrow \underset{(L)}{2FeCl_2} + SnCl_4$$

(*L*) is obtained from (*J*) by gas (*K*) (reducing agent) that can be H_2S.

$$\underset{(J)}{2FeCl_3} + H_2S \longrightarrow \underset{(L)}{2FeCl_2} + 2HCl + S$$

$$\underset{(J)}{FeCl_3} + K_4[Fe^{II}(CN)_6] \longrightarrow \underset{\substack{(M)\\ \text{(Prussian blue)}}}{KFe^{III}[Fe^{II}(CN)_6]} + 3KCl$$

$$\underset{(J)}{FeCl_3} + NH_4SCN \longrightarrow \underset{(N)}{[Fe(SCN)]Cl_2} + NH_4Cl$$

$$\underset{(L)}{FeCl_2} + K_3[Fe(CN)_6] \longrightarrow \underset{\substack{(O)\\ \text{(Turnbull's blue)}}}{KFe^{II}[Fe^{III}(CN)_6]} + 2KCl$$

4. (*C*) is hydrated double salt formed from (*B*) and $Al_2(SO_4)_3$

$\Rightarrow$ (*C*) can be an **Alum**

$M_2SO_4 \cdot Al_2(SO_4)_3 \cdot xH_2O$

(*B*) is obtained when S reacts with (*A*) is a sealed tube

$\Rightarrow$ (*A*) is also a source of oxygen that can convert S into SO_4^{2-}.

(*A*) has univalent cation M^+ and its compound with oxygen can be

M_2O (oxide)

M_2O_2 (peroxide)

MO_2 (superoxide)

Only MO_2 is paramagnetic.

$\Rightarrow A$ is MO_2 (binary given)

$$\underset{2(M+32)}{2MO_2} + \underset{32}{S} \longrightarrow \underset{(2M+96)}{M_2SO_4}$$

$2(M + 32)$ g MO_2 reacts with 32 of sulphur

$\therefore$ 1.422 g MO_2 reacts with $\dfrac{32}{2(M+32)} \times 1.422$ g sulphur

$$= \frac{32 \times 1.422}{2(M+32)} = 0.321 \text{ g sulphur}$$

$\Rightarrow \quad M = 39$ (K)

$\Rightarrow$ (*A*) is $\mathbf{KO_2}$

(*B*) is $\mathbf{K_2SO_4}$

(*C*) is $\mathbf{K_2SO_4 \cdot Al_2(SO_4)_3 \cdot 24H_2O}$ **(potash alum)**

[Value of *M* can also be evaluated by using weight of M_2SO_4 (*i.e.*, 1.743 g (given))]

$$\frac{\text{weight of S}}{\text{weight of } M_2SO_4} = \frac{32}{(2M+96)}$$

$$\frac{0.321}{1.743} = \frac{32}{2M+96}$$

5. (*A*) + $BaCl_2$ solution $\longrightarrow$ white ppt (*B*)

$\Rightarrow SO_4^{2-}$ is present

(*A*) + dil. HCl + H_2S $\longrightarrow$ black ppt soluble in 50% HNO_3

$\Rightarrow$ group II (copper subgroup) except Hg^{2+}

(*A*) + *aq* NH_3 (excess) $\longrightarrow$ deep blue solution (*C*)

$\Rightarrow Cu^{2+}$ is present.

Thus, (*A*) is $CuSO_4 \cdot xH_2O$

$$\underset{(159.5+18x)\text{ g}}{CuSO_4 \cdot xH_2O} + BaCl_2 \longrightarrow \underset{233\text{ g}}{BaSO_4}$$

$$\frac{159.5+18x}{233} = \frac{2.495}{2.33}$$

$\Rightarrow \quad x = 5$

Thus, (*A*) is $\mathbf{CuSO_4 \cdot 5H_2O}$

Explanation :

$$CuSO_4 + 4NH_3 \longrightarrow \underset{\text{deep blue } (C)}{[Cu(NH_3)_4]SO_4}$$

$$Na(NH_4)HPO_4 \xrightarrow{\Delta} NaPO_3 + H_2O + NH_3$$

$$NaPO_3 + CuSO_4 \xrightarrow{\Delta} \underset{\text{green}}{NaCuPO_4} + SO_3 \uparrow$$

$$2CuSO_4 + 4KI \longrightarrow Cu_2I_2 + 2K_2SO_4 + I_2$$

6. Let metal by (*A*).

(*A*) forms metal nitrate which gives red colouration to a non-luminous Bunsen burner.

$\Rightarrow$ (*A*) is Li (atomic weight 7)

metal nitrate is $LiNO_3$

metal oxide (*B*) is Li_2O

(*C*) is Li_3N

(*D*) is LiH

(*E*) is H_2

(*F*) is LiOH

There observations are confirmed quantitatively.

(i) $\underset{(2\times 7)\text{ g}}{2Li} \longrightarrow 2LiNO_3 \longrightarrow \underset{30\text{ g}}{Li_2O} + 2NO_2 + \frac{1}{2}O_2$

$$\frac{14}{30} = \frac{0.347}{0.747}$$

$0.467 = 0.464$ (very approximately equal)

hence Li.

(ii) $\underset{\substack{8\text{ g}\\0.1590\text{ g}}}{LiH} + H_2O \longrightarrow \underset{\substack{24\text{ g}\\x\text{ g}}}{LiOH} + H_2$

equivalent of LiOH ≡ equivalent of HCl

$$\equiv \frac{200\times 0.1}{1000} = 0.02$$

0.02 equivalent of LiOH ≡ 0.48 g

Thus, $\frac{8}{0.1590} = \frac{24}{x}$

$\Rightarrow$ $x = 0.477\text{ g} \approx 0.48\text{ g}$

which also confirms (A) is **Li**.

7. By (iii), it is clear that (A) has SO_4^{2-}

(A) decolourises MnO_4^-/H^+

(A) turns $Cr_2O_7^{2-}/H^+$ green

$\Rightarrow$ (A) has Fe^{2+}

$\Rightarrow$ (A) is $FeSO_4\cdot xH_2O$

(iii) $\underset{\substack{(152+18x)\\2.78}}{FeSO_4\cdot xH_2O} + BaCl_2 \longrightarrow \underset{\substack{233\\2.33}}{BaSO_4}$

$$\frac{152+18x}{233} = \frac{2.78}{2.33}$$

$\Rightarrow$ $x = 7$

$\Rightarrow$ (A) **is** $FeSO_4\cdot 7H_2O$

Explanation :

(i) $MnO_4^- + 8H^+ + 5Fe^{2+} \longrightarrow 5Fe^{3+} + \underset{\text{colourless}}{Mn^{2+}} + 4H_2O$

(ii) $Cr_2O_7^{2-} + 14H^+ + 6Fe^{2+} \longrightarrow 6Fe^{3+} + \underset{\text{green}}{2Cr^{3+}} + 7H_2O$

(iii) $FeSO_4 + 2H_2O \longrightarrow \underset{\text{green turbidity}}{Fe(OH)_2\downarrow} + H_2SO_4$

$Fe(OH)_2 + H_2SO_4 \longrightarrow \underset{\text{soluble}}{FeSO_4} + 2H_2O$

(iv) $2FeSO_4 \xrightarrow{\Delta} \underset{(B)}{Fe_2O_3} + \underset{(C)}{SO_2} + \underset{(D)}{SO_3}$

$\underset{(C)}{2SO_2} + O_2 \longrightarrow \underset{(D)}{2SO_3}$

$Cr_2O_7^{2-} + 3SO_2 + 2H^+ \longrightarrow \underset{\text{green}}{2Cr^{3+}} + 3SO_4^{2-} + H_2O$

$Fe_2O_3 + 6HCl \longrightarrow \underset{\text{yellow}}{2FeCl_3} + 3H_2O$

$FeCl_3 + KCNS \longrightarrow \underset{\text{red}}{[Fe(CNS)]Cl_2} + KCl$

Passage Comprehension Questions

Passage 1

1. (d) **2.** (a) **3.** (c) **4.** (b) **5.** (c) **6.** (d)

Passage 2

1. (a) **2.** (b)

Passage 3

1. (a) **2.** (b) **3.** (d) **4.** (c) **5.** (a)

Passage 4

1. (c) **2.** (b) **3.** (a) **4.** (c)

Passage 5

1. (c) **2.** (b) **3.** (a) **4.** (b) **5.** (d) **6.** (b)
7. (a)

Passage 6

1. (c) **2.** (a) **3.** (b) **4.** (a)

Passage 7

1. (b) **2.** (a)

Passage 8

1. (b) **2.** (a) **3.** (a)

Assertion & Reason

1. (e) **2.** (a) **3.** (b) **4.** (c) **5.** (b) **6.** (c)
7. (d) **8.** (a) **9.** (d) **10.** (a)

True & False

1. NH_4Cl is used alongwith NH_4OH in group III to decrease ionisation of NH_4OH by common-ion effect (of NH_4^+). $(NH_4)_2SO_4$ is also a source of NH_4^+ion but it can't be used since SO_4^{2-}ion will precipitate Zn^{2+}, Ca^{2+}, Ba^{2+} and Sr^{2+} ions as their sulphate; hence **false.**

2. HCl (common H^+ ion) decreases ionisation of H_2S giving less $[S^{2-}]$ that can precipitate group II as sulphides. But H_2SO_4, although a good source of H^+, will precipitate Pb^{2+}, Zn^{2+}, Ba^{2+}, Ca^{2+}, Sr^{2+} etc. as their sulphates, hence **false.**

3. $ZnCl_2 + H_2S \longrightarrow ZnS + 2HCl$

ZnS is formed but remains soluble due to formation of HCl. ZnS is precipitated when H_2S is passed in ammoniacal medium, hence **false.**

4. Reducing power of $I^- > Br^- > Cl^- > F^-$

$$2I^- + Cl_2 \longrightarrow 2Cl^- + I_2$$

$$2Br^- + Cl_2 \longrightarrow 2Cl^- + Br_2$$

Thus, I_2 is formed in preference of Br_2, and gives violet colour in $CHCl_3$, hence **true.**

5. $AB \rightleftharpoons A^+ + B^-$

AB is precipitated if $[A^+][B^-] > K_{sp}$

AgCl is precipitated if $[Ag^+][Cl^-] > K_{sp}$

PbI_2 is precipitate of $[Pb^{2+}][I^-]^2 > K_{sp}$, hence **true.**

6. $M(OH)_3$ is precipitated

$$[M^{3+}][OH^-]^3 > K_{sp}$$
$$[OH^-] > 1\times10^{-9}$$
$$[OH^-] > 1\times10^{-3}$$
$$\text{maximum pOH} = 3$$
$$\text{maximum pH} = 9$$

hence **true.**

7. Turnbull's blue colour is obtained when Fe^{2+} salts react with $K_3[Fe^{III}(CN)_6]$ and Prussian's blue colour is obtained when Fe^{3+} salts react with $K_4[Fe^{II}(CN)_6]$.

$$Fe^{II}Cl_2 + K_3[Fe^{III}(CN)_6] \longrightarrow \underset{\text{Turnbull's blue}}{KFe^{II}[Fe^{III}(CN)_6]} + 2KCl$$

$$Fe^{III}Cl_3 + K_4[Fe^{II}(CN)_6] \longrightarrow \underset{\text{Prussian's blue}}{KFe^{III}[Fe^{II}(CN)_6]} + 3KCl$$

hence **true.**

8.
$$BaBr_2 + AgNO_3 \longrightarrow \underset{\text{yellow ppt.}}{AgBr\downarrow} \text{ (due to } Br^-)$$

$$BaBr_2 + K_2CrO_4 \longrightarrow \underset{\text{yellow ppt.}}{BaCrO_4\downarrow} \text{ (due to } Ba^{2+})$$

hence **true.**

9.
$$AlCl_3 + NaOH \longrightarrow 3NaCl + Al(OH)_3\downarrow$$
$$Al(OH)_3 + NaOH \longrightarrow \underset{\text{soluble}}{Na[Al(OH)_4]}$$

hence **true.**

10. $Fe(OH)_3$ is insoluble while $Al(OH)_3$ is soluble in NaOH forming $NaAlO_2$ or $Na[Al(OH)_4]$

hence **true.**

11. SnS and As_2S_3 both are soluble in YAS,

hence **false.**

12.
$$CoCl_2 + 4NH_4SCN \longrightarrow \underset{\text{blue colour}}{(NH_4)_2[Co(SCN)_4]} + 2NH_4Cl$$

$$CuCl_2 + 2NH_4SCN \longrightarrow \underset{\text{black ppt.}}{Cu(SCN)_2\downarrow} + 2NH_4Cl$$

hence **true.**

13. NaOH can dissolve both $Al(OH)_3$ and $Zn(OH)_2$

$$Al(OH)_3 + NaOH \longrightarrow NaAlO_2 + H_2O \text{ or } Na[Al(OH)_4]$$
$$Zn(OH)_2 + NaOH \longrightarrow Na_2[Zn(OH)_4] \text{ or } Na_2ZnO_2 + H_2O$$

hence **false.**

14.
$$SbCl_3 + H_2O \longrightarrow \underset{\text{orange}}{SbOCl\downarrow} + 2HCl$$

$$BiCl_3 + H_2O \longrightarrow \underset{\text{white}}{BiOCl\downarrow} + 2HCl$$

white turbidity is due to BiOCl (while orange turbidity is due to SbOCl)

hence **false.**

15.
$$6Fe^{2+} + Cr_2O_7^{2-} + 14H^+ \longrightarrow 6Fe^{3+} + \underset{\text{(green)}}{2Cr^{3+}} + 7H_2O$$

$Cr_2O_7^{2-}/H^+$ can oxidise Fe^{2+} to Fe^{3+} giving green colouration at the end point

hence **true.**

16. White ppt. of $PbCl_2$ is soluble in hot water hence **false.**

17. White ppt. of Hg_2Cl_2 is blackened by NH_3 due to formation of $[Hg(NH_2)Cl + Hg]$

$$Hg_2Cl_2 + 2NH_4OH \longrightarrow \underbrace{Hg(NH_2)Cl + Hg}_{\text{black}} + NH_4Cl + 2H_2O$$

and not by iodide of Millon's base, hence **false.**

18. $K_4[Fe(CN)_6]$ gives different coloured ppt. with $Cu^{2+}, Fe^{3+}, Zn^{2+}, Cd^{2+}$

$$2Cu^{2+} + K_4[Fe(CN)_6] \longrightarrow \underset{\text{chocolate coloured}}{Cu_2[Fe(CN)_6]\downarrow} + 4K^+$$

$$Fe^{3+} + K_4[Fe(CN)_6] \longrightarrow \underset{\text{blue}}{KFe^{III}[Fe^{II}(CN)_6]} + 3K^+$$

hence **true.**

19. Yellow ppt. of $K_3[Co^{III}(NO_2)_6]$ and not of $K_4[Co(NO_2)_6]$ are obtained

$$Co^{2+} + 7NO_2^- + 2H^+ + 3K^+ \longrightarrow K_3[Co(NO_2)_6]\downarrow + NO\uparrow + H_2O$$

hence **false.**

20.
$$NH_4Cl + NaOH \longrightarrow NH_3 + NaCl + H_2O$$
$$3NaOH + NH_3 + 2K_2HgI_4 \longrightarrow 4KI + 3NaI + 2H_2O +$$

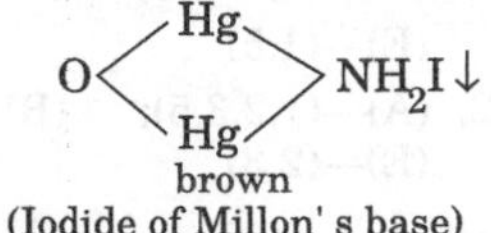

brown
(Iodide of Millon's base)

Fill in the Blanks

1. (a) common ion effect (b) K_{sp} values

2. (a) III (b) decrease (c) OH^- **3.** (a) dil. HCl (b) H_2S

4. (a) complex formation (b) KCN (c) $K_3[Cu(CN)_4]$ (d) $K_4[Cd(CN)_4]$ (e) CdS

5. (a) hot water (b) aq. NH_3 (c) blackened

6. (a) YAS or NaOH **7.** (a) SbOCl (b) Sb^{3+}

8. (a) NaOH (b) $Al(OH)_3$ (c) $NaAlO_2$ (d) $Fe(OH)_3$

9. (a) H_2O_2 (b) Na_2CrO_4 (c) yellow (d) $PbCrO_4$ (e) $(CH_3COO)_2Pb$

10. (a) dimethyl glyoxime

11. (a) Fe^{2+} (or $C_2O_4^{2-}$, SO_2 etc.)

12. (a) Fe^{2+} (b) $C_2O_4^{2-}$ **13.** (a) red

14. (a) Turnbull's blue (b) $K_3[Fe(CN)_6]$ (c) Prussian blue (d) $K_4[Fe(CN)_6]$

15. (a) $K_4[Fe(CN)_6]$

16. (a) CH_3COONH_4 (b) $(CH_3COO)_2Pb$

17. (a) $[Al^{3+}][OH^-]^3$ (b) greater

18. (a) KI (b) K_2HgI_4 (c) Nessler's reagent

19. (a) Fe^{3+} (b) $[Fe(SCN)]^{2+}$ **20.** (a) SO_2

21. (a) H_2S **22.** (a) $Na_3[Ag(S_2O_3)_2]$

23. (a) 'Ring' test (b) $[Fe(H_2O)_5NO]^{2+}$

24. (a) 'Chromyl chloride' test

25. (a) oxidise Fe^{2+} to Fe^{3+}

26. (a) NH_4SCN **27.** (a) $(C_2H_5O)_3B$

28. (a) $Ca(HSO_3)_2$, soluble

29. (a) Iodide of Millon's base $O\langle{}^{Hg}_{Hg}\rangle NH_2I$

30. (a) Oxidising agent (like SO_3^{2-}, NO_3^-)

31. (a) $K_3[Co(NO_2)_6]$ **32.** (a) K_2CrO_4 (b) $AgNO_3$

33. (a) 3 : 1 **34.** (a) $[Hg_2^{\ 2+}][Cl^-]^2$

35. (a) *aq.* NH_3 (b) cationic complex $[Ag(NH_3)_2^+]$

36. (a) Cu_2I_2 (b) KI (c) $[Cu(NH_3)_4]^{2+}$ (d) *aq.* NH_3 (excess)

37. (a) AsO_4^{3-} (b) PO_4^{3-} **38.** (a) $(NH_4)_3AsS_4$

39. (a) $[Fe_3(OH)_2(CH_3COO)_6]^+$

40. (a) white $(Ag_2S_2O_3)$ (b) black (Ag_2S)

Matrix-Match Type Questions

Only One Correct

1. (A)—(4); (B)—(3); (C)—(1); (D)—(5); (E)—(2)

One or More Correct

1. (A)—(1,2,3,4); (B)—(2,3); (C)—(2); (D)—(3,4) (E)—(1,5)
2. (A)—(1,2,3,5); (B)—(4); (C)—(5); (D)—(3,5); (E)—(2,3)
3.

List I	List II	List III
A	3	S
B	4	P
C	2	T
D	1	Q
E	5	R

4. (A)—(4); (B)—(4); (C)—(1); (D)—(2,3) (E)—(2,3,5)

Integer Answer Types

Set I

Questions →	1	2	3	4	5	6	7	8
Answers →	3	3	3	6	5	2	2	7

Set II

Questions →	1	2	3	4	5	6	7	8	9	10
Answers →	5	1	3	4	6	8	9	0	7	2

During a total eclipse of the sun on August 18, 1868, spectroscopic examination of the sun's chromosphere led to the discovery of helium, name derived from the Greek word 'helios' meaning sun."

20

Group 18 (VIIA—The Noble Gases)

Learning Framework of the Chapter ...

- Discovery
- Isolation of the Elements
- Position of Noble Gases in the Periodic Table
- Periodic Properties
- Study of Compounds of Group 18
- Structure of Xenon Compounds

$_{2}He$
$_{10}Ne$
$_{18}Ar$
$_{36}Kr$
$_{54}Xe$
$_{86}Rn$

Discovery

Discovery of the noble gases in the Periodic Table were based on the study of physical properties as spectral behaviour and density. Chemists associated with the discovery of inert gases were

1. Henry Cavendisch	1785	Mixture of noble gases (but could not be identified due to lack of facility of study)
2. Lockyer and Frankland	1868	Helium (Greek word helios — Sun) based on study of sodium D-lines in spectrum in solar eclipse
3. L. Palmieri	1881	As above
4. Ramsay	1895	Ar (meaning lazy)
5. Ramsay	1896	Assigned position to He and Ar (in Periodic Table) (zero group being inert). He also predicted that there should be other three inert gases to complete the table
6. Ramsay and Travers	1898	Isolated three new elements by low-temperature distillation of liquid air : neon (Greek-new) krypton (Greek-hidden) xenon (Greek-strange)
7. Rutherford an Soddy	1902	Isolated radon (Latin word radius—ray)

In 1904 Nobel Prize in physics was awarded to Lord Rayleigh for investigation on gas densities and for discovery of argon, Nobel Prize in Chemistry was awarded to W. Ramsay for discovery of the inert gas elements in air and their place in the periodic system.

Isolation of the Elements

- Main source of helium is natural gas. The hydrocarbons are liquefied at about 120 K and helium is pumped out.
- Liquid air is subjected to fractional distillation to isolate Ne, Ar, Kr and Xe. Separation depends on the boiling points.

Gas	Boiling points/K
He	4
Ne	27
Ar	87
Kr	120
Xe	166
N_2	77
O_2	90

- Rn (Radon) is produced by the α-decay of an aqueous solution of radium chloride ($RaCl_2$)

$$^{226}_{88}\text{Ra} \longrightarrow {}^{222}_{86}\text{Rn} + {}^{4}_{2}\text{He}\ (\alpha)$$

The gas is mainly diluted by ozone, hydrogen and oxygen (obtained by the α-irradiation of water) The mixture of gases is passed successively over heated copper and copper (II) oxide [Cu + CuO] and finally dried with phosphorus (V) pentoxide [P_2O_5] to get pure radon.

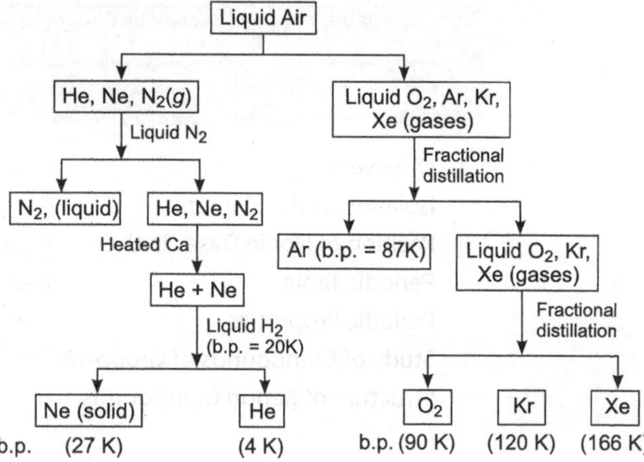

Position of Noble Gases in the Periodic Table

Group 17	Group 18	Group 1
Most electronegative	Inert	Most electropositive
Halogen	He_2	Li_3
F_9	Ne_{10}	Na_{11}
Cl_{17}	Ar_{18}	K_{19}
Br_{35}	Kr_{36}	Rb_{37}
I_{53}	Xe_{54}	Cs_{55}
At_{85}	Rn_{86}	Fr_{87}
ns^2np^5	ns^2np^6 (except He)	[Inert gas] ns^1

In Mendeleef's original Periodic Table there was a sharp transition from strongly electronegative elements of Group 17 (VIIA — halogen family) to the strongly electropositive elements of Group 1 (IA — alkali metals). It was suggested by Thomson that there should be a group between Group 17 and Group 1 which can make a smooth transition. When Ar and He were discovered they could be placed in this group easily. This was called zero group as the elements were found to be chemically inert and hence they may be assumed to have zero valency. Subsequently when neon krypton, xenon and radon were discovered, they fitted excellently with helium and argon in the same group.

They have electronic configuration $\boldsymbol{ns^2np^6}$ of the valence shell (except helium — $1s^2$) with complete octet in which the electrons are all paired. This accounts for the remarable stability of these gases and lack of chemical activity compared to other elements of the Periodic Table. According to the latest IUPAC convention, number 18 has been assigned to this group.

Molecular weight (by vapour density method) is same as atomic weight thus these are monoatomic gases.

Also, $$\frac{C_p}{C_V} = 1.667$$

which confirmed that they are monoatomic gases (C_p is the molar heat capacity at constant pressure and C_V is the molar heat capacity of constant volume).

Periodic Properties

- Because of highest ionisation energy of group 18 elements in their respective periods, they were given the name **'noble gases'** and thus inert. However, at least in the case of xenon, it is not so; it forms series of compounds.
- The noble gases are low boiling gases. Except for radon (Rn), they can be isolated by fractional distillation of liquefied air. Rn is collected from the radioactive disintegration :

$$^{226}_{88}\text{Ra} \longrightarrow {}^{222}_{86}\text{Rn} + {}^{4}_{2}\text{He}$$

- All isotopes of radon are radioactive and there is concern in the some areas of the world where poor ventilations in homes result in build-up of randon gas (which comes from the rocks upon which homes are built.

DEEP Focus

- The noble gases are colourless, tasteless and odourless. In the liquid and solid states the only forces of attraction among the atoms are **very weak London or van der Waals' forces. Polarisability and** interatomic interactions increase with increasing atomic number. The attractive forces among the atoms are so small that the remain liquid at 1 atmospheric pressure even at a temperature of 0.001 K.

- **Physical properties of Group 18 (the noble gases) have been summarised in Table 20.1 and Fig. 20.1**
- **Helium was first detected as a new element in the spectrum of the sun is 1868 (hence, its name is connected with helios, the Greek word for sun). Radon was discovered as the product of the radioactive decay of uranium and thorium.**

DEEP Focus

- **Helium has the lowest boiling point of any element, and its behaviour as a liquid is most unusual. It has no triple point that is there is no single combination of temperature and pressure at which the solid, the liquid and the gas all exist. At about 2K, helium undergoes a transition to a form of helium, labelled He_{II}. It will creep up the walls of the container in which it is placed and thus it exists in two phases in liquid helium, one of which has an extremely low viscosity. Uses have been summarised in Table 20.3**

Table 20.1 **Physical Properties of Group 18 Element**

Property	He	Ne	Ar	Kr	Xe	Rn
Atomic number	2	10	18	36	54	86
Electronic configuration	$1s^2$	[He] $2s^2 2p^6$	[Ne] $3s^2 3p^6$	[Ar] $3d^{10} 4s^2 4p^6$	[Kr] $4d^{10} 5s^2 5p^6$	[Xe] $4f^{14} 5d^{10} 6s^2 6p^6$
(IE)/kJ mol^{-1}	2372	2080	1521	1351	1170	1040
Melting point/°C	—	– 248.6	– 189.4	– 157.2	– 111.8	– 71
Boiling point/°C	– 269	– 246	– 186	– 153.6	– 108.1	– 62
Atomic radius/pm	120	160	191	200	220	—
Density at STP/ g/cm	1.8×10^{-4}	9.0×10^{-4}	1.8×10^{-3}	3.7×10^{-3}	5.9×10^{-3}	9.7×10^{-3}
Atmospheric content/% by volume	5.24×10^{-4}	1.82×10^{-3}	0.934	1.14×10^{-3}	8.7×10^{-6}	—
Principal oxidation no.	0	0	0	0	+2, +4, +6	0
Enthalpy of vaporisation/kJ mol^{-1}	0.08	1.7	6.5	9.1	12.7	18.1
Solubility in water at 20°C/cm^3 kg^{-1}	8.61	10.5	33.6	59.4	108.1	230
Discovery	1895 Sir William Ramsay (SWR)	1898 SWR & M.W. Travers (MWT)	1894 SWR & Lord Rayleigh	1898 SWR & MWT	1898 SWR & MWT	1902 Friedrich Dorn
Abundance in igneous rocks/ppm by weight	3×10^{-3}	7×10^{-5}	4×10^{-2}	—	—	1.7×10^{-10}

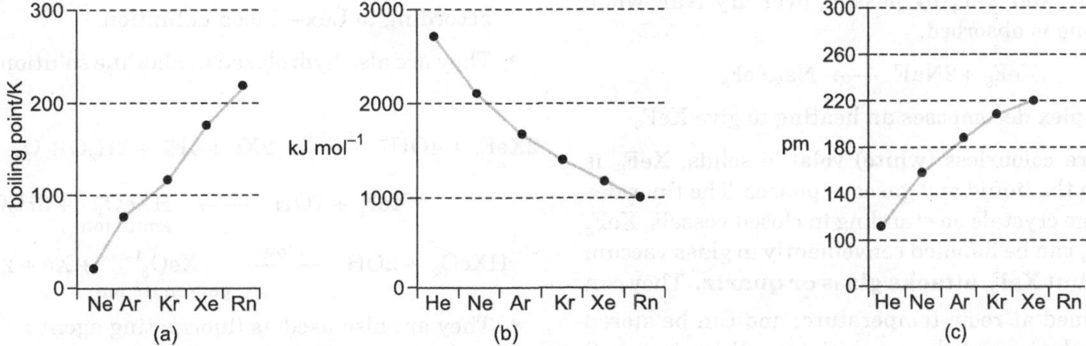

Fig. 20.1 Some group trends of the group 18 elements (a) Boiling point, (b) The first ionisation energy and (c) Atomic radius.

Study of Compounds of Group 18

- The noble gases were isolated and discovered because of their lack of reactivity. For a long time it was thought that they really were chemically inert. Before 1962, the only evidence for compound formation by the noble gases was some **molecular ions** formed in discharge tube and **clathrate** compounds.
- Several molecular ions such as He_2^+, HeH^+, HeH^{2+} and Ar^{2+} are formed under high energy conditions in discharge tubes.
- If an aqueous solution of quinol is crystallised under a pressure of 10–40 atm of Ar, Kr, or Xe, the gas becomes trapped in the cavities of about 4Å diameter in the β-quinol structure. These are called **clathrate compounds.** They do not form bonds in the cavities. When the clathrate is dissolved, the hydrogen bonded arrangement of β-quinol breaks down and the noble gas escapes. The smaller noble gases He and Ne do not form clathrate compounds because the gas atoms are small enough to escape from the cavities. Clathrate provides a convenient means of storing radioactive isotopes of Kr and Xe produced in nuclear reactors.

Xenon Fluorides

- Xenon reacts directly with fluorine when the gases are heated at 300–600°C in a sealed nickel vessel, and the products formed depend on the Xe/F_2 ratio.

$$Xe + F_2 \begin{cases} 2:1 \text{ mixture} \xrightarrow{400°C,\ 1\ atm} XeF_2 \\ 1:5 \text{ mixture} \xrightarrow{600°C,\ 7\ atm} XeF_4 \\ 1:20 \text{ mixture} \xrightarrow{300°C,\ 60\ atm} XeF_6 \end{cases}$$

- These conversions are usually quantitative under the given conditions. XeF_6 may be completely separated from XeF_2 and XeF_4 by passing over dry NaF when XeF_6 alone is absorbed.

$$XeF_6 + 2NaF \longrightarrow Na_2XeF_8$$

The complex decomposes an heating to give XeF_6

- These are colourless (white) volatile solids. XeF_6 is yellow in the liquid and gaseous phases. The fluorides form large crystals on standing in closed vessels. XeF_2 and XeF_4 can be handled conveniently in glass vaccum **system but XeF_6 attacks glass or quartz.** They can be sublimed at room temperature, and can be stored indefinitely in nickel or monel (an alloy of nickel) containers.
- H_2 reduces xenon fluorides to Xe :

$$XeF_2 + H_2 \longrightarrow Xe + 2HF$$

$$XeF_4 + 2H_2 \longrightarrow Xe + 4HF$$

$$XeF_6 + 3H_2 \longrightarrow Xe + 6HF$$

(at room temperature)

- Xenon fluorides oxidise Cl^- to Cl_2 and I^- to I_2 :
  Oxidising power increase in the order :

$$XeF_2 < XeF_4 < XeF_6$$

$$XeF_2 + 2HCl \longrightarrow 2HF + Xe + Cl_2$$

$$XeF_4 + 4KI \longrightarrow 4KF + Xe + 2I_2$$

This reaction is used to estimate XeF_4. I_2 formed is titrated using Hypo.

- They also oxidise Ce^{3+} salts to Ce^{4+} salts.
- XeF_2 is soluble in water but undergoes slow hydrolysis :

$$2XeF_2 + 2H_2O \longrightarrow 2Xe + 4HF + O_2$$

- XeF_4 and XeF_6 react violently with water giving XeO_3 :

$$3XeF_4 + 6H_2O \longrightarrow 2Xe + XeO_3 + 12HF + \frac{3}{2}O_2$$

$$XeF_6 + \underset{\text{(atmospheric moisture)}}{3H_2O} \longrightarrow XeO_3 + 6HF$$

This takes place in three stages :

$$XeF_6 + H_2O \longrightarrow XeOF_4 + 2HF$$

$$XeOF_4 + H_2O \longrightarrow XeO_2F_2 + 2HF$$

$$XeO_2F_2 + H_2O \longrightarrow XeO_3 + 2HF$$

- SiO_2 also converts XeF_6 into $XeOF_4$:

$$2XeF_6 + SiO_2 \longrightarrow SiF_4 + 2XeOF_4$$

$$2XeOF_4 + SiO_2 \longrightarrow 2XeO_2F_2 + SiF_4$$

$$2XeO_2F_2 + SiO_2 \longrightarrow 2XeO_3 + SiF_4$$

SiO_2 is part of glass or quartz. Hence, XeF_6 cannot be stored in glass or quartz vessel.

Here XeF_6 behaves as an acid (oxide ion acceptor) according to Lux—Flood definition.

- They are also hydrolysed in alkaline solution rapidly :

$$2XeF_2 + 4OH^- \xrightarrow{\text{rapid}} 2Xe + 4F^- + 2H_2O + O_2$$

$$XeF_6 + 7OH^- \longrightarrow \underset{\text{xenate ion}}{HXeO_4^-} + 3H_2O + 6F^-$$

$$HXeO_4^- + 2OH^- \xrightarrow{\text{slow}} \underset{\text{perxenate ion}}{XeO_6^{4-}} + Xe + 2H_2O + O_2$$

- They are also used as fluorinating agent :

$$2SF_4 + XeF_4 \longrightarrow 2SF_6 + Xe$$

$$Pt + XeF_4 \longrightarrow PtF_4 + Xe$$

$$C_6H_6 + XeF_2 \longrightarrow C_6H_5F + Xe + HF$$

- Xenon fluorides act as fluoride donor and forms complexes which are conducting due to formation of ions:

$$XeF_2 + MF_5 \longrightarrow [XeF]^+[MF_6]^-$$
$$(M = As^{3+}, Sb^{5+}, P^{5+}, \text{etc.})$$

$$XeF_6 + BF_3 \longrightarrow [XeF_5]^+[BF_4]^-$$

$$XeF_6 + HF \longrightarrow [XeF_5]^+[HF_2]^-$$

XeF_6 may also act as a fluoride acceptor:

$$XeF_6 + RbF \longrightarrow Rb^+ [XeF_7]^-$$

$$XeF_6 + CsF \longrightarrow Cs^+[XeF_7]^-$$

$$2Cs^+[XeF_7] \xrightarrow{\Delta} XeF_6 + Cs_2[XeF_8]$$

Structure of xenon fluorides are summarised in Table 20.3.

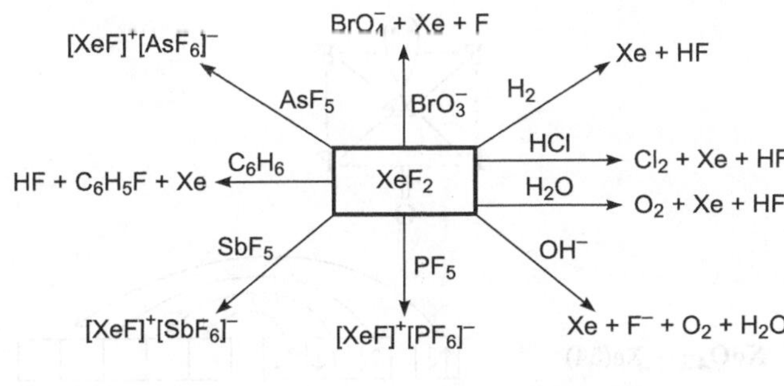

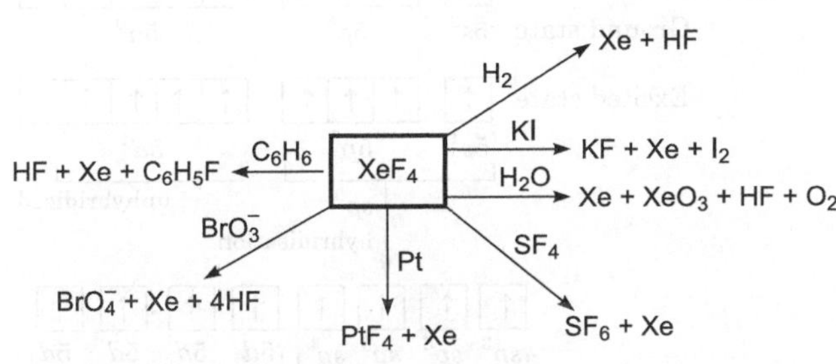

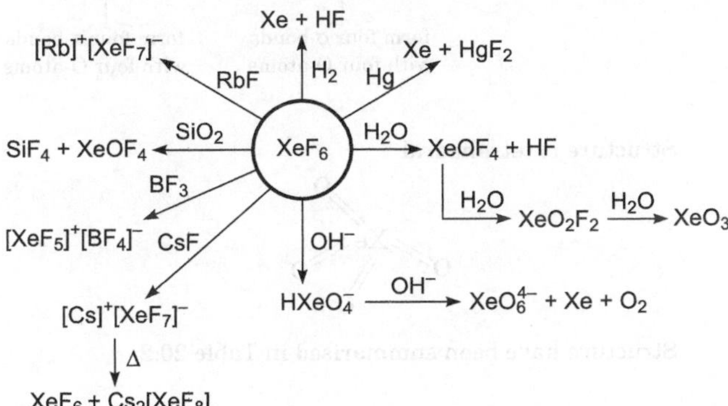

Structure of Xenon Compounds

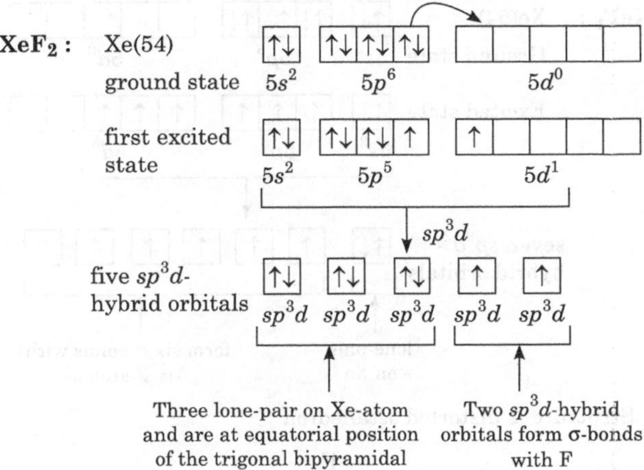

Thus, structure is linear

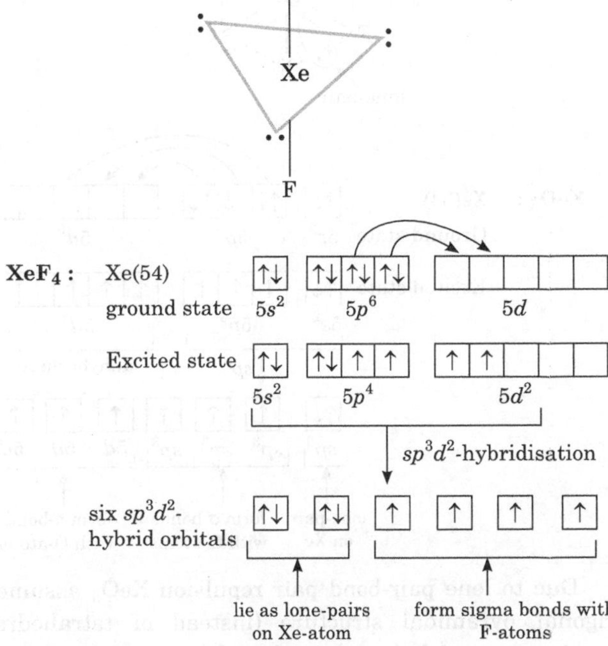

Thus, XeF_4 assumes square planar structure.

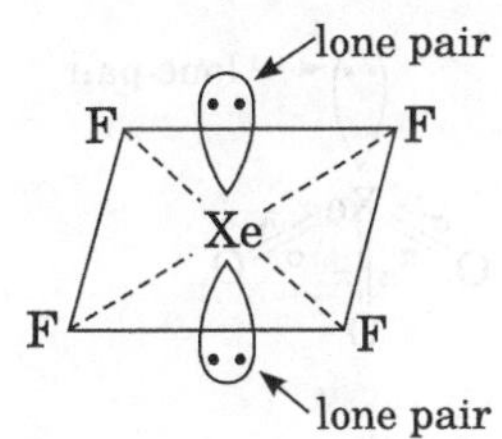

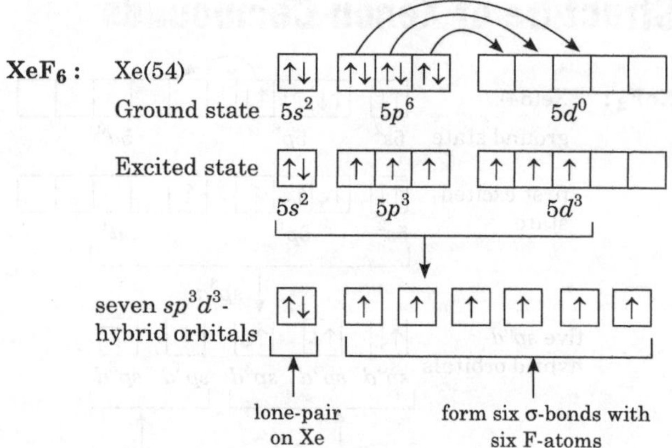

Structure is distorted octahedron

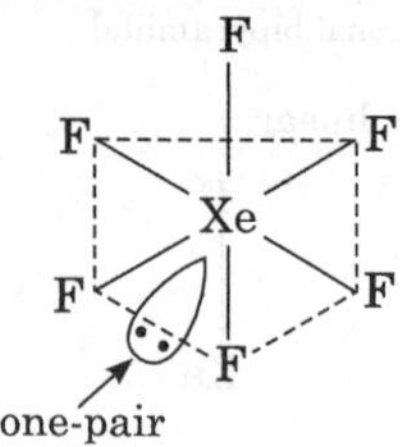

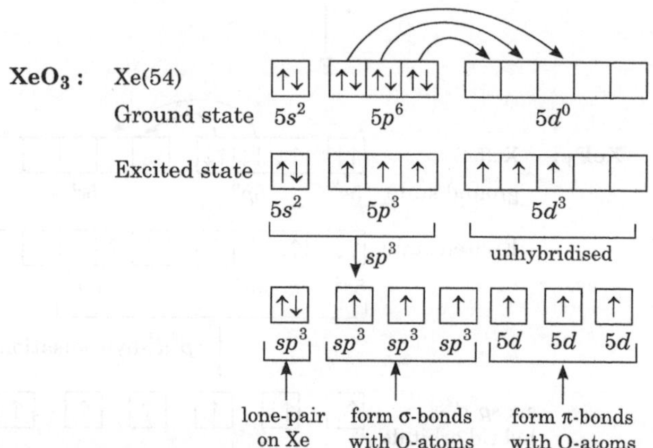

Due to lone pair-bond pair repulsion XeO_3 assumes trigonal pyramidal structure (instead of tetrahedral structure — sp^3 hybridisation) and bond angle is 103° (VSEPR Theory)

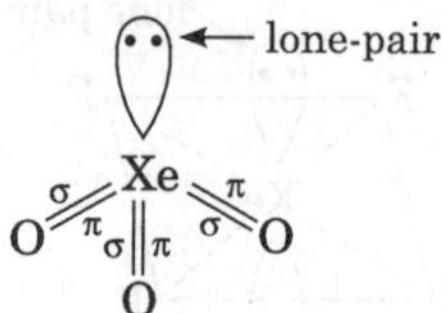

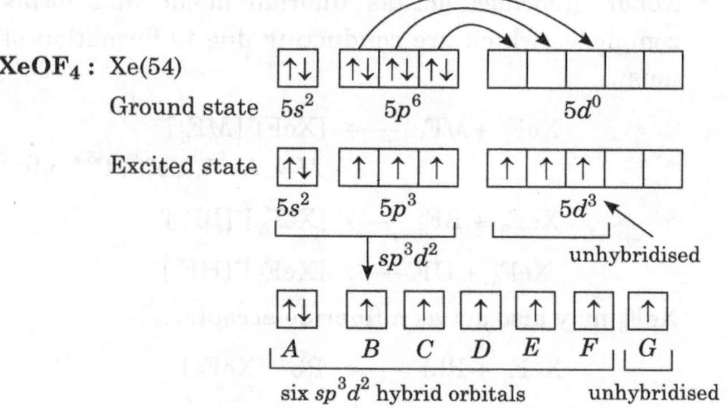

A — lone-pair on Xe

B to *E* — form four σ-bonds with four F-atoms

F — forms σ-bond with O-atom

G — forms π-bond with O-atom

Structure is square pyramidal

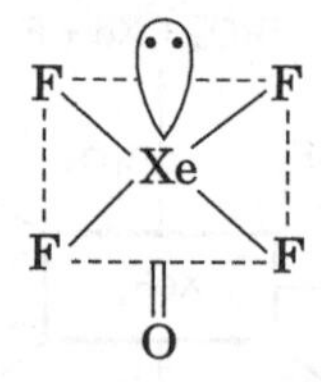

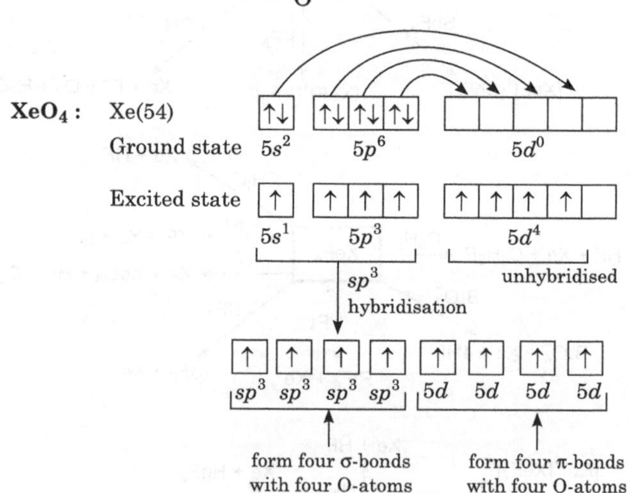

Structure is tetrahedral

$$O=Xe(=O)(=O)=O$$

Structure have been summarised in Table 20.2

Table 20.2 **Structures of Some Xenon Compounds**

Formula	(M.p.)/°C	Oxidation state	Number of bonding pairs	Number of lone pairs	Hybridisation of Xe	Structure	Form
XeF_2	129	+2	2	3	sp^3d	linear with three lone pairs in equatorial position.	colourless crystal
$XeF_2 \cdot 2SbF_3$	63	—	—	—	—	—	yellow solid crystals
XeF_4	117	+4	4	2	sp^3d^2	square planar with two positions occupied by lone pairs.	colourless crystals
$XeOF_2$	31	+4	4	2	sp^3d	*T*-shaped	colourless crystals
XeF_6	49.6	+6	6	1	sp^3d^3	distorted octahedron.	colourless crystals
XeO_3	—	+6	6	1	sp^3	trigonal pyramidal with three π-bonds and one lone pair.	colourless liquid
XeO_2F_2	—	+6	6	1	sp^3d^2	trigonal pyramidal with two π-bonds and one lone pair.	colourless liquid
$XeOF_4$	46	+6	6	1	sp^3d^2	square pyramidal with one π-bond and one lone pair.	colourless crystals
XeO_4	—	+8	4	0	sp^3	tetrahedral	colourless gas
XeO_6^{4-}	—	+8	4	0	sp^3d^2	octahedral	ions

Table 20.3 **Uses of the Noble Gases**

Noble gas	Use	Reasons
Helium	1. Filling of observation balloons and other lighter-than-air craft	Non-flammable; 93% lifting power as compared to flammable H_2
	2. He/O_2 mixture for deep-sea breathing, instead of N_2/O_2 mixture	Low solubility in blood; prevents nitrogen narcosis and "bends"
	3. Diluent for gaseous anaesthetics	Nonflammable, non-reactive
	4. He/O_2 mixtures for respiratory patients	Low density flows easily through restricted passages
	5. Heat transfer medium for nuclear reactions	Transfers heat readily; does not become radioactive; chemically inert
	6. Liquid He to maintain very low temperature in research (cryogenics)	Extremely low boiling point
Neon	Neon signs	Even at low Ne pressure, moderate electric current causes bright orange-red glow; can be modified by coloured glass or mixing with Ar or Hg vapour.
Argon	1. Inert atmosphere for welding	Chemically inert
	2. Filling incandescent light bulbs	Inert; prevents vaporisation of tungsten and blackening of bulbs
Krypton	Airport runway and approach lights	Gives longer life to incandescent lights than Ar, but more expensive
Xenon	Xe and Kr mixture in high-intensity, short-exposure photographic flash tubes	Both have fast response to electric current
Radon	Radiotherapy of cancerous-tissues	Radioactive

Target Practice 1

1. Arange the following in increasing s-character of the hybridised Xe atom.

$XeO_3, XeF_2, XeF_4, XeF_6$

....

....

2. (a) XeF_2 exists as .... in AsF_5.
(b) XeF_6 exists as .... in HF.
(c) XeF_6 exists as .... in RbF.

3. XeF_6 is hydrolysed to XeO_3 in following steps
(a) $XeF_6 + H_2O \longrightarrow$ ....(I)
(b) I + $H_2O \longrightarrow$ .... (II)
(c) II + $H_2O \longrightarrow XeO_3$

Answers

1.

			s-character	Increasing s-character
XeO_3	:	sp^3	25%	↑
XeF_2	:	sp^3d	20%	
XeF_4	:	sp^3d^2	16.7%	
XeF_6	:	sp^3d^3	14.3%	

2. (a) $[XeF]^+[AsF_6]^-$
(b) $[XeF_5]^+[HF_2]^-$
(c) $(Rb)^+[XeF_7]^-$

3. (a) I : $XeOF_4$
(b) II : XeO_2F_2
(c) III : XeO_3

Practice Exercise 1

1. He_2 does not exist, explain.
2. What is bond order of :
(a) He_2 (b) He_2^+
(c) He_2^{2+}
3. The bond energies of Cl_2 and F_2 are 243 and 155 kJ mol^{-1} respectively. Use these data to explain why XeF_2 is much more stable compound than $XeCl_2$.
4. Fill in the blanks :
(i) The most abundant noble gas in the atmosphere is ....
(ii) XeF_2 and .... can be kept in a dry pyrex vessels while .... attacks these.
(iii) XeO_3 has .... hybridised Xe with .... π-bonds and .... lone-pair (s).
5. Discuss the structure of XeF_2, XeF_4 and XeF_6.
6. Give the formula of the noble gas species that is isostructural with
(a) ICl_4^- (b) IBr_2^-
(c) BrO_3^-

Answers

1. He_2 has MO electronic configuration as

$$\underbrace{\sigma 1s^2}_{\substack{\uparrow \\ \text{Bonding}}} \quad \underbrace{\sigma^* 1s^2}_{\substack{\uparrow \\ \text{Anti-bonding MO}}}$$

Thus, electron in bonding MO = 2 = N_B

and in anti-bonding MO = 2 = N_A

$$\text{Bond-order} = \frac{N_B - N_A}{2} = 0$$

Since, bond-order is zero, hence He_2 does not exist.

2.

Species	MO Electronic configuration	N_B	N_B	Bond-order $= \frac{N_B - N_B}{2}$
(a) He_2 (4)	$\sigma 1s^2 \sigma^* 1s^2$	2	2	0
(b) He_2^+ (3)	$\sigma 1s^2 \sigma^* 1s^1$	2	1	0.5
(c) He_2^{2+} (2)	$\sigma 2s^2$	2	0	1.0

3.

	ΔH
$Xe(g) \longrightarrow Xe^+(g) + e^-$;	I (ionisation energy)
$X_2(g) \longrightarrow 2X(g)$;	D (Bond energy)
$2X(g) + 2e^- \longrightarrow 2X^-(g)$;	$2E$ (electron affinity)
$Xe^+(g) + 2X^-(g) \longrightarrow XeX_2(g)$;	$-U$ (lattice energy)
$Xe(g) + X_2(g) \longrightarrow XeX_2(g)$;	$\Delta H_f = I + D + 2F - U$

By taking values of I, D, E and U, we find that $\Delta H_f^\circ(XeF_2)$ is more negative than that of $XeCl_2$. Thus XeF_2 is much more stable.

4. (i) argon (ii) XeF_4, XeF_6
(iii) sp^3, three, one

5. (a) **XeF_2** : Three-lone-pairs are at three equatorial positions and two unpaired electrons from two σ-bonds with F-atoms in the trigonal bipyramidal.
Thus, it has linear structure.

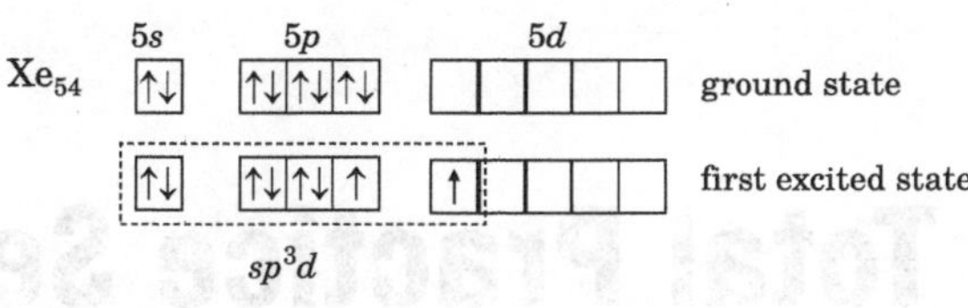

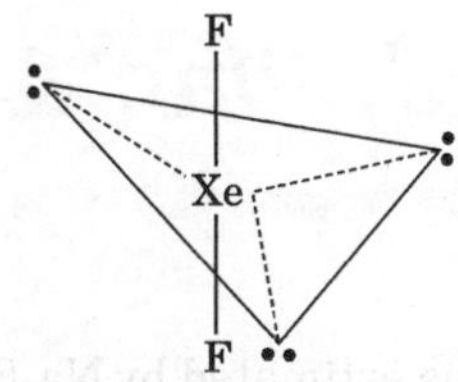

(b) **XeF_4 :** Four unpaired electrons form four σ-bonds with fluorine atoms and two lone-pairs are above and below the plane containing Xe and F atoms. The structure is square planar.

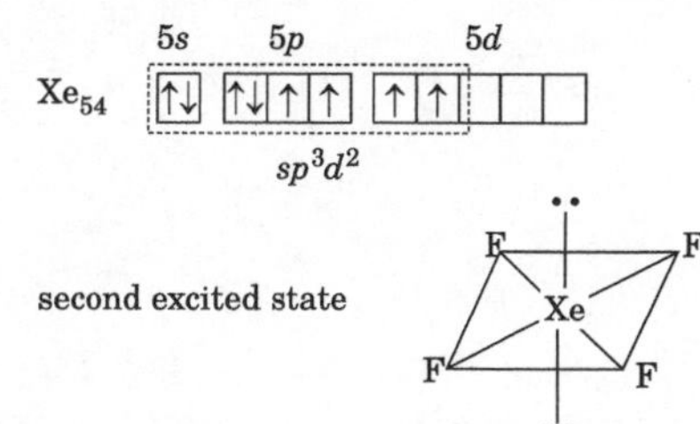

(c) **XeF_6 :** Six unpaired electrons form σ-bonds with F-atoms and one lone-pair at one corner in the distorted octahedron.

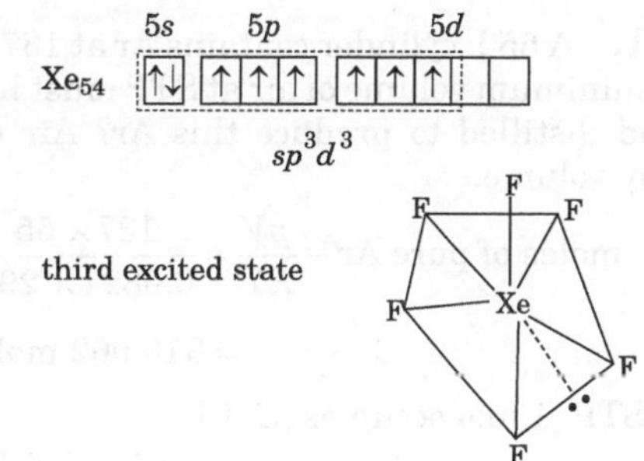

6. (a) XeF_4 (b) XeF_2 (c) XeO_3.

Total Practice Set

(Read, Plan & Solve!)

Problem 1. A 55 L cylinder contains Ar at 137 atm and 23°C. What minimum volume of air at STP must have been liquefied and distilled to produce this Ar? Air contains 0.934% Ar by volume.

Solution moles of pure Ar $= \frac{pV}{RT} = \frac{137 \times 55}{0.0821 \times 296}$

$$= 310.062 \text{ mol}$$

Since, at STP 1 mol occupies 22.4 L

Volume of Ar at STP = 6945.386 L

Volume of air to have this volume of argon

$$= \frac{6945.386}{0.934} \times 100$$

$$= \mathbf{743617.37\ L}$$

Problem 2. Complete the following reactions.

(a) $HeXO_4^- + OH^- \longrightarrow$

(b) $XeF_6 + OH^- \longrightarrow$

(c) $Ba_2XeO_6 + H_2SO_4 \longrightarrow$

(d) $XeF_4 + KI \longrightarrow$

followed by reaction with $Na_2S_2O_3$

Solution (a) $HXeO_4$ (xenates) slowly disproportionates in alkaline solution to form XeO_6^{2-} (perxenate) and Xe

$$2HeXO_4^- \longrightarrow 2OH^- \longrightarrow XeO_6^{4-} + Xe + O_2 + 2H_2O$$

(b) Alkaline hydrolysis of XeF_6 gives perxenates

$$2XeF_6 + 16OH^- \longrightarrow XeO_6^{4-} + Xe + O_2 + 12F^- + 8H_2O$$

(c) $BaSO_4$ is precipitated and XeO_4 (explosive) is formed

$$Ba_2XeO_6 + 2H_2SO_4 \longrightarrow 2BaSO_4 + XeO_4 + 2H_2O$$

(d) KI is oxidised to I_2 (and exists as KI_3)

$$XeF_4 + 4KI \longrightarrow Xe + 4KF + 2I_2$$

$$I_2 + KI \rightleftharpoons \underset{\text{(deep brown)}}{KI_3}$$

Liberated I_2 can be estimated by $Na_2S_2O_3$

$$I_2 + 2Na_2S_2O_3 \longrightarrow 2NaI + Na_2S_4O_6$$

Problem 3. Bartlett observed that PtF_6 changes colour on exposure to air. What conclusion is drawn regarding formation of Xe compounds?

Solution Bartlett observed that PtF_6 changes colour on exposure to air. Investigation concluded that PtF_6 is a very strong oxidising agent that can oxidise O_2 to O_2^+ and thus orange colour compound $O_2^+PtF_6^-$ is formed.

$$O_2 \longrightarrow O_2^+ + e^-$$

$$PtF_6 + e^- \longrightarrow PtF_6^-$$

$$O_2 + PtF_6^- \longrightarrow O_2^+PtF_6^-$$

Ionisation energy of Xe = 1169 kJ mol^{-1}

Ionisation energy of O_2 = 1175 kJ mol^{-1}

Since, ionisation energies of O_2 and Xe are very equal, thus oxidation of Xe by PtF_6 would also possible. Bartlett also isolated a compound $Xe^+PtF_6^-$ (orange). Later stage gave so many compound of Xe

Problem 4. Xe^{IV} disproportionates to Xe^0 and Xe^{VI} through Xe^{II} and Xe^{VIII}

$$3Xe^{IV} \longrightarrow 2Xe^{II} + Xe^{VIII}$$

$$\downarrow \qquad \downarrow$$

$$2Xe^0 \quad Xe^{VI}$$

Explain disproportionates of XeF_4 by H_2O.

Solution

$$\underset{(+4)}{3XeF_4} + 6H_2O \longrightarrow \underset{(+2)}{2XeO} + \underset{(+8)}{XeO_4} + 12HF$$

$$\downarrow \qquad \downarrow \qquad \downarrow$$

$$\underset{(0)}{2Xe} + \underset{(+6)}{XeO_3} + \frac{3}{2}O_2$$

Explanation

(a) XeF_6 *cannot be stored in a vessel made of glass or quartz.*

– XeF_6 produces an explosive XeO_3 on reaction with silica present in glass or quartz

$$2XeF_6 + SiO_2 \longrightarrow 2XeOF_4 + SiF_4$$

$$2XeOF_4 + SiO_2 \longrightarrow 2XeO_2F_2 + SiF_4$$

$$2XeO_2F_2 + SiO_2 \longrightarrow 2XeO_3 + SiF_4$$

Thus, XeF_6 cannot be handled in a vessel made of quartz or glass.

(b) XeF_6 *should be stored in a vessel free of moisture*

– XeF_6 is hydrolysed to XeO_3 (end product) which is very difficult to handle due to its explosive nature.

$$XeF_6 + H_2O \longrightarrow XeOF_4 + 2HF$$

$$XeOF_2 + H_2O \longrightarrow XeO_2F_2 + 2HF$$

$$XeO_2F_2 + H_2O \longrightarrow XeO_3 + 2HF$$

Hence, XeF_6 should be stored in moisture-free vessel.

(c) *Oxidising power of fluorides of* Xe *is in order*

$$XeF_2 < XeF_4 < XeF_6$$

– Greater the oxidation number of the central element, greater the tendency to be reduced to minimum state hence greater the oxidising power.

Fluorides	Oxidation number of Xe	Change in oxidation number if reduced to Xe(O)	Oxidising power
XeF_2	+2	2 units	↓
XeF_4	+4	4 units	Maximum
XeF_6	+6	6 units	

(d) XeO_3 *is* sp^3*-hybridised in* Xe *but it has pyramidal structure.*

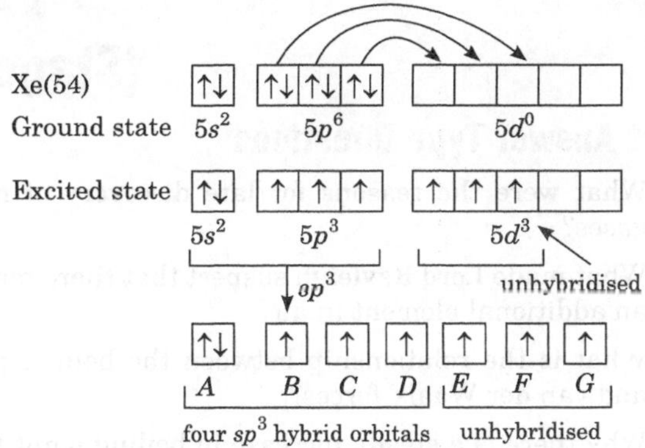

A — lone-pair on Xe
B, C, D — form three σ-bonds with O
E, F, G — form three π-bonds with O

Due to lone-pair there is lone-pair—bond-pair repulsion. Thus, geometry of the molecule is distorted to pyramidal instead of tetrahedral.

Master Exercises

Exercise 1
(Stage 1 : Learning)

Short Answer Type Questions

1. What were the reasons for late discovery of noble gases?
2. What made Lord Rayleigh suspect that there may be an additional element in air?
3. What is the relationship between the boiling point and van der Waals' forces?
4. Why there is a steady increase in boiling point from He to Rn?
5. Why no compounds of He and Ne are known?
6. Why noble gas compounds are formed only with O_2 and F_2 ?
7. Why does the tendency to form clathrates increase down the group?
8. How did the terrestrial helium originate?

Analytical Questions

1. What are the factors which led Bartlett to think that Xe can be oxidised by PtF_6?
2. Would you expect nitrogen to form compound of the type $N_2^+[PtF_6]$? Give reason for your answer (Ionisation potential of N_2 is 1403 kJ mol^{-1})
3. Would you expect radon to form more stable compounds than xenon.

Exercise 2
(Stage 2 : High Skill Problem Solving)

Only One Option Correct

1. Consider following properties of the noble gases:
   I : They readily from compounds which are colourless
   II : They generally do not form ionic compounds
   III : They have variable oxidation states in their compounds
   IV : Generally do not form covalent compounds.
   Select correct properties
   (a) I, II, III (b) II, III
   (c) I, III (d) I
2. A rare gas that was detected in the sun before its discovery on earth is
   (a) He (b) Ne
   (c) Ar (d) Kr
3. Which of the following is not known?
   (a) XeF_6 (b) XeF_4
   (c) KrF_6 (d) Cs_2XeF_6
4. End-product of the hydrolysis of XeF_6 is
   (a) XeF_4O (b) XeF_2O_2
   (c) XeO_3 (d) XeO_3^-
5. When water is allowed to freeze in the presence of Ar, Kr or Xe under pressure then
   (a) atoms of noble gas (G) trapped in the crystal lattice of ice giving clathrates corresponding to the composition $8G \cdot 26H_2O$
   (b) these are called noble gas hydrates
   (c) both (a) and (b) are correct
   (d) none of the above is correct
6. Select correct statements.
   (a) Helium has the lowest melting point and boiling point
   (b) Helium can diffuse through rubber, PVC and even glass
   (c) Ar, Kr and Xe form clathrate
   (d) All the above are correct statements

7. Consider following transformations

I. $XeF_6 + NaF \longrightarrow Na^+[XeF_7]^-$

II. $2PCl_5(s) \longrightarrow [PCl_4]^+[PCl_6]^-$

III. $[Al(H_2O)_6]^{3+} + H_2O \longrightarrow [Al(H_2O)_5OH]^{2+} + H_3O^+$

Possible transformations are

(a) I, II, III (b) I, III
(c) I, II (d) II, III

8. Match List I with List II and select the correct answer using the codes given below the list

List I	**List II**
A. XeF_4	1. Distorted octahedral
B. XeF_6	2. Tetrahedral
C. XeO_3	3. Square planar
D. XeO_4	4. Pyramidal

Codes

	A	B	C	D		A	B	C	D
(a)	1	2	3	4	(b)	3	1	4	2
(c)	1	3	2	4	(d)	2	4	1	3

9. Molecule(s) possessing two-centre-two electron bonds and three-centre-four electron bonds would include

(a) B_2H_6 and SiF_4 (b) XeF_6 and BF_6
(c) B_2H_6 and XeF_6 (d) B_2H_6 alone

10. Neon lamps are used in botanical gardens and in green houses as they

(a) provides oxygen
(b) provide better light
(c) stimulate growth of plants
(d) None of the above

11. Which of the following gas mixture is used by the divers inside the sea?

(a) $O_2 + He$ (b) $O_2 + Xe$
(c) $O_2 + Ar$ (d) $O_2 + N_2$

12. The gas not absorbed by coconut charcoal is

(a) He (b) Ne
(c) Ar (d) Kr

13. Which one of the following is square planar?

(a) XeF_4 (b) XeO_4
(c) XeO_3F (d) XeO_3F_2

14. For advertisement the coloured discharged tubes contain

(a) He (b) Ne
(c) Ar (d) Kr

15. The inert gas abundantly found in atmosphere is

(a) Kr (b) He
(c) Ar (d) Xe

16. The $[XeF]^+$ ions are attached to anion $[AsF_6]^-$

(a) with a linear F-bridge
(b) with an angular F-bridge
(c) with H-bond
(d) with tetrahdral F-bridge

17. Hydrates of helium and neon have not been prepared because of

(a) low polarisability (b) small size
(c) low boiling point (d) all of the above

18. Thermal stability of hydrates of group 18 elements

(a) increases down the group
(b) decreases down the group
(c) remains unchanged
(d) first increases up to Ar and then decreases

19. Which have distorted geometry based on VSEPR model?

(a) XeF_2, XeF_4, XeO_3 (b) $XeOF_2, XeO_3, XeF_2$
(c) NH_3, SO_2, H_2O (d) All of these

20. Second most abundant element in the universe (after hydrogen) is

(a) Al (b) Fe
(c) He (d) Ne

21. Helium accumulates in the earth's crust through

(a) α-decay of radioactive element
(b) thermonuclear reactions
(c) β-decay of radioactive element
(d) γ-decay of radioactive element

22. Most abundant noble gas in air is

(a) He (b) Ne
(c) Rn (d) Ar

23. Inert atmosphere for welding of metals is obtained by

(a) Ar, He (b) N_2, He
(c) O_2, N_2 (d) Ar, N_2

24. Xenon fluorides acts as fluoride donor with

(a) AsF_5 (b) BF_3
(c) HF (d) all of these

25. Xenon fluoride act as fluoride acceptor with

(a) RbF (b) CsF
(c) both (a) and (b) (d) None of these

26. XeF_6 is conducting in presence of

(a) BF_3 (b) RbF
(c) HF (d) all of these

27. Which is most basic fluoride?

(a) XeF_6
(b) XeF_2
(c) XeF_4
(d) None is basic, all are acids

28. Basic character of fluorides increases in the order

(a) $XeF_6 < XeF_4 < XeF_2$ (b) $XeF_2 < XeF_4 < XeF_6$
(c) $XeF_2 = XeF_4 < XeF_6$ (d) $XeF_2 > XeF_4 < XeF_6$

29. Which is the best fluorinating agent?

(a) XeF_2 (b) XeF_4
(c) XeF_6 (d) XeO_2F_2

30. XeF_2 forms ……….. with AsF_5.

(a) XeF^+ cation and AsF_4^+
(b) XeF_2^- anion and AsF_4^+
(c) AsF_4^+ cation and AsF_4^-
(d) XeF^+ cation and AsF_6^- anion

One or More Than One Options Correct

1. Select correct statement(s).
(a) The noble gases are present to the extent of about 1% in the earth's atmosphere, the chief component being argon
(b) Helium is the second most abundant element in the universe after hydrogen
(c) There are two stable isotopes of helium $_2^3He$, $_2^4He$
(d) $_2^4He$ is formed by thermonuclear fusion reaction in stars

$$4\,_1^1H \longrightarrow \,_2^4He + 2\,_1e^0$$

2. For inert gases $\dfrac{C_p}{C_V} = \phi$
(a) 1.33 (b) 1.40
(c) 1.66 (d) can't be predicted

3. Helium is used
(a) as a refrigerant (in liquid form) in low-temperature physics
(b) as a substituent for nitrogen in the breathing gas for deep-sea divers
(c) in radiation therapy
(d) in fluorescent tubes

4. Select correct statement(s) about XeF_4.
(a) They are thermodynamically stable
(b) Xe—F bond energy is very high
(c) They are endothermic compounds
(d) They are exothermic compounds

5. Which practical is/are supposed to be correct?
(a) Preparation of XeF_6 in glass or quartz apparatus
(b) Storing of HF in glass or quartz apparatus
(c) Storing of HNO_3 in Al vessel
(d) Storing of NaOH in Al vessel

6. In the following reaction,

$$XeF_2 + BrO_3^- + H_2O \longrightarrow BrO_4^- + Xe + 2HF$$

(a) XeF_2 is an oxidising agent
(b) XeF_2 is a reducing agent
(c) BrO_3^- is oxidised to BrO_4^-
(d) XeF_2 is reduced to HF

7. Select correct statement(s). Bond order of
(a) He_2 is zero (b) He_2^+ is 0.5
(c) He_2^{2+} is 1.0 (d) Be_2 is zero

8. Which is/are Lewis acid as well as Lewis base (amphoteric)?
(a) XeF_2 (b) XeF_4
(c) XeF_6 (d) XeF_6^{2-}

9. Oxidising agent(s) is/are
(a) XeF_2 (b) XeF_4
(c) XeF_6 (d) $XeOF_4$

10. According to the Lux-Flood definition XeF_6 behaves as
(a) an oxidising agent
(b) an acid (oxide ion acceptor)
(c) a base
(d) a reducing agent

Passage Comprehension Questions

Passage 1

XeO_3 is an oxidising agent and this can be used to estimate reducing agents.

$$I: XeO_3 + 6H^+ + 9I^- \longrightarrow Xe + 3H_2O + 3I_3^-$$
$$II: XeO_3 + 6H^+ + 6Fe^{2+} \longrightarrow Xe + 3H_2O + 6Fe^{3+}$$

Answer the following questions

1. Reaction can be studied by measuring pressure of the reaction mixture with time variation can be of the type

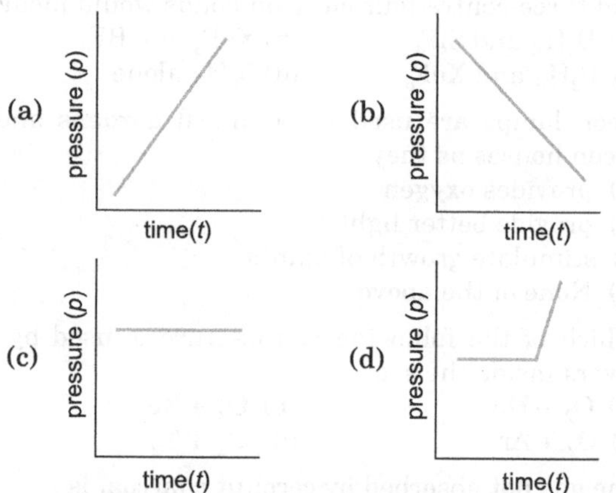

2. Select correct statement(s).
(a) Equivalent of XeO_3 in I and II is in the ratio 1 : 1
(b) Equivalent mass of $I_2 = \dfrac{\text{molar mass}}{9}$
(c) Equivalent mass of $Fe^{2+} = \dfrac{\text{molar mass}}{6}$
(d) All the above are correct statements

Passage 2

XeO_3 is one of the strong oxidising agents in aqueous media :

$$I: XeO_3 + 6H^+ + 6e^- \rightleftharpoons Xe + 3H_2O, \quad E° = 2.1\ V$$
$$II: HXeO_4^- + 3H_2O + 6e^- \rightleftharpoons Xe + 7OH^-, \quad E° = 1.2\ V$$

Answer the following questions

1. Which reaction is more spontaneous theoretically?
(a) I (b) II
(c) Both equally (d) Can't be predicted

2. Select correct statement(s).

(a) Equivalent mass of $XeO_3 = \frac{\text{molar mass}}{6}$ and equivalent mass of $HXeO_4^- = \frac{\text{molar mass}}{6}$

(b) As reaction I or II goes to forward side, pressure increases

(c) Both are correct statements

(d) None of the above is correct statement

3. Thermodynamically $HXeO_4^-$ should oxidise water, but reaction does not take place. It is due to the fact

(a) that there is a high activation energy barrier between oxidation state +6 to 0

(b) that standard emf is positive

(c) that $\Delta G° = +$ ve

(d) that $\Delta G° = -$ ve

Passage 3

XeF_n ($n = 2, 4, 6$) are oxidising agents and thus estimation of Xe salts can be determined by suitable methods.

$$XeF_4 + 4KI \longrightarrow Xe + 2I_2 + 4KF$$

Based on the above short write up, answer the following questions.

1. BrO_3^- is oxidised to BrO_4^- by XeF_2

$$\underset{\text{molar mass}\ m_1}{XeF_2} + \underset{m_2}{BrO_3^-} + H_2O \longrightarrow BrO_4^- + Xe + 2HF$$

Ratio of equivalent mass of XeF_2 and BrO_3^- is

(a) $\frac{m_1}{m_2}$ (b) $\frac{2m_1}{m_2}$

(c) $\frac{m_1}{2m_2}$ (d) $\frac{m_1}{3m_2}$

2. Oxidising power increases in the order

(a) $XeF_2 < XeF_4 < XeF_6$

(b) $XeF_2 < XeF_6 < XeF_4$

(c) $XeF_6 < XeF_4 < XeF_2$

(d) $XeF_6 = XeF_4 = XeF_2$

3. 1 mole of XeF_4 on reaction with excess of KI will liberate I_2 that requires

(a) 1 mole of $Na_2S_2O_3$ (b) 2 moles of $Na_2S_2O_3$

(c) 0.5 mole of $Na_2S_2O_3$ (d) 4 moles of $Na_2S_2O_3$

Assertion & Reason

Codes :

(a) Both **A** and **R** are true and **R** is the correct explanation of **A**.

(b) Both **A** and **R** are true but **R** is not the correct explanation of **A**.

(c) **A** is true but **R** is false.

(d) **A** is false but **R** is true.

1. Assertion (A) : XeF_6 cannot be handled in glass or quartz apparatus.

Reason (R) : Reaction between XeF_6 and SiO_2 of glass form explosive XeO_3.

2. Assertion (A) : He and Ne do not form clatharates.

Reason (R) : They are difficult to polarise due to smaller size.

3. Assertion (A) : Fluorides of Xe are thermodynamically stable.

Reason (R) : They are endothermic compounds.

4. Assertion (A) : XeF_4 may be estimated by titrating the I_2 by reaction with KI.

Reason (R) : Equivalent weight of $Na_2S_2O_3$ can be determined by iodimetric titration.

5. Assertion (A) : XeF_6 can be used as an oxidising agent.

Reason (R) : Oxidation state of Xe in XeF_6 is +6.

6. Assertion (A) : The reaction $XeO_3 + 6H^+ + 6e^- \rightleftharpoons Xe + 3H_2O, E° = 2.1$ V is not feasible thermodynamically to oxidise H_2O

Reason (R) : There is a high activation energy barrier between oxidation state +6 and 0 of Xe.

7. Assertion (A) : The reaction,

$$XeO_3 + 6H^+ + 9I^- \longrightarrow Xe + 3H_2O + 3I_3^-$$

can be studied kinetically by measurement of presure.

Reason (R) : Above reaction is called iodometric reaction.

True & False

1. Ionisation energy of O_2 and Xe are very similar and thus formation of cations O_2^+ and Xe^+ is possible.

2. $O_2^+PtF_6^-$ is an orange-coloured compound.

3. In 1904 Nobel Prize in Chemistry was awarded to Lord Rayleigh for investigation on gas densities and for discovery of argon.

4. Radon is produced by α-decay of radium compounds.

5. Group 18 elements have been placed in between most electronegative and most electropositive elements.

6. XeF_6 can be stored in glass and quartz vessels.

7. Xenon fluorides are Lewis acids and Lewis bases.

8. XeF_4 disproportionates to stable XeO and XeO_4.

9. Oxidising power of xenon fluorides increases in the order $XeF_6 > XeF_4 > XeF_2$.

10. He and Ne form stable hydrates.

Fill in the Blanks

1. $O_2^+PtF_6^-$ (an orange coloured compound) was prepared and thus formation of compound .... was feasible based on ....

2. XeF_6 cannot be stored in .... vessel NaOH cannot be stored in .... vessel and HF cannot be stored in ....

3. Xenon fluorides are strong ....

4. XeF_4 disproportionates to Xe(0) and Xe(VI) compounds through .... and ....

5. In 1904 Nobel prize in physics was awarded to .... for ....

6. In 1904, Nobel Prize in Chemistry was awarded to .... for ....

7. Radon is formed by ....

8. Structure of XeF_6 is .... with .... hybridisation of Xe.

9. XeF_2 is .... with .... lone pairs of electrons on .... position.

10. Bond-order of He_2^+ is .... but it is .... due to ....

Matrix-Match Type Questions

I. Only One Correct

1. Match the compounds (in List I) with the structures (in List II).

List I	List II
A. XeF_4	1. Distorted octahedral
B. XeF_6	2. Tetrahedral
C. XeO_3	3. Square planar
D. XeO_4	4. Pyramidal

2. Match the types of glass (in List I) with their characteristics (in List II).

List I	List II
A. Alumino-silicate glass	1. Very high transparency
B. Calcium-alkali silicate glass	2. Cheap laboratory glass-wares
C. Lead glass	3. Optical glass
D. Soda glass	4. Domestic glass for window
	5. Low coefficient of expansion

3. Match molecules/ions (in List I) with their shapes (in List II).

List I	List II
A. ClF_3	1. Square planar
B. $[PCl_4]^+$	2. T-shaped
C. XeF_4	3. Tetrahedral
D. XeO_3	4. Pyramidal

4. Match the compounds (in List I) with the structures (in List II).

List I	List II
A. P_4	1. Crown
B. S_8	2. Dimer
C. $AlCl_3$	3. Tetrahedral
D. PCl_5	4. Linear
	5. Trigonal bipyramid

5. Match the name of the compound (in List I) with its formula (in List II).

List I	List II
A. Borax	1. PH_3
B. Carborundum	2. $COCl_2$
C. Borazine	3. $Na_2B_4O_7 \cdot 10H_2O$
D. Water glass	4. SiC
E. Phosgene	5. $B_3N_3H_6$
	6. Na_2SiO_3

6. Match the acids (in List I) with number of acid salts formed by them (in List II).

List I	List II
A. Hypophosphorus acid	1. Two
B. Orthophosphorus acid	2. Zero
C. Orthophosphoric acid	3. One
D. Mellitic acid	4. Five

7. Match the reactions of metals with dilute HNO_3 (in List I) with the nitrogen compounds (obtained by oxidation/reduction (in List II).

List I	List II
A. Mg + dil. HNO_3	1. NO
B. Zn + dil. HNO_3	2. H_2
C. Sn + dil. HNO_3	3. N_2O
D. Pb + dil. HNO_3	4. NH_4NO_3

II. One or More Correct

1. Match manufacturing process (in Column I) with catalyst used (in Column II).

Column I	Column II
A. $Bi^{3+} \longrightarrow (BiO)^{+}$	1. Heat
B. $[AlO_2]^{-} \longrightarrow Al(OH)_3$	2. Hydrolysis
C. $SiO_4^{4-} \longrightarrow Si_2O_7^{6-}$	3. Acidification
D. $(B_4O_7^{2-}) \longrightarrow [B(OH)_3]$	4. Dilution by water

2. Match the species in Column I with their corresponding properties in Column II.

Column I	Column II
A. XeO_3	1. Hydrolysis product of XeF_6
B. $XeOF_4$	2. sp^3
C. XeO_2F_4	3. sp^3d^2
D. XeO_3F_2	4. Tetrahedral
E. XeO_4	5. Trigonal pyramidal
	6. Square pyramidal
	7. Trigonal bipyramidal

Integer Answer Types

This section contains 8 questions. The answer to each of the questions is a single digit integer, ranging from 0 to 9. The appropriate bubbles below the respectively question numbers in the *ORS* have to be darkened. For example, if the correct answers to question number *X, Y, Z* and *W* (say) are 6, 0, 9 and 2, respectively, then the correct darkening of bubbles will like the following

X	Y	Z	W
⓪	⓪	⓪	⓪
①	①	①	①
②	②	②	②
③	③	③	③
④	④	④	④
⑤	⑤	⑤	⑤
⑥	⑥	⑥	⑥
⑦	⑦	⑦	⑦
⑧	⑧	⑧	⑧
⑨	⑨	⑨	⑨

Set I

1. In XeF_4 there are .... lone pairs of electron.

2. There are total of .... π-bonds in XeO_3.

3. s-character in XeO_4 is 1/........... of total hybrid orbitals.

4. XeF_4 disproportionates in aqueous solution to Xe(0) and Xe(VI). Xe(VI) is formed from which oxidation state (...........)

5. For inert gases $\dfrac{C_p}{C_V} = \dfrac{5}{\cdots}$

6. Bond-order of He_2^{2+} is ....

7. In the following reaction

$$XeO_3 + 6H^+ + 6Fe^{2+} \longrightarrow Xe + 3H_2O + 6Fe^{2+}$$

equivalent mass of $XeO_3 = \dfrac{\text{molar mass}}{\text{...........}}$

8. In the following reaction

$$XeF_2 + BrO_3^- + H_2O \longrightarrow Xe + BrO_4^- + 2HF$$

equivalent mass of $BrO_3^- = \dfrac{\text{molar mass}}{\text{...........}}$

Set II

1. Electrons left in the valence shell when Xe is converted to XeF_6 is ....

2. Total number of hybrid orbitals in XeO_3 is ....

3. $A(s)$ changes to $B(s)$ in a nuclear reaction by

$${}_zA(s) \longrightarrow {}_{z-8}B(s)$$

emission of α-particles only. Pressure exerted per unit volume in a closed vessel at temperature T K after two half-life was $x\,RT$. What is value of x?

4. Number phases in the following equilibrium is

$$Ra(s) \longrightarrow Rn(g) + He(g)$$

5. XeF_6 reacts with quartz forming XeO_3

$$2XeF_6 + xSiO_2 \longrightarrow$$

What is value of x?

6. Out of F, Cl, Br, I, Xe, He, Ne, O, how many forms compounds with positive oxidation state?

7. 500 mL of chlorine contaminated water is treated with KI and mixture required 100 mL of 0.01 M hypo. What is concentration of Cl_2 in millimoles per litre?

8. Out of the following how many have Xe in sp^3d hybridisation? XeF_2, XeF_4, XeF_6, $XeOF_2$, XeO_3 ?

Test Yourself on the Periodic Table

1. Locate each of the following elements on the Periodic Table:

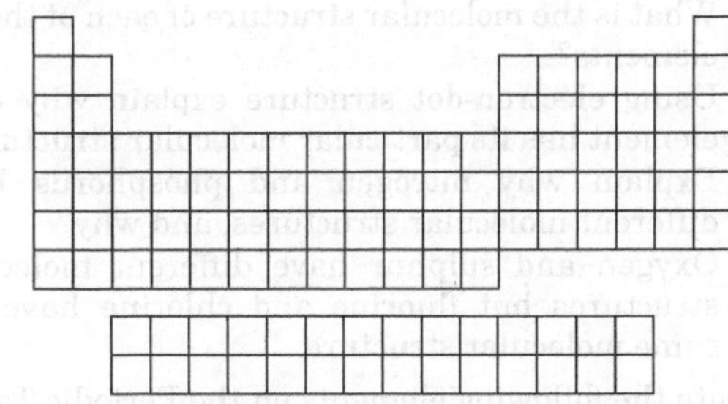

(a) Element with the lowest ionisation energy.

(b) Most electronegative element.

(c) Group 14 (IVA) element with the largest atomic radius.

(d) Group 16 (VIA) element with the smallest atomic radius.

(e) Group 13 (IIIA) element that is a semiconductor.

(f) Group 15 (VA) element that form the strongest π-bonds.

2. Locate the following element on the Periodic Table

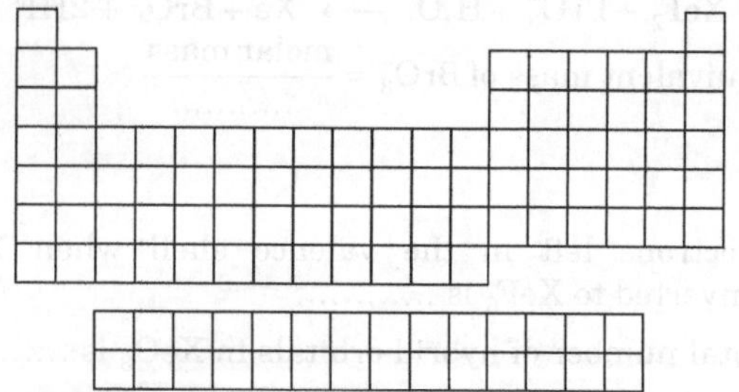

(a) Element that are gases at room temperature.
(b) A non-metal that is a liquid at 25°C.
(c) Non-metals that are solids at 25°C.
(d) Elements that exist as diatomic molecules at 25°C.

3. Consider the six, second and third-row elements in group 15–17 of the Periodic Table. Possible molecular structures for common allotropes of these elements are shown below:

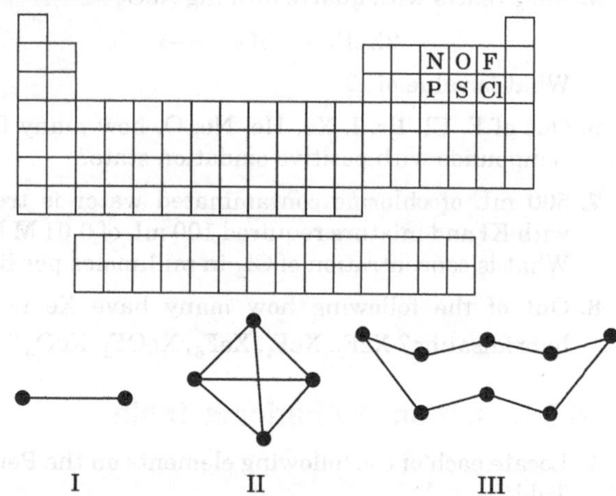

(a) What is the molecular structure of each of the six elements?
(b) Using electron-dot structure explain why each element has its particular molecular structure?
(c) Explain why nitrogen and phosphorus have different molecular structures, and why?
(d) Oxygen and sulphur have different molecular structures but fluorine and chlorine have the same molecular structure.

4. Locate the following elements on the Periodic Table

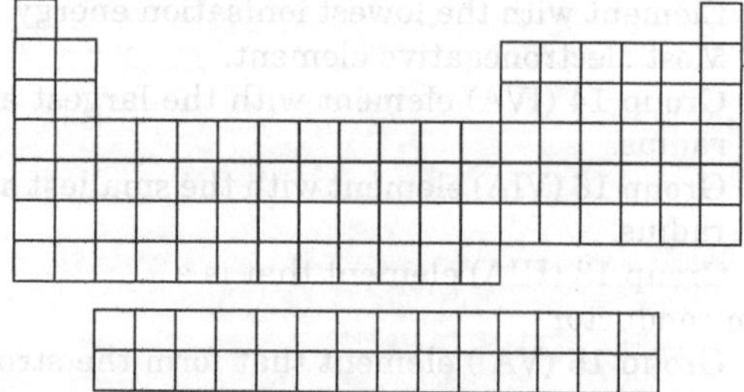

(a) Element *X* that form compound XN with nitrogen.
(b) Element with maximum electron-affinity.
(c) Element of group 14 (IVA) that form mixed oxide.
(d) Elements which together form artificial diamond.

5. Out of the elements *A*, *B*, *C*, *D*, *E*, *F* which element

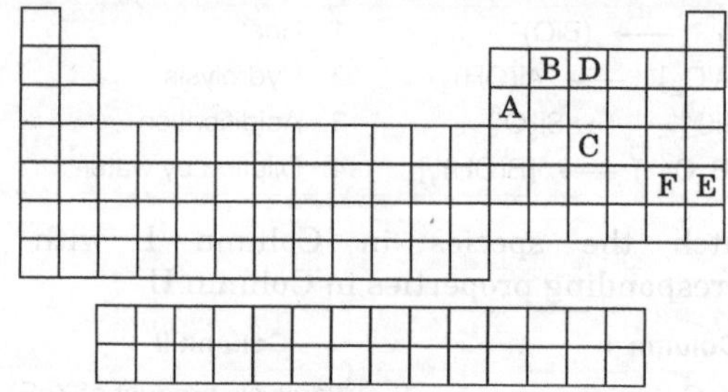

(a) is used for glass-cutter?
(b) shows inert-pair effect?
(c) whose carbide on reaction with H_2O forms CH_4?
(d) whose diatomic gas is used as inert medium?
(e) whose silver salt is used in artificial rain?
(f) whose tetrafluoride uses sp^3d^2-hybrid orbitals?

6. Out of the elements marked as *a*, *b*, *c*, *d* and *e*

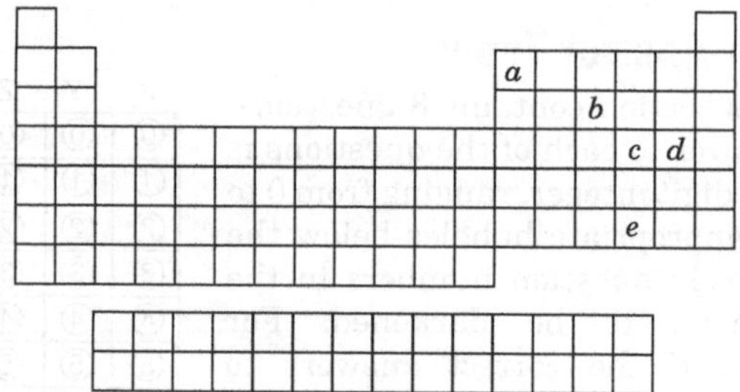

(i) Which element is a part of borax?
(ii) Which element (in diatomic state) is liquid at room temperature?
(iii) Which element is used in match-boxes?
(iv) Which element alongwith Cu used in red traffic-light?
(v) Which element is radioactive?

7. Consider the Periodic Table and locate the

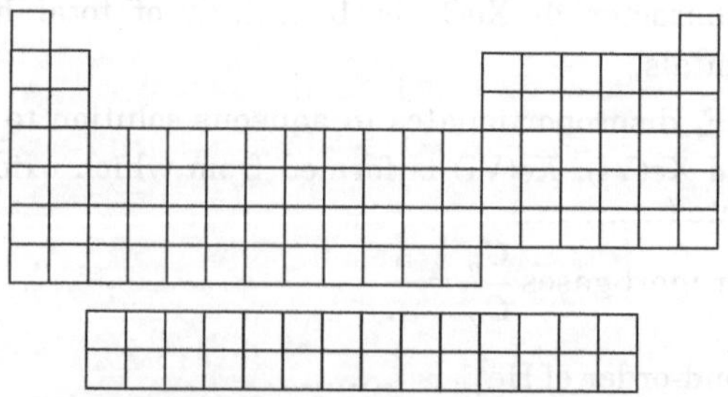

(a) elements with ns^2np^1 valence-shell electron configuration.
(b) element which has been recently (Aug, 2003) named as Ds by IUPAC
(c) elements which form positive and negative oxidation state in *p*-block.

Answers & Solutions
Master Exercises

Exercise 1

Short Answer Type Questions

1. Chemical inertness of lighter noble gases, *i.e.,* He, Ne and Ar, and rare abundance of comparatively reactive noble gases, *i.e.,* Kr, Xe and Rn were the reasons for the late discovery of noble gases.
2. Lord Rayleigh found that the density of nitrogen isolated from air was consistently about 0.5% higher than that of the nitrogen obtained chemically from ammonia. This observation made Lord Rayleigh suspect that there may be an additional gaseous element in air.
3. Boiling point is directly proportional to the van der Waals' forces between the atoms/molecules. More is the force of attraction, higher is the temperature required to overcome these binding force and higher will be the boiling point.
4. As we move down the group from He to Rn, the van der Waals' force of attraction between the noble gas atoms increases giving rise to a steady increase in boiling points.
5. Valence shells of He and Ne are completely filled, so they have no tendency to lose, gain or share electrons. Because of these reasons, compounds of He and Ne are not known.
6. Because the noble gases have extremely low reactivity, they need the strongest oxidising agents to form compounds.
7. In clathrates, the noble gases are held by weak van der Waals' forces which increase with the increase in the polarisability of the noble gas down the group. Therefore, the tendency to form clathrates increases down the group.
8. Terrestrial helium originated from the decay of radioactive elements in rocks.

Analytical Questions

1. The very fact that the first ionisation energy of xenon is slightly less than that of oxygen, led Bartlett to think that like oxygen, xenon can also be oxidised by PtF_6.
2. Oxygen and xenon which have ionisation energies, 1180 kJ mol^{-1} and 1170 kJ mol^{-1}, respectively, can be oxidised by PtF_6 while nitrogen having comparatively much higher ionisation energy, 1403 kJ mol^{-1} cannot be oxidised, so a compound of the type $[N_2]^+[PtF_6]^-$ cannot be formed.
3. On the basis of comparatively high polarisability and low ionisation energy, radon is expected to form more stable compounds. But because of very short half-life, its compounds ae highly unstable.

Exercise 2

Only One Option Correct

1. (b)	**2.** (a)	**3.** (c)	**4.** (c)	**5.** (c)	**6.** (d)	**7.** (a)	**8.** (b)	**9.** (d)	**10.** (c)
11. (a)	**12.** (a)	**13.** (a)	**14.** (b)	**15.** (b)	**16.** (a)	**17.** (d)	**18.** (a)	**19.** (d)	**20.** (c)
21. (a)	**22.** (d)	**23.** (a)	**24.** (d)	**25.** (c)	**26.** (d)	**27.** (b)	**28.** (a)	**29.** (c)	**30.** (d)

One or More Than One Options Correct

1. (a,b,c,d) **2.** (c) **3.** (a,b)
4. (a,b,d)
5. (c) **6.** (a,c) **7.** (a,b,c,d) **8.** (c)
9. (a,b,c,d) **10.** (b)

Passage Comprehension Questions

Passage 1

1. (a) **2.** (a)

Passage 2

1. (a) **2.** (c) **3.** (a)

Passage 3

1. (a) **2.** (a) **3.** (d)

Assertion & Reason

1. (a) **2.** (a) **3.** (c) **4.** (b) **5.** (b) **6.** (c)
7. (b)

True & False

1. T **2.** T **3.** F; Physics not in Chemistry

4. T **5.** T

6. F; XeF_6 is converted to explosive XeO_3

7. T

8. F; XeO further changes to Xe and XeO_4 further changes to XeO_3

9. F; $XeF_2 < XeF_4 < XeF_6$

10. F; do not form (due to smaller size and less polarising power)

Fill in the Blanks

1. $Xe^+PtF_6^-$, ionisation energy of Xe (which is very similar to that of O_2)

2. quartz or silica, Al, glass

3. oxidising agent

4. Xe(II), Xe(VIII)

5. Lord Rayleigh, discovery of argon

6. W. Ramsay, inert gas elements in air

7. α-decay of radium compounds

8. distorted octahedron, sp^3d^3

9. linear, three, equatorial

10. 0.5, paramagnetic, one unpaired electron

Matrix-Match Type Questions

I. Only One Correct

1. (A) — (3); (B) — (1); (C) — (4); (D) — (2)

2. (A) — (5); (B) — (4); (C) — (3); (D) — (2)

3. (A) — (2); (B) — (3); (C) — (1); (D) — (4)

4. (A) — (3); (B) — (1); (C) — (2); (D) — (5)

5. (A) — (3); (B) — (4); (C) — (5); (D) — (6); (E) — (2)

6. (A) — (2); (B) — (3); (C) — (1); (D) — (4)

7. (A) — (2); (B) — (3); (C) — (4); (D) — (1)

II. One or More Correct

1. (A) — (2,4); (B) — (3); (C) — (3); (D) — (2,3)

2. (A) — (1,2,5); (B) — (1,3,6); (C) — (1,3); (D) — (2,4); (E)—(7)

Integer Answer Types

Set I

Questions →	1	2	3	4	5	6	7	8
Answers →	2	3	4	8	3	1	6	2
	⓪	⓪	⓪	⓪	⓪	⓪	⓪	⓪
	①	①	①	①	①	①	①	①
	②	②	②	②	②	②	②	②
	③	③	③	③	③	③	③	③
	④	④	④	④	④	④	④	④
	⑤	⑤	⑤	⑤	⑤	⑤	⑤	⑤
	⑥	⑥	⑥	⑥	⑥	⑥	⑥	⑥
	⑦	⑦	⑦	⑦	⑦	⑦	⑦	⑦
	⑧	⑧	⑧	⑧	⑧	⑧	⑧	⑧
	⑨	⑨	⑨	⑨	⑨	⑨	⑨	⑨

Set II

Questions →	1	2	3	4	5	6	7	8
Answers →	0	4	3	2	3	5	1	2
	⓪	⓪	⓪	⓪	⓪	⓪	⓪	⓪
	①	①	①	①	①	①	①	①
	②	②	②	②	②	②	②	②
	③	③	③	③	③	③	③	③
	④	④	④	④	④	④	④	④
	⑤	⑤	⑤	⑤	⑤	⑤	⑤	⑤
	⑥	⑥	⑥	⑥	⑥	⑥	⑥	⑥
	⑦	⑦	⑦	⑦	⑦	⑦	⑦	⑦
	⑧	⑧	⑧	⑧	⑧	⑧	⑧	⑧
	⑨	⑨	⑨	⑨	⑨	⑨	⑨	⑨

Test Yourself on the Periodic Table

1.

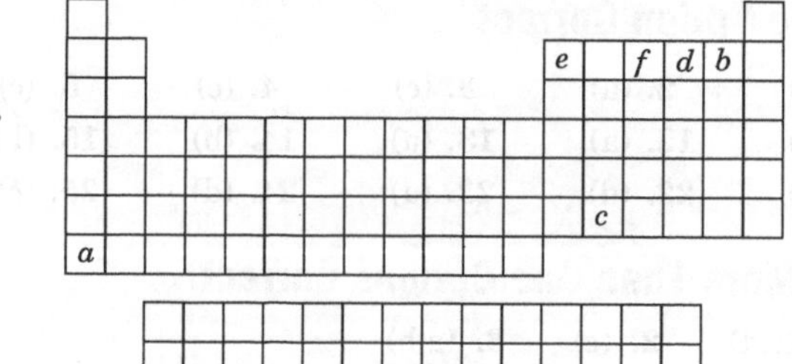

2.

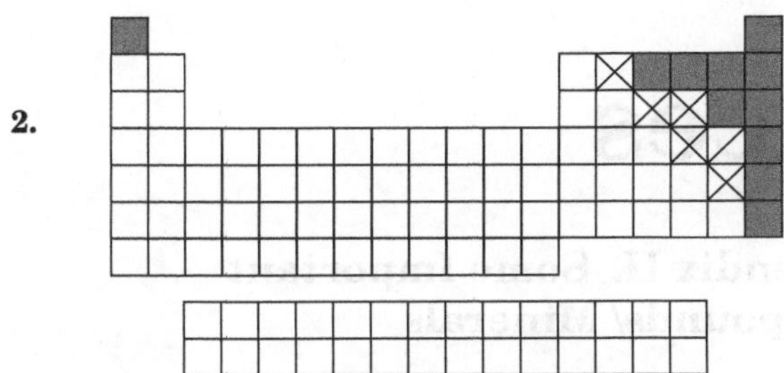

(a)

(b)

(c)

(d)

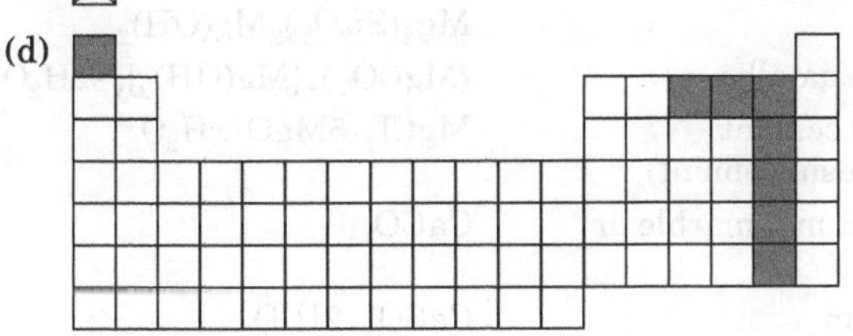

3. (a) I : N_2, O_2, F_2, Cl_2; II : P_4; III : S_8 (b) Refer text

(c) Refer text (d) Refer text

(a) Al, *X*N is AlN (b) Cl

(c) Pb_3O_4 : a mixed oxide $2PbO \cdot PbO_2$

(d) SiC is artificial diamond.

4.

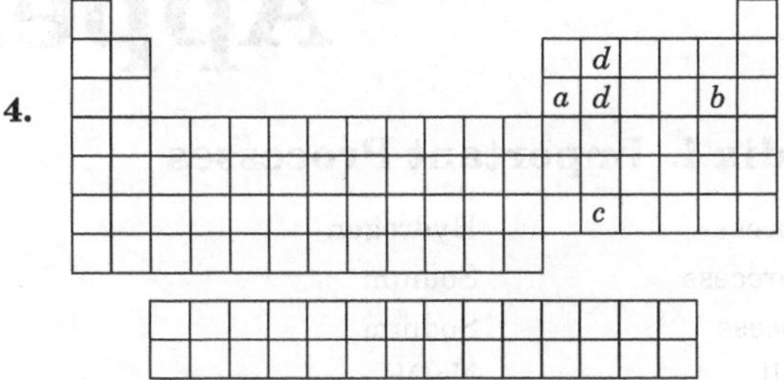

5. (a) B (b) C (c) A (d) D (e) F (f) E.

6. (i) a, (ii) d, (iii) b, (iv) c, (v) e.

7.

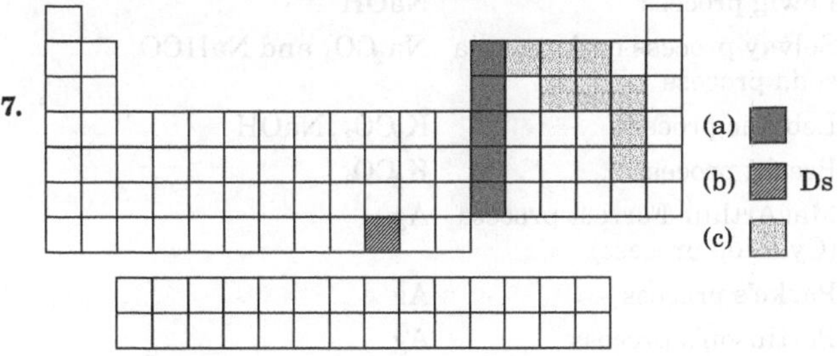

Appendices

Appendix I. Important Processes

Process	Product
Bosch process	Hydrogen
Castner process	Sodium
Down process	Sodium
Nelson cell	NaOH
Castner-Kellner cell	NaOH
Pidgeon process	Mg
Lowig process	NaOH
Solvay process or Ammonia soda process	Na_2CO_3 and $NaHCO_3$
Leblanc process	K_2CO_3, NaOH
Precht process	K_2CO_3
MacArthur Forrest process (Cyanide process)	Ag
Parke's process	Ag
Pattinson's process	Ag
Cupellation process	Ag (Purification)
Mond process	Ni
Baeyer's process	Al
Serpeck's process	Al
Hoop's process	Al (Purification)
Hall heroult process	Al
Goldschmidt process	Thermite welding
Carter process	Basic lead carbonate (White lead)
Haber's process	NH_3
Brikeland-Eyde process	NO, HNO_3
Ostwald process	NO, HNO_3
Deacon's process	Cl_2
Lead chamber process	H_2SO_4
Contact process	H_2SO_4
Kaldo and L.D. process	Steel
Corey-House	alkane
Oxo process	alcohol
Dow's sea water process	Mg
Cyanamide process	NH_3
Bessemer-Thomas process	Steel
Siemens process	Steel
Frasch process	Sulphur
IMI process	Ti
Wet process	H_3PO_4
Furnace process	H_3PO_4
Salt cake method	Cl_2
Hargreaves process	Cl_2

Appendix II. Some Important Compounds/ Minerals

Compound/Mineral	Formula
Magnesite	$MgCO_3$
Dolomite	$MgCO_3 \cdot CaCO_3$
Kieserite	$MgSO_4 \cdot H_2O$
Epsomite	$MgSO_4 \cdot 7H_2O$
Talc	$Mg_2(Si_2O_5)_2Mg(OH)_2$
Magnesia alba	$(MgCO_3)_x[Mg(OH)_2]_y \cdot zH_2O$
Sorel's cement (magnesia cement)	$MgCl_2 \cdot 5MgO \cdot xH_2O$
Lime stone, marble or chalk	$CaCO_3$
Gypsum	$CaSO_4 \cdot 2H_2O$
Plaster of Paris	$CaSO_4 \cdot \frac{1}{2} H_2O$
Anhydrite	$CaSO_4$
Dolomite	$CaCO_3 \cdot MgCO_3$
Fluorspar	CaF_2
Phosphorite	$Ca_3(PO_4)_2$
Quick lime	CaO
Slaked lime	$Ca(OH)_2$
Soda lime	$Ca(OH)_2 + NaOH$
Chlorapatite	$3Ca_3(PO_4)_2 \cdot CaCl_2$
Apatite (Fluorapatite)	$3Ca_3(PO_4)_2 \cdot CaF_2$
Pearl ash, Potash	K_2CO_3
Sylvite	KCl
Carnallite	$KCl \cdot MgCl_2 \cdot 6H_2O$
Felspar	$KAlSi_3O_8$
Salt petre (Indian salt petre)	KNO_3
Kainite	$KCl \cdot MgSO_4 \cdot 3H_2O$
Schonite	$K_2SO_4 \cdot MgSO_4 \cdot 6H_2O$
Permutit (Zeolite)	$Na_2Al_2SiO_8 \cdot xH_2O$
Calgon (Graham's salt)	$(NaPO_3)_n$, $Na_2[NH_4(PO_3)_6]$
Chile salt petre	$NaNO_3$
Salt cake	Na_2SO_4 (*anhydrous*)
Glauber's salt	$Na_2SO_4 \cdot 10H_2O$
Nitre salt	$NaHSO_4$
Borax	$Na_2B_4O_7 \cdot 10H_2O$
Soda ash (Crystal carbonate)	Na_2CO_3 (*anhydrous*)
Washing soda	$Na_2CO_3 \cdot 10H_2O$
Crystal carbonate	$Na_2CO_3 \cdot H_2O$
Baking soda	$NaHCO_3$

Coinage metals	Cu, Ag and Au
Caro's acid	H_2SO_5
Marshall's acid	$H_2S_2O_8$
Borazine (Inorganic benzene)	$B_3N_3H_6$
Boron nitride (Inorganic graphite)	$(BN)_x$
Sylvinite	(NaCl + KCl)
Bordeaux mixture (fungicides)	$CuSO_4/Ca(OH)_2$
Copper pyrites	$CuFeS_2$ or $Cu_2S \cdot Fe_2S_3$
Cuprite (Ruby copper)	Cu_2O
Copper glance	Cu_2S
Malachite	$Cu(OH)_2 \cdot CuCO_3$
Azurite	$Cu(OH)_2 \cdot 2CuCO_3$
Blue vitriol	$CuSO_4 \cdot 5H_2O$
Argentite (Silver glance)	Ag_2S
Horn silver (Chlorargyrite)	AgCl
Ruby silver (Pyrargyrite)	Ag_3SbS_3
Argentiferrous galena	PbS + Ag_2S
Lunar caustic	$AgNO_3$
Zinc blende	ZnS
Calamine	$ZnCO_3$
Zincite (Red ZnO)	ZnO
Philosopher's wool	ZnO
Willemite	Zn_2SiO_4
Bauxite	$Al_2O_3 \cdot 2H_2O$
Corundum (emarld)	Al_2O_3
Diaspore	$Al_2O_3 \cdot H_2O$
Cryolite	Na_3AlF_6
Kaolinite	$Al_2O_3 \cdot 2SiO_2 \cdot 2H_2O$
Mica	$K_2O \cdot 3Al_2O_3 \cdot 6SiO_2 \cdot 2H_2O$
Alum shale	$Al_2O_3 \cdot xSiO_2 + FeS_2$
Alum stone (Alunite)	$K_2SO_4 \cdot Al_2(SO_4)_3 \cdot 4Al(OH)_3$
Cassiterite (Tin stone)	SnO_2
Galena	PbS
Anglesite	$PbSO_4$
Cerussite	$PbCO_3$
Lanakite	$PbO \cdot PbSO_4$
Masscote (Litharge)	PbO
Haematite (Red haematite)	Fe_2O_3
Limonite (Brown haematite)	$Fe_2O_3 \cdot 3H_2O$
Magnetite	Fe_3O_4
Siderite (Spathic iron ore)	$FeCO_3$
Iron pyrites	FeS_2
Mohr's salt	$FeSO_4 \cdot (NH_4)_2SO_4 \cdot 6H_2O$
Green vitriol (copperas)	$FeSO_4 \cdot 7H_2O$
Asbestos	$CaMg_3(SiO_3)_4$
Aqua-regia	Conc. HNO_3 + Conc. HCl (1 : 3 part)
Anhdrone	$Mg(ClO_4)_2$
Barytes or barite	$BaSO_4$
Baryta water	$Ba(OH)_2$ solution
Bleaching powder	$CaOCl_2$
Boranes	Hydride of boron
Brine	NaCl solution
Baryta	BaO
Asbestos (Chrysotile)	$Mg_3Si_2O_2O_5(OH)_4$
Spinel	$MgAl_2O_4$
Carborundum	SiC
Cementite	FeC
Chinese white	ZnO
Caliche	$NaNO_3 + NaIO_3$
Caustic potash	KOH
Calomel	Hg_2Cl_2
Corrosive sublimate	$HgCl_2$
Dry ice (Drikold)	solid CO_2
Freon	CF_2Cl_2
Hydrolith	CaH_2
Hypo	$Na_2S_2O_3 \cdot 5H_2O$
Lead pencil	Graphite
Laughing gas	N_2O
Litharge (Masscote)	PbO
Lanakite	$PbO \cdot PbSO_4$
Mortar	Slaked lime + sand (1 : 3 in water)
Matte	Cu_2S + FeS
Minium	Pb_3O_4
Micro cosmic salt	$NaNH_4 \cdot HPO_4$
Milk of magnesia (Antacid)	Paste of $Mg(OH)_2$ in water
Marsh gas	CH_4
Nitro chalk	$[NH_4NO_3 + (NH_4)_2CO_3]$
Nitrolim	$CaCN_2$ (a fertilizer) + C
Nitrophos	$Ca(H_2PO_4)_2 + 2Ca(NO_3)_2$
Baking powder	$NaHCO_3 + Ca(H_2PO_4)_2$ + Starch
Oil of vitriol	Conc. H_2SO_4
Oleum	$H_2S_2O_7$
Phosgene	$COCl_2$
Pig iron	Impure form of iron
Producer gas	A mixture of CO + N_2 + H_2
Quartz (agate)	SiO_2
Quick silver	Hg
Refrigerant	CO_2, NH_3, CF_2Cl_2 etc.
Red lead	Pb_3O_4
Rochelle salt	Sodium-potassium tartrate
Rust	$Fe_2O_3 \cdot xH_2O$

TNT	Trinitro toluene (an explosive)
TNB	Trinitro benzene (an explosive)
Tincal	$Na_2B_4O_7 \cdot 10H_2O$
Water glass	Na_2SiO_3
Water gas	$CO + H_2$
Yperite	Mustard gas
Langbeinite	$K_2Mg(SO_4)_3$
Celestite	$SrSO_4$
Artificial explosive	Dynamite
First Noble Prize of chemistry was given to	vant Hoff
Some commonly used bleaching substances	H_2O_2, SO_2, SO_3, Cl_2
Neutral oxides of non-metals	CO, N_2O, NO, H_2O
Natural explosive	NCl_3
Some polymorphic elements	O, S, P
Wrought iron	Pure form of iron
Zinc white	ZnO
Olivine	$(Mg, Fe)_2SiO_4$
Strontianite	$SrCO_3$
Tincture iodine (antiseptic)	I_2 in alcohol
Some commonly used reductants	SO_2, H_2S, Cl_2
	Bleaching powder
Dry bleacher	O_3
Amphoteric oxides	$ZnO, PbO, Al_2O_3, SnO, BeO$
Some isomorphous substances	$FeSO_4 \cdot 7H_2O, MgSO_4 \cdot 7H_2O, ZnSO_4 \cdot 7H_2O, Na_2S_2O_3 \cdot 5H_2O, FeSO_4 \cdot 7H_2O$

Appendix III. Important Reagents

Reagents	Constituents	Uses
Aqua regia,	Conc. HCl and conc. HNO_3 (1 : 3 by volume)	To dissolve noble metals Au, Pt
Baeyer's reagent	Cold alkaline $KMnO_4$	To test unsaturation (alkene, alkyne) in organic compounds
Baryta water	5% $Ba(OH)_2$ solution	As a medicine to control digestive system
Barfoed reagent	$(CH_3COO)_2Cu$ and CH_3COOH	To make distinction between monosaccharides and disaccharides
Brown's catalyst	Nickel boride	To prepare stereoisomer from alkene by hydrogenation
Ceric ammonium nitrate (CAN)	CAN in HNO_3	To test alcohols
Candy water	$KMnO_4$ solution	For oxidation
Denige's reagent	Acidic solution of HgO	To test citrate
Etard's reagent	CrO_2Cl_2	Oxidising agent
Fehling's solution	A : $CuSO_4$ solution B : Sodium potassium tartrate	To test —CHO (only aliphatic) and reducing sugars
Fenton's reagent	$FeSO_4$ (or other Fe^{2+} ion) and H_2O_2	To oxidise polyhydric alcohols
Fusion mixture	Na_2CO_3 and K_2CO_3	In salt mixture analysis
Grignard's reagent	$RMgX$ (R = alkyl, aryl X = Cl, Br, I)	To synthesise alcohols, aldehydes, ketones acids, etc.
Hinsberg's reagent	Benzene sulphonyl chloride, $(C_6H_5SO_2Cl)$	To separate and to distinguish 1°, 2° and 3° amines
Lucas reagent	Conc. HCl and anhydrous $ZnCl_2$	To distinguish 1°, 2° and 3° alcohols
Millon's reagent	$Hg(NO_3)_2, Hg_2(NO_3)_2$ in HNO_3	To test proteins
Magnesia mixture	$MgCl_2 \cdot 6H_2O, NH_4Cl$, and NH_3	To test arsenate and phosphate
Molisch's reagent	Alcoholic solution of α-naphthol	To test carbohydrates
Nessler's reagent	K_2HgI_4 in NaOH	To test NH_4^+
Schiff's reagent	*p*-rosaniline hydrochloride solution decolourised by SO_2	To test —CHO group
Sodium nitroprusside	$Na_2[Fe(CN)_5NO]$	To test S^{2-} ion (sulphide)
Tollen's reagent	Ammoniacal $AgNO_3$	To test —CHO group and reducing sugars

Appendix IV. Scientists and Their Contributions

Scientist	Contribution
Anderson	Discovery of positron
Aspidin, J	Portand cement
Aston, F. William	Isolation of non-radioactive elements and discovery of mass-spectrograph
Bacon, R	Gun powder
Bohr	Atomic model, Long form of Periodic Table
Becquerrel H.	Radioactivity
Berzelius, J.J.,	Modern system of symbolising elements, relative atomic masses.
Bunsen R	Bunsen burner
Chadwick	Discovery of neutron
Curie I.J.	Radioactive elements
Formi E.	Nuclear reactor
Goldstein	Discovery of proton
Mendeleef	Periodic Table
Meyer L.	Classification of elements based upon atomic volume
Moseley	Modern Periodic Law
Pauling Linus C.	Pauling scale of electronegativity
Ramsay, Sir William	Discovery of noble gases Ar, Ne, He
Rontgen	Discovery of X-ray
Seaborg G.T.	Discovery of *trans*-uranium elements
Solvay, E	Solvay ammonia soda process for the manufacture of Na_2CO_3
Teller E	Hydrogen bomb
Thomson J.J.	Discovery of electron, e/m ratio of electron, photoelectric effect
Yukawa	Discovery of Mesons
Wohler F.	Synthesis of first organic compound (urea) from inorganic salt

Appendix V. Facts to Remember

Most reactive solid element	:	Li
Most reactive liquid element	:	Cs
Most reactive gaseous element	:	F
Highest electronegativity	:	F
Highest ionisation potential	:	He
Lowest ionisation potential	:	Cs
Lowest electron affinity	:	Noble gases (zero)
Highest electron affinity	:	Cl
Liquid element of radioactive nature	:	Francium (Fr)
Total number of radioactive elements in Periodic Table	:	25
Volatile *d*-block elements	:	Zn, Cd, Hg
Element containing no neutron	:	${}^1_1H$
Most abundant element of earth	:	Oxygen (O)
Rarest element of earth	:	Astatine (At)
Most abundant metal of earth	:	Al
Element having maximum tendency for catention	:	Carbon
Lightest element	:	H
Heaviest naturally occurring element	:	${}^{238}U$
Poorest conductor of current	:	Pb(metal), S(non-metal)
Amphoteric non-metal	:	Si
Elements showing diagonal relationship	:	Li—Mg; Be—Al; B—Si
Metalloids elements	:	B, Si, Ge, As, Sb, Te
Non-metals having metallic lustre	:	Iodine, graphite
Element sublime on heating	:	I
Noble metals	:	Au, Pt
Amphoteric metals	:	Zn, Al, Sn, Pb
Non-metal having highest m.p., b.p.	:	Diamond
Metals showing highest oxidation number.	:	Ru, Os
Element having highest tensile strength	:	Boron
Coolant in nuclear reactors	:	D_2O
Best electricity conductor among metals	:	Ag
Best conductor among non-metals	:	Graphite
Most poisonous element	:	Pu
Liquid non-metal	:	Br_2
Liquid metals	:	Hg, Ga, Cs, Fr, Eka
Element kept in water	:	P
Elements kept in kerosene oil	:	Na, K, I_2, Cs
Dry ice	:	CO_2
Most recently elements named by IUPAC	:	Ds (atomic number = 110)
'All purpose' grease	:	Lithium stearate
Old name of Astatine	:	Albamine
Most abundant gas in atmosphere	:	N_2
Rarest gas in atmosphere	:	Rn
Lightest gas in atmosphere	:	H_2
Metal with highest m.p.	:	W
Metal with lowest m.p.	:	Hg
Non-metal with highest refractive index	:	Diamond

Lowest refractive index	:	Air
Lowest b.p.	:	H_2
Group containing highest number of gaseous elements	:	Group 18 (Noble gases)
Most electrovalent compound	:	CsF
Heaviest solid metal	:	Os
Lightest solid metal	:	Li
Lightest solid non-metal	:	B
Hardest naturally occurring non-metal	:	Diamond
Hardest artificial substance good conductor of heat	:	B_4C (norbide)
but bad conductor of electricity	:	Mica
Strongest basic hydroxide	:	CsOH
Strongest basic oxide	:	CsO_2 (caesium peroxide)
Most stable metal carbonate	:	Cs_2CO_3
Element with highest radioactivity	:	Ra
Lighest radio-isotope	:	Tritium (^3_1H)
Compound with maximum covalent nature	:	H_2, N_2, O_2, Cl_2 (diatomic gases)
Strongest reducing agent	:	Azide (N_3^-)
Strongest oxidising agent	:	OF_2
Largest atomic size	:	Cs
Smallest atomic size	:	H
Smallest anion	:	H^- (hydrides)
First Noble Prize in Chemistry	:	van't Hoff
Most abundant noble gas	:	Ar
Most abundant element in human body	:	Oxygen
Most abundant gas in Sun	:	H_2
Most abundant gases in universe	:	H_2, He
Element with maximum number isotopes	:	Ag (46)
Element with minimum number of isotopes	:	H(3)
Element with maximum number of allotropes	:	Sn
Softest form of carbon	:	Lamp black
Latest allotrope of carbon	:	Fullerene or bucky ball
Strongest acid	:	$HSO_5F + 90\% SbF_5$ called magic acid
Most reactive form of P	:	White
Least reactive form of P	:	Red
Purest form of silica	:	Quartz
Most ductile metal	:	Gold

Appendix VI. Important Alloys

Alloy	Component	Alloy	Component
Aluminium bronze	Cu, Al	Electron	Mg, Zn
Alpax	Si, Al	Ferrochrome	Fe, Cr
Alnico	Al, Ni, Co, Fe, Cu	Ferromanga-nese	Fe, Mn
γ-alloy	Al, Cu, Mg, Ni	Ferrosilicon (silicon steel)	Fe, Si
Amalgam	Metal coated with Hg	German silver (Nickel silver)	Cu, Zn, Ni
Brass	Cu, Zn	Gun metal	Cu, Sn, Zn
Bronze	Cu, Sn	Hydrone	Na, Pb
Bell metal	Cu, Sn	Hastelloy C Nicrome	Ni, Cr
Britannia metal	Sn, Sb, Cu, Zn	Invar	Fe, Ni
Babbit metal	Sn, Sb, Cu	LA 141	Li, Al, Mg
Bearing metal	Cu, Zn, Sb	Magnalium	Mg, Al
Coin alloy (white)	Cu, Ag, Zn, Ni	Manganium	Ni, Cu, Mn
Coin alloy (red)	Cu, Zn, Sn	Manganese steel	Fe, C, Mn
Constantan	Cu, Ni	Monel metal	Cu, Ni, Fe, Mn
Cunico	Cu, Ni, Fe, Co	Nichrome	Ni, Cr, Fe, Mn
Delta metal	Cu, Zn, Fe	Newton metal	Sn, Pb
Dutch metal	Cu, Zn	Nickeloy	Ni, Cu, Al
Dental alloy	Sn, Hg, Ag	Nickel steel	Ni, Fe
Devarda's alloy	Cu, Zn, Al	Nimonic series	Ni, Cr, Co, Al, Ti
Duralumin	Al, Cu, Mg, Zn	Perm alloy	Ni, Fe, C
Phosphor bronze	Cu, Sn, P	Pewter	Pb, Sn
Rolled gold	Cu, Al	Platinite	Fe, Ni
Rose metal	Bi, Zn, Pb, Cd	Type metal	Pb, Sb, Sn
Soft silver (solder)	Sn, Pb	White metal	Cu, Sb, Sn
Stainless steel	Fe, Ni, Cr, C	Wood's metal	Bi, Pb, Sn, Cd

Appendix VII. Alloy Steels

Mechanical and other properties of steel are greatly modified by alloying with different metals and silicon. Some of these alloy steels are :

Type	Composition	Properties	Uses
Silicon steel	3.5% Si and very little carbon	magnetic	electromagnets and transformers.
Stainless steel	12-15% Cr	resists corrosion and action of acids	Stainless steel utensils
Manganese steel	12-13% Mn	resistant to wear	grinding machinery, safe, rock drill
Cobalt steel	up to 35% cobalt	magnetic	permanent magnets
Nickel steel	2-4% Ni	high tensile strength	cables, propeller shafts gears
Invar	up to 36% Ni	low coefficient of thermal expansion	scientific instruments
Chrome-nickel steel	1-4% Ni; 0.5-2% Cr	high tensile strength, great corrosion resistance	motor car and bicycle parts
Platinite	46% Ni	coefficient of thermal expansion same as that of glass	glassmetal-seals, electric bulbs
Chrome-vanadium steel	1-10% Cr, 0.14-0.55% V	high tensile strength, resistant to stress and torsion	automobile axles and springs
Tungsten steel	10-20% W, 3-8% Cr	hard, corrosion-resistant and retains temper at high temperature	high-speed cutting tools
Molybdenum steel	6-7% Mo	retains temper at high temperature	high-speed cutting tools
Misch-metal	70% Ce and traces of rare-earth metals.	produces spark on strike	to produce spark in gas and cigarette lighters.

Appendix VIII. Electronic Configurations of the Atoms of the Elements

Element	Atomic Number	Populations of Subshells										
		1s	*2s*	*2p*	*3s*	*3p*	*3d*	*4s*	*4p*	*4d*	*4f*	*5s*
H	1	1										
He	2	2										
Li	3	2	1									
Be	4	2	2									
B	5	2	2	1								
C	6	2	2	2								
N	7	2	2	3								
O	8	2	2	4								
F	9	2	2	5								
Ne	10	2	2	6								
Na	11				1							
Mg	12				2							
Al	13				2	1						
Si	14		Neon core		2	2						
P	15				2	3						
S	16				2	4						
Cl	17				2	5						
Ar	18	2	2	6	2	6						
K	19							1				
Ca	20							2				
Sc	21						1	2				
Ti	22						2	2				
V	23						3	2				
Cr	24						5	1				
Mn	25						5	2				
Fe	26						6	2				
Co	27			Argon core			7	2				
Ni	28						8	2				
Cu	29						10	1				
Zn	30						10	2				
Ga	31						10	2	1			
Ge	32						10	2	2			
As	33						10	2	3			
Se	34						10	2	4			
Br	35						10	2	5			
Kr	36	2	2	6	2	6	10	2	6			
Rb	37											1
Sr	38											2
Y	39									1		2
Zr	40									2		2
Nb	41									3		2
Mo	42									5		1
Tc	43					Krypton core				5		2
Ru	44									7		1
Rh	45									8		1
Pd	46									10		
Ag	47									10		1
Cd	48									10		2

Element	Atomic Number		Populations of Subshells									
			4*d*	4*f*	5*s*	5*p*	5*d*	5*f*	6*s*	6*p*	6*d*	7*s*
In	49		10		2	1						
Sn	50		10		2	2						
Sb	51		10		2	3						
Te	52		10		2	4						
I	53		10		2	5						
Xe	54		10		2	6						
Cs	55		10		2	6			1			
Ba	56		10		2	6			2			
La	57		10		2	6	1		2			
Ce	58		10	1	2	6	1		2			
Pr	59		10	3	2	6			2			
Nd	60		10	4	2	6			2			
Pm	61		10	5	2	6			2			
Sm	62		10	6	2	6			2			
Eu	63		10	7	2	6			2			
Gd	64		10	7	2	6	1		2			
Tb	65		10	9	2	6			2			
Dy	66		10	10	2	6			2			
Ho	67		10	11	2	6			2			
Er	68		10	12	2	6			2			
Tm	69		10	13	2	6			2			
Yb	70	Krypton core	10	14	2	6			2			
Lu	71		10	14	2	6	1		2			
Hf	72		10	14	2	6	2		2			
Ta	73		10	14	2	6	3		2			
W	74		10	14	2	6	4		2			
Re	75		10	14	2	6	5		2			
Os	76		10	14	2	6	6		2			
Ir	77		10	14	2	6	7		2			
Pt	78		10	14	2	6	9		1			
Au	79		10	14	2	6	10		1			
Hg	80		10	14	2	6	10		2			
Tl	81		10	14	2	6	10		2	1		
Pb	82		10	14	2	6	10		2	2		
Bi	83		10	14	2	6	10		2	3		
Po	84		10	14	2	6	10		2	4		
At	85		10	14	2	6	10		2	5		
Rn	86		10	14	2	6	10		2	6		
Fr	87		10	14	2	6	10		2	6		1
Ra	88		10	14	2	6	10		2	6		2
Ac	89		10	14	2	6	10		2	6	1	2
Th	90		10	14	2	6	10		2	6	2	2
Pa	91		10	14	2	6	10	2	2	6	1	2
U	92		10	14	2	6	10	3	2	6	1	2
Np	93		10	14	2	6	10	4	2	6	1	2
Pu	94		10	14	2	6	10	6	2	6		2
Am	95		10	14	2	6	10	7	2	6		2
Cm	96		10	14	2	6	10	7	2	6	1	2
Bk	97		10	14	2	6	10	9	2	6		2
Cf	98	Krypton core	10	14	2	6	10	10	2	6		2
Es	99		10	14	2	6	10	11	2	6		2
Fm	100		10	14	2	6	10	12	2	6		2
Md	101		10	14	2	6	10	13	2	6		2
No	102		10	14	2	6	10	14	2	6		2
Lr	103		10	14	2	6	10	14	2	6	1	2
Rf	104		10	14	2	6	10	14	2	6	2	2
Db	105		10	14	2	6	10	14	2	6	3	2
Sg	106		10	14	2	6	10	14	2	6	4	2
Bh	107		10	14	2	6	10	14	2	6	5	2
Hs	108		10	14	2	6	10	14	2	6	6	2
Mt	109		10	14	2	6	10	14	2	6	7	2
Ds	110		10	14	2	6	10	14	2	6	9	1

Appendix IX. The Elements : Discoverers, Their Names and Symbols

Element	Symbol	Atomic Number	Date of Discovery	Discoverer and Nationality
Actinium	Ac	89	1899	A. Debierne (Fr.)
Aluminium	Al	13	1827	F. Woehler (Ge.)
Americium	Am	95	1944	A. Ghiorso (USA), R.A. James (USA) G.T. Seaborg (USA), S.G. Thompson (USA)
Antimony	**Sb**	**51**	**Ancient**	
Argon	Ar	18	1894	Lord Rayleigh (GB), Sir William Ramsay (GB)
Arsenic	As	33	1250	Albertus Magnus (Ge)
Astatine	At	85	1940	D.R. Corson (USA), K.R. MacKenzie (USA) E. Segre (USA)
Barium	Ba	56	1808	Sir Humphry Davy (GB)
Berkelium	Bk	97	1950	G.T. Seaborg (USA), S.G. Thompson (USA) A. Ghiorso (USA)
Beryllium	Be	4	1828	F. Woehler (Ge.), A.A.B. Bussy (Fr.)
Bismuth	Bi	83	1753	Claude Geoffroy (Fr.)
Bohrium	Bh	107	1976	USA
Boron	B	5	1808	Sir Humphry Davy (GB), J.L. Gay-Lussac (Fr.) L.J. Thenard (Fr.)
Bromine	Br	35	1826	A.J. Balard (Fr.)
Cadmium	Cd	48	1817	Fr. Stromeyer (Ge.)
Calcium	Ca	20	1808	Sir Humphry Davy (GB)
Californium	Cf	98	1950	G.T. Seaborg (USA), S.G. Thompson (USA) A. Ghiorso (USA), K. Street, Jr. (USA)
Carbon	**C**	**6**	**Ancient**	
Cerium	Ce	58	1803	J.J. Berzelius (Swe.), William Hisinger (Swe.) M.H. Klaproth (Ge.)
Cesium	Cs	55	1860	R. Bunsen (Ge.), G.R. Kirchhoff (Ge.)
Chlorine	Cl	17	1774	K.W. Scheele (Swe.)
Chromium	Cr	24	1797	L.N. Vauquelin (Fr.)
Cobalt	Co	27	1735	G. Brandt (Ge.)
Copper	**Cu**	**29**	**Ancient**	
Curium	Cm	96	1944	G.T. Seaborg (USA), R.A. James (USA) A. Ghiorso (USA)
Darmstadtium	**Ds**	**110**	**2003**	**Sigurd Hofmann (Ge)**
Dubnium	Db	105	1970	
Dysprosium	Dy	66	1886	Locoq de Boisbaudran (Fr.)
Einsteinium	Es	99	1952	A. Ghiorso (USA)
Erbium	Er	68	1843	C.G. Mosander (Swe.)
Europium	Eu	63	1896	E. Demarcay (Fr.)
Fermium	Fm	100	1953	A. Ghiorso (USA)
Fluorine	F	9	1886	H. Moissan (Fr)
Francium	Fr	87	1939	Marguerite Perey (Fr.)
Gadolinium	Gd	64	1880	J.C. Marignac (Fr.)
Gallium	Ga	31	1875	Lecoq de Boisbaudran (Fr.)

Element	Symbol	Atomic Number	Date of Discovery	Discoverer and Nationality
Germanium	Ge	32	1886	Clemens Winkler (Ge.)
Gold	**Au**	**79**	**Ancient**	
Hafnium	Hf	72	1923	D. Coster (Du.), G. von Hevesey (H.)
Helium	He	2	1868	P. Janssen (spectr) (Fr.), Sir William Ramsay (isolated) (GB)
Hassium	Hs	108	1984	—
Holmium	Ho	67	1879	P.T. Cleve (Swe.)
Hydrogen	H	1	1766	Sir Henry Cavendish (GB)
Indium	In	49	1863	E. Reich (Ge.), T. Richter (Ge.)
Iodine	I	53	1811	B. Courtois (Fr.)
Iridium	Ir	77	1803	S. Tennant (GB)
Iron	**Fe**	**26**	**Ancient**	
Krypton	Kr	36	1898	Sir William Ramsay (GB), M.W. Travers (GB)
Lanthanum	La	57	1839	C.G. Mosander (Swe.)
Lawrencium	Lr	103	1961	A. Ghiorso (USA), T. Sikkeland (USA) A.E. Larsh (USA), R.M. Latimer (USA)
Lead	**Pb**	**82**	**Ancient**	
Lithium	Li	3	1817	A. Arfvedson (Swe)
Lutetium	Lu	71	1907	G. Urbain (Fr.), C.A. von Welsbach (Au.)
Magnesium	Mg	12	1808	Sir Humphry Davy (GB)
Manganese	Mn	25	1774	J.G. Gahn (Swe.)
Mendelevium	Md	101	1955	A. Ghiorso (USA), G.R. Choppin (USA) G.T. Seaborg (USA), B.G. Harvey (USA) S.G. Thompson (USA)
Meitnerium	Mt	109	1982	—
Mercury	**Hg**	**80**	**Ancient**	
Molybdenum	Mo	42	1778	G.W. Scheele (Swe.)
Neodymium	Nd	60	1885	C.A. von Welsbach (Au.)
Neon	Ne	10	1898	Sir William Ramsay (GB), M.W. Travers (GB)
Neptunium	Np	93	1940	E.M. McMillan (USA), P.H. Abelson (USA)
Nickel	Ni	28	1751	A.E. Cronstedt (Swe.)
Niobium	Nb	41	1801	Charles Hatchett (GB)
Nitrogen	N	7	1772	Daniel Rutherford (GB)
Nobelium	No	102	1958	A. Ghiorso (USA), T. Sikkeland (USA) J.R. Walton (USA), G.T. Seaborg (USA)
Osmium	Os	76	1803	S. Tennant (GB)
Oxygen	O	8	1774	Joseph Priestley (GB), C.W. Scheele (Swe.)
Palladium	Pd	46	1803	W.H. Wollaston (GB)
Phosphorus	P	15	1669	H. Brandt (Ge.)
Platinum	Pt	78	1735	A. De Ullao (Sp.)
			1741	Charles Wood (GB)
Plutonium	Pu	94	1940	G.T. Seaborg (USA), E.M. McMillan (USA) J.W. Kennedy (USA), A.C. Wahl (USA)
Polonium	Po	84	1898	Marie Curie (P.)
Potassium	K	19	1807	Sir Humphry Davy (GB)

Element	Symbol	Atomic Number	Date of Discovery	Discoverer and Nationality
Praseodymium	Pr	59	1885	C.A. von Welsbach (Au.)
Promethium	Pm	61	1945	J.A. Marinsky (USA), L.E. Glendenin (USA) C.D. Coryell (USA)
Protactinium	Pa	91	1917	O. Hahn (Ge.), L. Meitner (Au.)
Radium	Ra	88	1898	Pierre and Marie Curie, (Fr.; P)
Radon	Rn	86	1900	F.E. Dorn (Ge.)
Rhenium	Re	75	1925	W. Noddack (Ge.), I. Tacke (Ge) Otto Berg (Ge)
Rhodium	Rh	45	1804	W.H. Wollaston (GB)
Rubidium	Rb	37	1861	R.W. Bunsen (Ge.), G. Kirchhoff (Ge.)
Ruthenium	Ru	44	1844	K.K. Klaus (R.)
Rutherfordium	Rf	104	1965	USA
Samarium	Sm	62	1879	Lecoq de Boisbaudran (Fr.)
Scandium	Sc	21	1879	L.F. Nilson (Swe.)
Seaborgium	Sg	106	1976	—
Selenium	Se	34	1817	J.J. Berzelius (Swe.)
Silicon	Si	14	1824	J.J. Berzelius (Swe.)
Silver	**Ag**	**47**	**Ancient**	
Sodium	Na	11	1807	Sir Humphry Davy (GB)
Strontium	Sr	38	1808	Sir Humphry Davy (GB)
Sulphur	**S**	**16**	**Ancient**	
Tantalum	Ta	73	1802	A.G. Ekeberg (Swe.)
Technetium	Tc	43	1937	C. Perrier (I.)
Tellurium	Te	52	1782	E.J. Muller (Au.)
Terbium	Tb	65	1843	C.G. Mosander (Swe.)
Thallium	Tl	81	1861	Sir William Crookes (GB)
Thorium	Th	90	1828	J.J. Berzelius (Swe.)
Thulium	Tm	69	1879	P.T. Cleve (Swe.)
Tin	**Sn**	**50**	**Ancient**	
Titanium	Ti	22	1791	W. Gregor (GB)
Tungsten	W	74	1783	J.J. and F. de Elhuyer (Sp.)
Uranium	U	92	1789 1841	M.H. Klaproth (Ge.) E.M. Peligot (Fr.)
Vanadium	V	23	1801 1830	A.M. Del Rio (Sp.) N.G. Sefstrom (Swe.)
Xenon	Xe	54	1898	Sir William Ramsay (GB), M.W. Travers (GB)
Ytterbium	Yb	70	1907	G. Urbain (Fr.)
Yttrium	Y	39	1843	C.G. Mosander (Swe.)
Zinc	Zn	30	1746	A.S. Marggraf (Ge.)
Zirconium	Zr	40	1789	M.H. Klaproth (Ge.)

Appendix X. Minerals with Metallic Luster

Mineral (Formula)	Colour	Crystal System	Uses and Other Properties
Graphite (C)	black to grey	hexagonal	pencil lead, lubricants for locks, rods to control some small nuclear reactions, battery poles
Silver (Ag)	silvery white, tarnishes to black	cubic	coins, fillings for teeth, jewellery, silverplate, wires; malleable and ductile
Galena (PbS)	grey	cubic	source of lead, used in pipes, shields for X-rays, fishing equipment sinkers
Gold (Au)	pale to golden yellow	cubic	jewellery, money, gold leaf, fillings for teeth, medicines; does not tarnish
Bornite [Cu_5FeS_4]	bronze, tarnishes to dark blue, purple	tetragonal	source of copper; called "peakcock ore" because of the purple shine when it tarnishes
Copper (Cu)	copper red	cubic	coins, pipes, gutters, wire cooking utensils, jewellery, decorative plaques; malleable and ductile
Chalcopyrite ($CuFeS_2$)	brassy to golden yellow	tetragonal	main ore of copper
Chromite ($FeCr_2O_4$)	black or brown	cubic	ore of chromium, stainless steel, metallurgical bricks
Pyrrhotite (FeS)	bronze	hexagonal	often found with pentlandite, an ore of nickel; may be magnetic
Haematite (specular) (Fe_2O_3)	black or reddish brown	hexagonal	source of iron; roasted in a blast furnace, converted to "pig" iron, made into steel
Magnetite (Fe_3O_4)	black	cubic	source of iron, naturally magnetic, called lodestone
Pyrite (FeS_2)	light, brassy, yellow	cubic	source of iron, "fools's gold," alters to limonite
Fluorite (CaF_2)	colourless, white, blue, green, red, yellow, purple	cubic	used in the manufacture of optical equipment; glows under ultraviolet light
Limonite (hydrous iron oxides)	yellow, brown black	—	source of iron; weathers easily colouring matter of soils
Hornblende ($CaNa(Mg, Al, Fe)_5(Al, Si)_2Si_6O_{22}(OH)_2$)	green to black	monoclinic	will transmit light on thin edges; 6-sided cross section
Feldspar (orthoclase) ($KAlSi_3O_8$)	colourless, white grey green and yellow	monoclinic	insoluble in acids; used in the manufacture of porcelain
Feldspar (plagioclase) ($NaAlSi_3O_8$) ($CaAl_2Si_2O_8$)	grey, green white	triclinic	used in ceramics; striations present on some faces
Augite (Ca, Na) (Mg, Fe, Al) $(Al, Si)_2O_6$	black	monoclinic	square or 8-sided cross-section
Olivine ($(Mg, Fe)_2SiO_4$)	olive green	ortho-rhombic	gemstones, refractory sand
Quartz (SiO_2)	colourless, various colours	hexagonal	used in glass manufacture, electronic equipment, radios, computers, watches, gemstones
Garnet $(Mg, Fe, Ca)_3(Al_2Si_3O_{12})$	deep yellow-red green, black	cubic	used in jewellery, also used as an abrasive
Topaz $(Al_2SiO_4)(F, OH)_2$	white, pink yellow, pale blue, colourless	ortho-rhombic	valuable gemstone
Corundum (Al_2O_3)	colourless, blue, brown green white, pink, red	hexagonal	gemstones; ruby is red, sapphire is blue; industrial abrasive

Appendix XI. Mineral with Non-Metallic Lusture

Mineral (Formula)	Colour	Crystal System	Uses and Other Properties
Talc ($Mg_3(OH)_2Si_4O_{10}$)	white, greenish	monoclinic	easily cut with fingernail; used for talcum powder; soapstone is used in paper and for table tops
Bauxite (hydrous aluminium compound)	grey, red, white, brown	—	source of aluminium; used in paints, aluminium foil, and airplane parts
Kaolinite ($Al_4Si_2O_5(OH)_4$)	white, red, reddish brown, black	triclinic	clays; used in ceramics and in china dishes; common in most soils; often microscopic-sized particles
Gypsum ($CaSO_4 \cdot 2H_2O$)	colourless, grey, white, brown	monoclinic	used extensively in the preparation of plaster of Paris, alabaster, and dry wall for building construction
Sphalerite (ZnS)	brown	cubic	main ore of zinc; used in paints, dyes, and medicine
Sulphur (S)	yellow	orthorhombic	used in medicine, fungicides for plants, vulcanisation of rubber. Production of sulphuric acid
Muscovite ($KAl_3Si_3O_{10}(OH)_2$)	white, light grey, yellow, rose, green	monoclinic	occurs in large flexible plates, used as an insulator in electrical equipment, lubricant
Biotite ($K(Mg, Fe)_3AlSi_3O_{10}(OH)_2$)	black to dark brown	monoclinic	occurs in large flexible plates
Halite ($NaCl$)	colourless, red, white blue	cubic	salt; very soluble in water; a preservative
Calcite ($CaCO_3$)	colourless, white pale, blue	hexagonal	fizzes when HCl is added; used in cement and other building materials
Dolomite ($CaMg(CO_3)_2$)	colourless, white, pink, green, grey, black	hexagonal	concrete and cement, used as an ornamental building stone

JEE Main and Advanced (2017-10)

Previous Years' Questions

1. Fundamental Concepts of Inorganic Chemistry

1. A stream of electrons from a heated filament was passed between two charged plates kept at a potential difference V esu. If e and m are charge and mass of an electron, respectively, then the value of h/λ (where, λ is wavelength associated with electron wave) is given by **[JEE Main 2016]**

(a) $2\ meV$ (b) $\sqrt{meV}$

(c) $\sqrt{2meV}$ (d) meV

Ans. (c) As you can see in options, energy term is mentioned hence, we have to find out relation between $\frac{h}{\lambda}$ and energy. For this, we shall use de-Broglie wavelength and kinetic energy term in eV.

de-Broglie wavelength for an electron $(\lambda) = \frac{h}{p}$

$$\Rightarrow \quad p = \frac{h}{\lambda} \quad \ldots(i)$$

Kinetic energy of an electron = eV

As we know that, $KE = \frac{p^2}{2m}$

$$\therefore \quad eV = \frac{p^2}{2m}$$

$$\text{or} \quad p = \sqrt{2meV} \quad \ldots(ii)$$

From equations (i) and (ii), we get $\frac{h}{\lambda} = \sqrt{2meV}$

2. Chemical Bonding

1. The total number of lone pair of electrons in N_2O_2 is **[JEE Advanced 2015, Integer Type]**

Ans. (8) N_2O_3 has two proposed structures.

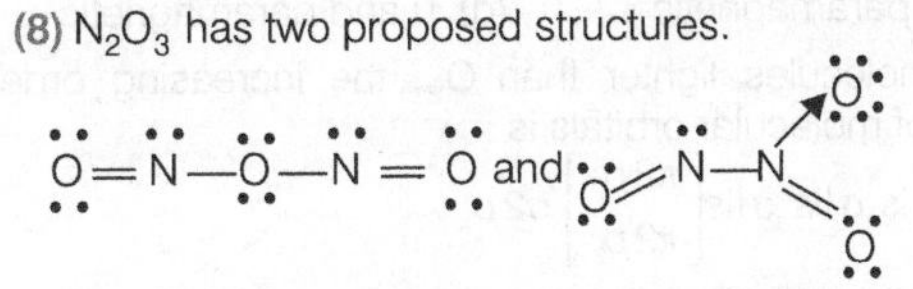

$$\ddot{O}=N-\ddot{O}-N=\ddot{O} \text{ and } O{=}N-N(\rightarrow \ddot{O}:)(=\ddot{O})$$

In both cases, number of lone pair of electrons are eight.

2. Among the triatomic molecules/ions $BeCl_2$, N_3^-, N_2O, NO_2^+, O_3, SCl_2, ICl_2^-, I_3^- and XeF, the total number of linear molecules(s)/ion(s) where the hybridisation of the central atom does not have contribution from the d-orbital(s) is [atomic number of S = 16, Cl = 17, I = 53 and Xe = 54] **[JEE Advanced 2015, Integer Type]**

Ans. (4) $Cl-\underset{\substack{\uparrow \\ sp}}{Be}-Cl$ $\quad N\equiv \underset{\substack{\uparrow \\ sp}}{N} \longrightarrow \ddot{N}:^-$ $\quad N\equiv \underset{\substack{\uparrow \\ sp}}{N} \rightarrow \ddot{O}:$ $\quad O=\overset{+}{\underset{\substack{\uparrow \\ sp}}{N}}=O$

All the above mentioned molecules/ions have sp-hybridised central atom and no one pair at central atom, hence linear also. Others are :

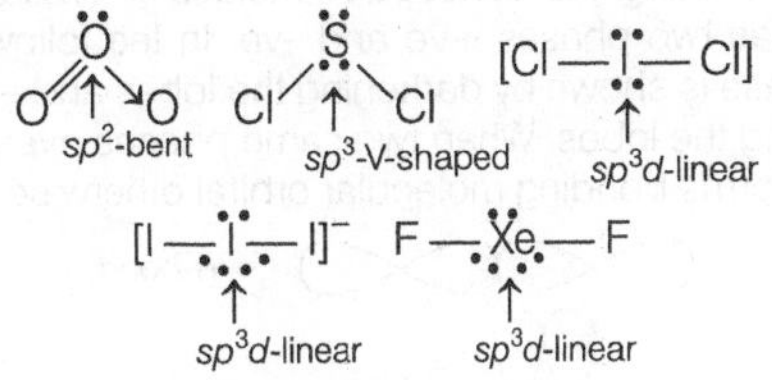

O_3: sp^2-bent $\quad SCl_2$: sp^3-V-shaped $\quad [Cl-I-Cl]^-$: sp^3d-linear

$[I-I-I]^-$: sp^3d-linear $\quad F-Xe-F$: sp^3d-linear

[Although ICl_2^-, I_3^- and XeF_2 all also are linear but in them d-orbital contribute in hybridisation.]

3. Assuming $2s$-$2p$ mixing is not operative, the paramagnetic species among the following is **[JEE Advanced 2014, Single Option]**

(a) Be_2 (b) B_2

(c) C_2 (d) N_2

Ans. (c) Assuming that no $2s$-$2p$ mixing takes place the molecular orbital electronic configuration can be written in the following sequence of energy levels of molecular orbitals

$$\sigma 1s, \overset{*}{\sigma} 1s, \sigma 2s, \overset{*}{\sigma} 2s, \sigma 2p_z, \pi 2p_x \equiv \pi 2p_y,$$
$$\overset{*}{\pi} 2p_x \equiv \overset{*}{\pi} 2p_y, \overset{*}{\sigma} 2p_z$$

(a) $Be_2 \rightarrow \sigma 1s^2, \overset{*}{\sigma} 1s^2, \sigma 2s^2, \overset{*}{\sigma} 2s^2$ (diamagnetic)

(b) $B_2 \rightarrow \sigma 1s^2, \overset{*}{\sigma} 1s^2, \sigma 2s^2, \overset{*}{\sigma} 2s^2, \sigma 2p_z^2$, (diamagnetic)

(c) $C_2 \rightarrow \sigma 1s^2, \overset{*}{\sigma} 1s^2, \sigma 2s^2, \overset{*}{\sigma} 2s^2, \sigma 2p_z^2, \begin{matrix}\pi 2p_x^1 \\ \pi 2p_y^1\end{matrix}, \begin{matrix}\overset{*}{\pi} 2p_x^0 \\ \overset{*}{\pi} 2p_y^0\end{matrix}$, (paramagnetic)

(d) $N_2 \rightarrow \sigma 1s^2, \overset{*}{\sigma} 1s^2, \sigma 2s^2, \overset{*}{\sigma} 2s^2, \sigma 2p_z^2, \begin{matrix}\pi 2p_x^2 \\ \pi 2p_y^2\end{matrix}$, (diamagnetic)

4. Match the orbital overlap figures shown in List I with the description given in List II and select the correct answer using the code given below the lists. **[JEE Advanced 2014, Matching Type]**

List I	List II
P.	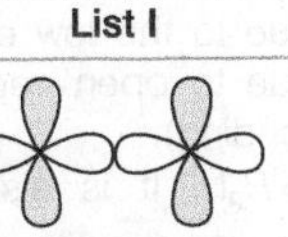 1. p-d π antibonding
Q.	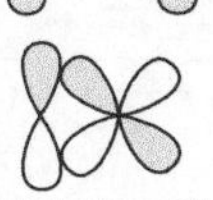 2. d-d σ bonding

	List I	List II
R.		3. p-d π bonding
S.		4. d-d σ antibonding

Codes

	P	Q	R	S		P	Q	R	S
(a)	2	1	3	4	(b)	4	3	1	2
(c)	2	3	1	4	(d)	4	1	3	2

Ans. **(c)** This problem includes basic concept of bonding. It can be solved by using the concept of molecular orbital theory. Any orbital has two phases +ve and –ve. In the following diagram +ve phase is shown by darkening the lobes and –ve by without darkening the lobes. When two same phases overlap with each other it forms bonding molecular orbital otherwise antibonding.

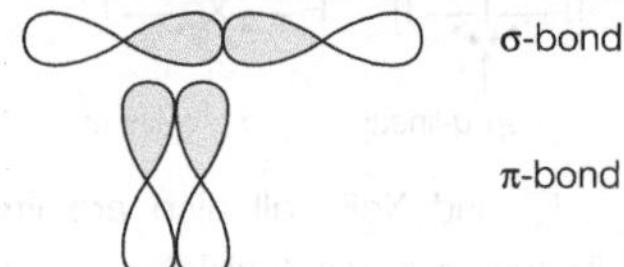

On the basis of above two concepts correct matching can be done as shown below

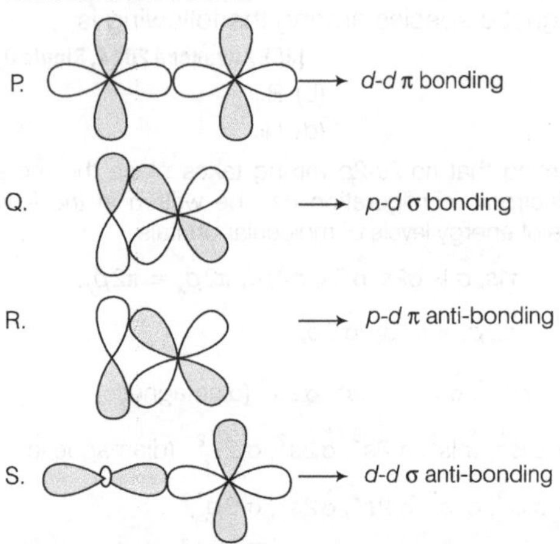

$\therefore$ P $\rightarrow$ 2, Q $\rightarrow$ 3, R $\rightarrow$ 1, S $\rightarrow$ 4

Hence, (c) is the correct option.

5. Hydrogen bonding plays a central role in the following phenomena **[JEE Advanced 2014, More than One Option]**

(a) ice floats in water

(b) higher Lewis basicity of primary amines than tertiary amines in aqueous solutions

(c) formic acid is more acidic than acetic acid

(d) dimerisation of acetic acid in benzene

Ans. **(a,b,c,d)** Ice floats in water due to the low density of ice as compare to water which is due to open cage like structure (formed by intermolecular H-bonding).

(b) Basic strength of $RNH_2 > R_3N$. It is also explained by hydrogen bonding.

(c)

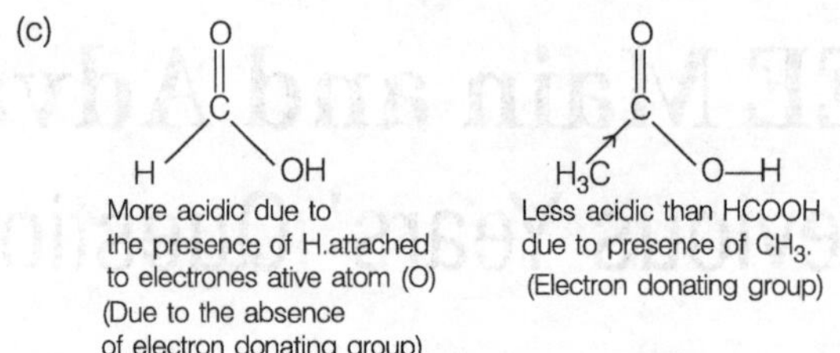

(d) Dimerisation of acetic acid in benzene is due to intermolecular hydrogen bonding.

H_3C—C(=O···H—O)(—O—H···O=)C—CH_3

6. A list of species having the formula XZ_4 is given below

XeF_4, SF_4, SiF_4, BF_4^-, BrF_4^-, $[Cu(NH_3)_4]^{2+}$, $[FeCl_4]^{2-}$, $[CoCl_4]^{2-}$ and $[PtCl_4]^{2-}$.

Defining shape on the basis of the location of X and Z atoms, the total number of species having a square planar shape is **[JEE Advanced 2014, Integer Type]**

Ans. **(4)** XeF_4, BrF_4^-, $[Cu(NH_3)_4]^{2+}$, $[PtCl_4]^{2-}$ are square planar as shown below

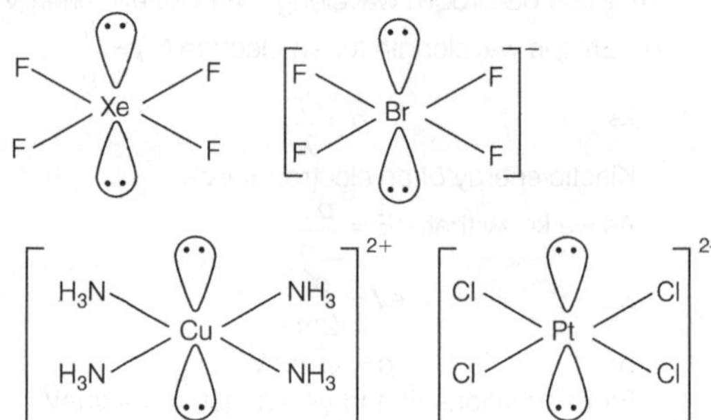

7. Assuming that Hund's rule is violated, the bond order and magnetic nature of the diatomic molecule B_2 is **[JEE Advanced 2010, Single Option]**

(a) 1 and diamagnetic (b) 0 and diamagnetic

(c) 1 and paramagnetic (d) 0 and paramagnetic

Ans. **(a)** For molecules lighter than O_2, the increasing order of energies of molecular orbitals is :

$$\sigma 1s\, \overset{*}{\sigma} 1s\, \sigma 2s\, \overset{*}{\sigma} 2s \begin{bmatrix} \pi 2p_y \\ \pi 2p_z \end{bmatrix} \sigma 2p_x$$

Where, $\pi 2p_y$ and $\pi 2p_z$ are degenerate molecular orbitals, first singly occupied and then pairing starts if Hund's rule is obeyed. If Hund's rule is violated in B_2, electronic arrangement would be : $\sigma 1s^2\, \overset{*}{\sigma} 1s^2\, \sigma 2s^2\, \overset{*}{\sigma} 2s^2 \begin{bmatrix} \pi 2p_y^2 \\ \pi 2p_z \end{bmatrix} \ldots$

No unpaired electron-diamagnetic.

Bond-order

$$= \frac{\text{bonding electrons} - \text{antibonding electrons}}{2} = \frac{6-4}{2} = 1$$

8. The species having pyramidal shape is **[IIT JEE 2010, Single Option]**

(a) SO_3 (b) BrF_3

(c) SiO_3^{2-} (d) OSF_2

Ans. (d)

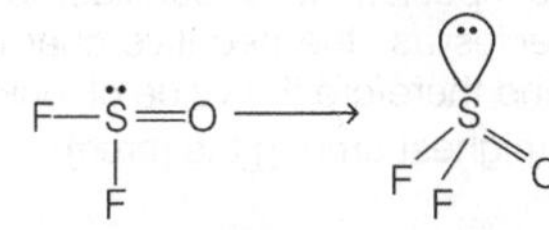

SO_3 is planar (S is sp^2 hybridised), BrF_3 is T-shaped and SiO_3^{2-} is planar (Si is sp^2 hybridised).

9. Based on VSEPR theory, the number of 90° F—Br—F angles in BrF_5 is **[IIT JEE 2010, Integer Type]**

(a) 0 (b) 1

(c) 2 (d) 3

Ans. (d) In BrF_5 number of electrons = 6 (1*lp* + 1*bp*)

So, the structure is supported to be square pyarmidal but will be distorted because of additional *lp*-*bp* interaction.

Additional *lp*-*bp* interaction reduced the all bond angle and do not let any angle to be 90°.

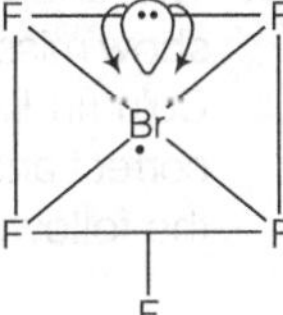

10. The correct statement for the molecule, CsI_3 is **[JEE Main 2014]**

(a) it is a covalent molecule

(b) it contains Cs^+ and I_3^- ions

(c) it contains Cs^{3+} and I^- ions

(d) it contains Cs^+, I^- and lattice I_2 molecule

Ans. (b) CsI_3 is an ionic compound. It consists of Cs^+ (cation) and I_3^- (anion) ions. I_3^- ion is made up of I_2 and I^- which is linear in shape.

11. Which one of the following molecules is expected to exhibit diamagnetic behaviour? **[JEE Main 2013]**

(a) C_2 (b) N_2

(c) O_2 (d) S_2

Ans. (a, b) MOEC of $C_2 (6+6=12) = \sigma 1s^2, \sigma^* 1s^2, \sigma 2s^2, \sigma^* 2s^2, \pi 2p_x^2 \approx \pi 2p_y^2$

Since, all the electrons are paired, it is a diamagnetic species.

MOEC of $N_2 (7+7=14) = \sigma 1s^2, \sigma^* 1s^2, \sigma 2s^2, \sigma^* 2s^2, \pi 2p_x^2 \approx \pi 2p_y^2, \sigma 2p_z^2$

It is also a diamagnetic species because of the absence of unpaired electrons.

MOEC of $O_2 (8+8=16)$

Or

$$= \sigma 1s^2, \overset{*}{\sigma} 1s^2, \sigma 2s^2, \overset{*}{\sigma} 2s^2, \sigma 2p_z^2, \pi 2p_x^2$$

$$\approx \pi 2p_y^2, \overset{*}{\pi} 2p_x^1 \approx \overset{*}{\pi} 2p_y^1$$

Due to the presence of two unpaired electrons O_2 and also S_2 both are paramagnetic molecules.

12. In which of the following pairs of molecules/ions both the species are not likely to exist? **[JEE Main 2013]**

(a) H_2^+, He_2^{2-} (b) H_2^-, He_2^{2-}

(c) H_2^{2+}, He_2 (d) H_2^-, He_2^{2+}

Ans. (c) Species having zero or negative bond order do not exist.

$H_2^{2+} (1+1-2=0) = \sigma 1s^0$, Bond order = 0

$He_2 (2+2=4) = \sigma 1s^2, \overset{*}{\sigma} 1s^2$

$$\text{Bond order} = \frac{N_b - N_a}{2} = \frac{2-2}{2} = 0$$

So, both H_2^{2+} and He_2 do not exist.

13. Stability of the species Li_2, Li_2^- and Li_2^+ increases in the order of **[JEE Main 2013]**

(a) $Li_2 < Li_2^+ < Li_2^-$ (b) $Li_2^- < Li_2^+ < Li_2$

(c) $Li_2 < Li_2^- < Li_2^+$ (d) $Li_2^- < Li_2 < Li_2^+$

Ans. (b) $Li_2 (3+3=6) = \sigma 1s^2, \sigma^* 1s^2, \sigma 2s^2$

$$\text{Bond order} = \frac{N_b - N_a}{2} = \frac{4-2}{2} = 1$$

$Li_2^- (3+3-1=5) = \sigma 1s^2, \sigma^* 1s^2, \sigma 2s^1$

$$\text{Bond order} = \frac{3-2}{2} = \frac{1}{2} = 0.5$$

$Li_2^- (3+3+1=7) = \sigma 1s^2, \sigma^* 1s^2, \sigma 2s^2, \sigma^* 2s^1$

$$\text{Bond order} = \frac{4-3}{2} = \frac{1}{2} = 0.5$$

Stability order is $Li_2 > Li_2^+ > Li_2^-$ (because Li_2^- have more number of electrons in antibonding orbital).

14. The molecule having smallest bond angle is **[AIEEE 2012]**

(a) NCl_3 (b) $AsCl_3$

(c) $SbCl_3$ (d) PCl_3

Ans. (c) $NCl_3, PCl_3, AsCl_3, SbCl_3, BiCl_3$

As electronegativity of central atom decreases and size of central atom increases therefore, bond angle decreases. According to question, $SbCl_3$ has lowest bond angle.

15. Which of the following pair is not isostructural? **[AIEEE 2012]**

(a) CO_3^{2-} and NO_3^- (b) PCl_3 and $SiCl_4$

(c) PF_5 and BrF_5 (d) AlF_6^{3-} and SF_6

Ans. (c) Compounds having same structure are known as isostructural. The structure of compound depends upon number of valence electrons. Except PF_5 and BrF_5 (where number of valence electrons are 40 and 42) all have same valence electrons and are isostructurals.

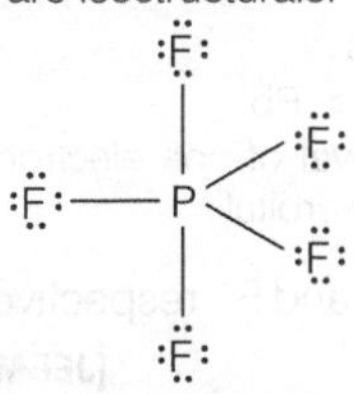

sp^3d hybridisation of P and triangular bipyramidal structural

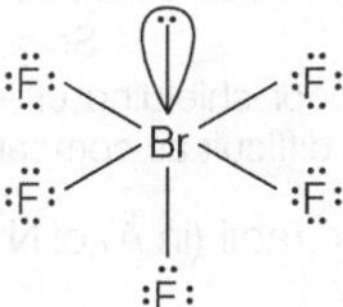

Six pairs of electrons around sp^3d^2 hybridisation of Br atom square pyramidal

16. Among the following the maximum covalent character is shown by the compound **[AIEEE 2011]**

(a) $FeCl_2$ (b) $SnCl_2$

(c) $AlCl_3$ (d) $MgCl_2$

Ans. (c) According to Fajan's rule, the polarising power and hence the covalent character of a molecule depends on

(i) Larger charge on ions (ii) Smaller size of cation

(iii) Larger size of anion

Here, all anions are same *i.e.*, Cl^- anion hence the covalent character will be decided by polarising power of cation.

Fe^{2+} Sn^{2+} Al^{3+} Mg^{2+}

↑ Maximum charge

Out of the above Al^{3+} has maximum charge and smallest radius and maximum polarising power, hence maximum covalent character.

17. The hybridisation of orbitals of N atom in NO_3^-, NO_2^+ and NH_4^+ respectively are **[AIEEE 2011]**

(a) sp^2, sp, sp^3 (b) sp, sp^3, sp^2

(c) sp^2, sp^3, sp (d) sp, sp^2, sp^3

Ans. (a)

sp^3 hybridised (structure of NO_3^-: N bonded to O^-, O, O)

sp hybridised $O{=}\overset{\oplus}{N}{=}O$

sp^3 hybridised (structure of NH_4^+: N bonded to four H)

18. The structure of IF_7 is **[AIEEE 2011]**

(a) square pyramid (b) trigonal bipyramid

(c) octahedral (d) pentagonal bipyramid

Ans. (d) $IF_7 = \frac{1}{2}(7+7) = 7$

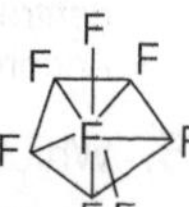

d^3sp^3 hybridisation

pentagonal bipyramidal

3. Periodicity of Elements

1. Which of the following atoms has the highest first ionisation energy? **[JEE Main 2016]**

(a) Na (b) K

(c) Sc (d) Rb

Ans. (c) Order of first ionisation energy is

Sc > Na > K > Rb.

Due to poor shielding effect, removal of one electron from 4*s* orbital is difficult as compared to 3*s*-orbital.

2. The ionic radii (in Å) of N^{3-}, O^{2-} and F^- respectively are **[JEE Main 2015]**

(a) 1.36, 1.40 and 1.71

(b) 1.36, 1.71 and 1.40

(c) 1.71, 1.40 and 1.36

(d) 1.71, 1.36 and 1.40

Ans. (c) Number of electrons in $N^{3-} = 7+3 = 10$

Number of electrons in $O^{2-} = 8+2 = 10$

Number of electrons in $F^- = 9+1 = 10$

Since, all the three species have each 10 electrons hence they are isoelectronic species. It is considered that, in case of isoelectronic species as the negative charge increases, ionic radii increases and therefore the value of ionic radii are

$N^{3-} = 1.71$ (highest among the three)

$O^{2-} = 1.40$

$F^- = 1.36$ (lowest among the three)

Time Saving Technique There is no need to mug up the radius values for different ions. This particular question can be solved through following time saving.

Trick The charges on the ions indicate the size as $N^{3-} > O^{2-} > F^-$. *Thus, you have to look for the option in which the above trend is followed. Option (c) is the only one in which this trend* is followed. Hence, it is the correct answer.

3. The unbalanced chemical reactions given in Column I show missing reagent or condition (?) which are provided in Column II. Match Column I with Column II and select the correct answer using the code given below the lists. Match the following **[JEE Advanced 2013, Matching Type]**

Column I	Column II
P. $PbO_2 + H_2SO_4 \xrightarrow{?} PbSO_4 + O_2$ + other product	1. NO
Q. $Na_2S_2O_3 + H_2O \xrightarrow{?} NaHSO_4$ + other + other product	2. I_2
R. $N_2H_4 \xrightarrow{?} N_2$ + other product	3. Warm
S. $XeF_2 \xrightarrow{?}$ Xe + other product	4. Cl_2

Codes

	P	Q	R	S		P	Q	R	S
(a)	4	2	3	1	(b)	3	2	1	4
(c)	1	4	2	3	(d)	3	4	2	1

Ans. (d) (P) $2PbO_2 + 2H_2SO_4 \xrightarrow{\text{Warm (3)}} 2PbSO_4 + O_2 + 2H_2O$

(Q) $Na_2S_2O_3 + H_2O \xrightarrow{Cl_2(4)} NaHSO_4 + HCl$

(R) $N_2H_4 \xrightarrow{I_2(2)} N_2 + HI$

(S) $XeF_2 \xrightarrow{NO(1)} Xe + NOF$

Thus, P—(3), Q—(4), R—(2), S—(1)

4. Which of the following represents the correct order of increasing first ionisation enthalpy for Ca, Ba, S, Se and Ar? **[JEE Main 2013]**

(a) Ca<S<Ba<Se<Ar

(b) S<Se<Ca<Ba<Ar

(c) Ba<Ca<Se<S<Ar

(d) Ca<Ba<S<Se<Ar

Ans. (c) Ionisation energy increases along a period from left to right and decreases down a group. The position of given elements in the periodic table is as group

group	2	16	18
	Ca	S	Ar
	Ba	Se	

The order of increasing ΔH_{IE}, is

Ba < Ca < Se < S < Ar.

5. The first ionisation potential of Na is 5.1 eV. The value of electron gain enthalpy of Na^+ will be **[JEE Main 2013]**

(a) –2.55 eV (b) –5.1eV

(c) –102 eV (d) +2.55eV

Ans. (b) $Na \longrightarrow Na^+ + e^-$ IE_1

$Na^+ + e^- \longrightarrow Na$

Electron gain enthalpy of Na^+

Because reaction is reverse so, $\Delta H_{e_g} = -5.1 eV$

6. The increasing order of the ionic radii of the given isoelectronic species is **[AIEEE 2012]**

(a) Cl^-, Ca^{2+}, K^+, S^{2-}

(b) S^{2-}, Cl^-, Ca^{2+}, K^+

(c) Ca^{2+}, K^+, Cl^-, S^{2-}

(d) K^+, S^{2-}, Ca^{2+}, Cl^-

Ans. (c) For the case of isoelectronic species, the size of ionic radii decreases as the atomic number increases. $\left(\Omega_n \propto \frac{1}{z}\right)$

7. The correct order of electron gain enthalpy with negative sign of F, Cl, Br and I, having atomic number 9, 17, 35 and 53 respectively, is **[AIEEE 2011]**

(a) I > Br > Cl > F

(b) F > Cl > Br > I

(c) Cl > F > Br > I

(d) Br > Cl > I > F

Ans. (c) As we move from top to bottom, size increases but force of attraction for the added extra electron decreases, thus electron gain enthalpy decreases.

The order should F < Cl < Br < I

but actual order is Cl > F > Br > I because extra e^- added to 2*p*-orbital of F is repelled to a greater extent than 3*p*-orbital of Cl atom due to shielding effect of 3d-orbital hence F has more electron affinity than Cl.

8. The correct sequence which shows decreasing order of the ionic radii of the elements is **[AIEEE 2010]**

(a) $Al^{3+} > Mg^{2+} > Na^+ > F^- > O^{2-}$

(b) $Na^+ > Mg^{2+} > Al^{3+} > O^{2-} > F^-$

(c) $Na^+ > F^- > Mg^{2+} > O^{2-} > Al^{3+}$

(d) $O^{2-} > F^- > Na^+ > Mg^{2+} > Al^{3+}$

Ans. (d) For isoelectronic species, higher the nuclear charge hence lower the size. $\left(\Omega_n \propto \frac{1}{z}\right)$

4. Oxidation-Reduction

1. In dilute aqueous H_2SO_4 the complex diaquadioxalatoferrate (II) is oxidised by MnO_4^-. For this reaction, the ratio of the rate of change of $[H^+]$ to the rate of change of $[MnO_4^-]$ is **[JEE Advanced 2015, Integer Type]**

Ans. (8) The balanced redox reaction is

$$MnO_4^- + [Fe(H_2O)_2(C_2O_4)_2]^{2-} + 8H^+ \longrightarrow Mn^{2+} + Fe^{3+} + 4CO_2 + 6H_2O$$

$$\Rightarrow \frac{r\,[H^+]}{r\,[MnO_4^-]} = \frac{8}{1} = 81111$$

2. The difference in the oxidation numbers of the two types of sulphur atoms in $Na_2S_4O_6$ is **[IIT JEE 2011, Integer Type]**

Ans. (5) $Na_2S_4O_6$ is a salt of $H_2S_4O_6$ which has the following structure

(5) (0) (0) (5)
HO—S(=O)(=O)—S—S—S(=O)(=O)—OH

⇒ Difference in oxidation number of two types of sulphur = 5

3. Among the following, the number of elements showing only one non-zero oxidation state is O, Cl, F, N, P, Sn, Tl, Na, Ti. **[IIT JEE 2010, Integer Type]**

Ans. (2) Only F and Na show only one non-zero oxidation state.

O = , –1, – 2, + 2; Cl = – 1 to +7

N = –3 to +5; P = – 3 to +5

Sn = + 2, +4; Tl = + 1, + 3 (rare but does exist)

Ti = + 2, + 3, +4

F, being the most electronegative element, exhibits only –1 state, and, Na being an alkali metal, exhibits only +1, state.

4. In which of the following reactions, H_2O_2 acts as a reducing agent? **[JEE Main 2014]**

I. $H_2O_2 + 2H^+ + 2e^- \longrightarrow 2H_2O$

II. $H_2O_2 - 2e^- \longrightarrow O_2 + 2H^+$

III. $H_2O_2 + 2e^- \longrightarrow 2OH^-$

IV. $H_2O_2 + 2OH^- - 2e^- \longrightarrow O_2 + 2H_2O$

(a) I and II (b) III and IV (c) I and III (d) II and IV

Ans. (d) In II and IV, H_2O_2 acts as a reducing agent as it is oxidised by losing two electrons in each.

5. Consider the following reaction,

$$x MnO_4^- + yC_2O_4^{2-} + zH^+ \longrightarrow xMn^{2+} + 2y\,CO_2 + \frac{z}{2}H_2O$$

The values of *x*, *y* and *z* in the reaction respectively are **[JEE Main 2013]**

(a) 5, 2 and 16 (b) 2, 5 and 8

(c) 2, 5 and 16 (d) 5, 2 and 8

Ans. (c) The half equations of the reaction are

$$MnO_4^- \longrightarrow Mn^{2+},\ C_2O_4^{2-} \longrightarrow CO_2$$

The balanced half equations are

$$MnO_4^- + 8H^+ + 5e^- \longrightarrow Mn^{2+} + 4H_2O$$

$$C_2O_4^{2-} \longrightarrow 2CO_2 + 2e^-$$

On equating number of electrons, we get

$$2MnO_4^- + 16H^+ + 10e^- \longrightarrow 2Mn^{2+} + 8H_2O$$

$$5C_2O_4^{2-} \longrightarrow 10CO_2 + 10e^-$$

On adding both the equations, we get

$$2MnO_4^- + 5C_2O_4^{2-} + 16H^+ \longrightarrow 2Mn^{2+} + 2 \times 5CO_2 + \frac{16}{2}H_2O$$

Thus, x, y and z are 2, 5 and 16 respectively.

5. Radioactivity

1. In the nuclear transmutation ${}^9_4Be + X \longrightarrow {}^8_4Be + Y$ (X, Y) is (are) **[JEE Advanced 2013, More than One Option]**

(a) (γ, n) (b) (p, D)
(c) (n, D) (d) (γ, p)

Ans. (a, b) Plan ${}^9_4Be + {}^b_aX \longrightarrow {}^8_4Be + {}^d_cY$

Atomic number same $4 + a = 4 + c$

$9 + b = 8 + d$

If $X = {}^0_0\gamma$ $a = 0; b = 0$

$Y = {}^1_0n$ $c = 0; d = 1$

If $X = {}^1_1p$ $a = 1; b = 1$

$Y = {}^2_1D$ $c = 1; d = 2$

$${}^9_4Be + {}^0_0\gamma \longrightarrow {}^8_4Br + {}^1_0n$$

$${}^9_4Be + {}^1_1p \longrightarrow {}^8_4Be + {}^2_1D$$

2. The Periodic Table consists of 18 groups. An isotope of copper, on bombardment with protons, undergoes a nuclear reaction yielding element X as shown below.

$${}^{63}_{29}Cu + {}^1_1H - 6{}^1_0n + {}^4_2a + 2{}^1_1H + X$$

To which group, element X belongs in the Periodic Table? **[JEE Advanced 2012, Integer Type]**

Ans. (8) Balancing the given nuclear reaction is terms of atomic number (charge) and mass number gives :

$${}_{29}Cu^{63} + {}_1H^1 \rightarrow 6{}_0n^1 + {}_2He^4(a) + 2{}_1H^1 + {}_2He^4 + {}_{26}X^{52}$$

The atomic number 26 correspond to transition metal "Fe" which belongs to 8th group of modern Periodic Table.

3. Bombardment of aluminium by α-particle leads to its artificial disintegration in two ways, (i) and (ii) as shown. Products X, Y and Z respectively are,

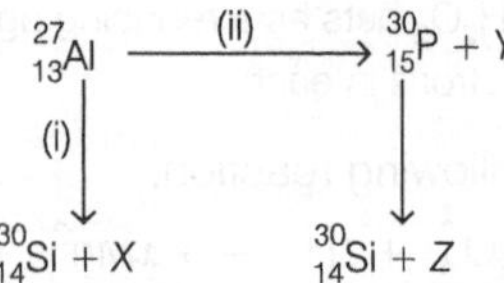

[IIT JEE 2011, Single Option]

(a) proton, neutron, positron (b) neutron, positron, proton
(c) proton, positron, neutron (d) positron, proton, neutron

Ans. (a) (i) ${}^{27}_{13}Al + {}^4_2He(a) \longrightarrow {}^{30}_{14}Si + {}^1_1X \Rightarrow X$ is proton ${}^1_1H$.

(ii) ${}^{27}_{13}Al + {}^4_2He(\alpha) \longrightarrow {}^{30}_{15}P + {}^1_0Y$

$\Rightarrow Y$ is neutron ${}^1_0n$. ${}^{30}_{15}P \longrightarrow {}^{30}_{14}Si + {}^0_1Z$

$\Rightarrow Z$ is positron ${}^0_{+1}e$.

4. The number of neutrons emitted when ${}^{235}_{92}U$ undergoes controlled nuclear fission to ${}^{142}_{54}Xe$ and ${}^{90}_{38}Sr$ is **[IIT JEE 2010, Integer Type]**

(a) 1 (b) 2
(c) 3 (d) 4

Ans. (3) ${}_{92}U^{235} \longrightarrow {}_{54}Xe^{142} + {}_{38}Sr^{90} + 3{}_0n^1$

6. Chemical Equations and Reactions

1. For the reaction, $I^- + ClO_3^- + H_2SO_4 \longrightarrow Cl^- + HSO_4^- + I_2$ the correct statement(s) in the balanced equation is/are **[JEE Advanced 2014, More than One Option]**

(a) stoichiometric coefficient of HSO_4^- is 6
(b) iodide is oxidised
(c) sulphur is reduced
(d) H_2O is one of the products

Ans. (a, b, d) Write both half-cell reactions, *i.e.*, oxidation half-cell reaction and reduction half-cell reaction.

Then balance both the equations.

Now determine the correct value of stoichiometry of H_2SO_4.

Oxidation half reaction, $2I^- \longrightarrow I_2 + 2e^-$ ...(i)

Here, I^- is converted into I_2. Oxidation number of I is increasing from –1 to 0 hence, this is a type of oxidation reaction.

Reduction half reaction

$$6H^+ + ClO_3^- + 6e^- \longrightarrow Cl^- + 3H_2O \quad ...(ii)$$

Here, H_2O releases as a product. Hence, option (d) is correct.

Multiplying Eq. (i) by 3 and adding in Eq. (ii)

$$6I^- + ClO_3^- + 6H^+ \longrightarrow Cl^- + 3I_2 + 3H_2O$$

$$6I^- + ClO_3^- + 6H_2SO_4 \longrightarrow Cl^- + 3I_2 + 3H_2O + 6HSO_4^-$$

Stoichiometric coefficient of HSO_4^- is 6.

7. Volumetric

1. A student performs a titration with different burettes and finds titre values of 25.2 mL, 25.25 mL, and 25.0 mL. The number of significant figures in the average titre value is **[IIT JEE 2010, Integer Type]**

Ans. (3) Average titre value is 25.15, but the number of significant figure cannot be greater than the same in either of them being manipulated. So, 25.15 ≈252, Thus, the number of significant figure is 3.

2. Among PbS, CuS, HgS, MnS, Ag_2S, NiS, CoS, Bi_2S_3 and SnS_2 the total number of black coloured sulphides is **[JEE Advanced 2014, Integer Type]**

Ans. (7) From qualitative analysis of the different metal ions it is found that

PbS, CuS, HgS, Ag_2S, NiS, CoS are black coloured

MnS—dirty pink/buff coloured, SnS_2—yellow coloured and Bi_2S_3

Hence, correct integer is (7).

3. Upon treatment with ammoniacal H_2S, the metal ion that precipitates as a sulphide is **[JEE Advanced 2013, More Than One Option]**

(a) Fe (III) (b) Al (III) (c) Mg (II) (d) Zn (II)

Ans. (d) K_{sp} (ZnS) is very high and Zn^{2+} is precipitated as ZnS by high concentration of S^{2-} formed when H_2S is passed in ammoniacal solution

$H_2S \rightleftharpoons Zn^+ + S^{2-}$ (I), $\quad H^+ + OH^- \rightleftharpoons H_2O$ (II)

Reaction (I) is favoured in forward side if H^+ is removed immediately by OH^- (NH_4OH)

$$Zn^{2+} + S^{2-} \longrightarrow \underset{\text{White ppt}}{ZnS\downarrow}$$

Fe^{3+} and Al^{3+} are precipitated as hydroxide.

4. An aqueous solution of a mixture of two inorganic salts,when treated with dilute HCl, gave a precipitate (P) and filtrate (Q). The precipitate P was found to dissolve in hot water. The filtrate (Q) remained unchanged, when treated with H_2S in a dilute mineral acid medium. However, it gave a precipitate (R) with H_2S in an ammoniacal medium. The precipitate R gave a coloured solution (S), when treated with H_2O_2 in an aqueous NaOH medium. **[JEE Advanced 2013, Passage Based]**

(i) The precipitate *P* contains

(a) Pb^{2+} (b) Hg_2^{2+}
(c) Ag^+ (d) Hg^{2+}

Ans. (a) In ammoniacal medium, cations of group III and IV may be precipitated as hydroxide or sulphide and dissolved in H_2O_2 due to oxidation.

Mixture of two inorganic salts $\xrightarrow[\text{(ii) Filter}]{\text{(i) HCl}}$ Filtrate ppt. (P)

$$\underset{\text{Yellow(S)}}{CrO_4^{2-}} \xleftarrow[\text{NaOH}]{H_2O_2} \underset{\text{ppt}}{Cr(OH)_3} \xleftarrow{NH_4OH/H_2S} \text{Filtrate ppt. (P)}$$

$PbCl_2$ is soluble in hot water.

(ii) The coloured solution *S* contains

(a) $Fe_2(SO_4)_3$ (b) $CuSO_4$
(c) $ZnSO_4$ (d) Na_2CrO_4

Ans. (d) Na_2CrO_4

5. For the given aqueous reaction which of the statement(s) is (are) true?

$$\underset{\text{Excess}}{KI} + K_3[Fe(CN)_6] \xrightarrow{\text{Dilute } H_2SO_4} \text{Brownish-yellow solution}$$
$$\downarrow ZnSO_4$$
$$\text{(White precipitate + Brownish-yellow filtrate)}$$
$$\downarrow Na_2S_2O_3$$
$$\text{Colourless solution}$$

[JEE Advanced 2012, More than One Option]

(a) The first reaction is a redox reaction
(b) White precipitate is $Zn_3[Fe(CN)_6]_2$
(c) Addition of filtrate to starch solution gives blue colour
(d) White precipitate is soluble in NaOH solution

Ans. (a, c, d)

$$K_3[\overset{+3}{Fe}(CN)_6] + KI \text{ (excess)} \longrightarrow K_4[\overset{+2}{Fe}(CN)_6] + \underset{\substack{\text{Brownish yellow}\\ \text{solution}}}{KI_3} \text{(Redox)}$$

$$K_4[Fe(CN)_6] + 2ZnSO_4 \longrightarrow \underset{\text{white ppt.}}{Zn_2[Fe(CN)_6]} + 2K_2SO_4$$

$$\underset{\substack{\text{Brownish}\\ \text{yellow filtrate}}}{I_3^-} + 2Na_2S_2O_3 \longrightarrow \underset{\substack{\text{Clear}\\ \text{solution}}}{Na_2S_4O_6} + 2NaI + \underset{\substack{\text{(turns starch}\\ \text{solution blue)}}}{I_2}$$

$K_2Zn[Fe(CN)_6]$ reacts with NaOH as

$$Zn_2[Fe(CN)_6] + NaOH \longrightarrow [Zn(OH)_4]^{2-} + [Fe(CN)_6]^{4-}$$

6. Passing H_2S gas into a mixture of Mn^{2+}, Ni^{2+}, Cu^{2+} and Hg^{2+} ions in an acidified aqueous solution precipitates **[IIT JEE 2011, Single Option]**

(a) CuS and HgS (b) MnS and CuS
(c) MnS and NiS (d) NiS and HgS

Ans. (a) In acidic medium, H_2S is very feebly ionized giving very small concentration of sulphide ion for precipitation. Therefore, the most insoluble salts CuS and HgS are precipitated only.

7. When a metal rod *M* is dipped into an aqueous colourless concentrated solution of compound *N*, the solution turns light blue. Addition of aqueous NaCl to the blue solution gives a white precipitate *O*. Addition of aqueous NH_3 dissolves *O* and gives an intense blue solution. **[IIT JEE 2011, Passage Based]**

(i) The metal rod *M* is

(a) Fe (b) Cu
(c) Ni (d) Co

Ans. (b) Appearance of blue colour on addition of the metal rod M to the solution of N is an indication that metal may be Cu. Later formation of white precipitate on addition of NaCl and dissolution of this precipitate in aqueous ammonia confirm that while precipitate is of AgCl. This implies that the solution N is of $AgNO_3$. This confirm that the metal M is only Cu.

(ii) The compound *N* is

(a) $AgNO_3$
(b) $Zn(NO_3)_2$
(c) $Al(NO_3)_3$
(d) $Pb(NO_3)_2$

Ans. (a) The compound N is $AgNO_3$, explained above.

(iii) The final solution contains

(a) $[Pb(NH_3)_4]^{2+}$ and $[CoCl_4]^{2-}$
(b) $[Al(NH_3)_4]^{3+}$ and $[Cu(NH_3)_4]^{2+}$
(c) $[Ag(NH_3)_2]^+$ and $[Cu(NH_3)_4]^{2+}$
(d) $[Ag(NH_3)_2]^+$ and $[Ni(NH_3)_6]^{2+}$

Ans. (c) Since in the beginning, concentrated $AgNO_3$ solution was taken, some $AgNO_3$ remain in solution which gives white precipitate of AgCl on addition of NaCl. The precipitate finally dissolve in ammonia and Cu^{2+} present in solution also forms complex with ammonia as

$$Cu(NO_3)_2 + 4NH_3(aq) \longrightarrow \underset{\text{Blue}}{[Cu(NH_3)_4]^{2+}} + 2NO_3^-$$

$$AgCl(s) + 2NH_3(aq) \longrightarrow [Ag(NH_3)_2]^+ + Cl^-$$

8. s-Block Elements - The Alkali Metals

1. Hydrogen peroxide in its reaction with KIO_4 and NH_2OH respectively, is acting as a **[JEE Advanced 2014, Single Option]**

(a) reducing agent, oxidising agent
(b) reducing agent, reducing agent
(c) oxidising agent, oxidising agent
(d) oxidising agent, reducing agent

Ans. **(a) Oxidant** Oxidant increases the oxidation number of the species with which it is reacted.

Reductant Reductant decreases the oxidation number of the species with which it is reacted.

H_2O_2 reacts with KIO_4 in the following manner

$$K\overset{+7}{I}O_4 + H_2O_2 \longrightarrow K\overset{+5}{I}O_3 + H_2O + O_2$$

On reaction of KIO_4 with H_2O_2, oxidation state of I changes from +7 to +5 *i.e.,* decreases, Thus, KIO_4 gets reduced hence, H_2O_2 is a reducing agent here.

With NH_2OH, it given following reaction

$$\overset{-1}{N}H_2OH + H_2O_2 \longrightarrow \overset{+3}{N_2}\, O_3 + H_2O$$

In the above reaction, oxidation state of N changes from –1 to +3. Here, oxidation number increases hence, H_2O_2 is acting as an oxidising agent here.

Hence, (a) is the correct choice.

9. s-Block Elements - The Alkaline Earth Metals

1. Which one of the following alkaline earth metal sulphates has its hydration enthalpy greater than its lattice enthalpy? **[JEE Main 2015]**

(a) $CaSO_4$ (b) $BeSO_4$
(c) $BaSO_4$ (d) $SrSO_4$

Ans. **(b)** As we move down the group, size of metal increases. Be has lower size while SO_4^{2-} has bigger size, that's why $BeSO_4$ breaks easily and lattice energy becomes smaller but due to lower size of Be, water molecules are gathered around and hence hydration energy increases.

On the other hand, rest metals i.e. Ca, Ba, Sr have bigger size and that's why lattice energy is greater than hydration energy.

Time Saving Technique In the question of finding hydration energy only check the size of atom. Smaller sized atom has more hydration energy.

Thus, in this question Be is placed upper most in the group has lesser size and not comparable with the size of sulphates. Hence, $BeSO_4$ is the right response.

2. The reagent(s) used for softening the temporary hardness of water is(+are) **[IIT JEE 2010, More than One Option]**

(a) $Ca_3(PO_4)_2$ (b) $Ca(OH)_2$
(c) Na_2CO_3 (d) NaOCl

Ans. **(b,c,d)** $Ca(OH)_2 + Ca(HCO_3)_2 \longrightarrow 2CaCO_3\downarrow + 2H_2O$ (Clark's method)

$$NaOCl + H_2O \longrightarrow NaOH + HOCl$$

$$HO^- + HCO_3^- \longrightarrow CO_3^{2-} + H_2O$$

$$Ca(HCO_3)_2 + Na_2CO_3 \longrightarrow CaCO_3\downarrow + 2NaHCO_3$$

3. Which of the following on thermal decomposition yields a basic as well as acidic oxide? **[AIEEE 2012]**

(a) $NaNO_3$ (b) $KClO_3$
(c) $CaCO_3$ (d) NH_4NO_3

Ans. **(c)** (a) $2NaNO_3 \xrightarrow{\Delta} 2NaNO_2 + O_2$

(b) $2KClO_3 \xrightarrow{\Delta} 2KCl + 3O_2$

(c) $CaCO_3 \xrightarrow{\Delta} \underset{\text{Basic oxide}}{CaO} + \underset{\text{Acidic oxide}}{CO_2}$

(d) $NH_4NO_3 \xrightarrow{\Delta} N_2O + H_2O$

10. Hydrogen and Its Compounds

1. Which of the following combination will produce H_2 gas? **[JEE Advanced 2017, Only One Correct Option]**

(a) Fe metal and conc. HNO_3
(b) Cu metal and conc. HNO_3
(c) Au metal and NaCN (*aq*) in the presence of air
(d) Zn metal and NaOH (*aq*)

Ans. **(d)** $\underset{\text{Amphoteric}}{Zn} + 2NaOH \longrightarrow Na_2ZnO_2 + H_2$

2. Which one of the following statements about water is false? **[JEE Main 2016]**

(a) Water can act both as an acid and as a base
(b) There is extensive intramolecular hydrogen bonding in the condensed phase
(c) Ice formed by heavy water sinks in normal water
(d) Water is oxidised to oxygen during photosynthesis

Ans. **(b)** There is extensive intermolecular H-bonding in the condensed phase.

3. From the following statements regarding H_2O_2, choose the incorrect statement. **[JEE Main 2015]**

(a) It can act only as an oxidising agent
(b) It decomposed on exposure to light
(c) It has to be stored in plastic or wax lined glass bottles in dark
(d) It has to be kept away from dust

Ans. **(a)** H_2O_2 acts as an oxidising as well as reducing agent, because oxidation number of oxygen in H_2O_2 is –1. So, it can be oxidised to oxidation state 0 or reduced to oxidation state–2.

H_2O_2 decomposes on exposure to light. So, it has to be stored in plastic or wax lined glass bottles in dark for the prevention of exposure. It also has to be kept away from dust.

4. Fe^{3+} is reduced to Fe by using **[JEE Advanced 2015, Only One Correct Option]**

(a) H_2O_2 in the presence of NaOH
(b) Na_2O_2 in water
(c) H_2O_2 in the presence of H_2SO_4
(d) Na_2O_2 in the presence of H_2SO_4

Ans. **(c)** H_2O_2 is alkaline medium acts as reducing agent, reduces Fe^{3+} to Fe^{2+}. In acidic medium the same H_2O_2 oxidises Fe^{2+} to Fe^{3+}.

5. Very pure hydrogen (99.9%) can be made by which of the following processes? **[AIEEE 2012]**
(a) Reaction of methane with steam
(b) Mixing natural hydrocarbons of high molecular weight
(c) Electrolysis of water
(d) Reaction of salts like hydrides with water

Ans. (c) On large scale, very pure hydrogen (>99.95%) can be obtained by electrolysis of water containing barium hydroxide in small amounts between nickel electrodes.

11. *p*-Block Elements - Boron Family [Group 13]

1. The increasing order of atomic radii of the following Group 13 elements is **[JEE Advanced 2016, Single Option Correct]**
(a) Al < Ga < In < Tl (b) Ga < Al < In < Tl
(c) Al < In < Ga < Tl (d) Al < Ga < Tl < In

Ans. (b) Due to poor shielding of *d*-orbital in Ga, atomic radius of Ga is smaller than that of Al. Thus,

Ga < Al < In < Tl.

2. The crystalline form of borax has **[JEE Adavanced 2016, One or More Than One Correct Option]**
(a) tetranuclear $[B_4O_5(OH)_4]^{2-}$ unit
(b) all boron atoms in the same plane
(c) equal number of sp^2 and sp^3 hybridised boron atoms
(d) one terminal hydroxide per boron atom

Ans. (a,c,d) $Na_2B_4O_7 \cdot 10H_2O$ (borax) is actually made of two tetrahedral and two triangular units, and is actually written as $Na_2[B_4O_5(OH)_4] \cdot 5H_2O$.

(a) Thus, correct.
(b) Boron atoms are in different planes thus, incorrect.
(c) Two sp^2 and two sp^3-hybridised B atoms thus, correct.
(d) Each boron has one —OH group thus, correct.

3. Three moles of B_2H_6 are completely reacted with methanol. The number of moles of boron containing product formed is **[JEE Advanced 2015, Integer Type]**

Ans. (6) $B_2H_6 + 6\,CH_3OH \longrightarrow 2[B(OCH_3)_3] + 6H_2$
Therefore, from 3 moles of B_2H_6, 6 moles of $B(OCH_3)_3$ will be

4. The correct statement(s) for orthoboric acid is/are **[JEE Advanced 2014, More than One Option]**
(a) it behaves as a weak acid in water due to self ionisation
(b) acidity of its aqueous solution increases upon addition of ethylene glycol
(c) it has a three dimensional structure due to hydrogen bonding
(d) it is a weak electrolyte in water

Ans. (b, d) For option (a) It does not undergo self ionisation in water but accepts an electron pair from water, so it behaves as weak monobasic acid

$$H_3BO_3 + H_2O \rightleftharpoons B(OH)_4^- + H^+$$

Hence, (a) is incorrect.

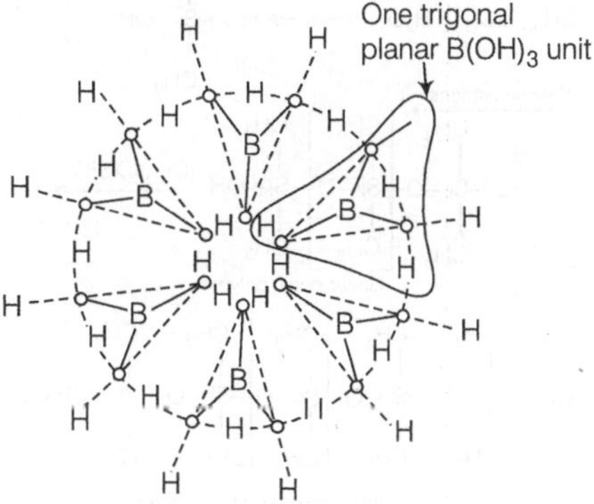

For option (b) When treated with 1, 2-dihydroxy or polyhydroxy compounds, they form chelate (ring complex) which effectively remove $[B(OH)_4]^-$ species from solution and thereby produce maximum number of H_3O^+ or H^+ ions, i.e. results in increased acidity. For option (c) Boric acid crystallises in a layer structure in which planar triangular BO_3^{3-} ions are bonded together through hydrogen bonds.

For option (d) In water the pK_a value of H_3BO_3 is 9.25.

$$H_3BO_3 + H_2O \rightleftharpoons B(OH)_4^- + H^+; \; pK_a = 9.25$$

So, it is a weak electrolyte in water.

5. The value of *n* in the molecular formula $Be_nAl_2Si_6O_{18}$ is **[IIT JEE 2010, Integer Type]**

Ans. (3) $Be_nAl_2Si_6O_{18}$, $2n + 6 + 24 - 36 = 0$
$\Rightarrow$ $n = 3$

6. Boron can't form which one of the following anions? **[AIEEE 2011]**
(a) BF_6^{3-} (b) BH_4^-
(c) $B(OH)_4^-$ (d) BO_2^-

Ans. (a)

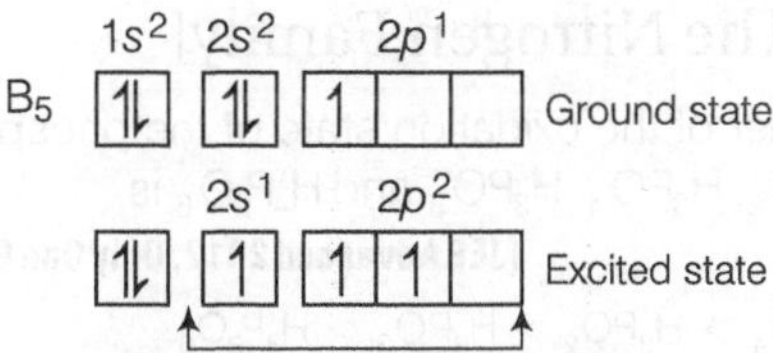

12. Group 14 [IVA-The Carbon Family]

1. Under hydrolysis conditions, the compounds used for preparation of linear polymer and for chain termination, respectively are **[JEE Advanced 2015, Single Option]**

(a) CH_3SiCl_3 and $Si(CH_3)_4$ (b) $(CH_3)_2SiCl_2$ and $(CH_3)_3SiCl$

(c) $(CH_3)SiCl_2$ and CH_3SiCl_3 (d) $SiCl_4$ and $(CH_3)_3SiCl$

Ans. (b)

$$(CH_3)_2SiCl_2 + H_2O \longrightarrow HO{-}Si(CH_3)_2{-}OH$$

$$\xrightarrow{\text{Polymerisations}} HO{-}Si(CH_3)_2{-}O{-}[Si(CH_3)_2{-}O]_n{-}Si(CH_3)_2{-}OH \xrightarrow{(CH_3)_3XSiCl}$$

Linear chain polymer

$$CH_3{-}Si(CH_3)_2{-}O{-}Si(CH_3)_2{-}O{-}[Si(CH_3)_2{-}O]_n{-}Si(CH_3)_2{-}O{-}Si(CH_3)_2{-}CH_3$$

Chain terminated linear chain silicone

2. With respect to graphite and diamond, which of the statement(s) given below is (are) correct? **[JEE Advanced 2012, More than One Option]**

(a) Graphite is harder than diamond.
(b) Graphite has higher electrical conductivity than diamond.
(c) Graphite has higher thermal conductivity than diamond.
(d) Graphite has higher C—C bond order than diamond.

Ans. (b,d) Diamond has a three dimensional network structure, a hard substance where graphite is soft due to layered structure.

In graphite, only three valence electrons are involved in bonding and one electron remains free giving electrical conductivity. In diamond, all the four valence electrons are covalently bonded hence, insulator. Diamond is better thermal conductor than graphite. Electrical conductivity is due to availability of free electrons, thermal conduction is due to transfer of thermal vibrational energy from one atom to another atom.

A compact and precisely aligned crystals like diamond thus facilitate better movement of heat.

In graphite C—C bond acquire some double bond character, hence, higher bond order than in diamond.

13. Group 15 [VA-The Nitrogen Family]

1. The order of the oxidation state of the phosphorus atom in H_3PO_2, H_3PO_4, H_3PO_3 and $H_4P_2O_6$ is **[JEE Advanced 2017, Only One Correct Option]**

(a) $H_3PO_4 > H_3PO_2 > H_3PO_3 > H_4P_2O_6$

(b) $H_3PO_4 > H_4P_2O_6 > H_3PO_3 > H_3PO_2$

(c) $H_3PO_2 > H_3PO_3 > H_4P_2O_6 > H_3PO_4$

(d) $H_3PO_3 > H_3PO_2 > H_3PO_4 > H_4P_2O_6$

Ans. (b) $H_3\overset{+5}{P}O_4 > H_4\overset{+4}{P_2}O_6 > H_3\overset{+3}{P}O_3 > H_3\overset{+1}{P}O_2$

Paragraph (Q. Nos. 2-3)

Upon heating $KClO_3$ in presence of catalytic amount of MnO_2, a gas *W* is formed. Excess amount of *W* reacts with white phosphorus to give *X*. The reaction of *X* with pure HNO_3 gives *Y* and *Z*.

2. *Y* and *Z* are, respectively

(a) N_2O_4 and HPO_3 (b) N_2O_4 and H_3PO_3
(c) N_2O_3 and H_3PO_4 (d) N_2O_5 and HPO_3

Ans. (d)

3. *W* and *X* are, respectively

(a) O_2 and P_4O_{10} (b) O_2 and P_4O_6
(c) O_3 and P_4O_6 (d) O_3 and P_4O_{10}

Ans. (a)

4. Assertion (A) Nitrogen and oxygen are the main components in the atmosphere but these do not react to form oxides of nitrogen.

Reason (R) The reaction between nitrogen and oxygen requires high temperature. **[JEE Main 2015]**

(a) Both Assertion and Reason are correct and the reason is the correct explanation for the Assertion
(b) Both Assertion and Reason are correct but the reason is not the correct explanation for the Assertion
(c) The Assertion is incorrect but the Reason is correct
(d) Both the Assertion and Reason are incorrect

Ans. (a) Nitrogen is an inert gas because of the presence of strong bond. That's why although there is 78% N_2 in the atmosphere but nitrogen oxide in not formed under ordinary conditions.

But when temperature is high enough i.e. ≈ 2000 K, it reacts with oxygen to form nitrogen oxide.

$$N_2 + O_2 \xrightarrow{\approx 2000\,K} 2NO$$

Thus, Assertion and Reason are true and Reason is the correct explanation of the Assertion.

5. Which ordering of compounds is according to the decreasing order of the oxidation state of nitrogen? **[JEE Advanced 2012, Single Option]**

(a) HNO_3, NO, NH_4Cl, N_2
(b) HNO_3, NO, N_2, NH_4Cl
(c) HNO_3, NH_4Cl, NO, N_2
(d) NO, HNO_3, NH_4Cl, N_2

Ans. (b) Let oxidation number of "N" be *x*.

In HNO_3, $+1 + x + 3(-2) = 0$

$\Rightarrow$ $x = +5$

In NO, $x - 2 = 0$

$\Rightarrow$ $x = +2$

In N_2, $x = 0$

In NH_4Cl, $x + 4 - 1 = 0$

$\Rightarrow$ $x = -3$

6. The reaction of white phosphorus with aqueous NaOH gives phosphine along with another phosphorus containing compound. The reaction type, the oxidation states of phosphorus in phosphine and the other product respectively are **[JEE Advanced 2012, Single Option]**

(a) redox reaction; – 3 and – 5
(b) redox reaction; 3 and + 5
(c) disproportionation reaction; – 3 and + 5
(d) disproportionation reaction; – 3 and + 3

Ans. **(c)** The reaction of white phosphorus with aqueous alkali is

$$P_4 + 3\,NaOH + 3H_2O \longrightarrow PH_3 + NaH_2PO_2$$

In the above reaction phosphorus is simultaneously oxidized $[P_4(0) \longrightarrow NaH_2\overset{+1}{P}O_2]$ as well as reduced $[P_4(0) \longrightarrow \overset{-3}{P}H_3]$.

Therefore, this is an example of disproportionation reaction. Oxidation number of phosphorus in PH_3 is – 3 and in NaH_2PO_2 is + 1. However, + 1 oxidation no. is not given in any option, one might think that NaH_2PO_2 has gone to further decomposition on heating.

$$NaH_2PO_2 + H_2O \longrightarrow NaOH + H_3PO_2$$

$$\underset{+1}{H_3PO_2} \longrightarrow \underset{+3}{H_3PO_3} \longrightarrow \underset{+5}{H_3PO_4}$$

7. Extra pure N_2 can be obtained by heating **[IIT JEE 2011, Single Option]**

(a) NH_3 with CuO (b) NH_4NO_3
(c) $(NH_4)_2Cr_2O_7$ (d) $Ba(N_3)_2$

Ans. **(d)** $Ba(N_3)_2 \xrightarrow{\text{Heat}} Ba(s) + 3N_2(g)$

Azide salt of barium can be obtained in purest form as well as the decomposition product contain solid Ba as byproduct alongwith gaseous nitrogen, hence no additional step of separation is required.

Other reactions are : $NH_4NO_3 \xrightarrow{\text{Heat}} N_2O + 2H_2O$

$$2NH_3 + 3CuO \xrightarrow{\text{Heat}} 3Cu + 3H_2O + N_2$$

$$(NH_4)_2Cr_2O_7 \xrightarrow{\text{Heat}} Cr_2O_3 + 4H_2O + N_2$$

8. Which of the following properties is not shown by NO? **[JEE Main 2014]**

(a) it is diamagnetic in gaseous state
(b) it is a neutral oxide
(c) it combines with oxygen to form nitrogen dioxide
(d) its bond order is 2.5

Ans. **(a)** NO is paramagnetic in gaseous sate because in gaseous state, it has one unpaired electron.

Total number of electron present = 7 + 8 = 15 e^-

Hence, there must be the presence of unpaired electron in gaseous state while in liquid state, it dimerises due to unpaired electron.

9. Which of the following is the wrong statement?

(a) ONCl and ONO^- are not isoelectronic **[JEE Main 2013]**
(b) O_3 molecule is bent
(c) ozone is violet black in solid state
(d) ozone is diamagnetic gas

Ans. **(a)** (a) ONCl = 8 + 7 + 17 = 32 electrons

ONO^- = 8 + 7 + 8 + 1 = 24 electrons (correct)

(b) O(central, with lone pair) bonded to two O atoms. O central atom O is sp^2 hybridised with one lone pair, so bent shape. (correct)

(c) In solid state, ozone is violet black (correct)

(d) O_3 has no unpaired electrons, so diamagnetic. (correct)

Hence, no option is wrong.

10. The molecule having smaller bond angle is **[AIEEE 2012]**

(a) NCl_3 (b) $AsCl_3$ (c) $SbCl_3$ (d) PCl_3

Ans. **(c)** N, P, As, Sb: Group = 15; Valence electrons = 5

MCl_3 has sp^3-hybridised M - elements with one lone pair. Lone pair and bond pair repulsion decreases bond angle.

However, the bond pair electrons are much farther away from the central atom than they are in PCl_3, $ASCl_3$ and $SbCl_3$. Thus, bond angle decreases from NCl_3 (maximum) to $SbCl_3$ (minimum).

11. Which of the following Statements is wrong? **[AIEEE 2011]**

(a) The stability of hydrides increases from NH_3 to BiH_3 in group 15 of the periodic table
(b) Nitrogen can't from $d\pi$-$p\pi$ bond.
(c) Single N—N bond is weaker than the single P—P bond
(d) N_2O_4 has two resonance structures

Ans. **(a)** (a) Thermal stability of the hydrides decreases as we go down the group in periodic table of group 15 (N-family)

$$\underset{\text{Least stable}}{BiH_3} < SbH_3 < AsH_3 < PH_3 < \underset{\text{Most stable}}{NH_3}$$

M—H—	255	247	322	391
Bond energy kJ mol^{-1}				

14. Group 16 [VI-A Oxygen family]

1. The correct statement(s) about O_3 is (are)
[JEE Advanced 2013, More than One Option]

(a) O—O bond lengths are equal
(b) thermal decomposition of O_3 is endothermic
(c) O_3 is diamagnetic in nature
(d) O_3 has a bent structure

Ans. **(a,c,d)** Process is endothermic if it takes place with absorption of heat.

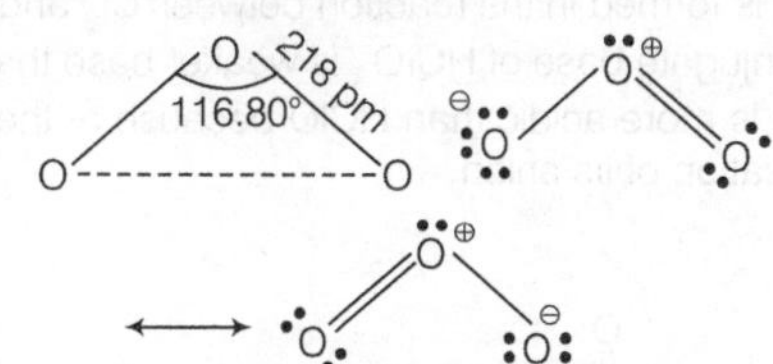

bent molecule all electrons paired thus, diamagnetic

$$2O_3 \longrightarrow 3O_2 \quad \Delta H° = -142 \text{ kJ mol}^{-1}$$
Exothermic

Thus, (b) is incorrect. (a, c, d) are correct.

2. Which of the following statements regarding sulphur is incorrect? **[AIEEE 2011]**

(a) S_2 molecule is paramagnetic.
(b) the vapour at 200°C consists mostly of S_8 rings.

(c) at 600°C, the gas mainly consists of S_2 molecules.
(d) the oxidation state of sulphur is never less than + 4 in its compounds.

Ans. (d)

(a) S_2 molecule is paramagnetic due to the presence of unpaired electron in *MO* and is blue-coloured compound-thus true.
(b) The vapour at 200°C consists mostly of S_8 rings, thus correct.
(c) At 600°C, the gas mainly consists of S_2 molecules—thus correct.
(d) Oxidation states of sulphur are

− 2	in	H_2S
0	in	S_8
+ 2	in	$S_2O_3^{2-}$
+ 4	in	SO_2
+ 6	in	SO_3

Thus, incorrect. (Valency can be less than 4).

3. Identify the incorrect statement from the following. **[AIEEE 2011]**

(a) Oxides of nitrogen in the atmosphere can cause the depletion of ozone layer
(b) Ozone absorbs the intense ultraviolet radiations of the sun
(c) Depletion of ozone layer is because of its chemical reactions with chlorofluorocarbons
(d) Ozone absorbs infrared radiations

Ans. (d) For healthy aquatic life, amount of dissolved oxygen should be 10 ppm.

15. Group 17 [VIIA-The Halogen Family]

1. The correct statement(s) about the oxoacids, $HClO_4$ and HClO, is (are)
[JEE Advanced 2017, One or More Than One Correct Option]

(a) The central atom in both $HClO_4$ and HClO is sp^3-hybridised
(b) $HClO_4$ is formed in the reaction between Cl_2 and H_2O
(c) The conjugate base of $HClO_4$ is weaker base than H_2O
(d) $HClO_4$ is more acidic than HClO because of the resonance stabilisation of its anion

Ans. (a, c, d)

(a)

HO—Cl(=O)(=O)=O (sp^3) HO—$\ddot{\text{Cl}}$: (sp^3)

(b) Incorrect : $Cl_2 + H_2O \longrightarrow HCl + HOCl$
(c) $HClO_4$ is stronger acid than H_2O.
(d) ClO_4^- is more stable than ClO^-.

2. The colour of the X_2 molecules of group 17 elements changes gradually from yellow to violet down the group. This is due to
[JEE Advanced 2017, One or More Than One Correct Option]

(a) decrease in $\pi^* - \sigma^*$ gap down the group
(b) decrease in ionisation energy down the group
(c) the physical state of X_2 at room temperature changes from gas to solid down the group
(d) decreases in HOMO-LUMO gap down the group

Ans. (a,d) Colour of halogen arises due to transition from HOMO to LUMO in the visible region. On moving down a group, the difference in energy between HOMO and LUMO decreases electronic transition occur more easily and colour intensity increases.

3. The pair in which phosphorus atoms have a formal oxidation state of +3 is **[JEE Main 2016]**

(a) pyrophosphorous and hypophosphoric acids
(b) orthophosphorous and hypophosphoric acids
(c) pyrophosphorous and pyrophosphoric acids
(d) orthophosphorous and pyrophosphorous acids

Ans. (d) Orthophosphorous acid, H_3PO_3 :

HO—P(=O)(H)—OH

$H_3\overset{x}{P}O_3 \quad 3 + x + 3(-2) = 0$
or $\quad x = +3$

Pyrophosphorous acid, $H_4P_2O_5$:

HO—P(=O)(H)—O—P(=O)(H)—OH

$H_4\overset{x}{P_2}O_5 = 4 + 2x + 5(-2) = 0$
$4 + 2x - 10 = 0$
$x = +3$

4. Which among the following is the most reactive? **[JEE Main 2015]**

(a) Cl_2 (b) Br_2 (c) I_2 (d) ICl

Ans. (d) Cl_2, Br_2 and I_2 are homonuclear diatomic molecule in which electronegativity of the combining atoms is same, so they are more stable and less reactive, whereas, I and Cl have different electronegativities and bond between them are polarised and hence, reactive. Therefore, interhalogen compounds are more reactive.

Time Saving Technique In this type of question of halogen, only go through the polarity of the molecule. As we know, diatomic molecule does not have polarity but molecules with dissimilar sizes have polarity resulting in more reactivity.

5. The correct statement(s) regarding, **[JEE Advanced 2015]**

(i) HClO,
(ii) $HClO_2$,
(iii) $HClO_3$ and
(iv) $HClO_4$ is (are)

(a) the number of Cl═ O bonds in (ii) and (iii) together is two
(b) the number of lone pair of electrons on Cl in (ii) and (iii) together is three
(c) the hybridisation of Cl in (iv) is sp^3
(d) amongst (i) to (iv), the strongest acid is (i)

***Ans.* (a)**

$$\text{H}-\ddot{\text{O}}-\ddot{\text{Cl}}: \qquad \text{H}-\ddot{\text{O}}-\ddot{\text{Cl}}=\ddot{\text{O}}$$
(i) (ii)

$$\text{H}-\ddot{\text{O}}-\overset{\text{O}}{\overset{\|}{\ddot{\text{Cl}}}}=\ddot{\text{O}} \qquad \text{H}-\ddot{\text{O}}-\overset{:\text{O}:}{\overset{\|}{\text{Cl}}}=\ddot{\text{O}}\ (\text{with } \|\ :\text{O}: \text{ below})$$
(iii) (iv)

(a) Number of $Cl^- = O$ bonds in (ii) and (iii) together is three. Hence, wrong.
(b) Number of lone pair on Cl in (ii) and (iii) together is three. Hence, correct.
(c) In (iv), Cl is sp^3-hybridised. Hence, correct.
(d) Amongst (i) to (iv), the strongest acid is (iv). Hence, wrong.

6. The product formed in the reaction of $SOCl_2$ with white phosphorus is **[JEE Advanced 2014, Single Option]**
(a) PCl_3 (b) SO_2Cl_2
(c) SCl_2 (d) $POCl_3$

***Ans.* (a)** White phosphorus on reaction with thionyl chloride ($SOCl_2$) produces phosphorus trichloride.

$$P_4(s) + 8SOCl_2(l) \longrightarrow 4PCl_3(l) + 4SO_2(g) + 2S_2Cl_2(g)$$

But if amount of thionyl chloride ($SOCl_2$) is in excess then it produces phosphorus pentachloride.

$$P_4 + 10SOCl_2(l) \longrightarrow 4PCl_5 + 10SO_2$$

Hence, (a) is the correct choice.

7. The reactions of Cl_2 gas with cold-dilute and hot-concentrated NaOH in water give sodium salts of two (different) oxoacids of chlorine, *P* and *Q*, respectively. The Cl_2 gas reacts with SO_2 gas, in the presence of charcoal, to give a product *R*. *R* reacts with white phosphorus to give a compound *S*. On hydrolysis, *S* gives an oxoacid of phosphorous *T*.
[JEE Advanced 2013, Passage Based]

P and *Q* respectively, are the sodium salts of
(a) hypochlorous and chloric acids
(b) hypochlorous and chlorous acids
(c) chloric and perchloric acids
(d) chloric and hypochlorous acids

***Ans.* (a)**

$$\underset{\text{Cold}}{2\text{NaOH}} + Cl_2 \longrightarrow NaCl + NaClO + H_2O$$

$$6NaOH + 3Cl_2 \longrightarrow 5NaCl + NaClO_3 + 3H_2O$$

$$\underset{\text{hypochlorous acid}}{\text{HOCl}} \xrightarrow{\text{NaCl}} \underset{P}{\text{NaOCl}}$$

$$\underset{\text{Chloric acid}}{HClO_3} \longrightarrow \underset{Q}{NaClO_3}$$

$$Cl_2 + SO_2 \longrightarrow \underset{S}{SO_2Cl_2}$$

$$SO_2Cl_2 + P_4 \longrightarrow 4PCl_5 + 10SO_2$$

$$PCl_5 + 4H_2O \longrightarrow \underset{T}{H_3PO_4} + 5HCl$$

8. Bleaching powder and bleach solution are produced on a large scale and used in several house hold products. The effectiveness of bleach solution is often measured by iodometry. 25 mL of household bleach solution was mixed with 30 mL of 0.50 M KI and 10 mL of 4 N acetic acid. In the titration of the liberated iodine, 48 mL of 0.25 N $Na_2S_2O_3$ was used to reach the end point. The molarity of the household bleach solution, is
[JEE Advanced 2012, Passage Based]
(a) 0.48 M (b) 0.96 M
(c) 0.24 M (d) 0.024 M

***Ans.* (c)** *The involved redox reactions are*

$$2H^+ + OCl^- + 2I^- \longrightarrow Cl^- + I_2 + H_2O \quad \ldots(i)$$

$$I_2 + 2S_2O_3^{2-} \longrightarrow 2I^- + S_4O_6^{2-} \quad \ldots(ii)$$

$$OCl^- \equiv I_2 \equiv 2I$$

$$\therefore \quad I = \frac{OCl^-}{2}$$

Thus, $xMOCl^- \equiv 2xNOCl^-$

$$2S_2O_3^{2-} \equiv I_2 \equiv 2I$$

$$\therefore \quad I \equiv S_2O_3^{2}$$

$$xMS_2O_3^{2-} \equiv 2xNS_2SO_3^{2-}$$

25 mL of bleach solution = 48 mL of 0.25 N $S_2O_3^{2-}$ Solution

$$\therefore \quad 25 \times N_1(\text{bleach}) = 48 \times 0.25$$

$$\therefore \quad N_1 = 0.48\text{ N}$$

$$\therefore \quad M(\text{bleach}) = \frac{0.48}{2} = 0.24\text{ M}$$

9. Bleaching powder contains a salt of an oxoacid as one of its components. The anhydride of that oxoacid, is
(a) Cl_2O (b) Cl_2O_7
(c) ClO_2 (d) Cl_2O_6

***Ans.* (a)** Bleaching powder is Ca(OCl) Cl. Therefore, the oxoacid whose salt is present in bleaching powder is HOCl. Anhydride of HOCl is —Cl_2O as

$$2\,HOCl \longrightarrow Cl_2O + H_2O$$

10. Reaction of Br_2 with Na_2CO_3 in aqueous solution gives sodium bromide and sodium bromate with evolution of CO_2 gas. The number of sodium bromide molecules involved in the balanced chemical equation is
[IIT JEE 2011, Integer Type]

***Ans.* (5)** Br_2 is disproportionated in basic medium as

$$3Br_2 + 3Na_2CO_3 \longrightarrow 5NaBr + NaBrO_3 + 3CO_2$$

11. Among the following, the number of compounds that can react with PCl_5 to give $POCl_3$ is O_2, CO_2, SO_2, H_2O, H_2SO_4, P_4O_{10}. **[IIT JEE 2011, Integer Type]**
(a) 1 (b) 2
(c) 3 (d) 4

***Ans.* (a)** PCl_5 produces $POCl_3$ with the following reagents.

16. The Transition and Inner Transition Elements

1. The colour of $KMnO_4$ is due to **[JEE Main 2015]**

(a) $M \to L$ charge transfer transition

(b) $d-d$ transition

(c) $L \to M$ charge transfer transition

(d) $\sigma - \sigma^*$ transition

Ans. **(c)** $KMnO_4 \longrightarrow K^+ + MnO_4^-$

$\therefore$ In MnO_4^-, Mn has +7 oxidation state having no electron in *d*-orbitals.

It is considered that higher the oxidation state of metal, greater is the tendency to occur $L \to M$ charge transfer, because ligand is able to donate the electrons into the vacant *d*- orbital of metal. Since, charge transfer is Laporte as well as spin allowed, therefore, it shows colour.

Time Saving Technique *There is no need to check all the four options. Just find out the oxidation state of metal ion. If oxidation state is highest and ligand present there is of electron donating nature, gives LMCT, which shows more intense colour.*

2. The correct statement(s) about Cr^{2+} and $Mn^{3+}$is/are [atomic number of Cr = 24 and Mn = 25] **[JEE Advanced 2015, More than One Option]**

(a) Cr^{2+} is a reducing agent

(b) Mn^{3+} is an oxidising agent

(c) both Cr^{2+} and Mn^{3+} exhibit d^4 electronic configuration

(d) when Cr^{2+} is used as a reducing agent, the chromium ion attains d^5 electronic configuration

Ans. **(a, b, c)** (a) In aqueous solution $Cr^{2+}(3d^4)$ acts as a reducing agent, oxidising itself to $Cr^{3+}(3d^3)$ that gives a completely half-filted t_{2g} level in octahedral ligand field of H_2O.

(b) $Mn^{3+}(3d^4)$ is an oxidising agent as it is reduced to $Mn^{2+}(3d^5)$, a completely half-filled stable configuration.

(c) Both Cr^{2+} and Mn^{3+} have d^4 configuration.

(d) $3d^4\ Cr^{2+}(aq) \xrightarrow{R.A} Cr^{3+}(aq) + e^-$

Hence (*d*) is wrong statement.

3. Consider the following list of reagents, acidified $K_2Cr_2O_7$, alkaline $KMnO_4$, $CuSO_4$, H_2O_2, Cl_2, O_3, $FeCl_3$, HNO_3 and $Na_2S_2O_3$. The total number of reagents that can oxidise aqueous iodide to iodine is **[JEE Advanced 2014, Integer Type]**

Ans. **(7)** Acidified $K_2Cr_2O_7$, $CuSO_4$, H_2O_2, Cl_2, O_3, $FeCl_3$ and HNO_3 oxidise *aq*. iodide to iodine. Alkaline $KMnO_4$ oxidise *aq*. iodide to IO_3^-. $Na_2S_2O_3$ is a strong reducing agent which on reaction with I_2 produces I^-.

$$Na_2S_2O_3 + I_2 \longrightarrow 2I^- + Na_2S_4O_6$$

Therefore, no reaction takes place between $Na_2S_2O_3$ and iodide ion. Hence, correct integer is (7).

4. The colour of light abosrbed by an aqueous solution of $CuSO_4$ is **[JEE Advanced 2012, Single Option]**

(a) orange-red (b) blue-green

(c) yellow (d) violet

Ans. **(a)** The aqueous solution of $CuSO_4$ consist of the complex $[Cu(H_2O)_4]^{2+}$ ion which absorbed in orange-red region and impart deep blue colouration to solution.

5. Reduction of the metal centre in aqueous permanganate ion involves **[IIT JEE 2011, More than One Option]**

(a) three electrons in neutral medium

(b) five electrons in neutral medium

(c) three electrons in alkaline medium

(d) five electrons in acidic medium

Ans. **(a,d)** In neutral medium : (Very dilute alkali solution)

$$MnO_4^- \longrightarrow MnO_2\ (Mn^{7+} + 3e^- \to Mn^{4+})$$

In strong alkaline medium $MnO_4^- \longrightarrow MnO_4^{2-}$

In acidic medium : $MnO_4^- \xrightarrow{+e^-} Mn^{2+}$ $(Mn^{7+} + 5e^- \to Mn^{2+})$

6. Which series of reactions correcly represents chemical relations related to iron and its compound? **[JEE Main 2014]**

(a) $Fe \xrightarrow{Dil.H_2SO_4} FeSO_4 \xrightarrow{H_2SO_4, O_2} Fe_2(SO_4)_3 \xrightarrow{Heat} Fe$

(b) $Fe \xrightarrow{O_2, heat} FeO \xrightarrow{Dil.\ H_2SO_4} FeSO_4 \xrightarrow{Heat} Fe$

(c) $Fe \xrightarrow{Cl_2, heat} FeCl_3 \xrightarrow{Heat, air} FeCl_2 \xrightarrow{Zn} Fe$

(d) $Fe \xrightarrow{O_2, heat} Fe_3O_4 \xrightarrow{CO, 600°C} FeO \xrightarrow{CO, 700°C} Fe$

Ans. **(d)** (a) $Fe + dil.H_2SO_4 \to FeSO_4 + H_2$

$$H_2SO_4 + 2FeSO_4 + \frac{1}{2}O_2 \longrightarrow Fe_2(SO_4)_3 + H_2O$$

$$Fe(SO_4)_3 \xrightarrow{\Delta} Fe_2O_3(s) + 3SO_3 \uparrow$$

The given reaction is incorrect in question.

(b) $Fe \xrightarrow{O_2} FeO$ (It could also be Fe_2O_3 or Fe_3O_4)

$FeO + H_2SO_4 \to FeSO_4 + H_2O$

$2FeSO_4 \xrightarrow{\Delta} Fe_2O_3 + SO_2 + SO_3$

(c) $Fe \xrightarrow[Cl_2]{\Delta} FeCl_3 \xrightarrow[air]{\Delta}$ No reaction, (It can not give $FeCl_2$)

(d) $Fe \xrightarrow[O_2]{\Delta} Fe_3O_4 \xrightarrow[600°C]{CO} FeO \xrightarrow[700°]{CO} Fe$

This is the correct equation.

7. Which of the following arrengements does not represent the correct order of the property stated against it? **[JEE Main 2013]**

(a) $V^{2+} < Cr^{2+} < Mn^{2+} < Fe^{2+}$ Paramagnetic behaviour

(b) $Ni^{2+} < Co^{2+} < Fe^{2+} < Mn^{2+}$ ionic size

(c) $Co^{3+} < Fe^{3+} < Cr^{3+} < Sc^{3+}$ stability in aqueous solution

(d) Sc < Ti < Cr < Mn number of oxidation states

Ans. **(d)** (a) $V^{2+} = 3$ unpaired e^-, $Cr^{2+} = 4$ unpaired e^-

$Mn^{2+} = 5$unpaired e^-, $Fe^{2+} = 4$unpaired e^-

Hence, the order of paramagnetic behaviour should be

$$V^{2+} < Cr^{2+} = Fe^{2+} < Mn^{2+}$$

(b) Ionic size decreases from left to right in the same period.

(c) $Co^{3+}/Co^{2+} = 1.97$; $Fe^{3+}/Fe^{2+} = 0.77$

$Cr^{3+}/Cr^{2+} = -0.41$ Sc^{3+} is highly stable because it does not show +2 oxidation state.

(d) The oxidation states increase as we go from group 3 to group 7 in the same period.

8. Four successive members of the first row transition elements listed below with atomic numbers. Which one of them is expected to have the highest $E^{\circ}_{M^{3+}/M^{2+}}$ value?

(a) Cr(Z = 24) (b) Mn(Z = 25) **[JEE Main 2013]**
(c) Fe(Z = 26) (d) Co(Z = 27)

Ans. **(d)** Standard reduction potential (SRP) value normally increases from left to right in the period of *d*-block elements. Some SRP value are exceptionally higher due to stability of product ion. *e.g.*,

$$E^{\circ}(Mn^{3+}/Mn^{2+}) = +1.57\text{ V}; E^{\circ}(Co^{3+}/Co^{2+}) = +1.97\text{ V}$$

Thus, $E^{\circ}(M^{3+}/M^{2+})$ is highest for Co.

9. Iron exhibits +2 and +3 oxidation states. Which of the following statements about iron is incorrect? **[AIEEE 2012]**

(a) Ferrous oxide is more basic in nature than the ferric oxide
(b) Ferrous compounds are relatively more ionic than the corresponding ferric compounds
(c) Ferrous compounds are less volatile than the corresponding ferric compounds
(d) Ferrous compounds are more easily hydrolysed than the corresponding ferric compounds

Ans. **(d)** (a) $FeO > Fe_2O_3$ (basic) correct.

(b) $FeCl_2 > FeCl_3$ (ionic), larger the charge, greater the polarising power thus, greater the covalent nature. Thus, correct.

(c) Fe^{2+} salts are more ionic thus, less volatile than Fe^{3+} salts. Thus, correct.

(d) Greater the covalent nature, more easily they are hydrolysed. Thus, $FeCl_3$ is more hydrolysed than $FeCl_2$. Thus, statement (d) is incorrect.

17. Metallurgical Extraction

1. Which one of the following ores is best concentrated by froth floatation method? **[JEE Main 2016]**

(a) Siderite
(b) Galena
(c) Malachite
(d) Magnetite

Ans. **(b)** Sulphide ores are concentrated by froth floatation method e.g. Galena (PbS)

2. Galvanisation is applying a coating of **[JEE Main 2016]**

(a) Cr (b) Cu
(c) Zn (d) Pb

Ans. **(c)** Zinc metal is the most stable metal to cover iron surfaces. The process of coating the iron surface by zinc is called galvanisation.

3. Extraction of copper from copper pyrite ($CuFeS_2$) involves **[JEE Advanced 2016, One or More Than One Option Correct]**

(a) crushing followed by concentration of the ore by froth-floatation
(b) removal of iron as slag
(c) self reduction step to produce 'blister copper' following evolution of SO_2
(d) refining of 'blister copper' by carbon reduction

Ans. **(a,b,c)** $CuFeS_2$ (copper pyrite) is converted into copper into following steps:

Step I Crushing (grinding) followed by concentration by froth-floatation process.

Step II Roasting of ore in the presence of SiO_2 which removes iron as slag ($FeSiO_3$).

$$2CuFeS_2 + O_2 \longrightarrow Cu_2S + 2FeS + SO_2$$
$$2FeS + 3O_2 \longrightarrow 2SO_2 + 2FeO$$
$$FeO + SiO_2 \longrightarrow FeSiO_3\text{(Slag)}$$

Step III Self-reduction in Bessemer converter

$$2Cu_2S + 3O_2 \longrightarrow 2Cu_2O + 2SO_2$$
$$2Cu_2O + Cu_2S \longrightarrow 6Cu + SO_2$$

Copper obtained is blister copper (98% pure).

Step IV Refining of blister copper is done by electrolysis

Impure copper — Anode
Pure copper — Cathode
At anode : $Cu \longrightarrow Cu^{2+} + 2e^-$
At cathode : $Cu^{2+} + 2e^- \longrightarrow Cu$

Carbon-reduction method is not used. Thus, (d) is incorrect.

4. In the context of the Hall-Heroult process for the extraction of Al, which of the following statements is false? **[JEE Main 2015]**

(a) CO and CO_2 are produced in this process
(b) Al_2O_3 is mixed with CaF_2 which lowers the melting point of the mixture and brings conductivity
(c) Al^{3+} is reduced at the cathode to form Al
(d) Na_3AlF_6 serves as the electrolyte

Ans. **(d)** (a) In Hall-Heroult process for extraction of Al, carbon anode is oxidised to CO and CO_2.

(b) When Al_2O_3 is mixed with CaF_2, it lowers the melting point of the mixture and brings conductivity.

(c) Al^{3+} is reduced at cathode to form Al.

(d) Here, Al_2O_3 is an electrolyte, undergoing the redox process. Na_3AlF_6 although is an electrolyte but serves as a solvent, not electrolyte.

5. Match the anionic species given in Column I that are present in the ore (s) given in Column II.

	Column I		Column II
A.	Carbonate	P.	Siderite
B.	Sulphide	Q.	Malachite
C.	Hydroxide	R.	Bauxite
D.	Oxide	S.	Calamine
		T.	Argentite

[JEE Advanced 2015, Matching Type]

Ans. **(d)** Siderite = $FeCO_3$, Malachite = $CuCO_3 \cdot Cu(OH)_2$
Bauxite = $Al_2O_3 \cdot 2H_2O_2$ consisting some $Al(OH)_3$
Calamine = $ZnCO_3$, Argentite = Ag_2S

6. Copper is purified by electrolytic refining of blister copper. The correct statement(s) about this process is/are **[JEE Advanced 2015, Single Option]**

(a) impure Cu strip is used as cathode
(b) acidified aqueous $CuSO_4$ is used as electrolyte
(c) pure Cu deposits at cathode
(d) impurities settle as anode-mud

Ans. (d)

(a) is wrong statement. Impure copper is set as anode where copper is oxidised to Cu^{2+} and goes into electrolytic solutions.

(b) $CuSO_4$ is used as an electrolyte in purification process.

(c) Pure copper is deposited at cathode as:

$$Cu^{2+} + 2e^- \longrightarrow Cu: \quad \text{(At cathode)}$$

(d) Less active metals like Ag, Au etc., settle down as anode mud.

7. Upon heating with Cu_2S, the reagent(s) that give copper metal is/are **[JEE Advanced 2014, More than One Option]**

(a) $CuFeS_2$ (b) CuO
(c) Cu_2O (d) $CuSO_4$

Ans. (b, c, d) (b) $4CuO \xrightarrow{1100°C} 2Cu_2O + O_2$

$$2Cu_2O + Cu_2S \xrightarrow{\Delta} 6Cu + SO_2$$

(c) $Cu_2S + 2Cu_2O \xrightarrow{\Delta} 6Cu + SO_2$

(d) $CuSO_4 \xrightarrow{720°C} CuO + SO_2 + \frac{1}{2}O_2$

$$4CuO \xrightarrow{1100°C} 2Cu_2O + O_2$$

$$2Cu_2O + Cu_2S \xrightarrow{\Delta} 6Cu + SO_2$$

Reaction is believed to proceed as

$$Cu_2S \rightleftharpoons 2Cu^+ + S^{2-}$$

$$2Cu_2O \rightleftharpoons 4Cu^+ + 2O^{2-}$$

$$S^{2-} + 2O^{2-} \longrightarrow SO_2 + 6e^-$$

$$6Cu^+ + 6e^- \longrightarrow 6Cu; \; E^\circ_{cell} = 0.52$$

Here, copper sulphide is reduced to copper metal. Solidified copper has blistered appearance due to evolution of SO_2 and thus obtained copper is known as blister copper.

Other compounds which give Cu are

(i) CuO as $4CuO \xrightarrow{1100°C} 2Cu_2O + O_2$

$$2Cu_2O + Cu_2S \xrightarrow{\Delta} 6Cu + SO_2$$

(ii) $CuSO_4$ as $CuSO_4 \xrightarrow{720°C} CuO + SO_2 + \frac{1}{2}O_2$

$$4CuO \xrightarrow{\Delta} 2Cu_2O + O_2$$

$$2Cu_2O + Cu_2S \xrightarrow{\Delta} 6Cu + SO_2$$

While $CuFeS_2$ will not give Cu on heating. The heating in the presence of O_2 gives Cu_2S and FeS with the evolution of SO_2.

8. Sulphide ores are common for the metals **[JEE Advanced 2013, Single Option]**

(a) Ag, Cu and Pb (b) Ag, Cu and Sn
(c) Ag, Mg and Pb (d) Al, Cu and Pb

Ans. (a)

Element	Ores	Name
Ag	Ag_2S	Argentite
Cu	$CuFeS_2$	Copper pyrites
Pb	PbS	Galena
Sn	SnO_2	Cassiterite
Mg	$MgCO_3 \cdot CaCO_3$	Dolomite
Al	$Al_2O_3 \cdot xH_2O$	Bauxite

9. The carbon-based reduction method is not used for the extraction of **[JEE Advanced 2013, More than One Option]**

(a) tin from SnO_2
(b) iron from Fe_2O_3
(c) aluminium from Al_2O_3
(d) magnesium from $MgCO_3$, $CaCO_3$

Ans. (c, d) Al has greater affinity for oxygen hence oxide is not reduced by carbon. MgO and CaO (formed in the calcination from carbonates) are stable species and not reduced by carbon.

During smelting $SnO_2 + C \xrightarrow{1300°C} Sn + CO$

$$2Fe_2O_3 + 3C \xrightarrow{\Delta} 4Fe + 3CO_2$$

10. In the cyanide extraction process of silver from argentite ore, the oxidizing and reducing agents used are **[JEE Advanced 2012, Single Option]**

(a) O_2 and CO respectively
(b) O_2 and Zn dust respectively
(c) HNO_3 and Zn dust respectively
(d) HNO_3 and CO respectively

Ans. (b) The reactions involved in extraction of silver by cyanide process are

$$Ag_2S + CN^- + O_2 \longrightarrow [Ag(CN)_2]^- + SO_2 \quad \ldots(i)$$

$$[Ag(CN)_2]^- + Zn \longrightarrow [Zn(CN)_4]^{2-} + Ag \quad \ldots(ii)$$

In reaction (i), sulphide is oxidized to SO_2 by oxygen. In the reaction (ii), silver ion (Ag^+) is reduced to Ag by Zn. Therefore, O_2 is oxidising agent and Zn is reducing agent.

11. Oxidation states of the metal in the minerals haematite and magnetite, respectively, are **[IIT JEE 2011, Single Option]**

(a) II, III in haematite and III in magnetite
(b) II, III in haematite and II in magnetite
(c) II in haematite and II, III in magnetite
(d) III in haematite and II, III in magnetite

Ans. (d) Haematite is Fe_2O_3, in which oxidation number of iron is III. Magnetite is Fe_3O_4 which is infact a mixed oxide ($FeO \cdot Fe_2O_3$), hence, iron is present in both II and III oxidation state.

12. Extraction of metal from the ore cassiterite involves **[IIT JEE 2011, More than One Option]**

(a) carbon reduction of an oxide ore
(b) self-reduction of a sulphide ore
(c) removal of copper impurity
(d) removal of iron impurity

Ans. (a, d) The important ore of tin is cassiterite (SnO_2). Tin is extracted from cassiterite ore by carbon reduction method in a blast furnace.

$$SnO_2 + 2C \longrightarrow Sn + 2CO$$

The product often contain traces of iron which is removed by blowing air through the melt to oxidise to FeO which then floats to the surface.

$$2Fe + O_2 \longrightarrow 2FeO$$

13. Copper is the most noble of the first row transition metals and occurs in small deposits in several countries. Ores of copper include chalcanthite ($CuSO_4 \cdot 5H_2O$), atacamite ($Cu_2Cl(OH)_3$), cuprite (Cu_2O), copper glance (Cu_2S) and malachite ($Cu_2(OH)_2CO_3$). However, 80% of the world copper production comes from the ore chalcopyrite ($CuFeS_2$). The extraction of copper from chalcopyrite involves partial roasting, removal of iron and self-reduction. **[IIT JEE 2010, Passage Based]**

(i) Partial roasting of chalcopyrite produces

(a) Cu_2S and FeO (b) Cu_2O and FeO
(c) CuS and Fe_2O_3 (d) Cu_2O and Fe_2O_3

Ans. (b) $2CuFeS_2 + O_2 \longrightarrow Cu_2S + 2FeS + SO_2 \uparrow$

$$2Cu_2S + 3O_2 \longrightarrow 2Cu_2O + 2SO_2 \uparrow$$
$$2FeS + 3O_2 \longrightarrow 2FeO + 2SO_2 \uparrow$$

(ii) Iron is removed from chalcopyrite as

(a) FeO (b) FeS
(c) Fe_2O_3 (d) $FeSiO_3$

Ans. (d) $FeO + SiO_2 \longrightarrow FeSiO_3$ (slag)

(iii) In self-reduction, the reducing species is

(a) S (b) O^{2-}
(c) S^{2-} (d) $3O_2$

Ans. (c) $Cu_2S + 2Cu_2O \longrightarrow 6Cu + SO_2$

In Cu_2S, sulphur is S^{2-} and in SO_2, sulphur is in +4 state.

Hence, S^{2-} is acting as reducing agent.

14. The metal that can not be obtained by electrolysis of an aqueous solution of its salts is **[JEE Main 2014]**

(a) Ag (b) Ca
(c) Cu (d) Cr

Ans. (b) Higher the position of element in the electrochemical series more difficult is the reduction of its cations. If $Ca^{2+}(aq)$ is electrolysed, water is reduced in preference to it. Hence, it can not be reduced electrolytically from their aqueous solutions.

$$Ca^{2+}(aq) + H_2O \longrightarrow Ca(OH)_2 + H_2 \uparrow$$

15. Which method of purification is represented by the following equation? **[AIEEE 2012]**

$$Ti(s) + 2I_2(g) \xrightarrow{523\text{ K}} TiI_4(g) \xrightarrow{1700\text{ K}} Ti(s) + 2I_2(g)$$

(a) Zon refining (b) Cupellation
(c) Poling (d) van-Arkel

Ans. (d) van-Arkel (and de-Boer) process is based on the thermal decomposition of a volatile compound like an iodide, which is first formed by direct combination of metal to be purified and iodine. Metal formed is an purest form. Titanium and zirconium are purified by this method.

18. Coordination Compounds

1. Addition of excess aqueous ammonia to a pink coloured aqueous solution of $MCl_2 \cdot 6H_2O(X)$ and NH_4Cl gives an octahedral complex Y in the presence of air. In aqueous solution, complex Y behaves as 1 : 3 electrolyte. The reaction of X with excess HCl at room temperature results in the formation of a blue colured complex Z. The calculated spin only magnetic moment of X and Z is 3.87 B.M., whersas it is zero for complex Y.

Among the following options, which statement(s) is (are) correct? **[JEE Advanced 2017, One or More Than Correct Option]**

(a) The hybridisation of the central metal ion in Y is d^2sp^3
(b) Addition of silver nitrate to Y given only two equivalents of silver chloride
(c) When X and Y are in equilibrium at 0°C, the colour of the solution is pink
(d) Z is a tetrahedral complex

Ans. (a, c, d) $\underset{\text{Pink }(X)}{[Co(H_2O)_6]Cl_2} \xrightarrow[O_2\text{(Air)}]{\text{Excess } NH_4OH/NH_4Cl} \underset{Y}{Co(NH_3)_6]Cl_3}$

$$\underset{X}{[Co(H_2O)_6]^{2+}} + \underset{\text{(Excess)}}{4Cl^-} \longrightarrow \underset{\text{blue } Z}{[CoCl_4]^{2-}}$$

(a) Since NH_3 is moderately strong ligand, hybridisation of cobalt in Y is d^2sp^3.
(b) Cobalt is sp^3-hybridised in $[CoCl_4]^{2-}$.
(c) $[Co(NH_3)_6]Cl_3 + 3AgNO_3(aq) \longrightarrow 3AgCl \downarrow$
(d) $\underset{\text{Blue}}{[CoCl_4]^{2-}} + 6H_2O \rightleftharpoons \underset{\text{Pink}}{[Co(H_2O)_6]^{2+}} + 4Cl^-;\ \Delta H < 0$

2. Which of the following compounds is metallic and ferromagnetic? **[JEE Main 2016]**

(a) CrO_2 (b) VO_2
(c) MnO_2 (d) TiO_2

Ans. (a) Only three elements iron (Fe), cobalt (Co) and nickel (Ni) show ferromagnetism at room temperature. CrO_2 is also a metallic and ferromagnetic compound which is used to make magnetic tapes for cassette recorders.

3. The pair having the same magnetic moment is [at. no. Cr = 24, Mn = 25, Fe = 26 and Co = 27] **[JEE Main 2016]**

(a) $[Cr(H_2O)_6]^{2+}$ and $[Fe(H_2O)_6]^{2+}$
(b) $[Mn(H_2O)_6]^{2+}$ and $[Cr(H_2O)_6]^{2+}$
(c) $[CoCl_4]^{2-}$ and $[Fe(H_2O)_6]^{2+}$
(d) $[Cr(H_2O)_6]^{2+}$ and $[CoCl_4]^{2-}$

Ans. (a)

Complex ion	Electronic configuration of metal ion	Number of unpaired electrons (n)
$[Cr(H_2O)_6]^{2+}$	Cr^{2+}; [Ar] $3d^4$	↑ ↑ ↑ ↑ □ ; 4
$[Fe(H_2O)_6]^{2+}$	Fe^{2+}; [Ar] $3d^6$	↑↓ ↑ ↑ ↑ ↑ ; 4
$[Mn(H_2O)_6]^{2+}$	Mn^{2+}; [Ar] $3d^5$	↑ ↑ ↑ ↑ ↑ ; 5
$[CoCl_4]^{2-}$	Co^{2+}; [Ar] $3d^7$	↑↓ ↑↓ ↑ ↑ ↑ ; 3

4. Among $[Ni(CO)_4]$, $[NiCl_4]^{2-}$, $[Co(NH_3)_4Cl_2]$ Cl, $Na_3[CoF_6]$, Na_2O_2 and CsO_2, the total number of paramagnetic compounds is **[JEE Advanced 2016, Single Option Correct]**

(a) 2 (b) 3
(c) 4 (d) 5

Ans. (b)

Compounds	Hybridisation	Unpaired electron(s)	Magnetic character
$Ni(CO)_4$	sp^3	No	Diamagnetic
$[NiCl_4]^{2-}$	sp^3	two	Paramagnetic
$[Co(NH_3)_4Cl_2]$ Cl	sp^3d^2	No	Diamagnetic
$Na_3[CoF_6]$	sp^3d^2	three	Paramagnetic
Na_2O_2	—	No	Diamagnetic (O_2^{2-})
CsO_2	—	One	Paramagnetic
			O_2^- (superoxide ion is paramagnetic)

5. The geometries of the ammonia complexes of Ni^{2+}, Pt^{2+} and Zn^{2+}, respectively, are **[JEE Advanced 2016, Single Option Correct]**

(a) octahedral, square planar and tetrahedral
(b) square planar, octahedral and tetrahedral
(c) tetrahedral, square planar and octahedral
(d) octahedral, tetrahedral and square planar

Ans. (a) $[Ni(NH_3)_6]^{2+}$ sp^3d^2 octahedral
$[Pt(NH_3)_4]^{2+}$ dsp^2 square planar
$[Zn(NH_3)_4]^{2+}$ sp^3 tetrahedral

6. Which of the following compounds is not coloured yellow? **[JEE Main 2015]**

(a) $Zn_2[Fe(CN)_6]$
(b) $K_3[Co(NO_2)_6]$
(c) $(NH_4)_3[As(Mo_3O_{10})_4]$
(d) $BaCrO_4$

Ans. (a) Zn_2 $[Fe (CN)_6]$, $K_3[Co (NO_2)_6]$ and $[(NH_4)_3$ As $(Mo_3O_{10})_4]$ show colour due to *d-d* transition while $BaCrO_4$ is coloured due to charge transfer phenomenon. Further, according to spectrochemical series the strong ligand possessing complex has higher energy and hence lower wavelength. Therefore, complexes containing NO_2, NH_4^+, O^{2-} etc ligands show yellow colour while CN^- forces the complex to impart white colour.

Spectrochemical series

$$I^- < Br^- < S^{2-} < SCN^- < Cl^- < NO_3^- < N_3^- < F^- < OH^- < C_2O_4^{2-} \approx H_2O < NCS^- < CH_3CN < py < NH_3 < en < bipy < Phen < NO_2^- < PPh_3 < CN^- \approx CO$$

7. The number of geometric isomers that can exist for square planar $[Pt (Cl) (py) (NH_3) (NH_2OH)]^+$ is (py = pyridine) **[JEE Main 2015]**

(a) 2 (b) 3
(c) 4 (d) 6

Ans. (b) $[Pt(Cl) (py) (NH_3) (NH_2OH)]^+$ is square planar complex. The structures are formed by fixing a group and then arranging all the groups.

8. If the freezing point of a 0.01 molal aqueous solution of a cobalt (III) chloride-ammonia complex (which behaves as a strong electrolyte) is $-$ 0.0558°C, the number of chloride(s) in the coordination sphere of the complex is [K_f of water = 1.86 Kkgmol^{-1}] **[JEE Advanced 2015, Integer Type]**

Ans. (1) + $\Delta T_f = iK_f m$

$$\Delta T_f = 0-(-0.0558°C) = 0.0558°C$$

$$\Rightarrow i \text{ (vant Hoff's factor)} = \frac{0.0558}{1.86 \times 0.01} = 3$$

This indicates that complex upon ionisation produces three ions as:

$$[Co(NH_3)_5Cl]Cl_2 \rightarrow [Co(NH_3)_5Cl]^{2+}(aq) + 2Cl^-(aq)$$

Thus, only one Cl is inside the coordination sphere.

9. For the octahedral complexes of Fe^{3+} in SCN^- (thiocyanato-S) and in CN^- ligand environments, the difference between the spin only magnetic moments in Bohr magnetons (when approximated to the nearest integer) is [atomic number of Fe = 26] **[JEE Advanced 2015, Integer Type]**

Ans. (4) When S is donor atom of SCN^-, it produces weak ligand field and forms high spin complex as
$[Fe(SCN)_6]^{3-}$: $Fe^{3+}(3d^5)$ =

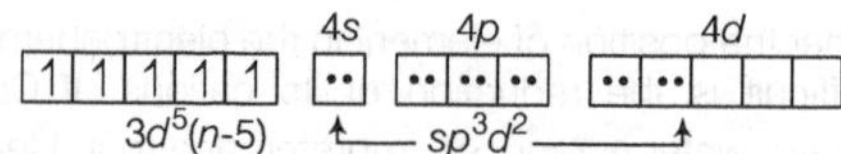

Spin only magnetic moment (μ_s) = $\sqrt{5(5+2)}$ BM = $\sqrt{35}$ BM

In case of CN^- ligand, carbon is the donor atom , it produces strong ligand field and forms low spin complex as $[Fe(CN)_6]^{3-}$: $Fe^{3+}(3d^5)$

3d 4s 4p

⇅ ⇅ 1 •• •• | •• | •• •• ••

d^2sp^3

Spin only magnetic moment (μ_s)

$$= \sqrt{1(1+2)} \text{ BM} = \sqrt{3} \text{ BM}$$

Hence, difference in spin only magnetic moment

$$= \sqrt{35} - \sqrt{3} \approx 4 \text{ BM}$$

10. In the complex acetylbromidodicarbonylbis (triethylphosphine) iron (II), the number of Fe—C bond (s) is **[JEE Advanced 2015]**

Ans. (b)

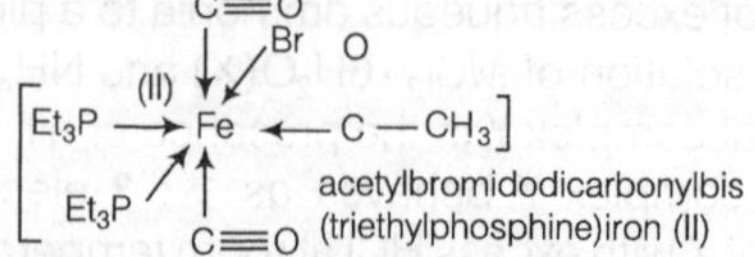

11. Among the complex ions, $[Co(NH_2—CH_2—CH_2—NH_2)_2Cl_2]^+$, $[CrCl_2(C_2O_4)_2]^{3-}$, $[Fe(H_2O)_4(OH)_2]^+$, $[Fe(NH_3)_2(CN)_4]^-$, $[Co(NH_2—CH_2—CH_2—NH_2)_2(NH_3)Cl]^{2+}$ and $[Co(NH_3)_4(H_2O)Cl]^{2+}$ the number of complex ion(s) that show(s) *cis-trans* isomerism is

[JEE Advanced 2015, Integer Type]

Ans. **(6)** All six complex will show *cis-trans* isomerism

Cis and *Trans*

and *Trans*

Cis and *Trans*

Cis and *Trans*

Trans and *Cis*

Cis and *Trans*

Cis and *Trans*

12. Consider the following complex ions, *P*, *Q* and *R*.

$P = [FeF_6]^{3-}$, $Q = [V(H_2O)_6]^{2+}$ and $R = [Fe(H_2O)_6]^{2+}$

The correct order of the complex ions, according to their spin-only magnetic moment values (in BM) is

[JEE Advanced 2013, Single Option]

(a) $R < Q < P$ (b) $Q < R < P$

(c) $R < P < Q$ (d) $Q < P < R$

Ans. **(b)** In presence of strong ligand (as CN^-, CO, NH_3, en) electrons are paired and electrons from ligands are filled in available inner orbitals

Complex	Atomic number of	O.N.	E.C.	Unpaired electrons	Magnetic moment
$P : [FeF_6]^{3-}$ weak ligand	26	+ 3	$[Ar]3d^5$	5	$\sqrt{35}$ BM
$Q : [V(H_2O)_6]^{2+}$ weak ligand	23	+2	[Ar]	3	$\sqrt{15}$ BM
$R : [Fe(H_2O)_6]^{2+}$	26	+2	$[Ar]3d^6$	4	$\sqrt{24}$ BM

Thus, order of spin-only magnetic moment = $Q < R < P$

13. The pair(s) of coordination complexes/ions exhibiting the same kind of isomerism is(are)

[JEE Advanced 2013, More than One Option]

(a) $[Cr(NH_3)_5Cl]Cl_2$ and $[Cr(NH_3)_4Cl_2]Cl$

(b) $[Co(NH_3)_4Cl_2]^+$ and $[Pt(NH_3)_2(H_2O)Cl]^+$

(c) $[CoBr_2Cl_2]^{2-}$ and $[PtBr_2Cl_2]^{2-}$

(d) $[Pt(NH_3)_3(NO_3)]Cl$ and $[Pt(NH_3)_3Cl]Br$

Ans. **(b,d)** Depending on the structure of the complex, different types of isomerism are shown.

	Complex	Isomerism
A.	$[Cr(NH_3)_5Cl]Cl_2$ $(C_2(NH_3)_4Cl_2)Cl$	Neither of structural nor stereo isomerism
B.	$[Co(NH_3)_4Cl_2]^+$	*trans* w.r.t. Cl; *cis* w.r.t. Cl

	Complex	Isomerism
	$[Pt(H_2O)\cdot(NH_3)_2Cl]^+$	H_2O, NH_3, Pt, Cl, NH_3 ; H_2O, NH_3, Pt, H_3N, Cl *trans*
C.	$[Co\,Br_2Cl_2]^{2-}$	sp^3tetrahedral
	$[PtBr_2Cl_2]^{2-}$	dsp^2-square planar
D.	$[Pt\,(NH_3)_3(NO_3)]Cl$	$[Pt\,(NH_3)_3(NO_3)]\,Cl$ / $[Pt\,(NH_3)_3\,Cl]\,NO_3$ ionisation
	$[Pt(NH_3)_3\,Cl]Br$	$[Pt\,(NH_3)_3\,Cl]\,Br$ / $[Pt\,(NH_3)_3\,Br]\,Cl$ ionisation

14. $EDTA^{4-}$ is ethylenediaminetetraacetate ion. The total number of N— Co — O bond angles in $[Co(EDTA)]^-$ complex ion is **[JEE Advanced 2013, Integer Type]**

Ans. (8) **EDTA** is a multidentate ligand as it can donate **six** pairs of electrons – two pair from the two nitrogen atoms and four pair from the four terminal oxygens of the $—COO^-$ groups.

$$\begin{matrix} \bar{O}OCCH_2 \diagdown & & \diagup CH_2COO^- \\ & \diagup \ddot{N} - CH_2CH_2 - \ddot{N} \diagdown & \\ \bar{O}OCCH_2 & & CH_2COO^- \end{matrix}$$

The structure of a chelate of a divalent Co^{2+} with EDTA is shown as

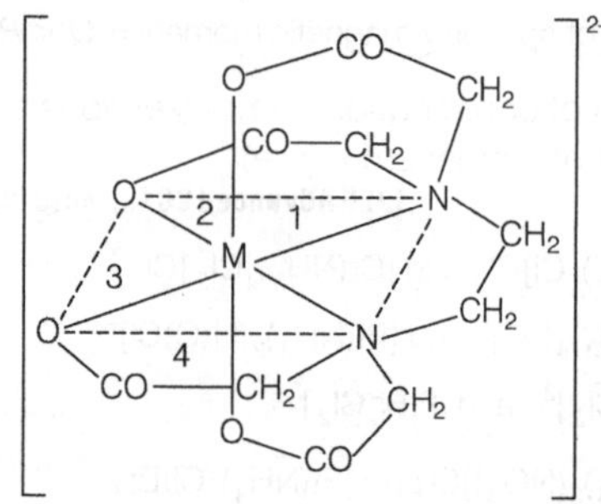

Each N has four N—Co—O bonds thus total eight N—Co—O bonds.

15. As per IUPAC nomenclature, the name of the complex $[Co\,(H_2O)_4(NH_3)_2]Cl_3$ is **[JEE Advanced 2012, Single Option]**

(a) tetraaquadiamminecobalt (III) chloride
(b) tetraaquadiamminecobalt (III) chloride
(c) diaminetetraaquacobalt (III) chloride
(d) diamminetetraaquacobalt (III) chloride

Ans. (d) First of all, the compound has complex positive part '$[Co(H_2O)_4(NH_3)_2]^{3+}$' therefore, according to IUPAC conventions, positive part will be named first. Secondly, in writing name of complex, ligands are named first in alphabetical order, irrespective of its charge, hence "ammine" will be written prior to "aqua".

16. $NiCl_2[P(C_2H_5)_2(C_6H_5)]_2$ exhibits temperature dependent magnetic behaviour (paramagnetic/diamagnetic) the coordination geometries of Ni^{2+} in the paramagnetic and diamagnetic states are respectively **[JEE Advanced 2012, Single Option]**

(a) tetrahedral and tetrahedral
(b) square planar and square planar
(c) tetrahedral and square planar
(d) square planar and tetrahedral

Ans. (c) In the given complex, $NiCl_2[P(C_2H_5)_2(C_6H_5)]_2$ nickel is in +2 oxidation state and the ground state electronic configuration of Ni^{2+} ion in free gaseous state is

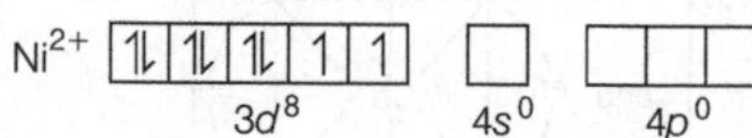

For the given four-coordinated complex to be paramagnetic, it must possess unpaired electrons in the valence shell. To satisfy this condition, four lone pairs from the four ligands occupies the four sp^3 hybrid orbitals as :

Ni^{2+} [1L|1L|1L|1|1] [••] [••|••|••] — sp^3

Therefore, geometry of paramagnetic complex must be tetrahedral. On the otherhand, for complex to be diamagnetic, there should not be any unpaired electrons in the valence shell. This condition can be fullfilled by pairing electrons of 3*d* orbitals against Hund's Rule as

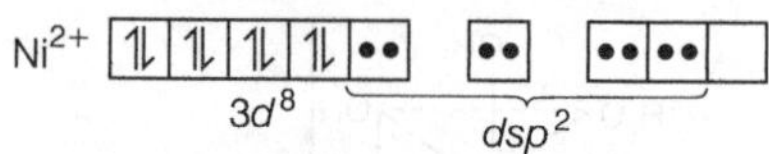

The above electronic arrangement gives dsp^2 hybridisation and therefore, **square planar** geometry to the complex.

17. Geometrical shapes of the complexes formed by the reaction of Ni^{2+} with Cl^-, CN^- and H_2O, respectively, are

(a) octahedral, tetrahedral and square planar **[IIT JEE 2011, Single Option]**
(b) tetrahedral, square planar and octahedral
(c) square planar, tetrahedral and octahedral
(d) octahedral, square planar and octahedral

Ans. (b) $Ni^{2+} + 4CN^- \rightarrow [Ni(CN)_4]^{2-}$

Here Ni^{2+} has d^8-configuration with CN^- as strong ligand.

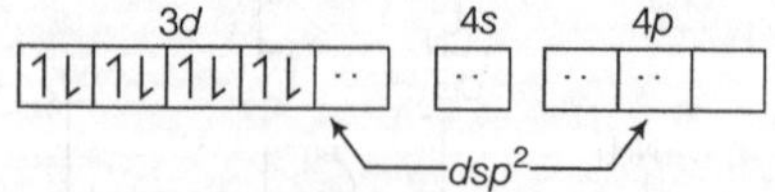

d^8-configuration in strong ligand field gives dsp^2hybridisation, hence square planar geometry.

$$Ni^{2+} + 4Cl^- \longrightarrow [NiCl_4]^{2-}$$

Here Ni^{2+} has d^8-configuration with Cl^- as weak ligand.

d^8-configuration in weak ligand field gives sp^3 hydridisation, hence tetrahedral geometry.

Ni^{2+} with H_2O forms $[Ni(H_2O)_6]^{2+}$complex and H_2O is a weak ligand.

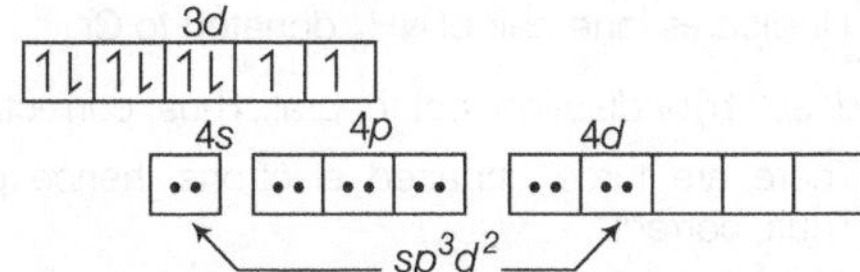

Therefore, $[Ni(H_2O)_6]^{2+}$ has octahedral geometry.

18. Among the following complexes *(K-P)*, $K_3[Fe(CN)_6]$ *(K)*, $[Co(NH_3)_6]Cl_3$ *(L)*, $Na_3[Co(ox)_3]$ *(M)*, $[Ni(H_2O)_6]Cl_2$ *(N)*, $K_2[Pt(CN)_4]$ *(O)* and $[Zn(H_2O)_6](NO_3)_2$ *(P)* the diamagnetic complexes are **[IIT JEE 2011, Single Option]**

(a) *K, L, M, N*
(b) *K, M, O, P*
(c) *L, M, O, P*(d) *L, M, N, O*

Ans. (c) For a diamagnetic complex, there should not be any unpaired electron in the valence shell of central metal.

In $K_3[Fe(CN)_6]$, Fe (III) has d^5-configuration (odd electrons), hence it is paramagnetic.

In $[Co(NH_3)_6]Cl_3$, Co (III) has d^6-configuration in a strong ligand field, hence all the electrons are paired and the complex is diamagnetic.

In $Na_3[Co(ox)_3]$, Co (III) has d^6-configuration and oxalate being a chelating ligand, very strong ligand and all the six electrons remains paired in lower t_{2g} level, diamagnetic.

In $[Ni(H_2O)_6]Cl_2$, Ni (II) has $3d^8$-configuration and H_2O is a weak ligand hence

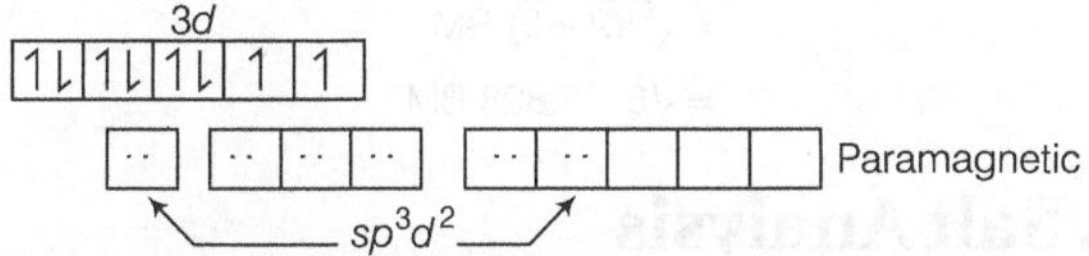

In $K_2[Pt(CN)_4]$, Pt(II) has d^8-configuration and CN^- is a strong ligand, hence all the eight electrons are spin-paired . Therefore, complex is diamagnetic.

In $[Zn(H_2O)_6](NO_3)_2$, Zn (II) has $_3d^{10}$configuration with all the ten electrons spin paired, hence diamagnetic.

19. The ionisation isomer of $[Cr(H_2O)_4Cl(NO_2)]Cl$ is **[IIT JEE 2010, Single Option]**

(a) $[Cr(H_2O)_4(O_2N)]Cl_2$
(b) $[Cr(H_2O)_4Cl_2](NO_2)$
(c) $[Cr(H_2O)_4Cl(ONO)]Cl$
(d) $[Cr(H_2O)_4Cl_2(NO_2)]\cdot H_2O$

Ans. (b) Ionisation isomers are the complexes that produces different ions in solution, *ie*, they have ions interchanged inside and outside the coordination sphere.

$[Cr(H_2O)_4Cl(NO_2)]Cl$ and $[Cr(H_2O)_4Cl_2](NO_2)$ have different ions inside and outside the coordinate sphere and they are isomers. Therefore, they are ionisation isomers.

20. The correct structure of ethylenediaminetetraacetic acid (EDTA) is **[IIT JEE 2010, Single Option]**

(a) $(HOOCCH_2)_2N-CH=CH-N(CH_2COOH)_2$

(b) $(HOOC)_2N-CH-CH-N(COOH)_2$

(c) $(HOOCCH_2)_2N-CH_2-CH_2-N(CH_2COOH)_2$

Ans. (c) $(HOOCH_2C)_2N-CH_2CH_2-N(CH_2COOH)_2$

21. The complex showing a spin-only magnetic moment of 2.82 BM is **[IIT JEE 2010, Single Option]**

(a) $Ni(CO)_4$ (b) $[NiCl_4]^{2-}$
(c) $Ni(PPh_3)_4$ (d) $[Ni(CN)_4]^{2-}$

Ans. (b) Magnetic moment = 2.83 BM indicates that there are two unpaired electrons.

$$u=\sqrt{n(n+2)}\text{ BM}$$
$$=\sqrt{8}\text{ BM}$$
$$=2.82\text{ BM}$$

In $[NiCl_4]^{2-}$, Ni has d^8 configuration and Cl^- is a weak ligand :

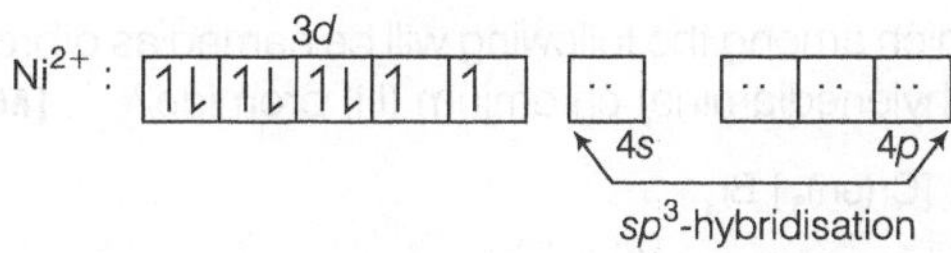

22. The octahedral complex of a metal ion M^{3+} with four monodentate ligands L_1, L_2, L_3 and L_4 absorb wavelengths in the region of red, green, yellow and blue respectively. The increasing order of ligand strength of the four ligands is **[JEE Main 2014]**

(a) $L_4 < L_3 < L_2 < L_1$
(b) $L_1 < L_3 < L_2 < L_4$
(c) $L_3 < L_2 < L_4 < L_1$
(d) $L_1 < L_2 < L_4 < L_3$

Ans. (b) Ligand field strength ∝ energy of light absorbed ∝1/ wavelength of light absorbed.

λ	L_1	L_2	L_3	L_4
Absorbed light	red	green	yellow	blue

Wavelength of absorbed light decreases.

Increasing order of energy of wavelengths absorbed reflect greater extent of crystal field splitting. Hence, higher field strength of the ligand.

Energy: blue (L_4) > green (L_2) > yellow (L_3)> red (L_1)

$\therefore L_4 > L_2 > L_3 > L_1 \rightarrow$ decreasing field strength of ligands.

23. The equation which is balanced and represents the correct product (s) is **[JEE Main 2014]**

(a) $Li_2O + 2KCl \rightarrow 2LiCl + K_2O$

(b) $[CoCl(NH_3)_5]^+ + 5H^+ \rightarrow Co^{2+} + 5NH_4^+ + Cl^-$

(c) $[Mg(H_2O)_6]^{2+} + (EDTA)^{4-} \xrightarrow[\text{NaOH}]{\text{Excess}} [Mg(EDTA)]^{2+} + 6H_2O$

(d) $CuSO_4 + 4KCN \longrightarrow K_2[Cu(CN)_4] + K_2SO_4$

Ans. **(b)** (a) $Li_2O + KCl \rightarrow 2LiCl + K_2O$ (wrong equation because a stronger base K_2O cannot be generated by a weaker base)

(b) $[CoCl(NH_3)_5]^+ + 5H^+ \rightarrow Co^{2+}(aq) + 5NH^{4+} + Cl^-$ (This is correct, all ammine complexes can be destroyed by adding H^+)

(c) $[Mg(H_2O)_6]^{2+} + EDTA^{4-} \xrightarrow[\text{excess}]{OH^-} [Mg(EDTA)]^{2+} + 6H_2O$

This is wrong equation. Correct equation is

$Mg^{2+} + (EDTA)^{4-}$ s $[Mg(EDTA)]^{2-}$

(d) The given reaction is incorrect. It can be correctly represented as

$$2CuSO_4 + 10KCN \rightarrow 2K_3[Cu(CN)_4] + 2K_2SO_4 + (CN)_2$$

24. Which of the following complex species is not expected to exhibit optical isomerism? **[JEE Main 2013]**

(a) $[Co(en)_3]^{3+}$ (b) $[Co(en)_2Cl_2]^+$

(c) $[Co(NH_3)_3Cl_3]$ (d) $[Co(en)(NH_3)Cl_2]^+$

Ans. **(c)** Optical isomerism is exhibited by only those complexes which lacks elements of symmetry. $[Co(NH_3)_3Cl_3]$ shows facial as well as meridional isomerism. But both form contain plane of symmetry. Thus, this complex does not exhibit optical isomerism.

25. Which among the following will be named as dibromidobis (ethylenediamine) chromium (III) bromide? **[AIEEE 2012]**

(a) $[Cr(en)_3]Br_3$

(b) $[Cr(en)_2Br_2]Br$

(c) $[Cr(en)Br_4]^-$

(d) $[Cr(en)Br_2]Br$

Ans. **(b)** Two Br, two (en) and one Cr are parts of complex. Charge on the complex is

$$\left.\begin{array}{l} 2(Br) = -2 \\ 2(en) = 0 \\ 1\ (Cr) = +3 \end{array}\right\} = +1$$

Thus, complex ion is $[Cr(en)_2Br_2]^+$

Since, anion is bromide thus, complex is $[Cr(en)_2Br_2]$ Br.

26. Which of the following facts about the complex $[Cr(NH_3)_6]Cl_3$ is wrong? **[AIEEE 2011]**

(a) The complex involves d^2sp^3 hybridisation and is octahedral in shape

(b) The complex is paramagnetic

(c) The complex is an outer orbital complex

(d) Complex gives white precipitate with silver nitrate solution

Ans. **(c)** $\underbrace{[Cr(NH_3)_6]}_{\text{Coordinate sphere}} \underbrace{Cl_3}_{\text{Ionisable}} \rightleftharpoons [Cr(NH_3)_6]^{3+} + 3Cl^-$

$$\xrightarrow{AgNO_3} \underset{\text{White precipitate}}{AgCl \downarrow}$$

$Cr(24) = [Ar]\,3d^5\,4s^1$

$Cr^{3+} = [Ar]\,3d^3\,4s^0$

$[G(NH_3)_6]^{3+}$ [1|1|1|☒|☒] [☒] [☒|☒|☒]

☒ Indicates lone pair of NH_3 donated to Cr.

(a) d^2sp^3-hybridisation, octahedral. Thus, correct.

(b) There are three unpaired electrons, hence paramagnetic. Thus, correct.

(c) d^2sp^3-inner orbital complex, thus statement (c) is incorrect.

(d) Due to ionisable Cl^- ions, white precipitate with $AgNO_3$, thus correct.

27. The magnetic moment (spin only) of $[NiCl_4]^{2-}$ is **[AIEEE 2011]**

(a) 1.82 BM (b) 5.46 BM

(c) 2.82 BM (d) 1.41 BM

Ans. **(c)** $[NiCl_4]^{2-}$; oxidation number of

Ni, $x - 4 = -2$

$\therefore$ $x = +2$

$Ni_{(28)} = [Ar]\,3d^8, 4s^2$

$Ni^{2+} = [Ar]$ $3d^8$ [⥮|⥮|⥮|1|1] $4s^0$ [ ] $4p$ [| |]

$[NiCl_4]^{2-} =$ $3d^8$ [⥮|⥮|⥮|1|1] $4s^0$ [☒] $4p$ [☒|☒|☒]

sp^3 $sp^3sp^3sp^3$ — Four sp^3- hybrid orbitals tetrahedral

Cl^- is a weak ligand and thus unpaired electrons are not paired.

Magnetic moment (spin only)

$$= \sqrt{n(n+2)}\ \text{BM}$$

$$= \sqrt{2(2+2)}\ \text{BM}$$

$$= \sqrt{8} = 2.828\ \text{BM}$$

19. Salt Analysis

1. The pair(s) of ions where both the ions are precipitated upon passing H_2S gas in the presence of dilute HCl, is (are) **[JEE Advanced 2015, Single Option]**

(a) Ba^{2+}, Zn^{2+} (b) Bi^{3+}, Fe^{3+}

(c) Cu^{2+}, Pb^{2+} (d) Hg^{2+}, Bi^{3+}

Ans. **(d)** Only radicals of I and II group of qualitative analysis get precipitated with H_2S in the presence of dilute HCl.

(c) $Cu^{2+} + H_2S \xrightarrow{H^+} \underset{\text{Black}}{CuS \downarrow}$

$Pb^{2+} + H_2S \xrightarrow{H^+} \underset{\text{Black}}{PbS \downarrow}$

(d) $Hg^{2+} + H_2S \xrightarrow{H^+} \underset{\text{Black}}{HgS \downarrow}$

$Bi^{3+} + H_2S \xrightarrow{H^+} \underset{\text{Brown ppt}}{Bi_2S_3 \downarrow}$

Ba^{2+}, Zn^{2+} and Fe^{3+} are not precipitated as sulphide.

20. Group 18 [VIIIA - The Noble Gases]

1. Which one has the highest boiling point? **[JEE Main 2015]**

(a) He
(b) Ne
(c) Kr
(d) Xe

Ans. **(d)** As we move down the group of noble gases , molecular mass increases by which dipole produced for a moment and hence London forces increases from He to Xe. Therefore more amount of energy is required to break these forces, thus boiling point also increases from He to Xe.

2. Under ambient conditions, the total number of gases released as products in the final step of the reaction scheme shown below is **[JEE Advanced 2014, Single Option]**

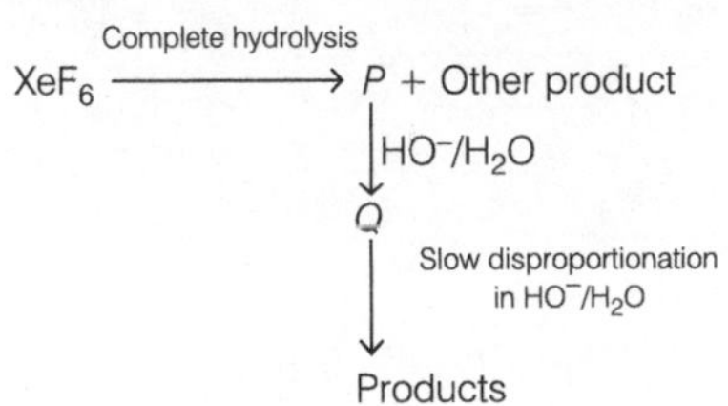

(a) 0
(b) 1
(c) 2
(d) 3

Ans. **(c)** XeF_6 on complete hydrolysis produces XeO_3.

$$XeF_6 + 3H_2O \longrightarrow XeO_3 + 3H_2F_2$$
$$\downarrow OH^-$$
$$HXeO_4^-$$
$$\downarrow OH^-/H_2O \text{ (disproportionation)}$$
$$XeO_6^{4-}(s) + Xe(g) + H_2O(l) + O_2(g)$$

Thus, (c) is the correct answer.

3. The shape of XeO_2F_2 molecule is **[JEE Advanced 2012, Single Option]**

(a) trigonal bipyramidal (b) square planar
(c) tetrahedral (d) *see-saw*

Ans. **(d)** In XeO_2F_2, the bonding arrangement around the central atom Xe is

O = Xe(••) = O, with two F atoms bonded to Xe

$$4\sigma \text{ bonds} + 1.0 \text{ lp} = 5$$

$\Rightarrow$ Hybridization of Xe = sp^3d

sp^3d hybridization corresponds to trigonal bipyramidal geometry.

Also, in trigonal bipyramidal geometry, lone pairs remain present on equatorial positions in order to give less electronic repulsion.

Remove lone-pair $\longrightarrow$

See-saw shape

JEE MAIN & ADVANCED
Solved Papers 2018*

JEE Main

1. According to molecular orbital theory, which of the following will not be a viable molecule?
(a) He_2^{2+} (b) He_2^+ (c) H_2^- (d) H_2^{2-}

2. The recommended concentration of fluoride ion in drinking water is up to 1 ppm as fluoride ion is required to make teeth enamel harder by converting $[3Ca_3(PO_4)_2 \cdot Ca(OH)_2]$ to :
(a) $[CaF_2]$ (b) $[3(CaF_2) \cdot Ca(OH)_2]$
(c) $[3Ca_3(PO_4)_2 \cdot CaF_2]$ (d) $[3\{Ca_3(PO_4)_2\} \cdot CaF_2]$

3. Which of the following compounds contain(s) no covalent bond(s)?

$$KCl, PH_3, O_2, B_2H_6, H_2SO_4$$

(a) KCl, B_2H_6, PH_3 (b) KCl, H_2SO_4
(c) KCl (d) KCl, B_2H_6

4. Which of the following are Lewis acids?
(a) PH_3 and BCl_3 (b) $AlCl_3$ and $SiCl_4$
(c) PH_3 and $SiCl_4$ (d) BCl_3 and $AlCl_3$

5. Total number of lone pair of electron in I_3^- ion is
(a) 3 (b) 6 (c) 9 (d) 12

6. Which of the following salts is the most basic in aqueous solution?
(a) $Al(CN)_3$ (b) CH_3COOK
(c) $FeCl_3$ (d) $Pb(CH_3COO)_2$

7. Hydrogen peroxide oxidises $[Fe(CN)_6]^{4-}$ to $[Fe(CN)_6]^{3-}$ in acidic medium but reduces $[Fe(CN)_6]^{3-}$ to $[Fe(CN)_6]^{4-}$ in alkaline medium. The other products formed are, respectively.
(a) $(H_2O + O_2)$ and H_2O
(b) $(H_2O + O_2)$ and $(H_2O + OH^-)$
(c) H_2O and $(H_2O + O_2)$
(d) H_2O and $(H_2O + OH^-)$

8. The oxidation states of Cr, in $[Cr(H_2O)_6]Cl_3$, $[Cr(C_6H_6)_2]$, and $K_2[Cr(CN)_2(O)_2(O_2)(NH_3)]$ respectively are
(a) +3, +4 and +6
(b) +3, +2 and +4
(c) +3, 0 and +6
(d) +3, 0 and +4

9. The compound that does not produce nitrogen gas by the thermal decomposition is
(a) $Ba(N_3)_2$ (b) $(NH_4)_2Cr_2O_7$
(c) NH_4NO_2 (d) $(NH_4)_2SO_4$

10. When metal '*M*' is treated with NaOH, a white gelatinous precipitate '*X*' is obtained, which is soluble in excess of NaOH. Compound '*X*' when heated strongly gives an oxide which is used in chromatography as an adsorbent. The metal '*M*' is
(a) Zn (b) Ca (c) Al (d) Fe

11. Consider the following reaction and statements :

$$[Co(NH_3)_4Br_2]^+ + Br^- \longrightarrow [Co(NH_3)_3Br_3] + NH_3$$

I. Two isomers are produces if the reactant complex ion is a *cis*-isomer.
II. Two isomers are produced if the reactant complex ion is a *trans*-isomer.
III. Only one isomer is produced if the reactant complex ion is a *trans*-isomer.
IV. Only one isomer is produced if the reactant complex ion is a *cis*-isomer.

The correct statements are
(a) (I) and (II) (b) (I) and (III)
(c) (III) and (IV) (d) (II) and (IV)

Note : *(*)Here we have given only Inorganic Chemistry related questions.*

JEE Advanced

Paper 1

1. The compound(s) which generate (s) N_2 gas upon thermal decomposition below 300 °C is (are)

(One or More than One Option Correct)

(a) NH_4NO_3 (b) $(NH_4)_2Cr_2O_7$
(c) $Ba(N_3)_2$ (d) Mg_3N_2

2. The correct statement (s) regarding the binary transition metal carbonyl compounds is (are) (Atomic numbers : Fe = 26, Ni = 28)

(One or More than One Option Correct)

(a) Total number of valence shell electrons at metal centre in $Fe(CO)_5$ or $Ni(CO)_4$ is 16
(b) These are predominantly low spin in nature
(c) Metal-carbon bond strengthens when the oxidation state of the metal is lowered
(d) The carbonyl C—O bond weakens when the oxidation state of the metal is increased

3. Based on the compounds of group 15 elements, the correct statement(s) is (are)

(One or More than One Option Correct)

(a) Bi_2O_5 is more basic than N_2O_5
(b) NF_3 is more covalent than BiF_3
(c) PH_3 boils at lower temperature than NH_3
(d) The N—N single bond is stronger than the P—P single bond

4. Among the species given below, the total number of diamagnetic species is____

H atom, NO_2 monomer, O_2^- (superoxide), dimeric sulphur in vapour phase,
Mn_3O_4, $(NH_4)_2[FeCl_4]$, $(NH_4)_2[NiCl_4]$, K_2MnO_4, K_2CrO_4

(Numerical Value Based)

Paper 2

5. The correct option(s) regarding the complex $[Co(en)(NH_3)_3(H_2O)]^{3+}$ (en = $H_2NCH_2CH_2NH_2$) is (are)

(One or More than One Option Correct)

(a) It has two geometrical isomers
(b) It will have three geometrical isomers, if bidentate 'en' is replaced by two cyanide ligands
(c) It is paramagnetic
(d) It absorbs light at longer wavelength as compared to $[Co(en)(NH_3)_4]^{3+}$

6. The correct option(s) to distinguish nitrate salts to Mn^{2+} and Cu^{2+} taken separately is (are)

(One or More than One Option Correct)

(a) Mn^{2+} shows the characteristic green colour in the flame test
(b) Only Cu^{2+} shows the formation of precipitate by passing H_2S in acidic medium
(c) Only Mn^{2+} shows the formation of precipitate by passing H_2S in faintly basic medium
(d) Cu^{2+}/Cu has higher reduction potential than Mn^{2+}/Mn (measured under similar conditions)

7. Match each set of hybrid orbitals from List–I with complexes given in List-II.

List–I	List–II
P. dsp^2	1. $[FeF_6]^{4-}$
Q. sp^3	2. $[Ti(H_2O)_3Cl_3]$
R. sp^3d^2	3. $[Cr(NH_3)_6]^{3+}$
S. d^2sp^3	4. $[FeCl_4]^{2-}$
	5. $[Ni(CO)_4]$
	6. $[Ni(CN)_4]^{2-}$

(Matching Type Question)

The correct option is

(a) $P \to 5$; $Q \to 4, 6$; $R \to 2, 3$; $S \to 1$
(b) $P \to 5,6$; $Q \to 4$; $R \to 3$; $S \to 1,2$
(c) $P \to 6$; $Q \to 4, 5$; $R \to 1$; $S \to 2, 3$
(d) $P \to 4,6$; $Q \to 5, 6$; $R \to 1,2$; $S \to 3$

Answer with Explanations

JEE Main

1. *(d)* **Key Idea** *According to M.O.T, the viability of any molecule can be judged through the calculation of bond order.*

Electronic	Configuration	Bond order
He_2^+	$\sigma_{1s^2}\sigma^*_{1s^1}$	$\frac{2-1}{2} = 0.5$
H_2^-	$\sigma_{1s^2}\sigma^*_{1s^1}$	$\frac{2-1}{2} = 0.5$
H_2^{2-}	$\sigma_{1s^2}\sigma^*_{1s^2}$	$\frac{2-2}{2} = 0$
He_2^{2+}	σ_{1s^2}	$\frac{2-0}{2} = 1$

The molecule having zero bond order will not be viable hence, H_2^{2-} (option d) is the correct answer.

2. *(c)* Fluoride ions help in making teeth enamel harder by converting $[3Ca_3(PO_4)_2 \cdot Ca(OH)_2]$ i.e. Hydroxy apatite to $[3Ca_3(PO_4)_2 \cdot CaF_2]$ i.e., Fluorapatite (Harder teeth enamel) *via* following reaction:

$$[3Ca_3(PO_4)_2 \cdot Ca(OH)_2] + \underset{\text{From drinking water}}{2F^-} \longrightarrow [3Ca_3(PO_4)_2 \cdot CaF_2] + 2OH^-$$

3. *(c)* KCl is the only ionic compound. The structure of PH_3, O_2, B_2H_6 and H_2SO_4 are given below

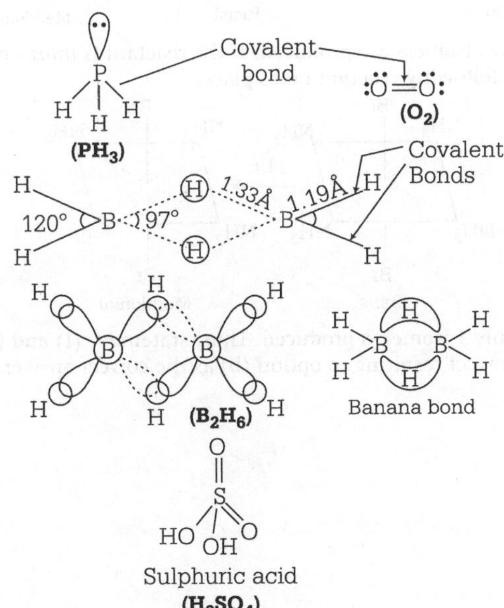

All bond between S and O atom are covalent bonds.

4. *(d)* **Key Idea** *Lewis acids are defined as,*

"Electron deficient compounds which have the ability to accept atleast one lone pair."

The compound given are

PH_3 -Octet complete although P has vacant 3*d*-orbital but does not have the tendency to accept lone pair in it. Hence, it cannot be considered as Lewis acid.

BCl_3 -Incomplete octet with following orbital picture.

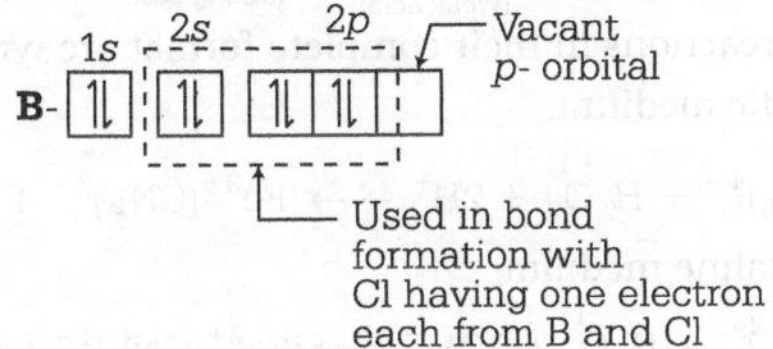

Hence, vacant *p*-orbital of B can accept one lone pair thus it can be considered as Lewis acid.

$AlCl_3$-Similar condition is visible in $AlCl_3$ as well i.e.

Al (Valence orbital only) =

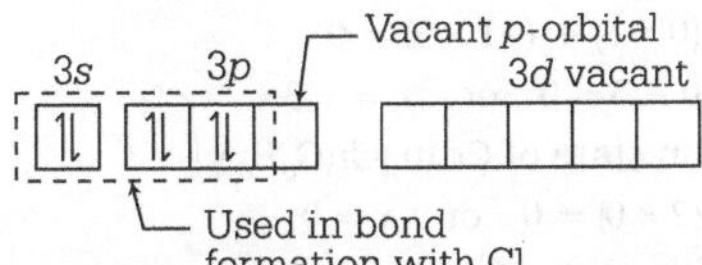

Hence this compound can also be considered as Lewis acid.

$SiCl_4$ - Although this compound does not have incomplete octet but it shows the tendency to accept lone pair of electrons in its vacant *d*-orbital. This tendency of $SiCl_4$ is visible in following reaction.

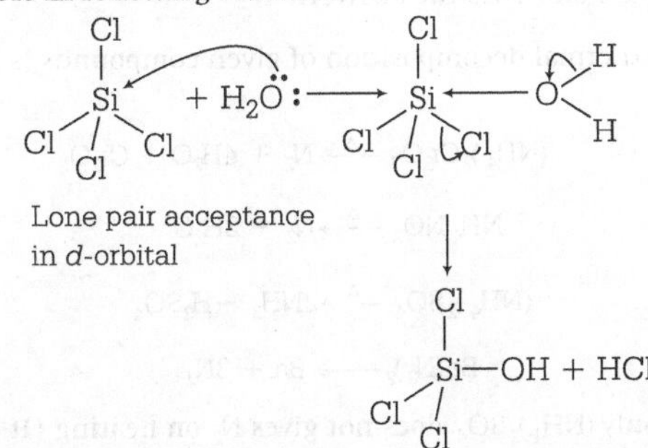

Thus option (b) and (d) both appear as correct but most suitable answer is (d) as the condition of a proper Lewis acid is more well defined in BCl_3 and $AlCl_3$.

5. *(c)* The structure of I_3^- ion is

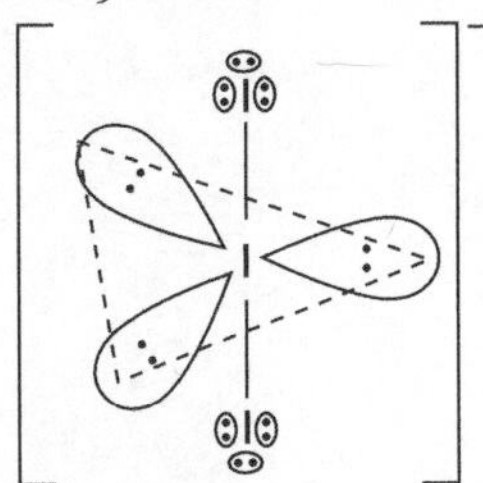

Hence, 9 is the correct answer.

6. *(b)* Among the given salts
$FeCl_3$ is acidic in nature i.e., have acidic solution as it is the salt of weak base and strong acid.
$Al(CN)_3$ and **$Pb(CH_3COO)_2$** are the salts of weak acid and weak base.
CH_3COOK is the salt of strong base and weak acid.
Hence, the solution of CH_3COOK will be most basic because of the following reaction.

$$CH_3COOK + H_2O \rightleftharpoons \underset{\text{(Weak acid)}}{CH_3COOH} + \underset{\text{(Strong base)}}{KOH}$$

7. *(c)* Both reactions in their complete format are written below
(i) In acidic medium,

$$[Fe^{2+}(CN)_6]^{4-} + H_2\overset{-1}{O_2} + 2H^+ \longrightarrow [Fe^{3+}(CN)_6]^{3-} + 2H_2\overset{-2}{O}$$

(ii) In alkaline medium,

$$[Fe^{3+}(CN)_6]^{3-} + H_2\overset{-1}{O_2} + 2OH^- \longrightarrow [Fe^{2+}(CN)_6]^{4-} + O_2 + 2H_2O$$

Hence, H_2O (for reaction (i)) and $O_2 + H_2O$ (for reaction (ii)) are produced as by product.

8. *(c)* Let the oxidation state of Cr in all cases is 'x'
(i) Oxidation state of Cr in $[Cr(H_2O)_6]Cl_3$

$$x + (0 \times 6) + (-1 \times 3) = 0$$

or $\quad x + 0 - 3 = 0 \quad$ or $\quad x = +3$

(ii) Oxidation state of Cr in $[Cr(C_6H_6)_2]$

$$x + (2 \times 0) = 0 \quad \text{or} \quad x = 0$$

(iii) Oxidation state of Cr in

$$K_2[Cr(CN)_2(O)_2(O_2)(NH_3)]$$

$$1 \times 2 + x + (-1 \times 2) + (-2 \times 2) + (-2) + 0 = 0$$

or $\quad 2 + x - 2 - 4 - 2 = 0 \quad$ or $\quad x - 6 = 0$

hence $x = +6$
Thus, +3, 0 and +6 is the answer.

9. *(d)* The thermal decomposition of given compounds is shown below

$$(NH_4)_2Cr_2O_7 \xrightarrow{\Delta} N_2 + 4H_2O + Cr_2O_3$$

$$NH_4NO_2 \xrightarrow{\Delta} N_2 + 2H_2O$$

$$(NH_4)_2SO_4 \xrightarrow{\Delta} 2NH_3 + H_2SO_4$$

$$Ba(N_3)_2 \longrightarrow Ba + 3N_2$$

Thus, only $(NH_4)_2SO_4$ does not gives N_2 on heating (It give NH_3). While rest of the given compounds gives N_2 on their thermal decomposition.

10. *(c)* Among the given metals Al forms white gelatinous ppt. with NaOH.
Hence, the probable metal can be Al. This ppt. is dissolved in excess of NaOH due to the formation of sodium metal Aluminate. Both the reactions are shown below.

$$Al^{3+} \xrightarrow{\text{NaOH}} \underset{\substack{\text{White gelatinous}\\ \text{ppt. }(X)\text{ of}\\ \text{aluminium}\\ \text{hydroxide}}}{Al(OH)_3} \xrightarrow{\text{Excess of NaOH}} \underset{\substack{\text{Sodium}\\ \text{metaaluminate)}\\ \text{soluble}}}{NaAlO_2}$$

Aluminium hydroxide on strong heating gives alumina (Al_2O_3) which is used as an adsorbent in chromatography. This reaction can be seen as :

$$2Al(OH)_3 \xrightarrow{\Delta} Al_2O_3 + 3H_2O$$

Thus, metal M is Al.
Ca, being below sodium in electrochemical reactivity series, cannot displaces Na from its aqueous solution.
Zn reacts with NaOH to form sodium zincate which is a soluble compound.
Fe reacts with sodium hydroxide to form tetrahydroferrate (II) sodium which is again a soluble complex.

11. *(b)* If the reactant is *cis* isomer than following reaction takes place.

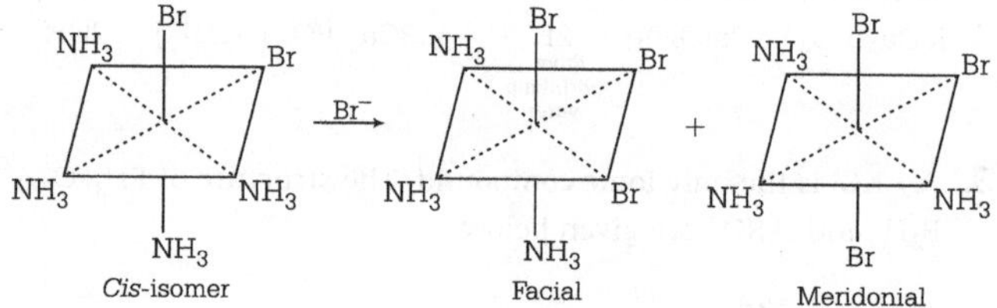

i.e. two isomers are produced. If the reactant is *trans* isomer than following reaction takes place.

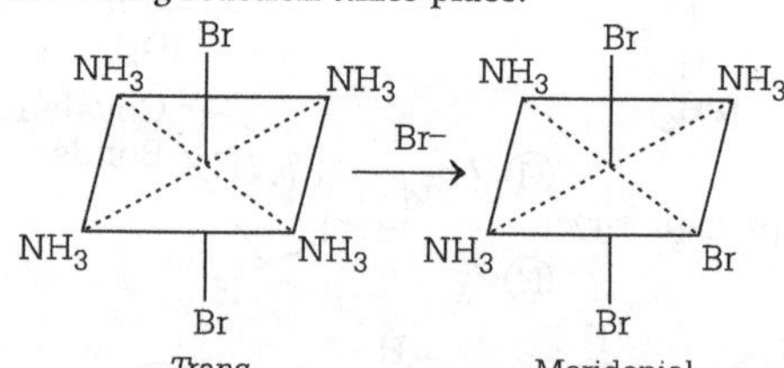

i.e. only 1 isomer is produced. Thus, statement (I) and (III) are correct resulting to option (b) as the correct answer. zx

JEE Advanced

1. *(b, c)* Among the given compounds, those which generate N_2 on thermal decomposition below 300°C are **ammonium dichromate** i.e., $(NH_4)_2Cr_2O_7$ and **barium azide** or nitride i.e., $Ba(N_3)_2$. Reactions of their thermal decomposition are given below

(i) $(NH_4)_2Cr_2O_7 \xrightarrow[\text{Below } 300°C]{\Delta} N_2 \uparrow + Cr_2O_3 + 4H_2O$

It is an exothemic reaction with

$\Delta H = -429.1 \pm 3$ kcal / mol.

(ii) $Ba(N_3)_2 \xrightarrow[\text{Around } 160° \text{ and above}]{\Delta} Ba + 3N_2 \uparrow$

Ammonium nitrate (NH_4NO_3) on heating below 300°C gives N_2O as

$$NH_4NO_3 \xrightarrow[\text{below } 300°C]{\Delta} N_2O + 2H_2O$$

However, on rapid heating or explosion (i.e. above 300°C) it gives off nitrogen as

$$2NH_4NO_3 \xrightarrow[\text{or explosion}]{\text{Rapid heating}} 2N_2 + O_2 + 4H_2O$$

Magnesium nitride (Mg_3N_2) does not decompose at lower temperatures being comparatively more stable. Its thermal decomposition requires a minimum temperature of 700°C and proceeds as

$$Mg_3N_2 \xrightarrow{700-1500°C} 3Mg + N_2 \uparrow$$

2. *(b, c)* Statement wise explanation is

(i) **Statement (a)** The total number of valence shell electrons at metal centre in $Fe(CO)_5$ or $Ni(CO)_4$ is 8 instead of 16 as shown below

Valence shell having 8 electrons in complex formation

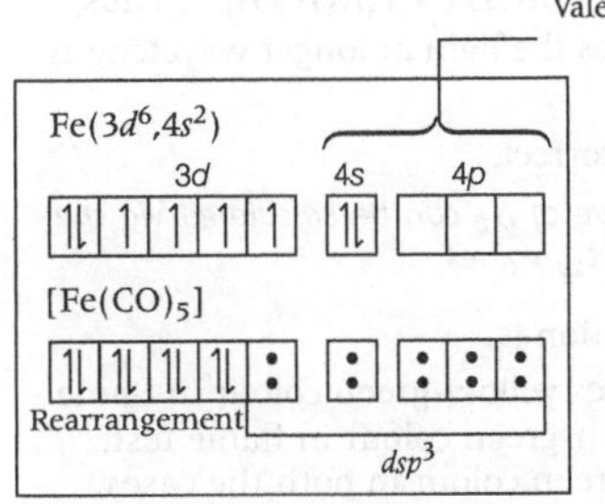

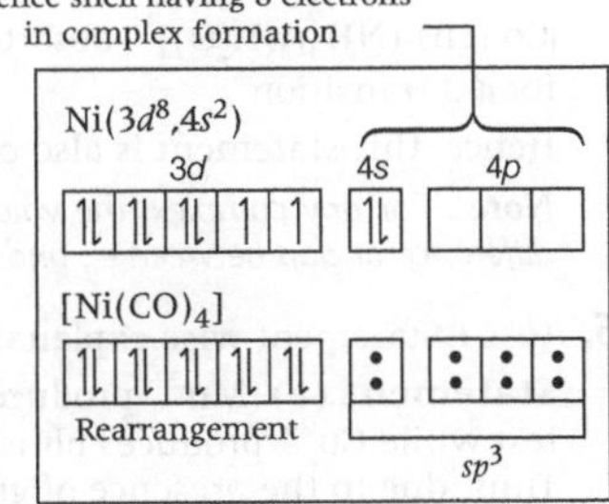

Hence, this statement is incorrect.

(ii) **Statement (b)** Carbonyl complexes are predominantly low spin complexes due to strong ligand fields. Hence, this statement is correct.

(iii) **Statement (c)** For central metal lowering of oxidation state results to increase in electron density on it. This in turn results to increase in extent of synergic bonding. Thus, we can say "metal carbonyl bond strengthens, when oxidation state of metal is lowered".

Hence, it is a correct statement.

(iv) **Statement(d)** Increase in positive charge on metal (i.e., increase in oxidation state) results to decrease in synergic bonding strength.

This in turn makes C—O bond stronger instead of weaker. Hence this statement is also incorrect.

3. ***(a,b,c)*** Statement wise explanation is

(i) **Statement(a)** Bi_2O_5 is a metallic oxide while N_2O_5 is a non-metallic oxide.

Metallic oxides being **ionic** are **basic in nature** while non metallic oxides being **covalent** are **acidic in nature**. This confirms more basic nature of Bi_2O_5 in comparison to N_2O_5. Hence, this is a correct statement.

(ii) **Statement (b)** The electronegativity difference between N(3) and F(4) is less as compared to the electronegativity difference between Bi(1.7) and F(4). More electronegativity difference leads to ionic compounds. Thus, NF_3 must be more covalent in nature as compared to BiF_3. Hence, this statement is also correct.

(iii) **Statement (c)** In NH_3 intermolecular hydrogen bonding is present, which is altogether absent in PH_3. Thus, PH_3 boils at lower temperature than NH_3.

Hence, this is also a correct statement.

(iv) **Statement (d)** Due to smaller size of N the lone pair-lone pair repulsion is more in N—N single bond as compared to P—P single bond. This results to weaker N—N single bond as compared to P—P single bond. Hence, this statement is incorrect.

4. *(1)* Among the given species only K_2CrO_4 is diamagnetic as central metal atom Cr in it has $[Ar]3d^0$ electronic configuration i.e., all paired electrons. The structure and oxidation state of central metal atom of this compound are as follows

Structure $K^+ [CrO_2(O^-)_2]$ (O=Cr with O, O⁻, O⁻), **Oxidation state** Cr^{6+}

Rest all the compounds are **paramagnetic**. Reasons for their paramagnetism are given below

(i) H-atom have $1s^1$ electronic configuration i.e., 1 unpaired electron.

(ii) NO_2 i.e. O=Ṅ–O in itself is **an odd electron species.**

(iii) O_2^- (Superoxide) has one unpaired electron in π^* molecular orbital.

(iv) S_2 in vapour phase has O_2 like electronic configuration i.e., have 2 unpaired electrons in π^* molecular orbitals.

(v) Mn_3O_4 has following structure

$\overset{+2}{Mn}(O)_2\overset{+4}{Mn}(O)_2\overset{+2}{Mn}$

Thus, Mn is showing +2 and +4 oxidation states. The outermost electronic configuration of elemental Mn is $3d^5 4s^2$. Hence, in both the above oxidation states it has unpaired electrons as

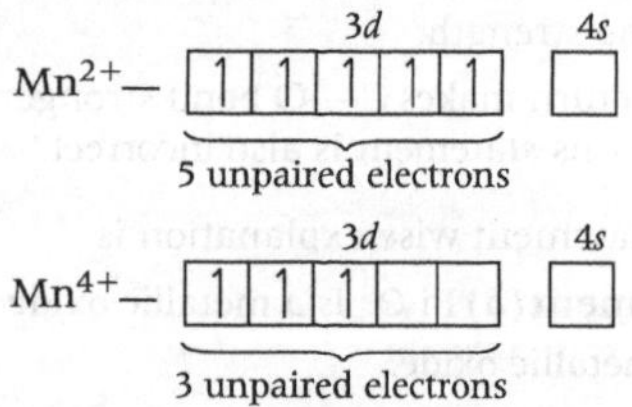

(vi) $(NH_4)_2FeCl_4$ has Fe as central metal atom with +2 oxidation state. The electronic configuration of Fe^{2+} in the complex is

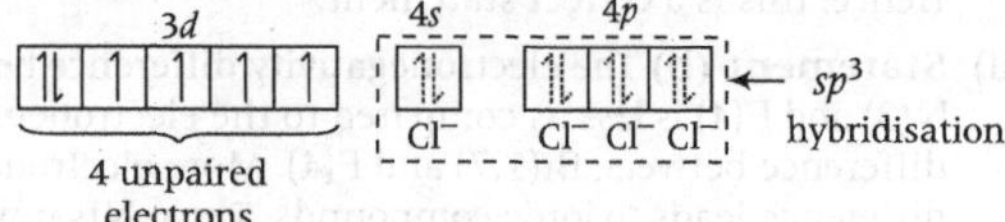

(vii) $(NH_4)_2NiCl_4$ has Ni as central metal atom with +2 oxidation state. The electronic configuration of Ni^{2+} in the complex is

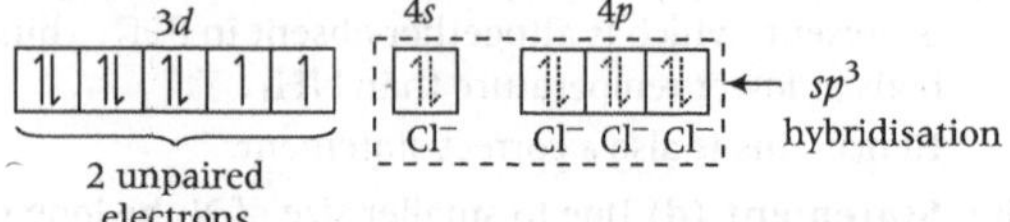

(viii) In K_2MnO_4 central metal atom Mn has +6 oxidation state with following structure

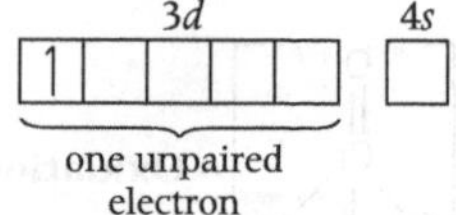

Electronic configuration of Mn^{6+} is

3d 4s

1						

one unpaired electron

5. *(a, b, d)* Statement wise explanation is

Statement (a) $[Co(en)(NH_3)_3H_2O]^{3+}$ have following 2 geometrical isomers.

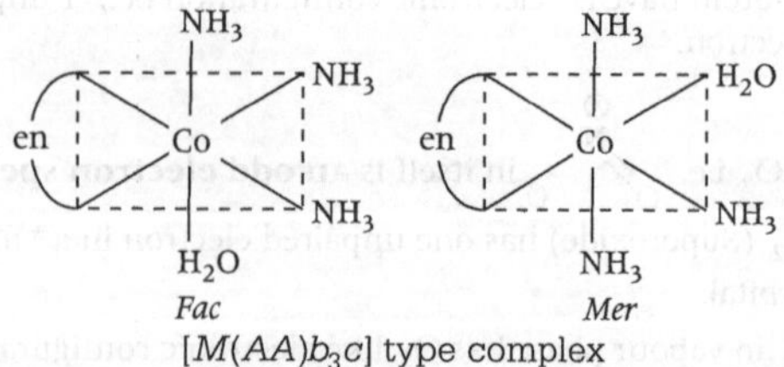

[M(AA)b_3c] type complex

Hence, this is correct statement.

Statement (b) If bidentate ligand 'en' is replaced by two cyanide ligands then $[Co(NH_3)_3(H_2O)(CN)_2]^+$ is formed.

It is [Ma_3b_2c] type complex which has following 3 geometrical isomers.

NH_3; $N{\equiv}C$, NH_3, Co, $N{\equiv}C$, NH_3; H_2O — (*Fac*)

NH_3; $N{\equiv}C$, NH_3, Co, H_2O, NH_3; CN — *Fac* with respect to NH_3 and *Mer* with respect to —CN

NH_3; $N{\equiv}C$, H_2O, Co, $N{\equiv}C$, NH_3; NH_3 — (*Mer*)

Hence, this statement is also correct.

Statement (c) Co metal has $[Ar]3d^7 4s^2$ configuration while in $[Co(en)(NH_3)_3(H_2O)]^{3+}$ it is in +3 oxidation state. Thus, Co^{3+} has $[Ar]3d^6$ configuration.

Co^{3+} = 3d: ⥮ 1 1 1 1 ; 4s: □

As *en* is a strong ligand, so pairing will occur

3d: ⥮ ⥮ ⥮ □ □ ; 4s: □

Due to the presence of all paired electrons it show diamagnetic behaviour rather than paramagnetic.

Hence, this statement is incorrect.

Statement (d) According to CFT, absorption of light by coordination complexes depends upon CFSE i.e., crystal field splitting energy (Δ_0)as

$$\Delta_0 \propto \frac{1}{\lambda}$$

Among the complexes given $[Co(en)(NH_3)_4]^{3+}$ has more Δ_0 value as compared to complex $[Co(en)(NH_3)_3(H_2O)]^{3+}$. Thus, $[Co(en)(NH_3)_3(H_2O)]^{3+}$ absorbs the light at longer wavelength for *d-d* transition.

Hence, this statement is also correct.

***Note**: For any complex, the value of Δ_0 can be calculated via the difference or gap between e_g and t_{2g} values.*

6. *(b,d)* Statement wise explanation is

Statement (a) Mn^{2+} produces yellow-green colour in flame test while Cu^{2+} produces bluish-green colour in flame test. Thus, due to the presence of green colour in both the cases, flame test is not the suitable method to distinguish between nitrate salts of Cu^{2+} and Mn^{2+}. Hence this statement is wrong.

Statement (b) Cu^{2+} belong to group II of cationic or basic radicals. It gives black ppt. of CuS if H_2S is passed through it in the presence of acid (e.g HCl). Mn^{2+} does not show this property hence this can be considered as a suitable method to distinguish between Mn^{2+} and Cu^{2+}.

Hence, this statement is correct.

Statement (c) In faintly basic medium when H_2S is passed both Cu^{2+} and Mn^{2+} forms precipitates. Thus, it is not suitable method to distinguish between them.

Hence, this statement is incorrect

Statement (d) The standard reduction potential of Cu^{2+}/Cu is +0.34 V while that of Mn^{2+}/Mn is –1.18V. This can be used to distinguish between Cu^{2+} and Mn^{2+}. In general less electropositive metals have higher SRP.

Hence, this statement is correct.

7. *(c)* **For, *P* i.e. dsp^2,** It is seen in $[Ni(CN)_4]^{2-}$

Ni—$[Ar]3d^8 4s^2$

Ni^{2+}—$[Ar]3d^8$

Ni^{2+}— 3d: ⥮ ⥮ ⥮ 1 1 | 4s: □ | 4p: □ □ □

as CN^- is a strong ligand so when it approaches towards central metal pairing of unpaired electrons takes place.

Thus, in $[Ni(CN)_4]^{2-}$

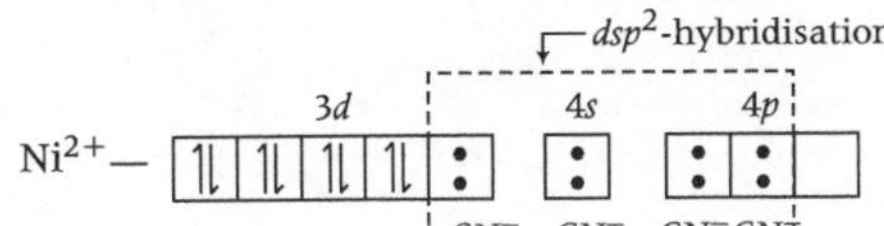

Structure : Square planar

So correct match for P is 6.

For *Q* i.e., sp^3

It is seen in $[FeCl_4]^{2-}$ and $Ni(CO)_4$

Fe – $[Ar]3d^6 4s^2$

Fe^{2+} – $[Ar]3d^6$

Fe^{2+}— 3d: ⥮ 1 1 1 1 | 4s: □ | 4p: □ □ □

As Cl^- is a weak ligand so when it approaches towards central metal pairing of unpaired electrons does not take place.

Thus, in $[FeCl_4]^{2-}$

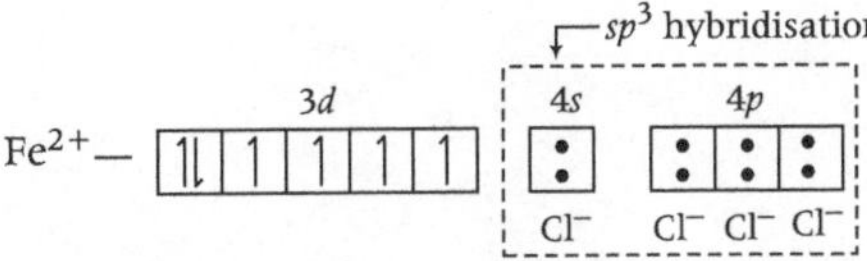

Structure-Tetrahedral

Likewise in $Ni(CO)_4$

Ni—$[Ar]3d^8 4s^2$

Ni — 3d: ⥮ ⥮ ⥮ 1 1 | 4s: ⥮ | 4p: □ □ □

As CO is a strong ligand, hence when it approaches towards central metal atom pairing of unpaired electron of central atom takes place.

Thus, in $Ni(CO)_4$

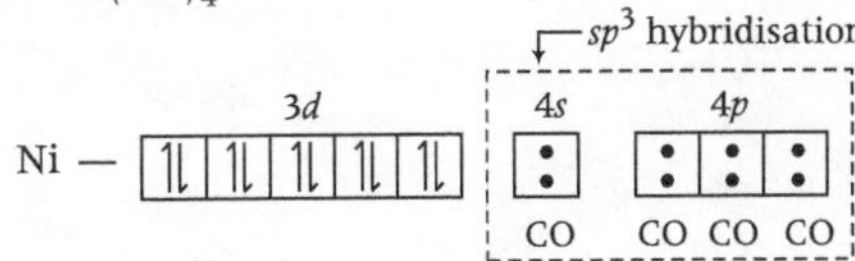

Structure Tetrahedral

So, for *Q*-4 and 5 are correct match.

For *R* i.e., sp^3d^2

It is seen in $[FeF_6]^{4-}$

Fe—$[Ar]3d^6 4s^2$

Fe^{2+}—$[Ar]3d^6$

Fe^{2+}— 3d: ⥮ 1 1 1 1 | 4s: □ | 4p: □ □ □ | 4d: □ □ □ □ □

As F^- is a weak field ligand hence, when it approaches towards central metal atom, pairing of its electrons does not take place.

Thus, in $[FeF_6]^{4-}$

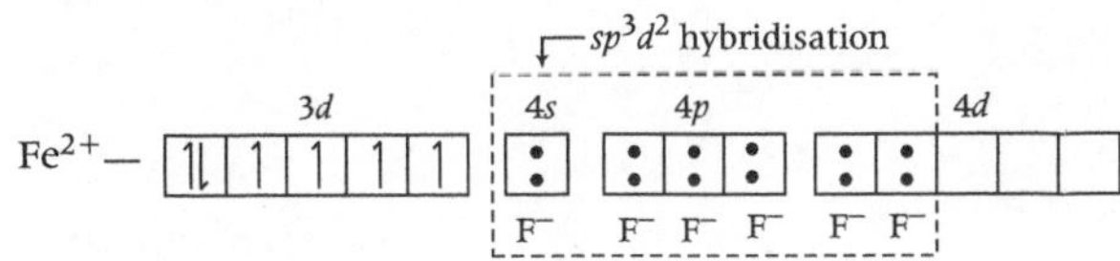

Structure : Octahedral

So, 1 is the correct match for *R*.

For *S* i.e., d^2sp^3

It is seen in $[Ti(H_2O)_3Cl_3]$ and $[Cr(NH_3)_6]^{3+}$

Ti—$[Ar]3d^2 4s^2$

Ti^{3+}—$[Ar]3d^1$

Ti^{3+} — 3d: 1 □ □ □ □ | 4s: □ | 4p: □ □ □

Here, both H_2O and Cl are weak ligands

So, in $[Ti(H_2O)_3Cl_3]$

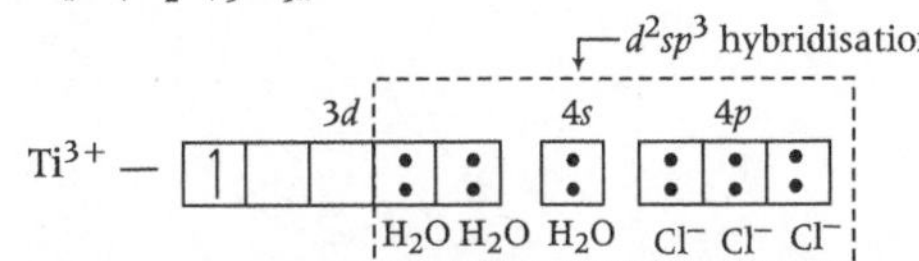

Structure Octahedral

Likewise in $[Cr(NH_3)_6]^{3+}$

Cr—$[Ar]3d^5 4s^1$

Cr^{3+}—$[Ar]3d^3 4s^0$

Cr^{3+} — 3d: 1 1 1 □ □ | 4s: □ | 4p: □ □ □

Here, NH_3 is also a weak field ligand so due to its approach no pairing takes place in Cr.

Thus, In $[Cr(NH_3)_6]^{3+}$

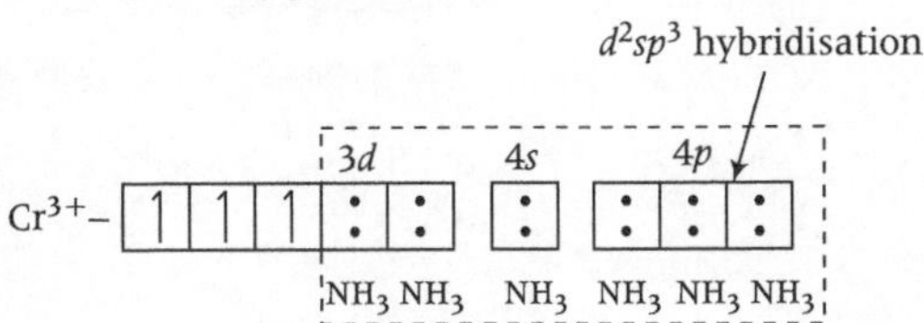

So for, S-2 and 3 are the correct match.